M000248785

CHEMICAL ENGINEERING IN THE PHARMACEUTICAL INDUSTRY

CHEMICAL ENGINEERING IN THE PHARMACEUTICAL INDUSTRY

R&D to Manufacturing

Edited by

DAVID J. AM ENDE

A JOHN WILEY & SONS, INC., PUBLICATION

Published by John Wiley & Sons, Inc., Hoboken, New Jersey
Published simultaneously in Canada

Library of Congress Cataloging-in-Publication Data:

Chemical engineering in the pharmaceutical industry : R&D to manufacturing / edited by David J. am Ende.
p. cm.
Includes index.
ISBN 978-0-470-42669-2 (cloth)
1. Pharmaceutical technology. 2. Biochemical engineering. 3. Chemical engineering. I. Ende, David J. am.
RS192.C525 2010
615′.19–dc22

2010013146

Printed in Singapore

10 9 8 7 6 5 4 3 2 1

CONTENTS

PREFACE

Chemical Engineering in the Pharmaceutical Industry is unique in many ways to what is traditionally taught in schools of chemical engineering. This book is therefore intended to cover many important concepts and applications of chemical engineering science that are particularly important to the Pharmaceutical Industry. There have been several excellent books written recently on the subjects of process chemistry in the pharmaceutical industry and separately on formulation development, but relatively little has been published specifically with a focus on chemical engineering.

The intention of the book is to highlight the importance and value of chemical engineering to the development and commercialization of pharmaceuticals covering active pharmaceutical ingredient (API) and drug product (DP) as well as analytical methods. It should serve as a resource for practicing chemical engineers as well as for chemists, analysts, technologists, and operations and management team members—all those who partner to bring pharmaceuticals successfully to market. The latter will benefit through an exposure to the mathematical and predictive approach and the broader capabilities of chemical engineers and also by the illustration of chemical engineering science as applied specifically to pharmaceutical problems. This book emphasizes both the need for scientific integration of chemical engineers with synthetic organic chemists within process R&D and the importance of the interface between R&D engineers and manufacturing engineers. The importance of analytical chemists and other scientific disciplines necessary to deliver pharmaceuticals to the market place is also emphasized with chapters dedicated to selected topics.

Although specific workflows for engineers in R&D depend on each company's specific organization, in general it is clear that, as part of a multidisciplinary team in R&D, chemical engineering practitioners offer value in many ways including API and DP process design, scale-up assessment from laboratory to plant, process modeling, process understanding, and general process development that ultimately reduces cost and ensures safe, robust, and environmentally friendly processes are transferred to manufacturing. How effective the teams leverage each of the various skill sets (i.e., via resource allocation) to arrive at an optimal process depends in part on the roles and responsibilities as determined within each organization and company. In general, it is clear that with increased cost pressures facing the pharmaceutical industry, including R&D and manufacturing, opportunities to leverage the field of chemical engineering science continue to increase. Indeed I have observed a significant increase in chemical engineering emphasis in API process development within Pfizer over the 15 years since I joined the company and especially in the past 5 years.

This book is divided into four main parts:

(1) Introduction
(2) Active Pharmaceutical Ingredient (API)
(3) Analytical Methods and Applied Statistics
(4) Drug Product (DP)

The introductory chapters span roles and opportunities for chemical engineering in small-molecule API, biologicals, drug products, as well as environmental sustainability and quality by design (QbD) concepts. The Active Pharmaceutical Ingredient part consists of 23 chapters covering chemical engineering principles applied to pharmaceutical specific unit operations (reaction engineering, crystallization, chromatography, filtration, drying, etc.) as well as pilot plant and scale-up manufacturing assessment chapters, including process safety. Process modeling promises to have significant payback as more *in silico* screening enables process design to be performed with fewer resources (for selection process conditions/optimization, solubility, distillation, and extraction design, etc.). Several chapters are devoted to process modeling with emphasis on several of the software tools currently available. The section on drug product includes formulation chapters as well as chapters highlighting unit operations specific to drug product (wet granulation, dry granulation, extrusion, controlled release, and lyophilization). In addition, process modeling within drug product chapter describes the various modeling approaches used to understand and predict performance of powder blending, mixing, tablet presses, tablet coating, and so on. The Analytical Methods and Applied Statistics part describes important topics on chemometrics, statistics, and analytical methods applied toward chemical engineering problems (e.g., material balance, kinetics, design of experiments, or quality by design for analytical methods).

The contributors were encouraged to provide worked out examples—so in most chapters a quantitative example is offered to illustrate key concepts and problem-solving approaches. In this way, the chapters will serve to help others solve similar problems.

There are many people to thank who made this work possible. First, I would like to thank all the contributors of this book. I also would like to thank my colleagues at Pfizer for writing many of the chapters and for my management (past and present) who encouraged and made this effort possible and who continue to encourage the role of chemical engineering in chemical R&D and pharmaceutical sciences.

Special thanks to my family (Mary, Nathan, Noah, and Brianna) for their support during the preparation of this book. Special thanks to Mary, not only for contributing two chapters to this book but also for her assistance in all phases of the project including the cover art. Finally, a special thanks to my parents for their encouragement to pursue chemical engineering in 1983 and their support through my attendance at the University of Iowa and Purdue University.

David J. am Ende, Ph.D.

Chemical R&D
Pfizer, Inc.
Groton, CT
January 2010

CONTRIBUTORS

Yuriy A. Abramov, Pfizer Global Research & Development, Pharmaceutical Sciences, Groton, CT, USA

Alberto Aliseda, Department of Mechanical Engineering, University of Washington, Seattle, WA, USA

David J. am Ende, Chemical R&D, Pfizer, Inc., Groton, CT, USA

Mary T. am Ende, Pharmaceutical Development, Pfizer Global Research & Development, Groton, CT, USA

Firoz D. Antia, Product Development, Palatin Technologies, Inc., Cranbury, NJ, USA

Leah Appel, Green Ridge Consulting, Bend, OR, USA

Simon J. Bale, Pfizer Global Research & Development, Sandwich, Kent, UK

Kimber L. Barnett, Pfizer Global Research & Development, Groton, CT, USA

Alfred Berchielli, Pharmaceutical Development, Pfizer worldwide Research & Development, Groton, CT, USA

Rahul Bharadwaj, Pharmaceutical Development, Pfizer Global Research & Development, Groton, CT, USA

Vivek Bhatnagar, Process Development, Amgen Inc., West Greenwich, RI, USA

Kevin J. Bittorf, Formulation Development, Vertex Pharmaceuticals Inc., Cambridge, MA, USA

Alan D. Braem, Process Research & Development, Bristol-Myers Squibb Co., New Brunswick, NJ, USA

Chau-Chyun Chen, Aspen Technology, Inc., Burlington, MA, USA

Jennifer Chu, Pharmaceutical Sciences, Neurogen Corporation, Branford, CT, USA

Paul C. Collins, Eli Lilly & Co., Indianapolis, IN, USA

Eric M. Cordi, Chemical Research and Development, Pfizer Inc., Groton, CT, USA

Philip C. Dell'Orco, Chemical Development, GlaxoSmithKline, King of Prussia, PA, USA

Wim Dermaut, Process Safety Center, API Small Molecule Development, Johnson and Johnson Pharmaceutical Research and Development, Beerse, Belgium

Pankaj Doshi, Chemical Engineering and Process Division, National Chemical Laboratory, Pune, India

Elizabeth Fisher, Merck & Co., Inc., Rahway, NJ, USA

Salvador García-Muñoz, Pharmaceutical Development, Pfizer Global Research & Development, Groton, CT, USA

Imogen Gill, Pfizer Global Research & Development, Sandwich, Kent, UK

Timothy W. Graul, Pfizer Global Research & Development, Groton, CT, USA

James R. Hagan, GlaxoSmithKline, Sustainability and Environment, 1 Franklin Plaza, Philadelphia, PA, USA

Jason Hamm, Process Research & Development, Bristol-Myers Squibb Co., New Brunswick, NJ, USA

Melissa Hanna-Brown, Pfizer Global Research & Development, Sandwich, Kent, UK

Robert E. Hannah, GlaxoSmithKline, Sustainability and Environment, Philadelphia, PA, USA

Joe Hannon, Scale-up Systems Limited, Dublin, Ireland

Karen P. Hapgood, Department of Chemical Engineering, Monash University, Clayton, VIC, Australia

José O. Jiménez, Intelligen, Inc., Amsterdam, The Netherlands

Concepción Jiménez-González, GlaxoSmithKline, Sustainability and Environment, Research Triangle Park, NC, USA

G. Scott Jones, Process Research & Development, Bristol-Myers Squibb Co., New Brunswick, NJ, USA

Matthew L. Jorgensen, Engineering Technologies, Chemical Research & Development, Pfizer Inc., Groton, CT, USA

Jeffrey P. Katstra, Formulation Development, Vertex Pharmaceuticals Inc., Cambridge, MA, USA

Stephen B. Kessler, Impact Technology Development, Lincoln, MA, USA

William Ketterhagen, Pharmaceutical Development, Pfizer Global Research & Development, Groton, CT, USA

Avinash R. Khopkar, Engineering Sciences, Dow Chemical International Private Ltd., Pune, Maharashtra, India

Andreas Klamt, COSMOlogic GmbH & Co. KG, Leverkusen, Germany and Institute of Physical and Theoretical Chemistry, University of Regensburg, Regensburg, Germany

Venkat Koganti, Pharmaceutical Development, Pfizer Global Research & Development, Pfizer, Inc., Groton, CT, USA

Alexandros Koulouris, A.T.E.I. of Thessaloniki, Thessaloniki, Greece

Theodora Kourti, GlaxoSmithKline, Pharma Launch and Global Supply, Global Functions

Joseph F. Krzyzaniak, Pfizer Global Research & Development, Pharmaceutical Sciences, Groton, CT, USA

Joseph L. Kukura, Global Science Technology and Commercialization, Merck & Co., Inc., Rahway, NJ, USA

Sourav Kundu, Process Development, Amgen Inc., West Greenwich, RI, USA

Pericles T. Lagonikos, Merck & Co., Union, NJ, USA

Thomas L. LaPorte, Process Research & Development, Bristol-Myers Squibb Co., New Brunswick, NJ, USA

Juan C. Lasheras, Jacobs School of Engineering, University of California, San Diego, La Jolla, CA, USA

Carl LeBlond, Department of Chemistry, Indiana University of Pennsylvania, Indiana, PA, USA

Li Li, Merck & Co., Inc., West Point, PA, USA

James D. Litster, School of Chemical Engineering and Industrial & Physical Pharmacy, Purdue University, West Layafette, IN, USA

Frederick H. Long, Spectroscopic Solutions, LLC, Randolph, NJ, USA

Mike Lowinger, Merck & Co., Inc., West Point, PA, USA

Sumit Luthra, Pharmaceutical Development, Pfizer Global Research & Development, Pfizer, Inc., Groton, CT, USA

Brian L. Marquez, Pfizer Global Research & Development, Groton, CT, USA

Francis X. McConville, Impact Technology Development, Lincoln, MA, USA

Craig McKelvey, Merck & Co., Inc., West Point, PA, USA

Robert Rahn McKeown, Chemical Development, GlaxoSmithKline, Research Triangle Park, NC, USA

Melanie Miller, Process Research & Development, Bristol-Myers Squibb Co., New Brunswick, NJ, USA

Saravanababu Murugesan, Process Research & Development, Bristol-Myers Squibb Co, New Brunswick, NJ, USA

Jason Mustakis, Chemical R&D, Pfizer, Inc., Groton, CT, USA

Roger Nosal, Global CMC, Pfizer, Inc., Groton, CT, USA

Charles J. Orella, Chemical Process Development and Commercialization, Merck & Co., Inc., Rahway, NJ, USA

Taeshin Park, RES Group, Inc., Cambridge, MA, USA

Naveen Pathak, Genzyme, Inc., Framingham, MA.

Edward L. Paul (Retired), Chemical Engineering R&D, Merck & Co., Inc., Rahway, NJ, USA

Klimentina Pencheva, Pfizer Global Research & Development, Pharmaceutical Sciences, Sandwich, Kent, UK

Demetri Petrides, Intelligen, Inc., Scotch Plains, NJ, USA

Michael J. Pikal, Pharmaceutics, School of Pharmacy, University of Connecticut, Storrs, CT, USA

Celia S. Ponder, GlaxoSmithKline, Sustainability and Environment, Research Triangle Park, NC, USA

Andrew Prpich, Pharmaceutical Development, Pfizer Global Research & Development, Groton, CT, USA

Tom Ramsey, Process Research & Development, Bristol-Myers Squibb Co., New Brunswick, NJ, USA

Vivek V. Ranade, Tridiagonal Solutions Private Ltd. and National Chemical Laboratory, Pune, Maharashtra, India

Tapan Sanghvi, Formulation Development, Vertex Pharmaceuticals Inc., Cambridge, MA, USA

Luke Schenck, Merck & Co., Inc., Rahway, NJ, USA

Richard L. Schild, Process Research & Development, Bristol-Myers Squibb Co., New Brunswick, NJ, USA

Kevin D. Seibert, Eli Lilly & Co., Indianapolis, IN, USA

Matthew Shaffer, Green Ridge Consulting, Bend, OR, USA

Praveen K. Sharma, Process Research & Development, Bristol-Myers Squibb Co, New Brunswick, NJ, USA

Joshua Shockey, Green Ridge Consulting, Bend, OR, USA

Charles Siletti, Intelligen, Inc., Mt. Laurel, NJ, USA

Brian Simpson, RES Group, Inc., Cambridge, MA, USA

Utpal K. Singh, Chemical Process R&D, Eli Lilly and Company, Chemical Product R&D, Indianapolis, IN, USA

Kamalesh K. Sirkar, Department of Chemical, Biological and Pharmaceutical Engineering, New Jersey Institute of Technology, Newark, NJ, USA

Omar L. Sprockel, Biopharmaceutics Research and Development, Bristol-Myers Squibb Co., New Brunswick, NJ, USA

Howard J. Stamato, Biopharmaceutics Research and Development, Bristol-Myers Squibb Co., New Brunswick, NJ, USA

Gregory S. Steeno, Pfizer Global Research & Development, Groton, CT, USA

Andrew Stewart, Process Research & Development, Bristol-Myers Squibb Co., New Brunswick, NJ, USA

Yongkui Sun, Department of Process Research, Merck & Co., Inc., Rahway, NJ, USA

Jason T. Sweeney, Process Research & Development, Bristol-Myers Squibb Co., New Brunswick, NJ, USA

Jose E. Tabora, Process Research & Development, Bristol-Myers Squibb Co, New Brunswick, NJ, USA

Michael Paul Thien, Global Science Technology and Commercialization, Merck & Co., Inc., Rahway, NJ, USA

Avinash G. Thombre, Pharmaceutical Development, Pfizer Global Research & Development, Groton, CT, USA

John E. Tolsma, RES Group, Inc., Cambridge, MA, USA

Jean W. Tom, Process Research & Development, Bristol-Myers Squibb Co., New Brunswick, NJ, USA

Gregory M. Troup, Merck & Co., Inc., West Point, PA, USA

Cenk Undey, Process Development, Amgen Inc., West Greenwich, RI, USA

Chenchi Wang, Process Research & Development, Bristol-Myers Squibb Co., New Brunswick, NJ, USA

Martin Warman, Analytical Development, Vertex Pharmaceuticals Inc., Cambridge, MA, USA

Timothy J. Watson, Global CMC, Pfizer, Inc., Groton, CT, USA

James T. Wertman, Chemical Development, GlaxoSmithKline, King of Prussia, PA, USA

Karin Wichmann, COSMOlogic GmbH & Co. KG, Leverkusen, Germany

R. Thomas Williamson, Roche Carolina, Inc., Florence, SC, USA

Xiao Yu (Shirley) Wu, Leslie Dan Faculty of Pharmacy, University of Toronto, Toronto, Ontario, Canada

Dimitrios Zarkadas, Chemical Process Development & Commercialization, Merck & Co., Inc., Union, NJ, USA

Mark Zell, Pfizer Global Research & Development, Pharmaceutical Sciences, Groton, CT, USA

CONVERSION TABLE

Quantity	Equivalent Values
Length	$1\ \text{m} = 100\ \text{cm} = 1000\ \text{mm} = 10^6\ \mu\text{m} = 10^{10}\ \text{Å}$ $= 39.37\ \text{in} = 3.2808\ \text{ft} = 1.0936\ \text{yards}$ $= 0.0006214\ \text{mile}$ $1\ \text{ft} = 12\ \text{in} = 0.3048\ \text{m} = 1/3\ \text{yard} = 30.48\ \text{cm}$
Area	$1\ \text{m}^2 = 10.76\ \text{ft}^2 = 1550\ \text{in}^2 = 10{,}000\ \text{cm}^2$ $1\ \text{in}^2 = 6.4516\ \text{cm}^2 = 645.16\ \text{mm}^2 = 0.00694\ \text{ft}^2$ $1\ \text{ft}^2 = 929.03\ \text{cm}^2 = 0.092903\ \text{m}^2$
Volume	$1\ \text{m}^3 = 1000\ \text{L} = 10^6\ \text{cm}^3\ (\text{mL}) = 1000\ \text{dm}^3$ $= 35.3145\ \text{ft}^3 = 220.83$ imperial gallons $= 264.17$ gal (U.S.) $1\ \text{ft}^3 = 1728\ \text{in}^3 = 7.4805$ U.S. gallons $= 0.028317\ \text{m}^3 = 28.317\ \text{L}$ 1 gal (U.S.) $= 3.785\ \text{L} = 0.1337\ \text{ft}^3 = 231\ \text{in}^3$
Mass	$1\ \text{kg} = 1000\ \text{g} = 0.001$ metric ton (MT) $= 2.20462\ \text{lb}_\text{m} = 35.27392\ \text{oz}$ $1\ \text{lb}_\text{m} = 16\ \text{oz} = 453.593\ \text{g} = 0.453593\ \text{kg}$ $1\ \text{MT} = 1000\ \text{kg} = 2204.6\ \text{lb}_\text{m}$
Pressure	$1\ \text{atm} = 1.01325\ \text{bar} = 1.01325 \times 10^5\ \text{N/m}^2$ (Pa) $= 0.101325\ \text{MPa}$ $= 101.325\ \text{kPa} = 1.01325 \times 10^6\ \text{dynes/cm}^2$ $= 760\ \text{mmHg}$ @ $0\,^\circ\text{C}$ (Torr) $= 10.333\ \text{m}\ H_2O$ @ $4\,^\circ\text{C}$ $= 14.696\ \text{lb}_\text{f}/\text{in}^2$ (psi) $= 33.9\ \text{ft}\ H_2O$ @ $4\,^\circ\text{C}$ $= 2116\ \text{lb}_\text{f}/\text{ft}^2$ $= 29.921$ in Hg @ $0\,^\circ\text{C}$
Temperature	$^\circ\text{C} = \frac{5}{9}(^\circ\text{F} - 32)$ $^\circ\text{F} = (9/5^\circ\text{C}) + 32$ $\text{K} = {}^\circ\text{C} + 273.15 = \frac{5}{9}{}^\circ\text{R}$ $^\circ\text{R} = {}^\circ\text{F} + 459.67$
Density	$1\ \text{g/cm}^3 = \text{kg/L} = 62.4\ \text{lb}_\text{m}/\text{ft}^3$ $1\ \text{lb}_\text{m}/\text{ft}^3 = 16.0185\ \text{kg/m}^3 = 0.01602\ \text{g/cm}^3$
Force	$1\ \text{N} = 1\ \text{kg-m/s}^2 = 10^5\ \text{dynes} = 10^5\ \text{g -cm/s}^2$ $= 0.22481\ \text{lb}_\text{f}$ $1\ \text{lb}_\text{f} = 32.174\ \text{lb}_\text{m}\ \text{ft/s}^2 = 4.4482\ \text{N}$ $= 4.4482 \times 10^5\ \text{dynes}$
Energy	Based on conventional thermochemical definitions of calorie and Btu, $1\ \text{J} = 1\ \text{N m} = 0.23901\ \text{cal} = 10^7\ \text{ergs}$ $= 10^7\ \text{dyne cm}$ $1\ \text{J} = 2.778 \times 10^{-7}\ \text{kW-h} = 0.7376\ \text{ft-lb}_\text{f}$ $= 0.00094845\ \text{Btu}$ $1\ \text{cal} = 4.184\ \text{J}$ (exact) $1\ \text{Btu} = 1054.35\ \text{J} = 1.054\ \text{kJ} = 251.9958\ \text{cal}$ $= 0.2930\ \text{W-h} = 10.406\ \text{L-atm}$ $1\ \text{kW-h} = 3.6\ \text{MJ}$ *Note*: The international steam table (IT) convention defines $(\text{calorie})_\text{IT} = 4.1868\ \text{J}$ and $(\text{Btu})_\text{IT} = 1055.056\ \text{J}$.
Heat generation	$1\ \text{Btu/lb}_\text{m}\text{-h} = 0.64612\ \text{W/kg}$
Heat transfer coefficient	$1\ \text{W/(m}^2\ \text{K)} = 0.1761\ \text{Btu/(h-ft}^2\ {}^\circ\text{F)}$ $1\ \text{Btu/(h-ft}^2\ {}^\circ\text{F)} = 5.678\ \text{W/(m}^2\ \text{K)}$ $= 4.886\ \text{kcal/(h-m}^2\ {}^\circ\text{C)}$
Latent heat	$1\ \text{Btu/lb}_\text{m} = 2.326\ \text{kJ/kg}$ $1\ \text{J/g} = 0.23901\ \text{cal/g}$
Power	$1\ \text{W} = 1\ \text{J/s} = 1\ \text{kg-m}^2/\text{s}^3 = 1\ \text{Nm/s}$ $= 0.23901\ \text{cal/s}$ $= 0.7376\ \text{ft-lb}_\text{f}/\text{s} = 0.0009486\ \text{Btu/s}$ $= 3.414\ \text{Btu/h} = 0.001341\ \text{hp}$
Power/volume	$1\ \text{W/L} = \text{kW/m}^3 = 0.03798\ \text{hp/ft}^3$ $= 96.67\ \text{Btu/h-ft}^3$ $= 12.9235\ \text{Btu/h-gal}$
Specific heat	$1\ \text{kJ/(kg-K)} = \text{J/(g-K)} = 0.2389\ \text{kcal/(kg}\ {}^\circ\text{C)}$ $= 0.2389\ \text{Btu/(lb}_\text{m}\ {}^\circ\text{F)}$ $1\ \text{Btu/(lb}_\text{m}\ {}^\circ\text{F)} = 1\ \text{cal/(g}\ {}^\circ\text{C)}$

(Continued)

Quantity	Equivalent Values
Thermal conductivity	1 Btu/(h-ft °F) = 1.7307 W/(m-K) = 0.00413 cal/(s cm-K) 1 W/(m-K) = 0.5779 Btu/(h-ft °F) = 0.85984 kcal/(h-m °C)
Throughput (continuous @ 365 days/year)	1 year = 365 days = 8760 h = 5.256×10^5 min 1 kg/h = 16.67 g/min = 24 kg/day = 8760 kg/year = 8.76 MT/year 10 MT/year = 10,000 kg/year = 27.4 kg/day = 1.14 kg/h = 19.03 g/min 1 Billion tablets/year = 2.74×10^6 tablets/day = 114,155 tablets/h = 31.7 tablets/s 10 MT API/year = 10,000 kg API/year formulated as a 10 mg dose API/tablet = 1.0 Billion tablets/year

Quantity	Equivalent Values
Diffusivity	1 m^2/s = 10.76 ft^2/s = 38749 ft^2/h 1 ft^2/s = 929.03 cm^2/s = 0.092903 m^2/s
Viscosity	1 centipose (cp) = 0.01 poise = 0.01 g/(cm-s) = 0.001 N-s/m^2 (Pa-s) = 3.6 kg/(m-h) = 0.001 kg/(m-s) = 2.419 lb_m/(ft-h) 1 centistoke (cs) = 1×10^{-6} m^2/s = 0.01 stoke = 0.0036 m^2/h = 0.0388 ft^2/h
Gas constant R	8.31451 J/(mol-K) = 1.987 cal/(mol-K) = 1.987 Btu/(lb-mol °R) 0.0820578 L-atm/(mol-K) = 82.057 atm-cm^3/(mol-K) = 10.73 psi-ft^3/(lb-mol °R)
Gravitational force	$g = 9.8066$ m/s^2 = 32.174 ft/s^2

PART I

INTRODUCTION

1

CHEMICAL ENGINEERING IN THE PHARMACEUTICAL INDUSTRY: AN INTRODUCTION

David J. am Ende
Chemical R&D, Pfizer, Inc., Groton, CT, USA

Although recently several excellent books have been published geared toward process chemistry [1–3] or formulation development in the pharmaceutical industry [4], relatively little has been published specifically with a chemical engineering (ChE) focus. This book, therefore, is about chemical engineering applied to the process research, development, and manufacture of pharmaceuticals. Across the pharmaceutical industry, chemical engineers are employed in R&D through to full-scale manufacturing in technical and management capacities. The following chapters provide an emphasis on the application of chemical engineering science to process development and scale-up for active pharmaceutical ingredients (APIs), drug products (DPs), and biologicals including sections on analytical methods and computational methods. This chapter briefly highlights a few industry facts and figures, in addition to some of the challenges facing the industry, and touches on how ChE can contribute to addressing those challenges. Chapter 2 by Kukura and Thien provides further perspective on the challenges and opportunities in the pharmaceutical industry and the role of chemical engineering.

In general, pharmaceuticals are drug delivery systems in which drug-containing products are designed and manufactured to deliver precise therapeutic responses [5]. The drug is considered the "active," that is, active pharmaceutical ingredient, whereas the formulated final drug is simply referred to as the drug product.

In the United States, federal and state laws exist to control the manufacture and distribution of pharmaceuticals. Specifically, the Food and Drug Administration (FDA) exists by the mandate of the U.S. Congress with the Food, Drug & Cosmetics Act as the principal law to enforce and constitutes the basis of the drug approval process [6]. Specifically in the United States, "The FDA is responsible for protecting the public health by assuring the safety, efficacy, and security of human and veterinary drugs, biological products, medical devices, our nation's food supply, cosmetics, and products that emit radiation. The FDA is also responsible for advancing the public health by helping to speed innovations that make medicines and foods more effective, safer, and more affordable; and helping the public get the accurate, science-based information they need to use medicines and foods to improve their health [7]."

A review of the structure within the FDA and the drug review process can be found in the cited references [8]. In Europe, the European Agency for the Evaluation of Medicinal Products (EMEA) is a decentralized body of the European Union with headquarters in London whose main responsibility is the protection and promotion of public and animal health, through the evaluation and supervision of medicines for human and veterinary use [9].

According to PhRMA statistics, more than 300 new medicines have been approved in the past 10 years that have contributed to increases in life expectancy. For example, since 1980, life expectancy for cancer patients has increased by about 3 years, and 83% of those gains are attributable to new treatments, including medicines. Death rates for cardiovascular disease fell a dramatic 26.4% between 1999 and 2005 [10]. The value of the biopharmaceutical industry to the American economy is substantial. In 2006, the industry employed over 680,000 people with each job indirectly supporting an additional 3.7 jobs. Thus, as an aggregate, the

Chemical Engineering in the Pharmaceutical Industry: R&D to Manufacturing Edited by David J. am Ende

industry supported 3.2 million jobs in 2006 contributing $88.5 billion in 2006 to the nation's gross domestic product [11]. In terms of the total value that the pharmaceutical sector outputs (sum of the direct value of goods produced, indirect value of goods and services that support the sector, and economic activity induced by the direct/indirect employees), it is estimated to be over $635 billion for 2006 [12].

As an industry, global pharmaceutical sales have steadily increased over the past decade and are now approaching an $800 billion industry based on 2009 revenues. Despite the slowing growth rate over the past decade (Figure 1.1), sales are still expected to grow at 4–7% per year to approach $975 billion by 2013 [13]. This is due, in part, from emerging market countries (China, Brazil, Russia, Mexico, India, Turkey, South Korea) where sales are expected to grow by 13–16% annually over the next 5 years (IMS Health). Amid the uncertainty in long-term growth, as an industry sector, the pharmaceutical industry still ranks near the top of most profitable industries with approximately 19% return on revenues according to Fortune 500 rankings [14]. The top 15 pharmaceutical companies are listed in Table 1.1 according to IMS Health.

The top 15 global selling drugs are shown in Table 1.2, with Lipitor/atorvastatin topping the list with 2008 global sales of $13.7 billion. The top 15 drugs total nearly $90 billion and comprise approximately 12% of the global market of $724 billion in 2008. Table 1.3 includes some of the top selling small-molecule APIs, including their formulation type and formulation ingredients.

With patent expirations and fewer blockbusters on the horizon, the pharmaceutical industry is undergoing a transformation in part through consolidation of drug portfolios via

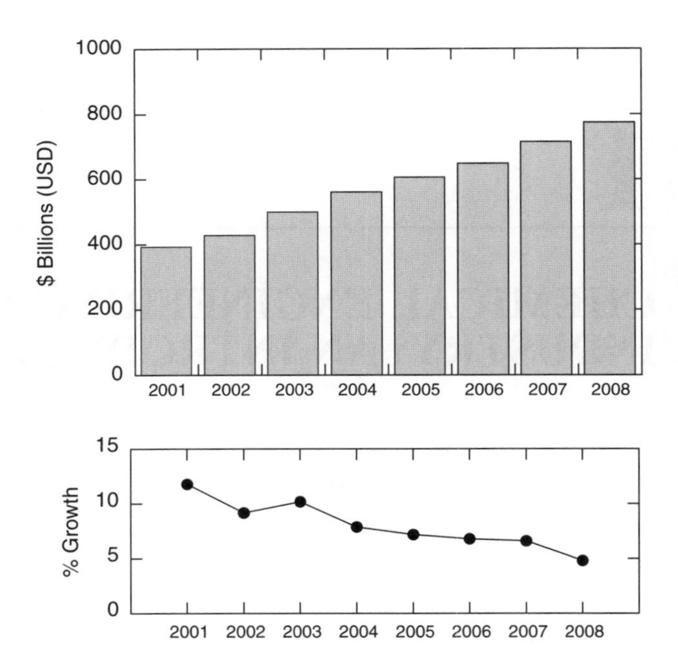

FIGURE 1.1 *Top*: Global pharmaceutical sales with worldwide pharmaceuticals sales approaching $725 billion for year ending 2008. *Bottom*: Declining growth rate based on global sales is defined as percentage change in global sales over the previous year. *Source*: Ref. 15.

mergers and acquisitions. At the time of this writing, further consolidation of the list in Table 1.1 includes Pfizer's acquiring Wyeth and Merck's acquisition of Schering-Plough in 2009. Patent expirations for branded pharmaceuticals create significant financial exposure to the industry. Specifically, products that generated $137 billion in sales face

TABLE 1.1 Top 15 Pharmaceutical Corporations in 2008 as Listed by IMS Health[15]

	2008 Rank (US$)	2008 Sales (US$ million)	2007 Sales (US$ million)	2006 Sales (US$ million)	2005 Sales (US$ million)	2004 Sales (US$ million)
Global market	0	724,465	673,043	612,013	572,659	530,909
Pfizer	1	43,363	44,651	45,622	45,869	49,401
GlaxoSmithKline	2	36,506	37,951	37,516	32,256	33,231
Novartis	3	36,172	34,409	31,560	29,616	26,404
Sanofi-Aventis	4	35,642	33,819	31,460	30,953	28,446
AstraZeneca	5	32,516	30,107	27,540	24,741	22,526
Roche	6	30,336	27,578	23,354	20,105	16,787
Johnson & Johnson	7	29,425	29,092	27,730	27,190	26,919
Merck & Co.	8	29,191	27,294	25,174	23,872	24,334
Abbott	9	19,466	17,587	16,065	14,849	13,310
Lilly	10	19,140	17,386	15,388	14,232	13,042
Amgen	11	15,794	16,536	16,270	13,435	10,944
Wyeth	12	15,682	15,965	14,695	14,469	14,019
Teva	13	15,274	13,547	12,001	10,053	8,675
Bayer	14	15,660	14,178	12,553	11,828	11,019
Takeda	15	13,819	12,778	11,880	11,370	10,707

Source: Ref. 15.

TABLE 1.2 Top 15 Global Pharmaceutical Products (in 2008)

Rank	Brand Name	Compound	Marketer	Indication	2008 Sales ($ Billion)
1	Lipitor	Atorvastatin	Pfizer	Hypercholesterolemia	13.655
2	Plavix	Clopidogrel	Bristol-Myers Squibb	Atherosclerotic events	8.634
3	Nexium	Esomeprazole	AstraZeneca	Acid reflux disease	7.842
4	Seretide/ Advair	Fluticasone and salmeterol	GlaxoSmithKline	Asthma	7.703
5	Enbrel	Etanercept	Amgen and Wyeth	Rheumatoid arthritis	5.703
6	Seroquel	Quetiapine	AstraZeneca	Bipolar, schizophrenia	5.404
7	Zyprexa	Olanzapine	Eli Lilly & Co.	Schizophrenia	5.023
8	Remicade	Infliximab	Centocor	Crohn's disease, rheumatoid arthritis	4.935
9	Singulair	Montelukast	Merck & Co.	Asthma, allergies	4.673
10	Lovenox	Enoxaparin	Sanofi-Aventis	Anticoagulant	4.435
11	MabThera	Rituximab	Roche	Lymphoma	4.321
12	Takepron/ Prevacid	Lansoprazole	Takeda	Antiulcer/gastric proton pump inhibitor	4.321
13	Effexor	Venlafaxine	Wyeth	Depression	4.263
14	Humira	Adalimumab	Abbott	Rheumatoid arthritis, Crohn's disease	4.075
15	Avastin	Bevacizumab	Genentech/Roche	Metastatic cancers	4.016

Source: Refs 13 and 15. Global sales figures are listed in US$ for 2008.

generic competition from 2009 to 2013 according to IMS Health [15], which represents approximately 17% of current global pharmaceutical sales. In addition, the United States is in the midst of U.S. health care reform (2010). It remains unclear whether the higher volume of prescription drugs that the program intends to ultimately provide coverage for, to the newly insured, will offset the lower price demands and how this will impact the industry as a whole.

Companies in general are broadly looking for ways to reduce costs to offset the exposure of patent expirations, rising generic competition, and current market pressures. The cost of advancing candidates and entire pharmaceutical portfolios in R&D is significant. In 2001, the average cost for an approved medicine was estimated to be $802 million as reported by Tufts Center for the Study of Drug Development. In 2008, the cost of advancing a drug through clinical trials and through FDA approval was estimated to range from $1 billion to $3.5 billion in 2008 dollars [16]. Although these figures clearly depend on the drug type, therapeutic area, and speed of development, the bottom line is that the upfront investments required to reach the market are massive especially when considering the uncertainty whether the upfront investment will pay back.

Given there might be 10 or more years of R&D costs without any revenue generated on a new drug, the gross margins of a successful drug need to cover prior R&D investments as well as cover the continuing marketing and production costs. Figure 1.2 shows the classic cash flow profile for a new drug developed and marketed. First, there is a period of negative cash flow during development. When the drug is approved and launched, only then are revenues generated, and the drug has to be priced high enough to recoup the investment and provide a return on the investment. The net present value (NPV) calculation is one way to assess return on investment; it considers the discounted revenue minus the discounted costs and is computed over the product development and marketing life cycle. These calculations are used to rationalize investment decisions. For example, a minimum threshold product price can be computed for which the NPV calculation hits a desired return on investment target. If this price is sustained by the market, then the investment can be considered viable. A discount rate of 10–12% is generally chosen in the pharmaceutical industry as the rate to which to value products or programs for investment decisions [17]. Patents typically have a validity of 20 years from the earliest application grant date based on applications filed after 1995, so it is in the company's best interest to ensure that the best patent protection strategy is in place to maximize the length of market exclusivity. Related to this is that patents and intellectual property in general need enforcement on a global basis to ensure fair competition and realize benefit in growth into emerging markets.

In some cases, time of market exclusivity can be extended through new indications, new formulations, devices, and so on, which are themselves patent protected. Once market exclusivity ends, generic competition is introduced, which will erode sales. It should be noted that independent of patent position or patent exclusivity, the FDA grants new drug product exclusivity (also known as Hatchman–Wax exclusivity) with specific periods of exclusivity. For example, the

TABLE 1.3 Top Selling Marketed Small-Molecule APIs and Dosage Formulations

Structure/Name/Company (2008 Sales)	Molecular Weight	Properties	Dosage Form	Formulation
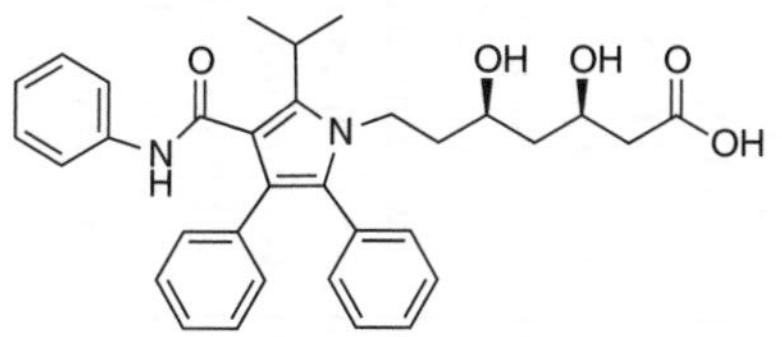 Lipitor / Atorvastatin Pfizer $12.4 Billion in sales in 2008	*Atorvastatin* Free acid: MW 558.64 Sodium salt: MW 580.62 Calcium salt trihydrate: $(C_{33}H_{34}FN_2O_5)_2Ca{\cdot}3H_2O$, MW 1209.39 *For the treatment of cardiovascular disease*	Calcium salt trihydrate: white to off-white crystalline powder. Freely soluble in methanol, slightly soluble in ethanol, very slightly soluble in acetonitrile, distilled water, and phosphate buffer (pH 7.4), and insoluble in aqueous solutions of pH 4 and below.	*Tablets* 10, 20, 40, or 80 mg	Lipitor tablets for oral administration contain atorvastatin calcium and the following inactives: calcium carbonate, USP; candelilla wax, FCC; croscarmellose sodium, NF; hydroxypropyl cellulose, NF; lactose monohydrate, NF; magnesium stearate, NF; microcrystalline cellulose, NF; Opadry White YS-1-7040 (hypromellose, polyethylene glycol, talc, titanium dioxide); polysorbate 80, NF; simethicone emulsion.
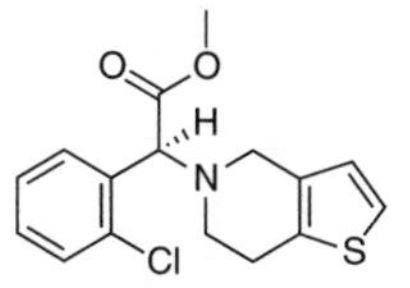 Plavix / Clopidogrel Bristol-Myers Squibb/Sanofi-Aventis $4.9 Billion BMS + €2.6 Billion S-A	*Clopidogrel* Free base: $C_{16}H_{16}ClNO_2S$, MW 321.82 Hydrogen sulfate: $C_{16}H_{16}ClNO_2S{\cdot}H_2SO_4$, MW 419.9 *Proton pump inhibitor; for the treatment of acid reflux disease*	Clopidogrel bisulfate is a white to off-white powder. It is practically insoluble in water at neutral pH but freely soluble at pH 1. It also dissolves freely in methanol, dissolves sparingly in methylene chloride, and is practically insoluble in ethyl ether.	*Tablets* 75 and 300 mg	Each tablet contains clopidogrel bisulfate and the following inactives: hydrogenated castor oil, hydroxypropylcellulose, mannitol, microcrystalline cellulose, and polyethylene glycol 6000. The pink film coating contains ferric oxide, hypromellose 2910, lactose monohydrate, titanium dioxide, and triacetin. The tablets are polished with carnauba wax.

Structure	Drug	Properties	Dosage forms	Formulation
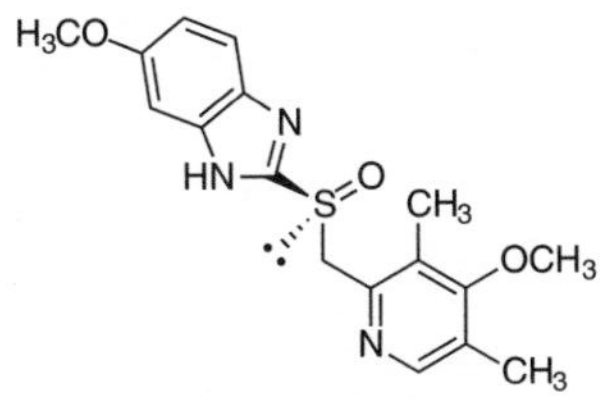 Nexium / Esomeprazole AstraZeneca $5.2 Billion in sales	*Esomeprazole* Free base: $C_{17}H_{19}N_3O_3S$, MW 345.42 Magnesium salt: $C_{34}H_{36}MgN_6O_6S_2$, MW 713.12 Magnesium salt trihydrate: $(C_{17}H_{18}N_3O_3S)_2Mg{\cdot}3H_2O$, MW 767.2 *For the treatment of acid reflux disease*	Esomeprazole magnesium trihydrate is a white to slightly colored crystalline powder. The solubility in water is 0.3 mg/mL, and the solubility in methanol is initially high, but followed by precipitation. The pK_a of the benzimidazole (omeprazole base) is 8.8, and that of the pyridinium ion is 4.0.	*Capsules (delayed release)* 20 and 40 mg *Sachet* 10 mg	Each delayed release capsule contains esomeprazole magnesium trihydrate in the form of enteric-coated granules with the following inactive ingredients: glyceryl monostearate 40–55, hydroxypropyl cellulose, hypromellose, magnesium stearate, methacrylic acid copolymer type C, polysorbate 80, sugar spheres, talc, and triethyl citrate. The capsule shells have the following inactive ingredients: gelatin, FD&C Blue #1, FD&C Red #40, D&C Red #28, titanium dioxide, shellac, ethyl alcohol, isopropyl alcohol, *n*-butyl alcohol, propylene glycol, sodium hydroxide, polyvinyl pyrrolidone, and D&C Yellow #10.
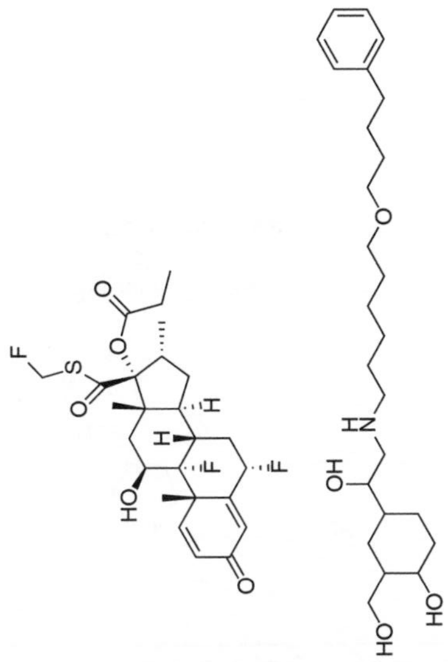 Advair (US)/Seretide(EU) fluticasone + salmeterol GlaxoSmithKline £4.137 Billion in sales	*Fluticasone propionate* $C_{25}H_{31}F_3O_5S$, MW 500.6 *Salmeterol xinafoate* $C_{25}H_{37}NO_4{\cdot}C_{11}H_8O_3$, MW 603.8 *For the treatment of asthma and chronic obstructive pulmonary disease*	Salmeterol xinafoate is white to off-white crystalline powder with a melting point ~123°C. *Solubility*: In water ~0.07 mg/mL (pH = 8) In methanol ~40 mg/mL In ethanol ~7 mg/mL In chloroform ~3 mg/mL In isopropanol ~2 mg/mL Fluticasone propionate is white to off-white powder. It is freely soluble in DMSO and DMF, sparingly soluble in acetone, dichloromethane, ethyl acetate, and chloroform, slightly soluble in methanol and 95% ethanol, and practically insoluble in water. Fluticasone propionate decomposes without melting.	*Dry powder inhaler* 50 µg salmeterol with 100, 250, or 500 µg fluticasone propionate/blister *Aerosol* 25 µg salmeterol with 50, 125, or 250 µg fluticasone propionate/ metered dose	*Dry powder inhaler* device containing a foil strip with 28 or 60 regularly placed blisters, each containing salmeterol (as the xinafoate salt) and fluticasone propionate. The inactives include lactose (milk sugar) and milk protein, which acts as the "carrier." *Aerosol* comprises a suspension of salmeterol and fluticasone propionate in the propellant HFA-134a (1,1,1,2-tetrafluoroethane). It contains no excipients.

(*continued*)

TABLE 1.3 *(Continued)*

Structure/Name/Company (2008 Sales)	Molecular Weight	Properties	Dosage Form	Formulation
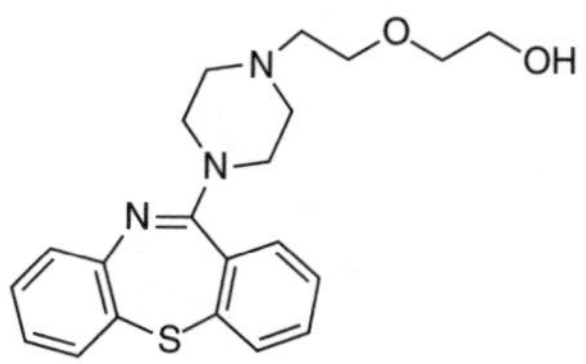 Seroquel/Quetiapine AstraZeneca $4.452 Billion in sales	*Quetiapine* $C_{21}H_{25}N_3O_2S$, MW 383.51 *Quetiapine fumarate* $(C_{21}H_{25}N_3O_2S)_2{\cdot}C_4H_4O_4$, MW 883.1 *For the treatment of schizophrenia and bipolar disorder*	Quetiapine fumarate is a white to off-white powder. It is only very slightly soluble in ether, slightly soluble in water, and soluble in 0.1 N HCl. Ionization constant: $pK_{a1} = 6.83$ in phosphate buffer at 22°C; $pK_{a2} = 3.32$ in formic buffer at 22°C. Partition coefficient: $\log P = 0.45$ (octanol/water). Melting point: 172.0–174°C.	*Immediate release tablet* 25, 100, 200, and 300 mg	Seroquel is available in four strengths containing 25, 100, 200, or 300 mg quetiapine per tablet (as quetiapine fumarate). The core of the tablet contains the following excipients: calcium hydrogen phosphate, lactose monohydrate, magnesium stearate, microcrystalline cellulose, povidone, and sodium starch glycolate type A. The coating of the tablet contains hydroxypropyl methylcellulose 2910, polyethylene glycol 400, red ferric oxide (25 mg tablets), titanium dioxide, and yellow ferric oxide (25 and 100 mg tablets).
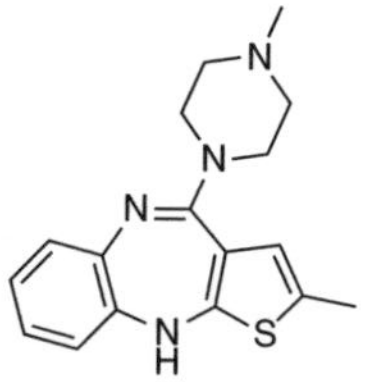 Zyprexa/olanzapine Eli Lilly & Co. $ 4.7 Billion in Sales	*Olanzapine* $C_{17}H_{20}N_4S$, MW 312.43 *For the treatment of schizophrenia and bipolar disorder*	Crystals from acetonitrile, mp 195°C. Practically insoluble in water.	*Tablets* 2.5, 5, 7.5, 10, 15, and 20 mg *Orally disintegrating tablets* zyprexa zydis 5, 10, 15, and 20 mg *Intramuscular injection* 10 mg vial	*Tablets*: inactive ingredients are carnauba wax, crospovidone, hydroxypropyl cellulose, hypromellose, lactose, magnesium stearate, microcrystalline cellulose, and other inactive ingredients. The color coating contains titanium dioxide (all strengths), FD&C Blue No. 2 aluminum lake (15 mg), or synthetic red iron oxide (20 mg). The 2.5, 5, 7.5, and 10 mg tablets are imprinted with edible ink that contains FD&C Blue No. 2 aluminum lake.

				Oral disintegrating tablets also contain the following inactives: gelatin, mannitol, aspartame, sodium methyl paraben, and sodium propyl paraben.
Singulair/montelukast Merck & Co. $4.337 Billion in Sales	*Montelukast* Free acid: $C_{35}H_{36}ClNO_3S$, MW 586.18 Montelukast sodium: $C_{35}H_{35}ClNNaO_3S$, MW 608.18 *For the treatment of asthma*	Montelukast sodium is a hygroscopic, optically active, white to off-white powder. Montelukast sodium is freely soluble in ethanol, methanol, and water and practically insoluble in acetonitrile.	*Tablet* 10 mg *Chewable tablets* 4 and 5 mg *Granules* 4 mg	Each 10 mg film-coated Singulair tablet contains montelukast sodium and the following inactive ingredients: microcrystalline cellulose, lactose monohydrate, croscarmellose sodium, hydroxypropyl cellulose, and magnesium stearate. The film coating consists of hydroxypropyl methylcellulose, hydroxypropyl cellulose, titanium dioxide, red ferric oxide, yellow ferric oxide, and carnauba wax. Each 4 and 5 mg chewable Singulair tablet contains montelukast sodium, with the following inactive ingredients: mannitol, microcrystalline cellulose, hydroxypropyl cellulose, red ferric oxide, croscarmellose sodium, cherry flavor, aspartame, and magnesium stearate.

(*continued*)

TABLE 1.3 *(Continued)*

Structure/Name/Company (2008 Sales)	Molecular Weight	Properties	Dosage Form	Formulation
Prevacid/Lansoprazole Takeda / Abbott $4.321 Billion in Sales (IMS Health)	*Lansoprazole* $C_{16}H_{14}F_3N_3O_2S$, MW 369.36 *For the treatment for peptic ulcer*	Lansoprazole is a white to brownish-white odorless, crystalline powder that melts with decomposition at approximately 166°C. Lansoprazole is freely soluble in DMF, slightly soluble in methanol, sparingly soluble in ethanol, slightly soluble in ethyl acetate, dichloromethane, and acetonitrile, very slightly soluble in ether, and practically insoluble in water and hexane. Octanol/water partition coefficient = 240 at pH 7.	*Capsules* Delayed release capsules contain enteric-coated granules and are available in two dosage strengths: 15 and 30 mg of lansoprazole per capsule. *Oral suspension sachets*	In addition to lansoprazole, each delayed release capsule contains the following inactive ingredients: cellulosic polymers, colloidal silicon dioxide, D&C Red No. 28, FD&C Blue No. 1, FD&C Green No. 3 (15 mg capsules only), FD&C Red No. 40, gelatin, magnesium carbonate, methacrylic acid copolymer, starch, talc, sugar spheres, sucrose, polyethylene glycol, polysorbate 80, and titanium dioxide. Oral suspension sachets include lansoprazole granules and inactive granules composed of the following ingredients: confectioner's sugar, mannitol, docusate sodium, ferric oxide, colloidal silicon dioxide, xanthan gum, crospovidone, citric acid, sodium citrate, magnesium stearate, and artificial strawberry flavor.
Effexor/venlafaxine Wyeth $3.928 Billion in sales	*Venlafaxine* $C_{17}H_{27}NO_2$, MW 277.40 Venlafaxine hydrochloride: $C_{17}H_{27}NO_2 \cdot HCl$, MW 313.86 *Antidepressant*	Venlafaxine HCl: white to off-white crystalline solid. *Solubility*: Water: 540, 542, 501, and 21.6 mg/mL at pH 1.0, 5.38, 7.09 and 7.97 Ethanol: 91.7 mg/mL Propylene glycol: 200 mg/mL Glycerin: 115 mg/mL pK_a value: 9.4	*Capsules* Effexor XR Hard gelatin capsule 37.5, 75, and 150 mg	*Composition*: venlafaxine hydrochloride, ethylcellulose, NF; gelatin, NF; hydroxypropyl methylcellulose, USP; iron oxide, NF; microcrystalline cellulose, NF 60; titanium dioxide, USP; White Tek SB-0007 and/or Opacode Red S-1-15034 ink; talc, USP.

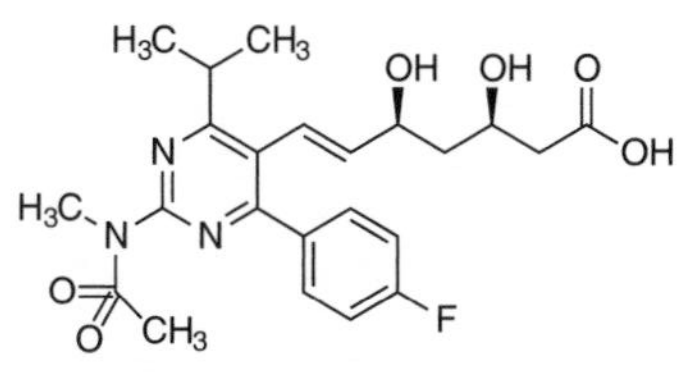 Crestor/Rosuvastatin AstraZeneca $3.597 Billion in sales	*Rosuvastatin* $C_{22}H_{28}FN_3O_6S$, MW 481.54 Rosuvastatin calcium salt: $(C_{22}H_{27}FN_3O_6S)_2Ca$, MW 1001.14 *For the treatment of high cholesterol*	Rosuvastatin calcium salt: white powder from water as the monohydrate; begins to melt at 155°C with no definitive melting point. Sparingly soluble in water, methanol and slightly soluble in ethanol.	*Tablets* 5, 10, 20, and 40 mg	*Composition*: each tablet contains 5, 10, 20, or 40 mg of rosuvastatin as rosuvastatin calcium. Each tablet also contains the following nonmedicinal ingredients: calcium phosphate, crospovidone, glycerol triacetate, hydroxypropyl methylcellulose, lactose monohydrate, microcrystalline cellulose, magnesium stearate, ferric oxide red, ferric oxide yellow, and titanium dioxide.
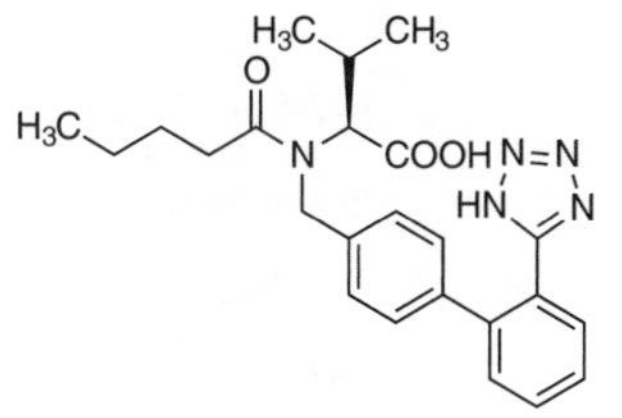 Diovan / valsartan Novartis $5.74 Billion in Sales	*Valsartan* $C_{24}H_{29}N_5O_3$, MW 435.52 *For the treatment of hypertension*	Valsartan is a white to practically white fine powder. It is soluble in ethanol and methanol and slightly soluble in water Crystals from diisopropyl ether, mp 116–117°C. Partition coefficient (*n*-octanol/aqueous phosphate buffer): 0.033. Soluble in water at 25°C.	*Tablets* 40, 80, 160, or 320 mg	The inactive ingredients of the tablets are colloidal silicon dioxide, crospovidone, hydroxypropyl methylcellulose, iron oxides (yellow, black, and/or red), magnesium stearate, microcrystalline cellulose, polyethylene glycol 8000, and titanium dioxide.
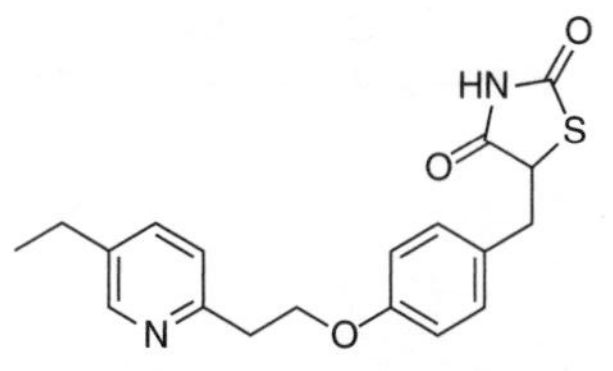 Actos / Pioglitazone Takeda Pharmaceuticals $3.87 Billion in sales	*Pioglitazone* $C_{19}H_{20}N_2O_3S$, MW 356.44 Pioglitazone hydrochloride: $C_{19}H_{20}N_2O_3S{\cdot}HCl$, MW 392.90 *For the treatment of diabetes mellitus type 2*	Pioglitazone hydrochloride: colorless prisms from ethanol, mp 193–194°C. Soluble in DMF, slightly soluble in ethanol, very slightly soluble in acetone and acetonitrile, practically insoluble in water, and insoluble in ether.	*Tablets* 15, 30, and 45 mg	Actos is available as a tablet for oral administration containing 15, 30, or 45 mg of pioglitazone (as the base) formulated with the following excipients: lactose monohydrate, NF; hydroxypropylcellulose, NF; carboxymethylcellulose calcium, NF; magnesium stearate, NF.

(*continued*)

TABLE 1.3 *(Continued)*

Structure/Name/Company (2008 Sales)	Molecular Weight	Properties	Dosage Form	Formulation
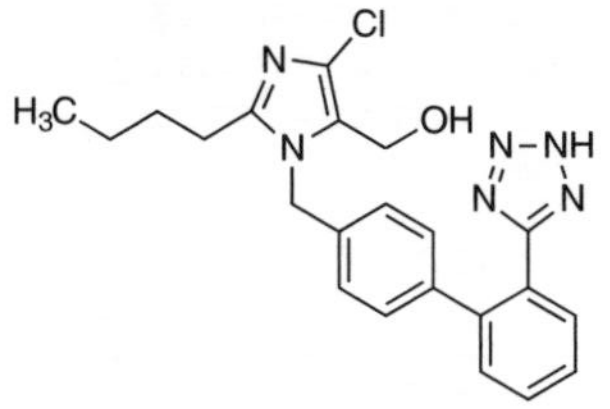 Cozaar (losartan potassium) Merck & Co. $3.558 Billion in sales	*Losartan* $C_{22}H_{23}ClN_6O$, MW 422.91 Losartan potassium: $C_{22}H_{22}ClKN_6O$, MW 461.00 *For the treatment of hypertension*	Losartan potassium is a white to off-white free-flowing crystalline powder. It is freely soluble in water, soluble in alcohols, and slightly soluble in common organic solvents, such as acetonitrile and methyl ethyl ketone.	*Tablets* 25, 50, or 100 mg	Cozaar contains losartan potassium and the following inactive ingredients: microcrystalline cellulose, lactose hydrous, pregelatinized starch, magnesium stearate, hydroxypropyl cellulose, hypromellose, titanium dioxide, D&C Yellow No. 10 aluminum lake, and FD&C Blue No. 2 aluminum lake.
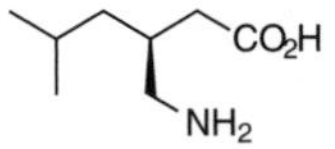 Lyrica / pregabalin Pfizer $2.6 Billion in sales	*Pregabalin* $C_8H_{17}NO_2$, MW 159.23 *For the treatment of neurologic pain*	Pregabalin is a white crystalline solid. It is soluble in water and in both basic and acidic aqueous solutions.	*Capsules* 25, 50, 75, 100, 150, 200, 225, and 300 mg	Each capsule of Lyrica contains pregabalin, lactose monohydrate, maize starch, and talc. The capsule shells contain gelatin and titanium dioxide. In addition, the orange capsule shells contain red iron oxide and the white capsule shells contain sodium lauryl sulfate and colloidal silicon dioxide.

Cymbalta/duloxetine HCl Eli Lilly & Co. $2.697 Billion in Sales	*Duloxetine* Free base: $C_{18}H_{19}NOS$, MW 297.41 Hydrochloride: $C_{18}H_{19}NOS{\cdot}HCl$, MW 333.88 *Antidepressant and analgesic*	Duloxetine hydrochloride is white to slightly brownish white solid. pK_a in DMF–water (66 : 34): 9.6. Slightly soluble in water. Freely soluble in methanol.	*Capsules* 30 and 60 mg	Each capsule contains enteric-coated pellets of duloxetine hydrochloride equivalent to 30 or 60 mg of duloxetine that are designed to prevent degradation of the drug in the acidic environment of the stomach. Nonmedicinal ingredients include FD&C Blue No.2, gelatin, hydroxypropyl methylcellulose, hydroxypropyl methylcellulose acetate succinate, sodium lauryl sulfate, sucrose, sugar spheres, talc, titanium dioxide, and triethyl citrate. The 60 mg capsules also contain yellow iron oxide.
Topamax /Topiramate Johnson & Johnson $2.731 Billion in Sales	*Topiramate* $C_{12}H_{21}NO_8S$, MW 339.36 *Anticonvulsant; antimigraine*	Topiramate is a white crystalline powder; bitter taste. Crystals from ethyl acetate + hexane, mp 125–126°C. Most soluble in alkaline solutions containing NaOH or sodium phosphate, pH 9–10. Freely soluble in acetone, chloroform, DMSO, and ethanol. Solubility in water: 9.8 mg/mL.	*Tablets* 25, 50, 100, and 200 mg *Capsules* 15 and 25 mg	Topiramate tablets contain the following inactive ingredients: lactose monohydrate, pregelatinized starch, microcrystalline cellulose, sodium starch glycolate, magnesium stearate, purified water, carnauba wax, hypromellose, titanium dioxide, polyethylene glycol, synthetic iron oxide (25, 50, 100, and 200 mg tablets) and polysorbate 80. (Topiramate capsules) Sprinkle capsules contain topiramate-coated beads in a hard gelatin capsule. The inactive ingredients are sugar spheres (sucrose and starch), povidone, cellulose acetate, gelatin, sorbitan monolaurate, sodium lauryl sulfate, titanium dioxide, and black pharmaceutical ink.

(*continued*)

TABLE 1.3 *(Continued)*

Structure/Name/Company (2008 Sales)	Molecular Weight	Properties	Dosage Form	Formulation
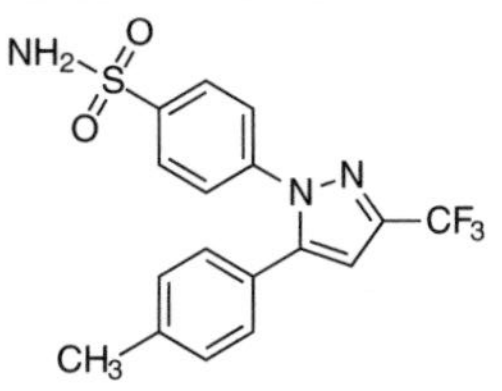 Celebrex/Celecoxib Pfizer, Inc $2.489 Billion in Sales	*Celecoxib* $C_{17}H_{14}F_3N_3O_2S$, MW 381.38 *Nonsteroidal anti-inflammatory drug (NSAID)*	Celecoxib is a white powder. mp 160–164°C. Celecoxib is a neutral molecule at physiologic pH. Celecoxib is "practically insoluble" in water (with an *n*-octanol/water partition coefficient of 10,000 at pH 7.0). Celecoxib is weakly acidic with a pK_a of 11.1.	*Capsules* 100 and 200 mg	The inactive ingredients in Celebrex 100 and 200 mg capsules include croscarmellose sodium, lactose monohydrate, magnesium stearate, povidone, and sodium lauryl sulfate. The capsules are made of gelatin and contain titanium dioxide (E171) and edible inks (ferric oxide (E172) for 200 mg capsules and indigotine (E132) for 100 mg capsules).
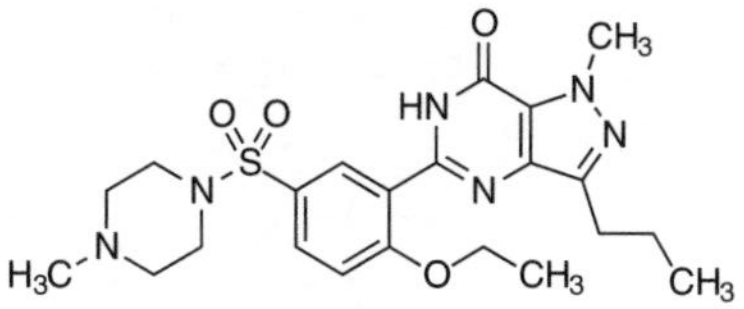 Viagra/Sildenafil Pfizer, Inc $1.934 Billion in Sales	*Sildenafil* Free base: $C_{22}H_{30}N_6O_4S$, MW 474.58 Citrate salt: $C_{22}H_{30}N_6O_4S{\cdot}C_6H_8O_7$, MW 666.70 *For the treatment of erectile dysfunction*	Sildenafil citrate is a white to off-white crystalline powder with a solubility of 3.5 mg/mL in water at 23°C, mp 187–189°C. pK_a (protonation of tertiary amine): 6.53. pK_a (deprotonation of pyrimidinone moiety): 9.17. Solubility at 23°C in 1 M HCl is 5.8 mg/mL and in 1 M NaOH is 42.3 mg/mL.	*Tablets* 25, 50, and 100 mg	Sildenafil citrate is formulated as blue, film-coated rounded-diamond-shaped tablets equivalent to 25, 50, and 100 mg of sildenafil for oral administration. In addition to the active ingredient, sildenafil citrate, each tablet contains the following inactive ingredients: microcrystalline cellulose, anhydrous dibasic calcium phosphate, croscarmellose sodium, magnesium stearate, hypromellose, titanium dioxide, lactose, triacetin, and FD&C Blue #2 aluminum lake.

Source: Merck Index (14th edition), Physicians Desk Reference, individual product monographs, and 2008 Annual Reports.

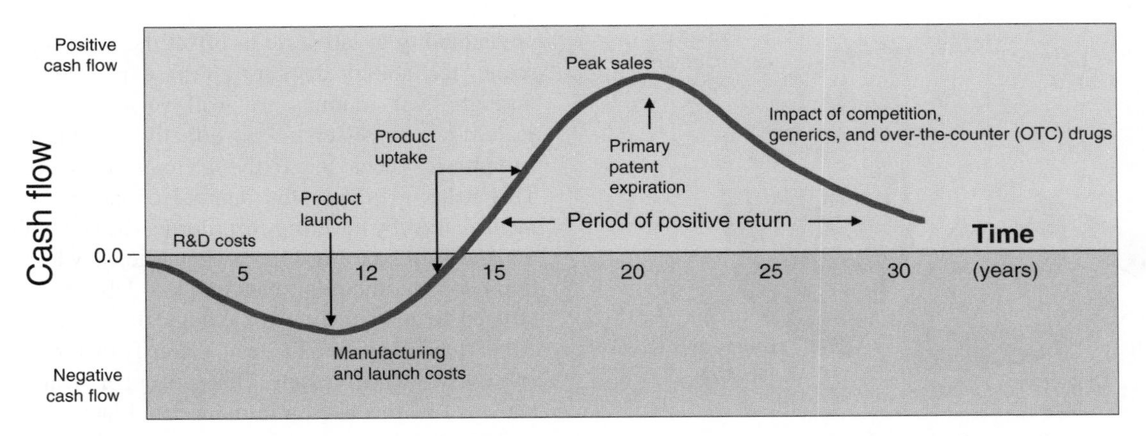

FIGURE 1.2 A hypothetical cash flow curve for a pharmaceutical product includes 10–15 years of negative cash flows of typically $1–3 billion. Reasonably high margins are needed, once the drug is on the market, if it is to recoup and provide a positive return on investment (ROI) over its life cycle.

following key points are quoted from the FDA on the subject of new drug product exclusivity.

> "Exclusivity provides the holder of an approved new drug application limited protection from new competition in the marketplace for the innovation represented by its approved drug product." "A 5-year period of exclusivity is granted to new drug applications for products containing chemical entities never previously approved by FDA either alone or in combination." "A 3-year period of exclusivity is granted for a drug product that contains an active moiety that has been previously approved, when the application contains reports of new clinical investigations (other than bioavailability studies) conducted or sponsored by the sponsor that were essential to approval of the application. For example, the changes in an approved drug product that affect its active ingredient(s), strength, dosage form, route of administration or conditions of use may be granted exclusivity if clinical investigations were essential to approval of the application containing those changes." "The Center for Drug Research and Evaluation (CDER) makes exclusivity determinations on all relevant applications. There is a procedure in CDER that provides review of all relevant applications, with or without a request from the applicant, for an exclusivity determination [18]."

The pharmaceutical industry invested approximately $60 billion into R&D in 2007. It now takes 10–15 years for a new medicine to go from the laboratory to the pharmacy. Figure 1.3 shows the typical development activity timeline from discovery to launch. From thousands of compounds evaluated for potential therapeutic effect, very few will clear all the safety, efficacy, and clinical hurdles to make it to approval. Figure 1.3 also shows how a general range of volunteers, and therefore clinical supplies, increases for clinical development through phases I to II with clinical development lasting 6–7 years. The cost of product development that includes the cost to manufacture clinical supplies is estimated to be in the range of 30–35% of the total

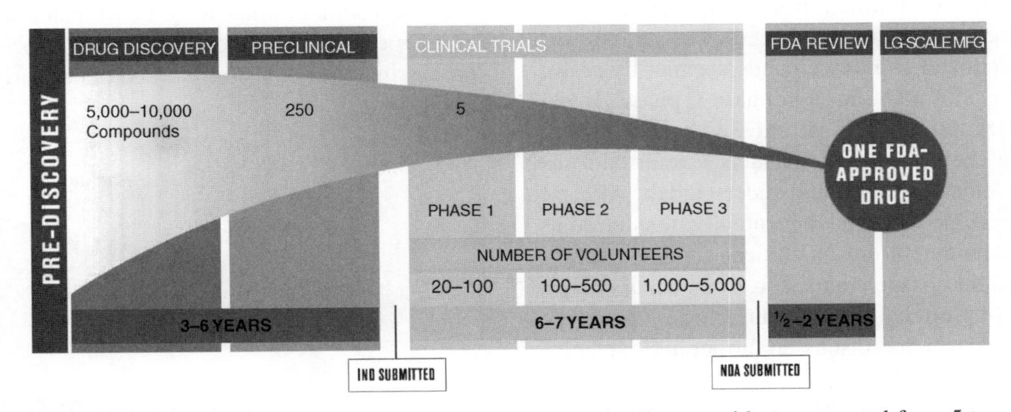

FIGURE 1.3 Drug research and development can take 10–15 years with one approval from 5 to 10,000 compounds in discovery. IND: investigational new drug; NDA: new drug application. *Source*: *Pharmaceutical Industry Profile 2009*, Pharmaceutical Research and Manufacturers of America (PhRMA) (www.phrma.org).

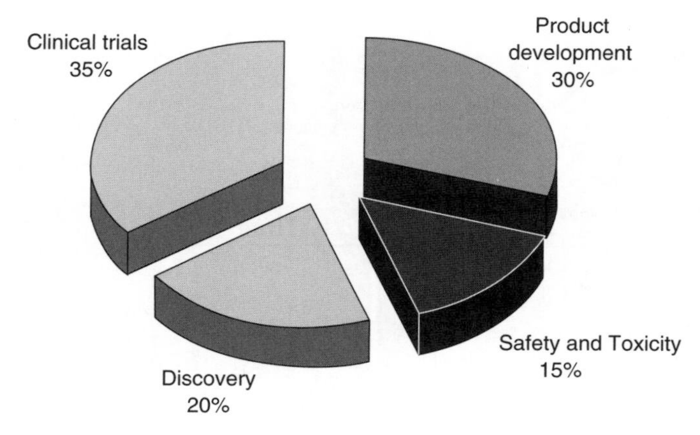

FIGURE 1.4 Estimated distribution of product development costs within R&D with the total cost to bring a new chemical entity to market in the range of $1–3.5 billion. *Source*: Ref. 16.

cost of bringing a new chemical entity to market with the following breakdown of development costs: discovery 20–25%, safety and toxicity 15–20%, product development 30–35%, and clinical trials 35–40%. The distribution is graphically displayed in Figure 1.4. Clearly, the distribution will depend on the specific drug, its therapeutic area, dose, and specific company [16].

1.1 PHARMACEUTICAL DEVELOPMENT

In general, pharmaceutical product development is different from most other research-intensive industries. Specifically in the pharmaceutical industry, there is a consistent need to ensure that clinical supplies are manufactured and delivered in a timely manner regardless of the current state of development or efficiency of the process. In other words, delivering clinical supplies when they are needed requires using technology that is good enough at the time even if it is not a fully optimized process. Further, process development, optimization, and scale-up historically tend to be an iterative approach [19]—clinical supply demands are met by scale-ups to kilo lab or pilot plant through phase I, phase II, and phase III and it is through this period R&D development teams (including chemist/engineers/analysts, referred to as the ACE model) refine, optimize, and understand the API and DP processes to enable them to be eventually transferred to manufacturing. Manufacturing of clinical supplies in kilo lab, pilot plant, solid dosage plants, and so on occurs under the constraints of current good manufacturing practices (cGMP) conditions, which is discussed further in Chapter 22 by Hamm et al. The pilot plant and kilo lab are also sometimes used to "test" the scalability of a process. In this way, they serve a dual purpose, which make them unique compared to nonpharmaceutical pilot plants. In terms of cost, however, large-scale experimentation in kilo lab or pilot plant can be significant, so there has been a shift toward greater predictability at lab scale to offset the need for pilot plant-scale "technology demonstration" experiments. Engineers through their training are well versed with scale-up or scale-down processes and can effectively model the chemical and physical behaviors in the lab to ensure success on scale. This helps to reduce the number of larger scale "experiments," thereby lowering costs during R&D. In this way, with the recent trend toward increasing efficiency and continuous improvement, the pilot plant and kilo labs are preferentially utilized to manufacture toxicological and clinical supplies rather than being used to "test" or verify that the chemistry or process will work on scale. This concept of "lean manufacturing" will be touched on in more detail later in this chapter.

The aim of process development is to drive down the cost contribution of the API to the final formulated pharmaceutical product cost, while at the same time optimizing process robustness. The impact of API costs on overall manufacturing costs is approximated in Figure 1.5. The cost contribution of API is expected to increase with increasing complexity of molecular structures of APIs. It is interesting to note that API molecular complexity can often impact API cost more than formulation or packaging costs. Federsel points out that, "Given the importance of 'time to market' which remains one of the highest priorities of pharmaceutical companies, the need to meet increasingly stretched targets for speed to best route has come to the forefront in process R&D" [20]. Recently, it was considered satisfactory to have a good enough synthetic route that was fit for purpose (i.e., could support the quantities of material needed) but not one having best or lowest cost ($/kg of API). The prevailing view was that the market would bear higher product pricing as compensation for higher cost of goods. Further cost reduction through new routes could be and was pursued post-launch with savings realized later in the life cycle. According to Federsel, and evidenced frequently in contemporary R&D organizations, this approach is no longer viable, at least not as a default position. Instead, the best synthetic route to API (i.e., route with ultimate lowest cost materials) coupled with best process design and engineering (process with lowest processing costs) must be worked out as early as possible in

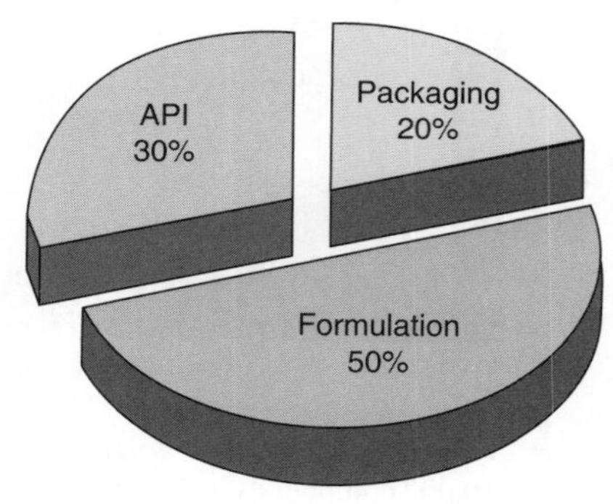

FIGURE 1.5 Average COG components in final dosage form across a large product portfolio, but for individual drugs this could vary widely (e.g., for API from 5% to 40%). *Source*: Ref. 19.

API process development [21]. The best API process developed by the time of launch is necessary to extract additional revenues and respond to reduced cost of goods margins. Achieving this requires continuous improvement in scientific and technical tools as well as multidisciplinary skill sets in the R&D labs, including chemical engineering science. Specific areas of opportunity for engineering are described in more detail in Chapter 2. The implementation of process design principles, drawing on the right skill sets, from both chemistry and engineering perspectives during clinical phase II is considered such an important step toward leaner more cost-effective processes readied for launch that several portions of this book will expand on this concept.

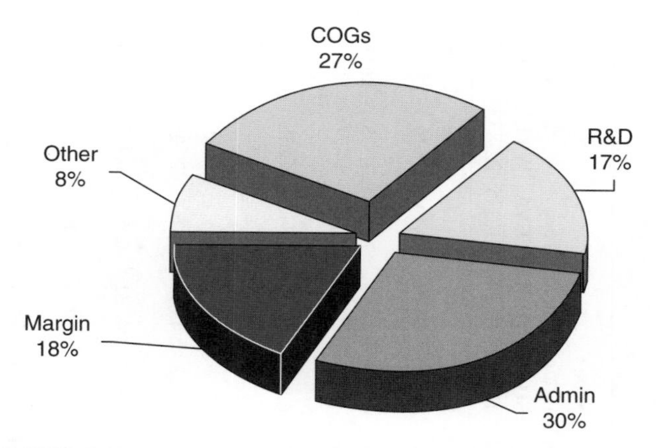

FIGURE 1.6 Distribution of revenue and expenses as a percentage of sales was averaged over 17 major pharmaceutical companies (listed in Table) based on 2008 annual reports.

1.2 MANUFACTURING

Pharmaceutical production plants of APIs and drug products can be generally characterized as primarily batch-operated multipurpose manufacturing plants. At these facilities, commercial supplies of API intermediates, APIs, and drug products are manufactured before being packaged, labeled, and distributed to various customers. Pharmaceutical production plants were typically designed to be flexible to allow a number of different products to be run in separate equipment trains, depending on the demand. Further, these facilities have various degrees of automation, relatively high levels of documentation, and change control to manage reconfigurations, with relatively long down times for cleanup and turnover of the plant between product changes [22]. Manufacturing often accounts for more than one-third of a company's human resources and a third of the total costs with expenses exceeding that of R&D [23]. Figure 1.6 shows the major components of revenues based on 2008 annual reports averaged over the 17 pharmaceutical companies shown in Table 1.4. Figure 1.6 shows that the manufacturing costs or costs of goods (COGs) are on average 27% of revenues where R&D represents 17% of revenues. The margin, shown in Figure 1.6, refers to the profit margin at 18%. As an industry with annual sales of over $700 billion, COGs of 27% represent close to $200 billion for the industry. For this reason, COGs have received considerable attention as an area of opportunity for potential savings [24].

It has been claimed that through adopting quality by design principles and principles of lean manufacturing, pharmaceutical companies, on average, could save in the range of $20–50 billion/year by eliminating inefficiencies in current manufacturing [25]. This translates to an improvement of 10–25% in reducing current COGs. Another critical factor in API cost determinations is the tax savings provided to companies who manufacture in selected countries. At present, for United States based pharmaceutical companies, significant tax savings are realized by locating production in tax-advantaged locations such as Ireland, Singapore, Puerto Rico, and Switzerland. Manufacturing in tax-advantaged locations can realize tax savings with tax rates of 2–12% versus the U.S. rate of 38% [26], so the cost advantages of any process or operational change need to be carefully determined and location-specific factors taken into consideration.

The principles of lean manufacturing are often cited as an approach to reduce COGs in pharmaceutical development and manufacturing. Lean manufacturing describes a management philosophy concerned with improving profitability through the systematic elimination of activities that contribute to waste; thus, the central theme to lean manufacturing is the elimination of waste where waste is considered the opposite of value. Based on the work of Taiichi Ohno, creator of the Toyota Production System, the following are considered wastes [27]:

- Overproduction
- Waiting
- Transportation
- Unnecessary processing
- Unnecessary inventory
- Unnecessary motion
- Defects

All of these wastes have the effect of increasing the proportion of nonvalue-added activities. Lean thinking is obviously applicable to many industries including pharmaceutical manufacturing as well as pharmaceutical development. Continuous processing, for pharmaceutical APIs and DPs, is one application of lean thinking applied to pharmaceutical manufacturing. The challenge that batch processing inherently leads to overproduction (e.g., inventory buildup of intermediates), leading to longer cycle times and excess inventory, is addressed through the concepts of continuous manufacturing.

According to Ohno, "The greatest waste of all is excess inventory" where in simplest terms excess inventory incurs

TABLE 1.4 Financial Data from 17 Pharmaceutical Companies Obtained from 2008 Year-End Reports

Stock ticker Revenues (in millions of USD)		Abbott ABT 29,526	Amgen AMGN 14,687	Astra AZN 31,601	BMS BMY 20,579	Boeh-ringer BOE[d] 16,139	GSK[a] GSK 35,592	J&J JNJ 63,747	Lilly LLY 20,378	Merck MRK 23,850	Novar-tis NVS 41,459	Pfizer PFE 48,296	Roche[b] RHHB-Y 33,692	Sanofi[c] SNY 38,372	Scher-Plough SGP 18,502	Takeda TKPHF 13,748	Teva TEVA 11,085	Wyeth WYE 22,833	Average AVG 28,476
Cost of sales		12,612	2,296	6,598	6,396	N/A	9,376	18,511	4,383	5,582	11,439	8,112	8397	10,212	7,307	2,786	5,117	6,248	7,836
Percent of revenues		42.71	15.63	20.88	31.08	N/A	26.34	29.04	21.51	23.40	27.59	16.80	24.92	26.61	39.49	20.26	46.16	27.36	27
R&D		2,687	3,030	5,179	3,585	2,936	5,380	7,577	3,841	4,805.30	7,217	7,945	7,405	6.368	3,529	2,757	786	3,373	4,612
Percent of revenues		9.10	20.63	16.39	17.42	18.19	15.12	11.89	18.85	20.15	17.41	16.45	21.98	16.60	19.07	20.05	7.09	14.77	17
Selling, informational, and administrative, expenses		8,436	3,789	10,913	4,792	N/A	11,190	12,490	6,626	7,377	11,852	14,537	7,746	9,977	6,823	6,730	1,842	6,838	8,247
Percent of revenues		28.57	25.80	34.53	23.29	N/A	31.44	19.59	32.52	30.93	28.59	30.10	22.99	26.00	36.88	48.95	16.62	29.95	29
Net income (margin)		4,881	4,196	6,130	5,247	1,983	6,887	12,949	−2,072	7,808.40	8,163	8,104	9,691	5,974	1,903	4,231	635	4,417	5,360
Percent of revenues		16.53	28.57	19.40	25.50	12.29	19.35	20.31	−10.17	32.74	19.69	16.78	28.76	15.57	10.29	30.78	5.73	19.34	18
Employees		68,000	17,500	67,000	41,000	39,800	99,003	119,200	40,600	59,800	96,717	81,800	78,604	100,000	55,000	15,717	38,000	50,527	62,839
Inventories	Finished goods	1,546	444	596	707	802	2.388	2,841	771	433	4,813	2,024	4,184	2,017	1,212	457	1,904	996	1,655
	Work-in-process	698	1,519	631	738	869	1,893	1,372	1,657	Incl. below	N/A	1,527	940	2,823	1,428	337	559	1,540	1,235
	Raw materials + consumables	532	112	409	320	487	1,647	839	236	2245.6	979	830	758	856	679	367	903	460	745
Total inventories		2,776	2,075	1,636	1,765	2,158	5,928	5,052	2,493	2,678	5,792	4,381	5,882	5,696	3,319	1,161	3,366	2,996	3,480

Source: Respective companies' 2008 annual reports. [Revenues, cost of sales, R&D net income, and inventory data are listed in the table. The averages listed in the last column were used for the distribution shown in Figure 1.6.] Cost of sales = cost of goods: Costs of sales include manufacturing costs such as raw materials, salaries of production labor, depreciation of equipment, production lines, infrastructure, utilities, maintenance costs, subcontractors, and other manufacturing costs. R&D: Research and development expenditures include raw materials, salaries of R&D, depreciation and infrastructure, and cost of utilities for R&D and professional services used in R&D. Selling, informational, and administrative expenses: They include sales, promotions, customer support and training, marketing, advertising campaigns, public relations, distribution, sponsorships, general corporate activities, and compensation of senior executives. The negative net income of Lilly is due to acquisitions including ImClone for $6.5 billion (See notes to consolidated statement—note 3,2008 Lilly financial report).

[a]British pounds converted to USD (1 GBP = 1.4619 USD) based on exchange rate as of December 31, 2008.

[b]Swiss francs converted to USD (1 CHF = 0.936944 USD) based on exchange rate as of December 31, 2008.

[c]Euros converted to USD (1 EUR = 1.3919 USD) based on exchange rate as of December 31, 2008.

[d]Abbreviated BOE—privately held company.

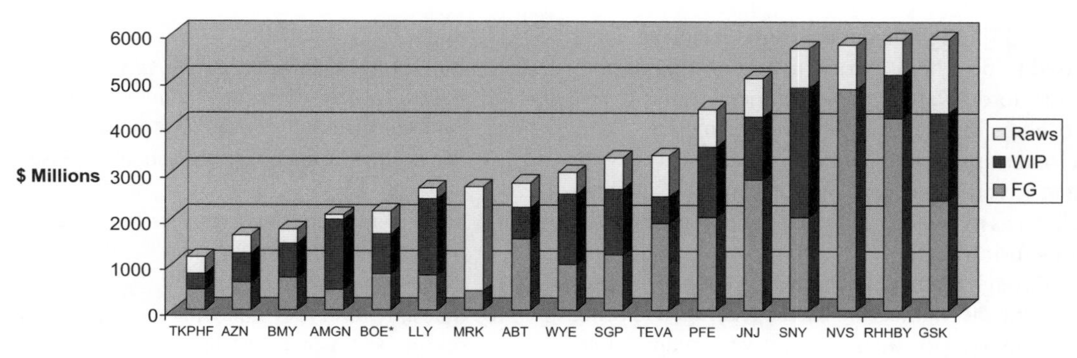

FIGURE 1.7 Inventory holdings for 17 pharmaceutical companies (numerical details shown in Table 1.4) based on inventories listed in each individual company's 2008 annual report. FG, Raws, and WIP refer to finished goods, starting materials, and work in process, respectively.

cost associated with managing, transporting, and storing inventories adding to the waste. Large inventories also tie up large amounts of capital. Implementation of lean manufacturing principles can be used to develop workflows and infrastructures to reduce inventories. One way to reduce inventories is through continuous processing. Chapter 23 by LaPorte et al. discusses the technical benefits of continuous manufacturing. A reliable steady delivery of product API and DP through small product-specific continuous plants could potentially reduce the level of inventory required in a dramatic way if the workflows were designed to ensure consistent delivery of product to packaging and distribution. The facilities of continuous production trains would likely be significantly smaller. Excess inventory represents an opportunity cost where capital is held up in the form of work in process (WIP), API finished goods, and formulated finished goods versus what could be invested elsewhere or back into R&D.

The costs of inventory holdings are significant, including both the carrying cost and the cash value of the inventory. Reductions in inventory equate to a one-time savings of the value of the inventory saved, which represents real savings and positive cash flow. The carrying costs of inventory include two main contributions: (1) weighted average cost of capital (WACC) and (2) overhead [28].

The weighted average cost of capital for the pharmaceutical industry is estimated to be 12% based on 2007 data and overhead costs are approximately 8% [29]. Estimates for the combined carrying cost of WACC and overhead range from 14% to 25%, which translates to approximately 20% return for every dollar of inventory reduction [28, 30]. In addition, every dollar of inventory reduced yields a one-time cash cost savings that can be invested to bolster the company's bottom line, for example, in R&D. Technology platforms and new workflows designed to minimize the need for stockpiling API and DP inventories across the industry therefore would seem to offer very rapid payback.

Figure 1.7 shows the range of inventories for the 17 pharmaceutical companies profiled in Table 1.4. For a large pharmaceutical company carrying $5 billion in inventories, the holding cost based on the combined WACC and overhead of 20% is approximately $1 billion/year. Considered another way, technologies that ensure a reliable and steady distribution of product with the result of eliminating the need to build and store massive inventories can return the company cost savings equivalent to a blockbuster drug (generating billions of dollars per year). Indeed, one of the three factors having the biggest impact on the profitability of a manufacturing organization is inventory with the other two being throughput and operating expense according to Goldratt and Cox [31]. Continuous processing if designed for reliable operations essentially year-round could potentially eliminate the need to accumulate significant inventories above and beyond 2–4 weeks of critical safety stocks of finished goods. Continuous manufacturing across API and DP integrated under one roof as a platform technology is one long-term approach to transforming the way the industry manages their commercial supply chain.

One textbook puts it, "Even for very small processes, continuous processes will prove to be less expensive in terms of equipment and operating costs. Dedicated continuous processes often put batch processes out of business" [32]. The real point here is that continuous manufacturing is one approach to lean manufacturing and to reducing inventories and costs but certainly not the only approach. Other lean systems can be devised that utilize the existing batch facilities as well.

1.3 SUMMARY

With current cost pressures on the pharmaceutical industry, there is an ever-increasing need for chemical engineering skill sets in process development and manufacturing. Chemical engineers are uniquely positioned to help address these needs in part derived from their ability to predict using mathematical models and their understanding of equipment and manufacturability. As Wu et al. highlighted, chemical engineers can help transform pharma from an industry focusing on inventing and testing to a process and product design industry [33]. Significant pressure exists on what historically used to be a high-margin nature of the pharmaceutical industry to deliver safe, environment-friendly, and economic processes in increasingly shorter timelines. This means fewer scale-ups at kilo and pilot plant scales, with expectation that a synthesis or formulation can be designed in the lab to perform as expected (and right the first time) at the desired manufacturing scale.

From R&D through manufacturing within the pharmaceutical industry, chemical engineering can be leveraged to bring competitive advantage to their respective organizations through process and predictive modeling that lead to process understanding, improving speed of development, and developing new technology platforms and leaner manufacturing methods. The chapters in this book are intended to provide examples of chemical engineering principles specifically applied toward relevant problems faced in the pharmaceutical sciences and manufacturing areas. Further, the broader goal of this work is to promote the role of chemical engineering within our industry, promote the breadth of skill sets therein, and showcase the critical synergy between this discipline and many other scientific disciplines that combine to bring pharmaceutical drugs and therapies to patients in need around the world.

ACKNOWLEDGMENTS

I would like to thank our chemical engineering co-op students Steven Modzelewski and Jamie Snopkowski from Northeastern University for their assistance in preparing this chapter.

REFERENCES

1. Abdel-Magid AF, Caron S, editors. *Fundamentals of Early Clinical Development*, Wiley, 2006.
2. Gadamasetti KG, editor. *Process Chemistry in the Pharmaceutical Industry*, Marcel Dekker, 1999.
3. Gadamasetti KG, Braish TF, editors. *Process Chemistry in the Pharmaceutical Industry*, Vol. 2, CRC Press, 2008.
4. Zheng J, editor. *Formulation and Analytical Development for Low Dose Oral Drug Products*, Wiley, 2009.
5. Janusz Rzeszotarski, W. *Pharmaceuticals Kirk-Othmer Encyclopedia of Science and Technology*, 2005.
6. Ibid.
7. http://www.fda.gov/AboutFDA/WhatWeDo/default.htm.
8. Janusz Rzeszotarski, W. *Pharmaceuticals Kirk-Othmer Encyclopedia of Science and Technology*, 2005, http://www.fda.gov/AboutFDA/WhatWeDo/default.htm.
9. http://www.ema.europa.eu/htms/aboutus/emeaoverview.htm.
10. Pharmaceutical Research and Manufacturers of America. *Pharmaceutical Industry Profile 2009*, PhRMA, Washington, DC, 2009.
11. Ibid.
12. Burns, LR. *The Biopharmaceutical Sector's Impact on the U.S. Economy: Analysis at the National, State, and Local Levels*, Archstone Consulting, LLC, Washington, DC, 2009.
13. Ainsworth SJ.Timely transformation, C&E News, December 7, 2009, pp. 13–21.
14. http://money.cnn.com/magazines/fortune/global500/2009/performers/industries/profits/.
15. http://www.imshealth.com.
16. Suresh P, Basu PK. Improving pharmaceutical product development and manufacturing: impact on cost of drug development and cost of goods sold of pharmaceuticals. *J. Pharm. Innov.* 2008;3:175–187.
17. Gregson N, Sparrowhawk K, Mauskopf J, Paul J. Pricing medicines: theory and practice, challenges and opportunities. *Nat. Rev. Drug Discov.* 2005;4:121–130.
18. www.fda.gov/Drugs/DevelopmentApprovalProcess/SmallBusinessAssistance/.
19. Dienemann E, Osifchin R. The role of chemical engineering in process development and optimization. *Curr. Opin. Drug Discov. Dev.* 2000;3(6):690–698.
20. Federsel H-J. In search of sustainability: process R&D in light of current pharmaceutical R&D challenges. *Drug Discov. Today* 2006;11 (21–22):966–974.
21. Ibid.
22. Behr A, Brehme VA, Ewers LJ, Gron H, Kimmel T, Kuppers S, Symietz I. New developments in chemical engineering for the production of drug substances. *Eng. Life Sci.* 2004;4(1):15–24.
23. Burns LR. *The Business of Healthcare Innovation*, Cambridge University Press.
24. The Gold Sheet, Pharmaceutical & Biotechnology Quality Control, Attention turns to the business case of quality by design, FDC Reports, January 2009.
25. Ibid.
26. Kager P, Williams D.How do you solve a problem like manufacturing? Pharmaceutical Executive, PharmExec.com, September 1, 2008.
27. Ohno T. *Toyota Production System Beyond Large Scale Production*, Productivity Press, 1988.
28. Cogdil RP, Knight TP, Anderson CA, Drennan JK. The financial returns on investments in process analytical technology and lean manufacturing: benchmarks and case study. *J. Pharm. Innov.* 2007;2:38–50.
29. Ibid.
30. Lewis NA. A tracking tool for lean solid-dose manufacturing. *Pharm. Technol.* 2006;30(10):94–108.
31. Goldratt EM, Cox J. *The Goal: A Process for Ongoing Improvement*, Gower Publishing, 1984.
32. Biegler LT, et al. *Systematic Methods of Chemical Process Design*, Prentice Hall, 1997.
33. Wu H, Khan MA, Hussain AS. Process control perspective for process analytical technology: integration of chemical engineering practice into semiconductor and pharmaceutical industries. *Chem. Eng. Commun.* 2007;194:760–779.

2

CURRENT CHALLENGES AND OPPORTUNITIES IN THE PHARMACEUTICAL INDUSTRY

JOSEPH L. KUKURA AND MICHAEL PAUL THIEN
Global Science Technology and Commercialization, Merck & Co., Inc., Rahway, NJ, USA

2.1 INTRODUCTION

The pharmaceutical industry bases its products, strategies, decisions, actions, and ultimately its very existence on the primary challenge of improving human health and the quality of life. The work of this industry uses a foundation of medical science to connect to the most basic struggle faced by all individuals and societies: the struggle for people to live healthy, productive lives. The industry has partnered with governments, health organizations, and society to achieve key successes in human history, including the eradication of smallpox, innumerable preventions of infection through the large-scale production and distribution of antibiotics such as penicillin, and significant reductions in cardiovascular events. Advances in pharmaceutics have contributed to the lower infant mortality rates and longer life spans observed over the past century. When considering the landscape of the pharmaceutical industry, one must retain the perspective that a challenge or opportunity that relates to the improvement of human health is at the core of challenges and opportunities shared by pharmaceutical companies.

2.2 INDUSTRY-WIDE CHALLENGES

Just as life and the state of human health often undergo significant changes, the pharmaceutical industry is profoundly changing. Fundamental elements that molded past business models are dynamically moving into new realms in a manner that will challenge the continued vitality of pharmaceutical companies much in the same way that changes in environments can influence the health of people. The parallels between diseases that the industry seeks to address and the pharmaceutical business climate are distinctly apparent. Illnesses such as HIV/AIDS are more complex and evolve at a faster pace than many previous diseases, requiring new approaches and advances. Similarly, economic, societal, and scientific forces are rapidly driving changes to the industry's business models. The forces challenging the industry align into four categories: increased costs and risks; revenue/price constraints; globalization of activities; and increasing complexity of pharmaceutical science.

2.2.1 Increased Costs and Risks

Bringing a new medicine to market involves a long, complex process in a highly regulated industry. Estimating accurate or typical costs for successfully launching new pharmaceutical products is difficult. The estimates are a strong function of the success rates assumed for moving a program through various clinical trial stages. Analysis shows that using different time periods to form assumptions for success rates can lead to variations in estimates of an average cost to launch a pharmaceutical product ranging from $900 million to $1.7 billion [1]. Across the industry, costs are trending upward in a manner that forces business practices to adjust.

Developing new medicines for unmet medical needs always involved significant costs and risks. For every product brought to market, pharmaceutical companies have typically invested in several thousands of compounds during the drug discovery stage, hundreds of compounds in preclinical testing, and many (7–12) unsuccessful clinical trials over a period

Chemical Engineering in the Pharmaceutical Industry: R&D to Manufacturing Edited by David J. am Ende

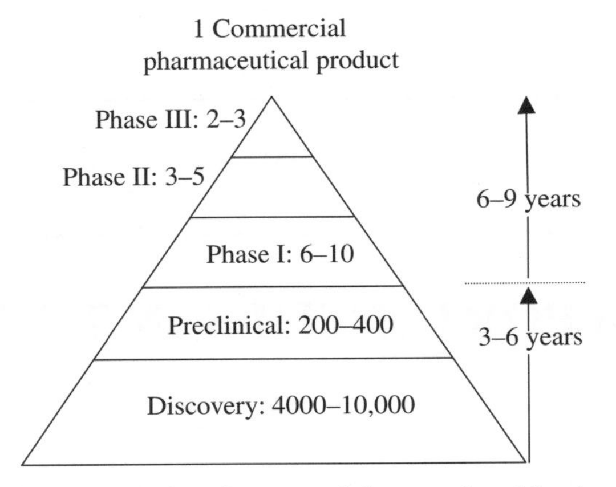

FIGURE 2.1 Number of compounds in research and development for every successful launch of a pharmaceutical product.

of 9–15 years, as depicted in Figure 2.1. Though the later stage phase II and phase III clinical trials are performed on the fewest number of compounds, they are also the most costly stages of development since they can require testing hundreds of patients in phase II and thousands in phase III in order to get statistically meaningful results. The cost of developing new medicines is therefore particularly sensitive to success rates of late-stage clinical trials. The industry has generally experienced a recent decline in the fraction of compounds proceeding through phase II and phase III clinical trials to a successful regulatory approval and commercial launch of a new product. This decline translates to expending more resources on programs that do not return value on their investment and greater overall spending on research and development.

The lower clinical trial success rates are due in part from the fact that pharmaceutical companies are attempting to treat more complex therapeutic targets. Many diseases with straightforward cause–effect relationships and less sophisticated biological mechanisms have already been addressed, leaving more challenging and intricate problems for the future. The setbacks and frustrations relating to the development of treatments and vaccines for HIV infection serve as a case in point. Numerous articles and presentations have reported that the HIV virus mutates, adapts, and changes features faster than predecessors that were studied for vaccine development. Similarly, the causes of many forms of cancer being addressed in clinical trials have more complex physiological traits in comparison to successfully treated conditions such as high cholesterol and high blood pressure. Treating more complicated illnesses leads to higher risks for clinical trial evaluations.

The pharmaceutical industry research and development costs are also increasing due to greater regulatory hurdles for getting approval of new medication. Many agencies have raised their requirements for approval in comparison to 5–10 years ago, leading to the need for larger and more comprehensive clinical trials and safety assessment testing. Government health agencies are showing a high level of caution with respect to side effects and risk–benefit assessments. This caution creates a need for outcomes data and in turn longer running trials and longer review periods that can delay introduction of a product to market. A more conservative regulatory approach ultimately forces greater spending on development and testing of new medicines to collect the data needed for the higher standards.

2.2.2 Revenue/Price Constraints

Beyond the challenge of increasing costs, the pharmaceutical industry is also facing constraints to income and product pricing. The patents of large revenue "blockbuster" drugs are expiring faster than they are being replaced by a comparable portfolio of new highly profitable products. The challenges of addressing more complex therapeutic targets previously described in the context of increasing cost also directly affect revenue in the industry. The greater level of complexity not only makes the research and development process more expensive, but also slows the realization of a return on investment in these areas. Many of these more sophisticated research efforts target a narrower patient base than preceding blockbusters. The largest sources of revenue for the industry over the past 20 years improved conditions that were widespread, such as depression, hypertension, and pain. Far fewer people have conditions that many products currently in development aim to improve, such as specific forms of cancer. With a smaller base of potential patients, these new products can be expected to generate less revenue than broadly used products already on the market.

Another strong influence on pharmaceutical sales relates to the means by which patients pay for medicine. Organizations responsible to pay for prescriptions, the payers such as insurance companies and health maintenance organizations (HMOs), are influencing the medical options for their membership. Pharmaceutical companies used to focus on physician–patient relationships when marketing products, but the decision-making process to select medicine now involves a more complex set of interactions between physicians, patients, and payers. The pharmaceutical industry must engage all three members of this collective to successfully bring products to those who need them. Payers acquired an increasingly important role in this process in the United States through consolidations that have allowed a few groups to represent a larger number of people. Single payers can control access to millions of patients [2]. Payers can exert their influence on the pharmaceutical industry in several ways. They cannot directly specify which medications a patient may use, but they can make copayments paid by patients much higher for some medications relative to others. If a payer wants to provide incentive for patients to request switching from a current treatment to a less expensive generic alternative, they can make the copayment for the generic version $50–100 per month less expensive. Similarly, payers can also choose to reimburse pharmacists at a higher rate for supplying generics and drive policies at pharmacies to favor the generic options.

In addition to consolidation of private payers, other events elevate the importance of payers to the pharmaceutical industry. The U.S. government became effectively the largest payer to the pharmaceutical industry in January 2006 with the implementation of Medicare Part D prescription plan, covering over 39 million people with that plan alone [3]. Even more people will be eligible for coverage benefits in the coming years. If the U.S. government alters its current policy to not negotiate medicine covered by Medicare or reimportation policies, the changes will create significant challenges to the business models of the pharmaceutical industry. The issue is certainly not limited to the United States. The changing demographics of the world will dramatically affect social medicine policies. The patients themselves in the patient–physician–payer relationship are changing in ways that will challenge the business models of the pharmaceutical industries. Across the world, the fraction of people above age 65 is growing as life expectancy increases. Never will the world have had this many people, this old. Along with economically challenged populations in emerging non-Western markets, this increasing fraction of the planet's population will generally have limited income available for health care, but they will have a disproportionally strong demand for pharmaceutical products. Ensuring access to medicine across the globe and across population sectors will require lower prices. The pharmaceutical industry must adapt to meet the needs of these large segments of customers.

Not only are the demographics of patients changing, but their behaviors and approaches to health care are also differing from the past. Survey results show that health care is a diverse consumer market with people seeking greater access to information to make their own choices with respect to health care needs [4]. With technology advances such as the Internet, patients can get more information to play a larger role in selecting treatment options. In some parts of the world, direct advertising to consumers is prevalent, raising new levels of their awareness of options. Patients are also willing to explore innovative techniques or travel outside their area and even their country to find options that best suit their preferences. To face the revenue and price constraints introduced by patients actions to the patient–physician–payer relationship, pharmaceutical companies need to understand the changing manner in which patient behaviors affect market demand and pricing.

2.2.3 Globalization of Activities

To address financial constraints and meet the global demands of emerging markets, the pharmaceutical industry is increasing the activity levels of its business in these regions, moving away from being primarily located and focused in the United States, Europe and Japan. Like many other industries, a greater fraction of manufacturing and research and development is shifting overseas from a Western base to countries such as China and India. Numerous clinical trials are conducted in these regions to achieve cost savings and to more quickly enroll patients who are not already undergoing another therapy. Development activities such as medicinal chemistry and process scale-up are being performed there as well, leading to an expansion of sophisticated laboratories in these countries. Manufacturing is becoming increasingly well established in regions outside the United States and Europe, supplying global medical needs from truly global locations. Like international efforts in other industries, the globalization of pharmaceutical activities increases challenges associated with logistics, language barriers, and cultural differences but pays dividends in cost and increases in the size of the talent pool.

2.2.4 Increasing Complexity of Pharmaceutical Science

As already mentioned in the context of rising costs and constrained revenues, current research and development of new medicine is attempting to address afflictions and therapeutic categories that are more complex than their predecessors. In order to understand and treat these more complex targets, the industry must use more complex and difficult science. The pharmaceutical industry has always employed a highly talented collection of several scientific disciplines, ranging from biologists and chemists to engineers and statisticians. All of these professions now face harder problems down to the molecular level of their fields to bring forward the next generation of medicines. The scientific challenges take many forms. For example, a larger fraction of compounds in development have low solubility and low permeability in human tissue, making drug delivery within the body more difficult. Highly potent compounds dictate that the amount of drug in the formulation be very small, sometimes in the submilligram ranges, and this also adds to the challenges of formulation development. Innovative and novel delivery systems are required to ensure that new medicines are effective. Advances in the academic understanding of the workings of human genetic code are creating especially challenging questions around how to translate this knowledge into practical improvements in human health. Employees of pharmaceutical companies must be prepared for a future with more difficult challenges.

2.3 OPPORTUNITIES FOR CHEMICAL ENGINEERS

The challenges faced by the pharmaceutical industry create several opportunities for its members, including chemical engineers. The pressures to reduce costs connect directly to engineering principles that seek economies of scale and the application of efficient technology. Technological innovation and engineering analysis also enhance products to create meaningful differentiation for patients, which provides value in the face of revenue constraints. The complexities of pharmaceutical science and constraints of approaches that need to be suitable for global use are interwoven with the application of chemical engineering tools to address cost issues and enable product value. Finally, the strategic management of technology used to meet industry challenges by chemical engineers is an additional overarching opportunity in the industry.

2.3.1 Reducing Costs with Engineering Principles

Owing to large margins, engineers in the nongeneric pharmaceutical industry have not had the same traditional focus on product cost as engineers in other businesses. With a renewed emphasis on cost, engineers are increasingly using a wide variety of engineering tools to improve costs and help maintain margins. These tools include modeling of unit operations, employment of efficient laboratory methods and design of experiments, combining the output of models and experiments to define advantageous processing options, and the use of standardized technology platforms. Some of these topics will be highlighted briefly here and discussed in greater detail in subsequent chapters.

Across many industries, engineers are employed to use process modeling and physical/chemical property estimation to maximize the yield and minimize the energy consumption and waste production associated with desired products. Using the broad applicability of this network of techniques is a continuing opportunity. Engineers in the pharmaceutical industry use computational tools originally created for oil refinery processes to optimize distillations and solvent recovery associated with the manufacture of active pharmaceutical ingredients (APIs) [5]. Similarly, thermodynamic solubility modeling can be applied to optimize crystallizations [6]. Computation Fluid Dynamics (CFD) has numerous applications to pharmaceutical flows [7]. The use of sound, fundamental chemical engineering science can eliminate bottlenecks, improve production, and unlock the full potential of biological, chemical, and formulation processes used to make medicine.

Chemical engineers can also use their training and expertise with technology to help reduce costs. In the R&D arena, the use of high-throughput screening tools and multireactor laboratory systems efficiently promotes the generation of data at faster rates. When modeling and estimation techniques cannot provide a complete picture, engineers can get the data they need quickly with high-efficiency technology. It is important to recognize that not all of the advanced laboratory technology works universally well in all situations. A miniature reactor system suitable for homogeneous reactions may have insufficient mixing for heterogeneous chemistry. Selection of the appropriate laboratory technology and proper interpretation of results produced by these laboratory tools benefit from the perspective of combined chemical engineering principles such as mass transfer, heat transfer, reaction kinetics, and fluid mechanics. With the appropriate equipment in hand, engineers can utilize a statistically driven design of experiments to maximize the value of data generated via the experimental methodology.

The process understanding that comes from combining models and experimental data is a key opportunity for chemical engineers. Changes in the regulatory environment contribute to this opportunity. As will be described later in this book, the advent of Quality by Design (QbD) principles (ICH Q8, Q9 and Q10) provides greater freedom after launching a product to modify operating parameters within a defined operating space. These changes can promote higher quality products and reduce process waste by applying the knowledge that comes with increased production experience once a medicine is commercialized. The modeling abilities of chemical engineers and their technology expertise will be able to provide crucial guidance to the definition and refinement of a QbD operating space. A thoughtful, well-conceived operating space will in turn lead to long-term gains in process efficiency and better results for consumers.

A primary method to achieve the benefits of chemical engineering principles at manufacturing scale comes from the development and application of technology platforms that can use a single set of equipment with common operating techniques across a portfolio of processes and products. The platforms not only reduce capital costs by allowing the purchase of a reduced amount of equipment for more applications, but development costs can be lowered as well through a streamlined approach that comes from having a deep understanding and expertise with a technology platform. The familiarity and data obtained from running multiple projects in a single platform will translate into benefits for future projects that share common features. Broad uses of standardized platforms also make processes more portable for global applications. The key challenges of platforms are (1) knowing for which compounds the platform will be applicable and (2) maintaining the knowledge gained about the platform and its underlying technology.

2.3.2 Improving Product Value

Chemical engineers also have opportunities to meet market demands for pharmaceutical products that deliver greater value to patients, payers, and physicians. These customers do not care about the manufacturing process, but they do care about product convenience, safety, and compliance. They are seeking meaningful differentiation in these areas among their options. The remainder of this section will discuss ways engineers can contribute to product value with two examples: drug delivery and diagnostics.

Contributions to improvements in drug delivery vehicles serve as an excellent example of how engineers can improve pharmaceutical product features. The application of particle engineering and convection modeling to inhalers can improve the consistency with which a dose is administered via the respiratory system independent of the strength of the patient's breath [8]. Greater consistency of delivery increases the associated compliance. In orally administered capsules and tablets, engineers can manipulate polymer properties and transport driving forces to afford a consistent extended release of an API [9]. A steady, slow release of medicine from a single delivery vehicle can reduce both the frequency with which the medicine needs to be taken and potential side effects, which can, in turn, improve conveniences for the patient and compliance with the dosing regimen. In order to realize the benefits of controlled release, the pure API particle size distribution usually must be kept consistent prior to formulation. A great deal of engineering effort has been applied to maintain control of crystal sizes during the crystallization, filtration, and drying unit operations for drug substances [10].

Engineering principles can also be used to improve diagnostic tools used to treat diseases. Diagnostics are especially important for payer organizations that want to utilize options that have the highest probability of success for the patient. A diagnostic tool that enables physicians to initially assign the best treatment without going through a trial and error approach reduces the costs charged to the payers through the preemptive elimination of ineffective options. The aforementioned chemical engineering skills that aid the process of making medicines also contribute to improvements in making diagnostic technology. The underlying governing equations that characterize the transport of medicine to a specific target in the body also have applications in the movement of a sample from a patient through a device to the analysis component. Beyond diagnostic effectiveness, the ability for chemical engineers to respond to patient preference and improve the convenience of diagnostics tools used in the home or other areas outside of hospitals and physician offices is a key opportunity with health care becoming an increasingly consumer-driven market [4]. Advances in polymer technology and manufacturing processes can lead to devices that are lighter, smaller, and more resilient to being dropped. Just like new models of an iPod™ garner increased use over their heavier, larger, and more delicate predecessors, delivery vehicles and diagnostics serve as examples of significant opportunities for chemical engineers in the pharmaceutical industry to meet consumer demands for improved products. As will be discussed in greater detail later in this book, the use of Quality by Design principles in the context of ICH Q8/Q9/Q10 guidelines will help ensure that features of the product best match the needs of the patient.

2.3.3 Strategic Technology Management

In addition to direct scientific contributions that reduce costs and improve value for the pharmaceutical industry, engineers have the opportunity to help direct strategic investments in technology. Companies cannot afford to individually develop, implement, and advance all technologies required for their business. Several case studies show how good and poor strategies relating technology to business considerations have affected multiple industries, including computer companies and international distributors [11]. The availability of global development and supply options creates relatively new decisions for the pharmaceutical industry. Technology investment must be managed through a careful balance of internal capabilities, strategic partnerships, and reliance on external vendors. In order for this balance to be established and maintained, a holistic definition and view of technology must first be established. Is any laboratory or production device such as a granulator or a blender considered technology or are they just pieces of equipment? In the context of strategic management, technology can be defined as a system comprising (1) technical knowledge, (2) processes and (3) equipment that is used to accomplish a specific goal. The knowledge encompasses the understanding of fundamental principles and relationships that provide the foundation of the technology. The processes are the procedures, techniques, and best practices associated with the technology. The equipment is the physical manifestation of the technology as devices, instruments, and machinery. The goal for strategic technology management is to make value-driven decisions around investments in the advancement, capacity, and capability with each of the technology components.

To make those investment decisions, industrially relevant technology can be assigned to categories. The concepts of core technology and noncore technology and associated subcategories are useful in this regard:

- *Core Technology*: Core technology is sophisticated technology that makes critical contributions to the core business. As such, it justifies investment in *all three* components of the technology (knowledge, process, equipment) to afford a competitive advantage. Options for core technology include maintaining internal capability for all three components as a primary technology or managing the technology via external capacity as a partnered technology. The distinguishing feature of all core technology is that internal capability in the knowledge component of the technology is typically required to ensure that the technology is adequately controlled to meet core business needs.
  - *Primary Technology:* Primary technology is a category of core technology that offers competitive advantage by maintaining internal capability for the complete technology system. The investment in

internal capacity does not need to meet all business uses of the technology but ensures that adequate resources can be provided for critical projects.

- *Partnered Technology:* Partnered technology is a technology that contributes to the core business, but the company can maintain a competitive advantage while relying on external partners to be primarily responsible for parts of the technology system. The company may invest in knowledge and process development for a partnered technology while utilizing external equipment capacity and potentially outside process expertise.

- *Noncore Technology*: Noncore technology does not warrant investment or control in all three components of the technology system. The subcategories of emerging technology and commoditized technology characterize the most relevant noncore technology.
  - *Emerging Technology:* Emerging technology has potential to contribute significant business value in the future but generally requires additional investment in the knowledge base before it can be applied in practice to the core business. Emerging technology is not necessarily brand new technology but its application to the core business may be atypical or speculative.
  - *Commoditized Technology:* Commoditized technology is mature technology that is reliable, well established within the industry, cost efficient, and available in the market such that little investment in the technology is required.

To determine if a technology of interest is core or noncore, the connection of the technology to business value must be assigned as well as risks associated with the technology's ability to meet business requirements. Business value (BV) can be determined by identifying the revenue enabled by products made via the technology. Risk (R) can be calculated or estimated by assessing the fraction of attempts that a technology fails to deliver intended results within predetermined specifications for both quality and efficiency. A minimum business value (BV_{min}) and minimum risk tolerance (R_{min}) for being a core technology should then be assigned based on a strategic business and financial perspective. If either the business value or risk associated with a technology is below the corresponding minimum, the technology should not be considered for core technology investment. When a technology meets the minimum risk and business value requirements, a threshold value (TV) can serve as the primary criteria for determining the core technology designation as follows:

$$\text{Core technology}: (BV - BV_{min}) \times (R - R_{min}) > TV$$

$$\text{Noncore technology}: (BV - BV_{min}) \times (R - R_{min}) < TV$$

The underlying principle of this approach is that the core technology investment ensures the value benefit of the technology to the core business and mitigates the risk of a severe failure in the application of the technology. This insurance and mitigation come from the direct investment and maintenance of expertise in all three components of the technology, whereas noncore technology takes more appropriate risks with lower investments. As technology evolves in importance and reliability, it can transition between the core and noncore regimes by regular assessment of the business value and risk associated with the technology.

Technology progression and the evolution of a concomitant investment approach can be illustrated graphically on a plot of business value versus risk. On such plots, core technologies fall into the upper right regions. Risk generally decreases as time progresses and experience with the technology increases. A life cycle thus moves from right to left on the value–risk plot, and two examples are shown in Figure 2.2. In both cases, the technologies start with a relatively high risk as emerging technologies and transition from noncore to core technology when the business value becomes sufficiently high. In case 1, the technology sustains business value long enough for the technology to become a low-risk commoditized technology. Though the business value remains high, the reduced risk drives the transition from core to noncore technology in this case. Technologies that make sterilized vials serve as an example here. At one point, the pharmaceutical industry needed to invest internal resources to ensure vials for vaccines would be sterile, but they are now readily available as a commodity made from reliable, established technology managed by vendors. In contrast, case 2 illustrates a decreased business value driving the transition from core to noncore technology, perhaps due to the introduction of a better replacement. Obsolete

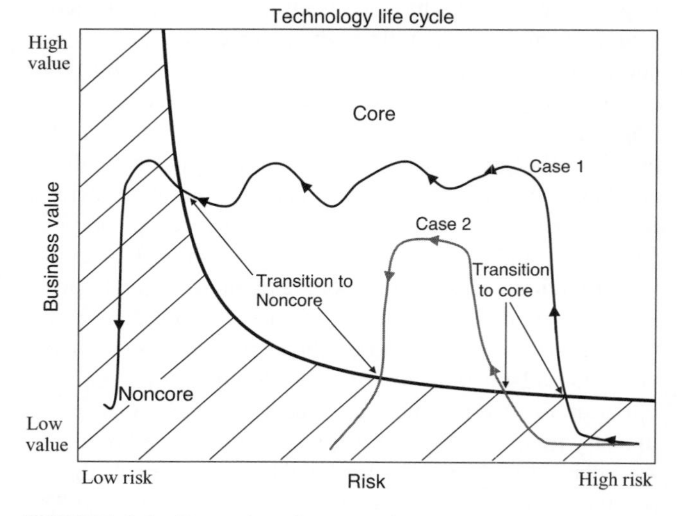

FIGURE 2.2 Examples of progression along a technology life cycle between core and noncore regimes.

open-top vacuum funnel filters that have been replaced by centrifuges and sealed filter dryers in manufacturing environments to improve industrial hygiene and efficiency provide an industrial example of case 2. Technologies relating to crystallization, spray drying, and roller compaction are representative of current chemical engineering core technology at several pharmaceutical companies.

Engineers have opportunities to substantially contribute to several facets of technology management as outlined above. Due to the complex nature of pharmaceutical processes, assigning a failure event to specific technology can be a challenging multivariant problem during risk assessment and risk management endeavors. Fundamental process understanding and technology expertise are vital to evaluating quantitative contributions to risk. Similar skills are useful for objectively determining whether value is enhanced through the use of internal capabilities versus external options. The understanding of a technology is important for determining the reliability of a prospective partner using that technology for critical business needs. Engineers can also contribute to investment choices among various emerging technologies with technical assessments of probabilities of success and potential applicability across a company's portfolio of products.

2.4 PROSPECTS FOR CHEMICAL ENGINEERS

Chemical engineers have made enabling contributions to health care that serve as a strong foundation for future success. However, the pharmaceutical industry is profoundly changing and the role of engineers must change with it. The industry challenges described in Section 2.2 translate into the opportunities for chemical engineers in Section 2.3. The use of modeling, standardized technology platforms and a sound technology strategy will allow engineers to help reduce costs. The platform technology will also assist with the challenges of making processes portable in an era of globalization. The need for greater product value to enable future revenues can partially be met by engineering enhancements to delivery devices and diagnostic tools. In addition, engineers may also be able to improve the stability of formulated products, thereby reducing the need for expensive cold storage and enhancing access options for patients in severe environments. As the underlying science and supporting academic chemical engineering research evolve toward an increasingly molecular basis, the perspectives and training of engineers must move from macroscopic and continuum foundations to a combined macroscopic, continuum, and molecular view. Chemical engineering will continue to integrate with the rest of scientific disciplines beyond a confined role in processing realms. The work of engineers must progress beyond connecting process contributions to production efforts to integrating the processes with the product itself. These efforts will be performed in the context of changing business models and pricing constraints on an increasingly global stage.

ACKNOWLEDGMENTS

The authors wish to acknowledge the contributions of the Merck Commercialization Technical Forum, most notably Anando Chowdhury, Louis Crocker, James Michaels, Lawrence Rosen, and Cindy Starbuck, to Section 2.3.3.

REFERENCES

1. Gilbert J, Henske P, Singh A. Rebuilding Big Pharma's business model, *In Vivo:* The Business and Medicine Report, Vol 21, No. 10, 2003.
2. Steiner M, Bugen D, Kazanchy B, Knox W, Prentice M, Goldfarb L. The continuing evolution of the pharmaceutical industry: career challenges and opportunities, RegentAtlantic Capital and Fiduciary Network, December 2007.
3. Medicare drug program enrollment hits 39 million, still open for advantage plans, Senior Journal.com, January 2007, http://www.seniorjournal.com.
4. Opportunities for life science companies in a consumer-driven market, Deloitte Center for Health Solutions, 2008, http://www.deloitte.com.
5. Li Y-E, Yang Y, Kathod V, Tyler S. Optimization of solvent chasing in API manufacturing process: constant volume distillation. *Org. Process Res. Dev.* 2009;13(1):73–77.
6. Kokitkar P, Plocharczyk E. Modeling drug molecule solubility to identify optimal solvent systems for crystallization. *Org. Process Res. Dev.* 2008;12(2):249–256.
7. Kukura J, Arratia P, Szalai E, Bittorf K, Muzzio FJ. Understanding pharmaceutical flows. *Pharm. Technol.* 2002;26 (1):48–72.
8. Finley WH. *The Mechanics of Inhaled Pharmaceutical Aerosols: An Introduction*, Academic Press, New York, 2001.
9. Wise D. *Handbook of Pharmaceutical Controlled Release Technology*, CRC Pres, New York, 2000.
10. Tung H-H, Paul EL, Midler M, McCauley J. *Crystallization of Organic Compounds: An Industrial Perspective*, Wiley, Hoboken, NJ, 2009.
11. Shimizu T, Carvalho MM, Laurindo FJB. Concepts and history of strategy in organizations. In: *Strategic Alignment Process and Decision Support Systems: Theory and Case Studies*, Idea Group Publishing, Hershey, PA, 2006.

3

CHEMICAL ENGINEERING PRINCIPLES IN BIOLOGICS: UNIQUE CHALLENGES AND APPLICATIONS

Sourav Kundu, Vivek Bhatnagar, Naveen Pathak,[1] and Cenk Undey
Process Development, Amgen Inc., West Greenwich, RI, USA

Therapeutic proteins now represent a major class of pharmaceuticals. These macromolecules, generally referred to as biologics, are either natural (e.g., protein fractions from donor human plasma) or are commonly made in bacterial or mammalian host systems by recombinant DNA technology. In the case of the latter, through modern molecular biology techniques, genes representing human proteins, protein fragments, or antibodies to therapeutic targets can be inserted into a host system such as the bacterium *Escherichia coli* or a mammalian host cell such as Chinese hamster ovary (CHO) cells. The host cells are then grown in fermenters or bioreactors using nutrient-rich media and highly controlled conditions. The host cells are internally programmed to synthesize the protein of interest. The synthesized protein may reside in high concentrations within the cell in inclusion bodies or secreted into the extracellular environment. The protein then can be harvested in large quantities and purified to the final therapeutic dosage form. This is of course an oversimplification of a very complex biological production process that brings a protein therapeutic to patients.

Chemical engineering principles apply widely in development and production of biologics. During discovery, biological drugs are first made in the laboratory in small scale (less than a milliliter to a few liters), quantities sufficient for testing *in vitro* or in an animal model. As the feasibility is proven, the process is refined and scaled up to manufacture quantities needed for extensive pharmacological, toxicological, and clinical testing. At this stage, the culture volume is typically increased from a few hundred to a few thousand liters. After the clinical safety and efficacy are proven and regulatory approval for commercialization is obtained, the drug may be manufactured at even larger scale. At each of these stages of product development lifecycle, chemical engineering principles such as mass transfer, heat transfer, fluid mechanics, chemical reaction kinetics, and so on are used. As the production process is scaled up during the product development, many of the process characteristics defined by dimensionless numbers widely used in chemical engineering (e.g., Reynolds number (*Re*), Nusselt number, Newton number which is also known as power number, etc.) are kept similar between scales. In fact, many of these dimensionless numbers provide the basis for scale continuum that is extremely important for consistency and comparability.

In the cell culture process for mammalian systems, chemical engineering principles are used to optimize transport of nutrients, supply of oxygen to the cells, facilitate mixing for homogeneity, removal of undesired metabolites and toxins, and collection of the protein of interest. In the downstream purification process, chemical engineering principles are used for separation of the protein drug from associated impurities, concentration of the protein drug, removal of undesired microorganisms, and stabilization of the protein drug to a state appropriate for storage, transport, and administration. Later in the chapter, we will describe in greater detail various unit operations that are typically involved in production of biologics and the specific fundamental principles associated with these unit operations.

[1]Currently at Genzyme, Inc.

Chemical Engineering in the Pharmaceutical Industry: R&D to Manufacturing Edited by David J. am Ende

3.1 WHY ARE BIOLOGICS UNIQUE FROM A MASS, HEAT, AND MOMENTUM TRANSFER STANDPOINT?

To fully appreciate why biologics are unique, it is important to discuss the complexity of the structure of these macromolecules. As stated earlier, biologics are protein molecules of therapeutic value most commonly administered as aqueous solution. The molecular weight of these macromolecules can range from 5–30 kDa (e.g., recombinant insulin, molecular weight approximately 6 kDa; recombinant human growth hormone, molecular weight approximately 22 kDa; etc.) to hundreds of kDa (e.g., recombinant factor VIII, molecular weight approximately 280 kDa). More complex and larger macromolecules are also being produced in recombinant forms such as recombinant von Willebrand factor that is made as a functional multimer with an approximate molecular weight reaching in the millions of daltons. Monoclonal antibodies represent an important class of protein drugs. Chimeric or humanized recombinant monoclonal antibodies have found wide applications as receptor blockers, receptor stimulators, delivery vehicles, or absorbents for undesirable antigens in various therapeutic, diagnostic, or imaging areas.

Protein molecules are comprised of amino acid sequences determined by the genetic code residing in the DNA. Protein synthesis takes place on ribosomes. The amino acids are connected to each other by peptide bonds forming polypeptide chains that provide the backbone of a protein molecule. Additional cross-links may be formed within a polypeptide chain or between adjacent branches of polypeptide chains through the disulfide bridges creating complex folds and a three-dimensional geometry. Protein architecture comprises of four levels of structure. The primary structure describes the amino acid sequence and the location of any disulfide bridges within the polypeptide chain and branches. The secondary structure describes the spatial arrangement of neighboring amino acids within the linear sequence. The tertiary structure of a protein describes the spatial arrangement of amino acids that are far apart in the linear sequence. If the protein molecule contains more than one polypeptide chain (also called the subunits), the spatial arrangements of these subunits are described in the quaternary structure. Additional complexity may arise when polypeptide chains fold into two or more compact globular regions joined by flexible portions. These domains resemble each other at times and vary in function at other times [1].

In addition to the characteristic three-dimensional peptide structures, many protein macromolecules also have covalently linked carbohydrate branches. These are known as glycoproteins and represent an important class of therapeutic molecules. The carbohydrate structure takes its initial shape in the endoplasmic reticulum during synthesis and is modified into its final form in the Golgi complex. The carbohydrates (also known as oligosaccharides) can be linked to the polypeptide chain of the glycoprotein through the side-chain oxygen atom on the amino acids serine or threonine residues by *O*-glycosidic linkages or to the side-chain nitrogen of asparagine residues by *N*-glycosidic linkages. In addition to these carbohydrate cores, there can be other polysaccharides attached to these cores in a variety of configurations causing a diverse and defining carbohydrate structure [1].

The three-dimensional structure of a protein is extremely important for its functionality. Many interactions between protein molecules or between a protein molecule and a receptor on a cell require a specific spatial geometry of the polypeptide chain. Any disturbance in the three-dimensional structure bears the risk of loss of the protein's biological activity, bioavailability, or therapeutic value. Environmental conditions that can alter the three-dimensional structure include pH, temperature, physical forces such as shear, specific concentration of chemicals such as chaotropic agents (chemicals that disrupt the protein' three-dimensional structure such as urea or guanidine hydrochloride at high concentrations), and organic solvents, and so on. The alteration of structure can be reversible and can actually be used for recovery of proteins. For example, highly aggregated recombinant protein housed in the inclusion bodies of *E. coli* can be solubilized in high concentration of urea or guanidine hydrochloride followed by a controlled removal of the chemicals. The process allows the protein to disaggregate and refold into its native desirable state. Under certain conditions, the alteration of structure becomes irreversible. The common environmental conditions causing permanent structural alteration of a protein are exposure to extreme heat, pH, or chemical concentration. In response to environmental conditions, such as heat, protein unfolds into a denatured state from its native state. Subsequent aggregation of the denatured molecules results in irreversible denaturation. Denaturation and aggregation during processing are of particular concern to a manufacturer of biologics. Thermal denaturation and unfolding of proteins can be studied by using differential scanning calorimetry (DSC), where heat flux is measured in a sample as the temperature is gradually increased. The protein sample provides a characteristic curve representative of glass transition and phase changes. Denaturation of a protein drug can lead to the loss of its biological activity. In one study [2], activities of natural and recombinant Protein C evaluated by DSC were observed to undergo rapid decline as temperature was increased from 20°C with a complete loss of activity at temperatures above 70°C. Narhi et al. observed irreversible thermal unfolding of recombinant human megakaryocyte growth factor upon heating, leading to formation of soluble aggregates ranging in size from tetramer to 14-mer [3]. Proteins can be affected by cold temperatures as well. Protein solutions may be exposed to temperatures lower than −50°C during freeze-drying to prepare a protein drug dosage form. Although less concerning than heat-induced structural damage, cold inactivation has been seen with enzymes such as phosphofructokinase [4] or with proteins such as β-lactoglobulin [5].

In addition to loss of biological activity, when administered to a patient, the structural change from the native state can trigger adverse cellular and immunological responses causing serious medical consequence. Therefore, manufacturing processes for biological drugs are designed to have appropriate controls such that the protein's active state is maintained. The likelihood of irreversible denaturation is high during operations involving elevated temperature, contact with excessively high or low pH, vigorous fluid movement (e.g., vortex formation), or contact with air interface (e.g., foaming). Thermal denaturation can occur during processing if the temperature is not maintained at the facility or the equipment level, or if unintended local heat buildup occurs in the processing equipment such as in a rotary lobe pump, a commonly used component of a bioprocess skid. Most biopharmaceutical plants have sophisticated environmental controls with high-efficiency HVAC systems to maintain the ambient temperature at a level that is far away from the denaturation temperature of the protein. Temperature is also controlled at the equipment level through water or ethylene glycol heat exchange fluids circulating through the jacket of a process vessel. When heating of a chilled protein solution (typically in between process steps) is needed, the jacket temperature is capped at 45–50°C to prevent excessive vessel skin temperature that may cause denaturation. Exposure to excessively high or low pH can occur during titration of a protein solution with acid or alkali during processing. It can be avoided by ensuring rapid mixing in the vessel through appropriately designed agitators, controlling the flow rate of the reagents, and by using lower titrant concentrations.

Appropriate handling of a protein solution during processing is extremely important. Abrupt and vigorous fluid movements such as those encountered during movement of tanks during shipping of liquid protein solutions can be detrimental and result in denaturation. Air entrapment and exposure of protein to an air interface create the bulk of the problem in these cases [6]. Process equipment and piping are designed such that the protein solution is not subjected to highly turbulent flow regions during transport. Effect of shear on proteins remains a controversial topic. Thomas provides an excellent overview of the issue of shear in bioprocessing [7]. While loss of activity has been seen with enzymes when exposed to high shear forces in a viscometer, denaturation of globular proteins is less likely just from high shear fields. As mentioned earlier, the primary mechanism of denaturation appears to be from the gas-liquid interface, especially if high velocity gradients are present. Lower concentration solutions are more susceptible to the damage. Design of equipment to minimize air entrapment (e.g., avoidance of pump cavitation) allows control of inactivation and protein denaturation in bioprocessing.

Cell culture fluids pose unique chemical engineering challenges as well. The cells require oxygen and respire carbon dioxide. Oxygen is sparged into the culture that produces bubbles and foam. Care needs to be taken to control the bubble size, velocity, and amount since the bubbles can carry cells to the surface and cause cell death upon bursting. Other chemical engineering challenges include the heat input (mammalian cells) or removal (microbial cells) requirements, mixing requirement, maintenance of appropriate pH conditions, nutrient delivery, and metabolite removal for large, industrial-scale cell culture systems. Further details of bioreactor design and operation are provided in a subsequent section.

The complexity of protein drugs and their biological production systems pose extraordinary challenges to the chemical engineers with responsibility for industrial-scale manufacturing of these products. Design, development, and operation of these processes require complete understanding and appreciation for the intricacies of proteins and living cells. In the subsequent sections, we will examine how concepts taught in traditional chemical engineering are applied to design equipment and develop processes that allow modern-day mass manufacture of biotechnology products.

3.2 SCALE-UP APPROACHES AND ASSOCIATED CHALLENGES IN BIOLOGICS MANUFACTURING

With the introduction of recombinant human insulin by Genentech (later licensed to Eli Lilly & Co.) in 1978, the commercial industrial biotechnology was born. It was clear that products created by recombinant DNA technology hold an immense potential for curing "uncurable" diseases such as cancer. The biotechnology products no longer were just research tools in the laboratory made in milligram quantities; they needed to be manufactured in bulk in commercial-scale manufacturing facilities. Scale-up and optimization of biologics manufacturing processes rapidly became critical for the success and growth of the industry.

Scale-up of biologics manufacturing processes utilizes similar engineering principles as scale-up of chemical manufacturing processes. In the laboratory, the cell culture may be performed in shaker flasks, roller bottles, or small glass stirred-tank bioreactors. Flask- or bottle-based cultures can be scaled up by simply increasing the number of flasks or bottles with some increase in the size. Bioreactors can be successfully scaled up linearly from 1 to 25,000 L or above by preserving the aspect ratios, impeller sizing ratios, impeller spacing ratios, and baffle geometries. Design of spargers for oxygen delivery to the cell culture may differ considerably between small-scale and industrial-scale bioreactors. While a small sintered sparger may be adequate for oxygen delivery in a small-scale bioreactor, much more elaborate sparging systems with one or more drilled pipes may be necessary to deliver the quantity of oxygen needed in a large-scale bioreactor. The oxygen concentration is maintained at the same level between the small and large scales by controlling to a specified dissolved oxygen setting determined during process development and preserved through scale-up. To achieve the level of dissolved oxygen during cell culture, an adequate amount of oxygen is fed to the bioreactor typically through a flow controller. For large-scale bioreactors, the piping and delivery system for gases can become rather massive. The delivery of oxygen to the cells depends on the mass transfer of oxygen from bubbles in the stirred tank to the liquid where the cells are situated. The bubble size is governed by the sparger type and whether the bubbles are broken up further with a Rushton (turbine)-type impeller at the bottom of the bioreactor. The mass transfer coefficient, $k_{\mathrm{L}}a$, is an important parameter that determines if adequate transport of oxygen to the cells can be achieved and must be calculated for each bioreactor where scale-up is being performed. The oxygen uptake rate (OUR) is the oxygen required for the optimum culture performance that must be matched with the oxygen transfer rate (OTR) that is determined by the mass transfer coefficient. The relationship between OUR and OTR at a constant dissolved oxygen level is shown by the following formula:

$$\mathrm{OUR} = \frac{\mu X}{Y_{X/\mathrm{O}_2}} = k_{\mathrm{L}}a(C_{\mathrm{sat}} - C_{\mathrm{L}}) = \mathrm{OTR}$$

where μ is the specific growth rate, X is the measured cell density, Y_{X/O_2} is the calculated cell yield per unit oxygen consumption, and C_{sat} and C_{L} are the dissolved oxygen concentrations at saturation and at any given time in the liquid phase, respectively.

Scaling up a suspension cell culture requires a thorough evaluation of mixing. Adequate mixing is necessary in the delivery of nutrients and oxygen to cells, removal of metabolites from the microenvironment, and dispersion of additives such as shear protectants and antifoam. Scale-up of mixing is often performed by maintaining similar power per unit volume (P/V) between the small-scale and large-scale bioreactors. P/V is calculated from the impeller geometry, agitation rate, and working volume of the stirred tank according to the following formula:

$$\frac{P}{V} = \frac{\rho n N_{\mathrm{p}} N^3 D_{\mathrm{i}}^5}{V}$$

Where ρ is density, n is the number of impellers, N_{p} is the impeller power number, N is the agitation rate, D_{i} is the impeller diameter, and V is the working volume of the tank. When P/V is preserved during scale-up, it is expected that with geometric similarity between the small- and large-scale bioreactors, circulation time, mixing time, and impeller tip speed increase but the size of the eddies does not change, hence ensuring mixing and mass transport [8]. P/V can be reported as an ungassed value or a gassed value. Typically, the measured ungassed value is slightly higher than the measured gassed value as a result of the loss of power upon introduction of gas sparging in the bioreactor. Impeller tip speed is determined by the following formula:

$$\text{Impeller Tip Speed} = \pi N D_{\mathrm{i}}$$

Maintaining impeller tip speed between scales can ensure similar shear-induced damage of cells between scales. It is important to minimize shear-induced cell damage to maintain cell viability and prevent release of unwanted enzymes, host cell DNA, and other impurities into the cell culture. The ability of cells to withstand shear occurring at the tip of the impeller blade depends on the type of cells. Typically, microbial cells can withstand more shear and, therefore, can be agitated more aggressively than mammalian cells. It is helpful to scale up in stages when transferring a process from a laboratory to a commercial manufacturing facility. The stepwise approach reduces the probability of failure and allows troubleshooting and understanding of scale-dependent product quality characteristics that are not uncommon. Beyond mixing, mass transfer, and shear characteristics, insight is needed with respect to bioreactor control parameters such as pH, temperature, dissolved oxygen, dissolved carbon dioxide, overlay pressure, and feed addition requirements for a successful bioreactor scale-up [8–11].

Scale-up of the downstream purification steps utilizes concepts similar to the upstream cell culture process. In the scale-up of the chromatography columns, the bed height and linear flow rate are generally preserved between the small-scale and the large-scale columns. The volumetric output needed for commercial manufacturing is achieved by increased diameter and bed volume of the large-scale column. Modern chromatography resins provide high protein loading capacity often exceeding 30 g/L. This helps reduce the total bed volume and size of the column necessary. In addition, modern chromatography resins can withstand higher pressures without getting crushed or deformed. This allows operation of the column at higher flow rates shortening the overall processing time. Column packing methods may differ significantly between the small columns used in the laboratory and large columns used in a commercial manufacturing facility. The complexity of packing increases significantly in larger columns just because of the size. In addition, column hardware components such as the screen, flow distribution system, and headplate play much more important role in large-scale columns. Subsequent chromatography section in this chapter provides more information on column packing.

Scale-up of tangential flow filtration (TFF) systems has become easier due to the availability of membrane cassettes of various sizes. For example, Pellicon™-2 TFF cassettes manufactured by Millipore Corporation (Billerica, MA) are available in 0.1, 0.5, and 2.5 m^2 membrane area configurations. Multiple 2.5 m^2 cassettes can be stacked together to provide the necessary membrane area for large-scale ultrafiltration operations. Other manufacturers also offer similar choice of sizes and molecular weight cutoffs to match scale and process needs. TFF can be scaled up by maintaining filtrate volume to membrane surface area ratios the same between laboratory-scale and commercial-scale systems. In addition, membrane material, molecular weight cutoff (determines retention characteristics), channel height and flow path type, and retentate and filtrate pressures are kept similar between the two scales [12, 13]. Other considerations for scale-up of a TFF system include the type of pump used, number of pump passes that the protein solution experiences during the operation, configuration and sizing of the piping, and process time. Rotary lobe pumps are commonly used in large scale, while a small-scale TFF operation may be performed with an air-driven diaphragm pump or a peristaltic pump. The number of pump passes is an important consideration for thermally labile proteins. The retentate protein solution may experience a rise in temperature if the concentration factor is high. Some large-scale retentate tanks are equipped with cooling jackets to prevent heat buildup and potential denaturation of the protein. The size of the piping and the configuration of the flow path in the large-scale equipment are carefully selected to ensure that excessive pressure drop does not occur or frictional forces do not become too large. Generally, these are not major concerns in the laboratory-scale equipment. It is a good idea to maintain similar process time between the small-scale and large-scale systems. This allows process consistency across scale and ensures predictable product quality during scale-up.

Microporous membrane filtration is another important type of unit operation employed in biopharmaceutical manufacturing. Hollow fiber cartridges can be used to produce filtrate in microfiltration. In this case, linear scaling can be readily accomplished by keeping the fiber length similar between scales. The size of the fiber bundle may increase depending on the volume of material to be processed.

Other unit operations that are typically used for bioprocessing are normal flow filtration (NFF), centrifugation, and various types of mixing. While we will not discuss scale-up of each one separately, it is clear from the discussion so far that similar approaches utilizing basic principles of chemical engineering can be successfully used in all these cases.

3.3 CHALLENGES IN LARGE-SCALE PROTEIN MANUFACTURING

Large-scale protein manufacturing utilizes a series of unit operations to grow cells, produce product, and isolate and purify product from the cell culture. In the following sections, we will examine these steps in more detail and discuss chemical engineering challenges associated with these steps.

3.3.1 Bioreactor

A bioreactor provides a well-controlled artificial environment to the protein producing host cells that promotes cell growth and product synthesis. Bioprocesses have multiple seed bioreactor steps to sequentially increase culture volume (and number of cells) that finally culminates in a production bioreactor where product expression occurs. The design criteria for production bioreactor and the seed bioreactors could differ substantially as the goals in the two systems are different. However, the concepts presented here are applicable for either system.

Mammalian cells lack an outer cell wall and are sensitive to shear forces and damage by bursting bubbles. Lack of a cell wall directly exposes a cell's plasma membrane to environmental stresses. Plasma membrane contains enzymes and structural proteins that play a key role in the communication between cell and its environment [14]. Microbial cells are enveloped by a cell wall that imparts much higher tolerance to shear. Growth rates for these microorganisms are also much higher. Thus, the design criteria for fermenters are geared toward providing adequate nutrients to the culture and less concerned with the shear introduced through agitation. Due to these conflicting requirements, very few dual-purpose vessels designed for both mammalian and microbial cell cultivations are in use.

Environmental conditions in the bioreactor such as osmolality, pH, and nutrient concentrations could significantly affect quantity of protein expressed as well as product quality attributes. By product quality attributes, we mean properties of the protein in terms of its structure, function, and stability. Complexity of maintaining live cells under optimal conditions dictates the design of the bioreactors. Over the years different designs of bioreactors have been developed. Most common is the stirred tank system that is widely used for free suspension cell culture. Other designs that are also used commercially are hollow fiber bioreactors, airlift bioreactors, and more recently developed various types of disposable plastic bag-based bioreactors. Each design provides some advantages and unique challenges. We will limit the scope of this discussion to the stirred tank bioreactor system.

Stirred tank bioreactors have emerged as a clear winner in the twenty-first century for culturing mammalian cells. One of the key reasons is that the mammalian cells have proved to be sturdier than initially thought of [15]. Stirred tank bioreactors are available in the widest capacity range (1–30,000 L) and commercially used in cultivating different cell lines (e.g., CHO, MDCK, BHK, NS/0, etc.). Fermenters used in commercial microbial and yeast fermentation processes are also primarily stirred tank vessels. Fermentation processes have been successfully scaled up to 250,000 L capacity.

Bioreactor and fermenter systems operate under aseptic conditions that require exclusion of undesired contaminating organisms. Bioreactors and associated piping are designed as pressure vessels such that these can withstand heat sterilization with saturated steam. All entry points to the vessel for adding and removing gases or liquids are designed to maintain aseptic conditions. The requirement for aseptic processing and heat sterilization of the system is specific to bioprocesses. Factors that determine success of any sterilization regimen are exposure time, temperature, and system conditions. Equipment is designed to avoid buildup of condensate that forms as the saturated steam transfers heat to the vessel interior or to the piping. Buildup of condensate can cause "cold spots" resulting in insufficient microorganism "kill" and failed sterilization. Piping connected to the bioreactor is equipped with steam traps to ensure aseptic removal of the condensate. Exposure to saturated steam causes coagulation of proteins in the microorganisms rendering them nonviable (unable to reproduce). The thermal death of microorganisms is a first-order process [16] and can be described by the following equation:

$$\frac{\mathrm{d}C_{\mathrm{v}}}{\mathrm{d}t} = -kC_{\mathrm{v}}$$

Where C_{v} is the concentration of viable microorganisms, t is the time, and k is the thermal death rate constant. Integration of this equation from a time 0 (t_0) when the viable microorganism concentration is C_{v0} to a finite time t provides the following expression:

$$\ln\left[\frac{C_{\mathrm{v0}}}{C_{\mathrm{v}}}\right] = k(t-t_0)$$

The equation can be used to calculate the time required to achieve a log reduction of microorganisms when the thermal death rate constant for a type of microorganism is known.

Bioreactor operation can be considered to be a two-step chemical reaction. The first step is the cell growth and the second step is the product synthesis by the cells. For growth associated products such as monoclonal antibodies, the two steps are not mutually exclusive and, thus, have to be considered as concurrent reactions. Protein expressed from genetically engineered mammalian cells is constitutive; that is, protein expression is independent of the growth phase of the culture. For nongrowth associated products such as expression of viral antigens, the two steps could be treated as different phases and may even require different operating conditions such as temperature shifts, pH shifts, and so on. Induction or infection of the culture for expression of these nongrowth associated products may even trigger cell destruction, for example, as seen in production of interferons. Similar to the chemical reaction kinetics, stoichiometric and kinetic equations for cell growth and product formation have been developed [17]. Modern laboratory systems are used to determine concentrations of cells, nutrients, metabolites, and product at different stages of the process. These key experimental data generated in the laboratory are used to develop temporal correlations for cell growth, nutrient uptake, and product synthesis. These correlations are used to calculate kinetic parameters such as specific growth rates or specific nutrient uptake rate. Table 3.1 lists the typical cell culture parameters related to growth and product formation along with their definitions.

TABLE 3.1 Parameters Commonly Used for Measurement of Cell Culture Growth and Product Formation

Parameter	Definition	Unit	Mathematical Expression
Growth Rate (G)	Rate of change in number of cells	No. of cells/(L-h)	$G = \frac{\mathrm{d}x}{\mathrm{d}t}$, x = viable cell density (cells/mL) at time t
Specific growth rate	Normalized growth rate	h^{-1}	$\mu = \left(\frac{1}{x}\right)\frac{\mathrm{d}x}{\mathrm{d}t}$
Specific nutrient consumption rate	Normalized uptake rate of a nutrient (e.g., glucose, glutamine, oxygen, etc.)	g nutrient/(g cell-h)	$q_s = -\left(\frac{1}{x}\right)\frac{\mathrm{d}s}{\mathrm{d}t}$, s = nutrient concentration
Specific product formation rate	Normalized rate of product synthesis	g product/(g cell-h)	$q_p = \left(\frac{1}{x}\right)\frac{\mathrm{d}p}{\mathrm{d}t}$, p = product concentration
Specific lactate formation	Lactate formed per unit glucose consumed	g lactate produced/ g glucose consumed	$q_{\mathrm{lac}} = \frac{\mathrm{d[lac]}}{\mathrm{d[glu]}}$, [lac] = lactate concentration and [glu] = glucose concentration

For other design aspects of a bioreactor, most chemical engineering principles for chemical reactor design remain applicable. Chemical engineers play an important role in establishing the design criteria for the bioreactors. These design criteria are established with the goal of providing the organism with optimal conditions for product synthesis. To prevent formation of localized environment, homogeneous mixing of the culture is essential. Furthermore, transfer of oxygen to the culture and removal of carbon dioxide from the system has to be achieved. The critical role for the process engineer is the identification of the design criteria that can meet these requirements. The process engineer analyzes the data generated in the laboratory using small-scale bioreactors and identifies key process operating parameters and their appropriate settings. These often include pH, dissolved oxygen, agitator speed, dissolved carbon dioxide, and initial viable cell density (VCD). The set points and normal ranges of these operating parameters are selected such that consistent and acceptable quality of the protein can be obtained run after run. Typically, the cell culture is optimized to yield the highest product concentration, commonly called titer. However, in many instances, as more titer is obtained from the culture, amounts of unwanted impurities such as fragmented or misfolded proteins, host cell DNA and cellular proteins, cell debris, and so on start to increase. Therefore, a careful balance has to be achieved between titer and amount of impurities. The capacity of the downstream purification steps for impurity removal is a major consideration. If the downstream steps are capable of removing large quantities of impurities coming from the production bioreactor, the cell culture can be forced to produce high titers accompanied by higher amounts of impurities that can be reproducibly cleared. During process development, process is scaled up from a small-scale to an intermediate-scale pilot laboratory bioreactor operated at the same set points as the small-scale bioreactor. Acceptable quality of the pilot-scale material confirms the validity of the scale-up. If successful, then the same criteria are used to scale up to the commercial-scale bioreactors. Otherwise, further characterization work is undertaken to better understand the cell culture process requirements and the process fit to the equipment, and the cycle is repeated.

Key characteristics of stirred tank bioreactors used for production of therapeutic protein molecules are described in the following sections.

3.3.1.1 Vessel Geometry The tanks used for mammalian cell culture were "short and fat." Original bioreactor systems had an aspect ratio of vessel height to inside diameter of 1:1. This geometry allowed proper mixing at low agitation speed. Currently, use of an aspect ratio of 2:1 is widely accepted although use of tanks with 3:1 aspect ratio has also been reported. Matching aspect ratios between the bench-scale and the commercial-scale bioreactors facilitate process scale-up but is seldom used as a strict scale-up criterion. Lower aspect ratio for small-scale bioreactors is acceptable, but for large-scale bioreactors, lower aspect ratio creates a larger footprint that becomes cost prohibitive to implement. Figure 3.1 shows a simplified schematic diagram of a bioreactor showing important internal components of the vessel. There are many possible variations for the configurations of the impellers, sparger assembly, and baffle system.

3.3.1.2 Mechanical Agitation Gentle mixing in the bioreactor is essential to maintain homogeneity, facilitate mass transfer, and support heat transfer. Three blade down-pumping axial flow impellers are the most common design used in bioreactors. Number of impellers used is governed by the aspect ratio of the vessel. A bioreactor with an aspect ratio of 2:1 is typically equipped with two impellers of diameter approximately equal to one-half of the tank diameter. If the impellers are too closely placed, then maximum power transfer is not achieved. For optimal results, space between the two impellers is between one and two impeller diameters. Actual placement of the impeller within this range is also governed by the operating volume in the tank so as to avoid the liquid levels where the rotating impeller is not fully immersed (also known as "splash zone") and may entrap air causing unnecessary and detrimental foaming.

Commercial-scale bioreactors have bottom mount drives. This is preferred because it allows a relatively shorter shaft. A shorter shaft provides structural integrity and absence of wobbling when the agitator is running. Removing the impellers for maintenance becomes easier with a bottom mount impeller system. In addition, headroom requirement within the production suite is also less compared to a top mount drive for which sufficient ceiling height is required to be able to remove the impeller without having to open the ceiling hatch and exposing the bioreactor suite to the outside environment. Shaft seal design for a bottom mount drive bioreactor poses significant challenges. Typically, a double mechanical seal is used that is constantly lubricated by clean steam condensate fed to the seal interface and a differential pressure is maintained across the seal so that bioreactor fluid does not enter the seal space reducing possibility of contamination. For smaller size seed bioreactors, a top mount drive could be considered to eliminate mechanical seal design issues. Design of the seal is often not given the same level of attention as some of the other parts of the system. Apart from the absolute requirement of maintaining aseptic conditions, the ease of maintenance of the seal should also be considered.

The *Re* in a bioreactor is expressed by the following equation:

$$\mathrm{Re} = \frac{\rho_{\mathrm{L}} N D^2}{\mu}$$

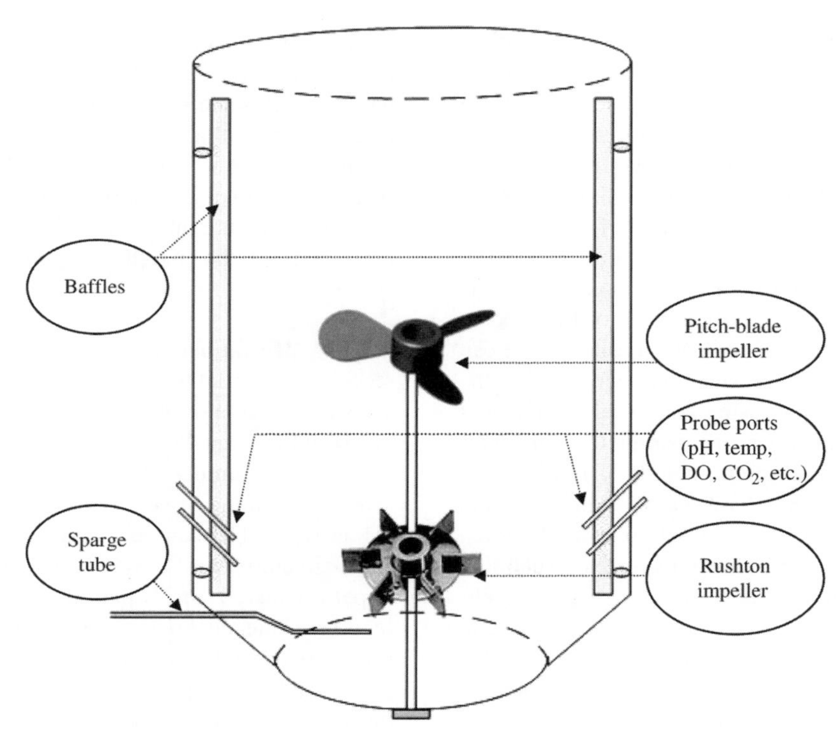

FIGURE 3.1 A simplistic schematic diagram of a cell culture bioreactor showing internal components. Many variations of sparge tube, impeller, and baffle configuration are possible.

where ρ_L is the density of the media (kg/m^3), μ is the viscosity of the media (Pa-s), D is the impeller diameter (m), and N is the impeller speed (revolution/s). The viscosity of the cell culture fluid is close to that of water. Thus, $Re > 10^4$ is often experienced in the bioreactor that allows utilization of turbulent flow theories to analyze the fluid mechanics in the bioreactor.

*3.3.1.3 **Mass Transfer*** Rate of oxygen transfer to the liquid in a bioreactor is governed by the difference between the oxygen concentration in the gas phase and the oxygen concentration in the liquid culture, fluid properties, and the contact area between the gas and the liquid. Pressurized air–oxygen mixture is supplied to the sparger that is either a ring or a tube with open end or with holes on the sidewall. Two spargers may be required in a production bioreactor, while the seed bioreactors may have only one sparger. The size of the bubbles and their distribution are key to the efficiency of mass transfer to the liquid. For a given air–oxygen mix, the higher the concentration of dissolved oxygen (DO) levels in the cell culture fluid, the lower is the efficiency of oxygen transfer to the fluid as the concentration difference is the driving force. To increase mass transfer in the existing equipment, composition of the gas (oxygen enrichment) may be altered or the gas flow rate may be increased. Higher sparge gas flows result in increased carbon dioxide stripping [15]. Thus, as a result of the selected strategy, carbon dioxide levels in the culture may vary. Above a certain concentration (usually partial pressure of >140 mmHg), carbon dioxide has a toxic effect on the cell culture, resulting in product quality issues. Therefore, a careful control of carbon dioxide concentration in the culture is essential for cell health and product quality. This is described later in this chapter in greater detail.

Respiratory quotient (RQ) is defined as the rate of carbon dioxide formation divided by the rate of oxygen consumption. RQ for mammalian cells is close to 1.0. This implies that for each mole of oxygen consumed, 1 mol of carbon dioxide is produced. Generally, bioreactors are maintained at DO levels of >30% of saturation level.

In the steady state, the oxygen transfer rate from the gas bubbles to the cells is matched by the oxygen uptake rate by the cells. The relationship between these two quantities has been described earlier. The mass transfer coefficient k_La in s^{-1} can be determined using the correlation developed by Cooper et al. [18] for a bioreactor with multiple impellers:

$$k_La = K \cdot \left(\frac{P_g}{V}\right)^{\alpha} (v_s)^{\beta}$$

where P_g is the gassed power input to the bioreactor agitator, V is the volume of liquid in the tank, v_s is the superficial velocity, and K is a function of number of impellers (N_i) used in the bioreactor given by the formula

$$K = A + B \cdot N_i$$

where A and B are positive constants.

EXAMPLE 3.1

Mass transfer coefficient, k_La, needs to be determined experimentally in a 10,000 L bioreactor to ensure that the OTR can match the OUR in a cell culture process.

Solution

OTR depends on the design of the bioreactor, impellers, sparging system, and operational parameters such as agitation speed. OUR is determined by the oxygen requirement of the cells for growth and functionality. For a successful cell culture operation, OUR must be, at minimum, met and preferably exceeded by OTR. This will ensure that adequate oxygen is available to the cells for biological functionality. OUR is generally determined from small-scale cell culture experiments. OTR of a large-scale bioreactor can be determined using the following relationship:

$$k_La(C_{sat} - C_L) = \text{OTR}$$

where k_L is the mass transfer coefficient that can be determined experimentally, a is the liquid–gas contact area or interfacial area of gas bubbles from the sparger, and $C_{sat} - C_L$ is the concentration gradient of oxygen between gas bubbles and the liquid.

To experimentally determine k_La for our 10,000 L bioreactor, the dynamic gassing technique is utilized. In this technique, the bioreactor is first filled to the working volume with a suitable pseudomedium or water, and oxygen is purged out of the liquid by equilibrating with nitrogen. Thereafter, oxygen is sparged in the bioreactor and the concentration of oxygen in the liquid (C_L) is measured as %DO at regular intervals using a fast response calibrated dissolved oxygen probe. Table 3.2 shows simulated data from such an experiment at a specific oxygen sparge rate and agitation speed.

The rate of change in oxygen concentration can be described by the following equation:

$$\frac{dC_L}{dt} = k_La(C_{sat} - C_L)$$

This equation can be integrated to yield

$$\ln\left[\frac{(C_{sat} - C_0)}{C_{sat} - C_t}\right] = k_Lat$$

TABLE 3.2 Simulated Dissolved Oxygen Measurements over Time at a Specific Oxygen Sparge Rate and Agitation Speed for a 10,000 L Bioreactor

Time (min)	%DO
0.0	18.61
1.0	19.82
2.0	24.22
3.0	29.12
3.5	31.35
4.0	33.49
4.5	36.12
5.0	39.5
5.5	41.6
6.0	43.26
6.5	45.14
7.0	48.16
7.5	49.94
8.0	51.52
8.5	53
9.0	55.21
9.5	57.29
10.0	58.82
10.5	60.59
11.0	61.94
11.5	63.9
12.0	65.15
12.5	66.45
13.0	67.97
13.5	69.04
14.0	70.68
14.5	71.67
15.0	73.3
15.5	74.3
16.0	75.53
16.5	76.21
17.0	77.51
17.5	79.31
18.0	80.18
18.5	80.43
19.0	81.03
20.0	83.47
21.0	84.72
22.0	86.14
23.0	87.8
24.0	89.24
25.0	90.5
26.0	91.52
27.0	92.95
28.0	93.65
29.0	94.81
30.0	95.7
31.0	96.6
32.0	97.08
33.0	98.33
34.0	98.5
35.0	99.84
36.0	100.25

where C_0 is the initial concentration of dissolved oxygen and C_t is the concentration of dissolved oxygen at time t. From Table 3.2, %DO at $t=0$ (or C_0) is 18.61 and %DO at saturation (C_{sat}) is 100.25. Substituting these values in the equation above and plotting over time, we get a straight line with a slope of 3.43. Therefore, k_La for our 10,000 L bioreactor at the experimental sparge rate and agitation speed settings is 3.43 h^{-1}. Substituting k_La in the earlier equation and assuming C_L to be a %DO set point typically between 20% and 60%, OTR can be calculated. If the OTR is not sufficient to match OUR, k_La can be raised by increasing the sparge flow rate and/or the agitation speed until sufficient OTR is obtained.

3.3.1.4 Heat Transfer In design of bioreactors for mammalian cell culture, heat transfer considerations are not as significant as that in design of fermenters for microbial culture. Mammalian cells have lower metabolic activity and generate less heat that needs to be removed during normal processing. The large-scale bioreactors are jacketed tanks and the bioreactor temperature is maintained by a temperature control module using chilled or hot water. The temperature of the culture generally remains between 30 and 40°C. However, the vessel is exposed to high temperatures during the sterilization by steam in place and requires an appropriate design to withstand thermal stresses.

3.3.1.5 Bioreactor Control Commercial-scale bioreactors are hooked up to a distributed control system (DCS) or, at the very least, have their own stand-alone programmable logic controllers (PLC). Parameters such as pH, dissolved oxygen, temperature, agitation, and gas flows rates are controlled in real time. The control of pH is achieved by adding carbon dioxide to lower pH and sodium carbonate solution to raise pH. A potentiometric pH probe measures the pH of the cell culture, and the signal is sent to a pH control module that then determines the necessity of the additions. Culture pH is maintained within a predefined range for optimum cell culture performance and typically controlled within a deadband around a set point. Use of a deadband prevents the constant interaction of the pH control loops and overuse of titrants. Similarly, the dissolved oxygen is typically measured by a polarimetric dissolved oxygen sensor. The control system determines the need for activation of gas flow to the sparger of the bioreactor through flow control devices. Feed solutions can be added to the bioreactor to supply enough nutrients to the growing and productive cells according to a predetermined regimen. In fed-batch cell cultures, product and metabolites remain in the culture until harvest. In perfusion cell cultures, the supernatant is periodically taken out of the culture to remove product and metabolites while the cells are returned to the culture. Periodic samples are drawn from the bioreactor for optical examination of the culture to assess culture health. Viable cell density, viability, nutrient and metabolite concentrations (e.g., glucose, glutamine, lactate, etc.), osmolality, off-line pH, and carbon dioxide concentration are determined at regular intervals for confirmation of culture health and process controls. Figure 3.2 provides an example of the time profile of some of these parameters. In a fed-batch culture, feed is added at regular interval stabilizing the nutrient supply to the cells. Significant effort has been made to mathematically model the mass transfer of oxygen and carbon dioxide in cell culture. This is described in greater detail later.

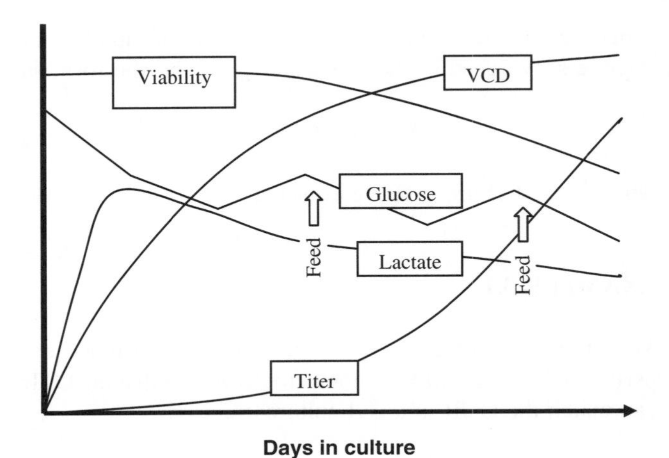

FIGURE 3.2 Example of cell culture parameter profiles that may be observed during the operation of a bioreactor in fed-batch mode (VCD is viable cell density). The arrows represent end of feed addition. Many other parameters not shown here may also be monitored. The profiles will change based on cell line, culture system, product, and selection of operating parameters.

3.3.2 Centrifugation

Initial challenge upon completion of the cell culture is to separate the cellular mass from the product stream. The protein of interest is generally soluble while the cells are suspended in the liquid culture media. In some cases with microbial fermentation, the protein of interest can be present at very high concentration within the cells in structures known as "inclusion bodies." Retrieval of the protein from these structures requires not only separation of the cells, but also rupture of cells and processing of the inclusion bodies. We will not discuss processing of inclusion bodies in this chapter. The cells also contain DNA, host–cell proteins, and proteolytic enzymes that have the potential to damage the protein of interest if released. Thus, to simplify the purification process, it is essential to remove the cellular mass while maintaining cellular integrity to keep the intracellular contents out of the process stream. To achieve this goal, various solid–liquid separation techniques have been used.

These techniques include centrifugation, microfiltration, depth filtration, membrane filtration, flocculation, and expanded bed chromatography [19]. These unit operations are used in a combination that depends on the process stream properties. Expanded bed chromatography has been developed as an integrated unit operation that combines harvest with product capture, but to date practical limitations have kept this technique from being adopted in large-scale operations [19].

The initial step of the harvest is either centrifugation or microfiltration for a large-scale operation. While not very common, processes can also have only a depth filtration step to separate cells from the culture fluid. This is known as the primary recovery step and is followed by a secondary clarification step that typically consists of a sequence of depth filtrations. Final secondary clarification step uses a sterilizing-grade membrane filter. It provides a bioburden-free process stream for the downstream purification operation.

In the quest for higher titers, the cell densities in the production bioreactor have been increased consistently along with processes being run for longer durations. This combination results in lower cell viability and increased cell debris at the harvest stage. In this scenario, centrifugation for primary recovery is becoming the method of choice as it can handle higher concentrations of insoluble material compared to the alternative techniques. Centrifugation uses the difference in densities of the suspended particles and the suspension medium to cause separation. A centrifuge utilizes the centrifugal force for accelerating the settling process of the insoluble particles. Cells in the harvest fluid can be approximated to be spherical. During centrifugation of this fluid, the cells are accelerated by the centrifugal force and at the same time experience the drag force that retards them.

From Stoke's law, a single spherical particle in a dilute solution experiences a drag force that is directly proportional to its velocity. A cell in suspension accelerates to a velocity where the force exerted by the centrifugal force is balanced by the drag force. Thus, based on the applied centrifugal force, cells achieve a terminal velocity in the centrifuge bowl. This property is exploited to effect the desired separation of the insoluble cellular mass from the process fluid. The terminal velocity of the cell (or other small insoluble spherical particles) can be calculated by the following equation [20]:

$$v_\omega = \frac{d^2}{18\mu}(\rho_s - \rho)\omega^2 r$$

where v_ω is the settling velocity; d is the diameter of the insoluble particle (cell); μ is the viscosity of the fluid; ρ and ρ_s are the densities of the fluid and the solids, respectively; ω is the angular rotation; and r is the radial distance from the center of the centrifuge to the sphere.

In commercial applications, continuous disk stack centrifuges are used. The design of these units is based on the above principle. Continuous disk stack centrifuges have multiple settling surfaces (disks) that yield high throughput and consistent cell-free filtrate (centrate). Settling velocity (v_s) is correlated to the operation and scale-up of the centrifuge by using the Σ (sigma) factor. The Σ factor relates the liquid flow rate through the centrifuge, Q, to the settling velocity of a particle according to the following equation:

$$v_s = \frac{Q}{\Sigma}$$

The ratio of flow rate to the Σ factor is held constant during scale-up of a centrifugal separation [20]. This ratio can be used as a scale-up factor even if two different models of centrifuges are involved. For a disk stack centrifuge, Σ factor is given by

$$\Sigma = \frac{2\pi n\omega^2}{3g}(R_0^3 - R_1^3)\cot\theta$$

where n is the number of disks; ω is the angular velocity; R_0 and R_1 are the distance from the center to the outer and inner edges of the disks, respectively; and θ is the angle at which the disks are tilted from the vertical. The Σ factor is expressed in the unit of $(\text{length})^2$. The calculation of the Σ factor varies for different types of centrifuges, while the measurement unit remains the same.

Disk stack centrifuges have a high upfront cost. Also, because of the complexity of the design, availability of equipment suitable for various scales of operation is very limited. Therefore, often harvesting at small scale is performed with a completely different type of centrifuge, or sometimes without a centrifuge. Even at pilot scale, representative equipment is not always available. This brings additional risk to the scale-up of the harvest process. Over the past decade, the two dominant continuous disk stack centrifuge manufacturers Alfa-Laval (Lund, Sweden) and GEA Westfalia (Oelde, Germany) have developed specific products for the biopharmaceutical application and have attempted making comparable pilot-scale equipment to address scale-up concerns. In addition to improvements in centrifuge design, depth filtration systems have also improved with filtration media capable of reliably removing cell debris, impurities, and other process contaminants. The combination has greatly improved the reliability and robustness of the harvest operation despite confronting more and more challenging source materials. With the recent improvements, harvest yields of >98% are being reported [19]. Harvest operation has been made robust to absorb the variability in the feed stream and yield a consistent output stream for the downstream chromatographic steps.

3.3.3 Chromatography

Liquid chromatography has been used in biotechnology in all phases of product development. In research, small-scale columns and systems have been the workhorse for purifying proteins that are used in several facets of drug development including high-throughput screening, elucidation of the three-dimensional structure using X-ray crystallography and NMR, and nonclinical studies. Although chemical engineering concepts apply universally to chromatography, they are neither as relevant nor as necessary to adhere to at the research stage. This is primarily because the cost of producing the protein and the amount of protein required are both small and success is primarily measured by being able to achieve the required purity in the shortest timeframe. The situation is quite different when a therapeutic protein is approaching commercialization. This now requires chromatography to be used as a key unit operation in the manufacturing process for the protein. Therapeutic proteins, including monoclonal antibodies, antibody fragments, fusion proteins, and hormones, derived from bacterial and mammalian cell culture processes, represent a sizeable portion of commercially available recombinant proteins. It is typical for the purification process for such molecules to include two to three chromatography steps. The column used in these processes can be up to two meters in diameter and can weigh several tons. To use chromatography successfully in large scale, it is important to ensure that the concepts of chemical engineering are incorporated early during process development and are carried forward through the development process and subsequently during routine manufacturing.

Fundamental to success of using chromatography is selecting the right ligand chemistry irrespective of scale or phase of development. This results in the required "selectivity." Selectivity is a measure of relative retention of two components on the chromatographic media. Most of the chromatography media used across scales consist of porous beads with an appropriate ligand (for the desired selectivity) immobilized throughout the surface area available. The beads may be compressible or rigid and can be made of a number of substances such as carbohydrate, methacrylate, porous glass, mineral, and so on. Although most beads are spherical, there are asymmetrical ones that are commercially available (e.g., PROSEP® media from Millipore Corp., Billerica, MA). For the porous beads, due to high porosity most of the ligands are immobilized in the area that is "internal" to the bead. The average bead diameter for most commercially used resins is between 40 to 90 μm. In order for the chemical interaction to occur, the protein of interest present in the mobile phase needs to be transported to the internal area of the beads via the pore structure and subsequently from the mobile phase to the ligands immobilized within the pores. Mass transfer theory, a key chemical engineering concept, plays a major role in understanding and predicting the behavior of such transport. Since the chromatography media are packed in relatively large diameter columns, appropriate packed bed stability and ensuring that the mobile phase is equally distributed across the column are critically important.

3.3.3.1 Mass Transfer in Chromatography Columns In the simplest case of a chromatography process, a mixture of a protein of interest (A) and an undesirable contaminant (B), referred to as the "load," is applied to a packed bed. As a result of convective and diffusive mass transfers, the load injected into the column as a bolus starts to assume a broader shape. As the load traverses the packed bed length, the peaks corresponding to the components A and B begin to separate based on the selectivity of the resin. At the same time, the peaks start to get broader as they move through the bed (Figure 3.3).

The peak broadening, which is a result of mass transfer resistances and diffusion, may lead to the overlap of the peaks of components A and B. The best separation can be achieved if there is no overlap between the peaks of components A and B by the time the whole packed bed length is traversed. Hence, the overlap caused in part by mass transfer plays a key role in achieving the desired level of purification. The characteristics of mass transfer in a packed bed under flow have been extensively studied [21]. The effect of mass transfer on separation of components and purification

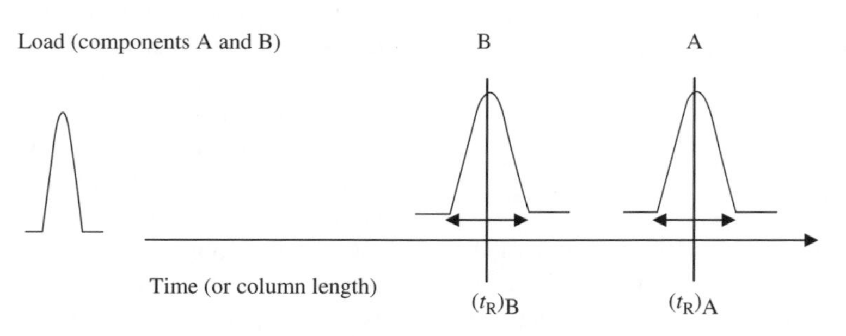

FIGURE 3.3 Schematic illustrating band broadening phenomenon.

efficiency is described by the term resolution. Resolution is described by the following equation:

$$R = \frac{2[(t_R)_B - (t_R)_A]}{W_A + W_B}$$

where $(t_R)_A$ is the retention time of early eluting peak, $(t_R)_B$ is the retention time of late eluting peak, W_A is the width of early eluting peak, and W_B is the width of late eluting peak.

The higher the separation efficiency, the better the resolution of the column. Concepts of height equivalent of theoretical plate (HETP) and residence time distribution have been successfully applied to quantify efficiency of a chromatography column. The van Deemter equation describes various mass transfer phenomena ongoing during chromatography, their impact on efficiency (as quantified by HETP), and the operating parameters that impact mass transfer. The resistance to mass transfer results in what is commonly referred to as "band broadening," that is, broadening of the peaks as they travel through the column as mentioned earlier. Higher HETP represents more "band broadening" and less separation efficiency. According to the van Deemter equation [22], the HETP is composed of three terms that describe three different mass transfer mechanisms.

$$H = A + \frac{B}{u} + Cu$$

where A represents eddy diffusion resulting from flow path inequality, B represents molecular diffusion, C represents all other resistances to mass transfer [21, 23], and u is the interstitial velocity defined by

$$u = \frac{\text{mobile phase flow rate}}{\text{porosity} \times \text{column cross-sectional area}}$$

HETP is expressed in the unit of length. As the flow rate increases beyond an optimum, the van Deemter equation predicts that for conventional media the efficiency of the column decreases (leading to increased plate height) (Figure 3.4). In addition to the van Deemter equation, many other mathematical relationships have been derived to model the band broadening phenomenon [21].

Having an understanding of how various parameters impact the efficiency of the column is critical for large-scale operation. A compromise in efficiency may lead to loss in yield, effective binding capacity, and product quality attributes such as purity. Acceptance criteria are generally established during process development and applied during commercial manufacturing to ensure that a chromatography column has the requisite separation efficiency (as measured by HETP) and appropriate peak shape during use. More recently, *transition analysis* is being used to get a more comprehensive understanding of column efficiency and

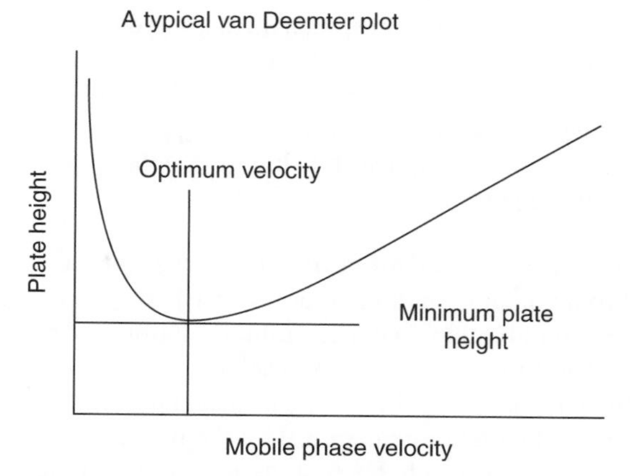

FIGURE 3.4 Van Deemter equation is the most commonly used model for "band broadening" in chromatography. For conventional media (plot shown), the resolving power decreases as the flow rate increases.

packed bed integrity. The van Deemter equation suggests that bead diameter and flow rate play a key role in mass transfer. Interestingly, the selection of these two parameters is based less on any mass transfer calculation and more on practical limitation set by acceptable pressure drop across the packed bed and capability of equipment in use. In addition, for columns in excess of 1 m diameter, time and cost sometimes determine the selection of normal operating range for flow rate. Despite these economic considerations, the scale-up principles are indeed based on the engineering fundamentals. Linear flow rate and bed height are generally held constant during scale-up to ensure that the mass transfer conditions under which the process was developed remain consistent across scale. As mentioned previously, this allows maintaining (or at least the best effort is made) comparable yield and purity from process development to commercial manufacturing.

Concepts of chemical engineering have also aided in the development and characterization of novel types of chromatographic media as highlighted by development of perfusion media. In perfusion media (e.g., POROS® from Applied Biosystems, Foster City, CA), the pore structure is controlled such that a balance is maintained between diffusive and convective mass transfers. This is achieved by having a certain percentage of "through pores" in the beads in addition to the network of smaller pores that branch from the "through pores." The "through pores" do not end within the beads and thus serve as transportation highways for solutes through the beads. The "through pores" with pore diameter greater than 5 nm maintain high degree of intraparticle mass transfer compared to diffusive (nonperfusive) media, whereas the smaller pores branching from them (diameters in the range of 300–700 Å) provide the adequate binding capacity [23]. Thus, such media can retain efficiency at significantly higher flow rates compared to conventional diffusive media, such as agarose-based beads. Tolerance to high flow rates and pressures enables the perfusion media to be the media of choice for high-throughput analytical chromatography.

3.3.3.2 Flow Distribution in Chromatography Columns

A chromatography unit operation consists of process steps such as equilibration, loading, washing, elution, and regeneration. The aim of these steps is to achieve the right condition for ligand binding, facilitate target/ligand interaction, collect the purified target protein, or restore the ligand to the initial state for the next batch. Each of these steps is carried out by flow of predetermined volumes of fluid through the column. For acceptable performance of these individual steps and the chromatography operation as a whole, uniformity of flow distribution across the column cross section is critical. Uniform flow distribution is a prerequisite for ensuring comparable mass transfer throughout a large commercial size column. Little attention is paid during bench-scale process development to flow distribution, since for small columns, typically of 1–3 cm in diameter, this is not a major concern. The column frit is sufficient to achieve adequate flow distribution. As the diameter of the column increases, achieving a uniform flow distribution becomes increasingly difficult. The design of the column headplate plays a key role in flow distribution and the efficiency of the packed bed [24]. Computational fluid dynamics (CFD) can be successfully utilized to assist in the design of chromatography column hardware [25]. CFD utilizes fundamental fluid mechanical and mass transport relationships, representative boundary conditions, and mathematical algorithms to predict fluid properties in a variety of fluid flow scenarios. Recently, the pharmaceutical industry has shown increasing interest in CFD for providing insight into flow characteristics and related phenomena that can help mitigate risks associated with scale-up as well as in troubleshooting [25]. Further discussion of CFD and its applications can be found in a subsequent chapter in this book. Frontal analysis techniques such as dye testing can be used to study the flow distribution. In brief, dye is injected into a packed and equilibrated column and allowed to flow for a predetermined period of time. Subsequently, the column is dismantled to expose the resin bed. Upon excavation, the dye profile in the bed reveals the quality of flow distribution. Alternatively, one can also look at the washout of the dye front from a column under flow and evaluate the effectiveness and uniformity of the column cleaning conditions. An appropriate cleaning procedure would be indicated when little or no residual dye is left, which can be confirmed by excavation of the resin bed after fluid flow through the column simulating the cleaning conditions.

It is ideal to use a complementary approach where CFD is used early in the hardware design process to ensure that predicted flow distribution is appropriate. Subsequent to the fabrication, dye testing can be performed to confirm the uniformity of flow distribution and flow conditions. This approach was utilized for characterization of flow dynamics in a process chromatography column [25]. A few highlights of this study are shown here. Figure 3.5 shows liquid velocity profile predicted by CFD within a packed bed under specific flow rates with clear areas of stagnation. Subsequently, when dye testing was performed, as shown in the lower panel of Figure 3.5, the same areas failed to show dye washout demonstrating zones of stagnation and confirmed the predictions from CFD modeling.

Why is flow distribution so critical for the success of large-scale chromatography? In addition to the issue of mass transfer and its relationship to purity and yield, there are many other factors that come into play for manufacture of biopharmaceuticals using chromatography. Without appropriate flow distribution, parts of the packed bed may remain unreachable for the process fluids. This reduces the total binding capacity of the bed and underutilizes expensive chromatography media such as Protein A media. The lack of flow uniformity and pockets of stagnation within the bed also increases the possibility of suboptimal cleaning and elevates the risk of microbial growth and protein carryover

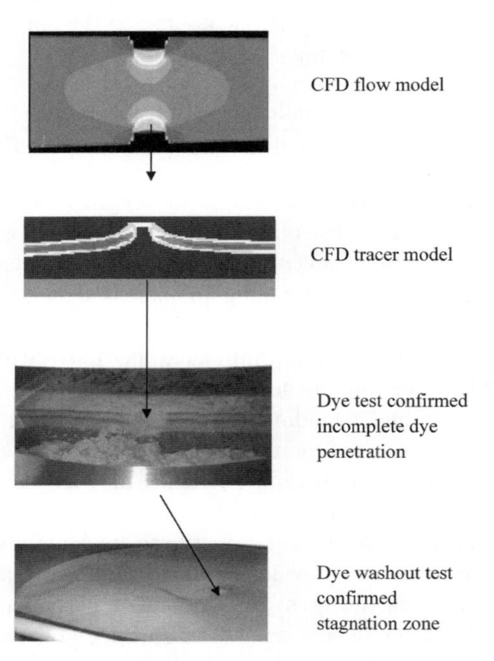

FIGURE 3.5 CFD flow and tracer modeling predicted stagnation zone beneath the chromatography resin introduction port, which was confirmed by dye test.

from one manufacturing batch to the next, creating significant safety and compliance concerns.

The chromatography system or "skid" design should ensure that appropriate flow can be delivered without excessive pressure drop from system components. Skid components (e.g., pumps) and piping should be selected appropriately to meet this requirement. The chromatography media used for biotechnology products are generally compressible, and hence, cannot withstand pressures in excess of 3 bar, thus putting practical limitation on flow rates and bed heights.

3.3.3.3 ***Column Packing and Packed Bed Stability*** Column packing for compressible resins is generally achieved by delivering a predetermined amount of media slurry to the column followed by packing the bed to a predetermined bed height. The slurry amount and the bed height are related such that the resin bed, upon achieving the target bed height, is under a target compression. The compression factor can be recommended by the manufacturer of the chromatography media or can be determined experimentally. In either case, it should be confirmed during packing development. Several engineering considerations are relevant during scale-up and commercial manufacturing to ensure that packed bed is fit for use. As mentioned earlier, due to the compressible nature of most of the chromatography media, operations are limited to lower pressures only. In a small-scale column with a smaller diameter, a significant part of the packed bed is supported by the frictional forces between the bed and the column wall. This is termed as the "wall effect." As the diameter of the column increases, the extent of the "wall effect" decreases. As a result, for the same bed height and flow rate, the pressure drop increases as the column diameter increases. Work performed by Stickel and Fotopoulos [26] can be used to predict pressure drop for larger columns once the data from the small scale are available. Additional work has been performed recently in an effort to improve the prediction models [27].

Column packing is time and resource intensive. For commercial manufacturing, the columns, once packed, are used for many cycles. This mandates that the packed bed remain stable and integral during multiple uses over long periods of time. Tools have been developed to measure and monitor bed stability. Qualification tests can be performed using these tools upon completion of packing. Traditionally, this has been done by injecting a tracer solution (e.g., salt solution or acetone, usually 1–2% of the column volume) and recording the output signal (either solution conductivity or absorbance of an ultraviolet light beam) at the column exit. Peak attributes are then used to calculate HETP and asymmetry (A_f) and compared to predetermined acceptance criteria. It should be noted that due to the significant resistance to protein mass transfer inside a bead, estimated plate heights for protein solutes are much greater than those obtained using a salt or an acetone solution. Hence, the results of column qualification are primarily indicative of packing consistency and integrity rather than the extent of protein separation [28]. Transition analysis, a noninvasive technique for monitoring the packed bed, is finding increasing utilization in process chromatography. Transition analysis is a quantitative evaluation of a chromatographic response at the column outlet to a step change at the column inlet [29]. It utilizes routine process data to calculate derived parameters that are indicative of the quality of the packed bed. All, or a subset of these parameters, can be used to qualify the column after packing, and subsequently during the lifetime of the column to assess bed integrity.

Moreover, a robust column packing procedure needs to ensure that the amount of resin packed in the bed is accurate and reproducible. This is achieved by effective mixing of media slurry, accurate measurement of media fraction in the slurry, and the slurry volume. Application of chemical engineering concepts can help ensure success of these measurements.

3.3.4 Filtration

Filtration is used across pharmaceutical and biopharmaceutical industries spanning all phases of drug development and commercialization. It provides a "quick" option to achieve a size-based separation. This is especially true when the difference in the molecular weight/size is significant (one order of magnitude or more). For this reason, research laboratories have used filtration as a workhorse in a variety of applications ranging from buffer exchange (dialysis) to removal of particulates. Similar to discussions presented in Section 3.3.3, yield is generally not a major performance parameter at the laboratory scale. Also, the filter and the system sizing are not rigorously performed. It is generally based on picking an "off-the-shelf" item that is judged to best suit the needs. In large-scale operation, yield is a major consideration, especially when the protein is expensive to produce. Appropriate filter sizing is also important as it determines the filtrate quality, time of operation, and cost.

At the heart of achieving separation by filtration are the membranes and filter media that provide the pore structure and size required to meet the separation performance requirement. Even though the primary mechanism of separation is based on size, in many cases that alone is not sufficient and charge interaction and/or adsorption are also exploited. Manufacturers of filter membranes do not always follow the same methodology to rate the membranes for pore size. Therefore, the rating provides only a first approximation of the retention capability of the membrane. Based on pore size and filter media structure, three broad categories exist as described in Figure 3.6. These categories and their application will be discussed later in this chapter.

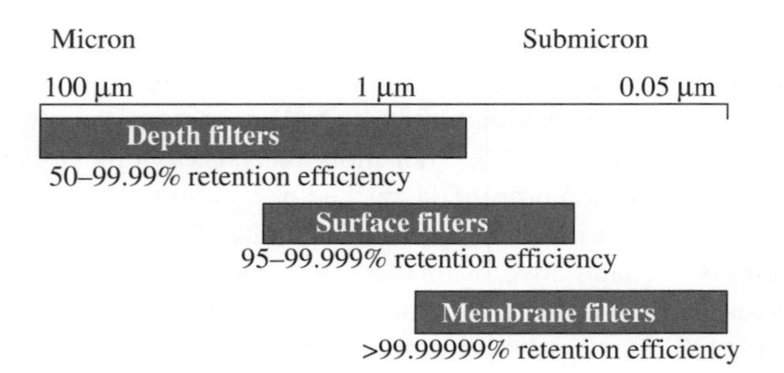

FIGURE 3.6 A comparison of filter media with respect to retention efficiency and pore size. (Figure courtesy of Millipore Corp.)

Filtration is performed primarily in two modes: normal flow and tangential flow (Figure 3.7). In NFF, the bulk flow on the retentate side is normal to the filter surface. During the course of filtration, particulates that are not allowed to pass through the filter can either plug the pores or build a residue cake on top of the filter. Both these events lead to reduced flux that may then lead to selection of larger filter surface area if process time is a major consideration. The theoretical models describing filtration are based on the relationship between pore dimensions and nature of components being filtered out and use pore plugging, cake buildup, or both as primary mechanism. Gradual pore plugging occurs when small deformable particles build up inside of the pores (Figure 3.8). The particles restrict flow through the pores and the flow decreases. At first, the flow decays relatively little, but as the effective diameter of the pores decreases further, the pace of the blockage rapidly increases and the resulting flow substantially decays (Figure 3.9). The gradual pore plugging model is recognized as most applicable to biological process streams.

The build up on the surface, referred to as formation of concentration polarization or gel layer, results in increased resistance to flow, which is undesirable. The TFF mode of operation attempts to mitigate this undesirable situation by having the flow parallel or tangential to the filter surface (Figure 3.7). This tangential flow, also called cross-flow, creates a "sweeping" action resulting in less gel layer and increased filtrate flux. Chemical engineering principles and empirical modeling have been extensively used to model membrane fouling and gel layer formation, and understand their relationship to operating parameters. This understanding leads to successful process development and scale-up of filtration. The flux through the filter can be expressed as the ratio of the driving force (i.e., transmembrane pressure) and the net resistance as shown in the following equation:

$$J = \frac{\text{TMP}}{(R_{\text{m}} + R_{\text{g}} + R_{\text{f}})}$$

where J is the flux through the membrane (commonly in L/(m^2 h)), TMP is the transmembrane pressure defined as the difference of average pressures on the retentate and permeate sides of the membrane, R_{m} is the resistance to flow through the membrane, R_{g} is the resistance to flow through the gel layer, and R_{f} is the resistance to flow due to membrane fouling. Empirical correlations relating flux to tangential flow rate under laminar and turbulent flow conditions have been developed [30].

Normal flow filters can be further classified as depth or membrane filters. Depth filters are used for clarification and prefiltration with higher solids content. They remove

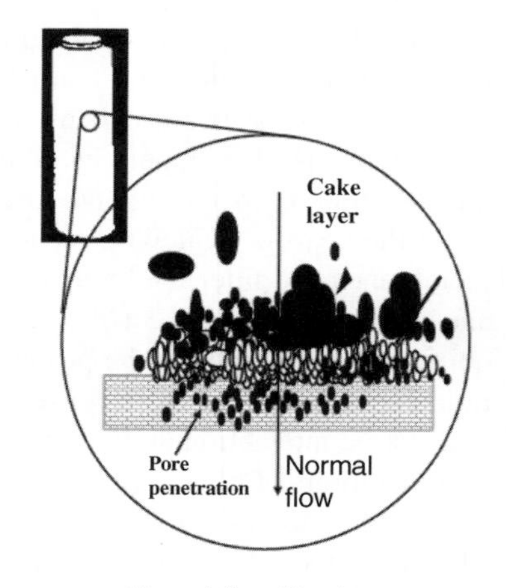

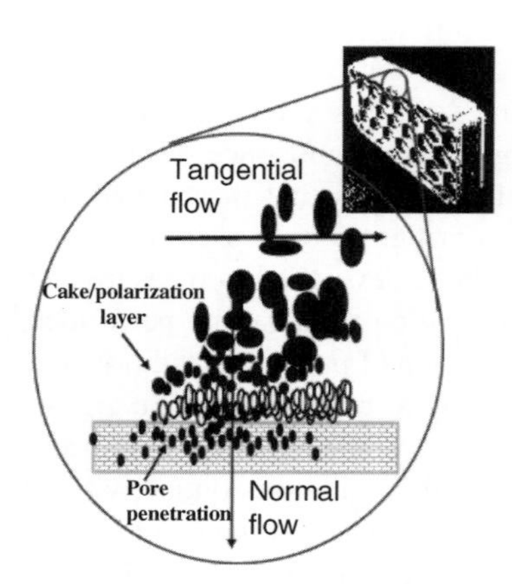

Normal flow filtration   Tangential flow filtration

FIGURE 3.7 An illustration of resistance to flux during the two modes of filtration. Components that penetrate the filter result in fouling. The components that build up on the surface result in the formation of cake and concentration polarization layer. Concentration polarization can be controlled by tangential flow. (Figure courtesy of Millipore Corp.)

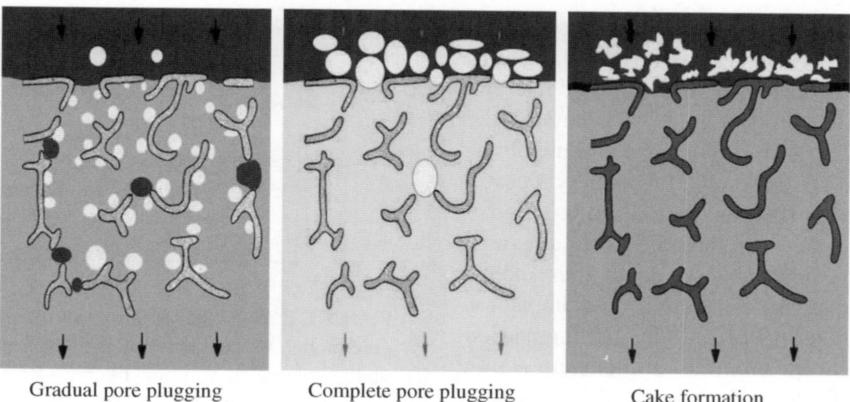

FIGURE 3.8 Mechanism and models for sizing normal flow filtration. Gradual pore plugging model is recognized as most applicable to biological processes. (Figure courtesy of Millipore Corp.)

particles via size exclusion, inertial impaction, and adsorption. They have high capacity for particulate matter and lower retention predictability. Membrane filters are generally used for prefiltration and sterilization. In membrane filtration, particles are removed via size exclusion. Membrane filters have very high retention predictability. Tangential flow filters are also membrane filters but with capability to retain various molecular sizes. Microfiltration devices are made of membranes that can retain cells, cell debris, and particulate matter of comparable size. Microfiltration can be utilized for separation of cells from the cell culture fluid at harvest. Ultrafiltration is used for concentration of protein solutions or for buffer removal or exchange. Most ultrafiltration operations in large scale are performed in the tangential flow mode to maximize throughput and minimize process time.

To successfully use filtration in a biopharmaceutical manufacturing process, the following three steps should be followed: (1) correct filter type and mode of operation should be selected, (2) filter size should be optimized for the type and quantity of product, and (3) system components (e.g., pumps, piping, housing, etc.) and size should be optimized for the application. Chemical engineering principles related to filtration play a key role in filter selection and determination of optimal size. Principles related to pumps and fluid flow through pipes must be considered to ensure that the system is sized appropriately.

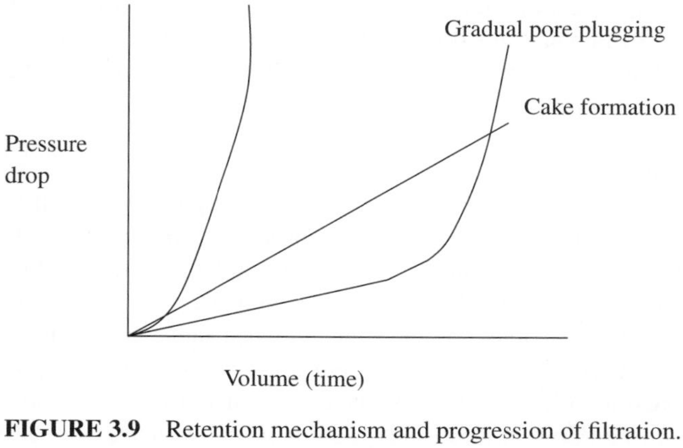

Complete pore plugging
Gradual pore plugging
Cake formation
Pressure drop
Volume (time)

FIGURE 3.9 Retention mechanism and progression of filtration.

3.3.4.1 Filter Selection and Sizing At a high level, filter selection and sizing is based on three primary elements: (1) chemical compatibility, (2) retention, and (3) economics.

3.3.4.1.1 Chemical Compatibility The process streams in biotechnology generally do not contain harsh chemicals, and therefore, chemical resistance to process streams is not a major concern. The filters operated in NFF mode are typically not reused and do not require cleaning. The filters operated in TFF mode are typically reused and are cleaned after every use. These membranes need to be compatible with cleaning agents such as sodium hydroxide, sodium hypochlorite, or detergent. Given that many biopharmaceuticals are injectables, the material of construction of a filter should be such that it does not add significant amounts of leachables and extractables into the product. In addition, material chosen should not result in high levels of adsorption of the product as it may lead to significant yield loss. Chemical engineers working in biopharmaceutical industry are responsible for selecting the appropriate material of construction for the filters used. Knowledge of materials sciences and polymer sciences are important in successful filter selection.

3.3.4.1.2 Retention At the initiation of the selection and sizing process, a well-defined design requirement is finalized. This requirement, at minimum, consists of (1) nature of the process fluid and components desired to be retained/not retained during the filtration step, (2) volume of process fluid that will undergo filtration, and (3) total time allowed for execution of the filtration operation. The required retention characteristic of a specified process fluid will generally steer the selection in the target range of pore size and membrane type as shown in Figure 3.6. Generally, for primary recovery steps such as separation of cells from the cell culture fluid, or removal of large particulates from the process stream, depth filters are used. Filtration mechanisms in depth filters include size-based exclusion, charged interaction, and impaction. Complete depth of the filter is utilized for catching particulates. Generally, for complex mixtures, a filter train is used where different types of filters may be placed in sequence. Alternatively, newer filtration products that have multiple layers of different filtration media assembled in one unit can also be utilized. Membrane filters are typically placed at the end of depth filters to obtain particulate-free process fluid. These can be nominally rated, absolute rated, or a combination of the two. Absolute rated filters follow a strict cutoff that allows retention of particles above a certain size. Unlike depth filters, only the topmost layer of a membrane filter is responsible for separation. The rest of the filter serves as a mechanical support that prevents the filtering layer from collapsing under pressure from the process fluid.

Ultrafiltration is utilized to remove or exchange soluble components from a protein solution. Its widespread uses in biopharmaceutical processes include removal of smaller molecular weight process additives, removal of small organic compounds, change of buffer system to impart special properties to the protein, and concentration of a protein solution. Ultrafiltration membrane filters have the smallest pore size. Filters composed of primarily polyether sulfone or regenerated cellulose in the range of 5–300 kDa nominal molecular weight cutoff are commercially available.

3.3.4.1.3 Filtration Economics Upon deciding the type of the filter that will be most suitable for the application, determination of appropriate size of the commercially available unit is the next step. Many choices exist in filtration with multiple suppliers providing an array of filtration products with respect to filter type, size, material of construction, pore size, mode of operation, and cost. Some filters are manufactured for specific applications, such as virus filters that are used to obtain validated levels of virus reduction in the process stream. Filter selection process requires determination of required filter size for these competing products so that a primary and a backup can be selected. The aim is to achieve design requirements using minimum area, that is, maximize flux during filtration. The filter selection and sizing is, in large part, an experimental process. However, chemical engineering principles, as they apply to filtration, have been used extensively to understand and model flow through the filters so that the experimental design is optimal. Sizing of filter is often done experimentally by determination of a filter's maximum "filterable" fluid volume before plugging (V_{max}) using the actual process fluid. A characteristic curve generated by such experimentation is shown in Figure 3.10. The V_{max} can be subsequently used to determine the minimum filter area required per unit volume of process fluid as shown in the following equation:

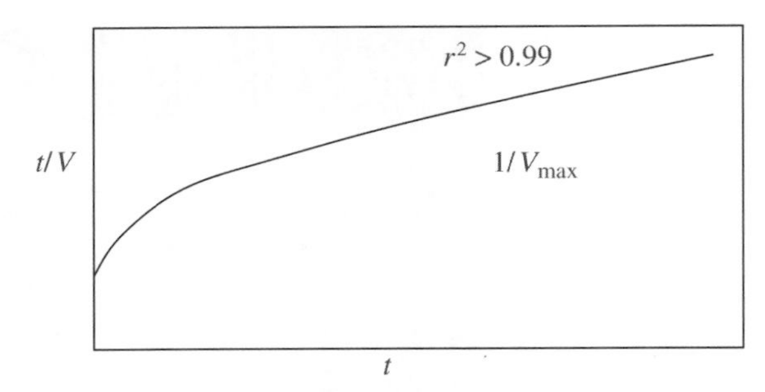

FIGURE 3.10 Filter sizing experiment using gradual pore plugging model. Process solution is filtered at a constant pressure at lab scale. Volume of solution filtered (V) with respect to time (t) is recorded and plotted as shown above. Filter with the highest V_{max} will result in least area requirement.

$$A_{min} = V_B\left(\frac{1}{V_{max}} + \frac{1}{J_i \times t_B}\right)$$

where A_{min} is the minimum filter area, V_B is the batch volume V_{max} is maximum volume (normalized) that can be filtered, t_B is the batch processing time, and J_i is the initial filtrate flux. The actual filter area recommended is always greater than the minimum area but generally is based on economic factors such as filter cost and processing time.

EXAMPLE 3.2

For a cell culture operation, 10,000 L of culture media is to be sterilized through a 0.2 μm absolute grade filter. The operational limitations in the plant require that the filtration be completed in about 2 h. The filter selected can withstand a maximum pressure drop of 20 psid. To design the appropriate filtration system for the plant, the filtration area needs to be calculated.

Solution

Filtration is a physical process of removing insoluble solids from a fluid stream by placing a porous filter in the flow path that retains the insoluble components and allows the fluid to pass through. As fluid passes through the filter, the filter medium becomes clogged and the resistance to the flow increases. Thus, either the driving force determined by the pressure differential across the filter is increased to maintain constant flow, or a constant pressure is maintained across the filter and the flow rate of the fluid is allowed to recede as the pores in the filter medium become plugged. We will assume that the differential pressure across the filter is kept constant for our application.

To estimate the minimum filter area and the filtration time, a small-scale flow decay study is executed using a capsule filter with 40 cm^2 (0.004 m^2) filtration area that is a scale-down version of the large-scale filter. Media prepared in the laboratory is passed through the capsule filter, and the filtrate volume is measured over time. The inlet and outlet pressures are monitored so as not to exceed the rated pressure drop. Table 3.3 provides simulated data of the study conducted at ambient temperature.

Since constant pressure differential is maintained in this case, the partial pore plugging model can be utilized. A relationship can be derived by plotting t/V against t and then performing a linear fit of the data. The plot resembles the curve shown in Figure 3.10. The slope of the line is $1/V_{max}$, where V_{max} is the maximum volume of media that can be filtered through the capsule. A linear fit of t/V against t from Table 3.3 (plot not shown) provides a slope of 0.0004 mL^{-1} with an R^2 of 0.93 indicating a good fit to the model. V_{max} can then be calculated to be 1/0.0004 or 2500 mL (2.5 L). The minimum filter area A_{min} can then be calculated as follows:

$$A_{min} = \frac{10,000\ \text{L}}{2.5\ \text{L}} \times 0.004\ \text{m}^2 = 16\ \text{m}^2$$

Finally, we check for the processing time requirement. Fluid flux is calculated for each data point and an average flux of 294 L/(m^2 h) is calculated from the individual fluxes. Using the calculated filter area above and the average flux determined from the data, we can calculate the estimated processing time to be 10,000 L/(16 m^2 × 294 L/(m^2 h)) = 2.12 h, close to our processing time requirement of 2 h. For the large-scale filter sizing, we would include a factor of safety of 50% over the calculated minimum filtration area to account for any scale-up issues and lot-to-lot variability in media components. Therefore, applying the safety factor, recommended filter area for this application would be 16 × 1.5 = 24 m^2.

The filtration cost also plays a role in determining if the mode of filtration will be NFF or TFF. Microfiltration operations can be mostly run successfully in a normal flow filtration mode (NFF) even though TFF has been used in cell culture harvest operation. The ultrafiltration step, however, is exclusively run in TFF mode for commercial operations. This is primarily due to the extent of concentration polarization layer on the filter surface under normal flow operations. The layer consisting of high concentrations of retained proteins acts as an additional source of resistance and significantly impedes the flux. In fact, in some cases the gel layer can also change the retention characteristics such that components smaller than the membrane pore size are also retained. Operation in a tangential flow mode helps reduce the buildup of the gel layer due to the sweeping action caused by the flow tangentially to the membrane. Concepts of chemical engineering have also shaped the understanding of how the gel layer impacts the flux. It has led to the understanding that the flow rate across the membrane and the pressure differential between the retentate and the permeate side can be varied to get two filtration regimes—one where pressure differential controls the flow, and the other where the gel layer controls the flow. This understanding helps form a strategy for operation of an ultrafiltration unit in a manner that minimizes the gel layer and maximizes the flux.

TABLE 3.3 Simulated Data from a Bench-Scale Filtration Study

Elapsed Time, t (min)	Volume Passed, V (mL)
0	0
3	91
6	272
12	496.6
17	683.5
22	607.5
27	896
32	964
37	1034
42	1123
47	1195.1
52	1246
57	1295
62	1346
67	1402
69	1416

3.3.4.2 System Sizing and Selection Similar to the chromatography system, the filtration system design should ensure that appropriate flow can be delivered without generating excessive pressure from system components. The process control is designed such that the normal flow through the membrane (flux) and tangential flow (retentate) can be monitored and controlled by the system. Flux in normal flow filtration is generally controlled by the differential pressure across the filter. In the case of tangential flow filtration, both transmembrane pressure, a measure of differential pressure across the membrane, and tangential flow are used to control the flux. The control system should ensure that maximum allowable filter pressure is not exceeded. The filters used in NFF mode are placed in housings generally of stainless steel construction. All stainless steel components are designed to be able to withstand clean-in-place and steam-in-place conditions. The housing design ensures that all entrapped air can be removed so that complete filter area comes in contact with process fluid. Many filters are self-contained capsules that do not require housings. These types of self-contained and single-use filter systems are gaining popularity rapidly since they provide significant advantages by eliminating complicated setup and dismantling, shortening process time and eliminating cleaning and sterilization.

3.4 SPECIALIZED APPLICATIONS OF CHEMICAL ENGINEERING CONCEPTS IN BIOLOGICS MANUFACTURING

In the preceding sections, we discussed how fundamental chemical engineering principles are used in process and equipment design for biologics manufacturing. In this section, we will examine how the area of process systems engineering in chemical engineering discipline offers unique solutions in biologics process development and manufacturing. Similar to a chemical plant, a biopharmaceutical plant utilizes the basic principles of mass, heat, and momentum transport to cultivate cells and produce product from raw materials and nutrients supplied to the cells, and then to purify the protein in the final dosage form. Also, similar to a chemical plant, a biopharmaceutical plant consumes energy and uses utilities such as water, steam, and compressed air. Because of these similarities, techniques for modeling of chemical plants can also be adapted to model a biopharmaceutical plant and the production processes.

Development of process models allows obtaining greater process understanding and predicting process behavior and is the first step toward process monitoring and control. Various levels of complexities can be incorporated into a process model. In biopharmaceutical manufacturing processes, it can vary between simple empirical models (data driven) to more sophisticated mathematical means such as metabolic flux and pathway models [31–33]. Comprehensive process understanding is imperative in achieving industry-wide guidelines set by regulatory agencies to promote quality by design (QbD), which is building quality in the process and product design. This can be achieved via correlative, causal, or mechanistic knowledge and at the highest level via first-principles models [34].

First-principles modeling involves in-depth understanding of the process dynamics that is typically defined in a set of material and energy balances via differential and algebraic equations, and depending on the modeling objectives, partial differential equations. As they demonstrate the process understanding at the highest level and provide opportunities for advanced process control, first-principles models are desirable and industry is encouraged by regulatory agencies as mentioned in various guidelines from U.S. Food and Drug Administration [35, 36] and International Conference on Harmonisation [37].

In this section, we will briefly review correlative (via multivariate modeling) and first-principles-based approaches by providing case studies to demonstrate their benefits and practical use.

3.4.1 A Case Study Using First-Principles Modeling: Mass Transfer Models in Cell Culture

Mathematical modeling of cell culture in bioreactors has received significant attention and has been successfully utilized for design and characterization. Most of the models developed included unstructured and unsegregated modeling approaches to describe the process at high level. In this section, we will demonstrate how first-principles modeling is used to develop representation of dissolved carbon dioxide (dCO_2) mass transfer in a bioreactor and practical use of the model in large-scale setting.

Excess accumulation of dCO_2 at high viable cell concentrations is known to have adverse effects to cellular growth and specific productivity in large-scale mammalian cell culture bioreactors. The accumulation can occur as a result of reduced surface-to-volume ratios and low CO_2 removal rates [38–40]. It is also known that high dCO_2 might be detrimental to protein structure and function (due to the alteration of glycosylation pattern of a therapeutic protein). On the other hand, excess stripping of CO_2 can alter bioreactor pH profile. Therefore, it is imperative to control dCO_2 levels.

There are three main sources of dCO_2 in cell culture: (1) CO_2 produced by cells during respiration, (2) CO_2 addition to control pH at its desired level, and (3) CO_2 produced by dissociation of sodium carbonate added to culture for pH control. Dissolved CO_2 mass balance in culture broth in relation to above-mentioned sources are provided in the following equations [40]:

$$CO_2(\text{gas}) \xleftrightarrow{k_La(CO_2)} CO_2(\text{liquid})[\text{also from cells}]$$
$$\xleftrightarrow{K_1} H_2CO_3 \xleftrightarrow{K_2} H^+ + HCO_3^- \leftrightarrow 2H^+ + CO_3{}^{2-}$$

and

$$\frac{dCO_2}{dt} = \frac{10^{-pH}}{10^{-pH} + K_1}\{\text{CER} + k_La([CO_2]^* - [CO_2])\}$$

where k_La is mass transfer coefficient, K_1 and K_2 are equilibrium constants, CER is CO_2 evolution rate, and $[CO_2]^*$ is the CO_2 concentration at the equilibrium.

Each of these sources can be mathematically described given the cell growth and lactate generation curves, as well as the bioreactor configuration and operating conditions as depicted in Figure 3.11. While more detailed generic mathematical description of the model can be found in the literature [38–40], we will focus on the industrial use of the model in this section.

The key inputs for model are the VCD and lactate concentration sampled over the time course of the bioreactor operation. In Figure 3.11, t_d designates discrete measurements performed, for example, daily measurements, and t

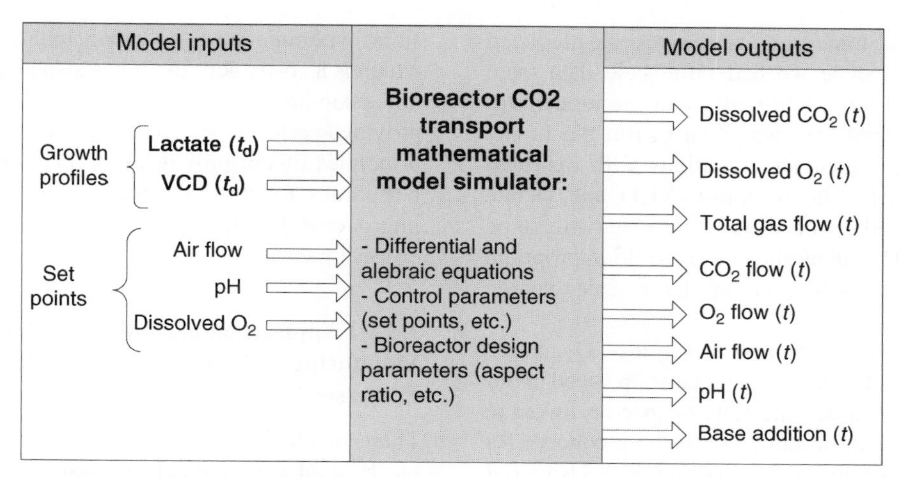

FIGURE 3.11 Schematic diagram representing the main input and output structures of CO_2 transport model.

designates continuous model outputs. The model receives a number of other values such as gas flow rates and base addition just for comparison purposes against model predictions for those. In other words, the model predicts those within its differential and algebraic equation set for every time instance that it solves and compares its outputs against the daily observations from the plant results. Model also allows studying the effect of pH control loop influence on the oxygen composition of the sparge gas so that controlling the levels of dCO_2 can be improved. Once a model is developed, it should be tested with existing process data and the output should be compared against the process outcome. In this case, the model was tested against large-scale production bioreactor data and found to be predictive of known performance. The model was able to provide a predicted relationship between CO_2 flow rate and dissolved CO_2 when the cell growth (VCD) and lactate concentration ([Lac]) time courses are provided and pH and aeration are at control set points. The predictive model can be utilized in a variety of applications. Perhaps the most beneficial application of such a model is in its ability to predict process behavior in a scale-up. During process development, a large body of small-scale process data is generated. These data can be used in the development of the model. Typically, the process is then scaled up to a pilot scale with perhaps a handful of process runs at that scale. Further refinement of the model can be performed with these data and scale comparability can also be assessed. The model can then be utilized to make predictions of large-scale process conditions.

In our specific scenario, the model was utilized to evaluate process behavior of a follow-on (second generation) process when a current process was already being operated at large scale. Therefore, we had access to process data at small, pilot, and large scales for the current (legacy) process. The follow-on process had small-scale and some pilot-scale process data. The scenario can be pictorially represented in Figure 3.12. The follow-on process was expected to have much higher cell densities and we were interested to know how we would select certain design criteria for large-scale bioreactors. One such criterion was the size of the mass flow controller that determines the amount of gas input to the culture. As can be seen from Figure 3.12, there were no process data available from a large-scale bioreactor (since no follow-on process had been run at this scale yet at that time). The approach undertaken was to first utilize the CO_2 first-principles model for the current process to ensure that the predictions are comparable to what was observed at both pilot and large scales from historical runs. Once that was confirmed, the

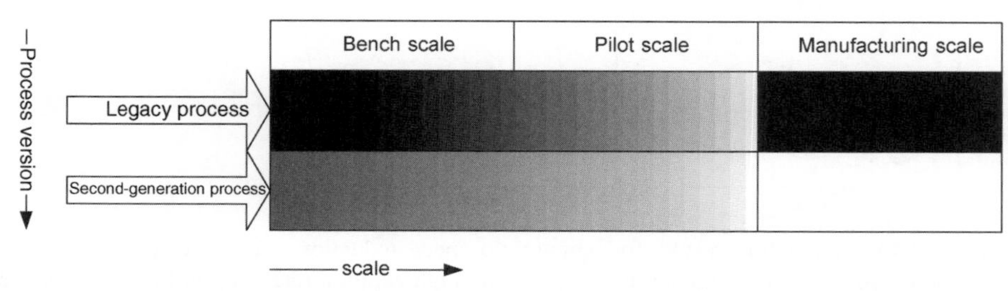

FIGURE 3.12 Color map representation of the available runs and data across scales (darker color means more manufacturing runs are available).

follow-on process conditions were used to generate predicted values at pilot scale. Since we had pilot-scale data from the follow-on process, we could verify that the model was still predictive. The final step was then to run the model using the large-scale bioreactor conditions with expected follow-on process metabolic response (VCD and lactate concentrations) to predict the dCO_2 levels that might be seen at large scale. This then allowed us to do appropriate calculations and size selection of large-scale sparge equipment.

In another application of the first-principles CO_2 model, we evaluated the impact of reducing agitator tip speed in our process. As mentioned earlier, the agitator speed is linked to mass transport in the culture but can also be a concern for shear or for the effect of air interface on the cells. Using our model, we were able to predict the culture properties at a lower agitator tip speed and determine whether our existing systems would be adequate for supporting the process requirement. Figure 3.13 shows some of the output of the model and demonstrates that for the most part key parameters such as pCO_2, pO_2, pH, and so on are comparable at different tip speeds. Despite the reduced agitation speed and aeration, dissolved oxygen was maintained at the required set point (mass transfer from gas to liquid is not rate limiting with respect to its consumption). Increase in sparged O_2 has allowed to attain the dissolved oxygen set point. Predicted increase in the demand on the sparged CO_2 under these conditions was well within the delivery capability of the existing mass flow controllers; and therefore, no resizing was required.

In these examples, the use of a first-principles model has shown its benefits by providing guidance for scale-up decisions without having to run actual experiments at scale, which is a costly activity. It has also helped increase process understanding of scale-up between pilot and large scales and provided performance comparability. Having a reliable first-principles model provides bioprocess engineers ample opportunities for process optimization, troubleshooting, and improvement and supports engineering decision-making process.

3.4.2 Application of Statistical Models in Process Monitoring and Control of Biologics Manufacturing Processes

There are challenges in modeling, monitoring, and control of batch biopharmaceutical processes due to their inherently complex biological and biochemical mechanisms and nonlinear time-variant process dynamics. There are also many variables measured during the course of a batch either off-line or online and at variable frequencies depending on the measurement system used. It is important to efficiently monitor and diagnose deviations from the in-control space for troubleshooting and process improvement purposes. One of the solutions successfully applied in chemical industry (both batch and continuous processing) is multivariate modeling and real-time statistical process monitoring [41–43]. These applications have been also successfully extended to pharmaceutical and biopharmaceutical cases [44–47].

In a typical industrial setting, data are generated by the process via various equipment online controls, off-line measurements, and assays that are all stored in various databases. Many process batches (I) are executed, where many variables (J) are measured at certain time intervals (K), forming a

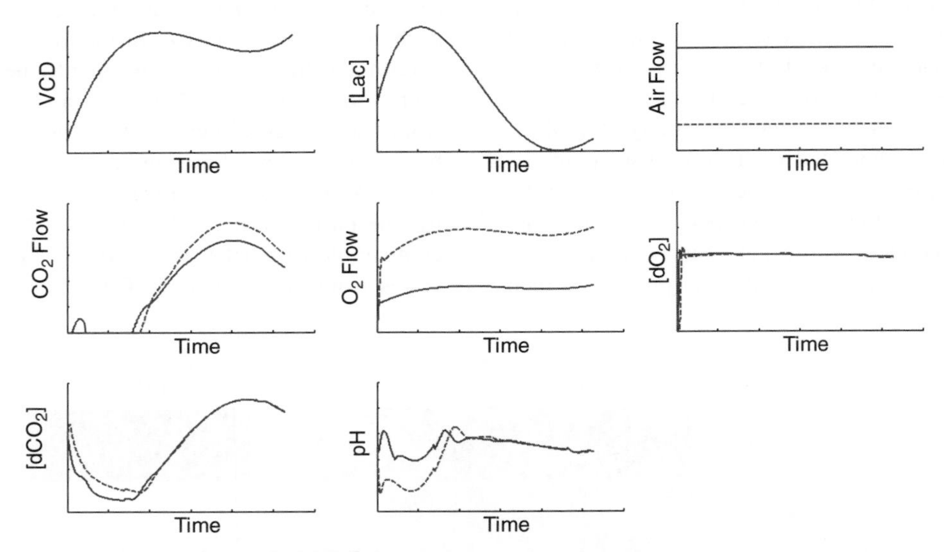

FIGURE 3.13 Simulation results for agitator tip speed reduction, and effect on CO_2 and O_2 requirements. Solid line (-), high agitation speed and high aeration rate; dashed line (- -), reduced agitation speed and aeration rate. VCD = viable cell density; [dCO_2] = concentration of dissolved CO_2; [Lac] = concentration of lactate; [dO_2] =concentration of dissolved oxygen.

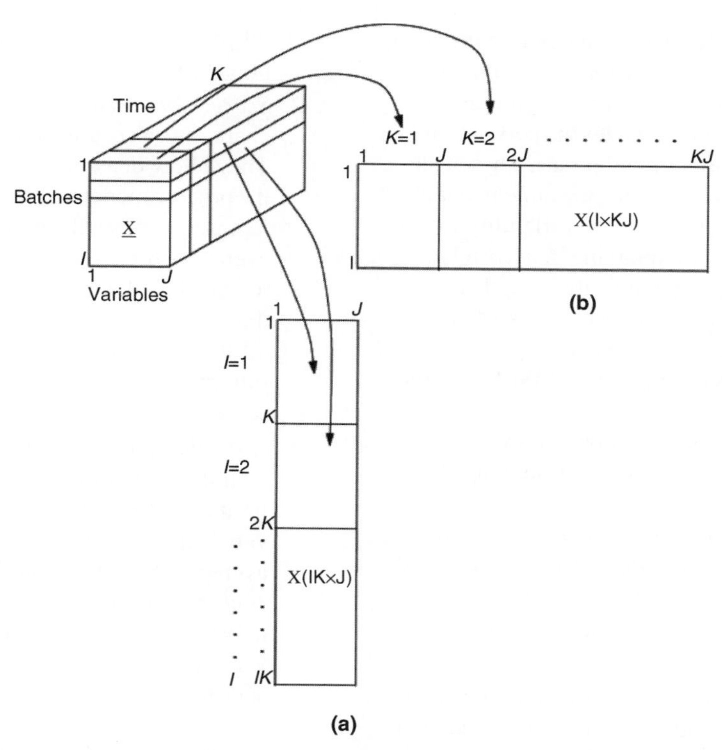

FIGURE 3.14 Unfolding of three-way batch data array: (a) observation level, preserves variable direction, and (b) batch level, preserves batch direction.

three-way data array ($\underline{\mathbf{X}}$) as depicted in Figure 3.14. Developing data-driven multivariate process models that define process variability has been shown to be beneficial in proactively monitoring the process consistency, performance, and for troubleshooting purposes. Typical process performance (as contained within data array $\underline{\mathbf{X}}$ from the process variables shown in Figure 3.14) can be modeled using multivariate techniques such as principal components analysis (PCA) and partial least squares (PLS). PCA is used when overall variability is to be modeled to find major variability dimension in a process (as shown for three variables in Figure 3.15a). This way the dimensionality problem is solved by reducing from many raw process variables to a few derived variables called principal components (PCs) that have linear contributions from raw variables. If the objective is also to correlate how changes in process variables impact-

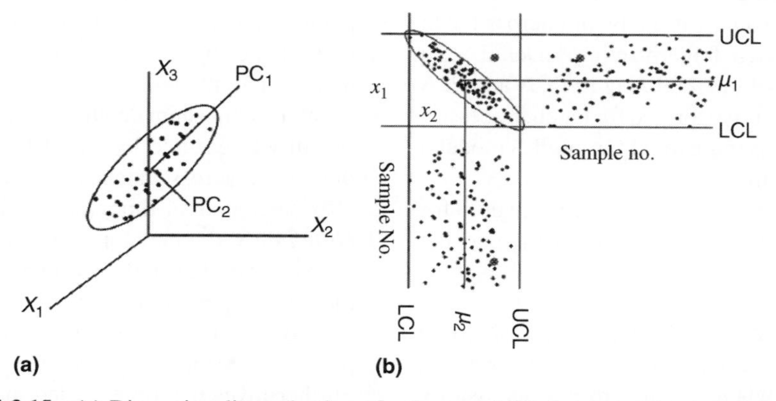

FIGURE 3.15 (a) Dimensionality reduction of a three-variable (x_1, x_2, and x_3) process. Overall variability of the process can be explained by using two principal components [33]. (b) Multivariate SPC versus univariate SPC [41], and a batch observation (depicted with a circled cross sign) that is in control in univariate charts is actually out of control in bivariate in-control region defined by the ellipsis (95% confidence region).

ing, say, a performance variable defining an end point, than PLS is the preferred choice. In that case, the multivariate regression model (PLS) is fitted to data from process variables (inputs) in such a way that maximizes the correlation to response variable (process output or end point). These techniques are extremely useful in (i) reducing the dimensionality problem (summarizing the overall process variability into a few variables from many), (ii) explaining the correlation structure of the variables, (iii) handling missing data, (iv) reducing the noise inherent to measurements so that the actual signal can be extracted, and (v) providing means for multivariate statistical process monitoring (MSPM) of the entire process (Figure 3.15).

MSPM for batch (bio)processes can be achieved in real time. Setting the framework up involves the following steps:

1. Collect historical batch data (from typically 20–30 batches of many process variables that are measured online) that define normal operating region
2. Detect and remove outliers in the data and apply preconditioning (autoscaling, etc.)
3. If the objective is only fault detection and diagnosis, develop a PCA model, and if the objective is also to predict process performance variables or critical quality attributes in real time, develop multivariate regression models such as PLS
4. Construct multivariate charts for real-time monitoring. Monitor a new batch by projecting its data (applying the same data conditioning) onto the model spaces that are developed for comparing to nominal performance.

PCA and/or PLS are performed on the unfolded array (**X**) in either direction as shown in Figure 3.14. Observation level models are suggested to avoid estimating the future portion of new batch trajectories [48]. These models are used in real-time monitoring of a new batch, whereas batch level models are used at the end of the batch for analyzing across batch trends, also known as "batch fingerprinting." Details and mathematical formulation of the modeling in PCA and PLS for batch processes (when it is performed for batch processes, they are usually referred to as multiway PCA or PLS models) can be found in the literature [46, 48].

A number of multivariate statistics and charts are used for monitoring new batches in real time. These include the following:

1. *Squared Prediction Error (SPE, also Known as Q Residuals or DModX) Chart*: SPE is used for process deviation detection where events are not necessarily explained by the model. When SPE control limit violation is observed, it is likely that a new event is observed that is not captured by the reference model (this can be triggered by a normal event that is part of inherent process variability that is not captured or a process upset).
2. *Score Time Series and Hotelling's T^2 Charts*: These charts are also used for process deviation detection. They allow detecting deviations that are explained by the process model and within the overall variability but represent unusually high variation compared to the average process behavior. Score time series allow monitoring the process performance at each model dimension separately, while T^2 allows monitoring all the model dimensions over the course of a batch run by using a single statistic calculated from all scores.
3. *Contribution Plots*: When the above detection charts identify a deviation (violation of multivariate statistical limits on SPE and/or T^2 charts) that indicates that some variables or a variable is deviating from the historical average behavior without diagnosing which variables are contributing the most, contribution plots are then used to delve into the original variable level to inspect which variable or variables are contributing to the inflated statistic.

3.4.2.1 Utilization of Real-Time Multivariate Statistical Process Monitoring in Bioprocessing Multivariate statistical process monitoring tools provide an important means to rapidly detect process anomalies as the process is running and take action to correct issues before the process drifts to an out-of-control region. In our example, an MPCA model was developed (steps are described above) for a train (multiple systems operated in an alternating fashion across batches) of a perfusion bioreactor system (bioreactor, tangential flow filtration skid, and media tank). The model used historical process data measured on 21 variables from a statistically relevant number of batches throughout the operation that typically spans several days yielding thousands of data points. Ninety-six percent of overall process variability could be explained by only using three principal components in this multivariate model.

When a transient decline ($\sim$3%) was observed in final day viability (measured by an off-line analytical system) in the bioreactor across batches (Figure 3.16a), deviations were also seen in real-time multivariate charts (Figure 3.16b, *Hotelling's T^2* chart for one of the low viability batches is shown). Variable contribution plots (Figure 3.16c) identified that higher perfusion feed and retentate temperature conditions occurred in the equipment in comparison to the normal process behavior demonstrated by the historical batches. Further investigation revealed that an equipment mismatch in the system caused the higher than normal temperature. The issue was corrected prior to the next batch and the temperature profile and the final day viability returned to their normal ranges (lower left corner of Figure 3.16b).

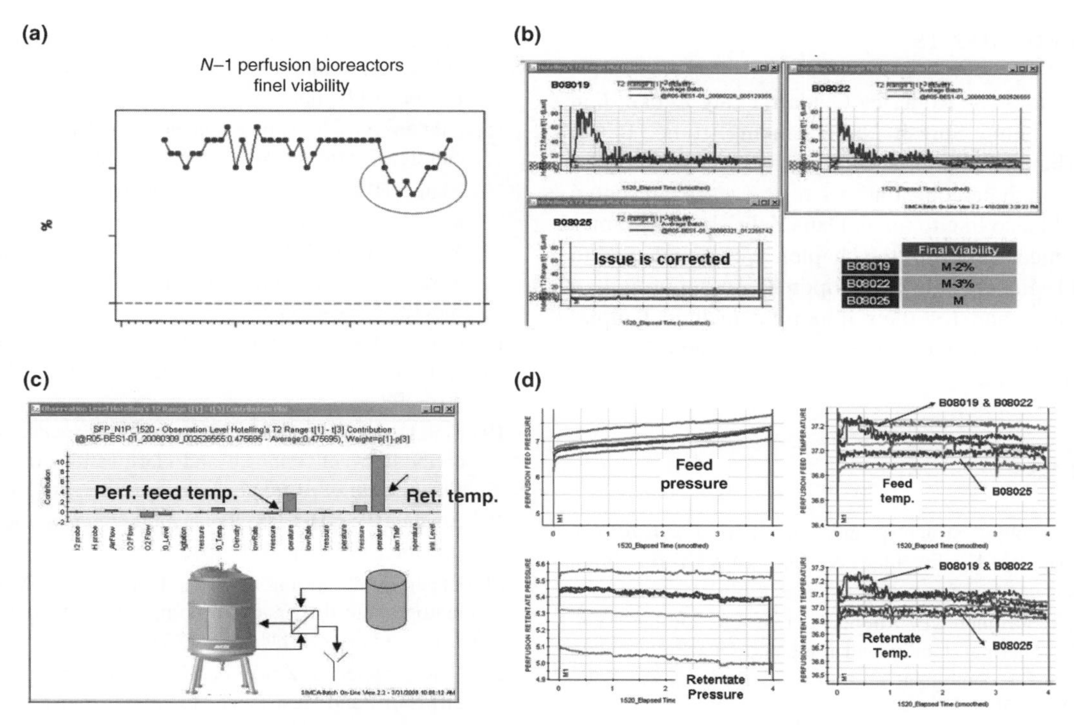

FIGURE 3.16 Real-time multivariate statistical process monitoring (RT-MSPM) steps involved in troubleshooting of a perfusion bioreactor system (shown in c): (a) offline observation of viability decline, (b) detection via T^2 charts, (c) contribution plot for diagnosing the potential root cause, and (d) inspection of the univariate variable profiles.

The example provided here demonstrates how multivariate statistical process monitoring can detect an abnormal process behavior, assist in determination of root cause, and confirm when a resolution is implemented and the process has returned to normal. The technology has the potential to increase operational success and reduce production costs in highly complex biopharmaceutical processing systems [49].

3.5 CONCLUSIONS

Since the time of synthesizing penicillin in milk bottles in the 1940s, bioprocesses have come a long way. Today, bioprocesses are used for commercial manufacturing of biologics that include therapeutic proteins, vaccines, enzymes, diagnostic reagents, and nonprotein complex macromolecules such as polysaccharides. Biologics are expected to be a major sector within the life sciences industry. The global market for biologics, currently at about $48 billion, is expected to rapidly grow to about $100 billion in the next few years. Currently, 6 out of 10 best selling drugs are biotechnology derived and over a third of all pipeline products in active development are biologics. The pipeline amounts to approximately 12,000 drug candidates in various stages of preclinical and clinical development covering about 150 disease states and promises to bring better lifesaving treatment to patients. These staggering statistics demonstrate that the fundamentals of biologics and bioprocess industry are strong.

Industrial biotechnology requires blending three key disciplines: biology, biochemistry, and chemical engineering. While our focus in this chapter has been on application of chemical engineering principles to biotechnology, we urge the readers to recognize that the principles of biology and biochemistry play equally (or more) important role in biologics and bioprocessing. Innovation is directly linked to the success of biotechnology products and processes. Innovation in protein design, genetic engineering, expression systems, and manufacturing technologies has allowed rapid commercialization of highly profitable biopharmaceutical products despite the presence of strict regulatory requirements. The innovation, however, has come at a considerable cost. Biologics are the most expensive among all medicinal products. As the biopharmaceutical industry proceeds through its evolution to maturity, it will be forced to undergo significant reinvention to include sound economic principles in addition to sound scientific and engineering principles that provide most cost-effective and value appropriate therapies to patients.

ACKNOWLEDGMENTS

We would like to thank Roderick Geldart, Craig Zupke, Jian Zhang, David Fang, Suresh Nulu, Sinem Ertunc, Bryan Looze and Thomas Mistretta, all currently or formerly of Amgen Inc. for their contributions to the work presented here. We would also like to thank Dane Zabriskie and James Stout for technical review of the chapter. We are indebted to Philip Lai and Michael Felo of Millipore Corp. for providing some the figures and for their important technical input. Finally, we would like to acknowledge Amgen Inc. for supporting this work.

REFERENCES

1. Stryer L. Protein structure and function. *Biochemistry*, 3rd edition, W. H. Freeman & Co., New York, 1988, pp. 15–42.
2. Medved LV, Orthner CL, Lubon H, Lee TK, Drohan WN, Ingham KC. Thermal stability and domain–domain interactions in natural and recombinant protein C. *J. Biol. Chem.* 1995; 270:13652–13659.
3. Narhi L, Philo J, Sun B, Chang B, Arakawa T. Reversibility of heat-induced denaturation of the recombinant human megakaryocyte growth and development factor. *Pharm. Res.* 1999;16:799–807.
4. Franks F. Conformational stability: denaturation and renaturation. In: Franks F, editor, *Characterization of Proteins,* 1st edition, Humana Press, New York, 1988, pp. 104–107.
5. Tang X, Pikal M. The effect of stabilizers and denaturants on the cold denaturation temperatures of proteins and implications for freeze-drying. *Pharm. Res.* 2005;22:1167–1175.
6. Maa Y-F, Hsu CC. Protein denaturation by combined effect of shear and air–liquid interface. *Biotechnol. Bioeng.* 1997;54:503–512.
7. Thomas CR. Problems of shear in biotechnology. In: Winkler MA, editor, *Chemical Engineering Problems in Biotechnology,* 1st edition, Elsevier, New York, 1990, pp 23–93.
8. Junker BH. Scale-up methodologies for *E. coli* and yeast fermentation processes. *J. Biosci. Bioeng.* 2004;97:347–364.
9. Xing Z, Kenty B, Jian Z, Lee SS. Scale-up analysis for a CHO cell culture process in large-scale bioreactors. *Biotechnol. Bioeng.* 2009;103:733–746.
10. Yang J. et al. Fed-batch bioreactor process scale-up from 3-L to 2500-L for monoclonal antibody production from cell culture. *Biotechnol. Bioeng.* 2007;98:141–154.
11. Ju L-K, Chase GG. Improved scale-up strategies of bioreactors. *Bioprocess Eng.* 1992;8:49–53.
12. Dosmar M, Meyeroltmanns F, Gohs M. Factors influencing ultrafiltration scale-up. *Bioprocess Int.* 2005;3:40–50.
13. Van Reiss R, Goodrich EM, Yson CL, Frautschy LN, Dzengeleski S, Lutz H. Linear scale ultrafiltration. *Biotechnol. Bioeng.* 1997;55:737–746.
14. Prokop A. Implications of cell biology in animal cell biotechnology. In: *Animal Cell Bioreactors*, Butterworth-Heinemann, Boston, 1991, pp. 21–58.
15. Nienow AW. Reactor engineering in large scale animal cell culture. *Cytotechnology* 2006;50:9–33.
16. Blanch HW, Clark DS. *Biochemical Engineering*. Marcel Dekker, Inc. 1997, pp. 415–426.
17. Xie L, Wang DIC. Stoichiometric analysis of animal cell growth and its application in medium design. *Biotechnol. Bioeng.* 1994;43:1164–1174.
18. Cooper CK, Fernstrom GA, Miller SA. Performance of agitated gas–liquid contactors. *Ind. Eng. Chem.* 1944;36: 504–509.
19. Shukla AA, KandulaJr., Harvest and recovery of monoclonal antibodies: cell removal and clarification. In: *Process Scale Purification of Antibodies*, Wiley, 2009, pp. 53–78.
20. Blanch HW, Clark DS. *Biochemical Engineering*. Marcel Dekker, Inc., 1997, pp. 461–467.
21. Usher KM, Simmons CR, Dorsey JG. Modeling chromatographic dispersion: a comparison of popular equations. *J. Chromatogr.* 2008;1200:122–128.
22. Van Deemter JJ, Zuiderweg FJ, Klinkenberg A. Longitudinal diffusion and resistance to mass transfer as causes of nonideality in chromatography. *Chem. Eng. Sci.* 1956;5: 271–289.
23. McCoy M, Kalghatgi K, Regnier FE, Afeyan, N. Perfusion chromatography—characterization of column packings for chromatography of proteins. *J. Chromatogr.* 1996;743: 221–229.
24. Moscariello J, Purdom G, Coffman J, Root TW, Lightfoot EN. Characterizing the performance of industrial-scale columns. *J. Chromatogr.* 2001;908:131–141.
25. Pathak N, Norman C, Kundu S, Nulu S, Durst M, Fang Z. Modeling flow distribution in large-scale chromatographic columns with computational fluid dynamics. *Bioprocess Int.* 2008;6:72–81.
26. Stickel JJ, Fotopoulos A. Pressure-flow relationships for packed beds of compressible chromatography media at laboratory and production scale. *Biotechnol. Prog.* 2001;17: 744–751.
27. McCue JT, Cecchini D, Chu C, Liu WH, Spann A. Application of a two-dimensional model for predicting the pressure-flow and compression properties during column packing scale-up. *J. Chromatogr.* 2007;1145:89–101.
28. Teeters MA, Quinones-Garcia I. Evaluating and monitoring the packing behavior of process-scale chromatography columns. *J. Chromatogr.* 2005;1069:53–64.
29. Larson TM, Davis J, Lam H, Cacia J. Use of process data to assess chromatographic performance in production-scale purification columns. *Biotechnol. Prog.* 2003;19 (2): 485–492.
30. Rudolph EA, MacDonald JH. Tangential flow filtration systems for clarification and concentration. In: Lydersen BK, D'Elia NA, Nelson KM, editors, *Bioprocess Engineering: Systems, Equipment and Facilities*, New York, John Wiley & Sons, 1994, pp.121–126.

31. Bailey JE. Mathematical modeling and analysis in biochemical engineering: past accomplishments and future opportunities. *Biotechnol. Prog.* 1998;14:8–20.
32. Bailey JE. Mathematical modeling and analysis in biochemical engineering: past accomplishments and future opportunities. *Biotechnol. Prog.* 1998;14:8–20.
33. Stephanopoulos G, Aristidou AA, Nielsen J. *Metabolic Engineering: Principles and Methodologies*, Academic Press, London, 1998, pp. 309–459.
34. Cinar A, Parulekar SJ, Undey C, Birol G. *Batch Fermentation: Modeling, Monitoring and Control*, CRC Press, New York, 2003, pp. 21–58.
35. Rathore, AS,and Winkle H. Quality by design for biopharmaceuticals. *Nat. Biotechnol.* 2009;27:26–34.
36. U.S. Food and Drug Administration. *Pharmaceutical cGMPs for the 21st Century: A Risk-Based Approach*, FDA, Rockville, MD, August 2002, available at http://www.fda.gov/oc/guidance/gmp.html (accessed date: July 23, 2009).
37. U.S. Food and Drug Administration. *Guidance for Industry: PAT—A Framework for Innovative Pharmaceutical Development, Manufacturing, and Quality Assurance*, FDA, Rockville, MD, September 2004, available at http://www.fda.gov/Cder/guidance/6419fnl.pdf (accessed date: July 23, 2009).
38. International Conference on Harmonisation (ICH). *Pharmaceutical Development Q8(R1)*, November 2008, available at http://www.ich.org/LOB/media/MEDIA4986.pdf (accessed date: July 23, 2009).
39. Gray DR, Chen S, Howarth W, Inlow D, Maiorella BL. CO_2 in large-scale and high-density CHO cell perfusion culture. *Cytotechnology* 1996;22:65–78.
40. Zupke C, Green J. Modeling of CO_2 concentration in small and large scale bioreactors. Presented at *Cell Culture Engineering* VI, 1998, San Diego, CA.
41. Mostafa S, Gu X. Strategies for improved dCO_2 removal in large-scale fed-batch cultures. *Biotechnol. Prog.* 2003;19:45–51.
42. MacGregor JF, Kourti T. Statistical process control of multivariate processes. *Control Eng. Pract.* 1995;3:403–414.
43. Neogi D, Schlags CE. Multivariate statistical analysis of an emulsion batch process. *Ind. Eng. Chem. Res.* 1998;37:3971–3979.
44. Undey C, Ertunc S, Tatara E, Teymour F, Cinar A. Batch process monitoring and its applications in polymerization systems. *Macromol. Symp.* 2004;206:121–134.
45. Westerhuis JA, Coenegracht PMJ, Lerk CF. Multivariate modeling of the tablet manufacturing process with wet granulation for tablet optimization and in-process control. *Int. J. Pharm.* 1997;156:109–117.
46. Albert S, Kinley RD. Multivariate statistical monitoring of batch processes: an industrial case study in fed-batch fermentation supervision. *Trends Biotechnol.* 2001;19:53–62.
47. Undey C, Ertunc S, Cinar A. Online batch/fed-batch process performance monitoring, quality prediction and variable contributions analysis for diagnosis. *Ind. Eng. Chem. Res.* 2003; 42:4645–4658.
48. Undey C, Tatara E, Cinar A. Intelligent real-time performance monitoring and quality prediction for batch/fed-batch cultivations. *J. Biotechnol.* 2004;108:61–77.
49. Wold S., Kettaneh N, Fridén H, Holmberg A. Modelling and diagnostics of batch processes and analogous kinetic experiments. *Chemom. Intell. Lab. Syst.* 1998;44:331–340.
50. Undey C, Ertunc S, Mistretta T, Pathak M. Applied advanced process analytics in biopharmaceutical manufacturing: challenges and prospects in real-time monitoring and control. *IFAC ADCHEM Proceedings*, Istanbul, Turkey, July 13–15, 2009.

4

DESIGNING A SUSTAINABLE PHARMACEUTICAL INDUSTRY: THE ROLE OF CHEMICAL ENGINEERS

CONCEPCIÓN JIMÉNEZ-GONZÁLEZ AND CELIA S. PONDER
GlaxoSmithKline, Sustainability and Environment, Research Triangle Park, NC, USA
ROBERT E. HANNAH AND JAMES R. HAGAN
GlaxoSmithKline, Sustainability and Environment, Philadelphia, PA, USA

4.1 INTRODUCTION

Pharmaceutical processing in general involves difficult and complex tasks, but is absolutely critical to improving and maintaining patient health and saving lives. In order to continue to create the life-saving medicines that society needs, it is necessary to discover and develop molecules that are very often complex (e.g, enantiomers with several chiral centers, high molecular weight, and high degree of functionality). Producing this type of molecule as a drug substance frequently requires complex chemistries and extensive purification processes. Bulk chemicals typically require between 1 and 2 chemical transformations, while pharmaceutical active ingredients usually require 6 or more. In addition to the technical complexities, the pharmaceutical industry needs to accommodate a high level of attrition, uncertainties in demand forecasts, and the relatively short length of the patent period, among other challenges. For every 5000 compounds evaluated in preclinical testing, only about 5 progress far enough to enter clinical trials and on average only 1 might actually gain approval by the US FDA for marketing [1]. Given an average of 12 years required to discover, develop, and deliver a new drug candidate to market, process development must be optimized in a relatively short amount of time under constraints where innovation and investment are difficult.

Given these historical challenges and complexities an enhanced approach to develop sustainable pharmaceutical processes is needed. A new approach will require innovation based on sustainability principles in order to improve and enhance the efficiency of our manufacturing processes. It must also have the potential to lower costs, which could in turn lower prices to widen global access to medicines in the marketplace.

Doing this effectively requires innovation, and of all the professions, engineering is in the best position to leverage innovative approaches [2]. The role of engineers in general, and chemical engineers in particular, is crucial to delivering more sustainable pharmaceutical processes.

4.2 A WORD ON SUSTAINABILITY

Since the 1980s sustainability was identified by the United Nations as the solution that would address the environmental issues that have affected the world. In a widely accepted definition, sustainability implies meeting the needs of today without compromising the ability of future generations to meet their own needs. This is translated into a triple-bottom line approach where environmental, economic, and social aspects are in balance. In other words, sustainable systems or process are the ones that

- minimize environmental impacts,
- are economically viable, and
- are socially responsible [3, 4].

However, implementing sustainability on an operational basis may be difficult due to the intrinsic interrelated, interdependent

Chemical Engineering in the Pharmaceutical Industry: R&D to Manufacturing Edited by David J. am Ende

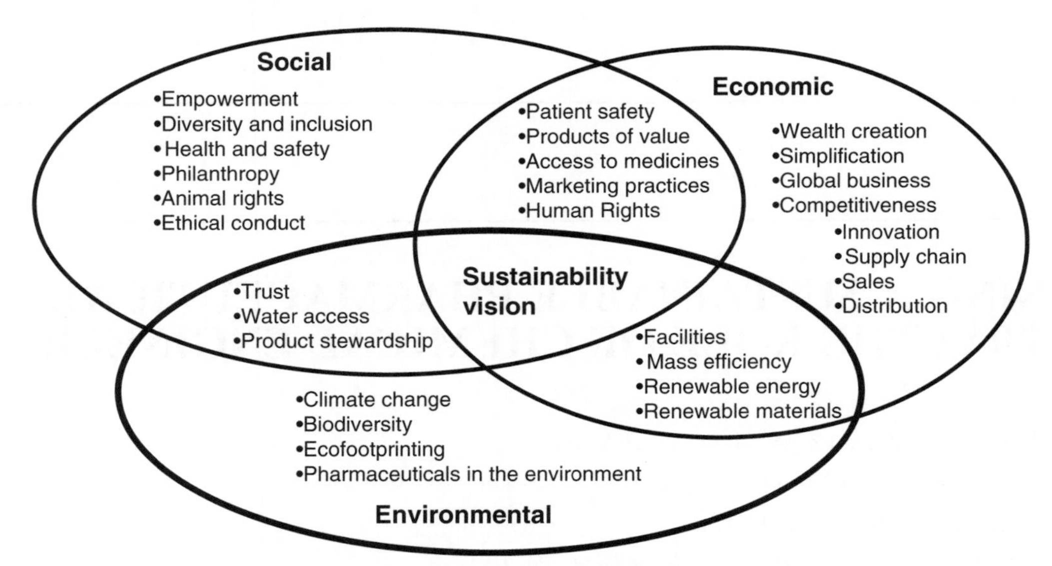

FIGURE 4.1 A triple-bottom approach to Sustainability, with some sustainability opportunities and issues for the pharmaceutical industry.

nature of the concept. For instance, just measuring the sustainability of pharmaceutical processes may is a very complex task more akin to a multivariable optimization. Chemical engineers have indeed proposed these approaches for measuring sustainability of processes [5–7]. In addition, sustainability is a dynamic system—it can only be measured with a long-term horizon and with extended boundaries (i.e., a life cycle approach). There have been many attempts to measure the sustainability or "greenness" of pharmaceutical synthesis through a series of "green metrics." Some approaches have focused on waste, such as the E-factor, the amount of waste generated in order to produce 1 kg of product. [8, 9] Approaches have focused on mass efficiency (or its inverse, mass intensity) trying to highlight the process optimization and innovation side of the equation through prevention instead of waste minimization approaches [10–12]. More evolved approaches try to include life cycle and process systems engineering in a more holistic view of the impacts. Figure 4.1 shows a triple-bottom approach to sustainability, with some sustainability opportunities and issues for the pharmaceutical industry.

At the end of the day, in order to design sustainable pharmaceutical processes, there is need to integrate sustainability criteria inherently in process development and optimization. Doing so is an essential part of anticipating, minimizing and solving problems that might arise in actual production, as well as embedding sustainable processes in manufacturing.

4.3 GREEN CHEMISTRY AND GREEN ENGINEERING PRINCIPLES

Following the publication of the green chemistry principles [13, 14], an initial attempt to capture principles of green engineering was made [15] with the publication of 12 proposed green engineering principles. These principles were aligned with the previous green chemistry principles, although one drawback is that the two lists were not integrated but were published separately. So in 2003, about 60 chemists and engineers from industry, government, and academia met in San Destin, Florida to discuss principles of green engineering. This group intended to appeal to the large engineering audience (beyond chemical industry), in addition to potentially broadening the scope of previous work to incorporate principles of sustainability [16]. These sets of principles are in general accepted by most people and have proven very powerful in disseminating the intent and guidelines of green chemistry and engineering. Green engineering and chemistry approaches are synergistic and need to be applied in parallel to realize the most potential benefit. However, these principles can be simplified [17]. So that when designing novel chemistry routes, selecting reactors or separations, designing chemical processes, building plants, and so on one should strive to

- maximize resource efficiency,
- eliminate and minimize EHS hazards, and
- design systems holistically and use life cycle thinking.

But is this simplified approach sufficient? The most commonly cited principles have been mapped to this simpler set of three [18], which seem to cover all the guidelines that have been postulated, as shown in Table 4.1.

With regard to pharmaceutical processes, there are efforts current efforts underway to integrate green chemistry and

TABLE 4.1 Mapping the Three Main Green Chemistry and Engineering Principles

Principles	Maximize Resource Efficiency	Eliminate and Minimize Hazards and Pollution	Design Systems Holistically and Using Life Cycle Thinking
Green chemistry (Anastas and Warner)	• Synthetic methods should be designed to maximize the incorporation of all materials used in the process into the final product • The use of auxiliary substances (e.g. solvents and separation agents) should be made unnecessary whenever possible and, innocuous when used • Energy requirements should recognized for their environmental and economic impacts and should be minimized. Synthetic methods should be conducted at ambient temperature and pressure • Unnecessary derivatization (blocking group, protection/deprotection, temporary modification of physical/chemical processes) should be avoided whenever possible • Catalytic reagents (as selective as possible) are superior to stoichiometric reagents	• It is better to prevent waste than to treat or clean up waste after it is formed • Wherever practicable, synthetic methodologies should be designed to use and generate substances that possess little or no toxicity to human health and the environment • Chemical products should be designed to preserve efficacy of function while reducing toxicity • Chemical products should be designed so that at the end of their function they do not persist in the environment and break down into innocuous degradation products • Analytical methodologies need to be further developed to allow for real-time in-process monitoring and control prior to the formation of hazardous substances • Substances and the form of a substance used in a chemical process should chosen so as to minimize the potential for chemical accidents, including releases, explosions, and fires	• A raw material feedstock should be renewable rather than depleting whenever technically and economically practical
• Green chemistry (Winterton)	• Identify and quantify by-products • Report conversions, selectivities, and productivities • Establish full mass balances for a process • Anticipate heat and mass transfer limitations • Quantify and minimize use of utilities	• Measure catalyst and solvent losses in aqueous effluent • Investigate basic thermochemistry • Recognize where safety and waste minimization are incompatible • Monitor, report, and minimize laboratory waste emitted	• Consult a chemical or process engineer • Consider effect of overall process on choice of chemistry • Help develop and apply sustainability measures • Recognize where safety and waste minimization are incompatible
Green engineering (Anastas and Zimmerman)	• Separation and purification operations should be designed to minimize energy consumption and materials use • Products, processes, and systems should be designed to maximize mass, energy, space, and time efficiency • Products, processes, and systems should be "output pulled" rather than "input pushed" through the use of energy and materials	• Designers need to strive to ensure that all material and energy inputs and outputs are as inherently nonhazardous as possible • It is better to prevent waste than to treat or clean up waste after it is formed	• Embedded entropy and complexity must be viewed as an investment when making design choices on recycle, reuse, or beneficial disposition • Targeted durability, not immortality, should be a design goal • Design for unnecessary capacity or capability (e.g., "one size fits all") solutions should be considered a design flaw

(*continued*)

TABLE 4.1 (*Continued*)

Principles	Maximize Resource Efficiency	Eliminate and Minimize Hazards and Pollution	Design Systems Holistically and Using Life Cycle Thinking
	• Design of products, processes, and systems must include integration and interconnectivity with available energy and materials flows		• Material diversity in multicomponent products should be minimized to promote disassembly and value retention • Products, processes, and systems should be designed for performance in a commercial "afterlife" • Material and energy inputs should be renewable rather than depleting
Green engineering (San Destin Declaration)	• Minimize depletion of natural resources • Strive to prevent waste • Conserve and improve natural ecosystems while protecting human health and well being	• Ensure that all material and energy inputs and outputs are as inherently safe and benign as possible • Strive to prevent waste • Conserve and improve natural ecosystems while protecting human health and well being	• Engineer processes and products holistically, use systems analysis, and integrate environmental impact assessment tools • Use life cycle thinking in all engineering activities • Develop and apply engineering solutions, while being cognizant of local geography, aspirations, and cultures • Create engineering solutions beyond current or dominant technologies; improve, innovate, and invent (technologies) to achieve sustainability • Actively engage communities and stakeholders in development of engineering solutions

green engineering principles. In 2005, the American Chemical Society (ACS), the Green Chemistry Institute (GCI), and several major pharmaceutical companies came together to form the ACS GCI Pharmaceutical Roundtable. The strategic priorities of the Roundtable are to inform and influence the green processing research needs of the industry, to identify innovations that will be required, to educate both pharmaceutical leaders as well as others in the benefits of this approach and to provide green processing expertise to global pharmaceutical operations.

Applying these three generic principles to process development strategies will allow us to design sustainable pharmaceutical processes. In other words, processes that are better, cheaper, faster, cleaner and are sustainable by design in that they

- optimize the use of material and energy resources;
- eliminate or minimize environment, health and safety hazards; and
- minimize life cycle impacts.

4.4 CHEMICAL ENGINEERS—DESIGNING SUSTAINABLE PHARMACEUTICAL PROCESSES

4.4.1 Resource Efficiency

Following the three general green principles above, let us start with one of the main challenges for chemical engineers in designing sustainable pharmaceutical processes: maximizing the use of material and energy resources.

This challenge has also been recognized by the ACS GCI PR, and as a result it has chosen process mass intensity (PMI) as a measure to drive efficiency improvements in pharmaceutical syntheses. The ACS GCI Pharmaceutical Roundtable members have used this common process mass intensity metric (total mass of materials per mass of product) to compare data from each company on an equitable basis.

This benchmarking, has allowed the group to drive innovation and improvements in terms of material utilization. For instance, during the 2008 benchmarking exercise of the ACS GCI PR it was found that the median mass intensity of the processes under different stages of development across all the stages was about 120 kg material/kg API according to the data provided by the seven-member companies at that time, with a maximum of 887 kg material/kg API and a minimum of 23 kg material/kg API. This benchmarking also found that most of the material requirements are solvent (about half) followed by water (about 30%), reactants (about 9%), and other materials as the balance [19]. This is generally aligned with other studies performed previously by GSK, both in the process and in the life cycle boundaries [20].

In general there is opportunity to improve the resource utilization during the development cycle, with the median moving from 185 kg material/kg API during preclinical to about 45 kg material/kg API (Figure 4.2) in the commercial phase. It is during this period of optimization that chemical engineers have the opportunity to collaborate with chemists and R&D scientists in improving the "sustainability profile" of pharmaceutical processes. Creating processes that are more material and energy efficient, are inherently safer, and that minimize the life cycle impacts on the environment.

This is part of the rationale that pharmaceutical companies have followed when they have set mass efficiency and energy reduction metrics to drive improvements. GlaxoSmithKline has set a target to double the average mass efficiency of processes for new products introduced between 2006 and 2010, roughly halving resource consumption and waste generation and aggressive targets for reduction of energy consumption and its impact on climate change [21]. Furthermore, more aggressive mass efficiency targets have been set beyond 2010 for R&D and manufacturing. In order to drive the changes required to achieve more sustainable processes, GlaxoSmithKline has established a Sustainable Processing Team drawn from discovery, development, and with strong links to manufacturing units to apply sustainability strategies developed within R&D and manufacturing [22]. The manufacturing unit also has established a Sustainable Manufacturing Center of Excellence to leverage innovative technologies and processes, to drive step change improvement in support of the ambition to deliver a sustainable and cost-effective active pharmaceutical ingredient supply base. In addition to sustainable processing initiatives, a Climate Change Program has been started and a central fund established to finance energy saving projects. In 2008, 171 projects were completed that are expected to result in a saving of more than 550,800 GJ of energy per year (more than 40,000 tons of carbon dioxide equivalents).

Recognizing the importance of the role that chemical engineers play in this area, GSK's Sustainable Processing Team has set up engineering working groups to improve and optimize processes by focusing primarily on green engineering and solvent optimization and recovery. The aim of these working groups is to interact and collaborate with the R&D chemists to achieve the design of more sustainable pharmaceutical processes.

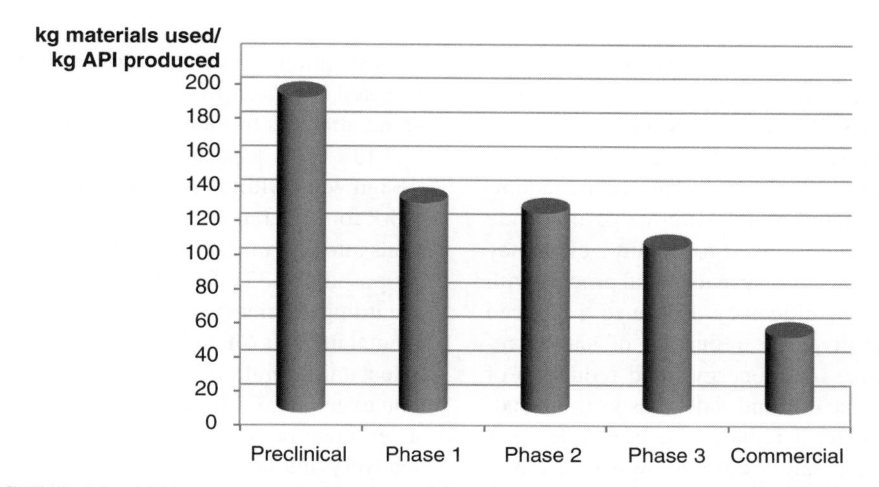

FIGURE 4.2 2008 process mass intensity benchmark of the ACS Green Chemistry Institute Pharmaceutical Roundtable. Medians by development phase are shown. Processes included in each phase: preclinical: 7; phase I: 5; phase II: 13; phase III: 16; commercial: 5. (Ref. 12.)

GSK has not been the only pharmaceutical company to leverage mass efficiency or mass intensity metrics to optimize processes. Merck has used PMI to drive improvements during the development phase with a primary focus on engineering. For Merck, this has meant optimizing the unit operations and processes with the assumption that the best chemistry is in place. Using this approach in a coupling reaction, the engineering group was able to reduce the PMI from 107 to 40 kg/kg API (more than doubling the mass efficiency, improve the cycle time and crystallization robustness, and replacing dimethyl chloride with either *iso*-propyl or ethyl acetate). This was accomplished by utilizing techniques of unit operation screening, solvent screening, polymorph screening, and process synthesis—without modifying the chemistry [23].

4.4.2 Beyond Resource Efficiency—Health, Safety, and Life Cycle Impacts

In the last section we saw good examples of how to drive improvements in terms of improving resource efficiency, but to complete the sustainability picture, one needs to integrate health and safety hazard and risk considerations, as well as life cycle impacts. The additional challenge that comes with this improve impacts that reach beyond manufacturing.

The importance of minimizing health and safety risks has been largely integrated into the work of developing pharmaceutical processes. However, much of these considerations have been made by the use of controls after a process is designed (e.g., globe boxes for charging potent materials, explosion suppression systems). The sustainability challenge at this point is to ensure that health and safety considerations are inherent to the process design as a built-in feature, not as a bolt on control. Chemical engineers have the remit to eliminate the need for materials of concern, designing systems where the remaining materials are contained by design and are inherently safer process through improved heat and mass transfer, among others.

The minimization of life cycle impacts has become one of the areas of work for chemical engineers, given our ethical responsibilities and the growing awareness and expectations of stakeholders. Ten years ago terms such as "carbon footprinting" or "ecofootprinting" were relegated to academic settings, and now are mainstream. Minimizing life cycle impacts however is intimately linked to resource efficiency and health and safety aspects. A pharmaceutical process with enhanced mass and energy efficiency will have a reduced environmental footprint given the reduction of natural resources consumption (mass and energy), and reduction of waste generated. When health and safety aspects are addressed in the design stage, it reduces the life cycle costs associated with health and safety controls, waste disposal, and reduces risks associated with improper control systems.

Life cycle assessment is also the framework that would allow us to address the wider aspects of sustainability, given the need to design process that addresses impacts that extend beyond our factory boundaries and over a longer term horizon. A process with a minimized environmental footprint is an obvious environmental benefit but it also provides societal and economic benefits that allows for providing medicines to patients globally at a lower price, with the potential business benefit of allowing the discovery company to be competitive after patent expiration. The sooner we can reduce our costs through process improvements the sooner the economic, social, and environmental goals can be realized. Using the life cycle assessment can highlight where in the process improvements can be focused to get the most impact on reducing energy, raw materials use, or emissions. For instance, a 10% reduction in solvent use may save more energy and reduce the environmental footprint than a 10% reduction in raw materials or improvement in yield. However, it can be challenging to acquire the information needed to assess the resource and energy consumption and emissions generated for all phases over the life of a product from cradle to grave, including extraction of raw materials, production, transport, use, and disposal.

Using life cycle thinking or taking a cradle-to-grave approach to design addresses the entire life of the product. For instance, the following questions are addressed:

- Are the raw materials from renewable or nonrenewable resources?
- How will the product or wastes be disposed?
- Can packaging be minimized?
- How can the product be designed to degrade at the end of life or for ease of recycling or reuse?
- How can it be designed to decrease impacts during the use of the product? (e.g., Can a propellant be removed from a spray?)

Are options (raw materials, unit operations, disposal, etc.) selected based on the life cycle impacts? There have been some attempts to facilitate the integration of health, safety and life cycle aspects as part of the chemical engineering design work within the pharmaceutical industry. One framework for integrating health, safety, and life cycle considerations into pharmaceutical technologies or processes has been proposed [24, 25]. This framework integrates considerations regarding efficiency, energy, health and safety, and environmental impacts (including life cycle impacts) to compare and select unit operations or processes from a "green" standpoint. For example, Table 4.2 shows the high-level comparative assessment performed between different options for solvent recovery and/or disposal within a particular pharmaceutical process. This framework has been utilized to develop GlaxoSmithKline's *Green Technology Review* and *Green Technology Guide* (Figure 4.3).

TABLE 4.2 Comparison of Technologies for Solvent Recovery or Disposal

	Environment	Safety	Efficiency	Energy
Pressure Swing Distillation, Vacuum	Yellow	Yellow	Yellow	Yellow
Pressure Swing Distillation, Atmospheric	Yellow	Yellow	Yellow	Yellow
Atmospheric Distillation plus Vapour Permeation	Yellow	Yellow	Yellow	Green
Incineration	Red	Yellow	Red	Red

Color Key:

Green	alternatives with significant advantages
Red	alternatives with significant disadvantages
Yellow	alternatives that do not exhibit significant advantages or disadvantages

* For the Environment category, mass indicators and life cycle indicators were considered. For the Energy category, energy requirements and life cycle energy were considered.

The safety column includes both health and safety considerations. For the environment category, mass indicators and life cycle indicators were considered. For the energy category, energy requirements and life cycle energy were considered.

However, there is much work that needs to be done in this regard, especially as related to the full integration of life cycle impacts within the decision-making processes of pharmaceutical companies.

4.4.3 Improving the Sustainability Profile

To realize sustainability aspirations, it is necessary to fully leverage the skill set of chemical engineers with an integrated sustainability mindset.

It all starts with innovation.

For the pharmaceutical industry, there are a series of key areas that chemical engineers need to integrate actively in designing more sustainable pharmaceutical processes. Some of these have been identified by the ACS GCI PR Green Engineering Subteam as a preliminary list, and include process intensification, continuous processing, bioprocessing, mass and energy integration, scale-up aspects, separation technologies, solvent selection, nanotechnology, life cycle assessment, and the integration of chemistry and engineering.

These areas represent different levels of development and different levels of innovation needed. Some of these areas such as process intensification, continuous processes, or bioprocessing are indeed not new, but need to be adapted and adopted into pharmaceutical processes. Some of them

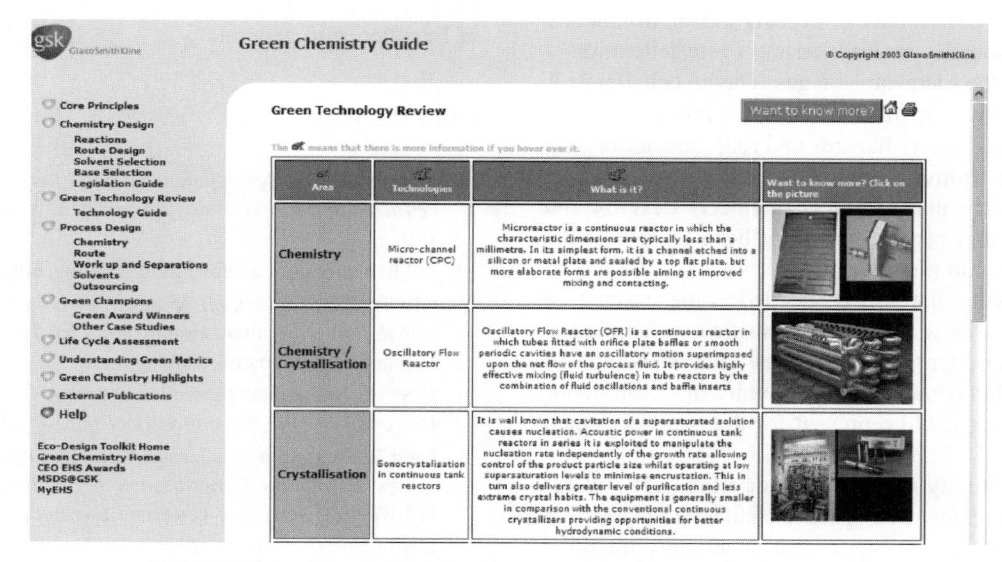

FIGURE 4.3 GlaxoSmithKline's *Green Technology Review* screenshot.

have not been explored in much detail, such as nanotechnology, some of them have been under development for sometime, such as separation technologies, solvent selection, and separations. Furthermore, areas such as life cycle assessment and the integration of chemistry and engineering will require implementation of different approaches in design, scale-up, and operation of manufacturing processes.

There are already examples of advances in these areas. Continuous microreactors from Corning [26] that have been utilized for API and intermediate production, continuous secondary processing alternatives explored by several companies, and bioprocesses currently established or being explored in our pharmaceutical plants [27, 28], process intensification has been demonstrated for some processes [29], life cycle assessment evaluations and tools of pharmaceutical processes [30, 31], to mention a few. However, a more widespread uptake is needed so these areas become best chemical engineering practices within the pharmaceutical industry.

In addition, these areas need to be addressed in a systematic, interrelated fashion and not in isolation. One can envision perhaps the possibility of an enzymatic process running continuously; or a heavily intensified process that utilizes hybrid reaction/separation unit operations to enhance mass and energy transfer. The extent to which these type of processes can be made operational will depend on how well innovation can be developed and applied.

4.5 FUTURE OUTLOOK

The challenge for chemical engineers will be to continue to advance the state-of-the-art of chemical engineering as it applies to the pharmaceutical industry in order to design more sustainable processes.

The aim would be to reduce both costs and environmental impact—both resource consumption and waste generation—while enhancing the social advantages. The first challenge is to improve the efficiency of the industry's processes while reducing health and safety hazards and risks and addressing life cycle impacts from a design perspective.

This will require that chemical engineers develop and utilize skill sets that perhaps have not been identified and intensively applied to pharmaceuticals.

One of the skills that will be needed is the mastery of continuous processes and process intensification applied to pharmaceuticals. Better scientific process understanding will be needed to fully leverage the opportunity that continuous flow manufacturing could represent for the pharmaceutical industry.

Another opportunity is the further development and extension of bioprocesses. One specific challenge may be that we could be dependent on the use of genetically modified microorganisms, which will require a increased dialogue with regulators and the public to ensure that the right controls are in place and that the public understands the risks and the benefits associated with genetically modified microorganisms. The development of new bioprocesses is also an exciting endeavor that will bring a particular need for process system engineers to develop quantitative decision-making tools and rapid simulation that will include both process design and sustainability principles.

Advances in process systems engineering will be dependent on development of better and more sophisticated tools (including property prediction packages and the development of a database for bio-based molecules).

Finally, to routinely assess sustainability of processes will require more robust and transparent life cycle inventory databases of pharmaceutical materials; as well as better modeling and understanding of the social and economic aspects of sustainability and their relationships.

The pharmaceutical industry is committed to discovering medicines that allow people around the world to live longer, healthier, and more productive lives. In pursuing a sustainable approach to process development and manufacturing, the industry may be in a better position to delivering on this promise.

ACKNOWLEDGMENTS

The authors want to thank and acknowledge all the individuals and teams working to develop and manufacture more sustainable processes within GlaxoSmithKline: Andrew Ross, Phil Dell'Orco, Tom Roper, Teresa Oliveira, Lucy Joyce, Joe Adams, Ted Chapman, Andrew Collis, Andrew Dywer, Richard Henderson, Rebecca De Leeuwe, and Lisa Cardo in particular; and all the colleagues in the Sustainable Processing Team, the Innovation and Sustainability Center of Excellence and the Sustainability and Environment Center of Excellence in general.

REFERENCES

1. Pisano GP. *The Development Factory. Lessons from Pharmaceuticals and Biotechnology*, Harvard Business School Press, 1997.
2. NAE, National Academy of Engineering, http://www.nae.edu/.
3. OECD. *The Application of Biotechnology to Industrial Sustainability—A Primer*. Organization for Economic Cooperation and Development, 2002.
4. U.S. Environmental Protection Agency. *Guides to Pollution Prevention in the Pharmaceutical Industry*, October, 1991.
5. Gani R, Jørgensen SB, Jensen N. *Design of Sustainable Processes: Systematic Generation & Evaluation of Alternatives. 7th World Congress of Chemical Engineering*, 2005.
6. Uerdingen E, Gani R, Fisher U, Hungerbuhler K. A new screening methodology for the identification of economically

beneficial retrofit options for chemical processes. *AIChE J.* 2003;49(9):2400–2418.

7. Carvalho A, Gani R, Matos H. Design of sustainable processes: systematic generation & evaluation of alternatives. In: Marquardt W, Pantelides C, editors. *16th European Symposium on Computer Aided Process Engineering and 9th International Symposium on Process Systems Engineering*, 2006. © 2006 Published by Elsevier B.V., Vol. 2, pp. 817–822.
8. Sheldon RA. Organic synthesis—past, present and future. *Chem. Ind.* 1992;23: 903–906.
9. Sheldon RA. The E factor: fifteen years on. *Green Chem.* 2007;9: 1273–1283.
10. Constable DJC, Curzons AD, Cunningham VL. Metrics to 'Green Chemistry'—which are the best? *Green Chem.* 2002;4: 521–527.
11. ACS GCI PR. American Chemical Society, Green Chemistry Institute, Pharmaceutical Roundtable Web site www.acs.org/greenchemistry, 2008.
12. Henderson RK, Kindervater J, Manley JB. *Lessons Learned Through Measuring Green Chemistry Performance—The Pharmaceutical Experience*, American Chemical Society, Green Chemistry Institute, Pharmaceutical Roundtable Web site www.acs.org/greenchemistry, 2008.
13. Anastas P, Warner J. *Green Chemistry: Theory and Practice*. Oxford University Press, USA, December 1998, 152 pp.
14. Winterton N. Twelve more green chemistry principles? *Green Chem.* 2001;3: G73–G75.
15. Anastas P, Zimmerman J, *Environ. Sci. Technol.* 2003;35(5): 98A–101A.
16. Abraham M, Nguyen N. *Environ. Prog.* 2003;22(4):233–236.
17. Beckman EJ. Using Principles of Sustainability to Design "Leap-Frog" Products. *Keynote Presentation during the 11th Annual Green Chemistry and Engineering Conference*, June 26–29, 2007.
18. Jiménez-González C, Constable DJC. *Green Chemistry and Engineering: A Practical Design Approach*, Wiley, 2010, ISBN: 978-0-470-17087-8.
19. ACS GCI PR. American Chemical Society, Green Chemistry Institute, Pharmaceutical Roundtable Web site www.acs.org/greenchemistry, 2008.
20. Jiménez-González C, Curzons AD, Constable DJC, Cunningham VL. Cradle-to-gate life cycle inventory and assessment of pharmaceutical compounds: a case-study. *Int. J. LCA* 2004;9(2):114–121.
21. GlaxoSmithKline. *Our Responsibility: Corporate Responsibility Report 2008*, 2008, 336 pp, www.gsk.com/responsibility.
22. Thayer AM. Sustainable syntheses. *Chem. Eng. News* 2009;87(23):13–22.
23. Cote AS, Dorwart JG, Fernandez PF, Hobbs DM, Massonneau V, Mohan MA, Moment AJ, Moses AW, Petrova RI, Wallace DJ, Wright TJ. Using Process Mass Intensity (PMI) to Guide Process Development and Design. *Presentation During the 13th Annual Green Chemistry & Engineering Conference*, College Park, MD, June 23–25, 2009.
24. Jiménez-González C, Curzons AD, Constable DJC, Overcash M, Cunningham VL. How do you select the 'greenest' technology? Development of guidance for the pharmaceutical industry. *Clean Products Processes* 2001;3(2001):35–41.
25. Jiménez-González C, Constable DJC, Curzons AD, Cunningham VL. Developing GSK's Green Technology Guidance: methodology for case-scenario comparison of technologies. *Clean Technol. Environ. Policy* 2002;4: 44–53.
26. Sutherland J. Efficient Processing with Corning® Advanced-Flow™ Glass Technology. *Presentation during the 13th Annual Green Chemistry & Engineering Conference*, College Park, MD, June 23–25, 2009.
27. Gebhard R. *Sustainable Production of Pharmaceutical Intermediates and API's—The Challenge for the Next Decade*, DSM Webminar, 2009.
28. Henderson RK, Jiménez-González C, Preston C, Constable DJC, Woodley JM. EHS & LCA assessment for 7-ACA synthesis: a case study for comparing biocatalytic & chemical synthesis. *Ind. Biotechnol.* 2008;4(2):180–192.
29. Poechlauer P, Braune S, Reintjens R. Continuous Processes in Small-Scaled Reactors Under cGMP Conditions: Towards Efficient Pharmaceutical Synthesis. *Presentation during the 13th Annual Green Chemistry & Engineering Conference*, College Park, MD, June 23–25, 2009.
30. Curzons AD, Jiménez-González C, Duncan A, Constable DJC, Cunningham VL. Fast life-cycle assessment of synthetic chemistry tool. *Int. J. LCA* 2007;12(4):272–280.
31. Jiménez-González C.Life cycle assessment in pharmaceutical applications, Ph.D. Thesis, North Carolina State University, Raleigh NC, 2000, 257 pp.

5

SCIENTIFIC OPPORTUNITIES THROUGH QUALITY BY DESIGN

Timothy J. Watson and Roger Nosal
Global CMC, Pfizer, Inc., Groton, CT, USA

Quality by design (QbD) is about understanding the relationships between the patient's needs and the desired product attributes by ensuring that all process attributes and parameters that are functionally related to safety and efficacy are consistently met. The value of prospectively developing enhanced product knowledge and process understanding can significantly minimize patient risk. The application of QbD principles also strengthens the balance between continued product improvements, technical innovation, business needs, and regulatory oversight. A QbD approach can serve as the foundation that links research and development, manufacturing, and regulatory conformance through a fundamental common language that is based on a science and risk-based principles.

For decades, much of the activity in quality and quality management focused on compliance rather than utilizing a fundamental understanding of the science behind process understanding. Business practices adapted to procedures and focused on minimizing regulatory risk. The implementation of new technology was not typically part of a strategy because oversight for novel technology was not precedented. Extremely risky and high attrition rates of research programs, unlike other industries, coupled with the lack of global regulatory harmonization fostered a minimalist paradigm and, significantly challenged investments in new technologies and using modern methodologies for development. In the 1990s the use of PAT started to gain interest in pharmaceutical manufacturing; however, it was primarily used for business purposes and not seriously considered for regulatory purposes. Describing to regulatory authorities a comprehensive view of process understanding was generally avoided for fear of being held accountable to increased scrutiny and higher standards.

In August 2002, the FDA launched their GMPs for the twenty-first century initiative in partial response to academics and consultants who criticized the pharmaceutical industry for not manufacturing to the highest standards. Companies were encouraged to use risk-based assessments, in particular when identifying product quality attributes, and adopt integrated quality systems that operated throughout the lifecycle of a product. This movement toward science-based regulations has not been limited to the United States as seen by the guidance provided by The International Conference on Harmonization (ICH). Thus, quality by design for the pharmaceutical industry evolved from a conceptual approach that envisioned an efficient, agile, flexible sector that reliably produces high-quality drug products without extensive oversight [1]. Guidance for QbD was crafted through the ICH process to what is now considered the "QbD trio"; ICH Q8, Q9, and Q10 (ICH Q11 for drug substance in progress). This movement away from prescriptive development programs has become an exciting and empowering platform for chemists, scientists, formulators, and engineers. While many elements associated with QbD, such as risk assessments, design of experiments (DoE), operational control strategies, etc., have been employed well before the adoption of the ICH guidelines, application was frequently not systematic, concerted or prospective, but rather retrospective in response to issues or problems encountered during development or after commercial launch. Consequently, companies were reluctant to pursue a QbD approach or introduce supplemental studies

Chemical Engineering in the Pharmaceutical Industry: R&D to Manufacturing Edited by David J. am Ende

on process capability for fear of unnecessarily increasing regulatory "requirements" and potentially delaying regulatory approvals.

QbD begins with a prospective vision that accepts and builds upon a science and risk-based platform with a commitment to maintain focus on the patient. It starts with the establishment of a quality target product profile (QTPP) that provides an inventory of expectations or "product attributes" required to ensure patient safety and efficacy and product quality. Using the QTPP, relationships between product attributes and the sources for meeting those attributes can be derived from drug product and drug substance platforms to establish a holistic understanding of how attributes are linked to patients needs, and how these attributes are functionally related through the entire manufacturing process. Ensuring patient safety and efficacy is not about "what measures we apply" to maintain the QTPP, it is about "how we develop" process understanding to establish the appropriate design and control elements of a process. The approach is predicated on executing a rigorous risk management exercise to determine "what we need and what we have" to demonstrate that quality is consistently met. It is about identifying the relationship between each attribute and its functional relationship to manufacturing variables and consistently controlling these relationships.

It is generally recognized that the three fundamental concepts of QbD are design space, control strategy, and criticality; where design space and control strategy are the deliverable outcomes from a systematic application of risk and science-based assessments, analyses, experiments, technical innovation, and control. The development of design spaces and control strategies is a *symbiotic relationship* that encompasses all of the concepts contained within the chapters of this book. In adopting a QbD approach and applying the science and risk-based principles to assess quality attributes and process parameters, design space can be created to describe the boundaries within which unit operations of a manufacturing process may operate. In essence, design space can demonstrate control of variables that may impact a critical quality attribute, and a control strategy can be established parametrically to as the resulting design modate design space. For example, a combination of well-space boundaries and real-time release testing can effectively demonstrate and confirm control and serve as the basis for release of the product without the need for specific end-product testing. Therefore, where the risk is understood and the severity and probability of impact are controllable, the demonstration of process control through the creation of design space could conceivably reduce the need to perform in-process testing as well. Continuous formal verification to demonstrate process capability in accordance with well-grounded design space criteria could serve as the basis for product release to a specification derived largely from critical quality attributes (Figure 5.1).

Scientists who embrace an enhance approach to development should consider these types of questions:

- How is prior knowledge substantiated, how can internal and external knowledge be used to leverage more accurate risk assessments?
- What level of detail is required to justify risk assessments?
- How should design space be presented and conveyed to demonstrate quality assurance?
- How can modeling be used to justify commercial manufacturing process changes?
- How should the control strategy connect drug product and drug substance quality attributes to process parameters and material attributes?
- Is there an attenuation of regulatory latitude for post-approval optimization and continual improvement?

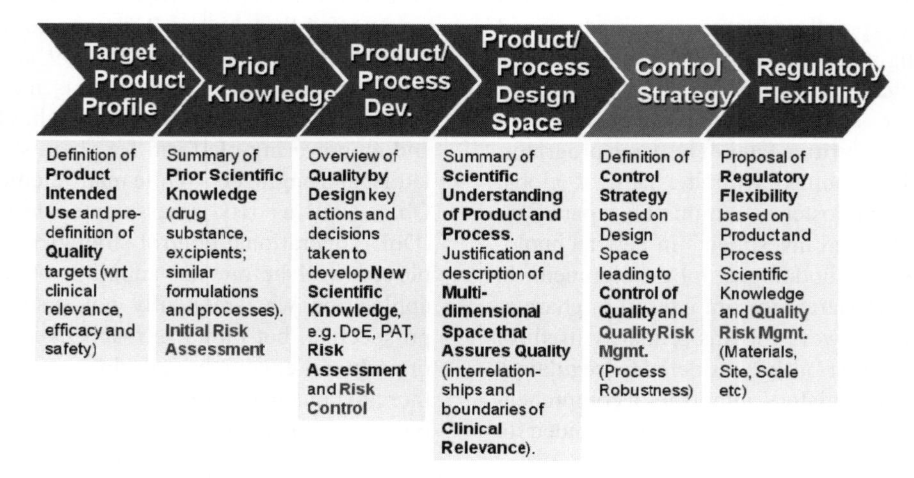

FIGURE 5.1 General outline of approach to application of quality by design. *Source:* EfPIA QbD WG.

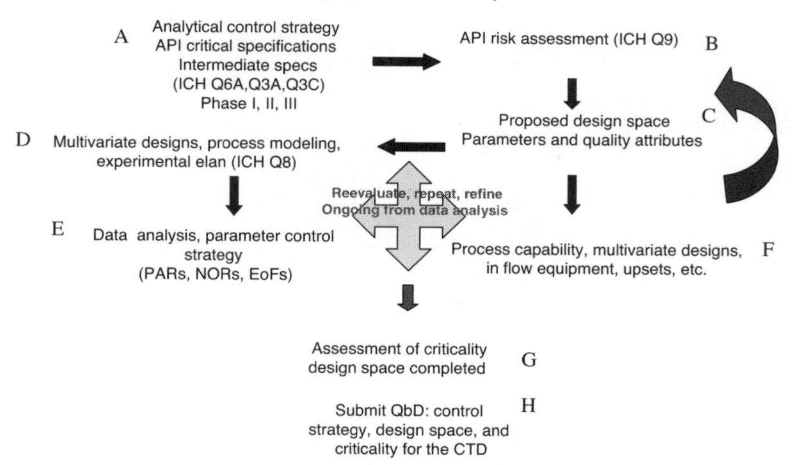

FIGURE 5.2 Small example of drug substance workflow for QbD.

In addition, there are many general processes that can be adapted to sketch out a general procedure for any team of subject matter experts to adapt a science and risk-based approach. One example is given are Figure 5.2. A Common thread that runs through all varieties of QbD applications is repeated risk assessment of process parameters and material attributes and their connectivity to the QTTP; adoption of an an iterative approach to risk and experimental data evaluation; creative experimental design to understand parameter interaction in a multivariate process; establishment of a well grounded design space and control strategy that ensures safety and quality. Finally, transparency in the interpretation and presentation of data and its justification for process design must meet the standards for peer review and "pass the red face test" for regulatory authorities. There are many options for implementing QbD. However the fundamental conceptual elements of the risk and science-based approach have emerged as relatively consistent within the industry. With appropriate scientific justification and consistent application most options are acceptable. Far from suppressing progress, the refinement of the meaning, application and implementation of QbD has stimulated regulatory authorities and industry to pursue clarification. As a result, subsequent progress has improved the consistent application and value of these concepts.

The intrinsic advantages of investing in enhanced process understanding increases confidence and assurance of product quality. Tangible benefits, for example, reductions in manufacturing costs associated with improved efficiencies and innovations, reduction in manufacturing recalls, and failures or extraneous investigations attributed to uncertainty, are largely realized over the lifecycle of a product.

The fundamental scientific premise derived from the application and implementation of Quality by Design principles that attracts scientific support from every discipline across this industry is driven by a common passion to develop improve process understanding and product knowledge. The movement away from prescriptive and in many cases retrospective development approaches has become an exciting and empowering platform for chemists, scientists, formulators, and engineers. QbD has also played an instrumental role in establishing the value and importance of cross-functional, scientific relationships in pharmaceutical development through proactively developing and understanding processes and formulations. Perhaps, most importantly, the application of a QbD approach and investment in robust Pharmaceutical Quality Systems are expected to reduce unexpected variability in manufacturing processes and unanticipated failures in product quality, thereby improving quality assurance of products.

ACKNOWLEDGMENT

Robert Baum from Pfizer for sharing his knowledge and thoughts on the historical perspective on QbD.

REFERENCES

1. Woodcock J,M.D. October 5, 2005.
2. Watson TJN, Nosal R, am Ende D, Bronk K, Mustakis J, O'Connor G, Santa Maria CL. *J. Pharmaceut. Innovation* 2007;2(3-4):71.

ACKNOWLEDGMENT

REFERENCES

PART II

ACTIVE PHARMACEUTICAL INGREDIENT (API)

6

THE ROLE OF CHEMICAL ENGINEERING IN PHARMACEUTICAL API PROCESS R&D

EDWARD L. PAUL*
Chemical Engineering R&D, Merck & Co., Inc., Rahway, NJ, USA

6.1 INTRODUCTION

The evolution of R&D and synthetic chemical processing in the pharmaceutical industry in the past century is remarkable in the accomplishments of bringing many important therapeutic products to the world markets.

These successes are illustrated by the proven ability of chemists and chemical engineers to overcome the many process problems that can be encountered in development and scale-up that are critical in bringing valuable therapeutic products to market. Failure to do so could have resulted in some of these products remaining as laboratory curiosities.

One of the outstanding examples of this type of success is illustrated by the penicillin process, in which penicillin could have remained a laboratory curiosity if not for a breakthrough using chemical engineering principles as applied to fermentation processing—further discussed below.

Cost to patients is, of course, a major source of concern for the industry. On the R&D side, minimizations of operating and capital costs are always key objectives in developmental programs. These efforts can at times be frustrated by the rather small impact some improvements can have in reducing costs. In other cases, a therapeutically important product can only be brought to market at a reasonable cost because of the chemical and engineering input in R&D. It is always interesting to speculate how critics decide something is far too expensive without knowing anything about the costs associated with discovering, developing, and making the product.

6.2 CHEMISTS AND CHEMICAL ENGINEERS

A fascinating aspect of the history of R&D and manufacturing is the relationship between chemists and chemical engineers and how this relationship has evolved over the years.

The accomplishments of chemists in the creation and development of complex syntheses provide the foundation of R&D in the Pharmaceutical Industry. It has always been a source of amazement to see how the synthesis of a complex molecule can be achieved from seemingly unrelated parts.

Chemists were unquestionably in charge of R&D through the 1950s and their role in process R&D was then viewed as dominant over chemical engineers. This latter point is well illustrated by noting that the then President of the Research Division, Merck & Co., Inc., Max Tishler, later to become president of the American Chemical Society, personally led the manufacturing plant start-up teams for new processes.

The question may be asked regarding what role chemical engineers have in making these complex syntheses commercially viable. Chemical engineers have played key roles in many processes and the remainder of this chapter will be focused on some aspects of the relationship. Some generalities

R&D has been dominated by chemists.
The role of chemical engineers is less defined.
The role has changed over the years
The role varies considerably between companies.

*Retired

Chemical Engineering in the Pharmaceutical Industry: R&D to Manufacturing Edited by David J. am Ende

The challenge to the chemical engineer is to recognize the following:

(1) when a process can just be made larger and therefore be readily scaled to manufacturing or
(2) when some changes requiring specialized equipment and/or process operations are needed for successful scale-up and/or economical operation.

The key to effective engineering input is the organization in R&D that allows for both early involvements in process design and process development. Also, continuing responsibility for plant design as well as the integration of laboratory and pilot plant programs with chemists—all leading to a manufacturing process and plant start-up. Management of these key functions must establish clear lines of responsibility that promote interdisciplinary cooperation. These roles are not difficult to state but can be difficult to accomplish. However, the benefits of this integrated effort can be realization of superior process design and operation.

6.3 PENICILLIN: A CHEMICAL ENGINEERING ACHIEVEMENT

The development of deep tank fermentation technology is cited as an achievement by chemical engineers, which has had a profound impact on both process technology and medicine. In the 1930s, after the discovery of penicillin, the only method to make the antibiotic was in a surface mold that was capable of making only gram quantities.

Surface culture was replaced by deep tank fermentation in the mid-1940s by collaboration between Abbott, Lederle, Squibb, Pfizer, and Merck with consultant Richard Wilhelm, Department Chair of Chemical Engineering at Princeton University. Implementation of this technology was achieved by application of chemical engineering principles developed in the 1930s and 1940s. Without this innovation, penicillin would have remained a laboratory curiosity for an unknown length of time. All subsequent antibiotic fermentation processes have utilized this technology.

In addition, the processing of large volumes of fermentation broth was achieved by continuous filtration and extraction. The following citation captures the extent of these accomplishments.

The American Chemical Society, in collaboration with the Royal Society of Chemistry, designated the development of penicillin as an International Historic Chemical Landmark on November 19, 1999. The text of the plaque commemorating the event reads

> In 1928, at St. Mary's Hospital, London, Alexander Fleming discovered penicillin. This discovery led to the introduction of antibiotics that greatly reduced the number of deaths from infection. Howard W. Florey, at the University of Oxford working with Ernst B. Chain, Norman G. Heatley and Edward P. Abraham, successfully took penicillin from the laboratory to the clinic as a medical treatment in 1941. The large-scale development of penicillin was undertaken in the United States of America during the 1939–1945 World War, led by scientists and engineers at the Northern Regional Research Laboratory of the US Department of Agriculture, Abbott Laboratories, Lederle Laboratories, Merck & Co., Inc., Chas. Pfizer & Co. Inc., and E.R. Squibb & Sons. The discovery and development of penicillin was a milestone in twentieth century pharmaceutical chemistry.
>
> *Source:* American Chemical Society [1].

The difficulties are summarized in the following quotation:

> Pfizer's John L. Smith captured the complexity and uncertainty facing these companies during the scale-up process: "The mold is as temperamental as an opera singer, the yields are low, the isolation is difficult, the extraction is murder, the purification invites disaster, and the assay is unsatisfactory." American Chemical Society [1].

6.4 BATCH AND CONTINUOUS PROCESSING

It is believed that some of the accomplishments in penicillin isolation were made possible in part because, in the 1940s, academic training in chemical engineering was focused on continuous processing with models from the petroleum industry. Thus, chemical engineers were ready to develop continuous processes since their training would lead them to think—continuous—first.

Batch and semi-batch operation will continue to predominate and are the methods of choice in many processes for readily documented reasons. Efficiency of these operations has been greatly enhanced in recent years by online computer control and analytical instrumentation. Start-up, operation, and shutdown of continuous operations have also been greatly facilitated by computer control. In addition, the small size of most in-line mixing devices can minimize or eliminate off-specification product from unsteady-state operation.

The development and manufacturing utilization of continuous processing in pharmaceutical manufacturing has been successfully accomplished in a variety of process applications. These process options are discussed by Paul and Rosas [9] along with the chemical engineer's role in the development and manufacturing implementation of safe and efficient processes with minimum capital and operating costs. The utilization of continuous operations is included in this developmental strategy, when appropriate to solve scale-up issues that may be encountered in some systems, including reactor selection and design, separation trains, and crystallization operations. In some of these systems, the inherent limitations of batch operation in mass transfer, mixing, and throughput require continuous operation for successful scale-up of selected steps.

Examples of these options are not limited to throughput efficiencies but, in some cases, include operations that cannot be successfully run batch wise. Examples include (1) fast, complex reactions that require more effective mixing and/or mass transfer than can be achieved in stirred vessels, (2) thermally hazardous reactions in which reaction volume must be minimized, and (3) some crystallizations including the direct resolution of optical isomers or control of particle size within tight limits .

Utilization of these types of process improvements has resulted in manufacturing efficiencies in many processes, as well as in evolution of strategies to integrate the R&D efforts of process chemists and chemical engineers starting early in the development cycle.

6.4.1 Literature

The design and utilization of continuous systems in the pharmaceutical industry has received little attention in the literature as opposed to the chemical industry in general. Since most of the applications are for fast reactions, large scale-up factors are not often encountered. The combination of high heats of reaction, high reactant concentrations, fast reaction rates, and simultaneous or consecutive reactions to undesired products presents an extreme challenge to chemists and development and design engineers. Heat transfer and micro-mixing requirements must be satisfied simultaneously.

The subject of mixing and fast chemical reactions has been extensively covered in the seminal work by Baldyga and Bourne [2], including the development of the test reactions that can be used in evaluation of various types of reactor systems for fast reactions. Mixing issues are also treated in the *Handbook of Industrial Mixing* [8], including several examples.

There are several issues that can lead to the decision to develop a semi-continuous or continuous process versus a batch or semi-batch operation. Each process requirement must be evaluated on the basis of manufacturing scale feasibility.

The most important reactor issue is achieving equal selectivity and by-product distribution on scale-up. For fast reactions, this may require in-line reactor configurations including tubular devices such as static mixers to achieve the necessary mixing intensity to minimize over-reaction in competitive–consecutive and parallel reactions.

Common reactor configurations such as continuous stirred tank reactors (CSTRs), packed beds, fluid beds, and trickle beds that are used in the chemical industry for high throughput find limited utility in the pharmaceutical industry because of reduced throughput requirements. In addition, this type of contacting can be achieved in standard batch or semi-batch reactors, including multiphase reactions, when mixing intensity is not an issue for selectivity on scale-up.

Factors that require evaluation include

- highly exothermic reactions,
- mass transfer limiting reactions,
- thermal hazard reactions,
- crystallization at nonequilibrium conditions and with narrow size distribution requirements, and
- high throughput requirements in reactions and separations.

The necessity to utilize a continuous reactor to achieve successful scale-up can, in some cases, be determined by analysis of data for a laboratory semi-batch reaction. For a classical consecutive–competitive reaction, the sensitivity to mixing can be evaluated by running the reaction under identical conditions of addition time, temperature, and feed point, but with increased mixing intensity. If the selectivity increases with increased mixing speed, thereby generating smaller amounts of by-products, the reaction is fast enough to be mixing sensitive. If increasing mixing speed does not result in achieving a plateau in selectivity, the reaction may not be scalable in a semi-batch configuration [8].

6.5 EXAMPLES

The following two examples, chosen as illustrations, are focused on R&D issues that required something more than direct scale-up either because

(1) the original laboratory process could not be successfully scaled up and achieve the desired yield or product quality and
(2) the process cannot be run successfully in a batch mode even in the laboratory.

Example 6.1 Alkylation Reaction with Continuous Liquid–Liquid Extraction and Crystallization

This example illustrates a combination of semi-batch and continuous operations in one step of a multistep synthesis (see Figure 6.1). The sequence is as follows

- semi-batch alkylation reaction,
- continuous multistage extraction (Karr extractor), and
- semi-continuous crystallization.

6.5.1 Reaction

The semi-batch reaction is a competitive–consecutive sequence, which is as follows

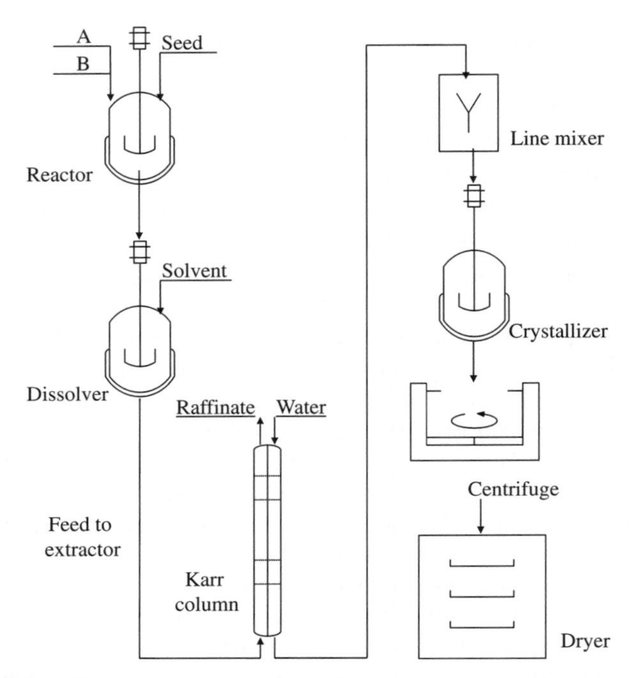

FIGURE 6.1 Flow sheet for alkylation step showing the configuration of reactor, extractor, crystallizer, and isolation.

$$\begin{aligned} A + B &\rightarrow R \\ R + B &\rightarrow S \end{aligned} \qquad (6.1)$$

where R is the product and S is the bis over-reaction product. When run as a homogeneous reaction system, the bis formation was excessive. A change to increase the concentration of the reaction mixture was possible to achieve increasing supersaturation of R as the reaction preceded. The addition of R seed at the start of the reaction was implemented to guarantee crystallization of R during the semi-batch addition of B. By separating the R as crystals in a second phase, the bis formation was minimized thereby raising the selectivity of R by >15%. Methods of increasing selectivity in heterogeneous systems are described by Sharma [10].

6.5.2 Extraction

On completion of the reaction, the R crystals are dissolved to prepare for purification by continuous solvent extraction using a multistage countercurrent Karr extraction column, which removes the aqueous-soluble components of the reaction mixture as illustrated in Figure 6.1.

6.5.3 Crystallization

The organic phase containing R and S is fed directly from the extractor to a crystallizer using a line mixer to induce nucleation of R in the feed stream. In this operation, the crystallizing product accumulates in the crystallizer. The required crystallization conditions, temperature, and mixing intensity for controlled nucleation and growth, are maintained in the line mixer between the extractor and the crystallizer. The slurry in the crystallizer is recycled around the crystallizer and through the line mixer throughout the operation.

The operation continues until the solution from the extractor feed tank is consumed and the product has accumulated in the crystallizer. The slurry is then cooled and centrifuged.

Example 6.2 Continuous Operation Required for Process Viability

The following example illustrates a case in which continuous processing is used because batch or semi-batch operation cannot produce the desired separations for the resolution of optical isomers.

This application operates at nonequilibrium conditions, therefore, continuous operation in separate R and S isomer crystallizers is the only feasible method to maintain optical purity in the crystallizers. Batch methods have been described in the literature [7] for laboratory operation but cannot be scaled-up because of the requirement to operate in the kinetic regime to prevent nucleation of the undesired isomer. The method applies only to racemic conglomerates and not to racemic compounds since the separation requires crystallization of each isomer separately, a condition that can only be met with racemic conglomerates.

The system has been described in patents [3–6, 11]. The system is shown in Figure 6.2. The fluidized bed provides the necessary solid–liquid separation between crystal beds as well as a controlled environment for crystal growth with minimum nucleation.

The key conditions for successful operation are (1) close control of supersaturation to prevent nucleation in the respective seed beds and (2) growth of seed crystals that are compatible with fluid bed operation . In addition, since the system is essentially an all-growth operation, the population balance must be maintained by some form of crystal fracture. This fracture is achieved by sonication as described in the Midler patents.

The fluidized bed technology can also be utilized in general for continuous crystallization and is particularly useful for the growth of large crystals for ease in downstream processing as well as for high throughput.

In both of these examples, the role of chemical engineers was critical in realizing the benefits of implementing changes in the original processes. For Example 6.1 (alkylation), the chemistry was established by the process chemists who then followed all aspects of the subsequent development in the

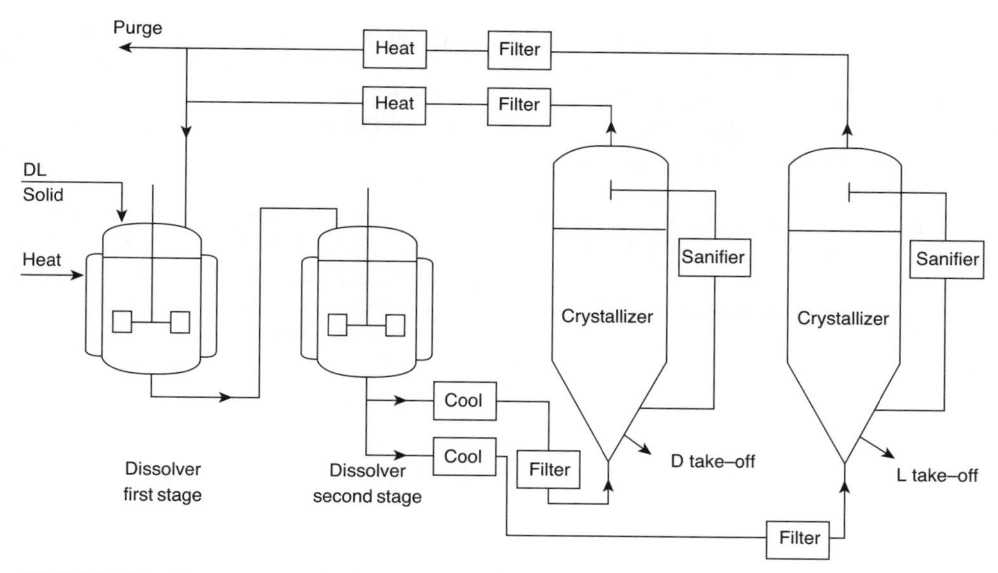

FIGURE 6.2 Flow sheet for resolution of optical isomers by direct crystallization (Source: Ref. 11, Fig. 7-27, p. 163. Reprinted with permission of Wiley).

engineering laboratory and pilot plant as well as participating in the manufacturing plant start-up. Engineering input came in three aspects of this step (1 of 17 steps for this synthesis) (1) selectivity improvement by crystallization to reduce bis formation, (2) process design to introduce continuous extraction and crystallization, and (3) crystallization equipment design to achieve crystal growth required for feasible filtration times on scale-up .

(1) Selectivity increase resulted in improvement of >15% with the resulting reduction in process cost in this late and critical step.
(2) Continuous operation was critical because of the high production volumes required where multistage batch extraction would be unfeasible.
(3) Crystallization by continuous in-line mixing and maintenance of a large seed bed was critical to both the high production volumes and to overcome nucleation of small poorly filtering crystals and achieve crystal growth.

For Example 6.2, the original processes for which this direct resolution technology was developed, accomplished the resolution through traditional diastereoisomer methods. These methods were successful but added costs through the use of resolving agents and their recovery as well as through multicrystallization and filtration steps.

Engineering input started in the laboratory, in first establishing feasibility for direct resolution—not a feasible option when the isomers form racemic compounds (solid solutions).

This development required the solution of several key issues including (1) solid–liquid separation after partial crystallization of each isomer, ultimately accomplished with fluidized bed technology, (2) development of a solvent system compatible with the solubility of the isomers, (3) establishment of supersaturation limits for operation with minimal nucleation (essentially all-growth), (4) a suitable method of crystal fracture to limit growth in the continuous operation, and (5) development of a control system to maintain the narrow limits of supersaturation and solid–liquid balance in the fluidized beds for continuous operation . The flow sheet is shown in Figure 6.2.

These engineering challenges were met in the evolution of robust processes that have been in large-scale operation for many years.

6.6 OTHER CHEMICAL ENGINEERING ACTIVITIES

The activities of chemical engineers in Pharma extend too many aspects of processing other than to reactions and separations as illustrated above. The range and importance of these activities is well demonstrated in the many topics addressed in this book.

REFERENCES

1. American Chemical Society. *Historic Chemical Landmark*, 2004.
2. Baldyga J, Bourne JR. *Turbulent Mixing and Chemical Reactions*. Wiley, Chichester, 1999.

3. *Chemical Engineering* magazine staff, Kirkpatrick Chemical Engineering Achievement Awards. *Chem. Eng.* 1965; 72 (23): 247.
4. Midler M. Production of crystals in a fluidized bed with ultrasonic vibrations. *U.S. Patent* 3,510,266, 1970.
5. Midler M. Process for the production of crystals in fluidized bed crystallizers. *U.S. Patent* 3,892,539, 1975.
6. Midler M. Crystallization system and method using crystal fracture external to a crystallizer column. *U.S. Patent* 3,996,018, 1976.
7. Mullin JW. *Crystallization*, 4th ed., Butterworth-Heinemann, Oxford, 2001.
8. Paul EL, Atiemo-Obeng V, Kresta SM, editors, *Handbook of Industrial Mixing: Science and Practice*, Wiley, New York, 2004.
9. Paul EL, Rosas CB. Challenges for chemical engineers in the pharmaceutical industry. *Chem. Eng. Progress* 1990; 86 (12): 17–25.
10. Sharma MM. Multiphase reactions in the manufacture of fine chemicals. *Chem. Eng. Sci.* 1988; 43 (8): 1749–1758.
11. Tung HH, Paul EL, Midler M, McCauley JA, *Crystallization of Organic Compounds: An Industrial Perspective*, Wiley, New York, 2009.

7

REACTION KINETICS AND CHARACTERIZATION

Utpal K. Singh
Chemical Process R&D, Eli Lilly and Company, Chemical Product R&D, Indianapolis, IN, USA
Charles J. Orella
Chemical Process Development and Commercialization, Merck & Co., Inc., Rahway, NJ, USA

7.1 INTRODUCTION

Characterization and understanding of reaction kinetics is an important part of chemical development in the pharmaceutical industry. The information gleaned from reaction kinetic studies are used for a range of different applications including process optimization, process safety evaluation, scale sensitivity understanding, and robustness testing. The use of reaction kinetics for each of these applications will be discussed in greater detail below with a particular emphasis on measuring reaction kinetics to probe the effect of scale and the effect of convolution of reaction kinetics with transport limitations.

This chapter focuses on some of the characteristics that make pharmaceutical processing unique from commodity and to a lesser degree from specialty chemicals. The literature has many excellent texts and articles devoted to a wide variety of perspectives on chemical reactions. Generally, these fall in three categories of mechanistic chemistry, reaction kinetics, or reactor design and operation. Far too frequently, these approaches are taken independently, without application of all three categories in a harmonized fashion. We have not recreated the full scope of excellent literature, and recommend the reader make good use of existing literature for chemistry or kinetics with or without reactor design and operation (reaction engineering) [1–6].

Our focus is on several aspects of these perspectives that offer room for increased application and impact by chemists and engineers working to successfully characterize, scale, and optimize chemical reactions within the pharmaceutical industry. The first differentiating feature of reactions within the pharmaceutical industry is the complexity and richness of the chemistry, with multiple reactive moieties present in molecules available for desired and more often undesired reactions. This often creates a barrier between chemists and engineers, which is only overcome by close collaboration to at least partially understand the chemistry and mechanism, the kinetic pathways and rates, and the most desirable reactor design model and operating mode. In order to address this complexity of the chemistry, it is necessary to make measurements that help uncover the underlying mechanisms, kinetic pathways, and rates. We seek to obtain unambiguous data about the reactions, which will bring together chemists and engineers to challenge and agree based on quantitative data.

The second feature that has differentiated the pharmaceutical industry to this point is the use of batch processing for the majority of operations, which results in scale up from the milliliter scale to the cubic meter scale (6 orders of magnitude). Over such a wide range of scale, the rate-limiting step is subject to vary with the scale, which poses challenges for understanding and controlling quality and consistency in the eventual manufacturing. In order to successfully scale up, it's necessary to apply or design the right equipment, and operate in such a manner to understand the rate-limiting step at every scale. For this reason, there needs to be a rational approach to characterizing the chemistry and the equipment used at every scale.

Chemical Engineering in the Pharmaceutical Industry: R&D to Manufacturing Edited by David J. am Ende

This incredible diversity in chemistry that is practiced in the pharmaceutical industry was recently captured by researchers at Pfizer. They reviewed the chemistry that was conducted at Pfizer over the course of 17 years and found that a small number of reaction classes contributed to a large fraction of their portfolio [7]. Better understanding of the mechanistic basis for these reaction classes by the practicing chemical engineer can aid process development by rationalizing and in-some cases predicting the effect of scale on the different reaction pathways. In light of the diversity of the chemistry described above, it is important to note that it is difficult, if not impossible, to make broad generalizations that can apply to the range of different chemistries in practice.

Some commonalities across the range of chemistries can be found from a report by which surveyed 22 different processes comprised of 86 different reactions; and classified them according to overall kinetics as well as homogeneous or heterogeneous nature of the reaction mixture [8]. They found that nearly 75% of the reactions were classified as either rapid or moderate and therefore have potential to be affected by scale. Additionally, a vast majority of the surveyed reactions were heterogeneous in nature due to the presence of multiple phases in the reaction mixture, that is, solids and gases. The combination of the heterogeneous nature of the reaction mixture along with the rapid nature of many of the reactions lends itself to the potential for scale sensitivity thereby underlying the need to understand and characterize reaction kinetics.

Characterization of reaction kinetics requires an understanding of the interplay of the rate of chemical transformation to that of the physical transformation. Scale sensitivity is exhibited when the rates of chemical transformation are faster than that for the physical transformation, that is, mixing, heat input, or removal. This chapter will outline some of the techniques that are available for characterization of chemical transformations as well as physical transformations.

7.2 EXPERIMENTAL APPROACHES

Several different equipment and technologies are available to aid in reaction kinetics measurements and this area is continuously evolving as the levels of automation and analyzer sophistication increases.

7.2.1 Calorimetry

Reaction calorimetry is a versatile and highly effective tool for reaction characterization in the pharmaceutical industry. The technique requires conducting an energy balance around the batch reactor yielding the following

$$MC_p\frac{dT_r}{dt} = UA(T_j - T_r) + R_{rxn}\Delta H_{rxn} + mC_p(T_{addn} - T_r) \tag{7.1}$$

where M, C_p, UA, T_j, T_r, R_{rxn}, ΔH_{rxn}, T_{addn} are reaction mixture mass, heat capacity of reaction mixture, heat transfer coefficient, jacket temperature, reactor temperature, reaction rate, heat of reaction, and temperature of added stream, respectively. The measurement can be conducted in an isothermal or nonisothermal mode which changes the relevant terms in the equation above. Since this techniques measure the total heat of reaction, it convolutes the heats associated with several chemical processes including heats of mixing, dissolution, crystallization as well as heats associated with all reactions including desired reaction and side reactions. For safety testing, this is ideal since such a measurement allows a lumped measurement of heat associated with all relevant chemical events in the process. For measurement of detailed reaction kinetics requiring deconvolution of different processes, reaction calorimetry offers the advantage of the fact that subtle changes in concentration profiles are magnified in heat flow measurements since the heat flow is directly proportional to the reaction rate. This methodology has been routinely highlighted in the works of Blackmond and coworkers as shown in Figure 7.1, which shows the rate profile during Pd/BINAP catalyzed cross coupling of bromobenzene with *N*-methylpiperazine. Conversion profiles measured using GC measurements (filled symbols) shows a potential inflection point during the first hour of reaction. The heat flow profile, which is effectively the reaction rate at a given time, accentuated this behavior clearly by showing an induction period during the first 100 min of reaction (approximately 50% conversion). This reaction was studied in great detail and systematic analysis has been highlighted in works of Blackmond and coworkers [9–12].

A systematic use of reaction progress kinetic analysis using *in situ* reaction calorimeter has been documented by Blackmond and coworkers; and several review articles articulate this approach is great detail [9, 10].

One important caveat should be applied when measuring rapid reaction kinetics especially when the process kinetics are of the same scale or faster than the equipment time constant, the measured rate constant can significantly vary. Table 7.1 shows a comparison of acetic anhydride hydrolysis from calorimetry with that from literature. As the reaction half-life is shortened to less than 1 min, the difference between the measured reaction rate and that from literature increases. A number of different algorithms are available for deconvoluting the equipment time constant from the measured kinetics [13]; however, this process can, depending on the vendor, be a black box. Nevertheless, these results indicate that reaction calorimetry can adequately measure reaction rates under

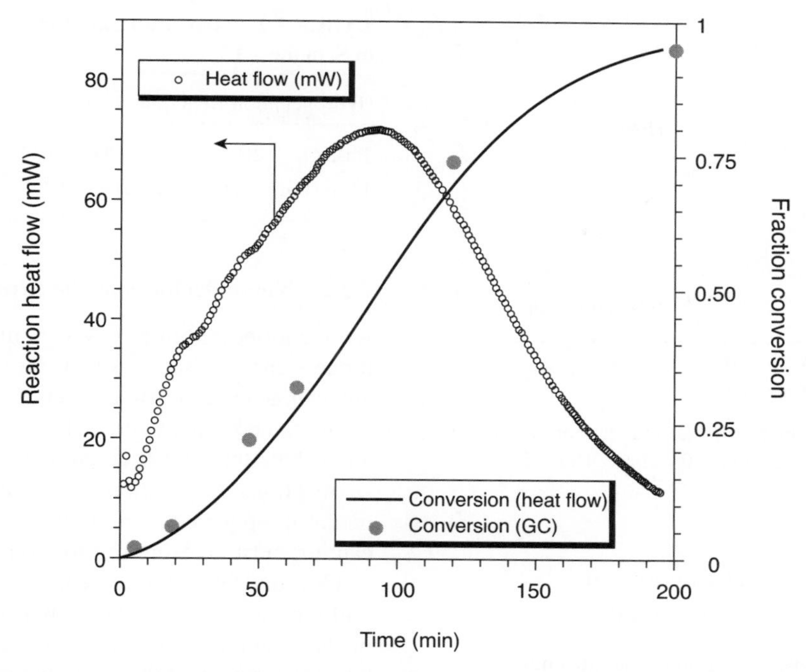

FIGURE 7.1 Reaction heat flow and fraction conversion versus time for the amination of bromobenzene (**1**, 0.71 M) with *N*-methylpiperazine (**2b**, 0.86 M) using NaO*t*Am (1.0 M) as base and a 0.5:1 mixture of $Pd_2(dba)_3$ and BINAP (2 mol% Pd based on [**1**]0) as catalyst. *Source*: Ref. 12.

TABLE 7.1 Comparison of Reaction Kinetics for Acetic Anhydride Hydrolysis Using an Omnical Z3 Calorimeter with that from Literature

Temperature (°C)	k_{obs} (1/s)	k_{lit} (1/s)	Measured Half-Life (s)	Expected Half-Life (s)
55	0.017	0.024	41	29
45	0.012	0.011	58	63
35	0.00585	0.00525	118	132

Source: Ref. 14.

synthetically relevant conditions with half-lives greater than 1 min.

Results from reaction calorimetry are further enhanced when orthogonal techniques are utilized in parallel. One such example of using orthogonal techniques is in the kinetic investigation of heterogeneous catalytic hydrogenation of nitro compounds shown in Scheme 7.1 [15]. The basic reaction network is described in Scheme 7.1.

Hydrogen uptake and reaction calorimetry data are shown in Figure 7.2 [15]; and similar temporal profiles are observed with both hydrogen uptake and reaction calorimetry. Concomitant LC sampling indicated that the zero order kinetics observed during the first 120 min, as evidenced by a flat temporal hydrogen uptake profile, is attributed to hydrogenation of the nitro moiety to the corresponding hydroxyl amine as shown in Scheme 7.2.

Taking the ratio of the two curves shown in Figure 7.2 yields the plot in Figure 7.3, which allow one to deconvolute the energetics for hydrogenation of the hydroxylamine with that to form the amine. The corresponding energetics extracted from the graph is shown in Table 7.2.

Such information and characterization is useful for safety assessment as well as reaction optimization. Understanding of reaction orders and energetics for each pathway in the

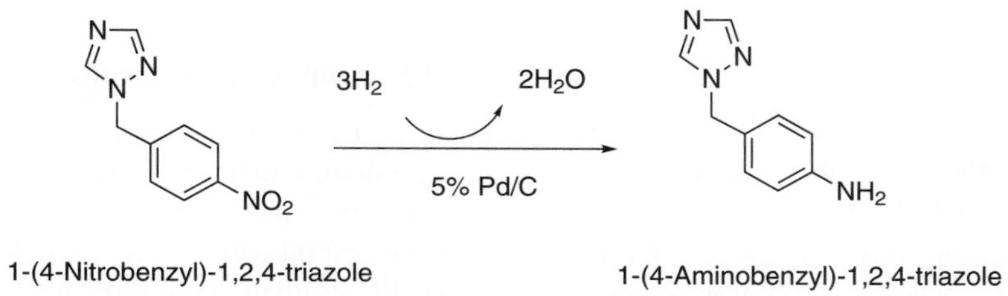

$3H_2$ $2H_2O$ 5% Pd/C

1-(4-Nitrobenzyl)-1,2,4-triazole → 1-(4-Aminobenzyl)-1,2,4-triazole

SCHEME 7.1 Hydrogenation of 1-(4-nitrobenzyl)-1,2,4-triazole.

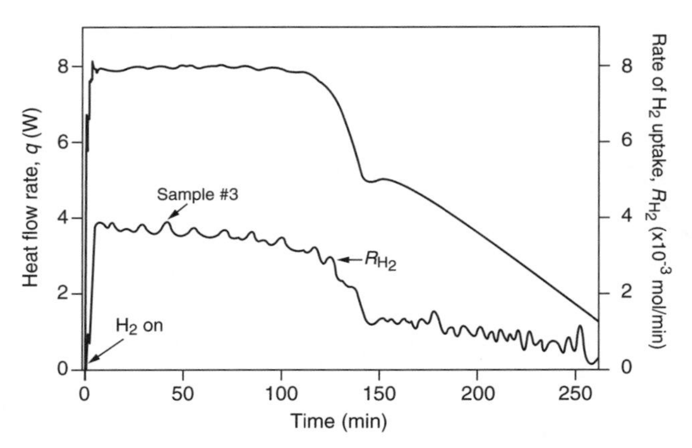

FIGURE 7.2 Temporal hydrogen uptake and reaction calorimetry for hydrogenation shown in Scheme 7.1. *Source*: Ref. 15.

$$Ph\text{-}NO_2 + 2H_2 \longrightarrow Ph\text{-}NHOH + H_2O$$
$$Ph\text{-}NHOH + H_2 \longrightarrow Ph\text{-}NH_2 + H_2O$$

SCHEME 7.2 Stepwise reduction of the nitro moiety.

reaction can be used to understand the operating design space. This example highlights the power of using orthogonal techniques to characterize reaction kinetics. Clearly the use of any one of the analytical techniques alone was not as powerful as the synergy of leveraging online hydrogen uptake, calorimetry, with off-line LC measurements.

Other calorimetry types, especially ARC (Accelerated Rate Calorimetry), are frequently used for process safety evaluation. Several other reviews have been written discussing the details of ARC testing and analysis [16].

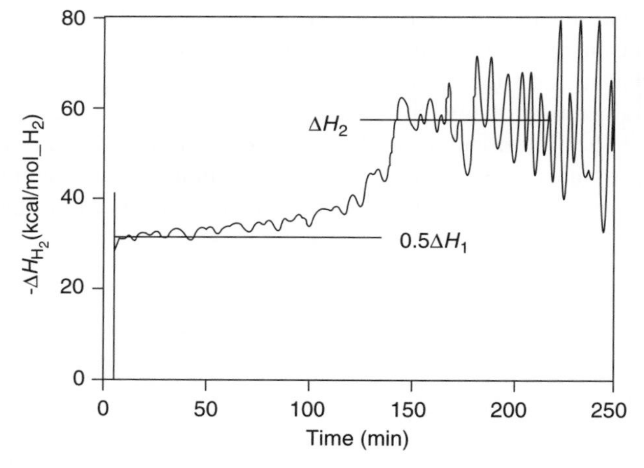

FIGURE 7.3 Ratio of temporal hydrogen uptake and calorimetry to elucidate the energetic of stepwise hydrogenation kinetics. *Source*: Ref. 15.

TABLE 7.2 Stepwise Heat of Hydrogenation of Nitro Group in Scheme 7.1

Hydrogenation Reaction	ΔH (kcal/mol)
$Ph\text{-}NO_2 + 2H_2 \rightarrow Ph\text{-}NHOH + H_2O$	−65
$Ph\text{-}NHOH + H_2 \rightarrow Ph\text{-}NH_2 + H_2O$	−58

7.2.2 Nonmolecule Specific Measurement

As mentioned above, physical measurements during the process can also serve as a means to track reaction progress and characterize reaction kinetics. These physical measurements can take many forms; however, temperature, gas flow, and pH are three more common measurements to characterize reactions. As mentioned with calorimetry, such measurements lump several different chemical events; and hence caution must be exercised for complex reaction systems.

Gas uptake measurements are particularly useful for multiphasic reaction such as hydrogenation as was outlined in the example above. As with calorimetry, care must be taken to ensure that the observed gas uptake measurement is correlated with the desired chemical transformation that is being tracked. Often times, side reactions such as over-reduction of desired products or catalyst reduction can mask the details of the chemical transformation that is to be tracked. Conversely, gas evolution measurements can also be used to track progress. This is frequently the case for decarboxylation reactions in which CO_2 evolutions can be used as a means to monitor and characterize decarboxylation kinetics.

Temperatures has been used for decades to track reaction progress; and it sometimes gets mistakenly neglected in favor of more complicated online sensors that are currently available. Tracking reaction progress with temperature, especially for exothermic reactions, that is, Grignards, are effective. Figure 7.4 shows the tracking of reaction progress at a 200 gallon scale during a benzyl Grignard formation. The initiation is evident during the time span of 150–200 min followed by formation of Grignard in a feed-limited manner up to approximately 330 min. The use of these physical measurements allow characterization and estimation of reaction rate constants both on lab and pilot plant scale which, in turn, can be used to understand scale sensitivity as will be discussed in subsequent sections.

7.2.3 Online Spectroscopy

The past 10 years has seen significant development of online technologies that have proven very effective for reaction characterization and measurement of reaction kinetics. While several online spectroscopic techniques are available, mid-IR, Raman, and NIR have proved to be the most valuable. A detailed review of each of these technologies is

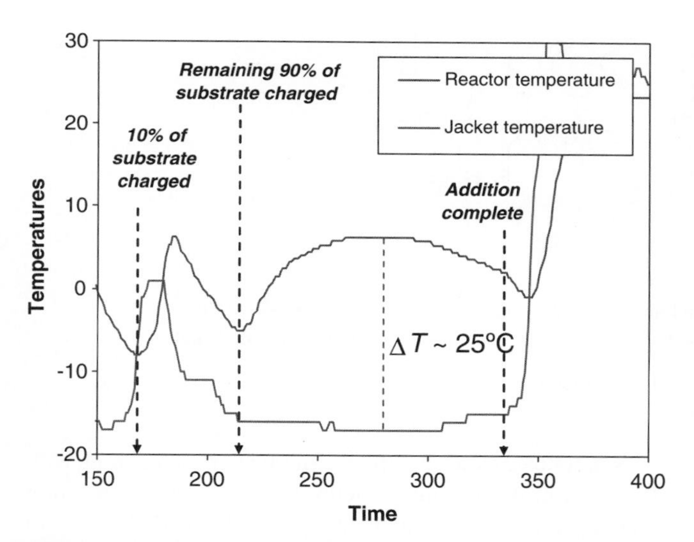

FIGURE 7.4 Reactor and jacket temperature profiles during the formation of a Grignard in a 200 gallon reactor. Both the initiation and the postinitiation reactive regimes are shown in the figure.

beyond the scope of this chapter; and several reviews have been written on this subject matter [17]. These technologies have been used routinely by the practicing chemist and engineering to extract detailed reaction kinetics and mechanistic information.

During the past few years, ReactNMR has also proved to be a valuable resource in understanding and characterizing reactions. *In situ* NMR has been widely used in academic environments; and lately, *in situ* NMR is also being used in industrial settings for probing reaction kinetics under synthetically relevant conditions. Use of different types of NMR (different nucleus) allows specific information to be gleaned that would otherwise not have been possible by conventional methods. This area will continue to garner more attention as additional applications show greater utility of this technique.

The above technologies are highly effective at measuring a vast majority of processes in the pharmaceutical industry; however, certain applications, such those requiring extreme reaction conditions and rapid kinetics, require specialized equipment such as stop-flow apparatus or tubular reactors.

7.2.4 Off-Line Concentration Measurements

There are several online and physical measurements that can be used to track reaction progress, as mentioned above; however, off-line concentration samples are a powerful means to track reaction progress especially with complex reaction networks and when tracking trace levels of impurities less than 0.5%. Newer technologies are being offered to allow online HPLC measurements that circumvent the time and discrepancies associated with manual off-line sampling. Sampling a minimum of 5–10 points across the reaction gives qualitative data regarding the overall reaction kinetics. Because of the separation capability and sensitivity of HPLC analysis applied to such samples, the kinetics of minor and major pathways leading to low level impurities as well as desired intermediates and products can be followed in this manner. In order to generate a richer set of data for quantitative analysis, more frequent sampling is required. This can be accomplished by means of integral data for concentrations of the major species using FTIR, or online or at-line HPLC with sampling taken at intervals of about 2–4% conversion. Recent developments in online LC have been reported in the open literature [18].

7.3 REACTION MODELING

Developing a model requires transformation of a reaction system into a discrete set of descriptions or elementary steps. Models of reaction systems can be developed in a number of different ways. Depending on the application, models can simply be a scale-down version of a pilot or commercial equipment that can be used to predict full-scale performance from laboratory measurements. Alternatively, models can be mathematical in nature in the form of dimensionless parameters or kinetics rate expressions that, when solved, can be used to predict performance. Regardless of the path chosen, models are developed to simplify and explain/rationalize an often complicated system into the most important/relevant elementary steps or rate-determining steps. Experimental data can be generated to test various steps. In such a way, the model guides the experiments, and the data from such experiments is used to support or refute the model. This allows a refining process for the model, which reflects the building of knowledge of the process. Accurate development and utilization of models enables us to have a high degree of confidence of the performance at different scales in batch processing or to apply the best design of a reactor configuration, be it batch or continuous mode. The model is an end goal, but is also helpful to the development of knowledge for any reaction system. As soon as a first draft of a model exists, it can be challenged with experimental data that helps improve or validate the draft of the current model.

Building a model requires measurements that "profile" the reaction from start to final conversion using any one or several techniques outlined above. This gives richer information for analysis than by simply analyzing the final conversion and selectivity or yield. This profiling of the reaction should capture concentration data of reagents, intermediates, and products plus selectivity for by-products along with direct measurements of the rate of reaction. Concentration-based data represent "integral" data, in that they are a direct measure of conversion, which comes from integration of the reaction rate, r.

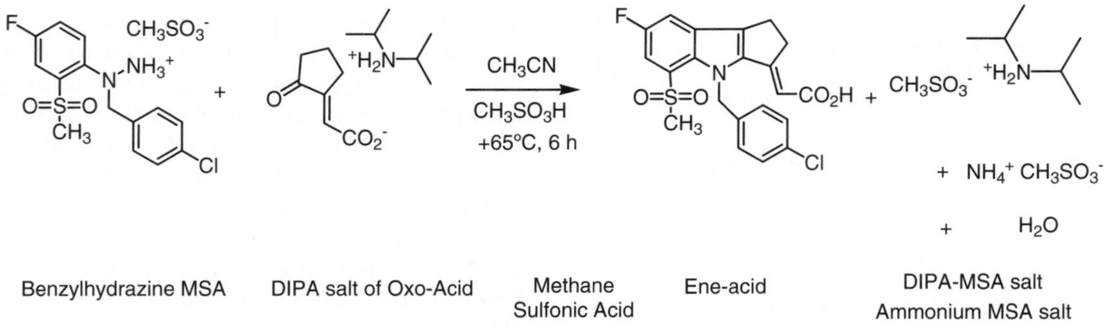

CH3CN
CH3SO3H
+65°C, 6 h

Benzylhydrazine MSA DIPA salt of Oxo-Acid Methane Sulfonic Acid Ene-acid DIPA-MSA salt Ammonium MSA salt

+ NH_4^+ $CH_3SO_3^-$
+ H_2O

SCHEME 7.3 Example of Fisher indole reaction.

$$X = C/C_0 = \int_0^t (r)dt \qquad (7.2)$$

Typically, samples are taken for off-line analysis that provides concentration data by HPLC, GC, FTIR, or equivalent methods, and the samples are taken at intervals of 5–20% conversion. This means that the frequency is a strong function of the half-life of the reaction(s).

Once the profiling data is obtained, it is often times necessary to visualize the data in different ways to characterize reaction kinetics. This has been done routinely for several catalyzed processes by Blackmond and coworkers. The example below adapts this technique for a Fisher indole reaction to understand the interplay of reaction kinetics with solid–liquid mass transfer (Scheme 7.3).

The Fisher indole reaction is expected to proceed through the hydrazone intermediate which exists a slurry with a solubility of 27 mg/mL before strong acid drives the cyclization to close the pyrrole ring and form the bicyclic indole as shown in Scheme 7.4 [19].

Small-scale experiments in a microcalorimeter along with RC-1 calorimeter were carried out to measure the reaction kinetics, with reaction calorimetry and off-line HPLC used to follow reaction progress. Multiple small (0.15 equiv) injections of methane sulfonic acid (MSA) was introduced into a slurry of hydrazone. Figure 7.5 shows calorimetry data modified to show the reaction rate data as a function of hydrazone concentration. This plot was obtained by integrating the heat flow data; and the conversion (or fractional heat evolution) was used to determine the hydrazone concentration during the course of the reaction. Each injection of MSA can be thought of and analyzed as an individual batch reaction. In contrast to conventional plots, the start of the reaction is shown on the right-hand side with higher concentrations of hydrazone. Reaction progresses by hydrazone consumption and movement to the left of the graph. Each peak observed in Figure 7.5 is due to an identical spike of MSA; and, as a result, the MSA concentration at the start of each peak is constant but the hydrazone concentration at the start of each peak is different allowing calculation of the reaction order in hydrazone during the course of the reaction.

+ H_2O
+ MSA

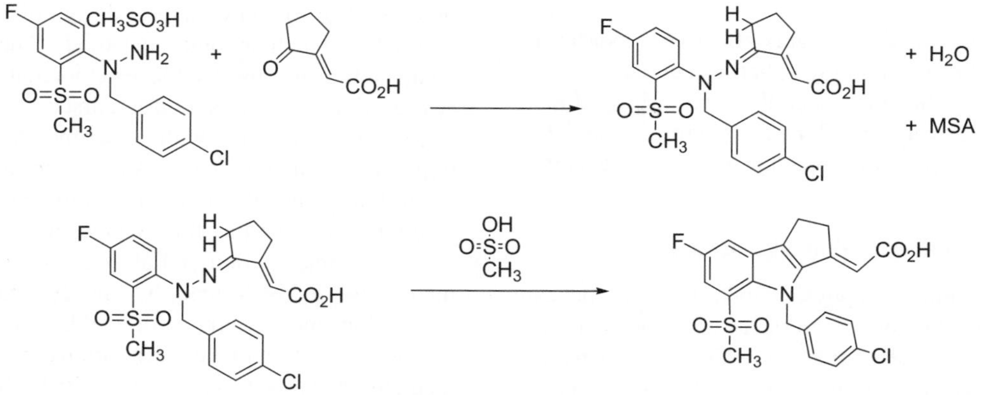

SCHEME 7.4 Proposed pathway for the Fisher indole reaction of Scheme 7.3.

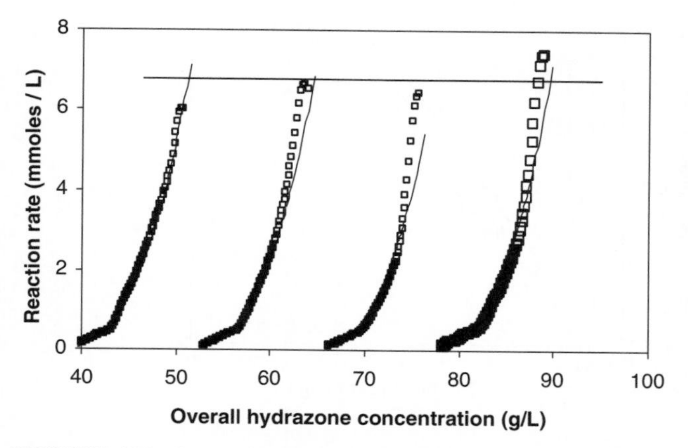

FIGURE 7.5 Rate of Fisher indole reaction as a function of overall hydrazone concentration.

Similarly, during the course of each spike, the MSA is effectively completely depleted whereas the hydrazone concentration would be expected to change little if at all allowing calculation of the order in MSA. The shape including slope and curvature of all four injections are nearly superimposable as shown in Figure 7.6 which shows the same data plotted as a function of MSA concentration. These results indicate that the order in MSA is not changing during the course of the reaction.

The combination of the results in Figures 7.5 and 7.6 indicate that the overall reaction rate is zero order in initial hydrazone concentration as shown by the solid line in Figure 7.5. Additionally, the reaction exhibits an overall third order behavior when plotted as a function of MSA concentration as shown in Figure 7.6 (solid line represents rate profile expected from third order kinetics). The initial zero order kinetics in hydrazone is consistent with the hydrazone being a slurry and the reaction kinetics being solubility limited in hydrazone. It was difficult to understand the chemical significance of the third order kinetics especially in the context of mechanistic data in the literature [19].

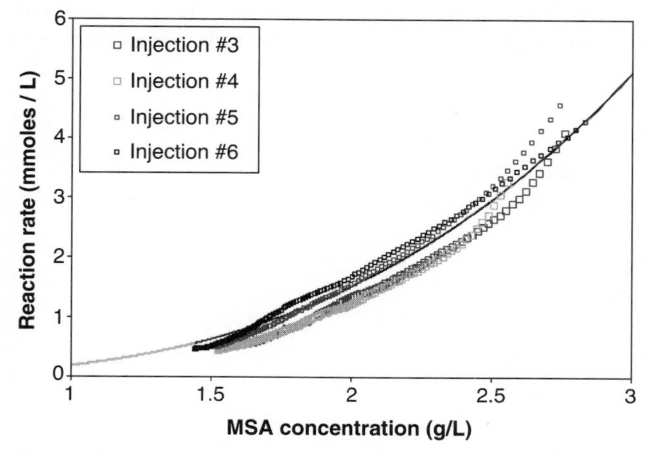

FIGURE 7.6 Rate of Fisher indole reaction as a function of the MSA concentration.

To better understand the underlying mechanism and effect of physical processes such as dissolution kinetics, the solids in the reaction mixture were filtered off and an identical study was conducted under homogeneous conditions using the dissolved hydrazone (saturated with a solubility of 27 mg/mL) in the filtrate; and the results are shown in Figure 7.7a and b. These results indicate a near first order dependence of reaction kinetics on hydrazone concentration and second order dependence on MSA concentration.

In the context of these results, the unusual third order kinetics in Figure 7.6 can be rationalized. The initial rate at the beginning of each MSA spike in Figure 7.5 indicated zero order kinetics in hydrazone since sufficient time was allowed for dissolution and equilibration resulting in identical solution phase concentration of hydrazone at the start of each peak in Figure 7.5. In contrast, first order kinetics with respect of hydrazone concentration was observed during the course of the reaction (each spike in Figure 7.7) since the dissolution rate was slower than the reaction rate resulting in a decrease in hydrazone concentration in each spike. As a result, the overall third order kinetics observed in Figure 7.6 is actually a convolution of second order dependency on MSA concentration and first order dependency on hydrazone concentration.

Independent dissolution rate measurements confirmed that the dissolution kinetics were occurring under the same timescale as that for the reaction. Understanding of this behavior led to a better phenomenological understanding and characterization of the reaction. This was particularly important since the solubility limited process (for both the starting material and the product) results in a slurry to slurry conversion resulting in significant occlusion of the starting material in the product depending on the addition mode and rate utilized.

As mentioned earlier, it is imperative when using online techniques such calorimetry or various spectroscopic tools to profile kinetics that one use orthogonal techniques to ensure the absence of artifacts affecting the measurements and hence the final conclusions. To that extent, Figure 7.8 indicates a comparison of the kinetic profile for hydrazone concentration as inferred from calorimetric measurements and that through quantitative IR measurements. Excellent agreement of data from different analytical techniques adds more credence to the measured kinetics.

The above example highlights an important point that will be examined in more detail in sections to follow, that is, convolution of reaction kinetics with physical rate processes. The next two examples highlight the importance of understanding the dynamics of reaction progress and the challenges associated with understanding a complicated reaction network with multiple pathways and by-products and

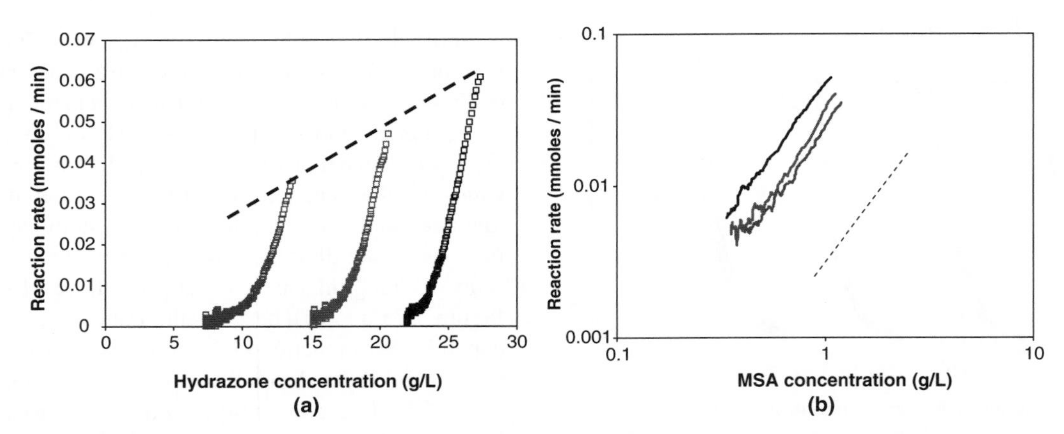

FIGURE 7.7 Plot of reaction rate as a function of (a) hydrazone and (b) MSA concentrations. Dashed lines in a and b represent first and second order curves, respectively.

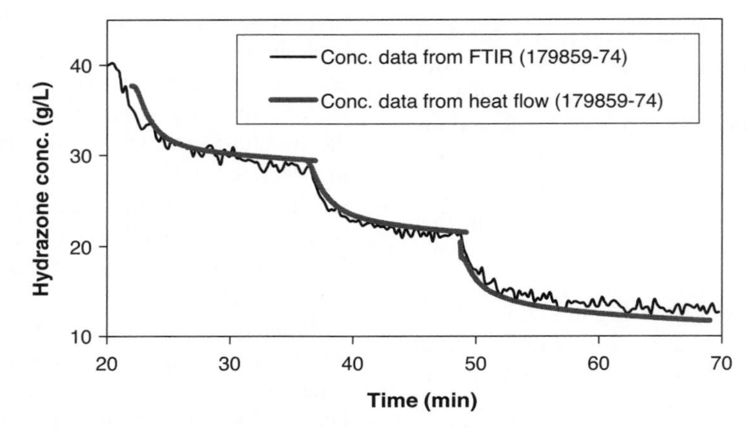

FIGURE 7.8 Comparison of temporal concentration profile obtained from heat flow and IR measurements.

changes in rate-determining step during the course of the reaction.

Let us start with an example from the literature for hydrogenation of ethyl pyruvate using modified Pt catalysts (Scheme 7.5) [20].

This reaction has been studied extensively in the literature to fully characterize the mechanism for enantioselective hydrogenation. The mechanistic understanding was linked to various reaction parameters including, but not limited to, solvent properties, modifier levels and interactions, and

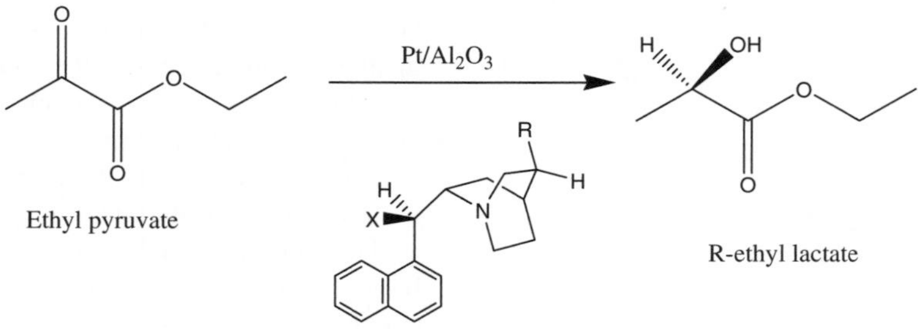

Ethyl pyruvate

Pt/Al_2O_3

R-ethyl lactate

	R	X
Cinchonidine	CH2CH	OH
Dihydrocinchonidine	CH3CH2	OH

SCHEME 7.5 Scheme for enantioselective reduction of ethyl pyruvate via heterogeneous catalysis.

catalyst dispersion and properties; and have helped the long-term strategic understanding. The effect of hydrogen transfer from the gas phase to the liquid phase significantly affects reaction kinetics. To that extent, Sun and coworkers examined the kinetics of the reaction and characterized the rate of hydrogen consumption as follows:

$$\frac{d[\mathrm{H_2}]}{dt} = k_{\mathrm{L}}a \times ([\mathrm{H_2}]^{\mathrm{sat}} - [\mathrm{H_2}]) - f\{[\mathrm{H_2}], [\text{catalyst}], [\text{substrate}]\} \quad (7.3)$$

where $k_{\mathrm{L}}a$, $[H_2]^{\mathrm{sat}}$, $[H_2]$, and $f([H_2]$, [catalyst], [substrate]) are mass transfer constant from gas phase to liquid phase, equilibrium concentration of hydrogen, hydrogen concentration at a given time, and intrinsic kinetic rate expression, respectively. Measurement of the mass transfer constant along with equilibrium hydrogen solubility and the kinetic rate expression allows one to numerically integrate the above expression to determine the hydrogen concentration during the course of the reaction. To that extent, the reaction rate was measured and the corresponding hydrogen concentration calculated under synthetically relevant conditions and the results are shown in Figure 7.9. Interestingly, there are two regimes evident in the kinetic profile. In the first regime, reaction kinetics is mass transfer limited. However, there is shift in the rate-determining step as the substrate is consumed and the intrinsic reaction kinetics slows sufficiently to become slower than the rate of hydrogen transfer from gas to liquid phase.

The shift in the rate-determining step is also associated with a shift in the selectivity as shown in Figure 7.9b [20]. The plot of cumulative selectivity (defined below) shows little to no change associated with the change in the rate-determining step; however, plotting the incremental selectivity (defined below) shows a marked change in incremental ee upon a change in the rate-determining step.

$$\text{Cumulative selectivity} = \frac{[\mathrm{R}]-[\mathrm{S}]}{[\mathrm{R}]+[\mathrm{S}]} \times 100 \quad (7.4)$$

$$\text{Incremental selectivity} = \frac{R_{\mathrm{R}}-R_{\mathrm{S}}}{R_{\mathrm{R}}+R_{\mathrm{S}}} \times 100 \quad (7.5)$$

[R], [S], R_{R}, and R_{S} are concentrations of R-enantiomer, S-enantiomer, and rates of formation of R- and S-enantiomers, respectively. The primary difference between the two modes of measuring selectivity is that the subtle changes in product distribution are masked and averaged out in the cumulative selectivity calculation whereas the incremental selectivity highlights incremental changes within two time points.

When modeling reaction kinetics, it must be understood that several competing reaction and equilibrium pathways can combine to yield complex nonintutive reaction kinetics. To this extent, apparent non-Arrhenius type behavior has been documented in literature in the form of activity minimum and maximum. Activity maximum has been observed during gas phase hydrogenation of benzene under initial rate conditions over supported transition metal catalysts; and the results have been rationalized by a conventional Langmuir–Hinshelwood mechanism for heterogeneous catalysis in which the increasing rate of elementary reaction steps

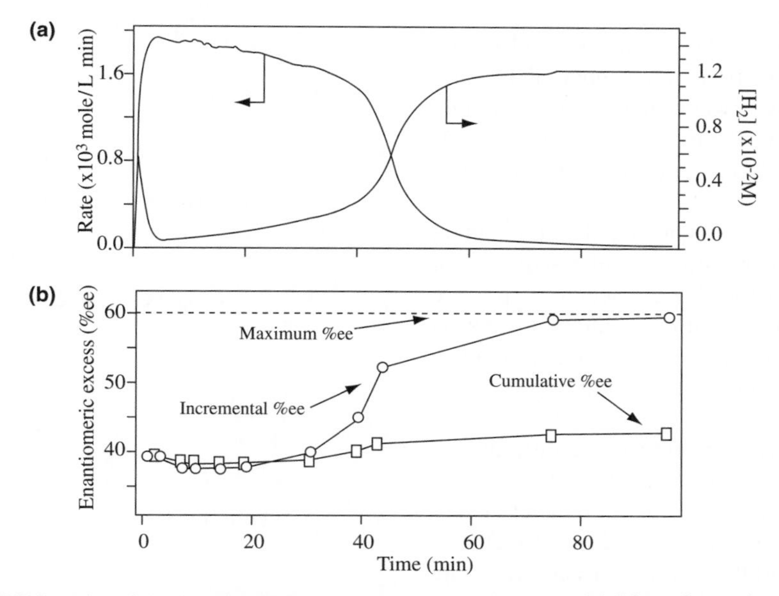

FIGURE 7.9 Plot of the kinetics of ethyl pyruvate hydrogenation. (a) Plot of reaction rate and solution phase hydrogen concentration; and (b) concomitant product distribution calculated during cumulative and incremental selectivities. *Source*: Ref. 20.

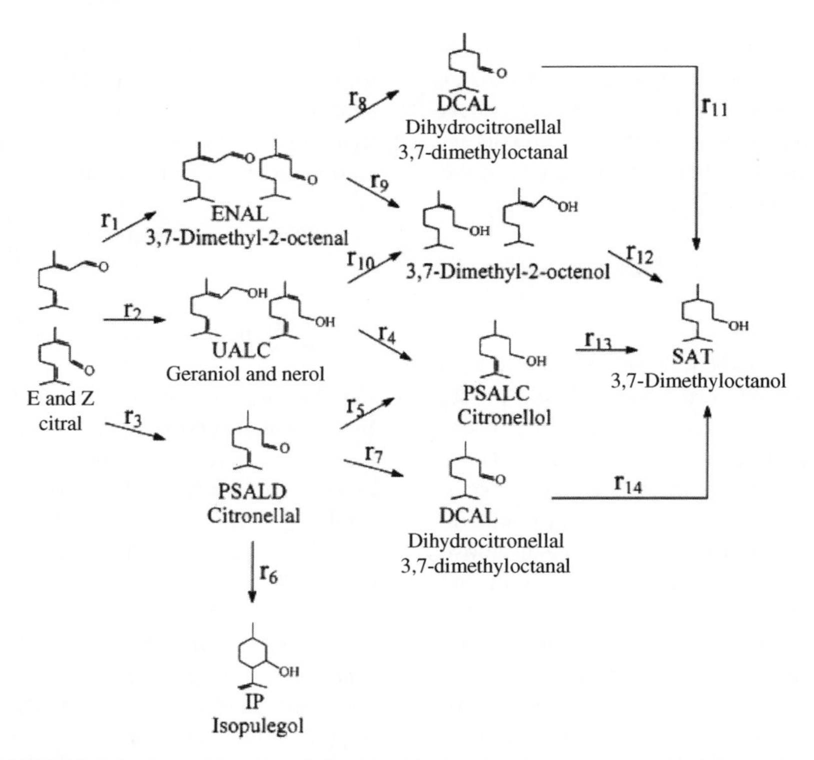

SCHEME 7.6 Reaction network for citral hydrogenation over supported Pt catalysts.

competes with decreasing equilibrium constant with increasing temperature. The opposing trends for these two parameters leads to an observation in which the reaction rate increases to a certain temperature after which the rate decreases with increasing temperature due to decreasing surface coverage of the substrate from decreasing adsorption equilibrium constant [21, 22].

Similarly, activity minimum has been observed during liquid phase citral hydrogenation over supported Pt catalysts [23]. The unusual behavior was rationalized by detailing a concurrent catalyst deactivation pathway occurring simultaneously with the reaction. The overall reaction pathway is shown in Scheme 7.6.

Focusing on pathways r_1, r_2, r_3, and r_5 and assuming that each of these pathways occurs with a similar sequence of elementary steps as outline below. The elementary steps were assumed to include preequilibrated dissociative adsorption of hydrogen and addition of the second hydrogen atom to the unsaturated moiety as the rate-determining step as shown in Scheme 7.7.

The corresponding deactivation pathway was described as follows, in the which the unsaturated alcohol (geraniol) adsorbs on the catalyst surface to make a metal-alkoxy species which then decomposes to form adsorbed CO. The adsorbed CO then desorbs off the catalyst surface with a rate constant k_D as shown in Scheme 7.8.

The rate expression has been derived in detail before [23] with the following result:

$$r_i = \frac{k'K'P_{H_2}C_{org}(1-\theta_{CO})^2}{(1+K_{citral}C_{citral})^2} \tag{7.6}$$

where θ_{CO} is the surface coverage of adsorbed CO and is defined by the following differential equation:

$$\frac{d\theta_{CO}}{et} = k'_{CO}C_{UALC}\theta_S^2 - k_D\theta_{CO} \tag{7.7}$$

$$H_2 + S \underset{}{\overset{K_H}{\rightleftharpoons}} 2H\text{–}S$$

$$Org + S \overset{K_C}{\rightleftharpoons} Org\text{–}S$$

$$Org - S + H\text{–}S \overset{K_U}{\rightleftharpoons} OrgH\text{–}S + S$$

$$OrgH\text{–}S + H\text{–}S \overset{K_i}{\longrightarrow} OrgH_2\text{–}S + S$$

$$OrgH_2\text{–}S \overset{1/K_P}{\rightleftharpoons} OrgH_2 + S$$

SCHEME 7.7 Elementary steps for hydrogenation of citral.

$$\text{UALC–S} + \text{S} \underset{}{\overset{K_{ET}}{\rightleftharpoons}} \text{Etoxy–S} + \text{H–S}$$

$$\text{Etoxy–S} + 2\text{S} \overset{K_{ACT}}{\rightleftharpoons} \text{Acyl–S} + 2\text{H–S}$$

$$\text{Acyl–S} + \text{S} \overset{K_{ACT}}{\longrightarrow} \text{CO–S} + \text{C}'\text{–S}$$

$$\text{CO–S} \overset{K_{D}}{\longrightarrow} \text{CO} + \text{S}$$

$$\text{C}'\text{–S} + \text{H–S} \overset{K_{HC}}{\longrightarrow} \text{H}_2\text{C} + 2\text{S}$$

SCHEME 7.8 Elementary steps for decomposition of geraniol to CO.

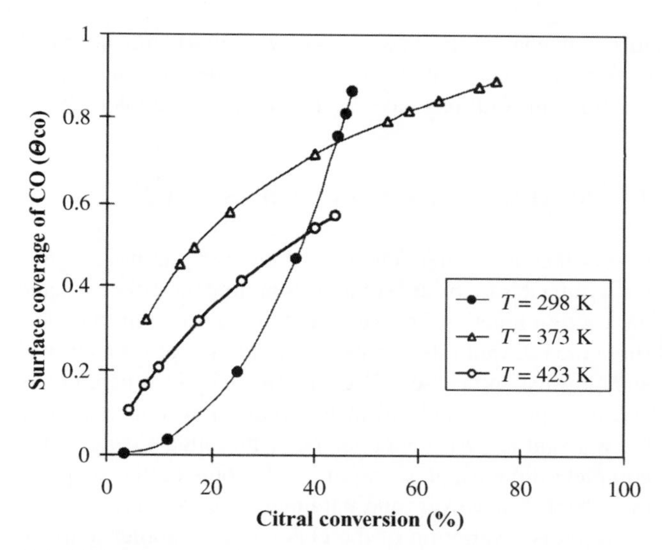

FIGURE 7.11 Calculated surface coverage of CO on supported Pt catalysts under synthetically relevant conditions during citral hydrogenation. *Source*: Ref. 23.

Representing the rate of CO formation via decomposition of the unsaturated alcohol (UALC) and subsequent desorption of CO from the catalyst surface. This deactivation pathway led to formation of CO, which poisoned the catalyst surface. Increasing the reaction rate led to increasing rate of catalyst deactivation and hence a reduction in rate with increasing temperature. After a certain critical temperature, CO starts to desorb off the catalyst surface leading to conventional Arrhenius-like behavior. The correlation of the model with experimental data is shown in Figure 7.10.

Measurement and quantification of the surface species in a solid–liquid interface for polycrystalline metal surfaces can be challenging. Kinetic modeling allows a means to understand the relative abundance of different reactive species and changes in the population of these species under different reaction conditions. For the case of citral hydrogenation, kinetic modeling was used to calculate the surface coverage of CO as shown in Figure 7.11. These relative trends in the surface coverage of CO with changes in temperature are consistent with the unusual trends in temperature that was described above.

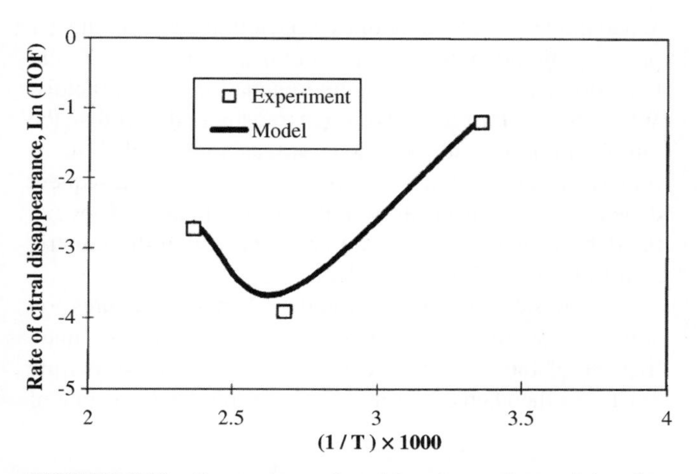

FIGURE 7.10 Comparison of model and experimental data for citral hydrogenation over supported Pt catalysts. H_2 (20 atm) and 1 M citral in hexane. Only the integral data is shown in the figure; however, actual curve fitting required regressing integral data at each temperature point shown. *Source*: Ref. 24.

Nonlinear regression of the model to the measured data allowed estimation of the adsorption equilibrium constant for the substrate along with the enthalpy and entropy of adsorption as well as the corresponding activation energy for the CO formation. Simply fitting of the data to a model is not justification alone for the validity of the model. Orthogonal techniques and independent measurements are necessary to add credibility to the results of curve fitting. Additionally, thermodynamic consistency of the fitted parameters should also be investigated. The standard states for the enthalpy and entropy of adsorption extracted from the temperature dependency of the adsorption equilibrium constant can be changed as shown in literature [23–25]; and the corresponding values have to satisfy additional thermodynamic constraints to ensure validity of the model and the nonlinear regression. For heterogeneous catalytic applications, two constraints have been discussed in the literature:

$$0 < \left|\Delta S^0_{ad}\right| < S^0_g \tag{7.8}$$

$$10 < \left|\Delta S^0_{ad}\right| < 12 - 0.0014(\Delta H^0_{ad}) \tag{7.9}$$

in which ΔS_{ad}, ΔH_{ad}, and S_g are the entropy of adsorption, enthalpy of adsorption, and standard entropy of the gas, respectively [26–28]. The constraint in equation 7.8 is stringent whereas that in equation 7.9 is a guideline suggested by Boudart and Vannice similar guidelines for elementary reactions in solutions have been detailed by Laidler [29].

7.4 SCALE SENSITIVITY ASSESSMENT

One of the areas where chemical engineers can have significant impact is in the area of understanding the effect of scale on reaction kinetics. To that extent, it is important to understand the relevant rate processes including that for chemical and physical processes. Use of dimensionless parameters can be especially helpful in this manner in understanding the relevant scale-up parameters. A thorough listing of the complete dimensionless parameters is outside the scope of this chapter; however, one parameter that deserves special attention is a variation of the classical Damkohler number. For the purposes of pharmaceutical process development Damkohler number can be applied as follows:

$$Da = \frac{\text{timescale for physical processes}}{\text{timescale for chemical transformation}} \qquad (7.10)$$

In which the rate of physical processes can include just about any process including those associated with mass transfer such as liquid–liquid mixing, gas absorption, gas desorption, solid suspension, as well as heat transfer including rate of heat generation, heat input/removal. Most, if not all, scale sensitivities can be understood in the context of the apparent Damkohler number. In general, no scale-sensitivities would be expected when the rate of the chemical transformation is slower than that for the physical process. In contrast, scale sensitivities are observed when the rate of physical process is slower than that for chemical transformations.

The above statement holds not only for the desired reaction pathway but also for all other chemical pathways that may result in impurity formation. One example of this behavior was the debenzylation of a fumarate salt of an amine to give the corresponding succinate salt of the secondary amine as shown in Scheme 7.9.

The hydrogenation process involves initial reduction of the fumaric acid to succinic acid followed by debenzylation to form the corresponding secondary amine succinate salt. The reaction rate profiles as function of hydrogen pressure is shown in Figure 7.12. These results along with concomitant sampling clearly indicate a positive order dependence of rate

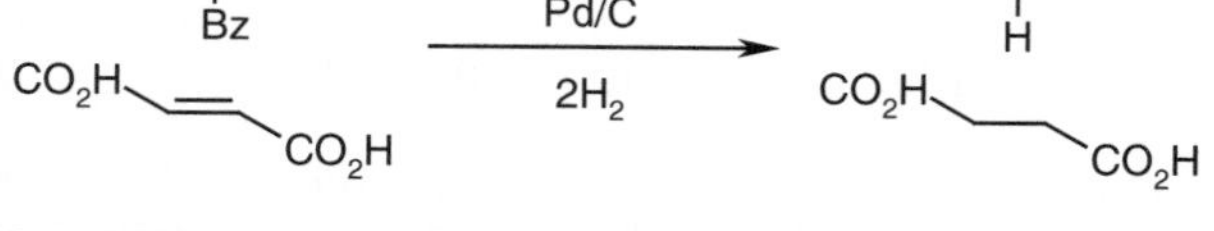

SCHEME 7.9 Debenzylation and corresponding fumaric acid reduction.

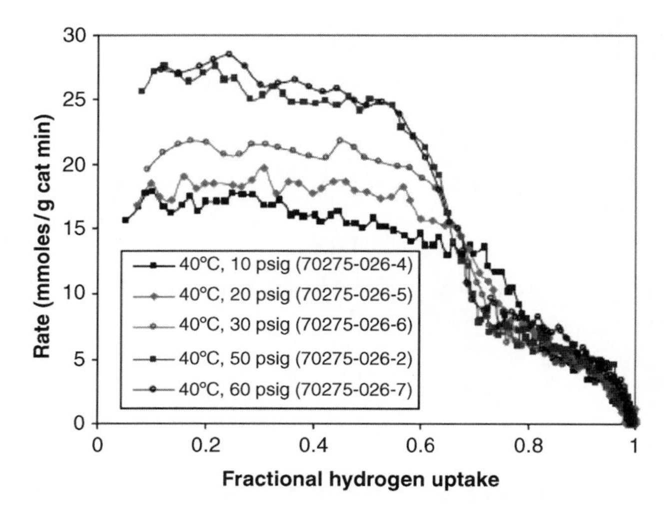

FIGURE 7.12 Temporal rate profile for concomitant debenzylation and fumaric acid reduction over Pd/C for reaction in Scheme 7.10.

of fumaric acid reduction on hydrogen pressure compared to a zero-order dependence for debenzylation. Hydrogen starvation resulted in significant decrease in the rate of fumaric acid reduction with little or no effect on the rate of debenzylation resulting in accumulation of the fumaric acid in the presence of a secondary amine thereby increasing the propensity for the formation of the Michael adduct as described in Scheme 7.10.

For such a process, the Damkohler number is defined as follows:

$$Da = \frac{\text{rate of fumaric acid hydrogenation}}{\text{rate of hydrogen transfer from gas phase to liquid phase}} \qquad (7.11)$$

When the $Da < 1$, the rate of hydrogen transfer from the gas phase to liquid phase is rapid compared to fumaric acid reduction and, as a result, its hydrogenation occurs rapidly. When $Da > 1$, the rate of hydrogen transfer is slower than the rate for fumaric acid reduction; and, as a result, the rate of fumaric acid reduction is slowed to the point that subsequent debenzylation can occur simultaneously thereby allowing the deprotected secondary amine to react with the fumaric acid to form the Michael adduct.

The physical process described in Damkohler expression above is usually related to some aspect of heat and mass transfer of the reactor process. In the case of exothermic reactions, the Damkohler number is expressed as the rate of

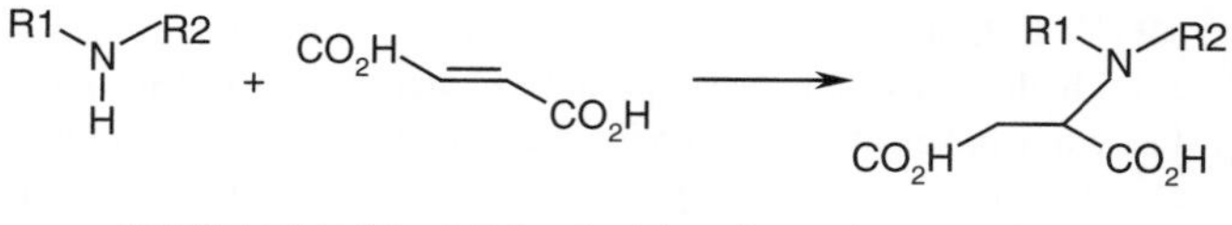

SCHEME 7.10 Michael adduct formation reaction.

heat formation to the heat removal by jacket services. In some cases, the rate of physical process is the rate of the relevant mixing mechanism to that for the reaction rate. Often times deconvoluting chemical and physical processes can be difficult and a more fundamental understanding of the relevant heat/mass transfer and chemical transformation is needed to understand the effect of scale.

It is difficult to make broad generalizations regarding how to measure and characterize reaction kinetics due to diversity of chemistry and the widely varying and highly complex chemical pathways that are known to occur especially when one focuses on pathways that lead to formation of impurities of the order of 0.1% (see discussion above on reaction optimization). Generalizations regarding the characterization of the relevant physical process can be made. To that extent, three relevant mass transfer processes will be discussed.

7.5 SOLID–LIQUID TRANSPORT

In many industrial applications, reactions involve reagents, catalysts, or intermediates that are heterogeneous in nature. In these cases, solid–liquid transport effects may need to be characterized and understood. In such a case, there are multiple ways to model the observed reaction. To do so, it is necessary to understand the rate of the solid–liquid transport, as well as the rate of the intrinsic chemical reactions. Deconvoluting the intrinsic rate of chemical kinetics from solid–liquid mass transfer rates can be complex. Calculations to estimate the transport from particles in heterogeneous reactions have been outlined by Zwietering [30]. Specifically, experimental data and mathematical correlations indicate that the rate of mass transfer changes appreciably up to the "just suspend" point for particles, at which point the particles no longer form a layer at the bottom of the vessel. Further increases in mixing intensity once solids have already been suspended give only marginal increases in the mass transport [31]. While this guidance provides an effective rule of thumb for guiding process development, detailed kinetic studies are occasionally necessary in order to decouple the different rate processes and obtain quantitative expressions for the intrinsic transport and reaction rates.

There are several ways to approach decoupling the transport rate from the chemical kinetics. A number of different mass transfer correlations are available in the literature [32]; and the different formulations can be used to estimate the rate of mass transfer across the solid–liquid interface and compare with the corresponding intrinsic reaction rate constant using a Damkohler number type approach articulated above. *A priori* determination of the mass transfer constant from correlations can be unreliable if one does not pay attention to the appropriate assumptions involved. Numerous correlations have been reported in the literature looking at functional relationship between the dimensional groups of Sherwood number, Reynolds number, and Schmidt number. Each of these dimensionless parameters is defined below

$$Re = \frac{\rho u d_p}{\mu} \text{(particle basis)} \quad \text{or} \quad Re = \frac{\rho u d_\mathrm{T}}{\mu} \text{(impeller basis)} \tag{7.12}$$

$$Sc = \frac{\mu/\rho}{D} \tag{7.13}$$

$$Sh = \frac{k_\mathrm{d} d_p}{D} \tag{7.14}$$

The term ρ represents the fluid density, u is the velocity, d is the characteristic length scale for particles (p) or turbine (T), μ is the fluid viscosity, D is the diffusion coefficient for the substrate or reagent of interest, and k_d is the mass transfer coefficient In general, these correlations have a functional form as shown below.

$$Sc = kRe^x Sc^y \tag{7.15}$$

where the constants k, x, and y vary depending on the system under consideration. One of the issues that arises when utilizing this correlation is the formulation of the Reynolds number. A number of different modified particles Reynolds number expressions are shown in the literature and care should be exercised to ensure that the assumptions are known and that the correct correlations are utilized [31]. Many studies have been carried out and published on the mass transfer to particles. Most studies are for transport of particles in stirred tanks although transfer in pipes was reported by Harriott [33–39]. It should be noted that there is wide spread dissolution measurements upon which these correlations have been made, and so the use of such correlations introduces a measurable level of uncertainty when applied in a new scenario.

Rather than using the correlations, it is preferable to explicitly measure the mass transfer constant across the solid–liquid interface using dissolution measurements. Several model systems have been evaluated and reviewed. This approach allows a direct measurement of the rate constant to compare with the corresponding reaction rate constant.

Alternatively, one could use the formulation of the rate equation that combines the rate constant for solid–liquid transport with the intrinsic reaction rate constant; and then use the relative activation energies as a means to deconvolute transport and reaction limited regimes. For a first order reaction, the rate expression that combines reaction rate constant and mass transfer across the boundary layer can be written as follows [40–41]:

$$R_G = \frac{k_\mathrm{d} k_\mathrm{r}}{k_\mathrm{r}(1-w_\infty)+k_\mathrm{d}} C \tag{7.16}$$

$$HCL + NaOH \xrightarrow{k_1} NaCl + H_2O$$

$$DMP + HCL + H_2O \xrightarrow{k_2} 2MeOH + Acetone + HCL$$

SCHEME 7.11 Reaction network for the fourth Bourne reaction. DMP stands for 2,2-dimethoxypropane.

where k_d and k_r are the rate constants for transport across the solid–liquid interface. In cases, where the mass transfer across the solid to liquid interface is rapid, that is, $k_d \gg k_r$ then the rate expression simplifies to $R = k_rC$ and chemical kinetics are rate controlling. In cases, where the mass transfer across the solid–liquid interface is slow, that is, $k_d \ll k_r$ then the rate expression simplifies to $R = k_dC$ and mass transfer across the boundary layer is the rate-limiting step.

The temperature dependency of the rate constant k_r and k_d allows deconvolution of the chemical kinetics with mass transfer kinetics. The influence of temperature on the transport rate is primarily through its influence on viscosity and/or diffusion coefficients. There is only a modest effect of temperature on these variables; and, as a result, the transport rates typically exhibit a weak dependency with temperature, that is, a low activation energy. As a rule of thumb, activation energies for transport limited processes are typically of the order of 10–20 kJ/mole compared to 40–60 kJ/mole for reaction processes. It should be stressed that this is a rule of thumb and some exceptions do exist in the literature especially in the context of the unusual temperature dependencies that were articulated in the section above. Such a technique has been routinely used in heterogeneous catalytic systems and has also been applied to crystallization processes [42].

7.6 LIQUID–LIQUID TRANSPORT

There have been several reviews written documenting the effect of liquid–liquid mixing in pharmaceutical applications; and a detailed discussion on the mechanistic aspects of liquid–liquid mixing is outside the scope of this review [43]. The focus instead will be on the convolution of reaction kinetics with liquid–liquid mixing either in a miscible system or in immiscible systems.

One of the powerful tools in characterizing liquid–liquid mixing is the Bourne reaction. Several Bourne reactions exist but the fundamental principle of the reactions is a competition between a sequence of slow and fast reactions. In the event of perfect mixings, only the fast reaction occurs; and the extent to which the slow reaction occurs is a measure of poor mixing. The known rate constants of the Bourne reaction can be used quantify mixing times to understand the interplay of mixing and chemical kinetics. The Bourne reactions are a great example in characterizing a reaction network and how the convolution of reaction kinetics with physical rate processes affects selectivity.

Johnson and Prud'homme [44] along with Mahajan and Kirwan [45] and Singh and coworkers [46] have used such reaction systems to characterize different mixing geometries to enhance mixing efficiency and reduce mixing times. A detailed description of the mixing characteristics and mechanisms is outside the scope of the present chapter, but rather the convolution of mixing with reaction kinetics will be discussed. This can be understood in the context of the fourth Bourne reaction as shown in Scheme 7.11.

The rate constants for k_1 and k_2 are 1.4×10^8 m^3/(mol s) and 0.6 m^3/(mol s), respectively. The extent to which MeOH (or acetone) is observed in the final stream is indicative of poor mixing. A plot of conversion of the slow reaction to Damkohler number, that is, reaction to mixing time, is shown in Figure 7.13 for conducting the process in a mixing elbow in which the two process streams containing 2,2-dimethoxypropane and NaOH were mixed with HCl in a 180° angle. The details of the formulation of the mixing time and Damkohler number can be found elsewhere [46]. It is interesting to note that in spite of 8 orders of magnitude difference in rate, significant conversion of the slow reaction to methanol is observed. The reaction time constant defined as follows

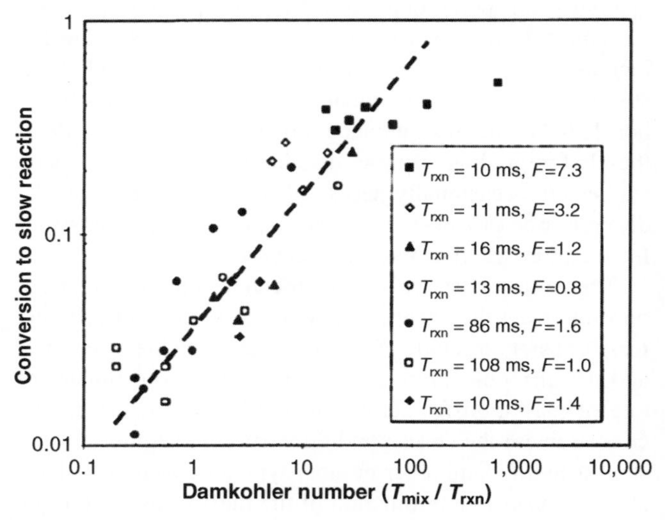

FIGURE 7.13 Plot of conversion of the 2,2-dimethoxypropane hydrolysis as a function of Damkohler number for time constants varying from 10 ms to 108 ms. *Source*: Ref. 46.

$$\tau_{rxn} = \frac{1}{k_2 C_0} \tag{7.17}$$

where k_2 and C_0 are the rate constant for the 2.2-dimethoxypropane hydrolysis and the concentration of the HCl in the final process stream, respectively. The reaction time constant was varied by 1 order of magnitude from 10 ms to 108 ms and the corresponding conversion to the slow reaction varied from 0.6 to 0.02. This application highlights the importance of understanding the convolution of physical processing parameters with reaction kinetics. Parameterization of the reaction kinetics allowed determination of the Damkohler number and developing a rationale for the significant changes in product distribution as a result of changes in operating parameters.

One of the challenges associated with scale-up is to determine whether liquid–liquid mixing under miscible conditions would be rate limiting and the effect on reaction rate and impurity formation. This is no trivial task and a number of diagnostic tests can be conducted, depending on the relevant reaction kinetics, to determine potential impact of mixing. Measurement of the rates of relevant chemical pathways (desired and undesired reactions) and comparison to the mixing timescales whether macromixing (bulk blending) or meso/micromixing. Correlations for mixing times under different mixing regimes for macro/meso/micromixing regimes have been articulated in the literature and will not be repeated in this report. Use of the Damkohler number type approach offers guidance on determining the effect of mixing on reaction performance.

Alternatively, one can also look at addition mode. Consider a case in which a stream of A is added to a stream of B to yield product C. Alternatively A and B could react to form D as shown in the scheme below

$$A + B \rightarrow C$$

$$A + B \rightarrow D$$

If the reverse addition is conducted, that is, a stream of B is added to a tank containing A, and no effect on rate on impurity formation is observed then one can fairly confidently conclude that mixing effects will be negligible. This is because the two addition modes mimic conditions in which you have either segregated high concentrations of A or B that would result from poor mixing at larger scales.

In contrast, if the order in B for formation of species C is greater than that for formation of D then the two different addition modes would give different ratios of species C and D; and hence mixing effects could potentially be pronounced at scale. There are several ways to manage the resulting mixing effect. Use of static mixers or auxiliary rapid mixing devices such as mixing elbows and vortex mixers can be used to enhance mixing while leaving the reaction kinetics unaffected thereby shifting the Damkohler number in the desirable direction. In contrast, the reaction kinetics itself can be slowed by either using dilution effects and taking advantage of differences in reaction order or temperature and leveraging the exponential dependence of temperature and/or differences in activation energy between the different chemical pathways.

7.7 GAS–LIQUID TRANSPORT

Gas–liquid mixing plays a central role in a number of commercialized synthetic processes. Transport of gas both into and out of solutions can drive reaction rates and selectivity.

One of the important issues that often arises when looking at gas–liquid phase reactions is the effect of solubility on reaction kinetics. In the presence of mass transfer limitations, gas solubility is clearly the driving force. However, the question of whether gas solubility affects reaction kinetics under conditions free of transport limitations can be more complex. Under ideal conditions, for multiphase reaction systems, that is, hydrogenations, the driving force, in the absence of transport limitations, is partial pressure of hydrogen and the rate in different solvents with varying hydrogen solubility would be independent of hydrogen solubility. There are unique situations in which nonidealities and interactions with the solvent can affect the driving force in a manner that hydrogen concentration becomes the driving force [24, 47, 48]. In the context of these results, care must be exercised when characterizing heterogeneous reaction system to ensure that the appropriate driving force is identified and used in the rate expressions.

A procedure for measuring the rate of mass transfer from the gas to liquid phases has been detailed previously [49]. The integral approach for measuring $k_L a$ is shown in the following equation:

$$k_L a \times t = \frac{P_f - P_o}{P_i - P_o} \ln \frac{P_i - P_f}{P - P_f} \tag{7.18}$$

where P_o, P_f, P_i, and P are solvent vapor pressure, final pressure, initial pressure, and pressure at a given time during the course of the experiment. Plotting the left-hand side of the above equation versus time yields a slope with units of 1/time; and it represents the mass transfer constant from gas phase to liquid phase. Alternatively, the initial slope of the pressure drop at the start of an uptake experiment to estimate the value of $k_L a$

$$k_L a \approx -\frac{dP}{dt} \frac{1}{P_i - P_f} \tag{7.19}$$

Note that for both large and small-scale measurements, it is important separately to understand the ramp up time for an agitator to reach full power. Experimental details for measuring k_La and factors that affect gas–liquid mixing efficiency have been captured elsewhere and will not be repeated here [50].

As was the case for solid–liquid, liquid–liquid systems, the convolution of reaction rate with mass transfer from gas phase to liquid phase can be described using the Damkohler number is defined in equation 7.20.

$$Da = \frac{R_{\text{rxn}}}{R_{\text{MT}}^{\text{max}}} = \frac{R_{\text{rxn}}}{k_La \times C_{\text{H}_2}^{\text{sat}}} \tag{7.20}$$

where R_{rxn} is the intrinsic reaction rate and $R_{\text{MT}}^{\text{max}}$ is the maximum rate of transfer from the gas phase to the liquid phase. A ratio of $Da > 1$ is indicative of mass transfer limitations whereas $Da < 0.1$ is indicative of a regime *free* of mass transport limitations.

Understanding of $k_{\text{L}}a$ is critical to understanding and characterizing reactions involving gas–liquid mixing. This has been routinely shown in earlier work of Sun and coworkers. One particularly effective case study was the enantioselective reduction over supported Pd catalysts as shown in Scheme 7.5. Results are plotted in Figure 7.14. The maximum rate of mass transfer was varied from 0.0035 mol/(L min) to 0.51 mol/(L min) resulting in ee varying from 24% to 60%. The maximum rate of mass transfer could be varied by either changing the rate of mass transfer, that is, k_La, at a given pressure or by changing hydrogen solubility through changing the pressure at a given stirring speed and mass transfer constant. The filled circles in Figure 7.14 were obtained with varying stirring speeds but at a constant pressure [51]. Manipulation of pressure at a constant stirring speed of 750 rpm was effectively able to mimic the trend observed by changing stirring speed at a constant pressure.

The discussion above pertains primarily for mass transfer from gas phase to liquid phase. Similar issues are encountered when gas is desorbed off from liquid phase to gas phase. This issue is routinely encountered during oxygen sensitive reactions such as asymmetric hydrogenations and cross couplings in which trace concentrations of oxygen can poison catalysts or decarboxylation reactions in which effective desorption of CO_2 is necessary prior to forward processing. Similar issues are also encountered when the desired chemical transformation is an equilibrium in which a gas, that is, CO_2 or HCl, is produced along with the product and efficient removal of the gas from the liquid phase is necessary to push the desired reaction forward to completion. Phenomenologically, this reaction network can be described as follows:

$$\text{A} + \text{B} \Longleftrightarrow \text{C} + \text{HCl (l)}$$

$$\text{HCl (l)} \Longleftrightarrow \text{HCl (g)}$$

The formulation of the fundamental rate expression that describes this driving force is similar to that for the rate of transfer from gas phase to liquid phase. Specifically, the rate can be described as follows:

$$\text{Desorption rate} = \frac{dC}{dt} = k_{\text{L}}a(C - C^*) \tag{7.21}$$

where C is the solution phase concentration of the gas at a given time and C^* is the equilibrium concentration of the gas described by Henry's law. It must be noted that the k_La describing the desorption rate constant is different from that for absorption processes. Depending on the measurement approach, the value of C^* may vary during the measurement process; and additional mass balance in the gas phase would be necessary. As before, a detailed account of the mechanism of transport from solution to gas phase is outside the scope of the present report. Rather, the convolution of this transport process with reaction kinetics is often encountered.

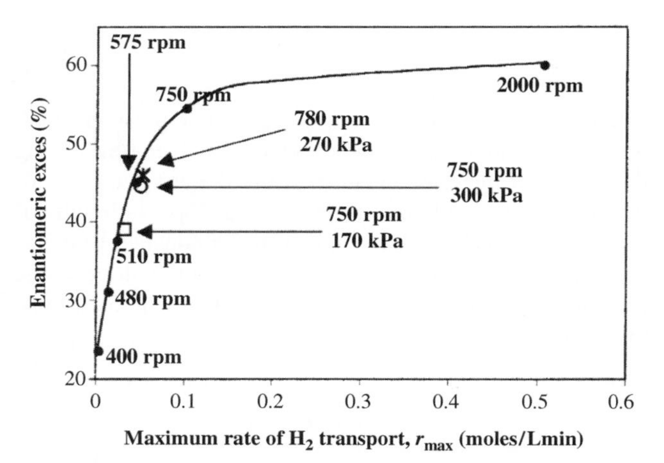

FIGURE 7.14 Effect of maximum rate of mass transfer on the final ee observed during enantioselective hydrogenation of supported Pd catalysts. *Source*: Ref. 24.

7.8 LEVERAGING THE ADVANTAGES OF CONTINUOUS REACTORS

Batch operating mode offers great flexibility in use of equipment, but can pose risks to successful scale-up. The most typical case of continuous reactors offering an advantage over batch operating conditions is for fast reactions with consecutive reactions that lead to "over-reaction." This is the case for either reactions that involve a reaction pathway with series of elementary reactions, or reactions in both series and parallel, and where the desired species is an intermediate [5]. This is the case for "unstable" reactions in quite a few instances in the pharmaceutical chemistry. Operating an unmixed (plug flow) continuous reactor or a continuous stirred reactor can offer advantages when such reactions are involved. In most other cases, batch reaction kinetics and that

TABLE 7.3 Opportunities for Continuous Reactions

Characteristic	Batch (8000 L Stirred Vessel)	Continuous (1 in. Pipe Reactor)	Advantage
Mixing time	>10 s (bulk blending)	>0.1 s	Rapid blending of reagents for fast reactions
Surface to volume available for heat transfer	~2 m^{-1}	~200 m^{-1}	Superior temperature uniformity with fewer "hot spots"
Typical temperature or pressure limit in absence of special designs	150°C; 10 Bar	200–250°C; 30–150 Bar	Ease of running reactions above the normal boiling point of solvents. Higher concentrations of gaseous reagents dissolved
Instantaneous amount reacting	100–1000 kg	1–5 kg	Lower energy potential and impact from runaway reaction

for plug flow reactions are identical when using the same conditions [5]. The use of continuous operation allows higher productivity than batch operating mode, but at the cost of more complex, less flexible, and capital intensive factories [52]. A list of general advantages is shown in Table 7.3.

The literature of the past 10–15 years has many instances of microreactors being investigated for the ability to improve performance, and highlighting issues such as those in Table 7.3, along with the ability to eliminate scale up by buying multiple reactor units. This sometimes includes a direct improvement, and other times involve subtle changes to the chemistry (Watts, Seeburger, and Jensen). Microreactors, such as larger scale continuous reactors, offer benefits in mixing and residence time control. At the same time, they present risks such as plugging that are not of the same level encountered in larger continuous reactors. They do offer great opportunity for lab-scale evaluation of rapid chemistry. The choice of whether to scale up in microreactors or traditional continuous reactors is linked to multiple considerations of economics that are beyond the scope of this effort.

Below, we present two industrial cases of using a continuous approach, one with a plug flow (unmixed) reactor and the second involving a stirred reactor.

In the original process to make imipenem, a desired inversion center in the backbone of the molecule was formed through a Mitsunobu inversion reaction followed by hydrolysis. Scheme 7.12 shows the original chemistry, which relied on hydrolysis under acidic conditions following the Mitsunobu inversion.

Because the inversion was carried out in water immiscible dichloromethane, the acidic hydrolysis required the addition of an equal volume of methanol prior to the addition of aqueous HCl. If scaled up directly, this would become a severe bottleneck to the volumetric productivity in manufacturing (Figure 7.15). The baseline yield for the inversion and hydrolysis was 80%, with the resulting intermediate **11** having a typical purity of about 99%. Importantly, the inversion and hydrolysis required over 40 L solvent per kilogram of **9**, and required significant use of water washing to remove methanol, then distillation to concentrate and dry **11** in DCM prior to crystallization.

Hence, hydrolysis at basic pH was investigated at lab scale as an alternative. In this case, intermediate **11** would be extracted into the aqueous phase as a sodium enolate. Batch experiments showed that the hydrolysis of the formate was feasible under basic conditions, but identified several challenges for successful processing. First, the hydrolysis was complete in less than 15 s. Second, the reaction tolerated up to 5 equiv of base, but only under cold conditions. Third, the beta-lactam began to undergo significant hydrolysis and ring opening after about 5 min of exposure to caustic. Fourth, the high solids loading prevented rapid (5 min) separation of the phases even at lab scale. Finally, neutralizing the excess

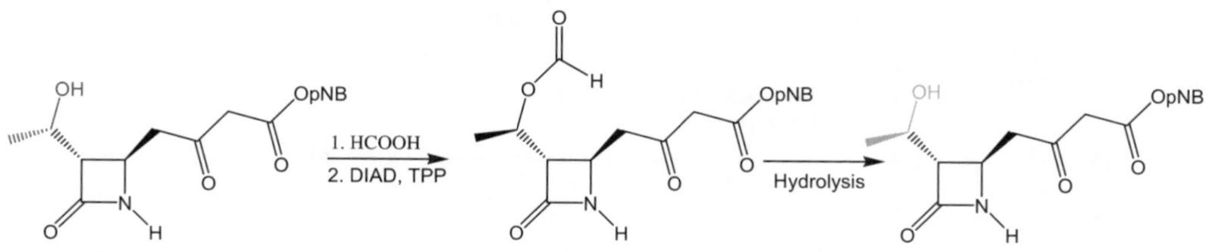

SCHEME 7.12 Extractive hydrolysis. *Source*: Ref. 53.

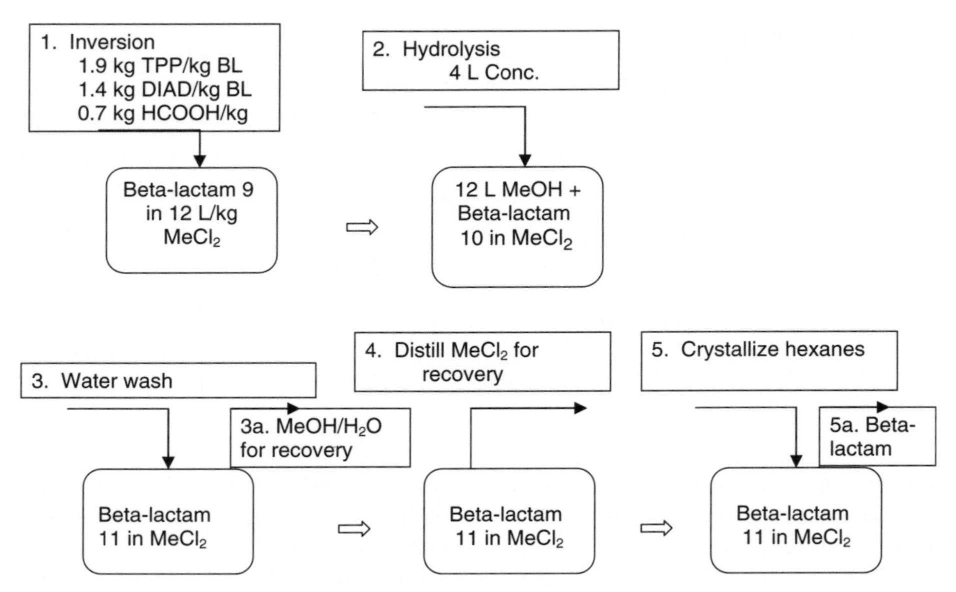

FIGURE 7.15 Flow sheet for acidic hydrolysis.

base prior to phase separation resulted in partitioning of **11** and spent reagents into the organic phase. This meant a very impure stream for crystallization.

From a reaction point of view, it was important to understand that the characteristic reaction time from laboratory studies indicated a half-life of no more than a few seconds, far less than the mixing time in a large stirred vessel. While the desired hydrolysis could be done in a stirred vessel, the half-life for degradation was not sufficiently longer than the mixing and batch wise phase separation time. So, scale-up in a batch process would result in 10–20% degradation by beta-lactam hydrolysis.

Thus, the poor separation required the use of enhanced separation of the liquid phases, and a centrifugal extractor was used to achieve the separation with a short contact time. In this case, the complete hydrolysis was straightforward, and the challenge became achieving a residence time and successful separation of spent reagents from **11** prior to neutralizing excess caustic. The equipment configuration ultimately implemented into the manufacturing process is shown in Figure 7.16.

This design implemented a rapid reaction of cold caustic with **10** followed by rapid separation and washing of the phases prior to neutralization. The productivity was more than 2.5 times that of the acidic hydrolysis route, exceeding the original goal. In addition, the yield was improved from 80% to more than 90%.

A second example is shown in Scheme 7.13, for the formation of an epoxide ring from an iodohydrin [53].

Initial batch experiments showed the reaction to take place very quickly, but require an excess of base to drive the reaction to completion, as seen in Figure 7.17.

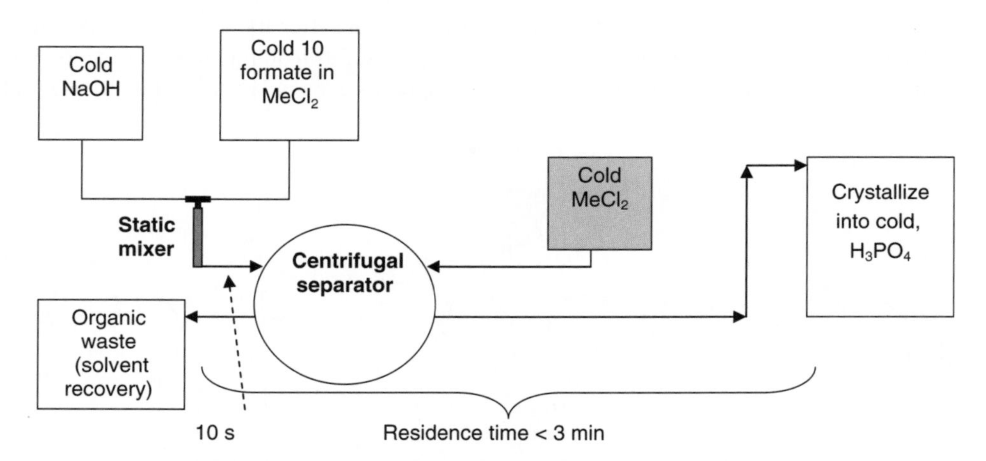

FIGURE 7.16 Flow sheet for continuous hydrolysis of **10** from Figure 7.15.

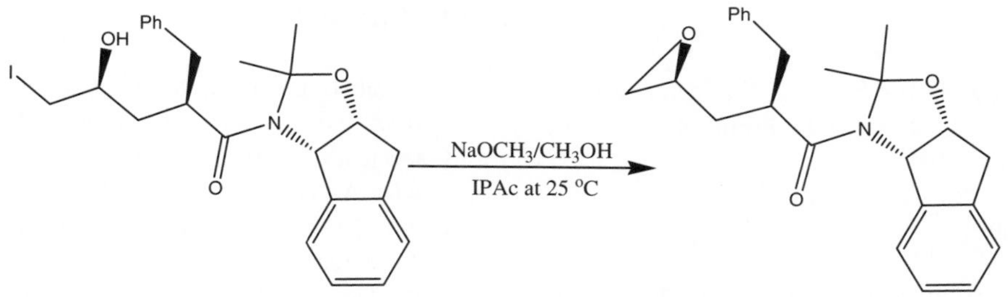

SCHEME 7.13 Epoxide formation from iodohydrin.

In this case, it was observed from batch experiments that back reaction could take place over several hours if sodium iodide remained in the reaction (Figure 7.18), and was sensitive to the pH. In particular, quenching the reaction only by adding water or IPAc would leave the resulting pH high and lead to enhanced back reaction, while neutralizing excess base would slow the back reaction kinetics.

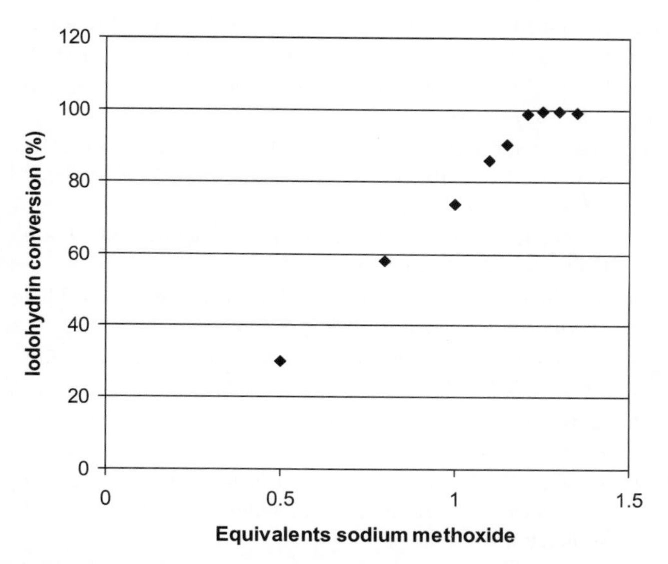

FIGURE 7.17 Impact of base charge on conversion in a "titration" experiment.

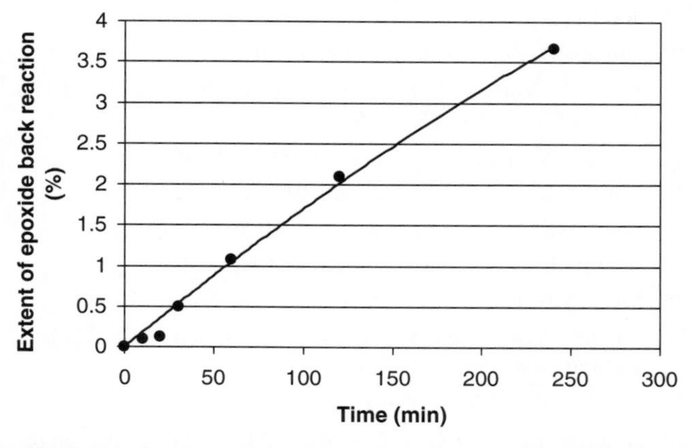

FIGURE 7.18 Extent of back reaction of epoxide to iodohydrin in an unquenched batch experiment (1.25 equiv of base).

Hence, it was desirable to limit the reaction time and exposure to basic conditions to avoid this reverse reaction. In order to avoid a loss of volumetric productivity that would accompany significant dilution to reduce back reaction, pH control would be desirable. This pH control also helped to minimize base catalyzed hydrolysis of IPAc by minimizing the exposure time of solvent at high pH. And finally, the neutralized salt layer was removed continuously from organic layer. Thus, the final process design was based on a continuous stirred reactor (CSTR) for the epoxidation followed by a CSTR with pH control to quench the excess base to prevent back reaction, and continuous separation of the aqueous phase to remove the aqueous salt layer (Figure 7.18). In addition to minimizing the risk and extent of back reaction, the short contact time also minimized IPAc hydrolysis, which improved yield in the solvent recycle step.

Today, there are many tools available to characterize such fast reactions at small scale, including stopped flow devices, microreactors and analytical technology such as described earlier in this chapter. All of this is useful to help characterize faster reactions in order to quantitatively understand the reaction timescale relative to that of mixing or transport rates of solids, gases, or even liquids to the reaction. As pointed out in Sections 7.5, 7.6, and 7.7, it is essential to quantitatively determine the timescale for reaction relative to the necessary mass or energy transport at every scale. Only this knowledge assures the process development expert that the chemistry is truly operating under the same limitations at each scale of operation, and with every equipment change.

7.9 CONCLUSIONS

In summary, the examples outlined above highlight the importance of understanding the interplay of chemical transformations with the physical rate processes, that is, heat and mass transfer resistances. Investigation of reaction kinetics and understanding of the elementary reactions in the pathway as well as quantification of the transport resistances is critical for ensuring successful process development and commercial scale-up and manufacture.

7.10 QUESTIONS

1. Derive the expression for $k_l a$ (gas–liquid). The mass transfer from the gas phase to the liquid phase can be described as shown below:

$$\text{Rate} = \frac{dC}{dt} = k_L a(C_t - C_{sat})$$

where C_t is the solution phase concentration of the gas at a given time and C_{sat} is the equilibrium concentration of the gas in solution. Integrating the above expression yield the following expression

$$\ln\left[\frac{(C_t - C_{sat})}{(C_0 - C_{sat})}\right] = \ln\left[\frac{(n_t - n_{sat})}{(n_0 - n_{sat})}\right] = k_L a \times t$$

where C_0 is the concentration of the gas in solution at $t = 0$ and n_0, n_t, and n_{sat} are the moles of gas initially in solution, moles of gas in solution at a given time t, and the moles of gas in solution at the saturation point. Mass balance for the gas yields the following:

$$n_t = n_0 + (P_0 - P_t)\frac{V_g}{RT}$$

$$n_{sat} = n_0 + \left(P_0 - P_f\right)\frac{V_g}{RT}$$

Substitution of the mass balance equation above yields the following expression for $k_L a$:

$$k_L a \times t = \ln\frac{P_f - P_t}{P_f - P_0}$$

2. For a reaction with a first order rate law, develop a rate expression that integrates the mass transfer constant for transport across the solid–liquid interface and the intrinsic reaction kinetics.
The rate of diffusion across the boundary layer is defined as follows:

$$R_{diff} = k_d(C_\infty - C_I)$$

where C_∞ and C_I are the bulk and interface concentrations, respectively. The reaction rate for a power law rate model can be described as follows:

$$R_{rxn} = k_r C_I$$

Equating the above two expressions and solving for C_I yields the following expression:

$$C_I = \frac{k_r k_d}{k_r + k_d} C_\infty$$

Substitution of this expression into the equation for R_{rxn} yields the following:

$$R_{rxn} = \frac{k_r k_d}{k_r + k_d} C_\infty$$

REFERENCES

1. March J. *Advanced Organic Chemistry: Reactions, Mechanisms, and Structure*, 4th edition, Wiley, New York, 1992.
2. Rylander PN. *Hydrogenation Methods (Best Synthetic Methods)*, Academic Press, New York, 1985.
3. Astarita G. *Mass Transfer with Chemical Reaction*, Elsevier, Amsterdam, 1967.
4. Dankwerts PV. *Gas–Liquid Reactions*, McGraw-Hill, New York, 1970.
5. Levenspiel O. *Chemical Reaction Engineering*, 2nd edition, Wiley, New York, 1972.
6. Froment GF, Bischoff KB. *Chemical Reactor Analysis and Design*, Wiley, New York, 1979.
7. Dugger RW, Ragan JA, Ripin DHB. *Org. Process Res. Dev.* 2005;9:253–258.
8. Roberge DM. *Org. Process Res. Dev.* 2004;8:1049–1053.
9. Blackmond DG, *Angew. Chem. Int. Ed.* 2005;44:4302–4320.
10. Mathew JS, Klussmann M, Iwamura H, Valera F, Futran A, Emanuelsson EAC, Blackmond DG. *J. Org. Chem.* 2005;71:4711–4722.
11. Shekhar S, Ryberg P, Hartwig JF, Mathew JS, Blackmond DG, Strieter ER, Buchwald SL. *J. Am. Chem. Soc.* 2006;128:3584–3591.
12. Singh UK, Strieter ER, Blackmond DG, Buchwald SL. *J. Am. Chem. Soc.* 2002;124:14104–14114.
13. Boddington T, Chia HA, Halford-Maw P, Hongtu F, Laye PG. *Thermochim. Acta.* 1992;195:365–372.
14. Zogg A, Fischer U, Hungerbuhler K. *Indus. Eng. Chem. Res.* 2003;42:767–776.
15. LeBlond C, Wang J, Larsen R, Orella C, Sun Y-K. *Top. Catal.* 1998;5:149–158.
16. Stoessel F. *Thermal Safety of Chemical Processes: Risk Assessment and Process Design*, Wiley-VCH Verlag GmbH, Weinheim, 2008.
17. Munson J, Stanfield CF, Gujral B. *Curr. Pharmaceut. Anal.* 2006;2:405–414.
18. Schafer WA, Hobbs S, Rehm J, Rakestraw DA, Orella C, McLaughlin M, Zhihong G, Welch C. *Org. Process Res. Dev.* 2007;11:870–986.
19. Hughes D, Zhao D. *J. Org. Chem.* 1993;58:232.
20. Sun Y-K, Wang J, LeBlond C, Landau RN, Blackmond DG. *J. Catal.* 1996;161:759–765.
21. Lin SD, Vannice MA. *J. Catal.* 1993;143:539.
22. Lin SD, Vannice MA. *J. Catal.* 1993;143:563.
23. Singh UK, Vannice MA. *J. Catal.* 2000;191:165–180.
24. Singh UK, Vannice MA. *AIChE J.* 1999;5:1059.
25. Mears DE, Boudart M. *AIChE J.* 1966;12:313.
26. Vannice MA, Hyun SH, Kalpakci B, Liauh WC. *J. Catal.* 1979;56:358.
27. Boudart M. *AIChE J.* 1972;18:465.
28. Boudart M, Mears DE, Vannice MA. *Ind. Chim. Belge.* 1967;32:281.

29. Laidler KJ. *Chemical Kinetics*, 3rd edition, HarperCollins Publisher, New York, 1987, pp. 183–220.
30. Zwietering ThN. *Chem. Eng. Sci.* 1958;8:244.
31. Paul EL, Atiemo-Obeng VA, Kresta SM. *Handbook of Industrial Mixing: Science and Practice*, Wiley, Hoboken, NJ, 2004.
32. Pangarkar VE., Yawalkar AA., Sharma MM., Beenackers AACM. *Ind. Eng. Chem. Res.* 2002;41:4141–4167.
33. Barker JJ, Treybel RE. *AIChE J.* 1959;6 (2): 295.
34. Harriott P. *AIChE J.* 1962;8 (1): p101ff.
35. Harriott P. *AIChE J.* 1962;8 (1): p93ff.
36. Brian PLT, et al. *AIChE J.* 1969;15 (5): 727.
37. Davies JR. *Chem. Eng. Process.* 1986;20 (4): 175.
38. Marrone GM, Kirwan DJ. *AIChE J.* 1986;32 (3): 523.
39. Armenante PM, Kirwan DJ. *Chem. Eng. Sci.* 1989;44 (12): 2781.
40. Garside J, Mersmann A, Nyvlt J. *Measurement of Crystal Growth and Nucleation Rates*, 2nd edition, Institution of Chemical Engineers, U.K., 2002.
41. *Perry Chemical Engineering Handbook*, 8th edition, McGraw-Hill, New York, 2007, pp. 7–19.
42. Singh UK, Pietz MA, Kopach M. *Org. Process Res. Dev.* 2009;13:276–279.
43. Bourne JR. *Org. Process Res. Dev.* 2003;7:471–508.
44. Johnson BK, Prud'homme RK. *AIChE J.* 2003;49 (9): 2264–2282.
45. Mahajan AJ, Kirwan DJ. *AIChE J.*, 1996;42 (7): 1801–1814.
46. Singh UK, Spencer G, Osifchin R, Tabora J, Davidson OA, Orella C. *Ind. Eng. Chem. Res.* 2005;44:4068–4074.
47. Madon RJ, O'Connell JP, Boudart M. *AIChE J.* 1978;24:904.
48. Gonzo RE, Boudart M. *J. Catal.* 1978;52:462.
49. Demling A, Karandikar BM, Shah YT, Carr NL. *Chem. Eng. J.* 1984;29:140.
50. Oldshue JY. *Chem. Eng. Progr.* 1980;76 (6): 60.
51. Singh UK, Vannice MA. *Appl. Catal. A Gen.* 2001;213: 1–24.
52. Khurana A. *Sloan. Mgmt. Rev.* 1999; *Winter*.
53. King ML, Forman AL, Orella CJ, Pines SH. *Chem. Eng. Progress* 1985; May: 36–39.
54. Maligres PE, et al. *Tetrahedron* 1996;52 (9): 3327–3338.

8

UNDERSTANDING RATE PROCESSES IN CATALYTIC HYDROGENATION REACTIONS

YONGKUI SUN

Department of Process Research, Merck & Co., Inc., Rahway, NJ, USA

CARL LEBLOND

Department of Chemistry, Indiana University of Pennsylvania, Indiana, PA, USA

8.1 INTRODUCTION

Hydrogenation is a powerful methodology in synthetic organic chemistry and has been broadly employed by organic chemists in drug synthesis. Heterogeneous-catalyzed hydrogenation has traditionally been very popular and continues to play a critical role in modern organic synthesis [1, 2]. Hydrogenation by homogeneous catalysts, particularly asymmetric hydrogenation, is gaining momentum in applications in the pharmaceutical industry. Since the pioneering work on the development of chiral catalysts for asymmetric hydrogenation by William Knowles in late 1960s and the first commercial application of asymmetric hydrogenation in the early 1970s by Monsanto in the production of the antiparkinsonian drug L-DOPA [3, 4], asymmetric hydrogenation has developed into a powerful chemical transformation, achieving enantioselectivities matching those previously seen only in enzymatic processes. Over the ensuing two to three decades, there has been rapid development in the science and technology of asymmetric hydrogenation. One of the key milestones in the development of this chiral technology was the discovery of BINAP in the 1980s by Ryoji Noyori and coworkers [5], which significantly broadened the scope of utility of asymmetric hydrogenation [6]. A recent special issue of the *Accounts of Chemical Research* documented the growing application of asymmetric hydrogenation in the pharmaceutical industry [7].

Hydrogenation reactions in the liquid phase are complex processes. Even for a homogeneously catalyzed hydrogenation for which the catalyst and the substrate are fully soluble in the solvent, the hydrogenation process is heterogeneous in nature since the dihydrogen reducing reagent H_2 is in a different phase. In addition to the catalytic reaction occurring on the catalyst, there are a number of mass transfer processes that can exert direct influence on the outcome of the catalytic reaction itself.

In the case of heterogeneous catalysis, the mass transfer processes include H_2 transport across the gas/liquid and the liquid/solid interfaces before the molecular H_2 chemisorbs dissociatively on the metal catalyst, as well as H_2 diffusion inside the pore structure of the catalyst particles. A schematic depicting the mass transfer processes (pore diffusion not displayed) is shown in Figure 8.1, along with the corresponding H_2 concentration profile. Among the mass transfer processes, the gas/liquid mass transfer needs significant attention in designing a hydrogenation process because it can have profound impact on the performance of the hydrogenation reaction [8], and it is greatly affected by agitation, reactor design and configuration, solvent properties such as viscosity, and solvent fill level. Mass transfer of H_2 across the liquid/solid interface and through the pore structure in the catalyst particles are often dominated by the physical characteristics of the catalyst, that is, particle size, shape and support material, and density and pore structures. While agitation intensity and reactor design can influence their

Chemical Engineering in the Pharmaceutical Industry: R&D to Manufacturing Edited by David J. am Ende

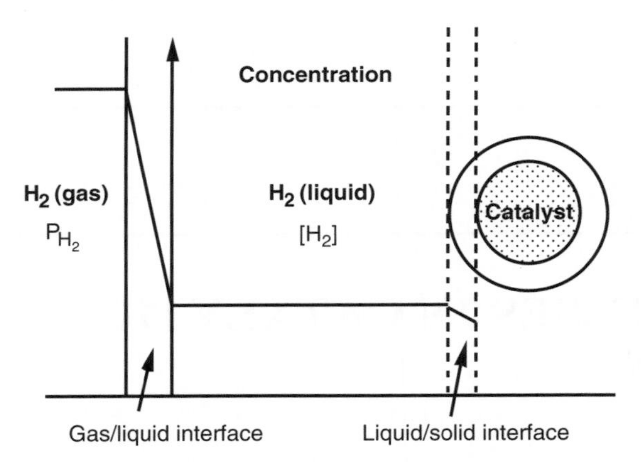

FIGURE 8.1 A schematic of hydrogen mass transfer processes across the gas/liquid and liquid/solid interfaces during a heterogeneously catalyzed hydrogenation reaction. Also plotted is a schematic of the hydrogen concentration profile.

kinetics, for catalytic system in which the diameter of the catalyst particle is <50 μm, the effects become minimal once a uniform catalyst suspension is achieved. All the mass transfer issues encountered in hydrogenation reactions carried out in slurry reactors are not reviewed in this chapter [9]. Instead, this chapter focuses only on the simple but crucial hydrogen gas/liquid mass transfer issue and its impact on the development of hydrogenation processes.

8.2 SOLUTION HYDROGEN CONCENTRATION DURING HYDROGENATION REACTIONS, $[H_2]$

When gas/liquid delivery is the dominant mass transfer step, the pathway followed by H_2 from the gas phase to its incorporation into the product in catalytic hydrogenation reactions using a homogeneous or a heterogeneous catalyst may be simply described as follows:

$$H_2(g) \xrightarrow{k_L a} H_2(l) \xrightarrow{k_r} \text{Hydrogenation product} \qquad (8.1)$$

where $k_L a$ is the mass transfer coefficient for the H_2 mass transfer across the gas/liquid interface, and k_r is the rate coefficient of the catalytic reaction. The intrinsic kinetics in the catalytic hydrogenation is a function of concentrations of the substrate, the catalyst, and the dissolved H_2 in addition to other factors such as the temperature. In developing a chemical process, one is normally demanding in the knowledge of the concentration of the substrate and try to follow [substrate] via various means during the course of the reaction. One is frequently, however, less demanding in the knowledge of the solution concentration of hydrogen, $[H_2]$, due, in most part, to an assumption that $[H_2]$ equals the equilibrium solubility of hydrogen, $[H_2]_{sat}$, at the temperature and pressure of the reaction.

The assumption holds, however, only when the rate of H_2 mass transfer is far greater than that of hydrogenation reaction itself. While $[H_2]_{sat}$ is fixed at a constant gas-phase H_2 pressure, $[H_2]$ may vary widely from nearly zero to nearly saturation, depending critically upon the relative magnitude of the rate of H_2 mass transfer from the gas to the liquid phase and the rate of reactive H_2 removal from the liquid phase due to the hydrogenation reaction. This is readily seen from a mass balance of the dissolved H_2 in the following equation:

$$\frac{d[H_2]}{dt} = k_L a([H_2]_{sat} - [H_2]) - k_r f([H_2], [\text{catalyst}], [\text{substrate}]) \qquad (8.2)$$

The first term to the right of equation 8.2 is the rate of H_2 mass transfer from the gas to the liquid phase, and the second term is the rate of reactive removal of the dissolved H_2 from the liquid phase by the catalytic hydrogenation.

The kinetics of H_2 mass transfer resembles that of the familiar first-order chemical reaction and may be characterized by a single parameter, the mass transfer coefficient $k_L a$. A comparison of the characteristics, including the kinetic expression and the maximum rate, between these two processes is given in Table 8.1. The significance of the mass transfer coefficient $k_L a$ is that it is the kinetic factor that dictates the maximum rate of mass transfer of hydrogen across a gas/liquid interface, that is,

$$R^{max}_{H_2,g/L} = k_L a[H_2]_{sat} \qquad (8.3)$$

in conjunction with the thermodynamic factor $[H_2]_{sat}$.

While k_r is an intrinsic kinetic property of the catalytic system, $k_L a$ is strongly affected by characteristics of the reactor vessel, including reactor type, configuration, liquid fill level, viscosity, and particularly agitation speed. At a constant H_2 pressure, the magnitude of $[H_2]$ during the reaction critically depends upon the relative magnitude of these two rate coefficients. In a situation where the kinetics of the mass transfer is much slower than the intrinsic reaction

TABLE 8.1 A Comparison of the Characteristics of Gas–Liquid Mass Transfer and a First-Order Chemical Reaction

	Gas–Liquid H_2 Mass Transfer	First-Order Reaction
Kinetic expression	$-\frac{d[H_2]}{dt} = k_L a([H_2] - [H_2]_{sat})$	$-\frac{d[C]}{dt} = k_r[C]$
First-order rate coefficient (s^{-1})	$k_L a$	k_r
Maximum rate	$R^{max}_{H_2,g/L} = k_L a[H_2]_{sat}$	$R^{max}_{rxn} = k_r[C]_0$

kinetics, that is,

$$R^{max}_{H_2,g/L} \ll R^{max}_{H_2,rxn}$$

or

$$k_La[H_2]_{sat} \ll k_r f([H_2]_{sat}, [catalyst], [substrate]) \quad (8.4)$$

$[H_2]$ deviates significantly from $[H_2]_{sat}$, and the solution is starved of H_2. In other words, the effective H_2 pressure, that is, the pressure that the catalyst experiences, is lower than the H_2 pressure in the gas phase. Under extreme hydrogen starved conditions, $[H_2]$ or the effective H_2 pressure may approach zero. In this case, the reaction becomes entirely limited by the mass transfer instead of by the catalytic processes on the catalyst, and the observed reaction rate equals the maximum rate of H_2 mass transfer. On the other hand, if the kinetics of mass transfer is much faster than the intrinsic reaction kinetics, that is,

$$R^{max}_{H_2,g/L} \gg R^{max}_{H_2,rxn}$$

or

$$k_La[H_2]_{sat} \gg k_r f([H_2]_{sat}, [catalyst], [substrate]), \quad (8.5)$$

$[H_2]$ approaches $[H_2]_{sat}$, and the effective H_2 pressure approaches the pressure in the gas phase.

Assuming that $[H_2]$ in equation 8.2 varies slowly with time, the lowest value of $[H_2]$ may be expressed as

$$[H_2] \approx [H_2]_{sat}\left(1-R^{max}_{H_2,rxn}/R^{max}_{H_2,g/L}\right) \quad (8.6)$$

To ensure that the solution is nearly saturated with H_2 throughout the entire course of the reaction and the observed rate is representative of the kinetics intrinsic to the catalytic system under the specified hydrogen pressure, a rule of thumb is that the maximum intrinsic reaction rate should be less than 10% of the maximum H_2 delivery rate:

$$R^{max}_{H_2,rxn}/R^{max}_{H_2,g/L} \leq 10\% \quad (8.7)$$

Equation 8.6 shows

$$[H_2] \geq 90\%[H_2]_{sat} \quad (8.8)$$

when the 10% rule of thumb is satisfied.

8.3 IMPACT OF k_La ON REACTION KINETICS AND SELECTIVITY

As shown in Section 8.2, even when the pressure in the gas phase is specified, the reaction conditions could in actuality be unspecified due to uncharacterized deviation of $[H_2]$ from $[H_2]_{sat}$ as a result of a lack of knowledge of the hydrogen mass transfer coefficient k_La. Under the unspecified conditions, rate measured reflects kinetics at an unknown $[H_2]$ instead of the intended constant $[H_2]_{sat}$. The kinetic data obtained under such conditions are not helpful and can even be harmful to the development of scalable processes. Irreproducibility in rate from reactor to reactor may be observed under seemingly identical reaction conditions due to different mass transfer coefficients in different reactors. A process developed under such conditions may both pose safety and selectivity problems in scale-up. In addition, the deviation of $[H_2]$ from $[H_2]_{sat}$ may cause irreversible alteration of the catalyst properties. The catalyst may be deactivated due to lack of hydrogen atoms on the catalytic surface, which not only reduces the catalytic activity, but also may change the selectivity of the catalyst in undesirable manners.

In addition, selectivity of the hydrogenation reactions may be strongly influenced by mass transfer by virtue of the intrinsic dependence of the selectivity on $[H_2]$, the effective pressure that the catalyst experiences. Many reactions exhibit strong dependence of selectivity on hydrogen pressure. Examples include the $[Rh(DIPAMP)]^+$-catalyzed asymmetric hydrogenation of α-acylaminoacrylic acid derivatives [10], asymmetric hydrogenation of γ-geraniol and geraniol catalyzed by Ru(BINAP) [11], and enantioselective hydrogenation of ethyl pyruvate over cinchonidine-modified Pt [12]. In the case of asymmetric hydrogenation of γ-geraniol catalyzed by Ru[(*S*)-BINAP], the enantioselectivity to (*R*)-β-citronellol decreases precipitously from 90% ee to nearly racemic with increasing effective H_2 pressure from nearly 0 to 100 psia. The case of asymmetric hydrogenation of geraniol is even more dramatic. The enantioselectivity flips from an ee of 93% (*R*) to 91% (*S*) when the H_2 effective pressure changes from 100 psia to nearly zero due to the presence of the competitive isomerization of geraniol to γ-geraniol under the hydrogenation conditions (Figure 8.2) [13]. In all these cases, even at constant H_2 gas-phase pressure, uncharacterized deviation of $[H_2]$ from $[H_2]_{sat}$ as a result of H_2 mass transfer limitations translates into unpredictable selectivity.

It is interesting to note that the mass transfer limitations may work for or against the desired selectivity, depending on how the selectivity is related to $[H_2]$. For instance, at a constant gas-phase pressure, mass transfer limitations help to enhance the enantioselectivity of the Ru[(*S*)-BINAP]-catalyzed asymmetric hydrogenation of γ-geraniol to (*S*)-citronellol, in which case the selectivity increases with decreasing pressure.

The effect of the interplay between the mass transfer and the intrinsic rate processes on kinetics and selectivity for reactions carried out at constant H_2 pressure is further demonstrated in Figure 8.3, using as an example the enantioselective hydrogenation of ethyl pyruvate over cinchonidine-modified Pt.[8(b)] The effects of the two extreme situations discussed above are graphically illustrated by the difference in the observed kinetics and enantioselectivity. At 400 rpm, the H_2 mass transfer is much slower than

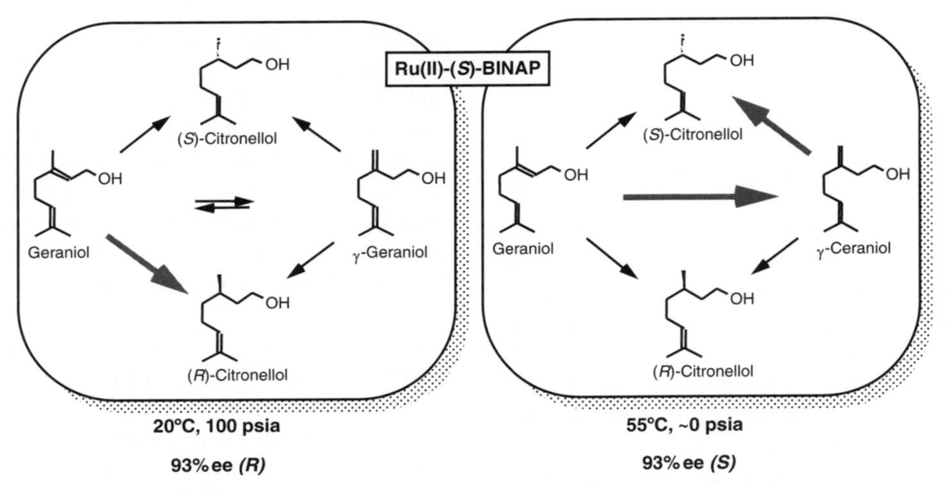

FIGURE 8.2 Striking dependence of enantioselectivity in the asymmetric hydrogenation of geraniol as a result of the interplay of rate processes in the isomerization/hydrogenation network.

the intrinsic hydrogenation rate ($k_La = 4 \times 10^{-3}\,s^{-1}$). [$H_2$] is virtually zero. The kinetics, being completely limited by the rate of the H_2 delivery across the gas/liquid interface instead of by the catalytic hydrogenation of ethyl pyruvate on the Pt surface, is independent of the substrate concentration and therefore exhibits zero-order kinetic behavior. A process developed under these conditions would potentially run into safety issues when scaled up as the hydrogenation rate would change greatly with differences in the mass transfer coefficient k_La associated with large hydrogenators. At 2000 rpm, the mass transfer is no longer limiting the rate ($k_La = 0.7\,s^{-1}$), the solution is saturated with H_2 at all times during the reaction, that is, [H_2] ≅ [H_2]$_{sat}$, and the rate is limited by the catalytic hydrogenation of ethyl pyruvate over Pt. Further increases in the mass transfer coefficient no longer change the rate profile. As a result, a picture of the intrinsic kinetics emerges. In addition to the kinetics, the mass transfer also influences selectivity by virtue of changing the availability of H_2 that the catalyst experiences. Figure 8.3 shows that the enantioselectivity increases from 23% to 60% ee due to a change in [H_2] from starvation to saturation upon increasing the agitation speed from 400 to 2000 rpm.

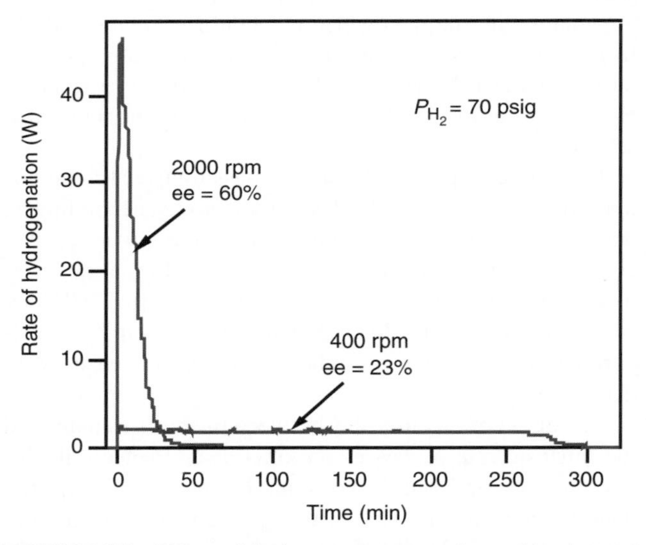

FIGURE 8.3 Effect of hydrogen mass transfer on kinetics and enantioselectivity in the chiral hydrogenation of ethyl pyruvate over cinchonidine-modified Pt. Except for the different stir rates, the two experiments were carried out under otherwise identical conditions (30°C, solvent: 1-propanol).

In developing a hydrogenation process, it is the intrinsic catalyst activity and selectivity, that is, the catalytic behaviors under well-defined conditions, that are of foremost concern, not those convoluted with the hydrogen mass transfer process. Obviously, it is not sufficient to specify H_2 pressure alone to unravel kinetics and selectivity of a catalytic hydrogenation reaction intrinsic to the catalytic system. It is necessary to characterize the mass transfer properties of the hydrogenator and conduct the hydrogenation experiments according to the 10% "rule of thumb" as described by equation 8.7.

8.4 CHARACTERIZATION OF GAS–LIQUID MASS TRANSFER PROCESS

Section 8.3 described how the gas/liquid mass transfer process, characterized by the simple parameter k_La, can exert significant influence on the outcome and robustness of the hydrogenation process. It is not uncommon in the pharmaceutical industry, however, to see chemists and chemical engineers employ hydrogenation vessels for screening, scaling-up, and commercialization of hydrogenation processes without characterizing their mass transfer coefficients.

In this section, a simple practical procedure for measuring k_La is reviewed and examples are given for measuring k_La in a variety of hydrogenators at scales ranging from 100 mL to 760 gal in volume. Given the importance of the mass transfer coefficient, it is recommended that all hydrogenators, including those used for screening, for process development and scaling-up, and for commercialization, be characterized in term of their mass transfer coefficients.

Among the methodologies available, the most straightforward one for measuring the gas/liquid mass transfer coefficient k_La is to measure directly the kinetics of nonreactive hydrogen uptake by the solution at various agitation speeds [14]. In this method, the solution is first degassed thoroughly by vacuum with the agitation on. The agitator is then turned off, and hydrogen gas is introduced to the headspace of the hydrogenator at a pressure close to the hydrogenation pressure at which point the hydrogen line valve to the hydrogenator is closed. When the agitation commences, the pressure in the reactor is recorded as a function of time using a fast-response pressure transducer. A schematic depicting the pressure drop in the headspace of the hydrogenator and the corresponding concentration rise of hydrogen in the liquid phase is shown in Figure 8.4. The rate at which the pressure decreases is directly related to the gas/liquid transfer rate, whereas the extent to which the pressure drops is related to the solubility of the gas in the liquid. The dependence of pressure P as a function of time t is governed by the simple first-order rate equation

$$\frac{P_f - P_0}{P_i - P_0} \ln\left(\frac{P_i - P_f}{P - P_f}\right) = (k_La)t \qquad (8.9)$$

where P_i is the initial pressure, P_f the final pressure, and P_0 the solvent vapor pressure.

Typical pressure–time curves in Metter Toledo's 1L RC1 MP10 reactor are displayed in Figure 8.5a. They graphically illustrate the difference in mass transfer rates at agitation rates of 400 and 1000 rpm. Figure 8.5b shows the pressure plot obtained using data in Figure 8.5a according to equation 8.9. The slope of the pressure plot yields k_La that is $3.8 \times 10^{-3}\,s^{-1}$ at 400 rpm and $0.24\,s^{-1}$ at 1000 rpm. Another way to look at the difference in the mass transfer rates is a comparison of the "half-life", that is, $t_{1/2} = \ln 2/k_La$. While reaching 50% of the hydrogen saturation concentration takes only ~3 s at 1000 rpm, it takes ~3 min at 400 rpm.

The mass transfer coefficient can be strongly influenced by a number of parameters including the agitation rate, the type of reactor, reactor configuration such as reactor size and shape, the type of agitator and its position in the reactor, fill level, temperature, solvent and solute, the use of baffle, and subsurface sparge line. The methodology described here is a convenient and fast way to measure the mass transfer characteristics of the hydrogenator relevant to the specific conditions of the hydrogenation process.

Using the simple methodology described in this section, the full mass transfer characteristics of three different laboratory hydrogenators, the 1L Mettler Toledo's RC1 MP10 reactor, the 250 mL Parr shaker, and a 5 gal stainless steel hydrogenator with mechanical agitation, have been conveniently and rapidly measured. The mass transfer coefficients and their dependence on agitation speed are shown in Figure 8.6.

Taking advantage of its design precision, for example, the reproducible agitation ramp from zero to the desired agitation rate in <0.2 s, we first used the 1L Mettler Toledo's RC1 MP10 system to study the effect of the hydrogen pressure on k_La using methanol as the solvent. The results show that the

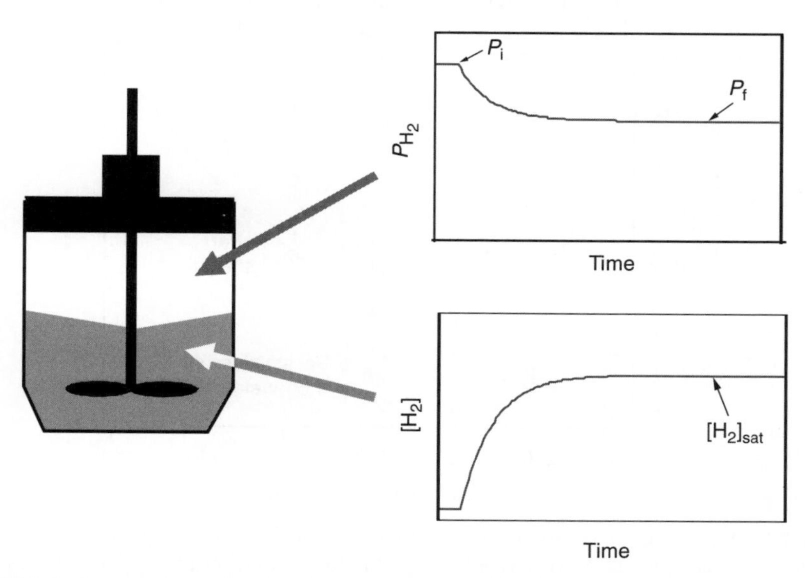

FIGURE 8.4 Schematics of the pressure drop and the corresponding concentration rise in a k_La measurement experiment.

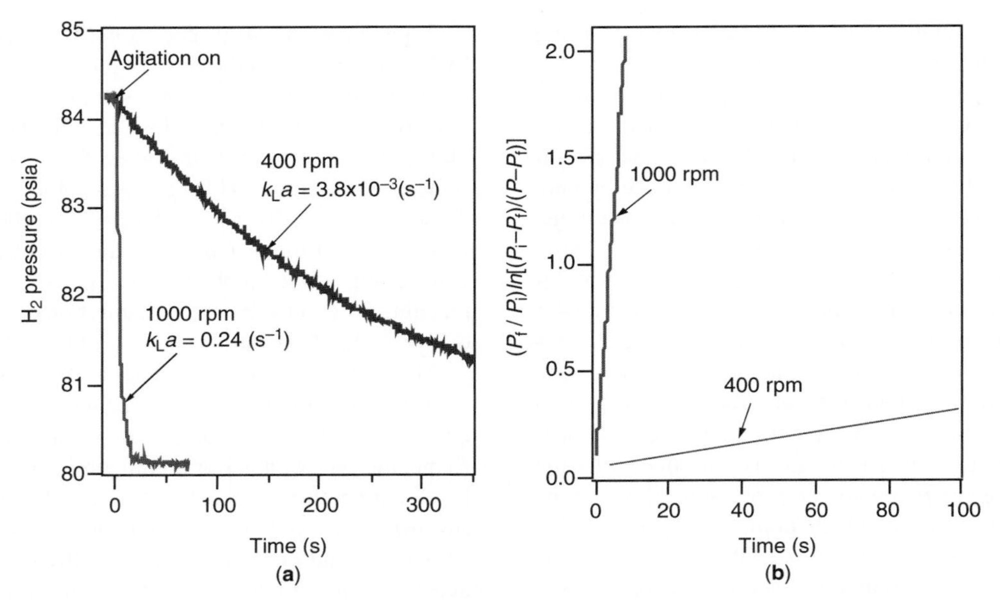

FIGURE 8.5 Measurements of the H_2 uptake kinetics in 1-propanol (0.5 L) in the Mettler Toledo's 1L RC1 MP10 reactor at 30°C. (a) Hydrogen pressure drop as a function of time. (b) Corresponding plots of the pressure function in equation 8.9 versus time. The slope is k_La.

k_La value is rather independent of the hydrogen pressure over the range of 15–100 psig. Given this observation, we simply choose any convenient hydrogen pressure in this pressure range for the k_La measurement.

To the best of our knowledge, mass transfer characteristics of the Parr shaker system, traditionally and frequently employed for development of hydrogenation processes, have not been reported. Figure 8.6 shows that the Parr shaker possesses surprisingly good mass transfer capability rivaling the best achievable in the RC1 system that is known for its good mass transfer capability. At the shaking frequency typically used for process development, the value of k_La is moderate, $0.1\ s^{-1}$ at 130 rpm. Higher k_La values are achievable at higher shaking frequencies. Figure 8.7 shows several pressure curves measured in the Parr shaker. Figure 8.6 also shows that the mass transfer capability of the specific 5 gal

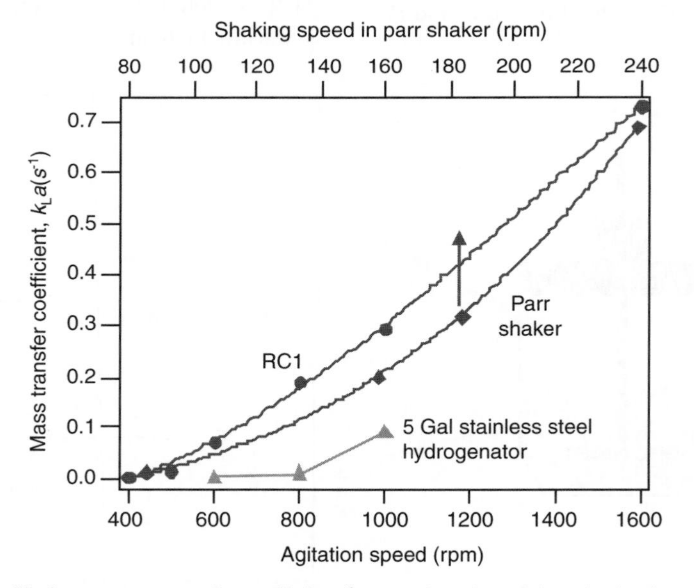

FIGURE 8.6 Hydrogen mass transfer coefficient k_La as a function of the agitation intensity for three types of hydrogenators: the 1L Mettler Toledo's RC1 MP10 reactor (500 mL MeOH), a 5 gal stainless steel hydrogenator with mechanical agitation (2.5 gal MeOH) and the 250 mL Parr shaker (90 mL MeOH).

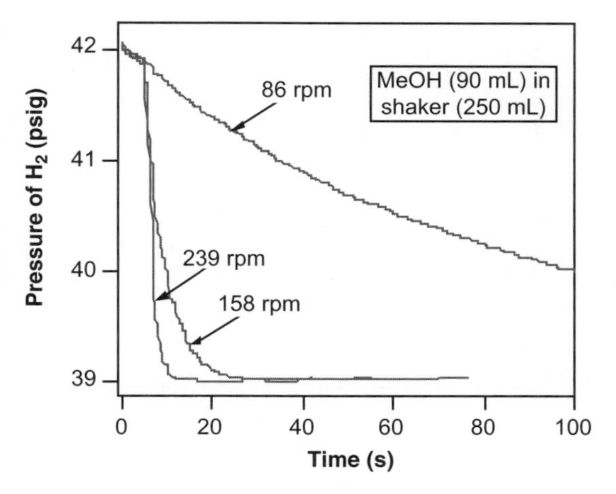

FIGURE 8.7 Decay of pressure in the headspace due to the nonreactive hydrogen uptake by methanol (90 mL) in the Parr shaker (250 mL) parametric in the shaking frequency.

hydrogenator studied even at the highest allowable rpm (1000 rpm) is somewhat limited ($k_{L}a = 0.1\ s^{-1}$).

The simple methodology is applicable to $k_{L}a$ measurement in large hydrogenators in manufacturing facilities. The mass transfer capability of a factory glass-lined hydrogenator with a nominal 750 gal volume (900 gal actual, with a retreat blade impeller, one baffle, subsurface sparging) was characterized. The results revealed a limited gas/liquid mass transfer capability of this specific hydrogenator even with 100% agitation power. For instance, at a 460 gal fill and with the full agitation power, the $k_{L}a$ value is $0.035\ s^{-1}$, a mass transfer coefficient equivalent to that in the RC1 reactor at 500 rpm only (Figure 8.6). The low mass transfer ability in this hydrogenator makes it unsuitable for running fast hydrogenation reactions that require $k_{L}a > 0.035\ s^{-1}$ at the existing configuration.

8.5 CHARACTERIZATION OF CATALYST REDUCTION PROCESS

For hydrogenation processes in a solvent with a solid-supported metal catalyst, an important process involved is the reduction of the catalyst itself to its metallic state. For example, the Pd in the typical Pd/C catalysts is either in the form of palladium hydroxide or in the form of metallic palladium particles with its surface layers oxidized. Reduction of the surface Pd to its metallic state is necessary for the catalytic hydrogenation processes. While reduction of Pd/C catalysts used in the gas–solid reactions can be conveniently studied, to the best of our knowledge, the kinetics of catalyst reduction under the gas–liquid–solid slurry hydrogenation conditions has not been reported. How long does it take to reduce the catalyst under the slurry hydrogenation conditions? What does the kinetics look like? Is it instantaneous upon pressurization of the hydrogenator? How do the solvent, the additives, and the substrate influence the kinetics of the catalyst reduction? How do the transient properties of the catalyst during the catalyst reduction process influence reactivity and selectivity of the hydrogenation of the substrate? These questions remain unanswered due to lack of research tools to characterize the kinetics of the catalyst reduction *in situ* under the slurry hydrogenation conditions. In this section, a simple procedure is described that allows one to measure the characteristics of the catalyst reduction process.

The procedure is an extension of the $k_{L}a$ measurement protocol described in Section 8.4. First, profile of the nonreactive uptake by the solvent is measured. Profile of the sum of the nonreactive and reactive hydrogen uptake due to the catalyst reduction is subsequently measured by repeating the same procedure (under the identical conditions) after addition of the heterogeneous catalysts to the solvent in the batch. The uptake profile due to the catalyst reduction can be extracted by taking the difference of the two uptake profiles.

The methodology is demonstrated in Figure 8.8 for measuring the reduction rate of a Pd/C catalyst in methanol at −10°C. The hydrogen uptake curves at 40 psig H_2 and at −10°C upon agitation (1000 rpm) for MeOH (500 mL) only and for MeOH (500 mL) plus the 5% Pd/C catalyst (10 g) were measured consecutively and the results are shown in Figure 8.8a. Because the two uptake curves were measured under the identical conditions except that one is with the Pd/C catalyst added, the nonreactive uptake curve can be directly subtracted from the sum curve to generate the uptake curve associated only with the Pd/C reduction. By subtracting the two curves in Figure 8.8a, the reduction profile of the Pd/C catalyst emerges and is shown in Figure 8.8b.

A few properties of the reduction process become apparent from Figure 8.8b. The reduction is a relatively fast process—it is nearly completed in 2 min at 40 psig H_2 and at −10°C. The reduction kinetics exhibits two distinct regimes and can be fitted nearly perfectly by a double exponential function, that is, uptake $= 2.55 - 1.65e^{-0.0177t} - 1.34e^{-0.0864t}$. The fast rate process has a half-life of about 8 s and is likely associated with the reduction of the outside layers of Pd catalyst particles, whereas the slower rate process has a half-life of about 39 s and is likely associated with the reduction of the bulk of the Pd catalyst particles and with the formation of bulk palladium hydrides. Figure 8.8b also shows that the overall stoichiometry of the reactive hydrogen update for this catalyst is H/Pd = 1.3.

A closer look at the uptake profile in Figure 8.8b shows that there is a short induction period (~4 s) in the catalyst reduction at −10°C; that is, there is no appreciable hydrogen uptake in the presence of 40 psig hydrogen for ~4 s. The induction period virtually disappears at 25°C for the catalyst reduction. This temperature effect is more evident when the reduction profiles at the two temperatures are placed in the same graph as shown in Figure 8.9. Interestingly, the kinetics in the fast rate regime at 25°C is similar to that at −10°C

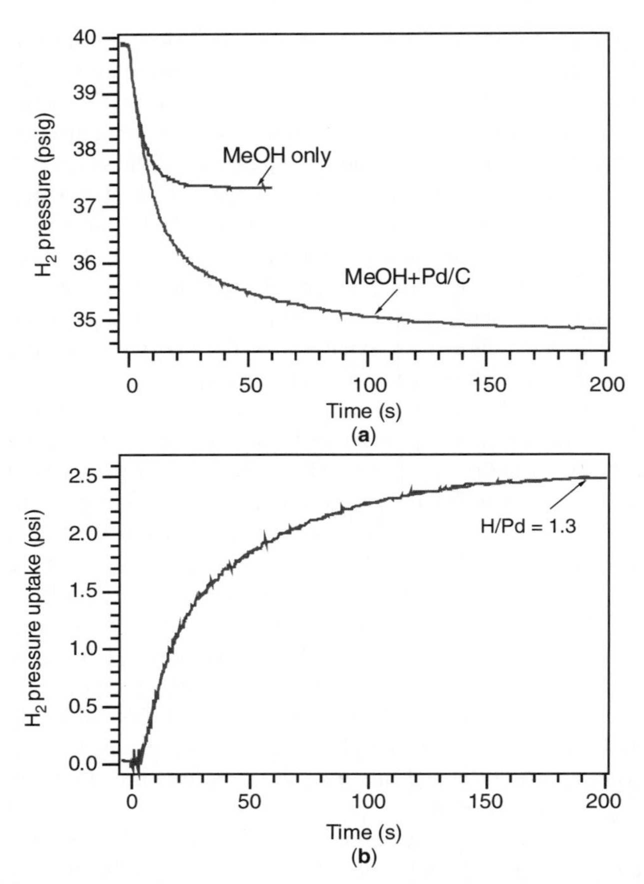

FIGURE 8.8 (a) Hydrogen uptake curves at 40 psig H_2 and at $-10°C$ upon agitation (1000 rpm) for MeOH (500 mL) only and for MeOH (500 mL) plus 5% Pd/C (10 g, Type 21, JM), following the standard k_La measurement procedure described in Section 8.4. Apparatus: Mettler-Toledo's RC1 with an MP10 reactor. (b) Difference between the two uptake curves in (a), representing the reactive hydrogen uptake by the reduction of the Pd/C catalyst only. The reactive hydrogen uptake curve can be fitted perfectly by a double exponential function. Uptake $= 2.55-1.65e^{-0.0177t} -1.34e^{-0.0864t}$.

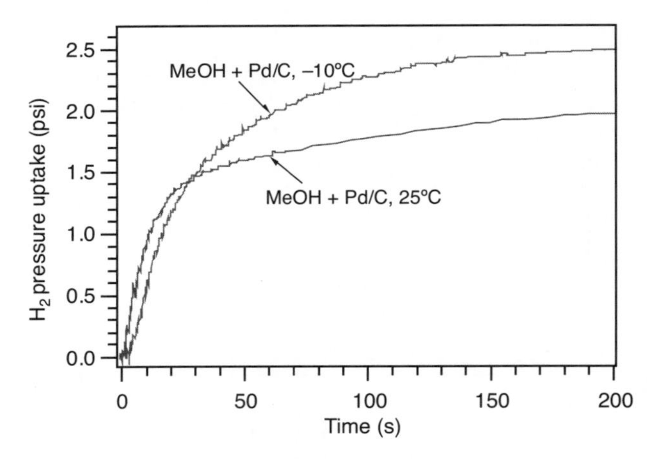

FIGURE 8.9 A comparison of the reduction kinetics of the 5% Pd/C catalyst (Type 21, JM) at -10 and 25°C. The experimental conditions are identical to those described in Figure 8.8.

(similar slopes of hydrogen uptake at the fast rate regimes). The total reduction in hydrogen uptake at 25°C is however less, presumably due to lower equilibrium surface coverage of the hydrogen atoms on Pd and the bulk concentration of the Pd hydride at higher temperatures.

The methodology provides a convenient way to measure the catalyst reduction kinetics in solvents under the hydrogenation conditions except that the substrate is not present. In the presence of the substrate, the reactive hydrogen uptake would originate from the catalyst reduction and from the substrate hydrogenation. The methodology described here cannot deconvolute the rate of the catalyst reduction from the rate of the substrate hydrogenation. How the presence of the substrate affects the reduction kinetics of the heterogeneous catalysts remains unclear. One thing is clear that the kinetics of the catalyst reduction can be altered by the substrate in the hydrogenation reaction. Figure 8.10 shows the effect of the addition of ammonia to the methanol solvent on the catalyst reduction at $-10°C$. In this experiment, a small amount of aqueous ammonia (0.78 mol) was added into methanol to a total volume of 500 mL, and the catalyst reduction kinetics was studied following the standard procedure. The two hydrogen uptake curves measured are shown in Figure 8.10a. The catalyst reduction kinetic profile derived from Figure 8.10a is plotted in Figure 8.10b along with the kinetic profile of the catalyst reduction in methanol alone as a reference.

The most striking feature of the reduction kinetics in the presence of ammonia is that there is a significant induction period, ~60 s, much longer than the 4 s induction period in MeOH without the ammonia addition. Over this long induction period, there is virtually no hydrogen uptake. At the end of the induction period, the rate of the Pd catalyst reduction accelerates. This induction period may result from strong chemisorption of ammonia on the surface of the Pd catalyst, inhibiting dissociative chemisorption of dihydrogen on Pd. Adsorption of a small amount of hydrogen on the catalyst surface through competitive adsorption process reduces a small fraction of the surface Pd that conversely facilitates dissociative chemisorption of additional hydrogen, leading to the rate acceleration. The maximum rate of the reduction, however, is slower than that in the absence of ammonia (Figure 8.10b). The long induction period is attributed to the presence of ammonia in the hydrogenation batch. Independent experiments showed that water as a result of the aqueous ammonia addition does not alter the catalyst reduction. Overall, the addition of ammonia extended the catalyst reduction timescale from about 2 to 7 min, suggesting that additives used in hydrogenation reactions and the substrate itself can influence the catalyst reduction kinetics.

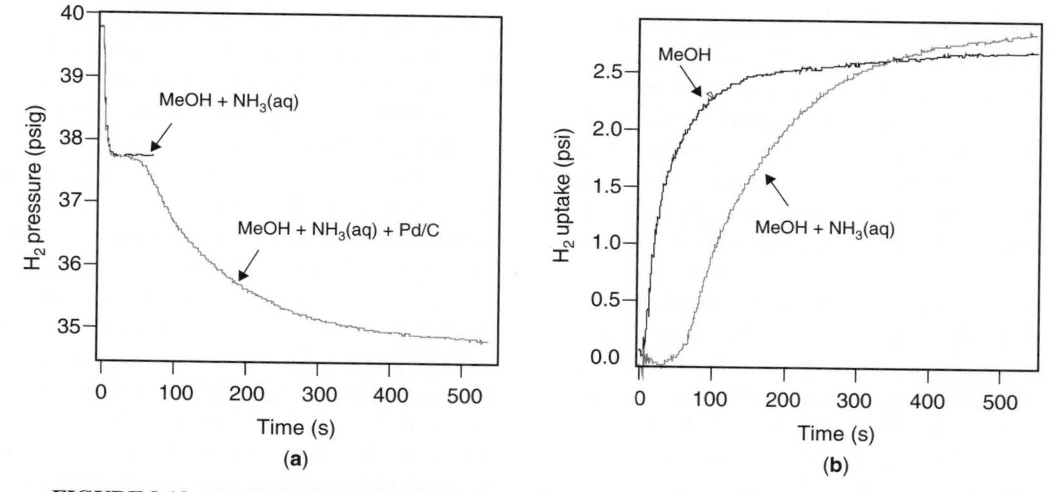

FIGURE 8.10 (a) Hydrogen uptake curves at 40 psig H_2, $-10°C$, 1000 rpm for MeOH + NH_3 (aq)) (500 mL) and for MeOH + NH_3(aq) (500 mL) plus 5% Pd/C (10 g, Type 21, JM), following the standard k_La measurement procedure described in Section 8.4. (b) Kinetics of the Pd/C catalyst reduction in MeOH + NH_3(aq)) in comparison with the kinetics in MeOH only (reproduced from Figure 8.8b).

8.6 BASIC SCALE-UP STRATEGY FOR HYDROGENATION PROCESSES

When a hydrogenation process is translated from the laboratory to the plant, chemists and chemical engineers sometimes encounter scale-up issues such as problems with reactivity and/or selectivity. Often the scale-up issues can be traced to a lack of characterization or understanding of the various rate processes and an ensemble of key factors that affect the reproducibility and robustness of the hydrogenation process. A successful scale-up necessitates that these factors be identified, measured, and controlled. This section discusses some of the fundamental factors that need to be considered for a successful scale-up. A basic strategy for scaling up hydrogenation processes is described in terms of several quantitative criteria.

The most basic requirement for a successful scale-up is for the chemists and engineers to have at the laboratory development stage a true understanding of reaction kinetics that is intrinsic to the catalytic system, that is, one that is not masked by H_2 mass transfer limitations but is obtained under well-defined conditions, in particular, with a known $[H_2]_{lab}$. The best strategy to achieve this is to characterize the mass transfer capability of the hydrogenation reactor and use the 10% rule of thumb (as described by equation 8.7) in the laboratory developmental work so that the condition

$$[H_2]_{lab} \approx [H_2]_{sat} \quad (8.10)$$

is also satisfied at all times. Availability of the intrinsic kinetic information allows one to make an intelligent choice of reactors and reaction conditions for reproducible and robust scale-ups.

The hydrogen mass transfer coefficient k_La of the hydrogenation reactor is one of the primary factors to consider in scale-up because it can have direct impact on kinetics, process safety, and selectivity. It is not uncommon, however, to find hydrogenation process descriptions specifying only the gas-phase H_2 pressure and agitation speed without specifying the requirement for the mass transfer coefficient k_La of the hydrogenator. To scale up a process in this manner subjects the process to the risk of running under ill-defined conditions; that is, undefined $[H_2]$ that is disengaged from the gas-phase hydrogen pressure. To reproduce in the plant runs the rate and selectivity observed in the laboratory, it is important that $[H_2]$ in the plant runs be the same as the $[H_2]$ used in the laboratory development work, that is,

$$[H_2]_{plant} = [H_2]_{lab} \approx [H_2]_{sat} \quad (8.11)$$

Using the same pressure as that employed in the laboratory development without specifying k_La does not necessarily guarantee that this condition (equation 8.11) is satisfied.

When one runs the plant process far from the hydrogen mass transfer-limited situations, the hydrogen concentration $[H_2]$ is known and constant (as long as the gas-phase pressure remains unchanged) over the entire course of reaction. This is particularly important for hydrogenation reactions with $[H_2]$-dependent selectivity. If this type of reactions is operated under hydrogen diffusion limitations, $[H_2]$ deviates from $[H_2]_{sat}$ and becomes disengaged from the gas-phase hydrogen pressure. The solution goes from hydrogen starved at the beginning stage of reaction when the intrinsic hydrogenation rate is fast to hydrogen saturated at the later stage of the reaction when the intrinsic hydrogenation slows down due to depletion of the starting material. The selectivity would vary throughout the course of the reaction as a result. In addition, rate of a process running under hydrogen mass transfer control is sensitive to slight changes in reaction conditions. For example, the rate can change significantly with changes in rpm due to the commonly observed exponential dependence of k_La on the agitation rate. The rate would not be affected by changes in the agitation rate, however, for processes running away from hydrogen mass transfer limitations.

To satisfy the scale-up requirement described by equation 8.11, the 10% rule of thumb (equation 8.7) provides a good general guideline for matching the mass transfer capability of the hydrogenation reactor to the kinetics of a process. Given a process with known intrinsic kinetics, the chosen reactor for scale-up needs to have a minimum mass transfer coefficient of

$$k_La \geq 10R^{\max}_{H_2,rxn}/[H_2]_{sat} \tag{8.12}$$

In addition, the heat transfer issue also needs to be considered for safe operations in scale-up. The following condition needs to be met

$$R^{\max}_{H_2,rxn} \leq q_r^{\max}/\Delta H_{H_2} \tag{8.13}$$

where $q_r^{\max}$ is the maximum heat removal capability of the reactor, and ΔH_{H_2} is the heat of hydrogenation per mole of H_2 reacted.

The inherent reaction kinetics and the conditions described by equations 8.12 and 8.13 form some of the basic requirements that need to be considered for successful scale-ups of hydrogenation processes.

8.7 SUMMARY

Dissolution of H_2 into the liquid phase is the first rate process along the H_2 pathway in catalytic hydrogenation reactions. The dissolution kinetics can play a key role in the kinetics and selectivity of catalytic reactions and the outcome of process scale-up. At a given gas-phase H_2 pressure, while thermodynamics determines the solubility $[H_2]_{sat}$, it is *kinetics* that determines the actual solution H_2 concentration $[H_2]$ or the "effective hydrogen pressure" that the catalyst experiences *during* hydrogenation reactions. Depending upon the relative magnitude of the rate of H_2 mass transfer across the gas/liquid interface versus the rate of the reactive removal of the dissolved H_2 from the liquid phase by hydrogenation, $[H_2]$ may vary greatly from saturation to nearly zero, even at a constant pressure in the gas phase. The influence of mass transfer on $[H_2]$ exerts a direct impact not only on rate but also on selectivity for reactions whose selectivity depends on $[H_2]$. It is also often the fundamental cause of irreproducibility in rate and selectivity observed from reactor to reactor (e.g., when scaling up a laboratory process in the manufacturing facility) when the same reaction is carried out under seemingly identical conditions. This is because different mass transfer capabilities of reactors of different types and scales can lead to different "effective pressures" even at a constant gas-phase pressure.

A catalytic hydrogenation process should be carried out under conditions where the intrinsic hydrogenation rate is at least 10 times lower than the maximum rate of H_2 mass transfer across the gas/liquid interface. This ensures that the observed kinetics and catalytic behaviors are not masked by the mass transfer limitations but are intrinsic to the catalytic system under well-defined conditions. This requirement may serve as a general guideline for designing scalable processes, since as long as the 10% rule of thumb is satisfied, the hydrogenation process is "portable"; that is, the same kinetics and catalytic behavior will be reproduced from reactor to reactor at any scale. Strategy for a successful scale-up of hydrogenation processes is formulated in terms of a set of quantitative criteria.

The process of reduction of a Pd/C catalyst under the hydrogenation conditions is characterized using a novel methodology. The results show that the catalyst reduction process can have an induction period, is relatively fast (on the order of <5 min) under typical hydrogenation conditions, and can be significantly influenced by the nature of the additives or the hydrogenation substrates.

ACKNOWLEDGMENTS

We thank Prof. Donna Blackmond for helpful discussions and Andy Newell, Charlie Bazaral, and Steve Conway for assistance in experiments.

REFERENCES

1. (a) Rylander PN. *Hydrogenation Methods*, Academic Press, 1990;(b)Rylander PN. *Catalytic Hydrogenation in Organic Synthesis*, Academic Press, 1979.
2. Augustine, R.L. *Heterogeneous Catalysis for the Synthetic Organic Chemists*, CRC Press, 1995.
3. Knowles, W.S., *Angew. Chem., Int. Ed.* 2002;41:1998.
4. Knowles, W. S., Noyori, R., *Acc. Chem. Res.* 2007;40:1238.
5. Noyori, R., *Angew. Chem., Int. Ed.* 2002;41:2008.
6. Noyori R. *Asymmetric Catalysis in Organic Synthesis*, Wiley, 1994, and references therein.
7. Krische MJ, Sun YK,guest editors. Hydrogenation and transfer hydrogenation, *Acc. Chem. Res.*, 40, 2007.
8. (a) See, for example, Sun, Y.K., Landau, R.N., Wang, J., LeBlond C., Blackmond, D.G., *J. Am. Chem. Soc.*, 1996;118:1348; (b)Sun, Y.K., Wang, J., Landau, R.N., LeBlond C., Blackmond, D.G., *J. Catal.* 1996;161:759.
9. (a) Ramachandran PA, Chaudhari, RV. Three-phase catalytic reactors. In: *Topics in Chemical Engineering 2*, Gordon and

Breach Science Publishers, 1983; (b) Beenackers, A.A.C.M., and Van Swaaij, Mass transfer in gas-liquid slurry reactors. *Chem. Eng. Sci.*, 1993;48:3109; (c) Hines AL Maddox RN. *Mass Transfer: Fundamentals and Applications*, Prentice-Hall, 1985.

10. (a) Landis, C.R.,and Halpern, J., *J. Am. Chem. Soc.* 1987;109:1746; (b) Sun, Y.K., Landau, R.N., Wang, J., Le-Blond, C.,and Blackmond, D.G., *J. Am. Chem. Soc.* 1996;118:1348.
11. Sun YK, LeBlond, C, Wang J, LeBlond C, Blackmond DG, Laquidara J, Sowa J., Jr., *J. Am. Chem. Soc.* 1995;117:12647.
12. Wang, J., Sun, Y.K., LeBlond, C., Landau. R, Blackmond, D.G., *J. Catal.* 1996;161:752.
13. Sun, Y.K., Wang, J., LeBlond, C., Reamer, R.A., Laquidara, J., Sowa, J.R., Blackmond, D.G., *J. Organomet. Chem.*, 1997;581:65.
14. (a) Matsumara, M, Masunaga, H.,and Kobayashi, J., *J. Ferment. Technol.*, 1979;57:107; (b) Deimling, A., Karandikar, B. M.,and Shah, Y.T., Carr, N.L., *Chem. Eng. J.* 1984;29:140; (c) Blaser, H.U., Garland, M., Jallet, H.P., *J. Catal.*, 1993;144: 569.

9

CHARACTERIZATION AND FIRST PRINCIPLES PREDICTION OF API REACTION SYSTEMS

Joe Hannon
Scale-up Systems Limited, Dublin, Ireland

9.1 INTRODUCTION

This chapter deals with how unit operations in API synthesis can be described, predicted, and scaled using classical chemical engineering principles. This allows design to be completed more quickly and with greater success than adopting a trial and error approach. Several common applications are worked through in detail for illustration. Examples and references to industry projects utilizing these concepts are included.

9.1.1 Rate Processes in API Unit Operations

Table 9.1 lists the rate processes involved in common unit operations in API synthesis. Characterization of each rate process is often feasible; the apparent complexity of unit operations arises from the combination of several rate processes in a single operation. In many cases, one rate process is limiting and dominates the others.

Other chapters in this book also address several of these rate processes and operations, including "Reaction Kinetics and Characterization," "Design of Distillation and Extraction Operations," "Design of Filtration and Drying Operations," "Kilo Lab and Pilot Plant Manufacturing."

9.1.2 Scale-Dependence and Scale-Independence

The intrinsic rate of a chemical reaction is independent of scale; in other words, if the reaction could be conducted without limitation by any other rate process, it would run at the same rate (per unit volume) at all scales; the reaction time, starting from the same initial composition and held at the same temperature, would be the same at all scales. This is true for reactions whose kinetics are slow compared to other rate processes. The choice of solvent, reagents and/or catalyst combination for reactions has been largely the domain of development chemists and these variables are taken here to be fixed already; in any case, those effects, including reagent solubility, are also independent of scale.

All other rate processes listed in Table 9.1 are scale-dependent. For example, the rate of heat transfer depends strongly on the ratio of surface area to volume, which reduces on scale-up. This means that heat removal on scale will be slower than in the laboratory, unless specific measures are taken to provide additional surface area or an increased temperature driving force. Therefore, the time required to heat or cool a batch of material tends to increase on scale. Similarly, the rate of mass (or phase) transfer is scale dependent; the agitation conditions inside the vessel determine the time required to reach equilibrium between the phases; it is easier to make this time short in the laboratory than on scale.

The rates of addition and removal of material also change with scale, normally taking longer at larger scales. For example, the rate of disengagement of gas evolved from a reaction depends again on the surface area to volume ratio, so reactions such as decarboxylation have to be conducted more slowly on scale to avoid partial loss of the reactor contents due to swelling of the batch.

The impact of scale dependence is to lengthen processing times on scale, reducing productivity somewhat but more

Chemical Engineering in the Pharmaceutical Industry: R&D to Manufacturing Edited by David J. am Ende

TABLE 9.1 Typical Incidence of Classical Chemical Engineering Rate Processes in Unit Operations

	Operation					
Rate Process	Reaction/Quench	Distillation	Extraction	Crystallization	Filtration	Drying
Chemical kinetics	☑			☑		
Mass transfer	☑	☑	☑	☑		☑
Heat transfer	☑	☑	☑	☑		☑
Addition	☑	☑	☑	☑	☑	☑
Removal	☑	☑	☑		☑	☑

importantly potentially allowing additional phenomena to occur that were not observed to the same extent in the laboratory, such as impurity-forming reactions, product degradation, catalyst poisoning or deactivation, unwanted crystal forms, nucleation, agglomeration or breakage. Fortunately, each of the rate processes that contribute to such problems can be characterized or estimated using classical chemical engineering methods, allowing scale-up problems to be anticipated and resolved in advance, or at least resolved quickly once they occur.

9.1.3 Scale-Up and Achievement of Similarity

Unit operations need to produce similar results and a similar quality product at each scale; historically the pharmaceutical industry has demonstrated this through process validation at each facility and operation thereafter with rigid limits on process parameters. Recent regulatory guidance [1] encourages adoption of "Quality by Design" and development of a "design space" in which the process may operate, with flexibility to move operating conditions around in this space without needing to obtain additional regulatory approval. The design space is proposed by the applicant and is intended to be applicable to the process at any scale. This implies a greater investment in developing process understanding during the development phase, balanced by flexibility, a reduced regulatory burden and opportunities for continuous improvement once the process is in full scale manufacturing.

There is some debate about how best to apply these principles while also satisfying business, process safety and environmental requirements and the underlying benefit of improved process understanding is underlined in this chapter. The operating window in which acceptable or similar results are obtained can be determined, demonstrated, and justified in terms of the chemical engineering rate processes involved in unit operations. For example, the design space for a chemical reaction whose intrinsic kinetics have been shown to be slow (compared to other rate processes involved) can be demonstrated to be scale independent, giving a flexible operating window that can be used at any scale. On the other hand, where specific equipment performance requirements exist, these can be framed in terms of the chemical engineering "rate constants" or time constants required by the process.

In the next section, we will see that these rate constants arise naturally in the chemical engineering rate equations, which provide a first principles basis for defining a scale-independent design space. The application of this approach is illustrated for several common types of chemical reaction and the same general principles and methods apply to other API synthesis operations. Examining the rate equations also leads to familiar scale-up rules.

9.1.4 Characterization of Reaction Systems

Guidance specific to each class of reaction is given below, but to avoid repetition for each class, the following general guidelines on reaction characterization are provided here. Guidelines on equipment characterization are given toward the end of the chapter.

Assuming that the solvent and reagents/catalyst have been selected, reactions may be characterized by a sequence of initial screening experiments to identify scale-dependent physical rate process limitations, followed by a more detailed kinetics study if necessary.

In all experiments aimed at characterizing a reaction, progress should be followed either by taking multiple samples or using an *in situ* analytical technique, not by relying on a single end point sample. If composition is analyzed using HPLC, determination of response factors will become more important as characterization proceeds. Other possibilities include infrared (IR), ultraviolet (UV), and Raman spectroscopy, or heat flow (Q_r).

Samples should be analyzed to indicate both reaction progress (e.g., product level, conversion or current yield) and product quality (e.g., impurity level). If either of these variables is sensitive to physical rates (e.g., agitation), the reaction is not limited by chemical kinetics, at least under typical processing conditions. There may be some scope to reduce the rates of the chemical reactions (relative to mixing) by operating at, for example, lower concentrations and/or temperatures; otherwise, if the process persists in its current form, the focus of characterization should be the physical capabilities of laboratory and larger scale equipment (see Section 9.6).

Under conditions where there is no effect of physical rate processes, a kinetics study may follow in which temperature, equivalents and/or other case-specific variables affecting the scale-independent chemical rates are varied experimentally. The choice of conditions and the sampling program for these experiments should be informed by the results of the previous experiments and an initial kinetic model [2] and in any event, multiple samples should be taken again to capture when the reactions are occurring as well as the final result.

Where possible, the parameters in equation 9.3, such as rate constant and activation energy should be fitted using a kinetic model to reliable results of experiments. Model development in this manner tends to be iterative, and the proposed reaction scheme may change a number of times before the model fits the data to an acceptable degree. Chemical knowledge that certain intermediate species are unlikely to exist for long may allow the rate constants of certain reactions to be set arbitrarily high so that the remaining parameters can be fitted with greater confidence. Model development should take place at the same time as experimentation and the design of remaining experiments may change as a result of indications from the model.

At the end of this phase, such a model may be used to find optimum conditions inside or outside the experimental region and to assist with definition of a scale-independent design space. Further experimentation should be focused on model verification, especially at the most forcing conditions over which it may be used.

9.2 BATCH PROCESSES WITH HOMOGENEOUS REACTIONS

Figure 9.1 shows a "process scheme" for a batch reaction. This consists of a simple representation of the (i) phases and (ii) rates involved in the operation. More or less detail may be added to such a schematic as required and this representation will be used below to introduce other operations.

9.2.1 Rate Equations

If for example the reaction involved is simply A + B → P, the rate equations for this system may be written as follows. The rate of change of concentration of A or B:

$$\frac{dC}{dt} = -r \tag{9.1}$$

where the rate expression on the right-hand side is of this or similar form:

$$r = kC_A^{\alpha}C_B^{\beta} \tag{9.2}$$

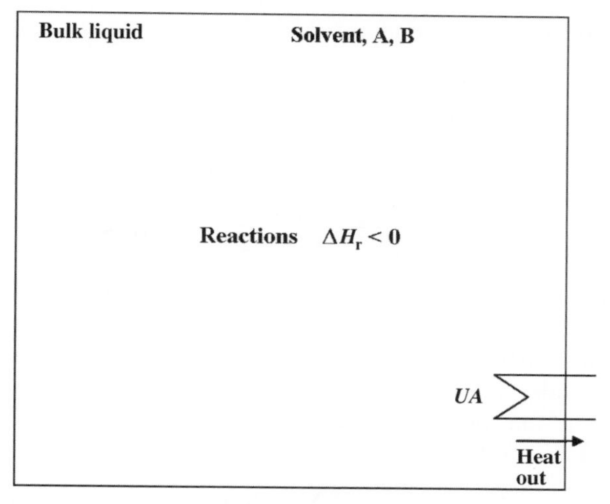

FIGURE 9.1 Process scheme for batch homogeneous reaction.

and the kinetic rate constant follows the Arrhenius relationship:

$$k = k_0 \exp\left(-\frac{E_A}{RT}\right) \tag{9.3}$$

More generally than (9.1), and when several reactions are occurring in the same mixture, the rate of change of concentration of any species in the mixture is given by

$$\frac{dC_i}{dt} = \sum_j -\nu_{ij} r_j \tag{9.4}$$

From equations 9.1 to 9.4, the rate constant depends on temperature and the rates of reaction depend on concentrations. This is the chemical engineering basis for the classical approach by chemists to study temperature and "equivalents" as two of the primary variables affecting the outcome of a reaction. In the above example, with 1 mol of A reacting with 1 mol of B, 1 equiv of A would mean 1 mol of A/mol of B; 2 equiv of A would be 2 mol of A/mol of B, and so on.

Integrating equations 9.4 in time, the final outcome (composition) of reaction will depend on the concentrations and temperature profiles followed during reaction and the time allowed for reaction. Temperature is often (but not always) held constant or nearly constant, in which case final concentrations at completion depend only on initial concentrations and initial temperature. Example 9.4 follows this behavior.

Temperature is an important influence and can change the relative rates of the reactions, depending on their activation energies. From a safety standpoint, temperature must be controlled to avoid thermal runaway and the batch reactor is not ideal for this purpose. The rate of change of reactor temperature is given by

$$\frac{dT}{dt} = -\frac{UA}{\rho VC_p}\Delta T_{LM} + \frac{Q_r}{\rho VC_p} \tag{9.5}$$

In which the log mean temperature difference is

$$\Delta T_{LM} = \frac{T_{jout} - T_{jin}}{\ln(T_{jout} - T)/(T_{jin} - T)} \cong (T - T_j) \tag{9.6}$$

The approximation on the right of equation 9.6 applies when the heat transfer fluid flow rate is high, that is, the heat transfer fluid temperature is almost constant between the jacket inlet and outlet.

The rate of heat evolution summed over all chemical reactions in the mixture is

$$Q_r = -\sum_j \Delta H_{rj} r_j V \tag{9.7}$$

Note that heat of reaction in equation 9.7 is the heat released per mole of reactant consumed when the stoichiometric coefficient of that reactant is 1 and is negative when the reaction is exothermic. Equation 9.5 shows that temperature can be controlled by balancing scale-dependent heat removal with scale-independent chemical kinetics. The former can be enhanced by increasing the surface area and/or increasing the temperature driving force, normally reducing the jacket inlet temperature. These measures can allow the reaction to run at the same temperature and concentration on scale, in the same reaction time as in the laboratory. There is an upper limit on the scale at which this will be feasible. On the other hand, reaction rate can be reduced to balance reduced heat removal by operating at lower concentration or lower temperature; the former reduces productivity; the latter could change the balance between desired and undesired reactions, lengthens reaction times, and also reduces the available temperature driving force for cooling.

For the common case where the temperature is held constant during reaction and the initial concentrations are the same at each scale, equation 9.5 with the left-hand side set to zero leads to the scale-up basis:

$$\frac{UA}{V}\Delta T_{LM} = \text{const} \tag{9.8}$$

Equations 9.4 and 9.8 taken together indicate that for scaling batch homogeneous reactions that are limited by chemical kinetics, a scale-independent design space can be expressed using temperature and equivalents as factors, as long as there is sufficient assurance that adequate heat transfer to maintain temperature control will be available at all scales. The latter condition amounts to a statement about the equipment capability at each location, specifically the heat removal capacity (in W/L) of the equipment operating at the required temperature.

All of the above effects can be calculated using mechanistic models in which equations 9.2–9.7 are solved by integration in time. In practice, batch homogeneous reactions may not be allowed to run to completion, because stopping the reaction earlier produces higher yield (yield peaks during the batch) and less impurity than waiting until the end. Similarly, batch homogeneous reactions may be run using temperature profiles, for example, charge at low temperature, heat to reaction temperature, then hold and possibly heat again. In these cases, the temperature profile followed by the reaction may be scale-dependent for the reasons given above, with, for example, longer heating ramps on scale; this will impact the relative rates of the reactions and could affect quality as well as rate.

If equation 9.3 is known for each reaction, these effects can be predicted easily, and operating conditions can be optimized to a high degree as shown in Example 9.4; good examples of this approach covering a variety of reaction types are available [3–7].

9.2.2 Characterization Tests for Batch Homogeneous Reactions

The general guidelines described in Section 9.1 should be followed with regard to experimental design.

The reaction should first be run in 2–3 otherwise identical (replicate) experiments in which stirrer speed is varied between, for example, 200 and 1000 rpm; four or more samples should be taken, biased toward the beginning of the reaction, for example, after 10, 30, 60 min and at the end.

Under conditions where there is no effect of physical rate processes, a kinetics study may follow in which temperature and equivalents are varied experimentally and the parameters in equation (9.3) fitted.

If agitation conditions affected the result of the initial screening experiments, the result of the reaction is influenced by the rate of mixing, or more specifically the rate of bulk blending of the vessel contents, known as "macromixing," This is unusual for a batch homogeneous reaction, but if it occurs, the scale-dependent macromixing time constant will be important as the outcome of reaction is influenced by this characteristic and not just chemical kinetics (see Section 9.6). A reaction such as this might be better run in fed-batch mode, with the fed reagent maintained at a low concentration. Alternatively, there may be some scope to reduce the rates of the chemical reactions (relative to mixing) by operating at lower concentrations and/or temperatures.

9.2.3 Achieving Similarity on Scale-Up

Achieving similarity on scale-up relies on having characterized both the reaction and the intended scale-up facility. Equations 9.2–9.7 provide the basis for a scale-independent design space.

When kinetics are rate limiting, other than agitation that is necessary to promote heat transfer, no particular additional agitation requirement exists for homogeneous reactions. As noted above, there may be an optimum temperature profile to adopt during reaction, and/or an optimum time at which to stop the reaction. These may be determined from a kinetic model, as in Example 9.4. A scale-independent design space may include factors such as equivalents, temperature, and reaction time.

When mixing is rate limiting, unusual for batch homogeneous reactions, agitation influences the outcome and similar macromixing time constants will be required at each scale; if the process persists in this form, the macromixing time constant should be a factor in design space definition.

9.3 MULTIPHASE BATCH PROCESSES WITH REACTIONS

Figure 9.2 shows a schematic of a multiphase batch reaction, in this case a hydrogenation.

These reactions are also referred to as "heterogeneous," "biphasic," or "slurry phase". Figure 9.2 shows the main liquid phase initially containing substrate, another phase (the headspace) containing reagent (hydrogen), reaction in the liquid phase between substrate and dissolving reagent, and removal of heat. (In Figure 9.2, the solid catalyst phase that is normally present for hydrogenation is omitted and the catalyst particles are assumed to follow the liquid. The headspace is continuously replenished with hydrogen as reaction proceeds.)

9.3.1 Rate Equations

Although relatively complex at first sight, this multiphase, chemically reacting, time-dependent problem can be broken into classical chemical engineering elements and rate processes just like the batch homogeneous reaction above. The rate equation for chemical species concentration in the liquid phase is given by combination of chemical kinetics with the film theory of scale-dependent mass transfer:

$$\frac{dC_i}{dt} = \sum_j -\nu_{ij} r_j + k_L a(C_i^* - C_i) \qquad (9.9)$$

Only the equations for components that transfer between phases contain the second term; for dissolving gases, solubility is given by Henry's Law:

$$C_i^* = \frac{p_i}{RTH_i} \qquad (9.10)$$

In other heterogeneous reactions, the solubility expression differs; for liquid–liquid (aqueous–organic) systems

$$C_i^* = \frac{C_{id}}{S_i} \qquad (9.11)$$

where S_i is the partition coefficient for component i between the phases. For solid–liquid systems

$$C_i^* \approx f(T, \text{composition}) \qquad (9.12)$$

where $f()$ is a function determined from phase equilibrium fundamentals and/or by experimental measurement (e.g., using NRTL equations or similar).

When the transferring component is a solvent (e.g., in a reactive distillation or solvent swap), the equilibrium is described using the Antoine or similar vapor pressure equation [8].

Examination of equation 9.9 indicates that when chemical kinetics are slow relative to mass transfer, the concentration of hydrogen (or any dissolving, reacting solute) will tend to

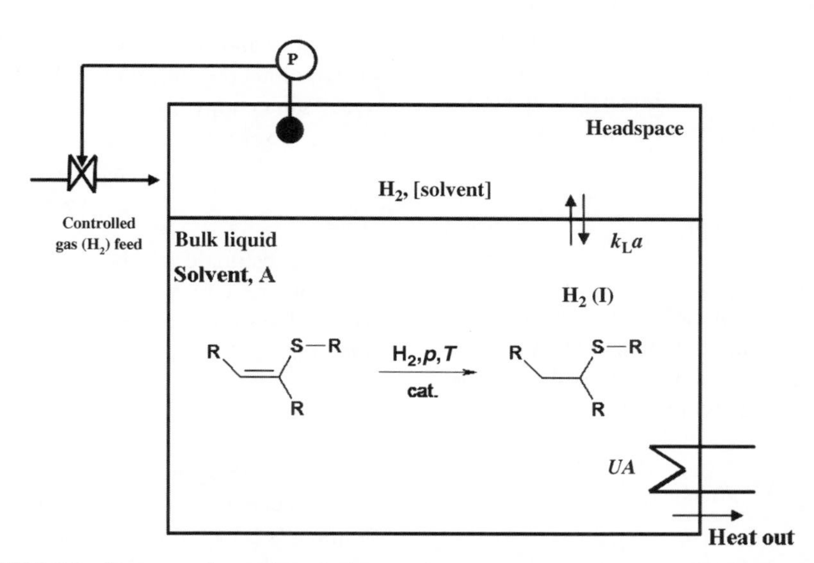

FIGURE 9.2 Process scheme for a hydrogenation, an example of batch multiphase reaction.

saturation with modest k_La values. On the other hand, if in equation 9.9, the kinetics of reactions consuming hydrogen are rapid compared to mass transfer, the concentration of hydrogen will tend toward zero during reaction. These two extremes of behavior change the outcome of reaction; slow mass transfer extends reaction time, may lead to increased impurities and can cause reaction to stall. The outcome of a batch multiphase reaction at constant temperature therefore depends on temperature, concentration, equivalents (including catalyst loading), pressure (if the solute is in the gas phase), and potentially k_La.

In reactions with three phases present, such as the additional solid catalyst phase in hydrogenation, that phase should be dispersed and its surface area made available for mass transfer in a similar way to the gas phase in the present example. Complete suspension of a solid phase is usually sufficient to prevent mass transfer to or from that phase being rate limiting; when the solid particles have the same characteristics at both scales, complete suspension is approximately equivalent to maintaining constant k_La for those particles (see Section 9.6).

For multiphase reactions, the rate equation for solution phase temperature is

$$\frac{dT}{dt} = -\frac{UA}{\rho V C_p}\Delta T_{LM} + \frac{Q_r}{\rho V C_p} - \frac{k_L a V (C_i^* - C_i)\Delta H_{\text{transfer}}}{\rho V C_p} \tag{9.13}$$

Equation 9.13 contains one additional term compared to equation 9.5, to account for any heat effect associated with mass transfer of dissolving material between phases; this effect is often negligible, but can be significant when solvent is vaporized (in distillation) or when a large quantity of solute is transferred quickly (e.g., when crystals come out of solution suddenly). The heat flow signature from the chemical reactions will be determined by kinetics when mass transfer is rapid and by mass transfer when kinetics are rapid. Neglecting the final term in equation 9.13, the scale-up relationship for maintaining constant temperature when kinetics limit is identical to the corresponding equation for homogeneous reactions:

$$\frac{UA}{V}\Delta T_{\text{LM}} = \text{const} \tag{9.14}$$

Equation 9.14 when applied to pressure reactions assumes constant pressure, that is, constant solubility. The scale-up relationship when kinetics are rapid relative to mass transfer can be obtained by noting that in this case, the rate of dissolution of the limiting solute dominates the rates of the main reactions:

$$Q_r \cong -\sum_j \Delta H_{rj} k_L a C^*_{\text{solute}} V \tag{9.15}$$

Substituting (9.15) into (9.13) leads to

$$\frac{UA}{k_L a V}\Delta T_{\text{LM}} = \text{const} \tag{9.16}$$

The above result may also be obtained by writing equation 9.13 in dimensionless form using $t^* = k_L a t$ and $T^* = T/T_0$, neglecting the last term and substituting equation 9.15 for Q_r:

$$\frac{dT^*}{dt^*} = -\frac{UA\Delta T_{\text{LM}}}{k_L a V \rho C_p T_0} + \frac{-\sum_j \Delta H_{rj} C^*_{\text{solute}}}{\rho C_p T_0} \tag{9.17}$$

The second term on the right-hand side is constant when pressure and initial temperature are fixed and when the left-hand side is set equal to zero (constant temperature), equation 9.16 is obtained.

Equations 9.16 and 9.17 indicate that when the solution concentration of solute is close to zero, the required rate of heat removal is directly proportional to the rate of mass transfer, that is, higher k_La implies the need for greater heat removal.

An additional constraint exists for pressure reactions, such as hydrogenation, which cannot be taken for granted in practice. The ability to maintain a desired or constant pressure in the headspace depends on the balance between supply of fresh gas to the headspace and removal of gas by mass transfer:

$$\frac{dp}{dt} = \frac{RT}{V_{\text{head}}}(N_{H_2} - k_L a V (C_i^* - C_i)) \tag{9.18}$$

Equation 9.18 shows that rapid mass transfer may cause the headspace pressure to reduce, while a temperature rise may cause it to increase. To maintain a constant pressure on scale-up requires adequate temperature control and adequate gas supply:

$$N_{H_2,\max} > k_L a V (C_i^* - C_i) \tag{9.19}$$

The right-hand side of equation 9.19 requires knowledge of the solution concentration of the dissolved gas. A conservative estimate may be made by setting the dissolved concentration to zero, leading to

$$\frac{N_{H_2,\max}}{k_L a V C_i^*} > 1 \tag{9.20}$$

A crude estimate of the required flow rate to maintain constant pressure may also be made from the number of moles of gas required for reaction and the intended reaction time, that is

$$N_{H_2,\max} \gg \frac{\text{moles\_required}}{\text{reaction\_time}} \tag{9.21}$$

Similar to homogeneous batch reactions, equations 9.9–9.13 and 9.18 can be easily incorporated into dynamic mechanistic models for multiphase reactions and the unknown parameters (e.g., rate constants, activation energies) regressed against experimental data. The effects of changing concentrations and equivalents (including catalyst loading), temperatures, pressures and the effects of mass transfer, heat transfer, and gas supply limitations can then be predicted. In many hydrogenations, for example, neither pressure nor temperature are maintained constant and the profiles they follow may be scale dependent; even the sequencing of nitrogen and hydrogen purges before reaction can affect the result. The impact of these changes can be predicted using classical chemical engineering rate equations with appropriate reaction and equipment characterization. Examples 9.1 and 9.2 illustrate this approach.

9.3.2 Characterization Tests for Multiphase Reactions

The general guidelines described in Section 9.1 should be followed with regard to experimental design.

For gas–liquid, gas–liquid–solid, liquid–liquid, and liquid–liquid–solid systems, the reaction should first be run in 2–3 otherwise identical (replicate) experiments in which stirrer speed is varied between, for example, 200 and 1000 rpm; four or more samples of the reacting phase should be taken during reaction, biased toward the beginning of the reaction, for example, after 10, 30, 60 minutes and at the end. With some reactions, additional profiling may be possible by following variables such as hydrogen uptake, pressure, and both pot and jacket temperature. The purpose of changing stirrer speed is to change the mass transfer contact area between the phases. In each of these experiments any solids present should be well suspended.

To check for mass transfer limitation due to dissolving solids, experiments should be run at an impeller speed that guarantees suspension, but either the particle size or the mass of solid reagent should be varied.

Under conditions where there is no effect of physical rate processes, a kinetics study may follow in which any of temperature, equivalents, pressure, catalyst loading, and/or other case-specific variables affecting the scale-independent chemical rates are varied experimentally.

If agitation conditions affected the result of the initial screening experiments, the outcome of the reaction is influenced by the rate of mass transfer. This is frequently the case for heterogeneous reactions and characterization of the scale-dependent equipment characteristics will be important. There may be some scope to reduce the rates of the chemical reactions (relative to mixing) by operating at lower concentrations, pressures, catalyst loadings, and/or temperatures, but this will also reduce volumetric productivity.

9.3.3 Achieving Similarity on Scale-Up

For batch multiphase reactions, achieving similarity on scale-up relies on having characterized both the reaction and the intended scale-up facility. Equations 9.9–9.13 and 9.18 provide the basis for a scale-independent design space.

When kinetics are slow (relative to mass transfer), adequate agitation is necessary to create a dispersion and to promote heat transfer; adequate gas supply is required for pressure reactions. Justifying the scale independence of the design space in this case, reduces to demonstrating that pressure and temperature can be maintained in the same range on scale as in the laboratory and that agitation is sufficient to disperse the phases.

When kinetics are fast, $k_{L}a$ should be included explicitly to make the design space scale independent.

For hydrogenations and other catalyzed pressure reactions, a design space could therefore contain concentrations, equivalents, temperature, pressure, catalyst loading, and time as factors; when kinetics are fast, $k_{L}a$ should also be a factor.

As noted above, there may be an optimum temperature or pressure profile to adopt during the reaction, and/or an optimum time at which to stop the reaction. These may be determined from a kinetic model, as used in the example described below [9]. Many examples of kinetic modeling of multiphase reactions are available, including a methanethiol producing reaction [10] and for other hydrogenations [11].

When kinetic models are used in reverse, to work backward from a desired end result to determine the possible operating conditions (or factors) that would produce this result, multiple acceptable combinations of factor settings may be found. For example, very poor mass transfer (low $k_{L}a$) can be partially compensated for by operating at higher pressure [12]; the effects of low $k_{L}a$ may also be partially mitigated by operating at lower concentrations of starting materials or catalyst. These combinations may be difficult to express in a design space definition, but can be found easily using a mechanistic model and justified if the model has been properly verified against experimental data. This has led to discussion of the use of models to more flexibly capture the definition of a design space and verify that a given set of operating conditions lie within it [13].

EXAMPLE 9.1

Figures 9.3 and 9.4 show examples of hydrogenation reactions with mass transfer limitations [9]. Figure 9.3 illustrates hydrogen heat flow and thermal conversion for a nitro reduction in a lab reactor, with reaction exhibiting apparent zeroth-order kinetics (constant rate, linear conversion profile) and taking over 6 h. The mass transfer capability ($k_{L}a$) of the lab reactor was known from previous equipment characterization to be rate limiting. The results of scale-up to the pilot plant are shown in Figure 9.4 (temperature, hydrogen uptake, and substrate level). The reaction time reduced to 1.5 h due to superior mass transfer in the pilot plant reactor.

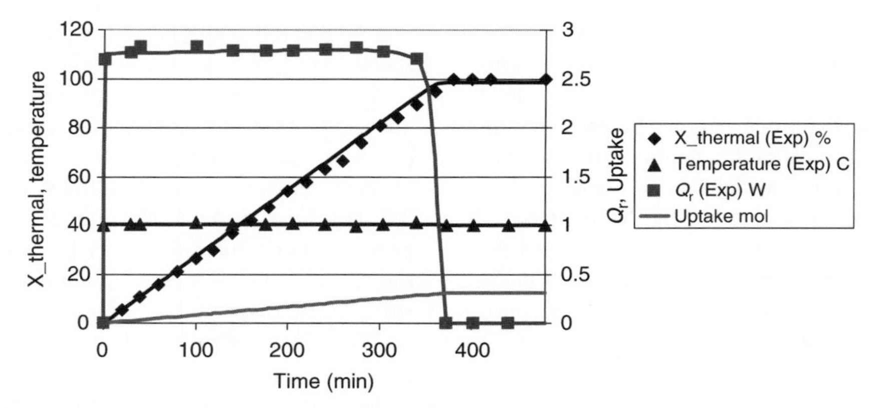

FIGURE 9.3 Laboratory results (discrete points) of this hydrogenation (reduction of a nitro group, Example 9.1) indicated a 6 h reaction time. Modeling (continuous curves) and vessel characterization revealed this was due to severe gas–liquid mass transport limitations ($k_L a$) in the laboratory equipment and that the process could perform much better.

This result was predicted and expected based on the reactor characteristics.

EXAMPLE 9.2

Figure 9.5 shows the reaction profile (hydrogen uptake) for a hydrogenation with dissolving solid substrate. On scale, the reaction had a long "tail" due to the bimodal size distribution of the substrate coming from the previous step. Larger particles dissolve more slowly than smaller particles and this caused the extended reaction time.

9.4 FED-BATCH PROCESSES WITH REACTIONS

Figure 9.6 shows a process scheme for a homogeneous fed-batch reaction.

These operations are common both in the laboratory and on scale and are also known as "exothermic additions" or "semi-batch reactions." Figure 9.6 shows the main reaction liquid phase initially containing solvent and one or more reactants, with another reactant initially in the feed tank. Liquid is added from the feed tank, reaction begins and the resulting heat is removed.

9.4.1 Rate Equations

The rate of change of concentration of each species in a homogeneous fed-batch reaction is

$$\frac{dC_i}{dt} = \sum_j -\nu_{ij} r_j + \frac{Q_f}{V}(C_{if} - C_i) \qquad (9.22)$$

Compared to equation 9.4, the additional term on the right-hand side represents addition of feed and the accompanying dilution effect.

When the kinetic rates of the reactions are slow compared to the rate of addition, reaction takes place after the addition and the result is very similar to a homogeneous batch

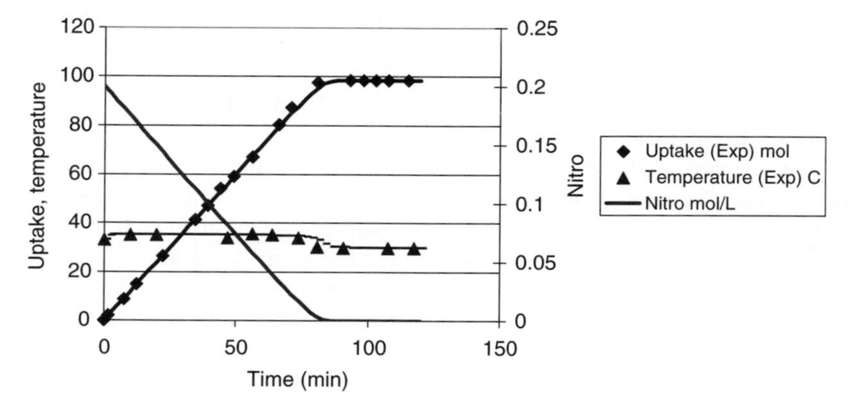

FIGURE 9.4 A reaction time of 1.5 h was predicted (curves) and achieved (symbols) in the pilot plant for the same reaction as in Figure 9.3 (Example 9.1). The mass transfer characteristics of the pilot plant reactor were superior to those of the lab reactor.

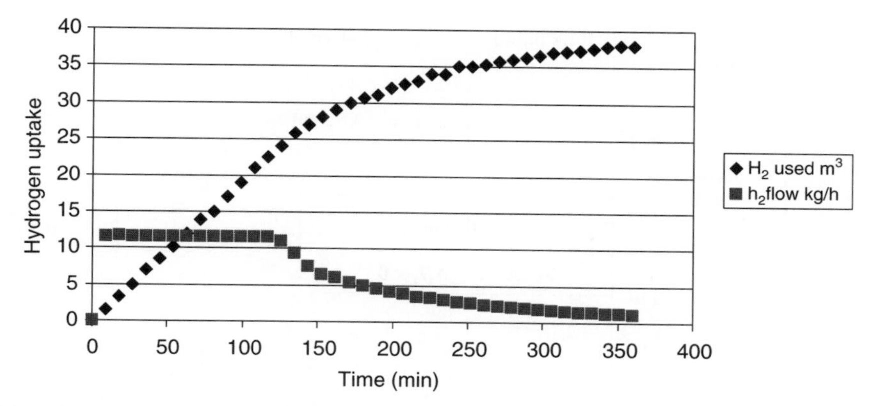

FIGURE 9.5 On scale-up of another hydrogenation (Example 9.2), reaction times extended to over 6 h, due to the appearance of a bimodal particle size distribution in the solid substrate, with about 50% of the solids mass having a much larger particle size; this reduces the rate of reaction, especially once the smaller solids have dissolved.

reaction, described above. When the kinetic rates are comparable to the addition rate, a significant amount of reaction occurs during the feed, which normally changes the outcome of reaction compared to the batch case. This is because the concentration of the fed reactant remains low during the addition.

When the kinetic rates are much faster than the addition rate, the concentration of the fed reactant tends to zero and in this case the reaction is highly localized around the feed region. Equation 9.22 in this case no longer fully describes the rates of change and a more detailed analysis involving phenomena known as mesomixing and micromixing is required [14, 15] for an accurate mathematical description. A process scheme for this situation is shown in Figure 9.7, with the notable addition of a "feed zone" or reaction zone in which most of the reaction takes place. The size of the feed zone and its composition are determined by micromixing and mesomixing rates. The outcome of the reaction is predominantly determined by these scale-dependent rates rather than by chemical kinetics.

The rates of mesomixing and micromixing can be thought of as somewhat analogous to the rate of mass transfer between two phases, which arises with multiphase systems as described above; for this reason, fed-batch reactions with rapid chemical kinetics are often referred to as "pseudohomogeneous," implying that the feed zone is like a separate phase.

The characteristics in Figure 9.7 arise in a significant fraction of applications, as reactions are often deliberately engineered to run with fast kinetics (e.g., by operating at high concentration, catalyst loading, and temperatures) because the heat output from such systems can then be halted in the event of a cooling failure by stopping the addition. Reactions such as these are also known as "feed controlled" and "dosing controlled." The existence of the feed zone does not necessarily change the outcome of the reaction signifi-

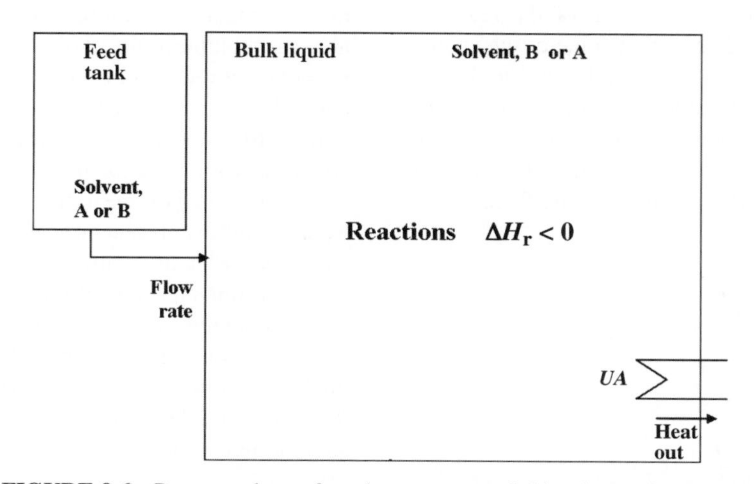

FIGURE 9.6 Process scheme for a homogeneous fed-batch reaction system.

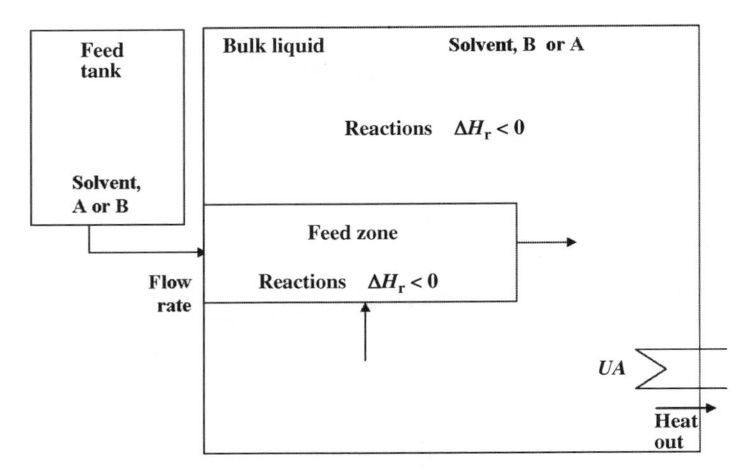

FIGURE 9.7 Process scheme required for a fed-batch reaction with rapid chemical kinetics.

cantly; this only occurs if the reaction scheme and kinetics are such that the concentration and temperature "hot spot" that exists in this zone causes the reactions to follow a different path compared to what they would follow if diluted fully throughout the bulk of the vessel. In a significant minority of cases, the feed zone has a major bearing on product quality and yield. Appropriate characterization experiments as described below can be used to determine whether a given reaction is subject to these effects.

Considering equation 9.22, when reaction kinetics are slow, such that all or almost all of the reaction takes place after the addition is complete, the solution tends toward that for a batch reaction with the same volume, as described by equation 9.4. Integrating equation 9.22 in time, the final outcome (composition) of the reaction will then depend on the concentrations and temperature profiles followed during the reaction and the time allowed for the reaction. Temperature is often (but not always) held constant or nearly constant, so final concentrations at completion depend only on initial concentrations and initial temperature. Example 9.4 follows this behavior.

When a significant amount, but not all of the reaction occurs during feeding, this indicates that some of the chemical kinetic rates are comparable to the addition rate. Equation 9.22 is more informative for this case when examined in dimensionless form; for a second-order reaction between A and B, using $C_i^* = C_i/C_{A0}$ and $t^* = kC_{A0}t$, the rate equation for the dimensionless A or B concentration is, from equation 9.22:

$$\frac{dC_i^*}{dt^*} = -\frac{C_A C_B}{C_{A0}^2} + \frac{Q_f}{VC_{A0}k}\frac{(C_{if}-C_i)}{C_{A0}} \tag{9.23}$$

Or, equivalently

$$\frac{dC_i^*}{dt^*} = -C_A^* C_B^* + \frac{Q_f}{VC_{A0}k}(C_{if}^*-C_i^*) \tag{9.24}$$

Defining the initial volume ratio of bulk solution to feed solution as

$$\alpha = \frac{V_{bulk}}{V_f} \tag{9.25}$$

and noting that feed time is given by

$$t_f = \frac{V_f}{Q_f} \tag{9.26}$$

Then, at the start of the addition[1]

$$\frac{Q_f}{V} = \frac{Q_f}{V_f}\frac{V_f}{V_{bulk}} = \frac{1}{t_f\alpha} \tag{9.27}$$

Substituting (9.27) into (9.24) leads to

$$\frac{dC_i^*}{dt^*} = -C_A^* C_B^* + \frac{1}{t_f\alpha C_{A0}k}(C_{if}^*-C_i^*) \tag{9.28}$$

Equation 9.28 shows that the outcome (or dimensionless concentration profile) of fed-batch reaction when a significant amount of reaction occurs during the addition depends on the initial concentrations in the bulk and feed vessels, the rate constant (temperature dependent), the addition time, and also the volume ratio of the reagents. This latter point is important as the volume ratio is independent of the number of equivalents, that is, reactions run at the same equivalents, temperature, and addition time may produce different results when the volume ratios mixed are different.

Apart from volume ratio, a further degree of freedom exists in fed-batch reactions that is not present in the batch case: the order of addition; when the order of addition is reversed, this can radically change the solution concentrations compared to the forward addition. Reverse addition is worth considering and testing when feasible and safe. From a quality point of view, order of addition should follow from the reaction scheme (expressed in equation 9.4 using the stoichiometric matrix ν_{ij}). For example, in the following case,

[1] By the end of the addition, the denominator contains $\alpha + 1$ instead of α

to maximize product it is more favorable to add A to B:

$$A + B \rightarrow \text{Product}$$
$$A + \text{Product} \rightarrow \text{Impurity}$$

In the next case, it is more favorable to add B to A:

$$A + B \rightarrow \text{Product}$$
$$B + \text{Product} \rightarrow \text{Impurity}$$

When essentially the entire reaction occurs during the addition, equations 9.22 or 9.28 are no longer directly applicable, as the reaction zone will be localized and some reactions will run to completion before the feed has been diluted into the bulk contents. In this case, the dominant effect on the composition of the reaction mixture is the rate of localized mixing near the addition point, a physical phenomenon somewhat analogous to mass transfer in the multiphase example above. The degree to which the reaction exhibits this characteristic may depend on the order of addition. The final outcome of this type of reaction will depend on the time constants for mesomixing or micromixing, whichever is slower and therefore rate determining; these influence the size and residence time of the feed zone, which in turn determines how much of the chemistry takes place in the concentration and temperature hot spot created by the feed. Both meso- and micromixing time constants are affected by the addition rate, the addition location, the diameter of the addition nozzle, the agitator configuration, and the agitator rotational speed [14]. Factors directly affecting the intrinsic kinetics have less influence in this scenario, such as reaction temperature.

The rate of change of temperature in fed-batch reaction systems is given by

$$\frac{dT}{dt} = -\frac{UA}{\rho V C_p}\Delta T_{LM} + \frac{Q_r}{\rho V C_p} + \frac{\rho_f C_{pf} Q_f}{\rho V C_p}(T_f - T) - \frac{\rho_f Q_f \Delta H_m}{\rho V C_p} \tag{9.29}$$

Compared to equation 9.5 for batch reactions, the two additional terms on the right-hand side represent contribution of sensible heat by the fed material (e.g., when warmer than the bulk) and any associated thermodynamic heat of mixing that also accompanies the addition. The heat flow signature from the chemical reactions will be determined by kinetics (kinetically controlled) when reaction occurs after the addition, by the rate of addition when the reactions occur entirely during the addition and by both kinetics and the rate of addition, when a significant amount of reaction, but not all, occurs during the addition.

From equation 9.29 with $Q_f = 0$, the scale-up relationship for maintaining constant temperature when kinetics limit is identical to the corresponding equation for homogeneous reactions

$$\frac{UA}{V}\Delta T_{LM} = \text{const} \tag{9.30}$$

At the other extreme, when kinetics are instantaneous relative to addition, the rate of heat output is primarily dependent on the rate of addition of limiting fed reactant:

$$Q_r \cong -\sum_j \Delta H_{rj} Q_f C_{fed} \tag{9.31}$$

Substitution of equation 9.31 into equation 9.29 and setting the right-hand side equal to zero leads to the widely used scale-up relationship:

$$\frac{UA}{Q_f}\Delta T_{LM} = \text{const} \tag{9.32}$$

Noting that the addition volumetric flow rate is the addition volume divided by the addition time, equation 9.32 may be rearranged in terms of feed time as

$$t_f = \text{const}\frac{V_f}{UA\Delta T_{LM}} \tag{9.33}$$

A more complete version of equation 9.33 may be obtained by writing equation 9.29 in dimensionless form using $T^* = T/T_0$ and $t^* = UAt/\rho V C_p$, then substituting equation 9.31 for Q_r and equation 9.27 for Q_f giving at the start of the addition:

$$\frac{dT^*}{dt^*} = -\Delta T^*_{LM} + \frac{V}{UA}\frac{1}{\alpha t_f}\frac{-\sum_j \Delta H_{rj} C_{fed}}{T_0} \tag{9.34}$$

Rearranging (9.34) with the right-hand side set equal to zero leads to

$$\frac{\alpha t_f UA\Delta T_{LM}}{-\sum_j \Delta H_{rj} C_{fed} V} = 1 \tag{9.35}$$

From equation 9.35, a longer feed time implies that a lower UA or ΔT may be tolerated; a higher volume ratio is most likely to be combined with a more concentrated feed and these effects cancel out.

From equations 9.33 and 9.35 it is clear that the addition time on larger scale will often need to be longer than on smaller scale in order to operate the reaction at the same constant temperature as at lab scale; the magnitude of the increase in addition time will depend on the degree to which the temperature driving force can be increased on scale-up.

Similarly to homogeneous batch reactions, the equations above can be easily incorporated into dynamic mechanistic models for fed-batch reactions and unknown parameters (e.g., rate constants, activation energies) regressed against experimental data. The effects of changing order of addition, concentrations (including catalyst loading), temperatures, volume ratio, and addition time and the effects of mixing and heat transfer can be predicted. In some fed-batch reactions, for example, temperature is deliberately profiled (e.g., using a heating ramp to drive reaction to completion) and the profile may be scale dependent. The impact of these changes can be predicted using classical chemical engineering rate equations. There may be an optimum time at which to stop the reactions, for example, after product concentration has peaked and before impurities are able to form. To understand these effects and take advantage of the opportunities, a detailed mechanistic model of the reaction is valuable. Several examples illustrating this approach are given below and more are included in Refs 7 and 11.

9.4.2 Characterization Tests for Fed-Batch Reactions

For fed-batch homogeneous reactions, if feasible the reaction should first be run using forward and reverse additions. If the desired process involves addition of A to B, then run this case followed by an otherwise identical case with B added to A. If the results of these experiments differ significantly, reaction occurs during the addition and is at least in the intermediate regime described above. Temperature or heat flow profiles when monitored during these experiments may indicate the degree of addition-controlled behavior and the amount of reaction happening during the feed versus after the feed. Likewise, *in situ* or sample data following the concentration of the bulk reactant both during and after the addition will be informative in this regard.

When a dosing-controlled reaction is suspected, a further two replicate experiments in which only stirrer speed is varied between, for example, 200 and 1000 rpm will further indicate the influence of agitation on the rate and outcome of reaction.

In each of the above initial screening experiments, four or more samples should be taken, for example, after 25%, 50%, 75%, and 100% of the feed have been added. With some reactions, additional profiling may be possible by following variables such as the rate of gas evolution and both pot and jacket temperature.

Under conditions where there is no effect of agitation on rate or quality, a kinetics study may follow in which addition time, volume ratio, temperature, equivalents, and, for example, catalyst loading are varied experimentally. The choice of conditions and the sampling program for these experiments should be informed by the results of the previous experiments and an initial kinetic model [2] and in any event, multiple samples should be taken again to capture when the reactions are occurring as well as the final result.

If rate or quality are sensitive to agitation, the reaction is not limited by chemical kinetics at current conditions but by localized mixing, at least under typical agitation conditions. The main focus of process design and characterization should be the mixing capabilities of laboratory and larger scale equipment (see Section 9.6). There may be some scope to reduce the rates of the chemical reactions (relative to mixing) by operating at lower concentrations and/or temperatures.

9.4.3 Achieving Similarity on Scale-Up

Achieving similarity on scale-up relies on having characterized both the reaction and the intended scale-up facility. Equations 9.22 and 9.29 provide the foundation for a scale-independent design space.

When kinetics are rate limiting, other than agitation that is necessary to promote heat transfer, no particular additional agitation requirement exists. As noted above, there may be an optimum temperature profile to adopt during reaction, and/or an optimum time at which to stop the reaction. These may be determined from a kinetic model, as in Example 9.4. A scale-independent design space may include factors such as equivalents, temperature, and reaction time.

When mixing is rate limiting (reaction is dosing-controlled and mixing affects quality), agitation influences the outcome and similar mesomixing or micromixing time constants will be required at each scale; in order to achieve a scale-independent design space, the mixing time constants should be factors in design space definition. Additional factors may include order of addition, addition time, volume ratio, equivalents, and temperature.

In the intermediate regime where a significant portion but not all of reaction occurs during the addition, reaction will tend to become more dosing-controlled on scale-up, as the addition time is increased to compensate for lower heat transfer area per unit volume. A scale-independent design space may include factors such as order of addition, addition time, volume ratio, equivalents, temperature (profile), and reaction time.

EXAMPLE 9.3

A reaction whose heat flow profile showed typical very different behavior when run with forward and reverse additions is shown below [16] in Figure 9.8. In this case, the forward addition produced the correct material but had an undesirable sudden exotherm at the end of the addition. The reverse addition was dosing controlled, as shown in Figure 9.9.

EXAMPLE 9.4

Many examples of design space mapping using chemical engineering rate equation-based models are available [5, 13, 17] and the results of one case are presented below [18, 19].

The production of *epi*-pleuromutilin in a homogeneous liquid-phase reaction was believed to follow the pathway shown in Figure 9.10.

To characterize and scale this reaction, six fed-batch experiments were carried out, in which acid was added to the other reactants and two factors were varied, temperature and equivalents of acid. The progress of reaction was fol-

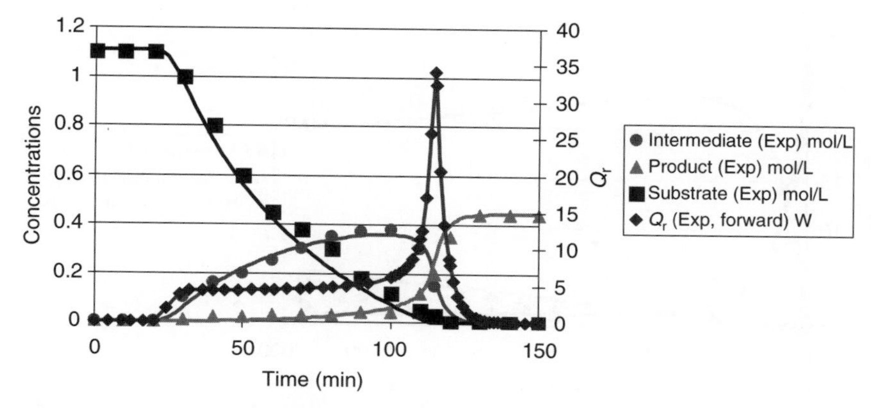

FIGURE 9.8 Concentration and heat flow profiles for Example 9.3: symbols are measured data and curves are model predictions. A large spike in heat flow occurred at the end of the forward addition.

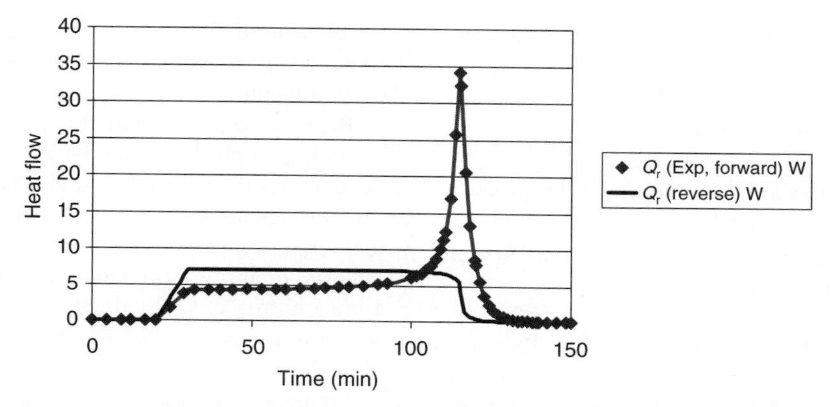

FIGURE 9.9 Heat flow profiles for Example 9.3, comparing forward and reverse addition: symbols are measured data (forward addition) and curves are model predictions (forward and reverse addition). The heat flow curve indicates dosing-controlled conditions when the addition is reversed.

lowed by taking multiple samples; a typical reaction profile is shown below (Figure 9.11).

There is a small temperature rise during the addition of acid and the majority of the reaction takes place after the addition is complete. The latter signals that the results of these experiments are determined by chemical kinetics. The fact that product increases to a peak and then reduces while the alkene increases lends support to the overall reaction scheme shown. The raw HPLC area percent data for each sample were converted with the aid of relative response factors to molar quantities for modeling. Kinetic fitting to the results of all six experiments led to the set of reactions and fitted parameters in Figure 9.12; the dissociation rate constant for sulfuric acid was not fitted but set to a high value.

During the kinetic fitting process, close attention was paid to the statistics relating to each parameter, each measured response and the model as a whole. In particular, confidence intervals on parameters of tens of percent or less were targeted. When comparing possible alternative reaction schemes/hypotheses, a low final sum of squares (quantifying the lack of fit) combined with a high value of the model discrimination statistic (the Akaike information criteri-

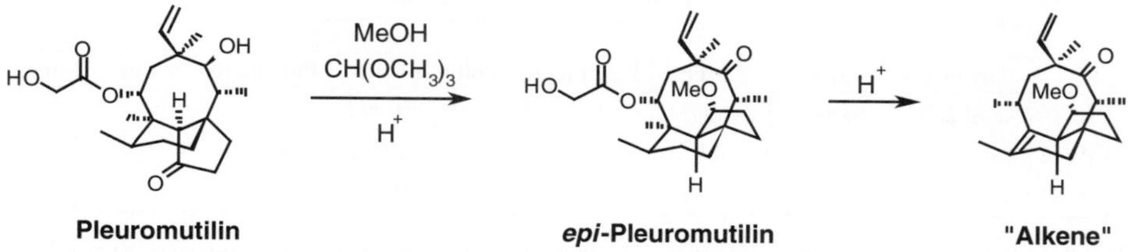

FIGURE 9.10 Overall reaction scheme for Example 9.4 (*epi*-pleuromutilin).

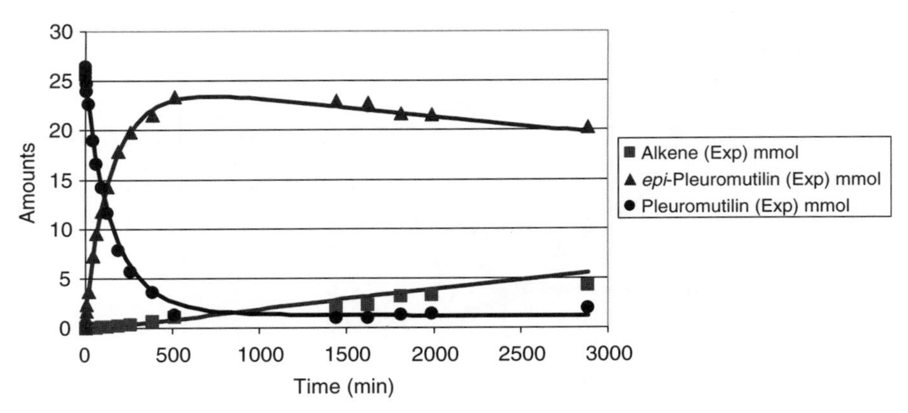

FIGURE 9.11 Typical reaction profiles for Example 9.4. Symbols are measured data (from HPLC area percent) and curves are model predictions.

on [20]) was achieved using the parameter values in Figure 9.12.

To leverage the model for process design and scale-up, two additional responses were defined:

- ProductMaxTime, that is, the time when the product concentration reaches its peak.
- QuenchWindow, that is, the time after that until 1% of product is lost to impurity (alkene).

The specification on these responses was that Product-MaxTime should be less than 8 h, for productivity and operational reasons, while QuenchWindow should be more than 2 h, to allow adequate time for sampling, analysis, and quenching the reaction before more than 1% of product was lost to alkene.

The design space was defined as the region of temperature and acid levels that simultaneous met both criteria. A series of 440 virtual experiments were carried out using the model to produce the response surfaces shown in Figures 9.13 and 9.14. The region of overlap of acceptable values of these responses is shown in Figure 9.15.

The original publication in 2007 [19] was based on 30 virtual simulations to which a polynomial equation was fitted for plotting; the above figures were reproduced in 2009 by running 440 virtual experiments in the same factor space with no interpolation.

Because kinetics limit this reaction at or near current operating conditions and because most of the reaction occurs after the addition, a scale-independent design space for these responses may be constructed using two variables: temperature and equivalents of acid. When all other input material quantities are held constant and adequate temperature control is maintained on scale, the result of this reaction at any point in this space would be independent of scale.

Figure 9.15 is quite typical in that there appear to be two "edges of failure" when the responses are overlapped [21]. One response relates to the quality attribute (amount of alkene in this case) and another relates in this case to a business attribute (reaction time). In a strictly QbD context, Figure 9.15 therefore has only one edge of failure: if the quench window is too short, the alkene level will exceed its limit; if the reaction time is too long, this will not directly impact quality but will slow productivity.

The design space shown in Figure 9.15 is approximately 10 mmol wide and 5° high. However if a rectangular region of proven acceptable ranges had to be defined inside this space,

Rxn1					H_2SO_4	⟶	2 H^+	+	SO_4^-	
Rxn2	TMOF	+	H^+	+	Pleuromutilin	⟶	*epi*-Pleuromutilin	+	By-product 1	+ H^+
Rxn3			H^+	+	*epi*-Pleuromutilin	⟶	Alkene	+	By-product 2	

	Parameter	Value	Units	Parameter	Value	Units	Parameter	Value	Units
Rxn1	*k*> at 25°C	10000	1/s	E_a>	0.00	kJ/mol			
Rxn2	*k*> at 25°C	1.4E-05	L^2/mol^2 s	E_a>	67.10	kJ/mol	K_{eq}	6.70	-
Rxn3	*k*> at 25°C	5.7E-07	L/mol s	E_a>	51.85	kJ/mol			

FIGURE 9.12 Final reaction scheme, rate constants and activation energies for Example 9.4 (*epi*-pleuromutilin).

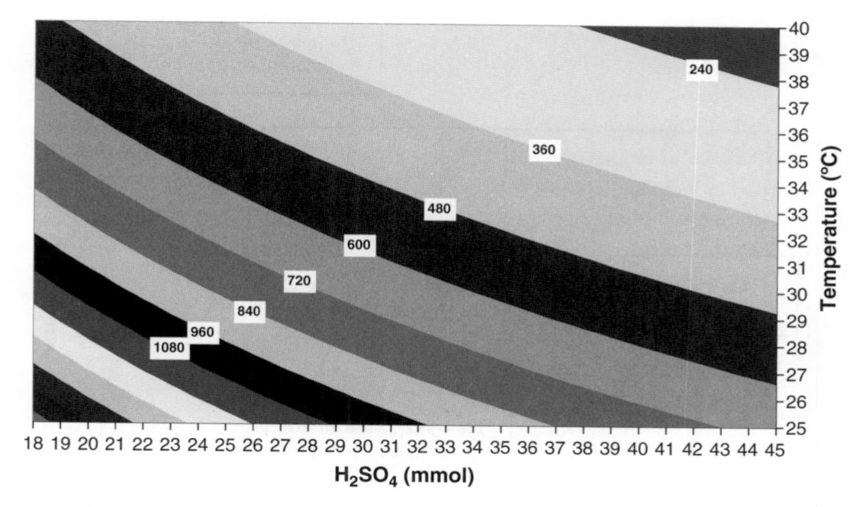

FIGURE 9.13 Response surface of ProductMaxTime (minutes) versus initial amount of acid and temperature (1 equiv = 26.5 mmol) for Example 9.4. Results produced using mechanistic model to run 440 virtual experiments.

the maximum width would be 7 mmol and the maximum height 2–3°. This illustrates how the design space has the potential to offer greater flexibility than rigidly defined proven acceptable ranges.

On the other hand, Figure 9.15 remains quite restrictive, in that all other factors (such as substrate concentration) have to be held constant for it to apply. This is one of the reasons why a more dynamic design space definition based on the full mechanistic model, rather than one set of response surfaces at otherwise fixed conditions, has been advocated by industry [13]. The model verification statistics associated with this example are discussed at the end of this chapter.

9.5 APPLICATION TO CONTINUOUS FLOW SYSTEMS

Continuous flow reactor systems are of interest in API synthesis for the potential benefits they offer in certain cases relative to batch or fed-batch reactors. These benefits may include greater process safety due to reduced holdup of hazardous material and the ability to quench a reaction more rapidly, improved containment of materials with low exposure limits, reductions in capital and/or operating costs, and volumetric productivity. In general, knowledge of the rate of reaction is even more important for design of continuous

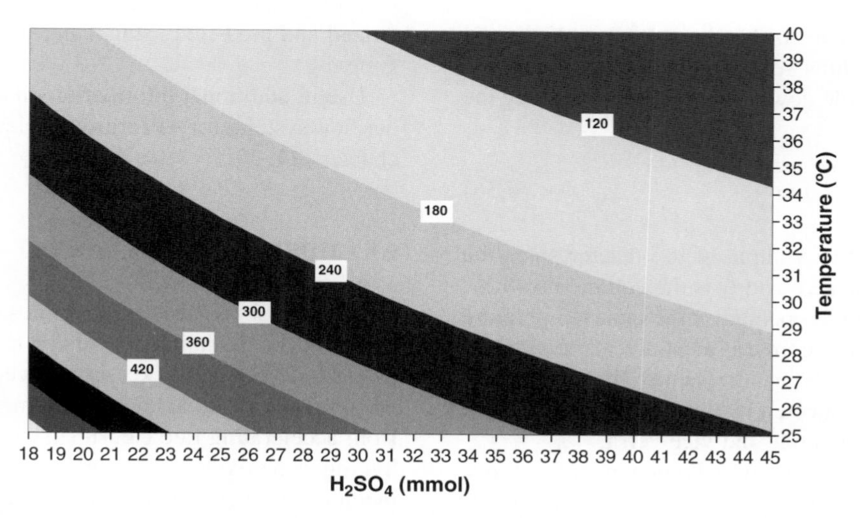

FIGURE 9.14 Response surface of QuenchWindow (minutes) versus initial amount of acid and temperature (1 equiv = 26.5 mmol) for Example 9.4. Results produced using mechanistic model to run 440 virtual experiments.

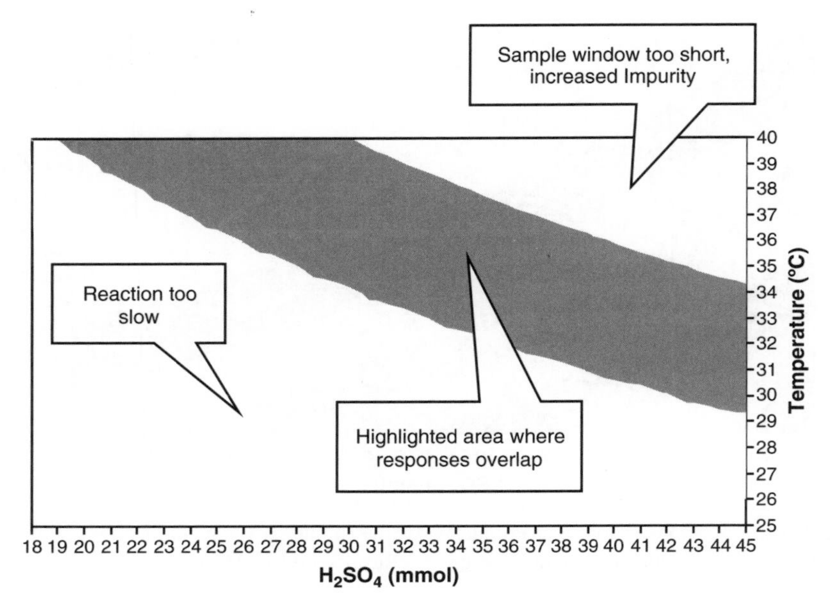

FIGURE 9.15 Region of overlapping response surfaces when ProductMaxTime and QuenchWindow are within specification, versus initial amount of acid and temperature (1 equiv = 26.5 mmol) for Example 9.4.

systems than for batch, as the residence time of the reactor is finite and it may be difficult to stop reaction sufficiently rapidly if problems arise. The economics of low volume production in a dedicated unit may limit the scope for continuous operation. Process issues specific to each application may also present challenges when reactors are operated for longer than typical batch cycle times, such as catalyst deactivation and difficulties handling slurries.

9.5.1 Plug Flow

Equations presented above for batch reactions may be applied directly to plug flow reactions after noting that the independent time variable now applies to position along the reactor [22]:

$$t = \frac{V}{Q} \tag{9.36}$$

Here, V is the cumulative volume of the reactor since the material entered at time zero and Q is the volume flow rate. When a plug flow reactor is operated at the same temperature and with the same residence time as a batch reactor with that cycle time, the end results are the same. This convenient result means that data collected in batch mode may be used for design in continuous mode and vice versa.

Plug flow reactors may be used for homogeneous systems, liquid–liquid and gas–liquid systems, with appropriate attention to ensuring the phases are dispersed and separated when required and that both phases travel along the tube without accumulation.

Reaction characterization experiments follow the same logic as the corresponding problems in batch systems described above [14]. Equipment performance characterization in terms of heat transfer is always required; mixing and mass transfer characterization will be required when kinetics are rapid and in multiphase systems [14, 23]. Design spaces are expressed by taking into account similar factors to batch reactors.

Plug flow reactors with multiple addition points along the reactor length are analogous to fed-batch systems with multiple sequential additions and likewise may be characterized and predicted using a similar approach to fed-batch systems.

Useful additional information on application of continuous flow systems for API are available in the references to this chapter [24, 25].

EXAMPLE 9.5

Figure 9.16 shows heat flow profiles measured in fed-batch mode for the oxidation conversion of a tertiary alcohol to a primary alcohol using excess hydrogen peroxide catalyzed in the presence of an acidic environment [26]. It was known from accelerating rate calorimetry (ARC) experiments that hydrogen peroxide in the presence of the acidic solution medium was thermally unstable at temperatures above 20°C. Carefully controlled fed-batch experiments were performed at two temperatures (5 and 15°C) with the same addition time; a reaction scheme and chemical kinetics were fitted to

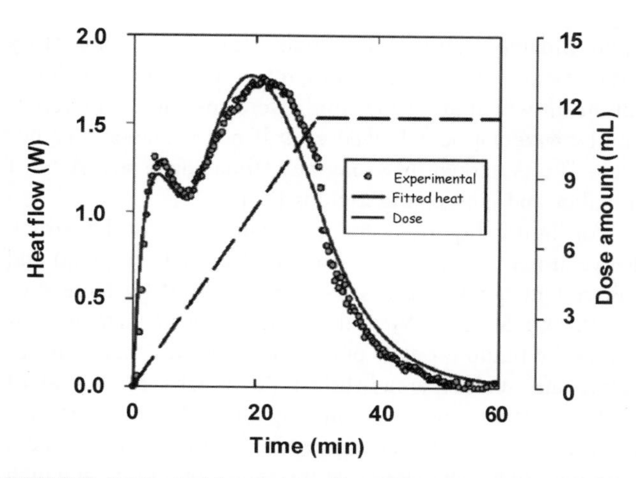

FIGURE 9.16 Fed-batch heat flow data (discrete points) for peroxide oxidation reaction in Example 9.5, showing double peak in heat output. Kinetic model results (curves) were used to design an intrinsically safer continuous reactor system.

the resulting heat flow data. Although not shown here, there was generally good agreement between model and experiments for both the heat and independently measured concentration profiles. An intrinsically safer continuous flow reactor was designed using these kinetics to produce kilograms of material at both pilot and manufacturing scales.

9.5.2 CSTRs in Series

Plug flow behavior may be approximated using a cascade of stirred reactors in series [23]; this can also ease the problem of running slurry reactions in continuous mode. The approach to reaction characterization and design for these systems is described above. Equipment characterization in terms of heat transfer is always required and batch characterization tests may be used for this purpose (see below).

Models developed for process prediction in batch or fed-batch systems may be reused to design continuous stirred tank reactor systems by adding appropriate feeding and removal between reactors in the cascade [27]. When the number of reactors is large, a simpler plug flow model will give equivalent results.

9.6 EQUIPMENT CHARACTERIZATION AND ASSESSMENT

Equipment performance characteristics play a major role in most operations in API synthesis including each of the reaction types described above. This section reviews methods to evaluate some of the key characteristics for reactions, quenches, extractions, distillation/solvent swap, and crystallization.

Equipment performance can in general be quantified using either or both of the following methods:

- Characterization tests, in which purpose-designed tests are carried out on the equipment and responses such as pressure, temperature, or concentration are monitored versus time.
- Assessment calculations, in which empirical correlations, often based on dimensionless groups, are used to estimate equipment performance as a function of dimensions, geometry, fluid properties, and the intended operating volume and recipe.

Pharma API development laboratories, kilo labs, pilot plants and full-scale manufacturing plants are dominated by relatively standard, multipurpose equipment into which each new process is fitted; this means that performance data can be reused many times over once generated. To support quality by design (QbD), equipment performance characteristics are stored in equipment databases that allow users at any location to retrieve performance data for equipment at both their own and other locations.

9.6.1 Heat Transfer

The product of overall heat transfer coefficient U and wetted area A appear throughout this chapter as equipment characteristics required for process scale-up.

The product UA is best evaluated using a solvent test in the intended process vessel, to which solvent is charged and the fill level and agitator speed are set to those of the intended process. The batch is heated and/or cooled over the range of pot and jacket temperatures required by the process. Pot and jacket inlet temperature are monitored continuously; jacket outlet temperature is monitored if available. In the absence of reaction and any other heat effects, equation 9.5 may be used to fit UA for both heating and cooling.

UA may also be estimated from chemical engineering correlations developed from measurements in similar types of equipment [28]. The coefficients in such heat transfer correlations may vary depending on the precise configuration and in some cases, new coefficients may be required to describe unusual vessel configurations. In mature applications of equipment characterization, those coefficients are stored with the equipment configuration data in an equipment database for future reuse.

9.6.2 Mass Transfer

The product of mass transfer coefficient, k_L and interfacial area, a (per unit volume) appear in this chapter whenever interphase transfer is involved.

The product k_La is best evaluated using a test in the intended process vessel, in which mass transfer is either the only phenomenon occurring, or is the rate-limiting phenomenon.

In one such test for headspace fed gas–liquid reactions, typically hydrogenations [29], solvent is charged and the fill level set to that of the intended process. The headspace is evacuated and then pressurized with gas and the agitator is turned on, with the speed increasing quickly to the intended process speed. Headspace pressure and both headspace and liquid temperature are monitored continuously. In the absence of reaction and any other phase transfer effects, equation 9.17 may be used to fit both gas–liquid (surface gassing) k_La and solubility. Alternatively, a reaction that consumes the dissolving gas may be run under conditions of high catalyst loading, such that mass transfer is rate limiting; in this case, equation 9.9 may be used to fit k_La to an indicator of reaction progress, such as hydrogen uptake. Note that (i) the k_La for surface gassing is very sensitive to the submergence of the top impeller [30] and (ii) sparging gas may be of little value in a "dead-end" system (i.e., unless gas is continuously removed from the headspace) and mass transfer due to surface gassing is more important.

The depressurization test described above must be done carefully in order to provide useful data; for example, the headspace and liquid must be at the same temperatures before the agitator is turned on, to avoid a pressure recovery due to heating of the gas by the liquid.

For solid–liquid and liquid–liquid reactions, analogous characterization tests in which the uptake of solute is monitored (e.g., by sampling the liquid) may be used; or a known fast chemical reaction may be run under conditions in which dissolution of solute is rate limiting. Equation 9.9 again provides the basis for fitting k_La to the monitored profiles.

Alternatively, empirical estimates may be made using chemical engineering correlations; for these, the molecular diffusion coefficient of the solute in the solvent is required, which limits applicability. A feature that makes solid–liquid systems somewhat easier to predict is that the wetted area does not change with stirrer speed once the solids are suspended; that is unless the particles are broken as a result of agitation. For liquid–liquid systems, the effect of minor components on the droplet size and the resulting surface area can be very significant, making accurate estimates difficult.

In all operations involving contact between multiple phases, a certain minimum level of agitation is required (even if kinetics are slow) in order to ensure that a dispersion of one phase exists in the other. In solid–liquid and liquid–liquid systems, the minimum stirrer speeds at which such a dispersion is created may be estimated with greater certainty than k_La.

For solid–liquid systems, this level of agitation, N_{JS}—the agitator speed that just suspends the solid particles, exposes the full surface area and k_La on scale may be taken as approximately equivalent to that applied in a laboratory experiment using the same raw materials at the same conditions in which all of the solids were suspended; therefore similar k_La can be achieved even if neither laboratory nor plant k_La is known. N_{JS} may be estimated for a variety of impeller and tank configurations [31].

For liquid–liquid systems, a balance must be struck between mass transfer rate (favored by small droplets) and subsequent quick phase separation (favored by large droplets). Operating at N_{JD}—the agitator speed that just disperses the liquid droplets of one phase in the other, exposes significant surface area while reducing the likelihood of forming a stable emulsion; once again the k_La on scale may be taken as similar to that applied in a laboratory experiment with the same raw material at the same conditions in which the liquids were just dispersed; therefore similar k_La can be achieved even if neither laboratory nor plant k_La is known. N_{JD} may be estimated for a variety of impeller and tank configurations [32, 35].

If solid–liquid or liquid–liquid k_La is a factor in definition of a design space, the proximity of agitator speed to N_{JS} or N_{JD}, respectively, is a reasonable surrogate variable for k_La; for example, a dimensionless agitator speed that is scale independent could be defined as a factor:

$$N^* = \frac{N}{N_{JS}} \tag{9.37}$$

N^* in equation 9.37 might for example in a particular application need to be above 1.0 in order to avoid mass transfer limitations.

9.6.3 Liquid Mixing

In batch homogeneous reactions there is the possibility of reaction rate limitation by macromixing, at very low agitation rates. A macromixing time constant of 30–60 s should eliminate this dependence and correlations are available to estimate this factor [33]. The liquid mixing characteristics relevant for fed-batch reactions are meso- and micromixing time constants; these appear in equations that describe the rate of localized mixing near the addition point in fed-batch and continuous reactors and may be rate determining when kinetics are fast, that is for pseudohomogeneous, dosing-controlled reactions as described above.

The time constants for meso- and micromixing may be characterized by running test reactions with known kinetics in the target equipment [14, 34]. The outcome of these reactions is mixing-sensitive under certain conditions and varies with factors such as impeller speed, addition rate and addition location. When the product distribution or selectivity from each experiment is combined with a mathematical model based on Figure 9.7, the time constants for meso- or micromixing for each set of conditions may be obtained.

Alternatively, formulas are available for these time constants [14] and if the spatial distribution of the local rate of energy dissipation, ε, is known, the time constants may be obtained from these. Specialized tools such as computational fluid dynamics or laser anemometry may be used to estimate the spatial distribution of ε. As a general rule, ε is a multiple (e.g., 10–100) of the vessel-averaged power input per unit mass (W/kg) near the impeller and a fraction (e.g., 1–10%) of the average near the liquid surface.

9.6.4 Phase Separation

Phase separation times, for example, after a liquid–liquid reaction or extraction, are longer on scale that in the laboratory [35]. As a rough indicator, the separation timescales with the liquid depth, so a separation time of 1 min in the laboratory can easily extend to 1 h on scale. Excessive separation times can be avoided by avoiding the formation of very fine droplets or stable emulsions; operating a liquid–liquid reaction at or near N_{JD} as described above represents a good balance between high mass transfer area and short phase separation time.

9.6.5 Gas Disengagement

When reactions evolve gas, the rate of gas evolution scales with the reaction volume, but the ability of the liquid surface to allow the gas to escape scales with the cross-sectional area of the vessel. The maximum velocity of gas escape through the liquid surface is approximately 0.1 m/s [35], allowing the maximum volume flow rate of gas evolution to be estimated:

$$Q_{gas} = 0.1\frac{\pi T^2}{4} \tag{9.38}$$

This can be converted into a molar rate using the ideal gas law:

$$N_{gas} = 0.1\frac{\pi T^2}{4}\frac{p}{RT} \tag{9.39}$$

If the rate of reaction exceeds the maximum rate of gas evolution, significant foaming will occur and in some cases material will be lost from the reaction vessel. This problem becomes more likely on scale, as the maximum rate of gas evolution per unit volume reduces

$$\frac{N_{gas}}{V} \approx 0.1\frac{1}{H}\frac{p}{RT} \tag{9.40}$$

Equation 9.40 indicates that the volumetric rate of reactions evolving gases may need to be reduced on scale (e.g., by slower feeding or otherwise) in order not to exceed the limitations imposed by the equipment.

For example, in a vessel of 2 L nominal volume with diameter 0.115 m, filled to a level of 1 L at 20°C, the maximum volumetric rate of a reaction evolving 1 mol of gas (e.g., CO_2) per mole of product without batch swelling is $0.1 \times (1/0.104) \times (1.01325 \times 10^5/8.314 \times (273.15 + 20)) = 39.97\ \text{mol/m}^3\ \text{s} \approx 0.04\ \text{mol/L s}$. The same process running in a 2000 L vessel with diameter 1.5 m at 1600 L batch size is limited to a rate of $0.1 \times (1/1)(1.01325 \times 10^5/8.314 \times (273.15 + 20)) = 4.16\ \text{mol/m}^3\ \text{s} \approx 0.004\ \text{mol/L s}$, that is, 10 times slower, to avoid batch swelling.

9.7 MODEL VERIFICATION STATISTICS

9.7.1 Parameter Uncertainty

All equations that are derived originally from experimental data have a degree of uncertainty associated with their calculations or predictions. This is true for calibration curves, regression lines, response surfaces generated by design of experiments, chemical engineering correlations for estimating equipment performance, models based on ordinary differential equations (such as many of the examples shown above) and those using partial differential equations (such as computational fluid dynamics, in which turbulence models, "wall laws" and other model parameters are ultimately based on experimental data). Unless these latter models involving differential equations are integrated over a sufficiently fine "grid," their uncertainty is further increased by lack of precision.

Once accepted as useful, models are often used without taking this uncertainty into account, but experienced practitioners will always allow a factor of safety to compensate for a margin of error, even if the size of that error is not known. When the original data and the model are both available, it is possible to use statistical methods to quantify the uncertainty level. This makes the potential deficiencies of the model more evident and explicit and may also focus further experimentation on reducing those uncertainties.

More details about how to calculate uncertainty are given elsewhere [36] and Figure 9.17 illustrates the typical situation for a linear model in which the slope and intercept have been fitted.

Confidence bands define an envelope within which there is a certain confidence level (typically 95%) of the true location of the best-fit line. Prediction bands (or intervals) define a wider envelope within which there is a certain confidence level (e.g., 95%) that all data points/observations will lie. The width of prediction bands relative to the model prediction indicates the likely relative error of the model predictions; this also reflects underlying variability or lack of reproducibility in the experimental data. For typical linear models such as that in Figure 9.17, the bandwidths are at a minimum at the center of the experimental data and increase in either direction from the center. This tallies with the belief that models are most applicable near the conditions where the

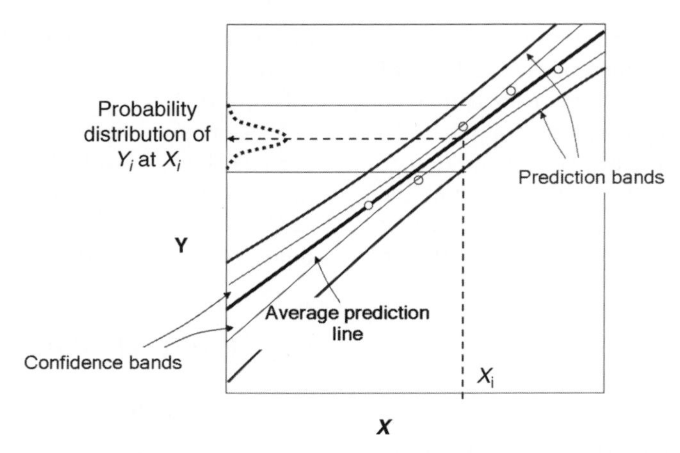

FIGURE 9.17 Schematic of confidence and prediction bands for a linear model in which the slope and intercept of the best-fit line have been fitted to data (symbols).

experiments were run and become less reliable at more extreme conditions.

Any such model should in the first instance be based on reliable, reproducible data; there should also be sufficient data to avoid "overfitting," that is, the number of observations should be significantly greater than the number of model parameters fitted. The degree to which such a model can be said to be verified depends on the width of its prediction bands compared to the accuracy needed in the intended application.

For example, if a model predicts an impurity level of 0.1%, with 95% prediction bandwidths of 0.05%, one could state with a high degree of confidence that the impurity level will be less than 1%; more formally, the probability of impurity exceeding 1% is almost zero, $p(\text{Impurity} > 1\%) \approx 0$. The same model is not sufficiently accurate to state with the same degree of confidence that the impurity level will be less than 0.2%; that is, $p(\text{Impurity} > 0.2\%) > 0$. However if the 95% prediction bandwidth was 0.01%, the model would be suitable for more confident predictions at lower impurity levels. There is therefore an element of fitness for purpose when judging whether a model has been verified sufficiently to apply it in a given situation.

Similarly, if the correlation used to calculate the stirrer speed required to suspend solids has a stated accuracy of 20% and the predicted $N_{JS} = 80$ rpm, this level of verification is sufficient to say that, for example, 40 rpm is too little—$p(\text{Suspended}) \approx 0$ and 120 rpm is too much—$p(\text{Suspended}) \approx 1$, but not whether 75 rpm is sufficient—$0 < p(\text{Suspended}) < 1$.

Returning to Example 9.4, confidence and prediction bands may be calculated for a dynamic model based on chemical engineering rate equations and typical results are shown in Figures 9.18 and 9.19.

Comparing Figures 9.18 and 9.19, the relative width of prediction bands is greater for alkene than for product. This

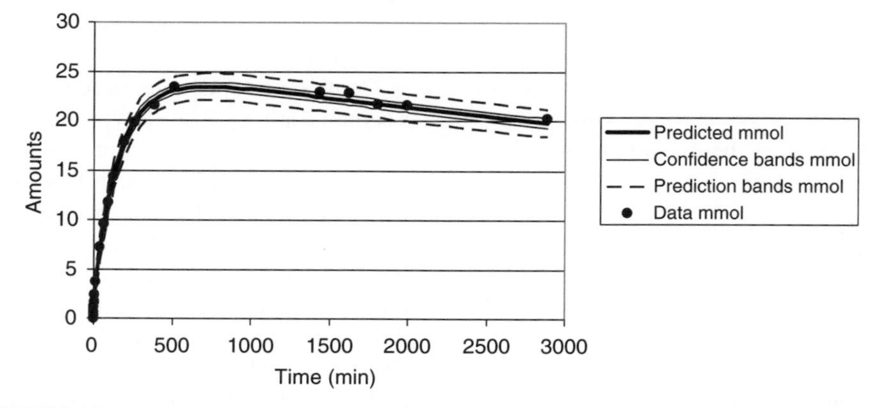

FIGURE 9.18 Model predictions with confidence and prediction bands/limits for the product profile in Example 9.4, compared with experimental measurements of that profile. All measured data lie within the envelope defined by the 95% prediction bands.

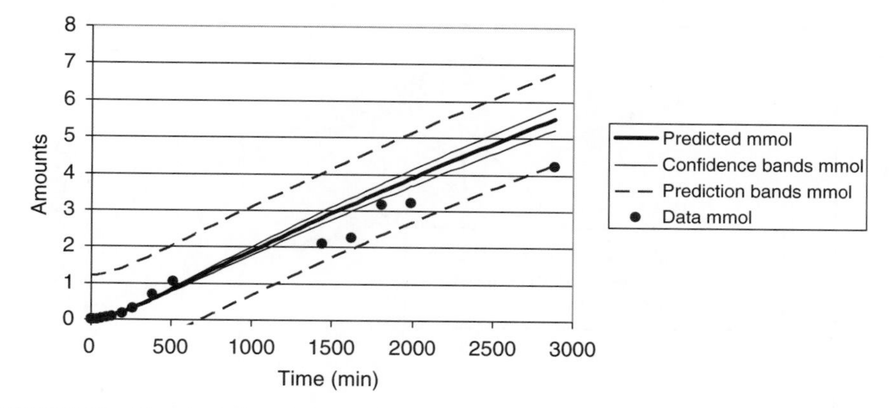

FIGURE 9.19 Model predictions with confidence and prediction bands/limits for the alkene profile in Example 9.4, compared with experimental measurements of that profile. All measured data lie within the envelope defined by the 95% prediction bands.

reflects greater uncertainty in the alkene measurements and predictions.

In quality by design work, the criticality of factors is sometimes judged according to their proximity to factor settings that define the "edge of failure," for example, crossing an impurity limit. Prediction bands should be taken into account when judging criticality in this way, as these will show that failure will occasionally occur at factor settings that are nearer to intended operating conditions.

9.7.2 Implications for Design Space Definition

The existence of uncertainty means that formal probability statements may be required to properly define a design space; while this implies some additional work, it has the benefit of quantifying the degree of assurance (or risk) that exists in relation to how the process will perform. In general, this approach will lead to design spaces that are smaller and more conservative than when uncertainty is ignored and which maximize the probability that the product quality will be in specification. Most popular statistical software packages at present do not take uncertainty into account in this way [37].

The relevant probability can be calculated in a variety of ways and Figure 9.20 illustrates application to Example 9.4. Recall from Figure 9.19 that the relative uncertainty of alkene is greater than that for product in this case; this leads to a relatively broader region of conditions in which the impurity has a significant probability of failing to meet specification. On the other hand, uncertainty in product predictions is low, leading to a narrow region separating success and failure. To produce Figure 9.20, the relative uncertainty in quench window was taken to be proportional to that of alkene and

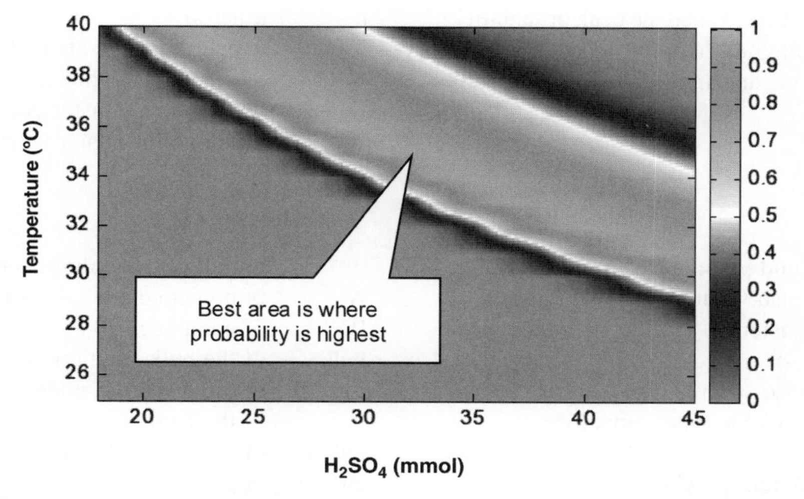

FIGURE 9.20 Response surface showing joint probability that responses ProductMaxTime and QuenchWindow will be in specification for Example 9.4.

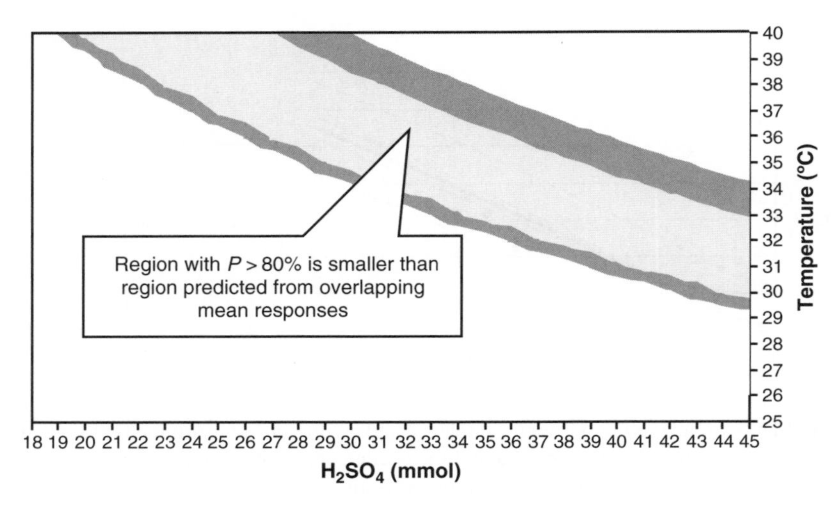

FIGURE 9.21 Comparison of design space for Example 9.4 defined using average responses and that defined using probability of success >80%.

the relative uncertainty of ProductMaxTime proportional to that of product.

Figure 9.20 illustrates the regions of the factor space with the highest probability of success; these are the most favorable regions in which to operate the process. In the highlighted area of Figure 9.20, the probability of success exceeds 80%. Figure 9.21 compares the design space defined on this basis ($p > 80\%$) with that obtained by overlapping the average responses; as expected, taking account of uncertainty reduces the size of the design space.

The above results show that for quality by design purposes, uncertainty in data or responses predicted by any model should be explicitly taken into account in defining the design space; probability is the natural way to do this. Uncertainty will tend to shrink the design space away from the edges of the area where average responses overlap, in line with good engineering practice. When the peak probability if far from 100%, this highlights the need to obtain greater process understanding, or an improved process, before proceeding.

SYMBOLS

a	area per unit liquid volume (m^2/m^3)
α	reaction order; also volume ratio (–)
A	area for heat transfer (m^2)
β	reaction order (–)
C	concentration; also specific heat capacity (mol/m^3 (concentration); J/kg K (heat capacity))
D	diameter (m)
ΔT	temperature difference (°C or K)
ΔH	heat of reaction (J/mol)
E	energy (J/mol)
H	Henry's law constant; also liquid depth (– (Henry constant); m (liquid depth)
k	rate constant; also mass transfer coefficient (m^3/mol s (kinetics); m/s (mass transfer))
N	impeller rotational speed; also hydrogen supply rate (1/s and rpm (impeller speed); mol/s (hydrogen supply rate))
ν	stoichiometric coefficient (–)
p	pressure; also probability (Pa (pressure); – (probability))
Q	volumetric flow rate; also heat flow rate (m^3/s (flow rate); W (heat flow rate))
r	rate of reaction (mol/m^3 s)
ρ	density (kg/m^3)
R	gas constant (J/mol K)
S	partition coefficient (–)
t	time (s)
T	temperature; also tank diameter (K or °C (temperature); m (tank diameter))
U	overall heat transfer coefficient (W/m^2 K)
V	liquid volume (m^3)

Subscripts

0	initial; also at infinite temperature
A	of component A; also of activation
B	of component B
bulk	of the bulk solution
d	of the dispersed phase
f	of the feed
fed	of limiting fed reactant
gas	of gas
head	of the headspace
H_2	of hydrogen

i	of the ith component
ij	of the ith component in the jth reaction
j	at the inlet to the cooling jacket or coil; also of the jth reaction
JD	just dispersed
JS	just suspended
L	in the liquid phase
LM	logarithmic mean
m	of mixing
max	maximum
p	at constant pressure
r	of reaction
solute	of limiting dissolving solute
transfer	of reaction

Superscripts

* at saturation, that is, equilibrium between the phases; also dimensionless

REFERENCES

1. International Conference on Harmonization (ICH) Q8, Q9, Q10 guidance documents. Available at http://www.ich.org/cache/compo/363-272-1.html.
2. Place D. Using DynoChem to Determine a Suitable Sampling Endpoint for Reaction Analysis in a DoE, *DynoChem User Meeting 2009*. Available at http://dcresources.scale-up.com/Publications/Default.aspx.
3. Vickery T. Scale-Up From RC1 and ARC Safety Tests Using DynoChem, *DynoChem User Meeting 2007*. Available at http://dcresources.scale-up.com/Publications/Default.aspx.
4. Bright R, Dale DJ, Dunn PJ, Hussain F, Kang Y, Mason C, Mitchell JC, Snowden MJ. Identification of new catalysts to promote imidazolide couplings and optimisation of reaction conditions using kinetic modelling. *Org. Proc. Res. Dev.* 2004; 8(6):1054–1058.
5. Jorgensen M. Modeling is the Easy Part!: Getting the Right Data and Getting the Data Right is the Challenging Part! *DynoChem User Meeting 2009*. Available at http://dcresources.scale-up.com/Publications/Default.aspx.
6. Eyley S. Why Study a Synthetically Useless Reaction?—Unravelling Sulphonate Ester Formation Using DynoChem, *DynoChem User Meeting 2009*. Available at http://dcresources.scale-up.com/Publications/Default.aspx.
7. Niemeier J. Using DynoChem to Scale Up Data from Various Calorimeters, *DynoChem User Meeting 2009*. Available at http://dcresources.scale-up.com/Publications/Default.aspx.
8. Nyrop J. Development of a High Performance, Company Specific DynoChem Front-End, *DynoChem User Meeting 2009*. Available at http://dcresources.scale-up.com/Publications/Default.aspx.
9. Hannon J, Hearn S, Brechtelsbauer C. Characterisation of the Scalability of Hydrogenation Reactions, *Scientific Update Scale-up Conference 2002*. Available at http://dcresources.scale-up.com/Publications/Default.aspx.
10. Remy B, Brueggemeier S, Marchut A, Lyngberg O, Lin D, Hobson L. Modeling-based approach towards on-scale implementation of a methanethiol-emitting reaction. *Org. Process Res. Dev.* 2008; 12: 381–391.
11. Richter S, Allian A. Process Safety Testing and Process Modeling in the PSL Using DynoChem, *3rd US Pharmaceutical Process Safety Forum*. Available at http://dcresources.scale-up.com/Publications/Default.aspx.
12. Hannon J. *Design Space for a Synthesis Reaction, Part 3*. Available at http://designspace-qbd.blogspot.com/search/label/Mass%20Transfer.
13. Stonestreet P, Hodnett N, Squires B, Escott R. Roles of Mechanistic and Empirical Modeling/DOE in Achieving Quality by Design, *DynoChem User Meeting 2009*. Available at http://dcresources.scale-up.com/Publications/Default.aspx.
14. Baldyga J, Bourne JR, Hearn SJ, Interaction between chemical reactions and mixing on various scales. *Chem. Eng. Sci.* 1997; 52(4):457–466.
15. Hoffmann W. DynoChem and Homogeneous Mixing: An Example, *DynoChem user Meeting 2007*. Available at http://dcresources.scale-up.com/Publications/Default.aspx.
16. Hoffmann W. *Workshop on Basics of Kinetics/Application of Software Tools, Introduction to the Determination of Kinetic Parameters*, Mettler Toledo RXE Forum, Lucerne, 2001.
17. Hallow D, Mudryk B, Braem A, Burt J, Rossano L, Tummala S. Application of DynoChem® Reaction Modeling to Quality by Design, *DynoChem User Meeting 2009*, Abstract available at http://dcresources.scaleup.com/Publications/Default.aspx.
18. Wertman J. GSK Approach to Enhancing Process Understanding Using DynoChem: Reaction Kinetics Examples, *DynoChem User Meeting 2007*. Available at http://dcresources.scaleup.com/Publications/Default.aspx.
19. Hannon J. Quality by Design for Drug Substance Scale-up, *Presented at Scientific Update Conference on Scale-up of Chemical Processes*, Vancouver, Canada, July 2009. Available at http://dcresources.scaleup.com/Publications/Default.aspx.
20. Akaike H. A new look at the statistical model identification. *IEEE Trans. Automatic Control* 1974; 19(6):716–723.
21. am Ende D, Bronk KS, Mustakis J, O'Connor G, Santa Maria CL, Nosal R, Watson TJN. API quality by design example from the torcetrapib manufacturing process. *J. Pharm. Innov.* 2007; 2: 71–86.
22. Levenspiel O. *Chemical Reaction Engineering*, 3rd Edition, John Wiley & Sons, 1999.
23. Zhu ZM, Hannon J, Green A. Use of high intensity gas–liquid mixers as reactors. *Chem. Eng. Sci.* 1992; 47: 2847–2852.
24. am Ende D. Lean and Green, The Value of API Process Design, *DynoChem User Meeting 2009*. Available at http://dcresources.scaleup.com/Publications/Default.aspx.

25. Roberge D. Microreactor Technology: Reaction Design and Scale-Up Concepts, *Scientific Update Scale-up Conference*, 2009.
26. Chan SH, Wang SSY, Kiang S. *Modeling and Alternative Reactor Design for a Highly Exothermic Reactive System, AIChE Annual Meeting*, 2005.
27. Erdman D. DynoChem Modelling of 3 Continuous Stirred Tank Reactors, *DynoChem User Meeting 2009*. Available at http://dcresources.scaleup.com/Publications/Default.aspx.
28. Kayode Coker A. *Modeling of Chemical Kinetics and Reactor Design*, Butterworth-Heinemann, 2001.
29. Machado R. *Fundamentals of Mass Transfer and Kinetics for the Hydrogenation of Nitrobenzene to Aniline*, Presented at RC User Forum USA, St Petersburg, 1994.
30. Lines PC. *Gas–Liquid Mass Transfer Using Surface-Aeration in Stirred Vessels, with Dual Impellers*, Trans IChemE, Vol. 78, Part A, 2000.
31. Zwietering TN. *Chem. Eng. Sci.* 1958; 8: 244.
32. Lines PC. *IChemE Symp. Ser.* 1990; 121: 167.
33. Paul EL, Atiemo-Obeng VA, Kresta SM. *Handbook of Industrial Mixing*, John Wiley & Sons, 2004.
34. Bourne JR, Yu S. Investigation of micromixing in stirred tank reactors using parallel reactions. *Ind. Eng. Chem. Res.* 1994; 33 (1): 41–55.
35. Atherton JH, Carpenter K. *Process Development—Physiochemical Concepts*, Oxford Science, Oxford, UK, 1999.
36. Box GEP, Hunter WG, Hunter JS. *Statistics for Experimenters: An Introduction to Design, Data Analysis, and Model Building*, John Wiley & Sons, 1978.
37. Peterson, JJ. A Bayesian approach to the ICH Q8 definition of design space. *J. Biopharmaceut. Stat.* 2008; 18 (5): 959–975. Available at http://dx.doi.org/10.1080/10543400802278197.

10

MODELING, OPTIMIZATION, AND APPLICATIONS OF KINETIC MECHANISMS WITH OPENCHEM

JOHN E. TOLSMA, BRIAN SIMPSON, AND TAESHIN PARK
RES Group, Inc., Cambridge, MA, USA

JASON MUSTAKIS
Pfizer, Inc., Groton, CT, USA

10.1 INTRODUCTION

Pharmaceutical manufacturing involves large-scale transformation of raw materials into drugs through various processes such as chemical synthesis, separation, and purification. Central to pharmaceutical manufacturing is the chemical synthesis of API (active pharmaceutical ingredient). The API is the active ingredient of the drug and its synthesis often involves complex chemical transformations of raw materials and strongly depends on operating conditions such as temperature, pressure, and agitation. As the mechanisms of API synthesis (e.g., relationship between operating conditions and yields of API and by-products) are better understood, the pharmaceutical manufacturing process can be optimized.

Current practice in the design and optimization of pharmaceutical manufacturing processes depends on the empirical knowledge obtained from experimental observations. Without a detailed mechanistic understanding of how APIs are synthesized, there is limited knowledge of the complex trade-offs that exist when developing the pharmaceutical manufacturing process. This is consistent with an observation made by the FDA in their Critical Pathway Initiative [1]. In pharmaceutical manufacturing, the FDA identified opportunities in adopting systematic methodology and modern science and technology into the manufacturing process to increase product yield, reduce waste, and improve process monitoring and control. Mechanistic understanding of API synthesis is one of the key components to realize these opportunities.

The API synthesis mechanism describes the chemical steps involved in the transformation of initial reagents into desired products, intermediates, and undesired by-products. Developing the API synthesis mechanism consistent with experimental observation is a complex and challenging process involving several activities. Some of these activities are described in Table 10.1 and Figure 10.1.

The goal of this chapter is to convey some of the challenges in developing API synthesis mechanisms and to describe several approaches for addressing these challenges. OpenChem software [2] will be used to illustrate some of these approaches for developing API synthesis mechanisms. OpenChem is a modeling and optimization software platform designed to help chemists to build, visualize, analyze, calibrate, and apply chemical reaction mechanisms.

The remainder of this chapter is organized as follows: first, we describe an example mechanism, involving a Buchwald–Hartwig amination reaction, which will be used throughout the chapter. Next, several approaches for building the mechanism and describing it on a computer are considered with their advantages and disadvantages. Mechanism building is followed by mechanism analysis with emphasis on extracting the most information from the model. In particular, multiple ways of visualizing the results of the simulation are described. Calibrating the mechanism with experimental data is discussed next with emphasis on a calibration workflow for

Chemical Engineering in the Pharmaceutical Industry: R&D to Manufacturing Edited by David J. am Ende

TABLE 10.1 Activities, Purpose, and Requirements in API Synthesis Mechanism Development

Activities	Purpose	Requirements
Building	Create one or more candidate mechanisms describing the API synthesis	Mechanism must capture correct reaction steps and reaction rates, including intermediate and unmeasured species
Analysis	Identify important reactions and parameters in mechanism influencing the predictions of interest	Analysis results should be valid even when values of parameters are highly uncertain
Calibration	Calibrate parameters such that model predictions best fit all available data	Calibration must be able to handle multiple data types (e.g., concentration, spectra, heat flow, hydrogen uptake, etc). Most methods will often converge to poor fits, but best fit should be ensured
Discrimination	Select, or discriminate, most likely mechanism from collection of candidate mechanisms	Correct mechanism should be selected with confidence from the set of proposed mechanisms. All available data should be used for this selection

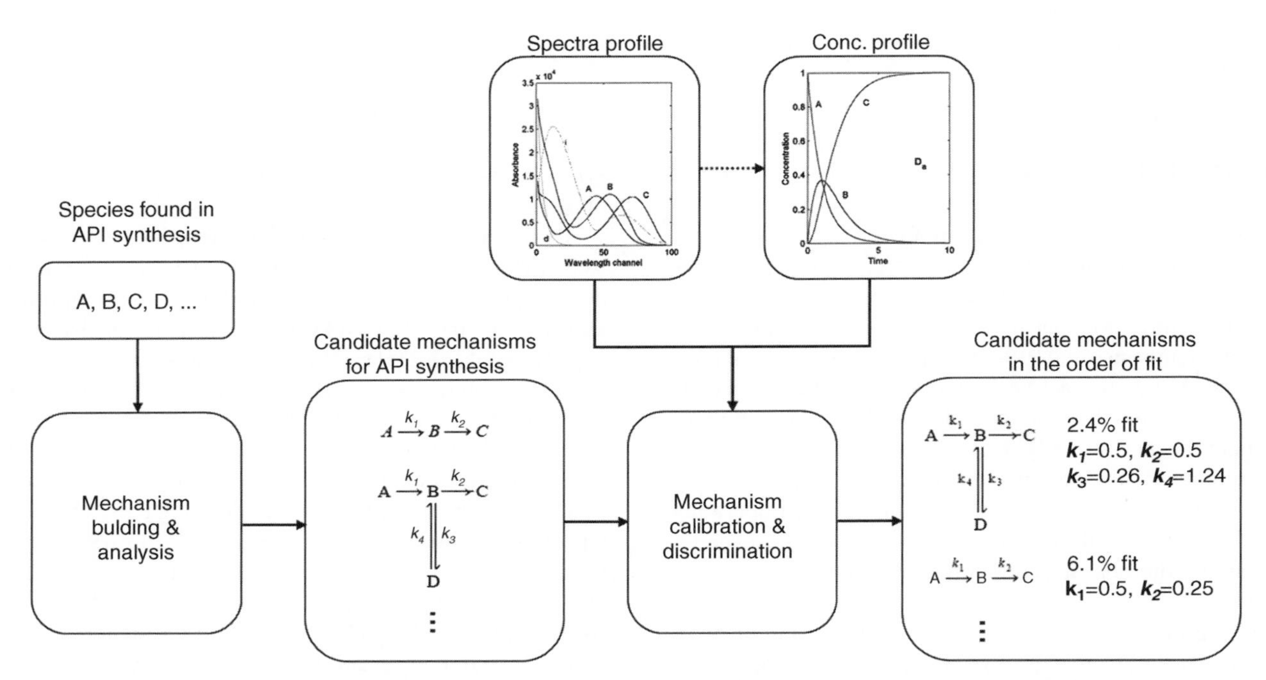

FIGURE 10.1 Activities in API synthesis mechanism development including mechanism building, analysis, calibration, and discrimination. Both concentration and spectra profile data are used for mechanism calibration and discrimination.

systematizing the complex task of model calibration. Finally, several applications of the calibrated model are considered.

10.2 DESCRIPTION OF THE EXAMPLE MECHANISM

The chemical synthesis mechanism describes the reaction steps and rates for producing the product, intermediates, and by-products from the initial reagents. Figure 10.2 shows a mechanism for a coupling reaction between an aryl halide (ArX) and an amine (Amine) in the presence of a base (Base) and a catalyst (Cat). This catalytic cycle offers a convenient example of API synthesis where a relatively complex mechanism is present but only a limited amount of observations are available. An additional complexity is present in this case as the base (potassium hydroxide, KOH) is heterogeneous requiring in addition to the standard chemical steps a mass transfer term. As can be seen in the upper right corner of Figure 10.2 the base appears to have an effect on both catalyst

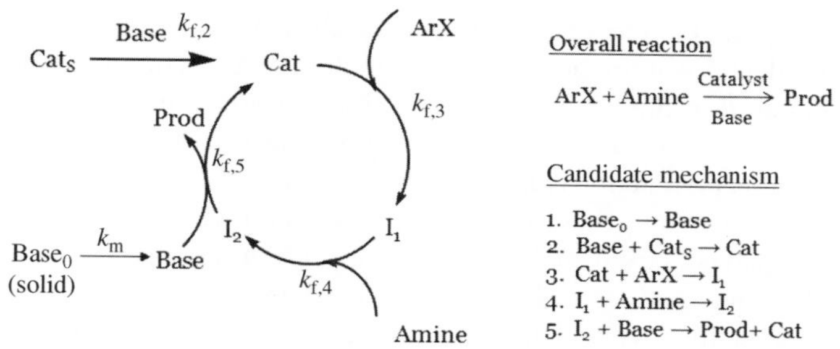

Overall reaction

$$ArX + Amine \xrightarrow[\text{Base}]{\text{Catalyst}} Prod$$

Candidate mechanism

1. $Base_0 \rightarrow Base$
2. $Base + Cat_S \rightarrow Cat$
3. $Cat + ArX \rightarrow I_1$
4. $I_1 + Amine \rightarrow I_2$
5. $I_2 + Base \rightarrow Prod + Cat$

FIGURE 10.2 Example API synthesis mechanism. Overall, an aryl halide reacts with an amine in the presence of a catalyst and strong base to form the desired aryl amine product. The schematic on the left shows the proposed mechanism steps, including formation of intermediate species. The corresponding candidate synthesis mechanism is shown on the right.

regeneration and initial catalyst activation. The proposed scheme and synthesis steps of the candidate mechanism are shown in Figure 10.2.

Overall, the aryl halide reacts with the amine in the presence of the catalyst and base to form the desired product, Prod. The solubility of the solid base is very low (e.g., KOH in toluene), so only a small concentration exists in solution to participate in the reaction. The equation used to describe the mass transfer term takes into account the effect of agitation for given reactor configurations. Please note that the formulation used for the base will still be valid if base reactions are happening on the surface. The first two steps in the proposed mechanism involve the solid base dissolving and the catalyst being reduced to the required oxidation state. This is followed in step 3 by the oxidative addition of the aryl halide to the modified catalyst, forming an intermediate I_1. In step 4, the halide (X) in the intermediate is replaced by the nitrogen in the amine to form an intermediate I_2, which then undergoes reductive elimination in the presence of the base to form the desired aryl amine (Prod) in step 5.

10.2.1 Reaction Formula

The reaction formulas of the candidate mechanism shown in Figure 10.2 specify the reaction stoichiometry. For example, in reaction step 4, one mole of intermediate I_1 reacts with one mole of amine to form one mole of intermediate I_2.

10.2.2 Rate Laws

To fully define the API synthesis mechanism, the rate of each reaction as a function of species concentrations, temperature, and other variables must also be specified. The reaction rates for the candidate mechanism shown in Figure 10.2 are the following:

$$r_1 = k_{m,1}\left(\frac{\text{RPM}}{\text{RPM}_{\text{ref}}}\right)^n([\text{Base}_0]-[\text{Base}])$$

$$r_2 = k_{f,2}[\text{Base}][\text{Cat}_s]$$

$$r_3 = k_{f,3}[\text{ArX}][\text{Cat}]$$

$$r_4 = k_{f,4}[\text{I}_1][\text{Amine}]$$

$$r_5 = k_{f,5}[\text{I}_2][\text{Base}]$$

where r_i denotes the rate of reaction i in moles per unit time per unit volume, $k_{m,1}$ and $k_{f,i}$ $(i=2, \ldots, 5)$ denote the rate constants, and $[\cdot]$ denotes the species concentration in moles per unit volume. Parameters RPM and RPM_{ref} are the agitator speed and reference speed, respectively. Parameter n indicates how the solid base dissolution rate scales with agitator speed. The reaction rate constant is usually a function of temperature, but may also depend on other quantities such as pressure and other species concentrations. Typically, temperature dependence of the reaction rate constant is given by the Arrhenius expression:

$$k(T) = AT^{\beta}\,\mathrm{e}^{-E_A/RT}$$

where $k(T)$ is the rate constant, A is the preexponential factor, β is the temperature exponent, E_A is the reaction activation energy, R is the gas constant, and T is the temperature. Parameters A, β, and E_A are typically adjusted in the synthesis mechanism to fit experimental data. This process is described in detail in Section 10.4. The expression defining the reaction rate is often referred to as the reaction rate law.

The mechanism is fully defined by the reaction formulas and rate laws. The following section describes how the mechanism can be specified and analyzed on a computer.

10.3 MECHANISM BUILDING AND ANALYSIS

10.3.1 Mechanism Building

When constructing a mechanism for the synthesis of an API, we usually know the reactants, main products, and some byproducts. From this we can postulate various reaction steps and develop a candidate mechanism. These steps can also be performed automatically with a computer, where standard chemical reactions or steps can be applied to create a graph connecting the reactants to the products. Automatic generation techniques have been applied for the creation of detailed homogeneous gas-phase reactions for many years [3] and have also been applied to more general reactions including API synthesis [4]. Automatic mechanism generation is outside the scope of this chapter and will not be covered here.

In practice, the experimentalist may only be able to measure a subset of the species actually involved in the reaction network. Also, the exact reaction steps that occur when transforming initial reagents into products, intermediates, and by-products are typically not known exactly. This uncertainty in reaction steps and limited observables creates a situation where several mechanisms may potentially explain the same API synthesis. For these reasons, mechanism building is an iterative process where several candidate mechanisms may be developed and tested. To facilitate this, the software tool used to describe the mechanism should be easy to use and allow rapid mechanism building and modifying. OpenChem supports two main approaches for specifying the synthesis mechanism: tabular and graphical. These are described below.

10.3.1.1 Tabular Input The API synthesis mechanism consists of a collection of one or more reaction formulas and associated rate laws. A natural input for this type of information is a table. Figure 10.3 shows the OpenChem tabular interface for specifying a reaction mechanism.

The user enters into the "New Reaction" input at the top of the form the reaction formula for the reaction to be added to the mechanism, for example, "Cat + ArX $\rightarrow$ I_1." OpenChem will parse this reaction formula to identify products and reactants. If new species are encountered, they are added to the mechanism. The new reaction is added to the mechanism and the user can add additional information, including a reaction identifier, the rate law type and expression, and annotation. The advantage of the tabular form is that it provides a compact representation of the mechanism. The table can also be sorted and queried. For example, only reactions involving a particular species can be listed. One disadvantage of the tabular form is that it can be difficult to visualize the relationships between various species in the mechanism. This drawback is addressed with the graphical view of the mechanism, described below.

10.3.1.2 Graphical Input The OpenChem Pathway Diagram provides a graphical interface for the mechanism. Figure 10.4 shows the Pathway Diagram for the mechanism shown in Figure 10.2. The round-cornered rectangles represent the species in the mechanism. The circles represent the reactions in the mechanism.

An arrow from a species node to a reaction node indicates the species is a reactant, or consumed in the reaction. An arrow from the reaction node to a species node indicates the species is a product, or produced in the reaction. Located on the top left of the OpenChem Pathway Diagram panel is a list of the species in the mechanism. Selecting one or more species in this list will highlight the corresponding species in the graph. Similarly, selecting one or more reactions in the "Reaction List," located below the "Species List," will highlight the corresponding reaction nodes in the graph. This feature is particularly useful when the mechanism is very large. An advantage of the graphical interface of the mechanism is that it provides an intuitive view of the system. Species interactions and reaction channels are more easily identified, providing greater insight into the mechanism. The graphical view also offers additional ways of viewing the simulation results, as described in Section 10.3.3.

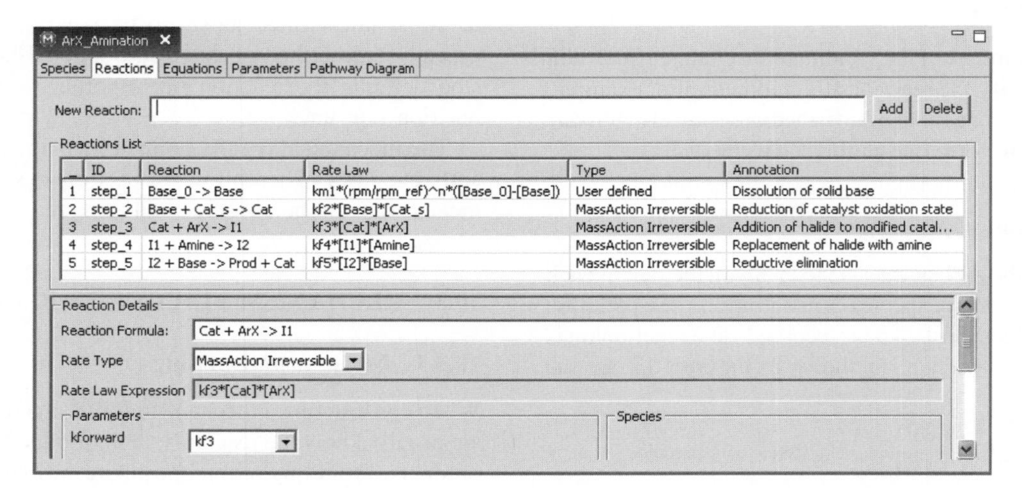

FIGURE 10.3 OpenChem form for specifying a mechanism in tabular format. The reaction formula, rate law type and expression, reaction identifier, and annotation are specified in this tab.

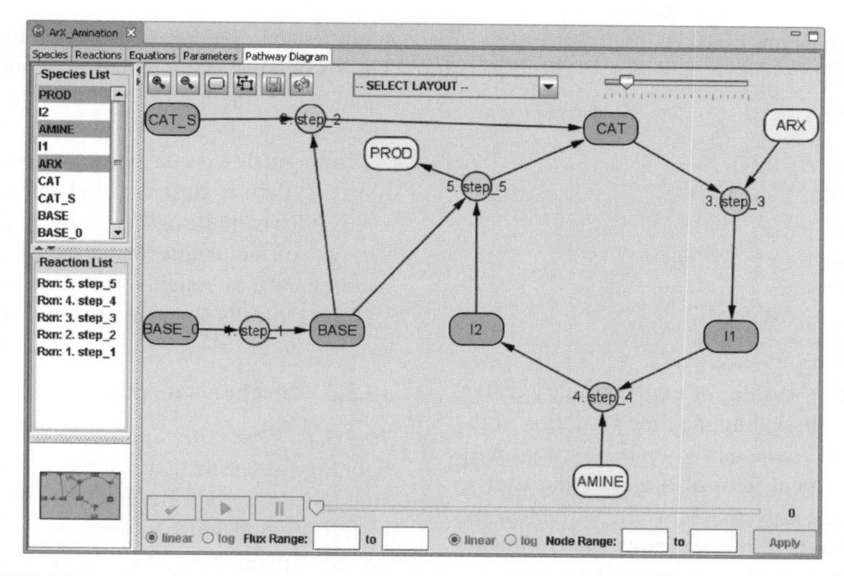

FIGURE 10.4 OpenChem form for specifying a mechanism in graphical format. Species are denoted by round-cornered rectangles and reactions are denoted by circles. Arrows point from the reactant nodes to the reaction nodes and from the reaction nodes to the product nodes. A list of species and reactions is shown on the left of the form. Selecting one or more species and/or reactions in these lists will highlight the corresponding nodes in the graph.

10.3.2 Connecting the API Synthesis Mechanism to the Reactor Model

The previous section describes how to specify the reaction stoichiometry and rates in the API synthesis mechanism. In this section, the mechanism will be coupled with a reactor model to simulate the behavior of the API synthesis. There are many reactor models one may choose to represent the laboratory experiment, pilot or commercial plant. OpenChem provides a library of reactor models, including

- isothermal, isobaric batch reactor for gas;
- isobaric batch reactor for gas;
- isothermal batch reactor for liquid;
- batch reactor for liquid;
- steady-state isothermal plug flow reactor;
- steady-state plug flow reactor;
- transient, isothermal plug flow reactor for gas with heterogeneous catalyst; and
- transient plug flow reactor for gas with heterogeneous catalyst.

OpenChem also provides an input language that can be used to customize the reactor models listed above or create entirely new reactor models. In the example that follows, the liquid-phase isothermal batch reactor will be used to simulate the laboratory experiments.

In general, the synthesis mechanism description is the same regardless of the reactor type. This important observation is exploited in OpenChem by separating mechanism creation from reactor selection. This allows the mechanism to be created once and then used with a variety of reactor types.

10.3.2.1 Mathematical Formulation for the Batch Reactor Model The reaction stoichiometry and rates specified in Section determine the molar production rate of each species, or the time rate of change of species concentration per unit volume. As an example, consider species Base in Figure 10.2: one mole is produced in step 1, one mole is consumed in step 2, and one mole is consumed in step 5. The net rate of change in Base is given by $\dot{\omega}_{\text{Base}} = r_1 - r_2 - r_5$. OpenChem constructs automatically the molar production rates using the reaction formulas and rate laws. The molar production rates for the candidate mechanism are shown below:

$$\begin{array}{lll} \dot{\omega}_{\text{Base}_0} = -r_1 & \dot{\omega}_{\text{Cat}} = r_2 - r_3 + r_5 & \dot{\omega}_{\text{I}_1} = r_3 - r_4 \\ \dot{\omega}_{\text{Base}} = r_1 - r_2 - r_5 & \dot{\omega}_{\text{ArX}} = -r_3 & \dot{\omega}_{\text{I}_2} = r_4 - r_5 \\ \dot{\omega}_{\text{Cat}_s} = -r_2 & \dot{\omega}_{\text{Amine}} = -r_4 & \dot{\omega}_{\text{Prod}} = r_5 \end{array}$$

The mechanism is connected to the reactor model by the molar production rates. From the user-specified mechanism and the selected reactor type, OpenChem will automatically construct the mathematical model describing the reactor. The isothermal, liquid-phase batch reactor equations combined with molar production rates are then

$$\frac{d[\mathrm{Base}]}{dt} = \dot{\omega}_{\mathrm{Base}} \qquad \frac{d[\mathrm{Amine}]}{dt} = \dot{\omega}_{\mathrm{Amine}}$$

$$\frac{d[\mathrm{Cat_s}]}{dt} = \dot{\omega}_{\mathrm{Cat_s}} \qquad \frac{d[\mathrm{I_1}]}{dt} = \dot{\omega}_{\mathrm{I_1}}$$

$$\frac{d[\mathrm{Cat}]}{dt} = \dot{\omega}_{\mathrm{Cat}} \qquad \frac{d[\mathrm{I_2}]}{dt} = \dot{\omega}_{\mathrm{I_2}}$$

$$\frac{d[\mathrm{ArX}]}{dt} = \dot{\omega}_{\mathrm{ArX}} \qquad \frac{d[\mathrm{Prod}]}{dt} = \dot{\omega}_{\mathrm{Prod}}$$

The above equations are a system of eight ordinary differential equations (ODEs) describing the time evolution of the species concentrations in a constant volume, constant temperature batch reactor. OpenChem also allows the user to define additional relationships to constrain variables and provide useful information. In this example, the following two algebraic equations are added to the model:

$$\mathrm{Base_0} = 1$$

$$y_{\mathrm{ArX}} = \frac{[\mathrm{ArX}]}{[\mathrm{Prod}] + [\mathrm{ArX}]}$$

Algebraic equations do not explicitly contain time derivatives of the state variables. The first algebraic equation specifies that the solid base concentration is fixed at unity. The second algebraic equation defines a new variable, y_{ArX}, equal to the fraction of reactant ArX that has reacted at any given time.

The eight differential equations and two algebraic equations form a system of differential algebraic equations (DAEs) [5]. DAEs offer more flexibility when formulating the problem and the cost of simulating (in terms of CPU time and memory requirements) is roughly the same as solving the stiff ODE systems typically found in chemical reaction simulations.

10.3.3 Mechanism Analysis

10.3.3.1 Dynamic Simulation One of the first tasks after defining the mechanism and selecting the reactor type is to simulate the model. Before doing this, however, three pieces of information must be provided: (1) values of the time-invariant parameters (e.g., rate constants), (2) initial conditions for the state or differential variables, and (3) the simulation duration. In this example, the initial concentrations of all the species are zero except reactants ArX and Amine, which have initial concentrations of 1.0 and 1.2, respectively. The dynamic system is simulated for 50,000 s (14 h) from the specified initial conditions. Figure 10.5 shows the OpenChem Simulation Exploring environment.

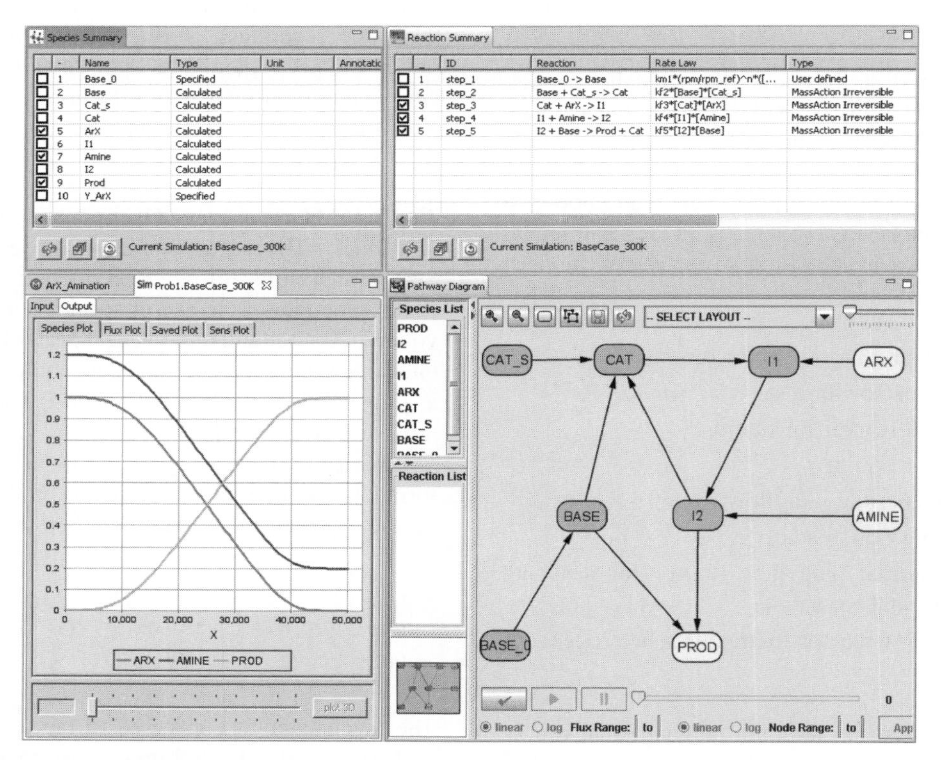

FIGURE 10.5 OpenChem interactive four-panel Simulation Explorer. The upper left and upper right panels contain a summary of the species and reactions, respectively. The lower left and lower right panels contain the Plotting Environment and Pathway Diagram, respectively. These panels interact, providing a convenient overview of the mechanism behavior.

The Simulation Exploring environment provides an overview of the mechanism and simulation results in four panels. The Species Summary panel is shown in the upper left. The Reaction Summary panel is shown in the upper right. The bottom left and bottom right panels contain the Plotting Environment and Pathway Diagram, respectively. The four panels of the Simulation Explorer are interactive. For example, one or more species may be selected in the Species Summary panel by clicking on the checkbox to the left of the species name. The Reaction Summary panel will be updated and show only the reactions that involve the selected species. The plotting panel (lower left) will be updated, showing the concentration profiles of the selected species. Similarly, if one or more reactions are selected in the Reaction Summary, all species participating in the selected reactions will be displayed in the Species Summary. Further, the Flux Plot tab in the Plotting Environment panel will display the flux profiles for all selected reactions.

Figure 10.6 shows more closely (a) concentration profiles for species ArX, Amine, Prod, I_1, and I_2 and (b) flux profiles for reactions step_1, step_2, and step_5. The species profiles indicate how the concentration changes in the reactor as a function of time. Reactants ArX and Amine monotonically decrease, product Prod monotonically increases, and intermediates I_1 and I_2 exhibit a maximum value at an intermediate time. Further, I_2 was produced in higher concentration than I_1. From this plot (Figure 10.6), we can see that reactant ArX is the limiting reagent and the reaction goes to completion. The flux profiles indicate how the reaction flux changes as a function of time. Visualizing the reaction fluxes allows the reactions to be compared to determine which reactions are most important. The reaction flux analysis described next provides an alternative method for identifying important reaction channels.

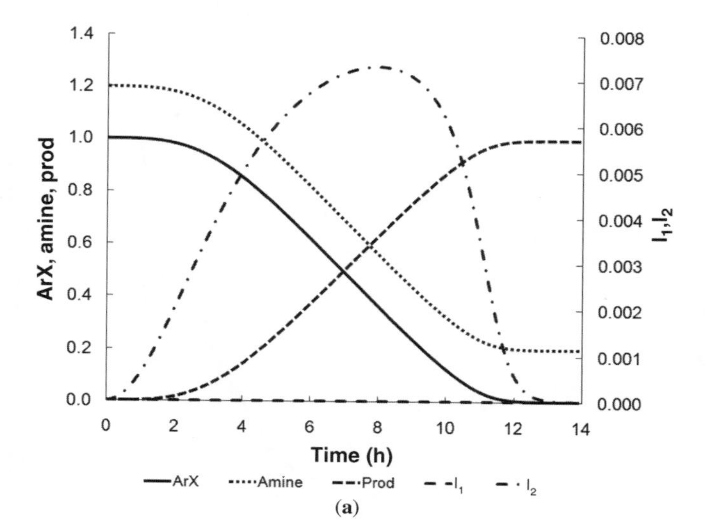

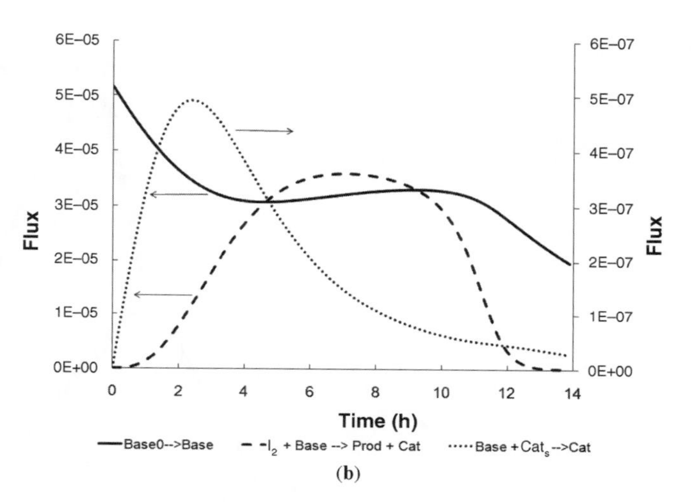

FIGURE 10.6 Selected species and flux profiles for the synthesis mechanism shown in Figure 10.2. Part (a) shows the concentration profiles for species ArX, Amine, Prod, I_1, and I_2. Part (b) shows the flux profiles for reactions $Base_0 \rightarrow Base$, $I_2 + Base \rightarrow Prod + Cat$, and $Base + Cat_s \rightarrow Cat$.

10.3.3.2 Reaction Flux Analysis The Pathway Diagram panel of the Simulation Explorer provides another way to visualize the results of a simulation. Selecting the checkmark button in the lower left corner of the Pathway Diagram panel enables the flux animation. Rather than viewing time profiles of the species and/or fluxes as shown in the previous section, the flux animation displays the reaction progress by changing the size of the species nodes and thickness of the reaction arrows. The size of the species node is proportional to the species concentration and the thickness of the arrow is proportional to magnitude of the flux. Figure 10.7 shows a snapshot of the flux animation at 1, 7, and 14 h of simulation time.

At 1 h of simulation time, the primary reaction occurring is the dissolution of the solid base ($Base_0 \rightarrow Base$). The arrow in the flux diagram for this reaction is thick and the nodes corresponding to the reactants ArX and Amine are large, indicating high concentration. At 7 h, sufficient solid base has dissolved and all reactions are proceeding. Finally, at the end of the simulation, after 14 h, reactants ArX and Amine have reduced and product Prod is formed. The animation can be played by selecting the run button in the lower left corner and stopped at any time using the stop button. The sliding bar at the bottom of the Pathway Diagram can be positioned to view a snapshot of the reaction progress at any time in the simulation.

When viewing the flux profiles as described in the previous section, individual reactions can be compared and important reactions identified. The reaction flux analysis, however, can be used to identify important reaction channels, that is, groups of reactions connecting reactants to products via intermediates. This is difficult to grasp when viewing individual reaction fluxes. This analysis is helpful for identifying important intermediates and pathways as well as finding bottlenecks or rate-determining pathways.

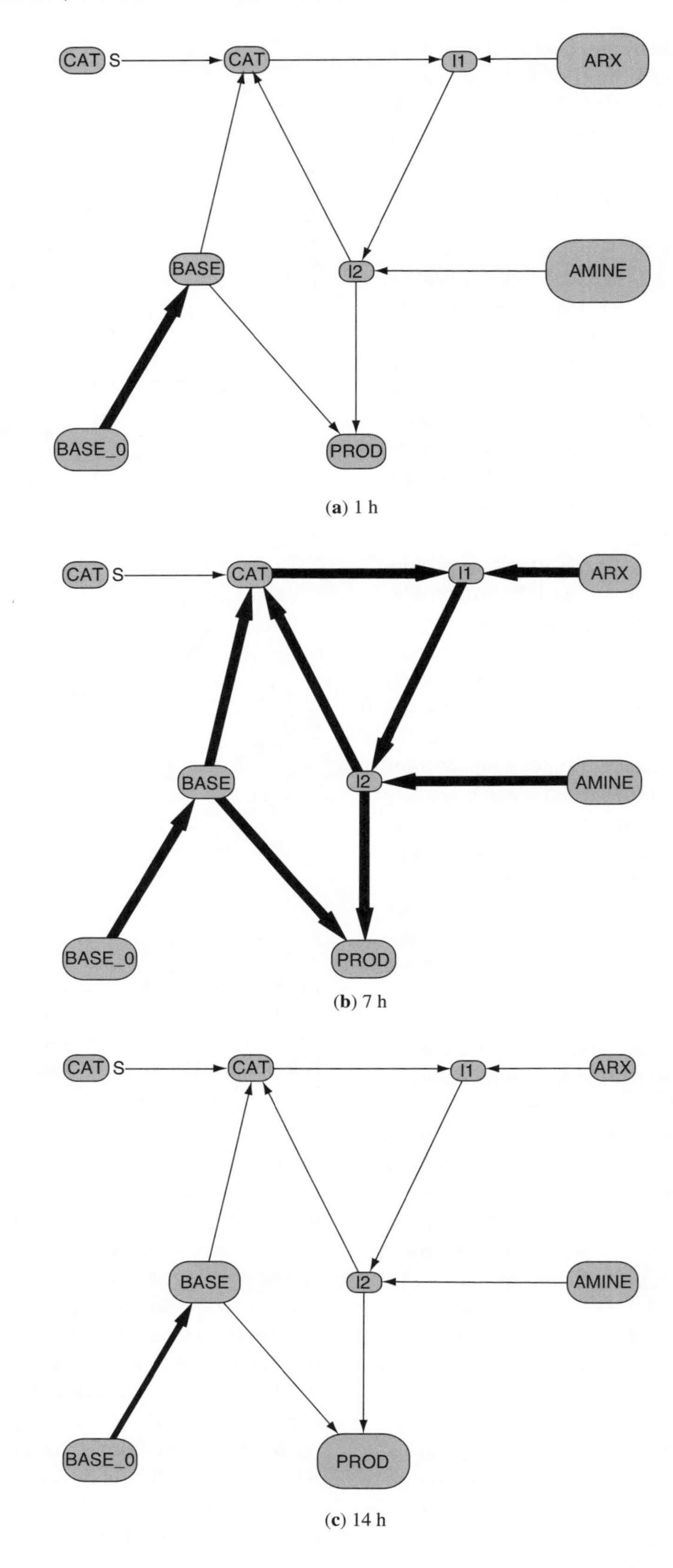

(**a**) 1 h

(**b**) 7 h

(**c**) 14 h

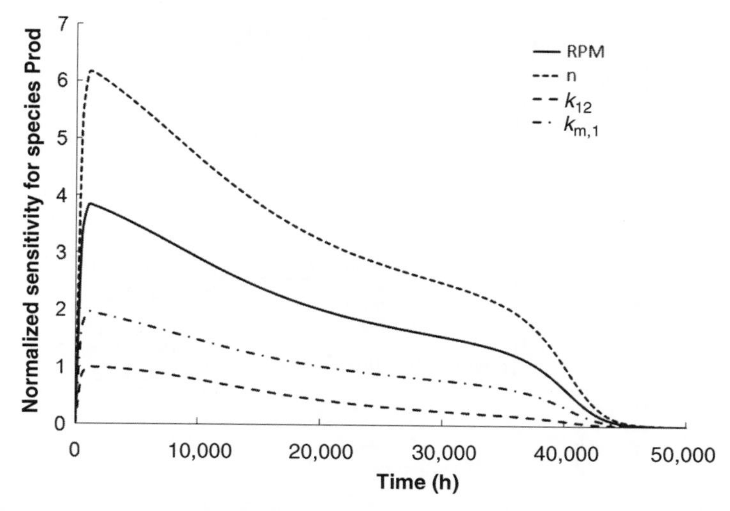

FIGURE 10.8 Normalized parametric sensitivities for species Prod with respect to several parameters.

10.3.3.3 Parametric Sensitivity Analysis In addition to computing species concentration and flux profiles, OpenChem can also compute the parametric sensitivities of the dynamic system. The parametric sensitivities are defined as

$$s_{ij}(t) = \frac{\partial y_i}{\partial p_j}$$

Here, y_i is a model output and p_j is a time-invariant parameter. This quantity is called the differential sensitivity. OpenChem can also compute the normalized sensitivities, defined as

$$\tilde{s}_{ij}(t) = \frac{\partial \ln y_i}{\partial \ln p_j} = \frac{p_j}{y_i}\frac{\partial y_i}{\partial p_j}$$

Sensitivity analysis has traditionally been used to determine how the parameters influence the model predictions [6]. This calculation involves solving a dynamic system (related to the original model) that computes the partial derivatives of the model outputs with respect to the time-invariant parameters as a function of time. The sensitivity trajectories can be compared to determine which parameters have the greatest influence on the outputs of interest and when during the simulation they are important. Figure 10.8 shows the normalized sensitivities for species Prod with respect to several parameters. Notice that the value of Prod is most sensitive to the parameter values early during the simulation. The sensitivity goes to zero as Prod reaches its final value.

10.4 MECHANISM CALIBRATION

When constructing a candidate API synthesis mechanism, there will be a number of time-invariant parameters with unknown values. For example, the values of rate constants in the mechanism shown in Figure 10.2, $k_{m,1}$, $k_{f,2}$, $k_{f,3}$, $k_{f,4}$, and $k_{f,5}$, will likely be unknown at the beginning of mechanism development. These parameters must be inferred from experimental data through the process of mechanism calibration. Mechanism calibration involves the following tasks:

- identifying the parameters in the model that strongly influence the model outputs of interest;
- collecting necessary experimental data through a series of experiments; and
- adjusting systematically the important parameters so that the model outputs best fit the experimental data.

FIGURE 10.7 OpenChem flux animation. This feature provides an additional view of the simulation results, showing how the reaction progresses in time by changing the relative sizes of the species nodes (to represent concentration) and thickness of the arrows (to represent flux magnitude) as an animation. Part (a) shows the reaction progress after 1 h of simulation time. Part (b) shows the reaction progress after 7 h of simulation time. Part (c) shows the reaction progress after 14 h of simulation time. For more complex mechanisms, this analysis is useful for identifying important intermediate species and reaction pathways.

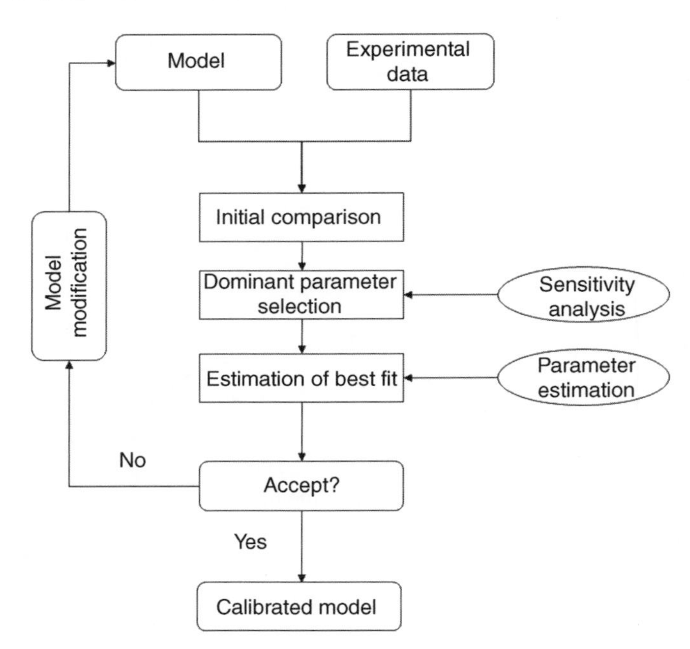

FIGURE 10.9 Calibration workflow. The major steps involved in model calibration are illustrated in this figure. Techniques that can be applied are shown in the ellipses. An important part of this workflow is model modification; deficiencies in the model and/or data are often uncovered during model calibration.

There are several challenges that must be addressed when calibrating a mechanism. The process of identifying the important parameters is complicated by the fact that the parameter values are unknown. An uncalibrated model can play an important role in determining what experiments should be performed and under what conditions, provided rough estimates for the parameter values and reasonable upper and lower bounds can be supplied. Mechanism calibration is further complicated by the fact that experimental data are often in a form that is not directly comparable to the model. For example, the model may involve species concentrations whereas the data may be spectral measurements (e.g., ultraviolet and infrared). The ability to effectively handle mixed data types is an important part of model calibration. The process of systematically adjusting the parameter values to best fit the data is referred to as parameter estimation. A well-known issue with parameter estimation is that the parameter estimates obtained often depend strongly on the initial values provided for the parameters being adjusted.

10.4.1 Calibration Workflow

An effective workflow for model calibration involves identification of the important parameters, parameter estimation with special provisions for handling mixed data and ensuring that the best possible fit is obtained, and unbiased discrimination between multiple calibrated candidate mechanisms. Figure 10.9 shows a schematic of one calibration workflow.

At the top of Figure 10.9 are the candidate mechanism and experimental data. The first step is to match the model outputs with the experimental data and perform an initial comparison. The next step is to identify the dominant parameters in the model. Only the parameters that influence the model outputs corresponding to the data can be effectively calibrated, so correctly identifying these parameters is an important step of model calibration. Computations such as sensitivity analysis assist the user with this selection. The next step is to adjust the parameters to best fit the data. If successful, the model and estimated parameters can be analyzed to determine the quality of the fit. In many cases, the process of model calibration can identify weaknesses in the model and/or experimental data. An important part of model calibration is to modify the model and/or collect new experimental data to ensure that the best possible model is obtained. The remainder of this section describes this workflow as implemented in OpenChem.

10.4.2 Parameter Identification

The time-invariant parameters in a model influence the model predictions in various ways. Some parameters strongly influence the model outputs of interest whereas others have a limited effect. This becomes particularly true as the size of the mechanism and the number of parameters increase. As described earlier, parametric sensitivity analysis can be used to quantitatively compare the influence of several parameters. Unfortunately, the approach described, referred to as a local sensitivity analysis, requires that the values of the parameters be specified prior to performing the calculation. Since these parameters are unknown or uncertain prior to calibration, the results of the sensitivity analysis depend on the initial estimates for the parameters.

To address the uncertainty in parameter values, OpenChem implements global sensitivity analysis (GSA) approach. This algorithm begins with the user selecting the parameters of interest and providing an appropriate range for the parameter values (e.g., upper and lower bound on each parameter). Even when the actual parameter values are not known, it is often possible to select reasonable values for the bounds. Next, values for the parameters are sampled from the parameter space defined by the bounds. This may be as simple as a uniform random sampling between the upper and lower bounds or a more complex sampling based on an *a priori* knowledge of the likelihood of the parameter values. A local sensitivity analysis is performed for each sample of parameter values. Each sensitivity profile is then converted to a scalar sensitivity metric, for example, maximum absolute value on the profile or integral of the absolute value of the profile:

$$\max_{0 \leq t \leq T} |s(t)| \quad \text{or} \quad \int_0^T |s(t)| \mathrm{d}t$$

Here $s(t)$ denotes the parametric sensitivity profile. This sampling is repeated many times and the scalar metric of each sensitivity trajectory is averaged over all samples. Figure 10.10 shows the results of GSA applied to the synthesis mechanism described in Figure 10.2.

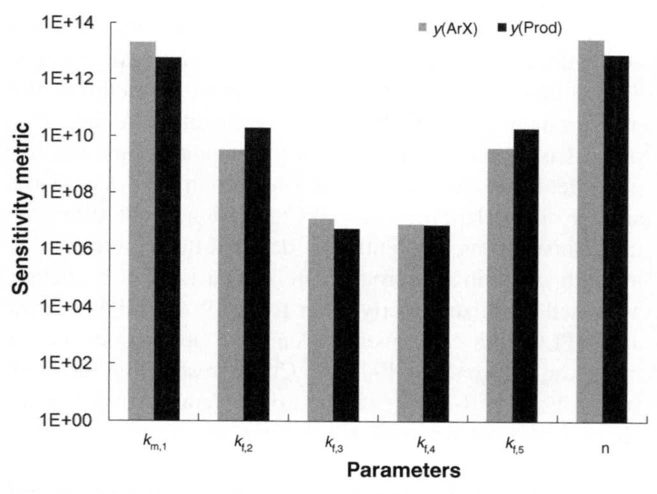

FIGURE 10.10 Global sensitivity analysis results for the synthesis mechanism shown in Figure 10.2 for species ArX and Prod. The importance of the parameters, as measured by the averaged sensitivity metric, is valid not for a single set of parameter values but rather a four-order magnitude range of parameter values.

In small mechanisms such as the example shown above, almost all parameters have a strong influence on the few model outputs. Figure 10.11 shows the results of GSA applied to a larger mechanism. In this case, there are clearly groups of parameters that have a greater effect on the selected model outputs. For example, the parameters k_1 and k_4 have strong influence on concentration $C(1)$ while the parameters k_2 and k_3 have no influence at all.

10.4.3 Parameter Estimation

10.4.3.1 Parameter Estimation with Concentration Data

Calibration with concentration data is relatively well established and there are a number of techniques available [7]. The most widely used are those based on a descent direction of the calibration objective function, such as the Levenberg–Marquardt method. OpenChem implements a control parameterization approach for solving parameter estimation problems for dynamic models. In this case, the discrepancy between the time series experimental data and the model predictions (i.e., the calibration objective function) is computed by numerically integrating the dynamic system. The gradient of the objective function is determined by computing the local parametric sensitivities with respect to the parameters being estimated. The scalar objective function is minimized by applying a nonlinear programming algorithm, such as successive quadratic programming [8].

One well-known problem of descent direction methods like those described above is that they often converge to locally optimal solutions. That is, parameters are estimated such that they are the best fit in some neighborhood, but a better fit might be obtained with parameters outside this neighborhood. As a result, the parameter fits obtained often depend on the initial values provided for the parameters. OpenChem addresses this dependence on initial parameter values by applying a multistart parameter estimation approach. Like the GSA method described above, multistart parameter estimation begins with the user providing appropriate lower and upper bounds for the parameters. Values for the parameter to be estimated are then sampled between the lower and upper bounds to provide the starting guess for a parameter estimation calculation. The sampling procedures can be as simple as a Cartesian grid search or more complex, like starting from a set of seed values generated by examining the objective function during GSA. Upon successful completion of each run, the final objective function and optimal parameter estimates are recorded. Figure 10.12 shows the final objective function values versus sample iteration number for 46,656 runs, sorted by decreasing value of objective function.

The iteration number on the x-axis corresponds to a different starting value for the parameters being estimated. Figure 10.12 shows a number of flat, plateau regions. These

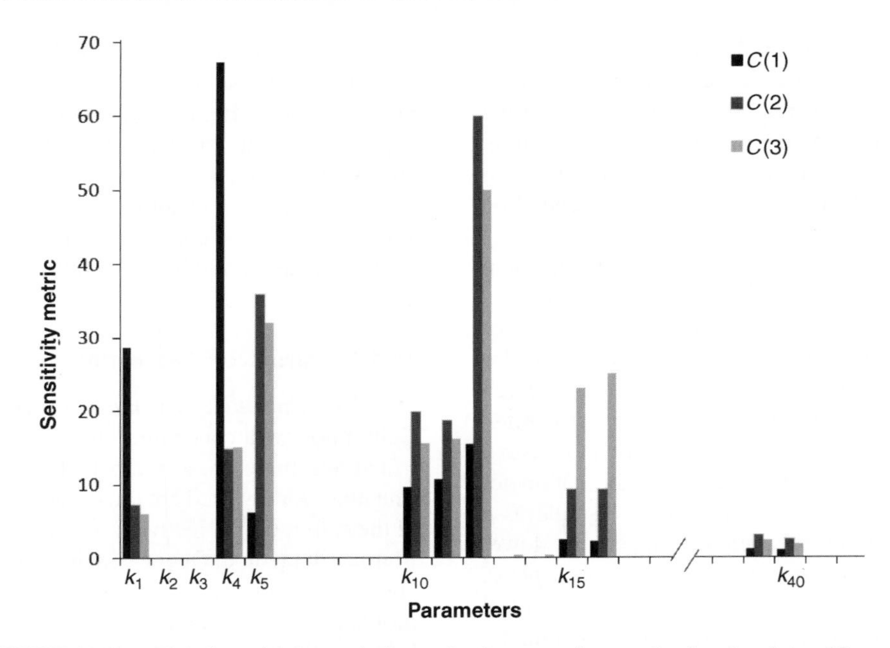

FIGURE 10.11 Global sensitivity analysis results for a reaction mechanism involving 30 species and 42 parameters with results shown for three key model outputs, $C(1)$, $C(2)$, and $C(3)$. The importance of the parameters, as measured by the averaged sensitivity metric, is valid not for a single set of parameter values but rather a four-order magnitude range of parameter values.

correspond to locally optimal parameter estimates that are obtained from several starting values for the parameters. This phenomenon is typical of many parameter estimations encountered when calibrating kinetic mechanisms. Two regions on Figure 10.12 have been marked A and B, corresponding to two distinct estimates for the parameter values. Overlay plots for these two estimates are shown in Figure 10.13.

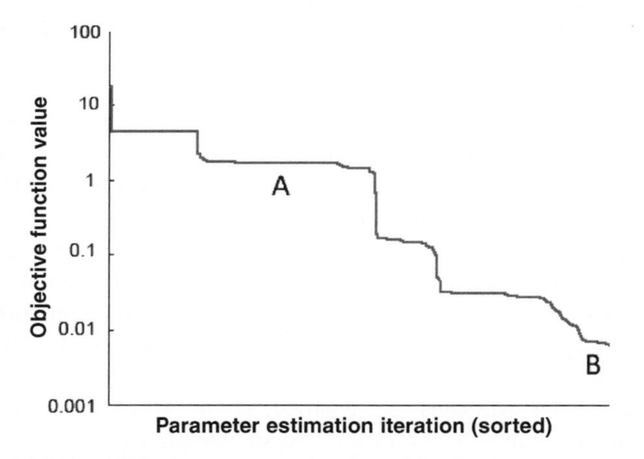

FIGURE 10.12 Parameter estimation objective function versus multistart iteration number (sorted) for 46,656 samples. The flat, plateau regions correspond to locally optimal solutions that were found from several initial parameter values. Overlay plots corresponding to two parameter estimates, marked A and B, are shown in Figure 10.13.

In addition to multistart parameter estimation, other optimization methods such as genetic and simulated annealing algorithms are also available in OpenChem.

10.4.3.2 Parameter Estimation with Mixed Concentration and Spectra Data The calibration method described above utilizes concentration data to compare model predictions directly to experimental values. However, most of the experimental data collected in practice are in a form other than concentration, such as spectra (e.g., ultraviolet and infrared) and HPLC (high-performance liquid chromatography). To use these data with traditional calibration methods, the measurements are first converted to concentration data. This process is time consuming, error prone, and sometimes not possible, for example, if the pure species in the system have similar or overlapping spectra. Calibration of the synthesis mechanism using concentration data is often referred to as a hard modeling approach. In contrast, soft modeling approaches utilize directly other forms of data, like spectra and HPLC, and can cope with data not applicable to hard modeling approaches [9, 10]. Curve resolution methods such as MCR-ALS (multivariate curve resolution–alternating least squares) are a category of soft modeling techniques.

The MCR-ALS algorithm represents a relatively recent advancement in calibrating the mechanism with spectra data. The basic idea is that the user provides experimentally measured spectra data (D) and initial guesses for the pure component spectra (S_0) and/or initial guesses for the species

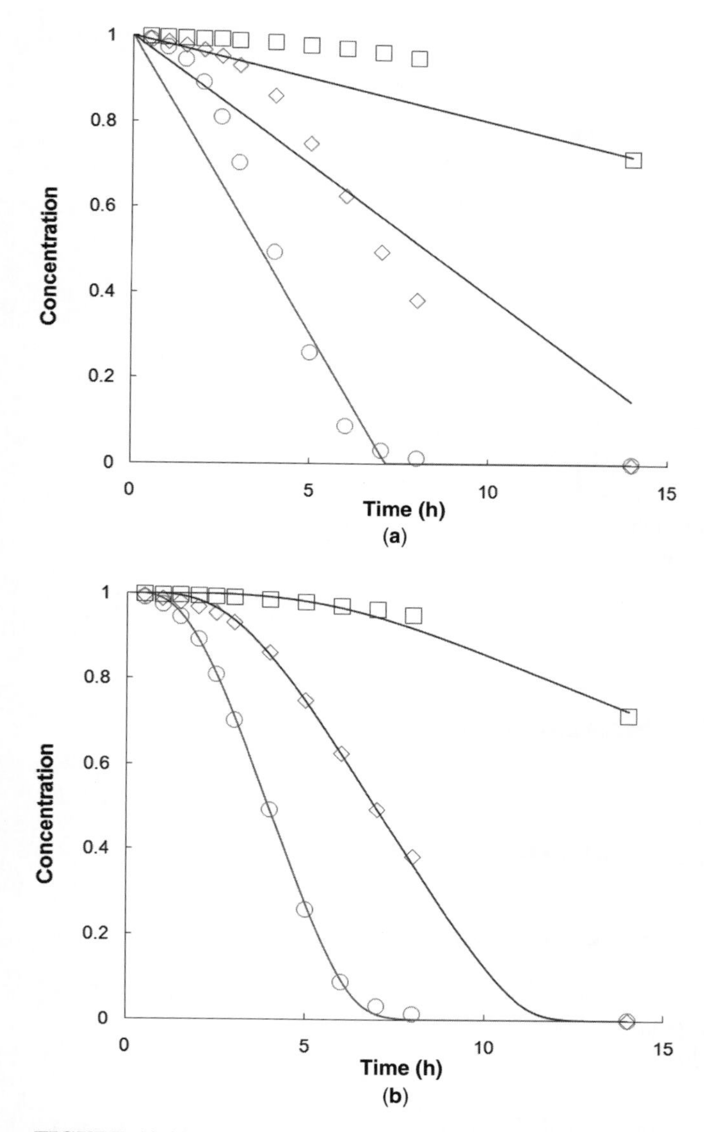

FIGURE 10.13 Results of the OpenChem multistart parameter estimation for the mechanism shown in Figure 10.2. Part (a) shows the overlay plot for a locally optimal, but poor fit. Part (b) shows the overlay plot for the best parameter fit obtained and is clearly superior to the other fit.

concentration profiles (C_0). The algorithm solves a sequence of linear least square problems, attempting to resolve the actual pure component spectra (S) and actual concentration profiles (C), which are related to the data according to the Beer–Lambert law [11]:

$$D = CS^{\mathrm{T}} + E$$

where E is the experimental error. At each iteration, constraints are applied to $C_{(k)}$ and $S_{(k)}$ (k is the current iteration number) to ensure that physically meaningful solutions are obtained. Also at each iteration, a standard parameter estimation is performed with the current $C_{(k)}$ to estimate the kinetic rate constants in the model.

When data from multiple sources are available, there are several advantages to using simultaneously all data sources when calibrating the mechanism, including

- more information is available leading to potentially better fits;
- estimated parameters will be consistent with all data; and
- combining all data sources in a single step simplifies the calibration and streamlines the calibration workflow.

OpenChem implements a modification of the MCR-ALS algorithm that simultaneously enables calibration with mixed spectra and concentration data [12]. The basic structure of this algorithm is similar to that of MCR-ALS described above, except that as part of one of the linear least squares problems, the concentration data are directly utilized in a hard modeling subproblem. In addition, ideas from the multistart parameter estimation approach described above are incorporated to ensure that the best possible fit is obtained.

The mixed data algorithm is illustrated with an example shown below involving three species and three reactions as shown in Figure 10.14.

Figure 10.15 shows the concentration and spectra data input used for calibration and optimal estimate output of the modified MCR-ALS algorithm. Figure 10.15a shows the concentration data for the three species in this mechanism. Figure 10.15b shows the spectra data for three measured wavelengths (this is the contents of matrix D described above). The concentration and spectra data are used simultaneously in the modified MCR-ALS algorithm to produce the fit shown in Figure 10.15c, where the simulated species profiles using the optimal estimates for the parameter values (solid and dashed lines) are plotted with the concentration profiles (square, circle, and triangle markers).

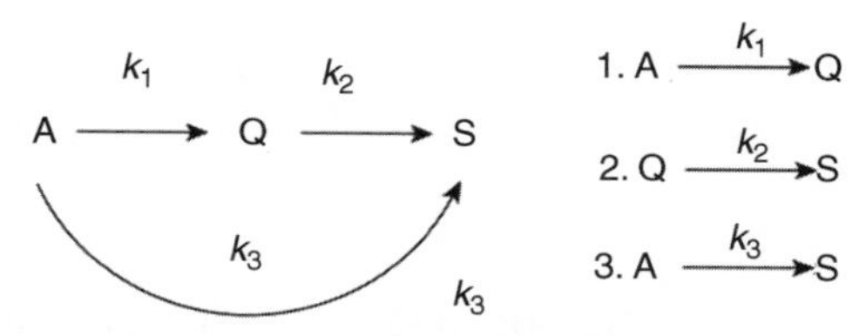

FIGURE 10.14 Schematic of a synthesis mechanism involving three species and three reactions. Reactant A is converted to main product, S, directly and through an intermediate species Q.

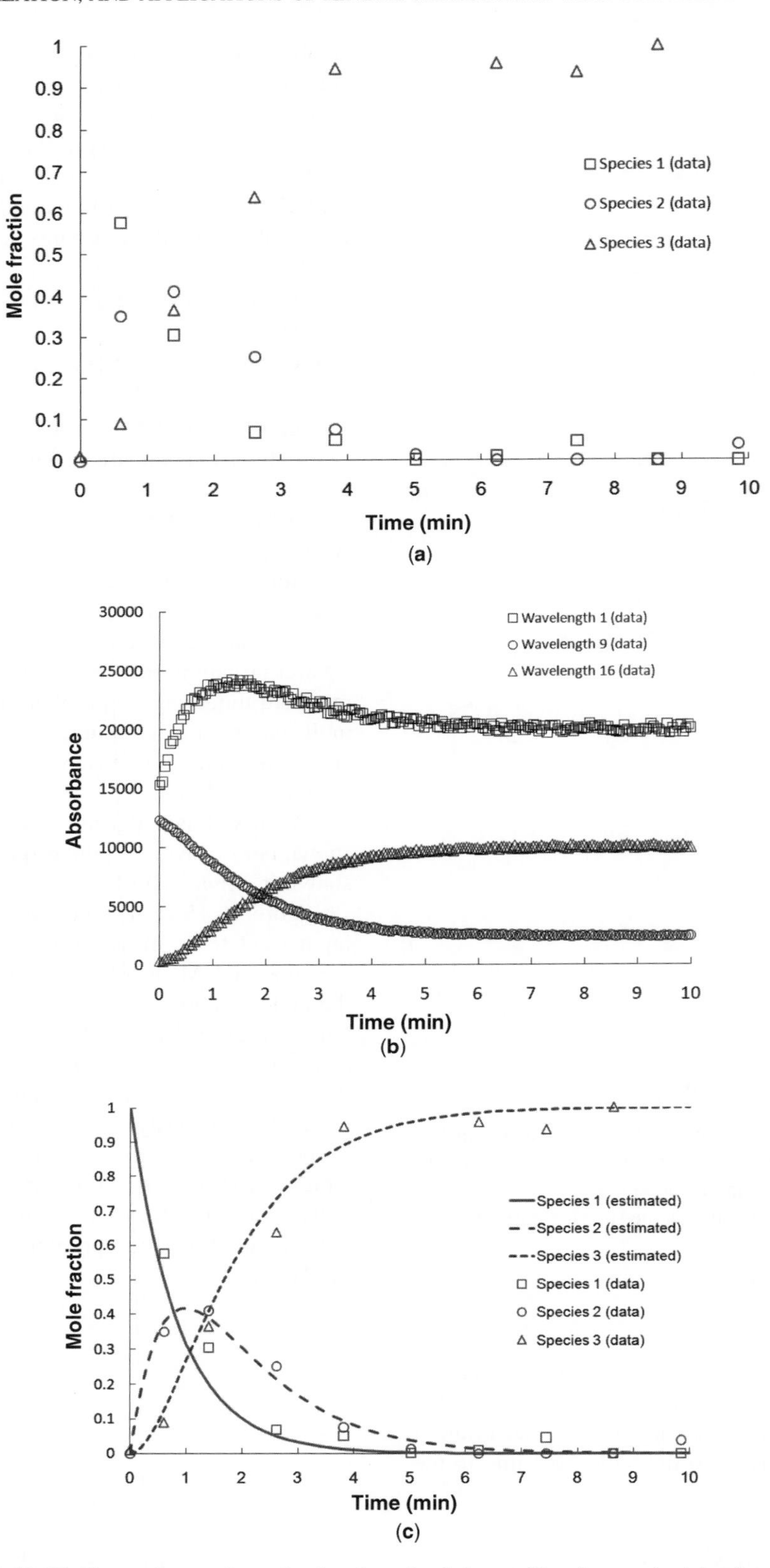

FIGURE 10.15 Input data and results for the mixed data calibration method implemented in OpenChem. Part (a) shows the measured concentration data (algorithm input). Part (b) shows the measured spectra data (algorithm input). Part (c) shows the overlay plot for the optimal solution (simulated profiles and measured concentration data).

10.5 MECHANISM APPLICATION

The above sections described how to build, analyze, and calibrate an API synthesis mechanism. Once calibrated, there are several potential applications of the synthesis mechanism in pharmaceutical manufacturing, including

- scale-up of laboratory procedures;
- detailed design of manufacturing equipment;
- optimization of process equipment and operation;
- design of process control strategies for safety and quality assurance; and
- process improvements.

Scale-up involves taking a chemist's recipe for synthesizing an API in a test tube and creating a manufacturing process that is able to mass produce the API. Effects such as mixing, heat transfer, and by-product formation that are trivial or neglected in the laboratory can have a significant effect in large-scale production. The calibrated API synthesis mechanism can help an engineer determine early in the development process whether it is economically feasible to mass produce the API. Once scale-up procedures have been developed, the synthesis mechanism can be used for detailed design of manufacturing equipment and process operation. For example, accurately determining how operating conditions such as temperature, pressure, and residence time affect reaction conversion and selectivity is critical for designing reactors and separators in the process. After the individual process equipment has been designed, a model of the process, utilizing the API synthesis mechanism, can be used to optimize the process operation and design control strategies necessary for safety and product quality assurance.

The remainder of this section describes three features in OpenChem that facilitate the application of the API synthesis mechanism:

- Operating map generation
- OpenChem scripting
- Application programming interface to third-party software tools

10.5.1 Operating Map Generation

OpenChem automates the task of running multiple simulations for more common operations such as operating map generation. An operating map shows visually how key model outputs depend on multiple operating conditions and parameters. Operating conditions may include reactor residence time, operating temperature, agitator speed, and so on. Model outputs may include product yield, product selectivity, maximum temperature rise, and so on. The operating map is useful for a variety of tasks, including

- visualizing trade-offs between multiple operating conditions;
- providing a visual, more intuitive view of parameter sensitivity than that provided by the sensitivity analysis described above; and
- identifying optimal operating conditions.

In OpenChem, the user is able to select two operating conditions or parameters and one model output. With these user-specified selections, OpenChem will automatically execute a series of simulations by varying the specified operating conditions/parameters and plotting the selected model output in a contour plot.

Figure 10.16 shows the operating map for a model of a batch reactor and the synthesis mechanism described in Figure 10.2. The operating conditions for the operating map in Figure 10.16 are agitator speed (x-axis) and catalyst loading (y-axis). The model output is the processing, or residence, time necessary to achieve 98% product yield.

10.5.2 OpenChem Scripting

Application of the API synthesis mechanism typically involves running the reactor simulations and/or optimizations many times under a variety of conditions. OpenChem provides a scripting language enabling this task. The language allows the user to write high-level commands for sweeping through various operating conditions and parameter values, executing multiple simulations, plotting results, and writing files summarizing the analysis in any customized manner. For example, operating map generation is enabled by this OpenChem scripting language.

10.5.3 Interface to Third-Party Software

Finally, OpenChem implements an application programming interface that enables the synthesis mechanisms to be utilized within other third-party software. One example of this is developing a synthesis mechanism in OpenChem, and then utilizing this mechanism in a detailed reactor design implemented in a CFD (computational fluid dynamics) software package.

10.6 CONCLUSION

The process of building, analyzing, calibrating, and applying API synthesis mechanisms is described. Building API synthesis mechanisms is a challenging task for several reasons, including (1) the reaction steps involved are usually not

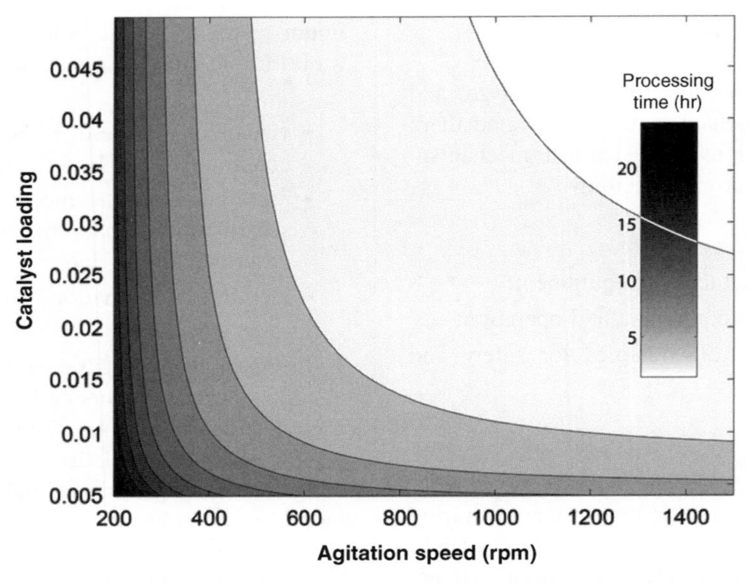

FIGURE 10.16 Operating map for reactor operation for the synthesis described in Figure 10.2. The x-axis is the agitator speed and the y-axis is the catalyst loading. The contour plot shows the residence time in hours required to achieve the desired product yield. This figure indicates that operation is most sensitive to agitator speed at low rpm.

known with certainty and (2) all intermediates and by-products are not measured (or even known). Because of this uncertainty, it is often possible to postulate multiple candidate mechanisms to describe a synthesis. These mechanisms contain several parameters, for example, reaction rate constants, with highly uncertain values. Further, these parameters have different influence on outputs of interest. We must decide which parameters are important and which experiments must be performed in order to calibrate the mechanism. Mechanism calibration is complicated by the fact that the data are often not in a form suitable for traditional calibration methods and the quality of the fit obtained is often dependent on the initial estimates of the uncertain parameters.

The OpenChem software has been designed to address the issues described above and facilitate the mechanism development process. OpenChem provides several ways for specifying the mechanism, tabular and graphical. These two mechanism views offer various advantages, including compact representation of the reactions and easy visualization of the interactions between species in the mechanism. A collection of reactor models is provided and new reactors can be easily added. The results of simulations can be viewed as regular two- and three-dimensional plots as well as a graphical flux analysis animation. The problem of identifying important parameters in the presence of uncertainty is addressed through global sensitivity analysis. In OpenChem, multiple disparate data sources (e.g., concentration and spectra) can be used simultaneously to calibrate the mechanism and special provisions are applied to increase the likelihood that the best possible parameter fit is obtained. Finally, several features are provided for mechanism application, including operating map generation, a scripting language for automating tasks, and an interface to third-party software.

NOTATION

A	Arrhenius expression pre-exponential factor
Amine	amine
ArX	aryl halide
Base	base
C	actual or estimated concentration profiles
Cat	catalyst
C_0	initial guesses for species concentration profiles
D	experimentally measured spectra data
E	experimental error matrix
E_A	Arrhenius expression reaction activation energy
I_1, I_2	intermediate species
$k_{m,1}$, $k_{f,i}$	reaction rate constants
$k(T)$	reaction rate constant
n	parameter indicating how the solid base dissolution rate scales with agitator speed
p_j	time-invariant parameter
Prod	product
R	gas constant
r_i	rate of reaction i in moles per unit time per unit volume
RPM	agitator speed
RPM_{ref}	reference agitator speed
S	actual or estimated pure component spectra

$s(t)$	parametric sensitivity profile
$s_{ij}(t)$	parametric sensitivity of output variable i with respect to parameter j
$\tilde{s}_{ij}(t)$	normalized parametric sensitivity of output variable i with respect to parameter j
S_0	initial guesses for the pure component spectra
T	temperature
X	halide
y_{ArX}	fraction of reactant ArX that has reacted at any given time
y_i	model output
$[\cdot]$	species concentration in moles per unit volume (e.g., [ArX] and [Cat])
β	Arrhenius expression temperature exponent
$\dot{\omega}_i$	molar production rate of species i in moles per unit volume per unit time

REFERENCES

1. Innovation or stagnation: challenge and opportunity on the critical path to new medical products, FDA White Paper, 2004.
2. RES Group, Inc. OpenChem: software for kinetic mechanism development and reactor design, RES Internal Report, June 2007.
3. Green WH, Barton PI, Matheu DM, Schwer DA, Song J, Sumathi R, Carstensen HH, Dean AM, Grenda JM. Computer construction of detailed chemical kinetic models for gas-phase reactors. *Ind. Eng. Chem. Res.*, 2001;40 (23):5362–5370.
4. Hopkinson GA.Computer-assisted organic synthesis design, Ph.D. thesis, University of Leeds, 1985.
5. Brenan KE, Campbell SL, Petzold LR. *Numerical Solution of Initial-Value Problems in Differential-Algebraic Equations*, SIAM, New York, 1996.
6. Feehery WF, Tolsma JE, Barton PI. Efficient sensitivity analysis of large-scale differential-algebraic systems, *Appl. Numer. Math.*, 1997;25 (1):41–54.
7. Bard J. *Nonlinear Parameter Estimation*, Academic Press, New York, 2001.
8. Bazaraa MS, Sherali HD, Shetty CM. *Nonlinear Programming: Theory and Algorithms*, 2nd edition, Wiley, 1993.
9. de Juan A, Maeder M, Martinez M, Tauler R. Combining hard- and soft-modeling to solve kinetic problems. *Chemometr. Intell. Lab. Syst.*, 2000;54: 123–141.
10. Bezemer E, Rutan SC. Multivariate curve resolution with non-linear fitting of kinetic profiles, *Chemometr. Intell. Lab. Syst.*, 2001;59: 19–31.
11. Ingle JDJ, Crouch SR. *Spectrochemical Analysis*, Prentice Hall, New Jersey, 1988.
12. RES Group, Inc. Calibration of kinetic models with combined concentration and spectra data, RES Internal Report, March 2008.

11

PROCESS SAFETY AND REACTION HAZARD ASSESSMENT

Wim Dermaut

Process Safety Center, API Small Molecule Development, Johnson and Johnson Pharmaceutical Research and Development, Beerse, Belgium

11.1 INTRODUCTION

When the issue of safety is raised in the context of pharmaceutical manufacturing, most of us might first think about issues of product and/or patient safety. There is another side of safety that might not get as much attention but that is also crucial to the production of pharmaceuticals: process safety. An often heard phrase in this context is "If the process can't be run safely, it shouldn't be run at all." Process safety should indeed be a concern, starting already in early development of a drug candidate. Running a small-scale synthesis in the lab only once is one thing, and running this process at metric ton scale on a routine basis in a chemical manufacturing plant is something completely different. Events such as exothermicity, gas generation, and stability of products might be relatively unimportant on a small scale, but they can pose tremendous challenges when this reaction is run at a larger scale.

This chapter provides an introduction to the field of process safety. The aim is to discuss some of the fundamentals of safety testing, in order to try to facilitate the communication between chemical engineers and development chemists. The focus will be on the interpretation and practical use of the different test results, rather than on the tests itself. The discussion will also focus on (semi-) batch reactors, since this is still the most commonly used type of reactors.

This chapter can be roughly divided in four main parts. We will start with a brief description of some general concepts such as the runaway scenario and the criticality classes. After that, we will consider some safety aspects of the desired synthesis reaction and how it can be studied on lab scale. The main focus will be on exothermicity (heat generation) and gas generation. We will then continue discussing how the data thus obtained can be used to scale up the reaction safely, with a large emphasis on the heat transfer at large scale. Finally, we will take a closer look at the undesired decomposition reactions that can take place in case of process deviations, how to study them at lab scale, and how to minimize the associated risks.

In the next section, the reader is offered some first insights into the domain of process safety and safety testing. It is by no means the intention of the author to give anything near an exhaustive overview of this field, but hopefully this introduction can provide some insight into the most common pitfalls of process safety. For a more in-depth review, the reader is referred to the widely available literature [1, 2, 10].

11.2 GENERAL CONCEPTS

11.2.1 Runaway Scenario

When discussing process safety, the cooling failure scenario is often used to illustrate the possibility of a runaway reaction in a reactor [3, 4]. In Figure 11.1, a possible cooling failure scenario is depicted. The normal process condition is indicated with the thin solid line: the reactants are being charged to the reactor (batch reaction), the reaction mixture is heated to the desired process temperature (T_p), the mixture is then

Chemical Engineering in the Pharmaceutical Industry: R&D to Manufacturing Edited by David J. am Ende

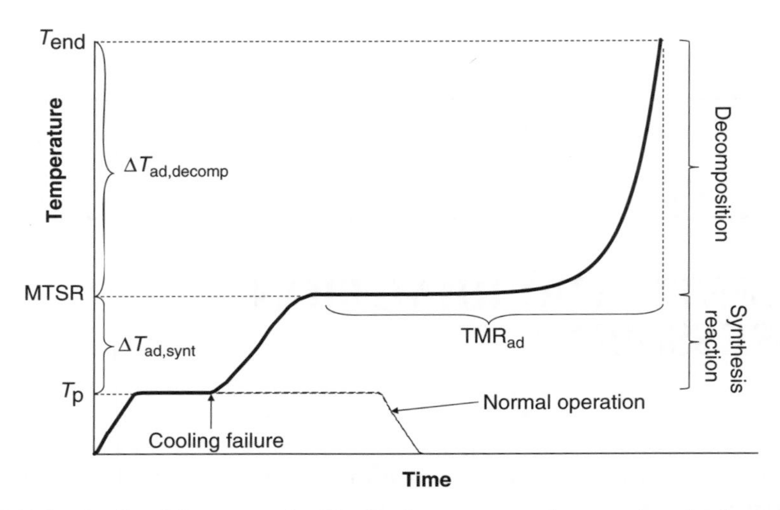

FIGURE 11.1 Cooling failure scenario. The thin line represents the normal mode of operation, and the thick line represents the possible consequences of a cooling failure. Reproduced with permission from Ref. 10. Copyright Wiley-VCH Verlag GmbH & Co. KGaA.

kept isothermally at this temperature with active jacket cooling (exothermic reaction), and when the reaction is finished, the mixture is brought back to room temperature for further workup. This is the process as it is intended to be run at both at small scale in the lab and at larger scale in the plant.

Let us now consider the possible consequences of a loss of cooling. We will assume that the loss of cooling power occurs relatively shortly after the desired process temperature has been reached, as indicated in the graph. From this point on, the exothermic reaction will proceed, but since the reaction heat is no longer removed by the jacket, the temperature in the reaction mass will start to increase. There is no heat exchange between the reactor and the surroundings, so the system is said to be adiabatic. After a certain time, the reaction has gone to completion, and hence a final temperature is reached that is called the maximum temperature of the synthesis reaction (MTSR). The total temperature increase from the process temperature to the MTSR is called the adiabatic temperature rise of the synthesis reaction ($\Delta T_{ad,synt}$). Up to this point in the cooling failure scenario, we are only dealing with the desired synthesis reaction. The study of the desired reaction will therefore be discussed first in the following paragraph.

When the MTSR is reached, a secondary exothermic reaction may take place, that is, a thermal decomposition of the reaction mixture or any of the ingredients. If such decomposition takes place, the temperature will increase further until the final temperature T_{end} has been reached. The time between reaching the MTSR and the point of the maximum rate of the decomposition reaction (i.e., thermal explosion) is called the time to maximum rate under adiabatic conditions (TMR_{ad}). It is generally accepted that a TMR_{ad} of 24 h or more can be considered as safe. The chance that a reactor would stay under adiabatic conditions for more than 24 h is low. A cooling failure should be noticed quite rapidly, and this leaves ample time to take corrective measures such as restoring the original cooling capacity, applying external emergency cooling, quenching the reaction mixture, or transferring it to another vessel or container with appropriate cooling. In analogy with the synthesis reaction, the total temperature increase from the MTSR to T_{end} is called the adiabatic temperature rise of the decomposition reaction ($\Delta T_{ad,decomp}$). How decomposition reactions are studied at lab scale and how they are dealt with during scale-up will be discussed later.

11.2.2 Criticality Classes

Starting from the cooling failure scenario, the criticality of any chemical process can be described in a relatively simple way by using the criticality classes as first introduced by Stoessel in 1993 [3]. In this method, the following four different temperatures need to be known to assess the possible consequences of a runaway reaction:

1. The process temperature under normal conditions (T_p).
2. The maximum temperature of the synthesis reaction.
3. The temperature at which the TMR is 24 h. In the description above, the time to maximum rate (TMR) concept was introduced. Since the reaction rate strongly depends on the temperature, the TMR_{ad} will vary with temperature as well. The importance of a TMR that is longer than 24 h was pointed out, and hence the third temperature we need to know is the temperature at which the TMR is 24 h (we will denote this temperature as $TMR_{ad,24h}$).
4. *The maximum temperature for technical reasons (MTT).* In an open system, this is the boiling point of

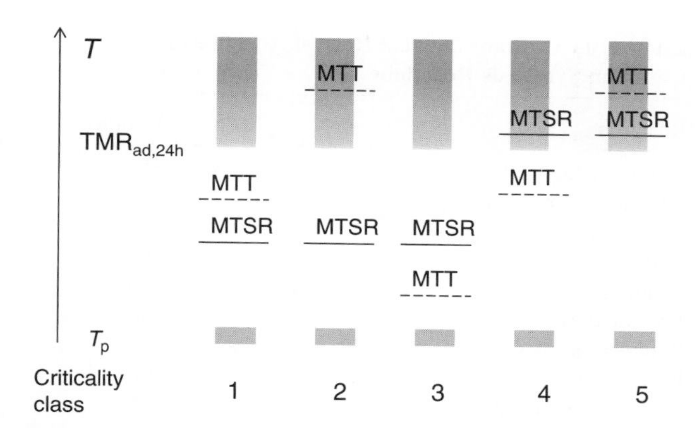

FIGURE 11.2 Criticality classes of a chemical process. In this classification, processes are divided into five different criticality classes, ranging from class 1 (intrinsically safe) to class 5 (high risk). Reproduced with permission from Ref. 10. Copyright Wiley-VCH Verlag GmbH & Co. KGaA.

the reaction mixture, and in a closed system, it is the temperature that corresponds to the bursting pressure of the safety relief system. This is the temperature that cannot be surpassed under normal process conditions and can therefore act as a safety barrier. Only when dealing with very rapid temperature rise rates a risk of over pressurization or flooding of the condenser lines might occur. This will be discussed later.

When these four temperatures are known for a given process, the criticality class can be determined according to Figure 11.2. Five different criticality classes are defined, ranging from the intrinsically safe class 1 processes to the critical class 5 processes.

Let us consider a process that corresponds to the class 1 type. In this case, the process is run at the process temperature T_p, and when a cooling failure takes place, the temperature will increase to the MTSR. This temperature is below the $TMR_{ad,24h}$, meaning that even in case the reaction mixture would remain at this temperature (under adiabatic conditions) for 24 h, there would be no serious consequences. Moreover, the MTT is situated between the MTSR and the $TMR_{ad,24h}$ giving an extra safety barrier for any possible further temperature increase. So even in case this process would run out of control due to a loss of cooling, there will be no real safety concerns.

The story is entirely different however when considering the class 5 process. In this case, a loss of cooling would raise the temperature inside the reactor to the MTSR, but here this temperature is higher than the $TMR_{ad,24h}$. This means that the secondary decomposition reaction will go to completion in less than 24 h if the reaction mixture remains under adiabatic conditions for a prolonged period of time. The MTT is higher than $TMR_{ad,24h}$, so there is a possibility that it will not be sufficient to prevent a true thermal explosion. This type of reactions are truly critical from a safety point of view and either a redesign of the process should be considered to bring it to a lower criticality class or appropriate safety measures should be taken.

The three other classes are intermediate cases and will not be described explicitly here, so the reader is referred to the original publication. The criticality index can be very useful to come to a unified risk assessment of a process. Some caution is needed, however, as this classification does not take pressure increase into account. As will be discussed later, pressure effects are at least as important as temperature effects in the assessment of process safety. This was addressed by the original author in a later publication [5], where a modified type of criticality index was proposed that does take pressure effects into account.

11.3 STUDYING THE DESIRED SYNTHESIS REACTION AT LAB SCALE

11.3.1 Compatibility

Before starting with any further safety assessment of a chemical process, it is crucial to evaluate the compatibility of all reagents being used. Ideally, the reagents should show no reactivity other than that leading to the desired reaction. Some of the incompatibilities are very obvious: developing a chlorination reaction with thionyl chloride in an aqueous solution simply does not make sense. Some other incompatibilities might be less known but can also have very serious consequences. The stability of hydroxylamine, for instance, is catastrophically influenced by the presence of several metal ions [6, 7], and even in the parts per million range this type of contamination can have severe consequences. The first starting point for any compatibility assessment should be "*Bretherick's Handbook of Reactive Chemical Hazards*," [8] a standard reference with a vast list of known stability and compatibility data on a wide range of chemicals.

Compatibility issues for several different conditions should be checked from the literature, or where the information is not available, the data should be generated experimentally.

1. Compatibility of all reagents used in combination with the other reagents present.
2. Compatibility of reagents with possible main contaminants in other reagents. Technical dichloromethane, for instance, is often stabilized with 0.1–0.3% of ethanol, which can turn out to be significant because of the large molar excess of the solvent in the reaction mixture.
3. Compatibility of the reagents with construction materials such as stainless steel (vessel wall) and sealings (Kalrez, Teflon, etc.). For example, the use

of disposable Teflon dip tubes may be appropriate for the handling of liquids that are very sensitive to contamination with metal ions such as hydroxylamine. Two questions need to be answered: will the product degrade when in contact with these materials and will the construction materials be affected by the product (corrosion, swelling of gaskets or sealings, etc.).

4. Compatibility of all products used with environmental factors such as light, oxygen, and water. If a product is incompatible with water, appropriate actions are needed to avoid contact with any source of water: containers should be closed under inert conditions in order to avoid contact with air humidity, containers should not be stored in open air in order to avoid water ingression due to rain, reactions should be run in a reactor where the heat transfer media (such as jacket cooling and condenser cooling) are water free, and so on. A first indication of possible compatibility issues with oxygen can be obtained from two DSC experiments in an open crucible, once under nitrogen atmosphere and once under air. If there is a pronounced difference between the outcomes of both experiments, the product is very likely to show some degree of reactivity with oxygen.

11.3.2 Exothermicity

Most chemical processes that run in pharmaceutical production plants are exothermic reactions. In general terms, a reaction is called exothermic when heat is generated during the course of the reaction. Reactions that absorb heat during their course are called endothermic reactions. Chemical processes in pharmaceutical production are in most cases designed as isothermal processes, so the heat that is generated during the course of reaction has to be removed effectively, usually through jacket cooling of the reactor. Intuitively, one can understand that an effective heat removal will become increasingly difficult when the scale of the process is increased from milliliter (lab) to cubic meter (production). Therefore, a correct assessment of the reaction heat becomes crucial when a process is being run at a larger scale.

From a thermodynamic point of view, the heat being released (or absorbed) by a reaction matches the difference in heat of formation between reactants and products. Hence, a first indication of the heat of reaction of any process can be obtained by making this calculation based on the tabulated literature data [9]. By convention, reaction enthalpies for exothermic reactions are negative values; for endothermic reactions, they are positive values.

The heat of reaction of a chemical process is usually expressed in the unit of energy per mole, for example, kcal/mol or kJ/mol. Some typical heats of reaction for common chemical processes are given in Table 11.1 [10].

TABLE 11.1 Some Typical Heat of Reactions for Common Synthesis Reactions

Reaction	ΔH_R (kJ/mol)
Neutralization (HCl)	−55
Neutralization (H_2SO_4)	−105
Diazotization	−65
Sulfonation	−150
Amination	−120
Epoxidation	−100
Polymerization (styrene)	−60
Polymerization (alkene)	−200
Hydrogenation (nitro)	−560
Nitration	−130

This table clearly shows that there is a big span in heats of reaction one can encounter in process chemistry, with the highest energies (and hence the highest risks) being related to the usual suspects such as hydrogenations of nitro compounds and polymerizations. When developing this type of reactions, extra care should be taken and a correct determination of the total reaction heat and the kinetics of the process by means of calorimetry is crucial.

The reaction heat of a chemical reaction can be determined by means of a reaction calorimeter. This is basically a small-scale reactor in which the reaction can be performed under controlled circumstances while recording any heat entering or leaving the system. Most used is the heat-flow calorimeter, where the reaction heat is measured by continuously monitoring the temperature difference between the reaction mixture and the cooling/heating fluid in the jacket.

$$Q_{\text{flow}} = U \times A \times (T_R - T_J) \tag{11.1}$$

where Q_{flow} is the heat flowing in or out the reaction mixture (W), U is the heat transfer coefficient (W/(m^2 K)), A is the heat exchange area (m^2), T_R is the reaction temperature (K), and T_J is the jacket temperature (K).

There are different heat-flow calorimeters available on the market, such as the RC1 (Mettler Toledo), the Calo (Systag), and the Simular (combined with power compensation calorimetry, HEL). Other systems offer reaction calorimetry based on a more direct measurement of the heat flux such as the Chemisens CPA (peltier based) and the Mettler Toledo RTCal. In our discussion, we will limit ourselves to heat-flow calorimetry, since it is the most widespread technique to date, but the interpretation of the data obtained with other types of calorimeters will be very comparable.

In principle, a heat-flow calorimeter can be considered as a scaled down jacketed reactor (usually in the range from 100 mL to 2 L), with a very accurate temperature control. Usually such a calorimeter is run in isothermal mode, so the temperature of the reaction mixture is kept constant during the course of the reaction. If the reaction is exothermic, the

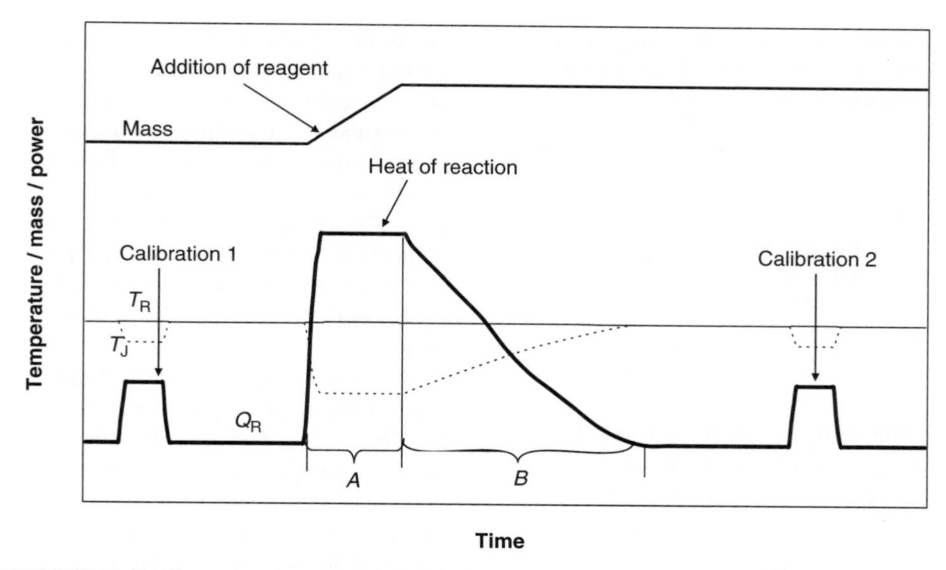

FIGURE 11.3 Example of a reaction calorimetry experiment. A calibration (with a calibration heater) is first performed, and then the reagent is dosed and the reaction takes place. When the reaction is finished, a final calibration is performed. The thick solid line represents the reaction heat.

jacket temperature will have to be lower than the reaction temperature in order to remove the reaction heat. As can be seen from equation 11.1, measuring T_J and T_R is not enough to obtain the reaction heat entering or leaving the reactor; we also need to know U and A. The heat transfer area A is usually easy to obtain: since the reactor geometry of the calorimeter is fixed, the heat exchange area as a function of the volume of the reaction mixture is known. The heat transfer coefficient U is most commonly obtained by the use of a calibration heater. During a certain period of time (typically 5 or 10 min), a calibration heater with a known heat output is switched on. The temperature of the jacket will be adjusted in such a way that the temperature in the reaction mixture remains unchanged. Since Q, A, T_R, and T_J from equation 11.1 are all known for this calibration period, U can be calculated. U is a function of a variety of factors such as viscosity, stirring speed, and temperature, as will be discussed later in greater detail. This means that U will be different in each calorimetry experiment, and it will even be different before and after the reaction, according to the physical properties of the reaction mixture. Therefore, the calibration is performed once before the reaction takes place and once after the reaction is finished, yielding the appropriate U values for the reaction mixture before and after the reaction.

An example of a semi-batch calorimetry experiment is shown in Figure 11.3. The reactor is filled with appropriate reagents and brought to the reaction temperature. After the temperature of both the reactor (T_R) and the jacket (T_J) has reached stable values, the calibration procedure as described above is started. When the calibration heater is switched on, it can be seen that the jacket temperature goes down almost immediately in order to maintain a constant reactor temperature. This is reflected in the Q_R signal that reaches a stable reading after a short period. After a couple of minutes, the calibration heater is switched off and reactor and jacket temperatures are again allowed to reach stable readings. The reaction is then started by a gradual dosing of the desired reagent, as can be read from the mass signal. Here, the response in heat profile is almost instantaneous as well, and a relatively stable heat signal is observed until all of the reagents have been dosed. At the end of the dosing, the heat signal does not drop to the baseline as rapidly as during the calibration procedure. This phenomenon where heat is being released after the addition of the reagent has been stopped is called thermal accumulation. The thermal accumulation at the end of the dosing can be calculated according to the following equation:

$$\text{Percentage of thermal accumulation} = 100 \times \frac{B}{A+B} \tag{11.2}$$

with A and B being the partial integrations of the heat signal as shown in Figure 11.3. If the heat signal dropped to zero immediately after the dosing had stopped, there would be no thermal accumulation. On the other hand, if the dosing was instant (which is the case in a batch reaction), there would be 100% thermal accumulation.

Let us now take a closer look at the key figures that can be extracted from a calorimetry experiment, and how they should be interpreted.

11.3.2.1 Reaction Heat, Adiabatic Temperature Rise, and MTSR The integration of the heat signal versus time gives

TABLE 11.2 Theoretical Example of the Resulting Adiabatic Temperature Rise for the Hydrogenation of Nitrobenzene Under Different Reaction Conditions

	Case 1	Case 2
Reaction heat	−560 kJ/mol	−560 kJ/mol
Concentration	2 M	0.5 M
Solvent	Chlorobenzene	Water
Density solvent	1.11 kg/L	1 kg/L
Specific heat solvent	1.3 kJ/(kg K)	4.2 kJ/(kg K)
ΔT_{ad} (equation 11.3)	776°C	67°C

us the total reaction heat, usually expressed in kJ or kcal. From this reaction heat, the adiabatic temperature rise of the synthesis reaction can be calculated according to equation 11.3.

$$\Delta T_{ad} = \frac{-\Delta H_R}{c_p} \tag{11.3}$$

where ΔH_R is the reaction enthalpy (kJ/kg) and c_p is the specific heat capacity of the reaction mixture (kJ/(kg K)).

The MTSR can be calculated merely by adding the adiabatic temperature rise to the reaction temperature (equation 11.4).

$$\text{MTSR} = T_{\text{Process}} + \Delta T_{ad} \tag{11.4}$$

Whereas the molar reaction enthalpy is an intrinsic property of a specific reaction, the adiabatic temperature rise is dependent on the reaction conditions. In the (hypothetical) example in Table 11.2, this difference is demonstrated.

This example clearly shows the importance of reaction conditions, with the adiabatic temperature rise being more than ten times higher in case 1. This dramatic difference can be fully attributed to the effect of the solvent acting as a heat sink. Working at higher dilution in a solvent with a higher heat capacity can drastically reduce the possible consequences of reaction that runs out of control. Unfortunately, running a process at higher dilution has an impact on the overall economy, so both aspects should be considered.

The adiabatic temperature rise is often used as a measure for the severity of a runaway reaction. A process with an adiabatic temperature rise of less than 50K is usually considered to pose no serious safety concerns, at least when there is no pressure increase associated with the reaction. When a process has an adiabatic temperature rise of more than 200K, a runaway reaction would most probably result in a true thermal explosion, and hence such processes require a very thorough safety study.

11.3.2.2 Thermal Accumulation According to equation 11.2, the thermal accumulation can be calculated by the partial integration of the heat signal. Thermal accumulation is an important parameter in the assessment of the safety of a process. If a problem occurs during a process (cooling failure, stirrer failure, etc.), it is common practice to stop the addition of chemicals immediately. In case there is no thermal accumulation, the reaction will also stop immediately and there will be no further heat generation that can lead to a temperature increase in the reactor. A reaction with 0% thermal accumulation is therefore called dosing controlled. If there is thermal accumulation, however, part of the reaction heat will still be set free after the dosing has been stopped, and hence the temperature in the reactor can increase.

Because of the importance of the thermal accumulation, the MTSR is often specified as being either $\text{MTSR}_{\text{batch}}$ or $\text{MTSR}_{\text{semi-batch}}$. For the calculation of the former, the total adiabatic temperature rise is added to the reaction temperature, whereas for the latter, the adiabatic temperature rise is first multiplied by the percentage of thermal accumulation. An example is given in Table 11.3.

This example shows the big difference in intrinsic safety of the process between the two cases. Should a cooling failure occur in case 1, the temperature would never be able to rise significantly above the process temperature, providing of course that the dosing is stopped as soon as the failure occurs. In case 2, on the other hand, the temperature would increase to 120°C without the possibility to cool, even when dosing is stopped immediately.

So, obviously, low thermal accumulation is to be preferred for any semi-batch process. A high degree of thermal accumulation is a sign that the reaction rate is low relative to the dosing rate. Two possible measures can be taken to decrease the thermal accumulation of a given process:

1. Increase the reaction rate. This can be done by increasing the reaction temperature. Increasing the reaction rate means also increasing the heat rate of the reaction, so a calorimetry experiment at this new (higher) process temperature is required to make sure that the cooling of the reactor can cope with the heat generation under normal process conditions. Obviously, the increased temperature will lead to a smaller

TABLE 11.3 Example of the Effect of the Thermal Accumulation on the $\text{MTSR}_{\text{semi-batch}}$

	Case 1	Case 2
Reaction temperature	60°C	60°C
Total reaction enthalpy	−200 kJ/kg	−200 kJ/kg
Thermal accumulation	2%	60%
Specific heat	2 J/(g K)	2 J/(g K)
$\Delta T_{\text{ad,batch}}$	100°C	100°C
$\text{MTSR}_{\text{batch}}$	160°C	160°C
$\Delta T_{\text{ad,semi-batch}}$	2°C	60°C
$\text{MTSR}_{\text{semi-batch}}$	62°C	120°C

safety margin between reaction temperature and possible decomposition temperature, and this should be dealt with appropriately.

2. Decrease the dosing rate.

11.3.2.3 ***Heat Rate*** Whereas a correct determination of the total reaction heat is important, the rate at which this heat is being liberated is at least equally important for a proper safety study. Where a process is run under identical conditions both at large scale and in the calorimeter, the heat rate (in W/kg) is scale independent. It should be kept in mind, however, that the cooling capacity of a reaction calorimeter is in most cases several orders of magnitude higher than that of a large-scale production vessel. A reaction calorimeter might still be able to keep a constant heat rate of 200 W/kg under control, while running this process at production scale will most certainly lead to a runaway reaction. In such a case, the process should be redesigned as to decrease the heat rate, and ideally the calorimetry experiment should be repeated under the new process conditions to make sure that no unwanted side effects occur (higher thermal accumulation, sudden crystallization, formation of extra impurities, etc.).

Not only the absolute value of the heat rate has to be considered, the duration of the heat evolution is also important. A peak in the heat evolution that surpasses the available cooling capacity but only lasts for a short period is not necessarily problematic. If such a peak is observed, one should calculate the corresponding adiabatic temperature rise and evaluate its consequences. For example, a heat rate of 200 W/kg for 2 min would give rise to a temperature increase of 12°C under adiabatic conditions, assuming a specific heat of 2 J/(g K). If a cooling capacity of 50 W/kg is available, this will only be 9°C. The issues one might encounter when scaling up a reaction to meet the heat removal capacities of the production vessel will be discussed in more detail later.

11.3.3 Gas Evolution

Up till now, we have focused only on the heat being generated by exothermic chemical processes. From a safety perspective, gas evolution and a resulting pressure buildup can have even more devastating consequences, so a proper knowledge of any gaseous products being formed during a process is crucial to ensuring a safe execution at production scale.

There are quite a few common reactions that do liberate considerable amounts of gas: chlorinations with thionyl chloride, BOC deprotections, quenching of excess hydride, and decarboxylations, to name a few. The most appropriate way to quantify the gas evolution during a reaction is to couple any type of gas flow measurement device to a reaction calorimeter and run the process under the same conditions as it will be run on scale.

There are several possibilities for the measurement of gas evolution at small scale.

1. *Thermal Mass Flow Meters*: This type of devices is probably the most widespread when a flow of gaseous products has to be measured. They are available in a large span of measuring ranges (from less than 1 mL/min to several thousand liters per minute), are relatively cheap, and deliver a signal that can be picked up easily as an input in the reaction calorimeter. However, this type of meters measures a mass flow (i.e., grams of gas per minute) but not a volumetric flow (milliliter per minute). When dealing with one known single type of gas, this is no problem as the volumetric flow can be easily calculated from the mass flow signal. When the gas to be measured is a mixture of different components, or when the composition of the gas stream is entirely unknown, the volumetric flow cannot be obtained reliably.
2. *Wet Drum Type Flow Meters*: The gas is led through a drum that is half submerged in inert oil, causing this drum to rotate. This rotation is recorded and is proportional to the volumetric flow (as opposed to the thermal mass flow meters). When using a unit that is entirely made of an inert material (e.g., Teflon), a very broad range of gaseous products can be studied. However, the dynamic range of this type of instruments is only modest, accuracy at the low end of the flow ranges (0–20 mL/min) is rather limited, and the fact that the drum rotates in a chamber filled with inert oil makes it susceptible to mechanical wear.
3. *Gas Burette*: This type of device measures the pressure increase in a burette that is filled with inert oil, releasing the overpressure at a predefined value, making an accurate determination of low gas flow rates possible. The signal is proportional to the volumetric flow, the setup is extremely simple without any moving parts, and it is fully corrosion resistant (only glass and silicon oil in contact with the gas). However, the output signal is difficult to integrate in any evaluation software (combination of pressure signal and count of the number of "trips"), and the flow range that can be measured is limited at the high end to approximately 50 mL/min (using the standard type of burette).
4. *Rotameters, Bubble Flow Meters, and so on*: There are other types of laboratory gas flow meters that will not be discussed here since they give only a visual readout and not a signal that can be incorporated electronically.

In the explanation above, it has been emphasized that the determination of a volumetric gas flow is of interest, rather than a mass flow. When scaling up the reaction to plant scale, we need to make sure that gas that is being produced can be removed safely from the vessel to the exhaust. This means that all of this gas will have to flow through piping with a certain diameter, and the limiting factor for a gas flowing through a pipe without causing pressure buildup is its volume, and not its mass. The maximum allowable gas rate for a specific process depends on the actual production plant layout, and this will be dealt with later.

Apart from the obvious importance of measuring the gas flow rate during a process, it might also be of interest to characterize the gas that is being emitted. Although there is no difference in possible pressure buildup, having a release of 50 m^3/h of carbon dioxide will obviously feel more comfortable for any chemist or operator than having a release of 50 m^3/h of hydrogen cyanide. When gas evolution comes into play, industrial hygiene, environmental emission limitations, and hazard classification (e.g., when hydrogen is being set free) should all be addressed appropriately.

Characterization of the gas being liberated during a process at lab scale is not an easy thing to do. Ideally, an online mass spectrometer can be used to quantify the exact composition of the gas stream at any time. Mass spectrometers with an appropriate measuring range (down to 28 Da when carbon monoxide is to be detected), low dead volume to eliminate unnecessary long holdup times, and high resolution (both nitrogen and carbon monoxide have a molecular weight of 28, so very high resolving power is needed to discriminate between them) do not come cheaply. Collecting the gas leaving the reactor in a gas sampling bag and subsequently injecting this gas into a regular mass spectrometer can be a viable alternative. Another widely used technique is to trap the gas in a wash bottle with an appropriate solvent in which the gas either dissolves or with which it reacts, and then to analyze this solution in a traditional way. Which technique is being used is irrelevant, but one should always try to know the composition of the gas stream leaving the reaction mixture.

11.4 SCALE-UP OF THE DESIRED REACTION

11.4.1 Heat Removal

11.4.1.1 Film Theory When designing an exothermic reaction for scale-up, it is important to know what the heat removal capacity of the reactor at production scale is. Unfortunately, this is easier said than done. The heat transfer between the heat transfer medium in the jacket and the reaction mixture is usually described in terms of a series of resistances, the so-called film theory. It considers three main factors governing the heat transfer in a stirred tank reactor: the resistance of the inner film (boundary reaction mixture – vessel wall), the resistance of the vessel wall, and the resistance of the outer film (boundary vessel wall – heat transfer fluid). This can be expressed numerically:

$$\frac{1}{U} = \frac{1}{h_r} + \frac{d}{\lambda} + \frac{1}{h_c} = \frac{1}{h_r} + \frac{1}{U_{max}} \qquad (11.5)$$

where U is the overall heat transfer coefficient (W/(m^2 K)), h_r is the inner film transfer coefficient (W/(m^2 K)), d is the thickness of the vessel wall (m), λ is the thermal conductivity of the vessel wall (W/(m K)), h_c is the inner film transfer coefficient (W/(m^2 K)), and U_{max} is the maximum heat transfer coefficient (W/(m^2 K)).

From this equation, it can be understood that there are two main contributions to the overall heat transfer coefficient: one that solely depends on the characteristics of the reaction mixture, and one that solely depends on the characteristics of the reactor. Indeed, the inner film transfer coefficient h_r is a measure for the resistance to heat transfer between the reaction mixture and the vessel wall, and is strongly correlated to the physicochemical properties of the reaction mixture (viscosity, density, heat capacity, etc.) and the stirring speed. U_{max}, on the other hand, can be interpreted as the maximum obtainable heat transfer coefficient in a certain reactor in the hypothetical case when the inner film resistance would approach zero. This term comprises of two contributing parts: one that is due to the thermal conductivity of the vessel wall and the other that is due to the outer film coefficient. These two terms solely depend on the characteristics of the reactor.

This explains why a correct scale-up of the heat transfer characteristics is so difficult: since U is dependent on both the process and the reactor, it should be ideally determined or calculated separately for each vessel–process combination. This is a rather time-consuming process, as can be seen from the list of actions that should be undertaken.

1. Determine the U_{max} of the calorimeter at the process temperature
2. Determine U for the reaction mixture under process conditions in the calorimeter
3. From 1 and 2, calculate h_r for the reaction mass in the calorimeter
4. Calculate h_r for the reaction mass in the production vessel (literature scale-up rules)
5. Determine the U_{max} of the production vessel at the intended jacket temperature
6. From 4 and 5, calculate the U value for the reaction mixture in the production vessel

For a thorough description of the theory behind this approach, the reader should refer to the literature [11–13]. In the next section, we will briefly discuss some major issues related to the determination of the heat transfer coefficient at production scale.

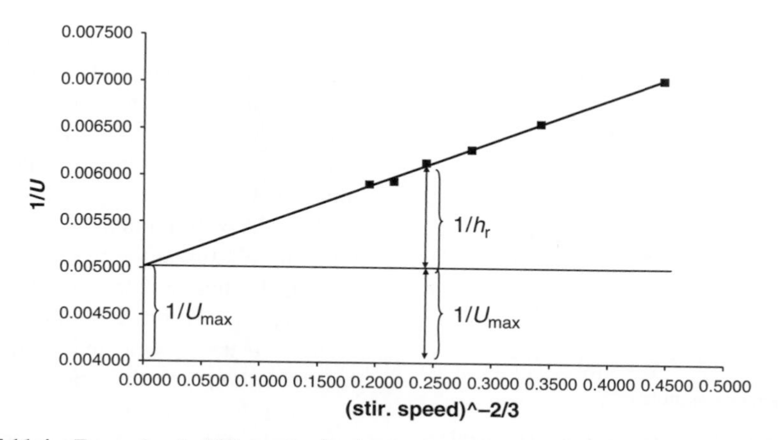

FIGURE 11.4 Example of a Wilson plot for the determination of the heat transfer characteristics of a reactor. The markers are experimentally obtained heat transfer coefficients at different stirring speeds. Through these points a straight line can be fitted that yields the contribution of both the maximum heat transfer coefficient and the inner film coefficient to the total heat transfer coefficient.

11.4.1.2 Determination of U The most widely used approach to determine the heat transfer characteristics of a reactor is by means of the Wilson plot [14]. The heat transfer coefficient is determined experimentally (by means of a cooling curve) at several different stirring speeds. When plotting the reciprocal heat transfer coefficient at a certain jacket temperature versus the stirring speed to the power −2/3 (equation 11.6), a straight line is obtained, with the intercept being equal to U_{max}^{-1} as given in the following equation:

$$\frac{1}{U} = c^{te} \times n^{-2/3} \qquad (11.6)$$

where n is the stirrer speed (in rpm).

An example of such a Wilson plot is shown in Figure 11.4.

This is intrinsically the most reliable way to determine the U_{max} for any reactor, since the result is independent of the solvent being used for the cooling experiment. When repeating the same experiments with a different solvent, the experimental points and the slope of the line will differ, but U_{max} will remain the same.

This method can be used for the characterization of a reaction calorimeter where a large number of either cooling curves or U determinations via the calibration heater can be run in automated way. Since U_{max} depends on the filling degree and jacket temperature, a vast range of U determinations are necessary for a proper description of the heat transfer properties of the reaction calorimeter. It is our experience that isopropanol is the most suitable solvent for obtaining good Wilson plots in the reaction calorimeter.

This method can be applied to the reaction calorimeter in programmed mode; however, it becomes rather cumbersome for use at plant scale. Since each measuring point in the curve has to be obtained from one cooling curve at one stirring speed, it becomes very time-consuming to gather the data needed for this plot. Therefore, another approach can be used for the estimation of U_{max} at production scale. As mentioned above, constructing the Wilson plot with different solvents will alter the slope but not the intercept. If water is used for the construction of the Wilson plot, the slope turns out to be very low; that is, the contribution of $1/h_r$ is low. This implies that determining the U value with a reactor filled with water at the highest possible stirring speed will yield a value that is close to U_{max}. In this way, a good approximation of the maximum heat transfer capacity of the vessel can be obtained from only one experiment. This approximation will obviously be less accurate, but the error made is always on the safe side: the U_{max} will be underestimated, and hence in reality there will be more cooling power available than anticipated. An example of this approach is shown in Figure 11.5, where the heat transfer coefficients for a typical stainless steel reactor and a glass-lined reactor of 6000 L are shown as a function of jacket temperature for a filling degree of 50% and 85%. This shows clearly the large range of U values that can be encountered in practice.

It is important to note that the U_{max} value is a function of the jacket temperature, rather than the reactor temperature because it is linked to the properties of the jacket and vessel wall, which is always at approximately the same temperature as the jacket. This implies that cooling curves should be determined with a constant temperature difference between jacket and reactor. If a cooling curve is recorded with the jacket constantly at its lowest temperature, the temperature dependence of the U_{max} is lost. Moreover, when calculating the final overall heat transfer coefficient of the reactor from its respective U_{max} value, the intended temperature offset between reactor and jacket should be kept in mind. This is less of an issue in the reaction calorimeter, since the observed temperature differences between jacket and reactor are usually a lot smaller than that in a large-scale reactor. Obviously, the values from Figure 11.5 apply only to the specific reactors in this specific plant, as different layouts in the cooling system, different materials of construction, different heat

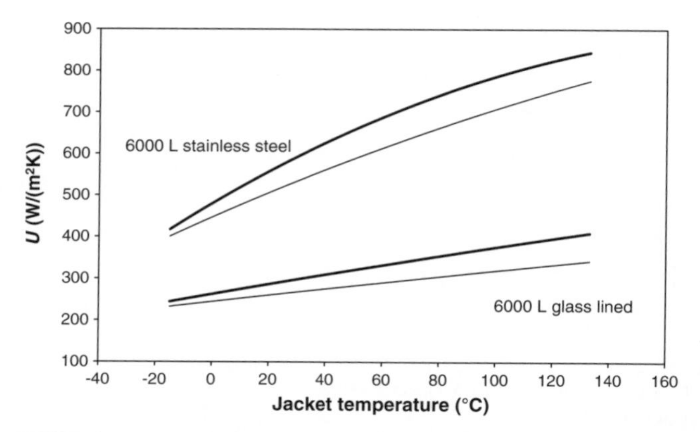

FIGURE 11.5 Heat transfer coefficients for a 6000 L stainless steel and a 6000 L glass-lined reactor. The reactor was filled with water for 50% of the nominal volume (thick line) or 85% (thin line), heated to the boiling point, and then cooled to room temperature with a constant temperature difference between jacket and reactor of 20°C at a high stirring speed. The heat transfer coefficient was determined as a function of the jacket temperature, and a curve was fitted through these data points to allow extrapolation to other temperatures.

transfer media, and different temperature control strategies can all have a large influence on the heat transfer characteristics of a vessel.

11.4.1.3 Influence of the Inner Film Coefficient Now that the U_{max} has been determined at both lab scale and production scale, let us turn our attention to the other term in equation 11.5, that is, the inner film coefficient h_r. This film coefficient is a measure for the resistance toward heat transfer between the reaction mass and the vessel wall. It is mainly governed by the stirring speed and the physical properties of the reaction mixture: a highly viscous reaction mixture will have more difficulties in dissipating reaction heat to the reactor wall than, for instance, pure water. Unfortunately, the influence of the inner film coefficient can be quite pronounced: if it were relatively small in comparison to the U_{max} term, it could be neglected and one single heat transfer coefficient could be used for each vessel, irrespective of the reaction mixture.

When the U_{max} of the reaction calorimeter is known at the (jacket) temperature and fill degree used, the inner film coefficient can be calculated from the overall heat transfer coefficient as determined in the calibration procedure according to equation 11.5.

As the inner film coefficient is dependent on the mixing characteristics of the vessel used, it is scale dependent and should be scaled up accordingly. This is usually done according to the following equation [12]:

$$h_{r(prod)} = h_{r(lab)} \times \left(\frac{D}{d}\right)^{1/3} \times \left(\frac{N}{n}\right)^{2/3} \times V_{i(prod)}{}^{0.14} \quad (11.7)$$

where $h_{r(prod)}$ is the inner film coefficient at large scale, $h_{r(lab)}$ is the inner film coefficient at calorimeter scale, D is the vessel diameter at large scale, d is the vessel diameter at calorimeter scale, N is the stirring speed at large scale, n is the stirring speed at calorimeter scale, and $V_{i(prod)}$ is the viscosity number at large scale.

The viscosity number is the ratio of the viscosity of the reaction mixture at the reaction temperature to its viscosity at the jacket temperature. When considering standard organic reactions in solution, this ratio is quite close to unity, so this factor is usually neglected. When studying polymerization reactions, however, this effect should be taken into account.

Using this equation, the inner film coefficient at production scale can be calculated, and hence the overall heat transfer coefficient is now known, according to equation 11.5. To get a more quantitative feeling for the influence of the different parameters on the overall heat transfer coefficient, let us take a look at a realistic numerical example.

EXAMPLE 11.1

A reaction is run in the reaction calorimeter at a temperature of 60°C. The reaction mixture is homogeneous and the solvent is methanol. The details of the reaction both in the calorimeter and in the production vessel (6000 L stainless steel) are given in Table 11.4.

The maximum heat transfer coefficient of the production vessel was estimated by running a cooling curve as follows: the reactor was filled with water up to 50% of its nominal volume, and the content was then heated to the boiling point and kept at that temperature for a while until both reactor and jacket temperatures have reached stable values. The reactor is then cooled to room temperature with a constant temperature offset between the reactor and the jacket. During the entire cooling cycle, rapid stirring is applied. The reactor temperature and the jacket temperature are recorded and put in a graph (temperature versus time). From the first derivative of this curve, the appropriate heat transfer coefficient is calculated. In this example, this yielded a value of 500 W/(m^2 K) at the intended jacket temperature (20°C). As described above, the U value for a reactor filled with water at high stirring speed is considered to be a good approximation of U_{max}.

The data from Table 11.4 are then inserted into equations 11.5 and 11.7.

First, h_r for the reaction calorimeter is calculated (equation 11.5):

$$\frac{1}{h_{r_{lab}}} = \frac{1}{U_{lab}} - \frac{1}{U_{max_{lab}}} = \frac{1}{180} - \frac{1}{215} \approx \frac{1}{1106} \frac{m^2K}{W}$$

From equation 11.7, we can now calculate h_r for the production vessel. We will assume that the viscosity of the reaction mixture at the reaction temperature is essentially

TABLE 11.4 Worked Example to Show the Influence of the Inner Film Coefficient on the Overall Heat Transfer Coefficient

	Calorimeter	Production Vessel	Calorimeter	Production Vessel
Reactor temperature (°C)	60	60	60	60
Jacket temperature (°C)	±60	20	±60	20
Filling degree (%)	50	50	50	50
U_{max} at jacket temperature (W/(m^2 K))	215	500	215	500
Diameter reactor (m)	0.12	2	0.12	2
Stirring speed (rpm)	450	100	450	100
U experimental (W/(m^2 K))	**180**		**120**	
U calculated (W/(m^2 K))		**337**		**169**

Note: The left part represents the case of a homogeneous nonviscous reaction mixture, and the right part that of strongly heterogeneous reaction mixture.

equal to the viscosity of the reaction mixture at the jacket temperature.

$$h_{r_{prod}} = h_{r_{lab}} \times \left(\frac{D}{d}\right)^{1/3} \times \left(\frac{N}{n}\right)^{2/3} = 1106\frac{W}{m^2K}$$

$$\times \left(\frac{2\,m}{0.12\,m}\right)^{1/3} \times \left(\frac{100\,rpm}{450\,rpm}\right)^{2/3} = 1036\frac{W}{m^2K}$$

Inserting this again in equation 11.5 yields the final U value for the production vessel.

$$\frac{1}{U_{prod}} = \frac{1}{h_{r_{prod}}} + \frac{1}{U_{max_{prod}}} = \frac{1}{1036} + \frac{1}{500} \approx 337\frac{W}{m^2K}$$

It is instructive to consider exactly the same reaction conditions, but this time with a strongly heterogeneous reaction mixture, where the experimentally obtained U value in the calorimeter is only 120 W/(m^2 K). If all other parameters are kept constant, a final U value at production scale of 169 W/(m^2 K) can be calculated. This clearly shows the importance of the inner film coefficient on the finally observed heat transfer coefficient: going from the maximum value of 500 W/(m^2 K) for pure water at the maximum stirring speed (U_{max}) to as low as 169 W/(m^2 K) for a heterogeneous reaction mixture at moderate stirring speed.

11.4.1.4 Shortcuts to U Value Determinations When the procedure for the determination of correct heat transfer data as described above is out of reach, there are other possibilities to make a rough estimation of the U values. If one chooses to go for these simplified estimation methods, care is needed to include a wide enough safety window when a batch is run for the first time in a certain reaction vessel.

The first possible estimation method is to simply use the heat transfer coefficient for a reactor filled with the neat solvent in which the reaction is to be performed. Cooling curves for some solvents are often readily available from cleaning campaigns. Since these cleaning cycles are usually repeated quite regularly, an indication of the evolution of the heat transfer characteristics of the reactor over time can also be obtained. It excludes the need for the separate determination of U_{max} with water, which is not often used as a cleaning solvent in a temperature cycle, and of the entire characterization of the U_{max} behavior of the reaction calorimeter. So if a reaction is to be run in a methanol solution at 30°C, one could simply calculate the U value at that temperature from a cooling curve with neat methanol. This approach will yield acceptable results, as long as the reaction mixture is not strongly heterogeneous or highly viscous.

Another possible approach is to use very conservative general heat transfer coefficients in the scale-up calculations. One could for instance record a cooling curve for methanol, calculate the U values from this curve, and then base all calculations on 50% of the heat transfer coefficient found. Although this method does not consider any specific effect of the physical properties of the reaction mixture on the heat transfer coefficient and should therefore only be used with great care and large safety margins, it can be an easy tool to give some guidance in the scale-up calculations.

Again, it should be stressed that when the correct U values are not known and can only be roughly estimated, broad safety margins should be incorporated into the process and the reactor data from the first batch should be checked carefully for any inconsistencies.

11.4.1.5 Practical Use of U Values So now the U value of the reaction mixture at production scale has been determined, but what can we do with it? The main use of heat transfer coefficients is to allow for a correct calculation of dosing times, making sure that all heat that is generated during the reaction can be safely removed. This is illustrated in the example below for a dosing controlled reaction. When dealing with a nondosing controlled reaction (i.e., with significant thermal accumulation), one should make sure that the available cooling capacity matches the heat release rate as observed in the calorimetry experiment at any time.

EXAMPLE 11.2

We will use the same reaction as in the previous example, with the U value at lab scale being 180 W/(m^2 K), and at production scale 337 W/(m^2 K). The reaction at 60°C is dosing controlled and has a total reaction heat of 100 kJ/kg reaction mass. The density of the reaction mixture is 0.8 kg/L. What dosing time is needed for a jacket temperature of 20°C? The reaction is performed in a 6000 L vessel with 3000 L of reaction mixture, and the heat exchange area is 8 m^2. We assume that the change in volume (and hence in heat exchange area) due to dosing is small.

Total heat to be removed = 3000 L × 0.8 kg/L × 100 kJ/kg = 240,000 kJ

Heat removal capacity (equation 11.1) = 337 W/(m^2 K) × 8 m^2 ×(60 − 20) K = 108 kW = 108 kJ/s

Dosing time needed = 240 000 kJ/108 kJ/s = 2222 s ≈ 37 min

If the same reaction were to be run in a glass-lined reactor at 20°C, with a U value of 150 W/(m^2 K) and a minimum obtainable jacket temperature of 5°C (water-cooled reactor), what would the dosing time be?

Total heat to be removed = 3000 L × 0.8 kg/L × 100 kJ/kg = 240,000 kJ

Heat removal capacity = 150 W/(m^2 K) × 8 m^2 ×(20 − 5) K = 18 kW = 18 kJ/s

Dosing time needed = 240,000 kJ/18 kJ/s = 13,333 s ≈ 4 h

A graphical representation of this example is given in Figure 11.6. It clearly shows the big influence the heat transfer characteristics of a vessel can have on the way in which a process can be run at production scale. In this context, our first concern is safety: is our cooling capacity sufficient to guarantee a safe operation at production scale? But there might be other consequences as well. In the above example, the recommended dosing time for the same process but in different equipment varies between 37 min and 4 h. Such a difference in dosing time can have serious consequences on the yield of the reaction, the impurity profile, thermal accumulation, and so on. It is therefore advisable to perform this type of scale-up calculations at a relatively early stage of development to avoid unpleasant surprises later on.

To conclude, some general trends about heat transfer in stirred tanks can be listed.

1. Stainless steel reactors generally have better heat transfer characteristics than glass-lined reactors.
2. For the same temperature difference between reactor and jacket, heat removal will be more efficient at higher reaction temperatures because of the higher U_{max} value at higher temperatures.
3. For the same heat transfer coefficient, increasing the temperature difference between jacket and reactor will make the heat removal more efficient. Care has to be taken, however, when going too low in jacket temperature so as to avoid crust formation.
4. The cooling circuit should be designed in accordance with the heat to be removed to make sure that the heat transfer medium returning to the jacket is sufficiently cooled.
5. Smaller vessels usually have better heat transfer capacities. This is due to the larger heat exchange area per volume of reaction mixture.
6. The physical properties of the reaction mixture have a profound influence on the overall heat transfer coefficient. Especially when dealing with highly heterogeneous or viscous reaction mixtures, heat transfer problems may occur.

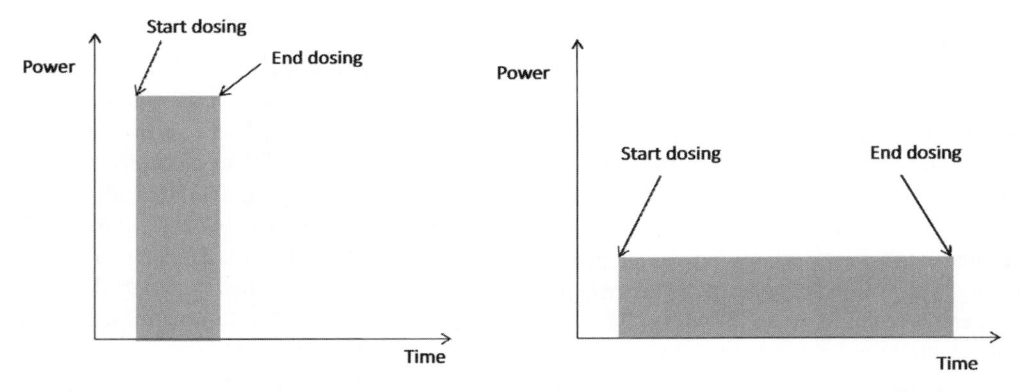

FIGURE 11.6 Influence of the heat transfer characteristics of a vessel on the dosing profile for an exothermic addition. The figure to the left shows the heat profile for a dosing controlled reaction in a vessel with a high heat transfer coefficient. The reaction mixture can be added relatively quickly, the cooling capacity is high enough to remove all the heat produced. In the figure to the right, the same addition is shown in a reactor with a lower heat transfer coefficient. Less heat can be removed in the same period of time, and hence the dosing should be slower in order to keep the reactor temperature constant. The overall reaction heat (shaded area) is the same in both profiles.

TABLE 11.5 Influence of the Limiting Piping Diameter in the Scrubber Lines on the Maximum Allowable Gas Flow at Production Scale (Left) and the Corresponding Gas Flows at Lab Scale (Right)

	Production Scale			Scale-Down from 6000 L		
	Diameter (cm)	Maximum Flow (L/min)	Maximum Flow (m^3/h)	2 L Reactor Maximum Flow (mL/min)	1 L Reactor Maximum Flow (mL/min)	100 mL Reactor Maximum Flow (mL/min)
DN25	2.5	295	18	98	49	5
DN50	5	1178	71	392	196	20
DN100	10	4712	283	1570	785	79
DN150	15	10603	636	3534	1767	177

These values assume that the installation is designed for a maximum gas speed of 5 m/s.

11.4.2 Gas Evolution

11.4.2.1 Gas Speed In the scale-up of a process in which gas is being liberated, it is of utmost importance to make sure that the gas that is set free can be evacuated from the reactor safely, without causing any pressure buildup. When a chemical plant is designed, a certain layout for the scrubber lines is worked out. This specific layout implies that there is a maximum gas speed in the piping in order to avoid pressure buildup, entrainment of powders, or unintended changes in flow pattern or even flow direction. The maximum gas speed that is often used is 5 m/s. If the maximum design gas speed and the minimum diameter through which the gas has to pass are known, the resulting maximum gas flow can be calculated. Some typical piping sizes and the corresponding maximum gas flow to meet the 5 m/s criterion are given in Table 11.5.

These figures clearly demonstrate the pronounced effect of the diameter of the narrowest piping the gas has to pass on its maximum flow rate: doubling the diameter allows a fourfold increase in gas flow.

The part of the Table 11.5 to the right shows some interesting scale-down data. In these columns, we have calculated what the corresponding gas flow at lab scale is. For instance, if the narrowest piping in the scrubber line is a DN50, a maximum flow rate of 71 m^3/h can be allowed at production scale. Using this maximum gas flow and assuming it concerns a 6000 L reactor, we can calculate the gas flow at lab scale. In this case, the corresponding gas flow in a 2 L reaction calorimeter would be 392 mL/min. This flow can be detected easily, but if the reaction is run in a 100 mL calorimeter, the corresponding gas flow is only 20 mL/min. One can imagine that such a low gas flow rate can be overseen easily during the process development work. This illustrates the importance of an accurate gas flow measurement combined with each reaction calorimetry experiment. Especially when small-scale reactors are used (less than 1 L), care should be taken in choosing the appropriate gas flow measuring device.

These values for the maximum allowed gas flow apply mainly to the desired synthesis reaction. Exceeding this flow to a limited extent might result in operational problems such as slight pressure buildup, process gases entering a neighboring reactor that is connected to the same scrubber line, or suboptimal condenser and scrubber performance, but will not necessarily lead to pronounced safety issues. When gas flow rates are considered that of an order of magnitude higher, vent sizing calculations come into play. This will be briefly discussed at the end of this chapter.

11.4.2.2 Reactive Gases In the previous discussion, the only parameter of concern was the gas flow rate. In many cases, however, the gas being emitted is reactive by itself, and this can cause particular safety problems. One example of having a very high yielding, but unfortunately enough undesired synthetic reaction, is depicted in Scheme 11.1.

These two processes were both run in the same plant. By coincidence, they were being run at exactly the same time in two neighboring reactors. The reaction depicted at the top resulted in the emission of hydrogen chloride, while reaction at the bottom was releasing ammonia. Both reactors were connected to the same scrubber lines, and they inevitably reacted with each other forming ammonium chloride in large amounts.

This solid material blocked the scrubber lines, and the reaction heat being evolved was large enough to partly melt the plastic scrubber lines. Fortunately, there were no serious consequences, but this demonstrates the need for a broad safety overview in any chemical plant.

11.4.2.3 Environmental Issues Although this factor is not related to process safety in the strict sense, the importance of the gas flow rate for environmental compliance should be mentioned here as well. It is important to know the layout of the gas treatment facility of the plant where the process is going to be run. If a carbon absorption bed is used, it is important to keep an overview of what the capacity of this bed is for the process gas being emitted: some gases are retained better than others, and some gases might even lead to dangerous hot spot formation in the bed. If a catalytic oxidation installation is used, it is important to know that

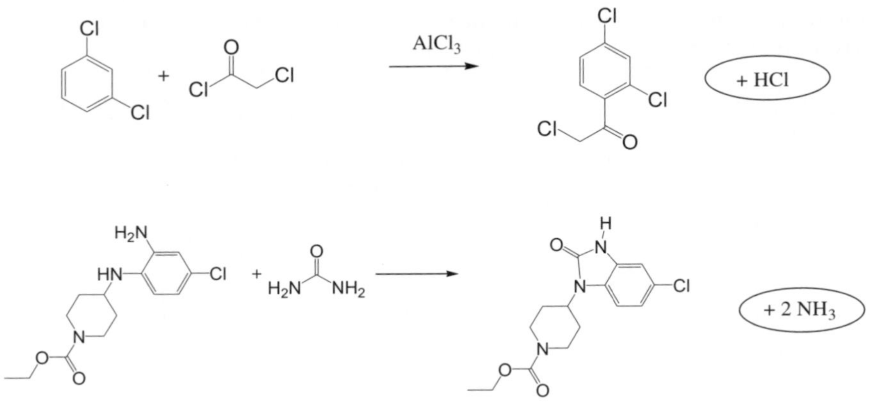

Scheme 11.1 Example of two synthetic reactions that generate reactive gases.

some compounds (such as hydrogen and alkenes) will lead to overheating in the installation if the flow is too high. Moreover, when no air treatment facility is installed at all, one should always make sure that the gas streams being emitted are within all environmental requirements. This assessment needs to be made for each production plant separately.

11.5 STUDYING THE DECOMPOSITION REACTION AT LAB SCALE

Having dealt with the study of the desired synthesis reaction (the first part of the cooling failure scenario), let us now turn our attention to the study of the undesired decomposition reaction. Decomposition reactions are of extreme importance for safety studies: in most cases, the energies being released in decomposition reactions are several orders of magnitude higher than those being released in the synthetic reaction, and hence the possible consequences of decomposition reactions can be catastrophic. The first prerequisite for any compound being used in a process is that it should be stable at the storage temperature for at least the time span anticipated for storage under normal operational conditions. Second, it should be at least sufficiently stable at the process temperature being used. Finally, it is important to assess its stability at the MTSR as well, since this is a temperature that can be attained in case of a cooling failure during the process (see Figure 11.1). We will discuss how the thermal stability of reagents and reaction mixtures can be studied at lab scale and how these data can be used to ensure a safe scale-up. But first we will start with another important characteristic of the stability of compounds, that is, shock sensitivity.

11.5.1 Shock Sensitivity

Some compounds are known to be prone to explosive decomposition when subjected to a sudden impact; therefore, they are called shock-sensitive compounds. Any compound that has at least one of the following characteristics should be considered as possibly shock sensitive:

1. The product has a very high decomposition energy (>1000 J/g)
2. The product has at least one so-called unstable functional group
3. The product is a mixture of an oxidant and a reductant

A list with some of the most common unstable functional groups that can make a product shock sensitive is given in Table 11.6. Please note that this list is not exhaustive, and when in doubt, the shock sensitivity of the compound should be tested [15].

When a compound is indeed shock sensitive, this may have serious consequences on the further development of the process, depending on the degree of shock sensitivity. There are restrictions for the transportation and storage of shock-sensitive compounds, so getting permission to purchase and store any of these products can be cumbersome. Therefore, it is vital to be aware of shock-sensitivity issues at an early stage.

When a reagent used in a synthesis is known to be shock sensitive, this does not necessarily exclude it from being used. For instance, hydroxybenzotriazole (HOBT) is known to be shock sensitive in its anhydrous form, but not in its hydrate form. Making sure that the appropriate grade of the chemical is used from an early stage can therefore avoid a lot of practical problems later on.

11.5.2 Screening of Thermal Stability with DSC

When evaluating the stability and risk potential of commonly used reagents, common literature and references such as Safety Data Sheets (SDS) can be a good starting point. More often than not in the development of active pharmaceutical ingredients, the compounds used are entirely new so the necessary safety data must be produced experimentally. It is

TABLE 11.6 Nonexhaustive List of Functional Groups That can be Shock Sensitive

Acetylenes	C≡C	Diazo	R=N=N
Nitroso	R-N=O	Nitro	R-NO_2
Nitrites	R-O-N=O	Nitrates	R-O-NO_2
Epoxides	(C–O–C ring)	Fulminates	C≡N-O
N-metal derivative	R-N-M	Dimercuryimmonium Salt	R-N=Hg=N-R
Nitroso	R-N-N=O	N-nitro	N-NO_2
Azo	R-N=N-R	Triazene	R-N=N-N-R
Peroxy acid	R-O-OH	Peroxides	R-O-O-R
Peroxide salts	R-O-O-M	Azide	R-N=N=N
Halo-aryl metals	Ar-M-X	N-halogen compounds	N-X
N–O compounds	N–O	X–O compounds	R-O-X

This list is extracted from an older version of Ref. 8 (version 4 and prior).

good practice to start with thermal stability screening of newly synthesized compounds at a very early stage (when the first gram of product becomes available), since changes in the process chemistry are still possible without too much impact.

The most widely used technique for thermal stability studies is the differential scanning calorimeter (DSC). In a DSC, a small cup with a few milligrams of product is heated at a predefined rate to a certain temperature. Typically, a sample could be heated from room temperature to 350°C at 5°C/min. During the heating phase, sensors detect any heat being generated (exothermic process) or absorbed (endothermic process) by the sample. The popularity of the DSC as a screening tool for thermal stability in process development is due to its low cost, wide availability of instruments from different suppliers, moderate experimental time (a typical run takes 1–2 h), appropriate sensitivity (1–10 W/kg), and small sample size (1–50 mg).

A typical DSC run is shown in Figure 11.7. As can be seen in the graph, at temperatures below 75°C, little thermal activity is observed. The first exothermic peak is observed at 110°C, followed by a much larger exotherm exhibiting two peaks at around 225°C and 270°C. Integration of the entire exothermic signal yields a reaction enthalpy of more than 1000 J/g, indicating that this particular compound has a very high decomposition energy.

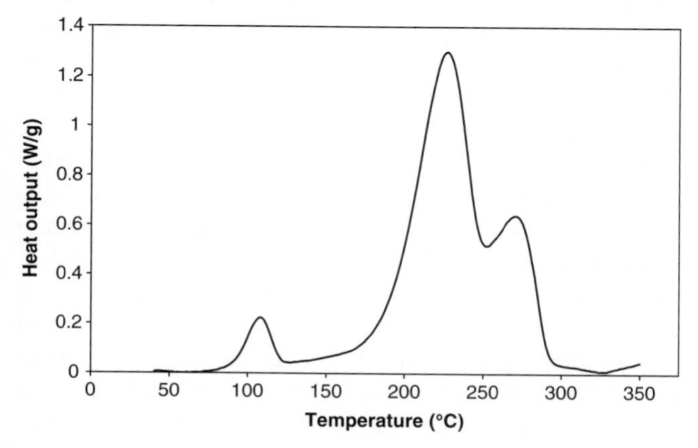

FIGURE 11.7 Example of a scanning DSC run of an unstable organic substance. The temperature was increased linearly from 30 to 350°C and the subsequent heat signal was recorded (exotherm is shown as a positive, upward, signal).

When evaluating a DSC run, there are three main parameters of interest.

11.5.2.1 Reaction Enthalpy Some typical decomposition energies for the most common thermally unstable groups are given in Table 11.7.

This table clearly shows that merely looking at a molecular structure can give an indication about the decomposition potential of a reagent. Note, however, that the data in this table are given in kJ/mol, whereas DSC data are usually reported in J/g. The latter gives, in fact, a better indication of the intrinsic energy potential of the product, since the influence of an unstable group will obviously be much larger in a small molecule than in a very large one. It is therefore no surprise that many of the most dangerous reagents (in terms of thermal stability) are indeed small molecules: hydroxylamine, nitromethane, cyanamide, methyl isocyanate, hydrogen peroxide, diazomethane, ammonium nitrate, and so on.

According to the observed decomposition enthalpy, a first assessment of the energy potential can be made. A decomposition energy of only 50 J/g is very unlikely to pose serious problems, even when the product would decompose entirely. On the other hand, decomposition energy in the order of magnitude of 1000 J/g should be considered as problematic, and the first choice should always be to avoid the use of such chemicals as much as possible.

As previously mentioned (equation 11.3), the reaction heat is directly proportional to the adiabatic temperature rise. For ΔT_{ad} to be known, the c_p of the compound or reaction mixture is needed. The heat capacity can be determined separately in the DSC, but to get a first indication an estimated value can be used as well. Usually a c_p of 2 J/(g K) can be used for organic solvents or dilute reaction mixtures, 3 J/(g K) for alcoholic reaction mixtures, 4 J/(g K) for aqueous solutions, and 1 J/(g K) for solids (conservative

TABLE 11.7 Typical Decomposition Energies for Some Common Unstable Functional Groups [2]

Functional Group	ΔH (kJ/mol)	Functional Group	ΔH (kJ/mol)
Diazo (-N=N-)	−100 to −180	Nitro (-NO_2)	−310 to −360
Diazonium salt (-N≡N^+)	−160 to −180	N-hydroxide (-N-OH)	−180 to −240
Epoxyde (C-O-C ring)	−70 to −100	Nitrate (-O-NO_2)	−400 to −480
Isocyanate (-N=C=O)	−50 to −75	Peroxide (-C-O-O-C)	−350

guess). A rough classification of severity of a decomposition reaction based on its reaction enthalpy and corresponding adiabatic temperature rise (in this case for $c_p = 2$ J/(g K)) is given in Table 11.8.

One final remark should be made here about endothermic decompositions. Some compounds decompose endothermically, and whereas one might consider them therefore to be harmless, special attention for these compounds is sometimes needed. Decomposition reactions in which gaseous products are formed (such as elimination reactions) are often endothermic. The intrinsic risk of this type of decompositions therefore lies not in the thermal consequences but in a possible pressure buildup. For this type of compounds, an evaluation of the possible pressure buildup associated with the decomposition is recommended.

11.5.2.2 Onset Temperature Even when a compound has very high decomposition energy, it can still be possible to use it safely in a process, provided that the difference between the process temperature and the decomposition temperature is sufficiently high. The term "onset temperature" is used quite frequently to denote the temperature at which a reaction or decomposition starts. In reality, however, there is no such thing as an onset temperature, since the temperature at which a reaction starts is strongly dependent on the experimental conditions. The point from which a deviation from the baseline signal can be observed is determined by the sensitivity of the instrument, the sample size, and the heating rate of the experiment. This implies that great care is needed when comparing "onset temperatures" obtained with different methods, but it does not completely rule out the use of this parameter in early safety assessment.

It should be mentioned here that the description above holds for the way in which the term "onset temperature" is usually interpreted in the process safety field. In other fields of research, the term onset is defined as the point where the tangent to the rising curve at the inclination point crosses the baseline. This definition is, for example, used in the calibration of a DSC by measuring the melting point of a suitable metal (mostly indium). In this text, however, the term "onset" will always refer to the temperature at which a first deviation from the baseline signal is observed.

TABLE 11.8 Classification of the Severity of Decomposition Reactions According to the Corresponding Adiabatic Temperature Rise (Assuming a c_p of 2 J/(g K))

ΔH (J/g)	ΔT_{ad} (°C)	Severity
Less than −500	>250	High
$-500 < \Delta H < -50$	$25 < \Delta T_{ad} < 250$	Medium
More than −50	<25	Low

A rule of thumb that has been used quite extensively in the past is the so-called 100 degrees rule. This rule states that a process can be run safely when the operating process temperature is at least 100° below the observed onset temperature. This rule has been shown to be invalid in certain cases, so it should certainly not be used as a basis of safety. This does not mean however that it is completely useless. If a process is to be run on a relatively small scale (50 L or less) in a well-stirred reaction vessel, natural heat losses to the environment are usually sufficiently large for the 100 degree rule to be valid. If the process is to be run a larger scale, a proper and more detailed study of the decomposition kinetics (and especially of the TMR_{ad}) should be made, as will be discussed further in this chapter. Here, it should be stressed that these remarks apply only to the thermal stability of the products, and extra care should be taken when gas evolution comes into play.

11.5.2.3 Reaction Type When dealing with DSC data of compounds or reaction mixtures, merely looking at the shape of the peak can give some clues about the type of reaction taking place. The DSC run shown in Figure 11.7 consists of several different overlapping peaks, indicating a rather complex reaction type. Very sharp peaks are indicative for autocatalytic reactions, and extra care with this type of decompositions is needed [16].

Autocatalytic reactions are reactions in which the reaction product acts as a catalyst for the primary reaction. This implies that the reaction might run slowly at a certain temperature for a while, but as time passes more catalyst for the reaction is formed and hence the reaction rate increases over time. Such reactions are therefore also called self-accelerating reactions. This is in contrast to the more classical behavior of reactions following Arrhenius kinetics where the reaction rate stays constant (zero order) or decreases (first order or higher) with time at a certain temperature. Having determined the decomposition reaction in such a case on a fresh sample will therefore always give the worst-case decomposition scenario; the initial rate measured

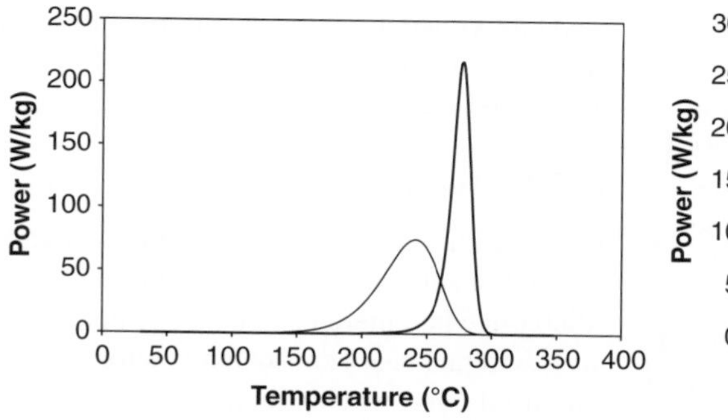

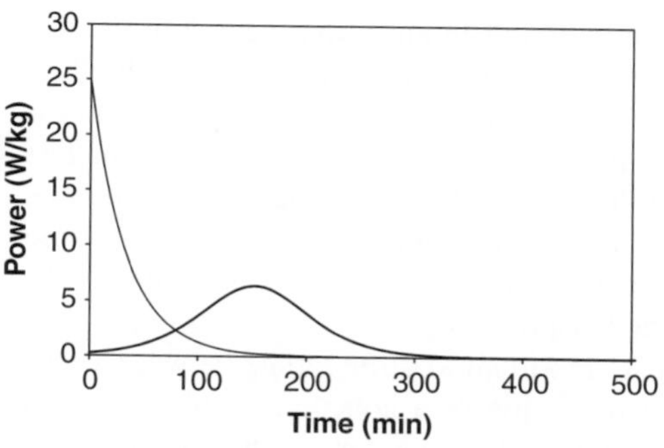

FIGURE 11.8 Example of a scanning DSC run (left) and an isothermal DSC run (right) of a reaction following *n*th-order kinetics (thin line) and of an autocatalytic reaction (thick line).

will be the maximum rate for that sample at that particular temperature. This is not the case for autocatalytic reactions, where the reaction rate is strongly dependent on the thermal history of the sample. Measuring the reaction rate for a pristine sample might lead to an underestimation of the risk associated with the decomposition of this sample when it has been subject to a certain thermal history (e.g., prolonged residence time at higher temperature due to a process deviation). An example of an isothermal and a scanning DSC run for both an autocatalytic and a first-order reaction are shown in Figure 11.8.

As pointed out, an autocatalytic reaction behavior is easily recognized by isothermal DSC experiments. If the temperature in the sample remains constant, the heat release over time will decrease in case of an *n*th-order reaction, but will show a distinct maximum in case of an autocatalytic reaction. Although not always as easily as in an isothermal DSC run, autocatalytic reactions can also be recognized in scanning DSC experiments where the peak shape is notably sharper than that for *n*th-order reactions. This is shown in Figure 11.8 to the left.

When a decomposition reaction is known to be autocatalytic, extra care is needed to avoid its triggering. For such compounds, any unnecessary residence time at elevated temperatures should be avoided. Extra testing might be appropriate to reflect the thermal history the product will experience at full production scale, since, for example, heating and cooling phases may take considerably more time than that in small-scale experiments. It should also be kept in mind that temperature alarms are not always an efficient basis of safety for this type of decomposition reactions: since the temperature rise can be very sudden, this type of alarm might simply be too slow to ensure that sufficient time is available to take corrective actions.

11.5.3 Screening of Thermal Stability: Pressure Buildup

As already mentioned earlier, in many cases the gas being released during a (decomposition) reaction can have greater safety consequences than merely the exothermicity. A proper testing method to determine whether or not a decomposition reaction is accompanied by the formation of a permanent (i.e., noncondensable) gas is therefore very important.

There are several instruments commercially available that are suited for this type of testing. Generally speaking, they consist of a sample cell of approximately 10 mL in which the sample is heated in a heating block or oven from ambient to around 300°C. During this heating stage, the temperature inside the sample is measured, as well as the pressure inside the sample cell. The main criteria these instruments should meet are an appropriate temperature range (ideally from (sub) ambient to 300°C), pressure range (up to 200 bar when measuring in metal test cells), and sample size (milligram to gram range). Some examples of such instruments available at the time of writing are the C80 from Setaram, the TSU from HEL, the RSD from THT, the miniautoclave from Kuhner, the Carius tube from Chilworth, and the Radex from Systag.

The most difficult aspect of interpreting the pressure data from this type of experiments lies in the differentiation between a pressure increase that is due to the formation of a permanent gas and a pressure increase due to an increased vapor pressure of the compounds at higher temperatures. When dealing with reaction mixtures in a solvent, the vapor pressure as a function of temperature can be calculated easily by means of the Antoine coefficients. These coefficients are readily available in the literature for most common solvents. If a plot of the vapor pressure as a function of the sample temperature matches the observed pressure profile, one can conclude that the observed pressure increase is due to the increased vapor pressure only. When in doubt or when the Antoine coefficients of the product are not known, for example, when dealing with a newly synthesized product that is an oil, it is advisable to run an isothermal experiment with pressure measurement at a temperature at which the decomposition is known or believed to occur at a considerable rate. If the pressure during this experiment remains constant, the pressure is due to vapor pressure, and if the pressure increases over time, it is due to the formation of a gaseous product. Alternatively, one could run two scanning experiments, each with a different filling degree in the test cell. If the pressure profile in the two runs match each other, the pressure is most likely to be due to vapor pressure, since it is not dependent on the free headspace available. If the test with the higher filling degree leads to a higher pressure, it is most likely to be due to the formation of a noncondensable gas, since less headspace is available for a larger amount of gas, leading to higher pressure.

But why is it so important to differentiate between vapor pressure and the formation of a noncondensable gas? If a certain pressure at elevated temperatures is due to only vapor pressure, it is relatively unlikely to pose problems. In such a case, a very rapid temperature increase is needed before the amount of vapor produced surpasses the amount that can be removed through the vent lines. When dealing with vent sizing calculations for serious runaway reactions, the effect of vapor pressure should definitely be taken into account. But when dealing with moderate temperature rise rates, or in isothermal operation, vapor pressure is unlikely to lead to major problems. The story is entirely different for the formation of a permanent gas due to a (decomposition) reaction. In this case, each process parameter that leads to an increased reaction rate will lead to an increased pressure rise rate, with possibly devastating effects. Obviously, a temperature rise will lead to an increased reaction rate, but other effects such as an increase in concentration due to the evaporation of the solvent (e.g., in case of a condenser failure) or sudden mixing of two previously separated layers (e.g., switching the stirrer back on after a failure) could also lead to an increased pressure rise rate in the vessel. This is also important for storage conditions: vapor pressure in a closed drum will reach an equilibrium at a certain pressure, whereas the formation of a permanent gas will lead to a pressure increase over time and the subsequent possibility of rupturing the drum.

EXAMPLE 11.3

In an isothermal stability test of a reaction mixture at 90°C, a gradual (linear) pressure increase is observed from 1.5 bar at the start of the experiment to 20 bar after 10 h. What is the gas release rate if the reaction is to be run at a 4000 L scale? We assume that the reaction behavior in an open system (production scale) is comparable to that in a closed system (lab test). In the test, 2 mL of reaction mixture was used in a system with an overall free headspace of 8 mL.

The pressure rise rate in the experiment is 18.5 bar in 10 h, that is, 0.031 bar/min.

The free headspace is 8 mL, so 8 mL of gas will lead to an increase of 1 bar.

The gas evolution rate is therefore 0.031 bar/min × 8 ml/bar = 0.248 mL/min.

Using the scale factor of 500,000 (4000 L/8 mL), this corresponds to 124 L/min.

At the production scale, a gas evolution rate of 124 L/min is expected. This corresponds to 7.4 m^3/h, which is only moderate (see Table 11.5).

11.5.4 Adiabatic Calorimetry

When discussing the cooling failure scenario previously, the concept of adiabaticity was introduced. A system is said to be adiabatic when there is no heat exchange with the surroundings. In a jacketed semi-batch reactor under normal process conditions, the reactor temperature is controlled by means of heat exchange between the reaction mixture and the heat transfer medium in the jacket. In case of a loss of cooling capacity (either because the heat transfer medium itself is no longer cooled or because it is no longer circulated), this heat exchange is no longer possible and the reactor will behave adiabatically. This is considered to be the worst-case situation in a reactor apart from a constant heat input (e.g., through an external fire), which will not be considered here.

For a lab chemist working on small-scale experiments only, the concept of adiabatic behavior in a large vessel is often hard to imagine. "I did it in the lab and I didn't notice any exothermicity" is an often heard statement. However, heat losses at small scale are a lot higher than at large scale, so the heat generation should already be relatively high before it is noticed during normal synthesis work at lab scale. This can be seen in Table 11.9, where some heat losses for different types of equipment are listed.

This table shows the vast difference in heat losses between small scale and large scale, and also the relevance of performing proper adiabatic tests. A 1 L Dewar calorimeter can be considered to be representative for other state-of-the-art adiabatic calorimeters, and it can be seen that its heat losses compare favorably to reactors in the cubic meter range.

Since this adiabatic behavior is considered to be the worst-case situation from a thermal point of view, it is of great interest to be able to mimic this situation in the lab under controlled conditions. An adiabatic calorimeter typically consists of a solid containment (to protect the operator against possible explosions that might take place inside the calorimeter) around a set of heaters in which the sample cell is placed. A thermocouple either inside the test cell or

TABLE 11.9 Typical Heat Losses for Different Types of Equipment [1]

	Heat loss (W/(kg K))	Time for 1°C loss at 80°C
5000 L reactor	0.027	43 min
2500 L reactor	0.054	21 min
100 mL beaker	3.68	17 s
10 mL test tube	5.91	11 s
1 L Dewar	0.018	62 min

attached to the outside of the test cell records the sample temperature, and the heaters are kept at exactly the same temperature at any time to obtain fully adiabatic conditions. During the entire experiment, the pressure inside the test cell is recorded, as well as the sample temperature. The criteria that an adiabatic calorimeter for safety studies should meet are obviously a high degree of adiabaticity (i.e., very low heat losses), an appropriate sample volume (typically between 5 and 50 mL), broad temperature range (ambient to 400°C), high pressure resistance or a pressure compensation system (up to 200 bar), and high speed of temperature tracking (>20°C/min). Some commercially available instruments are the ARC from Thermal Hazards Technologies, the Phi-Tec from HEL, the Dewar system from Chilworth, and the VSP from Fauske. Several pharmaceutical and chemical companies have developed their own adiabatic testing equipment, mainly based on a high-pressure Dewar vessel.

11.5.4.1 Adiabatic Temperature Profile Let us consider the situation as depicted in Figure 11.9. The reaction mixture is at a constant temperature of 120°C when a cooling failure takes place. At this temperature, the reaction starts relatively slowly and hence the temperature increases, albeit at a slow pace. Since most chemical reactions proceed faster at higher temperatures, the reaction rate (and thus the temperature rise rate) will increase as the reaction continues. This acceleration continues until finally the depletion of the reagents slows the reaction down again and a stable final temperature is achieved. The S-shaped temperature curve seen in the graph is very characteristic for an adiabatic runaway reaction. As indicated on the graph, there are three main parameters that describe the process of a runaway reaction, that is, the adiabatic temperature rise (ΔT_{ad}), the TMR, and the maximum self-heat rate (max SHR). The first two have been discussed previously, the maximum self-heat rate is a measure for the maximum speed with which the reaction occurs and can be directly correlated to the power output of the reaction: a SHR of 1°C/min corresponds to 33 W/kg reaction mixture for an organic medium with a c_p of 2, and to 70 W/kg in aqueous medium

$$1\frac{°\mathrm{C}}{\mathrm{min}} \times 2\frac{\mathrm{J}}{\mathrm{gk}} \times \frac{1}{60}\frac{\mathrm{min}}{\mathrm{s}} \times 1000\frac{\mathrm{g}}{\mathrm{kg}} = 33\frac{\mathrm{W}}{\mathrm{kg}}$$

This is important for vent sizing calculations and the assessment of using the boiling point as a safety barrier, as will be discussed later.

11.5.4.2 Heat-Wait-Search Procedure The most commonly applied method for adiabatic testing is the so-called heat-wait-search procedure, as shown in Figure 11.10. In this procedure, the sample is introduced to the instrument at room temperature and then heated (heating step) to the starting temperature of the test. The sample is then allowed to equilibrate at this temperature (waiting step), followed by the so-called search step. During this step (which usually takes between 5 and 30 min), the sample temperature is monitored to see if there is any sign of an exothermic reaction taking place. If the temperature rise rate under adiabatic conditions during this period is below the chosen detection threshold (typically 0.02 or 0.03°C/min), the temperature is increased with a couple of degrees and the cycle starts all over again until either an exotherm is detected or the preset final temperature has been reached. When an exotherm is detected, the instrument will track the sample temperature adiabatically until the temperature rise rate drops below the threshold value (end of the reaction), after which the heat-wait-search cycle starts again. Alternatively, the run is aborted during an exotherm if the upper temperature limit of the experiment has been surpassed.

Although adiabatic calorimeters can usually operate in other thermal modes as well, the heat-wait-search procedure is still the most widely used because it allows determining the

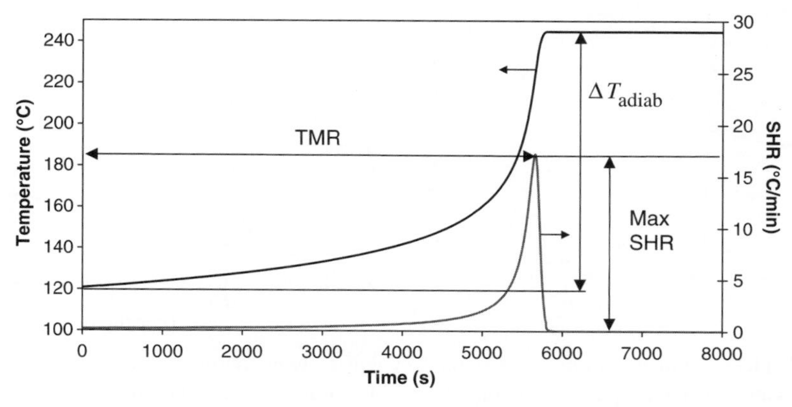

FIGURE 11.9 Typical adiabatic temperature versus time profile (left axis). The first derivative of the heat signal (self-heat rate) is shown on the right axis. The time to maximum rate (TMR), adiabatic temperature rise (ΔT_{ad}), and the maximum self-heat rate (max SHR) are indicated.

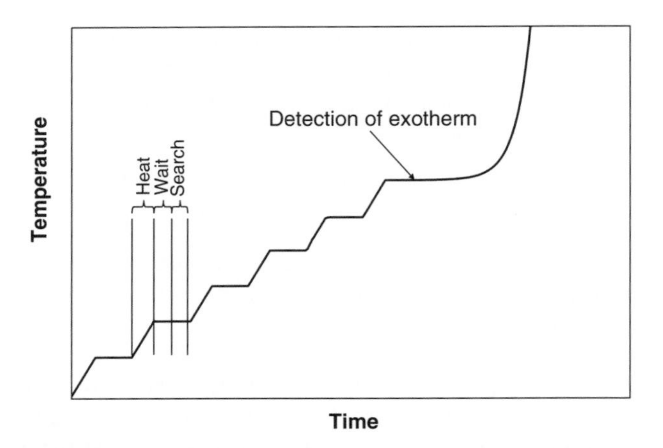

FIGURE 11.10 Typical representation of the heat-wait-search procedure in an adiabatic experiment. The sample is first heated to the desired temperature, then the temperature is allowed to stabilize during the wait period, and finally the temperature profile is checked for any sign of exothermicity during the search period. If exothermicity is detected, the temperature is adiabatically tracked until completion of the reaction or until the maximum experimental temperature has been reached. Otherwise the cycle is repeated.

onset temperature of the exotherm with great accuracy while keeping the experimental time acceptably short.

11.5.4.3 Thermal Inertia: φ Factor As mentioned above, one of the main reasons why it is hard to extrapolate adiabatic behavior at large scale from small-scale lab work is the dramatic difference in heat losses between these two working environments. There is another reason as well that plays a very important role in the interpretation of adiabatic calorimetry, that is, the thermal inertia or φ factor:

$$\varphi = \frac{m_c \times c_{p_c} + m_s \times c_{p_s}}{m_s \times c_{p_s}} \tag{11.8}$$

where m_c is the mass of the container (vessel or sample cell) (g), m_s is the reaction mass (g), c_{p_c} is the heat capacity of the container (J/(g K)), and c_{p_s} is the heat capacity of the reaction mass (J/(g K)).

When heat is generated in the reaction mixture, this heat will be used to increase the temperature of not only the reaction mixture itself, but also the container, being the vessel at large scale or the test cell at small scale. The φ factor is therefore a measure of which fraction of the thermal mass of the entire system is due to the thermal mass of the reaction mixture and which part is due to the container.

In large-scale equipment, the φ factor of a vessel during a runaway will be close to unity: that is, the thermal mass of the vessel itself (mainly the jacket) will be low compared to the thermal mass of the reaction mixture (i.e., $\varphi = 1$). In small-scale laboratory equipment, the φ factor is usually significantly higher than 1. To perform a lab-scale experiment at a φ factor that is close to unity, one would need to use a very light test cell that can accommodate a large amount of sample. The influence of the φ factor on the runaway behavior of a system is very pronounced [17, 21], as can be seen in Figure 11.11. In this figure to the left, the same adiabatic runaway profile is given for a sample being tested in two different test cells, one with a (hypothetical) φ factor of 1 and the other one with a φ factor of 2 (simulations). As can be seen, the curves differ drastically. In every aspect, the curve obtained with $\varphi = 1$ is by far more severe than the one obtained with $\varphi = 2$. The total adiabatic temperature rise scales linearly with φ; that is, the observed ΔT_{ad} in a $\varphi = 2$ experiment will be exactly half of the ΔT_{ad} in a $\varphi = 1$ experiment.

$$\Delta T_{ad,\varphi=1} = \Delta T_{ad,exp} \times \varphi \tag{11.9}$$

The TMR scales almost linearly with φ in most cases, but the max SHR scales far from linear with φ. The max SHR in an experiment with $\varphi = 1$ can easily be 10 times higher than

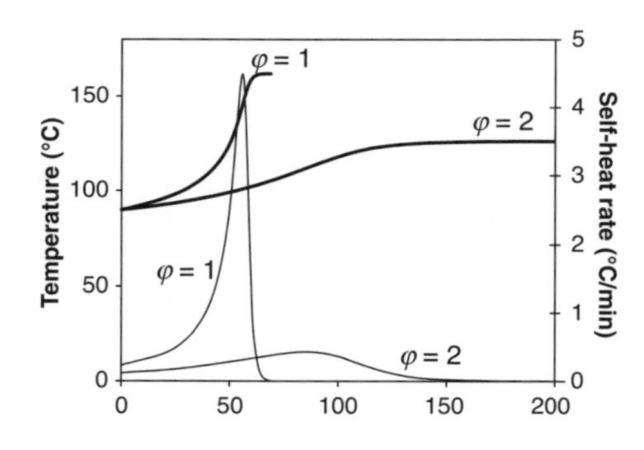

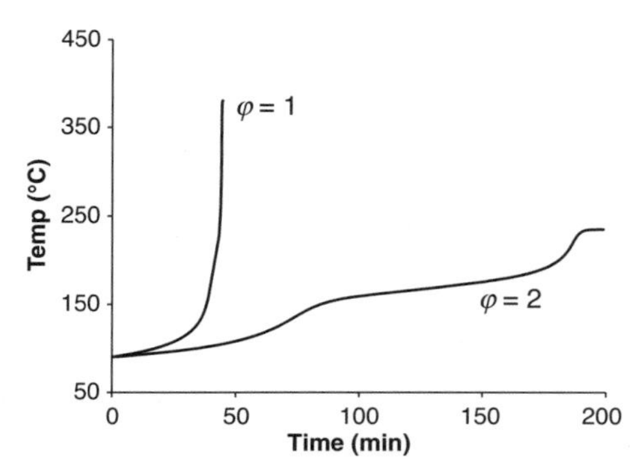

FIGURE 11.11 Influence of the φ factor on the adiabatic temperature profile. The figure to the left represents an adiabatic experiment on exactly the same sample, but once measured at a $\varphi = 1$ and once at $\varphi = 2$. The influence of the thermal inertia on the result is pronounced. This difference is even larger in the case of two consecutive reactions, as shown in the figure to the right. Reprinted with permission from Ref. 17. Copyright 2006, American Chemical Society.

TABLE 11.10 Key Figures for the Adiabatic Runaway Profile from Figure 11.11 to the Left

	$\Phi = 2$	$\Phi = 1$
ΔT_{ad}	70°C	140°C
TMR_{ad}	90 min	50 min
Max SHR	0.8°C/min	4.5°C/min

These data show that the adiabatic temperature rise scales linearly with φ. The TMR_{ad} scales approximately linearly with φ, whereas the max SHR does not.

in the $\varphi = 2$ experiment! The numerical data for the curves as depicted in Figure 11.11 are given in Table 11.10.

Figure 11.11 to the right also gives the runaway behavior of one reaction in a test cell with a (hypothetical) φ factor of 1 compared to the same run in test cell with a φ factor of 2. In this case, however, the difference between the two runs is even more pronounced. The reaction consists of two consecutive reactions and running this reaction at $\varphi = 1$ will result in a temperature profile where the first exotherm continues into the second one, leading to a very rapid temperature rise. In the run with $\varphi = 2$, the temperature rise from the first exotherm will be far less pronounced, and this will lead to a significant time interval between the two exotherms. Hence, the severity of this run will be significantly lower than that of the run with $\varphi = 1$.

These two examples show the importance of the φ factor on the experimental results. Ideally, one would try to perform the adiabatic measurement in a low φ test cell. In many cases, this is difficult to obtain experimentally, and a proper extrapolation to low-φ conditions is needed.

11.5.4.4 Interpretation of Adiabatic Experiments Adiabatic experiments can be performed for different reasons, but usually the main goal is to get a representative idea of what the temperature and pressure profile could be in a full-scale reactor in case of a runaway reaction. If the adiabatic experiment is performed at a φ factor close to unity, the match between the two will indeed be close. Let us take a look at the most important parameters to analyze.

1. *Adiabatic Temperature Rise*: This value can be directly extracted from the thermal profile. When dealing with very violent reactions, it is probably not possible to obtain the total adiabatic temperature rise as the maximum safety temperature or pressure of the equipment will be surpassed and the experiment will be automatically stopped. This is not necessarily a problem since ΔT_{ad} can be obtained from other experiments as well (e.g., from DSC), and for this type of violent reactions, the temperature profile close to the onset temperature is far more important. Whether the final temperature would be 1000 or 700°C is relatively unimportant, and it will be a full-blown thermal explosion anyhow. It should be kept in mind that the observed ΔT_{ad} should be multiplied with the φ factor to obtain the correct adiabatic temperature rise in case of a large-scale runaway (equation 11.9).
2. *Onset of the Exotherm*: Here as well, the term onset refers to the point where deviation from the baseline can be observed and is hence instrument dependent. In adiabatic calorimetry, a detection threshold of 0.02°C/min is often used, which corresponds to 1.4 W/kg in case of an aqueous reaction medium. Referring to Table 11.9, we know that the natural heat loss of a 5000 L reactor at 80°C and an ambient temperature of 20°C is 1.6 W/kg. These two figures match quite closely, so the temperature at which the exotherm is detected in the adiabatic calorimeter with this sensitivity is most likely to be the temperature at which exothermicity will be first noticed under adiabatic conditions at large scale (at least for temperatures higher than 80°C). If the onset temperature is well above the MTSR, the decomposition is unlikely to be triggered, even in case of a cooling failure during the synthesis reaction. If the onset temperature is close to the MTSR, a calculation of the TMR_{ad} should be made, as will be discussed later.
3. *Pressure Profile*: A careful analysis of the pressure profile should be made after each experiment. Very often, the pressure signal will be more sensitive to detect the start of a decomposition reaction than the temperature signal [18]. If a slow pressure increase during any of the search periods is observed, a decomposition reaction with gas evolution should be suspected. Some software packages allow a direct overlay of the vapor pressure of any chosen solvent related to the sample temperature. This can be very indicative to discern between permanent gas formation and vapor pressure. If this is not possible, looking at the pressure profile during the wait and search period should yield the same information: if the pressure remains constant during this stage (when the sample is at a constant temperature), vapor pressure is the most important contribution to the overall pressure. If the pressure rises during this stage, formation of a permanent gas is most likely to happen. If the pressure drops at a certain point, a leak of the test cell has most probably occurred.
4. *Self-Heat Rate (Temperature Rise Rate)*: The temperature rise rate of a runaway reaction can be calculated by taking the first derivative of the temperature versus time plot. Two things should be kept in mind when analyzing the SHR: first, the dramatic influence of the φ factor on the SHR, as discussed above, and second, the fact that the associated pressure rise rate can have far more serious consequences. So in case the

SHR is low at any time (e.g., below 2°/min) in a low-φ experiment ($\varphi < 1.2$) and there is no strong pressure increase, the consequences of the runaway reaction are unlikely to be severe.

5. *Pressure Rise Rate*: The analysis of the pressure rise rate data is far from trivial. First thing to keep in mind is the influence the free headspace volume in the adiabatic calorimeter has on the finally observed pressure profile. If a permanent gas is formed during the runaway reaction, the observed pressure increase will be considerably larger when the test has been performed with a small free headspace (e.g., a filling degree of 90% of the test cell) than in case of a large free headspace (e.g., 50% filling degree). Therefore, it is advisable to use a ratio reaction mixture versus free headspace that is comparable to the situation at large scale. As mentioned before, this is only relevant when dealing with the formation of a permanent gas, not when considering vapor pressure data. Second, as the pressure rise rate is in any case directly correlated to the temperature rise rate, the remarks made in the point above about the influence of the φ factor hold here as well. The pressure rise rate can be calculated back to a gas evolution rate (see also the previous worked example), and if the thus obtained gas flow is below the design limits of the installation under normal process conditions, no problems are to be expected. If this limit is exceeded to a limited extent, some operational issues can be suspected (limited condenser capacity, disturbed flow patterns in the venting line, etc.) without serious safety consequences. If the gas flow rate is considerably above this design limit, proper vent sizing calculations are needed to make sure that the emergency relief system is sufficient to cope with a runaway reaction. This point will be briefly discussed later.

Let us now consider a real-life example of such an adiabatic experiment. In Figure 11.12, a heat-wait-search experiment of a relatively concentrated solution of dibenzoyl peroxide in chlorobenzene is shown. The experiment was conducted in glass test cell with a φ factor of 1.5.

A close inspection of the temperature profile indicates that the thermal activity starts already at 45°C, but the detection threshold of 0.02°C/min is only reached at 58°C. At this temperature, the instrument goes into tracking mode and a maximum self-heat rate of more than 40°C/min is reached after 280 min. The observed adiabatic temperature rise is 140°C, but in reality it will be higher since the run was aborted at 200°C to prevent leakage of the silicone septa used to seal the glass test cell. An overlay of the vapor pressure curve with the pressure profile shows that a large part of the observed pressure is due to gas evolution of the decomposition reaction. The maximum pressure rise rate is very high, more than 200 bar/min. Keeping in mind that this experiment was conducted at a relatively high φ factor, it is clear that the severity of this runaway is totally unacceptable for introduction at a large scale. An obvious safety advice would be to investigate the use of a more dilute solution of this compound, or to turn to other, more stable, reagents.

11.5.4.5 Using the Boiling Point as a Safety Barrier

When a runaway reaction takes place in a reactor, the temperature of the reaction mixture can reach the boiling point. This can either be an extra risk that needs to be taken into account or it can act as an efficient safety barrier [19].

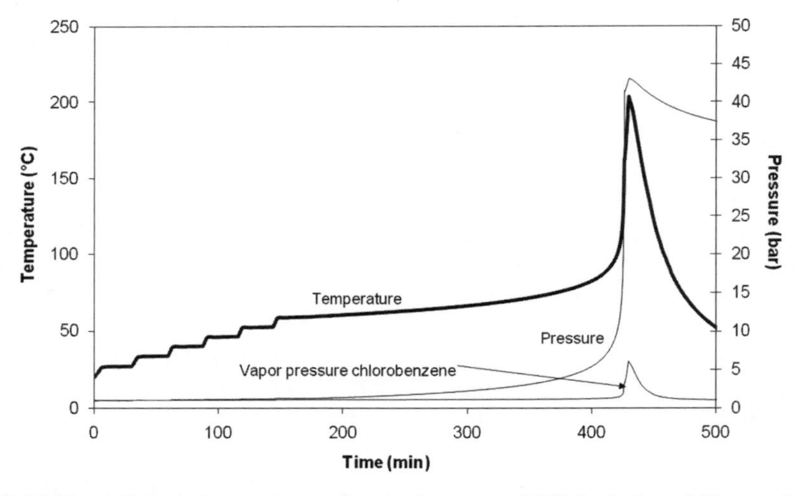

FIGURE 11.12 Adiabatic heat-wait-search experiment on a 0.75 M solution of dibenzoyl peroxide in chlorobenzene. The test was conducted in a glass test cell with a φ factor of 1.5. Exothermicity is observed at 58°C, and the run is aborted at 200°C to prevent leaking of the silicone septa used to close the test cell.

If the heat rate under adiabatic conditions at the boiling point is low, part of the solvent will be evaporated, but the temperature will remain constant and this will temper the runaway reaction. When the detected onset temperature in the adiabatic calorimeter is close to the boiling point, this can be a very effective safety barrier.

If the heat rate at the boiling point is relatively high, other effects might come into play.

1. *Evaporation of the Solvent*: If the boiling point is reached, part of the solvent will start to evaporate. This in itself will have a cooling effect, but when the vapor is no longer condensed and returned to the reactor, the reaction mixture will become more concentrated. This in turn will lead to an increased reaction rate and also to an increased boiling temperature. This should be taken into account when relying on the reflux barrier as a basis of safety.
2. *Swelling of the Reaction Mass*: If the reaction mixture starts to boil vigorously, a kind of "champagne effect" may take place, leading to an increase in the volume of the reaction mass. In such a case, the reactor content might even be forced out of the reactor into the condenser and scrubber lines.
3. *Flooding of the Vapor Line*: This effect will occur particularly when countercurrent condensers are used (i.e., the vapor and the condensate flow in opposite direction through the condenser). If the vapor flow through the condenser is too high, this flow will prevent the condensed liquid to flow back into the reactor and this liquid will be carried along with the vapor flow. Here as well, solvent will enter the scrubber lines with all possible problems associated with it.

Provided that all the above-mentioned factors are taken into account, the reflux barrier can be used as a very effective safety barrier. For a more quantitative description, the reader is referred to the literature.

11.5.5 TMR_{ad} Calculations

In the discussion of the cooling failure scenario, the importance of the temperature at which the TMR_{ad} is 24 h was pointed out. It was stated that if the TMR_{ad} at the MTSR is more than 24 h, the reaction can be considered as safe. This obviously implies a correct determination of this important parameter. Several approaches can be followed to do this, each with their merits and shortcomings.

11.5.5.1 Determination from One DSC Run A first approximation of the temperature at which the TMR_{ad} is 24 h (we will call this temperature $TMR_{ad,24h}$) can be made from one single DSC run. A full discussion of the theory behind this approach is out of scope here, so the reader is referred to the original publication by Keller et al [20]. We will however point out the basic concept with an example.

The idea behind this approach is quite simple: measure the heat release at one temperature and assume that the reaction follows zero-order reaction kinetics with a low activation energy of 50 kJ/mol. From this, the heat release at any temperature and hence the $TMR_{ad,24h}$ can be calculated. The assumptions on which this approach is based and the practical calculations are discussed below.

1. *Zero-Order Assumption*: A "classical" behavior for a reaction is that it follows *n*th-order Arrhenius kinetics. This means that the reaction rate increases with concentration and temperature. This implies that the reaction rate will decrease over time when the reaction temperature is constant (see Figure 11.8) as the concentration of the reagents drops with increasing conversion. In the reaction the follows zero-order kinetics, however, the reaction rate is independent of the concentration. This means that the reaction rate remains constant at a given temperature from the start (0% conversion) until the end (full conversion) of the reaction. Assuming this type of reaction kinetics leaves out the concentration dependence and makes the calculations a lot easier. It should be kept in mind, however, that this is an assumption, in reality the reaction is most likely not going to follow these kinetics. It is however a "safe" assumption, since the reaction rate will be overestimated as the decrease in reaction rate with decreasing reagent concentration is neglected. The fact that this is indeed a "worst-case" assumption is discussed thoroughly in the original publication.
2. *Low Activation Energy*: The dependence of the reaction rate on the temperature is dictated by the activation energy: the decrease in reaction rate when lowering the reaction temperature is more pronounced for a reaction with high activation energy than for a reaction with low activation energy. Since we will try to extrapolate the reaction rate at lower temperatures from one single measurement at one temperature, the activation energy needs to be known or estimated. A correct determination of the activation energy is possible by means of DSC, but not always straightforward. Therefore, the activation energy is assumed to be 50 kJ/mol, which is very low for organic reactions and decompositions. Choosing a low activation energy will again be on the safe side, since it will tend to overestimate the reaction rate at lower temperatures.
3. *Determination of the Reaction Rate (Heat Rate) at One Temperature*: A correct determination of the heat rate at one temperature is needed, preferably in a relatively early stage of the reaction since this will minimize the error introduced by the zero-order

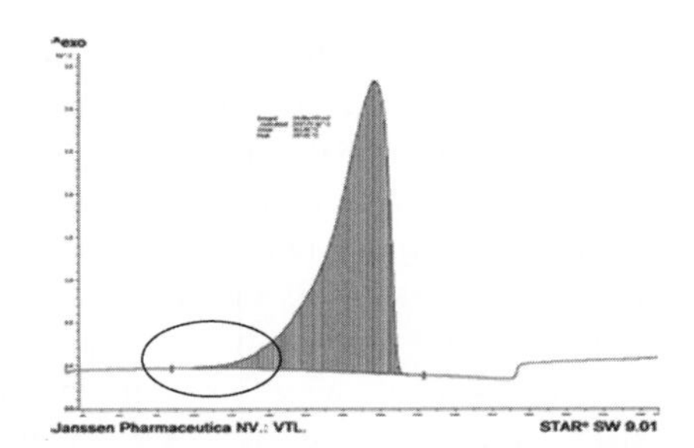

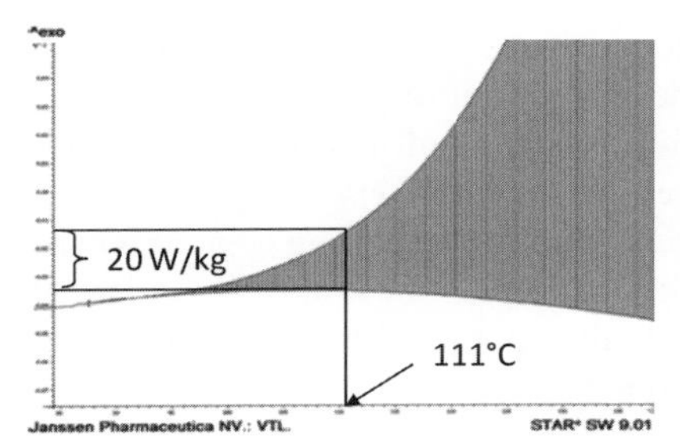

FIGURE 11.13 Example of a scanning DSC run of a highly unstable organic compound. The figure to the left shows the entire run, and the figure to the right is zoomed in on the start of the exotherm.

assumption. The heat signal should be well separated from the baseline, however, in order to obtain an accurate signal. Keller et al. suggest to search for the temperature at which the heat rate is 20 W/kg. This is a sensitivity that is well within reach of any decent DSC apparatus. In the example shown in Figure 11.13, this heat rate is observed at 111°C.

4. The heat rate at other temperatures can now be calculated according to the following equation:

$$q_0 = q_{\text{onset}} \times e^{\left(\frac{E_a}{R} \times \left(\frac{1}{T_{\text{onset}}} - \frac{1}{T_0}\right)\right)} \qquad (11.10)$$

where q_o is the heat rate at the new temperature, q_{onset} is the heat rate at the onset temperature (in this case 20 W/kg), E_a is the activation energy (50 kJ/mol), R is the universal gas constant (8.31 J/(mol K)), T_{onset} is the onset temperature (in this case 111°C), and T_0 is the new temperature at which the heat rate is to be calculated.

5. From this, the TMR_{ad} can be calculated for any temperature according to Equation 11.11

$$\text{TMR}_{\text{ad}} = \frac{c_p \times R \times T_0^2}{q_{T_0} \times E_a} \qquad (11.11)$$

If this calculation is performed for a number of temperatures, $\text{TMR}_{\text{ad,24h}}$ can be determined, as shown in Table 11.11.

TABLE 11.11 Extrapolation of the Heat Rate and TMR of the DSC Signal from Figure 11.13 to Lower Temperatures

T_0 (°C)	q_0 (W/kg)	TMR_{ad} (h)
111	20	0.7
101	13.2	1
91	8.5	1.4
81	5.3	2.2
71	3.2	3.4
61	1.9	5.4
51	1.1	8.8
41	0.6	14.9
31	0.3	26.2

In this example (using a c_p of 2 kJ/(kg K)), the $\text{TMR}_{\text{ad,24h}}$ is estimated to be 31°C. Another interesting point we can learn from this table is that a heat release of only 10 W/kg corresponds to a TRM_{ad} of roughly 1 hour!

This example shows how the $\text{TMR}_{\text{ad,24h}}$ can be extrapolated from only one DSC experiment. Because of the assumptions made, this will only be a rough estimate that can differ considerably from the true $\text{TMR}_{\text{ad,24h}}$. The merit of this method however lies in the fact that all assumptions are on the safe (conservative) side. If the thus obtained TMR_{ad} at the MTSR is longer than 24 h, no further testing is needed. If it is shorter than 24 h, a more accurate determination of the $\text{TMR}_{\text{ad,24h}}$ might be needed, as will be discussed below.

One final remark is needed about autocatalytic reactions. Since the thermal history of a sample is so important in the characterization of autocatalytic (decomposition) reactions, their TMR_{ad} is much harder to determine. The method described here should therefore not be used for this type of reactions; more elaborate adiabatic testing will be needed.

11.5.5.2 **TMR_{ad} from one Adiabatic Experiment** Probably the best way to determine the $\text{TMR}_{\text{ad,24h}}$ accurately is to perform a number of adiabatic experiments in a low-φ test cell, each at a different starting temperature, and then determine the TMR for each of these experiments until the temperature at which this TMR is 24 h is found. Obviously, this will be a very time-consuming procedure, and better alternatives are to be sought for. It would be beneficial if we could extract a reliable $\text{TMR}_{\text{ad,24h}}$ from one single adiabatic heat-wait-search experiment at a somewhat higher φ factor. This way, we would run an experiment under adiabatic conditions that is closer to the real situation in a vessel during a runaway reaction than a scanning DSC experiment. Also, if we can run in HWS mode, the experimental time will be reduced significantly, and running at a higher φ factor (e.g., 1.5) is experimentally easier than at a φ factor close to unity. Question is how to extrapolate data from this single adiabatic experiment to other temperatures and φ factors.

In one of the classical studies about adiabatic calorimetry, Townsend and Tou [21] evaluated in great detail the analysis of experimental adiabatic data. In this study, they present a method to extrapolate the experimental TMR to other temperatures and φ factors. A full description of the kinetic evaluation made in the original study is out of scope here, but the general concepts and the practical use of this approach are discussed below.

We start from an adiabatic experiment, where we can determine the TMR at the onset temperature directly from the temperature versus time plot. The first extrapolation to be made is from the experimental φ factor (i.e., >1) to the "ideal" case of $\varphi = 1$. Townsend and Tou state that for most relevant decomposition reaction with a high activation energy, the TMR scales linearly with φ (equation 11.12).

$$TMR_{\varphi=1} = \frac{TMR_{exp}}{\varphi} \qquad (11.12)$$

Second, a method is described to extrapolate TMR data to lower temperatures as well. Assuming the reaction follows zero-order kinetics, it can be shown that there is a linear correlation between the logarithm of the TMR and the inverse temperature according to equation 11.13.

$$\ln(TMR) = \frac{1}{T} \times \frac{E_a}{R} - \ln A \qquad (11.13)$$

Hence, plotting the logarithm of the TMR versus the inverse temperature will yield a straight line with a slope proportional to the activation energy and the intercept being equal to the logarithm of the frequency factor.

Using these two equations, the $TMR_{ad,24h}$ can be determined if the TMR_{ad} is known at a number of different temperatures. Since the approach is partly based on zero-order assumptions, it is important to focus on the early part of the exotherm, since the influence of a decrease in concentration due to conversion can be neglected there. This way, the experimentally obtained TMR_{ad} at the onset temperature and at a couple of temperatures that are slightly higher are determined. These values are then corrected for the experimental φ factor according to equation 11.12. Finally, a straight line is fitted through the plot of $\ln(TMR_{ad})$ versus $1/T$ and the point at which the TMR_{ad} is equal to 24 h can be read from the graph. This approach is illustrated in the following example.

Let us consider the adiabatic experiment as represented in Figure 11.12. The φ factor used in this experiment was 1.5. The onset of the exothermicity is detected at 58°C, 146 min after the start of the experiment. The maximum rate is reached after 429 min, and hence the TMR_{ad} at this temperature is 283 min. We will consider the TMR at five different points in the early part of the exotherm, that is, at 58, 60, 62, 64, and 66°C. The corresponding TMR_{ad} at the experimental φ factor and at $\varphi = 1$ are calculated, as shown in Table 11.12.

The according plot based on these data is shown in Figure 11.14. It can be seen that the correlation is indeed linear for both experimental and φ corrected data points. From the graph, the $TMR_{ad,24h}$ can be obtained directly. In this case, the TMR will be 24 h at 38.3°C.

TABLE 11.12 TMR at Different Temperatures from the Experimental Adiabatic Run as Shown in Figure 11.12

Temperature (°C)	Time (min)	TMR_{ad} (min), $\varphi = 1.5$	TMR_{ad} (min), $\varphi = 1$
58	146	283	189
60	189	240	160
62	230	199	133
64	265	164	109
66	294	134	90

The TMR_{ad} at $\varphi = 1$ is calculated according to equation 11.12.

The advantage of this method over the above method based on one DSC measurement is the increased accuracy. The reason for this can be found in different aspects of this approach.

1. This method is based on an adiabatic experiment that will be by definition more representative for the situation in a large-scale reactor during a runaway reaction than a DSC experiment.
2. The temperature range over which the extrapolation takes place is fairly limited: in our example the difference between the experimental onset temperature and the finally obtained $TMR_{ad,24h}$ is only 20°C (compared to 80°C in the DSC example).
3. The extrapolation for the φ correction is also limited (from $\varphi = 1.5$ to $\varphi = 1$).
4. By using only data points in the early part of the reaction, the error introduced by assuming zero-order kinetics is limited. Indeed, in this early stage with low conversion, the concentration of the reagents can be assumed to be constant.

Also, for this method of determining $TMR_{ad,24h}$, a word of caution is needed. Only the data points in the early part of the exotherm are used. Consequently, we are dealing with very low heat rates at that moment. This poses high demands on the quality of the experimental data: a small amount of drift in the temperature stability of the instrument (either positive or negative drift) can have a profound effect on the final result. It is our experience that a well-operated adiabatic calorimeter should be able to deliver reliable results, provided that regular drift checks on empty test cells are performed to confirm the stability of the instrument.

Here, extra attention is needed when dealing with strongly autocatalytic reactions, since this method might overestimate the TMR_{ad} for that kind of reactions, leading to unsafe extrapolations. When a very sudden and sharp temperature increase is noticed in an adiabatic experiment, autocatalysis should be suspected and more testing will be appropriate. When in doubt, an iso-aging experiment at a temperature

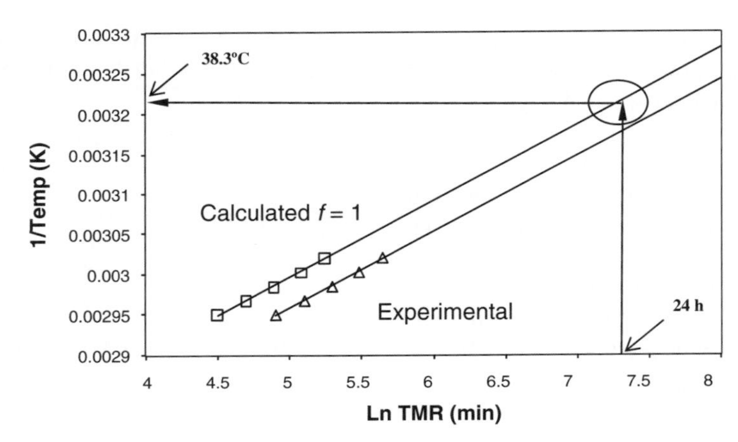

FIGURE 11.14 Extrapolation of the experimentally obtained TMR to lower temperatures and to $\varphi = 1$. The triangles represent the experimentally obtained TMR data at $\varphi = 1.5$, the squares are the corresponding calculated data at $\varphi = 1$, and the $TMR_{ad,24h}$ can be read from the graph as indicated by the arrows.

close to the calculated $TMR_{ad,24h}$ should be conducted to check the validity of the calculations.

11.5.5.3 Kinetic Modeling Both the above-mentioned methods make it possible to extract the $TMR_{ad,24h}$ from one single experiment, but with a limited accuracy due to the assumptions made. There are more advanced methods available as well that ask for a larger number of experiments and more advanced mathematical models, but lead to more accurate description of the reaction an allow for a broader range of simulations.

One possible approach to kinetic modeling is the fully mechanistic description of the reaction, which implies a complete understanding of the (decomposition) reaction at a molecular level. The reaction is therefore split up into its elementary reactions, and for each of those the frequency factor, reaction order, and activation energy are determined. It goes without saying that this method is quite elaborate from an experimental and computational point of view, but it also enables the widest range of process conditions that can be simulated (different concentrations, temperatures, φ factor, etc.) [22, 23].

Another possibility is the so-called nonparametric kinetic modeling. In this approach, a general kinetic model of the (decomposition) reaction is constructed based on a number (usually five) of DSC experiments with different heating rates [24, 25]. This kinetic description is said to be model free, meaning that there are no explicit assumptions being made about the reaction type. This type of modeling can lead to an accurate description of the reaction and enables the simulation of any temperature profile for the sample studied.

The reader is referred to the literature for a more detailed discussion of the different possibilities of kinetic modeling for both safety studies and process development.

11.6 OTHER POINTS TO CONSIDER

In the paragraphs above, we have discussed the fundamentals of process safety testing. Most of these techniques should be at least considered when developing any chemical process for scale-up. Some other techniques or practices should be brought to the attention of the reader, but, for lack of space, we will only touch upon them very briefly and refer to the literature for more details.

11.6.1 Flammability: Explosivity

Probably the single largest source of hazards in any chemical production plant is not due to intrinsic process safety, but to the risk of fire, especially when working with highly flammable organic solvents and reagents.

For solvents and liquid reagents, the flash point should be known to make sure that the instrumentation being used is suited for the job (hazardous area classification, zones, and Ex protection types). The flash point usually needs to be known for regulatory reasons for storage and transportation.

For solids, it might be necessary to determine the dust explosion characteristics. If a finely dispersed cloud of an organic solid finds an ignition source, a dust explosion can occur. Some products are more prone to this type of explosion hazards than others, so experimental testing is often needed. This will especially be the case in situations where a cloud of finely divided product can be formed in a noninert atmosphere (e.g., in fluidized bed driers, dry mills, or in a reactor during charging of a solid). Since these tests require relatively large amounts of product, they are usually conducted only at a late stage of development.

11.6.2 Static Electricity

Since a flammable atmosphere can be present in a chemical production plant, it is important to exclude any type of ignition source at any time. Especially when dealing with organic compounds, static electricity discharges can become very relevant. Many organic solvents and also a lot of the organic solids have a very low conductivity and can be charged easily. If the equipment being used is insufficiently grounded, a sudden discharge might occur, leading to a spark. This spark can in turn act as an ignition source for any flammable atmosphere, either of a vapor cloud or a dust cloud. Therefore, a proper understanding of static electricity is very important, and proper testing of the solvents and solids used might be needed (electrical conductivity measurements, charge decay measurements, etc.). The most obvious preventive measures against incidents related to static electricity are a proper grounding of any equipment being used (including the operator) and an appropriate inertization of reactors, driers, and so on whenever possible.

11.6.3 Vent Sizing

When a runaway reaction takes place in a reactor, the amount of gas (and vapor) being liberated can surpass the amount that can be removed through the conventional way, that is, through the condenser and scrubber lines. Therefore, vents are placed on the reactor. They usually consist of either a bursting disc or a pressure relief valve connected to a vent line of an appropriate diameter. In case of a serious runaway reaction or an external fire leading to an overpressure inside the reactor, the vents will open and the vent line will allow a safe depressurization of the reactor. This obviously implies a proper design of the vent system. The amount of gas that can be removed by such a system will obviously depend on the diameter of vent lines and also on the backpressure being generated by these lines. This backpressure is mainly a function of the amount and type of bends in the vent line and its total length. Another important factor one should consider is the composition of the gas flowing through the vent lines: if there is only gas flowing through the vent line, the minimum diameter needed will differ from the case when a mixture of liquid and gas is leaving the reactor. Proper vent sizing calculations are very complex and should only be undertaken by experts with the proper experience [26–28].

11.6.4 Safety Culture and Managerial Issues

Merely having a proper technical understanding of the process hazards present is not enough to guarantee the safety in a chemical production plant. The safety culture of the entire company, from the highest management level down to the shop floor, is of utmost importance. "Nothing is that important or urgent that it should not be done safely" should not be a hollow phrase but a natural part of the everyday work.

There are also a lot of managerial systems that are very important for reaching a high level of safety. Management of change should be taken very seriously: even minor changes made to a process can turn it from a safe process into an unsafe one, so safety should be kept in mind with every change made, and a proper hazard reevaluation might be needed. Adequate systems should be in place to establish the roles and responsibilities of all involved in process safety and to guarantee a standardized framework for safety assessments such as HAZOP, HAZAN, and PHA.

REFERENCES

1. Barton J, Rogers R. *Chemical Reaction Hazards*, Institution of chemical engineers UK, 1993.
2. Grewer T. *Thermal Hazards of Chemical Reactions, Industrial Safety Series*, Vol. 4, Elsevier, 1994.
3. Stoessel F. *Chem. Eng. Prog.*, 1993; 68.
4. Gygax R. *IUPAC, Safety in Chemical Production*, Blackwell scientific publications, 1991.
5. Stoessel F. *IChemE symposium series No. 153, 2007.*
6. Cisneros L, Rogers, WJ, Mannan MS. *Thermochim. Acta* 2004; 414:177.
7. Chervin, S, Bodman GT, Barnhart RW. *J. Hazard. Mater.* 2006; 130: 48–52.
8. Urben PG, editor, *Bretherick's Handbook of Reactive Chemical Hazards*, 7th edition, Elsevier, 2007.
9. Weisenburger G. et al., *Org. Process Res. Dev.* 2007;11(6): 1112–1125.
10. Stoessel F. *Thermal Safety of Chemical Processes: Risk Assessment And Process Design*, Wiley, 2008.
11. Burli M. Ueberprüfung einer nenen Methode zur Voraussage des ärmeüberganges in Rührkesseln. Dissertation ETHZ, 1979, 6479.
12. Choudhury S, Utiger L, Riesen R. Mettler-Toledo publication 00724218, 1990.
13. Zufferey B. Scale-down approch: chemical process optimisation using reaction calorimetry for the experimental simulation of industrial reactors dynamics. Dissertation No 3464, Ecole Polytechnique Federale de Lausanne, 2006.
14. Wilson EE. *Trans. Am. Soc. Mech. Eng.* 1915;37:47.
15. Yosida T. *Safety of Reactive Chemicals*, Elsevier, 1987.
16. Dien JM, Fierz H, Stoessel S, Kille G. *Chimia* 1994;48: 542–550.
17. Dermaut W. *Org. Process Res. Dev.* 2006;10:1251–1257.
18. McIntosh RD, Waldram SP. *J. Therm. Anal. Calorim.* 2003; 73: 35–52.
19. Wiss J, Stoessel F, Kille G. *Chimia* 1993; 47(11):417–423.
20. Keller A, Stark D, Fierz H, Heinzle E, Hungerbuhler K. *J. Loss Prev. Process Ind.*, 1997;10(1):31–41.
21. Townsend DI, Tou JC. *Thermochim. Acta* 1980; 37: 1–30.
22. Gigante L, Lunghi A, Martinelli S, Cardillo P, Picello L, Bortolaso R, Galvagni M, Rota R. *Org. Process Res. Dev.* 2003;7:1079–1082.

23. Remy B, Brueggemeier S, Marchut A, Lyngberg O, Lin D, Hobson L. *Org. Process Res. Dev.* 2008;12:381–391.
24. Dermaut W, Fannes C, Van Thienen J. *Org. Process Res. Dev.* 2007;11:1126–1130.
25. Roduit B, Dermaut W, Lunghi A, Folly P, Berger B, Sarbach A. *J. Therm. Anal. Calorim.* 2008; 93(1):163–173.
26. CCPS. *Guidelines for Pressure Relief and Effluent Handling Systems*, AICHE, 1998.
27. Fisher HG, Forrest HS, Grossel SS, Huff JE, Muller AR, Noronha JA, Shaw DA, Tilley BJ. *Emergency relief System Design Using DIERS Technology, the Design Institute for Emergency Relief Systems (DIERS) Project Manual*, AICHE, New York, 1992.
28. Etchells J, Wilday J. *Workbook for Chemical Reactor Relief System Sizing*, HSE, Norwich, 1998.

12

DESIGN OF DISTILLATION AND EXTRACTION OPERATIONS

ERIC M. CORDI

Chemical Research and Development, Pfizer Inc., Groton, CT, USA

12.1 INTRODUCTION TO SEPARATION DESIGN BY DISTILLATION AND EXTRACTION

Distillation and extraction are two important but often neglected unit operations in pharmaceutical process design and scale-up. Other than filtration following crystallization, these two processes represent the primary means of separating the important products of a complex manufacturing recipe from unwanted by-products, solvents, and other wastes. Successful design and execution of distillation and extraction operations have a direct impact on the purity and efficiency of downstream processing. These unit operations are material and cycle time intensive—often representing the point of maximum dilution and longest processing times in a synthetic manufacturing step. The cumulative effect of numerous distillation and extraction systems in active pharmaceutical ingredient (API) manufacture can be a determining factor in work center productivity in terms of kilograms of product per cubic meter of vessel space per week of operation. Beyond direct productivity, the overdesign of either unit operation leads to impacts on inventory and waste treatment systems—from thermal incinerators to solvent recovery to wastewater treatment facilities. Significant capital investments are often required to transport, store, recover, or treat liquids separated from products. The decision to waste or recover these liquids will have a direct impact on common green chemistry metrics such as the E-factor, which is defined as kilograms of waste per kilogram of product. Therefore, it is vital for process development chemists and their chemical engineering counterparts to jointly design the most efficient use of time and materials outside of traditional interests of bond-forming chemistry and product isolation.

12.1.1 Phase Equilibrium Thermodynamics

12.1.1.1 ***Phase Rule*** Distillation and extraction design is based on the fundamentals of phase equilibrium thermodynamics. Two phases in equilibrium must exist at the same temperature, pressure, and each species distributed between phases must possess identical partial molar Gibbs free energies in each phase.

$$T_\alpha = T_\beta$$

$$P_\alpha = P_\beta$$

$$G_i(T_\alpha, P_\alpha) = G_i(T_\beta, P_\beta)$$

The Gibbs phase rule for nonreactive components states that the number of degrees of freedom or the number of intensive state variables, available to fully specify a closed system at equilibrium is

$$F = C + 2 - P$$

where C is the number of distinct components, and P is the number of phases in equilibrium. Intensive variables are those that are independent of the size of a system as opposed to extensive variables that are a function of size. Mass and volume are examples of extensive variables, while temperature, pressure, and mass or mole fractions are intensive.

Chemical Engineering in the Pharmaceutical Industry: R&D to Manufacturing Edited by David J. am Ende

For example, a binary system in vapor–liquid equilibrium consists of two components and two phases in equilibrium, which results in two degrees of freedom ($F = 2 + 2 - 2 = 2$). Therefore, exactly two independent, intensive state variables (temperature, pressure, or mass/mole fraction of one species) may be specified to define the system at equilibrium.

12.1.1.2 Fugacity/Partial Fugacity As shown previously, partial molar Gibbs free energies are equivalent at equilibrium. Partial molar Gibbs free energy is also known as the chemical potential, μ_i, in phase equilibrium thermodynamics; and therefore, the chemical potentials of each component in each phase are equivalent at system equilibrium. Since chemical potential cannot be directly quantified, it can be conveniently redefined as a pseudopressure known as fugacity by the relationship

$$f_i = C\exp\left(\frac{\mu_i}{RT}\right)$$

where C is a temperature-dependent constant [1]. Partial molar fugacities of each component in each phase are therefore equivalent at equilibrium.

$$f_{i\alpha} = f_{i\beta}$$

In the vapor phase under ideal gas conditions, partial molar fugacity is equal to partial pressure (p_i), or

$$p_i = \frac{P}{y_i}$$

where P is total pressure and y_i is the mole fraction of component i. For ideal gas conditions or systems in which molecular interactions are absent in the vapor phase, can be assumed at moderate temperatures and at pressure of less than 10 bar, which holds true of most systems in pharmaceutical processing. Activity is defined as the partial molar fugacity of a component relative to its fugacity at some standard state. If the standard state is chosen as the fugacity of the pure component at the same temperature and pressure, then activity is

$$a_i = f_i/f_0 \text{ or } f_i = a_i f_0$$

Partial molar activity, a_i, can be further defined as the product of an activity coefficient, γ_i, and the mole fraction of the component, or for the liquid phase

$$a_i = \gamma_i x_i$$

where x_i is the mole fraction of a component in the liquid phase. Combining terms, the partial molar fugacity of a component in the liquid phase is therefore

$$f_i = \gamma_i x_i f_0$$

The standard state fugacity for a pure component at a given temperature is known as the vapor pressure, or $P^*(T)$.

12.1.1.3 Raoult's Law As stated previously, in vapor–liquid equilibrium the partial molar fugacities are equal, which forms the basis for Raoult's law for vapor–liquid equilibrium

$$f_{(i,v)} = f_{(i,l)} \text{ or } y_i P = \gamma_i x_i P_i^*(T)$$

For an ideal solution of components, the activity coefficient would be unity for all components across all possible compositions. Few multicomponent solutions exhibit ideal behavior upon mixing; and therefore, activity coefficient data are an essential part of phase equilibrium calculations. For noncondensable species with a mole fraction in solution approaching zero, Raoult's law can be restated by combining the activity coefficient and vapor pressure terms into what is known as a Henry's constant, H_i, and therefore Henry's law.

$$y_i P = x_i H_i$$

Liquid–liquid equilibrium follows the same thermodynamic principles; and therefore, equilibrium between phases can be described in a similar fashion and may include equilibrium with a vapor phase as well. That is, the partial molar fugacities of fugacities of each component of each phase must be equal, or

$$f_{i\alpha} = f_{i\beta} = f_{i\gamma}$$

where α is the vapor phase, β is the first liquid phase, and γ is the second liquid phase. Substituting know values for partial molar fugacities for each phase and assuming ideal gas behavior for the vapor phase, we have

$$y_i P = \gamma_{i\beta} x_{i\beta} P_i^*(T) = \gamma_{i\gamma} x_{i\gamma} P_i^*(T)$$

Since the pure component vapor pressure is independent of phase composition, this term can be eliminated from the equality to define vapor–liquid–liquid (VLL) equilibrium as

$$y_i P / P_i^*(T) = \gamma_{i\beta} x_{i\beta} = \gamma_{i\gamma} x_{i\gamma}$$

For noncondensable vapor, Henry's law for equilibrium between VLL phases would be

$$y_i P = H_{i\beta} x_{i\beta} = H_{i\gamma} x_{i\gamma}$$

12.1.1.4 Extended Antoine Equation Pure component vapor pressure data is well correlated as a function of temperature by an empirical relation known as the Antoine equation. In its extended form, the Antoine equation is

$$\ln \mathrm{P}_i^*(T) = C_{1i} + \frac{C_{2i}}{(T + C_{3i})} + C_{4i}T + C_{5i}\ln(T) + C_{6i}T^{C_{7i}}$$

where $P_i^*(T)$ is vapor pressure of component i and T is temperature. Parameter values are widely available in online data banks and within commercial properties and process simulation packages such as Aspen Properties and Aspen Plus.

12.1.1.5 K-Values Phase equilibrium is also represented frequently by ratios known as K values. In vapor–liquid equilibrium, this is simply the ratio of vapor to liquid mole fractions of a single component.

$$K_i = \frac{y_i}{x_i}$$

In relation to Raoult's law, this equilibrium constant can be restated as

$$K_i = \gamma_i P_i^*(T)/P$$

In the liquid–liquid equilibrium case, the K value is a ratio of mole fractions in each liquid phase and is known as a distribution coefficient

$$K_{iD} = \frac{x_{i\alpha}}{x_{i\beta}}$$

or through the phase equilibrium definition for the liquid–liquid system

$$K_{iD} = \frac{\gamma_{i\alpha}}{\gamma_{i\beta}}$$

Ratios of vapor–liquid equilibrium constants for a pair of components, i and j, are known as relative volatility, α, while the ratio of liquid–liquid equilibrium constants for a pair of components, i and j, are referred to as relative selectivity, β

$$\alpha = \frac{K_i}{K_j} = \frac{\gamma_i P_i^*(T)}{\gamma_j P_j^*(T)}$$

$$\beta = \frac{K_{iD}}{K_{jD}} = \frac{\gamma_{i\beta}\gamma_{j\alpha}}{\gamma_{i\alpha}\gamma_{j\beta}}$$

These ratios provide an alternative means of describing equilibrium between components although they remain temperature and composition dependent. Further, the ratios often provide a basis of comparison when selecting components in separation design. For instance, relative volatility may be used to determine the value of constant-level distillation as a means of separating a pair of components. Also, relative selectivity is useful in evaluating liquid–liquid extraction systems, where the choice of solvent is important to for efficient enrichment of the extract phase in the solute without undesirable carryover of feed solvent to the same.

12.1.2 Process Design from Laboratory to Manufacturing Plant

Accurate and efficient design of equilibrium stage separations requires fundamental physical property data for individual components and binary pairings as well as computational tools to store properties in data banks, to predict phase equilibrium, and to design new equipment or specify the operation of an existing plant. The range of activities leading to process design begins with component screening, phase diagramming, and single-stage equilibrium modeling. Most often in pharmaceutical manufacturing operations, a process design must fit the constraints of existing plant equipment, which is usually limited to relatively simple vessels with heat transfer utilities for establishing phase equilibrium, and in the case of vapor–liquid equilibrium an overhead condenser to separate vapor as distillate from the equilibrium vessel. Distillate is typically collected in a similar type of vessel and this by-product is sometimes sent to a separate facility on site for solvent recovery through further distillation. Liquid–liquid extractions are performed in similar equipment except that the receiver is used to collect the lower phase once equilibrium is established at the desired temperature and sufficient time is allowed for phases to separate. The separated phases in their respective vessels are then further processed with additional extractions using fresh solvent to achieve the desired design end point. For the reasons outlined here, the focus of subsequent sections will be single-stage equilibrium and separation design, although many of the early stage activities detailed here are applicable to selection and evaluation of solvents for relative volatility and selectivity as a basis for comparison of systems.

12.2 DESIGN OF DISTILLATION OPERATIONS

12.2.1 Vapor–Liquid Equilibrium Modeling

The graphical presentation of vapor–liquid equilibrium requires a collection of vapor pressure and activity coefficient data to calculate the mole fraction of each component in each equilibrium phase. Primary data from the laboratory are preferred as a starting point, but another valuable source of data is that contained within well-recognized data banks such as those of NIST, DIPPR, and DETHERM. In the absence of laboratory or a data bank source, pure component and binary properties may be estimated with the aid of property evaluation tools such as COSMOthermX or Aspen Properties. These tools use quantum chemical calculation of screening charge density or functional group contribution methods to provide pure component and binary data estimates.

12.2.1.1 Pure Component Vapor Pressure Data Fugacity of the liquid phase in Raoult's law is the product of an activity coefficient, liquid mole fraction, and the pure component vapor pressure. Vapor pressure is a function of temperature, and the data is often stored in tabular form or regressed into parameters of the extended Antoine equation for convenience and for use in modeling software. Figure 12.1 shows a collection of vapor pressure data plotted in semilog fashion for a set of common solvents.

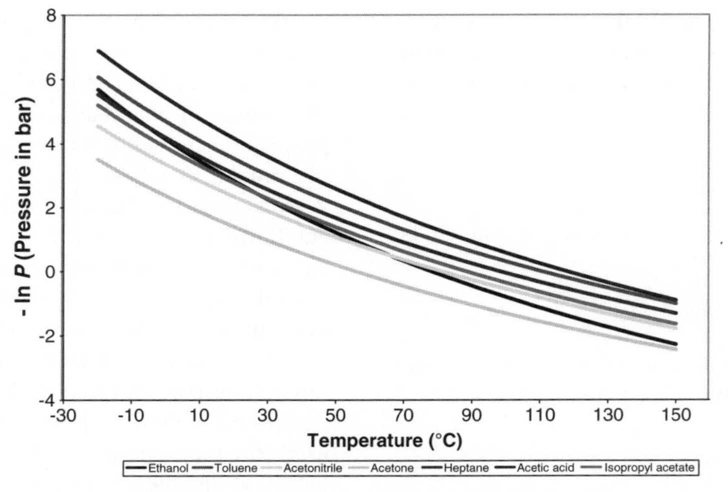

FIGURE 12.1 Vapor pressure of organic solvents.

12.2.1.2 Binary Activity Coefficient Data Activity coefficient data are vital to ensuring accurate nonideal solution vapor–liquid equilibrium modeling. Whether obtained from laboratory or estimated from equilibrium data, activity coefficients from binary pairs are typically regressed into NRTL binary interaction parameters, which may or may not be temperature dependent depending on the level of accuracy. The NRTL model is preferred for its ability to provide activity coefficient values for vapor–liquid and systems that display liquid–liquid immiscibility as well as predict maxima and minima in activity coefficients, where other models such as the Wilson equation are limited to vapor–liquid systems. The utility of this data is not limited to binary system, however, as the NRTL equation uses binary interaction parameters of possible parings of interest in a given system to calculate the liquid-phase activity coefficients for binary, ternary, and higher order systems. Figure 12.2 shows a set of binary activity coefficient plots for solvent pairs exhibiting a variety of behavior in solution. In ideal solutions, the activity coefficients of all components would be equal to unity. Strongly interacting solvents exhibit negative deviations from unity, while repulsive interactions are represented by positive deviations. In binary systems, the activity coefficient of a solvent is equal to unity at infinite dilution of all other components, which is shown in all plots at a mole fraction of unity.

12.2.1.3 Binary Phase Diagrams Simple binary phase diagrams contain a wealth of information about phase equilibrium behavior for a pair of solvents. Figure 12.3 shows examples of diagrams that combine vapor pressure and activity coefficient data of binary mixtures into a graphical representation of phase equilibrium for vapor–liquid and vapor–liquid–liquid systems. Generally, phase diagrams are divided into constant-pressure and constant-temperature plots. The constant-pressure phase equilibrium diagram, also known as a *T-xy* plot, is more common than the constant-temperature phase equilibrium diagram, or *P-xy* plot. The *T-xy* plot displays equilibrium temperature on the vertical axis versus liquid (x) and vapor (y) mole or mass fraction of one component on the horizontal axis. In a binary system, the mole or mass fraction of the other component in equilibrium is simply the difference between 1 and the plotted fraction. Several points of interest exist on either type of binary phase diagram including those that indicate the presence of an azeotrope or liquid–liquid immiscibility. On the left and right vertical axes, the liquid and vapor lines intersect at the boiling points of the pure components at the pressure or temperature specified by the plot. On a *P-xy* plot, these would be the pressure boiling points, while on a *T-xy* plot the pure component boiling point temperatures are indicated. Azeotropes are indicated by the intersection of the liquid and vapor curves between anywhere between 0 and 1 on the composition axis. In other words, vapor and liquid composition are equal at this point and are further defined by a temperature or pressure based on vertical position on the plot. On the *T-xy* plot, if this intersection occurs at a temperature less than the boiling temperature of either pure component (i.e., below the intersection of the liquid and vapor lines at either end of the composition axis), this is a minimum boiling azeotrope. If this temperature is greater than the boiling temperature of either pure component, this is a maximum boiling azeotrope. The reverse is true for less commonly used *P-xy* diagrams, since pressure is plotted on the vertical axis; that is, minimum boiling azeotropes appear above the boiling point pressures on either vertical axis. Azeotropes are divided into two classes heterogeneous and homogeneous. Homogeneous azeotropes are those that form in a single liquid phase, while heterogeneous azeotropes form in mixtures with liquid–liquid immiscibility. Further,

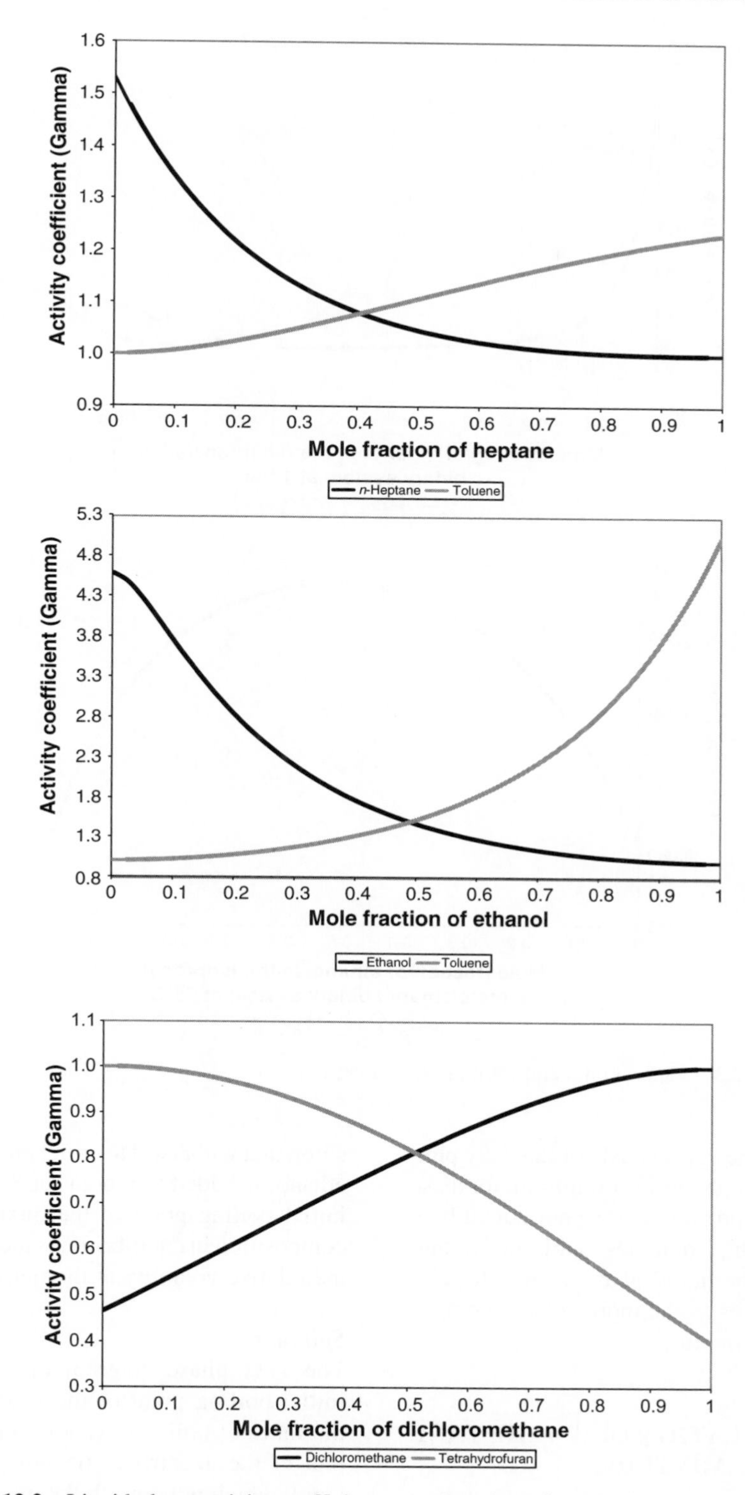

FIGURE 12.2 Liquid-phase activity coefficients as a function of composition.

heterogeneous azeotropes are always minimum boiling azeotropes, since a significant positive deviation in activity must occur for liquid-phase splitting to occur. On binary phase diagrams, liquid-phase immiscibility is indicated by a horizontal line for some portion of the composition range. A heterogeneous azeotrope occurs at the intersection of the vapor curve with this horizontal line. Heterogeneous liquid systems boil at a constant temperature indicated by the

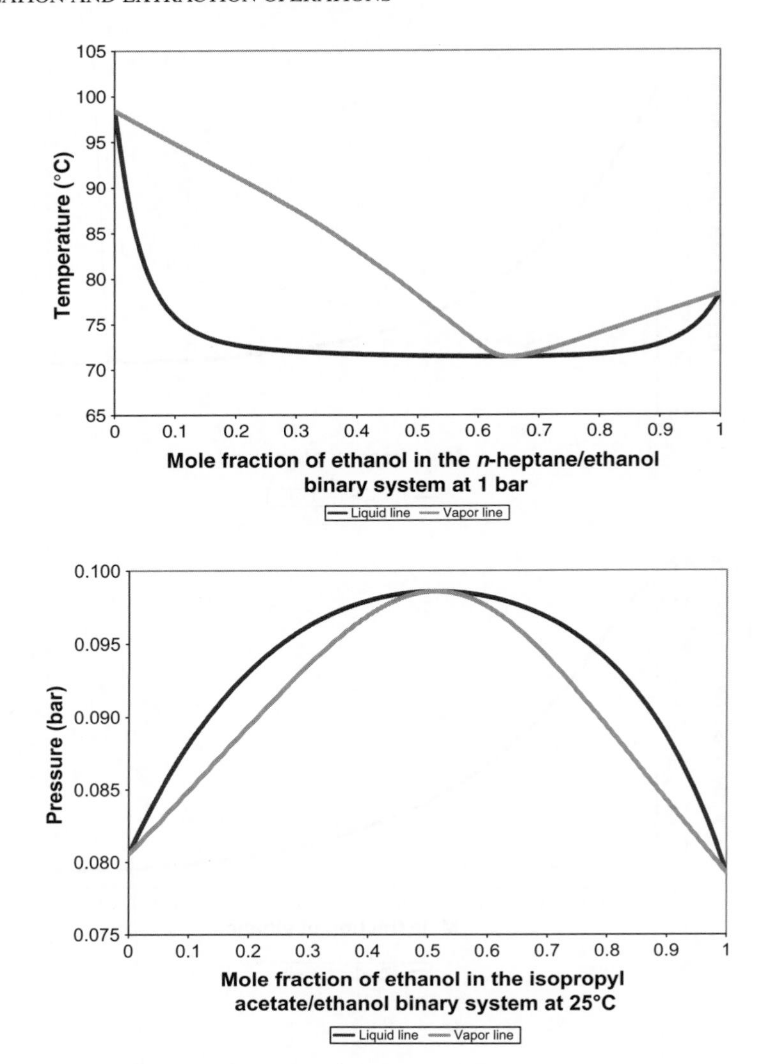

FIGURE 12.3 Vapor–liquid equilibrium-phase diagrams.

horizontal line position on the vertical axis of the *T-xy* plot. The composition of the liquid phases in equilibrium for these systems is indicated by the end points of the horizontal line. Constant temperature boiling continues until the system becomes homogeneous in the liquid phase, at which point the pot boiling point increases as the more volatile component is evaporated from the mixture.

EXAMPLE 12.1 DISTILLATION OF A *n*-HEPTANE AND ETHANOL BINARY MIXTURE

A 100 mole mixture of *n*-heptane (50 mol%) and ethanol (50 mol%) will be distilled to displace *n*-heptane from ethanol at ambient pressure. What is the initial boiling point of this mixture? As the mixture is distilled, which component will be enriched in the pot? If ethanol is not enriched in the pot, how much ethanol must be added to start at a composition that will distill to enrich the pot in ethanol? If sufficient ethanol is added to a point of 85 mol% ethanol, what is the initial boiling point of the mixture and what is the vapor composition in equilibrium at the start of distillation. What is the relative volatility at the initial mixture?

Solution

The *T-xy* phase diagram in Figure 12.3 shows that the initial boiling point of the mixture is about 72°C, and as the mixture boils the vapor composition is approximately that of the azeotrope (65 mol% ethanol, 35 mol% *n*-heptane), which will enrich the pot in *n*-heptane as the boiling point increases to that of 98°C. To begin at a point that will enrich the pot in ethanol, additional ethanol must be added to achieve a composition beyond the minimum-boiling azeotrope. To reach an initial mixture composition of 85 mol% ethanol and 15 mol% *n*-heptane, an ethanol material balance is used to determine the amount of extra ethanol to add:

$$100 \text{ moles} \times (50 \text{ mol\% ethanol}) + \text{A moles ethanol}$$
$$= \text{B moles} \times (85 \text{ mol\% ethanol})$$

$$100 \text{ moles} + \text{A moles} = \text{B moles}$$

Substituting the total mole balance for B into the first equation

$$50 \text{ moles ethanol} + \text{A moles ethanol} = (100 \text{ moles} + \text{A moles}) \times 85 \text{ mol\% ethanol})$$

$$\text{A} = 233 \text{ moles of ethanol added}$$

$$\text{B} = 333 \text{ moles total moles in new mixture}$$

The molar relative volatility of n-heptane to ethanol in the new mixture is

$$\alpha = \frac{y_1/x_1}{y_2/x_2}$$

where component 1 is n-heptane and component 2 is ethanol

$$= \frac{(0.30/0.15)}{(0.70/0.85)} = 2.4$$

The relative volatility of n-heptane over ethanol at ambient pressure is fairly low, which often results in inefficient batch strip-and-replace distillation. An alternative process in the form of constant-level distillation is a way to improve the time and material efficiency of batch distillation. Constant-level distillation will be discussed in detail in an upcoming section of the chapter.

12.2.1.4 Ternary Phase Diagrams Systems containing three components in vapor–liquid equilibrium are graphically represented in ternary phase diagrams, which are sometimes referred to residue plots. Unlike binary phase diagrams, these phase diagrams contain mostly information about the liquid-phase composition—hence the name residue plot—or in the case of vapor–liquid–liquid systems the liquid–liquid phase envelope with associated tie line end points defining the composition of each liquid phase. Examples of ternary-phase diagrams are shown in Figure 12.4 for a homogeneous azeotrope system and in Figure 12.5 a system with a heterogeneous azeotrope. Binary azeotropes are represented by points on the triangle edges labeled with the associated boiling point. Boiling points of individual components are included at the vertices of the triangle. When binary azeotropes are present, the ternary-phase diagram is divided into distillation regions separated by distillation boundaries. In the case of three or more binary azeotropes, ternary azeotropes may also form at the intersection of distillation boundaries. Pot composition trajectories are represented on the plot by residue curves that follow increasing temperature, either from one pure component vertex to another or between

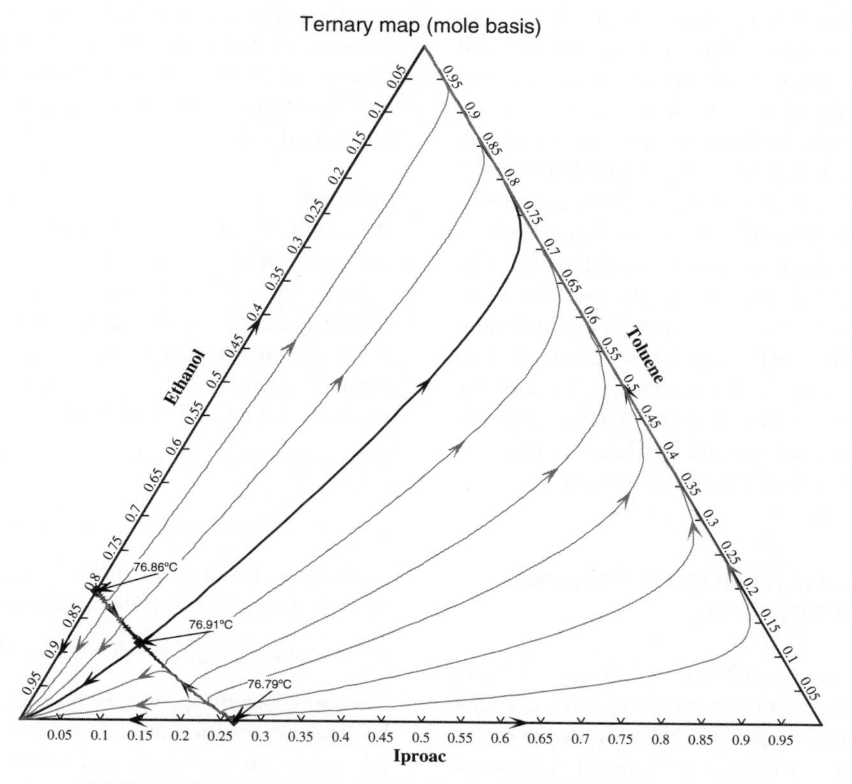

FIGURE 12.4 Ternary-phase diagram with homogeneous azeotrope.

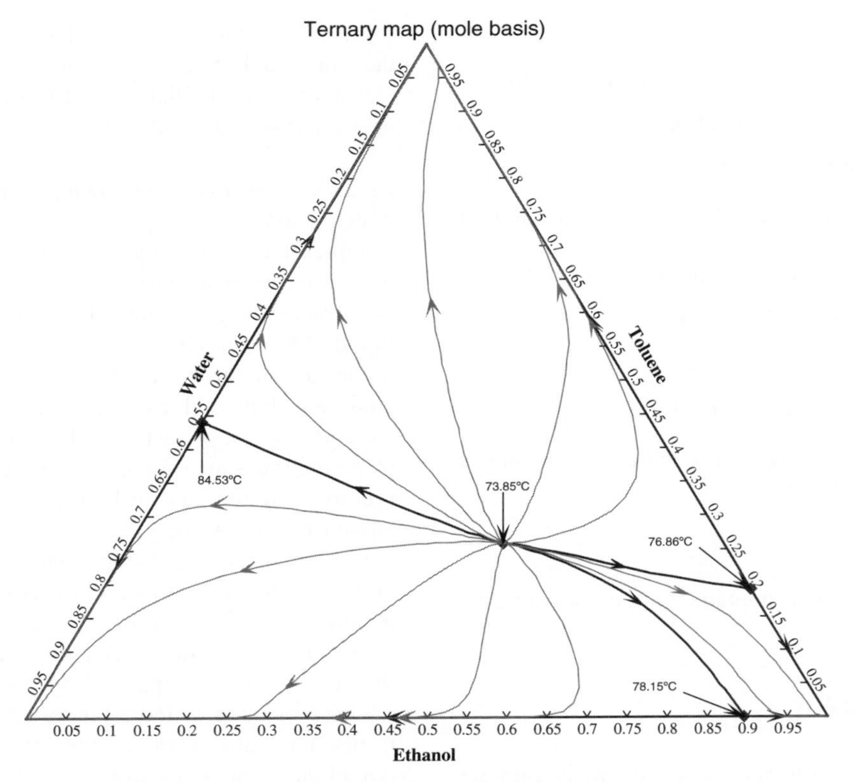

FIGURE 12.5 Ternary-phase diagram with heterogeneous azeotrope.

an azeotrope and a pure component. Analysis and design of distillation systems using ternary-phase diagrams can become complex with the number of possible scenarios posed by multiple binary and ternary azeotropes with and without liquid–liquid equilibrium phase envelopes, and there are many references available to consider for such in-depth analysis. As with binary diagrams, it is important for the chemical engineer to have access to the property data banks and plotting tools that allow quick access to system behavior in graphical form, which leads to a level of understanding for the purpose of process modeling and design that would be much more difficult and time consuming with the limitation of pure component boiling and perhaps binary azeotrope data at ambient pressure. These are the limitations imposed by tabulated data that can easily be overcome by computational tools such as Aspen Plus that provide flexibility in terms of depth of component choices and process conditions.

EXAMPLE 12.2 DISTILLATION OF ETHANOL AND WATER FROM TOLUENE

A 100 mole mixture of ethanol (63 mol%) and water (37 mol %) will be displaced with toluene at ambient pressure. Using the ternary VLE diagram in Figure 12.5, determine the moles of toluene that are added to reach the saddle ternary azeotrope near the center of the diagram. What is the significance of the mixture at this point? How much additional toluene is added to reach a starting distillation composition of 45 mol% toluene, 35 mol% ethanol, and 20 mol% water? Is this composition sufficiently high in toluene to ensure that distillation enriches the pot with toluene as distillate is collected from the condensed vapor?

Solution

A straight material balance line is drawn from the starting composition of 63 mol% ethanol and 37 mol% water to the toluene apex of the triangle. This line passes through the saddle azeotrope with a composition of 26 mol% toluene, 47 mol% ethanol, and 27 mol% water. A material balance on ethanol can be used to determine the total moles in the mixture at the saddle azeotrope

$$100 \text{ moles } (63 \text{ mol\% ethanol}) = \text{A moles } (47 \text{ mol\% ethanol})$$

$$\text{A} = 134 \text{ moles in total mixture at saddle azeotrope}$$

A material balance on toluene determines the amount of toluene added to reach the saddle azeotrope.

$$134 \text{ moles } (26 \text{ mol\% toluene}) = 34 \text{ moles of toluene added}$$

The saddle azeotrope represents the ternary point of constant boiling composition at ambient pressure. The distillate and pot compositions will be equivalent during the entire distillation process. This can only be overcome by adding additional solvent. In this case, additional toluene is added to achieve a composition within the distillation boundary that favors ethanol and water volatility over that of toluene. The material balance line for toluene lies on the starting point of interest for the distillation. A material balance on ethanol between the toluene-free mixture and the desired mixture distillation starting point will determine the amount of toluene added:

$$100\text{ moles }(63\text{ mol\% ethanol}) = \text{B moles }(35\text{ mol\% ethanol})$$

$$\text{B} = 180\text{ moles of mixture at distillation starting point}$$

The amount of toluene in this mixture is

$$180\text{ moles} - 100\text{ moles} = 80\text{ moles of toluene}$$

Therefore, the amount of toluene added past the saddle azeotrope is

$$80\text{ moles toluene} - 34\text{ moles toluene}$$

$$= 46\text{ moles toluene added}$$

The mixture pot composition will follow a residue contours interpolated from the figure toward the toluene apex of the ternary diagram.

12.2.2 Process Modeling and Case Studies

Chemical engineers designing distillation processes in the pharmaceuticals industry typically face the challenge of designing and optimizing single-stage equilibrium batch separations. These designs are further limited by the constraints of multi-purpose equipment also used for distillation. Absolute pressures in the equilibrium vessels and distillate receivers are limited to 0.05 bar–3 bar by the design limits of batch processing vessels and their associated vacuum pumping and pressure relief systems. Temperatures for the jacketed vessels and their associated condensers are often limited to the plant-wide heat transfer fluid service, or temperature control unit, service range of –20 to 150°C. Single-stage distillation is operated in two modes (1) strip and replace or (2) constant-level (or constant-volume). Strip-and-replace distillation is more common because it replicates the method used in the process development laboratory and does not require advanced control to operate on plant scale. However, constant-level distillation is beginning to gain favor for the benefits of improved solvent exchange efficiency for low relative volatility systems and for potential cycle time savings. The mathematical treatment of single-stage distillation is relatively easy for either mode, but speed of design is enhanced by the use of dynamic process simulators such as Aspen Batch Distillation, which is an adaptation of the more general dynamic modeler known as Aspen Custom Modeler. The scope of Aspen Batch Distillation goes beyond single-stage distillation, if needed, and also employs process control simulation, which is useful for implementing constant-level distillation models.

12.2.2.1 Strip and Replace Batch Distillation The more common mode of distillation in a pharmaceutical manufacturing facility is that of strip-and-replace distillation under ambient or vacuum conditions. This is achieved by stripping an initial mixture often containing a nonvolatile solute to a desired concentration or volume and then adding a replacement solvent to the equilibrium still or pot. Vapor is condensed into liquid distillate by an overhead condenser. Distillate is collected by gravity flow into a receiver that is usually similar in construction to the equilibrium still. The strip and replace procedure is repeated until a desired end point of composition and volume is obtained. Binary and ternary-phase diagrams are important references in the design of strip and replace batch distillation since the boiling point and azeotrope composition can have a significant impact on the efficiency of the operation. For instance, if a binary *T-xy* diagram is plotted with the composition of the undesirable solvent on the horizontal axis, the desired end point is toward the left-hand side of the plot. The pot residue will follow the path of increasing boiling point as the more volatile component is removed by vaporization and separated by condensation to the distillate receiver. If a minimum boiling azeotrope is present, then that path toward the higher boiling point may occur to the left or right of the azeotrope on the plot. In the standard plot with the undesirable solvent on the bottom axis, the desired direction is left; and therefore, the pot composition must be left of the azeotrope point for the pot to become enriched in the replacement solvent. Otherwise, on the right side of the azeotrope point, the replacement solvent will be lost through distillation. One solution to this problem is to add enough replacement solvent initially to reduce the mole or mass fraction to a point left of the binary azeotrope before beginning distillation to ensure the undesirable solvent has greater volatility than the replacement. Ambient pressure distillation is usually chosen for its simplicity, but vacuum distillation can be used to change the azeotrope composition of the system or reduce the operating temperature to preserve chemical stability of a solute.

Simple batch distillation is also known as differential distillation and has been described in mathematical terms by Lord Rayleigh and is derived as follows [2]. A charge of L_0 moles of liquid mixture is made to a still, which is heated to the boiling point at constant pressure. The initial composition for a given component of the liquid in the still has a mole fraction of x_1. As the liquid mixture boils at mole fraction x, the vapor V in equilibrium with the pot liquid is y, and this vapor is completely condensed to $y_d = x_d$. A material balance can be performed on this dynamic system for total moles.

$$\frac{dL}{dt} = -V \quad \text{(total material balance)}$$

or for a given component the material balance is

$$\frac{d(Lx)}{dt} = -Vy \quad \text{(component material balance)}$$

$$L\frac{dx}{dt} + x\frac{dL}{dt} = -Vy$$

If the above equation is combined with the total material balance

$$L\frac{dx}{dt} + x\frac{dL}{dt} = y\frac{dL}{dt}$$

Multiplying throughout by dt eliminates time dependence

$$Ldx + xdL = ydL$$

Rearranging L and mole fractions to either side

$$Ldx = (y-x)dL$$

or finally, the Rayleigh equation

$$\frac{dL}{L} = \frac{dx}{(y-x)}$$

A more useful form of the Rayleigh equation for numerical solution of time-independent batch distillation is

$$\frac{dx}{dL} = \frac{(y-x)}{L}$$

The values of y and x depend on knowledge of vapor–liquid equilibrium for a given system at temperature T and total pressure P.

This derivation describes composition trajectory of a solution during concentration in the batch still. If fresh replacement solvent is charged to the system, the new initial concentration can be determined by material balance and a composition new trajectory is then calculated again using the Rayleigh relationship. When a solute is present, one or more simplifying assumptions can be made when the nature of the solute in terms of mole fraction or binary interactions is unknown. The calculations may be performed ignoring the solute mole fraction and the results will be stated in terms of solvent fractions. This ignores the contribution of the solute as a mole fraction and presumes that the solute is nonvolatile. If solute is likely to impact boiling point temperature by the nature of its mole fraction in the solution, it may be included while still assuming the solute is nonvolatile, but this still ignores any binary interactions with solvent in the solution. Use of either assumption is reasonable for feasibility case studies or for initial design of the distillation process. The impact of a solute on vapor–liquid equilibrium could be assessed by estimation of binary interaction parameters for the NRTL model using COSMOthermX or group contribution methods such as UNIFAC, or ultimately measuring vapor–liquid equilibrium for the system in the laboratory with a range of compositions and concentrations. The latter is most time consuming and should only be used in the final stages of design of a commercial process.

EXAMPLE 12.3 BATCH STRIP-AND-REPLACE DISTILLATION WITH ETHANOL AND *n*-HEPTANE

A 100 L mixture of ethanol (65 wt%) and *n*-heptane (35 wt%) is added to a single-stage batch distillation vessel. Using a batch distillation simulator, determine the amount of ethanol that must be used in strip-and-replace distillation to achieve less than 0.1 wt% *n*-heptane at ambient pressure. The end point must be 100 L of final mixture, and the volume cannot exceed 200 L at any point in the distillation.

Solution

Using Aspen Batch Distillation, the simulation case is constructed for ambient pressure distillation of a single-stage system. The pot is filled with 100 L of 65 wt% ethanol and 35 wt% *n*-heptane. Next, 100 L of ethanol is added and distillation commences until a pot volume of 100 L is achieved at 77°C. An additional 100 L of fresh ethanol is added, and distillation resumes until 100 a pot volume of 100 L is achieved. After consuming 200 L of fresh ethanol in this operation, the pot contains 0.07 wt% *n*-heptane, which meets the requirement of achieving less than 0.1 wt% *n*-heptane by the end of distillation at a temperature of 78°C. In fact, the 0.1 wt% *n*-heptane specification was achieved at the 107 L mark. An additional 7 L of volume were distilled from the pot to achieve the final volume requirement of 100 L. *Note:* The volumes are approximate due to thermal expansion of solvents under ambient pressure distillation conditions.

12.2.2.2 Constant-Level Batch Distillation A variation of traditional batch distillation is constant-level distillation, which is sometimes referred to as constant-volume in this mode of operation. In this scenario, batch distillation takes place with a feed of fresh replacement solvent that matches the distillate rate in a fashion that maintains a constant liquid level in the equilibrium still. It is usually most efficient to concentrate the original solution to the greatest degree before adding replacement solvent, which minimizes the overall consumption of the added solvent to achieve the same end point. This effect is achieved my minimizing the amount of replacement solvent that is evaporated during the exchange. Analysis shows that systems with low relative volatility, $\alpha < 5$, enable replacement solvent savings of greater than 50% [3]. Constant-level distillation is an advance from traditional single-stage batch distillation efficiency when column rectification with reflux is unavailable. Distillation with partial reflux requires significant capital investment and results in increased batch cycle time and energy use. However, constant-level distillation may suffer inefficiencies without optimization of operating conditions. For instance, traditional batch strip-and-replace distillation uses a much larger fraction of the available heat transfer surface area in a vessel than constant-level distillation would under suboptimal conditions [4]. To maximize heat transfer area, the initially concentrated solution could be transferred by a suitably sized constant-level distillation vessel for batch processing in a full vessel. In a development or kilogram laboratory, constant-level distillation can be controlled by manual adjustment of replacement solvent flow that matches distillate collection. When volumes of solvent are small, the installation of advanced controls does not provide significant cost advantage. However, installation of level control would provide efficiencies in a pilot plant or commercial manufacturing facility under routine processing conditions. Automation enables programming of a defined end point in terms of replacement solvent volume charged. Level control can be maintained by a variety of inputs such as the still or distillate vessel level. Less direct due to differences in solvent density is the still or distillate mass measured by strain gauge as a control input. Beyond solvent exchange efficiency are the benefits of a reduced temperature profile over time experienced by the batch and reduced energy use by reduced solvent loading per batch. The reduced temperature profile may preserve chemical stability of solute leading to potentially higher yield and lower impurity loading in the isolated product.

EXAMPLE 12.4 CONSTANT-LEVEL DISTILLATION WITH ETHANOL AND *n*-HEPTANE

A 100 L mixture of ethanol (65 wt%) and *n*-heptane (35 wt%) is added to a single-stage batch distillation vessel. Using a constant-level distillation simulator, determine the amount of ethanol that must be used in constant-level distillation to achieve less than 0.1 wt% *n*-heptane at ambient pressure. The end point must be 100 L, which is identical to the starting mixture. Compare the amount of ethanol consumed in this example with that of the batch strip-and-replace distillation in Example 12.3, and determine which distillation method is more material efficient in this scenario.

Solution

Using Aspen Batch Distillation, the simulation case is constructed for ambient pressure distillation of a single-stage system. The pot is filled with 100 L of 65 wt% ethanol and 35 wt% *n*-heptane. The batch is heated to its initial boiling point of 72°C, and as distillate begins to collect, a level controller regulates the feed of fresh ethanol to the pot to maintain a 100 L volume. Distillation continues until 0.1 wt% *n*-heptane remains in the pot at a temperature of 78°C. The simulation shows that 150 L of fresh ethanol were consumed to achieve the *n*-heptane end point specification. The constant-level distillation processing scenario consumed 50 L less ethanol than the batch strip-and-replace distillation specified in Example 12.4. In the batch strip-and-replace scenario, 200 L of ethanol were consumed to achieve the 0.1 wt% *n*-heptane end point specification. Therefore, the constant-level distillation operation provided a savings of 25% of fresh ethanol used for strip-and-replace distillation. This level of savings is typical for scenarios in which the relative volatility of components is low.

12.2.3 Laboratory Investigation

12.2.3.1 Vapor–Liquid Equilibrium Data Collection As the number of possible binary combinations is quite large, a process developer or modeler is faced with screening and design without the benefit of binary interaction parameters for solvent pairs that are necessary to correlate activity coefficient for a given system as a function of composition and temperature. For instance, a selection of 100 solvents relevant to the pharmaceutical industry would provide 4950 binary combinations (100 × 99/2) among the set. Although the presence of a solute is often assumed to have negligible vapor pressure and minimal impact on the solution boiling point or vapor–liquid equilibrium of the system due to binary interactions, these assumptions may be ultimately tested in the laboratory. Accurate prediction requires careful measurement of important parameters in a laboratory equilibrium still. The experimentalist may choose to study binary systems for regression of collected data into binary interaction parameters, such as those for the NRTL model, that are temperature and composition dependent. Relying on Gibbs phase rule, the appropriate number of degrees of freedom is predetermined, and then the full condition of the system at equilibrium is recorded. Typically, in a binary system the pressure and total composition are predetermined and the batch is heated to its boiling point. At equilibrium, representative samples of vapor and liquid are extracted for analysis and system temperature is recorded. This data, converted to mole or mass fraction versus temperature, is then regressed within a convenient platform such as Aspen Properties into the a_{ij}, a_{ji}, b_{ij}, b_{ji}, and c_{ij} parameters (as subparameters of α_{ij} and τ_{ij}) of the NRTL model for later use in system phase diagramming and process modeling. Applications designed for regression of such data also conveniently perform thermodynamic consistency testing with the ability to exclude data points that are inconsistent for the system. Thermodynamic consistency testing is derived from the Gibbs–Duhem equation [5].

$$\sum x_i d \ln\gamma_i = 0$$

There are two derived forms for thermodynamic consistency testing based on the Gibbs–Duhem equation. First, the integral test is [6]

$$\int \ln\frac{\gamma_i}{\gamma_j} dx_i = 0$$

Graphically, the net area under a plot of $\ln(\gamma_i/\gamma_j)$ versus x_i must equal zero, or typically to some tolerance such as 10%.

Alternately, a differential or point test that tests consistency at a single composition may be constructed from the same equation and subjected to a tolerance similar to the integral test [7].

$$\frac{d\ln\gamma_j^{\circ}}{dx_j} = -\frac{x_i}{1-x_i} \times \frac{d\ln\gamma_i^{\circ}}{dx_j}$$

Graphically, the slopes of a combined plot of $\ln\gamma_i$ and $\ln\gamma_j$ versus x_i at a given composition must satisfy the equation to establish thermodynamic consistency. The results of consistency tests speak nothing about the accuracy of the data; only that it fits within the bounds imposed by partial molar Gibbs free energy of components, or chemical potential, within a system at equilibrium.

12.2.3.2 Batch Distillation Simulation in the Laboratory Performing batch distillations in the laboratory as part of process development is not a trivial exercise if reliable scale-up to manufacturing conditions is the goal. Lab-scale distillations should be studied in equipment representative of a single-stage still on scale, with the ability to collect samples of off-line analysis or insert analytical probes. The equipment must be adequately insulated from ambient conditions to prevent condensation and liquid holdup in places other than the condenser and distillate receiver, respectively. In either strip-and-replace or constant-level distillation, an accurate material balance and associate parameters should be used to record and describe the procedure. The total amount of replacement solvent consumed until the end point of the distillation is important as well as the liquid volumes and compositions in the still and the distillate receiver. A temperature profile should be noted as it may have a direct impact on product quality since distillation time on scale will be longer than in the laboratory. Vacuum distillations should be simulated by first applying vacuum to a desired set point that matches plant capacity, and then slowly applying heat to reach the solution boiling point to mimic common practice in manufacturing facilities. This practice also reduces the likelihood of bumping or boiling over of pot contents into the receiver. Unintended refluxing before the condenser will result in a distillation with higher efficiency than achievable in a single-stage on scale. During distillation development, other effects such as foaming or solute oiling should be noted. Uncontrolled crystallization of the solute may occur at any pressure during distillation, which may impact the quality of the isolated product by incorporating impurities or producing an undesired polymorph. Finally, an important simulation for batch distillation is that of extended time at a given temperature, whether it is the initial or final boiling point. The final boiling point will be higher that the start, so this will represent the most stressed condition in terms of chemical stability in the still. Stability at final reflux temperature is often tested for at least 24 h and sometimes longer to check for degradation that may occur on scale due to extended processing time or unexpected delays. The heat transfer efficiency of the plant will also determine the time to heat up and cool down the solution before and after distillation and this temperature history must be incorporated into the design. Similarly, there may also be heat and cool cycles associated with requirements around minimum temperature for sampling or solvent charging that should be considered.

12.2.4 Process Scale-Up

12.2.4.1 Equipment Selection The selection of equipment for batch distillation is normally limited in the pharmaceutical industry due to the design of synthesis plants around the strategy of maximum flexibility. Therefore, equipment selection is about choosing single-stage vessels appropriate for the task, which is to efficiently concentrate or exchange solvents without impacting the quality of pharmaceutical intermediates or active ingredients carried through. Efficiency is about choosing the volume of vessel that provides maximum heat transfer area per unit volume and with a minimum stir volume that meets the needs of the process. Further, a system capable of vacuum or pressure should be chosen carefully when necessary to achieve a distillation target. The vacuum and pressure limits should be well understood and the heat removal performance of the condenser in terms of operating temperature should be noted. If a constant-level distillation is planned, the replacement solvent should be added to the still in a fashion that allows rapid mixing with the batch. In other words, avoid adding the replacement solvent in such a way that it runs down the heated wall of the vessel and has opportunity to vaporize before mixing with the contents of the still.

12.2.4.2 Process Control and Unit Operation In operation the batch still requires temperature and perhaps vacuum control. In the mode of constant-level distillation, the replacement solvent feed and distillate rates must be matched with appropriate automation. For temperature control, the most beneficial mode of operation is temperature differential across the jacket to maintain vessel integrity and to preserve solute chemical stability by keeping the temperature at the vessel wall below a reasonable specification. A typical temperature differential set point is less than 30°C to achieve a desirable boil up rate at a given pressure. In this regime, the temperature of the jacket will increase with the boiling point to maintain the specified temperature differential through the distillation end point.

12.2.4.3 End Point Determination The desired end point distillation is usually based upon a desired solution composition or complete displacement of one solvent from the system. Detection of this end point is available through a variety of direct or indirect measurements that can be made on or off line. Sampling of the still contents for off-line solvent analysis is the most direct and most likely option in most pilot and manufacturing scale facilities. However, the turnaround time for results may be on the order of hours and sampling may require cooling of the batch to a safe temperature. Sometimes the batch temperature is a sufficient end point when a complete exchange of solvents is desired and the downstream process is insensitive to the presence of small amounts of the replaced solvent or water. In that case, the operator would continue distillation until the pot temperature reaches the boiling point of the pure solvent or that of the solution tested in the laboratory since the presence of a solute, which can raise the boiling point by $\geq 1°C$. If the distillation is run under vacuum, it is useful to provide boiling point data at the desired system pressure as a guide to distillation end point. Note that heterogeneous mixtures will exhibit a constant temperature boiling point until the mixture becomes homogeneous. From that point, there is often a rapid increase in temperature to the replacement solvent end point.

12.3 DESIGN OF EXTRACTION OPERATIONS

12.3.1 Liquid–Liquid Equilibrium Modeling

Equilibrium-phase models for the liquid–liquid case requires demonstration of equivalent fugacity in each phase for each component, and this case may be extended further by the inclusion of vapor-phase equilibrium. As with VLE, laboratory measured data is the preferred source of equilibrium data, and even without the facility to determine liquid–liquid equilibrium parameters directly, one may rely on well-populated data banks of NIST, DIPPR, and DETHERM for raw data. Or, as with VLE, the liquid–liquid equilibrium data may be estimated by computational tools such as COSMOthermX and activity coefficient data may be extracted for regression into nonideal solution models such as NRTL. With binary interaction parameters in hand, one may begin to plot liquid–liquid phase diagrams that are descriptive for binary or ternary systems. Higher order systems may be modeled easily with the binary interaction parameters even though graphical representation of equilibrium between four or more components is not as helpful.

12.3.1.1 Binary Activity Coefficient Data Focusing here on the liquid–liquid case alone, the important temperature and composition dependent element of equilibrium is the activity coefficient of each component of each phase. As discussed previously, a variety of nonideal solution models may be used to correlate activity coefficients for components as a function of temperature and composition, and in the case of liquid–liquid equilibrium the NRTL model is preferred because of the flexibility in the description of liquid-phase splitting as well as VLE. Equilibrium-phase splitting is only possible in nonideal solutions with significant deviations in the activity coefficient for each component away from unity.

12.3.1.2 Phase Diagrams The primary method of graphical representation of liquid–liquid equilibrium between three components is a ternary-phase diagram. There are several methods for plotting three-component liquid equilibrium, and the most common method is a triangular plot as shown in Figure 12.6. The plot may be formatted as an equilateral or isosceles triangle with the sides of either format divided as mole or mass fraction of each of the three components. Within the bounds of the triangle edges, a boundary line is drawn to divide single- and two-phase regions for a mixture of three components. Tie lines within the two-phase envelope determine the equilibrium mole or mass fractions of each component in each liquid phase at a given temperature. As liquid–liquid equilibrium is also temperature dependent, the size and shape of the two-phase region will be affected by changes in temperature. The point at which the tie lines become infinitely small is known as the plait point, where both liquid phases have identical compositions at their limit of immiscibility.

EXAMPLE 12.5 EXTRACTION OF ACETONE FROM WATER INTO TOLUENE

A 100 mole mixture of acetone (70 mol%) and water (30 mol %) is added to an extraction vessel. Determine the minimum number of moles of toluene must be added to the system to obtain a liquid–liquid system in equilibrium. Further, how much additional toluene must be added to the system to obtain a liquid–liquid system with the light phase containing 45 mol% acetone. What is the mole composition of the light and heavy phase in equilibrium at 20°C? Finally, determine the molar distribution coefficient of acetone and the selectivity of toluene for acetone relative to water in the final mixture.

Solution

Use the ternary-phase diagram in Figure 12.6 for the toluene, water, and acetone system to determine behavior of a ternary mixture of these solvents. A straight line drawn from the 70 mol% acetone and 30 mol% water intersection to the 100% toluene apex of the triangle determines the composition path for the addition of pure toluene to the system. The straight line intersects the phase boundary at 23 mol% toluene, 54% acetone, and 23% water. Performing a mole balance on acetone, we have

100 moles × (70 mol% acetone) − A mol × (54 mol% acetone) = 0; A = 130 moles of mixture after toluene

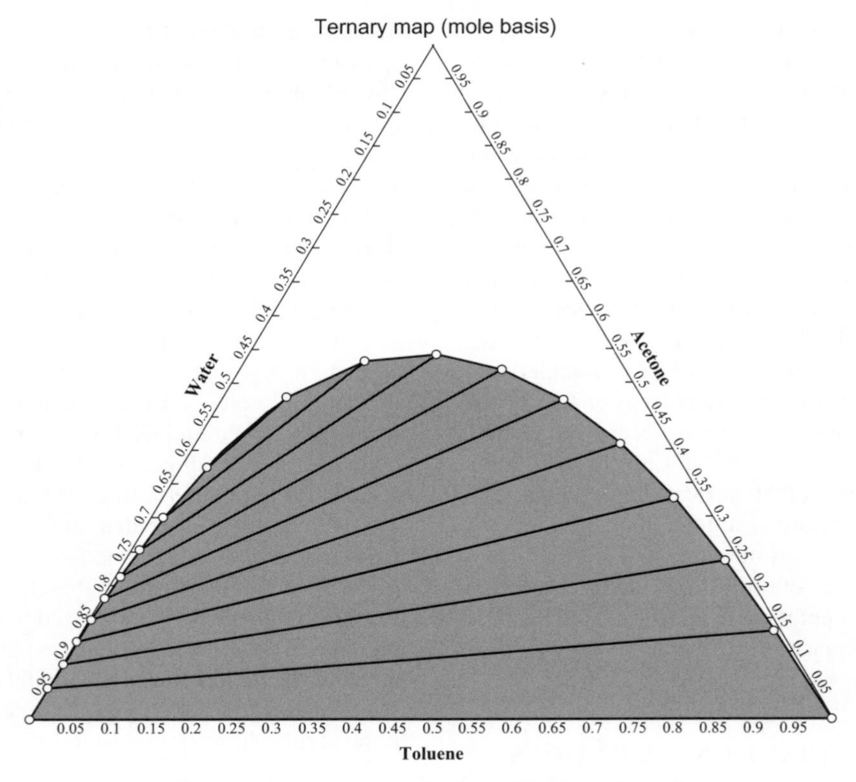

FIGURE 12.6 Ternary liquid–liquid equilibrium-phase diagram.

is added to the point of phase separation. A balance on toluene on the mixture shows

130 moles × (23 mol% toluene) = 30 moles of toluene added to reach the beginning of phase separation at 20°C.

A tie line is constructed at the intersection of 45 mol% acetone, 46 mol% toluene, and 9 mol% water to the other side of the phase boundary at 17 mol% acetone and 83 mol% water. This tie line intersects the composition path line at 40 mol% toluene, 42 mol% acetone, and 18 mol% water. Performing another acetone mole balance, we find

100 moles × (70 mol% acetone) – B mol × (42 mol% acetone) = 0; B = 167 mol total mixture.

The total amount of toluene in this second mixture is

167 moles (40 mol% toluene) = 67 mol toluene

The amount of extra toluene added beyond the initial phase split is

67 moles – 28 moles = 39 moles of toluene added in total

The compositions of the two phases in equilibrium based on the end points of the phase envelope tie line are

45 mol% acetone, 46 mol% toluene, and 9 mol% water (light phase)

17 mol% acetone, 83 mol% water, and trace toluene (heavy phase)

The molar distribution coefficient of acetone in terms of light-to-heavy phase compositions is the ratio of the mol% of acetone in the toluene phase divided by the mol% in the aqueous phase:

$$\frac{x_{a,l}}{x_{a,h}} = \frac{45\text{ mol\%}}{17\text{ mol\%}} = 2.65$$

where component 1 is acetone in the light phase and component 2 is acetone in the heavy phase.

A ratio greater than unity indicates the equilibrium distribution of acetone is favored in the light phase, which contains the most toluene. The selectivity of the light phase for acetone is equal to the distribution coefficient divided by the distribution coefficient of water between phases:

$$\beta = \frac{2.65}{(x_{w,l}/x_{w,h})} = \frac{2.65}{(9\text{ mol\%}/83\text{ mol\%})} = 24.4$$

This is a relative measure of the propensity of toluene to extract acetone while accounting for the amount of water taken by the light phase in the extraction. A higher selectivity is desirable to achieve efficient exchange of a cosolvent between liquid phases in equilibrium.

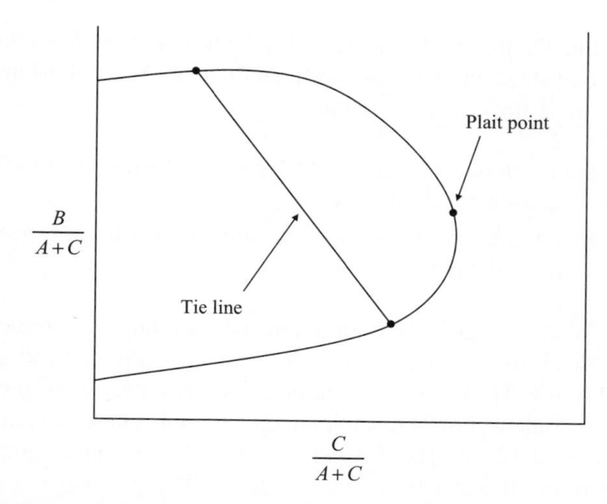

FIGURE 12.7 Ternary equilibrium phase diagram in Janecke format.

An alternative to triangular form for the ternary diagram is a Janecke plot on rectangular coordinates, as demonstrated in Figure 12.7. The rectangular plot contains the same amount of information as the triangular plot although not as direct, since each axis is a ratio of component fractions rather than pure component fractions. Another feature of interest on a ternary-phase diagram is the tie line that represents an equilibrium mixture in which each phase is of equal density, which may therefore present challenges in phase separation. This is referred to as the isopycnic line and is usually represented by a dashed line on the phase diagram.

Distribution and selectivity of solute between the extract and raffinate phases is of interest in the selection of solvents and design of extractive systems. Figure 12.8 is a demonstration of distribution plots of a solute between two liquid phases. The 45° line represents the boundary of unity in the distribution coefficient. The distribution curve ends at the plait point for the system. In some systems, the distribution curves may cross the 45° line before ending at the plait point. Such a system is referred to as solutropic because the distribution passes through unity. On a triangular ternary plot, the slope of tie lines changes sign in solutropic systems, which may also pose challenges in the design of a liquid–liquid extraction system.

12.3.2 Process Modeling and Case Studies

Similar to the case of batch distillation, liquid–liquid extraction processes in the pharmaceutical industry involving solutes of moderate manufacturing volume and high value are typically operated as single-stage operations in multipurpose manufacturing vessels. Multistage extractors present an opportunity for efficient separation of a solute from a given phase when the distribution coefficient with a given

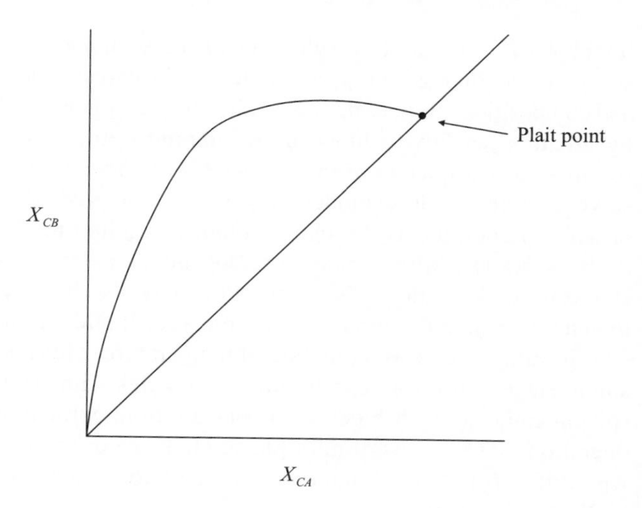

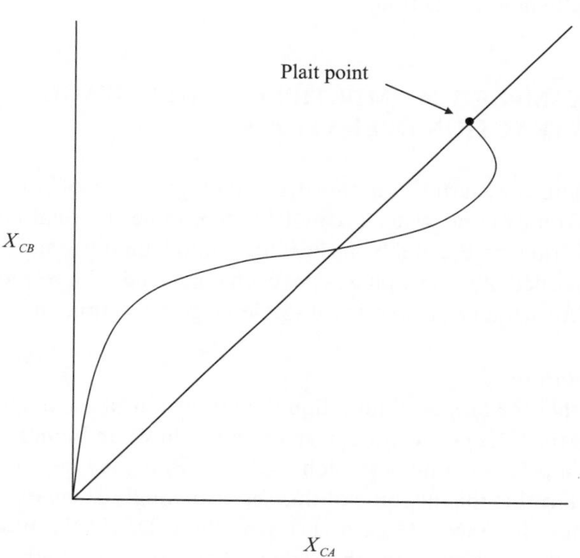

FIGURE 12.8 Liquid–liquid distribution curves.

solvent is small. However, significant investment in processing equipment and control systems may be required for multistage systems, whereas a single-stage extraction is easy to operate in flexible plant equipment at hand.

12.3.2.1 Single-Stage Extraction A single-stage liquid–liquid extraction may be performed in the same bulk vessel used for most other operations in the batch pharmaceutical plant. The temperature range of −15 to 150°C in these jacketed vessels provides plenty of flexibility in the design of liquid–liquid extraction. Phase equilibrium operations such as this are easily modeled in a process simulator such as Aspen Plus using a decanter (LLE) or three-phase flash (VLLE) model. Temperature and overall composition are required inputs, and results are posted in terms of mass or mole fraction plus total mass and volume for each resulting liquid phase. If liquid-phase splitting will not occur under the specified conditions, only a single liquid stream flow and composition is reported. Case studies are easily generated by creating sensitivity plots or by running optimization routines for parameters of interest such as temperature or solvent volumes. These models rely on accurate prediction of activity coefficients through a nonideal solution model such as NRTL, with temperature-dependent parameters stored in model data banks. Aspen Plus may also be used to generate triangular ternary-phase plots for liquid–liquid equilibrium systems as a function of temperature. Multiple single-stage contactors can be linked in series with fresh solvent input to each block to model multiple extraction operations. The results of multiple single-stage operations will differ from the countercurrent operation known as multistage extraction.

EXAMPLE 12.6 MULTIPLE SINGLE-STAGE EXTRACTION OPERATIONS

Using the starting mixture from Example 12.5 of acetone (70 mol%) and water (30 mol%), determine the final composition of the light phase if the initial mixture and two intermediate heavy phases are each contacted with 67 moles of fresh toluene in a series of single-stage extractions at 20°C.

Solution

Either the ternary liquid–liquid equilibrium-phase diagram for the toluene–acetone–water system shown in Figure 12.6 or a process simulator such as Aspen Plus may be used to determine the outcome of this series of single-stage extractions. In Aspen Plus, a series of three DECANT blocks would be linked by the heavy-phase streams, and each DECANT block. Equal amounts of fresh toluene are fed to each DECANT block. The first stage extraction mixes 100 moles of acetone (70 mol%) and water (30 mol%) with 67 moles of fresh toluene. At equilibrium, the first stage phases contain

Light phase: 146 moles (45.9 mol% toluene, 45.5 mol% acetone, 8.6 mol% water)

Heavy phase: 21 moles (0.4 mol% toluene, 17 mol% acetone, 82.6 mol% water)

The second stage receives the heavy phase from the first stage and 67 moles of fresh toluene. At equilibrium, the second stage phases contain

Light phase: 71 moles (95.0 mol% toluene, 4.7 mol% acetone, 0.3 mol% water)

Heavy phase: 18 moles (trace toluene, 1.7 mol% acetone, 98.3 mol% water)

Finally, the third stage receives the heavy phase from the second stage and 67 moles of fresh toluene. At equilibrium, the third stage phases contain

Light phase: 67 moles (99.4 mol% toluene, 0.4 mol% acetone, 0.02 mol% water)

Heavy phase: 18 moles (trace toluene, 0.14 mol% acetone, 99.86 mol% water)

In three single-stage extractions with a total of 201 moles of fresh toluene, the acetone in the water was reduced to 0.1 mol%. The heavy, aqueous phase contains only 0.002 moles of toluene, but the combined light, organic phases contain a total of 12.8 moles of water, which is 42% of the original 30 moles of water in the initial feed. Although toluene extracts acetone very well from water, this system is not very selective for acetone because a significant amount of water is extracted into the light, organic phase as well.

12.3.2.2 Multistage Countercurrent Extraction The use of multiple stage countercurrent extraction to improve the efficiency of liquid–liquid extractions is easily modeled by most process simulators. In multistage countercurrent extraction, feed and solvent streams enter opposite ends of the column. The product streams exiting the column are the extract and the raffinate. The extract stream contains the solvent with the desired component transferred between phases, whereas the raffinate is the stream remaining after the desired component is transferred to the extract. The internals of a multistage countercurrent extraction column can be quite varied. Stages are formed by zones of intense mixing between independent liquid phases, and this is accomplished by column packing in a variety of configurations (structured or random), by mechanical agitators such as those seen in Scheibel, Karr, or rotating disk columns, or by liquid dispersion with static sieve trays. Regardless of the liquid phase contacting technology, the important characteristic for a given column operating under specific temperature and flow rate conditions is the number of equivalent equilibrium stages the equipment achieves. If the equilibrium behavior of component distribution between phases is well understood and mapped with the aid of a ternary diagram or expressed in fundamental terms of activity coefficient model binary parameter terms, a process model may be used to explore operating conditions. The inputs for the model block are identical to that single-stage operation, except that the total number of equilibrium stages must also be specified. The results obtained from this countercurrent phase contacting scheme are reported in the same fashion as the single-stage model. Example 12.7 demonstrates the difference in efficiency based on choice of countercurrent multistage extraction in place of the multiple single-stage operations shown in Example 12.6.

EXAMPLE 12.7 MULTISTAGE COUNTERCURRENT EXTRACTION

Using an acetone (70 mol%) and water (30 mol%) feed equivalent to that of Example 12.6 (100 moles/h), determine how many moles of fresh toluene must be fed to a three-stage countercurrent extraction column at 20°C to achieve 0.14 mol% acetone in the heavy, aqueous phase at the outlet of the extraction column.

Solution

An Aspen Plus simulation is a convenient manner of solving this type of multistage extraction problem, although graphical methods using an equilibrium and operating line approach for staged operations may also be used to solve this type of problem. In Aspen Plus, the EXTRACT column model is used to simulate the countercurrent contact of the heavy feed stream entering the top of the column with a feed of fresh toluene entering the bottom of the column. Three equilibrium stages and adiabatic operation are specified for the model. The feed to the top of the column is 100 moles/h in total containing 70 mol% acetone and 30 mol% water at 20°C. The objective of the simulation is to add only enough fresh toluene to achieve 0.14 mol% acetone in the aqueous raffinate stream at the exit of the bottom of the column. This can be achieved by trial-and-error manipulation of the fresh toluene feed while monitoring the raffinate composition, or in Aspen Plus a design specification algorithm may be specified to manipulate the toluene feed rate with the objective set at the specified acetone composition in the raffinate. If design specification is used to solve the problem, the solution is to feed 48.5 moles/h of toluene to the column to achieve 0.14 mol% acetone in the aqueous raffinate. In this countercurrent extraction example, 20.5 mole/h of water are extracted by the fresh toluene compared with 12.8 moles of water in the previous example. This represents a loss of 68% of the total water fed to the column. The raffinate contains 0.001 moles/h of toluene, however, which is as low as the loss of toluene observed in Example 12.6. The most striking result, however, is that the countercurrent column requires only 48.5 mole/h of toluene compared with the multiple single-stage operation that requires 201 moles of fresh toluene. This represents 24% of the toluene requirement compared to the multiple single-stage operation.

12.3.3 Laboratory Investigation

The design of liquid–liquid extraction operations in the development laboratory may be divided generally into equilibrium data collection and operation stressing. Both require effective lab-scale equipment to provide robust data and scale-up assessment on the way to the design of an efficient process, whether single or multistage extraction is the objective.

12.3.3.1 Equilibrium Data Collection Liquid–liquid equilibrium data is often collected in a single-stage cell in which conditions such as temperature and total composition are altered to allow measurement of phase equilibrium across a wide operating space. An accurate representation of equilibrium data relies on accurate analytical measurement of individual phase compositions. The cell is loaded with components and agitated sufficiently to increase the surface area between the dispersed and continuous phases. The surface area between phases is the mass transfer area across which components distribute between phases to an equilibrium condition. Sufficient time must be given to ensure an equilibrium condition is met, and a time series of phase samples should be analyzed to ensure a constant value is obtained in the absence of some interfering form of component degradation. After equilibrium is achieved, agitation is stopped and the resulting phases are allowed to split into individual layers. In some cases only a single homogeneous solution may result, or sometimes relatively stable emulsion may form between phases. Both of these conditions signify a combination of components that is unsuitable for use in liquid–liquid extraction, since the goal of separation is not achievable at a given temperature and total composition. Data collected in the laboratory with the single-stage equilibrium cell is used to provide the basis for a ternary-phase diagram or for regression of binary interaction parameters of activity coefficient model such as NRTL. An early indication of unit operation behavior in terms of relative difficulty of phase separation by settling, the accumulation of interfacial contaminants, or potential component degradation can be observed and recorded during equilibrium data collection. That is, the experience obtained in equilibrium data collection may bring insight into the behavior of the system upon scale-up to manufacturing.

12.3.3.2 Extraction Laboratory Simulation Single or multistage laboratory models may be easily constructed to further assess behavior of system components under the conditions of interest and to refine those conditions to optimize the desired separation. Single-stage laboratory models are easiest to construct since they are close to the equilibrium cell described previously. In fact, there is no reason that a single-stage model could not be used for the purpose of gathering equilibrium data. The single-stage model incorporates several elements in terms of geometry and agitation conditions that mimic and which will be encountered on scale for the purpose of providing a close scale-down representation of manufacturing technology. In the pharmaceuticals industry, this is typically the same batch vessel used for other unit operations such as reaction, distillation, and crystallization. In addition to providing equilibrium data, the single-stage model should be used to examine and understand the dynamic approach to the equilibrium state. One can examine the impact of agitator choice and the power applied at the blade tip, where the highest shear forces are found, to determine the propensity for the system to irreversibly emulsify. The position of the agitator relative to vessel fill height is also an important factor in the extraction efficiency and approach to equilibrium. A system with a large difference in densities between phases may make efficient mixing difficult to achieve with an agitator set low in the vessel as is often the case in pharmaceutical manufacturing facilities for the purpose of enabling low stirring volumes. Multiple-stage models are scaled-down versions of the manufacturing scale in terms of the intended phase mixing technology. A direct scale-down of the contacting technology is necessary to ensure that the dynamic and equilibrium behavior of the components in the system behave in a desirable manner. These scaled-down systems are typically constructed of glass-walled columns to allow easy visibility of conditions within the operating space. This allows the user to observe the presence of emulsions or the accumulation of solid particulate or oily components within the equilibrium stages that could impact extraction efficiency or the mechanical operation of the system over an extended operating time.

12.3.3.3 Phase Split Efficiency The rate of phase separation is important to ensure an efficient operation of the extraction process. Since it is difficult or impossible to see into manufacturing-scale equipment to assess the quality of the liquid-phase splitting, the predicted time to phase separation on manufacturing scale must be known or inferred from laboratory behavior.

12.3.3.4 Impact of Missed Phase Splits and Desired Fate of Phase Interface The process designer must specify the desired fate of the phase interface to ensure proper operation of liquid–liquid extraction as a separation technique. In other words, the designer must state explicitly which phase must be disposed of entirely while keeping in mind that a small amount of one phase will always be carried with the other as a result of a phase split. Even after specifying the phase split strategy with regard to the interface after a batch liquid–liquid extraction operation, it is important to test the outcome of the reversed condition, or in other words a missed phase split on subsequent extractions, downstream operations, or perhaps process yield. Use the resulting product phase containing a small amount of the waste phase, and any associated rag material in downstream operations and note any impact on the process. If the impact is significant, then the phase split strategy or other specification may be enhanced to limit the extent of potential downstream problems.

12.3.3.5 Difficult Separations In some cases of liquid–liquid extraction, ease of dispersion is less of a problem than separating the resulting liquid phases after components are distributed in an equilibrium state. In the worst case, the dispersed phase remains stable as an emulsion, and this is a terminal condition to avoid. There are generally three physical properties of components and their mixtures that determine the relative ease of separation after equilibrium extraction and those are (1) liquid-phase density differences, (2) viscosity, and (3) interfacial tension. A large density difference between resulting phases enhances the rate of phase settling, while a low viscosity of the continuous phase also promotes separation. Moderate interfacial tension between the dispersed and continuous phase allows relatively easy separation to take place by coalescence of droplets but equally it does not hinder the dispersion of one liquid phase into another.

Occasionally an emulsion cannot be avoided due to the nature of the system or due to the batch variability in manufacturing. During process design or in a manufacturing environment, there are several methods of handling emulsified phases to recover the phase separation. Sometimes an interfacial particulate or rag layer may be responsible for stabilizing an emulsion, and its influence on the system may be tested by filtering the mixture to remove the material. The phases may separate as expected in the absence of this material. The other technique is to alter the physical properties of phases mentioned previously in a manner that benefits the rate of separation. For instance, density and interfacial tension of an aqueous layer can be increased by the addition of inorganic salts. These salts will also serve to reduce the solubility of water in the organic phase that is a common contributor to emulsification. An increase in temperature will typically reduce the continuous phase viscosity to further improve separation. Finally, a mechanical means of increasing the gravitational force, such as a centrifugal separator, can be used to increase the rate of sedimentation of the dispersed phase. Increasing the volume of the dispersed phase relative to the continuous phase may also accelerate coalescence of droplets to aid separation.

12.3.3.6 Process Scale-Up The selection of scale-up equipment and process controls should be considered as laboratory process development activities near completion. The primary objectives of the liquid extraction scale-up design should be to achieve time and material efficient separation and purification of components.

12.3.3.7 Equipment Selection Two factors for consideration are the agitator type and fill level of the vessel for efficient equilibrium mixing. Most general-purpose vessels have agitators placed relatively low to the maximum fill volume. If the densities of liquid phases are significantly different and the vessel is nearly full, the time to equilibrium during agitation may be extended. High-shear agitator types and vessel baffles make this operation more efficient under these conditions.

12.3.3.8 Process Control The most common parameter under adjustment and control in liquid–liquid extraction other than temperature is pH, which is often adjusted with aqueous acid or base. An adjustment in pH is used to convert the component of interest into a free acid, free base, or salt to enable it to distribute preferentially into a desirable liquid phase for further processing. The acid or base is added to the agitated solution and the pH is either measured by a probe installed in the tank or in a recirculation loop. Alternatively, a sample of the mixture or aqueous phase may be taken from the vessel for off-line analysis. The pH specification for a process should be carefully defined with regard to the state of agitation. The pH measurement of an aqueous and organic liquid-phase mixture results in a different pH value than measurement of the aqueous phase alone. Temperature also impacts solution pH and this effect must be accounted for when taking a sample for off-line analysis.

12.3.3.9 Unit Operation Maximum agitation is most often a benefit to liquid–liquid extraction operation since impeller shear creates interfacial surface area between phases. Greater the surface area between phases means increased mass transfer rates toward the equilibrium state. On a manufacturing scale in a batch pharmaceutical plant, the agitator is typically set very low to achieve stirring at very low volumes. This can hinder liquid–liquid extraction when the vessel is more than half full, since the upper portions of the fluid are less turbulent and coalescence between dispersed droplets is more likely. Using an impeller types with greater shear characteristics may assist the operation by ensuring the formation of minimum droplet size near the impeller. Where one impeller would be damaging in a crystallization process, the same impeller would typically improve liquid–liquid extraction performance. Once equilibrium is achieved, agitation is stopped, and the phases are allowed to settle into top and bottom layers.

12.3.3.10 Phase Split Determination Separation of phases from a single-stage extractor is usually achieved by draining the lower layer from the vessel through a bottom valve. The end point of the phase split is normally determined by visual inspection through a sight glass near the bottom valve. The sight glass allows detection of the interface between phases. When there is color similarity between phases, a conductivity probe may also be used in conjunction with the sight glass to detect the step change in conductance between phases as the split is nearing completion. The aqueous phase normally has greater conductivity than the organic phase and change between phases with the sensor is either indicated by the change in a signal light or displayed by digital indication of the conductivity measurement.

12.A APPENDIX

12.A.1 Guide to Generation of Binary Interaction Parameters for Solvent Pairs

12.A.1.1 Software COSMOthermX v.C21 and Aspen Properties V7.1 (or Aspen Plus V7.1).

12.A.1.2 Purpose This guide provides instruction on the generation of NRTL binary interaction parameters (BIPs) for liquid compounds pairs using COSMOthermX and Aspen Properties. Once the appropriate data are in Aspen Properties, they can be used as the basis for distillation or liquid–liquid extraction modeling. Estimation of BIPs through COSMOthermX is only required if the NRTL binary interaction parameters are not predefined in the Aspen Properties data bank. This guide assumes the user has basic skills in operating COSMOthermX and Aspen software packages.

12.A.1.3 Procedure The flowchart in Figure 12.9 describes the workflow for generating BIPs using COSMOthermX and Aspen software interfaces. The workflow demonstrates three paths to estimates for BIPs, and each is ranked by the accuracy of results based on the complexity of inputs used for estimation. The rankings for NRTL parameter estimation are summarized here:

- *Good*: Provides a set of *infinite dilution* activity coefficients for a pair of components at a *single temperature*. The NRTL parameters are estimated within Aspen.
- *Better*: Provides NRTL parameters directly from COSMOthermX, but the *parameters are insensitive to temperature* and are generally applicable to the pressure or isothermal conditions chosen in COSMOthermX.
- *Best*: Regression of activity coefficients from COSMOthermX is performed within Aspen to provide *temperature-dependent* NRTL parameters within Aspen.

Screenshots from both applications are shown after the flowchart as further guidance on the process. After determining the desired path through the workflow, the user should proceed to the matching workflow section of the guide for step-by-step instructions. A methanol–water system is used throughout the examples. A comparison of results for three binary solvent systems is also provided at the end of the section.

Estimation of NRTL BIPs from COSMOthermX infinite dilution activity coefficients

1. In COSMOthermX, choose the Activity Coefficient tab in the interface (Figure 12.10).
2. Select the temperature for the infinite dilution activity coefficient estimation (gamma).
3. Check the "Pure" box for each solvent and "Add" and "Run" the cases.
4. Examine the output from the COSMOthermX in terms of Ln(gamma) (Figure 12.11).
5. Convert the results to values of gamma for each solvent by calculating the exponential value of each.

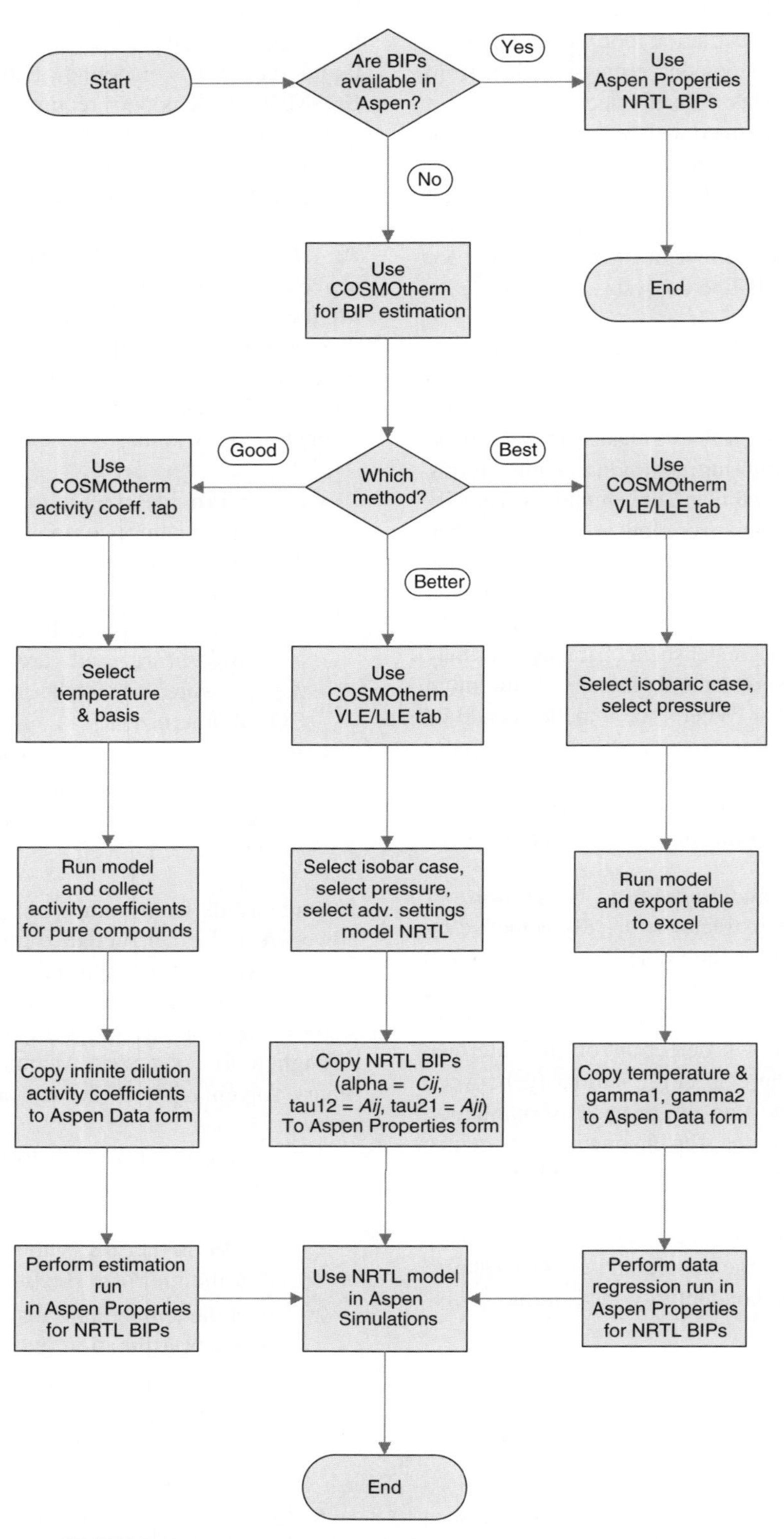

FIGURE 12.9 Binary interaction parameter estimation workflow.

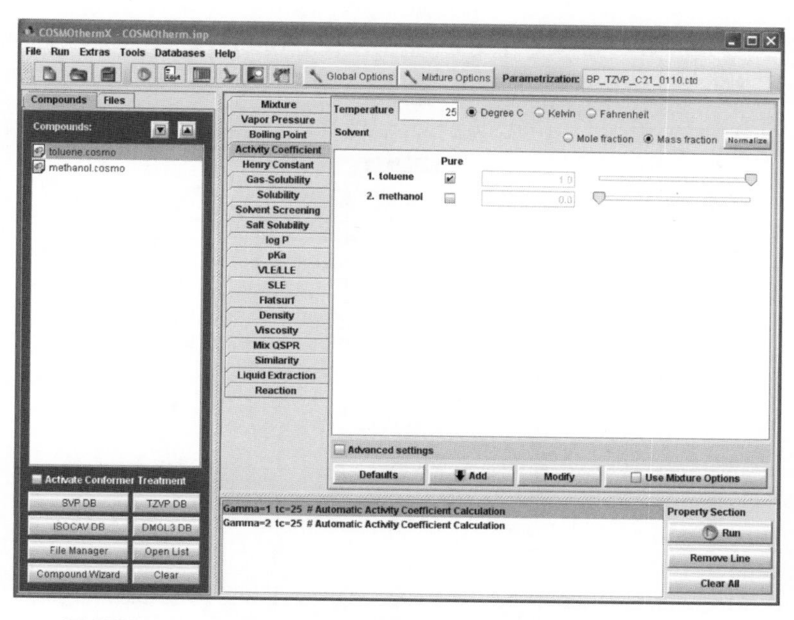

FIGURE 12.10 COSMOthermX activity coefficient user interface.

6. In the Aspen "Properties\Data" folder, create a data entry case.
7. Select "Mixture" as the Data type.
8. In the "Setup" tab, choose Category "For Estimation" and Data type "GAMINF", and then choose the pair of components for estimation.
9. Paste the temperature and gamma values in the appropriate locations on the "Data" tab (Figure 12.12).
10. In Aspen, select Tools/Estimation, and then request an estimation run within the Estimation folder.
11. Choose to "Estimate only the selected parameters."
12. On the Binary tab, specify Parameter "NRTL" and Method "Data" for "Component i" and "Component j" equal to "All."

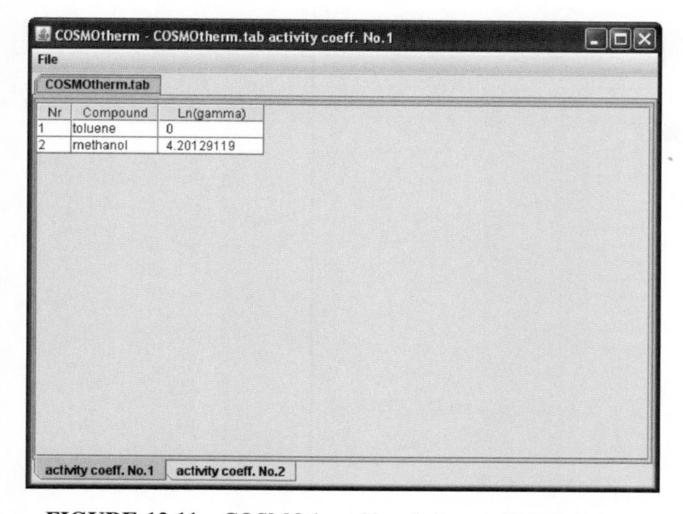

FIGURE 12.11 COSMOthermX activity coefficient output.

13. Run the estimation and examine the results in the "Parameters\Binary Parameters\NRTL-1" folder. Source should read "R-PCES," which is "Regression Property Constant Estimation System" (Figure 12.13).
14. The NRTL BIPs are in place and are ready for use in process modeling.
15. *Caution*: The BIPs are a result from an *isothermal* estimation of *infinite dilution* activity coefficients in COSMOthermX. The range of applicability is relatively narrow.

Estimation of Isothermal NRTL BIPs through COSMOthermX for export to Aspen

1. Choose the VLE/LLE tab in COSMOthermX (Figure 12.14).
2. Select an Isothermal or Isobar case and set the appropriate conditions.
3. Choose the components and make note of the order of entry.
4. Check "Advanced settings" and "Compute empirical activity coefficient models?" and select the "NRTL" model.
5. Add the case and press "Run" to perform the estimation.
6. Examine the bottom section of the text output from the run (Figure 12.15).
7. NRTL rms is an indication of parameter fit.
8. Copy each of Alpha, Tau12, and Tau21.
9. Within Aspen, under the "Properties\Parameters\Binary Interaction\NRTL-1," paste the values from

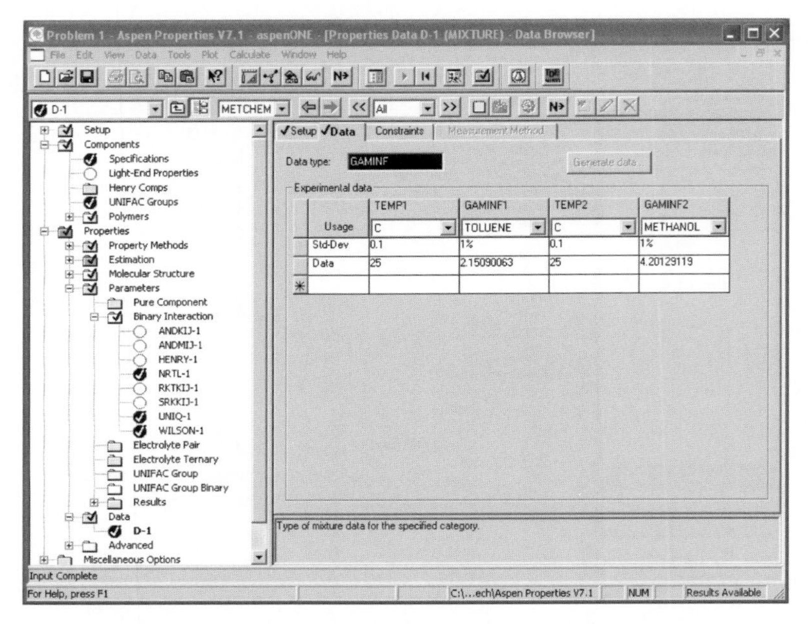

FIGURE 12.12 Aspen Properties activity coefficient data input.

COSMOthermX NRTL BIPs as follows in the column labeled with the solvent pair (Figure 12.16):

(a) CIJ = NRTL Alpha

(b) AIJ = NRTL Tau12

(c) AJI = NRTL Tau21

10. After pasting the parameters, the Source should be "USER."
11. The NRTL BIPs are in place and are ready for use in process modeling.
12. *Caution*: The resulting BIPs are *isothermal* parameters from COSMOthermX. The range of applicability is somewhat narrow since the parameters do not include coefficients for temperature-dependent NRTL terms.

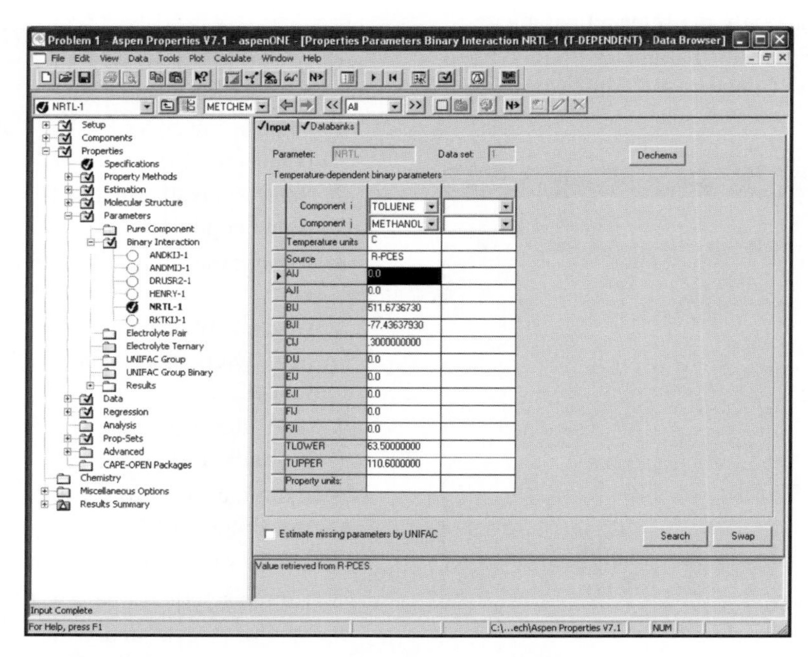

FIGURE 12.13 Aspen Properties estimated NRTL binary parameters.

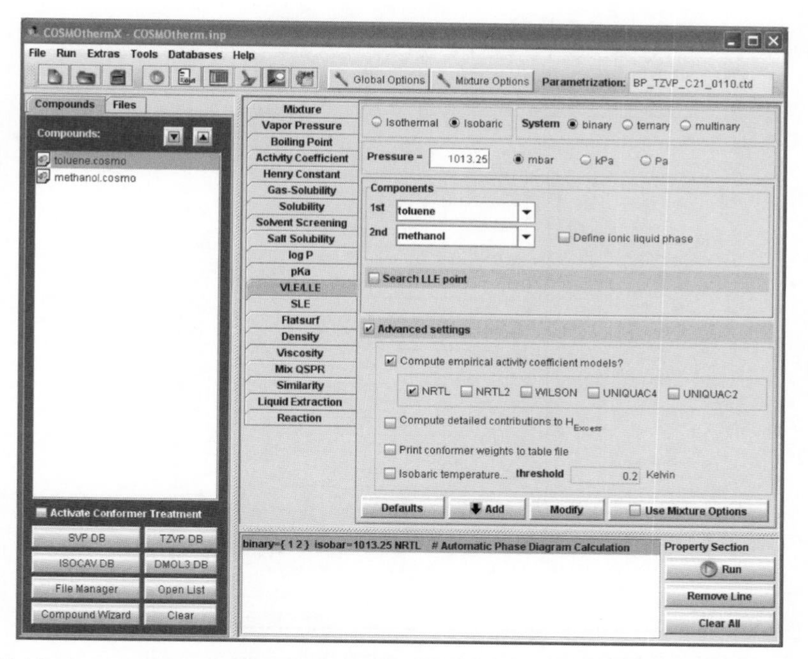

FIGURE 12.14 COSMOthermX VLE/LLE user interface for NRTL binary parameter estimation.

Estimation of NRTL BIPs from COSMOthermX liquid activity coefficients

1. In COSMOthermX, select the VLE/LLE tab (Figure 12.17).
2. Select an Isothermal or Isobar case and set the appropriate conditions.
3. Choose the components and make note of the order of entry.
4. Select "Search LLE point" with "default grid" algorithm.
5. Add the case and "Run" to perform the estimation.
6. Examine the tabular output from the run (Figure 12.18).
7. Choose "File\Save As" and save the file to a convenient location with a desirable filename.
8. In the Excel spreadsheet, insert columns after ln(γ1) and ln(γ2) as shown in Figure 12.19.
9. In the new columns, calculate "gamma1" and "gamma2" as exp(ln(γ)) for each (Figure 12.19).
10. In the Aspen "Properties\Data" folder, create a data entry case for a mixture.
11. In the "Setup" tab, choose Category "Thermodynamic" and Data type "GAMMA," and choose the

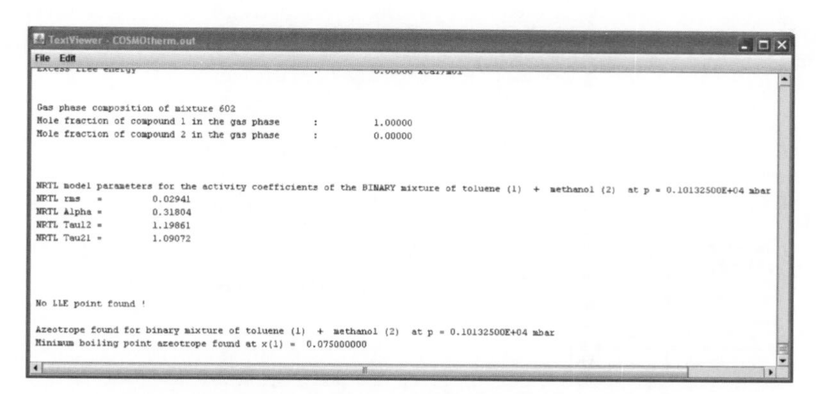

TextViewer - COSMOtherm.out

File Edit

```
Gas phase composition of mixture 602
Mole fraction of compound 1 in the gas phase     :          1.00000
Mole fraction of compound 2 in the gas phase     :          0.00000


NRTL model parameters for the activity coefficients of the BINARY mixture of toluene (1)  +  methanol (2)  at p = 0.10132500E+04 mbar
NRTL rms   =          0.02941
NRTL Alpha =          0.31804
NRTL Tau12 =          1.19861
NRTL Tau21 =          1.09072


No LLE point found !

Azeotrope found for binary mixture of toluene (1)  +  methanol (2)  at p = 0.10132500E+04 mbar
Minimum boiling point azeotrope found at x(1) =  0.075000000
```

FIGURE 12.15 COSMOthermX NRTL binary parameter output.

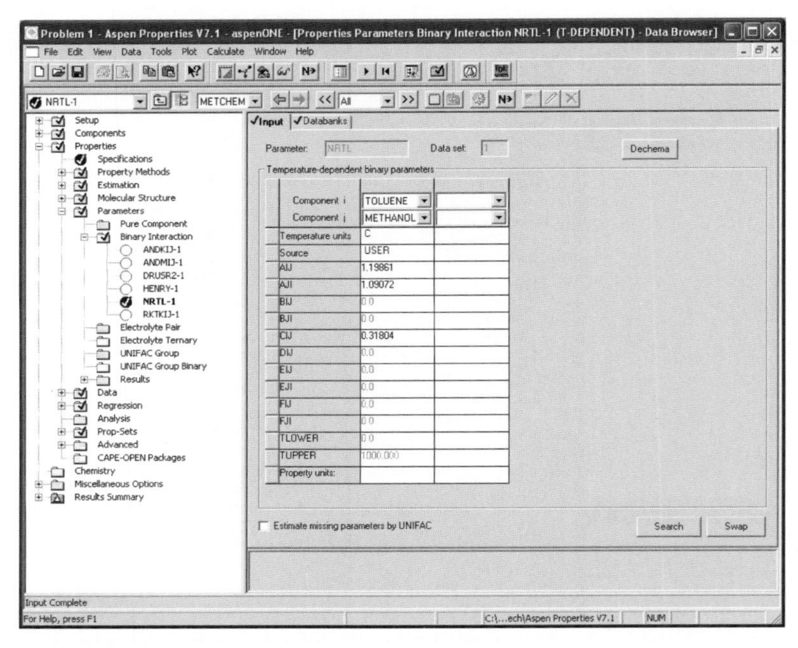

FIGURE 12.16 Aspen Properties NRTL binary parameter user input.

binary pair for estimation. Set the pressure at the bottom to that of the system.

12. Paste the temperatures and gamma values in the appropriate locations (Figure 12.20).
13. In Aspen, choose Tools\Regression, and then start a new case in the Regression folder. Choose the name of Data Set in the "Setup" section of Regression.
14. Under the Parameters tab, choose "Binary parameter" for type, select the Name as NRTL, and set Elements 1, 2, and 3 as shown in Figure 12.21.
15. For components, select the pair used in the COSMOthermX estimation.
16. Repeat the component pair for the same element but in reverse order, except for Element 3, which is

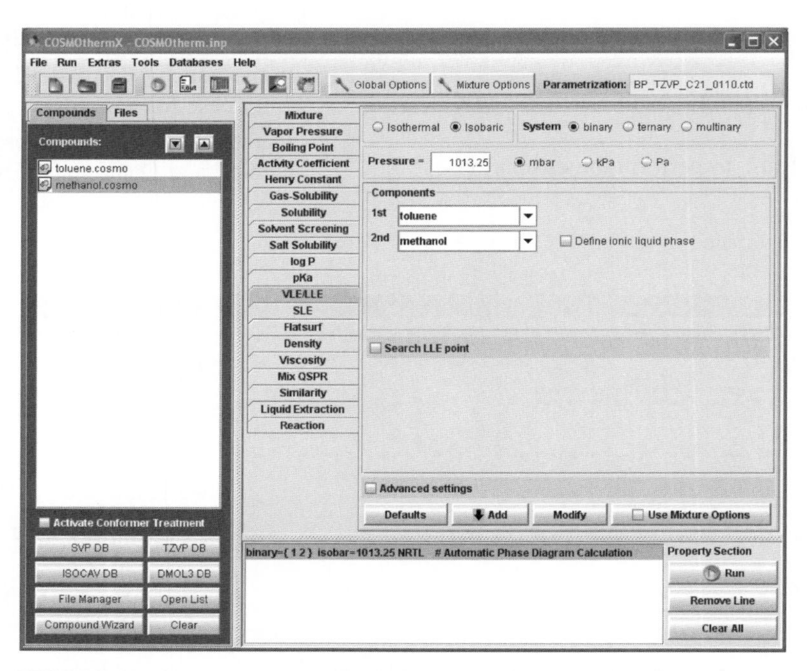

FIGURE 12.17 COSMOthermX VLE/LLE user interface for data estimation.

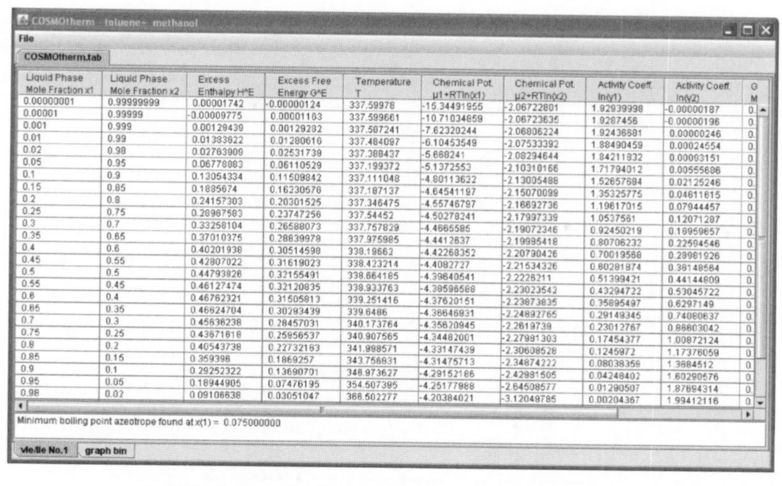

COSMOtherm - toluene+ methanol

COSMOtherm.tab

Liquid Phase Mole Fraction x1	Liquid Phase Mole Fraction x2	Excess Enthalpy H^E	Excess Free Energy G^E	Temperature T	Chemical Pot. μ1+RTln(x1)	Chemical Pot. μ2+RTln(x2)	Activity Coeff. ln(γ1)	Activity Coeff. ln(γ2)	G M
0.00000001	0.99999999	0.00001742	-0.00000124	337.59978	-15.34491955	-2.06722801	1.92939998	-0.00000187	0.
0.00001	0.99999	-0.00009775	0.00001163	337.599661	-10.71034659	-2.06723635	1.9287456	-0.00000196	0.
0.001	0.999	0.00129439	0.00129282	337.587241	-7.62320244	-2.06806224	1.92436681	0.00000246	0.
0.01	0.99	0.01383822	0.01280616	337.484097	-6.10453549	-2.07533392	1.88490459	0.00024554	0.
0.02	0.98	0.02763909	0.02531739	337.388437	-5.668241	-2.08294644	1.84211832	0.00093151	0.
0.05	0.95	0.06778883	0.06110529	337.199372	-5.1372553	-2.10318166	1.71794012	0.00555686	0.
0.1	0.9	0.13054334	0.11509842	337.111048	-4.80113622	-2.13005488	1.52657684	0.02125246	0.
0.15	0.85	0.1885674	0.16230576	337.187137	-4.64541187	-2.15070099	1.35325775	0.04611615	0.
0.2	0.8	0.24157303	0.20301525	337.346475	-4.55746797	-2.16692736	1.19617015	0.07944457	0.
0.25	0.75	0.28967583	0.23747256	337.54452	-4.50278241	-2.17997339	1.0537561	0.12071287	0.
0.3	0.7	0.33258104	0.26588073	337.757829	-4.4665585	-2.19072346	0.92450219	0.16959657	0.
0.35	0.65	0.37010375	0.28839978	337.975985	-4.4412637	-2.19985418	0.80706232	0.22594546	0.
0.4	0.6	0.40201938	0.30514598	338.19662	-4.42268352	-2.20790426	0.70019568	0.28981926	0.
0.45	0.55	0.42807022	0.31619023	338.423214	-4.4082727	-2.21534326	0.60281874	0.36148564	0.
0.5	0.5	0.44793826	0.32155491	338.664185	-4.39640541	-2.2226211	0.51399421	0.44144809	0.
0.55	0.45	0.46127474	0.32120835	338.933763	-4.38598588	-2.23023542	0.43294722	0.53045722	0.
0.6	0.4	0.46762321	0.31505813	339.251416	-4.37620151	-2.23873835	0.35895497	0.6297149	0.
0.65	0.35	0.46624704	0.30293439	339.6486	-4.36646931	-2.24892765	0.29149345	0.74080637	0.
0.7	0.3	0.45636238	0.28457031	340.173764	-4.35620945	-2.2619739	0.23012767	0.86603042	0.
0.75	0.25	0.43671618	0.25956537	340.907565	-4.34482001	-2.27981303	0.17454377	1.00872124	0.
0.8	0.2	0.40543738	0.22732163	341.998571	-4.33147439	-2.30608528	0.1245972	1.17376059	0.
0.85	0.15	0.359398	0.1869257	343.756931	-4.31475713	-2.34874222	0.08038359	1.3684512	0.
0.9	0.1	0.29252322	0.13690701	346.973627	-4.29152186	-2.42981505	0.04248402	1.60290576	0.
0.95	0.05	0.18944905	0.07476195	354.507395	-4.25177988	-2.64508577	0.01290507	1.87694314	0.
0.98	0.02	0.09108638	0.03051047	366.502277	-4.20384021	-3.12049785	0.00204367	1.99412116	0.

Minimum boiling point azeotrope found at x(1) = 0.075000000

vle/lle No.1 | graph bin

FIGURE 12.18 COSMOthermX VLE data output.

symmetric and only has one value. Optionally, "fix" the value of Element 3 under the "Usage" option.

17. Run the regression and examine the results under the Regression folder for closeness of fit. Predicted versus entered data can be plotted in the same area.
18. Examine the results in the "Parameters\Binary Parameters\NRTL-1" folder. Source should read "R-R-#," which means a result from regression case (Figure 12.22).
19. The NRTL BIPs are in place and are ready for use in process modeling.
20. *Caution*: The BIPs are a result of a regression across the boiling point temperature range of the liquid pair (i.e., an isobar case), which defines the range of applicability. To obtain NRTL BIPs at a different pressure, either *extend the first data set* by running a new case in COSMOthermX at a different isobaric condition, *or create a separate Regression case* in Aspen for the additional COSMOthermX isobaric data.

Examining the quality of liquid activity coefficients after estimation or regression

1. In Aspen, view the NRTL BIPs in the "Parameters \Binary Parameters\NRTL-1" folder and select the appropriate "Source."

Microsoft Excel - COSMOtherm.xls

A1 Results for binary mixture of toluene (1) + methanol (2) at p = 0.10132500E+04 mbar

Results for binary mixture of toluene (1) + methanol (2) at p =
energies are in kcal/mol temperature is in K
molecular weights (1) 92.1390 (2) 32.0420

Liquid Phase Mole Fraction x1	Liquid Phase Mole Fraction x2	Temperature T	Activity Coeff. ln(γ1)	gamma1	Activity Coeff. ln(γ2)	gamma2	Gas Phase Mole Fraction y1	Gas Phase Mole Fraction y2
0.00000001	0.99999999	337.6	1.9294	6.8854	0.0000	1.0000	0.0000	1.0000
0.00001	0.99999	337.6	1.9287	6.8809	0.0000	1.0000	0.0000	1.0000
0.001	0.999	337.6	1.9244	6.8508	0.0000	1.0000	0.0015	0.9985
0.01	0.99	337.5	1.8849	6.5857	0.0002	1.0002	0.0143	0.9857
0.02	0.98	337.4	1.8421	6.3099	0.0009	1.0009	0.0272	0.9728
0.05	0.95	337.2	1.7179	5.5730	0.0056	1.0056	0.0597	0.9403
0.1	0.9	337.1	1.5266	4.6024	0.0213	1.0215	0.0982	0.9018
0.15	0.85	337.2	1.3533	3.8700	0.0461	1.0472	0.1243	0.8757
0.2	0.8	337.3	1.1962	3.3074	0.0794	1.0827	0.1425	0.8575
0.25	0.75	337.5	1.0538	2.8684	0.1207	1.1283	0.1556	0.8444
0.3	0.7	337.8	0.9245	2.5206	0.1696	1.1848	0.1654	0.8346
0.35	0.65	338.0	0.8071	2.2413	0.2259	1.2535	0.1731	0.8269
0.4	0.6	338.2	0.7002	2.0141	0.2898	1.3362	0.1793	0.8207
0.45	0.55	338.4	0.6028	1.8273	0.3615	1.4355	0.1845	0.8155
0.5	0.5	338.7	0.5140	1.6720	0.4414	1.5550	0.1894	0.8106
0.55	0.45	338.9	0.4329	1.5418	0.5305	1.6997	0.1941	0.8059
0.6	0.4	339.3	0.3590	1.4318	0.6297	1.8771	0.1990	0.8010
0.65	0.35	339.6	0.2915	1.3384	0.7408	2.0976	0.2046	0.7954
0.7	0.3	340.2	0.2301	1.2588	0.8660	2.3775	0.2113	0.7887
0.75	0.25	340.9	0.1745	1.1907	1.0087	2.7421	0.2202	0.7798
0.8	0.2	342.0	0.1246	1.1327	1.1738	3.2341	0.2327	0.7673
0.85	0.15	343.8	0.0804	1.0837	1.3685	3.9293	0.2525	0.7475
0.9	0.1	347.0	0.0425	1.0434	1.6029	4.9674	0.2894	0.7106
0.95	0.05	354.5	0.0129	1.0130	1.8769	6.5335	0.3863	0.6137
0.98	0.02	366.5	0.0020	1.0020	1.9941	7.3457	0.5850	0.4150
0.99	0.01	373.9	0.0004	1.0004	1.9813	7.2523	0.7406	0.2594
0.999	0.001	382.7	0.0000	1.0000	1.9256	6.8595	0.9679	0.0321
0.99999	0.00001	383.8	0.0000	1.0000	1.9167	6.7987	0.9997	0.0003
0.99999999	0.00000001	383.8	0.0000	1.0000	1.9166	6.7981	1.0000	0.0000

Minimum boiling point azeotrope found at x(1) = 0.075000000

Results for binary mixture 1

FIGURE 12.19 COSMOthermX VLE data exported to Microsoft Excel.

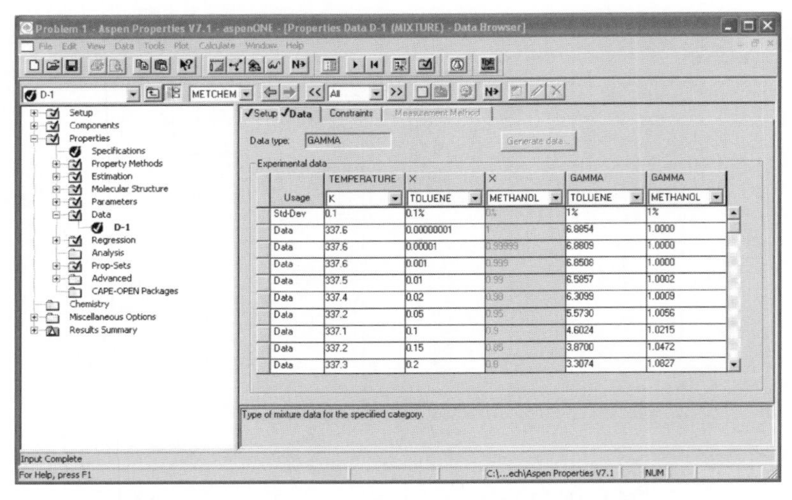

FIGURE 12.20 Aspen Properties user interface for VLE data input.

2. Under "Tools\Analysis\Binary," choose Type *T-xy* or *P-xy*, set other conditions to match those of the regressed or estimated BIPs, then run the analysis.
3. After seeing the resulting phase diagram, close the plot and choose the "Plot Wizard" from the table that appears.
4. In the Plot Wizard, choose "Gammas" as the desired plot.
5. An example of the resulting plot of liquid activity coefficients for each component is shown (Figure 12.23).

Comparison of T-xy diagrams and liquid activity coefficients by source

A graphical comparison of VLE and liquid activity coefficient data retrieved from various Aspen Properties data banks and generated through COSMOthermX is shown in Figures 12.24–12.29. Three pairs of binary systems is considered in these examples: (1) the zeotropic system of methanol and water (Figures 12.24 and 12.25), (2) the minimum boiling, homogeneous azeotropic system of acetonitrile and isopropanol (Figures 12.26 and 12.27), and (3) the minimum boiling, heterogeneous azeotropic system

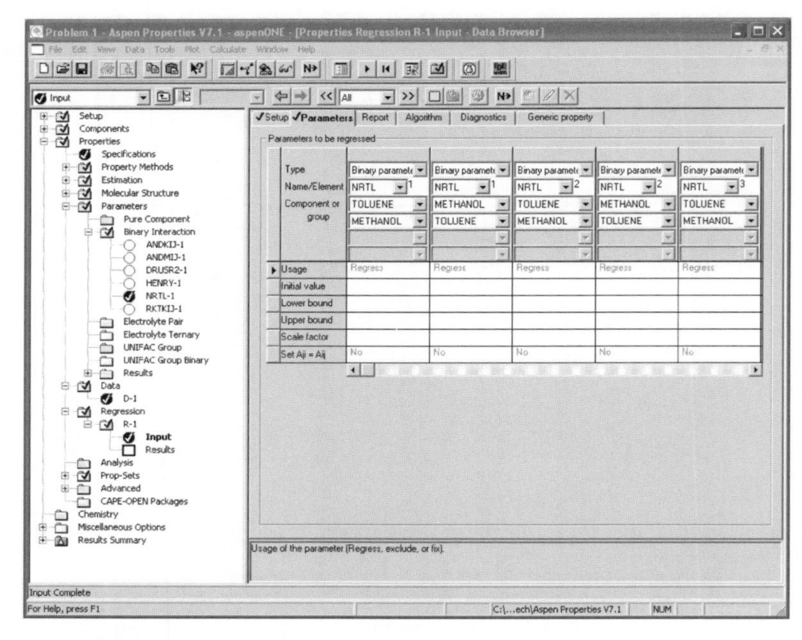

FIGURE 12.21 Aspen Properties data regression user interface.

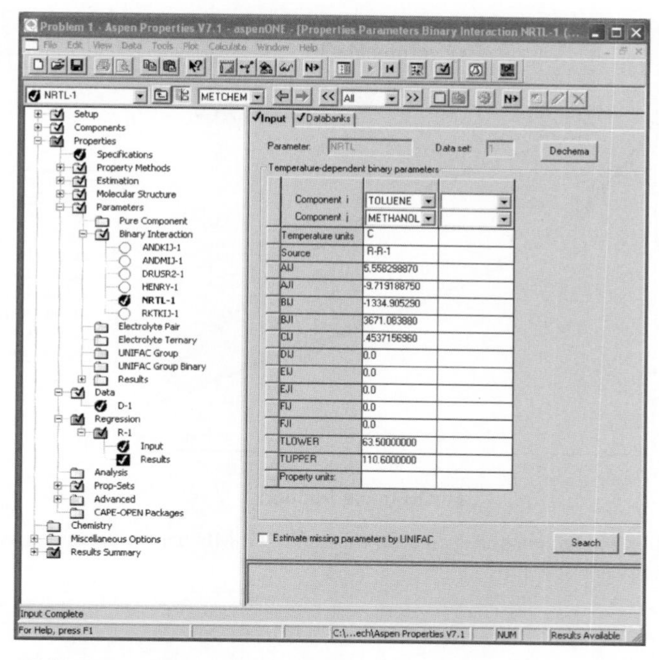

FIGURE 12.22 Aspen Properties VLE data regression results.

of *n*-butanol and water (Figures 12.28 and 12.29). In each case, the *T-xy* phase diagrams and liquid-phase activity coefficients plots were generated with the aid of COSMOthermX using the infinite dilution, NRTL parameter, and gamma regression protocols described in this guide. These three cases demonstrate that COSMOthermX is an exceptional method of obtaining estimated VLE data, and associated binary interaction parameter data, in the

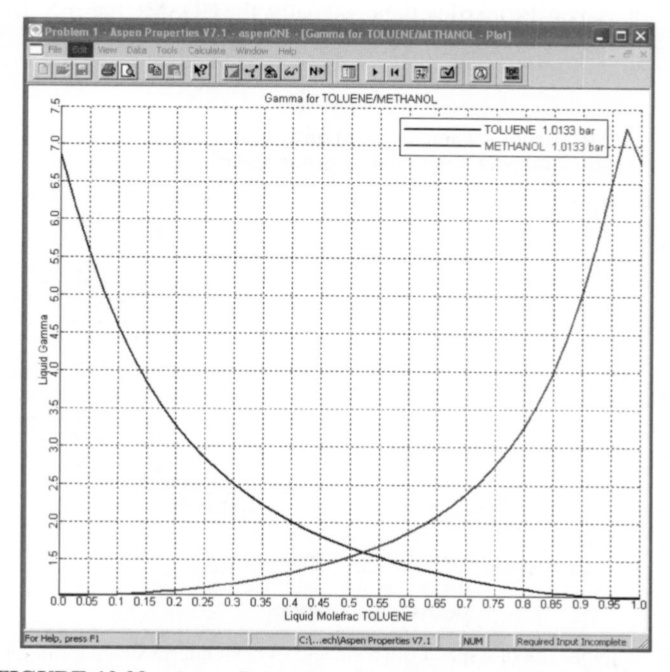

FIGURE 12.23 Aspen Properties plot of binary activity coefficients.

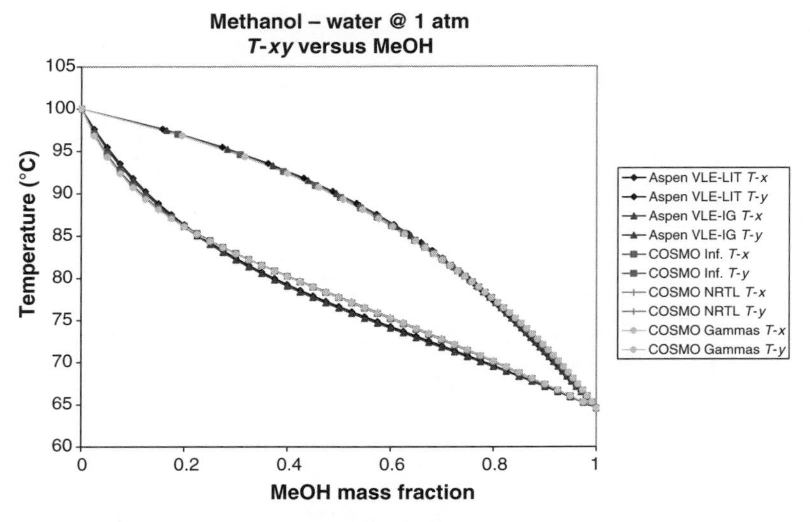

FIGURE 12.24 Comparison of Aspen Data bank and COSMOthermX data for *T-xy* phase diagram for methanol and water at 1 atm.

absence of laboratory or literature data. Aspen Properties is a convenient platform for the collection and regression of equilibrium data for further use in process modeling. The COSMOthermX estimation method provides immediate answers about the potential of azeotropic behavior, mixture boiling points, and relative volatility. As with any estimation technique, some deviation from real properties is expected, but the degree of deviation varies depending on the system. For these three cases, the deviations are not significant in all three categories of estimation. The results are plotted against two sources of Aspen Properties data bank data: VLE-IG (Dortmund Data Bank with ideal gas EOS) and VLE-LIT (other literature values including Dortmund Data Bank with ideal gas EOS). The three COSMOthermX estimation protocols are represented in the plots by this nomenclature: (1) "COSMO Inf" for estimation with infinite dilution activity coefficients, (2) "COSMO NRTL" for estimation from COSMOthermX generated NRTL binary parameters, and (3) "COSMO Gammas" for the regression of COSMOthermX liquid activity coefficient data within Aspen Properties. Generally, the COSMOthermX derived results are tightly grouped

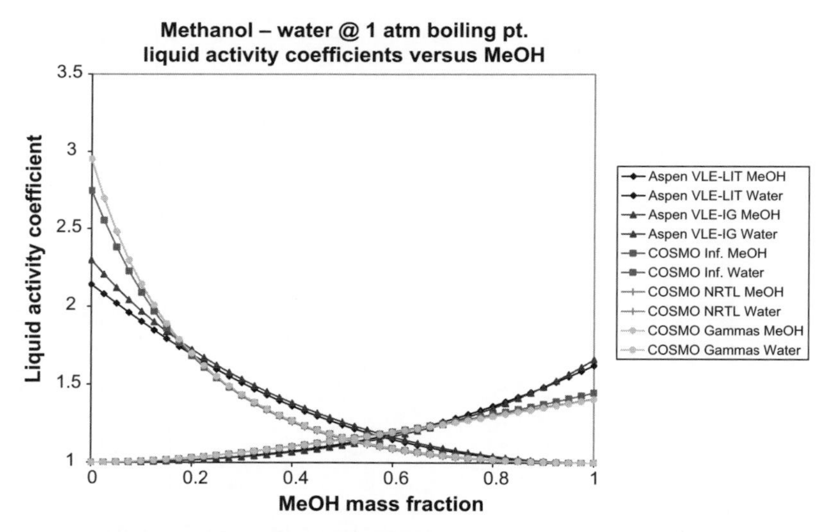

FIGURE 12.25 Comparison of Aspen Data bank and COSMOthermX data for liquid-phase activity coefficients for methanol and water at 1 atm.

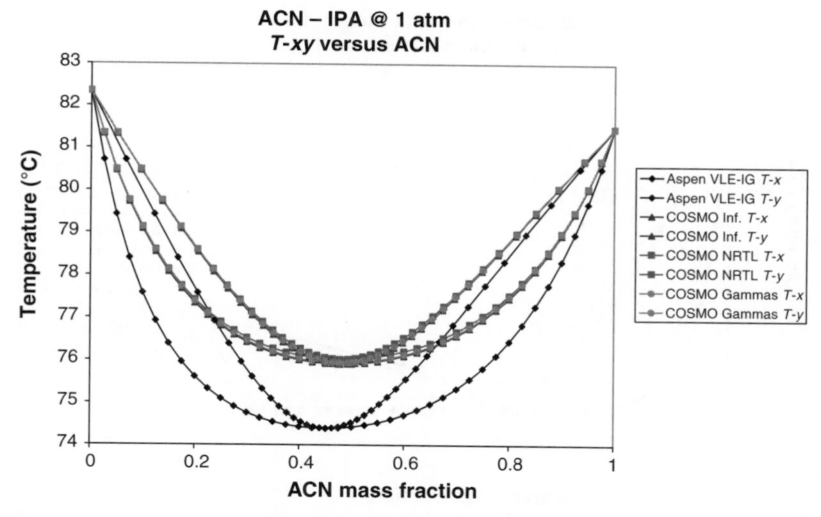

FIGURE 12.26 Comparison of Aspen Data bank and COSMOthermX data for *T-xy* phase diagram for acetonitrile and isopropanol at 1 atm with a homogeneous minimum boiling azeotrope.

in the resulting plots and in all cases the prediction of zeotropic and azeotropic behavior matches that from the Aspen Properties data banks. For the azeotropic systems, the boiling points of the mixtures deviates by less than 2°C and the composition of the binary azeotrope is offset by less than 0.1 mass fraction units. The activity coefficient plots show the greatest deviation from the data bank values at the end points approaching infinite dilution, although this deviation does not impact the *T-xy* phase diagram results to any great degree. As interest in any particular pair of components increases, the accuracy of the underlying physical properties can be improved with data from other sources and Aspen Properties can be used as a continuing storage and regression point for the information. COSMOthermX estimation techniques for binary interaction parameters have been used to successfully create custom data banks in Aspen Properties for thousands of binary systems not initially available from Aspen Properties data banks.

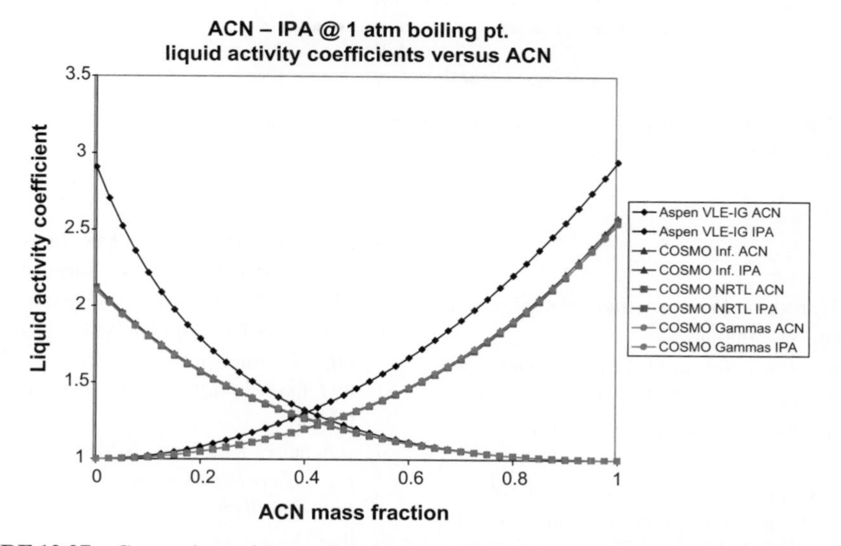

FIGURE 12.27 Comparison of Aspen Data bank and COSMOthermX data for liquid-phase activity coefficients for acetonitrile and isopropanol at 1 atm.

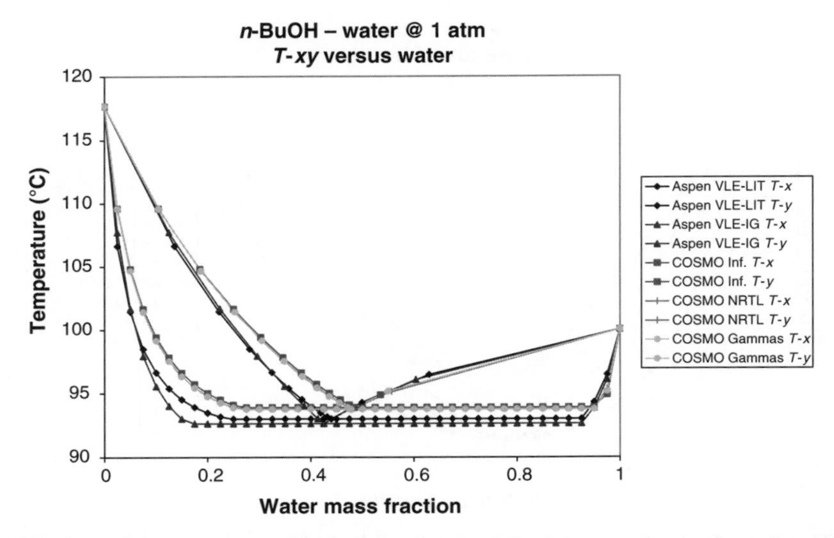

FIGURE 12.28 Comparison of Aspen Data bank and COSMOthermX data for *T-xy* phase diagram for *n*-butanol and water at 1 atm with a heterogeneous minimum boiling azeotrope.

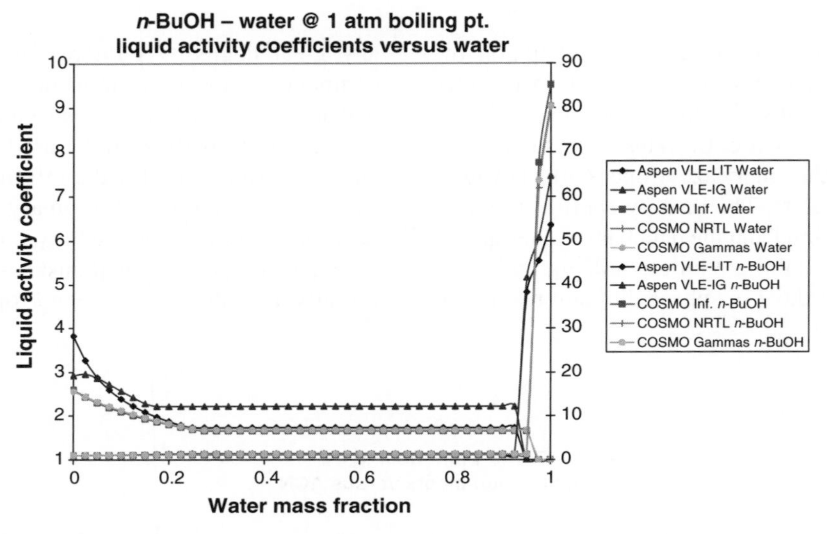

FIGURE 12.29 Comparison of Aspen Data bank and COSMOthermX data for liquid-phase activity coefficients for methanol and water at 1 atm.

REFERENCES

1. Seader JD, Henley EJ. *Separation Processes and Principles*, 2nd edition, Wiley, Hoboken, NJ, 2006, p. 31.
2. Geankoplis CJ. *Transport Processes and Unit Operations*, 2nd edition, Allyn and Bacon, Inc., Boston, MA, 1983, pp. 633–634.
3. Gentilcore MJ. Reduce solvent usage in batch distillation. *Chem. Eng. Prog.* 2002;56:56–59.
4. Li Y, Yang Y, Kalthod V, Tyler SM. Optimization of solvent chasing in API manufacturing process: constant volume distillation. *Org. Process Res. Dev.* 2009;13:73–77.
5. Balzheiser RE, Samuels MR, Eliassen JD. *Chemical Engineering Thermodynamics: The Study of Energy, Entropy, and Equilibrium*, Prentice-Hall, Englewood Cliffs, NJ, 1972, p. 387.
6. Balzheiser RE, Samuels MR, Eliassen JD. *Chemical Engineering Thermodynamics: The Study of Energy, Entropy, and Equilibrium*, Prentice-Hall, Englewood Cliffs, NJ, 1972, p. 452.
7. Balzheiser RE, Samuels MR, Eliassen JD. *Chemical Engineering Thermodynamics: The Study of Energy, Entropy, and Equilibrium*, Prentice-Hall, Englewood Cliffs, NJ, 1972, p. 453.

13

CRYSTALLIZATION DESIGN AND SCALE-UP

Robert Rahn McKeown
Chemical Development, GlaxoSmithKline, Research Triangle Park, NC, USA
James T. Wertman and Philip C. Dell'Orco
Chemical Development, GlaxoSmithKline, King of Prussia, PA, USA

13.1 INTRODUCTION

Crystallization can be defined as the formation of a solid crystalline phase of a chemical compound from a solution in which the compound is dissolved. In the synthesis of fine chemicals and pharmaceuticals, crystallization is extensively employed to achieve separation, purification, and product performance requirements. Despite its industrial relevance, an understanding of crystallization as a unit operation is often de-emphasized in engineering curricula and "learned on the job" in industrial settings.

In order to improve the knowledge and practice of crystallization science, several excellent volumes have been published that provide a comprehensive treatment of the subject [1–3]. The objective of this chapter is not to repeat these comprehensive overviews, but rather to provide a concise, basic understanding of crystallization design and scale-up principles, which can be applied toward common industrial problems. The focus is primarily on batch rather than continuous crystallization processes, as batch crystallization is the predominant processing method used in the pharmaceutical industry today.

The chapter begins with a discussion of crystallization design objectives and constraints on design, including a description of physical properties important to product performance. Thermodynamic principles of crystallization are then reviewed, followed by a discussion of crystallization kinetics. Crystallization design approaches are then presented, incorporating thermodynamic and kinetic considerations. Finally, the scale-up and scale-down of heat and mass transfer are discussed. Throughout the chapter, industrially relevant examples are used to illustrate the concepts presented.

13.2 CRYSTALLIZATION DESIGN OBJECTIVES AND CONSTRAINTS

Crystallization is used in pharmaceutical synthesis to accomplish the following two objectives: (1) separation and purification of organic compounds and (2) delivery of physical properties suitable for downstream processing and formulation. In achieving these objectives, a crystallization design is constrained by economic and manufacturing considerations, such as yield, throughput, environmental impact, and the ability to scale the process. An overview of these topics is presented in this section.

13.2.1 Separation and Purification

The synthesis of an active pharmaceutical ingredient (API) from raw materials involves a multistep synthetic procedure during which the raw materials undergo numerous chemical transformations and purification steps to ultimately prepare the desired molecular structure in high purity, (typically >99%). One of the first steps in overall process design is to understand where isolated intermediates are required to meet purification needs. An example from the literature

Chemical Engineering in the Pharmaceutical Industry: R&D to Manufacturing Edited by David J. am Ende

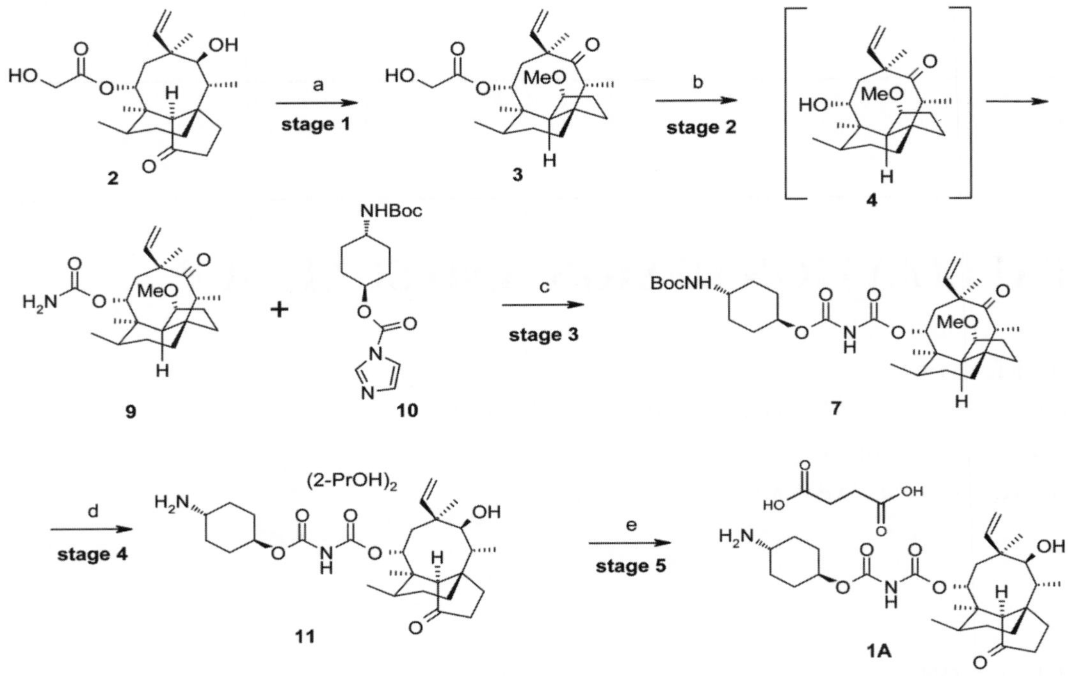

FIGURE 13.1 Schematic of a typical synthetic route to an active pharmaceutical ingredient [4]. This particular route uses six chemical transformations with five crystallizations to achieve the purity required to ensure product quality. Reprinted with permission from Ref. 4. Copyright (2009) American Chemical Society.

is used to exemplify a synthetic route and its separation/purification challenges. This example is illustrated in Figure 13.1 [4].

For the example in Figure 13.1, the API is produced in five stages from raw materials. Four of these five stages have crystallization steps to achieve purification, while the final stage uses a crystallization to control the composition and physical properties of the final molecule. The reference describes in detail the rationale for the placement of crystallization steps. Briefly, the stage 1 process was used to control key impurities in the process, as the input raw material 2 had approximately 35 impurities with the reaction producing additional impurities. This stage used a design space approach across the reaction and crystallization to ensure complete purging of raw material 2 at levels up to 3% and of an alkene impurity on the cyclohexyl ring of intermediate 3 at levels up to 4%. Crystallizations in stages 2, 3, and 4 increased the organic purity of the product from approximately 97% to greater than 99% prior to stage 5 so that this step could focus solely on the formation of the desired salt and control of resultant physical properties. Moreover, the use of crystallization in this synthesis allows intermediates to be "stabilized" by forming a less reactive solid phase, preventing solution phase side reactions (e.g., racemization), and allowing for material storage.

While organic impurities related to the molecular structure of the intermediates and products are one concern for product purification, inorganic and organic reagents also require separation. These include simple salts (e.g., NaCl, K_2CO_3, and NaOAc) and reagents (e.g., triethylamine, $Pd(OAc)_2$, and triphenylphosphine) commonly used in pharmaceutical synthesis. Crystallization is also used to control chiral purity, often through the use of chiral resolving agents [3].

Purification is enabled by selecting a solvent in which impurities are dissolved at the point where the desired product can be crystallized, or in which impurities remain undissolved and can be physically removed by filtration of the product solution. While thermodynamics are often the primary factor in achieving purification, kinetics may also impact the impurity content of a product by entrapment of impurities or solvents in the crystal lattice, often induced by rapid crystallization processes. This is discussed in additional detail in 13.5.3.

13.2.2 Product Performance

The second objective of crystallization is related primarily to API rather than to intermediate production. It concerns the delivery of the appropriate material physical properties to ensure acceptable downstream processing (e.g., isolation, drying, size reduction, and formulation unit operations) as well as the *in vivo/in vitro* performance of the formulated product. Of particular concern are the crystalline form of the compound, the particle size distribution of the active

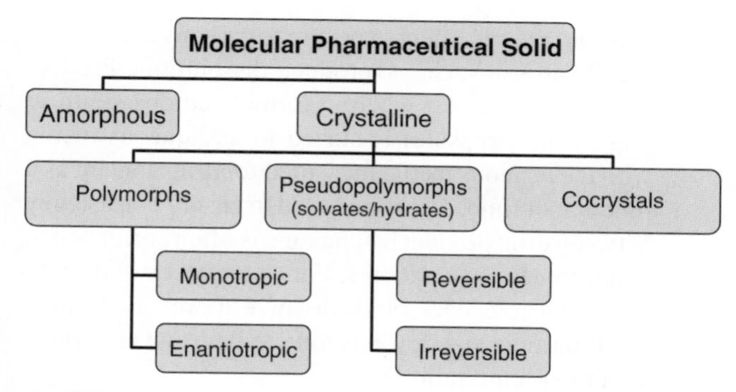

FIGURE 13.2 A map of the forms that a molecular pharmaceutical solid can exhibit. Crystalline materials are polymorphs, solvates/hydrates, and cocrystals. Cocrystals can be considered special cases of solvates, in which the "solvent" is instead an involatile compound that noncovalently bonds to the molecular solid in a regular, ordered manner. Irreversible solvates can convert either to polymorphs, different solvates, or amorphous materials upon desolvation/dehydration.

ingredient, and the morphology and flow properties of the product.

13.2.2.1 Crystalline Form Pharmaceutical solids are known for their ability to have multiple solid phases. A brief schematic of common solid phases is shown in Figure 13.2. Solid phases commonly exist as either "polymorphs" or "pseudopolymorphs." Pseudopolymorphs are also referred to as solvates and hydrates. Polymorphism occurs when a single compound exists in two or more solid-state forms that have identical chemical structures but different crystal lattice structures.

Polymorphs can have either "monotropic" or "enantiotropic" relationships [5]. This behavior is exhibited in Figure 13.3. When two polymorphs have a monotropic relationship they exhibit the same relative stability order up to the normal melting point of each polymorph. When an enantiotropic relationship exists, the polymorphs change stability order at a transition temperature below the normal melting point of either polymorph. In Figure 13.3, the Gibbs' free energy of two polymorphs is shown for both a monotropic case and an enantiotropic case. For the monotropic case (b), the Gibbs' free energy of polymorph I, G_I, is lowest across the temperature range until the melting point, where the liquid form of the material becomes most stable. In the enantiotropic case, polymorph II exhibits the lowest Gibbs' free energy, G_{II}, up to a transition temperature, IP_M at which point polymorph I exhibits the lowest Gibbs' free energy. As illustrated in Figure 13.3, the difference in enthalpy between polymorphs I and II (H_I and H_{II}) is approximately constant over the temperature range shown. Therefore, the change in their Gibbs' free energy relationship is predominately due to the entropy contribution, $T\Delta S$.

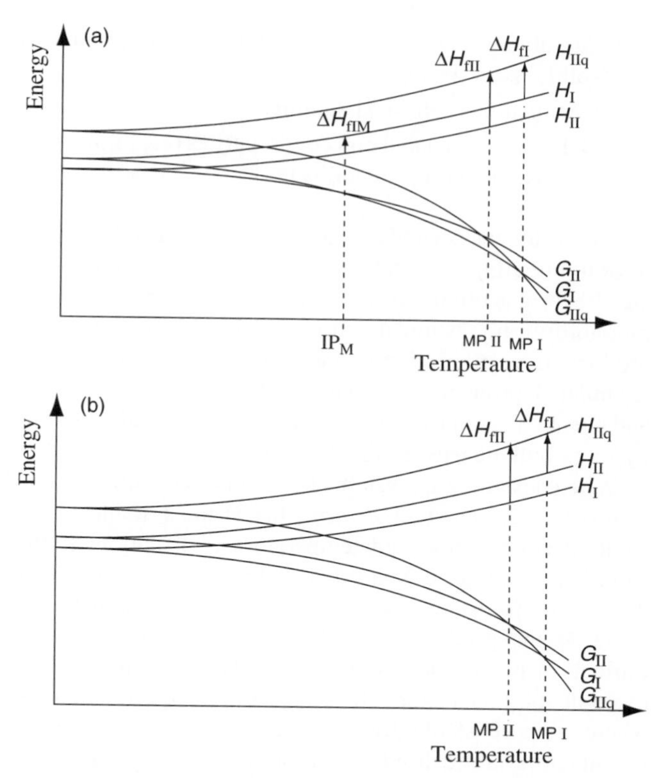

FIGURE 13.3 Thermodynamic description of polymorphism: enantiotropic (a) and monotropic (b) systems [6]. In the monotropic system, the stability order of forms is the same up to the melting point. For the enantiotropic system, a crossover temperature exists where the stability order changes. Reprinted with permission from Ref. 7. Copyright (1999) John Wiley & Sons, Inc.

Solvates and hydrates occur when a solvent or water molecule is integrated into the crystal lattice through a repeating, noncovalent bonding arrangement with the parent molecule. In the case of "reversible" solvates and hydrates, the molecule(s) of solvent or water can be removed from the pseudopolymorph without significantly affecting the crystallinity of the solid [8]. For "irreversible" pseudopolymorphs, removal of the solvent or water can lead to amorphous material. Amorphous material may also be produced through rapid precipitation or material comminution. Cocrystals are similar to solvates, except that the "solvent" or "water" molecule is instead an involatile solute (e.g., nicotinamide [9] and benzoic acid [10]), which forms a complex with the parent API via a repeating, noncovalent bonding pattern (typically through hydrogen bonding, π stacking, or van der Waals interactions between the parent and the solute). For a more in-depth discussion on crystalline forms, excellent texts are available [11, 12].

Crystalline forms are important because they may exhibit different properties, some of which can include

- Solubility
- Melting point

- Dissolution rate and bioavailability of a formulated solid dosage form
- Chemical and physical stability
- Habit and associated powder properties (e.g., flow, bulk density, and compressibility)

As a result, the desired crystalline form is typically defined prior to initiating a crystallization design. For an intermediate, the form is often selected based on ease of manufacture, filterability, and chemical and physical stability. For a final product, it is often chosen based on performance in the formulated product (e.g., bioavailability, dissolution rate, and chemical and physical stability) and on manufacturability (e.g., bulk density and melting point) [6, 13].

When designing a crystallization, it is frequently desired to produce the crystalline form that is most stable at the solution composition and temperature of isolation. If the form being produced is not thermodynamically stable, it is possible for form conversion to occur at some point in the life cycle of the product, with the unstable form becoming difficult or nearly impossible to manufacture. An excellent example of this is the oft-referenced Ritonavir example where a more stable form appeared during commercial manufacture and caused a disruption to supply while the form issue was resolved [14]. Because the conversion from an unstable to a stable form is a kinetic process, it can be affected by changes in impurities, equipment, concentration, and other process variables.

13.2.2.2 ***Particle Size*** The second product performance criteria of concern is often particle size, although other properties such as surface area may be of equal or greater importance. The focus on particle size, and its frequent specification for APIs, is due to its potential impact on the performance of solid dosage forms:

1. Particle size can affect exposure to patients and *in vitro* specifications such as product dissolution. Product dissolution is a test where a formulated dosage form (e.g., tablet, capsule) is stirred in an aqueous media with the aqueous media measured for drug content as a function of time. The test is used to ensure consistency between drug product batches and is often correlated to exposure levels in patients. For example, crystals with larger particle sizes often dissolve more slowly than small particle size crystals, due to their lower surface area to volume ratio [15].
2. Particle size can affect final dosage form production by impacting powder conveyance and mixing. Conveyance and mixing can impact granulation (the drug product unit operation through which API is mixed with excipients, lubricants, and disintegrants prior to preparing the final dosage form, e.g., tabletting or capsule filling). It can also affect the dose uniformity (i.e., the amount of drug in each dosage unit) and drug product appearance (e.g., color and shape).

For a detailed discussion of particle size and the nature of particle size distributions, thorough descriptions have been prepared [1]. Briefly, particle size distributions in pharmaceutical manufacture typically utilize volume- or mass-based distributions. Mass-based distributions indicate what percentage of product mass is distributed into size intervals. Volume-based distributions indicate what percentage of product volume is distributed into size intervals. Mass and volume distributions are related to each other by the crystal density. Figure 13.4 displays a typical volume-based distribution for a pharmaceutical, with references made to d_{90}, d_{50}, and d_{10}. These values represent the particle size below which more than 90%, 50%, and 10% of the total volume of the product lies, respectively. Specifications for APIs often

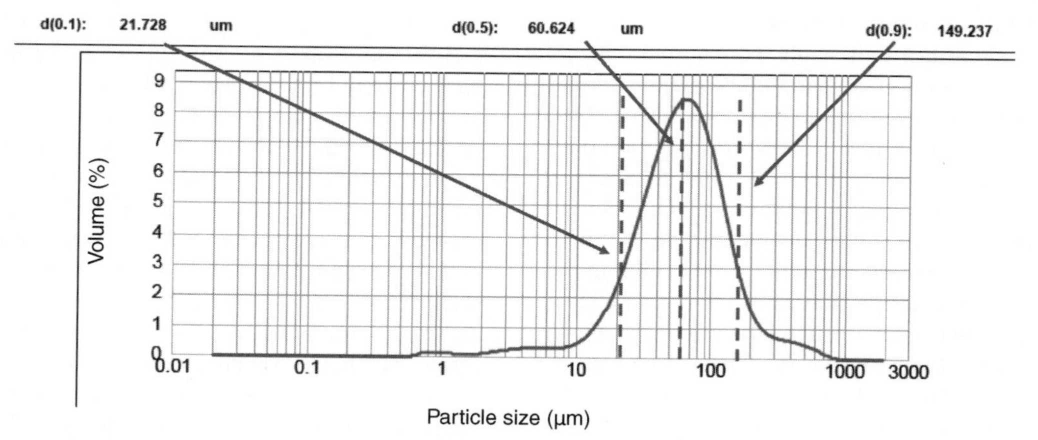

FIGURE 13.4 Particle size distribution (volume %) of a typical pharmaceutical product, measured by laser light diffraction. The $d(0.9)$ or d_{90} corresponds to the size at which 90% of the area of the curve is contained.

TABLE 13.1 Sieve Mass Fractions for an API

Sieve No.	Size Opening (μm)	Mass Retained on Sieve (g)	Sieve No.	Size Opening (μm)	Mass Retained on Sieve (g)
–	0	0.2	No. 30	595	12.2
No. 80	177	0.7	No. 25	707	8.7
No. 70	210	1.3	No. 20	841	5.3
No. 60	250	2.7	No. 18	1000	3.2
No. 50	297	5.3	No. 16	1190	2.4
No. 45	354	7.8	No. 14	1410	1.2
No. 40	420	10.9	No. 12	1680	0
No. 35	500	13.3	No. 10	2000	0

contain a range for particle size that includes one or more of d_{90}, d_{50}, and d_{10}. While the example in Figure 13.4 reflects a measurement obtained through a laser light scattering method for particle sizing, alternative methods such as sieving (which generates a mass distribution) are still routinely employed. Example 13.1 illustrates a mass particle size distribution calculation using sieves.

EXAMPLE 13.1 PARTICLE SIZE DISTRIBUTION CALCULATIONS

Seventy-five grams of API have been sieved using a cascade of 15 sieves. The mass on top of each sieve has been weighed, and is shown in Table 13.1. For mass distributions, instead of d_{90}, d_{50}, and d_{10}, the terms x_{90}, x_{50}, and x_{10} are used, where x represents mass fraction. Estimate the x_{90}, x_{50}, and x_{10} for the material.

For sieving, the first step is to assign a particle size to each interval. The amount on top of a sieve has a particle size that is between the size opening of that sieve and the next largest sieve size opening. The particle size of this interval is estimated by averaging these two sieve size openings. Once this is done for all size intervals, the mass amount is tabulated as a function of particle size. The mass amounts are then cumulatively added across sieve sizes with the amount at each interval divided by the total amount of material input to the sieve test. This gives a cumulative percentage of mass retained as a function of particle size. The mass retained and the cumulative percentage mass are then plotted as a function of particle size (Figure 13.5). The x_{90} can be estimated as the particle size at which the cumulative mass is 90%. A similar approach is used for x_{50} and x_{10}, giving values of $x_{90} = 890$ μm, $x_{50} = 520\,\mu$m, and $x_{10} = 300\,\mu$m.

13.2.2.3 Morphology and Powder Flow Properties A knowledge of crystal habits is important in understanding the particle size distribution and the behavior of powders and slurries, as different crystalline habits have different characteristic lengths and different flow characteristics. Flow

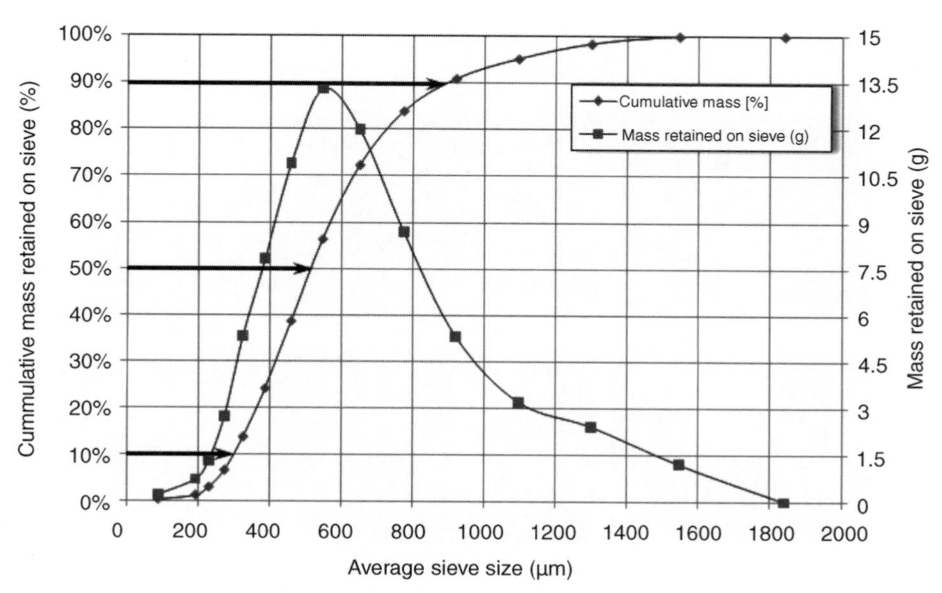

FIGURE 13.5 Size distribution estimated from sieve analysis data. Details of plot are described in Example 13.1.

characteristics are important because poor flowing powders can influence the ability of a powder to blend with excipients and can also impact the flow of powder in primary and secondary unit operations, causing undesirable phenomena such as core flow, also known as "rat-holing," in powder feed hoppers. The desired state for powder flow in a hopper is known as mass flow. In mass flow, the entire hopper contents are in motion. Mass flow is indicated by "first in–first out" flow of material, and a solid surface that sinks evenly. Core flow, on the contrary, is characterized by dormant zones near the walls of the hopper. Core flow is exemplified by a "last in–first out" flow of material and a solid surface that forms a core or rat-hole down the center of the hopper.

Figure 13.6 displays a summary of common habits observed in pharmaceutical crystallizations. Equant/block and bipyramidyl habits typically result in products that are easy to isolate, dry, and handle due to their relatively low surface area to volume ratio. Acicular and thin blade crystals, which are common in pharmaceuticals, tend to pose more processing difficulties such as long isolation times, agglomeration, and poor flow and handling properties. Despite processing difficulties, these habits may also result in high surface area materials, which can positively impact *in vivo* performance of a formulation. One of the key determinants of habit is the selected salt or hydration/solvation state of the parent molecule (i.e., the "version" of the molecule). The version and the selected form of that version, often selected early in development, can dictate the habit and constrain the ability to optimize downstream processing steps. The habit of a crystal may also be influenced by the crystallization solvent and the impurities present in the crystallization process.

The flow characteristics of a powder are frequently inferred from knowledge of powder densities, commonly the bulk and tapped densities. The bulk density of a powder is the density measured "as is," while the tapped density uses mechanical "taps" to facilitate further packing and settling of the powder. The ratio of the tapped to bulk density is referred to as the Hausner ratio, and provides an indication of the compressibility of powders and as a result the ease of powder conveyance. Materials with a Hausner ratio of less than 1.2 are generally considered to have acceptable flow properties, while those with a ratio greater than 1.4 are considered to have poor flow properties. The absolute values of density are also important, as they affect the level of fill for a piece of equipment. If material A has twice the bulk density of material B, the size of a batch can potentially be twice as large in the same equipment, resulting in increased throughput. Acicular materials routinely have Hausner ratios >1.3 and bulk densities <0.2 g/cm^3, while block and equant habits often have bulk densities >0.3 g/cm^3 and Hausner ratios <1.2 [16]. In addition to habit, particle size also affects material bulk densities, often proportionately (i.e., larger particle size generally corresponds to higher densities).

13.2.2.4 Manufacturability Common manufacturability criteria include yield, cycle and batch times, environmental impact, and processability. Theoretical process yields are calculated with equilibrium solubility data, as described in the next section. Environmental impacts are often dictated by the toxicity of the solvent (ICH Class 3 solvents being preferred), amount of solvent used (<10 L/kg solute preferred), ability to recover the solvent (single solvent systems preferred), and boiling point of the solvent (55–100°C preferred). Cycle and batch times are affected by the time it takes for the crystallization to proceed, and by the time spent in the isolation and drying unit operations. Generally, smaller

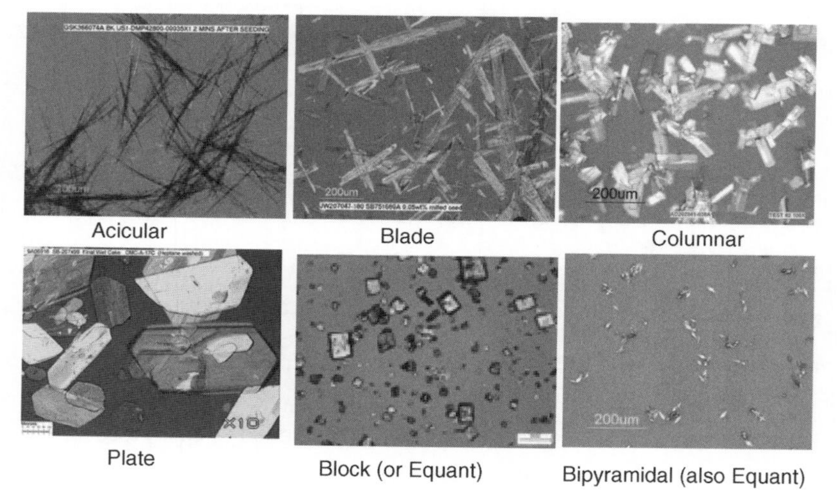

FIGURE 13.6 Commonly observed habits in pharmaceutical crystallization. For a more complete description of habits, please see Ref. 1.

particles will result in a higher filter cake resistance due to more efficient packing and reduced cake porosity [17]. As a result there is often a trade-off between time spent in each unit operation, as rapid crystallizations generate small particles (resulting in a slow isolation) and slow crystallizations generate larger particles (resulting in a rapid isolation). The integration of crystallization, isolation, and drying must be considered to deliver an optimum throughput for a product. An example of batch time analysis for a crystallization and isolation process is shown in Example 13.2.

EXAMPLE 13.2 ESTIMATION OF OPTIMAL CYCLE TIME FOR A CRYSTALLIZATION, ISOLATION, AND DRYING PROCESS

The API shown in Figure 13.7 is prepared by reactive crystallization where a base, ethylene diamine, is added to a molecular free acid. Two potential crystallization processes have been identified that meet the product performance needs.

Process 1 is performed in *tert*-butyl methyl ether (TBME), and requires 12 h to crystallize. Process 2 is performed in isopropyl alcohol (IPA), and requires 5 h to crystallize. Both crystallizations use 10 L solvent/kg product, the yields for both processes are the same, and both processes are isolated at 20°C. Either process is planned to use a 1 m^2 area filter with a mass loading of 100 kg and a 1 barg filtration pressure. Filter cake resistances, α, of the products were measured through small-scale tests to be 1.0×10^{11} and 0.5×10^{11} m/kg, respectively, where the cake resistance is estimated through equation 13.1, assuming a negligible media resistance. In equation 13.1, t is the filtration time, m_f is the mass of filtrate, ν is the kinematic viscosity, c is the mass of solids per mass of filtrate, A_F is the filtration area, and P is the pressure drop across the filter.

$$\frac{t}{m_f} = \frac{\alpha \nu c}{2 A_F{}^2 P} m_f \tag{13.1}$$

Which process offers the minimum batch time, and by how much (assume 1 bar atmospheric pressure)?

Solution

The filtration time for each material must be estimated at commercial scale. At 20°C, TBME has a kinematic viscosity of 4.7×10^{-7} m^2/s and a density of 0.740 kg/L, while IPA has a kinematic viscosity of 2.9×10^{-6} m^2/s and a density of 0.786 kg/L. From the equation provided, the isolation times for a 100 kg batch with 10 L solvent/kg product can be estimated as

$$c = \frac{1\,\text{kg}}{10\,\text{L}} \cdot \frac{1\,\text{L}}{0.740\,\text{kg}} = 0.135 \quad \text{and}$$

$$m_f = 10\,\text{L/kg} \cdot 100\,\text{kg} \cdot 0.740\,\text{kg/L} = 740\,\text{kg}$$

$$t = \frac{\alpha \nu c}{2 A_F{}^2 P} m_f^2$$

$$= \frac{1.0 \times 10^{11}\,\text{m/kg} \cdot 4.7 \times 10^{-7}\,\text{m}^2/\text{s} \cdot 0.135}{2 \cdot (1\,\text{m}^2)^2 \cdot (1 \times 10^5\,\text{kg/m s}^2)} \cdot (740\,\text{kg})^2$$

$$= 17390\,\text{s} \Rightarrow t = 4.8\,\text{hr for the TMBE process filtration}$$

$$c = \frac{1\,\text{kg}}{10\,\text{L}} \cdot \frac{1\,\text{L}}{0.786\,\text{kg}} = 0.127\ \text{and}$$

$$m_f = 10\,\text{L/kg} \cdot 100\,\text{kg} \cdot 0.786\,\text{kg/L} = 786\,\text{kg}$$

$$t = \frac{\alpha \nu c}{2 A_F{}^2 P} m_f^2$$

$$= \frac{0.5 \times 10^{11}\,\text{m/kg} \cdot 2.9 \times 10^{-6}\,\text{m}^2/\text{s} \cdot 0.127}{2 \cdot (1\,\text{m}^2)^2 \cdot (1 \times 10^5\,\text{kg/m s}^2)} \cdot (786\,\text{kg})^2$$

$$= 56985\,\text{s} \Rightarrow t = 15.8\,\text{h for the IPA process filtration}$$

So the total batch time using TBME is 12 + 4.8 = 16.8 h and using IPA is 5 + 15.8 h = 20.8 h. TBME offers a lower batch time by 4 h, mainly due to the lower kinematic viscosity of TBME compared with IPA. More detailed approaches optimizing crystallization parameters in-line with isolation times can be performed [18].

Example 13.2 illustrates a key aspect of crystallization design that is also illustrated in Figure 13.8. The crystallization

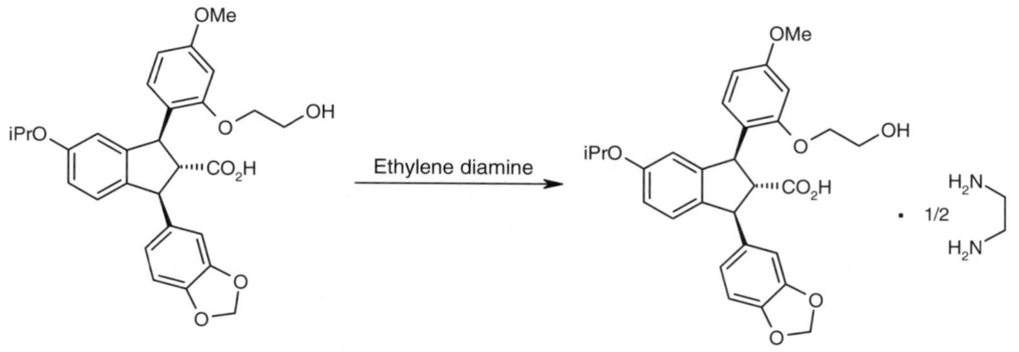

FIGURE 13.7 Example reactive API crystallization process. Details are described in Example 13.2.

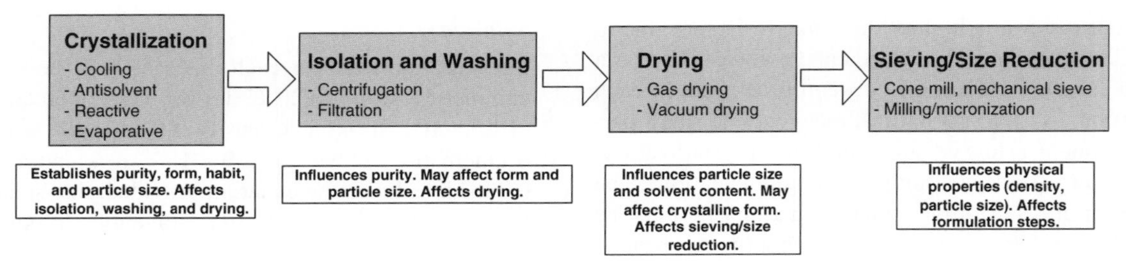

FIGURE 13.8 The relationships of crystallization with downstream processing steps. Crystallization typically needs to be studied in conjunction with downstream steps to understand the complete control of desired attributes.

in the example affects isolation performance, which in turn affects drying performance, which affects sieving and size reduction performance, which ultimately affects performance in the formulation process. As a result, it is often necessary to study crystallization in conjunction with several unit operations. As shown in Example 13.2, it is straightforward to assess the dependence of the crystallization on the isolation step and the drying step can further be integrated into structured studies. Each downstream unit operation must be investigated for its impact in ensuring the process, and not just the crystallization, achieves the desired performance objective.

13.3 SOLUBILITY ASSESSMENT AND PRELIMINARY SOLVENT SELECTION

An understanding of a compound's solubility is the starting point for crystallization design. The solubility of a compound determines the throughput and yield; it is the key measurement for selecting a solvent system and selecting a crystallization mode. Solubility is a thermodynamic property of a solute, which describes the equilibrium of a defined solid phase (e.g., a polymorph or pseudopolymorph) with a solution, as shown in equation 13.2. It is a dynamic equilibrium whereby the rate of dissolution is balanced by the rate of crystallization.

$$\text{drug(dissolved)} \underset{k_{\text{dissolutioin}}}{\overset{k_{\text{crystallize}}}{\longleftrightarrow}} \text{drug(s)} \tag{13.2}$$

Solubility is determined through equilibrium experiments in which a solid phase is slurried isothermally with a solvent until a constant concentration is achieved in the solution phase. Typical solubility data collection for a pharmaceutical compound is shown in Figure 13.9. These data illustrate the general approach to measurement: solid of a known phase is

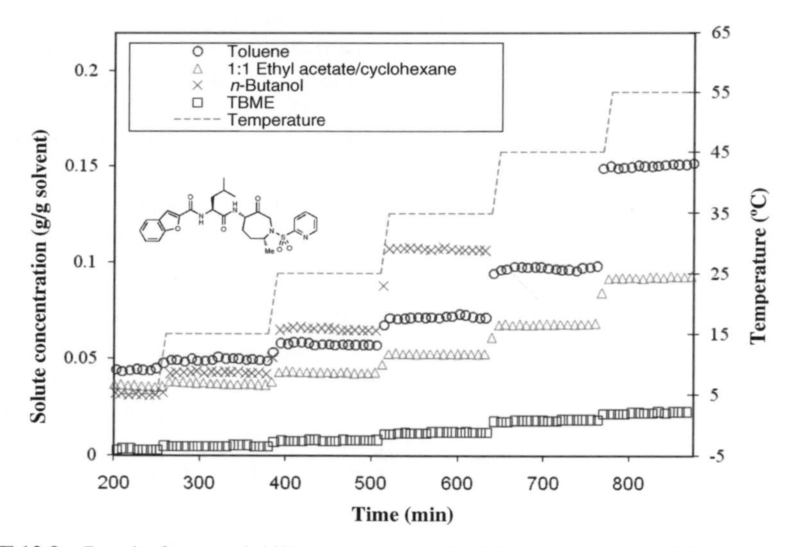

FIGURE 13.9 Results from a solubility experiment using Diamond Attenuated Total Reflectance IR spectroscopy to measure concentration *in situ* for a pharmaceutical active ingredient. At increasing temperatures, the solute achieves an equilibrium in the different solvent mixtures, indicated by the plateau in concentration upon achieving a new temperature. In this case, equilibrium is rapidly achieved.

added to a predefined amount of solvent at low temperature, an equilibrium concentration is achieved, and the temperature is then increased, while ensuring solids are still present. At the end of the experiment, the solid phase of the material is assessed to ensure the crystalline form has not changed, as different forms have different equilibrium concentrations. If binary or ternary solvents are being studied, the phase composition of the mixture should also be measured at the end of the experiment to ensure no changes. Each compound possesses its own timescale in which equilibrium is achieved, which typically ranges from minutes to hours, although days may be required in some circumstances. As illustrated in Figure 13.9, in-line methods of solute concentration measurement are useful in understanding time to equilibrium. When in-line methods are not practical, measurement of the filtered equilibrium solution by HPLC (or gravimetric analysis if the material is relatively free from impurities) is commonly used.

While Figure 13.9 displays solubility as a function of temperature, it may also be measured as a function of solvent composition. When temperature is used to generate solubility differences, the result is a "cooling" crystallization. When solvent composition is used, the result is often called an "antisolvent" crystallization. In all cases, the preferred units of solubility are mass solute per mass solvent. These units are convenient for many engineering calculations.

Once data are collected, they can be evaluated for use in process design. The thermodynamic description of two phases in equilibrium is

$$f_i^{\text{solid}} = f_i^{\text{solution}} \tag{13.3}$$

where f_i^{solid} is the fugacity of component i in the solid phase and f_i^{solution} is the fugacity of component i in the liquid or solution phase.

From this, one can derive an expression for solubility of the general form:

$$x_{\text{ideal}} = \frac{1}{\gamma_{\text{solute}}} \exp\left[\frac{\Delta H_{\text{tp}}}{R}\left(\frac{1}{T_{\text{tp}}} - \frac{1}{T}\right) - \frac{\Delta C_P}{R}\left(\ln\frac{T_{\text{tp}}}{T} - \frac{T_{\text{tp}}}{T} + 1\right) - \frac{\Delta V}{RT}(P - P_{\text{tp}})\right] \tag{13.4}$$

where x_{ideal} is the ideal solubility of the solute (mol solute/mol solution), γ_{drug} is the activity of the drug in solution, ΔH_{tp} is the enthalpy change for a liquid–solid solute transformation at the triple point, ΔC_P is constant pressure heat capacity difference between the liquid and solid solute phases, ΔV is the volume change, T is the temperature, T_{tp} the triple point temperature, P is the pressure, P_{tp} is the triple point pressure, and R is the universal gas constant.

In almost all situations, pressure has a little to no effect on solubility; therefore, the pressure term can be eliminated. The change in heat capacity is often assumed to be negligible, and the triple point is often replaced with the melting point of the solid to yield the approximation shown below:

$$x_{\text{ideal}} = \frac{1}{\gamma_{\text{solute}}} \exp\left[\frac{\Delta H_{\text{m}}}{R}\left(\frac{1}{T_{\text{m}}} - \frac{1}{T}\right)\right] \tag{13.5}$$

for an ideal solution, this can be reduced to a van't Hoff type expression and linearized

$$\ln x_{\text{ideal}}\gamma_{\text{solute}} = \ln(S_{\text{solute}}) = \frac{\Delta H_{\text{m}}}{RT_{\text{m}}} - \frac{\Delta H_{\text{m}}}{RT} = \frac{A_{\text{S}}}{T} + B_{\text{S}} \tag{13.6}$$

where S_{solute} is the observed solubility and A_{S} and B_{S} are constants obtained by regression. For ease of use in future calculations, the natural log of solubility is taken with solubility often having units of mass of solute per mass of solvent and temperature having units of Kelvin (Figure 13.10).

Almost all solutions containing a high fraction of API are nonideal; however this linearization technique (i.e., plotting $\ln(S_{\text{solute}})$ versus $1/T$) is a simple way to visualize and interpolate solubility from a few data points. For systems in which these plots are nonlinear, a correction can be added to equation 13.6 [1]:

$$\ln(S_{\text{solute}}) = \frac{A_{\text{S}}}{T} + B_{\text{S}} + C_{\text{S}}\ln(T) \tag{13.7}$$

where C_{S} is an additional constant. In other cases, polynomial or exponential functions may be used to represent data, but these expressions tend to be less representative when extrapolated beyond the range of temperature studied in the solubility experiment. When solubility is correlated as a function of both composition (solvent 1 and solvent 2) and temperature, A_{S} and B_{S} can often be empirically estimated as linear or quadratic functions of solvent 2 volume fraction to provide an adequate data fit.

Once solubility is measured, it can be used for a number of purposes. First, it can be used to select the crystallization solvent based on the design criteria: typically yield, throughput, and environmental constraints. The potential process yield, Y, is often estimated by equation 13.8 below:

$$Y = \frac{S_{\text{solute 1}}\, m_{\text{solvent 1}} - S_{\text{solute 2}}\, m_{\text{solvent 2}}}{S_{\text{solute 1}}\, m_{\text{solvent 1}}} \tag{13.8}$$

In equation 13.8, S_{solute1} and S_{solute2} are the solubility at the dissolution temperature and composition, and the solubility at isolation temperature and composition, respectively, and m_{solvent1} and m_{solvent2} are the mass of solvent at the dissolution temperature and composition, and the mass of solvent at isolation temperature and composition, respectively.

Another use of solubility data is the estimation of purification potential for a crystallization. To perform this assessment, the solubility of the impurity must also be known. Using equation 13.8, it is possible to calculate the mass of

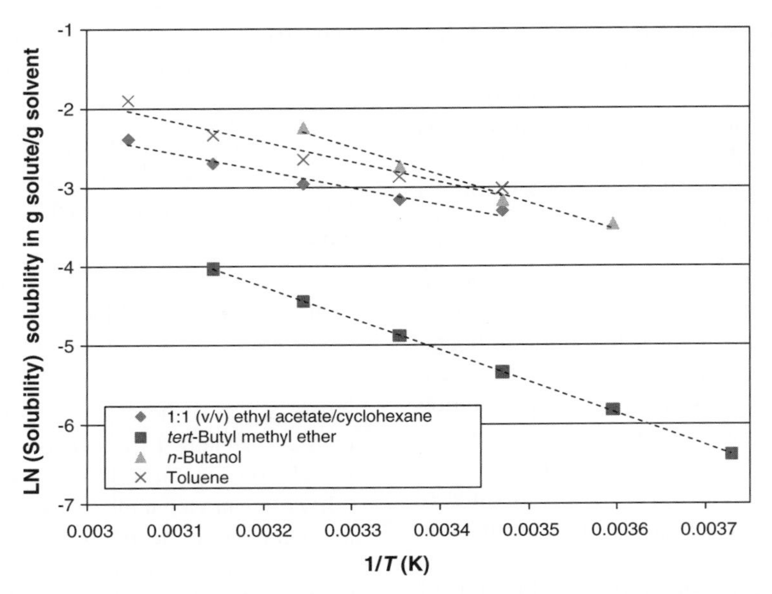

FIGURE 13.10 Solubility data from Figure 13.9 regressed using a simple van't Hoff relationship. For some solvents, the fit is very linear (e.g., *tert*-butyl methyl ether). Other solvents would benefit from an additional fitting parameter due to curvature (e.g., toluene) using the $C\ln(T)$ term, equation 13.7.

impurity and desired solute out of solution at the isolation temperature and composition, giving the potential product purity for the crystallization. The actual purification can be less than the calculated values due to impurity inclusion or occlusion in the product. The following example illustrates the use of solubility data to estimate the purification potential of a solvent.

EXAMPLE 13.3 USING SOLUBILITY TO PREDICT YIELD AND PURITY

(a) Given the solubility data in Table 13.2, calculate the maximum yield when the product is dissolved at 80°C and isolated at 10°C.

TABLE 13.2 API and Impurity Solubility Data for Example 13.3

Temperature (°C)	API Solubility (mg/mL solvent)	Impurity Solubility (mg/mL solvent)
0	1.8	0.9
10	3	1.2
20	6	1.7
30	10	2.2
40	15	2.7
50	22	3.3
60	33	4.1
70	50	5.0
80	78	6.2
90	120	7.7

(b) Given the additional data for the impurity, generate a graph of the yield and purity of the product versus dissolution temperature when isolating at 10°C with an input purity of 96%, assuming the impurity does not impact the API solubility.

Solution

(a) The yield can be calculated by performing a mass balance on the API in the solution phase. Equation 13.8 can be modified to equation 13.9, because the mass of solvent is the same at the beginning and end of the crystallization, and assuming the density of solvent does not change across the crystallization:

$$Y = \frac{S_{\text{solute 1}} - S_{\text{solute 2}}}{S_{\text{solute 1}}} \tag{13.9}$$

Therefore, the yield for a saturated solution at 80°C isolated at 10°C can be calculated as:

$$Y_{\text{API}} = \frac{S_{\text{API}}(80°\text{C}) - S_{\text{API}}(10°\text{C})}{S_{\text{API}}(80°\text{C})} = \frac{78-3}{78} = 96.2\%$$

The yield of API for each dissolution temperature can be calculated using equation 13.9. However, the level of impurity will always start at 4% of the initial API concentration, not the solubility limit. Therefore, the yield of impurity can be calculated as (Figure 13.11):

$$\text{Purity} = \frac{m_{\text{API}} Y_{\text{API}}}{m_{\text{API}} Y_{\text{API}} + m_{\text{impurity}} Y_{\text{impurity}}} = \frac{Y_{\text{API}}}{Y_{\text{API}} + (1 - \text{Purity}_{\text{input}}) Y_{\text{impurity}}} \tag{13.10}$$

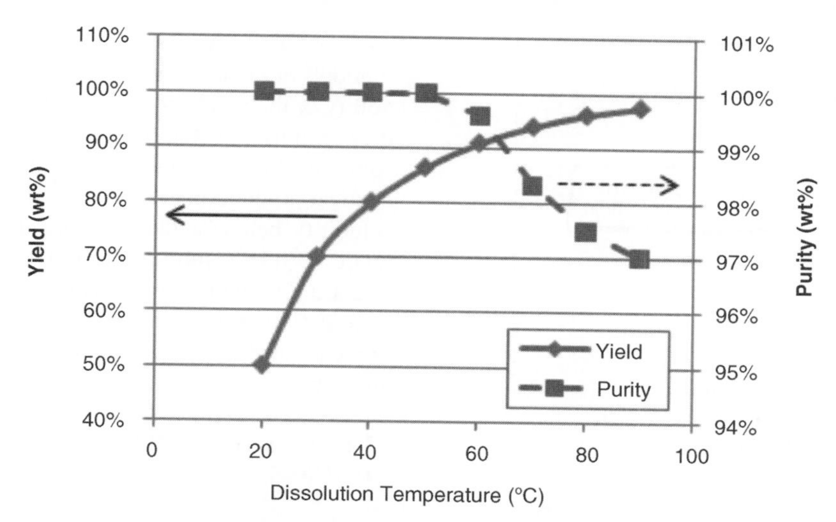

FIGURE 13.11 Illustration of the trade-off between API purity and yield varying the dissolution temperature (and solvent amount) with a constant isolation temperature. Details for the plot are described in Example 13.3.

An example of this calculation follows using the results from the first part, dissolution at 80°C and isolation at 10°C.

$$Y_{\text{impurity}} = \frac{0.04 \cdot S_{\text{API}}(80°\text{C}) - S_{\text{impurity}}(10°\text{C})}{0.04 \cdot S_{\text{API}}(80°\text{C})}$$

$$= \frac{0.04 \cdot 78 - 1.2}{0.04 \cdot 78} = 61.5\%$$

$$\text{Purity} = \frac{0.962}{0.962 + (1-0.96) \cdot 0.615} = 97.5\%$$

Solubility and related experiments are also essential in understanding the relative stability of crystalline forms. When slurrying solids at a constant composition and temperature, the crystalline form may stay the same or partially/fully convert to another form. If conversion occurs, the new form is more stable than the input form at that temperature and composition. The typical path for this type of process is shown in the scheme below for the formation of a hydrate from an anhydrate form:

$$\text{A(s)} + x\text{H}_2\text{O(l)} \Longleftrightarrow [\text{A} \cdot x\text{H}_2\text{O}](\text{l}) \Longleftrightarrow \text{A} \cdot x\text{H}_2\text{O(s)}$$

The anhydrate will initially dissolve, forming a dissolved and hydrated solute molecule. This hydrated species then spontaneously crystallizes as the more stable form, and the newly formed crystals continue to grow due to the higher solubility of the anhydrate relative to the hydrate. As the processes for spontaneous crystallization and growth can be slow, these experiments often take days or even weeks to achieve equilibrium. The process can be accelerated by performing the slurry experiment with both forms present initially or through minor fluctuations in temperature (e.g., ±5°C).

The "simple" phase diagram that can be constructed to understand the critical solvent activity or concentration required to facilitate a form change is illustrated in Figure 13.12a [19]. Such studies would also be conducted as a function of temperature to give a full view of the phase diagram.

An example of the use of solubility to understand the relative stability of two anhydrous forms is shown in Figure 13.12b. In Figure 13.12b, the two forms exhibit an enantiotropic relationship and a "crossover" temperature where the forms change in stability order. The "crossover" temperature is easily estimated through extrapolation of solubility data.

The solubility of a compound can be dramatically affected by the presence of impurities or residual solvents. When measuring the solubility for crystallization design purposes, it is recommended that the first measurement be made on relatively "pure" materials (>98%) in pure solvents. This gives a baseline understanding of behavior in the absence of nonidealities. Then, measurements of representative materials in representative process solvents should be taken. The values in actual systems should be used for forward design purposes; the differences between "ideal" and actual systems can often be narrowed through an adjustment of the actual system (e.g., removal of an impurity or better control of small quantities of undesired solvent).

Nonidealities caused by compositional differences can often be used advantageously. A specific instance involves the use of water as a cosolvent for poorly soluble intermediates and APIs. Water, when present as a minor component of a solvent system, often has a dramatic solubility enhancement

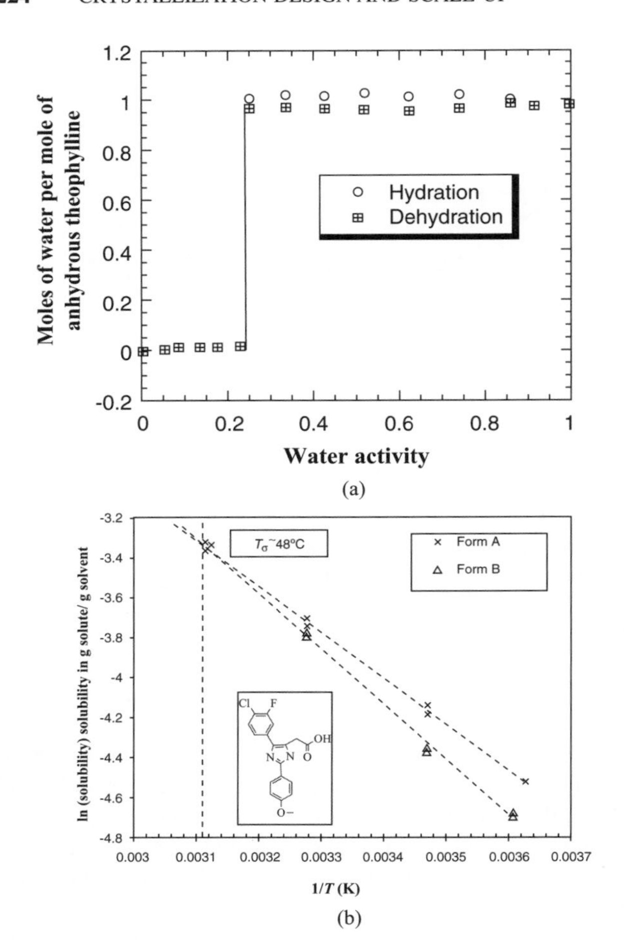

FIGURE 13.12 The use of equilibration and solubility experiments to understand form stability relationships. (a) Hydrate/anhydrate phase diagram that shows a critical water activity of ~0.25 needed to achieve full hydration. Below this activity, the anhydrate will be the most stable polymorph. The data was collected over 5 days starting from both 100% hydrate in one case (squares) and 100% anhydrate in the other case (circles) [19]. Reprinted with permission from Ref. 19. Copyright (1996) with permission from Elsevier.
(b) An enantiotropic system of an API. The solubility of Form B is less than the solubility of Form A at temperatures up to approximately 48°C (i.e., the crossover temperature), at which point Form A exhibits a lower solubility and becomes more stable.

effect that can be used advantageously, especially for compounds which exhibit poor solubility in neat solvents. This behavior is illustrated in Figure 13.13, in which a pharmaceutical compound exhibits a solubility maxima near 10% water content (158 mg API/g solvent), while exhibiting a low solubility in both neat solvent (28 mg API/g solvent) and pure water (0.1 mg API/g solvent).

Solvent evaluation using solubility data is essential to achieving many of the objectives for a crystallization: yield, throughput, and environmental impact (e.g., hazard of solvent and amount of solvent required). In addition, solubility data and assessment are the primary determinant of the "mode" of crystallization chosen. Without changes in solubility or the ability to increase solute concentrations above the equilibrium solubility, material will not crystallize from solutions. Frequently used crystallization modes in pharmaceutical production are shown in Table 13.3, along with the solubility behavior that typically leads to the use of each mode. While solubility is important to the initial mode selection, an understanding of kinetics, as described in the next section, is important in defining the parameters required to meet other crystallization objectives, such as particle size and crystalline form.

13.4 CRYSTALLIZATION KINETICS AND PROCESS SELECTION

Solubility, like any thermodynamic relationship, provides a start point and end point for a process. Knowledge of crystallization kinetics is critical in determining the path through which the beginning and end point are linked. For chemical reactions, kinetics are used to indicate the rate of change of molecular species; for crystallizations, kinetics are used to indicate the rate of solute mass transfer from solution phase to a solid phase. While solubility is often a primary control for achieving purification and separation objectives, the kinetic mechanisms of a crystallization are often the primary determinant for physical properties. The discussion below is a simplified description of crystallization kinetics, which are more comprehensively described in several references [1, 3].

Following the preliminary selection of solvent(s) from solubility data, the kinetics of the system must be understood in order to choose the conditions under which the crystallization operates and to validate the choice of solvent(s). If the chosen solvent system presents significant challenges related to kinetics that prevent the crystallization from achieving its design objectives, a new system is often sought.

Essential to understanding the kinetics of crystallization is the concept of supersaturation, which is the driving force for common crystallization mechanisms. Common expressions of supersaturation are shown by the equations presented below, where solubility and concentration units are adjusted for consistency:

$$\text{Supersaturation:} \quad \sigma_1 = C_{\text{solute}} - S_{\text{solute}} \tag{13.11}$$

$$\text{Supersaturation ratio:} \quad \sigma_2 = \frac{C_{\text{solute}}}{S_{\text{solute}}} \tag{13.12}$$

$$\text{Relative supersaturation:} \quad \sigma_3 = \frac{C_{\text{solute}} - S_{\text{solute}}}{S_{\text{solute}}} = \sigma_2 - 1 \tag{13.13}$$

In the equations above, C_{solute} is the actual concentration of solute at a temperature or composition condition, S_{solute} is

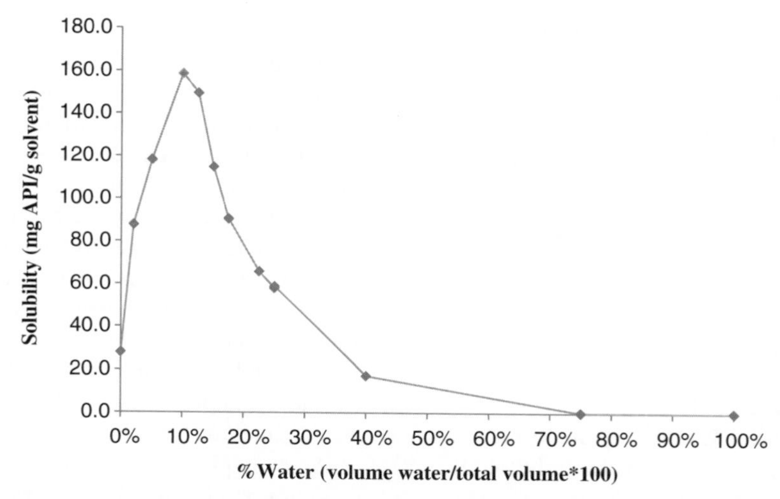

FIGURE 13.13 Solubility enhancement of an API using water as a cosolvent. Water often enhances solubility of pharmaceuticals at moderate concentrations. This particular example uses ethanol as a cosolvent. There is no form change across this water concentration range.

the equilibrium solubility at the same condition, σ_1 is the supersaturation in absolute terms (concentration), σ_2 is referred to as the supersaturation ratio, and σ_3 is referred to as the relative supersaturation. All three terms are frequently used in the analysis of crystallization processes.

Crystal mass formation can be achieved by either nucleation or growth. Nucleation can be described as the formation of new crystals from a solution or slurry, while growth can be defined as the deposition of solute mass on existing crystals of that solute. Nucleation can further be divided into two mechanisms: "primary" nucleation, which is the formation of new crystals from solutions devoid of crystals, and "secondary" nucleation, which is the formation of new crystals in the presence of existing crystals. Primary nucleation can occur within solutions (homogeneously) or at surfaces (heterogeneously), e.g., crystallizer walls and agitators.

Nucleation generates small crystals, which can be useful in preparing small particle size powders. However, nucleation can also lead to significant downstream processing problems such as long isolation times, significant agglomeration leading to poor performance in a formulation, and batch-to-batch variability.

TABLE 13.3 Common Crystallization Modes and the Influence of Solubility Behavior on the Selection of Modes

Crystallization Mode	Description	Solubility Behavior Leading to Mode Selection
Cooling	Crystallization is achieved by cooling solvent from a high temperature to a low temperature at constant solvent composition. Temperature is used to reduce solubility	Compound is soluble in a solvent at an elevated temperature below the normal boiling point of the solvent (e.g., >100 mg/g solvent), but relatively insoluble at a lower temperature (e.g., <20 mg/g solvent)
Antisolvent	Crystallization is achieved by adding an antisolvent to a solvent in which the solute is soluble. Composition is used to reduce solubility	Cooling crystallizations cannot achieve yield constraints (e.g., >90%) at reasonable dilutions (e.g., <20 L solvent/kg compound). The addition of an equal volume of antisolvent to a solvent reduces the solubility of a compound by more than 50%
Reactive	Crystallization is achieved by changing the compound ionically or structurally through reaction. The reactants are often soluble with the product being insoluble. Reaction is used to change the concentration of the product above the solubility limit	The product is completely insoluble in all potential solvents, and the precursors are readily soluble
Evaporative	Crystallization is achieved by the evaporation of solvent that increases the solute concentration above the solubility limit	Often used in combination with a cooling crystallization. For instance, if a cooling crystallization without evaporation can come close to meeting yield requirements, further concentration through distillation will allow additional mass to be recovered

Crystal growth is used to increase the size of the product, reduce batch-to-batch variability, and overcome downstream processing and handling issues [20]. It is also used to control the crystalline form of the compound being prepared, as multiple forms have the potential to simultaneously nucleate. As a result of the benefits afforded by crystal growth, it is generally preferred as the dominant mechanism in crystallization design, especially for controlling physical properties of materials.

13.4.1 Nucleation Kinetics and the Metastable Limit

When solutions are supersaturated, they are thermodynamically unstable. Like chemical reactions that do not react spontaneously, a thermodynamically unstable solution does not necessarily crystallize spontaneously as an energy barrier must be overcome to form a surface, analogous to the activation energy associated with a chemical reaction. Solutions that are supersaturated but do not spontaneously crystallize are referred to as "metastable." It is quite common for pharmaceutical intermediates and APIs to form metastable solutions at supersaturation ratios between 1 and 1.20. Eventually (i.e., weeks to years), many "metastable" systems might nucleate, but over the timescales associated with processing (i.e., minutes to hours) nucleation typically does not occur. A solute is said to be at its metastable limit when it is at the maximum supersaturation at which primary nucleation does not spontaneously occur.

After solubility, the metastable limit is the next critical measurement in crystallization design. It is measured by two primary methods, which are described in detail in Table 13.4.

An example of metastable limit determination and data reduction is provided by Example 13.4 for a cooling crystallization and an antisolvent crystallization.

TABLE 13.4 Description of Metastable Limit Measurement Techniques

Method	Description
Cooling rate (this method is primarily applicable to cooling crystallizations but can also be applied to evaporative crystallizations)	1. Using solubility data, a saturated solution of compound in solvent is prepared at a temperature close to the maximum temperature of the proposed crystallization in a reactor equipped with a particle measurement device (e.g., turbidity and Lasentec® FBRM®). 2. The solution is cooled at a slow rate (i.e., ~ 0.1°C/min) until particles are observed by the measurement device. The temperature at which crystallization is observed is recorded. 3. The experiment is repeated several times at faster cooling rates (e.g., 0.25, 0.5, 0.75, and 1°C/min). 4. A graph is prepared plotting crystallization temperature as a function of cooling rate. This plot should be linear; a linear fit will indicate the nucleation temperature at a 0°C/min cooling rate. 5. The supersaturation at the 0°C/min cooling rate is the metastable limit at this temperature. This limit is often expressed as a temperature difference between the saturation temperature and the temperature estimated at the 0°C/min cooling rate. 6. The experiment can be repeated at lower concentrations. It is best to get additional data at a concentration near the isolation condition. If there is little difference in the metastable limit in supersaturation terms relative to the high concentration point, additional data are not necessary. 7. It is recommended to perform duplicate experiments in the same and different equipment to understand the potential error and variability associated with the measurement. 8. For screening purposes, the observed crystallization temperature at the lowest cooling rate (i.e., 0.1–0.25°C/min) can be approximated as a 0°C/min cooling rate, and the metastable limit estimated from a single measurement.
Nucleation induction time method (this method is applicable to all crystallization types).	1. Saturated solutions of compound are prepared at conditions (temperature and composition) anticipated to be near the starting point of the crystallization. A particle detection probe is inserted (e.g., turbidity or Lasentec® FBRM®). 2. Supersaturation is generated by one of the following methods: (a) cooling the solution as rapidly as possible to a temperature at which the solution is supersaturated; (b) adding nonsolvent to a solvent composition at which the solution is supersaturated; (c) performing a partial reaction to generate supersaturation; or (d) rapidly evaporating a fraction of the solvent. 3. After the rapid generation of supersaturation in step 2, the solution is held isothermally until a particle detection device indicates that particles have formed. The time (i.e., nucleation induction time) to particle formation is recorded. 4. The experiment is performed at multiple conditions (e.g., different temperatures, different amounts of antisolvent added, different amounts of reactants used). 5. The nucleation induction time is plotted as a function of supersaturation. An asymptote will be observed. The supersaturation value at this asymptote is the metastable limit. 6. It is recommended to perform duplicate experiments in the same and different equipment to understand the potential error and variability associated with the measurement.

EXAMPLE 13.4 ESTIMATION OF METASTABLE ZONE WIDTH BY COOLING AND NUCLEATION INDUCTION TIME METHODS

Compound A is to be crystallized through a cooling crystallization in neat ethyl acetate. Compound B is to be crystallized using the addition of *n*-heptane (antisolvent) to a solution of tetrahydrofuran (THF, solvent). For Compound A, the "cooling rate" method is applied, and for Compound B, the nucleation induction time method is used. Here are the data for both studies.

Compound A: A solution of compound A is saturated in ethyl acetate at 65°C. The solubility follows a van't Hoff relationship, with $A_S = -3269.2$ and $B_S = 8.13$ (S_{solute} is in units of g solute/g solvent). The solution is heated to 70°C, and cooled at 0.25°C/min. The crystallization temperature, recorded by turbidity, is 50.4°C. After crystallization, the solution is reheated to 70°C to achieve dissolution, and cooled at 0.5°C/min, with a crystallization temperature of 48.9°C. The procedure is repeated at 0.67°C/min and twice at 1°C/min with crystallization temperatures of 47.3°C, 45.9°C, and 45.4°C, respectively. Estimate the metastable limit of the compound. Report results as supersaturation in units of g solute/g solvent.

Compound B: A solution of compound B is saturated in THF at 20°C. To the THF, different amounts of *n*-heptane (the antisolvent) are added, and the time to crystallization is noted. Relevant data are reported in Table 13.5. Estimate the metastable limit and report as relative supersaturation.

Solution

Compound A: With the data provided, a plot can be made of crystallization temperature as a function of cooling rate. This data can be linearly fit and extrapolated to a 0°C/min cooling rate, illustrated in Figure 13.14a. The extrapolation predicts a crystallization temperature of approximately 52°C at 0°C/min. Estimating the solubility at saturation (65°C) and at a 0°C/min cooling rate (52°C) using equation 13.6 gives

$$S_{solute}(65°C) = e^{\frac{-3269.2}{273.15+65} + 8.13} = 0.215 \text{ g/g solvent}$$

$$S_{solute}(52°C) = e^{\frac{-3269.2}{273.15+52} + 8.13} = 0.146 \text{ g/g solvent}$$

TABLE 13.5 Nucleation Induction Time Data for Compound B

Heptane Volume Fraction	Induction Time (min)	Solute Concentration (g/g solvent)	Solubility (g/g solvent)
0.055	180	0.104	0.0186
0.11	60	0.097	0.0097
0.22	15	0.085	0.0034
0.44	1	0.069	0.0009

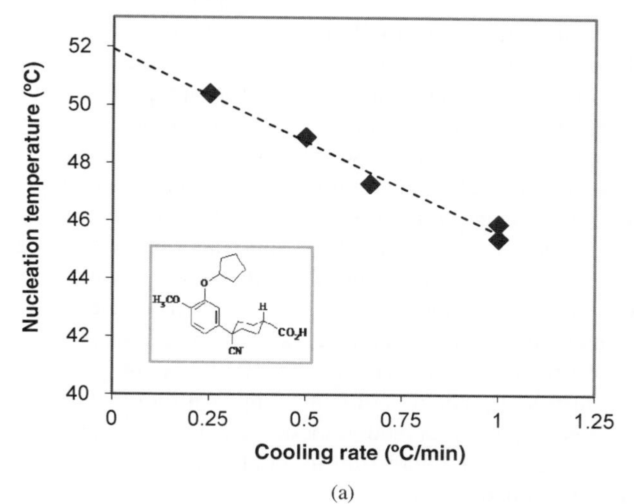

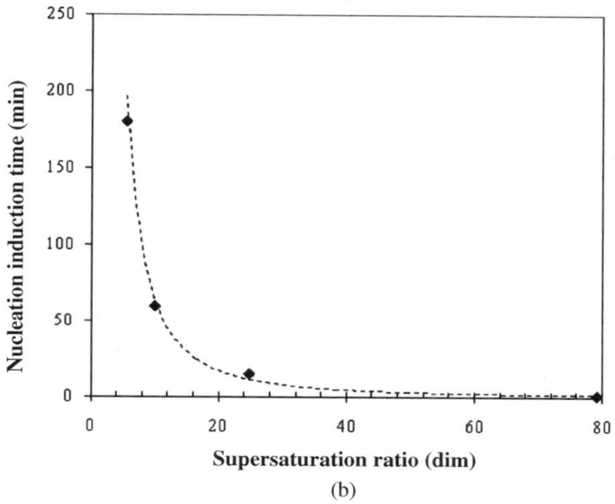

FIGURE 13.14 Metastable zone width measurement using the (a) cooling method for compound A and (b) nucleation induction time method for Compound B. Details are included in Example 13.4.

Then the supersaturation can be calculated for the metastable limit

$$\sigma_1 = 0.215 - 0.146 = 0.069 \text{ g/g solvent}$$

$$\sigma_2 = \frac{0.215}{0.146} = 1.47$$

Compound B: Induction time is plotted as a function of the supersaturation ratio, Figure 13.14b, as calculated from the data in Table 13.5. An asymptote is estimated from a power law fit to the data, giving a superaturation ratio σ_2 of approximately 5. As a result, the relative supersaturation σ_3 is approximately 4.

Of the two methods for estimating metastable zone width, the nucleation induction time method is often the most general, as it is broadly applicable to all crystallization modes. In addition, the nucleation induction time method gives an indication of the time window that is available to allow the addition and growth of seed materials and mix antisolvents/reactants with the main solution; as a result, it is necessary in understanding the scale-up implications of crystallizations that are described later in this chapter.

While the metastable limit is useful in understanding the conditions under which primary nucleation occurs, the potential for secondary nucleation must also be considered. Secondary nucleation occurs through several mechanisms, with the most common mechanism being contact nucleation. Contact nucleation is microattrition of crystals resulting in small crystalline fragments (i.e., $<10\,\mu m$) being present in the slurry [21]. The rate of contact nucleation is influenced by crystal–crystal, crystal–impeller, and crystal–wall collisions. A common expression for contact nucleation is indicated by equation 13.14:

$$B = k_N M^j N^k {\sigma_1}^b \tag{13.14}$$

where B is the nucleation rate (number per volume per time), k_N is the nucleation rate constant, M is the suspension density in number (mass per volume), and N is the agitation rate (a frequency or velocity). The variable b is the primary nucleation order and j and k are secondary nucleation orders. Secondary nucleation can occur either within or outside the metastable limit. A simplified version of equation 13.14 (j and $k = 0$) is used to represent primary nucleation kinetics. A thorough treatment of nucleation is provided by Kashchiev in Ref. 22.

13.4.2 Growth Kinetics

Crystal growth theory and mechanisms have been well described [1–3]. For the practicing engineer working on design and scale-up issues, this discussion is simplified below into the content that is typically required to analyze commonly used pharmaceutical crystallization modes.

Like nucleation, crystal growth is driven by a supersaturation driving force, with equation 13.15 illustrating a commonly used rate expression:

$$\frac{dm}{dt} = k_{GM} A_c \sigma_1^g. \tag{13.15}$$

In the equation above, k_{GM} is a temperature dependent growth rate constant, A_c is the surface area of crystals present in solution, g is the order of growth, and m is the mass of the solid solute. Equation 13.15 is a semiempirical equation merging the serial processes of diffusion of solute to the surface of a crystal and surface integration of solute onto the crystal. Therefore, the parameters k_{GM} and g may also be functions of mixing, as discussed in 13.6. Growth occurs either on material that has already been generated by nucleation or on material that has been purposefully added to the solution. Purposefully added material is referred to as "seed." Seeding is frequently employed in pharmaceutical crystallization processes to control crystalline form and physical properties, especially particle size. Seed material may be prepared from an alternative processing method, such as milling, or may be taken from one batch of material and added to a subsequent batch.

If a system exhibits growth as the primary mechanism for crystal mass formation, the amount and size of seed added control the particle size of the product. A common expression used to interpret the relationship of seed amount to particle size for batch or semi-batch crystallizations is expressed by the proportionality shown below:

$$\frac{m_s}{m_p} \propto \left(\frac{d_s}{d_p}\right)^n \tag{13.16}$$

In the proportionality, m_s represents the mass of seed, m_p represents the mass of product, and d_s and d_p represent sizes of the seed and product, respectively. Typically, d_{50} or d_{90} values from laser light diffraction measurement or sieve measurement are used for the size terms. The exponent term n is related to the habit of the crystal. For a perfectly spherical crystal, the exponent n would be equal to 3 and the proportionality would be an equality. In practice, the relationship between seed amount and particle size at the end of the crystallization is obtained by an empirical regression of the seed response curve (prepared by running the process at several seed loading (i.e., m_s/m_p) values and trending versus product size for several seed sizes). The preparation of seed response curves and subsequent analysis are listed in Table 13.6.

After regression of the seed response curve, the amount and size of seed required to deliver a desired particle size can be interpolated from the resultant correlations. When conducting these experiments, it is important that supersaturations remain well within the metastable limit, so that growth is the dominant mechanism. An example of a seed response curve is illustrated in Figure 13.15.

In performing seed response curve experiments, the kinetics of crystal growth can also be measured. Growth kinetics are valuable in crystallization design, as they can be used to determine the required rates of cooling, antisolvent addition, distillation, or reagent addition for the corresponding crystallization mode. Many methods for the determination of crystal growth kinetics can be used. A useful method in batch systems is through the measurement of solute concentration as a function of time following the addition of seed to a supersaturated solution. Varying the seed amount changes the "area" term in equation 13.15, allowing the determination of the growth rate constant. The area term is often indicated using a "specific surface area" measurement determined through nitrogen adsorption methods, or by understanding the size distribution of a material and the shape factor through which area of a material can be related to its mass or characteristic length. Measurement of the

TABLE 13.6 Method for Generation of Seed Response Curves

Step	Description of Step
1	Two to three different types of seed are generated. These may be generated by the following methods: (1) taking existing crystals from the as-is crystallization; (2) taking prepared crystals and performing a particle size reduction step, such as milling, micronization, or sonication (can also be done using a mortar and pestle or a blender); (3) crystallizing material in a different way (e.g., use of other solvents or other modes of crystallization). Each material is measured for particle size by an appropriate technique. The seed loading and seed sizes employed should vary by at least 1 order of magnitude
2	At least three experiments are performed with each seed at different seed loadings. The experiments are performed by seeding within the metastable limit allowing the solution to fully desupersaturate. For the remainder of the crystallization, supersaturation is generated very slowly until the crystallization has reached its completion (i.e., slow cooling rate or antisolvent addition rate). Supersaturation and particle size are monitored to ensure that the mechanism is growth throughout the crystallization
3	The product is isolated and weighed, giving the exact m_s/m_p value. The product is then measured for particle size
4	A seed response curve plots the size of the product versus the seed loading for each different seed size. The particle size as a function of seed loading can be fit through a variety of empirical functions

growth rate constant as a function of temperature enables a growth model over the entire path of a crystallization.

Figure 13.16 displays a typical crystal growth kinetic experiment and associated data fit. For this particular case, an online method (measurement of concentration by reflectance infrared spectroscopy) was used to measure solute concentration. The solute concentration data was used to calculate supersaturation at each time point. The initial conditions for the solution of equation 13.15 were $t = 0$, $C = 0.094$ g solute/g solvent, with the solubility at the experimental temperature (70°C) being 0.074 g solute/g solvent (1 L of solvent used with a density of 800 kg/m^3). A growth rate order of 1 was assumed. The data was adequately fit through numerical integration using a simple finite differences method and a sum of square residual minimization to estimate a value for $k_{GM} \cdot A_c$ of 1.6×10^{-6} m^3/s. With a seed area of 0.116 m^2, $k_{GM} = 1.4 \times 10^{-5}$ m/s. Due to the relatively small change in mass over the experiment, the value of A_c was approximated to be constant during this experiment. A more rigorous solution would relate mass change to area change through the use of a shape factor and a characteristic seed length, or alternatively an initial rates method would be used with a constant area approximation. Crystallization growth rate orders are often low, with values between 0.5 and 1. In the experience of the authors, a large number of pharmaceutical systems are modeled adequately when assuming first order in supersaturation (i.e., $g = 1$).

13.4.3 Controlling and Determining Crystallization Mechanisms

Figure 13.17 summarizes the kinetic discussions above by displaying regions of concentration in which different mechanisms are likely to occur for either cooling or nonsolvent crystallizations. To maximize the opportunity for crystal growth, operation in close proximity to the solubility is preferred. The nearer the solution concentration is to the metastable limit, the higher the likelihood of secondary nucleation, with primary nucleation possible at concentrations beyond the metastable limit. The principles in Figure 13.17 are also applicable to evaporative and reactive crystallizations, but in these situations supersaturation is typically generated by changes in concentration rather than solubility.

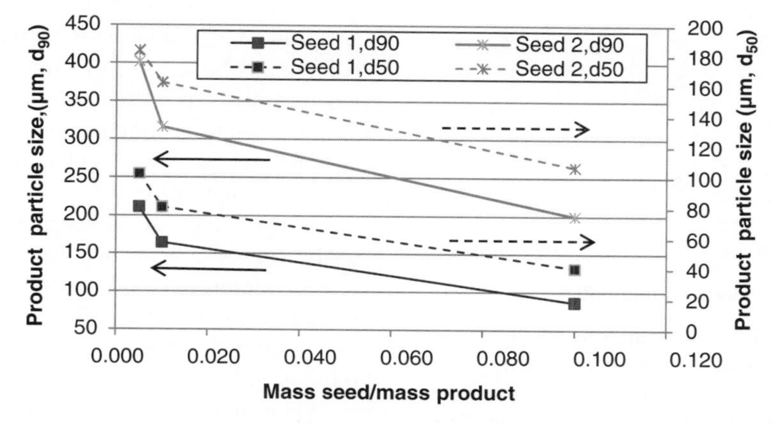

FIGURE 13.15 Typical seed response curve plot. Seed 1 in this case has a d_{90} value of 6.1 μm and a d_{50} of 2.2 μm, while Seed 2 has a d_{90} of 11.5 μm and a d_{50} of 4.4 μm. Both d_{90} and d_{50} values exhibit the anticipated response.

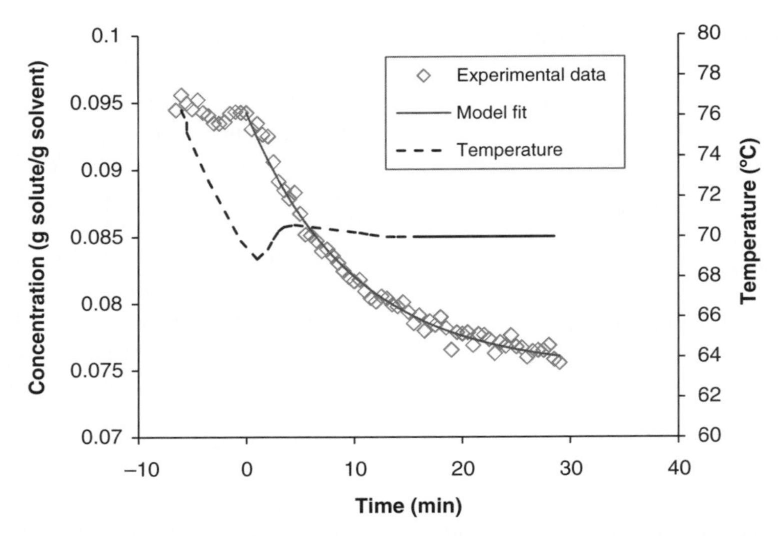

FIGURE 13.16 Crystal growth kinetics from concentration data. Crystallization seeded at time 0. Details are provided in the text. A finite differences approach was used to solve for the growth rate constant, $k_{GM} = 1.4 \times 10^{-5}$ m/s, with seed surface area $A_c = 0.116\,m^2$ and solvent volume $V_{solvent} = 1$ L.

Crystallization mechanisms are inferred from experimental data, especially microscopy and solute concentration data. From a design perspective, conditions are sought that provide the balance of mechanisms required to deliver the objectives of the crystallization. In Figure 13.18, successive micrographs are taken after seeding a pharmaceutical product within the metastable zone. As observed in the figure, the seed materials grow successively larger over time, with no new "fine" crystals observed. This is the type of behavior representative of a growth dominated system and indicates that the crystallization is operating sufficiently close to the solubility curve to minimize secondary nucleation [23].

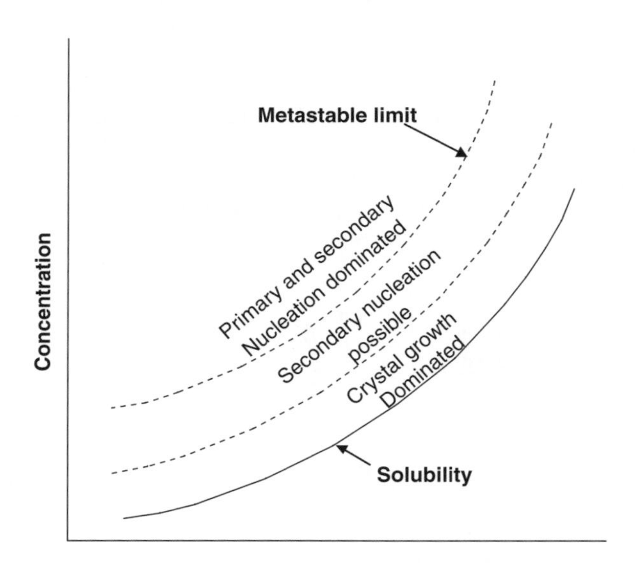

FIGURE 13.17 Mechanisms of crystallization and their relationship to solute concentration. Primary nucleation dominates when the solution is supersaturated beyond the metastable limit. Secondary nucleation can occur within the metastable zone, typically at supersaturations near the metastable limit. Growth frequently is the dominant mechanisms at relatively low supersaturations (i.e., concentration is close to the solubility).

In Figure 13.19, micrographs are overlaid with solute concentration data for a system that exhibits both growth and secondary nucleation. The solution was supersaturated within the metastable limit before adding seeds. Some crystal growth is observed immediately after seeding. After a growth period where seed material has clearly been enlarged through growth, a change is seen in the slope of the solute concentration curve, indicating a mechanism change from growth to nucleation. Upon further desupersaturation, the crystals have not grown larger and the habit has slightly changed from a columnar habit to more of a thin plate habit while maintaining the same crystalline form. The change in the rate of desupersaturation combined with the change in crystal habit and the lack of crystal enlargement are clear evidence of secondary nucleation.

Of particular use in mechanism detection are process analytical instruments that detect particulate matter, such as in-line microscopy or Lasentec® focused beam reflectance measurement (FBRM®). Overviews of the utility of FBRM® in crystallization mechanism inference can be found in other references [24].

When considering a design for a crystallization that favors growth, a recommended starting point for a design is to seed at a solution concentration no more than midway between the solubility curve and the metastable limit, and

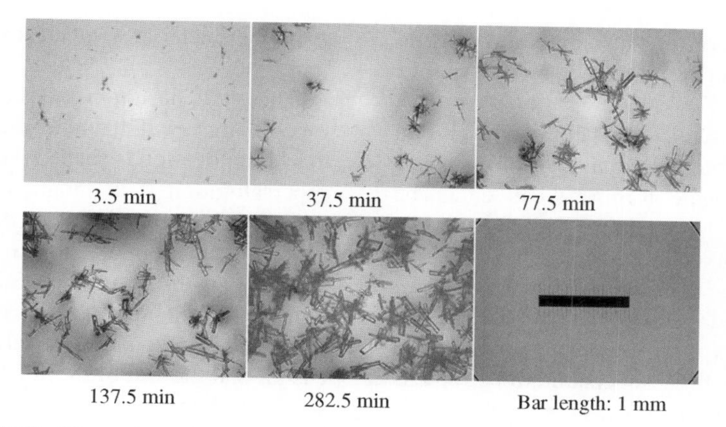

FIGURE 13.18 Illustration of crystal growth mechanism observed by optical microscopy. The process is seeded at time 0 and held isothermally for 282.5 min. Fine particles disappear as coarse particles appear, which grow larger throughout the duration of the experiment with no evidence of fines reappearance [18].

to maintain this minimum proximity to the solubility curve for the remainder of the crystallization process. Initial experiments are performed using microscopy, concentration, and online particle detection methods to ensure the selected conditions are indeed producing the desired mechanism. Conditions are then verified at larger scale to ensure that changes in mixing or equipment geometry and materials of construction have not altered the mechanistic behavior. Acicular habits, in particular, are highly prone to contact nucleation as they can easily be broken. Changes in scale can change the rate of secondary nucleation due to changes in agitation.

In addition to growth and nucleation, oiling, aggregation, and agglomeration are also potential phenomena in crystallization. In the case of oiling, supersaturation is typically in great excess of the metastable limit, or the solution is sufficiently concentrated with nucleation inhibiting impurities. As a result the solute forms a liquid phase consisting of a solvent–solute concentrate rather than a stable crystalline form. Oils are metastable and may crystallize spontaneously with sufficient holding times. Oiling is often prevented through seeding or by nucleation at low supersaturation; if this approach is unsuccessful, impurities must be individually

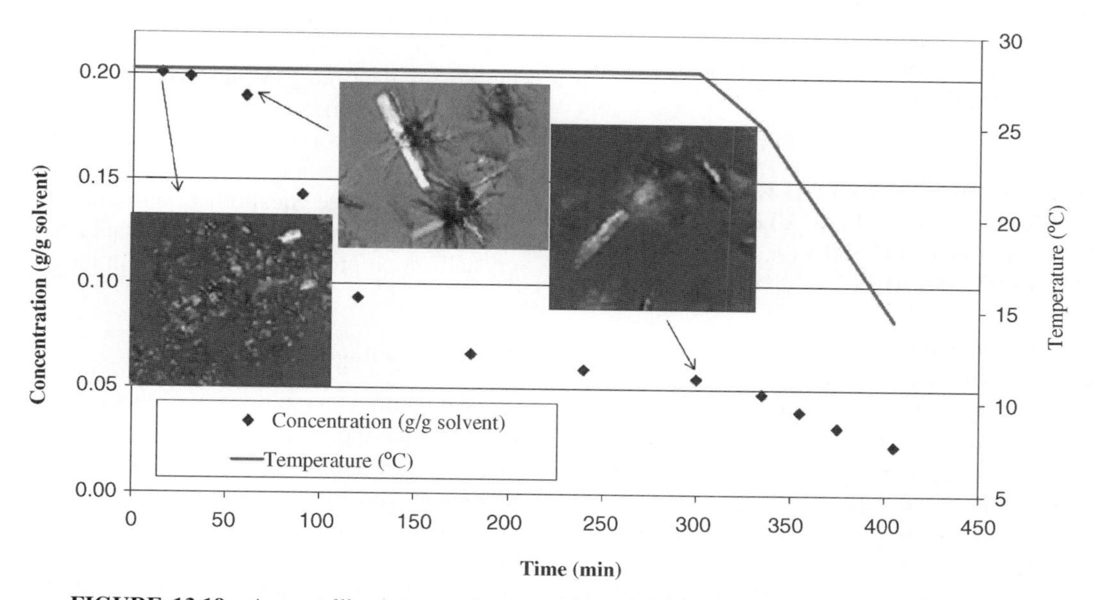

FIGURE 13.19 A crystallization experiment exhibiting both growth and secondary nucleation mechanisms. Initially, seeds grow as indicated by the presence of large crystals shortly after seeding. With additional time, a change in slope during desupersaturation is observed, indicating a change in mechanisms to secondary nucleation.

investigated for their inhibition of nucleation and growth, with the problematic impurity removed by alternative means.

Aggregation occurs when two crystals collide in solution and adhere to each other through favorable surface-surface interactions. Once an aggregate forms, it can either be "deaggregated" with crystals regaining their individual identity (this typically occurs through fluid shear), or the crystals may fuse together due to growth which links the two surfaces. When crystals fuse together, the result is called an agglomerate. Aggregation is often severe in processes in which nucleation occurs at high supersaturations, and is common for acicular habits or for crystals with high specific surface areas. Severe aggregation is often manifested as an immobile slurry that does not mix well and under certain conditions can form "shelves" of solid material on internal reactor equipment. As aggregation and agglomeration can affect processability, physical properties and mixing scale-up, the approach is often taken to minimize supersaturation to reduce the driving force for aggregate and agglomerate formation. When formed, agglomerates often cause bimodal size distributions and batch-to-batch variability in products. Because agglomeration is difficult to scale-up and control, it is often avoided as a selected mechanism.

In certain instances, aggregation or agglomeration are purposefully attempted to prepare a material that behaves like a small particle from a pharmaceutics perspective (e.g., rapid dissolution in a dosage form) but behaves like a larger particle from a manufacturability perspective (e.g., short filtration times). Examples of the favorable use of aggregates or agglomerates include spherical crystallizations [25]. The design of a spherical crystallization process is highly dependent on the properties of a molecule, and often must be established on a trial and error basis.

13.5 UNDERSTANDING CRYSTALLIZATION RATE PROCESSES: THE APPLICATION OF SOLUBILITY AND KINETICS DATA TO CRYSTALLIZATION MODES

The kinetic and thermodynamic principles described in the previous sections can be applied to common crystallization modes to complete a design. For most situations, it is recommended to pursue a design where growth is the dominant mechanism, as such processes are simpler to reproducibly scale-up from a heat and mass transfer perspective. To enable a growth basis for design, seeding is employed to provide the initial area required for growth. This is especially the case for active ingredients where control of particle size and crystalline form are key objectives. When tight control over physical properties is less of an issue, a preferred design approach may be to nucleate at a consistent, small supersaturation (i.e., just outside of the metastable limit), and then grow the nuclei at a supersaturation within the metastable limit.

These two design methods are illustrated in Figure 13.20 for common crystallization modes. In Figure 13.20b, d, and f, the crystallization is seeded within the metastable limit and supersaturation is maintained within the metastable limit until the crystallization has completed. This approach provides the best opportunity to maximize the potential for crystal growth. In Figure 13.20a, c, and e, nucleation is induced by increasing the supersaturation beyond the metastable limit, but supersaturation is then controlled to ensure that the rest of the crystallization occurs close to the solubility line.

A common error in crystallization design and scale-up is the generation of supersaturation to levels where nucleation is a dominant mechanism. This is caused by a lack of understanding of the timescales over which controlling process parameters need to be changed. This section addresses simple approaches to estimating the timescales for the controlling parameters in each major crystallization mode:

- Rate of temperature change (cooling crystallizations)
- Rate of addition of antisolvent (antisolvent crystallizations)
- Rate of reaction through control of temperature or addition of reactants (reactive crystallizations)
- Rate of solvent removal (evaporative crystallizations)

Only simple methods for the estimation of timescales and rate processes for batch or semi-batch crystallizations are described. There are many published examples of more rigorous approaches to solving similar problems, particularly the work of the Bratz [26] and Rawlings [27] research groups, which are recommended for further study. In particular, more detailed modeling approaches using partial differential equations and the method of moments as a solution of the population balance equation are encouraged. More detailed information regarding the use of population balances to model crystallization processes is provided in the literature [28].

13.5.1 Cooling Crystallization

Seeded, cooling crystallizations represent perhaps the simplest design approach for achieving consistent crystallizations while minimizing scale-up challenges. The design challenge is to balance the crystal growth rate with the rate of supersaturation generation. In order to understand the required rate of change of temperature to meet this objective, a knowledge of crystal growth kinetics for the seed material is required (Figure 13.16).

To estimate the rate of temperature change for a cooling crystallization design, the crystal growth rate is balanced with the rate of supersaturation generation through cooling

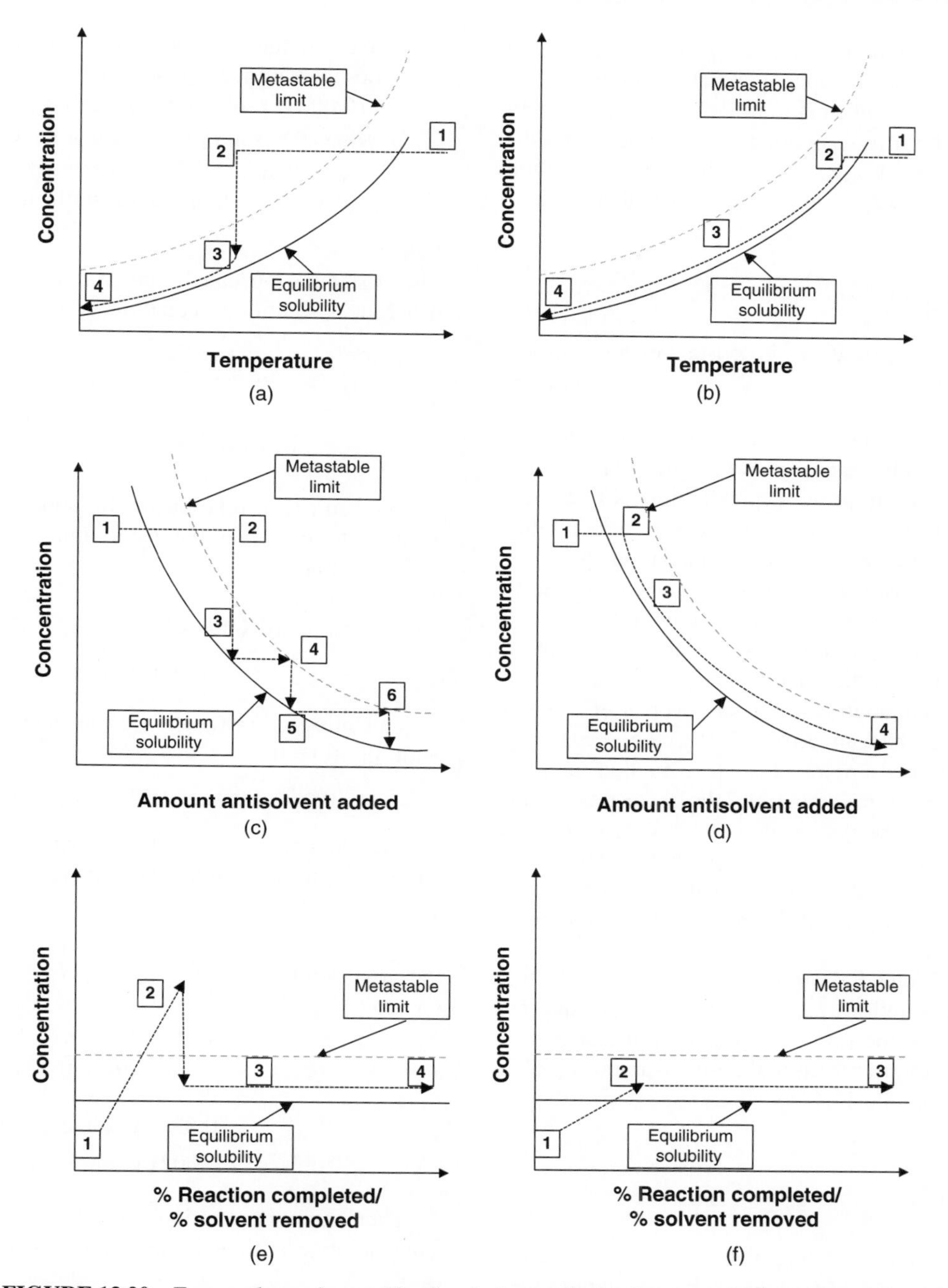

FIGURE 13.20 Frequently used crystallization design modes in pharmaceutical processes. (a) and (b) represent cooling crystallizations, (c) and (d) represent crystallizations induced by antisolvent addition, and (e) and (f) represent both reactive and evaporative crystallizations, which can be similar in terms of behavior. (a), (c), and (e) represent situations where initial crystal mass is generated by primary nucleation, with growth occurring after the nucleation event through controlled generation of supersaturation. The amount of nucleation is dependent on how much supersaturation is generated prior to crystallization. (b), (d), and (f) represent seeded crystallizations, where seed is added in the metastable zone. In each plot, [1] represents the starting point of the crystallization, where all material is dissolved [2] represents the point at which crystallization mass is first generated, either by seeding or by nucleation. The highest number on each plot represents the point at which the crystallization is complete.

such that the process remains within the metastable limit and near the solubility line. A simple approach to achieving this is through the evaluation of equation 13.17, where the growth rate is expressed both in terms of the solid phase (dm/dt) and the solution phase ($V_{solvent}(dC_{solute}/dt)$), and the solution phase concentration is constrained to a constant supersaturation, σ_1:

$$\frac{dm}{dt} = -V_{solvent}\frac{dC_{solute}}{dt} = k_{GM}A_c\sigma_1^g \cong -V_{solvent}\frac{\Delta C_{solute}}{\Delta T}\frac{\Delta T}{\Delta t}$$

where $C_{solute} = S_{solute} + \sigma_1$ and σ_1 is constant during cooling (13.17)

Through knowledge of the metastable limit and nucleation induction times, a supersaturation is first selected at which the material is desired to grow. Typically, this is a concentration within the metastable limit and close to the solubility line. The exact value selected is dependent on the compound and its propensity for secondary nucleation within the metastable limit. Care must be taken to not dissolve seed material due to process variability. For instance, if a solution is saturated at 70°C, and the selected seeding temperature is at 68.5°C, but the temperature probe has an error of 2°C and the solvent charge has an error of 1%, the seeds may dissolve. A commonly employed seeding method involves seeding at a supersaturation within the metastable limit but at a level where the typical processing errors that can cause seed dissolution are highly improbable. After seed growth desupersaturates the solution to an acceptable level, further supersaturation generation can be achieved through cooling.

Equation 13.17 and its variations can be solved with simple numerical solutions to determine both the time required for seed to desupersaturate a solution through growth and the cooling rates after the initial desupersaturation. This is illustrated in Example 13.5.

EXAMPLE 13.5

A cooling crystallization is being planned for an API in 1 L of solvent. The solubility (g solute/g solvent) of the compound in question can be modeled using a simple van't Hoff expression with $A_S = -3773.0$ and $B_S = 8.3930$. The starting concentration for the crystallization is 0.0952 g solute/g solvent, and the density of the solvent is 800 kg/m^3. A seeding temperature of 70°C is selected with the same area of seed as used in the kinetics experiment illustrated in Figure 13.16 (0.116 m^2). After the solution has reached a supersaturation $\sigma_1 = 0.002$ g solute/g solvent, the solution is cooled to 0°C at a constant supersaturation (0.002 g solute/g solvent). Using the kinetic constant and seed area from Figure 13.16, and assuming that $k_{GM} \cdot A_c$ is constant throughout the crystallization, answer the following questions:

1. What is the amount of time required for the seed to desupersaturate the solution to a supersaturation $\sigma_1 = 0.002$ g/g solvent at the seeding temperature, 70°C?
2. What is the minimum time it will take to cool the slurry to 0°C, maintaining a supersaturation $\sigma_1 = 0.002$ g/g solvent throughout the crystallization?

Solution

The first question can be answered using a simple finite differences solution to equation 13.17:

$$\frac{dm}{dt} = -V_{solvent}\frac{dC_{solute}}{dt}$$

$$= k_{GM}A_c\sigma_1^g \Rightarrow \Delta C_{solute} = -\frac{k_{GM}A_c\sigma_1^g}{V_{solvent}} \cdot \Delta t$$

The van't Hoff relationship for solubility, equation 13.6, is used to calculate the saturation concentration of the compound at 70°C:

$$S_{solute}(70°C) = e^{\frac{A_S}{T}+B_S} = e^{\frac{-3773.0}{70+273.15}+8.3930}$$

$$= 0.0741 \text{ g/g solvent}$$

This allows a calculation of supersaturation at the seed addition, time 0:

$$\sigma_1(0) = 0.0952 - 0.0741 = 0.0211 \text{ g/g solvent}$$

And the corresponding finite change in concentration at time 0, picking a suitable time step, here 1 min for illustration purposes:

$$\Delta C_{solute}(0\text{ min}) = -\frac{k_{GM}A_c}{V_{solvent}} \cdot \sigma_1^g \cdot \Delta t$$

$$= -\frac{1.40\times10^{-5}\text{ m/s}\cdot 0.116\text{ m}^2}{1.00\times10^{-3}\text{ m}^3}$$

$$\times (0.0211 \text{ g solute/g solvent})^1 \cdot 60\text{ s}$$

$$\Delta C_{solute}(0\text{ min}) = -0.0021 \text{ g/g solvent}$$

Then the concentration, supersaturation, and change in concentration at time 1 min can be calculated:

$$C_{solute}(1\text{ min}) = 0.0952 - 0.002 = 0.0930 \text{ g solute/g solvent}$$

$$\sigma_1(1\text{ min}) = 0.0930 - 0.0741 = 0.0189 \text{ g solute/g solvent}$$

$$\Delta C_{solute}(1\text{ min}) = -\frac{1.40\times10^{-5}\text{ m/s}\cdot 0.116\text{ m}^2}{1.00\times10^{-3}\text{ m}^3}$$

$$\times (0.0189 \text{ g solute/g solvent})^1 \cdot 60\text{ s}$$

$$= -0.00184 \text{ g solute/g solvent}$$

This sequence can be continued until reaching the desired supersaturation, $\sigma_1 = 0.002$ g/g solvent. This would most

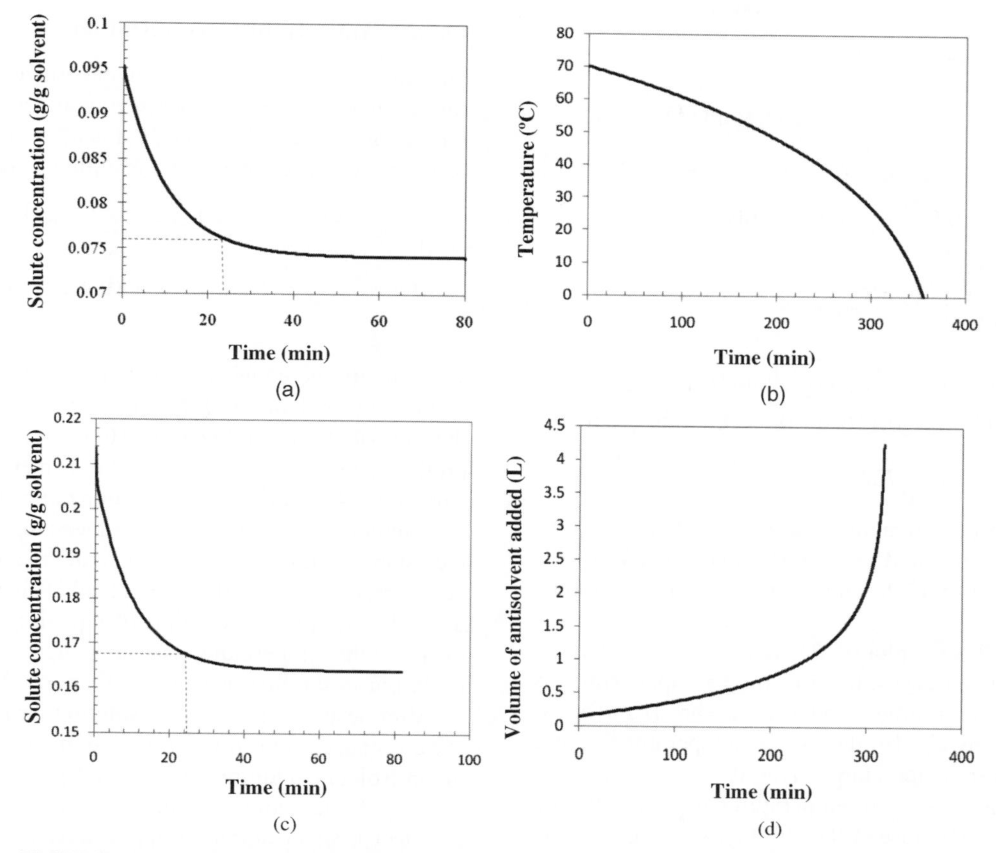

FIGURE 13.21 Calculated crystallization timescales and rate changes for a cooling crystallization (a) and (b) and an antisolvent addition (c) and (d). The cooling crystallization is described by Example 13.5, and the antisolvent example is described in the text. For (a) and (b), $k_{GM} \cdot A_c =$ constant $= 1.6 \times 10^{-6}$ m^3/s and $A_c = 0.116$ m^2. The solubility was represented by a simple van't Hoff correlation with $A = -3773.0$ and $B = 8.390$, and the solvent volume was 1 L. Seeding was performed at 70°C, with the criteria that a supersaturation, σ_1, of 0.002 g/g solvent be reached during isothermal seed growth (a), and with that value maintained during cooling (b). For (c) and (d), $k_{GM} \cdot A_c =$ constant $= 1 \times 10^{-5}$ m^3/s, with $A_c = 0.71$ m^2. The initial volume of methanol was 6 L, with 0.15 L water added to supersaturate. Seed was isothermally grown until a supersaturation of 0.004 g/g solvent was achieved (c), and then the balance of water, 4.1 L, was added isothermally while maintaining this supersaturation (d).

easily be completed in a spreadsheet, and smaller time steps would improve accuracy. The solution to this is illustrated in Figure 13.21a and the time required is approximately 24 min using a 0.2 min time step.

For the second question, the challenge is to crystallize at a constant supersaturation that constrains the growth rate to a constant value for the entire crystallization. Starting from the end of the last example, a finite differences approach is again applied. Using the starting point as a concentration of 0.0761 g/g solvent, and a constant growth rate as defined by $k_{GM} \cdot A_c \cdot \sigma_1$, the temperature is stepped in suitable ΔT increments to the end temperature, 0°C. The solubility is then calculated at each ΔT increment, and the supersaturation is added to the solubility at the ΔT increment to give the solution concentration at this time point, and a ΔC_{solute} from one temperature to the next. The time, Δt, is calculated for each ΔT interval by dividing ΔC_{solute} by $k_{GM} \cdot A_c \cdot \sigma_1 / V_{\text{solvent}}$. The time at each temperature is estimated by cumulatively adding the Δt values. This is illustrated below where $\Delta T = 1.0$°C:

$$T_1 = 70°\text{C};\; S_{\text{solute}}(70°\text{C}) = e^{\frac{A_S}{T} + B_S}$$

$$= e^{\frac{-3773.0}{70+273.15} + 8.3930} = 0.0741 \text{ g solute/g solvent}$$

$$C_{\text{solute } T_1} = S_{\text{solute } T_1} + \sigma_1 = 0.0741 + 0.002$$

$$= 0.0761 \text{ g solute/g solvent}$$

$$T_2 = 69°\text{C};\ S_{\text{solute}}(69°\text{C}) = e^{\frac{A_S}{T}+B_S}$$

$$= e^{\frac{-3773.0}{69+273.15}+8.3930} = 0.0718 \text{ g solute/g solvent}$$

$$C_{\text{solute } T_2} = S_{\text{solute } T_2} + \sigma_1 = 0.0718 + 0.002$$

$$= 0.0738 \text{ g solute/g solvent}$$

$$\Delta t_1 = \frac{[C_{\text{solute } T_1} - C_{\text{solute } T_2}]}{k_{\text{GM}} A_c \sigma_1} V_{\text{solvent}}$$

$$= \frac{(0.0761 - 0.0738)\text{ g solute/g solvent}}{1.40 \times 10^{-5}\text{ m/s} \cdot 0.116\text{ m}^2 \cdot 0.002\text{ g solute/g solvent}}$$

$$\cdot 1.00 \times 10^{-3}\text{m}^3 = 722\text{s}$$

For each subsequent temperature interval, a new time interval is then calculated in a similar manner and time is cumulatively added until the end crystallization temperature, 0°C, is reached.

From this analysis a plot of T versus t can be produced. The total calculated crystallization time is approximately 355 min, with the profile shown in Figure 13.21b. As observed in Figure 13.21b, the rate of temperature change increases at lower temperatures. This is because the solubility rate of change at high temperatures is greater than at low temperatures. Because of the constant $k_{\text{GM}} \cdot A_c$ assumption, this time is a conservative estimate. A more rigorous solution would update A_c as a function of the change in crystal mass, m, through the use of shape factors that relate the area of a crystal to its volume and a characteristic length.

In the event that seeding is not practical (e.g., for the crystallization of intermediates) an alternative approach is to cool to a temperature outside of the metastable limit, nucleate at a selected supersaturation, and then control the subsequent cooling in order to maximize crystal growth. The same methodology shown in Example 13.5 is applicable for the subsequent growth phase, except that the nucleated material must have an estimated area in order to apply a growth kinetic model.

An often used methodology in cooling crystallizations is the performance of a "ripening" step to increase particle size and improve downstream unit operations (e.g., isolation). Ripening involves reheating a slurry to partially dissolve some of the crystal mass (typically at least 10% of solids remain undissolved) and then recooling to promote growth on the remaining crystals (fines will typically dissolve first as dissolution rate is proportional to the surface area to volume ratio of a particle [6]). Ripening is frequently employed when nucleation is a significant mechanism due to slow crystal growth rates, and is a way to provide a large area for growth without initially adding large amounts of seed material (i.e., >10%).

13.5.2 Antisolvent Crystallization

The rate of antisolvent addition to maximize growth potential can be calculated through a similar approach to that described for a cooling crystallization. In this case, the volume is not constant but varies over time, resulting in equation 13.18:

$$\frac{dm}{dt} = -\frac{d\{(V_{\text{solvent}} + V_{\text{antisolvent}})C_{\text{solute}}\}}{dt} = k_{\text{GM}} A_c \sigma_1^g$$

$$\text{where } C_{\text{solute}} = S_{\text{solute}} + \sigma_1 \text{ during antisolventaddition} \tag{13.18}$$

To maintain the equality in equation 13.18, the antisolvent addition rate will often be slow initially and will increase near the end of the crystallization. However, the addition rate profile is highly dependent on the shape of the solubility curve [29]. As with a cooling crystallization, the first step is to add antisolvent to generate a supersaturation within the metastable limit, and then calculate the time needed to desupersaturate for a given seed load. This is accomplished through a numerical solution of equation 13.18, knowing $k_{\text{GM}} \cdot A_c$, the temperature, and the composition (the solution is similar to the first question in Example 13.5).

After desupersaturating the solution through seed growth at a constant composition, the next step is to calculate the rate of antisolvent addition. This can be achieved through a numerical integration of equation 13.18. It is first necessary to select a supersaturation appropriate to maintain crystal growth using metastable limit data. A simple finite difference approach involves starting at this supersaturation, σ_1, and calculating the growth rate, $k_{\text{GM}} \cdot A_c \cdot \sigma_1^g$. The volume is then incremented in small intervals, with the change in solute mass (Δm_{solute}) calculated over each volume interval ($V_{n+1} - V_n$) through the difference of the solubility times the mass of solvent across the volume interval:

$$\Delta m_{\text{solute}} = -\{(V_{\text{solvent}} + V_{\text{antisovlent}_{n+1}}) \cdot S_{\text{antisolvent}_{n+1}} - (V_{\text{solvent}} + V_{\text{antisovlent}_n}) \cdot S_{\text{antisolvent}_n}\}$$

The time required for each added increment of antisolvent, Δt, is calculated by dividing the change in solute mass by the calculated growth rate.

The results of antisolvent crystallization calculations are shown in Figure 13.21c and d for a system in which water is added to a methanol solution containing an API, assuming that $k_{\text{GM}} \cdot A_c$ is constant over the entire crystallization (i.e., 1.0×10^{-5} m^3/s) with a growth order of 1. For this particular scenario, the solvent is methanol and the antisolvent is water, and the solvents are assumed to mix ideally. The solubility of the binary system is described by an exponential function, $S_{\text{solute}} = 0.22 \exp(-12.04\, \phi_{\text{water}})$ g solute/g solvent, where ϕ_{water} is the water volume fraction. Initially, 1 kg of material is dissolved in 6 L of methanol, and 0.15 L of water is added to generate supersaturation at 20°C. The solution is then seeded and held until a supersaturation of 0.004 g/g solvent is achieved. Antisolvent is then added at a rate such that this supersaturation is maintained to achieve a total added volume of 4.25 L. The result is a seed hold time of approximately 24 min and a total addition time of approximately 317 min. The shape of the addition curve mirrors the solubility as a function of water volume fraction. As with the cooling crystallization, nucleation can also be used to initiate the crystallization, with growth rates calculated after estimating the area generated in the nucleation step.

13.5.3 Reactive Crystallization

In most cases, reactive crystallizations are of interest when the product of the reaction is sparingly soluble and the crystallization rate is slow compared with the reaction rate. For these systems, the challenge is to match the rate of product generation with the rate of crystal growth so that excess supersaturation is not generated. For a simple bimolecular reaction, where $\{A + B \rightarrow \text{Product}\}$, the following rate balance can be written:

$$\frac{\mathrm{d}m_{\text{product}}}{\mathrm{d}t} = \mathrm{MW}_{\text{product}} k_{\text{reaction}} [\mathrm{A}][\mathrm{B}] V_{\text{solvent}} = k_{\mathrm{GM}} A_{\mathrm{c}} \sigma_1^g \tag{13.19}$$

The rate of reaction for many pharmaceutical applications can be controlled by the rate of addition of reactant, resulting in the volume (V_{solvent}) being a function of addition rate and time. In the simplest instance, the reactant of interest is an ionizable compound that forms a salt with the pharmaceutical molecule, with the molecule then having a negligible solubility in the solvent (typically a solvent with low polarity). In this situation, the rate of reaction is essentially instantaneous with the addition of the reactant (B), and the left hand side of equation 13.19 becomes

$$\frac{dm_{\text{product}}}{\mathrm{d}t} = -\frac{d[V_{\mathrm{B,f}} C_{\mathrm{B,f}}]}{\mathrm{d}t} \frac{\mathrm{MW}_{\text{product}}}{\mathrm{MW}_B}$$

where f represents the feed solution in which reactant B is contained. This solution is similar to that shown for antisolvent crystallization.

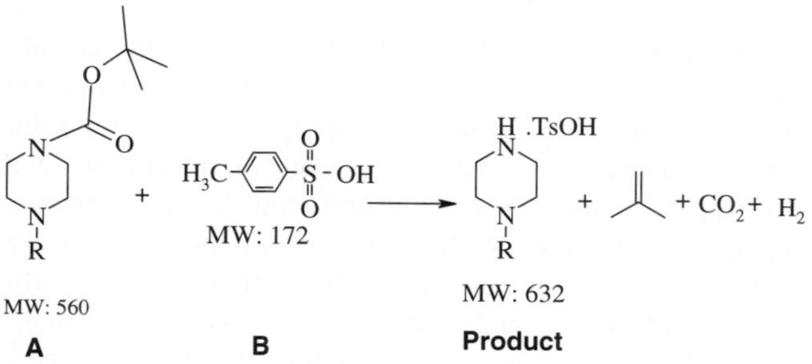

MW: 560 MW: 172 MW: 632

A B Product

FIGURE 13.22 Reaction scheme as an example of a reactive API crystallization process.

An example of a common pharmaceutical synthesis reaction is shown in Figure 13.22. This is a "Boc" deprotection reaction with an acid reactant. Upon deprotection, the resultant species is able to form a salt, which has limited solubility in the reaction solvent. A numerical solution to equation 13.19 is illustrated in Figure 13.23a for this example, which shows that the concentration of B must increase as a function of time to maintain constant supersaturation. This increase is required to maintain a constant reaction rate with [A] that is decreasing due to consumption. Conditions used are provided in Figure 13.23a.

A common problem in reactive crystallization is the presence of impurities in the final product at levels higher than predicted from solubility data. This can occur for two reasons. First, the impurity may be structurally similar to the desired molecule, and thus able to form some of the same bonding arrangements in the lattice. As a result, the impurity is integrated into the crystal. For reactive crystallizations, the reactant molecule is often similar in structure to the desired product, so integration is a distinct possibility. The second cause is referred to as inclusion, and occurs due to crystallization liquors being trapped into the crystal lattice as it forms or grows. Inclusion is more prone to occur when liquors are highly viscous or when crystal formation and growth kinetics are rapid. If impurities that are soluble in a wash solvent are not able to be effectively washed from surfaces, either inclusion or integration is the likely cause.

13.5.4 Evaporative Crystallization

In evaporative crystallization, the solubility is constant, so the rate of supersaturation generation is proportional to the rate of solvent mass removal ($\mathrm{d}m_{\text{solvent}}/\mathrm{d}t$)

$$\frac{\mathrm{d}m}{\mathrm{d}t} = S_{\text{solute}} \frac{\mathrm{d}m_{\text{solvent}}}{\mathrm{d}t} \cong S_{\text{solute}} \frac{U \cdot A_{\text{vessel}}(T_{\mathrm{j}} - T_{\mathrm{r}})}{\Delta H_{\mathrm{v}}} = k_{\mathrm{GM}} A_{\mathrm{c}} \sigma_1^g \tag{13.20}$$

In equation 13.20, U is the overall heat transfer coefficient of a batch reactor, A_{vessel} is the heat transfer area, and ΔH_{v} is the heat of vaporation of the solvent. The heat of crystallization is not included as it is often negligible relative to the heat transferred through vessel walls for batch crystallizations. A constant evaporation rate is necessary to maintain a constant crystal growth rate at a constant supersaturation and the evaporation rate is balanced with the crystal growth rate. The controlling variable in the determination of the evaporation rate is the jacket temperature of the reactor, T_{j}. As the volume decreases in the reactor, the effective heat transfer area decreases, thereby changing the temperature required on the jacket to maintain the constant evaporation rate. A sample calculation for T_{j} estimation during an evaporative crystallization is illustrated in Figure 13.23b. In practice, the T_{j} values shown in Figure 13.23b toward the

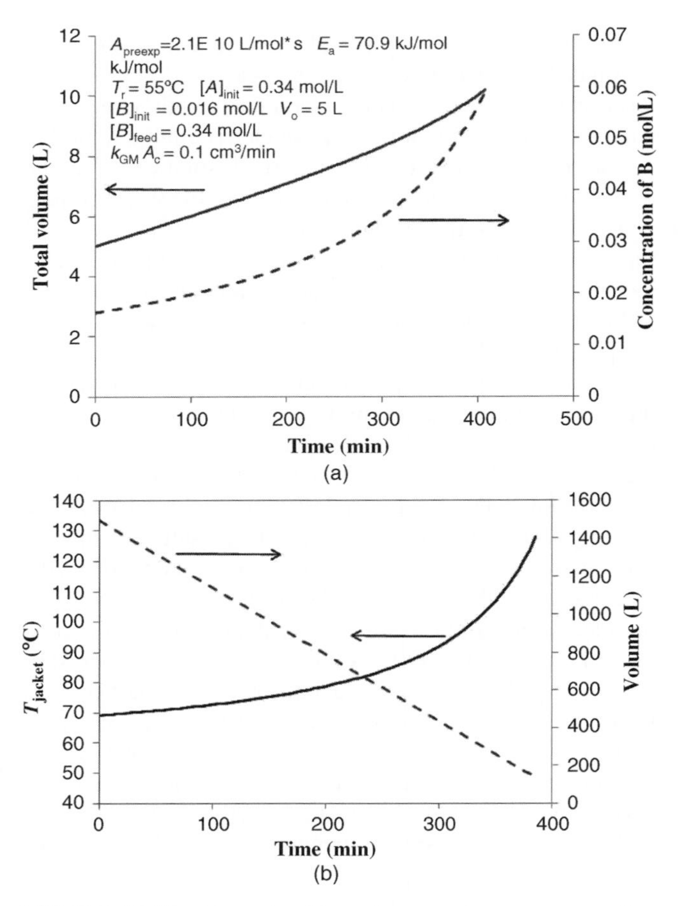

FIGURE 13.23 Examples illustrating constant supersaturation control for (a) reagent addition for a reactive crystallization and (b) jacket temperature for an evaporative crystallization. The reaction for (a) is shown in Figure 13.22, run isothermally at 55°C ($E_a = 70.9$ kJ/mol, $A_{preexp} = 2.1 \times 10^{10}$ L/mol s). Starting amount of compound A is 1 kg in 5 L solvent. Initially, the reaction was run to approximately 5% completion, through an addition of 5 mol% compound B, and a product concentration of 0.004 g/g solvent was achieved after seeding and seed growth. The balance of B was then added in a feed solution with a concentration of 0.336 mol/L. Compound B was added such that a supersaturation of 0.004 g/g solvent was maintained. $k_{GM} \cdot A_c = \text{constant} = 9.8 \times 10^{-6}$ m^3/s. For (b), the solvent was distilled from 1500 to 100 L; $k_{GM} \cdot A_c = \text{constant} = 2.9 \times 10^{-4}$ m^3/s. The vessel has a linear correlation between surface area and volume A_{vessel} (m^2) $= 0.00322 V_{solvent} + 0.53$. The overall heat transfer coefficient was 340 W/m^2 K, the solvent was methanol, and the supersaturation upon seeding was 0.005 g/g solvent; this was maintained throughout the crystallization.

end of the crystallization would not be used out of concern for material degradation. As a result, the crystallization would simply have a maximum allowed jacket temperature and proceed at a slower rate.

13.5.5 Continuous Crystallization

Continuous crystallization, though not historically used in pharmaceutical processing due to low production volumes and a batch manufacturing infrastructure, has recently gained attention as a way to reduce batch-to-batch variability and minimize solvent waste. A common "model" used for a continuous crystallizer is the mixed suspension, mixed product removal crystallizer, in which the same number of crystals entering the system leaves the system simultaneously. The crystals that leave the system will have grown commensurate with the growth rate, or nucleated if the system is sufficiently supersaturated. If, similar to the analysis above, the system is seeded immediately prior to entering the crystallizer, the resulting crystal growth can idealistically be modeled by the following combination of equations. Equation 13.21 relates crystal mass with its characteristic length [30]:

$$m = aL^3 \rho_M \tag{13.21}$$

where L is the characteristic crystal length, a is a proportionality constant relating the volume of a crystal to its characteristic length, and ρ_M is density of the crystal. Differentiating equation 13.21 and utilizing the growth rate structure introduced in equation 13.15, equation 13.22 provides a highly idealized approach to determine the growth of crystals:

$$\frac{dm}{dt} = 3aL^2 \rho_M \frac{dL}{dt} = k_{GM} A_c \sigma_1^g \tag{13.22}$$

Then over the average residence time of a particle in a continuous crystallizer, the growth can be approximated with equation 13.23:

$$\Delta L = \frac{k_{GM} A_c \sigma_1^g}{3aL^2 \rho_M} \tau \tag{13.23}$$

where τ is the mean residence time of a particle in the crystallizer, σ_1 represents the average supersaturation in the vessel, and ΔL represents the average change in the characteristic length for the particles across the vessel.

13.5.6 Statistical Experimental Design Methods

Factorial experimental designs have rightfully gathered substantial support recently in the pharmaceutical development field. As illustrated by the methodology provided for solubility measurement, seed response curve generation, and crystal growth rate estimation, a semi-empirical, mechanistic-based approach to crystallization design is preferred over a factorial design approach, due to the likelihood of crystallization mechanism change across wide experimental designs. Factorial designs are usefully applied to the robustness evaluation of process factors such as input quality, seed amount, and temperature variations. In robustness studies,

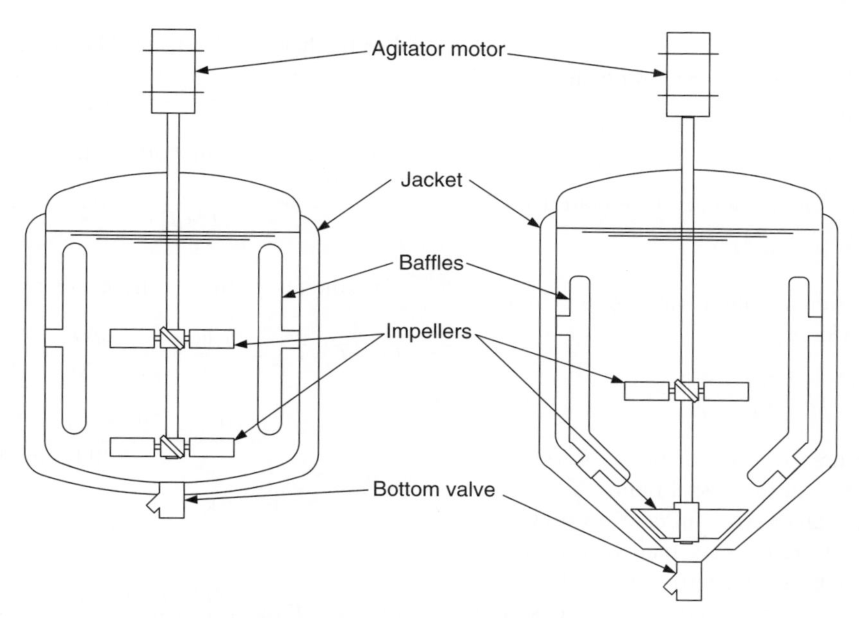

FIGURE 13.24 Schematics of typical vessel designs.

these variables are changed over intervals where the crystallization mechanism is likely to remain the same as the base design condition, so that the variance of a response to common scale-up perturbations can be well understood.

13.6 BATCH CRYSTALLIZATION SCALE-UP

In order to select appropriate scale-up parameters, crystallization scale-up involves understanding a system's relative sensitivities to heat and/or mass transfer [31]. Batch operation is the standard for much of the pharmaceutical industry, therefore stirred-tank reactors are the most commonly available equipment for crystallizations. Reactors are offered by many manufacturers and in several configurations. While they may be purpose-built for a process, they are usually built as multiuse reactors to offer the most flexibility for production. There are two common vessel designs for multipurpose batch reactors—both utilizing a cylindrical tank but having either a dish or conical bottom, as illustrated in Figure 13.24. Agitation designs, including impellers and baffles, vary widely. Independent of the reactor design, the two primary considerations for crystallization equipment scale-up are heat transfer and mixing.

13.6.1 Heat Transfer Considerations

Heat transfer for reactors is usually achieved by circulating heat transfer fluid (e.g., water, steam, or silicon-based fluids) through an external jacket or heating coils (Figure 13.25). External jackets are the most common for the multiuse reactors typically encountered, and for conventional vessel designs the reduced heating surface area to volume ratio on scale-up significantly reduces the heat transfer capacity, as illustrated by Example 13.6.

EXAMPLE 13.6 HEAT TRANSFER AREA

Approximate the relationship between the heat transfer area to volume ratio and the vessel radius, assuming the vessel is externally jacketed and cylindrical.

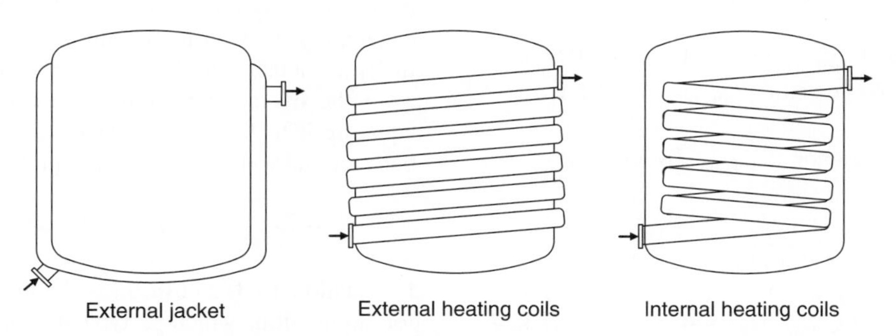

FIGURE 13.25 Schematics of common vessel heating arrangements.

Solution
The surface area of a cylinder is described by equation 13.24, neglecting the bottom surface:

$$SA = 2\pi rh \qquad (13.24)$$

The volume of a cylinder can be defined by equation 13.25:

$$V = \pi r^2 h \qquad (13.25)$$

As a result, the surface area to volume ratio for a cylinder is described by equation 13.26:

$$\frac{SA}{V} = \frac{2\pi rh}{\pi r^2 h} \propto \frac{1}{r} \qquad (13.26)$$

Generally, the surface area to volume ratio for vessels is proportional to the inverse of the vessel radius.

The obvious impact of an increase in scale is a slower rate for heating or cooling steps relative to small vessels, assuming similar overall heat transfer coefficients and $(T_j - T_r)$ values. This must be taken into account when designing crystallizations in the laboratory, so that unrealistic heating or cooling rates are not specified. For any specified cooling rate, the temperature difference (ΔT) between the vessel jacket and its contents $(T_j - T_r)$ will increase with vessel size to match the heating or cooling times specified in the small-scale design. One effect of this larger $T_j - T_r$ is that during cooling, the wall temperature can be much lower than experienced in the laboratory equipment, potentially leading to nucleation near the wall at lower bulk temperatures than expected. Conversely, during heating the wall can be much hotter at large scale than small scale. Solids attached to vessel walls will often be exposed to higher temperatures as a result, possibly affecting purity.

EXAMPLE 13.7 SOLUTION COOLING

What is the temperature difference between the solution and jacket necessary to achieve a cooling rate of 1 K/min for the following vessels? The solution is ethanol (MW = 46 g/mol; $\rho = 0.79$ g/mL; $C_p = 112$ J/mol K), the jackets each have an overall heat transfer coefficient $U = 200$ W/m^2 K, with no limitations from the jacket services.

(a) 1 L lab reactor: $V_{solvent} = 0.8$ L; $A_{vessel} = 0.038$ m^2
(b) 300 gal pilot plant reactor: $V_{solvent} = 900$ L; $A_{vessel} = 4.0$ m^2
(c) 3000 gal plant reactor: $V_{solvent} = 9000$ L; $A_{vessel} = 19$ m^2

Solution
For the solution, the required heat flow is

$$\dot{Q} = m_{solvent} \cdot C_p \cdot \frac{\Delta T}{\Delta t} \qquad (13.27)$$

For the heat flow between the jacket and solution:

$$\dot{Q} = U \cdot A_{vessel} \cdot (T_j - T_r) \qquad (13.28)$$

Therefore, the temperature difference is

$$(T_j - T_r) = \frac{m_{solvent} \cdot C_p}{U \cdot A_{vessel}} \cdot \frac{\Delta T}{\Delta t} \qquad (13.29)$$

And the temperature differences required are

$$\text{(a)}\ (T_j - T_r) = \frac{m_{solvent} \cdot C_p}{U \cdot A_{vessel}} \cdot \frac{\Delta T}{\Delta t} = \frac{(0.8\,\text{L} \cdot 790\,\text{g/L}) \cdot (112\,\text{J/mol K} \cdot 1\,\text{mol}/46\,\text{g})}{200\,\text{J/s m}^2\,\text{K} \cdot 0.038\,\text{m}^2} \cdot \frac{1\,\text{K}}{60\,\text{s}} = 3.4\,\text{K}$$

$$\text{(b)}\ (T_j - T_r) = \frac{(900\,\text{L} \cdot 790\,\text{g/L}) \cdot (112\,\text{J/mol K} \cdot 1\,\text{mol}/46\,\text{g})}{200\,\text{J/s m}^2\,\text{K} \cdot 4.0\,\text{m}^2} \cdot \frac{1\,\text{K}}{60\,\text{s}} = 36\,\text{K}$$

$$\text{(c)}\ (T_j - T_r) = \frac{(9000\,\text{L} \cdot 790\,\text{g/L}) \cdot (112\,\text{J/mol K} \cdot 1\,\text{mol}/46\,\text{g})}{200\,\text{J/s m}^2\,\text{K} \cdot 19\,\text{m}^2} \cdot \frac{1\,\text{K}}{60\,\text{s}} = 76\,\text{K}$$

What is the relationship between the temperature difference and the vessel radius, assuming it is cylindrical, externally jacketed, and neglecting the bottom surface? Building on the relationship shown in equation 13.26, it can be shown that the temperature difference between the jacket and solution is proportional to the vessel radius for a given heating or cooling rate, equation 13.30.

$$(T_j - T_r) \propto \frac{V}{A_{vessel}} \cdot \frac{\Delta T}{\Delta t} \propto r \cdot \frac{\Delta T}{\Delta t} \qquad (13.30)$$

Evaluating the jacket temperature may indicate potential issues for the scale-up of a process, but it is a conservative measure. A system can be evaluated in more detail by calculating the wall temperature, which the solution is actually in contact with. The wall temperature can be calculated from the overall heat transfer rate in equation 13.28 by equating it to the heat transfer rate across the liquid film between the wall and the bulk liquid:

$$U \cdot A_{vessel} \cdot (T_j - T_r) = h_i \cdot A_{vessel} \cdot (T_w - T_r) \qquad (13.31)$$

In equation 13.31 above, h_i is the heat transfer coefficient for liquid film, which is typically of the form shown in equation 13.32 for Newtonian liquids:

$$h_i = a \cdot \left(\frac{k}{T}\right)\left(\frac{C_p\mu}{k}\right)^{\frac{1}{3}}\left(\frac{D^2N\rho}{\mu}\right)^{b}\left(\frac{\mu}{\mu_w}\right)^{m} \qquad (13.32)$$

For equation 13.32, a, b, and m are empirical constants dependant on the mixing geometry, k is the thermal conductivity of the liquid, T is the tank diameter, C_p is the specific heat capacity of the liquid, μ is the liquid viscosity, D is the impeller diameter, N is the agitation rate, and ρ is the density of the liquid. All of the liquid physical properties are evaluated at the temperature of the bulk liquid, T_r, except μ_w which is evaluated at the wall temperature, T_w [32]. Since h_i is dependent on the wall temperature, it is common to iteratively solve for T_w, from an initial guess, until the calculated value from equation 13.31 equals the guess used for μ_w in equation 13.32.

Temperature control at larger scale involves a dynamic feedback loop to control the jacket services, often across several vessels. If tuning of the controller is inadequate, a temperature cycling of a couple degrees above and below a "constant" temperature set point can be observed. The mixing in the vessel may also lead to considerable temperature gradients through the vessel contents. These temperature fluctuations could promote ripening effects, which would not have been experienced on small scale.

Another primary issue associated with temperature is the thermal effect caused by added antisolvent or reactants, and the heat of crystallization. The effect of these energy terms is illustrated in equation 13.33 for a crystal growth only system with ideal mixing. For antisolvent additions, the dosing regimen (rate, temperature of addition) is designed into the crystallization often to minimize the $(T_{\text{antisolvent}} - T_r)$ term. Equation 13.33 can be used in conjunction with crystal growth kinetic equations to allow a more detailed process model to be constructed, and can be modified to incorporate additional energy balance terms (e.g., heat of reaction, heat of mixing, heat from nucleation processes) for reactive crystallizations and nucleation dominated crystallizations. In equation 13.33, m_r is the mass of reactor contents, ΔH_c is the enthalpy of crystallization, and $\dot{m}_{antisolvent}$ is the mass addition rate of antisolvent.

$$\frac{d\{m_r C_p T_r\}}{dt} = U \cdot A_{\text{vessel}}(T_j - T_r) + \Delta H_c k_{GM} A_c \sigma_1^g + C_{p,\text{antisolvent}}\,\dot{m}_{\text{antisolvent}}(T_{\text{antisolvent}} - T_r) \qquad (13.33)$$

13.6.2 Mixing Considerations

While the heat transfer considerations for crystallization scale-up are straightforward, those for mixing are increasingly complex, due to the wide variety of possible vessel configurations. Common impeller types include pitched-blade turbine, flat-blade turbine, curved-blade turbine, disc turbine, hydrofoil, retreat curve, propeller, and anchor, which are illustrated in Figure 13.26.

Mixing scale-up can be considered in three primary ways—geometrically, where the vessel and agitation configuration maintain the same shape and relative sizes, kinematically, where the relative velocities are maintained, and dynamically, where the relative forces are maintained [31]. The appropriate scale-up approach is dependent on the specific crystallization. The most common considerations for mixing scale-up for crystallization processes are provided in Table 13.7, including useful relationships for comparing geometric, kinematic, and dynamic similarities.

Vessel geometry (Figure 13.27) is a significant factor in mixing, but even when maintaining geometric similarity, on scale-up it is not possible to achieve both kinematic and dynamic similarity simultaneously, demonstrated illustratively in Figure 13.28.

Kinematic similarity is commonly compared using impeller tip speed or specific flow, but impeller, wall, or average

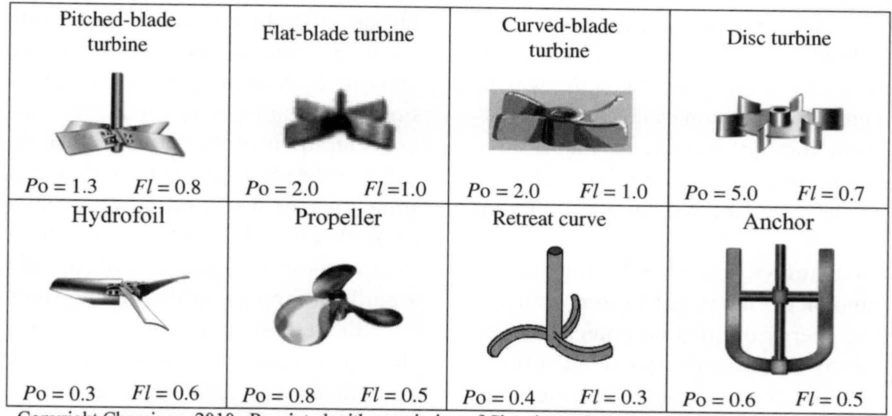

FIGURE 13.26 Common impeller types with typical power and flow numbers [17, 33–36].

TABLE 13.7 Useful Mixing Calculations [17, 33, 36]

Parameter	Formula	Description
Scale ratio (−)	$s = \frac{D_1}{D_2} = \frac{T_1}{T_2} = \cdots$	Ratio of geometric dimensions between two vessels (Figure 13.27)
Reynold's number (−)	$\text{Re} = \frac{\rho N D^2}{\mu}$	Ratio of inertial to viscous forces and describes flow regime: Laminar—$Re < \sim 10$ and $Po \propto Re^{-1}$ Transitional—$10 < Re < 10^4$ Turbulent—$Re > 10^4$; Po and $Fl =$ constant
Power number (−)	$Po = \frac{P}{\rho N^3 D^5}$	Characteristic impeller drag coefficient
Flow number (−)	$Fl = \frac{Q}{ND^3}$	Characteristic impeller discharge flow rate
Specific flow (1/s)	$\frac{Q}{V} = \frac{FlND^3}{V}$	Impeller discharge flow rate normalized by the fluid volume
Tip speed (m/s)	$v_T = \pi DN$	Impeller tip speed
Energy dissipation rate (W/m^3)	$\frac{P}{V} = \frac{Po \rho N^3 D^5}{V}$	Power dissipated by the impeller normalized by the fluid volume
Just suspension speed (rps)	$N_{js} = s\left(\frac{\mu}{\rho_l}\right)^{0.1}\left(\frac{g \cdot \Delta\rho}{\rho_l}\right)^{0.45} X^{0.13} d_p^{0.2} D^{-0.85}$	Minimum speed for complete suspension [37]
Mixing time (s)	$\theta_{\text{turbulent}} = C_1 \frac{T^{1.5} H^{0.5}}{Po^{1/3} N D^2}$	Time to achieve 95% homogeneity
	$\theta_{\text{transitional}} = C_2 \frac{T^{1.5} H^{0.5}}{Po^{2/3} \text{Re} N D^2}$	

D is the impeller diameter (m); T is the tank diameter (m); N is the impeller speed (1/s); ρ is the density (kg/m^3); μ is the viscosity (kg/m s); V is the fluid volume (m^3); s is the geometric constant; $\Delta\rho$ is the solid–liquid density difference (kg/m^3); X is the mass fraction solids (%); g is the gravitational acceleration (m^2/s); d_p is the particle diameter (m); H is the tank fill height (m); and C_1 and C_2 are the empirical constants (−); P is the impeller power (J/s).

shear rates can also be used. These can affect secondary nucleation during a crystallization process by causing crystal breakage, or attrition, through the crystal–impeller and crystal–wall impacts. Another kinematic consideration is solids suspension, typically evaluated with the just suspended agitation rate, or with cloud height prediction. Insufficient agitation could lead to stratification of the crystals, which could have the effect of growth dispersion by different crystal sizes residing in different mixing regimes of the vessel. Operationally, settling of product could cause issues transferring the slurry out of the vessel by blocking the bottom valve.

Dynamic similarity is commonly compared using energy dissipation rate, or the amount of energy transferred from the impeller to the fluid. The energy dissipation rate can affect the growth rate of crystals by changing the size of the diffuse boundary layer and, correspondingly, the mass transfer rate of bulk solute to the crystal surface. A lack of dynamic similarity could be manifested through a change in the crystal habit or aspect ratios, manipulating the independent crystal face growth rates. Additionally, secondary nucleation may be promoted if the energy dissipation rate is not sufficient to maintain the crystal growth rate, causing the system to maintain or increase its level of supersaturation. Depending on the mode of crystallization, another dynamic consideration for scale-up could be the mixing times, or degree of solution homogeneity. Crystallizations employing the addition of an antisolvent could be sensitive to the localized concentrations of the solvents since the solvent composition impacts the solubility and supersaturation. By extension, this principle can also apply to a reagent addition for a reactive crystallization. It should be noted that in these instances the addition point and geometry of the addition nozzle will have a significant impact on the mixing time.

This section covered conventional batch equipment and the most common considerations. However, specialized equipment and techniques can be utilized for modified crystallization mixing (e.g., to manipulate nucleation or attrition), including impinging jets, fluidized beds, sonication, and homogenizers [3].

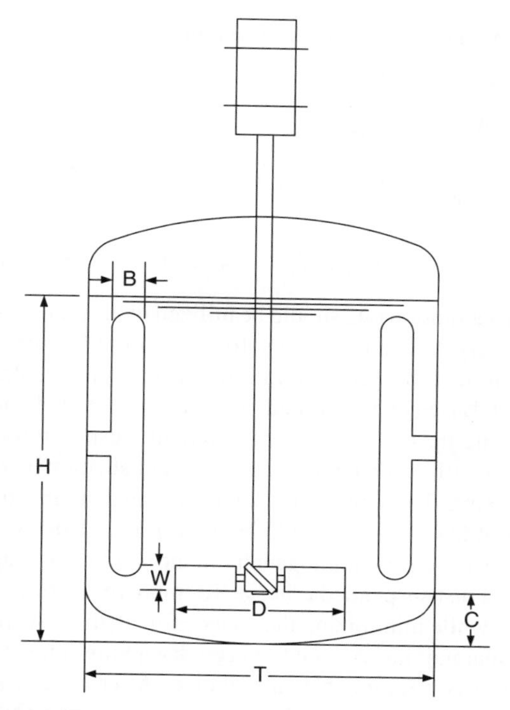

FIGURE 13.27 Illustration of geometric dimensions for an agitated vessel.

EXAMPLE 13.8 CRYSTALLIZATION MIXING

Given the product particle size distributions for a crystallization process evaluated at lab scale in Figure 13.29, what is the optimal agitation speed? And what are the likely explanations for the other two size distributions?

Solution

The narrowest particle size distribution was observed at 400 rpm, making that the most desirable of the three agitation rates. The lower agitation rate, 100 rpm, did not have sufficient suspension of the particles, leading to stratification and growth dispersion, evidenced by a broader distribution. The higher agitation rate, 700 rpm, led to attrition and breakage of the particles, evidenced by a bimodal distribution.

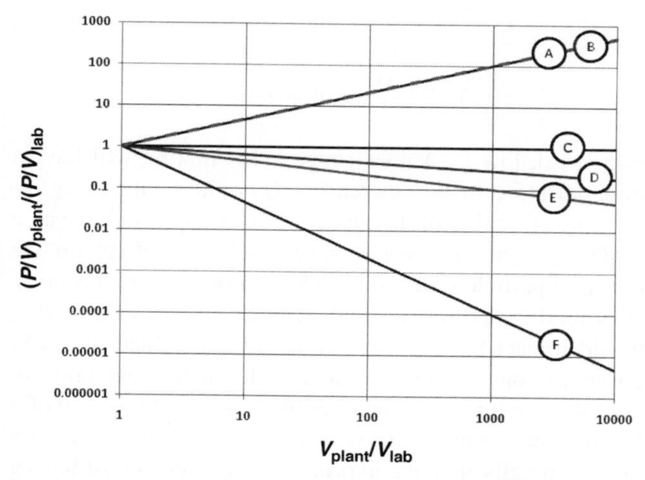

FIGURE 13.28 Power per unit volume scale-up comparison for constant: $A = Q/V$, $B = \theta_{turbulent}$, $C = P/V$, $D = N_{js}$, $E = v_T$, $F = Re$.

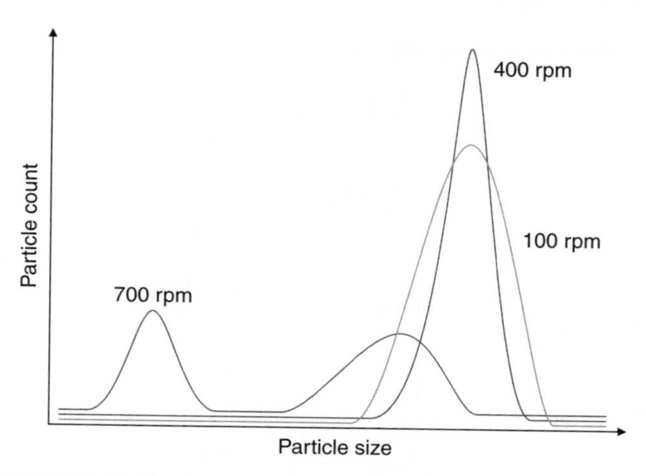

FIGURE 13.29 Lab-scale particle size distribution results. See Example 13.8 for additional details.

The lab impeller used to generate these results was 70 mm in diameter and the process is to be scaled to a plant vessel with a 700 mm diameter impeller. Assuming geometric similarity between the vessels and impellers, what agitation rate would be required to maintain the optimum conditions by

$$s = \frac{H_{lab}}{H_{plant}} = \frac{T_{lab}}{T_{plant}} = \frac{D_{lab}}{D_{plant}} = \frac{70\text{ mm}}{700\text{ mm}} = 0.1$$

$$V \approx \frac{\pi}{4}T^2H \Rightarrow \frac{V_{lab}}{V_{plant}} \approx s^3$$

(a) maintaining constant energy dissipation rate?

$$\left(\frac{P}{V}\right)_{lab} = \left(\frac{P}{V}\right)_{plant}$$

$$\frac{Po\rho N_{lab}^3 D_{lab}^5}{V_{lab}} = \frac{Po\rho N_{plant}^3 D_{plant}^5}{V_{plant}}$$

$$N_{plant}^3 = N_{lab}^3 \cdot \frac{D_{lab}^5}{D_{plant}^5} \cdot \frac{V_{plant}}{V_{lab}}$$

$$N_{plant} = N_{lab} \cdot \left(s^5 \frac{1}{s^3}\right)^{1/3}$$

$$N_{plant} = N_{lab} \cdot s^{2/3} = 200 \cdot 0.1^{2/3} = 86\text{ rpm}$$

(b) maintaining constant specific flow?

$$\left(\frac{Q}{V}\right)_{\text{lab}} = \left(\frac{Q}{V}\right)_{\text{plant}}$$

$$\frac{FlN_{\text{lab}}D_{\text{lab}}^3}{V_{\text{lab}}} = \frac{FlN_{\text{plant}}D_{\text{plant}}^3}{V_{\text{plant}}}$$

$$N_{\text{plant}} = N_{\text{lab}} \cdot \frac{D_{\text{lab}}^3}{D_{\text{plant}}^3} \cdot \frac{V_{\text{plant}}}{V_{\text{lab}}}$$

$$N_{\text{plant}} = N_{\text{lab}} \cdot s^3 \cdot \frac{1}{s^3}$$

$$N_{\text{plant}} = N_{\text{lab}} = 400\,\text{rpm}$$

(c) maintaining constant tip speed?

$$v_{T\text{lab}} = v_{T\text{plant}}$$

$$\pi D_{\text{lab}} N_{\text{lab}} = \pi D_{\text{plant}} N_{\text{plant}}$$

$$N_{\text{plant}} = N_{\text{lab}} \cdot \frac{D_{\text{lab}}}{D_{\text{plant}}}$$

$$N_{\text{plant}} = N_{\text{lab}} \cdot s = 400 \cdot 0.1 = 40\,\text{rpm}$$

(d) maintaining constant blend time?

$$\text{Re} = \frac{\rho N D^2}{\mu} > 10^4 \text{ for turbulent regime} \Rightarrow \frac{\rho}{\mu} > \frac{10^4}{ND^2}$$

$$\frac{\rho}{\mu} > 3 \times 10^5 \text{ for lab impeller}$$

$$\frac{\rho}{\mu} > 1 \times 10^5 \text{ for plant impeller assuming}$$

minimum rate of 10 rpm

$$\frac{\rho}{\mu} \approx \frac{1000}{0.001} = 1 \times 10^6 \text{ for common solvents,}$$

therefore the turbulent regime can be assumed

$$\theta_{\text{turbulent lab}} = \theta_{\text{turbulent plant}}$$

$$C_1 \frac{T_{\text{lab}}^{3/2} H_{\text{lab}}^{1/2}}{Po^{1/3} N_{\text{lab}} D_{\text{lab}}^2} = C_1 \frac{T_{\text{plant}}^{3/2} H_{\text{plant}}^{1/2}}{Po^{1/3} N_{\text{plant}} D_{\text{plant}}^2}$$

$$N_{\text{plant}} = N_{\text{lab}} \cdot \left(\frac{T_{\text{plant}}}{T_{\text{lab}}}\right)^{3/2} \cdot \left(\frac{H_{\text{plant}}}{H_{\text{lab}}}\right)^{1/2} \cdot \left(\frac{D_{\text{lab}}}{D_{\text{plant}}}\right)^2$$

$$N_{\text{plant}} = N_{\text{lab}} \cdot \left(\frac{1}{s}\right)^{3/2} \cdot \left(\frac{1}{s}\right)^{1/2} \cdot s^2$$

$$N_{\text{plant}} = N_{\text{lab}} = 400\,\text{rpm}$$

(e) maintaining solids suspension?

$$\frac{N_{\text{js plant}}}{N_{\text{js lab}}} = \frac{D_{\text{plant}}^{-0.85}}{D_{\text{lab}}^{-0.85}}$$

$$N_{\text{js plant}} = N_{\text{js lab}} \cdot \left(\frac{1}{s}\right)^{-0.85}$$

$$= N_{\text{js lab}} \cdot s^{0.85} = 400 \cdot 0.1^{0.85} = 57\,\text{rpm}$$

What agitation rate should be utilized in the plant to scale-up this crystallization? Even for this simplified example, selecting just the agitation rate for scale-up is not straightforward. Based on the limited information provided, the goal is to scale-up the process and maintain suspension of the particles with sufficient mixing, yet avoid attrition by excessive mixing. To maintain the solids suspension an agitation rate of at least 57 rpm would be appropriate. However, this rate does not ensure homogeneity in the solution as it applies only to a just suspended criteria. To improve mixture homogeneity while minimizing the chance for attrition, scaling up while maintaining constant energy dissipation rate (86 rpm on scale) is a reasonable first choice. After experience on scale, the approach would be revisited to confirm suitability.

13.6.3 Damköhler Numbers

A useful approach for evaluating crystallization scale-up and mixing sensitivity is the use of Damköhler numbers, *Da*[3]. The Damköhler number is a dimensionless group traditionally used to compare reaction timescales to other phenomena, such as mass transport or residence time. Depending on the mode of crystallization, applicable Damköhler numbers may include

$$Da_{\text{nucleation}} = \frac{\text{mixing time}}{\text{induction time}}$$

$$Da_{\text{growth}} = \frac{\text{mixing time}}{\text{crystal growth time}}$$

$$Da_{\text{reaction}} = \frac{\text{mixing time}}{\text{reaction time}}$$

In general, low *Da* values suggest that mixing will have a minimal effect, while increasing *Da* values suggesting increasing criticality of the mixing. For example, at low value of Da_{growth} mixing would have a minimal effect on the resulting particle size distribution. However, at high values, slow mixing and fast nucleation or crystal growth, mixing would impact the particle size distribution since localized concentrations would lead to variable nuclei generation or crystal growth rates throughout the solution. The value of a Damköhler number will depend on the specific definition used; generally though, an order of magnitude constitutes a low or high *Da* value (i.e., <0.1 and >10, respectively).

For reactive crystallization processes, Da_{reaction} has to be considered along with $Da_{\text{nucleation}}$ or Da_{growth} when analyzing the mixing criticality. For example, at high Da_{reaction} values, fast reaction relative to mixing, high-localized concentrations of product could be generated. Combined with a high $Da_{\text{nucleation}}$ or Da_{growth}, the fast nucleation or growth rates would make mixing critical to the product. However, if combined with a low $Da_{\text{nucleation}}$ or Da_{growth}, product generated by the reaction may be distributed evenly throughout the vessel before affecting nucleation or crystal growth. Alternatively, at low Da_{reaction} values, the reactants may be evenly distributed throughout the vessel before reaction, leading to a uniform product concentration and the mixing sensitivity solely due to the nucleation or growth.

The discussion above provides the concepts for utilizing Damköhler numbers to evaluate crystallizations, but the application does not have to be limited to these approaches. For example, one could also compare reaction rate and crystal growth rate directly for another Damköhler number applicable to reactive crystallizations.

13.6.4 Cooling Crystallization Considerations

Cooling crystallizations are commonly employed processes and the most typical considerations for their scale-up are as follows:

- Maintaining a cooling profile
- Managing the wall temperature
- Mixing that can impact both heat and mass transfer

As illustrated in the example regarding cooling a solution, scaling up a cooling profile for a batch process in typical equipment requires considerably larger temperature differentials when compared to lab scale. At scale, the wall temperature will be somewhat cooler than the bulk solution. Prior to seeding, the lower wall temperature could then lead to undesired nucleation by generating a higher localized supersaturation near the wall, which may exceed the metastable limit and induce nucleation. Alternatively, if the wall temperature is managed by controlling the solution–jacket temperature differential, the time required to cool would potentially be extended beyond the nucleation induction time, again leading to undesired nucleation. Often at scale there will be a limit beyond which the cooling rate cannot be maintained, a threshold rate. Design approaches must take into consideration the heat transfer characteristics of the manufacturing vessel so that practical cooling rates and timescales are incorporated into crystallization design. As no reagent additions or reactions are occurring, mixing considerations typically involve the assurance of adequate suspension without significant breakage; scaling by constant energy dissipation rate is often a good starting point for mixing calculations.

13.6.5 Antisolvent Crystallization Considerations

The most typical scale-up considerations for antisolvent crystallizations are as follows:

- Maintaining an antisolvent addition rate profile
- Managing temperature gradients due to the antisolvent addition
- Mixing that can impact both heat and mass transfer

Antisolvent crystallizations can be complicated by differences in temperature between the antisolvent and the solvent. If the antisolvent being added is at a different temperature than the crystallization solution, heat has to be added or removed to maintain the desired solution temperature (equation 13.33). Additionally, the equipment setup and mixing will significantly impact the distribution of the antisolvent and equilibration of heat. Therefore, antisolvent crystallizations have more potential to be sensitive to mixing than a common cooling crystallization. The simplest approach to minimize heat transfer or temperature considerations for an antisolvent addition is to adjust the antisolvent to the same temperature as the crystallization solution prior to addition. Addressing the distribution of antisolvent during an addition does not have a standard approach. Evaluating a Damköhler number provides a starting point, especially for comparing demonstrated laboratory conditions to proposed conditions at scale. Since mixing times generally increase with scale, it may be necessary to increase the antisolvent addition time accordingly to minimize impacts to the crystallization from localized concentrations. Effective mixing at the point of antisolvent introduction is especially critical to reduce the probability of nucleation due to the high local antisolvent concentration at that point. These effects are often investigated as sensitivity studies on small scale (i.e., $<20\,\text{L}$) by varying the addition geometry and rate and examining their sensitivity to particle size. For situations in which Da is small and a growth mechanism is employed, scaling up mixing by constant energy dissipation rate is a reasonable starting point, with increases in agitation rate toward a constant mixing time with increasing Da.

13.6.6 Reactive Crystallization Considerations

Reactive crystallizations can be more complex than the other modes of crystallization due to the addition of the reaction rate to the already present rates of mixing, mass transfer, and crystal growth. The most typical considerations for their scale-up are as follows:

- Maintaining a reagent addition rate profile
- Managing temperature gradients due to the reagent addition and heat of reaction
- Mixing that can impact both heat and mass transfer

Assuming the reaction requires the addition of a reagent, the heat transfer considerations for reactive crystallizations are similar to those of antisolvent crystallizations. However, it is more likely that the heat of reaction, and even the heat of addition in the case of acids or bases, will be significant and require significant jacket compensation for heat removal. Again, Damköhler numbers would provide a useful evaluation of a process and its scale-up. For reactive crystallizations the rate processes include the reagent addition, reaction, mixing, and nucleation, and/or growth. A thorough assessment requires the comparison of all of these components of the process. By comparing the reagent addition rate and the reaction rate it may be determined whether the reaction is addition rate controlled, where the reactant is consumed as fast as it is added, or if there is accumulation of the reagent. In the first case, it would be the addition rate that needs to be compared to the mixing times and nucleation and growth rates, but in the second it would be the reaction rate. This also highlights a potential method of adjusting a process for improved control—selecting a reagent addition rate slow enough to prevent reagent accumulation. Because of the number of simultaneous rate processes, reactive crystallizations tend to be the most problematic to scale-up.

13.6.7 Evaporative Crystallizations Considerations

The most typical scale-up considerations for evaporative crystallizations are as follows:

- Maintaining an evaporation rate profile
- Managing the wall temperature and encrustation
- Mixing that can impact both heat and mass transfer

Evaporative crystallizations encounter heat transfer limitations during scale-up that are similar to cooling crystallizations. With the reduced heat transfer area to volume on scale, larger temperature differences between the jacket and solution are required to maintain a desired evaporation rate. During scale-up, it is common to utilize higher jacket temperatures, since the solution temperature is fixed at the boiling point. Additional considerations due to the increased jacket temperature include chemical degradation, encrustation, nucleation, and foaming, each discussed briefly in this section. Chemical degradation of the product may be encountered with the higher localized temperatures at the vessel walls. The higher localized temperatures at the wall also promote faster evaporation of the solvent there, which combined with the reducing solution volume can lead to a buildup of product crystals on the wall above the solution surface. This encrustation may also exhibit increased levels of degradation since the solids are exposed to the heat from the jacket for the remainder of the process. If a process is initially designed as an evaporative crystallization at atmospheric pressure, reduction of the operating pressure can be the easiest approach to address the need for higher jacket temperatures. Reduced pressure alone will not address encrustation above the liquid level, and it increases the potential for foaming, which is commonly exacerbated by the presence of small particles. As a consequence of each of these potential issues, a reduced evaporation rate (using low $T_j - T_r$ values) and seeding at relatively high proportions may be required to successfully scale-up this mode of crystallization. Similar to cooling crystallizations, scaling by constant energy dissipation rate is a reasonable starting point for design.

REFERENCES

1. Mullin JW. *Crystallization*, 4th edition, Elsevier Butterworth-Heinemann, 2001.
2. Myerson AS. *Handbook of Industrial Crystallization*, 2nd edition, Butterworth-Heinemann, 2002.
3. Tung H, et al. *Crystallizations of Organic Compounds: An Industrial Perspective*, Wiley, Hoboken, 2009.
4. Sisko J, Oh LM, et al. Process development of a novel pleuromutilin-derived antibiotic. *Org. Process Res. Dev.* 2009; 13: 729–738.
5. Henck JO, Kuhnert-Brandstatter M. Demonstration of the terms enantiotropy and monotropy in polymorphism research exemplified by flurbiprofen. *J. Pharmaceut. Sci.* 1999; 88 (1): 103–108.
6. Byrn S, et al. Pharmaceutical solids: a strategic approach to regulatory considerations. *Pharmaceut. Res.* 1995; 12 (7): 945–954.
7. Henck J, Kuhnert-Brandstatter M. *J. Pharmaceut. Sci.* 1999; 88 (1): 103–108.
8. Vogt FG, et al. Physical, crystallographic, and spectroscopic characterization of a crystalline pharmaceutical hydrate: understanding the role of water. *Crystal Growth Design* 2006; 6 (10): 2333–2354.
9. Zaworotko MJ, et al. Crystal engineering of the composition of pharmaceutical phases: multiple-component crystalline solids involving carbamazepine. *Crystal Growth Design* 2003; 13 (6): 909–919.
10. Childs SL, et al. Crystal engineering approach to forming cocrystals of amine hydrochlorides with organic acids, molecular complexes of fluoxetine, hydrochloride with benzoic, succini, and fumaric acids. *J. Am. Chem. Soc.* 2004; 126 (41): 13335–13342.
11. Bernstein J. *Polymorphism in Molecular Crystals*, 1st edition, Oxford University Press, 2002.
12. Hilfiker R. *Polymorphism in the Pharmaceutical Industry*, 1st edition, Wiley-VCH, 2006.
13. Gibson M. *Pharmaceutical Preformulation and Formulation*, 1st edition, Interpharm/CRC, 2004.

14. Chemburkar SR, et al. Dealing with the impact of ritonavir polymorphs on the late stages of bulk. *Org. Process Res. Dev.* 2000; 4: 413–417.
15. Liu R. *Water-Insoluble Drug Formulation*, Interpharm Press, 2000.
16. Santomaso A, Lazzaro P, Canu P. Powder flowability and density ratios: the impact of granules packing. *Chem. Eng. Sci.* 2003; 58 (13): 2857–2874.
17. Perry RH, Green DW, Maloney JO. *Perry's Chemical Engineers' Handbook*, 7th edition, McGraw-Hill, 1997.
18. Togkalidou T, et al. Experimental design and inferential modeling in pharmaceutical crystallization. *AIChE J.* 2001; 47 (1): 160–168.
19. Zhu H, Yuen C, Grant DJW. Influence of water activity in organic solvent + water mixtures on the nature of the crystallizating drug phase 1-theophylline. *Int. J. Pharmaceut.* 1996; 135: 151–160.
20. Matthews HB, Rawlings JB. Batch crystallization of a photochemical: modeling, control, and filtration. *AIChE J.* 1998; 44 (5): 1119–1127.
21. Larson M, Khambaty S. Crystal regeneration and growth of small crystals in contact nucleation. *Ind. Eng. Chem. Fundam.* 1978; 17 (3): 160–165.
22. Kashchiev D. *Nucleation: Basic Theory with Applications*, 1st edition, Butterworth-Heinemann, 2000.
23. Dell'Orco P, Patience D, Rawlings J. Optimal operation of a seeded pharmaceutical crystallization with growth-dependent dispersion. *Org. Process Res. Dev.* 2004; 8 (4): 609–615.
24. Kougoulos E, et al. Use of focused beam reflectance measurement (FBRM) and process video imaging (PVI) in a modified mixed suspension mixed product removal (MSMPR) cooling crystallizer. *J. Crystal Growth* 2005; 273 (3–4): 529–534.
25. Nocent M, et al. Definition of a solvent system for spherical crystallization of salbutamol sulfate by quasi-emulsion solvent diffusion (QESD) method. *J. Pharmaceut. Sci.* 2001; 90 (10): 1620–1627.
26. Bratz. [Online] http://brahms.scs.uiuc.edu.
27. Rawlings. [Online] http://jbrwww.che.wisc.edu.
28. Ramkrishna D. *Population Balances: Theory and Applications to Particulate Systems in Engineering*, 1st edition, Academic Press, 2000.
29. Cote A, Zhou G, Stanik M. A novel crystallization methodology to ensure isolation of the most stable crystal form. *Org. Process Res. Dev.* 2009; 13: 1276–1283.
30. McCabe WL, Smith JC, Harriot P. *Unit Operations of Chemical Engineering*, 6th edition McGraw-Hill, 2001.
31. Schmidt B, et al. Application of process modelling tools in the scale-up of pharmaceutical. *Org. Process Res. Dev.* 2004; 8: 998–1008.
32. Geankoplis CJ. *Transport Processes and Unit Operations*, 3rd edition, Prentice Hall, 1993.
33. McConville FX. *The Pilot Plant Real Book: A Unique Handbook for the Chemical Process Industry*, 2nd edition, Fxm Engineering & Design, 2006.
34. Chemineer Impellers. [Online] http://www.chemineer.com/impellers.php.
35. Pfaudler Impellers. [Online] http://www.pfaudler.com/mixing_systems.php.
36. Kirk RE, et al. *Encylopedia of Chemical Technology*, 4th edition, Wiley, 1998.
37. Zwietering TN. Suspending of solid particles in liquids by agitators. *Chem. Eng. Sci.* 1958; 8: 244–253.

14

SCALE-UP OF MIXING PROCESSES: A PRIMER

Francis X. McConville and Stephen B. Kessler
Impact Technology Development, Lincoln, MA, USA

14.1 INTRODUCTION

The problems associated with the scale-up of mixing processes are universal. This is because the dynamics and mechanics of liquid agitation and blending are often poorly understood, yet these operations play a fundamental role in many aspects of the chemical and pharmaceutical industries. The success of homogeneous and heterogeneous chemical reactions, crystallizations, liquid–liquid extractions, and so many other operations critically depends on effective mixing and appropriately designed mixing systems. Unfortunately, as we shall see below, duplicating the energy and quality of mixing available in the laboratory at commercial scale can prove extremely difficult.

For example, the motor power required to turn agitators increases exponentially as the diameter of the agitators increases, making it prohibitively expensive to match, one to one, the mixing power input of small-scale reactors in large commercial vessels. This results in batch blend times, the time it takes for the contents of a batch reactor to become homogenized, sometimes orders of magnitude longer in commercial reactors than in the laboratory. This can have severe consequences for the results of many chemical operations.

Frequently, heterogeneous reactions such as catalytic hydrogenations fail to achieve expected reaction rates upon scale-up because there is insufficient mixing to fully suspend the catalyst particles. The catalyst settles to the bottom of the vessel where it is inaccessible to the reactants in solution and therefore cannot effectively catalyze the reaction.

Differences in local and average shear conditions due to differences in impeller diameter and impeller tip speeds in commercial vessels can have unexpected consequences for shear-sensitive processes such as fermentations using living cells or crystallization of materials that require a specific particle size distribution. High shear can also cause severe emulsification at large scale that might not have been experienced in the laboratory.

These are just a few of the types of problems often encountered at large scale due to the fact that mixing conditions differ so much from those available in the laboratory. Mixing scale-up often proves to be a compromise between cost and performance, between achieving the desired result and minimizing unexpected negative effects. The better the understanding of the fundamental principles of mixing and of the specific requirements of the process involved, the better the results of this compromise will be.

14.2 BASIC APPROACHES TO MIXING SCALE-UP

Over the years, scientists and engineers have considered many approaches to scaling up mixing processes, with the ultimate goal of successfully matching laboratory results at commercial scale at a reasonable cost. As a result, numerous scale-up parameters, equations, and principles have been developed, some of which work better or are more reliable than others depending on the specific application. No single method has been successful for all situations, and the characteristics of the system must be understood as well as possible to maximize the chances for success.

Chemical Engineering in the Pharmaceutical Industry: R&D to Manufacturing Edited by David J. am Ende

14.2.1 Principles of Similarity

Modeling theory considers two processes similar if they possess geometric, kinematic, and dynamic similarity. Geometric similarity requires that linear dimensions of two systems are scaled by the same ratios at different scales. Kinematic similarity requires geometric similarity and also that characteristic velocities scale by the same ratio. Dynamic similarity requires both geometric and kinematic similarity and adds the requirement that characteristic forces scale by the same ratio.

Rigorous application of modeling theory is rarely applied to scale-up of industrial mixing processes. One reason for this is that when more than two force properties are important in a mixing process, full dynamic similarity cannot be achieved. Since most mixing processes involve three or more force properties, a choice must be made among the possible properties to select one as a scaling factor. This choice is made by considering the nature of the process at hand and applying scaling factors that have been proven to work in similar processes. Some commonly used approaches to mixing scale-up and their utility in specific situations are described in the following sections.

14.2.2 Geometric Similarity

The concept of geometric similarity is illustrated in Figure 14.1. Adhering to geometric similarity can be extremely important in designing systems for scale-up, or for building small-scale experimental vessels designed to mimic the behavior of a larger system for research purposes. This latter approach, called scaling down or modeling, is an important aspect of mixing engineering and widely used to study the mixing behavior of commercial systems at a more convenient scale.

Figure 14.1 shows how certain key ratios would be held equal in two geometrically similar vessels of different sizes. Thus, the ratio of impeller diameter to tank diameter (*D*/*T*) is identical in both cases, as are the ratios of the liquid level (*Z*), the impeller bottom clearance (*C*), and the baffle width (*B*) to the tank diameter.

A number of practical issues limit the usefulness of this technique alone as a primary scale-up method. First, mechanical limitations may limit its utility in some cases. For example, marine impellers are often used in laboratory systems. However, in large-scale industrial mixing applications, these impellers are impractically heavy if scaled up by geometric similarity. Also, the shape of the vessel heads is usually not limited by mechanical considerations in the laboratory, but in most industrial applications, vessel head design is defined by codes that take mechanical stresses into account. These types of limitations can sometimes be overcome by anticipating large-scale design issues and creating scaled down laboratory vessels that match the large-scale geometry.

In addition to such limitations in the application of geometric similarity, there are limitations in what can be achieved when it is applied. Due to the rules of geometry, as a vessel doubles in diameter, its volume increases by a factor of 8 (2^3). Thus, when scaling up by a factor of 2, it is not possible to maintain certain key ratios such as surface area per unit volume or the volume/diameter ratio. It also proves impossible to operate these two systems in such a way that the intensity of mixing (as measured by power

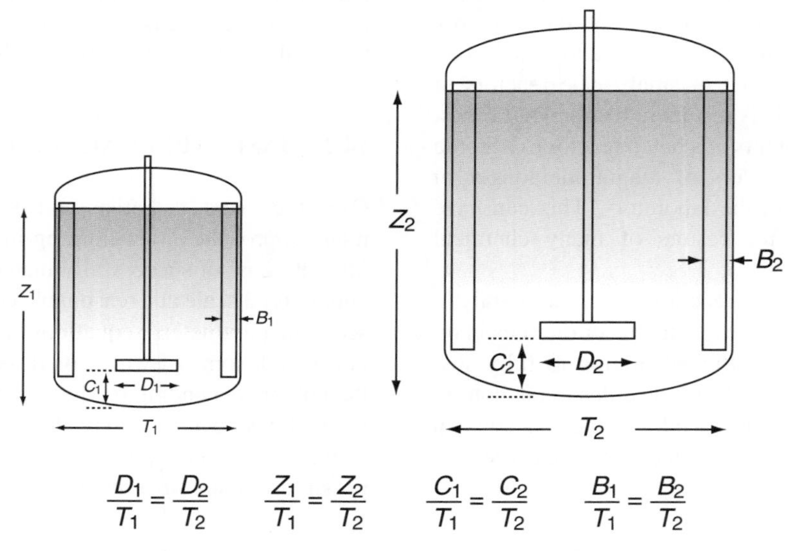

$$\frac{D_1}{T_1} = \frac{D_2}{T_2} \qquad \frac{Z_1}{T_1} = \frac{Z_2}{T_2} \qquad \frac{C_1}{T_1} = \frac{C_2}{T_2} \qquad \frac{B_1}{T_1} = \frac{B_2}{T_2}$$

FIGURE 14.1 The principle of geometric similarity for stirred tanks. Key ratios (*D*/*T*, *C*/*T*, *B*/*T*, *Z*/*T*) are held equal at both scales (*D* = impeller diameter, *T* = tank diameter, *B* = baffle width, *C* = impeller bottom clearance).

input per unit volume, P/V, for example) and the velocity of fluid circulation are *both* identical. It is possible to design and operate two systems of different sizes at an identical P/V, but the fluid circulation patterns, fluid velocities, degree of turbulence, and so on would likely be very different. If a successful process result relies on a particular fluid motion, it might not be achieved upon scale-up by simply maintaining geometric similarity and matching P/V. Such limitations are the source of much confusion and difficulty. In most cases, geometric similarity proves to be useful as a starting point for scale-up, but several other factors must be considered to ensure success.

Consequently, there are situations where deliberate deviation from geometric similarity is the best approach to scale down. Oldshue [1] uses the term "nongeometric similarity" to describe a situation where conventional concepts of similarity must be sacrificed so that certain factors can be controlled to achieve successful scale-up. For example, with geometric similarity observed, a scaled down vessel could be operated at the same tip speed as its full-scale counterpart, but this requires that the impeller in the scaled down vessel be operated at higher rpm. In this example, the maximum shear rate in the two vessels is the same, but the average shear rate in the impeller region is higher in the small vessel. If shear rate is one of the key process variables being modeled, the mismatch in maximum and average shear rates can be reduced by increasing the diameter of the impeller in the scaled down vessel relative to the vessel diameter (D/T). Tip speeds will now match at lower rpm in the small vessel, which would correlate with a smaller difference in impeller average shear rate.

More detailed information on system geometry and the application of geometric similarity to mixing processes can be found in Refs 2 and 3.

Figure 14.2 shows some typical "shape factors"—geometric ratios that have historically proven effective in systems designed for mixing processes, and can be used as a general guide to vessel design. For example, many mixing vessels employ agitators with diameters approximately one-third of the vessel diameter, located one impeller diameter off the bottom. Again, these values are typical but can vary significantly in equipment designed for specific applications.

14.2.3 Rate of Turbulent Energy Dissipation and P/V

A particularly useful and widely used approach to mixing scale-up involves maintaining a constant rate of turbulent energy dissipation ε across the various scales. ε, which is defined by equation 14.1, is usually expressed in units of W/kg:

$$\varepsilon = \frac{P}{\rho V} \tag{14.1}$$

where P is power input (W), ρ is liquid density (kg/m^3), and V is liquid volume (m^3).

ε is fundamental in describing the interrelationship between turbulence and mass transfer in mixing operations. This statement is illustrated by the Kolmogorov length scale, which characterizes the smallest eddies associated with turbulent mixing. The Kolmogorov eddy length η is defined

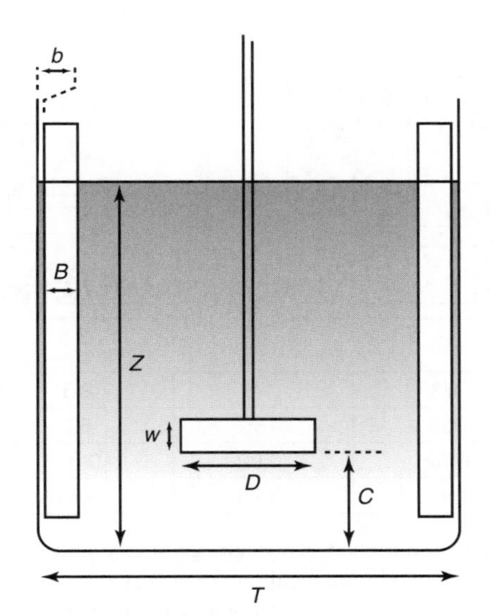

FIGURE 14.2 Typical shape factors, or geometric ratios, found useful for general mixing applications in stirred tanks.

by equation 14.2,, where ν is the kinematic viscosity:

$$\eta = \left(\frac{\nu^3}{\varepsilon}\right)^{1/4} \tag{14.2}$$

At the length scale represented by η, viscous forces in the eddy are equal to inertial forces due to turbulent velocity fluctuations. The Kolmogorov eddy length underlies and informs the use of ε as a scaling parameter. Kinematic viscosity ν is a liquid property that is scale independent; thus, constant ε is sufficient to fix a value for η over a range of scales. However, it remains to define the region for which ε is applicable. A mean value of ε can be calculated from the total power input and mass of liquid in the vessel. This overall mean value of ε is useful where an operation is governed by bulk flow characteristics. For geometrically similar vessels, it is sometimes assumed that holding overall mean ε constant is sufficient to provide accurate scaling in the impeller region. For more accurate scaling of local characteristics, a better approach is to define a volume based on the swept volume of the impeller instead of the total batch volume. A method for calculating impeller swept volume is given in Ref. 4.

Another parameter that is used to represent average mixing intensity in a vessel is power/unit volume (P/V). P/V is sometimes called power intensity, and is usually expressed in either W/L or HP/1000 Gal. In scaling equations that involve ratios of ε or P/V to represent different sizes of equipment, either ε or P/V works equally well. However, because of its units, P/V cannot be applied in fundamental equations that define turbulence and mass transfer in mixing systems. Also keep in mind that in large vessels, local values of ε may vary widely in different regions of the vessel.

When mean values of either ε or P/V are used for scale-up, it is important to also maintain geometric similarity. This is because in some mixing applications, a local value of ε may be of greater importance than the vessel average value. This point is well made in Figure 14.3 that shows three vessels all operated at the same P/V, but the fact that their geometries are very different (specifically impeller size) results in very different results in the suspension of solids. The effects shown are the result of calculations made with the commercial computational fluid dynamics program VisiMix® [5].

As shown in equations 14.3 and 14.4, ε and P/V can be expressed in terms of the impeller diameter D, its rotational speed N, liquid volume V, batch density ρ, and a parameter called the power number N_P that is explained in more detail below.

$$\varepsilon = \frac{N_P N^3 D^5}{V} \tag{14.3}$$

$$\frac{P}{V} = \frac{N_P \rho N^3 D^5}{V} \tag{14.4}$$

Power number, N_P, is a dimensionless number characteristic of a given impeller and vessel geometry. It is defined by equation 14.5.

$$N_P = \frac{P}{\rho N^3 D^5} \tag{14.5}$$

Figure 14.4 lists some typical values of power number for various types of impellers, but geometric factors such as impeller tip chord angle, number of blades, position of the

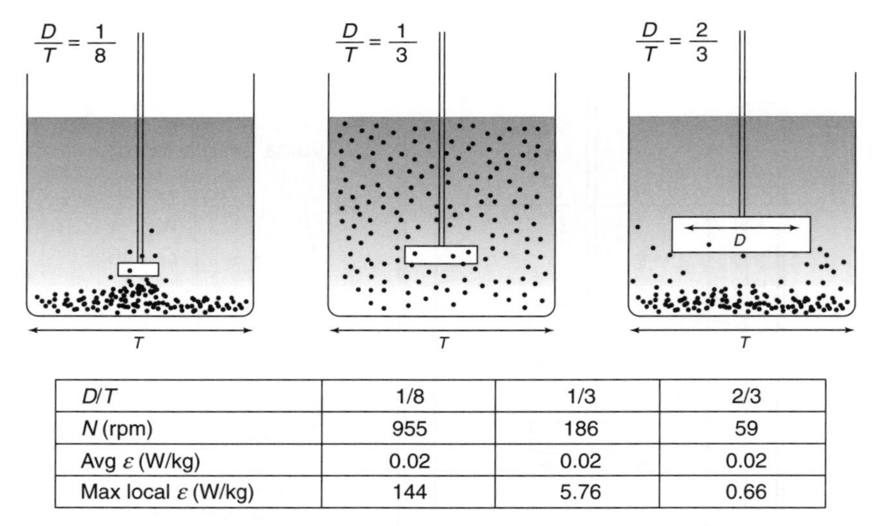

D/T	1/8	1/3	2/3
N (rpm)	955	186	59
Avg ε (W/kg)	0.02	0.02	0.02
Max local ε (W/kg)	144	5.76	0.66

FIGURE 14.3 Three cases illustrating the importance of system geometry and the distinction between mean rate of energy dissipation (ε) and maximum local rate of energy dissipation. While all three vessels are operating at the same average value of ε, differences in geometry result in very different fluid motion and mixing behavior, in this case manifested by differences in the suspension of solids as predicted by VisiMix [5].

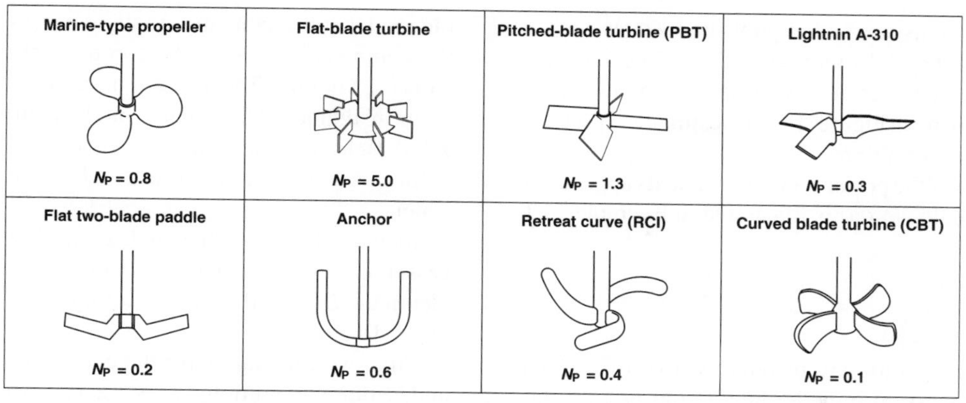

Marine-type propeller	Flat-blade turbine	Pitched-blade turbine (PBT)	Lightnin A-310
$N_P = 0.8$	$N_P = 5.0$	$N_P = 1.3$	$N_P = 0.3$
Flat two-blade paddle	**Anchor**	**Retreat curve (RCI)**	**Curved blade turbine (CBT)**
$N_P = 0.2$	$N_P = 0.6$	$N_P = 0.4$	$N_P = 0.1$

FIGURE 14.4 Typical power numbers (N_P) for various impeller types. These values are only approximate as the power number is significantly affected by number and pitch of blades, tip chord angle, position of the impeller within the vessel, baffle configuration, and other geometric factors.

impeller within the vessel, and the number and dimensions of baffles all affect the value of the power number. For this reason, accurate power number values for a particular system can only be obtained experimentally, by measuring power draw via a watt meter, or, more accurately, by directly measuring torque on the impeller shaft, under well-defined experimental conditions.

As in any type of fluid flow, fluid motion in mixing can be generally classified as either turbulent or laminar, depending on the velocity and other physical parameters. A common term for quantifying this is the impeller Reynolds number N_{Re}, a dimensionless parameter defined by equation 14.6. Values of N_{Re} can range from single digits for highly viscous flow to hundreds of thousands for very turbulent flow.

$$N_{Re} = \frac{\rho D^2 N}{\mu} \tag{14.6}$$

At impeller Reynolds numbers greater than about 10^4, fluid motion is considered turbulent, and under such conditions, the power number N_P assumes a constant value. Under laminar mixing conditions ($N_{Re} < 100$) and in the transitional regime between laminar and turbulent mixing, power number varies, typically increasing with decreasing Reynolds number as shown by the curves in Figure 14.5. The values of N_{Re} that delineate the transitional region are only approximate, and will vary depending on the system.

Note that the fluid viscosity term does not appear in the equations that define ε or P/V, but is captured indirectly in this relationship between power number and Reynolds

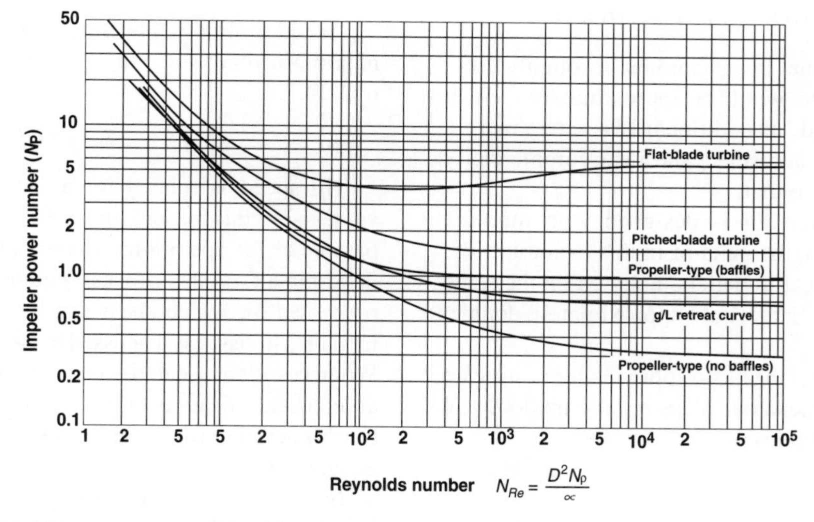

FIGURE 14.5 Relationship between impeller power number N_P and impeller Reynolds number N_{Re} for some typical impeller types. At $N_{Re} > 10^4$, flow is turbulent and N_P reaches a constant value. The power number generally increases under laminar conditions ($N_{Re} < 10$) and in the transitional regime between laminar and turbulent (after Hemrajani and Tatterson [6] and Rushton [7]).

number. It should be assumed that published values such as those given in Figure 14.4 represent turbulent power numbers. It is usually necessary to use empirical relationships such as those shown in Figure 14.5 to estimate N_P values under nonturbulent conditions.

The importance of the power number N_P and its application in typical mixing calculations is illustrated in Example 14.1.

EXAMPLE 14.1

Determine what size motor will be required to turn a 0.33 m diameter A-310 hydrofoil impeller at 120 rpm in a crystallizer with a working volume of 500 L. The process fluid has a density of 1150 kg/m^3 and a viscosity similar to water (approximately 0.001 Pa s).

The A-310 has a published turbulent power number of 0.3. To use this number, we must ensure that we are operating in the turbulent mixing regime ($N_{Re} > 10^4$). Apply equation 14.6 to calculate the Reynolds number. Note that the rotational speed must be expressed in rev/s.

$$N_{Re} = \frac{\rho D^2 N}{\mu} = 1150\,\text{kg/m}^3 \times (0.33\,\text{m})^2 \times 2\,\text{s}^{-1} \times \frac{1}{0.001\,\text{kg/(m s)}} = 250,470$$

This indicates that the mixing flow is clearly in the turbulent regime, so it is appropriate to use the published value of $N_P = 0.3$ for power draw estimation.

The power requirement is calculated by rearranging equation 14.5 as shown below:

$$P = N_P \rho N^3 D^5 = 0.3 \times 1150\,\text{kg/m}^3 \times (2\,\text{s}^{-1})^3 \times (0.33\,\text{m})^5 = 10.8\,\text{kg m}^2/\text{s}^3 = 10.8\,\text{W}$$

Estimating that frictional losses amount to roughly 20%, the total power requirement would be approximately 13 W. It is a common practice to add an additional 15% safety margin at the design stage and then select the next commercially available motor size above that.

In some cases, the agitator is designed with multiple impellers. If, for example, the agitator in this example were designed with two identical impellers mounted on the same shaft, the power requirements would approximately double.

Example 14.2 illustrates the basic approach for scaling up by maintaining constant mean rate of energy dissipation, ε, in two vessels of different scales.

EXAMPLE 14.2

A 2 L laboratory system is being designed to study the mixing characteristics of a commercial vessel. The goal is to operate the model at the same mean rate of energy dissipation (ε) as the commercial vessel. The commercial vessel is a 7500 L working volume (7.5 m^3) cylindrical vessel with $D = 2.0$ m, a 0.8 m diameter four-blade pitched turbine impeller ($D/T = 0.4$) that turns at a fixed speed of 68 rpm, and two vertical baffles. Assume that the process fluid has the properties of water.

In the interests of geometric similarity, the laboratory vessel is designed to have identical baffles and agitator, identical D/T and Z/T, resulting in $T = 12.85$ cm and $D = 5.14$ cm.

First, we calculate ε for the commercial vessel. The N_{Re} under these conditions is $\sim 7 \times 10^5$, so we can use the published turbulent N_P value of 1.3 in equation 14.3:

$$\varepsilon = \frac{N_P N^3 D^5}{V} = 1.3 \times (1.13\,\text{s})^3 \times (0.8\,\text{m})^5 \times \frac{1}{7.5\,\text{m}^3} = 0.082\,\text{m}^2/\text{s}^3 = 0.082\,\text{W/kg}$$

Now, determine the speed at which to operate the lab reactor to achieve the same mean value of ε by rearranging equation 14.3 and solving for N:

$$N^3 = \frac{\varepsilon V}{N_P D^5} = 0.082\,\text{m}^2/\text{s}^3 \times 0.002\,\text{m}^3 \times \frac{1}{1.3} \times \frac{1}{(0.0514\,\text{m})^5}$$

$$N = 7.06\,\text{s}^{-1} = 423\,\text{rpm}$$

Thus, we can match the commercial-scale mean rate of energy dissipation in the laboratory by operating the 5.14 cm impeller at 423 rpm.

14.2.4 Tip Speed

Tip speed is simply tangential velocity of the impeller at its maximum diameter and is calculated according to equation 14.7.

$$S_t = \pi D N \tag{14.7}$$

Tip speed is related to maximum shear rate in stirred vessels. For this reason, tip speed is often applied as a scaling parameter for operations where maximum shear is a critical determinant of the process outcome. This includes those processes for which shear can be either beneficial or detrimental. This issue is discussed in more detail in Section 14.2.6. When vessels are scaled according to geometric similarity and at constant mean energy dissipation rate, tip speed will be higher in the larger vessel, a fact supported by equation 14.7.

In addition to its relationship to maximum impeller shear, in geometrically similar vessels tip speed scaling corresponds exactly to scaling at constant torque per unit volume. In fully turbulent flow, that is, above $N_{Re} = 10^4$, all velocities scale with tip speed regardless of viscosity. Because of these relationships, tip speed or torque per unit volume is useful in scaling mixing processes that are controlled by flow such as blending of miscible liquids and suspension of solids in liquids.

14.2.5 Blend Time

Blend time is an empirical factor that describes the time it takes for the contents of a vessel to become homogenized, particularly important during chemical additions to a batch. It is usually determined experimentally by monitoring the dispersion of a dye or other tracer compound, either visually or by means of detection probes located at various points in the vessel.

Often, an acceptable blend time is established based on a practical, realistically achievable value such as 99% uniformity. Although somewhat subjective, blend time is a critical factor in the scale-up of many operations, particularly rapid chemical reactions that rely on rapid dispersion during controlled addition of a reagent. This is discussed in detail in the section on mixing-limited reactions.

Figure 14.6 illustrates that blend time increases rapidly when *P*/*V* is held constant, but vessel size increases. Holding blend time constant with increasing vessel size requires maintaining constant impeller speed in geometrically similar vessels. This approach leads to increasing *P*/*V* with vessel size and, ultimately, to unrealistically high power requirements. Values much higher than 1–2 W/L are difficult to achieve in standard stirred tanks at large scale as beyond that motors would become impractically large.

Impeller design and number of impellers will also have a significant effect on blend time. Some types of impellers, such as standard anchor-blade impellers, which are not designed for good bulk mixing, generally result in very long blend times, whereas a pitched-blade turbine operated at typical speeds in the same vessel would result in much shorter blend times.

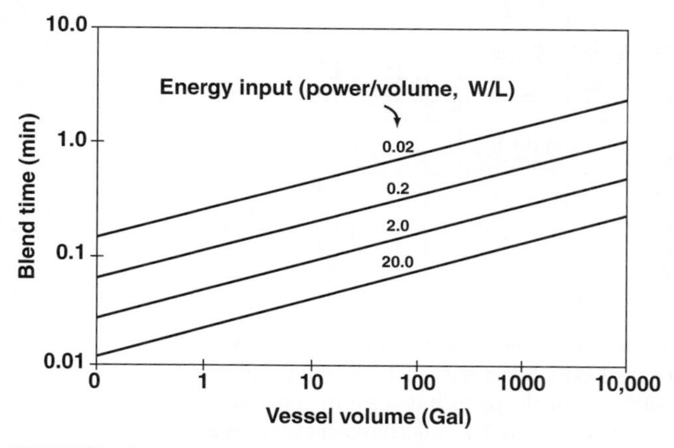

FIGURE 14.6 Relationship between blend time and vessel volume for various levels of power input (*P*/*V*).

Various correlations have been developed to help maintain constant blend time at different scales, such as the translation equations introduced below, but their success depends heavily on impeller design and other geometric factors. For standard vessel and impeller geometries, correlations are available that estimate blend times for the turbulent, transitional, and laminar regimes [8].

In mixing calculations, it is common to see a variable called "dimensionless blend time" that is essentially the product of the actual blend time and the impeller rotational speed, although often other geometric factors are included in equations used to calculate it.

14.2.6 Shear

As mentioned earlier, shear in a batch mixing operation can have desirable or undesirable effects, depending on the intended result of the operation. For example, maintaining sufficiently high shear rate in the impeller region may be required to rapidly disperse a reactant being fed into a vessel during a chemical reaction. However, the product of this same reaction may be a solid precipitate whose particles are shear sensitive and would suffer attrition, creating fines and complicating downstream recovery if high shear rates are maintained for too long.

Controlling shear rates when such a process is scaled up can become a complex undertaking. Various correlations presented in the literature to estimate shear rates in mixing vessels predict a broad range of shear rate values. Moreover, shear rates may be predicted to increase, decrease, or remain constant on scale-up, depending on the shear correlation and scale-up approach that are chosen. Some examples of the available correlations are described below.

One widely used correlation, the Metzner–Otto relationship, predicts average shear rate in the impeller region. This relationship, defined by equation 14.8, is valid for laminar, transitional, and moderately turbulent conditions.

$$\dot{\gamma} = k'N \tag{14.8}$$

where $\dot{\gamma}$ is the shear rate in s^{-1}, k' is a dimensionless Metzner–Otto coefficient characteristic of the impeller, and N is the impeller speed in rev/s.

For a Lightnin® A-310 hydrofoil impeller (see Figure 14.4), the value of k' is 8.6. Thus, the estimated average shear rate in the impeller region for an A-310 running at 100 rpm is

$$\dot{\gamma} = 8.6 \times 100/60 = 14\,s^{-1}$$

The value of shear rate predicted by the Metzner–Otto relationship depends only on impeller type and speed and is independent of impeller and vessel dimensions.

To estimate maximum shear rates produced in the flow near the impeller tip, an approach analogous to the

Metzner–Otto relationship is used. In this case, a single value of the coefficient, $k' = 150$, is applied regardless of the impeller type. For estimating maximum shear on the impeller surface, a value of $k' = 2000$ is sometimes applied.

As mentioned in the section on tip speed, this factor can be related to maximum shear rate near the impeller tip. While the $k' = 150$ rule described above applies for moderate Reynolds number (laminar and transitional) conditions, tip speed is recommended for higher N_{Re} conditions as a means of scaling on the basis of maximum shear rate. There is no general rule found in the literature that correlates tip speed with shear rate. When used for scaling purposes, tip speed is held constant as scale increases, which is assumed to provide constant maximum shear rate.

For estimates of shear rate averaged throughout a vessel under turbulent conditions, a vessel average shear rate can be calculated based on total energy dissipation. Equation 14.9 defines this correlation:

$$\dot{\gamma} = \frac{P}{V}\mu^{0.5} \tag{14.9}$$

As described above, the various shear rate correlations provide widely divergent values on scale-up. To illustrate this point, Figure 14.7 compares the shear correlations that are presented above. The graph presented in Figure 14.7 covers a range of 100:1 scale-up of impeller diameter under the condition of geometric similarity, which corresponds to a range of 10^6:1 in vessel volume. ε and P/V are held constant. Under these conditions, shear rates in the impeller region, in the flow near the tip, and on the tip surface, which are each defined by a constant multiplied by rpm, all decrease with increasing impeller diameter. Vessel average turbulent shear rate remains constant when P/V is held constant. Tip speed as an indicator of shear rate increases with increasing impeller diameter at constant P/V.

General guidelines in the literature indicate that Metzner–Otto-type correlations are best applied over the laminar and transitional Reynolds number ranges. Vessel average shear applies only under fully turbulent conditions. Tip speed can be used as a scaling factor for maximum shear under fully turbulent conditions. However, these guidelines should not be relied upon if a process is to be scaled up in which shear is an important consideration. In this case, lab- and pilot-scale experiments should be conducted to evaluate the effects of shear over a range of scales whereby a correlation can be selected for commercial scale-up.

14.2.7 Scaling (Translation) Equations

In keeping with the concept of similarity, a number of relationships, sometimes called translation equations, are used in an attempt to match operating conditions at two different scales. Various authors have developed different approaches for different situations.

For example, the equations below illustrate some relationships that have been proposed for maintaining *equal blend*

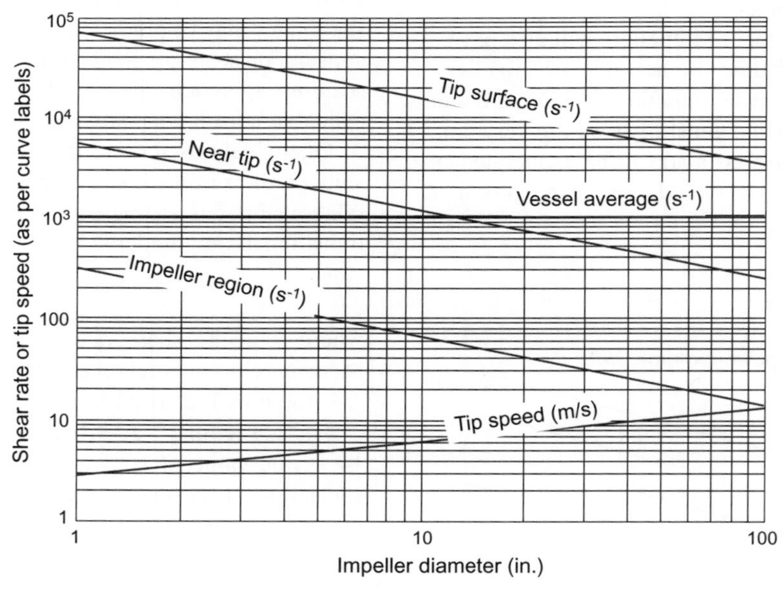

FIGURE 14.7 Theoretical effect of scale on various manifestations of shear in a mixing vessel, under conditions of geometric similarity and constant ε (and P/V). Shear rates in the region swept by the impeller, in the flow near the tip, and on the tip surface, which are each defined by a constant multiplied by rpm, all decrease with increasing impeller diameter. Vessel average turbulent shear rate remains constant under these conditions. Tip speed as an indicator of shear rate increases with increasing impeller diameter.

time between small-scale and large-scale vessels for batch mixing operations.

$$\frac{(P/V)_2}{(P/V)_1} \approx \left(\frac{D_2}{D_1}\right)^2 \quad [9] \qquad (14.10)$$

$$\frac{(T_Q/V)_2}{(T_Q/V)_1} \approx \left(\frac{D_2}{D_1}\right)^2 \quad [2] \qquad (14.11)$$

The application of translation equations such as these depends very much on the specific application and system geometry, and as always experimental validation at two different scales is strongly recommended when applying them to predict performance in commercial-scale operations. A wide array of translation equations used for various purposes under various conditions is examined by Uhl and Von Essen [2].

14.3 OTHER CONSIDERATIONS IN MIXING SCALE-UP

14.3.1 Importance of Fluid Rheology

While Reynolds number is seldom used as a mixing scale-up correlation *per se*, knowing whether a mixing operation is conducted under laminar, transitional, or turbulent flow conditions is vital to successful scale-up. For example, if the planned commercial-scale operation will be fully turbulent, then lab and pilot scale-down experiments must be designed to operate under turbulent conditions as well. Liquid viscosity is a primary determinant of the value of impeller Reynolds number, defined by equation 14.6, so knowledge of the viscosity that is characteristic of a given mixing operation is vital as well.

If an operation comprises blending of Newtonian liquids or suspending an immiscible solid in a Newtonian liquid, then obtaining the required viscosity data is straightforward. The liquids may have well-known viscosities that can be found in literature references. If not, then measurement of Newtonian viscosity is a simple matter that can be performed with inexpensive instruments. However, if the liquid being mixed contains macromolecular solutes or colloidal size particles, it may exhibit non-Newtonian characteristics. In this case, defining its rheology requires more than a single coefficient and measurements may require more sophisticated instruments and techniques.

The flow properties of a Newtonian liquid are defined by equation 14.12.

$$\tau = \mu\dot{\gamma} \qquad (14.12)$$

where τ is shear stress, μ is coefficient of viscosity, and $\dot{\gamma}$ is shear rate.

Shear thinning fluids are often encountered when dealing with macromolecular solutes or colloidal suspensions. Such fluids can be effectively modeled by a power law, shown in equation 14.13.

$$\tau = K\dot{\gamma}^n \qquad (14.13)$$

where K is a consistency index and n is a behavior index.

If a shear thinning fluid also exhibits a yield stress (i.e., there exists a shear stress below which no flow occurs), then the Herschel–Bulkley model, shown in equation 14.14, can be applied.

$$\tau = \tau_0 + K\dot{\gamma}^n \qquad (14.14)$$

where τ_0 is yield stress.

Mixing of yield stress fluids can prove particularly challenging to scale-up. If the yield stress is of sufficient magnitude, then use of a conventional turbine-style impeller may result in a well-mixed cavern of liquid surrounding the impeller with little or no liquid motion closer to the vessel walls. In this case, a close-clearance impeller that sweeps close to the walls of the vessel, such as an anchor or helical ribbon, may be required. Empirical correlations are available to assist with these scale-up problems. These correlations are specific to impeller geometrical factors and their efficacy will depend on choosing an appropriate rheological model and thorough characterization of the fluid.

Rheological models exist for many known types of fluid behavior. In addition to the non-Newtonian behaviors discussed above, additional levels of complexity such as time dependency and viscoelasticity can also be modeled. To support such models, measurement of the properties of non-Newtonian fluids requires the use of a rheometer. Rheometers are capable of controlling either shear stress or shear rate applied to a sample and are adaptable to multiple test geometries. A detailed discussion of non-Newtonian rheometry is beyond the scope of this chapter. A recommended reference in this regard is Ref. 10. A particularly powerful combination of techniques for scaling of mixing operations for non-Newtonian fluids is the use of appropriate rheological models in conjunction with computational fluid dynamics (CFD) as described in a later section.

14.3.2 The Role of Mixing in Heat Transfer

Heating and cooling batch vessels is a fundamental operation in any chemical processing endeavor. The rate and efficiency of heat transfer in and out of such vessels depend on many things, including the intrinsic heat transfer coefficient of the system, the temperature difference between the batch contents and the heat transfer medium (such as the fluid in the vessel heating jacket), and certain key properties of the batch itself (such as density, thermal conductivity, and heat capacity).

However, mixing also plays a major role in determining heat transfer efficiency. While a full treatment of heat transfer in agitated vessels is beyond the scope of this chapter, it is worth pointing out some fundamental principles.

A common dimensionless group that characterizes process-side heat transfer in stirred tanks is the Nusselt number Nu_L, which is a measure of the ratio of convective heat transfer to conductive heat transfer. It is defined in equation 14.15, where h_i is the process-side heat transfer coefficient, D is the vessel diameter, and k is the thermal conductivity of the batch.

$$Nu_L = \frac{h_i D}{k} \tag{14.15}$$

The value of Nu_L strongly depends on the mixing Reynolds number. Nu_L values close to unity indicate sluggish motion and heat transfer driven primarily by thermal conduction. Under highly turbulent conditions, Nu_L values can range from 100 to 1000, which indicates highly efficient convective heat transfer. Thus, providing a sufficient degree of mixing is an important factor in designing vessels that will be used for heating and cooling. Many very comprehensive texts on process heat transfer are available for additional information, such as Ref. 11.

14.3.3 Continuous Mixing Scale-Up

While a majority of mixing operations in the pharmaceutical industry are still performed in batch mixing vessels, there is a trend toward instituting continuous processing. This trend is being fostered by the FDA through elimination of regulatory constraints that previously limited most pharmaceutical processes to the batch approach. The advantages of continuous mixing include potentially much higher productivity, improved heat transfer, mass transfer, and mixing. The latter three advantages are attributable primarily to reduced mixing volume.

Current examples of continuous mixing processes in the pharmaceutical industry mainly involve mixing-sensitive chemical reactions, that is, fast consecutive reactions that occur on a timescale that is short compared to practical blend times for commercial-scale batch mixing vessels. Most of the reactors used for these operations are of tubular configuration, for example, in-line static mixers. The continuous stirred tank reactor (CSTR) is also used and may find new applications with the current regulatory environment favoring continuous processes.

The typical objective of batch mixing operations is to achieve a spatially homogeneous mixture within a fixed process volume, within a specified blend time. Continuous mixing operations are designed to produce a spatially and temporally homogeneous effluent stream within a specified residence time. While blend time is the key parameter characterizing batch mixing operations, residence time distribution (RTD) is the key parameter characterizing continuous operations. Figure 14.8 plots the response of a sensor at the outlet of a mixing vessel to a step input of tracer at the inlet. The two RTD curves illustrate ideal flow patterns that establish the bounds within which real stirred tanks operate: the plug flow reactor (PFR) and the CSTR. A PFR represents the unmixed limit or complete segregation, while the ideal CSTR represents perfect macromixing.

The y-axis parameter in Figure 14.8, $F_{CSTR} = A/A_0$, is the concentration of tracer measured at the outlet (A) divided by

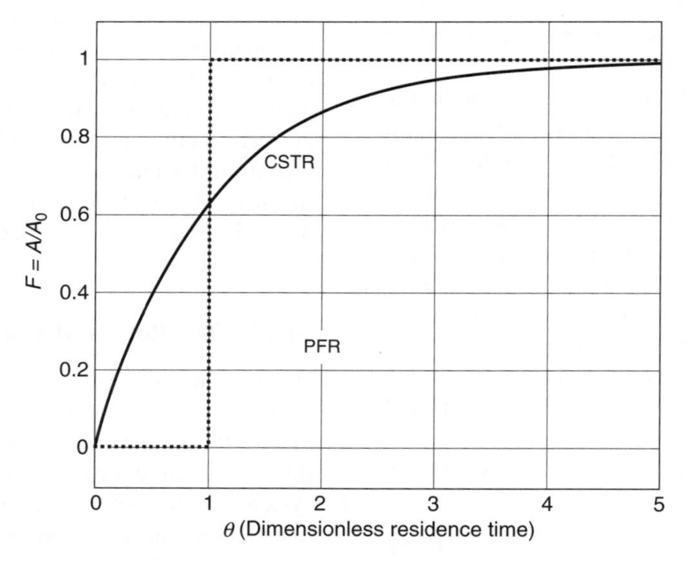

FIGURE 14.8 The response of a sensor at the outlet of a mixing vessel to a step input of tracer at the inlet for both plug flow reactors (PFR) and continuous stirred tank reactors (CSTR). The two ideal RTD curves illustrate the bounds within which real stirred tanks operate.

the concentration applied at the inlet (A_0). The parameter θ in Figure 14.8 is dimensionless residence time, defined by equation 14.16.

$$\theta = t/\bar{t} \tag{14.16}$$

where $\bar{t}$ is the mean residence time (vessel volume/flow rate) and t is the time elapsed following application of tracer.

In the case of plug flow (dotted curve), no tracer is detected at the outlet until $\theta = 1$, at which point the tracer concentration jumps to the value at the inlet. For the CSTR (solid curve), when tracer enters the vessel, some is detected instantly at the outlet and its concentration continues to rise exponentially, approaching asymptotically the inlet concentration (see equation 14.17).

$$F_{\mathrm{CSTR}}(\theta) = 1 - e^{-\theta} \tag{14.17}$$

Real stirred tanks are often assumed to behave as ideal CSTRs. However, some degree of nonideal flow is likely to occur due to channeling, recycling, stagnant regions, or a combination of these effects. In many cases, a real stirred tank may approximate the ideal CSTR closely enough that deviations from ideal flow have a negligible effect on the process. However, such deviations must be considered when scaling up. To quote Levenspiel [12], "The problems of nonideal flow are intimately tied to those of scale-up ... Often the uncontrolled factor in scale-up is the magnitude of the nonideality of flow, and unfortunately this very often differs widely between large and small units. Therefore, ignoring this factor may lead to gross errors in design."

For purposes of design and scale-up of continuous mixing operations, the ratio of the mean residence time in a CSTR divided by the batch blend time for the same vessel is defined by equation 14.18.

$$\alpha = \frac{V}{Q\Theta} \tag{14.18}$$

where V is mixed volume, Q is flow rate through the vessel, and Θ is the batch blend time for the vessel, which is either measured or estimated. To ensure continuous mixing that is near-ideal CSTR in character, a rule of thumb states that the ratio α should be >10. The basis for this ratio is discussed by Roussinova and Kresta [13].

In addition, the location of inlet and outlet ports must be considered. The rule to be followed in this regard is that a straight line drawn from the inlet port to the outlet port should pass through the impeller(s).

Scaling of the ratio of inlet flow to the impeller flow must also be considered. The simplest approach is to limit the average velocity of the liquid in the inlet port to be less than the tip speed of the impeller. Recommended ranges for this ratio can be found in Ref. 6. Ratios of momentum and specific energy dissipation between the entering liquid jet and the impeller flow are sometimes used as scaling factors instead of a velocity ratio. These scaling approaches are also frequently applied in semi-batch mixing, which is discussed in Section 14.5.

Scale-up of static mixers for use in continuous mixing processes is beyond the scope of this chapter. See Ref. 14 for further reading on this topic.

14.4 COMMON MIXING EQUIPMENT

Because of the wide variety of mixing processes encountered in the industry, a great number of mixing types and geometries have been developed, including fluidized beds, jet nozzles, and gas sparging. Here, however, we will focus on mechanically stirred tanks and examine the typical impeller types used in this application. Such stirred vessels are used for batch production of the vast majority of specialty chemicals and pharmaceuticals, for blending and homogenization, for creating dispersions, and for running chemical reactions.

14.4.1 Major Impeller Types Used in Batch Mixing

Batch vessels may employ a broad range of impeller designs, each optimized for a particular type of process duty. The impellers shown in Figure 14.4 are among the more common types used in agitated vessels in chemical processing. Some vessels use multiple impellers of different types on a single shaft to obtain better mixing results. For example, it is common to utilize a high-shear flat-blade turbine at the bottom and a high-flow pitched-blade impeller higher up the shaft in certain blending and dispersion operations.

Based on their design, impellers can be broadly categorized as generating an axial flow pattern or a radial flow pattern. In the case of a stirred tank, the axial flow pattern results in a pumping action, usually downward, that is very useful for preventing the settling of solids and generating good cross-mixing. Examples of axial flow impellers are marine propellers, pitched-blade turbines and hydrofoils such as the Lightnin® A-310. These impellers will be found in crystallizers, solid suspension applications, and the like. Radial flow impellers do not tend to generate a vertical flow field, but tend more to push the fluid outward radially from the impeller. Most high-shear impellers, such as flat-blade turbines or paddles, are radial flow styles. Figure 14.9 illustrates axial and radial flow patterns.

One design commonly seen in the industry is the retreat curve impeller (RCI), sometimes called a "crowfoot" impeller. Originally designed to prevent flexing and cracking of the enamel coating when mixing viscous polymers, this impeller has been ubiquitous in glass-lined chemical reactors for decades. Nowadays, it is being largely replaced in glass-lined reactors by the curved-blade turbine (CBT) for general process mixing.

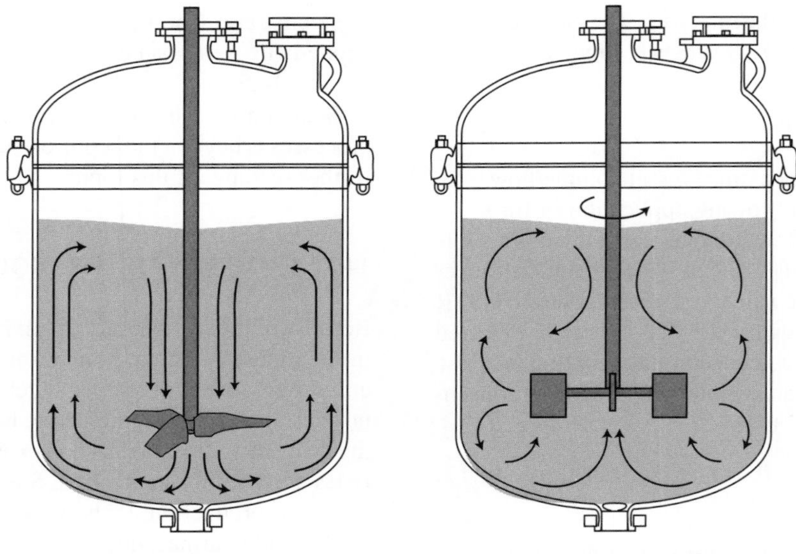

FIGURE 14.9 Typical stirred tank flow patterns. This figure shows an A-310 hydrofoil generating axial flow and a flat paddle impeller generating radial flow. Many impellers or combinations of impellers exhibit components of both types of flow patterns.

Close-clearance impellers, the so-called anchor styles, serve a rather specialized need in mixing highly viscous or non-Newtonian fluids, since a high-speed center-shaft impeller might simply rotate in the liquid without generating any movement at the vessel wall. The anchor provides this action near the wall that is critically important when heating or cooling the batch in a jacketed vessel. Often an anchor will be combined with a center mounted high-speed turbine to achieve sufficient heat transfer and good bulk mixing. Figure 14.10 illustrates two types of anchor blades. Pitched anchors and helical designs, albeit more expensive to construct than a flat anchor, can provide both wall motion and good bulk mixing.

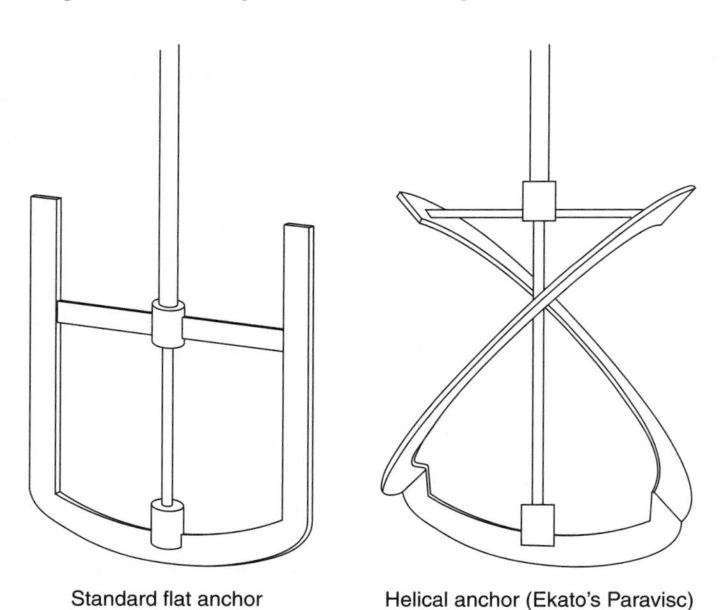

FIGURE 14.10 Anchor type impellers. The flat anchor generates motion at the wall, which is critical for heat transfer in mixing viscous liquids. Pitched anchor or helical styles (such as the "Paravisc" model designed by Ekato, Inc.) generate this motion at the wall and better bulk mixing throughout the rest of the vessel.

The very fact that so many types of impellers are used industrially further illustrates that there are many types of mixing duties, and no one impeller type is suitable for all. This is another source of difficulty in properly scaling up from the chemistry lab, where flat PTFE paddles are used almost exclusively for all overhead mixing service.

14.4.2 Mixing Baffles

Mixing baffles play a critical role in achieving efficient mixing in cylindrical vessels by preventing swirling and vortexing, increasing turbulence and cross-mixing, and providing better distribution of kinetic energy, especially for low-viscosity fluids (viscosity $< 5000\,\text{cP}$). Their use is limited to high-speed impeller applications, and would not be found in vessels utilizing close-clearance impellers such as anchor or helical impellers.

Many baffle designs exist, including those shown in Figure 14.11. The majority of those shown can be found in various glass-lined vessels, and are designed to be suspended from the vessel head. The rightmost baffle illustrated would be more typically used in stainless steel or other metal vessels where bolting directly to the wall is feasible. A space is normally left between the vessel wall and the baffle to allow flow and prevent collection of material there and simplify cleaning.

The introduction of baffles can actually have unwanted effects in certain cases, for example, tanks used for the dissolution of solids. Solids that are difficult to wet or that tend to float on the surface of the liquid may require the presence of a strong vortex to draw the material under the liquid surface. Baffles tend to reduce or eliminate this vortex, and can actually make this sometimes problematic processing step more difficult.

14.4.3 High-Shear Impellers

As part of the discussion on conventional impeller types in Section 14.4, impellers were described as axial flow (e.g., A-310), mixed flow (e.g., pitched-blade turbine), or radial flow (e.g., flat-blade turbine). With respect to shear in stirred tanks, axial, mixed, and radial flow impellers are considered to be low, medium, and high shear, respectively. As discussed in Section 14.2, there are various definitions for impeller shear. Average shear in the impeller region as defined by the Metzner–Otto relationship is one definition of shear that supports the categorization of impeller types given above.

While a radial flow impeller is considered high shear among conventional impellers, operations that are intended to create dispersions may require higher shear than can be produced by standard radial flow impellers. While gas/liquid dispersions are often created with flat-blade turbines, liquid/liquid and solid/liquid dispersions usually require higher shear to reduce droplet or particle sizes to desired levels. For dispersions that must be stable or settle slowly on standing, particle sizes of less than $10\,\mu\text{m}$ are usually required. When solid particles require deagglomeration or attrition to achieve the desired size, intense shear stresses must be generated at the length scale of single particles.

For the kinds of applications described above, the preferred dispersion devices are high-speed disperser blades or rotor/stator homogenizers. High-speed dispersers are simple devices that can be used in a stirred tank configuration, but are operated at much higher tip speeds than conventional impellers. Figure 14.12 shows two high-speed disperser blades of different designs. The blade on the right has a smaller number of teeth, but the teeth are larger than the standard Cowles design on the left. The blade with fewer, larger teeth will generate more flow than the Cowles blade,

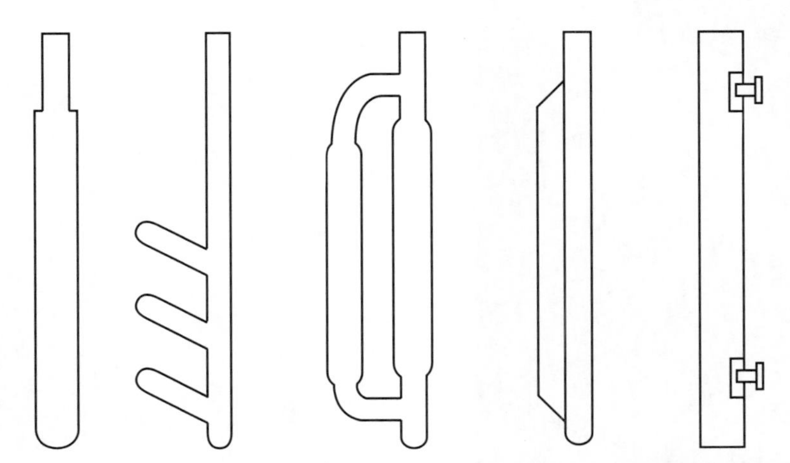

FIGURE 14.11 Various mixing baffle designs found in industrial tanks. From left to right: beaver tail, finger baffle, D-type baffle, fin baffle (all of which can be found in glass-lined vessels and do not attach to vessel side), and flat rectangular style for bolting directly to inner side wall of vessel.

FIGURE 14.12 Two high-speed disperser blades of different designs. The blade on the right has a smaller number of teeth, but the teeth are larger than the standard Cowles design on the left. The blade with fewer, larger teeth will generate more flow than the Cowles blade, but sacrifices some shear to achieve this.

but sacrifices some shear to achieve this. Scale-up of high-speed dispersers is typically done by tip speed. Commonly used tip speeds for these devices range from 2 to 25 m/s. The low end of this range is adequate for delumping of solids being introduced into a mixing tank, while the high end of the range is typical for producing fine particle dispersions.

Rotor/stator homogenizers provide a higher range of shear and energy dissipation than can be achieved by high-speed dispersers. A typical rotor/stator homogenizer is shown in Figure 14.13. While the usual tip speeds (5–50 m/s) are not that much higher than high-speed dispersers, much of the energy dissipation occurs within a small volume of liquid near the rotor and stator. This results in energy dissipation rates from 10^3 to 10^5 W/kg. This intense shear field results in very high shear stresses being transmitted to particles as they pass through the rotor/stator. Figure 14.14 shows the distribution of turbulent kinetic energy in a rotor/stator as predicted by CFD [15].

FIGURE 14.13 A typical rotor/stator homogenizer.

There are many design variations of rotor/stator units, which alter the balance between pumping and shear. Rotor/stator homogenizers can be used in batch mode within a stirred tank or in-line. In a stirred tank, it is best to provide an additional impeller to provide circulation and rely on the rotor/stator unit only to produce shear. In this way, the two effects can be decoupled, providing better control. This approach is essential when the fluid being mixed has significant yield stress. For yield stress fluids, multishaft mixers can offer both close-clearance impellers and rotor/stator homogenizers. While in-line rotor/stator homogenizers provide some pumping, it is best to use a separate pump so that flow and shear can be controlled independently.

Tip speed is the most common approach for scaling up rotor/stator homogenizers. Given the many design variations that are available and the complexity of some of the designs, successful scale-up depends on geometric similarity of the rotor/stator unit used at different scales. That being said, geometric similarity is not appropriate for the spacing between rotor/stator teeth. That gap must not be scaled up,

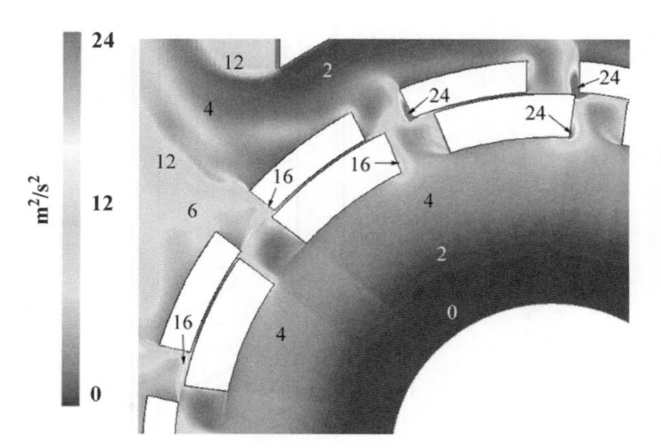

FIGURE 14.14 The distribution of turbulent kinetic energy in a rotor/stator as predicted by CFD. From Ref. 15. Reprinted with permission of John Wiley & Sons, Inc.

but must remain constant across scales for a given process to ensure the same intensity of shear [15].

14.5 SCALE-UP OF CHEMICAL REACTIONS

The scale-up of processes involving chemical reactions presents a special set of challenges, particularly in nonhomogeneous systems or in semi-batch reactions involving the controlled addition of reactive chemical reagents to a stirred vessel. Most chemical reactions are not 100% selective, that is to say that unwanted side reactions often accompany the main reaction. These reactions can reduce yield by consuming valuable starting materials, and the products of these side reactions can accumulate as contaminants that may be difficult or impossible to remove from the final product. These contaminants can also alter the crystal structure of some products, resulting in unexpected polymorphic crystal forms with poor solubility or other undesirable physical characteristics.

This can be a particularly vexing issue in an industry such as pharmaceutical manufacture, in which product quality is highly regulated and the presence of mere tenths of a percent of an unwanted impurity can result in an entire batch being rejected. Unfortunately, scaling up certain classes of reactions from a laboratory to commercial scale almost inevitably results in changes in selectivity, and often not for the better.

When a reaction is optimized in a laboratory setting, mixing is usually not an issue that comes into serious consideration, because laboratory stirrers provide very vigorous mixing and blend times in the 1–2 s range or less. However, upon scale-up, the reaction will be run in a system in which blend time may be on the order of 30 s or longer (see Figure 14.6).

Consequently, as the reactive material is added to the reactor, it may swirl around in a highly concentrated plume for sometime before it becomes dispersed throughout the reaction mixture. This localized zone of high concentration can cause an increase in side reactions that may not have been an issue in the laboratory, resulting in poor reaction selectivity and low batch quality. This is an extremely common problem in reaction scale-up, and below we discuss some possible solutions. First, some examples of reactions that are affected by this phenomenon (so-called "mixing-limited reactions") are in order.

14.5.1 Examples of Mixing-Limited Reactions

Consider the so-called Bourne reaction [16], in which trimethoxybenzene (TMB) is treated with bromine to produce monobromotrimethoxybenzene (see Scheme 14.1).

This reaction suffers from a consecutive competing reaction, in which the mono-Br product reacts with a second Br to form the di-Br product. The rate of the primary reaction (k_1) is about 1000× faster than that of the secondary reaction (k_2), so one would expect little of the di-Br to form. However, the rate of the secondary reaction is still fast enough that under typical mixing conditions, the mono-Br product is not swept away from the site of the reaction quickly enough and undergoes the second reaction. Figure 14.15 shows that reaction selectivity can be somewhat improved by increasing the intensity of agitation.

Another excellent example of the effect of mixing on reaction selectivity is the stereoselective enzymatic hydrolysis of a chiral organic ester (Scheme 14.2).

This is a biphasic reaction in which the enzyme is dissolved in the aqueous phase, and preferentially hydrolyzes only one enantiomer of the chiral ester (an insoluble organic liquid) as it slowly enters the aqueous phase by diffusion. Base is added to maintain a constant pH as the acid product is formed.

Figure 14.16 shows the results of the reaction under conditions of good mixing and poor mixing. Note that under conditions of rapid mixing, the product purity is on the order of 99%, a result of the intrinsic selectivity of the enzyme, until the conversion reaches roughly 50%, at which point the preferred enantiomer is essentially all consumed, and the enzyme begins to hydrolyze the other enantiomer. This results in reduced product purity at high conversions. However, with poor mixing, the addition of the base causes nonselective chemical hydrolysis at the point of addition, resulting in lower product purity even at very low conversions.

A final example will illustrate the importance of mixing in heterogeneous reacting systems due to its effect on mass transfer. Consider the reaction between phenol and benzoyl chloride shown in Scheme 14.3.

This reaction can be run in a biphasic system. The phenol is in aqueous solution, and the benzoyl chloride is a non-water-soluble organic liquid. In this process, the observed reaction rate is a function of both the intrinsic reaction kinetics and the rate at which the benzoyl chloride diffuses into the aqueous phase where it can react. As mixing speed increases, so does interfacial surface area (the dispersion

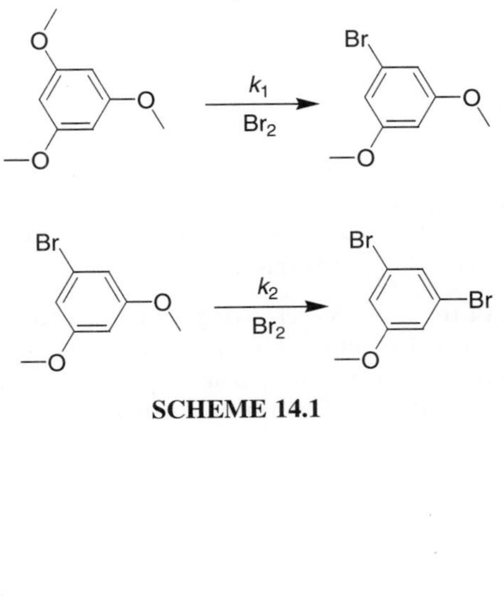

k_1, Br_2

Br

k_2, Br_2

Br Br Br

SCHEME 14.1

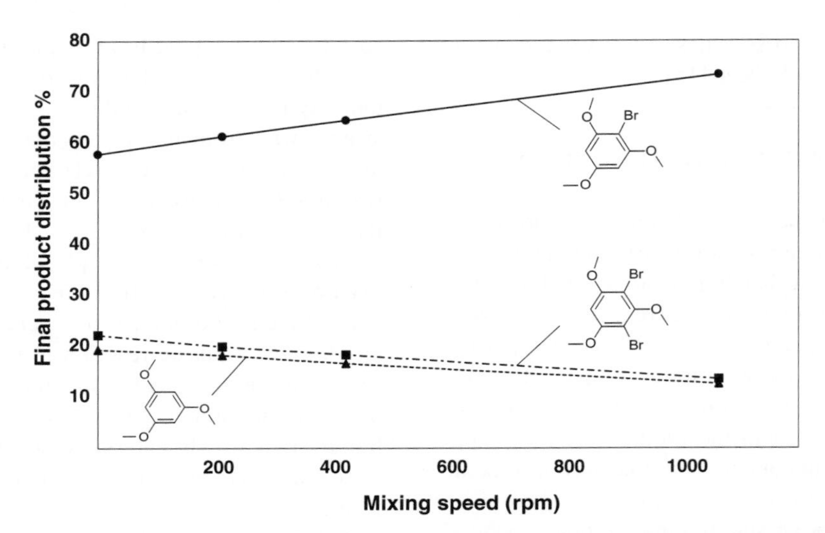

FIGURE 14.15 The effect of mixing on the selectivity of the trimethoxybenzene (Bourne) reaction [16].

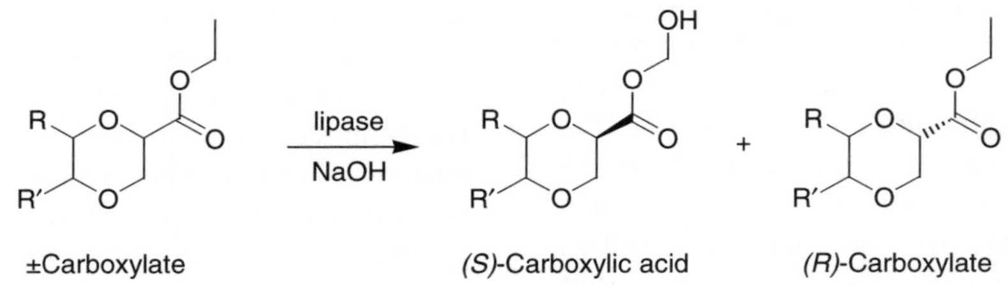

lipase
NaOH

±Carboxylate

(S)-Carboxylic acid

(R)-Carboxylate

SCHEME 14.2

droplets become smaller). This results in an observed increase in reaction rate because of the improved mass transfer, that is, the faster rate of transport of the benzoyl chloride into the aqueous phase (see Figure 14.17).

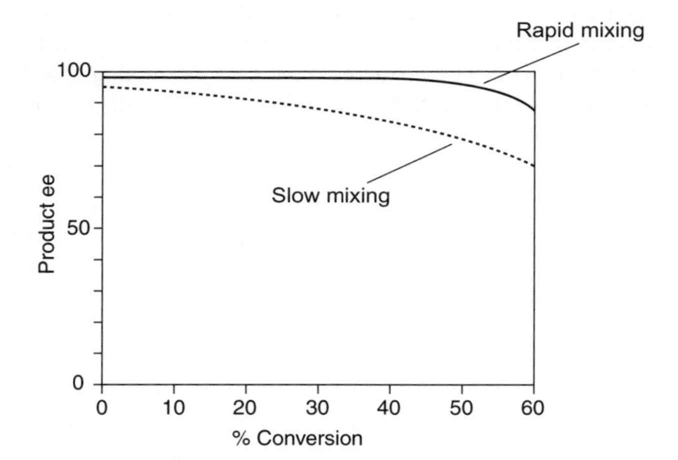

FIGURE 14.16 The effect of mixing intensity on stereoselectivity of the enzymatic hydrolysis of a chiral carboxylate ester. *Y*-axis is enantiomeric excess (ee) of the product, a measure of enantiomeric purity. *X*-axis is degree of enzymatic conversion.

14.5.2 Identifying Mixing-Limited Reactions

The types of reactions most likely to be affected by mixing upon scale-up are highly rapid reactions, such as acid–base neutralizations. It is wise to try to identify any mixing-dependent behavior of reactions in the laboratory prior to scale-up to minimize surprises and failed batches. Sometimes it is simply a matter of running the chemistry in the laboratory under conditions of intense, rapid mixing and slow, poor mixing. For example, for a controlled addition reaction, one could set up two side-by-side experiments. In one, the reagent is added slowly to a well-mixed flask; in the other, the reagent is added quickly to a flask with poor or no mixing. If there is a significant difference in product purity, then this system will likely experience issues at scale, and measures can be taken to minimize these effects prior to scale-up.

A more theoretical approach is to calculate the Damköhler number for the reacting system. The Damköhler number (Da) is a dimensionless reaction time that represents the dependence of a given chemical reaction on mixing. Da is a function of reaction rate constant, reaction order, and reactant concentrations, but in simple terms, for semi-batch reactions Da is generally defined as a ratio between some characteristic mixing timescale and the timescale of the reaction.

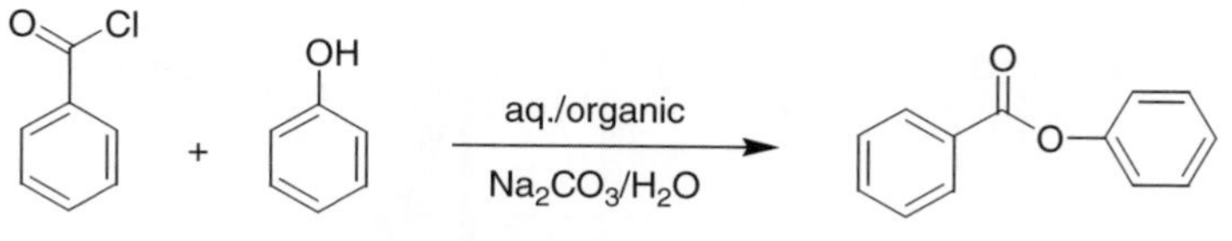

aq./organic

Na_2CO_3/H_2O

SCHEME 14.3

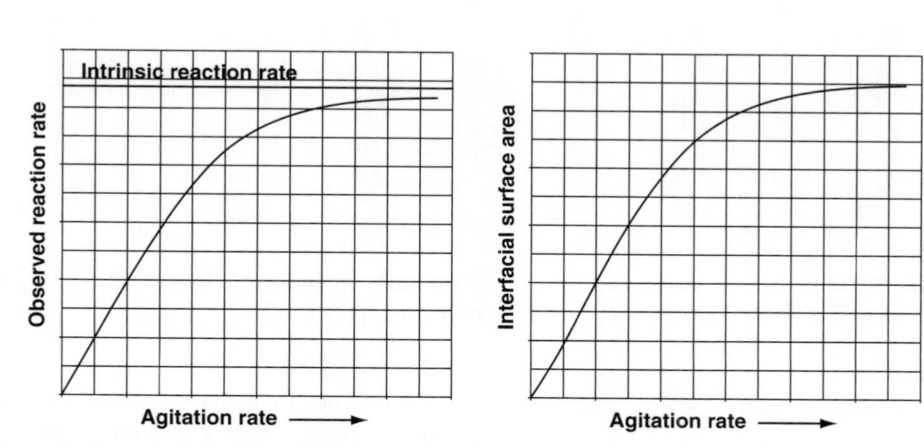

FIGURE 14.17 The effect of mixing speed on the reaction between benzoyl chloride and an aqueous solution of phenol. As agitation rate increases, so does interfacial surface area and diffusion rate, and consequently the observed reaction rate (after Ref. 17).

The higher the value of *Da*, the more susceptible is the reaction to mixing effects. Figure 14.18 shows that for a very rapid reaction, such as an acid–base neutralization, the Damköhler number is much larger than 1, whereas for slow reactions, such as hydrolysis of an ester, the value of *Da* is much smaller than 1. The higher the value of *Da*, the more likely it is that the reaction could suffer changes in selectivity upon scale-up.

14.5.3 Importance of Addition Point Design in Semi-Batch Reactions

Now that we understand one of the main causes behind poor reaction selectivity upon scale-up, we can examine some approaches to prevent it. In controlled addition reactions, selectivity can be improved by adding the reagent in such a way that it is dispersed and homogenized more rapidly. Therefore, rather than simply letting the reagent drip onto the surface of the batch or run down the side of the reactor, it should be injected at a zone of very high shear, such as right at the periphery of the rotating impeller using a delivery or "dip" tube.

For best results, the tube must be properly sized (i.e., small enough diameter and high enough flow velocity) to prevent backmixing in the tube that can lead to the same selectivity issues. Numerous setups are used to achieve rapid dispersion during chemical additions, some of which are shown in Figure 14.19. The perforated dispersion ring can be particularly useful in controlling pH by acid or base addition in biological or enzymatic systems that may be sensitive to high concentrations of these reagents. Some reactions have been significantly improved by spraying the reagent onto the surface of the batch by means of a "shower head"-type arrangement.

A number of other techniques are available for scaling up mixing sensitive reactions. One common approach is to install a static or mechanically agitated mixer in a forced recirculation loop. The reactive chemical reagent is then injected in a controlled fashion into the recirculation line just upstream of the in-line mixer. This can speed up dispersion and minimize the likelihood that a zone of very high concentration will exist in the vessel for any significant length of time. The static mixers, of which there are many designs, are particularly useful because they are generally well charac-

$$Da = \frac{\text{Characteristic timescale}}{\text{Reaction timescale}}$$

Acid-base neutralization	Base hydrolysis of ester
(rapid reaction)	(slow reaction)
$k \sim 10^{11}$ L/mol s	$k \sim 10^{-1}$ L/mol s
rxn timescale ~ 10^{-9} s	rxn timescale ~ 10^{3} s
$Da = \frac{10\ \text{s}}{10^{-9}\ \text{s}} = 10^8$	$Da = \frac{10\ \text{s}}{10^{3}\ \text{s}} = 0.01$

FIGURE 14.18 An example of Damköhler number (*Da*) calculations for a rapid reaction and a slow reaction. The higher the value of *Da*, the more susceptible the reaction is to changes in selectivity due to mixing effects. The characteristic mixing timescale in this example is the blend time, here set to a typical value of 10 s.

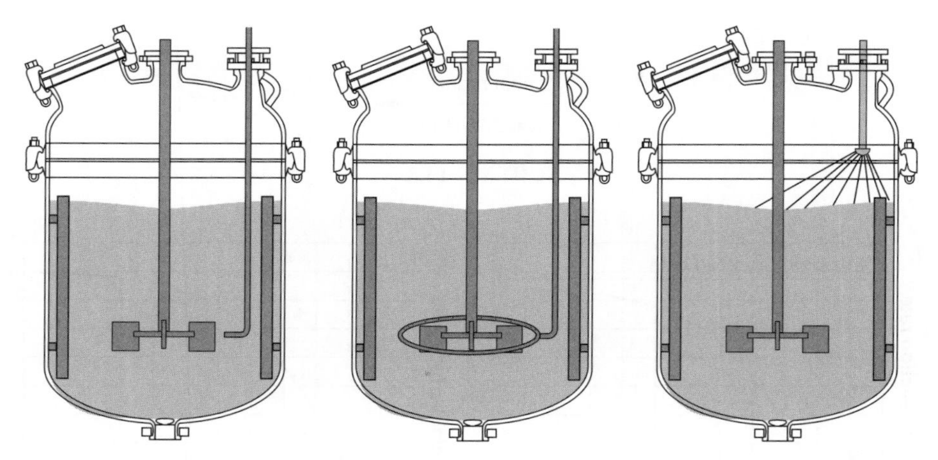

FIGURE 14.19 Some systems for improving performance of semi-batch or controlled addition reactions (from left to right: addition tube, dispersion ring, spray nozzle).

terized and have no moving parts, which minimizes maintenance.

14.6 CFD AND OTHER MODELING TECHNIQUES

One of the major tools for studying mixing and predicting mixing behavior in process equipment is CFD modeling. The advent of high-speed personal computers has made CFD widely available and it is finding use in many areas of technology, from plasma physics to the relatively simple liquid agitation we are concerned with here. Nonetheless, accurate modeling of fluid behavior requires the simultaneous calculation of huge numbers of equations and even the simplest of problems consumes considerable CPU time.

Basically, a mathematical model is constructed of the system of interest by dividing the fluid volume into hundreds of thousands or perhaps millions of contiguous cells (the model mesh; see, for example, Figure 14.20). The CFD software then tries to simultaneously solve the numerous momentum, velocity, force, heat transfer, and reaction mass balance equations associated with each of these cells in an attempt to converge on a single solution. When successful, these programs can accurately predict torque and mixing power requirements and can generate visual images or animations of fluid motion and circulation patterns that aid in identifying zones of high shear, stagnation, or other nonideal mixing behavior. Figure 14.14 is an example of this type of image.

Several commercial software platforms are available for CFD modeling, but they are all quite expensive and require considerable expertise to properly code the model, create the mesh, and run the simulations. Needless to say, the success of the model depends on the accuracy of the rheological, chemical, and geometric data that are used to build it.

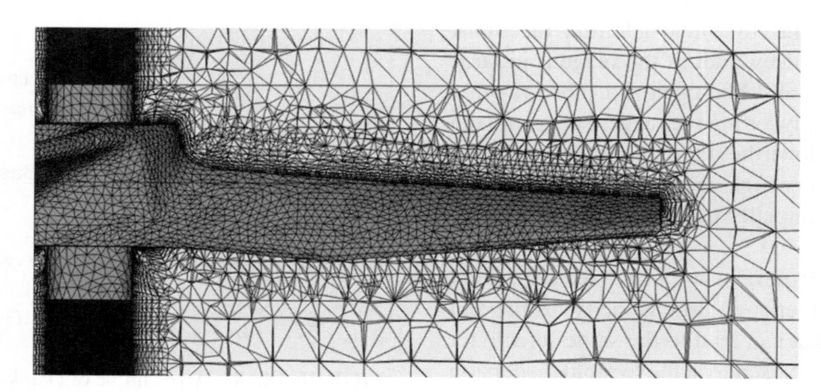

FIGURE 14.20 Showing the "wire mesh" for a portion of an A-310 impeller for a CFD simulation of a stirred tank. The CFD mesh can consist of millions of three-dimensional cells, usually with a finer grid size in the vicinity of the impeller where highest velocity and shear occur, and coarser in the bulk fluid to save CPU time.

Some uncertainty is inevitable when conducting mixing experiments or modeling studies using computer simulations. For this reason, most experts agree that for critical work, the CFD model should be validated by comparing predicted results with experimental measurements at at least two different scales. In a typical scenario in which an industrial mixing system is to be designed based on the results of CFD modeling, the best results will be obtained if experiments are conducted first at some laboratory scale, and then at a small pilot scale. CFD simulations of these two smaller scale operations would then be carried out, and once fine tuned to the point where predictions agree well with experiments, the model can be used for simulations to support the full-scale design.

SYMBOLS AND ABBREVIATIONS

A	tracer concentration (outlet)
A_0	tracer concentration (inlet)
B	tank baffle width
C	impeller bottom clearance
D	impeller diameter
Da	Damköhler number
F_{CSTR}	concentration ratio in CSTR
Gal	U.S. gallon
h_i	internal (process-side) heat transfer coefficient
HP	horsepower
k	thermal conductivity
k'	dimensionless Metzner–Otto constant
K	rheological consistency index
n	rheological behavior index
N	impeller rotational speed
N_{Re}	impeller Reynolds number
N_P	impeller power number
N_Q	impeller flow number
Nu_L	Nusselt number
P	mixing power
Q	flow rate
S_t	tip speed
t	time
$\bar{t}$	mean residence time
T	tank diameter
T_Q	torque applied to a mixer shaft
V	batch liquid volume
W	watt
Z	liquid height in batch vessel
α	ratio mean residence time/batch blend time
ε	rate of turbulent energy dissipation
ρ	density
η	Kolmogorov eddy length
ν	kinematic viscosity
$\dot{\gamma}$	shear rate
μ	viscosity
τ	shear stress
$\tau 0$	yield stress
θ	dimensionless residence time
Θ	batch blend time

REFERENCES

1. Oldshue JY.In: Ulbrecht JJ, Patterson GK,editors, Mixing of Liquids by Mechanical Agitation, Gordon and Breach, New York, 1985, Chapter 9.
2. Uhl VW, Von Essen JA. Scale-up of fluid mixing equipment. In: Uhl VW, Gray JB,editors, Mixing: Theory and Practice, Vol. 3, Academic Press, 1986, pp. 155–167.
3. Johnstone RE, Thring MW. Pilot Plants, Models, and Scale-Up Methods, McGraw-Hill, New York, 1957.
4. Kresta SM, Brodkey RS. Turbulence in mixing applications. In: Paul EL, Atiemo-Obeng VA, Kresta SM, editors, Handbook of Industrial Mixing, Wiley-Interscience, Hoboken, NJ, 2004, Chapter 2.
5. VisiMix Mixing Simulation Software, http://www.visimix.com/.
6. Hemrajani RR, Tatterson GB. Mechanically stirred vessels. In: Paul EL, Atiemo-Obeng VA, Kresta SM,editors, Handbook of Industrial Mixing, Wiley-Interscience, Hoboken, NJ, 2004, Chapter 6.
7. Rushton JH. et al. Power characteristics of mixing impellers, *Chem. Eng. Prog.* 1950;46(8):395.
8. Grenville RK, Nienow AW. Blending of miscible liquids. In: Paul EL, Atiemo-Obeng VA, Kresta SM,editors, Handbook of Industrial Mixing, Wiley-Interscience, Hoboken, NJ, 2004, Chapter 9.
9. Penney WR. *Chem. Eng.* 1971;78(7):86.
10. Macosko CW. Rheology: Principles, Measurements, and Applications, Wiley-VCH, New York, 1994.
11. Serth RW. Process Heat Transfer: Principles and Applications, Elsevier, 2007.
12. Levenspiel O. Chemical Reaction Engineering, 3rd edition, Wiley, Hoboken, NJ, 1999.
13. Roussinova VT, Kresta SM. Comparison of continuous blend time and residence time distribution models for a stirred tank, *Ind. Eng. Chem. Res.* 2008; 47 (10): 3532–3529.
14. Etchells AW, Meyer CF. Mixing in pipelines. In: Paul EL, Atiemo-Obeng VA, Kresta SM, editors, Handbook of Industrial Mixing, Wiley-Interscience, Hoboken, NJ, 2004, Chapter 7.
15. Atiemo-Obeng VA, Calabrese RV. Rotor-stator mixing devices. In: Paul EL, Atiemo-Obeng VA, Kresta SM, editors, Handbook of Industrial Mixing, Wiley-Interscience, Hoboken, NJ, 2004, Chapter 8.
16. Bourne J. Mixing and the selectivity of chemical reactions, *Org. Proc. Res. Dev.* 2003;7(4),471–508.
17. Atherton JH, Carpenter KJ. Process Development: Physicochemical Concepts, Oxford Science, 1999.

15

STIRRED VESSELS: COMPUTATIONAL MODELING OF MULTIPHASE FLOWS AND MIXING

AVINASH R. KHOPKAR

Engineering Sciences, Dow Chemical International Private Ltd., Pune, Maharashtra, India

VIVEK V. RANADE

Tridiagonal Solutions Private Ltd. and National Chemical Laboratory, Pune, Maharashtra, India

Stirred vessels are widely used in pharmaceutical industry to carry out a large number of multiphase applications (such as reactions, precipitations, and emulsions) and recipes. They offer unmatched flexibility in operation to manipulate the performance of the vessel. A skilled reactor engineer can use the offered flexibility to tailor the fluid dynamics, and therefore performance of a reactor by appropriately adjusting the reactor hardware and operating parameters. Performance of stirred vessels is influenced by variety of parameters such as the number, type, location and size of impellers, degree of baffling, sparger type, inlet/outlet locations, aspect ratio, and reactor shape. It is therefore essential to first translate the "wish list" of the reactor performance into a "wish list" of desired fluid dynamics. Despite the widespread use of these stirred vessels, the fluid dynamics in them, essentially for multiphase flows is not well understood. This lack of understanding and the knowledge of the underlying fluid dynamics have caused reliance on empirical information [1–3]. Available empirical information is usually described in an overall/global parametric form. This practice conceals detailed localized information, which may be crucial in the successful design of the process equipment. Reliability of such empirical information and, in particular, extrapolation beyond the range of parameters studied often remains questionable. It is therefore, essential to develop and apply new tools to enhance our understanding of the fluid dynamics prevailing in stirred vessels. Such understanding will be useful in devising cost-effective and reliable scale-up of stirred vessels.

In last two decades, with improvement in the knowledge of numerical techniques, turbulence models and the availability of fast computational resources have made it possible to develop models based on computational fluid dynamics and use them for *a priori* prediction of the flow field in chemical process equipment [4–6]. However, unlike single-phase flow, which is possible to predict with reasonable confidence [6], the computational models capable of predicting real-life turbulent multiphase flows involving complex geometries and with a wide range of space and timescales are yet to be established. The development of such models will be a significant step toward the prediction of local fluid dynamics. Such models will be useful for exploring the possibilities for performance enhancement of existing reactors, for evolving better reactor configurations and for reliable scale-up. In this chapter, we have critically reviewed the state of the art of computational modeling of multiphase flows in stirred vessels and demonstrated the extent of applicability of this with some examples and also provided our views on the path forward.

15.1 ENGINEERING OF MULTIPHASE STIRRED REACTORS

Multiphase stirred vessels are ubiquitous in pharmaceutical industry right from R&D to manufacturing. In many situations of practical interest, more than one phase need to be

Chemical Engineering in the Pharmaceutical Industry: R&D to Manufacturing Edited by David J. am Ende

contacted in a stirred vessel. In several cases, phase transitions such as generation of vapors by evaporation of volatile components, precipitation of solid particles via reactions or solidification and generation of liquid droplets via melting of solids or phase inversion occur in stirred vessels. Some examples of industrial multiphase processes carried out in stirred reactors are listed in Table 15.1. Engineering of these reactors begins with the analysis of the process requirements and evolving a preliminary configuration of the reactor. More often than not the reactor has to carry out several functions such as bringing reactants into intimate contact (to allow chemical reactions to occur), providing an appropriate environment (temperature and concentration fields by facilitating mixing, heat transfer and mass transfer) for adequate time, and allowing for removal of products. Naturally, successful reactor engineering requires expertise from various fields ranging from thermodynamics, chemistry, and catalysis to reaction engineering, fluid dynamics, mixing, and heat and mass transfer. Reactor engineer has to interact with chemists to understand intricacies of the considered chemistry. Based on such understanding and proposed performance targets, reactor engineer has to abstract the information relevant for identifying the characteristics of desired fluid dynamics of the reactor. Reactor engineer has to then conceive suitable reactor hardware and operating protocol to realize this desired fluid dynamics in practice.

The laboratory study and reactor engineering models, based on idealized fluid dynamics and mixing, help in this step. This step helps in defining performance targets of the reactor. The reactor engineer faces a major difficulty in translating the preliminary reactor configuration (laboratory scale or pilot scale) to the industrial reactor. Transformation of a preliminary reactor configuration to an industrial reactor proceeds through several steps. Some of these scale-up steps that are discussed in other chapters are highlighted below:

- *Scale-Down/Scale-Up Analysis*: It is essential to analyze the possible effects of scale of the reactor on the prevailing fluid dynamics and reactor performance. Conventionally, such an analysis is carried out with certain empirical rules (for example, equal power per unit volume and equal tip speed) and prior experience. However, it was observed that these rules do not guarantee the identical performance of reactor at two different scales. This can be explained by using the case of gas–liquid stirred reactor. A small-scale reactor provides a higher shear rate and more rapid circulations compared to a large-scale reactor. Gas dispersion, therefore, is often breakage (dispersion) controlled in a small-scale reactor but coalescence controlled in a large-scale reactor. The interfacial area per unit volume of reactor for gas–liquid interphase mass transfer decreases as the scale of the reactor increases for same specific power input.
- *Presence of Conflicting Process Requirements*: Presence of conflicting process requirements is also a major issue a reactor engineer needs to tackle in a multiphase stirred reactor. For example, the fluid dynamic characteristics required for better blending and heat transfer (flow controlled operations) are quite different from those required for better dispersion of a secondary phase and better mass transfer (shear controlled operations). Such conflicting process requirements make the task of evolving a "wish list" of desired fluid dynamics difficult. The reactor engineer needs to design for the conditions that achieve the most effective performance. It is therefore necessary to have a good understanding of the prevailing fluid dynamics and its relation with design parameters on one hand and with the processes of interest on other hand.
- *Designing New Reactor Concepts*: Development of reactor technologies relies on prior experience. Testing of new reactor concepts/designs are often sidelined due to lack of resources (experimental facilities, time, funding, and so on). Experimental studies have obvious limitations regarding the extent of parameter space that can be studied and regarding the extrapolation beyond the studied parameter space.

This brief review of the modeling of multiphase stirred reactors indicates that the detailed knowledge of the prevailing fluid dynamics will allow a reactor engineer to exploit the available degrees of freedom of stirred reactors. However, obtaining the detailed information on fluid dynamics in stirred reactors for multiphase flow is challenging. The complexity in modeling the fluid dynamics increases

TABLE 15.1 Some Industrial Applications of Multiphase Stirred Reactor

Phases Handled	Applications
Gas–liquid	Chlorination, oxidation, carbonylation, manufacture of adipic acid, oxamide, and so on
Gas–liquid–solid	Fermentation, hydrogenation, oxidation (*p*-xylene), waste water treatment, and so on
Liquid–liquid	Suspension and emulsion polymerization, oximations, methanolysis, extraction, and so on
Liquid–solid	Calcium hydroxide (from calcium oxide), anaerobic fermentation, regeneration of ion-exchange resins, leaching, and so on
Gas–liquid–liquid	Biphasic hydroformylation, carbonylation
Gas–solid	Stirred fluidized beds

significantly for multiphase flows. Till recent past, the complexity of fluid dynamics and multiphase processes occurring in stirred vessels was too overwhelming and most of the practical engineering decisions were based on empirical and semiempirical analysis. Several excellent reviews and books on such design procedures are available, for example, see Refs 1, 3, and 7). However, the information obtainable from these methods is usually described in an overall/global parametric form. This practice conceals detailed local information about turbulence and mixing, which may ultimately determine overall performance. The conventional approach essentially relies on prior experience and trial and error method to evolve suitable reactor hardware. This approach is being increasingly perceived to be expensive and time-consuming for developing better reactor technologies. It is necessary to adapt and develop better techniques and tools for relating reactor hardware with fluid dynamics and resultant transport processes.

In recent years, chemical engineers have started using the power of computational fluid dynamics (CFD) models to address some of these reactor engineering issues. CFD is a body of knowledge and techniques to solve mathematical models of fluid dynamics on digital computers. Considering the central role of stirred vessels in pharmaceutical industry, there is tremendous potential for applying these tools for better engineering of stirred vessels. Computational flow modeling can make substantial contributions to scale-up by providing quantitative information about the fluid dynamics at different scales. The computational model may offer a unique advantage for understanding the requirements of conflicting processes and their subsequent prioritization. The CFD model will allow a reactor engineer to switch on and off various processes and study interactions between different processes. Such numerical experiments can help to reduce and to resolve some of the challenges posed by conflicting demands made by different processes. CFD models can make valuable contributions to developing new reactor technologies by allowing *a priori* prediction of fluid dynamics for any configuration with just knowledge of reactor geometry, fluid properties, and operating parameters. These simulations allow detailed analysis, at an earlier stage in the design cycle, with less cost, with lower risk and in less time than experimental testing. It sounds almost too good to be true. Indeed, these advantages of CFD are conditional and may be realized only when the fluid dynamic equations are being solved accurately, which is quite difficult for most of the engineering flows of interest. It must be remembered that numerical simulations will always be approximate. There can be various reasons for differences between computed results and "reality" such as errors associated with fluid dynamic equations being solved, input data and boundary conditions, numerical methods and convergence, computational constraints, and interpretation of results.

It is indeed necessary to develop an appropriate methodology to harness the potential of CFD for better reactor engineering, design, and scale-up despite some of the limitations. This chapter is written with an intention of assisting practicing engineers and researchers to develop such methodology and approach.

Various aspects of computational flow modeling (CFM) and its application to multiphase stirred vessels are discussed and related in a coherent way. The emphasis is not on providing a complete review but is on equipping the reader with adequate information and tips to undertake a complex flow-modeling project. Modeling of single-phase flows and mixing in stirred vessels are not discussed and the reader is referred to Chapter 10 of Ref. 8. The scope of this chapter is restricted to multiphase flows and mixing. The basics of computational modeling and the extent of applicability of its applicability to simulating multiphase stirred reactors are discussed with examples of gas–liquid and solid–liquid flows. The insight obtained from these two examples may then be used to model the liquid–liquid and gas–liquid–solid stirred reactors. After describing these, possible applications to practical problems relevant to pharmaceutical industry are briefly discussed. Key conclusions and some suggestions for further work are outlined at the end of this chapter.

15.2 COMPUTATIONAL MODELING OF MULTIPHASE STIRRED REACTOR

The subject of modeling of multiphase flow processes is quite vast and covers a wide range of subtopics. It is virtually impossible to treat all the relevant issues in a single book, let alone in a single chapter. The scope of this chapter is restricted to modeling of dispersed multiphase flows in stirred reactors where the continuous phase is a liquid phase and a dispersed phase may be gas, liquid, or solid. There are mainly three approaches for modeling such dispersed multiphase flows:

- *VOF*: Volume of fluid approach (Eulerian framework for both the phases with reformulation of interface forces on volumetric basis);
- *EL*: Eulerian framework for the continuous phase and Lagrangian framework for all the dispersed phases; and
- *EE*: Eulerian framework for all the phases (without explicit accounting of interface between phases).

If the shape and flow processes occurring near the interface are of interest, VOF approach should be used. This approach is, however, naturally limited to modeling the motion of only a few dispersed-phase particles. The EL approach is suitable for simulating dispersed multiphase flows containing low volume fraction of the dispersed phases (motion of dispersed particles is not influenced by collisions). For denser dispersed-phase flows, it is usually necessary to use the EE approach. Considering that most of the pharmaceutical applications will involve dense dispersions, the scope of this chapter is restricted to EE approach. More information on modeling of other approaches may be found in Ref. 8 and references cited therein.

In the EE approach, the dispersed phases are also treated as continuum. All the phases "share" the computational domain and they may interpenetrate as they move within it. A concept of volume fraction of phase q, α_q, is used while deriving governing equations. Various averaging methods have been proposed (see Ref. 8 for more details). In this section, we will present a general form of governing equations for dispersed multiphase flows, which will be suitable for further numerical solution, without going into details of their derivation.'

15.2.1 Model Equations

For most of the operating regimes used in practice, flows in multiphase stirred vessels are turbulent. The statistical approach based on averaged equations is one of the most widely used approaches for modeling turbulent flows in practice. In this approach, an instantaneous value of any variable is decomposed into mean and fluctuating components. Two averaging approaches are commonly used for separating mean and fluctuating components, namely, Reynolds averaging and Favre averaging (see Ref. 8 for more details). Reynolds averaging is used for simple flows where instantaneous density exhibit very small turbulent fluctuations. Density-weighted averaging or Favre averaging is used when density fluctuations are important. The Favre averaging is now widely used in the commercial CFD packages. The mass and momentum balance equations governing such flows can be written as (Favre averaged equations for each phase without considering mass transfer):

$$\frac{\partial(\alpha_q\rho_q)}{\partial t} + \nabla.(\alpha_q\rho_q\vec{U}_{q,i}) = 0 \tag{15.1}$$

$$\frac{\partial(\alpha_q\rho_q\vec{U}_{q,i})}{\partial t} + \nabla.(\alpha_q\rho_q\vec{U}_{q,i} \times \vec{U}_{q,i}) =$$

$$-\alpha_q\nabla\vec{p} - \nabla.\left(\alpha_q\overline{\overline{\tau}}^{(lam)}_{q,ij}\right) - \nabla.\left(\alpha_q\overline{\overline{\tau}}^{(t)}_{q,ij}\right) + \alpha_q\rho_q g_i + \vec{T}_{fl} + \vec{F}_{12,i} \tag{15.2}$$

where α_q is the volume fraction (holdup) of phase q; ρ_q is the density of phase q; $\rightharpoonup U_q$ is the mean velocity of the phase q; p is the pressure shared by all the phases; and $\overline{\overline{\tau}}_{q,ij}{}^{(lam)}$ is the laminar shear stress. $\vec{T}_{fl}$ is the turbulent dispersion force accounting for the turbulent fluctuation in the phase volume fraction. It is modeled as

$$\vec{T}_{fl} = KV_{dr} \quad \text{where} \quad V_{dr} = -\left(\frac{D_p}{\sigma_{pq}\alpha_p}\nabla\alpha_p - \frac{D_q}{\sigma_{pq}\alpha_q}\nabla\alpha_p\right) \tag{15.3}$$

where V_{dr} is the drift velocity, D_p and D_q are the diffusivities of the continuous and dispersed phases, respectively, and σ_{pq} is the turbulent Prandtl number. The diffusivities, D_p and D_q can be calculated from the turbulent quantities following the work of Simonin and Viollet [9]. The turbulent Prandtl number σ_{pq} is usually set to 0.75–1.0. $\overline{\overline{\tau}}_{q,ij}{}^{(lam)}$ is the stress tensor in the phase q due to molecular viscosity and $\overline{\overline{\tau}}_{q,ij}{}^{(t)}$ is the Reynolds stress tensor of phase q (representing contributions of correlation of fluctuating velocities in momentum transfer). In 1877, Boussinesq postulated that the momentum transfer caused by turbulent eddies can be modeled based on an analogy between molecular and turbulent motions. Accordingly, the turbulent eddies are visualized like molecules, colliding and exchanging momentum obeying the laws similar to kinetic theory of gases. This approximation allows representation of the Reynolds stress tensor as

$$\overline{\overline{\tau}}^{(t)}_{q,ij} = \mu_{tq}\left(\left(\nabla\vec{U}_{q,i} + (\nabla\vec{U}_{q,i})^T\right) - \frac{2}{3}I(\nabla\vec{U}_{q,i})\right) \tag{15.4}$$

where, μ_{tq} is the turbulent viscosity of the phase q and I is the unit tensor. The turbulent viscosity, in contrast to the molecular viscosity, is not a fluid property but depends on local state of flow or turbulence.

The turbulent viscosity is rather a complex function of eddy length scale, turbulent kinetic energy, turbulent kinetic energy dissipation rate, turbulence timescale, velocity gradient, system geometry, wall roughness, fluid viscosity, and so on. Several turbulence models have been proposed to devise suitable methods/equations to estimate these characteristic length and velocity scales in order to close the set of equations. Despite the known deficiencies, the overall performance of the standard k-ε turbulence model for simulating flows in stirred vessels is adequate for many engineering applications [8]. Most of the modeling attempts of complex, turbulent multiphase flows mainly rely on the practices followed for the single-phase flows, with some *ad hoc* modifications to account for the presence of dispersed-phase particles. In this chapter, we present the standard k-ε turbulence model to estimate the turbulent viscosity of the liquid phase without going into critical review of different models and approaches. Additional details may be found in Ref. 8. The governing equations for turbulent kinetic energy, k, and turbulent energy dissipation rate, ε, are given below:

$$\frac{\partial}{\partial t}(\alpha_l\rho_l\phi_l) + \nabla.(\alpha_l\rho_l\vec{U}_{l,i}\phi_l) = -\nabla.\left(\alpha_l\frac{\mu_{tl}}{\sigma_{\phi l}}\nabla\phi_l\right) + S_{\phi l} \tag{15.5}$$

where ϕ_l is the turbulent kinetic energy or turbulent energy dissipation rate in the liquid phase, $\sigma_{\phi l}$ denotes the turbulent

Prandtl number for variable ϕ, and $S_{\phi l}$ is the corresponding source term for ϕ in liquid phase. Note that the turbulence equations are solved only for the continuous liquid phase. Source terms for turbulent kinetic energy and dissipation can be written as

$$S_{kl} = \alpha_l[(G_l + G_{el}) - \rho_l \varepsilon_l]$$
$$S_{\varepsilon,l} = \alpha_l \frac{\varepsilon_l}{k_l}[C_1(G_l + G_{el}) - C_2 \rho_l \varepsilon_l] \tag{15.6}$$

where G_l is generation in the liquid phase and G_{el} is extra generation (or dissipation) of turbulence in the liquid phase. Generation due to mean velocity gradients, G_l, and μ_{tl}, turbulent viscosity was calculated as

$$G_l = \frac{1}{2}\mu_{tl}\left(\nabla \vec{U}_{l,i} + (\nabla \vec{U}_{l,i})^T\right)^2 \quad \mu_{tl} = \frac{\rho_l C_\omega k_l^2}{\varepsilon_l} \tag{15.7}$$

Extra generation or damping of turbulence due to the presence of dispersed-phase particles is represented by G_{el}. Kataoka et al. [10] have analyzed the influence of the gas bubbles on liquid-phase turbulence. Motion of larger bubbles generates extra turbulence. However, their analysis indicates that the extra dissipation due to small-scale interfacial structures almost compensates for the extra generation of turbulence due to large bubbles. Numerical experiments on bubble columns also indicate that one may neglect the contribution of extra turbulence generation (see Ref. 11 for more details). Therefore, for stirred vessels, where impeller rotation generates significantly higher turbulence than that observed in bubble columns, the contribution of the additional turbulence generation due to bubbles can be neglected.

Following the general practice, the same values of parameters proposed for single-phase flow ($C_{1\varepsilon} = 1.44$, $C_{2\varepsilon} = 1.92$, $C_{3\varepsilon} = 1.3$, $C_\mu = 0.09$, $\sigma_k = 1.0$, and $\sigma_\varepsilon = 1.3$) may be used to simulate the turbulence in two-phase flow. In the dispersed k-ε turbulence model, no extra transport equations were solved for estimating the turbulent quantities for dispersed phase. Instead, a set of algebraic relations can be used to couple the dispersed-phase turbulence to continuous-phase turbulence using Tchen's theory [9, 12]. The turbulence of dispersed-phase depends mainly on three important timescales: characteristic time of turbulent eddy τ_1^t, bubble relaxation time τ_{12}^b, and eddy–bubble interaction time τ_{12}^t (see Ref. 13 for more details). This approach of modeling turbulent dispersed phase is computationally less expensive and can adequately simulate the turbulence in two-phase flow with low dispersed-phase holdup (<10%). In case of higher dispersed-phase holdup (>10%), simulating turbulence equations for individual phases may be required.

Interphase coupling terms make multiphase flows fundamentally different from single-phase flows. The formulation of time-averaged $\vec{F}_{12,i}$, therefore, must proceed carefully. The interphase momentum exchange term consists of four different interphase forces: Basset history force, lift force, virtual mass force, and drag force [14]. Basset force arises due to the development of a boundary layer around bubbles and is relevant only for unsteady flows. The Basset force involves a history integral, which is time-consuming to evaluate and in most cases, its magnitude is much smaller than the interphase drag force. Considering this, the Basset history force is usually not considered for simulating dispersed multiphase flows in stirred vessels. The interphase momentum exchange term, which included lift, virtual mass, and drag force terms, is written as

$$\vec{F}_{12,i} = \vec{F}_{D,i} + \vec{F}_{VM,i} + \vec{F}_{lift,i} \tag{15.8}$$

The virtual mass term in i-direction is given as

$$\vec{F}_{VM,i} = \alpha_2 \rho_1 C_{VM}\left(\frac{D\vec{U}_{2,i}}{Dt} - \frac{D\vec{U}_{1,i}}{Dt}\right) \tag{15.9}$$

where C_{VM} is virtual mass coefficient. In this chapter, the value of C_{VM} is set to 0.5.

The lift force in i-direction is given as

$$\vec{F}_{L,i} = -0.5\rho_l \alpha_l \alpha_g (V_{l,i} - V_{g,i})(\nabla \times V_{l,i}) \tag{15.10}$$

The interphase drag force exerted on phase 2 in i-direction is given by

$$\vec{F}_{D,i} = -\frac{3\alpha_1\alpha_2\rho_1 C_D\left(\sum(\vec{U}_{2,i} - \vec{U}_{1,i})^2\right)^{0.5}(\vec{U}_{2,i} - \vec{U}_{1,i})}{4d_b} \tag{15.11}$$

where C_D is a drag coefficient. This expression can be generalized to more than one dispersed phases in a straightforward way. It is necessary to correct the estimation of drag coefficient to account for the particle size distribution and nonspherical shapes of the particles, for the presence of other particles and for the presence of prevailing turbulence. Specific discussion of these as well as formulation of boundary conditions related to simulations of multiphase flows in stirred vessels is included in the following sections.

Denser dispersions of gas or liquid phases within a continuous liquid phase lead to issues such as coalescence and break-up. It is possible to extend the approach to incorporate population balance models to account for such processes. However, this may require significantly larger computational resources as well as input data on model parameters of coalescence and break-up kernels. Presence of dense solids suspension exhibits various additional complexities and requires substantial modifications of the governing equations. The governing equations for such cases are not included in this chapter for the sake of brevity and may be found in Ref. 8. Other conservation equations (enthalpy and species) for multiphase flows that can be written following the similar general format are also not included in this chapter.

It is important to formulate appropriate boundary conditions to solve these governing equations. Most of the turbulence models mentioned earlier are valid for flows away from the walls. Presence of wall alters turbulence in a nontrivial way. It damps turbulence in the region very close to the wall. At the outer part of the near-wall region, turbulence is rapidly generated due to the large gradient of mean velocity. The correct representation of influence of walls on turbulent flows is an important aspect of simulating wall-bounded flows such as stirred tanks. The "wall function" approach in which the viscosity affected inner regions (viscous and buffer layers) are not modeled is one of the most widely used approaches to represent wall boundary conditions. In this approach, semi-empirical formulas (wall functions) are used to bridge the viscosity-affected region between the wall and the fully turbulent region. The current practice is to use the standard wall functions even for multiphase flows, which is a reasonable approximation when dispersed-phase particles stay away from the walls. The top surface of the baffled stirred vessels is usually assumed to be flat and is usually modeled as a free slip wall. For the case of gas–liquid flows in stirred tanks, gas bubbles are allowed to escape from the top wall by implementing appropriate boundary conditions (see Ref. 8). In some cases, it may be important to quantify and to consider deformation of the top surface because of the rotating impeller and associated vortex (especially for unbaffled stirred tanks). Volume of fluid (VOF) approach may be used for such cases. Such cases are, however, not discussed in this chapter. The scope is restricted to turbulent multiphase flows in baffled vessels where the top surface may be assumed to be flat. Other boundary conditions, such as inlet and outlet, are specified by following standard practices for CFD simulations. Application of these equations to simulate multiphase flows in stirred vessels is discussed below.

15.2.2 Application to Simulate Gas–Liquid Flow in Stirred Reactor

Flow in baffled stirred reactors can be modeled by various approaches (see Chapter 10 of Ref. 8). These approaches differ in their handling of impeller rotation while solving the governing equations. Multiple reference frame (MRF) approach is one of the most widely used approaches and the scope of present discussion is restricted to this approach. In the MRF approach, a fictitious cylindrical zone with a radius more than that of impeller blade tips and less than inner edges of the baffles and height sufficient to include an entire impeller is defined. If there are more than one impellers, such inner zones are defined around each impeller. The flow characteristics of such inner regions are solved using rotating framework. These results are used to provide boundary conditions for the outer region (after azimuthal averaging), flow in which, is solved using a stationary framework. More information on other approaches are discussed in Ref. 8 and references cited therein.

After establishing the approach for handling impeller rotation, the most important step in the simulation of a gas–liquid stirred reactor is the appropriate selection of interphase force formulations. They play a very important role while simulating gas dispersion. Lane et al. [15] carried out order of magnitude analysis of all interphase forces. They observed that in the bulk region of the stirred reactor interphase drag force dominates the total magnitude of interphase forces and hence can determine the gas dispersion pattern. A few studies highlighted the influence of interphase drag force on the predicted gas holdup distribution (for example, see Refs 16 and 17). However, much information is not available in the literature on the virtual mass force and lift force and their effect on the predicted gas holdup distribution. To explain the influence of different interphase forces, we reproduce some of the results obtained by Khopkar and Ranade [17]. They have carried out simulations of gas–liquid flow in a stirred vessel for an experimental setup used by Bombac et al. [18]. All the relevant dimensions such as impeller diameter; impeller off-bottom clearance; reactor height and diameter; and sparger location and diameter were the same as used by Bombac et al. [18]. Considering the symmetry of the geometry, half of the reactor was considered as a solution domain (see Figure 15.1). The solution domain and details of the finite volume grid used was similar to those used by Khopkar and Ranade [17]. A QUICK discretization scheme with SUPERBEE limiter function (to avoid non-physical oscillations) was used. Standard wall functions were used to specify wall boundary conditions. The computational results are discussed in the following section.

15.2.2.1 Interphase Forces

Interphase Drag Force In stirred reactors, bubbles experience significantly higher turbulence generated by impellers. Unless, the influence of this prevailing turbulence on bubble drag coefficient is accounted, the CFD model was not

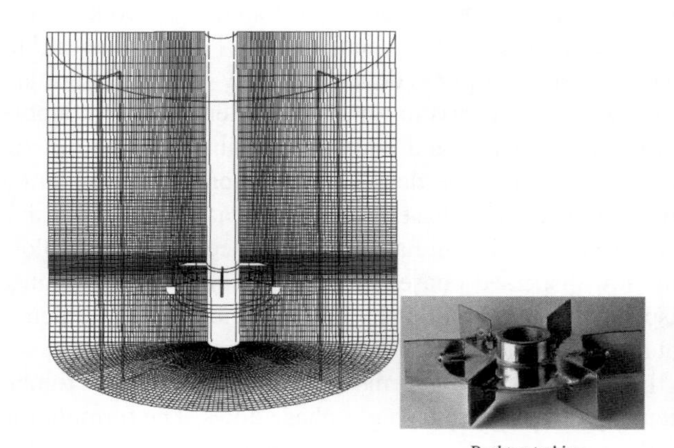

FIGURE 15.1 Computational grid and solution domain of stirred reactor. Grid details: $r \times \theta \times z$: $63 \times 98 \times 82$; Impeller blade: $18 \times 1 \times 19$; Inner region: $15 \leq k \leq 65, j \leq 45$.

able to predict the pattern of gas holdup distribution adequately. Relatively few attempts (experimental as well as numerical) have been made to understand the influence of prevailing turbulence on drag coefficient (see, for example, Refs 16 and 19–22). Bakker and van den Akker [20], Brucato et al. [21], and Lane et al. [16] attempt to relate the influence of turbulence on the drag coefficient to the characteristic spatiotemporal scales of the prevailing turbulence seem promising. Khopkar and Ranade [17] evaluated the three alternative proposals using a two-dimensional CFD-based model problem. They have observed that the predicted results deviate from the trends estimated by correlation of Lane et al. [16]. However, the predicted results show reasonable agreement with estimation based on correlation Bakker and van den Akker [20] (equation 15.12) and Brucato et al. [21] (equation 15.13), with 100 times lower correlation constant ($K = 6.5 \times 10^{-06}$). Interestingly, in both of these correlations they have used volume-averaged values of the turbulent viscosity and the Kolmogorov scale, respectively.

$$C_D = \frac{24}{Re^*}\left[1 + 0.15\left(Re^*\right)^{0.687}\right] \because Re^* = \frac{\rho_l U_{slip} d_b}{\mu_l + \frac{2}{9}\mu_t} \tag{15.12}$$

$$\frac{C_D - C_{D0}}{C_{D0}} = K\left(\frac{d_b}{\lambda}\right)^3$$

$$C_{D0} = \max\left\{\left(\frac{2.667 * Eo}{Eo + 4.0}\right), \left(\frac{24}{Re_b} * \left(1 + 0.15 * Re_b^{0.687}\right)\right)\right\} \tag{15.13}$$

where C_D is the drag coefficient in turbulent liquid, C_{D0} is the drag coefficient in a stagnant liquid, d_b is bubble/particle diameter, and λ is the Kolmogorov length scale (based on volume-averaged energy dissipation rate).

The gas–liquid flow in stirred reactor was simulated using the drag coefficients estimated with volume average values of Kolmogorov scale (equation 15.13) and turbulent viscosity (equation 15.12) for operating conditions of $Fl = 0.1114$ and $Fr = 0.3005$. This operating condition represents L33 flow regime (Large 33 cavities) in stirred reactor. In L33 flow regime, three large gas cavities and three small cavities coexist on alternate impeller blades (for the six-bladed Rushton turbine). The quantitative comparison of the predicted gas holdup distribution with the experimental data [18] is shown in Figure 15.2. It can be seen from Figure 15.2a,b that the gas holdup distribution predicted based on equation (15.12) shows fairly different gas distribution from the experimental data (shown in Figure 15.2a). The major disagreement was observed in the region below the impeller. The impeller-generated flow was not sufficient to circulate gas in a lower circulation loop. This leads to underprediction of the total gas holdup (fraction of volume occupied by gas in the gas–liquid dispersion). The predicted gas holdup was 2.55% compared to the experimental measurement of 3.3%. The predicted results based on equation (15.13) are closer to the experimental data (see Figure 15.2a,c). This model resulted in over prediction of total gas holdup (predicted holdup was 3.97% compared to the experimental measurement of 3.3%). Despite the over prediction, the predicted gas holdup distribution showed better agreement with the data than predicted by equation (15.12). The equation (15.13) can

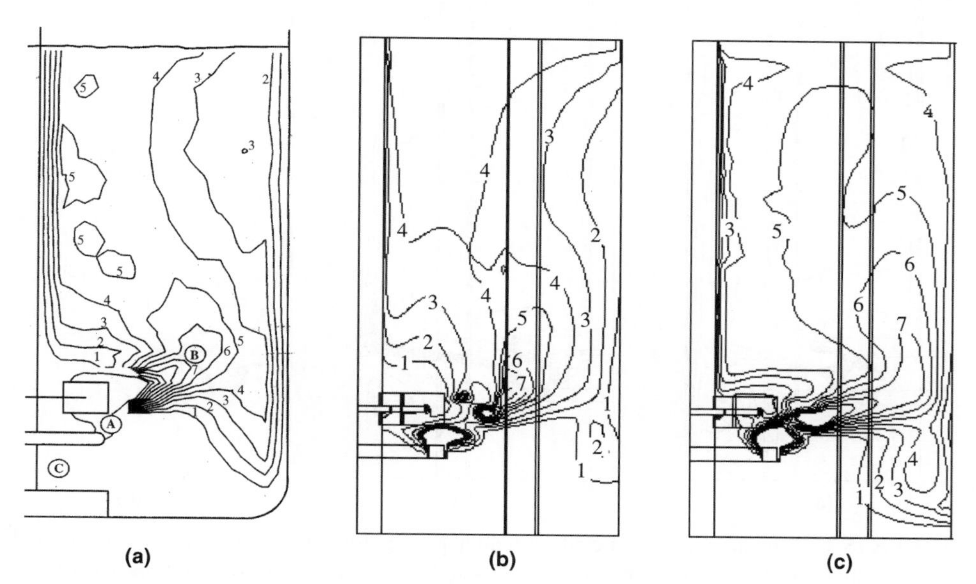

FIGURE 15.2 Comparison of experimental and predicted gas holdup distribution at mid-baffle plane for L33 flow regime, $Fl = 0.1114$ and $Fr = 0.3005$. (a) Experimental data of Bombac et al. [18]. (b) Predicted results with Bakker and van den Akker correlation (equation 15.12). (c) Predicted results with modified Brucato et al. correlation (equation 15.13). Contour labels denote the actual values of gas holdup (in percentage). Reprinted with permission from Ref. 17, copyright 2006, Wiley.

therefore be recommended for carrying out gas–liquid flow simulations in stirred tanks.

Virtual Mass and Lift Force The other two important interphase forces are virtual mass force and lift force. The virtual mass effect is significant when the secondary phase density is much smaller than the primary phase density. The effect of the virtual mass force was first studied. The predicted gas holdup distributions obtained with and without considering virtual mass force are shown in Figure 15.3a,b. It can be seen from Figure 15.3 that the influence of the virtual mass force on the predicted pattern of gas distribution was significant only in the impeller discharge stream. However, the influence of virtual mass force was not found to be significant in the bulk of the reactor. It should be noted that the value of virtual mass coefficient (C_{VM}) used in the present study (0.5) is valid for spherical bubble and may not be appropriate for wobbling bubbles. The reported value of virtual mass coefficient is somewhat higher than 0.5 (see, for example, Ref. 23). However, it should be noted that the predicted results are not very sensitive to the consideration of virtual mass terms. A comparison of the predicted results obtained with values of virtual mass coefficients as 0 and 0.5 did not show any significant differences (see Figure 15.3). Considering this, no specific effort was made to obtain the accurate value of the virtual mass coefficient.

Similarly, the simulations were carried out with and without considering lift force. The predicted gas holdup distribution obtained with considering lift force is shown in Figure 15.3c. It can be seen from Figure 15.3 that the influence of lift force on the predicted pattern of gas distribution was significant in the impeller discharge stream and below impeller region. The predicted results with lift force predict a lower gas holdup in the region below the impeller. In the upper impeller region of the reactor the influence of lift force was not found to be significant. Therefore, it can be said that the modeling of lift force and virtual mass force may not be essential while simulating gas–liquid flow in stirred vessels.

15.2.2.2 Modeling Bubble Size Distribution In a gas–liquid stirred reactor, gas bubbles of different sizes coexist. Very fine bubbles are observed in the impeller discharge stream (<1 mm) whereas bubbles of the size of few millimeter (~5 mm) are observed in the region away from the impeller [24]. The width of the bubble size distribution depends upon the turbulence level and prevailing flow regime. Appropriate selection of bubble sizes is very important for the correct prediction of the slip velocity and mass transfer area. Both the slip velocity and the mass transfer area can be more accurately estimated by modeling with local bubble size distributions. Local bubble size or gas–liquid mass transfer can be estimated more accurately from local bubble size distributions (BSDs) based on either a population balance [25] or by modeling the bubble number density function [16]. The bubble density function approach [16] is computationally less intensive and requires one additional

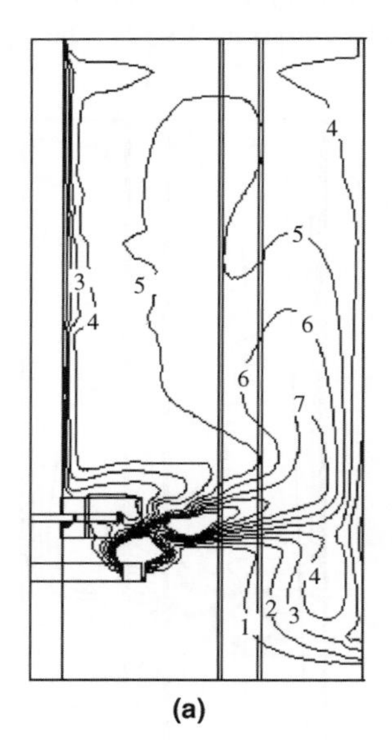

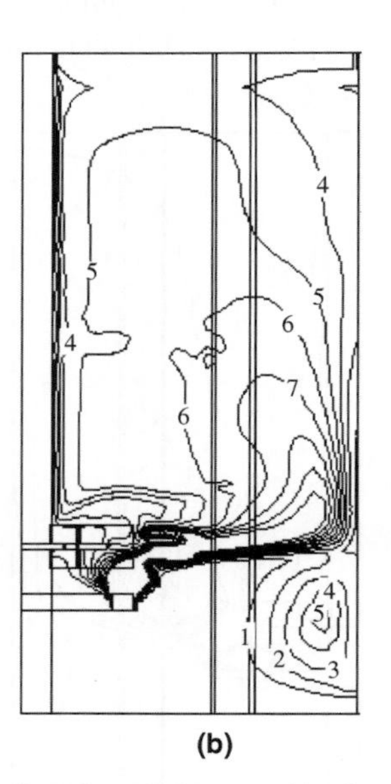

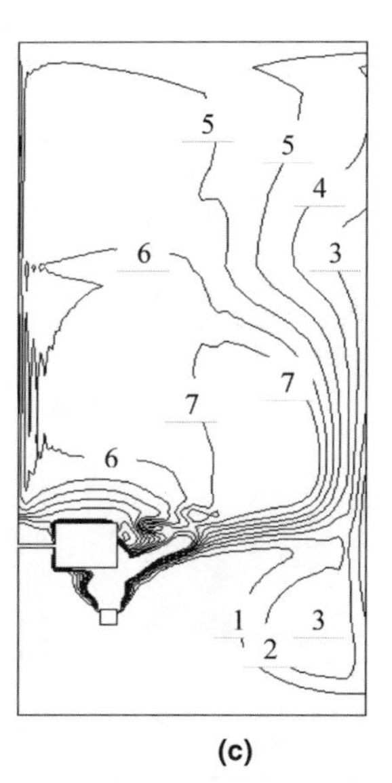

FIGURE 15.3 Comparison of predicted gas holdup profiles for with and without virtual mass and lift force effect for L33 flow regime, $Fl = 0.1114$ and $Fr = 0.3005$. (a) Predicted results without virtual mass and lift force effect: mid-baffle plane. (b) Predicted results with virtual mass effect: mid-baffle plane. (c) Predicted results with lift force effect: mid-baffle plane. Contour labels denote the actual values of gas holdup (in percentage). Reprinted with permission from Ref. 17, copyright 2006, Wiley.

equation to solve along with the two-fluid model. This approach predicts the Sauter mean diameter at every grid node point. The predicted results of Lane et al. [16] show reasonable agreement with the experimental data of Barigou and Greaves [24]. Laakkonen et al. [25] simulated the gas–liquid flows in a stirred reactor using population balance modeling. Their simulated results are discussed in this chapter to explain the need for modeling the bubble size distribution while simulating gas–liquid flow in stirred reactors. The details of the population balance formulation, bubble breakage, and coalescence model are not discussed in this chapter and can be found in Ref. 25. The influence of the prevailing turbulence on the interphase drag force was modeled with a slightly modified Bakker and van den Akker [20] correlation. The comparison of the predicted bubble size distribution and the mean bubble diameter with the experimental data is shown in Figure 15.4. The following conclusions can be drawn from the comparison between predicted results and experimental data:

- The parameters of the coalescence and breakage models were tuned to fit the experimental measurements. This limits the applicability of the model for different configurations of stirred reactor.
- The tails in the predicted volume BSD's are larger compared to the experimental measurements indicating underprediction of breakage process. The rate of breakage process is dependant on the predicted values of the turbulent energy dissipation rate. The CFD model underpredicts the turbulent kinetic energy dissipation rates and hence led to lower rate of bubble breakage process.
- The enormous requirement of computational requirement for multifluid model does not allow modeler to use fine mesh for simulating the turbulent multiphase flow. The use of a relatively coarse mesh significantly contributes to the underprediction of turbulent properties and hence influences the predicted breakage and coalescence rates.

The present state of understanding of the breakage and coalescence processes and the unavailability of experimental data for different reactor configurations suggest that it may not be advantageous to use population balance-based multifluid models while simulating industrial gas–liquid stirred reactors. It may be more effective to use effective combination of bubble diameter and interphase drag coefficient to get realistic results.

15.2.2.3 Gas Holdup Distribution in L33, S33, and VC Flow Regimes Gas–liquid flows generated by the Rushton turbine in a stirred vessel were simulated for two other gas flow regimes representing S33 ($Fl = 0.0788$; $Fr = 0.6$), and VC ($Fl = 0.026267$; $Fr = 0.6$). These flow regimes are defined based on the shapes of cavities observed behind blades under these regimes. In S33 flow regime, three large and three vortex clinging cavities were found behind impeller blades. Whereas, in VC flow regime, only clinging cavities were found behind impeller blades. More details on flow regimes in gas–liquid flows in stirred tanks may be found in Refs 1 and 3. As discussed previously, equation 15.13, based on volume-averaged dissipation rate and Kolmogorov scale (λ) was used to calculate effective drag coefficients. Comparisons of predicted gas holdup distributions with the

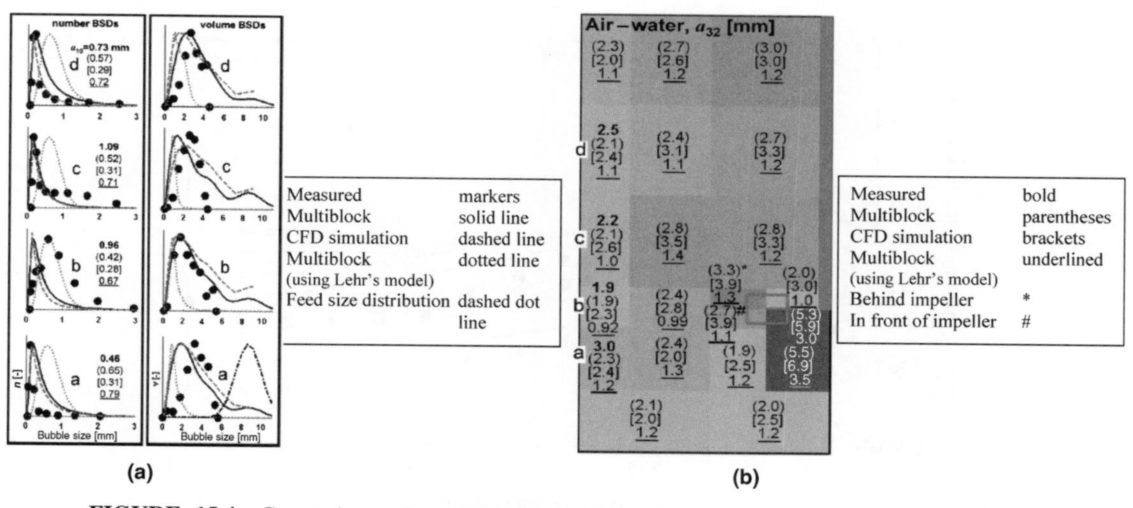

FIGURE 15.4 Comparison of predicted bubble size distribution with experimental data of Laakkonen et al. [25]. (a) Local bubble size distributions in the air–water dispersion, 14 L tank, $N = 700$ rpm and $Q = 0.7$ vvm. (b) Mean diameters (mm). Reprinted with permission from Ref. 25 Elsevier.

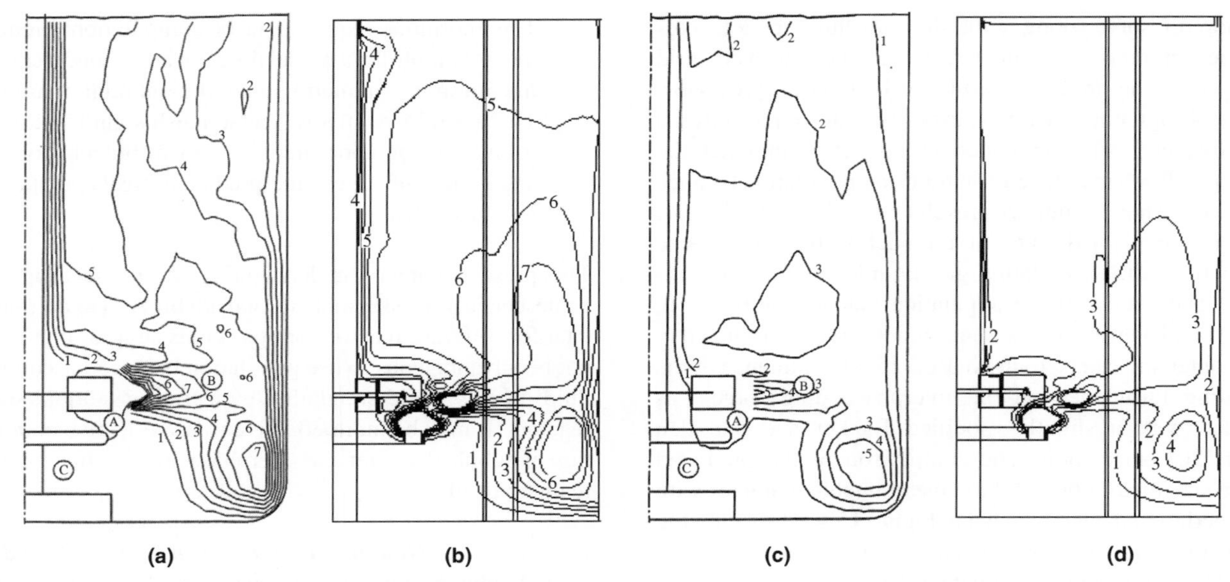

FIGURE 15.5 Comparison of experimental and predicted gas holdup distribution for S33 and VC flow regimes (Experimental data of Bombac et al. [18]). (a) Experimental data, S33 flow regime, $Fl = 0.0788$ and $Fr = 0.3005$ (mid-baffle). (b) Predicted data, S33 flow regime, $Fl = 0.0788$ and $Fr = 0.6$ (mid-baffle). (c) Experimental data, VC flow regime, $Fl = 0.026267$ and $Fr = 0.6$ (mid-baffle). (d) Predicted data, VC flow regime, $Fl = 0.026267$ and $Fr = 0.6$ (mid-baffle). Contour labels denote the actual values of gas holdup (in percentage). Reprinted with permission from Ref. 17, copyright 2006, Wiley.

experimental results at the mid-baffle plane are shown in Figure 15.5. It can be seen from these figures that the predicted gas holdup distributions for S33 and VC flow regimes are in reasonably good agreement with the experimental data. However, the computational model overpredicted the values of total gas holdup. The predicted value of total gas holdup (4.85%) was higher than the reported experimental value (4.2%) for the S33 flow regime. Similarly, the predicted value of total gas holdup (2.63%) was higher than the experimental data (2.2%) for the VC flow regime.

Comparisons of axial profiles of radially averaged gas holdup for all three regimes are shown in Figure 15.6. It can be seen from Figure 15.6 that the computational model overpredicts the values of gas holdup in the region above the impeller for all three regimes. The maximum value of predicted radially averaged gas holdup occurs at an axial

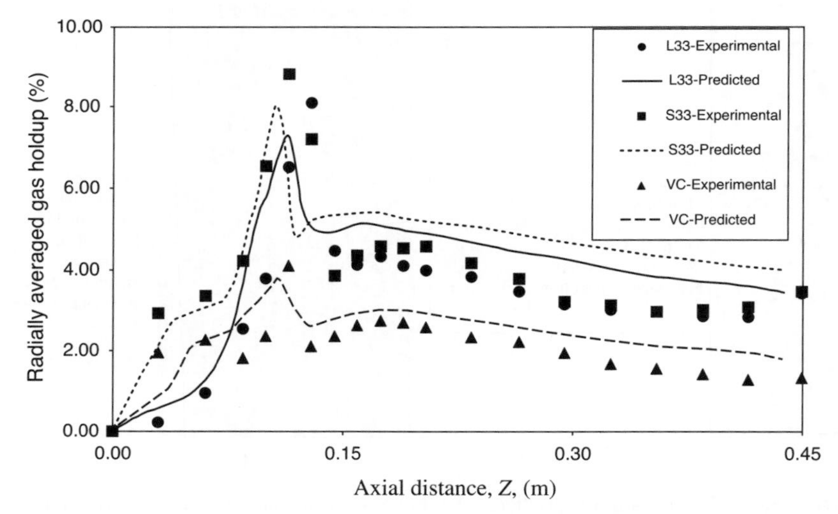

FIGURE 15.6 Comparison of predicted axial profile of radially averaged gas holdup with experimental data for L33, S33, and VC flow regimes. Symbol denotes the experimental data of Bombac et al. [18]. Reprinted with permission from Ref. 17, copyright 2006, Wiley.

distance of 0.117 m for L33 and 0.107 m for S33 as well as VC regimes compared to the experimentally observed distance of 0.13 m for L33 and 0.1125 m for S33 as well as VC regimes. The predicted values of gas holdups at this maximum are underpredicted (7.3% for L33, 7.94% for S33, and 3.82 for VC) compared with the experimental value (8.1% for L33, 8.8% for S33, and 4.1% for VC). Quantitative comparisons of angle-averaged values of predicted gas holdup and experimental data at three different axial locations for all three regimes are shown in Figure 15.7. It can be seen from Figure 15.7 that comparisons of the predicted values of gas holdup and experimental data are reasonably good for all three regimes. The computational model was thus able to simulate all three regimes reasonably well.

15.2.2.4 Gross Characteristics Predicted influence of the gas flow rate on gross characteristics, power and pumping numbers are also of interest. Pumping and power numbers were calculated from simulated results as

$$N_{\mathrm{Q}} = \frac{2\int_{-B/2}^{B/2}\int_{0}^{\pi} \alpha_{\mathrm{l}} r_{\mathrm{i}} U_{\mathrm{r}} \mathrm{d}\theta \mathrm{d}z}{N D_{\mathrm{i}}^{3}} \tag{15.14}$$

$$N_{\mathrm{P}} = \frac{2\int_{V} \alpha_{\mathrm{l}} \rho \varepsilon \mathrm{d}V}{\rho N^{3} D_{\mathrm{i}}^{5}} \tag{15.15}$$

where B is blade height, D_{i} is impeller diameter, N is impeller speed, r_{i} is impeller radius, and U_{r} is radial velocity. The calculated values of pumping and power number from the simulated results are listed in Table 15.2. As the gas flow rate increases, impeller pumping as well as power dissipation decreases. The extent of decrease increases with an increase in the gas flow rate (or in other words, as flow regime changes from VC to S33 and further to L33). Bombac et al. [18] have not reported their experimental values of power dissipation or pumping number. In the absence of such data, the predicted values were compared with the estimates of empirical correlations proposed by Calderbank [26], Hughmark [27], and Cui et al. [28]. While demonstrating the qualitative trend, the CFD model underpredicts the decrease in power dissipation in the presence of gas compared to the estimates of these correlations. CFD model, however, could correctly capture the overall gas holdup distribution and can therefore simulate different flow regimes of gas–liquid flow in stirred reactors.

15.2.3 Application to Simulate Solid–Liquid Flow in Stirred Reactor

Suspension of solid particles in a stirred reactor either in presence or in absence of gas is commonly encountered in chemical process industry (refer Table 15.1). In such reactors, the knowledge of the solid particles concentration distribution over the reactor (suspension quality) is an important parameter required for reliable design, optimum performance, and scale-up of the reactors. Despite significant

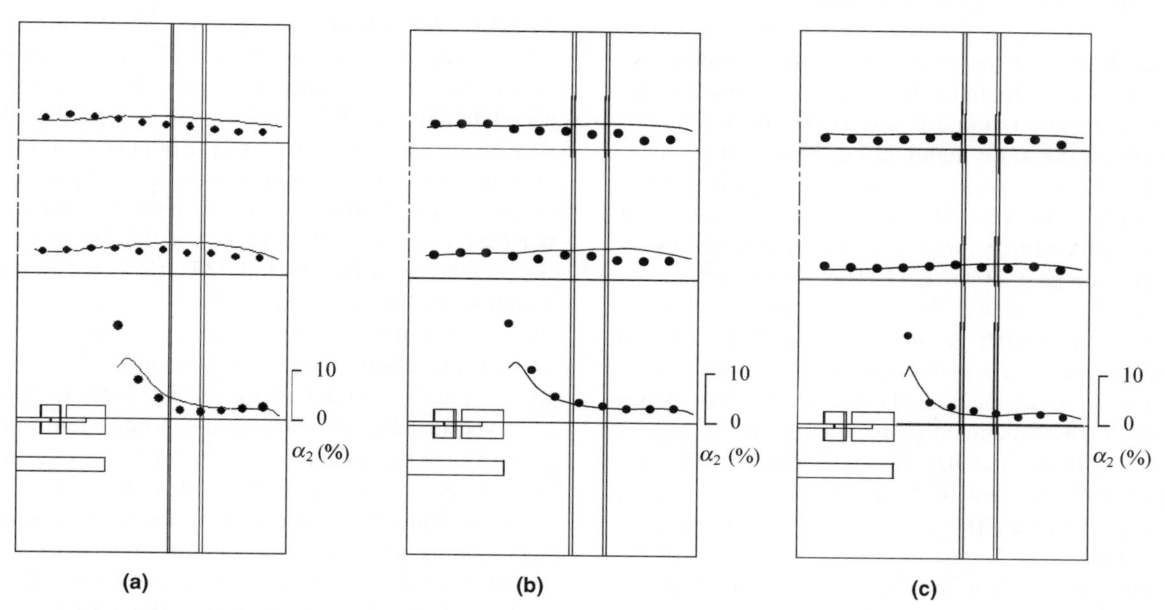

FIGURE 15.7 Comparison of predicted angle averaged values of gas holdup (α_2) with experimental data for L33, S33, and VC flow regimes. (a) L33 flow regime, $Fl = 0.1114$ and $Fr = 0.3005$. (b) S33 flow regime, $Fl = 0.0788$ and $Fr = 0.6$. (c) VC flow regime, $Fl = 0.026267$ and $Fr = 0.6$. ●: Experimental data of Bombac et al. [18]; —: Predicted results. Reprinted with permission from Ref. 17, copyright 2006, Wiley.

TABLE 15.2 Gross Characteristics of an Aerated Stirred Reactor. From Ref. 17

Operating Conditions	Total Gas Holdup (%)		Predicted Results		Influence of Gas on Power Number, N_{Pg}/N_P				Influence of Gas on Pumping Number, Predicted N_{Qg}/N_Q
						Predicted by Empirical Correlations			
	Predicted	Experimental (Bombac *et al.* 1997)	N_{Pg}	N_{Qg}	Predicted by CFD	Calderbank (1958)	Hughmark (1976)	Cui *et al.* (1996)	
Single-phase Flow	—	—	4.15	0.66	—	—	—	—	—
VC Flow Regime (Fl = 0.026267 &Fr = 0.6)	2.63	2.20	2.76	0.615	0.66	0.67	0.64	0.61	0.93
S33 Flow Regime (Fl = 0.0788 & Fr = 0.6)	4.85	4.20	2.196	0.6	0.53	0.47	0.49	0.41	0.9
L33 Flow Regime (Fl = 0.1114 & Fr = 0.3005)	3.97	3.30	1.66	0.49	0.4	0.41	0.51	0.41	0.74

research efforts, prediction of design parameters to ensure an adequate solid suspension is still an open problem for design engineers. Design of stirred slurry reactors relies on empirical correlations obtained from the experimental data. These correlations are prone to great uncertainty as one departs from the limited database that supports them. Moreover, for higher values of solid concentration, very few experimental data on local solid concentration is available because of the difficulties in the measurement techniques. Considering this, it would be most useful to develop computational models, which will allow *a priori* estimation of the solid concentration over the reactor volume.

The discussed two-fluid model is applied to simulate solid–liquid flow in a stirred reactor. In addition to interphase drag force, turbulent dispersion force plays an important role while simulating solid–liquid flows. There are few studies available in the literature highlighting the influence of interphase drag force on the predicted solid holdup distribution (see, for example, Refs 29–31). To explain the influence of different interphase forces, we reproduce some of the results obtained by Khopkar et al. [31]. They have carried out simulations of solid–liquid flow in a stirred vessel in an experimental setup used by Yamazaki et al. [32]. The system investigated consists of a cylindrical, flat-bottomed reactor (of diameter $T = 0.3$ m and liquid height $H = T$). Four baffles of width $0.1T$ were mounted perpendicular to the reactor wall. The shaft of the impeller was concentric with the axis of the reactor. A standard Rushton turbine with diameter, $D = T/3$ has been used. The impeller off-bottom clearance has been set as $C = T/3$, measured from the bottom of the reactor to the center of the impeller blade height. Water as liquid phase and glass beads (of density 2470 kg/m^3 and particle diameter $d_p = 264\,\mu$m) as solid phase were used in the simulations.

Considering geometrical symmetry, half of the reactor was considered as a solution domain. It is very important to use an adequate number of computational cells while numerically solving the governing equations over the solution domain. The prediction of the turbulence quantities is especially sensitive to the number of grid nodes and grid distribution within the solution domain. In the present work, the numerical simulations for solid–liquid flows in stirred reactor have been carried out with grid size of 298, 905 ($r \times \theta \times z$: 57 × 93 × 57). The details of computational grid used in the present work are shown in Figure 15.15. In this chapter, the standard wall functions were used to specify wall boundary conditions.

15.2.3.1 Interphase Drag Force Similar to the discussion included in the subsection 15.2.2.1, unless the influence of the prevailing turbulence on particle drag coefficient is accounted for the CFD model will not predict the solid suspension adequately. Recently, Khopkar et al. [31] evaluated the two alternative proposals [21, 33] using a two-dimensional CFD-based model problem. They have observed that the predicted results deviate from the trends estimated by the correlation of Pinelli et al. [33]. However, the predicted results show reasonable agreement with estimation based on the correlation by Brucato et al. [21]. They correlated the predicted results by considering the sole dependence of d_p/λ for a range of solid holdup values ($5 < \alpha < 25\%$). They observed that the predicted results require ten times lower proportionality constant ($K = 8.76 \times 10^{-5}$) in equation 15.13 as compared to that proposed by Brucato et al. [21].

Solid–liquid flow generated by the Rushton turbine has been simulated for a solid volume fraction of 10.0%, $d_p = 264\,\mu$m and at an impeller rotation speed $N = 20$ rps. Both of the formulations of drag coefficient, by Brucato et al. [21] ($K = 8.76 \times 10^{-4}$) and the modified Brucato correlation ($K = 8.76 \times 10^{-5}$) were used for the evaluation of the interphase drag force. The value of dispersion Prandtl number, σ_{pq}, has been set to the default value of 0.75. The predicted

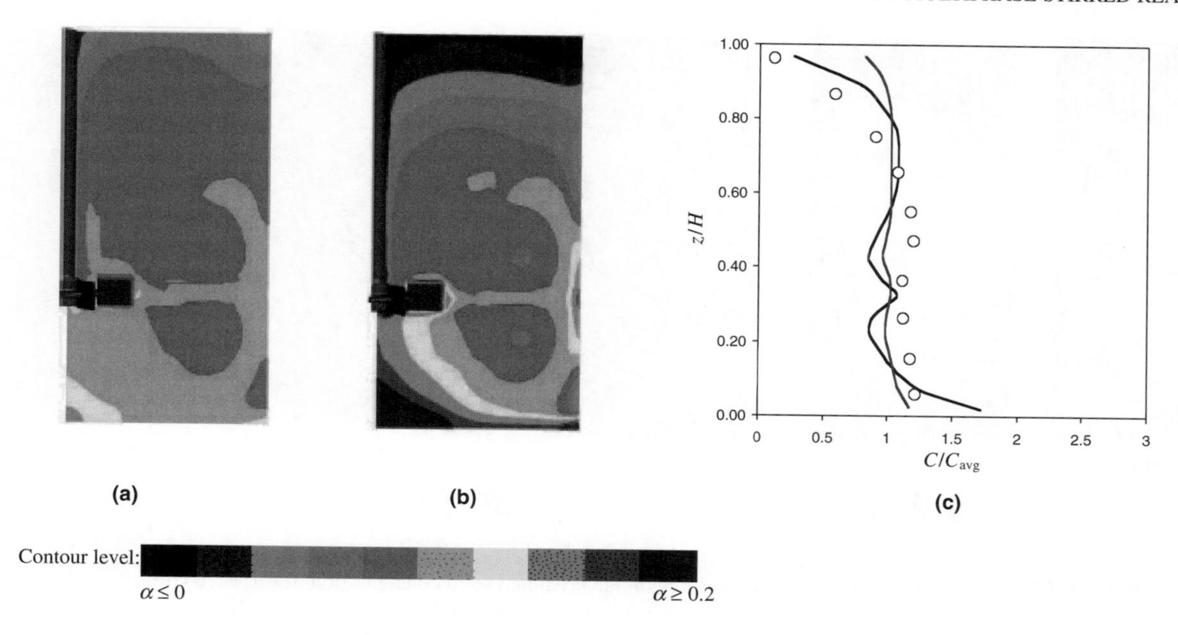

FIGURE 15.8 Simulated solid holdup distribution at mid-baffle plane for $d_p = 264\,\mu m$, $d_p/\lambda \approx 20$, $\alpha = 0.1$, $N = 20.0$ rps, and $U_{tip} = 6.283$ m/s. (a) $K = 8.76 \times 10^{-4}$. (b) $K = 8.76 \times 10^{-5}$. (c) Comparison of predicted results with experimental data. Circles: Experimental data [32]; Gray line: Predicted results (with [21]); Black line: Predicted results (with modified Brucato et al. correlation). Reprinted with permission from Ref. 31, copyright 2006, Wiley.

solid holdup distributions by using both drag coefficient formulations at the mid-baffle plane are shown in Figure 15.8a,b. It can be seen from Figure 15.8a that with the Brucato et al. [21] correlation, the computational model has predicted almost complete suspension of the solid particles. However, the simulated solid holdup distribution using the modified Brucato et al. [21] did not capture the complete suspension of solid particles in the stirred reactor (see Figure 15.8b). The simulated solid holdup distribution shows the presence of solid accumulation at the bottom and near the axis of the reactor. For quantitative comparison the predicted solid concentrations/holdups were compared with the experimental data of Yamazaki et al. [32]. The quantitative comparison of the azimuthally averaged axial profile of solid holdup at a radial location ($r/T = 0.35$) is shown in Figure 15.8c. It can be seen from Figure 15.8c that the computational model with drag coefficient formulation of Brucato et al. [21] has overpredicted the solid suspension height. However, the suspension height predicted by the modified Brucato correlation is in good agreement with the experimental data. It can also be seen from Figure 15.8 that solid holdup distribution predicted with the use of modified Brucato et al. [21] correlation has captured the presence of higher solid concentration in the impeller discharge stream (a bell shaped in the concentration profile), which is a characteristic of the solid–liquid flow generated by Rushton turbine. However, the prediction with Brucato et al. [21] correlation does not show any such characteristics. Overall, it can be said that the modified Brucato et al. [21] correlation predicted solid holdup distribution in stirred vessel more accurately.

15.2.3.2 Turbulent Dispersion Force The developed computational model was then extended to study the influence of the turbulent dispersion force on the suspension quality in the stirred reactor. The magnitude of the turbulent dispersion force was varied by varying the value of the dispersion Prandtl number, σpq, in the range 0.0375–3.75. Figure 15.9 shows the comparison of the predicted solid concentration distribution in the stirred reactor obtained with different values of turbulent dispersion force. It can be seen from Figure 15.9 that the turbulent dispersion has a significant effect on the predicted suspension quality in the stirred reactor. The computational model predicted a more uniform suspension with a decrease in the value of dispersion Prandtl number. This is expected as the drift velocity (or turbulent dispersion force) is inversely proportional to the dispersion Prandtl number (see equation 15.3). Decreasing the latter means increase in the turbulent dispersion force, which consequently results in more dispersion of the particles resulting in more uniform suspension. Overall, it can be said that the simulations carried out with $\sigma pq = 0.375$ and 0.0375 have over predicted the suspension quality. Whereas, for $\sigma pq = 3.75$ the computational model has underpredicted the suspension quality. Therefore, CFD simulation of solid–liquid stirred reactor are need to be

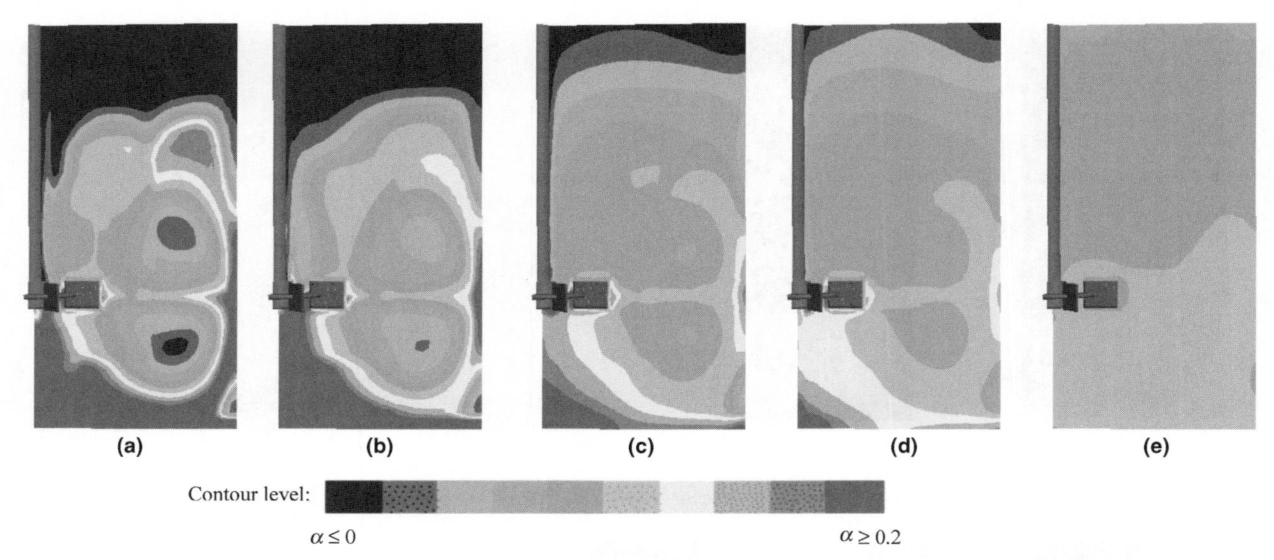

FIGURE 15.9 Effect of turbulent dispersion force on the predicted solid concentration, for $d_p = 264\,\mu$m, $d_p/\lambda \approx 20$, $\alpha = 0.1$, $N = 20$ rps, and $U_{tip} = 6.283$ m/s. (a) Without turbulent dispersion force. With turbulent dispersion force: (b) $\sigma_{pq} = 3.75$, (c) $\sigma_{pq} = 0.75$, (d) $\sigma_{pq} = 0.375$, and (e) $\sigma_{pq} = 0.0375$.

carried out with $\sigma pq = 0.75$ for adequate prediction of suspension quality.

15.2.4 Application to Simulate Gas–Liquid–Solid Flow in Stirred Reactor

Suspension of solid particles in presence of gas has various applications in the process industry. These applications include catalytic hydrogenations, oxidations, fermentations, evaporative crystallizations, and froth flotation. In a gas–liquid–solid system, the impeller plays a dual role of keeping the solids suspended in the liquid while dispersing the gas bubbles. Dylag and Talaga [34] have found that in a gas–liquid–solid stirred reactor, the gas phase is always uniformly dispersed before the solids are completely suspended. Therefore, the formation of a completely dispersed three-phase system depends on the condition under which the solids are suspended by the impeller action. Identification of these operating conditions is very important for operating a stirred reactor in an energy efficient mode. Some mass transfer studies (see, for example, Ref. 35) in two-phase solid–liquid mixing in stirred tanks have also shown that the particle–fluid mass transfer rate is comparable to the just off-bottom suspension (JS) point, irrespective of the power input level. Any incremental power input beyond this point for improving the mass transfer coefficient is often uneconomical. Attempts to extend the above hypothesis to three-phase systems introduces an additional complexity, as the impeller pumping efficiency changes in the presence of gas. It is also observed that the tank, the impeller, and the sparger geometry variations that have been proposed [36–38] are highly system specific with respect to gas–liquid and solid–liquid systems, and may not extend directly to gas–liquid–solid systems. It is therefore necessary to develop tools to examine the role of reactor hardware in meeting the demands associated with the simultaneous gas dispersion and solid suspension.

Critical analysis of available literature suggests that practically no information is available in the literature on the CFD simulation of three-phase gas–liquid–solid stirred reactor. Complex interactions between the solid particles, the gas bubbles, and the liquid phase make the fluid dynamics of three-phase stirred reactor very complex. Recently, Murthy et al. [39] made an attempt to simulate a three-phase stirred reactor. They used the approach proposed by Khopkar and Ranade [17] for modeling gas–liquid flow and approach of Pinelli et al. [33] for simulating solid–liquid flow. Murthy et al. [39] were able to predict the critical impeller speed required for solid suspension. However, their study was limited to very low solid loading (maximum solid loading is less than 10 wt%). The applicability of the same computational model to simulate solid suspension at higher solid loading (greater than 20 wt% or 10% by volume fraction) is not known. In this chapter, a CFD model was developed to simulate solid suspension in a three-phase stirred reactor. The approaches discussed in subsections 15.2.2 and 15.2.3 were used to model the gas–liquid and solid–liquid interactions in the three-phase stirred reactor.

Experimental setup of Pantula and Ahmed [40] was used to simulate gas–liquid–solid flows in a stirred reactor. The system investigated consists of a cylindrical, flat-bottomed reactor (of diameter $T = 0.4$ m and liquid height $H = T$). Four baffles of width $0.1T$ were mounted perpendicular to the reactor wall. The shaft of the impeller was concentric with the axis of the reactor. A standard Rushton turbine with diameter,

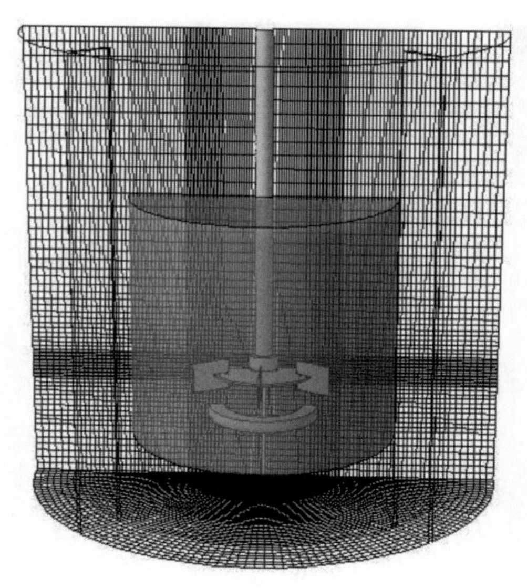

FIGURE 15.10 Computational grid and solution domain of stirred reactor. Grid details: $r \times \theta \times z$: $70 \times 93 \times 67$; Impeller blade: $20 \times 1 \times 15$; Inner region: $8 \leq k \leq 55$, $j \leq 51$.

$D = T/3$ has been used. The impeller off-bottom clearance has been set as $C = T/4$, measured from the bottom of the reactor to the center of the impeller blade height. A ring sparger of diameter, $D_s = 2D/3$ with evenly spaced holes at a clearance (C_s) of $T/6$ was provided for gas input. Water as liquid phase, air as gas phase and glass beads (of density $2500\,\text{kg/m}^3$ and particle diameter $d_p = 174\,\mu\text{m}$) as solid phase were used in the simulations. Simulations were carried out with solid-phase volume fraction of 12% (i.e., 30 wt%).

Considering the geometrical symmetry, half of the reactor was considered as a solution domain. In this chapter, the numerical simulations for gas–liquid–solid flows in stirred reactor have been carried out with grid size of 436,170 ($r \times \theta \times z$: $70 \times 93 \times 67$). The details of computational grid used in this chapter are shown in Figure 15.10. In this chapter, the standard wall functions were used to specify wall boundary conditions.

15.2.4.1 Solid Suspension in an Aerated Stirred Vessel

Minimum speed for just off-bottom suspension is a very important hydrodynamic parameter for designing gas–liquid–solid stirred reactor. Experimental studies so far on gas–liquid–sold suspensions have clearly indicated the requirement of increased suspension speed, thereby more power input, on the introduction of gas [34, 36–38]. This is because of a decrease in impeller pumping efficiency and power draw due to the formation of ventilated cavities behind the impeller blades on gassing [41]. Recently, Zhu and Wu [42] carried out experimental measurements in a three-phase stirred reactor to determine the just off-bottom suspension speed for a variety of solid sizes, solid loading, impeller sizes, and tank sizes. They suggested the possibility of relating relative just off-bottom suspension speed (RJSS) with just suspension aeration number (based on just suspension speed for solid–liquid system). They also observed that the proposed relation was independent of solid size, solid loading, and tank size and can be used to scale up laboratory data to full-scale vessel. The same definition (equation 15.16) is used in this chapter to identify the just off-bottom suspension speed for different gas flow rates.

$$\text{RJSS} = 1 + mNa_{\text{js}}^{n}$$

$$\text{RJSS} = {}^{N_{\text{jsg}}}\!/_{N_{\text{js}}} \quad \text{and} \quad Na_{\text{js}} = \frac{Qg}{N_{\text{js}}D^3} \tag{15.16}$$

where, m and n are constants. For the Rushton turbine, the values of m and n are 2.6 and 0.7, respectively. The simulations were carried out for three just suspension aeration number 0, 0.025, and 0.05. The impeller rotational speeds (N_{jsg}) for the three aeration numbers are 9.30, 11.16, and 12.27 rps, respectively.

The predicted solid holdup distribution at mid-baffle plane for all three aeration numbers is shown in Figure 15.11. It can be seen from Figure 15.11 that for three-phase system the computational model has predicted more accumulation of solids at the bottom of the reactor near the central axis in comparison to the two-phase system. The predicted cloud height values were also found to drop in presence of gas. The predicted solid volume fraction values were then used to describe the suspension quality in the reactor. The criterion based on the standard deviation value, calculated using equation 15.17, was used to describe suspension quality for all three cases. It was observed that the computational model predicted standard deviation value (σ) of 0.45 for two-phase flow. Whereas, for three-phase flow computational model predicted σ of 0.82 and 0.90 for Na_{js} of 0.025 and 0.05, respectively. Overall, it can be said that the computational model has predicted just off-bottom suspension condition for two-phase flow ($\sigma < 0.8$), whereas, incomplete suspension for three-phase system ($\sigma > 0.8$).

$$\sigma = \sqrt{\frac{1}{n}\sum_{i=1}^{n}\left(\frac{\alpha_i}{\alpha_{\text{avg}}} - 1\right)^2} \tag{15.17}$$

The predicted gas holdup distribution at mid-baffle plane for two suspension aeration numbers is shown in Figure 15.12. It can be seen from Figure 15.12 that for both conditions, the computational model has predicted higher values of gas holdup in both circulation loops of flow. This indicates that the computational model predicted complete dispersion condition of gas phase in the vessel. These predicted results also support the experimental observations made by Dylag and Talaga [34] on quality of gas dispersion.

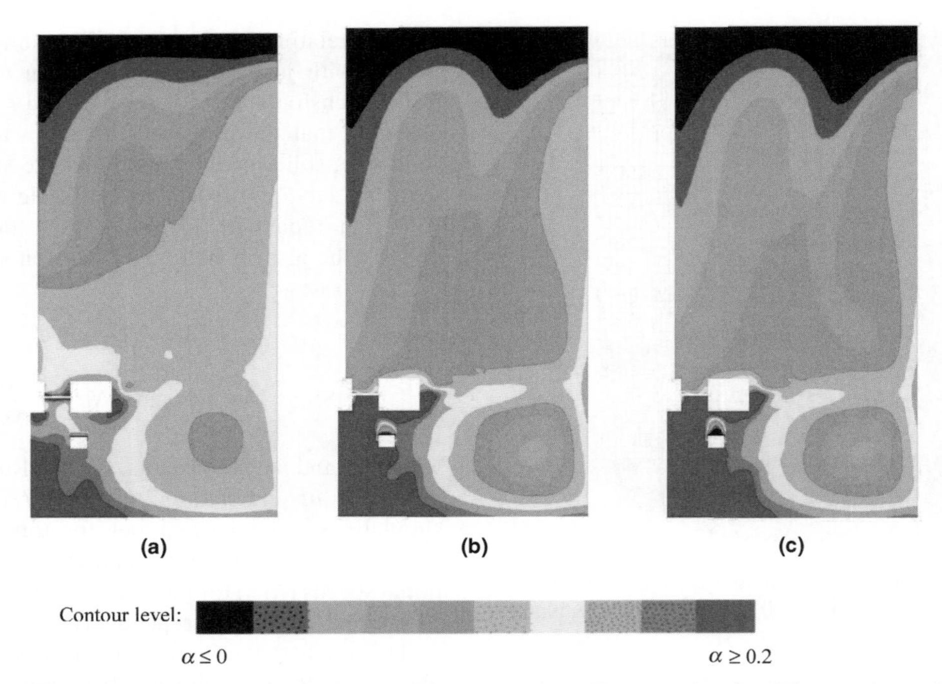

FIGURE 15.11 Simulated solid holdup distribution at mid-baffle plane for $d = 174\,\mu m$ and $\alpha_s = 0.12$. (a) $Na_{js} = 0$ and $N = 9.3$ rps. (b) $Na_{js} = 0.025$ and $N = 11.16$ rps. (c) $Na_{js} = 0.05$ and $N = 12.27$ rps.

15.2.4.2 Gross Characteristics The predicted influence of the gas flow rate on gross characteristics, power number, gas holdup, and suspension quality is also of interest. The calculated values of power number, gas holdup, and standard deviation from the simulated results are listed in Table 15.3. Few conclusions can be drawn from Table 15.3. First, the computational model has predicted the drop in impeller power number value in presence of gas. While demonstrating

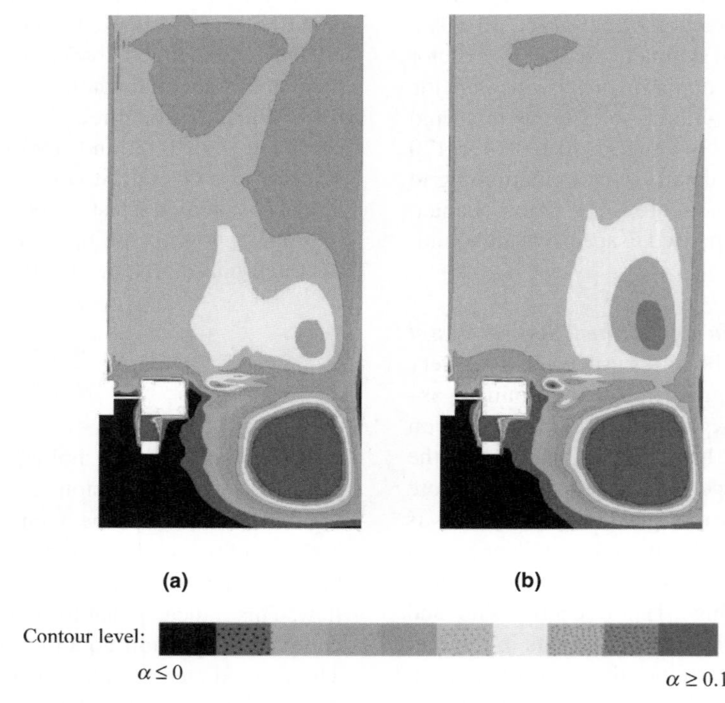

FIGURE 15.12 Simulated gas holdup distribution at mid-baffle plane for $d_p = 174\,\mu m$ and $\alpha_s = 0.12$. (a) $Na_{js} = 0.025$. (b) $Na_{js} = 0.05$.

TABLE 15.3 Gross Characteristics of a Gas–Liquid–Solid Stirred Reactor

Just Suspension Aeration number (Na_{js})	Predicted Total Gas Holdup (%) (ε_g)	Estimated Gas Holdup (%) [40]	Standard Deviation (σ)	Predicted Power Number(N_P)
0	—	—	0.45	3.95
0.025	5.33	4.80	0.81	2.61
0.05	6.45	9.55	0.90	2.54

the qualitative trend, the CFD model has underpredicted the actual power number value (predicted value of power number for single-phase flow of 3.85). Secondly, the CFD model was able to predict just suspension condition for solid–liquid flows. However, the model failed to predict just suspension condition in presence of gas. The standard deviation value (describing suspension quality) increases with an increase in gas flow rate. For lower aeration rate the model has predicted the total gas holdup values reasonably well. However, the model has underpredicted the total gas holdup value for higher aeration rate. Overall, the CFD model with the presented modeling approach, could reasonably predict gas–liquid–solid flow in a stirred reactor at low aeration rates. Further work is needed to develop adequately accurate model capable of simulating gas–liquid–solid flow in stirred reactor at higher aeration rates.

15.2.5 Application to Simulate Liquid–Liquid Flows in Stirred Reactor

Stirred tanks represent the most popular reactors and mixers, which are widely used in carrying out operations involving liquid–liquid dispersions. Drop size distributions and dynamics of their evolution are important characteristics of such dispersions as they are related to the rate of mass transfer and chemical reactions that may occur in a process. In some cases, the drops are stabilized against coalescence by the addition of stabilizers to have drops sized by agitation before chemical reaction begins (suspension polymerization). In other areas the break-up and coalescence processes can affect directly a reaction in the dispersed phase. It is well known that the other than physical chemistry, fluid dynamic interaction between the two phases plays a significant role in determining the features of the dispersion, but it is far from being fully understood. Average properties over the whole vessel are usually considered for system description and for scale-up. The following main aspects have been studied: minimum agitation speed for complete liquid–liquid dispersion [43]; correlation of mean drop size and DSD to energy dissipation rate and mixer geometric parameters [44] as well as to energy dissipation rate and flow in the vessel [45]; the influence of various impellers on the dispersion features [45–47]; and description of the interaction between the liquid phases in terms of intermittent turbulence [48–50].

CFD Modeling of these systems has also been attempted in recent years by using the Eulerian–Eulerian approach coupled with break-up and coalescence models (see, for example, Refs 51–55). All these efforts are analogous to the efforts made for simulating gas–liquid stirred reactor. In spite of the highly complex system and significant simplifications, the first results are encouraging [51]. To explain the CFD modeling of liquid–liquid stirred reactor, we have reproduced some of the results obtained by Alopaeus et al. [51] in this chapter.

Alopaeus et al. [51] simulated liquid–liquid dispersion in a stirred vessel coupled with population balance equations. For the working equations of population balance simulation, see Ref. 51. The general population balance equation call for the drop breakage and coalescence rate functions and convection terms before it can be used for simulating drop size distributions. In liquid–liquid dispersion the dispersed drops first deform and then break. The magnitude of deformation and breakage depends on the flow pattern around the drop. Most often, the systems characterized have low dispersed-phase viscosity. Such drops break-up provided that the local instantaneous turbulent stresses exceed the stabilizing forces due to the interfacial tension. Therefore, the earlier drop breakage models were only a function of the turbulence present in the continuous phase. In most practical applications, dispersion of high viscosity drops is commonly encountered. In such situations, contribution of the local turbulence on the drop breakage is not sufficient for modeling drop breakage. A viscous drop exposed to the pressure fluctuations causing its deformation tries to return to a spherical shape by the action of stabilizing stresses. Therefore, the stabilizing effect is found to be dominant in high viscosity dispersed phase. One can assume that for the breakage of a drop, the normal turbulent stress outside the drop has to be greater than the sum of viscous stresses developed within the drop due to deformation and stress due to interfacial tension. Narsimhan et al. [56] used both viscous and interfacial forces for estimating breakage frequency. Alopaeus et al. [51] used the same model for simulating breakage frequency in the population balance equation.

Coalescence of two drops depends on two subprocesses, namely, collision between two drops and drainage of film between two drops. Alopaeus et al. [51] used frequency of both these processes to estimate coalescence efficiency. They carried out preliminary simulations with multiblock model

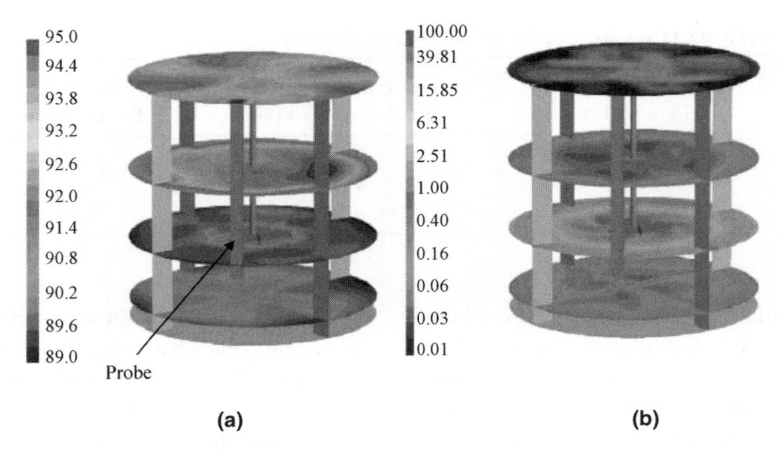

FIGURE 15.13 Predicted distribution of Sauter mean diameter and turbulent kinetic energy dissipation rate at heights of 0.03, 0.133, 0.25, and 0.4 m (from Ref. 51). (a) Sauter mean diameter (μm). (b) Turbulent kinetic energy dissipation rate (W/kg). Reprinted with permission from Ref. 51, copyright 2002, Elsevier.

(see Ref 51) for fitting the parameters of breakage and coalescence efficiency for dense dispersion. In such a multiblock model, the computational domain is divided into several blocks depending on the flow pattern. Each block is treated as a completely back-mixed zone. The flow field obtained from CFD simulation is used for quantifying mass and other exchanges between the two adjacent blocks. Alopaeus et al. [51] simulated dispersion of Exxsol in water in a 50 L stirred reactor equipped with Rushton turbine. Twenty drop size groups were used in the population balance model, with constant viscosity and density for both of the phases. Each group was introduced as mass fraction using user-defined scalars. The conservation equations for user-defined scalars are solved for each cell. Thus, only the source terms for drop breakage and coalescence had to be introduced. Alopaeus et al. [51] did not model the effects of drop size distribution and the volume fraction of the dispersed phase on the prevailing turbulence. They only modeled the effect of the population balance model on velocity and turbulence calculation is through density. The predicted distribution of Sauter mean drop diameter and turbulent kinetic energy distribution is shown in Figure 15.13. The comparison of the predicted local Sauter mean drop diameter with experimentally measured data at three different locations is shown in Table 15.4. It can be seen from the Table 15.4 that the CFD model was able to predict the local Sauter mean drop diameters reasonably well. Overall, the CFD model with the discussed modeling approach, could reasonably predict dense liquid–liquid flow in a stirred reactor. Further work is needed to evaluate CFD model for simulating dispersion of high viscosity drops in a stirred reactor.

TABLE 15.4 Comparison of Predicted Values of Sauter Mean Diameter with Experimental Data

Point	Measured Value (μm)	Predicted Tangential Distribution (μm)	Predicted Value (μm)
1	93.2	93.3–94.3	93.8
2	91.1	93.2–94.1	94.1
3	88.7	89.7–91.8	90.5

From Ref. 51.

15.3 APPLICATION TO ENGINEERING OF STIRRED VESSELS

Engineering of stirred vessels involves designing of vessel configuration and operating protocols to realize desired chemical and physical transformations. A reactor engineer has to ensure that the evolved reactor hardware and operating protocol satisfies various process demands without compromising safety, environment, and economics. Engineering of stirred reactors essentially begins with the analysis of process requirements. This step is usually based on laboratory study and on reactor models based on idealized fluid dynamics and mixing. In most of the industrial cases, this step itself may involve several iterations, especially for multiphase systems. Converting this understanding of process requirements to configuration and operating protocols for industrial reactor proceeds through several steps, such as examining the sensitivity of reactor performance with various flow and mixing related issues (such as short circuiting, bypass, and residence/circulation time distribution); resolving conflicting process requirements; and scale-up.

Not much progress can be made without better understanding of the underlying fluid dynamics of stirred reactors and its relation with the variety of design parameters on the one hand and with the processes of interest on the other hand. Experimental investigations have contributed significantly to the better understanding of the complex hydrodynamics of stirred vessels in the recent years. However, computational models offer unique advantages for understanding conflicting requirements of different processes and their subsequent prioritization. Using a computation model, one can switch on and switch off various processes, which otherwise is not possible while carrying out experiments. Such numerical experiments can give useful insight into interactions between different processes and can help to resolve the conflicting requirements.

It is essential to analyze possible influence of scale of reactor on its fluid dynamics and performance. It should be noted that small-scale reactor would invariably have higher shear and more rapid circulation than large-scale reactor. Multiphase processes, therefore, are often dispersion controlled in small scale and are coalescence controlled in large-scale reactor. The interfacial area per unit volume of reactor normally reduces as the scale of reactor increases for same specific power input. Scale-up/scale-down analysis is important to plan useful laboratory and pilot plant tests. It may often be necessary to use a pilot reactor configuration, which is not geometrically similar to the large-scale reactor in order to maintain the similarity of the desired process. Conventionally, such analysis is carried out based on certain empirical scaling rules and prior experience. Computational flow modeling can make substantial contributions to this step by providing quantitative information about the fluid dynamics. Computational flow models, which allow *a priori* predictions of the flow generated in a stirred reactor of any configuration (impellers of any shape), with just the knowledge of geometry and operating parameters, can make valuable contributions in evolving optimum reactor designs.

Recent advances in physics of flows, numerical methods, and computing resources open up new avenues of harnessing power of computational flow modeling for engineering of stirred vessels. It is, however, important to use this power judiciously. Conventional reaction engineering models and accumulated empirical knowledge about the hydrodynamics of stirred vessels must be used to get whatever useful information that can be obtained, before undertaking rigorous CFD modeling. Distinguishing the "simple" (keeping the essential aspects in tact and ignoring nonessential aspects) and "simpler" (ignoring some of the crucial issues along with the nonessential issues) formulations is a very important step toward finding useful solutions to practical problems. More often than not computational flow-modeling projects are likely to overrun the budget (of time and other resources) due to inadequate attention paid to this initial step of the overall project.

Another important point is it is beneficial and more efficient to develop computational flow models in several stages, rather than directly working with and developing a one-stage comprehensive model. For example, even if the objective is to simulate nonisothermal reactive multiphase flows, it is always useful to undertake a stagewise development. Such stages could (a) simulate isothermal single-phase flow, (b) evaluate isothermal, turbulent simulations, verify existence of key flow features, use the simulations to extract useful quantities such as circulation time distributions, (c) include nonisothermal effects (without reactions),(d) include multiphase models, and (e) include reactive mixing models. Such a multistage development process also greatly reduces various numerical problems, as the results from each stage serve as a convenient starting point for the next stage. The stagewise process also provides insight about relative importance of different processes, which helps to make judicious choice between "simple" and "simpler" representations.

More often than not, in many practical situations, models and results obtained at intermediate stages of such a stepwise process can provide useful support for decision making and continuous improvements without waiting for complete development of a comprehensive model. In this section, we illustrate application of computational flow models discussed in previous section to obtain useful information for some industrially relevant cases. It is possible to present actual case studies for various reasons. The presented examples are however useful to indicate the power and methodology of applying computational flow modeling to address industrially relevant multiphase mixing issues.

15.3.1 Tall Gas–Liquid Stirred Reactor: Flow and Mixing

In many industrial applications, tall vessels equipped with multiple impellers are used. The multiple impeller system provides better gas utilization, higher interfacial area, narrower residence time distribution in the flow system, lower initial cost, and take advantage of higher jacket wall area per volume for heat transfer as compared to a single impeller system. Also the multiple impeller systems are preferred in a bioreactor, as they offer lower average shear as compared to a single impeller system due to overall lower operational speed with nearly the same power input. Overall, the tall stirred vessel offers more degrees of freedom for controlling the gas dispersion as well as the bulk flow of liquid phase. Different fluid dynamic characteristics can be obtained in a tall vessel depending on the equipment and the operating parameters, such as impeller design, impeller spacing, rotational speed, and volumetric gas flow rates. These different fluid dynamic characteristics lead to different rates of transport and mixing processes (see, for example, Refs 31, 57–60). Khopkar et al. [31] and Khopkar and Tanguy [60] explained the influence of operating conditions on mixing and influence of reactor hardware on the prevailing local fluid dynamics, respectively. In this chapter, the case of gas–liquid flow generated by three down-pumping pitched blade turbines, studied by Khopkar et al. [31] was considered to explain the implications of prevailing flow patterns generated due to different flow regimes on the mixing process.

Shewale and Pandit [58] studied gas–liquid flows generated by three down-pumping pitched blade turbines in a stirred reactor. They varied impeller speed at a specific gas flow rate to realize different flow regimes ($Fl = 0.638$ and $Fr = 0.028$; $Fl = 0.438$ and $Fr = 0.0597$; and $Fl = 0.163$ and $Fr = 0.430$). Under these operating conditions, they had observed DFF, DDF, and DDL flow regimes, respectively; where D represents fully dispersed condition, L represents loading condition (impeller is able to disperse gas) and F represents flooding condition (impeller is not able to disperse gas). The DFF flow regime corresponds to upper impeller in the dispersed condition and the middle and bottom impellers in the flooded condition. The other two flow regimes can also be interpreted by analogy. Khopkar et al. [31] simulated these experiments using the EE approach.

The predicted liquid-phase velocity vectors for all the three operating conditions are shown in Figure 15.14. It can be seen from Figure 15.14 that the computational model captured the significantly different flow fields for all the three conditions. For DFF ($Fl = 0.638$ and $Fr = 0.028$) flow regime, the predicted velocity field shows the presence of two-loop structure. The predicted liquid-phase velocity field for DDF flow regime ($Fl = 0.438$ and $Fr = 0.0597$) also shows the two-loop structure (Figure 15.14b). However, the predicted two-loop structure for DDF flow regime was significantly different from the two-loop structure predicted for DFF flow regime. Along with these, two primary circulation loops, the computational model has also captured a secondary circulation loop, present between both circulation loops. For the DDL flow regime ($Fl = 0.163$ and $Fr = 0.430$), simulated results show (Figure 15.14c) three separate circulation loops for each impeller. The predicted velocity field for DDL condition also captured two secondary circulation loops, one at the bottom of the reactor and another between the lower and the middle impeller circulation loops. The complex interaction between the impeller-generated flow and the gas-generated flow was responsible for the formation of these two secondary circulation loops in the reactor.

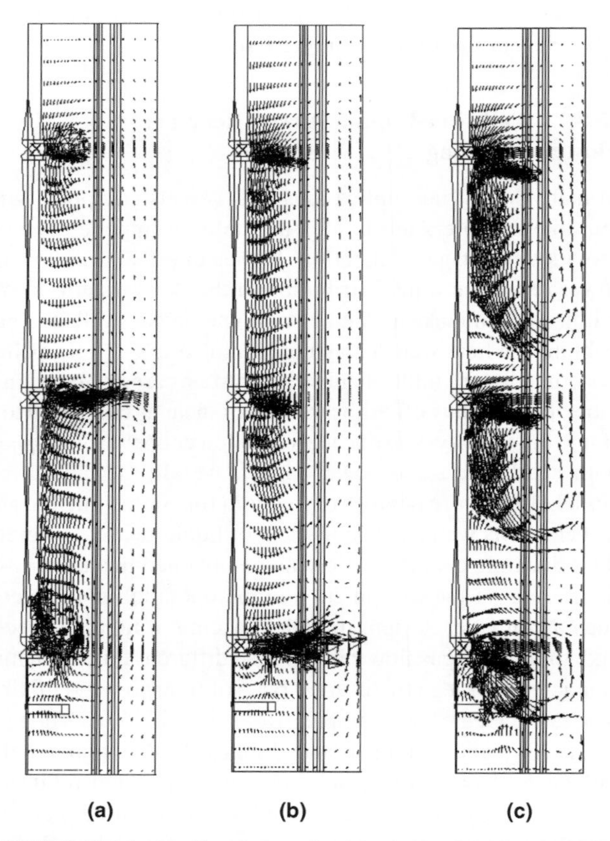

FIGURE 15.14 Predicted mean liquid velocity field at mid-baffle plane for DFF, DDF, and DDL flow regimes. (a) DFF flow regime, $Fl = 0.678$ and $Fr = 0.028$. (b) DDF flow regime, $Fl = 0.438$ and $Fr = 0.0597$. (c) DDL flow regime, $Fl = 0.163$ and $Fr = 0.430$. Reprinted with permission from Ref. 61, Copyright 2006, Elsevier.

The qualitative comparison of predicted gas holdup distributions for all the three operating conditions ($Fl = 0.638$ and $Fr = 0.028$ (DFF); $Fl = 0.438$ and $Fr = 0.0597$ (DDF); and $Fl = 0.163$ and $Fr = 0.430$ (DDL)) with experimental snapshots is shown in Figure 15.15. It can be seen from Figure 15.15a that similar to experimental condition, the simulation has captured the inefficient dispersion of gas at the bottom and middle impellers and dispersed condition of gas at the upper impeller for DFF flow regime. It can be seen from Figure 15.15b that the simulation has correctly captured the inefficient dispersion of gas by the bottom impeller and the complete dispersed conditions by the middle as well as upper impeller as observed in the case of DDF flow regime. For the DDL flow regime, the predicted gas holdup distribution shows the fully dispersed condition for upper and middle impeller and loading condition for the bottom impeller.

One of the major interests in developing such complex flow models is to gain insight into mixing. Mixing in the reactors is significantly influenced by the prevailing flow field, particularly flow regimes and interaction of internal circulation loops. Generally, mixing is characterized by "scale of segregation" and "intensity of segregation." The scale of segregation is a measure of the size of the unmixed lumps. An intensity of segregation is a measure of the difference in concentration between neighboring lumps of fluid. Lower the intensity of segregation, more is the extent of molecular mixing (for more detailed discussion, see Ref. 8 and references cited therein). Since most of the multiphase flows in industrial reactors will be turbulent, we will limit our discussion to turbulent mixing in this chapter. Convection and turbulent dispersion by large eddies lead to macroscale mixing and do not cause any small-scale mixing. Fluid motions in the inertial subrange reduce the scale of segregation via vortex stretching. Such a reduction in scale increases interfacial area between segregated lumps of the tracer fluid and the base fluid, which increases the rate of mixing by molecular diffusion. However, increase in inter-

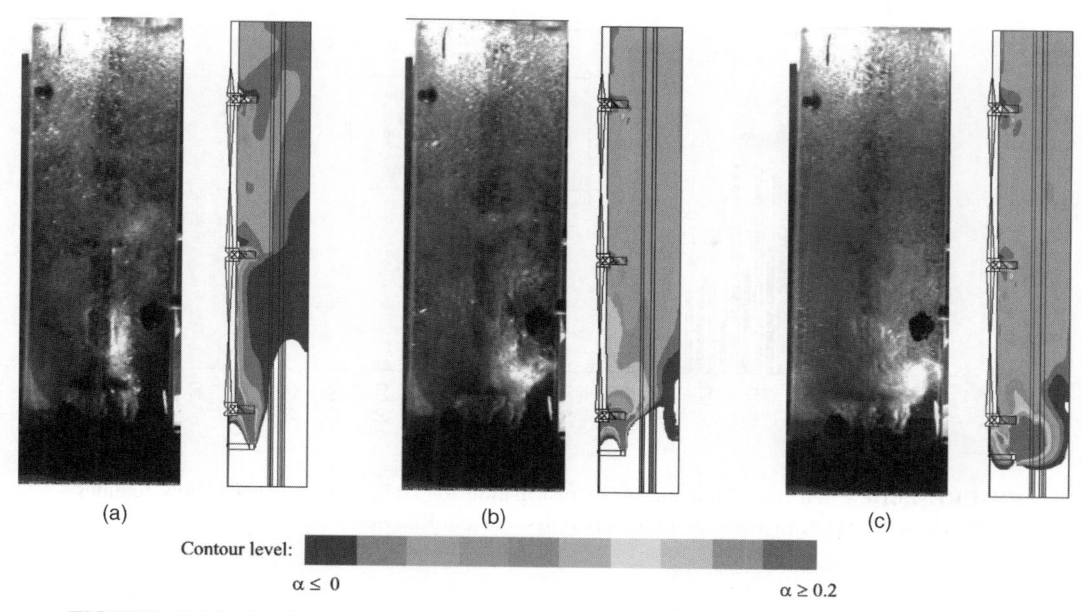

FIGURE 15.15 Qualitative comparison of experimental snapshot and predicted gas holdup distribution at mid-baffle plane for DFF, DDF, and DDL flow regimes. (a) DFF flow regime. (b) DDF flow regime. (c) DDL flow regime. Reprinted with permission from Ref. 61, copyright 2006, Elsevier.

facial area by the inertial subrange eddies may not be substantial. The mixing caused by this step is typically called as "meso-mixing." Meso-mixing reduces the scale of mixing substantially but does not affect intensity of mixing much. Engulfment and viscous stretching by Kolmogorov scale eddies lead to substantial increase in the interfacial area for molecular diffusion and therefore, contribute significantly to molecular mixing. The last step is diffusion process through such interfacial area between layers of different fluids accompanied by chemical reactions, if any. Molecular diffusion leads to complete mixing and dissipates concentration fluctuations. Comparison of the timescales of these mixing processes with the characteristic reaction timescales provides useful information about possible interaction of mixing and chemical reactions. For fast chemical reactions, effective reaction rate may not be controlled by reaction kinetics but may be controlled by rate of mixing. However, for most industrially relevant multiphase flow processes, fast reactions may often be controlled by interphase mass transfer rather than liquid-phase mixing. It is, however, often important to quantify characteristics timescale of "mixing" to understand interaction of interphase transport and mixing as well as possibility of short-circuiting and channeling. Usually, "mixing time" and '"circulation time" which essentially characterize macromixing in stirred tanks are used for this purpose.

Mixing time is the time required to achieve a certain degree of homogeneity [62], whereas circulation time is the time necessary for a fluid element to complete a one circulation within the vessel (time difference between an event of fluid element exiting from the impeller swept volume and an event of its re-entry into impeller swept volume). The circulation time distributions provide useful insight about possible short-circuiting and channeling. The mixing time is also usually related to mean circulation time [7]. In this example of tall gas–liquid stirred tanks, computational flow models are used to estimate mean circulation time to gain better understanding of macromixing process.

Using the Eulerian flow field obtained as discussed in the previous subsection, the particle trajectories were simulated for all the three operating conditions (DFF, DDF, and DDL). Based on the study of Rammohan et al. [63], neutrally buoyant particles of size less than 0.25 mm were released into the liquid at 10 randomly selected positions in the solution domain. The motion of particles in the liquid phase was simulated using the Lagrangian framework. The simulated particle trajectories were used to calculate the circulation time distribution.

The simulated circulation time distributions for all the three operating conditions are shown in Figure 15.16. It can be seen in Figure 15.16 that for DFF flow regime, significant fraction show circulation time higher than 16 s. These circulations were for particles following the upper circulation loop and may lead to slower mixing in the reactor. Almost no circulations with circulation times less than 4 s were found in the simulated circulation time distribution. For the DDF flow regime significant fraction show circulation time less than 6 s indicating faster mixing. For the DDL regime, not insignificant fraction show circulation times more than 30 s indicating slower mixing despite increase in the impeller speed. The predicted values of average circulation time and the experimental data are listed in Table 15.5. Figure 15.17

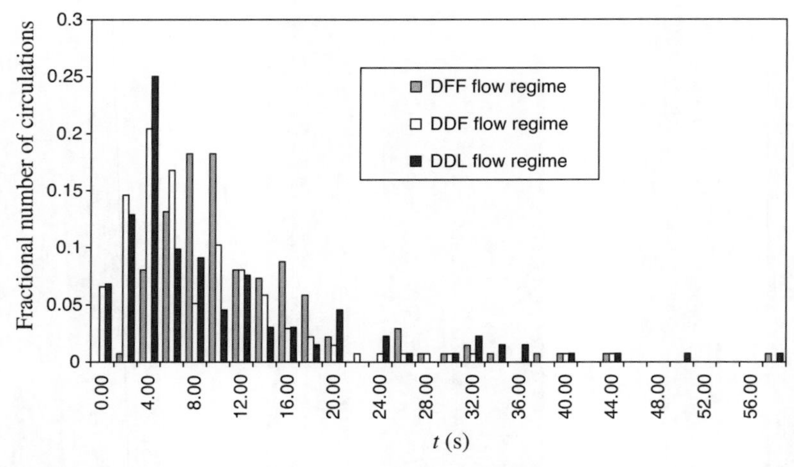

FIGURE 15.16 Predicted Circulation time distribution for DFF, DDF, and DDL flow regimes. Reprinted from [61], Copyright 2006, with permission from Elsevier.

TABLE 15.5 Gross Characteristics of a Tall Gas-Liquid Stirred Reactor. Experimental data from Ref. 61

Flow Regime	Total Gas Holdup (%)		Power Number, N_{Pg}		Average Circulation Time, t_c (Predicted)	Mixing Time, t_m (Experimental)	Percentage Change	
	Predicted	Experimental	Predicted	Experimental			$t_c/t_{c,min}$	$t_m/t_{m,min}$
DFF (Fl = 0.6328 & Fr = 0.028)	2.99	2.47	2.64	2.2	13.851	59	1.493	1.553
DDF (Fl = 0.438 & Fr = 0.0597)	3.43	2.79	2.98	2.55	9.277	38	1	1
DDL (Fl = 0.163 & Fr = 0.430)	5.58	3.65	4.05	3.45	11.234	45	1.211	1.184

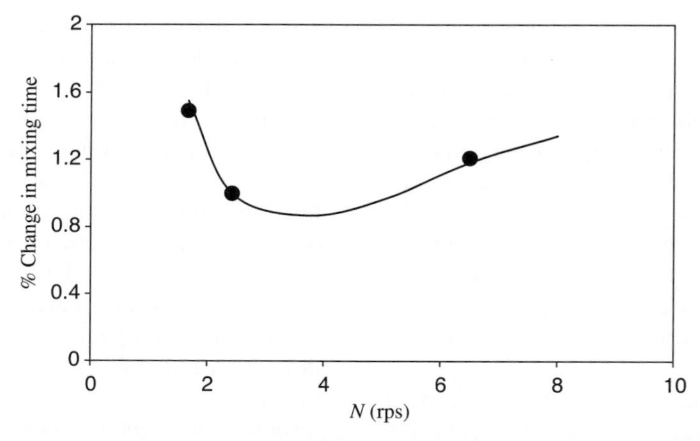

FIGURE 15.17 Comparison of the experimental data and predicted percentage change in mixing time as function of impeller speed. Dark circles: predicted results; —: Experimental data [58]. Reprinted with permission from Ref. 61, copyright 2006, Elsevier.

shows the variation in the mixing time with impeller speed as reported by Shewale and Pandit [58] and the time required for a fixed number of circulations as per the simulations in this chapter. It can be seen from Table 15.5 and Figure 15.17 that the predicted values of average circulation times have captured the apparently counterintuitive trend (increase in mixing time with increase in impeller speed) observed in the experimental study of Shewale and Pandit [58]. The developed computational model can thus be creatively used to address industrially important issues.

15.3.2 Solid Suspension and Mixing in Stirred Reactor

Liquid-phase mixing is a quite important in many solid–liquid reactions as well. It not only affects the selectivity of reactions but also controls the temperature distribution inside the reactor for exothermic reactions. In many cases, stirred slurry reactors are operated with high solid loading (solid volume fraction greater than 5.0%). In such situations, the liquid-phase mixing process was found to show a complex interaction with the suspension quality (see, for example, Ref. 64). A computational model, which is able to predict suspension quality and its influence on the liquid-phase mixing will definitely help reactor engineers to obtain optimum performance of stirred slurry reactors. Recently, Kasat et al. [65] simulated liquid-phase mixing in a stirred reactor for different operating conditions. Their simulations are reproduced in this chapter to explain the liquid-phase mixing in a stirred slurry reactor.

The simulations are carried out for the experimental setup of Yamazaki et al. [32] with solid volume fraction of 10.0% and particle diameter of 264 μm. The simulations are carried out for 10 different impeller rotational speeds starting from 2 to 40 rps. The completely converged solid–liquid flow simulations were used to simulate liquid-phase mixing. Mixing simulations were carried out with 1.0% (by volume) of tracer, with the same physical properties of liquid in the vessel. The tracer history was recorded at eight different locations. In stirred slurry reactor delayed mixing was usually observed near the top surface of the liquid. Therefore, tracer history was recorded at four different locations close to the top surface (for more details, see Ref. 65). The mixing time in this chapter is defined as the time required for the tracer concentration at these locations to lie within $\pm$5.0% of the final concentration, C_∞.

It will be very helpful to first shed light on the predicted suspension quality before discussing the influence of suspension quality on the mixing process. In a stirred slurry reactor, the critical impeller speed for complete off-bottom suspension Nc_s and complete suspension N_s are two very important design parameters. The concepts of a critical impeller speed was introduced more than 40 years ago and is the primary designed parameter used even today by reactor engineers for scale-up and design of stirred slurry reactor. The predicted suspension quality was analyzed to estimate the Nc_s and N_s. Several criteria are available in the literature to determine the values of Nc_s and N_s. However, those criteria are applicable for experimental measurements and cannot be extended directly to the CFD simulations with the EE approach. Bohnet and Niesmak [66] have proposed alternative criteria based on the standard deviation σ of solid concentration to describe the suspension quality (see equation 15.17). The same criterion is used in this chapter to describe the suspension quality. The decrease in standard deviation is manifested as an increase in the quality of the suspension. Based on the range of the standard deviation the quality of the suspension is broadly divided into three regimes: homogeneous suspension where the value of the standard deviation is smaller than 0.2 ($\sigma < 0.2$); complete off-bottom suspension—the value of the standard deviation lies between 0.2 and 0.8 ($0.2 < \sigma < 0.8$)—and incomplete suspension—the standard deviation value was found to be higher than 0.8 ($\sigma > 0.8$). This criterion enables the prediction of Nc_s and N_s and also gives the information on quality of suspension prevailing in the vessel.

The standard deviation values were estimated from the predicted solid volume fraction for all the 10 simulations carried out at different impeller rotational speeds. It must be noted that solid volume fraction values at all computational cells were used to estimate the standard deviation value. The predicted variation of standard deviation values with respect to impeller speed is shown in Figure 15.18. It can be seen from Figure 15.18 that three distinctly different suspension conditions, namely, incomplete suspension, complete off-bottom suspension and homogeneous suspension can be identified in the vessel. At a lower impeller speed, the computational model predicted very high values of the

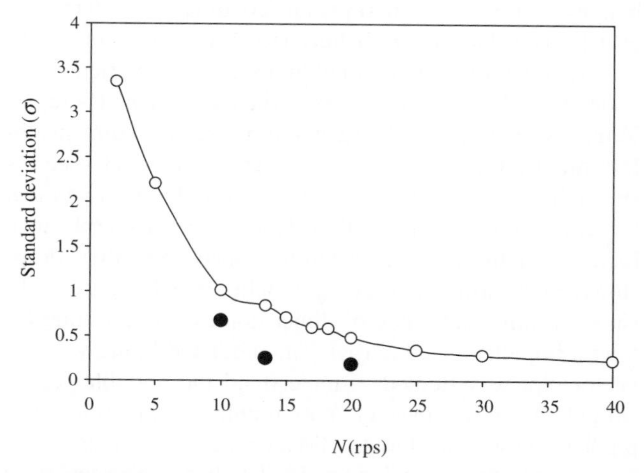

FIGURE 15.18 Predicted influence of impeller rotational speed on suspension quality for $d_p = 264\,\mu$m and $\alpha = 0.1$. Line with empty circles: With modified Brucato et al. [21] correlation. Dark circles: With Brucato et al. [21] correlation. Reprinted with permission from Ref. 65, copyright 2008, Elsevier.

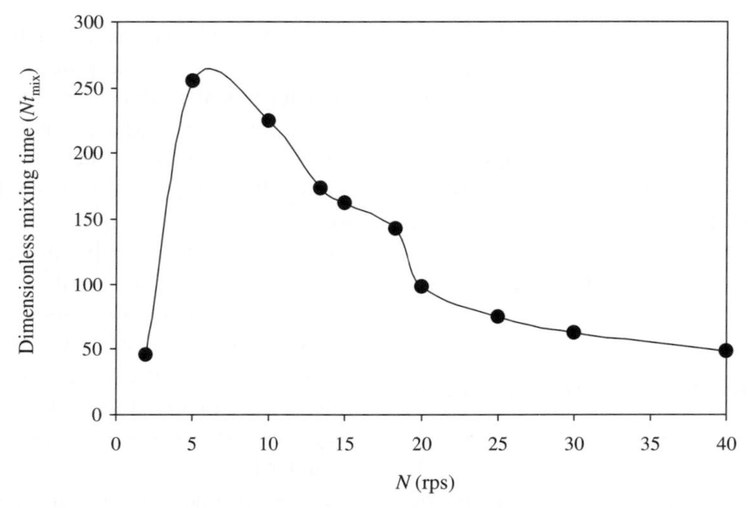

FIGURE 15.19 Predicted influence of impeller rotational speed on the dimensionless mixing time for $d_p = 264\,\mu m$ and $\alpha = 0.1$. Reprinted with permission from Ref. 65, copyright 2008, Elsevier.

standard deviation ($\sigma > 0.8$), indicating incomplete suspension in the vessel. It is also observed that the standard deviation values drop sharply with an increase in the impeller rotational speed until complete off-bottom condition is achieved. The computational model predicted standard deviation value of 0.7 for 15 rps. This indicates the presence of a critical impeller speed for complete off-bottom suspension ($\sigma = 0.8$) close to 15 rps. This is in good agreement with the Nc_s (= 13.4 rps) estimated using correlation proposed by Zwietering [67] for the experimental setup of Yamazaki et al. [32]. With further increase in the impeller rotational speed, the values of standard deviation drop slowly till the system achieves homogeneous suspension condition. The predicted results suggest that the homogeneous suspension condition for the experimental condition of Yamazaki et al. [32] is achieved at impeller speed N_s of 40 rps ($\sigma = 0.17$).

The species transport simulations are then carried out to understand the mixing process in the experimental setup of Yamazaki et al. [32]. The variation of predicted dimensionless mixing time (Nt_{mix}) with impeller rotational speed is shown in Figure 15.19. It can be seen from Figure 15.19 that the dimensionless mixing time first increases sharply with increase in the impeller rotational speed and then drops slowly with further increase in impeller speed. Figure 15.19 shows a minimum value of dimensionless mixing time for lowest impeller speed (2 rps). The predicted liquid velocity vector plot was studied to understand the possible reason behind the observance of a minimum mixing time. The predicted flow characteristics for an impeller rotational speed of 2 rps are shown in Figure 15.20a. It can be seen from Figure 15.20a that the computational model has predicted nearly a single loop velocity pattern in the vessel. It is possible that for such a low impeller speed, impeller action is not sufficient to lift the particles from the vessel bottom (see Figure 15.20a). The solid bed present at the bottom of the reactor might offer an apparent low off-bottom clearance to the impeller-generated flow and therefore lead to single loop flow pattern for the Rushton turbine. In such a scenario, all the energy dissipated by the impeller becomes available for generating liquid circulations in the vessel and for fluid mixing. Therefore, it is possible to have faster mixing in the reactor at low impeller rotational speed.

With increase in impeller rotational speed, the dimensionless mixing time increases, reaches a maximum and then drops slowly (see Figure 15.19). In this chapter, the maxi-

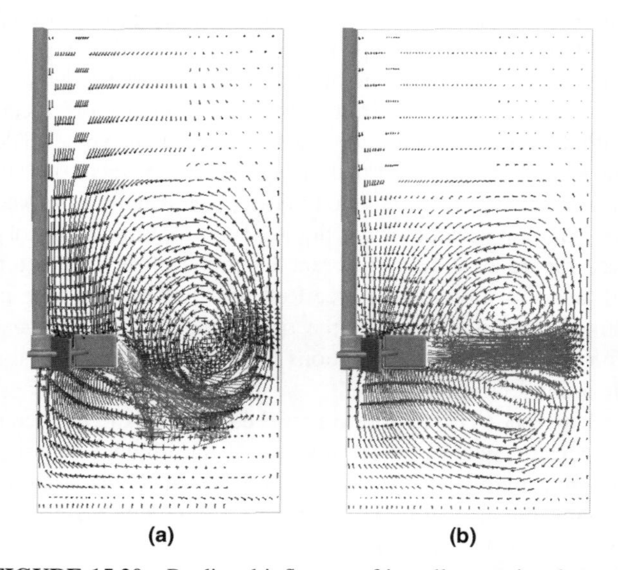

FIGURE 15.20 Predicted influence of impeller rotational speed on the liquid-phase flow field for $d_p = 264\,\mu m$ and $\alpha = 0.1$. (a) $N = 2$ rps. (b) $N = 5$ rps. Reprinted with permission from Ref. 65, copyright 2008, Elsevier.

mum in the mixing time was found to happen at around 5 rps. At 5 rps the impeller-generated flow becomes sufficient to start suspending solids into the bulk volume of the reactor. The energy dissipated by the impeller is now distributed for generating liquid circulations, fluid mixing and suspending the solids. The single-loop flow pattern changes to the classical two-loop structure for the Rushton turbine (see Figure 15.20b). The part of the energy dissipation for solid suspension and the rate of exchange between the two loops contribute to the slower mixing in the vessel for the 5 rps condition. A further increase in the impeller rotational speed leads to a reduction in the mixing time (increase in the mixing efficiency). This observed reduction in the mixing time continues till the system achieves the complete off-bottom suspension condition (i.e., $Nc_s = 15$ rps). The operating conditions (impeller rotational speed) after Nc_s show a gradual decrease in the mixing time with an increase in impeller rotational speed. The present simulations also supported the operating range at which a maximum of mixing time occurs, that is, $N = Nc_s/3$ [64].

The simulated results are further analyzed to understand the mixing in the stirred slurry reactor. The predicted tracer histories in the bulk volume of the reactor (closer to the impeller) and near the top surface of vessel were compared. The comparison of predicted tracer histories is shown in Figure 15.21. It can be seen from Figure 15.21 that the homogenization process is much faster in the region close to the impeller compared to the region near the top surface. It was also observed that the difference between the top region and the impeller region is strongly dependant on the suspension quality present in the vessel. Figure 15.21 shows that in incomplete suspension and in complete off-bottom suspension conditions, the time required for the homogenization near the top surface is significantly high compared to the time required for homogenization close to the impeller. The difference decreases as the system approaches the homogeneous suspension condition. Kasat et al. [65] showed that the lower liquid velocities present in the clear liquid layer above the solid cloud is responsible for the slower mixing process.

The subsections 15.3.1 and 15.3.2 discuss mixing issues in gas–liquid and solid–liquid systems. Similar approach can be used for addressing issues in other systems such as liquid–liquid and gas–liquid–solid stirred vessels. Apart from predicting dispersion or suspension quality and mixing time, CFD models allow estimation of circulation time distribution, different zones in stirred vessels with different prevailing shear rates, interaction of impeller stream with inlet and outlet nozzles, and so on. It is possible to creatively use information obtainable from CFD to gain better insight and support engineering decision making. For example, information on different shear zones and residence time distribution in these different zones often provides useful clues for quantifying influence of scale on "break-up" and '"coalescence" dominated zones. It is not possible to discuss actual industrial cases in this chapter for the sake of protecting confidential information. It is, however, hoped that information provided in this chapter will allow resourceful engineer to develop appropriate computational flow models and use the simulated results for addressing practical design and scale-up issues.

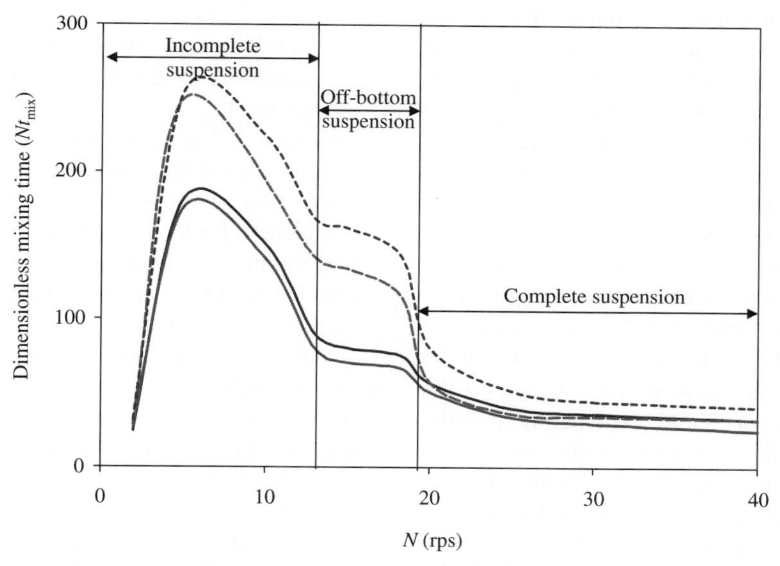

FIGURE 15.21 Predicted influence of suspension quality on delayed mixing in the top clear liquid layer for $d_p = 264$ μm and $\alpha = 0.1$. Dotted line: close to top surface, dashed line: close to top surface; black line: close to impeller; and gray line: close to impeller. Reprinted with permission from Ref. 65, copyright 2008, Elsevier.

15.4 SUMMARY AND PATH FORWARD

In this chapter, we have demonstrated the extent of applicability of computational models for simulating multiphase flows in stirred vessels with some examples. Role of turbulence, multiphase flow, interphase interactions (such as drag, lift, virtual mass, coalescence, and break-up) and flow regimes are critically analyzed for gas–liquid and solid–liquid flows. The presented computational models were found to capture key features of two-phase flows in stirred tank reasonably well. This chapter highlighted the limited applicability of direct extension of gas–liquid and solid–liquid modeling approaches for simulating three-phase flow. Despite some of the limitations, computational models were shown to provide useful information on important flow characteristics around the impeller blades as well as in the bulk. The computational models were able to predict the implications of reactor hardware, flow regimes and suspension quality on the transport and mixing process. Careful numerical experiments using these CFD models can be used for better understanding the characteristics of existing reactors, to enhance their performance, assess different configurations and greatly assist the engineering decision making process. The approach, the models, and the results discussed in this chapter will provide a useful basis both for practical applications and for further developments.

Although the models discussed in this chapter are capable of providing valuable and new insights, which hitherto were unavailable, there is still significant scope to improve the fidelity of these multiphase flow models. Some of the ways for improving the discussed models are listed in the following:

The results in this chapter have highlighted the importance of the correct modeling of interphase forces. The turbulent drag correction terms proposed by Khopkar and Ranade [17] for gas–liquid flows and Khopkar et al. [31] for solid–liquid flows were used in this chapter with reasonable success. Further improvements in these submodels to account for dispersed-phase holdup as well as particle Reynolds number may provide a more general framework to simulate industrial multiphase stirred vessels. Well-designed experiments and quantitative data (with error bars) are needed to validate some of these interphase drag models.

In a gas–liquid stirred reactor, the gas bubbles shear away from the tip of the gas cavities present behind the impeller blades. The size of the bubbles emanating from the cavity tip is controlled by the size of the cavity, breakage of cavity, and the turbulence level around the cavity. Unfortunately, no direct experimental data for turbulent kinetic energy dissipation rate are available for validating the available cavity breakage models. More experimental data in the region around impeller is needed to improve computational models.

All the simulations discussed in this chapter were carried out for laboratory-scale and pilot-scale reactors. For large-scale reactor, the ratio of characteristic length scales of impeller blades and the gas bubble are strikingly different from the corresponding value in a small reactor. Therefore, the interaction of gas bubbles with the trailing vortices and the structure of the cavities might be significantly different for industrial scale as compared to small reactor. Although some indirect evidence of this is available, no systematic study of the influence of scale on relative performance of different newly proposed impellers for dispersing a secondary phase is available.

Solid–liquid systems with polydispersed solid phase are encountered in process industry. However, there are no reports in the literature on the experimental measurements of the concentration profiles for the polydispersed system. Experimental and computational efforts are therefore needed to study the hydrodynamics of the solid–liquid stirred reactor with polydispersed solid phase.

In a three-phase stirred reactor, suspended solid particles will interact with the wake of gas bubble. This interaction will influence not only the drag experienced by both solid particles and gas bubble but also the lift experienced by solid particles. This might be a possible reason for the limited applicability of direct extensions of gas–liquid and solid–liquid modeling approaches for simulating gas–liquid–solid stirred reactor. Well-designed experiments and computational efforts need to be undertaken for the estimation of bubble and particle interaction.

The complexity of reactive flows may greatly expand the list of issues on which further research is required. Another area, which deserves mention in this chapter, is modeling of unsteady flows in stirred vessels. Most of the examples discussed in this chapter used a steady-state modeling approach for simulating flows in stirred vessels. In some conditions, the steady-state approach may not be appropriate and a full unsteady-state approach may be necessary. This is especially crucial when fast reactive mixing and interaction of nozzle and impeller stream are important. Fortunately for many multiphase stirred vessels, the overall performance is dominated by interphase transport rather than micromixing and therefore full unsteady simulations may not be necessary.

Adequate attention to key issues mentioned in this chapter and the creative use of computational flow modeling will hopefully make useful contributions to reactor engineering of multiphase stirred reactors. New advances made in modeling of multiphase flows in stirred vessels may be assimilated using the framework discussed in this chapter. We hope that this chapter will stimulate applications of computational flow modeling to reactor engineering in the pharmaceutical industry as well as the rest of the chemical process industry.

SYMBOLS

C	impeller off-bottom clearance (m)
C_1, C_2	model parameters (equation 15.5)
C_D	drag coefficient
C_{D0}	drag coefficient in stagnant water
C_{VM}	virtual mass coefficient
C_ω	model parameter (equation 15.7)
d_b	bubble diameter (m)
d_p	diameter of particle (m)
d_s	impeller shaft diameter (m)
d_{sp}	outer diameter of ring sparger (m)
D_i	impeller diameter (m)
D_{12}	turbulent diffusivity (m^2/s)
Eo	Evotos number
F_D	interphase drag force (N/m^3)
Fl	gas flow number
Fr	Froude number
F_L	lift force (N/m^3)
F_q	interphase momentum exchange term
F_{VM}	virtual mass force (N/m^3)
g	acceleration due to gravity (m/s^2)
H	vessel height (m)
k	turbulent kinetic energy (m^2/s^2)
K	constant (equation 15.13)
N	impeller rotational speed (rps)
N_P	power number
N_Q	pumping number
p	pressure (N/m^2)
Q_g	volumetric gas flow rate (m^3/s)
r	radial coordinate (m)
Re	impeller Reynolds number
Re_b	bubble Reynolds number
Re_p	particle Reynolds number
t	time (s)
t_{mix}	mixing time (s)
T	vessel diameter (m)
T_{fl}	turbulent dispersion force (N/m^3)
T_L	integral timescale of turbulence (s)
U	velocity (m/s)
U_{slip}	slip velocity (m/s)
V	volume of vessel (m^3)
V_{dr}	drift velocity (m/s)

Greek Symbols

α	secondary phase volume fraction
ε	turbulent kinetic energy dissipation rate (m^2/s^3)
θ	tangential coordinate
λ	Kolmogorov length scale (m)
μ	viscosity (kg/ms)
ρ	density (kg/m^3)
$\sigma_{\phi,1}$	model parameter (equation 15.4)
σ	standard deviation
σ_{pq}	dispersion Prandtl number
τ	shear stress (N/m^2)
τ_p	particle relaxation time (s)
ϕ	variable

Subscripts

1	liquid
2	secondary phase
g	gas
i	direction
l	liquid
s	solid particle
q	phase number
t	turbulent

Superscripts

—	time-averaged value
'	rms value

REFERENCES

1. Oldshue JY. *Fluid Mixing Technology*, McGraw Hill, New York, 1983.
2. Smith JM. In: Ulbrecht JJ, Patterson GK, editors. *Dispersion of gases in liquids in Mixing of Liquids by Mechanical Agitation*, Gordon and Breach, London, 1985.
3. Tatterson B. *Fluid Mixing and Gas Dispersion in Agitated Tanks*, McGraw Hill, London, 1991.
4. Ranade VV. Computational fluid dynamics for reactor engineering. *Rev. Chem. Eng.* 1995;11:225.
5. Kuipers JAM, van Swaij WPM. Application of computational fluid dynamics to chemical reaction engineering. *Rev. Chem. Eng.* 1997;13:1.
6. Joshi JB. Ranade VV. Computational fluid dynamics for designing process equipment: expectations, current status and path forward. *Ind. Eng. Chem. Res.* 2003;42:1115–1128.
7. Joshi JB, Pandit AB, Sharma MM. Mechanically agitated gas–liquid reactors. *Chem. Eng. Sci.* 1982;37:813–844.
8. Ranade VV. *Computational Flow Modelling for Chemical Reactor Engineering*, Academic Press, New York, 2002.
9. Simonin C, Viollet PL. Prediction of an oxygen droplet pulverization in a compressible subsonic conflowing hydrogen flow. *Numer. Methods Multi. Flows* 1998;1:65–82.
10. Kataoka I, Besnard DC. Serizawa A. Basic equation of turbulence and modelling of interfacial terms in gas–liquid two phase flows. *Chem. Eng. Commun.* 1992;118:221.
11. Ranade VV. Modeling of turbulent flow in a bubble column reactor. *Chem. Eng. Res. Des.* 1997;75:14.
12. Mudde R, Simonin O. Two- and three-dimensional simulations of a bubble plume using a two-fluid model. *Chem. Eng. Sci.* 1999;54:5061–5069.

13. Oye RS, Mudde R, van den Akker HEA. Sensitivity study on interfacial closure laws in two fluid bubbly flow simulations. *Am. Inst. Chem. Eng. J.* 2003;49:1621–1636.
14. Ranade VV. Numerical simulation of dispersed gas–liquid flows. *Sadhana* 1992;17:237–273.
15. Lane GL, Schwarz MP, Evans GM. Modelling of the interaction between gas and liquid in stirred vessels. Proceedings of the 10th European Conference on Mixing, Delft, The Netherlands, 2000, 197–204.
16. Lane GL, Schwarz MP, Evans GM. Computational modelling of gas–liquid flow in mechanically stirred tanks. *Chem. Eng. Sci.* 2005;60:2203–2214.
17. Khopkar AR, Ranade VV. CFD simulation of gas–liquid flow in stirred vessels: VC, S33 and L33 flow regimes, *Am. Inst. Chem. Eng. J.* 2006;52:1654–1671.
18. Bombac A, Zun I, Filipic B, Zumer M. Gas-filled cavity structure and local void fraction distribution in aerated stirred vessel. *Am. Inst. Chem. Eng. J.* 1997;43(11): 2921–2931.
19. Clift R, Gauvin WH. Motion of entrained particles in gas streams. *Can. J. Chem. Eng.* 1971;49:439–448.
20. Bakker A, van den Akker HEA. A computational model for the gas–liquid flow in stirred reactors. *Trans. Inst. Chem. Eng.* 1994;72;594–606.
21. Brucato A, Grisafi F, Montante G. Particle drag coefficient in turbulent fluids. *Chem. Eng. Sci.* 1998;45:3295–3314.
22. Pinelli D, Montante G, Magelli F. Dispersion coefficients and settling velocities of solids in slurry vessels stirred with different types of multiple impellers. *Chem. Eng. Sci.* 2004; 59:3081–3089.
23. Tomiyama A. Drag lift and virtual mass forces acting on a single bubble. 3rd International Symposium on two-phase flow modeling and experimentation, Pisa, Italy, September 22–24, 2004.
24. Barigou M, Greaves A. Bubble size distribution in a mechanically agitated gas–liquid contactor. *Chem. Eng. Sci.* 1992;47 (8):2009–2025.
25. Laakkonen M, Moilanen P, Alopeaus V, Aittamaa J. Modeling local bubble size distributions in agitated vessel. *Chem. Eng. Sci.* 2007;62:721–740.
26. Calderbank PH. Physical rate processes in industrial fermentation. Part I. The interfacial area in gas–liquid contacting with mechanical agitation. *Trans. Inst. Chem. Eng.* 1958;36,443.
27. Hughmark G. Power requirements and interfacial area in gas–liquid turbine agitated systems. *Ind. Eng. Chem. Proc. Des. Dev.* 1980;19:641–646.
28. Cui YQ, van der Lans RGJM, Luben KCAM. Local power uptake in gas–liquid systems with single and multiple Rushton turbines. *Chem. Eng. Sci.* 1996;51:2631–2636.
29. Angst R, Harnack E, Singh M, Kraume M. Grid and model dependency of the solid/liquid two-phase flow CFD simulation of stirred reactors. Proceedings of 11th European Conference of Mixing, Bamberg, Germany, 2003.
30. Montante G, Magelli F. Modeling of solids distribution in stirred tanks: analysis of simulation strategies and comparison with experimental data. *Int. J. Comp. Fluid Dyn.* 2005;19: 253–262.
31. Khopkar AR, Kasat GR, Pandit AB, Ranade VV. CFD simulation of solid suspension in stirred slurry reactor. *Ind. Eng. Chem. Res.* 2006;45:4416–4428.
32. Yamazaki H, Tojo K, Miyanami K. Concentration profiles of solids suspended in a stirred tank. *Powder Technol.* 1986;48: 205–216.
33. Pinelli D, Nocentini M, Magelli F. Solids distribution in stirred slurry reactors: influence of some mixer configurations and limits to the applicability of a simple model for predictions. *Chem. Eng. Comm.* 2001;188:91–107.
34. Dylag M, Talaga J. Hydrodynamics of mechanical mixing in a three-phase liquid–gas–solid system. *Int. Chem. Eng.* 1994;34 (4):539–551.
35. Nienow AW, Miles D. The effect of impeller/tank configurations on fluid-particle mass transfer. *Chem. Eng. J.* 1978;15:13–24.
36. Chapman CM, Nienow AW, Cook M, Middleton JC. Particle–gas–liquid mixing in stirred vessel. Part III. Three phase mixing. *Chem. Eng. Res. Des.* 1983;61:167–181.
37. Frijlink JJ, Bakker A, Smith JM. Suspension of solid particles with gassed impellers. *Chem. Eng. Sci.* 1990;45 (7):1703–1718.
38. Rewatkar VB, Raghava Rao KSMS, Joshi JB. Critical impeller speed for solid suspension in mechanically agitated three-phase reactors: experimental part. *Ind. Eng. Chem. Res.* 1991;30:1770–1784.
39. Murthy BN, Ghadge RS, Joshi JB. CFD Simulations of gas–liquid–solid stirred reactor: prediction of critical impeller speed for solid suspension. *Chem. Eng. Sci.* 2007;62: 7184–7195.
40. Pantula PRK, Ahmed N. Solid suspension and gas holdup in three phase mechanically agitated reactors. Presented at Chemeca 98, Port Douglas, Queensland, Australia, 1998. Paper No. 132.
41. Warmoeskerken MMCG, Smith JM, Flooding of disk turbines in gas–liquid dispersions: a new description of the phenomenon. *Chem. Eng. Sci.* 1985;40:2063.
42. Zhu Y, Wu J. Critical impeller speed for suspending solids in aerated agitation tanks. *Can. J. Chem. Eng.* 2002;80:1–6.
43. Armenante PM, Huang YT. Experimental determination of the minimum agitation speeds for complete liquid–liquid dispersion in mechanically agitated vessels. *Ind. Eng. Chem. Res.* 1992;31:1398–1406.
44. Calabrese RV, Chang TPK, Dang PT. Drop breakup in turbulent stirred-tank contactors. Part I. Effect of dispersed-phase viscosity. *Am. Inst. Chem. Eng. J.* 1986;32, 657–666.
45. Zhou G, Kresta SM. Correlation of mean drop size and minimum drop size with the turbulence energy dissipation and the flow in an agitated tank. *Chem. Eng. Sci.* 1998;53: 2063–2079.

46. Pacek AW, Man CC, Nienow AW. On the Sauter mean diameter and size distributions in turbulent liquid/liquid dispersions in a stirred vessel. *Chem. Eng. Sci.* 1998;53:2005–2011.

47. Giapos A, Pachatouridis C, Stamatoudis M. Effect of the number of impeller blades on the drop sizes in agitated dispersions. *Chem. Eng. Res. Des.* 2005;83 (A12): 1425–1430.

48. Bałdyga J, Bourne JR. Interpretation of turbulent mixing using fractals and multi-fractals. *Chem. Eng. Sci.* 1995;50: 381–400.

49. Bałdyga J, Podgorska W. Drop break-up in intermittent turbulence: maximum stable and transient sizes of drops. *Can. J. Chem. Eng.* 1998;76:456–470.

50. Bałdyga J, Bourne JR, Pacek AW, Amanullah A, Nienow AW. Effects of agitation and scale-up on drop size in turbulent dispersions: allowance for intermittency. *Chem. Eng. Sci.* 2001;56:3377–3385.

51. Alopaeus V, Koskinen J, Keskinen K, Majander J. Simulation of the population balances for liquid–liquid systems in a non-ideal stirred tank. Part 2. Parameter fitting and the use of the multi-block model for dense dispersions. *Chem. Eng. Sci.* 2002;57: 1815–1825.

52. Wang F, Mao Z, Wang Y, Yang C. Measurement of phase holdups in liquid–liquid–solid three-phase stirred tanks and CFD simulations. *Chem. Eng. Sci.* 2006;61: 7535–7550.

53. Zaccone A, Gabler A, Maaß S, Marchisio D, Kraume M. Drop breakup in liquid–liquid dispersions: modeling of single drop breakage. *Chem. Eng. Sci.* 2007;62:6297–6307.

54. Derksen J, van den Akker HEA. Multi-scale simulations of stirred liquid–liquid dispersions. *Chem. Eng. Res. Des.* 2007;85 (A2):169–179.

55. Laurenzi F, Coroneo M, Montante G, Paglianti A, Magelli F. Experimental and computational analysis of immiscible liquid–liquid dispersions in stirred vessels. *Chem. Eng. Res. Des.* 2009;87,507–514.

56. Narsimhan G, Gupta JP, Ramkrishna D. A model for transitional breakage probability of droplets in agitated lean liquid–liquid dispersions. *Chem. Eng. Sci.* 1979;34,257–265.

57. Hudcova V, Machon V, Nienow AW. Gas–liquid dispersion with dual Rushton turbine impellers. *Biotechnol. Bioeng.* 1989;34:617–628.

58. Shewale SD, Pandit AB. Studies in multiple impeller agitated gas–liquid contactors. *Chem. Eng. Sci.* 2006;61:489–504.

59. Kerdouss F, Bannari A, Proulx P. CFD modeling of gas dispersion and bubble size in double stirred tank. *Chem. Eng. Sci.* 2006;61:3313–3322.

60. Khopkar AR, Tanguy PA. CFD simulation of gas–liquid flows in a stirred vessel equipped with dual Rushton turbines: influence of parallel, merging and diverging flow configurations. *Chem. Eng. Sci.* 2008;63, 3810–3820.

61. Khopkar AR, Kasat GR, Pandit AB, Ranade VV. CFD simulation of mixing in tall gas–liquid stirred vessel: role of local flow patterns. *Chem. Eng. Sci.* 2006;61(9):2921–2929.

62. Ranade VV, Bourne JR, Joshi JB. Fluid mechanics and mixing in agitated tanks. *Chem. Eng. Sci.* 1991;46:1883–1893.

63. Rammohan AR, Dudukovic MP, Ranade VV. Eulerian flow field estimation from particle trajectories: numerical experiments for stirred tank type flows. *Ind. Eng. Chem. Res.* 2003;42:2589–2601.

64. Michelletti M, Nikiforaki L, Lee KC. Yianneskis M. Particle concentration and mixing characteristics of moderate-to-dense solid–liquid suspensions *Ind. Eng. Chem. Res.* 2003;42: 6236–6249.

65. Kasat GR, Khopkar AR, Ranade VV, Pandit AB. CFD simulation of liquid-phase mixing in solid–liquid stirred reactor. *Chem. Eng. Sci.* 2008;63:3877–3885.

66. Bohnet M, Niesmak G. Distribution of solids in stirred suspensions. *Ger. Chem. Eng.* 1980;3:57–65.

67. Zwietering TN. Suspending of solid particles in liquid by agitation. *Chem. Eng. Sci.* 1958;8:244–253.

16

MEMBRANE SYSTEMS FOR PHARMACEUTICAL APPLICATIONS

DIMITRIOS ZARKADAS
Chemical Process Development and Commercialization, Merck & Co., Inc., Union, NJ, USA
KAMALESH K. SIRKAR
Department of Chemical, Biological and Pharmaceutical Engineering, New Jersey Institute of Technology, Newark, NJ, USA

16.1 INTRODUCTION

Membrane separation technologies are being rapidly incorporated in a number of industries. There are a number of reasons: they are often cheaper, modular, athermal, and can achieve separations difficult to achieve otherwise. In specific industries, for example, desalination/water treatment they are becoming the dominant technology. In biopharmaceutical industry, the processes of dialysis for buffer adjustment, microfiltration for clarification, ultrafiltration, and membrane chromatography are widely used. An earlier brief review of applications of membrane technologies in pharmaceutical industry is available in Sirkar [1]. There have been, however, limited applications of membrane technologies in the pharmaceutical industry during active pharmaceutical ingredient (API) processing in the presence of organic solvents. The membrane technologies that are being used and/or explored in a more than cursory fashion in pharmaceutical processing are pervaporation and organic solvent nanofiltration. To that extent our focus in this chapter will be on pervaporation first and then on organic solvent nanofiltration. At the end, we will briefly focus on membrane solvent extraction.

16.2 PERVAPORATION IN THE PHARMACEUTICAL INDUSTRY

16.2.1 Introduction

Pervaporation is a process in which a feed liquid mixture at atmospheric or higher pressure is brought into contact with a membrane, which allows the selective removal of one or more components of the feed stream into a gaseous/vapor stream on the other side of the membrane (permeate side). Separation is achieved by maintaining a difference between the species partial pressure in equilibrium with the feed liquid and the permeate side partial pressure of the species in the feed to be removed. The partial pressure differential is commonly established by applying vacuum at the permeate side, flowing an inert gas or a combination of the above techniques. When the feed is a vapor stream the process is called vapor permeation. Some researchers treat the two processes as different. However, the operating principles and the membranes used in pervaporation or vapor permeation are similar and the two processes will be treated here as variations of the same technique.

The term pervaporation is a composite of the words permeation and evaporation. The evaporation heat required

Chemical Engineering in the Pharmaceutical Industry: R&D to Manufacturing Edited by David J. am Ende

to transfer the permeating component(s) from the liquid phase in the feed to the vapor phase in the permeate side is supplied by the sensible heat of the feed stream. Separation in pervaporation processes is the outcome of a sequence of three steps [2]:

- Preferential sorption of one or more components into the feed side of the membrane.
- Selective diffusion through the membrane.
- Desorption to the vapor phase at the permeate side.

It is apparent that complex mass and heat transfer phenomena occur during pervaporation. The membrane acts in two ways: first, as a physical barrier between the liquid and vapor phases and second as an additional component in the system, which alters its thermodynamics and allows the separation of its components. The last point is significant in understanding why pervaporation has been successfully used to break azeotropes (i.e., ethanol–water), for which conventional distillation is unsuccessful. Separation efficiency in distillation is governed by the vapor–liquid equilibria (VLE) of the system, which for an azeotrope cannot change. In pervaporation, separation is driven by differences in solubility and diffusivity of the components in the feed stream through the membrane used.

Pervaporation has found industrial applications in various fields including the dehydration of organic solvents [3, 4], the separation of organic–organic mixtures [5–7], the concentration/extraction of aroma compounds from water solutions in the food industry [8], the removal of VOCs from aqueous waste streams [9], and the enhancement of reaction conversion/rate by removal of water during condensation or esterification reactions [10]. In the above applications, pervaporation is applied either as a standalone technique or in a hybrid process combined with distillation. When the feed stream contains chemicals harmful to the membrane or solids, which would foul the membrane and reduce performance, vapor permeation can be applied. In this case, the stream is evaporated via distillation and while at the vapor state is passed through the pervaporation module. Vapor permeation has found extensive application in distillation–pervaporation hybrid units [11].

16.2.2 Process Description and Theory

A schematic of the process is shown in Figure 16.1. A liquid feed containing components 1 and 2 is entering the membrane device at a temperature $T_{\mathrm{f,in}}$ and pressure $P_{\mathrm{f,in}}$. A reduced pressure P_{p} is applied at the permeate side of the membrane. The membrane preferentially permeates component 1 over 2. Under these conditions, component 1 permeates through the membrane and appears in the vapor phase on the permeate side. The net outcome is the removal of component 1 from the feed stream. The heat of evaporation for the permeating component(s) is supplied by the sensible heat of the feed. The latter is, therefore, cooled to a temperature $T_{\mathrm{f,out}}$ at the outlet of the membrane.

Figure 16.2 shows a schematic of the two operating configurations most commonly used in pervaporation applications. Both of them utilize a pump to circulate the feed solution through the membrane module and optionally a heat exchanger to preheat the feed stream to the appropriate temperature, although an effort is always made to use heat available from upstream processing to minimize operating costs. At the permeate side a condenser is available to collect the permeating species. The two configurations shown differ only in the way the driving force across the membrane is established: in Figure 16.2a a vacuum pump is used, while in Figure 16.2b a carrier gas is used to reduce the mole fraction y_i of the permeating species and hence its partial pressure.

The quality of the separation is commonly expressed in terms of the separation factor, α, which for a binary system of species 1 and 2 is given by

$$\alpha = \frac{y_1}{x_1}\frac{x_2}{y_2} \tag{16.1}$$

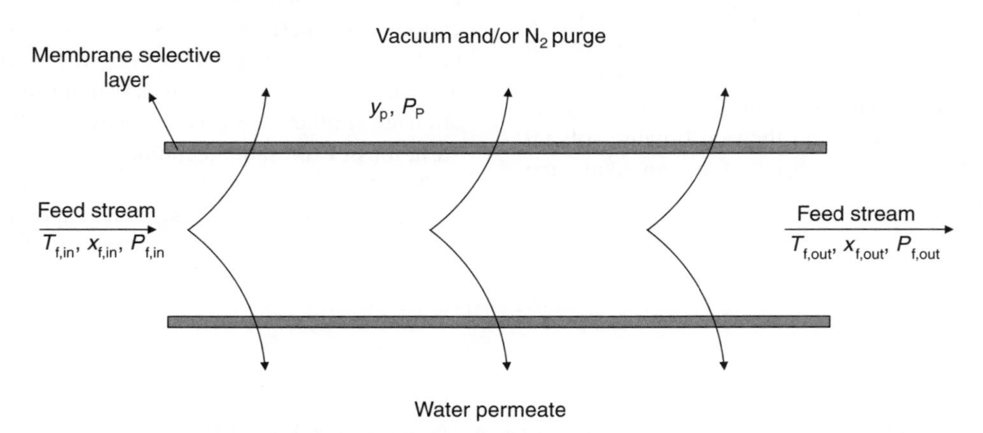

FIGURE 16.1 Operating principle of pervaporation.

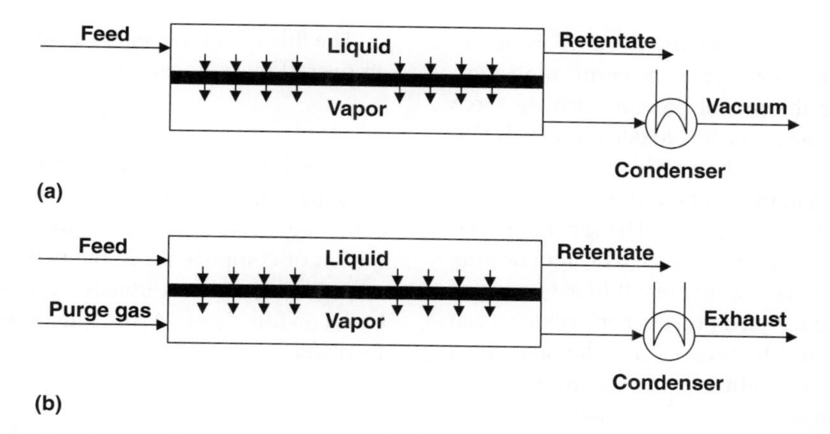

FIGURE 16.2 Schematic of different pervaporation operating schemes. (a) Vacuum on the permeate side and (b) inert carrier gas on the permeate side.

where y_i is the mole fraction of component i in the permeate, x_i is the mole fraction of component i in the feed stream.

The flux of component i through the membrane is given by the following expression:

$$J_i = \frac{Q_i}{l}\Delta p_i = \frac{Q_i}{l}(p_{i,\mathrm{f}} - p_{i,\mathrm{p}}) \tag{16.2}$$

where J_i is the flux of component i through the membrane, kg/(m^2 s); Q_i is the permeability of component i through the membrane, kg/(m s Pa); l is the effective membrane thickness, m; $p_{i,\mathrm{f}}$ is the partial pressure of component i in a gas stream in equilibrium with the feed liquid stream, Pa; $p_{i,\mathrm{p}}$ is the partial pressure of component i in the permeate, Pa.

The amount of component i that can be removed in time t from the feed stream is a function of the flux and the membrane area used to achieve the separation according to the following equation:

$$m_i = J_{i,\mathrm{ave}} A t \tag{16.3}$$

where m_i is the amount of component i removed, kg; A is the membrane surface area, m^2; t is the operating time, s; and an average value is used for the flux of component i to account for the fact that the flux declines as component i is removed from the feed stream to the permeate.

The permeability coefficient is dependent on the membrane material and varies with temperature and composition of the liquid feed stream. It is considered to be the product of the solubility and diffusivity of component i through the membrane

$$Q_i = S_i(T, C) D_i(T, C) \tag{16.4}$$

where S_i is the solubility of component i, kg/(m^3 Pa); D_i is the diffusivity of component i, m^2/s.

Although thermodynamic and transport models can be applied to calculate respectively the solubility and the diffusivity of a species through a membrane; in almost all practical applications, the permeability coefficient is estimated based on experimental data.

The partial pressure difference across the membrane is given by

$$\Delta p_i = x_i \gamma_i p_i^{\mathrm{sat}} - y_i P_{\mathrm{p}} \tag{16.5}$$

where γ_i is the activity coefficient of component i in the feed stream; p_i^{sat} is the vapor pressure of component i at the feed temperature, typically described by the Antoine equation.

Combining equations 16.2–16.5 a final expression for the flux of component i is obtained

$$J_i = \frac{S_i D_i}{l}(x_i \gamma_i p_i^{\mathrm{sat}} - y_i P_{\mathrm{p}}) \tag{16.6}$$

Equation 16.6 can provide an insight as to how the flux of the permeating species can be increased and hence the operation time needed to achieve a specified separation can be minimized. Four different cases and their combinations can be identified.

(a) *Increase the Permeability Coefficient*. Since the latter is dependent primarily on the membrane material, an increase of the flux can be achieved by selecting a membrane with a more open structure at the expense, however, of lower selectivity values. The allowable limits for this trade-off between flux and selectivity will generally depend on the intended application and can be decided by the process engineer during the process development stage.

(b) *Decrease the Effective Membrane Thickness*. This is also a membrane property, which has to be considered during the process development stage. Membranes with thin selective layers and open support layers to minimize diffusion limitations are the best choices.

(c) *Increase the Temperature of the Feed.* This action maximizes the vapor pressure of the permeating component and hence the partial pressure driving force across the membrane according to equation 16.5. Upper temperature limits are dictated by the boiling point of the liquid feed and the temperature stability of the active pharmaceutical ingredient. The second limitation is more severe in terms of process design. Boiling point limitations, which would result in cavitation of the feed pump and reduced process performance, can be partially overcome by pressurizing the feed. Such an action raises the boiling point according to the Clausius–Clayperon equation. This is the most effective and easy to implement modification to increase the flux of the permeating species due to the exponential dependence of vapor pressure on temperature.

(d) *Decrease the Partial Pressure of Component i in the Permeate Side.* This can be achieved by reducing the permeate's absolute pressure at the expense, however, of higher pumping costs. Alternatively, an inert gas can be pumped through the permeate side, which will reduce the mole fraction y_i to low levels. A combined approach, namely, starting by operating only the vacuum pump at the beginning of the process and then introducing a purge stream of an inert gas (i.e., N_2) toward the end of the process will be more effective since the partial pressure difference across the membrane is reduced throughout the process and becomes very small when most of the separation has been performed. This approach is similar to well established guidelines for conventional drying and is sometimes called the Combo Mode.

Equation 16.5 can be used to identify appropriate operating conditions with respect to the permeate pressure for a given separation. A positive flux and hence a separation is obtained if $\Delta p_i > 0$. For a given final concentration x_i in the feed stream, equation 16.7 then yields the maximum allowable operating pressure in the permeate side

$$P_{\mathrm{p,max}} = \frac{x_i \gamma_i p_i^{\mathrm{sat}}}{y_i} \tag{16.7}$$

Li et al. [12] used the above expression to calculate the permeate pressure for benzene dehydration at 70°C and select a vacuum pump appropriate for the desired levels of water removal. Such calculations require the estimation of the activity coefficient γ_i, which can be performed by group contribution or semi-empirical models such as UNIFAC and NRTL. Activity coefficient calculations are easy to perform for solvent–water systems but become more complicated for solvent–water–API/intermediate systems. The API or intermediate will alter the activity coefficient of water and will reduce or increase the water activity coefficient, depending on the nature of its interaction with water.

The flux of component *i* can also be expressed in terms of an overall mass transfer coefficient and concentrations

$$J_i = K_i (C_{i,\mathrm{f}} - C_{i,\mathrm{p}}) \tag{16.8}$$

where K_i is the overall mass transfer coefficient of component *i* through the membrane, m/s; $C_{i,\mathrm{f}}$ is the concentration of component *i* in the feed stream, kg/m^3; $C_{i,\mathrm{p}}$ is the concentration of component *i* in the permeate, kg/m^3.

The overall mass transfer coefficient can be expressed based on film theory as the sum of three resistances in series, feed side, membrane and permeate side:

$$\frac{1}{K_i} = \frac{1}{k_{i,\mathrm{f}}} + R_m + \frac{1}{k_{i,\mathrm{p}}} \tag{16.9}$$

where $k_{i,\mathrm{f}}$ is the mass transfer coefficient of component *i* at the feed side, m/s; R_m is the membrane mass transfer resistance, s/m; $k_{i,\mathrm{p}}$ is the mass transfer coefficient of component *i* at the feed side, m/s.

Equations 16.8 and 16.9 can be used to calculate fluxes based on film theory. In principle, mass transfer correlations can be used to calculate the feed and permeate side mass transfer coefficients with reasonable accuracy (see Ortiz et al. Ref. 13). Experimentation is needed to determine the membrane resistance. In practice, experimentation is performed and the overall coefficient is expressed as a function of the feed side mass transfer coefficient, since the membrane and permeate resistances are usually small compared to the feed side mass transfer resistance. The combined membrane and permeate resistance can then be obtained as the intercept in a Wilson plot [13].

The dependence of the flux of component *i* on temperature is usually expressed via an Arrhenius-type relationship [14]:

$$J = J_0 \exp\left(-\frac{E_\mathrm{a}}{RT}\right) \tag{16.10}$$

where E_a is the activation energy for permeation, J/mol; R is the universal gas constant, J/(mol K); T is the temperature, K.

The activation energy E_a in equation 16.10 is a compounded parameter accounting for the variation of permeation flux with both the membrane permeability and the permeation driving force Δp_i, as pointed out by Feng and Huang [14]. Both the preexponential term and the activation energy can be expressed as functions of temperature, feed, and permeate mole fractions of component *i* by curve fitting laboratory experimental data obtained for the intended application. These expressions can then be substituted in equation 16.10 and used to perform scale-up calculations, as detailed in Section 16.2.4.2 [15].

16.2.3 Pervaporation Membranes

The nature of the pharmaceutical industry poses certain limitations to the selection of membranes for manufacturing

TABLE 16.1 Commercially Available Inorganic Pervaporation Membranes

Membrane Type	Selective Layer	Support Layer	ID/OD (mm)	Selective Layer Location	Manufacturer
Zeolite	Zeolite (A-, T-, Y-)	Alumina	9/12	Outside of tube	Mitsui Engineering & Shipbuilding Ltd.
Ceramic	Amorphous silica	Alumina, Titania	8/14	Outside of tube	Sulzer Chemtech
Ceramic	Amorphous silica	Alumina, Titania	7/10	Inside of tube	Pervatech BV

API or intermediates. First, the membrane must be compatible with the pharmaceutical stream to be processed. The use of harsh solvents (i.e., DMF, THF) is still widespread in the industry, although a turn toward greener chemistry has been experienced in the last few years. In addition, many times the active ingredient will make the stream acidic or basic or it might increase the interaction between the stream and the membrane. The question of leachables into the pharmaceutical stream and its impact on the final drug substance (DS) quality must then be addressed appropriately. Membranes with increased chemical and thermal stability are preferable. A second characteristic of the pharmaceutical industry is the generally low production requirement compared to the chemical industry and the fact that only one in nine new chemical entities entering phase I clinical trials reaches commercialization [16]. Many times the intended application will have to be abandoned. Therefore, a membrane, which can be used for a variety of processes/streams, will be clearly preferable. Membranes tailored to a specific process would make sense only if the intended application enters commercial production. In summary, the three desirable characteristics for a pervaporation membrane in the pharmaceutical industry are the following:

1. Good chemical stability
2. Good thermal stability
3. Ability to handle a variety of streams and/or process conditions without significant loss in performance.

Three types of membranes are available for pervaporation: polymeric, inorganic, and mixed matrix* membranes [17]. Only the first two types of membranes are available commercially. Both of them are asymmetric membranes, namely, they consist of a porous support layer and a thin dense selective layer coated onto the support layer. The selective layer is the one in contact with the process stream and hence must have good chemical and thermal stability. It is also the layer that will start leaching first to the process stream.

Commercial polymeric hydrophilic membranes have a polyvinyl alcohol (PVA) selective layer. Sulzer Chemtech is the leader in the industry with probably more than 90% of market share. Spiral wound or hollow fiber modules are available. PVA membranes can tolerate temperatures of up to about 90°C and mild to relatively strong acidic and basic conditions. Fluxes reported for typical dehydration applications (water < 10 wt%) range between 1 and 2 kg/(m^2 h) [17, 18]. PVA membranes have a proven track record in a number of industrial dehydration applications. Their operating temperature limitations as well as stability issues when exposed to certain organic solvents should be taken into consideration. The above limitations make it unlikely that PVA membranes could be used as a general tool for dehydration problems in the pharmaceutical industry. However, they can still be useful in specialized applications. Recently, polymeric membranes with a fluoropolymer selective skin have been developed by Compact Membrane Systems (DE, USA) [19]. The selective layer can withstand temperatures of up to 200°C and a variety of different chemical conditions. It seems that this membrane can satisfy the performance criteria listed above and potentially be very useful for applications in the pharmaceutical industry. However, a suitable substrate must be selected for high temperature applications to avoid membrane damage due to thermal stresses.

Inorganic membranes can overcome the temperature limitations of polymeric membranes; certain types of inorganic membranes are resistant to a variety of chemical conditions also. Table 16.1 summarizes the commercially available inorganic membranes for dehydration applications. Two types of inorganic membranes exist: zeolite and ceramic. Zeolite membranes exhibit excellent selectivity and good fluxes [20–23], but they can only be used for a limited pH range, between 6 and 8 [1, 21]. These properties make zeolite membranes an excellent choice for solvent dehydration; however, in most cases zeolite membranes will be not able to withstand the presence of an API or an intermediate. Microporous silica membranes on the other hand can tolerate a much broader pH range, which makes them ideal for the dehydration of most pharmaceutical streams. The membranes manufactured by Pervatech BV can tolerate a pH down to 2–3, as has been confirmed with an actual pharmaceutical stream [24]. Silica membranes exhibit inferior separation factors compared to zeolite membranes but their flux

* Mixed matrix membranes consist of a polymeric base membrane impregnated with inorganic material.

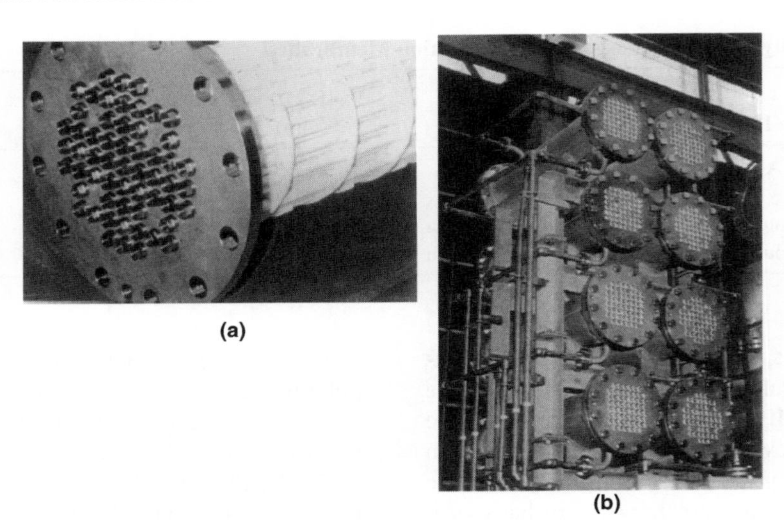

FIGURE 16.3 Zeolite NaA membranes from Mitsui Engineering & Shipbuilding Ltd. (a) Module configuration and (b) module layout (from Ref. 25, with permission).

is higher [20] and comparable if not better than polymeric membranes.

All commercially available inorganic membranes are of tubular geometry. Figures 16.3–16.5 show pictures of the membrane modules described in Table 16.1. The zeolite membrane modules from Mitsui have a typical shell-and-tube configuration [25]. The feed is introduced on the shell side and the baffles present force it to a path perpendicular to the tube length leading to higher feed side mass transfer coefficients. The permeate is collected in the tube side. The configuration of the Pervap® SMS module by Sulzer Chemtech is shown in Figure 16.4. The membrane tubes are placed inside the tubes of a heat exchanger to achieve isothermal operation and increase permeation rate. The feed flows in the annular space between the heat exchanger and the membrane tubes, while the permeate is collected inside the tubes. The membrane tubes can either be connected in series or in parallel. Both configurations yield acceptable pressure drops [26]. The modules by Pervatech BV have also a shell-and-tube configuration. They are made of 2 identical parts consisting of 54 tubes, 50 cm long, stacked one upon the other. The two parts are connected by a plate with machined channels connecting the individual tubes. The feed is introduced in the tube side and the permeate is collected at the shell side. The tubes are connected in series and the manufacturer recommends a linear velocity of 2 m/s through the tubes to minimize concentration polarization effects. This requirement results in a pressure drop of about 4 bar as measured for water by the manufacturer [27].

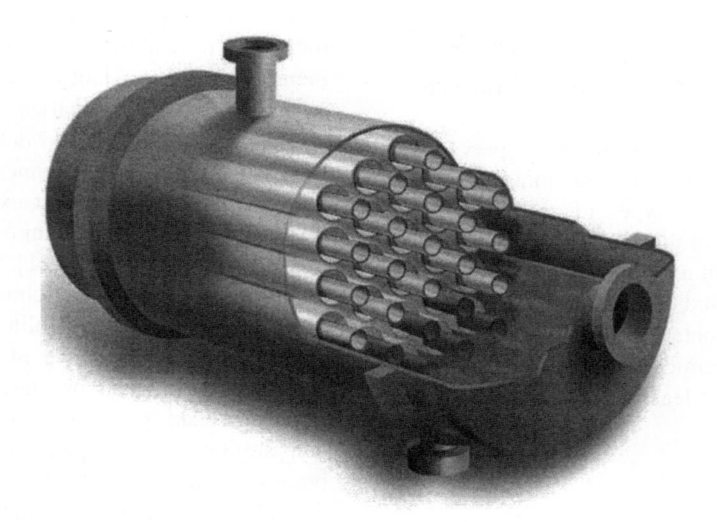

FIGURE 16.4 Pervap SMS silica membrane modules from Sulzer Chemtech (from Ref. 26, with permission).

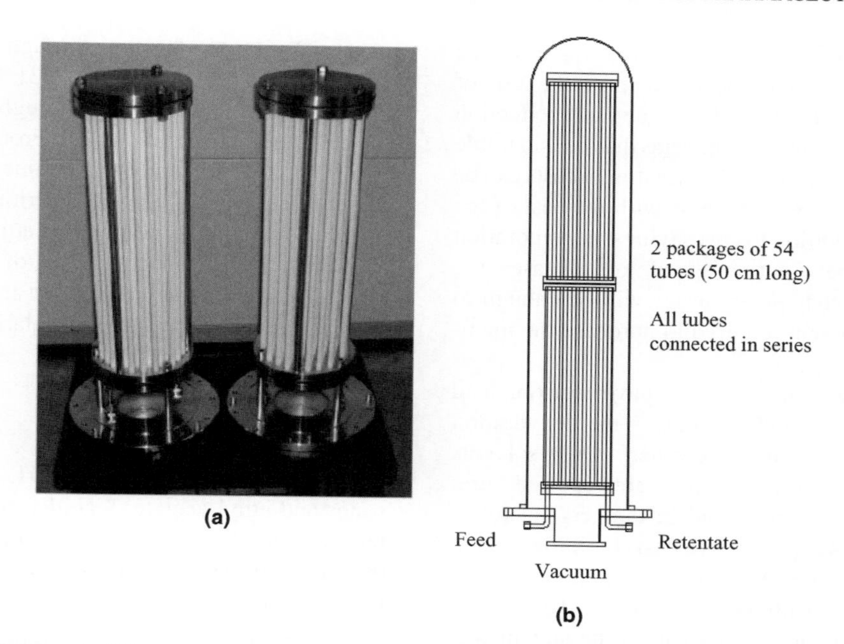

FIGURE 16.5 Silica membrane modules by Pervatech BV (a) Inner tube assembly and (b) module configuration.

16.2.4 Pervaporation Applications in the Pharmaceutical Industry

The majority of potential pervaporation applications in the pharmaceutical industry are related to dehydration of pharmaceutical streams. Pharmaceutical streams need to be dehydrated in the following cases:

1. *Removal of Moisture During Reactions.* Certain types of reactions, that is, esterification or condensation reactions, require the removal of water to increase conversion and yield. Such applications have already been demonstrated with nanofiltration membranes [28, 29]. In many cases, the presence of moisture is undesirable due to excess impurity formation and/or catalyst deactivation. The moisture can come from the API to be dissolved (hydrates or physically adsorbed water), the catalyst (in the form of adsorbed water) and/or the solvent itself.
2. *Removal of Moisture to Supersaturate Solutions in Crystallization Processes.* In many cases the solubility of pharmaceutical solids decreases with decreasing water content. If the nucleation kinetics is slow enough and a seeding step is envisioned for the process, these streams can be supersaturated by pervaporation and subsequently crystallized. Currently, batch distillation is used to achieve this goal.
3. *Dehydration of Waste Streams for Solvent Recovery.* These streams can originate from distillations or work-up procedures. Recovery of the solvent will make sense only in the case of large volume compounds and solvents with relatively high cost, for example, 2-MeTHF.

Another potential application of pervaporation is the removal of volatile organic compounds (VOCs) from pharmaceutical waste streams [30]. Since this application does not involve streams used in the production of an API, it will not be further explored here. The remainder of this section will explore in more detail the dehydration applications listed above.

16.2.4.1 Dehydration Applications of Pervaporation

This section will examine in more detail the work that has been performed up to now in the dehydration of organic solvents with ceramic membranes for two reasons. First, ceramic membranes are considered the most appropriate for applications in the pharmaceutical industry. Second, the work performed in pure solvents can serve as a basis for the design of applications with streams containing an API. The remainder of the section will focus on the operation modes that can be adopted in pharmaceutical manufacturing.

Dehydration in the pharmaceutical industry is primarily performed by distillation. Other means, for example, molecular sieves, can find only specialized applications and in many cases will not perform satisfactorily due to the presence of the pharmaceutical ingredient. The removal of water by azeotropic distillation can result in the intense use of solvents, even in the case where continuous** distillation is

** The term continuous is rather misleading; distillation is performed by continuously adding solvent and removing only distillate but not the batch, which is the heavy fraction.

used. One of the authors is aware of several examples where the removal of 100–200 kg of water requires the use of 2000–3000 L of solvent per batch. For an annual production of 20–50 batches, these numbers indicate that considerable savings in solvent cost as well as waste treatment costs can be achieved. The fiscal gains will increase with the cost of the solvent to be used. In addition to cost savings, pervaporation can lead to much greener solutions. In almost all cases, the cost savings from reduced solvent usage will be enough to repay the capital investment required to purchase the membrane modules.

Table 16.2 summarizes the results that have been obtained with commercial silica membranes during the dehydration of organic solvents by various researchers. Only solvents relevant to pharmaceutical applications are reported here; additional information can be found in the original references. The quoted fluxes can serve as a basis for initial design according to equation 16.3. A sample calculation is given in Example 16.1. As mentioned in Section 16.2.2, the presence of a pharmaceutical ingredient in the stream will alter its thermodynamics and especially the activity coefficient of water. According to equation 16.6, there will be an increase or decrease in the water flux depending on the nature of the interaction of the pharmaceutical ingredient with the rest of the stream components. The pressure required in the permeate side to effect a certain separation will also change accordingly. The presence of a pharmaceutical ingredient in a stream will also change its viscosity and hence the hydrodynamic conditions in the feed side. The latter can often negatively impact the flux through the membrane due to concentration polarization [26, 33]. Pervatech recommends a feed linear velocity of 2 m/s through the membrane to avoid concentration polarization [27]. From the above discussion, it follows that laboratory experimentation should be performed prior to scale-up to determine the actual flux and optimal operating conditions to adjust initial estimates of membrane surface area. As a rule of thumb, the flow should always be kept in the turbulent or at least in the transitional regime to avoid concentration polarization effects.

EXAMPLE 16.1

A pharmaceutical stream of 5000 L contains an API, 6 wt% water and ethyl acetate. Stability concerns dictate that the removal of water must be performed in 24 h. Calculate the membrane surface area that is needed for the separation. Rearrangement of equation 16.3 for membrane area

$$m_i = J_{i,\mathrm{ave}} A t$$

results in the following

$$A = \frac{m_i}{J_{i,\mathrm{ave}} t}$$

A water flux from ethyl acetate can be obtained from Table 16.2. Based on a value of 3.16 kg/(m^2 hr) and $t = 24$ h and substituting values yields a membrane area of 3.56 m^2.

TABLE 16.2 Results Obtained with Commercial Ceramic Membranes in the Dehydration of Organic Solvents

Solvent	Feed Water Content (wt%)	Operating Temperature (°C)	Water Flux (kg/(m^2 h))	Water in Permeate (wt%)	Separation Factor	Membrane	Reference
Methanol	10.4	60	1.87	58.84	10	Sulzer Chemtech[a)]	[20]
	10.5	60	0.39	71.69	20	Pervatech	
Ethanol	10.3	70	2.33	86.37	60	Sulzer Chemtech	[20]
	11.0	70	2.00	95.26	160	Pervatech	
IPA	4.5	80	1.86	98.10	1150	Sulzer Chemtech	[31]
	10.2	75	2.76	91.06	90	Sulzer Chemtech	[20]
	9.8	75	2.55	95.33	190	Pervatech	[20]
Acetone	10.0	70	0.52	>99.5	n/a	Sulzer Chemtech	[32]
	10	70	2.72	>99.5	n/a	Pervatech	[32]
Acetic acid	10.4	80	1.91	86.80	60	Sulzer Chemtech	[20]
Ethyl acetate	2.0	70	3.16	93.71	750	Sulzer Chemtech	[20]
THF	11.8	60	3.47	96.53	210	Sulzer Chemtech	[20]
	11.5	60	3.30	99.91	8400	Pervatech	
Acetonitrile	11.9	70	2.73	96.36	200	Sulzer Chemtech	[20]
	9.7	70	3.90	97.27	330	Pervatech	
DMF	10.2	80	1.53	92.28	100	Sulzer Chemtech	[20]
	9.1	80	1.14	92.19	120	Pervatech	

In the original text the membrane is quoted as ECN. This membrane has been commercialized by Sulzer Chemtech and is referenced by this name here.

$$A = \frac{m_i}{J_{i,\mathrm{ave}}t} = \frac{(5000\,\mathrm{L})(0.9\,\mathrm{kg/L})(0.06\,\mathrm{wt\%})}{(3.16\,\mathrm{kg/(m^2\,h)})(24\,\mathrm{h})}$$

$$= \frac{270\,\mathrm{kg}}{75.84\,\mathrm{kg/m^2}} = 3.56\,\mathrm{m^2}$$

This will give an initial estimate of capital cost to purchase the membrane modules. Additional laboratory experimentation with the actual pharmaceutical stream should be performed to obtain a more accurate estimate of water flux and hence membrane surface area.

Figure 16.6 shows a schematic for a typical pervaporation setup in the pharmaceutical industry. This setup will be useful when the pharmaceutical stream contains no solids. The batch is recycled between the reactor and the pervaporation module by means of a pump. A heat exchanger can be optionally used to ensure that the feed inlet temperature is maintained at appropriate levels. Alternatively, the reactor can be pressurized and its temperature increased to levels that will compensate any heat losses taking place between the reactor and the pervaporation module. A typical cartridge filter is installed prior to the pervaporation module to minimize membrane fouling and potential flux decline with time. The permeate side of the membrane module is connected to a condenser and a vacuum pump. In most cases, it will be practical to use chilled water (available at a temperature of about 6°C) for the condenser; however, glycol or Syltherm® condensers can also be used. It will usually be better to use a dedicated vacuum pump to ensure that the appropriate vacuum levels are maintained in the permeate side during operation. However, house vacuum can also be considered if the intended application does not require permeate pressures lower than 30–40 mmHg. An option to use a nitrogen purge stream is also depicted in Figure 16.6; its use can be decided based on laboratory process development. A portable skid arrangement for the setup depicted in Figure 16.6 will be preferable, since it gives the flexibility to use the unit in multiple applications. This is important in an industry where the majority of the products scaled up will not reach commercial scale production.

Figure 16.7 shows a typical setup for vapor permeation. This mode of operation will be useful when solids are present in the pharmaceutical stream (i.e., heterogeneous reactions, insoluble catalyst), which would foul the membrane and render it useless. This mode of operation is not as energy efficient as the one depicted in Figure 16.6; the heat of vaporization needs to be supplied to effect the separation. However, this is of relatively small concern for the pharmaceutical industry, whose end-products are of high value. The membrane module is positioned between the reactor and the condenser. Water is removed from the vapor phase and the rest of the vapor is condensed and returned as reflux to the reactor. Concentration polarization will be of smaller concern with this operation mode since the mass transfer resistance in the vapor phase will be considerably smaller compared to the liquid phase.

16.2.4.2 Process Modeling As Figures 16.6 and 16.7 illustrate, pervaporation processes in the pharmaceutical industry operate batchwise. During operation, part of the pharmaceutical stream is removed through the membrane and the stream mass decreases by the amount permeated. The temperature of the stream entering the membrane module in such applications can be considered constant, since it is regulated in the reactor containing the batch and the amount of stream present inside the module and the recirculation loop is small compared to the total batch volume. However, a temperature decrease is experienced inside the module, since the heat required for the removal of water is supplied by the sensible heat of the stream. Therefore, the temperature of the stream will vary with axial position inside the module. From the above discussion, it is apparent that both the mole fraction of the permeating species and the stream temperature are functions of the axial position inside the module.

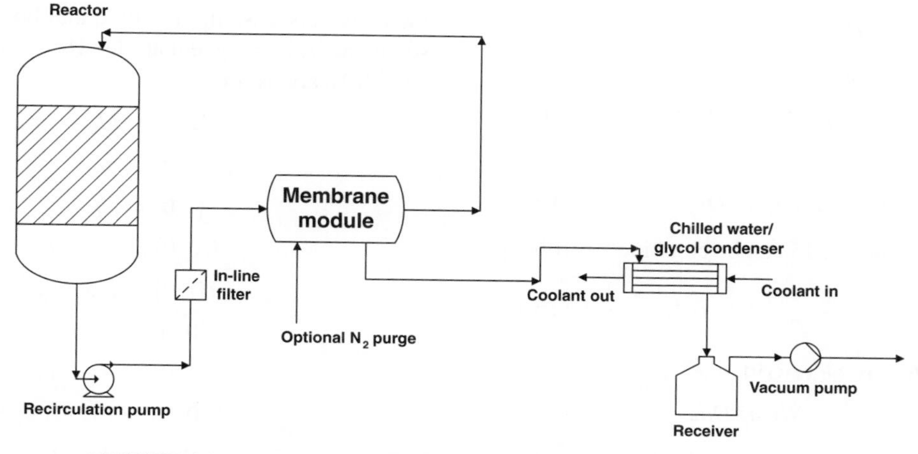

FIGURE 16.6 Typical pervaporation setup in the pharmaceutical industry.

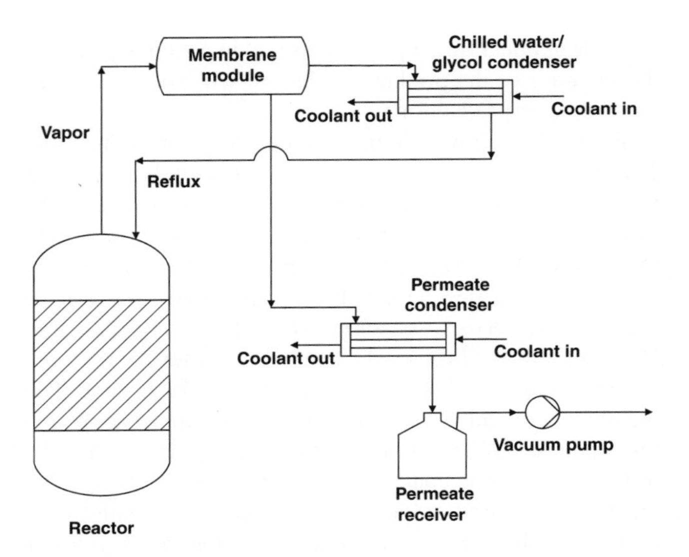

FIGURE 16.7 Typical vapor permeation setup in the pharmaceutical industry.

According to equations 16.6 and 16.10, the permeation flux also changes along the membrane module.

For a membrane module consisting of n tubes the following governing equations can be written [15]:

- *Overall system*

  Overall mass balance

$$dM = -J_t A_T dt \qquad (16.11)$$

  Mass balance for water

$$d(Mx_F) = -J_t A_T y_M dt \qquad (16.12)$$

  Combining equations 16.11 and 16.12, the following expression can be obtained

$$dx_F = \frac{(x_F - y_M) J_t A_T}{M} dt \qquad (16.13)$$

- *Membrane module*

  Overall mass balance

$$dF = -Jn\pi Ddz \qquad (16.14)$$

  Mass balance for water

$$d(Fx) = -Jyn\pi Ddz \qquad (16.15)$$

  Energy balance

$$d(FH_f) = -J\Delta H_v n\pi Ddz \qquad (16.16)$$

  Equations 16.14 and 16.15 can be combined as follows

$$dx = \frac{(x-y)}{F} J n\pi Ddz \qquad (16.17)$$

  Equation 16.16 may be rewritten as

$$FC_p dT = -J\Delta H_v n\pi Ddz \qquad (16.18)$$

  The following notations are used in the above equations: M is the mass of batch at time t, kg; J_t is the average flux through the membrane module for the time interval dt, kg/(m^2 s); A_T is the total membrane area, m^2; x_F is the water mole fraction in feed tank, dimensionless; y_M is the average water fraction in permeate for the time interval dt, dimensionless; F is the mass flow rate through membrane module at time t, kg/s; z is the axial displacement inside the membrane module from entrance, m; J is the local flux through a differential length dz of a membrane tube, kg/(m^2 s); D is the membrane tube diameter (inside or outside), m; H_f is the enthalpy of the stream entering the membrane module, J/kg; ΔH_v is the heat of vaporization of water, J/kg; C_p is the specific heat capacity of the stream entering the module, J/(kg K).

The model represented by equations 16.11, 16.13, 16.14, 16.17, and 16.18 neglects any heat losses due to vaporization of organic solvent. Its accuracy is not expected to deteriorate significantly due to this fact for the majority of organic solvents used with the exception of methanol, ethanol, and acetic acid, which as shown in Table 16.2 can permeate through the membrane to an appreciable extent. The system of equations 16.11, 16.13, 16.14, 16.17, and 16.18 can be solved by applying a finite difference scheme. Discretization is performed in both time and axial position inside the module domains. A prerequisite for this task is to have expressions of the flux J and the permeate water mole fraction as functions of feed water mole fraction and temperature

$$J = f(x, T) \qquad (16.19)$$

$$y_M = g(x, T) \qquad (16.20)$$

These expressions can only be obtained through laboratory experimentation. The latter should include permeation runs at different feed compositions and temperatures. Flux and permeate composition data will then need to be curve fitted against temperature and composition data. Equation 16.10 can be used for the flux, where the flux and/or activation energies are expressed as functions of feed composition; any expression for the water permeate fraction can be applied [15]. Based on the above discussion, the initial and boundary conditions to solve the following equations 16.11, 16.13, 16.14, 16.17 and 16.18 are as follows:

$$\begin{aligned} M(0) &= M_0 \\ x_F(0,0) &= x_0 \\ x_F(t,0) &= x_i \\ J(0,0) &= f(x_0, T_F) \\ J(t,0) &= f(x_i, T_F) \\ y(0,0) &= g(x_0, T_F) \\ y(t,0) &= g(x_i, T_F) \\ F(0,0) &= F(t,0) = F_0 \\ T(0,0) &= T(t,0) = T_F \end{aligned} \qquad (16.21)$$

A simpler but less accurate model can be obtained if the variation of flux inside the module is neglected. In this case, only equations 16.11 and 16.13 need to be solved with the aid of equations 16.19 and 16.20. This is an initial value problem that can be solved by numerical integration (i.e., fourth order Runge–Kutta scheme). The initial conditions are summarized below

$$\begin{aligned} M(0) &= M_0 \\ x_F(0) &= x_0 \\ J(0) &= f(x_0, T_F) \\ y(0) &= g(x_0, T_F) \\ T_F &= \text{constant} \end{aligned} \tag{16.22}$$

This model will probably be adequate for scale-up calculations without the need to face the complexity of the more rigorous model presented previously. The approximation will be better for modules where the tubes are arranged in parallel. When tubes are connected in series the variation of composition and temperature along the tube length will be significant and the simple model will tend to overpredict the flux and hence underpredict batch-processing time.

16.3 ORGANIC SOLVENT NANOFILTRATION IN PHARMACEUTICAL INDUSTRY

The API solutes of interest in pharmaceutical synthesis vary in molecular weight between 200 and 1000 Da. Nanofiltration (NF) membranes originally developed for treatment of aqueous solutions are particularly suited for rejecting solutes in such a size range and passing water through [34]; the solute molecular weight range is said to vary between 150 and 1000 Da. Organic solvent nanofiltration (OSN) is directed toward achieving the same goal with organic solvents.

16.3.1 Process Description and Principles

In nanofiltration, the feed solution is brought under pressure of as much as ≥ 35 bar to one side of the membrane (Figure 16.8). The solvent flows through the membrane to the other side at a lower pressure, generally atmospheric, called the permeate side. The liquid collected is called the permeate. The membrane is supposed to prevent the transmission of solutes of MW > 150~200 Da. If C_{ip} is the molar solute concentration in the permeate side, kgmol/m^3 and C_{if} is the molar solute concentration in the feed side, kgmol/m^3, then the extent of solute retention by the membrane is described by solute rejection, R_i,

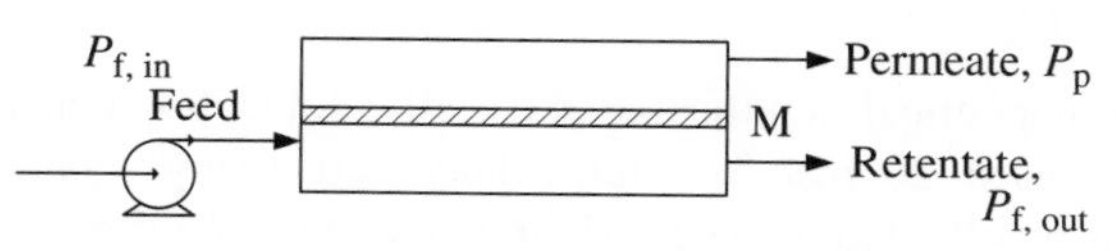

FIGURE 16.8 Schematic of a nanofiltration membrane unit for continuous operation.

$$R_i = 1 - \frac{C_{ip}}{C_{if}} \tag{16.23}$$

If the NF membrane retains the solute completely, then $C_{ip} = 0$ and $R_i = 1$. There are a variety of membranes; some membranes will reject higher molecular weight solutes/species/catalysts in the range of 200–1000 Da but not in the lower molecular weight range. The molecular weight cut off (MWCO) of the membrane is defined to be the solute molecular weight which will yield a value of $R_i = 0.95$; a smaller molecular weight will pass through the membrane more readily and have a lower R_i. An additional quantity of great interest is the solvent flux through the membrane N_i in gmol/(cm^2 s). It is reported often as a volume flux J_v in L/(m^2 h) (LMH); typical values at, say, 30 bar and 30°C may be between 10 and 150 LMH.

There are two general models for species transport through organic solvent nanofiltration membranes: (1) solution–-diffusion model and (2) pore–flow model. The molar flux expression J_i suggested [35] for solution–diffusion of a species through a nonswollen selective layer of the membrane (analogous to that for reverse osmosis by Wijmans and Baker [36]) is

$$J_i = Q_{i,m}\left[x_i - \frac{\gamma_{ip}}{\gamma_i} y_i \exp\left(-\frac{\overline{V}_i(P_f - P_p)}{RT}\right)\right] \text{mol}/(\text{m}^2\,\text{h}) \tag{16.24}$$

where $Q_{i,m}$ is the membrane permeability, mol/(m^2 h); γ_{ip} is the activity coefficient of component i in the permeate at pressure P_p; x_i is the mole fraction of component i in the feed; y_i is the mole fraction of component i in permeate; $\overline{V}_i$ is the partial molar volume of species i; P_f is the feed pressure, Pa; P_p is the permeate pressure, Pa.

According to this model, each component i gets dissolved in the membrane at the feed interface, then diffuses through the membrane and is desorbed at the other interface without any consideration of other species being transported through the membrane.

In the pore–flow model of transport for cylindrical pore models, the solvent flux (or volume flux) expression is

$$J_v = -\frac{\varepsilon d_{\text{pore}}^2}{32\mu\tau} \nabla P \ \text{m}^3/(\text{m}^2\,\text{s}) \tag{16.25}$$

where d_{pore} is the diameter of the pore, m; τ is the pore tortuosity dimensionless; μ is the viscosity of the solvent, kg/(m s); ε is the porosity of the microporous membrane, dimensionless; ∇P is the pressure gradient, Pa/m.

If the pore structure cannot be modeled as consisting of effectively cylindrical capillaries of diameter d_{pore}, the solvent volumetric flux is characterized as

$$J_v = -\frac{Q_{\text{sm}}}{\eta}\nabla P \tag{16.26}$$

where Q_{sm} is the solvent permeability through the membrane. The molar solute flux in the pore–flow model is described via

$$N_i = -\alpha'_i C_{im} J_v \tag{16.27}$$

where C_{im} is the molar concentration of solute i in the membrane and α'_i is a viscous flow characterization parameter [35]. Generally, membrane–solvent interactions occur and they would influence the parameters to be used in pore–flow models. The behavior of solute rejection R_i in such a model as a function of permeate volume flux (which increases essentially linearly with $-\nabla P$) has been illustrated for MPF-60 membrane in Whu et al. [28]. How to obtain various parameters for both models has been illustrated in Silva et al. [35] who have described the earlier literature. The local permeate solute concentration C_{ip} for component i may be described via

$$\frac{C_{ip}}{C_{sp}} = \frac{N_i}{N_s} \tag{16.28}$$

where N_s is the molar solvent flux and C_{sp} is the molar solvent concentration in the permeate. We could rewrite equation 16.28 for a dilute solution of component i as

$$C_{ip} = \frac{N_i}{J_v/\overline{V}_s} C_{sp} \cong \frac{N_i}{J_v} \tag{16.29}$$

where the partial molar volume of solvent is $\overline{V}_s$ yielding

$$R_i \cong 1 - \frac{N_i}{J_v C_{if}} \tag{16.30}$$

Most performance data in OSN reported in literature were obtained with small membrane samples. In larger units, there will be variation of feed concentration along the membrane. There is very little information in the literature on analysis of such a situation. However, in any situation, with small or larger membrane sample, one has to account for concentration polarization. Since OSN is used often to reject a solute, which is an API while the solvent is going through, the rejected solute concentration will increase on the membrane surface to C_{iw}, which is larger than the feed concentration C_{if}. The extent of this increase will depend on the solute mass transfer coefficient in the feed solution, k_{if} and the volume flux J_v as is observed in the processes of reverse osmosis and ultrafiltration

$$\frac{J_v}{k_{if}} = \ln\left(\frac{C_{iw} - C_{ip}}{C_{if} - C_{ip}}\right) \tag{16.31}$$

Therefore, when using the two models of membrane transport, C_{iw} should be used instead of C_{if} unless the ratio (J_v/k_{if}) is very small.

16.3.2 OSN Membranes

OSN membranes may be ceramic or polymeric in nature. There are no practical ceramic membranes whose MWCO value is less than 1000. Therefore, we will focus here on polymeric OSN membranes. Most polymeric membranes used in a variety of industries are prepared via the phase inversion technique in the first step of which the polymer is dissolved in a solvent. Therefore, most of those polymers are not suitable for OSN, since the membrane lacks solvent stability. Among those few that may be suitable, polyimide (PI) and polyacrylonitrile (PAN) stand out. A brief description of various aspects of the materials of OSN membranes is provided in Silva et al. [37].

The polymeric membranes of these materials may be of the asymmetric type or a composite membrane. In asymmetric membranes, a very thin skin at the top of the membrane is the solute-selective layer with the rest of the membrane providing a low transport resistance mechanical support; however, the whole membrane is of the same material and these are often described as integrally skinned. In composite membranes, the top selective skin is of a different material compared to the material of the porous support. This selective layer is cross-linked to impart substantial solvent stability. Cross-linked elastomeric barrier layers are often made out of polydimethylsiloxane on a PAN support.

Integrally skinned polyimide membranes of the STARMEM® type are available from W.R. Grace, Columbia, MD, USA and Membrane Extraction Technology Ltd., UK. The composite polymeric membranes with silicone top layers are identified as MPF types and are available from Koch membranes, Wilmington, MA, USA in small sizes as well as spiral-wound modules. These membranes are good with many organic solvents such as toluene, methanol, and ethyl acetate. However, polar aprotic solvents such as methylene chloride, dimethyl formamide (DMF), *n*-methyl pyrollidone (NMP), tetrahydrofuran (THF) are demanding; successful OSN has been reported with these solvents using integrally skinned Lenzing P84 polyimide membranes chemically cross-linked with aliphatic diamines [38].

16.3.3 Potential Applications of OSN in Pharmaceutical Industry

Pharmaceutical synthesis of small molecules generally involves 4–20 reaction steps. There are many separation steps involved in between where one could use OSN. It may involve separation of the catalyst and its reuse/recycle, removal of solvent, and its exchange with a different solvent,

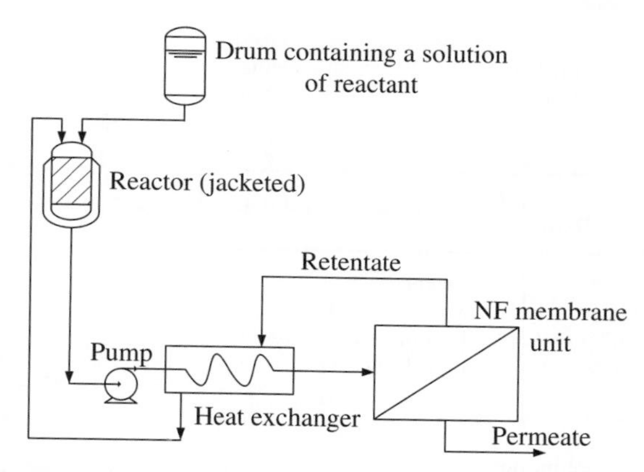

FIGURE 16.9 Schematic of a batch/semi-batch reactor coupled externally with an OSN unit (from Ref. 28, with permission).

concentration of a product/intermediate molecule rejected completely by the membrane, removal of smaller molecules/impurities from the reaction mixture along with the solvent through the membrane, recycling of resolving agents in chiral resolution processes, etc. These steps will most likely be implemented in a NF membrane unit external to the vessel/reactor where synthesis is being implemented (Figure 16.9) in a batch/semi-batch fashion with concentrate recycle.

There are two basic properties of successful membranes in OSN: (1) small molecular weight solvent and solutes pass through easily and the solvent flux is acceptable; and (2) molecules larger than the molecular weight cut off size of the membrane are essentially retained completely. The membrane MWCO may be 220/250, 400, 700 Da implying that species with molecular weight larger than the membrane MWCO is very likely to have a solute rejection (R_i) value ≥ 0.95. A basic expectation is also that the membrane performance will not change much with time either with respect to flux or solute rejection; therefore, membrane swelling in the solvents should be very low.

Scarpello et al. [39] demonstrated very high rejections of the following catalysts, Jacobsen catalyst (622 Da), Wilkinson catalyst (925 Da), and Pd-BINAP (849 Da) using STARMEM™ membranes of different kinds, STARMEM 122 (MWCO, 220 Da), STARMEM 120 (MWCO, 200 Da), and STARMEM 240 (MWCO, 400 Da) in the presence of different solvents, such as ethyl acetate and tetrahydrofuran. Usually STARMEM 240 showed higher rejection values, while STARMEM 120/122 showed very high rejection values >0.95 bordering onto 1. For smaller Pd-based catalysts used in Suzuki coupling, the leakage of catalyst through the STARMEM 122 membrane led to Pd levels in the product at a higher than the desired level [40]; Pd content was brought down to acceptable levels (<10 mg Pd kg/product) by using adsorbents on the permeate from OSN employed on the postreaction solution [41]. Adsorbents used alone for treating the postreaction solution were unsuccessful in achieving the desired level unless a very large amount was employed.

Homogeneous Heck catalysts were recycled from postreaction mixtures using OSN [42]. Recovery and reuse of ionic liquids used as solvents was also demonstrated during these studies with OSN for pharmaceutical synthesis processes [43]. Phase-transfer catalysts such as tetraoctylammonium bromide was separated and successfully reused via OSN [44, 45]. Recycle and reuse of organic acid resolving agents such as di-*p*-toluoyl-1-tartaric acid (DTTA) used for resolution of chiral bases such as racemic amines was demonstrated via OSN [46].

Solvent exchange is an important step in pharmaceutical synthesis. It is usually performed by batch distillation, which results in an intensive use of energy and solvent. In addition, the separation achieved by conventional distillation is not very satisfactory. The room temperature exchange of the solvent ethyl acetate with methanol in a solution containing the solute erythromycin (734 Da) was demonstrated via discontinuous and continuous diafiltration [47, 48] using MPF-50 and MPF-60 membranes (Figure 16.10). In Figure 16.10, E represents erythromycin. This process avoids distillation processes where thermally sensitive APIs are likely to be affected. Toluene as the solvent for solutes such as tetraoctylammonium bromide (547 Da) was exchanged with methanol in batch distillation using STARMEM 122 membrane with the solute retention exceeding 99% [29] compared to average solute rejection of around 96.37% in Sheth et al. [47, 48]. A continuous countercurrent cascade of three nanofiltration membrane cells for the feed solution and the exchange solvent entering at two ends of the cascade was demonstrated for a test solute such as tetraoctylammonium bromide (547 Da) and the solvents methanol and toluene [49].

Membrane-assisted organic synthesis wherein NF membranes are utilized to remove undesirable by-products/intermediates from the reaction mixture as the reaction proceeds and concentrate the reaction product has not received as much attention. Whu et al. [28] pointed out via modeling how OSN may facilitate organic synthesis. Consider the reaction

$$\underset{(\text{MW} \sim 400)}{\text{A}} + \underset{(\text{MW} \sim 50)}{\text{B}} = \underset{(\text{MW} \sim 400)}{\text{C}} + \underset{(\text{MW} \sim 50)}{\text{D}} \tag{16.32}$$

where the product D participates in a side reaction which consumes the reactant A to produce an undesired by-product E via

$$\text{A} + \text{D} = \text{E} \tag{16.33}$$

If we carry out OSN in an integrated fashion with a batch/semi-batch reactor, a few benefits are apparent. If reaction 16.32 is equilibrium limited, removal of D via OSN will shift the equilibrium to the right. Although reactant B is also removed through the membrane, one could continuously add reactant B from a drum in a solvent to replenish it. Removal of D reduces the extent of loss of A via side reaction 16.33. The concentration of C in the final reaction mixture can be

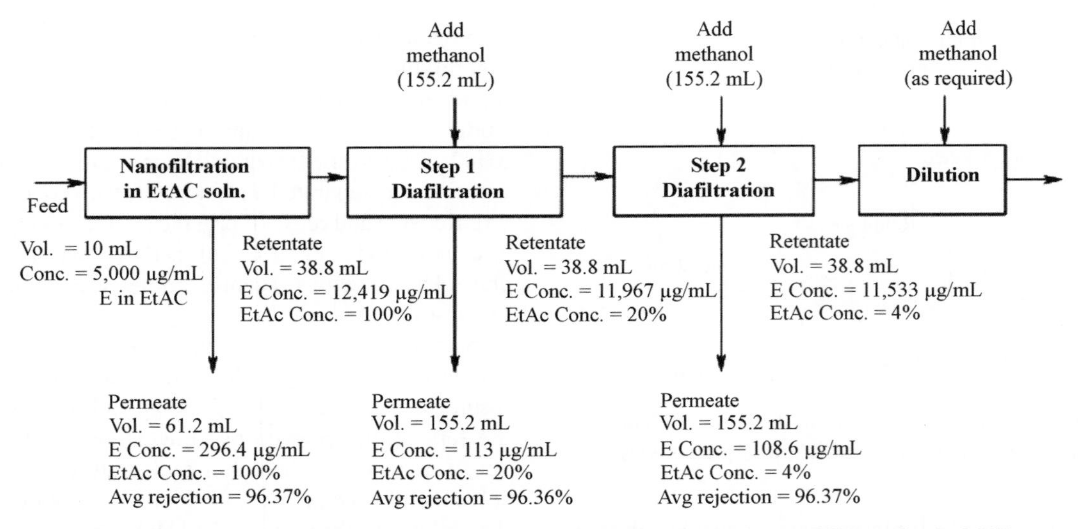

FIGURE 16.10 Mass balance during the preconcentration and discontinuous DF steps for the MPF-60 membrane in solvent exchange for erythromycin from ethyl acetate to methanol (from Ref. 48, with permission).

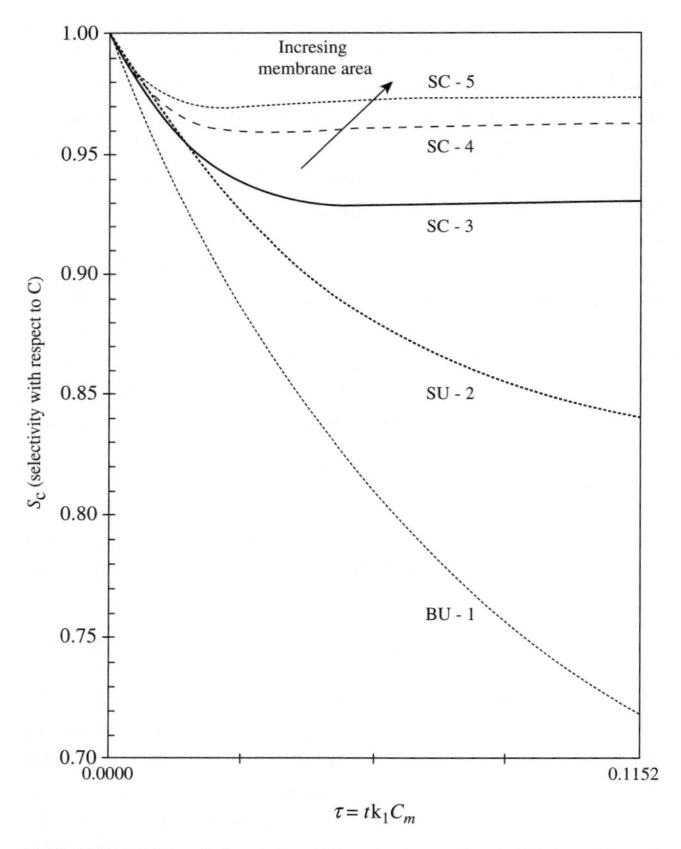

FIGURE 16.11 Selectivity with respect to the desired product C (S_c) as a function of time. See text for more details (from Ref. 28, with permission).

considerably enhanced. Further, the conversion time may be significantly reduced.

Figure 16.11 from Whu et al. [28] illustrates the enhancement of the selectivity with respect to species C as a function of dimensionless time for five modes of operation: BU-1 represents a batch reactor uncoupled from an OSN unit; SU-2 represents a semi-batch reactor uncoupled from an OSN unit; SC-3, SC-4, and SC-5 represent a semi-batch reactor coupled externally with an OSN unit with increasing membrane area and increasing concentration of species B in the drum as well as the volumetric rate of addition. This figure demonstrates that a much higher selectivity with respect to species C is achieved when the semi-batch reactor is coupled externally with an OSN unit to remove the solvent and the product D. Further, this enhanced selectivity is achieved in much less time.

16.4 NONDISPERSIVE MEMBRANE SOLVENT EXTRACTION

In pharmaceutical synthesis, sometimes one needs to extract a solute/API from an organic phase to an aqueous phase or vice versa. Much less frequently, one encounters extraction from a highly polar aqueous phase to a nonpolar immiscible organic phase or vice versa. Conventionally, these extractions are carried out in a mixer-settler type of extraction device by dispersing one phase as drops in another phase. Alternately, centrifugal extractors (Podbielniak) are employed. However, the possibility of a stable emulsion is a major problem in such processing. Nondispersive solvent extraction using microporous membranes has been developed to avoid such problems [50]. In this technique, the organic phase flows on one side of a porous hollow fiber membrane and the aqueous phase flows on the other side. One of the phases preferentially wets the membrane pores. Hydrophobic membrane pores are wetted by the organic phase. Hydrophilic membrane pores are wetted by the aqueous phase. However, the phase not wetting the pores is maintained at a pressure equal to or higher than the pressure of the phase present in the pores. There is no dispersion. One can have a wide range of phase flow rates on the two sides of the membrane. The aqueous–organic or organic–organic phase interfaces at the pore mouths are stable as long as the required relative phase pressure conditions are maintained.

Examples of commercial exploitation of such a technique are provided in Sirkar [51] for aqueous–organic systems. The commonly used porous hollow fiber membranes to this end are made of polypropylene with the tube-sheet made out of polyolefin resins. Therefore, only some of the more common and less aggressive solvents can be used at lower temperatures. Large modules are commercially available (Celgard/Membrana, Charlotte, NC). High performance ceramic membranes having a fluoropolymer coating is available from Kühni AG (Allschwill, Switzerland; Stanley, NC (Kühni, USA)). These modules can be used up to a temperature of 150°C and can be steam sterilized.

In such membrane solvent extraction techniques, porous hydrophobic membranes provide mass transfer advantages when extracting solutes/APIs from an aqueous into a solute-preferred organic phase. On the other hand, when the solute/API in an organic phase prefers the aqueous phase and an aqueous wash is desirable, a porous hydrophilic membrane is desirable. Porous nylon-based membranes are quite suitable for this goal [52].

16.5 CONCLUDING REMARKS

Pervaporation for removing water from organic reaction medium is an attractive opportunity in pharmaceutical processing. Large membrane modules are available. Organic solvent nanofiltration can provide multiple processing opportunities of great relevance; these include catalyst recovery/recycle, solvent exchange, product concentration, enhancement of reaction conversion/selectivity, and reduction of reaction time. Large modules for nondispersive membrane solvent extraction can facilitate pharmaceutical processing as long as the membrane module has acceptable solvent resistance.

REFERENCES

1. Sirkar, KK. Application of membrane technologies in the pharmaceutical industry. *Curr. Opin. Drug Discov. Dev.* 2000;3(6):714–722.
2. Mulder M. *Basic Principles of Membrane Technology*, Kluwer Academic Publishers, Dordrecht, 1996.
3. Sander, U, Jannsen, H. Industrial application of vapor permeation. *J. Membr. Sci.* 1991;61:113–129.
4. Jonquiéres, A, Clément, R, Lochon, P, Néel, J, Dresch, M, Chrétien, B. Industrial state-of-the-art of pervaporation and vapor permeation in the western countries. *J. Membr. Sci.* 2002;206:87–117.
5. Luo, GS, Niang, M, Schaetzel, P. Separation of ethyl *tert*-butyl ether–ethanol by combined pervaporation and distillation. *Chem. Eng. J.* 1997;68:139–143.
6. Hommerich, U, Rautenbach, R. Design and optimization of combined pervaporation/distillation processes for the production of MTBE. *J. Membr. Sci.* 1998;146:53–64.
7. Maus, E, Brüscke, HEA. Separation of methanol from methyl ethers by vapor permeation: experiences of industrial applications. *Desalination* 2002;149:315–319.
8. Lipinzki, F, Olson, J, and Trägård, G. Scale-up of pervaporation for recovery of natural aroma compounds in the food industry. *J. Food Eng.* 2002;54:197–205.
9. Abou-Nemeh, I, Majumdar, S, Saraf, A, Sirkar, KK, Vane, LM, Alvarez, FR, Hitchens, L. Demonstration of pilot-scale pervaporation systems for volatile organic compound removal from a surfactant enhanced aquifer remediation fluid. II: hollow fiber membrane modules. *Environ. Progr.* 2001;20 (1):64–73.
10. Lim, SY, Park, B, Hung, F, Sahimi, M, Tsotsis, TT. Design issues of pervaporation membrane reactors for esterification. *Chem. Eng. Sci.* 2002;57(22–23):4933–4946.
11. Wynn, N. Pervaporation comes of age. *Chem. Eng. Progr.* 2001;96:66–72.
12. Li, J, Chen, C, Han, B, Peng, Y, Zou, J, Jiang, W. Laboratory and pilot-scale study on dehydration of benzene by pervaporation. *J. Membr. Sci.* 2002;203:127–136.
13. Ortiz, I, Urtiaga, A, Ibanez, R, Gomez, P, Gorri, D. Laboratory and pilot plant scale study on dehydration of cyclohexane by pervaporation. *J. Chem. Technol. Biotechnol.* 2006;81:48–57.
14. Feng, X, Huang, RYM. Liquid separation by membrane pervaporation: a review. *Ind. Eng. Chem.* 1997;36:1048–1066.
15. Baig FU. Pervaporation. In: Li NN, Fane AG, Winston Ho WS, Matsuura T,editors, *Advanced Membrane Technology and Applications*, Wiley, Hoboken, NJ, 2008,pp.469–488.
16. Kennedy T. Strategic project management at the project level. In: Kennedy T,editor, *Drugs and the Pharmaceutical Sciences*, Vol. 182, Pharmaceutical Project Management, 2nd edition, Informa Healthcare, New York, NY, 2008.
17. Chapman, PD, Oliveira, T, Livingston, AC, Li, K. Membranes for the dehydration of solvents by pervaporation. *J. Membr. Sci.* 2008;318:5–37.
18. Kuzawski, W. Application of pervaporation and vapor permeation in environmental protection. *Pol. J. Environ. Stud.* 2000;9 (1):13–26.
19. Nemser SM, Majumdar S, Pennisi KJ. Removal of water and methanol from fluids. *U.S. Patent Application* 2008/0099400A1, 2008.

20. Sommer, S, Melin, T. Performance evaluation of microporous inorganic membranes in the dehydration of industrial solvents. *Chem. Eng. Process.* 2005;44:1138–1156.
21. Urtiaga, A, Gorri, ED, Casado, C, Ortiz, I. Pervaporative dehydration of industrial solvents using a zeolite NaA commercial membrane. *Separ. Purif. Tech.* 2003;32:207–213.
22. Van Hoof, V, Dotremont, C, Buekenhoudt, A. Performance of Mitsui Na type zeolite membranes for the dehydration of organic solvents in comparison with commercial polymeric pervaporation membranes. *Separ. Purif. Tech.* 2006;48:304–309.
23. Caro, J, Noack, M, Kölsch, P. Zeolite membranes: from the laboratory scale to technical applications. *Adsorption* 2005;11,215–227.
24. Zarkadas DM, Lekhal A.Unpublished data, 2009.
25. Morigami, Y, Kondo, M, Abe, J, Kita, H, Okamoto, K. The first large-scale pervaporation plant using tubular-type module with zeolite NaA membrane. *Separ. Purif. Tech.* 2001;25:251–260.
26. Sommer, S, Klinkhammer, B, Schleger, M, Melin, T. Performance efficiency of tubular inorganic membrane modules for pervaporation. *AIChE J.* 2005;51(1):162–177.
27. Velterop F. Personal communication with D. *Zarkadas* 2009.
28. Whu, J.A, Baltzis, BC, Sirkar, KK. Modeling of nanofiltration-assisted organic synthesis. *J. Membr. Sci.* 1999;163:319–331.
29. Livingston, A, Peeva, L, Han, S, Nair, D, Luthra, SS, White, LS, Freitas Dos Santos, LM. Membrane separation in green chemical processing. Solvent nanofiltration in liquid phase organic synthesis reactions. *Ann. N.Y. Acad. Sci.* 2003;984:123–141.
30. Shah, D, Bhattacharyya, D, Magnum, W, Ghorpade, A. Pervaporation of pharmaceutical waste streams and synthetic mixtures using water selective membranes. *Environ. Progr.* 1999;18:21–29.
31. Van Veen, HM, Van Delft, YC, Engelen, CWR, Pex, PPAC. Dewatering of organics by pervaporation with silica membranes. *Separ. Purif. Tech.* 2001;22–23:361–366.
32. Casado, C, Urtiaga, A, Gorri, D, Ortiz, I. Pervaporative dehydration of organic mixtures using a commercial scale silica membrane. Determination of kinetic parameters. *Separ. Purif. Tech.* 2005;42:39–45.
33. Cuperus, PF, Van Gemert, RW. Dehydration using ceramic silica pervaporation membranes-the influence of hydrodynamic conditions. *Separ. Purif. Tech.* 2002;27:225–229.
34. Schäfer AI, Fane AG, Waite TD. *Nanofiltration: Principles and Applications*, Elsevier, Oxford, UK, 2005.
35. Silva, P, Han, S, Livingston, AG. Solvent transport in organic solvent nanofiltration membranes. *J. Membr. Sci.* 2005;262:49–59.
36. Wijmans, JG, Baker, RW. The solution diffusion model: a review. *J. Membr. Sci.* 1995;107:1–21.
37. Silva P, Peeva LG, Livingston AG. Nanofiltration in organic solvents. In: Li NN, Fane AG, Winston Ho WS, Matsuura T, editors, *Advanced Membrane Technology and Applications*, Wiley, Hoboken, NJ, 2008, pp.451–467.
38. Toh, YHS, Lim, FW, Livingston, AG. Polymeric membranes for nanofiltration in polar aprotic solvents. *J. Membr. Sci.* 2007;301:3–10.
39. Scarpello, JT, Nair, D, Freitas dos Santos, LM, White, LS, Livingston, AG. The separation of homogeneous organometallic catalysts using solvent resistant nanofiltration. *J. Membr. Sci.* 2002;203:71–85.
40. Wong, HT, Pink, CJ, Ferreira, FC, Livingston, AG. Recovery and reuse of ionic liquids and palladium catalyst for Suzuki reactions using organic solvent nanofiltration. *Green Chem.* 2006a;8:373–399.
41. Pink, CJ, Wong, HT, Ferreira, FC, Livingston, AG. Organic solvent nanofiltration and adsorbents; A hybrid approach to achieve ultra low palladium contamination of post coupling reaction products. *Org. Process Res. Dev.* 2008;12:589–595.
42. Nair, D, Scarpello, JT, Vankelecom, IFJ, Freitas dos Santos, L, White, LS, Kloetzing, RJ, Welton, T, Livingston, AG. Increased catalytic productivity for nanofiltration-coupled Heck reactions using highly stable catalyst systems. *Green Chem.* 2002;4:319–324.
43. Wong, HT, Toh, YHS, Ferreira, FC, Cook, R, Livingston, AG. Organic solvent nanofiltration in asymmetric hydrogenation: enhancement of enantioselectivity and catalytic stability by ionic liquids. *Chem. Commun.* 2006b;19:2063–2065.
44. Luthra, SS, Yang, X, Freitas dos Santos, L, White, LS, Livingston, AG. Phase-transfer catalyst separation and re-use by solvent resistant nanofiltration membranes. *Chem. Commun.* 2001;1468–1469.
45. Luthra, SS, Yang, X, Freitas dos Santos, L, White, LS, Livingston, AG. Homogeneous phase transfer catalyst recovery and re-use using solvent resistant membranes. *J. Membr. Sci.* 2002;201:65–75.
46. Ferreira, FC, Macedo, H, Cocehini, U, Livingston, AG. Development of a liquid-phase process for recycling resolving agents within diastereomeric resolutions. *Org. Process Res. Dev.* 2006;10:784–793.
47. Sheth JC, Qin Y, Baltzis BC, Sirkar KK. *Nanofiltration-Based Diafiltration Process for Solvent Exchange in Pharmaceutical Manufacturing, 11th Annual Meeting of the North American Membrane Society*, Boulder, CO, May 27, 2000.
48. Sheth, JC, Qin, Y, Sirkar, KK, Baltzis, BC. Nanofiltration-based diafiltration process for solvent exchange in pharmaceutical manufacturing. *J. Membr. Sci.* 2003;211:251–261.
49. Lin, JC-T, Livingston, AG. Nanofiltration membrane cascade for continuous solvent exchange. *Chem. Eng. Sci.* 2007;62:2728–2736.
50. Prasad R, Sirkar KK. Membrane-based solvent extraction. In: Winston Ho WS, Sirkar KK,editors, *Membrane Handbook*, Chap. 41, Kluwer Academic Publishers, Boston, MA, 2001.
51. Sirkar, KK. Membranes, phase interfaces and separations: novel techniques and membranes-An overview. *Ind. Eng. Chem. Res.* 2008;47:5250–5266.
52. Kosaraju, P, Sirkar, KK. Novel solvent-resistant hydrophilic hollow fiber membranes for efficient membrane solvent back extraction. *J. Membr. Sci.* 2007;288(1–2):41–50.

17

DESIGN OF FILTRATION AND DRYING OPERATIONS

SARAVANABABU MURUGESAN, PRAVEEN K. SHARMA, AND JOSE E. TABORA
Process Research & Development, Bristol-Myers Squibb Co, New Brunswick, NJ, USA

17.1 INTRODUCTION

In this chapter, we consider two important unit operations in the pharmaceutical industry, filtration and drying. As part of the manufacturing process of active pharmaceutical ingredients (APIs), these two operations follow naturally from the main mechanism of isolation and purification used in this industry, that is, crystallization. It is in fact the dominance of crystallization as an isolation and purification technology that dictates the corresponding use of filtration and drying. As unit operations, both filtration and drying have wide applicability in the broad chemical and pharmaceutical industry; however, for brevity and clarity we will limit our discussion here to situations in the pharmaceutical industry in which these unit operations are a continuation of an isolation train that is preceded by a crystallization. In this context, the slurry resulting from the crystallization is a mixture of the suspended solids (valuable product) and the supernatant (waste solvent or mother liquor).

The goal of the filtration process is to separate the solids from the supernatant. In many cases, filtration operations are carried out to remove undesirable solid matter when the valuable material is present in the liquid phase. Waste filtration will not be discussed in this chapter, and we will concentrate on filtration as a unit operation to isolate the solid API from the solvent in which it was crystallized.

The process of filtration results in the separation of the solids as a wet cake, which requires subsequent washing followed by drying to yield a product of high purity that is subsequently packaged or processed further (e.g., milling) as a dry powder to complete the manufacturing of the drug substance. As discussed in previous chapters, the chemical manufacturing of an active pharmaceutical ingredient proceeds via a number of sequential chemical conversions during which some of the intermediates are isolated in the solid form. Although the isolation of all the intermediates may require the implementation of a filtration and drying sequence, these two unit operations acquire extreme importance in the final step when the API is isolated. Both filtration and drying operations play a significant role in the quality and the physical properties of the API.

17.2 FILTRATION

17.2.1 Background and Principles

In the pharmaceutical industry, filtration is the process of separating solids from a supernatant by exposing the slurry to a filter medium (filter cloth, screen, etc.) and applying a pressure gradient across the medium. The medium retains all or a large portion of the solids and the supernatant flows through the medium producing the separation. For the purpose of this discussion, it is assumed that we are interested in filtering a slurry stream generated from crystallization, to recover the solids (product) in high purity from the supernatant (waste, often called mother liquor). Typical methods to provide a pressure gradient are applied static pressure, applied vacuum downstream from the separating medium, and centrifugal force. The nature of the filtered slurry and the operating parameters of the isolation equipment dictate the filtration rate, and thereby the cycle time of this unit operation during the process, and the properties of the resulting wet cake.

Chemical Engineering in the Pharmaceutical Industry: R&D to Manufacturing Edited by David J. am Ende

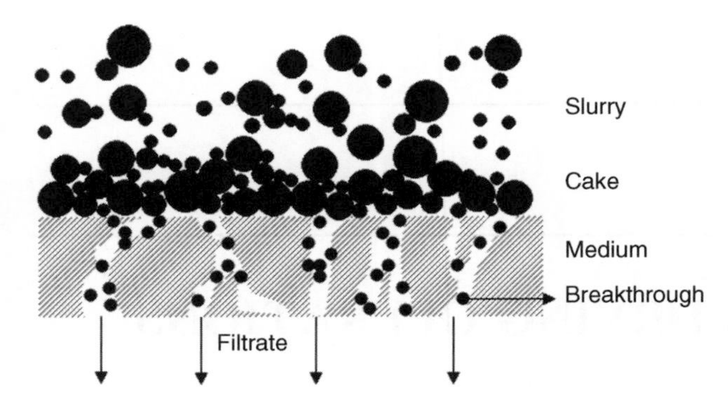

FIGURE 17.1 Cake filtration through porous medium.

17.2.1.1 Solid–Liquid Separation The typical solid–liquid separation performed in the pharmaceutical industry can be depicted as a cake filtration or filtration through a porous medium.

Cake filtration is performed by forcing slurry against a filter medium whose pore sizes are smaller than the average particle size of the solid particles present in the slurry so that the liquid phase passes through the medium while the solid phase is retained by the medium. During this process, cake is built up over time on the filter medium (Figure 17.1).

Filtration Principles In cake filtration, the resistance to flow increases with time as the cake builds on the filter medium. Initially, the only resistance to flow is provided by the filter medium; however, as the cake builds the cake itself becomes a resistance that must be overcome by the fluid being removed. If the filtration is performed at a constant pressure, the flow rate will decrease monotonically during the filtration. This is called a *constant pressure filtration*. In a less common *constant rate filtration*, the pressure drop is increased gradually to afford a constant flow rate.

In general, the rate of filtrate flow in a pressure-driven filtration can be depicted as shown in Figure 17.2. The rate of filtrate flow (q) is directly proportional to the driving force, pressure differential ($\Delta P = P_1 - P_2$) and inversely proportional to the length (L) of the cake (which in turn is a function of the amount of cake filtered).

For the same amount of filtered slurry, an increase in the filtration area (A) would provide more surface area for the filtrate flow and a reduction in the cake height L. An increase in the filtrate viscosity would adversely affect the filtration rate. Hence, the filtrate flow can be given as follows:

$$q = \frac{\mathrm{d}V}{\mathrm{d}t} \propto \frac{\Delta P_{\mathrm{Cake}} A}{\mu L} \tag{17.1}$$

where $q = \mathrm{d}V/\mathrm{d}t$ is the filtration flow rate (m^3/s), ΔP_{Cake} is the pressure drop across the cake (mbar), A is the filtration area (m^2), μ is the filtrate viscosity (Pa s), and L is the filtration length or cake height (m).

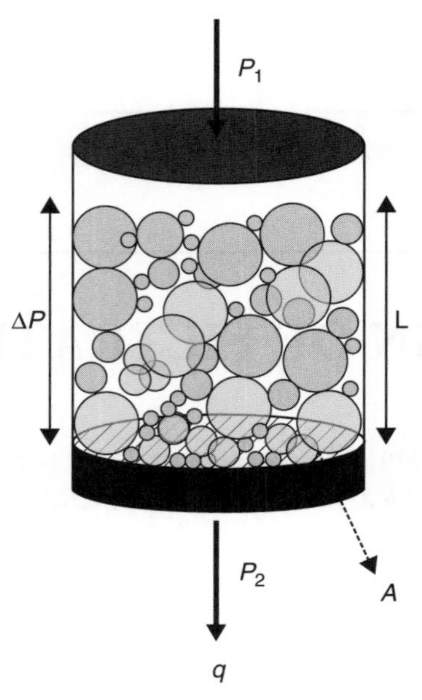

FIGURE 17.2 Schematic of solid–liquid separation.

Equation 17.1 is called Darcy's law [1] with a proportionality constant k defined as the cake permeability:

$$\frac{\mathrm{d}t}{\mathrm{d}V} = \frac{\mu}{\Delta P_{\mathrm{Cake}} A}\left(\frac{L}{k}\right) \tag{17.2}$$

where k is the permeability of the cake (m^2).

The term (L/k) is equivalent to the cake resistance (R_c) faced by the fluid flow during filtration.

$$\frac{\mathrm{d}t}{\mathrm{d}V} = \frac{\mu}{\Delta P_{\mathrm{Cake}} A}(R_c) \tag{17.3}$$

where R_c is the cake resistance (m^{-1}).

However, in addition to the cake resistance, the flow also faces the medium resistance ($R_m = L_m/k_m$). These resistances can be presented as resistances in series and determining the rate of filtration (rate at which volume of filtrate is collected) [2]

$$\frac{\mathrm{d}V}{\mathrm{d}t} = \frac{\Delta P A}{\mu}\frac{1}{(R_c + R_m)} \tag{17.4}$$

Where ΔP is the combined pressure drop (mbar) across the cake and the filter medium. Specific cake resistance (α) is then given by

$$\alpha = R_c A / CV \tag{17.5}$$

where C is the concentration of slurry (kg/m^3) and V is the volume of filtrate collected (m^3).

Specific cake resistance (m/kg) is an intrinsic property of the material being filtered and is a function of the particle size, cake porosity, particle density, and shape [3].

At time zero, the term ($\alpha CV/A$) is zero as the filtration has not been initiated ($V = 0$) and there is no cake built-up, and hence there is no resistance due to the cake ($R_c = 0$). The total resistance at this time is due to only the medium resistance (R_m). As the filtration proceeds ($V > 0$), the cake starts to build up and the value of R_c in equation 17.4 increases, thereby resulting in a corresponding increase in the overall resistance and further decreasing the rate of filtration. After a certain point of filtration, R_c dominates and R_m becomes negligible. In general terms, R_c originates from the packing of the particles as the cake forms, restricting the space available for the fluid to flow through. The fluid must then flow through the spaces between the particles with increasing friction, which in turn, for the same pressure drop, results in a decrease in the filtrate flow rate.

Combining equations 17.4 and 17.5 and rearranging

$$dt = \frac{\mu}{\Delta PA}\left(\frac{\alpha CV}{A} + R_m\right)dV \qquad (17.6)$$

Integrating equation 17.6 from the start of filtration ($t = 0$, $V = 0$) to an arbitrary time t at which a volume V of filtrate has been collected, we obtain

$$t = \frac{\mu}{\Delta PA}\left(\frac{\alpha CV^2}{2A} + R_m V\right) \qquad (17.7)$$

Equation 17.7 can be further rearranged as

$$\frac{V}{At} = \frac{\Delta P}{\mu}\left(\frac{1}{(\alpha CV/2A) + R_m}\right) \qquad (17.8)$$

The term V/At in equation 17.8 corresponds to the volume of the filtrate filtered through unit area in unit time and is denoted as the average filtration flux or simply filtrate flux. Note that the average filtration flux is an inverse function of the amount of the cake collected ($CV = M$).

In some cases the flow itself influences the packing of the solid particles which may result in an increased resistance. In a filtration process, this will be experienced as a change in the specific cake resistance with increased pressure (flow), and the cake is then said to be *compressible*. Specific cake resistance in the case of compressible cakes may be represented by an average specific cake resistance $\bar{\alpha}$ in the following equation:

$$\bar{\alpha} = \alpha_0 \Delta P^n \qquad (17.9)$$

where $\alpha_\circ$ and n are empirical constants, and n is called the compressibility index of the cake.

For incompressible cakes, α is independent of ΔP, and the compressibility index n is 0. If the cake is compressible (dependent of pressure), then n is greater than 0. Typical pharmaceutical cakes have compressibility index in the range of 0.2–1.

The average filtrate flux profiles of the compressible and incompressible cakes are illustrated in Figure 17.3 as a function of the pressure drop across the cake. As seen in the figure, in the case of the incompressible cakes ($n = 0$), an increase in the pressure drop results in a corresponding increased filtrate flow, while in moderately compressible cakes ($0 < n < 1$), the filtrate flow monotonically increases but with decreasing slope, and in highly compressible cakes ($n > 1$) the flow reaches a maximum. This is mainly because of the fact that in the case of highly compressible cakes, after a certain point an increase in pressure compresses the cake, filling the interstitial spaces between the particles, thereby reducing the space available for the filtrate to flow through the cake.

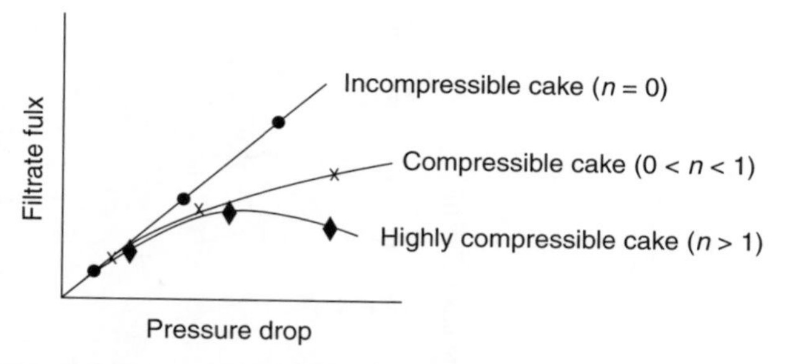

FIGURE 17.3 Filtrate flux versus pressure drop as a function of the compressibility index.

Derivation of equations 17.4–17.8 assumed that solid particles are deposited on the filter cake and there is no migration of particles across the surface of the cake.

Figure 17.4 shows the profile of a typical pressure drop seen across the cake and medium from the initiation ($t = t_0$) to the completion of filtration ($t = t_f$).

At $t = t_0$, there is no cake, and hence the overall pressure drop is only due to the medium resistance, and at $t = t_{1/2}$, the overall pressure drop is due to the combination of cake and medium resistances with cake resistance contributing to the majority of the pressure drop. At the end of the filtration, the overall pressure drop is mainly due to the cake resistance with minimal contribution from the medium. The linearity and the precise slopes of the profiles of the pressure drops depend on the compressibility of the cake.

From Figure 17.4, $\Delta P = P_a - P_f$

$$\text{At } t = t_f,\ \Delta P = (P_a - P_c) + (P_c - P_f)$$
$$= \Delta P_{cake} + \Delta P_{medium} \qquad (17.10)$$

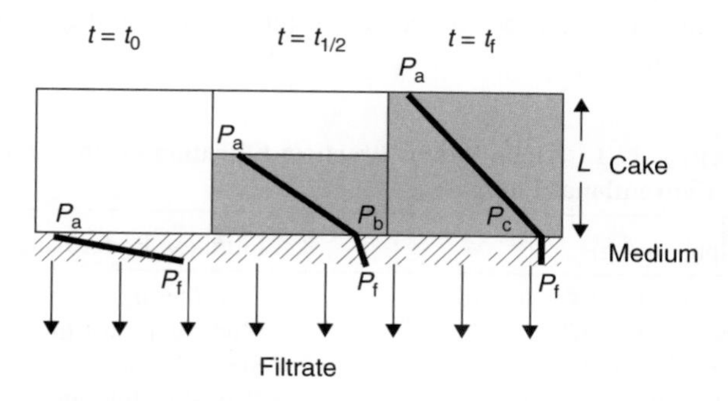

FIGURE 17.4 Pressure drop across the cake during filtration.

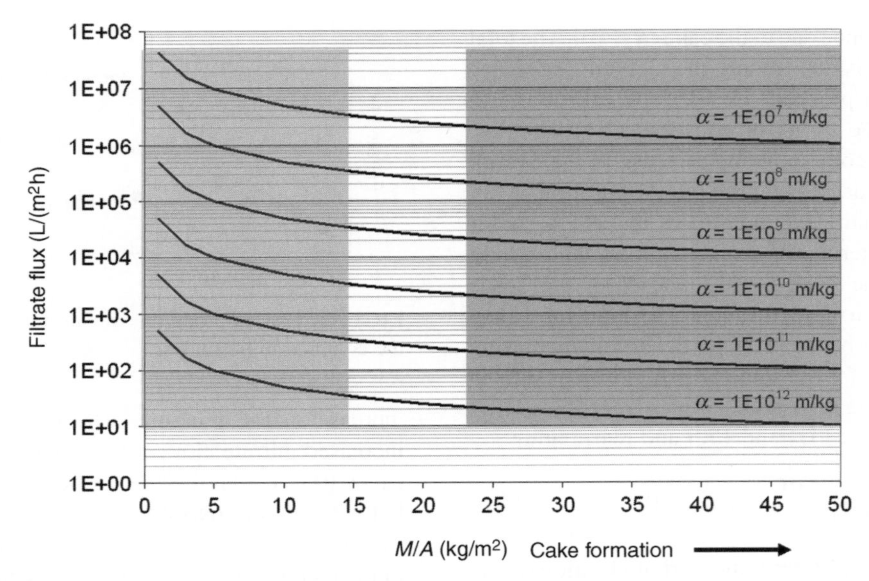

FIGURE 17.5 Filtrate flux profile versus M/A as a function of α (note the logarithmic scale on the Y axis).

As a result of the increasing resistance over time, the filtration flux decreases over time. The profile of the average filtration flux as a function of total cake formation (mass of cake over unit area of filtration, M/A) and specific cake resistance can be estimated for a given applied pressure differential and viscosity by using equation 17.8 as shown in Figure 17.5. A significant reduction in the flux occurs during the initial period of filtration as the cake resistance begins to dominate over the medium resistance. The profiles in Figure 17.5 are estimated for filtration at pressure 20 psi, viscosity of 1.2 cP and R_m $1 \times 10^6\ \text{m}^{-1}$ as an example case.

It is also evident from the figure that reporting the average filtration flux to compare the filtration performances of two different slurries could be misleading, as those two streams could result in the same filtration flux but for different levels of cake formation. A comparison of just the filtration fluxes (e.g., from Buchner funnel filtration) can be performed as long as the filtration pressure and the M/As are similar. Typical ranges for the ratio M/A seen in the laboratory (light gray) and pilot plant (dark gray) are also given in Figure 17.5 for better understanding. Table 17.1 gives a general guidance of the filtration performance in terms of specific cake resistance.

TABLE 17.1 Alpha Versus Practical Filtration Performance in Conventional Equipment

Alpha (m/kg)	Filtration Performance
1×10^7–1×10^8	Fast filtering
1×10^8–1×10^9	Moderately fast filtering
1×10^9–1×10^{10}	Slow filtering
$>1 \times 10^{10}$	Very slow filtering

EXAMPLE 17.1

A pilot plant filtration needs to be performed at 12 psi to separate 20 kg of API from 200 L of solvent. The area of the filter dryer is 0.5 m^2. Assume $\alpha = 1 \times 10^{10}$ m/kg for this API and $R_m = 1 \times 10^6\ \text{m}^{-1}$ for the filter cloth. Viscosity of the filtrate is 1.62 cP. Using equation 17.8, (a) generate a plot of filtration flux profile as a function of M/A (similar to Figure 17.5) and (b) calculate and compare the intrinsic cycle times to complete this filtration when the filtration was performed in (a) 1 load, (b) 3 loads, and (c) 5 loads.

Solution
The following are the M/A values:

(a) $M/A = 40\ \text{kg/m}^2$
(b) $M/A = 13\ \text{kg/m}^2$
(c) $M/A = 8\ \text{kg/m}^2$

(a) Using $\Delta P = 12$ psi, $\alpha = 1 \times 10^{10}$ m/kg, $\mu = 1.62$ cP, and $R_m = 1 \times 10^6\ \text{m}^{-1}$, a profile (Figure 17.6) of filtration flux could be generated as a function of M/A using equation 17.8 as shown below (similar to Figure 17.5).
(b) Plotting the above M/A values against the flux curve gives the average filtration flux for each case:

The corresponding filtration fluxes approximately are

(a) 920 L/(m^2 h)
(b) 2800 L/(m^2 h)
(c) 4600 L/(m^2 h)

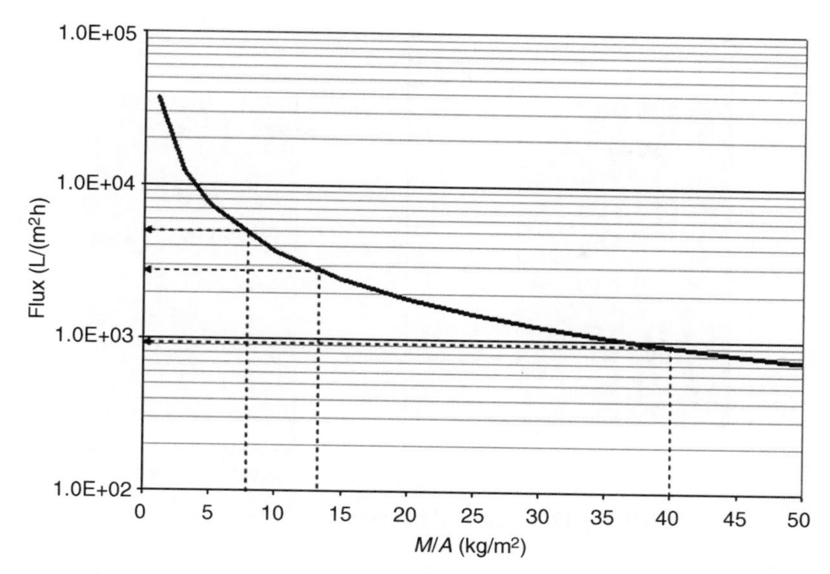

FIGURE 17.6 Filtrate flux profile versus M/A at $\alpha = 1 \times 10^{10}$ m/kg.

These filtration flux values are then used to calculate the cycle time to filter 200 L of solvent by using the following equation:

$$\text{Cycle time} = (\text{volume of filtrate}/(\text{filtration area} \times \text{filtration flux})) \times \text{number of loads}$$

The calculated cycle times are

(a) 0.435 h
(b) 0.143 h
(c) 0.087 h

It is worth noting that these cycle times do not include washing and discharging after the completion of filtering each load. Hence, in this case, splitting the batch to multiple loads to reduce the cake height (and to increase the filtration flux) may not reduce the *overall* cycle time for filtration.

17.2.1.2 Centrifugation Centrifugation is another type of filtration used to separate solids from liquids (or slurry) where the pressure drop across the filter medium is generated by using centrifugal force. Centrifugation is more common for larger batch sizes compared to the pressure filtration, as it can be performed in a relatively continuous fashion. In centrifugation, the filter medium supports the solid particles as they settle due to the centrifugal force (F_G) that can be orders of magnitude higher than standard gravitational force (F_g), and the liquid passes through the cake and then through the medium.

Centrifugation Principles In a small-scale centrifuge with a lower rotational speed, the liquid surface usually takes the shape of a paraboloid [4]. However, in large-scale industrial centrifuges, due to high rotational speed the centrifugal force dominates the force of gravity and the liquid surface takes a cylindrical shape, and it is coaxial with the axis of rotation (horizontal or vertical depending on the centrifuge). Figure 17.7 depicts a typical scenario inside a vertical axis centrifuge. As seen in the figure, the outer layer of cake (of radius r_i) and the outer layer of the liquid (of radius r_1) are cylindrical in shape and are coaxial with the axis of rotation of the basket (of radius r_2).

The pressure drop across the cake and liquid layer is [3]:

$$\Delta P = P_2 - P_1 = \frac{\omega^2 \varrho (r_2^2 - r_1^2)}{2} \tag{17.11}$$

where ω is the angular velocity of the basket (rad/s) and ϱ is the slurry density (kg/m^3).

Combining equations 17.6 and 17.11, and substituting for ΔP,

$$\frac{dt}{dV} = \frac{2\mu}{\omega^2 \varrho (r_2^2 - r_1^2) A} \left(\frac{\alpha C V}{A} + R_m \right) \tag{17.12}$$

Equation 17.12 has been derived with the following assumptions:

1. The filtration area does not change with the cake formation.
2. The liquid layer is on top of the cake layer at all times.
3. Gravitational force is negligible when compared to the generated centrifugal force.
4. Medium resistance is constant throughout the filtration.

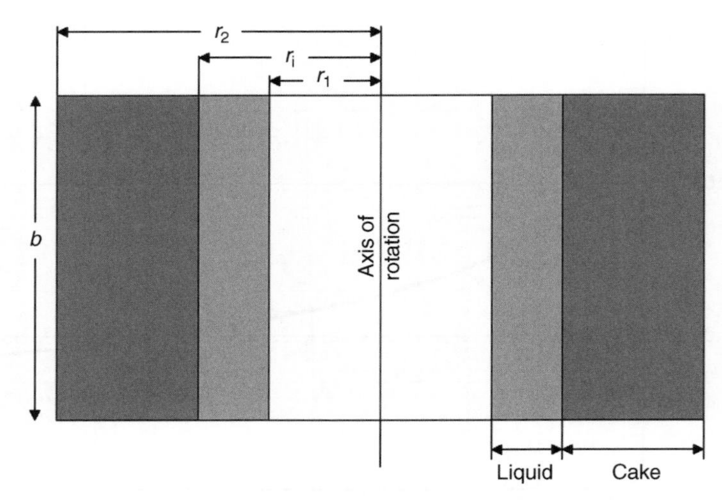

FIGURE 17.7 Cake filtration in a centrifugal filter.

Similar to the pressure filtration, for compressible cakes, α increases with an increase in the centrifugal force.

The centrifugal force F_G created by rotating the centrifuges is given by the following equation:

$$F_G/F_g = mG/mg \tag{17.13}$$

where F_g is the gravitational force of the earth, G is the centrifugal gravity $= \omega^2 \times r_2$, g is the earth's gravity.

G values can be determined using the radius of the bowl and the rotational speed as shown in Figure 17.8.

Filtration as a Multistage Process Filtration of a slurry to form a cake, either by pressure filtration or centrifugation, can also be modeled as a multistage process including stages such as the formation, the consolidation, and the immobilization of the cake bed. Discussion of this more sophisticated approach is beyond the scope of this chapter. More information about this approach is available in the literature [5, 6].

17.2.1.3 Effect of Slurry Properties As mentioned earlier in this chapter, the nature of the slurry as defined by particle size, concentration, and morphology heavily influences *filtration design* and *filtration performance*. *Filtration design* includes the choice of the equipment, wash solvents and filtration conditions, and *filtration performance* includes the purity/impurity profile of the resulting cake, its wetness, and cycle time for filtration. Hence, in several occasions, filtration performance may be improved by adjusting the crystallization conditions to yield a slurry with good filtering properties. Related crystallization concepts are described in detail in Chapter 13.

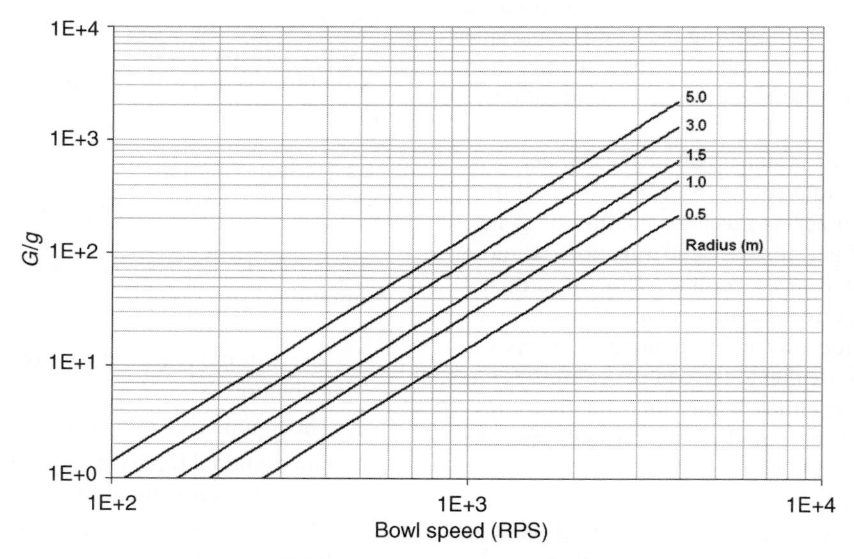

FIGURE 17.8 Centrifugal gravity versus bowl speed as a function of basket radius.

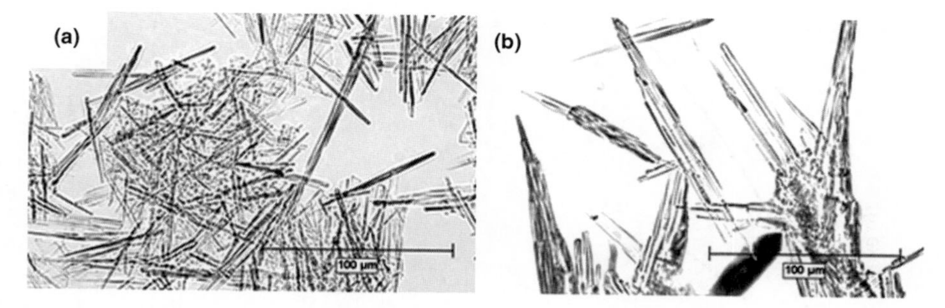

FIGURE 17.9 Optical microscopy images of the crystals obtained through (a) spontaneous nucleation and (b) seeded crystallization.

Here, we provide example cases in which minor adjustments in crystallization procedure resulted in significant improvements in filtration performance.

Spontaneous Nucleation Versus Seeded Crystallization Figure 17.9 shows the comparison of two crystals produced from the same pharmaceutical intermediate but through different crystallization conditions. The spontaneously nucleated batch (Figure 17.9a) has a significant portion of fines with a needle-like morphology and a high aspect ratio, while the seeded batch (Figure 17.9b) has a thicker rod-like morphology and comparatively low aspect ratio. In this case, the modification in crystallization reduced the filtration time significantly. The specific cake resistance measured for the needles (Figure 17.9a) was 6.9×10^{10} m/kg, while the specific cake resistance for the rods (Figure 17.9b) manufactured with an improved crystallization was to 9.0×10^{9} m/kg.

Distillative Crystallization Versus Antisolvent Crystallization In this example, the crystallization was changed from a distillative crystallization (DC) to an antisolvent crystallization (AS) with a different solvent system. This change in crystallization did not affect the crystal behavior significantly (Figure 17.10) and the estimated specific cake resistances were 1.37×10^{11} m/kg and 0.4×10^{11} m/kg respectively. However, a significant decrease in the viscosity of the mother liquor (from 2.4 to 0.45 cP) and in the solids concentration (from 126 to 27 mg/mL) between these two crystallizations played a major role in influencing the filtration performance in this example. The impact of these variables on the filtration time could be explained by equation 17.7 as below.

Since there is very minimal effect of R_m on filtration time, the medium resistance term in equation 17.7 could be ignored, and the filtration times (t_1 and t_2) for these two cases are given as follows:

$$t_1 = \frac{\mu_1}{\Delta P_1}\frac{C_1 V_1^2}{2A_1^2}\alpha_1, \quad t_2 = \frac{\mu_2}{\Delta P_2}\frac{C_2 V_2^2}{2A_2^2}\alpha_2$$

Since the values of ΔP and A were the same for both the cases, the ratio of t_1 and t_2 is

$$\frac{t_1}{t_2} = \frac{\mu_1}{\mu_2}\frac{C_1}{C_2}\frac{V_1^2}{V_2^2}\frac{\alpha_1}{\alpha_2}$$

By substituting M for CV, the ratio of the filtration times (t_1/t_2) of both filtrations to filter the same amount of cake ($M_1 = M_2$) is then given by

$$\frac{t_1}{t_2} = \frac{\mu_1}{\mu_2}\frac{C_2}{C_1}\frac{\alpha_1}{\alpha_2}$$

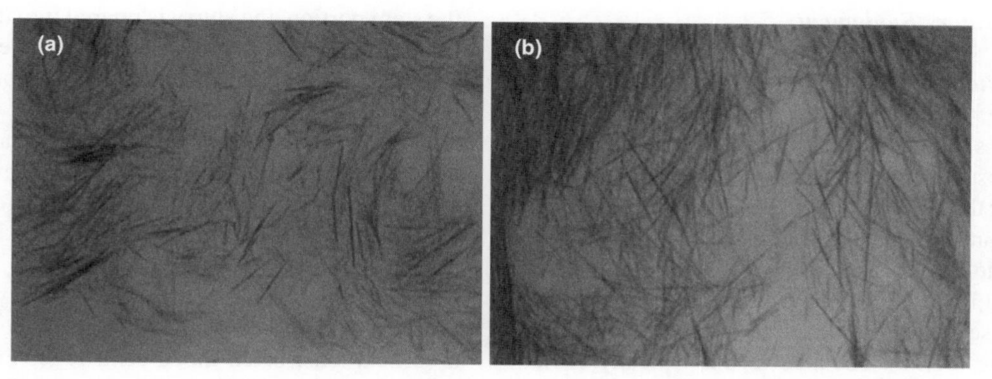

FIGURE 17.10 Optical microscopy images of the crystals through (a) distillative crystallization and (b) antisolvent crystallization.

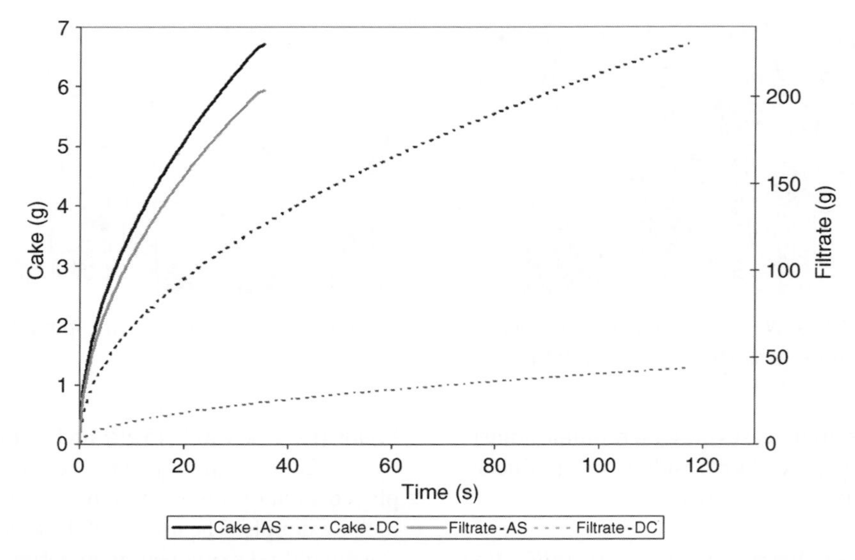

FIGURE 17.11 Profiles of cake formation and filtrate collection as a function of time for the two crystallizations.

Substitution of the values for viscosity, cake resistance, and filtrate volume (filtrate mass/filtrate density) in the above expression provides an estimate of the time ratio between the two filtrations:

$$\frac{t_1}{t_2} = \frac{2.4}{0.45}\frac{(27)}{(126)}\frac{1.37 \times 10^{11}}{0.4 \times 10^{11}} = 3.9$$

The profiles of the cake formation and the filtrate received over time for both filtrations are given in Figure 17.11. As seen in the figure, the slurry crystallized from distillative crystallization took ~117 s to filter while the one from antisolvent crystallization took only ~35 s to filter similar amount of cake.

17.2.2 Equipment

This section provides examples of common isolation equipment used in the laboratory and large scales.

17.2.2.1 Laboratory Equipment

Pressure Filtration

BUCHNER FUNNELS Buchner funnel filtration is one of the simplest and most widely used forms of laboratory filtration. It is named after the chemist Ernst Buchner who invented the Buchner funnel and Buchner flask. It is vacuum driven and can be used in a scale ranging from a few tens of milligrams to a few kilograms. Figure 17.12a shows a picture of a Buchner funnel filtration setup. The funnel is made of porcelain, glass, or plastic (Figure 17.12b). The filter media, typically a filter paper or a filter cloth, is kept on the perforated plate to retain the solid. In case of sintered glass funnel, the fritted surface acts as the filter medium. The usage of Buchner funnel is mainly driven to implement a separation and they are rarely used to collect scale-up parameters; however, with some care, they can provide a first approximation to the filtration properties of the cake. After completing the isolation, cake washes can be performed by adding the solvent on the wet cake and starting the filtration.

LEAF FILTERS Leaf filters are relatively larger scale filters compared to Buchner funnel filters. Their use is driven by the need to characterize slurry filtration properties and requires detailed measurement of the filtration profile, that is, rate of filtrate collection. Figure 17.13 shows a few examples of leaf filters of different materials of construction. Leaf filters consist of three parts: a bottom portion with flow valve, a middle stem portion, and a top lid portion. The bottom portion has the provision to place the filter medium, the middle stem portion holds the slurry, and the top portion closes the stem portion. After connecting the bottom portion (with filter paper of desired pore size) to the stem portion, the slurry is poured into the stem. The top lid portion is then connected to the stem. The chosen pressure set point is attained usually through compressed nitrogen/air. Cake washes can also be performed in a similar fashion. Filtration is performed at multiple pressures to fully characterize the cake properties such as the specific cake resistance and the compressibility index. Refer to the Example for clarity.

Centrifugation Pharmaceutical industry uses various types of centrifuges. This section provides an example of a vertical axis, batch mode centrifuge (Figure 17.14a). It uses a perforated basket (Figure 17.14b) rotating on a vertical axis to produce a centrifugal force on the slurry

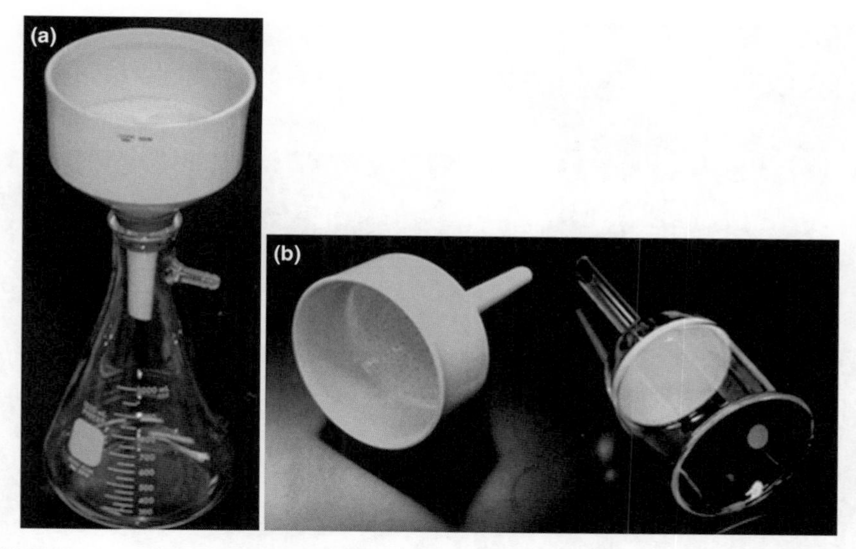

FIGURE 17.12 Pictures of Buchner funnel in (a) a filtration setup and (b) ceramics with perforated plate and glass with fritted filtration surface (ChemGlass Inc.).

that is fed to the basket through the feed pipe. The perforated basket is lined with a filter bag of desired pore size. This filter bag retains the solid particles as wet cake, while the filtrate passes through the filter medium. The filtrate flows out of the basket through the perforations and gets collected outside through the outlet valve. Cake washes can then be performed through similar steps by feeding the wash solvent to the basket.

*17.2.2.2 **Large-Scale Filtration Equipment*** The large-scale isolation equipment used in pilot and manufacturing plants have the same working principle as the laboratory filtration equipment. The operating conditions of the large-scale equipment, however, differ significantly from those of the laboratory equipment. Selection of appropriate isolation equipment for a specific process depends on various factors and this issue is discussed in detail in the next section.

Filter Dryer Filter dryers are one of the most robust isolation equipment used in the pharmaceutical industry. As the name implies, it encompasses the filtration and the drying unit operations into single equipment. Figure 17.15 shows the cross-sectional representation of a filter dryer. The mode of operation for filtration is similar to that of the leaf filters. The required differential pressure is generated by generated either pressure upstream of the filter medium

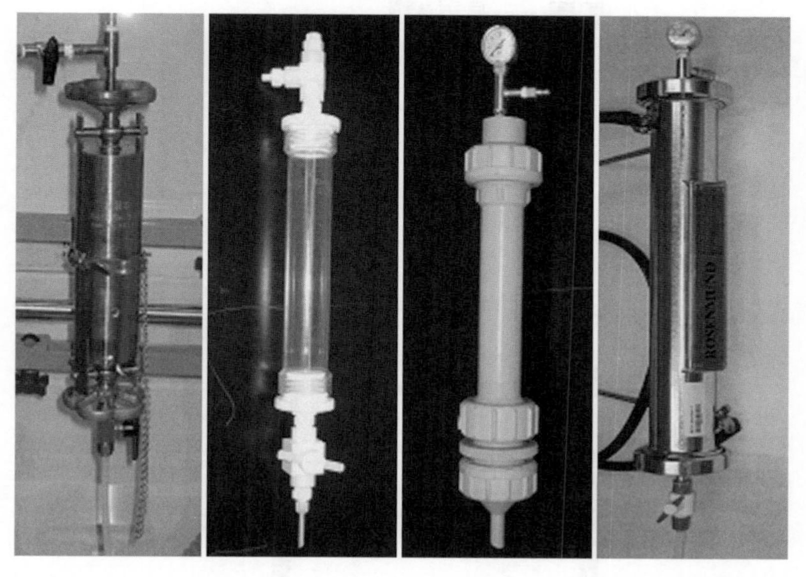

FIGURE 17.13 Examples of leaf filters of various materials of construction (BHS-Filtration Inc., Rosenmund, Inc.).

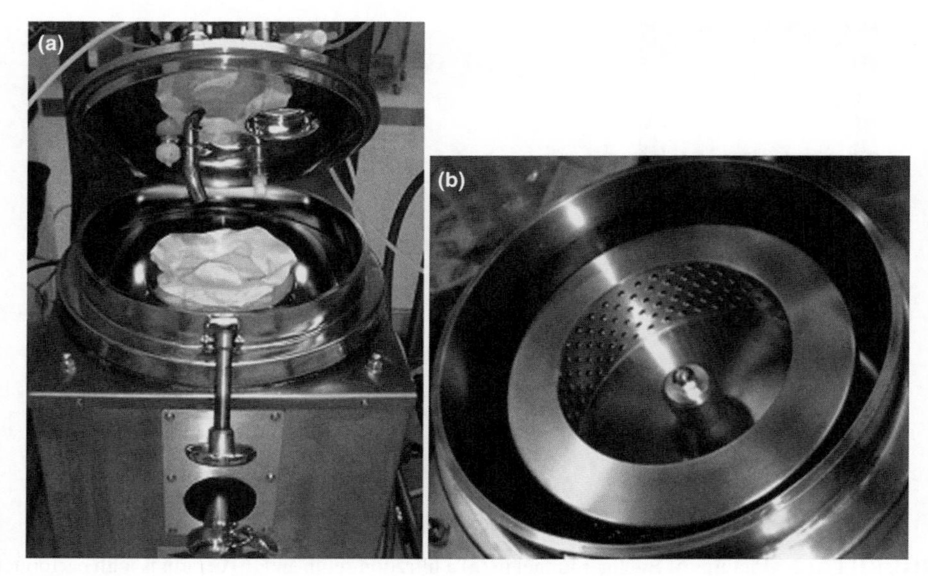

FIGURE 17.14 Pictures of (a) a vertical axis laboratory centrifuge with filter bag installed and (b) centrifuge basket (Rousselet Robatel).

or vacuum downstream. Typical operation would include equipment setup (fitting the filter cloth of appropriate pore size and material of construction and leak check), slurry load, deliquoring the mother liquor, wash load, deliquoring the wash, drying, and discharging. Filter dryers are usually fitted with an agitator that can be lowered or raised as needed. Presence of an agitator makes filtering and cake washing more efficient as the agitator aids in mixing the cake uniformly. Agitation also helps in smoothing the cake in case of crack formation that otherwise could cause channeling of the wash solvents resulting in an inefficient cake washing. Filter dryers are also fitted with a jacket to facilitate heating in the drying step. Provision of jacket also allows performing filtration at different temperatures. After

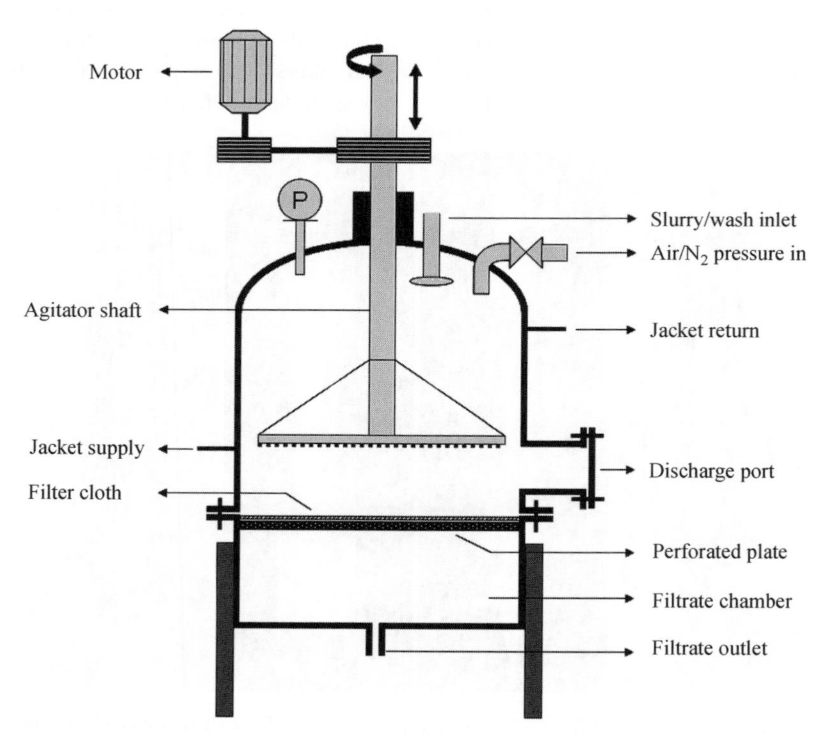

FIGURE 17.15 Cross-sectional representation of a filter dryer.

completing the filtration and washing steps, the cake can be dried in the same equipment and the dry cake can be discharged through the discharge port with the help of the agitator. There is typically an offset between the lowest position of the agitator and the filter cloth in order to avoid lifting or rupturing the cloth. This offset often results in residual cake (denoted as "heel") left on the cloth after the discharge. Heel could potentially slow down the filtration of the next batch. Heel removal is discussed in a later section of this chapter.

Centrifugal Filter Centrifuges can be used either continuously or in batch mode. Pharmaceutical industry prefers batch centrifuges from a regulatory standpoint as it gives an option of stopping the unit operation in case the product specification fails at some point. This section provides examples of batch centrifuge filters.

PEELER CENTRIFUGE Peeler centrifuges are available with horizontal or vertical axis of rotation. The mode of discharge is through a mechanical peeling action. Figure 17.16 shows the cross-sectional representation of a peeler centrifuge. Typical operation involves equipment setup (fitting the filter bag of appropriate pore size and material of construction, performing leak check), initiating the spinning, loading the slurry, deliquoring the mother liquor, loading the wash, deliquoring the wash, peeling, discharging and removing the heel before initiating the next load. Isolation starts as soon as the slurry is loaded as the basket is already spinning, yielding the necessary centrifugal force for the separation. After the liquid layer goes below the cake surface, cake washing can be initiated. After completely deliquoring the wash solvents, the cake is peeled off using a hydraulically activated peeler. The peeled solids are then discharged through a chute for the next unit operation. Often, there is an offset between the peeler's closest position and the cloth, resulting in a residual heel. Depending on the available equipment train, either the discharge chute could be connected to a dryer, and the material discharged directly to the dryer, or the cake could be discharged into drums and transported to the dryer for further processing. The operating parameters of the peeler centrifuge such as the load size and the spin speed are adjusted according to the process requirements. For example, a lower spin speed may be chosen to isolate a compressible cake, to prevent the formation of a highly compressed cake the heel of which might be difficult to remove.

INVERTING BAG CENTRIFUGE Inverting bag centrifuges are horizontally or vertically rotating centrifuges incorporating an automatic discharge. Figure 17.17 shows the schematic representation of a horizontally rotating, inverting bag centrifuge. The order of operation in this centrifuge is very similar to the peeler centrifuge except the mode of discharge. In this case, after completing the filtration and washing, the filter bag is stroked forward by a hydraulically controlled piston, turning the bag inside out, allowing the solids to be discharged. The filter bag is secured to the basket wall at the front end, preventing it from completely detaching from the basket. These centrifuges offer an efficient and faster cake removal as the bag is turned inside out.

Nonagitated Filter Dryers They are similar to filter dryers except that they are not fitted with agitators. Figure 17.18

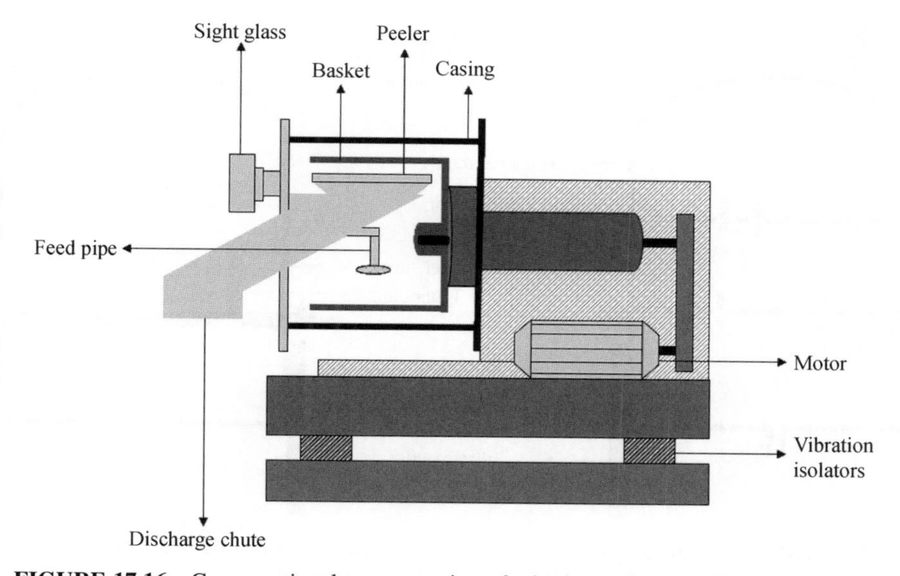

FIGURE 17.16 Cross-sectional representation of a horizontally driven peeler centrifuge.

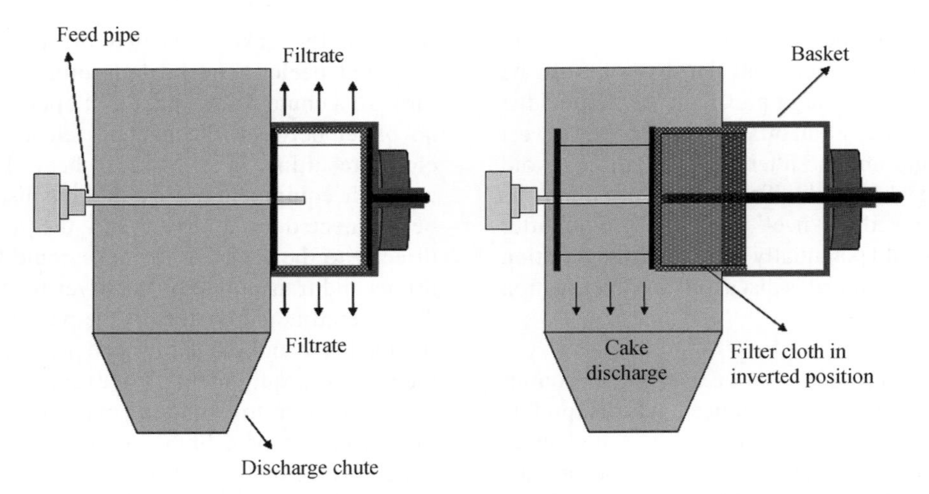

FIGURE 17.17 Schematic representation of an inverting bag centrifuge during (a) deliquoring phase and (b) discharge phase (derived from a model of Henkel Inc.).

shows a schematic representation of a non-agitated filter dryer in closed and cross-sectional views. The absence of an agitator may necessitate the manual mixing of the cake during washing and drying for content uniformity. In case of crack formation in the cake, the lid needs to be opened to manually smooth the cake. For this reason, operating a non-agitated filter dryer requires more personal protection (such as full suit, face shield, supplied breathing air, etc.) for the operator to avoid material exposure. Non-agitated filter dryer are usually fitted with a jacket.

17.2.2.3 Equipment Selection The specific equipment is selected for isolating a material based mainly on the nature of the slurry. The following are the factors that could affect the decision on the equipment selection:

1. Compressibility of the cake
2. Susceptibility of the cake to agglomerate upon agitation
3. Susceptibility of the cake to attrite upon agitation
4. Nature of the wet cake to crack during isolation/washing
5. Nature and availability of the equipment train
6. Environmental, health, and safety impact of the exposure of solids

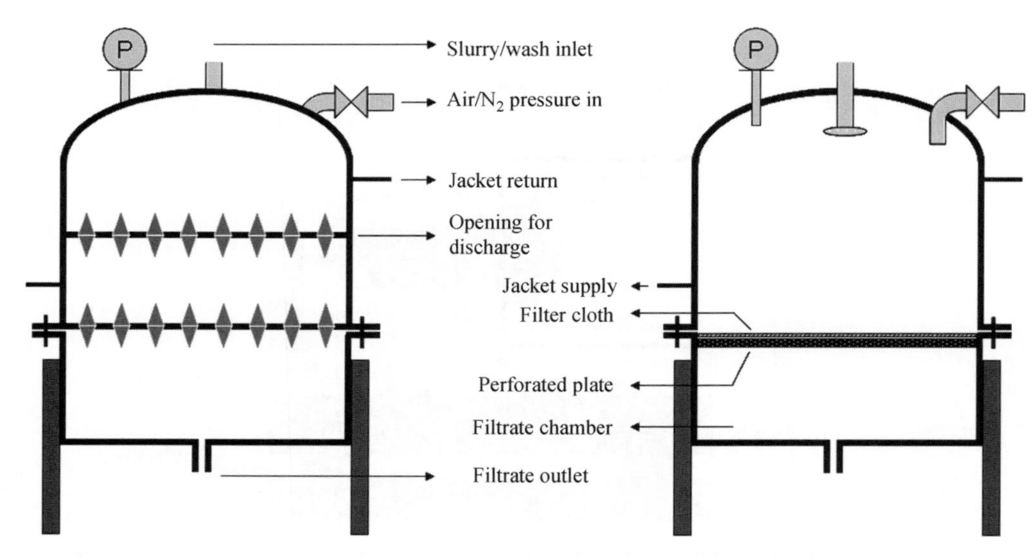

FIGURE 17.18 Non-agitated filter dryer in (a) closed position and (b) cross-sectional representation.

TABLE 17.2 Merits and Demerits of the Conventional Isolation Equipment

Filter	Agitated Filter Dryer	Non-agitated Filter Dryer	Centrifuge
Driving force	Pressure/vacuum	Pressure/vacuum	Centrifugal force
Filter medium	Filter cloth, sintered metal screens	Filter cloth, sintered metal screens	Filter bag
Operating parameters	Load size, applied pressure	Load size, applied pressure	Load size, spin speed, feed rate
Robustness	High	Medium	Medium
Washes possible[a]	Reslurry, displacement	Displacement	Displacement
Advantages	(1) Easy to operate	(1) Easy to operate	(1) Possibility to enhance the filtration performance for incompressible cakes by increasing the spin speed
	(2) High content uniformity due to agitation	(2) Possibility to continue to drying step in the same equipment	(2) Efficient heel removal in case of inverting bag centrifuge
	(3) Possibility to continue to drying step in the same equipment	(3) Heel removal is not needed as the material is discharged by manual operation without leaving any heel	
	(4) High containment of solids		
Disadvantages	Heel removal is difficult	(1) Low content uniformity due to the lack of agitation	(1) Possibility of making a highly compressed cake at high spin speeds
		(2) Greater chances of wash passing through the cracks as there is no agitator to smooth the cake	(2) Relatively difficult operation, need to look for vibration, basket balance, and splashing
		(3) Due to the lack of enough containment, extensive personal protective care is needed, especially when handling materials with high potency or toxicology profiles	(3) Greater chances of wash passing through the cracks as there is no provision to smooth the cake

[a]Reslurry and Displacement Washes are defined and discussed in the Cake Wash section of this chapter.

Table 17.2 compares the merits and demerits of the equipment discussed above.

17.2.3 Design

17.2.3.1 Data Collection and Modeling of Filtration Several data could be collected from the laboratory filtration experiments to assess the potential performance of filtration on-scale. These include data from Buchner funnel filtration, leaf filtration experiments at single and multiple pressures, laboratory-scale centrifugation experiments, and pilot plant-scale pressure and centrifugation filtrations.

Filtration modeling is performed (1) to use the laboratory data to estimate the cake properties and (2) to predict the filtration performance in the plant by using the estimated cake properties.

Buchner Funnel Filtration: First-Order Approximation As mentioned earlier, Buchner funnel filtration is performed when there is no requirement for extensive data collection. It is possible only to record the total amount of filtrate collected (V) and the total time (t) to complete the filtration at a single differential pressure (vacuum driven either through a vacuum pump or a house vacuum). An average filtrate flux could be calculated from V, t, and the filtration area. The average filtration flux (L/(m^2 h)) obtained from a Buchner funnel filtration is sufficient to estimate the average specific cake resistance ($\bar{\alpha}$) by using the equation 17.8 as a first-order approximation. An approximate R_m could be used for this calculation. Equation 17.8 can be rearranged as follows:

$$\text{Flux(av)} = \frac{V}{At} = \frac{\Delta P}{\mu}\left(\frac{1}{((\bar{\alpha}/2)(M/A)) + R_m}\right) \quad (17.14)$$

where $M = CV =$ mass of dry solids (kg).

The pilot plant filtration performance (filtration flux and cycle time) can then be calculated by using the estimated $\bar{\alpha}$ and the pilot plant operating parameters such as differential pressure (ΔP), filtration area (A), and batch size (M). Since, in typical pharmaceutical separations, the medium resistance is negligible when compared to the cake resistance, estimation of specific cake resistance through this approach by using an approximate low R_m is fairly reasonable. It should be noted that $\bar{\alpha}$ has been estimated from a single data point obtained by conducting filtration at a single pressure. Hence, it is not possible to calculate the compressibility index of the cake through this approach.

Leaf Filtration: Usage of t/V at a Single Pressure Data With the help of balances equipped with automated data collection, the instantaneous filtrate volume at a particular time interval could be collected during leaf filtration experiments at multiple pressures. However, in cases of limited material availability, data from filtration at a single pressure may be used to calculate the cake properties more accurately when compared to Buchner funnel filtration. Equation 17.7 can be rearranged as follows:

$$\frac{t}{V} = \frac{\mu\alpha C}{2A^2\Delta P}V + \frac{\mu R_m}{A\Delta P} \quad (17.15)$$

From the t versus V data, V can be plotted against t/V to obtain a straight line, the slope of which is $(\mu\alpha C/2A^2\Delta P)$and the intercept is $(\mu R_m/A\Delta P)$. With all the other parameters such as μ, C, A, ΔP known, α and R_m can be calculated. Pilot plant filtration performance can then be estimated using the similar approach described earlier. Compressibility index cannot be calculated through this set of data.

Leaf filtration: Usage of t/V at Multiple Pressures Data Inverse filtration rate (t/V) data at multiple pressures is the most complete data set required to calculate the cake properties including the compressibility index. Equation 17.9 can be modified as

$$\ln(\alpha) = \ln(\alpha_0) + (n)\ln(\Delta P) \quad (17.16)$$

After calculating α and R_m for the individual differential pressures as explained above, $\ln(\alpha)$ can be plotted against $\ln(\Delta P)$ to obtain a straight line. The slope of this straight line is the compressibility index (n) and the intercept is $\ln(\alpha_0)$.

Even though this approach is more complete and reliable, the previous two approaches can also be beneficial in cases of limited data.

Leaf Filtration: Filtration over Heel/Cake Washes Cake washes are typically performed when the cake is already deposited and the clean solvent is filtered through the deposited cake to displace the mother liquor. In often cases, crystal slurry is also filtered over the heel from the previous batch. Pilot plant performances of these cases can also be predicted by using similar equations. Equation 17.8 can be given as

$$\text{Flux(av)} = \frac{V}{At} = \frac{\Delta P}{\mu}\left(\frac{1}{((\bar{\alpha}/2)(M/A)) + R_m^{obs}}\right) \quad (17.17)$$

where $R_m^{obs} = R_m$ in case of slurry filtration over a clean filter cloth, $R_m^{obs} = R_m + (\alpha_{heel}m_{heel}/A)$ in case of slurry filtration over heel, and $R_m^{obs} = R_m + (\alpha_{cake}m_{cake}/A)$ in case of cake wash.

In case of cake wash, the term M ($=CV$) in equation 17.17 is zero as the concentration of API in wash solvent is zero.

Laboratory and Pilot Plant Centrifuge Filtration Similar set of data is collected from laboratory and pilot plant centrifuges as the pressure filtration. Spin speed (rpm) and feed rate (kg/h) are the only two parameters required in addition to the data set of t versus V collected from leaf filter at multiple pressures. Spin speed can be collected through the motor of the centrifuge, and the flow rate could be collected through mass flow meter, radar, or load cell.

T versus V data from leaf filtration is sufficient for estimating the specific cake resistance as discussed previously. In addition to α, for modeling a centrifuge operation in the pilot plant, the slurry feed rate and the centrifuge spin speed need to be optimized for the following constraints: (1) to minimize the centrifuge vibration, (2) to maintain uninterrupted filtration for completing a load, (3) to evenly distribute the cake, (4) to minimize the compression of the cake (in case of compressible cakes), and (5) to prevent the formation of the standard wave. In simplified terms, equation 17.12 can be used for this purpose in a similar fashion as described for the pressure filtration.

EXAMPLE 17.2

In an attempt to optimize the crystallization of an API, various crystallization parameters were studied that resulted in three different slurries. The three streams were then filtered separately in a leaf filter to assess the filtration performance of the streams. The following table provides the t versus V data for the streams filtered at 20 psi. Diameter of the leaf filter used was 5 cm. Viscosity of the mother liquor was 1.2 cP. Concentration of all the streams was 50 mg/mL. Calculate the specific cake resistances and the medium resistances for the three cakes.

Stream 1		Stream 2		Stream 3	
t (s)	V (mL)	t (s)	V (mL)	t (s)	V (mL)
0	0	0	0	0	0
3	22.48	9	29.69	3	27.06
6	40.88	18	47.70	6	49.32
9	55.72	27	61.15	9	67.06
12	68.44	36	72.48	12	82.34
15	79.74	45	82.50	15	95.92
18	90.03	54	91.62	18	108.27
21	99.52	63	100.05	21	119.65
24	108.38	72	107.95	24	130.28
27	116.72	81	115.58	27	140.28
30	125.13	90	122.68	30	149.76
33	132.63	99	129.47	33	159.37
36	139.64	108	135.95	36	167.95
48	166.57	117	142.03	–	–
–	–	126	148.07	–	–
–	–	135	153.89	–	–
–	–	144	159.48	–	–
–	–	155	166.12	–	–

Solution
From the given t versus V data, the following table was prepared with V versus t/V in SI units.

Stream 1		Stream 2		Stream 3	
V (m^3)	t/V (s/m^3)	V (m^3)	t/V (s/m^3)	V (m^3)	t/V (s/m^3)
0	–	0	–	0	–
2.25E-05	1.33E + 05	2.97E-05	3.03E + 05	2.71E-05	1.11E + 05
4.09E-05	1.47E + 05	4.77E-05	3.77E + 05	4.93E-05	1.22E + 05
5.57E-05	1.62E + 05	6.12E-05	4.42E + 05	6.71E-05	1.34E + 05
6.84E-05	1.75E + 05	7.25E-05	4.97E + 05	8.23E-05	1.46E + 05
7.97E-05	1.88E + 05	8.25E-05	5.45E + 05	9.59E-05	1.56E + 05
9.00E-05	2.00E + 05	9.16E-05	5.89E + 05	1.08E-04	1.66E + 05
9.95E-05	2.11E + 05	1.00E-04	6.30E + 05	1.20E-04	1.76E + 05
1.08E-04	2.21E + 05	1.08E-04	6.67E + 05	1.30E-04	1.84E + 05
1.17E-04	2.31E + 05	1.16E-04	7.01E + 05	1.40E-04	1.92E + 05
1.25E-04	2.40E + 05	1.23E-04	7.34E + 05	1.50E-04	2.00E + 05
1.33E-04	2.49E + 05	1.29E-04	7.65E + 05	1.59E-04	2.07E + 05
1.40E-04	2.58E + 05	1.36E-04	7.94E + 05	1.68E-04	2.14E + 05
1.67E-04	2.88E + 05	1.42E-04	8.24E + 05	–	–
–	–	1.48E-04	8.51E + 05	–	–
–	–	1.54E-04	8.77E + 05	–	–
–	–	1.59E-04	9.03E + 05	–	–
–	–	1.66E-04	9.33E + 05	–	–

By plotting t/V versus V the following graph (Figure 17.19) is obtained for all the three streams.

The slopes and the intercepts of the straight lines are obtained from the plot and are given below:

Streams	Slope	Intercept
Stream 1	1E + 09	102365
Stream 2	5E + 09	160315
Stream 3	8E + 08	85283

Calculation of α and R_m are then performed by using equation 17.15.

Slope = $(\mu\alpha C/2A^2\Delta P)$and the intercept = $(\mu R_m/A\Delta P)$.

By introducing the values for μ (1.2 cP, 0.0012 Pa s), A ($\pi \times (0.05/2)^2$, 1.96E − 03 m^2), ΔP (20 psi, 137,894 Pa) and C (50 mg/mL, 50 kg/m^3), the α and R_m values are calculated as follows:

For example, in the case of stream 1

$$\alpha = \frac{(1E+09)(2)(1.96E-03)^2(137,894)}{(0.0012)(50)} = 1.77E+10\,\text{m/kg}$$

$$R_m = \frac{(102,365)(1.96\text{E}-03)(137,894)}{(0.0012)} = 2.31\text{E}+10\,\text{m}^{-1}$$

The α and R_m values are calculated for all the three streams in a similar fashion.

Streams	Alpha (m/kg)	R_m (m^{-1})
Stream 1	1.77E + 10	2.31E + 10
Stream 2	8.83E + 10	3.61E + 10
Stream 3	1.41E + 10	1.92E + 10

By comparing the values of α, it seems that stream 2 is the one with the highest specific cake resistance, and stream 3 with the lowest specific cake resistance. Thus, the filtration performance of stream 3 is expected to be much better than that of stream 2 in the pilot plant.

Not surprisingly, R_m values are relatively comparable for all the streams as R_m depends mainly on the filter cloth and not on the cake properties.

17.2.3.2 Troubleshooting

Heel Removal Formation of heel may be an unavoidable problem in cake filtration in most of the equipments except in cases such as inverting bag centrifuges. As discussed earlier in this chapter, the offset between the agitator or peeler and the filter medium results in the formation of heel. The next load/batch could be filtered over the heel from the previous load/batch as long as it does not adversely affect the filtration performance, but the heel would become the primary filtration medium for the next load. For compounds with low toxicology profiles, a manual removal of heel is a viable option. However, for compounds with high potency/toxicology profiles, removal of heel in filter dryers/centrifuges needs an additional unit operation involving the dissolution of the

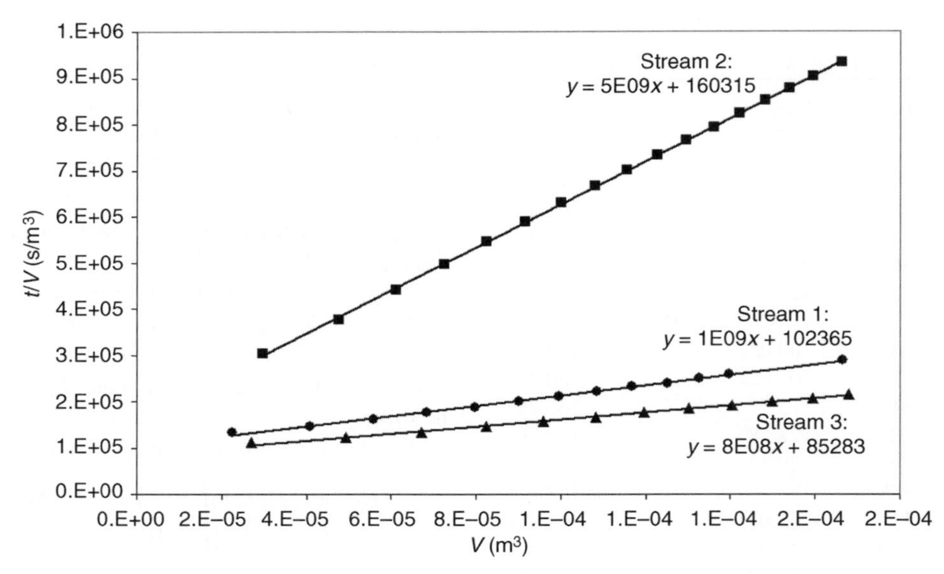

FIGURE 17.19 *t/V* versus *V* profiles of all the three crystallization streams.

cake in a solvent and passing the solution through the filter cloth. In modern-day centrifuges, there are provisions such as blowing compressed nitrogen from behind the cloth to remove the heel. Another possible solution in a centrifuge is to loosen the heel by rewetting it with wash solvent and discharging it.

Filter Medium Blinding Filter medium may play an important role in determining the filtration performance. At the beginning of the filtration process, filter medium retains the solid particles, while allowing the filtrate to pass through. The cake that is formed on the medium as a result then does the most of the filtration. However, in some cases, it has been observed that the filtration performance on a used filter medium is significantly different from that on a new one. This behavior could be observed even in the absence of heel over the medium. This is called "filter medium blinding." [7] It results from the clogging of the pores by the solid particles, thereby preventing the free movement of the filtrate through the pores as originally observed in a new medium. Due to this blinding, the initial medium resistance for the next batch/load might rise. Figure 17.20 shows the effect of filter cloth blinding on the medium resistance. From this figure, it is clear that the filter medium resistance increased gradually over the number of filtration cycles.

It is generally not possible to remove blinding through mechanical agitation or by passing compressed gas (nitrogen or air) through the medium, as it could result in permanently damaging the medium, rendering it unfit for use for the next batch/load. A typical solution for this problem is to use a solvent to dissolve off the entrained solid particles.

Cake Cracking In many cases, mostly after the deliquoring step, cake cracking is observed. In cases where a complete removal of the mother liquor (along with the unreacted starting material and reagents, by-products) is relied on washing the wet cake, cake cracking may pose a detrimental effect on the quality of the batch. Cracking of cake results in the formation of various channels. When washed with solvents without reslurrying the cake, the wash solvents tend to pass through these channels (paths of least resistances) leaving many parts of the cake unwashed. Cake cracking becomes more frequent when the average porosity of the cake is high [8], resulting in the formation of channels. Presence of cracks could be identified by (1) visual inspection through the sight glass, (2) monitoring the filtrate flow (an abnormally fast deliquoring could be a sign of cracking), and (3) measuring the purity of the cake, and the presence of a washable impurity in the cake could be another sign of an inefficient washing possibly due to channeling. This problem is commonly solved in the pharmaceutical industry by smoothing the cake before initiating the wash sequence. This smoothing action eliminates the channels allowing the wash solvent to filter/wash through the cake more uniformly.

Compressible Cake Compressible cakes may adversely affect the filtration performance if high pressures (pressure filtration) or high spin speeds (centrifugal filtration) are used. In addition to slow filtration, difficulty in discharging the wet cake and formation of hard agglomerates are other possible detrimental effects of a highly compressed cake.

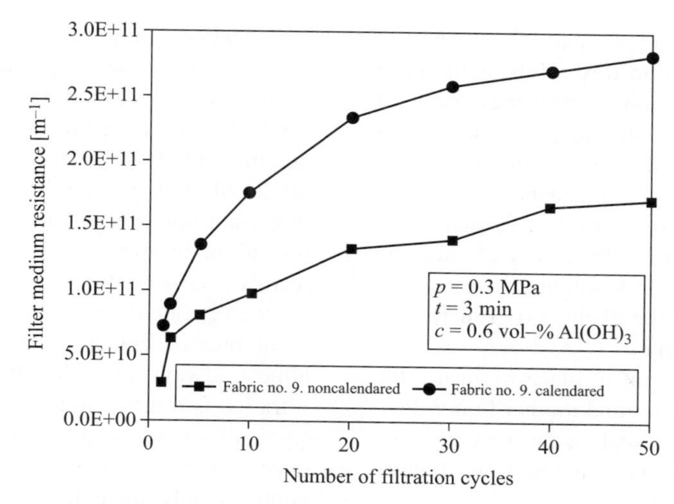

FIGURE 17.20 The effect of filter cloth binding on the filter medium resistance. Reprinted with permission from Ref. 7. Copyright 1997, with permission from Elsevier.

17.2.3.3 ***Cake Wash Design*** Proper design of cake washing is imperative to obtain the active pharmaceutical ingredient with acceptable purity/impurity specifications. Typically, the main objective of a cake wash is to displace/remove the mother liquor from the cake. Removal of the mother liquor facilitates the following:

1. Removal of unreacted/excess reagents, by-products, and impurities rejected by crystallization.
2. Removal of the crystallization solvent in case of significant product solubility at the anticipated drying temperature. This consideration is critical to prevent the redissolution of the product crystals in the crystallization solvent resulting in the potential formation of agglomerates upon the evaporation of the residual crystallization solvent during drying.
3. Removal of color of the cake if needed.

There are mainly two types of cake washes: reslurry and displacement. In reslurry wash, after introducing the solvent, the wet cake and the solvent are agitated to a slurry again and then filtered. In displacement wash, the solvent is introduced on top of the wet cake, and the filtration is started to initiate a plug-like flow of the wash solvent through the cake and displacing the mother liquor. When agitation is available, the overall cake wash is performed in the following sequence: displacement wash, reslurry wash, and displacement wash (DSD).

There are various factors that need to be considered to design and develop an efficient cake wash in order to attain the above-mentioned objectives. In typical cases, the crystallization solvent is the preferred solvent for cake wash and is generally used to displace the mother liquors even when a different solvent is implemented for other reasons. The following are the considerations while choosing a solvent for the cake wash.

- Identification of a suitable solvent that will have maximum solubility of the unwanted impurities. A low solubility might crystallize those impurities along with the already crystallized product.
- The solvent should have minimum solubility of the product to prevent product loss to the washes. However, a solubility drastically different from the mother liquors may result in the nucleation of material with high impurity level.
- The viscosity of the wash solvent should be minimal to afford a fast enough filtration such that the overall cycle time of the process is not extended.
- The thermal stability of the product in the presence of the residual wash solvent should be assessed under drying conditions.
- The thermal behavior (boiling point) of the wash solvent.

Volume Considerations for Cake Washes In principle, the volume of displacement wash required to displace the mother liquor should be estimated by the cake height and the coaxial mixing that occurs at the front end of the wash as it traverses the cake in a plug flow. Generally, however, the risk of channeling is too high, and as a rule of thumb, the solvent volume should be equal to at least thrice the volume of wet cake. In general terms, a reslurry wash requires more solvent than a displacement wash. If the isolation is performed in multiple loads, the cake wash volumes should be adjusted accordingly. This is true for both pressure filtration and centrifugation.

Uniform Washing To facilitate an efficient cake wash, care should be taken to ensure a uniform wash of the entire cake bed. Even though there are various approaches (such as spray balls, mists, or weirs) available to spray the solvent evenly across the surface of the cake, care should be taken to ensure that the cake surface is adequately smoothed to prevent (a) more solvent stagnating on troughs and (b) channeling of the wash solvent through cracks formed across the bed. These two factors might cause uneven washing of the cake, increasing the likelihood of inefficient washing of some parts of the cake.

In-process controls could be put in place to ensure the complete removal of the impurities and mother liquor. Possible controls include parameters such as purity, color, and pH. However, the accuracy of these methods is directly related to the uniformity of the wash.

17.2.4 Summary

A thorough understanding and a proper design of filtration unit operation are imperative in pharmaceutical industry to yield a high-quality product. Optimizing crystallization and filtration parameters to achieve an efficient filtration might also help in reducing the cycle time for the overall manufacturing process. Use of the laboratory filtration data in the filtration equations derived from Darcy's law provides a prediction of the filtration performance upon scale-up in pilot and manufacturing plants. There is an array of isolation equipment available with various sizes and operating parameters intended for laboratory, pilot plant, and manufacturing scales. An efficient cake wash design also aids in enhancing the product quality and the process yield.

17.3 DRYING

17.3.1 Background and Principles

Drying is an integral part of the isolation process and is important for the production of consistent and stable product. It is especially important in the case of the API where product properties can have a direct impact on product performance and the levels of residual solvent are considered a regulatory specification. Drying is basically a complex distillation carried out in a heterogeneous system involving solid, liquid, and vapor. Distillation involves vaporization of solvents from the liquid phase, the same phase transformation that in drying removes the solvents from "wetted" solids that were retained following the filtration process. One of the primary objectives of a drying operation is the removal of solvent to meet specifications in the product. In the case of API, the acceptable levels of residual solvent may be set according to safety (solvent toxicity), regulatory, or stability requirements (see Chapter 20). For earlier pharmaceutical intermediates, these specifications may be set according to the tolerance of solvents in chemistry steps that are immediately downstream of the drying operation. For APIs, another key objective of drying is to either achieve or maintain a specified crystallographic form. In most cases, the required form may have been produced during crystallization of the product and it is important not to effect a change in form during the drying operation. In some cases, the required form may not be readily accessible through direct crystallization and is more practically obtained during the drying operation.

In pharmaceutical processing, the drying operation is almost always preceded by filtration. The output from filtration is referred to as wet cake since it contains solvent (typically 5–50%) only and the output from a successful drying operation is referred to as dry cake since it usually contains only trace levels of residual solvent. A typical drying operation consists of the following general steps. The wet cake is charged/loaded into the dryer, which is heated in order to vaporize the solvent. The dryer is often equipped with a vacuum pump connected to the vessel's vent line and the vaporized solvent is removed through this line for downstream recovery that is typically done by means of a condenser. Alternatively, a continuous stream of inert carrier gas is passed through the dryer to sweep the vaporized solvent through the vent line. When the material is dry, it is discharged/unloaded from the dryer and packaged.

17.3.1.1 Phase Equilibria Since drying involves the vaporization of solvents, the vapor pressure, enthalpy of vaporization, and in some instances molecular diffusivity of the solvent are fundamental properties defining the drying operation. For some solvents, the boiling point will be beyond the operational range that can be achieved in a plant dryer, so the pressure in the dryer is reduced to effect a decrease in the boiling point. Often, pharmaceutical compounds are unstable at elevated temperatures and must be dried at lower temperatures that may also require reduced pressures. Hence, it is useful to understand the relationship between vapor pressure and temperature. The Antoine equation (shown below) provides a reasonable estimation of the vapor pressure of pure solvents.

$$\ln P^{\text{sat}} = A - \frac{B}{T+C} \tag{17.18}$$

where P^{sat} is the vapor pressure, T is the temperature, and A, B, and C are empirically determined constants that are readily available for many solvents. This relationship is used to estimate the pressure required to bring the drying operation within the limits of the dryer's capabilities (or the product's stability) for the particular solvent that is being removed. For more accurate representation of vapor pressure data over a wide temperature range, an equation of greater complexity such as the Riedel equation could be used [9]. However, for most applications in the

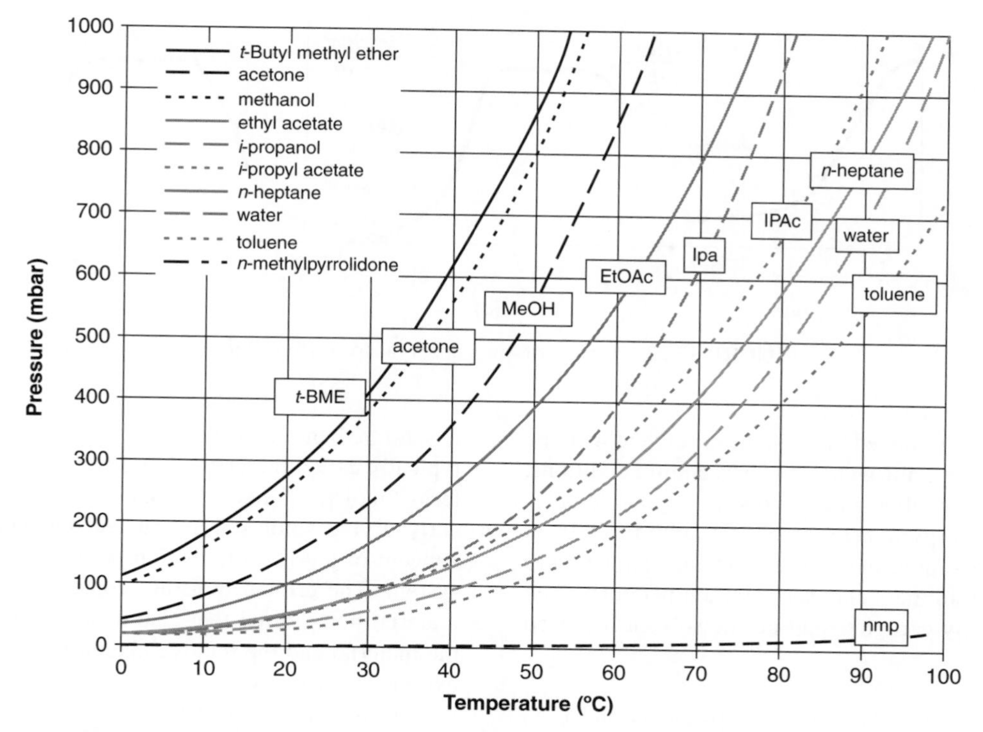

FIGURE 17.21 Example vapor pressure chart generated using Antoine's equation.

pharmaceutical industry, the Antoine equation provides a satisfactory approximation.

Figure 17.21 shows some examples of the vapor pressure versus temperature relationship for common solvents used in the pharmaceutical industry. Solvents with an equilibrium curve to the right of *n*-heptane are generally harder to remove in a drying operation. When the temperature at which drying is performed is above the boiling point of the solvent, the drying rate is typically increased significantly until mass transfer of the solvent becomes the rate-limiting step.

When a solid material is held in an environment with constant temperature and humidity, it will reach steady-state equilibrium with respect to its moisture content. The *equilibrium moisture content* is the limit of the water content that the material can retain under specific conditions of temperature and humidity. Materials encountered in drying operations can either be hygroscopic (e.g., porous materials) or nonhygroscopic (e.g., nonporous materials); however, typical pharmaceutical compounds are nonporous and do not exhibit significant moisture (or other solvent) movement unless they exist as hydrates (or solvates).

17.3.1.2 Drying Mechanisms/Periods Several mechanisms are involved during a drying operation and the drying process can be divided into different stages that reflect the mechanism that dominates the rate of solvent removal. Initially, drying is generally slow during the warm-up period as the system reaches steady state. Following this stage, drying may be faster during the heat transfer limited period, after which the rate of solvent removal may become considerably slower as the system becomes mass transfer limited near the end of the drying process. These mechanisms are described in more detail below.

The drying process can be monitored by following the solvent content versus time. The data can then be presented in several ways to determine when different drying mechanisms are dominant. These plots are called drying curves and an example schematic is shown in Figure 17.22. The first curve (a) shows a typical drying curve for solvent content, W versus time. In the second curve (b), the derivative of W with respect to time (i.e., the change in solvent content versus time, or the drying rate) is plotted against time. In the third curve (c), the drying rate is plotted against the solvent content. From each of these curves, distinct stages or periods can be identified and are denoted by points A through E in Figure 17.22 [9].

Jacket Warm-up Period The period between A and B is called the warm-up period and accounts for the time taken for the system to heat up toward the target set point of the jacket temperature controller. At this point, drying will proceed as solvent saturating the surface of the solids is vaporized. Under reduced pressure, this vapor-phase solvent would be continuously removed through the dryer vent. In the case where reduced pressure is not used, a continuous gas flow (i.e., sweep) through the dryer is utilized to ensure that

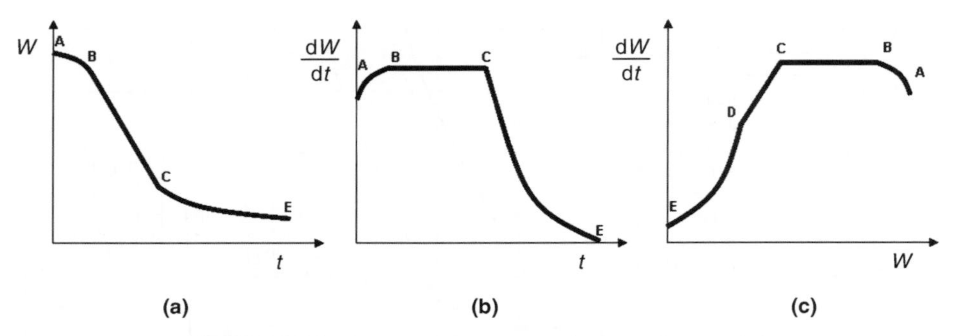

FIGURE 17.22 Typical drying curves (adapted from Ref. 9).

vaporized solvent is removed through the dryer vent with the inert carrier gas (e.g., nitrogen). Under vacuum conditions, the cake will often cool down to below its original temperature due to the energy transfer associated with the solvent enthalpy of vaporization. In this case, the cake temperature will increase back to the jacket set point temperature when drying is essentially complete and all the solvent has been vaporized.

Constant Rate Period The period between B and C is typically characterized by a constant drying rate (as can be seen from the second and third drying curves in Figure 17.22). The drying mechanism during this period is dominated by heat transfer and the drying rate is usually limited by the rate of heat transfer to the system. During this period, solvent that has saturated the surface of the solids is rapidly vaporized. The driving force during this constant rate period is the temperature difference between the dryer jacket set point and the temperature of the solvent being removed (at the system pressure). The drying rate can hence be increased by either increasing the jacket temperature set point or decreasing the pressure (i.e., to decrease the boiling point). It is important to note that the rate of solvent removal is also dependent on the size of the vacuum pump. If the pressure setting is too low, the rate of solvent vaporization may exceed the capacity of the vacuum pump to remove solvent from the system. In this case, the system settles at a higher pressure than the desired set point and the drying rate is determined by the vacuum pump flow rate rather than the heat transfer rate. For the drying rate to be controlled by the heat transfer rate, the vacuum pump must be appropriately sized for the system so that it imposes no limitations on solvent removal from the dryer. As would be expected, the drying rate is also affected significantly by the surface area available for heat transfer. As the cake dries, this area is reduced; however, agitation of the cake exposes "fresh" solvent to the heated surface and increases the area for heat transfer. Hence, agitation of the cake allows an increase in the area available for heat transfer.

Falling Rate Period At the point C, solvent is no longer abundant on the solid surface; that is, the surface becomes unsaturated and the drying rate begins to fall. This point is typically called the *critical moisture content*. Note that the term "moisture" derives from historical convention when early drying studies were carried out with water, but the concept of a critical transition point applies to any solvent system. The critical moisture content is a property of the material being dried that also depends on the thickness of the material and the relative magnitudes of the internal and external resistances to heat and mass transfer. In some cases, it could also be a function of the total surface area of the solid phase and hence the particle size [3]. It is important to take into account possible variations in the value of the critical moisture content during scale-up of a drying process.

During the period from C to E, heat transfer no longer becomes the limiting mechanism for drying. Instead, mass transfer becomes the dominant mechanism limiting the rate of solvent removal from the heated surface. As a result, the drying rate is typically observed to fall as depicted in the second and third drying curves of Figure 17.22. In some cases, the decrease in the rate of drying can become significant and lead to extremely long drying times. At some point during the falling rate period, the solvent present on the surface is depleted and solvent present inside the internal structure of the solids remains (either as a solvate or an occluded solvent that may not be removable). The drying rate is then determined by the movement of this internal solvent to the heated surface for vaporization. This transition is given at point D and can sometimes be seen by plotting the drying rate versus the solvent content as shown in the third drying curve of Figure 17.22. It is important to note that this solvent migration is extremely difficult to model and may not be a feature clearly visible on many drying curves.

17.3.1.3 Effect of Wet Cake Properties For any process operation, it is important to understand the properties of the input to the operation, and drying is not an exception. Careful analysis of the wet cake used in the drying process can provide insight into common problems faced during drying. There are properties of the wet cake that can have significant impact on the drying operation, such as the particle size distribution, the particle morphology/shape and the solvent content. For example, the particle size distribution and the particle morphology can have a direct impact on the surface area available for heat transfer, while the initial solvent content will clearly affect the overall drying cycle time. Furthermore, the solvent(s) that the cake is wet with can impact both the drying cycle time and the final particle properties. A higher initial solvent content can result from inadequate deliquoring (removal of solvent), practical equipment limitations during filtration, or a propensity of the material to retain solvent even in extreme deliquoring conditions. The interactions between solvent and solid phases can have an impact on the degree of particle attrition and/or agglomeration and hence the final particle size distribution. In addition, the composition of the solvents and the resulting vapor–liquid–solid equilibria can have an impact on the final particle properties (see Section 17.3.3.4).

The final particle properties of an API product are important characteristics that must be carefully controlled. The mechanical impact of the drying operation is exceedingly difficult to model and predict. Crystallization is a much better understood unit operation and control of particle size distribution and morphology is typically achieved through careful design and execution of this unit procedure. The typical goal and the general expectation for a drying process are therefore to have little to no impact on these particle properties and to attempt to maintain the same properties as the wet cake input.

17.3.2 Equipment

There are several types of dryers used in chemical plants that can be operated in both batch and continuous modes. Typical dryers can be classified as either *convective* or *conductive*. Convective dryers operate by passing heated air or gas over or through the material to be dried. Some common advantages of these types of dryers are that they tend to keep the product temperature relatively low and they can be scaled up to large sizes fairly easily. Some disadvantages are that they tend to have poor energy efficiency and require elaborate dust collection/capture units on the gas effluent side [10]. In pharmaceutical plants, conductive (or *contact*) dryers are by far the most commonly used type. In this type of dryer, the product is in direct contact with a heated surface. Contact dryers are mostly used in batch mode and are often more energy efficient than convective dryers [10]. Some common types include vacuum tray dryers, filter dryers, rotary cone dryers, and tumble dryers. The discussion here on large-scale equipment will focus on these types of dryers.

17.3.2.1 Laboratory Drying Equipment It is often difficult to accurately simulate the drying process in the laboratory; however, there are some common types of lab-scale equipment that can be helpful in providing guidance toward the selection of plant equipment, providing estimates of drying cycle time on scale and determining equipment settings necessary to achieve design objectives.

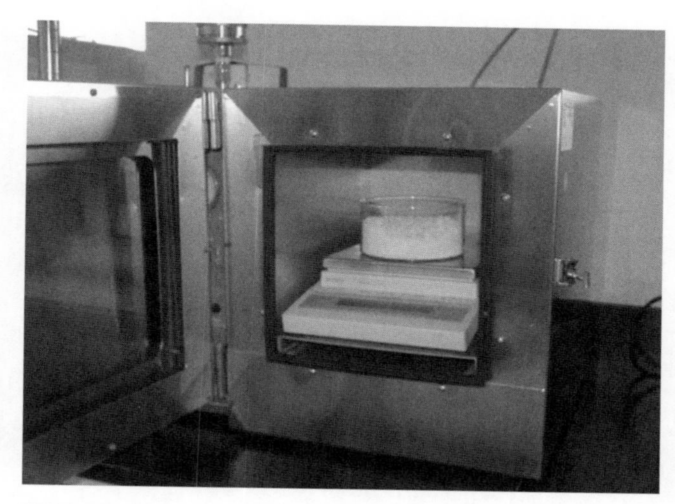

FIGURE 17.23 An oven balance used to generate drying curves.

Figure 17.23 shows an oven balance that can be used to determine a drying curve for a particular material. The setup consists of a temperature-resistant balance, placed inside a vacuum oven. The balance is connected to a computer outside of the oven that captures the mass of the product as it is dried. The solvent content at any time can be calculated from the product mass. The oven balance is therefore useful for generating drying curves such as those shown in Figure 17.22 and hence determining the critical moisture content. One advantage of using an oven balance is that it is not an *invasive* technique: that is to say that significant and discrete samples of the product are not required, hence data can be collected on a very small scale. One disadvantage is that drying in an oven is a purely static process and provides no information on the effect of agitation on the material.

The effect of agitation can sometimes be estimated using a pan dryer in the lab. A pan dryer is basically a closed jacketed vessel with an overhead motorized agitator with impeller blades. The jacket provides a heated surface to contact the product and the impeller blades allow the cake to be agitated (or turned over). The product and jacket temperatures can be recorded using thermocouples that provide the temperature differential (the driving mechanism for solvent evaporation in the constant rate period) and the agitator power can be recorded using a torque meter. The shape, type and size of the impellers in a pan dryer can be adjusted so that it is geometrically similar to the large-scale plant dryers; however, the shear force exerted on the cake may not be representative of a large-scale unit. The drying process is typically faster in a pan dryer than an oven due to the larger relative surface area in contact with the material from the effect of agitation. The use of a pan dryer is often helpful in determining the effect of agitation on powder properties of the product (e.g., particle

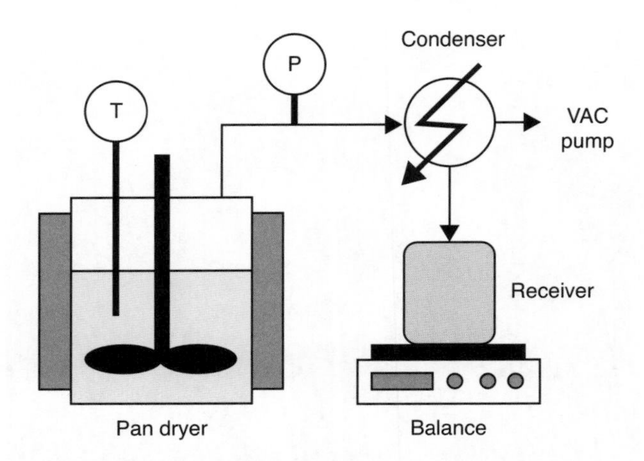

FIGURE 17.24 Typical laboratory setup.

size). The impact of different modes of agitation (i.e., turning over the cake at intermittent intervals or continuously agitating the cake throughout the process) can also be examined in a qualitative manner.

The typical setup for using a pan dryer in the lab is shown in Figure 17.24. Here, the solvent vapor removed from the product is condensed outside the dryer and collected in a vessel located on a balance. This allows an estimate of the solvent content to be determined (from a mass balance) at any time during the process without the need to take discrete samples, especially if the balance is connected directly to a computer or data recorder.

17.3.2.2 Large-Scale Drying Equipment The four most common types of dryers used for pharmaceutical products are vacuum tray dryers, filter dryers, rotary cone dryers (or conical dryers) and tumble dryers. These units are shown in Figures 17.25–17.28, respectively.

A tray dryer (see Figure 17.25) is basically a large oven connected to a vacuum pump and/or an inert gas supply. Product is loaded onto metal trays that are placed onto shelves in the dryer. Samples can be taken only by opening the oven. When the product is dry, the final dry cake is unloaded manually from the trays into product containers/drums.

Filter dryers (see Figure 17.26) are similar to pan dryers with the main difference being that they are used to both isolate and dry the product. After filtration and washing of the product, the jacket of the dryer provides a heated surface and the impeller is lowered into the cake to provide agitation. Filter dryers are usually also connected to a vacuum pump and typically have a sampling port on the side. The sampling port is comprised of a ball valve with a cup that is turned so that the orifice of the cup is facing the cake. The agitator is turned on to push a small sample of the cake into the cup. The ball valve is subsequently turned so that the cup faces the outside of the dryer and the contents of the cup are removed. Obtaining a significantly sized and representative sample of the cake can sometimes be challenging; however, this method provides a lower risk of exposure to the operator than attempting to access the cake directly.

A rotary cone dryer (also known as a conical dryer) is a jacketed dryer with an internal agitator (see Figure 17.27). The shape of the rotary cone dryer provides a larger surface area to volume ratio for contact drying than a filter dryer. A shaft that is permanently embedded within the cake provides the agitation and in some instances acts as a heat

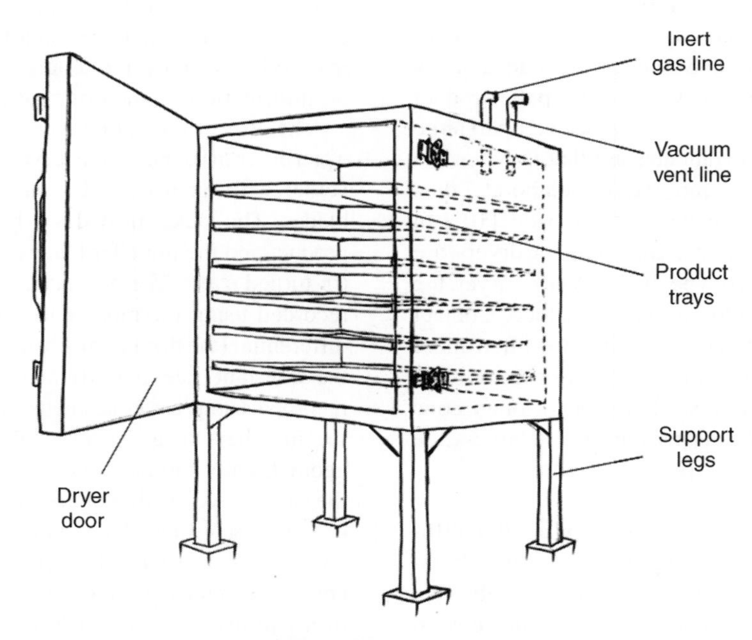

FIGURE 17.25 Tray dryer.

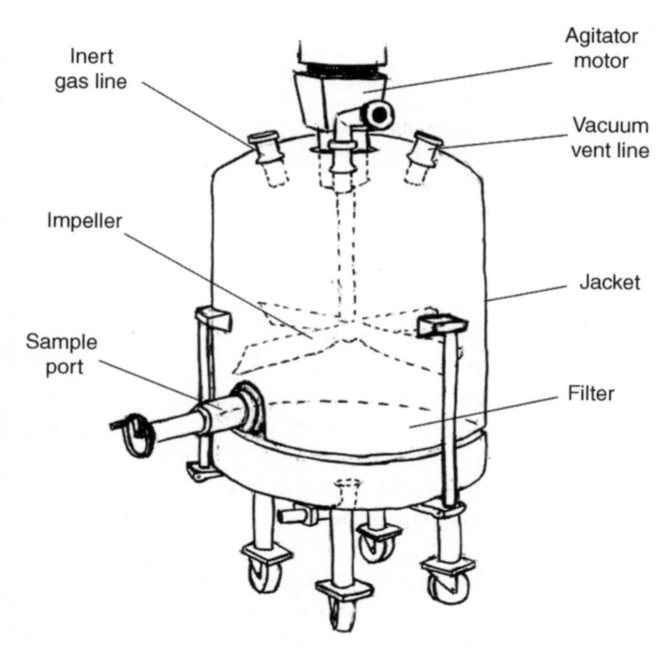

FIGURE 17.26 Filter dryer.

source as well. As the shaft turns, a screw built onto the shaft lifts the cake from the bottom of the cone to the top. In addition, the shaft itself orbits the vessel so that all parts of the cake can be turned over. Samples are typically obtained from the conical dryer via a sampling port similar to the one described for the filter dryer, and the temperature of the cake is measured by a temperature probe lance that extends down into the cake. Wet cake is charged into the top of the dryer

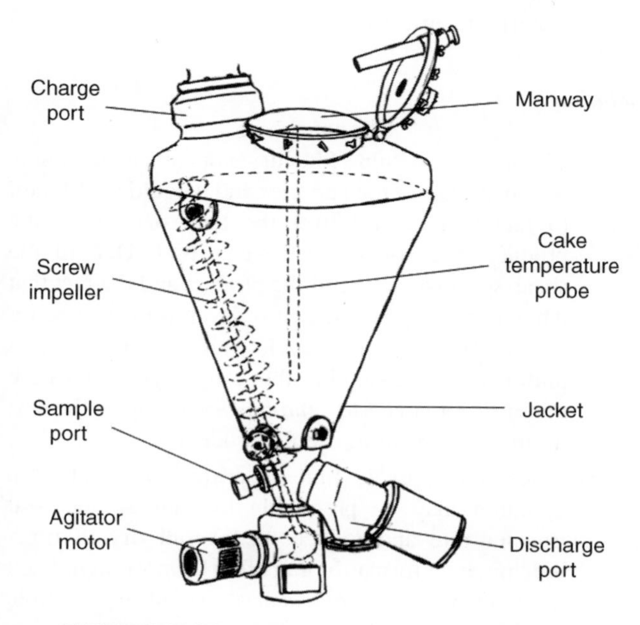

FIGURE 17.27 Rotary cone dryer or conical dryer.

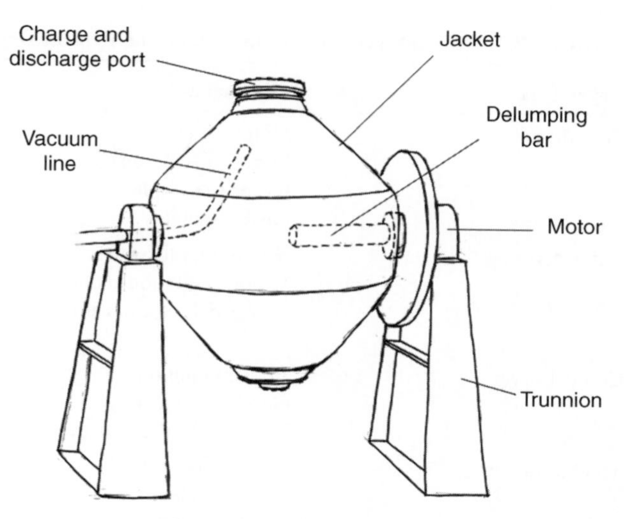

FIGURE 17.28 Tumble dryer.

and dry product is discharged from a special port built into the bottom of the cone.

A tumble dryer (see Figure 17.28) consists of a rotating or tumbling vessel and is most commonly found in a double-cone shape supported by two stationary trunnions. The vessel is surrounded by a heated jacket and a small vacuum line is installed within one of the trunnions and extends into the vessel (angled in the upward direction). The dryer can also be equipped with a delumping bar that extends into the vessel from one of the trunnions. The double-cone drying chamber rotates about the axis of the trunnions, causing the material to cascade inside. Through gentle tumbling and folding, the material is contacted with the heated wall to facilitate drying without a significant amount of shear being imparted to the material. Discharge from and cleaning of the tumble dryer is easier than other dryers since there are no internal shafts or agitators.

Other types of dryers that are available for use include fluid bed dryers and spray dryers. Fluid bed dryers are suitable for drying granular crystalline materials but not for very wet materials that have a pasty or liquid-like consistency.

In a fluid bed dryer, a hot gas stream is introduced into the bottom of a chamber filled with the material to be dried. The gas stream expands the bed of material to create turbulence and the solid particles attain a fluid-like state—a phenomenon known as fluidization. Heat transfer is extremely efficient and uniform since the solid particles are surrounded by hot gas, leading to fast drying times. The exhaust to the chamber is equipped with particulate filters to prevent the product from escaping the chamber.

Spray dryers are used to dry product that is either dissolved in a solvent or suspended as slurry. The liquid stream is dispersed into a stream of hot gas and sprayed via a nozzle into a cylindrical chamber (often with a conical bottom) as a mist of fine droplets. Solvent is vaporized rapidly to leave

TABLE 17.3 Advantages and Disadvantages of Typical Dryers

Dryer Type	Advantages	Disadvantages
Tray dryer	Simple operation Easier to scale up Easier to sample Easier to clean	Labor intensive Long drying times Poor uniformity Operator exposure
Filter dryer	No product loss after isolation Various agitation modes Lower risk of exposure	Difficult to sample Agglomerate formation Particle attrition Difficult to scale up
Conical dryer	Active agitation Good homogeneity Lower risk of exposure	Difficult to sample High particle attrition Difficult to scale up
Tumble dryer	Simple operation Easier to clean Lower capital cost Suitable for shear-sensitive materials	Lower efficiency Material must be free flowing Long drying times
Fluid bed dryer	Good homogeneity Lower risk for agglomeration Short drying times	Unsuitable for pasty or liquid materials Product recovery may be required from exhaust High volumetric consumption of inert gas
Spray dryer	Suitable for heat-sensitive materials Good product uniformity Combines crystallization and drying	Liquid or slurry feed required Difficult to control bulk density High volumetric consumption of inert gas

a residue of dry solid product particles in the chamber. It is important to ensure that particles of product are not wet with solvent when they touch the walls of the chamber and hence spray drying chambers tend to have large diameters. The heating period is very short, hence functional damage to the product is usually not an issue.

*17.3.2.3 **Equipment Selection*** Selection of the appropriate type of dryer to be used is usually based on the objectives of the drying operation. Some typical objectives include minimization of drying time, achieving or maintaining powder properties (i.e., prevention for particle agglomeration), and maintaining crystallographic form. Table 17.3 shows some advantages and disadvantages for each of the dryer types discussed that may be used for equipment selection.

EXAMPLE 17.3

Select and recommend appropriate drying equipment for the following three products (note there may be several dryers that are suitable for these systems):

(a) Crystals of product A are needles that tend to break easily and it is desirable to maintain the integrity of both their particle size distribution and the bulk density of the material while achieving a uniform powder. There are no exposure issues associated with product A.

(b) Product B is a potent compound that may require careful control of exposure hazards. A low cycle time for the drying operation is of critical importance and the formation of agglomerates causes issues in the downstream formulation process.

(c) Product C is in the early stages of development and requires rapid scale-up. The crystallization and isolation procedures have not been finalized yet and the material is unstable at higher temperatures.

Solution

(a) The agitation within a filter dryer or conical dryer may be too strong for the particles and it could be difficult to maintain control over the bulk density of the product if a spray dryer were used. Due to the shear-sensitive nature of the product and the fact that it has no exposure issues, a tray dryer would be a better option. However, this could lead to poor product uniformity. So assuming that long drying times are not of major concern, a *tumble dryer* may be the best drying equipment to use for product A.

(b) Since a low cycle time is critical, a dryer with agitation may be preferable to increase the heat transfer rate and reduce the overall drying time. Agglomerate formation can often be more significant in a filter dryer, so a *conical dryer* would be preferable for this material. A conical dryer would also provide the necessary containment to handle the exposure issues associated with a potent compound. Alternatively, a *fluid bed dryer* would also provide a short drying time and lower risk for agglomeration; however, recovery of material from the exhaust may be required and could lead to containment issues.

(c) A *spray dryer* would provide a combined crystallization and drying operation and would alleviate the temperature sensitivity issue. However, rapid and simple scale-up may not be possible as the operation would require more detailed development in order to implement successfully. For relatively fast, simple and low-cost scale-up, a *tray dryer* or *tumble dryer* could be used, especially since the isolation has not been finalized and there may be an opportunity to influence which solvent the product input will be "wetted" with. The operating pressure would need to be designed in conjunction with this to ensure that the product temperature does not exceed the stability limit.

17.3.3 Design

The two primary concerns during scale-up are powder properties and drying cycle time. Powder property control is important for APIs since particle size, particle morphology and solvate form can all be impacted by the drying operation. For intermediates, scale-up is generally only a concern if there is a tendency to form hard macroscopic lumps during drying.

Scale-up of drying operations is not well understood and similar performance between lab and plant equipment is often difficult to achieve, making the selection of the correct scale-up parameters challenging. Data collection and modeling are important aspects to designing a drying operation and are discussed below.

*17.3.3.1 **Data Collection*** The drying process is typically monitored using temperature and pressure since most dryers already have existing product temperature probes and pressure sensors installed. In the case of temperature, correct placement of the temperature probe in the drying cake is critical to obtaining reliable information from the data. It should be noted that during static drying, a temperature probe may not provide a good representation of the average temperature of the cake. Agitation of the cake may be required to get a representative measure of the bulk temperature. Alternatively, it can be helpful to have multiple temperature probes situated at different locations within the cake. This can be useful for both static and agitated drying.

Pressure and an estimate of the solvent flow from the dryer can help to monitor the stage of the drying operation and determine the end point. One possible way to identify when solvent is being vaporized is to monitor the position of the valve controlling the pressure of the dryer. As solvent is vaporized, it causes an increase in pressure in the dryer and in order to maintain vacuum, the vacuum control valve in the dryer vent opens. So while solvent is being removed, the vacuum control valve remains open. When there is no longer a significant amount of solvent to create enough pressure in the dryer, the vacuum control valve begins to close until it reaches a position corresponding to the leak rate of the dryer setup. This movement of the vacuum control valve can be used as an indication of the end of the heat transfer limited period of the drying process. Figure 17.29 shows an example of how the vacuum control valve position starts to decrease from fully open (a value of 100%) as the pressure in the dryer approaches the vacuum set point. In this case, the position corresponding to the leak is approximately 35%.

Drying is also typically monitored by estimating the residual solvent content of the drying solids and is accomplished by either a direct measurement or by inference from surrogate process parameters. Direct measurement methods include loss on drying (LOD), gas chromatography (GC), and Karl Fischer coulometric or volumetric titration (KF).

An LOD instrument rapidly heats up a sample of the solids to a temperature high enough to vaporize most solvents at atmospheric pressure (typically 120°C is used). The sample is then held at this temperature for approximately 30 s to ensure that any solvent is removed before cooling back to ambient temperature. The mass of the sample before and after heating is recorded and the change in mass is expressed as a fraction of the starting mass to obtain the final LOD result.

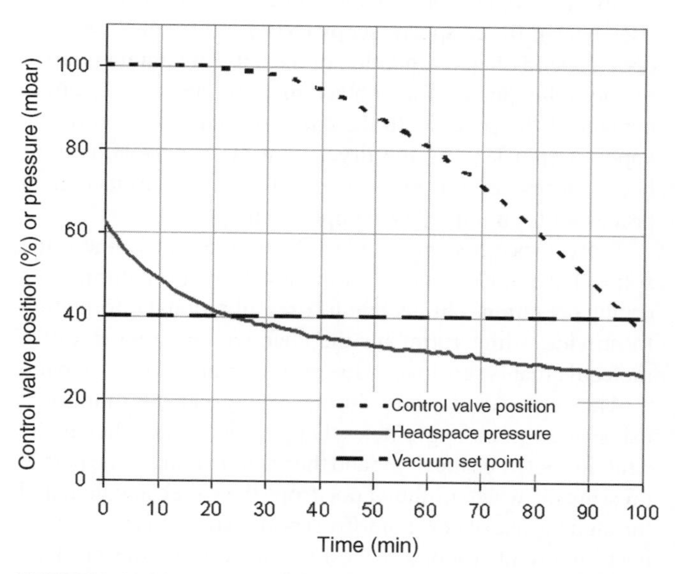

FIGURE 17.29 Using pressure and vacuum control valve position to determine the end point of drying.

GC also directly measures the content of volatile materials in a sample. *A priori* calibration of the GC measurements with solvent standards is typically required.

KF titration is a technique that is used specifically to determine the water content of a sample. Residual water down to the parts per million (ppm) range can often be detected using a KF instrument.

Estimating the solvent content using offline methods such as LOD, GC and KF can have potential drawbacks. First, these methods are "obtrusive" as they may require significantly sized samples to be taken to obtain a representative and reproducible result: this may be a problem when carrying out drying studies in the lab in a small-scale dryer. Furthermore, samples taken from a plant dryer may not be representative of the bulk material, depending on the location and size of the sampler. Another important disadvantage of taking samples is the risk of exposure to operators and analysts. This is especially the case when the product being dried is toxic, as can be the case for many pharmaceutical intermediates. It is therefore desirable to utilize online monitoring techniques to minimize or even eliminate the need to take samples from a dryer. Online techniques can also reduce the analysis time associated with many offline methods used to determine drying end point. Brief introductions to some of these online methods are given below.

Mass spectrometry has gained popularity recently and is particularly useful when different solvents are being removed from the cake. The vent of the dryer is sampled for the vapor content and molecular masses of ionized species in the vapor are determined. The intensity of identified peaks obtained from a mass spectrum enables the determination of solvent ratios and through the use of standards and appropriate calibration, solvent amounts can also be determined.

Spectroscopic methods including infrared spectroscopy (IR), near-infrared spectroscopy (NIR) and Raman spectroscopy depend on quantification of the solvent content relative to the solid phase (for contact measurements, i.e., direct contact of the probe with the cake) or quantification of the vapor stream leaving the dryer vent. These methods have been successfully implemented in many pharmaceutical processes to monitor drying operations [11].

Raman spectroscopy can also be used to determine if the solids have undergone a form transformation during the drying operation. This technique uses vibrational information to provide a fingerprint for the chemical environment of a molecule that is sensitive to its crystallographic arrangement.

The removal of water during drying can be monitored using a dew point hygrometer that is typically installed in the vent line between the dryer and the vacuum pump. The partial pressure of water in the vapor from the dryer is calculated through the use of dew point/frost point tables. In cases where it is important to control the hydrate form of the product, the partial pressure of water can then be used to calculate the relative humidity of the vapor stream [12].

An offline technique frequently used to characterize the crystal structure and detect changes in morphology or hydrate/solvate forms is X-ray powder diffraction (XRPD). Peaks associated with different crystallographic faces can be identified in the pattern from an XRPD measurement. The existence of different polymorphic forms will result in the appearance of new or shifted peaks in the pattern and a fingerprint pattern can be established for each particular form. As with other offline techniques, the disadvantage of XRPD is that it is an analytical technique that must be carried out on samples that have been taken of the drying cake that may not be representative of the bulk material in the dryer.

17.3.3.2 Modeling of Drying Modeling the rate at which solvent is removed from the drying cake can be useful to determine optimal parameter settings for the dryer or to choose the appropriate equipment. To model the drying process, mass and energy balances must be carried out on the system. By coupling these two balances, it is possible to calculate the solvent content of the cake at any time and hence the drying rate. A simple static drying model [3] is presented here that should enable simulations of the constant rate period and falling rate period to be run, given some basic information about the system. For more complex drying models, the reader is referred to more detailed treatments proposed in the literature [13, 14].

Consider a bed of product wet cake in a static dryer where the cake is in contact with a wall heated by a jacket service. The heat transfer rate from the jacket to the cake ($\dot{q}$) is given by the heat transfer coefficient (U), the area available for heat transfer (A_{ht}) and the difference between the dryer jacket temperature (T_j) and the vaporized solvent temperature (T) which is related to the operating pressure by Antoine's equation:

$$\dot{q} = UA_{ht}(T_j - T) \tag{17.19}$$

At steady state, the heat transfer rate from the wall into the cake is completely consumed by solvent vaporization, giving the following equation for the mass transfer rate of solvent during the constant rate period ($\dot{m}$):

$$\dot{m} = \frac{\dot{q}}{\Delta H_{vap}} \tag{17.20}$$

where ΔH_{vap} is the heat of vaporization of the solvent. The mass transfer rate of the vapor allows calculation of the rate of change of solvent content in the cake versus time.

$$-\frac{dX}{dt} = \frac{\dot{m}}{m_{drycake}} \tag{17.21}$$

In equation 17.21, X is the solvent content in the cake per unit mass of dry cake and $m_{drycake}$ is the final mass of dry product cake obtained. Substituting equations 17.19 and

17.20 into equation 17.21 and integrating allows calculation of the drying time during the constant rate period (t_C):

$$t_C = \frac{m_{drycake}\Delta H_{vap}}{A_{ht}} \frac{(X_1 - X_C)}{U(T_j - T)} \quad (17.22)$$

where X_1 is the initial solvent content and X_C is the critical moisture content.

Below the critical moisture content, the model must be modified to take into account the fact that the drying rate is no longer controlled by the heat transfer rate. As an approximation, the mass transfer rate of the solvent can be assumed to be proportional to the solvent content in the falling rate period [3].

$$\dot{m} = aX \quad (17.23)$$

where a is a proportionality constant that may be related to the molecular diffusivity of the solvent. By substituting equation 17.23 into equation 17.21 and integrating, the drying time during the falling rate period (t_F) can be estimated:

$$t_F = \frac{m_{drycake}}{a} \ln\left(\frac{X_C}{X_2}\right) \quad (17.24)$$

where X_2 is the final solvent content at the end of drying. Equations 17.22 and 17.24 can be combined to calculate the total drying time (t_T) for the process.

$$t_T = t_C + t_F \quad (17.25)$$

Given the pressure set point, jacket temperature, dryer heat transfer coefficient, enthalpy of vaporization, and critical moisture content, the equations can be used to generate solvent content profiles as a drying operation proceeds. Alternatively, experimental data on the solvent content profile versus time can be used to determine the critical moisture content and/or the value of the constant a through a data fitting approach. Example simulations based on these equations are shown in Figure 17.30, where the DynoChem® software application was used to solve the equations in conjunction with complete material and energy balances. The simulations were carried out under identical operating conditions ($T_j = 60°C$, $T_{initial} = 20°C$, $P = 150$ mbar, $UA = 10$ W/K, $X_1 = 1$ kg/kg, $X_C = 0.15$ kg/kg, $X_2 = 0.0101$ kg/kg, $m_{drycake} = 5$ kg, and $a = 0.1$ min^{-1}) for five typical solvents. The results illustrate the impact of solvent properties on drying performance and hence wash solvent choice during upstream isolation.

Charts (a), (b), and (c) in Figure 17.30 are typical drying curves that can be obtained from the model equations and the transition from the constant rate period to the falling rate period is clearly visible. The deviation from expected behavior seen for some of the simulations at the start and near the critical moisture content is related to the choice of numerical integration parameters in DynoChem®. Comparison of the profiles for each of the solvents shows that more volatile solvents such as acetone and ethyl acetate are removed very rapidly relative to less volatile solvents such as water that can take a prohibitively longer time to dry (in fact the water simulation does not reach the critical moisture content until approximately 84 h). Chart (d) shows that the cake temperature reaches a steady-state value corresponding to the boiling point during the constant rate period. At the transition point, the temperature starts to increase toward the jacket set point (in this case 60°C). When the boiling point of a volatile solvent is below the initial system temperature, the cake will be cooler than the initial temperature for the duration of the constant rate period. According to Antoine's equation, this behavior should be expected when the system pressure is set low enough that the boiling point of the solvent is below the initial conditions. In this case, the saturation temperature of acetone is just below 10°C at a pressure of 150 mbar, so the cake is seen to cool down from an initial temperature of 20°C until the end of the constant rate period.

An important point to note is that the model does not take into account the effect of agitation that would lead to the redistribution of "wet" and "dry" portions of the cake effectively changing the area available for heat transfer (above the critical moisture content). An approach to describe the effect of agitation on heat transfer during drying has been proposed in the literature [15]. Random particle motion during agitation is represented by an empirical mixing parameter known as the mixing number. This parameter essentially represents how many revolutions of the agitator are necessary to achieve full turnover or redistribution of the wet and dry zones in the cake so that they can no longer be distinguished as separate zones on either side of a drying front moving through the cake. The larger the value of the mixing number, the less efficient the mixing provided by the agitator. The drying rate was shown to be sensitive to the mixing number in the case of finely grained particles and the reader is referred to the original reference for a more detailed description of how to use this approach for agitated systems [15].

EXAMPLE 17.4

A total of 10 kg of wet cake that is initially 50 wt% wet with isopropanol is to be dried at 100 mbar pressure and 60°C jacket temperature in a dryer that has 0.125 m^2 surface area available for heat transfer and a heat transfer coefficient of 40 W/(m^2 K). The enthalpy of vaporization of isopropanol is 662.789 kJ/kg and the Antoine coefficients are $A = 18.6929$, $B = 3640.2$, and $C = 219.61$, where T is in units of °C and P^{sat} is in units of mmHg. If the critical moisture content is expected to be 0.1 kg solvent/kg dry cake and the target solvent content is 0.02 kg solvent/kg dry cake, provide

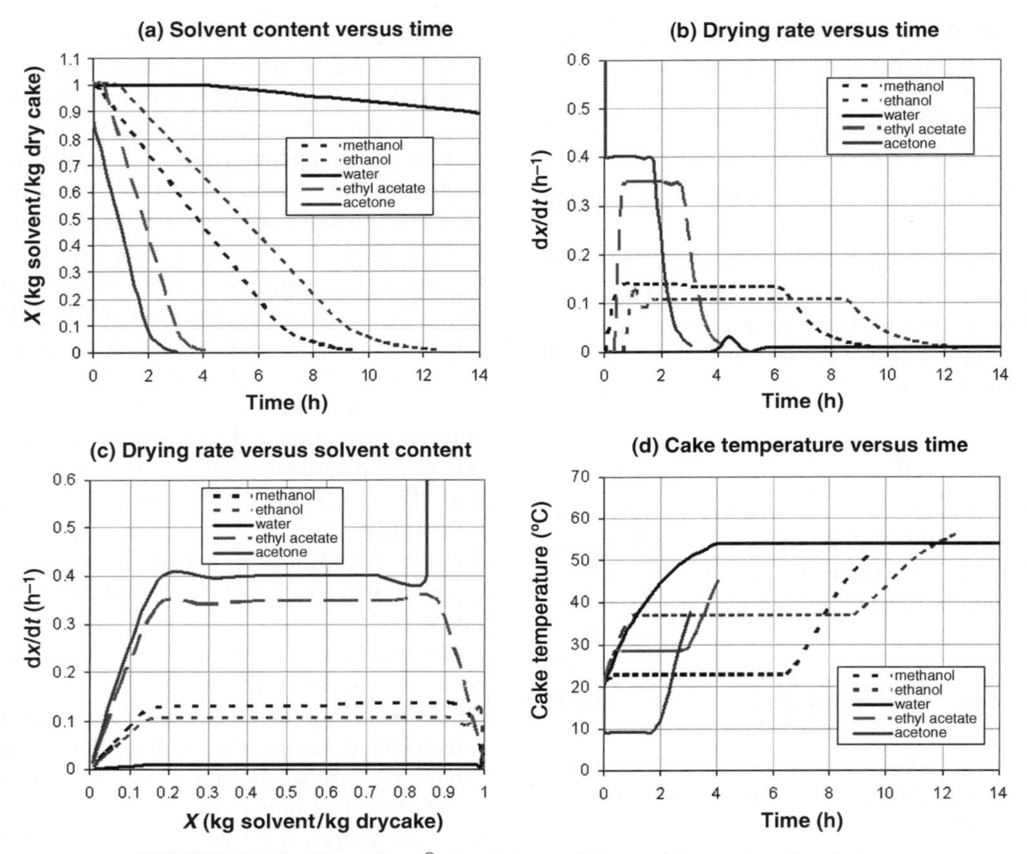

FIGURE 17.30 DynoChem® simulations of the model equations for drying.

a first-order estimate for the total time it will take to complete the drying operation.

Solution

The portion of the total time spent in the constant rate period is given by equation 17.22, the portion of the total time spent in the falling rate period is given by equation 17.24, and the total drying time is the sum of these as given by equation 17.25. Hence,

$$t_T = t_C + t_F = \frac{m_{\text{drycake}}\Delta H_{\text{vap}}}{A_{ht}} \frac{(X_1 - X_C)}{U(T_j - T)} + \frac{m_{\text{drycake}}}{a} \ln\left(\frac{X_C}{X_2}\right).$$

At the critical moisture content, X_C the final mass transfer rate of solvent from the constant rate period must be equal to the initial mass transfer rate of solvent from the falling rate period. Hence, from equations 17.19, 17.20, and 17.23, we can write

$$\frac{UA_{ht}(T_j - T)}{\Delta H_{\text{vap}}} = aX_C$$

This gives us an expression for the proportionality constant a:

$$a = \frac{UA_{ht}(T_j - T)}{X_C \Delta H_{\text{vap}}}$$

Substituting into the expression for total drying time and simplifying

$$t_T = \frac{m_{\text{drycake}}\Delta H_{\text{vap}}}{UA_{ht}(T_j - T)} \left(X_1 - X_C + X_C \ln\left(\frac{X_C}{X_2}\right)\right)$$

Therefore, the total time for the drying operation can be estimated as follows:

$$m_{\text{solvent}} = 0.5 \times m_{\text{wetcake}} = 0.5 \times 10\,\text{kg} = 5\,\text{kg}$$

Hence, $m_{\text{drycake}} = 5\,\text{kg}$

$$X_1 = m_{\text{solvent}}/m_{\text{drycake}} = 1\,\text{kg solvent/kg dry cake}$$

T can be calculated from the system pressure and a rearrangement of Antoine's equation:

$$T = \frac{B}{A - \ln P^{\text{sat}}} - C$$

For isopropanol, $A = 18.6929$, $B = 3640.2$, and $C = 219.61$, where T is in units of °C and P^{sat} is in units of mmHg.

Hence, using $P^{\text{sat}} = 100\,\text{mbar} \times (760\,\text{mmHg}/1013.25\,\text{mbar}) = 75.0\,\text{mmHg}$ in the equation, $T = 33.6°\text{C}$.

Now, substituting this into the expression, we obtain the following total drying time:

$$t_T = \frac{(5)(663 \times 1000)}{(40)(0.125)(60-34)}\left(1-0.1 + (0.1)\ln\left(\frac{0.1}{0.02}\right)\right) = 27{,}100\ \text{s}$$

So the estimated total drying time for this operation is 7–8 h. Note that ΔH_{vap} and T were approximated for the calculation since the objective was simply a first-order estimate of the drying time.

*17.3.3.3 **Form Control*** Hydrates, solvates, or other forms may require careful control of humidity or other drying parameters. In the case where the product is a hydrate and the hydration level needs to be maintained during drying (i.e., to avoid the loss or gain of water from the crystals), the relative humidity in the dryer must be controlled carefully. The composition of wash solvents prior to drying can be critical to ensure that there is sufficient water in the cake during drying to avoid dehydration. The use of a dew point hygrometer in the vent line of a dryer can provide information on the relative humidity around the cake and can be used to monitor the removal of water during drying and ensure that the cake is not overdried [12]. In some cases, it may be helpful to charge additional water prior to or during drying to ensure that there is sufficient water present. The development of phase diagrams based on water activity can be used to design the appropriate conditions for maintaining a specific hydrate [16].

The same principles can be applied to maintain a certain solvate where a solvent other than water is bound within the crystal lattice of the product. In these cases, MS or NIR can be used to monitor the removal of critical solvents during drying.

Sometimes, the product crystals can undergo polymorph transformation during drying. In this case, offline techniques such as XRPD can be helpful to determine this after drying is complete. Online methods such as Raman spectroscopy can be used to monitor the cake while the cake is still drying to ensure that the product form is not changing. Polymorph transformation can be a result of shear and agitation, so this must be carefully controlled.

EXAMPLE 17.5

During a drying operation in which both isopropanol (IPA) and water are removed, it is desirable to preserve a monohydrate that is formed with the product. The product has also been shown to form an undesirable solvate with IPA, which must be avoided. Use the data given in Figure 17.31 to suggest a strategy for maintaining the desired monohydrate while avoiding the undesired IPA solvate. Given that the composition of the azeotrope formed between IPA and water is 16 wt% water at 1 mmHg, would the strategy be successful at any operating pressure?

Solution
From the solubility plot, it is clear that the solubility of the IPA solvate and monohydrate intersect at approximately 8 wt % water. At any given condition, the polymorph with the lower solubility will be the thermodynamically stable crystal form, and hence the monohydrate will be favored over the IPA solvate if the water content is maintained above 8 wt%.

In order to form and preserve the monohydrate, it is necessary to maintain a solvent composition in the drying cake of more than 8 wt% water. From the *x-y* diagram, there is a minimum boiling azeotrope between IPA and water at 12 wt% water. If the final cake wash (i.e., the starting point for the drying operation) contains less than 12 wt% water, the liquid phase will shift to lower water concentrations as evaporation takes place, and it will be possible to reach

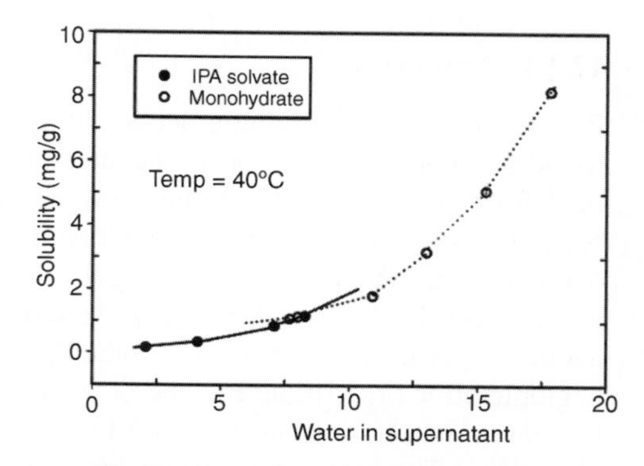

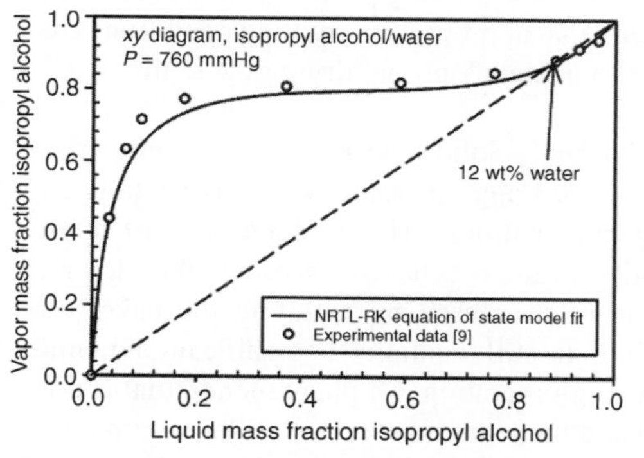

FIGURE 17.31 Solubility plot (on the left) showing the relative solubility of the IPA solvate and the monohydrate at different water contents (although the data was collected at 40°C, the results are applicable at all temperatures—see monotropism), and an *x–y* diagram (on the right) for IPA/water showing the existence of a minimum boiling azeotrope at 12 wt% water. Reprinted with permission from Ref. 12. Copyright 2004, American Chemical Society.

a point during drying where the solvent composition in the cake is less than 8 wt% water. In this case, the monohydrate will not be stable and the IPA solvate could be formed instead. If the final cake wash composition is greater than 12 wt% water, the solvent composition in the dryer will never be below the solvent composition transition point due to the presence of the azeotrope and the monohydrate will remain the more stable form over the drying process. A suggested strategy is therefore to use a final wash composition of say 20 wt% water to ensure that the solvent composition always remains above 8 wt% water during the drying operation.

Using any final wash composition containing more than 16 wt% water should result in a successful strategy for obtaining the monohydrate, down to an operating pressure of 1 mmHg.

17.3.3.4 Troubleshooting There are some common challenges faced during drying that we discuss below.

Long Drying Times Long drying times can cause significant bottlenecks in the overall process and is most often a result of a prolonged period under the mass transfer limited mechanism. Drying times can be improved by ensuring that most of the drying operation is carried out under heat transfer limited conditions and that both the driving force and the heat transfer area available are maximized as much as possible (within the constraints of product stability and equipment limitations). From equation 17.19, it can be seen that the difference between the jacket temperature and the product temperature provides the driving force for heat transfer and the heat transfer coefficient is equally important. Static dryers such as a vacuum oven often do not provide enough surface area for heat transfer, whereas drying in agitated dryers can provide the benefits of increased surface area for heat transfer. Note that certain solvents such as water can be difficult to remove regardless of the equipment used, leading to longer drying times. Estimation of the critical moisture content in the lab and simulation of the drying process with a model can be helpful to ensure that appropriate operating conditions are used in the plant to achieve optimal drying cycle times.

Agglomeration/Balling Solid particles can sometimes aggregate to form hard agglomerates with solvent trapped inside (also known as balling). This problem is most often seen in agitated dryers and is generally worse in filter dryers. Agglomeration is often caused by agitating the cake too aggressively while it still contains a significant amount of solvent. Since agglomeration is a phenomenon that is not well understood and there are no specific guidelines on how to avoid the issue during drying, a general approach is to avoid agitation in the initial stages of the drying. One factor that may be important for avoiding the formation of agglomerates is the solubility of the product in the individual solvents that are being removed. Sometimes the cake has been washed with a mixture of solvents prior to drying and if there is enough solubility in one of these solvents by itself, the product could become partially dissolved that could lead to fusion of the crystals. One approach to avoiding agglomeration is therefore to adjust the solvent composition of the final cake wash to avoid the potential enrichment of highly soluble solvents during the drying operation. Agglomeration has been found to occur when the drying rate is high and the shear rate is relatively low [17, 18], indicating that there is a higher risk for agglomeration if agitation is applied during the initial stages of drying when the process is heat transfer controlled as opposed to mass transfer controlled.

Particle Attrition If the product crystals are sensitive to shear, agitation of the cake can lead to particle breakage and attrition. In these cases, maintaining a specified particle size during drying can then become a challenging aspect of the drying operation. Equipment selection and careful consideration of how the agitation will be operated are critical for ensuring particle size integrity. Based on the geometry of the equipment and how the agitator interacts with the cake, the force exerted on the particles should be considered. Particles are likely to experience less force in a tumble dryer as opposed to a filter dryer or conical dryer. When the use of a tumble dryer is not an option, lab studies to determine the maximum shear that the particles can endure before breakage becomes a problem can be carried out. The plant agitation protocol could then be scaled up based on maintaining a lower tip speed than the critical value. However, it is important to note that the ranges of tip speeds achievable in the lab often do not overlap with those achievable in the plant. Particle attrition has been shown to occur when the drying rate is low and the shear rate is high [17, 18], indicating that there is a higher risk for particle attrition if continuous agitation is applied during the later stages of drying when the drying rate is limited by mass transfer.

17.3.4 Summary

Proper control of the drying process is important especially for API compounds whose solvent content, particle size, and final form need to meet specifications. Depending on the solvent system and the equipment being used, a drying operation can often have a long cycle time and become a bottleneck in the overall process. Agitation is helpful to avoid this issue but can lead to other problems such as agglomeration or attrition. It is therefore important to select the appropriate equipment and carefully design the operating conditions. The scale-up of drying processes can be challenging due to the lack of understanding with some of the issues encountered; however, there are many tools and methods available for simulation, monitoring, and data collection that can be applied for design purposes.

REFERENCES

1. Darcy H. *Les Fontaines Publiques de la Ville de Dijon* (The Public Fountains of the Town of Dijon), Dalmont, Paris, 1856.
2. Ruth, BF, Montillo, GH, Montonna, RE. Studies in filtration: I. Critical analysis of filtration theory; II. Fundamentals of constant pressure filtration. *Ind. Eng. Chem.* 1933;25:76–82,153–161.
3. McCabe WL, Smith JC, Harriott P. *Unit Operations of Chemical Engineering*. 7th edition, McGraw-Hill, 2005.
4. Bird RB, Stewart WE, Lightfoot EN. *Transport Phenomena*, 2nd edition, Wiley, 2001.
5. Landman, KA, White, LR, Eberl, M. Pressure filtration of flocculated suspensions. *AIChE J.* 1995;41:1687–1700.
6. Barr, JD, White, LR. Centrifugal drum filtration: I. A compression rheology model of cake formation. *AIChE J.* 2006; 52: 545–556.
7. Weigert, T, Ripperger, S. Effect of filter fabric blinding on cake filtration. *Filtr. Sep.* 1997;507–510.
8. Rideal, G. Roundup. *Filtr. Sep.* 2006, 36–37.
9. Perry RH. *Perry's Chemical Engineer's Handbook*, 6th edition, McGraw-Hill, 1984.
10. McConville FX. The Pilot Plant Real Book: A Unique Handbook for the Chemical Process Industry. *FXM Engineering and Design* 2004.
11. Burgbacher, J, Wiss, J. Industrial applications of online monitoring of drying processes of drug substances using NIR. *Org. Process Res. Dev.* 2008;12:235–242.
12. Cypes, SH, Wenslow, RM, Thomas, SM, Chen, AM, Dorwart, JG, Corte, JR, Kaba, M. Drying an organic monohydrate: crystal form instabilities and a factory-scale drying scheme to ensure monohydrate preservation. *Org. Process Res. Dev.* 2004;8:576–582.
13. Kohout, M, Collier, AP, Štěpánek, F. Mathematical modeling of solvent drying from a static particle bed. *Chem. Eng. Sci.* 2006;61:3674–3685.
14. Michaud, A, Peczalski, R, Andrieu, J. Modeling of vacuum contact drying of crystalline powder packed beds. *Chem. Eng. Process.* 2008;47(4):722–730.
15. Schlünder, E-U, Mollekopf, N. Vacuum contact drying of free flowing mechanically agitated particulate material. *Chem. Eng. Process.* 1984;18:93–111.
16. Variankaval, N, Lee, C, Xu, J, Calabria, R, Tsou, N, Ball, R. Water activity-mediated control of crystalline phases of an active pharmaceutical ingredient. *Org. Process Res. Dev.* 2007;11:229–236.
17. Lekhal, A, Girard, KP, Brown, MA, Kiang, S, Glasser, BJ, Khinast, JG. Impact of agitated drying on crystal morphology: KCl-water system. *Powder Technol.* 2003;132: 119–130.
18. Lekhal, A, Girard, KP, Brown, MA, Kiang, S, Khinast, JG, Glasser, BJ. The effect of agitated drying on the morphology of L-threonine (needle-like) crystals. *Int. J. Pharm.* 2004;270: 263–277.

18

THE DESIGN AND ECONOMICS OF LARGE-SCALE CHROMATOGRAPHIC SEPARATIONS

FIROZ D. ANTIA
Product Development, Palatin Technologies, Inc., Cranbury, NJ, USA

18.1 INTRODUCTION

Chromatography ("color writing") was the fanciful name given in 1906 by the botanist Mikhail Tswett to the adsorptive separation method he employed to separate plant pigments on columns of calcium carbonate using petroleum ether as the eluting solvent [1]. Chromatography is a differential migration process; components of a mixture that distribute differently between a particulate[1] adsorbent and a fluid separate because they move at different rates through the fluid-perfused[2] adsorbent bed. If conditions are chosen correctly, the individual species in the mixture emerge at the outlet of the chromatographic column in pure bands.

The broad array of sorbent and fluid combinations available today make chromatography a versatile and widely applied analytical and preparative purification tool at the laboratory, pilot, or industrial scale for virtually any class of pharmaceutical compound, particularly when gentle, but nonetheless highly selective, separation conditions are required. Chromatography is an enabling technology in biotechnology—practically all industrial biopharmaceutical purification processes contain one or more chromatography steps. While employed less frequently in small-molecule drug (API) manufacturing processes, it is nevertheless heavily used in early stages of API development, and does find important industrial applications in natural product and chiral separations. In addition, it is indispensable for the large-scale purification of synthetic peptides. In this chapter, the focus is on chromatography as it is practiced in the purification of small-molecule drugs and peptides.

Chromatography is considered by many to be an expensive step; a useful tool to obtain from milligrams to a few hundred grams of intermediates or drug substances for deliveries early in development, but ultimately a step that must be superseded in favor of more cost-effective methods (extractions and crystallizations, or improved synthesis routes) before commercialization of the process. However, done properly, chromatography is economically competitive; indeed with modern equipment, method optimization, and solvent-sparing technologies—including solvent recycling and continuous multicolumn chromatography techniques—chromatography is both effective and efficient for industrial scale purifications.

This chapter will begin with an outline of the key design elements in chromatography and a discussion on fundamental chromatographic relationships such as retention and selectivity. The various available chromatographic chemistries will then be discussed. This will be followed by brief sections on chromatographic operating parameters, the choice of the mode of operation including multicolumn systems, choice of equipment, scale up, a short section on parametric design space, and a discussion on chromatographic economics.

18.2 KEY DESIGN ELEMENTS

Choice of the combination of the adsorbent (also known as chromatographic media or the stationary phase) and the fluid

[1] Monolithic adsorbents are becoming increasingly popular in the literature, but casting monoliths at the industrial scale remains a challenge.
[2] The fluid may be a gas, a supercritical fluid, or a liquid; for this chapter, the discussion will be confined to liquid chromatography.

Chemical Engineering in the Pharmaceutical Industry: R&D to Manufacturing Edited by David J. am Ende

(or mobile phase) is the first design element in chromatography. This choice sets the chromatographic "chemistry," dictating the type of physicochemical interactions that take place within the system as well as the means to manipulate their intensity to achieve the desired separation goals. Many types of adsorbents with a wide array of surface chemistries are available today, and several are prepared with large-scale applications in mind, that is, they are manufactured under controlled conditions in large batch sizes. The mobile phase, an aqueous or solvent-based mixture that could also contain other components such as acids, bases, buffers, or salts, is selected to be compatible with the adsorbent, and acts to mediate adsorption and release of the separating species from the stationary phase. Manipulation of the mobile phase composition is the primary means to control retentivity—a measure of the strength of adsorption of the separating species on the adsorbent—and obtain selectivity—a measure of the difference in retentivities of separating species, which is the key to effecting the desired separation. Solubility of the separating species in the mobile phase is also an important consideration; high solubility is desirable to achieve high productivity. A classification of the various adsorbents and compatible mobile phases on the basis of the underlying physicochemical factors governing retention is provided in Section 18.3.

Another important design element is the choice of operating mode. Most chromatographic separations are carried out in the elution mode, where the separating species move through the system in the presence of all components of the mobile phase. However, there are other means to carry out chromatography. In the displacement mode, one or more mobile phase components (introduced after feeding the separating species) binds tightly to the adsorbent, swamping available binding sites so the separating species move ahead of it. Continuous chromatographic techniques are operating modes in their own right; simulated moving bed (SMB) chromatography and the multicolumn solvent gradient process (MCSGP) have become more popular in recent years as applications for large-scale chromatographic processes increase. All of these are discussed in more detail in Section 18.5.

Once the choice of media and operating mode is determined, the chromatographic process is defined by its operating parameters such as the column dimensions, the adsorbent particle size, mobile phase flow rate, the operating temperature, and other factors pertinent to the operating mode (e.g., cycle and column switching times in simulated moving bed systems). Variation of these operating parameters would form the basis for process and economic optimization as well as the regulatory design space for a chromatographic process. Operating parameters and design space are also discussed later in this chapter.

18.3 FUNDAMENTAL CHROMATOGRAPHIC RELATIONSHIPS

18.3.1 Chromatographic Velocity

Chromatographic operations take place in a column packed with particles. The volume fraction of the interstitial space between the particles in a randomly packed bed, or the interstitial porosity, ε_e, typically has a value of about 0.4. The totally porous sorbent particles commonly used in chromatographic applications often have an internal void fraction, ε_i, of approximately 50%, so that the packed bed has a total porosity, ε_T $(=\varepsilon_e + (1-\varepsilon_e)\varepsilon_i)$, of roughly 0.7. For a fluid flow rate F in a bed of cross-section area A, three flow velocities[3] may be defined and are related as follows

$$\frac{F}{A} = u_s = \varepsilon_e u_e = \varepsilon_T u_0 \tag{18.1}$$

Here, u_s is the superficial velocity, u_e is the interstitial velocity, and u_0, the average velocity of an unretained molecule that explores the entire void space, is known as the chromatographic velocity. For modeling purposes, it is assumed that the mobile phase migrates through the system at the chromatographic velocity.[4]

18.3.2 Operating Pressure

Chromatographic operations are mostly carried out at a fixed velocity or flow rate. The pressure drop, ΔP, across a packed column is related to the chromatographic velocity, u_0, the column length, L, the adsorbent particle diameter, d_p, and the mobile phase viscosity, η, via Darcy's law as

$$\Delta P = \frac{L u_0 \eta \varphi}{d_p^2} \tag{18.2}$$

Where the factor φ is a proportionality factor related to porosity given by the Cozeny–Karman equation

$$\varphi = 180 \frac{(1-\varepsilon_e)^2}{\varepsilon_e^2} \frac{\varepsilon_T}{\varepsilon_e} \tag{18.3}$$

For the typical values of the porosities given above, this factor has a numerical value of approximately 710, while in practice it may vary from 600 to 1000 depending on the exact porosities, as well as the regularity and roughness of the particles. In equation 18.3, the variables can be expressed in any consistent units; the equation is also consistent without the need to add any correcting factors if the pressure drop is

[3] In this analysis, we assume one-dimensional axial flow in a cylindrical column; radial flow systems exist but will not be considered here.

[4] Strictly speaking, in the surface layer near the sorbent, the composition of a multicomponent mobile phase is often different from that in the bulk and changes in mobile phase composition can lead to difference in migration rates of the solvent components themselves, but this subtlety can often be ignored in practice.

expressed in bars, the velocity in centimeters per second, the column length in centimeters, the particle size in micrometers, and the viscosity in centipoise.

18.3.3 Mass Balance Equation

A differential one-dimensional mass balance for a retained species in a chromatographic system can be written in simplified form as

$$\frac{\partial c}{\partial t} + \phi \frac{\partial q}{\partial t} + u_0 \frac{\partial c}{\partial x} = D_e \frac{\partial^2 c}{\partial x^2} \qquad (18.4)$$

Here, the parameters t and x are time and distance along the column, respectively, c is the concentration of the species in the mobile phase, q is its concentration in the stationary phase, and ϕ is the ratio of stationary to mobile phase volumes (i.e., $(1 - \varepsilon_T)/\varepsilon_T$). In this simplified view, all dispersive effects, including molecular diffusion, transport of the species to and within the particle and any dispersive influence of slow adsorption kinetics are lumped into the effective dispersion coefficient, D_e.

18.3.4 Retention and Retention Factor

A system of mass balance equations (one for each migrating species), with suitable initial and boundary conditions, serves as an adequate model for chromatography [2]. The adsorbed concentration q is related to the mobile phase concentration via an equilibrium relationship known as the adsorption isotherm. In general (and very often in the practical case of preparative chromatography that is carried out at high concentrations of the migrating species), the adsorption isotherm of each species is dependent not only on its own concentration in the mobile phase but also on that of all the other species present. Consequently, migration through the system is concentration dependent; the system of equations is nonlinear, and must be solved numerically. On the other hand at low concentrations, it can be assumed that the distribution coefficient ($K = q/c$) for each species is an independent constant. In this circumstance, in an ideal system without dispersion, the mass balance equation reduces to

$$\frac{\partial c}{\partial t} + \frac{u_0}{(1 + \phi K)} \frac{\partial c}{\partial x} = 0 \qquad (18.5)$$

This is a one-dimensional wave equation with propagation velocity—or the migration velocity of a species through the system—of $u = u_0/(1 + \phi K)$. The product ϕK is known as the retention factor, k'. Despite being defined only at low concentrations in limited circumstances, k' is a key factor in the understanding and characterization of chromatographic behavior.

Conditions where the distribution coefficient K of each migrating species is independent of concentration (the Henry's law region) are known as "linear" (or "analytical" conditions, as these are conditions under which chromatographic analyses are carried out). A system in which the mobile phase composition is kept constant over time is termed "isocratic." In a linear isocratic system, if the mixture to be separated is injected to approximate a δ-function, the retention factor k' of each separating species can be determined from its retention time, t_R, (the time of elution of the center of gravity of the migrating component) and the dwell time, t_0 (the time of elution of an unretained component), as follows [3]

$$k' = \phi K = \frac{t_R - t_0}{t_0} \left(= \frac{V_R - V_0}{V_0} \right) \qquad (18.6)$$

(V_R and V_0 are the corresponding elution and dwell volumes; $V_R = Ft_R$, where F is the mobile phase flow rate, and $V_0 = Ft_0 = \varepsilon_T V_C$, where V_C is the total column volume.)

The distribution coefficient K is a thermodynamic property, related to the free energy of adsorption, and as a result k' is a function of temperature via the relationship[3]

$$\ln k' = -\frac{\Delta H}{RT} + \frac{\Delta S}{R} + \ln \phi \qquad (18.7)$$

Here, ΔH and ΔS are the enthalpy and entropy of adsorption, T is the temperature in degrees Kelvin, and R is the universal gas constant. Values of k' can increase or decrease with increasing temperature depending on the sign of the enthalpy of adsorption; in most instances the latter is true as adsorption is often enthalpically favored (i.e., ΔH is negative), but there are important exceptions, such as in hydrophobic interaction chromatography of proteins, and in other individual cases, where the opposite can hold.

18.3.5 Selectivity

The power of a chromatographic system to discriminate between two species is quantified by the ratio of their retention factors, termed the "selectivity," $\alpha (= k'_1/k'_2$, where the subscripts refer to the two species and $k'_1 > k'_2$, so that $\alpha > 1)$. The type of system chemistry, the mobile phase composition, and the temperature are major factors that influence selectivity. In industrial applications, concentrations are usually high, so that Henry's law no longer applies and distribution coefficients as well as selectivities can be nonlinear functions of the concentration of all the locally present separating species. Nevertheless, the first step in designing a chromatographic separation is to operate in the Henry's law regime to find conditions (i.e., the right stationary and mobile phase composition) that maximize selectivity. To achieve this, one must understand how retention and selectivity can be manipulated in the context of the various chromatographic chemistries. A discussion on some of the more popular chemistries is given in Section 18.4.

18.3.6 Efficiency

During chromatography, dispersive effects counteract the effectiveness of the separation; chromatographic efficiency thus increases when dispersion is decreased. Contributions to dispersion arise from flow anastomosis, molecular dispersion, transport in and out of the particle, and slow adsorption kinetics. Efficiency is characterized by the so-called plate number, N, which arises from a model that treats the chromatography column as a series of stirred cells containing equal amounts of stationary phase. Based on this model applied in the Henry's law adsorption regime, the plate number is related to the width of a chromatographic peak[5] by the relationship[3]

$$N = 5.54\left(\frac{t_R}{w_{1/2}}\right)^2 \tag{18.8}$$

The plate number is related to the lumped dispersion coefficient in equation 18.4 as follows [4]

$$N = \frac{u_0 L}{2D_e} \tag{18.9}$$

The larger the number of plates, the more closely the system approximates plug flow, and the more efficient it is.

The height equivalent of a theoretical plate, $H(=L/N)$, is a related measure of the efficiency. The plate height is a function of the particle size and flow velocity, as elucidated first by van Deemter and his colleagues [5]

$$H = Ad_p + \frac{BD_m}{u_0} + \frac{Cd_p^2}{D_m}u_0 \tag{18.10}$$

Here, A, B, and C are constants, with typical numerical values of about 1.5, 0.8, and 0.3, respectively [6] and D_m is the molecular diffusivity of the separating species in the mobile phase. The first (or A) term of this equation is independent of flow rate and arises from flow nonuniformity within the packed bed. The second (or B) term is a result of molecular dispersion and rapidly becomes insignificant as flow velocity increases. The third (or C) term is a consequence of diffusion of the separating molecule in and out of the stagnant fluid within the particle pore space. It increases linearly with flow rate and is usually the most significant factor related to band spreading. Contributing factors to the plate height, such as diffusion through the boundary layer around the particle, and the effect of slow adsorption kinetics, are usually of less importance and have thus been ignored in the simplified equation presented above.

[5] The peak is theoretically the output of a δ-function input to the system; in practice a small analytical size injection is employed. The peak approximates a Gaussian distribution. The width at half height, $w1/2$, is more convenient to use than the width at the baseline.

The van Deemter equation can be written in dimensionless form by introducing the reduced plate height, $h(=H/d_p)$, and the reduced velocity, $\nu(=u_0 d_p/D_m)$ so that

$$h = A + \frac{B}{\nu} + C\nu \tag{18.11}$$

Figure 18.1 shows the van Deemter curve and indicates the approximate practical range for the reduced plate height for small molecules (of 15–80, assuming diffusivities in the $2–5 \times 10^{-6}\ cm^2/s$ range, flow velocities from 0.1 to 0.3 cm/s, and a 10 μm particle diameter). With similar flow velocities and particle size, peptides have a higher reduced velocity range because of their lower molecular diffusivities. This shows that in practice, the B term reduces to zero and that band spreading is dominated by diffusion through the sorbent particle.

Since in practical cases diffusion through the particle is the major contributing factor to the plate height (and thus to N), another useful scaling factor related to efficiency, termed the Lightfoot number, Li, can be constructed by taking the ratio of the characteristic diffusion time across a particle ($t_{diff} = d_p^2/D_m$) and the dwell time in the system ($t_0 = L/u_0$) so that

$$\mathrm{Li} \equiv \frac{t_0}{t_{diff}} = \frac{LD_m}{u_0 d_p^2} \tag{18.12}$$

18.3.7 Operation at High Concentration

As mentioned earlier, at high concentration Henry's law no longer applies, and the adsorption isotherm of each species is generally a nonlinear function of the concentration of all the species present. Migration of the components under these conditions is concentration dependent.

While many forms of the adsorption isotherm are possible depending on the molecular properties of the mixture and the sorbent, in many cases the sorbent displays a finite maximum capacity for the adsorbing species. In this situation, compounds vie for a limited number of sites on the sorbent and the adsorption is competitive. At elevated concentrations then, relatively fewer sites are available and the slope of the adsorption isotherm decreases, implying that regions of higher concentration migrate faster. In this circumstance, the low concentration at the leading edge of a concentration front tends to move more slowly than the high concentration of its trailing edge; this has a self-sharpening effect, countering the dilutive effect of dispersion and leading to a sharp shock wave. The opposite is true for the rear, producing a long dilutive tail. Peaks that at low concentration appear symmetrical become triangle shaped at higher concentrations. This effect is illustrated in Figure 18.2. In multicomponent systems, which are of more interest in practice,

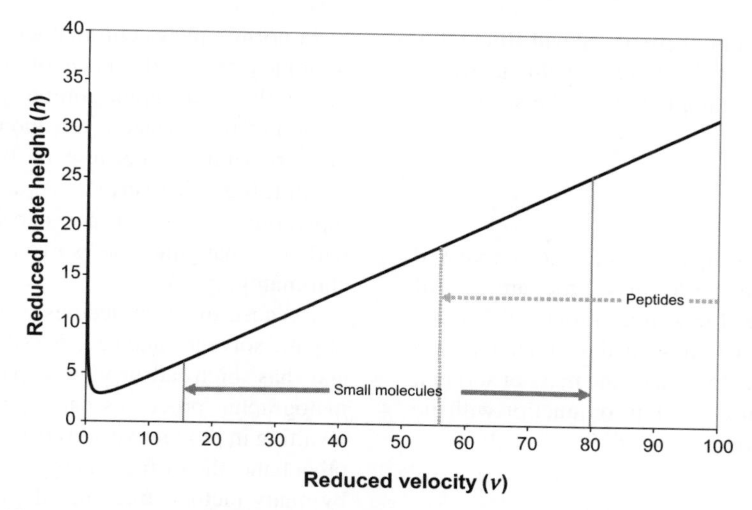

FIGURE 18.1 van Deemter plot of reduced plate height versus reduced velocity showing typical operating regimes for small molecules and peptides as described in the text.

strongly bound compounds displace and thus suppress the binding of more weakly adsorbed species. Under the right conditions in a chromatographic system, this results in more weakly bound components being pushed ahead of more strongly bound ones; some concentration can occur as a result. This so-called "displacement effect" is beneficial if an early eluting component is the target of the purification. Unfortunately, competitive adsorption also leads to a perverse "tag-along" effect; weakly bound components act to suppress somewhat the binding of strongly bound components, accelerating their motion and dragging them ahead, reducing the extent of separation. The effect is shown for two components eluting individually and together in Figure 18.3.

It is worth noting that not all systems display competitive binding, and the displacement and tag-along effects are by no means universal. Some systems display cooperative binding; the chromatographic effects in such circumstances are the inverse of those discussed above. Some systems show a combination of these types of binding, and the resulting peak shapes in chromatography can be quite complex.

One practical consequence of operating at high concentration is that the neat symmetric separated peaks one is used to seeing in analytical separations disappear. Nonspecific detection at the end of the column (usually by UV light absorption), often shows an undifferentiated blob. Once properly characterized, such detection can indeed be used to govern collection of pure fractions, but at least in early

FIGURE 18.2 Schematic showing overlaid elution profiles resulting from injections of increasing concentration of a single species when adsorption is competitive. The elution profiles appear as a series of nested near-triangular shapes, with sharp fronts and extended tails. The end of each tail coincides with the retention time of the species at infinite dilution.

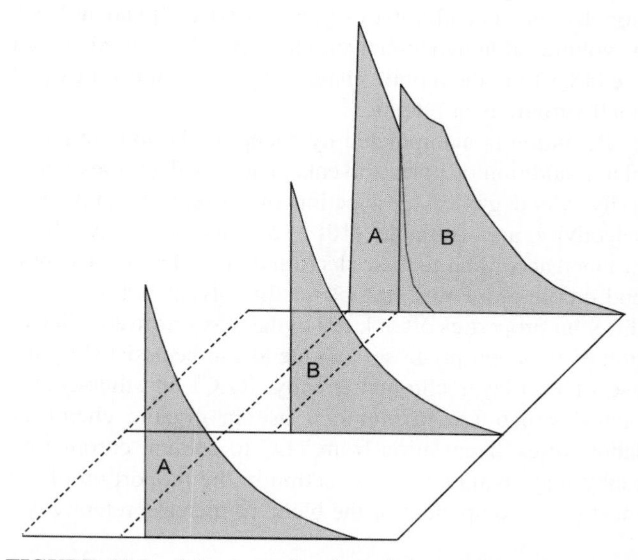

FIGURE 18.3 Schematic showing elution profiles of separate injections of two species A and B, and of a mixture of A and B. In the mixed injection, the peak shape of each species is altered by the presence of the other: A is compressed and concentrated by the displacement effect, and the front end of B elutes a little earlier than in the single component case because of the tag-along effect.

stages of process development there is no substitute for collecting multiple fractions of the emerging eluent stream and analyzing these to understand how well the separation has progressed.

18.3.8 Load

The load on a column is the quantity of feed, expressed conveniently either as total solids or as the amount of the desired product in the feed mixture, introduced into the system per injection. A dimensionless load parameter, Γ, can be defined as the column load divided by the mass of sorbent, and this can be used as a scaling factor in conjunction with the plate number, as discussed later.

18.4 CHROMATOGRAPHIC ADSORBENT CHEMISTRIES AND BASIS OF RETENTION

18.4.1 Normal Phase Chromatography

"Normal phase" chromatography implies use of a polar stationary phase, usually unmodified porous macroreticular silica, with an organic solvent blend as the mobile phase. Mobile phase solvents covering the spectrum from the very nonpolar (heptane), to the very polar (methanol) are used. The silica surface is populated with silanol groups with a distribution of activities that depend on their structure [7], and water can bind strongly at the most polar sites. As a consequence, minor fluctuations in the water content of nominally dry solvents can have a profound effect on retention. Deliberate addition of some water (usually about 0.5% by volume, although higher amounts up to 7% have also been used [8, 9],) to the mobile phase can provide uniformity and quell variations in retention.

Retention is manipulated by changing the mobile phase blend; addition of polar solvents generally decreases retentivity. Good guides to selection of solvents to maximize selectivity are available [10]. Solvents used have been grouped according to their electron donor, electron acceptor and dipole properties, and blends of solvents with broadly different properties often leads to the best selectivity. Selection of the appropriate solvent blend can be assisted by the use of thin-layer chromatography (TLC) on silica-coated plates, which are in common use in organic chemistry laboratories. Translation from TLC to column chromatography may require some reduction in the proportion of the most polar component of the blend to increase retentivity.[6]

[6] In TLC, practitioners like to elute components so that they migrate to about half the length of the plate, so that the "retardation factor" R_F (distance migrated/plate length) is ~0.5. Retention factor and retardation factors are related ($k' = 1/R_F - 1$), so the corresponding k' is ~1. For column chromatography, $k' > 3$ is often desirable, hence the need to increase retention while translating from TLC to LC.

In normal phase chromatography with bare silica as the stationary phase, the layer of solvent closest to the silica surface has a different composition from the bulk. Equilibration of the surface layer upon changes in the bulk takes time, presumably because of the heterogeneity of active sites on the silica surface. As a result, most normal phase separations on silica are carried out isocratically (i.e., without changing the solvent composition during the chromatography).

Because many molecules of interest are very soluble in organic solvents, and because silica is relatively inexpensive and has high adsorptive capacity, normal phase chromatographic processes can be run at high concentrations, resulting in productive and cost-effective separations. On the other hand, the surface energy of bare silica can be influenced by many factors (trace metals, for instance), and a normal phase system can suffer from variability for a variety of reasons, including variations in silica manufacturing, or fluctuations in the crude feedstock which may contain components that bind practically irreversibly to the silica and can slowly change the character of the surface. Bonded phases with more homogenous surfaces are available that reduce or eliminate these sensitivities, and these can also be used in a gradient mode (i.e., where the composition of the mobile phase is changed during the chromatographic run). Phases with diol and amino functions are available that have niche applications (for instance, for separation of molecules containing polyols) with selectivities that differ considerably from bare silica. Bonding chemistry adds to the cost of the adsorbent but this may be offset by a longer column lifetime under the right conditions.

18.4.2 Reversed-Phase Chromatography

So-called "reversed-phase" chromatography is carried out with nonpolar stationary phases using a mixture of water with a miscible solvent as the mobile phase (the polarity of the system is reversed compared to the "normal" phase described above). Molecules separate from each other on the basis of their hydrophobicity; more hydrophobic species are more highly retained. Porous macroreticular silica-based stationary phases whose surfaces have been modified by bonding an alkyl group, such as octadecyl(C18), octyl (C8), or butyl(C4), are most commonly used, although a wide variety of others, including macroreticular polymer-based adsorbents, are available. Commonly employed water-miscible solvents used in the mobile phase are acetonitrile, methanol, ethanol, isopropanol, and more rarely, tetrahydrofuran. Mobile phases are often buffered or acidified to influence retention of ionizable species. Organic or inorganic acids (e.g., acetic, trifluoroacetic, methanesulfonic, and phosphoric acids) can interact with basic groups on the separating species, forming ion-pairs and altering their hydrophobicity, and simultaneously suppress ionization of residual (unbonded) silanols on the stationary phase surface, reducing ionic interactions. For separation of ionizable species, manipulation of the nature of the ion-pairing agent and the pH can have a profound effect on selectivity.

For a given pH and ion-pairing agent, the retention factor in reversed-phase chromatography decreases with the increasing volume percent of miscible organic solvent in the mobile phase, ψ. In such a system, the relationship between k' and ψ can be adequately described over a broad range of ψ by the expression

$$\ln k' = \ln k'_0 - S\psi \quad (18.13)$$

Here, k'_0 is the retention factor extrapolated to zero organic content. The slope of the plot, S, correlates roughly with molecular weight, large molecular weight species having larger S values, although there is often considerable variation for species with similar molecular weight. The relationship breaks down at high organic content, where retention can sometimes increase with increasing ψ. Figure 18.4 shows the k' versus ψ relationship for several compounds in a hypothetical mixture.

Reversed-phase is the most widely used form of chromatography on an analytical scale and is also used extensively in industrial applications. Practically every peptide manufacturing process, for instance, includes a reversed-phase chromatography step.

18.4.3 Chiral Chromatography

Perhaps the most important development in the field of chromatography over the past two decades has been the introduction of stationary phases with chiral selectivity. Such phases have seen increasing use on an industrial scale for the separation of enantiomers, providing an alternative to stereoselective synthesis or classical resolution techniques. The phases contain a chiral selector (for instance, substituted cellulose or amylose) bonded or coated onto a chromatographic particle (usually porous macroreticular silica). A broad variety of chiral stationary phases (CSPs) are available utilizing different selectors and these can be operated in either the reversed- or normal-phase modes, depending on their design and solvent compatibility. Choice of the best phase for an application is typically a process of screening a broad range of CSPs under analytical conditions for the ones with the best selectivity. Since selective adsorption capacity can be limited on CSPs, it is worth choosing a few different CSPs to see which one works best under the higher-concentration conditions intended for the application.

CSPs are expensive relative to normal- or reversed-phase sorbents, and separation design efforts often focus on minimizing the amount of stationary phase. In chiral separation applications, the intent usually is to separate enantiomers; presence of other impurities is incidental. For this reason, simulated moving bed chromatography and related multicolumn technologies, which are ideally suited to binary separations, have seen increasing use in the pharmaceutical industry.

18.4.4 Other Chemistries

While most applications of large-scale chromatography for separation of small molecules employ the types of phases discussed above, other types of chemistries are also sometimes employed. Notable among these are phases containing weak or strong anionic or cationic ionized groups bonded on macroreticular silica or polymeric particles, commonly

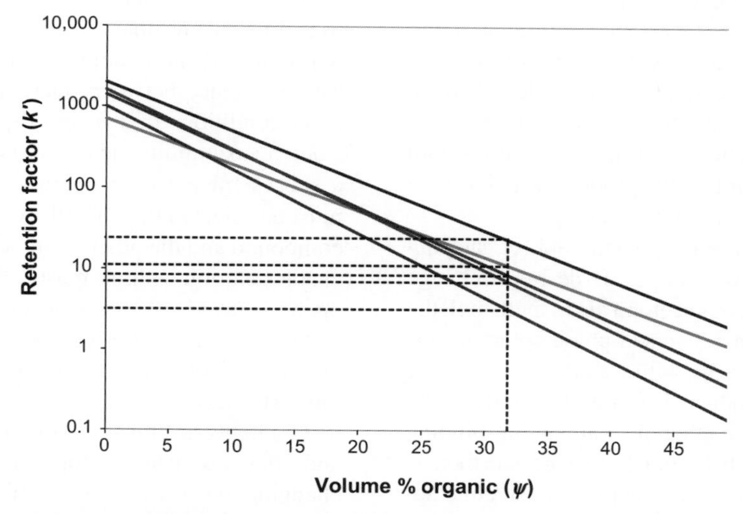

FIGURE 18.4 Plots of retention factor of various components of a mixture versus the organic solvent volume percentage in the mobile phase in a reversed-phase chromatography system, where there is typically a straight line relationship between log k' and ψ.

referred to as ion exchange or electrostatic interaction phases. These are usually operated with aqueous or hydro-organic mobile phases containing salt and buffers for pH control. Salt screens electrostatic interactions, so retention on these phases decreases with increasing mobile phase salt content. pH effects ionization of the separating molecules as well as the ionizable groups on the stationary phase surface and is thus a useful means to manipulate retention.

Hydrophilic interaction chromatography (HILIC) uses polar stationary phases and water-miscible solvents; retention increases with increasing polarity and increases in the water content of the mobile phase result in a decrease in retention. Other types of chemistries, such as those used for hydrophobic interaction chromatography (HIC) where weakly hydrophobic surfaces are employed, or for size exclusion chromatography (SEC), where molecules are not retained but separate on the basis of size, are usually applied only in large molecule separations.

18.5 OPERATIONAL ASPECTS

18.5.1 Chromatographic Applications

Chromatographic systems can be used in several ways in a chemical process. One simple application is for removal of impurities—a process stream is run through a chromatographic column and one or more impurities are adsorbed on the bed. Once the bed is saturated, it is either discarded, or regenerated, reequilibrated, and returned to use. This requires conditions where the impurities are tightly bound and the product is minimally affected by the sorbent. Conditions for this simple operation can often be found if the product and the impurities are not very closely related; separating neutral impurities from a charged product, for instance.

Another application is the inverse operation; a dilute process stream is run through a chromatographic column to capture the product. The product is then desorbed in a concentrated form with a suitable eluent. For example, a clarified aqueous fermentation broth may be run through a reversed phase system to capture the product, which is then eluted in concentrated form in methanol.

A chromatographic system is convenient tool for carrying out certain solvent switches. For example, if it desired to change the product solvent from aqueous acetonitrile to methanol, the product could be captured on a reversed-phase column. This may require dilution of the stream with water to create strong binding conditions for the product. The column could then be washed with dilute aqueous methanol (maintaining binding conditions) to remove acetonitrile, and the product can then be eluted in methanol. A similar product-capture/wash/elute approach can also be used to desalt a process stream. Yet another variation is the use of a reversed-phase system to effect a counterion switch, an approach often used after the purification of peptides. For example, a peptide purified by reversed phase chromatography using trifluoroacetic acid (TFA) as the ion-pairing agent (often a good choice to obtain selectivity for reverse-phase separation, but usually not desired in the final product) can be switched to another counterion form, say acetate, by (a) capturing it (after suitable aqueous dilution) on a reversed-phase column, (b) washing under binding conditions with streams containing first ammonium acetate to remove the TFA and then acetic acid to wash away excess ammonium ion, and finally (c) eluting the peptide by increasing the organic solvent content of the acetic acid containing mobile phase.

By far the most useful application of chromatography, of course, is the purification of a product from closely related impurities; for instance, separation of a desired peptide from a dozen or more synthesis impurities. Such impurities can have differences as subtle as, say, a beta-aspartic acid substituted for aspartic acid in the original sequence, resulting in an isomer with an extra methylene group in the peptide backbone. The following operational modes are employed for such applications.

18.5.2 Elution Chromatography

Elution is the most familiar mode of operation for chromatography; species migrate through the column and separate in the presence of a mobile phase that directly mediates retention. When the mobile phase composition is held unchanged, this is known as "isocratic" elution. When the mobile phase composition is varied during the migration, this is known as "gradient" elution.

The motivation for changing the mobile phase composition in a gradient elution process lies in the potentially wide range of retentivities of all the components in the mixture to be separated. An increase in the strength of the mobile phase over time ensures that the most strongly retained component will elute off the column within a practicable time period. The difference between operating under isocratic and gradient conditions in a reversed-phase system can be understood by examining the $\ln k'$ versus ψ plot in Figure 18.4. For a successful isocratic operation, a single solvent strength must be selected that affords selectivity between the desired compound and the other components of the mixture and also enables elution within a reasonable time frame. For instance, operating at 32 vol% organic, the first and last components to elute would have k's of about 3 and 23, respectively. A gradient in solvent strength would hasten the elution of the more strongly absorbed compounds.

Gradients are most commonly either linear or stepwise, and they are usually formed by blending two streams, changing the blend composition appropriately over time. Some equipment allows formation of convex or concave gradients, but this is not recommended if transfer between different brands of equipment is anticipated.

When the mobile phase strength increases over time, the front and rear of a migrating peak are often in different solvent environments, so that retention is lower at the rear of the peak than the front. This results in a peak "compression" that acts counter to and somewhat mitigates the dilution that is always observed under isocratic conditions. Reducing dilution can have significant impact on solvent volumes and processing of the collected purified material.

18.5.3 Displacement Chromatography

Under conditions of competitive binding and at high column loads, the displacement effect mentioned earlier leads to concentration of early eluting, weakly bound species displaced by more strongly bound components. Displacement chromatography, introduced in the 1940s by Tiselius [11], and developed further in more recent years by the academic groups of Horváth [12] and Cramer [13], among others, exploits this effect to the fullest extent possible. A mixture introduced into the column under strong binding conditions (e.g., low salt content in an ion-exchange system, where k' is high), is pushed through the column not by the customary means of changing the eluent strength, but by introducing a solution containing a species that is more strongly bound than any of the mixture components, called the displacer. Velocity of the displacer front can be manipulated by adjusting its concentration. The displacer swamps the binding sites on the sorbent, acting almost like a piston to move the other species ahead of it. The mixture species, concentrated and moved forward by the action of the displacer, then act to displace each other in order of binding strength, until—after development of the displacement train over a sufficient column length—they separate into contiguous concentrated bands of individual components. The displacer does not mix with the separating components except by dispersion at the end of the separation train. The displacer must be removed from the column before it can be reused. This can be accomplished by changing the eluent strength or other conditions such as the pH to facilitate washing the displacer from the column.

This mode of separation is ideally suited for economical preparative separations, since it perforce runs at high concentration and makes efficient use of the capacity of the stationary phase. However, relatively few displacement separations have been implemented. One reason may be that development of a displacement separation is often time-consuming and can require expert attention. Some headway in widening the use of this technique is being made, particularly in ion-exchange chromatography applications [14].

18.5.4 Multicolumn Chromatography

It has long been recognized by chemical engineers that the most effective mass transfer between two phases takes place when they are contacted in a continuous, countercurrent manner. In a chromatographic system with a particulate solid sorbent, actual movement of the solid would disrupt the packing structure negating any potential efficiency gains. Nevertheless, many of the benefits of countercurrent solid motion can be achieved by simulating the solid motion by appropriately switching inlet and outlet positions in a multicolumn system. When the inlet and outlet ports are simultaneously switched in the direction of fluid flow, as shown in Figure 18.5, the sorbent appears to flow in the opposite

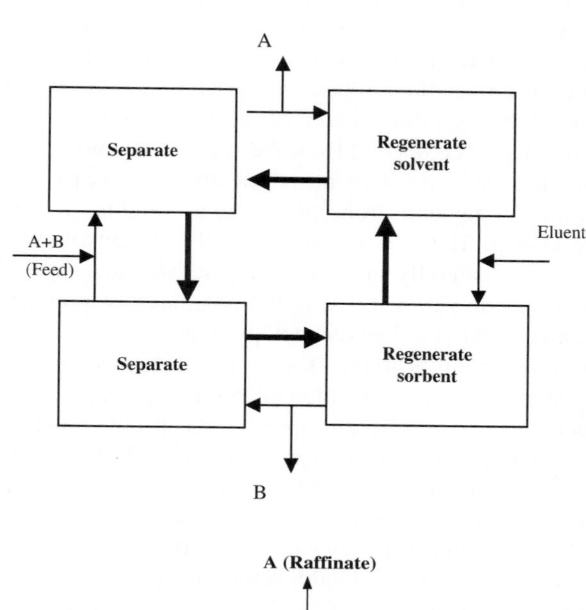

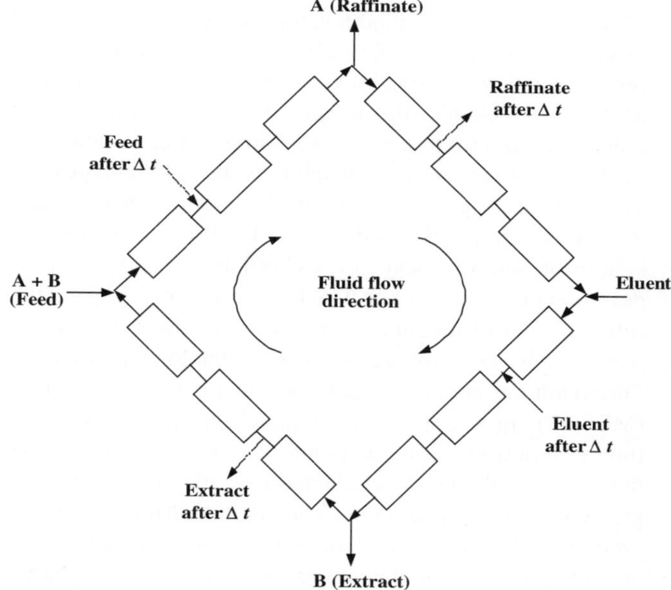

FIGURE 18.5 The four-zone simulated moving bed chromatograph. *Top*: Shows a schematic of the four-zone moving bed—dark arrows show the desired direction of solid flow. *Bottom*: Shows how simulation of the solid movement is carried out in a 12 columns system with 3 columns in each zone. The position of each of the external flow switches by one column after each switching period Δt. Flow within the column ring is driven by a pump.

direction. Simulated moving beds of this kind have been used in multiton industrial operations such as the purification of *p*-xylene from C8 fractions since the early 1960s, with installed capacity exceeding 10^7 tons/year [15]. Over the last two decades, simulated moving beds and related multi-column chromatography (MCC) technologies, have been increasingly employed in pharmaceutical applications.

The configuration shown in Figure 18.5 shows two inlet streams (one for the feed mixture, the other for the eluent) and two outlet streams (one for the separated early eluting compound(s), known as the raffinate, the other for the separated late eluting compound(s), known as the extract). The system has four zones, two separation zones on either side of the feed inlet, a solvent recycle zone that strips the early eluting compound from the solvent, sending it toward the raffinate, and a sorbent recycle zone that strips the late eluting compound from the sorbent, sending it toward the extract. The presence of two outlet streams makes the technique naturally suited to binary separations, and it thus comes as no surprise that in the pharmaceutical arena the simulated moving bed technology is used most widely for enantiomer purifications. There are three major operating advantages over conventional chromatography: (a) the stationary phase is used to maximum capacity, minimizing sorbent requirements; (b) operation can be adjusted to obtain close to quantitative yields at the targeted purity, minimizing loss of precious product, and (c) internal recycle of the solvent minimizes the solvent consumption [16].

Various efficiency modifications have been introduced recently into the repertoire of multicolumn operations. A process employing asynchronous column switching, known as "Varicol," enables the assignment of a fractional part of a physical column to a given separation zone, reducing the number of columns required and thus the column hardware needed [17]. Appropriately timed flow rate changes within a switching period (termed "power feed") have also led to significant improvement in MCC efficiency [18]. A recent development that is expected to significantly impact the future design of nonbinary large-scale gradient elution separations is the introduction of a technology termed the "multicolumn countercurrent solvent gradient purification (MCSGP)" process [19]. This process can employ as few as three columns in a semicontinuous countercurrent operation that in principle could be adapted to work for any existing gradient elution separation. The major advantage of the process is that the semicontinuous operation enables near quantitative yield of the desired compound at the target purity. Unlike the more traditional four-zone binary systems, the MCGSP does not afford significant solvent savings. Nevertheless, yield considerations can be a sufficient driver to implement such a scheme.

Many of the advantages of the simulated moving bed can also be realized using a single or dual column multi-injection steady recycle process invented by Charles Grill in the mid-1990s [20]. This process has now been commercialized [21].

18.6 EQUIPMENT

18.6.1 Columns

Until the introduction of large-scale high performance chromatographic equipment in the 1980s, industrial chromatographic separations were carried out in large columns packed literally with tons of relatively large sorbent particles (c. 100 μm diameter or larger). One consequence of large particle size was the need for significant column lengths (several meters) to achieve separations. Poor distribution in such columns created inlet and outlet flow development zones, creating serious inefficiencies and making column aspect ratio an important factor in scale-up. Modern high performance columns, now available from several vendors, have excellent flow distributors, so that performance is virtually independent of column diameter. Sorbent particle diameters used today are usually between 10 and 20 μm. While these provide high efficiency so that column lengths of less than 1 m (and usually $\ll$50 cm) are sufficient for many separations, they demand proper packing techniques and relatively high operating pressure, of up to about 100 bar. One technology that has solved these issues and so has dominated the large-scale column market is the so-called dynamic axial compression (DAC) column. A hydraulic piston inside the column is used initially to compress a slurry of sorbent particles into a firmly packed bed, and pressure is maintained on the bed during column operation, ensuring that any holes or pockets that may arise from bed subsidence over time are eliminated, ensuring efficient long-term operation. It is not unusual for a DAC column of 1 m diameter to show the same efficiency as an analytical column packed with the same sorbent. This makes scale up facile, as the only important scaling parameter is the cross-section area of the column. Figure 18.6 shows an example of a large-scale DAC column. The tall gray assembly is part of the hydraulic system that transmits pressure to the column piston. For unpacking, the bottom flange of the column is opened and the piston is driven downwards to force the sorbent out. Column packing and unpacking of the few tens of kilos of sorbent is thus straightforward, taking only a few hours (including all preparation time) in an industrial setting.

18.6.2 Pumps, Detectors, and Controllers

High-performance chromatography requires pumping of the mobile phase at a fixed flow rate through the system at pressures up to 100 bar. Industrial chromatography pumping systems almost universally employ double (or sandwich) diaphragm positive displacement pumps. The two diaphragm

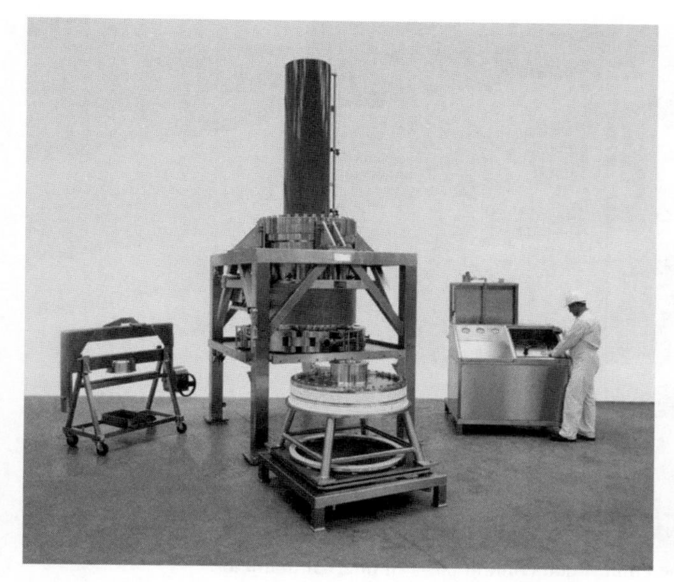

FIGURE 18.6 A 1 m internal diameter dynamic axial compression column for large scale high-performance liquid chromatography. The gray cylinder above the column is part of the hydraulic system to maintain pressure on the piston during packing an operation. Components of the piston assembly are seen in the foreground. Photograph courtesy of Novasep Inc.

design assures that failure of a single diaphragm will not contaminate the process side with oil or fluid from the pump. Continuous monitoring of the space between the diaphragms for rupture provides assurance that any failure will be detected immediately.

It is advisable to use separate pumps for feed and mobile phase streams, although some pumping designs use the same pump for both.

A common feature in pumping system designs is a gradient forming system that enables blending of two (or more) inlet streams to vary mobile phase composition for gradient elution operation. Valves and mass flow meters along with electronic controllers are often used to achieve the gradient. Gradient composition monitoring and control using PAT systems, such as near IR probes with feedback control, are available commercially.

The effluent stream is usually monitored by a detector; signals from the detector can be used to decide when to collect fractions. The most commonly employed detector is the variable wavelength UV detector, although a plethora of others (conductivity detectors, for instance) can be used depending on the application. Some systems take a small slipstream to pass through a detector, others use a full-flow detector cell. Under the high concentration conditions typical in preparative chromatography, it is common for the UV absorption signal to saturate at the wavelengths and cell path lengths commonly used in analytical chromatography. Often sensitivity needs to be dampened to properly capture the emerging peak and make useful fraction collection decisions. One simple way to accomplish this in practice is to use a UV wavelength far from the absorption maxima of the compound.

Fraction collection is accomplished by directing flow to different collecting vessels. Decision making on fraction collection can be triggered by the detector signal. As mentioned earlier, UV signals from industrial chromatographs do not look quite as simple as those in analytical systems, so some logical rules based on time, flow volume, and thresholds of UV signal and/or UV slope can be employed to start and end fraction collection periods. Most commercial equipment manufacturers supply software with sophisticated fraction collection algorithms.

Pump valves and detectors are often mounted in one assembly, called a pumping skid. Control of the skid is accomplished with the aid of a programmable logic controller (PLC) with appropriate software running from a PC. Fully automatic operation is possible. Electronic controls are either located remotely from the skid or in an appropriately purged box to ensure the unit is explosion proof. In explosion proof equipment, valves are activated pneumatically.

Equipment for multicolumn units likewise involves a combination of columns, pumps, valves, and process analytics specific to the design of the units. Figure 18.7 shows a series of columns intended for use in a Varicol MCC at industrial scale.

18.7 SCALE-UP

With good radial flow distribution, inlet and outlet flow development zones are practically eliminated and chromatography can be genuinely described by a one-dimensional (axial) mass balance such as that in equation 18.4. The simplest possible scale-up paradigm under these circumstances is to maintain the same sorbent particle size, column length, and chromatographic velocity across scales, setting all other parameters proportional to the column cross-section area. This will ensure practically identical pressure drop, column efficiency (plate number), separation quality, and specific production rate (production rate normalized to sorbent mass) across scales. An example showing such a scale up from a standard 0.46 cm internal diameter (ID) column to a 45 cm ID column is shown in Table 18.1.

If identical particle sizes are not employed, similar quality of separation can be obtained by keeping the dimensionless number Li and the normalized feed load constant across scales. Rearranging equation 18.2 to solve for the chromatographic velocity u_0 and substituting the result into the expression of Li from equation 18.12 one obtains

$$\mathrm{Li} = \frac{\left(L/d_{\mathrm{p}}^{2}\right)^{2}}{\Delta P}\left(\frac{\eta\varphi}{D_{\mathrm{m}}}\right) \tag{18.14}$$

FIGURE 18.7 A six-column simulated moving bed unit. Each column has 1 m internal diameter. The unit was designed to operate in a Varicol process. Photograph courtesy of Novasep Inc.

TABLE 18.1 Example of Scale-Up Keeping Particle Size and Column Length Fixed

Parameters	Small Scale	Scale Factor	Large Scale	Comments
Key Scale Up Parameters				
Column diameter	0.46 cm		45.00 cm	
Column cross section	0.17 cm^2	9,570	1590.43 cm^2	Basis for scale up 9569.94
Parameters that must be Fixed to Enable Scale-Up based on Cross-Section Area				
Particle size	10.00 μm		10.00 μm	
Column length	25.00 cm		25.00 cm	
Parameters that Scale with Cross-Section Area				
Mass sorbent in column	2.53 g	9,570	24.25 kg	Packed density of 0.61 g/cm^3
Flow rate	1.50 mL/min	9,570	14.35 L/min	
Feed injected per run	25.00 mg	9,570	239.25 g	Load factor held constant
Important Parameters that Remain Unchanged in this Scale-Up Paradigm				
Chromatographic velocity	0.215 cm/s		0.215 cm/s	
Pressure drop across column[a]	40 bar		40 bar	Equation 18.2: $\varphi = 710$; $\eta = 1.05$ cP
Feed composition	–		–	
Mobile phase composition	–		–	
Run time	25 min		25 min	
Gradient program	–		–	
Detector settings	–		–	
Temperature	30°C		30°C	
Performance Attributes				
Yield	85%		85%	
Production rate	1.22 g/day		11.71 kg/day	Feed injected/day × Yield
Specific production rate	0.48		0.48	kkd = kg produced/kg sorbent/day
Mobile phase used per day	2.16 L		20,671 L	
Specific solvent consumption 1	1.76 L/g		1.76 L/g	No recycle
Specific solvent consumption 2	N/A		176 L/kg	90% recycle

[a] There will be additional pressure drop caused by flow in piping, not accounted for here.

TABLE 18.2 Productivity Comparison for Columns Packed with 10 and 60 Particles

	Case 1	Case 2	Comments/Calculations
Scaling Calculations			
Particle size [μm]	10	60	
Pressure drop [bar]	40	3	3 bar typical of low pressure equipment
Column length [cm]	25	246	Calculation based on scaling parameter
$(L/dp^2)^2/\Delta P$ [dyne^{-1}]	15,625,000	15,625,000	Scaling parameter
Productivity Calculations: Columns Packed with 1 kg of Sorbent in Both Cases			
Linear flow velocity, u_0 [cm/s]	0.215	0.059	Equation 18.2: $\varphi = 710$; $\eta = 1.05$ cP
Dwell time, t_0 [min]	1.94	69.89	$t_0 = L/u_0$
Run time, t_R [min]	25	900	Case 2 based on ratio of dwell times
Column ID that contains 1 kg sorbent [cm]	9.14	2.91	Packed density of 0.61 g/cm^3
Feed injected per run [g]	9.9	9.9	From Table 18.1, keeping feed to sorbent mass constant
Yield	85%	85%	Based on Example Table 18.1
Production rate [g/day]	484.70	13.46	A 36-fold difference
Column Dimensions and Sorbent Quantities Required to Obtain the Same Productivity			
Column diameter [cm]	45	86	
Column volume [L]	40	1,431	Must have 36 × volume to maintain equal productivity
Sorbent mass to pack column [kg]	24.25	873.15	
Feed injected per run [g]	239.25	8613	
Number of injections/day	57.6	1.6	# = 24 × 60/t_R
Production rate [kg/day]	11.71	11.71	= # × feed injected per inj* Yield
Specific productivity	0.48	0.013	kg produced/kg sorbent/day

The viscosity η and diffusivity D_m are constants across scales as long as temperature is maintained constant. The bed permeability φ should remain constant across scales as long as similar sorbent particle geometry (e.g., spherical particles) is maintained. Thus, the pertinent invariant scale factor when sorbent particle size is not held constant is $(L/d_p^2)^2/\Delta P$.

Table 18.2 shows an example of two systems with different particle sizes, 10 and 60 μm, respectively, where the factor $(L/d_p^2)^2/\Delta P$ is held constant. Separation quality (not quantified in the table) and important performance characteristics such as solvent consumption are expected to be identical in both cases, but specific production rates are dramatically different. The smaller particle size sorbent has an increased specific productivity proportional to the square of the ratio of particle sizes (a factor of 36 in the example). If the same production rate was required in the two cases, the quantity of 60 μm sorbent required would be 36-fold that in the 10 μm system. Dimensions of identically performing columns packed with the two different particles are also given in Table 18.2. This example illustrates the incentive for using smaller particles and higher pressures for chromatographic operations.

18.7.1 Scaling of Gradients

Solvent gradients should be expressed in scalable terms. For example, in reversed-phase chromatography, varying solvent composition from X% to Y% in a straight-line manner over several column volumes (CVs) of flow (rather than expressing it as A% of solvent A to B% of solvent B over 20 min) enables seamless scale up.

18.8 DESIGN SPACE

The so-called design space is the window within the operating parameter space in which acceptable process performance is achieved. The goal of this section is to provide the reader with some appreciation for the various operating parameters and the sensitivity of conventional chromatographic operations to these parameters. Multicolumn systems are not discussed. The manner of defining the design space for regulatory agencies—carrying out design of experiments, etc—is not within the scope of this discussion.

Figure 18.8 shows a view of a chromatographic process illustrating the various process parameters as well as salient product quality and process performance attributes. Design issues associated with some of the parameters, for instance sorbent selection, column dimensions and efficiency, have been discussed earlier in this chapter. Some of the other parameters are discussed below.

18.8.1 Process Quality Attributes

Critical quality attributes of the product stream from a chromatographic process are product purity, which must be above a minimum target (e.g., >98.5%), with key individual

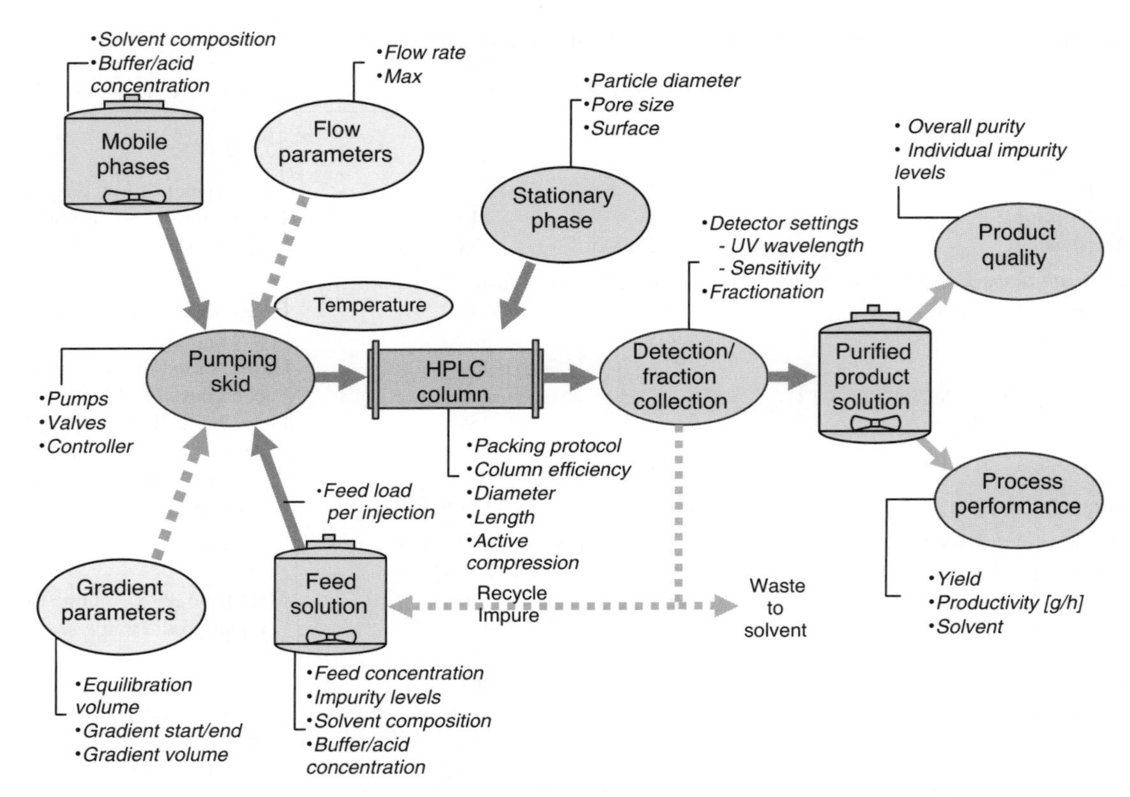

FIGURE 18.8 Overview of chromatographic operating parameters, product quality attributes, and process performance measures.

impurities maintained below prespecified maximum limits. The concentration in the product stream is a quality attribute, but a wide range is usually acceptable and thus it is not critical.

18.8.2 Process Performance Attributes

Key process performance (and indeed process economic) measures are (a) yield, calculated as the moles of desired product recovered at or above the target quality as a percentage of the moles introduced in the crude feed, (b) process productivity, which is simply a mass rate of production, expressed in gram or kilogram/hour (a specific production rate, normalized to the mass of sorbent, is sometimes a useful measure), and (c) solvent consumption, most usefully expressed in terms of each solvent species consumed per unit mass of product. These parameters are discussed further in Section 18.9.

18.8.3 Feed Parameters

As shown in Figure 18.8, feed concentration, impurity levels and the feed solution characteristics, including solvent composition, buffer, or counterion concentration, and pH are important factors. The rule of thumb is that for the same feed load, better results are obtained the higher the feed concentration. Nonetheless, feed concentration is usually more critical in isocratic elution than in gradient elution chromatography. This is because in gradient elution, feed introduction is usually under relatively strong binding conditions, enabling concentration of the feed at the column inlet regardless of its concentration. There is usually no such opportunity under isocratic conditions, unless the feed solvent composition is manipulated to afford somewhat stronger binding conditions than the eluent itself. (In reversed-phase chromatography, this would imply using lower organic modifier in the feed than in the eluent.) Feed solubility is of course a strong limiting factor influencing concentration; some practitioners using isocratic elution chromatography manipulate feed solvent composition to provide more solubility at the expense of strong binding (e.g., raising the methanol content in normal phase chromatography) [9]. The pH can also play a powerful role; there is at least one example where a pH mismatch between feed and eluent has been exploited to enhance separation [22]. Criticality and sensitivity of the process to these parameters needs to be experimentally explored during process design.

18.8.4 Mobile Phase Parameters

Solvent composition (i.e., the blend of different solvents employed), buffer concentration and pH are key mobile phase parameters. In systems, where two or more mobile phase streams are blended by the gradient-forming system in the pumping skid, a wide composition range may be permissible depending on whether or not appropriate process analytical technology is employed to monitor or control the gradient. Heats of mixing and gas evolution upon solvent blending (gas solubility may change upon blending and solvents that are not degassed may outgas as a consequence) need to be accounted for. Typically outgassing occurs only at low pressures at the column outlet and may or may not interfere with detection. Prior degassing of solvent streams may not be required if the system has sufficient pressure at the detector cell to prevent outgassing.

18.8.5 Fraction Collection

Issues with UV or other detection of product in the column effluent have been discussed above. Proper collection and combining of fractions containing purified material is critical to the success of the chromatographic process. Ideally, a process should be designed to collect one purified fraction and perhaps one or two lower purity fractions that could be recycled. The remainder of the process stream can be sent to solvent recovery units. Collection of a single pure fraction is possible only after all the nuances of the chromatographic operation have been fully understood and a fail-safe rich cut collection algorithm is devised that can handle all possible variation in feedstock purity and any minor drifts in retention caused by operating within a parameter ranges. Until sufficient process history is available, it is prudent to collect multiple fractions; combination of *contiguous* fractions at different purity levels *in their entirety* to achieve a rich cut at the desired purity is operationally the same as taking a single rich fraction, and thus does not carry the stigma associated with blending of poor and high quality materials.

18.8.6 Column Lifetime

An important design and economic parameter is column lifetime. Some minimum lifetime should be specified before commercial implementation of a process, and this can be obtained from process history or from multiple small-scale injections on a small scale. Minimum plate count and selectivity criteria, measured with a test tracer solution (either a small injection of product or other marker compounds), should be in place to define acceptable column performance, and these should be tied to process performance. For small-molecule applications, it should be feasible to extend the use of a column beyond the existing historic lifetime based on a tracer test. Since column performance may diminish for mechanical reasons (poor maintenance of bed integrity owing to lack of sufficient axial pressure, for instance), the sorbent may be unpacked and repacked on multiple occasions. For regulatory purposes, some limits for such operations must be defined. It is also useful to establish whether or not the feed components can eventually poison the sorbent (by irreversible binding, for instance). In addition, depending on nature of further downstream operations, it may be important to develop tests for column leachables and demonstrate absence or establish acceptable limits of these compounds in the process stream.

18.8.7 Temperature

Column and solvent temperature is an important parameter. Ideally isothermal conditions should be maintained, as retentivity is temperature dependent as shown in equation 18.7. However, during operation of larger columns, a small increase (1–3°C) between the column skin temperature and the solvent inlet temperature is useful to overcome frictional heating effects and within the column and related wall effects [23].

18.9 ECONOMICS

Given the many parameters associated with design of a chromatographic process, rigorous economic assessment and optimization of chromatography is a complex task. Key economic parameters, such as return on capital and Lang factors (the ratio of total installation cost to cost of equipment), vary from company to company, so that outcomes of net present value (NPV) assessments of the same separation problem may lead to different conclusions at different locations. Nevertheless, cost drivers for chromatographic operations are common to most other processes; in an industrial manufacturing operation the overall cost, expressed per kilogram or ton of product purified, is the sum of costs of amortized capital, labor, consumables, waste, and product loss. Intangibles not usually taken into account because they are hard to quantify are development opportunity costs. For instance, a chiral separation is typically easier and quicker to implement and scale up than an asymmetric synthesis. If the unwanted enantiomer can be racemized, the major objection to a separation approach (automatic 50% yield loss associated with the unwanted enantiomer) may be mitigated. With commensurate resources for optimization and engineering, ultimate process costs of the chiral chromatography and a more elegant asymmetric synthesis may not be significant, but the time talented chemists spent on the synthesis development could have been used more fruitfully elsewhere. Only organizations that develop sufficient competence in chromatographic operations would have the confidence to exploit the technology to its full potential.

Capital and labor costs are usually specific to the site of operation. In this author's opinion, cost for large-scale high performance chromatographic equipment and installation is not prohibitive compared to new installation of other process equipment. There is sufficient competition among vendors and important equipment designs are now off patent. The key economic concerns for the chromatography development engineer should be the minimization of cost of consumables and product loss. Several good literature references for quantitative optimization are available [24,25].

18.9.1 Stationary Phase

During development, cost of the stationary phase can be high. Sorbent usually is dedicated to a particular product and, since product failure rates are high, there is limited opportunity to spread the cost of the phase over a long period of time. On the other hand, during commercial operation, sorbent costs, while significant, are not usually a limiting factor since column lifetime can be quite long. For instance, chiral stationary phase lifetimes of several years have been reported. Unpublished reports suggest stationary phase costs in such applications can be held below $\$10\,kg^{-1}$ purified [26].

18.9.2 Solvent Consumption

Solvent volumes employed in chromatography are indeed high. Operating a 45 cm ID column at 10 L a minute involves handling 4000 gal a day. However, measures can be taken to drastically limit solvent *consumption* and costs by implementing solvent recycle procedures. At full production, it is reasonable to expect upwards of 90% solvent recycle (on the organic component; water is rarely recycled; in some instances this may give organic solvent rich mobile phases the economic edge, since disposal costs for aqueous waste must be considered).

For separations where multicolumn techniques, such as simulated moving bed technology or multiinjection steady state recycle technology can be employed (mostly for chiral separations) solvent consumption can be drastically reduced compared to column chromatography. Coupled with solvent recycle, use of SMB technology may consume less solvent than a conventional three-step classical chiral resolution by crystallization with a tartrate salt, and the total cost for a simulated moving bed operation can be competitive with or lower than such a process [27]. New chiral stationary phases with high adsorption capacity are becoming available in the market; use of these phases is expected to improve the economics of chromatographic separation of enantiomers even further.

18.9.3 Product Loss

For high value products, yield loss can have a significant economic impact. In conventional chromatography (e.g., a gradient elution separation of peptides), yield losses can be minimized by recycling impure fractions. Regulatory agencies expect defined criteria for choosing fractions for recycling, and defined limits on the number of times recycle is allowed. Multicolumn techniques have a distinct advantage since high yield is an inherent attribute of continuous processes. An example showing yield increase from 85% in a conventional gradient peptide purification to >95% in an MCSGP process, with a concomitant 25-fold productivity increase, has been published [27].

18.10 CONCLUSIONS

With modern high performance equipment and improvements in operating strategies, including the increasing use of multicolumn technologies, robust and economic large-scale chromatographic processes can be designed for purification of a wide variety of pharmaceutical compounds.

ACKNOWLEDGMENTS

I am grateful to Dr. Olivier Dapremont, of Ampac Fine Chemicals, for insightful discussion on sorbent costs and economics of simulated moving bed chromatography operations. I also thank Dr. Henri Colin, of Ulysse Consult, Nancy, as well as Professor Anita Katti, Purdue University, Calumet, and Professor Georges Guiochon, University of Tennessee, for useful discussions. Also, I am grateful to Novasep Inc. for providing photographs of their equipment used in Figures 18.6 and 18.7.

REFERENCES

1. Twett MS. *Ber. Dtsch. Botan. Ges.* 1906;24:316–323; Twett MS. *Ber. Dtsch. Botan. Ges.* 1906;24:384–393.
2. Antia FD, Horváth Cs. *Ber Bunsenges. Phys. Chem.* 1989; 93: 961–968.
3. Horváth Cs, Melander WR. In: Heftmann E, editor, *Chromatography: Fundamentals and Applications of Chromatographic and Electrophoretic Methods, Part A*, Elsevier, Amsterdam, 1983, pp. 27–135.
4. Antia FD, Horváth Cs. Gradient elution in non-linear preparative liquid chromatography. *J. Chromatogr.* 1989; 484: 1–27.
5. van Deemter JJ, Zuiderweg FJ, Klinkenberg A. *Chem. Eng. Sci.* 1956; 5: 271–289.
6. Antia FD, Horváth Cs. *J. Chromatogr.* 1988; 435: 1–15.
7. Unger KK. *Porous Silica*, Elsevier, Amsterdam, 1978.
8. Roush DJ, Antia FD, Göklen KE. *J. Chromatogr. A* 1998; 827: 373–389.
9. Nti-Gyabaah J, Antia FD, Dahlgren ME, Göklen KE. *Biotechnol. Prog.* 2006; 22: 538–546.
10. Meyer V. *Practical High-Performance Liquid Chromatography*, 5th edition, Wiley, New York, 2010.

11. Tiselius A. *Kolloid Z.* 1943; 105: 101.
12. Frenz J, Horváth Cs. *AIChE J.* 1985; 31: 400–409.
13. Tugcu N, Bae S, Moore JA, Cramer SM. *J. Chromatogr. A* 2002; 954: 127–135.
14. Liu J, Hilton Z, Cramer SM. *Anal. Chem.* 2008; 80 (9): 3357–3364.
15. Gattuso MJ. *Chim. Oggi. (Chem. Today)* 1995; 13: 18–22.
16. Antia FD. In: Rathore AS, Velayudhan A,editors, *Scale-Up and Optimization in Preparative Chromatography: Principles and Biopharmaceutical Applications*, Marcel Dekker, New York, 2002, pp. 173–201.
17. Ludemann-Hombourger O, Nicoud RM, Bailly M. The "VARICOL" process: a new multicolumn continuous chromatographic process. *Separ. Sci. Technol.* 2000; 35: 1829–1862.
18. Zhanga Z, Mazzotti M, Morbidelli M. *J. Chromatogr. A* 2003; 1006: 87–99.
19. Aumann L, Morbidelli M. *Biotechnol. Bioeng.* 2007; 98: 1043–1055.
20. Grill CM. *U.S. Patent* 5,630,943, 1997.
21. http://www.novasep.com/upload/news/PDF/31Press%20 Release%20Process%20Cyclojet.pdf.
22. Vailaya A, Sajonz P, Sudah O, Capodanno V, Helmy R Antia FD. *J. Chromatogr. A* 2005; 1079: 85–91.
23. Dapremont O, Cox GB, Martin M, Hilaireau P, Colin H, *J. Chromatogr. A* 1988; 796: 81–99.
24. Guiochon G, Shirazi DG, Felinger A, Katti AM. *Fundamentals of Preparative Chromatography*, 2nd edition, Elsevier, Amsterdam, 2002, Chap. 18.
25. Katti A.In: Valkó K, editor, *Handbook of Analytical Separations*, Vol. 1, Elsevier, Amsterdam, 2000, pp. 213–291.
26. Dapremont O. Ampac Fine Chemicals, personal communication.
27. Auman L, Stroehlein G, Schenkel B, Morbidelli M. *BioPharm Int.* 2009; 22 (1).

19

MILLING OPERATIONS IN THE PHARMACEUTICAL INDUSTRY

KEVIN D. SEIBERT AND PAUL C. COLLINS
Eli Lilly & Co., Indianapolis, IN, USA
ELIZABETH FISHER
Merck & Co., Inc., Rahway, NJ, USA

19.1 INTRODUCTION

Two aspects of a successful formulation are that it produces consistent results *in vivo* and it can be manufactured reproducibly. One of an engineer's main goals in pharmaceutical development is to design a process that results in a high level of consistency and control of final product performance from research through manufacturing. Particle size of ingredients in a formulation, especially the active pharmaceutical ingredient (API), greatly impacts bioperformance and process capability, and as such it is an important parameter to understand and control. Milling, or mechanical size reduction of solids, is frequently used to achieve API or formulated intermediate particle size control, and can also be used for reagent and excipient size control [1]. Size reduction can be performed with enough energy to break individual particles or with less energy, to break granules or agglomerates during formulation; it can be performed on dry solids, partially wet solids, or in slurry mode; it can be performed on reagents and excipients as well as final API. While there are other processing options that can be used to achieve these goals, milling is a powerful tool that is frequently chosen as a relatively straightforward way not only to evaluate whether particle size control will provide the desired performance and consistency, but also to effect the desired process or product performance control from early development through commercial-scale manufacturing.

Milling can increase the surface area of solids to increase their dissolution rate. This can be extremely important for improving bioavailability of a formulated drug product. The Biopharmaceutics Classification system [2–4] is one way of characterizing whether increasing API dissolution rate may improve its bioperformance; in this system, compounds are classified as BCS II or IV when their solubility is low, based on a set of standard dissolution test conditions. While clearly particle size does not affect compound equilibrium solubility, dissolution rate is directly proportional to particle size (see Noyes–Whitney equation, Fick's first law).

Control of dissolution rate can also be important for facilitating or improving chemical reaction performance when a reagent is charged in solid form. Increasing surface area of the API can also increase its effectiveness as seed for crystallization. As discussed in previous chapters, providing enough available surface of the correct crystal form is important for controlling API crystal size and form uniformity.

Size reduction of API and/or excipients can also be important to ensure consistency of formulation processing. This processing frequently involves dry blending of API with one or more excipients, and the mean size and particle size distribution of the API relative to the excipients can influence the tendency of the API to segregate, or not mix uniformly with the other ingredients. Formulations where relatively small (mean size less than 10 μm) API is required include small unit dose formulations (submilligram) to ensure blend uniformity. Milled API results can aid in obtaining proportionately tight distributions and small API mean size, which

Chemical Engineering in the Pharmaceutical Industry: R&D to Manufacturing Edited by David J. am Ende

may be necessary to facilitate meeting these specifications [5]. Suspension or controlled release formulations may also have tight requirements on mean and top (largest) size in the API particle size distribution. For example, API that is used in dry powder or metered dose inhalers must have sizes less than 5–6 μm and more than 0.5–1 μm to reach deep lung and not be exhaled [6, 7].

In a milling operation, mechanical energy imparts stress to particles, and they are strained and deformed. When strained to the point of failure, cracks are formed and can propagate through the particle and result in breakage. Crystalline materials tend to break along crystal planes, while amorphous materials break randomly, and flawed particles will be easier to break than those with fewer internal weaknesses. When sufficient force is rapidly applied normal to a particle surface and directed toward its center, mass fracture, or breakage into a few large fragments, can occur. When force is applied more slowly, compression will be the main cause of particle breakage. If force is applied parallel to the surface of the solid, over time the particle can break into many fine particles, which is usually described as attrition [1]. Mill design and operation play a large role in determining which mode of breakage occurs.

There are several techniques available to characterize the distribution of particle and granule sizes produced by milling. Optical microscopy or image analysis is useful when little material is available, and observations of particle/granule morphology can be invaluable in troubleshooting milling problems or interpreting other particle size analysis results and powder flow behaviors. Laser light scattering particle size analysis requires more sample, but is commonly used to evaluate API particle size distributions. Solids to be analyzed can be dispersed in a carrier gas or liquid, and some force (air pressure or sonication) is applied to disrupt agglomeration and permit analysis of individual particles (Figure 19.1). This technique can also be used online as milling takes place and thereby permit additional process understanding and control [8–10]. Sieve analysis, or determining the amount of material that passes through or is retained on standard mesh screens is also used to assess size distribution of granules in a formulation process. Perry's *Chemical Engineer's Handbook* [11] summarizes several size analysis techniques and provides additional references on this topic.

It is important to note that in certain situations milling can be challenging and alternative size reduction techniques should be considered. If the milling operation results in material that contains a high proportion of fines (particles much smaller than the average size), processing problems can occur in subsequent unit operations. Large proportions of fine particles can increase cohesive behavior of the bulk solid, resulting in poorly flowing powder as the API is handled during formulation [12]. When solids are milled in a slurry and then need to be isolated from the liquors, fines can pass through the filter media and reduce overall yield. Fines can also reduce cake and filter media permeability and reduce filtration productivity. At pilot and manufacturing scale, formulation usually involves mechanical solids feeding, which requires that the solids flow in a predictable manner. This "flowability" can be assessed in different ways [13–15] and there is no standard industry practice for this testing in pharmaceutical manufacturing. However, it is clear that solids that tend to demonstrate cohesive behaviors and clump or stick to equipment surfaces can be challenging to process reproducibly and efficiently in formulation without additional effort and safeguards.

Applying mechanical energy via milling does more than break crystals. As the applied energy increases to be sufficient to make individual crystals smaller, the force and stress that cause fracture can also induce changes in the crystal. These changes may manifest themselves in a variety of ways, from increasing numbers of surface cracks and flaws that may increase specific surface area and surface roughness to changes in crystal form, including conversion to an amorphous form. Both types of changes can affect performance in formulation/*in vivo* and can affect physical and chemical stability. This observation is more common for dry milling operations that break primary particles than it is for delumping/granule breaking operations or for wet milling. While milling-induced disorder likely occurs during a wet milling

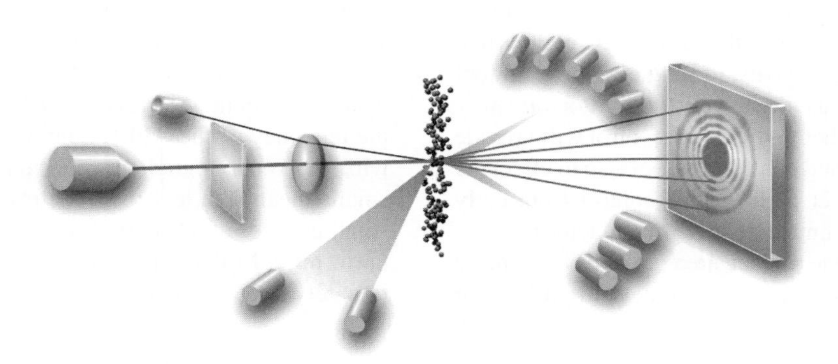

FIGURE 19.1 Laser light scattering device—Photo courtesy of Malvern.

process, contact of the crystal with a liquid medium facilitates surface annealing via dissolution and subsequent redeposition, and the liquid present can transfer heat more readily than a gas stream. As understanding of solid surface chemistry and crystal structure has increased, and analytical detection and quantification capabilities have improved, these consequences of milling have become more frequently reported and studied. Ongoing work in this area continues to show that while milling can be a desirable way to control particle size, it is important to fully characterize the quality of the milled product. Some analytical techniques to assess amorphous content or crystal form include X-ray powder diffraction and differential scanning calorimetry, and techniques to assess surface energy include atomic force microscopy and inverse gas chromatography [16–18].

Each type of milling equipment is designed to impart different amounts of energy to the solids and therefore can produce different mean sizes and size distributions. Product requirements and process limitations will inform what target size and/or largest size is required. The following sections provide some guidance on how to use that information to select milling equipment and the key parameters that can influence its performance. The resulting mean particle size and particle size distribution depend not only on the mill but also on physical properties of the feed solids such as brittleness and hardness [19], as well as its initial size distribution. Generally speaking, the most efficient way to assess milling feasibility is to perform trial experiments. Even when combined with a size classification operation, it is not possible to use a mill to precisely "dial in" a mean size or particle size distribution with any reliability without the use of feedback control from in-line particle size analysis equipment. However, engineering understanding of mill operation and key process parameters permits effective scale-up of milling, so that the particle size to be achieved at large scale can be predicted accurately from carefully conducted gram scale bench experiments.

19.2 TYPES OF MILLING AND MILL EQUIPMENT

The two primary means of reducing particle size of a solid product are wet and dry milling as defined by the media in which the engineer chooses to carry out the milling activity. Selecting a milling strategy depends on a variety of factors including particle physical and chemical properties, chemical stability, and safety related issues.

API is most often isolated by crystallization, followed by filtration of the solids and a terminal drying step. Based on the thermodynamic and kinetic properties of the system, the physical properties of the solids of interest and the dynamics of the environment within the crystallizer, several outcomes are possible. Crystallizations that end in final isolated particles at the desired particle size distribution (PSD) specification require no further processing to be carried into the downstream processing train, and are often the result of targeted particle engineering efforts [20–22] Particles below the target PSD can either be tested for processability in the drug product manufacturing, or the size increased through modified crystallization strategies to yield larger particles of the desired PSD [15, 23]. Most often, and generally by design, the isolated crystals are larger than the target PSD as established through formulation development or bioavailability testing. In these situations, the resulting material must be milled to reduce the PSD to the target.

Deliberately growing particles larger than the target PSD may offer the advantage of cycle-time savings from reduced filtration time during isolation. While the filtration time can be reduced through increased pressure, equipment limitations and nonlinearity associated with compressible solids may result in diminishing returns. As a result, it is often advantageous to attempt to grow, isolate, and dry larger particles in the final processing steps of an API synthesis. In situations with solution instability, this may be the only suitable processing option as the time associated with milling a slurry could negatively impact API purity in these cases. Therefore, in considering the overall manufacturing efficiency associated with making API, the purposeful generation of large particles, which are later reduced in size, is a viable option to consider. In these scenarios, dry milling is the preferred technology for micronization.

Dry milling may also be preferred over wet milling due to incompatibility between the processing solvent and the available wet milling equipment, such as when the process stream is corrosive to stainless steel. Additionally, dry milling may be a desirable unit operation, if an isolation lends itself to some process advantage. Motivating factors driving an isolation prior to reducing particle size may include impurity removal or rejection resulting from a selective crystallization and rapid removal of the liquors.

Wet milling offers an advantage in that milling can be combined with terminal isolation such as selective crystallization of the desired product. When impurity removal is still of principle concern, techniques such as decantation, cross flow microfiltration with diafiltration, or filtration with reslurry to remove crystallization, mother liquors may be employed as a way to remove rejected impurities or selectively switch to a desired solvent system.

An additional process design benefit is potentially realized with wet milling, as it allows for greater flexibility in terms of how a crystallization process is developed. As an example, as crystallization progresses the mean size of the crystals can be simultaneously reduced by milling during the crystallization process. This is time sparing and also presents crystallization benefits by continually exposing new surface area for use in the ongoing crystallization. Wet mills are also preferred if the material being milled exhibits undesirable physical properties or phase changes at higher temperatures.

The increased heat capacity of the liquid carrier media allows for smaller temperature fluctuations during milling. This can be especially important for materials that have either a low melting point, or are susceptible to crystal form conversion at lower temperature.

Wet milling may offer significant operating advantages over dry milling when the solvent system used for particle size reduction is directly compatible with the downstream drug product processing operation. Drug product unit operations including spray coating and wet granulation require the blend of the milled drug substance with liquid and may therefore be amenable to wet milling followed by direct use of the slurry stream in the formulation processing.

19.2.1 Dry Milling Setup

A general dry milling setup is shown in Figure 19.2. Unmilled solids are charged to the feed hopper via several methods, generally designed to contain any particle dusting during the charge operation. Solids are fed from the feed hopper through a feed device that will allow a constant mass flow to the milling unit. They are conveyed by either gravitational flow through the milling apparatus, screw conveyance, or by conveyance with a carrier gas such as nitrogen. Typically, rotary valves and screw feeders are used to ensure constant feed rates and prevent back flow of solids due to the slightly positive gas pressures on the mill itself. Feed hoppers are generally designed with sloped sides to enable proper solids flow into the rotary valve and screw feeder system. For poorly flowing solids, "bridge-breakers" or small agitators designed to facilitate solids flow from the hopper into the screw feeder are also installed.

Gas is fed to the mill along with the solids. Depending on the type of mill being used, some of the gas may be introduced to the mill separately from the entering solids, such as in the design of a loop or spiral jet mill, or the gas may be used as a carrier fluid to convey the solids to the mill as is the case in a hammer or pin mill design. While air may be used to convey and mill the solids, nitrogen is most often used to ensure adequate inertion of the environment around the solid phase. For solids with a low melting point, or a propensity to undergo a pressure-induced phase transition to an amorphous solid, liquid nitrogen may be supplied via an orifice upstream of the milling operation to maintain cryogenic temperatures within the mill internals.

Depending on the mechanism of grinding within the mill, the residence time of the solids may vary from one to several seconds of residence time. Once milled, particles will exit the grinding chamber and be carried to product collection. When tighter particle size control is desired, a classifier may be installed in line, which allows for larger particles to be returned to the feed hopper while particles that have been adequately reduced will be routed to the collection area. Product collection from the gas stream will frequently use a cyclone to separate most of the larger solids from the gas stream, and then send the gas containing finer particles to a bag filter and HEPA filter for final dust removal.

In most size reduction milling equipment, residence time in the mill is a key parameter determining outlet particle size. As a result, the feed rate becomes a significant part of the

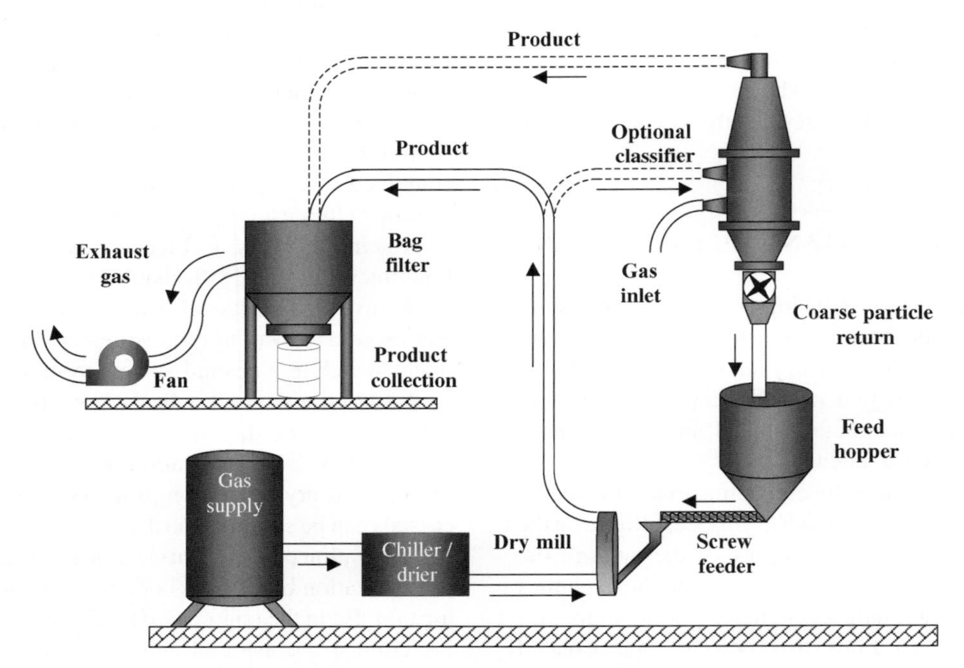

FIGURE 19.2 General setup for dry milling operations.

control strategy and must be carefully considered when determining scale-up parameters for milling equipment.

19.2.2 Dry Milling Equipment

Dry milling may be accomplished by grinding or high force collisions of particles with a moving pin or hammer. Alternatively, it may be accomplished by high energy particle--particle collisions as is the case with jet mills. The following sections briefly describe different types of dry mills. Note that for reduction of particle size to submicron, dry milling strategies are not usually adequate, and therefore wet milling strategies are typically employed. The table below summarizes some different types of dry mills, their key scale-up parameters, and typical minimum particle sizes they can produce. As noted above, solids residence time in the mill is always an important scale-up parameter for dry milling. This can be easily measured experimentally and can be controlled by limiting solid and gas feed rates to match mill efficiency.

Mill Type	Typical Minimum Milled Size (μm)	Key Parameters
Hammer mill	20–60	Mill speed, solids residence time, hammer type, screen size/type, feed size
Universal/pin mill	15–30	Mill speed, solids residence time, head type, feed size
Jet mill	2–10	Gas pressure, solids residence time

19.2.2.1 Hammer Mill Hammer mills involve feeding solids through a series of spinning hammers contained within a casing that may also contain breaker plates. Attrition of particles is accomplished through their impact with the hammers and the mill internals. A sieve screen at the mill outlet is used to limit the size of the particle that can exit the system. Hammer mills come in a wide range of motor speeds, and this is the most important parameter to investigate to ensure reliable scale-up. Screen design is another important parameter, in that size and shape of perforations in the screen will affect mill residence time or how easily milled particles escape the chamber. The hammer mill is not ideal for milling very abrasive materials because significant metal erosion/mill wear can result over time. It is also not preferred for milling highly elastic materials, which could blind the screen, reduce gas flow through the mill, and lead to overheating.

19.2.2.2 Universal/Pin Mill The term "universal" mill usually refers to a mill configuration where multiple milling heads can be used. These mills often can be fitted with pin, turbo-rotor, and hammer-type heads. Pin mills operate similarly to hammer mills, but with typically faster rotor speeds and small clearances between rotating and stationary pins. Solids are fed to a milling chamber in a conveyance gas stream. The milling chamber contains a high-speed rotor–stator configuration of pins, which impact the particles as solids are directed from the center of the pin disc out through all the rows intermeshing pins (Figure 19.3). Control parameters to adjust and vary the output PSD include pin gap or pin spacing, the rotational speed of the rotor, solids feed rate, size of the mill, and velocity of the carrier gas used to convey the solids out of the mill.

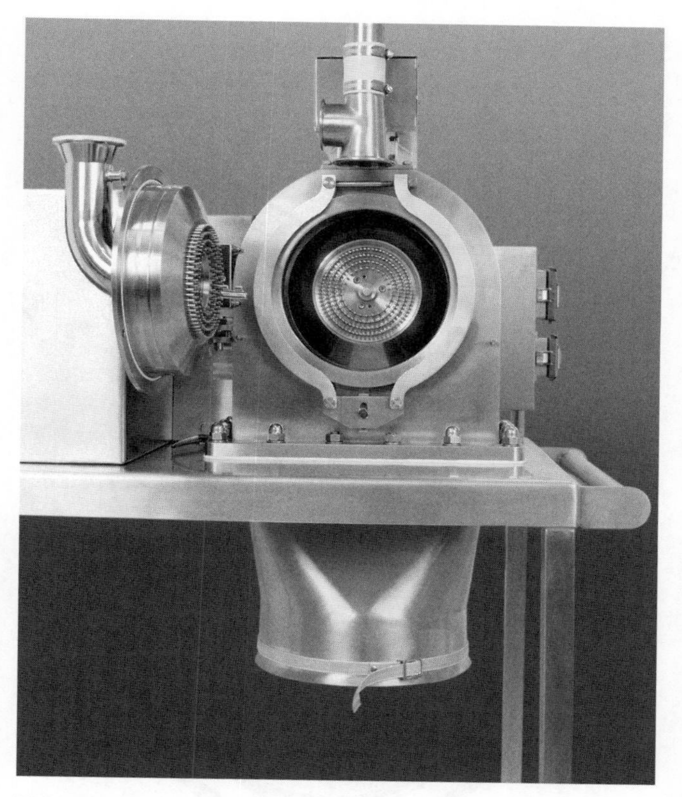

FIGURE 19.3 Pin mill internals—Photo courtesy of Hosokawa Micron Powder Systems.

19.2.2.3 Jet Mill Jet mills are an alternative to hammer or pin milling, where the primary mode of action is mechanical impact of the mill with the particle. With jet milling, micronization is accomplished mainly through particle–particle collisions caused by high velocity gas streams. Spiral jet mills, loop jet mills, and fluidized bed jet mills are examples of jet or fluid energy mills. These jet mills use the same general operating principle. Through the introduction of high-pressure air or other carrier gas (i.e., nitrogen) into specially designed nozzles, the potential energy of the compressed gas is converted into a grinding stream at sonic or supersonic velocities.

Differences in the various jet mills are predominantly in the geometry of the grinding chamber itself. Spiral mills

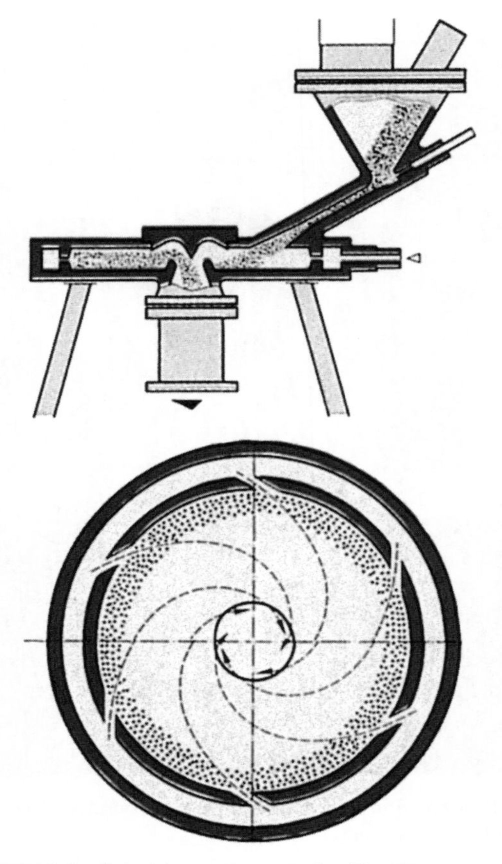

FIGURE 19.4 Spiral jet or "pancake" mill—Photo courtesy of Hosokawa Micron Powder Systems.

create a high velocity helix of gas that rotates around the center of the circular jet mill. Solids are introduced via a venturi feed eductor (Figure 19.4) and become entrained in the turbulent helical flow. The resulting high-energy collisions between particles as well as between the particles and the mill internals fracture particles to micron and submicron size. For loop jet mills, air or carrier gas is injected into an oval grinding loop or "race track" through specially designed nozzles (Figure 19.5). Solids are introduced in a manner similar to the spiral jet mill. In both designs, particles will stratify based on their relative inertia toward the outlet of the mill, resulting in larger particles remaining in the grinding chamber longer while smaller particles are carried out of the mill with the overall gas exhaust to the collection cyclone or chamber. This permits some internal classification and makes the spiral or loop mill performance somewhat less dependent on feed particle size, compared with hammer and pin mills. Key scale-up parameters are the grinder nozzle pressure and the mill residence time, typically determined as particle density in the mill (ratio of solid mass flow to volumetric gas flow through the mill).

In fluidized bed jet mills, the grinding chamber is oriented as a fluidized bed, with specially designed nozzles introducing the grinding gas at the bottom of the mill chamber, creating high intensity collisions between particles (Figure 19.6). Net vertical gas flow out of the mill fluidizes the milled material in the grinding chamber. Fluidized bed jet mills are usually fitted with a classification wheel. Based on the rotational speed of the classifier wheel, large particles gain radial momentum and are returned to the grinding zone of the

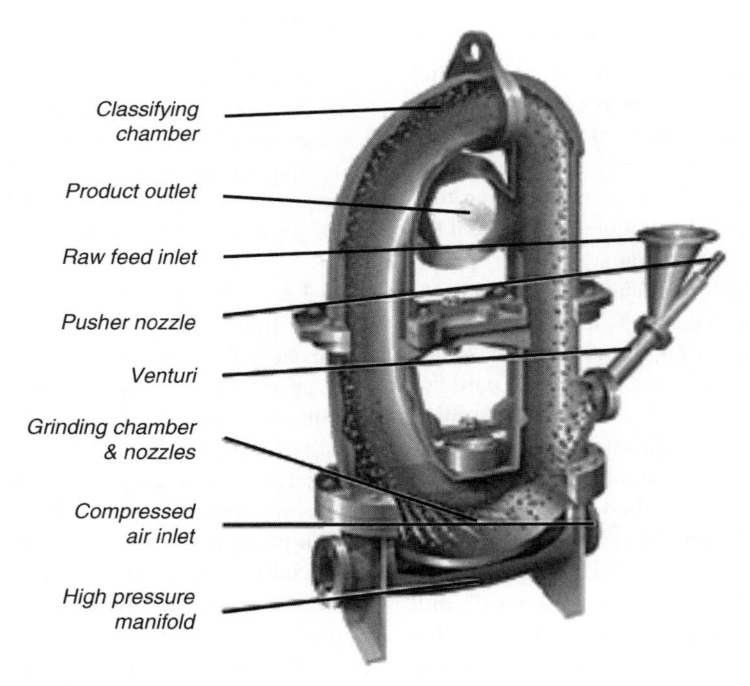

FIGURE 19.5 Loop jet mill—Photo courtesy of Fluid Energy & Co.

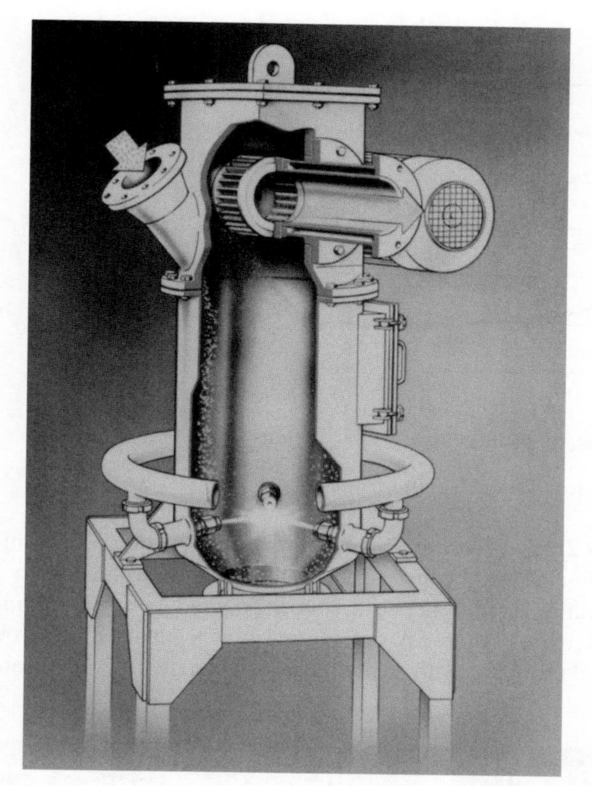

FIGURE 19.6 Fluidized bed jet mill with classification—Photo courtesy of Hosokawa Micron Powder Systems.

fluidized bed. Lighter particles can escape the mill through vanes in the classifier wheel, carried by the main gas exhaust. This type of arrangement can lead to much narrower particle size distributions than other types of jet milling equipment. In addition to grinder pressure, classifier speed is clearly an important scale-up parameter. To control the solids residence time in the mill, nozzle diameter, total gas flow rate, and mill chamber pressure can be varied with solids feed rate, and these variables must be controlled on scale-up as well.

19.2.3 Wet Milling Setup

Milling a solid suspended in liquid is referred to as either wet or slurry milling. Figure 19.7 shows a generalized setup for a wet milling operation. Wet milling typically occurs in a recirculation loop from a well-mixed holding vessel. Solids are suspended in an appropriate solvent or solvent mixture, either through selective crystallization from that solvent mixture, or the reslurry of previously isolated solids. The system is agitated to maintain a well-mixed system and prevent plugging of the vessel outlet. The slurry is circulated through the wet mill, and cycled back to the holding vessel. The recirculation is continued until slurry samples from the vessel show adequate reduction in particle size. Because of the nature of the recirculation operation, there exists a residence time distribution for particles within the slurry vessel. Where some particles will have traveled through the recirculation loop in multiple passes, some will have not yet passed at all. As a result, the distribution of particle sizes can tend to broaden over time. Operating repeatedly in single pass or "once-through" mode may tighten the distribution, depending on solid properties.

Generally, a lower limit exists for the obtainable particle size. Figure 19.8 shows the d_{90} (90% of the particles have a diameter less than or equal to the d_{90}) of a slurry of drug substance as a function of the mill passes (defined as the ratio

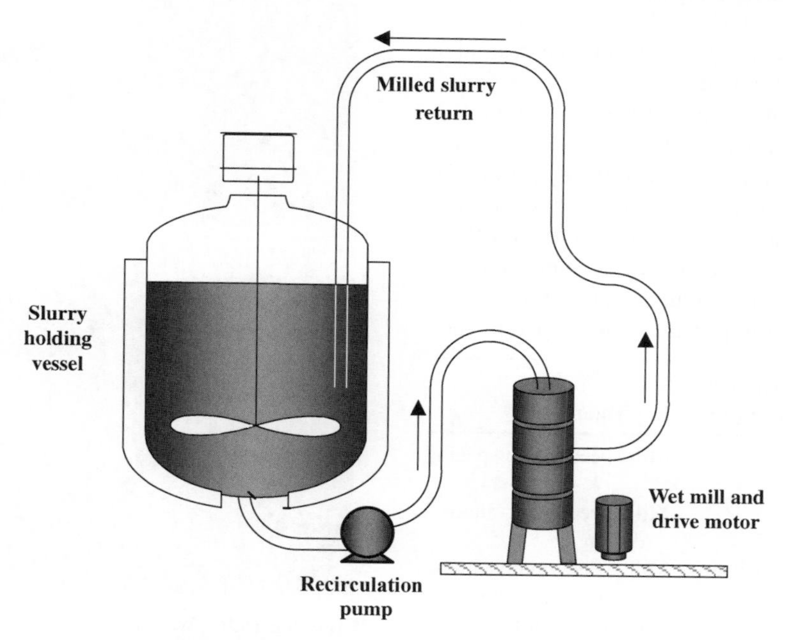

FIGURE 19.7 General setup for wet milling operations.

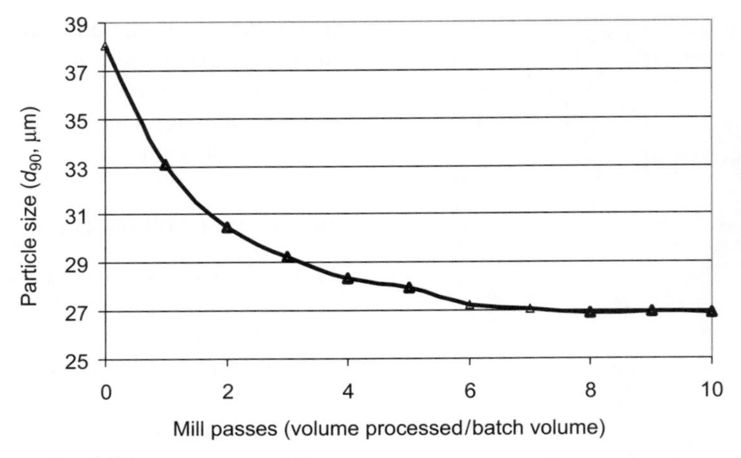

FIGURE 19.8 Particle size versus mill passes in a slurry milling operation.

of the volume of material processed to the volume of the slurry). As can be seen by the figure, a point of diminishing returns is often reached where further circulation through the mill does very little to further reduce the particle size, but will continue to increase the levels of fines produced by chipping rather than mass fracture.

19.2.4 Wet Milling Equipment

Wet milling offers many advantages to dry milling, although there are some limitations. One significant advantage is the ability to eliminate a separate unit operation of dry milling. Wet milling can be carried out as part of the final crystallization–isolation sequence. This approach eliminates the cycle time and cost associated with an extra unit operation, which can be particularly significant if special containment is necessary (see next section). Wet mills also provide the ability to protect the product from the heat input from some dry milling equipment.

As with dry mills, multiple types of wet mills are available. The three most commonly used in pharmaceutical manufacturing are toothed rotor–stator mills, colloid mills, and media mills. The table below summarizes typical minimum particle sizes and main scale-up parameters for these mills. If the mills are operated in recycle mode but not allowed to reach "steady state" milled particle size, feed particle size will also be an important scale-up parameter.

Mill Type	Typical Minimum Milled Size (μm)	Key Parameters
Toothed rotor–stator mill	20–30	Mill speed, tooth spacing
Colloid mill	1–10	Mill speed, rotor–stator design
Media mill	Submicrometer–1 μm	Mill speed, composition, amount, and size of media in mill

19.2.4.1 ***Toothed Rotor–Stator Mills*** Rotor–stator mills (Figure 19.9) consist of a rotating shaft (rotor), with an axially fixed concentric stator. Toothed rotor–stator mills have one or more rows of intermeshing teeth on both the rotor and the stator with a small gap between the rotor and stator.

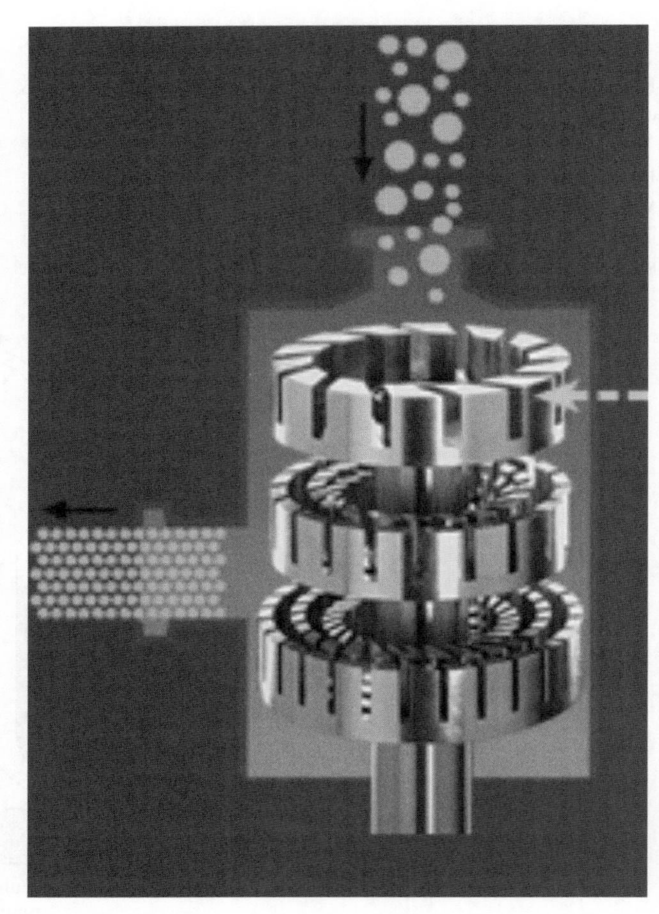

FIGURE 19.9 Wet mill internals—Photo courtesy of IKA Works Inc.

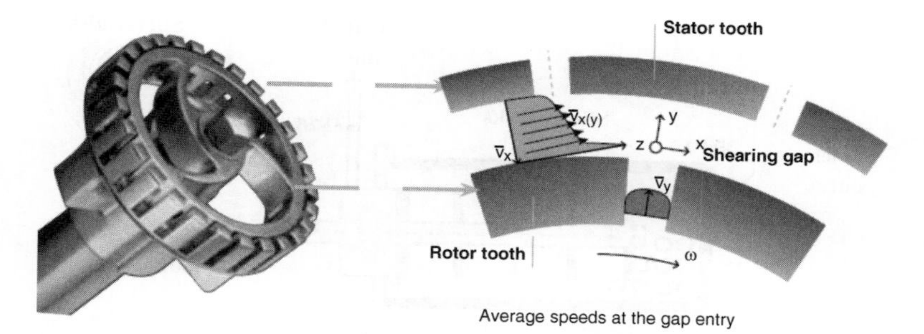

FIGURE 19.10 Rotor–stator working principle—Photo courtesy of IKA Works Inc.

Variations in the number of teeth, teeth spacing, angle of incidence, etc., all impact the milling efficiency of toothed rotor–stator mills [24–26].

The differential speed between the rotor and the stator imparts extremely high shear and turbulent energy in the gap between the rotor and the stator. In this configuration, particle size is reduced by both the high shear created in the annular region between the teeth and by the collisions of particles on the leading edge of the teeth (Figure 19.10). Size is affected by selecting rotor–stator pairs with different gap thickness, or by operating at different rotational rates (or tip speeds) of the rotor. Tip speed is a very important factor when considering the amount of shear input into the product. Tip speed (v_{ts}) is determined according to the following equation

$$v_{ts} = \pi D \omega \tag{19.1}$$

where D is the rotor diameter and ϖ is the rotation rate (rev/min).

The shear created between the rotor and the stator is a function of the tip speed and the gap thickness. The shear rate $\dot{\partial}$ is given by

$$\dot{\partial} = \frac{v_{ts}}{h} \tag{19.2}$$

where h is the gap distance.

Another important factor is the number of occurrences that rotor and stator openings mesh. This is known as the shear frequency (f_s) and is calculated as

$$f_s = N_r N_s \omega \tag{19.3}$$

where N_r and N_s are number of teeth on the rotor and the stator, respectively.

The shear number (N_{sh}) is given by

$$N_{sh} = f_s \dot{\partial} \tag{19.4}$$

Depending upon the rotor–stator design, the number of rows should be accounted for when calculating the shear number. Scale-up of toothed rotor–stator mills is accomplished by determining for the material in question whether shear frequency or tip speed has greater control over milled size, and then working with rotor–stator design options and mill speed to maintain the key parameter across scales. If milling will be stopped before steady state/minimum size is achieved, the process must also be scaled on batch turnovers or passes through the mill.

19.2.4.2 Colloid Mills Colloid mills are another form of rotor–stator mill. A colloid mill is composed of a conical rotor rotating in a conical stator (Figure 19.11). The surface of the rotor and the stator can be smooth, rough, or slotted. The spacing between the rotor and the stator is adjustable by varying the axial location of the rotor to the stator. The gap can be as little as a few hundred micrometers or as large as a few millimeters [17]. Varying the gap varies not only the shear imparted to the particles but also the mill residence time and the power density applied. Particle size is affected by adjusting the gap and the rotation rate, and these are the key parameters to maintain at increased scale.

19.2.4.3 Media Mills Media mills, also referred to as pearl or bead mills, are much different in operation than a

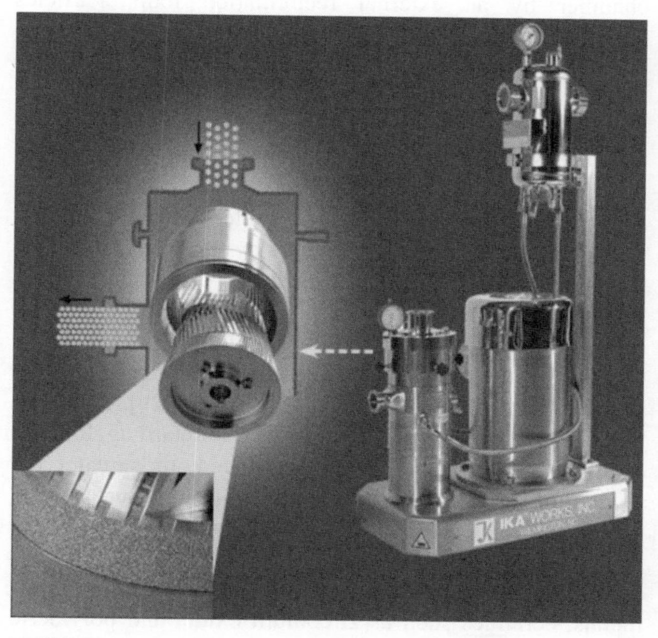

FIGURE 19.11 Colloid mill—Photo courtesy of IKA Works Inc.

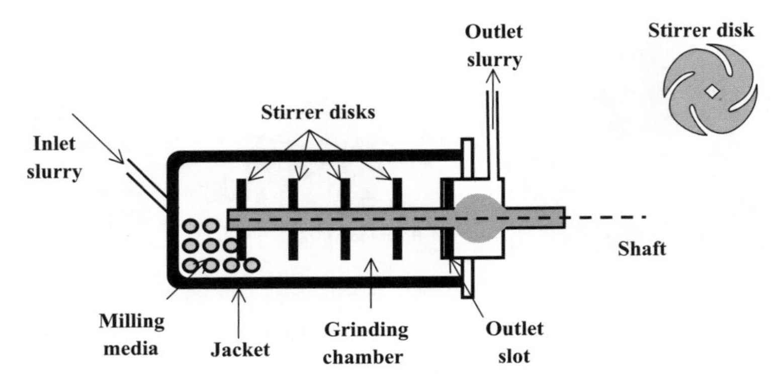

FIGURE 19.12 Schematic diagram of a media mill.

rotor–stator mill. The mill is composed of a milling chamber, milling shaft, and product recirculation chamber (Figure 19.12). The milling shaft extends the length of the chamber. A shaft can have either radial protrusions or fingers extending into the milling chamber, a series of discs located along the length of the chamber, or a relatively thin annular gap between the shaft and the mill chamber. The chamber is filled with spherical milling media usually less than 2 mm in diameter and typically 1 mm in diameter or less. Media are retained in the mill by a screen located at the exit of the mill. The rotation of the shaft causes the protrusions to move the milling media, creating high shear forces [27]. Scale up of media mills is accomplished by maintaining residence time in the mill, keeping the milling media size constant, and holding energy input constant.

The high energy and shear that result from the movement of the milling media is imparted to the particles as the material is held in the mill or recirculated through the milling chamber by an external recirculation loop. Particles are ground by a combination of particle–media, particle–particle, and particle–wall collisions. The result is the ability to create submicron particles. Thermally labile material is easily handled as the milling chamber is jacketed. By utilizing smaller media (less than 100 μm), nano-sized (20 nm) particles are achievable.

At the small size scales achievable by media milling, particle–particle interactions caused by van der Waals forces can begin to dominate [28]. By inclusion of certain additives to the dispersion fluid, the possible agglomeration and resulting reduced efficiency and reduced effectiveness of the mill can be mitigated. Surfactants can inhibit agglomeration by both electrostatic and steric stabilization [29]. Similarly, polymeric stabilizers can also be used to retard agglomeration. The smaller the particle produced, the greater the amount of surfactant needed, since the specific surface area of the particle increase with decreasing one over the radius of the particles.

Milling media are available as several materials glass, metals, ceramics such as zirconium oxide, and polymeric such as a highly cross-linked polystyrene resin. The proper selection of the milling media is an important criterion in the milling process. Materials such as glass and metals are typically not used for APIs since some abrasion of the milling media can occur, potentially arising in extraneous material concerns. As a result, either polymeric or ceramic media are usually used.

19.3 SAFETY AND QUALITY CONCERNS

Chemical engineers must proactively address process safety concerns and quality risks associated with wet and dry milling processes. While these operations are not unique to the pharmaceutical industry, performing these operations on potent organic molecules can pose some unique challenges.

The main quality concern for both wet and dry milling is to minimize introduction of extraneous matter into the milled material. The most common material of construction of milling equipment is stainless steel, although ceramic, Hastelloy C, and various types of Teflon are also frequently used depending on the material to be milled. Abrasion of the mill surfaces by API or excipient particles can result in erosion of the mill material, and levels of residual metals such as Cr and Ni must be minimized relative to the human dose of API [30]. Assessing the hardness of the API during milling process development, either empirically through small-scale milling experiments or fundamentally through experiments such as indentation testing [19], will help prevent unexpected outcomes on scale-up.

The most obvious safety risk associated with dry milling is personnel exposure to highly potent airborne dust. This is especially important to consider during drug development, before full API safety testing has been completed, and a conservative equipment setup is recommended at early development stages. While most mill equipment is or can be enclosed during operation, minimizing this exposure risk, solids recovery/packaging operations and equipment cleaning should not be overlooked as opportunities for exposure. Typical methods to address this risk include clean-in-place systems and performing the entire milling operation in a negative pressure isolator with HEPA filtered exhaust. Multiple levels of exposure control, beginning with these engineering controls and continuing through personal protective equipment (PPE), are often required to ensure adequate operator protection.

Perhaps less apparent but no less serious is the risk of dust explosion during dry milling or postmilling activities [31, 32]. There have been numerous reported incidents of explosions occurring in the chemical, food, and pharmaceutical industry, in situations where fine powders are dispersed in air and a spark may be present. The most straightforward way to address this risk in development is to mill in an inert atmosphere (no/limited oxygen), using nitrogen or argon, and maintaining a scrupulously clean processing area. When sufficient API is available (tens of grams), the explosive properties of the dust should be quantified to better understand dust explosion risks. Typical measurements include determining the minimum ignition energy (MIE) (smallest amount of energy required to ignite a dust cloud of optimal concentration) and minimum ignition temperature (minimum temperature at which a dust cloud will ignite) [33, 34]. If an explosive dust (MIE < 20 mJ) will be dry milled at pilot or manufacturing scale, engineering controls must be employed to minimize risk of explosion [35]. The force of a possible explosive event can be quantified by measuring K_{st} (expression of the burning dust's rate of pressure increase) and this parameter can be used in equipment/facility design. Engineering or design solutions include inertion of the process train and use of rupture disks or automatic shutoff valves that quickly close when pressure spikes are detected.

The most common risk associated with wet milling is that of accumulating a static charge in a recirculating loop, especially when a nonconductive solvent such as hexane, heptane, or toluene is used. Electrostatic charge accumulation results from frictional contact between flowing liquids and solids in recirculating loops and the surfaces of equipment (pipes, vessels, agitators). The resulting potential difference between batch and processing surfaces could lead to a static discharge, which could ignite a fire or explosion, especially while processing with flammable organic solvents. Conductivity of organic solvents can range from 10^7 to <1 pS/m. Another way to interpret these values is to estimate the time required for an accumulated charge to decay (a multiple of the relaxation time). For 99% charge decay this may range from milliseconds for water to several minutes for *n*-hexane [36]. Slurry conductivities of <100 pS/m are considered at high risk for electrostatic discharge if a mechanism for charge formation is present (such as a recycle loop or wet mill). Charges can be encouraged to dissipate by use of conductive equipment (stainless steel in place of Teflon-lined pipe) and grounding and bonding the equipment. Additional precautions include operating under nitrogen and keeping linear velocity through the recirculation loop as low as is practical (while maintaining adequate flow rates to ensure that solids do not settle within the recirculation piping and low points in the loop).

19.3.1 Example Approach to Milling Scale-Up

Throughout the development cycle, different PSDs may be studied to determine the optimum size range for an API or excipient. Determining milling parameters to achieve these profiles has often been done in an empirical fashion through the use of test milling—portions of a material lot are subjected to different milling conditions, and the resultant output tested to determine particle size distribution. While useful in establishing milling ranges for one particular lot of material, test mill runs do not easily allow extrapolation to different milling input conditions. As Quality by Design expectations continue to increase within the pharmaceutical processing world, an understanding of milling parameters, as well as input PSD and their effect on final API attributes is desirable.

As an example, a simple model for particle breakage, like the one shown in equation 19.5, would likely predict an output parameter, such as the d_{50} of the PSD, as a function of one or more input parameters.

$$d_{50,\text{milled}} = \mathit{function}(\text{input size, mill energy, residence time}) \tag{19.5}$$

To utilize this type of model, the input properties must be chosen and parameters simulating residence time and energy input must be determined. As a first approach, a static parameter such as the unmilled d_{50} of the slurry could be chosen as the input size, the mass feed rate (m) to the mill chosen as a surrogate for residence time, and rotor tip speed (ν) chosen for energy input.

$$d_{50,\text{milled}} = K(d_{50,\text{unmilled}})^{\varphi} m^{\alpha} \nu^{\beta} \tag{19.6}$$

Figure 19.13 shows the relationship between predicted and actual milled values using such a model following a calibration exercise to fit the exponents and the constant K.

While a rough trend may be observed between predicted and actual, the accuracy of this model is not sufficient for predicting actual output PSD. Examining further the choice of input parameters for the model, one can observe that while the d_{50} of an in-process sample from the crystallizer may describe the slurry PSD well prior to drying (Figure 19.14), this parameter may be insufficient to capture variations in actual mill input material following the final drying unit operation (Figure 19.15). The relatively unimodal peaks from the slurry samples clearly change to a wide variety of average sizes and modalities after drying in an agitated filter dryer. It is this variation in size entering the milling step that makes prediction difficult for a simple model such as that of equation 19.6.

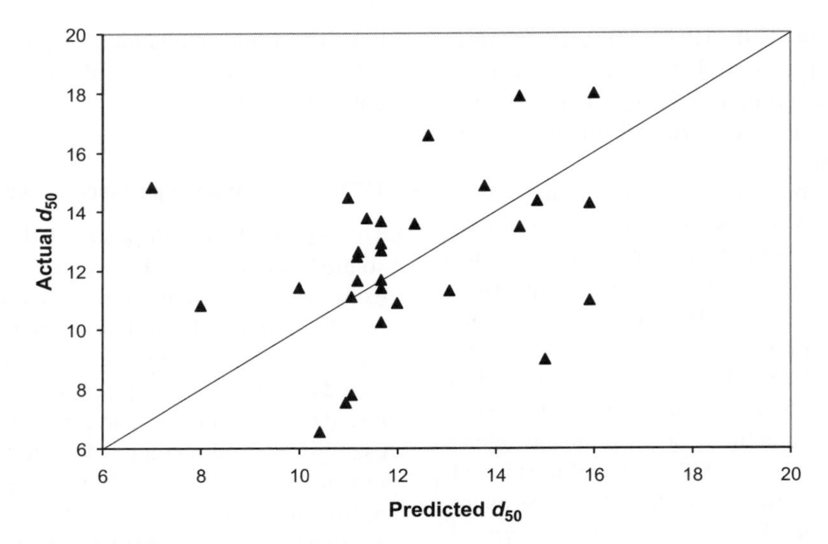

FIGURE 19.13 Predictive ability of a milling model utilizing unmilled slurry d_{50} as an input parameter.

To account for the variability seen in the dried material prior to milling, a simple summary statistic cannot be used. Rather, a measure of the entire PSD should be utilized to inform the model of the distribution of particle sizes moving into the breakage operation. Most particle size analyzers generate histograms with multiple channels of data describing the overall distribution of particle sizes in the sample. Chemometric methods such as principal component analysis (PCA) allow the full shape of the PSD to be reduced from a very large number of histogram components to 2–3 principal components describing the breadth, shape, and skewedness of the distribution. Using PCA in this manner maintains model inputs to a minimum number without significant loss in ability to characterize the full PSD spectrum.

After adding the principal components to the model to accurately represent the shape of the distribution, the equation becomes

$$d_{50,\text{milled}} = K(\text{PCA1})^{\varphi}(\text{PCA2})^{\mu}(\text{PCA3})^{\partial}m^{\alpha}v^{\beta} \qquad (19.7)$$

where PCA1, PCA2, and PCA3 represent the three principal components of the PSD. This enhancement to the model, following a short number of calibration runs, allows much better prediction of output d_{50}, as shown in Figure 19.16.

The final predictive model is still limited to the API that was used to generate the data. However, the model allows for prediction as needed, and the process of constructing the model may provide valuable information regarding the API,

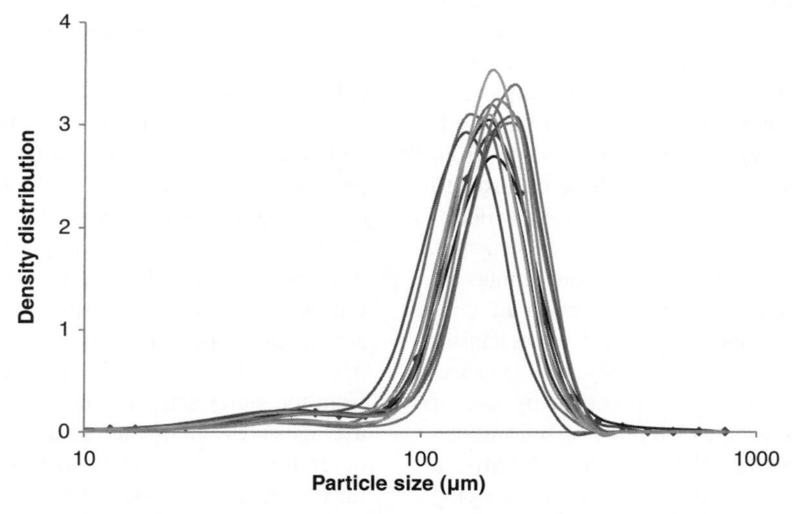

FIGURE 19.14 Particle size of the API coming out of the crystallizer.

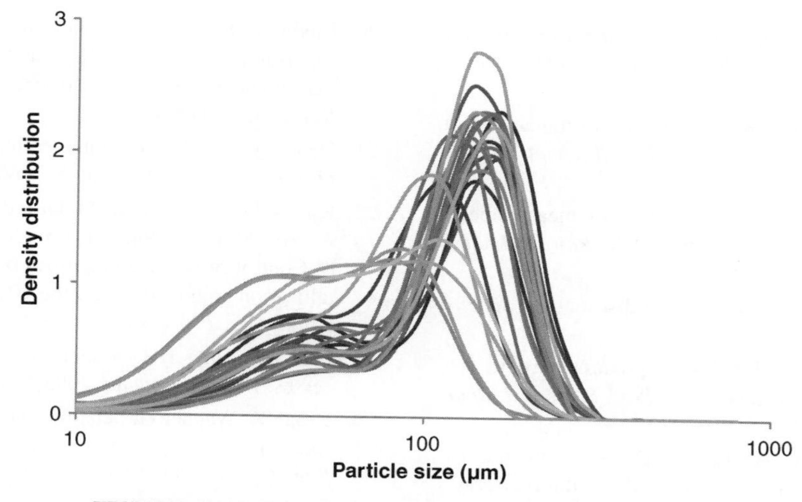

FIGURE 19.15 Particle size of API following agitated drying.

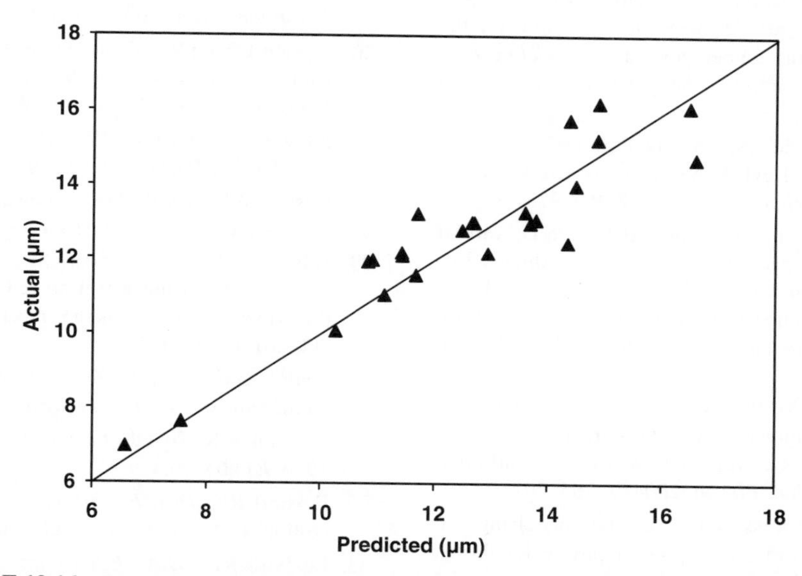

FIGURE 19.16 Predictive ability of a milling model incorporating full PSD spectrum as an input.

including the relative importance of inclusion of the entire PSD spectrum as well as the impact that filtration and drying technique may have upon milling performance.

REFERENCES

1. Parrott EL. Milling of pharmaceutical solids. *J. Pharm. Sci.* 1974;63(6):813–829.
2. Amidon GL, Lennernas H, Shah VP, Crison JR. A theoretical basis for a biopharmaceutic drug classification: the correlation of *in vitro* drug product dissolution and *in vivo* bioavailability. *Pharmaceut. Res.* 1995;12:413–420.
3. Oh DM, Curl R, Amidon G. Estimating the fraction dose absorbed from suspensions of poorly soluble compounds in humans: a mathematical model. *Pharmaceut. Res.* 1993;10: 264–270.
4. *Waiver of in vivo bioavailability and bioequivalence studies for immediate-release solid oral dosage forms based on a Biopharmaceutics Classification System.* U.S. Department of Health and Human Services Food and Drug Administration Center for Drug Evaluation and Research (CDER), August 2000.
5. Rohrs BR, Amidon GE, Meury RH, Secreast PJ, King HM, Skoug CJ. Particle size limits to meet USP content uniformity criteria for tablets and capsules. *J. Pharm. Sci.* 1996;95 (5):1049–1059.

6. Shoyele SA, Cawthorne S. Particle engineering techniques for inhaled biopharmaceuticals. *Adv. Drug Deliv. Rev.* 2006;58 (9–10):1009–1029.
7. Malcolmson RJ, Embleton JK. Dry powder formulations for pulmonary delivery. *Pharmaceut. Sci. Technol. Today* 1998;1:394–398.
8. Holve DJ, Harvill TL. Particle size distribution measurements for in-process monitoring and control. *Adv. Powder Metall. Particul. Mater.* 1996;1:4.81–4.93.
9. Collin A. Online control of particle size distribution. *Spectra Analyse* 1997;26(196):31–33.
10. Harvill TL, Hoog JH, Holve DJ. In-process particle size distribution measurements and control. *Part. Part. Syst. Char.* 1995;12(6):309–313.
11. Snow RH, Allen T, Ennis BJ, Litster JD. Size reduction and size enlargement. In: Perry RH, Green DW, editors, *Perry's Chemical Engineers' Handbook*, 7th edition, Chap. 20, McGraw Hill, New York, New York, 1997.
12. Popov KI, Krstic SB, Obradovic MC, Pavlovic MG, Irvanovic ER. The effect of the particle shape and structure on the flowability of electrolytic copper powder. *I. Modeling of a representative powder particle. J. Serb. Chem. Soc.* 2003;68 (10):771–777.
13. Brittain HG, Bogdanowich SJ, Bugay DE, DeVincentis J, Lewen G, Newman AW. Review: physical characterization of pharmaceutical solids. *Pharm. Res.* 1991;8:963–973.
14. Gabaude CMD, Gautier JC, Saudemon P, Chulia D. Validation of a new pertinent packing coefficient to estimate flow properties of pharmaceutical powders at a very early development stage, by comparison with mercury intrusion and classical flowability methods. *J. Mater. Sci.* 2001;36: 1763–1773.
15. Lindberg N-O, Palsson M, Pihl A-C, Freeman R, Freeman T, Zetzener H, Enstad G. Flowability measurements of pharmaceutical powder mixtures with poor flow using five different techniques. *Drug Dev. Ind. Pharm.* 2004;30:785–791.
16. Ward GH, Shultz RK. Process-induced crystallinity changes in albuterol sulfate and its effect on powder physical stability. *Pharm. Res.* 1995;12:773–779.
17. Buckton G. Characterization of small changes in the physical properties of powders of significance for dry powder inhaler formulations. *Adv. Drug Deliv. Rev.* 1997;26:17–27.
18. Price R, Young PM. On the physical transformations of processed pharmaceutical solids. *Micron* 2005;36:519–524.
19. Ghadiri M, Kwan C, Ding Y. Analysis of milling and the role of feed properties. In: *Handbook of Powder Technology*, Vol. 12, Chap. 14, Elsevier, 2007.
20. Liotta V, Sabesan V. Monitoring and feedback control of supersaturation using ATR-FTIR to produce and active pharmaceutical ingredient of a desired crystal size. *Org. Process Res. Dev.* 2004;8:488–494.
21. Dennehy RD. Particle engineering using power ultrasound. *Org. Process Res. Dev.* 2003;7:1002–1006.
22. Kim S, Lotz B, Lindrud M, Girard K, Moore T, Nagarajan K, Alvarez M, Lee T, Nikfar F, Davidovich M, Srivastava S, Kiang S. Control of particle properties of a drug substance by crystallization engineering and the effect on drug product formulation. *Org. Process Res. Dev.* 2005;9:894–901.
23. Paul EL, Tung HH, Midler M. Organic crystallization processes. *Powder Technol.* 2005;150(2):133–143.
24. Urban K, Wagner G, Schaffner D, Röglin D, Ulrich J. Rotor–stator and disc systems for emulsification processes. *Chem. Eng. Technol.* 2006;29(1):24–31.
25. Lee I, Variankaval N, Lindemann C, Starbuck C. Rotor–stator milling of apis—empirical scale up parameters and theoretical relationships between the morphology and breakage of crystals. *Am. Pharm. Rev.* 2004;7(5):120–123.
26. Atiemo-Obeng V, Calabrese R. Rotor stator mixing devices. In: Paul EL, Atiemo-Obeng V, Kresta S.editors, *Handbook of Industrial Mixing*, 1st edition, Wiley, New York, 2003.
27. Kwade A, Schwedes J. Wet grinding in stirred media mills. In: *Handbook of Powder Technology*, 2007, Elsevier, pp. 229–249.
28. Russell WB, Saville DA, Schowalter WR. *Colloidal Dispersions*, Cambridge University Press, New York, 1991.
29. Bilgili E, Hamey R, Scarlett B. Nano-milling of pigment agglomerates using a wet stirred media mill: elucidation of the kinetics and breakage mechanisms. *Chem. Eng. Sci.* 2006;61(1):149–157.
30. EMEA *Guideline on the specification limits for residues of metal catalysts or metal reagents*, 21 Feb 08.
31. Cashdollar K. Coal dust explosivity. *J. Loss Prev. Process Ind.* 1996;9(1):65–76.
32. Eckhoff RK. *Dust Explosions in the Process Industries*, 3rd edition, Gulf Professional Publishing, Burlington, MA, 2003.
33. Eckhoff RK. *Dust Explosions in the Process Industries*, Butterworth-Heinemann Ltd., Oxford, 1991.
34. Jaeger N, Siwek R. Prevent explosions of combustible dusts. *Chem. Eng. Progress* 1999;June:25–37.
35. Stevenson B. Preventing disaster: analyzing your plant's dust explosion risks. *Powder Bulk Eng.* 2001;January:19–27.
36. Britton LG. *Avoiding Static Ignition Hazards in Chemical Operations*. Center for Chemical Process Safety of the AIChE, New York, 1999.

20

PROCESS SCALE-UP AND ASSESSMENT

ALAN D. BRAEM, JASON T. SWEENEY, AND JEAN W. TOM
Process Research and Development, Bristol-Myers Squibb Co., New Brunswick, NJ, USA

20.1 INTRODUCTION

One of the key roles of chemical engineers in drug substance process development is transforming an active pharmaceutical ingredient (API) synthesis route into a scalable commercial process. In this chapter, we present an approach to process assessment and scale-up focused on risks to safety, quality, and business that takes into consideration the product's stage of development and the magnitude of scale-up. Emphasis is placed on understanding common unit operations' scale-up factors, and use of this knowledge in the assessment and definition of a development strategy.

The initial chemical synthesis of a small molecule active pharmaceutical ingredient is typically developed by an exploratory or discovery chemistry group. The goal of the initial synthesis is to quickly enable production of milligrams to grams of the API to support exploratory studies and to confirm the biological activity of the molecule. This initial route is designed to be divergent and allow access to a variety of targets and is not designed for further scale-up to kilogram scale, much less to a manufacturing process. It is the role of process chemists to design a synthesis that can be developed into a scalable process to deliver sufficient API quantity and quality to support clinical, toxicology assessment, and downstream formulations. It is a collaboration of process chemical engineers with the process chemists that shapes the synthesis from a procedure to a plant-scale process.

20.1.1 Phases of Development

The focus of the chemical engineer in transforming a synthesis into a scalable process changes as the compound goes through the various stages of development. The evolution of the synthesis route is tied to the clinical timeline. Table 20.1 provides a simplified overview of the evolution of the synthesis and product requirements as the stage of development progresses.

In reality, the progression of development is likely not as clearly defined, and overlap of these categories will occur, depending on the compound's potency, synthesis and molecular complexity, the therapeutic class of the compound, and the infrastructure and organization of the Process R&D group. However, the process development goals can be generally delineated within these milestones in terms of the magnitude of scale-up, and the process safety, business, and quality risks. This, in turn, guides the allocation of engineering resources and determines the level of risk assessment needed.

20.1.1.1 Early Development In early development, the synthesis milestone is the selection of an appropriate route for the initial scale-up. The key considerations are process safety, chemical hygiene, number of synthetic steps, availability of reagents, raw materials, and intermediates, and ability of the synthesis to address API quality. The top process assessment priority is to identify hazardous reactions and reagents, and to evaluate safe operating limits and material exposure limits. Altering the reagents and conditions and/or developing engineering controls to ensure safe operation are the main engineering foci. The secondary process assessment priority is to meet targeted process development goals. These goals would be (1) to obtain sufficient process knowledge to support at-scale operations (first-order scale effects, stability over duration of unit

Chemical Engineering in the Pharmaceutical Industry: R&D to Manufacturing Edited by David J. am Ende

TABLE 20.1 Process Research and Development Requirements During Stages of Development [1,2]

Stage of Development	Discovery	Early Development	Full Development	Launch
Clinical stage	IND toxicology	Phase I	Phase II	Phase III/launch
Type of synthesis	Expedient	Practical	Efficient	Optimal
Synthesis milestone	Enabling synthesis	Route (intermediates) selected	Sequence of unit operations finalized	Process parameter ranges finalized
Amount of compound	10 mg–10 g	100 g–10 kg	10–100 kg	>100 kg
Site for preparation	Laboratory	Kilo laboratory	Pilot plant	Manufacturing plant
Number of batches	1–5	1–10	10–100	10–1000
Probability of success to next stage of development [3–6]	40–60%, preclinical to clinical	40–60%, phase I to phase II	40–60%, phase II to phase III	80–100%, phase III to NDA filing and launch

operations, heat/mass transfer, and hold points) and (2) understanding, but not optimization, of the process conditions (stoichiometry, temperature, concentration, filtration) to enable appropriate equipment usage. Process knowledge may be limited to single point information rather than a design range. The API product quality must meet an initial set of specifications, which may include form, purity, stability, and impurities.

At this stage of development, the probability of the molecule achieving success for a New Drug Application (NDA) is still low (<10%), and there is a high likelihood that the chemistry will not be used beyond this stage of development. Thus, fewer engineering resources will generally be allocated.

20.1.1.2 Full Development In this stage of development, the compound has demonstrated some key human safety milestones (phase I) and/or some evidence of clinical response or efficacy. A clinical timeline can be projected for the compound that will lead to a New Drug Application.

By this stage, the process chemists will have evaluated alternate synthetic routes and will have determined the desired sequence of intermediates. The development will focus on finalizing the chemistry and reagents and defining the API crystal form and powder attributes. The process scale-up assessment will continue to have strong process safety and quality focus. In addition, an evaluation of the business risks will guide the efforts to develop a robust, efficient, and economical manufacturing process. Key aspects to evaluate for this assessment include yield, cycle time, equipment usage, waste output, and need for analytical support such as in-process assays and process analytical technology.

At this phase of development, the probability of success for the molecule to be filed for approval has increased significantly. The optimized process must consistently meet the quality requirements at the pilot plant scale before the next stage of development.

20.1.1.3 Launch This stage of development will focus on the final process optimization and providing a full understanding of the chemistry, manufacturing, and controls (CMC) for the New Drug Application. Detailed information on process parameter ranges and fundamental understanding of the key unit operations (e.g., reactions, crystallizations) will be required to support the process validation at the manufacturing site and the submission of the NDA.

20.1.2 Process Safety and Risk Assessments

Risk is often defined as the combination of the probability and the severity of a harmful occurrence. Regulatory guidance to the pharmaceutical industry on evaluating risk is that the protection of the patients is of prime importance [7]. Protection of workers and maximizing business objectives are also key goals in assessing risk. Thus, the assessment of a chemical process must cover three key aspects: safety, product quality, and business parameters. Safety and quality risk assessments are inherent in all phases of development, while the emphasis on business risk assessment increases in the later stages of development. Process safety is tied to standards and regulations governed by government safety agencies and acceptable and appropriate product quality is guided by the various ICH (International Conference on Harmonization) guidelines for registration of pharmaceuticals for human use [8].

20.1.2.1 Process Safety Assessment Process safety assessment is one of the main chemical engineering concerns throughout process development. As Chapter 11 covers this topic in more depth, this chapter will briefly review some of the key elements of such an assessment and its impact on process scale-up.

For any scale-up, the first step is a review of process safety to (1) examine the reaction issues such as impact of chemical mixing, rate, sequence, and mode of charges, parameters for self-heat, potential gas evolution, corrosivity; (2) evaluate the thermal and chemical stability; and (3) evaluate the electrostatic and dust hazards. Issues raised in this review require identification of safe limits by understanding the mechanisms of the hazards, and an engineering evaluation to provide appropriate controls to meet such safe limits. This review is typically followed by a safety assessment against the scale-up implementation plan. An example would be a Process Hazards Analysis (PHA) [9] that uses a standard methodology to evaluate potential process and equipment hazards that could cause the catastrophic release of hazardous materials or other significant safety impacts and ensures that appropriate safeguards are in place to prevent, detect, or mitigate these occurrences.

As an example, for a highly exothermic reaction, chemical engineers can (1) evaluate the equipment capabilities (i.e., heat transfer or mass transfer rates), (2) optimize the chemistry via reactants and reaction conditions (i.e., changing the mode, sequence, or rate of a reagent charge, which change the kinetic stoichiometry), and (3) adjust the process setup to mitigate the hazard (i.e., emergency quench vessel, addition of external heat exchanger to increase cooling).

20.1.2.2 Process Risk Assessment There are various goals in a process risk assessment, which may take many forms. Regulatory guidance sets the primary goal of a risk assessment as identification of issues resulting in drug product quality that adversely impacts the patient's health. For API synthesis, this means identifying the process parameters that can cause the drug substance to fail its critical quality attributes. Such critical quality attributes may include product potency, crystal form, impurity levels, and physical properties such as particle size distribution. Process issues that impact overall productivity, capacity, or process greenness are considered business risks, because they have little or no impact on the quality of the drug substance.

Some risk assessment goals are short term such as an evaluation to ensure that the chemistry is scalable in the proposed equipment. This usually involves having sufficient prior experience to maximize the probability of success to safely produce material of desired quality. There are also longer term risk assessment goals such as evaluating the process design and synthesis against the combination of desired safety, quality, and business criteria.

The risk management process may be informal, using relatively simple empirical tools such as flow charts, check lists, questionnaires, process mapping, and cause and effect diagrams to organize the data and facilitate decisions. Risk evaluation may also utilize formal processes and methodologies. Two recognized tools are listed as follows:

1. Failure mode evaluation and analysis (FMEA) [7] to score and quantify risks by identifying potential failure modes and the impact on product quality.
2. Kepner–Tregoe analysis [10,11] to provide a quantitative assessment of the synthesis, taking into consideration both quality and business risks.

Other tools, adapted from safety risk analysis, such as PHA, Hazard Operability Analysis (HAZOP), and Hazard Analysis and Critical Control Points (HAACP) [7], apply a failure analysis to meet set criteria of safety, product quality, and/or business deliverables. These approaches are particularly valuable when performing a systematic review of the processability and scalability of the chemistry for commercial manufacturing.

20.1.3 Manufacturing Considerations

The short-term focus of process scale-up and assessment is on glass plant or pilot plant processing; however, as a project moves through development, the focus will shift to address manufacturing scale concerns. There are a number of differences to consider when assessing the process for either the pilot plant or manufacturing scale-up. Some of these differences are generalized and tabulated in Table 20.2.

These differences can have significant implications for the process scale-up assessment. The flexibility and technical support in a pilot plant environment may allow equipment setups and tighter control of process parameters than is typical or possible in a manufacturing plant. These factors must be understood to design a robust commercial process.

20.2 DRIVERS FOR DEVELOPMENT/RISK ASSESSMENT

The process development strategy is defined by risk management across three main areas: safety, quality, and business. Both safety and quality are necessary attributes of a scalable synthesis at any phase of development. Therefore, at a minimum, there must be sufficient development to manage the risk to safety and quality. In the absence of safety and/or quality concerns, business considerations define the development strategy. The challenge is defining the level and timing of development work as projects transition from early

TABLE 20.2 Pilot Plant Versus Manufacturing Differences

Consideration	Pilot Plant Scale	Manufacturing Scale
Equipment size	50–4000 L	400–12,000 L
Operating hours	5 days (≤24 h/day)	7 days (usually 24 h/day)
Technical support	Process engineer/ chemist	Plant engineer
Automation	Manual to fully automated	More likely automated
Analytical support	Short turnaround (<1 h)	Long turnaround
Equipment setup	• More likely to utilize unique equipment with flexible setup • Mostly sequential operations	• More likely to utilize only standardized equipment • More likely to have parallel operations

to full development. Throughout the stages of development, it will be impossible to understand and eliminate all the risks. The key is to eliminate enough risk to ensure that the differences upon scale-up do not impact the development goals.

By understanding the goal of the scale-up and performing an assessment against the goal, the development needs can be prioritized. In early development, the goal will include safety and quality aspects. While the business factors for process optimization are not key drivers, there are still scenarios to consider such optimization. These drivers will be based on the complexity of the synthesis route, the long-term synthesis strategy, the project timeline, and the facilities' constraints. Some examples of such drivers are outlined in Table 20.3. As the project moves forward through development, business goals will become a higher priority.

The initial process scale-up assessment is typically performed as a paper exercise guided by prior experience. Then, a laboratory assessment is usually necessary to evaluate unknown risks. Evaluation of such unknown risks can usually be performed in a few well-planned experiments. For example, the impact of processing time is typically unknown in early development and is highly likely to change as the process is scaled. During laboratory runs of the process, sampling at key points and aging the samples at the processing conditions will provide sufficient information about the process stability over the plant-scale time frame. Example 20.1 is given to illustrate an initial process assessment for an early development project. The impact of each unit operation on safety, quality, and business is evaluated (Table 20.4).

TABLE 20.3 Drivers for Development Outside of Safety/ Quality Issues in Early Development Programs

Issue	Driver for Development	Potential Development Activities
Length of synthesis	Availability of plant time to prepare the chemistry to meet API needs	Alternate route development
Low yield in synthesis		Increased throughput by • higher yields • smaller maximum volume (V_{max}) • decreased cycle time
Sourcing of reagents	Key reagent not available in time for scale-up activities	Alternate route development Replacement of reagent
Intellectual property	Some key elements of synthesis under patent protection	Alternate route development to replace this element of synthesis

EXAMPLE 20.1

Process Description

1. Compound **1** is mixed with solvent and base, heated to 40°C, and aged for 1 h to activate the amide hydrogen (**1**) (Scheme 20.1).

$$R_2NHR_1\ (\mathbf{1}) + \text{Base} \xrightarrow[\text{(2) Isopropyl alcohol}]{\text{(1) } ClC(O)R_3\ (\mathbf{2})} R_1N(R_2)C(O)R_3\ (\mathbf{3}) + HCl$$

2. The solution is then cooled to 0°C and excess acid chloride (**2**) is added, initiating the coupling reaction to form amide (**3**). The reaction is exothermic and the temperature must be maintained at ≤5°C.
3. The reaction is then quenched at ≤20°C by addition of isopropyl alcohol. The quench is exothermic. Several process impurities are formed during the reaction and quench, with impurity A found to be a suspected carcinogenic compound.
4. Compound **3** is then crystallized at 20°C by addition of an antisolvent. Impurity A cocrystallizes with compound **3**.
5. The slurry is then filtered, the wet cake is washed to remove impurity A, and solid compound **3** is dried under vacuum at 50°C.

In this assessment (Table 20.4), the initial activation is shown as low priority and the crystallization and stability are both shown as medium priority. In early development, the absence of information regarding the activation is not considered an issue since it is likely that the process will not deviate significantly from the lab procedure. Similarly, the crystallization and stability should be manageable by keeping as close to the lab procedure as possible. A medium risk was assigned since mixing during the crystallization and time are key scale factors that could result in differences between the lab and glass plant procedures. If the same process was assessed for later stage development, all of the unit operations would receive a medium to high priority since the process knowledge is limited.

The subsequent sections will discuss in more detail areas that should be considered when evaluating the development needs for a program. It is important to keep in mind that these factors are not independent and therefore a compilation of factors may in itself be a driver for development.

TABLE 20.4 Step Assessment for Example 20.1

	Risk to			
Unit Operation	Safety	Quality	Business	Priority
Activation	Unknown—no apparent exotherm	Unknown—no observed degradation	Incomplete activation leading to low yield	Low
Reaction and quench	Highly exothermic, HCl liberated	Key impurities form at higher temperature	None	High—impurity formation and exotherm must be understood and controlled
Crystallization	Highly acidic	Key impurity rejection	None	Medium—key impurity rejection is sufficient but should be understood
Isolation	None	Impurity A is removed during the cake wash	None	High—process impurity A must be controlled
Drying	None	None	None	Low
Impact of time	None	Unknown		Medium—effect of age time should be understood

20.2.1 Process Safety

In general, safety is a risk to be understood and then managed. For most risks, mitigation strategies can be developed, once the risk is understood and acceptable risks have been defined. The definition of acceptable risk will likely change as the scale of operation increases and this could drive process development as the project moves forward along its timelines. For example, the safety analysis for a $\leq$20 L scale-up may be limited to a paper review when no sign of exotherm has been observed, whereas at larger scale thermal and corrosion testing would be required.

20.2.1.1 Personnel Safety The safety issues related to personnel may include exposure to highly potent and toxic compounds (i.e., teratogens, mutagens), sensitizers, and genotoxic and cytotoxic intermediates. Highly potent and toxic compounds are usually characterized by having an exposure limit $<1\,\mu g/m^3$. The majority of pharmaceutical intermediates have exposure limits in the 10–100 $\mu g/m^3$ range. Compounds with exposure limits of 1–10 $\mu g/m^3$ are considered to have medium to high potency/toxicity [12]. Personnel risk can typically be managed by a combination of engineering controls such as closed isolation and handling equipment, personnel protective equipment (e.g., breathing apparatus and chemically resistant clothing, gloves, and face/eye protection), and administrative controls such as restricted access and specialized training. With advances in containment technology, exposure levels down to $<1.0\,\mu g/m^3$ can be achieved. This level is typical of exposure guidelines for compounds considered to be highly potent and toxic. However, the cost associated with purchasing and maintaining the appropriate high containment equipment for highly potent and toxic compounds can drive the decision to look for an alternative chemistry route when possible.

20.2.1.2 Exceptional Process Hazards The assessment of a process should include investigating exceptional process hazards. In general, pilot-scale and manufacturing facilities are set up to handle a "typical process," so variables such as high or low temperature (-20 to 110°C), solids charging, and slight exotherms ($\Delta T_{ad} < 5$°C) would not be considered unusual when performing a hazards assessment. However, many processes have one or more steps that include additional hazards such as gas evolution, dust explosion, and static and/or significant exotherms (Table 20.5). Such hazards need to be addressed with further development or by appropriate equipment selection, additional engineering controls, and personnel training.

In the case of gas evolution, understanding the chemistry and particularly the source and identity of the gas is important. Generation of CO_2 as a by-product in a reaction may not pose a significant risk, if the total gas generated is too low to result in a significant pressure increase for the processing equipment or if the rate of CO_2 evolution can be controlled by adjusting reaction rates or reagent addition times. However, if the amount, gas composition, or generation rate are not understood, further development is necessary to ensure that the gas evolution does not pose a significant hazard. Release of a flammable gas such as hydrogen poses an additional challenge since it has a wide flammability range (4–75% in air) and low minimum ignition energy [16].

20.2.1.3 Material Compatibility Material compatibility refers to the ability of the materials in a given equipment train to withstand exposure to the process streams (i.e., to maintain mechanical integrity at the temperature and timescale for the process). Most lab development occurs in glassware using Teflon accessories (agitators, seals, etc.), which ensures compatibility with the exception of hydrogen fluoride. In the pilot plant and manufacturing facilities, the range of

TABLE 20.5 Exceptional Process Hazards

Hazard	Safety Limit	Risk	Mitigation Strategy
Gas evolution	Dependent on equipment vent capacity and gas properties (i.e., minimum ignition energy, flammability)	Equipment overpressure, hazardous or combustible gas release	Understand gas formation mechanism to develop control strategy
High exotherm [13]	$\Delta T_{ad} > 50°C$, $TMR_{ad} \leq 24$ h. *Note:* Proximity of th operating temperature to the initiation temperature for secondary decomposition exotherms should also be considered	Runaway reaction and potential thermal explosion	Control reaction by addition method and cooling, maintain reaction sufficiently, maintain distance from decomposition exotherm. Training and awareness
Dust explosivity [14]	KST $\geq$ 1 bar m/s	Potential explosion	Based on explosion risk, consider not isolating, alternative form, or alternative intermediate. Inert handling, containment, explosion suppression, blow out panel. Training and awareness
Static [15]	<100 pS/m nonconductive	Arching from static charge buildup leading to risk to personnel, equipment damage (glass or Teflon), fire, and/or explosion	Bonding and grounding, inert handling, antistatic additives, solvent changes, conductive components (pumps, antistatic bags), appropriate hold times to match relaxation time for solvent. Training and awareness
Hydrogenation [16]	Concentration <4% or >75%	Potential fire and/or explosion, hydrogen embrittlement	Pressure testing, inert handling, grounding, explosion protection systems, hydrogen/LEL monitoring, equipment selection (motor and equipment rating), H_2 rated flash arrestors, open to air venting handling procedures
Flammable liquids [16]	Flash point <60.5°C (closed cup)	Potential fire and/or explosion	Inert handling, bonding, and grounding

physical equipment (glass lined, stainless steel, hastelloy C, tantalum) will likely provide the flexibility to ensure a compatible fit for any process. However, an initial assessment is usually necessary to ensure that the equipment is properly selected. This understanding of materials and process stream compatibility is required to ensure that the right set of equipment is selected for any scale-up. It is also important to be aware of not just the raw materials (solvents, reagents), the intermediates, and the products, but also by-products in the process. Table 20.6 describes some common equipment materials of construction and materials to avoid. Though some plastics and elastomers are compatible with many solvents, leaching must be considered since it would be difficult to find an impurity leached from the polymer in the final API. In certain cases, compatibility issues may conflict. For example, the use of heptane in a highly acidic environment may present a glass (static) and hastelloy (corrosion) concern. Also, material concerns should be extended to include interactions of the process stream with jacket and condenser fluids. This is of particular concern with water-sensitive process steams.

Material compatibility will likely not play a significant role in driving process development for early and mid-stage processes unless the equipment available is limited. Even then, a development effort based on incompatibility is not necessary until the program progresses to full development. When transferring a process to manufacturing, it becomes highly desirable to reduce or eliminate materials' compatibility issues to allow easy movement of a process between facilities.

20.2.1.4 Hazardous Reagent Handling Highly hazardous reagents are materials that warrant special consideration as the general safety hazards are well known throughout the chemical industry. In general, these materials should be limited when developing a commercial process since the complexity associated with risk mitigation can be costly and difficult to manage. However, most highly hazardous materials have been well-studied and methods to mitigate the risk have been developed. Despite the added cost, it is not uncommon to use a hazardous material in early development when the scale-up is still limited and the risks are well-known

TABLE 20.6 Examples of Incompatible Materials

Equipment Material of Construction	Incompatible Material
Carbone (condensers)	Bromine, NMP
Glass	Hydrogen fluoride, inorganic base at high temperature
Stainless steel	Acids, acid salts, chlorinating reagents
Polypropylene (filter media)	Some solvents (i.e., methylene chloride, heptane, toluene)
Hastelloy B	Ferric and cupric salts
Elastomers (seals) [17]	Some solvents
EPDM	Organic chlorides, cyclohexane
Neoprene	Ethers, acetates, acids
Viton	Acetone, amines, ammonia, acetates, ethers, ketones, caustics

and can be mitigated. The key is risk awareness and communication to ensure that all parties involved are aware of the hazards and the necessary controls used to address them. Table 20.7 lists some examples of hazardous reagents used in API syntheses.

20.2.2 API Quality

Prior to use in human clinical studies, impurities and other foreign contaminants within the API must be controlled at a level specified according to regulatory guidelines. Quality in pharmaceuticals refers to adhering to these regulatory rules as well as understanding the impact of process variations. Similar to safety, product quality is a key development driver. In this section, we focus on discussion of key aspects of API quality: (1) critical quality attributes, (2) genotoxic risk, and (3) process robustness.

20.2.2.1 Critical Quality Attributes Critical quality attributes (CQAs) are quantifiable properties of an intermediate or final product that are considered critical for establishing the intended purity, efficacy, and safety of the product. For API, CQAs must meet specifications prior to release for formulation and eventual use in clinical trials or commercial production. These may include overall purity, levels of impurities, form, color, metals, solvent content, and powder properties (Tables 20.8–20.12). The selection of CQAs will be based on ensuring that the API does not pose a significant risk to patients.

For impurities, the target levels are conservatively set assuming a high dose of 1 g without consideration of the actual clinical trial dosage or therapy duration. It is therefore possible that if the risk is sufficiently understood and can be managed, these target properties can be adjusted to less conservative values for compound used at low dosage and/or in a short duration clinical study. The inability to meet a purity CQA will drive additional development. At early stages of development, this may mean developing a rework strategy for the API, and subsequently in later development, may entail obtaining a detailed and complete understanding of the mechanism of formation for a key impurity.

The powder properties of the API, such as particle size and morphology, can impact the design of the formulation as well as the formulation process performance, and need to be addressed in concert with drug product development. Particle size becomes more of an issue depending on the Biopharmaceutics Classification System (BCS) since the higher the solubility and permeability the less likely that a change in the particle size will have an impact (Table 20.13). The crystal form of the API will impact the compound's chemical and physical stability as well as its pharmaceutical properties (solubility, permeability). Most APIs are crystalline and can exist as different polymorphs, solvates, or salts, or as co-crystals with other organic compounds. An optimal crystal form needs to be selected based on its stability and pharmaceutical properties. Identification of the various forms and selection of the most appropriate form are primary objectives in early development of the compound.

Changes to CQAs of the API that significantly impact the drug product could impact the program's clinical timeline, if additional clinical studies are needed to demonstrate equivalency of the drug products.

20.2.2.2 Genotoxic Impurity Risk Genotoxic compounds have the potential to impact cells in a mutagenic or carcinogenic manner. All intermediates and known impurities present in the API need to be analyzed to assess their genotoxicity. This is typically done first by computerized structural analysis against a database of known genotoxic structural moieties (*in silico*) and then followed up with tests on bacteria (Ames test) to verify positive results. Genotoxic impurities present a significant challenge for drug development since they must be controlled to levels much lower than the standard HPLC detectability limit. The limits for these impurities are set based on daily intake (Table 20.14). Therefore, at high drug load, the limit may be so low that the impurity cannot be detected with standard analytical methods.

As with API quality attributes, the primary risk mitigation for genotoxic compounds is sufficient removal or prevention of its formation. The added cost associated with development of a control strategy and appropriate analytical testing methods makes the presence of genotoxic compounds a formidable development challenge [27]. The need to develop a control strategy for genotoxic impurities will often force a reexamination of the synthetic route to either eliminate the formation of the compound from the synthesis or move its formation earlier in the synthetic sequence, allowing the subsequent reaction, workup, and isolation steps to more effectively remove the impurity prior to the API step. If an intermediate is genotoxic, a new synthetic route that avoids the intermediate may be designed. In the case of an impurity,

TABLE 20.7 Examples of Hazardous Reagents

Chemical	Hazard	Risk Mitigation
t-Butyl lithium [16]	Pyrophoric, air sensitive	Proper handling and inert handling to prevent exposure to air
Azides (e.g., sodium azide) [16]	1. Under acidic conditions, sodium azide forms highly toxic and explosive hydrazoic acid gas 2. Sodium azide and hydrazoic acid are extremely toxic and can be absorbed through the skin 3. Heavy metal azides are shock sensitive	1. Proper use and waste handling techniques. Avoid contact with water or acid 2. When handling use impervious protective clothing and proper PPE (Personal Protective Equipment) 3. Plastic materials are preferred for handling or nonsparking metals
Hydroxybenzotriazole (HOBt, anhydrous) [18]	Explosive, flammable solid	Proper handling and storage. Avoid heat and mechanical shock. Consider use of the HOBt hydrate due to reduced explosivity risk
Alkali and alkaline hydrides (e.g., sodium hydride) [16]	Sodium hydride reacts violently with water. The heat given off is sufficient to ignite the hydrogen decomposition product. Sodium hydride is spontaneously flammable in air	Proper handling and storage. Avoid contact with water
Alkali metal cyanides (e.g., sodium cyanide) [16]	Highly toxic	Use only in well-ventilated area. *Note:* Hydrogen cyanide can be absorbed through the skin. Avoid contact with water or acid. Communicate the hazards prior to use
Hydrazine [16]	Highly toxic, highly explosive (limits in air 4.7–100%, flash point 52°C)	Proper PPE and communication of hazards prior to use. Maintain inert environment and avoid contact with oxidants and catalyst
Chlorinating agents (e.g., oxalyl chloride, $SOCl_2$, $POCl_3$, PCl_3) [19]	Highly toxic. React violently with water and release HCl gas. In addition, thionyl chloride ($SOCl_2$) and oxalyl chloride ($(COCl)_2$) release SO_2 and CO gas, respectively	Proper use and waste handling techniques. Avoid contact with water or acid. When CO and SO_2 gas release are anticipated, ensure proper ventilation and consider monitoring
Metal and supported metal catalyst (e.g., Pd or Pt on carbon, Raney nickel) [19]	Catalysts such as Raney nickel or activated heterogeneous catalyst such as supported Pt and Pd are pyrophoric. The risk of fire is especially high in the presence of flammable liquids	Proper use and waste handling techniques. Catalyst in contact with flammables should be kept under nitrogen inertion. When kept wet with water, Pt and Pd supported catalyst can be handled in air
Grignard reagents [20,21]	Grignard reagent formation and reactions are highly exothermic. The exotherm is particularly hazardous since the reaction initiation can be delayed. Grignard reagents react exothermically with air, water, and CO_2	Proper transportation, handling, and storage. Avoid contact with air and water. Evaluate the reaction exotherm and the equipment cooling capacity to ensure that proper control can be achieved. Charge no more than 20% of the reagent prior to reaction initiation. Consider online techniques such as FTIR to monitor the reaction progress. Evaluate the potential for water content including jacket services. Care should be taken when disposing of residual magnesium turnings
Halogenating agents (Cl_2, Br_2, I_2) [16]	Highly toxic vapors. Water reactive	Proper handling and storage to address inhalation hazards
Hydrogen flouride [19]	Highly toxic. Fluoride ion can penetrate skin and bind with ions such as calcium, potassium, and magnesium in the body	Select proper PPE to prevent inhalation or direct exposure. HF is incompatible with glass and most metals [16,17]
Phosgene [16]	Highly toxic gas. Reacts in lungs reducing lung function leading suffocation	1. Thorough plant controls to prevent emission 2. Leak detection devices and personnel monitors 3. Escape breathing mask or respirator

TABLE 20.8 Typical API Properties Analyzed Prior to Releasing the Material for Drug Product Formulation [22]

Property	Purpose
Purity/impurity profile	The weight percentage can be measured by HPLC or titration. An HPLC (High Performance Liquid Chromatography) purity profile can also show the impurity concentration and indicate whether there is a significant amount of unknown present. The purity of an API is regulated by ICH guidelines (Table 20.9)
Chiral purity	Chiral purity for single chiral center compounds is defined by the enantiomeric excess (ee) and is derived from HPLC using chiral columns. ee is defined as $(R - S)/(R + S)$, where R and S are the fractions of the enantiomers and $R + S = 1$. Chiral purity for diastereomers (multiple chiral centers) is also derived from HPLC
Crystal form (polymorph, solvate, salt)	Form is usually verified by X-ray powder diffraction (XRD), solid-state NMR (Nuclear Magnetic Resonance), or spectroscopic methods (i.e., Raman). The appropriate form ensures that the compound has good physical and chemical stability as well as pharmaceutical properties
Color	Color can be an indicator of an unidentified impurity or degradate. It is also important for ensuring a uniform tablet color. Color can be assessed visually or quantitatively with UV
Inorganic impurities (including metals)	Inorganic impurities can be quantified by residue on ignition or atomic adsorption spectroscopy. A generic heavy metals test is performed by ICPMS (Inductively Coupled Plasma Mass Spectrometry). In addition, individual metals known to be present in the process streams such as Pd and Pt are monitored and need to be controlled based on dose. A typical target for heavy metals is $\leq$10 ppm [23]
Solvent content (including water)	A GC (Gas Chromatography) analysis of the final API is performed specifically looking for any solvent present within the final two API synthesis steps. Solvents have differing level of toxicity and therefore different target limits (Tables 20.10-20.12). Residual water is important for compounds that are hygroscopic, degradable by moisture, or known hydrates. Standard methods include Karl Fisher titration or loss on drying
Powder properties	Typically, particle size measured by laser diffraction, crystal habit assessed by microscopy, and form measured by XRD or Raman are of primary concern. Other measures such as surface area and density may also be appropriate. Powder properties can have a significant impact on the formulation process and the API's bioavailability
Microbial limits	Such assays include total count of aerobic microorganism, yeast or molds, and absence of specific bacteria. The need for such testing is based on nature of drug substance and intended use of drug product (e.g., endotoxin testing for drug substance to be formulated into injectable drug product)

a detailed understanding of how the impurity forms may afford a method to limit or prevent the formation. Typical control strategies for the impurity might include reaction conditions and/or extraction and crystallization design.

20.2.2.3 Process Robustness The ability of a process to demonstrate acceptable quality and performance while tolerating variability in inputs and process parameters is referred to as robustness [28]. Robustness is a function of both the process design (synthesis route selected, the equipment capabilities and settings, and environmental conditions) and the process inputs (quality of raw materials). Process robustness and therefore process understanding is of critical importance to enabling commercialization of a drug. The use of in-process controls and assays ensures that processing activities produce API with the required quality. Understanding of the process variability is critical to ensuring that the API quality will be consistently achieved.

As a project moves into full development and toward commercialization, increased emphasis is placed on under-

TABLE 20.9 International Conference on Harmonization Reporting Guidelines for Impurities Present in an API [8]

Maximum Daily Dose[a]	Reporting Threshold[b]	Identification Threshold[c]	Qualification Threshold[d]
$\leq$2 g/day	0.05%	0.10% or 1.0 mg/day intake (whichever is lower)	0.15% or 1.0 mg/day intake (whichever is lower)
>2 g/day	0.03%	0.05%	0.05%

[a] The amount of drug substance administered per day.
[b] The reporting threshold is a limit above which an impurity should be reported. Higher reporting thresholds should be scientifically justified.
[c] Identification threshold is a limit above which an impurity should be structurally identified.
[d] Qualification threshold is a limit above which an impurity should be qualified in clinical studies. Lower thresholds can be appropriate if the impurity is unusually toxic.

TABLE 20.10 ICH Class 1 Solvents Should be Avoided for Use in Drug Substance Synthesis [8]

Solvent	Concentration Limit (ppm)	Concern
Benzene	2	Carcinogen
Carbon tetrachloride	4	Toxic and environmental hazard
1,2-Dichloroethane	5	Toxic
1,1-Dichloroethene	8	Toxic
1,1,1-Trichloroethane	1500	Environmental hazard

TABLE 20.12 ICH Class 3 Solvents May be Regarded as Less Toxic and of Lower Risk to Human Health [8]

Solvent	PDE (mg/day), Concentration Limit (ppm)
Acetic acid	Class 3 solvents may be regarded as less toxic to human health and need to be controlled to <0.5% or 50 mg/day. This specification does not require specific testing as long as the product loss on drying test is less than 0.5 wt%
Acetone	
Ethanol	
Ethyl acetate	
Heptane	
Isopropyl acetate	
Methyl ethyl ketone	
Methyl isobutyl ketone	
Isopropanol	

standing which step and process parameters have the potential to impact an API CQA. A design range for these parameters can then be defined to ensure that the API quality is consistently achieved. Within the design range, a target set point is selected and the process and equipment capability are then used to define a normal operating range. Critical process parameters (CPPs) are parameters that have a direct and significant influence on a CQA when varied beyond a limited range. Failure to operate within the defined range leads to a high likelihood of failing a CQA specification. Numerous approaches have been presented on defining this range to distinguish a CPP from non-critical process parameters. One approach is to evaluate whether material of acceptable quality can be made within 6σ of the normal operating range, where σ is the equipment specific operational variability [29].

Building process knowledge is typically a significant undertaking in the later stages of development since each unit operation may have up to 10 process parameters, and it is unlikely that each of these process variables has been studied. For example, for a given reaction, process parameters could include reaction temperatures, time to ramp up to the temperature, reaction time at temperature, agitation, variability of charge equivalents, sequence of charges, hold times between charges, and concentrations. An earlier stage assessment would likely focus the development effort only on a subset of these parameters with the largest impact on process robustness. In addition, the potential for multivariable interaction such as time and temperature must be evaluated at later stages.

TABLE 20.11 ICH Class 2 Solvents Should be Limited Because of Their Inherent Toxicity [8]

Solvent	PDE (mg/day)	Concentration Limit (ppm)
Acetonitrile	4.1	410
N,N-Dimethylacetamide	10.9	1090
N,N-Dimethylformamide	8.8	880
Methanol	30	3000
N-Methylpyrrolidone	5.3	530
Tetrahydrofuran	7.2	720
Toluene	8.9	890

An important component of process robustness for a given step is the understanding of process impurity generation and rejection. It is then necessary to determine the "fate" of the impurity in later processing steps; more specifically, is it inert or transformed into other process impurities and is the original impurity or new impurity removed during an extraction or crystallization. By carrying this analysis forward through the API step, the impurity "tolerance" or limit can be established. Target purity profiles for each intermediate can then be defined through similar analysis with all process impurities. This assessment is often completed to support the establishment of the appropriate critical quality attributes for the drug substance.

20.2.3 Business Optimization

In the absence of quality or safety issues, process development is driven by optimization of parameters to improve the business of manufacturing drug substance (e.g., productivity, flexibility, or throughput). Often, the majority of this development effort can be deferred until there is a high probability that the compound will be commercialized. A key milestone for any compound is the achievement of proof of concept in

TABLE 20.13 Biopharmaceutical Classification System [24]

Class I	High permeability,[a] high solubility[b]
Class II	High permeability, low solubility
Class III	Low permeability, high solubility
Class IV	Low permeability, low solubility

[a] A drug substance is considered *highly permeable* when the extent of absorption in humans is determined to be >90% of an administered dose, based on mass balance or in comparison to an intravenous reference dose.
[b] A drug substance is considered *highly soluble* when the highest dose strength is soluble in <250 mL water over a pH range of 1–7.5.

TABLE 20.14 FDA Draft Guidance on Genotoxic Impurities [25, 26]

Duration of clinical trial exposure	<14 days	14 days to 1 month	1–3 months	3–6 months	6–12 months	>12 months
Allowable daily intake	120 μg	60 μg	20 μg	10 μg	5 μg	1.5 μg

the therapeutic hypothesis, typically successful completion of phase IIA clinical trials. However, the complexity of the process and the duration of the overall clinical program will play a role in assessing the timing of process optimization.

There are two key business drivers for API process development: (1) meeting the project timeline and (2) reducing the cost of manufacturing. The first driver applies to both early- and late-stage products. The second priority, reducing the cost to manufacture, is not usually considered in the early stage of development unless there are specific issues that will impact scale-up to generate the required quantity of API for the program's development. Reducing the cost of manufacturing involves both synthesis design to reduce the material cost and number of steps as well as process optimization to address productivity and capacity through improvements in yield and volume efficiency.

As part of the business drivers for process development, we examine the impact of project timelines, process fit and ease of manufacturing, and process greenness. Evaluation of a process' productivity and fit into a manufacturing plant through time cycles, yield, and mass balance will also provide insight into setting the direction for process development. The use of process metrics is an important tool that will enable a common platform to evaluate the evolution of a process.

20.2.3.1 Project Timeline Hierarchically, the project timeline does not drive development but rather the development strategy. Generally, this timeline will define both immediate and long-term API needs. The immediate needs are driven by API required for the clinical studies, the drug safety studies, and drug product development. All of these activities are on the critical path to bringing a drug to market. The highest business priority is delivering sufficient API to meet the clinical and drug safety study timelines. The short-term needs may drive changes, which will be covered in the next few sections. In the long term, the project timeline will drive development decisions. With an extended project timeline due to long clinical trials (10–14 years), deferring process development focused on optimization allows resources to be used on programs with shorter timelines. Alternatively, an accelerated program (5–7 years) leaves little time for process development and may drive parallel development efforts to meet short-term needs as well as to develop the manufacturing process.

20.2.3.2 Process Cycle Time At commercial scale, an optimized cycle time is critical to control cost and manufacturing capacity utilization. The process cycle time can be thought of at multiple levels including the time to complete (1) the entire synthetic sequence, (2) one isolated intermediate or process step, and (3) an individual unit operation. In early development, optimization of the cycle time is less critical. Time cycle optimization would only be considered in the rare case when the program timeline cannot be met. Even then, the first choice would be to use alternative equipment such as a larger vessels or filters to accelerate the timeline. At the transition to full development, a significant effort will be placed on improving the overall process cycle time. This timing will vary depending on the severity of the bottleneck and the potential for changes to impact the API quality. As a project moves through development, emphasis will shift from individual step and unit operation optimization to debottlenecking of the whole synthetic sequence. In development, individual batches are typically run in sequence, so a reduction in time anywhere will lead to a shorter overall delivery time. At commercial scale, the process unit operations and process steps will likely be run in parallel, so resources should be more strategically placed on true bottlenecks. These bottlenecks will not likely be known until the commercial process fit is identified since multiple unit operations may be planned for the same equipment. This is most often the case with isolation and drying where filtering on a centrifuge and drying in a conical dryer can process in parallel whereas filtration and drying on a filter dryer must occur in series.

As an example, a simple process timeline involving a reaction, aqueous workup, solvent swap, crystallization, isolation, and drying steps is depicted in Figure 20.1. For ease of discussion, each of these steps is assumed to have an 8 h cycle time. The first timeline illustrated is a process fit with two vessels and a filter dryer. During the isolation step, both the crystallizer and the filter dryer are active, which extends the time cycle in the crystallizer. In the second timeline, the fit is the same, but the equipment downtime has been minimized by running processing steps in parallel. This provides a significant improvement in time cycle allowing the third batch to be completed in a similar time frame as the second development batch. The bottleneck, however, in this process is vessel 2, as it requires 24 h versus the 16 h for all other equipments. If you add an additional vessel to hold the slurry during isolation, as shown in the third timeline, the parallel processing timeline is reduced by 8 h. An alternative approach to debottleneck is a focused development effort on reducing the total time cycle to complete the solvent swap, crystallization, and isolation. It is therefore important to

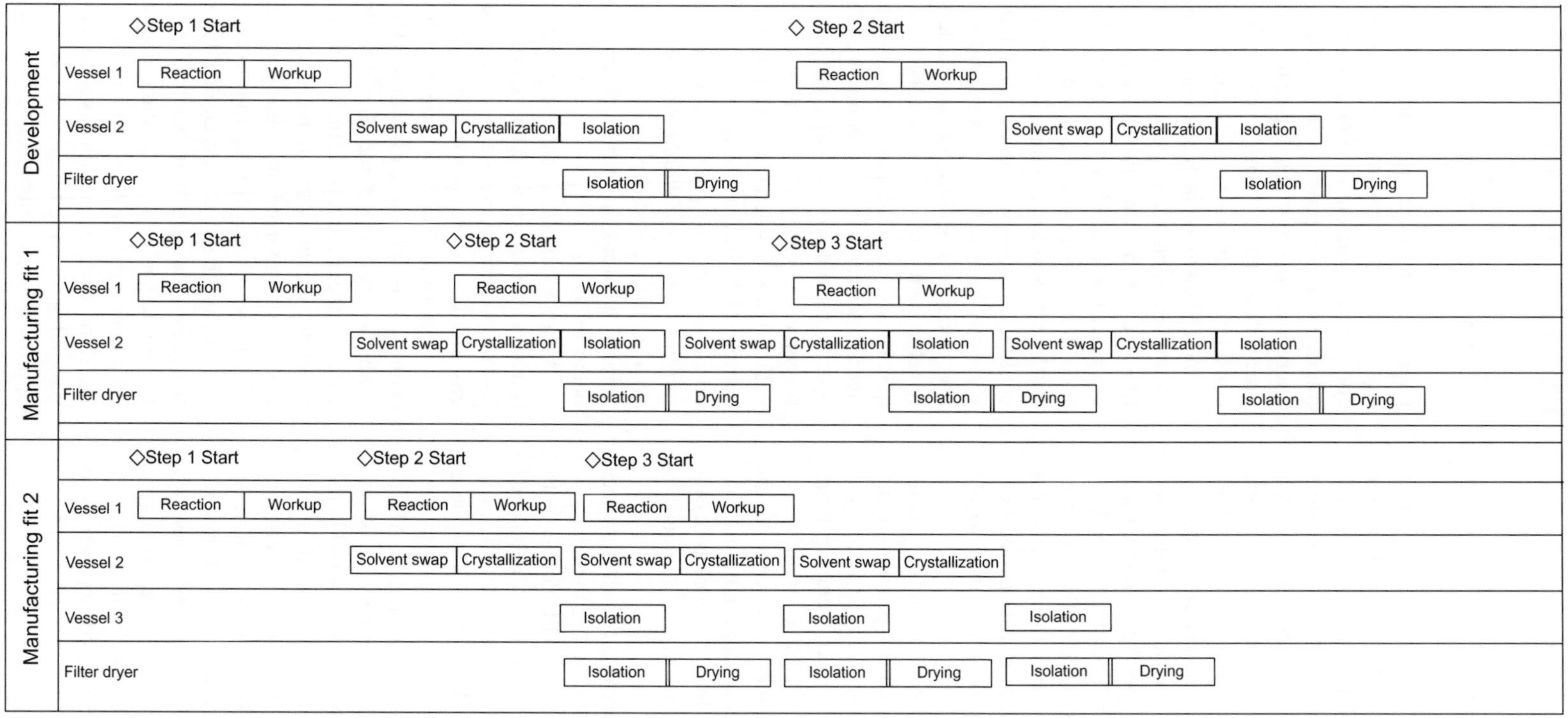

FIGURE 20.1 Single-step processing timeline.

evaluate the process as a whole. In this case, the only equipment change that would optimize the process time cycle is debottlenecking vessel 2. Similarly, the process as a whole should be considered when evaluating where to focus the development effort. In this example, all steps were assumed to take the same amount of time, whereas in reality the processing time for each unit operation will vary widely and should be considered in the evaluation.

*20.2.3.3 **Process Fit and Ease of Manufacture*** The process design can have a significant impact on the manufacturing cost and flexibility as well as the process portability. Ideally, a process is flexible enough to fit into any facility, regardless of the equipment available. However, there are often process constraints such as volume requirements as well as specialized processing equipment needs such as hydrogenators, cryogenic reactors, or continuous processing that impact the selection of manufacturing equipment. In early development, process fit is of little concern since the equipment flexibility is built into glass plant and pilot-scale facilities. However, as a project moves toward full development, the flexibility of the process becomes a significant development driver. In general, development would focus on improving the process flexibility and reducing the process complexity. Therefore, as a project moves toward full development, the desire to reduce the equipment cost and operational complexity may mean minimizing the use of specialized technology.

The first step in improving process fit and ease of manufacture is minimizing the number of unit operations by reducing the need for solvent exchanges, extractions, and isolations. In early development, the process may not be well understood and process steps are added to ensure that a quality material is achieved. An increase in process knowledge around impurity formation and control will allow optimized solvent usage and may allow for the elimination of extractions and isolations. Once the process steps are defined, a high priority is placed on reducing the maximum process volume as well as providing a wider volume range. For a given process train, the maximum volume will dictate the maximum batch size and therefore manufacturing efficiency. Outside of volume reduction, the goal is to crystallize product that can be isolated and dried in either a filter dryer or a centrifuge and conical dryer. Flexibility in crystallization, isolation, and drying are typically linked. Regarding reactions, flexibility in scale-up is typically associated with mixing in heterogeneous systems. It is often not possible to eliminate heterogeneous reactions, so the focus is placed more on understanding and minimizing the impact at scale. Similarly, while hydrogenations can impact the process fit, it is often a very efficient chemical transformation and in the case of asymmetric hydrogenation can significantly increase the overall process yield.

*20.2.3.4 **Process Greenness*** Process greenness is typically considered as part of the overall development strategy to select the final synthetic route and is rarely the main driver for process development. This is especially true in early development where the waste treatment cost and environmental impact are minimal. There are many aspects to consider when discussing process greenness. The most general methods such as the *E*-factor or the process mass intensity (PMI) account for the total waste or mass used relative to the product mass. These factors align quite well with business priorities since it would drive development toward lower cost by reducing material requirements, volatile organic carbon emissions, and chemical wastes. Though these factors give a quick guide to compare the efficiency of materials used, they fail to account for safety and environmental risks posed by specific reagents and solvents. Therefore, the *E*-factor and PMI are typically used only as an early guide. As a program moves through to full development, a more comprehensive evaluation is performed and includes reagents' and solvents' risks. An example of such a tool is the process greenness scorecard, developed by Bristol–Myers Squibb, which tracks about 15 parameters for each step in process and uses green chemistry and engineering principles to assign values that are weighted into an overall score [30]. It is important to note that the definition of process greenness is continually evolving toward a more holistic evaluation. Some proposals include factors such as the impact of operating temperature and certain inefficient unit operations such as classical chromatography [31,32].

*20.2.3.5 **Yield and Mass Balance*** The yield and mass balance are key indicators for the process and, with the exception of early development, drive the team toward further development. The two key measures of yield and mass balance are the absolute number and the batch-to-batch variation. The target yield and mass balance will vary based on the step complexity; however, a target yield of 80–90% and mass balance of >95% are typically acceptable. Though a low yield and/or mass balance are of concern as a process moves through development, a focused development effort to improve yield and mass balance is likely not justified if the process is consistent. However, significant batch-to-batch variability in both yield and mass balance is an indication that a key parameter in the process is not well understood. This lack of knowledge is a critical issue that should be considered even in early development since the quantity or quality of the API synthesized in a given scale-up campaign is at risk. Therefore, even in early development, an effort should be made to understand significant inconsistencies in yield or mass balance.

*20.2.3.6 **Process Metrics*** In the previous sections of this chapter, numerous factors to assess a given process have been discussed. Process metrics can be a powerful tool to evaluate

TABLE 20.15 Process Metrics for a Process Step

Productivity metrics	Yield (mol%)
	kg intermediate/kg API
	Number of chemical transformations
	Longest reaction time (h)
	Number of workups (count of below total)
	• Distillations
	• Extractions
	• Waste filtrations
	• Chromatography
	Peak V_{max} (L/kg)
	V_{max}/V_{min} (V_{max} swings)
Material usage and waste generation	kg starting material/kg product
	kg reagents/kg product
	kg aqueous charges/kg product
	kg solvents/kg product
Quality metrics	Purity (wt%), normalize for salts, solvates
	Purity (% A)
	Potential GTIs
	Impurities above ICH identification threshold
	Number of unknown impurities

the quality and business drivers for process development as well as to track the process evolution. Table 20.15 lists example of process metrics to consider for a given process step. Such process metrics can then be tabulated for a given synthesis step or summarized for the entire synthetic sequence (Table 20.16).

20.3 UNIT OPERATIONS

In assessing the suitability of a process to run at a given scale, there needs to be an assessment of both the overall characteristics of the process such as cycle time, cost of goods, and yield (as described in Section 20.2) and the characteristics of each individual unit operation. This section will examine the most common unit operations and enumerate factors that contribute to the scalability of each operation. These factors should be considered in a process scale-up assessment.

TABLE 20.16 Process Metrics for an Overall Synthesis

Overall Yield
Total kg intermediates/1 kg API
Total number of workup operations
Total number of isolated intermediates
Total number of potential GTIs
Total kg solvents/kg API
Total kg aqueous/kg API

20.3.1 Introduction to Evaluation

One of the hallmarks of a readily scalable process is that it can be run in a standard facility, using standard equipment, with an ordinary degree of control over the process parameters. Therefore, it is critically important for an engineer to understand how processes are generally run on pilot and manufacturing scale.

The vast majority of pharmaceutical processes are run as a batch operation, rather than as a continuous or semicontinuous operation. The process train typically consists of multiple stirred vessels, pumps and lines for liquid charges/transfers, waste receiver vessels for distillate, mother liquors, and waste streams, product isolation equipment (pressure filters, centrifuges, or filter dryers), and dryers (tray, conical, rotary, filter dryers). The batch reactors are generally equipped with ports for charges/feeds/probes, a bottom valve for discharge, a fixed agitator type and fixed baffling configuration, and overhead piping system for providing venting, vacuum, and emergency pressure relief, typically with a condenser on the main vent path. Flexible lines and a manifold system are commonly used to allow transfers from vessel to vessel, or from vessels to the isolation equipment. Common instrumentation on the equipment includes the temperature and pressure of the equipment's contents, the temperature of the equipment's jacket, and product stream's density.

Given the standardized nature of the equipment, standard unit operations are preferred to achieve the process goals. A typical sequence of unit operations includes solution preparation, reaction, separation (extraction, distillation), crystallization, isolation, and drying.

20.3.1.1 Selection of Unit Operations Before the sequence of unit operations for a given step can be determined, an understanding of the objectives for the step is needed. The objectives of each step in the synthetic sequence should be considered collectively, since there are likely trade-offs between steps in the sequence with respect to yield, quality, process cycle time, and the need for specialized equipment. Key to assessing these trade-offs are well-established API quality requirements (including powder property requirements), and knowledge of the material value for a given step (e.g., what is the value of an additional 5% yield). The intermediate quality requirements can then be defined after considering trade-offs between the steps. For example, one may tighten the quality specification in an early step at the expense of step yield, in exchange for eliminating the need for difficult or costly purification downstream.

Once the objectives for the overall step are established, the objectives of each individual unit operation should be understood. An optimized process will involve no additional operations (or more complicated operations) than needed to safely, reliably, and robustly meet the process objectives.

Prior practice at smaller scales may dictate the initial choice of unit operations, but as the process is optimized the number and type of operations are expected to change. For example, a prior iteration of a process may involve multiple liquid–liquid extractions, designed to remove a key process impurity. If subsequent improvement to the reaction conditions reduces the number or extent of side reactions, fewer or no extractions may be needed. The process optimization to reduce the number and complexity of unit operations is a key process development objective.

20.3.1.2 Process Fit Another core process engineering activity is understanding the process fit. Engineers are frequently tasked with fitting a process in an existing facility in such a way to minimize capital expenditure (modifications to existing equipment or purchase of new equipment) and to minimize the deployment of shared resources (portable equipment). To accomplish this task, the engineer must clearly understand the capabilities and limitations of the plant. Specifically, vessel configurations (minimum and maximum volumes, baffles, number and type of agitators), vacuum and temperature control capabilities, heat and mass transfer coefficients, filtration capabilities (e.g., centrifuge versus pressure filter, filter area, filter porosity), and drying capabilities (e.g., agitated versus nonagitated, heat transfer, vacuum control) will need to be considered. The engineer will then be able to assess if the process as designed can operate in the plant without modification and, if necessary, modify the process to fit existing equipment.

20.3.1.3 Common Scale-Up Factors There are many scale-up factors that are not specific to any one particular unit operation. Time is a particularly important example. Nearly every activity requires more time to accomplish at manufacturing scale compared to the lab scale. The ramifications of this will be discussed in the individual unit operations section. One concern that is common to all the unit operations is stability. The stability of the reaction mixture with respect to undesired side reactions (degradation) must be assessed for each unit operation on timescales relevant to the plant scale.

Another issue common to most unit operations is the potential for residual material in process lines and dead legs to interact with material being charged or discharged. For example, a single charge line may be reused for multiple reagent charges, with a solvent flush in between each charge. If the flush is inadequate (or not done), and the materials are not compatible with each other, a deleterious reaction may occur. It is critically important for both safety and quality reasons for the engineer to be cognizant of what lines are being used for what purpose, and to systematically consider what residues may be left behind as process fluids are transferred throughout the equipment train.

For all unit operations where heat transfer is important, the surface area to volume ratio will be a common issue. As scale (vessel size) increases, the surface area to volume ratio decreases. Since the rate of heat flow is proportional to the heat transfer area, and the overall heat capacity of the system is proportional to the mass of the batch (and thus the volume), heat transfer will be significantly slower as a process is scaled up.

For the reaction, extraction, and crystallization unit operations, mixing is a common scale-up factor. Generally speaking, the mixing power is much greater in the plant than in the laboratory. It can be challenging to simulate the mixing behavior that will be obtained on scale in the laboratory, since there are many variables one can choose to hold constant between the experiment and the plant run. These variables include power, power per volume, tip speed, rotational speed, flow per volume, torque per volume, Reynolds number, blend time, and geometric similarity (ratio of impeller diameter to vessel diameter). Different phenomena scale with different variables, and it is not always well understood which variable is the best choice for scale-up and scale-down. Some case studies and rules of thumb are available in the literature [33, 34].

A final consideration that applies to several unit operations is the issue of dip tube depth. For any operation that involves sampling, the engineer must consider whether or not the dip tube is below the liquid level, to allow a sample to be taken. In this case, the minimum volume for a unit operation may need to be increased.

The following sections discuss the common individual unit operations. Detailed treatment of the chemical engineering theories of heat transfer, mass transfer, thermodynamics, chemical kinetics, etc. and their application to batch reactors is available elsewhere, and is outside the scope of this chapter. Instead, a brief discussion of the factors an engineer needs to consider is presented.

20.3.2 Reaction

The objective of the reaction unit operation is to convert a starting material or materials into the desired product, with maximum yield and minimum degree of by-product (impurity) formation. In the laboratory, reagent selection, solvent selection, stoichiometry, sequence of addition, and temperature are generally established. This list of process variables is unlikely to change upon scale-up, since, as Caygill et al. [35] state, "chemical rate constants are scale independent, whereas physical parameters are not." The many physical parameters that play a role in the outcome of the reaction that are scale dependent (see Table 20.17) are the main cause of scale-up problems.

A key consideration is whether the reaction is homogeneous (single phase) or heterogeneous (multiphase). Generally speaking, a standard batch reactor is configured such that

TABLE 20.17 Scale-Up Factors for Reactions

Factor	In Lab	At Scale	Impact	Means to Evaluate
Time to charge reagents	1 min or less	Between 5 and 60 min	Different stoichiometry profiles with time may impact reaction kinetics	Simulate longer additions at lab scale
Charge method	Pouring, pump, addition funnel	Pump from drum, pressure from vessel, vacuum from drum	Choice of charge method may impact rate. Vacuum charges may cause volatilization of components	Simulate charge method
Charge port	Generally above surface	Above-surface, subsurface, sprayball	Backmixing may occur during subsurface charges. Use of above-surface ports may leave material on the vessel walls. Use of sprayball can help rinse solids from sides of the vessel	Backmixing calculation from engineering correlations [36]
Sequence of addition	Based purely on chemistry/convenience	Limited number of lines and ports may necessitate different order of addition (e.g., to avoid incompatibles in the same line). Also, order of solids versus liquids may differ in the plant based on considerations such as inert handling	Can affect the kinetics of main and side reactions	Test different orders of addition in lab experiments
Mixing time	Can vary over wide range	Varies, max agitation likely affords longer mixing time than lab maximum	If reaction time is fast compared to mixing time, undesired reactions may occur	Experiments to determine reaction kinetics + blend time calculation. Can evaluate Damköhler number. If large, mixing is an issue
Solids suspension	Typically not an issue	May be an issue	Insufficient suspension equals lower effective surface area of solids	Njs (agitator speed to just suspend) calculation [37]
Mass transfer	$k_{La} = 0.02–2\ s^{-1}$	$k_{La} = 0.02–0.2\ s^{-1}$ (batch reactor), $k_{La} = 1–3\ s^{-1}$ (Buss Loop)	Either mass transfer or chemical kinetics may be rate limiting at different k_{La}. This will impact reaction profile	Gas uptake experiments in lab and at scale to determine k_{La}. k_{La} predictions by engineering correlations
Heat transfer	Excellent, high area/volume	Lower area/volume as scale increases	Safety (runaway reaction), excursion from acceptable temperature range	UA evaluation at scale (mock batch/solvent trial heating trend data may be used) and in the lab

reagents in a homogeneous reaction can be sufficiently well mixed to avoid the need for detailed consideration of mixing and mass transfer. There are, of course, exceptions, such as highly exothermic reactions, where temporary hot spots can cause a high level of impurity formation before reagents are well-mixed. In contrast to homogeneous reactions, heterogeneous reactions (reactions with separate liquid–liquid, liquid–solid, or liquid–gas phases that participate in the reaction) are likely to be highly dependent on mass transfer considerations.

Table 20.17 enumerates several scale-up factors for reactions.

TABLE 20.18 Scale-Up Factors for Extractions

Factor	In Lab	At Scale	Impact	Means to Evaluate
Tendency to form emulsions	May be seen	Additional factors cause emulsions to be seen even if not seen in lab (e.g., increased agitation power)	Stable emulsions must be broken before processing can continue	Test extremes of composition. Maximize mixing to stress. See text for means to break emulsions
Settling time (settling velocity)	Variable	Variable	Settling *velocity* should be similar between lab and plant. Settling time therefore will be much longer in a plant vessel	Measure settling time and height of phase boundary in the lab. Estimate time on plant scale using constant velocity
Mixing	Shaking in sep funnel, stir bar, overhead stirring. Generally easy to mix the phases, but low power	Various agitator, vessel, and baffle configurations. Much greater power. For fixed equipment operated within normal volume ranges, mixing is typically good	Adequate mixing needed to properly mix the phases and equilibrate composition. Excessive agitation may promote stable emulsion formation	Determine at-scale blend time to ensure that phases will be adequately mixed. Stress the process by mixing as vigorously as possible in the lab
Ability to catch the split	Generally easy	May be difficult to see through small sight glass. Use of mass meters to detect density differences to determine split	Missing the phase boundary can cause repeated operations (waste of time) or loss of yield	Note when the phase boundary is more difficult to see (e.g., phases have similar phase color and opacity) and inform the plant staff of the need for caution. Before plant run, calculate expected phase volumes, so that the rate of discharge can be adjusted depending on how close the phase boundary is
Rag layer	Minimal or not seen	Rag layer of significant volume may be present	Must determine disposition of rag layer in advance—inclusion can affect purity, exclusion can affect yield	Difficult to assess in lab. Establish rag layer disposition in advance of processing based on quality/yield requirements. Typically, rag layer is kept with the product phase, except for the final phase split

20.3.3 Separation

20.3.3.1 Extraction The objective of an extraction unit operation is to remove undesired components (organic impurities, inorganic salts) from the product solution, and in some cases to quench the reaction. This is achieved by adding a liquid that is immiscible with the reaction mixture. Typically, the reaction mixture is organic and the added liquid is water or an aqueous salt solution, but the reverse situation is possible. In the laboratory, the liquid is added to the vessel and stirred, or the liquids are combined in a separatory funnel and shaken together. The agitation is stopped and the phases are allowed to settle, followed by separation. The relative densities of the phases are a key parameter in determining how quickly the phases will settle. Table 20.18 enumerates several scale-up factors for extraction:

Emulsions Several differences between the lab and the plant scale can contribute to emulsion formation. One is the mixing power per volume. Most often the plant-scale agitation is high power per volume, and thus there may be a greater tendency to form emulsions. Another factor is the likelihood to precipitate either product or salts during the extraction. In the laboratory, the midpoints of acceptable temperature and solvent composition (distillation end points, charge ranges) are often studied, whereas in the plant the parameters may be near the upper or lower part of the range. If one of the phases is near the solubility limit for a component, tiny particles that have the potential to stabilize an emulsion may form. Also, at scale the reagents may introduce tiny particulates or impurities that affect solubility. A final consideration is the position of the agitator blade relative to the phase boundary. This can influence which phase is dispersed in which, potentially affecting the stability of the dispersed phase. These factors may be proactively investigated in the laboratory to determine if an emulsion is likely.

If an emulsion is formed, methods to break the emulsion should be studied. If the emulsion is seen for the first time in the plant, such a study may be undertaken with a batch sample. Typical means of breaking an emulsion include addition of either solvent or water to change the composition, heating, filtration (to remove stabilizing entities such as tiny particles), pH adjustment, salt addition, and in rare cases, addition of a demulsifier. Some case studies and rules of thumb are available in the literature [38].

20.3.3.2 Distillation Generally, the objective of a distillation operation is to change the solvent composition of the system to facilitate downstream processing. This is generally performed in a semi-batch mode by either continually adding the new solvent at a constant volume or sequentially adding the new solvent and then distilling down to the original volume, sometimes repeatedly (put/take). More rarely, reactive distillation may be used in cases where a volatile component must be removed to drive the reaction to completion. Distillation is also occasionally used to change the solvent composition to drive crystallization of the product (distillative crystallization).

A good first step in understanding a distillation operation is to obtain thermodynamic vapor–liquid equilibrium (VLE) data for the solvent system in question. The effect of pressure, the presence or absence of azeotropes, and the difference in vapor compositions across the liquid composition space are all easily visualized (for two solvent systems) with a x–y or T–x–y diagram (or several diagrams for different pressures). Several software packages (e.g., DynoChem™, Aspen™) are available to perform VLE calculations and distillation simulations. Typically, calculations and simulations based on pure solvents (ignoring the presence of the product or starting material) provide sufficiently accurate estimates.

Table 20.19 enumerates several scale-up factors for distillations:

20.3.3.3 Color/Metal Removal The objective of a color or metal removal unit operation is to purify the process stream with respect to color bodies or metals. Typically, this is accomplished through the use of an adsorbent material. Common examples include activated carbon, functionalized silica, or functionalized polymeric materials. Use of this unit operation at scale is not desirable since color and metal removal requires special materials and often special equipment. If other means of meeting product specifications are available, they should be considered.

There are two typical ways that an adsorption step is scaled up: (1) slurry of loose adsorbent followed by filtration and (2) filtering the process stream through a cartridge or a filtration equipment (sparkler, Nutsche) containing the adsorbent. If the cartridge option is available, it is preferred, since the loose materials are often challenging to filter from the process stream and are difficult to clean from process equipment.

In any investigation of absorbents, there are two key criteria for absorbent selection: (1) degree of removal of the color or metal (as a function of percent loading of the adsorbent) and (2) loss of product to the adsorbent. Secondary considerations include cost of the adsorbent and availability (lead time) of the adsorbent. As a general rule, activated carbons are cheaper and more readily available compared to functionalized materials. A typical protocol for studying the adsorption unit operation is described in Example 20.2.

EXAMPLE 20.2

The final intermediate in the synthesis of an API is received from a vendor and found to have a dark brown color. The intermediate (designated compound A) is used in a laboratory run to produce API, which is found to also have a brown color. The specification for the API is off-white, so color will

TABLE 20.19 Scale-Up Factors for Distillation

Factor	In Lab	At Scale	Impact	Means to Evaluate
Time	Can vary from very short to very long, depending on ΔT (difference between boiling point temperature and jacket temperature)	Typically longer than that in the lab (hours)	Instability of the stream at elevated temperature can cause a quality issue	Establish stability by refluxing the process stream at the highest anticipated pressure, and at both extremes of composition
Vacuum control	Generally excellent	Depends on equipment, may be poor. May be difficult to achieve pressures lower than 50 mmHg	Fluctuations in vacuum = fluctuations in boiling point. Some process streams have a tendency to foam, or to "bump"	Carefully test the effect of decreased pressure on distillation
Jacket temperature	Choice of ΔT between jacket and boiling point drives distillation. Lower jacket temperatures might be adequate in the lab to achieve a reasonable distillation rate	Due to lower heat transfer area per volume, greater jacket temperatures may be desirable	Higher jacket temperatures may result in decomposition of any solids that are deposited onto the vessel walls during distillation	Assess the stability of the product at high temperature to see if high jacket temperature is an issue
Minimum volume	Often very low in terms of liters per kg of input (ability to agitate small volumes) or to use a rotary evaporator	Usually larger in terms of liters per kg of input. Some conical bottom vessels can have low minimum agitatable volumes. There may be safety issues associated with highly concentrated process streams on scale	More solvent is needed to achieve the end point as the minimum volume increases	Use the plant's minimum volume (in L/kg) in the lab study. Perform safety evaluation of concentrated process streams
End point volume	Easy to mark a volume end point on glassware	May be difficult to see volume landmarks in the vessel, radar level sensor may be present, but may not be very accurate (±5% typical, sometimes ±10%). Foaming can result in inaccurate radar measurement	If no in-process control is established (e.g., quantitation of the product concentration), the concentration going into the next unit operation may be off-target	Assess the sensitivity of downstream unit operations to variations in the concentration representing both under- and overdistillation

(*Continued*)

TABLE 20.19 (*Continued*)

Factor	In Lab	At Scale	Impact	Means to Evaluate
End point composition	In early development, the process stream may be rotovapped to an oil followed by dissolution in the next solvent. In later development, a put/take or constant volume distillation is done, and the process stream is analyzed for solvent composition by GC	Can choose between sampling the product stream for solvent composition, analyzing the distillate composition, PAT monitoring of the distillate, use product temperature as a guide, or fixed distillation protocol with no in-process control	If no in-process control is established (e.g., GC on the process stream), the composition going into the next unit operation may be off-target	Assess the sensitivity of downstream unit operations to variations in the composition representing both under- and over-distillation. Develop reliable control scheme as needed
Constant volume distillation	Usually requires full-time supervision	Usually requires full-time supervision	Constant volume distillation is usually much more solvent efficient than put/take distillation; however, the need for supervision may make it a less desirable choice	Evaluate both constant volume and put/take distillation modes
Sampling	Can draw a sample through a septum	Sampling often requires the distillation be temporarily stopped to draw the sample due to limitations on the sample temperature, or the possible need to pressurize the vessel to draw a sample	Sampling can increase cycle time	To minimize the number of samples, establish vessel landmarks, end point temperature (as a function of pressure), or if the distillation is a critical operation, PAT monitoring

TABLE 20.20 Example 20.2: Screening Results

Carbon Type	Color (by Visual Inspection)	Recovery of Compound A (%)
Carbon 1	Brown	97
Carbon 2	Brown	95
Carbon 3	Very light yellow	93
Carbon 4	Brown	98
Carbon 5	Clear	82

need to be removed. This could be done either via rework of the intermediate or as a processing step in the API step.

At the beginning of the API step, compound A is dissolved in 20 L of methanol per kg of compound A. The project team decides to pursue color removal by carbon filtration after this dissolution step.

The first step is to screen various potential adsorbents. The team has five common carbons available for scale-up. One hundred milligrams of each carbon is placed in a vial along with 4 mL of compound A solution in methanol. Since the solvent quantity in the solution is 20 L/kg (or 20 mL/g), the 4 mL solution contains 200 mg of compound A. Thus, the loading of carbon in the screening experiment is 50% (100 mg of carbon to 200 mg of compound A). The samples placed in a shaker block for 60 min, and then filtered. The color is inspected visually, and the concentration of the filtrate is analyzed by HPLC for wt%. The recovery of compound A is calculated based on the HPLC quantitation and results are shown in Table 20.20.

Carbons 3 and 5 are the only adsorbents that afford color removal. Carbon 5 results in the best color; however, too much of the desired compound is lost to the carbon. The team decides to use carbon 3 for scale-up.

Since the pilot plant will use carbon cartridges, a breakthrough study is performed in the laboratory to simulate the plant operation and to determine what carbon area is needed per liter of process stream to be decolorized. A 47 mm carbon disk is set up in a filter housing. This disk is known from vendor literature to have an effective carbon surface area of 0.0135 ft^2. A fluid reservoir is connected to a pump, which is subsequently connected to the carbon disk. Downstream from the carbon disk is a filter and a UV/Vis detector. First, methanol is flushed through the system for 20 min at 5 mL/min. Then, the feed is switched to a reservoir of 300 mL of compound A solution. UV/Vis monitoring is started, and continues until all the solution is passed through the pad. The solution that has passed through the pad is collected in 5 mL fractions. A plot of the absorbance at 310 nm versus time is presented in Figure 20.2. The color begins to breakthrough at about 30.5 min, and after 35 min the color breakthrough is increasing rapidly. Judging the breakthrough point to be 35 min, the team pools all the fractions from 20 to 35 min, and proceeds with the API chemistry. The resulting material is found to be white, so 35 min is verified to be an acceptable breakthrough point.

Given the 20 min of flush and the 5 mL/min flow rate, the breakthrough point is calculated to be 75 mL (35 − 20 min = 15 min × 5 mL/min = 75 mL). Given the 0.0135 ft^2 carbon area of the 47 mm pad, the carbon "life" is 5.56 L/ft^2. The flux, or flow per area, was 0.37 L/(min ft^2).

For the scale-up to 5 kg API batch, the batch volume will be 100 L at the point of dissolution. Given the life of 5.56 L/ft^2, the needed carbon area to remove color in this batch is 18 ft^2 (100 L/5.56 L/ft^2 = 18 ft^2). Based on the flux of the experiment (0.37 L/(min ft^2)), a total minimum time of 15 min is required for the operation (100 L/0.37 LPM/ft^2/18 ft^2). This could also be expressed as a flow rate of 6.7 L/min.

20.3.4 Crystallization

The objective of the crystallization operation is to isolate the product as a solid, purify by leaving impurities in the liquid

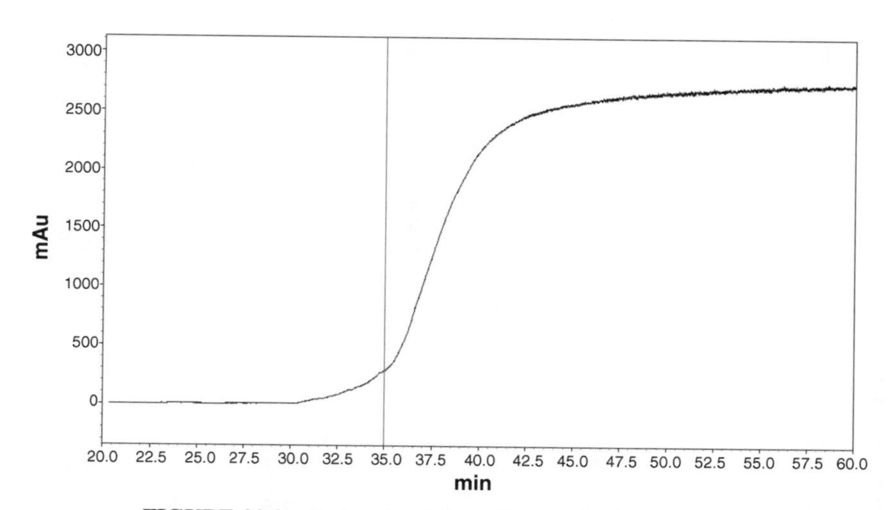

FIGURE 20.2 Carbon breakthrough curve for Example 20.2.

TABLE 20.21 Scale-Up Factors for Crystallization

Factor	In Lab	At Scale	Impact	Means to Evaluate
Metastable zone width (see text for explanation)	Generally broader	Generally narrower	Narrower metastable zone width may result in spontaneous nucleation prior to seeding, and thus uncontrolled crystallization (lack of polymorph control, lack of control of particle size distribution)	Test metastable zone width in the lab as a function of agitation, cooling rate (if applicable), and antisolvent addition rate (if applicable). Generally, to grow large particles of a desired polymorph, it is best to seed the batch under conditions of very low supersaturation [39]
Seeded/unseeded	Either is possible—easy	Either is possible—for some equipment, solids charge is difficult and the seeds are transferred as a slurry. Seed charges can be small in quantity, and it may be challenging to ensure that the seed charge reaches the batch	Polymorph control and particle size distribution control	Evaluate effect of amount of seeds (loading), seeding point, and dry versus slurry transfer
Mixing	Stir bar, or overhead stirring. Generally easy to achieve good mixing	Various agitator, vessel, and baffle configurations. For fixed equipment operated within normal volume ranges, mixing is typically good	Inadequate mixing can result in hot spots of high supersaturation, and thus uncontrolled nucleation. High agitation may reduce metastable zone width	Evaluate the impact of mixing in scale-down experiment. One means of eliminating scale dependence is by performing the mixing outside the vessel, for example, with a T-mixer or a jet mixer
Attrition	Variable	Variable	Some particles are easily broken. This process may be sensitive to agitation parameters (including time). This may affect particle size distribution, filterability, and ultimate powder properties	Explore the particle size distribution as a function of agitation using overhead stirring, or stress the process by using a high-shear mixer
Measurement of temperature	By thermocouple, not typically a problem	Temperature probes can sometimes be coated with crystals and provide false information	If jacket temperature control is used, this may not be a processing issue, but accurate temperature data will not be collected. If operating in batch temperature control mode, the control system may adjust the jacket temperature and thus move the batch temperature outside the normal operating range	Observe the tendency for coating during the crystallization. Specify jacket temperature control rather than batch temperature control

phase, and create particles of the correct form and desired physical properties (i.e., size distribution, density, surface area). As a brief review of fundamentals, crystallization consists of several physical phenomena, the most important of which are nucleation and growth (others include attrition and aggregation). Nucleation refers to the formation of very tiny crystals from the solution, and growth refers to the increase in size of the nuclei by transfer of product from the solution to the crystal faces. The balance of the rate of nucleation and growth is a key determinant of the particle size distribution. If the nucleation rate is dominant throughout the crystallization, small particles with a nonuniform distribution will form. If growth is dominant throughout the crystallization, large particles with a more monodisperse distribution will form.

Supersaturation (i.e., the state where the product concentration is above the equilibrium solubility) is required for nucleation and growth. Supersaturation is often induced by the addition of antisolvent or by lowering the batch temperature. Generally, very high supersaturation favors nucleation over growth, and low supersaturation favors growth over nucleation. In a system with no nuclei present (added seeds or foreign matter that can act as nuclei), spontaneous nucleation does not happen immediately at the onset of supersaturation. The region in the parameter space (concentration and solvent composition, or concentration and temperature) in which the solubility is exceeded but spontaneous nucleation does not occur is referred to as the metastable zone. The width of the metastable zone depends not only on the inherent characteristics of a given system but also on physical parameters such as agitation, rate of cooling or rate of antisolvent addition, and the presence of other nuclei (foreign matter or seeds). For this reason, metastable zone width depends on scale.

Table 20.21 enumerates several scale-up factors for crystallizations.

TABLE 20.22 Scale-Up Factors for Isolation

Factor	In Lab	At Scale	Impact	Means to Evaluate
Filtration flux	Up to 10× greater, depending on lab versus plant cake thickness and cake compression	Often up to 10× slower than lab, or longer	Longer cycle time	Measure filtration flux as a function of cake height or mass of cake. Use engineering correlations to predict at-scale performance (see Chapter 17)
Filter media	Typically done with filter paper, 6–25 μm	Limited choices of pore sizes and material of construction	Potential for filter media to blind or pass through of product	Evaluate plant-scale filter media in lab-scale experiments
Compressibility	Low ΔP compared to the plant, so effect of compressibility is less of a factor	Higher ΔP, so effect of compressibility is more of a factor	Can greatly slow down the filtration at high ΔP or high centrifugation spin speeds	Evaluate compressibility with leaf filter studies (pressure filtration measurements of rate versus ΔP)
Cake wash	Able to smooth cracks in the wet cake	Sometimes not able to smooth cracks in the wet cake (this can be done in a filter dryer)	Channeling of cake wash through cracks results in poor washing of the cake, affecting impurity profile and solvent content	Evaluate propensity to crack by allowing cake to deliquor completely between filtration and each wash
Extent of deliquoring	Typically easy to achieve low solvent content	Solvent content after isolation may be much greater	Greater solvent content impacts cake wash efficiency and drying operations. Stability of the product may be an issue	Study stability of wet cake under very wet conditions (e.g., 50% wash solvent)
Discharge	Easy—by scooping wet cake from Büchner funnel into a drying dish	May be challenging depending on equipment. Safety considerations such as electrostatic buildup from nonconductive washes may dictate need to delay (for relaxation of charge)	Wet cake properties needed to select appropriate parameters on equipment (i.e., peeler centrifuges—LOD, wet cake density). Longer discharge requires additional product stability under wet conditions	Study stability of wet cake under very wet conditions (e.g., 50% wash solvent)

20.3.5 Isolation

The objective of the isolation operation is to separate the solids (product or waste) from the mother liquors as rapidly as possible, and efficiently wash nondesired components (organic impurities, inorganic salts, solvents, or product) from the isolated material. Table 20.22 enumerates several scale-up factors for isolations.

20.3.6 Drying

The objective of the drying operation is to remove solvents to achieve a final product solvent specification, and to maintain or create desired powder properties. A typical drying target is set to remove solvent below a maximum allowable concentration. When drying a solvate crystalline form, minimum and maximum solvent content criteria will be set. Table 20.23 enumerates several scale-up factors for drying.

20.3.7 Particle Size Reduction (Milling)

An active pharmaceutical ingredient typically has a specification related to the powder properties. Particle size control may also be critical for process intermediate seeds to ensure sufficient impurity rejection or to improve filterability. The most common specification is related to final particle size distribution and often given as a single number that characterizes the particle size distribution. Example specifications include the mean (volume or mass based), or a D "number" (i.e., D_{50}, D_{90}, D_{97}), which refers to a value on the distribution such that "number" % (by mass) of the particles have a diameter of this value or less. Different moments of the particle size distribution as well as surface area and bulk density may also be chosen as a specification.

Development scientists can attempt to address the powder property requirement by several means, including crystallization engineering, wet milling, and dry milling. Each of these technologies is addressed in more detail elsewhere, and crystallization scale-up factors are discussed above. A requirement to make amorphous API would entail consideration of additional technologies such as spray drying. Issues related to scale-up of milling processes depend on (1) the equipment for the specific milling technology and (2) the physical properties of the compound (bulk density,

TABLE 20.23 Scale-Up Factors for Drying

Factor	In Lab	At Scale	Impact	Means to Evaluate
Agitated drying	Not always evaluated	Agitated filter dryers, rotary tumble, and conical dryers are most common drying methods. The LOD in which agitation begins is important parameter for determining powder properties	Agitation can promote lump/ball/boulder formation in cohesive powders. Agitation can influence all of the key final powder properties through breakage or attrition of particles—bulk density, particle size distribution, flowability, electrostatics	Lab-scale agitated dryer units are available. Scale-down of agitation drying experiments is not straightforward, so laboratory data may only provide trends or insights into tendencies of the system, not quantitative prediction of scale behavior
Bulk density	Easy to adapt to low bulk density	Bulk density dictates needed dryer size	Can greatly affect the choice of equipment or number of dryer loads. May affect formulation performance	See above
Sampling	Scoop/spatula	Sampling configurations differ between different dryers. The operation may be difficult, and samples may not be fully representative. Multiple samples generally taken	Too frequent sampling adversely affects cycle time. Nonrepresentative samples can result in false passing results from in-process controls	Establish tolerance for solvents/water in downstream processing (or API release). PAT methods for monitoring drying may sometimes be implemented if drying is a critical operation. PAT may be especially useful if attempting to maintain a solvate (to prevent overdrying)
Discharge	Scoop/spatula	Depends on dryer—discharge may occur through a small port and may not be trivial. Often, a significant heel is left behind after discharge	Poorly-flowing powders can be very difficult to discharge and may require excessive time/operator intervention	Measure the flow characteristics of the powder after agitated drying

TABLE 20.24 Summary of Milling Technologies

Milling Technology	Key Advantages [40]	Key Parameters and Issues for Evaluation and Scale-Up
Air attrition milling (jet or loop mills)	• Capable of attrition down to D_{97} of 2–10 μm • No heat generation—ideal for heat-sensitive compounds • Easy maintenance—no moving parts • Inert milling	• Pressures (pusher and grinding) • Mass of solids/gas flow rate ratio • Tendency of material to compact and stick to raceway surface
Fluidized bed air attrition mills with classifiers	• Capable of attrition down to D_{97} of 2–10 μm • No heat generation—ideal for heat-sensitive compounds • Steeper particle size distributions are achievable • Inert milling	• Pressure • Classifier speed • Nitrogen flow to achieve fluidization • Product feed rate/product removal rate
Impact milling (hammer, pin)	• Capable of attrition down to D_{97} of 30–50 μm • Large industrial-scale units for very high throughput	• Pin or hammer speed • Product feed rate • Sensitivity of compound to temperature
High-shear rotor–stator wet milling [41, 42]	• Capable of attrition down to 10–30 μm as a mean • Technique can be set up as a recycle of the crystallized slurry • No exposure to dry powders • More suited for "needle" morphologies to reach lower end of attrition	• Rotor–stator configuration (number of teeth, gap width) • Shear frequency, shear rate • Slurry concentration • Batch turnovers • Point of wet milling initiation during the crystallization time cycle • Product filterability
Media and ball milling	• Capable of attrition down to $D_{97} < 1$ μm • Technique can be set up as a recycle of the crystallized slurry • No exposure to dry powders	• Media size • Media material compatibility • Duration of the milling run • Product filterability

flowability, morphology, tendency for compaction, fragility). The parameters to consider for scale-up will vary with milling technology since the mechanisms for attrition are different. For milling scale-down, laboratory-sized units are available for experiments in the 10–100 g scale. While scale factors and empirical rules are used to determine the initial parameters for scaling up milling operations, a small test batch is often run to verify the physical properties (PSD, etc.) prior to milling the entire batch. PAT monitoring of the particle size distribution through online particle size analysis (Insitec, FBRM) is a prudent means of ensuring that the correct particle size is achieved. Some of the key advantages and parameters/issues to consider for the various milling technologies are described in Table 20.24.

20.4 SUMMARY

Understanding process scale-up and assessment is a core activity for process chemical engineers in the pharmaceutical industry. It enables transformation of a chemical synthesis to a scalable pilot plant process and then to a robust manufacturing process. The numerous factors to consider in the scale-up and assessment encompass addressing the specific risks to safety, quality, and manufacturing productivity as well as the more general strategic risks in managing a portfolio of projects that span different stages of development. In this chapter, we have discussed many drivers for development, including the requisite process and personnel safety, product quality, and business optimization. Understanding these drivers is the key to both efficiently prioritizing development activities for a given project's stage of development and ensuring that resources are appropriately prioritized across the portfolio. We have also discussed the unit operations that constitute a typical process. Understanding of the process fit and scale-up factors for these unit operations is critical to defining and executing a process development strategy. By applying these concepts, along with more detailed insights from the other chapters in this book, the process engineer will be well-prepared to meet the challenges of API process development.

REFERENCES

1. Anderson N. *Practical Process Research & Development*, Academic Press, 2000.
2. Rubin, E., Tummala, S., Both, D., Wang, C. and Delaney, E. Emerging technologies supporting chemical process R&D and their increasing impact on productivity in the pharmaceutical industry, *Chem. Rev.*, 2006;106(7):2794–2810.
3. DiMasi, J.A., Hansen, R. W., Grabowski, H.G., and Lasagna, L. Research and development costs for new drugs by therapeutic category, *PharmacoEconomics*, 1995;7(2):152–169.
4. Pisano GP. *The Development Factory*, Harvard Business School Press, 1997, p.99.
5. Gilbert, J., Henske, P. and Singh, A. Rebuilding Big Pharma's business model, *In Vivo* Business and Medicine Report 21(10), November 2003.
6. Adams, C.P. and Brantner, V.V. New drug development: estimating entry from human clinical trials, Federal Trade Commission, July 7, 2003, www.ftc.gov/be/workpapers/wp262.pdf
7. Guidance for Industry Q9—Quality Risk Management, U.S. Department of HHS, FDA, CDER and CBER, June 2006.
8. International Conference on Harmonization of Technical Requirements for Registration of Pharmaceuticals for Human Use. Q3A: Impurities in New Drug Substances; Q3C: Impurities—Residual Solvents; Q8: Pharmaceutical Development.
9. Noren, A., Naik, N, Wieczorek, G, Basu P. Keep pharmaceutical pilot plants safe, *Chem. Eng. Prog.*, 1999;95(3):39–43.
10. Parker, J.S. and Moseley, J.D. Kepner–Tregoe decision analysis as a tool to aid route selection. Part 1. *Org. Process. Res. Dev.*, 2008;12:1041–1043.
11. Moseley, J. et al. Kepner–Tregoe decision analysis as a tool to aid route selection. Part 2. Application to AZD7545, a PDK inhibitor. *Org. Process. Res. Dev.*, 2008;12:1044–1059.
12. Naumann, B. D., Sargent, E. V., Starkman, B. S., Fraser, W. J., Becker, G. T., Kirk, G. D., Performance-based exposure control limits for pharmaceutical active ingredients, *Am. Ind. Hyg. Assoc. J.*, 1996;57:33–42.
13. Stoessel, F, What is your thermal risk? *Chem. Eng. Prog.*, 1993;89(10):68–75.
14. Ebadant, V. and Pilkington, G., Assessing dust explosion hazards in powder handling operations, *Chem. Process.*, 1995;74–80.
15. Gritton, L.G. Avoiding static ignition hazards in chemical operations, CCPS-AIChE, 1999.
16. *Ullmann's Encyclopedia of Industrial Chemistry*, Wiley, 2009.
17. Cole-Parmer Chemical Compatibility Charts, www.coleparmer.com/techinfo.
18. Wehrstedt, K.D., Wandrey, P.A., Heitkamp, D. Explosive properties of 1-hydroxybenzotriazoles, *J. Hazard. Mater.*, 2005;126 (1–3):1–7.
19. Sigma–Aldrich MSDS Database.
20. Rakita, P.E., Aultman, J.F., Stapleton, L. Handling commercial Grignard reagents. *Chem. Eng.*, 1990: 97(3):110–113.
21. Silverman GS, Rakita PE. *Handbook of Grignard Reagents*, Marcel Dekker, 1996.
22. International Conference on Harmonization Topic Q6A, Specifications: Test Procedures and Acceptance Criteria for New Drug Substances and New Drug Products: Chemical Substances (CPMP/ICH/367/96), May 2000.
23. EMEA Committee for Human Medicinal Products, Draft Guideline on the Specifications Limits for Residues of Metal Catalyst, January 2007.
24. http://www.fda.gov/AboutFDA/CentersOffices/CDER/ucm128219.htm
25. Draft Guidance for Industry. Genotoxic and Carcinogenic Impurities in Drug Substances and Products: Recommended Approaches, Center for Drug Evaluation and Research, FDA, December 2008.
26. McGovern, T. and Jacobson-Kram, D. Regulation of genotoxic and carcinogenic impurities in drug substances and products. *Trends Anal. Chem.*, 2006;25:790–795.
27. Pierson, D.A., Olsen, B.A., Robbins, D. K., DeVries, K.M., Varie, D.L. Approaches to assessment, testing decisions, and analytical determination of genotoxic impurities in drug substances. *Org. Process. Res. Dev.*, 2009;12:285–291.
28. Glodek et al. Process robustness: a PQRI White Paper. *Pharm. Eng.*, 2006;25:1–11.
29. Seibert, K., Sethuraman, S., Mitchell, J., Griffiths, K, McGarvey, B. The use of routine process capability for the determination of process parameter criticality in small-molecule API synthesis. *J. Pharm. Innov.*, 2008;2:105–112.
30. Thayer, A. Sustainable synthesis, *C&E News* 2009;87 (23):12–22.
31. Van Aken, K., Strekowski, L., and Patiny, L. EcoScale, a semi-quantitative tool to select an organic preparation based on economical and ecological parameters, *Beilstein J. Org. Chem.* 2006;2(3).
32. Ritter, S. K. Chemists and chemical engineers will be providing the thousands of technologies needed to achieve a more sustainable world, *C&E News*, 2008;86(33):59–68.
33. Paul EL, Atiemo-Obeng V, Kresta S. *Handbook of Industrial Mixing*, Wiley, 2004.
34. Oldshue JY, Herbst NR. *A Guide to Fluid Mixing*, Mixing Equipment Co., 1990.
35. Caygill, G., Zanfir, M., Gavriilidis, A. Scalable reactor design for pharmaceuticals and fine chemicals production. Part 1. Potential scale-up obstacles. *Org. Process. Res. Dev.*, 2006;10:539–552.
36. Fasano, J., Penney, R., Xu, B. Feedpipe backmixing in agitated vessels, *AIChE Symp. Ser.*, 1992;293:1–7.
37. Zweitering, T.N. Suspension of solid particles in liquid by agitators. *Chem. Eng. Sci.*, 1958;8:244–253.
38. *The Nalco Water Handbook,* McGraw-Hill, 1988, Chapter 11.
39. Singh, U.K, Pietz, M.A., Kopach, M.E. Identifying scale sensitivity for API crystallizations from desupersaturation measurements. *Org. Process. Res. Dev.*, 2009;13:276–279.
40. www.hmicronpowder.com.

41. Lee, I., Variankaval, N, Lindemann, C, Starbuck, C. Rotor–stator milling of APIs: empirical scale-up parameters and theoretical relationships between the morphology and breakage of crystals. *Am. Pharm. Rev.*, 2004;7(5): 120–128.

42. Kamahara, T., Takasuga, M., Tung, H.H., Hanaki, K., Fukunaka, T., Izzo, B., Nakada, J., Yabuki, Y., Kato, Y. Generation of fine armaceutical particles via controlled secondary nucleation under high shear environment during crystallization: process development and scale-up. *Org. Process. Res. Dev.*, 2007;11:699–703.

21

SCALE-UP DOS AND DON'TS

Francis X. McConville
Impact Technology Development, Lincoln, MA, USA

21.1 INTRODUCTION

One of the chemical engineer's primary responsibilities in the pharmaceutical industry is to assist in scaling up laboratory or development-stage processes for commercialization. An unfortunate fact is that so much of the practical information that can make scale-up more efficient and ultimately more successful is generally acquired in the workplace only through years of on-the-job training. Most university engineering curricula are simply not geared toward teaching students about the many real-world issues that can complicate scale-up or lead to unexpected results, unsuccessful campaigns, and potentially dangerous situations.

21.2 LEARNING THE HARD WAY

I learned many of the important things about technology transfer and process scale-up on the floor of the pilot plant long after I left school. Suffice it to say that some of these lessons were hard won, sometimes at the cost of out-of-spec batches, close calls, and unnecessary delays. Lucky is the new graduate in a position to be mentored by someone who has learned about scale-up through experience by putting time in "in the trenches."

It sometimes seems that scale-up is simply a lesson in learning to expect the unexpected. But over time, one realizes that many of the surprises encountered during scale-up could have been anticipated and often prevented by paying attention to the appropriate details, conducting some relatively simple laboratory studies, or collecting the right quantitative data during early process development. That is why I have always been a strong proponent of appropriate process engineering studies early on in new process design. There are countless simple laboratory measurements that allow the process chemist or engineer to characterize and quantify the behavior of the reactants and other materials used in the process, and this can go a long way toward streamlining scale-up.

21.3 TYPICAL SCALE-UP ISSUES

One of the most common effects of scale-up is a change in reaction selectivity, especially in semi-batch reactions. This results mainly from differences in mixing between the laboratory or kilo-lab scale and the commercial scale. This is discussed in more detail in Chapter 14. Changes in selectivity can lead to lower yields and higher levels of impurities in the final product, or changes in the impurity profile that in turn can alter the physical form or polymorph of crystalline products. The appearance of previously unseen polymorphs of pharmaceutical solids upon scale-up is all too common, and this change in impurity profile is one of the major underlying reasons.

Product isolation can also lead to unexpected results upon scale-up. Despite the industry's best efforts to design better and more efficient filters and centrifuges for the recovery of solid products, the fact remains that the removal of impurities in the cake washing step is often not as complete or as efficient at large scale as in the laboratory. This is, in part, simply due to the difficulty of ensuring even distribution of the cake and cake wash at large scale.

Chemical Engineering in the Pharmaceutical Industry: R&D to Manufacturing Edited by David J. am Ende

Other unexpected consequences of scale-up result from the very long times required to complete many processing steps at large scale. Operations as fundamental as charging raw materials can take many hours; likewise for transfers, distillations, and product isolations. It is critical to conduct the necessary laboratory stability studies to ensure that the various process streams do not undergo degradation during these prolonged processing times.

Sometimes, problems result from poor communication at the technology transfer stage. One of my more uncomfortable memories involves an API process being transferred to the 8000 L scale at a foreign CMO. It was only upon arrival at the manufacturing site that our team learned that the designated reactor train was not equipped for vacuum distillation—all stripping operations were to be conducted at atmospheric pressure. However, all previous development work for this process had utilized vacuum distillation. We somewhat unwisely proceeded with the campaign based on the results of a quick laboratory test, and wound up cooking our final product stream until it was dark brown and failed spec, because of the very elevated temperature and the many hours that the distillation took at that scale.

21.4 PURPOSE OF THIS CHAPTER

Throughout the long years as a process engineer and working in kilo labs and pilot plants, I came to develop a list of things that should be followed—things that made pilot operations more efficient and safer—and a list of things that should be avoided, or operations that could not be scaled up effectively. At times, I worked with scientists with little scale-up experience, and often used these lists to help educate them about the types of procedures that might work well in a plant setting and those that would probably not. These lists evolved into a set of scale-up "dos and don't" for professionals involved in technology transfer, operating pilot plants, or laboratory personnel developing processes for eventual scale-up that I first published in 2002 [1] and that I expand upon in this chapter. This is a very subjective list, representing my own personal take on scale-up. Many of the activities or approaches I recommend are considered *de rigor* for companies involved in the GMP manufacture of pharmaceutical materials to comply with ICH guidelines, but even for small companies and start-ups, these practices often just make good sense and can be beneficial for long term safety and success.

21.5 THINGS TO DO DURING SCALE-UP

21.5.1 Make It a Team Effort

One thing that has helped me greatly in my career is establishing strong channels of communication with the many process development chemists with whom I have worked. I feel strongly about the important role that process engineering plays in process development. There have been those few inexperienced chemists who tended to practice "over the wall" chemistry—that is they would develop a new process and then toss it "over the wall" to the engineering/pilot plant group and expect to never see it again.

We know this cannot happen in the real world. It is only through close cooperation between the various disciplines that a successful, scaleable process can be obtained. The best process chemists understand this. While the chemist has a deeper understanding of the effects of the various process variables on product quality, the engineer may have a better appreciation for the physical limitations of pilot equipment, or in short, what operations can or cannot be conducted safely in a pilot or commercial plant. Each has his or her own set of priorities, and they must be communicated clearly to each other, and the earlier in the development cycle, the better.

For example, the engineer can inform the chemist that his or her choice of solvents will not be acceptable in the plant, or that certain laboratory operations such as evaporating to dryness cannot be scaled up. Likewise, the chemist can communicate the need for tight temperature control during a certain critical step, or the need for a certain degree of agitation, and so on.

Engineers bring many valuable skills to the table to assess and improve the scalability of novel processes. Some of the areas in which process engineers can make important contributions to the development effort are as follows:

- Identify and determine limits for critical process parameters
- Identify process hazards and conduct the necessary hazards analysis, including isothermal and adiabatic calorimetry and explosivity studies.
- Compare projected commercial costs of alternate routes (COGS)
- Complete mixing and heat transfer calculations for scale-up and assist with "scale-down" experimental design
- Conduct reactor design calculations
- Size and specify equipment
- Complete material and energy balances
- Investigate opportunities for continuous or other alternate processing

However, as mentioned above, there are many simple experiments and measurements that fall under the category of process engineering that any laboratory scientist can carry out easily, such as solids drying studies; distillation stability studies; and simply measuring and recording operating pressures, temperatures, densities, and other physical properties of the various process streams. This type of information will be a great aid in speeding scale-up.

21.5.2 Develop an Operating Philosophy

No matter how large or how small the facility is, it is very important to lay down some ground rules for the safe transfer and scale-up of laboratory processes to the kilo lab or pilot plant. For example, at one company, we made a key decision of not operating the new kilo-lab under cGMP. This eliminated the need for strict compliance to regulations imposed by outside agencies, streamlined scale-up, and allowed more operating flexibility. This did not mean, however, that we threw caution to the wind and worked haphazardly.

We established clear, strict requirements for all processes to be transferred to the kilo lab. With management support, we were able to adhere to these requirements even in the face of pressure to meet aggressive timelines. For example, we required a written batch record for all processes, and that the batch record be proven by using it to run a minimum of three laboratory-scale batches (see below). One of the three laboratory runs would typically serve as the raw material use test, conducted using the actual kilo-scale raw material lots.

We also called for strict cleaning and rinse-test protocols for all equipment to minimize the possibility of cross-contamination between batches. Another important requirement was the completion of a Haz-Op study, conducted by a team of at least three individuals representing the kilo-lab staff, process chemistry, and when possible a company safety officer.

Although it may appear that these requirements would hinder efficiency, it was not the case. Once it was clear that the rules would be strictly enforced, everyone on the development team adjusted his or her thinking accordingly with the end result that kilo lab experienced no serious accidents and virtually no failed batches over many years of operation.

21.5.3 Establish Use-and-Maintenance Files

Both good engineering practice and good common sense are required in a cGMP environment No matter how small the operation is, developing a sound recordkeeping system to capture how each piece of equipment in one's kilo lab or plant has been used and maintained, starting from the moment of installation, can pay big dividends in extending the life of the equipment, minimizing life cycle cost, and providing traceability for important clinical or preclinical materials.

Established companies will have detailed IQ–OQ (installation qualification and operational qualification) protocols and preventive maintenance (PM) systems. Smaller companies should, at a minimum, establish a logbook (a laboratory notebook works well) in which they can record manufacturer data and wiring and engineering diagrams, in addition to information about each batch of material processed, the results of cleaning operations or rinse tests, maintenance performed, calibrations or other measurements made such as heat transfer coefficients, volume calibrations, and so on. Such records should be kept over the life of each reactor or mixing vessel, each filter or centrifuge, and even reusable transfer hoses and other portable components.

21.5.4 Establish a Sample Database

Another recordkeeping practice that will prove invaluable is maintaining a sample database in the kilo-lab or pilot plant. Keep a permanent record of each and every sample or process stream that is collected for analysis or observation during your operations. That includes any in-process batch samples, distillates, wet cakes, final products, and waste streams. Every sample should be given a unique sample number and the batch, time, date, amount, reason for its collection and any other important observations should be recorded. Larger companies may have a fully integrated digital laboratory information management system (LIMS) but for kilo-scale operations, any kind of logbook is sufficient. Provide columns to make sure that all the pertinent data are recorded.

In one kilo-lab, we used a laboratory notebook for the purpose, and simply assigned unique sequential numbers to all process and retain samples. We frequently needed to go back and reconfirm the identity of past samples for research or regulatory purposes, for conducting mass balances and the like, and the information we needed was readily available.

21.5.5 Collect Retain Samples

Along with a sample database, it is important to institute a retain sample system. Samples of all dried isolated intermediates and final products should be kept in appropriate sealed containers stored in a cool, dry, dark place, well organized, and readily accessible should the need arise. The exact sample size, storage container, and conditions will of course depend on the specifics of the material being stored, but the important things are that the containers be clearly, permanently labeled and that the storage system be carefully thought out and enable easy retrieval of samples when needed.

21.5.6 Fix the Process Before Scaling

This point goes hand in hand with developing a consistent operating philosophy. Last minute changes to a process being scaled up can lead to serious unexpected consequences, and possibly unsafe situations. It is important to minimize last minute changes by fixing the process well in advance of scale-up and ensure that laboratory demonstration runs of the actual final process have been conducted prior to scale-up.

The cases of processes that have failed due to on-the-fly changes in the plant are legion. One example that comes to mind is a failed selective diastereomer crystallization that was "bumped" into a 12,000 L stainless steel reactor because the usual glass-lined reactor was occupied by another process. The expected enantiomeric enrichment did not occur and the batch failed miserably, most likely due to material surface effects or differences in the nature of mixing in the two reactors. This was a harmless enough—albeit tremendously expensive—lesson but last minute changes can, and often have led to the creation of very serious hazards.

21.5.7 Conduct a Haz-Op Review

There are two terms commonly heard in the chemical process industries, "Haz-An" (for hazards analysis) and "Haz-Op" (for hazards and operability study). The former is a detailed examination of a particular hazard, such as a potential decomposition reaction, in which one might conduct calorimetry studies to determine onset temperature, time to maximum rate, maximum adiabatic temperature rise, and so on. The Haz-Op, on the other hand, is a more general study conducted by a team in an attempt to identify all the potential hazards of a process prior to scale-up.

Most companies insist on some level of Haz-Op study before scaling up new processes, even to the kilo-scale, and of course a detailed study should be absolutely required for larger scale operations. The first step is assembling the team, which should consist of chemists and process and safety experts from within the company who are most familiar with the process and the plant. This team approach eliminates potential oversights by individuals working in isolation. The next step is the preparatory work and assembling the necessary documentation (batch record, plant P&ID's, process flow diagrams, etc.) to facilitate the study. And then finally there are the Haz-Op meetings themselves, wherein the team leader should encourage free expression of ideas and uninhibited "what-if" thinking in an effort to develop a list of potential dangers and "mal-operations" for each and every process step. Clearly documenting the findings and proceedings of the meetings is also a very important part of the exercise.

There is much excellent information on including Haz-Ops as part of a safety management program, and the reader is encouraged to take advantage of these and other resources [2, 3].

21.5.8 Quantify Reaction Energetics

Underestimating or failing to recognize the potential hazards of exothermic reactions is perhaps the single most common cause of harmful and destructive process industry accidents. One of the reasons for this is a failure to understand the very limited heat transfer (cooling) available in large chemical reactors, a consequence of the low surface area per unit volume.

Thus, an important part of safe reaction scale up should be calorimetry or similar studies designed to quantify the exotherm, identify the potential maximum rate of reaction, predict the adiabatic temperature rise, and so on. For companies that cannot conduct such studies in house, many safety laboratories offer these services on a contract basis. A number of innovative reaction classification systems have been proposed to help categorize the potential hazards of chemical reactions based on calorimetric parameters and to better ensure safe scale up [4, 5].

The same can be said for determining the explosion potential of dusts and powders, along with minimum ignition energy (MIE), limiting oxygen concentration (LOC) for ignition, maximum attainable pressure, shock sensitivity, and so on. Standard test methods exist for conducting and interpreting all these characteristics, and the information from these studies can help design safer procedures to prevent explosions and to engineer improved mitigative equipment [6].

Of course, engineering judgment and chemical wisdom must be applied to interpretation and use of the data obtained from such studies, as the data often do not tell the whole story. Gustin [7] reports the case of a nitration reaction that counterintuitively exploded after the reaction was completed and the reactor was being cooled. The reason appears to have been the crystallization of a highly nitrated shock-sensitive sodium salt that came out of solution upon cooling. A piece of this material may have collected on the impeller shaft, broken free and impacted the impeller or baffle resulting in the explosion. Calorimetry or other studies might not have predicted this event, but an examination of the chemistry by someone experienced with these compounds might have identified the potential for this compound to form. Then more exacting safety studies could have been conducted to determine the risk and potential consequences of this happening.

21.5.9 Create a Written Batch Record

Variously called a batch log sheet, batch ticket, or batch record, this is an approved document (either paper or electronic) to be filled out by the operating staff as the batch is conducted. The batch record is based on a detailed process description prepared by the project chemist or process engineer and formally approved by his or her direct supervisors or others as necessary. It is a step-by-step recipe sheet, if you will, with spaces for recording pertinent data such as raw materials charges, processing times, and temperatures, and so on, and with spaces for the initials or signatures of the individuals completing each task and checkers where necessary.

The batch record should be a controlled document. There is no need to point out the importance of ensuring that most current version of the record is in use, and that it is as free of errors as possible. The tremendous convenience and speed of modern word processing has led to more than one pilot plant mishap due to careless cut-and-paste errors and poor proofreading. That is why a laboratory shake-down run is so important in ensuring the correctness of the batch record.

21.5.10 Understand the Raw Material Grades

Many common raw materials and chemical reagents are available from a variety of sources and in a variety of qualities, purities, or grades. By convention, some terminology has arisen around these various grades, such as reagent grade, technical grade, but the precise meaning of these terms is anything but consistent from manufacturer to manufacturer, and the global supply chain creates even more confusion.

What one company may call spectrophotometric grade, another may call reagent grade and vice versa. Many supply companies have devised their own systems of nomenclature and provide various proprietary grades or purity levels that are impossible to directly compare to those of other suppliers. That is why it is important to understand the critical quality attributes of the raw materials used in a process and ensure that the chosen supplier can consistently provide the quality needed. If there a particular impurity in a raw material that can adversely affect a reaction, then its effects must be quantified and the specification for that impurity must be set based on the resulting data.

Much laboratory work uses reagent grade solvents and materials, but beware that the commercial grades available at large-scale may not match the purity of many of these substances. It is wise to work with commercial grade materials as early in development as practical to better anticipate the results upon scale-up. This is also why it is so critical to conduct the laboratory-scale raw material use test prior to the first scale-up batch (see below).

Chemical purity is not the only characteristic to be concerned about. The physical form, such as particle size in the case of solids, can have a major effect on results of the process. A well-known example of this is the use of K_2CO_3 as an acid-sequestering agent in many alkylations and in other organic reactions. Moseley [8] tells a tale of woe about their experiences in scaling up just such a process, due to the fact that a granular form of the carbonate with relatively low specific surface area was used during some large-scale runs. Their problems were exacerbated by the fact the mixing conditions in the large-scale vessel were insufficient to suspend these granular solids and so the carbonate lay unmoving and inaccessible on the bottom of the reactor, a very common phenomenon in large vessels.

21.5.11 Conduct a Raw Material Use Test

It should be considered an absolute requirement that a laboratory-scale experiment be conducted, ideally following the written batch record, that uses the very same raw material lots or batches of intermediates that will be used in the scale-up campaign. This demonstration batch should be carried to completion; the product should be isolated and analyzed in full just as the pilot batch would be. The pilot batch should not proceed until all analytical results are reported and it has been demonstrated that the batch passes all specifications. In this manner, if and when there are quality or processing issues in the scale-up batch, the raw materials can for the most part be eliminated as a cause. This can save a great amount of work and head-scratching and can prevent an investigation of a problem from being focused in the wrong area.

21.5.12 Make the Most of the Opportunity

A tremendous amount of time, effort and money is expended in preparing for and conducting pilot-scale batches. These batches can consume alarming amounts of precious raw materials and labor. Consequently, in most cases only a limited number of pilot-scale batches can be conducted. That is why it is important to try to learn as much as possible from each batch. For example, a well planned sampling and analytical plan will allow you to complete a mass balance, identify unexpected side products and otherwise troubleshoot batches that have not gone as expected. Basically every process stream, including wasted streams, should be weighed and sampled. There may never be another opportunity to collect many of the samples generated in a pilot batch.

All observations should be carefully noted and retain samples of isolated intermediates and final products should be saved for future reference if necessary. Putting in a nutshell, one should make the best use of the opportunity to gather as much scale-up data as possible and clearly document the results of the batch in a comprehensive campaign report.

21.6 THINGS TO AVOID DURING SCALE-UP

21.6.1 Avoid Complexity

We have no doubt heard of the famous “KISS” principle. There is certainly much to be said for keeping it simple, particularly in chemical process development and scale-up. The less complexity, the less opportunity for processing errors, operator slips and unforeseen complications.

Commercial processes are carried out by a chemical operations staff who are well trained, but often not educated as chemists or engineers, and certainly not as familiar with the idiosyncrasies and hazards of new processes as their developers are. These operators will carry out their jobs only as well as their training, experience, and operating instructions allow them. A clearly written batch record is supremely important, and this is much more difficult to achieve if a process is overly complex. A simple example would be making an extra effort in development to find a single solvent or mixture of solvents that can be used throughout the reaction, workup, and crystallization steps so as to eliminate the need for time-consuming and wasteful solvent switches between each operation.

Of course, most development chemists understand the importance of simplicity, not only for improved safety and efficiency but also for minimizing processing time, minimizing waste, and so on. However, it is not always possible to keep things simple when the process involves chemistry that requires sophisticated controls or specialized equipment such as hydrogenation reactions, nitrations, aminations, or other types of reactions that could potentially be hazardous. It is always an option to contract out these particular steps to manufacturers who have expertise in those types of reaction and the equipment to carry them out.

21.6.2 Avoid "All-in-and-Heat" Reactions

One of the most dangerous practices in chemical processing, and one that is frowned upon at all scales, is to charge all reactants to a batch vessel and then begin heating it up. The danger is that once the mixture reaches the onset temperature of reaction, and the reaction starts, there will be no way of controlling it. Many reactions are highly exothermic, and once they "kick-off" they will continue to heat themselves to higher and higher temperatures, possibly exceeding the boiling point of the mixture and erupting. The mixture could also begin to decompose at higher temperatures. Many times, the decomposition itself is self-accelerating and more exothermic than the process reaction. Explosive gases can be evolved and a highly energetic explosion could ensue.

Of course, when the chemistry is well understood and known to be safe, this type of all-in operation may be acceptable. But when scaling up new processes for the first time, it should be forbidden.

A number of factors make carrying out exothermic reactions at large scale much more dangerous than at laboratory scale. Of course, the consequences of a large explosion are more devastating than a small one, but the key difference at large scale is the limited heat transfer area characteristic of large reactors, and the long response time if cooling is suddenly required. It can take many hours to cool a very large chemical reactor, by which point it may be too late.

The recommended approach for scaling up exothermic reactions is to maintain some degree of control. The most common approach for this is to use a "controlled addition" scheme in which the nonreactive components of the batch are charged to the reactor, and then the reactive (controlling) reagent is slowly added with agitation, allowing the reaction to proceed at a controllable pace. In the event of a temperature excursion, addition of the reagent can quickly be stopped, halting the reaction.

It is important in this method to ensure that the reaction is in fact proceeding and that the reactive agent is being consumed as it is added. If, for some reason, this is not happening, the reaction "stalls" for instance, then the reagent will accumulate in the reactor and may suddenly react all at once, putting us back where we started. Many methods are available for monitoring the progress of a reaction, but the simplest is to monitor the batch temperature and ensure that the anticipated rise is observed.

21.6.3 Do Not Apply Heat Without Agitation

A former colleague of mine would attest to the practical nature of this advice. He was performing a toluene/aqueous extraction experiment at about 70°C in a round-bottom flask with a heating mantle. He stopped the agitator briefly to let the phases separate, collected his sample, and then restarted the agitator. The entire contents of the flask instantly erupted out the top of the reflux condenser. Luckily no one was hurt. We surmised that while the agitator was stopped, the glass surface must have exceeded the 85°C boiling point of the toluene–water azeotrope. As soon as the agitator was restarted, the mixture boiled violently.

In my opinion it is never acceptable to apply heat to a reactor without agitation. Heat transfer is severely limited in large reactors to begin with, and what little heat transfer does occur is highly dependent on the degree of agitation in the vessel and the convection that it creates.

In addition to the incidents such as the one described above, countless other undesirable effects can also result. Without agitation, there can be no accurate reading of the internal reactor temperature. Because the reactor wall temperature is quite often much hotter than the bulk batch temperature, product can easily become overheated and "baked" against the side of the vessel, which can lower yield and lead to dangerous degradation reactions. Temperature gradients created by insufficient agitation can result in poor reaction selectivity and out-of-spec products.

Countless cases of violent reactions and explosions have been attributed to "agitation issues" of one kind or another. For example, one incident occurred during a highly exothermic reaction between the sulfuric acid and an organic amine. This biphasic reaction was normally carried out by slow addition of the amine to the hot acid with vigorous agitation. One fateful day, at shift change, the agitator was inadvertently left off when amine addition was started. The amine pooled in the bottom of the reactor but did not react. Much later, the second shift noted that the agitator was off and proceeded to turn it on, at which point the reactor exploded as all materials reacted instantaneously. A neighborhood in Frankfurt, Germany was dusted with a yellow coating of *o*-nitroanisole in a similar incident that eventually resulted in the company going out of business.

21.6.4 Do Not Ignore Potential Decomposition Reactions

This recommendation goes hand in hand with Section 21.6.3. Not only must the necessary calorimetry be conducted on exothermic reactions but also the possibility of self-heating decomposition reactions must be considered. This may require additional testing, such as adiabatic reaction calorimetry (ARC). Such testing should be considered for all process streams if it is believed that the particular chemistry involved has the potential to create unstable decomposition products.

One of the difficulties is that these self-heating decomposition reactions may happen very slowly that they may not be identified in routine testing. Even below the onset temperature, exothermic reactions are still happening at some finite rate. In one case, a reactor exploded many hours after the reaction was completed, the services shut off and the reactor left to cool on its own. A previously unknown decomposition reaction was occurring so slowly that no one noticed the temperature rising in the reactor. Eventually the temperature reached the onset temperature and the self-accelerating reaction kicked in, resulting in an explosion (it is generally safer to cool a reactor to a safe temperature using the jacket rather than allow it to simply cool down on its own). Waste stills have been known to explode days after being charged with a waste stream that underwent a slow decomposition and unanticipated increase in temperature.

21.6.5 Avoid Adding Solid Reactants to Reacting Mixtures

Another common laboratory technique that is not easy to scale (without some advance planning) is the portion-wise addition of solid reagents to a reacting mixture. Laboratory scientists routinely use this technique to avoid the "all-in-and-heat" approach, but in the laboratory it is a simple matter to remove a glass stopper from the flask in a fume hood and add a small spatula full of the solid and close it back. This is repeated until the addition is complete.

There are a number of reasons that this becomes much more difficult at scale. First, it is most inadvisable to open the manway of a large reactor containing flammable solvents because an ignitable atmosphere may form as vapors exit the vessel or air enters it. This is especially true if the contents of the reactor are being heated. Second, it is quite possible that the reagent may react quickly, causing the eruption or ejection of material out of the manway, endangering the personnel (sadly, operators have died because of this very thing). Therefore, the solids addition must somehow be accomplished with the reactor sealed.

One possibility is to reconsider the order of addition. It is always much easier to charge solids to a reactor first, and then the solvents and liquid ingredients. Of course, changing the order of the addition may change the selectivity of the reaction, or worse, may remove an important component of exothermic control. Another possibility is to make a solution of the solid, which can be conveniently charged by pump or other methods, or even a slurry, although charging slurries at a consistent rate is somewhat more difficult than charging a solution.

Where modifying the process is not possible, number of solids charging apparati do exist that enable the controlled addition of solids to a reacting mixture. An excellent review of available options was recently published, which discusses the advantages and weaknesses of each approach for a number of given situations [9]. Many of these devices are also very helpful for improving the dissolution of hard-to-wet solids or solids that tend to float or form large lumps, which can become major processing issues in commercial-scale operations.

21.6.6 Avoid Evaporating to Dryness

Generally speaking, the common laboratory technique of evaporating a process stream to dryness by removing all solvent on a rotary evaporator or similar piece of equipment simply will not fly in pilot-scale equipment. First, toward the end of distillation, the liquid level will fall below the agitator (minimum stir volume) in most standard reactors. As discussed above, it is inadvisable to continue applying heat without adequate agitation.

One of the consequences of heating without agitation, or stripping to dryness, could be the decomposition of the product in contact with the hot reactor walls. This presents not only quality but also safety issues. When complete removal of the solvent is necessary so that a different solvent can be introduced for the next processing step, the standard approach is to conduct a "solvent exchange." In this operation, the solvent is first partially distilled down to a safe (but mixable) level; then the second solvent is added and distillation is continued. This process is repeated until the concentration of the first solvent is as low as required. Of course, the specifics of the operation and its success depend on the relative volatilities of the solvents and whether or not they form an azeotrope. It is also well known that the so-called constant volume solvent exchanges are much more efficient than the add-and-distill approach [10, 11].

21.6.7 Do Not Underestimate Processing Times at Scale

I have often said that one of the biggest surprises that an R&D chemist experiences when bringing a new process to the pilot plant for the first time is how long everything takes. My first scale-up campaign at a CMO was no exception.

I remember arriving early on the first day eager to get the process underway. Step 1, charging a major raw material, a solid, took the entire 8 h first shift. This included time for completing the release paperwork, transporting the material from the warehouse, the operators suiting up, the laborious act of charging the material manually, and coffee breaks. Later in the process, a distillation step that we routinely accomplished in about 30 min in our kilo-lab took over 12 h, including heating up and cooling down the reactor. The final product isolation, which used a product centrifuge, required seven or eight separate centrifuge loads; the entire recovery operation took over 24 h. Again, this was something I was used to completing in 1 h, even at the pilot scale, where one filter load could accommodate the entire batch.

Extended processing times are a fact of life in plant-scale operations, but for the unprepared, it will be a test of their patience. More importantly, serious quality and safety issues might arise if the question of process stream stability has not been considered prior to the scale-up. As a case in point, Dunn [12] describes a deprotection step involving trofluoroacetic acid (TFA) that was subsequently distilled off. This technique was quite successful in the laboratory, but in the very first scale-up batch, that distillation step took many hours, and when it was completed only about 5% of the product remained! No one had recognized that the product was not stable in the presence of TFA at elevated temperature because the stripping only took a few minutes in a laboratory rotovap.

A simple stability experiment in which the product stream is cooked at distillation temperature for a time would have given the researchers a heads-up that there could be a problem. Every step of a new process should be considered from this perspective since even under the best of conditions, not withstanding equipment failures or scheduling delays, every operation will take much longer in the plant than in the laboratory or kilo-lab.

21.6.8 Do Not Ignore Plant-Scale Solvent Issues

Chemists often develop processes using their favorite solvents, those that provide solubility for a wide range of substances, or those that are most easily removed in distillation and drying. Unfortunately, some of these very solvents may need to be avoided in pilot or commercial operations due to safety concerns or environmental issues.

Hexane comes to mind as an excellent organic solvent for running many types of reactions and for crystallizing a broad range of organic compounds. However, with a flash point of only −23°C, a relatively low enthalpy of vaporization, and very low conductivity (making it prone to electrostatic buildup), many processing plants will simply not allow it as a production solvent. It is also toxic, as are a number of other common organic solvents.

Ethyl ether, methylene chloride, chloroform, methyl isobutyl ketone (MIBK), methyl ethyl ketone (MEK), *n*-methyl pyrrolidone (NMP) are but a few of the solvents that have been widely used in industry for decades, but which suffer from either flammability and safety issues, high water solubility and the accompanying environmental concerns, toxicity or other reasons that make their large-scale use unfavorable, or in many cases prevented by law. There are also those solvents that, because of health concerns are not allowed by the FDA and other regulatory bodies to be present in the final formulation of human drug products, and this list of solvents is constantly evolving. Finally, there may be waste-disposal concerns for certain solvents in particular locales.

Thus, it is important for those working in process development to understand as early as possible what limitations on solvents they will have to deal with in their companies, or the plants or countries they will be operating in. Sometimes, it is not easy to find replacement solvents with a better safety profile, and the process may need to be entirely redesigned, but the earlier this is recognized the better.

21.6.9 Avoid Hot Filtration Operations

A common processing step is the "polish filtration," in which a final product solution is filtered through a small-pore cartridge filter or a filter coated with celite or other filter aid to remove any particulates or undissolved contaminants as a final polishing step. This is usually carried out just prior to isolation of a final product by crystallization.

At laboratory scale, this step is often ignored, but plant operators know that small amount of dust and dirt and other undissolved solids often wind up in a product batch after multiple processing steps, and that this material needs to be removed prior to crystallization. Unfortunately, the way many crystallization processes are designed, the product solution is supersaturated to help maximize crystal yield. Thus, the polish filtration step must be carried out at elevated temperatures to ensure that no product crystallizes out in the filter or the pipes connecting it.

This can present some difficulties and potentially unsafe situations on scale-up. In order to prevent material crystallizing in the filter, the filter and all lines leading to it must be heated, perhaps by steam tracing, which complicates the operation. If the temperature of the pipes falls too low, the lines and filter may become plugged with solids. Also the handling and transferring of heated flammable solvents is not a particularly safe practice.

For these reasons, it is best to avoid filtering heated, supersaturated solutions. The most obvious way to do this is to dilute the product stream so that it is not supersaturated at ambient temperature and then distill off the solvent prior to crystallization. This, of course, can be time-consuming and could potentially affect product quality in other ways, illustrating again that process scale-up always involves compromises and trade-offs, and that processing decisions must be made while keeping the whole process in mind.

21.6.10 Do Not Underestimate the Quench/Extraction Step

Most reactions carried out in organic solvents are conveniently "quenched" by adding an aqueous solution of an acid, base, buffer, or other quenching agent. This rapidly stops the reaction by neutralizing the reactive species, and via the phase separation step that follows, provides a convenient way to extract and remove unreacted starting materials, side products and other impurities from the reaction mixture. However, often this important step is taken for granted and simply tacked on at the end of a clever new synthesis with the attitude of "work up as usual." This is unfortunate because the quench and extraction (or "work-up") step is a source of countless problems during scale-up and as much care should go into the design of this step as goes into the chemistry that precedes it.

For one thing, the quench usually represents the highest volume step, and to maximize volumetric productivity, the workup should be designed to minimize the use of extract phase while still accomplishing the necessary goals. This also minimizes waste disposal. Also, the settling and phase separation part of the operation can take much longer than in the laboratory due to a number of factors, including finer dispersions and more of a tendency to emulsify due to the higher tip speeds and higher shear associated with commercial-scale impellers. Inexperienced operators may operate agitators at speeds much higher than necessary during extraction steps and create emulsions that can take many hours to separate.

There is also the issue of phase continuity. Every dispersion consists of a continuous phase and a dispersed phase. Depending on the relative ratios of the solvents, their surface tensions, densities and other factors, the aqueous phase may be dispersed in the organic, or the organic phase may be dispersed in the aqueous. These two systems can often show drastically different behavior, sometimes creating an intractable emulsion in one case and readily separating into two phases in the other. These differences should be studied in the laboratory in order to minimize difficulties on scale-up. This phenomenon is described in better detail in Atherton [13].

Unfortunately, emulsions are sometimes unavoidable in batch chemical scale-up, but an awareness of how easily a mixture can become emulsified and a familiarity with the ways to minimize or deal with emulsions will make the scale-up less problematic.

21.6.11 Avoid Routine Reliance on Flash Chromatography

Chromatographic separation techniques are important tools in process development, and in certain sectors of the industry, biomolecule and protein production, for example, they play a major role in commercial production. Certain specialized types of chromatography such as simulated moving bed (SMB) are also used in full-scale production of small molecule products for the separation of chemical enantiomers.

But then there is so-called "flash" chromatography, the purification of a product stream by means of a single pass through a packed silica-gel column, and then eluting out the various bands with solvents. This is a favorite technique of many laboratory chemists, especially early in development where the goal is to simply isolate a small amount of product for analysis and characterization. Unfortunately, this technique is not very amenable to scale up for commercialization.

This type of chromatography can use very large amounts of solvent. Solvents are generally responsible for the majority of the environmental impact of chemical processes in the pharma industry, and chromatography utilizes a disproportionately large amount of solvent per unit product. This solvent needs to be disposed of or recovered and purified for reuse. It is also difficult to design, manufacture and pack very large chromatography columns so that they will operate without short-cutting and backmixing. There will be higher pressure drops through larger columns, necessitating larger pumps, and the temperature and flow control becomes more difficult for very large columns.

For these reasons, it is best for the process developer to recognize that flash chromatography is generally a method for the laboratory only, and that a reliable, scaleable purification process, such as crystallization, will be much better for commercialization. There are numerous ways to approach this, either by crystallizing the product directly from a solvent or mixture of solvents or, if necessary, by forming a crystalline salt form of the molecule with an appropriate anion or cation.

21.6.12 Play It Safe

While safety always has to be the number one priority, what I specifically refer to here is minimizing the risk of losing all of your valuable raw materials or intermediates in a single batch, especially when scaling up for the first time. Avoid the temptation to go for the home run and convert all of that hard-earned feedstock to final product at once. No matter how careful you are, and how much time you spend going over the details of the process, there will always be unexpected occurrences the first time a new process is scaled up.

An instruction might be misinterpreted by an operator, processing steps will take longer than expected possibly resulting in product degradation, a key piece of equipment may not operate as anticipated. Better to play it safe by running two or more smaller batches to ensure that the program is not stopped in its tracks because all key intermediate or custom raw material is used up.

Running smaller batches has other advantages. There is improved heat transfer area per unit volume, and there will be fewer issues with mixing and chemical transfers, and so on if the scale-up factor is smaller. One group described scaling up a hydrogen-generating reaction, and their plans to dilute the hydrogen in the off gas with nitrogen to keep it below its lower flammability limit. Halfway through the first batch an unexpected alarm went off, indicating that the liquid nitrogen tank that supplied the whole building was empty. They had completely drained it; to complete the campaign they split the remaining work into a number of smaller batches to eliminate this from recurring.

21.7 CONCLUSIONS AND FINAL THOUGHTS

It should be clear from the above that experience plays a major role in the speed and success of any scale-up campaign. There is also a tremendous amount of valuable published information about scale-up and process safety available and the process engineer should certainly make an effort to access it and learn from it. Although it is impossible to anticipate every possible mishap during first-time scale-ups, I hope that the above list of dos and don'ts will provide food for thought and a better appreciation for the fact that making cross-disciplinary communication and cooperation high priorities (i.e., keep it a team effort) will maximize your chances for success.

REFERENCES

1. McConville F. *The Pilot Plant Real Book*, 2nd edition, FXM Engineering & Design, Worcester, MA, 2007.
2. Kletz T. *Hazop and Hazan*, 4th edition, Institute of Chemical Engineers, Rugby, UK, 1999.
3. Kletz T. How we changed the safety culture. *Org. Process Res. Dev.* 2007;11(6):1091.
4. Bollyn M. DMSO can be more than a solvent: thermal analysis of its chemical interactions with certain chemicals at different process stages. *Org. Process Res. Dev.* 2006;10 (6):1299.
5. Saraf SR, et al. Classifying reactive chemicals. *Chem. Eng. Prog.* 2004; 100 (3): 34.
6. Zalosh R, et al. Safely handle powdered solids. *Chem. Eng. Prog.* 2005;101(12):23.
7. Gustin J. Runaway reaction hazards in processing organic nitro compounds. *Org. Process Res. Dev.* 1998;2(1):27.
8. Moseley J, et al. Trouble with potassium carbonate and centrifuges: mass transfer and scale-up effects in the manufacture of ZD9331 POM quinacetate, *Org. Process Res. Dev.* 2006;10: 153.
9. Glor M. Prevent explosions during transfer of powders into flammable solvents. *Chem. Eng.* 2007;114(10):88.
10. Chung J. Solvent exchange in bulk pharmaceutical manufacturing. *Pharm. Technol.* 1996;20(6):38–48.
11. Li Y.E. et al. Optimization of solvent chasing in API manufacturing process: constant volume distillation, *Org. Process Res. Dev.* 2009;13(1):73.
12. Dunn PJ, et al. The chemical development and scale-up of sampatrilat, *Org. Process Res. Dev.* 2003;7:244–253.
13. Atherton JH, Carpenter KJ. Process Development: Physicochemical Concepts, Oxford Science, Oxford, 1999.

22

KILO LAB AND PILOT PLANT MANUFACTURING

JASON C. HAMM, MELANIE M. MILLER, THOMAS RAMSEY, RICHARD L. SCHILD, ANDREW STEWART, AND JEAN W. TOM

Process Research and Development, Bristol-Myers Squibb Co., New Brunswick, NJ, USA

22.1 INTRODUCTION

The pharmaceutical industry has traditionally divided its scientific efforts to bring a drug to market into three stages: drug discovery, process development, and manufacturing. Process development is the link between the worlds of the laboratory and the commercial manufacturing plant. To accomplish this mission, chemical engineers in process development groups must be knowledgeable and capable in both arenas—laboratory experimentation and plant scale-up.

The two primary deliverables from the development efforts are the supply of drug substance or active pharmaceutical ingredient (API) and the process knowledge generated. The API fuels several key activities including clinical studies, toxicological studies, and formulation development. The process knowledge forms the basis of a sustainable commercial manufacturing process. Figure 22.1 shows the work process and the organizational structure metaphorically represented as a bridge. In this case the bridge is Process Development physically linking Drug Discovery to Manufacturing. At the foundation of the bridge is knowledge, both process knowledge and regulatory knowledge as required by agencies such as the Food and Drug Administration (FDA). Also shown are several steps of the bridge representing the progression and deliverables of the development process (toxicological study supplies, clinical supplies, and process optimization). Two important tools available to process development groups to carry out their mission are kilo labs and pilot plants.

There can be significant differences across the pharmaceutical industry in what is meant by the terms kilo lab and pilot plant. Some companies differentiate kilo labs from pilot plants by equipment size, and may refer to kilo labs as glass plants because of the extensive use of glass equipment typically found in these facilities. Other companies designate all of their scale-up facilities as pilot plants making no distinction between equipment size or scale of the facility.

Kilo labs are typically utilized for the first scale-up to produce sufficient quantities for initial drug testing, i.e., toxicological studies or phase I clinical trials. Typical chemistry used is less developed than at later stages and is often a modification of the longer, less efficient discovery chemistry route. Kilo labs are designed to yield quantities ranging from 100 g to under 10 kg of materials with typical batch sizes of 2–3 kg (hence the historic reference to kilo lab). Reactor sizes usually range from 20 to 100 L and are made from glass. The kilo lab represents the first step out of the laboratory and the equipment design is typically closer to a laboratory (larger glassware) than a manufacturing plant (industrial equipment).

Kilo labs typically have a high degree of flexibility to accommodate a range of complex chemistries. Portable equipment setups are configured to meet specific process needs and disposable components may be used to decrease turnaround time for process areas. Ideally, the kilo lab infrastructure is sufficient to support "mini-piloting" or scale-down studies to explore process changes prior to scale-up in the pilot plant. Equipment similar in design to the pilot plant equipment (e.g., filter dryers) is often available to accelerate identification of scalability issues. While many kilo labs are manually operated, kilo labs with robust data collection systems can provide key process

Chemical Engineering in the Pharmaceutical Industry: R&D to Manufacturing Edited by David J. am Ende

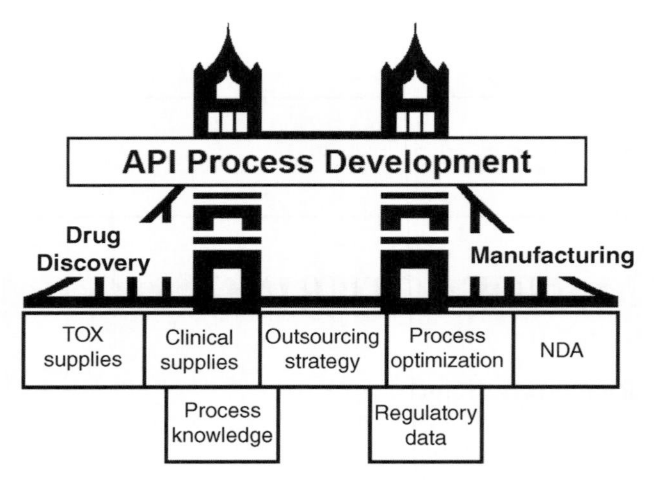

FIGURE 22.1 Process Development links Drug Discovery and Manufacturing.

information to further development, making the kilo lab more than a place to generate kilogram deliveries.

Pilot plants in contrast are typically larger than kilo labs with designs that more closely resemble the commercial manufacturing plant. Supplies generated in pilot plants often support phase II and later clinical or chronic toxicology studies. At the pilot plant stage, demonstration of potential commercial routes and generation of process knowledge through data collection can be key objectives. Though design standards and size differ among companies, pilot plant reactors typically range in scale from 200 to 4000 L.

The pilot plant, like the kilo lab, must operate with a degree of flexibility and absorb process and schedule uncertainties. However, pilot plants typically have more substantial infrastructure and regulatory requirements for quality, safety, and environmental concerns. Pilot plant operations often run 16 hours per day and 5 days per week and can expand to a 24 hours and/or 7 day schedule as needed to manage process chemistry requirements and changes in campaign objectives (e.g., timing or quantity). Portable equipment will also supplement the fixed equipment trains to support process-specific needs. Examples of such equipment include portable tanks, filters, pumps, and process analytical technology (PAT) instruments (Figure 22.2).

One element that differentiates facilities in the pharmaceutical industry from those in the chemical process industry is the requirement to follow current good manufacturing practices or CGMPs when preparing clinical supplies. Good manufacturing practices are governed by regulatory health authorities such as the Food and Drug Administration and European Medicines Agency (EMEA), and are defined within six quality systems: quality, production, facilities and equipment, laboratory controls, materials, and packaging and labeling. More information is provided in Section 22.3.

Figure 22.3 shows a graphical representation that matches the facilities and scale to the phase of clinical development. Many pharmaceutical companies have laboratories, kilo labs, and pilot plant facilities available as an internal capability. Other companies leverage vendor facilities to supplement or replace internal capabilities at each scale of development. It

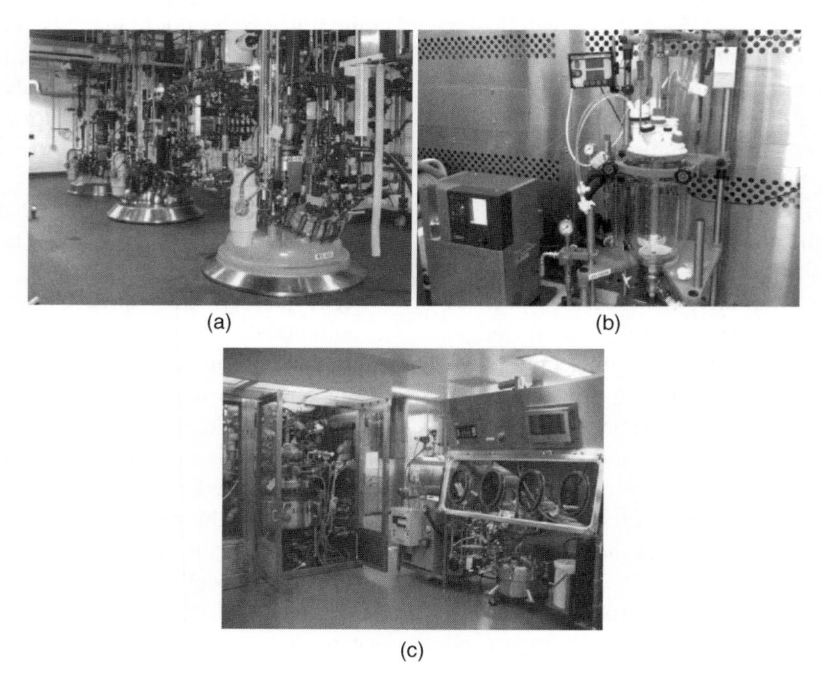

FIGURE 22.2 Examples of pilot plants and kilo labs: (a) pilot plant process area showing three reactors (1000–4000 L) and associated overhead piping; (b) a typical kilo lab with small glass reactor (10 L); and (c) kilo lab process area showing larger reactors (left) in down flow booth and Nutsche filter and tray dryer inside isolator (right) with filtrate receiver underneath.

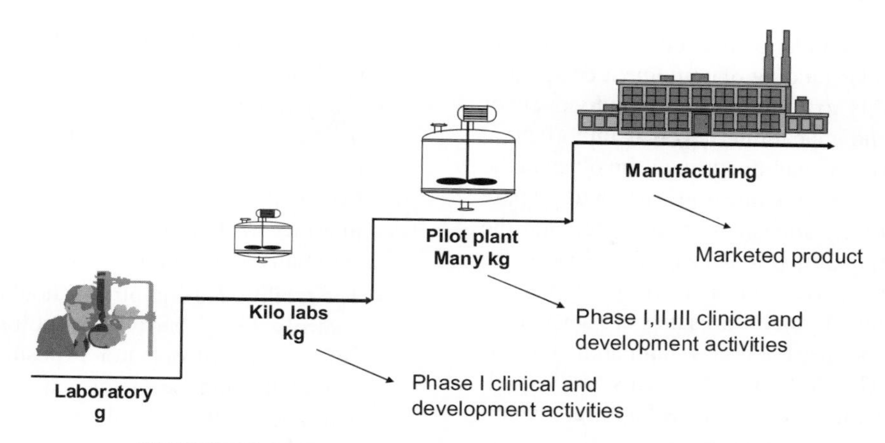

FIGURE 22.3 Progression of scale-up of chemical processes.

is common for companies to outsource several of the early steps in the synthesis to vendors while executing only the last two to four steps internally depending upon business and regulatory drivers. Each company develops its strategy through assessment of the cost and benefit of maintaining internal capabilities and to what scale and capacity.

The basis of an operating philosophy for a kilo lab or pilot plant is safe production of high-quality product while meeting all necessary regulatory requirements. Process safety is constant throughout process development and a primary consideration for process design and execution decisions. Appropriate systems and procedures regarding plant operation and material flow following CGMP provide the basis for the operating philosophy to support generation of high-quality supplies for clinical studies. Processing activities in the facility are the source of process data and knowledge, and the plant's capability to generate such information is a key part of the operating philosophy for the facility. Business objectives for cost- and time-efficient operation of the facilities must also be met. As a result, there is a complex balance of timely and cost-effective manufacture of chemical inventory and obtaining sufficient processing experience to advance knowledge and development of the process.

22.1.1 Operating Staff

The personnel overseeing and working in the facility are key components of a kilo lab or pilot plant facility's ability to operate in a manner demanded by safety and regulatory requirements. They play a critical role in the workflows to operate the facility as well as compliance with the regulatory drivers. The size and makeup of the facility staff will also determine how the facility is run.

In the kilo lab, operating staff may include chemists or chemical engineers. The kilo lab scientist supplements the project scientists, and brings an understanding of scalability issues and process safety risks as well as strong process troubleshooting skills. In some kilo lab facilities, a laboratory manager may oversee the facility and its maintenance while coordinating the project scientists who execute process activities in the facility.

In a pilot plant, many of the named roles are similar to those in a manufacturing plant: plant manager/area head, operations or process engineer/supervisor, chemical operator, and facility support staff, such as maintenance, process automation, and supply chain. The process safety group can be part of the facility staff, but in some companies it is part of the process development group. Analytical support for in-process assays and quality assurance groups are important links to ensuring overall operation of the pilot plant, but typically are not part of the plant operations staff. The number of staff in each role is determined in large part by the planned capacity utilization of the pilot plant, the size and complexity of the infrastructure, and the operational cost base. For the plant manager, the operations engineer, and the chemical operator, there are significant differences in the skill set required between R&D and manufacturing.

The plant manager/area head of a pilot plant is at the intersection of process development activities, campaign execution, and facility support with a broad range of responsibilities from resource management, safety, CGMP, and environmental compliance to readiness for regulatory inspection. It is essential that the manager have sufficient knowledge to understand the various customer needs and manage the dynamic environment of new process implementations.

The operations engineer serves an important function in taking the process description and engineering development data into a set of process instructions for execution in a plant environment. In partnership with the process development team, the operations engineer identifies scalability issues, drives equipment decisions, and determines processing schedules. During execution, the operations engineer may work as a team with the chemical operator to ensure successful implementation. A solid foundation in chemical engineering fundamentals, good understanding of equipment design and operation, and the ability to communicate to individuals of diverse backgrounds and educational levels is important.

The chemical operator in manufacturing is involved in the execution of fully developed chemical processes to prepare marketed products; process robustness is expected and the operator is trained and qualified to execute the process. By contrast, the chemical operator in an R&D pilot plant is involved in the execution of processes as they are developed; process variability is routine. Each new campaign is a new process introduction. The R&D operator trains and is qualified on process equipment, process troubleshooting, and unit operations independent of a specific process.

22.2 KILO LAB AND PILOT PLANT FACILITY DESIGN

For both the kilo lab and the pilot plant, the key objective is to provide for as much flexibility and agility as possible to manage a dynamic portfolio of potential products while meeting environmental, health, and safety (EHS), quality and business requirements. Though they may appear separate, a well-designed facility acknowledges the benefits of integrated quality and safety systems. Below are several characteristics common to R&D pilot plant and kilo lab design:

- The materials of construction (MOC) of all equipment and associated piping systems provide corrosion resistance to a range of process conditions. This reduces concerns of material incompatibility between equipment components and process streams and allows a wide variety of chemistries to be run. The most common choices are glass or glass-lined equipment, Teflon®, stainless steel, and high end alloys such as Hastelloy®.
- The equipment is designed to cover broad temperature and pressure ranges to support safe execution of all typical unit operations.
- Pilot plant facilities should incorporate a blend of fixed and portable equipment in a complementary fashion to increase the diversity of available equipment configurations while managing the amount of time required to change over from one product to a different product. The equipment can be setup relatively fast and inexpensively through use of standard equipment designs and utilization of quick connect hoses.
- The equipment and facility should be designed for cleanability and to mitigate the risk of cross-contamination between products. Kilo labs and pilot plants often execute multiple processes simultaneously. It is important to keep individual process areas clean and to design building airflows to minimize the chance of product contamination between process areas.
- The building and process automation systems enable the capture of data to further the understanding of the process and to satisfy regulatory requirements. Scientists will benefit from process control systems that interface with common engineering analysis tools to summarize and analyze data.
- The facility and equipment should include engineering controls (e.g., barrier technology) as the primary defense against operator exposure to chemicals, drug candidates, and their intermediates. Personal protective equipment (PPE) should not be used as the primary control to mitigate the risk of worker exposure.
- The facility should be able to execute products at both early and late stages of the development cycle. A kilo lab may be called upon to execute a low volume product at a late stage in development while a pilot plant may be the right venue for early stage products with high volume requirements.
- The facility must support the ability to safely produce clinical material being consumed by humans with the proper employment of CGMP for the stage of development. Adherence of key aspects of the facility to CGMP needs to be considered. These include raw material and personnel flow, heating/ventilation/air conditioning (HVAC) design and air balance, maintainability, and cleanability.
- The facility must meet all relevant safety codes as required by regional, federal, and state regulations.

Each chemical process can be broken down into a series of unit operations. Unit operations for a typical process are described below.

- *Charging Components*: The starting molecule is typically a solid and is charged to the reactor through an open system, such as scooping material in through the reactor manway, or through a closed system. Closed charging systems protect the worker and the material from contamination and include split butterfly valves, powder transfer pumps, and glovebox technology. Liquids are charged through reactor nozzles by pump, vacuum or pressure transfer.
- *Reaction*: The reaction is where the molecule is being synthesized within each batch step progressing to the preparation of the API. It normally involves heating, cooling, controlled addition rates of reagents, and agitation. Thermal chemistry studies should be completed prior to execution to ensure safe processing.
- *Work-up*: A work-up can be comprised of multiple unit operations including extractions, polish filtrations, and distillations.

- *Isolation*: The isolation typically includes volume reduction to a predetermined product concentration and subsequent crystallization. The crystallization is one of the most complex and important steps in controlling product quality and downstream processing characteristics.
- *Filtration*: The isolated solids are separated from the carrying liquids (the mother liquors) using filtration equipment. Filter dryers, pressure filters, and centrifuges are common filtration devices found in pilot plants and kilo labs.
- *Drying*: The solids are dried to remove residual solvent. Commonly found drying equipment includes tray dryers, filter dryers, conical dryers, tumble dryers, and pan dryers.
- *Dry Powder Finishing*: Delumping, milling, or blending is performed as needed based on the powder properties of the API and the requirements for successful formulation into the drug product.

22.2.1 Kilo Lab Design

Kilo labs are typically installed within a laboratory building and are generally not found as stand-alone facilities. Kilo labs in close proximity to process development laboratories encourage interactions between scientists allowing for greater synergy between laboratory and scale-up operations. Kilo lab workflow as an extension of laboratory practices can improve efficiency and decrease the time needed to execute a process. However, the shared infrastructure can present challenges to the control of material and personnel workflow. Requirements for controlled personnel access and separate physical areas to manage segregated storage of raw materials and material subdivision apart from laboratory operations will increase the complexity of the area management.

Laboratory buildings typically employ central corridors for transport of materials and personnel. Air locks are often employed as physical separation of the kilo lab process areas from the rest of the building and reduce unnecessary personnel flow into the areas. The air locks also help to maintain proper air balance and pressure differentiation between the process areas and the corridor (see Section 22.2.3.1).

Most of the processing for a single intermediate or API is performed in one room or area. This includes all the material charging, reactions, workup, isolation, and filtration. The most common exceptions are drying and milling, which may be performed in separate locations specifically designed for handling dry powders.

Potential for cross-contamination between products should be considered at each phase of facility and process design. Contamination occurs primarily through airborne particulates and/or insufficient cleaning of product-contact surfaces. Many approaches designed to minimize contamination also have benefit in protecting the kilo lab scientist from exposure to the compound. The use of down flow booths, fume hoods, barrier technology, and other engineering controls to achieve closed processing are good examples (see Section 22.2.5). Placement of reactor systems into down flow booths or fume hoods, as shown in Figure 22.4, can increase process segregation for safety and quality. The same approach can be employed to segregate isolation and drying equipment. Where kilo labs are an extension of laboratory operations, specific laboratories can be designated for preparation of supplies destined for clinical use to further segregate process operations.

FIGURE 22.4 Typical kilo lab reactor shown in a down flow booth.

Equipment cleanability should be a consideration in the initial design of the equipment as well as in the design of the process equipment configuration for a specific product. Cleaning procedures should be developed to address a wide range of chemistries and process conditions recognizing that little is known about the cleanability of the product at the early stages of development.

22.2.2 Pilot Plant Design

Unlike kilo labs, pilot plants are commonly designed as stand-alone facilities. As the scale of the equipment increases, the set of equipment, often called the equipment train, designed to work together for producing a single intermediate or product will expand into multiple rooms. Since the 1990s, it has been common for new pilot plants to be constructed with a "gravity" design where the process will flow from upper floors to lower floors, in the direction of gravity (Figure 22.5). In such a design, smaller equipment is typically located in the upper floors and the larger vessels and isolation equipment are located in the lower floors.

Another design aspect is the use of closed space for a single train of equipment versus an open floor space without walls or physical barriers between multiple equipment trains. The latter is common to designs of older pilot plant facilities where product segregation is supported by procedural

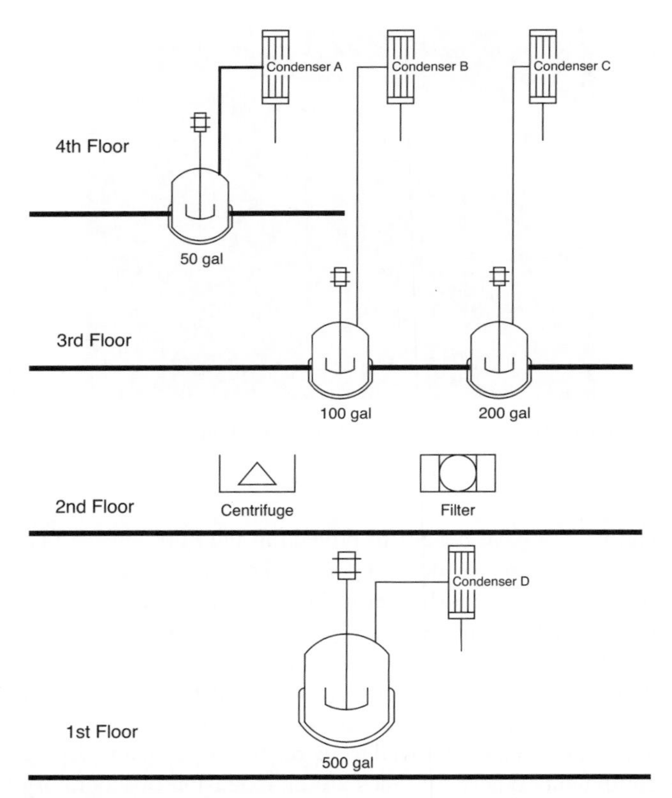

FIGURE 22.5 Schematic of gravity feed flow in a pilot plant facility.

controls rather than physical separation between the individual equipment trains for each product.

In stand-alone pilot plants, loading dock areas, proximity of elevators, corridor layout, storage space, air locks, utilities, and processing areas can be addressed in a complimentary design. Material flow, personnel flow, and HVAC design can be engineered soundly with good anticipation of the end user's needs. Dryer discharge areas and milling areas may employ temperature and humidity monitoring and/or control. As in kilo labs, pressure differentials are utilized to mitigate risk of cross-contamination from one area to another. Open floor designs increase the difficulty of maintaining pressurization, airflows, and system cleanability and are not preferred.

Pilot plants utilize a vast array of different surface finishes designed to promote compliance with CGMP requirements and a cost effective building maintenance life cycle. There is substantial emphasis on corrosion resistant finishes in wet chemistry areas where surfaces are subject to heavy traffic and potential chemical spills. Isolation and drying areas are designed with less concern about corrosive chemical exposure and greater emphasis on finishes that are easy to clean, including floor, ceiling and wall surface areas.

Pollution control is an area of emphasis in pilot plant design. While kilo lab facilities can control the small amounts of process emissions through standard ventilation systems, pilot plants typically employ additional emission control technologies to minimize emissions to the atmosphere and ensure compliance with regulatory permits. Technologies used include condensers, thermal oxidizers, carbon absorption, gas scrubbers, and cryogenic condensers. It is common to employ primary and secondary control systems in series to minimize the impact of system failure. Regulations governing the reporting and controlling of process emissions vary by country or region (e.g., European Union) and in the United States, by state.

Kilo labs and pilot plants following a standard design may not be suitable to run all types of chemistry. Safe execution of some chemical processes requires special facility design features that are not generally part of a standard design. Considerations include electrical classification, fire suppression systems, and wall construction (fire-rated, damage limiting, blast resistant). Once the type of chemistry and list of chemicals is identified for a new facility or facility renovation, a full code review can be performed by the project architects and engineers to identify specific design requirements and/or gaps against the existing design. Electrical classifications or ratings determine the list and quantity of chemicals (flammable liquids and dry powders) that a facility can safely process within that design. The electrical rating influences the design of all electrical systems and components within the facility. Chemistries, such as hydrogenation reactions or those involving pyrophoric materials, generally require specially designed areas or dedicated facilities to better mitigate the process safety risks.

22.2.3 Utility Systems

Utility systems generally fall into two groups: building systems and process support systems. Building utilities include electrical systems, potable water systems, utility steam/condensate systems, compressed air systems, fuel oil/natural gas systems, and control systems for process emissions (e.g., thermal oxidizer). Building utility systems do not come into direct contact with the product and generally do not impact its quality. Process support utilities include process gases (e.g., nitrogen, hydrogen), process solvents, process water, and temperature control fluids (e.g., glycol, thermal oils). These utilities influence the process environment through either direct process contact or their effect on process conditions. The impact of the utility on the product quality and the ability to meet CGMP production requirements determines the commissioning, qualification, and life cycle management for the system. Information on commissioning and qualification of equipment and systems is provided in Section 22.2.7.

In a chemical processing facility, process safety and environmental systems can critically impact operations and employee safety, yet do not directly impact the quality of the product. The impact of system failure against environmental, health, and safety criteria should be a factor in the design and maintenance of the system.

The centralization or decentralization of utility systems has significant impact on CGMP operations as well as business economics. Centralization can be more cost-effective and improve the maintainability of the equipment [1]. Some utilities are typically distributed to the pilot plants and kilo labs from a central source on the manufacturing site (e.g., steam). Kilo labs may share building systems and some process support systems when situated in multipurpose buildings. However, some process utilities should be dedicated to the kilo lab area to mitigate the risk of system disruption due to general use or to segregate CGMP controlled systems (e.g., temperature control system, process nitrogen, vacuum). Both building and process support utilities are typically dedicated to a single pilot plant or group of pilot plants. Backup power systems or uninterrupted power supply to support key parts of the facility, such as a designated equipment train or utility, is another design feature that mitigates the risks posed by hazardous chemistry during power disruptions in the building or manufacturing site. Some utilities with an important role in the safe execution of chemical processes while meeting CGMP production requirements are HVAC, process nitrogen, process water, and temperature control systems.

22.2.3.1 Heating/Ventilation/Air Conditioning The HVAC system includes equipment for the control of air pressure, airflow, particulates, humidity, and temperature. Proper design prevents the spread of contamination throughout a facility, which in turn provides a layer of protection to the worker from airborne compound exposure as well as protection of the product from airborne contamination of other products (cross-contamination). The location of the facility and CGMP requirements are important considerations in determing the needed HVAC capacity and control strategies. Air change requirements are typically 10–15 air changes per hour requiring large air movement equipment (air handlers and exhaust fans). Considerations for design include the requirements for air balancing within and between manufacturing spaces, directional airflow, environmental control in process areas, and the potential for compound exposure to the environment. In general, the process area will have a negative pressure differential relative to the air lock, the air lock is typically positive relative to both the process area and the corridor, and the corridor is a relatively neutral area (Figure 22.6). Air locks may also be designed to be negative relative to the corridor depending on the room function and company design practices.

22.2.3.2 Process Water The design of the process water system is driven by the quality requirements for the water produced, which are based on its intended use. Several factors influence those requirements including where the water is used in the synthesis (intermediate or API, early or late) and any microbial specifications for the API (sterile or low endotoxin). Water quality standards for pharmaceutical processing range from potable water, to endotoxin reduced, and to water for injection (sterile API). There are several guidelines available to guide design of pharmaceutical water systems [2, 3]. Some considerations include the water quality and temperature, routine sampling and monitoring requirements, system sanitization and cleaning, and maintainability. Process water systems should be designed for high reliability and availability; system efficiency and cost-effectiveness are other important considerations.

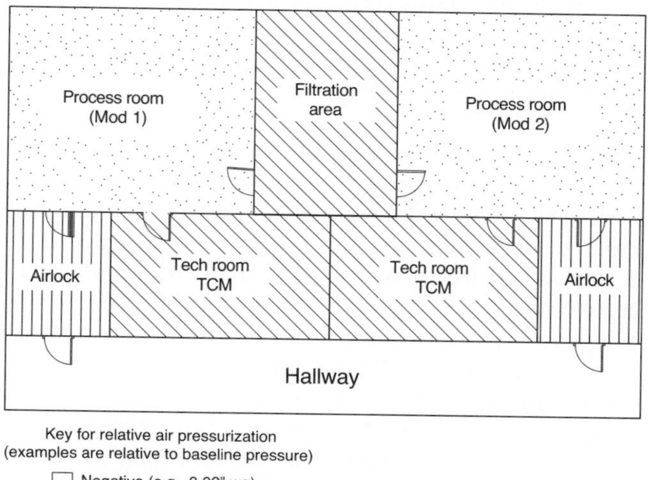

FIGURE 22.6 Example of design pressure in a pilot plant facility.

22.2.3.3 Process Nitrogen Gaseous process nitrogen has multiple uses in chemical processing. As a critical process safety system, nitrogen is used to inert process equipment used with flammable solvents and explosible dry powders, effectively reducing the potential for fires by displacing air as a source of oxygen from the system (one leg of the fire triangle) [4]. As a process support utility, nitrogen is used to assist several process operations including pressurizing equipment to perform liquid transfers and liquid sparging to remove unwanted dissolved gases. Liquid nitrogen can also be used on the jackets of equipment to achieve low reaction temperatures (<-15 to -30°C).

The nitrogen is filtered prior to equipment entry to avoid introduction of particulates into the process equipment. Kilo lab requirements may be sourced from cryogenic bulk systems or individual cylinders, depending upon the available source and capacity needs. Pilot plant requirements are often sourced from bulk cryogenic nitrogen tanks and distributed at high purity to the process equipment at sufficient pressure to support pressurized operations. System design should consider assuring the integrity of the system and equipment while minimizing any safety risks.

22.2.3.4 Process Temperature Control System As pilot plant facilities have evolved from designs utilizing multifluid heating and cooling systems (e.g., pressurized steam, water, brine) to single fluid systems, the design of the temperature control system has become more complex. The temperature control module (TCM) has become an integral component of equipment heat transfer as a means to ensure robust process temperature control. TCM is a term given to a collection of piping, instrumentation, valves, pumps, and heat exchangers designed to act as a unit to provide control of temperature to process equipment; temperature control within one or two degrees from a desired set point. A typical operating range requirement is −29°C to 120°C. Heat transfer fluids are chosen based on their range of heat transfer capability, cost, and maintainability. Two frequently used fluids are glycol and Syltherm XLT™.

For TCMs to function properly, a facility must have a cold loop as part of its infrastructure and an ability to heat (usually via heat exchanger) using electricity or steam. When the equipment requires heating, the TCM pump recirculates the heat transfer fluid through a heat exchanger or electric heater in a closed loop until the desired temperature is reached. When the equipment requires cooling, the system bleeds in fresh fluid from the facility cold loop until the target temperature is reached. Cooling performance is adequate with these types of systems; however, heating performance can be slow, particularly when using electric heaters. An alternate design is the use of a hot loop and mixing control of feeds from the hot and cold loop to achieve the desired temperature.

22.2.4 Equipment Design

The design of equipment differs for kilo lab and pilot plant applications yet there are several common key attributes. As engineering control technologies evolve, the design of kilo lab equipment to address containment has become more similar to those found in pilot plants.

22.2.4.1 Reactors/Vessels Kilo lab reactors and vessels are typically made of borosilicate glass or glass on steel. Pilot plant reactors are typically made of glass on carbon steel. Where metal materials of construction are required, Hastelloy alloys offer a corrosion resistant but more costly option to stainless steel. Both kilo lab and pilot plant applications require the ability to heat, cool, and agitate the contents. Agitator system design can be simple or complex with a multitude of options available. Most reactors are equipped with agitators made of the same materials of construction as the reactor they support. Pilot plant reactors are typically designed to achieve higher pressures and temperature applications than their kilo lab counterparts. Metal reactors are often used for cryogenic temperature application (−30°C to −100°C), however, glass on stainless steel vessels can also be used for milder cryogenic applications (>−30°C). Some pilot plants will have reactor systems designed to specifically support gas–liquid reactions, such as hydrogenations, utilizing specific agitator system configurations and pressure ratings.

An often used strategy to increase process configuration flexibility is to standardize reactor nozzle piping designs (e.g., number, type, and size of connections) within a facility or plant. If implemented correctly, the approach will significantly reduce customization of process equipment trains for each new process introduction. A thorough review of the common nozzle requirements for charging, material transfers, sampling, workup (extraction, distillation, concentration), recirculation loops, and PAT instruments. This information can be incorporated into a standard design that meets the general requirements of most processes.

22.2.4.2 Isolation Equipment (Filtration and Drying) As described in Section 22.1, the manufacture of an API proceeds through a number of chemical steps during which some intermediates are isolated in solid forms. While the filtration and drying sequence occurs in each case, the operations have increased importance in the final isolation of the API. Both the filtration and drying equipment design and operation play a significant role in the quality and physical properties of the isolated API (see Chapter 17). Filtration and drying operations can have a substantial impact on overall plant throughput and manufacturing costs. Slow filtrations or drying cycles can become the rate limiting steps for a process, increasing cycle time and reducing availability of the equipment for subsequent batches.

The filtration and drying operations are dependent upon the characteristics of the crystal slurry and are best developed as a coordinated effort with the crystallization process. Kilo lab equipment design is often an extension of the laboratory and should facilitate process troubleshooting and manage a broad range of slurry characteristics. Pilot plant scale equipment is closer to that found in manufacturing providing the ability to demonstrate the performance of the operations with different equipment designs.

Filters Considerations for filter selection include compressibility of the solid cake, susceptibility of the solid cake to agglomeration or attrition upon agitation, compound containment requirements and occupational health concerns, the design of the equipment train, and development objectives to inform equipment selection for manufacturing. Commonly used filters include filter dryers, centrifuges, and Nutsche filters.

The filter dryer is found in both the kilo lab and the pilot plant and combines filtration and drying unit operations within a single unit. As a single plate filter, it is fitted with an agitator, which can be raised and lowered as needed, and a jacket to facilitate the drying operation. In some designs, the agitator can also be heated increasing the heated contact surface area. The agitator is used to mix and smooth the cake during the filtration minimizing crack formation and increasing the efficiency of the filtration and wash operations. During drying, the agitator is used to increase the uniformity of the cake temperature and efficiency of the solvent removal. The agitator also assists in the removal of the cake during discharge through the discharge port. Engineering controls can be added to the discharge configuration to avoid worker exposure to the product. The closed system design from slurry entry through discharge has made this a frequent choice for potent and highly potent compounds.

Two batch centrifuge designs commonly found in pilot plants are the peeler centrifuge and the inverted bag centrifuge. Both have the same sequence of operations with the exception of the discharge. The peeler centrifuge is defined by its mechanical peeling action during discharge. The inverted bag centrifuge incorporates an automatic discharge achieved by inverting the bag (turning the bag inside out) allowing the solids to discharge. The inverted bag centrifuge offers a more efficient and faster heel removal than the peeler centrifuge. Centrifuges are commonly found in the pilot plant but are used less frequently in the kilo lab. The most often used centrifuge in the kilo lab is a basket centrifuge. As the name implies, the slurry is fed into the top of the "basket" and manually discharged.

Nutsche filters are jacketed, single plate filters without agitators and are commonly used in both pilot plant and kilo lab operations (Figure 22.7). The absence of the agitator may require the operator to manually mix and smooth the cake during filtration. The potential for operator exposure during use is increased and additional engineering controls and/or PPE are needed to provide adequate worker protection. Older designs also require manual removal of the cake upon discharge. More recent designs include engineering controls to reduce the risk of operator exposure during filtration and discharge. Nutsche filters do not offer the same level of containment as filter dryers and are typically not used with potent compounds.

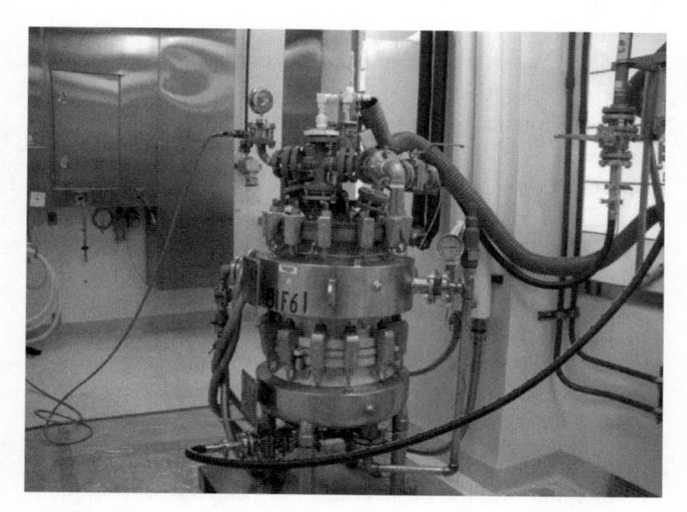

FIGURE 22.7 Nutsche filter for kilo lab or pilot plant use.

Dryers Drying equipment used in the kilo lab and pilot plant are generally contact dryers (solid in contact with heated surface) and operated in batch mode. Considerations for dryer selection include the acceptable drying time, the impact on powder properties (i.e., prevention of particle agglomeration or crystallographic form), and the explosibility of the dry powder. Commonly found types of dryers include tray dryers, filter dryers, tumble dryers, and rotary cone dryers. Drying equipment used in kilo labs is generally limited to tray dryers and filter dryers. Pilot plants typically standardize on a few different types of dryers, harmonizing where possible with Manufacturing.

A tray dryer is a large oven connected to a vacuum pump or inert gas supply. Product is loaded onto trays (metal or glass), which are placed onto shelves in the dryer. Sampling is achieved by interrupting the drying cycle and removing material from the trays. The dry product is manually unloaded from the trays into product containers. The tray dryer design offers little containment and additional engineering controls are needed to provide worker protection. Like Nutsche filters, tray dryers are typically not used for other than laboratory amounts of potent compounds unless additional containment is included in the equipment design.

The filter dryer is also usually connected to a vacuum pump during the drying operation and typically has a sampling port on the side of the filter wall that facilitates contained sampling operations. A standard agitation protocol (intermittent versus continuous, agitation rate) can be deployed and then varied as needed for individual products. The sampling port is typically a ball valve design; during sampling, the cup faces the cake and the agitator pushes a small sample of the cake into the cup. The valve is then turned facing the cup to the outside of the dryer and the sample can be removed.

22.2.5 Engineering Control Equipment

Engineering control equipment provides the primary level of protection for the worker from the hazards associated with chemical exposure during process operations. The engineering controls should provide protection for the specific operations where the operator exposure levels (OEL) are above a threshold of concern for an individual chemical, intermediate, or API (see Section 22.3). Typically the operations with the greatest potential for worker exposure are charging, discharging, sampling, milling, and cleaning operations. The standard designs for equipment used in these operations do not provide sufficient containment on their own and another layer of protection is needed. Common types of engineering controls found within pilot plants and kilo labs are flow hoods, barrier isolation technology, solids transfer technology, and sampling technology.

Flow hoods employ various technologies to achieve their containment target. Common types of flow hoods found in the pharmaceutical industry are down flow hoods and chemical fume hoods. Down flow hoods provide a high level of solids containment by directing the flow of particulates in the air away from the worker and into the hood exhaust system and particulate filtration system. Chemical fume hoods provide a constant flow of air from the front of the hoods that is exhausted to the atmosphere directly or via a filter. Fume hoods are effective for reducing exposure levels from chemical vapors and tend to be less effective for solids containment compared to down flow hoods. The small scale of equipment used in kilo labs often enables installation of the reactor vessels within the fume hood or down flow hood allowing for sufficient worker protection during operations presenting the greatest risk of exposure. Figure 22.8 shows an open filter design often found in kilo labs. The filter design enables manual manipulation of the wet cake and facilitates troubleshooting during filtration, but offers no protection to the worker from the product. The filter must be placed in a down flow booth or used with other engineering controls to ensure adequate worker protection.

Barrier isolation technology directly separates the worker from the chemical hazards via a temporary or permanent wall. The barrier isolator may resemble a box and enables the worker to perform all of the necessary standard operations (e.g., discharging, sampling) through the use of glove ports and pass through ports through the walls of the box. This technology is incorporated into standard types of equipment such as reactors, filters, and mills at both the pilot plant and kilo lab scale to perform operations involving highly potent compounds (see Figure 22.2c). Further information is provided in Section 22.3. Barrier isolators contain the particulates from the outside environment through the use of high efficiency particulate air (HEPA) filters on the exhaust. Where needed, the degree of containment achieved can be increased by maintaining the internal environment in the isolator under a negative pressure differential. Isolator technology is commonly used today where primary equipment design is insufficient to provide sufficient worker protection during operation.

FIGURE 22.8 Open filter often found in kilo labs.

Solids transfer technology significantly reduces the worker exposure levels resulting from the charging/discharging operation. The technology allows material charging and discharging operations to be performed in a closed system preventing direct worker exposure to the compound as well as the room contamination that results from open handling. Some common solids charging and discharging technologies used in the pharmaceutical industry are barrier isolators, split butterfly valve technology, and continuous liners. The barrier isolator can be installed as a hard wall or soft wall isolator at the equipment charge or discharge ports to provide the desired level of containment. Split butterfly valve technology supports low operator exposure through the use of active and passive valves. The active and passive valves are mounted separately to the charge/discharge vessel nozzle and the equipment as shown in Figures 22.9 and 22.10. The solids transfer operation proceeds only when the active and passive valves are joined or docked together. Continuous liners made of pliable material are used to increase the flexibility of solids containment. A continuous liner mounted at the receiving dryer can be crimped to isolate a portion of the solids and then cut to form a separate bag containing those solids without breaking containment.

22.2.5.1 Sampling Engineering controls used during sampling operations are designed to minimize operator exposure potential during the direct sampling of reactors via pressure or caused by sample spills. This is of increased importance when the APIs are highly potent or have substantial toxicological concerns. To minimize operator exposure during the sampling operation, barrier technology, closed vent sampling devices, or similar technologies may be installed within the scale-up facilities as shown in Figure 22.11.

22.2.6 Process Automation

Today, most pilot plant facilities have modern process control systems that provide basic control of primary process parameters such as temperature and pressure. Less common to

(a)

(b)

FIGURE 22.9 Solids handling using split butterfly valve technology: (a) material charge into a reactor using a continuous liner and (b) discharging the wet cake from a centrifuge into a flexible liner.

pilot plant facilities are the recipe driven process automation systems (recipe control) found in manufacturing, which may feature advanced process control and business systems interfaces. Most pilot plants will also have data historians to capture process data (e.g., pressure, temperature) as well as events (e.g., valve opening, changed set points). Kilo labs that are an extension of laboratory systems may have basic control for reactor temperature and pressure, while others deploy process control systems for regulatory control. Few kilo labs have the extensive process control systems found in pilot plant facilities.

The approach to process control and automation in kilo labs and pilot plants differs from that in manufacturing facilities. Commercial products are manufactured using validated processes supported by knowledge of the process design space. Goals of the process control system are to ensure efficient and consistent execution of the chemical process minimizing process excursions and reducing batch to batch variability. Integration of the process control system with other manufacturing execution systems allows for greater efficiency in product release and inventory management leading to reduced operating costs. Process data is collected to support batch release and continuous process improvement. In contrast, processes in pilot plant facilities span early to late development and are typically run only a few times. Process changes are expected between campaigns! The primary goal of the data collection system is to generate process knowledge and provide development scientists access to process data. The process control system should be designed for flexibility and robustness to accommodate changing requirements from each process, to facilitate process troubleshooting, and to enable chemical operators to focus on the process.

Kilo lab and pilot plant process control system have the following general goals:

- Provide control of process and utility parameters to maintain safe operation and reduce variability of the process. This will reduce the risks encountered when scaling process chemistry from one scale to the next.
- Notify the operator when the process is out of range.

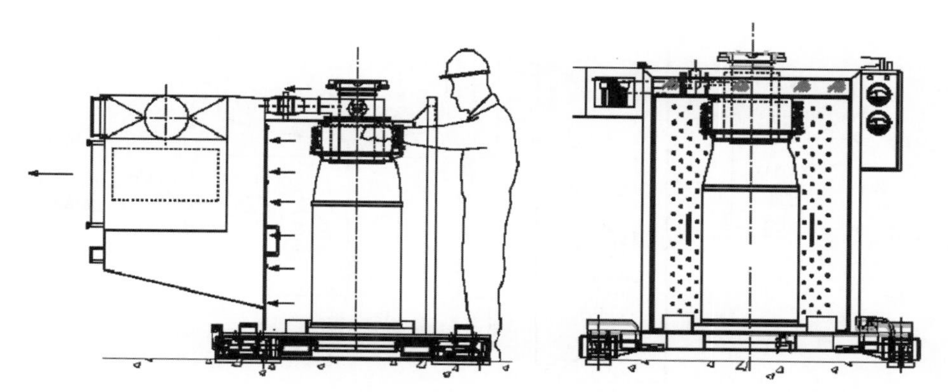

FIGURE 22.10 Schematic of closed discharge system consisting of discharge drum in laminar flow booth connected by continuous liner.

FIGURE 22.11 Sampling device contained within a glovebox, shown open on the left and closed on the right. Sample bottles are removed via continuous liner sleeve on left port, crimping and cutting the liner for each bottle.

- Document what has occurred during the batch to support regulatory requirements and review. This may range from a simple trend of temperature to a log of all continuous process data and discrete events that have occurred.
- Help reduce development time by enabling process understanding and process robustness through knowledge capture.

22.2.6.1 Control Systems A typical batch reactor control system with temperature and pressure control loops is shown in Figure 22.12. The control system relies upon primary instruments, such as thermocouples and pressure sensors, and control elements, such as control valves, that are wired back to stand-alone controllers or to a computerized process control system such as a distributed control system where hundreds of control loops are executed. These sensors also

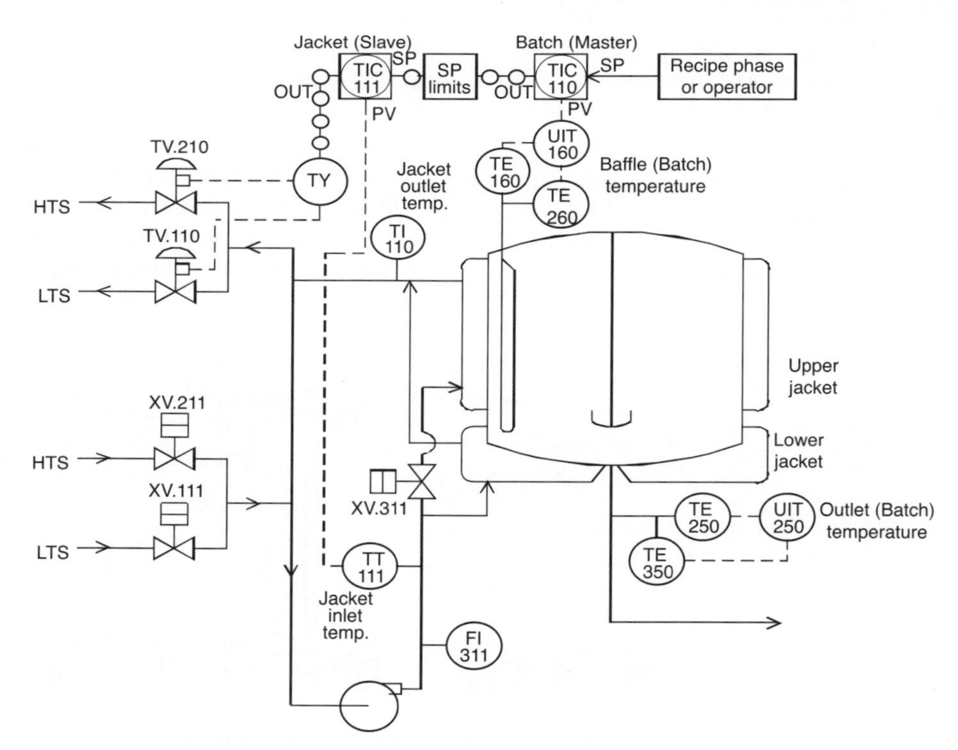

FIGURE 22.12 Batch reactor with typical split-range configuration to control batch temperature.

facilitate capture of continuous data (e.g., temperature) and discrete, event data (e.g., when does a valve open and close during a charge).

The basic regulatory control process, PID (proportional-integral-derivative) controller is still used in most pilot plants though there are more advanced approaches such as model-predictive control. In modern pilot plants, the flexibility inherent in these digital systems not only allows for plug and play of various unit operations equipment but also allows the control strategies to be easily changed to meet the needs of each product, facilitates regulatory compliance, and supports generation of process knowledge.

22.2.6.2 Recipe Driven Batch Control A "batch" consists of a sequence of unit operations to produce a product. For example, to carry out a reaction, a typical sequence of events is shown in Table 22.1.

The instructions or recipe to execute the sequence is written in the control software and includes the parameters specific to the process. The recipe will contain set points and alarm limits for each step. The recipe can also select the desired control strategy, such as jacket temperature or batch temperature control.

In 1995, the ISA-S88 Batch Manufacturing Standard was issued describing the definition and control of batch processes [5]. At the core of the standard is the separation of the product knowledge from equipment capability, the "S88" model.

"S88" provides a logical, consistent structure for building batch recipes as shown in Figure 22.13, which combine these elements. The "recipe" sequences the different "Unit Procedures" such as those described in Table 22.1. Each Unit Procedure consists of a series of "Operations" (e.g., heat, hold, sample, cool) that contain parameters (e.g., set points and limits) specific to that step. Each Operation then consists of a series of "Phases" that instruct the equipment-specific elements to carry out that operation (open valve X-101 on reactor R-101). During the execution of the recipe, the control system captures what was done (operations and phases), when it was done, who did it (by electronic signatures) and how it was done (process data capture of continuous, temperature and discrete data, weight of a drum charge).

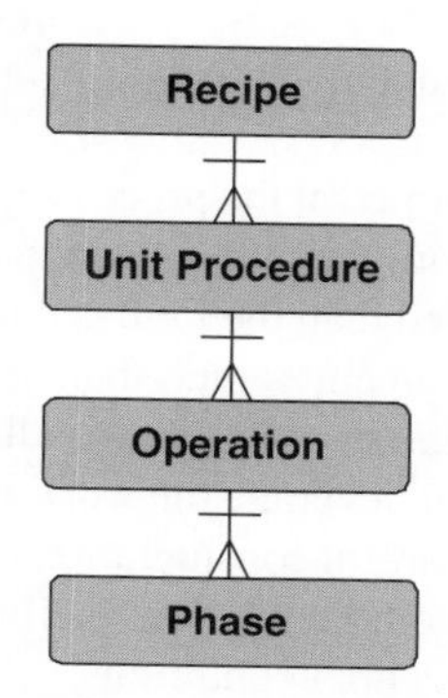

FIGURE 22.13 S88 procedural model.

TABLE 22.1 Typical Sequence of Events

Sequence	Event
1	Inert the vessel with nitrogen
2	Charge solvents, substrates, and reagents to vessel
3	Turn on agitator
4	Heat to reaction temperature
5	Hold until reaction is complete as determined by sampling
6	Sample by pressurizing vessel, withdrawing sample, and depressurizing
7	Cool the vessel and crystallize product
8	Transfer the product to filter
9	Dry the product

With a control system using an S88 recipe approach and a process and event data historian, a complete record of the batch operations can be obtained to support development, quality and regulatory requirements. Phases and operations are general purpose and become process specific when the parameters for the process (e.g., reaction temperature, hold time) are added.

22.2.7 Equipment Commissioning and Start-Up

When starting up a new or renovated facility or new equipment in an existing facility, the building systems and process equipment must be tested to ensure that the design specifications are met. This is required for adherence to regulations (CGMP and EHS) and good engineering practices (GEP). The standard processes to ensure that a system or piece of equipment is started up, qualified, and rendered ready for use in a pharmaceutical environment consists of commissioning, qualification, and validation (CQV). The CQV plan includes acceptance test criteria and an impact analysis to evaluate the impact of the system or equipment on the quality of the product produced (direct, indirect, or no impact) [6]. The outcome of the impact assessment determines the extent of testing required for an individual system or equipment.

The Factory Acceptance Test (FAT) entails sending a team to the equipment manufacture's location to execute an approved protocol geared toward assessing that the equipment can perform as specified. The goal is to ensure that the equipment is suitable to be shipped and installed.

Commissioning is an engineering approach of bringing a facility and/or equipment into an operational state. For qualified systems, commissioning activities can overflow into the Site Acceptance Testing (SAT). For nonqualified system, the commissioning is usually more extensive since there is no subsequent activity to ensure operational readiness.

The SAT is performed at the processing site (pilot plant or kilo lab) after the equipment has been installed. The SAT is usually more involved than the FAT as it needs to thoroughly check out the equipment rather than just ensure that the equipment is complete enough to allow for shipment. While it is possible to leverage the work at the FAT to reduce the SAT activities, careful consideration must be given to the installation work and how the final facility conditions differ from the factory (vendor) conditions.

The next phase of the process is the qualification, which is captured in a series of protocols to confirm the installation, operation, and performance of the equipment. Installation qualification (IQ) confirms that the correct equipment has been installed as specified in the design documentation. It is a documentation review comparing nameplate data (serial number, model number, pressure rating, etc.) with the appropriate specification to ensure compliance with the design. Additional information is also collected for analysis and future use including materials of construction, construction techniques, testing, and similar information. The operational qualification (OQ) is a protocol to systematically test the equipment to ensure that it can perform according to the user requirements for the equipment. Equipment classes tend to have similar checkout requirements for individual equipment within that class. Examples of reactor class testing include pressure, temperature control, and agitation testing.

The performance qualification (PQ) is perhaps the most detailed manner of checkout due to the inclusion of a test specification. Water and nitrogen system are the two most common systems that require PQs in the pilot plant and kilo lab. This is to ensure that materials charged to the batch (e.g., water) or in contact with the batch (e.g., nitrogen) meet the required quality specifications.

Successful start-up and qualification of equipment and facility systems requires a systematic approach. The employment of FAT, SAT, IQ, OQ, and PQs provides a consistent, methodical way to prove facility integrity. Subsequent changes to the facility are managed using a change management program to ensure that facility and equipment integrity is maintained.

22.3 OPERATING PRINCIPLES AND REGULATORY DRIVERS

The design and operation of scale-up facilities in the pharmaceutical industry is influenced by a matrix of regulatory guidelines. Current good manufacturing practices, established to ensure the quality of the drug delivered to the patient, overlay EHS regulations at the local, state and federal level established to protect the worker, the community and the environment. CGMPs must be followed when producing API for clinical studies to ensure that the quality and purity of the drug substance is adequately controlled. Guidelines for CGMPs are governed by the regulatory health authorities, such as FDA and the EMEA. The guidelines of each agency may differ and substantial effort has gone into establishing a consistent set of practices across agencies through the International Conference on Harmonization (ICH) [7]. Scale-up facilities located in the United States are governed by EHS agencies that include the Environmental Protection Agency (EPA) [8], Occupational Safety Health Authority (OSHA) [9], and equivalent state agencies. Equivalent agencies govern chemical process operations outside of the United States. In many cases, the design of a compliant EHS programs can be structured around the scale of operations. Kilo labs are commonly classified as laboratories, which enables these facilities to fall under the local, state, and federal regulations for laboratories. In most cases, pilot plants operate at a scale that precludes laboratory classification and must comply with regulations established for plant scale of operation and purpose. Additional EHS regulations, such as New Jersey's Toxic Catastrophe Prevention Act (TCPA), apply where the quantity of listed hazardous chemicals exceeds a threshold value.

The expectations set by CGMP and EHS guidelines are shared in many areas. Most companies develop an integrated approach to operating procedures, work practices, and staffing to more effectively meet safety and quality requirements. Once established, robust training, documentation, and monitoring programs ensure compliance against regulations.

22.3.1 Process Safety

Prior to the implementation of an intermediate or API process step into a kilo lab or pilot plant, the chemistry and process details must undergo a process safety evaluation. The safety evaluation should provide data to determine the thermal hazards of the solids, key process streams and any relevant off-gassing data. The safety evaluation should also examine the physical properties, toxicity hazards, fire and explosion hazards, and hazardous interactions between different chemicals to determine the safety risks associated with running the process in a scale-up facility. For projects that are further in development, a safety evaluation on the powder properties of any isolated intermediates and the API should be completed. The complete evaluation should be used to determine the intrinsic safety of the process design and whether a process can be safely implemented at the desired scale.

During the transfer of process knowledge from the process development group to the plant operations group, an assessment of the process hazards should be completed following a standard methodology to identify the process safety risks of execution into the specific scale-up facility. Some common methodologies used for the hazard assessment are a what-if analysis, checklist of hazards, process hazard analysis

(PHA), hazard and operability study (HAZOP), failure mode and effect analysis (FMEA), or fault tree analysis. The data from the safety evaluation, plant equipment details, plant procedures, and the intended process batch instructions should be used to develop the safety assessment. The intent of the assessment is to address hazards from the process, consequences of failure from engineering and administrative controls within the plant, and potential human errors. Any process changes proposed after the initial hazard assessment is completed should be examined prior to introduction to the scale-up facility using the facility management of change policy to ensure that all safety precautions are taken. Further information is available in Chapter 11 and related references [10, 11].

22.3.2 Current Good Manufacturing Practices

The Quality Management section of ICH Q7A [7], specifically states "quality should be the responsibility of all persons involved in manufacturing" and that each individual manufacturer is responsible for establishing and documenting effective quality systems to manage quality of products produced both internally and through external partners. CGMP can be classified into six quality systems: facilities and equipment, production, packaging and labeling, material systems, and laboratory controls. Each requirement listed in ICH Q7A will map to one of the six systems as shown in Figure 22.14.

These systems address review of completed manufacturing records for critical process steps prior to the release of API, making sure that equipment used for manufacturing is maintained, calibrated and fit for its intended use, and that the proper level of documentation for all activities is reviewed,

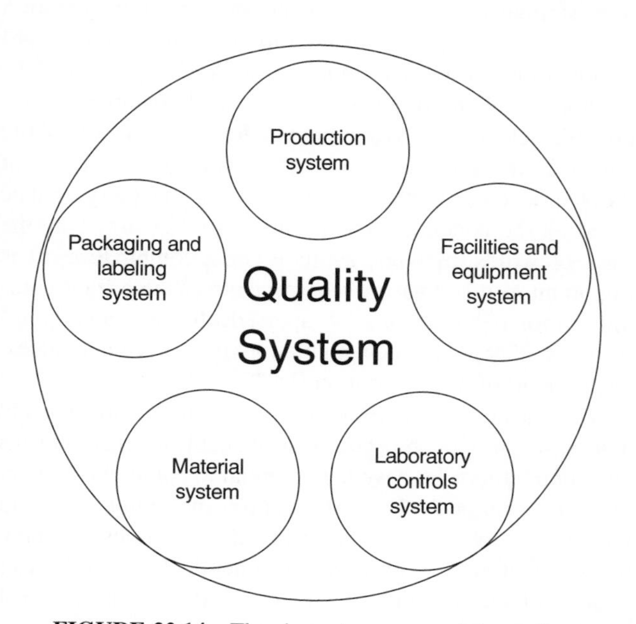

FIGURE 22.14 The six-system approach to quality.

complete, and accurate. An internal audit program designed to ensure that a state of compliance is maintained must be established, including a defined audit frequency with documented audit findings and corrective and preventative actions (CAPAs) to address identified quality risks.

A sound personnel training and qualification program is an important element of the CGMPs. The roles and responsibilities for employees associated with pharmaceutical manufacturing facilities and operations should be defined, documented, and maintained. Requirements for education, experience, general training, job function training, proper hygiene and sanitation habits, and suitable clothing worn during manufacturing or laboratory operations are included. A comprehensive and well-documented employee qualification program assists operations managers in assigning the right employee with the right skill set to the right task in compliance with CGMP.

CGMP considerations for scale-up facilities and equipment are described in Section 22.2 and include design, construction, and start-up activities. Facility and equipment design and construction should facilitate maintenance and cleaning, minimize contamination of the product, and provide adequate control systems. Materials of construction for product-contact equipment are also an important consideration. Materials that are incompatible with process streams can contribute to quality issues potentially rendering the product unusable. Process safety issues may also result where the introduction of contaminants into the process stream causes an unexpected reaction. Commissioning, qualification, and validation activities are completed to demonstrate that the facilities, equipment and systems were constructed as designed and are fit for their intended use in the manufacture of APIs and chemical intermediates. Additional facility considerations include material management such as material receipt, identification, sampling, storage, and separation of released and unreleased input materials, intermediates, and APIs.

Proper qualification should be maintained through a change management program for systems and equipment that impact the quality of the API and/or chemical intermediates. Written procedures for facility and equipment maintenance and cleaning operations should be designed to promote consistent execution over time independent of an individual operator.

Documentation and records is one of the most visible CGMP requirements. This is the primary evidence of "what, when, who, and how" as it relates to an operation or task. Documentation requirements vary by task and may include what task was performed, who performed it, when it was performed, how it was performed, and reconciliation of any deviation from the expected procedure. The documentation is the link to ensure that an operation or task was completed as intended, and that the impact of any deviation on the quality of the outcome is evaluated and understood. CGMP documentation includes equipment cleaning records, laboratory (analytical) controls, training records, and batch records, which document the execution of the manufacturing process. Batch records contain input material information, a description of executed unit operations, analytical control results, and process data. They should also contain an adequate explanation of any process deviations that occurred during processing. Any changes to batch records during execution should follow a change management program that includes appropriate comment to reconcile the change with the original entry.

22.3.3 Health and Safety

Kilo lab and pilot plant operations include the use of hazardous chemicals that may be toxic, flammable, corrosive, and/or explosive. The health and safety of the employees must be the highest priority and worker protection programs should be in place before hazardous chemicals are introduced into the facilities. Regulations governing employee and workplace safety can vary substantially across the world and many companies have established a minimum standard for all of their facilities to supplement country, regional, and local regulations. In the United States, OSHA sets standards to ensure worker and workplace safety; the standards can be found under the Regulation Standard 29 CFR 1910 Occupational Safety and Health Standards [12]. Table 22.2 lists some of the regulations under 29 CFR 1910. A chemical hygiene plan (CHP) or similar program is typically developed for kilo lab and pilot plant facilities to ensure compliance with OSHA standards. The plan or program consists of policies and procedures around specific topics that define the operational safety framework, including hazard communication, occupational exposure, PPE, and the procurement.

TABLE 22.2 Some U.S. Health and Safety Regulations for Kilo Labs and Pilot Plants

Topic	Regulation Standard (29 CFR)
Emergency action plans	1910.38
Occupational noise exposure	1910.95
Process safety management of highly hazardous chemicals	1910.119
Hazardous waste operations and emergency response	1910.120
Permit-required confined spaces	1910.146
The control of hazardous energy (lockout/tagout)	1910.147
Hazardous (classified) locations	1910.307
Air contaminants	1910.1000
Hazards communication	1910.1200
Occupation exposure to hazardous chemicals in laboratories	1910.1450
Hazardous materials	1910 Subpart H
Personal protective equipment	1910 Subpart I
Fire protection	1910 Subpart L

22.3.3.1 ***Hazards Communications*** A well-defined hazard communication procedure is essential to understanding, accessing, and communicating the hazards associated with chemicals used in the kilo labs and pilot plants. Required elements include a hazard determination process, a procedure for management and use of material safety data sheets (MSDS), proper container labeling requirements, and effective information and training on hazardous chemicals.

The hazard determination process should involve the evaluation of chemicals purchased for use in and produced by the kilo lab and pilot plant facilities to determine if they are hazardous. Hazard information for chemicals that are purchased is contained in the MSDS provided by the vendor. For compounds produced as part of the research and development process, the company producing the compound is responsible for generating the MSDS. MSDS information for all hazardous chemicals used in the scale-up facilities should be made available to the operating staff.

A MSDS contains the chemical name of the compound as found on the label and other common names in addition to any available hazard information. If the material is a mixture, the MSDS will list the major components and nominal composition. The types of hazard information commonly found are related to the properties of the material, such as toxicological information, stability and reactivity, exposure control, environmental impact (i.e., ecological concerns and disposal procedures), and procedures to handle the material (e.g., storage, accidental release measures and transportation information). There are also additional sources for hazard information beyond that found on the MSDS [12, 14].

Proper labeling of hazardous chemical containers is another vehicle for hazard communications. The identity of the chemical, the appropriate hazard warning, and the name of the chemical manufacturer or other responsible party must be listed on the container. The hazard warnings may take the form of words, pictures, symbols, or a combination. It is important to note that the requirements of the hazard communication regulation do not supersede the labeling requirements of other government regulations (e.g., in United States: Department of Transportation (DOT), EPA).

The hazard information for hazardous chemicals and appropriate training must be made available to all employees and contractors who have the potential to come into contact with the chemical. The information and training should include the details of the hazard communications program, the physical and health hazard of chemicals in the labs or plant, and the methods of protection to prevent chemical exposure.

22.3.3.2 Occupational Exposure for Pharmaceutical Compounds Companies are responsible for developing the hazard information for research and development compounds. There is often limited information on the compound early in development. Therefore, a variety of methods, including toxicology studies and chemical structure-based assessments, are often used to develop the information needed to evaluate the potential hazards to employees within the laboratories and scale-up facilities. An evaluation of the available data is typically performed by an industrial hygienist, who assigns an exposure control limit (ECL) to the compound if sufficient data is available. It is often not feasible to assign a single number ECL early in pharmaceutical development and most companies use a performance-based exposure control band strategy assigning the compound into a control band. The exposure control limits for each control band are set with consideration of engineering control technologies, PPE, procedural controls, and other available tools to mitigate the risks of worker exposure as required [13]. The data generated from the toxicological evaluation and testing is typically added to the MSDS for the API or intermediates.

There is not one standard control band classification used across the pharmaceutical industry through the concept is widely accepted. Based on the control band assignment, the handling procedures for a compound in the laboratory, kilo lab, and pilot plant are designed to ensure that the potential worker exposure is controlled or contained within acceptable limits. Engineering controls should be used as the primary containment strategy (Section 22.2.5) supplemented by PPE as necessary to provide redundancy. PPE should not be used as the primary control.

22.3.3.3 Personal Protective Equipment The proper selection and use of PPE is important to worker safety in the kilo lab and pilot plant. Various types of PPE are available and should be used as appropriate to mitigate risk of chemical exposure. A standard policy defining the minimum requirements for all workers and visitors is typically developed for each facility based on facility operations and construction. For kilo labs and pilot plants, minimum requirements generally consist of the items found in Table 22.3.

There may be additional requirements for PPE for use when handling specific chemicals or performing certain operations where the PPE is applicable to the hazards involved. The selection criteria for protective face shields, glove type, respiratory protection, and outer safety garments should include potential hazards as well as known hazards.

Gloves provide a barrier between the worker and chemical or physical hazard. When choosing a glove, the kilo lab or pilot plant operator should select one that provides the type of protection necessary for the operation being performed and also provides the dexterity necessary to complete the task. There is no single type of glove that provides protection against all chemicals and it is important to carefully select the appropriate glove for the operation. The respiratory protection used in a kilo lab or pilot plant must provide protection against gas, vapors, mists, and solid particulates. The type of respiratory protection selected must be able to maintain the employee's exposure levels below permissible levels. The types of respirators currently used in pharmaceutical scale-up facilities include air-purifying respirators, powered air-purifying respirators, and supplied air respirators. The MSDSs for the chemicals to be used and additional literature sources should be consulted to determine the appropriate glove and respirator selection.

TABLE 22.3 Common Minimum PPE Requirements for Kilo Labs and Pilot Plants

Type of PPE	Common Requirements
Eye protection	Safety glasses with side shields. Protection should meet the ANSI Z87 standard for Occupational and Education Eye and Face Protection
Foot protection	Steel-toed shoes that are static dissipating
Protective apparel	Fire retardant uniform or laboratory coat
Head protection (if applicable to facility)	Protection should meet the ANSI S89 standard for Protective Headwear for Industrial Workers Requirements

22.3.3.4 Management of Hazardous Chemicals Procedures should be established for the procurement, distribution, and storage of hazardous chemicals that provide for adequate control over chemical inventory to mitigate the risk of safety-related issues. Consideration of interactions and incompatibilities between different classes of chemicals should be included at all phases of the management process. The design and operation of chemical storage and staging areas must comply with regulations for segregation of certain hazardous chemicals.

22.3.4 Environmental

Pharmaceutical kilo labs and pilot plants handle larger quantities of hazardous and toxic chemicals than standard laboratories and have a responsibility to prevent accidental release of substances into the environment. Local, state, and federal governments have established laws and regulations to prevent the accidental release of hazardous substances into the environment. These laws and regulations typically require that the facilities have procedures and safeguards for air pollution and controls, site management of toxic substances, and waste disposal management. Examples of U.S. environmental regulations are the Clean Air Act, Resource Conservation and Recovery Act (RCRA) and Toxic Substances Control Act.

The government regulations for air pollution are set to provide protections for the public health against hazardous pollutants. The regulations aim to reduce the overall emissions by setting limits on certain pollutants, setting performance levels for pollutant controls, managing environmental permit programs, and providing enforcement powers. Facilities which produce above certain quantities of pollutants, as defined by local, state, or federal regulations, are required to obtain operating permits that limit air pollutants. The permit will typically require the facilities to report the actual air pollutants generated through the use of theoretical models. A typical emissions report provides the actual emissions produced categorized by type of pollutant and documents that the emissions do not exceed the allowable permitted quantity for each type of pollutant. Scale-up facilities will typically reduce or limit air pollutants with air emission controls, such as condensers, cold traps, gas scrubbers, and thermal oxidizers, connected to the facility or equipment train. When properly engineered, these devices can significantly reduce or eliminate specific pollutants.

22.3.4.1 ***Waste Disposal*** Scale-up facilities must have a waste management program for waste that is generated during processing. There are local, state, and federal government laws and regulations for the management of hazardous waste and nonhazardous waste. Hazardous waste is considered any waste with properties that makes it dangerous or capable of having a harmful effect on human health or the environment. Nonhazardous solid waste is considered to be any waste that is generated by industrial process or nonprocess sources that does not meet the classification of hazardous waste. The regulations surrounding hazardous waste requires control from "cradle-to-grave". Procedures for the generation, storage, treatment, transportation, and disposal of hazardous waste must be developed to ensure compliance with the regulations.

22.3.4.2 ***Emergency Response for Spills and Accidents*** An emergency response plan should be established for kilo labs and pilot plants to enable rapid and appropriate response. The response plan should include environmental, health, or safety hazards associated with hazardous material spills, fires, and explosion emergencies, proper notification details, the containment and control of the hazard, cleanup and restoration of the area, a list of the emergency response groups, the evacuation plan, and reporting requirements. Some companies train operating staff in kilo labs and pilot plants to provide the first response to spill cleanup. Emergency response procedures should be used when there are properties of the hazardous substances, circumstances of the release, and other factors in the work area that require more that an incidental cleanup.

22.3.5 Programs and Procedures

Procedures and programs developed for kilo labs and pilot plants should be structured to promote compliance with CGMP, health and safety, and environmental regulations while enabling process development. Operating procedures, employee training, housekeeping, maintenance, and inspection programs provide a foundation for operation of the kilo lab and pilot plant.

22.3.5.1 ***Operating Procedures*** Operating procedures and other written instructions for kilo labs and pilot plants should provide process operators with clear instructions on how to safely operate equipment and conduct routine plant or kilo lab operations. Equipment instructions should describe start-up operations, normal operations, shutdown operations, and emergency operations for the specified equipment. Additional information about equipment operating limits and an interlock or safety system list may be included. Other topics for written instructions include industrial hygiene testing, hazardous and/or toxic materials handling, hazardous waste handling, and energy control procedures to manage the multiple sources of hazardous energy during maintenance and operations (see Section 22.3.5.3).

22.3.5.2 ***Employee Training*** A well-defined and documented employee training program is essential to ensuring compliance with procedures and program expectations. The training plan is role-based and assigned to an individual based on his role within the kilo lab or pilot plant operations. The training objective may range from awareness of the topic to hands-on application within daily operations. Awareness training will usually apply to policies or procedures where the knowledge of the general context is important but the individual is not responsible for performing the operations. Application training is used when the knowledge is necessary to perform an assigned task. Between the two is comprehension-based training, which applies to employees that need to understand the specifics of the procedures due to their role in the organization. An essential element of a good training program is the requirement for refresher training at a specified time interval to ensure that the operating staff maintains a competent skill level and knowledge needed to be compliant with regulatory requirements.

22.3.5.3 ***Housekeeping, Maintenance, and Inspection Program*** Preventative programs are designed to proactively address workplace safety and ensure CGMP compliance of the facility. A housekeeping policy, preventative maintenance (PM) program, and inspection practices allow for early identification of issues and corrective actions before problems arise. A housekeeping policy for kilo labs or pilot plants will include keeping process areas clear of obstructions, ensuring clean and sanitary work areas, and the proper storage of hazardous chemicals.

Preventative maintenance programs (PM) help to ensure that the systems installed within the facility are functioning properly and within the design specifications. A well-designed PM program enables determination of whether equipment replacement or calibration is necessary and allows the tasks to be scheduled with the least disruption to processing activities. The program includes examination of instruments, such as pressure gauges, transmitters, and relief valves, evaluation of wear on equipment train components such as agitator seals, pumps, and valves, and testing of facility-related equipment such as safety showers, fire extinguishers, and engineering controls (e.g., laminar flow hoods, isolators).

A procedure for the control of hazardous energy in a kilo lab or pilot plant is necessary when performing maintenance activities on equipment with multiple energy sources (e.g., electricity, pressure, mechanical). Maintenance on the equipment is defined as any activity performed on the equipment, including any setup, adjustment, or inspections. The procedure is necessary to prevent the unexpected energization of equipment while the maintenance is being performed causing serious injury to the employees. The basis of the procedure should contain the identification of the energy sources, proper training of the affected personnel, and periodic inspections of the energy sources. To control the hazardous energy, lockout or tagout devices should be installed on all energy sources to prevent the accidental energization on the equipment or machine. The use of locks is the preferred method for any energy control procedures as tags are only a warning against hazardous conditions. It is critical for the individual or groups of individual who are performing the maintenance operations to confirm the de-energization of all equipment being serviced.

A safety inspection program acts to alert the staff to any safety risks associated with inadequate housekeeping and deficiencies within the PM program. It can also provide a real-time check on compliance with safety procedures such as labeling, hazardous chemical handling, and improper use of personal protective equipment. Completing the integration of safety and quality, the inspection program can also supplement the CGMP internal audit through identification of potential quality risks.

22.4 SUMMARY

Kilo labs and pilot plants are important tools supporting process development in the pharmaceutical industry. Designed for flexibility and responsiveness to change, they are an extension of the scientist's laboratory supporting process knowledge generation and successful introduction of new chemistries at all stages of process development. Governed by CGMP and EHS regulations, they are designed and operated in compliance with a multitude of regulations. Protection of the patient, the worker, and the environment are top priorities. Joined together, the kilo lab and the pilot plant can provide a competitive edge within process development in the pharmaceutical industry.

EXERCISE

GMP regulations require investigation of unplanned events that occur within scale-up facilities to determine the affect on the quality of the product manufactured. Regulations define certain elements that must be included within a compliant "Quality Events program," however, individual company programs will vary. At the core of a quality events program is the investigation of the unplanned event to enable identification of the root cause and determination of the impact to the quality of the product. Corrective actions are then implemented that address the root cause mitigating the risk of a future occurrence. Facility operations at the time of the event, the state of the equipment, input materials, and the chemical process are some factors to consider during the investigation. Investigations in kilo labs and pilot plants can be further complicated where there are limitations on the process knowledge available for the product. A commonly used methodology for investigation of such events is the Kepner-Tregoe process (www.kepner-tregoe.com). A situational analysis is typically performed to determine if an investigation is needed based on the potential for impact to the quality of the product. As with any investigation, constructing a timeline and collecting "forensic evidence" are important first steps in determining if the product quality is affected.

You have been asked to lead the situational appraisal for an unplanned event that occurred in the manufacture of an API. Below is a summary of information provided to you.

Where: Pilot plant

What: An API slurry was transferred from a crystallizer into a filter dryer using nitrogen pressure. The mother liquor was collected in a receiver downstream of the filter. After the slurry transfer was complete, the API was washed with fresh solvent charged into the filter from solvent drums to remove any residual mother liquor from the cake. The process flow diagram (PFD) for the operation in depicted in Figure 22.15. During the washing step, black particulates were observed by the operator on the top of the wet cake in the filter dryer.

Background

- The API is designated for use in a clinical phase II study.
- The crystallizer and filter dryer are used for processing of both intermediates and API.
- After each campaign, the equipment is cleaned and inspected.
- The filter dryer has been used in the manufacture of multiple products since the last preventative maintenance, both intermediates and API.

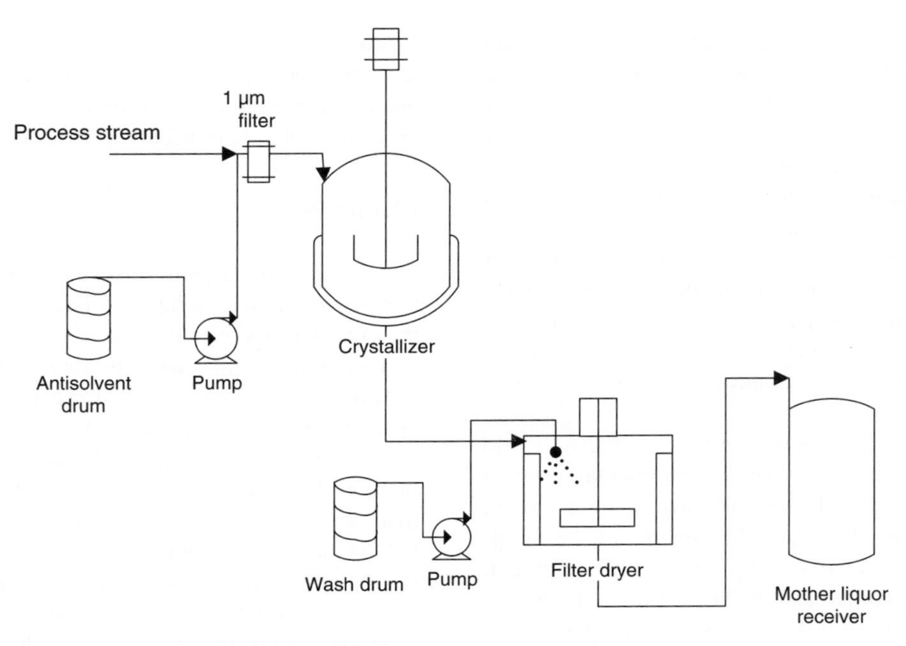

FIGURE 22.15 Process flow diagram for exercise.

- The crystallizer has not been used since the last preventative maintenance was completed. It was verified as clean prior to use in this batch.
- The process stream and the crystallization solvent were filtered through a 1 μm filter before they were charged to the crystallizer.

(A) Construct the timeline for the event.

(B) What is your next step? Should an investigation be conducted? Explain your answer.

(C) Would your conclusion change if the particulates were found before the manufacture of the processing began?

Additional data

- The black particulates were eventually determined to be a polymer frequently found in *o*-rings, which are used to seal equipment to seal against the escape of gas or fluid.
- There are several *o*-rings in the filter dryer assembly that come into contact with the process stream.
- The polymer is listed as incompatible with the crystallization solvent in the available literature and is known to swell after a short period of time when immersed in the solvent.

(D) Consider the additional data and review the PFD. What is the likely source of the black particulates observed on the filter cake?

(E) Is there another obvious potential source?

(F) Should the investigation be expanded to include other products? Why or why not?

REFERENCES

1. *Pharmaceutical Engineering Guides for New and Renovated Facilities*, Biopharmaceutical Manufacturing Facilities, Vol. 6, ISPE, 2004.
2. *United States Pharmacopeia Monograph: Purified Water*, USP32–NF27, 3872.
3. *Pharmaceutical Engineering Guides for New and Renovated Facilities*, Water and Steam Systems, Vol. 4, ISPE, 2001.
4. Brauer RL. *Safety and Health for Engineers*, Wiley, 2006.
5. ANSI/ISA88.01-1995 (IEC61512). The Instrument, Systems and Automation Society sponsored the creation of the S88 standard and supports the expansion and development of industrial instrumentation and control expertise for process and other industries.
6. Pharmaceutical Engineering Guides for New and Renovated Facilities, Vol. 5, Commissioning and Qualification, ISPE, 2001.
7. *International Conference on Harmonization of Technical Requirements for Registration of Pharmaceuticals for Human Use*, Q7A Good Manufacturing Practice Guidance for Active Pharmaceutical Ingredients, ICH, 2001.
8. www.epa.gov.
9. www.osha.gov.
10. *Guidelines for Process Safety in Batch Reaction Systems*, AICHE Center for Chemical Process Safety, 1999.
11. *Guidelines for Process Safety Fundamentals in Plant Operations*, AICHE Center for Chemical Process Safety, 1995.
12. 29 CFR 1920, Occupational Safety and Health Standards.
13. Naumann BD, et al. Performance based exposure control limits for pharmaceutical active ingredients. *Am. Ind. Hyg. Assoc. J.* 1996; 57: 33–42.
14. *Ullmann's Encyclopedia of Industrial Chemistry*, Wiley, 2009.

23

PROCESS DEVELOPMENT AND CASE STUDIES OF CONTINUOUS REACTOR SYSTEMS FOR PRODUCTION OF API AND PHARMACEUTICAL INTERMEDIATES

THOMAS L. LAPORTE, CHENCHI WANG, AND G. SCOTT JONES
Process Research and Development, Bristol-Myers Squibb Co., New Brunswick, NJ, USA

23.1 INTRODUCTION

Batch processing in stirred tank reactors is the default mode of operation for production of process intermediates and active pharmaceutical ingredients in the pharmaceutical industry. This is true for both homogeneous and heterogeneous reactions, as well as the subsequent workup unit operations and final crystallization. While commonplace in the commodity chemical industry, continuous processes are somewhat rare in the pharmaceutical industry. However, the potential advantages of organic synthesis reactions operated via a continuous mode include enhanced safety, improved quality, reduced energy costs, and greater cycle efficiencies [1]. These benefits are largely the result of smaller active reaction volumes and superior mass and heat transfer. Recently in the pharmaceutical industry, there has been renewed emphasis on holistic continuous processing where not only the reaction, but downstream extractions, solvent exchanges, and crystallizations are performed continuously as well [2]. However, most examples of continuous processing in the pharmaceutical industry are reaction only at this point, and this chapter will primarily focus on implementation of continuous reactions.

Part of the appeal of stirred tank batch reactors is their general versatility and the fact that a single piece of capital equipment can serve as a reactor, an extractor, a still, or a crystallizer depending on the needs of the process. This versatility enables a wide array of unit operation combinations and therefore, a single plant with multiple stirred tank reactors can manufacture a large number of products, with different processes. Additionally, batch processing on scale is similar to how a process chemist typically works in the laboratory. For example, charge ingredients, heat to reaction conditions, react for a specified time and sample for reaction completion. This systematic approach affords a simple and reproducible methodology for processing. However, laboratory and manufacturing scale batch reactors typically have vastly different heat and mass transfer characteristics. For chemistries in which heat or mass transfer controls selectivity, a direct scale-up of a laboratory batch process may be problematic in manufacturing. Similarly, limitations in heat transfer in stirred tank reactors may render some energetic laboratory processes unsafe at manufacturing scale. Finally, the versatility provided by general purpose stirred tanks comes at an efficiency cost when compared to continuous equipment designed for a specific unit operation. In each of these instances, continuous processing can offer advantages over traditional batch processing.

This chapter discusses opportunities for continuous processing of pharmaceutical intermediates and API, review some considerations for developing and implementing continuous processes, present two brief case studies from the authors' experience, and consider some of the barriers to widespread use of continuous processes. Since the engineering design equations for continuous reactors are covered extensively in undergraduate Chemical Engineering curricula, that level of detail is not presented here.

Chemical Engineering in the Pharmaceutical Industry: R&D to Manufacturing Edited by David J. am Ende

23.2 BENEFITS OF CONTINUOUS PROCESSING

23.2.1 Safety

Process safety is probably the greatest driver for development of continuous processes within the pharmaceutical industry. The two attributes of continuous processes that facilitate improved safety are a reduced inventory of reactive species and improved heat transfer. For a given throughput, continuous reactors are relatively small when compared to batch reactors. Additionally, continuous reactors are often operated at higher temperatures than batch reactors, resulting in higher rates of conversion. Both of these factors reduce the potential heat release contained within the reactor volume, by reducing the inventory of reactive species. The reduced chemical inventory greatly reduces the severity of failure and also allows for a rapid emergency quench of the entire reactor contents in the case of potential runaway reaction. The improved heat transfer rates of continuous reactors also help to reduce safety concerns when scaling exothermic reactions. This characteristic results in dramatically improved temperature control and enables operation within a safe operating window. In some cases, continuous processing may be the only practical means of scaling a highly exothermic process. Some examples employing continuous processing to mitigate safety concerns are given below.

Many pharmaceutical syntheses involve reactions with short half-lives and high heats of reaction, and thereby pose thermal runaway potential. Some examples include nitrations, oxidations, and other reactions involving energetic compounds such as peroxides, azides, and diazo compounds [3, 4]. Nitrations are highly exothermic, involve explosive or hazardous nitrating agents, and continuous processes have been developed to implement this chemistry more safely. In one example, the nitration of a pharmaceutical intermediate utilized a continuous reactor to enable high chemoselectivity while mitigating temperature control and decomposition concerns that existed in the batch process [4]. The continuous process operated at 90°C with a 35 min residence time in a microreactor. In contrast, the batch process operated for 8 h at 50°C and required very precise addition control for nitrating reagents.

23.2.2 Product Quality

The selectivity of organic reactions is determined by the amount of time molecules are exposed to a given set of conditions, that is, stoichiometry and temperature. In batch processing, spacial gradients exist for temperature and reactant concentration due to the mixing times achievable with conventional batch reactors. Restated, in batch reactors, the reaction conditions vary with location in the reactor. That nonuniform reaction environment can lead to undesirable side products and the extent of their formation depends upon the mixing characteristics of the reactor and the rate laws for both desired and undesired reactions. The increased heat and mass transfer capability of continuous reactors can result in improved reaction impurity profiles since conditions can be controlled more uniformly than with batch reactors. Improvements in impurity profiles at the reaction stage lessen the burden of downstream unit operations designed to remove impurities. This can allow for yield improvements due to optimization of downstream workup and crystallization. The improved control of reaction conditions should also help to minimize batch-to-batch variability that sometimes exists with batch processes.

There are additional consequences of the inferior heat and mass transfer properties of conventional batch reactors. Often reagents must be added over extended periods of time, and this means that there is a wide distribution in the amount of time that substrate molecules, starting material or product, are exposed to reaction conditions. While this increases cycle times, it also affects product quality and choice of operating conditions. These temporal gradients necessitate that conditions are defined to accommodate those molecules exposed to process conditions for the longest periods of time. Mean residence times are reduced in continuous reactors and molecules experience reaction conditions for more uniform periods of time. Additionally, the increased heat and mass transfer rates also mean that reaction conditions can be manipulated more rapidly than with batch reactors. The minimization of temporal gradients, coupled with the ability to rapidly manipulate reaction conditions, allows the process development engineer to consider operating conditions that would lead to unacceptable impurity profiles in batch processes. One example of this benefit is in the case where a relatively unstable intermediate is produced. Consider the time-temperature stability envelope displayed for a hypothetical first-order decomposition in Figure 23.1. The stability of a chemical intermediate increases at lower temperature, decreases with time, and these parameters are coupled. This fact means that batch reactions requiring low temperatures and long reaction times for stability reasons can possibly be converted to high temperature continuous processes when operated for a much shorter period of time. This same concept applies to all reactions, desired and undesired, and by understanding the rate laws governing them, continuous processing conditions that improve reaction selectivity can often be identified.

Many examples exist where continuous processing led to improved product quality [5, 6]. For example, the biphasic BOC-protection of an amine was investigated with continuous flow reactors due to its high heat of reaction, −213 kJ/mol, and the propensity to form dimeric impurities [5]. The dimeric impurities were reduced and the overall selectivity was improved from 97% to 99.9% in the continuous process. The improvements were attributed to the

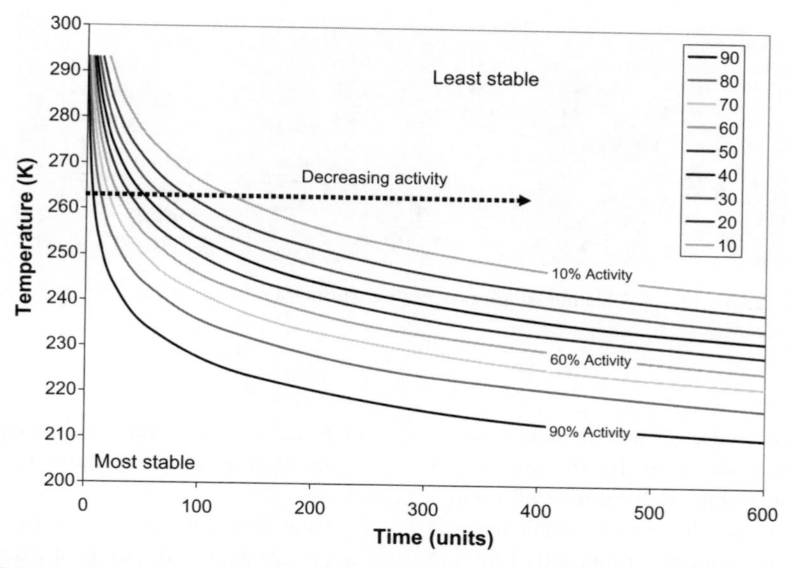

FIGURE 23.1 Operating chemistry envelope for a hypothetical first order degradation mechanism. The stability (activity) of an intermediate is a function of time and temperature. The rate of degradation increases with higher temperatures.

reduction of spacial and temporal gradients in reaction conditions.

23.3 CONTINUOUS REACTOR AND ANCILLARY SYSTEMS CONSIDERATIONS

The three main components of a continuous reaction process include the feed solutions, the reactor, and the quench [7] (Figure 23.2). We will first consider the reactor followed by the ancillary systems for the feed solutions and quench.

23.3.1 Continuous Reactors

23.3.1.1 Plug Flow Reactors Ideal plug flow reactors (PFRs) have flow with minimal back mixing along the flow path, no radial concentration or temperature gradients and a precise residence time for all flowing material. In the case of

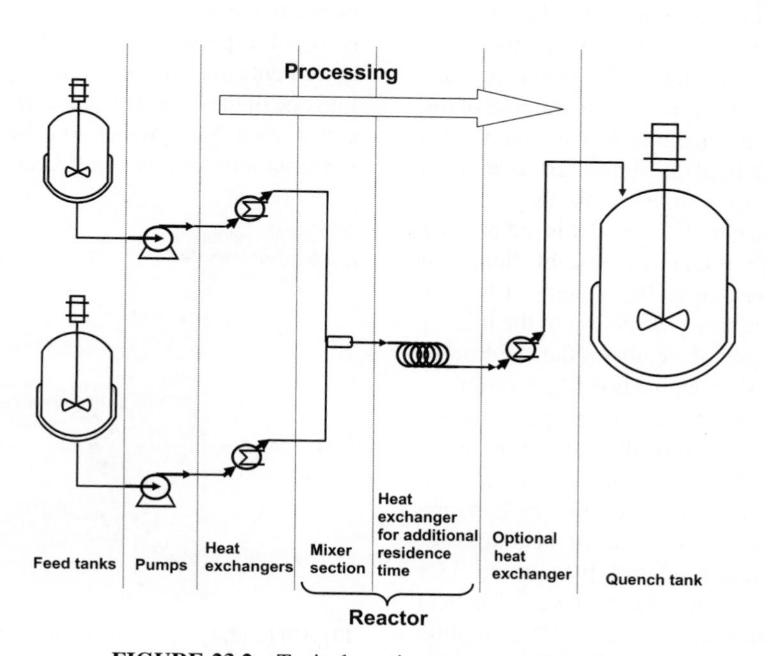

FIGURE 23.2 Typical continuous processing scheme.

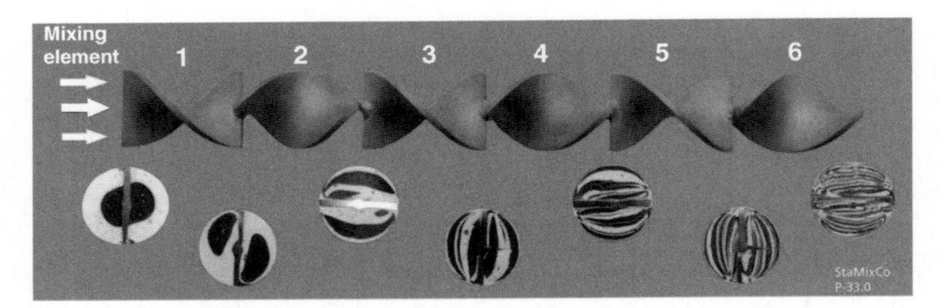

FIGURE 23.3 Cartoon demonstrating operating principle of a static mixer. Courtesy of StaMixCo LLC.

laminar flow, radial gradients may exist and there may not be plug flow in the truest sense. However, for the rest of this chapter, the term plug flow reactor will refer to all tubular flow reactors, regardless of the degree of turbulence and radial gradients. Plug flow reactors are composed of mixing zones for mass transfer and heat exchangers for heat transfer, and often these components are present in a single device. The static mixers and heat exchangers commonly utilized for these purposes in the pharmaceutical industry are described below and can be used in any combination required to meet the demands of the particular process.

In-line static mixers are commonly utilized in plug flow reactor systems to efficiently mix multiple feed streams. Tubular static mixers have characteristic mixing times of a few seconds or less depending on the degree of turbulence and provide efficient mixing even in the case of laminar flow. Static mixers consist of sequential static, often helical, mixing elements housed in a tube. The mixing elements typically alternate between left and right handed torsion and simultaneously produce flow division and efficient radial mixing, minimizing radial gradients in velocity, temperature, and concentration. A cartoon demonstrating the operating concept of a static mixer is given in Figure 23.3. In the case of the common Kenics® static mixer, each mixing element divides the process fluid in half. Each fluid division is further divided by subsequent mixing elements and the number of fluid striations is theoretically equal to 2^N, where N is the number of individual elements. In this simple way, miscible fluids can be thoroughly mixed within a very short length of tubing, even under laminar flow conditions. Selection of the mixing inserts and number of elements depends on the fluid properties and the specific processing application [8]. A variety of vendors (Kenics, Komax, Sulzer, etc.) manufacture static mixers and can aid in the selection of the most appropriate mixing elements. Static mixers are typically jacketed to control temperature when used as a reactor, and in one example, the mixing elements are made of heat transfer tubes for improved temperature control [9]. Figure 23.4 shows an example of a lab static mixer where 27 helical mixing elements are contained within 7 in. of 1/4 in. tubing, equating to a theoretical 134 million striations. Static mixers offer advantages over mechanical agitators such as more rapid mixing, ease of maintenance, and lower operating costs.

Heat removal and temperature control in plug flow reactors are achieved with heat exchangers. The versatility of heat exchangers is demonstrated in the case studies of the latter sections of this chapter, where heat exchangers are utilized to adjust the temperature of feedstocks prior to reaction, to control the temperature in the mixing zone of the reactor, to provide additional residence time for complete conversion of the reaction, and to thermally quench reactions. They are an essential component of plug flow systems. The key advantage of heat exchangers, versus conventional stirred tanks, is their improved heat transfer rates that result from much higher surface area to volume ratios. Most commonly used by the authors are concentric tube and shell and tube heat exchangers that are readily available from vendors and easily constructed in house as well. Schematics of several variants of these are shown in Figure 23.5. These heat exchangers are readily available at low prices, have reasonable pressure drops, and meet the heat transfer requirements for most reactions. By inserting mixing elements into one of these heat exchangers (Figure 23.5), a PFR can be constructed that provides good heat and mass transfer. Due to the simplicity of construction and lack of moving parts and

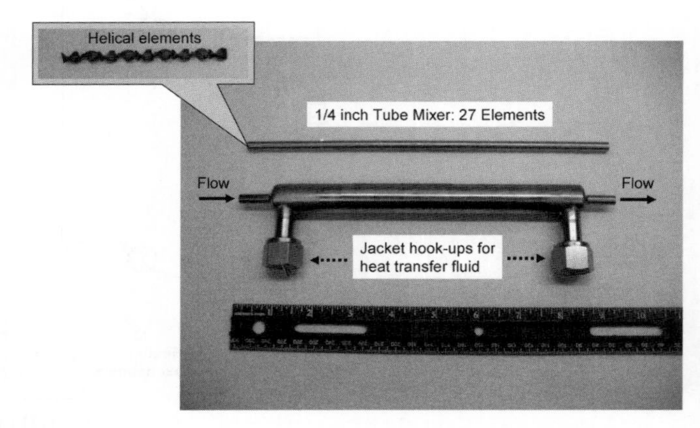

FIGURE 23.4 1/4 in. tube mixer with 27 elements and a 1/2 in. jacket for added temperature control.

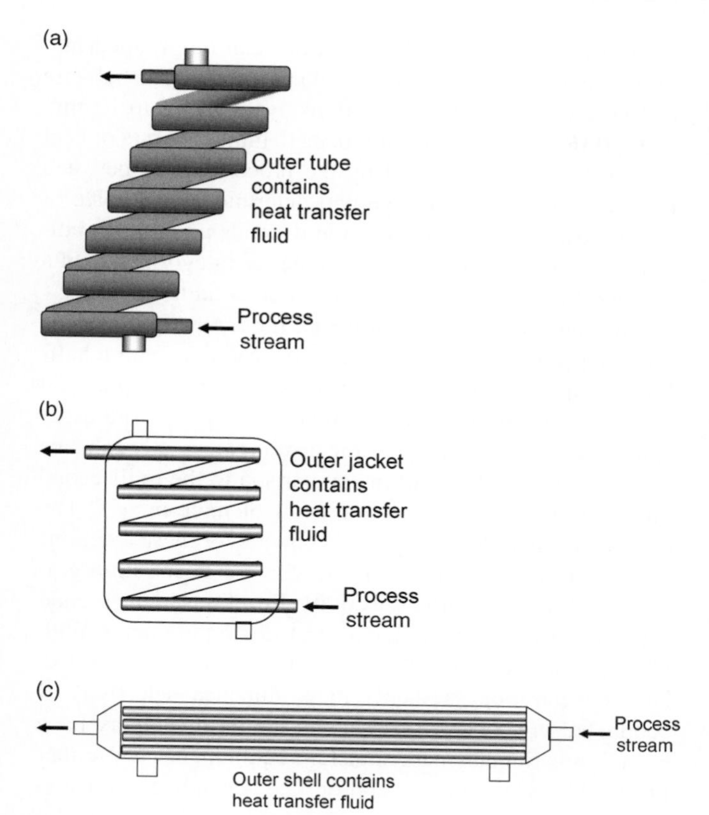

FIGURE 23.5 Heat exchangers commonly employed in continuous processing: (a) concentric tube, (b) jacketed coil, (c) shell and tube.

associated seals, a PFR provides a cost efficient reactor that is easy to construct and operate. Although they have better heat transfer properties than batch tank reactors, PFRs often operate nonisothermally for exothermic reactions.

23.3.1.2 ***Microreactors*** Microreactors are another type of flow reactor that have been increasingly studied and applied as laboratory tools for process screening and scale-up studies. The term microreactor typically implies a single unit integrating a static micromixer and heat exchanger combined with an additional heat exchanger that provides time for reaction conversion beyond the mixing zone. Other more specialized reactors such as spinning tube-in-tube [10] and spinning disk reactors [11] are less wide spread and will not be discussed here. Laboratory microreactors fabricated by glass or metals are available with an internal volume of less than 1 mL. As an example, a standard microreactor from Micronit Microfluidics [12] includes a preheating section for each input stream, a mixing section, and a quenching section from a third input. The total volume of this borosilicate reactor is 3.4 mL of which the mixing zone is 2.4 mL.

Microreactors are suitable tools to employ with fast reactions that require extremely efficient mixing and the reaction requires only low flow rates. Micromixers have internal microchannels that typically lie in the range of 50–500 µm. In these microfluidic devices, molecular diffusion is the governing mixing mechanism within the laminar flow domain, unlike turbulent mixing created in a pipe or mechanically agitated vessels. Many micromixers can maximize the interfacial surface contact of fluid lamination and efficiently minimize concentration gradients. The internal microstructures promote multiple flow divisions and recombinations and are designed for specific flow arrangements and fluid types [13]. For example, T-mixers and interdigital mixers (Figure 23.6) are routinely used in microreactors for obtaining efficient liquid–liquid mixing. Microreactors also possess extremely high surface to volume ratios for enhanced heat transfer and can therefore operate isothermally even with exothermic reactions. This expands processing

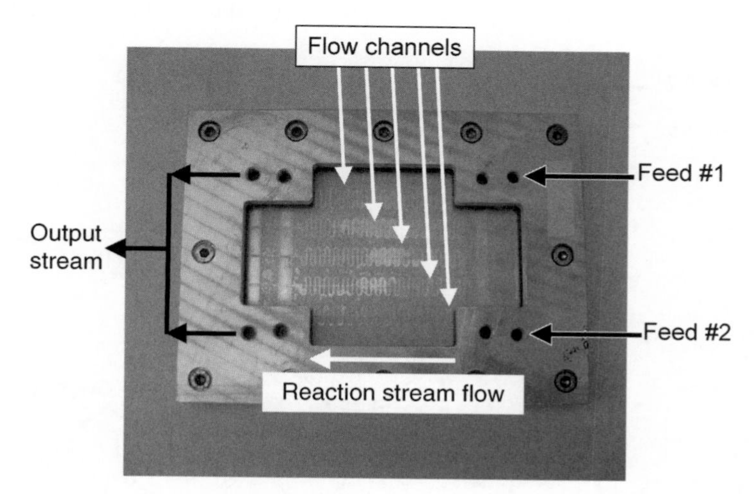

FIGURE 23.6 Microreactor from Mikroglas Chemtech GmbH: interdigital mixer with heat exchanger, five channels with a width of 500 µm and a depth of 250 µm.

opportunities for managing hazardous or highly energetic chemistries with enhanced safety.

23.3.1.3 Continuous Stirred Tank Reactors Continuous stirred tank reactors (CSTRs) are presented last because they are less commonly used in the pharmaceutical industry. They are essentially batch tank reactors that are operated continuously by simultaneously flowing reactants in and product out. Since they are tank reactors, their heat and mass transfer characteristics are equivalent to similarly sized batch tank reactors. Additionally, single CSTRs have broad residence time distributions and low conversion rates per unit volume. Some of these characteristics of CSTRs can be improved by using a series of smaller reactors cascaded together as shown in Figure 23.7. CSTRs cascaded in this way have been used for several processes at AMPAC Fine Chemicals LLC for the production of hazardous or energetic chemicals [14]. In one facility, they utilize up to seven cascaded reactors from 0.25 to 1 L in volume for a continuous process [15].

23.3.2 Choosing Between CSTRs, PFRs, and Microreactors

As demonstrated earlier, continuous processes have the potential to deliver higher throughput, improved heat and mass transfer, and improved impurity profile through control of precise reaction conditions. The ability of the continuous process to deliver on this potential largely depends on the type and size of reactor chosen. In this section, we will make some general comparisons between PFRs, microreactors, and CSTRs. A qualitative comparison of key attributes for these reactors is given in Table 23.1. The case will be made that PFRs and microreactors are generally preferred over CSTRs. PFRs are preferred over microreactors when they are capable of meeting the heat and mass transfer demands of the reaction of interest.

As shown in Table 23.2, reactions can be grouped into three general kinetic categories: (1) very fast with a half-life of less than 1 s, (2) rapid reactions, typically 1 s to 10 min, and (3) slow reactions greater than 10 min. The rate of heat and mass transfer required by the process varies between these categories and in large part determines the choice of reactor. Since the rate of reaction depends upon the conditions chosen, it is sometimes possible for categorization of a reaction to change based upon reaction conditions.

The mass transfer requirements for a reaction depends upon the reaction categorization. For reactions with a half-life less than 1 s, microreactors may be the only practical choice due to mass transfer limitations of PFRs, and especially CSTRs. Even though static mixers can greatly enhance mixing in PFRs, they pail in comparison to the millisecond mixing times that are characteristics of micromixers. The degree to which PFRs may be acceptable for these reactions depends in part on the extent to which concentration gradients influence reaction selectivity. Reactions in the second category have less stringent mass transfer requirements. With PFRs or microreactors, they are likely kinetically controlled, but concentration gradients may influence selectivity if conducted in CSTRs. Reactions in the third category have even less stringent mass transfer requirements and either PFRs and CSTRs may be appropriate depending upon specific process needs.

Another major factor in reactor selection is heat transfer requirements of the process. Since the heat generated by a reaction is proportional to reactor volume and the heat removal is proportional to reactor surface area, the ratio of surface area to volume provides an easy means of comparing a reactor's ability to remove heat. Table 23.3 shows this ratio for a 2000 L batch reactor compared to a smaller CSTR, tubular PFR, and a system of microreactors capable of similar throughputs. Obviously conversion of an existing batch reactor to a CSTR does nothing to improve the heat transfer characteristics. Although a smaller CSTR represents a great

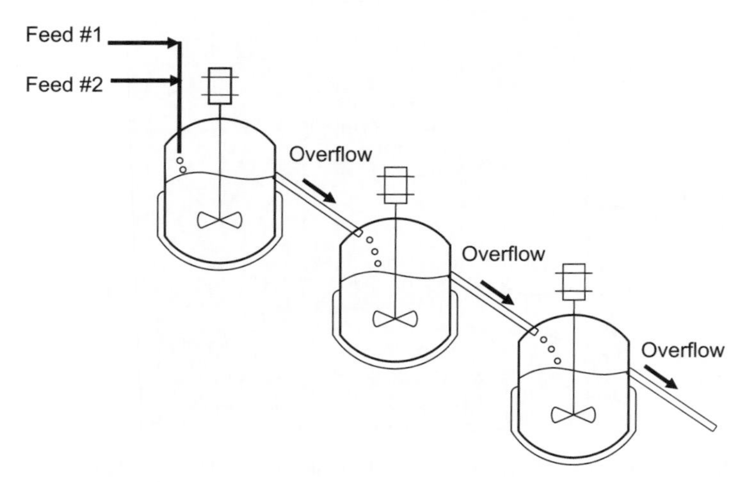

FIGURE 23.7 Cascaded continuous stirred tank reactors in series.

TABLE 23.1 Attributes of CSTRs, PFR and Microreactors

Reactor Mode	Multiple CSTRs	PFR	Multiple Microreactors
Handling of solids	+ +	−	− −
Gas evolution	+ +	−	− −
Slow reaction kinetics	+ +	−	− −
Quickly achieve steady state	−	+ +	+ +
High conversions per volume	−	+ +	+ +
Narrow residence time distribution	−	+ +	+ +
Mitigates product reacting with starting material	−	+ +	+ +
Initial large heat sink	+	−	−
Low operational complexity	−	+ +	− −
Low level of equipment intensity	+	+ +	− −
Enhanced heat transfer	−	+	+ +
Enhanced mass transfer	−	+	+ +
Low cost	+	+	− −

+ +: strong positive characteristic; +: positive characteristic; − −: strong negative characteristic; −: negative characteristic.

TABLE 23.2 Categorization of Reactions for Continuous Process Fit [16]

	Reaction Rate	Characteristics
1	Very fast	Reaction half-life of less than 1 s. Reaction is mixing sensitive since rate is faster than mixing. Most of the reaction occurs in the mixing stage for a continuous reaction. Heat management can be an issue
2	Rapid	Reaction half-life is 1 s to 10 min. Reaction may be mixing sensitive, but typically kinetically controlled. Heat management may be an issue
3	Slow	Reaction half-life greater than 10 min. Implemented in a continuous process mainly for hazardous chemistries

improvement over batch reactors in terms of surface to volume ratios, it cannot compete with tubular PFRs or integrated microreactors. Reactions in category 1 are often highly energetic and the rapid generation of heat may require a microreactor if a high degree of temperature control is required. PFRs however meet the heat removal requirements of many common reactions, and can be used for more energetic reactions if nonisothermal operation is acceptable. The nonisothermal characterization refers primarily to temperature gradients in the axial rather than radial direction.

In order to realize all of the benefits described earlier, a narrow residence time distribution is often required. For an ideal PFR, all molecules have the same residence time and the distribution is represented by a Dirac delta function. While real PFRs are less perfect, they have very narrow residence time distributions. Microreactors operate in laminar flow and axial dispersion models have been used to model the residence time distribution [17]. This deviation from ideal plug flow is less important for microreactors since they are typically operated with shorter mean residence times, and it is the absolute value of the residence time at the upper boundary of the distribution that impacts impurity profiles, not the percent deviation. CSTRs have a broad residence time distribution and some molecules spend considerably longer in the reactor than others. In fact, the standard deviation of the CSTR residence time distribution is equal to the mean residence time. For processes requiring exposure to reaction conditions for a precise period of time, PFRs are preferred. Furthermore, the residence time distribution in CSTRs results in longer transient periods, typically three to four residence times, prior to reaching steady state. The longer transient periods for CSTRs result in larger amounts of wasted product and are an additional drawback of CSTRs.

For processes that are not constrained by heat and mass transfer or residence time distribution considerations,

TABLE 23.3 The High Surface Area to Volume Ratios for Continuous Reactors are Due to the Small Characteristic Reactor Dimension

Reactor	Characteristic Dimension (mm)	Surface Area/ Volume (cm^{-1})
2000 L tank	680	2.9×10^{-2}
50 L CSTR	200	0.10
Tubular PFR (500 mL)	3.2	3.1
Microreactor[a]	0.05	400

[a] 50+ units would be needed to meet throughput requirements.

throughput considerations may be important when choosing between CSTRs or PFRs. Continuous reactors will always offer a higher throughput than batch processing. Conceptually this is quite simple since in continuous processing reactants are constantly fed to the reaction and reactors operate at a constant volume, usually full. For positive order reactions with simple kinetic rate laws, the design equations for CSTRs and PFRs dictate that PFRs deliver a given conversion with smaller reactor volumes than CSTRs. The extent of divergence between CSTR and PFR volume depends upon the rate law and the conversion required in the reaction, and it is especially pronounced when high conversions are required. Conversions in the pharmaceutical industry are nearly always greater than 95% and quite frequently are greater than 99%. Table 23.4 compares the CSTR reactor volume, relative to a PFR, required for a first- or second-order reaction to achieve 99% conversion. For a single CSTR to reach 99% for a second-order reaction, it would need to be 100 times larger than its PFR counter part. The conversion efficiency of CSTRs is improved by cascading several in series, and in the limit of an infinite number of CSTRs in series performance equals that of a PFR. Based upon conversion, or throughput per unit volume, PFRs are clearly superior to CSTRs.

Considerations for ease of operation and reactor costs may factor into choice of reactor as well. Based upon the authors' experience, PFRs represent a good balance between cost, heat, and mass transfer efficiency, and ease of operation, and they are preferred when they meet the demands of the process. To the extent possible, attempts are made to modify reaction conditions to allow the use of PFRs. While microreactors have superior heat and mass transfer rates they are significantly more expensive than PFRs due to the fine machining required to construct their microchannel flow paths. PFRs on the other hand are simple jacketed tubes with mixing elements and their cost reflects this simplicity. Another practical drawback of microreactors is their relative inability to handle even small amounts of solids. Individual particles may be sufficient to block flow and interrupt operation. By comparison, the larger diameters of most PFRs allows slurries with low solids density to flow. Slurries with higher solids loading likely require CSTRs for operation, or may not lend themselves to continuous processing at all. Additionally, CSTRs can be better suited for handling reactions that involve large amounts of gas evolution. In microreactors, generation of large amounts of gas can serve to reduce the residence time by forcing the process stream through the reactor more rapidly than intended. Non-CSTR reactors can be designed to handle gases and one of the authors has developed and implemented a continuous trickle bed oxidation column for production of a pharmaceutical intermediate [18]. A final instance where CSTRs might be preferred is the case of slower reactions requiring longer residence times for complete conversions. In such circumstances, the length of a PFR required to accommodate the longer residence time may result in impractical pressure drops.

TABLE 23.4 Comparison of Reactor Volume for Multiple Stirred Tank Reactors in Series Versus a Plug Flow Reactor Based on 99% Conversion of Starting Material

# of Stirred Tank Reactors in Series	Volume Relative to a Plug Flow Reactor	
	First-Order Kinetics	Second-Order Kinetics
1	22	100
2	4	8
3	2.4	4
4	1.8	2.6
6	1.5	2
∞	1	1

23.3.3 Ancillary Systems

23.3.3.1 Feed Solutions All of the reactants for a continuous process must be in a form that is easily transported by pumps or pressure transfer. Since a higher number of feed solutions requires a proportional number of tanks and feed control systems it is generally desirable to combine several solvents and reactants when possible. Of course, species that react with one another should be prepared in separate feed streams. Typically, one feed solution will contain the starting material and the bulk solvent while a second feed contains the reagent. In some cases, a third feed may contain a second reagent, a catalyst, or possibly a second compound in the case of a coupling reaction. Ideally, the feed solutions should be homogeneous to avoid reactor plugging or fouling, and knowledge of substrate solubility in all process streams is desirable. Additionally, knowledge of the chemical stability of each feedstock is imperative for successful operation.

23.3.3.2 Quench The quench brings the reaction mixture to a nonreactive and stable condition for downstream processing. The quench can be chemical or thermal in nature and the choice depends on the reactivity of the processing stream and downstream processing needs. Examples of both are given in the two case studies sections. The three predominant quench modes utilized in continuous processing are demonstrated in Figure 23.8. While these modes are depicted for chemical quenches, slight variants can be envisioned for thermal quenches as well. The first is a reverse batch quench where the reaction stream flows into a reactor containing the quench material. Depending upon processing needs, parallel quench vessels can be setup to alternately receive the continuous reaction stream and allow uninterrupted operation of the reactor. Under this scenario, the contents of the off-line

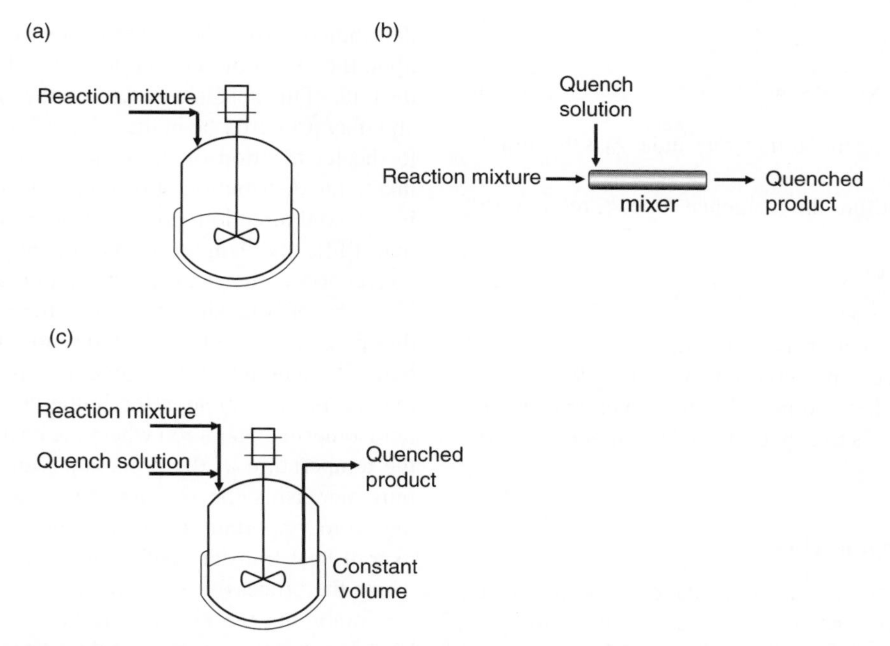

FIGURE 23.8 Continuous reaction stream quench scenarios. (a) Reverse quench; (b) continuous quench with in-line mixer; (c) continuous quench with CSTR.

quench tank are worked up, while the second quench tank continues to receive the reaction stream. This operating mode is the least complex, ensures an excess of quench solution, may provide a good heat sink for exothermic quenches, and provides a well-defined delineation of batches from a GMP perspective. The second mode of quenching utilizes a static mixer, typically jacketed, to introduce the quench solution. Assuming downstream processing is conducted batchwise, the quenched solution would be collected in stirred tank reactors. The third mode of quenching is similar to the second, but uses a CSTR for the continuous quench. Since this quench mode would require an additional reactor as a collection vessel, it is relatively impractical unless subsequent processing is conducted continuously.

23.4 PROCESS DEVELOPMENT OF THE CONTINUOUS REACTION

23.4.1 Reaction kinetics

A prerequisite to designing a continuous process is to understand the rate laws governing the kinetics of the desired and undesired reactions. The level of knowledge required depends upon the complexity of the process, but where possible a complex reaction should be broken down into its elementary reaction steps and the rate laws for each step established. In some cases, the development of an overall apparent rate law may be sufficient. In either case, the activation energy, and effects of reactant concentration should be established for both desired and undesired reactions. It should be noted that the kinetic experiments need not be conducted in a continuous reactor since the reaction kinetics do not depend upon mode of operation. However, in some instances such as fast reactions, flow reactors may offer a practical means of studying reaction rates. With rate laws established, an overall kinetic model can be constructed to help identify operating conditions—temperature, concentration, and time—that promote high rates of conversion and selectivity toward the desired product. In this way, the process development engineer can realize the full potential of the improved heat and mass transfer rates and precise residence times of plug flow reactors and microreactors.

EXAMPLE 23.1

Show how residence time varies with conversion in a PFR for a constant-density first-order reaction. Generate a table of conversion versus residence time. Assume the first-order rate constant is 0.01 s^{-1}. Compare the residence times required to reach 90% conversion, 99%, 99.9%, and 99.99%.

Starting with the design equation for a PFR:

$$\frac{V}{F_{A0}} = \int_0^X \frac{dX}{-r_A} \tag{23.1}$$

Substituting the rate equation

$$-r_A = kC_A \tag{23.2}$$

$$C_A = C_{A0}(1-X) \tag{23.3}$$

$$\frac{V}{F_{A0}} = \frac{1}{kC_{A0}} \int_0^X \frac{dX}{1-X} \tag{23.4}$$

where F_{A0} is the entering molar flow rate and C_{A0} is the initial molar concentration.

Integrate and substitute the residence time, τ, relation to obtain

$$\tau = \frac{VC_{A0}}{F_{A0}} = -\frac{1}{k}\ln(1-X) \tag{23.5}$$

Create a table for X versus τ using $k = 0.01\ s^{-1}$.

Thus, the residence time required to achieve 90% conversion is 230.3 s. It takes another 230 s to convert from 90% to 99% and another 230 s to convert from 99% to 99.9%, and so on.

23.4.2 Reaction Engineering

In addition to the kinetics, an understanding of the heat generated by the process needs to be understood in order to design an appropriate reactor for the process. The two main sources of heat generation are the heat of mixing and heat of reaction. The heat of mixing refers to heat generated upon mixing of the feed streams, including heats of dilution. The heat of reaction is proportional to reaction conversion and is distributed across the length of the plug flow reactor based upon the extent of conversion. For PFRs and microreactors, the bulk of the heat is generated in the first part of the reactor, and may primarily be in the mixing stage. This is in part due to the localization of the heat of mixing, but is primarily due to the distribution of reaction conversion, and thus heat, in the axial direction. For a first-order reaction in an isothermal PFR, the length of reactor required to reach 90% conversion is the same length required to go from 90% to 99% conversion. The first half of the reactor would need to dissipate an order of magnitude more heat than the second half. The amount of heat generated in the early part of the reactor is even greater for higher order reactions and in nonisothermal operation where the heat of reaction increases the temperature of the process stream early in the reactor, thus increasing the reaction rate and heat generated. It is therefore important to understand the intended reactor's overall heat transfer coefficient, or to design a reactor that meets the process's requirements.

Combining the kinetic rate laws, heats of reaction, and knowledge of the reactor's heat transfer coefficients provides a powerful means to model expected outcomes. A combined experimental and modeling approach is essential for rapid process development since so many parameters depend on one another. Figure 23.9 shows a generic workflow for

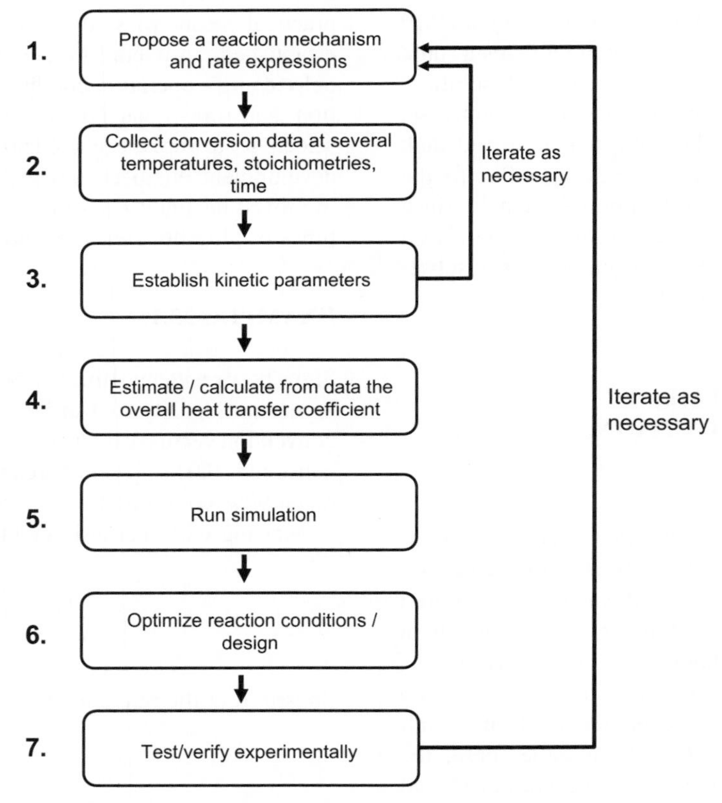

FIGURE 23.9 Workflow for process modeling.

continuous reaction development from a reaction engineering perspective. An understanding of the factors influencing heat generation is established in steps 1 through 3 by combining the kinetic rate laws and heats of reaction. The reactor's heat transfer properties and ability to remove heat are established in step 4. Simulations can then be conducted to determine reaction conditions at different positions along the reactor and ultimately product quality. Such simulations can be used to evaluate different reactor types and configurations, as well as changes in flow rates, stoichiometry, and temperature. With a good model, much of the process development can be facilitated by virtual experiments. The final step is to experimentally verify the optimized conditions, or redesign the reactor.

Examples of this type of methodology exist in the literature [3, 19] and an example is also given in the case studies section of this chapter. Bogaert-Alvarez et al. [3] undertook a good example of this approach. They solved the rate laws and nonisothermal heat transfer equations as two ordinary differential equations for a plug flow reactor. They assumed a constant temperature for the heat transfer medium although an energy balance of it could also be included. The model enabled them to evaluate the effects of various parameters including jacket temperature, reactor length, flow rate, and heat transfer coefficients on reaction conversion and peak reaction stream temperatures.

As discussed earlier, microreactors and PFRs can combine any number of mixing zones and heat exchangers to accommodate the needs of a process. Combining this flexibility with the predictive models described earlier can lead to improved reactor design and influence conversion rates and product quality. Since the reactant and product concentrations vary along the length of the reactor, different stages may benefit from different operating temperatures. In the case of moderately or highly exothermic reactions, the reaction temperature may spike above the desired control point during the initial portion of the reactor (Figure 23.10a). Because the greatest amount of heat is generated at the entrance of the plug flow reactor, a two-zone jacket temperature may facilitate greater reaction temperature control (Figure 23.10b). In this example, the two zones consist of a lower initial jacket temperature of 63°C versus the 80°C on the remaining portion of the reactor. Alternatively higher temperatures can be utilized later in the reactor to improve conversion rates. Similarly reactors can be easily designed to accommodate multiple feed points at different stages along the reactor to further manipulate reaction conditions if required. In this way reactors can be specifically designed for maximum throughput and product quality.

While the reaction engineering discussion thus far has focused on product quality, the same concepts can be utilized for process safety evaluations. Since most of the reactions employed in the pharmaceutical industry are exothermic, this safety aspect is an important consideration. This is especially true of nonisothermal PFR operation where rates of heat generation and temperature vary along the length of the reactor. Reactions should be evaluated in combination with proposed reaction conditions to avoid potential runaway reactions and ensure a sufficiently large safe operating window.

23.5 SCALE-UP: VOLUMETRIC VERSUS NUMBERING-UP

Classical scale-up of a batch process consists of increasing the volume of the batch reactor. As a result of poorer heat and mass transfer in larger reactors, many of the common operations of batch processing take significantly longer at larger scales. Activities that frequently take longer at scale are charging of reagents, batch heat-up or cool-down, and reaction quench. The improved heat and mass transfer capabilities of continuous reactors means that lab- and plant-scale processing times are much better aligned. For example, the reaction time does not change with scale since it is the design criterion for the continuous process. Additionally, the reaction stream is quenched immediately upon completion of the reaction at any scale.

Scale-up of most continuous processes occurs by increasing the total reactor volume and the flow rate to maintain the same residence time established during development. However, an alternative approach in continuous processing scale-up, particularly for using microreactors, is to number up. Here the reactor system is duplicated numerous times, with all running in parallel [20]. At DSM, multiple parallel microreactors were utilized for pilot-scale production of a nitration reaction of a pharmaceutical intermediate [21]. In this scenario, the replication of the same geometries and flow rates for each unit provides the higher overall process flow rates, and thus avoids any scale-up effects. The logistics, complexity, and capital investment of such systems may limit widespread implementation for high-volume products. Examples where processes have been numbered up using microreactors for commercial manufacturing are rare and this approach may not be amenable for most processes without additional technological advances particularly in automated flow stream division and control.

LaPorte et al. [18] demonstrated a less intensive example of numbering-up of a gas–liquid reaction. They replicated a trickle bed column and housed the set of four in a common baffled jacketed tube for temperature control. An enolate stream was split equally into four streams using rotameters and each stream flowed into one of four trickle bed columns. This scale-up facilitated a 4× numbering-up by maintaining the same fluid dynamics, mass transfer and heat transfer characteristics in each tube. This operation did require constant monitoring since the flow splitting was not

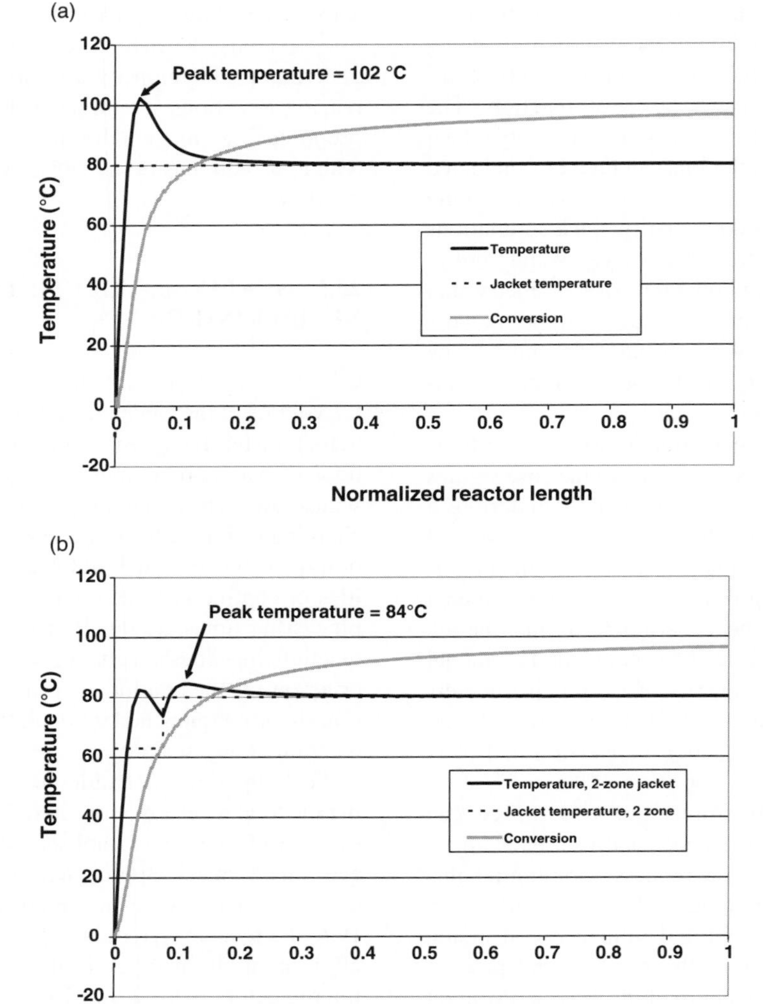

FIGURE 23.10 Simulated reaction temperature profile for a second-order reaction in a plug flow reactor. A constant jacket temperature is assumed for a single jacket temperature zone (a) and a two-zone jacket (b).

automated. Commercial processes would require a high degree of automation to ensure the proper flow at all times. Other approaches to scale-up, particularly for a microreactor, include adding to the volume of the reactor with serial addition of reactor plates. Hence, a large range of flow rates from milliliters to several hundred milliliters per minute is possible for a specific reactor platform [22].

23.6 PLANT OPERATIONS

23.6.1 Flow Control

Flow control is a critical parameter for a continuous process. The total flow of the streams ensures the proper residence time for reaction, and the ratio of the individual streams ensures the proper stoichiometry of the reagents. The feed streams must be accurate and consistent, and the constraints on those parameters depend on the tolerance of the process. One approach to minimize pulsating flows is to utilize pressured feed tanks along with flow meters and control valves to control the flow rate. Another possibility is to use metering pumps for each of the feed streams. For pumps that pulsate (piston, diaphragm, etc.), synchronization, dampening devices or multiple pistons that are sequenced and positioned on one pump may be required. Pulseless pumps with an integrated mass flow meter and feedback control system are ideal. These systems provide precise metering for processes with tight flow tolerances.

23.6.2 Process Analytical Technology

Process analytical technology (PAT) is an important part of most continuous processes as it provides a useful means of monitoring the state of the reaction. Indeed, one of the stated goals of the FDA's PAT initiative is "Facilitating continuous processing to improve efficiency and manage variability" [23]. Typical PAT tools include Raman, FTIR, NIR, and UV-Vis spectroscopy [24, 25] or other noninvasive monitoring techniques that can be adapted to a flow cell or tube reactor. These tools can be used during both transient and steady-state operations. As an example, FTIR was used to determine the proper ratio of reagent feed rate to starting material feed rate during the start-up of a continuous process to make an active pharmaceutical ingredient [18]. For this particular process, the same PAT equipment could have been used to monitor the product quality. In a well-defined continuous process it is envisioned that feedback controllers could adjust operating parameters based on input signals from PAT analyzers. Even in the absence of feedback control, PAT can provide valuable information to plant operators who can modify the operation if necessary. If PAT analyzers indicate that product quality is suspect, flow can be diverted to alternative holding tanks for further analysis. In the absence of spectroscopic analyzers, simple temperature measurements at various reactor positions can provide a wealth of information regarding reaction performance.

23.7 CASE STUDY: CONTINUOUS DEPROTECTION REACTION—LAB TO KILO LAB SCALE-UP

A batch process to carry out an acidolysis and deprotection chemistry for a pharmaceutical intermediate involved adding the substrate solution to triflouroacetic acid (TFA) at approximately 0°C. The complete reaction mixture was immediately quenched into a biphasic mixture of aqueous base and ethyl acetate. An amide impurity was formed at high levels of >2%. The longer quench times anticipated upon scale-up were expected to further increase the level of the amide impurity.

A continuous processing approach was undertaken to minimize impurity formation through improved control of reaction time and reduced quenching time. The continuous reaction was assessed in the laboratory by mixing two feed streams, one for the substrate and the other TFA, in a glass microreactor with an overall volume less than 10 mL. Experiments varying temperature and residence time identified process conditions, 25°C and a minimum residence time of 4 min, which provided complete conversion and a significantly lower level of amide impurity, approximately 1%.

The preliminary reaction kinetics obtained from the small-scale continuous reactions paved the way for a rapid process scale-up. A 100-fold increase in flow rate in the substrate and TFA streams was used to process approximately 5 kg of starting material using the setup shown in Figure 23.11. The starting material solution was not stable at room temperature and required storage at −10°C. A preconditioning heat exchanger was used to continuously heat up the starting material feed stream to the reaction temperature, 25°C, just prior to reaction. A PFR, constructed of a jacketed static mixer for mixing the two feeds and three sequential concentric tube heat exchangers, operated with an overall residence time of 5 min. The reaction stream was continuously quenched in a jacketed static mixer and the quenched mixture flowed into a receiver for subsequent processing. The use of a static mixer for the quench ensured effective mixing of the biphasic process stream while rapidly quenching the reactive species.

This particular batch process was relatively simple to convert to a continuous process. However, it is a good

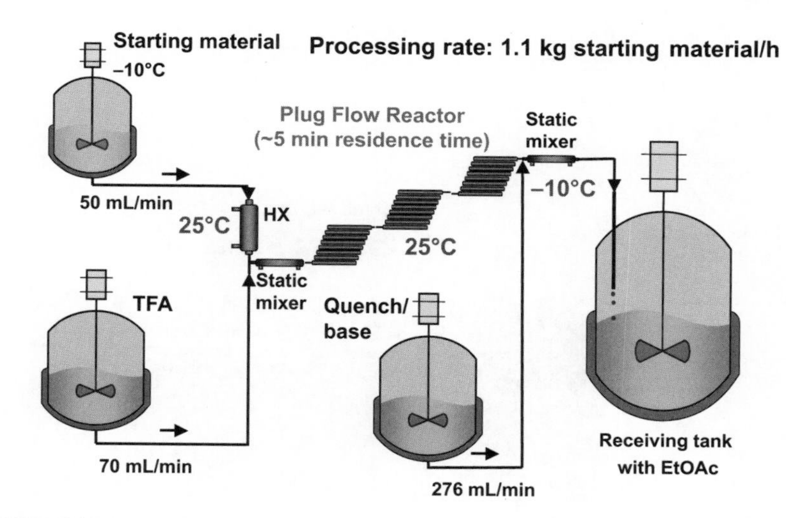

FIGURE 23.11 Kilo lab continuous flow setup for acidolysis and deprotection process.

example to demonstrate the key components and strategies behind the development process. This includes the use of stable feeds, preconditioning of a feed, and combined in-line jacketed static mixer and heat exchangers as the reactor.

23.8 CASE STUDY: CONTINUOUS PRODUCTION OF A CYCLOPROPONATING REAGENT

23.8.1 Introduction

The Simmons–Smith cyclopropanation is a well-known reaction to form cyclopropanes from olefins utilizing zinc and an alkyl iodide. The structure of the reactive zinc carbenoid species is the subject of numerous papers [26,27]. Formation of the active species is relatively exothermic with an adiabatic temperature rise above 120°C. Additionally, the complexes are known to be unstable for extended periods of time above 0°C. The exothermic nature of the reaction, combined with the complexity and incomplete understanding of the mechanism, and relative instability of the active species made scale-up very challenging in a batch process. One solution to the scale-up was the development of a continuous process for formation of the cyclopropanating reagent. The process was demonstrated at lab scale, scaled-up to pilot plant scale, and was used to make launch supplies for the starting material of a commercial API. The development and implementation of this process are discussed here.

23.8.2 Process Development

The strategy for developing a continuous process was to operate a PFR with a short residence time and higher temperatures, followed by a rapid thermal quench. A short reaction time was required to minimize the size of the plug flow reactor as well as reagent degradation. Initial screening work utilized a coiled 1/8 in. stainless steel jacketed tube as the reactor. Later in development, multiple 26 mL shell and tube heat exchangers containing up to 19 1/8 in. stainless steel tubes (Figure 23.12) were utilized. The heat exchangers were sequenced end to end to form the plug flow reactor. The reactor was operated with a short 50 s residence time. Due to the short residence time and the large amount of heat generated early in the reactor, isothermal operation was not possible. Details of the process are described below.

The laboratory setup used for development of the continuous process is shown in Figure 23.13. Two feed streams, one containing diethyl zinc (13.5 wt%) and dimethoxyethane in toluene and the other containing diiodomethane in dichloromethane, were held at ambient temperature. These streams were fed to the reactor with gear pumps and mass flow meters to ensure proper stoichiometry and residence time. Both streams passed through independent heat exchangers with a 30°C jacket temperature prior to mixing in a nonjacketed static mixer containing 27 helical mixing elements. The feeds entered the static mixer at about 29°C and exited at about 48°C. The reaction mixture flowed through a series of three shell and tube heat exchangers with 30°C jacket temperature to facilitate formation of the reagent. The process stream exit temperature was 52°C after the first heat exchanger and 32°C after the third. The 2°C temperature difference observed between process and jacket sides of the third heat exchanger indicates that the reaction was nearly complete by that stage. Additional calorimetric laboratory experiments that evaluated residual heat generation of the reaction mixture confirmed this observation. Finally, the reaction was thermally quenched to less than −10°C, again with a shell and tube heat exchanger. The process attained a steady state within three residence times based on multiple temperature measurements at various points along the plug flow reactor.

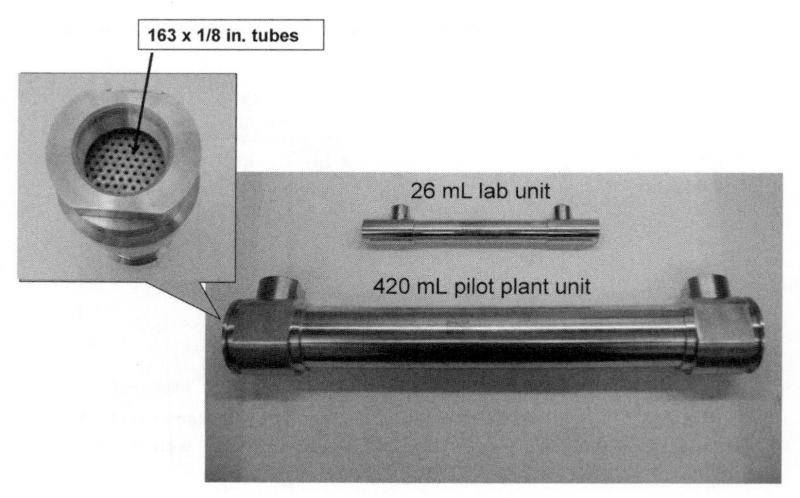

FIGURE 23.12 Mini shell and tube heat exchangers for laboratory or pilot plant use.

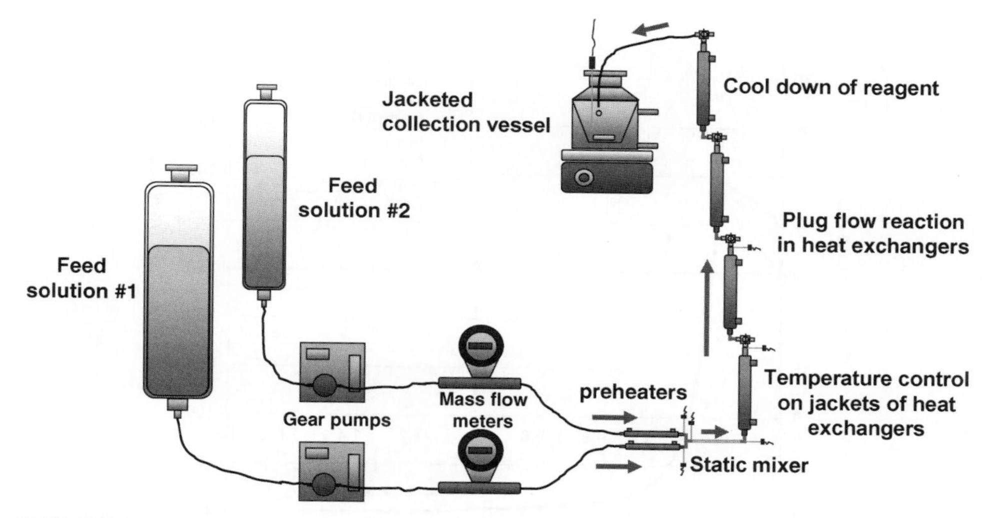

FIGURE 23.13 Laboratory setup for development of the Simmons–Smith reagent continuous process.

23.8.3 Modeling and Simulation

A reaction engineering approach similar to that described earlier was employed here to gain further insight into the continuous process. The proposed reactions for the model are formation of the Furukawa complex and Wittig complex and are shown in Figure 23.14. A proposed kinetic model and energy balance equation governing the reaction are shown as follows:

Nonisothermal plug flow reaction model

Assumptions:

- Completely mixed in radial direction
- No diffusion in flow direction (axial)
- Constant shell side temperature
- Constant stream density
- A two step reaction mechanism producing first the Furukawa complex followed by the Wittig complex
- Modeled on a per tube basis in the heat exchanger
- Overall heat transfer coefficient is independent of the number of tubes in the heat exchanger (constant shell side temperature)

Reaction Mechanism:

$$\mathrm{A}+\mathrm{B}\rightarrow\mathrm{F} \quad \Delta H_1=-99\,\mathrm{kJ/mol} \tag{23.6}$$

$$\mathrm{F}+\mathrm{B}\rightarrow\mathrm{W} \quad \Delta H_2=-94\,\mathrm{kJ/mol} \tag{23.7}$$

where A is the diethyl zinc/dimethoxyethane, B is the diiodomethane, F is the Furukawa complex, and W is the Wittig complex.

Rate Expressions:

$$k=A\,\mathrm{e}^{-E/RT} \tag{23.8}$$

$$r_1=-k_1C_\mathrm{A}C_\mathrm{B} \tag{23.9}$$

$$r_2=-k_2C_\mathrm{F}C_\mathrm{B} \tag{23.10}$$

Simplified Mass/Energy Balance:

$$u\rho C_\mathrm{p}\frac{dT}{dz}=\Delta H_1r_1+\Delta H_2r_2-UA_V(T-T_c) \tag{23.11}$$

where u is the reaction stream velocity, ρ is the reaction stream density, C_p is the reaction stream heat capacity, T is the reaction stream temperature, z is the axial position in plug flow reactor, ΔH_i is the heat of reaction, r_i is the rate of reaction, U is the overall heat transfer coefficient, A_V is the specific heat transfer area (area/unit volume), and T_c is the temperature of jacket coolant.

In this example, the reaction kinetics were not studied in separate detailed studies. Rather, the activation energies and frequency factors were fitted using the process stream temperatures at numerous reactor locations, a calculated overall heat transfer coefficient, and information on complex formation. Several assumptions were made in the modeling of

Et_2Zn + DME + CH_2I_2	----▶	$EtZnCH_2I$ * DME + EtI	Furukawa
$EtZnCH_2I$ * DME + CH_2I_2	----▶	$Zn(CH_2I)_2$ * DME + EtI	Wittig

FIGURE 23.14 Modeled formation of complexes for the cyclopropanating reagent.

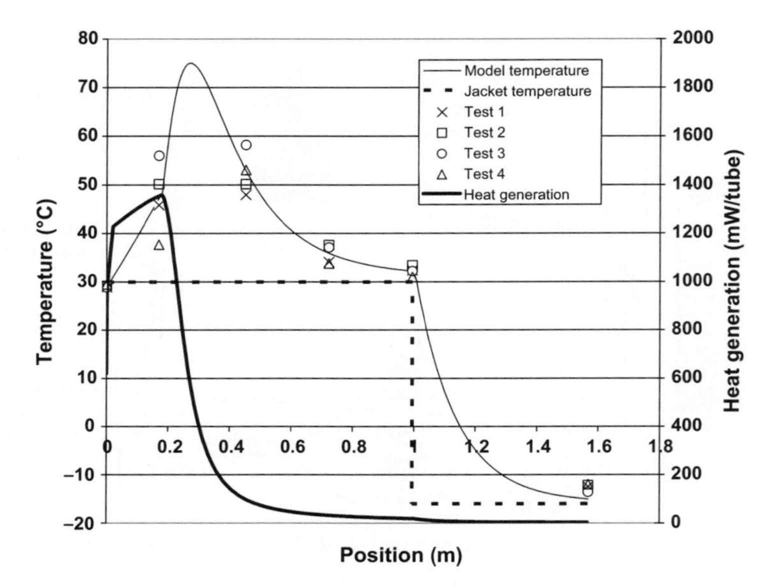

FIGURE 23.15 Test and model results for lab-scale plug flow reactor. Note that the axial velocity is higher in the static mixer due to a lower cross-sectional area relative to the heat exchangers. As a result, the shape of the temperature curve is influenced when plotted against position.

this process. Ideal plug flow was assumed although the Reynolds number was low and suggested laminar flow. The surface temperature of the heat exchanger tubes was assumed constant. Calorimetric studies provided the heat of reaction data. The kinetic model was fitted to the temperature profile with an estimated overall heat transfer coefficient [28]. Using the kinetic parameters, the reaction was simulated as it progressed through the reactor. Several local minima were determined during the model fitting exercise, requiring additional data to refine the model. The simulation results are shown in Figure 23.15, which plots temperature and heat generated versus axial position in the reactor. Although the residence time in the static mixer was only about 1.5 s, the maximum rate of heat generation was experienced there, at a reactor position of 0.18 m. As a consequence, the temperature of the reaction stream increased by 18°C since the static mixer was not jacketed. The static mixer could have been jacketed for additional temperature control, but it was not necessary in this case. The simulated reaction shows a maximum temperature of 75°C at about 0.27 m down the reactor. At this point, the heat generation is equal to the heat removal by the coolant flow. Although the predicted maximum temperature was not measured experimentally due to limited thermocouples in the PFR, the results seem reasonable and are consistent with the proposed reaction mechanism and experimental observations. The reaction reached 88% yield prior to being thermally quenched to less than −10°C for complex stability. Overall, the simulation does an adequate job in modeling the observed behavior and results from the laboratory. Despite limited knowledge of the reaction kinetics prior to modeling, the exercise demonstrates the utility of simulating a nonisothermal plug flow reactor. This type of process knowledge could be used to modify reaction conditions if necessary, but in this case was primarily used to guide the design and operation of pilot plant and commercial manufacturing reactors.

23.8.4 Process Scale-Up to Pilot Plant

With little additional development work, the process described in the previous sections was scaled up in a pilot plant to generate 700 kg of the cyclopropanating reagent solution. The feed tanks were pressurized and an actuated diaphragm valve coupled with a mass flow meter controlled the flow rate of each feed. The process was scaled by maintaining a similar residence time as in the laboratory, and essentially numbering up the laboratory setup by having a larger number of tubes in each heat exchanger, while maintaining the tube diameter. Unlike the batch process, this approach ensured similar heat and mass transfer characteristics and little change in reaction conditions when moving from the laboratory to the pilot plant. The PFR was constructed from a static mixer and five shell and tube heat exchangers, each containing 163 1/8 in. tubes. The total residence time was 65 s, similar to the 51 s residence time utilized in the laboratory. In the pilot plant, a spiral heat exchanger was used to facilitate the thermal quench. The design of the pilot plant system was otherwise similar to the previously described laboratory system.

Table 23.5 shows the different specifications for the laboratory and the pilot plant setups. The operation on pilot scale was similar to the laboratory process with slight differences in measured peak process temperatures most

TABLE 23.5 Laboratory to Pilot Plant PFR Specifications

	Laboratory	Pilot Plant
DME/DEZ (g/min)	70	1358
Diiodomethane/DCM (g/min)	48	940
Total mass flow rate (g/min)	118	2298
Volumetric flow rate (mL/min)	94	1868
PFR residence time (s)	50	65
Fluid velocity (cm/s)	1.7	3.85
Number of tubes (per shell)	19	163
Tube size (ID, cm)	0.254	0.254
Re number	53	120
Reactor length (m)	1.0	2.55
Heat load (W)	240	4700

likely due to differences in heat transfer characteristics. The maximum reaction stream temperature, measured after the first heat exchanger in the PFR, was 56–62°C. As a result, the PFR was operated with a higher jacket temperature relative to the laboratory reactor. After exiting the PFR, the reaction stream was thermally quenched and collected in a jacketed 2000 L reactor for later use. The process ran until all the feed solutions were consumed. The implemented continuous process facilitated production of 700 kg of reagent of consistent quality under reproducible conditions.

23.9 INTEGRATED CONTINUOUS PROCESSING IN PHARMA

While implementing continuous reactions can lead to improved safety and product quality, the increased manufacturing efficiencies experienced in the commodity chemical industry are largely unrealized when the downstream processing is conducted in a semi-batch fashion. This is a result of equipment downtime in such a scenario. Coupling multiple unit operations into a continuous process train has the potential to accelerate introduction of new drugs through more efficient production processes, and decrease the costs of production with smaller facilities, minimization of waste, lower energy consumption, and decreased raw material use [29]. Post-reaction processing in the pharmaceutical industry typically involves extractions, solvent exchanges, crystallizations, and drying and technologies currently exist to perform many of these unit operations continuously. For example, traditional chemical processing equipment such as Podbielniak centrifugal extractors, wiped film evaporators, and continuous crystallizers can perform extractions, solvent exchanges, and crystallizations continuously. These devices offer not only higher throughput but can also increase efficiencies as well, resulting in yield improvements and less waste. Additionally, parallel drying trains can be setup to alternately receive material from upstream continuous process trains. While integrated continuous processing of API is in its infancy, some companies have efforts underway [30] and others are collaborating with academia to develop new technologies for such purposes [2]. An integrated continuous processing plant may become more common in the pharmaceutical industry as technologies develop and as cost pressures rise. Whether these exist as smaller plants dedicated to a single product or modular multiproduct plants remains to be seen. Either way, the evolution will likely be slow given the entrenchment of existing batch processing plants and the real, or perceived, barriers to widespread acceptance of continuous processes.

23.10 BARRIERS TO IMPLEMENTATION OF CONTINUOUS PROCESSING IN PHARMA

The barriers to continuous processing in pharma are largely historical and involve GMP documentation concerns, lack of experience and understanding, and an existing infrastructure designed for batch processing. The pharmaceutical industry has traditionally preferred batch processing largely because of GMP documentation and traceability purposes. By having obviously defined discrete batches or lots of material it is straightforward to meet GMP requirements to document and verify the processing activities, parameters, and raw materials that go into each batch. Since continuous processes do not have well defined and frequent beginning and end points, there is a perception that the definition of a batch is less obvious. This is a misperception however, since the FDA's own guidance states "In the case of continuous production, a batch may correspond to a defined fraction of the production" [31]. Clearly, the FDA is willing to work with industry to adapt traditional GMP approaches to work with continuous processing. The prolonged absence of continuous processing in pharma has led to a dearth of continuous processing know how, both in development and manufacturing. That barrier has largely been reduced, especially on the development side, over the last decade as the regulatory hurdles to continuous processing have lessened and chemical engineers bring their skill sets to bear on the industry. With regulatory acceptance and development capabilities in place, the question then becomes one of economics. Where continuous processes enable improvements in safety or product quality they are currently being utilized on a case-by-case basis. The large investments that the pharmaceutical industry has in existing batch plants represent a significant hurdle to widespread adoption of continuous processing. Furthermore, given the high rates of attrition during development, companies may be hesitant to invest in less familiar processing technologies. Transitioning to a continuous process post-NDA also represents a significant cost and regulatory burden. While many scientists and engineers recognize the benefits of continuous processing, the transition from a batch industry to a continuous industry will likely be very slow.

23.11 SUMMARY

In this chapter, we have discussed how to implement a continuous processing paradigm for organic synthesis reactions. Converting processes from batch to continuous has the advantages of improved intermediate stability, enhanced safety, greater risk management, and enhanced mass and heat transfer. A wide range of reactors, including traditional PFRs and CSTRs, as well as novel microreactors are available for continuous processing. Knowledge of the kinetics and heats of reactions is a prerequisite for the development of a continuous process, and modeling helps to guide reactor choice and identify operating conditions. Continuous operation may not be appropriate for all processes. When looking for development opportunities, the initial focus should be on processes that have safety issues, followed by issues of quality, and lastly economics. The economic considerations are difficult to realize during process development and may not be substantial in manufacturing in the absence of integrated continuous processing. Chemical engineers can take the lead in helping the pharmaceutical industry realize all the benefits of continuous processing.

REFERENCES

1. Higgins S. Are more fine specialty chemicals being moved into continuous process plants? *CHIMICA OFFI/Chemistry Today*, September 1998.
2. Pellek A, Van Arnum P. Continuous processing: moving with or against the manufacturing flow, *PharmTech* 2008;9(32): 52–58.
3. Bogaert-Alvarez RJ, Demena P, Kodersha G, Polomski RE, Soundararajan S, Wang SS. Continuous Processing to control a potentially hazardous process: conversion of aryl 1,1-dimethylpropargyl ethers to 2,2-dimethylchromens (2,2-dimethyl-2*H*-1-benzopyrans). *Org. Process Res. Dev.* 2001;5:636.
4. Panke G, Schwalbe T, Stirner W, Taghavi-Moghadam S, Wille G. A practical approach of continuous processing to high energetic nitration reactions in microreactors. *Synthesis* 2003; 18:2827.
5. Brechtelsbauer C, Ricard F. Reaction engineering evaluation and utilization of static mixer technology for the synthesis of pharmaceuticals. *Org. Process Res. Dev.* 2001;5:646.
6. Roberge DM, Bieler N, Thalmann M. Microreactor technology and continuous processes in the fine chemical and pharmaceutical industries. *Fine Chem.* 2006; 14.
7. LaPorte TL, Wang C. Continuous processes for the production of pharmaceutical intermediates and active pharmaceutical ingredients. *Curr. Opin. Drug Discov. Dev.* 2007;10(6): 738–745.
8. Thakur RK, Vial C, Nigam KDP, Nauman EB, Djelveh G. Static mixers in the process industries—a review. *Trans IChemE* 2003;81(A):787.
9. Stankiewicz A, Drinkenburg AAH. Process intensification: history, philosophy, principles. *Chem. Ind. (Dekker)* 2004; 98:1–32.
10. Ritter SK. *A New Spin on Reactor Design—Refined Rotor-Stator System has the Potential to Boost Chemical Process Intensification Efforts*, C&EN, July 29, 2002.
11. Brechtelsbauer C, Lewis N, Oxley P, Richard F. Evaluation of a spinning disc reactor for continuous processing. *Org. Process Res. Dev.* 2001;5:65–68.
12. Micronit Microfluidics BV http://www.micronit.com, 2009.
13. Ehrfeld W, Hessel V, Löwe H. *Microreactors, New Technology for Modern Chemistry*, 1st edition, Wiley-VCH, Weinheim 2000, Chapter 3.
14. Dapremont O, Zeagler L, DuBay W. Reducing costs through continuous processing. *Sp2*, June 22–24, 2007.
15. *Reuters: AMPAC Fine Chemicals Announces the Inauguration of the New cGMP Continuous Processing Development Facility*, March 9, 2009.
16. Roberge DM, Ducry L, Bieler N, Cretton P, Zimmermann B. Microreactor technology: a revolution for the fine chemical and pharmaceutical industries? *Chem. Eng. Technol.* 2005;28(3): 318–323.
17. Günther M, Scheider S, Wagner J, Gorges R, Henkel T, Kielpinski M, Bierbaum R, Köhler JM. Characterisation of residence time and residence time distribution in chip reactors with modular arrangements by integrated optical detection. *Chem. Eng. J.* 2004;101:373–378.
18. LaPorte TL, Hamedi M, DePue JS, Shen L, Watson D, Hsieh D. Development and scale-up of three consecutive continuous reactions for production of 6-hydroxybuspirone. *Org. Process Res. Dev.* 2008;12:956–966.
19. Choe J, Kim Y, Song KH. Continuous synthesis of an intermediate of quinolone antibiotic drug using static mixers. *Org. Process Res. Dev.* 2003;7(2):187–190.
20. Schwalbe T, Kursawe A, Sommer J. Application report on operating cellular process chemistry plants in fine chemical and contract manufacturing industries. *Chem. Eng. Technol.* 2005; 28(4):408–419.
21. Thayer A. Handle with care. *Chem. Eng. News* 2009, March 16.
22. Lonza News Release: *Lonza secures important manufacturing contract using its proprietary microreactor technology,* May 26, 2009.
23. US Food and Drug Administration, Center for Drug Evaluation and Research: *Guidance for Industry PAT—A Framework for Innovative Pharmaceutical Manufacturing and Quality Assurance,* September 2004.
24. Lobbecke S, Ferstl W, Panic S, Turcke T. Concepts for modularization and automation of microreaction technology. *Chem. Eng. Technol.* 2005;28(4):484–493.
25. Ferstl W, Klahn T, Schweikert W, Billeb G, Schwarzer M, Loebbecke S. Inline analysis in microreaction technology: a suitable tool for process screening and optimization. *Chem. Eng. Technol.* 2007;30(3):370–378.
26. Charette A, Marcoux JF. Spectroscopic characterization of (iodomethyl)zinc reagents involved in stereoselective

reactions: spectroscopic evidence that $IZnCH_2I$ is not An $(CH_2I)_2 + ZnI_2$ in the presence of an ether. *J. Am. Chem. Soc.* 1996;118:4539–4549.

27. Davies S, Ling K, Roberts P, Russell A, Thomson J. Diastereoselective Simmons–Smith cyclopropanation of allylic amines and carbamates. *Chem. Commun.* 2007;4029–4031.
28. Pitts DR, Sissom LE. *Heat Transfer*, McGraw-Hill, 1977.
29. Mullin R. Cover story—breaking down barriers: drugmakers are paving the way to more streamlined manufacturing via culture change in R&D. *Chem. Eng. News* 2007;85(04): 11–17.
30. Berry M. After a Century of Batch Manufacturing API, What Does Continuous Processing Offer? PhRMA API Workshop, 2009.
31. ICH Harmonised Tripartite Guideline Q7: *Good Manufacturing Practice Guideline for Active Pharmaceutical Ingredients*, 2000.

24

DRUG SOLUBILITY AND REACTION THERMODYNAMICS

KARIN WICHMANN

COSMOlogic GmbH & Co. KG, Leverkusen, Germany

ANDREAS KLAMT

COSMOlogic GmbH & Co. KG, Leverkusen, Germany, and Institute of Physical and Theoretical Chemistry, University of Regensburg, Regensburg, Germany

24.1 INTRODUCTION

24.1.1 Methods for Compounds in Solution

There is a variety of computational methods for the treatment of compounds in solution. The scope of this chapter is not to give an overview of them, but to concentrate on applications of COSMO-RS, a young and very efficient method for the *a priori* prediction of thermophysical data of liquids. COSMO-RS combines unimolecular quantum chemical calculations that provide the necessary information for the evaluation of molecular interaction in the fluid phase, with a very fast and accurate statistical thermodynamic procedure. It has established itself as an alternative to structure-based group contribution methods (GCMs) on the one hand and to force field-based simulation methods on the other hand. Because of its special approach, COSMO-RS is a generally applicable method for compounds in solution. It has been applied successfully in such diverse areas as solvent screening, partitioning behavior, liquid–liquid and vapor–liquid equilibria, and ADME property prediction, and for such diverse compound types as drugs, pesticides, common organic compounds, halocarbons, and ionic liquids. COSMO-RS is used in chemical, pharmaceutical, agrochemical and petrochemical industry.

In this contribution, two application fields important in drug development and drug production will be considered: solubility prediction and prediction of free energy of reaction in solution. Solubility prediction methods are important during the drug design and development process, because in the early drug design phase compounds are often only virtually considered by computational drug design methods, or the synthesized amount of substance is insufficient for experiments. In both cases the only tools for the selection of promising drug candidates with adequate solubility are computational methods that predict the solubility with sufficient accuracy just from the chemical structure of the compound. A method requiring experimental data for solubility prediction is unfeasible in this situation, since such data will not be available.

Prediction of thermodynamic properties of compounds in solution is also important in industrial process development. Here, specifically reaction energies and equilibrium constants of reaction in solution are of particular interest when a new process is developed or alternative pathways for existing processes are explored. The Gibbs free energy of reaction varies with the choice of the solvent or solvent mixture, and hence the chosen solvent system can strongly influence the process in solution. Generally, experimental data for the equilibrium constant or the free energy of reaction in solution are rare, but are available relatively straightforward from a computational approach. Prediction of thermochemical data like heat of reaction and heat of vaporization furthermore helps designing a chemical process such that process hazards can be prevented.

24.1.2 COSMO

In conventional quantum chemistry, molecules are treated as isolated particles at a temperature of 0K. In physical reality

Chemical Engineering in the Pharmaceutical Industry: R&D to Manufacturing Edited by David J. am Ende

however, the major part of reactions takes place in solution and at higher temperatures. Since direct treatment of a large number of molecules is computationally very demanding, solvent effects are often treated indirectly by continuum solvation models, where the solute is embedded in a dielectric continuum and the solvent is represented by a mean interaction with a surrounding dielectric medium. The interaction of the solute with such a dielectric solvent is taken into account in the quantum chemical calculation by polarization charges that arise from the dielectric boundary condition.

The "COnductor-like Screening MOdel" (COSMO) is an efficient variant of dielectric continuum solvation methods [1]. In quantum chemical COSMO calculations the solute molecules are calculated in a virtual scaled conductor environment, that is, the scaled boundary condition of a conducting medium is used, where the molecule is ideally screened, and not the exact dielectric boundary condition. In such a conducting environment the solute molecule induces a polarization charge density σ on the interface between the molecule and the conductor, that is, on the molecular surface. These charges act back on the solute and generate a more polarized electron density than in vacuum. During the quantum chemical self-consistency algorithm, the solute molecule is thus converged to its energetically optimal state in a conductor with respect to electron density. Due to the analytic gradients available for the COSMO energy contributions, the molecular geometry can be optimized using conventional methods for calculations in vacuum. The quantum chemical calculation has to be performed once for each molecule of interest.

24.1.3 COSMO-RS

As discussed in more detail elsewhere the simple dielectric continuum models suffer from a number of insufficiencies [2,3]. The polarization charge density σ resulting from unscaled COSMO calculations (also called screening charge density σ), which is a good local descriptor of the molecular surface polarity, is used to extent the model toward "Real Solvents" (COSMO-RS).

In COSMO-RS, a liquid is considered to be an ensemble of closely packed ideally screened molecules, as shown in Figure 24.1. In this figure, each piece of surface has one direct contact partner, but is still separated from its partner by a thin film of conductor. Since the conducting medium that was assumed to surround the molecules in the COSMO calculation is not existent in reality, the energy difference between the pairwise contacts and the ideally screened situation has to be defined as a local electrostatic interaction energy that results from the removal of the conductor film between the molecules. Considering a contact on a region of molecular surface of area a_{eff} (effective contact area), and considering that the two contacting pieces of molecular surface have average ideal screening charge densities σ and σ' in the conductor, it is possible to calculate this interaction energy as the energy that is necessary to remove the residual screening charge density $\sigma + \sigma'$ from the contact. In the special case of $\sigma = -\sigma'$, the contact is an "ideal electrostatic contact" and the interaction energy is zero, because the two molecules screen each other as well as the conductor did. In the general case however $\sigma + \sigma'$ does not vanish and the arising electrostatic interaction energy is

$$E_{\text{misfit}}(\sigma,\sigma') = a_{\text{eff}} e_{\text{misfit}}(\sigma,\sigma') = a_{\text{eff}} \frac{\alpha'}{2} (\sigma+\sigma')^2, \quad (24.1)$$

where $e_{\text{misfit}}(\sigma,\sigma')$ is the misfit energy density on the contact surface and α' is a general constant that can be calculated approximately, but in COSMO-RS is fitted to experimental data as fine-tuning. The misfit term (equation 24.1) subsumes the polarization response of the molecules to the electrostatic misfit quite well [4, 5].

Hydrogen bonding interactions are to some extent already covered by the description of electrostatic interactions, but we still have to parameterize the additional hydrogen bonding energy resulting from interpenetration of the atomic electron densities in some reasonable way. This energy should only be relevant if two sufficiently polar pieces of surface of opposite polarity are in contact, and it should be the more important, the more polar both surface pieces are. Taking the screening charge density σ as a local

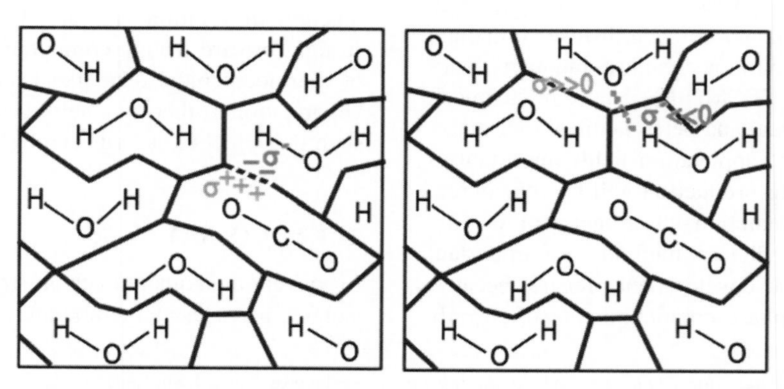

FIGURE 24.1 Schematic illustration of contacting molecular cavities and contact interactions.

measure of polarity, the following function realizes such behavior:

$$E_{hb}(\sigma, \sigma') = a_{eff} e_{hb}(\sigma, \sigma') = a_{eff} c_{hb} \min\{0, \min(0, \sigma_{don} + \sigma_{hb}) \max(0, \sigma_{acc} - \sigma_{hb})\} \quad (24.2)$$

with $\sigma_{don} = \min(\sigma, \sigma')$ and $\sigma_{acc} = \max(\sigma, \sigma')$. General parameters are σ_{hb}, the threshold for hydrogen bonding, and c_{hb}, the coefficient for the hydrogen bond strength. Both parameters have to be adjusted to experimental data. With equation 24.2, the hydrogen bond interaction energy is zero, unless the more negative of the two screening charge densities is less than the threshold $-\sigma_{hb}$, and the more positive exceeds σ_{hb}. Because positive molecular regions have negative screening charge, the negative σ now is the donor part of the hydrogen bond and the positive is the acceptor. In this case, the hydrogen bonding energy is proportional to the product of the excess screening charge densities $(\sigma_{don} + \sigma_{hb})(\sigma_{acc} - \sigma_{hb})$.

van der Waals interactions are described by element-specific parameters τ in COSMO-RS. The τ parameters have to be fitted to experimental data. Then, the vdW energy gain of a molecule X during the transfer from the gas phase to any solvent is given by

$$E^X_{vdW} = \sum_{\alpha \in X} a^X_\alpha \tau_{vdW}(e(\alpha)) \quad (24.3)$$

The vdW term is spatially nonspecific. Because E_{vdW} is independent of any neighboring relations, it is not really an interaction energy, but may be considered as an additional contribution to the energy of the reference state in solution. Currently nine of the vdW parameters (for elements H, C, N, O, F, S, Cl, Br, and I) have been optimized. For the majority of the remaining elements reasonable estimates are available. Nonadditive vdW corrections are used for a few element pairs, but they are of minor importance for the topics of this contribution.

The transition from microscopic surface interaction energies to macroscopic thermodynamic properties of a liquid is possible via a statistical thermodynamics procedure. The exact solution of the thermodynamic problem would require sampling of all different arrangements of all molecules of the systems, weighting the contribution of each arrangement by its Boltzmann factor. This direct approach, which is used in the molecular dynamics and Monte Carlo type methods, is very time-consuming and requires compromises regarding sampling and regarding the accuracy of the energy evaluations. COSMO-RS follows a different concept. The basic approximation is that the ensemble of interacting molecules may be replaced by the corresponding ensemble of independent, pairwise, interacting surface pieces. This approximation implies the neglect of any neighborhood information of surface pieces on the molecular surface and the loss of steric information. The advantage of this approximation is the extreme reduction of the complexity of the problem, which allows for a fast and exact solution. It should be noted that GCMs as UNIFAC are also based on the assumption of independent pairwise interacting surfaces.

Since the screening charge density σ is the only descriptor determining the interaction energy terms in equations 24.1 and 24.2, the ensemble of surface pieces characterizing an ensemble S is sufficiently described by its composition with respect to σ. For this purpose we introduce the molecular σ-profile $p^X(\sigma)$, which is a histogram of the screening charge densities σ on the surface of a molecule X (Figure 24.2). The σ-profile can easily be derived from the COSMO files produced as output of the quantum chemical COSMO calculation for molecule X, applying a local averaging algorithm in order to take into account that only screening charge densities σ averaged over an effective contact area are of physical meaning in COSMO-RS [5].

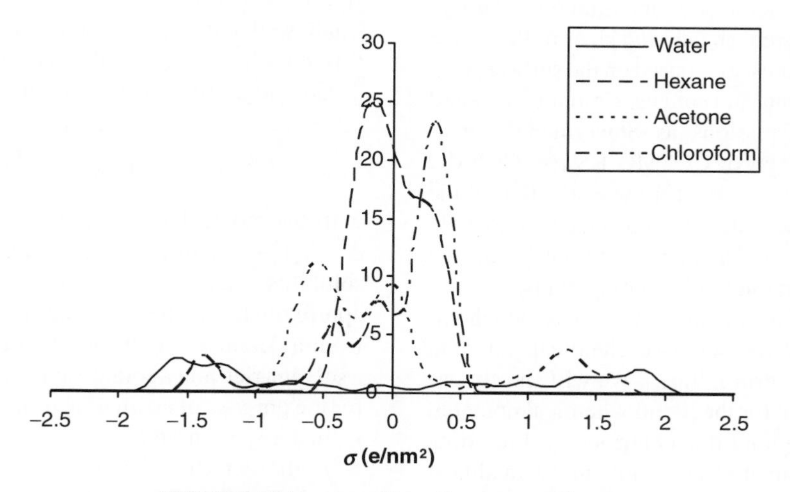

FIGURE 24.2 σ-profiles of common solvents.

The σ-profile for the entire solvent of interest S, which might be a mixture of several compounds, $p_S(\sigma)$, is given by the weighted, surface area normalized sum of the σ-profiles of the components X_i:

$$p'_S(\sigma) = p_S(\sigma)/A_S = \sum_{i \in S} x^i p^{X_i}(\sigma) / \sum_{i \in S} x^i A^{X_i} \qquad (24.4)$$

where A^{X_i} is the COSMO surface of a compound X_i in the system.

Under the condition that there is no free surface in the bulk of the liquid, that is, each piece of molecular surface has a direct contact partner, the statistical thermodynamics of the system can be solved using the exact equation:

$$\mu_S(\sigma) = -RT \ln\left\{ \int p'_S(\sigma') \exp\left(\frac{\mu_S(\sigma') - a_{\text{eff}} e(\sigma, \sigma')}{RT} \right) d\sigma' \right\} \qquad (24.5)$$

In this equation, $\mu_S(\sigma)$ is the chemical potential of an average molecular contact segment of size a_{eff} in the ensemble S at temperature T, and $e(\sigma,\sigma')$ is the interaction energy functional $e(\sigma,\sigma') = e_{\text{misfit}}(\sigma,\sigma') + e_{\text{hb}}(\sigma,\sigma')$. Since $\mu_S(\sigma)$ appears on both sides of equation 24.5, it must be solved by iteration, starting with $\mu_S(\sigma') = 0$ on the right-hand side. Fortunately, the solution converges rapidly and $\mu_S(\sigma)$ can be computed up to numerical precision within milliseconds on a personal computer. For a formal derivation of equation 24.5, we refer to Ref. 4.

Now it is straightforward to define the chemical potential of a solute X in the ensemble S by

$$\mu_S^X = \mu_{\text{res},S}^X + \mu_{\text{comb},S}^X = a_{\text{eff}}^{-1} \int p^X(\sigma) \mu_S(\sigma) d\sigma + \mu_{\text{comb},S}^X \qquad (24.6)$$

where the residual part, that is, the part resulting from the interactions of the surfaces in the liquid, is given by the surface integral of function $\mu_S(\sigma)$ over the solute surface, which is expressed using the σ-profile of the solute in equation 24.6. The second part is the combinatorial contribution, which arises from the different shapes and sizes of the solute and solvent molecules. Expressions based on the surface areas and volume ratios of solvents and solutes, similar to standard chemical engineering expressions as Staverman-Guggenheim, are used in the context of COSMO-RS [6]. COSMO surface areas and volumes are used for the evaluation of the combinatorial term. Hence, equation 24.6 can be completely evaluated based on the information resulting from the COSMO calculations of the individual compounds.

The chemical potential of equation 24.6 is a pseudochemical potential [7], that is, the standard chemical potential without the concentration term $RT \ln x_i$. We will shortly use the term chemical potential for the pseudochemical potential from equation 24.6 throughout this contribution. Providing the chemical potential of an arbitrary compound X in almost arbitrary solvents and mixtures as a function of temperature and concentration, equation 24.6 allows for the prediction of almost all thermodynamic properties of compounds or mixtures, such as activity coefficients, partition coefficients, or solubility, as shown in the flowchart for a COSMO-RS property prediction in Figure 24.3.

As mentioned above, the COSMO-RS method depends on a small number of adjustable parameters. Some of the parameters are predetermined from physics, while others are determined from selected properties of mixtures. The parameters are not specific to functional groups or types of molecule. As a result, COSMO-RS is the least parameterized of all quantitative methods for the prediction of chemical properties in the liquid phase [8].

24.1.4 Treatment of Conformers in COSMO-RS

Many molecules can adopt more than one conformation, and different conformers of one molecule can have different σ-profiles. The chemical potentials of the individual conformers and hence the conformer distribution as well as the chemical potential of the compound represented by an ensemble of conformers depend on the composition of the system and the temperature. Thus, it is essential for property prediction with COSMO-RS to take conformers with different σ-profiles into account, each described by individual quantum chemical COSMO calculations. The relative contributions of the conformers are determined by an iterative procedure using the Boltzmann-weight of the free energies of the conformers in the liquid phase. This results in a thermodynamically fully consistent treatment of multiple molecular conformations.

24.2 SOLUBILITY PREDICTION WITH COSMO-RS

For the calculation of the solubility S_S^X of a liquid compound X in a solvent S we require the chemical potentials of X in S and in its pure liquid state, μ_S^X and μ_X^X. If S_S^X is sufficiently small, so that the solvent behavior of the X-saturated solvent S is not significantly influenced by the solute X, then the decadic logarithm of the solubility is given by

$$\log S_S^X = \log\left(\frac{\text{MW}^X \rho_S}{\text{MW}_S} \right) - \frac{\ln(10)}{kT} \Delta_S^X \qquad (24.7)$$

with the molecular weight MW, the solvent density ρ, and $\Delta_S^X = \mu_S^X - \mu_X^X$. In the case of high solubility (typically for solubility greater than 10 wt%), equation 24.7 becomes approximate and the true solubility would have to be derived from a detailed search for a thermodynamic equilibrium of a solvent-rich and a solute-rich phase. But, in general, at least for the purpose of estimating drug solubility, equation 24.7 is sufficiently accurate.

If the zeroth order $S_S^{X_0}$ as initially provided by equation 24.7, using infinite dilution of X in S, is resubstituted

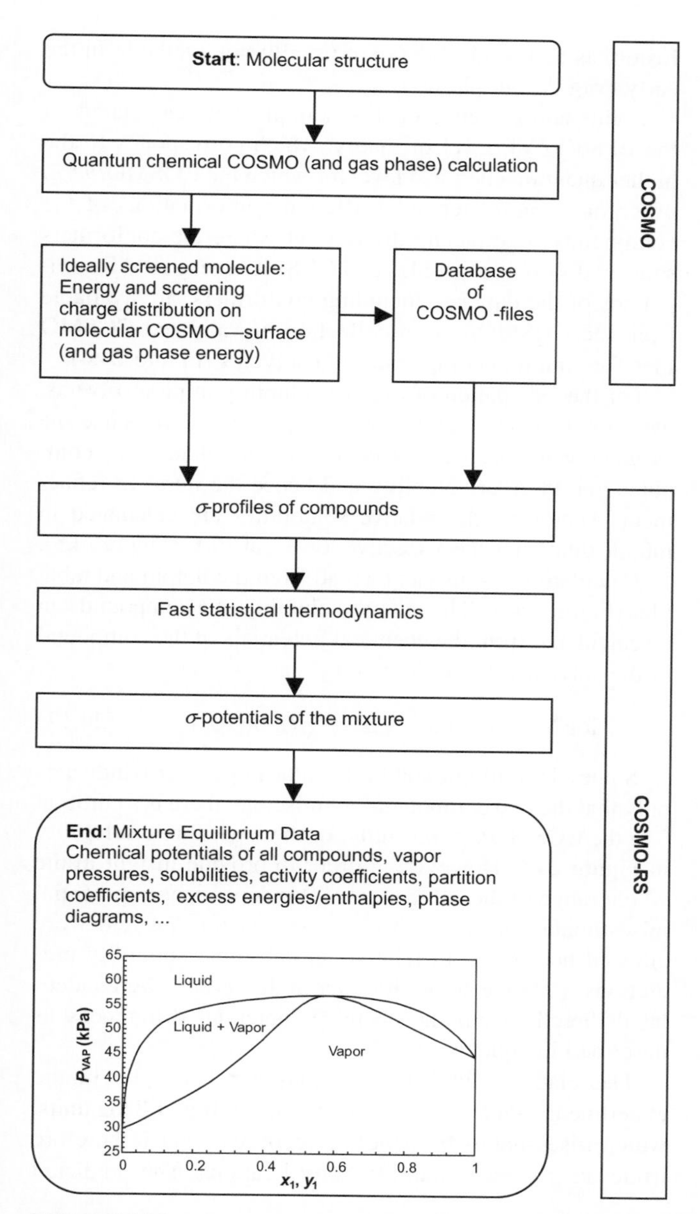

FIGURE 24.3 Flowchart of a property prediction procedure with COSMO and COSMO-RS.

into the solubility calculation via $\Delta_S^{X_1} = \mu_{S(X_0)}^X - \mu_X^X$ a better approximation for S_S^X is achieved. In other words, the solute chemical potential μ_S^X is computed for the solvent–solute mixture with the finite mole fraction of X in S that was predicted by the zeroth order S_S^X. Then, using equation 24.7 with the new $\mu_{S(X_0)}^X$ and the resulting values, an improved solubility $S_S^{X_1}$ is computed. Iterating this process to convergence an iterative solubility can be achieved, which is also implemented in our COSMO-RS program COSMO*therm* and allows for the accurate prediction of solubility values even for cases of high solubility (solubility up to 50 wt%). Thus, except for rare cases of very high solubility, a complicated search for a multiphase thermodynamic equilibrium of a solvent-rich and a solute-rich phase can be avoided, but instead equation 24.7 and its iterative refinement can be used.

Drugs are mostly solid at room temperature. Because the solid state of a compound X is related to its liquid state by the free energy difference ΔG_{fus}^X, which is negative in the case of solids, a more general expression for solubility reads

$$\log S_S^X = \log\left(\frac{\text{MW}^X \rho_S}{\text{MW}_S}\right) + \frac{\ln(10)}{kT}\left[-\Delta_S^X + \min(0, \Delta G_{\text{fus}}^X)\right] \tag{24.8}$$

For liquid compounds, ΔG_{fus}^X is positive and equation 24.8 reduces to equation 24.7.

If melting point temperature T_{melt} and heat of fusion ΔH_{fus} or entropy of fusion ΔS_{fus} are known experimentally for a solid compound, the free energy of fusion ΔG_{fus} in equation 24.8 can be estimated from

$$\Delta G_{\text{fus}}(T) = -\Delta H_{\text{fus}}\left(1 - \frac{T}{T_{\text{melt}}}\right) \tag{24.9}$$

or

$$\Delta G_{\text{fus}}(T) = -\Delta S_{\text{fus}}(T_{\text{melt}} - T) \tag{24.10}$$

Equations 24.9 and 24.10 can be complemented by an additional temperature-dependent term using the heat capacity of fusion ΔCp_{fus} in order to obtain good absolute predictions, but data for ΔCp_{fus} are rarely available from experiment.

The free energy of fusion of new compounds is often not known, because experimental measurements can be cumbersome and substance may be scarce. Computational prediction of ΔG_{fus}^X requires evaluation of the free energy of a molecule of compound X in its crystal, that is, the crystal structure needs to be known or predicted. In general however, crystal structure prediction for drugs has to be considered as an unsolved problem. Thus, there is no viable way to a fundamental model. As an alternative, a QSPR approximation for ΔG_{fus} can be used in COSMO*therm*, which is based on a few rather obvious factors that should influence crystallization. Larger molecules should have larger ΔG_{fus} than smaller ones, compounds with more polarity and hydrogen bonding ability should have larger ΔG_{fus} than less polar ones, and also rigidity should give rise to larger ΔG_{fus}. We found that a good regression equation for ΔG_{fus}^X can be achieved by a combination of the descriptors V^X, the cavity volume from the COSMO calculation as size descriptor, $N_{\text{ring atom}}^X$, the number of ring atoms in X as a descriptor of the compounds rigidity, and μ_W^X, the compounds chemical potential in water as a combined measure of polarity and hydrogen bonding: [8]

$$\Delta G_{\text{fus}}^X = c_0 + c_1\mu_W^X + c_2 N_{\text{ring atom}}^X + c_3 V^X \tag{24.11}$$

24.2.1 Relative Solubility and Solubility Screening

The computational prediction of the relative solubility of a drug candidate in a variety of solvents with COSMO-RS is straightforward and can be done without wasting any of the substance in this step. The required DFT/COSMO calculations can be done even before the compound comes to the development laboratory, and a COSMO-RS solubility screening can be already completed when the work in the development department starts.

Experimental data for melting point and free energy of fusion can often be obtained through differential scanning calorimetry. If melting point temperature and heat of fusion or entropy of fusion are known for the compound in question, the free energy of fusion $\Delta G_{\rm fus}$ can be calculated according to equations 24.9 or 24.10 and a solubility screening for absolute solubilities can be done. If data for $\Delta G_{\rm fus}$ are not known, an estimated $\Delta G_{\rm fus}$ may be used, either from the QSPR model implemented in COSMO*therm* or from an external model.

EXAMPLE 24.1 RELATIVE AND ABSOLUTE SOLUBILITY OF ACETAMINOPHEN IN PURE SOLVENTS AT 30°C

Experimental data for the solubility of acetaminophen in pure solvents were reported by Granberg and Rasmuson [9]. We use this data set to validate the calculated acetaminophen solubilities. Furthermore, a melting point temperature of $T_{\rm melt} = 441.2$K, a heat of fusion of $\Delta H_{\rm fus} = 26.0$ kJ/mol and an entropy of fusion of $\Delta S_{\rm fus} = 59.0$ J/mol have been reported for acetaminophen [10]. These data will be used to compute absolute solubility predictions for acetaminophen in the solvent data set.

With the case study of acetaminophen solubility we want to show first the prediction of relative solubility. For relative solubility calculations, we do not make use of any experimental data like melting point temperature or enthalpy of fusion, as that kind of data are usually not available in the early drug design phase.

Solute and solvents for this example were calculated on the BP86/TZVP level of theory, which corresponds to the higher quantum chemical level for which the COSMO*therm* program is parameterized. Different conformations of the compounds were taken into account where the conformers showed different σ-profiles and COSMO energies. All compounds of the data set, including conformers, are available from the COSMO*base*, a collection of validated COSMO files for common compounds and solvents (Figure 24.4).

For the calculation of relative solubility in pure solvents, we do not use the iterative refinement procedure, since the assumed value of $\Delta G_{\rm fus} = 0$ kcal/mol will influence the computed zeroth-order solubility and hence the iterative refinement. Therefore, the relative solubilities are calculated in infinite dilution in the respective solvent at 30°C (Figure 24.5).

Calculation results can be read from the output and table files (Figure 24.6). The relative solubility of a compound can be calculated from the chemical potentials of the compound in the solvent $\mu_i^{(j)}$ and in its pure state $\mu_i^{(P)}$ as

$$\log S_{\rm rel}(x) = (\mu_i^{(P)} - \mu_i^{(j)})/(RT\ln(10)) \qquad (24.12)$$

Since a logarithmic solubility value larger than 0 indicates only that the two compounds are miscible, there is a cutoff at 0 in the COSMO*therm* results for the logarithmic solubility in Figure 24.6. However, in order to provide insight in the whole range of the solvent data set independent of potential misestimates of $\Delta G_{\rm fus}$, positive values for $\log S_{\rm rel}(x)$ are allowed here for both relative and absolute solubility predictions. Relative solubility data in Table 24.1 were calculated directly from the chemical potential differences as described in equation 24.12.

The relative solubility predictions correlate well with the experimental data, revealing an overall shift of 1.8 log units, which arises mainly from the neglect of $\Delta G_{\rm fus}$ and is therefore irrelevant for real solubility considerations. The predicted

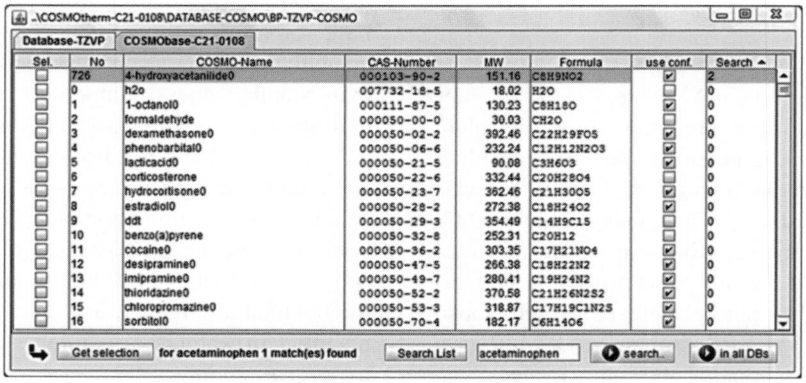

..\COSMOtherm-C21-0108\DATABASE-COSMO\BP-TZVP-COSMO

Database-TZVP | COSMObase-C21-0108

Sel.	No	COSMO-Name	CAS-Number	MW	Formula	use conf.	Search
☐	726	4-hydroxyacetanilide0	000103-90-2	151.16	C8H9NO2	☑	2
☐	0	h2o	007732-18-5	18.02	H2O	☐	0
☐	1	1-octanol0	000111-87-5	130.23	C8H18O	☑	0
☐	2	formaldehyde	000050-00-0	30.03	CH2O	☐	0
☐	3	dexamethasone0	000050-02-2	392.46	C22H29FO5	☑	0
☐	4	phenobarbital0	000050-06-6	232.24	C12H12N2O3	☑	0
☐	5	lacticacid0	000050-21-5	90.08	C3H6O3	☑	0
☐	6	corticosterone	000050-22-6	332.44	C20H28O4	☐	0
☐	7	hydrocortisone0	000050-23-7	362.46	C21H30O5	☑	0
☐	8	estradiol0	000050-28-2	272.38	C18H24O2	☑	0
☐	9	ddt	000050-29-3	354.49	C14H9Cl5	☐	0
☐	10	benzo(a)pyrene	000050-32-8	252.31	C20H12	☐	0
☐	11	cocaine0	000050-36-2	303.35	C17H21NO4	☑	0
☐	12	desipramine0	000050-47-5	266.38	C18H22N2	☑	0
☐	13	imipramine0	000050-49-7	280.41	C19H24N2	☑	0
☐	14	thioridazine0	000050-52-2	370.58	C21H26N2S2	☑	0
☐	15	chloropromazine0	000050-53-3	318.87	C17H19ClN2S	☑	0
☐	16	sorbitol0	000050-70-4	182.17	C6H14O6	☑	0

Get selection | for acetaminophen 1 match(es) found | Search List | acetaminophen | search.. | in all DBs

FIGURE 24.4 Database view in COSMO*therm*X. Databases can be searched and columns are sortable.

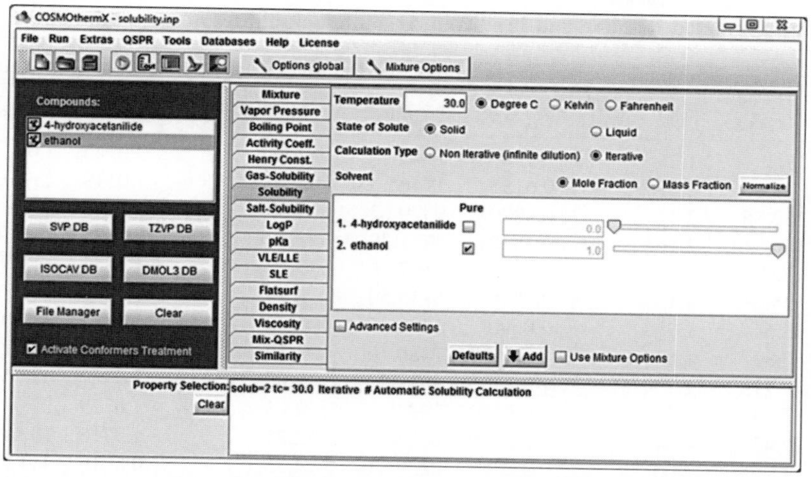

FIGURE 24.5 Overview of COSMO*therm*X with compound list, solubility panel and input section. When the solute state is set to liquid, $\Delta G_{fus} = 0$ is used, while with solid solute state, given or estimated values for ΔG_{fus} are used. The iterative refinement can also be set in the solubility panel. In the solvent frame, the solvent composition is set to pure for the respective solvent. Pictured here are settings for absolute solubility using the iterative refinement procedure.

relative solubility data apparently fall into two groups, as can be seen from Figure 24.7. Solubility data in alcoholic solvents are grouped together on a rather straight line with a slope of 0.7 and a relative shift of 1.4 log units compared to the experimental data, while the second, more scattered group has a larger shift ($\geq$2 log units), but a similar slope. There is one severe outlier in the data set, carbon tetrachloride, where the acetaminophen solubility is, in contrast to the trend observed in the other solvents, underestimated by 1.5 log units. Since the experimental value for the solubility in carbon tetrachloride comes from a single measurement, and the other solvents appear to be described reasonably by the model prediction apart from a general overestimation due to the missing free energy of fusion term, we tend to consider this experimental value as questionable.

Using the experimental data for T_{melt} and ΔH_{fus}, absolute solubilities of acetaminophen in the solvent data set are also computed. The absolute predictions are in good quantitative agreement with experimental data, as shown in Figure 24.8. Of the 26 solvents, 4 are predicted with a positive $\log S(x)$:

```
 Solubility at T =  303.15 K  in compound    2 (h2o) -  energies are in
kcal/mol volume is in A^3  - Solvent Density =  995.363 [g/l]

 Nr Compound              log10(x_solub)     mu(self)       mu(solv)       DG_fus
  1 4-hydroxyacetanilide  -1.68194563  -5.05034328  -2.71690337  0.00000000
  2 h2o                    0.00000000  -2.90487525  -2.90487525  0.00000000
...
 Solubility at T =  303.15 K  in compound   25 (chcl3) -  energies are in
kcal/mol volume is in A^3  - Solvent Density = 1478.286 [g/l]

 Nr Compound              log10(x_solub)     mu(self)       mu(solv)       DG_fus
  1 4-hydroxyacetanilide  -1.88836005  -5.05034328  -2.43053523  0.00000000
 25 chcl3                  0.00000000  -5.29052995  -5.29052995  0.00000000
...
 Solubility at T =  303.15 K  in compound   26 (ccl4) -  energies are in
kcal/mol volume is in A^3  - Solvent Density = 1572.230 [g/l]

 Nr Compound              log10(x_solub)     mu(self)       mu(solv)       DG_fus
  1 4-hydroxyacetanilide  -4.51493557  -5.05034328   1.21343282  0.00000000
 26 ccl4                   0.00000000  -7.50124553  -7.50124553  0.00000000
...
 Solubility at T =  303.15 K  in compound   27 (toluene) -  energies are in
kcal/mol volume is in A^3  - Solvent Density =  860.666 [g/l]

 Nr Compound              log10(x_solub)     mu(self)       mu(solv)       DG_fus
  1 4-hydroxyacetanilide  -3.05977298  -5.05034328  -0.80538062  0.00000000
 27 toluene                0.00000000  -4.62709909  -4.62709909  0.00000000
```

FIGURE 24.6 Excerpt from the COSMO*therm* table file for the solubility calculation of acetaminophen in pure solvents. log 10(x_solub) indicates the logarithmic solubility in mole fractions.

TABLE 24.1 Experimental, Predicted Relative and Predicted Absolute Solubilities of Acetaminophen in Pure Solvents

Solvent	Experimental		Predicted Relative			Predicted Absolute			Error
	c_S (g/kg)	log S (mg/g)	log $S(x)$	S (mg/g)	log S (mg/g)	log S (x)	S (mg/g)	log S (mg/g)	
Water	17.39	1.24	−1.6819	174.53	2.24	−3.0745	7.07	0.85	−0.39
Methanol	371.61	2.57	0.6155	19465.33	4.29	−1.0119	459.02	2.66	0.09
Ethanol	232.75	2.37	0.4809	9929.62	4.00	−1.0687	280.11	2.45	0.08
1,2-Ethanediol	144.3	2.16	0.1829	3711.29	3.57	−1.2535	135.86	2.13	−0.03
1-Propanol	132.77	2.12	0.2182	4157.36	3.62	−1.2441	143.37	2.16	0.03
2-Propanol	135.01	2.13	0.3858	6115.50	3.79	−1.1268	187.86	2.27	0.14
1-Butanol	93.64	1.97	0.0690	2390.35	3.38	−1.3675	87.50	1.94	−0.03
1-Pentanol	67.82	1.83	−0.0639	1480.10	3.17	−1.4866	55.92	1.75	−0.08
1-Hexanol	49.71	1.70	−0.1756	987.32	2.99	−1.5932	37.75	1.58	−0.12
1-Heptanol	37.43	1.57	−0.2630	709.92	2.85	−1.6779	27.31	1.44	−0.14
1-Octanol	27.47	1.44	−0.3589	507.95	2.71	−1.7716	19.64	1.29	−0.15
Acetone	111.65	2.05	0.9328	22297.15	4.35	−0.8016	411.01	2.61	0.57
2-Butanone	69.99	1.85	0.6508	9380.65	3.97	−0.9153	254.79	2.41	0.56
4-Methyl-2-pentanone	17.81	1.25	0.1126	1956.04	3.29	−1.2844	78.41	1.89	0.64
Tetrahydrofuran	155.37	2.19	1.6842	101306.53	5.01	0.0527	2366.72	3.37	1.18
1,4-Dioxane	17.08	1.23	1.0644	19898.20	4.30	−0.7332	317.14	2.50	1.27
Ethyl acetate	10.73	1.03	0.0872	2097.21	3.32	−1.2415	98.38	1.99	0.96
Acetonitrile	32.83	1.52	0.1079	4720.92	3.67	−1.2689	198.25	2.30	0.78
Diethylamine	1316.9	3.12	3.5201	6845852.30	6.84	1.4457	57683.45	4.76	1.64
N,N-Dimethylformamide	1012.02	3.01	2.2018	276105.51	5.44	0.4567	4965.64	3.70	0.69
Dimethyl sulfoxide	1132.56	3.05	3.3062	3915389.90	6.59	0.2699	3601.53	3.56	0.50
Acetic acid	82.72	1.92	0.3232	5298.62	3.72	−1.1632	172.87	2.24	0.32
Dichloromethane	0.32	−0.49	−1.8354	26.00	1.41	−3.1820	1.17	0.07	0.56
Chloroform	1.54	0.19	−1.8884	16.37	1.21	−3.2647	0.69	−0.16	−0.35
Carbon tetrachloride	0.89	−0.05	−4.5149	0.03	−1.52	−5.9289	0.00	−2.94	−2.89
Toluene	0.34	−0.47	−3.0598	1.43	0.16	−4.4573	0.06	−1.24	−0.77

log $S(x)$ indicates the logarithmic solubility in mole fractions. Experimental data are taken from Ref. 12.

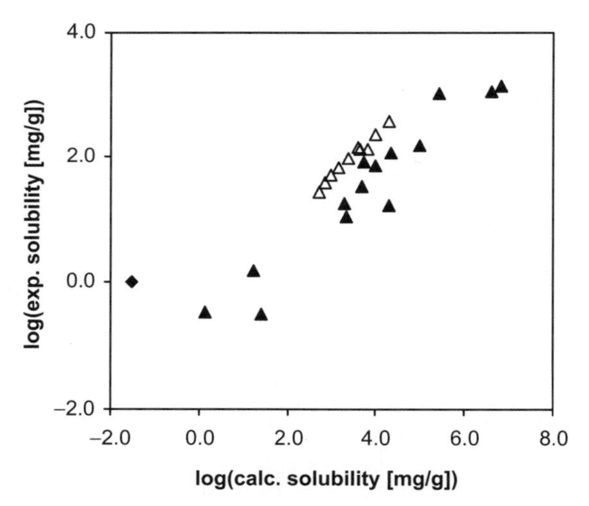

FIGURE 24.7 Predicted relative solubility of acetaminophen versus experimental data in pure solvents at 303.15K. Triangles represent relative solubility data, with empty triangles (△) representing alcoholic solvents and water and solid triangles (▲) representing the remainder of the solvent data set. One outlier (carbon tetrachloride) is represented by a solid diamond (♦).

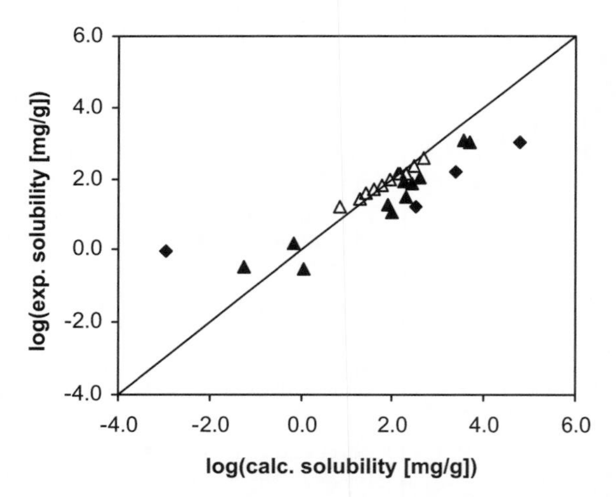

FIGURE 24.8 Predicted absolute solubility of acetaminophen versus experimental data in pure solvents at 303.15 K. Empty triangles (△) represent absolute solubility data of alcoholic solvents and water, solid triangles (▲) represent the remainder of the solvent data set. Four outliers (carbon tetrachloride, diethylamine, 1,4-dioxane, tetrahydrofuran) represented by solid diamonds (♦).

tetrahydrofuran, diethylamine, *N,N*-diethylformamide, and dimethyl sulfoxide. The rmse for all solvents is 0.77 log units. There are, however, four significant outliers in the data set, one of them, carbon tetrachloride, deviates by almost 3 log units from the experimental solubility and has already been discussed above. While the solubility of carbon tetrachloride is severely underestimated by the model prediction, the other ouliers are overpredicted solubilities. For diethylamine, the predicted solubility is 1.64 log units too high. We attribute this error to the known systematic error of COSMO and COSMO-RS for secondary and tertiary aliphatic amines [5]. Another outlier is the predicted solubility of acetaminophen in 1,4-dioxane (317.14 mg/g), which deviates by 1.27 log units from the value reported by Granberg and Rasmuson (17.1 mg/g). It is noteworthy that the deviation is much less (0.56 log units) if compared to the experimental solubility reported by Romero et al. (86.9 mg/g) [11]. The fourth outlier in the data set is in tetrahydrofuran, where the predicted solubility deviates by 1.13 log units from the experimental data. Here, we do not have an explanation for the error of the model prediction. However, since tetrahydrofuran and 1,4-dioxane are rather similar solvents, the question arises whether the uncertainty of the experimental data for tetrahydrofuran might be comparable to the case of 1,4-dioxane, where there is a deviation between the published experimental data from the different sources.

With the four outliers removed, the overall rmse reduces to 0.46 log units. While the rmse is very small for alcohols (0.10 log units), the calculated solubilities in polar aprotic solvents like acetone or hydrophobic solvents like toluene are systematically overpredicted. The rmse for the remainder of the data set without the alcoholic solvents is 0.86 log units. It should be noted that the estimate of ΔG_{fus} based on ΔH_{fus} and T_{melt} itself may have an error of the order of 0.5 log units, making these absolute deviations uncertain.

EXAMPLE 24.2 SOLUBILITY OF ACETAMINOPHEN IN BINARY MIXTURES OF WATER–ACETONE AND TOLUENE–ACETONE AT 25°C

Experimental data for acetaminophen solubility in water–acetone and acetone–toluene binary solvents were also reported by Granberg and Rasmuson [12] and are used here for comparison with the model prediction.

As in the previous example, the COSMO files of the compounds were calculated on the BP86/TZVP level of theory. Absolute solubility predictions are calculated using the experimental data for ΔG_{fus} of acetaminophen and employing the iterative refinement procedure for the solubility. The calculations are done for the compositions that were measured by Granberg and Rasmuson (Figure 24.9).

The predicted solubilities of acetaminophen in the binary solvent system can be extracted from the COSMO*therm* table file shown in Figure 24.10. Table 24.2 lists the predicted solubilities together with the experimental data.

Figures 24.11 and 24.12 show the prediction results for the acetaminophen solubility in water–acetone and acetone–toluene binary solvent mixtures. The solubility of acetaminophen in the water–acetone binary mixture is nonideal, with a solubility peak at ~70% mass fraction of acetone. The nonideal solubility behavior is also captured by the model prediction, with the solubility peak slightly shifted to higher acetone content of the binary solvent. Since the solubility in pure acetone is overpredicted by 0.59 log units, the model prediction for the binary mixture does not show the strong decrease in solubility for very high acetone content of the solvent mixture that is found in the experiment.

The prediction results for the acetone–toluene system (Figure 24.12) are consistent with the trends exhibited by the experimental data. Again, we see effects of the

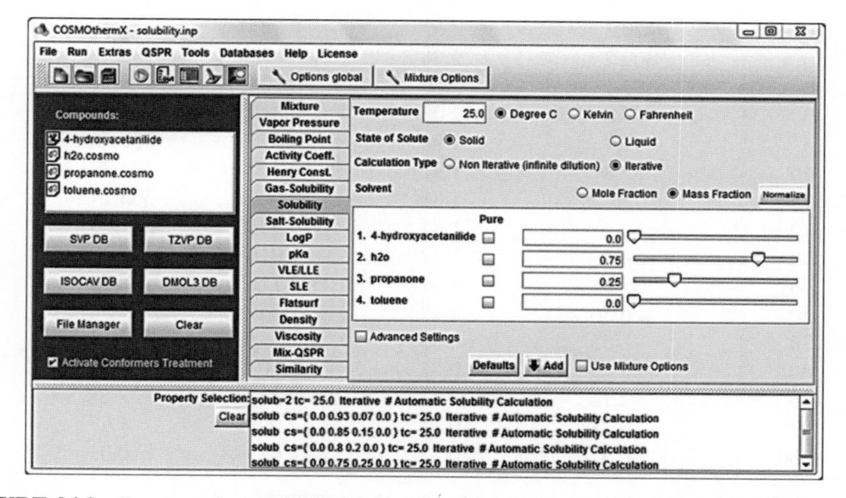

FIGURE 24.9 Input section of COSMO*therm*X with entries for a list of solvent compositions.

```
Solubility at T =  298.15 K  at given concentration CS={ 0.0 85 15 0 }  -
energies are in kcal/mol volume is in A^3

 Nr Compound              log10(x_solub)    mu(self)      mu(solv)       DG_fus
  1 4-hydroxyacetanilide   -2.39941309   -5.22146393   -3.98117592   2.03362348
  2 h2o                     0.00000000   -2.98262140   -2.98262140   0.00000000
  3 propanone               0.00000000   -1.72077455   -1.72077628   0.00000000
  4 toluene                -3.08354307   -4.65191005   -0.44452822   0.00000000
```

FIGURE 24.10 Excerpt from the COSMO*therm* table file for the solubility calculation of acetaminophen in a binary solvent system.

overpredicted solubility in pure acetone and the lower predicted solubility in pure toluene, but the relatively ideal solubility behavior of acetaminophen in the acetone–toluene binary mixture is found also by the model prediction.

24.3 CHEMICAL REACTIONS IN SOLUTION

Calculation of reaction energies in the gas phase is a standard application in quantum chemistry. The computational prediction of free energies of reaction in solution is more involved, but still a rather straightforward procedure. Generally, the free energy of reaction is the difference of the total free energies of the reactants and the products of the reaction. For a reaction

$$aA + bB \rightarrow cC + dD$$

where A and B are the reactants with stoichiometric coefficients a and b, and C and D are the reaction products with

TABLE 24.2 Experimental and Predicted Solubilities of Acetaminophen in Water–Acetone and Acetone–Toluene Binary Solvent Mixtures

% Mass Fraction			Experimental			Predicted		Error
Water	Acetone	Toluene	c_S (g/kg)	log S (mg/g)	log S (x)	S (mg/g)	log S (mg/g)	
100	0	0	14.90	1.17	−3.1508	5.93	0.77	−0.40
93	7	0	28.18	1.45	−2.7658	13.69	1.14	−0.31
85	15	0	53.0	1.72	−2.3994	29.99	1.48	−0.25
80	20	0			−2.1978	45.87	1.66	1.66
75	25	0			−2.0125	67.47	1.83	1.83
70	30	0	150.0	2.18	−1.8424	95.66	1.98	−0.20
65	35	0			−1.6879	130.58	2.12	2.12
60	40	0			−1.5500	171.24	2.23	2.23
55	45	0			−1.4290	215.47	2.33	2.33
50	50	0	327.0	2.51	−1.3242	260.58	2.42	−0.10
30	70	0	454.6	2.66	−1.0260	408.68	2.61	−0.05
15	85	0	420.3	2.62	−0.8851	452.18	2.66	0.03
7	93	0	302.2	2.48	−0.8364	438.41	2.64	0.16
3	97	0	197.1	2.29	−0.8247	415.72	2.62	0.32
0	100	0	99.8	2.00	−0.8285	386.32	2.59	0.59
0	95	5	91.7	1.96	−0.8517	359.46	2.56	0.59
0	90	10	82.4	1.92	−0.8774	332.39	2.52	0.61
0	85	15	75.7	1.88	−0.9062	305.14	2.48	0.61
0	80	20	66.4	1.82	−0.9386	277.63	2.44	0.62
0	70	30	52.8	1.72	−1.0165	222.79	2.35	0.63
0	60	40	37.08	1.57	−1.1186	168.79	2.23	0.66
0	50	50	26.56	1.42	−1.2563	117.58	2.07	0.65
0	30	70	8.55	0.93	−1.7149	37.19	1.57	0.64
0	20	80	3.39	0.53	−2.1018	14.50	1.16	0.63
0	15	85	2.12	0.33	−2.3663	7.68	0.89	0.56
0	7	93	0.78	−0.11	−2.9938	1.73	0.24	0.35
0	0	100	0.37	−0.43	−4.6734	0.03	−1.46	−1.03

log $S(x)$ indicates the logarithmic solubility in mole fractions. Experimental data are taken from Ref. 15.

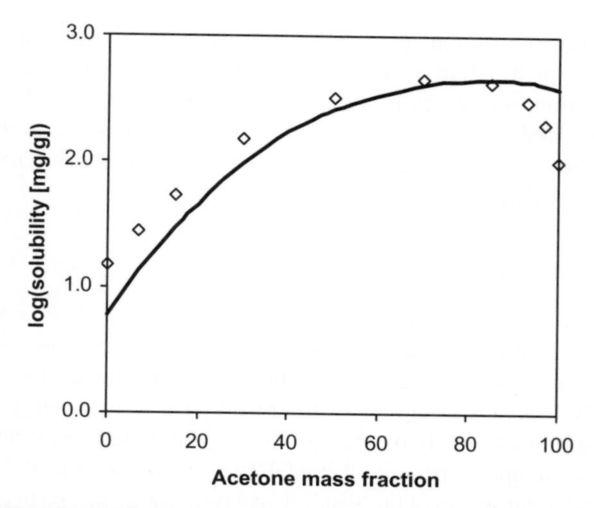

FIGURE 24.11 Experimental and predicted solubility of acetaminophen in acetone–water binary mixtures at 298.15K. Diamonds (◊) are experimental data, the solid line is from the model prediction.

stoichiometric coefficients c and d, the free energy of reaction can be calculated from the difference of the sums of free energies on both sides of the reaction:

$$\Delta G_r = [c \cdot G(C) + d \cdot G(D)] - [a \cdot G(A) + b \cdot G(B)] \tag{24.13}$$

The free energies of the reactants and products, and thus the free energy of reaction, depends on the conditions under which the reaction takes place. The free energy of reaction in the gas phase differs from the free energy of reaction in solution, and it is different in each specific solvent.

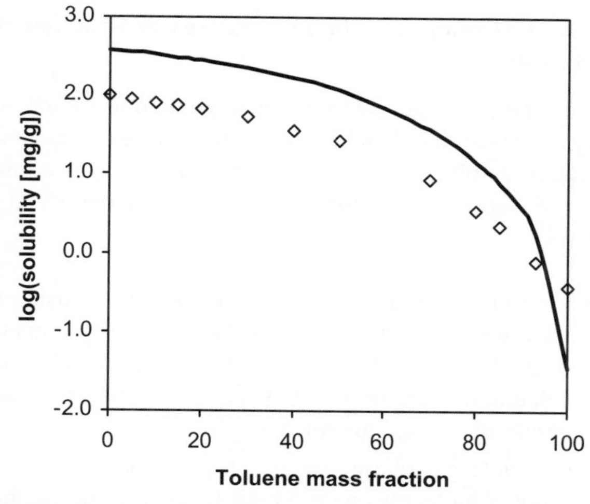

FIGURE 24.12 Experimental and predicted solubility of acetaminophen in toluene–acetone binary mixtures at 298.15K. Diamonds (◊) are experimental data, the solid line is from the model prediction.

In solution, the Gibbs free energy of a species is

$$G(i) = E_{gas}(i) + \Delta G_{solv}(i) \tag{24.14}$$

$E_{gas}(i)$ is the gas-phase energy of the compound, and for computational predictions of the reaction free energy it should be taken from an adequate quantum chemical (DFT or post-Hartree Fock) level. $\Delta G_{solv}(i)$, the free energy of solvation, describes the change of the free energy that occurs when the compound is dissolved from the gas phase into the liquid phase. This contribution to the total free energy of a compound can be computed using COSMO-RS.

Using the gas-phase energies of the compounds and the free energies of solvation of the compounds, the free energy of reaction in solution can be calculated according to a thermodynamic cycle as depicted in Figure 24.13.

In order to compute the lower horizontal leg of the cycle, corresponding to the reaction in solution, we have to take the appropriate sums and differences of the upper horizontal leg, that is, the gas-phase reaction, and the vertical legs, that is, the solvation energies of the compounds:

$$\begin{aligned}\Delta G_r(\text{sol}) = \Delta G_r(\text{gas}) &+ [c\Delta G_{solv}(C) + d\Delta G_{solv}(D)] \\ &- [a\Delta G_{solv}(A) + b\Delta G_{solv}(B)]\end{aligned} \tag{24.15}$$

$\Delta G_r(\text{gas})$ can be calculated from the chemical potential of the compounds in the gas phase. As already mentioned, the quantum chemical gas-phase energy E_{gas} of a compound is computed in vacuum at absolute zero. Furthermore, E_{gas} does not account for vibrational motion that is present even at $T = 0$K. The so-called zero-point energy or zero-point vibrational energy (ZPE) can be computed quantum chemically from the vibrational frequencies of the compound and is a standard correction to E_{gas}. Using the ZPE, the free energy of a compound can be calculated as

$$G(i) = E_{gas}(i) + \text{ZPE}(i) + \Delta G_{solv}(i) \tag{24.16}$$

As a further refinement for the gas-phase energies of the compounds and the resulting reaction energy, the temperature dependent thermal contributions to the free energy μ_{vib} of the molecule can be calculated. From vibrational frequencies the molecular translational, rotational, and vibrational partition functions, q_{trans}, q_{rot} and q_{vib}, can be calculated,

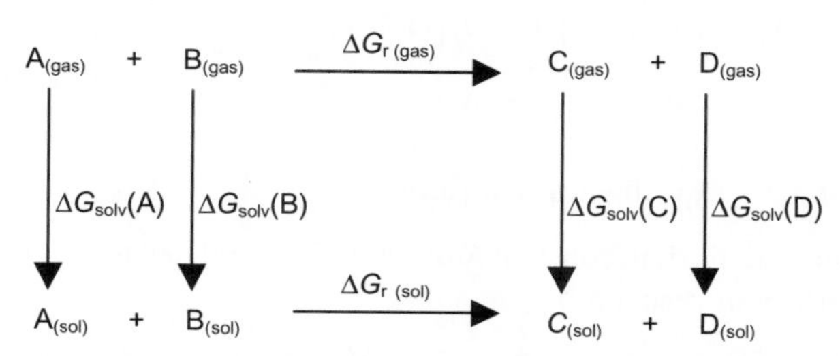

FIGURE 24.13 Cycle for computation of a free energy change in solution.

thus enabling prediction of thermodynamic functions at temperatures other than 0K and finite pressure.

$$\begin{aligned} G(i) &= E_{gas}(i) + \mathrm{ZPE}(i) - RT\ln(q_{trans}\cdot q_{rot}\cdot q_{vib}) \\ &\quad + \Delta G_{solv}(i) \\ &= E_{gas}(i) + \mu_{vib} + \Delta G_{solv} \end{aligned} \quad (24.17)$$

The other terms required for equation 24.15, the free energies of solvation ΔG_{solv}, can be obtained from a COSMO*therm* prediction of the reverse process, that is, from a vapor pressure prediction. The partial vapor pressure *P*(*i*) that is calculated by COSMO*therm* corresponds to the pure compound vapor pressure times the activity coefficient and is related to ΔG_{solv} by

$$\Delta G_{solv}(i) = RT\ln(10)[\log_{10}(P(i)) - \log_{10}(P)] \quad (24.18)$$

with *P* being the reference pressure at which the reaction takes place.

Note that if the reactant or product compounds are present in the mixture at a finite concentration with a mole fraction *x* (*i*) (e.g., if the reaction takes places in bulk reactant liquid), an entropic contribution $RT\ln(x(i))$ of the compound has to be added to the compounds free energy *G*(*i*).

24.3.1 Heat of Reaction

The heat of reaction or reaction enthalpy in solution can be calculated by a procedure similar to the free energy of reaction. Instead of the free energy of solvation of the compounds we make use of the heat of vaporization ΔH_{vap}. Since ΔH_{vap} is the enthalpy that is needed to transfer the compound from the liquid phase to the gas phase, it has to be substracted from the gas-phase energy to obtain the enthalpy of the compound in solution:

$$H(i) = E_{gas}(i) - \Delta H_{vap}(i) \quad (24.19)$$

Zero-point vibrational energy corrections or thermal correction terms for the enthalpy in the gas phase can also be used as corrections to the gas-phase energies of the compounds for the calculation of the heat of reaction.

Similarly to equation 24.13 for the free energy of reaction ΔG_r, the heat of reaction ΔH_r can be calculated from the enthalpy of the compounds:

$$\Delta H_r = [c\cdot H(\mathrm{C}) + d\cdot H(\mathrm{D})] - [a\cdot H(\mathrm{A}) + b\cdot H(\mathrm{B})] \quad (24.20)$$

24.3.2 Equilibrium Constants

The equilibrium constant *K* of a reaction is related to the free energy of reaction by

$$\ln K = -\frac{\Delta G^\circ}{RT} \quad (24.21)$$

For a reaction in an ideal solution, that is, in infinite dilution, the equilibrium constant can be calculated using the reaction free energy in solution according to equation 24.15. The free energies of the individual compounds can be computed using different quantum chemical correction terms as described above.

24.3.3 Accuracy

In the described procedure the free energies of the compounds are calculated from two main contributions, the quantum chemical gas-phase energy and the free energy of solvation. The accuracy of the resulting reaction energy is determined mainly by the accuracy of the underlying quantum chemical method. With DFT methods like the BP86 functional, errors of the absolute reaction energy can be in the range of 10 kcal/mol or more [13, 14]. However, for relative reaction energies of one reaction in different solvents in a solvent screening application, that is, if we are looking at the variation of the solvation energy only, calculated as the COSMO*therm* contribution in the liquid phase, the accuracy is much higher. For such relative reaction energy predictions considering ΔG_{solv} or ΔH_{vap} from the COSMO*therm* vapor pressure prediction only, an accuracy of 0.5 kcal/mol can be expected.

For a higher accuracy of absolute predictions of the reaction energy or enthalpy, it follows that a more accurate quantum chemical method should be applied for the calculation of the gas-phase energy, for example, the MP2 or Coupled-Cluster methods combined with adequate basis sets. Quantitative improvement of the total free energy of a compound *G*(*i*) can also be achieved by the ZPE and thermal corrections to the gas-phase energy of the compound.

24.3.4 Calculation of the Free Energy of Reaction and Heat of Reaction

A procedure for the computational prediction of the free energy of reaction ΔG_r or heat of reaction ΔH_r using quantum chemical gas-phase energies and free energies of solvation or heats of vaporization from COSMO-RS is described as follows:

- Compute the reactant and product molecules using the DFT methods for which COSMO*therm* is parameterized. Both COSMO and gas-phase quantum chemical calculations are required. Then use COSMO*therm* to obtain the ΔG_{solv} and/or ΔH_{vap} values.
- Compute gas-phase energies E_{gas} of the reactant and product molecules with a high-level *ab initio* method, for example, the Coupled-Cluster method.
- Compute vibrational frequencies for reactant and product molecules to obtain the ZPE correction or the

thermal corrections to the gas-phase energies of the compounds. Vibrational frequency calculations at a DFT level of theory are usually sufficiently accurate.

- Combine E_{gas}, ZPE or thermal corrections and ΔG_{solv} to compute the compounds $G(i)$ according to equations 24.16 or 24.17, or E_{gas}, ZPE or thermal corrections and ΔH_{vap} to compute the compounds $H(i)$.
- Calculate the free energy of reaction ΔG_r or the heat of reaction ΔH_r from the difference of the sums of free energies or enthalpies on both sides of the reaction as described by equations 24.13 and 24.20.

EXAMPLE 24.3 ESTIMATE THE HEAT OF REACTION ΔH_R FOR THE REDUCTION OF NITROBENZENE TO ANILINE IN THE LIQUID PHASE IN THF AT 25°C

$$PhNO_2(\mathbf{1}) + 3H_2(\mathbf{2}) \rightarrow PhNH_2(\mathbf{3}) + 2H_2O(\mathbf{4})$$

Following the procedure described above, we calculate the gas-phase energies and the heats of vaporization. In this example, we calculate the gas-phase energies of the compounds on the MP2/TZVPP quantum chemical level. Furthermore, we employ the ZPE and thermal corrections to the enthalpy in the gas phase to refine the calculated heat of reaction.

For the calculation of the heat of vaporization with COSMO*therm* we have to provide COSMO files for all compounds involved in the reaction, including the solvent, and the gas-phase energies for the reactants and products. For the compounds involved here, the COSMO files and gas-phase energies are available from the database included in the COSMO*therm* package. However, to exemplify the procedure, we will give a short overview of the required quantum chemical calculations.

Since we require the gas-phase energy as well as the screening charge surface of the compounds for the calculation of the heat of vaporization, we need to do geometry optimizations both in the gas phase and in the conductor, using the COSMO model with ideal screening. COSMO*therm* is parameterized for the BP86/TZVP and the BP86/SVP//AM1 quantum chemical levels. Here, we will use the higher one of the two levels, that is, the BP86 functional and the TZVP basis set. The QC calculations are performed using the TURBOMOLE [15] quantum chemical program suite.

For the QC calculation we have to provide 3D structures of the compounds, which can be generated by an external tool or built with the molecular builder of the TURBOMOLE graphical user interface. With the starting geometries, the following QC calculations are performed for each compound

- A gas-phase geometry optimization on the BP86/TZVP quantum chemical level.
- A geometry optimization on the BP86/TZVP/COSMO quantum chemical level.
- A gas-phase geometry optimization on the MP2/TZVPP quantum chemical level.
- A vibrational frequency calculation with the optimized BP86/TZVP molecular gas-phase structure to obtain the zero-point vibrational energy.
- Using the results from the vibrational frequency calculations, we also compute the thermal contributions with the corresponding interactive tool of the TURBOMOLE suite.

These steps are described in more detail in Appendix 24. A. Further information about how the quantum chemical calculations are carried out can also be found in the documentation of the TURBOMOLE program suite and the TmoleX documentation.

The heats of vaporization H_{vap} are calculated with COSMO*therm* at a temperature of 25°C and a solvent composition of pure tetrahydrofuran. For the COSMO-RS vapor pressure prediction, quantum chemical gas-phase energies from the BP/TZVP level are used to calculate the chemical potential of the compound in the gas phase.

The calculated data for the heats of vaporization of the compounds can be extracted from the COSMO*therm* output file (Figure 24.14). In Table 24.3, the results for the individual energy contributions of the compounds both from the QC calculations and the COSMO*therm* calculation are tabulated together with the resulting data for the heat of reaction. The data for the heat of reaction were calculated from $\Delta H_r = [H(3) + 2 \cdot H(4)] - [H(1) + 3 \cdot H(2)]$, taking into account the stoichiometry of the reaction (equation 24.20). For the enthalpy values of the compounds H(i), different correction terms for the quantum chemical gas-phase energies were employed. For comparison, heats of reaction using the quantum chemical gas-phase energies from the MP2/TZVPP level and the BP86/TZVP level are tabulated.

The catalytic reduction of aromatic nitro compounds in the gas phase is known to be a highly exothermic process. For the gas-phase reduction of nitrobenzene, a heat of reaction of $\Delta H_r = -131 \pm 3$ kcal/mol was published [16]. Absolute values for the heat of reaction of the reduction of nitrobenzene to aniline in tetrahydrofuran solution could not be found in the literature, but heats of reaction for the reduction of other R-NO_2 compounds in solution have been found independent of R or solvent to be in the range of -125 to -130 kcal/mol [17].

Using MP2/TZVPP gas-phase energies for the compounds, thermal corrections, and the heat of vaporization from the COSMO-RS prediction, the calculated heat of reaction in solution is $\Delta H_r = -127$ kcal/mol. With ZPE correction only, the calculated heat of reaction in solution

```
Results for mixture   1
-----------------------
Temperature            :    298.150 K

Compound Nr.           :         1          2          3          4          5
Compound               :  nitrobenz    aniline         h2        h2o        thf
Mole Fraction          :     0.0000     0.0000     0.0000     0.0000     1.0000

Compound:  1  (nitrobenzene)
Chemical potential of the compound in the mixture :      -3.54480 kcal/mol
Log10(partial pressure [mbar])                    :      -0.32317
Free energy of molecule in mix (E_COSMO+dE+Mu)    : -274197.04529 kcal/mol
Total mean interaction energy in the mix (H_int)  :      -5.23132 kcal/mol
Misfit interaction energy in the mix (H_MF)       :       2.46378 kcal/mol
H-Bond interaction energy in the mix (H_HB)       :      -0.27976 kcal/mol
VdW interaction energy in the mix (H_vdW)         :      -7.41534 kcal/mol
Ring correction                                   :      -1.14821 kcal/mol
Vapor pressure of compound over the mixture       :       0.00000 mbar
Chemical potential of compound in the gas phase   :       0.98955 kcal/mol
Heat of vaporization                              :      12.02087 kcal/mol
```

FIGURE 24.14 Excerpt from the mixture output section of the COSMO*therm* output file for a vapor pressure calculation of compounds in pure tetrahydrofuran at 25°C.

is $\Delta H_r = -125$ kcal/mol. Both predicted values are well inside the range of the experimental data for comparable reactions. In contrast, when ZPE and thermal corrections are ignored, the heat of reaction is overestimated by several kcal/mol with a value of $\Delta H_r = -141$ kcal/mol. Table 24.3 also shows heats of reaction calculated from the BP86/TZVP gas-phase energies. The heat of reaction without ZPE or thermal corrections is $\Delta H_r = -125$ kcal/mol, which agrees well with experimental data, but when the correction terms, which should in general lead to a better prediction, are included, the heat of reaction in solution is significantly underestimated with predicted values of $\Delta H_r = -109$ kcal/mol and $\Delta H_r = -111$ kcal/mol, respectively. However, it should be noted that absolute errors in the range of 10–20 kcal/mol are not unusual for pure DFT functionals like BP86 [13,14].

EXAMPLE 24.4 ESTIMATE REACTION FREE ENERGY, EQUILIBRIUM CONSTANT, AND EQUILIBRIUM COMPOSITION FOR THE REACTION OF 1-METHOXY-2-PROPANONE 1 AND ISOPROPYLAMINE 2

First, the gas-phase energies of the reactants and products of the transamination reaction (Figure 24.15) are calculated quantum chemically. In this example, we also employ the MP2 level of theory and the TZVPP basis set. As starting structures for the gas-phase geometry optimizations, the 3D structures from the BP86/TZVP level are used, which are available from the COSMO*base*, a database of validated COSMO files and gas-phase structures. With these structures we perform gas-phase geometry optimizations using the MP2/TZVPP method and basis set combination.

TABLE 24.3 Gas-Phase Energies, Heat of Vaporization, Zero-Point Vibrational Energies, Thermal Corrections, Total Enthalpies of Compounds, and Heat of Reaction for the Hydrogenation of Nitrobenzene

	1	2	3	4	ΔH_r	
					MP2/TZVPP	(BP86/TZVP)
H_{vap}	12.02	1.35	14.54	10.24		
E_{gas} (BP86/TZVP) (Hartree)	−436.944122	−1.177446	−287.715215	−76.465165		
E_{gas} (MP2/TZVPP) (Hartree)	−435.958485	−1.164647	−287.000283	−76.323461		
ZPE (BP86/TZVP) (Hartree)	0.099295	0.009850	0.113236	0.020640		
ΔH_T (thermal corrections)	67.36	8.26	75.45	15.32		
$H = E_{gas} - H_{vap}$	−273580.11	−732.17	−180109.95	−47903.93	−141.19	(−125.06)
$H = E_{gas} + \text{ZPE} - H_{vap}$	−273517.80	−725.99	−180038.89	−47890.98	−125.08	(−108.95)
$H = E_{gas} + \Delta H_T - H_{vap}$	−273512.75	−723.92	−180034.50	−47888.61	−127.22	(−111.09)

Enthalpy terms are in kcal/mol, quantum chemical gas-phase energies and ZPE are in Hartree.

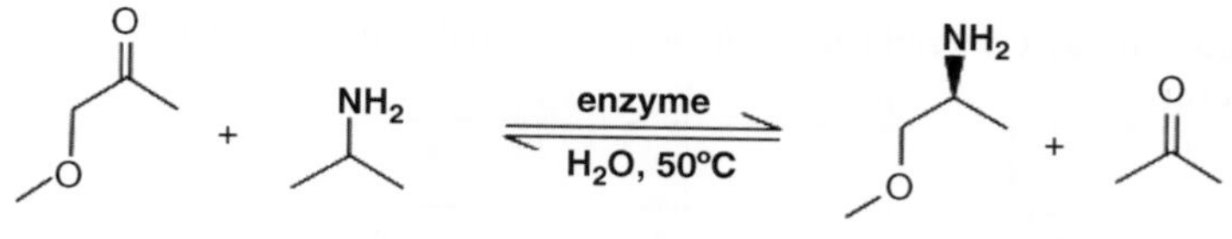

FIGURE 24.15 Catalytic transamination of 1-methoxy-2-propanone and isopropylamine.

As a further refinement for the gas-phase energies, vibrational frequency calculations for the compounds are carried out, and the thermal contributions from the vibrational frequencies to the gas-phase energy are computed using the corresponding tool of the TURBOMOLE suite. These calculations are carried out on the BP86/TZVP level.

In our next step for the calculation of the reaction free energy, the free energy of solvation of each compound is calculated from equation 24.18, using the partial pressures from the vapor pressure prediction of the COSMO*therm* program. Reactants, products, and solvent for the reaction are taken from the COSMO*base*. The conditions for the vapor pressure calculation are set to a temperature of 50°C and the solvent composition is set to pure water. Quantum chemical gas-phase energies of the compounds are used to calculate the chemical potentials of the compounds in the gas phase. The calculated partial pressures can be extracted from the COSMO*therm* output file (Figure 24.16).

Energy terms and corrections for the reactants and products from this procedure are tabulated in Table 24.4. Calculated data for the free energy of reaction ΔG_r are also tabulated, using different QC correction terms. ΔG_r is calculated according to the stoichiometry of the reaction as $\Delta G_r = [G(3)+G(4)]-[G(1)+G(2)]$.

With the MP2/TZVPP gas-phase energies and the solvation free energies of the compound, the free energy of reaction in solution is $\Delta G_r = -3.27$ kcal/mol, corresponding to an equilibrium constant of $\ln K = 5.10$. The zero-point vibrational energy corrections for the reactants and products have very little influence on the overall reaction energy ($\Delta G_r = -3.29$ kcal/mol, $\ln K = 5.12$ kcal/mol), but when thermal correction are included in the free energies $G(i)$ of the compounds, the free energy of reaction decreases to $\Delta G_r = -2.20$ kcal/mol and the equilibrium constant decreases to $\ln K = 3.43$. The equilibrium constant indicates that the equilibrium position of the reaction lies on the right-hand side of the reaction equation. The relative amount of reactants at equilibrium is 0.08 each, and the relative amount of products at equilibrium is 0.42 each. The experimental equilibrium constant for the reaction of 1-methoxy-2-propanone with isopropylamine is $K = 7.8 (\ln K = 2.05)$ [18]. Thus, the calculated free energy of reaction and equilibrium constant are in excellent agreement with the experimental data.

24.4 CONCLUSION AND OUTLOOK

In this chapter, we presented an overview of two different applications of COSMO-RS in the drug development process. The power and main benefit of the COSMO-RS model is that properties in solution can be obtained from *ab initio* calculation without any experimental input. It does not require external data for modeling and can also be used when empirical models are not parameterized. Complex multifunctional molecules and new chemical functionalities are treated on the same footing as simple organic molecules.

```
Results for mixture   1
 -----------------------
 Temperature               :     323.150 K

 Compound Nr.              :          1         2         3         4         5
 Compound                  :  isopropyl 1-methoxy 1-methoxy propanone       h2o
 Mole Fraction             :     0.0000    0.0000    0.0000    0.0000    1.0000

 Compound:  1  (isopropylamine)
 Chemical potential of the compound in the mixture :       -1.17237 kcal/mol
 Log10(partial pressure [mbar])                    :        4.11202
 Free energy of molecule in mix (E_COSMO+dE+Mu)    :  -109542.55539 kcal/mol
 Total mean interaction energy in the mix (H_int)  :       -7.42349 kcal/mol
 Misfit interaction energy in the mix (H_MF)       :        2.15679 kcal/mol
 H-Bond interaction energy in the mix (H_HB)       :       -4.65492 kcal/mol
 VdW interaction energy in the mix (H_vdW)         :       -4.92535 kcal/mol
 Ring correction                                   :        0.00000 kcal/mol
 Vapor pressure of compound over the mixture       :        0.00000 mbar
 Chemical potential of compound in the gas phase   :       -2.81690 kcal/mol
 Heat of vaporization                              :       10.89292 kcal/mol
```

FIGURE 24.16 Excerpt from the mixture output section of the COSMO*therm* output file for a vapor pressure calculation of compounds in pure water at 50°C.

TABLE 24.4 Energy Terms and Energy Corrections for the Reactants and Products for reaction 2 and Free Energies of Reaction with the Various Correction Terms

	1	2	3	4	ΔG_r	ln *K*
log 10($P(i)$) (mbar)	2.72032	4.11202	3.12339	3.75518		
G_{solv}	−0.41	1.64	0.18	1.12		
E_{gas} (MP2/TZVPP) (Hartree)	−307.100753	−174.113883	−288.440644	−192.779317		
ZPE (BP86/TZVP) (Hartree)	0.112968	0.117222	0.149297	0.080872		
μ_{vib} (thermal corrections)	54.44	227.770	71.54	32.03		
$G = E_{gas}$ (MP2) + G_{solv}	−192709.05	−109256.47	−180999.06	−120969.74	−3.27	5.10
$G = E_{gas}$ (MP2) + G_{solv} + ZPE	−192638.16	−109182.91	−180905.38	−120918.99	−3.29	5.12
$G = E_{gas}$ (MP2) + G_{solv} + μ_{vib}	−192660.99	−109202.03	−180927.52	−120937.70	−2.20	3.43
Relative amount of compound in equilibrium	0.08	0.08	0.42	0.42		

The relative amount of the compounds in equilibrium has been calculated from the equilibrium constant ln $K = 3.43$. Free energy terms and chemical potential are in kcal/mol, quantum chemical gas-phase energies and ZPE are in Hartree.

Prediction of relative drug solubility with COSMO-RS is based on a consistent thermodynamic modeling of interactions in the solvent and the supercooled state of the drug. Solvent mixtures can be treated in the same way as pure solvents and with similar accuracy. Absolute solubility prediction is limited by the availability of free energy of fusion data.

Although COSMO-RS in its present state cannot be proven to be more accurate than more empirical models with many adjusted parameters, its strength is the essential independency from experimental data. This allows for independent modeling and avoids errors resulting from erroneous experimental data on which empirical models rely. Potential improvement to the current COSMO-RS solubility prediction model include a more accurate fusion term for absolute solubility prediction, and improvement of the COSMO-RS interaction terms themselves, especially for the chemically important group of secondary and tertiary amines, but requires more reliable experimental data as are available at present.

With COSMO-RS, solubility prediction is also possible for salts. This is important as many drugs are formulated as salts. Solvent systems involving ionic liquids can also be treated with very good accuracy [19]. Furthermore, different conformational forms of molecules can be used for solubility screening, and the relative weight of conformers in different solvents can be determined. This feature basically allows for examination of conditions influencing the crystallization process and solvent screening for conformational polymorphism and pseudopolymorphism [20, 21].

Another application of COSMO-RS frequently used in pharmaceutical and agrochemical industry deals with reaction modeling. In principle, reaction equilibrium, mechanism, rate and by-product formation may be solvent dependent. Here, we investigated the influence of solvation on the free energy and heat of reaction and the reaction equilibrium only. A straightforward procedure for the computational prediction of the free energy of reaction and heat of reaction has been shown, and the effect of the employed quantum chemical level on the absolute heat of reaction has been demonstrated. Elsewhere, it has been shown that although, depending on the quantum chemical level, absolute values for the free energy of reaction may differ substantially from experimental data, general trends are predicted correctly [22].

All applications of COSMO-RS require quantum chemical calculations of compounds, taking into account the various molecular conformations of the compounds. This constitutes the computationally most demanding part of the procedure, but has to be done only once per compound. The COSMO files of the involved compounds can be reused for other projects and all types of properties. Thus, if combined with a database of precalculated COSMO files for common compounds, thermodynamic property calculations with COSMO*therm* can be carried out quite fast and efficiently.

24.A APPENDIX

24.A.1 Details of COSMO and Gas-Phase Calculations

For later use of the COSMO files in COSMO*therm*, the details of the quantum chemical COSMO calculation should be consistent with one of the parameterizations of COSMO*therm*. There are two levels of different quality mainly used in COSMO*therm*, a lower, computationally less expensive level, and a higher level, which is computationally more time-consuming but better suited for chemical engineering applications. The higher level has also been used throughout in this contribution and requires

- BP86 DFT geometry optimization with a TZVP quality basis set and the RI approximation applied in the gas phase and in the conductor.

- For COSMO calculations only: COSMO applied in the conductor limit ($\varepsilon = \infty$) using optimized element-specific COSMO radii for the cavity construction.
- If more than one conformation is considered to be potentially relevant for a compound, QC calculations have to be done for all conformations.

24.A.2 Steps for Calculating the Free Energy or Enthalpy of a Compound

Apart from the heat of vaporization or free energy of solvation that are calculated by COSMO*therm*, the free energies and enthalpies of compounds involved in the reaction examples are composed from several quantum chemical contributions.

24.A.2.1 Gas-Phase Energy The BP86/TZVP gas-phase energy is obtained from a gas-phase geometry optimization of the compound using the settings described in Appendix 24.A.1. The resulting energy value can be found in output files of the TURBOMOLE program suite. The protocol file job.last comprises the output of the last complete cycle and information about the settings that were applied in the program run, for example, convergence criteria. The gas-phase energy can be read from "total energy" line in the output section displayed in Figure 24.17. If the graphical user interface TmoleX is used, the gas-phase energy can also be taken from the Energy block of the Job-Results panel. This panel also displays information about the run of the job (Figure 24.18).

The MP2/TZVPP gas-phase energy is obtained from an MP2 geometry optimization employing the TZVPP basis set (also called def-TZVPP) and the RI approximation (Figure 24.19). The module used for this type of calculation is ricc2. In TmoleX, the method has to be set to RI-MP2 in the "Level" section of the "Level of Theory" panel. In the MP2 calculations for this contribution, the 1s orbitals of elements C, N, and O were kept frozen. Details about settings for frozen core orbitals can be found in the TURBOMOLE documentation.

24.A.2.2 Correction Terms from Vibrational Frequency Calculations For the zero-point vibrational energy contribution, a vibrational frequency calculation is required. This type of calculation can be done with the aoforce module of TURBOMOLE. The ZPE can be taken either directly from the aoforce.out file or from the interactive module freeh, which also allows for calculation of the molecular partition functions at temperatures other than 0K and finite pressure. Results of the freeh module are printed to standard I/O (Figure 24.20). Note that vibrational frequency calculations are based on the assumption of an harmonic oscillator and partition functions are computed within the assumption of an ideal gas and no coupling between degrees of freedom.

Vibrational frequency calculations can also be started from TmoleX. Start the job as a Single-Point calculation with the "Frequency Analysis" radio button ticked in the "Job Selection" section of the Single-Point calculation panel.

The thermal enthalpy contribution ΔH_T can be calculated from the thermal energy contribution printed in the freeh output. This is done via $\Delta H_T = \Delta G_T + RT$, where ΔG_T is the value in the "energy" column of the freeh output, T is the temperature, and R is the gas constant $R = 8.314472$ J/(K mol).

For the calculation of the thermal contribution to the free energy of a compound, μ_{vib}, the freeh value for "chem. pot." should be used and not the value from the "energy" column.

After converting the contributions from the individual steps to consistent units, the total enthalpy H of a compound can be calculated as the sum of the individual contributions. For compound nitrobenzene on the MP2/TZVPP level of theory, this involves the sum $H = E_{gas}(\text{MP2}) + \Delta H_T - H_{vap}$ consisting of the following contributions listed in Table 24.3

- E_{gas} (MP2) = −435.958485 Hartree = −273568.09 kcal/mol from the MP2 gas-phase geometry optimization.
- $\Delta H_T = \Delta G_T + RT = 279.36$ kJ/mol + 8.314472 J/(K mol) · 298.15K = 281.84 kJ/mol = 67.36 kcal/mol

```
***************************************************************************
    nitrobenzene

***************************************************************************


                         ----------------------------------------
                        |  total energy      =    -436.94412249634  |
                         ----------------------------------------
                        :  kinetic energy    =     435.14545720492  :
                        :  potential energy  =    -872.08957970126  :
                        :  virial theorem    =       1.99588353476  :
                        :  wavefunction norm =       1.00000000000  :
                         ........................................
```

FIGURE 24.17 Excerpt from the TURBOMOLE output file job.last from a gas-phase geometry optimization of nitrobenzene.

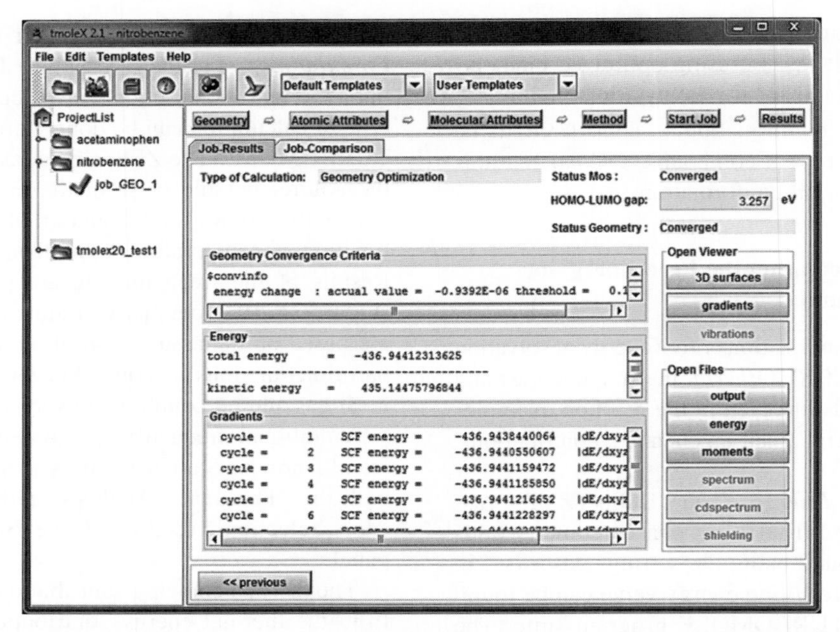

FIGURE 24.18 Job-results panel of the graphical user interface TmoleX of the TURBOMOLE program suite.

from the vibrational frequency calculation and subsequent computation of partition functions with the freeh module.

- $H_{\text{vap}} = 12.02$ kcal/mol from the COSMO-RS vapor pressure prediction.

Note that quantum chemical energies are usually expressed in Hartree, which is the atomic unit of energy. The conversion factor to the kcal/mol unit system is 627.5095 kcal/mol.

The free energy G of a compound can similarly be calculated from

$$G = E_{\text{gas}}(\text{MP 2}) + G_{\text{solv}} + \mu_{\text{vib}}$$

For compound isopropylamine, this requires the following energy terms, listed in Table 24.4:

- $E_{\text{gas}}(\text{MP2}) = -174.113883$ Hartree $= -109258.16$ kcal/mol from the MP2 gas-phase geometry optimization.
- $\mu_{\text{vib}} = 227.77$ kJ/mol, the chemical potential value ("chem. pot.") from a vibrational frequency calculation and subsequent computation of partition functions with the freeh module.
- $G_{\text{solv}} = 1.64$ kcal/mol, calculated from $\Delta G_{\text{solv}}(i) = RT\ln(10)[\log_{10}(P(i)) - \log_{10}(P)]$ (equation 24.18). The partial pressure of isopropylamine, $\log_{10}(P(i)) = 4.11202$ mbar, was be taken from a COSMO*therm*

```
****************************************************************
 *                                                             *
 *<<<<<<<<<<  GROUND STATE FIRST-ORDER PROPERTIES  >>>>>>>>>>>*
 *                                                             *
 ****************************************************************



        ----------------------------------------------
           Method          :  MP2
           Total Energy    :    -435.9584853137
        ----------------------------------------------
```

FIGURE 24.19 Excerpt from the TURBOMOLE output file job.last from an MP2 gas-phase geometry optimization of nitrobenzene using the ricc2 module of TURBOMOLE.

```
enter the range of temperatures (K) and pressures (MPa)
at which you want to calculate partition sums and free enthalpies :

tstart=(real) tend=(real) numt=(integer) pstart=(real) pend=(real) nump=(integer)

default values are  :
tstart=298.15 tend298.15 numt=1 pstart=0.1 pend=0.1 nump=1

or enter q or * to quit


         -----------------
          your wishes are :
         -----------------

 pstart=  0.1000      pend=  0.1000      nump=   1

 tstart=   298.1      tend=   298.1      numt=   1

           zero point vibrational energy
           ----------------------------
           zpe=   260.7      kJ/mol

  T         p       ln(qtrans) ln(qrot) ln(qvib) chem.pot.   energy    entropy
 (K)       (MPa)                                 (kJ/mol)  (kJ/mol) (kJ/mol/K)

298.15   0.1000000      17.82     12.90     3.49   175.88    279.36   0.35539

  T         P             Cv              Cp
 (K)      (MPa)       (kJ/mol-K)      (kJ/mol-K)
298.15   0.1000000     0.1128303       0.1211446
```

FIGURE 24.20 Excerpt from the interactive output of the freeh module of TURBOMOLE.

vapor pressure prediction (Figure 24.16) and the reference pressure P was assumed to be 1 bar.

24.6 ABBREVIATIONS

BP86	approximate DFT functional, consisting of Becke's exchange functional [23] and Perdew's correlation functional [24]
COSMO	conductor-like screening model
COSMO-RS	conductor-like screening model for realistic solvation
DFT	density functional theory
GCM	group contribution method
MC	Monte Carlo method
MD	molecular dynamics method
MP2	second-order Møller-Plesset perturbation theory
QC	quantum chemistry/quantum chemical
rmse	root mean square error
TZVP, TZVPP	Ahlrich's triple-zeta valence polarization basis sets [25]
vdW	van der Waals interaction

24.7 SYMBOLS

E_{gas}	gas-phase energy
ΔG_{fus}	free energy of fusion
ΔG_{solv}	free energy of solvation
ΔH_{vap}	heat of vaporization
μ	chemical potential
σ	screening charge density

REFERENCES

1. Klamt A, Schüürmann G. COSMO: a new approach to dielectric screening in solvents with explicit expressions for the screening energy and its gradient. *J. Chem. Soc. Perkin Trans. 2* 1993;799–805.
2. Klamt A. Conductor-like screening model for real solvents: a new approach to the quantitative calculation of solvation phenomena. *J. Phys. Chem.* 1995;99:2224–2235.
3. Klamt A. *COSMO-RS: From Quantum Chemistry to Fluid Phase Thermodynamics and Drug Design*, Elsevier, Amsterdam, 2005.
4. Gmehling J. Present status of group-contribution methods for the synthesis and design of chemical processes. *Fluid Phase Equilib.* 1998;144:37–47.
5. Klamt A, Jonas V, Bürger T, Lohrenz JCW. Refinement and parameterization of COSMO-RS. *J. Phys. Chem. A* 1998; 102: 5074–5085.
6. Klamt A, Eckert F. COSMO-RS: a novel and efficient method for the *a priori* prediction of thermophysical data of liquids. *Fluid Phase Equilib.* 2000; 172: 43–72.
7. Ben-Naim A. *Solvation Thermodynamics*, Plenum Press, New York and London, 1987.

8. Klamt A, Eckert F, Hornig M, Beck ME, Bürger T. Prediction of aqueous solubility of drugs and pesticides with COSMO-RS. *J. Comput. Chem.* 2002;23:275–281.
9. Granberg RA, Rasmuson AC. Solubility of paracetamol in pure solvents. *J. Chem. Eng. Data* 1999;44:1391–1395.
10. Manzo RH, Ahumada AA. Effects of solvent medium on solubility. V: Enthalpic and entropic contributions to the free energy changes of di-substituted benzene derivatives in ethanol:water and ethanol:cyclohexane mixtures. *J. Pharm. Sci.* 1990;79:1109–1115.
11. Romero S, Reillo A, Escalera B, Bustamante P. The behavior of paracetamol in mixtures of amphiprotic and amphiprotic–aprotic solvents. relationship of solubility curves to specific and non-specific interactions. *Chem. Pharm. Bull.* 1996;44:1061–1064.
12. Granberg RA, Rasmuson AC. Solubility of paracetamol in binary and ternary mixtures of water + acetone + toluene. *J. Chem. Eng. Data* 2000; 45: 478–483.
13. Koch W, Holthausen MC. *A Chemist's Guide to Density Functional Theory*, Wiley-VCH, Weinheim, 2001.
14. Cramer CJ. *Essentials of Computational Chemistry*, Wiley, Chichester, 2002.
15. TURBOMOLE V6.0 2009, a development of University of Karlsruhe and Forschungszentrum Karlsruhe GmbH, 1989–2007, TURBOMOLE GmbH, since 2007. Available from http://www.turbomole.com.
16. Macnab JI. The role of thermochemistry in chemical process hazards: catalytic nitro reduction processes. *I. Chem. E. Symp. Ser.* 1981;68(3/S):1–15.
17. am Ende DJ.Private communication, 2009.
18. Matcham G, Bhatia M, Lang W, Lewis C, Nelson R, Wang A, Wu W. Enzyme and reaction engineering in biocatalysis: synthesis of (*S*)-methoxyisopropylamine (=(*S*)-1-methoxypropan-2-amine). *Chimia* 1999;53:584–589.
19. Diedenhofen M, Eckert F, Klamt A. Prediction of infinite dilution activity coefficients in ionic liquids using COSMO-RS. *J. Chem. Eng. Data* 2003;48:475–479.
20. Abramov Y, Mustakis J, am Ende DJ.Private communication, 2009.
21. Geertman R.Private communication and presentation at *COSMO-RS Symposium 2009*, http://www.cosmologic.de/index.php?cosId=1901&crId=5.
22. Peters M, Greiner L, Leonhard K. Illustrating computational solvent screening: prediction of standard Gibbs energies of reaction in solution *AICHE J.* 2008;54: 2729–2734.
23. Becke AD. Density-functional exchange-energy approximation with correct asymptotic behaviour. *Phys. Rev. A* 1988;38: 3098–3100.
24. Perdew JP. Density-functional approximation for the correlation-energy of the inhomogenous electron gas. *Phys. Rev. B* 1986;33:8822–8824.
25. Schäfer A, Huber C, Ahlrichs R. Fully optimized contracted Gaussian basis sets of triple zeta valence quality for atoms Li to Kr. *J. Chem. Phys.* 1994;100:5829–5835.

25

THERMODYNAMICS AND RELATIVE SOLUBILITY PREDICTION OF POLYMORPHIC SYSTEMS

YURIY A. ABRAMOV

Pfizer Global Research & Development, Pharmaceutical Sciences, Groton, CT, USA

KLIMENTINA PENCHEVA

Pfizer Global Research & Development, Pharmaceutical Sciences, Sandwich, Kent, UK

25.1 INTRODUCTION

Polymorphism of the crystalline state of pharmaceutical compounds is quite a common phenomenon, which has been the subject of intense investigation for more than 40 years [1]. Polymorphs may significantly differ from each other in variety of physical properties such as melting point, enthalpy and entropy of fusion, heat capacity, density, dissolution rate, and intrinsic solubility. These differences are dictated by the differences in the free energies of the forms, which in turn determine their relative stability at specific temperatures. Two polymorphs are monotropically related to each other if their relative stability remains the same up to their melting points. Otherwise, the forms are related to each other enantiotropically and may display a solid–solid transition at a temperature below the melting point. In practice, monotropic and enantiotropic behavior are usually differentiated by several simple rules based on the experimental heats of fusion, entropies of fusion, heat of solid–solid transition, heat capacities, and densities [2, 3].

In the pharmaceutical industry, drug polymorphism can be a critical problem, and is the subject of various regulatory considerations [4, 5]. One of the principal concerns is based on an effect that polymorphism may have on a drug's bioavailability due to change of its solubility and dissolution rate [6]. A famous example of a polymorphism-induced impact is the anti-HIV drug Norvir (also known as Ritonavir) [7]. Abbot Laboratories had to stop sales of the drug in 1998 due to a failure in a dissolution test, which was caused by the precipitation of a more stable form II [8]. As a result, Abbot lost an estimated $250 million in the sales of Norvir in 1998 [9, 10].

A large number of studies have been focused on the polymorphism effect on solubility, many of which were summarized by Pudipeddi and Serajuddin [11]. Several-fold solubility decrease was observed for many polymorphic systems. Therefore, in pharmaceutical industry, it is quite crucial to get comprehensive experimental information on the available drug polymorphs and their relative stability and solubility. Beyond that, it is important to be able to perform an estimation of the potential impact of an unknown, more stable form on a drug's solubility. Knowledge of such an impact should be considered in a risk assessment of the API solid form nomination for commercial development.

There have been a number of studies considering the quantitative models to estimate the solubility ratio of two polymorphs based on the thermal properties of both forms [11–15]. One of the major objective of this work is to determine the potential impact of an unknown more stable form on the drug solubility. This is accomplished by reevaluating those models and paying a special attention to errors, which may be introduced by the most common assumptions with the hope of producing a new more accurate model. Such model should satisfy the following two conditions. It should

Chemical Engineering in the Pharmaceutical Industry: R&D to Manufacturing Edited by David J. am Ende

require a smaller number of input parameters, predominately relying on the thermal properties of only one (the known) form. When applied to a pair of observed polymorphs, the accuracy of the solubility ratio prediction by this equation should be at least as accurate as any currently known model.

25.2 METHODS

Methods used in this work are based on a combination of purely theoretical considerations and statistical analysis of available experimental data. A theoretical analysis of all popular approaches for prediction of absolute and relative solubilities of crystalline forms was performed. Special attention was paid to errors that are introduced by each of the approximations. Literature reports were carefully reviewed for solubility and thermal data of the organic crystals, with focus on drug-like molecules. In order to increase the statistical significance of the analysis, a comprehensive compilation was made of available polymorph solubility ratio data. However, only low solubility data (dilute solutions) for nonsolvated polymorphs was considered.

25.3 RESULTS AND DISCUSSION

25.3.1 Solubility of a Crystalline Form

The solubility, X_i, of a crystal form i in a solution can be presented as

$$\ln X_{i_i} = \ln\left(\frac{X_i^{id}}{\gamma_i}\right) = -\frac{\Delta G_i}{RT} - \ln\gamma_i \qquad (25.1)$$

where X_i^{id} is an ideal solubility; γ_i is an activity coefficient, which accounts for deviations from the ideal behavior in a mixture of liquid solute and solvent; ΔG_i is a free energy difference between the liquid and solid solute at the temperature of interest, T, and R is the universal gas constant. In case no additional phase transition takes place in the temperature range between the temperature of interest, T, and the melting point, T_m, the ΔG_i can be presented as

$$\Delta G_i = \Delta H_{fus}\left(1-\frac{T}{T_m}\right) + \int_{T_m}^{T} \Delta C_p dT - T\int_{T_m}^{T} \frac{\Delta C_p}{T} dT \qquad (25.2)$$

Here, ΔH_{fus} is the heat of fusion of the polymorph i at its melting point, T_m; ΔC_p is a difference between the heat capacities of the liquid and solid states of the form i, which is always positive. For practical reasons, it is usually assumed that ΔC_p is constant and equal to one estimated at the T_m, ΔC_{pm}. In that case the free energy difference, ΔG_i, can be presented as

$$\Delta G_i = \Delta H_{fus}\left(1-\frac{T}{T_m}\right) - \Delta C_{pm}(T_m - T) + \Delta C_{pm} T \ln\frac{T_m}{T} \qquad (25.3)$$

However, as a rule, the ΔC_{pm} property is not available and further approximations should be taken. The most popular assumptions which are used in the literature are $\Delta C_{pm} = 0$ (Assumption A) and $\Delta C_{pm} = \Delta S_{fus}$ (Assumption B), where ΔS_{fus} is entropy of fusion at the melting point, $\Delta S_{fus} = \Delta H_{fus}/T_m$. Equation 25.3 is simplified upon these assumptions to equations 25.4 and 25.5, respectively.

$$\Delta G_i = \Delta H_{fus}\left(1-\frac{T}{T_m}\right) \qquad (25.4)$$

$$\Delta G_i = \Delta H_{fus}\frac{T}{T_m}\ln\frac{T_m}{T} = \Delta S_{fus} T \ln\frac{T_m}{T} \qquad (25.5)$$

While the first Assumption A is usually justified by negligibly low value of the ΔC_{pm} (which is not always true), the latter one (B) is based on the observation by Hildebrand and Scott that $\ln X_i^{id}$ is linearly related to $\ln T$ [16].

In order to understand the errors introduced by both assumptions, they were mathematically derived below from equation 25.3 based on a first-order Taylor expansion of $\ln(T_m/T) \approx ((T_m/T)-1)$, which is correct only in case of T_m/T close to 1 (Table 25.1).

Assumption A

Transformation of $\ln(T_m/T)$ to $((T_m/T)-1)$ in the last term of the equation 25.3, results in the complete cancellation of the last two terms of the equation

$$\Delta G_i \approx \Delta H_{fus}(1-(T/T_m)) - \Delta C_{pm}(T_m - T) + \Delta C_{pm} T((T_m/T)-1) = \Delta H_{fus}(1-(T/T_m))$$

Thus, the applied transformation is equivalent to neglecting the ΔC_{pm} term ($\Delta C_{pm} = 0$, equation 25.4). Since at $T < T_m$, $((T_m/T)-1)$ is always larger than $\ln(T_m/T)$ (Table 25.1) and ΔC_{pm} is always positive, Assumption A leads to the systematic overestimation of the ΔG_i resulting into underestimation of the solubility relative to the predictions based on the equation 25.3. The ΔG_i error introduced by the Assumption A relative to equation 25.3 is related to the error of the first-order Taylor series expansion and can be presented as

$$\Delta G^{A}_{i,\mathrm{error}} = \Delta C_{pm} T\left\{\left(\frac{T_m}{T}-1\right) - \ln\frac{T_m}{T}\right\} \qquad (25.6)$$

This error is proportional to the ΔC_{pm} property, and increases with T_m/T due to an increasing inaccuracy in the first-order Taylor expansion (Table 25.1).

Assumption B

In an attempt to counterbalance the error introduced by the direct Taylor expansion transformation used in the

TABLE 25.1 Relative Errors of the First-Order ln (T_m / T) Expansions for Different T_m / T Values

T_m/T[a]	$\ln(T_m/T)$	$(T_m/T)-1$[b]	Relative Error (%)[b]	$2((T_m/T)-1)/((T_m/T)+1)$[c]	Relative Error (%)[c]
1.1 (330/300)	0.095	0.1	4.9	0.095	−0.1
1.2 (360/300)	0.182	0.2	9.7	0.182	−0.3
1.3 (390/300)	0.262	0.3	14.3	0.261	−0.6
1.4 (420/300)	0.337	0.4	18.9	0.333	−0.9
1.5 (450/300)	0.406	0.5	23.3	0.400	−1.3
1.6 (480/300)	0.470	0.6	27.7	0.462	−1.8

[a] Examples of T_m and T values in K are presented in the parenthesis. T is chosen to be close to the room temperature.
[b] First-order Taylor series expansion: $\ln(T_m/T) \approx ((T_m/T)-1)$.
[c] First-order expansion adopted by Hoffman [22] $\ln(T_m/T) \approx 2((T_m/T)-1)/((T_m/T)+1)$. This expansion is significantly more accurate than the first-order Taylor expansion.

Assumption A, one may apply a reverse transformation, $((T_m/T)-1) \approx \ln(T_m/T)$, to equation 25.4

$$\Delta G_i = \Delta H_{fus}(1-(T/T_m)) = \Delta H_{fus}(T/T_m)((T_m/T)-1)$$
$$\approx \Delta H_{fus}(T/T_m)\ln(T_m/T) = \Delta S_{fus}T\ln(T_m/T)$$

The result is equivalent to equation 25.5, which was derived under assumption of $\Delta C_{pm} = \Delta S_{fus}$. Since ΔH_{fus} (and ΔS_{fus}) is always positive and the error introduced by the reverse transformation is opposite to the one introduced by the direct transformation (Assumption A), a cancellation of errors should take place. The resulting ΔG_i error introduced by the Assumption B relative to equation 25.3 is equal to

$$\Delta G^{B}_{i,error} = \Delta G^{A}_{i,error} + \Delta S_{fus}T\left\{\ln\frac{T_m}{T} - \left(\frac{T_m}{T}-1\right)\right\}$$
$$= (\Delta C_{pm}-\Delta S_{fus})T\left\{\left(\frac{T_m}{T}-1\right)-\ln\frac{T_m}{T}\right\} \quad (25.7)$$

It is apparent from this equation that the error will change sign in case of $\Delta S_{fus} > \Delta C_{pm}$, resulting in underestimation of the ΔG_i and overestimation of solubility relative to the predictions based on the equation 25.3. In the case where ΔS_{fus} is more than twice as large as ΔC_{pm}, an absolute error introduced by Assumption B will exceed the error introduced by Assumption A. This phenomenon may have resulted in contradicting results of the relative accuracy of Assumptions A and B in the literature [17–20]. It was shown recently [21] that a relation between ΔS_{fus} and ΔC_{pm} properties is dependent on a chemical class of organic compounds. A ratio of the absolute ΔG_i errors introduced by Assumptions B (equation 25.7) and A (equation 25.6), $\Delta G^{B}_{i,error}/\Delta G^{A}_{i,error} = (|\Delta C_{pm}-\Delta S_{fus}|/\Delta C_{pm})$, is presented for 68 organic compounds in Figure 25.1. Only 12 nondrug-like compounds out of total 68 displayed a relative error of more than 1, indicating that Assumption B introduces a higher absolute error than Assumption A. A majority of these compounds can be characterized by the low value of their differential heat capacities, $\Delta C_{pm} < 40$ J/(mol K) (Figure 25.1). All of these considerations provide justification for the application of Assumption B over Assumption A for drug-like compounds.

Assumption C

Another valuable approximation of equation 25.2, was proposed by Hoffman [22] based on a significantly more accurate series expansion of $\ln(T_m/T)$ than the first-order Taylor expansion applied above (Table 25.1)

$$\Delta G_i = \Delta H_{fus}(T_m-T)\frac{T}{T_m^2} \quad (25.8)$$

The differential heat capacity, ΔC_p, is assumed to be both not negligible and independent of temperature. The lack of significant errors introduced by the $\ln(T_m/T)$ expansion,

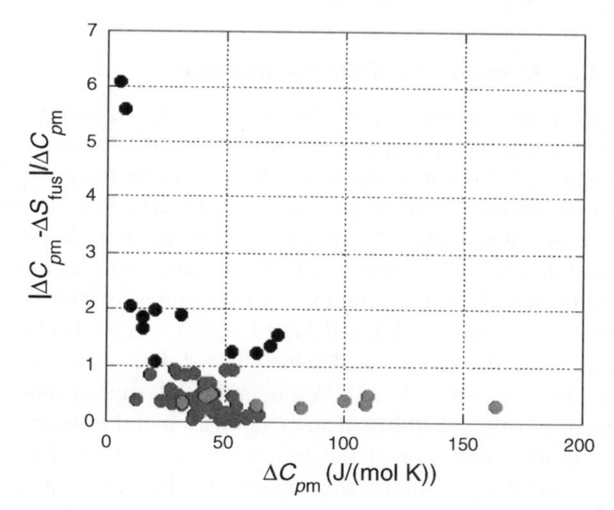

FIGURE 25.1 A ratio of the absolute ΔG_i errors introduced by the Assumptions B and A relative to equation 25.3, $|\Delta C_{pm}-\Delta S_{fus}|/\Delta C_{pm}$, versus differential heat capacity values ΔC_{pm} for 68 organic compounds. The closer this ratio is to zero, the lower is the error introduced by the Assumption B relative to equation 25.3. The compounds for which the Assumption B introduces higher absolute error than the Assumption A are highlighted in black. The drug compounds (Paracetamol [20], Anisic acid [20], Diethylstilbestrol [20], Mannitol [20], Naproxen [20], Caffeine I [25], Carbamazepine I [25], Progesterone I [25], Acetamide [27]) are highlighted in light gray. All other data are taken from Pappa et al. [21]

TABLE 25.2 Absolute Errors of the ΔG_i Predictions at the Room Temperature According to the Equation 25.3 and Assumptions A (Equation 25.4), B (Equation 25.5), and C (Equation 25.8) Relative to the Results Obtained Utilizing Temperature-Dependent ΔC_p Values (Equation 25.9)[a]

Name	T_m (K)	ΔH_{fus} (kJ/mol)	ΔC_{pm} (J/(mol K))	ΔC_p (T=298.2 K) (J/(mol K))	Error Relative to Equation 25.9 (kJ/mol)			
					Equation 25.3	Equation 25.4	Equation 25.5	Equation 25.8
Carbamazepine I [25]	463.7	26.3	109.8	164.6	0.7	4.4	2.5	1.0
Carbamazepine III [25]	452.4	27.2	111.3	184.3	1.0	4.3	2.5	1.1
Paracetamol [24]	442.2	28.1	99.6	165.8	0.6	3.3	1.6	0.3
Anisic acid [20]	455.4	27.8	81.4	150.6	0.8	3.3	1.4	0.0
Diethylstilbestrol [20]	441.8	28.8	43.8	262.3	2.0	3.2	1.5	0.2
Mannitol [20]	438.7	50.6	163.8	290.3	1.1	5.3	2.4	0.1
Naproxen [20]	428.5	31.5	108.6	220.3	0.9	3.3	1.7	0.4
MAE[b] (kJ/mol)					*1.0*	*3.9*	*1.9*	*0.4*

[a] The ΔA_2 term (equation 25.9) is different from zero only in the case of carbamazepine.
[b] Mean absolute error is calculated as an arithmetic average of the absolute errors of the predictions performed by the corresponding approach.

and perhaps a more justified approximation for the ΔC_p, make Assumption C a generally more thermodynamically sound model than Assumptions A and B. Additionally, equation 25.8 can be seen as equivalent to equation 25.4 (Assumption A) scaled down by a factor of T/T_m. This effectively introduces a correction for the overestimation of the ΔG_i by Assumption A.

25.3.2 Comparison of the Assumptions

A rigorous comparison of Assumptions A, B, and C is complicated by the fact that the differential heat capacity in equation 25.2 is temperature dependent, and for the general case increases as temperatures decrease [18, 20, 23]. Even for the cases when the ΔC_p at the melting point is known, equation 25.3 might not produce a reliable reference for comparison. An accurate temperature dependence of the heat capacities of both solid and liquid states in a polynomial form, $C_p = A_0 + A_1 T + A_2 T^2$, has limited data available in the literature [20, 24, 25]. Applicability of such a model depends on the reliability of an extrapolation of the observed temperature behavior of the heat capacities above (liquid state) and below (solid state, supercooled liquid) T_m at the temperature of interest. The difference between the coefficients (A_i) of the liquid and solid forms reflects a temperature dependence of the differential heat capacity. In such a case, the free energy difference between the liquid and solid solutes can be presented as

$$\Delta G_i = \Delta H_{fus}\left(1-\frac{T}{T_m}\right) - \Delta A_0(T_m - T) + \Delta A_0 T \ln\frac{T_m}{T} - \Delta A_1 \frac{(T_m-T)^2}{2} - \Delta A_2 \frac{2T_m^3 + T^3 - 3TT_m^2}{6} \quad (25.9)$$

where $\Delta A_i = A_i$ (liquid) $- A_i$ (solid). In Table 25.2, ΔG_i predictions at room temperature using equation 25.3 and Assumptions A, B, and C are compared with the results based on equation 25.9. The differential heat capacities of all the compounds increase significantly at room temperature relative to the values at their melting points (Table 25.2). The temperature dependence of the ΔC_p leads to a decrease of the predicted ΔG_i values at the room temperature relative to the predictions based on the differential heat capacities at T_m (equation 25.3). A resulting mean absolute error (MAE) of equation 25.3 predictions is 1.0 kJ/mol (Table 25.2). This ΔG_i error corresponds to an average underestimation of the ideal solubilities at room temperature by 33%. The MAE values of the ΔG_i predictions based on Assumptions A, B, and C for the same compounds relative to results obtained by the equation 25.9 are 3.9, 1.9, and 0.4 kJ/mol, respectively. The corresponding errors of the ideal solubility predictions at the room temperature are 79%, 54%, and 15%, respectively. Thus, the presented results demonstrate that the Hoffman approximation significantly outperforms Assumptions A and B. The largest error of ideal solubility prediction at ambient temperature is made using Assumption A which introduces a large ΔG_i error.

25.3.3 Application to Polymorphs Solubility Ratio

The solubility ratio of polymorphs can be presented by equation 25.10, and seems to be an optimal test for validation of the different ΔG_i models considered in the Section 25.3.1.

$$\frac{X_i}{X_j} = \frac{X_i^{id}\gamma_j}{X_j^{id}\gamma_i} = \frac{\gamma_j}{\gamma_i}\exp\left(-\frac{\Delta G_i - \Delta G_j}{RT}\right) \equiv \frac{\gamma_j}{\gamma_i}\exp\left(-\frac{\Delta\Delta G_{ij}}{RT}\right) \quad (25.10)$$

Recently, evaluations of different models for polymorph solubility ratio prediction based on thermal properties of the polymorphs were reported [11, 12]. Pudipeddi and Serajuddin have found that for 10 polymorphic pairs, predictions based on Assumption C were "slightly closer" to the experimental data than the results obtained by Assumption A [11]. Mao et al. have considered calculations based on the Assumption A [12]. A validation of this approach on nine polymorphic systems led to the conclusion that the utilization of Assumption A typically leads to an error of only 10% or less. An obvious drawback of these two studies is that the very limited datasets of the polymorph pairs were adopted for the testing of only selected assumptions. Therefore, to increase statistical significance of the results, further side-by-side verification of all three assumptions using larger experimental data sets could prove to be very important.

Two different data sets were selected for the models validation in this study, which contains 10 monotropically related (Table 25.3) and 18 enantiotropically related (Table 25.4) pairs of nonsolvated polymorphs. Each data point in these sets contains information on experimental properties such as solubility, X_i, melting point, T_m, and heats of fusion, ΔH_{fus}. The following considerations were taken into account during the data selection. There is quite a common misperception that the polymorph solubility ratio is solvent independent. However, according to equation 25.10, this is only true when the activity coefficients for the two polymorphs are identical to each other in any solvent [12, 26]. This takes place in the case of an infinite solubility limit (dilute solution). In such a case, each polymorph in the liquid state is not a significant part of the solvent system in which the actual solubility is measured. Thus, whenever possible, solubility data was chosen for polymorphs approximately tens of milligram per milliliter or less. Moreover, at these low concentrations, there is no need to convert milligram per milliliter or microgram per milliliter units to mol fractions, in which equation 25.10 is presented. One drawback of the selection of very low solubility data is a higher standard deviation of the experimental polymorph solubility ratios (Appendix 25.A).

25.3.3.1 Monotropic Case The initial validation of the solubility ratio models was performed using monotropically related polymorphs. In the following discussions, notations 1 and 2 will refer to the higher and lower soluble polymorphs. Equations used for the solubility ratio predictions in this section are explicitly listed in Appendix 25.B. Given that low solubility experimental data was selected for the test, it seems reasonable to expect that cancellation will not only take place between the activity coefficients of both polymorphs in the solution, but also between the errors introduced by the ΔG_i assumptions. Results of the relative solubility predictions utilizing Assumptions A, B, and C (Appendix 25.B) for each polymorph are presented in Table 25.3. The corresponding MAE values relative to the experimental X_1/X_2 observations are 1.01, 0.50, and 0.32, respectively. These observations disagree with previous reports that Assumption A results in an error of only 10% or less [12], and that Assumption C is just slightly closer to the experimental data than the results obtained by Assumption A [11]. The obtained MAE values demonstrate that a complete cancellation of errors does not take place, and, as a result, Assumption C remains significantly more accurate than the others. According to the error analysis presented in the Section 25.3.1 (equations 25.6 and 25.7), the lack of error cancellation in the $\Delta\Delta G_{12}$ prediction can be accounted for by nonnegligible differences of the ΔC_{pm} (($\Delta C_{pm} - \Delta S_{fus}$) in case of the Assumption B) and/or T_m properties between the two polymorphs. For example, in case of Assumption A, the error of the $\Delta\Delta G_{12}$ prediction relative to the one based on equation 25.3 can be presented as a difference of $\Delta G^{A}_{i,error}$ errors (equation 25.6) between two polymorphs

$$\Delta\Delta G_{error} = T\left\{\Delta C_{pm1}\left[\left(\frac{T_{m1}}{T}-1\right)-\ln\frac{T_{m1}}{T}\right] - \Delta C_{pm2}\left[\left(\frac{T_{m2}}{T}-1\right)-\ln\frac{T_{m2}}{T}\right]\right\} \quad (25.11)$$

The following two limiting cases can be derived from equation 25.11. In the case of relatively insignificant variations of the ΔC_{pm} terms, the ΔG_{error} is proportional to $T\{[((T_{m1}/T)-1)-\ln(T_{m1}/T)]-[((T_{m2}/T)-1)-\ln(T_{m2}/T)]\}$. In the case where variations of ΔC_{pm} are noticeably more significant than the variations of T_m, the $\Delta\Delta G_{error}$ is proportional to $(\Delta C_{pm1} - \Delta C_{pm2})$. The ΔC_{pm} values should be replaced by the $(\Delta C_{pm} - \Delta S_{fus})$ differences for the error estimation of the $\Delta\Delta G_{12}$ prediction based on the Assumption B.

It is easy to show from equation 25.10 that for dilute solutions, the difference between the natural logarithms of polymorph solubility ratios as predicted by Assumptions A or B, and equation 25.3, should be proportional to the $\Delta\Delta G_{error}$

$$\ln\left(\frac{X_1}{X_2}\right)_{A,B} - \ln\left(\frac{X_1}{X_2}\right)_{\text{equation 25.3}} = -\frac{\Delta\Delta G_{error}}{RT} \quad (25.12)$$

According to equation 25.11 in case of the Assumption A this difference will be equal to $\{\Delta C_{pm2}[((T_{m2}/T)-1)-\ln(T_{m2}/T)]-\Delta C_{pm1}[((T_{m1}/T)-1)-\ln(T_{m1}/T)]\}/R$. It is reasonable to propose that the difference between natural logarithms of polymorph solubility ratios as predicted by Assumptions A or B, and those observed experimentally, may be described by the similar factors as presented in equations 25.11 and 25.12. In the absence of the ΔC_{pm} values, a correlation was tested between the $\ln(X_1/X_2)$

TABLE 25.3 Comparison of the Experimental and Predicted Solubility Ratios for Monotropically Related Polymorphic Pairs

Compound	T_{m1} (K)	ΔH_{fus1} (kJ/mol)	T_{m2} (K)	ΔH_{fus2} (kJ/mol)	T (K)	$S_1/S_{2,exp}$	Assumption A	Assumption B	Assumption C	Equation 25.13	Equation 25.14	Equation 25.15	Equation 25.16
Chloramphenicol palmitate (A/B) [27, 28]	362	41.9	368	64.1	303	4.2	5.95	4.93	4.19	5.14	4.74	4.10	3.60
Tolbutamide (I/III) [27, 29]	379	18.5	400	24.5	303	1.22	2.42	2.08	1.84	1.29	1.78	1.65	1.55
Ritonavir (I/II) [9]	395.2	56.4	398.2	63.3	298	2.39[a]	2.29	2.01	1.80	2.17	2.02	1.83	1.69
MK571 (II/I) [30]	425.2	49.0	437.2	54.0	309	1.64	2.59	2.08	1.77	1.56	1.77	1.61	1.50
Cyclopenthiazide (II/I) [31][b]	496.2	98.42	512.5	105.5	310	1.78	6.32	3.40	2.30	1.72	2.96	2.31	1.93
E2101 (II/I) [32]	413.0	35.2	421.3	38.2	298	1.25	1.74	1.54	1.40	1.32	1.43	1.35	1.28
Indomethacine (α/γ) [15]	429.2	36.14	435.2	36.49	318	1.1	1.19	1.14	1.10	1.11	1.04	1.03	1.03
Acemetacin (II/I) [3]	423.2	48.4	423.7	50.7	293	1.67	1.36	1.29	1.23	1.52	1.34	1.27	1.22
Torasemide (II/I) [33]	430.2	29	434.7	37.2	293	2.74	3.26	2.58	2.16	2.51	3.00	2.45	2.10
Cimetidine (A/B) [34, 35]	413.5	44.03	413.7	44.08	298	1.15	1.01	1.01	1.01	1.21	1.01	1.00	1.00
MAE							*1.01*	*0.50*	*0.32*	*0.19*	*0.38*	*0.27*	*0.33*

Results of application of the equations 25.14–25.16 adopting $T_m = T_{m2}$ are presented. The explicit equations for solubility ratio predictions are presented in Appendix 25.B.

[a] Polymorph solubility data in ethyl acetate:heptanes (2:1) mixture is adopted for solubility ratio estimation.

[b] An enantiotropic relationship between forms I and II with a very low transition temperature was proposed in the literature [14].

$-\ln(X_1/X_2)_{\text{exp}}$ predictions based on the different assumptions and $\{[((T_{m2}/T)-1)-\ln(T_{m2}/T)]-[((T_{m1}/T)-1)-\ln(T_{m1}/T)]\}$ property (Figure 25.2). High linear correlation coefficients, R^2, of 0.92 and 0.91 were found for the predictions based on Assumptions A and B, respectively (Figure 25.2a and b). This observation suggests a higher and a more systematic contribution to the $\Delta\Delta G_{\text{error}}$ by the differences in the T_m values, rather than by the differences in the ΔC_{pm} or $(\Delta C_{pm}-\Delta S_{\text{fus}})$ properties. A noticeably weaker correlation (having an R^2 of 0.72, Figure 25.2c) was observed for the predictions based on Assumption C.

Found simple linear regressions (Figure 25.2) can be used for estimations of likely errors in the solubility ratio predictions of monotropically related polymorphs based on the different assumptions. The MAE values of the prediction (0.24, 0.22, and 0.19, respectively) are based on Assumptions A, C, and B after the errors are corrected by using simple functions of the melting points. The latter result corresponds to the best agreement with the experimental observations using the approaches presented in Table 25.3. This suggests that the polymorph solubility ratio of the monotropically related polymorphs can be best predicted through the following relationship

$$\ln\frac{X_1}{X_2}=\ln\left(\frac{X_1}{X_2}\right)_{\text{B}}+0.188-43.096\left\{\left[\left(\frac{T_{m2}}{T}-1\right)-\ln\frac{T_{m2}}{T}\right]-\left[\left(\frac{T_{m1}}{T}-1\right)-\ln\frac{T_{m1}}{T}\right]\right\} \quad (25.13)$$

Based on the above observation of the high contribution to the $\Delta\Delta G_{\text{error}}$ by the differences in the T_m values, an alternative approach can be suggested. In order to better counterbalance the prediction errors, it was proposed to adopt a single T_m value for both polymorphs used in the solubility ratio predictions. In this case, an improvement of the predictions should take place through the increase of ΔG_1 ($T_m=T_{m2}$), or the decrease of ΔG_2 ($T_m=T_{m1}$). The following simplifications of the $\Delta\Delta G_{12}$ calculation based on Assumptions A, B, and C are proposed

$$\Delta\Delta G_{12}=(\Delta H_{\text{fus1}}-\Delta H_{\text{fus2}})\left(1-\frac{T}{T_m}\right) \quad (25.14)$$

$$\Delta\Delta G_{12}=(\Delta H_{\text{fus1}}-\Delta H_{\text{fus2}})\frac{T}{T_m}\ln\frac{T_m}{T} \quad (25.15)$$

$$\Delta\Delta G_{12}=(\Delta H_{\text{fus1}}-\Delta H_{\text{fus2}})(T_m-T)\frac{T}{T_m^2} \quad (25.16)$$

Besides a possible improvement of the polymorph solubility prediction, the proposed equations more importantly depend on only two input parameters: T_m of one of the forms,

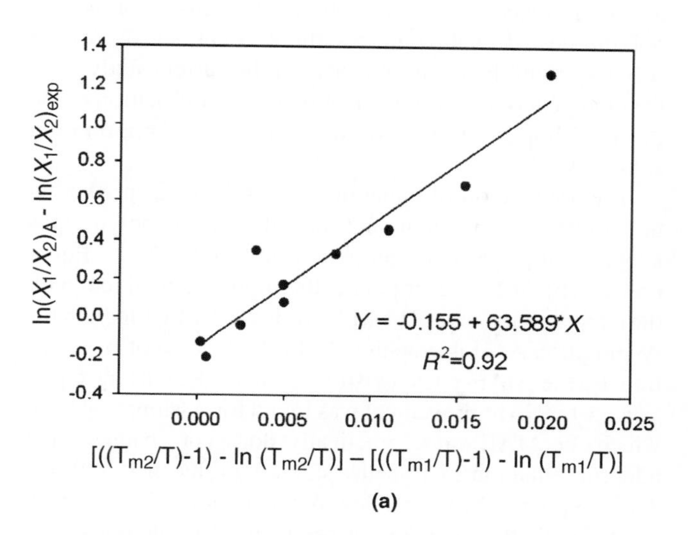

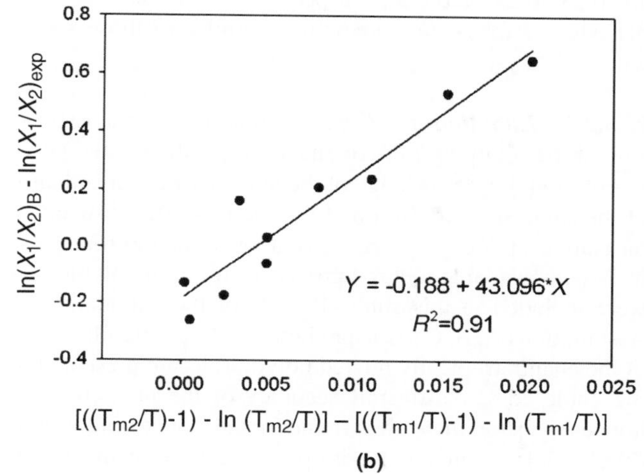

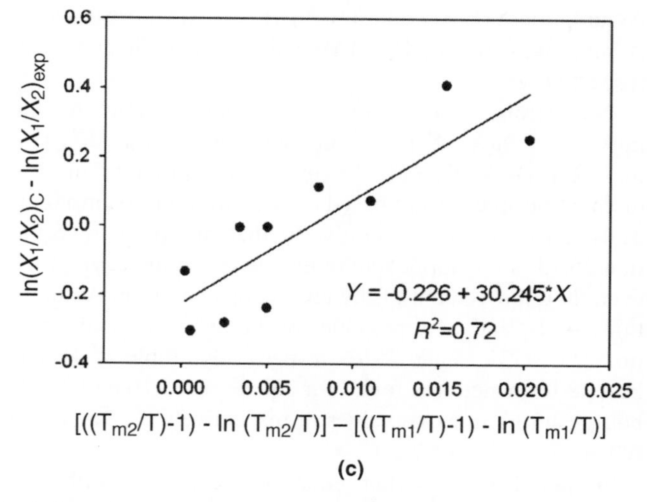

FIGURE 25.2 A correlation between the $\ln(X_1/X_2)-\ln(X_1/X_2)_{\text{exp}}$ values based on the Assumptions A (a), B (b), and C (c) and the $\{[((T_{m2}/T)-1)-\ln(T_{m2}/T)]-[((T_{m1}/T)-1)-\ln(T_{m1}/T)]\}$ property.

and a difference between the enthalpies of fusion of the two polymorphs. This fact makes these equations useful for solving one of the major objective of the current study—the development of a working equation for the estimation of the potential impact of an unknown more stable form on drug solubility.

The application of equations 25.14–25.16 in predicting the solubility ratio of monotropically related polymorphs adopting $T_m = T_{m2}$ is presented in Table 25.3. Equations 25.14 and 25.15 dramatically improve agreement with the experimental data. The MAE drops from 1.01 to 0.38 for Assumption A using equation 25.14. In the case of Assumption B, the MAE changes from 0.50 to 0.27 using equation 25.15. No improvement was found for Assumption C, in which the MAE value practically does not change when adopting equation 25.16 with a single T_m value of T_{m2}. When $T_m = T_{m1}$, the MAE values for equations 25.14–25.16 are 0.28, 0.29, and 0.35, respectively. This demonstrates behavior of the X_1/X_2 predictions similar to those found with $T_m = T_{m2}$.

25.3.3.2 Enantiotropic Case A thermodynamic expression of the solubility ratio of enantiotropically related polymorphs requires knowledge of the temperature and enthalpy of the solid–solid transition [12], which is often difficult to measure accurately. For this reason, only enantiotropic systems with available melting properties for both polymorphs were included in this study. Results of the application of Assumptions A, B, C to the predictions of the solubility ratio of the enantiotropically related polymorphs are presented in the Table 25.4. An overall accuracy of the predictions is noticeably better than it was found for the monotropic system (Table 25.3). As in the monotropic case, the agreement with the experimental data is worse for the calculations based on Assumption A (MAE value is 0.34), relative to those based on Assumptions B and C (MAE values are 0.28 and 0.25, respectively).

No strong correlation was found between the $\ln(X_1/X_2) - \ln(X_1/X_2)_{exp}$ values and the $\{[((T_{m2}/T) - 1) - \ln(T_{m2}/T)]\} - [((T_{m1}/T) - 1) - \ln(T_{m1}/T)]$ property in case of enantiotropic system based on the different assumptions. Thus, error correction similar to that proposed by equation 25.13 is not applicable to the enantiotropic case. However, an improvement of the predictions based on Assumptions A, B, and C is possible by the application of equations 25.14, 25.15 and 25.16, respectively (Table 25.4). The best performance was found for equations 25.16 and 25.15, both resulting in MAE values of respectively 0.22 and 0.24 (where $T_m = T_{m2}$ or T_{m1}).

It should be noted that equations cannot describe the change of the relative stability of the enantiotropically related polymorphs with temperature. To do so would result in the $\Delta\Delta G_{12}$ property having the wrong sign above the solid–solid transition, T_t, ($\Delta\Delta G_{12}(T_t) = 0$). Thus, the application of these equations to enantiotropic polymorphs is limited to systems with temperatures below T_t.

From the above considerations which are based on the analysis of the largest reported experimental data set of both monotropic and enantiotropic systems, an application of the original (equation 25.8, Appendix 25.B) and modified (equation 25.16) Hoffman approaches as well as of equation 25.15 are recommended for an accurate solubility ratio prediction for both monotropic and enantiotropic polymorphic systems. Since the latter two approaches utilize the melting temperature measurements of only one form, T_{m2} or T_{m1}, they can be used in combination with the statistical analysis of the differences of the heat of fusions of polymorphs, for an estimation of the potential impact of an unknown more stable form on drug solubilities (see Section 25.4 for more details).

EXAMPLE 25.1 PREDICTION OF IDEAL SOLUBILITY RATIO BETWEEN FORMS II AND I OF ACEMETACIN AT 293 K BASED ON THE ASSUMPTIONS A, B, AND C AND REGRESSION EQUATION 25.13

The thermal data for both forms of Acemetacin are presented in Table 25.3. Initially ΔG_{II} and ΔG_I properties should be calculated for each form adopting equations corresponding to Assumptions A (equation 25.4), B (equation 25.5), and C (equation 25.8). The resulting values at 293K are listed in Table 25.5. At the next step, differences between ΔG_{II} and ΔG_I properties should be calculated to obtain $\Delta\Delta G$ values. In order to calculate $\ln(X_{II}^{id}/X_I^{id})$ values, the negative of the $\Delta\Delta G$ predictions should be divided by RT factor. RT at 293K is equal to 8.314×10^{-3} (kJ/(mol K)) $\times$ 293 (K) = 2.436 kJ/mol. All the above steps are combined in the explicit equations presented in Table 25.B1 in Appendix 25.B. The $\ln(X_{II}^{id}/X_I^{id})$ value calculated based on the Assumption B is used in combination with $\{[((T_{m2}/T) - 1) \quad - \ln(T_{m2}/T)] - [((T_{m1}/T) - 1) - \ln(T_{m1}/T)]\}$ property for $\ln(X_{II}^{id}/X_I^{id})$ prediction based on the regression equation 25.13. Results of all the intermediate calculations are summarized in Table 25.5. Finally, exponent of $\ln(X_{II}^{id}/X_I^{id})$ results gives the polymorphs solubility ratio predictions based on all four methods. For this particular example, the best and the worst agreement with the experimental value of 1.67 is obtained by the equation 25.13 and Assumption C, respectively.

25.4 APPLICATION TO AN ESTIMATION OF LIKELY IMPACT ON DRUG SOLUBILITY BY UNKNOWN MORE STABLE FORM

Below we present two approaches to predict a likely change of drug solubility due to form change. The first thermal data approach is based on a combination of

TABLE 25.4 Comparison of the Experimental and Predicted Solubility Ratios for Enantiotropically Related Polymorphs

Compound	T_{m1} (K)	ΔH_{fus1} (kJ/mol)	T_{m2} (K)	ΔH_{fus2} (kJ/mol)	T (K)	$S_1/S_{2,exp}$	Assumption A	Assumption B	Assumption C	Equation 25.14	Equation 25.15	Equation 25.16
Axinitib (IV/VI) [36–38]	491.90	47.15	484.80	51.79	310	1.25	1.62	1.53	1.45	1.91	1.67	1.51
Axinitib (IV/XXV) [36–38]	491.90	47.15	490.40	50.43	310	1.25	1.54	1.42	1.33	1.60	1.45	1.34
Paracetamol (II/I) [27, 39]	429	26.9	442	28.1	303	1.3	1.45	1.30	1.21	1.16	1.13	1.11
Buspirone–HCl (II/I) [40]	476.8	42.24	463.0	47.45	303	1.7	1.49	1.49	1.46	2.04	1.78	1.60
Carbamazepine (I/III) [41]	462	26.4	448	29.3	299	1.20	1.19	1.21	1.21	1.47	1.37	1.30
F2692 [27]	453	27.17	445	29.32	303	1.8	1.15	1.16	1.15	1.31	1.25	1.20
Gepirone hydrochloride (II/I) [42]	485	41.6	453	47.1	303	2.01	0.99	1.19	1.31	2.06	1.80	1.62
Indiplon [43]	465.51	41.63	463.08	45.89	298	1.1	1.74	1.58	1.46	1.85	1.63	1.48
Nimodipine (I/II) [3]	397.2	39	389.2	46	298	2	1.52	1.49	1.46	1.94	1.78	1.66
Phenylbutazone (II/III) [27]	370	21.9	368	24.4	303	1.1	1.15	1.14	1.13	1.19	1.17	1.16
Piroxicam (I/II) [44]	475.8	36.54	472.9	37.43	310	0.99	1.06	1.07	1.06	1.13	1.10	1.08
Propranolol hydrochloride (I/II) [45]	436.2	31.35	435.0	36.62	293	1.34	1.98	1.75	1.60	2.03	1.78	1.61
Retinoic acid (II/I) [46]	456.3	36.8	456.9	37.1	310	1.32	1.05	1.04	1.03	1.04	1.03	1.03
RG12525 (II/I) [47]	431.0	43.10	427.8	46.86	304	1.26	1.41	1.35	1.31	1.54	1.43	1.36
Sulfathiazole (I/III) [27]	474.2	27.75	446.8	29.47	303	1.68	0.81	0.93	1.01	1.25	1.20	1.16
WIN63843 (I/III) [48]	337.7	28.87	334.4	31.88	296	1.04	1.04	1.04	1.05	1.15	1.14	1.13
Cimetidine (C/B) [34, 35]	417.5	43.13	413.7	44.08	298	1.23	0.99	1.01	1.03	1.11	1.09	1.08
Cimetidine (C/A) [34, 35]	417.5	43.13	413.5	44.03	298	1.07	0.98	1.01	1.02	1.11	1.09	1.08
MAE							*0.34*	*0.28*	*0.25*	*0.29*	*0.24*	*0.22*

Results of application of the equations 25.14–25.16 adopting $T_m = T_{m2}$ are presented. The explicit equations for solubility ratio predictions are presented in Appendix 25.B.

TABLE 25.5 Prediction of Ideal Solubility Ratio Between Forms II and I of Acemetacin at 293K Based on the Assumptions A, B, C and the Regression Equation 25.13

Approach	ΔG_{II} (kJ/mol)	ΔG_{I} (kJ/mol)	$\Delta\Delta G$ (kJ/mol)	$\{[((T_{m\mathrm{I}}/T)-1)-\ln(T_{m\mathrm{I}}/T)]$ $-[((T_{m\mathrm{II}}/T)-1)-\ln(T_{m\mathrm{II}}/T)]\}$	$\ln(X_{\mathrm{II}}^{id}/X_{\mathrm{I}}^{id})$	$X_{\mathrm{II}}^{id}/X_{\mathrm{I}}^{id}$
Assumption A	14.891	15.640	−0.749		0.308	1.36
Assumption B	12.321	12.932	−0.612		0.251	1.29
Assumption C	10.309	10.815	−0.506		0.208	1.23
Equation 25.13				5.257E-04	0.416	1.52

statistical analysis of the experimental heat of fusion differences between polymorphic pairs and the one proposed in the current work equation 25.16 for the ideal solubility ratio prediction. The second, solubility ratio approach is based on statistical results from experimental solubility ratio observations.

25.4.1 Thermal Data Approach

This approach is based on application of the modified Hoffman equation 25.16 coupled with statistical analysis of experimentally determined heat of fusion differences between polymorphs. The ideal solubility ratio predictions can be carried out for a known form with available melting temperature and likely changes in heat of fusion, $\Delta\Delta H_{\mathrm{fus}}$. In order to do that, a survey of thermal data for 101 polymorphic pairs was carried out, where most of the data where found in one literature source [27] and the rest were taken from the Tables 25.3 and 25.4 of the current study. Trends in heat of fusion changes between polymorphs were presented in the form of the cumulative relative frequency distribution in Figure 25.3. The cumulative relative frequency distribution is particularly useful for describing the likelihood that a variable (heat of fusion difference) will not exceed a certain value. It was found that there is a 50% probability that the change in heat of fusion for a polymorphic pair is less or equal to 3.0 kJ/mol (Figure 25.3). The probability of heat of fusion difference between a pair of polymorphs not exceeding values of 6.2 kJ/mol and 16.7 kJ/mol is respectively 80% and 95%. Combining these $\Delta\Delta H_{\mathrm{fus}}$ values with equation 25.16 allows estimation at the different probability levels of maximum impact on ideal solubility by a new more stable polymorph. Although the thermal data approach relies on the statistical analysis (of $\Delta\Delta H_{\mathrm{fus}}$), it introduces some degree of dependence on the thermal properties (T_m) of the reference form through equation 25.16. Therefore, predictions based on this method are form-specific.

25.4.2 Solubility Ratio Approach

An alternative approach is based on the statistical analysis of the polymorph solubility ratio observations. A survey of solubility changes for 153 polymorphic pairs was carried out, where most of data where found in open literature sources [11], and some were extracted from in-house Pfizer data or provided by company associated institutions. A

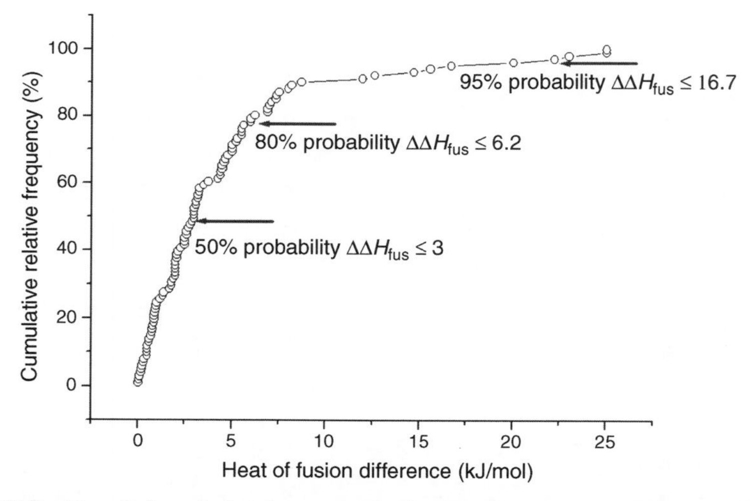

FIGURE 25.3 Cumulative relative frequency distribution of experimental differences of heats of fusion, $\Delta\Delta H_{\mathrm{fus}}$, for 101 polymorphic pairs. Data points corresponding to 50%, 80%, and 95% probabilities of $\Delta\Delta H_{\mathrm{fus}}$ is not exceeding a certain threshold are indicated by arrows.

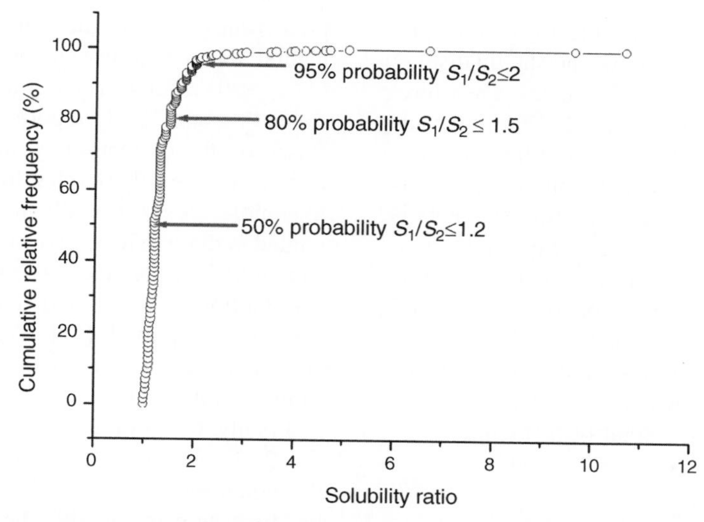

FIGURE 25.4 Cumulative relative frequency distribution of experimental solubility ratios, X_1/X_2, for 153 polymorphic pairs. Data points corresponding to 50%, 80%, and 95% probabilities of solubility ratio is not exceeding a certain threshold are indicated by arrows.

statistical analysis of the experimental data was performed on the basis of cumulative relative frequency distribution, presented in Figure 25.4. It was found that there is a 50% probability that the change in solubility is less than 1.2-fold for a polymorphic switch (Figure 25.3). The probability of the solubility ratio between a pair of polymorphs not exceeding value of 1.5 is 80%. It was also shown that only in 5% of the studied cases the change in solubility for polymorphic pairs would be more than twofold (Figure 25.4).

The presented trend of probabilities of the relative solubility changes is purely statistical and does not provide any direct dependence on thermal data of the current form. Therefore, this approach may be considered as form nonspecific one. In addition, majority of the solubility ratio measurements are performed at different temperatures in the range of 20–40°C [11], rather than at the room temperature, which may introduce some level of noise in the predictions based on this method.

EXAMPLE 25.2 ESTIMATION OF A LIKELY IMPACT ON SOLUBILITY BY A NEW FORM OF RITONAVIR

Analysis of a possible impact on solubility by a new form should be defined by a selected probability limit of the X_1/X_2 (the solubility ratio approach) and $\Delta\Delta H_{fus}$ (the thermal data approach) changes. In this example, the 80% probability was selected to provide a reasonably high level of confidence of predictions by both methods. In case more than one polymorphic form exists, a probability of further increase of the polymorphs solubility ratio as well as of the $\Delta\Delta H_{fus}$ should be estimated relative to the least stable form. In case of Ritonavir, the most unstable and soluble form is I. The 80% probability of heat of fusion increase according to the thermal data approach is not exceeding 6.2 kJ/mol (Figure 25.3). This $\Delta\Delta H_{fus}$ value together with the melting point of form I, $T_{mI} = 395.2$K (Table 25.3), are used to predict a likely change of solubility of form I at the room temperature by equation 25.16. The estimated impact is not exceeding 1.59-fold. The 80% probability of the change in solubility between two polymorphs according to the solubility ratio approach is less or equal to 1.5-fold. The two methods provide similar estimated change of the ideal solubility with respect to the least stable form for a probability level of 80%. It is known that a more stable form II was discovered later for Ritonavir. The observed heat of fusion difference and solubility ratio between forms II and I are respectively 6.9 kJ/mol and 2.39 (Table 25.3). These values correspond to respectively 82% and 98% probability limits of the $\Delta\Delta H_{fus}$ and X_1/X_2 changes, and exceed the thresholds predicted within the probability limit of 80%. Therefore, both approaches suggest a quite low probability of further impact on solubility by a hypothetical new stable form.

25.4.3 Qualification/Quantification of Impact of Likely Form Change on Drug Absorption

A significant solubility difference between two polymorphs can result in difference in oral absorption and may affect bioavailability [49]. Orally administrated immediate-release drug products are categorized in the Biopharmaceutics Classification System (BCS) according to their aqueous solubility and permeability [50]. These properties together with dissolution rate control drug absorption. Absolute bioavailability of a drug is also affected by first-pass intestinal and hepatic metabolism [51]. It is reasonable to assume that polymorphic forms of a particular compound should display similar permeabilities and first-pass clearances. Therefore, the differences in fraction absorbed and absolute bioavailability between oral products (based on different polymorphs) is controlled by solubility and dissolution rate. This assumes that the polymorphs interaction with excipients is negligible. While the dissolution rate can be generally controlled by changing the particle size, a thermodynamic aqueous solubility is a fundamental property of the polymorphic form which cannot be modified.

The classification of drug form solubility is based on dimensionless dose number D_0, which is a function of maximum dose strength D (mg) and solubility, S (mg/mL)

$$D_0 = \frac{D}{V_0 S}. \tag{25.17}$$

Here, V_0 is volume of water taken with the dose which is generally set to 250 mL. Solid forms of drugs with D_0 equal or less than 1 are considered being highly soluble [52]. According to the BCS system such forms are characterized as Classes I and III. An estimation of a likely change in solubility due to transformation to a new more stable form allows prediction of the potential impact on D_0 that a new form could present. The potential risk associated with the late discovery of a new stable form can be accessed based on a degree of probability of solubility (and D_0) change, as discussed above, and projected change of drug absorption. A qualitative analysis of the impact of form change on absorption can be based on the BCS system; here, we classify risk as associated with a potential change of the drug class from I to II or from III to IV. In addition, computational simulations (e.g., GastroPlus, Simulationsplus, Inc., Lancaster, CA) may be adopted for a (semi) quantitative analysis of sensitivity of drug absorption to a potential form change.

25.5 CONCLUSION

One of the main purposes of this study is to develop valid methods for the estimation of a potential impact of an unknown more stable form on drug solubility. This information has a crucial practical application in the pharmaceutical industry by supporting the risk assessment of an API solid form selection for commercial development of an oral drug. Two independent approaches to predict a likely change of drug solubility due to the form change were suggested in the current study. One of them is based on the modified Hoffman equation 25.16, which was found through a consistent theoretical consideration of the errors introduced by the different popular assumptions used for absolute and relative polymorph solubility predictions.

In addition, the first side-by-side validation of all three popular assumptions for the relative polymorph solubility prediction was performed on the largest up-to-date experimental dataset. It was demonstrated that Assumption A ($\Delta C_{pm} = 0$) results in noticeable errors which significantly exceed the previously reported values of 10% or less [12]. Based on the current study, this assumption is not recommended for the polymorph solubility ratio prediction of drug-like molecules, especially in case of the monotropically related systems. The superiority of Assumption C (Hoffman equation) over the other assumptions, and in particular over Assumption A was found to be much stronger than was previously reported [11]. Assumption B ($\Delta C_{pm} = \Delta S_{fus}$) demonstrated an intermediate performance between Assumptions A and C.

Finally, based on the error analysis, a new model, equation 25.13, was proposed for the solubility ratio prediction of the monotropically related polymorphs. This model provided the best agreement with the experimental dataset of 10 polymorphic pairs.

25.A APPENDIX

25.A.1 Propagation of Errors of the Solubility Ratio Measurements

Assuming independence of the solubility measurements of two polymorphs, X_1 and X_2, the standard deviation of the solubility ratio, $k = X_1/X_2$, can be expressed as [53]

$$\sigma(\kappa) = \left[\left(\sigma(X_1)\frac{\partial k}{\partial X_1}\right)^2 + \left(\sigma(X_2)\frac{\partial k}{\partial X_2}\right)^2\right]^{1/2} \tag{25.A.1}$$

In case of $\sigma(X_1) \approx \sigma(X_2) = \sigma(X)$, the following resulting equation can be obtained

$$\sigma(\kappa) = \left[\left(\frac{\sigma(X)}{X_2}\right)^2 + \left(\frac{k\sigma(X)}{X_2}\right)^2\right]^{1/2} = \frac{\sigma(X)}{X_2}(1+k^2)^{1/2} \tag{25.A.2}$$

Equation (25.A.2) demonstrates that the error of the solubility ratio measurements increases with increase of the k value, and with decrease of the polymorph solubility, X_2. For example, for the solubility X_2 of 0.2 μg/mL, $\sigma(X)$ equal to 0.02 μg/mL, and k value of 2, the $\sigma(k)$ is equal to 0.22.

25.B APPENDIX

25.8 ACKNOWLEDGMENTS

The authors would like to thank Mr. Brian Samas, Dr. Neil Feeder, Dr. Paul Meenan, Dr. Robert Docherty, and Dr. Bruno Hancock for the valuable comments and discussions. The authors also wish to thank Mr. Anthony M. Campeta for providing the experimental thermal and solubility data on the

TABLE 25.B1 Explicit Equations Used for Predictions of Polymorphs Solubility Ratio

Based on	Equation	Comments
Assumption A	$\ln\frac{X_1}{X_2}=\frac{\Delta H_{fus2}(1-\frac{T}{T_{m2}})-\Delta H_{fus1}(1-\frac{T}{T_{m1}})}{RT}$	
Assumption B	$\ln\frac{X_1}{X_2}=\frac{\Delta H_{fus2}\frac{T}{T_{m2}}\ln\frac{T_{m2}}{T}-\Delta H_{fus1}\frac{T}{T_{m1}}\ln\frac{T_{m1}}{T}}{RT}$	
Assumption C	$\ln\frac{X_1}{X_2}=\frac{\Delta H_{fus2}(T_{m2}-T)\frac{T}{T_{m2}^2}-\Delta H_{fus1}(T_{m1}-T)\frac{T}{T_{m1}^2}}{RT}$	
Equation 25.13	$\ln\frac{X_1}{X_2}=\ln\left(\frac{X_1}{X_2}\right)_B+0.188-43.096\left\{\left[\left(\frac{T_{m2}}{T}-1\right)-\ln\frac{T_{m2}}{T}\right]-\left[\left(\frac{T_{m1}}{T}-1\right)-\ln\frac{T_{m1}}{T}\right]\right\}$	Only for the monotropic system
Equation 25.14	$\ln\frac{X_1}{X_2}=\frac{(\Delta H_{fus2}-\Delta H_{fus1})(1-\frac{T}{T_m})}{RT}$	$T_m=T_{m2}$ or T_{m1}
Equation 25.15	$\ln\frac{X_1}{X_2}=\frac{(\Delta H_{fus2}-\Delta H_{fus1})\frac{T}{T_m}\ln\frac{T_m}{T}}{RT}$	$T_m=T_{m2}$ or T_{m1}
Equation 25.16	$\ln\frac{X_1}{X_2}=\frac{(\Delta H_{fus2}-\Delta H_{fus1})(T_m-T)\frac{T}{T_m^2}}{RT}$	$T_m=T_{m2}$ or T_{m1}

axitinib polymorphs. YAA is thankful to Mr. Brian D. Bissett for a thorough review of the manuscript.

REFERENCES

1. Brittain H.G.editor. *Polymorphism in Pharmaceutical Solids*, Marcel Dekker, New York 1999.
2. Burger A, Ramberger R. On the polymorphism of pharmaceuticals and other organic molecular crystals. I: theory of thermodynamic rules. *Mikrochim Acta* 1979; II: 259–271.
3. Grunenberg A, Henck J-O, Siesler HW. Theoretical derivation and practical application of energy/temperature diagrams as an instrument in preformulation studies of polymorphic drug substances. *Int. J. Pharm.* 1996; 129: 147–158.
4. Byrn S, Pfeiffer R, Ganey M, Hoiberg C, Poochikian G. Pharmaceutical solids: strategic approach to regulatory considerations. *Pharm. Res.* 1995; 12: 945–954.
5. Yu LX, Furness MS, Raw A, Woodland Outlaw KP, Nashed NE, Ramos E, Miller SPF, Adams RC, Fang F, Patel RM, Holcombe FO Jr., Chiu Y, Hussain AS. Scientific considerations of pharmaceutical solid polymorphs in abbreviated new drug applications. *Pharm. Res.* 2003; 20: 531–536.
6. Brittain HG, Grant DJW. Effects of polymorphism and solid–state solvation on solubility and dissolution rate. In: Brittain HG,editor, *Polymorphism in Pharmaceutical Solids*, Marcel Dekker, New York, 1999, pp. 279–330.
7. Kempf DJ, Marsh KC, Denissen JF, McDonald E, Vasavanonda S, Flentge CA, Green BE, Fino L, Park CH, Kong XP, Wideburg NE, Saldivar A, Ruiz L, Kati WM, Sham HL, Robins T, Stewart KD, Hsu A, Plattner JJ, Leonard JM, Norbeck DW. ABT-538 is a potent inhibitor of human immunodeficiency virus protease and has high oral bioavailability in humans. *Proc. Natl. Acad. Sci. U.S.A.* 1995; 92: 2484–2488.
8. Bauer J, Spanton S, Henry R, Quick J, Dziki W, Porter W, Morris J. Ritonavir: an extraordinary example of conformational polymorphism. *Pharm. Res.* 2001; 18: 859–866.
9. Chemburkar SR, Bauer J, Deming K, Spiwek H, Patel K, Morris J, Henry R, Spanton S, Dziki W, Porter W, Quick J, Bauer P, Donaubauer J, Narayanan BA, Soldani M, Riley D, McFarland K. Dealing with the impact of ritonavir polymorphs on the late stages of bulk drug process development. *Org. Process Res. Dev.* 2000; 4: 413–417.
10. Morissette SL, Soukasene S, Levinson D, Cima MG, Almarsson O. Elucidation of crystal form diversity of the HIV protease inhibitor Ritonavir by high-throughput crystallization. *Proc. Natl. Acad. Sci. U.S.A.* 2003; 100: 2180–2184.
11. Pudipeddi M, Serajuddin ATM. Trends in solubility of polymorphs. *J. Pharm. Sci.* 2005; 94: 929–939.
12. Mao C, Pinal R, Morris KR. A quantitative model to evaluate solubility of polymorphs from their thermodynamic properties. *Pharm. Res.* 2005; 22: 1149–1157.
13. Grant DJW, Higuchi T, *Solubility Behavior of Organic Compounds*, Wiley, New York, 1990.
14. Gu C-H, Grant DJW. Estimating the relative stability of polymorphs and hydrates from heats of solution and solubility data. *J. Pharm. Sci.* 2001; 90: 1277–1287.
15. Hancock BC, Parks M. What is the true solubility advantage for amorphous pharmaceuticals? *Pharm. Res.* 2000; 17: 397–403.
16. Hildebrand JH, Scott RL. *Regular Solutions*, Prentice-Hall, Englewood Cliffs, NJ, 1962.
17. Mishra DS, Yalkowsky SH. Ideal solubility of a solid solute: effect of heat capacity assumptions. *Pharm. Res.* 1992; 9: 958–959.
18. Neau SH, Flynn GL. Solid and liquid heat capacities of *n*-alkyl *para*-aminobenzoates near the melting point. *Pharm. Res.* 1990; 7: 1157–1162.
19. Neau SH, Flynn GL, Yalkowsky SH. The influence of heat capacity assumptions on the estimation of solubility parameters from solubility data. *Int. J. Pharm.* 1989; 49: 223–229.

20. Neau SH, Bhandarkar SV, Hellmuth EW. Differential molar heat capacity to test ideal solubility estimations. *Pharm. Res.* 1997; 14: 601–605.

21. Pappa GD, Voutsas EC, Magoulas K, Tassios DP. Estimation of the differential molar heat capacities of organic compounds at their melting point. *Ind. Eng. Chem. Res.* 2005; 44: 3799–3806.

22. Hoffman JD. Thermodynamic driving force in nucleation and growth processes. *J. Chem. Phys.* 1958; 29: 1192–1193.

23. Gracin S, Brinck T, Rasmuson AC. Prediction of solubility of solid organic compounds in solvents by UNIFAC. *Ind. Eng. Chem. Res.* 2002; 41: 5114–5124.

24. Hojjati H, Rohani S. Measurement and prediction of solubility of paracetamol in water–isopropanol solution. Part 2. Prediction. *Org. Process Res. Dev.* 2006; 10: 1110–1118.

25. Defossemont G, Randzio SL, Legendre B. Contributions of calorimetry for C_p determination and of scanning transitiometry for the study of polymorphism. *Cryst. Grow. Des.* 2004; 4: 1169–1174.

26. Higuchi WI, Lau PK, Higuchi T, Shell JW. Solubility relationship in the methylprednisolone system. *J. Pharm. Sci.* 1963; 52: 150–153.

27. Yu L. Inferring thermodynamic stability relationship of polymorphs from melting data. *J. Pharm. Sci.* 1995; 84: 966–974.

28. Aguiar AJ, Krc J. Jr., Kinkel AW, Samyn JC. Effect of polymorphism on the absorption of chloramphenicol from chloramphenicol palmitate. *J. Pharm. Sci.* 1967; 56: 847–853.

29. Rowe EL, Anderson BD. Thermodynamic studies of tolbutamide polymorphs. *J. Pharm. Sci.* 1984; 73: 1673–1675.

30. Ghodbane S, McCauley JA. Study of the polymorphism of 3-(((3-(2-(7-chloro-2-quinolinyl)-(*E*)-ethenyl)phenyl)((3-(dimethylamino-3-oxopropyl)thio)methyl)-thio)propanoic acid (MK571) by DSC, TG, XRPD and solubility measurements. *Int. J. Pharm.* 1990; 59: 281–286.

31. Gerber JJ, vander Watt JG, Lötter AP. Physical characterization of solid forms of cyclopenthiazide. *Int. J. Pharm.* 1991; 73: 137–145.

32. Kushida I, Ashizawa K. Solid state characterization of E2101, a novel antispastic drug. *J. Pharm. Sci.* 2002; 91: 2193–2202.

33. Rollinger JM, Gstrein EM, Burger A. Crystal forms of torasemide: new insights. *Eur. J. Pharm. Biopharm.* 2002; 53: 75–86.

34. Shibata M, Kokubo H, Morimoto K, Morisaka K, Ishida T, Inoue M. X-ray structural studies and physicochemical properties of cimetidine polymorphs. *J. Pharm. Sci.* 1983; 72: 1436–1442.

35. Crafts PA. The role of solubility modelling and crystallization in the design of active pharmaceutical ingredients. In: Ng KM, Gani R, Dam-Johansen K, editors, *Chemical Product Design: Toward a Perspective through Case Studies*, Elsevier, Dordrecht, The Netherlands, 2007, pp. 23–85.

36. Ye Q, Hart RM, Kania R, Ouellette M, Wu ZP, Zook SE. Polymorphic forms of 6-[2-(-(methylcarbomoyl) phenylsulfanyl]-3-*E*-[2-pyridin-2-yl)ethenyl]indazole. U.S. Patent 0094763, 2006.

37. Campeta AM, Chekal BP, McLaughlin RW, Singer RA. Novel crystalline forms of a VEGF-R inhibitor. *PCT Int. Appl.* 2008; WO 2008122858.

38. Chekal B, Campeta AM, Abramov YA, Feeder N, Glynn P, McLaughlin R, Meenan P, Singer R. Facing the challenges of developing an API crystallization process for a complex polymorphic and highly-solvating system. Part I. *Org. Process Res. Dev.* 2009; 13: 1327–1337.

39. Sohn YT. Study on the polymorphism of acetaminophen. *J. Korean Pharm. Sci.* 1990; 20: 97–104.

40. Sheikhzadeh M, Rohani S, Traffish M, Murad S. Solubility analysis of buspirone hydrochloride polymorphs: measurements and prediction. *Int. J. Pharm.* 2007; 338: 55–63.

41. Behme RL, Brooke D. Heat of fusion measurement of a low melting polymorph of carbamazepine that undergoes multiple-phase changes during differential scanning calorimetry analysis. *J. Pharm. Sci.* 1990; 80: 986–990.

42. Behme RJ, Brooke D, Farney RF, Kensler TT. Research article characterization of polymorphism of gepirone hydrochloride. *J. Pharm. Sci.* 1985; 74: 1041–1046.

43. Collman B, private communication.

44. Vrečer F, Vrbinc M, Meden A. Characterization of piroxicam crystal modifications. *Int. J. Pharm.* 2003; 256: 3–15.

45. Bartolomei M, Bertocchi P, Ramusino MC, Santucci N, Valvo L. Physico-chemical characterization of the modifications I and II of (R, S) propranolol hydrochloride: solubility and dissolution. *J. Pharm. Biomed. Anal.* 1999; 21: 299–309.

46. Caviglioli C, Pani M, Gatti P, Parodi B, Cafaggi S, Bignargi G. Study of retinoic acid polymorphism. *J. Pharm. Sci.* 2006; 95: 2207–2221.

47. Carlton RA, Difeo TJ, Powner TH, Santos I, Thompson MD. Preparation and characterization of polymorphs for an LTD_4 antagonist, RG 12525. *J. Pharm. Sci.* 1996; 85: 461–467.

48. Rocco WL, Swanson JR. WIN 63843 polymorphs: prediction of enantiotropy. *Int. J. Pharm.* 1995; 117: 231–236.

49. Singhal D, Curatolo W. Drug polymorphism and dosage form design: a practical perspective. *Adv. Drug Del. Rev.* 2004; 56: 335–347.

50. Amidon GL, Lennernas H, Shah VP, Crison JRA. Theoretical basis for a biopharmaceutic drug classification: the correlation of *in vitro* drug product dissolution and *in vivo* bioavailability. *Pharm. Res.* 1995; 12: 413–420.

51. Varma MVS, Obach RS, Rotter C, Miller HR, Chang G, Steyn SJ, El-Kattan A, Troutman MD. Physicochemical space for optimum oral bioavailability: contribution of human intestinal absorption and first-pass elimination. *J. Med. Chem.* 2010; 53: 1098–1108.

52. Kasim NA, Whitehouse M, Ramachandran C, Bermejo M, Lennernäs H, Hussain AS, Junginger HE, Stavchansky SA, Midha KK, Shah VP, Amidon GL. Molecular properties of WHO essential drugs and provisional biopharmaceutical classification. *Mol. Pharm.* 2004; 1: 85–96.

53. Taylor JR. *An Introduction to Error Analysis*. University Science Books, Mill Valley, CA, 1982.

26

TOWARD A RATIONAL SOLVENT SELECTION FOR CONFORMATIONAL POLYMORPH SCREENING

Yuriy A. Abramov, Mark Zell, and Joseph F. Krzyzaniak
Pfizer Global Research & Development, Pharmaceutical Sciences, Groton, CT, USA

26.1 INTRODUCTION

Crystalline solids with the same chemical composition but different molecular arrangements in the crystal lattice are known as polymorphs [1]. Changes in polymorphic form during pharmaceutical development can have a negative impact on a drug's performance, that is, solubility and bioavailability (Chapter 25), chemical and physical stability, and mechanical properties. Therefore, it is necessary to identify the stable crystal form under normal manufacturing and storage conditions to ensure that this form does not change during the life cycle of the drug product.

Polymorph screens are conducted early in drug development to identify unique crystal forms of the active pharmaceutical ingredient (API). Each crystal form discovered is characterized to identify whether the crystalline phase is an anhydrous form or a solvate. The polymorphic lattice can also consist of either the same or the different molecular conformations. Conformational polymorphism describes the latter case when different conformations of the same molecule occur in different crystal forms [2]. Solid-state characterization studies are then conducted to develop an understanding of the stability relationship between all crystalline phases since the thermodynamically stable form is directly related to conditions (crystallization, environmental, and manufacturing) in which the API is exposed to during the drug development process [1].

During the preparation of the desired polymorphic form, the science of crystallization has shown to be a very complex phenomenon that is dictated by interplay between different thermodynamic and kinetic factors. The presence of different molecular conformations in saturated solution introduces an additional degree of complexity allowing crystallization of polymorphs different not only in the packing arrangement but also in the molecular geometry as seen in conformational polymorphs [1–3]. Crystallization is believed to be a multiple stage process in which molecules associate into prenucleation molecular clusters followed by their assembling into crystal nuclei leading to crystal growth (Figure 26.1) [3, 4]. It was assumed [5] and later demonstrated [6, 7] that a saturated phase contains clusters of molecules displaying packing of all possible polymorphs. Final growth of a specific polymorph can be achieved by altering crystallization conditions, such as degree of supersaturation, type of solvent, and additives [8–11].

From a thermodynamic viewpoint, a primary factor for conformational polymorph formation is stabilization of the conformer free energy in the crystalline environment relative to that in saturated solution. That consideration defines the type of solvent as one of the major factors in polymorphic selectivity. Solvent selection for polymorph crystallization is usually based on achieving a reasonably high API solubility [12] to facilitate crystal growth during drowning-out, evaporative, cooling, or slurry crystallization techniques. With this, it is reasonable to assume that a higher population of a specific molecular conformation is needed to feed a crystallization of a corresponding conformational polymorph. A higher conformer population should contribute to increased nucleation of the corresponding conformational polymorph, structural organization of which is most readily

Chemical Engineering in the Pharmaceutical Industry: R&D to Manufacturing Edited by David J. am Ende

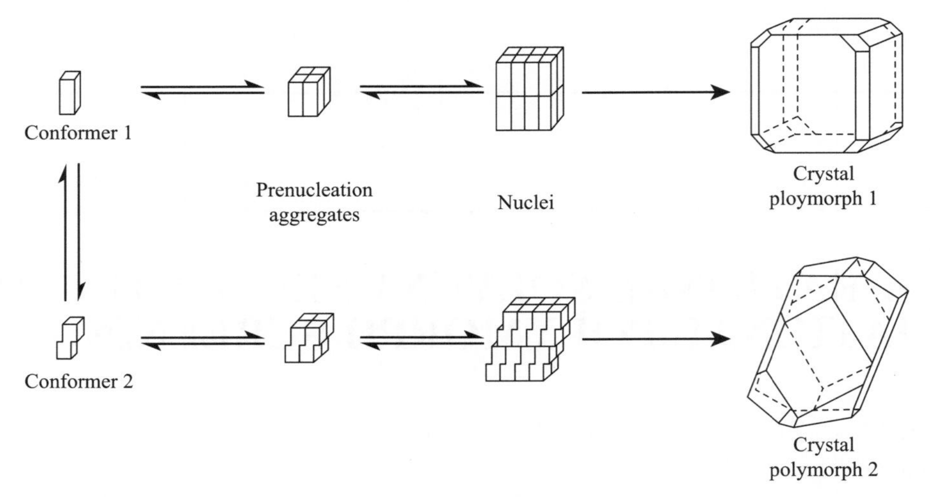

FIGURE 26.1 An illustration of the crystallization of conformationally flexible molecules. Reproduced from Ref. 3 with permission. Copyright (2010) American Chemical Society.

derived from the preferred conformations in solution [13, 14]. The focus of this chapter is testing a computational approach for conformational population prediction in different solvent media in order to explore diversification of conformational populations in solution. A working hypothesis is that controlling the selective conformer's population should facilitate a rational solvent selection for conformational polymorph screening. In addition, prediction of conformer population in a supercooled liquid (self-media) will also be performed and should mimic stabilization of molecular conformations in an amorphous solid state. An assumption is made that while the conformer distribution in solution is important for nucleating a conformational polymorph, a high conformation population in the supercooled liquid should reflect stability of the conformation in a solid state neglecting long-order contributions. The interplay between conformer distribution in solution and self-media should reflect a driving force for crystallization of the different molecular conformations.

In addition, the proposed conformational population analysis technique will be discussed for preferable conformation selection for crystal structure prediction (CSP). CSP is an important computational tool that is valuable not only in guiding polymorph screening [15–17] but also in performing a polymorphic risk assessment for solid form nomination during pharmaceutical development [18]. However, the currently available CSP methods heavily rely on a correct selection of the starting molecular conformations that are typically held rigid at least during initial crystal packing generation. That is why a rational selection of the preferred starting molecular conformations for CSP of complex flexible molecules is of great importance [19–21].

Although important, the following topics and considerations are out of scope of this chapter.

1. Due to the effect of different crystalline environments, molecular conformations may be distributed around local minima of the potential energy surface as defined in the reference media (e.g., gas phase) [22]. Molecular conformations considered in this chapter are defined strictly by a local minimum in the aqueous or gas media neglecting crystal packing effects. Therefore, we will be considering different types of conformations representing possible conformers with slightly different geometries. This is especially true for the freely rotatable bonds, rotation of which is defined by a flat potential energy surface in the gas phase or in the implicit solution, while torsion angle of such bonds may be fixed in a crystalline environment by a specific hydrogen-bonding interaction (e.g., hydroxyl and carboxyl groups).
2. Relative stability of polymorphs is described by an equilibrium-phase diagram and cannot be adequately determined by the methods developed in this chapter. Moreover, the intention of this chapter is to facilitate crystallization of a diverse set of conformational polymorphs including those that are metastable.
3. Rotational barriers that are important for kinetics of conformational interconversion will not be considered. Therefore, only thermodynamic factors will be used in theoretical evaluation of the conformational distributions.
4. In addition, no considerations of kinetic factors that may affect selectivity of the polymorph crystallization will be addressed. Among such factors are the degree of supersaturation, cooling and stirring rates, impurities and additives that affect the polymorphic form, morphology, and crystal growth [8–11].

26.2 METHODS

26.2.1 Theoretical

The conformational population study was performed in two steps. During the first step, a stochastic conformational search was performed using the MOE 2008.10 software package [23] with the following parameters: MMFF94× force field with distance-dependent dielectric, chiral constrains, allowing amide-bond rotation and energy cutoff of 7 Kcal/mol. After the conformational search was complete, the conformations generated were further optimized using aqueous media at PBE/DNP/COSMO level of theory (defined by theoretical method and a basis set) as implemented in $DMol^3$ [24–26]. This utilizes density functional theory (DFT) PBE approximation [27] with all-electron double-numeric-polarized basis set. The effect of bulk water is estimated by conductor-like screening model (COSMO) as implemented in $DMol^3$ [28]. It was demonstrated recently [29] that PBE is one of the best DFT functionals in prediction of energies of hydrogen-bonding (HB) systems. The *cosmo* files generated by the PBE/DNP/COSMO calculations for each conformer are further used for solvation free energy calculations, ΔG_{solv}, and for prediction of conformational distribution in different solvents adopting COSMO-RS theory [30] as implemented in COSMOtherm software [31].

An obvious advantage of the application of the PBE/DNP/COSMO calculations is that the COSMOtherm parameters are specifically optimized for this level of theory. However, the DFT is known to underestimate strong electron correlation effects (such as dispersion energy) [32]. Therefore, in order to assure the quality of the PBE/DNP calculations, we have also performed calculations at a combined level of theory according to the following procedure (Chapter 24). The conformer free energy in solution is presented as

$$G(i) = E_{\text{gas}}(i) + \text{ZPE}(i) + \Delta G_{\text{solv}}(i) \tag{26.1}$$

where $E_{\text{gas}}(i)$ is the gas-phase energy of the conformer and ZPE(i) is its zero-point vibrational energy. The value of $\Delta G_{solv}(i)$ is calculated at the PBE/DNP/COSMO level as described above, while the gas free energy of the conformer was calculated at the RI-MP2/TZVPP level [33] adopting TURBOMOLE software [34]. The ZPE(i) contribution was estimated at the $DMol^3$ PBE/DND level of theory.

In the following discussions, the two theoretical approaches described above for simplicity will be referred to as PBE and RI-MP2 levels, respectively.

The equilibrium conformer population, $p(i)$, was calculated according to the following equation [30]:

$$p(i) = \frac{\varpi(i)\exp(-G(i)/RT)}{\sum_j \varpi(j)\exp(-G(j)/RT)} \tag{26.2}$$

where $\varpi(i)$is a multiplicity of the conformer i, which is based on the geometrical degeneration factors, R is the universal gas constant and T is the temperature in Kelvin. A challenge in an accurate prediction of $p(i)$ is introduced by an exponential dependence on the calculated $G(i)$ values, so that a relatively small error in conformer's free energy transforms into significant errors in conformer populations. For example, in case of a simple system with two conformations displaying similar energies and multiplicities, the true population of each conformer is 50%. However, an error in predicted relative $G(i)$ values, $G(2) - G(1)$, of 0.4 kCal/mol would result in $p(i)$ error at the room temperature of ±16.3%. The predicted conformer populations would be 66.3% and 33.7%.

26.2.2 Experimental

26.2.2.1 NMR Measurements ^{1}H, gHMBC, and G-BIRD$_{R,X}$-CPMG-HSQMBC experiments were performed on a Bruker Avance DRX 600 spectrometer equipped with a 5 mm BBO probe with z-axis gradient. All experiments were performed at a temperature of 298 K.

Samples were prepared at concentrations of 10, 100, and 300 mg/mL in acetone-d_6, acetonitrile-d_3, and methanol-d_4 with tetramethylsilane as an internal standard. All experiments were performed using Bruker standard pulse sequences, except for G-BIRD$_{R,X}$-CPMG-HSQMBC, which was written and implemented by a staff member in our laboratory.

For ^{1}H NMR analysis, typically one transient was acquired with a 1s relaxation delay using 32 K data points. The 90° pulse was 10.5 μs and a spectral width of 10775 Hz was used.

The gHMBC spectrum was acquired with 4096 data points for $F2$ and 128 $F1$ increments. gHMBC data was acquired with 4K data points in $F2$, 128 increments for $F1$ (16 scans per increment) and $F2 \times F1$ spectral window of 5200 × 23800 Hz. Data were processed with 4K data points zero-filled to 8K in $F2$ and 128 data points zero-filled to 1K in $F1$.

The G-BIRD$_{R,X}$-CPMG-HSQMBC spectra were acquired in approximately 14 h with 4K data points in $F2$, 128 increments for $F1$ with 256 scans per increment and $F2 \times F1$ spectral window of 5200 × 23800 Hz. Data were processed with 4K data points zero-filled to 8K in $F2$ and 128 data points zero-filled to 1K in $F1$. A sine squared window function was applied to the $F1$ dimension before Fourier transformation and no apodization was applied in the $F2$ dimension. The gradient ratios for G-BIRD$_{R,X}$-CPMG-HSQMBC were G1:G2:G3:G4:G5 = 2.5:2.5:8:1: +/−2. The delay for long-range polarization transfer was set to 63 ms. The delay used for (delta) was set to 200 μs. All measured coupling constants are believed to be within +/−0.5 Hz. The values of $^3J_{\text{CH}}$

were determined from direct measurement and subsequent manual peak fitting analysis [35].

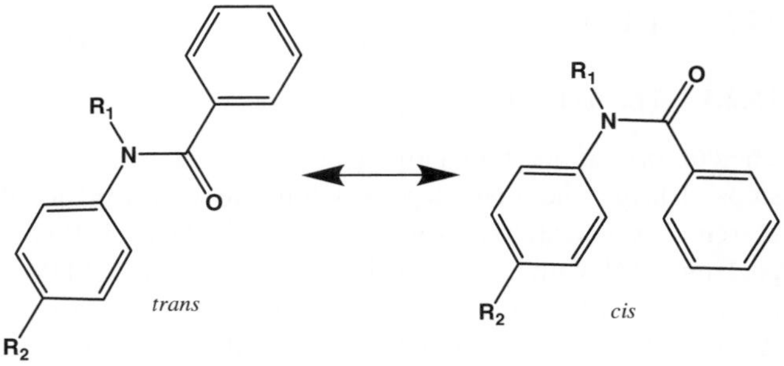

1. $R_1 = H, R_2 = H$
2. $R_1 = CH_3, R_2 = H$
3. $R_1 = OH, R_2 = H$
4. $R_1 = OH, R_2 = CH_3$
6. $R_1 = NH_2, R_2 = H$
7. $R_1 = OCH_3, R_2 = H$

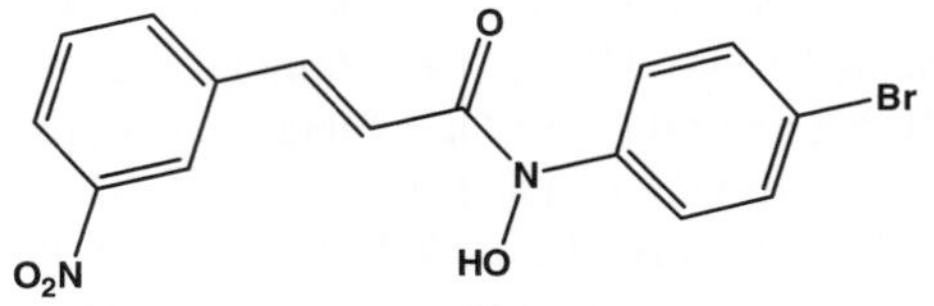

FIGURE 26.2 Chemical structures of *N*-substituted amides series [36].

26.3 RESULTS AND DISCUSSION

26.3.1 Test of Accuracy of Conformational Population Predictions

The accuracy of conformational population prediction for flexible organic molecules was tested at different levels of theory. For this, three test cases were selected consisting of *N*-substituted amides series [36], *N*-(pyridin-2-yl)benzamides chemical series [37], and *S*-ibuprofen. While accurate experimental conformational populations based on the solution NMR experiments were available for the two first cases, a separate experimental work was performed for *S*-ibuprofen.

26.3.1.1 N-Substituted Amides Series Yamasaki et al. [36] reported a detailed study of the *cis–trans* ratio of conformer population of a series of *N*-substituted amides (Figure 26.2) in DCM-d_2, methanol-d_4, and acetone-d_6 at 183 K. It was demonstrated that the compounds **1** and **2** display *trans* and *cis* conformations in all three solvents, respectively. A very weak solvent dependence was found for the compounds **6** and **7**, the former being preferably in the *cis*-conformations while the latter displaying close to a uniform *cis–trans*-distribution. The most remarkable result of the study was an observation of a pronounced solvent dependent conformational switching of the phenylhydroxamic acids (**3–5**) (Figure 26.3) (Table 26.1) from the predominantly *cis*-conformations in DCM to preferably *trans*-conformations in acetone.

Theoretical predictions at the DFT (PBE) and RI-MP2 levels are in agreement with the experimental observations (Table 26.1). In particular, the strong solvent dependence of conformations of the compounds **3–5** was reproduced correctly, although the absolute values of *cis*-populations in methanol and acetone are underestimated for the compounds **3** and **4**, especially by the DFT method (Figure 26.3). At the same time, both levels of the predictions overestimated *cis*-populations of the compound **7**. The overestimation is more pronounced in the case of the predictions using the RI-MP2 level of theory.

26.3.1.2 N-(Pyridin-2-yl)Benzamides Series Populations of *cis* conformations of a series of *N*-(pyridin-2-yl) benzamides (**8–11**) (Figure 26.4) as well as of *N*-(2,6-dimethylphenyl)acetamide (**12**) (Figure 26.4) were studied by means of solution NMR spectroscopy in three solvents: chloroform-d_1, methanol-d_4, and acetone-d_6 at 243 K (Table 26.2) [37]. It was demonstrated that the compounds **8** and **10** display *trans*- and *cis*-conformations in all three solvents, respectively. These observations are analogous to the results reported by Yamasaki et al. [36] for the compounds **1** and **2** (Figure 26.2) (Table 26.1). Solvent-dependent conformational behavior was observed for the compounds **9** and **12**. Theoretical predictions at the DFT PBE level are in a quite good agreement with the experimental observations (Table 26.2). In particular, the solvent dependence of the compounds **9** and **12** conformations was reproduced correctly, although the absolute values of *cis*-populations in methanol and acetone are somewhat overestimated (Figure 26.5). The overestimation is more pronounced in case of the predictions at the RI-MP2 level of theory.

It is found for both *N*-substituted amides and *N*-(pyridin-2-yl)benzamides series that a combined approach with RI-MP2/TZVPP gas-phase calculations does not demonstrate an advantage over PBE/DNP/COSMO level of theory in predicting conformational distributions in different solvents. Therefore, a less demanding PBE/DNP/COSMO analysis was adopted for further calculations.

EXAMPLE 26.1 *S*-IBUPROFEN CONFORMATIONAL POPULATION IN METHANOL, ACETONE, AND ACETONITRILE

Ibuprofen is a nonsteroidal anti-inflammatory drug in which activity is usually associated to the $S(+)$-isomer. As shown in Figure 26.6, the ibuprofen molecule displays four flexible

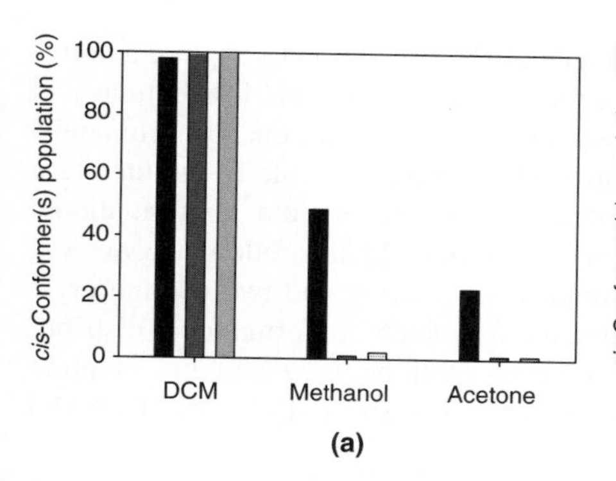

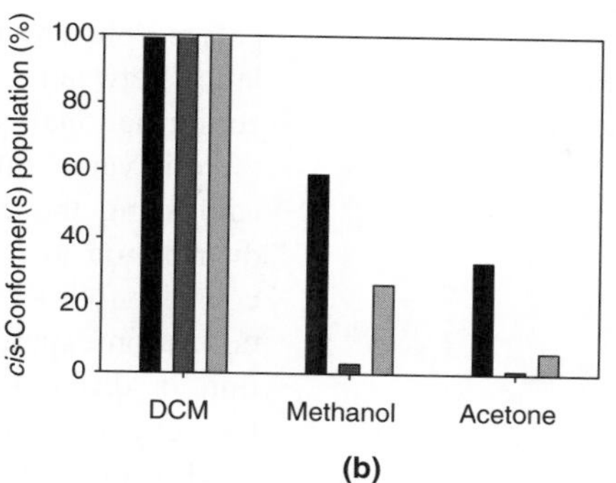

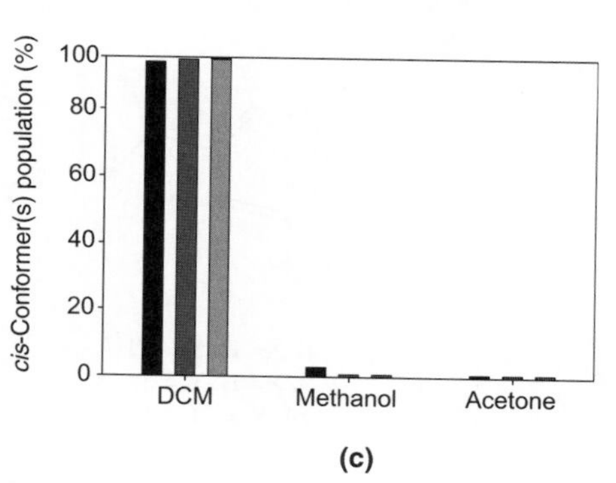

FIGURE 26.3 *cis*-Conformer(s) populations of the compounds **3–5** ((a)–(c)) of *N*-substituted amides series (Table 26.1) (Figure 26.2) [36]. Experimental, PBE, and RI-MP2 results are represented by black, dark gray, and gray bars, respectively.

torsion angles. The initial stochastic conformational search was performed using the MOE software package [23] adopting default parameters (MMFF94× force field with distance-dependent dielectric, chiral constrain, and energy cutoff of 7 Kcal/mol). The 25 lowest energy conformations generated were further optimized in water at the PBE/DNP/COSMO level of theory. The optimized *S*-ibuprofen conformations are aligned and presented in Figure 26.7. The *cosmo* files generated were used for the conformational distribution study in methanol, acetone, and acetonitrile solvents at 25°C, adopting COSMOtherm software [31]. No significant solvent effect on conformer distribution was found. Population of four conformers (Figure 26.8) was found to be in the range of 12–23% in all solvents under consideration, which significantly exceeded the population of any other conformation. The torsion angles of the preferred conformations are listed in Table 26.3. However, it should be expected that in a crystalline environment, the angle τ_1 may change (e.g., switch by ∼180°) in order to accommodate intermolecular hydrogen-bonding interaction.

The solvent and concentration dependences of *S*-ibuprofen conformations were investigated by G-BIRD$_{R,X}$-CPMG-HSQMBC experiment. For this, the $^3J_{CH}$ heteronuclear coupling constants that define the flexible torsion angles (τ_1–τ_4) in the *S*-ibuprofen molecule were extracted (Table 26.4). Due the fast conformational interconversion at the room temperature, the NMR observations are averaged over specific conformational distributions.

The $^3J_{CH}$ coupling constant data presented in Table 26.4 suggest that there is no change in the molecular conformation of *S*-ibuprofen in solution, within the limits of the NMR measurements (> +/−0.5 Hz) as a function of either solvent composition or concentration. This data further support the computational results for *S*-ibuprofen presented above that also suggest no change in conformational populations as a function of solvent composition. An exercise of converting the measured $^3J_{CH}$ coupling constant values into their respective torsion angles was not carried out.

26.3.2 Conformational Distribution and Polymorph Crystallization

Polymorph crystallization may be performed by utilizing different crystallization techniques. Typically a slurry experiment is considered to be the best to induce solvent-mediated transformation to the most stable form [38, 39]. The system is

TABLE 26.1 Experimental [36] and Predicted *cis*-Conformers Populations (%) of *N*-Substituted Amides (Figure 26.2) at −90°C in Three Solvents

	DCM			Methanol			Acetone				
Compound	Experimental	PBE	RI-MP2	Experimental	PBE	RI-MP2	Experimental	PBE	RI-MP2	Self PBE	*T* (°C)
1	<1	<1	<1	<1	<1	<1	<1	<1	<1	<1	−90
2	98	98.8	>99	>99	95.4	>99	>99	98.8	>99	97.0	−90
3	98	>99	>99	49	<1	2.2	23	<1	<1	98.1	−90
4	>99	>99	>99	59	3.0	26.4	33	<1	6.4	98.6	−90
5	>99	>99	99.6	3	<1	<1	<1	<1	<1	79.6	−90
6	98	>99	>99	95	86.5	>99	85	60.2	>99	97.2	−90
7	50	88.1	>99	63	88.7	>99	55	52.2	>99	45.5	−90

Details of the theoretical approaches based on PBE and RI-MP2 levels of theory are described in Section 26.2.1.

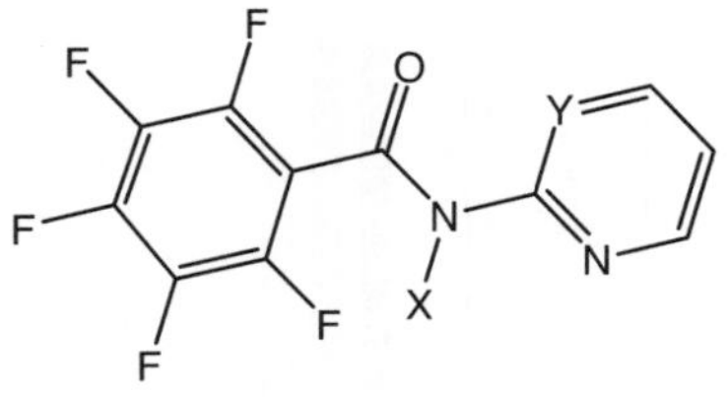

8. X= H, Y = CH
9. X= H, Y = N
10. X = CH_3, Y = N

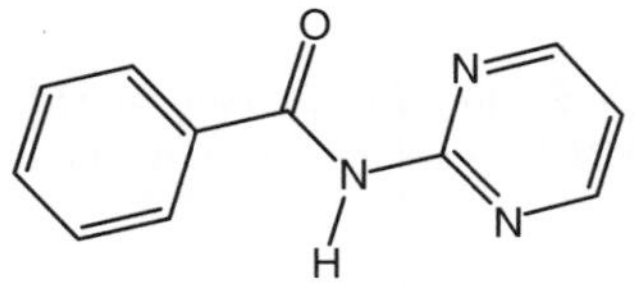

11

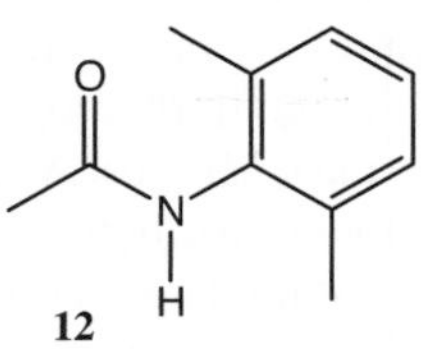

12

FIGURE 26.4 A series of *N*-(pyridin-2-yl)benzamides (**8–11**) and *N*-(2,6-dimethylphenyl)acetamide (**12**) [37].

preferably under thermodynamic control and the solution is saturated with respect to a metastable form and supersaturated with respect to a more stable form. It has been shown that a solvent that gives high solubility provides faster transformation to the most stable form and usually 8 mM solubility threshold has been adopted when designing the slurry experiment [39]. At the same time, cooling and evaporation crystallization in general produces a supersaturated solution with respect to all possible forms, and the system is preferably under kinetic control. In that case, polymorph crystallization typically follows Ostwald law of stages [40] and crystallization of the least stable form is expected first, followed by transformation to a more stable polymorph. The focus of this chapter is to explore a correlation between conformational polymorph distributions and conformational polymorph crystallization. Unfortunately, there is very limited information available in the literature addressing the effect of solvent on both conformational distribution and crystallization. In the following, we will consider four examples—two slurry and two cooling crystallization experiments, for which conformational distribution in different solvents will be predicted by adopting theoretical approaches described above (PBE/DNP/COSMO level of theory).

26.3.2.1 ***N-Phenylhydroxamic Acids*** Experimental and conformational distributions of *N*-phenylhydroxamic acids in three different solvents were reported by Yamasaki et al. [36] and discussed above (Table 26.1, compounds **3–5**). Both NMR spectroscopy and theoretical calculations (Table 26.1, compounds **3–5**) demonstrated that all *N*-phenylhydroxamic acids display switching from the *cis*-conformations in dichloromethane to the *trans*-conformations in methanol and especially in acetone (183 K). Yamasaki et al. recrystallized compound **3** from DCM and acetone producing polymorphs with *cis* (crystal A)- and *trans* (crystal B)-molecular conformations, respectively (Figure 26.9). It appears from differential scanning calorimetry (DSC) profile [36] that an enantiotropic relationship exists between A and B forms so that at room temperature the crystal B is presumably more stable. The stability assignment is opposite to the one that may be based on the predicted population distribution in the self-media (or an amorphous solid, Table 26.1). This reflects the importance of the long-range order contributions in the *N*-phenylhydroxamic acid crystals. An important conclusion from the results of the Yamasaki et al. study [36] is that the polymorph crystallization in different solvents may follow the conformational population trend rather than the Ostwald rule of stages.

26.3.2.2 ***Taltirelin*** Taltirelin ((4*S*)-*N*-[(2*S*)-1-[(2*S*)-2-carbamoylpyrrolidin-1-yl]-3-(3*H*-imidazol-4-yl)-1-oxopropan-2-yl]-1-methyl-2,6-dioxo-1,3-diazinane-4-carboxamide),

TABLE 26.2 Experimental [37] and Predicted *cis*-Conformers Populations (%) of *N*-(Pyridin-2-yl)Benzamides (8–11) (Figure 26.4) and *N*-(2,6-Dimethylphenyl)Acetamide (Figures 26.4 and 26.5) at −30°C in three solvents

	$CHCl_3$			Methanol			Acetone			Self	
Compound	Experimental	PBE	RI-MP2	Experimental	PBE	RI-MP2	Experimental	PBE	RI-MP2	PBE	T (°C)
8	<1	<1	<1	<1	<1	<1	—	<1	<1	<1	−30
9	60.6	51.7	66.5	7.4	26.1	56.3	50.3	78.2	96.4	80.0	−30
10	>99	>99	>99	—	>99	>99	—	>99	>99	>99	−30
11	<1	<1	8.4	—	<1	1.3	—	<1	32.4	<1	−30
12	20.6	25.4	35.9	—	7.0	7.8	2.9	3.3	4.2	8.3	−30

Details of the theoretical approaches based on PBE and RI-MP2 levels of theory are described in the Section 26.2.1.

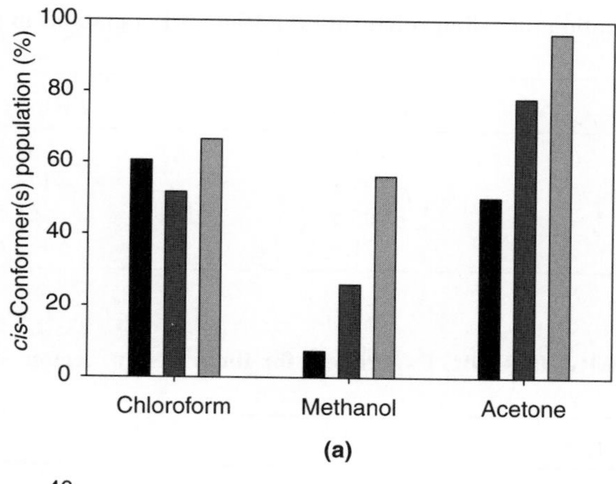

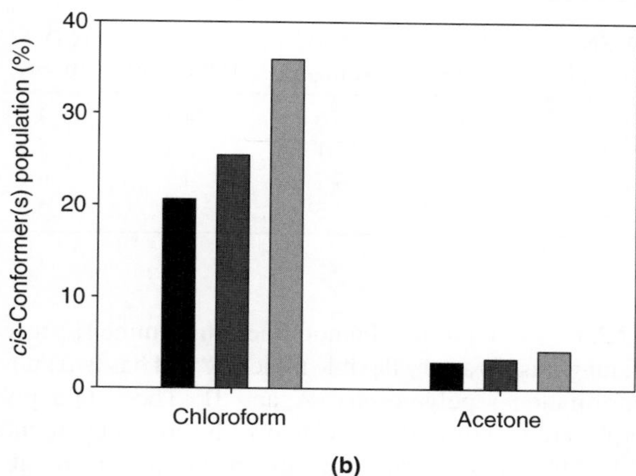

FIGURE 26.5 *cis*-Conformer(s) populations of (a) 2,3,4,5,6-pentafluoro-*N*-(pyrimidine-2-yl)benzamide (**9**) (Table 26.2) (Figure 26.4) and (b) *N*-(2,6-dimethylphenyl)acetamide (**12**) (Table 26.2) (Figure 26.4) [37]. Experimental, PBE, and RI-MP2 results are represented by black, dark gray, and gray bars, respectively.

a central nervous system activating agent, was reported to have two crystalline tetrahydrate forms: a metastable α-form and stable β-form (Figure 26.10) [41, 42]. It was found that the solvent-mediated transformation to the β-form occurring in the water slurry can be significantly promoted by adding a small amount (10 wt%) of MeOH. It was demonstrated that although the polymorph solubility had little affect with the added MeOH (it is actually slightly decreasing), an increase in methanol concentration causes an induction period of transformation to become shorter. In addition, Shoji Maruyama and Hiroshi Ooshima [42] demonstrated by means of nOe NMR analysis that MeOH causes the conformation change of taltirelin from the α-form to the β-form conformers through the solute-MeOH interaction. That observation supports the importance that the conformer population has on driving the corresponding conformational polymorph crystallization.

FIGURE 26.6 Structure of ibuprofen molecule with four flexible torsion angles. τ_1: $O-C_1-C_2-C_4$; τ_2: $C_1-C_2-C_4-C_9$; τ_3: $C_8-C_7-C_{10}-C_{11}$; τ_4: $C_7-C_{10}-C_{11}-C_{13}$.

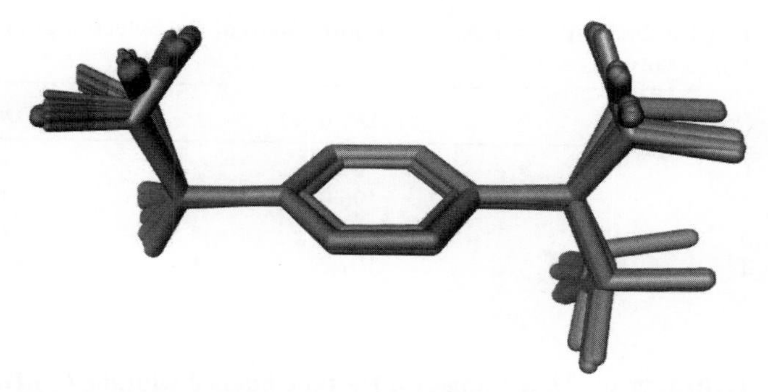

FIGURE 26.7 Aligned 25 *S*-ibuprofen conformations optimized in water at PBE/DNP/COSMO level of theory. All hydrogens are omitted.

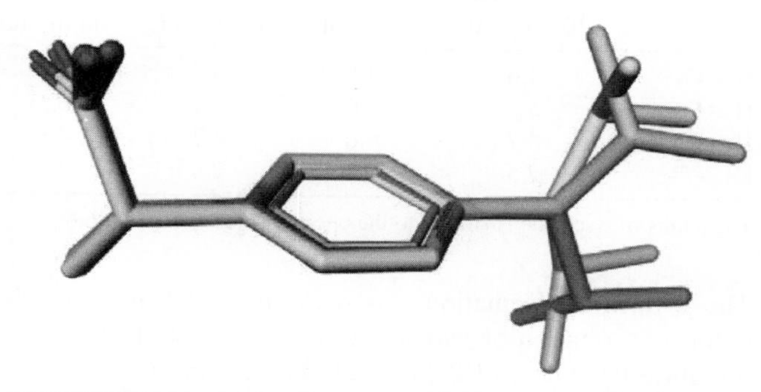

FIGURE 26.8 Aligned four *S*-Iibuprofen conformers that displayed the highest populations in methanol, acetone, and acetonitrile solvents at 25°C.

Taltirelin is a very flexible molecule with eight rotatable bonds, which makes a reliable conformation search a very challenging task. In order to test whether the MeOH effect on conformational populations can be predicted, the following calculations were performed. The α-form conformation was taken from the crystal structure available in the Cambridge Structure Database [43] (CSD, reference code REPLIH[1]).

[1] REPLIH conformer represents a mirror image ((4R)-*N*-[(2*R*)-1-[(2*R*)-2-carbamoylpyrrolidin-1-yl]-3-(3*H*-imidazol-4-yl)-1-oxopropan-2-yl]-1-methyl-2,6-dioxo-1,3-diazinane-4-carboxamide tetrahydrate) of taltireline α-form as it is described in Refs 41 and 42.

TABLE 26.3 Torsion Angles (Figure 26.6) of the Selected Four Conformations of *S*-ibuprofen with the Highest Populations in the Three Solvents

Conformer	τ_1 (Degrees)	τ_2 (Degrees)	τ_3 (Degrees)	τ_4 (Degrees)
1	94.7	−65.6	−73.1	−63.2
2	96.0	−62.7	−105.3	−173.2
3	95.8	−64.2	103.6	−65.5
4	107.3	−65.7	71.8	−172.8

TABLE 26.4 $^3J_{CH}$ Values (in Hertz) Obtained with the G-BIRD$_{R,X}$-CPMG-HSQMBC Experiment for Ibuprofen in Acetone-d_6, Acetonitrile-d_3, and Methanol-d_4.

Correlation	$^3J_{CH}$ (Hz)								
	CD_3OD 300 mg/mL	CD_3OD 100 mg/mL	CD_3OD 10 mg/mL	CD_3CN 300 mg/mL	CD_3CN 100 mg/mL	CD_3CN 10 mg/mL	$(CD_3)_2CO$ 300 mg/mL	$(CD_3)_2CO$ 100 mg/mL	$(CD_3)_2CO$ 10 mg/mL
H_3−C_4	4.6	4.6	4.5	4.4	4.5	4.5	4.6	4.7	4.7
H_8−C_1	5.1	5.0	5.1	5.1	5.0	5.1	5.1	5.1	5.1
$H_{12(13)}$−C_{10}	4.7	4.6	4.7	4.8	4.7	4.6	4.7	4.7	4.7
H_{11}−C_7	2.5	2.4	2.4	2.5	2.4	2.4	2.5	2.5	2.3

The numbering scheme is similar to the one presented in Figure 26.6.

The β-form conformation was reconstructed from that of α-form by rotation of two single bonds as described in the literature (Figure 26.10) [41, 42]. The α- and the β-form conformations were further adopted for the conformational population analysis at the PBE/DNP/COSMO level of theory. Although no significant effect was reproduced at 10%wt MeOH concentration, the calculations demonstrated a general qualitative increase of the β-conformer population when switching from aqueous to methanol solution (Figure 26.11).

26.3.2.3 Famotidine Famotidine, a histamine H2-receptor antagonist, is a very flexible molecule that has two known conformational polymorphs: A and B. These two polymorphs are monotropically related with form A being more stable [44]. Since form B is the metastable form, it is kinetically favored and according to the Ostwald law should crystallize first when cooling a saturated solution. Selective cooling crystallization of Famotidine polymorphs was reported by Lu et al. [44]. It was found that the form prepared

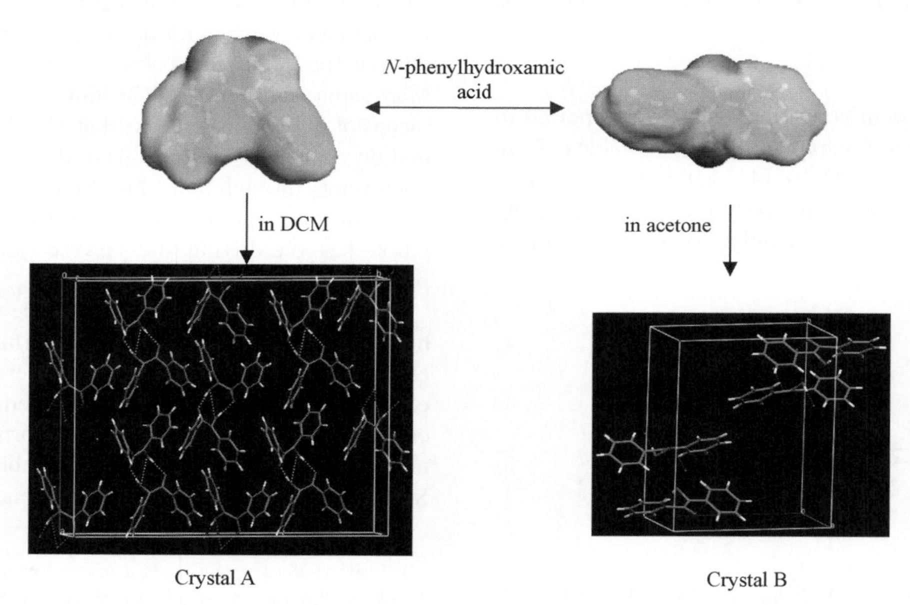

FIGURE 26.9 Conformational equilibrium of *N*-Phenylhydroxamic acid (**3**) (Table 26.1) and crystal structures from recrystallization in DCM (crystal A) and acetone (crystal B) [36].

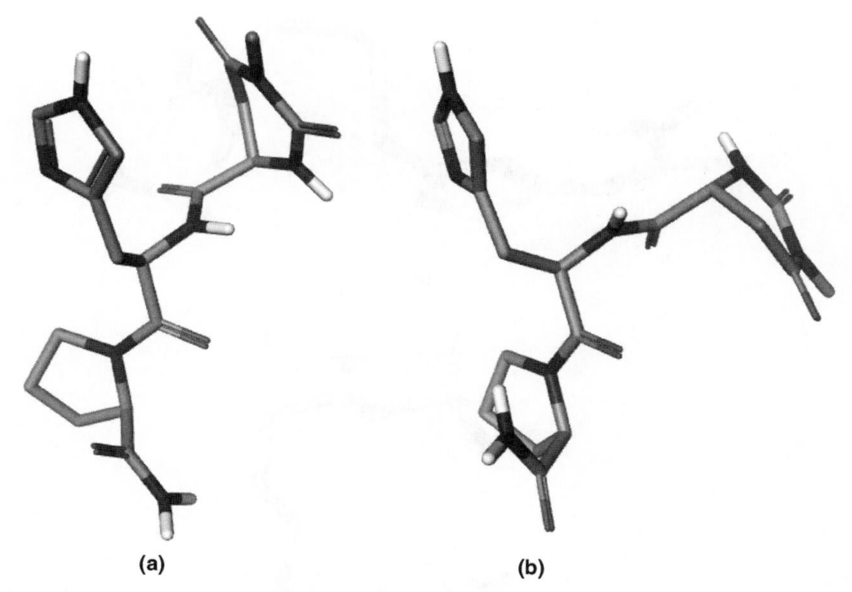

FIGURE 26.10 Conformations of taltireline α-form (a) and β-form (b). Only polar hydrogens are shown.

was not only influenced by the cooling rate but also affected by the solvent of crystallization. For example, form B was crystallized from a water solution with high initial drug concentration independent of the rate of cooling. Additionally, the stable form A was crystallized from methanol and acetonitrile solutions at low initial drug concentrations. No conformational population study in solutions was reported by the authors.

Conformational populations of forms A and B were predicted at a crystallization temperature of 50°C in three solvents at the PBE/DNP/COSMO level of theory. Due to the very high flexibility of the drug, crystallographic conformations (Figure 26.12) were adopted for the calculations with no conformational search. The resulting conformer population in three solvents is presented in Figure 26.13. It

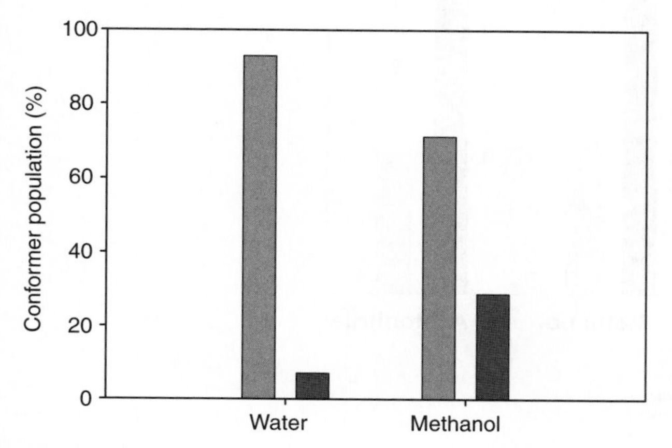

FIGURE 26.11 Predicted populations of the α-form (gray) and β-form (black) taltireline conformers in water and methanol.

is demonstrated that in addition to any solubility and kinetic factors, the preference of crystallization of B or A conformers in different solvents can be accounted for by a trend in a conformational distribution. The conformer A displays the highest population in methanol and acetonitrile from which it was preferably crystallized. The population of the conformer B increases and becomes the highest in the water solution, from which it was crystallized.

26.3.2.4 Ritanovir Ritanovir, a HIV-protease inhibitor, is a well-known example of the impact polymorphism has on drug development [45]. Currently, two polymorphic forms, I and II, are known with form II being the most stable at room temperature. A solvent-mediated polymorphic conversion study was reported by Miller et al [39]. It was demonstrated that, although the slurry crystallization is preferentially thermodynamically controlled, relatively high drug solubility (>8 mM but <200 mM) is needed to ensure solvent-mediated conversion to the most stable form. These conclusions were supported by polymorph screening in 13 solvents. Conformational populations of forms I and II in 13 solvents were predicted at room temperature at the PBE/DNP/COSMO level of theory. Due to the extremely high flexibility of the ritanovir drug, the crystallographic conformations were adopted (Figure 26.14) for the calculations. The conformations were taken from CSD database (reference codes are YIGPIO and YIGPIO01). The resulting conformer populations together with the polymorph screening results are presented in Table 26.5. It is demonstrated that except for MTBE, the results of 2-week slurry crystallization follow the trend of the preferred conformer population in the corresponding solvent.

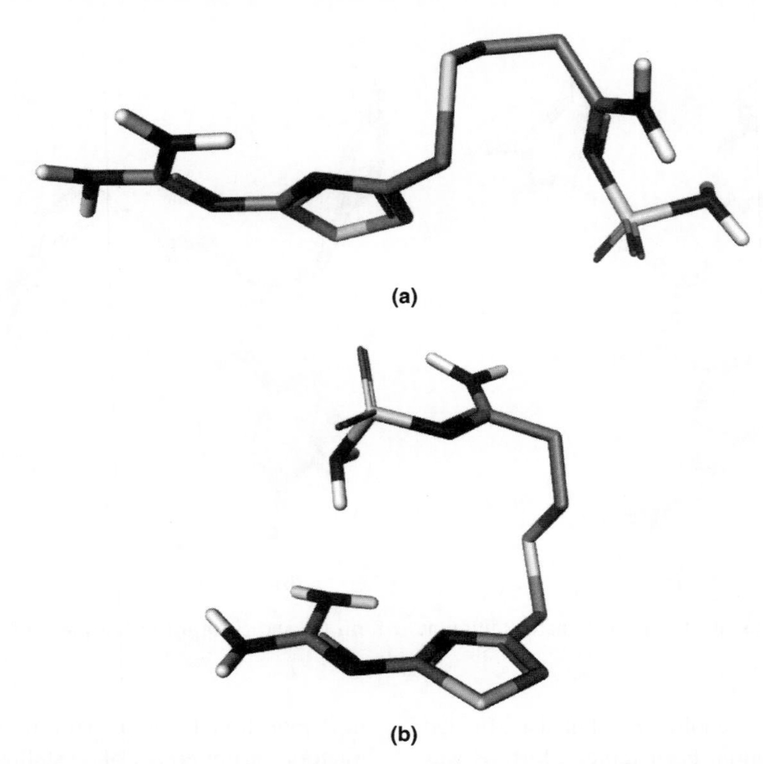

FIGURE 26.12 Molecular conformations of famotidine polymorphs A (a) and B (b). The crystal structures were taken from CSD, reference codes: FOGVIG04 (form A) and FOGVIG05 (form B). Only polar hydrogens are shown.

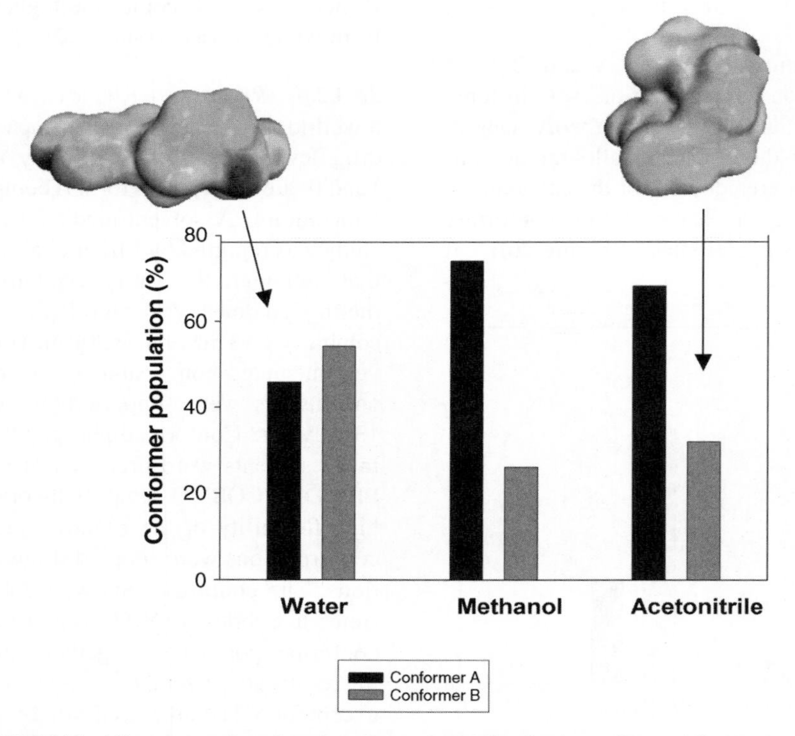

FIGURE 26.13 Histogram of predicted conformational distributions of famotidine in three solvents at 50°C. COSMO surfaces of the two conformers are shown on the top.

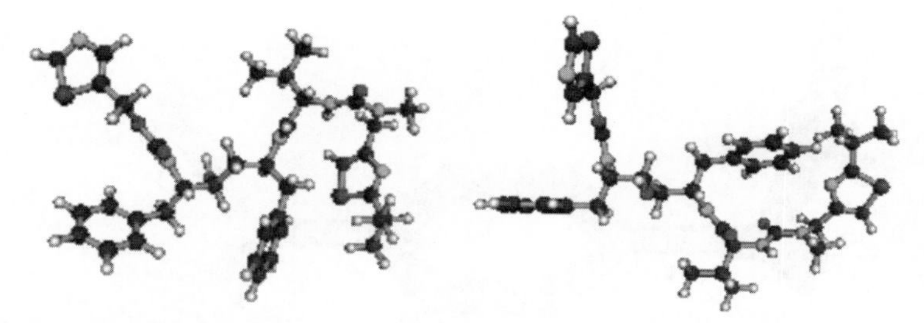

FIGURE 26.14 Crystallographic conformations of ritanovir forms I and II. The crystal structures were taken from CSD, reference codes: YIGPIO and YIGPIO01.

26.3.3 Implication to Crystal Structure Prediction

Crystal structure prediction is becoming a useful tool conducted in parallel to polymorph screening as well as being able to assess the risk of discovering a more stable form [15–18]. A starting point for CSP is the selection of molecular conformations that are as a rule held rigid during the generation of potential packing diagrams. Typically, the conformations are generated in the gas phase [19–21]. Since crystallization in the pharmaceutical industry is not performed from the gas phase but rather from different solvents, we recommend adopting the method of conformer distribution analysis, considered in this chapter, be applied to the conformations selection for virtual polymorph screening. For this, a diverse set of solvents should be considered in order to determine whether the solvent induces conformational switching of the active pharmaceutical. An indication that a molecule may switch conformations is the presence of intramolecular hydrogen bonding or a noticeable variation of molecular hydrophobic and hydrophilic surfaces. We propose a small diverse set of solvents to be considered for selecting molecular conformations used in CSP: polar protic—water (both HB donor and acceptor capabilities) and diethyl amine (HB donor capabilities); polar aprotic—acetone (HB acceptor capabilities); non-polar—hexane; and self-media to mimic solid amorphous. A combined set of conformations, which displayed the highest populations in any of the above solvents, can be recommended for the virtual conformational polymorph screening.

In order to illustrate this point, a conformer selection in support of CSP analysis was performed using a theoretical conformational population study of the flexible *S*-ibuprofen molecule. This study allowed for the selection of only four preferred conformations (Figure 26.8) with the highest populations using the following three solvents: methanol, acetone, and acetonitrile. Theoretical conformational distribution analysis in water, diethylamine, hexane, and self-media at the PBE/DNP/COSMO level of theory discover the same four favorable conformations. Thus, the selected four conformations can be used as a starting point for CSP. In addition, we should take into account that the acid group may rotate in the crystalline environment to participate in hydrogen-bonding interactions. This adds at least another four

TABLE 26.5 Results of Polymorph Screen of Ritanovir [39] and Predicted Conformer Populations at Room Temperature

Solvent	Solid Form (2 days)[a]	Solid Form (2 weeks)[a]	Conformer I Population[b] (%)	Conformer II Population[b] (%)
Water	I	II	21	79
Hexane	I	I,II mixture	85	15
Methyl-*t*-butyl ether	I	I	39	61
1,2-Xylene	I	I	70	30
Toluene	I	I	68	32
Nitromethane	II	II	31	69
Ethyl acetate	II	II	35	65
Acetonitrile	II	II	26	74
2-Propanol	II	II	32	68
2-Butanone	II	II	34	66
Acetone	II	II	27	73
1,2-Dimethoxyethane	II	II	28	72
Ethanol	II	II	30	70

[a] Results of stable polymorph screen reported by Miller et al. [39].
[b] Predictions at PBE/DNP/COSMO level of theory.

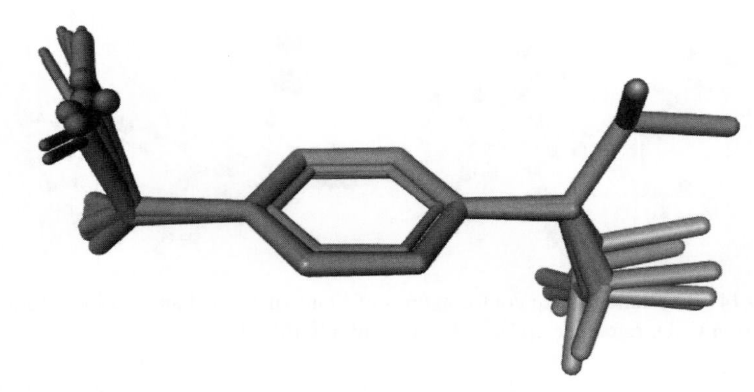

FIGURE 26.15 Alignment of the crystallographically observed *S*-ibuprofen conformations (CSD reference codes are IBPRAC, JEKNOC10, JEKNOC11, HUPPAJ, and RONWOG). All hydrogens are omitted.

conformations that are different from the initial set only by rotation of the acidic group by ~180°. In order to validate this selection process, the generated favorable conformers for *S*-ibuprofen (Figure 26.8) were compared with the observed molecular conformations from the Cambridge Structural Database (reference codes: IBPRAC, JEKNOC10, JEKNOC11, HUPPAJ, and RONWOG). Aligned crystallographic conformations are shown in Figure 26.15. An accuracy of the best alignment of the four selected favorable conformations (hydrogens are removed for the alignment) with those crystallographically observed is presented in Table 26.6. Quite low RMSD values demonstrate a good quality of theoretical selection of the conformations. It is interesting that while all crystallographic conformations were represented in the selected set of the favorable conformations of *S*-ibuprofen, one of the preferred conformations (conformer 3 in Table 26.3; Figure 26.8) has not yet been crystallographically observed (Figure 26.15).

26.4 CONCLUSIONS

Molecular crystallization is a very complex phenomenon, which is dependent on a combination of multiple thermodynamic and kinetic factors. Added complexity occurs when attempting to crystallize a conformational polymorph that is selected for drug development. In this study, we have shown the use of one parameter, such as relative conformer population, in exploring the possibility of controlling crystallization of conformational polymorphs. We have demonstrated that relative conformational population behavior in different solvents can be reasonably accurately predicted by theoretical methods. A correlation between trends of predicted (and, in some cases, also observed) conformational distribution changes in different solvents and crystallization results for conformational polymorphs were found in the four examples considered in this chapter. Although more extended testing is required, the following applications may be proposed.

TABLE 26.6 Best Alignments of the Calculated Favorable Conformations of *S*-Ibuprofen with the Similar Crystallographic Conformations from CSD

Conformer	RMSD (Å)	Reference Code
1	0.27	RONWOG
2	0.23	IBPRAC, JEKNOC10, JEKNOC11
4	0.12	HUPPAJ

1. The solvent selection strategy may focus on an attempt to diversify predicted conformer distribution in general, adopting crystallographic and/or theoretically generated conformations. This may facilitate crystallization of new form(s) that are both stable and metastable.
2. Another application of solvent selection is based on predicted change (increase or decrease) of the population(s) of the known conformer(s). This may allow tuning (promotion or inhibition) of the corresponding polymorph(s) crystallization.

In all cases, the final selection of solvents should also satisfy the criteria of relatively high (predicted or measured) solubility of the compound under consideration. In addition, the theoretical prediction of conformer distribution in different solvents, including self-media, was proposed for selection of preferable starting conformation(s) for virtual polymorph screening via crystal structure prediction.

ACKNOWLEDGMENTS

The authors would like to thank Mr. Brian Samas and Dr. Brian Marquez for the valuable comments and discussions.

Yuriy A. Abramov is thankful to Dr. Alex Goldberg for the valuable suggestions in running DMol^3 applications. YAA is grateful to Dr. Frank Eckert for the valuable consultations on COSMOtherm implementation of conformational population analysis.

REFERENCES

1. Bernstein J. *Polymorphism in Molecular Crystals*, Clarendon, Oxford, 2002.
2. Nangia A. Conformational polymorphism in organic crystals. *Acc. Chem. Res.* 2008;41:595–604.
3. Yu L, Reutzel-Edens SM, Mitchell CA. Crystallization and polymorphism of conformationally flexible molecules: problems, patterns, and strategies. *Org. Proc. Res. Dev.* 2000;4: 396–402.
4. Weissbuch I, Kuzmenko I, Vaida M, Zait S, Leiserowitz L, Lahav M. Twinned crystals of enantiomorphous morphology of racemic alanine induced by optically resolved α-amino acids: a stereochemical probe for the early stages of crystal nucleation. *Chem. Mater.* 1994;6:1258–1268.
5. Weissbuch I, Popoviz-Biro R, Leiswerowitz L, Lahav M. Lock-and-key processes at crystalline interfaces: relevance to the spontaneous generation of chirality. In: Behr J-P, editor. *The Lock-and-Key Principle: The State of the Art—100 Years On*, John Wiley & Sons, New York, 1994, pp. 173–246.
6. Lee AY, Lee IS, Dette SS, Boerner J, Myerson AS. Crystallization on confined engineered surfaces: a method to control crystal size and generate different polymorphs. *J. Am. Chem. Soc.* 2005;127:14982–14983.
7. Lee AY, Lee IS, Myerson AS. Factors affecting the polymorphic outcome of glycine crystals constrained on patterned substrates. *Chem. Eng. Tech.* 2006;29:281–285.
8. Weissbuch I, Addadi L, Leiswerowitz L. Molecular recognition at crystal interfaces. *Science* 1991;253:637–645.
9. Datta S, Grant DJW. Effect of supersaturation on the crystallization of phenylbutazone polymorphs. *Cryst. Res. Technol.* 2005;40:233–242.
10. Davey RJ, Blagden N, Righini S, Alison H, Ferrari ES. Nucleation control in solution mediated polymorphic phase transformations: the case of 2,6-dihydroxybenzoic acid. *J. Phys. Chem. B* 2002;106:1954–1959.
11. Davey RJ, Blagden N, Potts GD, Docherty R. Polymorphism in molecular crystals: Stabilization of a metastable form by conformational mimicry. *J. Am. Chem. Soc.* 1997;119: 1767–1772.
12. Mullin JW. *Crystallization*, Linacre House, Jordan Hill, Oxford, 1992.
13. Threlfall T. Structural and thermodynamic explanations of Ostwald's Rule. *Org. Proc. Res. Dev.* 2003;7:1017–1027.
14. Hursthouse MB, Huth S, Threlfall TL. Why do organic compounds crystallise well or badly or ever so slowly? Why is crystallization nevertheless such a good purification technique? *Org. Proc. Res. Dev.* 2009;13:1231–1240.
15. Blagden N, Davey RJ. Polymorph selection: challenges for the future? *Cryst. Growth Des.* 2003;3:873–885.
16. Cross WI, Blagden N, Davey, RJ, Pritchard, RG, Neumann, MA, Roberts, RJ, Rowe RC. A whole output strategy for polymorph screening: combining crystal structure prediction, graph set analysis, and targeted crystallization experiments in the case of diflunisal. *Cryst. Growth Des.* 2003;3:151–158.
17. Price SL. From crystal structure prediction to polymorph prediction: interpreting the crystal energy landscape. *Phys. Chem. Chem. Phys.* 2008;10:1996–2009.
18. Price SL. The computational prediction of pharmaceutical crystal structures and polymorphism. *Adv. Drug Del. Rev.* 2004;56:301–319.
19. Ouvrard C, Price SL. Toward crystal structure prediction for conformationally flexible molecules: the headaches illustrated by aspirin. *Cryst. Growth Des.* 2004;4:1119–1127.
20. Day GM, Motherwell WDS, Jones W. A strategy for predicting the crystal structures of flexible molecules: the polymorphism of phenobarbital. *Phys. Chem. Chem. Phys.* 2008;9: 1693–1704.
21. Cooper TG, Hejczyk KE, Jones W, Day GM. Molecular polarization effects on the relative energies of the real and putative crystal structures of valine. *J. Chem. The. Comput.* 2008;4:1795–1805.
22. Weng ZF, Motherwell WDS, Allen FH, Cole JM. Conformational variability of molecules in different crystal environments: a database study. *Acta Cryst.* 2008; *B64*:348–362.
23. MOE 2008.10, Chemical Computing Group, Inc., 1010 Sherbrooke Street West, Suite 910, Montréal, Québec, Canada. http://www.chemcomp.com
24. Delley B. An all-electron numerical method for solving the local density functional for polyatomic molecules. *J. Chem. Phys.* 1990;92:508–517.
25. Delley B. From molecules to solids with the DMol^3 approach. *J. Chem. Phys.* 2000;113:7756–7764.
26. Accelrys Software. Material Studio 4.4 DMol^3; Accelrys Software, Inc., San Diego, CA, 2008.
27. Perdew JP, Burke K, Ernzerhof M. Generalized gradient approximation made simple. *Phys. Rev. Lett.* 1996;77: 3865–3868.
28. Andzelm J, Kölmel C, Klamt A. Incorporation of solvent effects into density functional calculations of molecular energies and geometries. *J. Chem. Phys.* 1995;103:9312–9320.
29. Zhao Y, Truhlar DG. Benchmark databases of nonbonded interactions and their use to test density functional theory. *J. Chem. Theory Comput.* 2005;1:415–432.
30. Klampt A. *COSMO-RS: From Quantum Chemistry to Fluid-Phase Thermodynamics and Drug Design*, Elsevier, Amsterdam, 2005.
31. COSMOtherm, Version C2.1_0109, COSMOLogic GmbH, Leverkusen, Germany.
32. Lein M, Dobson JF, Gross EKU. Towards the description of van der Waals interactions within density functional theory. *J. Comput. Chem.* 1999;20:12–22.

33. Weigend F, Häser M, Patzelt H, Ahlrichs R. RI-MP2: optimized auxiliary basis sets and demonstration of efficiency. *Chem. Phys. Lett.* 1998;294:143–152.
34. TURBOMOLE, V6.0 2009, a development of University of Karlsruhe and Forschungszentrum Karlsruhe GmbH, 1989–2007, TURBOMOLE GmbH, since 2007.
35. Keeler J, Neuhaus D, Titman JJ. A convenient technique for the measurement and assignment of long-range carbon-13 proton coupling constants. *Chem. Phys. Lett.* 1988;146:545–548.
36. Yamasaki R, Tanatani A, Azumaya I, Masu H, Yamaguchi K, Kagechika H. Solvent-dependent conformational switching of *N*-phenylhydroxamic acid and its application in crystal engineering. *Cryst. Growth Des.* 2006;9:2007–2010.
37. Forbes CC, Beatty AM, Smith BD. Using pentafluorophenyl as a Lewis acid to stabilize a *cis* secondary amide conformation. *Org. Lett.* 2001;3:3595–3598.
38. Gu C, Young JrV, Grant DJW. Polymorphs screening: influence of solvents on the rate of solvent-mediated polymorphic transformation. *J. Pharm. Sci.* 2001;90:1878–1890.
39. Miller JM, Collman BM, Greene LR, Grant DJW, Blackburn AC. Identifying the stable polymorph early in the drug discovery--development process. *Pharm. Dev. Technol.* 2005;10:291–297.
40. Ostwald W. Studien Uber Die Bildung und Umwandlung Fester Korper. *Z. Physik. Chem.* 1897;22:289–302.
41. Maruyama S, Ooshima H, Kato J. Crystal structures and solvent-mediated transformation of taltireline polymorphs. *Chem. Eng. J.* 1999;75:193–200.
42. Maruyama S, Ooshima H. Mechanism of the solvent-mediated transformation of taltirelin polymorphs promoted by methanol. *Chem. Eng. J.* 2001;81:1–7.
43. Allen FH, Bellard S, Brice MD, Cartwright BA, Doubleday A, Higgs H, Hummelink TWA, Hummelink-Peters BG, Kennard O, Motherwell WDS, Rodgers JR, Watson DG. The Cambridge Structural Database: a quarter of a million crystal structures and rising. *Acta Cryst.* 1979;B35:2331–2339.
44. Lu J, Wang X-J, Yand X, Ching C-B. Characterization and selective crystallization of famotidine polymorphs. *J. Pharm. Sci.* 2007;96:2457–2468.
45. Chemburkar SR, Bauer J, Deming K, Spiwek H, Patel K, Morris J, Henry R, Spanton S, Dziki W, Porter W, Quick J, Bauer P, Donaubauer J, Narayanan BA, Soldani M, Riley D, McFarland K. Dealing with the impact of ritonavir polymorphs on the late stages of bulk drug process development. *Org. Proc. Res. Dev.* 2000;4:413–417.

27

MOLECULAR THERMODYNAMICS FOR PHARMACEUTICAL PROCESS MODELING AND SIMULATION

CHAU-CHYUN CHEN

Aspen Technology, Inc., Burlington, MA, USA

27.1 INTRODUCTION

In this increasingly demanding pharmaceutical market, chemical engineers face major challenges in maintaining a competitive advantage in process development: (1) driving speed and efficiency into the active pharmaceutical ingredient (API) process development workflow, (2) managing information flow throughout the process development stages to improve collaboration, and (3) enabling improved process design that delivers quality assured product and lower cost of goods in commercial-scale manufacturing operations. In addition, U.S. FDA and other agencies are increasing their emphasis on process understanding. Quality by design (QbD) is an evolving initiative from FDA emphasizing that quality should be built into a product, with a thorough understanding of the product and the process by which it is developed and manufactured, along with a knowledge of the risks involved in manufacturing the product and how best to mitigate those risks. As part of QbD, first-principles modeling, statistically designed experiments, and scale-up correlations will all be considered in determining the approved design space of acceptable manufacturing conditions. To meet these challenges, chemical engineers must strive to apply modern-day first-principles modeling and simulation technology and advance science-based, mechanistic understanding of pharmaceutical manufacturing processes. Without such understanding, pharmaceutical process development would remain "lagging behind potato chip and laundry detergent makers in the use of modern manufacturing systems [1]." However, to date, the use of first-principles modeling and simulation technology in the pharmaceutical industry remains very limited, if any.

27.2 PROCESS SIMULATION AND MOLECULAR THERMODYNAMICS

> Process simulation, which emerged in the 1960s, has become one of the great success stories in the use of computing in the chemical industry. For instance, steady-state simulation has largely replaced experimentation and pilot plant testing in process development for commodity chemicals, except in the case of reactions having new mechanisms or requiring new separation technologies. Tools for steady-state process simulation are nowadays universally available to aid in the decisions for design, operation, and debottlenecking; they are part of every process engineer's toolkit. Their accuracy and predictive ability for decision-making is widely accepted to make routine plant trials and most experimental scale-up obsolete in the commodity chemicals industry [2].

While the foundation of quality by design is the availability of intrinsic (kinetic and mechanistic) process knowledge collected through the use of the various (modeling) tools [1], the scientific foundation of process modeling and simulation technology are the molecular thermodynamic models that provide thermodynamically consistent descriptions of thermophysical properties and

Chemical Engineering in the Pharmaceutical Industry: R&D to Manufacturing Edited by David J. am Ende

phase behavior of chemical systems that are being investigated.

> Indeed, the development of process modeling technologies and software tools follows and evolves around the development of molecular thermodynamics. For example, advances in equations-of-state and activity coefficient models since the 1970s set the stage for widespread applications of process modeling tools in the oil and gas industry and the petrochemical industry. For example, recent advances in molecular thermodynamics have made it possible to develop and apply process modeling technologies for complex chemical systems, i.e., processes involving synthetic fuels, aqueous electrolytes, polymer manufacturing processes, etc. Without these advances in molecular thermodynamics, chemical engineers would not have been able to develop first-principles-based, high-fidelity process models that are instrumental for process and product design and optimization [3].

The key technical barrier to the successful application of process modeling and simulation technology in the pharmaceutical industry has been the lack of accurate and robust molecular thermodynamic models that meet the unique challenges of the pharmaceutical industry. The process modeling and simulation challenges faced by the pharmaceutical industry are not so much with the mathematical formulation or simulation software of various unit operations, batch or continuous. Such mathematical formulation and simulation software have been well advanced and widely applied in the petrochemical, chemical, and specialty chemical industries and they are ready to be used in the pharmaceutical industry. Rather, without robust molecular thermodynamic models that can accurately describe the thermophysical properties and phase behavior of systems with drug molecules, process modeling and simulation technology will not provide the accuracy and predictive capability required to simulate real performance of pharmaceutical manufacturing processes and little process knowledge and few benefits, if any, could ever be derived from it.

27.2.1 Activity Coefficient as Key Thermophysical Property

Models for wide varieties of thermophysical properties and phase behavior are required in process modeling and simulation. Among the various thermophysical properties of concern to pharmaceutical process modeling and simulation, activity coefficient stands out clearly as the single most critical property. Activity coefficient plays the central role in determining the solubility and related phase behavior of pharmaceutical molecules in major pharmaceutical unit operations, for example, crystallization, chromatography, extraction, distillation, reaction, and so on.

As summarized by Lipinski et al. [4], "the knowledge of the thermodynamic solubility of drug candidates is of paramount importance in assisting the discovery, as well as the development, of new drug entities at later stages." Given a solid polymorph and a fixed temperature, equation 27.1 shows that the solubility of drug candidate is only a function of its activity coefficient in solution as the solvent composition changes.

$$\ln x_I^{\mathrm{sat}}\gamma_I^{\mathrm{sat}} = \frac{\Delta H_{\mathrm{fus}}}{R}\left(\frac{1}{T_{\mathrm{m}}} - \frac{1}{T}\right) \tag{27.1}$$

where x_I^{sat} is the mole fraction of solute I at saturation, ΔH_{fus} is the enthalpy of fusion for the solid polymorph, R is the ideal gas constant, T is the temperature, T_{m} is the melting point, and γ_I^{sat} is the activity coefficient of solute I at saturation. ΔH_{fus} and T_{m} vary with polymorphic forms of the solute.

Equation 27.2 shows that the magnitude of solute retention in chromatography under isocratic condition (i.e., constant mobile-phase solvent composition) is also related to solute activity coefficients in the mobile phase and the stationary phase.

$$k_I = \frac{x_{\mathrm{s}} V_{\mathrm{s}}}{x_{\mathrm{m}} V_{\mathrm{m}}} = K_I \Phi = \frac{\gamma_{\mathrm{m}}^{\infty}}{\gamma_{\mathrm{s}}^{\infty}} \Phi \tag{27.2}$$

where k_I is the retention factor for solute I; K_I is the partition coefficient; Φ is the phase ratio, that is, the ratio of the volume of the stationary phase V_{s} to that of the mobile phase V_{m}; x_{s} and x_{m} are the solute concentrations in the stationary phase and the mobile phase; and $\gamma_{\mathrm{s}}^{\infty}$ and $\gamma_{\mathrm{m}}^{\infty}$ are the solute infinite dilution activity coefficients in the stationary phase and the mobile phase, respectively.

Equation 27.3 shows the isoactivity relationship for extraction:

$$x_I^{\mathrm{L1}}\gamma_I^{\mathrm{L1}} = x_I^{\mathrm{L2}}\gamma_I^{\mathrm{L2}} \tag{27.3}$$

where x_I^{L1} and x_I^{L2} are the concentrations for solute I in the first liquid phase and the second liquid phase, respectively. γ_I^{L1} and γ_I^{L2} are the solute activity coefficients in the two liquid phases.

Equation 27.4 shows the isofugacity relationship for distillation:

$$x_I \gamma_I p_I^{\mathrm{o}} = y_I \phi_I^{\mathrm{V}} P \tag{27.4}$$

where x_I and y_I are the concentrations for solute I in the liquid phase and the vapor phase, respectively, γ_I is the solute liquid-phase activity coefficient, ϕ_I^{V} is the vapor-phase fugacity coefficient, p_I^{o} is the solute vapor pressure, and P is the system pressure.

As seen by the inclusion of activity coefficients in the above equations, a prerequisite to executing meaningful process modeling and simulation of any pharmaceutical manufacturing processes is having robust and thermodynamically consistent models that can accurately describe activity coefficients of various components in the system of interest. Availability of such activity coefficient models is a prerequisite to meaningful first-principles process modeling and simulation in the pharmaceutical industry.

27.2.2 Thermodynamic Activity Coefficient Models

Numerous molecular thermodynamic models have been proposed in the literature to correlate or predict activity coefficients [5]. Many of them have been incorporated into process simulators [6]. Popular semiempirical correlative models, such as NRTL and UNIQUAC, are the gold standard activity coefficient models for process modeling and simulation of the petrochemical and chemical industries. However, these correlative models require identification of binary interaction parameters from phase equilibrium data for each of the solvent–solvent, solvent–solute, and solute–solute binary mixtures. While such solvent–solvent, solvent–solute, and solute–solute binary phase equilibrium data are often available for commodity chemicals, they are rarely available for new chemical entities and reaction intermediates encountered in the pharmaceutical industry. Consequently, these correlative models find very limited use in pharmaceutical process modeling, simulation, and design except in very limited solvent recovery applications.

The predictive, group contribution-based UNIFAC activity coefficient model requires only chemical structure information for the solvents and solutes [7]. Unfortunately, UNIFAC fails for complex pharmaceutical molecules for which either the UNIFAC functional groups are undefined or the functional group additivity rule becomes invalid for rigid molecular structure [8]. Recent developments in computational chemistry yielded the COSMO-RS [9] and COSMO-SAC [10] predictive models that represent promising alternatives to UNIFAC. However, the predictive powers of UNIFAC- and COSMO-based models are still inadequate [11] and their usability has been limited to nonelectrolytes.

The Hansen solubility parameter model has been the most widely used activity coefficient model in the pharmaceutical industry [12]. Incorporating the "like dissolves like" concept, the model is useful as a guide to help chemists and engineers explain API solubility behavior. However, due to its oversimplistic assumptions, the model has very limited practical use in the quantitative estimation of drug molecule solubility [11].

27.2.3 NRTL Segment Activity Coefficient (NRTL-SAC) Model

Designed to overcome the gap between molecular thermodynamics and process modeling and simulation technology for the pharmaceutical industry, the NRTL-SAC model is a very interesting and promising new development [11]. As an extension of the NRTL model [13] and the polymer NRTL [14] model for systems with solvents, solutes, oligomers, and polymers, NRTL-SAC computes activity coefficients from a combinatorial term and a residual term.

$$\ln \gamma_I = \ln \gamma_I^{C} + \ln \gamma_I^{R} \tag{27.5}$$

The combinatorial term γ_I^C is calculated from the Flory–Huggins approximation for the combinatorial entropy of mixing. The residual term γ_I^R is calculated from the local composition (lc) contribution γ_I^{lc} of the polymer NRTL model. Incorporating the segment interaction concept, the equation computes the activity coefficient for component I in solution by summing up contributions to the activity coefficient from all segments that make up component I.

$$\ln \gamma_I^{R} = \ln \gamma_I^{lc} = \sum_i r_{i,I}\left[\ln\Gamma_i^{lc} - \ln\Gamma_{i,I}^{lc}\right] \tag{27.6}$$

$$\ln\Gamma_i^{lc} = \frac{\sum_j x_j G_{ji}\tau_{ji}}{\sum_k x_k G_{ki}} + \sum_m \frac{x_m G_{im}}{\sum_k x_k G_{km}}\left(\tau_{im} - \frac{\sum_j x_j G_{jm}\tau_{jm}}{\sum_k x_k G_{km}}\right) \tag{27.7}$$

$$\ln\Gamma_{i,I}^{lc} = \frac{\sum_j x_{j,I} G_{ji}\tau_{ji}}{\sum_k x_{k,I} G_{ki}} + \sum_m \frac{x_{m,I} G_{im}}{\sum_k x_{k,I} G_{km}}\left(\tau_{im} - \frac{\sum_j x_{j,I} G_{jm}\tau_{jm}}{\sum_k x_{k,I} G_{km}}\right) \tag{27.8}$$

where I is the component index, i,j,k,m are the segment species index, $r_{i,I}$ is the number of segment species i contained only in component I, x_j is the segment-based mole fraction of segment species j, $x_{j,I}$ is the segment-based mole fraction of segment species j in component I, Γ_i^{lc} is the activity coefficient of segment species i, and $\Gamma_{i,I}^{lc}$ is the activity coefficient of segment species i contained only in component I. G and τ in equations 27.7 and 27.8 are binary quantities related to each other by α $\left(\text{i.e., } G = \exp(-\alpha\tau)\right)$. α and τ are the nonrandomness factor parameter and the segment–segment binary interaction energy parameter, respectively.

Chen and Song identified four unique "conceptual" segments that broadly characterize surface interaction characteristics of molecules, solvents, or solutes [11]. These four conceptual segments, together with their corresponding nonrandomness factor and segment–segment binary interaction parameters (i.e., α and τ), are capable of qualitatively describing the various solvent–solvent, solvent–solute, and solute–solute molecular interactions and the resulting phase behavior of mixtures of solvents and solutes. Specifically, Chen and Song proposed to describe the molecular surface interactions of all solvents and solutes in solution with four types of conceptual segments: hydrophobic segment, electrostatic solvation segment, electrostatic polar segment, and hydrophilic segment. The conceptual segment numbers for each molecule, solvents or solutes, are measures of the effective molecular surface areas that exhibit surface interaction characteristics of hydrophobicity, solvation, polarity, and hydrophilicity. The hydrophilic segment simulates molecular surfaces that are "hydrogen bond donor or acceptor." The hydrophobic segment simulates molecular surfaces that show aversion to forming a hydrogen bond. The polar and solvation segments simulate molecular surfaces that are "electron pair donor or acceptor." The solvation segment is attractive to the hydrophilic segment while the polar segment is repulsive to the hydrophilic segment. The molecule-specific conceptual segment numbers correspond to $r_{i,I}$ in equation 27.6.

Also proposed are "reference compounds" for the conceptual segments. They are used to identify the segment–segment nonrandomness factor and binary interaction energy parameters for the conceptual segments from regression of available experimental vapor–liquid and liquid–liquid equilibrium data associated with these reference compounds. Chen and Song further identified the conceptual segment numbers for solvents commonly used in the pharmaceutical industry.

To determine the conceptual segment numbers of a solute molecule, solubility data or equivalent activity coefficient data in at least four solvents of varied surface interaction characteristics are needed. The parameterization is improved if a range of hydrophilic solvents, polar solvents, solvation solvents, and hydrophobic solvents are used. Once the segment numbers of the solute molecule are determined, the NRTL-SAC model can then provide robust, qualitative prediction for the solute activity coefficient and the corresponding solubility in pure solvents and solvent mixtures.

It is estimated that half of all the drug molecules used in medicinal therapy are administered as salts [15]. This conceptual segment methodology has also been successfully extended for activity coefficient modeling of organic salts [16, 17].

27.2.4 Acetaminophen: An Example

Figure 27.1 shows NRTL-SAC predictions versus experimental data for acetaminophen solubility in 23 pure solvents at 303.15K [18]. As representatives of hydrophilic (water and ethanol), solvation (DMSO), polar (acetone, acetonitrile, and THF), and hydrophobic (chloroform and toluene) solvents, eight pure solvent solubility data points (shown as solid squares) were used to identify the acetaminophen parameters. Empty diamonds represent the predictions for the remaining 15 pure solvents.

Figures 27.2–27.5 show robust predictions (as solid lines) versus experimental data (as solid squares) for acetaminophen solubility in mixed solvents at 298.15K [18]. Figure 27.2 shows the prediction for a mixed solvent of two hydrophilic solvents, for example, ethanol–water binary. The acetaminophen solubility in this binary is nonideal but without significant peak solubility. Figure 27.3 shows the prediction for a mixed solvent of one polar solvent and one hydrophilic solvent, for example, acetone–water binary. The solubility behavior of acetaminophen in this binary is extremely nonideal, with "bell"-shaped solubility behavior as a function of solvent composition and a four- to fivefold solubility increase. Figure 27.4 shows the prediction for

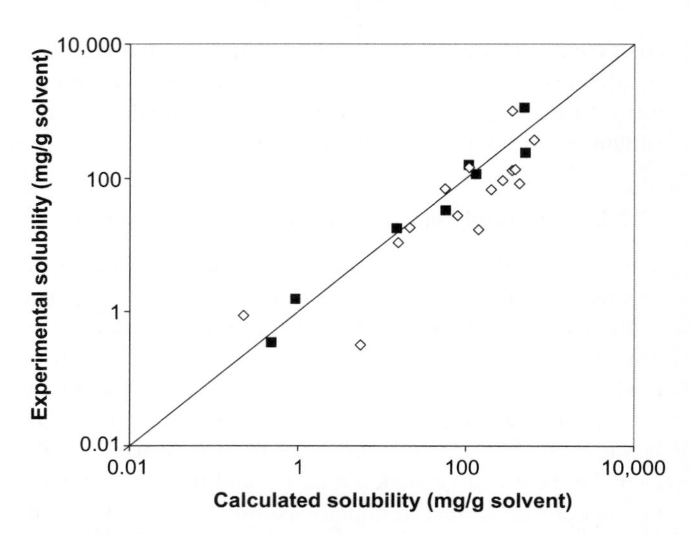

FIGURE 27.1 Model prediction versus experimental data for acetaminophen solubility in pure solvents at 303.15K (solid squares represent the eight pure solvent solubility data points that were used to identify the solute parameters; empty diamonds represent pure solvent solubility data, excluding the eight pure solvents). Reprinted with permission from Ref. 18. Copyright 2006, American Chemical Society.

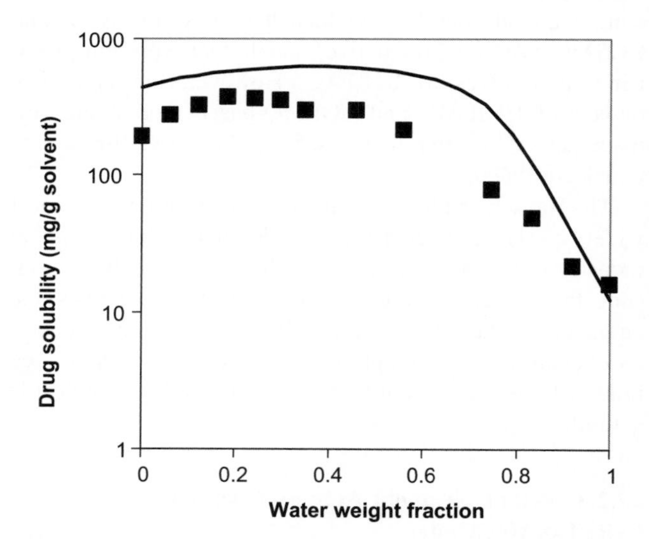

FIGURE 27.2 Model prediction versus experimental data for acetaminophen solubility in ethanol–water binary solvents at 298.15K (solid squares are experimental data and solid line represents model predictions). Reprinted with permission from Ref. 18. Copyright 2006, American Chemical Society.

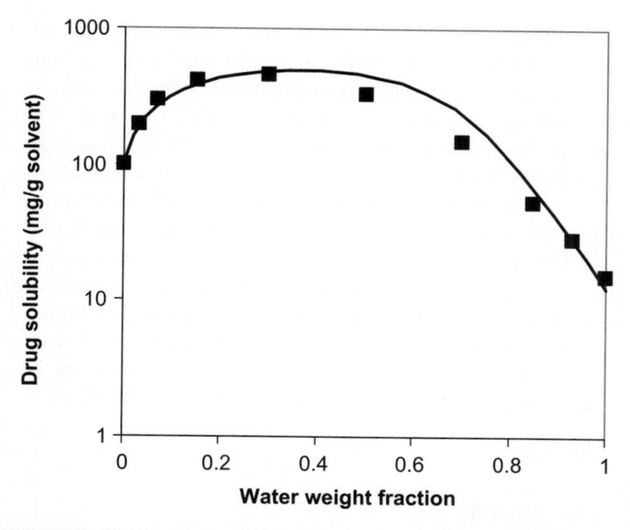

FIGURE 27.3 Model prediction versus experimental data for acetaminophen solubility in acetone–water binary solvents at 298.15K (solid squares are experimental data and solid line represents model predictions). Reprinted with permission from Ref. 18. Copyright 2006, American Chemical Society.

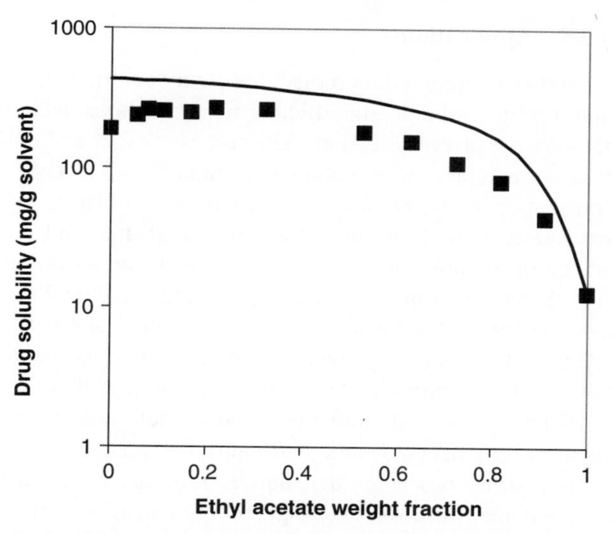

FIGURE 27.5 Model prediction versus experimental data for acetaminophen solubility in ethanol–ethyl acetate binary solvents at 298.15K (solid squares are experimental data and solid line represents model predictions). Reprinted with permission from Ref. 18. Copyright 2006, American Chemical Society.

a mixed solvent of one polar solvent and one hydrophobic solvent, for example, acetone–toluene binary. The acetaminophen solubility in this binary is relatively ideal. Figure 27.5 shows the prediction for a mixed solvent of one hydrophilic solvent and one hydrophobic solvent, for example, ethanol–ethyl acetate binary. Again, the model predicts nonideal acetaminophen solubility in the binary, consistent with the trend exhibited by the experimental data.

While the quality of the NRTL-SAC model predictions depends on the quality of the experimental data used to

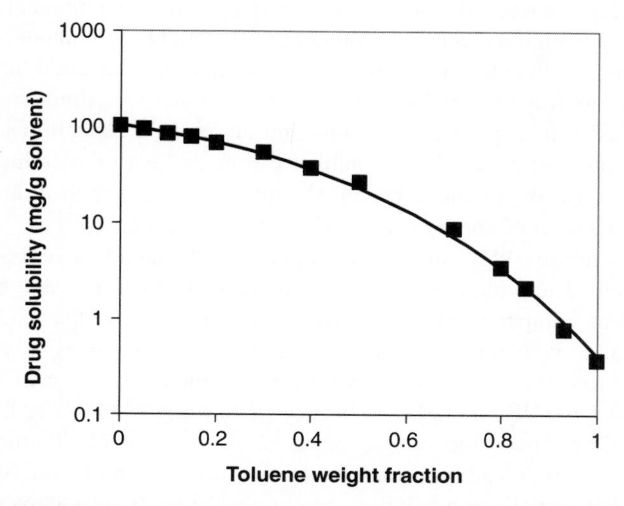

FIGURE 27.4 Model prediction versus experimental data for acetaminophen solubility in acetone–toluene binary solvents at 298.15K (solid squares are experimental data and solid line represents model predictions). Reprinted with permission from Ref. 18. Copyright 2006, American Chemical Society.

identify the solute parameters and there is no guarantee that the model will always yield correct, quantitative predictions, the predictive capability of NRTL-SAC has been successfully demonstrated with hundreds of drug molecules. For example, in a study with six Merck compounds [19], Merck researchers showed that NRTL-SAC offers solubility prediction accuracy within the range of ±50% and meets the needs of solvent selection and API process design. In contrast, the prediction accuracies from UNIFAC are in the range of ±500%, while the correlation accuracies with the Hansen model are in the range of ±200%. In a subsequent study [20], Merck researchers reported application of NRTL-SAC and COSMO-SAC in the solubility estimation of four Merck compounds: lovastatin, simvastatin, rofecoxib, and etoricoxib. They concluded that NRTL-SAC offered superior performance over COSMO-SAC. The maximum average log square error for NRTL-SAC was 0.10 (i.e., prediction accuracy of ±30%) while the maximum average log square error for COSMO-SAC was 0.32 (i.e., prediction accuracy of ±100%).

A very recent evaluation of NRTL-SAC also found a satisfactory agreement between experimental and calculated values for four drugs: paracetamol, allopurinol, furosemide, and budesonide [21]. The solubility data in pure organic solvents were used to regress the solute model parameters that were used afterward for the prediction of solubility of these compounds in water and in mixed solvent systems. The absolute average deviation was 68% for the correlation in the organic solvents and 38% for the prediction in water. The model was shown to be an appropriate tool to represent and predict the solubility of these compounds.

27.2.5 Applications

The ability to predict drug molecule activity coefficients and solubility in a reliable and efficient manner is invaluable to the tasks of solvent selection, API process design and optimization, and process modeling and simulation. NRTL-SAC represents a quantum leap forward in the ability to first correlate a limited number of experimental data and then predict drug molecule activity coefficients and resulting phase behavior in pure solvents and mixed solvents. When integrated with Microsoft® Excel and process simulators, NRTL-SAC offers a rigorous and practical thermodynamic framework that provides robust predictions of API activity coefficients across all unit operation models and enables chemical engineers design and optimize pharmaceutical manufacturing processes that deliver required drug purity and yield, minimize solvent usage, reduce hazardous solvent waste, consume less energy, and lower overall cost.

Some successful industrial applications of NRTL-SAC have started to emerge in the public domain. A few examples are summarized here:

Design of crystallization processes for the manufacture of API is a significant technical challenge to process research and development groups. AstraZeneca researchers examined the role of activity coefficient modeling and its application within the crystallization process design framework [22]. NRTL-SAC has been demonstrated, through the case study on cimetidine, to be a valuable aid in solubility data assessment and targeted solvent selection for crystallization process design.

Eli Lilly scientists applied NRTL-SAC to screen solvents for a crystallization medium with the goal to maximize API solubility and to minimize solvent usage [23].

> The NRTL-SAC model parameters for the molecule in development are first identified from a minimal set of solubility experiments in selected solvents. We then perform numerous *in silico* virtual experiments to explore the solubility behavior of the molecule in other pure solvents and mixed solvents. The modeling results suggested optimal solvent systems for the crystallization medium which are validated in physical laboratories and chosen for process scale-up. This study demonstrated the effectiveness of the NRTL-SAC model and supports its use as a tool in drug development.

Using models, Bristol-Myers Squibb researchers demonstrated an efficient approach to identify optimal solvent compositions during conceptual design of an API process [24].

> A ternary solvent system was considered for a reaction, extraction, distillation, and crystallization sequence. Two thermodynamic models, NRTL-SAC and NRTL, as well as Aspen modeling tools, were employed to predict the liquid-liquid, vapor-liquid, and solid-liquid phase behaviors. We used these modeling tools to identify a solvent composition space for the reaction that allows for reasonable reaction volume while continuously removing a byproduct into a second aqueous phase. This composition also reduces API loss during subsequent aqueous extractions. Furthermore, the composition of the organic phase allows for an efficient azeotropic distillation during solvent exchange, resulting in a shorter cycle time needed to achieve the desired composition for final crystallization. Overall solvent usage for the process is also significantly reduced. This approach was applied retrospectively to a late-stage API process under experimental development and was validated with the production of API of excellent quality at the pilot scale with solvent compositions of the process in agreement with those predicted by the models.

27.2.6 Benefits

At the highest level, and for any R&D centric industry, the sooner improvements to new products (or the quality of decisions surrounding them) can be made, the greater the overall value potential. Value potential here is measured not only by value delivered to a product or a process, but also by "redundant cost avoidance." It is this that hits pharma's "value sweet spot" square on because pharmaceutical research is fundamentally much more "selection" than "instruction" in its nature, and so the biggest value impact may be felt in cost avoidance. This is where process modeling and simulation comes into its own and can yield millions of dollars of accrued value in the course of subsequent R&D and throughout the life cycle of the new drug thereafter. This is illustrated in Figure 27.6: a value plot against timeline to launch a new drug. Here, areas of application for modeling and simulation include lead optimization, API process development, and Drug Product (DP) process development.

Given the recent development of the NRTL-SAC model, and the long timelines for product development required for a new drug, value benefit can only be a qualitative estimation, based on current applications and anticipated capabilities. Notwithstanding the inevitably "estimated" nature of value benefit, any potential benefit should also be seen against the backdrop of apparently ever-increasing R&D costs.

In the 1980s and 1990s, relative R&D spending represented approximately 15–17% of revenue for the average drug company. Today, that average is approaching 20%, and for some companies may exceed that level. Estimates have placed the cost of bringing a new chemical entity (NCE) successfully to market to be anywhere from $700 to $1200 million over the course of 9–12 years of R&D. Some companies estimate that getting as far as completion of lead optimization requires spending some $300 million over the first 4 or 5 years of research. With the discovery and development of high-value medicines becoming harder and harder, and new drug application (NDA) annual submissions on the decline, the time is fast approaching where dramatic

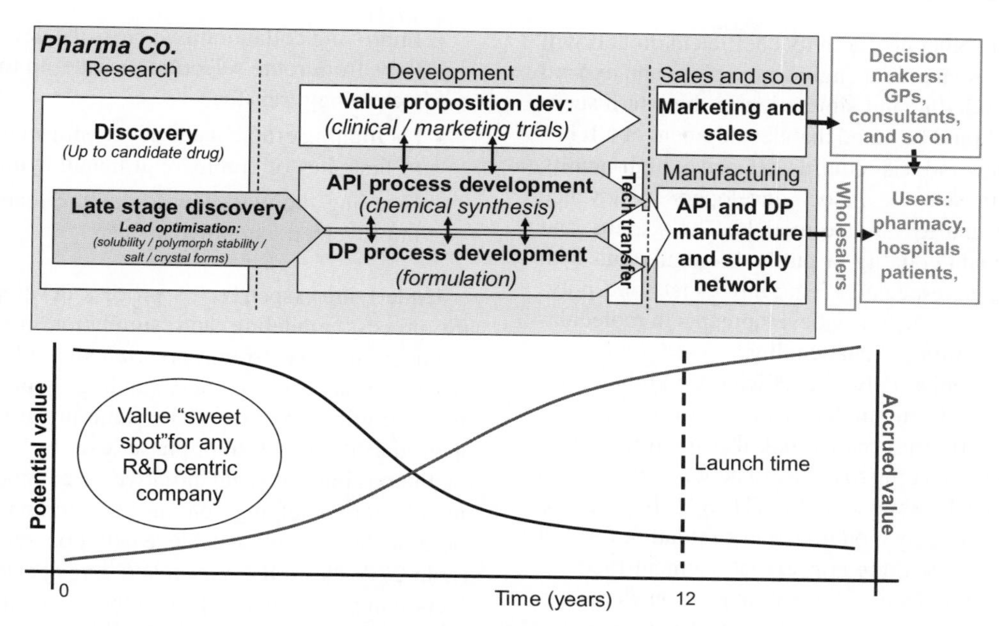

FIGURE 27.6 Value potential and benefit plot against timeline for a new drug.

operational efficiency improvements in R&D will be as much a central plank for competitive advantage in pharmaceutical industry as it already is for manufacturing operations.

With the above in mind, the application of process modeling and simulation should add value in four major ways across any pharmaceutical R&D organization:

1. *Efficiency Improvement*: by driving up the efficiency with which NCE solubility can be fully characterized, and all the potentially advantageous effects that this can confer in the lead optimization space. Literally hundreds of "experimental hours" could be reduced to just a few through the use of modeling and prediction software. This value may manifest as a reduction in cost through headcount reduction or an increase in throughput rate of NCEs in late discovery/early development. The latter is the likelier benefit route for companies with healthy pipelines of NCEs.
2. *Risk Management/Better Decision Making*: by exploiting the predictive power of the model to drive more informed and earlier decisions relating to selecting the candidate drug to best progress with respect to its "processability" downstream in API and DP manufacture. This ultimately enables a more informed and better investment focus. Delaying or dropping candidate drugs exhibiting very significant process challenges could save time, money, and resource or direct attention to solving "knockout" issues first, before devoting more investment. In addition, this should augment an "eyes open" approach to portfolio management of NCEs in early development with respect to their risk profile for manufacturability.
3. *Speed to Market Launch/Continuity of Supply After Launch*: by enabling aspects of process development activity (often delayed owing to insufficient NCE material) to proceed earlier. This can be achieved by using the modeling and predictive power of the software to sidestep this common cause of delay by using prediction to replace what would otherwise be experimentally derived process design data. This may translate into earlier clinical trials, and just possibly faster to market. Further value may manifest by avoiding or reducing the emergence of unforeseen disasters downstream that may severely compromise launch times or continuity of supply after launch. (Late emerging crystal polymorphs are a good example here, in which solubility characteristic of the active drug are permanently changed and can impact dose form stability, and even bioavailability.)
4. *API and Drug Product Manufacturing Process Performance and Cost Profiles*: by enabling the informed design of many aspects of the API and DP manufacturing processes, such that the final developed process is better characterized, optimized, greener, and higher in yield, thereby reducing cost of goods from the outset of launch.

27.3 FUTURE DEVELOPMENTS

NRTL-SAC represents one recent successful molecular thermodynamic model that makes it possible to carry out meaningful first-principles process modeling and simulation for pharmaceutical manufacturing processes. While the development of NRTL-SAC is new, further developments of NRTL-SAC and other new activity coefficient models will certainly emerge as chemical engineers explore and expand use of process modeling and simulation in the pharmaceutical industry. It should be noted that development of NRTL-SAC and other new models requires extensive, high-quality experimental data sets for model developers to advance physical insights and to validate models. The scarcity and the often questionable quality of public literature data with drug-like molecules, especially regarding exact solid polymorphs, indeed makes such new developments in molecular thermodynamics extremely challenging.

The ability to predict activity coefficients and the corresponding solubility of drug molecules is important not only for API manufacturing processes, but also for formulation design and delivery of a drug to its site of action in the human body [4]. While the development of NRTL-SAC has focused on applications in process modeling and simulation for API manufacturing processes, the conceptual segment methodology of describing molecular surface interaction characteristics is equally applicable to surfactants and polymers that are often used as excipients in drug formulation. Another example of particular interest, although not obvious and under investigation, is solubility modeling and prediction of biologically derived or engineered macromolecules, such as monoclonal antibodies and genetically engineered proteins. This could be particularly exciting if further research shows that the segmentation nature of the NRTL-SAC model lends itself well to more and more complex chemical/biochemical entities. With further experimentation, it may also be possible to characterize human organs, body fluids, and tissues in a similar manner to study drug bioavailability, pharmacokinetics, and toxicity. Application of NRTL-SAC in drug formulation and other pertinent areas should be actively pursued.

27.4 ASPEN TECHNOLOGY'S PROCESS DEVELOPMENT SOLUTION

aspenONE Engineering, AspenTech's model-based process development solution for the pharmaceutical industry, delivers "design for manufacture" (DfM) capability by enabling people to model, simulate, design, and optimize API manufacturing processes. The solution is designed to transform the pharmaceutical process development workflow into a productive and efficient process by

- Enabling development of robust and thermodynamically consistent thermophysical property models including NRTL-SAC
- Providing the modeling and simulation framework for first-principles-based process models, both batch and continuous
- Improving collaboration across the development workflow, from route selection to scale-up to completion of technology transfer
- Facilitating efficient and successful technology transfer to first sites of commercial manufacture
- Enabling the workflow to capture learning and apply improvements iteratively

Underlying AspenTech's process development solution are process modeling and simulation tools that enable engineers to develop first-principles-based process models to achieve mechanistic understanding of the pharmaceutical manufacturing processes. The same process models developed during the design phase can be used to support continuous improvement initiatives in commercial-scale API manufacture—helping engineers eliminate bottlenecks, increase throughput, or achieve better operational efficiency. Table 27.1 shows the *aspenONE Engineering* suite of products comprised of independently deployable components with specific modeling functionalities.

Clearly, one cannot overemphasize the importance of applying exactly the same thermophysical property methods, models, and data across all process modeling and simulation tools and activities to ensure rigor and consistency of results. Therefore, a key common requirement for the various process modeling and simulation products is the library of rigorous, thermodynamically consistent models for thermophysical properties and phase behavior. Aspen

TABLE 27.1 aspenONE Engineering suite of products

Product Name	Description
Aspen Properties	Physical property modeling system with comprehensive chemical database and estimation capability
Aspen Solubility Modeler	Modeling solubility of drug molecules and predicting drug molecule solubilities in solvents and solvent mixtures
Aspen Plus	First-principles rigorous modeling for continuous steady-state processes (including single and multistage separations such as distillation and extraction, reactors, heat exchangers, pumps, compressors, etc.)
Aspen Batch Distillation	First-principles rigorous modeling for batch distillation processes
Aspen Reaction Modeler	First-principles rigorous reaction modeling, including kinetic data fitting
Aspen Custom Modeler	Custom environment for detailed modeling of other unit operations
Aspen Batch Process Developer	Recipe-based process modeling designed for route selection, recipe development, process scale-up, scheduling, and technology transfer

Properties provides such a common technology component and delivers best-in-class thermophysical property methods, models, and data. It includes extensive databases of pure component and phase equilibrium data and libraries of estimation methods. Also included are the activity coefficient models such as NRTL, electrolyte NRTL, UNIQUAC, UNIFAC, Hansen, COSMO-SAC, NRTL-SAC, electrolyte NRTL-SAC, and so on. Ongoing collaboration with the U.S. National Institute of Standards and Technology (NIST) ensures continuing access to the newly available methods, models, and data.

There are three modeling tools in Table 27.1 that deserve special attention due to their particular pertinence to the process modeling and simulation of pharmaceutical manufacturing processes: Aspen Solubility Modeler (ASM), Aspen Batch Distillation, and Aspen Reaction Modeler. Brief summaries are given below.

Aspen Solubility Modeler is a tool with an Excel-based front-end that provides users with the capability to define drug properties and identify NRTL-SAC parameters for drug molecules by regressing user-specified experimental solubility data in pure solvents or mixed solvents. Also available is the NRTL-SAC databank that provides NRTL-SAC parameters for over 150 solvents and excipients commonly used in the pharmaceutical industry. The Data Analysis Excel package allows users to predict phase equilibria (vapor–liquid equilibrium and vapor–liquid–liquid equilibrium) and solubility behavior of drugs under various operating conditions and solvent compositions. Appendix A shows how to set up Aspen Solubility Modeler, how to use ASM to regress solubility data to identify NRTL-SAC parameters for drug molecules (i.e., caffeine), and how to use ASM to perform solubility calculations based on the regressed NRTL-SAC parameters.

Aspen Batch Distillation is a comprehensive simulation tool for conceptual design, analysis, and optimization of batch distillation processes. Key features include the following:

- Intuitive interface designed specifically for simulating batch distillation
- Interoperability of Aspen Batch Distillation models inside the industry-leading Aspen Plus process simulation environment
- Optimization tool enabling rigorous identification of optimum operating steps to minimize cycle time while maintaining operating and performance constraints
- Equation-oriented architecture allowing timely and robust dynamic simulation of complex columns
- Rigorous equipment modeling including flexible, configurable controller models; pressure drop correlations; multiphase, azeotropic, and reactive distillation; and options to start from dry or total reflux conditions

Aspen Reaction Modeler enables users to identify reaction kinetics models using experimental measurements from reaction calorimeters. Key features include the following:

- Easy to use user interface that fully supports the model identification workflow and enables easy copying of experimental data from Excel
- Powerful numerical solvers for finding the best fit to experimental data
- Comprehensive kinetics include power law and Langmuir–Hinshelwood reaction kinetics, reversible reactions, and mass transfer effects
- Kinetic models consistent with other AspenTech products such as Aspen Plus, Aspen Plus Dynamics, and Aspen Batch Distillation, enabling users to apply the fitted parameters directly within these products
- Use of Aspen Properties for estimating physical properties required for the model identification process

27.5 CONCLUSIONS

Modern-day first-principles process modeling and simulation is the enabling technology to advance science-based, mechanistic understanding of pharmaceutical manufacturing processes and to succeed in quality by design. Molecular thermodynamics provides the scientific foundation for process modeling and simulation technology, and the lack of suitable molecular thermodynamic models for systems with complex pharmaceutical molecules has been the primary technical barrier for productive practice of process modeling and simulation in the pharmaceutical industry. The recent development of the NRTL-SAC activity coefficient model brought about a long-awaited breakthrough in molecular thermodynamics and it opened the door for meaningful application of process modeling and simulation technology in the pharmaceutical industry. Incorporating NRTL-SAC and other pertinent molecular thermodynamic models, AspenTech's model-based process development solution for the pharmaceutical industry ensures thermodynamic consistency in modeling of thermophysical properties and phase behavior and enables development of intrinsic process knowledge through the use of first-principles process modeling and simulation technology.

ACKNOWLEDGMENT

C.-C. Chen is grateful for extensive inputs, discussions, and assistance in generating the examples from his colleagues including Phil Norris, Russell Schofield, Catherine Jablonsky, Gary O'Neill, Jack Vinson, David Tremblay, Yuhua Song, Emmanuel Lejeune, Joseph DeVincentis, Jiangchu Liu, and

Zhiming Huang. The Benefits section was published previously in *Pharmaceutical Manufacturing* (October 2009, pp. 30–32) and the author thanks Agnes Shanley of Pharmaceutical Manufacturing for permission to reprint.

APPENDIX A SOLUBILITY MODELING WITH ASPEN SOLUBILITY MODELER

I. Setting Up Aspen Solubility Modeler

Objective

This workshop shows how to set up Aspen Solubility Modeler on your computer.

Description

Task 1: Set up the Aspen Properties Add-Ins in Excel

- ❑ Start Excel
- ❑ Go to the Tools menu, Add-Ins... (Figure 27.7)
- ❑ Click the Browse button (Figure 27.8)
- ❑ Navigate to C:\Program Files\AspenTech\Aspen Properties v7.1\Engine\Xeq
- ❑ Select the file Aspen Properties.xla, and then click OK (Figure 27.9)

Click OK on any message box that may be displayed. Click the OK button back on the Add-ins window.

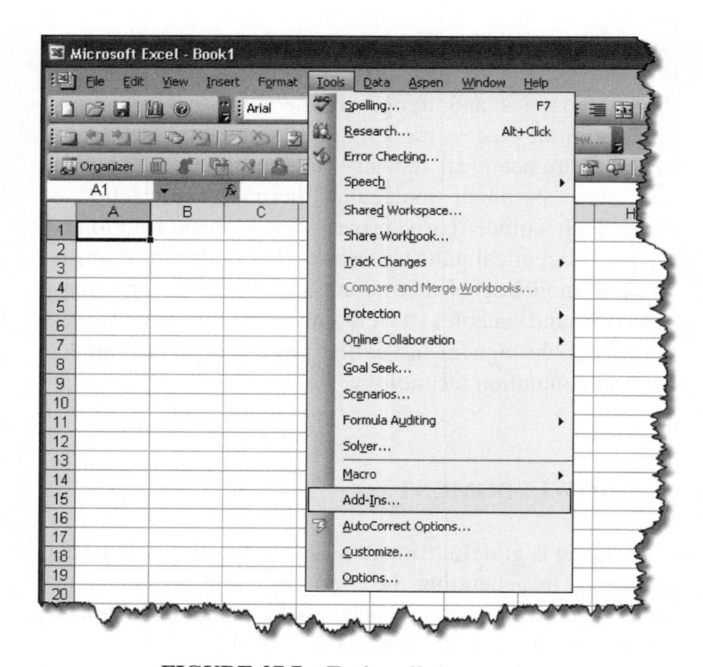

FIGURE 27.7 Tools pull-down menu.

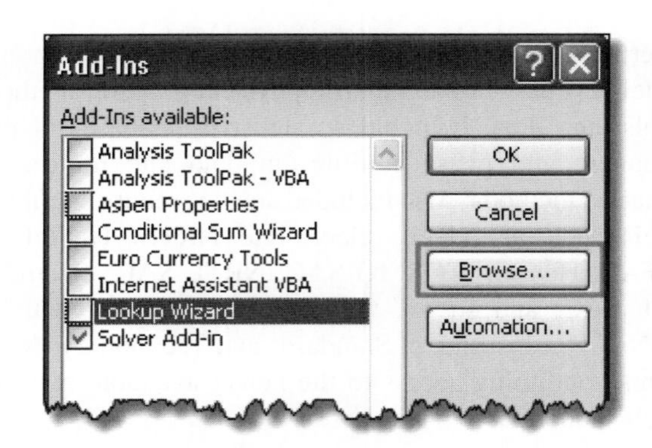

FIGURE 27.8 Add-Ins menu.

Task 2: Allow Macros

- ❑ In Tools, Macros, Security, select the "High" security level (or Medium or Low)
- ❑ Close Excel
- ❑ Go to the Start menu of Windows, Programs, AspenTech, Process Development v7.1, Aspen Solubility Modeler
- ❑ Open the spreadsheet "Regression.xls"
- ❑ When prompted to allow macros signed by AspenTech Inc., click "Enable macros"
- ❑ Close Excel

Note: This procedure needs to be done only once.

II. Caffeine Solubility Data Regression

Objective

This workshop shows how to regress solubility data.

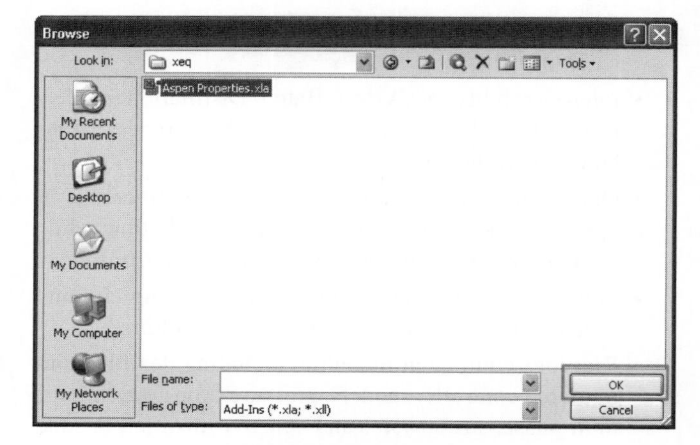

FIGURE 27.9 Browse file.

Description

The solubility of caffeine in different solvents is given in the following table.

Solvent	T (°C)	x (Solubility) (g Drug/g Solvent)	Standard Deviation (%)
N-HEXANE	25	8.8802E–06	40
2-ETHOXYETHANOL	25	1.4707E–02	10
1-OCTANOL	30	3.6693E–03	10
1,4-DIOXANE	25	1.8241E–02	10
1,4-DIOXANE	25	1.8873E–02	10
N,N-DIMETHYLFORMAMIDE	25	3.3787E–02	10
WATER	25	2.4274E–02	10
WATER	25	2.4741E–02	10
WATER	25	2.1667E–02	10
WATER	25	2.3316E–02	10
WATER	30	2.0436E–02	10

The caffeine pure properties are given in the following table:

Property	Value
MW	194.19 kg/kmol
Melting point	512.15K
Enthalpy of fusion	21600 kJ/kmol

Source: Ref. 25.

Task 1: Copy the NRTL-SAC Folder

- From the Windows Start menu, go to Programs, AspenTech, Process Development v7.1, Aspen Solubility Modeler
- Select the folder NRTL-SAC, and then copy it into any location that is convenient for you (e.g., the Desktop)

Note: We recommend copying the NRTL-SAC folder instead of working on the original files. This is because some data specific to your project will be stored in the physical property package, that is, the regressed parameters.

Task 2: Specify the Property Package

- Open the file Regression.xls with Excel
- On the worksheet, make sure the option "Pure Solvents" is selected, and then click the OK button

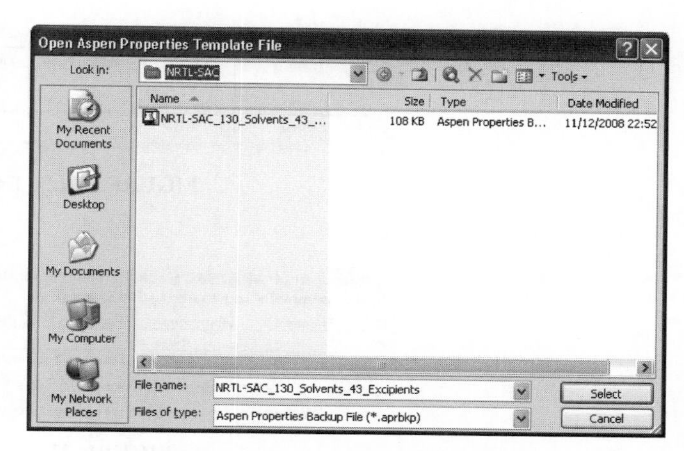

FIGURE 27.11 Open Aspen Properties template file.

- On the "Pure Solvent" sheet, click the "Execute Step 1" button (Figure 27.10)
- Select the file shown, and then click Select (Figure 27.11)

This will launch Aspen Properties as a hidden application, so there will be a delay before you can continue.

Task 3: Define the Drug Parameters

- Change the name to "Caffeine" (this is only for reporting purposes)
- Set the MW to 194.19
- Set the melting point to 512.15K
- Set the enthalpy of fusion to 21600 kJ/kmol
- Clear the entropy of fusion value
- Click the button "Execute Step 2" (Figure 27.12)

The entropy of fusion is now calculated (enthalpy of fusion divided by melting point).

Task 4: Enter NRTL-SAC Model Parameter

- Click the button "Calculate Ksp A & Ksp B" (Figure 27.13)

You should see the values of Ksp A and B are updated. The values of Ksp A and Ksp B can be calculated from the enthalpy and entropy of fusion (Ksp A = entropy of fusion/

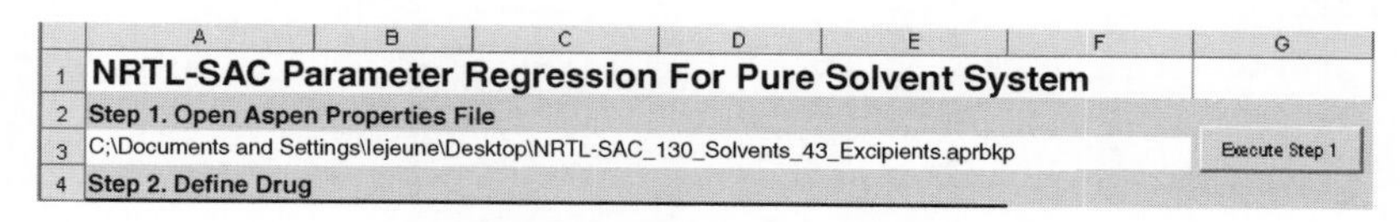

FIGURE 27.10 Properties file.

4	Step 2. Define Drug					
5	Name	MW	Melting Point	Enthalpy of	Entropy of	
6				Fusion	Fusion	
7		(Kg/Kgmole)	(K)	(KJ/Kmole)	(J/Kmole-K)	Execute Step 2
8	Caffeine	194.19	512.15	21600	42175.144	
9	Either the entropy or the enthalpy can be used here. If both are entered the entropy will be used.					

FIGURE 27.12 Enter data for caffeine.

10	Step 3. Enter NRTL SAC Model Parameters for Drug						
11	(For parameter regression, go to Step 4)						
12	Parameter X	Parameter Y-	Parameter Y+	Parameter Z	Ksp A	Ksp B	Ksp C
13	1	1	1	1	5.072542096	-2597.902434	0
14	The values of Ksp A and Ksp B can be either calculated				Calculate Ksp A & Ksp B		
15	by the programe or designated by the user.						Execute Step 3
16	Default value of Ksp C is 0, which can be overrided by the user						

FIGURE 27.13 Select parameters.

gas constant; Ksp B = −enthalpy of fusion/gas constant). The other parameters (X, Y−, Y +, Z) may be left at the value of 1 or specified to a positive value (e.g., $X = 0$ if you believe the component does not manifest any hydrophobic behavior).

- Click the Execute Step 3 button (Figure 27.13)

This will copy the parameters from Excel to the Aspen Properties file.

Task 5: Enter the Experimental Data

- Enter the experimental data
- For each data row, select the solvent using the pull-down list and enter the temperature, the solubility, and the standard deviation
- Delete the data on the rows you are not using

Note: You can copy and paste the data from the spreadsheet "Solubility Data.xls."

- For the parameter Ksp B, select the option "EXCLUDE"

Note: We exclude the parameter Ksp B because the experimental temperature range is not very large.

See Figure 27.14 for how the spreadsheet should look.

Task 6: Run the Regression and Review the Results

- Click the button "Execute Step 4"

The following window will be displayed. Click the OK button (Figure 27.15).The Aspen Properties application will be made visible.

- Click the OK button when this window is displayed (Figure 27.16)

17	Step 4. Perform Data Regression to Compute Model Parameters						
18	(At least four data points are required)						
19	Parameter X	Parameter Y-	Parameter Y+	Parameter Z	Ksp A	Ksp B	Ksp C
20	REGRESS	REGRESS	REGRESS	REGRESS	REGRESS	EXCLUDE	EXCLUDE
21							
22	Solvents	Temp (Exp)	Solubility (Exp.)	Std-Dev	Data to Regress		
23		(C)	(g drug/g solvents)	Solubility(%)			
24	N-HEXANE	25	8.88E-06	40	Yes		
25	2-ETHOXYETHANC	25	1.47E-02	10	Yes		
26	1-OCTANOL	25	3.67E-03	10	Yes		
27	1,4-DIOXANE	25	1.82E-02	10	Yes		
28	1,4-DIOXANE	25	1.89E-02	10	Yes		
29	N,N-DIMETHYLFOF	25	3.38E-02	10	Yes		
30	WATER	25	2.43E-02	10	Yes		
31	WATER	25	2.47E-02	10	Yes		
32	WATER	25	2.17E-02	10	Yes		
33	WATER	25	2.33E-02	10	Yes		
34	WATER	30	2.04E-02	10	Yes		
35	ETHYLENE-GLYCC				Yes		

FIGURE 27.14 Perform data degression.

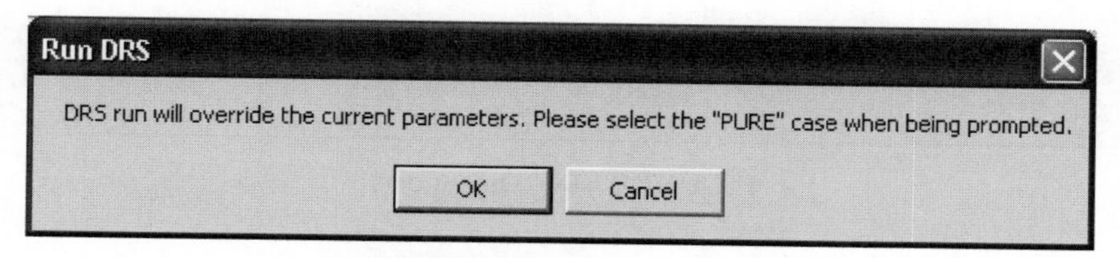

FIGURE 27.15 Dialog box.

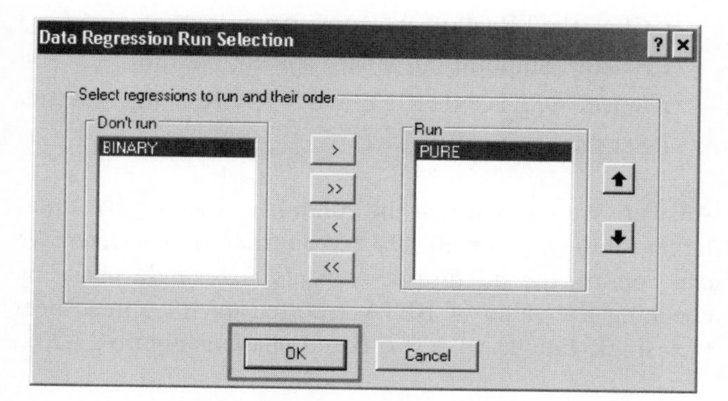

FIGURE 27.16 Dialog box.

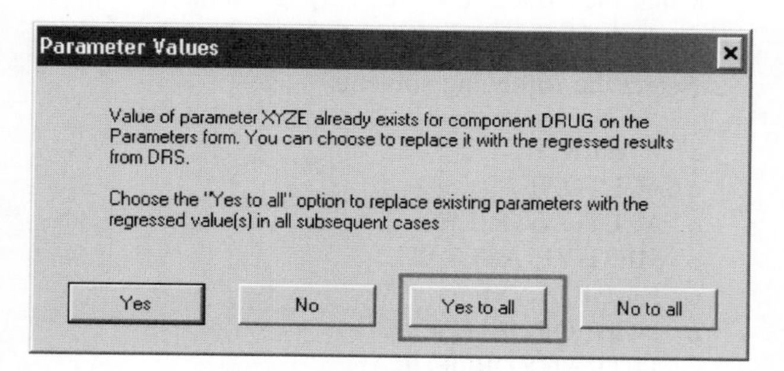

FIGURE 27.17 Dialog box.

Aspen Properties is now processing the data regression. When the calculations are complete, you will see the following window (Figure 27.17).

Click the "Yes to all" button

Back in Excel you can now inspect the regression results.

Solution

The following results are obtained.

	Value	Standard Deviation
Parameter X	0.012959095	0.013825285
Parameter Y−	0.509289197	0.06132528
Parameter Y+	0.920405633	0.060143778
Parameter Z	0.422110828	0.050881132
Ksp A	3.90390191	0.056377518
Ksp B	−2597.902434	0
Ksp C	0	0

SSQ	22.5715669
R^2	0.915104372
R^2 (log)	0.988690968
RMSE	0.0026638
RMSE (log)	0.104301944

The regression looks good: the values of the parameters are reasonable, the standard deviations are smaller than the regressed parameter, and the parity plot shows a good correlation.

III. Caffeine Solubility Calculation

Objective

This workshop shows how to use the calculation spreadsheet.

Description

Task 1: Open Calculation Spreadsheet

- ❑ Navigate to the folder where the workshop files are stored, in the folder asm-calc-caffeine
- ❑ Open the spreadsheet "Calculation.xls"

Note: When Excel displays a warning window about ActiveX controls, click the "Yes" button to allow them. These are required for the ternary diagram plots used on some calculation sheets.

Task 2: Pure Solvent Solubility

- ❑ Select the option "Solubility in Solvents"
- ❑ On the "Solubility in Solvents" sheet, select the following solvents and enter the experimental solubility (Figures 27.18 and 27.19)

Temperature: 25°C
Pressure: 1 bar

Solvent	Solubility (g/100 g Solvent)	Calculated Solubility (g/100 g Solvent)
N-HEXANE	8.88E−4	
2-ETHOXYETHANOL	1.47	
1-OCTANOL	0.366	
WATER	2.42	

			Input x		Calculated x		Calculated w		Solubility (g solute/100 g solvent)		
Solvent Name	Solute	Solvent ID	Solute	Solvent	Solute	Solvent	Solute	Solvent	Calculated	perimental	% Error
N-HEXANE	DRUG	HEXANE	0.5	0.5	7.62E-06	0.999992	1.72E-05	0.99998284	1.716E-03	8.88E-04	93.28102
2-ETHOXYETHANOL	DRUG	2ETHOXYE	0.5	0.5	0.009147	0.990853	0.019502	0.9804976	1.989E+00	1.47	35.30822
1-OCTANOL	DRUG	OCTANOL	0.5	0.5	0.002148	0.997852	0.0032	0.9967999	3.210E-01	0.366	-12.2849
WATER	DRUG	WATER	0.5	0.5	0.002051	0.997949	0.021675	0.97832495	2.216E+00	2.42	-8.44929

FIGURE 27.18 Output data.

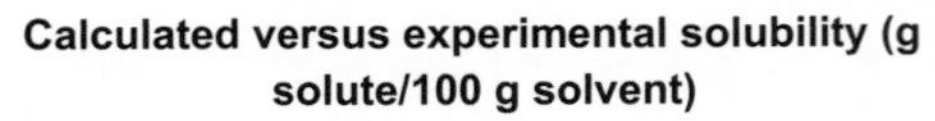

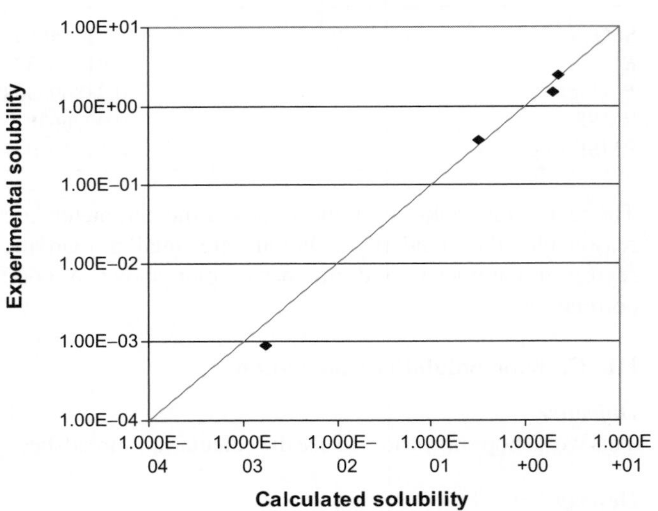

FIGURE 27.19 Comparison of calculated versus experimental solubility.

Task 3: Find the Solubility in Acetonitrile

Using the same sheet, change one solvent to Acetonitrile. This will report the solubility of caffeine in acetonitrile.

Answer: 2.79 g/100 g solvent.

Task 4: Find the Solubility in Binary Mixture

- ❑ Click the "Back to Welcome page" button
- ❑ Select "Solubility in binary solvent mixture"
- ❑ Select ACETONITRILE and WATER as the solvents
- ❑ Set the temperature to 25°C

Answer: We can see that the solubility is about five times larger for the mixture 60 wt% acetonitrile/40 wt% water. We can confirm the mixture of solvents is a single phase by checking the value of BETA reported on the spreadsheet (1 = single liquid, <1 = two liquid phases) (Figure 27.20).

Task 5: Use the "High Throughput" Sheet

- ❑ Click the "Back to Welcome page" button
- ❑ Select "High Throughput Prediction" option
- ❑ Select the following solvents:
  - o WATER
  - o ACETONE
  - o ACETONITRILE
  - o METHYL-ACETATE
  - o 1,4-DIOXANE
  - o 1-CHLOROBUTANE
  - o TETRAHYDROFURAN
  - o ETHYL-ACETATE

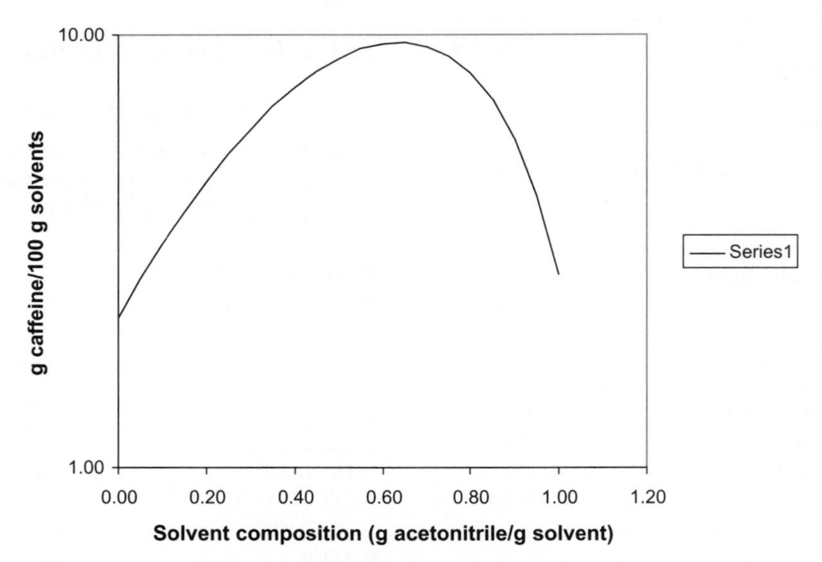

FIGURE 27.20 Graphical answer to Task 4.

- ETHANOL
- METHANOL

Note: When prompted "Do you want to continue," click "No" except when selecting the last solvent. This will prevent Excel recalculating the sheet while you are still setting up the list of solvents.

Is there another binary mixture in which the caffeine solubility becomes larger than in the pure solvents?

Answer: Yes, essentially all binary mixtures with water, especially water/ethanol where the solubility is even higher than that in water/acetonitrile.

REFERENCES

1. McKenzie P, Kiang S, Tom J, Rubin AE, Futran M. *AIChE J.* 2006;52:3990–3994.
2. *Beyond the Molecular Frontier: Challenges for Chemistry and Chemical Engineering*, Committee on Challenges for the Chemical Sciences in the 21st Century, Board on Chemical Sciences and Technology, National Research Council, National Academics Press, Washington, DC, 2003, p. 86.
3. Chen C-C. Fluid Phase Equilib. 2006;241:103–112.
4. Lipinski CA, Lombardo F, Dominy BW, Feency PJ. *Adv. Drug Deliv. Rev.* 1997;23:3–25.
5. Prausnitz JM, Tavares FW. *AIChE J.* 2004;50:739–761.
6. Chen C-C, Mathias PM. *AIChE J.* 2002;48:194–200.
7. Fredenslund A, Jones RL, Prausnitz JM. *AIChE J.* 1975;21: 1086–1099.
8. Gracin S, Brinck T, Rasmuson AC. *Ind. Eng. Chem. Res.* 2002; 41:5114–5124.
9. Eckert F, Klamt A. *AIChE J.* 2002;48:369–385.
10. Lin S-T, Sandler SI. *Ind. Eng. Chem. Res.* 2002;41:899–913.
11. Chen C-C, Song Y. *Ind. Eng. Chem. Res.* 2004;43:8354–8362.
12. Hansen CM. *Hansen Solubility Parameters: A User's Handbook*, CRC Press, 2000.
13. Renon H, Prausnitz JM. *AIChE J.* 1968;14:135–144.
14. Chen C-C. *Fluid Phase Equilib.* 1993;83:301–312.
15. Stahl PH, Wermuth CG, editors. *Handbook of Pharmaceutical Salts: Properties, Selection and Use*, Verlag Helvetica Chimica Acta, Zürich, 2002.
16. Chen C-C, Song Y. *Ind. Eng. Chem. Res.* 2005;44:8909–8921.
17. Song Y, Chen C-C. *Ind. Eng. Chem. Res.* 2009;48:5522–5529.
18. Chen C-C, Crafts PA. *Ind. Eng. Chem. Res.* 2006;45: 4816–4824.
19. Bakken D, et al. Solubility Modeling in Pharmaceutical Process Design, paper presented at AspenTech User Group Meeting, New Orleans, LA, October 5–8, 2003.
20. Tung H-H, Tabora J, Variankaval N, Bakken D, Chen C-C. *J. Pharm. Sci.* 2008;97:1813–1820.
21. Mota FL, Carneiro AP, Queimada AJ, Pinho SP, Macedo EA. *Eur. J. Pharm. Sci.* 2009;37:499–507.
22. Crafts P. The role of solubility modeling and crystallization in the design of active pharmaceutical ingredients. Ng KM, Gani R, Dam-Johanson K, *Chemical Product Design: Toward a Perspective through Case Studies*, Elsevier, 2007.
23. Kokitkar PB, Plocharczyk E, Chen C-C. *Org. Process Res. Dev.* 2008;12:249–256.
24. Hsieh D, Marchut AJ, Wei C-K, Zheng B, Wang SSY, Kiang S. *Org. Process Res. Dev.* 2009;13:690–697.
25. Marrero J, Abildskov J. *Solubility and Related Properties of Large Complex Chemicals, Part 1: Organic Solutes Ranging from C_4 to C_{40}*, Chemistry Data Series, Vol. 15, DECHEMA 2003.

28

THE ROLE OF SIMULATION AND SCHEDULING TOOLS IN THE DEVELOPMENT AND MANUFACTURING OF ACTIVE PHARMACEUTICAL INGREDIENTS

DEMETRI PETRIDES
Intelligen, Inc., Scotch Plains, NJ, USA

ALEXANDROS KOULOURIS
A.T.E.I. of Thessaloniki, Thessaloniki, Greece

CHARLES SILETTI
Intelligen, Inc., Mt. Laurel, NJ, USA

JOSÉ O. JIMÉNEZ
Intelligen, Inc., Amsterdam, The Netherlands

PERICLES T. LAGONIKOS
Merck & Co., Union, NJ, USA

28.1 INTRODUCTION

The global competition in the pharmaceutical industry and the increasing demands by governments and citizens for affordable medicines have driven the industry's attention toward manufacturing efficiency. In this new era, improvements in process and product development approaches and streamlining of manufacturing operations can have a profound impact on the bottom line. Process simulation and scheduling tools can play an important role in this endeavor. The role of such tools in the development and manufacturing of pharmaceutical products has already been reviewed in the past [1]. This chapter focuses on the role of these tools in the development and manufacturing of active pharmaceutical ingredients (APIs), and specifically on small-molecule APIs that are produced through organic synthesis. Information on the role of such tools in the development and manufacturing of biologics is also available in Ref. 2.

Process simulation and scheduling tools serve a variety of purposes throughout the life cycle of product development and commercialization in the pharmaceutical industry [2–6]. During process development, process simulators are used to facilitate the following tasks:

- Represent the entire process on the computer
- Perform material and energy balances
- Estimate the size of equipment
- Calculate demand for labor and utilities as a function of time
- Estimate the cycle time of the process
- Perform cost analysis
- Assess features such as environmental impact

The availability of a good computer-based model improves the understanding of the entire process by the team members and facilitates communication. What-if and sensitivity

Chemical Engineering in the Pharmaceutical Industry: R&D to Manufacturing Edited by David J. am Ende

analyses are greatly facilitated by such tools. The objective of such studies is to evaluate the impact of critical parameters on various key performance indicators (KPIs), such as production cost, cycle times, and plant throughput. If there is uncertainty for certain input parameters, sensitivity analysis can be supplemented with Monte Carlo simulation to quantify the impact of uncertainty. Cost analysis, especially capital cost estimation, facilitates decisions related to in-house manufacturing versus outsourcing. Estimation of the cost of goods identifies the expensive processing steps and the information generated is used to guide R&D work in a judicious way.

When a process is ready to move from development to manufacturing, process simulation facilitates technology transfer and process fitting. A detailed computer model provides a thorough description of a process in a way that can be readily understood and adjusted by the recipients. Process adjustments are commonly required when a new process is moved into an existing facility whose equipment is not ideally sized for the new process. The simulation model is then used to adjust batch sizes, figure out cycling of certain steps (for equipment that cannot handle a batch in one cycle), estimate recipe cycle times, and so on.

Production scheduling tools play an important role in manufacturing (large scale as well as clinical). They are used to generate production schedules on an ongoing basis in a way that does not violate constraints related to the limited availability of equipment, labor resources, utilities, inventories of materials, and so on. Production scheduling tools close the gap between ERP/MRP-II tools and the plant floor [7, 8]. Production schedules generated by ERP (enterprise resource planning) and MRP-II (manufacturing resource planning) tools are typically based on coarse process representations and approximate plant capacities and, as a result, solutions generated by these tools may not be feasible, especially for multiproduct facilities that operate at high capacity utilization. That often leads to late orders that require expediting and/or to large inventories in order to maintain customer responsiveness. "Lean manufacturing" principles, such as just-in-time production, low work in progress (WIP), and low product inventories cannot be implemented without good production scheduling tools that can accurately estimate capacity.

28.2 COMMERCIALLY AVAILABLE SIMULATION AND SCHEDULING TOOLS

Computer-aided process design and simulation tools have been used in the chemical and petrochemical industries since the early 1960s. Simulators for these industries have been designed to model continuous processes and their transient behavior for process control purposes. Most APIs, however, are produced in batch and semicontinuous modes. Such processes are best modeled with batch process simulators that account for time-dependency and sequencing of events. *Batches* from Batch Process Technologies, Inc. (West Lafayette, IN) was the first simulator specific to batch processing. It was commercialized in the mid-1980s. All of its operation models are dynamic and simulation always involves integration of differential equations over a period of time. In the mid-1990s, Aspen Technology (Burlington, MA) introduced *Batch Plus*, a recipe-driven simulator that targeted batch pharmaceutical processes. Around the same time, Intelligen, Inc. (Scotch Plains, NJ) introduced *SuperPro Designer*. The initial focus of SuperPro was on bioprocessing. Over the years, its scope has been expanded to include modeling of small-molecule API and secondary pharmaceutical manufacturing processes.

Discrete-event simulators have also found applications in the pharmaceutical industry, especially in the modeling of secondary pharmaceutical manufacturing processes. Established tools of this type include *ProModel* from ProModel Corporation (Orem, UT), Arena and Witness from Rockwell Automation, Inc. (Milwaukee, WI), and Extend from Imagine That, Inc. (San Jose, CA). The focus of models developed with such tools is usually on the minute-by-minute time-dependency of events and the animation of the process. Material balances, equipment sizing, and cost analysis tasks are usually out of the scope of such models. Some of these tools are quite customizable and third-party companies occasionally use them as platforms to create industry-specific modules. For instance, BioPharm Services, Ltd. (Bucks, UK) have created an extend-based module with emphasis on biopharmaceutical processes.

Microsoft Excel is another common platform for creating models for pharmaceutical processes that focus on material balances, equipment sizing, and cost analysis. Some companies have even developed models in Excel that capture the time-dependency of batch processes. This is typically done by writing extensive code (in the form of macros and subroutines) in Visual Basic for Applications (VBA) that comes with Excel. K-TOPS from Biokinetics, Inc. (Philadelphia, PA) belongs to this category.

In terms of production scheduling, established tools include *Optiflex* from i2 Technologies, Inc. (Irving, TX), *SAP APO* from SAP AG (Walldorf, Germany), *ILOG* Plant PowerOps from ILOG SA (Gentilly, France), and *Aspen SCM* (formerly Aspen MIMI) from Aspen Technology, Inc. (Burlington, MA). Their success in the pharmaceutical industry, however, has been rather limited so far. Their primary focus on discrete manufacturing (as opposed to batch chemical manufacturing) and their approach to scheduling from a mathematical optimization viewpoint are some of the reasons of the limited market penetration.

SchedulePro from Intelligen, Inc. (Scotch Plains, NJ) is a finite capacity scheduling tool that focuses on scheduling of batch and semicontinuous chemical and related processes. It is a recipe-driven tool with emphasis on generation of feasible solutions that can be readily improved by the user in an interactive manner.

28.3 MODELING AND ANALYSIS OF AN API MANUFACTURING PROCESS

The steps involved during the development of a model will be illustrated with a simple process that represents the manufacturing of an active compound for skin care applications.

The first step in building a simulation model is always the collection of information about the process. Engineers rely on draft versions of process descriptions, block flow diagrams, and batch sheets from past runs, which contain information on material inputs and operating conditions, among others. Reasonable assumptions are then made for missing data.

The steps of building a batch process model are generally the same for all batch process simulation tools. The best practice is to build the model step by step, gradually checking the functionality of its parts. The registration of materials (pure components and mixtures) is usually the first step. Next, the flow diagram (see Figure 28.1) is developed by putting together the required unit procedures and joining them with material flow streams. Operations are then added to unit procedures (see the following paragraph for explanation) and their operating conditions and performance parameters are specified.

In SuperPro Designer, the representation of a batch process model is loosely based on the ISA S-88 standards for batch recipe representation [9]. A batch process model is in essence a batch recipe that describes how to make a certain quantity of a specific product. The set of operations that comprise a processing step is called a "unit procedure" (as opposed to a unit operation that is a term used for continuous processes). The individual tasks contained in a procedure are called "operations." A unit procedure is represented on the screen with a single equipment-looking icon. Figure 28.2 displays the dialogue through which operations are added to a vessel unit procedure. On the left-hand side of that dialogue, the program displays the operations that are available in the context of a vessel procedure; on the right-hand side, it displays the registered operations (Charge Quinaldine, Charge Chlorine, Charge Na2CO3, Agitate, etc.). The two-level representation of operations in the context of unit procedures enables users to describe and model batch processes in detail.

For every operation within a unit procedure, the simulator includes a mathematical model that performs material and energy balance calculations. Based on the material balances, it performs equipment-sizing calculations. If multiple operations within a unit procedure dictate different sizes for a certain piece of equipment, the software reconciles the different demands and selects an equipment size that is appropriate for all operations. The equipment is sized so that it is large enough and, hence, not overfilled during any operation, but it is no larger than necessary (in order to minimize capital costs). If the equipment size is specified by the user, the simulator checks to make sure that the vessel is not overfilled. In addition, the tool checks to ensure that the vessel contents do not fall below a user-specified minimum volume (e.g., a minimum stirring volume) for applicable operations.

In addition to material balances, equipment sizing, and cycle time analysis, the simulator can be used to carry out cost-of-goods analysis and project economic evaluation. The following sections provide illustrative examples for these features.

Having developed a good model using a process simulator, the user may begin experimenting on the simulator with alternative process setups and operating conditions. This has the potential of reducing the costly and time-consuming laboratory and pilot plant effort. Of course, the GIGO (garbage in, garbage out) principle applies to all computer models. If critical assumptions and input data are incorrect, so will be the outcome of the simulation.

When modeling an existing process, input data required by the model can be extracted from the data recorded by the actual process. A communication channel must, therefore, be established between the modeler and the operations department. The application of some data mining technique is usually required to transform the process data to the form required by the model. When designing a new plant, experience from similar projects can be used to fill in the information gaps. In all cases, a certain level of model verification is necessary after the model is developed. In its simplest form, a review of the results by an experienced engineer can play the role of verification. Running a sensitivity analysis on key input variables can reveal the parameters with the greatest impact on the model's most important outputs. These parameters would then constitute the focal points in the data acquisition effort in an attempt to estimate their values and uncertainty limits with the best possible accuracy.

28.3.1 Design Basis and Process Description

A simple batch process is used to illustrate the steps involved in building a model with SuperPro Designer. It is assumed that the process has been developed at the pilot plant and it is ready to be moved to large-scale manufacturing. Based on input from the marketing department, the objective is to produce at least 27,000 kg of active ingredient per year at a cost of no more than $330 per kilogram. A production suite can be dedicated to this process that includes two 3800 L reactors (R-101 and R-102), one 2.5 m^2 Nutsche filter (NFD-101), and a 10 m^2 tray dryer (TDR-101).

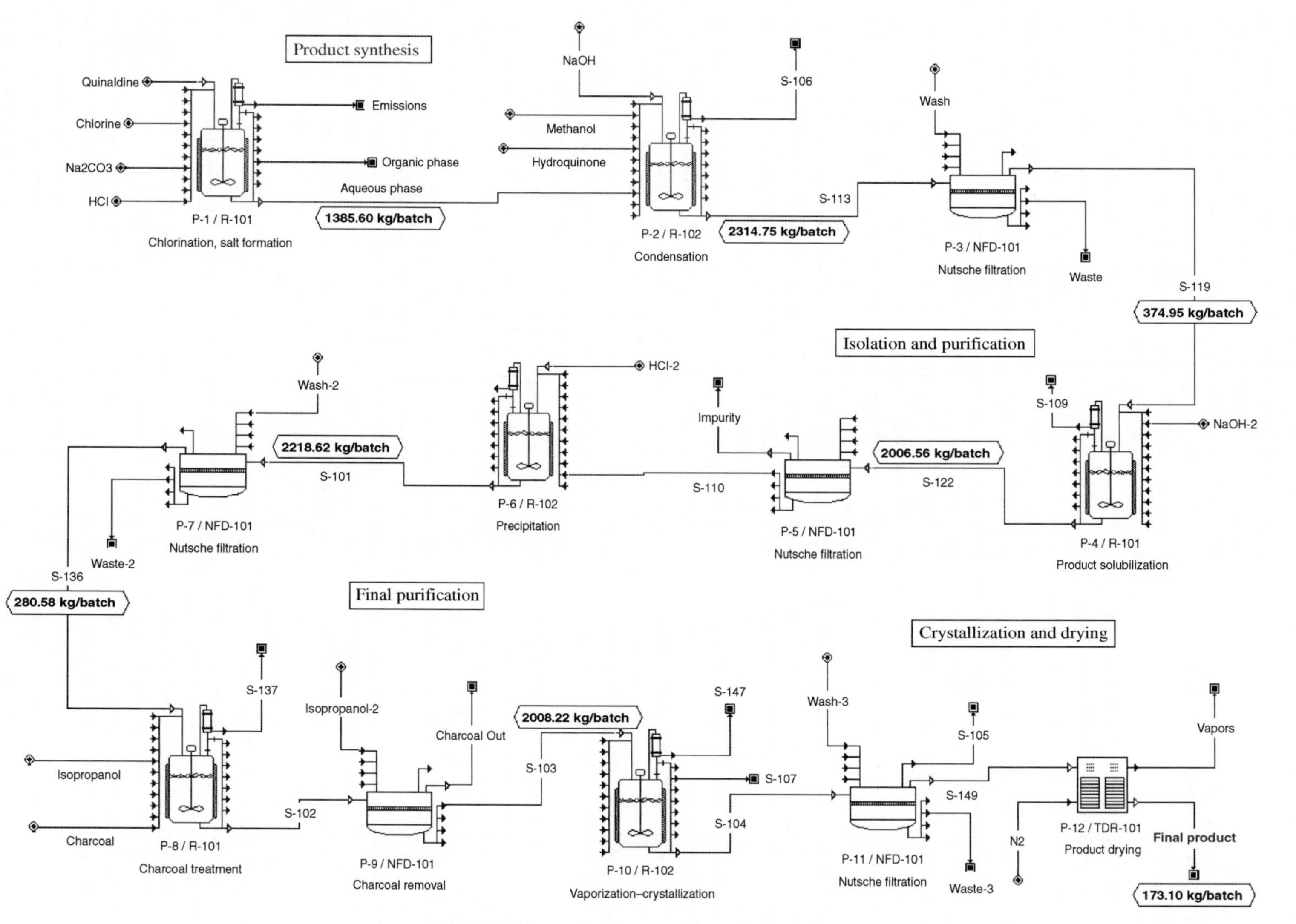

FIGURE 28.1 Flow diagram of the API process.

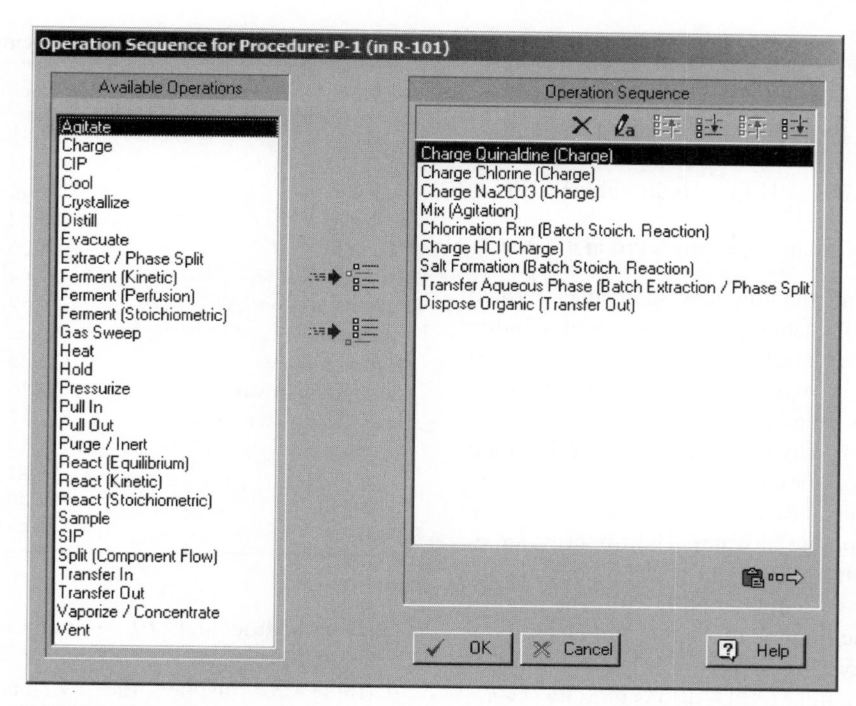

FIGURE 28.2 The operations associated with the first unit procedure in Figure 28.1.

The entire flow sheet of the batch process is shown in Figure 28.1. It is divided into four sections: (1) Product synthesis; (2) Isolation and purification; (3) Final purification; and (4) Crystallization and drying. A flow sheet section in SuperPro Designer is simply a group of unit procedures (processing steps).

The formation of the final product in this example involves 12 unit procedures. The first reaction step (procedure P-1) involves the chlorination of quinaldine. Quinaldine is dissolved in carbon tetrachloride (CCl_4) and reacts with gaseous Cl_2 to form chloroquinaldine[1]. The conversion of the reaction is around 98% (based on amount of quanaldine fed). The generated HCl is then neutralized using Na_2CO_3. The stoichiometry of these reactions is as follows:

$$\text{Quinaldine} + Cl_2 \rightarrow \text{Chloroquinaldine} + HCl$$

$$Na_2CO_3 + HCl \rightarrow NaHCO_3 + NaCl$$

$$NaHCO_3 + HCl \rightarrow NaCl + H_2O + CO_2$$

Small amounts of unreacted Cl_2, generated CO_2, and volatilized CCl_4 are vented. The above three reactions occur sequentially in the first reactor vessel (R-101). Next, HCl is added in order to produce chloroquinaldine-HCl. The HCl first neutralizes the remaining $NaHCO_3$ and then reacts with chloroquinaldine to form its salt, according to the following stoichiometries:

$$NaHCO_3 + HCl \rightarrow NaCl + H_2O + CO_2$$

$$\text{Chloroquinaldine} + HCl \rightarrow \text{Chloroquinaldine.HCl}$$

Small amounts of generated CO_2 and volatilized CCl_4 are vented. The presence of water (added with HCl as hydrochloric acid solution) and CCl_4 leads to the formation of two liquid phases. Then, small amounts of unreacted quinaldine and chloroquinaldine are removed with the organic phase. The chloroquinaldine-HCl remains in the aqueous phase. This sequence of operations (including all charges and transfers) requires about 14.5 h.

After removal of the unreacted quinaldine, the condensation of chloroquinaldine and hydroquinone takes place in reactor R-102 (procedure P-2). First, the salt chloroquinaldine-HCl is converted back to chloroquinaldine using NaOH. Then, hydroquinone reacts with NaOH and yields hydroquinone-Na. Finally, chloroquinaldine and hydroquinone-Na react and yield the desired intermediate product. Along with product formation, roughly 2% of chloroquinaldine dimerizes and forms an undesirable by-product impurity. This series of reactions and transfers takes roughly 13.3 h. The stoichiometry of these reactions is as follows:

[1] Note that carbon tetrachloride is an ideal solvent for this specific reaction from a chemistry perspective, but this solvent is considered highly undesirable from an environmental, health, and safety perspective.

$$\text{Chloroquinaldine-HCl} + \text{NaOH} \rightarrow \text{NaCl} + H_2O + \text{Chloroquinaldine}$$

$$2\text{Chloroquinaldine} + 2\text{NaOH} \rightarrow 2H_2O + 2\text{NaCl} + \text{Impurity}$$

$$\text{Hydroquinone} + \text{NaOH} \rightarrow H_2O + \text{Hydroquinone-Na}$$

$$\text{Chloroquinaldine} + \text{Hydroquinone-Na} \rightarrow \text{Product} + \text{NaCl}$$

Both the product and the impurity molecules formed during the condensation reaction precipitate out of solution and are recovered using a Nutsche filter (procedure P-3, filter NFD-101). The product recovery yield is 90%. The filtration, wash, and cake transfer time is 6.4 h.

Next, the product/impurity cake recovered by filtration is added into a NaOH solution in reactor R-101 (procedure P-4). The product molecules react with NaOH to form product-Na, which is soluble in water. The Impurity molecules remain in the solid phase, and are subsequently removed during procedure P-5 in filter NFD-101. The product remains dissolved in the liquors. Procedure P-4 takes about 10 h, and procedure P-5 takes approximately 4 h.

Notice that the single filter (NFD-101) is used by several different procedures. The two reactors are also used for multiple procedures during each batch. Please note that the equipment icons in Figure 28.1 represent unit procedures (processing steps), as opposed to unique pieces of equipment. The procedure names (P-1, P-3, etc.) below the icons refer to the unit procedures, whereas the equipment tag names (R-101, R-102, etc.) refer to the actual physical pieces of equipment. The process flow diagram in SuperPro designer is essentially a graphical representation of the batch "recipe" that displays the execution sequence of the various steps.

After the filtration in procedure P-5, the excess NaOH is neutralized using HCl and the product-Na salt is converted back to product in reactor R-102 (procedure P-6). Since the product is insoluble in water, it precipitates out of solution. The product is then recovered using another filtration step in (procedure P-7). The product recovery yield is 90%. The precipitation procedure takes roughly 10.7 h, and the filtration takes about 5.7 h. The recovered product cake is then dissolved in isopropanol and treated with charcoal to remove coloration. This takes place in reactor R-101 under procedure P-8. After charcoal treatment, the solid carbon particles are removed using another filtration step in (procedure P-9). The time required for charcoal treatment and filtration is 15.9 h and 5 h, respectively.

In the next step (procedure P-10), the solvent is distilled off until the solution is half its original volume. The product is then crystallized in the same vessel with a yield of 97%. The crystalline product is recovered with a 90% yield using a final filtration step (procedure P-11). The distillation and crystallization steps take approximately 18.3 h, and the filtration requires roughly 3.3 h. The recovered product crystals are then dried in a tray dryer (procedure P-12, TDR-101). This takes an additional 15.6 h. The amount of purified product generated per batch is 173.1 kg.

TABLE 28.1 Raw Material Requirements

Material	kg/batch	kg/kg MP
Carbon	497.31	2.87
Quinaldine	148.63	0.86
Water	3621.44	20.92
Chlorine	89.52	0.52
Na_2CO_3	105.06	0.61
HCl (20% w/w)	357.44	2.07
NaOH (50% w/w)	204.52	1.18
Methanol	553.26	3.20
Hydroquinone	171.45	0.99
Sodium hydroxid	74.16	0.43
HCl (37% w/w)	217.57	1.26
Isopropanol	2232.14	12.90
Charcoal	15.85	0.09
Nitrogen	1111.49	6.42
Total	9399.84	54.30

Table 28.1 displays the raw material requirements in kilogram per batch and per kilogram of main product (MP = purified product) that correspond to the maximum batch size achievable with the available equipment. Note that around 54.3 kg of raw materials (solvents, reagents, etc.) are used per kilogram of main product produced. Thus, the product to raw material ratio is only 1.84%, an indication that large amounts of waste are generated by this process. A more detailed description of this process along with information on how the pilot plant process is transferred to the large-scale manufacturing facility is available in Ref. 10.

28.3.2 Process Scheduling and Cycle Time Reduction

Figure 28.3 displays the equipment occupancy chart for three consecutive batches (each color represents a different batch). The process batch time is approximately 92 h. This is the total time between the start of the first step of a batch and the end of the last step of that batch. However, since most of the equipment items are utilized for shorter periods within a batch, a new batch can be initiated every 62 h, which is known as the minimum cycle time of the process. Multiple bars on the same line (e.g., for R-101, R-102, and NFD-101) represent reuse (sharing) of equipment by multiple procedures. If the cycle times of procedures that share the same equipment overlap, scheduling with the assumed equipment designation is infeasible. White space between the bars represents idle time. The equipment with the least idle time between the consecutive batches is the *time (or scheduling) bottleneck* (R-102 in this case) that determines the maximum number of batches per year. Its occupancy time (approximately 62 h) is the minimum possible time between the consecutive batches.

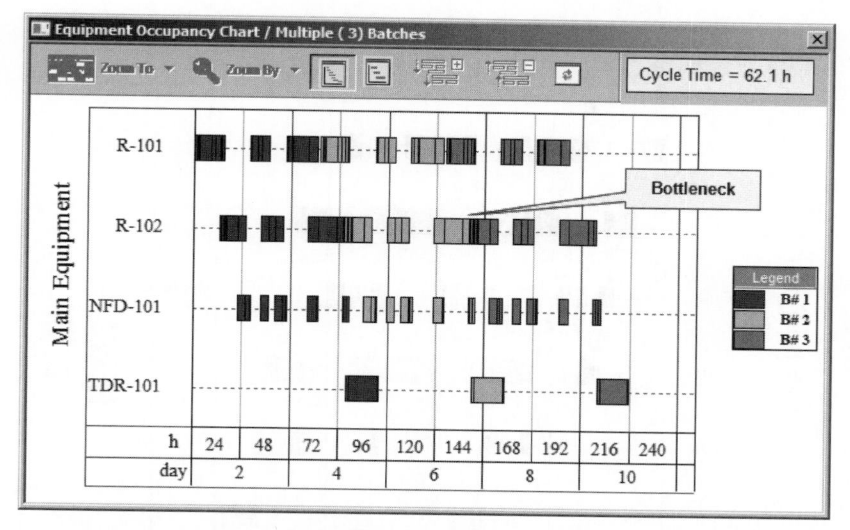

FIGURE 28.3 Equipment occupancy chart for three consecutive batches.

Scheduling in the context of a simulator is fully process driven and the impact of process changes can be analyzed in a matter of seconds. For instance, the impact of an increase in batch size (that affects the duration of charge, transfer, filtration, distillation, and other scale-dependent operations) on the plant batch time and the maximum number of batches can be seen instantly. Due to the many interacting factors involved with even a relatively simple process, simulation tools that allow users to describe their processes in detail, and to quickly perform what-if analyses, can be extremely useful.

If this production line operated around the clock for 330 days a year (7920 h) with its minimum cycle time of 62 h, its maximum annual number of batches would be 126, leading to an annual production of 21,810 kg of API (126 batches × 173.1 kg/batch), which is less than the project's objective of 27,000 kg. And since the process operates at its maximum possible batch size, the only way to increase production is by reducing the process cycle time and thus increasing the number of batches per year. The cycle time can be reduced through process changes or by addition of extra equipment. However, major process changes in GMP manufacturing usually require regulatory approval and are avoided in practice. Addition of extra equipment is the practical way for cycle time reduction. Since R-102 is the current bottleneck, addition of an extra reactor can shift the bottleneck to another unit. Figure 28.4 displays the effect of the addition of an extra reactor (R-103). Please note that under the new conditions, each reactor handles two procedures instead of three.

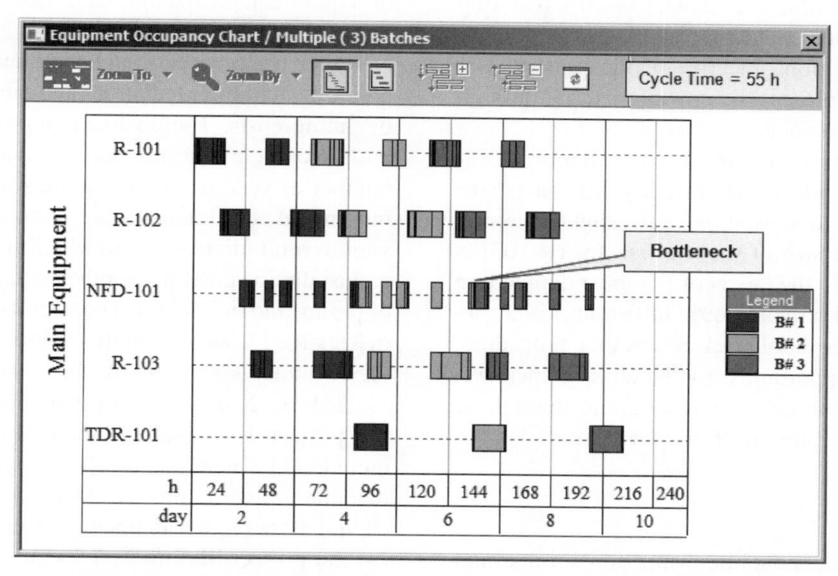

FIGURE 28.4 Equipment occupancy chart for the case with three reactors.

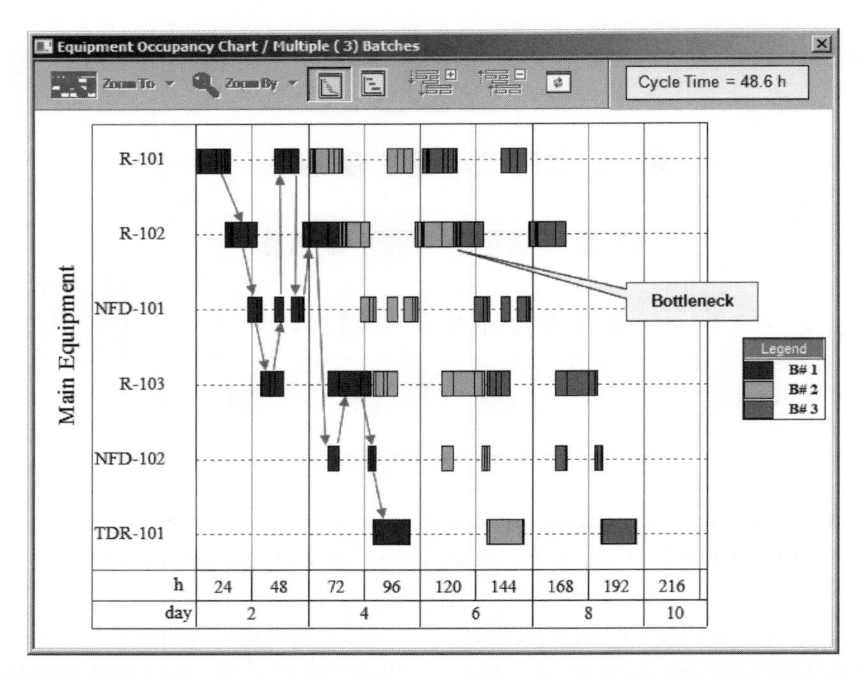

FIGURE 28.5 Equipment occupancy chart for the case with three reactors and two filters.

The addition of R-103 reduces the cycle time of the process to 55 h, resulting in 143 batches per year and annual throughput of 24,753 kg. Under these conditions, the bottleneck shifts to NFD-101. Since the annual throughput is still below the desired amount of 27,000 kg/year, addition of an extra Nutsche filter to eliminate the current bottleneck is the next logical step. Figure 28.5 shows the results of that scenario. In this scenario, the first Nutsche filter (NFD-101) is used for the first three filtration procedures (P-3, P-5, and P-7) and the second filter (NFD-102) handles the last two filtration procedures (P-9 and P-11). Under these conditions, the process cycle time goes down to 48.6 h, resulting in 162 batches per year and annual throughput of 28,042 kg, which meets the production objective of the project. The arrows in Figure 28.5 represent the flow of material through the equipment for the first batch.

Debottlenecking projects that involve installation of additional equipment provide an opportunity for batch size increases that can lead to substantial throughput increase. More specifically, if the size of the new reactor (R-103) is selected to accommodate the needs of the most demanding vessel procedure (based on volumetric utilization) in a way that shifts the batch size bottleneck to another procedure, then, that creates an opportunity for batch size increase. Additional information on debottlenecking and throughput increase options can be found in Refs 11,12.

28.3.3 Cost Analysis

Cost analysis and project economic evaluation is important for a number of reasons. If a company lacks a suitable manufacturing facility with available capacity to accommodate a new product, it must decide whether to build a new plant or outsource the production. Building a new plant is a major capital expenditure and a lengthy process. To make a decision, management must have information on capital investment required and time to complete the facility. To outsource the production, one must still do a cost analysis and use it as a basis for negotiation with contract manufacturers. A sufficiently detailed computer model can be used as the basis for the discussion and negotiation of the terms. Contract manufacturers usually base their estimates on requirements of equipment utilization and labor per batch, which is information that is provided by a good model. SuperPro Designer performs thorough cost analysis and project economic evaluation calculations and estimates capital as well as operating costs. The cost of equipment is estimated using built-in cost correlations that are based on data derived from a number of vendors and literature sources. The fixed capital investment is estimated based on total equipment cost using various multipliers, some of which are equipment specific (e. g., installation cost) while others are plant specific (e.g., cost of piping and buildings). The approach is described in detail in Refs 10, 13. The rest of this section provides a summary of the cost analysis results for this example process.

Table 28.2 shows the key economic evaluation results for this project. Key assumptions for the economic evaluations include (1) a new plant will be built and dedicated to the manufacturing of this product (2) the entire direct fixed capital is depreciated linearly over a period of 12 years; (3) the project lifetime is 15 years, and 27,000 kg of final product will be produced per year.

TABLE 28.2 Key Economic Evaluation Results

Total capital investment	$19.5 million
Plant throughput	27,000 kg/year
Manufacturing cost	$8.6 million/year
Unit production cost	$318/kg
Selling price	$450/kg
Revenues	$12.2 million/year
Gross margin	29.3%
Taxes (40%)	$1.1 million/year
IRR (after taxes)	14.0%
NPV (for 7% discount interest)	$8.5 million

For a plant of this capacity, the total capital investment is around $19.5 million. The unit production cost is $318/kg of product, which satisfies the project's objective for a unit cost of under $330/kg. Assuming a selling price of $450/kg, the project yields an after-tax internal rate of return (IRR) of 14% and a net present value (NPV) of $8.5 million (assuming a discount interest of 7%).

Figure 28.6 breaks down the manufacturing cost. The facility-dependent cost, which primarily accounts for the depreciation and maintenance of the plant, is the most important item accounting for 35.74% of the overall cost. This is common for high-value products that are produced in small facilities. This cost can be reduced by manufacturing the product at a facility whose equipment has already been depreciated. Raw material is the second most important cost item accounting for 32.12% of the total manufacturing cost. Furthermore, if we look more closely at the raw material cost breakdown, it becomes evident that quinaldine, hydroquinone, and isopropanol make up more than 80% of this cost (see Table 28.3). If a lower priced quinaldine vendor could be found, the overall manufacturing cost would be reduced significantly.

Labor is the third important cost item accounting for 18.8% of the overall cost. The program estimates that 12 operators are required to run the plant around the clock supported by three QC/QA scientists. This cost can be reduced by increasing automation or by locating the facility in a region of low labor cost.

28.4 UNCERTAINTY AND VARIABILITY ANALYSIS

Process simulation tools typically used for batch process design, debottlenecking, and cost estimation employ deterministic models. They model the "average" or "expected" situation commonly referred to as the base case or most likely scenario. Modeling a variety of cases can help determine the range of performance with respect to key process parameters. However, such an approach does not account for the relative likelihood of the various cases. Monte Carlo simulation is a practical means of quantifying the risk associated with uncertainty in process parameters [14]. In a Monte Carlo simulation, uncertain input variables are represented with probability distributions. A simulation calculates numerous scenarios of a model by repeatedly picking values from a user defined probability distribution for the uncertain variables. It then uses those values in the model to calculate and analyze the outputs in a statistical way in order to quantify risk. The outcome of this analysis is the estimation of the confidence by which desired values of key performance indicators can be achieved. Inversely, the analysis can help identify the input parameters with the greatest effect on the bottom line and the input value ranges that minimize output uncertainty.

In batch pharmaceutical processing, uncertainty can emerge in operation or market-related parameters. Process times, equipment sizes, material purchasing, and product

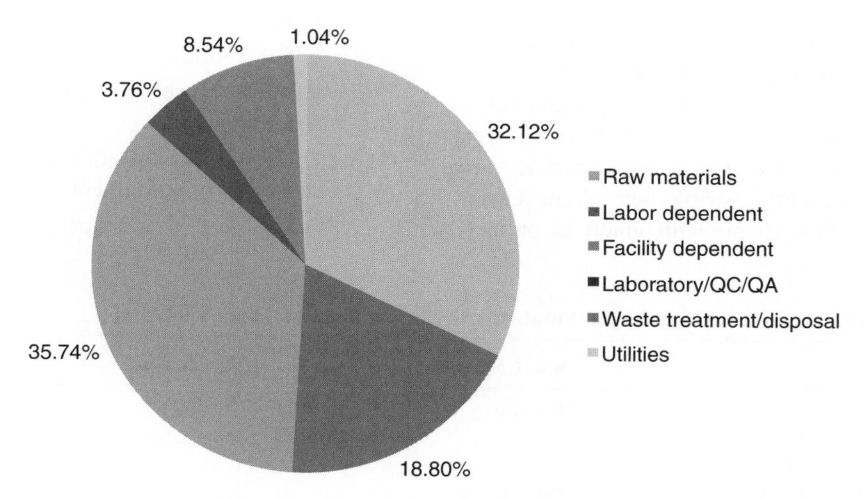

FIGURE 28.6 Manufacturing cost breakdown.

TABLE 28.3 Raw Material Requirements and Costs

Bulk Material	Unit Cost ($/kg)	Annual Amount (kg)	Annual Cost ($)	%
Carbon tetrachloride	0.80	77,581	62,065	2.25
Quinaldine	60.00	23,187	1,391,215	50.40
Water	0.10	564,944	56,494	2.05
Chlorine	3.30	13,965	46,083	1.67
Na_2CO_3	6.50	16,389	106,528	3.86
NaOH (50% w/w)	0.15	31,905	4,786	0.17
Methanol	0.24	86,308	20,714	0.75
Hydroquinone	18.00	26,746	481,427	17.44
Sodium hydroxide	2.00	11,569	23,138	0.84
HCl (37% w/w)	0.17	33,942	5,770	0.21
Isopropanol	1.10	348,214	383,035	13.88
Charcoal	2.20	2,473	5,440	0.20
Nitrogen	1.00	173,393	173,393	6.28
TOTAL		1,466,376	2,760,088	100.00

selling prices are common uncertain variables. Performing a stochastic analysis early on in the design phase increases the model's robustness and minimizes the risk of encountering unpleasant surprises later on.

For models developed in SuperPro Designer, Monte Carlo simulation can be performed by combining SuperPro Designer with *Crystal Ball* from Decissioneering, Inc. (Denver, Colorado). Crystal Ball is an Excel add-in application that facilitates Monte Carlo simulation. It enables the user to designate the uncertain input variables, specify their probability distributions and select the output (decision) variables whose values are recorded and analyzed during the simulation. For each simulation trial (scenario), Crystal Ball generates random values for the uncertain input variables selected in frequency dictated by their probability distributions using the Monte Carlo method. Crystal Ball also calculates the uncertainty involved in the outputs in terms of their statistical properties, mean, median, mode, variance, standard deviation, and frequency distribution.

Section 28.3.3 discusses the production and cost objectives of the project (27,000 kg/year of API for less than $330/kg) based on the assumed operating parameters and material unit costs. If the variability related to process parameters and uncertainty related to cost parameters can be represented with probability distributions, Monte Carlo simulation can estimate the certainty with which the project objectives can be met. For this exercise, a normal distribution was assumed for the price of quinaldine, which is the most expensive raw material, with a mean value equal to that of the base case ($60/kg).

The annual throughput (or number of batches per year) is determined by the process cycle time. Since procedure P-8 that utilizes vessel R-102 is the time bottleneck, any variability in the completion of P-8 leads to uncertainty in the annual throughput. Variability in the completion of P-8 can be caused by variability in the operations of P-8 as well as by variability in the operations of procedures upstream of P-8. Common sources of process time variability in chemical manufacturing are as follows:

(1). Fouling of heat transfer areas that affect duration of heating and reaction operations
(2). Fouling of filters that affect duration of filtration operations
(3). Presence of impurities in raw materials that affect reaction rates
(4). Off-spec materials that require rework
(5). Random power outages and equipment or utility failures
(6). Differences in skills of operators that affect setup and operation of equipment
(7). Availability of operators

TABLE 28.4 The Input Parameters Used for the Monte Carlo Simulation and Their Variation

Variable	Base Case Value	Distribution	Variation and Range
Quinaldine cost	60 ($/kg)	Normal	S.D. = 10 (30–90)
Chlorination reaction time (in P-1)	6 h	Triangular	(4–8)
Condensation reaction time (in P-2)	6 h	Triangular	(4–8)
Cloth filtration flux in P3, P5, P7, P9	200 (L/m^2-h)	Triangular	(150–250)

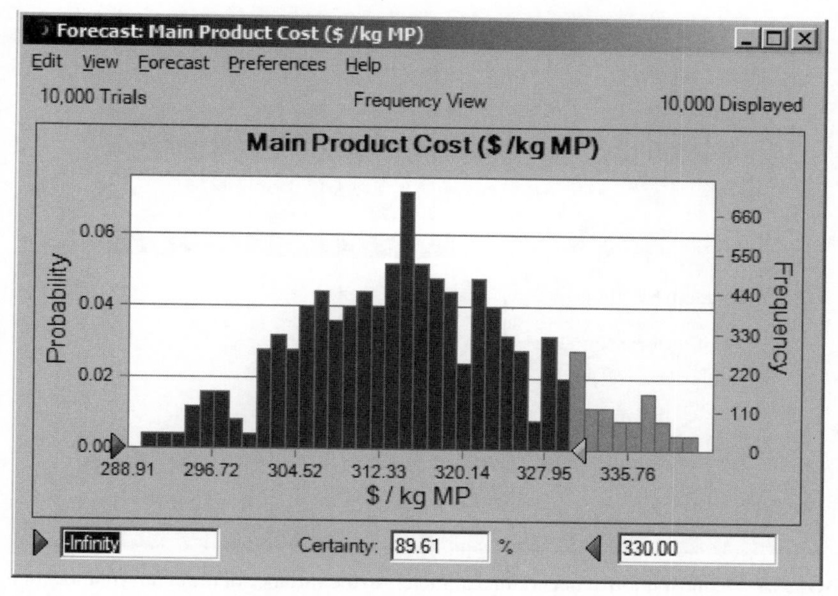

FIGURE 28.7 Probability distribution of the unit production cost (10,000 trials) (mean = 315.52; median = 315.03; S.D. = 10.49; range = 290.19–342.25).

Triangular probability distributions were assumed for the duration of the two main reaction operations and the filtration steps that precede P-8 (Table 28.4). Even though variability distributions were assigned to specific operations, it may be deemed more accurate to assume that they account for the composite variability of their procedures. If this type of analysis is done for an existing facility, historical data should be used to derive the probability distributions. Crystal Ball has the capability to fit experimental data.

The two decision variables considered in this study are the number of batches that can be processed per year and the unit production cost. These are key performance indicators important for production planning and project economics. The output variables of the combined SuperPro Designer–Crystal Ball simulation are quantified in terms of their mean, median, mode, variance, and standard deviation. These results are shown in Figures 28.7 and 28.8 for the "unit production cost" and the "number of batches," respectively. Based on the

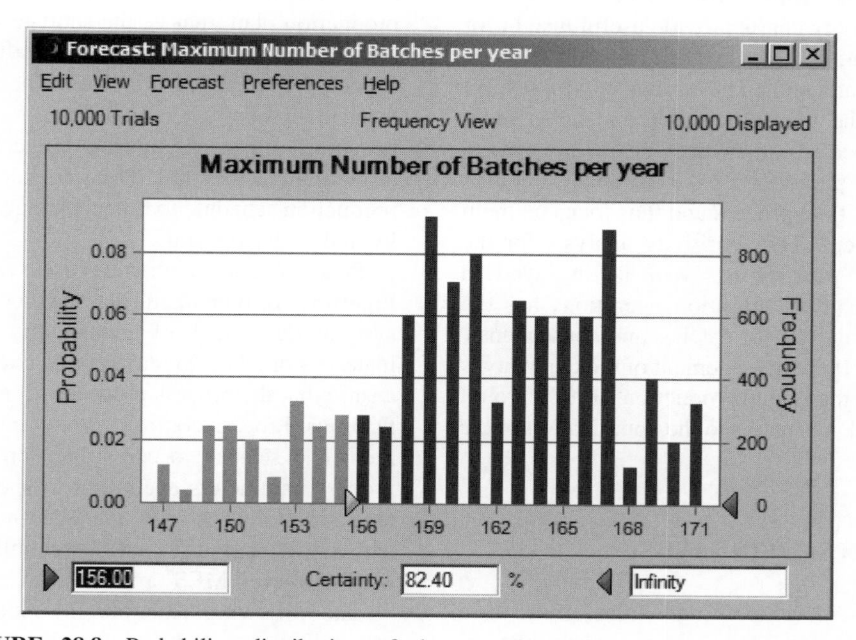

FIGURE 28.8 Probability distribution of the annual number of batches (10,000 trials) (mean = 161.0; median = 161, mode = 159, S.D. = 5.72, range = 147–171).

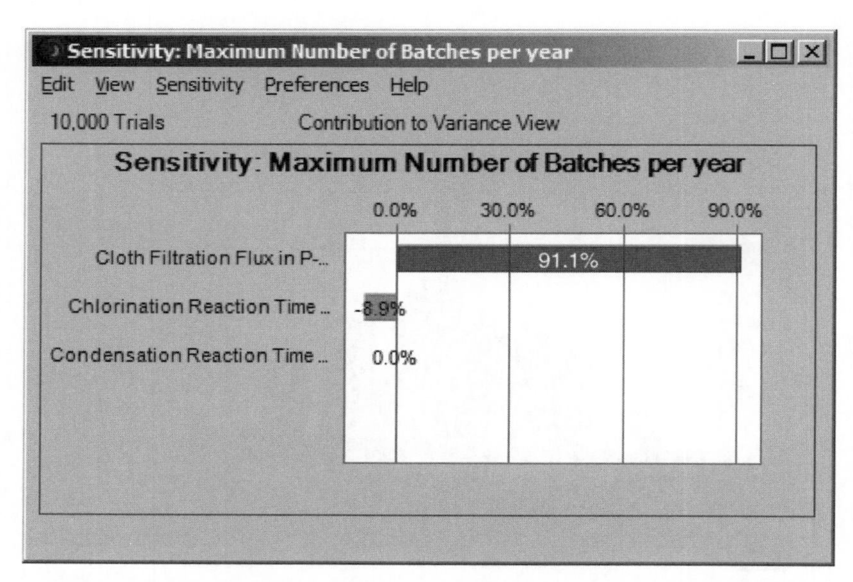

FIGURE 28.9 Contribution of uncertain parameters to the variance of the annual number of batches.

assumptions for the variation of the input variables we note that average values (mean/median/mode) calculated for the decision variables satisfy the objective. The certainty analysis reveals that we can meet the unit production cost goal (unit cost of under \$330/kg) with a certainty of 89.6% (black area in Figure 28.7). The certainty of meeting the production volume goal (of 27,000 kg or 156 batches) is only 82.4% (black area in Figure 28.8). Such findings constitute a quantification of the risk associated with a process and can assist the management of a company in making decisions on whether to proceed or not with a project idea.

The dynamic sensitivity charts provide useful insight for understanding the variation of the process. They illustrate the impact of the input parameters on the variance (with respect to the base case) of the final process output, when these parameters are perturbed simultaneously. This allows us to identify which process parameters have the greatest contribution to the variance of the process and thus focus on them for process improvement. The sensitivity analysis for the *maximum number of batches per year* is displayed in Figure 28.9. The flux of the filtration operations has the greatest impact on the number of batches and consequently the annual throughput. If the management of the company is seriously committed to the annual production target, it would be wise to allocate R&D resources to the optimization of the filtration operations.

28.5 PRODUCTION SCHEDULING

After the process is developed and transferred to a manufacturing facility for clinical or commercial production, it becomes the job of the scheduler to ensure that all the activities are correctly sequenced and that the necessary labor, materials, and equipment are available when needed. The short-term schedule includes the upcoming production campaigns and may span from a week to several months. The general workflow begins with the long-term plan that describes how much of each product should be made over the planning period. The long-term plan, which is described in the next section, is based on approximate batch or campaign starts and does not include details about process activities. The scheduler uses the plan and knowledge about the process and available equipment and resources to generate a detailed production plan, that is, the short-term schedule, and communicate it to the appropriate staff. As the schedule is executed, there may be deviations between the schedule and the actual process execution. Tests, for example, may need to be redone, operations may take longer than the time assumed, or equipment may fail. The scheduler must recalculate the production schedule to reflect changes in resource availability and notify the staff.

Pharmaceutical companies use a variety of plant systems. Enterprise or manufacturing resource planning (ERP/MRP) systems keep track of the quantity of resources, such as materials or labor. Manufacturing execution systems (MES) ensure that the process proceeds according to precise specifications. Process control systems interface with the equipment and sensors to carry out steps and to maintain the process parameters according to specification [15]. Short-term scheduling is often managed manually or with standalone systems, but it could potentially interface with ERP/MRP and even MES programs.

The following example introduces SchedulePro as a scheduling tool. SchedulePro does not model the process itself with respect to its material and energy balances; it is mainly concerned with the time and resources that tasks consume. If a user is interested in both process modeling and scheduling, he/she can generate the process model in SuperPro Designer, perform the material and energy balances there, and then export it as a recipe to SchedulePro for a thorough capacity planning or scheduling analysis in the context of a multiproduct facility. Within SchedulePro, capacity/scheduling information imported from SuperPro related to processing tasks can be expanded in the following ways:

- For every procedure, an equipment pool can be defined representing the list of alternative equipment that could potentially host that procedure.
- Auxiliary equipment can be assigned, possibly through pools, to operations.
- Materials supplied or generated through operations can be linked to supply, deposit or intermediate storage units.
- The rigidity in recipe execution is relaxed with the introduction of the ability to delay the start or break the execution of an operation (if the resources it requires are not available).

The inclusion in the production model of this additional information is motivated mainly by the needs of the pharmaceutical/biotech industry where it is known that quite frequently the bottlenecks exist in the use of auxiliary equipment (e.g., CIP skids, transfer panels) or are related to support activities (e.g., cleaning, buffer preparation) that tend to have flexible execution.

With the resources and facilities in place, simulation of the production activity in SchedulePro can proceed through the definition and scheduling of campaigns. A *campaign* is defined as a series of batches of a given recipe leading to the production of a given quantity of product. A series of campaigns organized in a priority list constitute the production plan that needs to be realized. As a finite capacity tool, SchedulePro attempts to schedule production of campaigns while respecting capacity constraints stemming from resource unavailability (e.g., facility or equipment outages) or availability limitations (e.g., equipment can only be used by only one procedure at a time). Conflicts (i.e., violations of constraints) can be resolved by exploiting alternative resources declared as candidates in pools, introducing delays or breaks if this flexibility has been declared in the corresponding operations, or moving the start of a campaign or batch at a time where the required resources are available. The automatically generated schedule can subsequently be interactively modified by the user through local or global interventions in every scheduling decision. Through a mix of automated and manual scheduling, users can formulate a production plan that is feasible and satisfies their production objectives.

28.5.1 Illustrative Example

This example uses the optimized version of the pharmaceutical intermediate process described above. The objective in this example is to create a schedule for the month of October. Specifically, the process is the three-reactor, two-filter case outlined in Figure 28.5. SchedulePro serves as the scheduling tool. The scheduling model or recipe captures the step-by-step timing and the use of equipment, materials, utilities and labor. Table 28.5 shows a recipe representation for the product dissolution step (procedure P-4).

Pharmaceutical process scheduling is unlike scheduling general work activities because tasks are generally assumed to progress one after the other without delay or interruption. Due to chemical stability limitations, delays in the process are defined and limited. The recipe representation of an allowed delay or safe-hold is the *flexible shift*. In this example, when the product is in a solid form, it may be held for up to 6 h.

The plant scheduler must create a schedule that meets product demand and respects the resource limitations of the facility. The target, plan is for 15 batches with an average

TABLE 28.5 Sample Scheduling Recipe

Operation	Description	Scheduling and Timing	Operators
Cake charge	Transfer in for 233 min from NFD-101 (in P-3) to R-103 (in P-4).	Starts concurrently with TRANSFER-OUT-1 in P-3. The duration matches the duration from TRANSFER-OUT-1 in P-3 to TRANSFER-OUT-2 in P-3	2.0
NaOH charge	Charge 1740 kg of mixture to R-103 (in P-4), using stream NaOH-2.	Starts at the end of cake charge in P-4. Duration is 3.233 h	2.0
Product solubilization	React in batch mode for 2 h., at 50°C and pressure of 2.5 bar.	Starts at the end of NaOH in P-4. Duration is 2 h	1.0
Transfer to filter	Transfer 100% from R-103 (in P-4) to NFD-101 (in P-5) for 253 min.	Starts at the end of product solubilization in P-4. The duration matches the duration of impurity removal in P-5	2.0

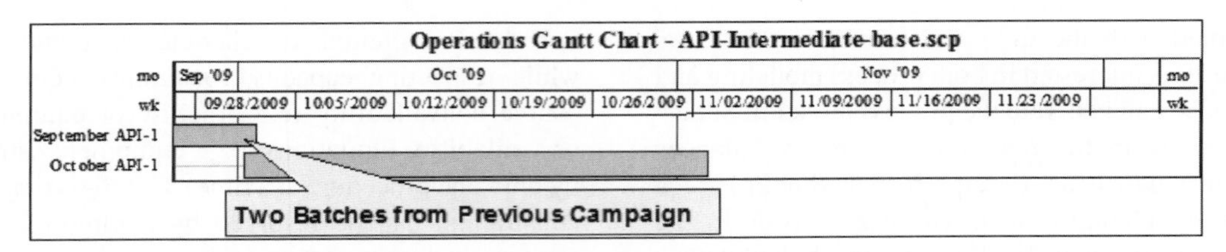

FIGURE 28.10 Campaign to produce 15 batches in October.

cycle time of 50 h. The suite has a crew of seven operators, three reactors, two Nutsche filters and a tray dryer. The scheduler creates a campaign of 15 batches starting at 8:00 on October 1. The last two batches of the October campaign finish in November. This is balanced by the October completion of the final two batches of the September campaign. The two campaigns are shown in the Gantt chart in Figure 28.10.

28.5.1.1 Labor Shortages Under normal circumstances, seven operators are required. Figure 28.11 shows the typical operator demand for 1 week. The horizontal line indicates the limit of seven operators. The thin line shows the average labor requirement.

Short-term schedulers often need to account for labor availability. In this case, a training program during the week of October 12 effectively reduces the crew size to six. The scheduler must decide whether or not to request overtime operators. Rescheduling with the new temporary limit produces the result shown in Figure 28.12. The scheduling tool manages the temporary labor constraint by delaying the start times of the two batches that begin during the week of October 12.

The revised schedule still meets the 15-batch goal for the month of October; however, the completion of the final batch is delayed by about 1 day.

28.5.1.2 Maintaining the Schedule Time does not always specify the completion of an operation in pharmaceutical processing. The concentration of a key component may, for example, be the primary specification. The durations of actual operations may therefore vary from those in the scheduling recipe. The scheduler must, therefore, regularly update the schedule based on new information about the status of the batches. For example, suppose the scheduler updates the schedule on Tuesday, October 27 at 5:00 and learns that batches 12 and 13 are in progress. Furthermore in batch 12, the evaporation step (P-10 in R-103) was delayed by 3 h due to some mechanical issues. The scheduler sets the current time in the schedule and updates the status by entering the actual duration for the vaporization in batch 12. The scheduling tool predicts that R-103 will be over-allocated by 11:30.

Figure 28.13 shows the conflicted schedule. The diagonal hatch indicates activities that are in progress. Over-allocated or conflicted equipment is shown as an additional line on the chart and the conflicted procedures are outlined.

To resolve the conflict, the scheduler has the scheduling tool attempt. The scheduling tool takes advantage of the safe hold point in the process to delay the transfer of batch 13 material to R-103. Figure 28.14 shows the result. This solution does not affect the production target.

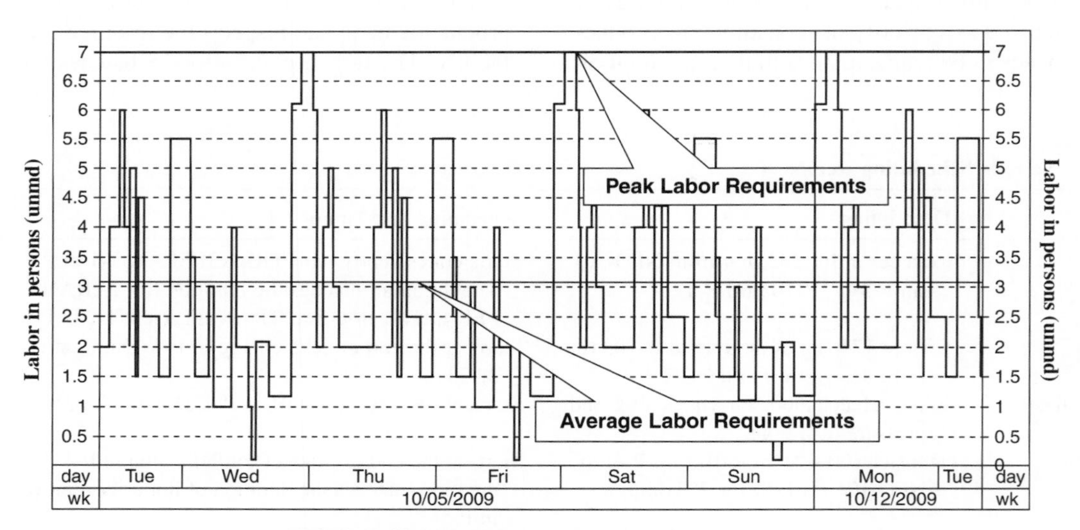

FIGURE 28.11 Typical labor demand for 1 week.

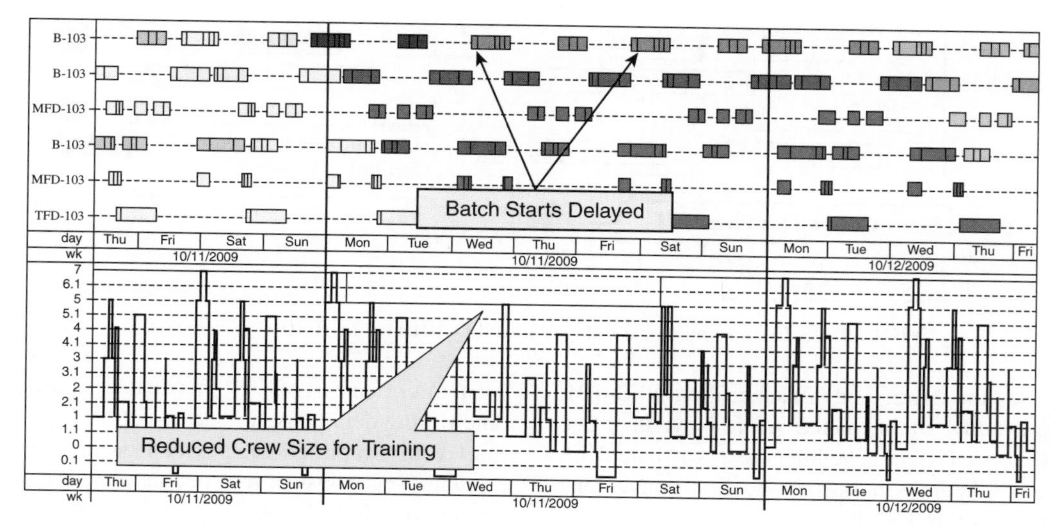

FIGURE 28.12 Equipment occupancy (top) and labor demand (bottom).

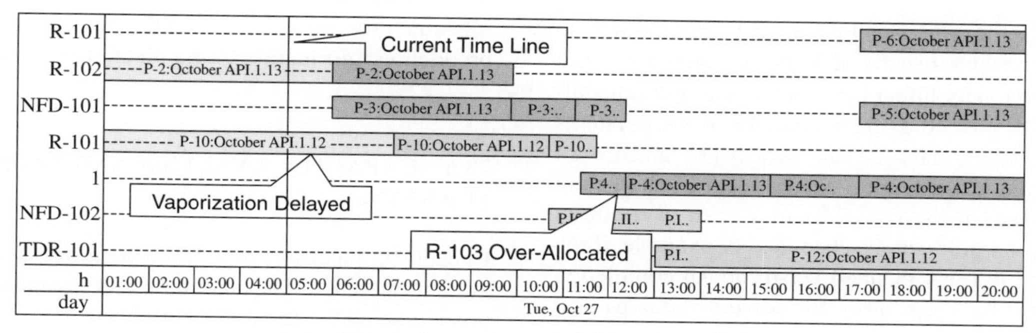

FIGURE 28.13 Conflict with R-103 in batch 13.

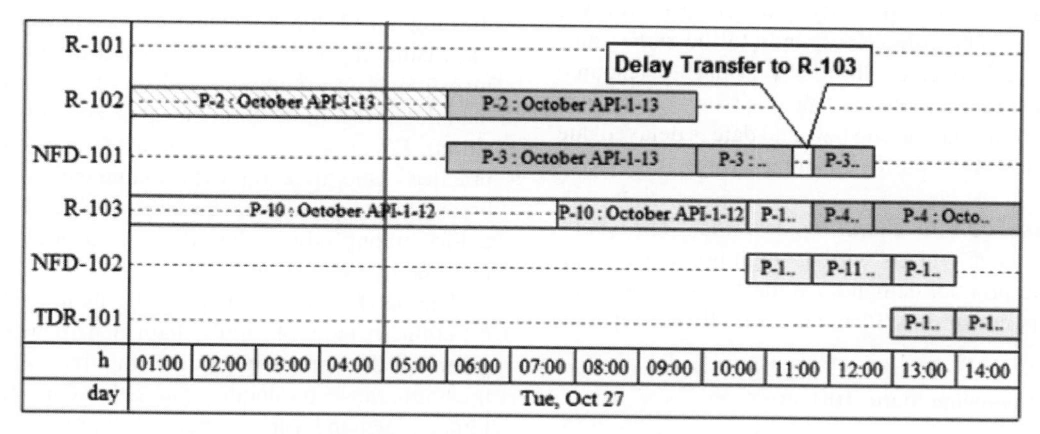

FIGURE 28.14 Conflict resolved with a hold in P-3.

28.5.1.3 Accounting for Equipment Outages Scheduling around equipment maintenance is one of the scheduler's routine tasks. From a scheduling standpoint, preventative maintenance (PM) represents periods of unavailability. Maintenance may be fixed to a particular date or it may be floating. Figure 28.15 illustrates each type. The dryer, TDR-101, has a firm maintenance outage on Monday, October 19. The scheduling tool "plans around" the outage by delaying

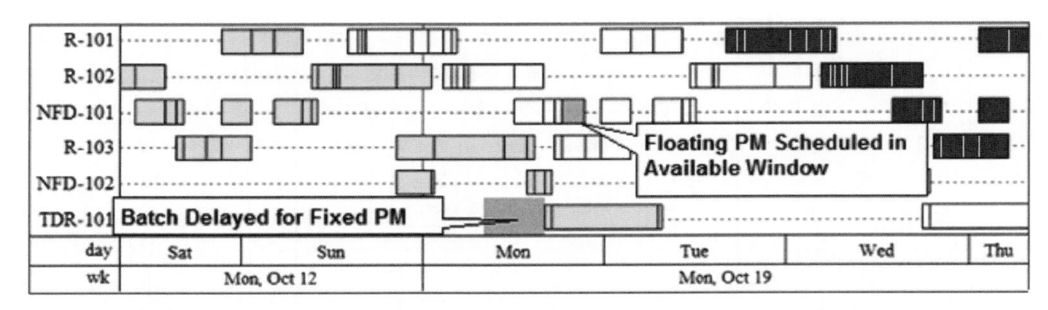

FIGURE 28.15 Maintenance outages for NFD-101 and TDR-101 (gray areas).

the batch that uses the dryer. The Nutsche filter, NFD-101 requires a 4 h preventive maintenance before November but after 10:00 on October 19. The scheduler creates a "batch" of maintenance and schedules it during the first convenient window of opportunity in October.

28.5.2 Tracking

The scheduler's focus is usually limited to the immediate future. The scheduler usually deletes completed batches because they no longer affect the current or future scheduling. Electronic batch records are maintained in other systems that are focused on permanence and security [16] and are not well suited to the fast-changing environment of short-term scheduling. However, the scheduler may wish to recall an earlier version of a production schedule. For example, if the scheduler wants to track planned versus actual completion dates, there must be a repository for scheduled campaigns. The schedule changes as batches are made, so the repository may store multiple versions of the schedule.

Table 28.6 displays the results of a simple report. The first row corresponds to the originally planned date for the campaign. The second row corresponds to the same campaign just before it begins. The delay is due to resource constraints from other campaigns. The third row corresponds to the completed campaign. The new end date is delayed due to constraints that arose during manufacturing.

28.5.2.1 Connection to Planning Systems The short-term scheduler usually starts with a rough production plan that is based on product demand, estimated plant capacity, and/or inventory constraints. Planners generally use separate systems, often part of ERP, that do not require all the resource and timing details. As the next section shows, a scheduling tool with simplified recipes can be an effective planning system. Regardless of the specific planning system used, an automated link between planning and scheduling, can streamline both the planning and the scheduling processes. For example, the scheduler may download campaign information with estimated dates from the planning system. As the schedule progresses, the scheduler uploads revised date and production information to the planning system.

TABLE 28.6 Campaign Status History

Campaign	Start Date	End Date	Entry Date	Comment
October API-1	10/01/2009	11/04/2009	09/15/2009	Original Plan
October API-1	10/05/2009	11/06/2009	10/18/2009	Before Start
October API-1	10/05/2009	11/07/2009	11/26/2009	Completed

28.6 CAPACITY ANALYSIS AND PRODUCTION PLANNING

Capacity is a measure of how much product a manufacturing system can make. The amount of product manufactured in a given time period (hour, day, week, etc.) or the time required to produce a given quantity of product are the most intuitive and commonly used measures of capacity. The capacity of a manufacturing system should exceed demand at least over the long run. On the other hand, excess capacity is costly [17]. Increasing capacity to meet demand might require capital investments in equipment and buildings or extending the manufacturing time (through labor overtime or additional shifts). *Effective capacity* is the actual capacity achieved in practice. Due to equipment maintenance or unexpected breakdowns, scheduling inefficiencies, and labor unavailability among others, the effective capacity is usually less than the nominal plant capacity.

The need for an estimate of the plant capacity arises in different activities of supply chain management. In aggregate planning, the objective is to generate feasible long-range or medium- range production plans that can satisfy expected lumped demand for a range of aggregate products. The validity of these plans depends on the accuracy of the aggregate plant capacity estimates. If an MRP-II approach is used to create a master production plan and more detailed production orders, the feasibility of the generated schedules should be checked against the plant capacity, this time measured with greater detail and for each product separately. Inventory management, batch sizing, and operation scheduling are other examples of activities that relate to capacity analysis.

Depending on the complexity of the production system, the range of different products produced and the diversification of their routings (recipes), the level of difficulty in estimating a plant's capacity can vary from trivial to formidable. The capacity of a single-product batch plant depends only on the batch size, the cycle time, and the allocation of production time. If greater capacity is required, either the production time should be extended or the cycle time should be reduced by removing bottlenecks. In multiproduct or multipurpose facilities, however, with complex material flows, multiple equipment used in parallel, shared resources and sequence-dependent changeover and cleaning times, the estimation of the capacity is far from trivial. In fact, in these cases, capacity estimates emerge through the same activities that capacity analysis is supposed to serve, that is, planning and scheduling. In other words, only after specific production planning and scheduling scenarios have been laid out, can capacity be estimated. Capacity analysis is, therefore, interlinked with the production planning and scheduling activities providing important data to carry out these activities and simultaneously emerging as their outcome. This is the reason why in this section capacity analysis and production planning are treated simultaneously.

Both production planning and capacity analysis, in different contexts, have been the subject of intense research and industrial activity for many years. It is now recognized that there is no solution to these problems that can fit all cases; there is too much variability in the problem structure for a single solution to cover all aspects. The differences between process industries and discrete manufacturing industries have also been investigated and the applicability in the process industries, of the methods developed mainly for discrete manufacturing, has been questioned (see, for example, Refs [18–20].)

Pharmaceutical manufacturing facilities are typically multipurpose plants equipped with multiple production lines that share utilities, labor resources, and auxiliary equipment, such as CIP skids, transfer panels, delivery lines, and occasionally main equipment. Production is typically campaigned. Considerable changeover time is often required between campaigns of different products. API synthesis, in particular, is characterized by complex material flows and the need to handle and store a variety of required intermediates.

Simulation is an appropriate tool to cope with the complexity of production planning and capacity analysis in pharmaceutical manufacturing. Rather than attempting to formalize a single model and come up with a single solution as optimization-based methods do, simulation allows the planner to formulate and analyze different scenarios and select the one that best fits the objectives and constraints of the problem. Such "what-if" analyses can generate feasible production plans utilizing the available capacity or provide justifications for facility expansions and/or outsourcing of production. The types of capacity analysis questions that can be answered using simulation will be demonstrated in this section with the use of the software tool SchedulePro.

28.6.1 Simulating the Production Process

Production planning is the activity of assigning facility resources to processing tasks. This makes a scheduling tool appropriate because it manages timing and resources without the necessity of engineering calculations.

A simulation-based approach can be used to support both planning/capacity analysis and scheduling activities. The level of detail included in the simulation model is the only difference between the two. In planning, the recipe representations are coarse, products could be lumped in aggregates with similar production recipes and only the most basic resources are considered. In scheduling, recipes are expanded to their fullest detail, products are differentiated and all potentially limiting resources are included. The following example will demonstrate the use of simulation for planning and the types of what-if scenarios that can be investigated under different assumptions and objectives.

28.6.2 Capacity Analysis Example

In this example, we will consider the last three steps required for the production of a small-molecule API assuming that any required raw or intermediate material is supplied from external sources. Based on detailed process analysis done in a process simulator such as SuperPro Designer, the amount of raw material and the amount of product produced in each step have been calculated. By considering the main plant resources, the cycle time of each step has also been estimated. This is shown in Figure 28.16. The material denoted as SM in the figure is the raw material supplied externally (e.g., by a contract manufacturer); Int-1 and Int-2 are the two stable isolated intermediates, which can be stored until turned into the final product FP. Step-1 takes 975 kg of SM as input per batch and generates 1500 kg of Int-1 as output. Step-2 takes

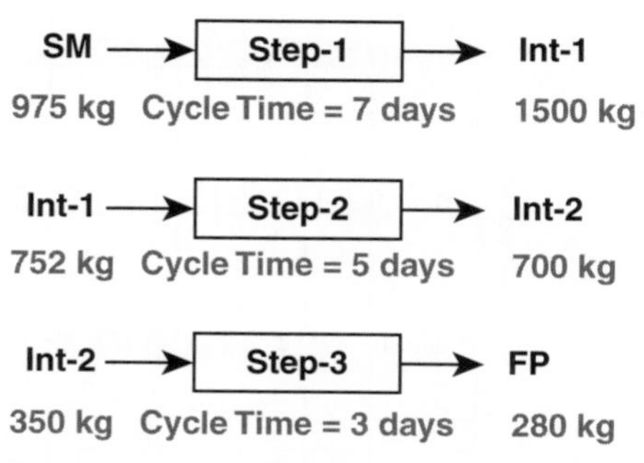

FIGURE 28.16 Cycle time, input amount, and output amount for the three chemical steps.

752 kg of Int-1 per batch and generates 700 kg of Int-2. Finally, Step-3 takes 350 kg of Int-2 and generates 280 kg of FP per batch. The cycle times for the three steps are 7, 5, and 3 days, respectively.

The objective of this study is to find out whether 35,000 kg of FP (corresponding to the anticipated demand) can be produced per year if three independent production areas ("Train-1," "Train-2," and "Train-FP") can be made available to this synthesis route. Since the output of Step-3 is 280 kg of FP per batch, the required number of batches of Step-3 per year is 125 (= 35,000/280). For Step-2, the required amount of produced Int-2 is 43,750 kg (= 35,000*350/280) and the corresponding number of batches per year is 63 (= 43,750/700). Finally, for Step-1 the required amount of produced Int-1 is 47000 (= 43,750*752/700) and the corresponding number of batches per year is 32 (= 47,000/1500).

In SchedulePro's terminology, each step corresponds to a separate recipe. For the purposes of this long-term planning study, it suffices to represent each step as a single-procedure recipe that utilizes one of the available production trains. In other words, the entire recipe is abstracted to a single processing task and all resources are represented through a single resource corresponding to each plant. Assuming that the capacity of each plant to execute each detailed step recipe has been checked, the above simplification comes at no loss of generality. The advantages of faster implementation and production plan development exceed by far the effects of possible inaccuracies (such as end effects in the planning horizon's beginning or end) caused by the simplified representation.

In this representation, each procedure is assigned a duration equal to the step's cycle time (as reported earlier) and a pool of equipment representing each of the three available plants. It should be noted that the reported cycle times are a bit longer than the minimum (optimal) cycle times calculated by the step's detailed analysis. Operating at a cycle time that is somewhat larger than the minimum enables the schedule to absorb any delays without long deviations from the original plan.

Under the above assumptions, it is quite easy to calculate the required capacity (measured in production days) for the required quantity of the final product. For 32 batches of Step-1, 224 days (= 32 × 7) are required. Similarly, for Step-2 and Step-3 the corresponding duration is 315 (= 63 × 5) and 375 (= 125 × 3). Adding the time required to produce the first batch of Int-1 so that Step-2 can start and the time to produce the first batch of Int-2 so that Step-3 can start, brings the total campaign make span beyond the desired 365-day completion horizon.

The above simple calculations can be easily verified with a simulation of this case scenario in SchedulePro. The equipment occupancy chart in Figure 28.17 is generated under the assumption that each step is executed independently in the three separate lines named (Train-1 for Step-1, Train-2 for Step-2, and Train-FP for Step-3). The total make span of the production schedule is approximately 56 weeks.

The implementation of the above schedule requires large inventories for Int-1 and Int-2 since they are produced at a much faster pace than they can be consumed by the subsequent steps. This is clearly demonstrated in Figure 28.18a,b where the inventories of Int-1 and Int-2 are shown. Storage capacity of over 15,000 kg for Int-1 and 8000 kg for Int-2 will be required to implement this production plan.

On the other hand, the capacity of the available trains dedicated to the production of intermediates is underutilized (see Figure 28.17) and the objective of completing the production campaign in less than a year is not satisfied. Modifications on the above basic scenario can be driven by two different objectives: reduction of storage capacity and reduction of total make span.

The key to satisfy the make span objective is better plant capacity utilization. The underutilization of capacity in Train-1 and Train-2 creates the possibility of exploiting that excess capacity for Step-3 that is responsible for the delay. Please recall the initial assumption that each production train can be used interchangeably for every step. The base case can therefore be modified by inserting Train-1 (with the lowest utilization) into the pool of candidate trains for executing

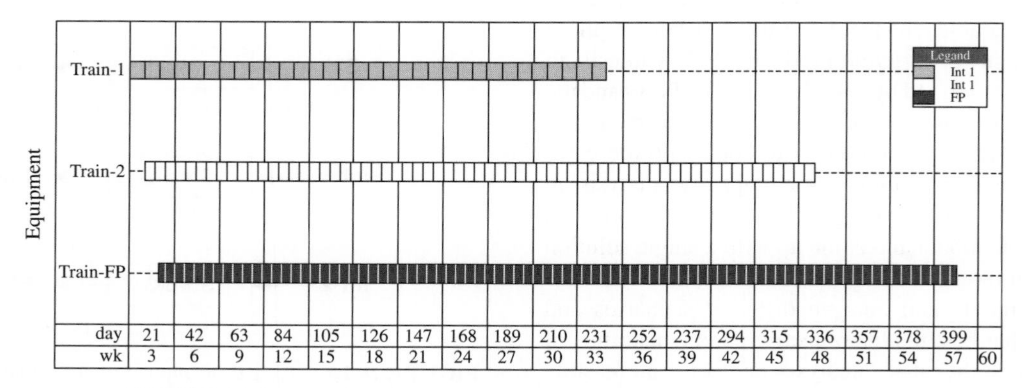

FIGURE 28.17 Equipment occupancy chart for the base production scenario.

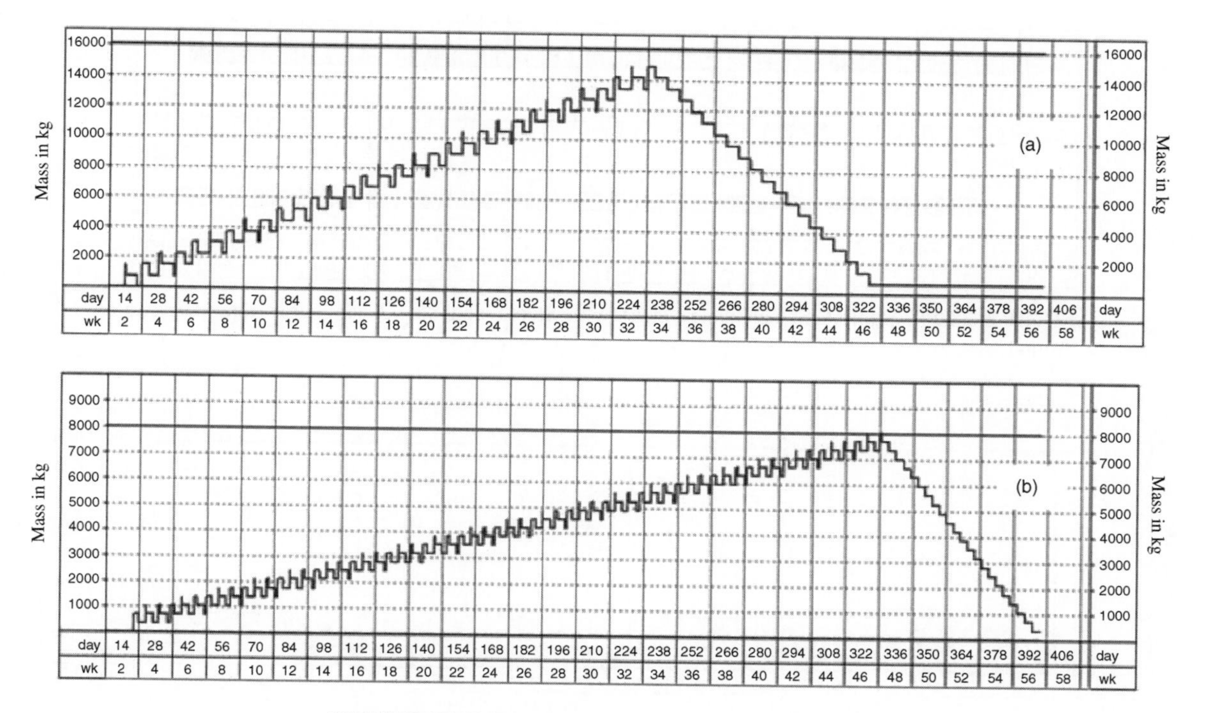

FIGURE 28.18 Inventories of Int-1 (a) and Int-2 (b).

Step-3. Initially, it will be assumed that an extended (5-week period) is required for a thorough cleaning/changeover when switching campaigns within a train—a constraint that makes alternation of batches of different steps within the same train prohibitive. The generated production plan under these assumptions is shown in Figure 28.19.

In this case, the excess Train-1 capacity is utilized to host some of the Step-3 campaign batches after campaign Step-1 is completed and following the extensive cleaning of the line. As a result, the make span is significantly reduced to less than 47 weeks and the production objective is satisfied. This change has not affected the demand for intermediate storage, though, which remains high at the levels previously shown in Figure 28.18.

The potential to satisfy the production make span objective while minimizing inventories exist only if it is possible to alternate batches of different steps within a single line. If we assume that the cleaning required when switching products is not as extensive as before, it is possible to break the long 32-batch campaign of Step-1 in multiple shorter campaigns which can be spread throughout the year and interject batches of Step-3 in the time gaps. With this strategy, it is expected that both objectives can be met.

To implement this scenario, it is assumed that the Step-1 campaign is split into four 8-batch campaigns released every 11 weeks and 3 days of cleaning are required before and after switching products within a plant. Figure 28.20 shows the updated schedule. The total make span is again shorter than a

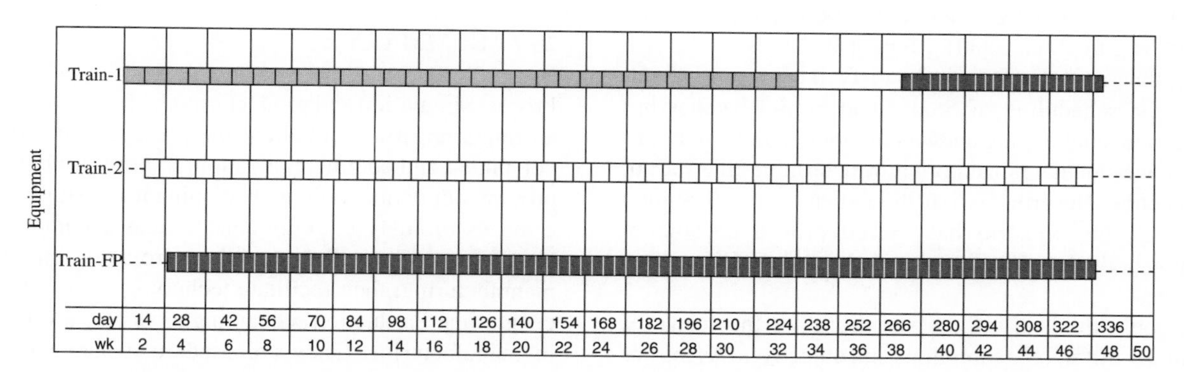

FIGURE 28.19 Equipment occupancy chart with Step-3 batches following Step-1 batches in Train-1.

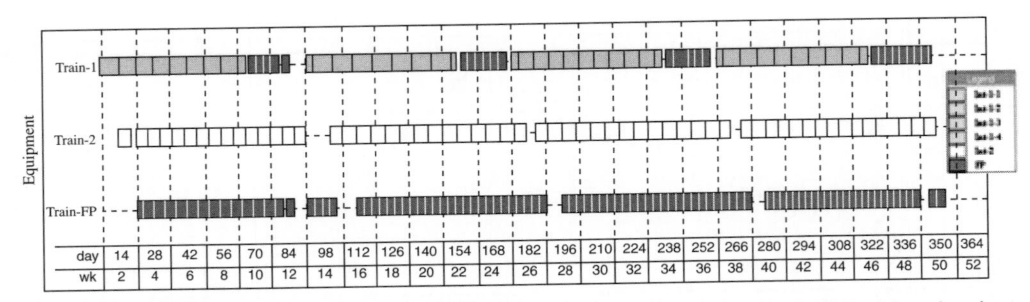

FIGURE 28.20 Equipment occupancy chart with Step-1 and Step-3 campaigns alternating in Train-1.

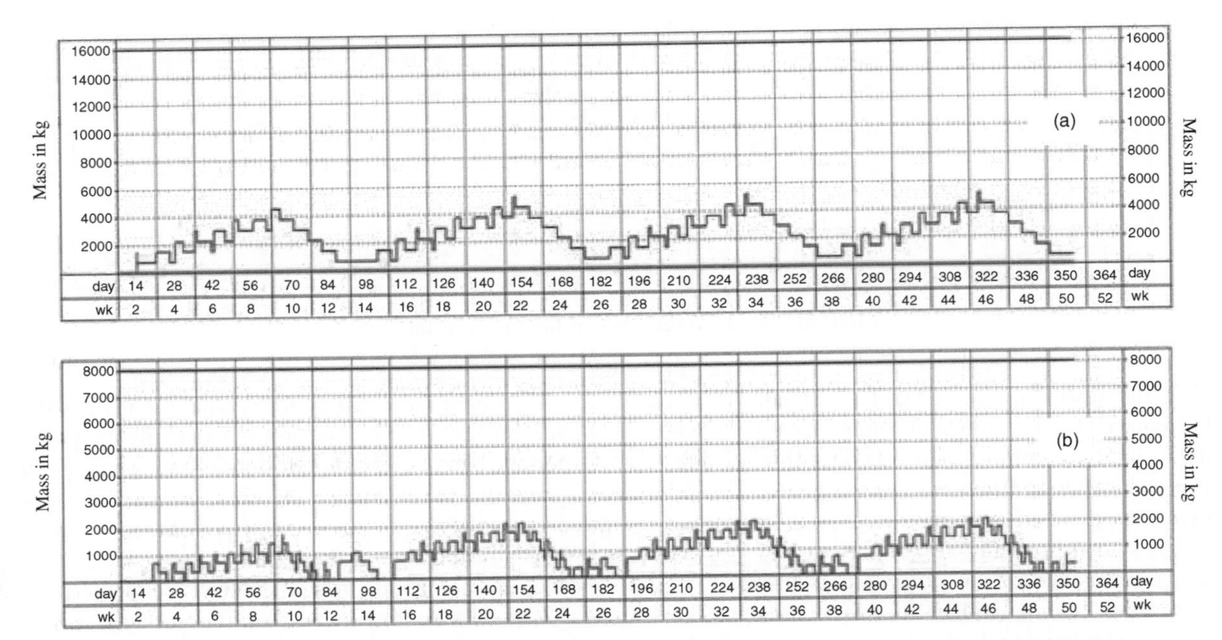

FIGURE 28.21 Inventories of Int-1(a) and Int-2 (b) with Step-1 and Step-3 campaigns alternating in Train-1.

year, but, unlike the previous scenario, it is now possible to reduce considerably the inventory of intermediates. As shown in Figure 28.21, the required storage capacity has dropped from 16,000 kg to less than 5000 kg for Int-1 and from 8000 kg to about 2000 kg for Int-2.

Note that in the attempt to satisfy the inventory constraints, the scheduling of Step-2 and Step-3 batches has become more challenging and less obvious. The existence of adequate intermediate inventory now determines the start of these batches. Nevertheless, at the expense of less regular scheduling, this scenario has indeed proved capable of satisfying both the make span and the reduced inventory objectives.

One could think of infinite variations for the scenarios above under different assumptions and objectives. As long as the capacity and planning constraints can be intuitively captured, formulating and developing feasible and satisfactory solutions under variable assumptions and objectives can be easily performed in a simulation environment.

28.7 SUMMARY

Process simulation and production scheduling tools can play an important role throughout the life cycle of product development and commercialization. In process development, process simulation tools are becoming increasingly useful as a means to analyze, communicate, and document process changes. During the transition from development to manufacturing, they facilitate technology transfer and process fitting. Production scheduling tools play a valuable role in manufacturing. They are used to generate production schedules based on the accurate estimation of plant capacity, thus minimizing late orders and reducing inventories. Such tools also facilitate capacity analysis and debottlenecking tasks. The pharmaceutical industry has only recently begun making significant use of process simulation and scheduling tools. Increasingly, universities are incorporating the use of such tools in their curricula. In the future, we can expect to see increased use of these technologies and tighter integration with other enabling IT technologies, such as supply chain tools, MES, batch process control systems, process analytics tools (PAT), and so on. The result will be more robust processes and efficient manufacturing leading to more affordable medicines.

REFERENCES

1. Papavasileiou V, Koulouris A, Siletti CA, Petrides D. Optimize manufacturing of pharmaceutical products with process simulation and production scheduling tools. *Chem. Eng. Res. Des.* 2007;85(A7):1–12.
2. Petrides DP, Calandranis J, Cooney CL. Bioprocess optimization via CAPD and simulation for product commercialization. *Genet. Eng. News* 1996;16(16):24–40.
3. Petrides DP, Koulouris A, Lagonikos PT. The role of process simulation in pharmaceutical process development and product commercialization. *Pharm. Eng.* 2002;22(1).
4. Petrides DP, Calandranis J, Cooney CL. Bioprocess optimization via CAPD and simulation for product commercialization. *Genet. Eng. News* 1996;16:24–40.
5. Hwang F. Batch pharmaceutical process design and simulation. *Pharm. Eng.* 1997;17(1):28–43.
6. Thomas CJ. A design approach to biotech process simulations. *BioProcess International*, October, 2003, pp. 2–9.
7. Petrides DP, Siletti CA. The role of process simulation and scheduling tools in the development and manufacturing of biopharmaceutical. In: Ingalls RG, Rossetti MD, Smith JS, Peters BA, editors, *Proceedings of the 2004 Winter Simulation Conference*, 2004, pp. 2046–2051.
8. Plenert G, Kirchmier B. *Finite Capacity Scheduling: Management, Selection, and Implementation*, John Wiley & Sons, 2000.
9. Parshall J, Lamb L. *Applying S88: Batch Control From a User's Perspective*, ISA, 2000.
10. Intelligen. User's guide of SuperPro Designer, Chapter 2. http://www.intelligen.com/demo. 2007.
11. Petrides D, Koulouris A, Siletti C. Throughput analysis and debottlenecking of biomanufacturing facilities: a job for process simulators, *BioPharm*, August 2002.
12. Tan J, Foo DCY, Kumaresan S, Aziz RA. Debottlenecking of a batch pharmaceutical cream production. *Pharm. Eng.* 2006;26(4):72–82.
13. Harrison RG, Todd P, Rudge SR, Petrides DP. *Bioseparations Science and Engineering*, Oxford University Press, 2003.
14. Achilleos EC, Calandranis JC, Petrides DP. Quantifying the impact of uncertain parameters in the batch manufacturing of active pharmaceutical ingredients. *Pharm. Eng.* 2006; 26(4):34–40.
15. Abel J. Enterprise knowledge management for operational excellence. *Pharma. Eng.* 2008;28:1–6.
16. Abel J, LeBlanc L. Specifying a batch management system for electronic records and signatures: a checklist for compliance with 21-CFR Part 11. *Pharm. Eng.* 1999;19(4):1–9.
17. Sipper D, Bulfin R. *Production: Planning, Control and Integration*, McGraw-Hill International Editions, Singapore, 1997.
18. Crama Y, Pochet Y, Wera Y.A discussion of production planning approaches in the process industry, Université Catholique de Louvain, Center for Operations Research and Econometrics, CORE discussion papers No. 2001042, 2001.
19. Kallrath J. Planning and scheduling in the process industry. *OR Spectrum* 2002;24; 219–250.
20. Schuster EW, Allen SJ. A new framework for production planning in the process industries. *38th APICS Conf. Proc.* 1995: 31–35.

PART III

ANALYTICAL METHODS AND APPLIED STATISTICS

29

QUALITY BY DESIGN FOR ANALYTICAL METHODS

TIMOTHY W. GRAUL AND KIMBER L. BARNETT
Pfizer Global Research & Development, Groton, CT, USA
SIMON J. BALE, IMOGEN GILL, AND MELISSA HANNA-BROWN
Pfizer Global Research & Development, Sandwich, Kent, UK

29.1 INTRODUCTION

For any analytical measurement used in pharmaceutical development or product quality control, it is essential that the integrity of that measurement is understood. The "rules" that assure measurement integrity were derived during the 1980s through cross-industry consensus and with the regulators via the International Conference on Harmonization [1] These rules were complimentary to the good manufacturing practices (GMPs) enshrined in regulatory law.

The creation of ICH standards followed a period during which instrument-dependant separation techniques (e.g., HPLC/GC) displaced the earlier wet chemistry and manual techniques (e.g., titration/TLC/gravity column chromatography) and during which quality assurance of pharmaceutical product was the concern of governments due to some high profile public health disasters [2]. The pharmaceutical quality context that dominated the creation of these "rules" were as follows:

- Quality assurance of product can best be achieved by following a set of instructions (compliance) that are shown to repeatedly give the same product (product validation) and product quality assurance is supported by analytical testing (quality control).
    - Similarly, assurance of *measurement integrity* can best be achieved by following a set of instructions (compliance) that are shown to repeatedly give the same result (analytical method validation).
- In addition to compliance in following instructions (standard operating procedures, SOPs) and completing validation, it is essential to have an underlying Quality System that enforces training, equipment validation, maintenance, calibration, facilities, and so on.

The benefits of these international agreements have been clarity of regulatory expectations for developing pharmaceutical products and associated test methods worldwide across all regulatory authorities. Although specific variations are required for some countries, in most cases product and methods are expected to receive global market approval. The pharmaceutical company knows what regulators expect to see in a submission for the description of the test method, the control and validation of the method and data sets related to these. The impact on global health should not be underestimated as ease of global registration correlates directly to the rapid access to medicines for the world population. However, this approach of compliance to a strict set of testing instructions has a significant disadvantage. The regulatory control of postapproval changes is a barrier to introducing technology advances during product lifecycle and also drives conservatism in initial product development. The 10–15-year gestation time of a product in R&D followed by the 20–30-year life of marketed product means this barrier has a real effect on the availability of affordable, effective highest quality medicines. The FDA recognized this in 2004 and produced a white paper (FDA PAT Team and Manufacturing Science White Paper—*Innovation and Continuous*

Chemical Engineering in the Pharmaceutical Industry: R&D to Manufacturing Edited by David J. am Ende

Improvement in Pharmaceutical Manufacturing) [3] calling for an evaluation of development practice to create better products for patients. The white paper led to a dialogue between the industry and the regulators on how product development practices could be improved, and how regulatory submissions and quality management systems could enable innovation while assuring the product integrity for the patient. The use of "Quality by Design" (QbD) as a framework for product development has been widespread in engineering for several years, and was already described in guidance by the FDA for development of Medical Devices [4]. It was recognized that this methodology could also apply to pharmaceutical product development. The subsequent evolution of the Quality by Design concept for development of pharmaceutical formulations and manufacturing processes and practices led to three new ICH papers launched in 2006; Q8, Q9, and Q10 [5–7]. These outline QbD concepts through discussions on pharmaceutical development, quality risk management, and a pharmaceutical quality system. The industry engaged in developing the first products using QbD via pilot programs with close communication with the FDA. These pilots have led to successful approvals and product launches for Pfizer Inc., Astra-Zeneca, GlaxoSmithKline, Merck, Wyeth and others. The QbD approach to product development is now widely adopted throughout the industry.

QbD for product requires a deep understanding of the critical quality attributes (CQAs) that impact final product quality. To gain that understanding, analytical techniques will be applied. This has reinvigorated attention to online analysis and consideration of in-process testing as an alternative to, or complementary to end-product testing in quality assurance. The appropriate use of in-process or PAT techniques is well documented but the development, validation, transfer between laboratories, the control and the life cycle use of methods has remained unchanged, and this is the subject of this chapter on QbD. This chapter describes QbD as a system for analytical method development and lifecycle management. It describes how the concepts of enhanced scientific understanding, the use of quality management systems and structured risk assessments may be applied to analytical methods and how the concept of defining method factors and attributes can be used to define the control strategy for the method to ensure it is robust and rugged.

29.1.1 Criticisms of Current Practices

The FDA PAT Team and Manufacturing Science White Paper—Innovation and Continuous Improvement in Pharmaceutical Manufacturing made the following criticisms of the industry arising from poor lifecycle management of product:

- Pharmaceutical manufacturing operations are inefficient and costly.
- Processes are not robust
  - Out of specification (OOS) observations can occur frequently.
- Measurement systems are not good enough
  - Variability and/or uncertainty in a measurement system can pose significant challenges when OOS results are observed.
  - Measurement system variability can be a significant part of total variability.
- Knowledge Management is poor
  - Information needed for process improvement can be in a different organization and often not available at the right time.
  - Similar and repeating OOS observations for different products across the industry and a less than optimal understanding of variability.
- Continuous improvement is difficult, if not impossible.

These criticisms apply also to analytical measurement for quality control; the methods can be inefficient and costly, the robustness of methods is a frequent cause of OOS results, variability may be poorly understood or is not fit for purpose once product manufacture reaches more exacting efficiency. Knowledge management overly relies on experts and recall. Continuous improvement of methods is stifled by the cost and inconvenience of postapproval changes when conducted on a global scale with multiple regulatory agencies. So the principles that QbD enables for manufacturing products should also give benefit when applied to analytical test methodology.

Beginning in 2007, analytical scientists began to consider how applying QbD principles to the method lifecycle can lead to better methods; methods that work more reliably and give information that not only supports product quality but can support manufacturing process improvements. To do this, the application of QbD principles would need to overcome the barrier that is stifling innovation in analytical technology. The industry bodies, PhRMA and EFPIA each set up subgroups to explore the subject; these groups have collaborated on a concept paper to enable dialogue across the industry and with the regulators.

QbD necessitates a rigorous evaluation of the intended purpose of a measurement, followed by development of a method and routine use built upon a thorough understanding of the science underpinning the analytical methodology selected. There is a need to improve the reliability of the analytical method by understanding, reducing, and controlling all sources of variability. QbD facilitates adopting new technology, particularly where it enhances understanding of the analyte and so enables continuous improvement. QbD also uses knowledge management systems to improve application of the method and understanding of the data.

There are two key concepts in QbD for analytical methods. The first (the analytical target profile, ATP) addresses the *purpose* of the measurement and forms the basis for development of the initial method. It is also the basis for substitution of subsequent methods as technology develops. While the product control strategy defines which attributes will be routinely measured to assure the product is of the desired quality, each measurement requirement for each *attribute* is formally defined in the analytical target profile. The ATP is proposed as a new mechanism for describing analytical methods in regulatory submissions that would reduce the burden of postapproval variation. At present, method changes typically involve comparisons of data sets from a common sample pool generated using the original method and the proposed new method. This traditional approach of comparison biases changes to those where the new technology delivers results that are *the same as* the original method, and hence this can stifle continuous improvement by preventing adoption of a technique that *enhances* understanding of the analyte. It is proposed that when an ATP is registered, subsequent method changes would be referenced to the ATP. A significant advantage in the new approach is that it enables introduction of methods with improved reliability and enhanced accuracy providing they meet the ATP descriptors. For example, a method for routine manufacturing use may have been developed to ensure a given attribute lies within specification limits. Modern pharmaceutical production may require knowledge about variability of that attribute within a batch in order to improve batch yield and lower product cost. Introduction of a new technique, or an enhanced method would generate information for both purposes without jeopardizing the quality control application of the method.

The ATP would first be generated at the time a need for a method is identified, in pharmaceutical product development this would relate to the needs generated from the QbD for the product as shown in Figure 29.1. Once the product control strategy is established, the ATP for each CQA of the product or process should be reviewed to ensure it is appropriate and the ATP is then suitable for a filing.

Figure 29.1 compares QbD for analytical methods with QbD for product and illustrates how variations in operating conditions could be managed. QbD for analytical methods starts with a description of the requirements of a measurement, the ATP, which may be derived from a critical quality attribute of a product. In QbD for product, flexibility is gained by the opportunity for varying operating parameters and is based on the design space that has been documented around a CQA of the product or process. In QbD for analytical methods, flexibility is gained by the opportunity for varying *method factors* (defined as *any factor that forms part of the method definition* [8], *e.g., machines, materials, people, processes, measurements, and environments*) and is based on the design space that has been documented around the ATP. The ATP is in effect a representation of the critical quality attributes of the measurement. In recent years, analytical scientists have sought to minimize detail in regulatory methods in order to allow flexibility in subsequent application of the method. In doing so, pharmaceutical companies focus only on the method factors that they believe are critical to obtaining a true and accurate result. The rationale for why these method factors are critical, and more importantly, the functional relationships between these factors and method performance is typically not included in regulatory filings (nor are they rigorously studied during method development). The discipline imposed by a QbD approach will ensure the essential elements of the method are recognized. Furthermore, as the ATP relates to the product CQA, the benefits of this systematic approach to design and development are realized consistently throughout the quality assurance and quality control of product development, manufacture, and lifecycle management.

The second concept addresses how QbD steps, tools and approaches can be applied to design and development of an analytical method and can be used for implementation and lifecycle management of analytical methods in a manner analogous to those described for pharmaceutical

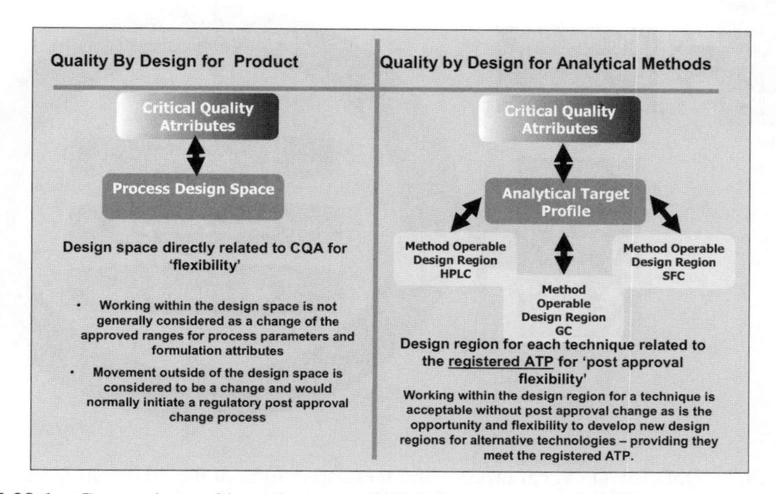

FIGURE 29.1 Comparison of key elements of QbD for product and QbD for analytical methods.

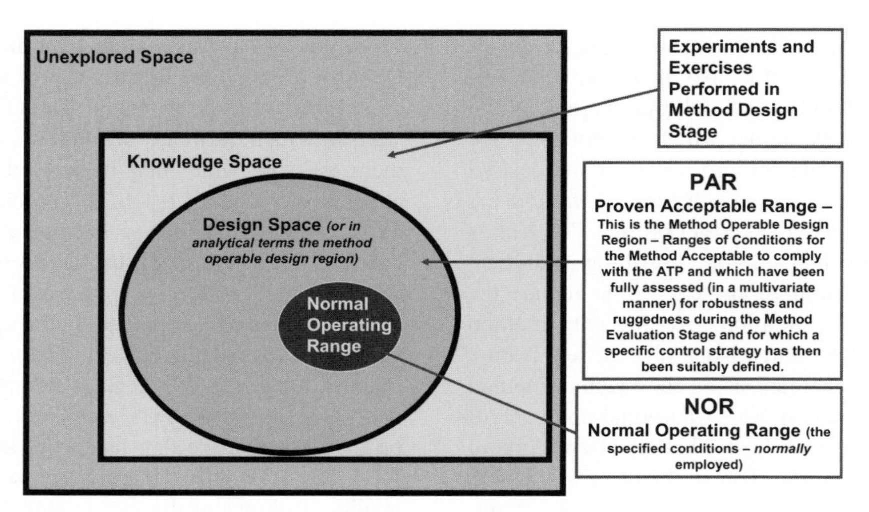

FIGURE 29.2 Description of how the QbD approach to development of an analytical method can be considered in "traditional" QbD terms.

manufacturing in ICH Q8, Q9, Q10. In this concept, the stepwise process of method development, method validation, and method transfer are superseded by a new paradigm with a design phase followed by a comprehensive method development and evaluation. This yields a robust and rugged method that is described in terms of the "method operable design region" where all of the factors that may affect the method have been evaluated or explored through experimentation. The method operable design region is essentially the design space (or proven acceptable range in QbD for product) over which the robustness and ruggedness experimentation has shown the method can meet the requirements of the ATP. For convenience, the method may typically run with more restricted conditions for business operational reasons (i.e., the normal operating range). Of course, in determining the method operable design region some experiments will identify conditions that do not meet the ATP, this is illustrated as the knowledge space beyond the design space. For some techniques there may be conditions that remain unexplored as they are unlikely to yield usable conditions. These terms are illustrated in Figure 29.2.

Figure 29.3 gives an overview of the concepts discussed and illustrates the key components of the process for QbD of analytical methods. The adoption of a well-researched ATP and the implementation of the new approach to change control will result in methods that are exquisitely matched to the requirements for the measurement and that allow changes to facilitate adoption of new technologies that augment continuous improvement. The robustness and ruggedness evaluation will enable changes to method factors to allow flexibility

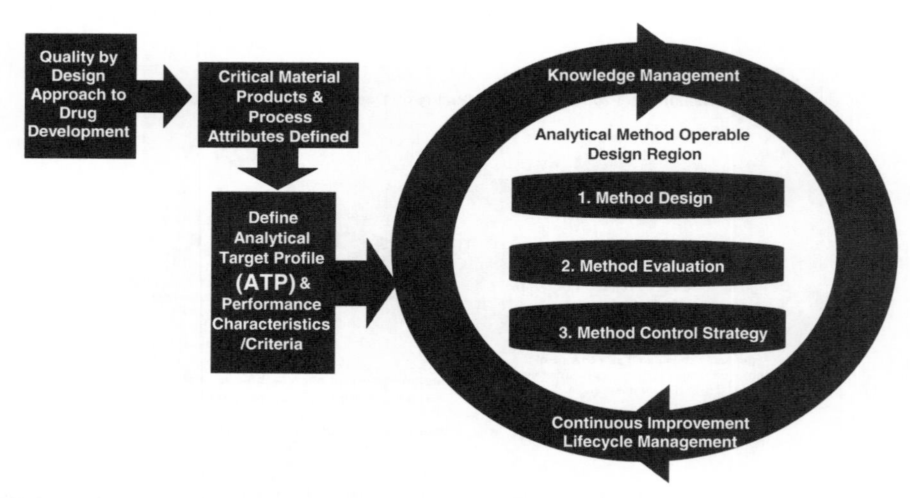

FIGURE 29.3 QbD for analytical process and its relation to both the ATP concept and the QbD approach to drug development.

TABLE 29.1 Example ATP for Measurement of Impurities Present in a Drug Product

Analytical Target Profile: The procedure must be able to quantify specified and unspecified impurities (degradation products) in the presence of API (active pharmaceutical ingredient), excipients and other potential impurities and degradation products over a range of 0.05–0.5% relative to the drug substance. The accuracy and precision of the method must be such that measurements fall within the range of $100\% \pm 15\%$ for levels $\leq 0.15\%$ and $100 \pm 10\%$ for impurity levels $>0.15\%$ with 90% probability

In this example, the impurities are controlled at 0.2% and have a reporting threshold of 0.05%.

during the method lifecycle with an enhanced method control regime. In the sections below the key components as given in Figure 29.3 are described in more detail, giving the general case for application of these components, and illustrated throughout by an example of an HPLC impurity assay for a drug product tablet formulation.

29.2 ANALYTICAL TARGET PROFILE

Creation of an analytical target profile is the initial step for QbD for analytical methods. The ATP is a description of the key measurement system requirements that must be satisfied by the analytical method and will take into account the nature and purpose of the measurement, whether it is being carried out to give process understanding, in-process control or as part of finished API or product release testing. A typical ATP may cite the analytes to be measured, the level at which they are to be measured, the required sensitivity, specificity and/or allowable uncertainty (precision and accuracy) in the measurement. The ATP is not intended to be method or technique dependent in that any analytical method or technique may be deployed if it is demonstrated that the method meets the ATP criteria.

An example ATP for control of impurities in a drug product is given in Table 29.1. At first glance, drafting an ATP may appear to be a simple process. However, creation of a meaningful ATP requires a good understanding of the manufacturing process capability, the effect of the API and/or drug product quality attributes [9] on patient safety and efficacy, and how the generated data will be used, interpreted, and reported. A strong connection between the ATP and the manufacturing processes is important. The ATP should be aligned with critical process parameters and critical quality attributes (definitions provided in Table 29.2) identified during the QbD for product process [10] that are impactful to patient safety and efficacy. A strong connection between the ATP and manufacturing process also enables the design of methods providing the appropriate level of feedback for process development, optimization, and ultimately control. If data will be used to support regulatory filings, appropriate agency and regulatory guidelines should be consulted when developing an ATP. For example, the ICH guidelines for impurity identification and qualification should be consulted when creating an ATP for an impurity method [11,12].

ATP criteria will vary depending on the measurement to be made, the type of sample to be analyzed, and the intended use of the data (whether the measurement/result is required for process understanding, in-process testing or final API or drug product release testing). Several examples are described as follows:

- A limits test for a drug product impurity will have different ATP criteria when compared to a quantitative test for an impurity.
- The requirements for measuring the unwanted enantiomer in an API or drug product could be very different from the requirements for measuring low-level genotoxic impurities where adherence to guidelines and regulatory requirements is critical [13].
- An ATP for water content of drug product tablet blends tested during manufacturing as in-process control testing may be different when compared to water content testing performed for finished goods testing of tablets.

Some considerations when creating an ATP are discussed below:

Identify the Quality Attribute to be Measured: The quality attribute to be measured may be, for example, assay, impurity levels, water content, residual solvents, dissolution, identity, sterility, endotoxins, and particle size. Each quality attribute is likely to result in a unique ATP, although on occasions more that one quality attribute may be combined into a single ATP, such as the quantitation of multiple impurities.

TABLE 29.2 Definitions of Critical Process Parameters and Critical Quality Attributes [14]

Critical Process Parameter: A process parameter whose variability has an impact on a critical quality attribute and therefore should be monitored or controlled to ensure the process produces the desired quality

Critical Quality Attribute: A physical, chemical, biological, or microbiological property or characteristic that should be within an appropriate limit, range, or distribution to ensure the desired product quality

Identify the Levels and Required Range of Values to be Measured by the Method: When determining the targeted levels to be measured, consider the analyte and the levels that might be expected to be present and consider the levels that may affect safety and efficacy, and also regulatory guidelines. The range of values to be covered by the methodology will direct technique selection, method development and eventually method evaluation. Ranges may be quite wide for some applications, such as methods used to gather data to further understanding of manufacturing processes or more narrow for measurements used to support release testing of finished goods.

Identify the Allowable Overall Uncertainty in the Measurement and Appropriate Probability: An approach that captures the acceptable overall uncertainty allowed in the measurement can be used. The overall uncertainty is the difference between the true value and the measured value and contains contributions from both precision and accuracy. This approach is preferred compared to explicitly stating the individual requirements for both accuracy and precision because it allows for some variability in both the accuracy and precision of the analytical method while still ensuring the value provided by the measurement is within an allowable distance from the true value. In the example provided in Table 29.1, the overall allowable uncertainty varies depending on the levels that are measured, with levels near the reporting limit allowing a larger uncertainty ($\pm 15\%$) compared to levels at or near the specification limit ($\pm 10\%$).

Identify the Required Method Specificity: The required specificity should be stated in the ATP. Method specificity and accuracy are linked in that interferences with the analyte of interest can affect measurement uncertainty depending on the magnitude of interference. Using the example of a drug product impurity, the ATP states that the impurity should be accurately quantitated in the presence of API, excipients, and other degradation products and process-related impurities.

Once an ATP is established, any method that meets the requirements may be implemented provided it is demonstrated to meet the requirements of the ATP and the method is developed following a QbD for analytical methods approach. Having an ATP in place should also facilitate changing from one analytical method to another as shown in Figure 29.1. When changing to a new method, method equivalency is demonstrated by showing the new method satisfies the criteria outlined in the ATP. Business requirements such as efficiency, cost, or improvements to process understanding drive the need to change methodology. Therefore, typically only one method would be in use at any time for a repeat situation such as product release testing. Furthermore, when the requirement for trending is important, for example when trending for stability testing during product development, changing to a new method should be approached with due caution. The requirements for establishing method equivalency are currently being debated by the USP [15] and the pharmaceutical industry.

29.3 METHOD DESIGN

Once an ATP has been created, the criteria contained within it will help guide selection of an appropriate analytical technique. As an example, for the ATP cited in Table 29.1, a chromatographic method could be considered since it has the potential to meet the required selectivity, measurement uncertainty, range, and sensitivity. A spectroscopic method, such as NIR, will likely not have the appropriate sensitivity and selectivity.

The overall objectives of the method design phase are to:

- achieve a set of "starting" method conditions for the selected analytical technique through technique selection and initial experimental screening;
- achieve a list of method factors associated with each unit operation of the method that have been thoroughly assessed with respect to "potential risk" to method "failure";
- achieve a thorough understanding of which method factors will be controlled and how they will be controlled; and
- achieve a list of method factors that require further evaluation to assess robustness and ruggedness and ultimately describe the boundaries of the method operable design region.

The process of method design starts with technique selection. Once the appropriate technique has been selected a series of experiments will be performed to identify a suitable starting set of method conditions that can be further assessed using QbD principles. The separate unit operations of the method are then identified, and an exercise is carried out where all method factors associated with each unit operation are identified. The final step of the process involves a risk assessment exercise where each method factor is categorized and prioritized according to potential "risk." The outcome of this exercise is a list of fixed method factors plus an experimental plan derived to evaluate method robustness and ruggedness using the remaining noise factors and nonfixed method parameters. The following discussion outlines each component of method design in more detail.

29.3.1 Technique Selection

This involves selection of an appropriate analytical technique that will be capable of achieving the desired measurement of the material, product, or process attribute defined in the ATP.

A simple prioritization exercise can be performed to build the rationale for choice of technique. For example, in Table 29.3, rationalization of a suitable technique to meet the ATP requirements is presented. The ATP requirements were listed as the primary technique selection driver and any business drivers were listed as secondary factors that could influence the final choice of technique. It can be seen that more than one technique, if rationalized purely from the ATP performance requirements alone, would be "suitable" from a scientific perspective to comply with the requirements defined in the ATP. However, if the business operational considerations are also accounted for, then it soon becomes clear why uHPLC-UV (DAD) would be a sensible choice in terms of analysis time and costs drivers. HPLC-UV (DAD) would also be sensible, but possibly to a lesser extent due to analysis time. Of course, this selection rationale is highly dependent upon the stage of development of the product for which the ATP is designed. This sort of process is useful to document when designing a method as it captures the rationale behind technique selection clearly—which could well be valuable later in the method lifecycle—or when, for example, a new innovative technology appears on the analytical landscape that could be assessed in the same way against the originally defined criteria.

29.3.2 Initial Experimental Screening to Develop "Initial" Method Conditions

Having identified a suitable technique (HPLC-UV) to meet the ATP and business drivers, the method design process continues with initial experimentation to build a set of starting method conditions for more intensive evaluation. If one were to consider the experimentation process like a funnel that at the top is wide and encompasses a variety of method factors (which are ideally considered in a multivariate way so that interactions and interdependencies are understood) then this stage is the very top of the funnel. All experiments and exercises performed herein contribute to the definition of the knowledge space description for the method (Figure 29.2) and of course ultimately to the definition of conditions that will be extensively interrogated in the method evaluation stage of the process.

The timing of this knowledge gathering within the drug development lifecycle is important to ponder further. Regardless of the development stage in which the QbD for analytical methods process might be initiated—any previous knowledge is certainly valuable to capture in this knowledge space description.

There are of course two possible approaches that companies might wish to adopt here with respect to timing. The first approach might be where a company develops method conditions in a "traditional" one factor at a time (OFAT) manner (these experiments would define the "knowledge space" of the method) and retrospectively applies QbD thinking to that knowledge gathered at a later point during development. At the point of application of the QbD approach, the area around the normal operating range of the method would be interrogated so that method operable design region could be mapped. In defining this method operable design region, the company would be afforded the opportunity to optimize the normal operating range further from the critical "edges" of the method operable design region than it may have originally resided.

The second approach involves application of the QbD process at an earlier stage of development. Here, the systematic application of orthogonal platform screening strategies to thoroughly map the knowledge space and define a suitable set of starting method conditions would be generically applied to all compounds in development regardless of their developmental stage. The next step would then be to perform multivariate experiments to map the likely method operable design region and nominate the normal operating range for ongoing developmental support. The employment of a multivariate approach here would be key to achieving a thorough understanding both of the interdependencies of the method factors and the criticality of each method factor to the method success. As the compound progresses in development (the synthetic route of the API is modified, formulations are developed, degradation mechanisms and structures are understood, etc.), the knowledge space may well be redefined or expanded leading to a slightly different method operable design region. The application of scientifically rationalized orthogonal platform screening approaches in combination with multivariate experimentation allows for a consistent response to building such a description. Once the commercial API route and/or commercial product formulation has been nominated, then a more thorough evaluation of the method in concert with the receiving manufacturing laboratories would be pursued so that the robustness and ruggedness of the method is thoroughly understood and that the optimal normal operating range is selected for routine laboratory use (this is the topic of method evaluation—described later).

29.3.3 Risk Assessment

Whatever the route to defining the "knowledge space," be it though a traditional OFAT approach or through orthogonal platform screening approaches and multivariate experimentation, once a "knowledge space" for a method has been established—an initial set of conditions should have been arrived at for successful operation of the method in order to achieve compliance with the ATP. Now, there is a need to thoroughly interrogate these conditions to ensure that they are indeed robust and rugged for testing laboratories to operate on a routine basis. Historically, little was done at this stage to test the performance of the method in a receiving laboratory. Development laboratories would validate conditions according to ICH Q2(R1) [16] and then eventually

TABLE 29.3 Method Performance and Operational Requirements Influencing Analytical Measurement Technique Selection of an Impurity Method for a Tablet Formulation

	1. Method Performance Requirements			2. Business Drivers				
Technique	Specificity	Sensitivity	Capability to Meet Accuracy and Precision Requirements	Capability for At-line or Online Measurement	Technology Available to Customer Lab?	Analysis Time (H, M, L)	Analysis Cost (H, M, L)	Need for Sample Preparation
uHPLC-UV-MS	×	×	×	×		L-M	M	Yes
uHPLC-UV (DAD)	×	×	×	×	×	L	L-M	Yes
HPLC-UV (DAD)	×	×	×	×	×	M	L-M	Yes
HPLC-UV-MS	×	×	×	×		M-H	M	Yes
TLC	×	×			×	M	L	Yes
CE-UV (DAD)	×	×	×			L	L	Yes
UV			×	×	×	L	L	Yes
GC-MS or GC-FID	×		×		×	L-M	L	Yes
NIR				×		L	L	No

complete a formal exercise to transfer the methodology to a receiving laboratory for release or stability testing laboratory within a company's manufacturing unit using comparative testing. Only if issues occurred during the transfer testing would there be indication of future robustness problems. Ownership of the method transferred to the testing laboratory with little knowledge of how the method was developed or understanding of the operating or design space. In QbD for analytical methods this becomes significantly more of a partnership, with receiving analytical laboratories actively contributing to the method design and method operable design region creation.

A structured risk assessment is conducted to identify all potential sources of variation in the practice of the intended method. The goal is to consider potential functional relationships between each method factor and the performance characteristics/criteria defined in the ATP or those that would be specific to the employed technique, and assign a risk ranking to these method factors through a scoring system. Essentially, all of the method factors that could impact method success are discussed and their relative significance is considered. Before a risk assessment exercise is conducted, several preparatory activities should take place. The receiving/testing laboratory should review the normal operating range conditions for the method and ideally run the method to provide initial feedback, but this could extended to completing testing on materials identified as appropriate for purposes such as formulation development or clinical supply release. Although most experienced analysts should have the capability of participating in a risk assessment exercise by reviewing the method, hands-on experience in preparing samples and operating the instrumentation will likely provide additional perspective. The risk assessment should therefore proceed with representation from both the developing analysts and receiving analysts with the requisite method familiarity. The exercise itself is extremely valuable for understanding different strategies adopted by each site—as even the simplest and seemingly most innocuous method factor can be one that could contribute to method failure. These conversations therefore often highlight and circumvent the common "assumptions" made in method transfers between one laboratory and another that contribute to unsuccessful operation of the method in a receiving laboratory. The three components of a risk assessment are defined below.

The risk assessment process proceeds by first defining the unit operations of the method. For an HPLC method to profile impurities in a drug product formulation, the unit operations may look something like the depiction in Figure 29.4, which also breaks out the subunit operations associated with the tablet sample preparation step.

Once the unit operations have been defined, all method factors that can influence the performance of the method can be mapped to each unit operation appropriately. Method factors can fall into multiple categories including those associated with machines, materials, people, processes, measurements, and environmental conditions. Examples of method factors at the bulk sampling stage could include the batch homogeneity, the integrity/identity of the batch, the sample size being taken, the sampling strategy (e.g., size, % of batch, thief), and the "human" contribution to variability (training/skill/experience); examples for a sonication step in standard preparation could be sonication time, bath temperature, and bath fill volume; and for a sample preparation mixing and extraction step, factors could include shaker type, shaker time, vessel orientation, and shaker speed. Examples demonstrating method factors associated with unit operations are presented in Figures 29.5 and 29.6 for sample preparation of a drug product tablet for an impurity assay and the subsequent HPLC impurity assay.

Now that each factor has been mapped to its appropriate unit operation, the risk assessment exercise may continue either through application of appropriate risk assessment tools or using experience (prior knowledge) or a combination of both. There are several approaches that could be employed to carry out this assessment and the most commonly used are failure mode effects analysis (FMEA) [17,18] and cause-and-effects matrix (C&E) [19]. In an FMEA, potential failure modes of the method (e.g., analytical balance out of

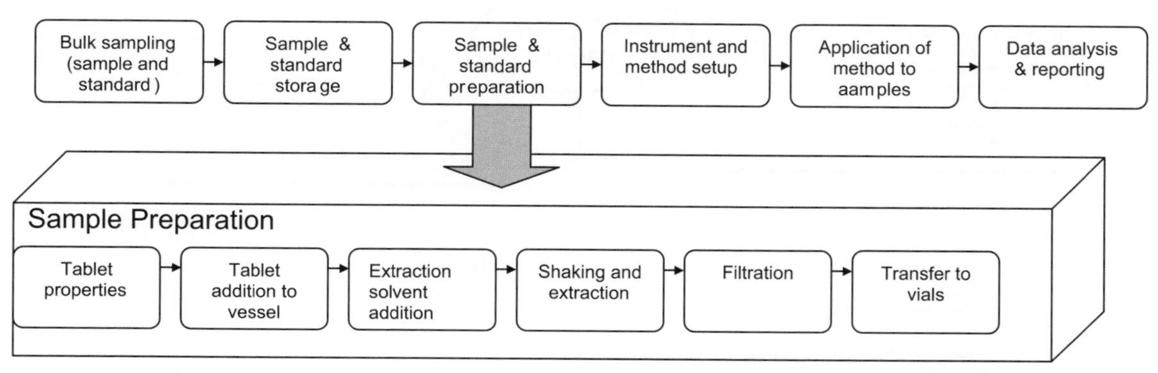

FIGURE 29.4 Unit operations of a tablet sample preparation unit operation within an HPLC impurity method.

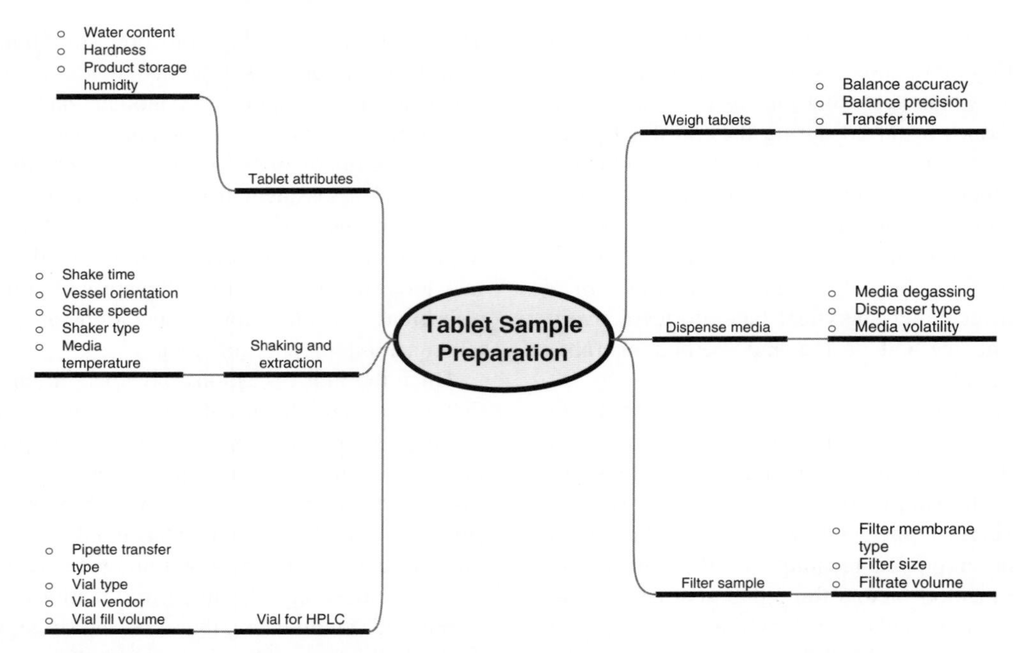

FIGURE 29.5 Mapping example of method factors to the sample preparation unit operation of an HPLC impurity method for a drug product tablet formulation.

calibration) are brainstormed by the group and each of those modes is scored for impact against the performance characteristics/criteria defined in the ATP or those that would be specific to the employed technique. High ranking failure modes are then addressed through experimentation plans or other means (e.g., fixing/controlling factors to specified levels/criteria) that attempt to lessen the risk. In a C&E matrix, the method process is mapped out into individual activities and the factors of each activity; then during the scoring exercise each method factor is evaluated with respect to any attribute that could be considered significant in affecting the performance criteria as defined in the ATP. Note that these attributes may not be specifically listed in the ATP, but do indeed strongly influence the capability of the

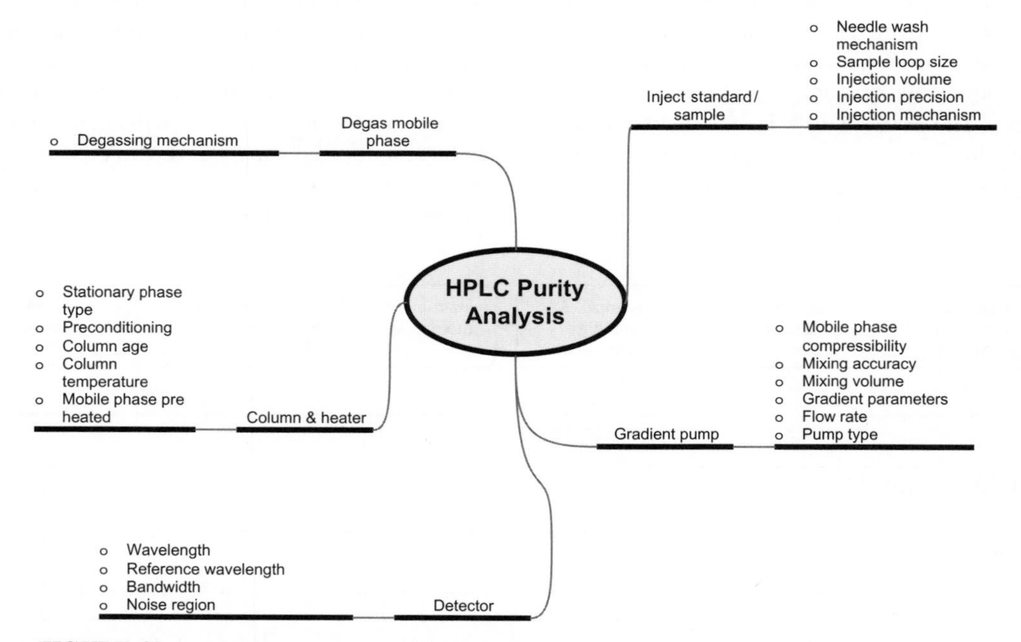

FIGURE 29.6 Mapping example of method factors to the HPLC unit operation of an HPLC impurity method for a drug product tablet formulation.

TABLE 29.4 Definitions of C, N, and X Assignments for Method Factors in Risk Assessment Exercise

C	Analytical Method Factors that form part of the method definition and can be specified at **C**ontrollable unique levels These variables do not require experimentation to "optimize"—they are variables that are fixed and should be clearly stated. During the group work between developing and receiving laboratory analysts—these variables are often the important ones to define carefully as clearly communicating how these variables are fixed often circumvents unsuccessful operation of the methodology in the receiving laboratory
N	Analytical Method **N**oise Factors (factors that cannot be controlled or are allowed to vary randomly from a specified population) and if identified as potentially critical may require ruggedness testing. Examples of these variables in an HPLC method could include the column batch or the instrument make in a FTIR method for identification
X	Analytical Method Factors that form part of the method definition and can be varied continuously and if potentially *critical* may require e**X**perimentation to optimize the method operable design region These variables are important as they will define the starting point for multivariate DoE's in the robustness assessment of the method. Examples of these variables for an HPLC method would include the temperature of the separation or the pH of the mobile phase

method to meet the criteria of the ATP, for example, sample solution stability of an HPLC impurity method. Thereby, a risk factor is assigned to each parameter. As the C&E matrix provides a more thorough understanding of an analytical method, all further references to a risk assessment will reference the application of a C&E matrix.

When a method factor is believed to have a strong relationship to an attribute, it should be scored higher. These exercises tend to bring forth a lot of additional information as any prior experience with similar methodologies or differing laboratory practices can be taken into account to determine whether a method factor should be ranked higher or lower. After each method factor is scored, the influence (score) of a method factor against each attribute is summed. This value represents the risk of a single method factor on impacting the performance of a method where the higher the value, the higher the risk. The team performing the risk assessment has to make a decision on what level of risk is deemed appropriate for designating as "potentially critical" to the success of the method meeting the ATP criteria.

Also during this exercise, as each method factor is being scored, an assignment of method factor "type" is decided. Method factors are designated C, N, or X according to the definitions in Table 29.4

Examples from the outcome of prioritized C&E matrices where the potential critical factors are highlighted and classified as C, N, or X are shown in Figures 29.7 and 29.8.

Factor Parent	Factor Name	Sample Solution Stability	Sample Accuracy	Sample Precision	Final Score	C, N, X	Experimental Strategy
		10	5	5			
Shaking and extraction	Media temperature	9	9	9	180	N	DOE 1
Shaking and extraction	Vessel orientation	9	9	9	180	C	Method controlled
Shaking and extraction	Shake time	5	10	10	150	X	DOE 1
Filter	Filter membrane type	5	10	10	150	X	OFAT 1
Tablet testing	Product storage humidity	9	1	5	120	N	Ruggedness 1
Shaking and extraction	Shake speed	1	10	10	110	X	DOE 1
Vial for HPLC	Vial type	5	5	5	100	N	Ruggedness 1
Filter	Filter size	1	9	9	100	X	OFAT 1
Vial for HPLC	Vial vendor	5	5	5	100	N	Ruggedness 1
Dispense media	Dispenser type	5	5	5	100	N	Ruggedness 1
Filter	Filtrate volume	1	9	9	100	X	OFAT 1
Shaking and extraction	Shaker type	1	5	5	60	C	Method controlled
Tablet testing	Water content	1	5	5	60	N	Ruggedness 1
Vial for HPLC	Vial fill volume	1	5	5	60	N	Ruggedness 1
Tablet Testing	Hardness	1	5	5	60	N	Ruggedness 1
Weigh tablets	Transfer time	1	1	1	20		
Weigh tablets	Balance accuracy	1	1	1	20		
Weigh tablets	Balance precision	1	1	1	20		
Vial for HPLC	Pipette transfer type	1	1	1	20		
Dispense media	Media volatility	1	1	1	20		
Dispense media	Media degassing	1	1	1	20		

FIGURE 29.7 Example from the outcome of a prioritized C&E matrix for the sample preparation unit operation of an HPLC impurity method for a drug product tablet formulation.

Factor Parent	Factor Name	Individual & Total Impurity (area %)	Peak Retention Time	Peak Resolution	Signal/ Noise LOQ	Baseline Quality	Linearity	Peak Area	Peak Plates	System Precision	Peak Tailing	Final Score	C, N, X	Experiment Strategy
		10	10	10	10	10	10	5	5	5	5			
Column & heater	Preconditioning	10	10	10	10	10	5	10	10	5	10	725	N	Ruggedness-1
Column & heater	Column Age - # of injections	10	10	10	10	10	5	10	10	5	10	725	N	Ruggedness-2
Pump (gradient)	Gradient parameters	10	10	10	10	5	5	9	10	5	9	665	X	DOE-1
Column & heater	Stationary phase type	10	10	10	5	10	5	10	10	1	10	655	X	DOE-1
Column & heater	Column temperature	5	10	10	9	5	5	5	10	5	1	545	X	DOE-1
Column & heater	Mobile phase preheated	5	9	9	5	5	5	5	9	5	5	500	N	OFAT-1
Detector	Detector wavelength	10	1	5	10	5	5	10	5	5	1	465	X	DOE-1
Pump (gradient)	Mixing accuracy	5	10	10	5	5	1	1	1	5	1	400	X	DOE-1
Detector	Reference wavelength	9	1	1	9	5	5	9	1	1	5	380	N	OFAT-2
Auto-sampler	HPLC injection precision	5	1	1	10	1	1	10	1	10	5	320	N	DOE-1
Auto-sampler	Injection volume	5	1	1	10	1	1	10	1	1	10	300	X	DOE-1
Auto-Sampler	Injection mechanism (fixed/partial)	5	1	5	5	1	1	5	9	1	5	280	N	OFAT-3
Pump (gradient)	MP compressibility	5	5	5	1	5	1	1	5	1	1	260	N	None
Detector	Detector bandwidth	5	1	1	5	5	1	9	1	1	5	260	N	None
Pump (gradient)	Pump type	5	5	5	1	1	1	1	5	1	1	220	N	None
Detector	Noise region for S/N	1	1	1	10	1	1	1	1	1	1	170	N	None
Pump (gradient)	Mixing volume	1	5	1	1	1	1	1	1	1	1	120	N	None
Auto-sampler	Needle wash mechanism	1	1	1	1	1	1	1	1	1	1	80	N	None
Auto-sampler	Sample loop size	1	1	1	1	1	1	1	1	1	1	80	N	None
Degas	Degasing mechanism	1	1	1	1	1	1	1	1	1	1	80	N	None

FIGURE 29.8 Example from the outcome of a prioritized C&E matrix for the HPLC unit operation of an impurity method for a drug product tablet formulation.

29.4 METHOD EVALUATION

Having established a prioritized ranking of potential influences on method performance, an experimental plan is developed to understand the impact of high-risk method factors that are labeled N or X. A method factor labeled as C might also be at higher risk, but is controlled through execution of the method (possibly understood through previous method development experiments) or by establishing agreed/fixed practices that indicate that the method factor is not allowed to influence the method's performance.

Robustness studies can be designed from method factors labeled X. Often these studies can be completed through a DoE (design of experiments) and input from statistical expertise is particularly helpful at this point. For an HPLC method, an example would be to further refine the method operable design region around resolution of a critical pair of impurities by evaluating chromatographic factors such as temperature, % mobile phase modifier, buffer concentration, gradient conditions, and flow rate. Chromatographic factors in a DoE have been frequently discussed in the literature [20–23], but this approach should be applied to other focus areas of an impurity profile method (factors such as shake time, speed, extraction solvent composition for tablet preparation) or other types of methodology such as dissolution, water content, and spectroscopy. Experimental plans can use an OFAT approach to evaluate a single factor through a univariate study or alternatively explore the interdependence of a number of factors in a multivariate design. Results from these robustness studies could either lead to the control of certain method factors within the methodology or to an understanding of how each can be varied while still meeting the analytical target profile. This is explored a little more in the discussion on method control below.

An example DoE is presented below for a gradient HPLC impurity assay with an inflection point in the gradient profile nearly halfway through the program. The design of this study is consistent with the rankings from the C&E matrix presented in Figure 29.7. Table 29.5 shows the method factors that were investigated, along with their ranges. The study was a fractionated factorial design (2^{6-2}). The advantage of using a fractionated design is that it reduces the number of analyses needed while still incorporating interactions into the results. Additionally, two center points were run.

TABLE 29.5 Method Factors Analyzed in a DoE to Determine Method Operable Design Region for an HPLC Impurity Method of a Drug Product Tablet Formulation

Factor	Center	Low	High
Temperature	30°C	25°C	35°C
Flow rate	1.0 mL/min	0.9 mL/min	1.1 mL/min
Buffer concentration	0.05%	0.025%	0.075%
Gradient time 1	1 min	0 min	3 min
Gradient time 2	18 min	15 min	21 min
Gradient time 3	40 min	37 min	43 min

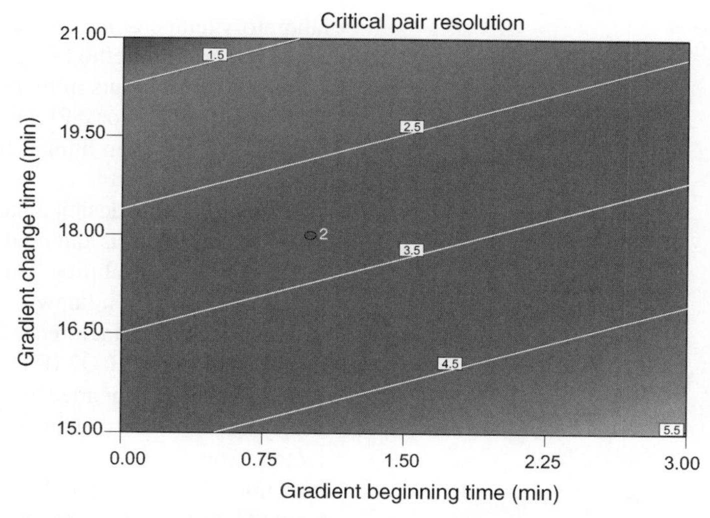

FIGURE 29.9 Influence of gradient parameters on resolution between closest eluting impurities in an HPLC impurity method. All other parameters are held at the center point.

For the above design, attributes evaluated were resolution of critical impurities, limit of quantification, efficiency (theoretical plates), peak symmetry (tailing), main band retention time, and last peak retention time. Based on the results, the attribute that was monitored closest was resolution of two closely eluted peaks. Additionally, main band retention time was looked at in detail because of its relationship to resolution between the main band and several key impurities. Figures 29.9 and 29.10 demonstrate the influence of this design region on the attribute of critical pair resolution. As shown by the area in the statistical contour plot in which the resolution was greater than 1.5, a rather large method operable design region was determined through this study. Although not demonstrated in this design, another factor that should be evaluated to determine the most robust conditions is column type. Finding multiple columns that can offer adequate method performance can afford greater method flexibility as an analytical laboratory may find one vendor preferable over the other for the reasons such as cost, availability, or future changes/discontinuation of the column.

Method robustness for an HPLC impurity assay has been typically described in the literature by understanding the effects of the chromatographic factors. However, understanding of the standard and sample preparation factors can be every bit as critical in the development of the assay. Preparation conditions that either do not extract the API and/or impurities from the sample matrix or do not reproducibly do so can lead to poor analytical results. In

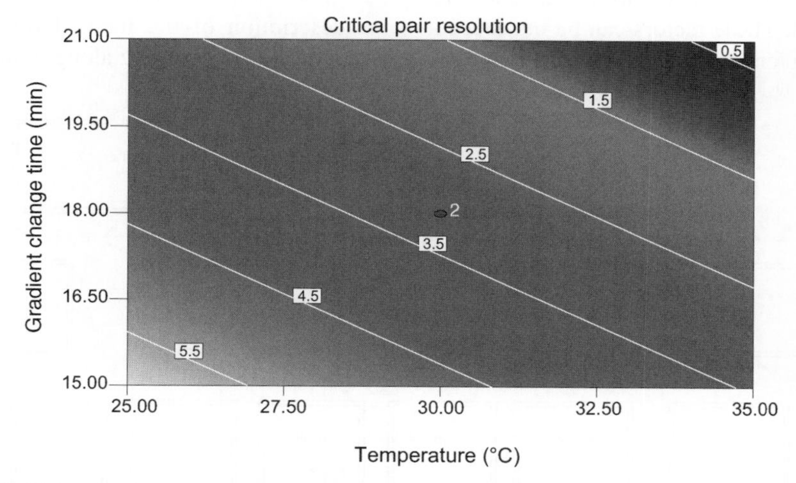

FIGURE 29.10 Influence of gradient parameters and temperature on resolution between closest eluting impurities in an HPLC impurity method. Note that gradient time 1 and all other parameters are held at the center point.

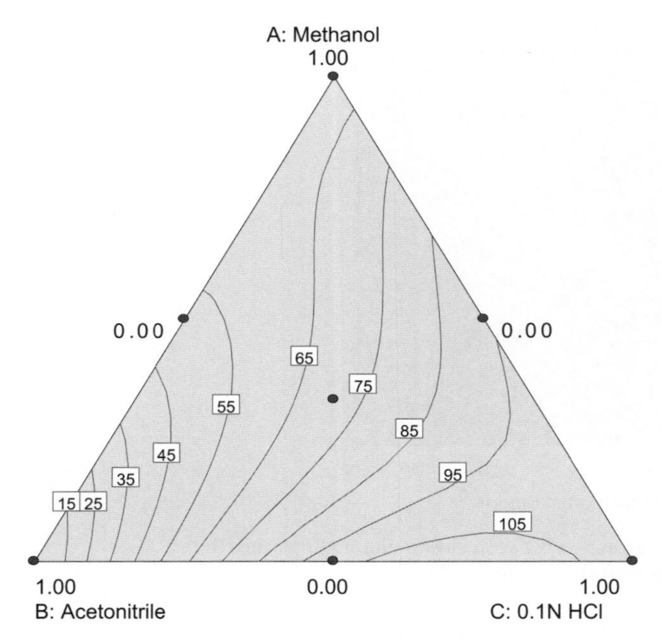

FIGURE 29.11 Contour plot from sample preparation DOE on solvent composition. Design evaluated influence of various solvents on API recovery after 15 min of shaking on a reciprocal shaker. Contours correspond to percent of API recovered from the tablet.

Figure 29.11, a statistical contour plot is provided to demonstrate the most robust solvent composition for sample preparation. For this study, a DoE was run to understand the impact of methanol, acetonitrile, and several aqueous solutions on the extraction of API from an immediate release tablet formulation. Similar studies could also be designed to understand the optimal device for sample extraction (e.g., reciprocating shaker, homogenizer, stir plate), extraction time, and filter type.

Noise factors should be evaluated to understand the ruggedness of the method. These factors can be introduced into the method through environmental conditions (e.g., laboratory temperature, humidity), inputs from the materials described in the method (e.g., filter material variability or reagent source), inputs from the process (e.g., particle size of API on NIR assay for API, tablet hardness on sample tablet preparation), or even human factors like the training/experience of the analyst.

The ruggedness design detailed in Figure 29.12 evaluates noise factors such as the analyst (training/experience) and HPLC column variability. This is done by running an analytical sample with a known value or samples from a well-characterized manufactured lot. This design is very similar to that outlined in ICH Q2 (R1) for an intermediate precision study that is recommended by Japanese regulators. While it is nearly impossible to gain a large understanding of noise factors from such a small study and that this type of study does not afford regulatory flexibility, a variability assessment from the results of this design can point to areas of significant risk. Compiling larger databases to evaluate the impact of noise factors through the lifetime of the method is discussed further in Section 29.5.

Again, statistically designed studies can be completed to understand the impact of the noise factors. To monitor the impact of process parameters on method ruggedness a DOE can be constructed. As part of a NIR tablet assay development and validation, evaluation of process parameters such as tablet hardness and API particle size can be intentionally varied along with percent assay of the API. This type of study demonstrates the variability of assay results when there is noise in process parameters.

As with the experimental factors, OFAT designs can be completed to evaluate noise factors. An example would be a HPLC column lifetime study where multiple injections are made to understand how long a column will last.

The goal at the end of this phase is to have a full understanding of how the method factors influence the method's ability to comply with the ATP. This will include a description of the method operable design region and normal operating range along with a plan for which factors

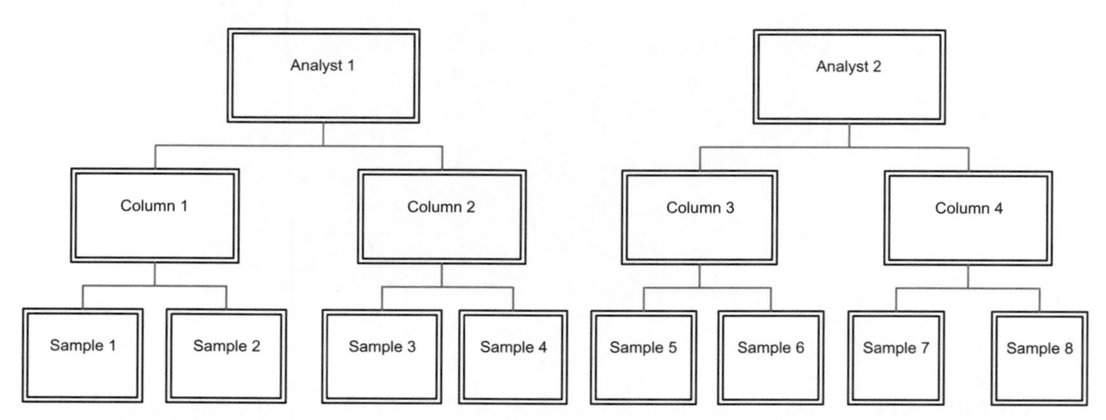

FIGURE 29.12 Ruggedness design to study noise factors in an HPLC impurity method for a drug product tablet formulation.

will be fixed/controlled. Developing sufficient understanding within the design region of the method serves to avoid conventional validation testing as a stand-alone activity. Increased communication, properly designed studies, and critical evaluation of the data set the stage for a relevant and efficient control strategy and continuous improvement.

29.5 METHOD CONTROL AND LIFECYCLE MANAGEMENT

The discussion so far has described how method conditions can be identified such that a good understanding has been developed of the factors that have a significant influence on the performance of the method—this knowledge is the method operable design region of the method. The method operable design region consists of method factors that must be known by the operating analyst in order to obtain accurate results. These factors may be fixed to a single point or two (e.g., light protective glassware used in sample preparation for a photosensitive product) when tightly controlling a factor is critical to meet method performance criteria outlined in the ATP. Alternatively, factors may be described by a range (e.g., column temperature 25–35°C) when it has been demonstrated that the criteria of the ATP are met throughout the range. Sample preparation factors such as extraction time or apparatus can be expressed broadly when demonstrated that multiple techniques can be used to sufficiently recover API from a tablet. An example of a design region for an HPLC impurity method for a drug product tablet formulation is described in Table 29.6. The current expectation for the routine practice of methodologies developed through the QbD for analytical concept is that a target set of conditions will be established within the design region and these will constitute the method that all laboratories utilize for a particular measurement. These target conditions should be seen as equivalent to the normal operating range as defined from QbD for product in Figure 29.1 but are not expected in reality to be practiced as a range. Also, it is not intended that a variety of methodologies or conditions based on individual laboratories or analysts' preferences would be employed for making a given measurement, even if they were all within the design region.

When practicing analytical QbD principles, the current vehicles for demonstrating control of an analytical method and understanding of the influences on its performance—namely method validation and system suitability, may be rendered redundant given the extra information generated in creating the design region. As much of the method operable design region is constructed in a multivariate way, it should be possible to develop significantly more knowledge than is gained from conventional validation experiments for characteristics (or attributes) such as linearity, accuracy, and precision for the entire design region. Additionally, rather than adhering to rigid expectations for these attributes, method performance would be judged directly against the ATP. Other attributes that are currently associated with method validation such as specificity—determined in the chromatographic arena through indicators like resolution, should be less important as the work done in creating the design region will have ultimately determined how varying the chromatographic conditions affect the final result. Specificity should therefore be seen as a way of defining how the chromatography should look when the knowledge of its impact on the final result is not known or well understood. The requirements for these attributes are often set high on the

TABLE 29.6 Example Method Operable Design Region for an HPLC Impurity Method of a Tablet Formulation

Method Factor	Design Region			
HPLC column	Brand X, Y, or Z C18 4.6 × 150 mm, 3 μm particle size			
Mobile phase concentration	Mobile phase A: methanol (15–30%): buffer (70–85%)			
	Mobile phase B: methanol (40–50%): buffer (50–60%)			
	Mobile phase C: methanol (55–65%): buffer (35–45%)			
Gradient parameters	Time (min)	%A	%B	%C
	0	100	0	0
	0–3	100	0	0
	16–20	0	100	0
	37–43	0	0	100
Buffer concentration	0.025–0.075%			
Column temperature	25–35°C			
Flow rate	0.9–1.1 mL/min			
Detector wavelength	250–258 nm			
Sample preparation	Tablets should be prepared at a concentration of 0.1 mg/mL in mobile phase A in light-protective glassware. API should be sufficiently extracted from tablet matrix using a proven process			

assumption that in so doing, the method will be operating well away from any edge of failure without establishing the edges of failure or understanding the impact on the result of allowing a lower requirement. Dictating a high number of theoretical plates, a low tailing factor or a minimum peak width are other chromatographic examples. Current method validation requirements around intermediate precision, ruggedness and robustness again represent relatively crude attempts to assess method performance. Intermediate precision testing represents a very narrow review of the impact of different analysts, equipment and possibly laboratory/site performing the methodology and focuses on establishing that these factors do not impact the results obtained for a particular measurement. This testing is carried out in a very short period, with relatively few measurements being made and compared and is therefore not that relevant to the long-term use of the method. Likewise, robustness testing, as described in current validation guidelines, focuses on establishing that minor deliberate perturbations to the target conditions do not have any effect on the validation rather than understanding the impact on the true result of the measurement.

It is recognized that extracting this information from the DoE/method evaluation exercises and presenting it in a way that would give regulators the same level of confidence that they get from conventional validation reports is work that still needs to be progressed before full adoption of QbD for analytical would be acceptable in all areas. One approach, and that used within the practice of QbD for product, would be to validate the target conditions to be run routinely in the conventional manner (e.g., ICH Q2(R)) and provide analysis results from method evaluation studies that clearly demonstrate the influence of method factor ranges on performance. Any desired change to the target conditions is then subsequently assessed for the potential impact on the validation "status" of the method. Proposed changes might be rationalized as to having little or no impact on the validation status given the knowledge associated with the design region, alternatively some revalidation may be considered appropriate. Changes outside the design region would require more work to justify.

In addition to the current practice of carrying out validation to establish limited knowledge of method performance, it is expected that system suitability requirements are established prior to running a method. System suitability is a further blunt tool that, in the absence of knowledge, seeks to ensure that a method is performing within tight constraints. These tests often focus entirely on the instrumental elements of a method and are in many cases derived from an era when the engineering and instrumentation employed in making analytical measurements were highly variable, with a large potential impact on the result. As instrumentation is much more precisely engineered today, the checking of system suitability attributes such as injection precision for a method employed for impurities (especially with wide requirements for measurement uncertainty) often yields no useful information while there are many parameters associated with determining an accurate result that go unchecked—the extraction procedure being the most obvious. Depending on the design region achieved for a method it may be that some system suitability should be included—particularly if the design region is small and the method conditions are close to the edge—a chiral separation may be an example where in many instances the separation of an enantiomeric pair is achieved only with a specific column chemistry and mobile phase composition with relatively little toleration for variation or the need for injection precision for an assay method with narrow requirements for measurement uncertainty.

Lifecycle management can be thought of in two terms; first is the lifecycle management of the method operable design region for a specific technique. The more progressive definition is lifecycle management of the ATP such that changes from one analytical technique to another may be applied. The following discussion addresses the former term for lifecycle management. Understanding the uncertainty in measurement systems throughout the lifetime of a method is critical for determining the variability of the manufacturing process from batch to batch, and to know when a change is needed to the measurement system. During the lifetime of the methodology, numerous issues could influence the desire or need to make a change within the method operable design region. These issues may be attributed to manufacturing changes or to changes within the analytical method factors. Performance of the methodology could be reviewed through collection and monitoring of a variety of performance indicators such as some of the critical method factors required for system suitability. Data acquisition systems could be set to do this automatically and flag either significant change or a trend away from the initial value. The change could then be assessed against the knowledge within the design region to determine what, if anything should be done. Another approach to monitor long-term performance is that each time the method is run, a reference sample for which the true result is known is run alongside the samples being analyzed by the methodology and the results for the reference material are compared to the true value for that sample. This could be useful in cases where the method operable design region is narrow or sensitive to method factors. Alternatively, a less proactive process could be that the method performance is reviewed only when failures are encountered. The reasons for failure, either out of specification measurements or failure to meet some other requirement that would drive a laboratory investigation would be reviewed in the context of design space. Corrective action could be taken to alter the method conditions within the design region or create more knowledge to expand the design region. Catastrophic failure of the method brought about by changes to components like the column or reagents (change in characteristics or cessation of supply) would also be dealt with in this way. Further, when the need to change a method is identified, all experiences associated with running the method would be collated and considered to determine a new set of target conditions to be adopted by all analysts making that measurement. The flexibility to make changes would also allow improvements to the overall efficiency of making the measurement or improve the actual quality of the measurement. For example, a simpler impurity profile (impurities no longer observed and therefore do not need to be tracked) brought about by changes to the process or with routine manufacturing experience would provide rationale to make modifications to the method (e.g., shorter run times, fewer criteria needed before running the method, etc.). The presence of these impurities will have had an influence on the design of the original method conditions—their absence represents an opportunity to review the method conditions and potentially speed up the run time to achieve the measurement while still meeting the requirements of the ATP. This change would not only improve efficiency but would ultimately lead to reducing the cost of testing, and therefore, the cost of the product. However, in the current environment, the cost of facilitating the change through refiling and approval processes would outweigh the benefits of making the change for all but the highest cost or highest volume products. The following paragraph proposes an option to address this cost-benefit concern.

In the more progressive and optimal definition of lifecycle management, the ATP will be seen as the regulatory commitment for making the measurement and so variations to the conditions within the method operable design region should be allowed without impacting that commitment. Changes that are recognized as being outside the existing design region would be expected to require further experimental work to redefine the design region such that the new conditions were within it. This might be true of major changes to conditions associated with a particular analytical technique, say HPLC—but would certainly be true if there were a desire to switch to a different technique or technology where a whole new method operable design region would need to be created. Changing from one technique to another but one with the same or similar fundamental principles (e.g., chromatography—and a potential change from HPLC to GC, or vice versa) might be seen as a change of lesser magnitude to a proposed change to a technique with fundamentally different principles, such as a change from a chromatographic technique to a spectroscopic technique. It is the long-term vision that this change would still be "allowed" by regulators, provided the requirements of and commitment to the ATP were unchanged. At the current time, the preliminary feedback from regulators who have commented on QbD for analytical methods suggests that it should be relatively straightforward to achieve the desired freedom for changes within the design region and where the full knowledge set has been laid out. However, changes to a technique or method where a new method operable design region was needed, even though the requirements of the ATP were still met, will likely require significantly more dialogue and may be seen as changes requiring prior approval.

29.6 CONCLUSION

Quality by Design for analytical methods is an evolutionary step in pharmaceutical development as analytical scientists are in full partnership with API and drug product process engineers to understand the requirements of a manufacturing process and align them with the requirements of the measurement technique and technologies capable of making those measurements. There are significant benefits to adopting a QbD paradigm for the development of analytical methodology, irrespective of any freedom to change methodology that might be agreed with regulators. The practice will result in more robust methodology as there will be a greater knowledge about the factors that influence its performance and there will be improved clarity of understanding between those developing the methodology and the community of analysts that will ultimately run it routinely. There will be enhanced focus on both generic laboratory practices as well as the specifics of operating a given analytical method. The other, perhaps more significant benefit, which may take longer to realize, is that this concept will ultimately allow companies to make changes to methodology for a variety of reasons without the need for time-consuming and costly refiling activities required to secure prior approval, provided the changes were within the regulatory commitment represented by the ATP.

ACKNOWLEDGMENTS

The authors would like to acknowledge the following colleagues for their thoughtful discussions and contributions to this chapter: Jackson Pellett, James Morgado, Gregory Steeno, Sadia Abid, Zena Smith, and Dawn Hertz.

REFERENCES

1. *International Conference on Harmonisation Meeting*, Brussels, April 1990.
2. Agalloco J. Validation: an unconventional review and reinvention. *PDA J. Pharm. Sci. Technol.* 1995;49(4):175–179.
3. *Innovation and Continuous Improvement in Pharmaceutical Manufacturing*, The PAT Team and Manufacturing Science Working Group Report, Sept. 2004.
4. FDA Guidance *Design Control Guidance for Medical Device Manufacturers,* FDA Center for Devices and Radiological Health, March 11, 1997.

5. ICH Guideline Q8 (R2), *Pharmaceutical Development,* Nov. 2008.
6. ICH Guideline Q9, Quality Risk Management, Nov. 2005.
7. ICH Guideline Q10, Pharmaceutical Quality System, June 2008.
8. Schweitzer M, Pohl M, Hanna-Brown M, Nethercote P, Borman P, Hanson G, Smith K, Larew J. *Implications and opportunities of applying QbD principles to analytical measurements*, Position Paper: QbD Analytics, Pharmaceutical Technology, February 2010, 52–59.
9. Schweitzer M, Pohl M, Hanna-Brown M, Nethercote P, Borman P, Hanson G, Smith K, Larew J. *Implications and opportunities of applying QbD principles to analytical measurements*, Position Paper: QbD Analytics, Pharmaceutical Technology, February 2010, 52–59. The definition of a quality attribute is a physical, chemical or microbiological property or characteristic that directly or indirectly relates to pre-defined product quality (e.g., potency, purity, identity, dissolution, etc.).
10. am Ende D, Bronk-Karen S, Mustakis J, O-Connor G, Santa-Maria CL, Nosal R, Watson-Timothy JN. *J. Pharmaceut. Innov.* 2007;2(3–4):71–86.
11. ICH Guideline Q3A (R2), Impurities in New Drug Substances, Oct. 2006.
12. ICH Guideline Q3B (R2), Impurities in New Drug Products, June 2006.
13. CHMP Guideline, Guideline on the Limits of Genotoxic Impurities, June 28, 2006.
14. ICH Guideline Q8(R2), *Pharmaceutical Development,* Nov. 2008.
15. Hauck WW, DeStefano AJ, Cecil TL, Abernethy DR, Koch WF, Williams RL. *Pharmacopeial Forum* 2009;35(3):772.
16. ICH Guideline Q2 (R1), *Validation of Analytical Procedures: Text and Methodology,* Nov. 2006.
17. *Procedures for Performing a Failure Mode, Effects, and Criticality Analysis,* MIL-STD-1629A, United States of America Department of Defense Military Standard, November 24, 1980.
18. Chrysler LLC, Ford Motor Co. and General Motors, *Potential Failure Mode and Effects Analysis (FMEA) Reference Manual*, fourth edition, AIAG, 2008.
19. Ishikawa K. *What is Total Quality Control?: The Japanese Way*, Prentice-Hall, Englewood Cliffs, NJ, 1985.
20. Srinubabu G, Raju ChAI, Sarath N, Kiran Kumar P, Seshagiri Rao JVLN. *Talanta*, 2007;71:1424–1429.
21. Müller A, Flottmann D, Schulz W, Seitz W, Weber WH. *Anal. Bioanal. Chem.* 2008;390:1317–1326.
22. Ye C, Liu J, Ren F, Okafo N. *J. Pharmaceut. Biomed. Anal.* 2001;23:581–589.
23. Li W, Rasmussen HT. *J. Chromatogr. A* 2003;1016:165–180.

30

ANALYTICAL CHEMISTRY FOR API PROCESS ENGINEERING

Matthew L. Jorgensen

Engineering Technologies, Chemical Research & Development, Pfizer Inc., Groton, CT, USA

30.1 INTRODUCTION

The role of the analytical chemist in API process development is critically important in the pharmaceutical industry. The analysis and the analytical data they provide are the "eyes" on the process. Without accurate analytical results, the process would be running blind. Often the process engineer and chemist know what to expect. But without reliable analytical data, it is impossible to know if the processes have quantitatively met expectations.

The level of importance placed on the analytical data highlights how critical it is that the data be sound and truly representative of the process.

Occasionally the analytical results may be confounded with unquantified or unseparated components or simply may be nonrepresentative due to oversight on the part of the chemist, engineer, analyst, or a combination of the three. This breakdown in the quality of the analytical results is traced back to a breakdown in the communication between the parties involved. Information that one or all parties are unaware of can directly impact the quality of analytical results. The entire process team needs to be cognizant of information such as the stability of reaction components, composition of samples (in addition to starting materials, intermediates, and products), and what level of precision is required of the results.

This chapter will deal directly with what a process engineer should know about the analytical data. This includes information around what is required to insure that the data that are produced, be it by an analyst or engineer, is of the highest quality needed for a particular study. Details around what each analytical technique is tracking and what are its limitations, common mistakes that may confound analytical results, and coupling analytical methods to overcome these limitations will all be covered in this chapter. Finally, it will be shown through examples how this level of understanding of the analytical techniques can be leveraged by the engineer to solve the problems of mass balance and estimate kinetic parameters.

High-quality analytical data are paramount if one wishes to accurately know how a process is truly performing. In most cases, certain assumptions are made during the application of the analytical data and understanding the validity of the assumptions is important. Information in this chapter will help the engineer be aware of these typical assumptions and their applicability.

30.2 USE OF ANALYTICAL METHODS APPLIED TO ENGINEERING

Occasionally the analyst and the engineer can feel that the other is speaking different languages. For example, the terms potency and purity are commonly used and can be a source of confusion without clarification around what these numbers mean and how their values were arrived at. Both potency and purity refer to a measure of the active or desired ingredient relative to the sample. The details of how purity and potency are actually determined are important to understand and are the subject of the next section.

Chemical Engineering in the Pharmaceutical Industry: R&D to Manufacturing Edited by David J. am Ende

30.2.1 Purity

A strict definition of percent purity would require qualifying what the purity basis is, that is, purity percent by weight or purity percent by HPLC (high pressure/performance liquid chromatography) area at 254 nm. Often in the pharmaceutical industry purity percent by HPLC area is shortened to just purity and when the more rigorous definition is applied, purity percent is stated as purity by wt%.

The term purity typically is based on area percent values alone.

$$\text{Purity} = \left[\frac{\text{Active}_{\text{area}}}{\text{Active}_{\text{area}} + \text{Other}_{\text{area}}}\right] \times 100\% \qquad (30.1)$$

where "$\text{Other}_{\text{area}}$" refers to the peak areas of all the other peaks in the chromatogram. Thus, any impurity is assumed to have the same response factor as that of the main component. The reason area percent purity is reported is one of timing. In early development of a new chemical entity, there are usually no standards. As area percent purity is something that can be reported from the first injection, a meaningful metric can be generated without a lot of work to develop standards. The area percent purity values can be used to compare the historical samples with each other to compare different chemical approaches to the project. Later on in the project when standards have been made and characterized, percent purity by mass values can also be reported. This percent purity by mass relative to the standard (also referred to as potency) taken with the percent purity by area value is a good indicator of how well the standards are characterized. For the remainder of this chapter, the percent purity by area will be referred to as just purity.

30.2.2 Potency

$$\text{Potency of A}^{\text{s}} = \left[\frac{\text{Mass}_{\text{A}}^{\text{S}}}{\text{Mass}_{\text{total}}^{\text{S}}}\right] \times 100\% \qquad (30.2)$$

where

$$\text{Mass}_{\text{A}}^{\text{S}} = \text{Area}_{\text{A}}^{\text{S}} R_{\text{f}_{\text{A}}} \qquad (30.3)$$

and

$$R_{\text{f}_{\text{A}}} = \frac{\text{Mass}_{\text{A}}^{\text{STD}}}{\text{Area}_{\text{A}}^{\text{STD}}} \qquad (30.4)$$

Here, S denotes sample, STD denotes standard, and A denotes material A.

The term potency in equation 30.2 is a bit deceiving at first glance as it appears that when samples are reported at a given percent potency, this is a percent by mass intrinsic to the test sample. This is not the case. Actually, this value is a percent active compared to an external standard as shown in equation 30.3 through the use of a proportionality constant called a response factor designated by R_{f}. This response factor is generated from a reference standard as shown in equation 30.4 by taking the ratio of the response, HPLC area in this case, to the mass of the sample. The reference standard is typically a well-characterized purified sample of the desired material used to calibrate HPLC peak area to mass of the sample. In most cases in the pharmaceutical industry, the reference standard is not commercially available and has to be purified through crystallization or preparative-scale chromatography.

Response factor can be simplified as the ratio of the output response to the input material and is used in most analytical techniques with linear responses such as mid-IR spectroscopy, mass spectroscopy, or in this case UV spectroscopy. After close inspection of equations , it becomes apparent that the accuracy of the potency value hinges on the quality of the standard used to generate the response factor. As all reported values are relative to the standard, it is possible to have a test result indicating a potency greater than 100%, showing the sample is more potent than the standard.

Early in development, this standard may be nothing more than the most pure obtained sample to date. The limited characterization of the standard consists of analysis for residual solvents including water and residue on ignition testing (ash). Anything that is not ash or solvent is then attributed to the material of interest. So to reiterate, potency refers to the % active component in a given sample relative to an external reference standard for that active component.

EXAMPLE 30.1

(A) An isolated sample is submitted for the typical purity and potency analysis. The results reported were the following:

Purity: 99.5%
Potency: 97.3%

What do these results tell us about the sample?

Solution
A typical mass balance for the reference standard is the following:

$$\text{Mass}_{\text{total}}^{\#} = \text{Mass}_{\text{A}}^{\#} + \text{Mass}_{\text{residual solvent}}^{\#} + \text{Mass}_{\text{ASH}}^{\#} + \text{Mass}_{\text{impurities}}^{\#} \qquad (30.5)$$

Where the superscript # can be either STD or S if the mass balance is for the standard or the sample, respectively. The term $\text{Mass}_{\text{A}}^{\#}$ refers to the mass of the desired compound of interest in the standard or the sample. Writing equation 30.5 in terms of the sample and solving for $\text{Mass}_{\text{A}}^{\#}$ results in equation 30.6.

$$\text{Mass}_A^S = \text{Mass}_{\text{total}}^S - \text{Mass}_{\text{residual solvent}}^S - \text{Mass}_{\text{ASH}}^S - \text{Mass}_{\text{impurities}}^S \quad (30.6)$$

Substituting equation 30.6 into equation 30.2 results in the following equation:

$$\text{Potency of}A^S = \left[\frac{\text{Mass}_{\text{total}}^S - \text{Mass}_{\text{residual solvent}}^S - \text{Mass}_{\text{ASH}}^S - \text{Mass}_{\text{impurities}}^S}{\text{Mass}_{\text{total}}^S}\right] \times 100\% = 97.3\% \quad (30.7)$$

From equation 30.7, it become apparent that the terms for residual solvent, ash, and impurities are why the potency is less than 100%. At first glance, it would be easy to assume that because the purity value is 99.5%, the $\text{Mass}_{\text{impurities}}^S$ term is low. Remember though that the percent purity is actually percent purity by area percent. If there are any impurities that have a drastically larger response factor than the desired material, then they will be underreported. At best, the purity value of 99.5% infers what are the dominate terms that are reducing the potency of the sample. Submitting the sample for further analysis for residual solvents and ash is the only way to identify if the missing 2.7% is due to residual solvents and ash or underreported impurities.

(B) Consider the example where the magnitudes of purity and potency are reversed; that is, the sample has a higher potency than purity:

Purity: 95.3%

Potency: 103.2%

What does this tell us about the product and, more importantly, about the standard?

Solution

The potency value of 103.2% means that the sample is more potent than the standard. Because potency is a relative activity of a sample compared to the standard, it is possible to have values greater than 100%. What this indicates is that the mass balance around the standard is not fully closed. From equation 30.5, the mass of the active material in the standard is determined by difference, so

$$\text{Mass}_A^{STD} = \text{Mass}_{\text{total}}^{STD} - \text{Mass}_{\text{residual solvent}}^{STD} - \text{Mass}_{\text{ASH}}^{STD} - \text{Mass}_{\text{impurities}}^{STD} \quad (30.8)$$

Substituting equations 30.3, 30.4 and 30.8 into equation 30.2 results in the following equation:

If the areas and total mass of both the sample and the standard are accurate, then from equation 30.8 to have potency greater than 100% the only way is that the characterization of the reference standard around ash, solvent, or impurities is off. The source of the failure to close the mass balance of the standard is most likely due to the impurities not being fully characterized, as residual solvent and ash are standard analysis. If the purity of the standard (UV area percent) was near 100%, then there may be impurities that are not showing up at the wavelength that the detector is set at, or they may not be UV active. In such a case, further purification of the standard by chromatography or recrystallization is needed to better close the mass balance of the standard and gain an accurate R_f.

To further complicate the issue, a sample purity value of 95.3% indicates that the R_f is too high due to a poorly characterized standard, and the samples' total impurities of 4.7% indicate that some of these impurities have higher molar absorption coefficient relative to the desired and thus will appear to be present in higher concentration. The assumption with area percent values is that everything has the same R_f as the main peak. If any of the impurities have a lower R_f than the main peak, then the impurities will be overreported by the area percent value. The various scenarios discussed above have been summarized in Table 30.1:

For comparing processes with each other based solely on isolated yield and relative potency, a less than fully characterized standard still allows relative comparison; that is, 103% potent material is better than 95% potent material. For work that would require a more stringent mass balance, kinetics, or process understanding, the mass balance should be closed by utilizing a combination of complementary analytical techniques such as quantitative H^1 NMR and HPLC. Two areas where analytical data are most frequently needed by the API process development engineer are data to close the mass balance and data to develop kinetic models. These two utilizations of the data are not independent of each other, as it is necessary to have a reasonable mass balance before attempting to develop a kinetic model. As such, it is imperative to have analytical techniques available that can both "see" what needs to be tracked and give values of concentrations that are needed for both the mass balance and the kinetic model.

30.3 METHODS USED AND BACKGROUND

What follows is a brief overview of the most common analytical techniques used and some concepts that need to

$$\text{Potency of } A^S = \left[\text{Area}_A^S \frac{\text{Mass}_{\text{total}}^{STD} - \text{Mass}_{\text{residual solvent}}^{STD} - \text{Mass}_{\text{ASH}}^{STD} - \text{Mass}_{\text{impurities}}^{STD}}{\text{Area}_A^{STD}} / \text{Mass}_{\text{total}}^S\right] = 103.2\% \quad (30.9)$$

TABLE 30.1 Possible Scenarios of Purity and Potency Values

	Purity: $\text{purity(area\%)} = \frac{(A_{\text{area}})}{(\sum \text{UVactive}_{\text{area}})} \times 100$ Potency: $\text{potency(wt\%)} = \frac{(A_{\text{area}})(R_{\text{fA}})}{(\text{Sample}_{\text{mass}})} \times 100$
Purity = potency	Can occur if the response factors of all the UV active components including impurities are very similar so that an area percent of A is equivalent to a wt%. It also requires that the reference standard for the potency determination be highly accurate
Purity < potency	If we assume the reference standard is accurate, then this situation can arise if the impurity peaks have a higher extinction coefficient and higher absorbance than the desired component A. This translates to artificially high impurity count and lower purity by area percent
Purity > potency	If we assume the reference standard is accurate, then purity will exceed potency when there are non-UV components present that are not being detected by HPLC. This will contribute to lower potency values and higher HPLC area percent—for example, if the sample has high salt content (ash). This will look pure by HPLC because the ash is not detected
Potency > 100%	Reference standard likely not well characterized with respect to wt% ash, residual solvent, or impurities

be kept in mind when attempting to analysis the data generated. A more through discussion about each technique can be found elsewhere [1].

All methods of column chromatography rely on the same basic principles. First, there is a sample that is made up of a mixture of components. This mixture is loaded onto a column that separates the individual components as they partition between two phases, the mobile and stationary phases. In liquid chromatography (LC), the partitioning is driven by the polarity of the components and the differing polarity of the mobile phase versus the stationary phase, absorbing and deabsorbing onto the stationary phase down the length of the column. In gas chromatography, the partitioning is driven by the relative volatility of the components as it alternates between the gas phase and dissolution into the stationary phase. The net effect of any chromatographic system is to separate the components of the sample mixture. It is the detector attached to the outlet of the chromatographic system that allows one to see the relative concentrations of each species in the sample. As such, the type of detector used will dictate what is "seen" by the analytical method. Table 30.2 lists the types of detectors available, what type of chromatographic system they are most often paired with, and what they are capable of detecting.

The underling similarity in all these methods of detection excluding FID is that the resulting signal is proportional to the concentration. The important thing to remember is that for every component of a sample that is being analyzed, there is a proportionality constant that is unique to that compound. So in the example of the UV detector, the most common detector for different LC methods, this proportionality constant is the molar absorptivity ε. The relationship of ε to concentration and absorption is described by Beer–Lambert law as shown in the following equation:

$$A = \varepsilon b c \tag{30.10}$$

where A is the absorption (dimensionless), ε is the molar absorptivity (L/(mol cm)), c is the concentration (mol/L), and b is the detector path length (cm).

In the case of mass spectrum detectors (MS), the proportionality constant is the ionization potential, and in electrochemical detection, it is the redox potential. Even in the case of nonchromatographic methods, the idea of proportionality constants should always be remembered. As an example, quantitative NMR has relaxation times that can be thought of as proportionality factors. So for every sample analyzed, be it with chromatography or not, the individual components will have a unique proportionality constant that may or may not be similar to other components in that sample. This is why most detectors are not universal detectors; all species that are chemically different will have different proportionality constants. In many instances, if the components are all structurally similar, then their proportionality constants may be very similar as well, but this is not always the case. This is why taking area percent values as direct replacements for concentration can lead to erroneous results. At best, these area percent values can be used to indicate relative abundances, but care around the possibility of different response factors must be taken if area percent values are used as replacements for concentration values for calculating mass balances or kinetic profiles. The following example illustrates this point.

EXAMPLE 30.2 RESPONSE FACTORS VERSUS AREA PERCENT

A high-temperature coupling reaction was evaluated with potassium hydroxide in a high boiling solvent. Initial reaction completion HPLC looked promising with apparent conversion of >80% although long reaction times were required (Figure 30.1). However, the isolated yields were

TABLE 30.2 Properties of Most Common Detector Types [2]

	Chromatography Method	Detection	Sensitivity	Notes
Ultraviolet (UV)	LC	Absorption of UV light by pi–pi bonds, that is, conjugation	Dependent on ε of the analyte	Variation of ε on the order of 100- to 1000-fold is possible
Mass spectrometer (MS)	LC/GC	Charged particles	μM concentrations	Does not detect mass but rather mass to charge (*m/z*) ratio
Flame ionization detector (FID)	GC	Ionized particles from combustion of organic compounds	ppm–ppb	Signal proportional to the number of carbon atoms, that is, signal proportional to mass not concentration
Conductivity	LC	Charged ions by measuring resistance in detection cell	5×10^{-9} g/mL	Most often used for ion exchange chromatography
Electrochemical	LC	Current generated by oxidation or reduction of sample	Order of magnitude more sensitive than UV	More selective and sensitive than UV, but detector not as rugged as UV
Refractive index (RI)	LC	Variations in refractive index	0.1×10^{-7} g/mL	Universal detector, poor detection limit, and sensitivity to external condition (temperature, dissolved gas, etc.) limit practical use
Evaporative light scattering detector (ELSD)	LC	Nonvolatile particles of analyte scattering light	0.1×10^{-7} g/mL	Analyte needs to be nonvolatile whereas mobile phase needs to be volatile
Florescence	LC		Low ng/mL range	Typically require derivatization with fluorophore reagents

low (<50%), but this was attributed to a laborious workup. The workup involved two extractions followed by distillation and then crystallization. An extensive amount of time was spent trying to optimize the reaction with an eye toward fixing the workup and increasing isolated yield after the reaction was optimized.

The "conversion" was calculated in the lab as:

$$\text{Conversion} = \frac{\text{Area\% product}}{(\text{area\% starting material} + \text{area\% product})} \tag{30.11}$$

There was some concern around this approach, but the argument against pulling samples for quantitative HPLC was that the reaction was very thick and heterogeneous, making it hard to sample representatively. Provide a solution to the approach.

Solution

To get accurate quantitative HPLC data and potency values, the sampling limitation was avoided by not sampling. A reaction was run and the quantitative HPLC sample made by using the entire reaction in a volumetric flask. By doing this, there would be no sampling error as the entire reaction would be used.

In this instance, the quantitative HPLC conversion was calculated as

$$\text{Conversion} = \frac{\text{moles product}}{(\text{moles starting material at } T_0)} \tag{30.12}$$

where moles of product are calculated as

$$\text{moles product} = \text{Area}_{\text{product}} \times R_{\text{f}}^{\text{mole}}{}_{\text{product}} \tag{30.13}$$

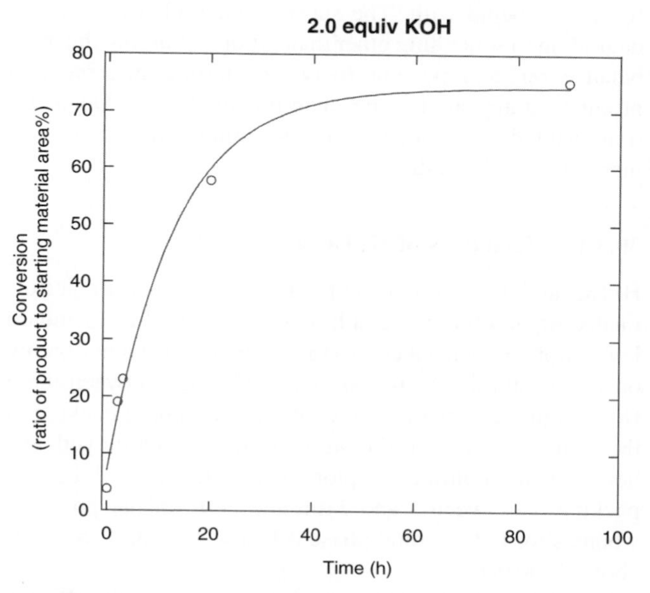

FIGURE 30.1 Conversion as calculated by the chemist in the lab by equation 30.11 only taking the ratio of starting material and product area percents into account.

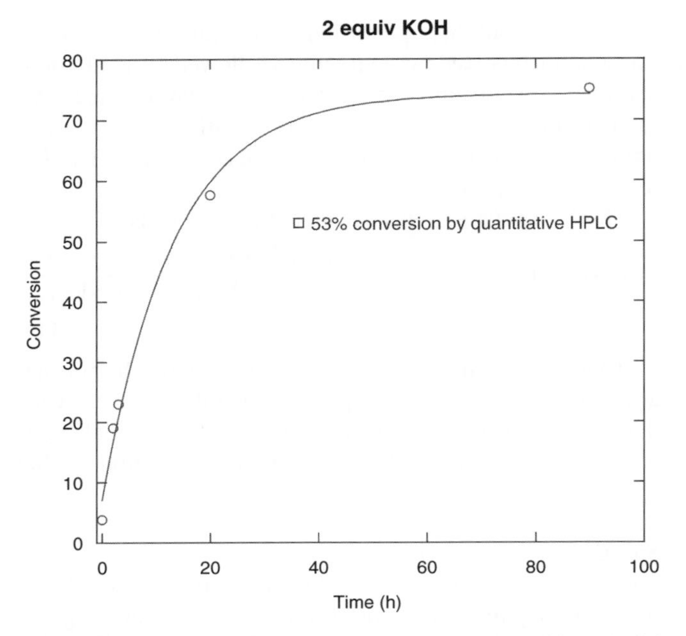

FIGURE 30.2 Conversion calculated by equation 30.11 (solid line) compared with conversion calculated by equation 30.12 (empty box). This discrepancy between the two values at the 48 h time point indicates that the forced mass balance of equation 30.11 elevated the product concentration by not taking into account a possible side reaction of the starting material that did not result in product.

R_f^{mole} is calculated from equation 30.4 on a mole basis.

Quantitative HPLC showed there to be much less product after 45 h than originally assumed (Figure 30.2). Low yield was not due to product loss in workup, but rather was never formed to begin with. The starting material was reacting/degrading to something other than product. Forcing the mass balance on area percent (between starting material and product), it appeared higher than it actually was. Quantization resulted in the decision to discontinue further development on these conditions.

30.3.1 Mechanics of HPLC and UPLC

HPLC and the more recent UPLC (ultrapressure/performance liquid chromatography) are considered the standard lab equipment when it comes to understanding what is going on in a synthesis or process. The difference between these two techniques lies in the size of the solid phase packing in the columns as well as the pressures that are employed, and hence the high/ultra descriptors. For HPLC, the solid-phase packing is between 5 and 3 μm and 200–400 bar pressure, whereas for UPLC, solid phase is below 3 μm and pressures above 1000 bar.

A quick aside about the equation that governs the efficacy of both techniques, as well as any other column chromatography, the van Deemter equation in its simplified version (equation 30.14) [3].

$$H = A + \frac{B}{u} + Cu \tag{30.14}$$

where H (sometimes shown as HETP) is the variance per unit length, also referred to as height equivalent to a theoretical plate; u is the volumetric flow rate; A is the term describing the multipaths in the packed bed; B is the term describing longitudinal diffusion; and C is the term describing resistance to mass transfer.

This hyperbolic function relates the variance per unit length to particle size, mass transfer between the stationary and mobile phases, and the linear velocity of the mobile phase. This relationship was the first result from applying rate theory to the chromatography process and was originally developed to describe gas chromatography. It has been extended to describe liquid chromatography as well with modifications to the lumped parameters terms A, B, and C in equation 30.14. A typical van Deemter plot is shown in Figure 30.3.

The van Deemter equation is useful in describing the theory and mechanism of the chromatography process, not only for the small analytical chromatography used in analysis, but also for large-scale separations done on large pilot plant and commercial scale. This equation explains why the problem of unresolved peaks cannot be solved by just going to a longer column at the same flow rate. The increased resolving power of more packing in a longer column is lost to the increase of the B term in equation 30.14 (increase of eddy and longitudinal diffusion) due to increased time spent on the column. Thus, the number of theoretical plates is less for the longer column (when held at the same flow rate) even though it is longer with more packing because the height of the plates, the H term in equation 30.14, is larger as well. This is where UPLC comes into its own. By decreasing the packing size and increasing the pressure, the linear velocity is kept high and the increased resolving power of more packing is

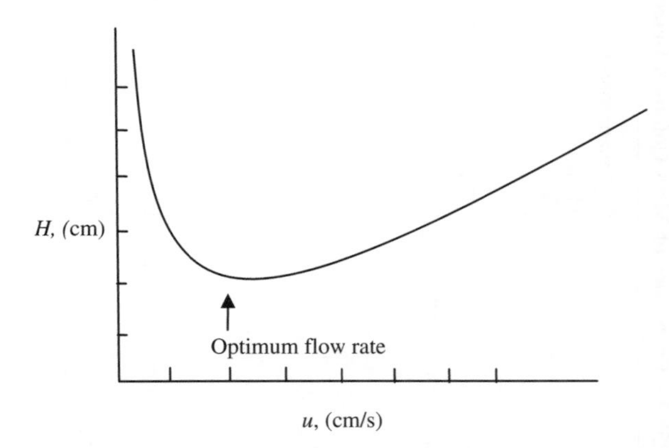

FIGURE 30.3 Characteristic van Deemter plot shape illustrating the presence of an optimum flow rate to maximize column efficiency.

not lost to increased diffusion, thus giving higher plate counts for a given time on the column. The net result is increased resolution and throughput for the analyses.

Both methods, HPLC and UPLC, are only a tool for separating individual components from a mixture and feeding them to a detector will give a response that is proportional to concentration. It is concentration data that are typically the most applicable to the engineer and their accuracy is of foremost importance.

30.4 THINGS TO WATCH OUT FOR IN LC AND GC

30.4.1 Injections have Everything in Them, Not Just the Desired Reactants

The first thing that must be communicated to the analyst, or kept in mind for those who are acquiring their own data, is to account for what is in the reaction mixture. The most common mistake in LC that everyone makes once and hopefully only once is the toluene mistake. This is what happens when people forget that toluene, unlike most organic solvents, has a chromophore and is retained on most LC columns. I cannot tell you how many bright analytical chemists have come running down in a panic telling everyone that there is this major new impurity only to find out that the project has switched to toluene in the process and had not notified the analyst. Worse yet are the chemists who report that they have excellent *in situ* yield only to be looking at a nonexistent reaction because they assumed that large peak that was not starting material was product when in reality it was the toluene peak. This can lead to wasted development time chasing a nonexistent reaction.

In GC, the major concern is nonvolatiles, that is, salts. If a lot of reaction mixtures are injected that contain a large percentage of salts, then the injector may become plugged, necessitating the cleaning of the injector before accurate analysis can resume. What is more complicated is when the product or reactants are salts, that is, charged species. These will not "fly" on the GC and will require some sort of quench to run on the GC. Most often this is a neutralization of the reaction mixture to quench the charge on the desired compounds so as to facilitate GC analysis.

30.4.2 Unplanned/Planned Modification of Stationary Phase

If running one's own analysis, one must be cognizant of possible changes to the HPLC column due to history. Depending on the nature of the mobile phase being used, "conditioning" of the column may take place such that the results may not be repeatable or representative of differing HPLC systems. This opens the possibility of analysis that cannot be duplicated, leading to confusion around what results are accurate. A prime example of conditioning of the column is ion-pairing mobile phase, such as sodium dodecyl sulfate, or a weak ion-paring agent such as perchloric acid. In the case of ion-pairing mobile phases, the stationary phase is modified or conditioned over time to be more retentive of polar species such as primary amines due to the stationary phase being modified by the mobile phase containing the ion-pairing agents. There is a memory effect now for this column that will still maintain the effect even if the mobile phase is switched to a more traditional acidic mobile phase. This has the greatest impact when someone develops a method with such a conditioned column as it will be impossible to replicate these results without this preconditioned column.

The other extreme is when the column is conditioned negatively or destroyed by running samples of reaction mixtures that destroy the resolving power of the stationary phase. This is most often seen with samples from reactions such as hydrogenations that contain metal species that bind to the stationary phase resulting in reduced resolving power. If a column is suspected of being conditioned either negatively or positively, then the only option available is to replace the column and see if the previous analysis is replicated. Thus, it is always good to periodically run a system suitability test to check analytics with a reference mixture to confirm retention times/peak shapes.

30.4.3 Product Stability/Compatibility with Analysis Method

Stability of the reaction mixture or products to the chromatographic conditions is another major concern that needs to be addressed before a strategy for analyses can be agreed upon. The majority of aqueous mobile phases utilized in UPLC and HPLC are acidic. This regularity of acid mobile phases is due to two major factors. The first factor is that until recently the silicon support for the column mobile phase was not stable to high pH values as silicon is soluble at pH levels above 11. The second factor is that if the mobile phase pH is near the pK_a of any of the sample components, then slight variations in the pH of the mobile phase can change the polarity of the components. This change in polarity will then change retention time and order of elution of the components.

Because of these two factors, nearly 80% of the mobile phases are acidic ($pH < 1$) to both maintain the stability of the mobile phase and prevent any change in the analysis due to pH variations. The idea is to protonate everything and prevent pH gradients from forming on the column that may cause chromatographic artifacts.

This proclivity of mobile phases to be acidic makes stability to acid aqueous conditions one of the main stability concerns. As the amount of sample that will be loaded on the column for each injection is so miniscule, on the order of microliters, the sample gets swamped by the mobile phase. If the sample is not stable to aqueous/acid conditions, then degradation will be taking place as the sample travels and elutes from the column. The net effect is that the sample that was representative of the reaction at a given time or point in the process is scrambled by the analysis method, rendering the results no longer representative. This is unfortunate if one lab-scale reaction is lost because of this, devastating if three weeks of DOE experimentation is rendered useless because of this, and both have happened.

In GC, the major issue with stability of the reaction mixture is that of thermal stability. Remembering that the standard injector temperature for GC is 280°C, this is the temperature that the reaction samples have to "endure" just to get on the column.

If stability is a problem in LC or GC, then quenching the reaction samples (if this improves stability) or some sort of derivatization method may be required.

30.4.4 Derivatization

Derivatization is the process by which a reaction sample is further reacted to form a new compound as part of the sample preparation. This may be done for various reasons such as increasing reactant stability to the analysis method, modifying the components of a sample to make them detectable such as attaching a chromophore, or increasing volatility for GC analysis [4]. The major issue with derivatization is that this is a second reaction that is in series with the desired reaction. The net effect of this is that if the reaction of interest is to be accurately characterized, then the derivatization reaction needs to be quantitative in reaction completion and at a reaction rate that is orders of magnitude faster than the desired so as not to skew the analysis. For a system that a kinetic model is being developed and derivatization is required, then a quench of the reaction should be used before derivatization, or a derivatization that also quenches the reaction. This is to insure that the analysis is representative of the time the sample was taken, not the time the sample was analyzed.

30.5 USE OF MULTIPLE ANALYTICAL TECHNIQUES

Oftentimes when trying to understand a process by developing both a mass balance and a kinetic model, it is best to start at the beginning. In most cases, the beginning is a full characterization of the feedstocks going into the process. It will be very difficult to close the mass balance if one is not aware of what is going into the process. The use of two or more complementary analytical techniques can greatly aid in fully understanding the inputs for a process. The following Case Study 1 illustrates this point.

30.5.1 Case Study 1: Mass Balance Around Starting Materials to Develop a Kinetic Model

A process involves coupling secondary aniline with a volatile chiral epoxide utilizing an ytterbium catalyst in isopropyl acetate at 60°C. The secondary aniline was synthesized from the primary aniline as shown in Scheme 30.1:

The secondary aniline was telescoped into the reaction with the chiral epoxide with some residual primary aniline present. The observation was that when the process to coupling the secondary aniline and epoxide was first scaled up in a kilo-scale facility, a second charge of the epoxide reagent was required to drive the reaction to completion. The reaction had been run in a sealed reactor so as to limit losses of the epoxide with its very high vapor pressure. Owing to the sealed reactor configuration, the time for the reaction to complete was believed to be 16 h but had not been confirmed as reaction completion samples were not taken during this initial 16 h over fears of venting the epoxide during sampling. In addition, the incoming secondary aniline reagent was in a solution of isopropyl acetate that was known to have residual primary aniline from the first reaction. The question was what the impact of residual primary aniline was on the desired reaction of the secondary aniline with the epoxide.

It was decided to undertake a kinetic study of this reaction to identify the answers to the following questions:

1. How long does the reaction take?
2. Why did the first scale-up require the second charge of epoxide to drive the reaction to completion?
3. What is the impact of residual primary aniline?

The first step to developing a kinetic model was to fully characterize the two incoming reagent streams. The

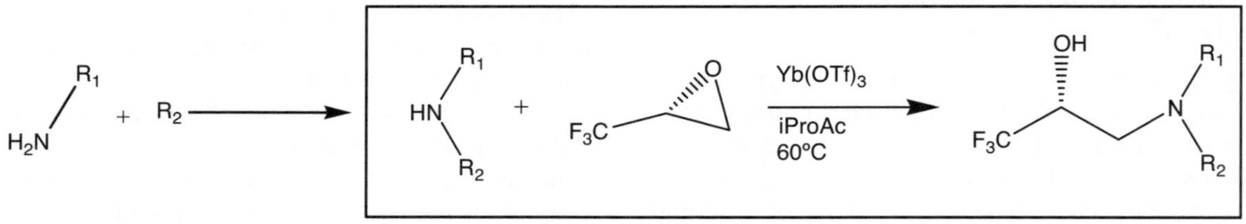

Scheme 30.1 The desired reaction of the secondary aniline (the reaction in the box) that is synthesized from the primary aniline.

aniline reagent stream had an unknown potency as standards were not available for quantitative HPLC. The epoxide reagent had a certificate of analysis (COA) from the vendor, but its potency would be reevaluated to confirm these numbers.

To gain a handle on the composition of the aniline reagent stream, a H^1 NMR was taken that resolved the isopropyl acetate from the aniline compounds. Figure 30.4 shows the H^1 NMR with the peaks assigned to the structure. From this figure, it becomes apparent that there was not enough resolution between the primary and secondary aniline compounds with NMR to decouple their individual concentrations. Using the HPLC area percent, the concentration of the different anilines was calculated.

The assumption in this approach was that the only other components in the stream besides the secondary aniline were the isopropyl acetate and the primary aniline. The second assumption was that the NMR relaxation times for all the components were on the same timescale. The third assumption was that the response factors for the two anilines were similar enough to be able to use the area percent values directly. The calculation for the composition of the starting material is shown below:

From Figure 30.4, the ratio of isopropyl acetate to aniline compounds

isopropyl acetate $\frac{71.1}{3\text{H}} = 23.7$

where H is the proton

aniline compounds $\frac{28.9}{3\text{H}} = 9.63$

The values 23.7 and 9.63 represent the relative number of moles of IPAC and aniline compounds. So the mol% isopropyl acetate is easily calculated

$$\frac{23.7}{23.7 + 9.63} = 71.1\%$$

and the mol% aniline compounds is

$$\frac{9.63}{23.7 + 9.63} = 28.2\%$$

The step of taking the ratio of the area to the number of protons in this case is redundant as both peaks in the NMR being compared are for methyl groups (three protons), but this is an important step that can often be missed.

The area percent values from the HPLC in Figure 30.5 were used to decouple the concentration of aniline compounds.

Secondary aniline 84.6%

Primary aniline 15.4%

So on a mol% the composition is

$0.846 \times 0.282 = 0.239 \times 100 = 23.9\%$ secondary aniline

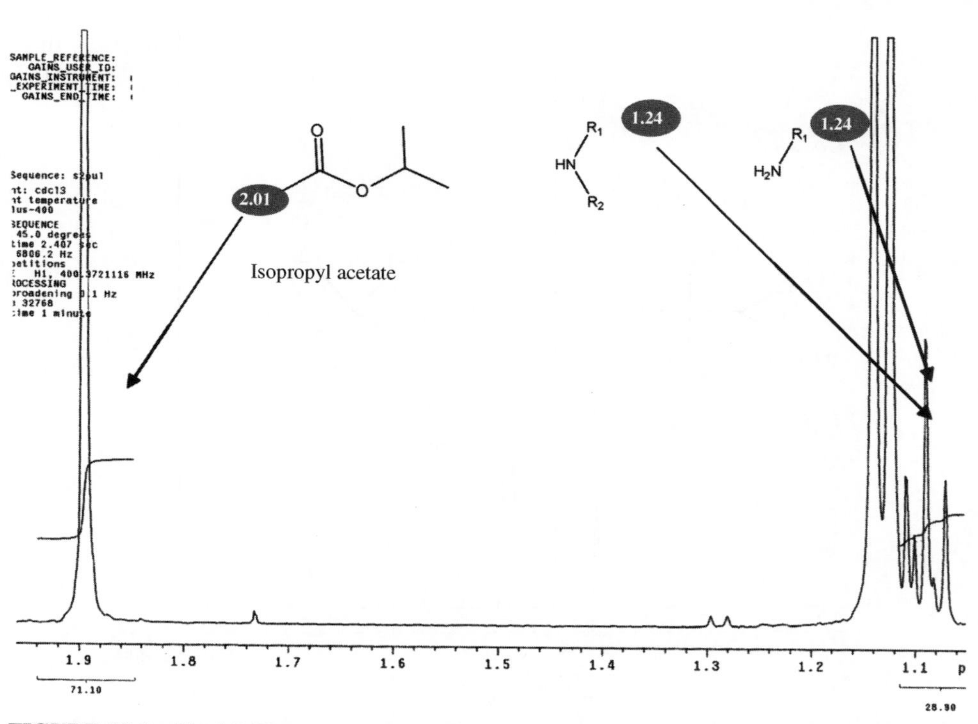

FIGURE 30.4 The NMR scan of the secondary aniline solution with the methyl group protons integrated for both the primary and secondary aniline compounds, as well as the methyl protons for the isopropyl acetate.

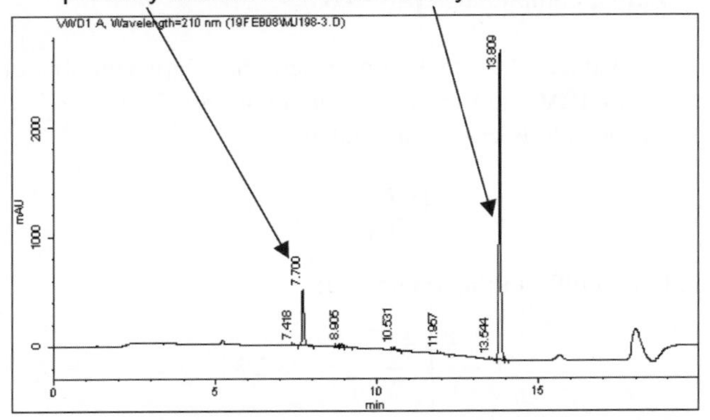

FIGURE 30.5 HLPC of the secondary aniline starting solution showing the primary aniline present at 15% by area.

$$0.154 \times 0.282 = 0.043 \times 100 = 4.3\% \text{ primary aniline}$$

Using molecular weights and assuming a basis of 1 mol

$$0.711 \text{ mol} \times 102.13 \text{ g/mol} = 72.61 \text{ g isopropyl acetate}$$

$$0.239 \text{ mol} \times 453.86 \text{ g/mol} = 108.47 \text{ g secondary aniline}$$

$$0.043 \text{ mol} \times 247.72 \text{ g/mol} = 10.65 \text{ g primary aniline}$$

$$72.61 \text{ g} + 108.47 \text{ g} + 10.65 \text{ g} = 191.73 \text{ g total}$$

Weight percent

$$\frac{72.61 \text{ g}}{191.73 \text{ g}} \times 100 = 37.87 \text{ wt\% isopropyl acetate}$$

$$\frac{108.47 \text{ g}}{191.73 \text{ g}} \times 100 = 56.57 \text{ wt\% secondary aniline}$$

$$\frac{10.65 \text{ g}}{191.73 \text{ g}} \times 100 = 5.55 \text{ wt\% primary aniline}$$

This characterizes the incoming aniline reagent stream, and now the same process is repeated with the epoxide reagent stream. By the certificate of analysis from the vender the epoxide, which is a liquid, it is known to have methyl *tert*-butyl ether present at 12% by weight. As this compound cannot be detected with HPLC with a UV detector, H^1 NMR was again used to characterize the material as shown in Figure 30.6. Taking the area of the MTBE compared to the area of the epoxide peaks, we are able to calculate the mol%

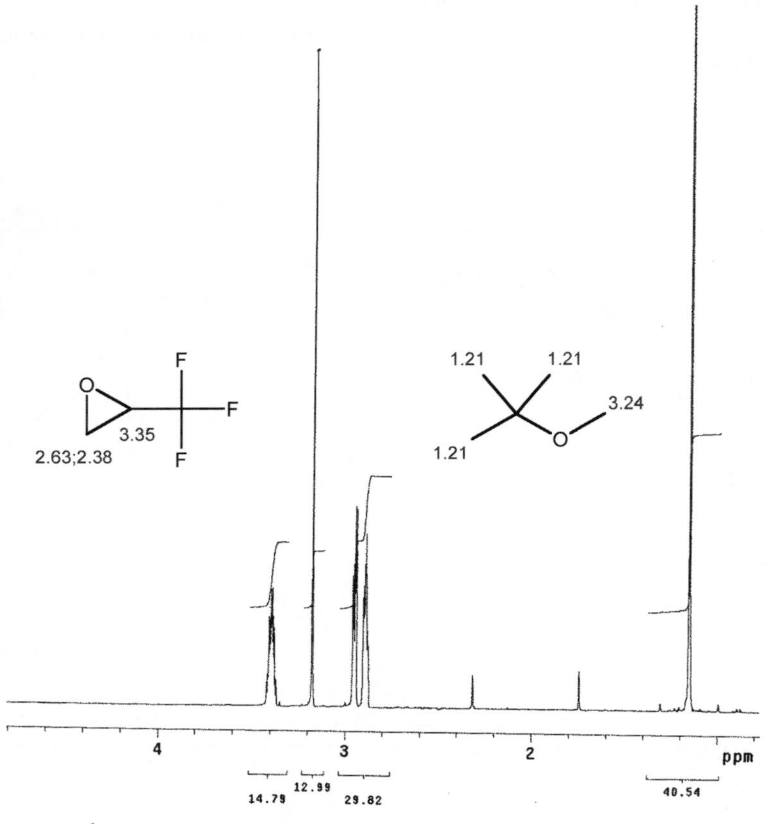

FIGURE 30.6 H^1 NMR of epoxide starting material with three separate peaks, each representing one proton, while the two singlets at 1.21 ppm (nine protons) and 3.24 ppm (three protons) indicate the presence of MTBE.

of each as the following:

$$\frac{12.99 + 40.54}{12\text{H}} = \frac{4.46\%}{\text{H}}$$

$$\frac{14.79 + 29.82}{3\text{H}} = \frac{14.87\%}{\text{H}}$$

$$\frac{14.87}{4.46 + 14.87} \times 100 = 76.93\% \text{ epoxide}$$

$$\frac{4.46}{4.46 + 14.87} \times 100 = 23.07\% \text{ MTBE}$$

Using molecular weights and assuming 1 mol total solution to convert to weight percent

$$0.7693 \text{ mol} \times 112.05 \text{ g/mol} = 86.20 \text{ g}$$

$$0.2307 \text{ mol} \times 88.15 \text{ g/mol} = 20.34 \text{ g}$$

$$20.34 \text{ g} + 86.20 \text{ g} = 106.54 \text{ g total}$$

Weight percent

$$\frac{86.20 \text{ g}}{106.54 \text{ g}} \times 100 = 80.91\% \text{ epoxide}$$

$$\frac{20.34 \text{ g}}{106.54 \text{ g}} \times 100 = 19.09\% \text{ MTBE}$$

Remembering that from the COA at the time the epoxide was received, it was 12 wt% MTBE, but due to the volatile nature of the epoxide, every time the container was opened, it had been concentrating the MTBE to nearly 20% by evaporation of the epoxide. This is most likely why when it was first scaled up, an additional amount of epoxide had to be charged as the first charge was effectively an undercharge due to lower than expected potency of the epoxide.

With these characterized reagent streams, a kinetic model could now be developed for the system. Reactions were set up in small septum-capped vials at three temperatures and two catalyst loadings. The septum caps allowed sampling of the reaction mixtures without venting the epoxide. The HPLC of these IPC samples showed the fate of the aniline species when reacted with the chiral epoxide. Over time, the secondary aniline reacts with the epoxide as the desired reaction, but so does the primary aniline according to Scheme 30.2:.

The primary aniline reacts with the epoxide to form impurity 1 that then reacts with a second mole of the epoxide to form impurity 2. This reaction progression is shown in the HPLC traces in Figure 30.7. The shoulder peak on impurity 2 is in fact the diastereomer that is formed when the second chiral epoxide is added. Standard reverse phase HPLC column packing is not capable of separating enantiomers but can separate diastereomers.

To calculate the concentration over time, the area percent values from the HPLC were used. Each reaction system (Schemes 30.1 and 30.2) had the area percent values normalized only for that system. The normalized values were then used to calculate the concentration at that time point by multiplying by the initial starting concentrations. To illustrate this process, Table 30.3 lists the area and normalized values for each reaction system for the 60°C reaction using the standard catalyst loading.

The data from Table 30.3 were then transformed into concentration data by multiplying the normalized area values for each reaction system with the starting concentrations. The secondary aniline starting concentration of 0.797 M was used for the desired reaction system of Scheme 30.1 and the concentration of the primary aniline of 0.145 M for the

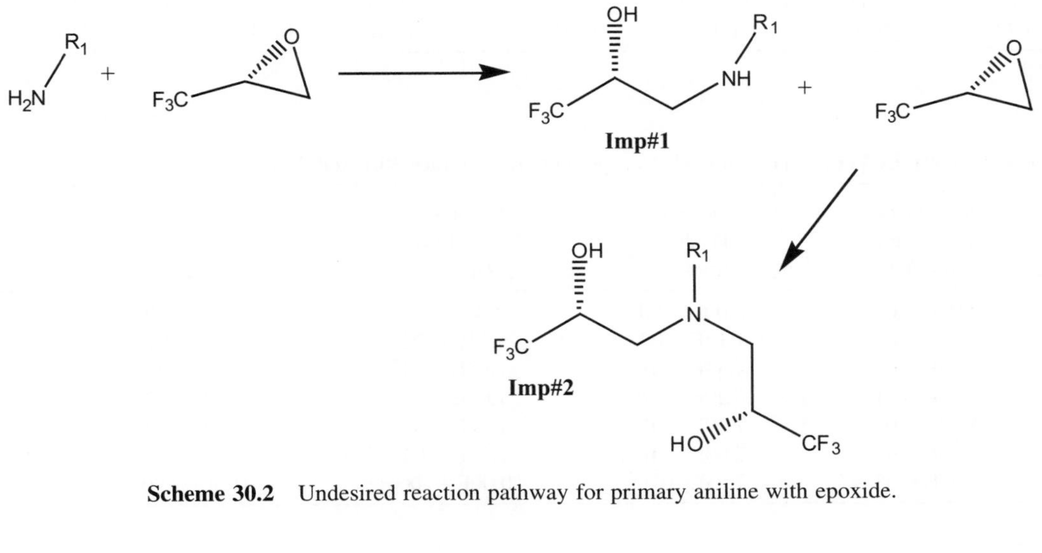

R_1, H_2N, +, F_3C, O, OH, NH, Imp#1, N, Imp#2, HO, CF_3

Scheme 30.2 Undesired reaction pathway for primary aniline with epoxide.

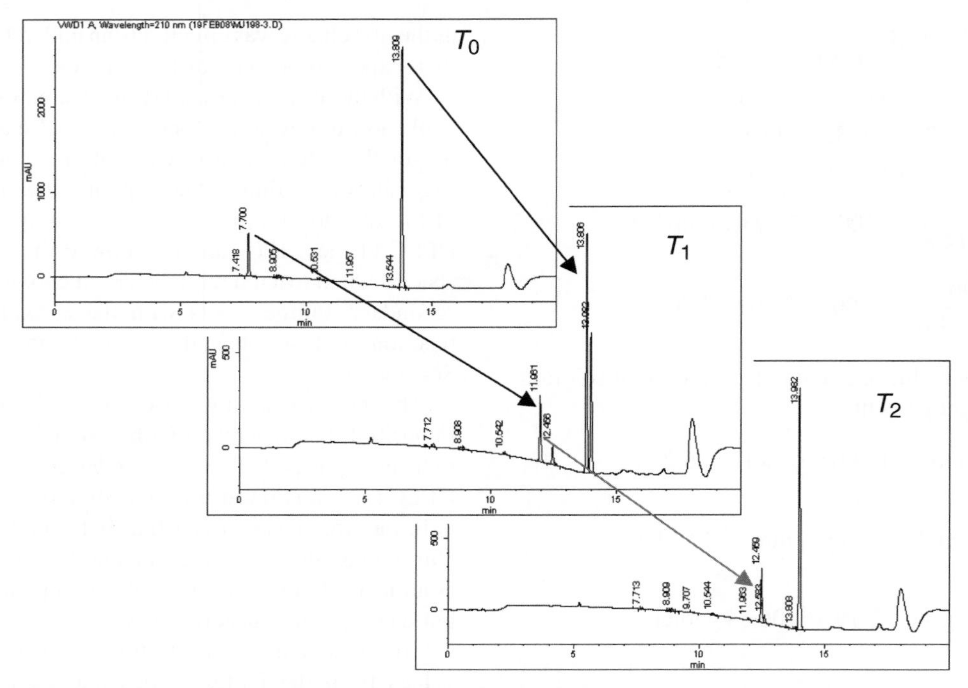

FIGURE 30.7 HPLCs over time showing both desired and undesired reactions. The desired reaction is secondary aniline at 13.8 min going to product at 13.9 min, while the undesired reaction is primary aniline at 7.7 min going to impurity 1 at 11.9 min, which further reacts to impurity 2 at 12.4 min.

TABLE 30.3 HPLC Area and Normalized to Each Reaction System

Time (s)	Secondary Aniline Area	Desired Product Area	Secondary Aniline Area	IM1 Area	IM2 Area	Normalized Area for System of Scheme 30.1		Normalized Area for System of Scheme 30.2		
0	11763.66	0.00	2138.86	0.00	0.00	1.00	0.00	1.00	0.00	0.00
720	7701.32	755.84	636.18	1281.50	40.90	0.91	0.09	0.32	0.65	0.02
4680	3884.43	4631.85	23.46	1202.17	864.00	0.46	0.54	0.01	0.58	0.41
9420	1771.80	6652.65	15.72	639.57	1432.61	0.21	0.79	0.01	0.31	0.69
16380	619.75	7895.18	25.21	289.47	1830.73	0.07	0.93	0.01	0.13	0.85
19380	292.27	6106.68	25.17	160.02	1451.73	0.05	0.95	0.02	0.10	0.89
78240	15.82	6365.85	0.00	0.00	1626.31	0.00	1.00	0.00	0.00	1.00

TABLE 30.4 Concentrations Versus Time for All Components in Schemes 30.1 and 30.2

Time (s)	Secondary Aniline (mol/L)	Desired Product (mol/L)	Secondary Aniline (mol/L)	Imp#1 (mol/L)	Imp#2 (mol/L)
0	7.97E − 01	0.00E + 00	1.45E−01	0.00E + 00	0.00E + 00
720	7.25E − 01	7.12E − 02	4.71E − 02	9.48E − 02	3.02E − 03
4680	3.63E − 01	4.33E − 01	1.63E − 03	8.33E − 02	5.99E − 02
9420	1.68E − 01	6.29E − 01	1.09E − 03	4.44E − 02	9.94E − 02
16,380	5.80E − 02	7.39E − 01	1.70E − 03	1.95E − 02	1.24E − 01
19,380	3.64E − 02	7.60E − 01	2.23E − 03	1.42E − 02	1.28E − 01
78,240	1.97E − 03	7.95E − 01	0.00E + 00	0.00E + 00	1.45E − 01

undesired reaction system of Scheme 30.2. This resulted in the concentration versus time data as shown in Table 30.4.

With this understanding of the reactions involved, the temperature-dependent kinetic model could be developed using DynoChem software as shown in equation 30.15.

$$\mathrm{Yb}(OTf)_3 + \text{secondary aniline} + \text{epoxide} \xrightarrow{k_2} \text{product} + \mathrm{Yb}(OTf)_3$$

$$\mathrm{Yb}(OTf)_3 + \text{primary aniline} + \text{epoxide} \xrightarrow{k_2} \text{imp 1} + \mathrm{Yb}(OTf)_3$$

$$\mathrm{Yb}(OTf)_3 + \text{imp 1} + \text{epoxide} \xrightarrow{k_3} \text{imp 2} + \mathrm{Yb}(OTf)_3 \tag{30.15}$$

where the temperature-dependant rate constants $k_{\#}$ are defined as follows:

$$k = k_{\text{ref}} \times \exp^{-E_aR}(1/T - 1/T_{\text{ref}}) \tag{30.16}$$

The fit in DynoChem resulted in the values given in Table 30.5.

TABLE 30.5 The Output Values for the Kinetic Model with the Confidence Interval (CI)

	Final Value	Units	CI (%)
$k_{1\text{ref}}$	0.0031	$L^2/(mol^2\ s)$	11.157
$k_{2\text{ref}}$	0.0173	$L^2/(mol^2\ s)$	7.745
$k_{1\text{ref}}$	0.0029	$L^2/(mol^2\ s)$	9.346
E_{a1}	59.554	kJ/mol	10.953
E_{a2}	57.586	kJ/mol	7.521
E_{a3}	56.99	kJ/mol	9.932

Equation 30.15 with the values from Table 30.5 results in the predicted versus actual plots of reaction progression as shown in Figure 30.8. The first reaction of the primary aniline with the epoxide was found to have a rate constant ($k_{2\text{ref}}$) that was an order of magnitude faster than the desired reaction rate constant ($k_{1\text{ref}}$), further explaining why additional epoxide was needed to consume all the secondary aniline starting materials.

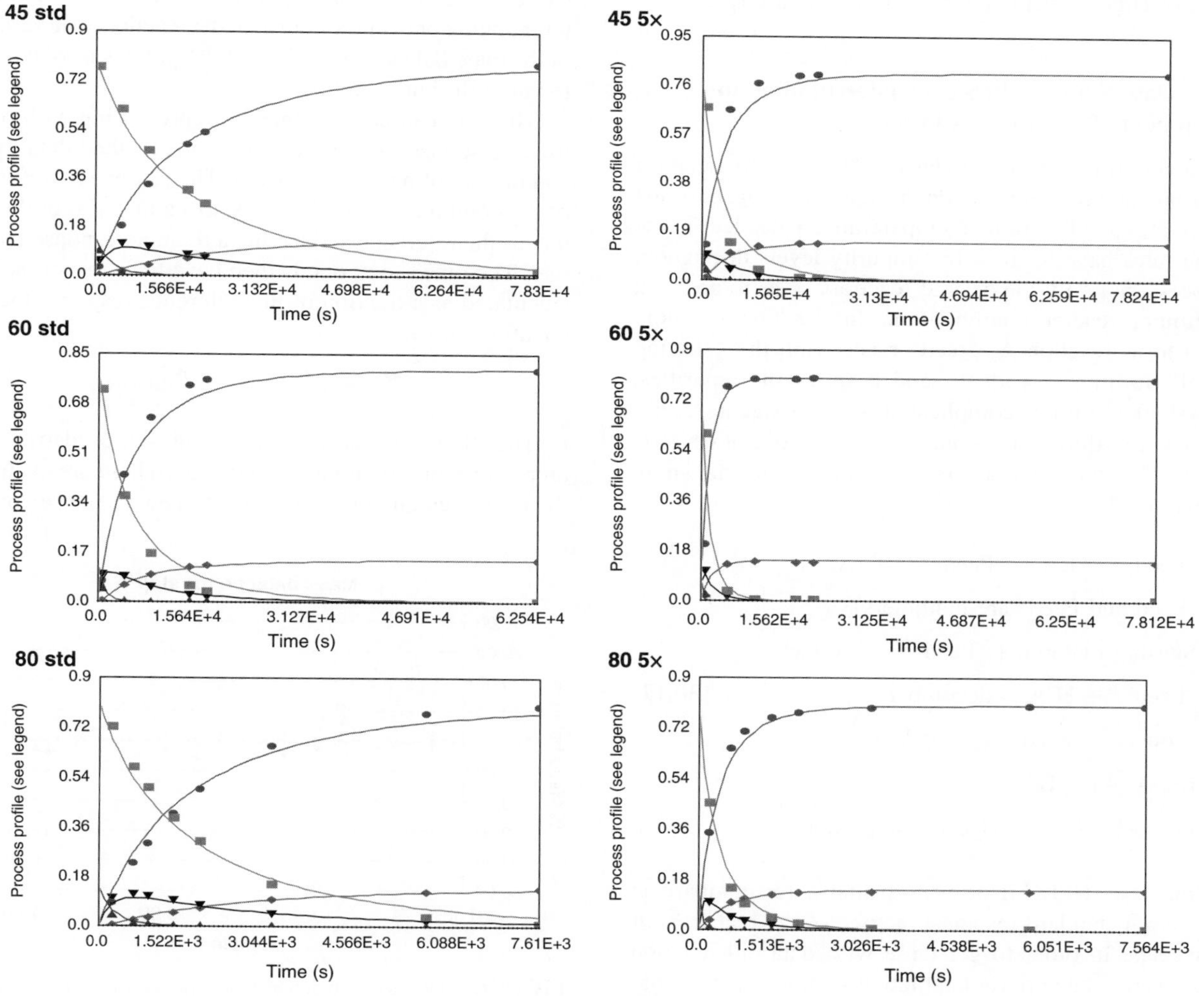

FIGURE 30.8 Predicted versus actual values for reaction progression from kinetic model.

From Case Study 1, we see that using the complementary analytical techniques of NMR and HPLC-UV, it was possible to fully characterize the starting materials. Once this was done, the area percent values were used to calculate the concentration over time that was then used to develop the kinetic model that gave the necessary process understanding. The assumption in this case was that all the species in a reacting system, that is, secondary aniline to the desired product for one system and primary aniline to impurity 1 onto impurity 2 for the other reactive system, had the same response factor and that area percent could be used without response factors. This is a reasonable assumption to make as the epoxide that was being added to the molecules did not have a chromophore and the electronics of the UV chromophore between starting materials and products were not changing much; that is, the responds of primary aniline differs slightly from that of impurity 1 and 2, as did the responds of secondary aniline to the desired product. But what if this assumption about starting materials and products having the same response cannot be made, what is the course of action? This problem is explored in Case Study 2.

30.5.2 Case Study 2: Process Understanding for Development of Continuous Process

Two reactions in series need to have CSTR reactors sized for a given annual throughput. To do this reactor sizing, absolute reaction rates as a function of temperature are needed. These reaction rates have to track the impurity levels throughout the process, not only the desired reaction, so as to arrive at an optimum reactor configuration. In the first reaction, referred to as reaction A, "feed" reacts with the "starting material" forming "product" and a series of impurities. This system is further complicated as the reagent "feed" can exist as two different tautomers, with only one of which is reactive. The six reactions in this system are shown in equation 30.17.

$$\begin{aligned}&\text{“Feed”} + \text{TEA} = \text{“Feed”*} + \text{TEA}\\&\text{(Feed and Feed* are tautomers)}\\&\text{Starting material} + \text{“Feed”*} \rightarrow \text{Product}\\&\text{“Feed”*} + \text{TEA} \rightarrow \text{decomp}\\&\text{Product} + \text{“Feed”*} \rightarrow \text{ImpA}\\&\text{ImpA} \rightarrow \text{ImpB}\\&\text{Product} + H_2O \rightarrow \text{Hydrolysis product}\end{aligned} \tag{30.17}$$

In this case, HPLC data were available for comparison with external standard response factors to arrive at wt% of each species. In order to get these wt% data, all reaction samples were made up as quantitative samples, that is, mg/mg reaction in samples in volumetric flasks.

With these data it is now possible to close the mass balance around the incoming limiting reagent "starting material." This mass balance at time t was calculated by comparing to initial starting material (SM_0) with equation 30.18.

$$\frac{SM_t + \text{prod}_t + \text{impA}_t + \text{impB}_t + \text{Hydrolysis}_{\text{prod}_t}}{SM_0 * 100} = \%\text{measured\_mass\_balance} \tag{30.18}$$

Calculations using equation 30.18 were performed for every sample taken from the reaction over the course of the reaction to give the mass balance. Mass balance for step A around the starting material including the impurities that are being tracked is shown in Figure 30.9. This mass balance illustrates that for three reactions at three different temperatures and over the course of each reaction, the mass balance fluctuates near 100%. In this case, the mass balance is not being forced to 100% by taking ratios but is calculated from external standards. What we can conclude from this data is that there is not an unaccounted for reaction as this would cause a systematic drain on the system as a function of temperature or over the course of the reaction. The variability of the mass balance around the 100% point is most likely due to variability in sampling.

The data can be smoothed by reprocessing with relative molar response factors, resulting in smoothed data that has had the sampling error removed. This is done by setting one of the compounds, usually the starting material or the product, as the reference and having a relative response factor of one. The other compounds then have their response factors calculated as a fraction of this reference response factor by equation 30.19.

$$R_{\text{f comp}}/R_{\text{f product}} = R_{\text{f relative}} \tag{30.19}$$

The relative response factors can now be used to adjust the area values of individual components. These area values are then summed and used to calculate new response corrected

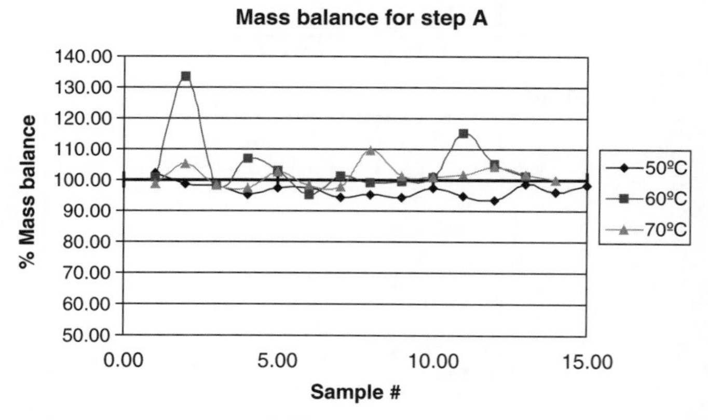

FIGURE 30.9 Mass balance from quantitative HPLC over the course of the reaction at three temperatures.

TABLE 30.6 Relative Response Factors Used to Smooth the Data by Converting Area into Fraction T_0 Concentration

Component	A	B	C	D	E
Relative response factor	1	0.4415	0.2628	0.4173	1.9011
Area	980,582.93	138,507.25	118,161.23	1125.98	2110.25
RRF corrected area	980,582.93	313,730.71	449,665.07	2698.10	1110.04
Fraction of $[T_0]$	0.755	0.242	0.314	0.002	0.001

area percent values. These area percent values are then used in conjunction with the starting material concentration at time zero to calculate individual component concentrations.

An example of these calculations for one time point are shown below for the reaction where B and C are reacted together, with C in excess to give product A and impurities D and E shown in Table 30.6.

The fraction of T_0 concentration is calculated by taking the relative response factor corrected areas summed together resulting in 174,7786.85 total area counts. The subtle point now is to make sure that the reagent in excess is not double counted. So product and impurities fraction of T_0 concentration values are calculated as shown in equation 30.20:

$$\frac{\text{Area}_{\text{A,B,D,E}}}{(\text{Area}_{\text{total}} - \text{Area}_{\text{C}})} = \text{Fraction\_}T_0 \qquad (30.20)$$

where the reagent in excess, reagent C, is calculated by equation 30.21:

$$\frac{\text{Area}_{\text{C}}}{(\text{Area}_{\text{total}} - \text{Area}_{\text{B}})} = \text{Fraction\_}T_0 \qquad (30.21)$$

These fractions of T_0 values can now be multiplied by the T_0 concentrations (limiting reagent concentration for all but the reagent in excess, which is multiplied by its T_0 concentration). The results of these calculations are shown in Table 30.7.

The profiles of the starting material and the product calculated with external standards versus calculated with relative response factors are shown in Figure 30.10. These smoothed data were then used to develop the kinetic model with DynoChem software package. The model versus predicted data from this model is shown in Figure 30.11.

TABLE 30.7 Final Conversion of T_0 Concentration to Concentration at this Time Point

Component	A	B	C	D	E
$[T_0]$	0	0.512	0.603	0	0
Fraction of $[T_0]$	0.755	0.242	0.314	0.002	0.001
Concentration at this time point	3.869E − 01	1.238E − 01	1.890E − 01	1.065E − 03	4.380E − 04

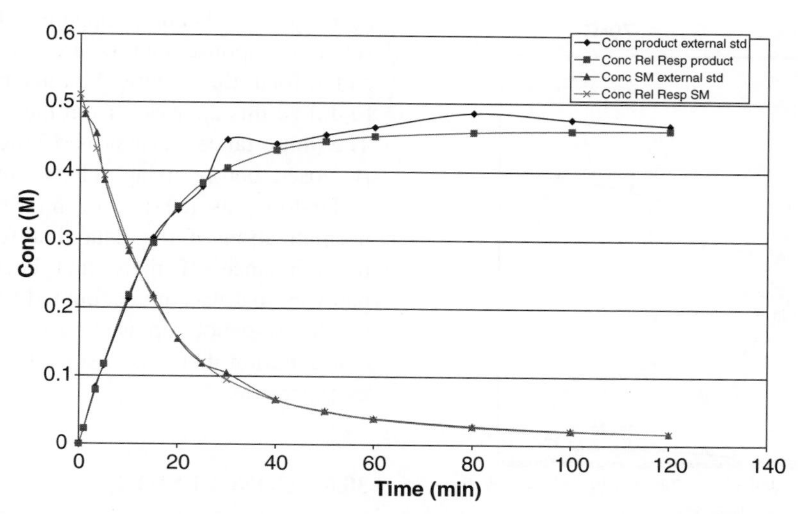

FIGURE 30.10 Starting material and product profiles comparing external standards and relative responds factors. The effect of smoothing the data and indicating where possible errors in sampling may have occurred is relatively straightforward once displayed graphically.

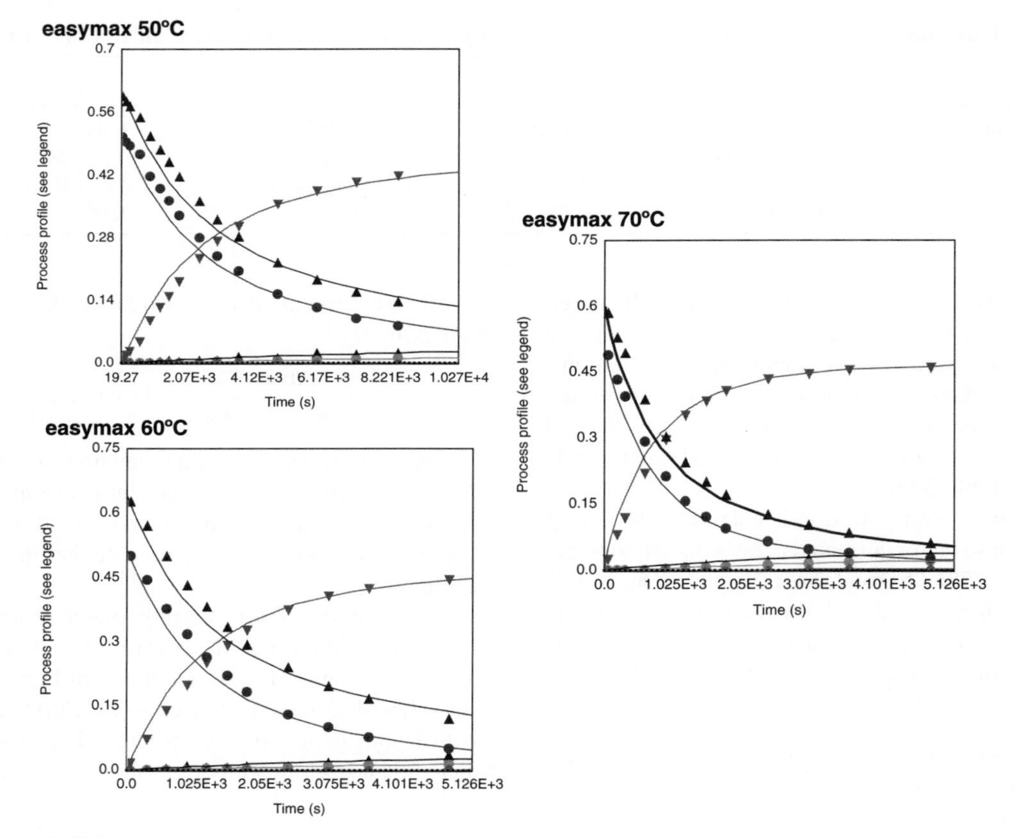

FIGURE 30.11 Kinetic model predicted versus actual concentrations at three different temperatures.

The model for step A was then validated by comparing the model to a semi-batch reaction done in a Mettler 0.5L RC-1 reactor in which the data were not calculated with relative response factors. These data are shown in Figure 30.12.

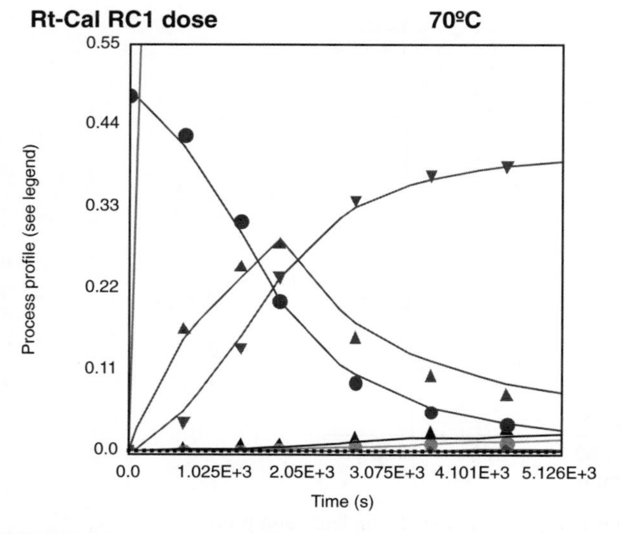

FIGURE 30.12 Validation semibatch reaction with 30 additions done in Mettler RC-1 reactor.

This process was repeated with the second reaction, reaction B, to develop a kinetic model. These models for both reactions A and B were then used to optimize a design for a series of CSTR reactors that would allow for the appropriate annual production.

It quickly become apparent that the data from the relative response factors are much smoother and better suited for fitting kinetic parameters. One could be tempted to utilize this approach from the beginning of the analysis. The importance of first verifying that the mass balance is closed before using relative response factors must be understood, as the use of relative response factors is a normalization of the data that forces the closure of the mass balance. If there had been a secondary reaction pathway that was not accounted for, then the kinetic model would have not represented the process and any reactor configuration that was designed would not have preformed as expected.

30.6 CONCLUSION

The use of analytical methods to elucidate process parameters, be it mass balance or ultimately kinetic information, can be full of assumptions. Oftentimes, in the pharmaceutical industry tight timelines prevent the investigation into every assumption. Being aware of the assumptions is the only way that one is ever going to be able to test the ones that will have the biggest impact on the data. The key to understanding what assumptions are being made is being aware of what each analytical method is looking at, what it is proportional to, and understanding the complementary test methods that can give a clearer picture of the problem.

REFERENCES

1. Meloan C. *Chemical Separations Principles, Techniques, and Experiments*, Wiley, New York, 1999.
2. Snyder LR, Kirkland JJ, Glajch JL. *Practical HPLC Method Development*, 2nd edition, Wiley, New York, 1997.
3. Skoog DA, Holler FJ, Nieman TA. *Principles of Instrumental Analysis*, 5th edition, Harcourt Brace & Co., Orlando, FL, 1998.
4. Little JL. Artifacts in trimethylsilyl derivatization reactions and ways to avoid them. *J. Chromatogr. A*, 1999;844:1–22.

31

QUANTITATIVE APPLICATIONS OF NMR SPECTROSCOPY

Brian L. Marquez
Pfizer Global Research & Development, Groton, CT, USA
R. Thomas Williamson
Roche Carolina, Inc., Florence, SC, USA

31.1 INTRODUCTION

31.1.1 General Principles of NMR

Nuclear magnetic resonance is an analytical method that takes advantage of the magnetic properties of certain atomic nuclei. This approach is similar to other types of spectroscopy in which the absorption or emission of electromagnetic energy at characteristic frequencies provides analytical information. However, NMR differs from other types of spectroscopy in which the discrete energy levels and the transitions between them are created by placing the samples in a strong magnetic field (B_0).

When an atom is placed in a magnetic field, its electrons circulate about the direction of the applied magnetic field. The circulation of these nuclei generates a very small magnetic field that is generally on the order of 1–20 ppm of the total applied magnetic field for ^{1}H and 1–200 ppm for ^{13}C. This field opposes the applied magnetic field and can be detected through the same R_f coil that is used to excite the nuclei of interest. When nuclei spin about the axis of this externally applied magnetic field, they possess an angular momentum. This angular momentum can be expressed as a function of a proportionality constant, **I**, that can be either an integer or a half-integer. **I** is referred to as the spin quantum number, or more simply as the nuclear spin. It is possible for some isotopes to have a spin quantum number $\mathbf{I} = 0$. These nuclei are not considered magnetic and cannot be detected by NMR. In order for a nuclei to have a spin quantum $\mathbf{I} = 0$ it must have an even atomic number and even mass. Commonly occurring non-NMR active nuclei include ^{12}C, ^{16}O, and ^{32}S. The group of nuclei most commonly observed by NMR methods is nuclei with a spin of 1/2. These include ^{1}H, ^{13}C, ^{19}F, ^{31}P, and ^{15}N. Other spins with $\mathbf{I} = 1$ or $\mathbf{I} > 1$ nuclei can be observed with slightly more difficulty. These nuclei include ^{2}H and ^{14}N ($\mathbf{I} = 1$) and ^{10}B, ^{11}B, ^{17}O, and ^{23}Na ($\mathbf{I} > 1$).

The rate at which a particular nuclei spins in a particular magnetic field is known as its precession frequency. This frequency is both a function of the externally applied field, the nucleus of interest, and the environment in which it resides (Figure 31.1). For a proton (^{1}H), in an applied 2.35 Tesla (T) magnetic field, the reference precession frequency is ~100 MHz. In the same externally applied field, other nuclei will have different gyromagnetic ratios. For example, ^{13}C with a gyromagnetic ratio of 4 will precess at 25 MHz in the same magnetic field. This characteristic precession frequency is known as the Larmor frequency of the nucleus. An NMR sample may contain many different magnetization components, each with its own Larmor frequency. Therefore, an NMR spectrum may be made up of many different frequency lines.

Nuclei aligned with the axis of the externally applied magnetic field will be in the lowest possible energy state. Thermal processes oppose this tendency, such that there are

Chemical Engineering in the Pharmaceutical Industry: R&D to Manufacturing Edited by David J. am Ende

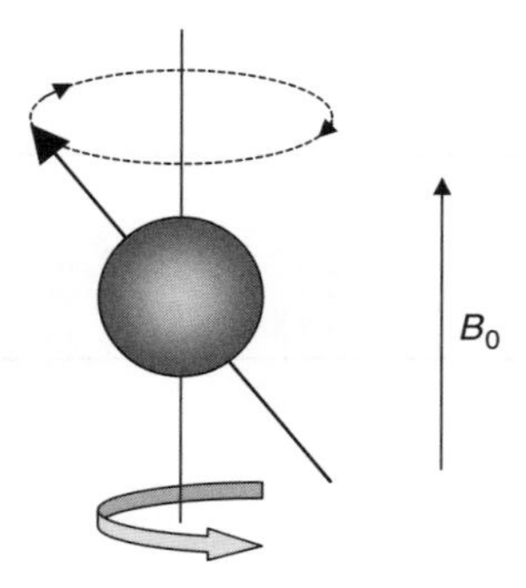

FIGURE 31.1 A spinning magnetic nuclei in an externally applied magnetic field with its axis precessing around the direction of the applied field (B_0) analogous to a gyroscope.

two populations of nuclei in an externally applied magnetic field. One is aligned with the axis of the field and another, which is only slightly smaller, is aligned opposite to the direction of the applied field. The distribution of spins between these two energy levels is referred to as the Boltzman distribution, and it is this population difference between the two levels that provides the observable collection of spins in an NMR experiment. This difference is very small compared to the total number of spins present, which leads to the comparative inherent insensitivity of NMR.

In the modern NMR experiment, pulsed radiofrequency (R_f) energy is used to excite all frequencies at once. In a simplistic sense, a certain amount of energy from an R_f pulse is absorbed by each nuclei. As these nuclei relax back from the excited state to the ground state, a corresponding amount of R_f energy is emitted. The frequency and the amplitude of this emitted energy contain important information about the nucleus from where it originated. In other words, the application of a radiofrequency pulse of energy orthogonal to the axis of the applied magnetic field perturbs the Boltzmann distribution thereby producing an observable event that is governed by the Bloch equations. When the application of that pulse is completed, the vector that has been rotated into the xy-plane will continue to precess about the axis of the externally applied field (z-axis), generating an oscillating signal in a receiver coil as the vector rotates about the z-axis at its characteristic Larmor frequency. The signal from the magnetization vector will decay back to an equilibrium condition along the B_0 axis as a function of two time constants, the spin–spin or transverse relaxation time, T_2, and the spin–lattice relaxation time, T_1. Once equilibrium is reestablished after a short delay, the process can be repeated and multiple acquisitions can be added to increase S/N. Immediately after the original R_f pulse, a receiver is turned on and a signal known as a time domain interferogram is acquired via an R_f receiver. With the help of an analogue to digital converter, these data are saved to a computer. This so-called interferogram contains information on all signals emitted by the sample at various NMR frequencies. This information is present as a sum of all damped-sinusoid signals emitted by the sample at various nuclear resonance frequencies. The specific resonance frequencies of these signals vary by the strength of the corresponding magnetic field. For instance, at 11.7 T the ^1H resonance frequency is ~500 MHz and at 18.8 T it is 800 MHz. The collected interferogram is called a free induction decay (FID). An example of free induction decay is shown in Figure 31.2.

Once acquired and saved, these data are converted from the time domain to frequency data to make them more meaningful and easier to interpret for the end user. This conversion is typically done through a mathematical manipulation known as the Fourier transform named for the mathematician Jean-Baptiste Joseph Fourier. This mathematical transformation also provides signal intensity information that is a key to the usefulness of NMR data. Typically, in common practice, a derivation of the Fourier transform is done by computer using the Cooley–Tukey fast Fourier algorithm otherwise known as the fast Fourier transform (FFT).

The frequency of the R_f energy absorbed by a particular nucleus is strongly affected by its chemical environment. These variables in the chemical environment can lead to changes in its so-called chemical shift. Three basic components of an NMR spectrum reveal its extreme usefulness. These include (1) the chemical shift, (2) the amplitude of the signal at that chemical shift, and (3) the splitting of the signal in response to its interaction with neighboring nuclei.

The location of an NMR signal in a spectrum is known as the signal's chemical shift. The location of this chemical shift is a function of the chemical environment of the sampled nuclei and is designated with a scale value referenced to an internal standard. For solution state, ^1H and ^{13}C NMR experiments, the reference standard is tetramethylsilane, or TMS, which has an accepted chemical shift of 0.00 ppm. Most proton signals appear to the left or "downfield" of TMS. Aliphatic hydrocarbon signals will generally be grouped nearer the position of TMS and are said to be "shielded" relative to vinyl or aromatic signals that are as a group referred to as "deshielded" (Figure 31.3). Chemical moieties involving heteroatoms, for example, $-OCH_3$ and $-NCH_2-$, typically will be located in a region of the NMR spectrum between the aliphatic and vinyl/aromatic signals.

In addition to chemical shift information, an NMR spectrum may also contain scalar coupling information. For NMR of liquids, scalar (J) couplings in proton spectra provide information about the local chemical environment of a given proton resonance. Proton resonances are split into multiplets related to the number of neighboring protons. For example, an ethyl fragment will be represented by a triplet with relative peak intensities of 1:2:1 for the methyl group, the splitting due to the two neighboring methylene protons, and a 1:3:3:1 quartet for the methylene group, with the splitting due to the three equivalent methyl protons. This concept is easily

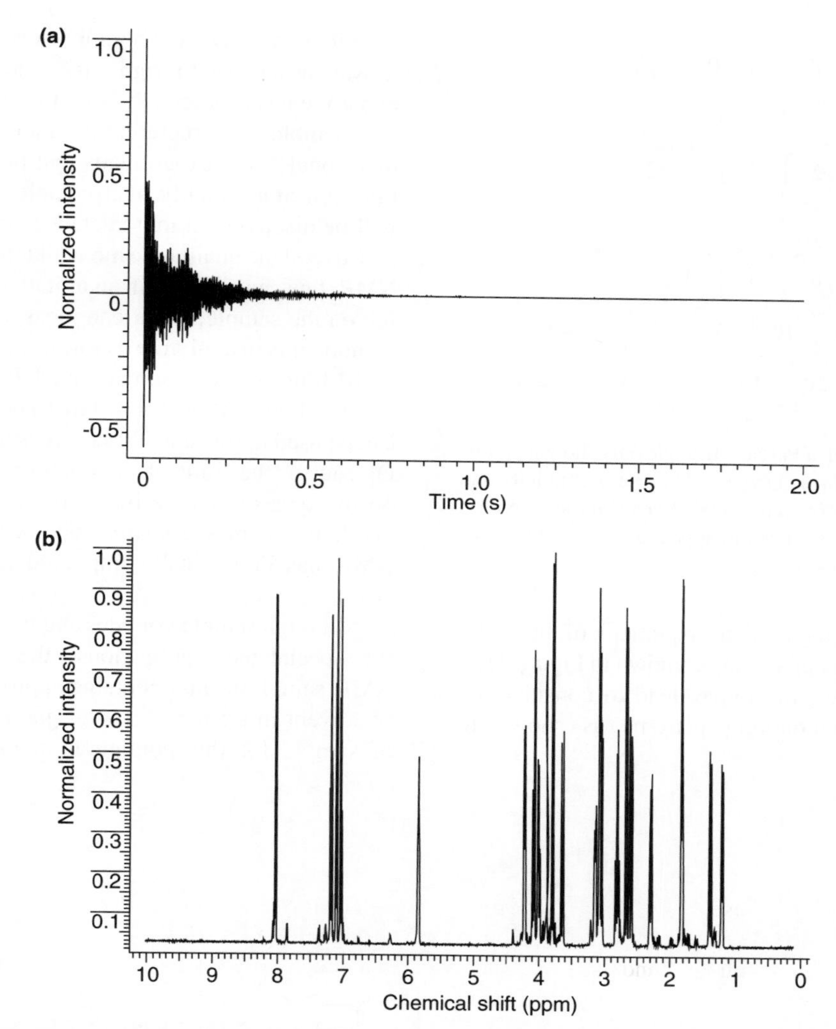

FIGURE 31.2 (a) The FID acquired for a sample of strychnine (**3**) at an observation frequency of 500 MHz. The spectrum was digitized with 16 K points and an acquisition time of ~ 2 s. Fourier transforming the data from the time domain to the frequency domain yields the spectrum of strychnine presented as intensity versus frequency shown (b).

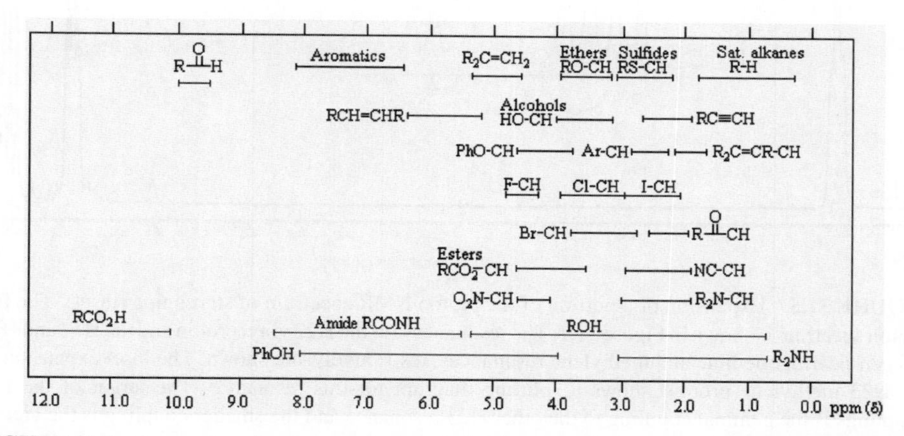

FIGURE 31.3 Image showing the proton chemical shift range depending on chemical environment.

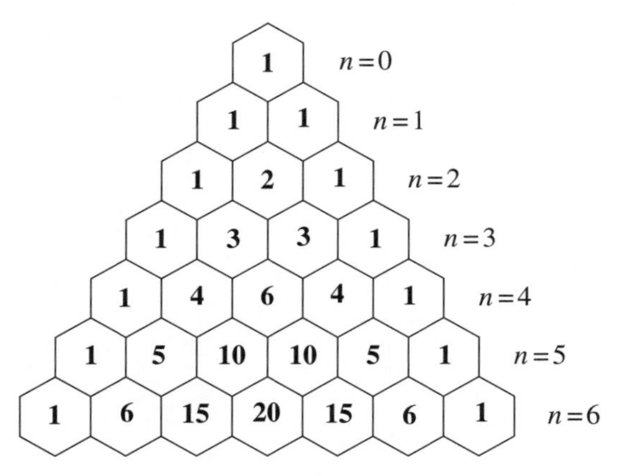

FIGURE 31.4 Illustration of Pascal's triangle only showing ratios to $n = 6$ according to $M = (n + 1)$ where M is the multiplicity and n is the number of scalar-coupled nuclei. For example, a proton adjacent to three protons ($n = 3$) would appear as a quartet ($M = 4$) with relative peak intensities of 1:3:3:1.

illustrated with the geometrical arrangement of binomial coefficients known as Pascal's triangle shown in Figure 31.4. More complex molecules, of course, lead to considerably more complicated spin-coupling patterns as shown in Figure 31.5.

Prior to the advent of homonuclear 2D NMR experiments, it was necessary to rigorously interpret a proton NMR experiment and identify all of the homonuclear couplings to assemble the structure. Alternatively, there are multidimensional NMR experiments that provide similar information in a more readily interpretable way. These techniques will be discussed in more detail later.

Beyond the qualitative molecular information afforded by NMR, one can also obtain quantitative information. Depending on the sample, NMR can measure relative quantities of components in a mixture as low as 0.1–1% in the solid state. NMR limits of detection are much lower in the liquid state, often as low as 1000:1 down to 10,000:1. Internal standards can be used to translate these values into absolute quantities. Of course, the limit of quantitation is dependent not only on the type of sample but also on the amount of sample. While not as mass sensitive as other analytical techniques, NMR has dramatically improved in sensitivity in recent years.

Although sample volumes and tube sizes can vary greatly for specialized applications, the most often analyzed NMR sample format contains approximately 500–1000 μL of solvent in a 5 mm diameter glass tube. Typical amounts of sample for this configuration range from 1 to 20 mg

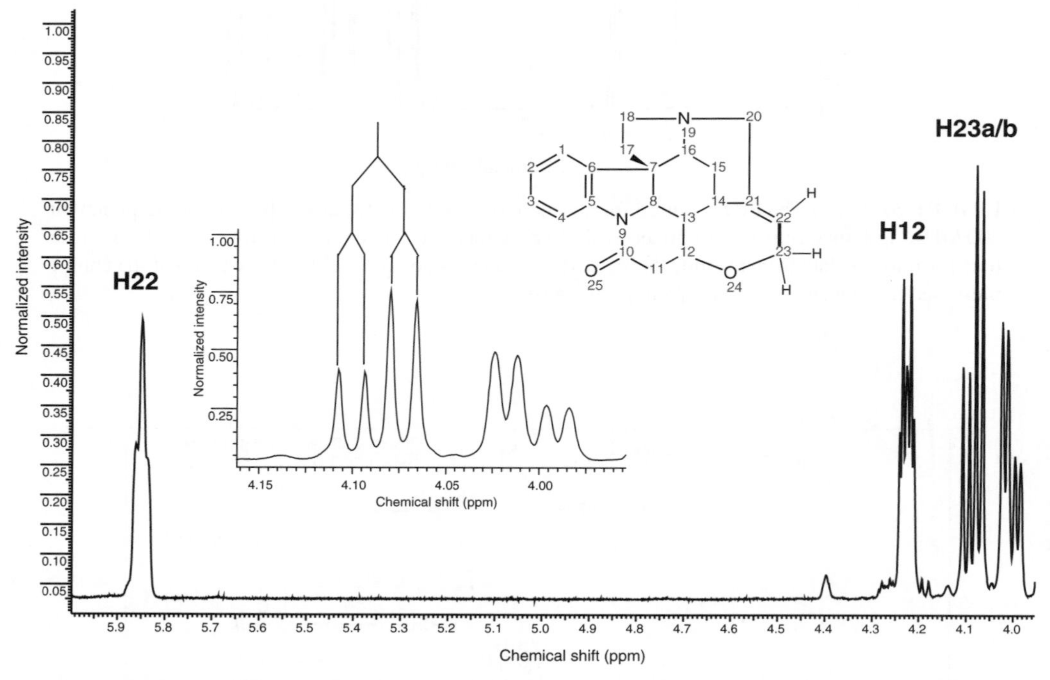

FIGURE 31.5 Expansion of a portion of the proton NMR spectrum of strychnine (inset). The full proton spectrum is shown in Figure 31.1. The resonances for the H22 vinyl proton and the H12 and H23 oxygen-bearing methine and methylene resonances, respectively, are shown. The inset expansion of the H23 methylene protons shows a splitting diagram for this resonance. The larger of the two couplings is the geminal coupling to the other H23 resonance and the smaller coupling is the vicinal coupling to the H22 vinyl proton.

depending on the amount and solubility of the sample available. NMR hardware accommodating smaller diameter sample tubes enable the detection of much smaller samples. Common commercially available liquid NMR tubes range from 1 to 11 mm in diameter. Amounts of detectable sample for these configurations can be as low as hundreds of nanograms for ^{1}H NMR detection. Carbon sensitivity is on the order of 100 times worse, so detection limits are usually limited to tens of micrograms.

Essentially all magnetic nuclei can be observed by NMR but the sensitivity of each is determined by the relative sensitivity and the natural abundance of the nuclei. For example, ^{1}H (with a spin of 1/2) is not only one of the most sensitive nuclei but also has a natural abundance of 99.985%. Alternatively, ^{13}C is fairly sensitive but only has a natural abundance of 1.108%. Other nuclei, such as ^{15}N (spin $\mathbf{I} = -1/2$), have both low natural abundance (0.37%) and low sensitivity making them doubly difficult to detect. These types of nuclei are very difficult to observe directly and are most commonly detected using what is known as "inverse detection" through a more sensitive nucleus. This technique can dramatically increase sensitivity for insensitive nuclei and can make available chemical shift and coupling information available that otherwise would have never been possible with existing technologies.

For the NMR analysis of solid samples, the amount of sample required is much greater in part because the apparent signal-to-noise ratio is significantly reduced. This is due to much broader line shapes, easily an order of magnitude wider than those observed in equivalent spectra of liquids. A standard solids NMR sample is a powder packed tightly into a small zirconia rotor and sealed with end caps. As for liquids, solid sample configurations are described in terms of their diameter, in this case the diameter of the sample rotor. Common, commercially available rotors range from 2.5 to 11 mm in diameter. Amounts of sample for these configurations depend on the sample and its density and typically range from 30 to 500 mg.

In common practice, most ^{1}H NMR spectra are recorded in solutions prepared with deuterated solvents. The advantages of this customary practice of sample preparation are twofold. Firstly, the lack of protons in the solvent enables observation of the protons of interest without interference from solute protons. Secondly, the presence of deuterium in the sample allows the NMR instrument to "lock" onto a reference frequency outside of those frequencies being observed. The ability to lock on a signal as the magnetic field drifts slightly over time can substantially improve the quality of spectra in experiments with long acquisitions. This can be very important when observing small amounts of sample or when the highest quality spectra with narrow lines are required. For a deeper understanding of the theory, application, and experimental setup of many common place experiments, see Refs 1–9.

31.2 ONE-DIMENSIONAL NMR METHODS

31.2.1 1D Proton NMR Methods

31.2.1.1 Magnetically Equivalent Nuclei A critical element to one of the fundamental phenomenon that is used for structure characterization is that of *J*-coupling (spin–spin coupling, scalar coupling, or often referred to as simply "coupling"). Of particular importance is the understanding of magnetic equivalency and this coupling interaction. Magnetically equivalent nuclei are defined as nuclei having both the same resonance frequency and spin–spin interaction with neighboring atoms. The spin–spin interaction does not appear in the resonance signal observed in the spectrum. It should be noted that magnetically equivalent nuclei are inherently chemically equivalent, but the reverse is not necessarily true. An example of magnetic equivalence is the set of three protons within a methyl group. All three protons attached to the carbon of a methyl group have the same resonance frequency and encounter the same spin–spin interaction with its vicinal neighbors. As a result, the resonance signal will integrate for three protons split into the appropriate coupling pattern (see next section) to adjacent nuclei.

31.2.1.2 NOE Experiments The acronym NOE is derived from the nuclear Overhauser effect. This phenomenon was first predicted by Overhauser in 1953 [10]. It was later experimentally observed by Solomon in 1955 [11]. The importance of the NOE in the world of small molecule structure elucidation cannot be overstated. As opposed to scalar coupling described above, NOE allows the analysis of dipolar coupling. Dipolar coupling is often referred to as through-space coupling and is most often used to explore the spatial relationship between two atoms experiencing zero scalar coupling. The spatial relationship of atoms within a molecule can provide an immense amount of information about a molecule, ranging from the regiochemistry of an olefin to the three-dimensional solution structure.

A simplified explanation of the NOE is the magnetic perturbations induced on a neighboring atom resulting in a change in intensity. This is usually an increase in intensity but may alternatively be zero or even negative. One can envision saturating a particular resonance of interest for a time t. During the saturation process, a population transfer occurs to all spins that undergo an induced magnetization.

There are two types of NOE experiments that can be performed. These are referred to as the steady-state NOE and the transient NOE. The steady-state NOE experiment is exemplified by the classic NOE difference experiment [12]. Steady-state NOE experiments allow one to quantitate relative atomic distances. However, there are many issues that can complicate their measurement, and a qualitative interpretation is more reliable [13]. Spectral artifacts can be observed from imperfect subtraction of spectra. In addition, this experiment is extremely susceptible to inhomogeneity issues and temperature fluctuations.

1D transient NOE experiments employing gradient selection are more robust and therefore are more reliable for measuring dipolar-coupling interactions [14]. Shaka et al. published one such 1D transient NOE experiment that has, in most cases, replaced the traditional NOE difference experiment [15]. The sequence dubbed the double-pulsed field gradient spin echo (DPFGSE) NOE employs selective excitation through the DPFGSE portion of the sequence [16]. Magnetization is initially created with a 90° ^{1}H pulse. Following this pulse are two gradient echoes employing selective 180° pulses. The flanking gradient pulses are used to dephase and recover the desired magnetization as described above. This selection mechanism provides very efficient selection of the resonance of interest prior to the mixing time where dipolar coupling is allowed to build up.

31.2.1.3 Relaxation Measurements Relaxation is an inherent property of all nuclear spins. There are two predominant types of relaxation processes in NMR of liquids. These relaxation processes are denoted by the longitudinal (T_1) and transverse (T_2) relaxation time constants. When a sample is excited from its thermal equilibrium with an RF pulse, its tendency is to relax back to its Boltzmann distribution. The amount of time to reequilibrate is typically on the order of seconds to minutes. T_1 and T_2 relaxation processes operate simultaneously. The recovery of magnetization to the equilibrium state along the z-axis is longitudinal or the T_1 relaxation time. The loss of coherence of the ensemble of excited spins (uniform distribution) in the xy-plane following the completion of a pulse is transverse or T_2 relaxation. The duration of the T_1 relaxation time is a very important feature as it allows us to manipulate spins through a series of RF pulses and delays. Transverse relaxation is governed by the loss of phase coherence of the precessing spins when removed from thermal equilibrium (e.g., an RF pulse). The transverse or T_2 relaxation time is visibly manifest in an NMR spectrum in the linewidth of resonances; the linewidth at half height is the reciprocal of the T_2 relaxation time. These two relaxation mechanisms can provide very important information concerning the physical properties of the molecule under study, tumbling in solution, binding or interaction with other molecules, and so on.

T_1 relaxation measurements provide information concerning the time constant for the return of excited spins to thermal equilibrium. For spins to completely relax, it is necessary to wait for a period of five times T_1. To accelerate data collection, in most cases one can perform smaller flip angles than 90° and wait for a shorter time before repeating the pulse sequence. Knowing the value of T_1 proves to be very useful in some instances, and it is quite simple to measure. The pulse sequence used to perform this measurement is an inversion recovery sequence [17]. The basic linear sequence of RF pulses (an NMR pulse sequence) consists of a 180-τ-90-acquire. Knowledge of the T_1 is of paramount importance when looking to increase the accuracy of quantitative NMR (qNMR) experiments. This will be described in detail in Section 31.4.

31.2.2 1D Carbon NMR Methods

31.2.2.1 Proton Decoupling Carbon-13, or ^{13}C, is a rare isotope of carbon with a natural abundance of 1.13% and a gyromagnetic ratio, γ_C, that is approximately one-quarter that of ^{1}H. Early efforts to observe ^{13}C NMR signals were hampered by several factors. First, the 100% abundance of ^{1}H and the heteronuclear spin coupling, $^{n}J_{CH}$ where $n = 1$–4, split the ^{13}C signals into multiplets, thereby making them more difficult to observe. The original efforts to observe ^{13}C spectra were further hampered by attempts to record them in the swept mode, necessitating long acquisition times and computer averaging of scans. These limitations were circumvented, however, with the advent of pulsed Fourier transform NMR spectrometers with broadband proton decoupling capabilities [18].

Broadband ^{1}H decoupling, in which the entire proton spectral window is irradiated, collapses all of the ^{13}C multiplets to singlets, vastly simplifying the ^{13}C spectrum. An added benefit of broadband proton decoupling is NOE enhancement of protonated ^{13}C signals by as much as a factor of 3.

Early broadband proton decoupling was accomplished by noise modulation that required considerable power, typically 10 W or more, and thus caused significant sample heating. Over the years since the advent of broadband proton decoupling methods, more efficient decoupling methods have been developed including GARP, WURST, and others [19]. The net result is that ^{13}C spectra can now be acquired when needed with low power pulsed decoupling methods and almost no sample heating.

31.2.2.2 Standard 1D Experiments Of the multitude of 1D ^{13}C NMR experiments that can be performed, the two most common experiments are a simple broadband proton-decoupled ^{13}C reference spectrum, and a DEPT sequence of experiments [20]. The latter, through addition and subtraction of data subsets, allows the presentation of the data as a series of "edited" experiments containing only methine, methylene, and methyl resonances as separate subspectra. Quaternary carbons are excluded in the DEPT experiment and can only be observed in the ^{13}C reference spectrum or by using another editing sequence such as APT [21]. The individual DEPT subspectra for CH, CH_2,

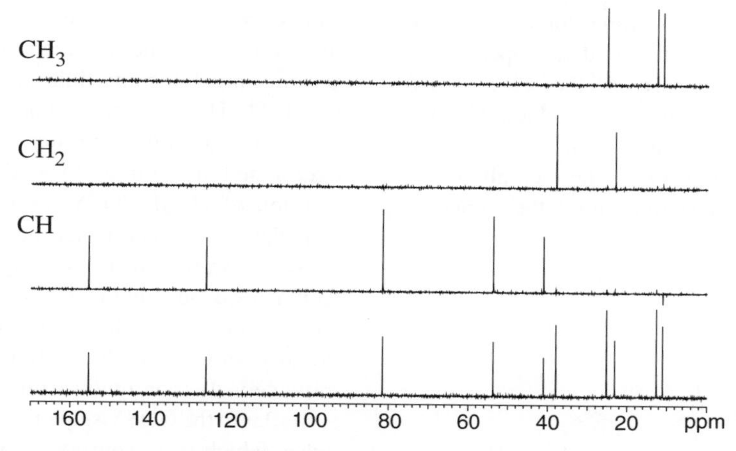

FIGURE 31.6 Multiplicity-edited DEPT traces for the methine, methylene, and methyl resonances of santonin (1). Quaternary carbons are excluded in the DEPT experiment and must be observed in the ^{13}C reference spectrum or through the use of another multiplicity editing experiment such as APT.

and CH_3 resonances of santonin (**1**) are presented in Figure 31.6.

1 Santonin

31.3 TWO-DIMENSIONAL NMR METHODS

31.3.1 Basic Principles of 2D

2D NMR methods are highly useful for structure elucidation. Jeener described the first 2D NMR experiment in 1971 [22]. In standard NMR nomenclature, a data set is referred to by one less than the total number of actual dimensions, since the intensity dimension is implied. The 2D data matrix therefore can be described as a plot containing two frequency dimensions. The inherent third dimension is the intensity of the correlations within the data matrix. This is the case in "1D" NMR data as well. The implied second dimension actually reflects the intensity of the peaks of a certain resonance frequency (which is the first dimension). The basic two-dimensional experiment consists of a series of 1D spectra. These data are run in sequence and have a variable delay built into the pulse program that supplies the means for the second dimension. The variable delay period is referred to as t_1.

All 2D pulse sequences consist of four basic building blocks: preparation, evolution (t_1), mixing, and acquisition (t_2). The preparation and mixing times are periods typically used to manipulate the magnetization (also known as "coherence pathways") through the use of RF pulses. The evolution period is a variable time component of the pulse sequence. Successive incrementing of the evolution time introduces a new time domain. This time increment is typically referred to as a t_1 increment and is used to create the second dimension. The acquisition period is commonly referred to as t_2. The first dimension, generally referred to as F_2, is the result of Fourier transformation of t_2 relative to each t_1 increment. This creates a series of interferograms with one axis being F_2 and the other the modulation in t_1. The second dimension, termed F_1, is then transformed with respect to the t_1 modulation. The resultant is the two frequency dimensions correlating the desired magnetization interaction, most typically scalar or dipolar coupling. Also keep in mind that there is the "third dimension" that shows the intensity of the correlations.

While performing 2D NMR experiments, one must keep in mind that the second frequency dimension (F_1) is digitized by the number of t_1 increments. Therefore, it is important to consider the amount of spectral resolution that is needed to resolve the correlations of interest. In the first dimension (F_2), the resolution is independent of time relative to F_1. The only requirement for F_2 is that the necessary number of scans is obtained to allow appropriate signal averaging to obtain the desired S/N. These two parameters, the number of scans acquired per t_1 increment and the total number of t_1 increments, are what dictate the amount of time required to acquire the full 2D data matrix. 2D homonuclear spectroscopy can be summarized by three different interactions, namely, scalar coupling, dipolar-coupling, and exchange processes.

31.3.2 Homonuclear 2D Methods

31.3.2.1 Scalar-Coupled Experiments: COSY and TOCSY

The correlated spectroscopy (COSY) experiment is one of the simplest 2D NMR pulse sequences in terms of the number of

RF pulses it requires [23]. Once the time domain data are collected and Fourier transformed, the data appear as a diagonal in the spectrum that consists of the ^{1}H chemical shift centered at each proton's resonance frequency. The off-diagonal peaks are a result of scalar coupling evolution during t_1 between neighboring protons. The data allow one to visualize contiguous spin systems within the molecule under study.

2 Astemizole

In addition to the basic COSY experiment, there are phase-sensitive variants that allow one to discriminate the active from the passive couplings allowing clearer measurement of the former. Active couplings give rise to the off-diagonal cross-peak. However, the multiplicity of the correlation has couplings inherent to additional coupled spins. These additional couplings are referred to as passive couplings. One such experiment is the double quantum filtered (DQF) COSY experiment [24]. Homonuclear couplings can be measured in this experiment between two protons isolated in a single spin system. Additional experiments have been developed that allow the measurement of more complicated spin systems involving multiple protons in the same spin system [25]. The 2D representation of the scalar-coupled experiment is useful when identifying coupled spins that are overlapped or are in a crowded region of the spectrum. An example of a DQF COSY spectrum is shown in Figure 31.7. This data set was collected on astemizole (**2**).

Total correlation spectroscopy (TOCSY) is similar to the COSY sequence, that is, it allows observation of contiguous spin systems [26]. However, the TOCSY experiment additionally will allow observation of many coupled spins simultaneously (contiguous spin system). The basic sequence is similar to the COSY sequence with the exception of the last pulse, which is a "spin-lock" pulse train.

31.3.2.2 Scalar-Coupled Experiments: INADEQUATE 2D homonuclear correlation experiments are typically run using ^{1}H as the nucleus in both dimensions. This is advantageous, as the sensitivity for proton is quite high. However, there are 2D homonuclear techniques that detect other nuclei. One such experiment is the INADEQUATE experiment [27]. Its insensitive nature arises from the low natural abundance of ^{13}C and the fact that one is trying to detect an interaction between two adjacent ^{13}C atoms. The chance of observing this correlation is 1 in every 10,000 molecules. Given sufficient time or isotope-enriched molecules, the INADEQUATE provides highly valuable information. The data

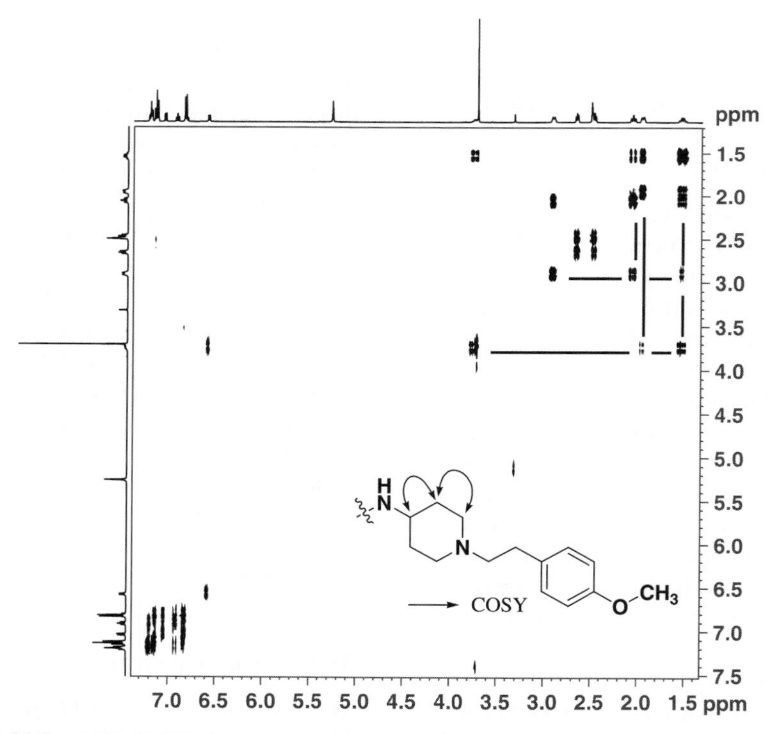

FIGURE 31.7 DQF COSY data of astemizole (**2**). The black bars indicate the contiguous spin system drawn with arrows on the relevant portion of astemizole.

generated from this experiment allow one to map out the entire carbon skeleton of the molecular structure. The only missing structural features occur where there are intervening heteroatoms (e.g., O and N).

31.3.2.3 Dipolar-Coupled Experiments: NOESY The 2D experiments described thus far rely solely on the presence of scalar coupling. There are other sequences that allow one to capitalize on the chemical shift dispersion gained with the second dimension for dipolar-coupling experiments as well. One example in this category is the 2D NOESY pulse sequence [28]. The pulse sequence for the 2D NOESY experiment is essentially identical to that of the DQF COSY experiment with the exception of an element that allows the buildup of dipolar couplings. The processed data is reminiscent of a COSY spectrum in that there is a diagonal represented by the 1D ^{1}H spectrum in both frequency dimensions. In sharp contrast to the scalar-coupled cross-peaks in the COSY experiment, NOESY provides off-diagonal responses that correlate spins through space. This sequence is used extensively in the structure characterization of small molecules for the same reason as its 1D counterparts. The spatial relationship of ^{1}H atoms is an invaluable tool.

The similarity of the NOESY to the COSY also causes some artifacts to arise in the 2D data matrix of a NOESY spectrum. The artifacts arise from residual scalar coupling contributions that survive throughout the NOESY pulse sequence. These artifacts are usually quite straightforward to identify as they have a similar antiphase behavior as can be seen for the DQF COSY data (Figure 31.8).

31.3.3 Heteronuclear 2D Methods

31.3.3.1 Direct Heteronuclear Chemical Shift Correlation There are many numbers of heteronuclear correlation experiments that reach into the spin systems of many different chemical environments. This chapter will focus on the two major types of experiments that are used for structure elucidation. These are the $^1J_{CH}$ scalar-coupled experiments and the long-range ($^nJ_{CH}$, where $n > 1$) scalar-coupled experiments. The two predominant experiments for $^1J_{CH}$ are the heteronuclear multiple quantum coherence (HMQC) and the heteronuclear single quantum coherence (HSQC) methods. Both of these methods rely on the sensitive and time-efficient proton or so-called "inverse"-detected heteronuclear chemical shift correlation experiments are preferable [29]. For molecules with highly congested ^{13}C spectra, ^{13}C rather than ^{1}H detection is desirable due to high resolution in the F_2 dimension [30].

Using strychnine as a model compound, a pair of HSQC spectra are shown in Figure 31.9. Figure 31.9a shows the HSQC spectrum of strychnine without multiplicity editing. All resonances have positive phase. In contrast, in the opinion of the authors, the much more useful multiplicity-edited variant of the experiment is shown Figure 31.9b. The multiplicity-editing feature allows one to phase the data so that the correlations representing methyl and methine groups are

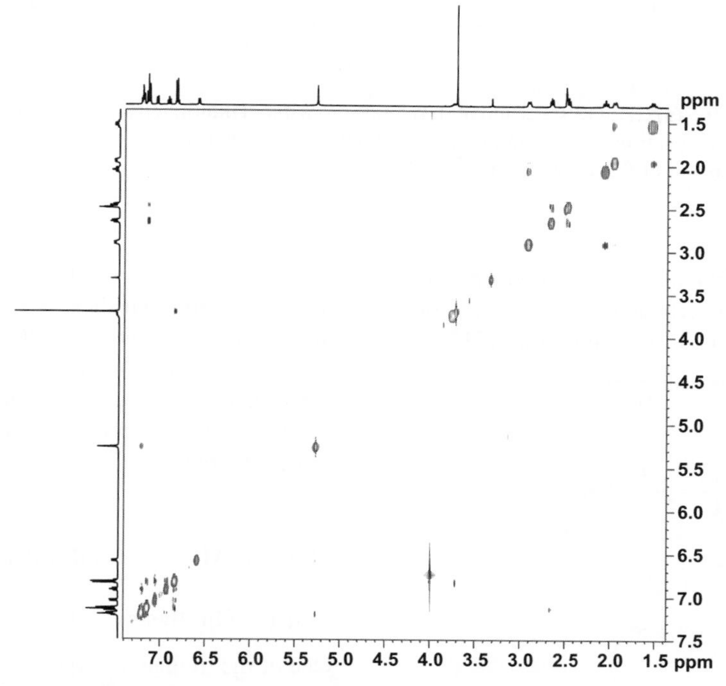

FIGURE 31.8 2D NOESY data of astemizole (**2**). The mixed phase correlations in the aromatic region are examples of the artifacts described in the text.

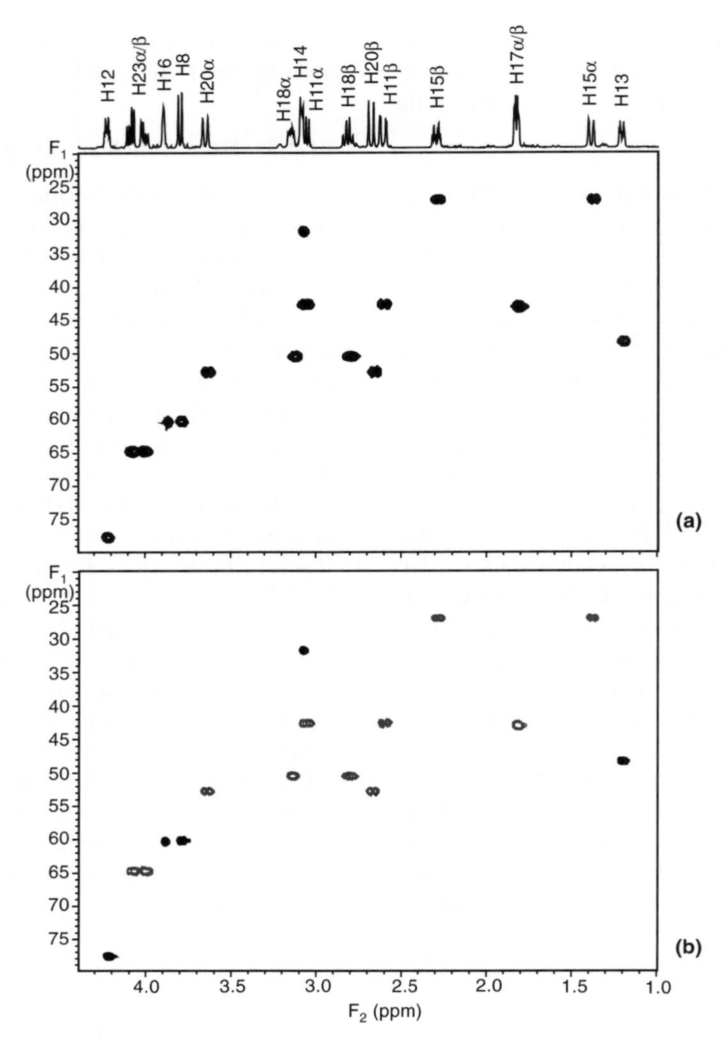

FIGURE 31.9 (a) GHSQC spectrum of strychnine (**3**) without multiplicity editing. (b) Multiplicity-edited GHSQC spectrum of strychinine showing methylene resonances (grey contours) inverted with methine resonances (black contours) with positive phase. (Strychnine has no methyl resonances.)

in the same direction and methylene's are the opposite phase. Other less common direct heteronuclear shift correlation experiments have been described in the literature [31].

3 Strychnine

31.3.3.2 Long-Range Heteronuclear Shift Correlation Methods There are numerous ^{13}C detected long-range heteronuclear shift correlation methods developed [32]. The primary reason that these methods have largely fallen into disuse is because of the HMBC (heteronuclear multiple bond correlation) experiment [33]. The proton-detected HMBC experiment and its gradient enhanced variant offer considerably greater sensitivity than the original heteronucleus-detected methods.

31.4 qNMR SPECTROSCOPY

31.4.1 The Basics

One of the tremendous qualities of NMR spectroscopy is its intrinsic quantitative attributes. A few recent publications have done an excellent job of detailing the basic principles and examples of the use of qNMR as well as pertinent parameters and reference standards [34–36]. The ability to quantitatively determine relative concentrations of multiple components in a mixture, determine potency, and product recovery, to name a few, have led this technique to gain tremendous popularity in the last 15 years. As with all things, popularity can lend itself to usage of the technique without carefully considering the physical phenomenon that allow these properties to be measured. This chapter will deal with the typical uses of qNMR, including the key parameters to consider, as well as examples of applications.

With respect to qNMR, the most relevant relationship is that between the integrated signals to the number of nuclei responsible for the resonance:

$$I_X = K_S + N_X \quad (31.1)$$

where I represents the integrated resonance that is directly correlated to the number of nuclei responsible for the integrated resonance while K is a spectrometer constant. The factor K cancels as the substances are undergoing the same experimental conditions, assuming proper care is taken to ensure appropriate parameters are in place [34]. The utilization of NMR for quantitative explanations is also described in the USP-NF 28, Chapter 761 [37]. There are two primary ways in which qNMR is used. These are a relative approach and an absolute method.

31.4.2 Relative and Absolute Methods

The most frequently used method, often in a pseudoqualitative mode (not optimal acquisition parameters), is the relative approach (Figure 31.10). For a quick analysis of two components of similar chemical properties (e.g., similar molecular weight and atomic composition) then one can select two resonances of similar hybridization and integrate and get a rough estimate of molar ratios $\left(\frac{n_x}{n_y}\right)$ using equation 31.2. One can solve this equation to provide the mole fraction of either of the two components.

$$\frac{n_x}{n_y} = \frac{I_x N_y}{I_y N_x} \quad (31.2)$$

If the chemical composition of both components are known then the weight percent is inherent. However, one must be careful when using this "crude" method with components of an unknown chemical composition as there are many factors that can lead to ever increasing erroneous results. One of the most important of these is the intrinsic relaxation rates of the compounds under study. If little is known of the compound being measured then large differences in this chemical property can lead to a poor result. To make a rudimentary comparison it would be equivalent to measuring the relative response of two compounds by UV

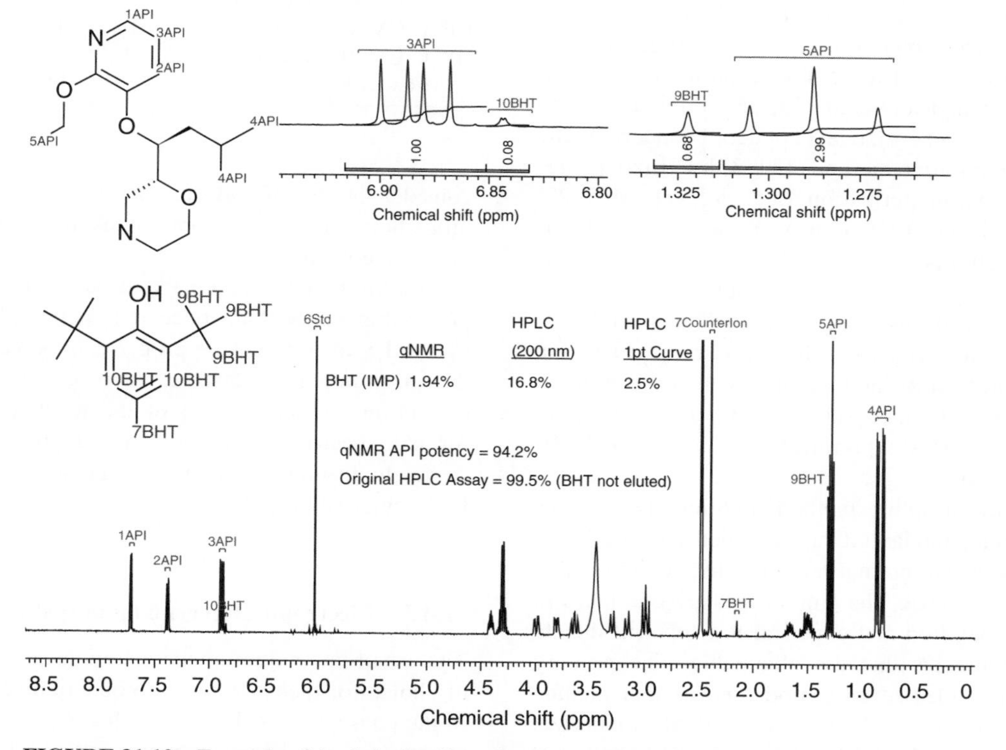

FIGURE 31.10 Example of the "absolute" method for qNMR experiments. In this case, BHT was used as an internal reference with a known active pharmaceutical (API). See Table 31.2 to observe the impact of using NMR as opposed to UV for quantitation.

TABLE 31.1 Summary of the Universal Spectrometer Parameters for qNMR

Parameter	Bruker	JEOL	Variant	Value
90° Pulse strength	Pl1	x_atn	tpwr	Instrument specific
90° Pulse length	P1	X_90_width	pw90	Instrument specific
Spin rotation		Spin_set	spin	Optional
Measurement temperature	TE	Temp_get	temp	300 K
Frequency of excitation	o1	Irr_freq	tof	Middle of spectrum
Pulse angle		X_angle	pw	30°
Preacquisition delay	DE	Initial_wait	alfa	5 μs
Acquisition time	AQ	X_acq_time	at	3.41 s
Relaxation delay	D1	Relaxation_delay	D1	$\geq(7/3)x$ longest T_1
Sweep width	SW	X_sweep	sw	16 ppm
Filter width	FW	Filter_width	fb	≥20 ppm
Number of FID points	TD	X_points	np	32 K
Number of scans	ns	Scans	nt	Declined of reached S/N
Signal-to-noise ratio	S/N	Sn_ratio	dsn	≥150
Line broadening (em)	lb	Width	lb	0.3 Hz
Number of frequency points	SI	X_points	fn	64 K

Taken from Ref. 34

at a single wavelength with no knowledge of the extinction coefficient/relative response factors.

The above description provides a very fast approach yet it is not a rigorous quantitative method. To make this a rigorous relative quantitative method one needs to measure a few critical NMR parameters that are intrinsic to the compounds in the mixture. In addition, there have been a few papers that have been published that detail general sets of parameters to be used to achieve a particular level of confidence in the measurement [35, 36]. Publications by Malz et al. go into a very good level of detail concerning the validation of quantitative methods [34, 35]. Utilizing careful experiment parameterization one can get ~0.5–2% accuracies in their quantitative measurements. A "validated" parameter setup is shown in Table 31.1. This table has been taken directly from the paper published by Malz et al [35]. Malz et al. also deal with attributes related to specificity and selectivity as well as accuracy, precision, measurement uncertainty, and sensitivity. It is the author's opinion that this reference should serve as the primary reference to begin ones exploration into the use of NMR as a quantitative tool.

The use of the absolute method traditionally involves the use of internal standards that are added into the NMR sample and used as an internal reference to use the relative method described above. The amount of care one takes in both sample preparation and experiment setup will dictate the level of accuracies that are obtainable. The primary literature also provides several good reviews that include a vast number of potential internal standards to utilize. Standards should be selected for the sample at hand based on several factors including, chemical shift, similar molecular weight, relaxation rates, and so on. One should also consider the use of solvents that allow for reasonable manipulation without evaporation as this can also lead to inaccurate measurements.

TABLE 31.2 Data Showing the Variation of Potency Determination with Variable Wavelength and NMR

Method	% BHT	% API
HPLC-UV Area% (200 nm)	16.8	83.2
HPLC-UV Area% (214 nm)	11.2	88.8
HPLC-UV Area% (224 nm)	2.7	97.3
HPLC-UV Area% (282 nm)	0.98	99.0
HPLC-UV RF wt/wt%	2.5	
qNMR wt/wt%	1.94	94.2

One thing that has not been discussed is the criticality of processing of the data once it is acquired. Several steps should be taken to ensure proper data processing. Three of the most important features involve the use of baseline correction, proper phasing of the resultant Fourier transformed spectrum, and choosing appropriate integration ranges to ensure the majority of the signal is encompassed in the integration [34–36].

31.4.3 Electronic Referencing in qNMR

The use of internal reference materials has been the method of choice for many years to do quantitative measurements. In the pharmaceutical industry, this is particularly true as it relates to potency measurements, mass balance, and so on. An alternative approach has been gaining traction for those that work in this area. This approach utilizes an electronic signal that is generated artificially and is calibrated to a particular value and subsequently inserted into the collected NMR spectrum and used as a "standard" for comparison. The method was published under the acronym of ERETIC [38]. The original application was developed for imaging. It was later elaborated for use in high-resolution NMR [39]. In effect, an electronic signal is routed through a "spare" coil in the probe. The amplitude of the signal is set by the operator and this signal can be inserted at a user defined resonance frequency so as to not interfere with any of the signals related to the compound under study. This signal can then be used with a reference standard independently related back to the sample of interest using the same amplitude.

Subsequent to the ERETIC method there have been alternate approaches developed. One such method uses software to simulate a signal that is incorporated into a reference spectrum [40]. The simulated signal is then integrated relative to the reference. As with the ERETIC, the signal is then placed in the spectrum containing the compound/chemistry under investigation as a reference. The simulated signal can be placed anywhere in the spectrum and can be crafted to have similar lineshape and intensity to the resonance/species of interest.

31.4.4 Quantitation in Flow NMR

The use of quantitation under flow-NMR conditions requires a bit more consideration than a static NMR tube arrangement. The theory and practicality of flow-NMR has been described thoroughly in the literature, including several comprehensive reviews [41–44]. Unlike static tube based experiments where all spins in the sample are uniformly experiencing B_0 and are therefore at Boltzmann distribution prior to excitation with a radiofrequency pulse, flow experiments require the scientist to adjust the flow rate in such a manner to ensure that the sample flowing through has enough residence time within the B_0 field that they reach Boltzmann distribution prior to the active region of the flow cell for excitation. To ensure quantitation, this condition must be met.

$$\frac{1}{T_{i,\text{flow}}} = \frac{1}{T_{i,\text{static}}} + \frac{1}{\tau} \quad \text{with } \tau = \frac{V_{\text{active}}}{V_{\text{flow}}} \tag{31.3}$$

Equation 31.3 shows the interdependency of the flow of the system to the intrinsic relaxation processes as well as the geometry of the probe ensuring quantitative conditions, where T_i represents either T_1 or T_2^* and τ is flow dependant [34]. Ensuring the sample is at or near Boltzmann's conditions prior to excitation is critical. The typical amount of time that the sample must reside in the magnetic field prior to irradiation is approximately five times T_1. It should be noted that this time should be calculated based on the entity with the slowest T_1. The experimental parameters employed for the measurement of T_1 values are described in detail elsewhere [9]. Calculation of the maximum flow rate while accomplishing optimal premagnetization of the spins within the sample can be done through equation 31.4.

$$V_{\text{flow,maximum}} = \frac{V_{\text{premag}}}{5T_{1,\text{max}}} \tag{31.4}$$

Within a standard 1D proton NMR experiment, there are several factors that must be taken into account. Two of these are the acquisition time (to ensure proper digitization), t_{aq}, and the relaxation time to ensure the vast majority of spins "reachieve" Boltzmann distribution, t_{d}. Both of these parameters (see Table 31.1) together are known as the pulse repetition time t_{p}. One can envision the interdependency of the flow and the repetition time in equation 31.5.

$$t_{\text{p}} = t_{\text{aq}} + t_{\text{d}} = \frac{V_{\text{active}}}{V_{\text{flow,max}}} \tag{31.5}$$

An example would be a compound with a T_1 of 5 s and a 100 μL premagnetization volume would have a maximum flow rate ($V_{\text{flow,maximum}}$) of ~ 0.24 mL/min (Figure 31.11).

31.4.5 Reaction Kinetics

One important, sometimes overlooked, application of quantitative NMR is the ability to monitor the progress and kinetics of a chemical reaction directly by NMR. A recent publication does a good job of reviewing this topic and provides many references to the primary literature. This section will give highlights of the methodology and a few examples [45]. A simplistic example of the use of NMR to

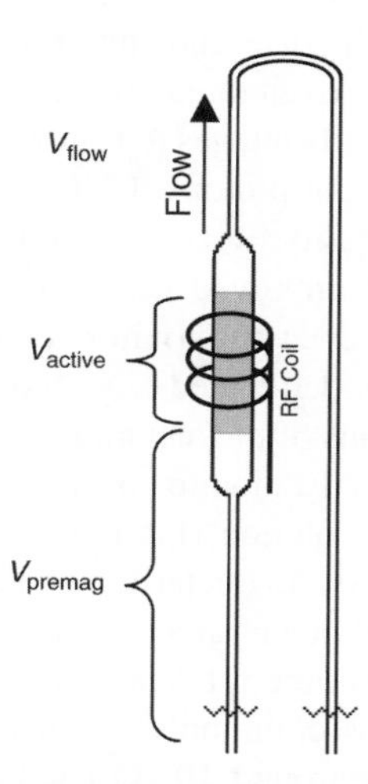

FIGURE 31.11 Diagram showing the key components within the flow path of the flow cell located in the NMR probe. This diagram is showing the key elements along the flow path that must be considered when designing flow-NMR experiments under quantitative condition.

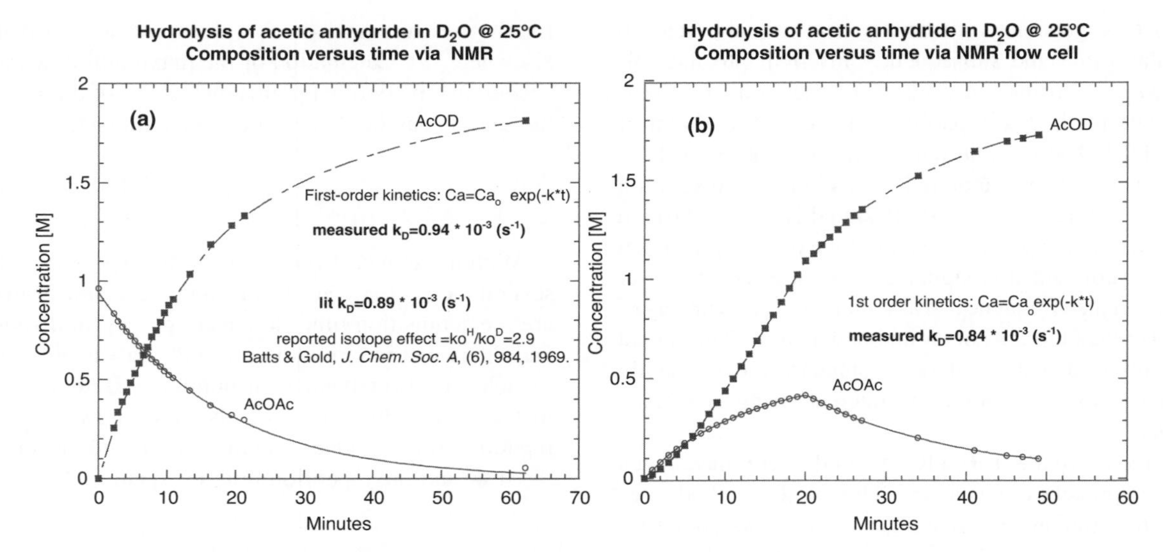

FIGURE 31.12 (a) NMR tube kinetics; the reagent is injected, NMR tube is shaken and placed in the magnet, and spectra are recorded. (b) Flow NMR kinetics; the reactor is integrated to flow-NMR and run in SemiBatch mode. The conditions of the reaction were 20 mL AcOAc dosed over 20 min into 200 mL of D_2O. The flow rate through the probe was 3 mL/min.

monitor reactions is shown for the hydrolysis of acetic anhydride to acetic acid in the presence of D_2O. Figure 31.12 shows the reaction progression under two different experimental conditions. One is static in an NMR tube and the other is measure in real time through the application of flow-NMR. Both of these methods give good accuracy in the measurement of this first-order rate constant.

The practice of monitoring reactions in NMR tubes has been used since the inception of NMR and involves preparing reactions on a small scale and initiating them in an NMR tube with a suitable deuterated solvent. Although there are certain and obvious advantages to following this practice, the use of solvents lacking protons ($CDCl_3$, D_2O, CCl_4, etc.) is by no means a necessity. Recently, with the advent of more stable magnets and improved instrument electronics, it has become routine practice to monitor reactions by collecting NMR spectra in non-deuterated (e.g. "no-D NMR") solvents [46]. The applications of the technique are wide ranging and can be used to directly monitor or assay most any reaction mixture or reagent solution (Figures 31.13 and 31.14). As mentioned previously, no deuterium solvents are used so the spectra are recorded in an "unlocked mode." One factor that facilitates the acquisition of NMR data in this way is that the concentration of most commonly used neat organic solvents is usually somewhere around 10 M. The concentration of the reactants in most reaction solutions is in the range of 0.1–1 M (and of solutions of most commercial reagents about 0.5–2.5 M). Thus, the ratio of solvent to solute/analyte molecules is usually between 100:1 and 10:1 in most solutions of interest. It is a straightforward matter for most NMR spectrometer hardware to handle dynamic ranges of proton intensities of greater than four orders of magnitude. If needed, techniques can be applied that use special "pulse sequences" to suppress unwanted signals from the solvent [34]. By doing so, the dynamic range and spectral quality of the reactions species can be improved.

Other more advanced techniques such as flow-NMR can be utilized to monitor larger scale reactions in real time. The use of flow-NMR has been there for many years and has been used to monitor everything from reaction completion, kinetics, combinatorial chemistry, and so on [44]. Of late, there has been a great deal of activity in the area of monitoring process chemistry using flow-NMR [45, 47–51].

There have been many different designs (instrument setup) to accomplish real-time reaction monitoring. However, one that works very well is that published by Maiwald et al. and is used by many others in the field [45]. The basic premise is to have a fast loop that carries the reaction mixture from a reactor to ensure that the loop is just a "real-time" extension of the reactor. From this loop, there is a split (many ways to accomplish this) that allows a slow loop to flow into the NMR flow cell. One must keep in mind that the criteria must be met as described in Section to be under quantitative conditions if quantitation is desired. Obviously if one is interested in strictly observing the reaction for gross features (e.g., reaction completion and/or reaction optimization) the quantitative conditions are not needed and a less rigorous approach can be taken. A general schematic is described in Ref. 45.

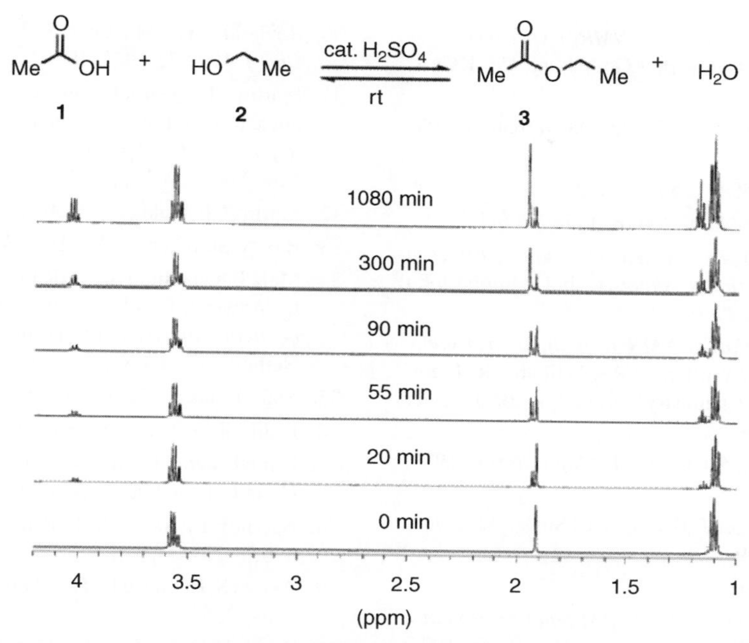

FIGURE 31.13 Example of the No-D NMR method Fischer esterification of acetic acid in ethanol (1:4 molar ratio). Reprinted with permission from Ref. 46.

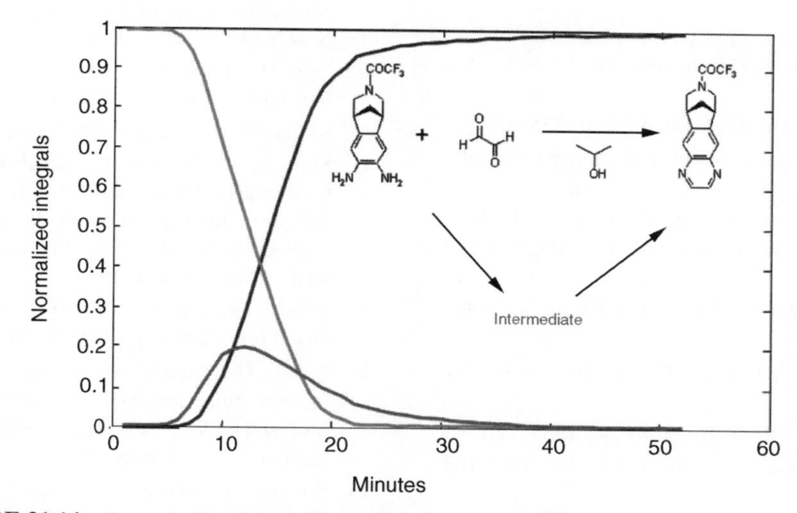

FIGURE 31.14 A graph showing the presence of an intermediate in the formation of varenicline. This intermediate was readily observed by NMR.

REFERENCES

1. Friebolin H. *Basic One- and Two-Dimensional NMR Spectroscopy*, VCH Publishers, New York, 1993.
2. Gunther H. *NMR Spectroscopy*, 2nd edition, John Wiley & Sons Ltd, New York, 1995.
3. Bloch F, Hensen WW, Packard M, *Phys. Rev.* 1946;69:127.
4. Derome AE. *Modern NMR Techniques for Chemistry Research*, Pergamon Press Ltd, Tarrytown, NY, 1987.
5. Claridge TDW. *High Resolution NMR Techniques in Organic Chemistry*, Pergamon Press Ltd, Tarrytown, NY, 1999.
6. Keeler J. *Understanding NMR Spectroscopy*, John Wiley & Sons Ltd, New York, 2006.
7. Levitt MH. *Spin Dynamics: Basics of Magnetic Resonance*, John Wiley & Sons Ltd, New York, 2002.
8. Croasmun WR, Carlson RMK. *Two-Dimensional NMR Spectroscopy: Applications for Chemists and Biochemists*, 2nd edition, VCH Publishers, NY, 1994.

9. Berger S, Braun S. *200 and More NMR Experiments: A Practical Course*, Wiley-VCH Verlag GmbH and Co. KGaA, Weinheim, Germany, 2004.
10. Overhauser AW. *Phys. Rev.* 1953;89:689;Overhauser AW. *Phys. Rev.* 1953;92:411.
11. Solomon I. *Phys. Rev.* 1955;99:559.
12. Richarz R, Wuthrich K. *J. Magn. Reson.* 1978;30:147.
13. Neuhaus D, Williamson M. *The Nuclear Overhauser Effect in Structural and Conformational Analysis*, 2nd edition, John Wiley & Sons, New York, 2000.
14. Claridge TDW. *High-resolution NMR techniques in organic chemistry*, In: Baldwin JE, Williams FRS, Williams RM, editors, Tetrahedron Organic Chemistry Series, Vol. 19, Elsevier Science, Oxford, UK, 1999.
15. Stott K, Keeler J, Van QN, Shaka AJ. *J. Magn. Reson.* 1997; 1125:302.
16. Stott K, Stonehouse J, Keeler J, Hwang T-L, Shaka AJ. *J. Am. Chem. Soc.* 1995;117:4199.
17. Hahn EL. *Phys. Rev.* 1949;76:145.
18. Levy GC, Nelson GL. *Carbon-13 Nuclear Magnetic Resonance for Organic Chemists*, Wiley-Interscience, New York, 1972; Stothers JB. *Carbon-13 NMR Spectroscopy*, Academic Press, New York, 1972.
19. Shaka AJ, Keeler J, *Prog. Nucl. Magn. Reson. Spectrosc.* 1987; 19:47–129.
20. Doddrell DM, Pegg DT, Bendall MR. *J. Magn. Reson.* 1982;48: 323–327; Doddrell DM, Pegg DT, Bendall MR, *J. Chem. Phys.* 1982;77:2745–2752.
21. Patt SL, Shoolery JN. *J. Magn. Reson.* 1982;46:535–539.
22. Jeener J,*Ampere International Summer School*, Basko Polje, 1971, proposal.
23. Aue WP, Bartholdi E, Ernst RR. *J. Chem. Phys.* 1976;64: 2229–2246; Bax A, Freeman R, Morris GA. *J. Magn. Reson.* 1981;42:164–168.
24. Piantini U, Sorensen OW, Ernst RR. *J. Am. Chem. Soc.* 1982; 104:6800.
25. Griesinger C, Sorensen OW, Ernst RR. *J. Am. Chem. Soc.* 1985;107:6394.
26. Braunschweiler L, Ernst RR. *J. Magn. Reson.* 1983;53:521.
27. Bax A, Freeman R, Frenkiel TA. *J. Am. Chem. Soc.* 1981;103: 2102.
28. Jeener J, Meier BH, Bachmann P, Ernst RR. *J. Chem. Phys.* 1979;71:4546. Please note this is the first appearance of the NOSY sequence, however, this publication deals with chemical exchange. For a quite comprehensive text on NOE, please see Ref. 13.
29. Müller L. *J. Am. Chem. Soc.* 1979;101:4481;Bodenhausen G, Ruben DJ. *Chem. Phys. Lett.* 1980;69:185–189.
30. Reynolds WF, MacLean S, Jacobs H, Harding WW. *Can. J. Chem.* 1999;77:1922–1930.
31. Martin GE. Qualitative and quantitative exploitation of heteronuclear coupling constants. In: Webb GA, editor, *Annual Report NMR Spectroscopy*, Vol. 46, Academic Press, New York, 2002, pp. 37–100.
32. Martin GE, Zektzer AS. *Magn. Reson. Chem.* 1988;26:631.
33. Bax A, Summers MF. *J. Am. Chem. Soc.* 1986;108:2093–2094.
34. Malz F. Quantitative NMR in the solution state. In: Holzgrabe U, Wawer I, Diehl B, editors, *NMR Spectroscopy in Pharmaceutical Analysis*, 1st edition, Elsevier, Amsterdam, The Netherlands, 2008, pp. 43–60.
35. Malz F, Jancke H. *J. Pharm. Biomed. Anal.* 2005;38:813.
36. Pauli GF, Jaki BU, Lankin DC. *J. Nat. Prod.* 2005;68:133.
37. United States Pharmacopoeia (USP) 30 (2007), No 761, The United States Pharmacopoeia Convention, Rockville, MD.
38. Barantin L, Akoka S, LePape A.French Patent CNRS No. 95 07651, 1995.
39. Akoka S, Barantin L, Trierweiler M. *Anal. Chem.* 1999;71(13): 2554–2557.
40. Wider G, Drefer L. *J. Am. Chem. Soc.* 2006;128 (8):2571–2576.
41. Albert K. *On-line LC-NMR and Related Techniques*, John Wiley & Sons, England, 2002.
42. Jones DW, Child TF. NMR in following systems. In: Waugh JS, editor, *Advances in Magnetic Resonance*, Academic Press, New York, 1978, Chapter 3.
43. Dorn HC. Flow NMR. In: *Encyclopedia of Nuclear Magnetic Resonance*, Wiley, New York, 1996, pp. 2026–2037.
44. Keifer PA. *Ann. Rep. NMR Spect.* 2007;62:1–47.
45. Maiwald M, Steinhof O, Sleigh C, Bernstein M, Hasse, H. Quantitative NMR in the solution State. In: Holzgrabe U, Wawer I, Diehl B, editors, *NMR Spectroscopy in Pharmaceutical Analysis*, 1st edition, Elsevier, Amsterdam, The Netherlands, 2008, pp. 471–491.
46. Thomas RH, Brian ME, Troy DR, Mikhail V, Letitia JY. *Org. Lett.* 2004; (6): pp. 953–956.
47. Zennie TM, Rothhaar R, Kaerner A.Reaction NMR: a method for real time reaction monitoring. Presented at the Small Molecules Are Still Hot NMR Conference, Burlington, Vermont, 2006, Poster 45.
48. Kaerner A, Marquez BL.RxnNMR: real time monitoring of chemical reactions and processes using high-field NMR and PAT. Presented at the Small Molecules are Still Hot NMR Conference, Santa Fe, NM, 2008, Conference workshop.
49. Maiwald M, et al. *J. Mag. Resonance*, 2004;166:135–146.
50. Hasse H, Albert K, et al. *Chem. Eng. Process.* 2005;44: 653–660.
51. Horvath IT, et al. *Chem. Rev.* 1991;91:1339–1351.

32

EXPERIMENTAL DESIGN FOR PHARMACEUTICAL DEVELOPMENT

GREGORY S. STEENO
Pfizer Global Research & Development, Groton, CT, USA

32.1 INTRODUCTION

In the pharmaceutical industry and in today's regulatory environment, process understanding in terms of characterization or optimization is critical in developing and manufacturing new medicines that ensure patient safety and drug efficacy. Experimentation is the key component in building that knowledge base, whether those activities are physical experiments or trials run *in silico*. This understanding comes from relating changes in observed response(s) back to changes, both intended and observational, in independent factors. If the factor changes are deliberate, then the scientist is testing hypotheses about factor effects and qualifying or quantifying their impact. However, what can be often neglected is how important the *quality* of the experimental plan is, and how that connects to making inferences from those data. This chapter focuses on both of these aspects, the experimental design and the data modeling.

For engineers characterizing a chemical reaction, experimental inputs span catalyst load, reaction concentration, jacket temperature, reagent amount, and other factors that are continuous in nature, as well as types of solvents, bases, catalysts, and other factors that are discrete in nature. These are examples of *controllable factors* and are represented as $x_1, x_2, \ldots, x_p$. Additional elements that can influence reaction outputs, such as analysts, instruments, and laboratory humidity, are examples of *uncontrollable factors* and are represented as $z_1, z_2, \ldots, z_q$. Figure 32.1, as shown by Montgomery [1], depicts a general process where both factors types impact the output.

The set of trials to develop relationships between factors and responses plus the structure of how the trials are executed, comprise the *experimental design*. What follows are strategies for sound statistical design and analysis.

As a motivating example, consider a process where the output is yield (g) and is expected to be a function of two controllable factors, reaction time (min) and reaction temperature (°C). That is, yield $= f(\text{time, temperature})$. As a first step in understanding and optimizing this process, experiments were executed by fixing time at 30 min and observing yield across a range of reaction temperatures. Figure 32.2 shows the data and it is concluded that 35°C is a reasonable choice as an optimal temperature. For the second step, experiments were executed by now fixing temperature at 35°C and observing yield across a range of reaction times. Figure 32.3 shows the data and it is concluded that the optimal time is about 40 min. The combined information from the totality of experiments produces an optimal setting of (temperature, time) = (35°C, 40 min) with a predicted yield around 70 g.

This experiment was conducted using a *one factor at a time* (OFAAT) approach—while easy to implement and instinctively sensible, there are a few shortcomings when compared to a statistically designed experiment. In general, OFAAT studies (1) are not as precise in estimating individual factor effects, (2) cannot estimate multivariate factor effects, such as linear × linear interactions, and (3) as a by-product are not as efficient at locating an optimum. Figure 32.4 illustrates the experimental path (• · · · •), the chosen optimum ("*XX*"), and the actual underlying relationship between time

Chemical Engineering in the Pharmaceutical Industry: R&D to Manufacturing Edited by David J. am Ende

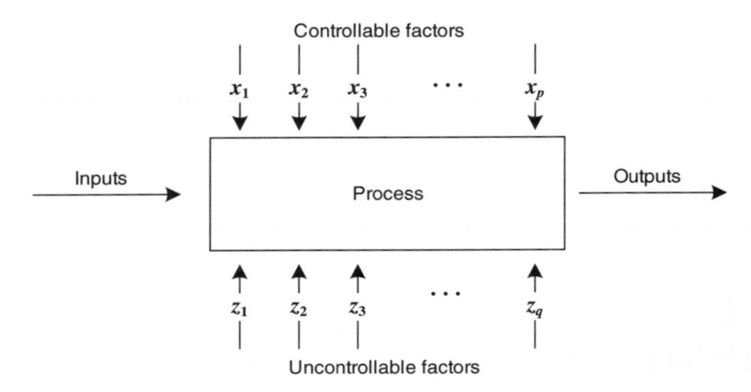

FIGURE 32.1 General process diagram.

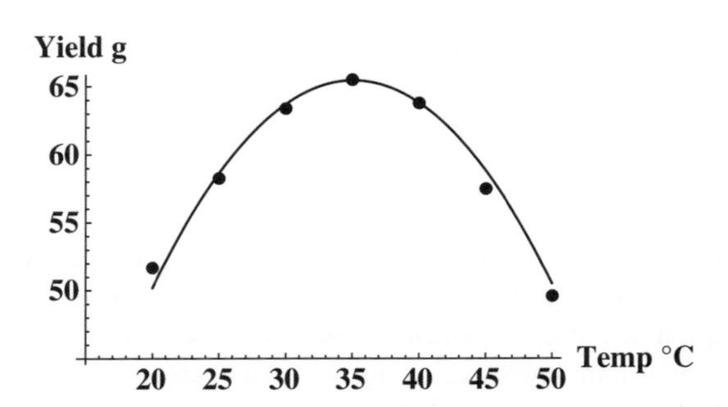

FIGURE 32.2 Scatterplot of yield against temperature with empirical fit.

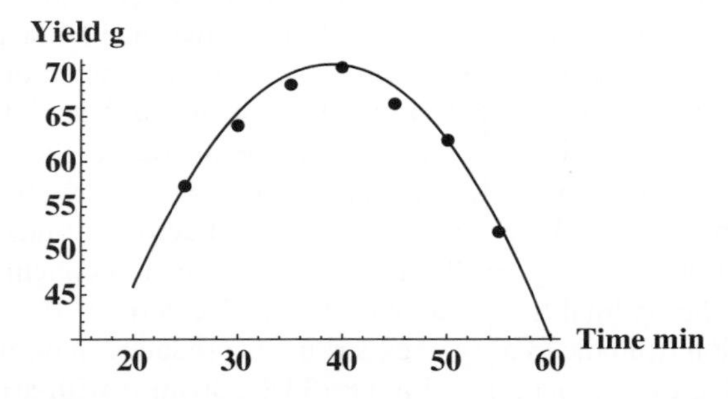

FIGURE 32.3 Scatterplot of yield versus time with empirical fit.

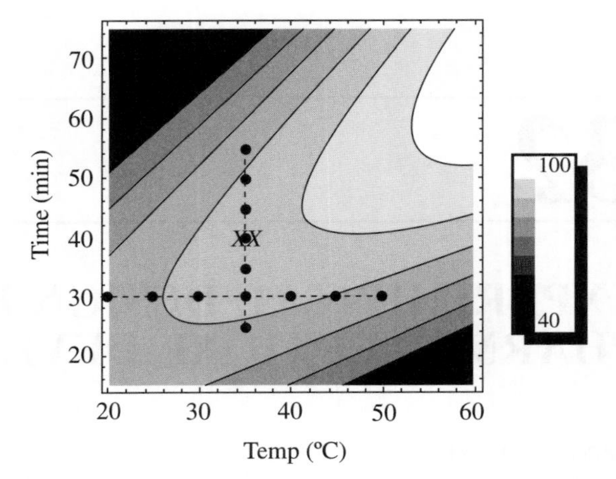

FIGURE 32.4 Contour plot of underlying mechanistic relationship of temperature and time on yield, along with experimental path (● · · · ●) and chosen optimum (*XX*).

and temperature on yield. Since the joint relationship between time and temperature was not fully investigated, the true optimum around (temperature, time) ≈ (60°C, 60 min) was missed.

This example illustrates that even with only two variables, the underlying mechanistic relationship between the factors and response(s) can be complex enough to easily misjudge. This is especially true of many pharmaceutical processes, where functional relationships are dynamic and nonlinear. Because of this complexity, experiments that produce the most complete information in the least amount of resource (time, material, and cost) are vital. This is a major aspect of *statistical experimental design*, which is an efficient method for evaluating process inputs in a systematic and multivariate way. The integration of experimental design and model building is generally known as *response surface methodology*, first introduced by Box and Wilson [2]. Data resulting from such a structured experimental plan coupled with regression analysis easily lend to deeper understanding of where process sensitivities exist, as well as how to improve process performance in terms of speed, quality, or optimality.

The experimental design and analysis procedure is straightforward and intuitive, but is described below for completeness to ensure that all information is collected and effectively used.

1. Formulate a research plan with purpose and scope
2. Brainstorm explanatory factors denoted $X_1, X_2, X_3, \ldots$, that could impact the response(s). Discuss ranges and omit factors that have little scientific value.
3. Determine the responses to be measured, denoted Y_1, Y_2, Y_3, ..., and consider resource implications.
4. Select appropriate experimental design in conjunction with purpose and scope. Consider the randomization sequence. Hypothesize process models.
5. Execute the experiment. Measurement systems should be accurate and precise. The randomization sequence is key to balancing effects of random influences and propagation of error.
6. Appropriately analyze the data
7. Draw inference and formulate next steps

The statistical experimental designs discussed in this chapter are used to help estimate approximating response functions for chemical process modeling. To begin, the functional relationship between the response, y, and input variables, $\xi_1, \xi_2, \ldots, \xi_k$, is expressed as

$$y = f(\xi_1, \xi_2, \ldots, \xi_k) + \varepsilon$$

which is unknown and potentially intricate. The inputs, $\xi_1, \xi_2, \ldots, \xi_k$, are called the natural variables, as they represent the actual values and units of each input factor. The ε term represents the variability not explicitly accounted for in the model, which could include the analytical component, the laboratory environment, and other natural sources of noise. For mathematical convenience, the natural variables are centered and scaled so that coded variables, $x_1, x_2, \ldots, x_k$ have mean zero and standard deviation one. This does not change the response function, but it is now expressed as

$$y = f(x_1, x_2, \ldots, x_k) + \varepsilon$$

If the experimental region is small enough, $f(\cdot)$ can be empirically estimated by lower order polynomials. The motivation comes from Taylor's theorem that asserts any sufficiently smooth function can locally be approximated by polynomials. In particular, first-order and second-order polynomials are heavily utilized in response modeling from designed experiments.

A first-order polynomial is referred to as a *main effects model*, due to containing only the primary factors in the model. A two-factor main effects model is expressed as

$$y = \beta_0 + \beta_1 x_1 + \beta_2 x_2 + \varepsilon$$

where β_1 and β_2 are coefficients for each factor and β_0 is the overall intercept, and represents a plane through the (x_1, x_2) space. As an example consider an estimated model

$$\hat{y} = 100 - 10x_1 + 5x_2$$

Figure 32.5 shows a 3D view of that planar response function, also called a surface plot. Figure 32.6 represents the 2D analogue called a contour plot. The contour plots are often easier to read and interpret since the response function height

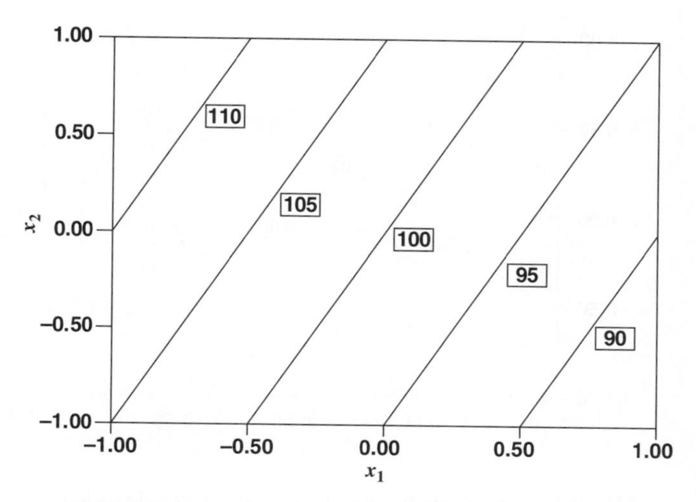

FIGURE 32.6 Contour plot of $\hat{y} = 100 - 10x_1 + 5x_2$.

is projected down onto the (x_1, x_2) space. If there is an interaction between the factors, it is easily added to the model as follows:

$$y = \beta_0 + \beta_1 x_1 + \beta_2 x_2 + \beta_{12} x_1 x_2 + \varepsilon$$

This is called a first-order model with interaction. To continue with the example, let the estimated model be

$$\hat{y} = 100 - 10x_1 + 5x_2 - 5x_1 x_2$$

The additional term $-5x_1x_2$ introduces curvature in the response function, which is displayed on the surface plot in Figure 32.7 and the corresponding contour plot in Figure 32.8. Occasionally, the curvature in the true underlying response function is strong enough that a first-order plus interaction model is inadequate for prediction. In this case, a second-order (quadratic) model would be useful to approximate $f(\cdot)$ and takes the form

$$y = \beta_0 + \beta_1 x_1 + \beta_2 x_2 + \beta_{12} x_1 x_2 + \beta_{11} x_1^2 + \beta_{22} x_2^2 + \varepsilon$$

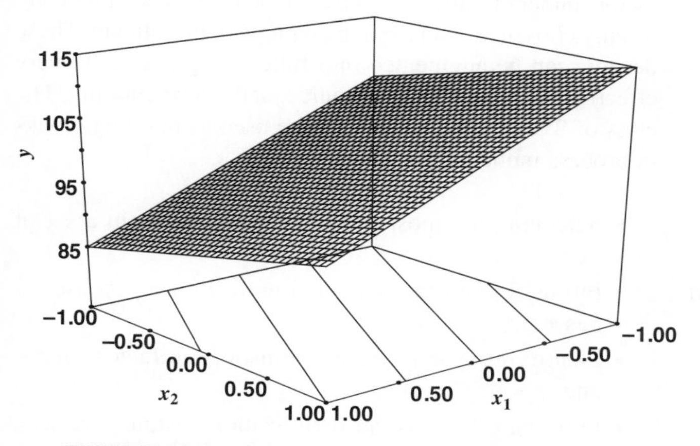

FIGURE 32.5 Surface plot of $\hat{y} = 100 - 10x_1 + 5x_2$.

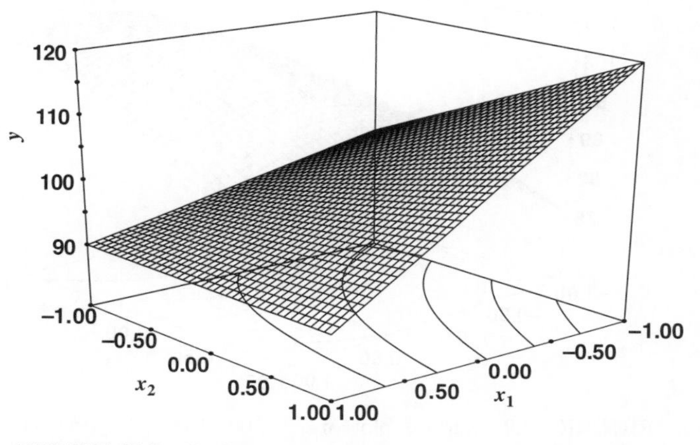

FIGURE 32.7 Surface plot of $\hat{y} = 100 - 10x_1 + 5x_2 - 5x_1x_2$.

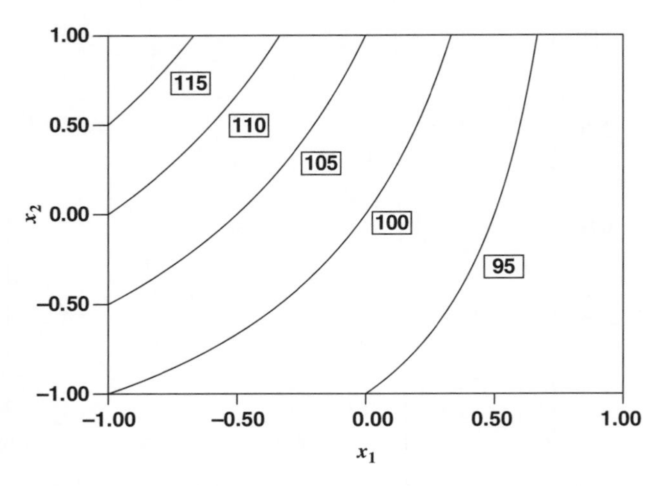

FIGURE 32.8 Contour plot of $\hat{y} = 100 - 10x_1 + 5x_2 - 5x_1x_2$.

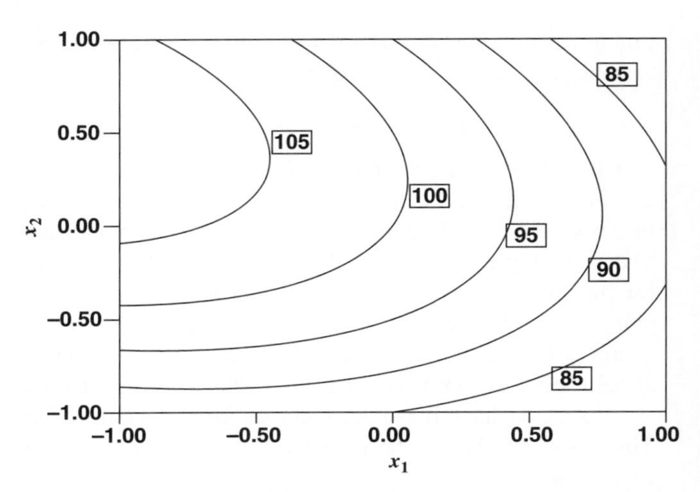

FIGURE 32.10 Contour plot of $\hat{y} = 100 - 10x_1 + 5x_2 - 5x_1x_2 - 4x_1^2 - 10x_2^2$.

To finish the example, let the estimated model be

$$\hat{y} = 100 - 10x_1 + 5x_2 - 5x_1x_2 - 4x_1^2 - 10x_2^2$$

Figure 32.9 shows a parabolic relationship between y and x_1, x_2, while Figure 32.10 displays the typical elliptical contours generated by this model.

There is an iterative, sequential nature to understanding and optimizing the performance of chemical processes. If the goal is to first identify the most important factors for further study, a screening design may be carried out. This is sometimes referred to as *phase zero* of the study. Once this is complete, the next objective is to determine if the optimum lies within current experimental region, or if the factors need adjustment to locate a more desirable one, say by using methods of steepest ascent/descent. This is referred to as *phase one* of the study, also known as *region seeking*. Finally, once the region of desirable response is established, the goal becomes to precisely model that area and identify optimal factor settings. For this case, higher order models are employed to capture likely curvature about the optimum point.

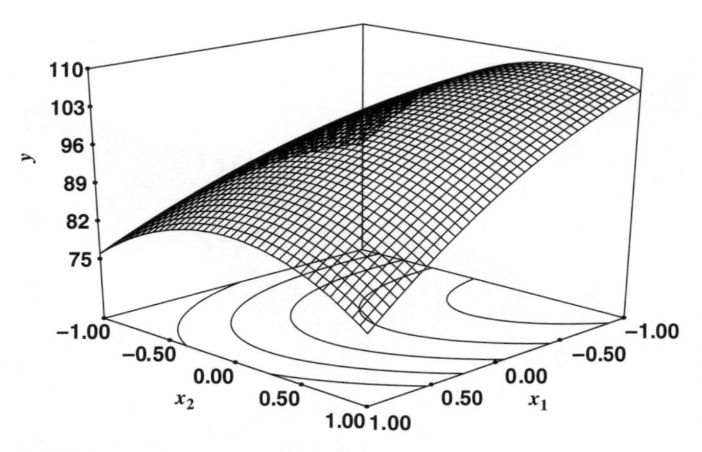

FIGURE 32.9 Surface plot of $\hat{y} = 100 - 10x_1 + 5x_2 - 5x_1x_2 - 4x_1^2 - 10x_2^2$.

32.2 THE TWO-LEVEL FACTORIAL DESIGN

Factorial designs are experimental plans that consist of all possible combinations of factor settings. As an example, a factorial design with three different catalysts, two different solvents, and four different temperatures produces a design with $3 \times 2 \times 4 = 24$ unique experimental conditions. The advantage of these designs is that all joint effects of factors can be investigated. The disadvantage is that these designs become prohibitively large and impractical when factors contain more than just a few levels or the number of factors under investigation is extensive.

The simplest and most widely used factorial designs for industrial experiments are those that contain two levels per factor, called 2^k factorial designs, where k is the number of factors under investigation. The two levels for each factor are usually chosen to span a practical range to investigate. These designs can be augmented into fuller designs and are very effective in terms of time, resource, and interpretability. The class of 2^k factorial designs can be used as building blocks in process modeling by:

- Screening the most important variables from a set of many;
- Fitting a first-order equation used for steepest ascent/descent;
- Identifying synergistic/antagonistic multifactor effects; and
- Forming a base for an optimization design, such as a central composite (to be introduced).

TABLE 32.1 Example 2^2 Experimental Design with $n=2$ Replicates Per Design Point

Treatment	Design: Temperature (A)	Design: Time (B)	Temperature × Time (AB)	Response: Rep 1	Response: Rep 2
(1)	−1	−1	1	y_{11}	y_{12}
a	1	−1	−1	y_{21}	y_{22}
b	−1	1	−1	y_{31}	y_{32}
ab	1	1	1	y_{41}	y_{42}

Factors can either be continuous in nature or discrete. For the rest of the chapter it is assumed that the factors are continuous. This allows for predictive model building and regression analysis that includes the linear and interaction terms, and subsequently the quadratic/second-order terms.

To illustrate a two-level factorial design, consider the previous case where there are $k=2$ factors, temperature (Factor A) and time (Factor B), each having an initial range under investigation. In coded units, the low level of the range for each factor is scaled to -1, and the high level of the range is scaled to $+1$. The 2^2 experimental design with all four treatment combinations is shown in Table 32.1 and graphically depicted via circles in Figure 32.11. Notice that each treatment condition occurs at a vertex of the experimental space. For notation, the four treatment combinations are usually represented by lowercase letters. Specifically, a represents the combination of factor levels with A at the high level and B at the low level, b represents A at the low level and B at the high level and ab represents both factors being run at the high level. By convention, (1) is used to denote A and B each run at the low level.

Two-level designs are used to estimate two types of effects, main effects and interaction effects, and these are estimated by a single degree of freedom *contrast* that

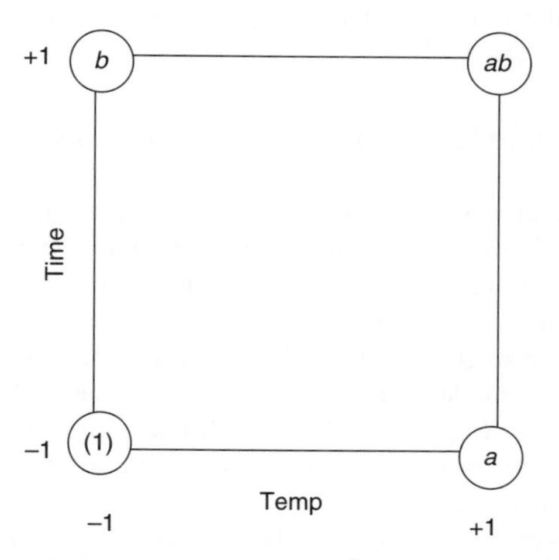

FIGURE 32.11 Factor space of 2^2 experimental design.

partitions the design points into two groups: the low level (-1) and the high level $(+1)$. The contrast coefficients are shown in Table 32.1 for each of the main factors of temperature and time, as well as the temperature × time interaction obtained through pairwise multiplication of the main factor contrast coefficients.

A main effect of a factor is defined as the average change in response over the range of that factor and is calculated from the average difference between data collected at the high level $(+1)$ and data collected at the low level (-1). For the 2^2 design above and using the contrast coefficients in Table 32.1, the temperature (A) main effect is estimated as

$$A=\bar{y}_{\text{Temp}+}-\bar{y}_{\text{Temp}-}=\frac{ab+a}{2n}-\frac{b+(1)}{2n}=\frac{1}{2n}[ab+a-b-(1)]$$

where (1), a, b, and ab are the respective sum total of responses across the n replicates at each design point ($n=2$ in Table 32.1). Geometrically this is a comparison of data on average from the right side to the left side of the experimental space in Figure 32.11. If the estimated effect is positive, the interpretation is that average response increases as the factor level increases. Similarly, the time (B) main effect is estimated as

$$B=\bar{y}_{\text{Time}+}-\bar{y}_{\text{Time}-}=\frac{ab+b}{2n}-\frac{a+(1)}{2n}=\frac{1}{2n}[ab+b-a-(1)]$$

which is a comparison of data on average from the top side to the bottom side in Figure 32.11.

An interaction between factors implies that the individual factor effects are not additive and that the effect of one factor depends on the level of another factor(s). As with the main effects, the interaction is estimated by partitioning the data into two groups and comparing the average difference. The contrast coefficients in Table 32.1 show that the temperature × time (AB) interaction effect is estimated as

$$AB=\bar{y}_{\text{Temp}\times\text{Time}+}-\bar{y}_{\text{Temp}\times\text{Time}-}=\frac{ab+(1)}{2n}-\frac{a+b}{2n}$$
$$=\frac{1}{2n}[ab+(1)-a-b]$$

which is a comparison on average of data on the right diagonal against the left diagonal in Figure 32.11.

The sum of squares for each effect are mathematically related to their corresponding contrast. Specifically, the sum of squares for an effect is calculated by the squared contrast divided by the total number of observations in that contrast. For the example above, the sums of squares for temperature, time, and the temperature × time interaction are

$$SS_{\text{Temp}}=\frac{[a+ab-b-(1)]^2}{4n}$$

$$SS_{\text{Time}}=\frac{[b+ab-a-(1)]^2}{4n}$$

TABLE 32.2 ANOVA Table for Completely Randomized 2^2 Design with n Replicates Per Design Point

Source	SS	DF	MS	F	p-value
Temp	SS_{Temp}	1	MS_{Temp}	MS_{Temp}/MS_E	p_{Temp}
Time	SS_{Time}	1	MS_{Time}	MS_{Time}/MS_E	p_{Time}
Temp × Time	$SS_{Temp\times Time}$	1	$MS_{Temp\times Time}$	$MS_{Temp\times Time}/MS_E$	$p_{Temp\times Time}$
Error	SS_E	$4(n-1)$	MS_E		
Total	SS_T	$4n-1$			

$$SS_{Temp\times Time} = \frac{[ab + (1) - a - b]^2}{4n}$$

In 2^k designs, the contrasts are orthogonal, thus additive. The total sum of squares, SS_T, is the usual sum of squared deviations of each observation from the overall mean of the data set. Because the contrasts are orthogonal, the error sum of squares, denoted SS_E, can be calculated as the difference between the total sum of squares, SS_T, and the sums of squares of all effects. For the 2^2 example, $SS_E = SS_T - SS_{Temp} - SS_{Time} - SS_{Temp\times Time}$. With this information, the analysis of variance (ANOVA) table is constructed, as shown in Table 32.2.

The ANOVA table contains all numerical information in determining which factor effects are important in modeling the response. The hypothesis test on individual factor effects is conducted through the F-ratio of MS_{Factor} against MS_{Error}. If this ratio of "signal" against "noise" is large, this implies that the factor explains some of the observed variation in response across the experimental design region and should be included in the process model. If the ratio is not large, then the inference is that the factor is unimportant and should be deleted from the model. All statistical evidence of model inclusion comes via the p-value. Large F-ratios imply low p-values, and a common cutoff for model inclusion of a factor is $p \leq 0.05$, although this should be appropriately tailored with experimental objectives, such as factor screening where the critical p-value is normally a little higher. Another interpretation of the p-value is in terms of the confidence level, equal to $(1-p) \times 100\%$. Thus, a factor with a p-value less than 0.05 implies there is greater than 95% confidence that the observed factor effect is real and not due to noise.

Clearly it is important to identify all significant factors for modeling change in response back to change in factor level. An *underspecified* model, one that does not contain all the important variables, could lead to bias in regression coefficients and bias in prediction. One approach to mitigate this issue is to not model edit but rather incorporate all factor terms in the process model, including those that contribute very little or nothing of value in predicting. However, an *overspecified* model, one that contains insignificant terms, produces results that lead to higher variances in coefficients and in prediction. Thus, a proper model will be a compromise of the two. This can be completed manually, say, by investigating the full model ANOVA and then deleting insignificant effects one at a time, all while updating the ANOVA after every step. Yet for models that could contain many effects, this exercise becomes cumbersome. There are several variable selection procedures that can aid in helping identify smaller sized candidate models. The most common algorithms used in standard software packages entail either sequentially bringing in significant factors to build the model up (called forward selection), sequentially eliminating regressors from a full model (called backward elimination), or a hybrid of the two (called stepwise regression). The procedures typically involve defining a critical p-value for factor inclusion/exclusion in the model building process. Once a term enters or leaves, factor significance is recalculated and the process is repeated for the next step. The engineer should use these tools not as a panacea to the model building process, but rather as an exercise to see how various models perform. For more information on the process and issues of model selection, the reader is instructed to see Myers [3].

Once the significant factors are selected, the estimated regression coefficients in the linear predictive model are functionally derived from the factor's effect size. To estimate the regression coefficient β_i for factor i, its effect is divided by two. The rationale being that by definition the regression coefficient represents the change in y per unit change in x. Since each factor effect is calculated as a change in response over a span two coded units (-1 to 1), division by 2 is needed to obtain the per-unit basis. Finally, the model intercept, β_0, is calculated as the grand average of all the data.

EXAMPLE 32.1 2^3 FACTORIAL DESIGN

A factorial experiment is carried out to investigate the effect of three factors on percent reaction conversion, catalyst load (Factor A), ligand load (Factor B), and temperature (Factor C). Each experimental condition is completely randomized and independently replicated ($n = 2$). The design in coded units, full model, and data are listed in Table 32.3, and depicted in Figure 32.12. Note that the run sequence in Table 32.3 is in *standard order*, as opposed to a randomized order for the actual experiment.

As previously remarked, the estimated effects and associated sums of squares are functions of their respective

TABLE 32.3 2^3 Reaction Conversion Experimental Design, Model, and Data

Treatment	Design			*AB*	*AC*	*BC*	*ABC*	Data	
	Catalyst *A*	Ligand *B*	Temperature *C*					Conversion	
								Rep 1	Rep 2
(1)	−1	−1	−1	1	1	1	−1	63.8	64.1
a	1	−1	−1	−1	−1	1	1	76.4	74.9
b	−1	1	−1	−1	1	−1	1	66.5	63.7
ab	1	1	−1	1	−1	−1	−1	76	76.3
c	−1	−1	1	1	−1	−1	1	78.7	77.7
ac	1	−1	1	−1	1	−1	−1	77.6	80.4
bc	−1	1	1	−1	−1	1	−1	78	81.3
abc	1	1	1	1	1	1	1	79	77.2

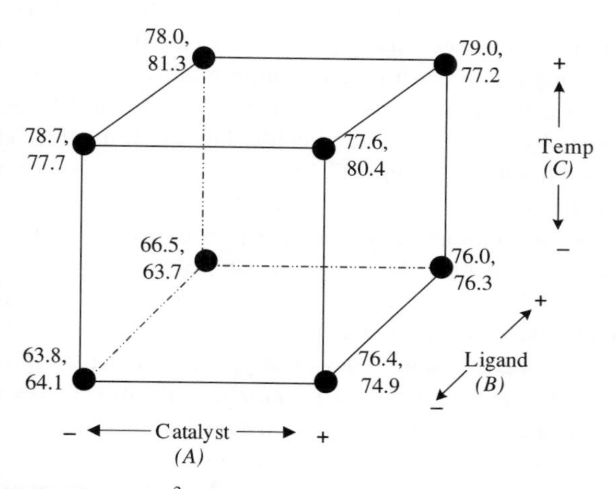

FIGURE 32.12 2^3 reaction conversion experimental space and corresponding data.

contrasts. Using catalyst load (*A*) as an example, the contrast coefficients shown in Table 32.3 represent the comparison between data on the right side of the cube (+) in Figure 32.12 to the data on the left side of the cube (−). The data where catalyst load is high sum to

$$a + ab + ac + abc = 76.4 + 74.9 + 76.3 + 76.0 + 77.6 + 80.4 + 79.0 + 77.2 = 617.8$$

while the data where catalyst load is low sum to

$$b + c + bc + (1) = 66.5 + 63.7 + 78.7 + 77.7 + 78.0 + 81.3 + 63.8 + 64.1 = 573.8$$

The estimated effect of catalyst load is calculated as

$$A = \bar{y}_{\text{Cat}+} - \bar{y}_{\text{Cat}-} = \frac{a + ab + ac + abc}{4n} - \frac{b + c + bc + (1)}{4n} = \frac{617.8}{4 \times 2} - \frac{573.8}{4 \times 2} = 5.5$$

TABLE 32.4 Reaction Conversion Experiment ANOVA Table with Full Model

Source	SS	DF	MS	*F*	*p*-value
A (catalyst)	121.00	1	121.00	58.24	<0.0001
B (ligand)	1.21	1	1.21	0.58	0.4673
C (temperature)	290.70	1	290.70	139.93	<0.0001
AB	2.25	1	2.25	1.08	0.3284
AC	138.06	1	138.06	66.46	<0.0001
BC	0.30	1	0.30	0.15	0.7127
ABC	0.72	1	0.72	0.35	0.5717
Error	16.62	8	2.08		
Total	570.87	15			

The interpretation is that the average yield increases by 5.5% as the catalyst load increases from low to high. The corresponding sum of squares for the catalyst load effect is calculated as

$$\text{SS}_A = \frac{[a + ab + ac + abc - b - c - bc - (1)]^2}{8n} = \frac{(617.8 - 573.8)^2}{8 \times 2} = 121$$

The other estimated factor effects and sums of squares follow the same logic as above and are trivial to calculate. The full model ANOVA table is shown in Table 32.4. Based on the information, there are three highly significant effects ($p < 0.0001$): catalyst load, temperature, and their corresponding interaction. All other effects are insignificant at the 95% confidence level ($p > 0.05$). The final model and ANOVA table after sequential model editing are shown in Table 32.5.

This model accounts for ~96.3% of the observed variability in reaction completion, as determined by the coefficient of determination, R^2.

TABLE 32.5 Reaction Conversion Experiment ANOVA Table After Model Editing

Source	SS	DF	MS	F	p-value
A (catalyst)	121.00	1	121.00	68.80	<0.0001
C (temperature)	290.70	1	290.70	165.29	<0.0001
AC	138.06	1	138.06	78.50	<0.0001
Error	21.10	12	1.76		
Total	570.87	15			

$$R^2 = \frac{SS_{Model}}{SS_{Total}} = \frac{SS_A + SS_C + SS_{AC}}{SS_{Total}} = \frac{549.76}{570.87} = 0.9630$$

The final regression model expressed in coded units is estimated as

$$\hat{y} = 74.48 + \left(\frac{5.5}{2}\right)x_1 + \left(\frac{8.26}{2}\right)x_3 + \left(\frac{-5.87}{2}\right)x_1x_3$$

$$\Rightarrow \hat{y} = 74.48 + 2.75x_1 + 4.26x_3 - 2.94x_1x_3$$

where x_1 and x_3 represent catalyst load and temperature, respectively.

Because the factors are all centered and scaled and all the effects are orthogonal, one can compare which effects are the most dominant by the size of the coefficient. Using the reaction conversion model, temperature has the largest effect, followed by the catalyst × temperature interaction and then catalyst. However, because of the interaction between catalyst and temperature, the main effects of those individual factors have lost some interpretability. Specifically, the conclusion from the temperature effect is that for every unit change in temperature, the conversion increases 4.26% via the coefficient on the temperature term. But this estimate is a pooled average over the other factors. That is, it smoothes over the significant joint effect between temperature and catalyst. The information contained in the interaction will need to be visually explored in greater detail.

After obtaining the final equation, residual analysis and other model diagnostics are carried out. This is critical step in validating the process model and having trust in its ability to accurately predict over the experimental region. Standard residual analyses consist of inspecting the normality assumption, checking for constant variance, identifying outliers relative to the model, and observing patterns in residuals over time. There are many flavors of model diagnostic information, both numerical and graphical. With the aid of Design Expert [4] software package, highlighted below are two visuals that, in this author's view, capture a significant snapshot of model performance. First, the normal probability plot is an effective graphical tool to verify the normality assumption of the errors as well as for outlier detection. Figure 32.13 shows this plot for the reaction conversion residuals using the final regression model. If the residuals fall along a straight line, then the normality assumption is valid. Any large errors or outliers from the fitted model would be visually apparent by significantly falling off the line. The interpretation of Figure 32.13 is that there is no severe problem with the normality assumption.

Second, another very effective plot is the model-based predictions against the observed data, often referred to as "Predicted versus Actual." This reveals how well the model predicts back the original data and is a graphical depiction of the calculated R^2 value. Additionally, it will aid in identifying (sets of) data that are not well captured by the model, as well as indicate any trends of nonconstant variance across the prediction range. Figure 32.14 is an example using the reaction conversion data and final model. The interpretation from this graph is that the regression model is performing well and the spread of about the 45° line is relatively constant across the range.

Once the engineer is satisfied with the diagnostics information, visualizing the change in response across the

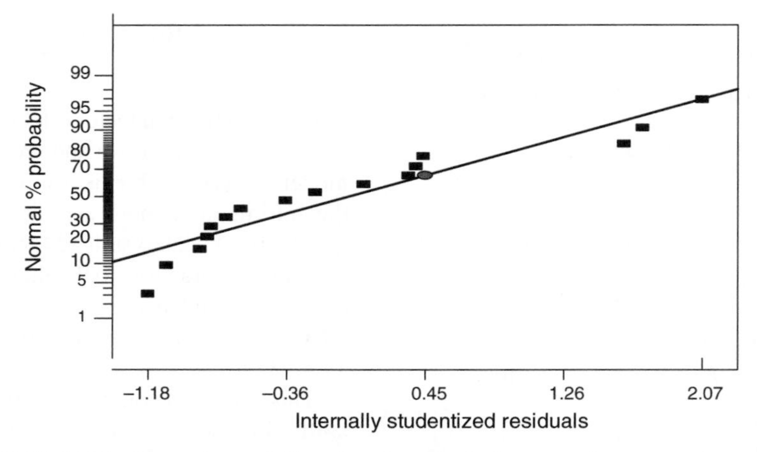

FIGURE 32.13 Normal probability plot of residuals from reaction conversion model.

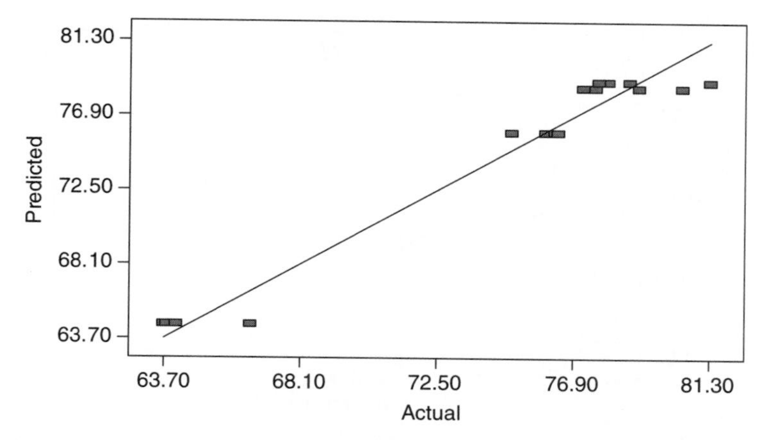

FIGURE 32.14 Predicted versus actual plot using reaction conversion example.

subspace of significant factors is the next step. This is usually accomplished by main effect and interaction plots. If factors are continuous, then contour and surface plots are very informative and more descriptive in illustrating bivariate factor effects on response. As previously shown with the reaction conversion example, the interaction between catalyst and temperature is significant and therefore the synergistic relationship between these two factors needs further inspection. Figure 32.15 displays an interaction plot for these two factors. Notice that when temperature is at the high (+ 1) level, there is no observed effect of catalyst load on the reaction conversion. That is, it is robust to changes in catalyst. However, when temperature is at the low (−1) level, reaction conversion is now a function of catalyst load, and higher load leads to higher predicted conversion. Figure 32.16 shows the corresponding contour plot that also illustrates the predicted change in conversion across the temperature and catalyst levels, but takes advantage of the continuous nature of the factors. Regardless, the inference is the same: At high temperatures, the prediction is virtually constant across catalyst (~78% conversion), while at lower temperatures, the catalyst effect is present.

Two-level designs are very intuitive, comprehensive, and powerful in identifying main and interaction effects on responses of interest. However, even at two levels they can become impractically large as the number of factors increases. For example, if $k = 6$ factors are under investigation, then a full factorial experiment with no replication would consist of $2^6 = 64$ runs, which is not a cost-effective experiment. If this design were executed, the full regression model with all possible main effects and interactions would leave zero degrees of freedom to estimate variability for statistical inferences on factor effects. There are strategies and tools that combat the issues of unreplicated designs and variance estimation. One such strategy takes advantage of the *sparsity-of-effects principle* (or similarly, the Pareto principle) that states that a process is usually dominated by only the vital few effects, such as main effects and two-factor interactions, from the trivially many. More specifically, observing effects from higher order terms such

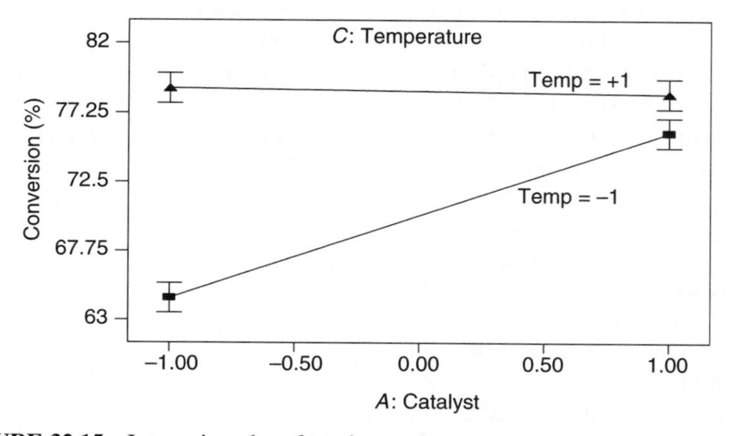

FIGURE 32.15 Interaction plot of catalyst and temperature on reaction conversion.

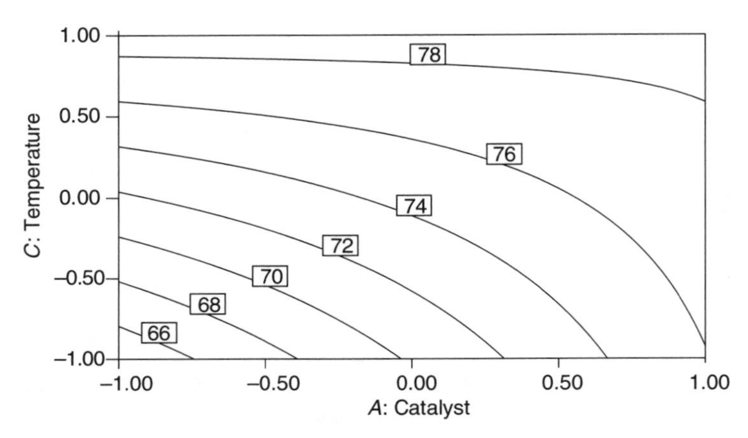

FIGURE 32.16 Contour plot of predicted reaction conversion across catalyst and temperature levels.

as three-factor interactions and beyond is rare in practice. If it is reasonable to assume that effects from higher order interactions are negligible, then those terms can be pooled to estimate variability through the mean squared error. Another approach is to use a normal probability plot or half-normal probability plot to determine the significant effects. Proposed by Daniel [5], these are effective graphical tools that use the estimated effects to highlight which factors are important in explaining response variation relative to those that do not. Negligible effects are assumed to be normally distributed with mean zero and variance σ^2, while significant effects are assumed normally distributed at their true effect size different from zero with variance σ^2. Normal and half-normal effect plots are standard in most statistical software packages.

32.3 BLOCKING

There are situations that call for the design to be run in groups or clusters of experiments. This often occurs when the number of factors is large, in which case the design would be broken down into smaller *blocks* of more homogeneous experimental units. These are referred to as incomplete blocks, as not all treatment combinations occur in these smaller sets. This situation also occurs with equipment set up or limitations, where groups of experiments are executed at the same time. As one example, consider a replicated 2^2 design that investigates temperature and solvent concentration on reaction completion. The equipment chosen for the experiment is a conventional process chemistry workstation that has four independent reactors. A natural and appropriate way to execute this design is to run each replicate of the 2^2 design on the workstation with random assignment of the reactor vessels to each treatment combination. By doing this, any block-to-block variability is accounted for and does not influence the analysis on factor effects. As another example, consider a 2^3 design using the same four-reactor workstation that investigates temperature, solvent concentration, and catalyst loading on reaction completion. To execute this study, the design should be split into two groups of four, since it is not possible to execute all experiments in one block and impractical to execute the experiments one at a time.

Both of these examples result in observations within the same block to be more homogeneous than those in another block. For situations where the experiment is subdivided, appropriate blocking can help in optimally constructing a design based on the assumption or knowledge that certain (higher order) interactions are negligible. This design technique is called *confounding* or *aliasing*, where information on treatment effects is indistinguishable from information on block effects. The number of blocks for the two-level design is usually a multiple of two, implying designs are run in blocks of two, four, eight, and so on.

To illustrate, recall the previous 2^3 design with temperature (A), solvent concentration (B), and catalyst loading (C) as the factors, with the design executed in two blocks of four on the workstation. The design and full model with associated contrast coefficients are shown in Table 32.6.

One candidate design partitioned into two blocks of four is to group all combinations where ABC is at the low level (-1) into one block, and all combinations where ABC is at the high

TABLE 32.6 2^3 **Design Full Model Contrast Coefficients**

Treatment	A	B	C	AB	AC	BC	ABC
(1)	−1	−1	−1	1	1	1	−1
a	1	−1	−1	−1	−1	1	1
b	−1	1	−1	−1	1	−1	1
ab	1	1	−1	1	−1	−1	−1
c	−1	−1	1	1	−1	−1	1
ac	1	−1	1	−1	1	−1	−1
bc	−1	1	1	−1	−1	1	−1
abc	1	1	1	1	1	1	1

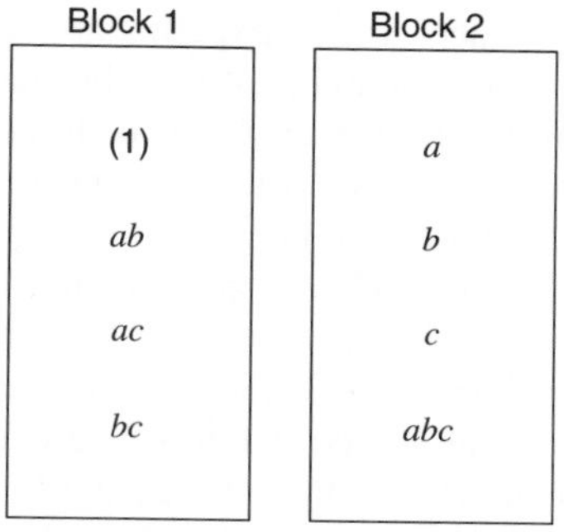

Block 1	Block 2
(1)	a
ab	b
ac	c
bc	abc

FIGURE 32.17 Schematic of 2^3 factorial divided into two blocks of size four.

level (+ 1) into the other block. Schematically, each block would appear as in Figure 32.17.

The contrast to estimate the effect of A (temperature) assuming n=1 replicate is written as

$$A = \frac{1}{4}[a + ab + ac + abc - (1) - b - c - bc]$$

Any block effect between experiments conducted in Block 1 and those conducted in Block 2 is canceled out in this contrast, as the overall effect of A is actually a pooled sum of the *simple within-block effects* of A. That is, let A_1 be the comparison of A at the high level versus A at the low level within Block 1, and A_2 be the corresponding comparison within Block 2. The overall effect of A is calculated as

$$\begin{aligned} A_1 &= [ab + ac - (1) - bc] \\ A_2 &= [a + abc - b - c] \\ \Rightarrow A &= \tfrac{1}{4}[A_1 + A_2] \end{aligned}$$

This holds for all other effects except the ABC effect, where its corresponding contrast from Table 32.6 is

$$ABC = \frac{1}{4}[a + b + c + abc - (1) - ab - ac - bc]$$

By design, the difference in those treatment combinations corresponds exactly to the design partition into Blocks 1 and 2. That is, the effect of ABC is not estimable; it is confounded with blocks. Assuming the ABC effect is negligible, this is an optimally constructed design, as the ABC effect was intentionally confounded to preserve inferences on lower order effects in the presence of any block-to-block variation.

The last example showed that for a design partitioned into two blocks, one effect (ABC) is chosen to be confounded with the block effect. Another way of stating this is that the block effect and the ABC effect *share a degree of freedom* in the analysis. For the case of four blocks that use three degrees of freedom in the analysis, the general procedure is to independently select two effects to be confounded with blocks, and then a third confounded effect is determined by the generalized interaction. This will be described in a more detail with fractional factorial designs. For more information on constructing blocks in 2^k designs, see Ref. 6.

32.4 FRACTIONAL FACTORIALS

The unreplicated 2^6 design contains 64 unique treatment combinations of the 6 factors and therefore 63 degrees of freedom for effects. Of those 63 degrees of freedom, only 6 are used for main effects (Factors $A, B, \ldots, F$) and 15 for two-factor interactions ($AB, AC, \ldots, EF$). Assuming the sparsity-of-effects principle holds, only a subset of the total degrees of freedom are used for the vital few effects that should adequately model the process. This discrepancy gets bigger as k gets larger, making full factorial designs an inefficient choice for experimentation. As with blocking, it is possible to optimally construct designs based on the assumption or knowledge that higher order interactions are negligible, which are smaller in size yet preserve critical information about likely effects of interest. These are called *fractional factorial* designs and they are widely utilized for any study involving, say, five or more factors. In particular, these are highly effective plans for *factor screening*, the exercise to whittle down to only the crucial process factors to be subsequently studied in greater detail. As with the full factorial designs, the fractional factorials are balanced and the estimated effects are orthogonal. Two-level fractional factorial designs are denoted 2^{k-p}, where k still represents the number factors in the study, and p represents fraction level. A 2^{k-1} design is called a one-half fraction of the 2^k, a 2^{k-2} design is called a one-quarter fraction of the 2^k, and so on. A 2^{k-p} design is a study in k factors, but executed with 2^{k-p} unique treatment combinations.

Consider the 2^3 design, but due to limited resources only four of the eight treatment combinations can be studied. The candidate design is a one-half fraction of a 2^3 factorial, denoted 2^{3-1}. The primary questions in design construction are similar to those encountered in blocking, which center on how the four treatment combinations should be chosen and what information is contained in those experiments. Under the sparsity-of-effects principle, the ABC effect is negligible. Thus, one choice of design is to choose those treatment combinations that are all positive in the ABC contrast coefficients. Equivalently one could choose the set that are all negative in ABC. Table 32.7 shows the experimental design for those coefficients positive in ABC.

TABLE 32.7 One-Half Fraction of the 2^3 Full Factorial Design

Treatment	A	B	C	AB	AC	BC	ABC
a	1	−1	−1	−1	−1	1	1
b	−1	1	−1	−1	1	−1	1
c	−1	−1	1	1	−1	−1	1
abc	1	1	1	1	1	1	1

It should be clear to the reader from the contrast coefficients in Table 32.7 that (1) all information for the ABC term is sacrificed in creating this design and (2) the contrast coefficients that estimate one effect exactly match those for another. For instance, the contrast to estimate the A effect simultaneously estimates the BC effect. That is,

$$\tfrac{1}{2}[a + abc - b - c] = A + BC$$

As previously discussed with the blocks, the effect of A is said to be confounded (aliased) with BC. Similarly, the B effect is confounded with the AC effect, and the C effect is confounded with the AB effect. There is no way to individually estimate those effects, only their linear combinations. This pooling of effect information is a by-product of fractional factorial designs.

In the previous example, the ABC term was assumed the least important effect and used as the basis for constructing the 2^{3-1} experimental design. Formally, ABC is called the *design generator* and is algebraically expressed in the relation

$$I = ABC$$

This is known as the defining relation, where I stands for identity, and implies that the ABC effect is confounded with the overall mean. Knowing this relationship helps determine details about the alias structure. This is accomplished by multiplying each side of the defining relation by an effect of interest and deleting any letter raised to the power 2 (i.e., via modulo 2 arithmetic). Any effect multiplied by I gives the effect back. Below demonstrates how to determine which effects are confounded with the main effect of A.

$$\begin{aligned} A \cdot I &= A \cdot ABC \\ \Rightarrow A &= A^2BC \\ \Rightarrow A &= BC \end{aligned}$$

The interpretation is that the estimated A effect is confounded with the BC effect ($A = BC$), which was previously observed via the contrast coefficients in Table 32.7. Likewise it is trivial to show $B = AC$ and $C = AB$.

The defining relation for the chosen fraction above is more descriptively expressed as $I = +ABC$, since all contrast coefficients were positive in ABC. This is called the principle fraction, and will always contain the treatment combination with all levels at their high setting. Alternatively, the complementary fraction could have been selected such that the defining relation would be expressed as $I = -ABC$. As a consequence, the linear contrasts would estimate $A - BC$, $B - AC$, and $C - AB$. Irrespective of sign, both fractions are statistically equivalent as main effects are confounded with two-factor interactions, although there may be a practical difference between the two.

Appropriately, one-half fraction designs are always constructed with the highest order interaction in the defining relation. For instance, a 2^{5-1} experiment uses $I = ABCDE$ to create the fraction and to investigate the alias structure, as this interaction is assumed the least likely effect to significantly explaining response variation. However, the one-half fraction may still be too large to feasibly execute and therefore fractions of higher degrees should be considered. The quarter-fraction design, denoted 2^{k-2}, is the next highest degree fraction from the half-factorials and comprises of a fourth of the original 2^k factorial runs. These designs require two defining relations, call them $I = E_1$ and $I = E_2$, where the first designates the half-fraction based on the " + " or "−" sign on the E_1 interaction, and the second divides it further into a quarter fraction based on the " + " or "−" sign on the E_2 interaction. Note that all four possible fractions using $\pm E_1$ and $\pm E_2$ are statistically equivalent, with the principle fraction corresponding to choosing $I = +E_1$ and $I = +E_2$ in the defining relation. In addition, the generalized interaction $E_3 = E_1 \cdot E_2$ using modulo 2 arithmetic is also included. To investigate the alias structure for a 2^{k-2} fractional factorial design, the complete defining relation is written as $I = E_1 = E_2 = E_3$. These interactions need to be chosen carefully to obtain a reasonable alias structure.

As an example, consider the 2^{6-2} design for factors A through F, and let $E_1 = ABCE$ and $E_2 = BCDF$. The generalized interaction, E_3, is computed by multiplying the two interactions together and deleting any letter with a power of 2. That is,

$$E_3 = (ABCE) \cdot (BCDF) = AB^2C^2DEF = ADEF$$

and hence the complete defining relation is expressed as

$$I = ABCE = BCDF = ADEF$$

It is easy to show for this design that main effects are confounded with three-factor interactions and higher (e.g., $A = BCE = ABCDF = DEF$) and that two-factor interactions are confounded with two-factor interactions and higher ($AB = CE = ACDF = BDEF$). Again, individual effects are not estimable, only linear combinations.

To succinctly describe the alias structure of fractional factorials, *design resolution* is introduced. The resolution of a fractional factorial design is summarized by the length of the shortest effect (often referred to as the *shortest word*) in the defining relation and is represented by a Roman numeral subscript. The one-half fraction of a 2^5 factorial with defining relation $I = ABCDE$ is called a resolution V design and is formally denoted 2_{V}^{5-1}. Similarly, the one-quarter fraction of a 2^6 factorial with defining relation $I = ABCE = BCDF = ADEF$ is a resolution IV design and is denoted 2_{IV}^{6-2}. The design resolutions of greatest interest are described below.

- *Resolution III*: There exist main effects aliased with two-factor interactions. These designs are primarily used for screening many factors to identify which are the most influential in process modeling.

- *Resolution IV*: There exist main effects confounded with three-factor interactions and two-factor interactions confounded with each other. Assuming the sparsity-of-effects principle, main effects are said to be estimated free and clear, since three-factor interactions and higher are considered negligible.
- *Resolution V*: There exist main effects confounded with four-factor interaction and two-factor interactions confounded with three-factor interactions. Assuming the sparsity-of-effects principle, main effects and two-factor interactions are said to be estimated free and clear, since three-factor interactions and higher are considered negligible.

32.5 DESIGN PROJECTION

One of the major benefits to using the two-level factorials and fractional factorials is to take advantage of the *design projection property*. This states that factorial and fractional factorial designs can be projected into stronger designs in a subset of the significant factors. In the case of unreplicated full factorials, those designs project into full factorials with replicates. For example, by disregarding one insignificant factor from a 2^4 full factorial the design becomes a full $2^{4-1} = 2^3$ factorial with 2^1 replicates at each point. In the case of fractional factorial designs of resolution R, those designs project into full factorials in any of the $R-1$ factors, possibly with replicates. It may be possible to project to a fuller design with more parameters than the $R-1$ rule dictates, but this is not guaranteed.

As a specific example, consider the 2_{III}^{3-1} fractional factorial design with defining relation $I = +ABC$ that investigates catalyst equivalents (A), ligand equivalents (B), and solvent volume (C). The four treatment combinations are depicted on the cube in the middle of Figure 32.18. As this is a resolution III design, it projects into a full factorial in any two of the three factors, also displayed in Figure 32.18. The projection property of two-level designs is very important, simply due to its usefulness in obtaining full modeling information on a subset of factors, and its implicit use in sequential experimentation.

32.6 STEEPEST ASCENT

The content so far has focused on employing experimental designs with the sole purpose of identifying magnitude of effects plus two-parameter synergies. Often though, the process models are used for optimization or improvement. The combination of experimental design, model building, and *sequential experimentation* used in searching for a

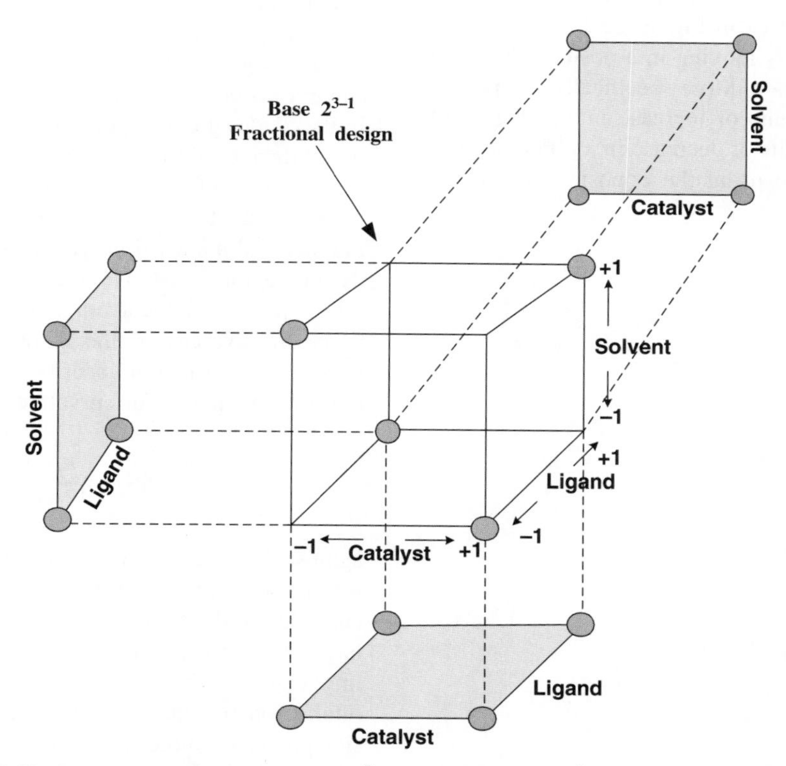

FIGURE 32.18 Illustration of projecting a 2^3 fraction onto a full 2^2 factorial in a subspace of the experimental region.

region of improved response is called the method of steepest ascent (or "descent" if the goal is to explicitly minimize). The goal is to effectively and efficiently move from one region in the factor space to another. Thus, model simplicity as well as design economy are very important. The general algorithm consists of the following:

- Fitting a first-order (main effects) model with an efficient two-level design.
- Computing the path of steepest ascent (descent), where there is an expected maximum increase (decrease) in response.
- Conduct experiments along the path. Eventually response improvement will slow or start to decline.
- Carry out another factorial/fractional design.
- Recompute new path, or augment into optimization design.

Constructing the path of steepest ascent is straightforward. Consider all points that are a fixed distance from the design region center (i.e., radius r) with the desire to seek the parameter combination that maximizes the response. Mathematically, one uses the method of Lagrange multipliers to find where the maximum response lies, constrained to the radius r. Intuitively, the path of steepest ascent is proportional to the size and sign of the coefficients for the first-order model in coded units. For example, let fitted equation be $2+3x_1-1.5x_2$. As shown in Figure 32.19, the path of steepest ascent will have x_1 moving in a positive direction and x_2 in a negative direction. More specifically, the path is such that for every 3.0 units of increase in x_1, there will correspondingly be 1.5 units of decrease in x_2. For steepest decent, the path is chosen using the opposite sign of the coefficients.

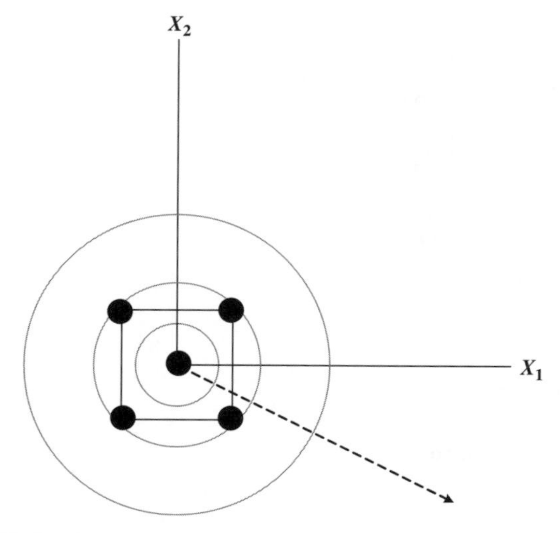

FIGURE 32.19 Path of steepest ascent for the model $\hat{y}=2+3x_1-1.5x_2$.

The success of the steepest ascent method rests on whether the region where the path is constructed is main-effect driven. Steepest ascent should still be successful in the presence of curvature (interaction or quadratic), as long as it is small relative to size of the main effects. If curvature is large, then this exercise is self-defeating. In addition, the success of the path is dependent on the overall process model. Models that are poor and have high uncertainty lead to paths with high uncertainty. Finally, modifying the steepest ascent path with linear constraints are both mathematically and practically easy to incorporate. For more information on process improvement with steepest ascent, see Ref. 6.

32.7 CENTER RUNS

One of the model assumptions in using a two-level design is linearity across the experimental region. If the region is small enough, this is a fair assumption. If the region spans a somewhat broader space and/or the region contains the optimal process condition, then it would not be surprising if nonlinearity exists. Unfortunately, two-level designs by themselves cannot even detect any curvilinear relationship across the design region, much less model it. A cost-effective strategy to initially identify curvature and also have an independent estimate of variance is to add *center runs* to the experimental design. This second point is critical as in practice most 2^k designs are unreplicated. Using the standard ±1 scaling of factor levels, center runs are replicated n_c times at the design point $x_i=0$, $i=1, 2, \ldots, k$. Note that adding center runs produces no impact on the usual effect estimates in the 2^k design. The pure error variance is estimated with n_c-1 degrees of freedom and the test for nonlinearity is via a single degree of freedom contrast that compares the average response at the center to the average response from the factorial points. If nonlinearity is nonexistent across the design region, these averages should be comparable. Specifically, let $\bar{y}_c$ be the average of the n_c center points and let $\bar{y}_f$ be the average of the n_f factorial points. The formal hypothesis test of nonlinearity is conducted by comparing the sum of squares for curvature,

$$SS_C=\frac{n_f n_c(\bar{y}_f-\bar{y}_c)^2}{n_f+n_c}$$

against the mean square error. This test does not give any information on which factors contribute the sources of curvature, only whether curvature exists or not. If $\bar{y}_f-\bar{y}_c$ is large, then curvature is present across the design region. The implication is that the linear model with main effects and interactions is inadequate for prediction and additional design points or a more advanced design is necessary to identify which specific factors are contributing to the nonlinearity in order to accurately predict across the experimental region.

32.8 RESPONSE SURFACE DESIGNS

Previous discussion has centered on fitting first-order and first-order plus interaction models. However, a higher order model is necessary when in the neighborhood of optimal response. In this case, second-order models are very good approximations to the true underlying functional relationship when curvature exists. These take the form

$$\begin{aligned} y = \beta_0 + \beta_1 x_1 + \beta_2 x_2 + \cdots + \beta_k x_k \\ + \beta_{12} x_1 x_2 + \beta_{13} x_1 x_3 + \cdots + \beta_{k-1,k} x_{k-1} x_k \\ + \beta_{11} x_1^2 + \beta_{22} x_2^2 + \cdots + \beta_{kk} x_k^2 + \varepsilon \end{aligned}$$

To estimate the second-order model each factor must have at least three levels. Implicitly there has to be at least as many unique design points as model terms. Many efficient designs are available that accommodate the above model. The most common is the central composite design, abbreviated CCD [2]. The k-factor CCD is comprised of three components, (1) a full 2^k factorial or resolution V fraction, (2) center runs, and (3) $2 \cdot k$ axial points. One compelling feature of CCD designs is that the axial points are a natural augmentation to the standard 2^k or 2_{V}^{k-p} plus center run designs.

As the name suggests, the axial points lie on the axes of each factor in the experimental space. In coded units they are set at a distance $\pm\alpha$ from the center of the design region. The axial value, α, can take on any value, which speaks to the flexibility of these designs. In practice they are usually taken at either $\alpha = 1$ for a face-centered design, $\alpha = \sqrt{k}$ for a spherical design, or $\alpha = f^{1/4}$ for a rotatable design, where f is the size of factorial or fractional used in the CCD. Table 32.8 is an example of a two-factor CCD, and Figure 32.20 displays the experimental region.

Referring to Figure 32.20, if the axial value is set at $\alpha = 1$, then all of the experimental conditions except the center runs lie on the surface of the cube. Similarly, if the axial value is set at $\alpha = \sqrt{k}$, all of the experimental conditions except the center runs lie on the surface of a sphere.

TABLE 32.8 General Two-Factor Central Composite Design

X_1	X_2	
−1	−1	Factorial runs
1	−1	
−1	1	
1	1	
0	*0*	Center runs (≥1)
$-\alpha$	0	Axial runs
α	0	
0	$-\alpha$	
0	α	

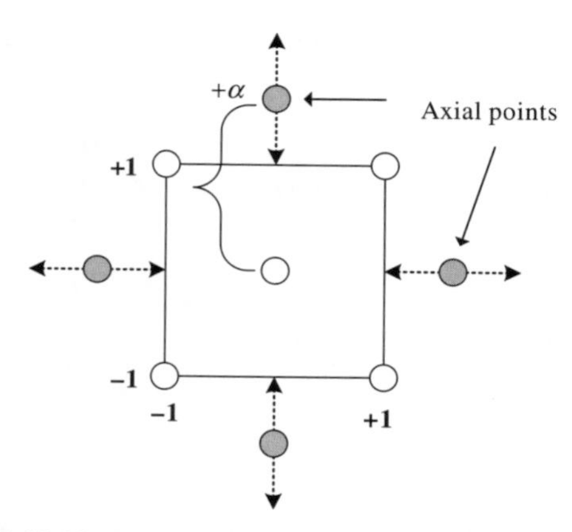

FIGURE 32.20 Illustration of general two-factor central composite design.

For rotatable designs, the precision on the model prediction is a function of only the distance from the design center and the error variance, σ^2. This is illustrated through the two-factor CCD in Table 32.8. The number of factorial points is $n_f = 4$, which yields $\alpha = 4^{1/4} = \sqrt{2}$ as the axial value for rotatability. Assume the design contains $n_c = 3$ center runs. Using this experimental plan, Figure 32.21 displays how the scaled standard error of prediction, a quantity proportional to the size of the confidence interval on the model prediction, varies across the design region. First, the prediction error increases toward the boundary of the design region, a behavior typically seen with confidence bands about a simple linear regression fit. Second, as the design is rotatable, the standard error of prediction is constant on spheres of radius r. Consider two model predictions in Figure 32.21, $\hat{y}(\underline{x}_1)$ and $\hat{y}(\underline{x}_2)$, that are at different coordinates in the experimental region. While the model-based predictions should be different, the precision of $\hat{y}(\underline{x}_1)$ and $\hat{y}(\underline{x}_2)$ is the same as both are equidistant from the center of the design region.

An alternative to the class of central composite designs are the Box–Behnken [7] designs (BBD). These experimental plans are very efficient in fitting second-order models, are nearly rotatable, and have a potentially practical advantage of experimenting with three equally spaced levels over the experimental region. These designs are constructed by incorporating 2^2 or 2^3 factorial arrays in a balanced incomplete block fashion, with the other factors set at their center value. Table 32.9 is an example of a three-factor BBD and Figure 32.22 displays the design in graphical form. The BBDs are spherical designs and there are no factorial "corner points" or face points. For the $k = 3$ design shown in Figure 32.22, all conditions except the center runs are at $\sqrt{2}$ distance from the center. This should not deter the engineer from using this design, especially if predicting at the corners is not of interest, potentially due to cost, impracticality, feasibility, or other issues.

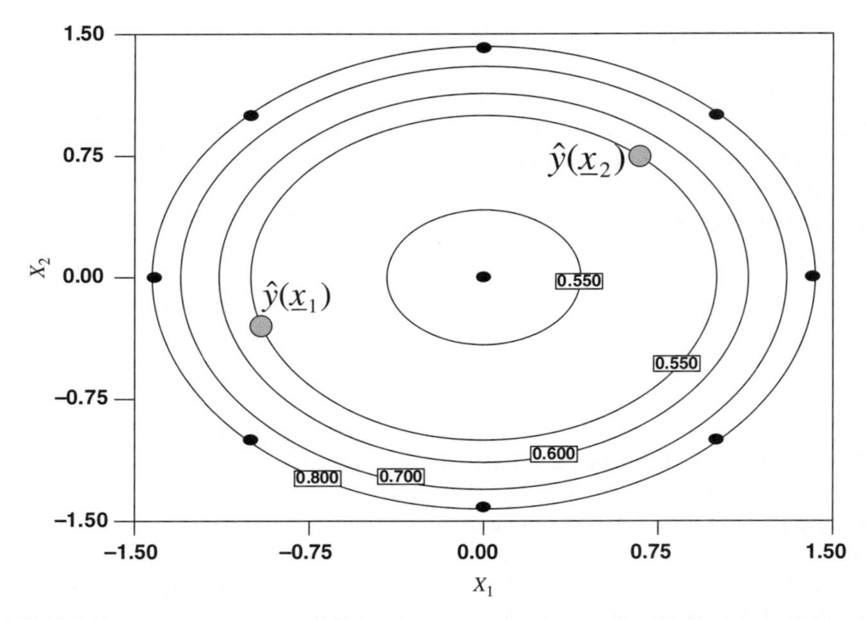

FIGURE 32.21 Illustration of rotatability using the standard error of prediction across factor space.

TABLE 32.9 Three-factor Box–Behnken Design

X_1	X_2	X_3
−1	−1	0
1	−1	0
−1	1	0
1	1	0
−1	0	−1
1	0	−1
−1	0	1
1	0	1
0	−1	−1
0	1	−1
0	−1	1
0	1	1
0	*0*	*0*

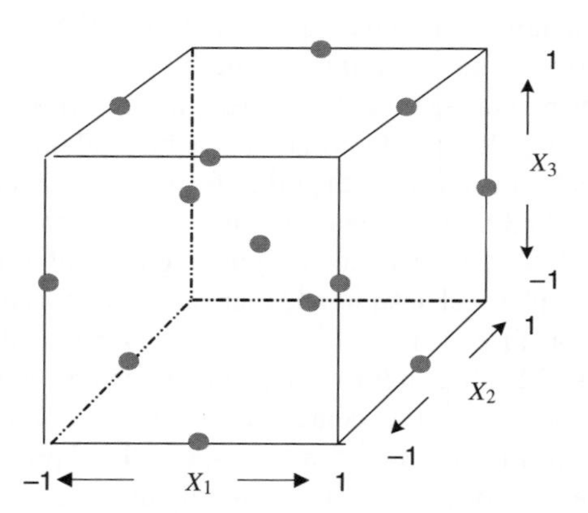

FIGURE 32.22 Illustration of three-factor Box–Behnken design.

32.9 COMPUTER GENERATED DESIGNS

Classical experimental designs such as those discussed so far, may not be appropriate for a practical situation, due to one or more constraints. These could include:

- Sample size limitations due to time, budget, or material, which may yield a nonstandard number of runs, say 11 or 13;
- Nonviable factor settings, thereby impacting the experimental region's geometry. Examples include solubility limitations, safety concerns, and small-scale mixing sensitivities;
- Desired factor levels are nonstandard, say 4 or 6;
- Factors are both qualitative and quantitative;
- Proposed model may be more complicated than first- or second-order polynomial, either higher in order or nonlinear.

For such cases, experimental designs should be tailored to accommodate any constraints, yet still preserve properties that the more classical designs typically possess, such as those based on model precision and prediction precision. The design construction is accomplished with computer assistance and falls under the class of *computer-generated designs*. The computer is a vital tool to construct appropriate designs that meet certain objectives, but unfortunately can be viewed as a black box and misused because of gaps in understanding exactly what the computer algorithms are doing.

Computer-generated designs, and the area of optimal design theory, can be attributed to Kiefer [8, 9]· and

Kiefer and Wolfowitz [10]. The general algorithm consists of the engineer providing an objective function that reflects the design property of interest, a hypothesized model, the sample size for the study, and other design elements potentially corresponding to blocks, center runs, and lack-of-fit. The algorithm then uses a fine grid of candidate experimental conditions and searches for the design that optimizes the objective function.

There are several objective functions that speak to design properties of interest and are referred to as alphabetic optimality criteria. As background to the objective functions, consider the linear model

$$\underline{y} = X\underline{\beta} + \underline{\varepsilon}$$

where $\underline{y}$ is the $(N \times 1)$ vector of observations, X is the $(N \times p)$ model matrix, $\underline{\beta}$ is the $(p \times 1)$ vector of model coefficients, and $\underline{\varepsilon}$ is the $(N \times 1)$ vector of errors assumed to be independent and normally distributed with mean zero and variance σ^2. The ordinary least squares estimate of model coefficients is

$$\hat{\underline{\beta}} = (X'X)^{-1}X'\underline{y}$$

and the covariance matrix of those estimates is given by

$$Var(\hat{\underline{\beta}}) = (X'X)^{-1}\sigma^2 \tag{32.1}$$

In addition, the variance of a predicted mean response, $\hat{y}_0$, at coordinates $\underline{x}_0$ is given by

$$\text{Var}(\hat{y}_0) = \underline{x}'_0(X'X)^{-1}\underline{x}_0\sigma^2 \tag{32.2}$$

It should be apparent to the reader from (32.1) and (32.2) the importance of good experimental design on the model and prediction precision, as demonstrated through the $(X'X)^{-1}$ matrix embedded in both of those quantities.

The most common optimality criterion is D-optimality, which *minimizes the joint confidence region* on the regression model coefficients. A-optimality is a similar and common criterion that *minimizes the average size* of a confidence interval on the regression coefficients. Both D- and A-optimality use functions of (32.1) to obtain the appropriate design, and are defined by the scaled moment matrix, M, expressed as

$$M = \frac{X'X}{N}$$

for completely randomized designs. The scaling takes away any dependence on σ^2, a constant independent of the design, and the sample size, N, which allows for comparisons across designs of different size. The algorithm finds that set of design points that maximizes the determinant of M for D-optimality, and minimizes the trace of M^{-1} for A-optimality.

Two other criteria are G-optimality and IV-optimality, which use functions of the prediction variance in (32.2) to obtain a candidate design. A G-optimal design minimizes *the maximum size* of a confidence interval on a prediction over the entire experimental region, whereas IV-optimality minimizes *the average size* of a confidence interval on a prediction over the entire experimental region. Similar to the scaling done with the D-criterion above, these criteria are defined by the scaled prediction variance given by $\upsilon(\underline{x}) = N\underline{x}'(X'X)^{-1}\underline{x}$

Research and software application in the area of optimal design theory has grown immensely over the recent past. Clearly the flexibility is appealing and at times invaluable. It allows the engineer to generate an experimental design for any sample size, number of factors (both discrete and continuous), type of model (linear or nonlinear), and randomization restrictions. Here are some additional notes and cautions regarding computer-generated designs:

- They are *model-dependent* optimal. Occasionally, the engineer will proposed a mechanistic model that represents the true relationship between y and $\underline{x}$. Often though, empirical models are proposed and inevitably edited after collecting data. Optimal designs constructed for one model could be fairly suboptimal with respect to an edited model, thereby impacting process modeling performance. There are strategies and graphical tools available to help generate and assess model robustness, such as those proposed by Heredia-Langner [11].
- The optimal design for one criterion is usually robust/near optimal across other criteria [12]. This is not overly surprising due to the importance of the $(X'X)^{-1}$ matrix. However, this is not guaranteed as it is possible to generate a D-optimal design that has poor prediction variance.
- Slight variations in algorithms and software packages could lead to generated designs that are statistically equivalent but different experimentally. As a matter of good scientific practice, the engineer needs to scrutinize the design for merit and practicality.
- Two-level designs for main effects only and main effects plus two-factor interaction models are all A-, D-, G-, and IV-optimal.
- Classical CCD and BBD designs for second-order models are near optimal. That is, they are highly efficient relative to the most optimal designs.

32.10 MULTIPLE RESPONSES

Up until now any discussion involving effect identification and regression analysis has focused on a single response, and the process model for that response can be used to hone in on a region of *desirability* in the factor space, as defined by the engineer. However, rarely in practice are experiments conducted when only a single response is collected. With chemical reaction trials, natural outputs could include various impurities levels, completion time, and yield, just to name a few. Standard analysis practice would involve (1) modeling each of those outcomes separately, (2) defining their respective region of acceptable response over the factor space, and (3) identifying the intersection of individual regions where all responses are deemed acceptable. Some software packages include this feature of *overlaying* contour plots, which is an effective approach in locating an optimal process-operating region. However, when the number of responses and/or factors gets somewhat large, this exercise can become quite cumbersome. In addition, it is not surprising to have competing responses, meaning the optimal region for one response is suboptimal for one or more of the other responses. This commonly occurs with crystallization processes, where maximum impurity purge is often at the sacrifice of higher yield, due to similar solubility properties of the chemical species. The question then becomes how to effectively merge process model information to identify conditions that are optimally balanced across multiple criteria.

The idea of desirability functions introduced by Derringer and Suich [13] addresses this problem. This is a formula scaled between [0,1] inclusively, where the researchers own priorities and requirements are built into the optimization procedure. To illustrate, consider minimizing the total impurity level (%) from a chemical reaction and assume that any reaction that produces $\leq 0.5\%$ is highly desirable. On the other hand, assume a reaction with $>3\%$ is unacceptable. The desirability function, d, for that response is expressed as

$$d = \begin{cases} 1, & \hat{y} \leq 0.5 \\ \left(\dfrac{3.0-\hat{y}}{3.0-0.5}\right)^S, & 0.5 < \hat{y} < 3.0 \\ 0, & \hat{y} \geq 3.0 \end{cases}$$

Model predictions less than 0.5% get the highest desirability score of 1, whereas model predictions higher than 3.0% get the lowest desirability score of 0. For cases in between those levels, there exist a gradient of desirability scores that are a function of both the model prediction and a weight, S. This weight is chosen by the engineer and it determines the severity of not achieving the most desirable goal, which in this case is 0.5% or less. Figure 32.23 gives a visual of how that desirability function behaves for various values of S.

For a given factor setting, each of the m responses has its own desirability score. That is, response i gets desirability, d_i, $i = 1, 2, \ldots, m$. To obtain the overall desirability across m responses at any experimental condition, the overall desirability score, D, is calculated as the geometric mean of each individual d_i, expressed as $D = \{d_1, d_2, \ldots, d_m\}^{1/m}$. This overall score is easily modified when responses vary in importance. For example, impurity responses that affect drug product quality (and therefore affect the patient) are considered more important versus, say, yield that affects a sponsor's bottom line. The final objective is to locate parameter conditions that make D largest. This is normally accomplished via response surface modeling of D across experimental space and/or numerical techniques. Many software packages include this functionality as part of optimization. Note that any identified conditions should be confirmed for acceptability.

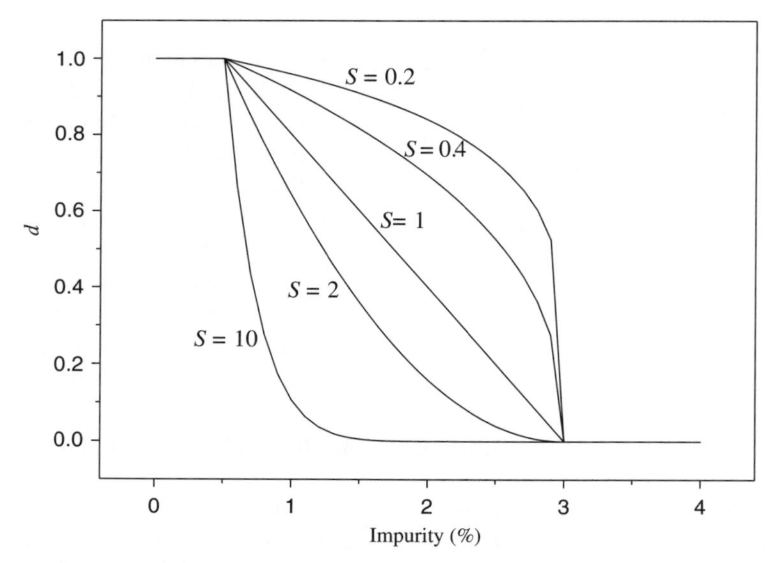

FIGURE 32.23 Example total impurity desirability function across various weights of S.

EXAMPLE 32.2 THREE-FACTOR CENTRAL COMPOSITE DESIGN

A three-factor face-centered central composite design is carried out to identify factor combinations that simultaneously minimize total impurities (%) and maximize yield (g). The three factors are catalyst (Factor A), concentration (Factor B), and temperature (Factor C). Each experimental condition is completely randomized and the center runs are replicated three times ($n_c = 3$). Total size of the study is 17 runs. The randomized design in coded units and data are listed in Table 32.10, and depicted in Figure 32.24. A full second-order model was fit to each response. The final ANOVA, fit statistics, predicted model equation, and relevant plots are presented below.

TABLE 32.10 Three-Factor CCD Design to Optimize Total Impurities and Yield

Design			Data	
Catalyst	Concentration	Temperature	Total Impurities	Yield
−1	1	−1	23.54	33.95
−1	−1	−1	10.57	15.27
0	0	0	12.20	35.10
1	1	−1	19.74	15.43
−1	1	1	5.20	18.68
1	−1	−1	8.61	2.50
0	0	0	11.10	37.20
1	−1	1	0.79	24.00
0	0	1	8.63	33.40
−1	−1	1	2.59	8.30
0	0	0	10.10	42.20
1	0	0	11.96	32.16
1	1	1	4.40	42.10
0	−1	0	5.47	31.90
0	0	−1	15.98	21.60
0	1	0	17.07	47.80
−1	0	0	12.82	31.17

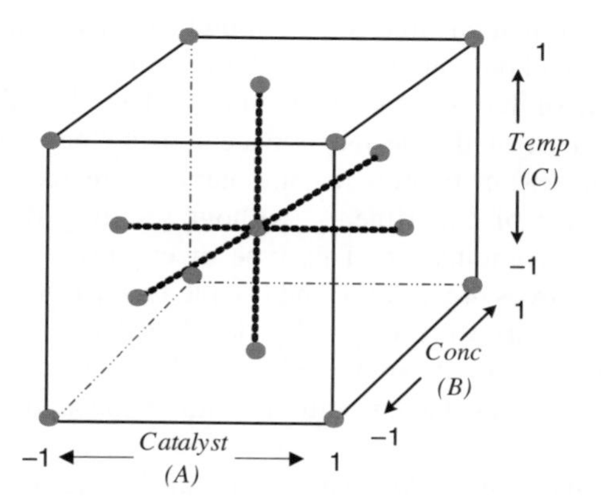

FIGURE 32.24 Illustration of three-factor face-centered CCD design.

Figure 32.25 gives an example snapshot of all relevant information in modeling total impurities. The table shows that total impurities are jointly impacted by concentration and temperature, as shown by the significant p-value on the interaction term, and independent of catalyst level. The model explains approximately 90.7% of the variation in the data as calculated by the R^2 value, and the final predicted model equation in coded units is embedded as well. Two previously highlighted diagnostic plots are included in this snapshot, although there could be others of interest that highlight or confirm aspects of the data/model. The normal probability plot of residuals shows no deviation from the normality assumption, and the predicted versus actual plot indicates that the model is performing well with constant variability across the range of prediction. Finally, a model-based contour plot of predicted total impurities across concentration and temperature is shown along with its 3D analogue surface plot. From this graphs, total impurities are minimized at the combination lower concentration and higher temperatures levels, independent of catalyst. That is, (concentration, temperature) = (−1, 1).

Figure 32.26 shows all relevant information in modeling yield. The embedded table shows that yield is impacted by all three factors, including second-order catalyst and temperature effects. Notice also that the main effect of catalyst is included even though the p-value is insignificant. This is to preserve *model hierarchy*, as catalyst does explain some of the variation in response, but in conjunction with higher order terms (either interactions and/or as a second-order effect). The model explains approximately 97.4% of the variation in the data as calculated by the R^2 value, and the final predicted model equation in coded units is shown. The normal probability plot of residuals demonstrates that the residuals can be assumed normally distributed, and the predicted versus actual plot indicates that the model is performing quite well with no departures from the constant variability assumption. And again, a model-based contour plot of predicted yield across catalyst and temperature levels at the high concentration is displayed along with its corresponding surface plot. The concentration level is set to high because that factor comes in the model *only* as a positive main effect, implying higher concentration predicts higher yield. Therefore setting concentration at its high level is in the optimal direction. From this information, yield is maximized at (catalyst, concentration, temperature) ≈ (0.25, 1.0, 0.25) in coded units.

In identifying a candidate process-operating region that is optimal over both responses, criteria that define acceptable performance are established. For this example, assume that the process is acceptable if the predicted total impurities are below 10% and the yield is above 40 g. Figure 32.27 displays

Source	SS	DF	MS	F	p-value
B - Conc	175.7286	1	175.73	41.46	< 0.0001
C- Temperature	322.9649	1	322.96	76.19	< 0.0001
BC	39.9618	1	39.96	9.43	0.0089
Error	55.10706	13	4.24		
Total	593.7624	16			

$R^2 = 90.7\%$ $\qquad \hat{y}_{\text{imp}} = 10.63 + 4.19(\text{Conc}) - 5.68(\text{Temp}) - 2.24(\text{Conc})(\text{Temp})$

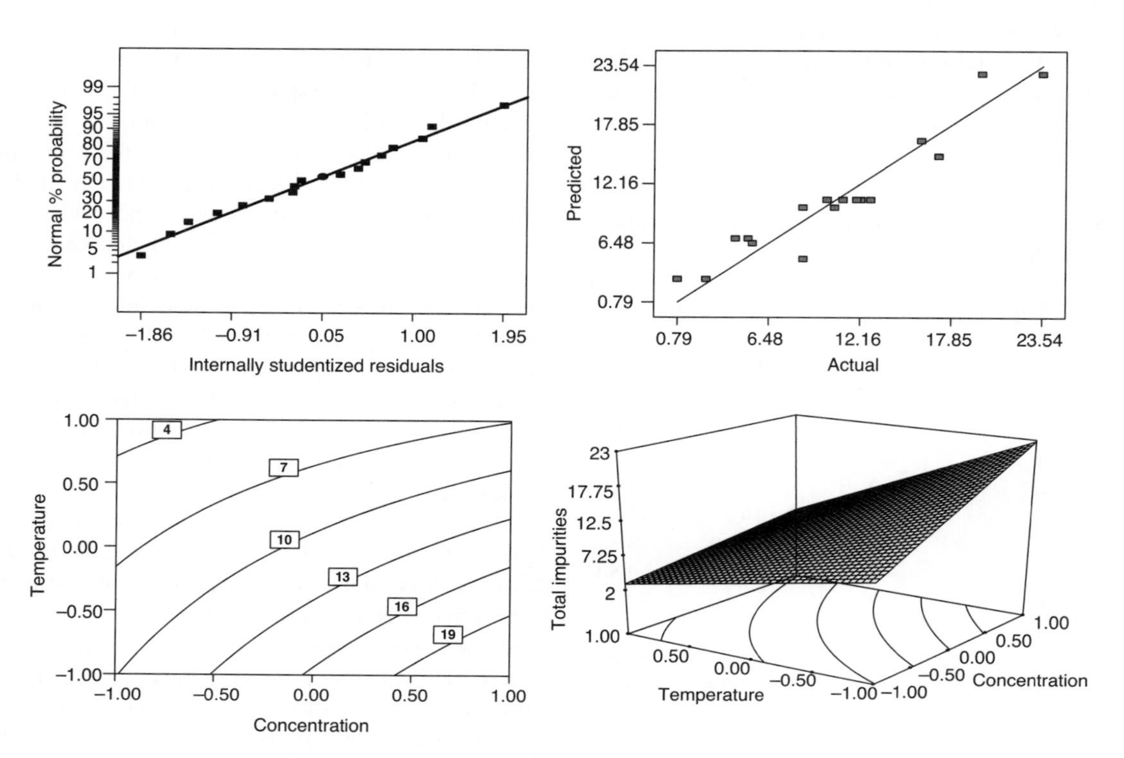

FIGURE 32.25 Summary of total impurities modeling.

at the high (+ 1) concentration level the subspace (in white) of catalyst–temperature combinations where simultaneously both responses meet the specified process performance criteria. Similarly, one solution from using the desirability function predicts a candidate optimal setting at (catalyst, concentration, temperature) $\approx (0.67, 1, 1)$. This is denoted in the upper part of the white area by (•). This condition would be verified experimentally.

32.11 ADVANCED TOPICS

32.11.1 Industrial Split-Plot Designs

One assumption made throughout the chapter is that all experimental designs have complete randomization of the treatment combinations, which are generally called *completely randomized designs*. However, for experiments run at larger scale and/or with equipment limitations, complete randomization of the experiments is arduous. This happens with factors that are difficult to independently change for each design run, or impractical when certain factors can and should be held constant for some duration of the experiment due to resource or budget constraints. Temperature and pressure are examples that immediately come to mind of hard-to-change factors. When this situation occurs, part of the design is executed in "batch-mode." That is, certain treatment combinations are fixed across a sequence of experiments, without resetting the actual treatment combination. This type of execution results in nested sources of variation and is called a *split-plot design*. These designs were originally developed for agronomic experiments, but its applicability easily spans all fields of science, even as the agricultural naming conventions have endured.

The basic split-plot experiment can be viewed as two experiments that are superimposed on each other. The first corresponds to a randomization of hard-to-change factors to

Source	SS	DF	MS	F	p-value
A-Catalyst	7.78	1	7.8	1.2	0.3047
B - Conc	577.45	1	577.4	86.9	< 0.0001
C-Temp	142.36	1	142.4	21.4	0.0009
AC	619.70	1	619.7	93.2	< 0.0001
A^2	162.79	1	162.8	24.5	0.0006
C^2	400.29	1	400.3	60.2	< 0.0001
Error	66.46	10	6.6		
Total	2536.27	16			

$R^2 = 97.4\%$

$$\hat{y}_{\text{yield}} = 38.88 + 0.88(\text{Catalyst}) + 7.6(\text{Conc}) + 3.77(\text{Temp}) + 8.8(\text{Catalyst})(\text{Temp}) - 7.33(\text{Catalyst}^2) - 11.50(\text{Temp}^2)$$

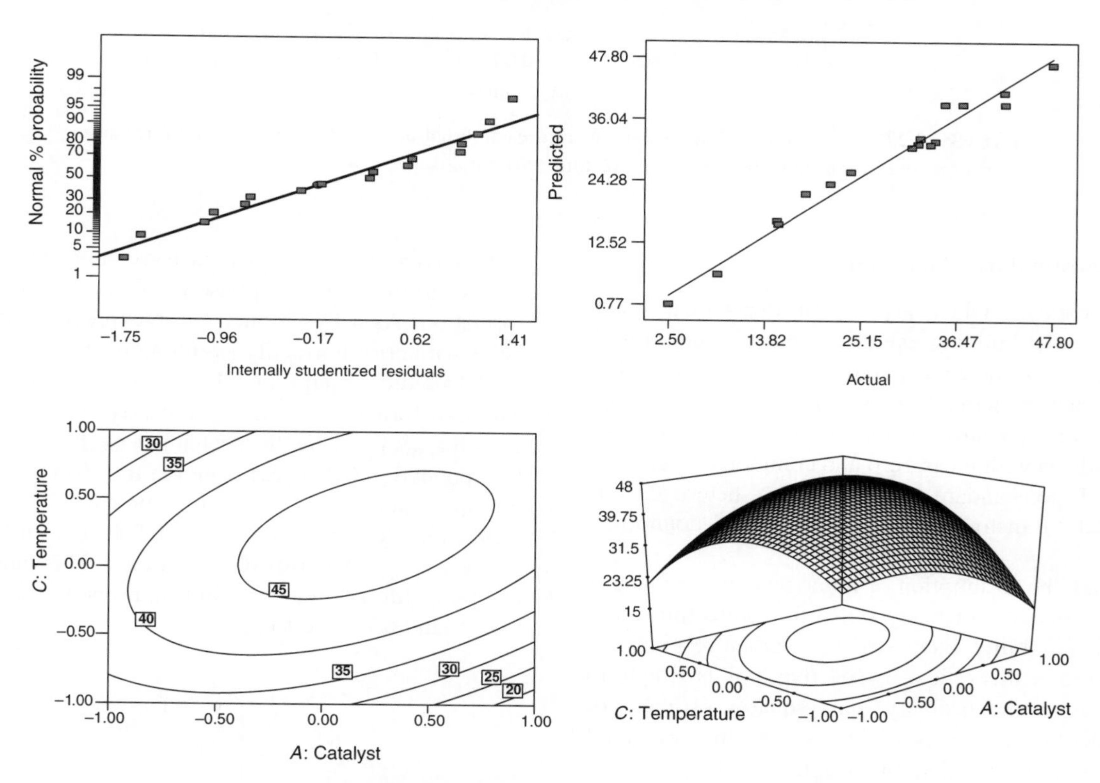

FIGURE 32.26 Summary of yield modeling.

experimental units called *whole plots*, while the second corresponds to a separate randomization of the easy-to-change factors within each whole plot, called *subplots*. This is different than completely randomized designs that have unrestricted randomization factor combinations across all experimental units. The blocking of the hard-to-change factors along with the two separate randomization sequences creates a correlation structure among data collected within the same whole plot. Inferentially, the associated ANOVA needs to reflect that the experiment was executed in a split-plot fashion. If data from a split-plot structure were analyzed as if the experiments were completely randomized, then it is possible to erroneously conclude significance of hard-to-change effects when they are not, while conclude insignificance of easy-to-change effects when they are. A good discussion on classical split-plot designs can be found in Hinkelmann and Kempthorne [14], as well as in Box and Jones [15] regarding response surface methodology.

In the recent past considerable attention has been given to constructing and evaluating optimal split-plot designs, especially with the computational horsepower of today's computers. Topics span algorithms for D-optimal split-plot designs [16, 17], to comparing the performance between classical response surface designs in a split-plot structure [18], to graphical techniques for comparing competing split-plot designs [19]. This is an important area of research as many industrial experiments are conducted with restricted randomization, whether deliberately planned or not. For those cases when it is planned, more software tools are becoming widely available so that the split-plot experiment are both powered sufficiently and analyzed appropriately.

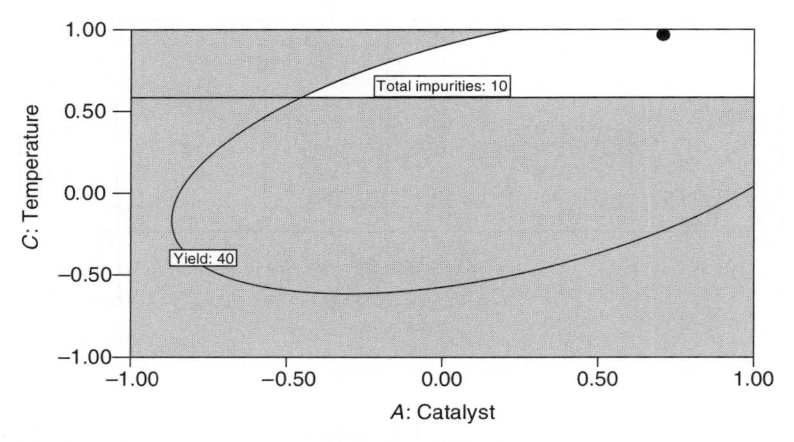

FIGURE 32.27 Overlay plot of catalyst–temperature combinations that meet specified performance criteria for total impurities and yield, with candidate optimal setting (●).

32.11.2 Nonstandard Conditions

Model diagnostics and the analysis of residuals are a crucial part of any model building exercise and are standard with any software package with a regression component. The information can numerically or visually identify any conditions from the standard assumption of normal and independent residuals with mean zero and common variance σ^2. Examples of nonstandard conditions are heterogeneous variance, data transformations, outliers, and nonnormal errors.

In practice, the assumption of homogeneous variance is most likely violated. For many scientific applications, it is natural that variability increases with either response or regressor. From a theoretical perspective, this is simple to accommodate as the ordinary least squares estimate of regression coefficients, $\hat{\beta}$, is slightly altered to a weighted least squares estimate, where the weights are the inverse of the variances. The implication is that residuals with large variances (small weights) should not count as much in the model fit as those with small variances (large weights). For a given matrix of weights, W, the expression for the weighted least squares estimate of β is given by

$$\underline{\hat{\beta}}_{WLS} = (X'WX)^{-1}X'W\underline{y}$$

The immediate practical issue is how to obtain the weights. One approach is to collect multiple data at each experimental condition and use the corresponding estimate of variance. However, this is problematic if the sample variances are based on a small number of data, and therefore potentially unreliable as a solution. A loose rule of thumb is that any estimated weights should be based on no less than a sample of nine [20]. Less than that can result in very poor performance of the weighted least squares solution, and ignoring the weights would be the better course of action.

One effective approach to the issue of increasing variance relative to increasing response is through a response transformation. As a by-product, the reexpression of data also helps with error normality assumption, as well as with both model prediction and model selection. The most common data transformation is the log transformation ($y^* = \ln(y)$ or $y^* = \log_{10}(y)$), where the mechanism of change in y across $\underline{x}$ is exponential. Other less common transformations include the square root ($y^* = \sqrt{y}$) and reciprocal ($y^* = y^{-1}$). All of these examples fall under the class of power transformations, where $y^* = y^{\lambda}$. The Box–Cox method [21] is one particular and powerful technique that aids in properly transforming y into y^* and takes the form

$$y^*(\lambda) = \begin{cases} \dfrac{y^{\lambda} - 1}{\lambda \dot{y}^{\lambda-1}} & \lambda \neq 0 \\ \dot{y}\ln(y) & \lambda = 0 \end{cases}$$

where $\dot{y}$ is the geometric mean of the original data. The assignment of ln(y) when $\lambda = 0$ comes from the limit of $(y^{\lambda} - 1)/\lambda$ as λ approaches zero, which allows for continuity in λ. The use of $\dot{y}$ rescales the response to the same units so that the error sum of squares can be comparable across different values of λ. As mentioned, common power transformation values of λ are -1, 0, 0.5, and 1, the last corresponding to no response transformation. In practice, after an appropriate model is fit to the data, an estimate of λ that minimizes the error sum of squares is estimated. If this value is significantly different than 1.0, a transformation on the data is recommended to produce a better fit, with potentially a different model. Most software packages that contain the Box–Cox procedure as part of model diagnostics provide a confidence interval on λ that simultaneously tests the hypothesis of no response transformation needed and puts a bound on a recommended one.

32.11.3 Designs for Nonlinear Models

Except for a brief mention in the section on computer generated designs, all the discussion and examples in this chapter have only considered models that are linear in their parameters, which are then (successfully) used to approximate underlying nonlinear behavior over a small region. By definition, a linear model is one where the partial derivatives with respect to each model parameter are only a function of the factor/regressor. A simple example is the two-factor interaction model

$$y = \beta_0 + \beta_1 x_1 + \beta_2 x_2 + \beta_{12} x_1 x_2$$

It is trivial to show that $\partial y/\partial \beta_i$ ($i = 0$, 1, 2, and 12), does not depend on β_i. Conversely, a nonlinear model is defined as one where at least one of the partial derivatives is a function of a model parameter. An example of a nonlinear model would be

$$y = \beta_0 \, e^{-\beta_1 x}$$

where both $\partial y/\partial \beta_0$ and $\partial y/\partial \beta_1$ are still functions of the unknown β parameters. Of course, chemical engineers are fully aware that the majority of kinetic models are nonlinear and not even closed form but expressed as differentials.

As opposed to linear models, in most cases classical experimental designs are not appropriate or optimal for nonlinear models. (The notable exception is in some applications of generalized linear models, as discussed by Myers et al. [22].) Thus, practitioners typically rely on computer generated designs that are optimal in some respect, such as by D- or G-optimality criteria previously described. However, this is problematic and circular in terms of design construction, since *the optimal design for a nonlinear model is a function of the unknown parameters*. To circumvent that issue, initial parameter estimates are used that are often the results of previous studies and/or scientific knowledge. These designs are called *locally optimal*, and Box and Lucas [23] discuss locally D-optimal designs for nonlinear models. Intuitively, if the initial model parameter estimates are poor, then the locally optimal design will suffer in performance. One avenue that does not rely on initial parameter point-estimates is to use Bayesian approach to experimental design [24], where the scientist would postulate a *prior distribution* on the parameters plus specify a utility function, which is similar in spirit to the objective functions discussed in classical alphabetic optimal criteria. Regardless, optimal designs for nonlinear models are inherently sequential, which may be an obstacle in adoption. In addition, commonly available technology has not caught up to the contemporary thinking in this field, adding another very tangible barrier. Nevertheless, independent of model-type it should be apparent that the process model fit is only as good as the design behind the data.

REFERENCES

1. Montgomery DC. *Design and Analysis of Experiments*, 6th edition, Wiley, New York, 2005.
2. Box GEP, Wilson KB. On the experimental attainment of optimum conditions. *J. Royal Stat. Soc., Ser. B* 1951;13:1–45.
3. Myers RH. *Classical and Modern Regression with Applications*, 2nd edition, Duxbury Press, Belmont, CA, 1990.
4. *Design Expert version 7.1.6*, Stat-Ease, Inc, Minneapolis, MN November, 2008.
5. Daniel C. Use of half-normal plots in interpreting factorial two-level experiments. *Technometrics*, 1959;1:311–342.
6. Myers RH, Montgomery DC. *Response Surface Methodology: Process and Product Optimization Using Designed Experiments*, 2nd edition, Wiley, New York, 2002.
7. Box GEP, Behnken DW. Some new three-level designs for the study of quantitative variables. *Technometrics* 1960;2:455–475.
8. Kiefer J. Optimum experimental designs. *J. Royal Stat. Soc., Ser. B* 1959;21:272–304.
9. Kiefer J. Optimum designs in regression problems. *Ann. Math. Stat.* 1961;32:298–325.
10. Kiefer J, Wolfowitz J. Optimum designs in regression problems. *Ann. Math. Stat.* 1959;30:271–294.
11. Heredia-Langner A, Montgomery DC, Carlyle WM, Borror CM. Model robust designs: a genetic algorithm approach. *J. Quality Technol.* 2004;36:263–279.
12. Cornell JA. *Experiments with Mixtures: Designs, Models, and the Analysis of Mixture Data*, 3rd edition, Wiley, New York, 2002.
13. Derringer G, Suich R. Simultaneous optimization of several response variables. *J. Quality Technol.* 1980;12:214–219.
14. Hinkelmann K, Kempthorne O. *Design and Analysis of Experiments Volume I: Introduction to Experimental Design*, Wiley, New York, 1994.
15. Box GEP, Jones S. Split-plot designs for robust product experimentation. *J. Appl. Stat.* 1992;19:3–26.
16. Goos P, Vandebroek M. Optimal split plot designs. *J. Quality Technol.* 2001;33:436–450.
17. Goos P, Vandebroek M. D-optimal split plot designs with given numbers and sizes of whole plots. *Technometrics* 2003;45:235–245.
18. Letsinger JD, Myers RH, Lentner M. Response surface methods for bi-randomization structure. *J. Quality Technol.* 1996;28:381–397.
19. Liang L, Anderson-Cook CM, Robinson TJ, Myers RH. Three dimensional variance dispersion graphs for split-plot designs. *J. Comput. Graph. Stat.* 2006;15:757–778.

20. Deaton ML, Reynolds MR, Jr., Myers RH. Estimation and hypothesis testing in regression in the presence of non-homogenous error variances. *Commun. Stat.* 1983;B12(1): 45–66.

21. Box GEP, Cox DR. An analysis of transformations (with discussion). *J. Royal Stat. Soc., Ser. B* 1964;26: 211–246.

22. Myers RH, Montgomery DC, Vining GG. *Generalized Linear Models With Applications in Engineering and the Sciences*, Wiley, New York, 2002.

23. Box GEP, Lucas HL. Design of experiments in non-linear situations. *Biometrika* 1959;46:77–90.

24. Chaloner K, Verdinelli I. Bayesian experimental design: a review. *Stat. Sci.* 1995;10:273–304.

33

MULTIVARIATE ANALYSIS FOR PHARMACEUTICAL DEVELOPMENT

FREDERICK H. LONG
Spectroscopic Solutions, LLC, Randolph, NJ, USA

Multivariate analysis (MVA) is the statistical analysis of many variables at once. Many problems in the pharmaceutical industry are multivariate in nature. The importance of MVA has been recognized by the U.S. FDA in the recent guidance on process analytical technology [1]. MVA has been made much easier with the development of inexpensive, fast computers, and powerful analytical software. Chemometrics is the statistical analysis of chemical data, which is an important area of MVA. Spectral data from modern instruments is fundamentally multivariate in character. Typically pharmaceutical process monitoring requires more than one variable. Furthermore, the powerful statistical methods of chemometrics are essential for the analysis and application of spectral data including NIR and Raman. In this chapter, we will review the subject of chemometrics and MVA and its application in the pharmaceutical industry.

With spectral data, it is not uncommon to measure several thousand variables at one time. However, it is often hard to conceptualize so many variables; therefore, we will begin our discussion of MVA with a few simple examples that illustrate important statistical concepts which are essential in chemometrics. The first problem is a set of pharmaceutical quality data. Measurements of density and assay have been measured for 43 lots of material. The data is shown in Table 33.1. Inspection of the data reveals that the density values are near 1.0, while the assay values are closer to 100. A goal of the data analysis is to understand the variation within the data set. It will be advantageous to have the two variables in the data set with similar magnitudes; therefore, we will scale each of the two variables by its own standard derivation. The standard deviation, s, of a set of measurements ($x_1, \ldots, x_n$) is given by

$$\sigma = \left(\frac{\sum (x_i - \overline{x})^2}{n-1} \right)^{1/2} \tag{33.1}$$

where $\overline{x}$ is the average value of the n measurements. The denominator in equation 33.1 is $n-1$, because once the average is calculated, there are $n-1$ degrees of freedom. We note that the standard deviation has the same units as the variable of interest.

A plot of the scaled data is shown in Figure 33.1. The x-axis is the scaled density and the y-axis is the scaled assay values. Each point represents 1 of the 43 lots of material. From the plot in Figure 33.1, one data point is far away from all of the others. Statisticians call data points that do not belong to the data set outliers. Outliers are important to identify and remove from the analysis of the data set, because a single outlier can greatly influence the statistical analysis and obscure underlying trends in the data. We note that while outliers are often removed in a research and development environment during method development, great caution must be used in removing outliers during validation or use in actual production.

The scaled data are replotted in Figure 33.2, with the outlier point removed. The reader will also note that the origin of the graph has been moved to the center point of the data set. This operation is called *mean centering*, when the average of the overall data set is subtracted from the data. As mentioned earlier, in MVA we are concerned with

Chemical Engineering in the Pharmaceutical Industry: R&D to Manufacturing Edited by David J. am Ende

TABLE 33.1 Pharmaceutical Quality Data Example

Density (g/cm^3)	Assay (mg)
0.801	121.410
0.824	127.700
0.841	129.200
0.816	131.800
0.840	135.100
0.842	131.500
0.820	126.700
0.802	115.100
0.828	130.800
0.819	124.600
0.826	118.310
0.802	114.200
0.810	120.300
0.802	115.700
0.832	117.510
0.796	109.810
0.759	109.100
0.770	115.100
0.759	118.310
0.772	112.600
0.806	116.200
0.803	118.000
0.845	131.000
0.822	125.700
0.971	126.100
0.816	125.800
0.836	125.500
0.815	127.800
0.822	130.500
0.822	127.900
0.843	123.900
0.824	124.100
0.788	120.800
0.782	107.400
0.795	120.700
0.805	121.910
0.836	122.310
0.788	110.600
0.772	103.510
0.776	110.710
0.758	113.800

investigation of the variation within the data set. The average values of the data set are not of primary importance. Two arrows in the figure illustrate the two directions of variation within the data set. P1 is the largest direction of variation and P2 is the second direction of variation. It is important to note that P1 and P2 are perpendicular to each other. In MVA, P1 and P2 are the first and second principal components of the data set, respectively.

For each one of the data points, the projection of the data point onto the P1 or P2 vector is called a score value. Plots of score values for different principal components, typically P1 versus P2 are called score plots. Score plots provide

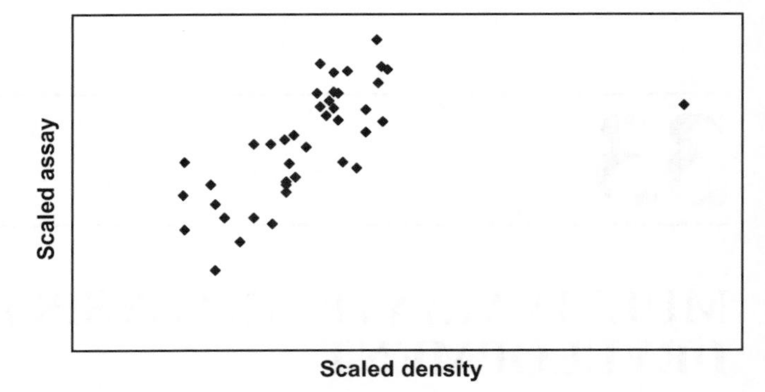

FIGURE 33.1 Scaled pharmaceutical quality data. Both the density and assay are scaled by the standard deviation of the data for each variable. Because the variables are scaled by the standard deviation, they are dimensionless.

important information about how different samples are related to each other. Principal component plots, also called loading plots, provide information about how different variables are related to each other. Because we are working with scaled variables, the PCs and scores are dimensionless variables.

The mathematics of PCA can be clearly described using linear algebra [2]. An excellent discussion of linear algebra can be found in the references [3]. By convention, the data matrix, **X**, has p columns and n rows, and each column represents another variable and new rows for each observation or sample. The average data matrix, $\overline{\mathbf{X}}$, is the average of each individual column (i.e., variable) in the data set. Mean centering is written as

$$(\mathbf{X}-\overline{\mathbf{X}}) \tag{33.2}$$

The covariance matrix is written as

$$\mathbf{C} = (\mathbf{X}-\overline{\mathbf{X}})^{\mathrm{T}}(\mathbf{X}-\overline{\mathbf{X}}) \tag{33.3}$$

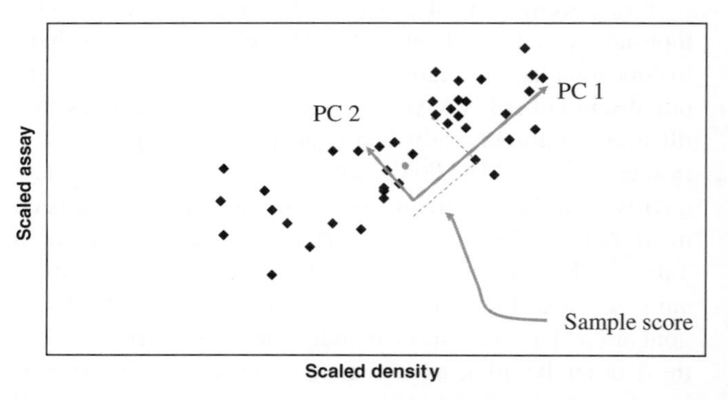

FIGURE 33.2 Scaled pharmaceutical quality data showing both the first and the second principal components for the data set. The first principal component is the direction of the maximum variation within the data set. The second principal component is perpendicular to the first PC. The scores for each sample point are given by the projection of the data point onto the principal component vector.

where an upper script **T** represents a matrix transpose. The covariance matrix is a square, symmetric, $p \times p$ matrix. The covariance matrix provides information about the relationship between different variables. For example, the i,j element of the covariance matrix quantifies the relative change between the i,j variables. If an element of the covariance matrix is zero, there is no relationship (correlation) between the two variables.

Related to the covariance matrix is the correlation matrix where all the variables have been scaled for their standard deviations. The correlation matrix is useful when one or more of the variables have much higher numerical values than the other variables. The scaling of the variables means that all variables will contribute to the analysis in roughly the same way. Mathematically the correlation matrix, **R**, is written as

$$r_{ik} = \frac{s_{ik}}{\sqrt{s_{ii}}\sqrt{s_{jj}}} = \frac{\sum_{i=1}^{n}(x_{ji}-\bar{x}_i)(x_{jk}-\bar{x}_k)}{\sqrt{(x_{ji}-\bar{x}_i)^2}\sqrt{(x_{jk}-\bar{x}_k)^2}} \qquad (33.4)$$
$$\mathbf{R} = \begin{pmatrix} 1 & & r_{1p} \\ & 1 & \\ r_{p1} & & 1 \end{pmatrix}$$

where the elements of the correlation matrix are given by r_{ij}. R is a square $p \times p$ matrix, where p is the number of variables. The diagonal elements of R are equal to one.

PCA is the systematic analysis of the covariance or correlation matrix. It can be shown that the eigenvalues are positive and the eigenvectors are orthogonal for both matrices [4]. The eigenvector equation for **C** is

$$\mathbf{C}u_i = \lambda_i u_i \qquad (33.5)$$

where u_i is the ith eigenvector and λ_i is the corresponding eigenvalue. By convention, the eigenvalues are placed in descending order, where λ_1 is the largest eigenvalue. In PCA, the eigenvectors are also called principal components. It can be shown that the first PC represents the largest source of variance in the data set. The percentage variation explained by the ith PC is given by

$$100 \times \frac{\lambda_1}{\sum_i \lambda_i} \qquad (33.6)$$

It is common with spectral data that the data set can be well approximated by a few principal components. As explained earlier, score values provide information about the relationship between different observations. The PCs form a basic set which can be used to approximate the original data set. For a single mean-centered observation, x_j,

$$x_j = \sum_{i=1}^{p} t_{ji}\mathrm{PC}_i = \sum_{i=1}^{A} t_{ji}\mathrm{PC}_i + E \qquad (33.7)$$

where t_{ji} are the score values, A is the number of principal components, E is the error when the number of principal components is less than the number of variables. Because the PCs are orthogonal, a direct expression for the score values can be given by the following equation.

$$t_{ji} = (\mathbf{x}_j - \overline{\mathbf{X}}) \bullet \mathrm{PC}_i \qquad (33.8)$$

Equation 33.7 is derivable from equation 33.6 by taking a dot product of both sides and exploiting the orthogonality of the principal components. The previous example is somewhat trivial because only two variables were involved.

Let us now consider another example with more variables. In Table 33.2, a set of data describing the properties of 43 raw materials is shown. The variables that describe the raw materials are labeled QV1–QV8. The variables QV1–QV8 describe different properties of the raw material such as moisture, assay, and particle size. Using commercial software, we can do a PCA analysis of the data set using the same approach that was used for the first data set, that is, scaling by standard deviation and mean centering. A few of the critical results are shown in Figures 33.3 and 33.4. The loading (principle component) plot shows some results that are clearly interpretable, Figure 33.3. The principle component plot shows how different variables relate to each other. In the plot the reader can observe that QV5 and QV8 are close to each other and therefore are well correlated to each other. QV1 and QV7 are also correlated. A plot of the score values for each 1 of the 43 raw materials is shown in Figure 33.4. The origin of the score plot corresponds to the average of the entire data set. The samples that are farther away from the origin are more likely to be possible outliers. The ellipse in Figure 33.4 is called the Hotelling T^2 ellipse and is showing the 95% probability level for outliers. The Hotelling T^2 ellipse is based on scaled, squared score values [2]. The T^2 value for observation i given below.

$$T_i^2 = \sum_{a=1}^{A} \frac{t_{ia}^2}{S_{ta}^2}$$
$$S_{ta}^2 = \frac{\sum_{i=1}^{N} t_{ia}^2}{N} \qquad (33.9)$$

where A is the number of principal components and t_{ia} is the ath principal component score value for the ith sample. S_{ta}^2 is the variance of t_a, because the average of the score values is zero [2]. T^2 is closely related to the often-used parameter Mahalanobis distance. An important property of the T^2 statistic is that it is directly proportional to an F value, which is a statistical parameter that is rigorously related to a probability value.[1] The numerical value of the F value is dependent on the number of samples, principal components,

[1] $T_i^2 \frac{(N-A)N}{A(N^2-1)}$ is approximately F-distributed, see Ref. 2.

TABLE 33.2 Multivariable Quality Data Set

Primary ID	QV1	QV2	QV3	QV4	QV5	QV6	QV7	QV8
1	110	2	2	180	1.5	10.5	10	70
2	110	6	2	290	2	17	1	105
3	110	1	1	180	0	12	13	55
4	110	1	1	180	0	12	13	65
5	110	1	1	280	0	15	9	45
6	110	3	1	250	1.5	11.5	10	90
7	110	2	1	260	0	21	3	40
8	110	2	1	180	0	12	12	55
9	100	2	1	220	2	15	6	90
10	130	3	2	170	1.5	13.5	10	120
11	100	3	2	140	2.5	8	140	m
12	110	2	1	200	0	21	3	35
13	140	3	1	190	4	15	14	230
14	100	3	1	200	3	16	3	110
15	110	1	1	140	0	13	12	25
16	100	3	1	200	3	17	3	110
17	110	2	1	200	1	16	8	60
18	70	4	1	260	9	7	5	320
19	110	2	0	125	1	11	14	30
20	100	2	0	290	1	21	2	35
21	110	1	0	90	1	13	12	20
22	110	3	3	140	4	10	7	160
23	110	2	0	220	1	21	3	30
24	110	2	1	125	1	11	13	30
25	110	1	0	200	1	14	11	25
26	100	3	0	0	3	14	7	100
27	120	3	0	240	5	14	12	190
28	110	2	1	170	1	17	6	60
29	160	3	2	150	3	17	13	160
30	120	2	1	190	0	15	9	40
31	140	3	2	220	3	21	7	130
32	90	3	0	170	3	18	2	90
33	100	3	0	320	1	20	3	45
34	120	3	1	210	5	14	12	240
35	110	2	0	290	0	22	3	35
36	110	2	1	70	1	9	15	40
37	110	6	0	230	1	16	3	55
38	120	1	2	220	0	12	12	35
39	120	1	2	220	1	12	11	45
40	100	4	2	150	2	12	6	95
41	50	1	0	0	0	13	0	15
42	50	2	0	0	1	10	0	50
43	100	5	2	0	2.7	1	1	110

and probability level desired, α. Examination of equation 33.9 for two PCs shows that

$$F(\alpha) = C\left(\frac{t_1^2}{S_1^2} + \frac{t_2^2}{S_2^2}\right) \tag{33.10}$$

where C is a constant. Equation 33.9 is an equation for an ellipse in the t_1, t_2 space. By convention, the Hotelling T^2 ellipse is usually drawn at the 95% probability level.

PCA can be viewed as a method for approximating the original data set. The approximation is based on a linear combination of the principle components where the amplitude coefficients are the previously described scores. The approximation is exact when the number of principle components equals the number of variables in the data set. For most spectral data sets, a small number of principle components (also called factors) can be used to approximate the spectral data set very well. The determination of the correct

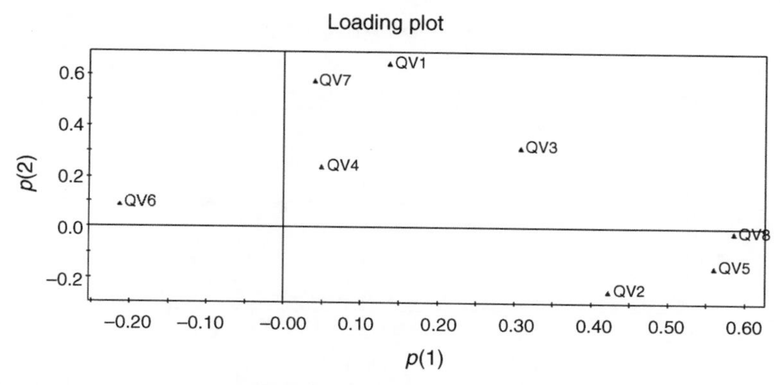

FIGURE 33.3 Loading plot for the data set in Table 33.2. The first principal component is plotted on the x-axis and the second principal component is on the y-axis. Variables that are close to each other are highly correlated.

number of factors can be done by a variety of numerical methods. Too many factors in the PCA model will over fit the data and the model will not predict reliably. Most multivariate analysis software packages will suggest a suitable number of principle components. The suggested number is usually a good starting point; however, it is best practice to verify the optimum number of principal components with additional independent test data.

Classification is an important application of chemometrics. Classification is the sorting of data into different groups. These groups can be quite diverse such as different sources or different quality grades of the same raw material. Chemometrics methods for raw material identification using NIR or Raman spectra as important but are relatively simple and are discussed elsewhere [5]. In this chapter, we will discuss soft independent modeling of class analogies (SIMCA) [6]. A method for classification of similar classes using multivariate analysis. PCA score plots sometimes show data sets to consist of several subgroups. For example, Figure 33.5 shows the score plot for the mid-IR spectra for a series of oils. Color coding the score plot clearly illustrates the differences between the four oils (olive, corn, safflower, and corn margarine).

SIMCA is designed to improve on this separation of classes by using the residuals from the PCA analysis. Residuals are the difference between the PCA model and the data. In the SIMCA analysis, a separate PCA model is built for each class in the training set. The average residual value for each class (S_0) is also calculated. Test or validation data are then fit to each PCA class model. The correct class is the class that has the best fit to the PCA model. The comparison is quantified by the use of the scaled residual S_0 (Dmod$X$) values. The equations are given below

$$S_i = \sqrt{\frac{\sum_{k=1}^{K} e_{ik}^2}{(K-A)}}$$

$$S_0 = \left(\frac{\sum_{i,j} e_{ij}^2}{(N-A-1)(K-A)} \right)^{1/2} \tag{33.11}$$

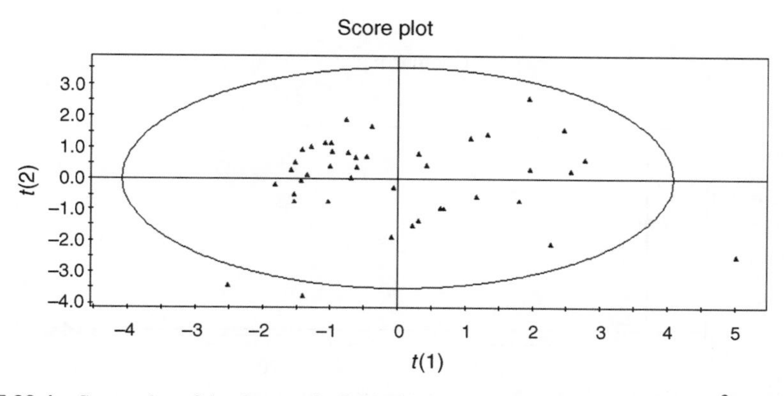

FIGURE 33.4 Score plot of the data set in Table 33.2. The ellipse is the Hotelling T^2 ellipse at 95% probability level. Samples outside the ellipse have a probability of greater than 95% of being statistical outliers.

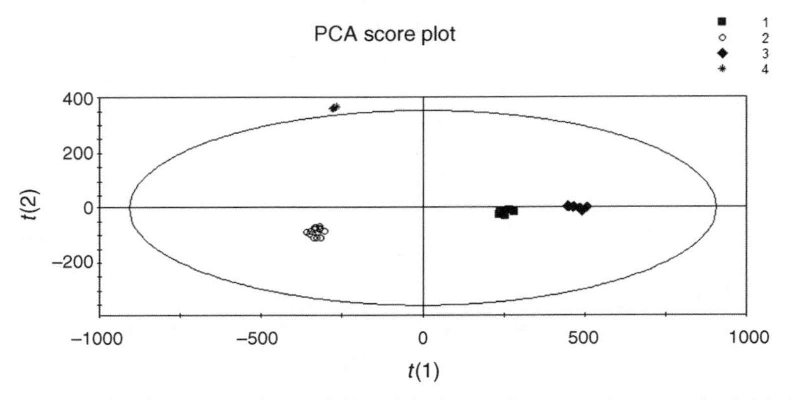

FIGURE 33.5 Score plot of mid-IR spectral data for a series of oils. Legend group 1: corn oil, group 2: olive, group 3: safflower, and group 4: corn margarine.

N is the number of samples, A is the number of principal components, K is the number of variables, s_i is the root mean square residual value for the ith sample, and e_{ij} is the spectral residual, that is, the difference between the spectra and the PCA model for observation i and variable j. If the test sample residual is close to the average residual for the entire class, then the sample has a high probability of belonging to the class. The relationship to actual probability values is possible because the scaled residual values $(S_i/S_0)^2$ in equation 33.11 are described by an F-distribution. The results of a SIMCA analysis are often displayed in a Cooman's plot. In a Cooman's plot, two classes are compared as shown in Figure 33.6.

A typical Cooman's plot is shown below. PCA models for corn oil and olive oil are used to predict the classification of a set of test samples. The test samples include olive, corn, corn margarine, safflower, and walnut oils. The different classes are color coded as shown in the legend. The x-axis on the Cooman's plot is the Dmod$X$ (distance to model) value for the corn oil; the y-axis is the same for olive oil. The red vertical line is the 5% probability level for the corn oil model, samples to the right of this line are probable outliers for the corn oil models. The red vertical line is the same for olive oil. Note most olive and corn oil test samples are correctly classified. Test samples form other classes are well separated from the oil and corn oil groups.

In many cases, spectral data requires mathematical transformations before multivariate analysis is performed [7]. The mathematical transformations are collectively referred to as spectral preprocessing. Derivative preprocessing is the most common form of spectra preprocessing with NIR spectra. Derivative preprocessing will eliminate or at least minimize the background variation associated with the NIR spectra of many pharmaceutical materials. The effects of a first derivative preprocessing on a typical NIR spectra are shown in Figure 33.7 (top and bottom). The first derivative removes the slowly varying baseline typical of NIR spectra of powders, positive and negative peaks correspond to regions where the slope of the raw spectrum has a positive or negative value. There are several methods for the calculation of spectral derivatives; however, they all start with the definition of first or second derivative from elementary calculus.

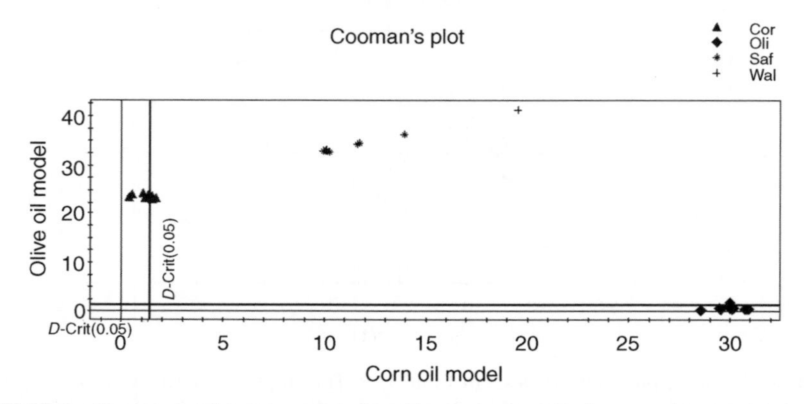

FIGURE 33.6 Cooman's plot comparing the olive and corn oil classes using a test set. Legend triangle: corn oil, diamond: olive oil, asterisk: safflower oil, and plus: walnut.

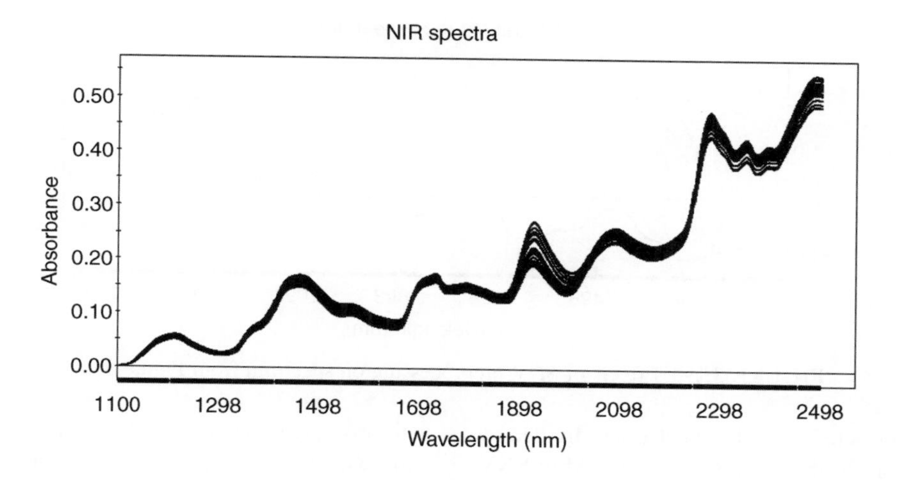

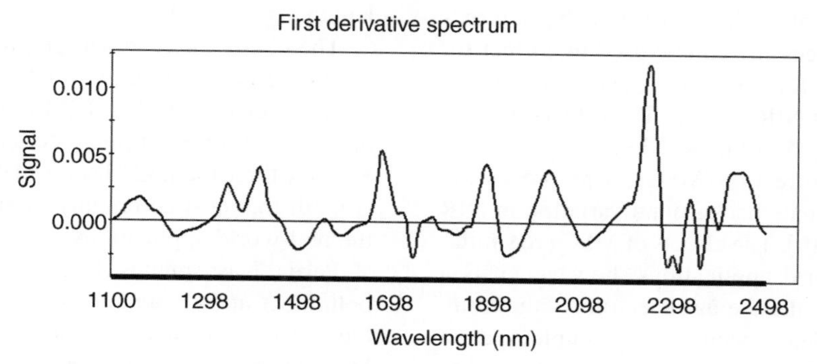

FIGURE 33.7 Effects of first derivative spectral preprocessing. *Top*: Several raw NIR spectra. *Bottom*: First derivative spectrum.

$$f'(x) = \frac{f(x+\Delta x) - f(x)}{\Delta x}$$

$$f''(x) = \frac{f(x+2\Delta x) - 2f(x+\Delta x) + f(x)}{(\Delta x)^2} \quad (33.12)$$

$$f''(x+\Delta x) \approx f''(x) + (\Delta x) f^{(3)}(x)$$

The two most common approaches for the calculation of derivatives are gap or Savitzky–Golay derivatives [8]. A gap derivative is based on the calculation of a running average for n points to the right and the left of a center point. The average values for the right- and left-hand sides are then used to calculate a finite difference derivative. There is an optional gap or separation between the right- and left-hand sides. Gap derivatives are described in Figure 33.8. A first-order gap derivative uses an n-point average to calculate a finite difference first derivative. Commonly the gap is set to zero in many applications.

Savitzky–Golay derivatives are based on fitting N points of the data to either a quadratic or cubic polynomial. The derivative is found by differentiation of the polynomial. For both methods of numerical differentiation, it is important to properly determine the number of points used in the averaging. Too few points can compromise the signal to noise; too many points will filter out important high frequency components of the data.

Standard normal variant (SNV) is a preprocessing method that is used to autoscale individual spectra [9]. The equation for SNV is given below.

$$\frac{x-\mu}{\sigma} \quad (33.13)$$

where μ is the average value for the spectrum of interest and σ is the standard derivation of the numbers, which make up the spectrum. During SNV preprocessing, the average value

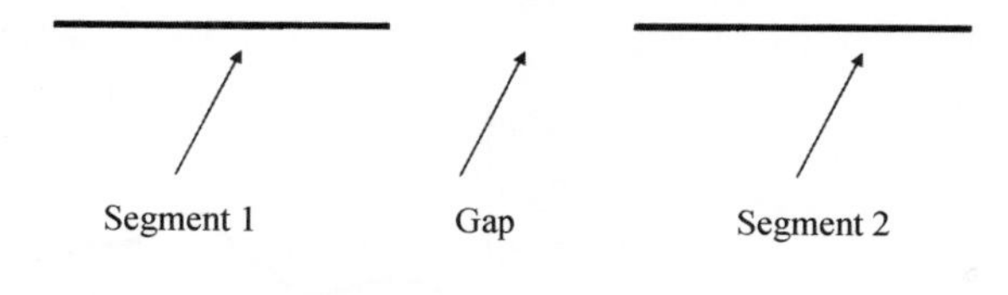

$$f'(x) = \frac{\text{Segment2} - \text{Segment1}}{\text{Length}}$$

FIGURE 33.8 Illustration of gap derivative algorithm.

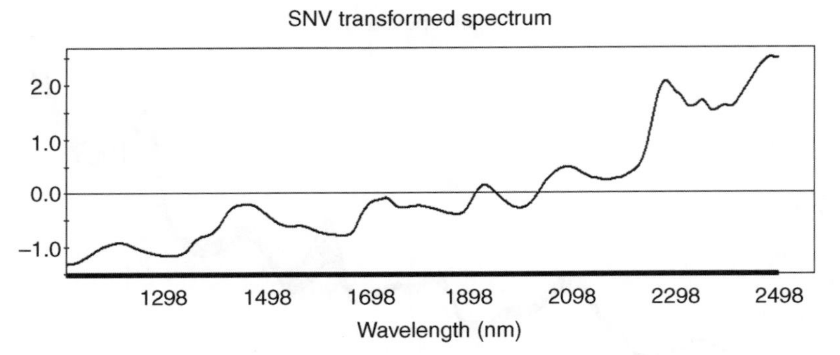

FIGURE 33.9 Effects of SNV preprocessing on spectrum from Figure 33.7.

for each spectra is subtracted, then the spectrum is divided by the standard derivation for the sample spectrum. After SNV preprocessing, the range of each spectra will be approximately −2 to 2. SNV processing can be used to correct for laser intensity variation in Raman spectra, and several kinds of path length variation in NIR spectra. The effects of SNV preprocessing are illustrated in Figure 33.9.

Multiplicative scatter correction (MSC) is a preprocessing method designed to eliminate background variation in NIR spectra due to scattering [10]. The effects of MSC are similar to SNV in many real-world applications; however, it is a distinct method. MSC uses the average spectrum of the entire data set and not individual spectra. The sample spectra are then regressed against the average spectrum producing slope and offset values at each wavelength for all samples in the data set. The slope and offset values are then used to correct the data set. Results of MSC preprocessing are illustrated in Figure 33.10. MSC preprocessing will remove the variation due to scattering in the data set but not change the average spectral value as SNV preprocessing does. The equations for MSC preprocessing are given below.

$$x_j = a_j\mathbf{1} + b_j\boldsymbol{\mu} + \boldsymbol{\varepsilon}_j$$
$$x'_j = (x_j - a_j)/b_j \tag{33.14}$$

$\boldsymbol{\mu}$ is the average spectrum, b_j is the slope, and a_j is the offset values for each wavelength. An important advantage of MSC preprocessing is that it can be used on filter wheel data, where the wavelength spacing is irregular and only a few wavelengths are typically measured.

There are many other preprocessing method used in chemometrics such as wavelets, orthogonal signal correction, and extended MSC (EMSC) [3, 11]. In practice, combinations of different preprocessing can also be used. However, the three methods discussed derivatives, SNV, and MSC are still the most commonly used preprocessed methods in the real-world applications.

Partial least squares (PLS) is an extension of PCA where both the *X* and *Y* data are considered [12, 13]. In PCA, only the *X* data is considered. The goal of the PLS analysis is to build an equation that predicts *Y* values (laboratory data) based on *X* (spectral) data. The PLS equation or calibration is based on decomposing both the *X* and *Y* data into a set of scores and loadings, similar to PCA. However, the scores for both the *X* and *Y* data are not selected based on the direction of maximum variation but are selected in order to maximize the correlation between the scores for both the *X* and *Y* variables. As with PCA, in the PLS regression development the number of components or factors is an important practical consideration. A short description of the PLS algorithm is given below, a more detailed discussion of the PLS algorithm can be found elsewhere [12, 13]. Commercial software can used to construct and optimize both PCA and PLS calibration models.

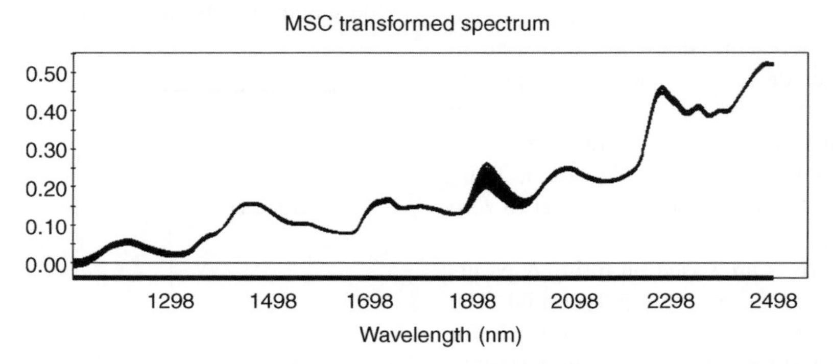

FIGURE 33.10 Effects of MSC preprocessing on spectrum from Figure 33.7. Note difference in *y*-axis from Figure 33.9.

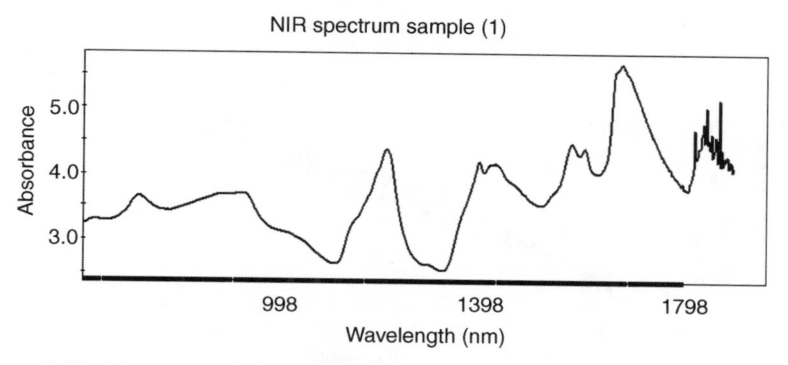

FIGURE 33.11 NIR transmission spectrum of a pharmaceutical tablet.

PLS decomposition of both X and Y data into scores and loadings is given in equation 33.15.

$$\mathbf{X} = \mathbf{T}\mathbf{P}^{\mathrm{T}} + \mathbf{E}$$
$$\mathbf{Y} = \mathbf{U}\mathbf{Q}^{\mathrm{T}} + \mathbf{f} \qquad (33.15)$$

The score matrices for **X** and **Y**, that is, **T** and **U**, are calculated together. This self-consistent approach allows for a set of scores and loadings that represent the variation in the Y data set. Therefore, the scores and loadings are much better than PCA scores and loadings for quantitative prediction. The algorithm proceeds by mean centering the data and then finding the first loading spectrum and first component scores. The prediction of a PLS method is summarized in the regression vector or coefficient, **B**. The predictions are related to the **x** sample data by

$$y = \mathbf{B} \cdot \mathbf{x} \qquad (33.16)$$

We will now consider an example of a PLS calibration using NIR data. NIR transmission spectra from 155 tablets have been measured [14]. The tablet calibration set included samples with a range of assay values and lots of production samples in order to capture the typical variations seen in the tablets. After scanning with the NIR instrument, the amount of active ingredient in each tablet was measured by HPLC. The weight of the tablet was about 800 mg and the target value for the drug content was 200 mg. We will use chemometrics to develop a model for the amount of active. This model could be used to monitor the stability of tablets over time in a nondestructive manner. For brevity, we will only outline the analysis procedure. Typical NIR transmission spectra for the pharmaceutical tablet are shown in Figure 33.11. The broad, overlapping spectra with a considerable background is typical of NIR spectra. Derivative preprocessing can be used to remove the unnecessary background and elucidate the underlying peaks in the spectra. A first derivative spectrum is shown in Figure 33.12.

A calibration curve showing the predictions of the PLS model versus the laboratory data is shown in Figure 33.13. The clear quality of the calibration curve is evident. The calibration curve can be evaluated by several methods including outlier detection and removal and optimization of the spectral range used for PLS calibration. A detailed discussion of these issues can be found in the references [12, 13]. Common examples of quantitative methods done with NIR data and PLS regression are moisture, particle size, and assay [7].

Method validation for NIR or Raman spectroscopic methods using chemometrics is outlined in United States Pharmacopoeia (USP) Chapter ⟨1119⟩ [15]. The criteria for method validation are the same as other quantitative analytical methods, such as accuracy, precision, intermediate precision, linearity, specificity, and robustness. Since these methods are statistical in nature and are based on a previously

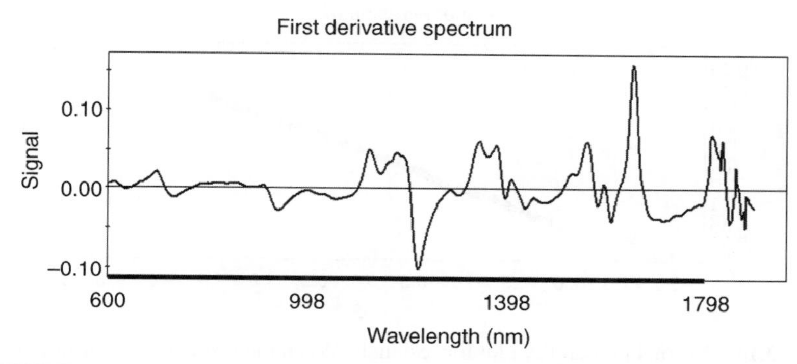

FIGURE 33.12 Spectrum from Figure 33.11 after first-derivative preprocessing.

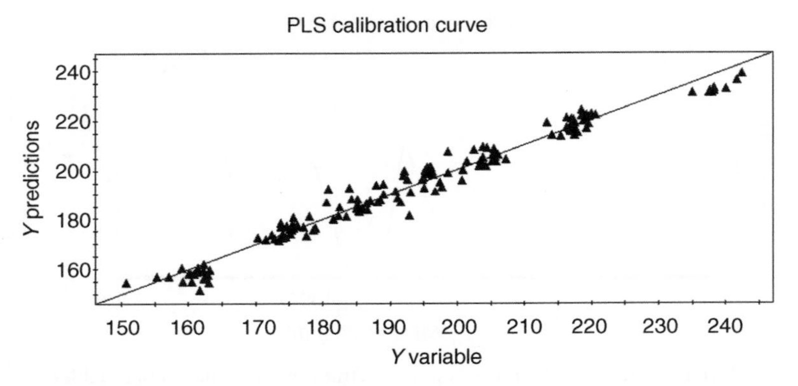

FIGURE 33.13 Calibration curve for PLS method for tablet assay value.

validated analytical method, the validation of MVA methods is somewhat different than traditional analytical methods. In this chapter, we will briefly discuss chemometric method validation, a more detailed discussion can be found elsewhere.

Accuracy of the MVA method refers to how closely the MVA method and the original laboratory method compare. The accuracy of a chemometric method is evaluated by comparing the predictions of the MVA model with the actual laboratory data for a set of validation samples. The validation samples should be from lots of material not used in the original calibration set. There are several mathematical ways to express the accuracy. The most commonly used approach is the standard error of prediction (SEP). The SEP is defined in equation 33.14.

$$\mathrm{SEP} = \sqrt{\sum \frac{(\mathrm{NIR}-\mathrm{LAB})^2}{n}} \qquad (33.17)$$

where n is the number of validation samples. The SEP value should be close to the actual error of the original laboratory method. The actual error of the laboratory method should include normal sources of variation such as different analysts, different instruments, different materials analyzed on different days.

The *linearity* of a multivariate method is an important topic. Typically the linearity of a chromatographic method is evaluated by the R^2 (coefficient of determination) value of a recovery measurement. R^2 is the fraction of variation in the y-variable explained by the linear fit; r is the correlation coefficient that quantifies the correlation between the x and y variables [16]. R^2 is often used in the analysis of chromatography recovery studies [16]. In contrast, R^2 is not a good statistical parameter for multivariate methods. The linearity of a multivariate method is evaluated by the inspection of the residual values, that is, the difference between the predictions of the multivariate model and the actual laboratory data. A linear model will have residuals that are random, that is, normally distributed. A nonlinear model will have residuals that are not normally distributed. The USP Chapter ⟨1119⟩ states that the linearity should be evaluated by examination of the residuals, but no specific threshold or criteria are given. In the opinion of this author, visual inspection of the residuals using a normality plot is recommended. In Figure 33.14, a normality plot of residuals is shown. The data points in Figure 33.14 do follow a straight line, indicating a normal distribution of residuals, consistent with a linear model or

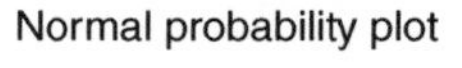

FIGURE 33.14 Normal probability plot for residuals. When the residuals fall on straight line, the calibration under consideration is linear.

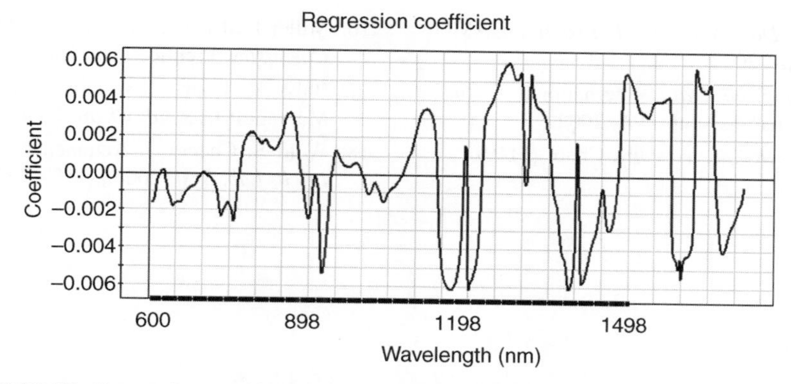

FIGURE 33.15 Regression coefficient from the PLS model for tablet assay described earlier. The regression coefficient is a method for documenting and examination of which wavelengths are most important for the PLS calibration.

calibration [13]. In some cases, the linearity of the model can be improved by removing some of the points in the normality graph, which are probable outliers.

Method *specificity* is the extent the multivariate calibration is specific to the analyte of interest. With a PLS calibration, the specificity is documented by the regression coefficient of the calibration. The regression coefficient shows which wavelengths are most important for the PLS calibration. Important wavelengths may have either positive or negative regression coefficient values. The most important wavelengths should correspond to the absorption peaks of the analyte of interest. For example, the regression coefficient for a moisture model will have peaks at the known water absorbance band locations. In practice, the regression coefficient is often documented in the method development report. A regression coefficient from the PLS calibration for tablet assay described earlier in this chapter is shown in Figure 33.15.

The *range* of a multivariate calibration method is determined by the range of laboratory values in the calibration and validation data sets. A method is validated over the range of laboratory values of the samples used in the independent validation set. The range of the validation samples can also depend upon the application of the method. For example, in-process testing or testing where a limited number of samples are available may require a fairly small range of values because samples outside of a small range are not available or do not exist.

This chapter has briefly summarized the essential principles of chemometrics and their application to spectral data. There are many applications of chemometrics that have not been discussed here due to space limitations. Two important examples of this are chemical imaging and batch monitoring. Raman and NIR chemical imaging have been applied to pharmaceutical products including tablets and drug-coated stents [17]. Both Raman and NIR chemical imaging methods typically require chemometrics for the creation of useful images. Batch monitoring involves the use of multivariate control charts based on score plots developed from a collection of good batches [18]. Batch monitoring can be used with spectral or process data. Batch monitoring has been used to monitor a variety of complex pharmaceutical products to improve yields and provide improved process understanding [2]. In summary, chemometrics is a vital part of process analytical technology, quality by design, and the overall future of both pharmaceutical development and manufacturing.

REFERENCES

1. US Food and Drug Administration. *Guidance for Industry Process Analytical Technology*, 2004.
2. Eriksson L, Johansson E, Kettaneh-Wold N, Trygg J, Wikström C, Wold S. *Multi- and Megavariate Data Analysis*, Umetrics AB, Umeå, Sweden, 2006.
3. Strang, G. *Computational Science and Engineering*, Wellesley-Cambridge Press, Wellesley, MA, 2007.
4. Johnson RA, Wichern DW. *Applied Multivariate Statistics*, Prentice Hall, Upper Saddle River, NJ, 2002.
5. Long F. *Am. Pharm. Rev.* 2008; Sept/Oct.
6. Wold S. *Pattern Recogn.* 1976;8:127–139.
7. Siesler H, Ozaki Y, Kawata Y, Heise H. *Near-Infrared Spectroscopy: Principles, Instruments, Applications*, Wiley-VCH, Weinheim, 2002.
8. Savitzky A, Golay MJE. *Anal. Chem.* 1962;36:1627–1639.
9. Barnes RJ, Dhanoa MS, Lister SJ. *Appl. Spec.* 1989;43: 772–777.
10. Martens H, Jensen SA, Geladi P. *Proceedings from Nordic Symposium on Applied Statistics,* Stokkand Forlag Publishers, Norway, 1983, pp. 205–233.
11. Martens H, Stark EJ. *Pharmaceut. Biomed. Anal.* 1991;9: 625–635.
12. Wold S, Josefson M. Multivariate analysis of analytical data. In: Meyers R, editor, *Encyclopedia of Analytical Chemistry*, Wiley, New York, 2000, pp. 9710–9736.

13. Esbensen K. *Multivariate Data Analysis—In Practice*, CAMO Process AS, Oslo, Norway, 2002.
14. Ritchie G. *Software Shootout Data,* International Diffuse Reflection Conference, Chambersburg, PA, 2002.
15. US Pharmacopoeia. *Near IR Spectroscopy*, Chap. ⟨1119⟩.
16. Miller J, Miller J. *Statistics and Chemometrics for Analytical Chemistry*, Prentice Hall, Harrow, UK, 2000, p. 137.
17. Balss K, Long F, Veselov V, Akerman E, Papandreou G, Maryanoff C. *Anal. Chem.* 2008;80:4853–4859.
18. Wold S, Cheney J, Kettaneh N, McCready C. *Chemometr. Intell. Lab. Syst.* 2006;84:159–163.

PART IV

DRUG PRODUCTS

34

PROCESS MODELING TECHNIQUES AND APPLICATIONS FOR SOLID ORAL DRUG PRODUCTS

Mary T. am Ende, Rahul Bharadwaj, Salvador García-Muñoz, William Ketterhagen, and Andrew Prpich

Pharmaceutical Development, Pfizer Global Research & Development, Groton, CT, USA

Pankaj Doshi

Chemical Engineering and Process Division, National Chemical Laboratory, Pune, India

34.1 INTRODUCTION

The budget-restricted pharmaceutical environment is countered by the heightened expectations for drug products to be developed with more intensive use of science-based principles. The issues that arise during drug product development are often attributed to the impact of changing batch size, and equipment type or scale on the formulation and the process. In the commercial arena, one of the more prevalent causes for batch failures, or product not meeting specifications, is property shifts in the excipients or active pharmaceutical ingredient (API). During development and commercialization of a drug product, it is important to design a robust dosage form that is minimally sensitive to raw material variations and equipment scale.

The fundamental principles taught in the chemical engineering curriculum equip the chemical engineer with the skill base that allows them to address these complicated process operations and the understanding to connect the material properties to the processing equipment and design. Courses on the fundamental laws of heat transfer, mass transfer, momentum transfer, transport phenomena, and physical chemistry allow the chemical engineer to mathematically describe the process. For example, an impeller used to mix API and excipients imparts energy and momentum on the material to achieve uniformity of the blend. In this chapter, an example of powder discharging from a bin is monitored through computational methods to predict segregation. Utilizing the basics of the process calculations course enables students to break down the system into a control volume and solve mass and energy balances to determine the solution. An example of this type will be illustrated in this chapter during the derivation of the thermodynamic film-coating model. Process control principles allow the chemical engineer to develop models to accommodate variations in the inlet stream properties and can adjust the process through feedforward or feedback control to produce consistent quality product. In this chapter, this approach is used in reverse to set specifications on the raw material properties to ensure product quality using empirical models. The undergraduate curriculum, including many other courses not specifically highlighted in this chapter, provide a well-rounded understanding of how to approach solving process problems and how to break down a problem into its fundamental parts. In addition, it provides the science-based hypothesis testing principles that are important to understanding the solution of the problem. Chemical engineers are skilled at writing in mathematical terms the driving forces affecting processes, and are capable of modeling a process using first principles. They are able to construct a control volume for engineering balance determination across inlet and outlet streams, and use empirical methods, such as traditional regression polynomials, neural networks, or multivariate latent variable models (LVMs) to understand complex processes. The two

Chemical Engineering in the Pharmaceutical Industry: R&D to Manufacturing Edited by David J. am Ende

opposing external environment factors of setting expectations to reduce costs from consumers and the heightened scientific expectations from the regulators have created a crucial opportunity for chemical engineering principles to be applied and implemented across the industry.

In this chapter, modeling techniques applied to formulation and processing operations are discussed as support to the design, development, and scale-up for solid oral drug products. These process modeling techniques are discussed and exemplified with case studies ranging from raw material specifications to process parameter predictions. In general, the main unit operations utilized to produce tablets include blending and other powder processing, dry or wet granulation, tablet compression (powder compaction), and film coating. Specific consideration is given in this chapter to transfer and scale-up issues along with general process design related challenges to pharmaceutical process R&D.

34.1.1 Benefits of Using Modeling Tools to Design, Develop, and Optimize Drug Products

Over the last decade, pharmaceutical companies, in an effort to reduce costs, have embarked on bulk conserving methods for drug product design and development. These efforts have resulted in significant advances in process scale-up that utilize science of scale tools and predictive models. The major benefits of using modeling during development of pharmaceutical products have been previously highlighted by Wassgren and Curtis [1] who illustrated how employing reliable models can improve understanding of critical processes that may rapidly accelerate process improvements. An economic analysis of one specific engineering company utilizing computational fluid dynamics (CFD) modeling for a 6-year period revealed a sixfold return on investment (ROI) [2].

A more comprehensive economic analysis was conducted by Louie et al. [3] for the modeling of API and material science properties within the pharmaceutical development. Examples of the modeling capabilities considered in this analysis included API material properties (such as crystal morphology, surface area, and powder X-ray diffraction patterns), solubility, polymorphism, breakage planes, refractive index, molecular and solvent interactions. The benefits considered in the Louie analysis [3] included improved experimental effectiveness, broader/deeper understanding in the exploration of solution to a problem, improved productivity by employing knowledge-based reasons for moving forward, reduced time to market for new products (IP/exclusivity), and fewer unknowns with a reduced risk for failures.

The analysis indicated that the use of modeling and simulation tools in pharmaceutical development is producing an ROI of \$4 to \$10 per dollar invested for an occasional user to a superuser, respectively. The greatest impact on ROI was found to originate from employing superusers (or subject matter experts) in material sciences and API development. A similar analysis would be beneficial for the modeling of drug product processes during development and commercialization. The modeling capabilities available and applied to solid oral drug products consist of a balance between fundamental models, engineering-based models, and empirical models with intentional focus on applied use (Table 34.1). There

TABLE 34.1 Comparison of Modeling and Simulation Capabilities for API Material Properties Versus Solid Oral Drug Product Properties and Processes

Materials Science Modeling and Simulation of API [3]	Solid Drug Product Modeling and Simulation
Structural properties (crystal morphologies, orientations, attachment energy, surface energy, PXRD, possible API crystal forms	Raw material properties (particle size distribution, particle shape, density, material properties) [1, 4]
Physical properties (solubility, hydration, predicting preliminary physical data)	Blend properties (flow, velocity, segregation) [1, 5–10]
Molecular interactions (hydrogen bonding, solvent interactions)	Agglomeration properties (population balances of wet granulation [11])
Purity (polymorphism, impurities, predicting stability of crystal forms)	Breakage properties (population balances of milling [12])
Mechanical properties (shear strength, hardness)	Fluid dynamics (turbulent fluid flows in spray dryers, dry powder inhalers) [1]; (agitated vessels, fluidized beds) [13–15]
Thermal properties	Solid mechanics (stress analysis and density distribution for tablet and tooling) [13, 16–19]
Optical properties (refractive index, spectral absorption, circular dichroism)	Mass and energy balances (film-coating pan, fluid bed dryer) [20]
Electrical properties (conductivity, resistivity, dielectric behavior)	
Empirical models [21, 22]	Empirical models [21, 22]

appears to be a similar level of modeling and simulation capabilities for materials science of API and solid oral drug products; therefore, the authors would anticipate the return on investment for drug products to be positive also.

34.1.2 Summary of Modeling Approaches for Solid Oral Drug Products

Kremer and Hancock [13] were first to review modeling in the pharmaceutical industry and point out that process modeling can be considered as numerical simulations of the underlying physical processes. The process models in this category, which are based on first principles, can be expressed by the governing equations that are solved either analytically or numerically. The primary modeling examples presented in this chapter for the physics-based models include fluid dynamics, solids mechanics for tooling design, and particle-based models of powder discharging from hoppers.

Fluid dynamics is the study of flowing media such as gases, liquids, and certain types of solids such as dense, rapidly flowing powders. Pharmaceutical researchers have utilized commercially available CFD software packages to simulate a variety of applications including spray drying, inhalation, mixing in agitated vessels and flow of granular material. The performance of several unit operations has been investigated and optimized using CFD.

Discrete element method (DEM) is a particle-scale modeling approach in which the motion and forces associated with each particle are tracked individually. Commercial DEM software packages have only recently become available, and a specialist is often required to develop particle-based modeling using DEM. The primary disadvantage for DEM is the significant computational resources required to compute and track the wealth of particle-level information produced: particle velocities, forces, residence times, and stresses. The maximum number of particles, N, which can be modeled for reasonable simulation times is typically on the order of $N \sim 10^5$.

Engineering models can be considered a subclass of physics-based models because they are based on first principles; however, they are applied to a defined control volume typically encompassing a unit operation. These types of models are often built upon mass, momentum, and/or energy balances across the control volume, or derived from nondimensional analysis of the driving forces involved in a certain process. One example of the former (tablet film coating) and two examples of the latter approaches (wet granulation and fluid bed drying) will be discussed in this chapter. The thermodynamic film-coating model is used to scale-up the tablet film-coating process based on matching exhaust air temperature and humidity as a representation of the tablets in the coating pan (environmental similarity). Nondimensional analysis (e.g., Froude number and Reynolds number) can be employed to blending, milling and wet granulation processes to examine the specific driving forces. In addition, momentum transfer of drying air to the wet granules can be analyzed for corresponding fluidization conditions based on the granule particle size as an applied engineering approach to predict acceptable drying airflow rates.

Tablet film-coating models [20, 23–25] have been used for process scale-up based on environmental similarity in the coating pan and maintaining constant droplet size from the spray guns. The thermodynamic film-coating model has been validated across lab to production scales, and can be used to predict the temperature and relative humidity of outlet air stream, determine process set points based on desired exhaust air conditions, and minimize or eliminate the need for scale-up trials. The film-coating atomization model is a validated theoretical model that describes the film coater atomizer performance, and allows for the prediction of droplet diameter at the spray guns. The model requirements for the thermodynamic model include previous process data for the coating pan in question to generate the heat loss factor; and for the atomization model include the nozzle specifications of the spray guns and rheological properties of coating solution (viscosity, surface tension, and density).

Empirical modeling approaches are typically based on existing process data, which can yield a set of parameters that can then be interpreted from a fundamental deterministic knowledge of the process. Multivariate LVM is one specific empirical approach that has been proven to deliver a deep process understanding for unit operations that are particularly difficult to describe in a first principles model, and in situations where there is a wealth of data from the process under study [22].

The application of LVM can be better appreciated when it is mapped along the life cycle of a product/process. Although the methods can be widely applied, their application is limited by the availability of data. LVM can be applied at early stages of development, where materials (and ratios of) are being selected along with processing conditions [26]. These methods can also be applied during process design and scale-up [27] to minimize experimental work at the larger scale. For a commercial process, LVM can be used to troubleshoot/diagnosis issues [28–30], to optimize [31, 32] and to control [33–35] the operation. Other applications of LVM include the analysis of images [36, 37] and the establishment of multivariate specifications for incoming materials [38].

The first stage in the general strategy for these applications is to (i) fit a model to data, which is considered relevant to the particular application (e.g., for a monitoring application, the data will correspond to normal operating conditions, which will serve as a basis); and (ii) once the model is fitted and deemed valid, the parameters of the model are interpreted (if possible) from a fundamental perspective, trying to associate each of the identified principal components (or latent variables) with a driving force acting upon a system.

A model that is considered valid and representative of a given system can be used in either passive mode or active mode. In passive mode, the model does not influence the process directly (e.g., troubleshooting or monitoring applications). In active mode, the model is actually influencing a decision either through a feedback control or by a design exercise.

Multivariate latent variable modeling is an alternative approach when fundamental modeling is not an option due to timing or investment constraints or lack of available fundamentals. The main difference between these types of latent variable models and other empirical approaches (such as neural networks or traditional ordinary least squares fitting) is the capabilities embedded in the method to handle massive amounts of incomplete and ill-conditioned data (which is common from an industrial process).

34.1.3 Overview of Modeling Approaches Mapped onto Unit Operations

In order to design dosage forms effectively, the engineer must be aware of the potential issues that might occur and the driving forces associated with these issues. For solid oral dosage forms produced through a dry granulation process (Figure 34.1) [39], the potential issues that can occur during process scale-up can be consider from an engineering perspective. Blending operations are intended to evenly disperse API within the excipient powder by transferring mass and energy from the mixer or impeller to the powder. Between unit operations, there often exists a transfer step that can create an opportunity for the uniform blend to segregate resulting from mass transfer. Dry granulation and milling processing is intended to reduce this segregation potential by altering the particle size distributions of active powder blend through a combination of mass and energy transfer. The goal of the tablet compression process is to produce tablets with the target product properties such as tablet weight, hardness, potency, and dissolution. This process can also be susceptible to tablet weight variations or powder segregation issues, all of which are related to the differential mass transfer, energy transfer, and momentum transfer of the materials. Film coating of tablets can exhibit issues of overdrying, overwetting, or attrition that cause defects in the final product. These issues are related to their associated driving forces of energy transfer (high drying flow rates), mass transfer (accumulation of moisture and coating from high spray rates), and momentum transfer (impacting low-density regions of tablet surfaces that attrite). The dosage form design criteria must ensure the stability, performance, and manufacturing capability of the final product. The source of the energy and/or momentum imparted onto the drug product formulation is through the processing equipment. Therefore, engineers can use energy/momentum transfer analysis as the design levers to adjust the product to the desired result in the quality attributes (e.g., blend and granulation content uniformity, tablet potency, dissolution, and stability profile).

34.2 FORMULATION MODELING

34.2.1 Empirical/Statistical Models for Raw Materials

Material properties are inherently variable, and therefore understanding the impact of these variations is an important factor in the formulation and process development of a drug product. A well designed, robust product will be minimally

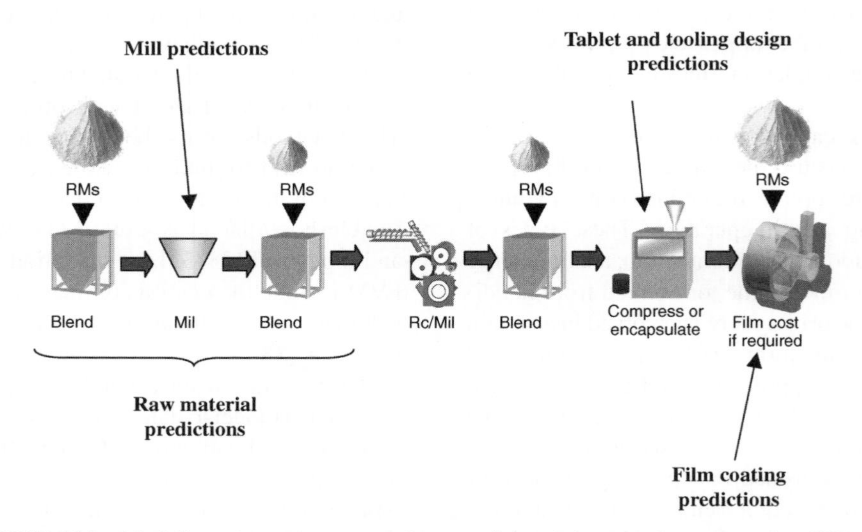

FIGURE 34.1 Modeling approaches mapped onto solid oral drug product processing [39]. Reprinted from Ref. 39 with permission of John Wiley & Sons, Inc.

sensitive to ingoing raw material changes. This section will address the relationship between the raw material variability and the drug product quality attributes, demonstrated both conceptually and through an example.

Quality should not be qualitative but quantitative, and even more, should be multidimensionally quantitative. As Duchesne and MacGregor write, “Quality is a multivariate property requiring the correct combination of all measured characteristics” [38]. So from this point and beyond, the concept of quality will not refer to a single numeric value representing a measured attribute of a product, but rather to a vector of multiple values (of multiple attributes of a product) that represent quality as a set.

Recent guidance documents from the U.S. Food and Drug Administration emphasize that quality should be built into, rather than tested on, the product [40–43]. This should be pursued from the design stages all the way to commercialization to varying degrees. Quality is defined on a case-specific basis, and in principle should be guided by the ultimate effect on product safety and efficacy and performance [44]. This guidance is referred to as the “quality by design” paradigm, and has established a science and risk based approach to pharmaceutical development and commercialization.

A process-driven design exercise is executed downstream and in sequence. Such a scenario may start with the effect of the starting material onto the first unit operation and will continue downstream with emphasis remaining on the process. It is only at the end of the process (the design of the last few unit operations of the train) that the product quality will be considered.

A quality by design exercise implies upstream design, where the design starts with a focus on the product quality and how it is impacted by the process. It continues upstream, with emphasis on the product, and ends with the analysis of the effect of the starting materials on the complete manufacturing train. This analysis will eventually end in the establishment of an acceptance criterion for each of the raw materials. These acceptance criteria will be mostly based on statistical diagnostics.

Typical statistical tools focus on testing one variable at a time in order to accept/reject a given lot of raw material. Recent trends in data analysis suggest performing this diagnosis on multiple variables simultaneously. First, a mathematical description of each of the statistical tools is required, followed by an illustration of these concepts highlighting the benefits of each technique.

34.2.2 Traditional Statistical Tools

As a well-established field, statistical diagnostics are now well accepted and widely applied in practically all areas of engineering. This section is not intended to serve as a reference in statistics; the reader is referred to other texts for this purpose [31]. What is included is a high level description of some of the concepts mentioned in statistics books so that the reader can better interpret the information presented in these.

34.2.3 A High Level View of Hypothesis Testing and Significance Levels

A set of experiments should always be carried out with a purpose in mind. The purpose is usually to obtain information from a given system and learn from it. And this learning usually comes from proving (or disproving) a preconceived idea about the system. This is referred to as hypothesis testing. The type of test used will depend on the particular hypothesis being tested, and the available data.

In engineering, hypothesis tests are usually of the quantitative nature. An inequality test—for example, is the density > 1.5 g/mL?—would require a one-sided hypothesis test, whereas a range (is the density greater than 1.5 and lower than 1.8) would require a two-sided test, and so on.

The reader should be aware that in statistics all statements involve a probability, usually referred to the level of significance, and quantified as $100(1-\alpha)$ for a given conclusion. Typical values chosen for $100(1-\alpha)$ are 95%, 99%, or 99.73% (which is the probability associated with six standard deviations or six-sigma in a normal distribution). The choice of the level of significance for a given test usually depends on the consequences and the implications (some times legal) of drawing the wrong conclusion (99% confidence on the conclusion that a plane engine will not fail means 1 in 100 times it might!). Obvious to mention is that a 95% significance level implies a value of 0.05 for α. If the test to be used is a single sided test, α is taken as is. For a double-sided test, α is usually divided by two to allow the test to be centered on the 50% probability.

Often in the establishment of specifications for materials, the hypothesis to be tested is a double-sided one. In these cases, the value of property A for a new material is compared against a reference set of values for property A to verify that it lies within a given range. The reference values for property A are usually chosen from materials used in the past due to desirability of those materials in terms of quality of the product or cost of manufacturing.

34.2.4 The Estimate of a Mean Value and Its Confidence Intervals

A simple tool to establish a double sided diagnostic to test property A for new materials is to estimate the mean value of property A and estimate the upper and lower confidence intervals for this mean value. This estimation will yield lower and upper bounds for the mean value of property A in the reference set. A new lot of raw material will be tested to determine its value of property A, and then decided if the mean value of property A for this new lot is also a plausible (probable) mean value for property A in the reference materials.

This confidence interval is a function of the degrees of freedom (number of samples n used to estimate the mean of property A minus 1), the standard deviation of the reference values for property A and the t value from statistical tables. This t value is a function of the desired significance level (α) and a number of degrees of freedom ($n-1$). Although this t value is a strong function of α, the α value is often the less questioned parameter. For all practical purposes, α can easily be fixed (e.g., 99.73%) for the sake of testing if the mean value of a given property of a new material is the same as the one in the reference set.

Equation (34.1) describes the calculation of the confidence intervals for the mean value of a normally distributed population with unknown (only estimated) variance α^2. In practice, the factor that has the greatest impact on this calculation is the number of samples (n). The range of acceptance (upper minus lower bound) will be large for a small number of samples, and will asymptotically narrow as the number of samples increases.

$$\bar{x}-t_{\alpha/2,n-1}\sigma/\sqrt{n} \leq \mu \leq \bar{x}+t_{\alpha/2,n-1}\sigma/\sqrt{n} \qquad (34.1)$$

In this formula, σ is the calculated standard deviation, $\bar{x}$ is the calculated average, n is the number of samples considered, and $t_{\alpha/2,n-1}$ refers to the value of t for a significance level of $100(1-\alpha)$ and $n-1$ degrees of freedom; this value is taken from statistical tables.

For example, consider three sample sets taken from the same population. All sample sets with an average of 4.25, and standard deviations of 0.0470, 0.0565, and 0.0895, respectively, which were calculated using 30, 10, and 3 samples. At a 95% level of significance the values of t for $n=30$, 10 and 3 are 2.042, 2.26, and 4.3, respectively. With these values, the confidence intervals on the means for each population are given in Table 34.2. Notice the dramatic increase in the range, just due to the number of samples considered in the calculations. The data for this example are from a pharmaceutical grade polymeric material and are a real illustration of how acceptance limits can vary in an application. The practitioner is encouraged to sample properly to avoid artificially large acceptance regions simply due to a limited number of samples available.

TABLE 34.2 Confidence Intervals for the Mean Value of Three Sample Sets of the Same Population

Mean Value	Sample Sets of Same Population		
	$n=30$	$n=10$	$n=3$
Mean lower bound	4.234	4.214	4.027
Mean upper bound	4.268	4.295	4.472
Range	0.03397	0.08088	0.44471
% Change from $n=30$	0	138.12	1209.27

34.2.4.1 Emerging Multivariate Techniques

A natural implication of the evolution of analytical technology is the fact that a given material can be characterized by a large number of attributes. It is the duty of the engineer to determine which of these attributes are relevant to the product/process. The answer to this question for a pharmaceutical product is rarely a single property (a scalar), and more often a set of properties (a vector) that will impact the product or process. The challenge now is how to establish a specification for multiple quantities.

The simple solution to this challenge is to establish multiple univariate diagnostics using traditional statistical tools described previously. This practice, however, implies that all the measured characteristics for the new material can be tested and assessed independently of each other. This assumption falls apart quite easily for complex materials where a large number of properties are related. For example, for a polymeric material the molecular weight distributions (or compositional distributions for a copolymer) are not independent of viscosity or density, which can also be linked to cross-linking.

In such a situation, there is a need for a tool that will enable the establishment of acceptance criteria for multiple properties simultaneously, accounting for their correlated nature. It is no longer enough to know the desired level (mean) and tolerance for each property. Additional information is necessary to account for the dependencies across the multiple properties being tested.

Multivariate LVMs have been proposed in literature to address this need. LVMs will empower the user to establish diagnostics and acceptance criteria based on a model of the data [38].

34.2.4.2 Using a Model to Establish a Test of Acceptance

Before describing the calculations behind a multivariate specification, it is important to discuss the overall strategy of using a model (any model!) to establish acceptance regions. To illustrate this point, consider the case of a material that is characterized by two properties A and B (plotted in Figure 34.2 where each dot is plotting the numerical values of property A versus property B for a given lot of raw material). Assume also that there is enough data to conclude that the lots of material represented by gray markers are desirable, and those represented by black dots correspond to undesirable material.

The challenge is to somehow delineate the region spanned by the gray markers. If univariate measures are taken (define a lower and upper bound for property A and B separately) one would end with a region equivalent to the smallest square that fits in the "gray marker" zone (drawn with a dotted line square). Although feasible and simple, this may constrain the practitioner to a very small region of acceptance and will end in large amounts of rejected material.

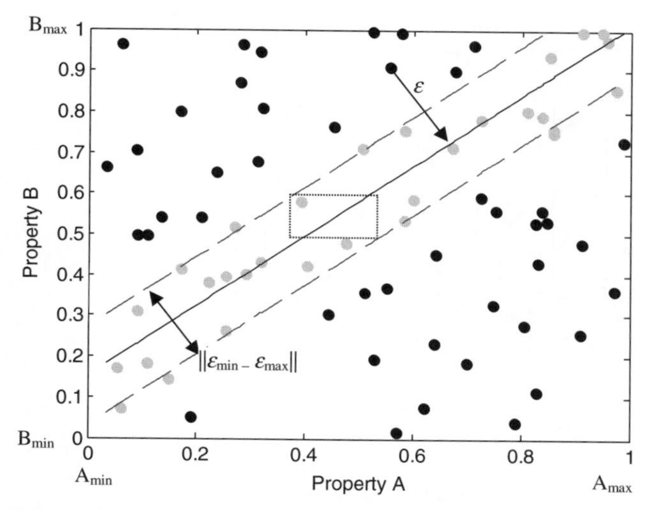

FIGURE 34.2 Conceptual illustration of a case where a multivariate specification is required.

The alternative is to use mathematics to delineate this region. Taking all the data that correspond to the gray markers, a regression model can be fitted. The regression line given by $A_{gray} = mB_{gray} + \delta$ is illustrated with a black line in Figure 34.2 (where m and δ are fitted parameters from the data).

This simple model (a line) can then be used to quantify the perpendicular distance from any given point, to the regression line (illustrated as ε in Figure 34.2).

The set of distances (ε) for all the gray points can be used to determine a maximum and minimum ε (ε_{min} and ε_{max}). These limits (ε_{min} and ε_{max}) on the perpendicular distance and the regression model and the upper–lower boundaries (A_{min}, A_{max}, B_{min}, and B_{max}) can be used to establish a multivariate specification that will ensure properties A and B for a new lot are within the region spanned by the gray markers. There is still a delicate statistical exercise, that is, to determine the perpendicular distance to tolerate (given by $||\varepsilon_{min} - \varepsilon_{max}||$), this will determine the width of the acceptance region (region bounded by the dashed lines in Figure 34.2) and the risk associated with the test.

For this conceptual case, the steps to accept a new lot of material (with properties A and B equal to a_{new} and b_{new}, respectively) using the overall bounds, a simple regression line as a model and an acceptable perpendicular distance (ε) would be as follows:

(i) If $A_{min} < a_{new} < A_{max}$ and $B_{min} < b_{new} < B_{max}$ continue, otherwise reject a_{new} and b_{new}.
(ii) Calculate $\varepsilon = a_{new} - (mb_{new} + \delta)$.
(iii) If $\varepsilon_{min} \le \varepsilon \le \varepsilon_{max}$, then the new lot is not rejected (there is no statistical evidence to prove that this lot of material is any different from the population represented by the gray markers in Figure 34.2).

Step (i) will ensure that the values of a_{new} and b_{new} are at least within range, step (ii) uses the model to determine the perpendicular distance to the line that runs in the middle of the acceptance region, step (iii) determines if this perpendicular distance is within tolerance. Notice that the model is *not* being used for predictive purposes.

Both properties (A and B) still need to be measured and none of them are being predicted from the other. The model in this case is just a mere geometrical tool to delineate a region that is one degree of complexity beyond a simple squared region (which is the result of two univariate specifications together). Also notice that there is still an exercise of probability and risk analysis associated with determining the upper and lower bounds for the key diagnostic(s) involved (in this case ε_{min} and ε_{max}). The strategy proposed here is multivariate in the sense that it handles more than one variable, but more important, it is multivariate simultaneous, which means it handles more than two properties at the same time.

For the conceptual case illustrated in Figure 34.2, it is easy to see how a line can be used as a model in the specification since the data are composed of two properties. As the dimensionality of the problem increases (the amount of variables to consider simultaneously) so does the need to have a model that considers all variables, their uncertainty levels and correlation simultaneously. And for that, principal component analysis is suggested.

34.2.4.3 Principal Components Analysis Principal component analysis (PCA) is a well-established technique to project or compress data to a lower number of new variables called principal components. Geometrically, the PCA works by identifying a new coordinate system within the data (see Figure 34.3), so that each point can be referred to by its coordinates with respect to this new system. Hopefully, the number of coordinates needed to span data well enough will be dramatically less than the original number of variables. The example in Figure 34.3 illustrates a data set with three variables (X_1, X_2, and X_3) that can significantly be represented with a two-dimensional coordinate system, assuming the deviation from this plane (see Figure 34.3b) is negligible.

Many software packages[1] are available in the market to fit a PCA model, and therefore such calculation is not discussed here. A PCA model is quite powerful as a tool to establish specifications due to the diagnostics provided by the model. This application is extensively discussed in literature [4, 38] and only summarized in this chapter.

Once a PCA model is fitted, each of the observations used in the model can be summarized by two overall diagnostics, the squared prediction error (SPE) and the Hotelling's T^2 statistic.

[1] www.umetrics.com; www.prosensus.ca; www.camo.com; www.eigenvector.com

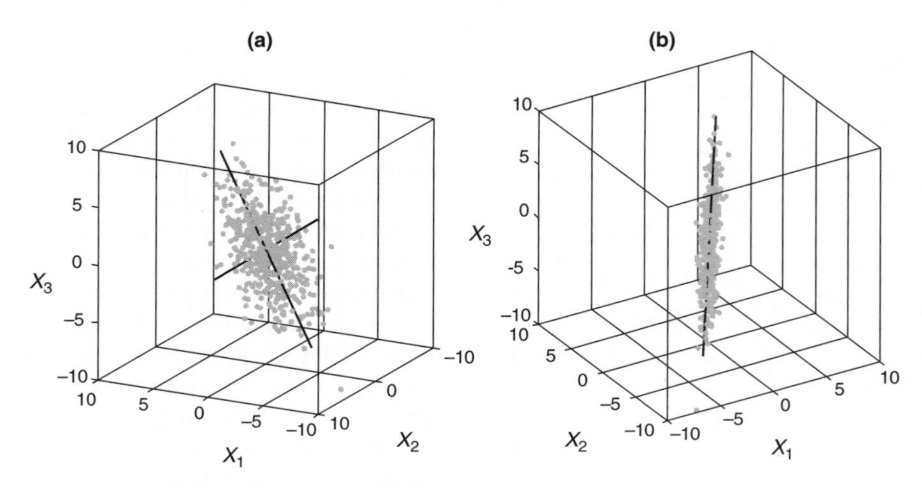

FIGURE 34.3 Dimension reduction of a 3D data set to a 2D plane by PCA.

34.2.4.4 ***The Squared Prediction Error*** This diagnostic is identical in interpretation to the perpendicular distance (ε) mentioned in the conceptual problem presented previously in this section. This is a quantitative measure on how well the new sample adheres to the structure of the reference data. This is a positive number and is well accepted to follow a chi-squared distribution [45], which means that confidence intervals for a given level of significance can be computed.

34.2.4.5 ***The Hotelling's T^2 Statistic*** This diagnostic is a measure of the squared distance from the origin of the data (the mean values) to the expected conditional value for the properties of the new lot. In the conceptual problem, this would be equivalent to the square of the distance along the black line (illustrated with an arrow), from the center of the box to the intersection where b_{new} meets $\hat{a}$, which is given by the regression model (Figure 34.4). Since the model prediction is being used here, it is imperative to first assess that the SPE is within tolerance. Notice that imposing a bound on this diagnostic, implicitly imposes a bound on the magnitude of properties A and B, and hence, step (i) in the suggested sequence, could be replaced by a simple one-sided test on the Hotelling's T^2 diagnostic.

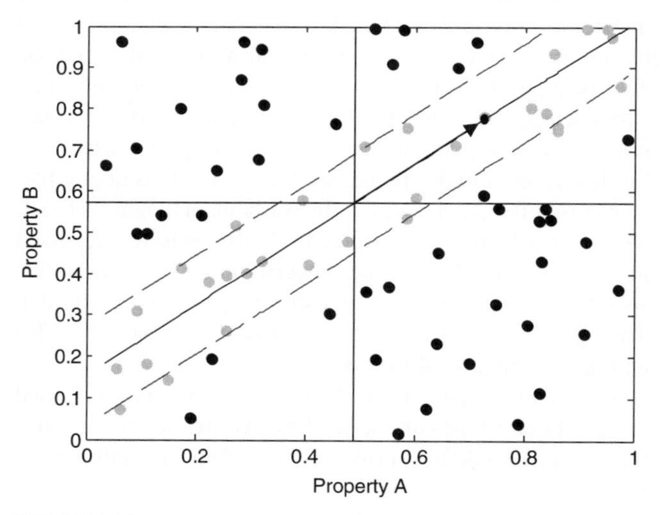

FIGURE 34.4 Conceptual problem illustrating the distance being diagnosed by the Hotelling's T^2 in a two-dimensional problem.

The use of these multivariate diagnostics is illustrated with an example taken from a real scenario in the pharmaceutical sector.

EXAMPLE 34.1 SETTING MULTIVARIATE RAW MATERIAL SPECIFICATIONS USING PCA

Consider a polymeric ingredient that is characterized by three descriptors of its particle size: D_{10}, D_{50}, and D_{90} (each number represents the average size at the tenth, fiftieth, and ninetieth quantile from the distribution). Each quantity is reported in logarithmic scale in Table 34.3. This table also contains the lower and upper limits for the mean, as calculated from this table. The fundamental concept to illustrate is that these three descriptors for particle size are not independent of each other, and a change in one of them will imply a change in the others (Figure 34.5).

Consider now six new lots of material (data provided in Table 34.4). If the three independent specifications are used to decide weather to accept or not these six lots of material, it is quite obvious that the lots 1, 5, and 6 (marked by a ●, ▼, and ★, respectively) will be rejected. In contrast, a principal component analysis model was fitted to the data using one significant component, the Hotelling's T^2 and the SPE diagnostic, and finally a multivariate specification was built (Figure 34.6 where the new lots are colored in gray). Notice that this specification also rejects lots 1, 5, and 6, however, lots 3 and 4 (marked by ▲ and ◇, respectively) are rejected and only lot 2 (marked by a ■) is accepted. The reason is

TABLE 34.3 Example of Data from a Polymeric Pharmaceutical Excipient Used in Example 34.1

Sample	$\log_{10}(D_{10})$	$\log_{10}(D_{50})$	$\log_{10}(D_{90})$
Lot 1	4.17	4.57	5.42
Lot 2	4.17	4.59	5.48
Lot 3	4.19	4.57	5.43
Lot 4	4.19	4.57	5.37
Lot 5	4.20	4.53	5.40
Lot 6	4.20	4.54	5.37
Lot 7	4.22	4.58	5.40
Lot 8	4.22	4.58	5.40
Lot 9	4.22	4.59	5.39
Lot 10	4.23	4.60	5.44
Lot 11	4.23	4.60	5.40
Lot 12	4.23	4.59	5.44
Lot 13	4.23	4.57	5.44
Lot 14	4.23	4.59	5.40
Lot 15	4.24	4.60	5.41
Lot 16	4.25	4.59	5.38
Lot 17	4.25	4.61	5.40
Lot 18	4.25	4.62	5.46
Lot 19	4.25	4.61	5.44
Lot 20	4.26	4.58	5.46
Lot 21	4.26	4.59	5.45
Lot 22	4.27	4.58	5.43
Lot 23	4.27	4.62	5.44
Lot 24	4.28	4.62	5.41
Lot 25	4.29	4.61	5.47
Lot 26	4.31	4.67	5.51
Lot 27	4.31	4.64	5.46
Lot 28	4.31	4.65	5.47
Lot 29	4.32	4.63	5.47
Lot 30	4.32	4.63	5.49
Lot 31	4.32	4.63	5.52
Lot 32	4.35	4.67	5.50
Mean	4.25	4.60	5.44
Standard deviation	0.047	0.032	0.041
n	32.00	32.00	32.00
$t(\alpha/2),n$ [$\alpha = 0.05$]	2.042	2.042	2.042
Mean low limit	4.23	4.59	5.42
Mean upper limit	4.27	4.61	5.45

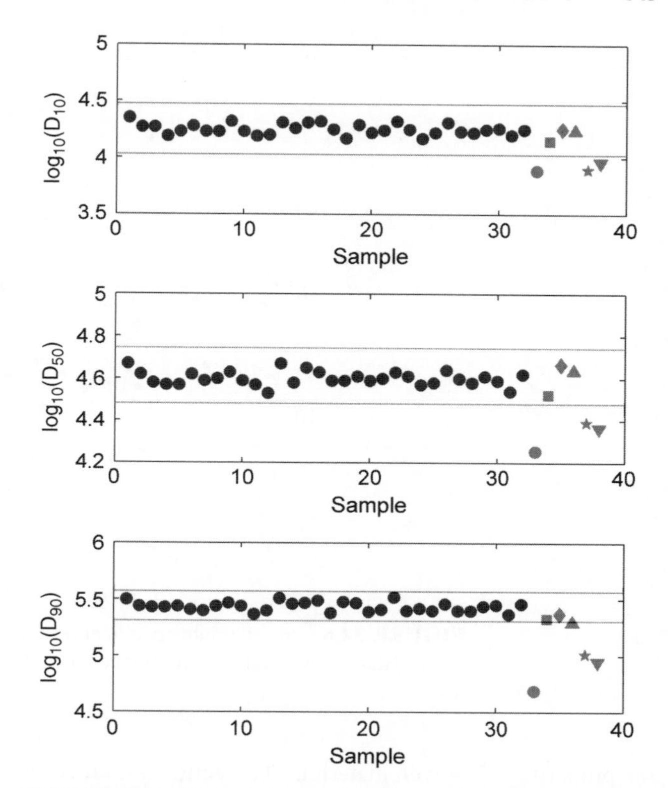

FIGURE 34.5 Univariate specifications and acceptance criteria for particle size descriptors. Black markers denote reference lots and gray markers denote new lots to be tested for acceptance.

TABLE 34.4 Data for Six New Lots of Polymer for Example 34.1

New lot	$\log_{10}(D_{10})$	$\log_{10}(D_{50})$	$\log_{10}(D_{90})$
A	3.9	4.2	4.7
B	4.1	4.5	5.3
C	4.2	4.7	5.4
D	4.2	4.6	5.3
E	3.9	4.4	5.0
F	4.0	4.4	4.9

simple to understand by looking at the three descriptors simultaneously, as shown in Figure 34.7.

In the top plot of Figure 34.7, lots 1, 5, and 6 were rejected because all three particle size descriptors are clearly different from the lots used as a reference. The bottom plot (which is a rotated version of the top) illustrates why lot 2 (marked by a ■) was accepted while lots 3 and 4 (marked by ▲ and ◇) were rejected. The particle size descriptors for lot 2 exhibit the same expected proportions between the $\log_{10}(D_{10})$, the $\log_{10}(D_{50})$, and the $\log_{10}(D_{90})$ (referred to as covariance structure) as in the reference set and hence it is safer to accept this material than other materials where the proportions between these properties is different. An added advantage of the multivariate specification (Figure 34.6) is that a single plot can be used to impose specifications on multiple properties simultaneously. For this case, it was possible to visualize the three particle size variables in a three dimensional plot; a real case scenario may consist of several hundreds of variables, and then the power of a multivariate approach is the ability to still monitor the multivariate proportion of the all properties, simultaneously in a couple of plots.

Ultimately, the impact of the raw material physical and chemical properties on the final drug product depends on the manufacturing process (e.g., wet granulation will be affected by properties that don't affect dry granulation). This section exemplified a general method to establish specifications on

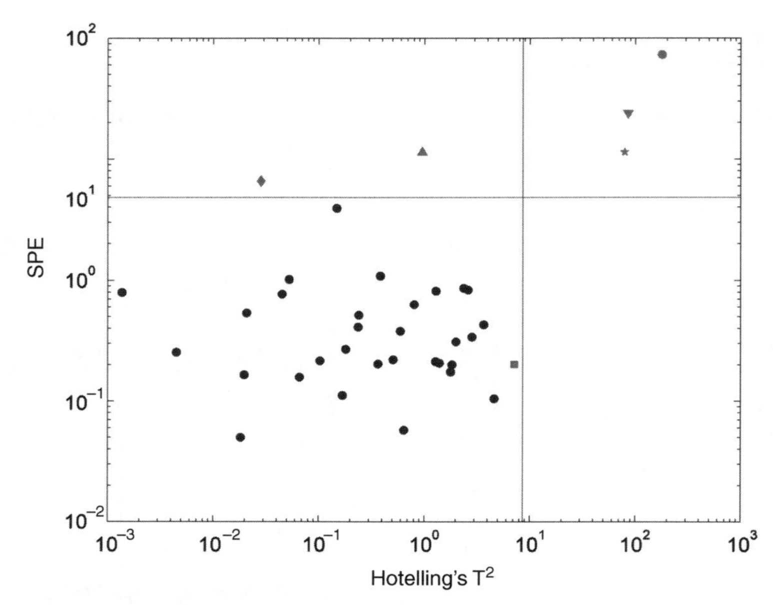

FIGURE 34.6 Multivariate specification for particle size data in Example 34.1 with acceptance limits. Black markers denote reference lots, gray markers denote new lots to be tested for acceptance.

the properties of a given material. The Section 34.3 focuses on unit-operation specific details where the relevance of certain physical/chemical properties of the material are discussed in the context of the processing route.

34.3 PROCESS MODELING FOR SOLID ORAL DRUG PRODUCT PROCESSES

34.3.1 Powder Flow Models

34.3.1.1 Model Development for Powder Processes Using Discrete Element Method The DEM is one possible approach to model powder flow in processing operations. DEM models are a computational approach whereby the state of each particle in the system is tracked at each instant in time. These models produce a wealth of particle-level data including particle positions and velocities as well as the forces acting on each particle. These data can then be used to calculate many other useful quantities such as velocity profiles, stress tensors, solid fractions, and local mass fractions. This wealth of data is a key advantage of the DEM approach; many of these quantities are expensive and difficult, if not impossible, to measure experimentally. However, acquiring these data via DEM does have significant costs, primarily in long computational times. Depending on the number of particles modeled, the complexity of the simulation domain, and the length of time modeled, simulation times may range from a few hours to well over a month of computing time. Thus, high-speed computers and efficient software algorithms are quite important to obtain results in reasonable times.

The algorithm of a typical DEM model is shown in Figure 34.8. The simulation is initiated by defining the computational domain and creating particles within it. Each particle is given a size, mass, and density and assigned a position and velocity. The simulation is started and all contacting pairs (both particle–particle and particle–wall) are identified. This step is among the most time consuming aspect of DEM programs. A brute force algorithm that searches between all possible pairs scales with N^2, where N is the number of particles in the simulation. However, using techniques such as a neighbor searching algorithm can reduce the time to the order of N Ln (N). Once each contacting pair is identified, force–displacement models are used to determine the contact forces acting on each of the particles. While several such models can be used, most models specify the normal and tangential forces as a function of the overlap distance between particles (the overlap approximates particle deformation during contact and is typically constrained to a small value (<1% of particle diameter)). For example, one such model for the normal force is based on the theoretical work by Hertz [46] in 1882. This model describes the elastic contact of a sphere, and gives the normal force, $\mathbf{F}_N$, as

$$\mathbf{F}_N = k_N \delta^{3/2} \hat{\mathbf{n}} \tag{34.2}$$

where k_N is a stiffness related to the radii and material properties of the contacting spheres, δ is the overlap distance between the spheres and $\hat{\mathbf{n}}$ is the unit normal vector. Other, more complex, models build upon this and other theories to include dissipative effects for modeling inelastic contacts. In

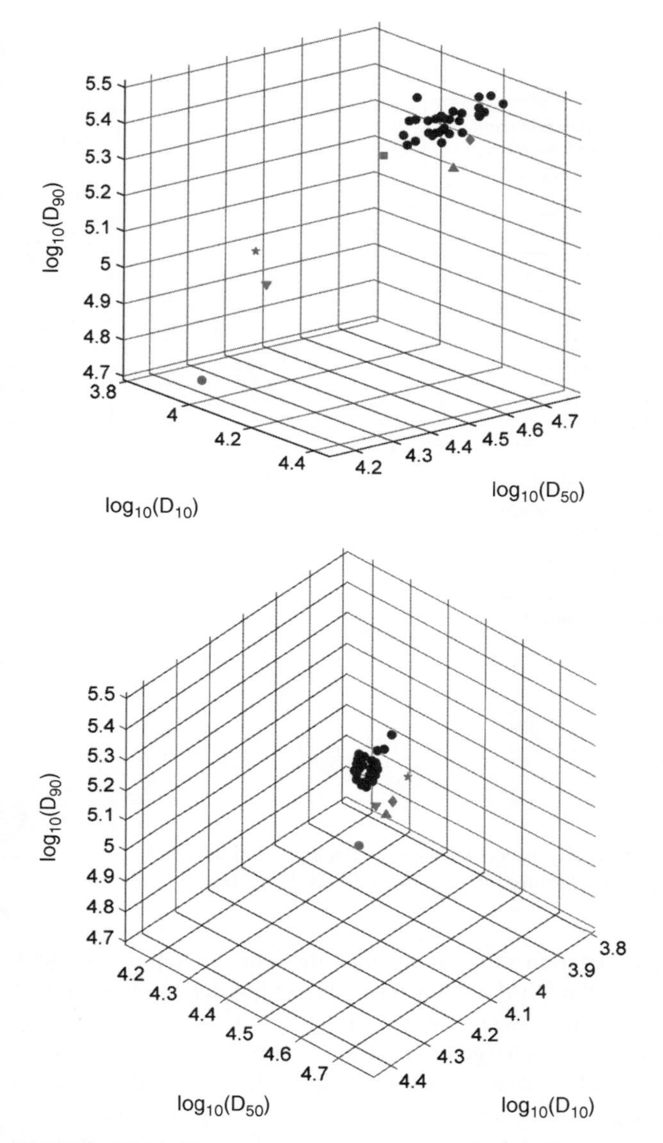

FIGURE 34.7 Three particle size descriptors plotted simultaneously black markers denote reference lots, gray markers denote new lots to be tested for acceptance. Bottom plot is a rotation of the top plot.

addition to the normal and tangential contact forces, body forces—such as the acceleration due to gravity, electrostatics, and magnetic fields – can also be included. The resultant contact and body forces acting on each particle are summed. Newton's second law is then used to calculate the translational and rotational accelerations, respectively:

$$\mathbf{F}_{N,\mathrm{Total}} = m\frac{\partial^2 \mathbf{x}}{\partial t^2} \tag{34.3}$$

$$\mathbf{M}_{\mathrm{Total}} = I\frac{\partial^2 \theta}{\partial t^2} \tag{34.4}$$

where $\mathbf{F}_{N,\mathrm{Total}}$ is the total, resultant normal force acting on a given particle, m is the particle mass, $\mathbf{x}$ is the particle position vector, t is time, $\mathbf{M}_{\mathrm{Total}}$ is the total, resulting moment due to the tangential forces acting on a given particle, I is the particle moment of inertia, and θ is the particle orientation vector. Subsequently, these accelerations are integrated in time to determine updated particle velocities and positions. At this point, virtually any quantity of interest may be measured and the simulation then proceeds to the next iteration by incrementing the time step and repeating the necessary contact detection calculations. This procedure is repeated until the simulation has reached the desired end point, such as when a hopper is completely discharged.

Often times, certain assumptions are made in DEM models to simplify the computational demands. For example, assumptions of spherical particles, cohesionless particles, and negligible interstitial fluid effects are often made. Each of these assumptions help to not only make the simulations faster but also make the modeled system less representative of the real system of interest. Ongoing research efforts currently are working toward relaxing these assumptions from the models, and many recent research papers describe work where one or more of these assumptions have been removed. Interested readers are referred to review papers [6, 47, 48] and the references therein for more detailed information.

34.3.1.2 Powder Discharge and Segregation Modeling Using DEM The drug product manufacturing process typically involves several powder handling operations that are used to create the final product—a dosage form such as a tablet or capsule—from several raw materials—typically powders with varying physical and chemical properties. In all cases, content uniformity is a critical quality attribute of the final dosage form. The drug loading in each dosage form is important because if this were to vary considerably, patients would receive doses that might be ineffective or possibly result in undesired side effects or worse. Detailed guidelines issued by the United States Pharmacopeia (USP) state the acceptance limits of variability in drug loading and prescribe the testing procedure used to determine the variability of a particular batch [49].

There are many powder processing unit operations used in the manufacture of tablets and capsules. These operations include blending, hopper filling and discharge, and flow through various feeding or dispensing devices, to name but a few. The object of the blending process is, as the name suggests, to combine the raw material powders into a well-mixed blend containing a uniform distribution of all materials. This is not a trivial task, as each raw material consists of particles with a range of sizes and morphologies. Differences in these properties will cause particles to segregate or de-mix. If this occurs in a blender, it may be difficult or impossible to achieve a well-mixed blend. Assuming that a uniform

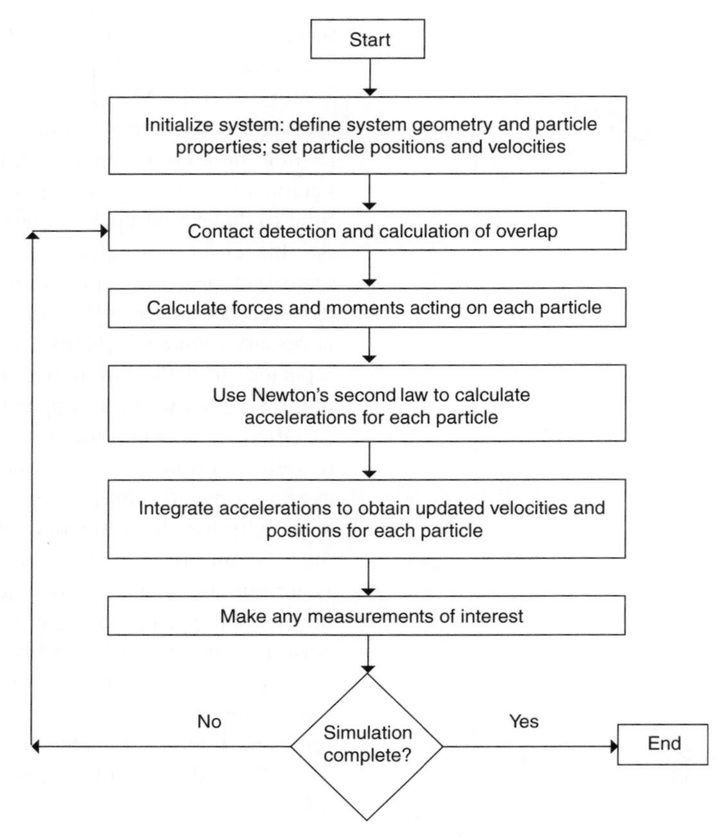

FIGURE 34.8 A flowchart showing the algorithm typically used in Discrete Element Method simulations of powder flow [6]. Reprinted from Ref. 6 with permission of John Wiley & Sons, Inc.

mixture can be obtained in a blender, segregation in subsequent operations can lead to content uniformity that is out of specification. It should be noted here that blend uniformity is not the sole concern of a formulation scientist. Other properties such as flowability and compressibility may present similar constraints on the development and processing of a drug product formulation. Nevertheless, in the remainder of this section, we will focus on content uniformity.

Hopper filling and discharge is one common operation that occurs downstream from blending operations. Therefore, it is critical that the hoppers are filled in a manner that does not cause the uniform blend to segregate. Additionally, it is important that the hopper design and raw materials—to the extent that their properties can be modified or selected—be designed to minimize segregation during the hopper discharge process. If these hopper flows and other subsequent operations such as flow through a tablet press feeder do not induce segregation of the blend, favorable content uniformity results should be obtained.

Here, we show an example of how the DEM as described earlier can be used to computationally model powder flow and gain a better understanding of the powder dynamics. While DEM models can be useful for discerning a wide variety of data from powder flows, we will focus on segregation of a binary mixture. This example will show how the DEM approach can be used to determine the effect of particle size ratio on the extent of segregation during hopper discharge. Consider a binary mixture of spherical particles—one species with a large diameter and the second species with a smaller diameter. While typical pharmaceutical formulations contain several different components, we presently assume a binary mixture where the small species represents the API and the larger species represents all of the excipients (diluents, binders, disintegrants, lubricants, etc.). Typically, the API particle size is much smaller than most of the excipients, and the ratio of these size differences can affect the degree of segregation during hopper discharge.

The model hopper is filled with the binary mixture containing 5% by mass of the smaller species. The initial state is well mixed, which permits analysis of the segregation during the discharge event only. A similar model could be developed to model the combined filling and discharge processes. Figure 34.9 shows an image taken from the simulation after the hopper is filled and discharge has been initiated. This model uses periodic boundaries to enable modeling just a thin slice of a larger hopper with a rectangular cross section.

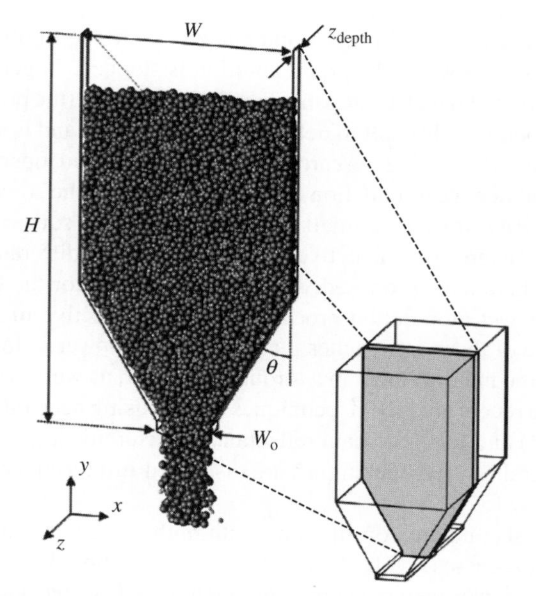

FIGURE 34.9 A simulation image showing a DEM model of hopper discharge [8]. Reprinted from Ref. 8 with permission of John Wiley & Sons, Inc.

As the mixture is discharged, the simulation program tracks the time at which each particle is discharged. These data are analyzed and the fines mass fraction is calculated. Figure 34.10 shows the plots of the normalized fines mass fraction, x_i/x_f, where x_i is the fines mass fraction of a given sample of the discharge stream and x_f is the fines mass fraction of the initial blend charged to the hopper as a function of the fractional mass discharged M/M_{Total} where M is the cumulative mass discharged and M_{Total} is the mass of material initial charged to the hopper. For all size ratios, the

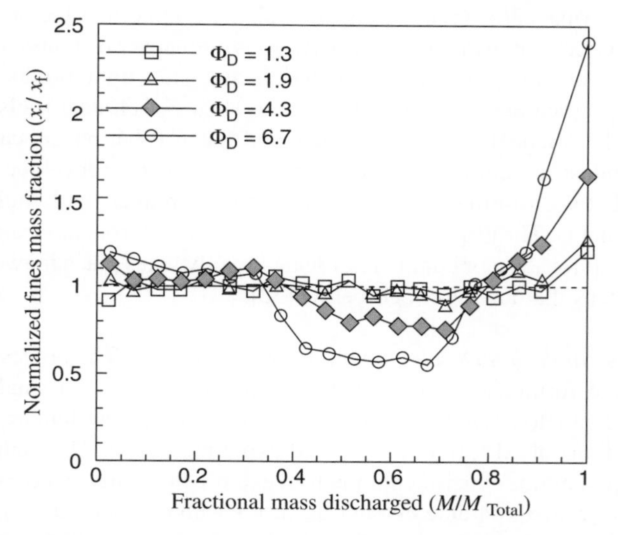

FIGURE 34.10 Segregation of a binary mixture of the given particle size ratios during discharge from a hopper [8]. Reprinted from Ref. 8 with permission of John Wiley & Sons, Inc.

discharge stream is relatively well-mixed for the first 30% of the discharge process. Then, for the cases with larger particle diameter ratios, Φ_D, a depletion in the smaller species is observed. This is followed by a spike in the fines mass fractions near the end of hopper discharge ($M/M_{Total} > 80\%$). These results show that discharge of a well-mixed pharmaceutical blend from a hopper may result in significant content uniformity issues if the particle size ratio is greater than ~2. For $\Phi_D = 6.7$, tablets produced during the middle of the batch may only have ~50% of the expected API, while tablets produced at the end of the batch could have over 200% of the expected API.

The simulation animations also provide insight into the flow dynamics that might be difficult to observe experimentally. In fact, in such an animation, one can observe the small particles segregating via the sifting or percolation mechanism [50]. During this process, small particles preferentially move in the direction of gravitational acceleration through a matrix of larger particles. This process causes an accumulation of fine particles near the hopper walls that gets discharged last. The material from which the fines sifted can now considered to be depleted in fines. This fines-depleted material generally exits the hopper after 40% of discharge but less than 70% of discharge according to Figure 34.10.

In practice, these potency variations within a batch can occur. However, they typically do not reach these extremes. Due to several assumptions in this example model, most predominantly that of cohesionless particles, the model predicts the worst case scenario in terms of segregation potential. The inclusion of cohesive forces in the model would tend to reduce the extent of segregation as the particles, once in a well-mixed state, will have reduced freedom to move relative to one another and segregate. Nevertheless, the model described here with cohesionless particles helps to improve process understanding and guide process development and scale-up decisions.

This example has illustrated how the DEM approach can be used to gauge the segregation potential during hopper discharge for a range of particle size ratios. The effects of other particle properties—such as density, shape, and surface roughness—and hopper geometries—such as the hopper wall angle, diameter, outlet diameter, and wall roughness can also be modeled using the same approach [8]. Similarly, other processes such as blending and hopper filling can also be modeled with DEM.

In this section, we have reviewed the importance of maintaining content uniformity during hopper discharge and highlighted one of the methods by which segregation of materials can be modeled. In the next section, we discuss modeling of wet granulation, a process that helps to bind particles of different materials together, thereby reducing the potential for segregation in subsequent processing and handling operations and improving the likelihood of good content uniformity in the final dosage form.

34.3.2 Wet Granulation Process Models

34.3.2.1 Model Development for Wet Granulation Using Engineering Principles Wet granulation is a particle size enlargement process that is commonly used in the manufacture of drug product dosage forms. There are several reasons to wet granulate pharmaceutical blends. Increasing the particle size will tend to improve flow. Fine powders usually have significant cohesive forces between the constituent particles that act to retard flow. By enlarging the particle size, these cohesive forces become less significant compared with the particle mass, thereby improving the flowability of the bulk powder. Another reason to granulate includes reducing the potential for segregation of the API. The wet granulation process physically binds the blended particles together thereby reducing the likelihood that one species will segregate and cause potential content uniformity problems.

Wet granulation processes are often carried out as batch processes using high-shear mixers as shown in Figure 34.11. While wet granulation can also be conducted in other equipment such as planetary mixers, fluidized beds, or extruders, high-shear mixers are the most common in the pharmaceutical industry and will be our focus in this section. In the high-shear mixer, a centrally located impeller (in this case, a top-driven impeller is shown although some high-shear mixers utilize a bottom-driven impeller) is used to mix and consolidate the granulation. A chopper (located on the left side of this schematic) spins at a high speed and helps to break up very large granules. Finally, a spray nozzle (not shown in Figure 34.11) is used to add a liquid granulating agent. This liquid may contain a liquid binder or may consist only of water when a dry binder has already been added to the formulation.

Granulation in a high-shear mixer begins with the addition of the dry powder blend and dry mixing with the impeller for a short period of time. With both the impeller and the chopper rotating, the liquid addition phase begins. After the desired amount of liquid has been added, the "wet massing" phase begins where the impeller and chopper continue to mix the granulation while the liquid addition is stopped. In general, the point at which to stop the granulation process (the process end point) is difficult to determine scientifically and is still a matter of ongoing research. In the past, skilled operators would deem a granulation complete if it passed the so-called *squeeze test* where a small amount of material is squeezed in one's hand and subjectively observed. Many different researchers have proposed various ways to monitor the high-shear wet granulation process in a more objective manner. Some of these approaches include impeller power or torque, off-line measurement in a torque rheometer, as well as some more recent analytical techniques such as using near-infrared (NIR) and focused beam reflectance measurement (FBRM). Several of these techniques are discussed further in a recent review [51].

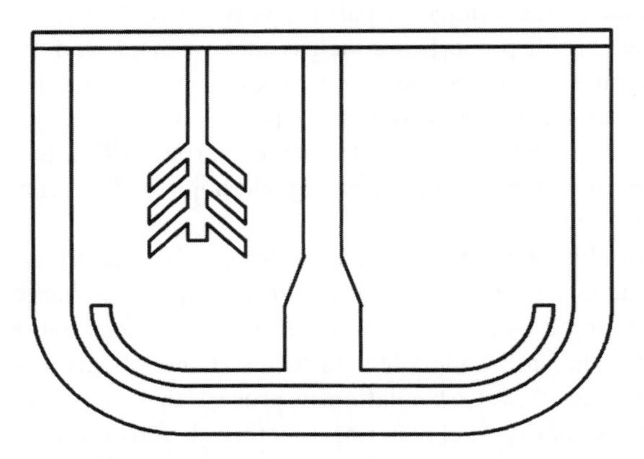

FIGURE 34.11 A schematic of a high shear wet granulator [57]. Reprinted from Ref. 57, Copyright (1999), with permission from Elsevier.

A second area of difficulty with high-shear wet granulation processes revolves around process scale-up. Many of the process parameters such as impeller speed, water addition amount, water addition rate, and wet massing time are determined through experimental design of experiments (DOEs). However, the process dynamics change significantly as larger scale granulators are used. Hence, similar DOEs are conducted at each of the scales during the process scale-up effort. These experiments consume significant labor resources and also incur large raw material costs if granulations of proprietary APIs in limited quantities are being conducted. This is especially true at the largest of scales where batch sizes may be on the order of 1000 L. Thus, the use of models to (1) predict process parameters for scaling-up and (2) determine when to stop the wet granulation process (end point) can be extremely useful.

34.3.2.2 Wet Granulation Scale-Up Process scale-up in pharmaceutical industry is driven by two important factors: (1) cost of API, which usually runs into several thousand dollars per kilogram of material, and (2) tight product specifications as desired by various stages of clinical trials. The practical considerations demand that in a pharmaceutical industrial setting, a wet granulation process can be developed that is cost-effective, robust, and deliver products with high quality. The process development and scale-up from lab scale to pilot scale or commercial scale broadly takes place in two steps that are described below [52].

Formulation Development and Optimization The process and formulations are developed and optimized in a small granulator, typically at one L lab scale using few hundred grams of API, through detailed experimentation. The main goal of this experimentation is to explore the design space of the process parameters such as the total amount of binder and its rate of addition, impeller speed, total processing time, dry powder fill height, and end point. This exploration should result in a process that yields granules with desired size and porosity distribution, which can be compressed into tablets with the desired quality attributes including hardness–compression behavior and dissolution profiles.

Process Development and Scale-Up After the design space is explored and the formulation is optimized at lab scale, it is scaled up to pilot or production scale. Essentially, a process template is developed at lab scale and it has to be replicated at a much larger scale that results in product with similar quality attributes. Several routes to process scale-up have been demonstrated by various groups, mostly applicable to a specific set of granulators and limited formulations [51]. Almost all the procedures of scale-up are based on the similarity between the two granulators of interest. Formally, the similarity principal for a process is established using dimensional analysis. The application of the similarity principal begins with the recognition that any physical process can be represented by a dimensional relationship between n process variables and constants as shown below.

$$F(X_0, X_1, X_2, \ldots, X_n) = 0 \qquad (34.5)$$

The above relation can be reduced by applying Buckingham Π theorem, which simply states between $m = n - r$ mutually independent dimensionless groups, where r is the number of dimensional units, that is, fundamental units (rank of the dimensional matrix). The equation (34.5) can be reduced to the following relationship.

$$F(\Pi_0, \Pi_1, \ldots, \Pi_n) = 0 \qquad (34.6)$$

The above relationship can be rewritten by expressing first dimensionless group in terms of the rest of them as shown below.

$$\Pi_0 = f(\Pi_1, \Pi_2, \ldots, \Pi_n) \qquad (34.7)$$

It must be noted that the similarity analysis should be applied to processes where a clear understanding of process is established.

Scale-Up Approach 1

The earliest application of similarity analysis for the scale up of wet granulation process was demonstrated by Leunberger and coworkers at University of Basel and Sandoz AG [53–56]. The physical relationship used to describe the granulation process can be written as

$$P = f(\rho, D, \Omega, q, t_p, V_b, H, g) \qquad (34.8)$$

The description of various physical quantities is shown in Table 34.5.

The above relationship is nondimensionalized using d, $1/\omega$, and ρd^3, as length, time, and mass scales, respectively. The dimensionless quantities are shown in Table 34.6.

These investigators performed wet granulation experiments using a placebo formulation (86% w/w lactose, 10% w/w corn starch, and 4% w/w polyvinylpyrrolidone as binder) in mixers of planetary type (e.g., Dominici, Glen, and Molteni). The batch size ranged from 3.75 up to 60 kg. The impeller speed was scaled using a constant Froude number ($d_1\omega_1^2 = d_2\omega_2^2$). The volume fraction and geometric ratio were also kept constant. It was seen that the power profile measured during the granulation can be divided into five different phases (S_1–S_5) as shown in Figure 34.12. It was found that the amount of binder liquid added during the process varies linearly with the batch as shown in Figure 34.13. Therefore, the functional relationship between dimensionless groups is as follows:

$$\Pi_0 = f(\Pi_1) \qquad (34.9)$$

From these findings, one can conclude that the correct amount of granulating liquid per amount of particles to be granulated is a scale-up variable. It is necessary, however, to mention that during this scale-up exercise only a low-viscous granulating liquid was used. The exact behavior of a granulation process using high-viscous binders and different batch sizes is unknown. It is shown that the first

TABLE 34.5 List of Important Process Variables and Parameters That Define a Wet Granulation Process in Scale-Up Approach 1

No.	Quantity	Symbol	Units	Dimension
1	Power consumption	P	Watt	ML^2T^{-3}
2	Specific density	ρ	kg/m^3	ML^{-3}
3	Impeller diameter	D	m	L
4	Revolution speed	Ω	rev/s	T^{-1}
5	Binder flow rate	q	kg/s	MT^{-1}
6	Bowl volume	V_b	m^3	L^3
7	Gravitational constant	g	m/s^2	LT^{-2}
8	Bowl height	H	m	L
9	Process time	t_p	s	T

TABLE 34.6 Dimensionless Numbers and Groups Used in Scale-Up Approach 1

Number	Symbol	Dimensionless Group	Description
1	Π_0	$P/(d^5\omega^3\rho)$	Power number
2	Π_1	$qt_p/V_b\rho$	Specific amount of liquid binder
3	Π_2	V/d^3	Volume fraction of dry powder
4	Π_3	$(d\,\omega^2)/g$	Froude number (centrifugal force/gravitational force)
5	Π_4	H/d	Geometric ratio

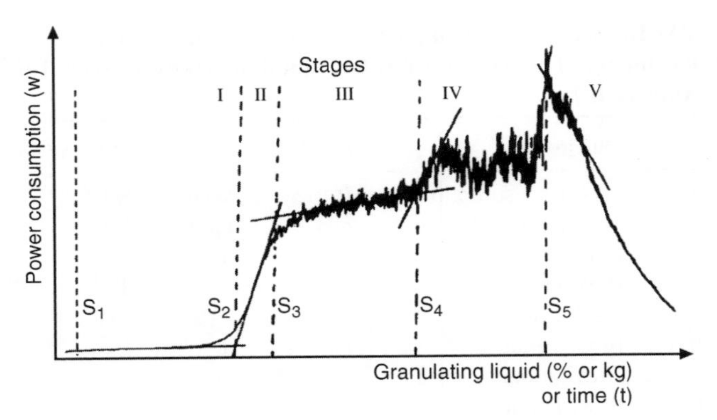

FIGURE 34.12 Division of a power consumption curve [55]. Reprinted from Ref. 55, Copyright (2001), with permission from Elsevier.

derivative of the power consumption curve is a scale-up invariant and it is proposed that it can be used as an in-process control or a fine-tuning of the correct amount of granulating liquid.

Scale-Up Approach 2

One of the key assumptions made in this study that the viscosity of the wet mass is unimportant, may not hold true for many formulations and viscous binders. Rowe and coworkers [57, 58] developed a different approach to scale up wet granulation process. These authors defined the process by following relationship.

$$\Delta P = f(\rho, R, \Omega, \mu, R_b, g, m) \tag{34.10}$$

The description of the various physical quantities is shown in Table 34.7. The above relationship can be nondimensionalized using d, $1/\omega$, and ρd^3, as length, time, and mass scales, respectively. The dimensionless quantities are

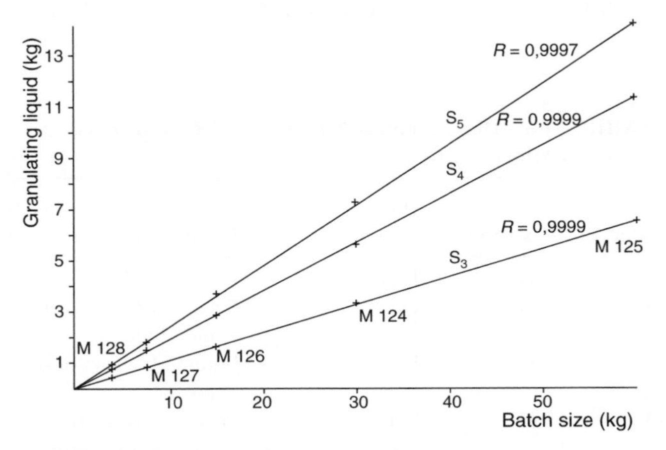

FIGURE 34.13 Scale-up precision measurements with identical charges [55]. Reprinted from Ref. 55, Copyright (2001), with permission from Elsevier.

TABLE 34.7 List of Important Process Variables and Parameters That Define a Wet Granulation Process in Scale-Up Approach 2

No.	Quantity	Symbol	Units	Dimension
1	Net power consumption	ΔP	Watt	ML^2T^{-3}
2	Wet mass density	ρ	kg/m^3	ML^{-3}
3	Impeller radius	R	m	L
4	Revolution speed	Ω	rev/s	T^{-1}
5	Wet mass consistency/ viscosity	μ	Nm	ML^2T^{-2}
6	Bowl radius	R_b	m	L
7	Gravitational constant	G	m/s^2	LT^{-2}
8	Amount of wet mass	M	kg	M

shown in Table 34.8.The power number is expressed as a function of the other dimensionless quantities as follows:

$$\log_{10}(N_P) = a \cdot \log_{10}(\psi Re \cdot Fr \cdot \text{fill ratio}) + b \tag{34.11}$$

where a and b are regression constants. Faure et al. [57] carried out wet granulation experiments of lactose and maize starch-based placebo formulations in a series of Collette Gral Granulators with sizes 8, 25, 75, and 600 L. They fitted the experimental data with equation 34.11 and found that the regression coefficient was $r^2 > 0.88$ using the data from the 8, 25, and 75 L bowls with PTFE lining, and the 600 L bowl that did not require the lining (see Figure 34.14). The slope was found to be $a = -0.926$, and the intercept $b = 3.758$. This work shows that a nonlinear scale-up relationship exists between wet granulation carried out in geometrically similar granulators of different sizes. Moreover, this relationship can be effectively used when scaling up the process from one size to another size. In the next section, a case study is presented, which demonstrates the application of above-mentioned scale-up approaches.

TABLE 34.8 Dimensionless Numbers and Groups Used in Scale-Up Approach 2

No.	Symbol	Dimensionless Group	Description
1	N_P	$\Delta P/(d^5\omega^3\rho)$	Power number
2	ψRe	$\rho R^2\omega/\mu$	Pseudo Reynolds number (inertial force/viscous force)
3	Fr	$(R\omega^2)/g$	Froude number (centrifugal force/gravitational force)
4	Fill ratio	$\rho R_b^3/m$	Fill ratio (granulator volume/wet mass)

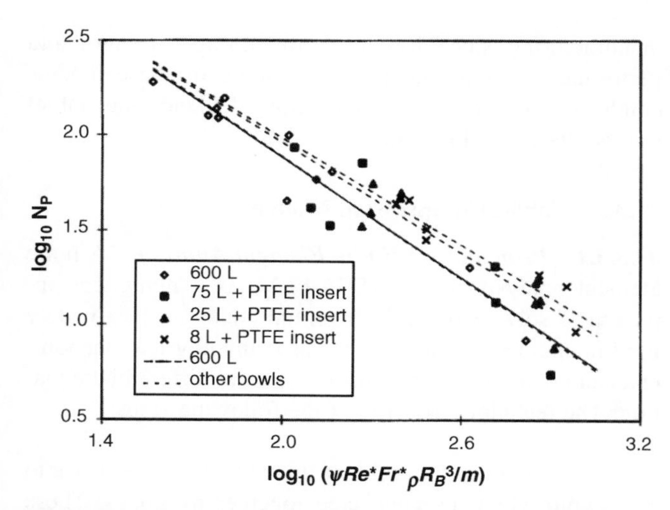

FIGURE 34.14 Dimensionless power relationship of the 600 l Collette Gral mixer-granulator [57]. Reprinted from Ref. 57, Copyright (1999), with permission from Elsevier.

EXAMPLE 34.2 SCALE-UP OF WET GRANULATION USING ENGINEERING MODELS

A wet granulation process is carried out for a placebo formulation with water as a binding liquid, in an 8 L Collette Gral granulator. The process parameters are optimized for desirable end point through a design of experiments study. These parameters are listed in Table 34.9. This process is to be scaled-up to a granulator with 75 L capacity such that the final product is of the same quality. In this, case the product quality is identified by the specific density and wet mass consistency of granules at the end point. Hence, desirable end point is one that gives product with essentially the same specific density and wet mass consistency or viscosity obtained at 8 L scale. There are three main process parameters that are to be determined at 75 L scale, (a) impeller rotation speed, (b) amount of water used, and (c) power consumption near the end point. All these quantities will be calculated using above discussed scale-up approaches.

Solution

(a) The impeller rpm is scaled using constant impeller tip speed:

$$\Omega_2 = \Omega_1\left(\frac{r_1}{r_2}\right)$$
$$\Omega_2 = 350\left(\frac{0.12}{0.25}\right) = 164\,\text{rpm} \qquad (34.12)$$

(b) The amount of water used at 75 L scale is determined by assuming water to dry powder weight ratio is invariant across the scales.

TABLE 34.9 Process Variables and Parameters for Two Different Collete-Gral Bowl Sizes Used in the Example 34.2

Process Parameters and Variables	Units	8 L Bowl	75 L Bowl
Impeller radius	m	0.119	0.254
Bowl radius	m	0.123	0.262
Revolution speed	1/s	5.83	2.73
Revolution speed	1/min	350.0	**164.0**
Dry powder weight	kg	1.50	14.00
Total water added	kg	0.50	**4.67**
Gravitational constant	N/m^2	9.81	9.81
Bulk density	kg/m^3	400.0	400.0
Viscosity	N/m	0.30	0.30
Power consumption	Watt	127.7	**3503**
Froud number		0.41	0.19
Pseudo Reynolds number		110.1	235.1
Fill ratio		0.37	0.38
Power number		67.41	405.8
$\log(N_P)$		1.83	2.61
$\log(Re^* Fr^*$fill ratio$)$		1.23	1.24
Scale-up constant a		−0.93	−0.93
Scale-up constant b		3.76	3.76

Bolded values are calculated by scale-up rules.

Water used at 75 L

$$= \frac{(\text{amount of water used at 8 L}) \times (\text{dry powder mass at 75 L})}{(\text{dry powder mass at 8 L})}$$

$$\text{Water used at 75 L} = \frac{(0.5\,\text{Kg}) * (14\,\text{Kg})}{(1.5\,\text{Kg})} = 4.67\,\text{Kg} \qquad (34.13)$$

(c) Finally, the power consumption at the end point is determined using scaling relationship given by equation 34.11 as follows:

$$\begin{aligned} Fr &= 0.19 \\ \psi Re &= 235.1 \\ \text{Fill ratio} &= 0.38 \\ \log_{10}(N_P) &= -0.926 * \log_{10}(0.19 * 235.1 * 0.38) \\ &\quad + 3.758 \\ N_P &= 405.8 \\ \Delta P &= N_P * (\rho\Omega^3 R^5) = 3503\,\text{W} \end{aligned} \qquad (34.14)$$

The process parameters for both the scales are listed in Table 34.9. The power calculated using the above scale-up approach is used to guide determination of the wet granulation end point. However, it must be kept in mind that this scale-up approach, like any other approach based on dimensional analysis, is semiempirical in nature and needs some experimental work to achieve optimum scale-up and process design. This approach should be contrasted from a process model based on fundamental principles, which requires material properties and process parameters to achieve optimum design.

34.3.2.3 Fluidization Regime During Granule Drying

In the area of fluid bed drying of granules produced by wet granulation processing, there is an engineering approach to assess fluidization regimes *a priori* if the particle size distributions and equipment airflows are known. A semiempirical approach in that empirical heat transfer data is required for the simulations to compare favorably with experiment, has been used to predict process parameters for fluidization properties in a fluid bed dryer. Based on the mean diameter of the granule distribution, the process map shows the proposed equipment is adequate to fluidize the granules in the desired bubbling regime (Figure 34.15). Granule characteristics, such as moisture content and granule size distribution, are known for a particular product entering the fluid bed dryer. The range of volumetric flow rates of the drying gas and the dimensions of the air inlet were obtained from the equipment manufacturer. Finally, a review of the literature indicated that researchers had constructed models to predict both the minimum fluidization velocity and the transitions to turbulent and fast fluidization.

The granules produced through wet granulation and dried through fluid bed drying are finally compressed into tablets (Figure 34.1). The focus of this section was on the use of modeling to achieve consistent granulation properties on scale-up to assure consistent input to the tableting process. Tableting is an important unit operation for solid dosage

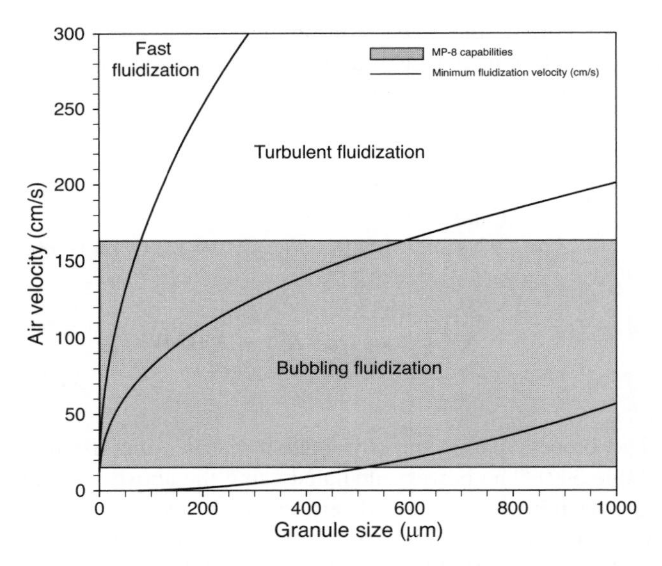

FIGURE 34.15 Engineering model of granule fluidization using empirical relationships found in literature based on granule distribution and equipment airflow [14, 15]. (Courtesy of D.M. Kremer.)

manufacturing since it defines the dosage strength and performance. In the next section, some of these models employed to understand compaction of blends into tablet will be discussed in detail.

34.3.3 Tablet Compression Models

34.3.3.1 Principles of Finite Element Analysis A finite element analysis/method (FEA/FEM) is a numerical approach to solve a partial differential equation. It is widely used in engineering and science as many physical phenomena can be described in terms of a partial differential equation. The technique consists of the following steps:

1. Subdividing the problem domain (or geometry) into finite elements connected together by nodes. These finite elements are commonly termed as a *mesh*.
2. Development of equations (such as force and mass balance) for each element and then assembling them for the entire domain or system of elements.
3. Solving the resulting system of equations.
4. Analyzing quantities of interest such as stresses and strains and obtain visualizations of the response of the system to the applied loadings.

Powder Compaction Tablet compression is an important unit operation in the pharmaceutical industry as it significantly affects the mechanical strength and relative density of the drug product. Finite element analyses are a common method that is employed to study the tableting process where the formulation powder is assumed to be a continuum material [59–62]. The approach is based on the following components:

- Continuity equations (e.g., conservation of mass)
- Equilibrium equations (e.g., force balance on the material)
- Initial and boundary conditions of the problem
- Dynamics of the loading and the geometry of the problem
- Constitutive behavior of the powder (e.g., stress–strain relationships)

Due to the availability of powerful, inexpensive computers and commercial finite element software, the continuity and equilibrium equations can be solved accurately and quickly after the appropriate boundary conditions are defined. Moreover, it is possible to define a complex sequence of loading and unloading steps such as the compression, decompression and ejection during model setup. However, it is a challenge to obtain model inputs to the FEA solver [60–64] such as the constitutive relationships of the formulation powder (e.g., stress–strain relationships) and friction between powder and die wall to get accurate estimation of the powder stress levels during tableting. The following section highlights some of the commonly used constitutive relationships for the powder continuum.

34.3.3.2 Tablet Finite Element Analysis

Powder Material Models The stress–strain relationships for powders were originally developed for classical soil mechanics applications and were used to simulate compaction of ceramic powders. They were assumed to be elastic-plastic materials and appropriate relationships were developed to describe the yield surface of the material. There are a number of phenomenological models to describe the yield surface of the powder materials such as the Gurson model [16], Cam Clay model [17] and the Drucker–Prager cap plasticity (DPC) model [18, 19]. The DPC model has been widely used [59, 62, 64–67] in comparison with the other models for two main reasons: firstly, it can efficiently capture the shear failure during the decompression and ejection phase of the tableting process and secondly, because experiments on real powders can be designed to efficiently characterize its parameters. The stress–strain relationship and the yield loci used in the DPC model are shown in Figure 34.16. The yield surface consists of three segments: a shear failure surface, a "cap" surface that represents plastic compaction or inelastic hardening and a transition surface between them. The transition surface is introduced for smooth numerical implementation. For a detailed description of the equations of the different yield surfaces of the DPC model and the experimental procedure to obtain its parameters, refer to Han et al. [60] or Cuningham et al. [59].

Applications FEA analyses performed using the DPC material model has been used to study the elastic recovery or "springback" of material during the compression and ejection phases of tableting that leads to capping incidence in tablets [64, 67]. The relative density distribution after ejection [62, 66] and temperature distribution in the compact and tooling during compaction [68] have also been investigated using FEA methods. Moreover, the stresses during the different tableting stages (see Figure 34.17) have been analyzed and correlated to possible tablet failure mechanics [60, 67].

This section summarized the concept of applying a continuum based finite element model to predict the stresses on a tablet during tableting operations. Further investigation is necessary to establish if these predicted stresses and density distributions by FEA have an implication in understanding possible failures that might occur in other unit operations, such as during tablet film coating. The next section explains the tablet-coating operation and available engineering models used to predict the coating process parameters.

34.3.4 Tablet Film-Coating Models

34.3.4.1 Model Development for Film Coating Using First Law of Thermodynamics Tablet film coating is a widely used unit operation within the pharmaceutical industry for applying both aesthetic and functional coatings on tablets. Color coating is often used in combination with tablet shape to enable manufacturers, pharmacists, and patients to distinguish between not only different products but also different dosage strengths within the same product. From a physical standpoint, a thin film-coat layer can improve the mechanical integrity of the tablets and also make them smoother, which improves tablet flowability, enhances packaging efficiency, and increases palatability for patients. Film coating can improve functionality by providing a barrier against environmental exposure to moisture, light, or air. This can enhance product stability and reduce the requirement for more expensive packaging materials. In some cases, the color

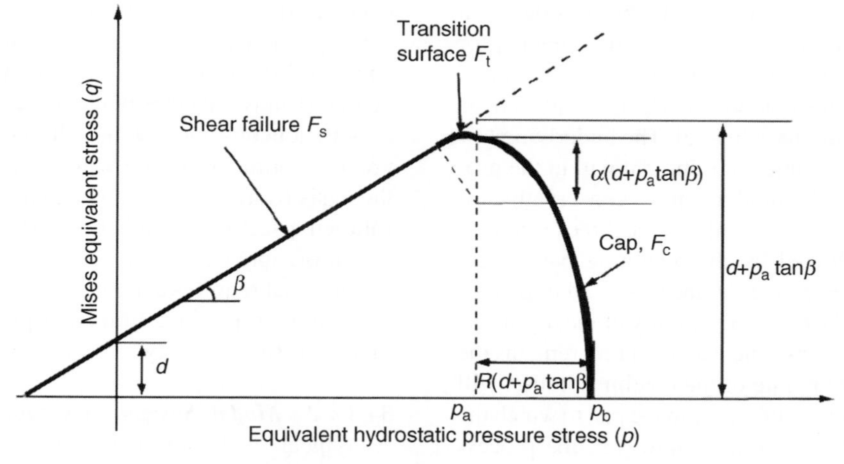

FIGURE 34.16 Drucker–Prager cap model. Yield surface in the *pq*-plane [60]. Reprinted from Ref. 60, Copyright (2008), with permission from Elsevier.

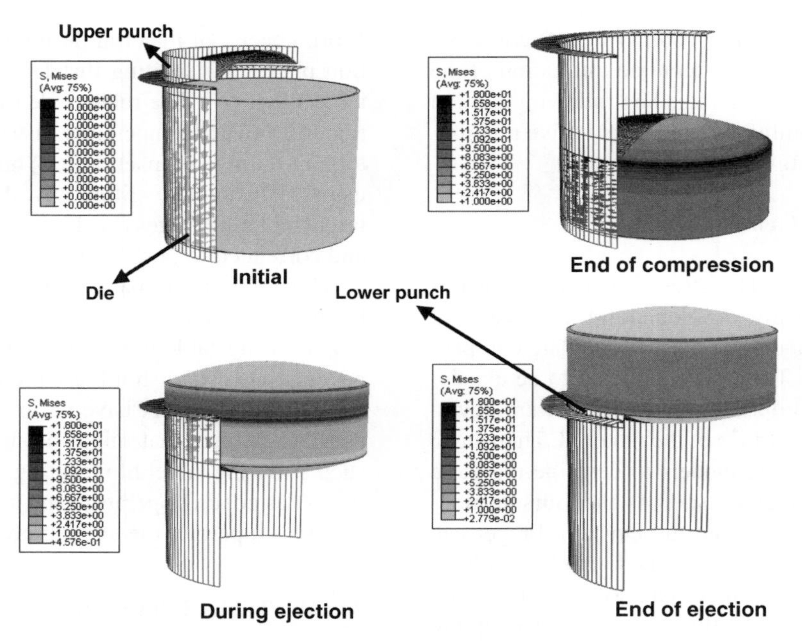

FIGURE 34.17 Stress distributions on the tablet during compaction using a standard round concave (SRC) punch.

of a tablet can be an important part of brand recognition and can even be used to trademark.

Typically coating formulations can be either aqueous or solvent-based systems. While aqueous coatings are rapidly becoming the preferred method for many applications, solvent-based coatings are still used to apply many functional coatings onto tablet cores for controlled drug delivery (e.g., semipermeable or delayed/sustained-release membranes) [69]. Although solvent-based systems hold a number of advantages in terms of application and flexibility, the move to aqueous coatings has largely been driven by factors such as cost, safety concerns, more stringent regulations on effluent discharge, as well as the broad variety of film formulations that have been developed recently for aqueous application [70].

The film coating process can essentially be considered an adiabatic evaporative cooling process. The driving energy for fluid evaporation is a combination of the airflow volume, temperature, and moisture content of the air. This can be considered as the bulk gas phase in total. The underlying law controlling the thermodynamic environment within the process is the first law of thermodynamics (conservation of energy). It is important to understand that the three principles of driving energy are linked by the operating parameters within the process. For example, an increase in the inlet air temperature will lower the relative humidity of the drying air into the pan (although the absolute water content remains the same). Increasing the spray rate of the coating solution will increase the moisture content of the air in the pan to which the tablets are exposed. Once stabilized, however, the process will remain in equilibrium unless disturbed by the alteration of a process parameter or external condition. Since the quality of the overall coating is greatly influenced by the thermodynamic conditions inside the pan, it is of great importance to understand these relationships and how to control them.

Thermodynamic models utilizing material and energy balances have been used in the past with some success to model the aqueous film coating process [23–25, 70, 71]. These models are used to predict the key process parameters that impact the quality of the film coat; mainly exhaust air temperature and exhaust air relative humidity. Film-coating models are particularly important in pharmaceutical development where process conditions vary greatly for the purpose of design of experiment, scale-up, and coating formulation changes. Most of the previous models have been restricted to aqueous film coatings, and one coating pan type or scale, making them limited in their scope and applicability. More recently, am Ende and Berchielli developed a universal thermodynamic model that is applicable to both aqueous and organic film-coating systems to aid in process optimization and scale-up [20]. This model will be the basis for the discussion and calculations laid out in the following section. Whether the system in question is aqueous or organic based, the film-coating process is a delicate balance that requires a high level of control over the process conditions to produce films that provide the required aesthetics or functionality.

34.3.4.2 Model System, Assumptions, and Limitations

A typical tablet film coater schematic is shown in Figure 34.18 for a perforated coating pan. The drying air is heated by an external heating source to a target temperature,

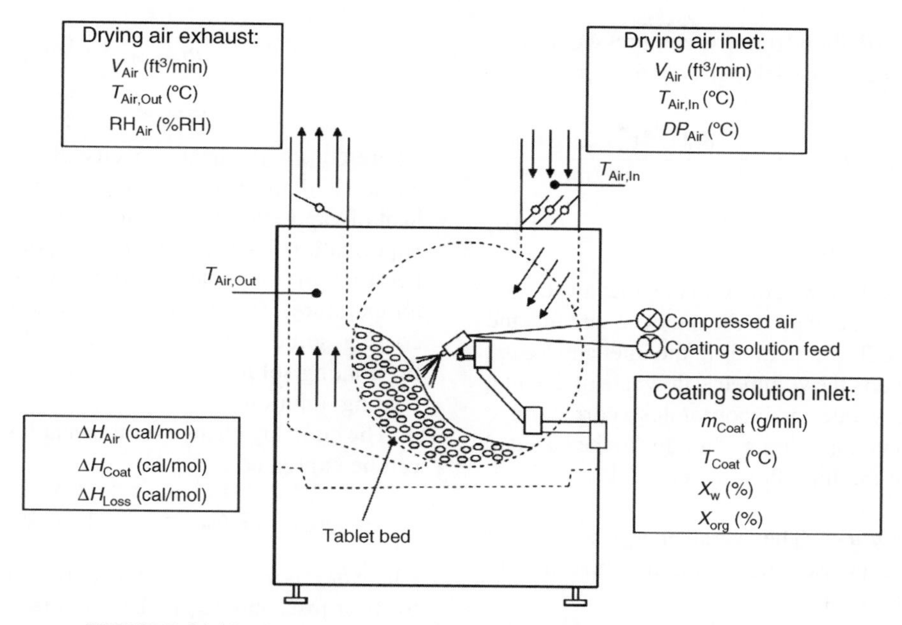

FIGURE 34.18 Schematic of a typical perforated tablet film-coating pan.

$T_{\text{air,in}}$, which is typically controlled by the operator to target a desired exhaust air temperature, $T_{\text{air,out}}$. The air flows through the tablet bed where it serves to dry the damp tablets as the coating solution droplets impact and spread on the tablet face. The air exits the pan through the exhaust air duct at a lower temperature due to the evaporative cooling effect from the volatile components (aqueous or organic) in the coating solution. The coating solution is supplied to the spray nozzle(s) by a pump where it is atomized by a compressed air stream into a pattern of tiny droplets that are propelled from the nozzle toward the tablet bed.

The rotating tablet bed defines the control volume for the material and energy balances. The model applies to steady-state conditions where the heat, temperature and mass do not change with time:

$$\frac{dq}{dt} = \frac{dT}{dt} = \frac{dm}{dt} = 0 \tag{34.15}$$

The inlet streams to the system include the drying air, and the film-coating solution. The compressed air stream is neglected in the overall airflow through the pan since it is only a minor component compared to the drying air. The outlet stream consists of the drying air exhaust and it is assumed that the volatile components in the coating solution exit the coater through this stream as vapor (i.e., the tablets do not retain any moisture). This is a reasonable assumption since coating pans are typically overdesigned with regards to drying capacity. The model was developed for a closed but not isolated system since energy exchange occurs as heat loss from the pan during operation and sampling. The model also neglects the humidity from the compressed air as well as the sensible heat term for the solid components of the coating solution.

This film-coating model is a macroscopic analysis of the process and, therefore, the physical operating conditions such as the spray gun-to-bed distance, pan speed, pan load, and spray zone coverage are not considered. It is well known that these parameters also influence key tablet attributes such as coating uniformity and elegance, and, therefore, they should be monitored carefully during process optimization.

*34.3.4.3 **Material Balance*** The material balance for the tablet bed control volume can be expressed in terms of each of the three components involved (e.g., water, organic solvent, and air). Assuming no reaction or accumulation the total mass entering the system should be equal to the total mass exiting the system:

$$\sum m_{\text{in}} + \sum m_{\text{coat}} = \sum m_{\text{out}} \tag{34.16}$$

The material balance for water takes into account the humidity from the inlet drying air stream as well as the water in the coating solution:

$$m_{\text{w,in}} + m_{\text{w,coat}} = m_{\text{w,out}} \tag{34.17}$$

where $m_{\text{w,in}}$ is the mass flow rate of water in the inlet air stream, $m_{\text{w,coat}}$ is the mass flow rate of water in the coating solution, and $\text{m}_{\text{w,out}}$ is the mass flow rate of water in the outlet air stream.

For coating formulations with organic components, the material balance can be expressed as

$$m_{\text{org,coat}} = m_{\text{org,out}} \tag{34.18}$$

where $m_{\text{org,coat}}$ is the mass flow rate of organic in the coating solution and $m_{\text{org,out}}$ is the mass flow rate of organic in the outlet air stream.

Neglecting the contribution from the compressed air line to the spray nozzle the material balance for air can be expressed as follows:

$$m_{\text{air,in}} = m_{\text{air,out}} = V_{\text{air,in}}\,\text{ft}^3/\text{min} \times \left(\frac{28.3\,\text{L/ft}^3 \times 29\,\text{g/mol}}{22.4\,\text{L/mol}}\right) \times \left(\frac{273\,\text{K}}{273\,\text{K} + T_{\text{air,in}}}\right) \quad (34.19)$$

where $m_{\text{air,in}}$ is the mass flow rate of air in the inlet air stream, $m_{\text{air,out}}$ is the mass flow rate of air in the outlet air stream, and $V_{\text{air,in}}$ is the volumetric flow rate of air in the inlet air stream.

This equation converts the volumetric flow rate of the inlet air stream to a mass flow rate and incorporates a correction to account for change in molar volume of air due to the elevated temperature, $T_{\text{air,in}}$, at the inlet flow meter.

34.3.4.4 ***Energy Balance*** The overall energy balance for the system can be expressed based on the first law of thermodynamics as follows:

$$\Delta H = \Delta H_{\text{air}} + \Delta H_{\text{coat}} + \Delta H_{\text{loss}} = 0 \quad (34.20)$$

where ΔH is the overall enthalpy change across the control volume and individual terms, ΔH_{air}, ΔH_{coat}, and ΔH_{loss}, represent the enthalpy change across the control volume due to the drying airflow, the coating solution, and heat loss, respectively. The individual enthalpy terms can be expressed in terms of a sensible heat term and, where applicable, a latent heat of vaporization.

$$\Delta H = mC_p\Delta T + m\Delta\hat{H}_{\text{vap}} \quad (34.21)$$

The enthalpy change for the airflow then becomes

$$\Delta H_{\text{air}} = m_{\text{air,in}}C_{\text{p,air}}\left(T_{\text{air,out}} - T_{\text{air,in}}\right) \quad (34.22)$$

where $C_{\text{p,air}}$ is the heat capacity of air, $T_{\text{air,out}}$ is the temperature of the exhaust air, and $T_{\text{air,in}}$ is the temperature of the inlet air.

Since the tablet bed temperature, T_{tablet}, typically is not measured during normal operation it is assumed to be the same or similar as the exhaust air temperature. Therefore, the sensible heat term in the coating solution enthalpy change is defined in terms of the temperature difference between the exhaust air and the coating solution temperature, T_{coat}. Thus, the enthalpy change for the coating solution can be written as

$$\Delta H_{\text{coat}} = m_{\text{w,coat}}C_{\text{p,w}}\left(T_{\text{air,out}} - T_{\text{coat}}\right) + m_{\text{w,coat}}\Delta\hat{H}_{\text{vap,w}} + m_{\text{org,coat}}C_{\text{p,org}}\left(T_{\text{air,out}} - T_{\text{coat}}\right) + m_{\text{org,coat}}\Delta\hat{H}_{\text{vap,org}} \quad (34.23)$$

$$m_{\text{w,coat}} = x_{\text{w}}m_{\text{coat}} \quad (34.24)$$

$$m_{\text{org,coat}} = x_{\text{org}}m_{\text{coat}} \quad (34.25)$$

where $C_{\text{p,w}}$ is the heat capacity of water, $C_{\text{p,org}}$ is the heat capacity of the organic component, $\Delta\hat{H}_{\text{vap,w}}$ is the latent heat of vaporization for water, $\Delta\hat{H}_{\text{vap,org}}$ is the latent heat of vaporization for the organic component, T_{coat} is the temperature of the coating solution (assumed to be room temperature), x_{w} is the mass fraction of water in the coating solution, x_{org} is the mass fraction of organic in the coating solution, and m_{coat} is the mass flow rate of the coating solution.

The enthalpy change for the heat loss to the surroundings can be expressed as

$$\Delta H_{\text{loss}} = h_{\text{loss}}A\left(T_{\text{air,out}} - T_{\text{RT}}\right) \quad (34.26)$$

where h_{loss} is the heat transfer coefficient, A is the surface area for heat loss, and T_{RT} is the room temperature.

Since the heat transfer coefficient and surface area for heat loss can differ greatly between coating pans, these two terms are lumped together into an empirically determined heat loss factor (HLF):

$$\Delta H_{\text{loss}} = \text{HLF}\left(T_{\text{air,out}} - T_{\text{RT}}\right) \quad (34.27)$$

The overall energy balance can be obtained by substituting the individual enthalpy terms into equation (34.20) as follows:

$$\Delta H = m_{\text{air,in}}C_{\text{p,air}}\left(T_{\text{air,out}} - T_{\text{air,in}}\right) + x_{\text{w}}m_{\text{coat}}C_{\text{p,w}}\left(T_{\text{air,out}} - T_{\text{coat}}\right) + x_{\text{w}}m_{\text{coat}}\Delta\hat{H}_{\text{vap,w}} + x_{\text{org}}m_{\text{coat}}C_{\text{p,org}}\left(T_{\text{air,out}} - T_{\text{coat}}\right) + x_{\text{org}}m_{\text{coat}}\Delta\hat{H}_{\text{vap,org}} + \text{HLF}\left(T_{\text{air,out}} - T_{\text{RT}}\right) = 0 \quad (34.28)$$

The energy balance equation can then be rearranged to solve for the unknown exhaust air temperature:

$$T_{\text{air,out}} = \frac{m_{\text{air,in}}C_{\text{p,air}}T_{\text{air,in}} + x_{\text{w}}m_{\text{coat}}C_{\text{p,w}}T_{\text{coat}} - x_{\text{w}}m_{\text{coat}}\Delta\hat{H}_{\text{vap,w}} + x_{\text{org}}m_{\text{coat}}C_{\text{p,org}}T_{\text{coat}} - x_{\text{org}}m_{\text{coat}}\Delta\hat{H}_{\text{vap,org}} + \text{HLF} \times T_{\text{RT}}}{m_{\text{air,in}}C_{\text{p,air}} + x_{\text{w}}m_{\text{coat}}C_{\text{p,w}} + x_{\text{org}}m_{\text{coat}}C_{\text{p,org}} + \text{HLF}} \quad (34.29)$$

The thermodynamic film-coating model detailed above provides a direct relationship between inlet air temperature, drying airflow rate, coating solution spray rate, and composition to the temperature of the exhaust air stream. Once the HLF is determined for a specific coating pan, the model can be used to predict the exhaust air temperature based on the operating conditions of the coater. The percent relative humidity of the exhaust air stream (%RH_{out}) can be calculated based on the material balance for water around the control volume and then taking the ratio of the partial pressure of water vapor in the exhaust air ($P_{w,out}$) to the vapor pressure of water at the exhaust air temperature ($P^*_{w,Tair,out}$):

$$\%RH_{out} = \frac{P_{w,out}}{P^*_w(@T_{air,out})} \times 100\% \quad (34.30)$$

34.3.4.5 Prediction of Target Film-Coating Parameters Using Thermodynamic Model Before the predictive capabilities of the thermodynamic film-coating model can be utilized a value for the HLF must be determined for the specific coating pan in question. The model outlined in equation (34.29) has two unknowns, the HLF and the exhaust air temperature, $T_{air,out}$. The HLF can be determined empirically by comparing equation (34.29) to a set of experimental data where the exhaust air temperature has been measured. The HLF is used as a variable fitting parameter to minimize residual sum of squared error between the experimental data and the predicted exhaust air temperature from the model. This can be done with any simple optimization function such as Solver in Microsoft Excel®. A sample data set with an optimized heat loss factor calculation for a Vector HCT-30 model film coater is shown in Table 34.10. The data include both aqueous and organic coating formulations for a wide range of operating conditions (inlet temperature, airflow, and spray rate). By minimizing the sum of squared error between the actual and the predicted exhaust air temperature the HLF was determined to be 150 cal/min°C.

Once the HLF is determined the model can be used for process optimization and scale-up predictions for that specific pan. For example, an operator can see how a change in spray rate or drying airflow will affect the exhaust air temperature of the process. The operator can also determine what inlet air temperature set point will be required to target a specific exhaust air temperature in the coater.

EXAMPLE 34.3 THERMODYNAMIC FILM-COATING MODEL

A small-scale Vector LDCS-5 film coater has a HLF of 282 cal/min°C. (a) Determine the required inlet air temperature set point to achieve a target an exhaust air temperature of 50°C at an airflow of 40 ft^3/min, and a spray rate of 4 g/min. The coating solution is an aqueous formulation with 20 wt% solids and the room temperature is 22°C. The heat capacity of air and water, and the latent heat of vaporization of water can be readily found in any physical chemistry textbook. (b) What is the relative humidity of the exhaust air at these conditions if the dew point of the inlet air is 10°C?

Solution

(a) Equation (34.29) cannot easily be rearranged to solve for the inlet air temperature, $T_{air,in}$, since this term is embedded in the denominator of equation (34.19) to calculate the mass flow rate of air in the coater. Therefore, the equation must be solved iteratively by guessing a value of $T_{air,in}$ and solving for $T_{air,out}$. Since we know $T_{air,in}$ must be higher than $T_{air,out}$ a good initial guess might be 20°C higher than the target exhaust temperature (i.e., $T_{air,in} = 70°C$). Thus, from equation (34.19)

$$m_{air,in} = 40\ \text{ft}^3/\text{min} \times \left(\frac{28.3\ L/\text{ft}^3 \times 29\ \text{g/mol}}{22.4\ \text{L/mol}}\right) \times \left(\frac{273\ \text{K}}{(273 + 70°C)\text{K}}\right) = 1166.45\ \text{g/min}$$

TABLE 34.10 Sample Data Set with Heat Loss Factor Determination for an HCT-30 Film Coater

Trial	Acetone (%)	Water (%)	Room Temperature (°C)	Inlet Temperature (°C)	Drying Airflow (ft^3/min)	Spray Rate (g/min)	Actual $T_{air,out}$ (°C)	Predicted $T_{air,out}$ (°C)	Difference Predicted - Actual
1	85	5	22	41	32	20	27.6	26.3	−1.3
2	85	5	18	61	24	22	30.3	31.1	0.8
3	85	5	18	50	37	21	31.4	31.0	−0.4
4	85	5	18	49	37	21	30.3	30.9	0.6
5	0	94	18	74	38	7	45.2	44.2	−1.0
6	0	94	18	77	38	8	44.9	45.5	0.6
7	0	94	18	75	38	8	44.6	43.7	−0.9
8	0	94	18	71	38	7	41.6	42.6	1.0
9	0	94	21	78	35	8	44.1	44.6	0.5
10	0	94	18	79	35	8	45.0	44.4	−0.6
							Group average		−0.06
							Standard Deviation		0.85
							Sum of squares		6.58
							HLF (cal/min°C)		**150**

and from equation (34.29)

$$T_{\text{air,out}} = \frac{(1166.45\text{g/min})(0.238\ \text{cal/g°C})(70°\text{C}) + 0.8(4\text{g/min})(1.0\ \text{cal/g°C})(22°\text{C})}{(1166.45\text{g/min})(0.238\ \text{cal/g°C}) + 0.8(4\ \text{g/min})(1.0\ \text{cal/g°C}) + 282\ \text{cal/min°C}}$$
$$+ \frac{-0.8(4\text{g/min})(540\ \text{cal/g}) + (282\ \text{cal/min°C})(22°\text{C})}{(1166.45\ \text{g/min})(0.238\ \text{cal/g°C}) + 0.8(4\text{g/min})(1.0\ \text{cal/g°C}) + 282\ \text{cal/min°C}}$$

$T_{\text{air,out}} = 42.6°C$

Based on this result, the inlet air temperature must be higher than 70°C. Increasing the estimate to $T_{\text{air,in}} = 85°\text{C}$ gives

$$m_{\text{air,in}} = 1117.57\ g/\text{min}$$

and

$$T_{\text{air,out}} = 49.3°C$$

This could be iterated further to get an exhaust air temperature closer to 50°C, but this estimate is well within the accuracy of the model and the variability of the actual equipment. Therefore, the inlet air temperature set point required to achieve a target exhaust air temperature of 50°C is approximately 85°C.

(b) To determine the relative humidity of the exhaust air, we need to determine the molar flow rate of water vapor exiting the coater. Based on the material balance for water around the control volume, the molar flow rate of water vapor in the exhaust stream should be equal to amount of water vapor entering in the drying air plus the amount of water added through the coating solution. First, calculate the water vapor partial pressure in the inlet air, which is, by definition, the vapor pressure at the dew point temperature. This can be determined using the Arden Buck equation.

$$P_{\text{w,in}} = P^*_{\text{w}}(@T_{\text{dew}}) = 0.012\ \text{atm}$$

Assuming an ideal gas mixture the mole fraction of water vapor in the inlet stream can be calculated by the ratio of the partial pressure to the total pressure (which is assumed to be atmospheric).

$$y_{\text{w,in}} = \frac{P_{\text{w,in}}}{P_{\text{total}}} = \frac{0.012\ \text{atm}}{1\ \text{atm}} = 0.012$$

and, thus, the mole fraction of dry air is

$$y_{\text{air,in}} = 1 - y_{\text{w,in}} = 0.988$$

The overall molar flow rate of the inlet air stream is the mass flow rate divided by the average molecular weight of the inlet air:

$$\dot{n}_{\text{inlet}} = \frac{m_{\text{air}}}{MW_{\text{inlet}}} = \frac{m_{\text{air}}}{y_{\text{w,in}}MW_w + y_{\text{air,in}}MW_{\text{air}}}$$
$$= \frac{1117.57\ \text{g/min}}{0.012 \times 18\ \text{g/mol} + 0.988 \times 29\ \text{g/mol}}$$
$$= 38.713\ \text{mol/min}$$

and the molar flow rate of water vapor in the inlet air is

$$\dot{n}_{\text{w,inlet}} = y_{\text{w,in}} \times \dot{n}_{\text{inlet}} = 0.012 \times 38.713$$
$$= 0.4646\ \text{mol/min}$$

The contribution of water vapor from the coating solution is

$$\dot{n}_{\text{w,spray}} = m_{\text{coat}}\left(\frac{x_{\text{w}}}{MW_{\text{w}}}\right)$$
$$= 4\ \text{g solution/min} \times \left(\frac{0.8\ \text{g Water/g Solution}}{18\ \text{g Water/mol}}\right)$$
$$= 0.1778\ \text{mol/min}$$

By conservation of mass, the molar flow rate of water vapor in the exhaust air stream is

$$\dot{n}_{\text{w,outlet}} = \dot{n}_{\text{w,inlet}} + \dot{n}_{\text{w,spray}} = 0.4646 + 0.1778$$
$$= 0.6424\ \text{mol/min}$$

and the overall molar flow rate of the exhaust air stream is

$$\dot{n}_{\text{outlet}} = \dot{n}_{\text{inlet}} + \dot{n}_{\text{w,spray}} = 38.713 + 0.1778$$
$$= 38.891\ \text{mol/min}$$

Finally, the partial pressure and vapor pressure of water in the exhaust air stream can be calculated as follows:

$$\text{P}_{\text{w,out}} = y_{\text{w,out}}P_{\text{total}} = \frac{\dot{n}_{\text{w,outlet}}}{\dot{n}_{\text{outlet}}}P_{\text{total}}$$
$$= \frac{0.6424\ \text{mol/min}}{38.891\ \text{mol/min}} \times 1\ \text{atm} = 0.0165\ \text{atm}$$

$$P^*_w(@T_{\text{air,out}}) = 0.1177\ \text{atm}$$

and thus the relativity humidity of the exhaust air stream is

$$\%\text{RH}_{\text{out}} = \frac{P_{\text{w,out}}}{P^*_{\text{w}}(@T_{\text{air,out}})} \times 100\%$$
$$= \frac{0.0165\ \text{atm}}{0.1177\ \text{atm}} \times 100\% = 14\%.$$

34.3.4.6 Process Scale-Up Based on Model Predictions

An important feature of the thermodynamic film-coating model is in its predictive capabilities, which can be used for process simulation and to assist with scale-up from one film coater to another. Typically, when a film-coating process is scaled to a larger unit it is desirable to maintain the same thermodynamic conditions in the pan that the tablets are exposed to during coating. This can be accomplished by matching the temperature and relative humidity of the exhaust air stream across the two coaters when the HLF for each coater is known. The thermodynamic model can be used to determine the process parameters on the new coater that will result in similar exhaust air conditions to the proven operation of the original coater. Ideally, this will reduce the number of trials required during scale-up and minimize any failed batches during validation.

This concept can be best illustrated through a simple example. Consider a commercial tablet film-coating process that is currently being executed on a Glatt GC-750 model film coater. The heat loss factor for the GC-750 has been determined to be 1080 cal/min°C. The process is well defined on this coater and the proved design space includes inlet air temperatures ranging 60–70°C, spray rates ranging 30–80 g/min, and a drying airflow rate of 300 ft^3/min. The coating is an aqueous formulation consisting of 15 wt% solids. Due to high product demand the commercial site is considering scaling the process to a larger Glatt GC-1000 model film coater, which is typically operated at spray rates ranging 120–250 g/min and a drying airflow rate of 900 ft^3/min. Use the thermodynamic film-coating model to determine the inlet air temperature range in the GC-1000 that will give similar exhaust air temperature and relative humidity across the two coaters.

First, the exhaust air temperature and relative humidity in the GC-750 at each point of the design space conditions outlined above can be calculated using the thermodynamic model. The resulting four data points define the corner points for the operating space of the GC-750 as outlined in Figure 34.19. This operating space represents the proven acceptable range of thermodynamic conditions that will be used for scale-up to the larger coater.

Since the spray rate range and drying airflow are known on the GC-1000, the thermodynamic model can be used directly to calculate inlet air temperature required to match the exhaust air temperature at each of the four points on the GC-750 operating space. Once the exhaust air temperature is known the corresponding relative humidity can be calculated based on the material balance equations. This procedure is similar to the solution of Example 34.3. The resulting predicted exhaust air temperature/relative humidity operating space of the GC-1000, along with the GC-750 operating space for comparison, is shown in Figure 34.20. The input parameters and model predictions for both the GC-750 and the GC-1000 are shown in Table 34.11. According to the model predictions, the inlet air temperature required in the GC-1000

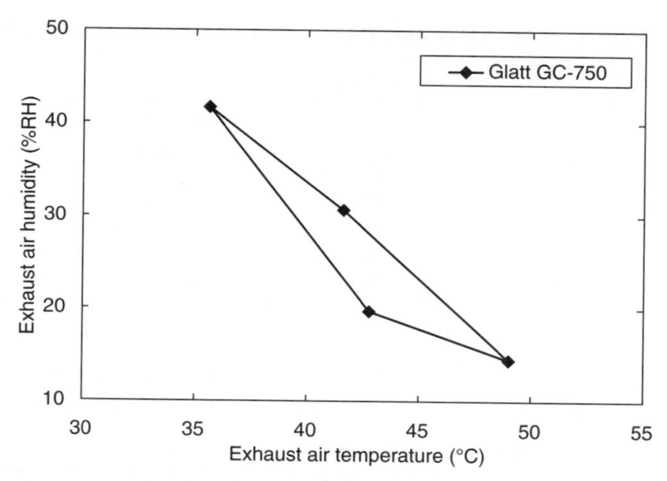

FIGURE 34.19 Exhaust air temperature/relative humidity operating space for the Glatt GC-750 based on inlet air temperatures of 60–70°C, spray rates of 30–80 g/min, and drying airflow rate of 300 ft^3/min.

to match the exhaust air temperature of the GC-750 ranges approximately 57–65°C. It is evident from Figure 34.20 that at these conditions the exhaust air temperature/relative humidity operating space of the GC-1000 matches very well with the proven space of the GC-750.

By matching the thermodynamic conditions across the two coaters we have defined the potential design space for this product on the new coater without running any trials at scale. The operating space can be verified with as many or as few confirmation batches as the operators and formulators deem necessary. We can also map out a design space plot that allows one to visualize the relationship between changes in spray rate, inlet air temperature, and exhaust air temperature

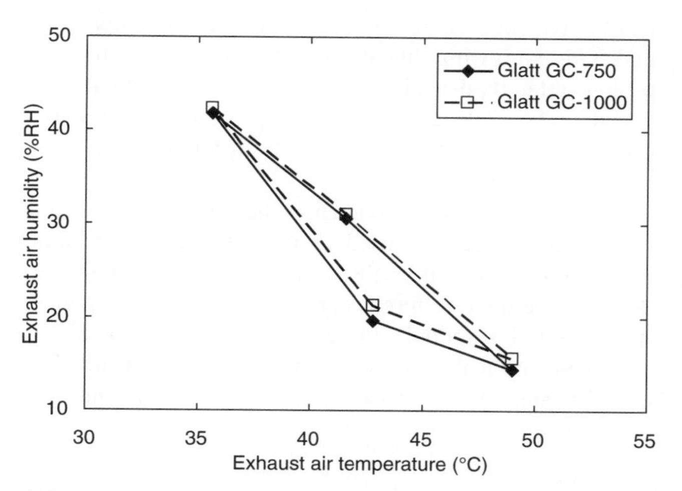

FIGURE 34.20 Exhaust air temperature/relative humidity operating space for the Glatt GC-1000 based on inlet air temperatures of approximately 57–65°C, spray rates of 120–250 g/min, and drying airflow rate of 900 ft^3/min. The GC-750 operating space is shown for comparison.

TABLE 34.11 Operating Conditions and Model Predictions for GC-750/GC-1000 Scale-Up

Glatt GC-750				Glatt GC-1000			
Inlet Temperature (°C)	Spray Rate (g/min)	Exhaust Temperature (°C)	%RH	Inlet Temperature (°C)	Spray Rate (g/min)	Exhaust Temperature (°C)	%RH
60	30	42.8	19.6	56.6	120	42.8	21.3
60	80	35.6	41.7	57	250	35.6	42.3
70	30	49	14.4	64.7	120	49	15.6
70	80	41.6	30.6	65.2	250	41.6	31

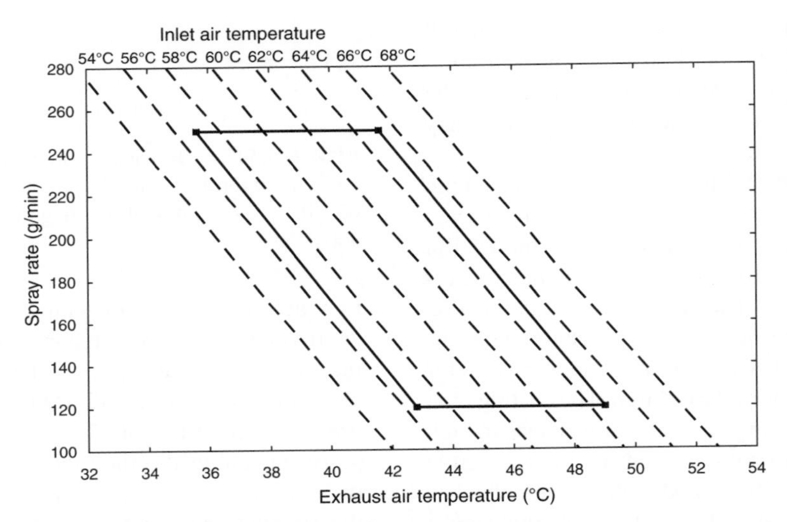

FIGURE 34.21 Design space contour plot for the Glatt GC-1000 film coater for a drying airflow rate of 900 ft^3/min.

for a given drying airflow rate. This type of plot for the GC-1000 film coater is shown in Figure 34.21. The contour plot allows one to map out any combination of operating conditions (inlet air temperature, exhaust air temperature, and spray rate) within the design space for the given drying airflow rate. It also allows operators and formulators to visualize how the exhaust air temperature changes as a function of both spray rate and inlet air temperature across the entire design space.

This exercise has illustrated how the thermodynamic film-coating model can be used to predict specific operating conditions, assist with scale-up from one coater to another, and also simulate different operation scenarios on a given film coater. The model is an extremely versatile predictive tool that is applicable to a wide range of coating formulations on units ranging from lab scale coaters to commercial sized equipment.

34.4 SUMMARY

This chapter highlighted several modeling techniques applied to the design, development, and scale-up for solid oral drug products. These process modeling techniques were discussed and exemplified with case studies ranging from raw material specifications to process parameter predictions in wet granulation and film coating. Specific consideration was given to transfer and scale-up issues along with general process design related challenges to pharmaceutical process R&D. There are many other modeling approaches available to formulation and process scientists that were not covered in this chapter. However, the purpose was to demonstrate that the fundamental principles taught in the chemical engineering curriculum ensure the chemical engineer is well poised to apply and implement modeling techniques to solve challenging issues in the pharmaceutical industry.

REFERENCES

1. Wassgren C, Curtis JS. The application of computational modeling to pharmaceutical materials science. *MRS Bull.* 2006;31(11):900–904.
2. Davidson DL. The enterprise-wide application of computational fluid dynamics in the chemicals industry. 6th World Congress of Chemical Engineering Melbourne, Australia, 2001.

3. Louie AS, Brown MS, Kim A.Measuring the return on modeling and simulation tools in pharmaceutical development. White paper No, HI204892, Health Industry Insights, 2007.
4. Garcia-Munoz S. Establishing multivariate specifications for incoming materials using data from multiple scales. *Chemometr. Intell. Lab. Syst.* 2009;98(1):51–57.
5. Ketterhagen WR. Modeling granular segregation during hopper discharge. 2006.
6. Ketterhagen WR, am Ende MT, Hancock BC. Process modeling in the pharmaceutical industry using the discrete element method. *J. Pharm. Sci.* 2009;98(2):442–470.
7. Ketterhagen WR, Curtis JS, Wassgren CR. Stress results from two-dimensional granular shear flow simulations using various collision models. *Phys. Rev. E* 2005;71(6-1):061307/1–061307/11.
8. Ketterhagen WR, Curtis JS, Wassgren CR, Hancock BC. Modeling granular segregation in flow from quasi-three-dimensional, wedge-shaped hoppers. *Powder Technol.* 2008; 179(3):126–143.
9. Ketterhagen WR, Curtis JS, Wassgren CR, Hancock BC. Predicting the flow mode from hoppers using the discrete element method. *Powder Technol.* 2009;195(1):1–10.
10. Ketterhagen WR, Curtis JS, Wassgren CR, Kong A, Narayan PJ, Hancock BC. Granular segregation in discharging cylindrical hoppers: a discrete element and experimental study. *Chem. Eng. Sci.* 2007;62(22):6423–6439.
11. Iveson SM, Litster JD, Hapgood K, Ennis BJ. Nucleation, growth and breakage phenomena in agitated wet granulation processes: a review. *Powder Technol.* 2001;117(1–2):3–39.
12. Bilgili E, Scarlett B. Population balance modeling of non-linear effects in milling processes. *Powder Technol.* 2005;153(1):59–71.
13. Kremer DM, Hancock BC. Process simulation in the pharmaceutical industry: a review of some basic physical models. *J. Pharm. Sci.* 2006;95(3):517–529.
14. Kunii D, Levenspiel O. *Fluidization Engineering*, Butterworth-Heinemann, Newton, MA, 1991.
15. Bi HT, Ellis N, Abba IA, Grace JR. A state-of-the-art review of gas–solid turbulent fluidization. *Chem. Eng. Sci.* 2000; 55(21):4789–4825.
16. Gurson AL. Continuum theory of ductile rupture by void nucleation and growth. Part I. Yield criteria and flow rules for porous ductile media. *J. Eng. Mater. Technol.* 1977;9:2–15.
17. Schofield AN, Wroth CP. *Critical State Soil Mechanics*, McGraw-Hill, London, 1968.
18. DiMaggio FL, Sandler IS. Material model for granular soils. *J. Eng. Mech. Div.* 1971;97:935–950.
19. Drucker DC, Gibson RE, Henkel DJ. Soil mechanics and work hardening theories of plasticity. *Trans. Am. Soc. Civil Eng.* 1957;122:338–346.
20. am Ende MT, Berchielli A. A thermodynamic model for organic and aqueous tablet film coating. *Pharm. Dev. Technol.* 2005;10(1):47–58.
21. Box GEP, Hunter WG, Hunter JS. *Statistics for Experimenters: An Introduction to Design, Data Analysis, and Model Building*, Wiley, 1978.
22. MacGregor JF, Yu H, Garcia Munoz S, Flores-Cerrillo J. Data-based latent variable methods for process analysis, monitoring and control. *Comput. Chem. Eng.* 2005; 29(6):1217–1223.
23. Page S, Baumann KH, Kleinebudde P. Mathematical modeling of an aqueous film coating process in a Bohle Lab-Coater. Part 1. Development of the model. *AAPS Pharm. Sci. Tech.* 2006; 7(2):42.
24. Page S, Baumann K-H. Mathematical modeling of an aqueous film coating process in a Bohle Lab-Coater. Part 2. Application of the model. *AAPS Pharm. Sci. Tech.* 2006; 7(2):43.
25. Stetsko G, Banker GS, Peck GE. Mathematical modeling of an aqueous film coating process. *Pharm. Technol.* 1983;7(11):50, 52–53, 56, 58, 60, 62.
26. Muteki K, MacGregor JF, Multi-block PLS modeling for L-shape data structures with applications to mixture modeling. *Chemometr. Intell. Lab. Syst.* 2007;85:186.
27. Garcia Munoz S, MacGregor JF, Kourti T. Product transfer between sites using Joint-Y PLS. *Chemometr. Intell. Lab. Syst.* 2005;79(1–2):101–114.
28. Garcia-Munoz S, Kourti T, MacGregor JF, Mateos AG, Murphy G. Troubleshooting of an industrial batch process using multivariate methods. *Ind. Eng. Chem. Res.* 2003;42(15):3592–3601.
29. Kourti T, Lee J, MacGregor JF. Experiences with industrial applications of projection methods for multivariate statistical process control. *Comput. Chem. Eng.* 1996; *20S*:S745.
30. Neogi D, Schlags CE. Multivariate statistical analysis of an emulsion batch process. *Ind. Eng. Chem. Res.* 1998; 37(10):3971–3979.
31. Garcia-Munoz S, Kourti T, MacGregor, JF, Apruzzese F, Champagne M. Optimization of batch operating policies. Part I. Handling multiple solutions. *Ind. Eng. Chem. Res.* 2006; 45(23):7856–7866.
32. Garcia-Munoz S, MacGregor JF, Neogi D, Latshaw BE, Mehta S. Optimization of batch operating policies. Part II. Incorporating process constraints and industrial applications. *Ind. Eng. Chem. Res.* 2008;47(12):4202–4208.
33. Doyle FJ, Harrison CA, Crowley TJ. Hybrid model-based approach to batch-to-batch control of particle size distribution in emulsion polymerization. *Comput. Chem. Eng.* 2003;27:1153–1163.
34. Flores-Cerrillo J, MacGregor JF. Within-batch and batch-to-batch inferential-adaptive control of semi-batch reactors: a partial least squares approach. *Ind. Eng. Chem. Res.* 2003; 42(14):3334–3345.
35. Flores-Cerrillo J, MacGregor JF. Latent variable MPC for trajectory tracking in batch processes. *J. Process Control* 2005;15(6):651–663.
36. Bharati MH, Liu JJ, MacGregor JF. Image texture analysis: methods and comparisons. *Chemometr. Intell. Lab. Syst.* 2004;72(1):57–71.
37. Yu H, MacGregor JF. Multivariate image analysis and regression for prediction of coating content and distribution in the production of snack foods. *Chemometr. Intell. Lab. Syst.* 2003;67(2):125–144.

38. Duchesne C, MacGregor JF. Establishing multivariate specification regions for incoming materials. *J. Qual. Technol.* 2004;36(1):78–94.
39. am Ende MT, Blackwood DO, Gierer DS, Neu CP. Challenges in development and scale-up of low dose dry granulation products: a case study. In: Zheng J, editor. *Analytical and Formulation Development For Low-Dose Oral Drug Products*, Wiley, New York, 2009; pp.117–157.
40. U.S. FDA. Draft Guidance for Industry: Quality Systems Approach to Pharmaceutical CGMP Regulations, 2006.
41. U.S. FDA. International Conference on Harmonisation: Guidance on Q8(R1) Pharmaceutical Development. *Fed. Regist.* 2009;74(109):27325–27326.
42. U.S., FDA, International Conference on Harmonisation: Guidance on Q9 Quality Risk Management. *Fed. Regist.* 2006;71(106):32105–32106.
43. U.S. FDA, International Conference on Harmonisation: Guidance on Q10 Pharmaceutical Quality System. *Fed. Regist.* 2009;74(66);15990–15991.
44. Nosal R, Schultz T. PQLI definition of criticality. *J. Pharm. Innov.* 2008;3:79–87.
45. Nomikos P, MacGregor JF. Multivariate SPC charts for monitoring batch processes. *Technometrics* 1995;37(1):41–58.
46. Hertz H. Über die Berührung fester elastischer Körper. *J. Reine Angew. Math.* 1882;92:136.
47. Zhu HP, Zhou ZY, Yang RY, Yu AB. Discrete particle simulation of particulate systems: theoretical developments. *Chem. Eng. Sci.* 2007;62(13):3378–3396.
48. Zhu HP, Zhou ZY, Yang RY, Yu AB. Discrete particle simulation of particulate systems: a review of major applications and findings. *Chem. Eng. Sci.* 2008;63(23):5728–5770.
49. United States Pharmacopeia. Uniformity of dosage units. *Pharmacopeia Forum* 2006;32(6):1653–1659.
50. Johanson JR. Predicting segregation of bimodal particle mixtures using the flow properties of bulk solids. *Pharm. Technol. Europe* 1996;8(1):38–44.
51. Levin M. Wet granulation: end-point determination and scale-up. In: Swarbrick J, editor. *Encyclopedia of Pharmaceutical Technology*, Taylor & Francis, Boca Raton, FL, 2006, pp. 4078–4098.
52. He X, Lunday KA, Li L-C, Sacchetti MJ. Formulation development and process scale up of a high shear wet granulation formulation containing a poorly wettable drug. *J. Pharm. Sci.* 2008;97(12):5274–5289.
53. Bier HP, Leuenberger H, Sucker H. Determination of the uncritical quantity of granulating liquid by power measurements on planetary mixers. *Pharm. Ind.* 1979;41(4);375–380.
54. Leuenberger H. Granulation, new techniques. *Pharma. Acta Helv.* 1982;57(3):72–82.
55. Leuenberger H. New trends in the production of pharmaceutical granules: the classical batch concept and the problem of scale-up. *Eur. J. Pharm. Biopharm.* 2001;52(3):279–288.
56. Leuenberger H, Puchkov M, Krausbauer E, Betz G. Manufacturing pharmaceutical granules: is the granulation end-point a myth? *Powder Technol.* 2009;189(2):141–148.
57. Faure A, Grimsey IM, Rowe RC, York P, Cliff MJ. Applicability of a scale-up methodology for wet granulation processes in Collette Gral high shear mixer-granulators. *Eur. J. Pharm. Sci.* 1999;8(2):85–93.
58. Landin M, York P, Cliff MJ, Rowe RC, Wigmore AJ. Scale-up of a pharmaceutical granulation in fixed bowl mixer-granulators. *Int. J. Pharm.* 1996;133(1–2):127–131.
59. Cunningham JC, Sinka IC, Zavaliangos A. Analysis of tablet compaction. I. Characterization of mechanical behavior of powder and powder/tooling friction. *J. Pharm. Sci.* 2004;93:2022–2039.
60. Han L, Elliott J, Bentham A, Mills A, Amidon G, Hancock B. A modified Drucker–Prager cap model for die compaction simulation of pharmaceutical powders. *Int. J. Solids Struct.* 2008;45(10):3088–3106.
61. Sinka IC. Modelling powder compaction. *KONA* 2007;25:4–22.
62. Sinka IC, Cunningham JC, Zavaliangos A. Analysis of tablet compaction. II. Finite element analysis of density distributions in convex tablets. *J. Pharm. Sci.* 2004;93(8):2040–2053.
63. Sinka IC, Cunningham JC, Zavaliangos A. The effect of wall friction in the compaction of pharmaceutical tablets with curved faces: a validation study of the Drucker–Prager cap model. *Powder Technol.* 2003;133(1–3):33–43.
64. Wu CY, Ruddy OM, Bentham AC, Hancock BC, Best SM, Elliott JA. Modelling the mechanical behaviour of pharmaceutical powders during compaction. *Powder Technology* 2005;152(1–3):107–117.
65. Frenning G. Analysis of pharmaceutical powder compaction using multiplicative hyperelasto-plastic theory. *Powder Technology* 2007;172(2):103–112.
66. Michrafy A, Ringenbacher D, Tchoreloff P. Modeling the compaction behavior of powders: application to pharmaceutical powders. *Powder Technol.* 2002;127(3):257–266.
67. Wu CY, Hancock BC, Mills A, Bentham AC, Best SM, Elliott JA. Numerical and experimental investigation of capping mechanisms during pharmaceutical tablet compaction. *Powder Technol.* 2008;181(2):121–129.
68. Zavaliangos A, Galen S, Cunningham J, Winstead D. Temperature evolution during compaction of pharmaceutical powders. *J. Pharma. Sci.* 2008;97(8):3291–3304.
69. am Ende MT, Herbig SM, Korsmeyer RW, Chidlaw MB. Osmotic drug delivery from asymmetric membrane film-coated dosage forms. In: Wise DL,editor. *Handbook of Pharmaceutical Controlled Release Technology*, Marcel Dekker, Inc., New York, 2000, pp.751–785.
70. Ebey GC. A thermodynamic model for aqueous film-coating. *Pharma. Technol.* 1987;11(4):40–50.
71. Rodriguez L, Grecchi R, et al. Variation of operational parameters and process optimization in aqueous film coating. *Pharm. Technol.* 1996;10:76–86.

35

PROCESS DESIGN AND DEVELOPMENT FOR NOVEL PHARMACEUTICAL DOSAGE FORMS

LEAH APPEL, JOSHUA SHOCKEY, AND MATTHEW SHAFFER
Green Ridge Consulting, Bend, OR, USA
JENNIFER CHU
Pharmaceutical Sciences, Neurogen Corporation, Branford, CT, USA

35.1 INTRODUCTION

Strong fundamental knowledge of the formulation and process that is used for a pharmaceutical product is critical to ensuring efficacy, safety, and robust product quality. The design and development approach that is generally advocated in the pharmaceutical industry is called "quality by design" (QbD). This is the application of a scientifically logical approach to developing a formulation and process that is robust, well understood, and well characterized. Knowledge of science and engineering principles and how to apply them are imperative to this product development process.

This approach is especially important in novel dosage forms that are used to produce a drug product that may have even tighter tolerances for performance, stability, and/or manufacturability than a standard dosage form. For example, if the release rate of drug is governed by a functional coating, where the coating thickness and morphology impact the rate of release, it becomes critical to control the coating process such that it consistently provides the same coating quality. While this is important for cosmetic coatings too, the range of coating thickness that yields acceptable performance and appearance is much broader than that for a functional coating. This chapter demonstrates the application of energy and mass transport principles to both the dosage form mechanism of release and the manufacturing process for a novel pharmaceutical formulation.

The formulation and process utilized to make a drug product are coupled and each must be examined in the context of the other. For example, in choosing the materials that are used in the formulation it is important to understand the process implications of the selected materials. Similarly, understanding the mechanism of drug release is key to understanding which product attributes are most critical in achieving the target release profile. This knowledge can be used to help guide process design and development.

In this chapter, an osmotic rupturing multiparticulate formulation manufactured by fluid bed coating is used as a model to demonstrate the application of engineering principles to develop a formulation and process for a novel dosage form. The rupturing multiparticulate is designed to provide a burst of drug release at a specific point in time after dosing. The primary use for this type of dosage form would be where a delayed release of drug is required; quite commonly, this is in combination with an immediate release dose of drug in a single unit dose. This approach allows the combination of multiparticulates with different release profiles in a capsule to provide the overall target release for the product.

In order to identify a process suitable for manufacture of a formulation, it is important to first understand the critical attributes of the formulation. Second, it is important to have a general understanding of the process, in this case fluid bed coating, to understand the effect of the process equipment

Chemical Engineering in the Pharmaceutical Industry: R&D to Manufacturing Edited by David J. am Ende

and key process parameters on the formulation. Finally, these bodies of knowledge should be combined to design the most appropriate process for manufacture of a specific formulation. This chapter is organized accordingly. We will first introduce the formulation and its mechanism of release in order to understand the critical product attributes. Next, we will discuss the fluid bed coating process from a general perspective. Finally, we will combine knowledge of the formulation and mechanism of release with this general process understanding to discuss specific fluid bed coating process considerations for the rupturing multiparticulate formulation.

35.2 ARCHITECTURE AND FORMULATION

The architecture of the multiparticulate system referred to in this chapter is modeled after formulations described by Ueda et al. [1, 2] and Dashevsky and Mohamad [3]. The release of drug from a rupturing multiparticulate occurs when water passes through a delayed release functional coating into the multiparticulate, builds pressure, and eventually ruptures the delayed release coating, allowing the drug in the multiparticulate to be released. The rupturing multiparticulate is composed of a seed core, surrounded by drug and sweller layers. The final layer is composed of a semipermeable polymer that controls the rate of water ingress. This "delayed release" layer is considered a functional coating since it governs the drug release rate from the multiparticulate. Figure 35.1 illustrates the multiparticulate architecture.

1. Seed core
2. Drug layer
3. Sweller layer
4. Delayed release layer

FIGURE 35.1 Osmotic rupturing multiparticulate architecture.

The various layers of the multiparticulate must have certain attributes in order to achieve the target release profile. In addition, all components that contact the drug must be chemically compatible with it. The general attributes of each component are described below, followed by the specific components used in the model system.

The drug in the model system is adipiplon, a small molecule that has relatively high solubility over a physiologically relevant pH range (>2 mg/mL from pH 1 to 8). For the rupturing multiparticulate formulation, the desired dose was low (<10 mg). The desired release profile was an immediate release dose followed by a pulse of drug 1–3 h after administration. Formulations for delayed release multiparticulates with 1, 2, and 3 h lag times were identified. The osmotic rupturing multiparticulate was selected as the lead approach for pulsatile release. The exact composition of the adipiplon multiparticulate formulation that was developed is shown in Table 35.1. The rationale for choosing the specific components for each layer is presented below.

The seed core material should provide an inert, durable, and smooth substrate for coating. Spherical microcrystalline cellulose (Celphere CP-708) with a mean particle size of 700 μm was used in the model system.

The drug layer should provide immediate drug release once the delayed release coating has ruptured. A water-soluble binder, HPMC E5 premium, was used in the model system with a high drug loading (75% A). The relatively high drug loading was selected to reduce processing time.

The sweller layer should provide sufficient driving force in the form of water activity at a sufficiently rapid rate to hydrate the core up to the point of rupture. The swelling component should be uniformly distributed to provide a smooth coating surface. Milled croscarmellose sodium (Ac-Di-Sol) with a water-soluble binder (HPC, Klucel EF) was used in the model system.

TABLE 35.1 Example Adipiplon Multiparticulate Formulation Composition

Layer	Amount (mg/g Final Multiparticulates)	Component (% in Layer)	Name	Function
Seed core	391.5	700 μm Celphere	Microcrystalline cellulose (Celphere CP-708)	Substrate
Drug layer	14.6	HPMC (25%)	Hypromellose (E5 premium)	Binder
	43.9	API (75%)	API (milled)	Active
Sweller layer	128.6	HPC (28.6%)	HPC (Klucel EF)	Binder
	321.4	Ac-Di-Sol (71.4%)	Croscarmellose sodium (Ac-Di-Sol, milled)	Sweller
Delayed release layer	50	Talc (50%)	Talc (IMP-1889L)	Coating strength modifier
	50	Ethylcellulose (50%)	Ethylcellulose (Ethocel STD 10 cP)	Semipermeable polymer

The delayed release layer has several properties that are critical to the performance of the system. The coating contains a semipermeable polymer that controls water ingress. This makes both the morphology and thickness of the coating critical to the performance. The coating in this layer must also fail via fracture or rupture; ideally the coating is brittle enough that it does not require a large degree of swelling prior to rupture. The semipermeable polymer selected was ethylcellulose. This polymer is commonly used in osmotic systems. A coating strength modifier, talc, was also used to increase the brittleness of the coating. Figure 35.2 shows the effect of two levels of talc on performance. Increasing the level of talc resulted in a much sharper burst, indicating more of the multiparticulates burst in a narrower time window. Figure 35.3 shows the performance of the 1, 2, and 3 h delayed release formulations. All formulations had the same composition differing only in the thickness or coating weight of the delayed release layer. To further understand the mechanism of drug release from this system, a more detailed explanation of the mechanism of release is given in the next section.

35.3 MECHANISM OF RELEASE

The physical model for the mechanism of release from rupturing multiparticulates is shown in Figure 35.4. The target release profile for a rupturing multiparticulate formulation is shown in Figure 35.5. The release profile has two primary components: the lag time, defined as the time between aqueous exposure of the multiparticulates and when rupture of the functional coating is initiated, and the duration of release, defined as the time between when rupturing begins and when drug release from the multiparticulates is substantially complete.

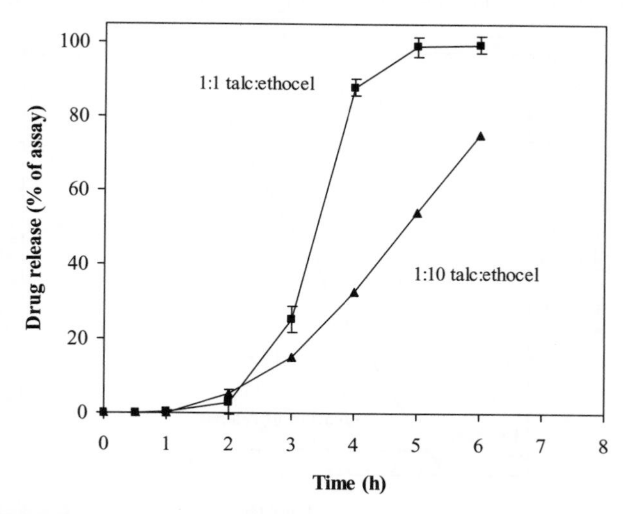

FIGURE 35.2 Effect of talc level in delayed release layer on the performance of adipiplon multiparticulates.

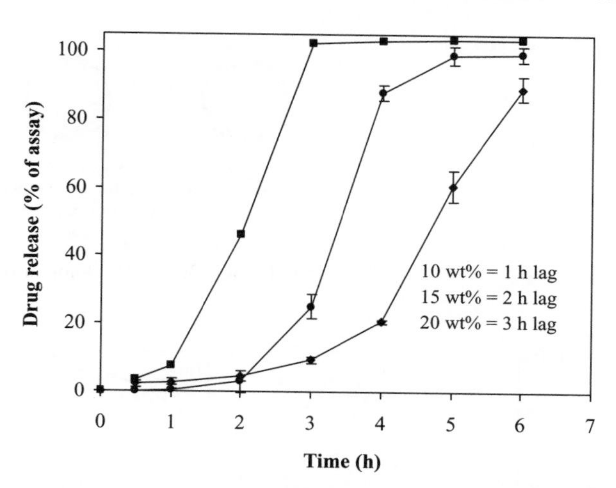

FIGURE 35.3 Effect of delayed release layer coating weight on the performance of adipiplon multiparticulates.

If a dissolution profile on a single multiparticulate were obtained, it would show a lag time dependent on the thickness and composition of the delayed release coating, and would have an immediate release profile once it ruptured (see Figure 35.6). However, since a dosage form consists of hundreds or thousands of multiparticulates, not one single multiparticulate, there is a distribution of final multiparticulate size, with small cores with thin coatings on one end of the distribution and large cores with thick coatings on the other (see Figure 35.7). If this distribution is then translated to predicted performance, multiparticulates with different coating thicknesses would be expected to rupture at different times. If many of these multiparticulates are in a dosage form, then the overall dissolution profile is the composite of many individual dissolution profiles and the overall dissolution profile will have a much broader duration of release (see Figure 35.8). The lag time for the dosage form is then defined as the amount of time prior to the first multiparticulate rupturing and the duration of release is proportional to the breadth of the coating weight distribution.

The mechanism of release from rupturing multiparticulates can be presented mathematically. The water uptake by the multiparticulates can be represented by equation 35.1, as demonstrated for osmotic systems by Theeuwes [4]:

$$\frac{dV_w}{dt} = \frac{A}{h} L_p(\sigma \Delta\Pi - \Delta P) \quad (35.1)$$

where V_w is the volume of water in the multiparticulate, t is the time, A is the cross-sectional area of the coating, h is the coating thickness, L_p is the mechanical permeability of the coating to water, σ is the reflection coefficient, $\Delta\Pi$ is the osmotic pressure difference across the coating, and ΔP is the hydrostatic pressure difference across the coating.

From a practical formulation approach, using this equation to guide formulation selection ensures the multiparticulate is formulated such that the osmotic pumping term is

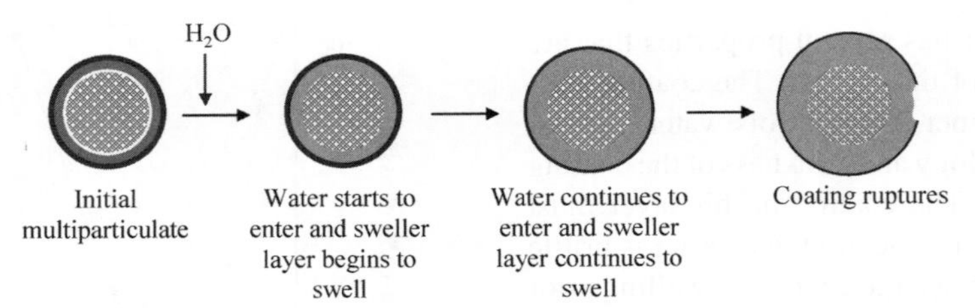

FIGURE 35.4 Mechanism of release for rupturing multiparticulates.

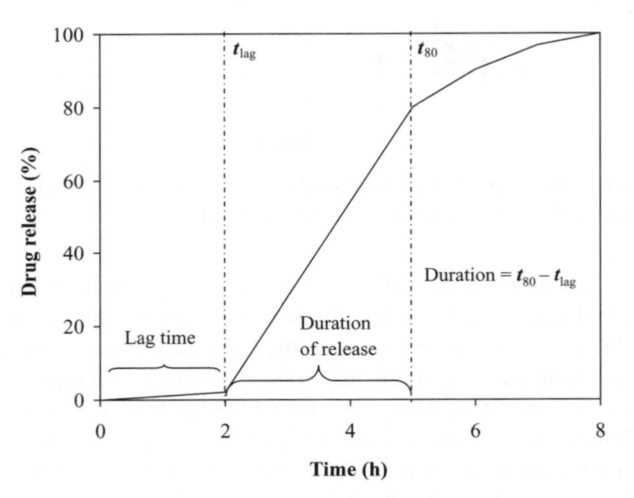

FIGURE 35.5 Example release profile of rupturing multiparticulates.

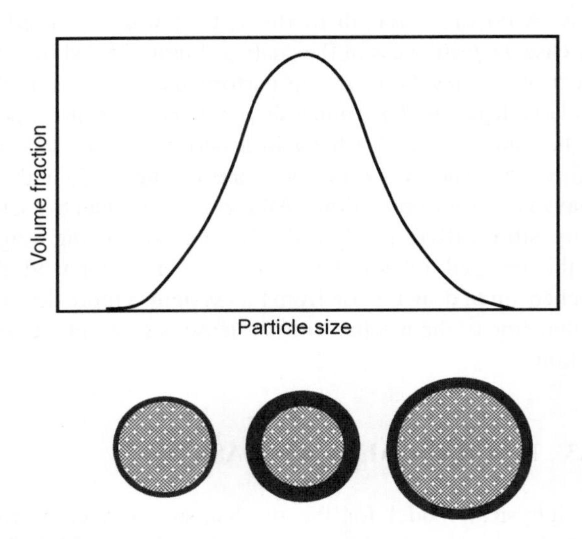

FIGURE 35.7 Coated particle size distribution and bead population.

always significantly greater than the hydrostatic pressure resistance term. This will ensure that the direction of water flow is into the multiparticulates (e.g., $dV_w/dt > 0$) and that the multiparticulates will ultimately rupture.

To examine the hydrostatic pressure difference, ΔP, it is useful to use Laplace's law, as shown in equation 35.2. This law describes the relationship between pressure of a sphere and wall tension or stress.

$$\Delta P = \frac{2E\varepsilon_e h}{r} \tag{35.2}$$

where E is the modulus of elasticity for the semipermeable coating, ε_e is the engineering strain of the system

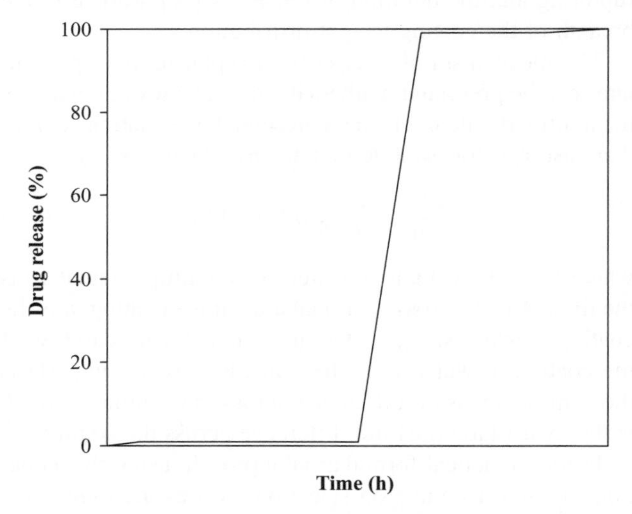

FIGURE 35.6 Single bead dissolution.

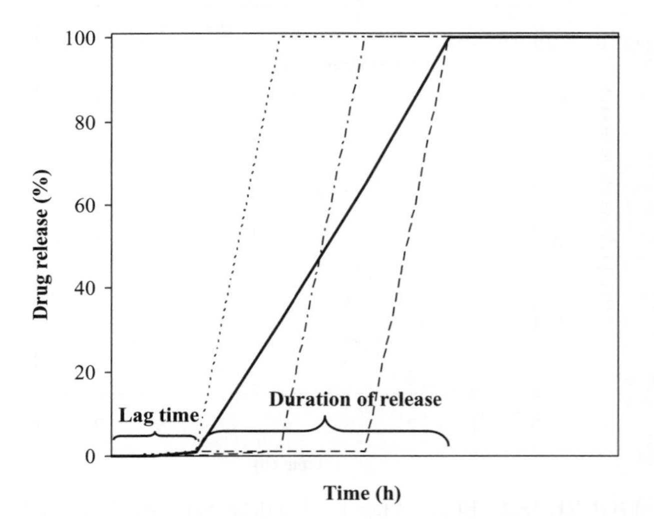

FIGURE 35.8 Composite dissolution profile.

(deformation of the system), and r is the radius of the multiparticulate.

Equation 35.2 assumes that the coating is perfectly elastic and does not yield (i.e., it displays Hookean behavior, where the strain is directly proportional to the stress) while it expands and that the multiparticulates are spherical with the coating thickness significantly smaller than the radius of the multiparticulate. For this system, these are reasonable simplifying assumptions. For $dV_w^*/dt > 0$ (where * denotes the value at rupture), we can combine equations 35.1 and 35.2 to obtain equation 35.3.

$$\sigma\Delta\Pi > \frac{2E\varepsilon_e^* h^*}{r^*} \quad (35.3)$$

Practically, this equation tells us that the pressure required to rupture the coating is directly proportional to the coating's modulus of elasticity E, the engineering strain ε_e (a measure of how much the coating has changed in size), and the thickness of the semipermeable coating. Also, the pressure required to rupture the coating is inversely related to the radius of the multiparticulate; that is, smaller multiparticulates are harder to rupture.

To summarize, from the preceding discussions of the formulation architecture and mechanism of release, critical product attributes include the size and distribution of seed cores, the uniformity of all coating layers, the potency of the drug and sweller layers, and the thickness and morphology of the delayed release layer. Now that the factors governing the release of the drug from the multiparticulates have been discussed, the next section covers understanding the fluid bed coating process both generally and specifically as it pertains to manufacturing this dosage form.

35.4 PROCESS

The primary process used to manufacture the rupturing multiparticulates is Wurster fluid bed coating. In this chapter, when we refer to fluid bed coating it is always Wurster fluid bed coating. To obtain the target performance and have a robust and well-characterized process, it is important to have a good understanding of the fluid bed coating unit operation.

A schematic of a bottom-spray fluid bed coater with Wurster column is shown in Figure 35.9. The fluid bed coater consists of an air distributor plate, a nozzle, a Wurster column, an expansion chamber, and a downbed. Atomizing air and fluidizing air flow from the bottom to the top of the coater.

The fluid bed coating process can be envisioned as controlled circulation of particles through the Wurster column, where coating is applied via coating solution droplets from a two-fluid atomizer. Hot drying gas is introduced at the bottom of the fluid bed where the design of the distributor

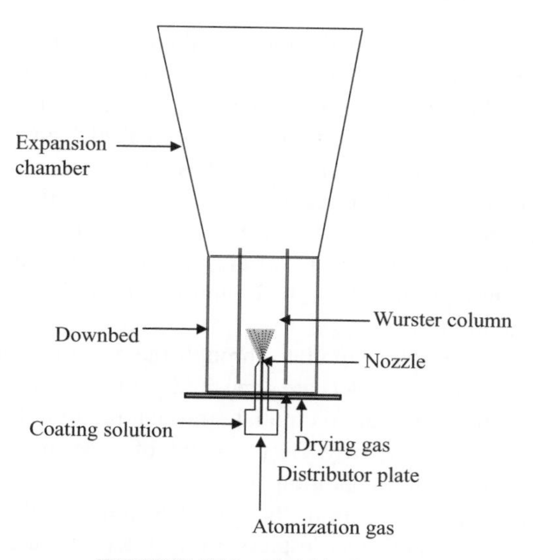

FIGURE 35.9 Fluid bed coater.

plate causes the majority of the gas to go up through the Wurster column. As a result, the velocity in the column is much higher than that in the downbed and there is a pressure difference between the column and downbed, causing particles to move from the downbed into the column. These entrained particles are transferred upward through the column where coating is deposited and partially dried. As the particles leave the column, they enter the expansion chamber where the particles continue to dry and the gas velocity drops (due to the equipment geometry) below the minimum entrainment velocity and the particles disengage from the gas stream and fall back to the downbed to start the cycle over again.

There are many subprocesses that occur in a Wurster fluid bed coating process. All of these subprocesses must work appropriately to successfully coat particles. However, in the context of this chapter, the main phenomena that need to be considered in the fluid bed coating process are (1) the frequency of coating and circulation of individual particles as they affect the uniformity of the coating and (2) coating deposition and drying as it directly affects the coating quality. The first phenomenon will be discussed qualitatively in this chapter, while the second phenomenon will be addressed in more detail.

It is important to have a qualitative understanding of the frequency of coating individual particles as it can have a direct impact on the coating uniformity. First of all, it is important to understand that when particles pass through the Wurster column, only a small percentage is actually coated on each pass [5, 6]. The consequence is that in order to uniformly coat all particles, the coating process needs to be long enough to ensure a sufficient number of passes through the spray zone for each particle. In some cases, the coating solution solids content may need to be adjusted to ensure a sufficiently long coating time.

Particle circulation through the fluid bed must also be consistent for all particles in order to achieve uniform coating. Preferential entrainment due to static or particle size, or uneven fluidization of the downbed, can lead to nonuniform coating across particles during the process. Practically, particle circulation can be assessed through visual observations of fluidization, and can be improved through use of a narrower particle size distribution of cores and/or through increasing bed humidity to minimize static accumulation.

As the particles circulate through the Wurster column, coating solution droplets are deposited onto the particles. As the particles continue to circulate through the fluid bed, the coating droplets then dry. The properties of these droplets and the environmental conditions such as temperature and solvent concentration within the fluid bed directly affect the drying rate of both the droplets prior to deposition and the deposited coating. The drying rate can have a significant impact on the coating morphology and therefore the drug release rate if it is a functional coating. At faster drying rates, the coating droplets contain less solvent and produce more porous coatings. These more porous coatings can be both more permeable and mechanically weaker than coatings applied under wetter conditions. Thus, it is important to choose an appropriate drying rate for a particular coating process and ensure that it is maintained throughout the process.

Droplet size can also have a significant effect on the coating properties, since it impacts the drying rate. There are four variables that affect the droplet size: atomization gas flow rate (commonly controlled by pressure), atomizer design, spray rate, and solution properties. From a practical standpoint, the last three variables are usually fixed for a given process and equipment train, leaving the atomization gas flow rate as the most common process variable. For most solutions, once there is sufficient atomizing gas flow to fully atomize the solution, there is generally a small effect of atomizing gas on droplet size (see Figure 35.10) [7].

The three process parameters that affect the droplet drying/coating formation conditions are drying gas flow rate, drying gas temperature, and spray rate. Practically, drying gas flow rate is constrained as it is coupled to particle circulation, leaving spray rate and inlet temperature as variables. It is often helpful to evaluate the effect of these parameters on the driving forces for heat and mass transfer, namely, the dependent variables of bed temperature and solvent concentration. Figure 35.11 shows such a plot for an aqueous coating system with constant bed temperature and humidity lines. Such a plot can be constructed for any solvent/drying gas coating system using mass and energy balances and is demonstrated later in this chapter.

In the upper left-hand corner of the plot, there is a low driving force for mass transfer and thus slower drying rates. Moving to the lower right-hand corner, the solvent concentration decreases and temperature increases, resulting in much faster drying rates and thus decreasing coating efficiency and increasing coating porosity.

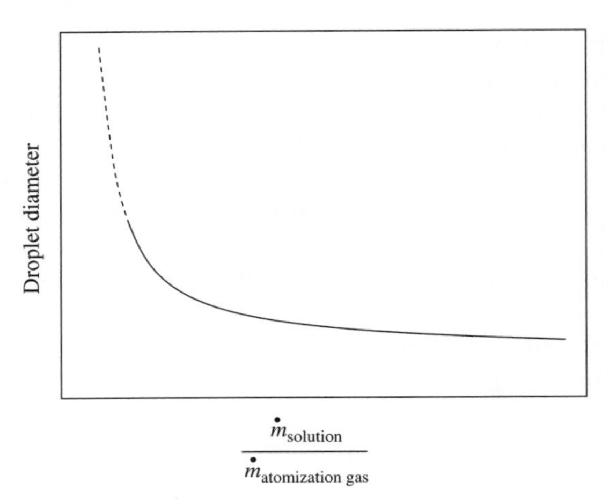

FIGURE 35.10 Droplet size as a function of $\dot{m}_{solution}/\dot{m}_{atomization\ gas}$.

This plot can be overlaid with the practical limitations of the coating process (some of which are determined experimentally) as shown in Figure 35.12 to define the acceptable processing space. This process map can be used as a guide to understand the effect of process variables on product properties within the process space.

35.4.1 Mass and Energy Balance

To understand the overall process on a macroscopic scale and draw the process plots, it is necessary to evaluate the overall mass and energy balance for the unit operation. The objective is to evaluate the overall system when operating at steady state with a control volume drawn around the entire system, as shown in Figure 35.13. Using the principles of the

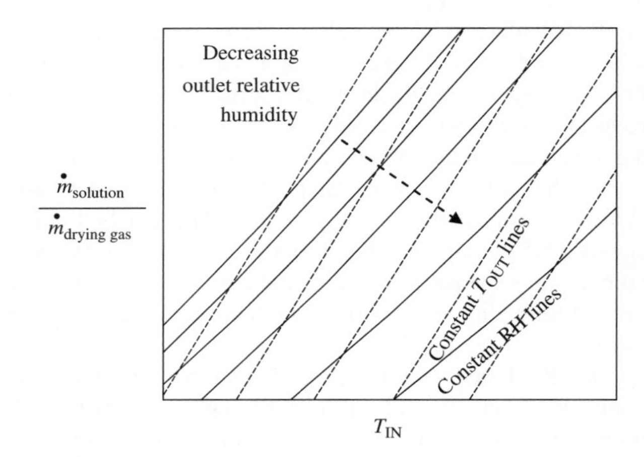

FIGURE 35.11 Outlet temperature and relative humidity as a function of $\dot{m}_{solution}/\dot{m}_{drying\ gas}$ versus T_{IN}.

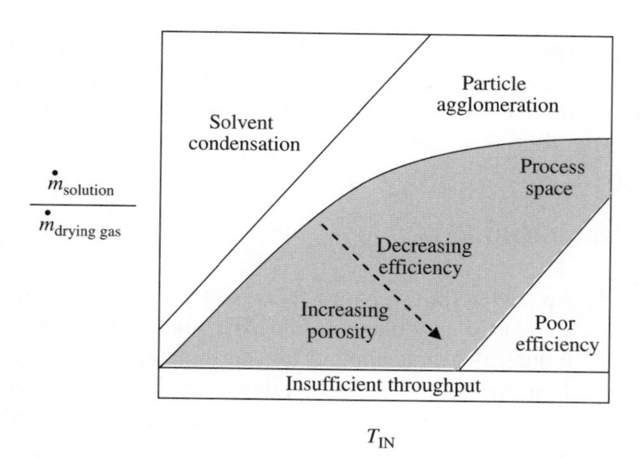

FIGURE 35.12 Conceptual process space.

conservation of mass and the first law of thermodynamics, the inputs of this control volume must be equal to the outputs plus any accumulation that occurs in the control volume.

The inputs into the control volume are

- mass flow rate, temperature, and solids content of the coating solution,
- mass flow rate, inlet temperature, and relative humidity of the drying gas, and
- mass flow rate, temperature, and relative humidity of the atomization gas.

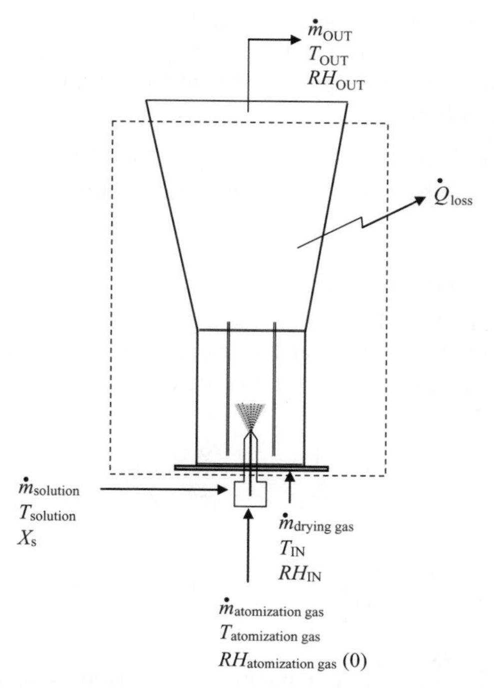

FIGURE 35.13 Control volume for fluid bed coater.

The outputs from the process are

- mass flow rate, outlet temperature, and relative humidity of the exiting gas stream, and
- the heat loss from the system.

If an energy balance on the system at steady state is done and it is assumed that potential and kinetic energy changes are negligible, the following equation can be written:

$$\dot{m}\,\Delta H = \dot{Q} + \dot{W} \qquad (35.4)$$

where $\dot{m}\,\Delta H$ is the change in enthalpy for the system, $\dot{Q}$ is the rate of heat flow into the system, and $\dot{W}$ is the rate of work done on the system.

It is assumed that there is negligible work done on the system and that the enthalpy contributions of the spray solids and atomizing gas are negligible. Therefore, only the enthalpy contributions of the drying gas and the solution are significant. Furthermore, it is assumed that the heat of vaporization is constant with respect to temperature and the spray solvent is completely evaporated. Under these conditions, the energy balance can be written as follows:

$$\dot{m}_{\text{dgas}} C_{p\ \text{dgas}} \Delta T_{\text{dgas}} + \dot{m}_{\text{solution}}(1-x_{\text{S}}) C_{p\ \text{solution}} \Delta T_{\text{solution}} + \dot{m}_{\text{solution}}(1-x_{\text{S}})\lambda_{\text{v}} = \dot{Q} \qquad (35.5)$$

where $\dot{m}$ is a mass flow rate, C_p is specific heat, x_{S} is the solid fraction in the coating solution, and λ_{v} is the enthalpy of vaporization.

This equation can be rearranged to solve for the drying gas outlet temperature to construct the lines in Figure 35.11. Similarly, the lines of constant humidity (or solvent) can be constructed using psychometric principles [8].

Now that the formulation, the mechanism of release, and the general fluid bed coating process have been discussed, we can use this information to understand critical process parameters for manufacture of rupturing multiparticulates.

35.4.2 Process Considerations Specific to Rupturing Multiparticulates

As was discussed in Section 35.3, critical product attributes for rupturing multiparticulates are particle size and distribution of the seed core, uniformity of the coating for all layers, and the coating morphology and thickness for the delayed release layer. Specific processing considerations for rupturing multiparticulates that impact the critical product attributes are discussed below.

35.4.2.1 Seed Core Many types of particles with a wide range of sizes can be coated in a fluid bed coater. In general, with smaller particles, minimizing static, ensuring uniform coating thickness particle to particle, and running the process with high efficiency is more challenging than for large particles. Large particles present their own set of process challenges such as maintaining good fluidization while minimizing attrition.

The key properties of the seed core that are important to achieving a uniform coating for all layers are the particle size and size distribution. The seed core size and distribution serves as the basis for the overall final multiparticulate size and distribution. As discussed in Section 35.3, this is important because the thickness of the coating on each particle affects when that multiparticulate bursts.

Based on this understanding, it is very important in cases where a sharp delayed release pulse is desired to start the process with as narrow a particle size distribution of seed cores as possible. Furthermore, it is critical to run the process to minimize the breadth of the size distribution for the final coated multiparticulates. In some cases, sieving the starting seed cores to narrow the distribution may be appropriate.

35.4.2.2 Uniformity of Coatings: Drug Layer and Sweller Layer The critical product attributes for the drug and sweller layer are uniformity of the coating and achieving target potency of either drug or sweller. To ensure uniformity of the coating, good atomization, sufficiently long process time, and good particle circulation are required during the process. To ensure that the potency of the drug and sweller is as desired, the coating efficiency should be high (i.e., efficiency as a function of process conditions should be understood and maximized).

Particle size can be a factor for materials applied as suspensions. A rule of thumb for coating multiparticulates is that particles in suspension should be at least one order of magnitude smaller in their longest dimension than the size of the core being coated. This aids in obtaining high coating efficiency.

35.4.2.3 Delayed Release Layer In addition to uniformity of the coating being critical for the delayed release layer, as we have seen in our discussion of mechanism of release, the morphology of this coating (e.g., porosity) is also critical to performance. As discussed, the drying rate of droplets has significant impact on coating porosity. Based on this, the acceptable process space for coating this layer is further constrained not only by efficiency but also by product performance. Practically, once the process map introduced in Section 35.4 is established for this process, it can be used to define a range of processing parameters that result in acceptable product performance.

35.5 SUMMARY

The key to developing a successful novel dosage form from a processing perspective is understanding the underlying mechanism of release of the dosage form, the unit operations used in the process, and how the key process variables impact the product properties. While this can be more complex with novel dosage forms, the principles remain the same for all dosage form development.

35.6 PROBLEMS

1. An ethylcellulose-based delay release coating, having an elastic modulus of 500 MPa and elongation at failure of 3%, is applied to a formulated core that is 1 mm in diameter. The coating is applied to achieve 25 wt% coating weight. Assuming the coating and formulated core have similar densities, and the final dosage form has a displaceable volume of 0.2 mm^3,
   (a) What is the hydrostatic pressure required to rupture the coating?
   (b) What is the volume of water required to rupture the coating?
2. Given the information from the previous problem, and assuming a mechanical permeability of the membrane of 5×10^{-7} cm^2/(atm h), a reflection coefficient of 1, and a constant osmotic pressure difference of 50 atm between the media and the core, what is the approximate time to hydrate the core?
3. An aqueous suspension of ethylcellulose, TEC, and talc at 20 wt% total solids is being coated onto formulated multiparticulates in a fluid bed coater. The drying gas is conditioned to an inlet temperature of 55°C and a measured dew point of 10°C. The air flow rate is 600 cfm and the solution spray rate is 15 kg/h. Assuming all the solvent is evaporated, the solution is at 20°C, and the rate of heat loss is approximately 20 kJ/min,
   (a) Estimate the temperature of the bed.
   (b) What is the relative humidity of the exhaust gas?

35.7 PROBLEM SOLUTIONS

1. (a) Based on the applied coating, first the coating thickness and the starting particle radius must be determined. Assume that $h \ll r$ and the coating and core are of similar density.

$$h = \frac{1}{3}\frac{\rho_{\text{bead}}}{\rho_{\text{coat}}}\left(\frac{X_{\text{coat}}}{1-X_{\text{coat}}}\right)r$$

$$h_0 = \frac{1}{3}(1)\left(\frac{0.25}{1-0.25}\right)\left(\frac{1.00\text{ mm}}{2}\right) = 0.056\text{ mm} = 56\ \mu\text{m}$$

$$r_0 = \left(\frac{1.00\text{ mm}}{2} + 0.056\text{ mm}\right) = 0.556\text{ mm} = 0.56\text{ mm}$$

Plug these values into the equation for the hydrostatic pressure difference term at the point of failure indicated below:

$$\Delta P^* = 2E\frac{h_0}{r_0}\frac{\varepsilon_e^*}{(1+\varepsilon_e^*)^3}$$

$$\Delta P^* = 2(500\text{ MPa})\left(\frac{0.056\text{ mm}}{0.556\text{ mm}}\right)\left(\frac{0.03}{(1.03)^3}\right)$$

$$= 2.8\text{ MPa} = 27\text{ atm}$$

(b) The amount of water is equal to the displaceable volume plus the change in volume of the expanding particle (based on its strain). Use the definition of strain to define what the volume of the particle will be at the point of coating failure indicated below:

$$V_{\text{bead}}^* = V_0(1+\varepsilon_e^*)^3$$

$$V_w^* = V_d + V_0\left[(1+\varepsilon_e^*)^3 - 1\right]$$

$$V_w^* = 0.2\text{ mm}^3 + \frac{4\pi}{3}(0.556\text{ mm})^3(1.03^3-1) = 0.27\text{ mm}^3$$

---

2. Use the differential equation for the volumetric flow rate of water through the semipermeable membrane (equation 35.1). By assuming that the coating thickness and surface area do not change during the time of hydrating the core, the equation can be simplified since the hydrostatic pressure difference is negligible, and integrated to result in the following relationship:

$$\int_0^{V_d} dV_w = \frac{4\pi r_0^2}{h_0} L_p\sigma\Delta\Pi \cdot \int_0^t dt$$

Solving for time,

$$t_{\text{hydrate}} = \frac{V_d h_0}{4\pi r_0^2 L_p \sigma \Delta\Pi}$$

$$t_{\text{hydrate}} = \frac{2\times 10^{-4}\text{ cm}^3(0.0056\text{ cm})}{4\pi(0.0556\text{ cm})^2(5\times 10^{-7}\text{ cm}^2/(\text{atm h}))(1)(50\text{ atm})} = 1.2\text{ h}$$

---

3. (a) Based on the defined inputs of inlet temperature and dew point for the drying gas, a psychrometric chart [9] can be used to define the properties of the drying gas, namely, the specific volume and humidity, determined to be 0.94 m^3/kg dry air (DA) and 0.0076 kg/kg DA, respectively. The heat capacity of water was taken to be 4.186 kJ/(kg °C), and enthalpy of vaporization 2390 kJ/kg (for 310 K). The specific heat for the moist air was interpolated to be 1.026 kJ/(kg °C) based on the individual values for air and water.

First determine the mass flow rates of both dry and moist air.

$$\dot{m}_{DA} = (600\text{ ft}^3/\text{min})(\text{m}^3/35.315\text{ ft}^3)(\text{kgDA}/0.94\text{ m}^3)$$
$$= 18.07\text{ kg/min}$$

$$\dot{m}_{\text{dgas}} = 18.07\text{ kg/min} + 18.07\text{kg DA/min}(0.0076\text{ kg/kg DA})$$
$$= 18.21\text{ kg/min}$$

Rearrange equation 35.5, and solve for the outlet temperature using units on a per minute basis (i.e., 15 kg/h is 0.25 kg/min).

$$T_{\text{OUT}} = \frac{\dot{m}_{\text{dgas}} C_{p\text{ dgas}} T_{\text{IN}} + \dot{m}_{\text{solution}}(1-x_S)C_{p\text{ solution}}T_{\text{solution}} + \dot{Q} - \dot{m}_{\text{solution}}(1-x_S)\lambda_v}{\dot{m}_{\text{dgas}} C_{p\text{ dgas}} + \dot{m}_{\text{solution}}(1-x_S)C_{p\text{ solution}}}$$

$$T_{\text{OUT}} = \frac{18.21(1.026)(55) + 0.25(1-0.20)(4.186)(20) + (-20) - 0.25(1-0.20)(2390)}{18.21(1.026) + 0.25(1-0.20)(4.186)} = 28°\text{C}$$

(b) Evaluate the mass balance for water in the system and solve for the outlet humidity assuming complete evaporation of solvent.

$$H_{\text{OUT}} = H_{\text{IN}} + \frac{\dot{m}_{\text{solution}}(1-x_S)}{\dot{m}_{\text{DA}}}$$

$$H_{\text{OUT}} = 0.0076 + \frac{0.25(1-0.20)}{18.07} = 0.0187\text{ kg/kg DA}$$

The psychometric chart can be used to interpolate the relative humidity of the exhaust gas at the calculated outlet temperature of 28°C, or alternatively the humidity can be converted into a partial pressure of water vapor, and Antoine's equation used to calculate the saturation pressure for the solvent at the calculated outlet temperature.

$$\text{RH}_{\text{OUT}} = 78\%$$

ACKNOWLEDGMENT

The authors wish to acknowledge Neurogen Corporation for allowing us to use adipiplon osmotic rupturing multiparticulate data in the presentation of the material in this chapter.

REFERENCES

1. Ueda, S., et al. *J. Drug Target.*, 1994;2:35–44.
2. Ueda, S., et al. *Chem. Pharm. Bull.* 1994;42(2):359–363.
3. Dashevsky, A., Mohamad, A. *Int. J. Pharm.* 2006;318:124–131.
4. Theeuwes, F. *J. Pharm. Sci.* 1975;64(12):1987–1991.
5. Cheng, X. X. Turton, R., *Pharm. Dev. Technol.*, 2000;5(3): 311–322.
6. Cheng, X. X. Turton, R., *Pharm. Dev. Technol.*, 2000;5(3): 323–332.
7. Masters K. *Spray Drying Handbook*, 4th edition, Wiley, New York, 1985, pp. 236–252.
8. McCabe W, Smith J, Harriott P. *Unit Operations of Chemical Engineering*, 6th edition, McGraw-Hill, New York, 2001, pp. 596–621.
9. Moyers C, Baldwin G. Psychrometry, evaporative cooling, and solids drying. In: Perry R, Green D, Maloney J, editors, *Perry's Chemical Engineers' Handbook*, 7th edition, McGraw-Hill, New York, 1997.

36

DESIGN OF SOLID DOSAGE FORMULATIONS

KEVIN J. BITTORF, TAPAN SANGHVI, AND JEFFREY P. KATSTRA
Formulation Development, Vertex Pharmaceuticals Inc., Cambridge, MA, USA

36.1 INTRODUCTION

The oral route is the most common way of administering drugs. It not only represents a convenient (self-administered) and safe way of drug administration but is also more profitable to manufacture than the parenteral dosage forms that must be administered, in most cases, by trained personnel. This is reflected by the fact that well over 80% of the drugs in the United States that are formulated to produce systematic effects are marketed as oral dosage forms. Among the oral dosage forms (Table 36.1), tablets of various different types are the most common because of their low cost of manufacture (including packaging and shipping), increased stability, and virtual temper resistance.

Following oral administration of tablets, the delivery of the drug to the systemic circulation requires initial transport through the gastrointestinal (GI) membrane. The drug absorption from the GI tract requires that the drug is brought into solution in the GI fluids and that it is capable of crossing the intestinal membrane into the systemic circulation; therefore, the rate of dissolution of the drug in the GI lumen can be a rate-limiting step in the absorption of drugs given orally. Particles of drugs, for example, insoluble crystalline forms or specific delivery systems such as liposomes, are generally found to be absorbed to a very small extent. The cascade of events from release of the drug from tablet, that is, disintegration of tablet into granules or aggregates followed by dissolution of the drug in the gut lumen, interactions and/or degradation within the lumen, and the absorption of the drug across the intestinal membrane into the systemic circulation, is schematically shown in Figure 36.1. The slowest of these events (dissolution and/or absorption) determines the rate of availability of the drug from the tablet formulation. Many factors in each step influence the rate and extent of availability of the drug. Physical, chemical, and biopharmaceutical properties of the drug, as well as the design and production of the tablet, play a very important role in its bioavailability after oral administration. These considerations make the seemingly simple tablet formulation approach complex to formulate in reality. These realizations have resulted in a change in philosophy of tablet formulation design in the last decade or more, wherein it is no longer considered an art but well-defined science.

The single greatest challenge to the tablet formulator is in the definition of the purpose of the formulation and the identification of the suitable materials to meet development objectives. A good formulation must not only be bioavailable but also be manufacturable, and chemically and physically stable from manufacturing through the end of shelf life. In addition, many quality standards and requirements must be met to ensure the efficacy and safety of the product.

All these formulation goals can be described as the target product profile (TPP). A TPP is a summary of characteristics that if achieved will provide optimal efficacy, patient compliance, and marketability. A TPP (Table 36.2) often includes attributes such as pharmacokinetic information (e.g., immediate release (IR) versus extended release (ER)), dosage form (e.g., tablet versus injectable), and shelf life information (e.g., 2 years at 25°C/60% relative humidity (RH)). There are also many other potential inputs for drug development that a formulator may or may not need, such as warnings, and precautions, adverse reactions, drug interactions, use in specific populations, drug abuse and dependence, clinical studies, and patient counseling information.

Chemical Engineering in the Pharmaceutical Industry: R&D to Manufacturing Edited by David J. am Ende

TABLE 36.1 Types of Solid Oral Dosage Forms

Type of Oral Dosage Form	Characteristics
Immediate release tablets	Disintegrate in stomach after taken orally
Delayed release tablets	Enteric-coated tablets to keep tablets intact in stomach and disintegrate in intestine for absorption
Sustained/controlled release tablets	Release drug slowly over a period of time to decrease the frequency of administration
Chewable tablets	Tablets are broken by chewing before swallowing with water
Orally disintegrating tablets	Disintegrate in oral cavity without drinking water to form a suspension for ease of swallowing
Hard gelatin capsules	Two-piece capsule shells filled with granules, powders, pellets, sprinkles, semisolids, oils
Soft gelatin capsules	One-piece capsule filled with oily liquid
Sachets	Single-dose unit bag containing granules

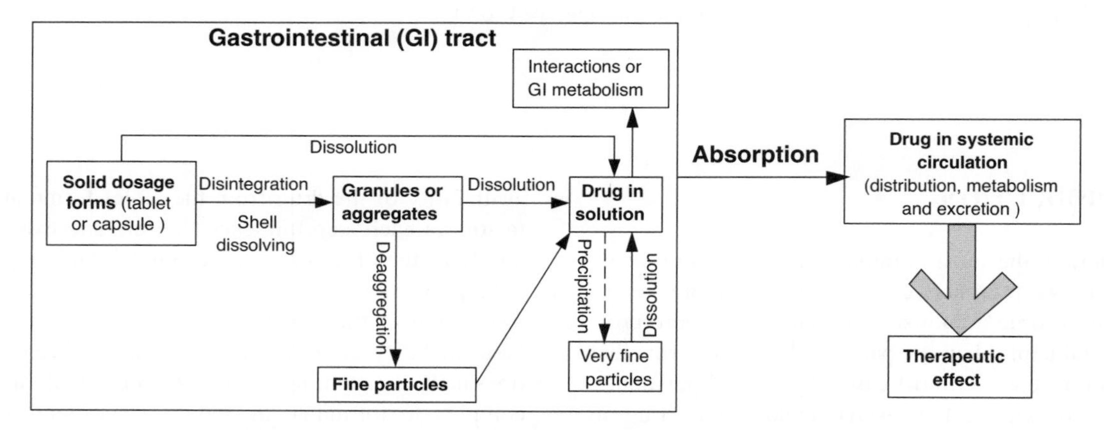

FIGURE 36.1 Fate of solid dosage form following oral administration.

It is important to establish the TPP so that the formulation effort can be effective and focused. When the formulation requirements are defined by the TPP, a strategy must be established to facilitate effective formulation development. To establish a formulation strategy, one must consider the physical, chemical, and biopharmaceutical characteristics of the drug; optimal technologies to achieve formulation goals; and the manufacturing capabilities to support the product.

This chapter examines tablet formulation design and development of an immediate release oral solid dosage form using a mix of pharmaceutical science, statistical, and engineering approaches. The chapter is aimed toward providing

TABLE 36.2 Typical Target Product Profile for an Immediate Release Tablet

Characteristic	How Used by a Formulator	Typical for IR Tablet
Indications and usage	Examine other products in the same class: examine improvements	Once a day (QD) Twice a day (BID) Three times a day (TID)
Dosage and administration	Good to know what is expected before one starts formulating	Oral tablet
Dosage forms and strengths	Multiple strengths may be needed depending on the population being targeted (adults versus children)	Dependent on drug, typically 10–500 mg
Overdosage	Useful if designing an extended release dosage, in which overdose (dose dumping) is a possibility	Dependent on drug
Description	This is up to the formulator and marketing: shape, size, and color of the tablet	A tablet with markings and color
Clinical pharmacology	Helps determine where the drug is absorbed and how fast the drug must get into solution	Dependent on drug
How supplied/stored/handled	Important as most people do not like refrigerated dosage forms	Two years room temperature shelf life

engineers an overview of the key physicochemical, mechanical, and biopharmaceutical properties of the drug and their influence on the selection of formulation process platform. Subsequently, critical tablet characteristics that affect the stability and bioavailability of the drug product are discussed. Finally, strategy for tablet process optimization and scale-up is defined to select proper equipment and to define operational design space. A systematic scientific approach to tablet formulation and process development along with practical examples is discussed to expedite the drug product development.

36.2 UNDERSTANDING DRUG SUBSTANCE

Integration of physicochemical, mechanical, and biopharmaceutical properties of a drug candidate is a prerequisite in developing a robust and bioavailable drug product that has optimal therapeutic efficacy. The measurement of physical, mechanical, and chemical properties not only helps guide the selection of dosage form but also provides an insight into their processability and storage to ensure optimal drug product quality. Figure 36.2 lists the critical physicochemical, mechanical, and biopharmaceutical properties that need to be understood to aid in design of tablet formulation.

36.2.1 Physicochemical Properties

Prior to the development of tablet dosage form, it is essential to determine certain fundamental physical and chemical properties of the drug molecule along with other derived properties. This information dictates many of the subsequent approaches in tablet formulation development and is known as preformulation. It should be kept in mind that many of these properties are dependent on the solid form, and complete characterization of each of the most relevant solid forms is needed to provide a complete physicochemical picture.

36.2.1.1 Solubility and Drug Dissolution Solubility of a drug candidate may be the critical factor in determining its usefulness, since aqueous solubility dictates the amount of compound that dissolves, and therefore, the amount available for absorption. A compound with low aqueous solubility could be subject to dissolution rate-limited absorption within the GI residence time.

Dissolution is the dynamic process by which a material is dissolved in a solvent that is characterized by a rate (amount dissolved per time unit), while *solubility* is the amount of material dissolved per unit volume of a certain solvent that is characterized by a concentration. Solubility is often used as a short form for "saturation solubility," which is the

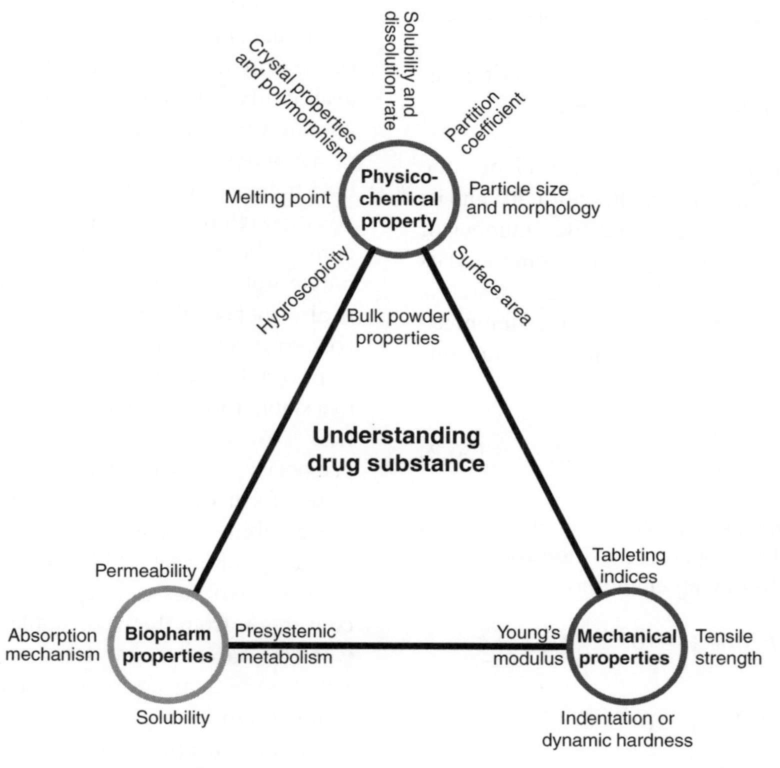

FIGURE 36.2 Understanding drug substance properties.

maximum amount of drug dissolved at equilibrium conditions. Finally, *intrinsic solubility* is the solubility of the neutral form of an ionizable drug.

Dissolution rate is directly proportional to the aqueous solubility, C_s, and the surface area, A, of drug exposed to the dissolution medium. It is a common, when developing an immediate release dosage form of poorly soluble drug, to increase drug dissolution rate by increasing the surface area of a drug through particle size reduction.

The dissolution rate of a solute from a solution is described by the Noyes–Whitney equation as follows [1]:

$$\frac{dC}{dt} = \left(\frac{D \times A}{h}\right) \times (C_s - C_t) \qquad (36.1)$$

where D is the diffusion coefficient of the drug substance in a stagnant water layer around each drug particle with a thickness h, A is the drug particle surface area, C_s is the saturation solubility, and C_t is the drug concentration in the bulk solution at a given time.

The dissolution rate, rather than the saturation solubility, is most often the primary determinant in the absorption process of a sparingly low soluble drug. Determining the dissolution rate is critical. The main areas for dissolution rate studies are evaluations of different solid forms of a drug (e.g., salts, solvates, polymorphs, amorphous, stereoisomers) or different particle sizes of the drug. The dissolution rate can be determined either for a constant surface area of the drug in a rotating disk apparatus [2] or as a dispersed powder in a beaker with agitation (as detailed in pharmacopoeias such as U.S. Pharmacopoeia, etc.).

The impact of solubility and dissolution rate on formulation selection is discussed later in the chapter.

36.2.1.2 Partition Coefficient Partition coefficient is the relationship between chemical structure, lipophilicity, and its disposition *in vivo* and has been reviewed by a number of authors [3]. The lipophilicity of an organic compound is described in terms of a partition coefficient $\log P$, which is defined as the ratio of the concentration of the unionized compound, at equilibrium, between organic and aqueous phases:

$$\log P = \frac{[A]_{organic}}{[A]_{aqueous}} \qquad (36.2)$$

For ionizable drugs, the ionized species does not partition into the organic phase, and the apparent partition coefficient, D, is calculated from the following equations:

$$\text{Acids}: \log D = \log P - \log[1 + 10^{(pH - pK_a)}] \qquad (36.3)$$

$$\text{Bases}: \log D = \log P - \log[1 + 10^{(pK_a - pH)}] \qquad (36.4)$$

pK_a is the dissociation constant.

Since it is virtually impossible to determine $\log P$ in a realistic biological medium, the octanol/water system has been widely adopted as a model of the lipid phase [4]. There has been much debate about the suitability of this system [5], but it remains the most widely used in pharmaceutical studies.

Generally, compounds with $\log P$ values between 3 and 6 show good passive absorption, whereas those with $\log P$'s of less than 3 or greater than 6 often have poor passive transport characteristics. The role of $\log P$ in absorption processes occurring after oral administration has been discussed by Navia and Chaturvedi [6].

36.2.1.3 Crystal Properties and Polymorphism Most drug substances appear in more than one polymorphic form. Polymorphs differ in molecular packing (crystal structure), but share the same chemical composition [7]. Hydrates or solvates are often called "pseudopolymorphs" because in addition to containing the same given drug molecule, they also contain molecules of solvents that are incorporated into the crystal lattice. Amorphous forms are characterized by absence of long-range order.

Polymorphism has a profound implication on formulation development and biopharmaceutical properties because polymorphs may exhibit significantly different solubility, dissolution rate, compactibility, hygroscopicity, physical stability, and chemical stability [7]. Figure 36.3 provides a detailed list of physical properties that can differ among the polymorphs.

Higher solubility and faster dissolution rates of the metastable polymorph may lead to significantly better oral bioavailability. Chloramphenicol palmitate [9] (bacteriostatic antimicrobial) and ampicillin [10] (antibiotic) are examples of the anhydrous form that gave higher blood serum levels than the less soluble trihydrate form.

Although use of a faster dissolving polymorph may have clinical benefit, it is important to keep in mind that a polymorph with a higher solubility or faster dissolution rate is also metastable (i.e., a higher energy form) and tends to convert to a thermodynamically more stable form over time or in certain conditions. Conversion from a metastable form to a stable form could lower a drug's oral bioavailability and lead to inconsistent product quality. From a formulating perspective, it is desirable to use the thermodynamically stable form of the API; however, biopharmaceutical and processability considerations may dictate the deliberate selections of a metastable form for processing.

It is important to keep in mind that polymorphic form conversion from the most stable form may still occur, even when a stable crystal form is chosen for development. Polymorphic transformations can take place during pharmaceutical processing, such as particle size reduction, wet granulation, drying, and even during the compaction process and compression process [11, 12], as each of these processes

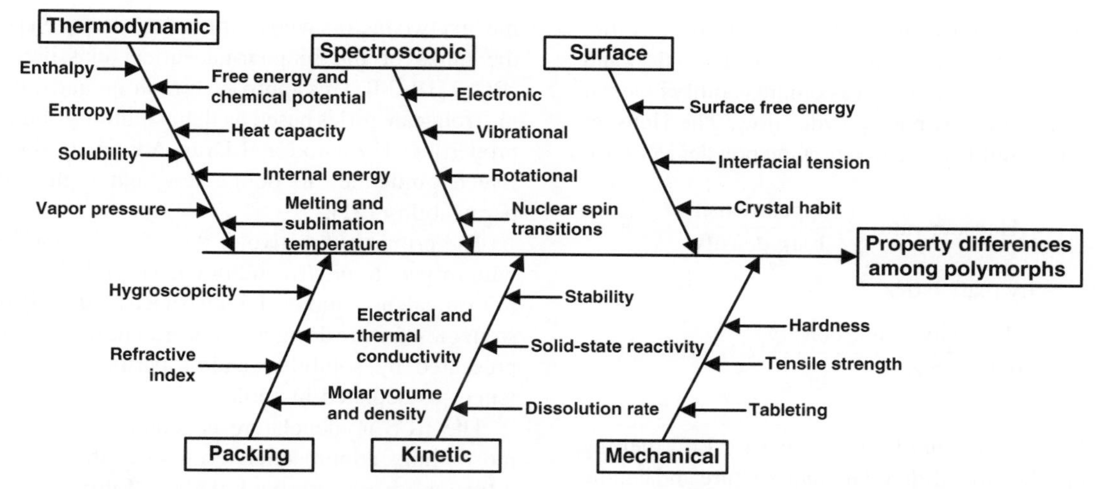

FIGURE 36.3 Fishbone schematic of physical properties differences among polymorphs (adapted from Ref. 8).

may add the energy required to move the drug to the unstable form.

36.2.1.4 Particle Size, Particle Morphology, and Surface Area Bulk flow, compactability, formulation homogeneity, and surface area control dissolution and chemical reactivity, which are directly affected by size, shape, and surface morphology of the drug/API (active pharmaceutical ingredient).

Spherical particles have the least contact surface area and exhibit good flow, whereas acicular particles tend to have poor flow [13]. Milling of long acicular (or needle) crystals can enhance flow properties; however, excessively small particles tend to be cohesive and aggravate flow problems.

In addition to the flow properties, crystal shape and size has been demonstrated to impact mixing and tabletability. L-Lysine monohydrate with plate-shaped crystals exhibited greater tabletability than the prism-shaped crystals [14]. Kaerger et al. [15] studied the effect of paracetamol particle size and shape on the compactibility of binary mixture with microcrystalline cellulose (MCC), showing that compressibility increased with particle size and irregular crystals whereas compactibility increased with decrease in particle size.

Particle size affects drug content uniformity (CU). For low dose direct compression formulations, where drug content uniformity is of particular concern, the particle size of the drug substance has to be small enough to meet the U.S. Pharmacopoeia requirement on content uniformity [16]. For example, Zhang and Johnson [17] showed that low dose blends containing a larger drug particle size (18.5 μm) failed to meet the USP requirement, whereas a blend containing smaller particle sizes (6.5 μm) passed.

Surface areas of drug particles are important because dissolution is a function of this parameter (as predicted by the Noyes–Whitney equation (36.1)). This is particularly true in those cases where the drug is poorly soluble. Such drugs are likely to exhibit dissolution rate-limited absorption. For such drugs, particle size reduction (e.g., micronization) is often utilized to increase the surface area that enhances the dissolution rate; for example, micronization enhanced the bioavailability of felodipine when administered as an extended release tablet [18].

Methods to determine particle size and shape include light diffraction, scanning electron microscopy (SEM), sieve analysis, and various electronic sensing zone particle counters. Methods available for surface area measurement include air permeability and various gas adsorption techniques.

36.2.1.5 Bulk Powder Properties Density and porosity are two important pharmaceutical properties that are derived from the information on particle size, particle shape, and surface area. A comparison of true particle density, apparent particle density, and bulk density can provide information on total porosity, interparticle porosity, and intraparticle porosity. Generally, porous granules dissolve faster than dense granules, since pores allow water to penetrate more readily.

$$\text{Interparticle (interspace) porosity} = 1 - \frac{\text{bulk density}}{\text{apparent particle density}} \tag{36.5}$$

$$\text{Intraparticle porosity} = 1 - \frac{\text{apparent particle density}}{\text{true particle density}} \tag{36.6}$$

$$\text{Total porosity} = 1 - \frac{\text{bulk density}}{\text{true particle density}} \tag{36.7}$$

The increase in bulk density of a powder is related to the cohesivity of a powder. Bulk density and tapped density are used to calculate compressibility index and Hausner ratio, which are measures of the propensity of a powder to flow and to be compressed. *A rule of thumb*: a compressibility index of higher than 30% indicates poor powder flow. The Hausner ratio varies from about 1.2 for a free flowing powder to 1.6 for cohesive powders.

$$\text{Hausner ratio} = \frac{\text{tapped density}}{\text{bulk density}} \quad (36.8)$$

$$\text{Compressibility (carr index)} = \frac{100 \times (\text{tapped density} - \text{bulk density})}{\text{bulk density}} \quad (36.9)$$

36.2.1.6 *Melting Point and Hygroscopicity*

Low melting materials tend to be more difficult to manufacture and handle in conventional solid dosage forms. *A rule of thumb*: melting points below 60°C are considered to be problematic. Temperatures in conventional manufacturing equipment, such as fluid bed dryers and tablet presses, can exceed 50°C. During the milling process, hot spots in the milling chamber may have much higher temperatures.

Moisture uptake is a concern for pharmaceutical powders and is known to affect a wide range of properties, such as powder flow, compactibility, and stability. On the other hand, moisture may improve powder flow and uniformity of the bulk density, as well as appropriate amount of moisture may act as a binder to aid compaction. Thus, knowledge of the type and level of moisture is critical for understanding its impact not only on deformation behavior but also on the attributes of the final product.

36.2.2 Biopharmaceutical Properties[a]

Complete oral absorption occurs when the drug has a maximum permeability coefficient and maximum solubility at the site of absorption, which results in rapid and uniform pharmacological response. Based on this premise, a key objective in designing a rational oral dosage form is having sound understanding of multiple factors, including physicochemical properties of the drug and dosage form components, and physiological aspects of GI tract.

Generating formulations with relevant oral bioavailability depends on a number of factors including solubility, permeability,[b] and metabolic stability.[c] Absorbability is related to the first two factors whose importance has been recognized in the guise of the biopharmaceutical classification system (BCS) [19, 20]. This approach bins drugs and drug candidates into four categories based on their solubility and permeability properties. The Food and Drug Administration (FDA) has issued guidelines to define low and high solubility and permeability [21].

The primary objective of the BCS[d] is to guide decisions with respect to *in vivo* and *in vitro* correlations and need for bioequivalence studies; it is also used to identify dosage form strategies that are designed at overcoming absorption barriers presented by solubility and/or permeability related challenges as depicted in Table 36.3.

The BCS nomenclature is centered on the premise that most orally administered drugs are absorbed via passive diffusion[e] process through the small intestine and excludes other important factors such as the drug absorption mechanism (carrier-mediated,[f] P-glycoprotein efflux,[g] etc.) and presystemic degradation or complexation that may enhance or limit oral bioavailability.

36.2.3 Mechanical Properties

Material mechanical properties play a role in manufacturing drug product. Particle properties influence the true areas of contact between particles and can affect unit operations, such as compression, milling, and granulation. Characterization of mechanical properties of drug substance is important in three areas: choosing a processing method, such as granulation or direct compression; selecting excipients with properties that mask the poor properties of the drug; and helping to document what went wrong, that is, when a tableting process is being scaled up or when a new bulk drug process is being tested. Since all these can influence the quality of the final product, it is to the formulator's advantage to quantify and understand the importance of the mechanical properties of the active and inactive ingredients and their combinations.

Pharmaceutical materials are elastic, plastic, viscoelastic, hard, tough, or brittle in the same sense that metals, plastics, or wood have similar properties. The same concepts that

[a] Biopharmaceutics is defined as the study of the relationships between physicochemical properties, dosage forms, and routes of administration of drugs and its effect on the rate and extent of absorption in the living body.
[b] Permeability determines the ability of drug to move across the lipophilic intestinal membrane in gastrointestinal tract (GIT). Permeability of a drug may be predicted using computational (*in silico*) models or measured using both physicochemical and biological methods (*in vitro*, *in situ*, or *in vivo*).
[c] Metabolic stability refers to ability of a drug to withstand metabolism or degradation in the gut wall and the liver.

[d] BCS (Biopharmaceutics Classification System) is a guidance for predicting the intestinal drug absorption using solubility and permeability as defined by the U.S. Food and Drug Administration.
[e] Passive diffusion is a transport process, wherein drug molecules pass across the lipoidal intestinal membrane from a region of higher concentration in the lumen (GIT) to a region of lower concentration in the blood (systemic circulation). Mathematically, it is described by Fick's first law of diffusion.
[f] Carrier-mediated transport may be subdivided into active transport and facilitated diffusion or transport. Active transport is a process whereby drug is bound to a carrier or membrane transporter and is transported against the concentration gradient across a cell membrane. Facilitated diffusion differs from active transport in that it cannot transport a substance against a concentration gradient of that substance.
[g] P-glycoprotein is one of the key countertransport efflux proteins that expel specific drugs back into the lumen of the GIT after they have been absorbed.

TABLE 36.3 Dosage Form Options Based on Biopharmaceutical Classification System[a]

Class I: high solubility, high permeability	Class II: low solubility, high permeability
Class I: high solubility, high permeability • No major challenges for immediate release dosage form • Controlled release dosage forms may be needed to limit rapid absorption	Class II: low solubility, high permeability • Formulation are designed to overcome solubility • Salt formation • Precipitation inhibitors • Metastable forms • Solid dispersions • Lipid technologies • Particle size reduction
Class III: high solubility, low permeability • Prodrugs • Permeation enhancers • Ion pairing • Bioadhesives	Class IV: low solubility, low permeability • Formulation would have to use a combination of the approaches identified in class II and III • Strategies for oral administration are not really viable. Often use alternative delivery methods such as intravenous administration

[a]A drug is considered to be highly soluble when the highest dose is soluble in 250 mL or less of aqueous media over the pH range 1–8. A drug is considered to be highly permeable when the extent of absorption in humans is expected to be greater than 90% of the administered dose.

materials/mechanical engineers use to explain/characterize tensile, compressive, or shear strength are relevant to pharmaceutical materials. A number of characterization tools as outlined in Table 36.4 are available for understanding the mechanical property of the material.

Based upon the analysis of the physicochemical, mechanical, and biopharmaceutical properties of the drug substance, selection of excipients and the formulation process is performed. The next section discusses excipients, their types, and the selection procedure based upon their effect on the drug substance properties.

36.3 EXCIPIENTS

Excipients facilitate formulation design to perform a wide range of functions to obtain desired properties for the finished drug product. Historically, pharmaceutical excipients have been regarded as inert additives, but this is no longer the case. Each additive must have a clear justification for inclusion in the formulation and must perform a defined function in the presence of the active and any other excipients included in the formulation. Excipients may function, for example, as an antimicrobial preservative, a solubility enhancer, a stability enhancer, or a taste masker, to name a few.

Excipients are selected based on their chemical/physical compatibility with drugs, regulatory acceptance, and processability. First, excipients shall be chemically compatible with drug substances. Second, at the time of globalization, excipients are to meet the requirements of not only the FDA or EMEA but also the regulatory agencies of other potential marketing countries. Third, excipients impact the properties of a powder mixture, such as flowability, density, compactibility, and adhesiveness. For example, different fillers are selected carefully to balance the plasticity, elasticity, and brittleness of the precompaction powder mixture, in order to make large-scale production feasible.

For tablets, excipients are needed both for the facilitation of the tableting process (e.g., glidants) and for the formulation (e.g., disintegrants). Except for diluents, which may be present in large quantity, the level of excipient use is usually limited to only a few percent and some lubricants are required at $<1\%$. Details of the types, uses, and mechanisms of action of various excipients for tablet production have been discussed at length in multiple of articles and books. The types and functions of excipients for tablet production are summarized in Table 36.5.

It is worth noting that some of these tableting excipients may exert effects in opposition to each other. For example, binders and lubricants, because of their respective bonding and waterproofing properties, may hinder the disintegration action of the disintegrants. In addition, some of these tableting excipients may possess more than one function that may be similar (e.g., talc as lubricant and glidant) or opposite (e.g., starch as binder and disintegrant) to each other.

Furthermore, the sequence of adding the excipients during tablet production depends on the function of the excipient. Whereas the diluents and the binders are to be mixed with the active ingredient early on for making granules, disintegrants may be added before granulation (i.e., inside the granules) and/or during the lubrication step (i.e., outside the granules) before tablet compression.

36.4 DRUG–EXCIPIENT COMPATIBILITY STUDIES

Excipient compatibility testing provides a preliminary evaluation of the physical and chemical interactions that can occur. Testing is carried under stressed temperature and humidity conditions between a drug and potential excipients. This helps excipient selection, particularly for tablet formulations in order to minimize unexpected formulation stability problems during product development.

Traditionally, a binary mixture of drug with the excipient being investigated is intimately mixed, and the ratio of drug to excipient is often 1:1; however, other mixtures may also be investigated. These blends are stored at various

TABLE 36.4 Characterization Tools for Understanding Mechanical Properties of Materials

	Quasistatic Testing	Dynamic Testing
API required	1–100 g	2–10 g
Advantages	"Independently" dissect out and investigate various mechanical properties	Understand the mechanics of materials at speeds representative of production tablet compaction
Limitations	Cannot determine properties at representative production scales	Difficult to factor out the individual mechanical property "component"
Characterization tests	Tensile strength • Describes the global strength of the material • Measured using traditional tablet hardness tester [22] or transverse compression in tensile tester [23] • Typical desired value greater than 1 MPa Indentation/dynamic hardness • Describes the "local" plasticity of the material • Measured using pendulum impact device or free falling indenter [24] Young's modulus • Describes stiffness and toughness of the material • Measured using both four- and three-point beam bending, flexure testing [25] Tableting indices • Dimensionless numbers that integrate above described tests Bonding index (BI) • Defines the tendency of the material to remain intact after compression • Desired value >0.01 Brittle fracture index (BFI) • Measure of brittleness of a material • BFI = 1 represents very brittle material and BFI < 0.3 is relatively nonbrittle material Strain index (SI) • Indirect measure of elastic strain	Force–displacement profiles • Indicator of tablet-forming ability of powder • Assessment of the elastic properties • Thermodynamic analysis of the process of compact formation Tablet volume–applied pressure profiles • Measured using hydraulic press, rotary press, compaction simulator, and compaction emulator Heckel equation • Tablet porosity–applied pressure function

temperatures and humidity and analyzed for potential degradation products.

More recently, the use of a model formulation approach to excipient screening has become much more widespread across the industry. Model formulations include commonly used excipients in each functional category such as fillers, binders, disintegrants, and lubricants and those with different chemical structures, namely, celluloses, starches, and sugars. Both wet and dry model formulations may be prepared for stability testing. It is recommended that a DOE be used to assist in the development and interpretation of results for these types of studies. Table 36.6 contains an example of the model formulation approach. It lists excipients and their approximate composition that would be found in a typical tablet formulation.

Powders are physically mixed and may be granulated or compacted to accelerate any possible interaction. Samples may be exposed in open pans or scaled in bottles/vials to mimic product packaging. The storage conditions used widely vary in terms of temperature and humidity, but a temperature of 40°C for storage of compatibility samples is considered appropriate. Some compounds may require higher temperatures to make reactions proceed at a rate that is measured over a convenient period. Methods of analysis also vary widely, ranging from thermal techniques (DSC) to chromatographic techniques (TLC, HPLC) to microcalorimetry.

An example of an excipient compatibility study utilizing partial factorial design (2^{7-3}) is illustrated in Table 36.7. In this study, a model compound (BCS class II) is blended with excipients (shown in Table 36.6) to make 16 formulations and stationed on open dish stability at 25°C/60% RH, 40°C dry, and 40°C/75% RH. The study duration is 3 months, which is analyzed for physical and chemical stability.

Figure 36.4 shows a regression model that is defined for assessing the effect of formulation and time on degradation

TABLE 36.5 Types and Functions of Tableting Excipients

Excipient	Function	Some Examples of Excipients
Diluents	Act as bulking/filling material	Sugars, lactose, mannitol, sorbitol, sucrose, calcium salts, microcrystalline celluloses (MCC),
Binders and adhesives	Holds powder together	Sugars, glucose, polymers, starch, gelatin,
Disintegrants	To facilitate the breakup of the tablet in the gastrointestinal tract	Croscarmellose sodium (CCS), sodium starch glycolate (SSG), crospovidone
Glidants	Improve the flow of granules, needed for compression	Silica, magnesium stearate (MgSt), talc
Lubricants	Reduce friction between granules and the compression equipment	Magnesium stearate, stearic acid, talc, sodium lauryl sulfate (SLS),
Antiadherents	To minimize the problems if sticking to the tablet punch head	Talc, cornstarch, SLS, MgSt
Colorants	For identification and marketing	Natural pigments and synthetic dyes
Flavors and sweeteners	To improve the taste of chewable tablets	Mannitol, aspartame

TABLE 36.6 Typical Excipients Selected for a Model Formulation Study

Excipient Type	% Composition	Level 1	Level 2
API	10	–	–
Filler 1	38–40	MCC	Mannitol
Filler 2	38–40	Dicalcium phosphate (ATAB)	Spray-dried lactose
Surfactant	0–4	None	Sodium lauryl sulfate (SLS)
Binder	4	Polyvinylpyrrolidone (PVP)	Hydroxy propyl cellulose (HPC)
Disintegrant	5	Sodium starch glycolate (SSG)	Croscarmellose sodium (CCS)
Lubricant	1	Magnesium stearate (MgSt)	Sodium stearyl fumarate (SSF)
Wet granulation	20% w/w water	No	Yes

TABLE 36.7 Formulation Composition for Excipient Compatibility Study

Number	Formulation Composition							
	10%	38–40%	38–40%	0–4%	4%	1%	5%	20% w/w water
1	API	MCC	ATab	None	PVP	MgSt	SSG	Dry (no)
2	API	MCC	ATab	None	HPC	MgSt	CCS	Wet
3	API	MCC	ATab	SLS	PVP	SSF	CCS	Wet
4	API	MCC	ATab	SLS	HPC	SSF	SSG	Dry (no)
5	API	MCC	Lactose	None	PVP	SSF	CCS	Dry (no)
6	API	MCC	Lactose	None	HPC	SSF	SSG	Wet
7	API	MCC	Lactose	SLS	PVP	MgSt	SSG	Wet
8	API	MCC	Lactose	SLS	HPC	MgSt	CCS	Dry (no)
9	API	Mannitol	ATab	None	PVP	SSF	SSG	Wet
10	API	Mannitol	ATab	None	HPC	SSF	CCS	Dry (no)
11	API	Mannitol	ATab	SLS	PVP	MgSt	CCS	Dry (no)
12	API	Mannitol	ATab	SLS	HPC	MgSt	SSG	Wet
13	API	Mannitol	Lactose	None	PVP	MgSt	CCS	Wet
14	API	Mannitol	Lactose	None	HPC	MgSt	SSG	Dry (no)
15	API	Mannitol	Lactose	SLS	PVP	SSF	SSG	Dry (no)
16	API	Mannitol	Lactose	SLS	HPC	SSF	CCS	Wet

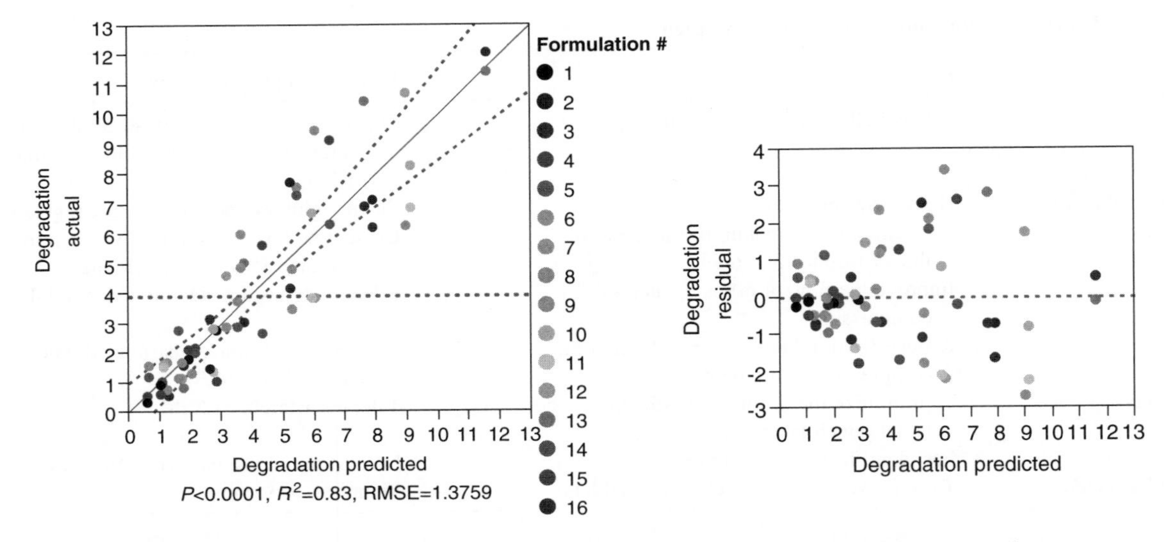

FIGURE 36.4 Degradation actual versus predicted plot and degradation residuals versus degradation predicted plot. The residuals are evenly distributed, indicating that there is no bias in the model. The symbol represents the formulation described in Table 36.7.

growth at storage condition of 40°C/75% RH. A regression analysis is completed for data at 40°C/75% RH to determine which excipient affects the growth of degradation products. From the analysis (Table 36.8), it is found that time, filler 1, disintegrant, and granulation have effect on degradation, as well there are some interactions between time and filler 1, time and disintegrant, and time and granulation (borderline as p value ≈ 0.05).

The parameters analyzed that did not show significance were filler 2, surfactant, binder, and lubricant and were subsequently removed from the model during stepwise regression.

The prediction profiler and the interaction profiles (Figure 36.5) provide information on the specific excipient within a significant class (from Table 36.7) and the sensitivity of each of the variables on the degradation growth. As seen from the prediction profiler, within filler, mannitol causes more degradation compared to MCC. Similarly, SSG is better than CCS among disintegrant and dry blend is better than wet granulation as the latter causes more degradation.

These results suggest that both mannitol and CCS could be detrimental to the stability of the API and are not being assessed for formulation development. Also, wet granulation is to be avoided to increase the shelf life.

36.5 PROCESSING OF FORMULATIONS

The properties of a drug substance dictate the design of formulation composition and the choice of formulation processing platform technology. The most commonly used processing platforms for solid oral dosage form include direct compression and granulation (wet and dry).

Direct compression is the term used to define the process where powder blends of the drug substance and excipients are compressed on a tablet machine. There is no mechanical treatment of the powder apart from a mixing process.

Granulation is a generic term for particle enlargement, whereby powders are formed into permanent aggregates. The purpose of granulating tablet formulations is to improve

TABLE 36.8 Regression Results from the Excipient Compatibility Experiments

Term	Estimate	Std Error	t Ratio	Prob > |t|
Intercept	−0.029267	0.338094	−0.09	0.9313
Time (months)	2.4060773	0.179126	13.43	<.0001*
Filler 1[mannitol]	0.9593669	0.171987	5.58	<.0001*
Disintegrant [CCS]	0.6043247	0.171987	3.51	0.0009*
Granulation [Dry]	−0.717924	0.171987	−4.17	0.0001*
(Time (months) − 1.625) × filler 1 [Mannitol]	0.640884	0.179126	3.58	0.0007*
(Time (months) − 1.625) × Disintegrant [CCS]	0.5068487	0.179126	2.83	0.0065*
(Time (months) − 1.625) × Granulation [Dry]	−0.358106	0.179126	−2.00	0.0505

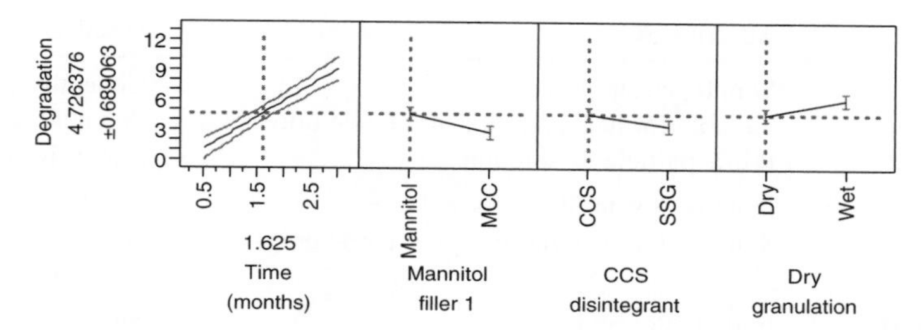

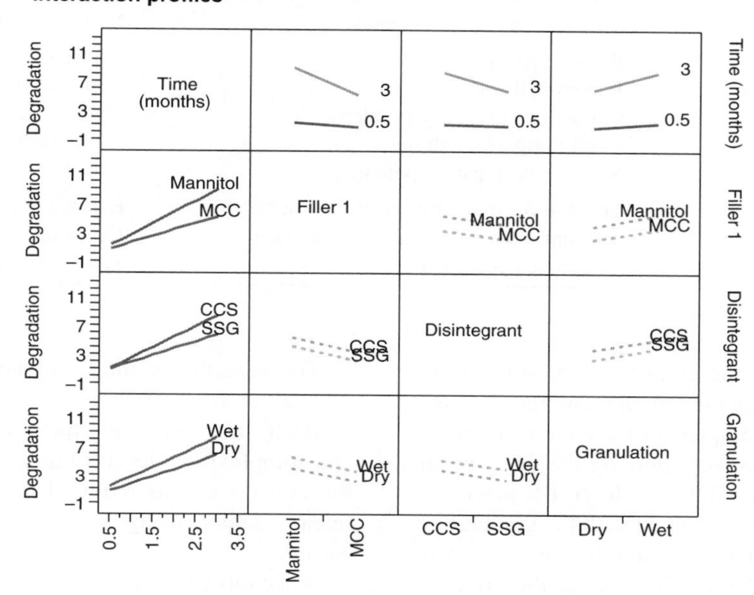

FIGURE 36.5 Prediction profiler and interaction profiles for drug–excipient compatibility studies.

the flow and compaction properties. A number of methods are used to achieve agglomeration or aggregation; these are normally classified as either wet granulation, where a liquid is used to aid the agglomeration process, or dry granulation, where no liquid is used.

36.5.1 Dry Granulation

In the dry methods of granulation, the primary powder particles are aggregated under high pressure. There are two main processes. Either a large tablet (known as a "slug") is produced in a heavy-duty tableting press (a process known as "slugging") or the powder is squeezed between two rollers to produce a sheet of material ("roller compaction"). In both cases, these intermediate products are broken using a suitable milling technique to produce granular material that is usually sieved to separate the desired size fraction. The unused fine material may be reworked to avoid waste. This dry method may be used for drugs that do not compress well after wet granulation, or those that are sensitive to moisture.

36.5.2 Wet Granulation

Wet granulation involves the massing of a mix of dry primary powder particles using a granulating fluid. The fluid contains a solvent that must be volatile so that it is removed by drying. Typical liquids include water, ethanol, and isopropanol, either alone or in combination. The granulation liquid may be used alone or, more usually, as a solvent containing a dissolved *adhesive* (also referred to as a *binder* or *binding agent*) that is used to ensure particle adhesion once the granule is dry.

The three main methods of producing pharmaceutical granulates are low shear granulation, high shear granulation, and fluid bed granulation. Low shear mixers encompass machines such as Z-blade mixers and planetary mixers that, as their name suggests, impart relatively low shear stresses onto the granulate.

High shear granulators are closed vessels that normally have two agitators: an impeller that normally covers the diameter of the mixing vessel and a small chopper positioned perpendicular to the impeller. The powders are dry mixed

TABLE 36.9 Processing Platforms: Advantages and Disadvantages

Processing Platform	Advantages	Disadvantages
Direct compression	Simple, cheap process Suitable for heat and moisture labile drugs Prime particle dissolution	Generally limited to low dose compounds Potential to segregation Expensive excipients
Dry granulation (slugging)	Imparts flowability to formulation Suitable for heat and moisture labile drugs	Dusty process Not suitable for all compounds Slow process
Dry granulation (roller compaction)	Imparts flowability Suitable for heat and moisture labile drugs Limits segregation tendency	Slow process Loss of compactibility for tableting No hydrophilization of surfaces
Wet granulation (aqueous)	Robust process Improves flowability Can reduce elasticity problems Can improve wettability Reduces segregation potential	Expensive Specialized equipment Stability concerns for moisture sensitive, thermolabile, and metastable drugs with aqueous granulation
Wet granulation (nonaqueous)	Suitable for moisture-sensitive drugs Vacuum drying techniques can reduce/ remove need for heat	Expensive equipment Explosion proof Solvent recovery

using the impeller, and then the granulating fluid is added. Wet massing takes place using the impeller and the chopper, and granulation is usually completed in a span of minutes.

Fluid bed granulation involves spraying the dry powder with a granulating fluid inside a fluid bed dryer. The powder is fluidized in heated air and then sprayed with the granulating fluid. When all the granulating liquid has been added, the fluidization of the powder continues until the granules are dry.

Seager et al. [26] produced a detailed analysis of the influence of manufacturing method on the tableting performance of paracetamol granulated with hydrolyzed gelatin. The main difference in the granules produced by different methods is their final density: high shear mixers producing denser granules than low shear granulators that in turn produced denser granules than fluid bed granulations. Disintegration times were greater for tablets produced from the denser granulates. A detailed description of granulation process development and scale-up is found in the literature [27].

The advantages and disadvantages of each process are detailed in Table 36.9.

Each processing platform has unique characteristics and complexity in terms of unit operations. Table 36.10 lists the unit operations required for manufacturing immediate release tablet using the processing platform discussed earlier.

Since more than one platform technology may be used to manufacture a drug product, selection of the most appropriate processing platform is affected by many factors as shown in Figure 36.6.

36.6 TABLET FORMULATION DESIGN

Having decided on a formulation design strategy, the process of preparing and screening initial formulation possibilities begins. It is important to appreciate that the goal is to develop a "robust" formulation, and this objective facilitates identification of the factors that influence the selection of a design

TABLE 36.10 Unit Operations Required for Various Processing Platforms

Unit Operation	Direct Compression	Dry Granulation	Wet Granulation
Raw materials (weighing and sieving)	✓	✓	✓
Blending	✓	✓	✓
Compaction		✓	
Wet granulation			✓
Wet screening			✓
Drying			✓
Milling		✓	✓
Tablet compression	✓	✓	✓

process as depicted in Figure 36.6. The first major design criterion is the nature of the API and in particular the possible dosage level (described in preformulation report and TPP). The knowledge of biopharmaceutical class to which the API belongs helps in deciding the formulation rationale. In particular, the implications of low permeability and low solubility must be carefully considered prior to the selection of the processing platform. For example, a poorly soluble drug often tends to be poorly wettable, too. If the objective is to obtain a fast dissolving and dispersing dosage form, inclusion of a wetting agent such as sodium lauryl sulfate or polysorbate 80 may be appropriate or even necessary.

Processing methods may also significantly impact dosage form performance. For example, it may not be appropriate to wet granulate amorphous drug because water may lower the glass transition temperature and facilitate recrystallization during or after processing. In other situations, wet granulation can be used to avoid potential segregation and content uniformity problems where there is a significant difference in particle size or bulk density between the drug and excipients.

Another major consideration must be the anticipated dosage level. It is worth emphasizing that in the case of a high dose active form, a major proportion of the processing difficulties are traced to the physicochemical and mechanical properties of the API. Unfortunately, the key properties of the API may change during scale-up of the synthetic API process, or from lot to lot when outsourced. It follows that continuous monitoring of critical quality attributes (CQAs) of API that affect the process is an essential policy. Figure 36.7 depicts a decision-guiding flowchart for selection of the processing platform.

36.7 TABLET CHARACTERISTICS

There are two important classes of tablet characteristics. The first set examines the tablet immediately after manufacturing and the second class examines what happens to the tablet over time.

Immediately after manufacturing and during the formulation process of a tablet, the release of the tablet is of utmost importance. If the tablet does not disintegrate or dissolve in the body, then the efficacious effect desired is likely not going to occor. There are many factors that can affect this, from excipient choice to manufacturing.

After manufacturing a tablet must maintain consistency over time. Similarly to drug release, excipients and processing can affect the shelf life of a tablet.

36.7.1 Release Profile: Factors That Affect *In Vivo* Performance

Release profile of a tablet can affect *in vivo* drug performance, and as this is the case, it is important to measure this characteristic during development. The FDA guidance, Dissolution of Immediate Release Solid Oral Dosage Forms, states the dissolution requirements for an immediate release drug. Dissolution testing is useful in development to determine how processing and formulations can potentially affect *in vivo* performance. What is a dissolution test?

> Dissolution is a test that provides some assurance of tablet performance by an indication of the mass transfer the drug into solution.

There are many stages in the development of a dissolution method. The final quality control (QC) form of the method is

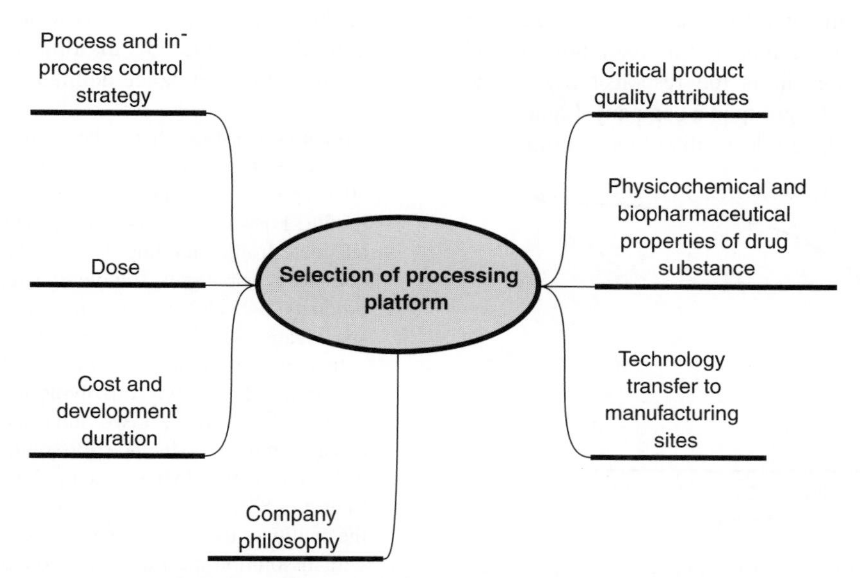

FIGURE 36.6 Factors affecting selection of processing platform.

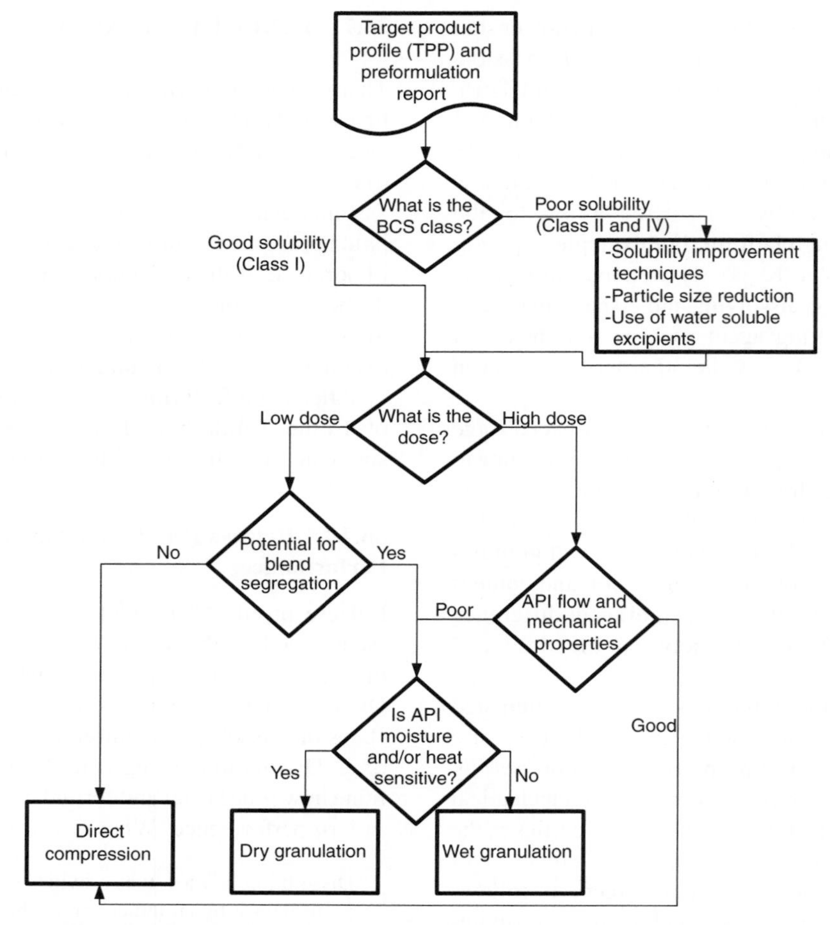

FIGURE 36.7 Flowchart for selection of adequate processing platform.

used in day-to-day production to ensure consistency of the tablets produced. In early development, dissolution testing is useful in screening formulations, but this dissolution test may not be or even resemble the final QC test used when the drug has been approved. The development of a dissolution method at each stage of development is the responsibility of the analytical development (AD) group in a company. Figure 36.8 shows a "typical" immediate release dissolution profile.

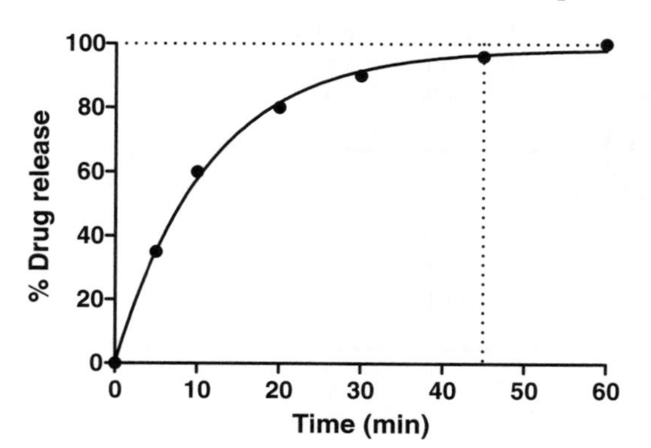

FIGURE 36.8 Typical drug release profile; very fast initial release with a leveling off.

Even though it is generally the responsibility of AD group to develop the dissolution method, it is critical for the drug developer to understand the final QC requirements from a regulation perspective so as to aid in developing a final drug product. A final QC dissolution method is carried out according to the guidance defined as:

> Dissolution testing should be carried out under mild test conditions, basket method at 50/100 rpm or paddle method at 50/75 rpm, at 15-minute intervals, to generate a dissolution profile. For rapidly dissolving products, generation of an adequate profile sampling at 5- or 10-minute intervals may be necessary. For highly soluble and rapidly dissolving drug products (BCS classes 1 and 3), a single-point dissolution test specification of NLT 85% (Q = 80%) in 60 minutes or less is sufficient as a routine quality control test for batch-to-batch uniformity. For slowly dissolving or poorly water soluble drugs (BCS class 2), a two-point dissolution specification, one at 15 minutes to include a dissolution range (a dissolution window) and the other at a later point (30, 45, or 60 minutes) to ensure 85% dissolution, is recommended to characterize the quality of the product. The product is expected to comply with dissolution specifications throughout its shelf life. If the dissolution characteristics of the drug product change with

> time, whether or not the specifications should be altered will depend on demonstrating bioequivalence of the changed product to the original biobatch or pivotal batch. To ensure continuous batch-to-batch equivalence of the product after scale-up and post approval changes in the marketplace, dissolution profiles should remain comparable to those of the approved biobatch or pivotal clinical trial batch(es) [28].

This is important knowledge to ensure compliance when developing and changing formulations. The QC method described above is not always the best method to use during development to assess potential impact on bioavailability; alternate media or methods may provide additional insight.

36.7.2 Problems and Troubleshooting Dissolution Testing

Beyond compliance, dissolution is used to determine performance of the tablet. Assuming a well-developed dissolution method, there are many things that can affect the dissolution of the tablet:

- Processing conditions: compressing the tablet too hard and/or overblending the lubricant
- Excipients: choice and amount
- API physical properties
- Storage: over time the tablet dissolution may slow down due to excipient interactions with the drug and excipient reaction with each other

A discriminating dissolution method is useful in developing a tablet formula and manufacturing process; however, a proper method may take time for the AD group to develop, just as it takes a while to develop a reliable process.

36.7.2.1 Problems with Dissolution: Nonengineered Mixing Vessels and Troubleshooting Assuming a good dissolution method may not be the best assumption. Dissolution is a QC test required for regulatory compliance; however, there are many problems with the dissolution test.

Dissolution apparatus 1 (Figure 36.9) is a paddle mixer in a cylindrical vessel; from an engineering standpoint, this does not provide good mixing. If an engineer is designing this, he/she would have put a baffle or two in there to promote top to bottom mixing. As is imagined, there may be problems with bottom settling and coning with tablets that disintegrate into large particles having a high density. In this case, the dissolution results have significant variation as how the drug settles and the percentage of the drug setting has an effect on the results.

Apparatus 2 is a basket mixer in a cylindrical vessel; again from an engineering standpoint, this does not provide good mixing. There is little mixing power associated with the method; if the powder flows out of the basket, the powder settles, floats, or suspends depending on the buoyancy of powders. If the powder stays in the basket, the method has a high probability to be reliable [29].

FIGURE 36.9 Typical dissolution apparatuses.

When examining dissolution results method, there are five considerations to determine if results are method biased.

1. What is the media used in the dissolution bath? What is the solubility of the drug in the media? This determines the mass transfer driving force for the drug to go into solution.
2. Does my drug change forms in the dissolution media? If it does, the form it changes into may not have the same solubility. Form conversion is a stochastic event and affects the consistency of the results.
3. Are the particles suspended and flowing? This also affects the mass transfer of the drug into the media.
4. Is the tablet submerged in the media? Often a floating tablet provides many problems and inconsistent results.
5. What is the dissolution medium comprised of? The media may react with the API or excipients used in the tablet.

When analyzing a change in dissolution profiles ensure that the changes made are due to the process and formula versus problems with the method. It is always a good idea to observe the dissolution testing so as to see what is actually occurring.

36.8 USING DISSOLUTION TO DETERMINE CQAs

Assuming an acceptable dissolution method has been developed, dissolution is a useful tool to determine CQAs for the

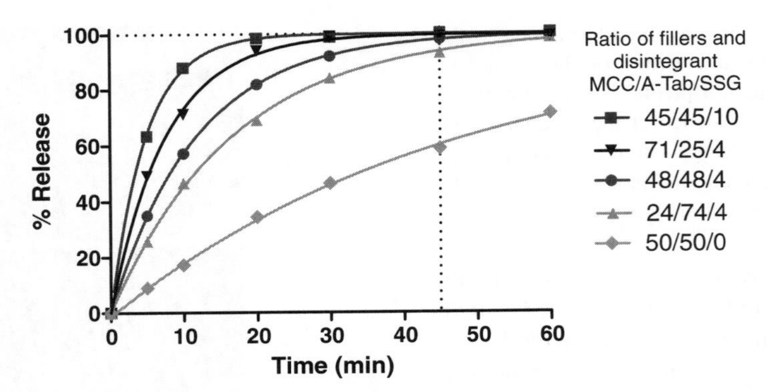

FIGURE 36.10 Dissolution comparison between different excipient ratios and a constant drug load of 20%.

tablet. Dissolution can help determine the maximum tablet hardness, the optimal drug substance particle size and/or density, and the proper ratio or the amount of excipients.

36.8.1 Using Dissolution to Determine the Ratio of Excipients

A tablet formulation can affect the dissolution profile. A tablet often contains a mixture of water-soluble and insoluble fillers/binders and disintegrants that all have the potential to affect the dissolution profile. Determining the optimal loading of excipients is a difficult task even after the compatible excipients have been chosen.

Let's examine excipient optimization of a BCS class II tablet based on dissolution performance. For example, compressing a tablet consisting of 20% API with a particle size of 29 μm at a hardness of ~10 kP. The remaining 80% of the tablet consisting different ratios of filler, binder, and disintegrant. Two commonly used fillers MCC and calcium dibasic phosphate (A-Tab), and a commonly used disintegrant SSG, are used based on excipient compatability example. These are in five different compression ratios, and dissolution results are shown in Figure 36.10.

As is seen in the Figure 36.10, different excipient ratios can affect tablet performance. From this example, it looks like 71/25/4 MCC/A-Tab/SSG has most optimal performance without putting an excess amount of disintegrant in the tablet (Table 36.11).

36.8.2 Using Dissolution to Determine the Optimal API Particle Size and Tablet Hardness

The next properties that can affect dissolution are API particle size and tablet hardness. API particle size has the potential to affect dissolution based on different surface area or particle morphology and the tablet hardness can affect how fast the tablet disintegrates into primary particles

TABLE 36.11 Percent Release Data from Figure 36.10

	Release Data of Tablets with Different MCC/A-Tab/SSG Ratios				
Time (min)	48/48/4	45/45/10	25/71/4	71/25/4	50/50/0
0	0	0	0	0	0
5	35.0	63.4	25.8	49.1	8.8
10	57.1	87.9	46.5	71.3	17.1
20	81.8	98.5	69.0	94.0	34.2
30	91.9	98.8	84.0	99.1	46.2
45	98.1	100.3	93.0	99.1	58.6
60	99.6	100.4	98.6	99.1	71.3

Note: Due to method variability, it is common to see tablet performance slightly above or below 100%.

enabling the API to dissolve. As a rule of thumb about particle size:

> There is never an instance where bigger particles will improve the immediate release performance but there are many instances where it will not change the performance.

In optimizing the release of the drug, first a target CQA must be defined, which is determined from IVIVC[h] or good scientific reasoning. A hypothetical CQA could be NLT (not less than) 70% release at 30 min to ensure proper absorbance in the body; 30 min is chosen as it is the approximate gastric emptying time of an empty stomach [30].

Continuing with the example, for determining the optimal hardness and API particle size range, dissolution is chosen at the CQA at $t = 30$ min. Starting with the "optimal" formulation from the example (71/25/4 MCC/A-Tab/SSG), the material is compressed at five hardnesses, ranging from ~10 to 30 kP, and four different API average particle sizes (d_{50}), ranging from 29 to 73 μm. Table 36.12 indicates the

[h] IVIVC: *In vitro–in vivo* correlation, by which benchtop data accurately correlate to human bioavailability.

TABLE 36.12 Effect of API Particle Size μ and Tablet Hardness (kP) at the 30 min Dissolution Time Point

%Dissolved	d_{50}	Hardness	%Dissolved	d_{50}	Hardness (kP)
99.1	29	9.7	87.18	50	10.3
85.17	29	15.6	75.91	50	14.4
77.45	29	20.8	65.03	50	20.4
64.3	29	25.6	52.42	50	24.7
54.15	29	30.2	40	50	30.2
89.04	42	10.7	72.65	73	9.6
84.35	42	14.7	60.13	73	15.3
69.17	42	20.9	50.84	73	20.9
61.31	42	24.1	43.62	73	25.4
46.61	42	29.4	26.97	73	29.6

results attained, and from observation there is an effect of both hardness and API particle size.

On examining Figure 36.11, it is found that the data have a linear relationship between % release and hardness; moreover, there is a relationship between release and particle size. It is noted that 8 of 20 experiments met the CQA requirement of NLT 70% release at 30 min. From this point, a design equation is developed to mathematically describe the design space.

Regression is completed providing an expression for the relationship of acceptable hardness and API particle size combinations. The expression is used to describe the design space (Tables 36.13 and 36.14).

Based on this information, the relationship between hardness, particle size, and % release at 30 min is

$$\%R_{@30\ \text{min}} = 139.2 - 0.59 \times d_{50} - 2.25 \times \text{hardness} \quad (36.10)$$

This is not an ideal form of the equation as >100% release is predicted at some values; however, it is used to determine the maximum range of hardness and particle size to attain release >80%. The model is further developed to attain the curvature but more data above 30 kP and smaller particle sizes are required. Determining tablet and API properties there is sufficient information for control.

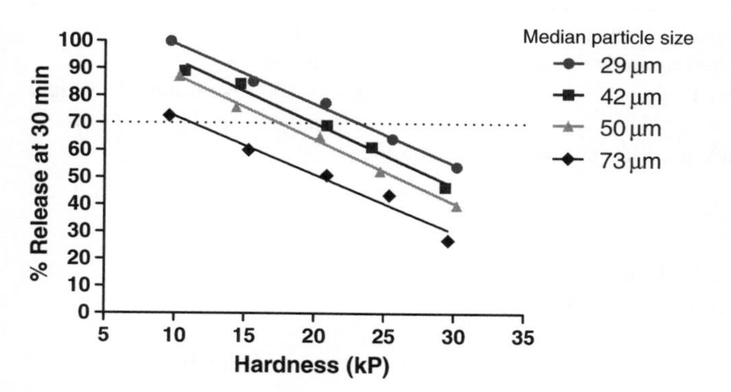

FIGURE 36.11 Comparison of hardness and API particle size to dissolution release at the 30 min time point.

TABLE 36.13 Results from the Linear Regression

Term	Estimate	Std Error	t Ratio	Prob > |t|
Intercept	139.2	1.99	69.9	<0.0001*
d_{50}	−0.592	0.0289	−20.5	<0.0001*
Hardness	−2.247	0.0659	−34.0	<0.0001*

The asterisk indicates that the variable is significant.

TABLE 36.14 Summary of Fit of the Regression

R^2	0.989
R^2 adj	0.988
Root mean square error	2.070
Mean of response	65.31
Observations	20

To determine the acceptable combinations of hardness and particle size to maintain the CQA of NLT 70% release at 30 min, equation 36.10 is rearranged.

$$69.3 \geq 0.59 \times d_{50} + 2.25 \times \text{hardness} \quad (36.11)$$

As long as this equation is satisfied, the CQA is maintained. The design space is described in Figure 36.12.

The last check is examining the residuals to ensure there is no systematic error. Shown in Figure 36.13 are the randomly distributed data, indicating the regression does not have a systematic error. Another way is to confirm that the model residuals are normally distributed by using a goodness of fit.

36.8.3 Physical Tablet Characteristics

The physical attributes of the tablet are important for processing and ensuring that a consistent quality drug product is delivered to the customer. Physical attributes include tablet hardness, thickness, friability, disintegration, and weight.

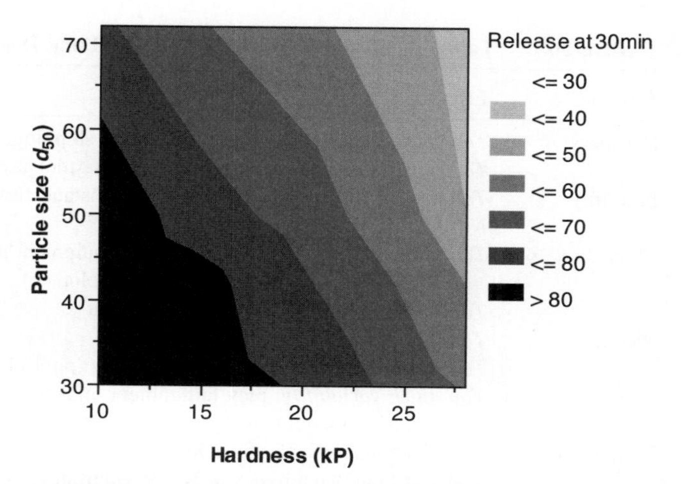

FIGURE 36.12 Contour plot showing dissolution as a function of particle size and hardness.

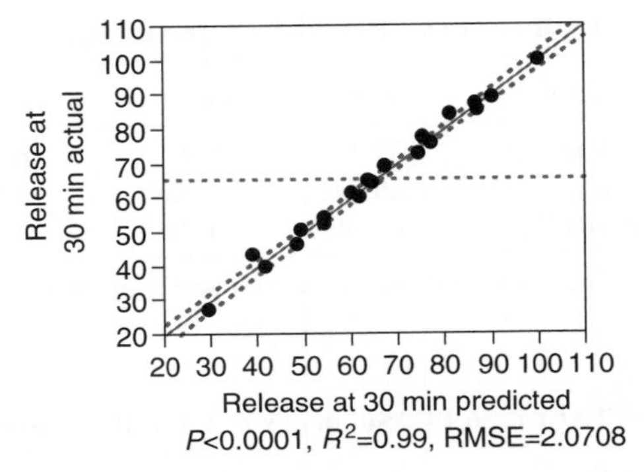

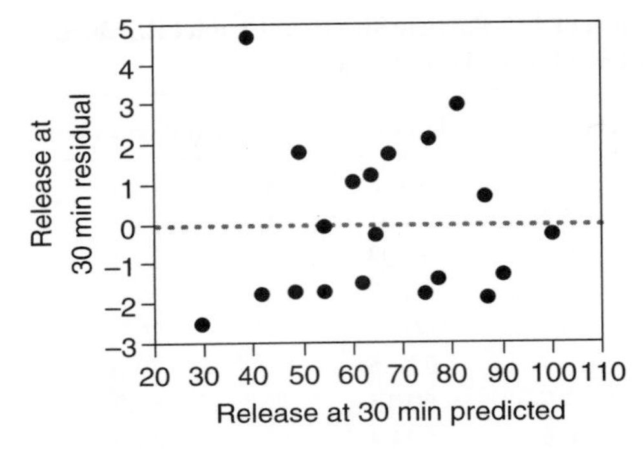

FIGURE 36.13 Residual plot from the regression. The data are randomly distributed and nonsystematic.

When determining tablet characteristics, consider how the material is to be handled after compression (Table 36.15).

36.9 DRUG PRODUCT STABILITY

Stability is critical in all drug product/formulation design; without stability, there is no commercial product. Stability is examined in two different manners: meeting the minimum regulatory requirement needed to launch a drug, and/or examining the root causes of degradation. From a scientific/engineering perspective, it is important to determine what affects the stability of the drug to design a drug product process around these factors.

36.9.1 Regulatory Requirements for Drug Product Stability

Regulatory requirements for stability need to be done in the intended primary commercial package. There is ICH guidance that governs the expectations of pivotal stability studies—see "Stability Testing of New Drug Substances and Products Q1A(R2)." The minimum required for submission is shown in Table 36.16; real-time data are needed for shelf life dating over 2 years.

This is an excellent guide for the regulatory requirements once a package(s) has been chosen for clinical trial, registration, and commercial distribution. Registration batch minimum is three lots of at least 100,000 tablets and at least 1/10 of the expected commercial batch size that is packaged into the intended commercial package. But there is much work required before selecting the primary package.

36.9.2 What Affects Stability and How to Predict Shelf life?

In a QbD world, the minimum is generally not sufficient to launch a product—the more a scientist determines what affects stability, the better engineered is the product.

Packaging is usually not known in early development and it can range from blister packaging to bottles to pouches. Each packaging type can vary significantly in the materials used. Different materials can protect from light, moisture,

TABLE 36.15 Tablet Attributes and Their Effect on Final Dosage Form

Attribute	Effect	Measurement
Hardness	*Too soft*: the tablet can break in storage shipping, coating, packaging *Too hard*: the tablet cannot dissolve and may not have the required clinical effect	Hardness tester Typical units: kP, N
Friability	*Too friable*: tablet cannot be able to withstand further testing *Not friable*: nothing wrong	*Friabilator*: 100 revolutions; % weight loss; if capping/lamination occurs
Thickness	*Too thick*: may not fit into packaging equipment/package. May not be able to swallow (poor marketing compliance) *Too thin*: may clog packaging equipment	Caliper
Weight	*Too heavy*: the drug may be overpotent *Too light*: there may not be enough drug (poor clinical efficacy) *Too much variability*: may fail content uniformity. Too much yield loss during manufacturing	Scale
Disintegration	*Too slow*: may not be efficacious *Too fast*: may have issues in humid environments and coating	Disintegration, dissolution bath
Elegance	*Nonelegant*: shows inconsistent production and may turn off customers	Acceptable quality limit (AQL)

TABLE 36.16 Minimum Guideline from the ICH Q1A(R2) for Room Temperature Product

Study	Storage condition	Minimum time period covered by data at submission
Long term[a]	25°C ± 2°C/60% RH ± 5% RH or 30°C ± 2°C/65% RH ± 5% RH	12 months
Intermediate[b]	30°C ± 2°C/65% RH ± 5% RH	6 months
Accelerated	40°C ± 2°C/75% RH ± 5% RH	6 months

[a] It is up to the applicant to decide whether long-term stability studies are performed at 25 ± 2°C/60% RH ± 5% RH or 30°C ± 2°C/65% RH ± 5% RH.
[b] If 30°C ± 2°C/65% RH ± 5% RH is the long-term condition, there is no intermediate condition.

oxygen, and other environmental factors. Early research on the effect of heat, moisture, oxygen, and light enables primary and secondary packaging selection. Though these studies are comprehensive, studies still need to be completed with the primary package to ensure that no reaction occurred between the packaging material and the drug product. The primary packaging may have leachables, extractables, or antistatic properties that may react with the drug product.

A structured approach helps determine the conditions under which a drug converts into a degradation product. This occurs given there is enough time in certain temperature, relative humidity, or light conditions. Packaging is often used to prevent such occurrence; the proper choice and storage conditions are critical depending on the stability of the product.

Determining package type is as easy as answering a few questions: what is the drug product sensitive to—temperature, moisture, light? Is the drug sensitive to impurities/components in the packaging, impurities/components in the excipients, or starting impurities in the drug substance?

This section provides a framework to determine what affects stability and how. The simplest experiments are placing the product at open dish conditions, and an example of this is provided in the excipient compatibility section, though a more integrated set of experiments is used to create a predictive model on how the tablet can degrade. From the initial readout of stability, a more extensive experimentation is completed to model the stability of the drug.

36.9.3 Open Dish Experiments

These are the easiest experiments to get a quick read on how the drug product can degrade and what changes in formulation affect degradation. Using conditions of 40°C dry and 40/75% RH provides immediate (i.e., 1–4 weeks) information of how the drug reacts with both temperature and humidity and the degradants or form change to expect upon stability.

Open dish experiments are used to test specification of excipients. Possible effects on stability can occur from changing excipient vendors or lot-to-lot variation within a vendor. For example, MCC, a commonly used binder, has often a residue on ignition (ROI[i]) specification of not more than (NMT) 0.050%, so it is possible to receive material with ROI of 0.040%, 0.005%, and 0.020%. A tablet is compressed with these different lots of MCC and placed on open dish stability and depending on the drug, the results could affect the stability. Figures 36.14 and 36.15 show the results from this example.

Both Figures 36.14 and 36.15 show a relationship between ROI and impurity growth. A regression analysis is completed and shown in Table 36.17. The regression shows that both time and ROI affect stability, but relative humidity (RH) does not affect stability.

Another manner to examine the data is plotting the slopes from Figures 36.14 and 36.15 (degradation rate) against ROI of the MCC. Figure 36.16 shows how the ROI affects the growth rate of impurities; this could be important and may provide justification in setting excipient specifications.

Open dish studies are useful in determining what can degrade the drug, but these are harsh conditions and do not simulate what would happen upon shelf life. However, they do give an indication of what to look for on stability.

36.9.4 Modeling and Predicting Shelf Life.

Using information gained from open dish studies, a more elegant study is then conducted to determine the drug shelf life. Experiment on the effect of temperature and tablet moisture on impurity growth is used to develop a model to predict shelf life. This type of study is called the TRH study that models the effects temperature (*T*) and RH have on tablet shelf life.

Setting up this study requires tablets, RH equilibration chambers, foil pouches, and a heat sealer for the pouches. Tablets are equilibrated at different RH conditions, and then packaged in foil pouches to ensure the moisture content of the tablet remains constant throughout the time material is on stability. As well, every time point and condition should be individually packaged to maintain the tablet moisture, as opening and closing packages could adulterate the tablets. The idea is to equilibrate separate tablets to a minimum of three different groups of RH (i.e., 15%, 25%, and 45%); equilibration may take up to 2–7 days. The last three steps are as follows:

1. Measure the tablet moisture content (Karl Fisher (KF) is one of the more effective measurements) and

[i] The ROI test measures the amount of residual substance not volatilized from a sample when ignited in the presence of sulfuric acid. The test determines the content of inorganic impurities. USP <281>.

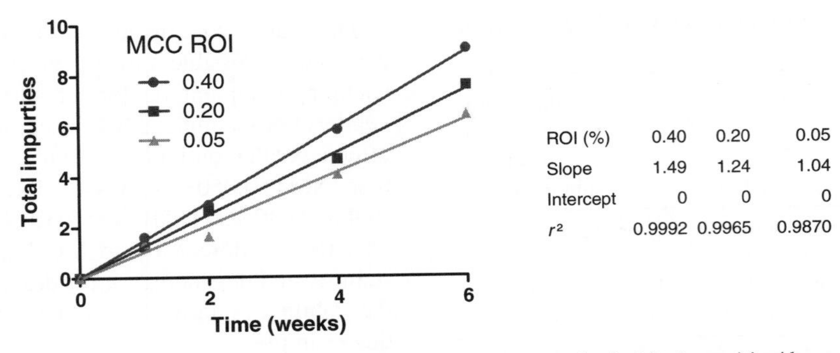

FIGURE 36.14 A hypothetical effect of MCC ROI and the growth of tablet impurities/degradants at 40°C/75% RH open dish conditions.

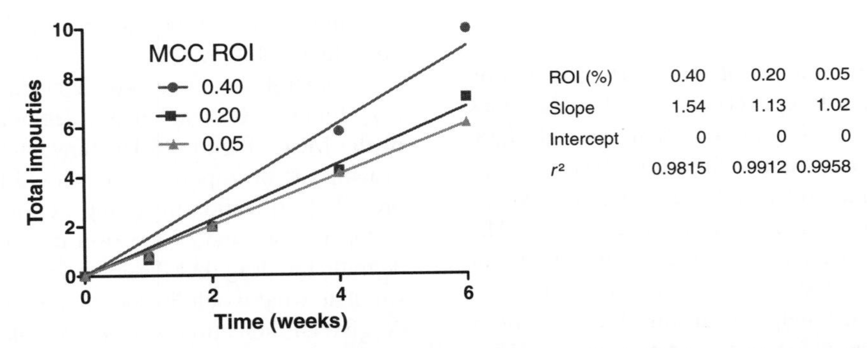

FIGURE 36.15 A hypothetical effect of MCC ROI and the growth of tablet impurities/degradants at 40°C dry open dish conditions.

separate the tablets into three moisture categories (i.e., 1%, 2%, and 5.5%)

2. Determine the amount of time pulls required (i.e., 1, 3, 6, 9, 12, 18, 24 months)
3. Determine the storage temperatures (T) to place the tablets at; a minimum of three is recommended (i.e., 25, 30, and 40°C), and these are typical ICH temperatures.

The study described requires 63 foil pouches to cover each time point and condition. This is an extensive study but does not account for different lots of API or excipients. Much is learned from this study about packaging protection requirements. To expand the study, excipient ROI is examined as a factor, which increases the study samples by three times.

In the experimental results shown in Figure 36.17, the data set is extensive, but it is important to analyze interim data and guide the packaging decisions. At the end of the study, a complete predictive model for temperature, moisture, time, and excipient ROI is attained to guide decisions made around storage temperature, shelf life, and excipient specifications.

A regression analysis is completed to provide a prediction equation to determine what and how much each of the

TABLE 36.17 Regression Results for the Material Stored at 40°C Dry and 40°C/75% RH

Term	Estimate	Std Error	t Ratio	Prob > |t|
Intercept	−0.902499	0.136978	−6.59	<0.0001*
Time (weeks)	1.2841237	0.029206	43.97	<0.0001*
ROI (%)	3.29304	0.438798	7.50	<0.0001*
RH conditions[75]	0.1093065	0.062911	1.74	0.0946
(Time (weeks) − 2.6) × (ROI − 0.21667)	1.5833776	0.203707	7.77	<0.0001*

Asterisk indicates that the term is statistically significant (Prob < 0.05).

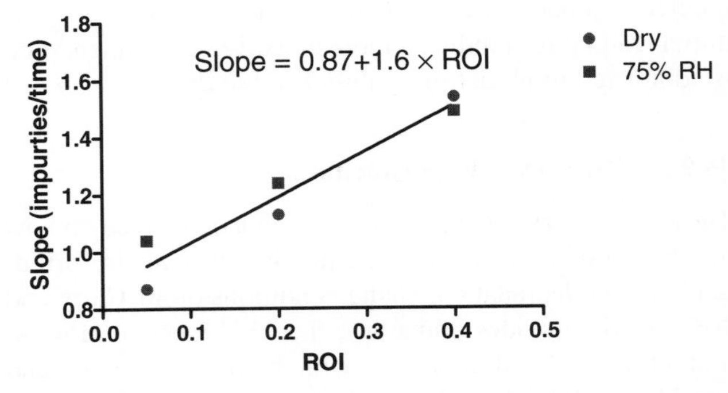

FIGURE 36.16 Comparison between impurity growth rate and ROI of excipient.

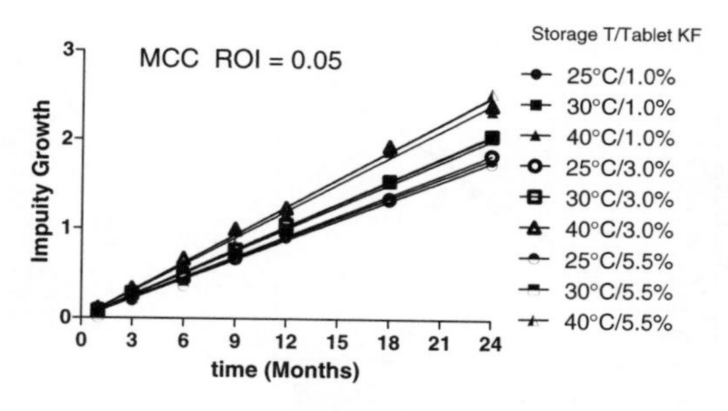

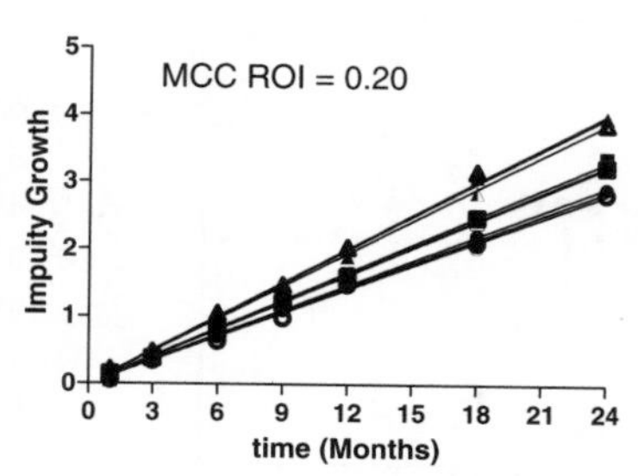

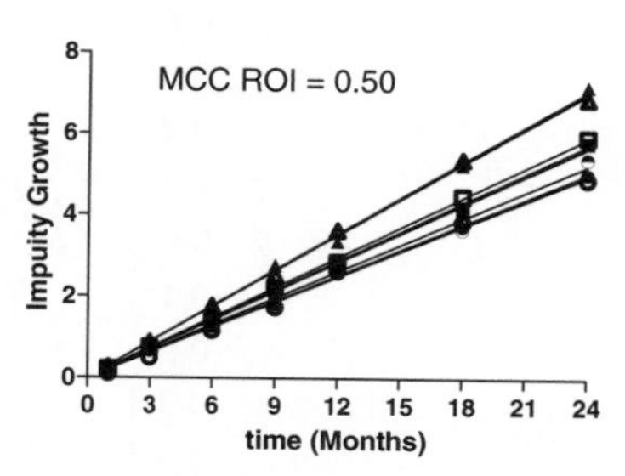

FIGURE 36.17 Result from the temperature moisture study completed by examining the effect of MCC ROI at three different levels 0.05%, 0.20%, 0.50%. The figure seems to indicate that temperature, time, and ROI have the largest effect on impurities.

variables affects impurity growth. The analysis indicates that time, temperature, and ROI have effect on degradation; in addition, there are some interactions between time and temperature, time and ROI, and temperature and ROI (Table 36.18). As expected, tablet moisture content did not affect stability as is seen from Figure 36.17.

Figure 36.18 shows that the regression is not biased and the residuals are evenly distributed.

$$\text{Impurities} = 0.633 - 0.041 \cdot T - 0.020 \cdot t + 1.17 \cdot \text{ROI} + 0.00035 \cdot (t \cdot T) + 0.360 \cdot (t \cdot \text{ROI}) + 0.080 \cdot (T \cdot \text{ROI}) \quad (36.12)$$

The above equation is the expression for expected impurities at any given time, storage temperature, and MCC ROI. This is used to test different scenarios such as what would the ROI need to be to attain room temperature (25°C) storage condition with acceptable amount of impurities or what would the shelf life be in warmer climates 30°C. It must be noted that impurity levels are set by a combination of toxicity and process capability.

Contour plots (Figure 36.19) are useful as the sensitivity of each variable is more easily visualized.

TABLE 36.18 Regression Results from the TRH Experiments

Term	Estimate	Std Error	t Ratio	Prob > $\|t\|$
Intercept	−2.11	0.043	−49.12	<0.0001*
Time, T (months)	0.160	0.00102	156.82	<0.0001*
Temperature, t (°C)	0.037	0.00125	29.60	<0.0001*
MCC ROI	3.71	0.042	89.00	<0.0001*
(Time (months) − 10.43) × (temperature (°C) − 31.67)	0.0035	0.00016	21.60	<0.0001*
(Time (months) − 10.43) × (MCC ROI − 0.25)	0.360	0.0055	65.99	<0.0001*
(Temp (°C) − 31.67) × (MCC ROI − 0.25)	0.080	0.0067	11.91	<0.0001*

All parameters shown are significant. The parameters analyzed that did not show significance were tablet moisture and all tablet moisture interactions.

Stability is important and knowing what can predict stability is valuable in determining packaging and excipient grade selection.

36.10 PROCESS OPERATIONS AND SCALABILITY OF DOSAGE FORM

There are many considerations in scaling up unit operations that manufacture solid dosage forms. Scaling up through preclinical → early clinical (phase I and phase II) → late clinical (phase IIb and phase III) → registration → engineering/validation batches has many challenges (Table 36.19). Scale-up usually takes the course of laboratory experiments, pilot scale tests, and finally commercial-scale operation and continuous improvement [31,32].

Beyond development, scale-up or scale-down also occurs after approval, in which case changes are governed by Post-Approval Changes (SUPAC[j]) guidelines as specified by the Center for Drug Evaluation and Research (CDER). Finally, tech transfer (TT) is needed if multiple plants or CMOs are required.

Limited and costly API or drug substance (DS) and resources may hinder the experimental understanding that could be gained; therefore, know-how prior to manufacturing is extremely valuable. Understanding and using engineering first principles, dimensional analysis, and design of experiments (DOE) improves the likelihood that the process(es) and drug product (DP) will succeed.

Pharmaceutical process scale-up shall consider formulation, process development, and marketing needs. A risk-based approach is to examine how the TPP of the drug is affected by CQAs of the final dosage form and the design space of the process. Quality by design (QbD) principles are used to ensure a safe and efficacious product. Design space, controls, and specifications are continuously improved through continuous learning.

[j] http://www.fda.gov/Drugs/default.htm.

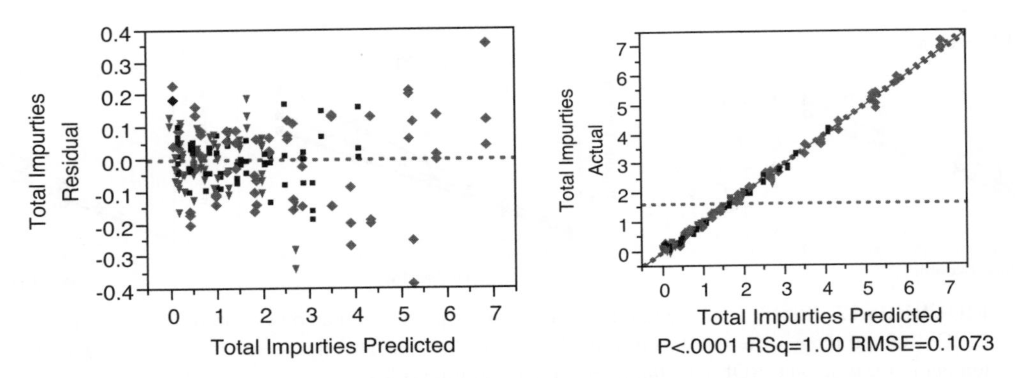

FIGURE 36.18 The residuals plot of predicted versus actual. The residuals are evenly distributed, indicating there is no bias in the model. ▼ is 0.05 ROI, □ is 0.20 ROI, and ◇ is 0.50 ROI.

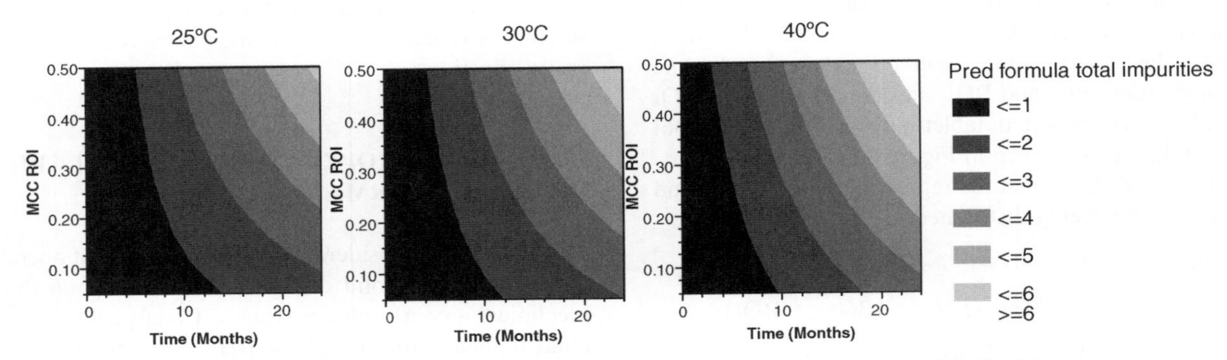

FIGURE 36.19 Contour plots examining total impurities versus time and MCC ROI.

Definitions referring to pharmaceutical manufacturing are given as follows (adapted from PQRI) [33]:

Critical Quality Attribute: A quantifiable property of an intermediate or final product that is considered critical for establishing the intended purity, efficacy, and safety of the product. That is, the property must be within a predetermined range to ensure final product quality.

Target Product Profile: A summary of characteristics that if achieved provides optimal efficacy, patient compliance, and marketability. A TPP often includes attributes such as pharmacokinetic information (e.g., immediate release versus extended release), dosage form (e.g., tablet versus injectable), and shelf life information (e.g., 2 years at 25°C/60% RH).

Design Space: The design space is the established range of process parameters that has been demonstrated to provide assurance of quality. In some cases, design space can also be applicable to formulation attributes.

Critical Process Parameter (CPP): A CPP is a process input that, when varied beyond a limited range, has a direct and significant influence on a CQA.

Critical Material Attribute (CMA): A CMA is an attribute that has direct impact on the processability or CQA of the drug product. CMAs could include impurities from the excipient or raw materials.

Normal Operating Range (NOR): A defined range, within the proven acceptable range (PAR), specified in the manufacturing instructions as the target and range at which a process parameter is to be controlled, while

TABLE 36.19 Typical Scale-Up

Stage	Typical Material Required	Reason
Preclinical	0.05–1 kg	Early toxicology testing
Phase I and II	0.2–50 kg	Healthy volunteers and early proof of concept
Phase IIb and III	10–1000 kg	Proof of concept and verification trials
Registration	>100 kg and >100,000 unit dosages and minimum 1/10th commercial batch size	From FDA guidance
Engineering/validation	Based on registration and expected product demand	Final process testing and process confirmation runs

producing unit operation material or final product meeting release criteria and CQAs.

Proven Acceptable Range: A characterized range at which a process parameter may be operated within, while producing unit operation material or final product meeting release criteria and CQAs.

Another useful tool during scale-up is process failure modes effects analysis (pFMEA) that is used to understand the failure modes of the CQAs to help mitigate risks in unit operations [34, 35]. This is done by understanding the severity, occurrence, and detection of any or all potential failure modes. pFMEA is different from root cause analysis (RCA) as a RCA is performed after deviations have already occurred. Risk prioritization numbers (RPNs; scores from 1 to 100) are calculated using pFMEA (RPN = severity [1 − 10] × occurrence [1 − 10] × detection [1 − 10]) and point the engineer toward corrections that are implemented to reduce risk to the drug product. Usually, an engineer starts to look to ameliorate problems with high RPN numbers.

Besides CQAs, critical business attributes (CBAs) are also considered. Business decisions that involve scale-up can relate to choice of CMO, batch size, operators needed, equipment purchases, use of PAT tools, and so on.

It is imperative to learn the CPPs of the unit operation at hand. This is done through an evolution of understanding the engineering principles and the processing knobs at the engineer's disposal. These CPPs affect any or all the CQAs of the DP [34].

Unit Operation	CQAs	CPPs	Potential Failure Mode
Roller compaction	Ribbon density, degradants, downstream dissolution	Roll speed feed screw speeds, roll force/pressure, roll separation/gap, room temperature/humidity	Ribbon density variation, high degradation
Slugging	Hardness, dissolution	Slugging force	Too little or too much
Wet granulation	Particle size, powder density, degradants, downstream dissolution	Granulation fluid mixing time, granulation fluid mixing speed, granulating fluid amount, granulating fluid addition rate, granulating fluid temperature, spray nozzle air volume, dry mixing time, wet mixing time, impeller speed, chopper speed, power consumption	Too little or too much
Fluid bed granulation	Particle size, powder density, powder wetness, degradants, downstream dissolution	Granulation fluid mixing time, granulation fluid mixing speed, granulating fluid amount, granulating fluid addition rate, granulating fluid temperature, spray nozzle air volume, bed mixing time, supply air flow rate, temperature, dew point, product bed temperature, exhaust air temperature, dew point, filter shaking intervals	Loss of yield, powder degradation
Milling	Particle size, degradants	Impeller speed, feed rate, room temperature, humidity	Undesired particle size, degradation
Lyophilization	Degradants, physical form, product wetness	Pretreatment, freezing, drying, temperature, cycle times, chamber pressure	Degradation, loss of stability, yield loss
Blending	Blend uniformity, content uniformity (CU)	Blend time (pre- and postlube), rotation rate, agitator speed, room temperature, humidity	Underblending may lead to bad CU, overblending may lead to poor compressibility
Encapsulation	Powder density, downstream dissolution, weight	Speed, dosing	Improper weight, broken capsules, too much dense powder in capsule
Tableting	Hardness, thickness, weight, dissolution, degradants, content uniformity	Tablet weight, press (turret) speed, main compression force, precompression force, feeder speed, upper punch entry, room temperature, humidity	Capping if dwell time is too low, low weights or high weight variability if powder flow is bad
Tablet coating	Appearance, dissolution	Coating suspension mixing time, coating suspension mixing speed, coating suspension solids load, atomization pressure, preheat time, jog time number, type of guns, gun to bed distance	Twinning if tablet shape is not round, spray drying of coating suspension if temperature is too high, nonuniform coating if pan speed is too slow, tablet defects if pan speed is too fast
Tablet printing	Appearance, degradants	Ink dosage amount, force, location	Ink degrades product

There are many unit operations that are used for drug product manufacture; the easiest and most economical is direct compression (DC). In oral solid dosage manufacturing, direct compression process technology is the most effective and efficient way to make powder materials suitable for tableting or encapsulation without a step to increase the particle size [36]. In the example, the TPP is a DC tablet that focuses on the unit operations of (1) blending, (2) compression, and (3) coating.

36.10.1 Blending Scale-Up

Blending is a critical operation that determines how well the product is to perform in the next phases. Achieving and maintaining homogeneous mixing of powders is critical, especially in formulations involving small amounts of high-potency components. Lack of blend uniformity at the blending stage may result in the lack of CU in the finished product dosage forms.

Tumbling blenders are typically used. The most common types of blenders are in-bin and V-shell blenders. In-bin blenders are typically used for high drug load blends and are good for storage of said blend. V-shell blenders are used in intermediate drug load blends (Figure 36.20). The main difference in these blenders is the geometry.

There are three mechanisms of particle mixing: convection, dispersion, and shearing [37]. In tumbling blenders, convective and dispersive mixing are dominant, unless intensifier bars or chopper blades are added to cause shear mixing. For example, within a V-shell blender, convective blending occurs within each shell side during tumbling, and dispersive mixing happens between shells.

Blending in a DC case consists of a prelubricant and a postlubricant blend ahead of compression. Lubricants such as sodium starch fumarate (SSF) and magnesium stearate are normally used.

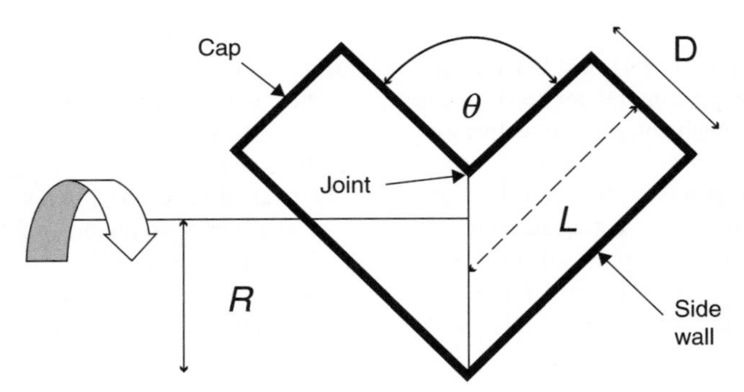

FIGURE 36.20 V-shell blender schematic.

Important parameters are as follows:

CMA/CPP	CPP/CQA/CBA It Can Affect
Particle shape and size	Blend uniformity, content uniformity, compressibility
Powder density	Blend uniformity, content uniformity, compressibility
Fill level	Blend uniformity, content uniformity
Loading procedure	Blend uniformity, content uniformity
Number of rotations (pre- and postlube addition)	Blend uniformity, content uniformity
Rotation speed	Blend uniformity, content uniformity
Blender size	Throughput
Room humidity	Degradants, compressibility

There are many ways to determine if a blend is well blended. Three simple ways are the following:

1. Use online process analytical technology (PAT) of near-infrared (NIR) technology
2. Perform thief sampling over blending time and test
3. Simply compress the blend material and access CU

The NIR region spans the wavelength range 780–2526 nm, in which absorption bands mainly correspond to overtones and combinations of fundamental vibrations. NIR spectroscopy is a fast and nondestructive technique that provides multiconstituent analysis of virtually any matrix. As NIR absorption bands are typically broad and overlapping, chemometric data processing is used to relate spectral information to sample properties.

The left graph in Figure 36.21 shows the second derivative of spectral data gathered from the API and the other components/excipients in the blend. The right graph shows when the API spectra reach a $<1\%$ RSD distribution within the blend, which is the blend end point. Commonly, 1 min after $<1\%$ RSD for the API is accomplished, called as the blend end point, but the engineer can see the asymptote of the line over time, sample number. Other determinations of blend end point are used as well, and method development is to be used for particular blends. This tool is useful as the engineer receives online data without sampling bias.

If PAT tools cannot be used, a more traditional sample method is used. Samples are commonly pulled from many locations (Figure 36.22) within the V-shell blender in order to understand if there is any location bias versus blend uniformity.

Usually, blenders are scaled from V-shell (laboratory, pilot scale, commercial scales); however, depending on the product and manufacturing needs, the blending operation may be transferred to an in-bin blender (pilot, commercial scales). Changing geometric characteristics of tumbling blenders may lead to different mixing behaviors;

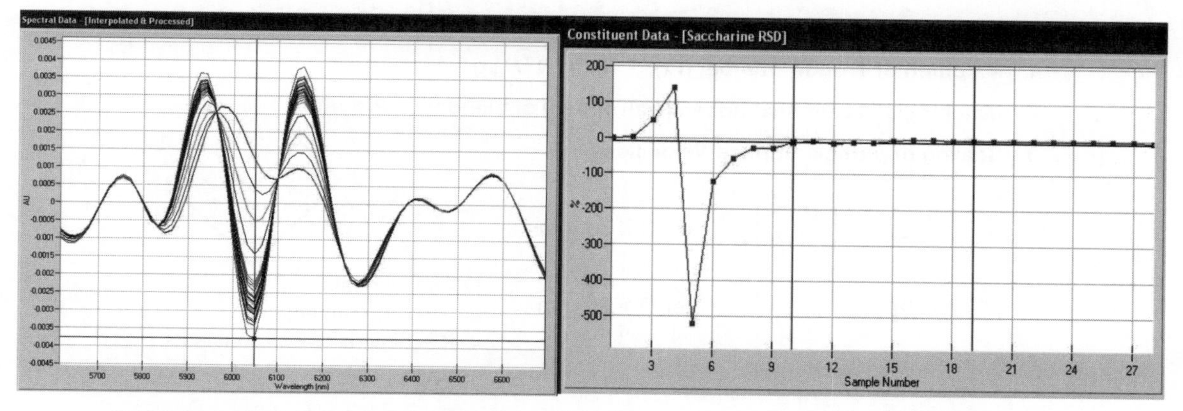

FIGURE 36.21 NIR spectral and constituent data for a blend containing saccharine as a model API[k].

therefore, a straightforward transition cannot be accomplished unless engineering principles are used. Some scale-up approaches are matching Froude (Fr) number, matching tangential/wall speed, or scaling particle surface velocities (Figure 36.23) [38–40].

36.10.2 Compression Scale-Up

Compression is important to make robust tablets. Tablets that are too soft cannot withstand the downstream coating or packaging processes without chipping or breaking and losing tablet weight/active component. Tablets that are too hard

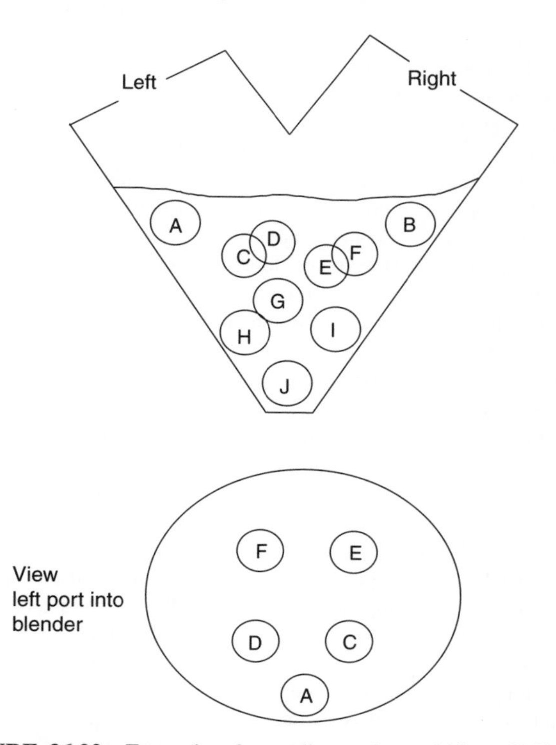

FIGURE 36.22 Example of sampling points within a V-Shell blender.

[k]NIR technology from CDI Pharma

cannot dissolve effectively and therefore also cannot be efficacious when considering the TPP.

Compression is typically used to make solid oral dosage forms of core tablets. Many types of equipment are manufactured; some include single station, rotary presses. Typical manufactures are Korsch, Elizabeth Hata, SMI, GEA Courtoy, and Manesty. Tablet presses are capable of using tooling of various sizes such as A, B, and D.

Parameters that may be critical in tablet production are as follows:

CPP	CQA It Affects
Incoming blend	Tablet weight (flowability), compressibility in general
Feeder speed	Tablet weight
Fill depth	Tablet weight
Press speed (dwell time)	Appearance (defects via capping or lamination)
Precompression	Tablet hardness
Main compression	Tablet hardness
Upper punch entry	Tablet hardness
Room humidity	Compressibility in general, degradants, tablet water content
Press temperature over time	Tablet hardness, degradants, possible change in physical form

To access compressibility of the drug product DOEs are performed to evaluate precompression force, main compression force, and press speed (Figure 36.24). Tests such as tablet weight, thickness, hardness, friability, and dissolution are performed to understand the processing affects on the CQAs. These data are used to determine processing targets, NORs, and PARs.

Tablet dies and tooling may be the same from laboratory to pilot to commercial scale; the change is in tooling dwell time.

$$\text{DwellTime} = \frac{60,000 \times \text{Punch HeadFlatDiameter}}{\pi \times \text{PitchCircle Diameter} \times \text{PressSpeed}} \qquad (36.13)$$

1. **Matching of froude number (Fr),** $Fr = [\Omega^2 R]/g$
2. **Matching of tangential speed (wall speed) of blender,** $2\pi\Omega R$
3. **Scaling of particle surface velocities,**

$$V = kR\Omega^{2/3}\left(\frac{g}{d}\right)^{1/6} \quad \text{for } \Omega \le 30 \text{ rpm}$$

$$V = KR\Omega^{1/2}\left(\frac{g}{d}\right)^{1/4} \quad \text{for } \Omega > 30 \text{ rpm}$$

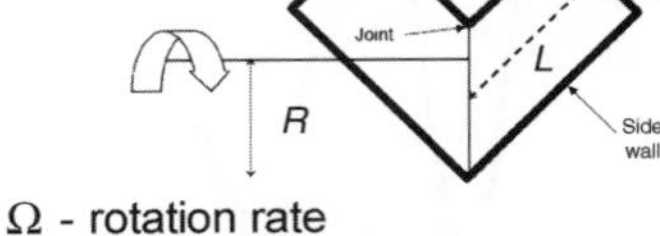

Ω - rotation rate
d - vessel diameter
k, K – dimensionless constants

FIGURE 36.23 Common scale-up techniques for the process of blending.

Different dwell times can cause problems such as tablet capping or lamination. Tools such as compaction simulators could be used early on to save both time and money. Scaling up based upon mechanical similarity and quality attributes of the product is important, but sometime also is scaling down. Analytical techniques such as shear cells to understand powder flow and compaction simulators to understand compressibility behavior have been developed with the mindset of scaling down.

36.10.3 Coating Scale-Up

Tablet coating is the unit operation consisting of spray coating functional or nonfunctional/aesthetic coating onto the surface of the already compressed tablets. There are various sizes of tablet coaters, ranging up to ≥60 in. coating pans. Coating pans are either perforated or nonperforated. PAT tools may be implemented, for example, NIR for water content.

Important parameters are as follows:

CPP	CQA It Affects	Potential Problems
Pan load	Appearance, tablet water content	Improper pan loading for the scale being used
Spray gun to bed distance	Appearance	Improper spray to tablet bed
Number of spray guns	Appearance	
Exhaust temperature	Degradants, tablet water content	Spray drying of coating suspension
Atomization air flow rate	Appearance, tablet water content	Improper spray
Pattern air flow rate	Appearance, tablet water content	Improper spray
Spray rate	Appearance, tablet water content	Improper spray
Spray formulation	Appearance, dissolution	May impede tablet dissolution
Weight gain	Appearance, dissolution	Too high may impede tablet dissolution, too low may not cover tablets/ appearance
Pan speed	Appearance	Too high of pan speed
Jogging	Appearance	Too much or too little jogging of the tablet bed
Incoming tablets	Appearance, dissolution	Too soft tablets, too much disintegrant, especially on the surface of the tablets

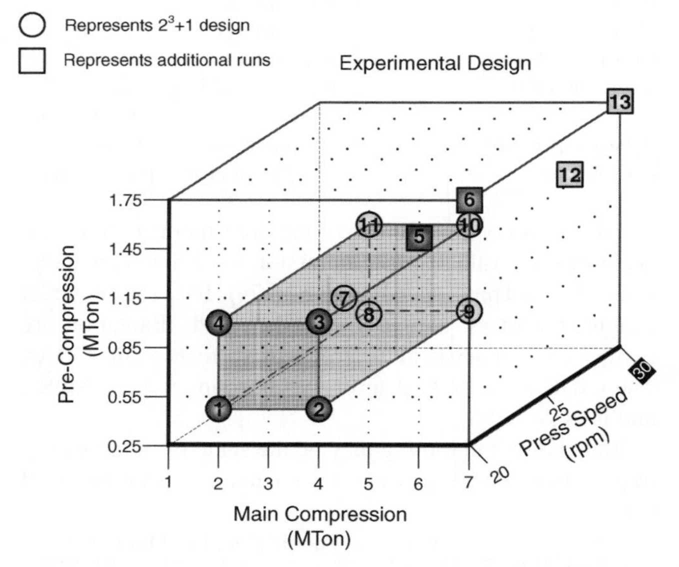

FIGURE 36.24 Fractional factorial experimental design for tableting.

TABLE 36.20 Spray Coating Design and Number of Defects Observed in an 800 Tablet Sample

Experiment	Spray Rate (g/min)	Exhaust Temperature (°C)	Pan Speed (RPM)	Suspension Concentration (wt%)	Defects
1	350	50	5	18	4
2	350	50	10	22	1
3	350	60	5	22	7
4	350	60	10	18	0
5	450	50	5	22	12
6	450	50	10	18	2
7	450	60	5	18	9
8	450	60	10	22	2
9	300	55	7.5	20	1
10	300	55	7.5	20	0

Postcoated tablets are examined by AQL for the number and type of defects (minor, major, or critical) (ANSI/ASQ Z1.4-2008). Common reasons for defects stemming from film coating include the following:

- Improper EEF, thermodynamic conditions
- Incoming raw material including tablets
- Operating conditions (nonthermodynamic)

Thomas Engineering Inc. provides a thermodynamic analysis of aqueous coating (TAAC) model for coating scale-up that uses thermodynamic heat and mass transfer equations to characterize the environmental conditions inside a coating pan during a steady-state film coating process [41, 42].

The environmental equivalency factor (EEF) is the most important piece of data output by the TAAC program. It is a dimensionless number proportional to the ratio of the dry area of the tablet bed to the wetted area and, as such, is indicative of the drying rate of the film being applied. The dimensionless EEF lumps together all thermodynamic terms for ease of modeling or scaling up.

If there is a concern with water content increase, changing parameter values to increase the EEF helps; however, too high an EEF may cause unwanted spray drying of the coating suspension, leading to undesired tablet defects. A balance is usually found empirically. *A good rule of thumb*: use an EEF of 2–5, with 3.3 being a typical production value.

A spray coating half factorial design around the parameters of spray rate, exhaust temperature, pan speed, and suspension concentration is executed to better define the coating processing design space with respect to tablet defects.

Table 36.20 shows the experiment and the number of defects seen in a sample size of 800. Stepwise linear regression of the data yields the model shown in Figure 36.25 and Tables 36.21 and 36.22.

Spray rate, pan speed, and suspension concentration were all seen to be significant parameters on the response of tablet defects. Exhaust temperature is removed from the model as it is insignificant in the range studied. Two interaction terms were also found to be significant: spray rate × pan speed and

TABLE 36.21 Summary of Fit for Defects Model

R^2	0.990806
R^2 adj	0.979314
Root mean square error	0.598029
Mean of response	3.8
Observations (or sum wgts)	10

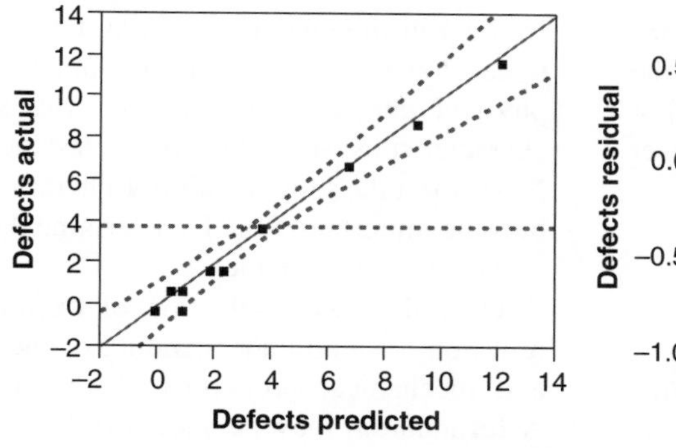

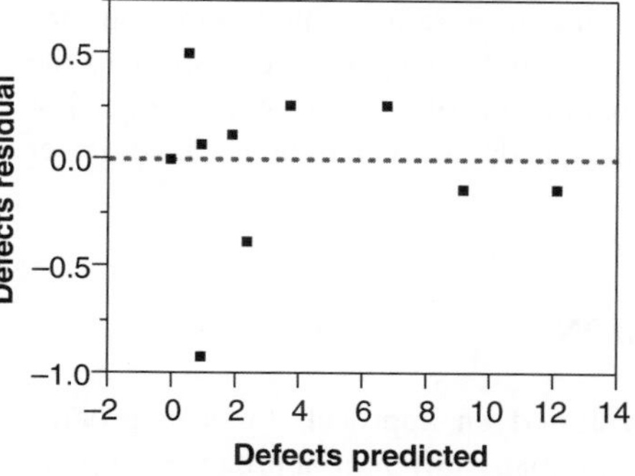

FIGURE 36.25 Defects actual versus predicted and defects residuals versus predicted.

TABLE 36.22 Parameter Estimates for Defects Model

Term	Estimate	Std Error	t Ratio	Prob > $\|t\|$
Intercept	−9.702778	2.531304	−3.83	0.0186*
Spray rate (g/min)	0.0363889	0.003152	11.55	0.0003*
Pan speed (RPM)	−1.21	0.091089	−13.28	0.0002*
Suspension concentration (wt%)	0.4375	0.105718	4.14	0.0144*
(Spray rate (g/min) − 380) × (pan speed (RPM) − 7.5)	−0.007	0.001691	−4.14	0.0144*
(Pan speed (RPM) − 7.5) × (suspension concentration (wt%) − 20)	−0.125	0.042287	−2.96	0.0417*

The asterisk indicates that the variable is significant.

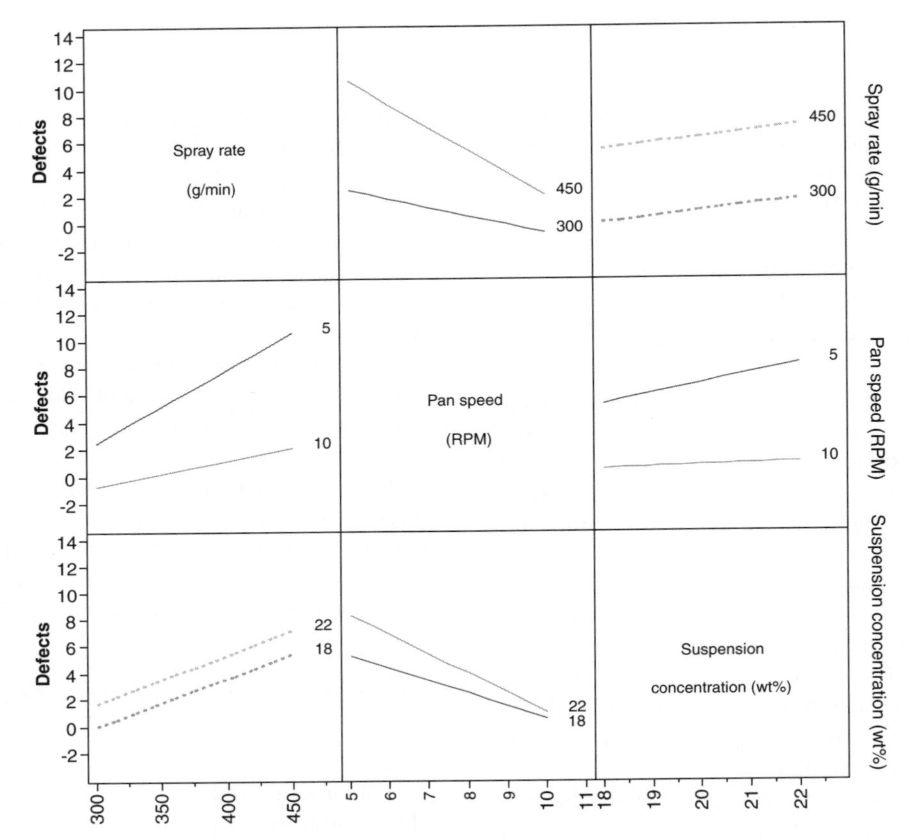

FIGURE 36.26 Interaction profiler for the tablet defects model.

pan speed × suspension concentration. The interaction profiles are shown in Figure 36.26. The engineer optimizes the process by using the parameters of the strongest leverage. For example, higher pan speeds were shown to have fewer defects.

36.11 CONCLUSION

This chapter demonstrated an approach to drug product development. There are many different approaches; a good engineer examines the best approach for the situation. Defined in the chapter was a systematic manner in which to develop an immediate release tablet. First, account for the physical characteristics of the drug substance: particles characteristics, solubility, BCS classification, and stability. Next, the release and stability characteristics of the tablet become important. Finally, determine the processability and scaleability of the tablet.

Overall, a wise selection of excipients and processes relies on a sound understanding of the physical, chemical, and mechanical properties of the drug and excipients. A formulation may be successfully scaled up and consistently meet performance and manufacturing requirements only when one fully understands the complex relationship between the drug, excipients, processing, and desired dosage form performance criteria.

When formulating any pharmaceutical dosage form, it is important to remember that there is equilibrium between the bioavailability of the product, its chemical and physical stability, and the technical feasibility of producing it. Any changes made to a formulation in an attempt to optimize one of these properties are likely to have an effect on the other two parameters that must be considered. This is especially true of immediate release solid dosage forms.

REFERENCES

1. Martin A. *Physical Pharmacy*, 4th edition, Lea & Febiger, Philadelphia, PA, 1993, p. 223.
2. Niklasson M, Brodin A, Sundelöf L-O. Studies of some characteristics of molecular dissolution kinetics from rotating discs. *Int. J. Pharm.* 1985;23:97–108.
3. Koehler MG, Grigoras S, Dunn WJ, III. The relationship between chemical structure and the logarithm of the partition coefficient. *Quant. Struct.-Act. Relat.* 1988;7:150–159.
4. Leo A, Hansch C, Elkins D. Partition coefficients and their uses. *Chem. Rev.* 1971;71:525–616.
5. Dearden JC, Bresnen GM. The measurement of partition coefficients. *Quant. Struct.-Act. Relat.* 1988;7:133–144.
6. Navia MA, Chaturvedi PR. Design principles for orally bioavailable drugs. *DDT* 1996;1:179–189.
7. Byrn SR, Pfeiffer RR, Stowell JG. *Solid State Chemistry of Drugs*, 2nd edition, SSCI, West Lafayette, IN, 1999.
8. Grant DJW. Theory and origin of polymorphism. In: Britain HG, editor, *Polymorphism in Pharmaceutical Solids*, Marcel Decker, New York, 1999, pp. 1–34.
9. Aguiar AJ, Krc J, Kinkel AW, Samyn JC. Effect of polymorphism on the absorption of chloramphenicol from chloramphenicol plamitate. *J. Pharm. Sci.* 1967;56(7):847–883.
10. Poole JW, Owen G, Silverio J, Freyhof JN, Rosenman SB. Physicochemical factors influencing the absorption of the anhydrous and trihydrate forms of ampicillin. *Current Therapeutic Research,* 1968;10:292-303
11. Matsumoto T, Nobuyoshi K, Iliguchi S, Otsuka MJ. Effect of temperature and pressure during compression on polymorphic transformation and crushing strength of chlorpropamide tablets. *J. Pharm. Pharmacol.* 1991, 43, 74.
12. Zhang GZ, Law D, Schmitt EA, Qiu Y. Phase transformation considerations during process development and manufacture of solid oral dosage forms. *Adv. Drug Deliv. Rev.* 2004;56(3): 371–390.
13. Venables HJ, Wells JJ. Powder mixing. *Drug Dev. Ind. Pharm.* 2001;27:599–612.
14. Sun C, Grant PJW. Influence of crystal shape on the tableting performance of L-lysine monohydrochloride dihydrate. *J. Pharm. Sci.* 2001;90:569–579.
15. Kaerger JS, Edge S, Price R. Influence of panicle size and shape on flowability and compactibility of binary mixture of paracetamol and microcrystalline cellulose. *Eur. J. Pharm. Sci.* 2004;22(2–3):113–119.
16. U.S. Pharmacopeia USP29/NF24. *U.S. Pharmacopeial Convention*, Rockville, MD, 2006, General Chapter<905>, pp. 2778–2784.
17. Zhang Y, Johnson KC. Effect of drug particle size on content uniformity of low-dose solid dosage forms. *Int. J. Pharm.* 1997;154(2):179–183.
18. Johansson D, Abrahamsson B. *In vivo* evaluation of two different dissolution enhancement principles for a sparingly soluble drug administered as extended-release (ER) tablet. Proceedings of the 24th International Symposium on Controlled Release of Bioactive Materials, 1997, pp. 363–364.
19. Amidon GL, Lennernäs H, Shah VP, Crison JR. A theoretical basis for a biopharmaceutical drug classification: the correlation of *in vitro* drug product dissolution and *in vivo* bioavailability. *Pharm. Res.* 1995;12:413–420.
20. Dressman JB, Amidon GL, Reppas C, Shah VP. Dissolution testing as a prognostic tool for oral drug absorption: immediate release dosage forms. *Pharm. Res.* 1998;15:11–22.
21. Food and Drug Administration guidance for industry (2000). *Waiver of In Vivo Bioavailability and Bioequivalence Studies for Immediate-Release Solid Oral Dosage Forms Based on a Biopharmaceutics Classification System.* Food and Drug Administration, Center for Drug Evaluation and Research, Rockville, MD.
22. Fell JT, Newton JM. Determination of tablet strength by the diametral-compression test. *J. Pharm. Sci.* 1970;59(5):688–691.
23. Hiestand E, Wells JE, Peot CB, Ochs JF. Physical processes of tableting. *J. Pharm. Sci.* 1977;66(4):510–519.
24. Hiestand EN, Bane JM, Strzelinski EP. Impact test for hardness of compressed powder compacts. *J. Pharm. Sci.* 1971;60 (5):758–763.
25. Ho R, Bagster DF, Crooks MJ. Flow studies on directly compressible tablet vehicles. *Drug Dev. Ind. Pharm.* 1977;3:475.
26. Seager H, Rue PJ, Burt I, Ryder J, Warrack NK, Gamlen MJ. Choice of method for the manufacture of tablets suitable for film coating. *Int. J. Pharm. Technol. Prod. Manuf.* 1985;6: 1–20.
27. Parikh D,editor. *Handbook of Pharmaceutical Granulation Technology*, 2nd edition, Marcel Dekker, New York, 2005.
28. *Guidance for Industry, Dissolution Testing of Immediate Release Solid Oral Dosage Forms,* U.S. Department of Health and Human Services, Food and Drug Administration, Center for Drug Evaluation and Research (CDER), August 1997.
29. Kukra J, Arratia PC, Szalai ES, Bittorf KJ, Muzzio F, Understanding Pharmaceutical Flows, Pharmaceutical Technology, 2002, p. 48–72.
30. Ewe K. Press AG, Bollen S, Schuhn I. Gastric emptying of indigestible tablets in relation to composition and time of ingestion of meals studied by metal detector. *Dig. Dis. Sci.*, 1991;36(2):146–I52.
31. Monkhouse DC, Rhodes CT,editors. *Drug Products for Clinical Trials: An International Guide to Formulation,*

Production, Quality Control, Drugs and the Pharmaceutical Sciences, Vol. 87, Marcel Dekker, 1998.

32. Levin M,editor. *Pharmaceutical Process Scale-Up, 2nd edition, Drugs and the Pharmaceutical Sciences*, Vol. 157, Marcel Dekker, 2006.
33. Process robustness—a PQRI white paper. PQRI Workgroup Members. November/December 2006. Pharmaceutical Engineering On-Line Exclusive.
34. Marder R, Sheff RA. *The Step-by-Step Guide to Failure Modes and Effects Analysis*, Opus Communications, Marblehead, MA, 2002.
35. Stamatis DH. *Failure Mode and Effect Analysis; FMEA from Theory to Execution*, ASQC Quality Press, Milwaukee, WI, 1995.
36. Parikh DM. *Handbook of Pharmaceutical Granulation Technology, Drugs and the Pharmaceutical Sciences*, Vol. 81, Informa Healthcare, 1997.
37. Lacey PM Developments in the theory of particle mixing. *J. Appl. Chem.* 1954;4:257–268.
38. Swarbrick J. *Encyclopedia of Pharmaceutical Technology*, 3rd edition Vol. 5, Taylor & Francis, 2006.
39. Alexander A, Shinbrot T, Muzzio FJ. Scaling surface velocities in rotating cylinders as a function of vessel radius, rotation rate, and particle size. *Powder Technol.* 2002;126: 174–190.
40. Alexander AW, Muzzio FJ. *Batch size increase in dry blending and mixing. In: Pharmaceutical Process Scale-up*, 2nd edition, Marcel Dekker, New York, 2001.
41. Strong JC.Psychrometric analysis of the environmental equivalency factor for aqueous tablet coating. *AAPS PharmSciTech*, Vol. 10, No. 1, 2009.
42. Ebey GC. A thermodynamic model for aqueous film-coating. *Pharm. Technol.* 1987;11:40–50.

37

CONTROLLED RELEASE TECHNOLOGY AND DESIGN OF ORAL CONTROLLED RELEASE DOSAGE FORMS

Avinash G. Thombre and Mary T. am Ende
Pharmaceutical Development, Pfizer Global Research & Development, Groton, CT, USA
Xiao Yu (Shirley) Wu
University of Toronto, Toronto, Ontario, Canada

37.1 INTRODUCTION

An oral controlled release drug delivery system is designed to deliver a drug in a controlled and predictable manner over a period of time or at a predetermined position in the gastrointestinal tract. There are several other terms used interchangeably to describe controlled release dosage forms. The U.S. Food and Drug Administration defines modified release dosage forms as those whose drug release characteristics of time course and/or location are chosen to accomplish therapeutic or convenience objectives not offered by conventional dosage forms such as a solution or an immediate release dosage form [1]. Modified release oral dosage forms include extended release, that is, dosage forms designed to make the drug available over an extended period of time after ingestion, and delayed release, that is, dosage forms designed to provide a delay before drug release. Additionally, terms such as sustained release, prolonged release, pulsatile release, and targeted release have also been used in the literature. Orally disintegrating tablets that are designed to disintegrate more rapidly than an immediate release tablet can also considered being controlled release dosage forms. They disintegrate on contact with saliva, thus eliminating the need to chew the tablet, swallow an intact tablet, or take the tablet with liquids [2].

Over the past five decades, oral drug delivery systems have matured and currently are a dominant segment of the pharmaceutical market. Oral dosage forms are preferred because of their convenience and cost-effectiveness. Although they were once considered quite exotic, oral controlled release systems have now become commonplace and their advantages accepted both in the development of new molecular entities and in the product enhancement. The controlled release market was estimated to be worth over U.S. \$17 billion globally in 2007 with a +2% year-on-year growth [3]. Some top-selling controlled release products in the U.S. market are listed in Table 37.1.

There are several reasons for pursuing the development of controlled release dosage forms. Controlled release formulations can reduce the dosing frequency and minimize side effects. Drugs with short biological half-lives (i.e., those where the drug is metabolized or rapidly eliminated from the blood stream) have to be dosed frequently in order to maintain efficacious levels in the blood. By slowing the rate at which the drug is released, a controlled release dosage form can increase the apparent half-life and maintain efficacious levels for a longer duration, thereby reducing the need for frequent dosing. Reducing the dosing frequency to once daily assures patient convenience and compliance and a reduction in the peak to trough blood concentrations of the drug results in a more uniform therapeutic effect and can potentially lead to a lower total dose. Controlled release dosage forms can reduce undesirable side effects that are related to high and rapidly rising drug peak blood levels. In some cases, the undesirable side effects are related to a local irritation of the upper part of the gastrointestinal tract by the

Chemical Engineering in the Pharmaceutical Industry: R&D to Manufacturing Edited by David J. am Ende

TABLE 37.1 Some Top-Selling Oral Controlled Release Products in the United States

Name	Drug	Indication	Company	Type of Controlled Release Formulation	US Sales in 2008[a] (in million dollars)
Effexor XR	Venlafaxine HCl	Antidepressant	Wyeth Pharmaceuticals (Pfizer, Inc.)	Diffusion through a coating membrane on spheroids	2.87
Oxycontin	Oxycodone	Opioid agonist for pain management	Purdue Pharma	Diffusion through a matrix tablet	2.16
Adderall XR	Amphetamine and Dextroamphetamine	Attention deficit hyperactivity disorder (ADHD)	Shire Pharmaceuticals	Capsule containing two types of drug-containing beads designed to give a double-pulsed delivery	1.34
Concerta	Methylphenidate	ADHD	Ortho-McNeil-Janssen Pharmaceuticals	Trilayer capsule shaped tablet with two distinct drug layers and a push layer (osmotic technology) and a drug overcoat layer	1.00
Niaspan	Niacin (nicotinic acid)	Antihyperlipidemic agent	Abbott Laboratories	Diffusion through the gel that forms by hydration of the matrix tablet	0.81
Stilnox	Zolpidem tartrate	Hypnotic for the treatment of insomnia characterized by difficulties with sleep onset and/or sleep maintenance	Sanofi-aventis	Coated two-layer tablet with an immediate release and extended release layer	0.88
Detrusitol/ Detrol LA	Tolterodine L-tartrate	Treatment of overactive bladder	Pfizer, Inc.	Coated drug layer beads filled in a gelatin capsule	0.84

[a]Data from IMS Health.

drug. In such cases, a delayed release dosage form can help bypass the upper part of the gastrointestinal tract and reduce the frequency and intensity of these side effects.

Orally disintegrating tablets have the advantage that they can be taken without water. This can be very important to pediatric and geriatric patients, and to patients who have difficulty swallowing tablets or capsules. A controlled release dosage form intended to avoid degradation of acid-labile drugs is typically an enteric-coated dosage form (delayed release). The enteric coat prevents drug release in the acidic environment of the stomach, and, at a higher intestinal pH, the coating dissolves to enable drug release.

The rapid advance in the field of controlled release occurred because of two main reasons: (1) Interdisciplinary teams worked together on novel concepts and designs for drug delivery devices and (2) advances in many fields that could be related to controlled release. Chemical engineers and the science of chemical engineering played a major role by introducing concepts of mass transfer and drug diffusion through matrices and membranes, material properties of excipients, thermodynamics, and kinetics of drug release. The science of biopharmaceutics provided the understanding of gastrointestinal physiology and its relationship to controlled release dosage forms, with respect to both the transit of dosage forms and the absorption of drug as a function of position in the gastrointestinal tract. It also provided preclinical *in vivo* models such as beagle dogs that led to an increased understanding of the *in vivo* performance of controlled release dosage forms and their *in vitro–in vivo* relationships. Polymer science and engineering provided novel materials with a range of properties, which could be tailored to suit a particular application, for example, polymers that eroded with time, thereby releasing the drug. Advances in the understanding of the pharmacokinetics and pharmacodynamics of drugs allowed controlled release dosage forms to be designed in a rationale manner. Finally, advances in manufacturing science and engineering were important, for example, advances in the ability to manufacture precise laser drilled orifice in osmotic tablets at rates suitable for commercial production.

This chapter focuses on the design of oral controlled drug release dosage forms. However, the field of controlled release is much broader. It spans other pharmaceutical dosage forms such as long-acting injections and implants, transdermal patches, ocular devices, and targeted drug delivery systems. Furthermore, controlled release is also used in veterinary applications [4] and diverse fields such as the sustained release of fertilizers, insecticides, herbicides, fragrances, and the food industry [5].

37.2 DEVELOPMENT OF CONTROLLED RELEASE FORMULATIONS IN AN INDUSTRIAL SETTING

The rational development of controlled release formulations in the setting of a large multinational pharmaceutical company with discovery and development operations typically starts with establishing the rationale for modifying the release rate and the desired product profile, that is, defining the medical need. The next steps involve selection of the dose, delivery duration, and release kinetics based on the known or the assumed target blood levels. It is highly recommended that prior to initiating a development program, an assessment of the feasibility of developing a controlled release formulation based on the physicochemical and biopharmaceutical properties of the drug candidate be conducted and the most appropriate technology be selected based on the attributes of the technology and manufacturing considerations such as availability of commercial scale equipment, operator expertise, and prior experience with the technology [6].

Many compounds fail to become drugs because of their poor physicochemical and/or poor biopharmaceutical properties [7]. The physicochemical properties that have an impact on the feasibility of a controlled release formulation include molecular weight, partition coefficient, solubility, pH-solubility profile, potential for solubilization, salt forms, polymorphs, particle size distribution, and stability. The biopharmaceutical and pharmacokinetic properties that have an impact on the feasibility of a controlled release formulation include gastrointestinal transit of the dosage form, fed/fasted state, permeability, efflux, and extent of gut wall/first-pass metabolism. Good absorption throughout the length of the gastrointestinal tract is important in the successful development of controlled release formulations [8].

37.3 CONTROLLED RELEASE PROFILES AND MECHANISMS

37.3.1 Types of Controlled Release Profiles

Different drug release profiles and release rates may be required based on the pharmacokinetic and pharmacodynamic need of the medication. Commonly used drug release profiles are illustrated in Figure 37.1. The first four release profiles are based on the time dependence of their release rates, while the last one is based on the onset of drug release. In the case of zero-order release, the release rate

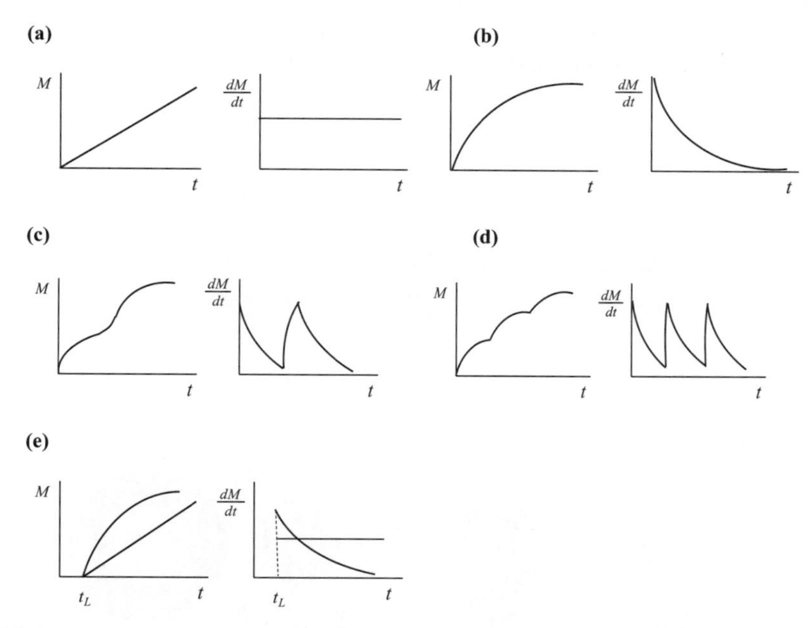

FIGURE 37.1 Schematic illustration of various types of release profiles commonly seen in oral controlled release dosage forms. (a) Zero-order release, (b) first-order release, (c) biphasic release, (d) pulsatile release, and (e) delayed or timed release.

remains constant, reflected by a linear relation between the amount of drug released and time. In the case of first-order release, the release rate decreases exponentially with time. Biphasic, multiphasic, and pulsatile release are typified by two or more modes of release. When there is no drug released until a lag time, t_L, the release profile is called delayed (or timed) release. After t_L, the release profile can be in any shape, such as zero-order or first-order release. It should be noted that there may be differences between the *in vitro* and the *in vivo* controlled release profiles because conditions of pH, hydrodynamics, fluid volume, and presence of enzymes and bacteria vary in the gastrointestinal tract.

37.3.2 Controlled Release Mechanisms and Structure of Controlled Release Systems

Various drug release profiles can be obtained by utilizing different drug release mechanisms, device geometry and structure, and materials. The following five major release mechanisms have been utilized alone or in combinations to design oral controlled release dosage forms: diffusion, erosion/degradation, ion exchange, swelling, and osmotic pressure.

The drug delivery systems are frequently referred to by the mechanism that dominates the drug release rate. Corresponding mathematical models are then derived based on the dominating drug release mechanism, the geometry of the delivery system, and the boundary conditions which the delivery systems are exposed to. The following sections describe the major drug release mechanisms, delivery systems, and their associated release profiles, and present essential mathematical equations of analytical or semianalytical solutions derived from mechanistic models. Interested readers are referred to specialized books [9–14] and original papers for the derivation of the equations.

37.3.3 Controlled Release Via Diffusion

Drug release from a device is considered diffusion-controlled when diffusion of drug molecules through the device is the rate-determining step. Depending on the structure of the delivery system, diffusion-controlled systems can be classified as membrane–reservoir (Figure 37.2a) or monolithic (matrix) systems (Figure 37.2b).

37.3.3.1 Membrane–Reservoir Systems In membrane–reservoir systems, there is a drug-rich core (drug reservoir) enclosed by a membrane, which may or may not contain drug initially. Drug diffusion from the reservoir through the membrane is the rate-limiting step. Each delivery system can be made into various geometries. Figure 37.3 shows membrane–reservoir systems of four basic geometries—slab, cylinder, sphere and disk, which are commonly used for drug delivery. Irrespective of the geometry, membrane–reservoir systems should result in a zero-order release profile as long as the drug core provides a constant drug supply. This is true when an excess amount of solid drug is loaded in the core and drug dissolution is much faster than drug diffusion through the membrane. In this case, drug solution at the inner side of the membrane is maintained at a constant concentration that normally equals the drug saturation solubility. Once the excess drug is dissolved, the drug core can no long provide a constant supply, resulting in a decrease in the release rate.

37.3.3.2 Monolithic (Matrix) Systems In the monolithic (or matrix) systems, uniformly distributed drug is released

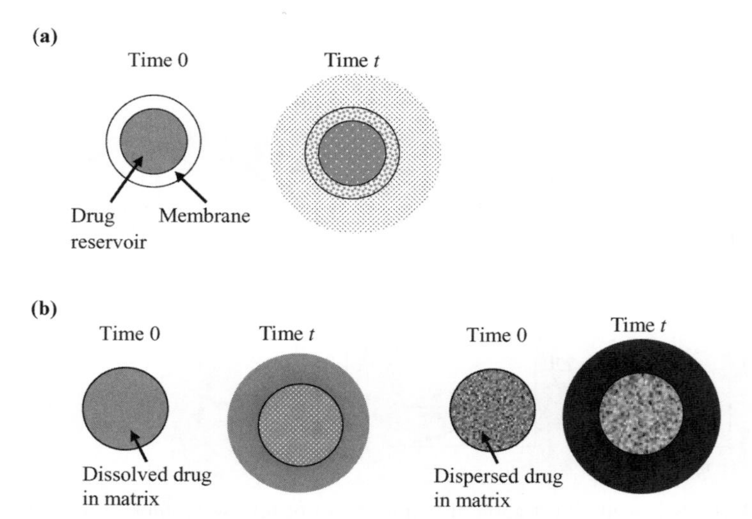

FIGURE 37.2 Schematic illustration of diffusion-controlled systems. (a) Membrane–reservoir system and (b) matrix system.

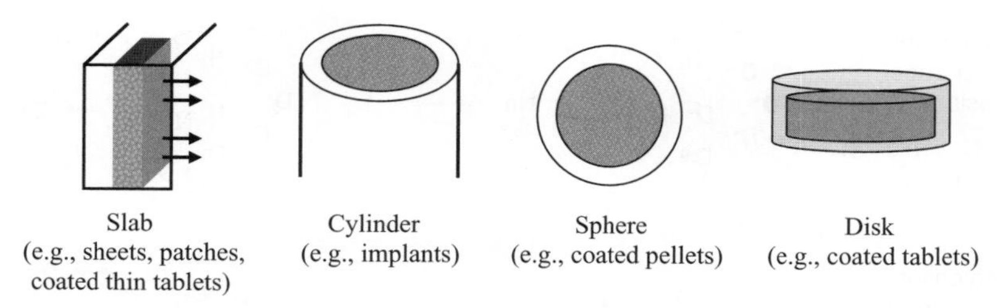

FIGURE 37.3 Membrane–reservoir systems of various geometries.

by diffusion through the matrix. Depending on the loading level and drug solubility in the matrix, the drug may exist as a molecular solution (dissolved drug) or a particle dispersion (dispersed drug). The drug loading level (C_0) relative to the drug solubility (C_s) and initial drug distribution in the matrix can influence the release profiles, as do the volume of release medium and stirring conditions.

A monolithic system usually provides first-order release profiles because the drug concentration within the matrix decreases with time and the diffusional distance increases with time. The nonlinearity of the release curve increases as the device is changed from slab to cylinder and from cylinder to sphere. Figure 37.4 compares release profiles of fractional drug release from one-dimensional slab, cylinder, and sphere with the same characteristic dimension ($a = 0.2$ cm) into a perfect sink, computed using AP-CAD® software.

37.3.4 Controlled Release Via Erosion or Degradation

Erosion- or degradation-controlled systems are special cases of matrix systems, in which matrix erosion or degradation is the rate-limiting step of drug release. Thus, while the rate of drug release for dissolution-based systems depends mostly on the drug solubility, erosion-based systems limit drug release by dissolution (erosion) or degradation of the materials that form the matrix. Pure erosion/degradation-controlled release is hard to find in oral controlled release dosage forms. Matrix erosion and degradation are often concurrent with other release mechanisms such as drug diffusion and dissolution. Biodegradable polymers typically used in erosion-controlled systems have been reviewed elsewhere [15].

37.3.4.1 Heterogeneous Erosion When the matrix is rigid and hydrophobic with minimal hydration in the release medium, entrapped drug is released mainly by matrix surface erosion, that is, heterogeneous erosion (Figure 37.5a). If drug solubility in the medium is very low, drug release rate may still be dictated by matrix erosion (solution) even if the matrix is hydrophilic. The released drug particles may dissolve following release from the matrix. An ideal heterogeneous erosion-controlled system should give a zero-order release if it is a planar shape, or nonlinear release if it is a cylinder or a sphere. Again, the nonlinearity is higher for sphere than cylinder due to more dramatic reduction in the area toward the center of the device.

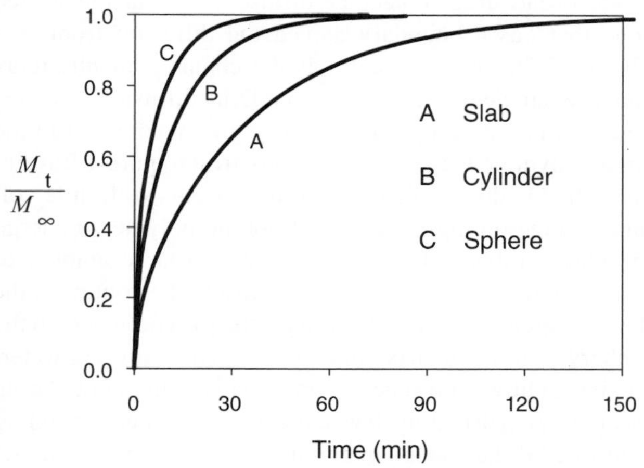

FIGURE 37.4 Comparison of release profiles of fractional drug release from one-dimensional slab, cylinder, and sphere with the same characteristic dimension ($a = 0.2$ cm) into a perfect sink, computed using AP-CAD software.

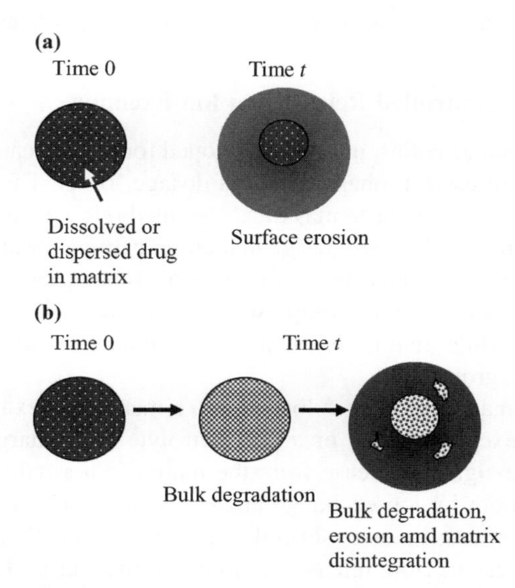

FIGURE 37.5 Schematic illustrations of erosion and degradation-controlled release. (a) Heterogeneous (surface) erosion and (b) homogeneous (bulk) degradation and erosion.

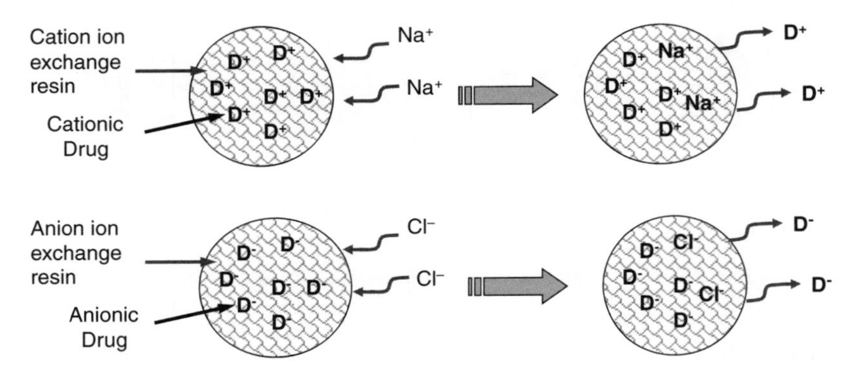

FIGURE 37.6 Schematic of ionic drug release from ion-exchange resin that involves steps of counterion diffusion into the resin, exchange with bound drug, and diffusion of dissociated drug out of the resin.

37.3.4.2 Homogeneous Degradation and Erosion When the matrix undergoes bulk degradation, the molecular weight of the matrix polymer decreases gradually, resulting in a higher drug diffusion coefficient in the matrix with time. Eventually the matrix may disintegrate and dissolve, releasing the remaining drug. This process is named homogeneous degradation and erosion (Figure 37.5b). Usually a first-order release curve is seen for a period of time, followed by an accelerated release when disintegration of the device occurs. The rate of degradation of conventional biodegradable polymers, such as polylactides, is too slow to be suitable for oral controlled release dosage forms that are retained in the gastrointestinal tract for a maximum of about 24 h. Instead, microbially degradable polymers, especially azo cross-linked polymers, which are degraded specifically by colonic bacteria, have been investigated for colonic drug delivery. The release profile of such delivery system is rather complex and can vary from near zero-order to first-order release.

37.3.5 Controlled Release Via Ion Exchange

Ion-exchange resins, initially developed for water treatment, have been used in pharmaceutical dosage forms. Their applications include taste masking, sustained release, and gastric retention. They are designated either cationic or anionic based on the counterions. Cationic ion-exchange resins are comprised of anionic groups such as $-COO^-$ and $-SO_3^-$ groups, while anionic exchange resins contain $-NR_2^+$ or $-NR_3^+$ groups.

When an ionic drug is loaded into a matrix, for example, an ion-exchange resin or a polyeletrolyte with charges of opposite sign, its release from the matrix is normally controlled both by ion-exchange and by diffusion because the release process involves several essential steps: (1) Diffusion of counterions from release medium into the matrix; (2) exchange of counterions with bound drug molecules in the matrix; and (3) diffusion of free drug molecules out from the matrix into the medium.

These steps are depicted in Figure 37.6. In case of hydrophobic polyeletrolytes, matrix swelling may also play a role in the release kinetics. In general, a first-order release profile is seen in ion exchange-controlled delivery systems.

37.3.6 Controlled Release Via Swelling

When a swellable glassy polymer matrix is placed in a thermodynamically compatible solvent, it undergoes an abrupt transition from the glassy state to the rubbery state. Because the polymer chains at the glassy state are rigid, drug diffusion in the glassy region is negligible as compared to that in the rubbery region. Pharmaceutical dosage forms are usually made from swellable hydrophilic polymers such as hydroypropylmethylcellulose and polyethylene oxide. When a matrix tablet made from such polymer and loaded with drug is introduced into an aqueous medium, water penetrates into the matrix, wets the polymer and drug particles therein and fills the pores. In the hydrated layer, drug particles start to dissolve and drug molecules diffuse out from the wetted zone that has a boundary named the diffusion front (see Figure 37.7). The hydrated polymer chains gradually relax and disentangle forming a gel layer. Drug diffusion in the gel layer is much faster than in the dry glassy core and in the slightly hydrated layer as well. In contrast to pure diffusion-controlled hydrophobic matrix drug systems with little volume change during release, hydrophilic polymers undergo the glassy–rubbery transition and absorb large amount of water due to osmotic pressure. As a result, the volume of the device increases, so does the drug diffusion coefficient in the rubbery zone and matrix porosity if high quantities of water-soluble additives are added, or in the case of high initial drug loadings. A matrix drug device is classified as being swelling controlled if the change in polymer morphology by interaction with the external release medium controls or alters the drug release rate. Note that noncross-linked hydrophilic polymers may dissolve before all payload is released, which is often seen in pharmaceutical hydrophilic matrices.

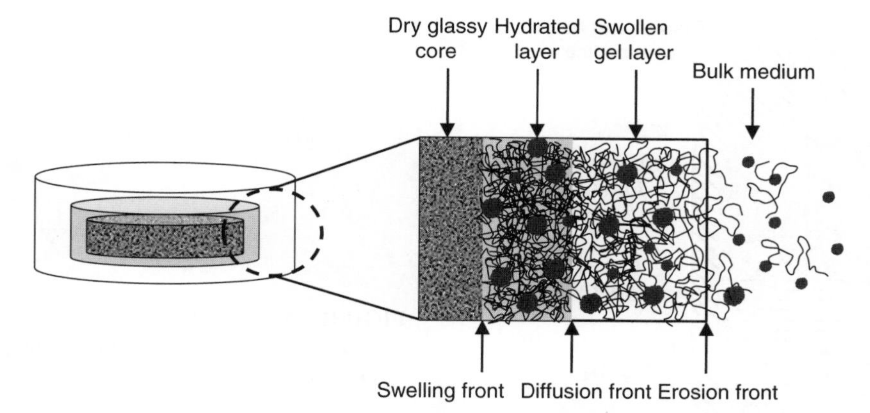

FIGURE 37.7 Illustration of drug release from a hydrophilic matrix tablet by swelling mechanism.

37.3.7 Controlled Release Via Osmotic Pressure

Osmotic-controlled release of drug molecules involves the regulation of osmotic permeation of water through the use of a semipermeable membrane. The diffusion of water across the semipermeable membrane induced by an existing chemical potential gradient between the dissolution medium and the tablet core creates a hydrostatic pressure. The hydrostatic pressure generated by the influx of water forces the release of a saturated solution of the drug through delivery ports in the device. In addition to the mechanism of osmotic pumping, drug release can also take place through the membrane as a result of the solution–diffusion mechanism. Since the device volume is constant, the volume of drug solution delivered will be equal to the volume of osmotic water uptake within a given time interval. Therefore, the rate of drug delivery will be constant as long as a constant osmotic pressure gradient is maintained across the membrane. Prolonged zero-order release can then be achieved with this system. However, as the reservoir concentration falls below saturation, the rate declines asymptotically. It is also conceivable for osmotic systems to achieve release rates much higher than systems that solely involve solution–diffusion mechanism.

Osmotic devices can be manually activated or self-activated. Manually activated devices have to be stored empty and loaded with water prior to use. Other versions have an impermeable seal between the semipermeable membrane and the water chamber, allowing the devices to be stored fully loaded with water. The osmotic pump then becomes activated when the seal is broken. Self-activated devices are activated by water imbibed from the gastrointestinal tract or the dissolution vessel medium driven by the device itself.

37.3.7.1 Rose–Nelson Pump The Rose–Nelson pump [16] shown in Figure 37.8 consists of a drug chamber, a salt chamber containing excess solid salt, and a water chamber. The drug and the water chambers are separated by a rigid semipermeable membrane. Water moves from the water chamber into the salt chamber as a result of the difference in osmotic pressure across the membrane. The increase in volume of the salt chamber as a result of water uptake moves the piston and causes drug to be pumped out of the device.

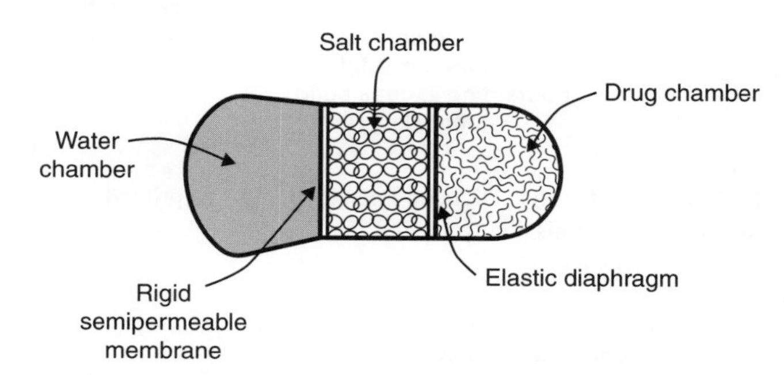

FIGURE 37.8 The three-chamber Rose–Nelson osmotic pump. Reprinted from Ref. 16, Copyright (1995), with permission from Elsevier.

37.3.7.2 Higuchi–Leeper Pump The Higuchi–Leeper pump differs from the Rose–Nelson pump in that the water chamber is absent (Figure 37.9). The Higuchi–Leeper pump usually consists of a salt chamber that contains a fluid solution with excess solid, and a rigid housing with the semipermeable membrane supported on a perforated frame.

37.3.7.3 Higuchi–Theeuwes Pump The semipermeable membrane in the Higuchi–Theeuwes pump acts as the outer casing of the pump. As shown in Figure 37.10, the pump is comprised of a rigid rate-controlling outer semipermeable membrane surrounding a solid layer of salt coated on the inside by an elastic diaphragm and on the outside by the membrane. During its operation, water is osmotically drawn by the salt through the semipermeable membrane. This water increases the volume of the salt chamber, forcing the drug release from the chamber.

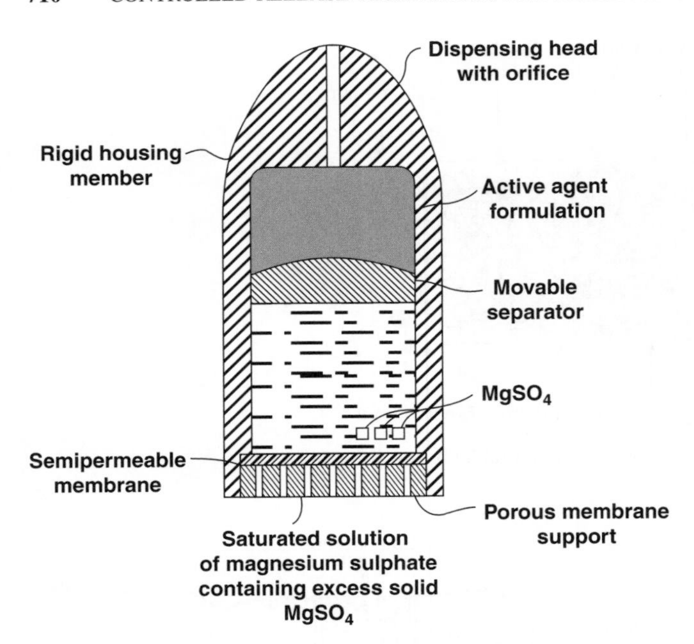

FIGURE 37.9 The Higuchi–Leeper pump. Reprinted from Ref. 16, with permission from Elsevier.

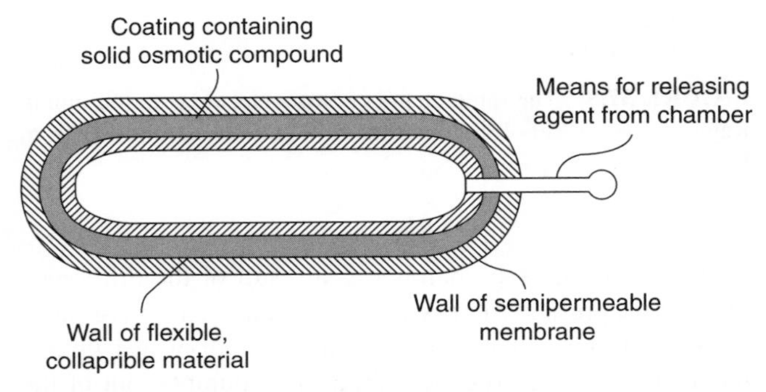

FIGURE 37.10 The Higuchi–Theuwes Pump. Reprinted from Ref. 16, with permission from Elsevier.

*37.3.7.4 **Elementary Osmotic Pump*** In these systems, a semipermeable membrane with a delivery orifice surrounds an osmotic core that contains the drug. The delivery rate from these devices is regulated by the osmotic pressure of the osmotic agent of the core formulation and by the water permeability of the semipermeable membrane (Figure 37.11). For example, the OROS® system developed by ALZA Corporation is used to deliver Acutrim, an over-the-counter appetite suppressant, at a controlled rate [17]. Similarly, Elan Corporation of Ireland has developed MODAS (multi-directional oral absorption system). This system differs from OROS in that it has a multitude of small pores through which the drug can exit.

*37.3.7.5 **Push–Pull Osmotic Pump*** The OROS system described in the previous section is somewhat limited

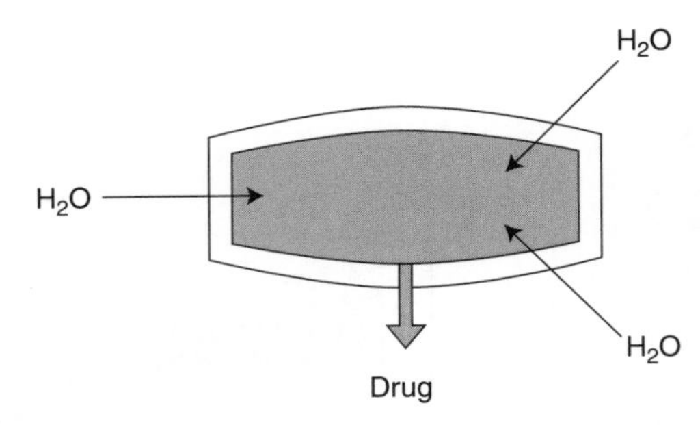

FIGURE 37.11 The elementary osmotic pump. Reprinted from Ref. 29, Copyright (2000), with permission from Marcel Dekker, Inc.

because it can only deliver drugs with good aqueous solubility. The push–pull osmotic pump, which delivers a suspension of drug, was an advancement over the elementary osmotic pump because it could be used for the delivery of low solubility drugs and it could be manufactured using conventional pharmaceutical equipment [18]. For this system, the core consisted of a bilayer tablet with one layer containing a swelling agent and other layer containing the drug formulation. The swelling agent functioned to push a suspension of drug from the orifice (Figure 37.12). ALZA Corporation, which was acquired by Johnson & Johnson in 2001, developed the gastrointestinal therapeutic system (GITS), using this dosage form to deliver Nifedipine to provide once-a-day dosing for hypertension [19].

*37.3.7.6 **Semipermeable Membranes Containing Micropores*** Innovations in the osmotic drug delivery field continued in the 1990s by the development of the controlled porosity osmotic pump tablet (CP-OPT) by Zentner and others [20–22]. The main advancement of the CP-OPT compared to the OROS system was the new design of the semipermeable membrane to contain pores sufficient in size to eliminate the need for laser drilling an orifice. The CP-OPT membrane also contains a pore former and plasticizer. This osmotic dosage form is designed to deliver a drug solution by an osmotic mechanism; therefore, limited in application

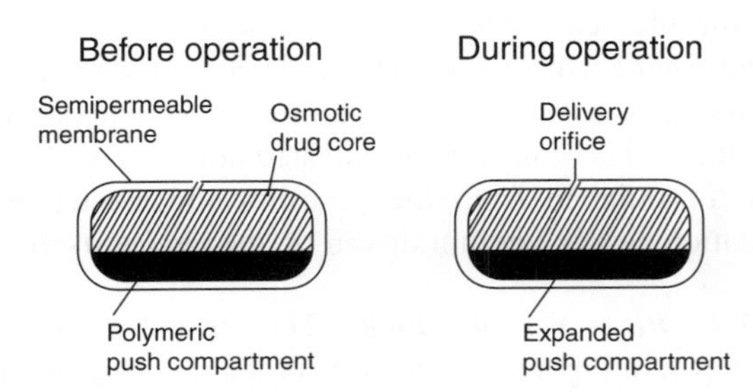

FIGURE 37.12 Illustration of the push–pull osmotic pump known as the gastrointestinal therapeutic systems. Reprinted from Ref. 19, Copyright (1987), with permission from Elsevier.

to soluble compounds. In the late 1990s, Okimoto and Stella [23–27] advanced the CP-OPT technology to encompass poorly water soluble compounds that could be solubilized by Captisol™ (sulfobutyl ether-β-cyclodextrin or $(SBE)_{7m}$-β-CD) that serves as both a solubility enhancing agent and an osmotic agent. The use of $(SBE)_{7m}$-β-CD enabled the osmotic release from CP-OPT of low solubility drugs such as prednisolone, chlorpromazine, and testosterone.

The asymmetric membrane (AM) film-coated tablet is a unique embodiment within the field of osmotic drug delivery. The membrane is formed by a phase inversion process and is composed of a several layers of polymer with a network of interconnecting pores [28]. The polymer acts as a semipermeable barrier while the interconnected pores provide a path for dissolved core components to exit. A laser-drilled orifice is not necessary in the AM system as required for the OROS technology, and similar to the CP-OPT. In fact, the entire AM film coating acts as hundreds of preformed delivery orifices. Therefore, the drug release can be adjusted by varying the type and concentration of the pore former present in the semipermeable membrane as well as the membrane thickness [29]. Unlike the CP-OPT, in the AM tablet design the porous, semipermeable membrane contains polyethylene glycol in a dual role, serving as plasticizer and pore former. The holes through which drug is released are pores created in the tablet coating as a result of the method of coating and polymer solution used or occur when the water-soluble component of the tablet coating is leached out after the tablet is swallowed [16]. As with Theeuwes's elementary osmotic pump, a porous membrane surrounds an osmotic core containing the drug. It has been demonstrated that the mechanism of drug release from spherical beads consisting mainly of phenylpropanolamine hydrochloride and sucrose that were coated with a porous ethylcellulose film is predominantly osmotic, irrespective of film porosity [30]. It has been shown that high water fluxes can be achieved with asymmetric membrane tablets [28, 29]. The asymmetric coating consists of a porous substrate with a thin outer skin. The high water fluxes from these asymmetric coatings permits the osmotic delivery of drugs with lower solubilities [29] (Figure 37.13).

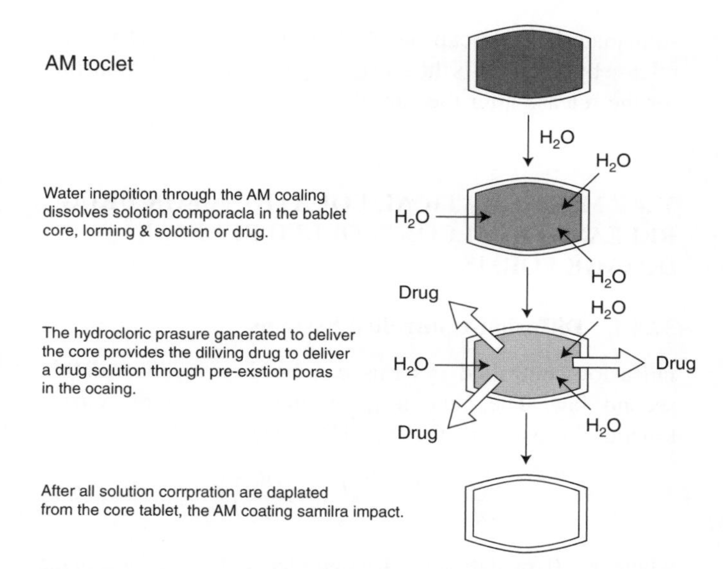

FIGURE 37.13 Semipermeable membrane containing micropores. Reprinted from Ref. 29, Copyright (2000), with permission from Marcel Dekker, Inc.

37.3.7.7 ***Polymer Drug Matrix Systems*** Polymer drug matrix systems are comprised of polymer-encapsulated drug particles dispersed within a polymer matrix (Figure 37.14). Several researchers have postulated different phenomena accounting for drug release. For example, Wright et al. [31] have postulated that drug release occurs as soon as water drawn osmotically in through the encapsulating polymer causes the coating to rupture. An osmotic pressure gradient is then believed to pump the dissolved drug to the surface through fractures created via interconnected pores. In other words, after rupturing, osmotic pressure driven convection is believed to be responsible for the release of the remaining

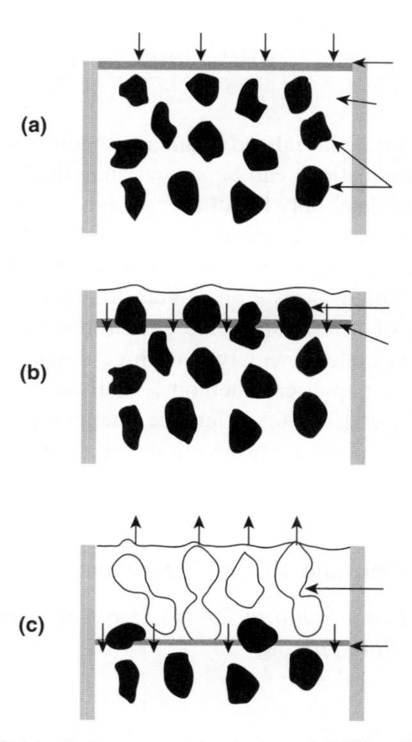

FIGURE 37.14 Polymer matrix system. (a) The diffusion of water to first layer of encapsulated particles. (b) Water imbibition into encapsulated particles. (c) Zone of interconnected capsules, imbibing capsules, and intact capsules. Reprinted from Ref. 32, Copyright (1994), with permission from Elsevier.

solid material in the capsule. According to Amsden et al. [32], release by diffusion is the most likely phenomena responsible for the release after the capsules rupture.

37.4 MATHEMATICAL EQUATIONS FOR DRUG RELEASE FROM CONTROLLED RELEASE DOSAGE FORMS

37.4.1 Diffusion-Controlled Systems

Diffusion-controlled systems can be described by Fick's second law. The general governing equation of release kinetics for one-dimensional (1D) release is

$$\frac{\partial C}{\partial t} = \frac{1}{x^{\alpha}}\left[\frac{\partial}{\partial x}\left(x^{\alpha} D \frac{\partial C}{\partial x}\right)\right] \tag{37.1}$$

where $\alpha = 0$ for slab, $\alpha = 1$ for cylinder, and $\alpha = 2$ for sphere; D is the drug diffusion coefficient in the device; c is the drug concentration as a function of time t and distance x. The 1D release model is applicable to infinite large slab or infinite long cylinder, where drug release from the edge of the slab or the ends of the cylinder is negligible.

For general multidimensional problems, the governing equation is

$$\frac{\partial C}{\partial t} = D\left[\frac{\partial^2 C}{\partial x^2} + \frac{\partial^2 C}{\partial y^2} + \frac{\partial^2 C}{\partial z^2}\right] \tag{37.2}$$

With appropriate initial and boundary conditions, the partial differential equations can be solved analytically or numerically. Up to date, only handful explicit exact solutions and approximate analytical solutions for simple geometries have been obtained. The final expressions are presented below without detailed derivation. Interested readers can find procedures of derivation in the cited references. For complex delivery systems with two- or three-dimensional (2D or 3D) release, numerical methods, such as finite element method [33–35], finite difference method [36, 37], and Monte Carlo method can be employed, which will not be elaborated here.

37.4.2 Membrane–Reservoir Systems

37.4.2.1 Exact Solution for 1D Slab with Constant Reservoir in a Sink [10, 38]

$$M = \frac{DKC_r t}{\delta} + \frac{2KC_r\delta}{\pi^2}\sum_{n=1}^{\infty}\frac{(-1)^n}{n^2}\left[1-\exp\left(\frac{-Dn^2\pi^2 t}{\delta^2}\right)\right] + \frac{4C_m\delta}{\pi^2}\sum_{m=0}^{\infty}\frac{1-\exp\left(\frac{-D(2m+1)^2}{\delta^2}\pi^2 t\right)}{(2m+1)^2} \tag{37.3}$$

where M is the cumulative amount of drug released from unit area; δ is the membrane thickness; C_r is the drug concentration in the reservoir, which is normally taken as the drug solubility in the presence of excess dispersed drug; C_m is the initial drug concentration in the membrane; K is the partition coefficient of drug between the membrane and the reservoir; and D is the drug diffusion coefficient in the membrane. This equation predicts drug released from unit area until all dispersed drug is exhausted. By letting $C_m = 0$ and $t \to \infty$ in equation 37.3, drug released at steady state and time lag, t_l, can be obtained as

$$M = \frac{DKC_r}{\delta}(t - t_l) \tag{37.4}$$

Similarly by letting $C_m = C_r$ and $t \to \infty$, the initial burst time t_b is found from

$$M = \frac{DKC_r}{\delta}(t + t_b) \tag{37.5}$$

where $t_l = \delta^2/(6D)$ and $t_b = \delta^2/(3D)$

37.4.2.2 Analytical Solution for 1D Slab with Nonconstant Reservoir in a Sink [38]

After all solid drug is dissolved, the drug concentration in the reservoir decreases with time. The amount of drug released from unit area is then described by

$$M = C_s V\left[1-\exp\left(-\frac{DK}{\delta V}t\right)\right] \tag{37.6}$$

where V is the volume of the reservoir with unit area. This equation is based on pseudosteady state assumption and mass balance.

Using equation 37.3 for time up to t^*, the time at which all dispersed drug is depleted, and equation 37.6 after t^* one can obtain a release profile covering the entire release course from constant reservoir to nonconstant reservoir. To find t^*, let $M = (C_r - C_s) \times V$; substitute it into the left-hand side of equation 37.3, and then solve for t^*.

37.4.2.3 Exact Solution for 1D Cylinder with Constant Reservoir in a Sink [38]

$$M = \frac{2\pi KC_r Dt}{\ln\left(\frac{b}{a}\right)} + 4\pi\sum_{n=1}^{\infty}\left(\frac{KC_r J_0(b\alpha_n)}{J_0(a\alpha_n) - J_0(b\alpha_n)} + C_m\right) \times \frac{J_0(a\alpha_n)[1-\exp(-D\alpha_n^2 t)]}{\alpha_n^2[J_0(a\alpha_n) + J_0(b\alpha_n)]} \tag{37.7}$$

where M is the cumulative amount of drug released from a cylinder of constant reservoir through a membrane of unit length; a and b, respectively, are the internal and the external radius of the cylindrical membrane, which defines the membrane thickness $\delta = b - a$. α_n values are the positive roots of $J_0(a\alpha_n)Y_0(b\alpha_n) - J_0(b\alpha_n)Y_0(a\alpha_n)$, J_0 and Y_0 are Bessel function of the first and the second kind of order zero.

Similar to slab case, the corresponding steady-state drug released with lag time or initial burst are

$$M = \frac{2\pi DKC_r}{\ln\left(\frac{b}{a}\right)}(t - t_l) \tag{37.8}$$

$$t_l = \frac{2\ln\left(\frac{b}{a}\right)}{D}\sum_{n=1}^{\infty}\frac{J_0(a\alpha_n)J_0(b\alpha_n)}{\alpha_n^2[J_0^2(b\alpha_n) - J_0^2(a\alpha_n)]} \tag{37.9}$$

$$M = \frac{2\pi DKC_r}{\ln\left(\frac{b}{a}\right)}(t + t_b) \tag{37.10}$$

$$t_b = \frac{2\ln\left(\frac{b}{a}\right)}{D}\sum_{n=1}^{\infty}\frac{J_0^2(a\alpha_n)}{\alpha_n^2[J_0^2(b\alpha_n) - J_0^2(a\alpha_n)]} \tag{37.11}$$

37.4.2.4 Analytical Solution for 1D Cylinder with Nonconstant Reservoir in a Sink [38]

$$M = C_s V\left[1 - \exp\left(-\frac{2DK}{a^2\ln\left(\frac{b}{a}\right)}t\right)\right] \tag{37.12}$$

where V is the volume of the cylindrical reservoir with unit length.

37.4.2.5 Exact Solution for Sphere with Constant Reservoir in a Sink [38]

$$M = \frac{4\pi abDKC_r t}{\delta} + \frac{8ab\delta(KC_r - C_m)}{\pi}\sum_{n=1}^{\infty}\frac{(-1)^n}{n^2} \times\left[1 - \exp\left(\frac{-Dn^2\pi^2 t}{\delta^2}\right)\right] + \frac{8b^2\delta C_m}{\pi}\sum_{n=0}^{\infty}\frac{1}{n^2}\left[1 - \exp\left(\frac{-Dn^2\pi^2 t}{\delta^2}\right)\right] \tag{37.13}$$

where a and b, respectively, are the internal and external radius of the spherical membrane, which defines the membrane thickness $\delta = b - a$. This equation describes the drug release from constant reservoir through a spherical membrane. Similar to slab case, the corresponding steady-state drug released with lag time or initial burst are

$$M = \frac{4\pi abDKC_r}{\delta}(t - t_l) \tag{37.14}$$

$$M = \frac{4\pi abDKC_r}{\delta}(t + t_b) \tag{37.15}$$

where $t_l = \delta^2/(6D)$ and $t_b = b\delta^2/(3aD)$.

37.4.2.6 Analytical Solution for Sphere with Nonconstant Reservoir in a Sink [38]

$$M = C_s V\left[1 - \exp\left(-\frac{3bDK}{\delta a^2}t\right)\right] \tag{37.16}$$

where V is the volume of the spherical reservoir.

37.4.2.7 Analytical Solution for 2D Tablet with Constant Reservoir in a Sink [39]

$$M = 2[\pi a^2 M_a + (H - \delta_a)M_r + M_c] \tag{37.17}$$

where M is the cumulative amount released from the axial direction and radial direction, M_a and M_r, are given in equations 37.3 and 37.7, respectively, and M_c is expressed below

$$M_c = \frac{\pi KC_r t}{\ln\left(\frac{a}{b}\right)}\left\{\frac{D_a}{\delta_a}\left[\frac{a^2 - b^2}{2} - a^2\ln\left(\frac{a}{b}\right)\right] - D_r\delta_a\right\} \tag{37.18}$$

where a and b are internal and external radius of the tablet, δ_a is axial membrane thickness, and the radial membrane thickness $\delta_r = b - a$. H is the half-thickness of the tablet. D_a and D_r are the drug diffusion coefficients in the axial and the radial directions, respectively. The parameters for radial and axial directions can be identical for symmetric coating, or different for asymmetric coating. The amount of drug released at the steady state with time lag or initial burst are

$$M = 2\pi KC_r\left(\frac{a^2 D_a}{\delta_a} + \frac{2(H - \delta_a)D_r}{\ln\left(\frac{b}{a}\right)}\right)(t - t_l) \tag{37.19}$$

$$t_l = \frac{\frac{a^2 h}{6} + 4(H - \delta_a)\sum_{n=1}^{\infty}\frac{J_0(a\alpha_n)J_0(b\alpha_n)}{\alpha_n^2[J_0^2(b\alpha_n) - J_0^2(a\alpha_n)]}}{\frac{a^2 D_a}{\delta_a} + \frac{2(H - \delta_a)D_r}{\ln\left(\frac{b}{a}\right)}} \tag{37.20}$$

$$M = 2\pi KC_r\left(\frac{a^2 D_a}{\delta_a} + \frac{2(H - \delta_a)D_r}{\ln\left(\frac{b}{a}\right)}\right)(t + t_b) \tag{37.21}$$

$$t_b = \frac{\frac{a^2 h}{3} + 4(H - \delta_a)\sum_{n=1}^{\infty}\frac{J_0^2(a\alpha_n)}{\alpha_n^2[J_0^2(a\alpha_n) - J_0^2(b\alpha_n)]}}{\frac{a^2 D_a}{\delta_a} + \frac{2(H - \delta_a)D_r}{\ln\left(\frac{b}{a}\right)}} \tag{37.22}$$

37.4.2.8 Analytical Solution for 2D Tablet with Nonconstant Reservoir in a Sink [39]

$$M = C_s V \left\{ 1 - \exp\left[\frac{-2KD_r}{a^2 \ln \frac{b}{a}} - \frac{KD_a}{h(H-\delta_a)} - \frac{\pi K}{V}\left(\frac{D_a}{\delta_a}\left(\frac{a^2-b^2}{2\ln\left(\frac{a}{b}\right)} - a^2 \right) - \frac{D_r \delta_a}{\ln\left(\frac{a}{b}\right)} \right) \right] t \right\} \quad (37.23)$$

where V is the volume of the tablet reservoir.

Although the equations for describing drug release kinetics for membrane–reservoir systems are presented above separately for constant and nonconstant reservoir, the total amount of drug released during the entire course can be combined seamlessly based on mass balance as outlined by Zhou et al. [39].

37.4.3 Monolithic (Matrix) Systems Containing Dissolved Drug ($C_0 \leq C_s$)

37.4.3.1 Exact Solution for 1D Slab with Dissolved Drug ($C_0 \leq C_s$) in a Sink [10]

$$\frac{M_t}{M_0} = 1 - \sum_{n=0}^{\infty} \frac{8}{(2n+1)^2 \pi^2} \exp\left[\frac{-D(2n+1)^2 \pi^2 t}{4l^2} \right] \quad (37.24)$$

where l is the half thickness of the slab, $M_0 = 2C_0 l$, C_0 is the initial drug loading, and D is the drug diffusion coefficient. If l is defined as the thickness of the slab, the equation is

$$\frac{M_t}{M_0} = 1 - \sum_{n=0}^{\infty} \frac{8}{(2n+1)^2 \pi^2} \exp\left[\frac{-D(2n+1)^2 \pi^2 t}{l^2} \right] \quad (37.25)$$

37.4.3.2 Exact Solution for 1D Slab with Dissolved Drug ($C_0 \leq C_s$) in a Well-Stirred Finite Volume [10]

$$\frac{M_t}{M_\infty} = 1 - \sum_{n=1}^{\infty} \frac{2\alpha(1+\alpha)}{1+\alpha+\alpha^2 q_n^2} \exp\left(\frac{-Dq_n^2 t}{l^2} \right) \quad (37.26)$$

where the q_n values are the nonzero positive roots of $\tan q_n = -\alpha q_n$ and effective volume ratio, $\alpha = L/Kl$. K is the partition factor between solute in the slab in equilibrium and that in the solution. L is the thickness of the external volume, on one side of the slab, excluding the space occupied by the half thickness of the slab. Note that in a finite volume, the amount of drug released at equilibrium, M_∞, may be smaller than the initial payload, M_0, because the external medium may be saturated by the released drug before all payload is released. Should the saturation occur, the time for $\frac{M_t}{M_\infty}$ to reach unity is shorter than $\frac{M_t}{M_0}$ and the difference between these two increases as the effective volume ratio decreases. This phenomenon is seen in all geometries [33–35].

37.4.3.3 Exact Solution for 1D Cylinder with Dissolved Drug ($C_0 \leq C_s$) in a Sink [10]

$$\frac{M_t}{M_0} = 1 - \sum_{n=1}^{\infty} \frac{4}{R^2 q_n^2} \exp(-Dq_n^2 t) \quad (37.27)$$

where the q_n values are the roots of $J_0(Rq_n) = 0$ and $J_0(x)$ is the Bessel function of the first kind of order zero, R is the radius of the cylinder, and $M_0 = C_0 \pi R^2$.

37.4.3.4 Exact Solution for 1D Cylinder with Dissolved Drug ($C_0 \leq C_s$) in a Well-Stirred Finite Volume [10]

$$\frac{M_t}{M_\infty} = 1 - \sum_{n=1}^{\infty} \frac{4\alpha(1+\alpha)}{4+4\alpha+\alpha^2 q_n^2} \exp\left(\frac{-Dq_n^2 t}{R^2} \right) \quad (37.28)$$

where the q_n values are the nonzero positive roots of $\alpha q_n J_0(q_n) + 2J_1(q_n) = 0$, $J_1(x)$ is the Bessel function of the first order, $\alpha = V/(\pi R^2 K)$,K is the partition factor between solute in the cylinder and that in the medium at equilibrium, and V is the external volume excluding the space occupied by the cylinder.

37.4.3.5 Exact Solution for Sphere with Dissolved Drug ($C_0 \leq C_s$) in a Sink [10]

$$\frac{M_t}{M_0} = 1 - \frac{6}{\pi^2} \sum_{n=1}^{\infty} \frac{1}{n^2} \exp\left(\frac{-Dn^2 \pi^2 t}{R^2} \right) \quad (37.29)$$

where R is the radius of the sphere and $M_0 = 4\pi R^3 C_0/3$.

37.4.3.6 Exact Solution for Sphere with Dissolved Drug ($C_0 \leq C_s$) in a Well-Stirred Finite Volume [10]

$$\frac{M_t}{M_\infty} = 1 - \sum_{n=1}^{\infty} \frac{6\alpha(1+\alpha)}{9+9\alpha+\alpha^2 q_n^2} \exp\left(\frac{-Dq_n^2 t}{R^2} \right) \quad (37.30)$$

where q_n values are the nonzero positive roots of $\tan q_n = 3q_n/(3+\alpha q_n^2)$ and $\alpha = 3V/(4\pi R^3 K)$. K is the partition factor between solute in the sphere in equilibrium and that in the solution. V is the external volume excluding the space occupied by the sphere.

37.4.3.7 Exact Solution for 2D Tablet with Dissolved Drug ($C_0 \leq C_s$) in a Sink [40]

Considering drug release from all surfaces of a matrix tablet with symmetric properties, Fu et al. derived an exact solution:

$$\frac{M_t}{M_0} = 1 - \frac{8}{H^2 R^2} \sum_{m=1}^{\infty} \exp(-D\alpha_m^2 t)(\alpha_m^{-2}) \sum_{n=1}^{\infty} \exp(-D\beta_n^2 t)(\beta_n^{-2}) \quad (37.31)$$

where α_m values are the roots of $J_0(R\alpha)=0, \beta_n=(2n+1)\pi/2H, R$ is the radius, and H is the half-thickness of a tablet.

37.4.4 Monolithic (Matrix) Systems Containing Dispersed Drug ($C_0 > C_s$)

37.4.4.1 Exact Solution for 1D Slab with Dispersed Solute ($C_0 > C_s$) in a Sink [41]

$$M_t = \frac{2C_s}{\mathrm{erf}(\beta^*)}\sqrt{\frac{Dt}{\pi}} \qquad (37.32)$$

β^* can be found from the following equation

$$\sqrt{\pi}\beta^*\exp(\beta^*)\mathrm{erf}(\beta^*) = \frac{C_s}{C_0 - C_s} \qquad (37.33)$$

where C_0 is the initial drug concentration and C_s is drug solubility in the matrix. This equation predicts a linear plot of M_t versus $\sqrt{t}$, that is, a square root relationship. It is only applicable when the excess dispersed drug is present.

37.4.4.2 Analytical Solution for 1D Slab with Dispersed Solute ($C_0 > C_s$) in a Sink [42]

$$M_t = \sqrt{DC_s(2C_0 - C_s)t} \qquad (37.34)$$

This solution calculates drug released from unit area based on pseudosteady state assumption and is generally applicable for $C_0 > 3C_s$.

37.4.4.3 Analytical Solution for 1D Cylinder with Dispersed Solute ($C_0 > C_s$) in a Sink [43]

$$M_t = \pi C_0(R_0^2 - r^2) \qquad (37.35)$$

$$\frac{r^2}{2}\ln\left(\frac{r}{R_0}\right) + \frac{R_0^2 - r^2}{4} = \frac{C_s Dt}{C_0} \qquad (37.36)$$

where R_0 is the radius of the cylinder and r is the moving front of dispersed drug. For a given series of r such as $r_1, r_2, \ldots, r_n$, solve for $M_{t1}, M_{t2}, \ldots M_{tn}$ from equation 37.35 and $t_1, t_2, \ldots, t_n$ from equation 37.36, and then correlate M_t and t to get a release profile This solution is based on pseudosteady state assumption and is generally applicable for $C_0 > 3C_s$.

37.4.4.4 Analytical Solution for a Sphere with Dispersed Solute ($C_0 > C_s$) in a Sink [42]

$$M_t = \frac{4}{3}\pi R_0^3 C_0 - 4\pi\left[\frac{r^3 C_0}{3} + \frac{C_s r}{6}(R_0^2 + rR_0 - 2r^2)\right] \qquad (37.37)$$

$$t = \frac{1}{6DC_sR_0}\left[C_0(R_0^3 + 2r^3 - 3R_0r^2) + C_s\left(4r^2R_0 + R_0^3\ln\frac{R_0}{r} - R_0^3 - R_0^2r - 2r^3\right)\right] \qquad (37.38)$$

where R_0 is the radius of the sphere, r is dispersed drug moving front. Using the same approach given in the case of 1D cylinder a correlation between M_t and t is obtained. This solution is based on pseudosteady state assumption and is generally applicable for $C_0 > 3C_s$.

37.4.4.5 Analytical Solution for a 2D Tablet with Dispersed Solute ($C_0 > C_s$) in a Sink [44]

$$M_t = 2C_0\pi[H(R^2 - r^2) + zr^2] \qquad (37.39)$$

where R is radius and H is half thickness, C_0 is initial drug loading, and r and z are the moving front of dispersed drug in the radial and the axial directions, respectively. They are given as follows

$$z = \sqrt{\frac{2D_a}{D_r}\left[\frac{r^2}{2}\ln\frac{r}{R} + \frac{R^2 - r^2}{4}\right]} \qquad (37.40)$$

$$t(r) = \frac{C_0}{C_sD_r}\left[\frac{r^2}{2}\ln\frac{r}{R} + \frac{R^2 - r^2}{4}\right] \qquad (37.41)$$

For a given series of r such as $r = 0, r_1, \ldots, R$, solve for corresponding t and z from equations 37.40 and 37.41, and then substitute r and z into equation 37.39 to calculate M_t and correlate M_t with t. This solution is based on pseudosteady state assumption and is generally applicable for $C_0 \gg C_s$.

37.4.4.6 Assumptions, Applications, and Implementations of Models for Diffusion-Controlled Systems Certain important assumptions were used in the derivation of the models presented above for membrane reservoir and matrix systems such as dissolution much faster than diffusion, constant material properties and no dimensional change during the complete release process. If these assumptions can be justified for a given delivery system and release process, the mechanistic models for diffusion-controlled release can be applied for prediction of release kinetics, sensitivity tests of formulation variables, parameter identification of dosage forms, and *in vitro–in vivo* correlation. Compared with regression models such as $M_t = kt^n$, mechanistic models can reveal more information about effects of important formulation variables on drug release kinetics, such as dimension (R, l, δ), geometry, material properties (D, C_s, K), and initial loading (C_0, C_m). It is noticed that transcendental expression and nonlinear equations are involved in the mechanistic models, which is cumbersome for daily usage. However several computer software packages for dosage form design such as AP-CAD and Simulation Plus have been developed to implement the computation tasks.

Mathematical models aforementioned for diffusion-controlled systems describe the general trends of the drug release process. However, one major assumption made is that drug dissolution is much faster than drug diffusion, which means all or part of the drug, depending on the drug solubility, has dissolved in the beginning of the release process. As a phenomenological approximation, it is acceptable for quick dissolving drugs. While it may not be suitable for poorly water-soluble drugs and thus the drug dissolution process needs to be taken into account. Improved models have been proposed to embrace both drug diffusion and dissolution processes [45–49]. The governing equations describe diffusion- and dissolution-controlled drug release processes are presented as follows for a one-dimensional slab problem. The second term on the right-hand side of equation 37.42 depicts the change rate of concentration of dispersed drug due to drug dissolution that is described by equation 37.43 as an example.

$$\frac{\partial C_{\mathrm{d}}}{\partial t} = D\frac{\partial^2 C_{\mathrm{d}}}{\partial x^2} - \frac{\partial C_{\mathrm{sd}}}{\partial t} \quad (37.42)$$

$$\frac{\partial C_{\mathrm{sd}}}{\partial t} = -K_{\mathrm{d}}(C_{\mathrm{s}} - C_{\mathrm{d}}) \quad (37.43)$$

where C_{d} is the concentration of dissolved drug, C_{sd} is the concentration of dispersed drug, and K_{d} is the dissolution rate coefficient of drug.

For the coupled partial differential equations of diffusion and dissolution, explicit exact or analytical solutions, such as those for diffusion-controlled systems, have not been found yet. Hence, numerical approaches such as finite element [47, 48] and finite difference [49] methods have been used to solve this mathematical problem.

37.4.5 Erosion-Controlled Systems

37.4.5.1 Surface (Heterogeneous) Erosion

Analytical Solutions for a 1D Slab, Cylinder, and Sphere [50]

$$\frac{M_t}{M_\infty} = 1 - \left(1 - \frac{k_0 t}{aC_0}\right)^n \quad (37.44)$$

where k_0 is surface erosion constant (mg/(hr-cm^2)), a is the radius of a sphere or a cylinder or the half thickness of a slab, and $n = 1, 2$, and 3 for slab, cylinder, and sphere, respectively.

Analytical Solutions for a 2D Tablet [14]

$$\frac{M_t}{M_\infty} = 1 - \left(1 - \frac{k_0 t}{r_0 C_0}\right)^2 \left(1 - \frac{2k_0 t}{l_0 C_0}\right) \quad (37.45)$$

where r_0 and l_0, respectively, are the initial radius and initial thickness of the tablet.

37.4.5.2 Bulk (Homogeneous) Erosion There are few explicit analytical solutions available for bulk erosion problems. Lee developed a model for drug release from an erodible slab with consideration of simultaneous diffusion and erosion processes [51]. A more comprehensive model including erosion, diffusion, and chemical reaction was developed by Thombre and Himmelstein [52–54] for a slab in a sink. The model considered water, drug, acid generator, and acid with partial differential equations as follows:

$$\frac{\partial C_i}{\partial t} = \frac{\partial}{\partial x}\left(D_i(x,t)\frac{\partial C_i}{\partial x}\right) + v_i \quad i = A, B, C, E \quad (37.46)$$

where C_i and D_i, respectively, are the concentration and diffusion coefficient of the diffusing species and v_i is the net sum of synthesis and degradation rate of species. A, B, C, and E are water, acid generator, acid, and drug respectively. Concentration-dependent diffusion coefficient is expressed as

$$D_i = D_i^0 \exp\left[\frac{\mu(C_D^0 - C_D)}{C_D^0}\right], \quad i = A, B, C, E \quad (37.47)$$

Finite difference method was used to solve the equations with various initial and boundary conditions.

37.4.6 Ion Exchange-Controlled Systems

37.4.6.1 Drug Loading Onto Ion-Exchange Spheres [55]

$$F = \frac{3}{\lambda\theta_0} B_1 \left[\sqrt{\tau}\exp\left(-\frac{1}{4\tau}\right) + \sqrt{\pi}\tau\,\mathrm{erf}\left(\frac{1}{2\sqrt{\tau}}\right) - 2\sqrt{\tau}\right] \quad (37.48)$$

where F is the fraction of drug loaded onto the sphere, $\lambda = 3V/(4\pi R^3)$, $\theta_0 = C_0/C_{\max}$, V is the external fluid volume, R is the radius of sphere, C_0 is the initial solute concentration in the external solution, and $C_{\max}$ is the maximum solute binding capacity of the ion-exchange spheres, $\tau = Dt/R^2$, D is diffusion coefficient of polymer, and B_1 is obtained by solving equation 37.49.

$$\frac{B_1}{\sqrt{\tau}} = (1-\alpha)B_1\gamma - \beta[1 + B_1\gamma] \left\{\theta_0 - \frac{3B_1}{\lambda}\left[\sqrt{\tau}\exp\left(-\frac{1}{4\tau}\right) + \tau\gamma - 2\sqrt{\tau}\right]\right\} \quad (37.49)$$

where $\alpha = VRK_{\mathrm{des}}/(DA)$, $\beta = VRK_{\mathrm{ads}}C_{\max}/(DA)$, and $\gamma = \sqrt{\pi}\,\mathrm{erf}[1/(2\sqrt{\tau})]$, A is the surface area of sphere, K_{des} is the dissociation rate constant, and K_{ads} is the association rate constant.

37.4.6.2 Drug Release from Ion-Exchange Spheres [56]

$$\frac{M_\tau}{M_0} = 1 - \frac{3}{\theta_{\mathrm{RS}}^0}\int_0^1 \left(\theta_{\mathrm{s}^+} + \frac{K\theta_{\mathrm{s}^+}}{\theta_{\mathrm{Na}^+} + K\theta_{\mathrm{s}^+}}\right) x^2 dx \quad (37.50)$$

$$\frac{M_\tau}{M_0} = 1 - \frac{3}{\theta_{\mathrm{RS}}^0}\int_0^1 \left(\theta_{\mathrm{s}^+} + \frac{\sqrt{K^2\theta_{\mathrm{s}^+}^4 + 8K\theta_{\mathrm{s}^+}^2\theta_{\mathrm{Ca}^{2+}}} - K\theta_{\mathrm{s}^+}^2}{4\theta_{\mathrm{Ca}^{2+}}}\right) x^2 dx \quad (37.51)$$

where $M_0 = 4\pi R^3 C_{RS}^0/3$, C_{RS}^0 is the initial concentration of the drug in the sphere which is bound with binding sites of the ion-exchange polymer, K is Langmuir isotherm constant, $x = r/R$, C_m is maximum solute binding capacity of the ion-exchange sphere. $\theta_{s^+} = C_{s^+}/C_m$, $\theta_{Na^+} = C_{Na^+}/C_m$, and $\theta_{Ca^{2+}} = C_{Ca^{2+}}/C_m$, detailed numerical procedure can be found from the original reference [56].

37.4.7 Swelling-Controlled Systems

Swelling-controlled release involving solvent penetration into and drug release out from a polymeric matrix system, such as 1D planar and cylindrical devices, beads or 2D tablets, have been modeled and solved numerically [57–59]. The following presents governing equations for a 2D tablet in a perfect sink. Equations 37.52–37.54 describe the rates of solvent penetration, drug diffusion and dissolution, respectively. Dimensional change due to swelling and matrix erosion is expressed by equation 37.55, where the first term on the right-hand side represents the dimensional increase from swelling and the second term for polymer dissolution. Water concentration-dependent diffusion coefficients are expressed by equations 37.56 and 37.57.

$$\frac{\partial C_w}{\partial t} = \frac{1}{r}\frac{\partial}{\partial r}\left(rD_w\frac{\partial C_w}{\partial r}\right) + \frac{\partial}{\partial z}\left(Dw\frac{\partial C_w}{\partial z}\right) \quad (37.52)$$

$$\frac{\partial C_d}{\partial t} = \frac{1}{r}\frac{\partial}{\partial r}\left(rD_d\frac{\partial C_d}{\partial r}\right) + \frac{\partial}{\partial z}\left(D_d\frac{\partial C_d}{\partial z}\right) - \frac{\partial C_{sd}}{\partial t} \quad (37.53)$$

$$\frac{\partial C_{sd}}{\partial t} = -K(C_s - C_d) \quad (37.54)$$

$$z_t r_t^2 = 2\int_0^{z_t}\int_0^{r_t}\left[\frac{C_w(r,z,t)}{\rho_w} + \frac{C_d(r,z,t)}{\rho_d}\right] r\,\mathrm{d}r\,\mathrm{d}z + \frac{1}{2\pi\rho_p}\left(m_{p,0} - \int_0^t K_p A_s \mathrm{d}t\right) \quad (37.55)$$

$$D_w = D_w^{eq}\exp\left[-\beta_w\left(1 - \frac{C_w}{C_w^{eq}}\right)\right] \quad (37.56)$$

$$D_d = D_d^{eq}\exp\left[-\beta_d\left(1 - \frac{C_w}{C_w^{eq}}\right)\right] \quad (37.57)$$

where C_d is the concentration of dissolved drug, C_{sd} is the concentration of dispersed drug, C_s is the drug solubility, K is the dissolution rate coefficient of drug, C_w is the solvent concentration, D_d and D_w are the diffusion coefficients of drug and solvent, respectively, D_d^{eq} and D_d^{eq} are equivalent coefficients of drug and solvent at saturated solvent state, β_d and β_w are characteristic constants for drug and solvent, respectively. For initial drug loading below the drug solubility, the dissolution term in equation 37.53 can be omitted.

37.4.8 Osmotic Pressure-Controlled Systems

The basic equations for the osmotic component of drug release versus time from osmotic pressure-controlled systems are obtained by expressing the mass delivery rate ($\mathrm{d}m/\mathrm{d}t$) from the dosage form as a product of the total volumetric flow rate ($\mathrm{d}V/\mathrm{d}t$) of water into the interior of the device and the concentration of drug, C, in the solution or suspension being released. For several osmotic dosage forms, the expression for the volumetric flow rate is derived from irreversible thermodynamics.

37.4.8.1 Miniosmotic Pump The pumping rate is given by the following equation:

$$\frac{\mathrm{d}m_t}{\mathrm{d}t} = \frac{A\theta\Delta\pi C}{h} \quad (37.58)$$

where $\frac{\mathrm{d}m_t}{\mathrm{d}t}$ is the drug release rate, C is the concentration of the drug in the chamber, A is the surface area of the membrane, θ is the osmotic permeability, h is the membrane thickness, and $\Delta\pi$ is the osmotic pressure difference between the two solutions on either side of the membrane.

37.4.8.2 Elementary Osmotic Pump The release of drug from this system is controlled by the solvent influx (water) across the semipermeable membrane whereby this influx of water carries the drug to the outside via the orifice. According to Theeuwes [60], the general expression for the solute delivery rate, $\frac{\mathrm{d}m}{\mathrm{d}t}$, obtained by pumping through the orifice can be described by

$$\frac{\mathrm{d}m}{\mathrm{d}t} = \frac{\mathrm{d}V}{\mathrm{d}t}C \quad (37.59)$$

where C is the concentration of the compound in the dispensed fluid expressed per unit volume of the solution, and $\frac{\mathrm{d}V}{\mathrm{d}t}$ is the volume flux across the semipermeable membrane. The volume flux is described as follows:

$$\frac{\mathrm{d}V}{\mathrm{d}t} = \frac{A}{h}L_p(\sigma\Delta\pi - \Delta P) \quad (37.60)$$

where $\Delta\pi$ and ΔP are the osmotic and hydrostatic pressure differences, respectively, between the inside and the outside of the device; L_p is the hydraulic permeability; σ is the reflection coefficient; A is the membrane area; and h is the membrane thickness. Since $\Delta\pi \gg \Delta P$ and the hydrostatic pressure inside the device is minimized as the delivery orifice increases, ΔP can be omitted from equation 37.60. Furthermore, when the osmotic pressure of the core, π, is significantly larger than the osmotic pressure of the dissolution fluid, equation 37.58 can be written as

$$\frac{\mathrm{d}m}{\mathrm{d}t} = \frac{A}{h}\theta\pi C \quad (37.61)$$

where θ equals $L_p\sigma$.

Theeuwes [60] characterized the mode of release mathematically, namely, zero-order delivery rate and nonzero-order release, over the entire life of the system.

$$\left(\frac{dm}{dt}\right)_z = \frac{A}{h}\theta\pi_s S \quad (37.62)$$

Equation 37.62 defines the zero-order release rate from $t=0$ until a time t_z, where S is the solubility, and π_s is the osmotic pressure at saturation. The solubility in equation 37.62 replaces the concentration term, C, from time $t=0$ to $t=t_z$ by assuming the rate of dissolution of a single compound within the system is much larger than the rate of pumping.

The nonzero-order release rate, as defined by Theeuwes [60], as a function of time, indicates a parabolic decline:

$$\frac{dm}{dt} = \frac{F_s S}{\left[1+\frac{F_s}{V}(t-t_z)\right]^2} \quad (37.63)$$

where F_s is the flux during the zero-order time and is related to the volume flux, F, into the device during nonzero release by

$$\frac{F_s}{F} = \frac{\pi_s}{\pi} = \frac{S}{C} \quad (37.64)$$

Moreover, the nonzero release rate can be further written as a fraction of the zero-order rate:

$$\frac{dm}{dt} = \frac{\left(\frac{dm}{dt}\right)_z}{\left[1+\frac{1}{SV}\left(\frac{dm}{dt}\right)_z(t-t_z)\right]^2} \quad (37.65)$$

In order for the aforementioned equations defining the mode of drug release from an elementary osmotic device to be applicable, osmotic pumping has to be the sole mechanism of release. Therefore, the size of the orifice must be such that it is smaller than a maximum size to minimize the solute diffusion through the orifice. It is also imperative to have the orifice larger than a minimum size to reduce hydrostatic pressure inside the system. Hydrostatic pressure within the system will decrease the osmotic influx as well as it may cause an increase in the volume of the system.

The equations for the elementary osmotic pump represent the mass delivered per unit time due to the mechanism of osmotic pumping. In fact, the total mass delivered per unit time from such systems results from osmotic pumping, diffusion through the orifice, and diffusion through the membrane itself [60]. If diffusion through the orifice is negligible, we have

$$\left(\frac{dm}{dt}\right)_t = \left(\frac{dm}{dt}\right)_o + \left(\frac{dm}{dt}\right)_d \quad (37.66)$$

where $\left(\frac{dm}{dt}\right)_0$ is the rate of release due to osmotic pumping, and $\left(\frac{dm}{dt}\right)_d$ is the release rate resulting from diffusion, as demonstrated by Zentner et al., for KCl release rates from controlled porosity osmotic tablet [61].

The total zero-order release rate during the steady state portion can then be expressed by

$$\left(\frac{dm}{dt}\right)_{t,z} = \frac{A}{h}(\theta\pi_s S + PS) \quad (37.67)$$

where P is the permeability coefficient of the drug in the polymer. In a similar fashion, the total nonzero-order rate can be given as

$$\frac{dm}{dt} = \frac{F_s}{S}C^2 + \frac{A}{h}PC \quad (37.68)$$

However, to express $\frac{dm}{dt}$ as a function of time, the concentration, C, inside the system must be expressed as a function of time.

37.4.8.3 Semipermeable Membrane containing Micropores Many researchers have shown that besides simple diffusion, osmotic pumping mechanism contributes significantly to the release of drugs from film-coated preparations [30, 62]. The zero-order steady state release for such systems under the influence of zero hydrostatic pressure can be expressed by equation 37.67. According to Lindstedt et al. [63], the equations presented by Theeuwes cannot describe osmotic pumping as contributing to the mechanism of drug release. They argue that since during zero-order release a steady state is maintained with no volume expansion of the tablets, the net bulk volume flux through the membrane must be zero. It therefore follows that release of drugs would be independent of osmotic pressure and be exclusively diffusive. Therefore, they presented the solute flux, F_s, as follows:

$$F_s = CF_v(1-\sigma) + \frac{A}{h}P_s\Delta C \quad (37.69)$$

where

$$F_v = \frac{A}{h}L_p(\Delta P - \Delta\pi) \quad (37.70)$$

In order to remove the limitation of zero net bulk volume flux through the membrane, the membrane was considered to consist of two areas with different reflectivity. The release rate, Q, is the sum of the solute fluxes in areas 1 and 2:

$$Q = F_{s1} + F_{s2} = (1-\sigma_1)F_{v1}C_1 + (1-\sigma_2)F_{v2}C_2 + D_s \quad (37.71)$$

where $F_{v1} + F_{v2} = 0$ at steady state, since they are equal and in opposite direction, and

$$D_s = \left(\frac{A_1}{h_1} P_{s1} + \frac{A_2}{h_2} P_{s2}\right) \Delta C \qquad (37.72)$$

is the diffusional release through areas 1 and 2. Equation 37.70 reduces to

$$Q = (1-\sigma_2)(-F_{v1})C_s + D_s \qquad (37.73)$$

The assumptions were that the bulk volume flux in area 1 is directed into the tablet and C_1 is zero and C_2 is equal to the concentration in the core, C_s, and $-F_{v1} = F_{v2}$. Therefore,

$$Q = (1-\sigma_2)\frac{A_1}{h_1} L_{p1}(\sigma_1 \Delta\pi - \Delta P)C_s + D_s \qquad (37.74)$$

It was further inferred that the low-reflective area is very small compared to the total area. From this inference, it was concluded that the release rate is calculated from

$$Q = (1-\sigma_2)\frac{A}{h} L_{p1}(\sigma_1 \Delta\pi - \Delta P)C_s + D_s \qquad (37.75)$$

If ΔP is assumed negligible and $\sigma_2 < \frac{\Delta P}{\Delta \pi}$ since the volume flux is directed out of the tablet and $F_{v2} > 0$, equation 37.74 is reduced to

$$Q = \frac{A}{h} L_{p1}\sigma_1 \Delta\pi C_s + D_s \qquad (37.76)$$

37.4.8.4 Push–Pull Osmotic Pump Drug release from this system is controlled by the solvent influx (water) across the semipermeable membrane into the tablet core and resulting simultaneous push action from the swelling layer. According to Swanson et al. [19], the general expression for the solute delivery rate, $\frac{dm}{dt}$, obtained by pumping through the orifice can be simply modified from equation 37.59, is given as

$$\frac{dm}{dt} = \frac{dV}{dt} C_s \qquad (37.77)$$

where C_s is the concentration of the drug in suspension in the dispensed fluid expressed per unit volume of the solution.

The osmotic volume flow into the osmotic compartment is described as

$$(dV/dt)_o = k/h A_p(H)\, \pi_p(H) \qquad (37.78)$$

where k is the osmotic membrane permeability coefficient, h is the membrane thickness, A_p is the area of the push compartment, and π_p is the imbibition pressure of the push compartment. An additional consideration for the push–pull system is the osmotic volume imbibition flow into the drug compartment is described as

$$(dV/dt)_D = k/h(A - A_p(H))\, \pi_D(H) \qquad (37.79)$$

where A is the total area of the dosage form and π_D is the imbibition pressure in the drug compartment.

The total volume flow from the dosage form is the summation of the osmotic flow into the osmotic compartment and the osmotic imbibition flow into the drug compartment, as described below:

$$dV/dt = (dV/dt)_o + (dV/dt)_D \qquad (37.80)$$

The concentration of dispensed drug from the dosage form can be expressed as

$$C_s = F_D\, C_o \qquad (37.81)$$

where F_D is the fraction of drug in the drug compartment and C_o is the concentration of solids dispensed from the dosage form.

Substituting equations 37.78 and 37.79 into equation 37.80, and substituting equations 37.80 and 37.81 into equations 37.77, the total drug release is given as

$$dm/dt = [k/h A_p(H)\, \pi_p(H)] + [k/h\,(A - A_p(H))\, \pi_D(H)]\, F_D\, C_o \qquad (37.82)$$

37.4.8.5 Polymer Drug Matrix Systems The osmotic release mechanism from the polymer matrix-type of device is expressed by the model developed by Wright et al. [31]:

$$\frac{dm}{dt} = \frac{3\alpha\rho\phi\Delta\Pi S_o L_p}{d(\lambda_b^3 - 1)\left[1 - \left(\frac{6\phi}{\pi}\right)^{1/3}\right]} \qquad (37.83)$$

where $\frac{dm}{dt}$ is the zero-order release rate, α is a constant of proportionality, ρ is the solid density of the drug, ϕ is the volumetric loading of the drug, $\Delta\Pi$ is the osmotic pressure difference between the capsule solution and the external medium, S_o is the surface area of the device, L_p is the polymer hydraulic permeability, d is the particle size, and λ_b is the polymer extension ratio at rupture. In deriving this model, it was assumed that the drug particles were spherical and were released by osmotic rupturing. Also, the matrix was considered to consist of two zones, a ruptured capsule zone separated from a zone of water imbibing capsules by a moving water front.

Schirrer et al. [64] developed the following model by assuming that the water flow into a capsule per unit time per unit area at the capsule–polymer interface is constant for a given osmotic agent and is directly proportional to the volumetric loading:

$$\frac{1}{Q_o}\frac{dQ}{dt} = \frac{S_o \Phi}{V_o \phi^{1/3} 3(\lambda_b - 1)} \qquad (37.84)$$

where Φ is the water flow into a capsule per unit time per unit area at the capsule–polymer interface, Q is the volume of salt released, Q_o is the initial volume of salt in the matrix, and V_o is the initial volume of the device. The rate of mass drug release is defined as

$$\frac{dm}{dt} = \rho \frac{dQ}{dt} \tag{37.85}$$

According to Amsden et al. [65], these models are limited because of their dependence on drug density and undefined constant of proportionality. Further, it is believed that at low drug loading a portion of the drug will be released by dissolution and diffusion. Based on these premises, a model was developed. Paramount to this model is the assumption that the solutes that remain in a capsule after it ruptures were released by diffusion and not convection. Also, the model accounts for capsule swelling and that not all capsules in the monolith were ruptured. The model is expressed as follows:

$$\frac{dm}{dt} = \frac{8\pi\{f(1-F_D)\phi\}^{2/3} S_o L_p E \omega C_{sat}(\lambda_b^3 - 1)}{3dt_b^*} \tag{37.86}$$

where f is the mass fraction of material remaining after the initial burst that is released by osmotic pressure induced polymer rupturing, $f(1-F_D)$ is the mass fraction of particles in the monolith released by rupturing, F_D represents the mass fraction of particles released by dissolution and diffusion, S_o is as defined previously, and E is the Young's modulus of elasticity.

$$f = 1 - \exp\left(-\frac{\pi}{4}\left(\frac{h_c}{h}\right)^2\right) \tag{37.87}$$

where h_c and h are the critical wall thickness for rupturing to occur and the average wall thickness, respectively.

$$C_s\omega = C_s \exp\left[\frac{-DA_o(t_b)}{V_f h}\right] \tag{37.88}$$

where D is the drug diffusivity, A_o is the cross-sectional area of the channel, C_s is the agent saturation concentration, V_f is the capsule volume after rupture, t_b is the time require for sufficient water to flow into a capsule to induce rupture, and

$$t_b^* = \frac{4L_p E t}{r_o^2} \tag{37.89}$$

where r_o is the initial particle radius.

37.5 CASE STUDY

37.5.1 Background for Example 37.1

A controlled release formulation was required for a drug to reduce the high dosing frequency related to its short half-life (2.5 – 3.8 h) and to reduce C_{max}-related side effects. The projected dose strengths were 2 and 10 mg. The drug solubility was greater than 100 mg/mL at pH 4. The dose–solubility map indicated that an osmotic dosage form based on asymmetric membrane technology would be suitable for this drug [8].

EXAMPLE 37.1

An asymmetric membrane tablet core is composed of 10 mg of a highly soluble drug and mannitol ($\pi = 38$ atm) as osmogen. This core tablet is coated with a 15% w/w semipermeable coating composed of cellulose acetate, polyethylene glycol, acetone, and water. The drug release into various media (distilled water, sucrose solutions, and saturated drug solutions) is listed in Tables 37.2 and 37.3.

(a) What are the osmotic pressures of the sucrose solutions in Table 37.2?
(b) What osmotic pressure will shut down the osmotic release mechanism for this AM tablet?
(c) What percentage of the release mechanism is due to diffusion?

Solution

(a) The osmotic pressures for sucrose solutions can be located in the *Handbook of Chemistry and*

TABLE 37.2 Drug Release from Asymmetric Membrane Tablet into Media Containing Varying Sucrose Concentrations

	% Drug Dissolved				
Time (h)	Distilled Water	216.2 g/L Sucrose	363.7 g/L Sucrose	470.6 g/L Sucrose	601 g/L Sucrose
0	0	0	0	0	0
2	10.5	7.1	5.0	2.1	0.95
6	32.1	21.5	14.8	6.2	3.7
12	60.8	45.2	30.0	12.4	5.4
18	81.7	63.7	44.9	18.4	7.1
24	99.9	88.3	59.6	24.7	6.4

TABLE 37.3 Drug Release from Asymmetric Membrane Tablet into Saturated Drug Solution Media

Time (h)	Π Release Shut-Off	Diffusional Release Shut-Off (D)	Cumulative Release (Π + D)	Actual Drug Release
0	0	0	0	0
2	5.1	13.1	18.2	12.8
6	3.7	40.9	44.6	—
8	—	—	—	52.2
16	—	—	—	84.5
18	7.1	86.1	93.2	—
24	6.4	91.8	98.2	93.9

% Drug dissolved = milligram released/total milligram in tablet * 100%.

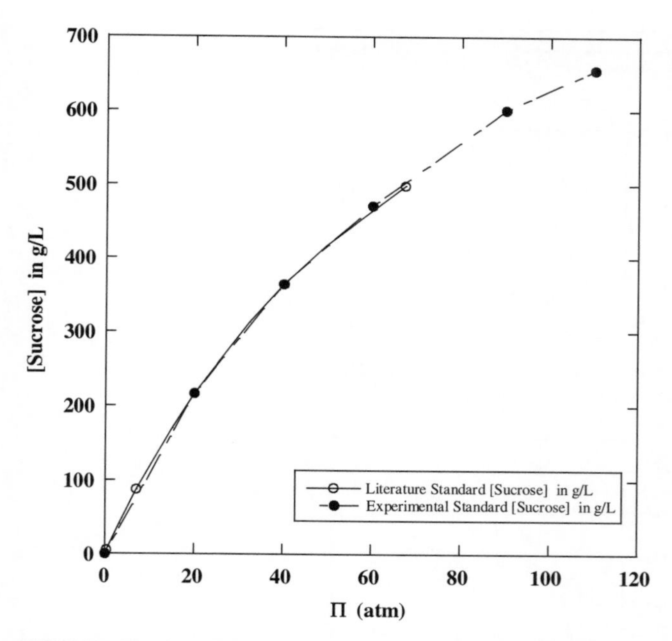

FIGURE 37.15 Calibration curve for sucrose solutions prepared to produce osmotic pressures ranging from 0 to 100 atm.

Physics [66]. These literature standard values are plotted in Figure 37.15, which are similar to experimental results reported by am Ende and Miller [67], and demonstrate consistency with accepted standards. The sucrose solution concentrations of 0, 216, 364, 471, and 601 g/L have osmotic pressures of 0, 20, 40, 60, and 90 atm, respectively.

(b) The drug release profiles from Table 37.2 are plotted in Figure 37.16, and slopes for the initial 0–60% release are

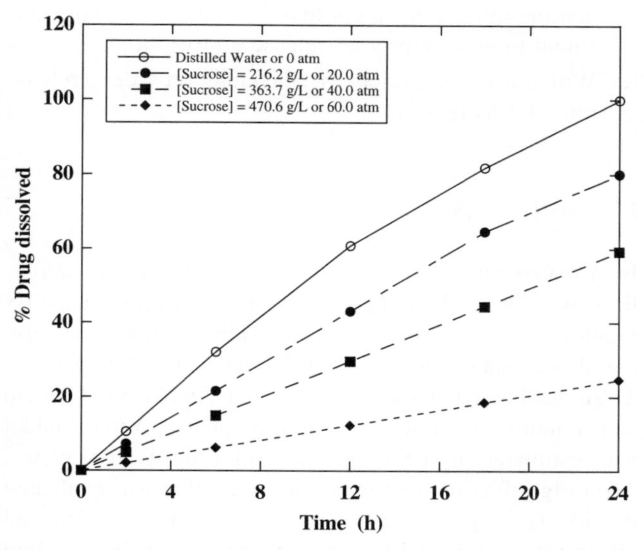

FIGURE 37.16 Drug release from AM tablets into varying concentrations of sucrose solutions, and associated osmotic pressures determined in list item (a).

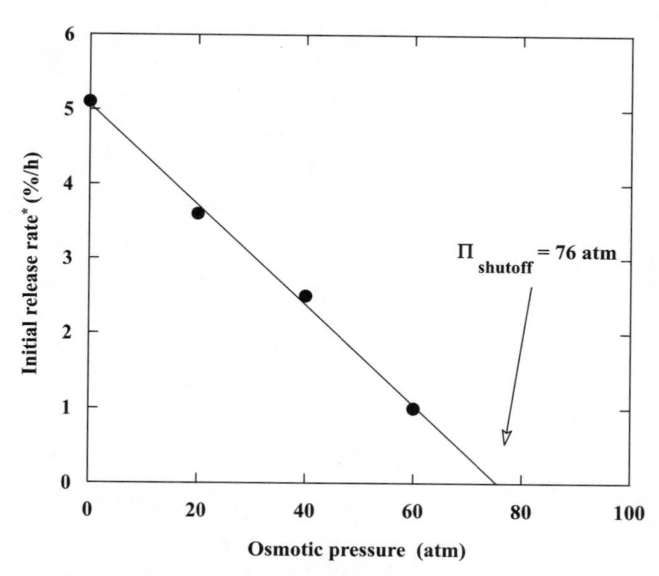

FIGURE 37.17 Initial drug release rates from AM tablets as a function of media osmotic pressure.

calculated using linear fits. The initial release rates as a function of sucrose media osmotic pressure are 5.1, 3.6, 2.5, 1.0 %/h for 0, 20, 40, and 60 atm, respectively. These resulting initial release rates are then plotted as a function of media osmotic pressure, as shown in Figure 37.17. The

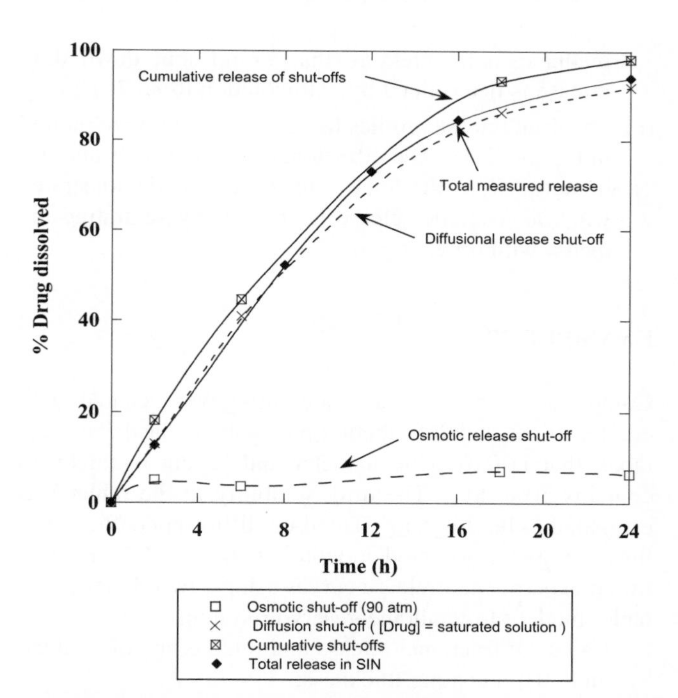

FIGURE 37.18 Drug release from AM tablets into media with osmotic pressure in excess of the determined shut-off, for example, 90 atm, and into media saturated with drug to determine diffusional shut-off. [54] Reprinted from Ref. 67, Copyright (2007), with permission from Springer.

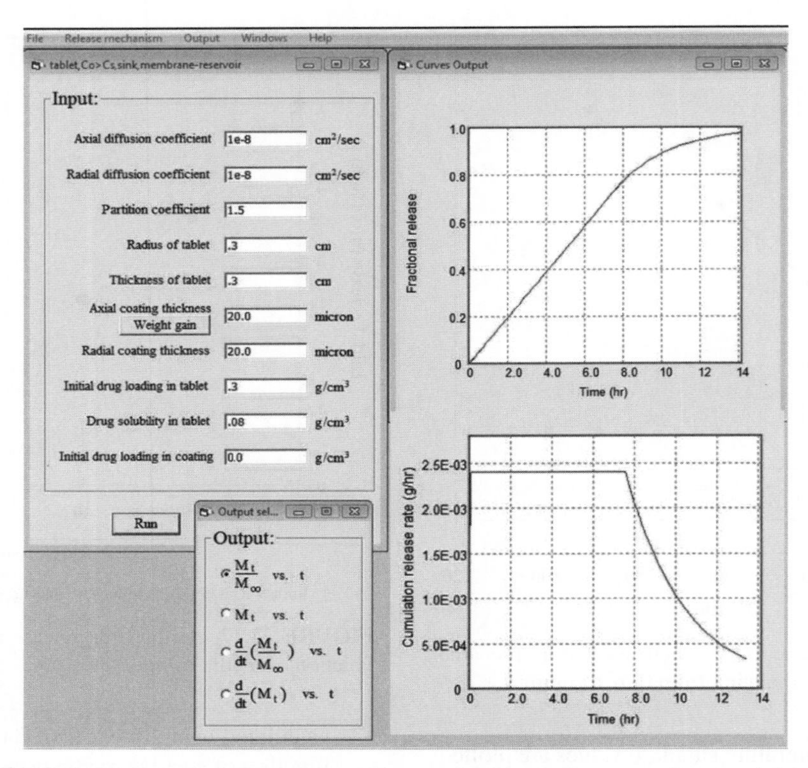

FIGURE 37.19 Fractional drug release profile (top right) and release rate (bottom right) of a 2D membrane-coated tablet predicted by equations 37.17 and 37.23 and computed by AP-CAD software package. In this example, identical axial and radial coating thickness and diffusion coefficients were used (see the input parameter values on the left panel).

media osmotic pressure that would shut down drug release is determined by extrapolation to be 76 atm.

(c) The drug release profiles listed in Table 37.3 are plotted in Figure 37.18. The diffusional contribution is approximately 5% of the total drug release, as demonstrated when all osmotic release is shut-off by saturating the media with drug.

EXAMPLE 37.2

Compressed tablets of a model drug were coated with ethylcellulose and 20% diethyl phthalate, a plasticizer. The tablet that is 0.6 cm in diameter and 0.3 cm in thickness contains 30% drug. The drug solubility in the tablet was estimated to be 0.1 g/cm^3. The drug diffusion coefficient in the coating was evaluated previously to be 1×10^{-8} cm^2/s by fitting experimental release curves using a two-dimensional tablet model of membrane–reservoir system.

(a) Calculate fractional release and release rate of a tablet, and 20 μm coating thickness;

(b) If a zero-order release up to 12 h is desirable, what should be the coating thickness?

(c) What is the release rate at this coating thickness?

Solution

(a) The fractional release and release rate were computed based on equations (37.17), (37.18), and (37.23) and plotted in Figure 37.19.

(b) By implementing computer simulation using various coating thickness, an optimal thickness of 31.5 μm was found to give zero-order release up to 12 h.

(c) With this coating thickness, the tablet provides a release rate of 1.5 mg/h.

EXAMPLE 37.3

Inert matrix tablets of two model drugs of different solubilities were made by compression of drug-excipient matrix granules. The granules were prepared by using acrylic polymer dispersion (Eudragit® FS 30D) as the granulating agent. The tablet has a diameter of 0.6 cm and a thickness of 0.3 cm and contains 30% drug. The solubility of drug A in the tablet was estimated to be 0.1 g/cm^3 and drug B 0.01 g/cm^3. The drug diffusion coefficient in the coating was evaluated previously to be 8×10^{-7} cm^2/s by fitting experimental release curves of tablets containing a low initial drug loading ($C_0 < C_s$) using the two-dimensional tablet model

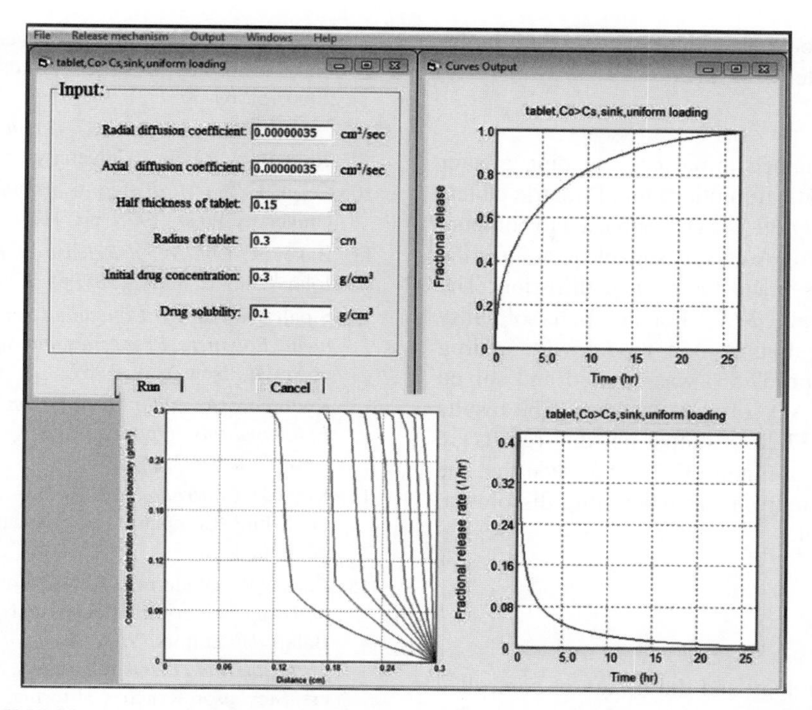

FIGURE 37.20 Fractional release (top right), release rate (bottom right), and concentration profiles at various times (bottom left) of a matrix tablet containing a water-soluble drug with the dimension, initial drug loading, and diffusion coefficient indicated in the left panel.

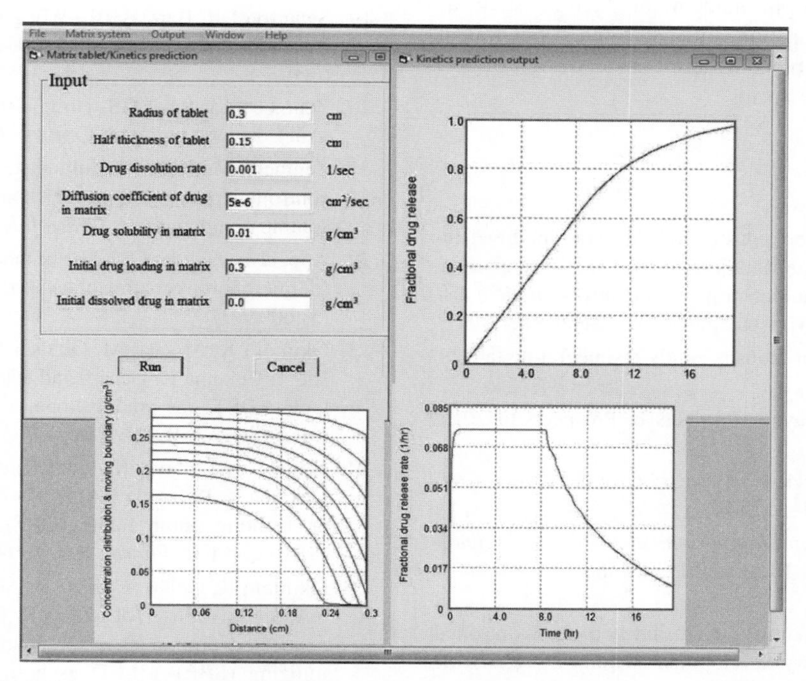

FIGURE 37.21 Fractional release (top right), release rate (bottom right), and concentration profiles at various times (bottom left) of a matrix tablet containing a poorly water-soluble drug with the dimension, initial drug loading, diffusion coefficient, and drug dissolution rate constant indicated in the left panel.

(equation 37.31). How would drug solubility and dissolution influence the rate and release profiles?

Solution

Because the tablets contained initial drug loading greater than drug solubility, analytical solutions for 2D matrix tablets with $C_0 > C_s$ (equations 37.39–37.41) were used to compute drug release profiles of drug A. It is reasonable to assume the dissolution of drug A is much faster than diffusion. The results are presented in Figure 37.20. For drug B, its solubility is one-tenth of that of drug A, the model with a drug dissolution term (equation 37.53) was applied and solved numerically using AP-CAD software package. The results are shown in Figure 37.21. Comparing the results in Figure 37.20 with those in Figure 37.21, it is seen that the release profile becomes more linear when drug dissolution becomes more significant.

37.6 CONCLUSIONS

Controlled release technology and the design of controlled release dosage forms were discussed in this chapter. The equations governing the rate of drug release have been derived based on the dominant mechanism of drug release, for example, Fick's second law of diffusion for nonerodible monolithic and reservoir devices and irreversible thermodynamics for certain osmotic systems. In addition to physicochemical factors, biopharmaceutical factors such as mechanism of drug absorption and gastrointestinal transit of the dosage form must be considered in the design of oral controlled release dosage forms.

REFERENCES

1. U.S. FDA. Guidance for industry SUPAC-MR: modified release solid oral dosage forms scale-up and postapproval changes: chemistry, manufacturing, and controls; *in vitro* dissolution testing and *in vivo* bioequivalence. September 1997.
2. U.S. FDA. Guidance for industry orally disintegrating tablets. December 2008.
3. Controlled Release: Players, Products & Prospects to 2015. Epsicon Report.
4. Rathbone MJ, Bowersock T. Veterinary drug delivery. *J. Control. Release* 2002;85(1–3):284.
5. Lakkis JM, editor. *Encapsulation Controlled Release Technologies in Food Systems*, Blackwell Publishing, Oxford, 2007, pp. 239.
6. Thombre AG. Assessment of the feasibility of oral controlled release in an exploratory development setting. *Drug Discov. Today* 2005;10(17):1159–1166.
7. Lipper RA. How can we optimize selection of drug development candidates from many compounds at the discovery stage? *Mod. Drug Discov.* 1999;2:55–60.
8. Thombre AG. Feasibility assessment and rapid development of oral controlled release prototypes. *ACS Symp. Ser.* 2000;752:69–77.
9. Carslaw HS, Jaeger JC. *Conduction of Heat in Solids*, 2nd edition, Oxford University Press, 1959, pp. 510.
10. Crank J. *The Mathematics of Diffusion*, 2nd edition, Oxford University Press, 1975, pp. 414.
11. Baker R. *Controlled Release of Biologically Active Agents*, John Wiley & Sons, 1987, pp. 279.
12. Singh SK, Fan LT. *Controlled Release: A Quantitative Treatment: Polymers, Properties and Applications*, Vol. 13, Springer-Verlag, New York, 1989.
13. Kydonieus A, editor. *Treatise on Controlled Drug Delivery: Fundamentals, Optimization, Applications*, Marcel Dekker, New York, 1992, pp. 553.
14. Kim C-J. *Controlled Release Dosage Form Design*, Technomic Publishing Company, Inc., Lancaster, PA 17604 USA 2000, pp. 310.
15. Carrier RL, Waterman KC. Use of biodegradable polymers in oral drug delivery: challenges and opportunities. In: Surya KM, Balaji N, editors, *Handbook of Biodegradable Polymeric Materials and Their Applications*, American Scientific Publishers, Stevenson Ranch, California, 2006, pp. 33–56.
16. Santus G, Baker RW. Osmotic drug delivery: a review of the patent literature. *J. Control. Release* 1995;35(1):1–21.
17. Ranade VV, Hollinger MA. *Drug Delivery Systems*, 2nd edition, CRC Press, Boca Raton, FL, 1996, pp. 16–39.
18. CardinalJr., Controlled release osmotic drug delivery systems for oral applications. *Drugs Pharm. Sci.* 2000;102:411–444.
19. Swanson DR, Barclay BL, Wong PS, Theeuwes F. Nifedipine gastrointestinal therapeutic system. *Am J Med* 1987;83 (6B):3–9.
20. Zentner GM, Rork GS, Himmelstein KJ. The controlled porosity osmotic pump. *J. Control. Release* 1985;1(4):269–282.
21. Zentner GM, Rork GS, Himmelstein KJ. Osmotic flow through controlled porosity films: an approach to delivery of water soluble compounds. *J. Control. Release* 1985;2:217–229.
22. Appel LE, Zentner GM. Use of modified ethyl cellulose latexes for microporous coating of osmotic tablets. *Pharm. Res.* 1991;8 (5):600–604.
23. Okimoto K, Miyake M, Ohnishi N, Rajewski RA, Stella VJ, Irie T, Uekama K. Design and evaluation of an osmotic pump tablet (OPT) for prednisolone, a poorly water soluble drug, using $(SBE)_{7m}$-β-CD. *Pharm. Res.* 1998;15(10);1562–1568.
24. Okimoto K, Ohike A, Ibuki R, Aoki O, Ohnishi N, Irie T, Uekama K, Rajewski RA, Stella VJ. Design and evaluation of an osmotic pump tablet (OPT) for chlorpromazine using $(SBE)_{7m}$-β-CD. *Pharm. Res.* 1999;16(4):549–554.
25. Okimoto K, Ohike A, Ibuki R, Aoki O, Ohnishi N, Rajewski RA, Stella VJ, Irie T, Uekama K. Factors affecting membrane-controlled drug release for an osmotic pump tablet (OPT) utilizing $(SBE)_{7m}$-β-CD as both a solubilizer and osmotic agent. *J. Control. Release* 1999;60(2–3):311–319.
26. Okimoto K, Rajewski RA, Stella VJ. Release of testosterone from an osmotic pump tablet utilizing $(SBE)_{7m}$-β-cyclodextrin

as both a solubilizing and an osmotic pump agent. *J. Control. Release* 1999;58(1):29–38.

27. Stella VJ, Rao VM, Zannou EA. The pharmaceutical use of Captisol: some surprising observations. *J. Inclusion Phenom. Macrocycl. Chem.* 2003;44(1–4):29–33.
28. Herbig SM, Cardinal JR, Korsmeyer RW, Smith KL. Asymmetric-membrane tablet coatings for osmotic drug delivery. *J. Control. Release* 1995;35(2–3):127–136.
29. am Ende MT, Herbig SM, Korsmeyer RW, Chidlaw MB. Osmotic drug delivery from asymmetric membrane film-coated dosage forms. In: Wise DL, editor. *Handbook of Pharmaceutical Controlled Release Technology*, Marcel Dekker, Inc., New York, 2000, pp. 751–785.
30. Narisawa S, Nagata M, Hirakawa Y, Kobayashi M, Yoshino H. An organic acid-induced sigmoidal release system for oral controlled-release preparations. III. Elucidation of the anomalous drug release behavior through osmotic pumping mechanism. *Int. J. Pharm.* 1997;148(1):85–91.
31. Wright J, Chandrasekaran SK, Gale R, Swanson D. A model for the release of osmotically active agents from monolithic polymeric matrixes. *AIChE Symp. Ser.* 1981;77(206):62–68.
32. Amsden BG, Cheng Y-L, Goosen MFA. A mechanistic study of the release of osmotic agents from polymeric monoliths. *J. Control. Release* 1994;30(1);45–56.
33. Zhou Y, Wu XY. Finite element analysis of diffusional drug release from complex matrix systems. I. Complex geometries and composite structures. *J. Control. Release* 1997;49(2–3):277–288.
34. Wu XY, Zhou Y. Finite element analysis of diffusional drug release from complex matrix systems. II. Factors influencing release kinetics. *J. Control. Release* 1998;51(1):57–71.
35. Wu XY, Zhou Y. Studies of diffusional release of a dispersed solute from polymeric matrixes by finite element method. *J. Pharm. Sci.* 1999;88(10):1050–1057.
36. Siegel RA. Theoretical analysis of inward hemispheric release above and below drug solubility. *J. Control. Release* 2000;69 (1):109–126.
37. Siepmann J, Streubel A, Peppas NA. Understanding and predicting drug delivery from hydrophilic matrix tablets using the "sequential layer" model. *Pharm. Res.* 2002;19(3);306–314.
38. Good WR, Lee PI. Membrane-controlled reservoir drug delivery systems. *Med. Appl. Control. Release* 1984;1:1–39.
39. Zhou Y, Chu JS, Li JX, Wu XY. Theoretical analysis of release kinetics of coated tablets containing constant and non-constant drug reservoirs. *Int. J. Pharm.* 2010;385(1–2):98–103.
40. Fu JC, Hagemeir C, Moyer DL. A unified mathematical model for diffusion from drug–polymer composite tablets. *J. Biomed. Mater. Res.* 1976;10(5);743–758.
41. Paul DR, McSpadden SK. Diffusional release of a solute from a polymer matrix. *J. Membr. Sci.* 1976;1(1):33–48.
42. Higuchi T. Mechanism of sustained-action medication. Theoretical analysis of rate of release of solid drugs dispersed in solid matrices. *J. Pharm. Sci.* 1963;52(12):1145–1149.
43. Roseman TJ, Higuchi WI. Release of medroxyprogesterone acetate from a silicone polymer. *J. Pharm. Sci.* 1970;59(3): 353–357.
44. Zhou Y, Chu JS, Zhou T, Wu XY. Modeling of dispersed-drug release from two-dimensional matrix tablets. *Biomaterials* 2005;26(8):945–952.
45. Chandrasekaran SK, Paul DR. Dissolution-controlled transport from dispersed matrixes. *J Pharm Sci* 1982;71(12): 1399–1402.
46. Harland RS, Dubernet C, Benoit JP, Peppas NA. A model of dissolution-controlled, diffusional drug release from non-swellable polymeric microspheres. *J. Control. Release* 1988; 7(3):207–215.
47. Frenning G, Brohede U, Stromme M. Finite element analysis of the release of slowly dissolving drugs from cylindrical matrix systems. *J. Control. Release* 2005;107(2):320–329.
48. Zhou Y, Li JX, Wu XY. QbD of oral controlled release dosage forms by computational simulation. Abstract #537, 35th Annual Meeting & Exposition of the Controlled Release Society, New York, NY, 2008.
49. Chang NJ, Himmelstein KJ. Dissolution–diffusion controlled constant-rate release from heterogeneously loaded drug-containing materials. *J. Control. Release* 1990;12(3):201–212.
50. Hopfenberg HB. Controlled release from erodible slabs, cylinders, and spheres. *Pap. Meet. Am. Chem. Soc. Div. Org. Coat. Plast. Chem.* 1976;36(1):229–234.
51. Lee PI. Diffusional release of a solute from a polymeric matrix: approximate analytical solutions. *J. Membr. Sci.* 1980;7(3): 255–275.
52. Thombre AG, Himmelstein KJ. A simultaneous transport-reaction model for controlled drug delivery from catalyzed bioerodible polymer matrixes. *AIChE J.* 1985;31(5): 759–766.
53. Thombre AG. Theoretical aspects of polymer biodegradation: mathematical modeling of drug release and acid-catalyzed poly (ortho-ester) biodegradation. *Spec. Publ. R. Soc. Chem.* 1992;109:214–225.
54. Thombre AG, Himmelstein KJ. Modeling of drug release kinetics from a laminated device having an erodible drug reservoir. *Biomaterials* 1984;5(5):250–254.
55. Abdekhodaie MJ, Wu XY. Drug loading onto ion-exchange microspheres: modeling study and experimental verification. *Biomaterials* 2006;27(19):3652–3662.
56. Abdekhodaie MJ, Wu XY. Drug release from ion-exchange microspheres: mathematical modeling and experimental verification. *Biomaterials* 2008;29(11):1654–1663.
57. Siepmann J, Peppas NA. Hydrophilic matrixes for controlled drug delivery: an improved mathematical model to predict the resulting drug release kinetics (the "sequential layer" model). *Pharm. Res.* 2000;17(10):1290–1298.
58. Wu N, Wang L-S, Tan DC-W, Moochhala SM, Yang Y-Y. Mathematical modeling and *in vitro* study of controlled drug release via a highly swellable and dissoluble polymer matrix: polyethylene oxide with high molecular weights. *J. Control. Release* 2005;102(3):569–581.
59. Wu XY, Zhou Y. Numerical simulation of controlled drug release from matrix tablets involving swelling, erosion and diffusion. *Pharm. Res.* 1997;14:S716.

60. Theeuwes F. Elementary osmotic pump. *J. Pharm. Sci.* 1975;64 (12):1987–1991.
61. Zentner GM, Rork GS, Himmelstein KJ. Osmotic flow through controlled porosity films: an approach to delivery of water soluble compounds. In: Anderson JM, Kim SW, editor. *Advances in Drug Delivery Systems*, Elsevier, New York, 1986, pp. 217–229.
62. Rekhi GS, Porter SC, Jambhekar SS. Factors affecting the release of propranolol hydrochloride from beads coated with aqueous polymeric dispersions. *Drug Dev. Ind. Pharm.* 1995;21(6):709–729.
63. Lindstedt B, Ragnarsson G, Hjaertstam J. Osmotic pumping as a release mechanism for membrane-coated drug formulations. *Int. J. Pharm.* 1989;56(3):261–268.
64. Schirrer R, Thepin P, Torres G. Water absorption, swelling, rupture and salt release in salt–silicone rubber compounds. *J. Mater. Sci.* 1992;27(13):3424–3434.
65. Amsden BG, Cheng Y-L, Goosen MFA. A mechanistic study of the release of osmotic agents from polymeric monoliths. *J. Control. Release* 1996;38(2–3):275. Erratum to document cited in CA121:17834.
66. Weast RC, Selby SM. *Handbook of Chemistry and Physics*, 55th edition, CRC Press, Cleveland, OH, 1974, pp. 2304.
67. am Ende MT, Miller LA. Mechanistic investigation of drug release from asymmetric membrane tablets: effect of media gradients (osmotic pressure and concentration), and potential coating failures on *in vitro* release. *Pharm. Res.* 2007;24(2): 288–297.

38

DESIGN AND SCALE-UP OF DRY GRANULATION PROCESSES

OMAR L. SPROCKEL AND HOWARD J. STAMATO

Biopharmaceutics Research and Development, Bristol-Myers Squibb Co., New Brunswick, NJ, USA

38.1 OVERVIEW OF THE DRY GRANULATION PROCESS

Granulation offers the opportunity to alter the properties of solid particulate material. An increase of particle size reduces potential hazards or nuisance from dust. The granule structure can improve the flow and compression properties of the powder and helps to assure that the composition remains uniform throughout the powder mass if a mixture of materials is used.

Nishii and Horio [1] have described how dry granulation can be achieved during mixing, or as a powder is passed through a screen. As reported by Horio [2] particles can be dry granulated by a fluidization technique where the particle size of the starting materials is small, for example, lactose at a particle size of approximately 3 μm. However, granulation without a solvent to help with the binding is usually accomplished by subjecting powders to pressure.

Dry granulation by slugging is achieved by compressing powders in a cavity formed by a set of tools and a die, similar to the way tablets are made. Pietsch [3], in a review of granulation technology, described dry granulation by slugging among the methods and referenced a patent for a tabletting machine from 1843. Slugging and tabletting use similar technology and the patent shows how long slugging may have been available. When comparing slugging with tabletting it is apparent that tablets are typically much smaller than slugging compacts and can be the final product, whereas slugging compacts are reduced to granules feeding a subsequent tabletting or capsule filling operation. As the purpose of the slugging operation is to form granules with larger particle size, better flow and improved compression properties, the feed to a slugging press often consists of very fine powder with poor flow properties when compared to a tabletting operation.

Johnson [4] described dry granulation by roller compaction noting that passing powders through the nip of two rollers in order to produce larger sized material was originally used in coal processing. One of the earliest patents for a roller compaction system, “A Method for Converting Fine Coal into Lumps,” was issued in 1848. The largest volume applications of roller compaction are in the coal, soda ash, potash, calcined lime, and magnesium oxide industries. Bakele [5] gives an example of soda ash production of 160,000 ton/year. In pharmaceutical applications, Kleinebudde [6] has reviewed the advantages of granulation and the application of roller compaction to form granules used in tabletting and filling of hard gelatin capsules. Mouro et al. [7] showed how granulation via roller compaction improved the processing of low bulk density materials in a capsule filling operation.

The roller compaction unit operation is actually comprised of several subprocesses. Once the components and composition have been selected, the powder to be compacted must be mixed so that the feed to the compactor is relatively homogeneous. The container with the prepared powder is placed in position to feed the roller compaction machine and the powder must be metered to the area in front of the compaction rollers. A variety of devices and configurations such as valves, stirring devices, and feed screws are used in order to convey the powder from the holding container, to the roller compaction unit, and inside the machine to the area in front of the rollers. The resulting powder bed is dragged by

Chemical Engineering in the Pharmaceutical Industry: R&D to Manufacturing Edited by David J. am Ende

the rollers to the nip area where it is compacted. During the compaction the powders are deformed to a degree that causes the formation of a continuous sheet. The compacted output is commonly referred to as a ribbon but has also been referred to as a compact, flake, or tape.

The last step of the roller compaction unit operation is the reduction of the ribbon to granules. This can happen in one or more steps where mechanical elements impact the ribbon, break it into pieces, and force the broken pieces through a screen. The size reduction subprocess of roller compaction may be integral to the roller compaction machine or performed as a separate operation. In either case the size reduction typically yields particles with a distribution of sizes. Because one of the objectives of the compaction process is to increase size, a large fraction of granules below a minimum size may be undesirable. In some cases fines are separated from the granulation and returned for another pass through the compaction step.

The sections of this chapter will give an overview of the roller compaction unit operation, followed by consideration of material behavior and measurement. A step-by-step review of the subprocesses that comprise the roller compaction unit operation including various types of models and control strategies is offered next. In the final section, case studies including scale-up, one for a parametric-based scale-up approach and one for an attribute-based scale-up approach are presented.

38.2 GENERAL CONSIDERATIONS FOR ROLLER COMPACTION OPERATIONS AND EQUIPMENT

Although the roller compaction process is used to improve the flow of material, the incoming feed needs to have sufficient flow properties in order to be delivered with as consistent a composition, flow, and density as possible. The requirements may include limitations on the micromeritic properties of the active ingredient. Excipients are usually selected to dilute the active ingredient to an appropriate concentration and enhance various aspects of the powder properties. Some of the flow properties to help feed the roller compactor and compression properties to assist in ribbon formation can be provided by the addition of materials called binders because of their compression properties (e.g., microcrystalline cellulose (MCC), and lactose). Including these materials is often necessary to achieve the desired granulation properties. Flow aids (e.g., silicon dioxide) may be needed to achieve the desired mixture flow properties. A lubricant (e.g., magnesium stearate) typically added to modify interaction with the equipment surfaces, may also be necessary to achieve the desired performance of the roller compactor. Other components necessary for performance of the final product, rather than performance in the roller compactor, such as disintegrants (e.g., croscarmellose sodium, cross-linked polyvinyl pyrrolidone, or sodium starch glycolate) may also be used.

The next selection for the overall roller compaction operation is the unit operation that will prepare the powder blend for roller compaction. A diffusion mixer is one of the more popular choices for the mixing subprocess of roller compaction used to prepare the roller compactor feed. The batch size, sequence of loading, number and speed of revolutions, and the shape of the blender may be considered when designing the process. Multiple steps involving a geometric dilution strategy or intermediate milling between blending steps may be necessary to distribute the ingredients and achieve the appropriate level of uniformity and handling properties. Although a homogeneous powder feed will reduce variability in the roller compaction operation, content uniformity of the powder blend is not required at the same limits that would normally be applied to pharmaceutical dosage forms. Issues with blocking, bridging, segregation, or adhesion to equipment surfaces may result from poor selection of materials or insufficient blending prior to roller compaction.

The arrangement of the blender, bin, or other container relative to the roller compactor is also to be considered. Powder needs to flow evenly from the container to the equipment without bridging, flooding, or segregating between the bin and the roller compactor feed hopper. A general arrangement of the elements of a roller compaction process is shown in Figure 38.1.

Once the powder reaches the integrated roller compaction machinery there will be various means of conveying the powder toward the rollers depending on the manufacturer, model, and user requirements. Typically there is a feed hopper to receive material from the main powder container. The hopper may have a device to break any powder bridges and help move the material for further processing. The hopper may feed directly to the rollers by gravity, but a majority of designs also use one or more feed screws to deliver the powder and consolidate it in front of the rollers. In all cases the feed screw is oriented in-line with the roller nip. The orientation of the screw will depend on the orientation of the rollers.

The powder bed must be contained to assure it travels to the rollers and through the gap. The most typical system involves plates mounted to cover the powder bed before the gap, which extend over the gap itself, preventing powder migration outside the compaction area. It is important to assure proper mounting of the plates and sufficient maintenance to prevent the leakage of powder from the compaction area. Unprocessed fines could join the product stream if powder leaks from the seals. A different type of sealing system also uses ridges on the roller edges such that one roller fits into a channel in the surface of the other in order to form a tighter seal.

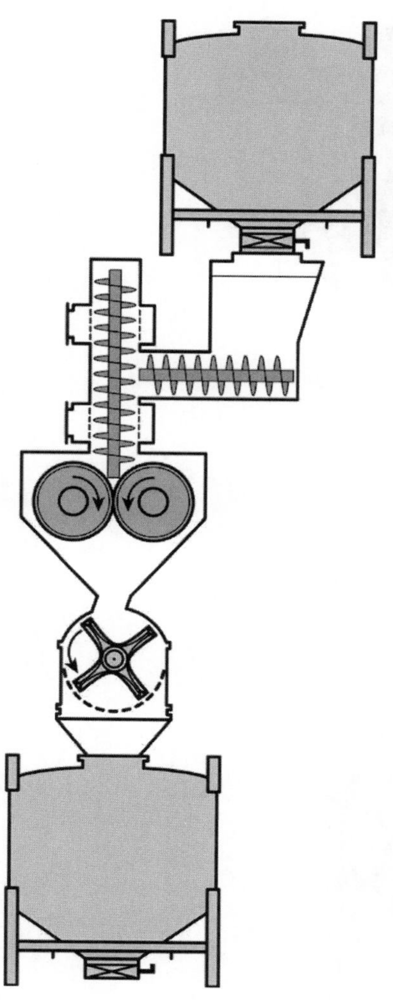

FIGURE 38.1 General arrangement of the roller compaction process. (*Courtesy*: The Fitzpatrick Company.)

Kleinebudde [6] surveyed various manufacturers with regard to their roller and screw configurations, offering a list of suppliers correlated to references focused on the various designs. Some machines have the powder conveyed to the nip vertically such that the open space between the rollers—the gap—is vertical and a line drawn between the roller axes is horizontal. A system with this configuration may lose a small amount of powder through the nip as conditions are stabilized in the beginning of the run. Other roller compaction machines have this configuration turned on its side so that the powder feed and the nip opening are oriented horizontally and a line drawn between the roller axes is vertical. This reduces the possibility of powder flooding the machine at start-up and depends entirely on the action of the feed screw to move powder to the rollers. A third configuration, used in other types of roller compactors, orients the powder flow and nip at an angle to the vertical giving some of the advantages of both the horizontal and vertical configurations. An illustration of the various configurations is shown in Figure 38.2. Typical roller compactors from laboratory to production scale from different manufacturers are shown in Figure 38.3. A unit shown in an expanded view so that all of the parts are visible is shown in Figure 38.4.

During roller compaction, the porosity of the powder is reduced in the compaction step to form the ribbon. The air filling the pores must escape through the forming compact and adjacent powder bed. In some types of roller compaction machines the powder is deaerated before reaching the nip by passing over a porous plate with vacuum applied underneath. The deaeration is intended to reduce any disruption from air moving through the powder as the compact is formed. Miller [8, 9] described the effects of air entrainment and

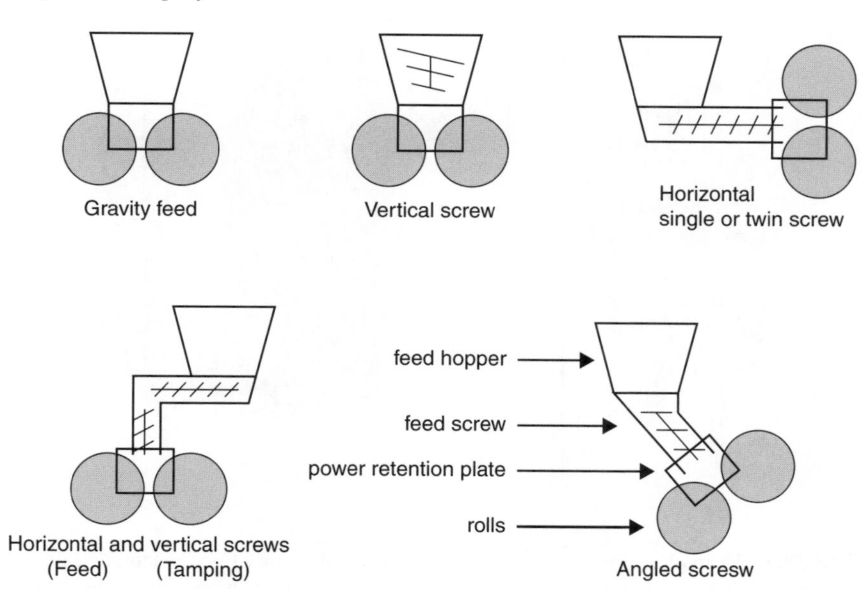

FIGURE 38.2 Common feed, feed screw, and roller configurations.

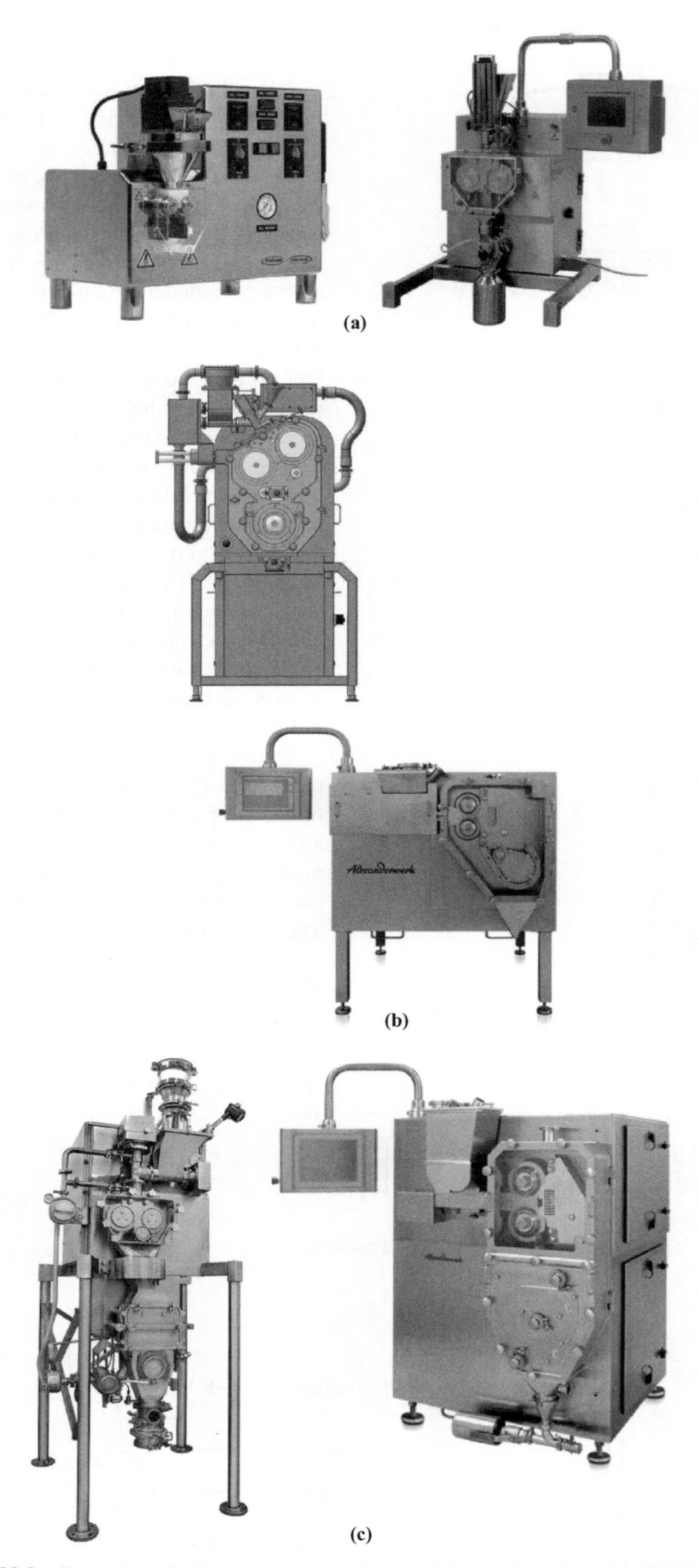

FIGURE 38.3 Examples of roller compactor units from laboratory to commercial scale. (a) Lab-scale roller compactors; (b) pilot-scale roller compactors; (c) production-scale roller compactors. (*Courtesy*: Vector Corporation, The Fitzpatrick Company, Gerteis Maschinen + Processengineering AG, Alexanderwerk AG.)

FIGURE 38.4 Expanded view of a roller compactor showing the components. (*Courtesy*: The Fitzpatrick Co.)

improvements in the roller compaction operation due to deaeration of the powder feed.

In the simplest case the roller faces in contact with the powder are smooth. However, other types of roller finishes, such as grooves or other inscribed patterns, are used in an attempt to change the interaction with powder and the surface of the rollers are also available. Rollers with different surface finishes and the resulting ribbons are shown in Figures 38.5 and 38.6. Daugherity and Chu [10] studied the effect of roller surfaces on the compacted ribbon properties. Rambali et al. [11] also noted differences attributed to the use of smooth versus grooved rollers. Pietsch [12] described a briquetting application in which the faces of the rollers are indexed so that shaped indentations in each roller face form the desired final product shape during the compaction step. Rollers with internal channels for cooling liquid may be used for temperature sensitive products that might melt from the energy of compaction as mentioned by Pietsch [3].

Even on a relatively small compaction machine the nominal roller pressure may be set to a value of 50–70 bar. This pressure is typically applied by use of a hydraulic cylinder pushing on the roller axle. The pressures involved are in a

(a)

(b)

FIGURE 38.5 Rollers with different surface treatments. (*Courtesy*: The Fitzpatrick Company and Gerteis Maschinen + Processengineering AG.)

FIGURE 38.6 Examples of product produced from rollers with different surfaces. (*Courtesy*: The Fitzpatrick Company.)

range where deflection of the axle can actually affect the operation of the machine. Most manufacturers of roller compaction machinery have carefully considered the design of the hydraulic system and axle deflection so that the width of the gap is consistent across the roller width. It should be noted that the pressure on the powder may be different than the nominal pressure set on the hydraulic system although the two values are related.

Guangsheng et al. [13] described an application for roller compaction of magnesium alloy powder to form sheets. In this case, after some posttreatment, the roller compacted ribbon is the final product. However, in most roller compaction operations the ribbon itself is not the useful product and is broken into granules that have the desired properties for further processing. Although used in some high volume applications, briquetting rollers, as described by Pietsch [12], are not used to form the final product in pharmaceuticals due to the need for tight uniformity of finished product. Different types of output from roller compaction operations including broken ribbons with different surfaces, granules, and long briquettes are shown in Figure 38.7.

The size reduction step for the ribbon typically consists of an impeller without a screen to start the process followed by one or two stages of milling, this time with an impeller mounted in proximity to the screen. The mill configuration may take the form of an oscillating screener, hammer mill, or conical mill. The screening mill may or may not be integrated with the roller compaction unit.

Fines may be present in the resulting granulation from several sources. The sealing mechanism at the edges of the rollers may not contain all of the powder or powder may not be completely incorporated into the ribbon and is carried over into the final product. The milling operation itself can create fines. In some cases the amount of fines are considered unacceptable for the final granulated material.

For operations where more fines are present the equipment may be fitted with mechanisms to separate the fine fraction. Some fines reduction can be accomplished by collecting powder which did not get properly compacted by use of a chute into which the fines fall while the ribbon is collected separately prior to milling. However, this method will not completely assure a low fines fraction as some fines are generated during the milling process. In other cases the entire output after the milling operation is size separated to remove fines from the granulation caused by the milling operation along with any carry over of uncompacted fines. In both cases the undersized material may be returned to the

FIGURE 38.7 Examples of roller compactor output. (*Courtesy*: The Fitzpatrick Company.)

roller compactor to improve yield of granules with the desired size.

It should be noted that repeating the compaction step can reduce the performance of the granules in subsequent processing. If reprocessing is necessary the number of passes through the compaction should be limited and the combination of reprocessed material with the main granulation should be monitored. Material behavior and the characterization of raw material, and finished granules, and compacted ribbon are discussed in the next section.

38.3 MATERIAL BEHAVIOR, ATTRIBUTE TESTING, AND PROCESS SENSORS

The performance of the roller compaction unit process and the resulting products (e.g., ribbon, granules, tablet, or capsules) depend on the properties of the powder being processed. Some considerations of the choices for the unit operation related to material behavior are described in this section.

38.3.1 Material Behavior During the Roller Compaction Process

38.3.1.1 Powder Handling of Feed and Granulated Product Powder handling is important in the feed of the roller compaction unit and is also applied to the granules produced by the roller compaction operation. The particle size distributions, densities, and flow properties of the powders going into the roller compactor must be matched, manipulated, or modified by additives (e.g., lubricants and flow aids) and preprocessing (e.g., milling and blending) in order to feed material evenly and with the appropriate levels of the components to the compaction machine. The powder mix will have to maintain these properties throughout the compaction run that may be an extended time because roller compaction is typically performed as a semicontinuous operation. The powder to be compacted will be subjected to various conveying and consolidation operations that may induce segregation due to material or air motion, and vibration. Bacher et al. [14] reported that the shape and size distribution of calcium carbonate and sorbitol used to prepare roller compacted granules affected the granule content as a function of granule size. Similarly the output granules must have appropriate flow, resist segregation, and have sufficient strength to withstand handling when being conveyed to various downstream process or packaging into the final product presentation.

38.3.1.2 Compaction In addition to the feeding properties, the raw material properties can also have an effect on the compaction of the powder into ribbons. Material properties were considered an input for the earliest mathematical models, such as Johanson [15] who suggested measuring the internal and surface friction of the powder, to more recent studies including process analytical technology, such as Soh et al. [16, 17], who studied twenty material parameters and suggested particle size, span (a measure of size distribution), and angle of fall as the most interesting measurements to characterize the input material.

Many studies have focused specifically on the behavior of an excipient or class of excipients in the roller compaction operation and can give insights to assist in selection of excipients. In many studies a formulation including a model active ingredient was used to simulate the response of the excipient under typical use conditions. Because the material in a roller compactor undergoes some of the same physical processes as in a tablet press many of the same considerations of material behavior apply. For example, Sheskey and Dasbach [18] noted that slower roller speeds, giving a longer dwell time under pressure, allowed plastically deforming materials to perform better as binders. Falzone et al. [19] noted differences in behavior between plastically deforming microcrystalline cellulose and lactose that exhibited behavior influenced by brittle fracture.

Some roller compaction studies of specific materials are summarized in Table 38.1.

TABLE 38.1 Selected References Reporting Material Behavior in Roller Compaction

Material	Reference
Microcrystalline cellulose	Inghelbrecht and Remon [20]
Lactose	Riepma et al. [21]
	Inghelbrecht and Remon [22]
Hydroxypropylmethylcellulose	Sheskey et al. [23]
Hydroxypropylcellulose, methylcellulose, polyvinylpyrrolidone, starch, and microcrystalline cellulose	Sheskey et al. [18]
Magnesium carbonate	Freitag et al. [24]
Magnesium carbonate in combination with powdered cellulose	Freitag et al. [25]
Hydroxypropylmethylcellulose, hydroxypropylcellulose, microcrystalline cellulose, and polyvinylpyrrolidone	Herting et al. [26]
Calcium carbonate and sorbitol	Bacher et al. [14, 27]

The addition of small amount of components to influence the powder properties is common practice, but may have an effect on the compaction operation. He et al. [28] studied the effect of lubricating the roller compactor feed with magnesium stearate on the strength of tablets compressed from the resulting granules. The authors noted that lubricant is often added to roller compaction feed to prevent sticking to the rollers. One caution offered was that the lubricant can change the interaction of the powder and rollers resulting in a smaller nip angle and reduced process efficiency. It was found that tablets produced from roll compacted granules of unlubricated microcrystalline cellulose did not show a significant decrease in strength at two of the three levels of ribbon densification studied. At the highest level of ribbon densification there was approximately a 30% drop in tablet strength. For microcrystalline cellulose lubricated with 0.5% magnesium stearate the drop in tablet strength was 90% over a much broader range of ribbon densities.

Many excipients can absorb water from the environment under typical processing conditions. The moisture can affect the material properties, perhaps causing unwanted effects in the roller compaction operation. The effect of moisture on the behavior of microcrystalline cellulose in roller compaction with ambient relative humidity ranging from 15% to 75% was studied by Gupta et al. [29]. It was found that as water content increased; the powder yield strength decreased, indicating better powder rearrangement, while tensile strength of the resulting ribbons decreased, indicating poorer bonding. Inghelbrecht and Remon [30] intentionally added water to blends of lactose, microcrystalline cellulose and hydroxypropylmethylcellulose, and other ingredients in order to reduce the amount of fines in the roller compacted granules. They found that the fines produced during the compaction step were negligible and the fines fraction produced during the milling operation were reduced. The resulting granules and tablets had lower friability and the tablets had higher strength.

A more typical strategy to reduce fines is the recycle of the fine fraction to the inlet of the roller compaction step. However, the change in material properties when subjected to roller compaction can be cumulative and has been reported by many investigators. Bultmann [31] studied this phenomenon as a function of the number of repeated roller compactions for microcrystalline cellulose and found the repeated compactions decreased the amount of fines but also decreased the compressibility of the resulting granules. Up to 10 cycles of roller compaction were studied with most of the losses in material properties seen in the first and second processing by the roller compactor. Shesky et al. [23] found a similar reduction in compressibility for hydroxypropylmethylcellulose which decreased with increasing roller pressure after a single pass through the machine. In contrast Riepma et al. [21] found that for material displaying brittle fracture characteristics, such as lactose, dry granulation had minimal influence on the compatibility during subsequent tabletting.

A review of specific material attributes and their effect on the products from a roller compaction process are discussed in the following sections.

38.3.2 Incoming Powder and Outgoing Granule Properties

The performance of the roller compaction process and the quality of the output materials (granules) depends on material characteristics as well as on process parameters. The types of attributes and measurements that may be considered for roller compaction operations are considered in this section. For example, Miguelez-Moran [32] showed that the size of the three flow regions in the roller compaction zone and the transition from one to the other depends on the properties of the input powder. Examples of powder properties that affect the powder flow behavior in the compaction, zone are internal friction, cohesion and friction between the powder and the rollers and side shields. Particle size distribution is an attribute that is commonly measured for the input and output materials. Density and flow are characteristics that are influenced by particle size distribution, and often monitored. Offline characterization techniques are covered in the respective property section below, but online methods are covered in Section 38.3.4.

38.3.2.1 ***Particle Size Distribution*** Much of the knowledge gained on the influence of input material properties from the tablet compaction field can be applied to roller compaction. The influence of particle size and compression characteristics (plastic or brittle nature) is dominant in ribbon and granule quality.

Several authors have reported on the importance of selecting the proper diluent particle size. Herting et al. [33, 34] reported on the effect of microcrystalline cellulose particle size on granule and tablet properties. Under similar processing parameters, a reduction in the particle size of the input material (MCC and theophylline) resulted in larger mean granule size and higher compactability. Inghelbrecht and Remon [20] evaluated the effect of MCC particle size in ibuprofen/MCC drug mixtures. They found that smaller MCC particles produced stronger granules. The irregularity of the MCC particles was ascribed a secondary role for the differences seen in granule strength.

In evaluating the effect of particle size of sorbitol on granules properties, Bacher et al. [27] found that smaller sized sorbitol produced granules that had higher compactability due to the increase in surface area with smaller particles of sorbitol.

Inghelbrecht and Remon [22] evaluated the influence of lactose particle size as well as the type on granule and tablet properties. As with MCC and sorbitol, they found that reducing the lactose particle size (regardless of type) improved the granule quality (less friable). They showed that anhydrous lactose produced granules that were less friable, because the crystals were more compactable and less elastic.

In summary, the work cited above shows that selecting a smaller particle for the input material (MCC, sorbitol, or lactose) results in granules that are larger, stronger and less friable. The effects were ascribed to the increased surface area.

As with input materials, particle size of the resulting granules is often used as a metric to determine input or parameter effects. The granule size is controlled by the roller compaction process as well as downstream processing. The initial granule size is a result of the material properties and the process parameters. Granule attrition during postroller compaction processing (blending) will determine the final particle size distribution.

The quality of the ribbons (density and strength) produced has a direct impact on the granule size distribution. Ribbons with higher solid fractions (lower porosity) produced granules with larger mean sizes that resulted in a better flowing powder (Herting and Kleinebudde [33, 34]). Ribbon solid fraction is correlated with ribbon strength, which is sometimes used as an alternate measure of ribbon quality and the effect on granule size. Under similar milling conditions (mill speed and screens) ribbons with higher strength produced larger granules compared to ribbons with lower strength (Shesky and Hendren [35], Inghelbrecht and Remon [20, 22], Weyenberg et al. [36]). Farber et al. [37] hypothesized that during roller compaction the particles deform under load, causing them to interlock. Upon milling, the break in the ribbon occurs at the weakest interlocking junction, but the deformed particles remain intact in the resulting granules. Rambali et al. [11] showed that the mean granule size produced from thicker ribbons was smaller that that for thinner ribbons. The ribbon thickness effect on granule size was marginal.

Methods for particle size determination fall into two general categories: laser light scattering (LLS) and sieve analysis.

A secondary way in which the particle size of the granules is affected is in processing downstream from the roller compaction operation. This is primarily governed by the strength of the granules. This is assessed by determining the friability of these granules by measuring the change in particle size distribution under stress (e.g., due to additional mixing). There are two ways by which the friability of the granules can be determined. An indirect way is to compare the particle size distribution of the granules after milling with that of the final blend after additional mixing. After accounting for any extragranular materials added, the change is size distribution would be indicative of granule friability.

Inghelbrecht and Remon [20] quantified granule friability directly by tumbling granules of a particular particle size range with glass beads for certain duration, and then determined the change in particle size. The reduction in mean particle size was ascribed to particle attrition. Patel et al. [38] used stress–strain analysis on single particles using a 2 mm flat probe to determine the particle fracture potential. Inghelbrecht and Remon [22] ranked the process parameters for their effect on granule friability in order of decreasing influence: roller pressure, roller speed, and feed rate.

38.3.2.2 Density Soh et al. [16] and Freitag et al. [24] investigated the effect of raw material attributes on their performance in roller compaction. One attribute tracked was material density. Since, material density was confounded with other materials properties, such as particle size or morphology; a clear relationship between input density and output (ribbon or granules) properties was difficult.

Herting and Kleinebudde [34] characterized the hardening of granules postroller compaction by measuring the yield pressure. They showed an increasing relationship between the applied roller pressure and the apparent yield pressure. They ascribed the increased resistance to deformation as granule hardening. Traditional density measurement techniques typically used include bulk density, tap density, and true density by helium pycnometry. Derived parameters, such as porosity, Carr Index, Hausner ratio, also have been used (Soh [16]).

38.3.2.3 Flow For processing downstream from the roller compaction operation powder flow dominates process stability. Flow of the powder blend into the compression operation controls the variability in tablet weight. Weyenberg et al. [36] reported that the fastest powder flow was obtained with a combination of low roller speed and high roller pressure. These conditions yielded ribbons with high strength that resisted attrition during the milling operation. The rate of powder feeding had a minor impact of granule flow properties. These conditions also produced the largest and strongest granules with the lowest friability. Granule size increased with higher roller pressure, lower roller speed and higher powder feed in order of importance.

There are several techniques used to gauge the ability of a powder to flow. A mass flow determination can be obtained by measuring the flow time of a certain mass of powder passing through a certain orifice. An alternative method would be to determine the minimum orifice opening that would support continuous flow (Flow index). Other techniques rely on indirect means to assess flow, such as Carr Index, shear cell, and other measures of powder rheology.

38.3.2.4 Compactability One of the common objectives of a roller compaction process is to improve the performance of granules in a downstream capsule or tablet operation. Compactability, the ability to form tablets of a desired strength at an acceptable pressure, is one of the main attributes to consider for the roller compacted granules. Compactability is typically studied by experiments on a tablet press or compaction simulator to assure that the desired tablet properties can be achieved.

Several articles report a reduction in powder compactability of the roller compacted material (Sun and Himmilspach [39], Herting and Kleinebudde [26], Herting and Kleinebudde [30], Malkowska and Khan [40], Shesky and Cabelka [41]). Investigators have identified two possible causes for this reduction: work hardening and size growth. Malkowska and Khan [40] describe the effect as a loss of bonding capacity between the particles. Plastically deforming materials are more susceptible to work hardening.

Sun and Himmilspach [39] ascribed the reduction in compactability to an enlargement of granule size relative to the input particle size. The growth in particle size reduces the area available for bonding. Herting and Kleinebudde [30] concluded from their investigation that the reduction in compactability is due to both size enlargement and hardening of the granules. They observed that work hardening could be countered by producing smaller granules.

38.3.3 Ribbon Properties

38.3.3.1 Density and Porosity The packing in a ribbon after roller compaction can be stated by three related terms: density, solid fraction (density relative to the true density), and porosity (measured or calculated from the solid fraction). Ribbon solid fraction is an attribute that indicates the degree to which the powder has been compressed (Zinchuk et al. [42]). The density across the width of a ribbon can vary (Miguelez-Moran et al. [43]).

The density of the ribbon is often highest in the center and lowest at the edge. This density gradient is caused by the friction between the powder and the face plates covering the rollers (Guigon and Simon [44]) and can be reduced by the inclusion of lubricant. Funakoshi et al. [45] showed that density distributions are related to the force distribution across the ribbon. Funakoshi et al. [45] and Parrott [46] evaluated an alternate roller design (concave–convex roller pair) to address the nonuniform distribution of pressure on the rollers during roller compaction. Several incline angles were evaluated and a 65° angle was shown to be optimum for a uniform pressure distribution over the roller surface (Funakoshi et al. [45]).

The force distribution is related to the flow patterns of the powder passing between the rollers (Miguelez-Moran et al. [32]). This effect is accentuated at high roller speeds (Funakoshi et al. [45], Miguelez-Moran et al. [32]). The mean ribbon density is higher at a narrower gap setting compared to a wider gap setting. Guigon and Simon [44] also showed that if the screw feeding in a roller compactor is nonuniform in time or space, and this produces ribbons whose solid fraction varies over the width of the ribbon.

The solid fraction of the ribbons produced is a result of powder properties, process parameters, and equipment geometry factors. The increase in solid fraction affects the mechanical properties of the granules and the material behavior. However, when comparing ribbons made from different materials, solid fraction is insufficient as a sole descriptor of quality. Ribbons with similar solid fractions could have dissimilar mechanical properties such as strength. Ribbon strength is an attribute that is indicative of performance during the milling operation postribbon production. Solid fraction can be used as a surrogate for ribbon strength within a single composition.

Two physical techniques were reported by various authors for determining ribbon density (sectioning, enveloping). Additional methods using NIR or ultrasound are described in Section 38.3.4. With the sectioning method, a portion of the ribbon is removed and the volume and mass measured to yield the density. The dimensions of the section must be carefully measured. With the enveloping method a volume displacement approach is used so that the precise dimensions of the section are not critical. In determining the ribbon solid fraction, Soh et al. [16] argued that envelope volume is required for higher precision rather than using sectioning. The reason stated for this is the imprecise nature of the edges during sectioning.

In place of the techniques cited above that provide the average overall density of a ribbon specimen, Miguelez-Moran et al. [43] used X-ray microcomputed tomography to obtain the distribution of the densities in a given ribbon sample. This enabled the investigators to not only determine the effect of roller compaction process parameters on ribbon density, but also on the density distribution.

38.3.3.2 Thickness Investigators use a micrometer to measure the thickness of the ribbons produced (Miguelez-Moran et al. [32, 43]). This measurement has to be repeated at several places across the ribbon sample to account for variation in the ribbon thickness. This average thickness then is used for density calculations using the sectioning technique or for feedback control of the roller compaction process. In lieu of measuring the ribbon thickness, some investigators use the roller gap as a surrogate. This approach ignores the relaxation that may occur postconsolidation.

38.3.3.3 Strength Many investigators use ribbon density as a metric for comparing ribbons made under different conditions to determine equivalency. Zinchuck et al. [42] argued that a ribbon's resistance to milling postribbon production is a better metric. They determined the tensile strength of ribbons using a three-point beam bending analysis. Miguelez-Moran et al. [43] used a microindentation technique to determine the hardness of ribbons at a microscale. The size of the indentation made depends on the shape of the indentor, the force used to make the indentation and the hardness of the ribbon.

Within a composition the ribbon tensile strength or hardness varies directly with the solid fraction of the ribbons. Similar to tablets, the ribbon tensile strength or hardness is directly proportional to the pressure used to make the ribbons. Farber [37] hypothesized that during roller compaction the particles, deform under load, causing them to interlock. This is the reason for the increased strength of ribbons postcompaction. The strength of ribbon varies across the width of the ribbon, due to the variation in ribbon solid fraction across the width of the ribbons. This variation is due to a nonuniform distribution of stress across the roller width. Guigon and Simon [47] used a series of pressure transducers on the rollers to obtain the distribution of the pressure on the rollers during roller compaction. They observed that the pressure varied with a period that coincides with the screw feeder pushing the powder into the nip region. Guigon and Simon [47] showed this variation in applied force across the rollers visually by adding charcoal particles to lactose. The areas of high stress were identified by the large number of broken charcoal particles.

38.3.4 Online or At-Line Process Sensors

38.3.4.1 Particle Size Distribution There are several examples in the literature of technologies for determining the particle size of powders online, but Zhang and Yan [48] made the point that development of cost-effective online particle size instruments is challenging. Zhang and Yan [48] used an electrostatic sensor combined with digital signal processing to determine particle size distribution. A slip stream of particles was carried on an air stream toward the sensor causing the particles to acquire a charge; the magnitude of the charge was size dependent. Frake et al. [49], Rantanen et al. [50] and Gupta et al. [29, 51] used a near infrared technique to quantify the particle size of granules. The near infrared spectrum was affected by physical properties such as particle size, particle surface and density (Rantanen et al. [52]).

Two additional means of ascertaining the particle size distribution of the granules as they exit the milling chamber postribbon production are laser light scattering and high-speed image analysis. For the LLS method a slip stream sampling of the falling granules is needed. Traditional LLS techniques can be used on the sample of granules (Bordes et al. [53]). Image analysis would require high-speed image capture that is processed against a predictive model of pixels versus volume or diameter. Liao and Tarng [54] developed a high-speed optical inspection system to determine particle size. A CCD or CMOS camera was used to acquire the image that was processed and analyzed against a reference.

38.3.4.2 Ribbon Density Two general methods were reported by various authors: physical (sectioning, enveloping) and associative (NIR, ultrasound). The physical methods were described previously and can be used only at-line. Several investigators have reported on the associative methods (Sprockel et al. [55], Gupta et al. [29], Feng et al. [56]). Even though the fundamental principles of the associative methods differ, both rely on a predictive model that correlates density to the underlying measurement. The associative methods lend themselves to online or at-line measurements, whereas the physical methods can only be used at-line.

Ghorab et al. [57] reported on their evaluation of the relationship between ribbon physical attributes (such as density and strength) and thermal effusivity. They found strong correlations between thermal effusivity and ribbon density or strength. The relationships were first- or second-order polynomials depending on the composition. Even though these relationships are composition dependent, the utility of this method is intriguing. Herting and Kleinebudde [33] reported a means of calculating the in-gap porosity by calculating the volume of ribbon produced per unit time. The mass of granules corresponding to this time unit was used to calculate the ribbon porosity.

38.3.4.3 Composition and Uniformity Bacher et al. [13] investigated the cause for the nonhomogeneous distribution of calcium carbonate in roller compacted granules containing sorbitol. They showed that the particle size of the sorbitol diluent was the main contributing factor. Using the smaller particle size sorbitol produced granule fractions with near theoretical mean calcium carbonate content. When sorbitol with larger particle sizes was used a higher content of calcium carbonate was seen in the fines.

They postulated that the weakest interparticulate bond in the granule was the calcium carbonate–calcium carbonate bond. It is at this juncture that the ribbons fractured during the milling process. This rupturing of the ribbons at the weakest point exposed the calcium carbonate to attrition. Gupta et al. [29] investigated near-infrared coupled with multivariate analysis to relate the spectral data to content uniformity of the ribbon.

The material properties of the raw materials and the resulting products determine the type of equipment selections for the roller compaction unit operation. The various tests of input, output, and intermediate material properties help to understand the operation of the equipment and selection of parameters described in the next section for each subprocess of the roller compaction operation.

38.4 PRINCIPLES OF OPERATION

Roller compaction is a unit operation composed of several subprocesses the principles of, and considerations for operations are described for each subprocess in the sections below.

38.4.1 Roller Compactor Feed Preparation and Delivery

The preparation and handling of powder fed to the roller compactor has challenges common to similar operations in other equipment trains. Powder and handling are specialized areas with adequate sources from which to assemble information for the design of the mixing and handling subprocesses of a roller compaction process. Blending and blend batch size increase was discussed by Alexander and Muzzio [58]. Prescott [59] presented aspects of powder handling, and the metering and dispensing of powders was reviewed by Yang and Evans [60].

Some common themes from powder handling technology are important to the powder handling for roller compaction. For example, it is commonly held that powder flow can be affected by particle size and shape, factors which can be important considerations in the selection of material for a roller compaction process. Yang and Evans [60] mentioned how humidity and electrostatic charge can affect powder flow, knowledge that can also be applied to the conveying of powder to the roller compactor. Pietsch [12] has described how air entrainment can influence powder handling, including the densification step and the postcompaction recovery in the roller compaction process. A study by Miller et al. [8] showed that the leakage of uncompacted powder was reduced from 20–30% to <2% of the material by the use of a vacuum deaeration system fitted to the roller compactor. The throughput of the roller compactor was also increased by 20–40% with the deaeration system activated.

Most roller compaction equipment configurations move the powder toward the rollers via a screw feeder. Sander and Schonert [61] showed that the delivery of powder from an unconstricted feed screw follows a linear relation with the screw speed. It was found that the screw feed needed to have a minimum speed in order to feed sufficient material to the rollers at a given roller speed assuring the roller compactor operated properly. At higher screw speeds the screw throughput was less than predicted by the unconstricted delivery rate indicating that the screw exerts a pressure on the powder before the rollers (Figure 38.8). The screw feed increased the throughput of the roller compactor by causing consolidation of the powders before the nip. Similar results showing the behavior of the screw feeder and its effect the roller compaction throughput were also reported by Guigon and Simon [44].

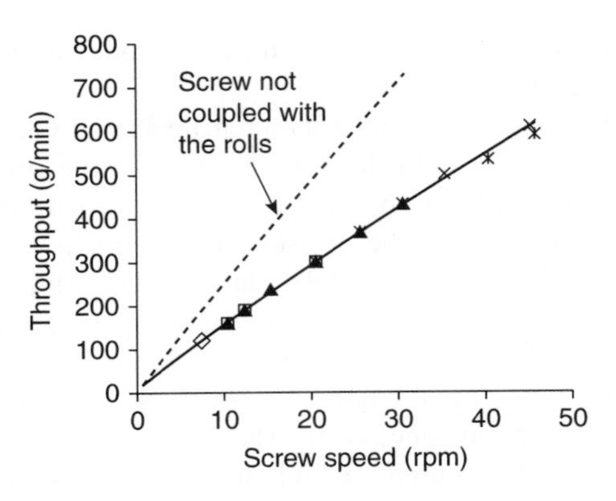

FIGURE 38.8 Screw throughput as a function of screw speed with and without the influence of the rollers. Reprinted from Ref. 44, Copyright (2003), with permission from Elsevier.

The design and rotation of the screw can have unintended effects on the compacted ribbon. Patterns of color or differences in ribbon properties oriented along the main axis of the ribbon in a sinusoidal pattern were reported by Simon and Guigon [62] for operations with a single screw (Figure 38.9). They used piezoelectric sensors in the rollers to study the pressure during compaction, and image analysis of the ribbons to determine the density of the ribbon. Both sets of data showed a periodic variation that could be correlated with the frequency of the feed screw rotation. Experiments with a piston device to feed the powder did not show periodic variations confirming that the screw feeder was causing the powder to consolidate differently depending on the screw rotation. It was postulated that the screw is preferentially applying pressure where the clearance between the flight and the nip are at the minimum, a position corresponding to the screw flight terminus. Miller [9] reported that a dual screw feed design in combination with vacuum deaeration of the powder minimized any influence of the screw rotation on the ribbon.

Lecompte et al. [63] did not find screw-related variations in similar experiments with an instrumented roller compactor. It was proposed that because the screw terminated further

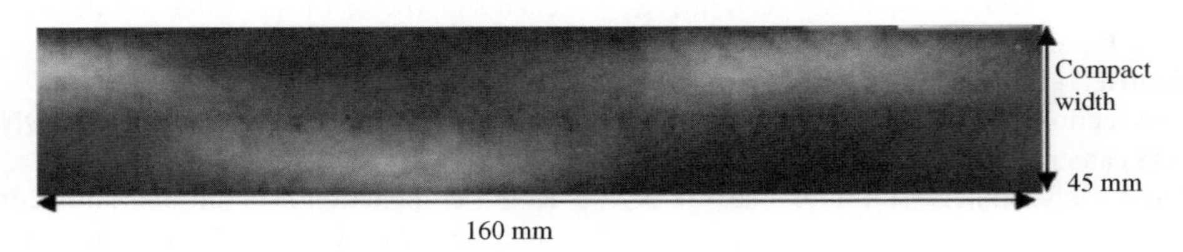

FIGURE 38.9 Variation in ribbon color due to variation in ribbon compaction induced by the motion of the feed screw. Reprinted from Ref. 44, Copyright (2003), with permission from Elsevier.

from the rollers in the experimental setup used by Lecompte, variations due to screw rotation were not carried forward into the ribbon. The authors also examined how various combinations of parameters such as feed screw speed, roller speed, and gap could be adjusted to assure that the powder spread evenly across the rollers and was converted to a ribbon with consistent properties. Settings which increased the amount of powder entering the nip region, such as high screw speed, low roller speed, and a narrow gap setting, promoted the distribution of powder across the roller width.

The sealing plates can have an influence on the distribution of powder and compaction of the ribbon. This phenomenon was studied by Miguelez-Moran et al. [32, 43] using an instrumented laboratory roller compactor, and several techniques to characterize ribbon density. A distribution of density was found across the ribbon (parallel to the roller axes, transverse to the ribbon motion) with lower density found at the ribbon edges. The effect was less pronounced for slower roller rates, smaller gaps and powders that slipped along the seal surfaces more readily. The authors suggested that the lower density at the ribbon edges was attributed to drag induced in the powder feed by the sealing plates.

From the end of the screw feed to the gap, the powder is typically contained by plates that may also be considered part of the feeding system. Some leakage of powder from the seals may contribute to the amount of fines in the granulation. If not properly installed so that the leakage is at a minimum, and sufficiently maintained so that the performance of the seal plates is consistent, the sealing plates may cause an isolated batch-to-batch variation or trend in the granule properties over time.

To improve the powder sealing and process performance some manufacturers offer rollers that interlock. One roller has a rim on the edge such that the edge of the second roller fits the channel and seals the powder into the nip. A system of this type was explored by Funakoshi, et al. [45]. Several roller designs with different modifications to the rim geometry were explored. The amount of leaked powder was reduced from ~20% to ~5% by use of the interlocking rollers. When examining the resulting ribbons it was found that without the interlocking roller, the ribbon experienced the highest pressure and achieved the highest density in the center. With the interlocking roller system in place, the pressure experienced by the ribbon and resulting density was even across the ribbon width.

38.4.2 Consolidation and Compaction in Between the Rollers

Once powder is delivered to the area before the nip it begins its interaction with the rollers, moves forward and becomes compacted into ribbon as it passes through the gap between the rollers. Johanson [15] developed one of the most referenced descriptions and mathematical models of the roller compaction process. The roller compaction operation was described based on the machine geometry and assumptions of powder behavior.

Johanson took the pressure exerted on the powder bed from the feed screw as an input to the mathematical model but did not otherwise discuss the powder motion up to the rollers. From the feed area the powder moves forward and begins to be influenced by the roller motion. This region is typically referred to as the slip region because the rollers move faster than the powder with a boundary condition of slip between the powder and the roller surface. As the powder is dragged forward the space in between the rollers narrows so that the powder bed consolidates and the pressure between the roller surfaces and the powder increases. A schematic diagram of the powder and roller interactions is presented in Figure 38.10.

To describe the process in the slip region Johanson built upon earlier work describing steady-state powder flow originally developed by Jenike in 1961. The Jenike model and other powder flow concepts, developed more recently, has been summarized by Podczeck [64].

It was proposed that for a cohesive, compressible, isotropic powder, a shear test could give information about the

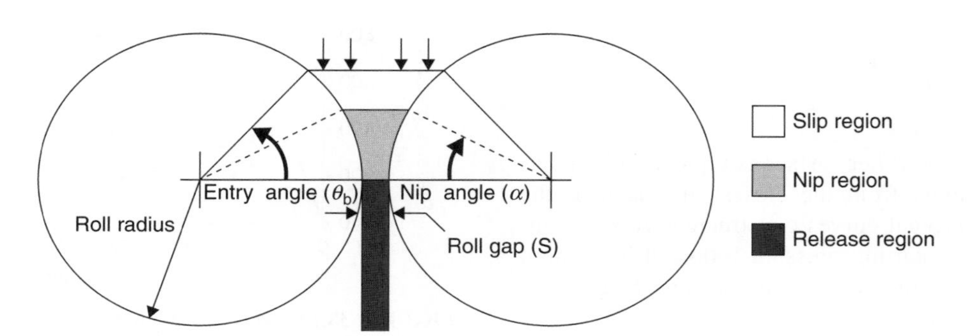

FIGURE 38.10 Schematic diagram of the roller and powder interactions. Reprinted from Ref. 65, Copyright (2005), with permission from Elsevier.

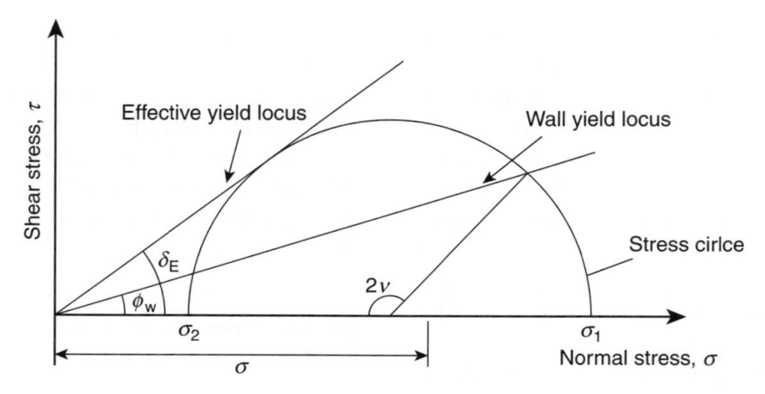

FIGURE 38.11 Jenike-Shield yield criterion for the slip region. Reprinted from Ref. 65, Copyright (2005), with permission from Elsevier.

effective angle of friction and the surface friction angle. This information could then be used to plot yield loci describing the response of the powder to shear and normal stresses (Figure 38.11). The intersection of the wall yield locus and the stress circle gives the resolution of shear and normal stresses at the roller surface. The angle between the normal stress and the tangent to the roller is then described by equation 38.1 (equations were prepared from Ref. [65] Copyright (2005) with permission from Elsevier).

$$2\nu = \pi - \arcsin\frac{\sin\phi_\omega}{\sin\delta} - \phi_\omega \tag{38.1}$$

Johanson used the combination of the incoming pressure, roller geometry, and powder properties to predict a pressure distribution as a function of position as shown in equation 38.2.

$$\frac{d\sigma}{d\chi_{\text{Slip}}} = \frac{4\sigma((\pi/2-\vartheta-\nu)\tan\delta_E}{(D/2)[1+(S/D)-\cos\vartheta][\cot(A-\mu)-\cot(A+\mu]} \tag{38.2}$$

where A is given by

$$A = \frac{\vartheta+\nu+(\pi/2)}{2}$$

The pressure at the roller surface is typically plotted as pressure versus angle from the closest approach of the rollers, the gap. Typical curves constructed with this approach show a low starting pressure followed by a rapid nonlinear increase of pressure moving toward the gap (see Figure 38.12, line a).

At some point the pressure on the rollers increases such that the powder no longer slips along the roller surface but moves with the roller surface until the powder exits the gap as compacted ribbon. This region close to the gap is referred to as the nip region. In order to predict the pressure in this region, Johanson considered the densification of a cross section of powder moving between the rolls. This section of powder is compressed as if in uniaxial compression with density increasing as the distance between the rollers decreases. Powder property measurements from compression experiments with a die can determine the functional relationship between pressure and density. Johanson developed the expression shown in equation 38.3 to describe the pressure as a function of position in this region.

$$\left(\frac{d\sigma}{d\chi}\right)_{Nip} = \frac{K\sigma\vartheta(2\cos\vartheta-1-(S/D))\tan\vartheta}{(D/2)[(1+(S/D)-\cos\vartheta)\cos\vartheta]} \tag{38.3}$$

This function can be used to construct a plot of pressure as a position from the gap in a similar fashion to the plot generated for the case where the powder slips along the roller surface. For the case of no powder slip the resulting pressure versus position curve typically has a different shape with a higher starting pressure and a more linear increase as

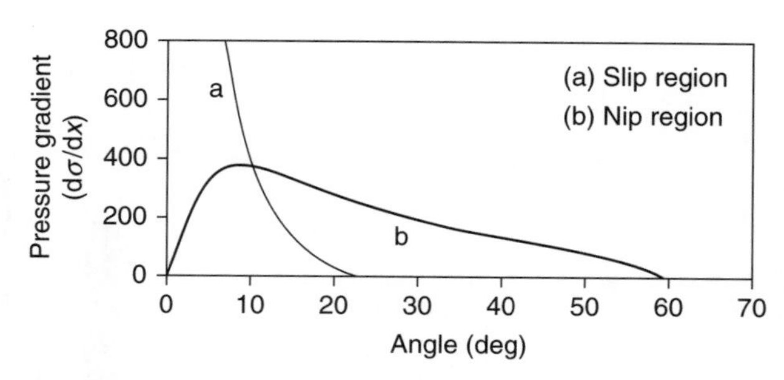

FIGURE 38.12 Pressure gradient versus angle from the nip. Reprinted from Ref. 65, Copyright (2005), with permission from Elsevier.

the gap is approached than the plot for the slip condition (see Figure 38.12, line b).

The difficulty is predicting when the powder behavior will switch from slip to no-slip at the roller surface, thereby transitioning from the slip to the nip region. Since the pressure is most often plotted as an angle from the gap, the position of this transition was referred to by Johanson, and is commonly called, the nip angle. The lowest pressure predicted by the slip and no-slip assumptions for the roller surface boundary condition is followed. At the nip angle the pressure predictions from the slip and no-slip condition are equal and the powder no longer slips at the roller surface from this point onward as no-slip is the lowest pressure needed to induce powder motion. This relationship is represented in equation 38.4.

$$\left(\frac{d\sigma}{d\chi}\right)_{\text{Slip}} = \left(\frac{d\sigma}{d\chi}\right)_{\text{Nip}} \tag{38.4}$$

or

$$\frac{4((\pi/2)-\alpha-v)\tan\delta_{\text{E}}}{[\cot(A-\mu)-\cot(A+\mu)]} = \frac{K(2\cos\alpha-1-(S/D))\tan\alpha}{\cos\alpha}$$

A schematic illustration of the concepts from the Johanson model showing the two pressure predictions as a function of angle from the gap and the intersection of line a (slip condition) and line b (no-slip condition) defining the nip angle is shown in Figure 38.12. A graphic representation of the stress on the rollers is shown in Figure 38.13.

Forty years later, Bindhumadhavan et al. [65] compared the results of calculations using Johanson's model with experiments. Microcrystalline cellulose was roller compacted with the benefit of better instruments to measure powder properties, pressure sensors mounted in the roller

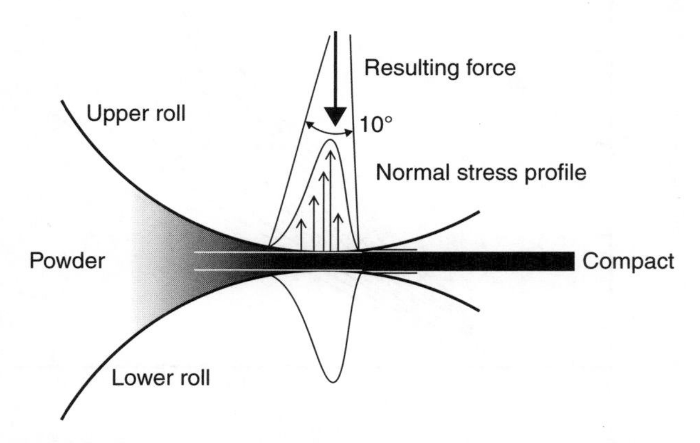

FIGURE 38.13 Graphic representation of the stress profile in the roller compactor nip. Reprinted from Ref. 44, Copyright (2003), with permission from Elsevier.

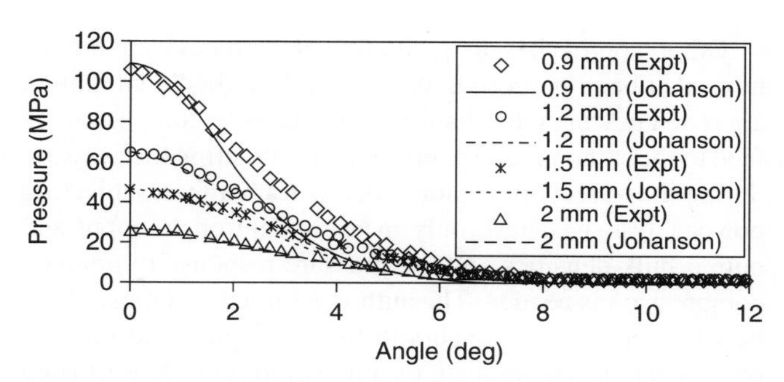

FIGURE 38.14 Predicted and experimental roller surface pressure as a function of gap. Reprinted from Ref. 65, Copyright (2005), with permission from Elsevier.

faces to measure the compaction forces, and better computing power to calculate the model and evaluate the data. Good agreement was found between the predictions using the method of Johanson and the experimental values as shown in Figure 38.14 relating the pressure as a function of angle from the nip for different gap settings. Yusof et al. [66] also reviewed Johanson's model and conducted experiments with maize powder to compare the model and experimental values. Once again good agreement between the model and the experiments was found.

Sommer and Hauser [67] reviewed the Johanson model and found that predictions were useful given the limited number of inputs. However, they examined the assumptions, which Johanson used and concluded that there could be limitations from the assumptions, which might cause the model to deviate from the results found in practice. One example presented was the model's possible sensitivity due to the method of choosing the boundary condition for the point at which the rollers engage the powder feed. The boundary condition assumption could cause an unrealistic sensitivity to the feed pressure in the model results. Simple material models were used for the description of the nip region, and it was proposed that the limitations in these models could also lead to inaccuracies.

An additional mechanism of slip has been explored by Schonert and Sander [68]. They reviewed several theoretical models and concluded that there could be slip between the compacted ribbon from the point of maximum stress to the exit of the gap. Instrumentation embedded in the rollers capable of resolving the normal and tangential stresses at the roller surface confirmed that the powder does begin to move with the rollers at some level of consolidation. However, the maximum stress occurred slightly before the line of centers between the two rollers. From the point of maximum stress it was found that the compacted ribbon accelerates, moving faster than the rollers, toward the gap exit. A similar measurement showing the maximum pressure before the line of centers and an acceleration of the ribbon was obtained by Lecompte et al. [63].

Other approaches to predicting the process conditions have also been proposed. Yehia [69] described a simplified approach in which the change in bulk density going from the feed to the ribbon is considered when estimating the pressure during the roller compaction process. Rather than subjecting powder to a shear test, only measurement of the input and output bulk densities and the pressure response to uniaxial compression is needed. The author assumed that most of the powder densification occurs in the nip region and that the pressure at the beginning of the nip region could be estimated by a density measurement and information about the density–pressure relationship for the material studied. The speed and geometry of the rollers and gap width is used to estimate the change in volume, and therefore, density of the material as it is processed. It was proposed that the performance at scale could be predicted from the limited material testing required for the calculation and information obtained on a lab-scale roller compactor. The lab-scale machine would need to be adjusted to a variety of geometries and equipped with different roller finishes to determine an appropriate regime to predict the performance at scale.

Dec et al. [70] reviewed various models including a method of estimating roller compaction conditions. The "slab method," originally developed for the compaction of metal powders, considers a section of material or "slab" passing as a single element through the nip region. The stepwise calculations assume a condition at the inlet of the nip and uses experimental data to determine the nip angle. Pressure is predicted by successive iterations of the calculation until the prediction matches the measured properties of the ribbon.

The mechanistic understanding of roller compaction developed by Johanson and others is useful to understand the processing history of the powder as it is compacted into ribbon. The approach of these investigators grew of an interest in designing roller compaction equipment. For most applications, the machine geometry, speeds, and pressure ratings have been predetermined by an equipment manufacturer for a range of materials similar to the proposed process. However, the preselected machine capabilities still leave the choice of the various processing parameters to be made in designing the roller compaction operation.

Other investigators reported studies designed to understand the interaction of the process variables and particular materials or combinations of materials to determine the most effective processing parameters. Most of the studies of this type used a statistical design of experiments to evaluate the affect of parameter changes on the ribbon, granule, and finished tablet properties but may also include raw material changes, composition or other variables as part of the study.

Falzone et al [19] studied the effect of the horizontal and vertical feed screw speeds, and roller speed on granule size and compactability for microcrystalline cellulose, lactose, and a model active blend consisting of 60% acetominophen. The results for microcrystalline cellulose granules and tablets were successfully modeled by a quadratic regression that included the horizontal feed speed and the roller speed with high feeder speed and low roller speed giving the highest values for both granule size and compactability. For lactose the vertical feed speed also had to be used to successfully model the results. The acetaminophen blend results showed granule size equal to or less than the starting material due to fracture of the acetominophen crystals. The results for granule size could not be modeled by a quadratic fit of the data, but the acetominophen granulation compactability could be described by the quadratic regression and showed a dependence on both the feed screw speeds and the roller speed.

Hervieu et al. [71] used a model powder to study the effect of feeder speed, roller speed, and compaction force on granule properties and the hardness and friability of the final tablet. The Box Wilson experimental design required 15 batches to complete. It was observed that at low feeder speeds compared to the roller speed, the powder could not be effectively compacted and had high friability. If the feed speed was too fast in comparison to the roller speed the material temperature increased and the roller compactor jammed. Compaction force only had a secondary effect on the results. Similar results showing dependence on the feeder to roller speed ratio were reported by Guigon and Simon [44]. At higher feed speeds, overfeeding resulted in a poor quality compact with loss of uncompacted powder while at lower feeder speeds underfeeding resulted in no compact being formed. The effect of feeder and roller speed selections at constant pressure on the gap opening is shown in Figure 38.15 with lower feed resulting in a smaller gap at constant roller

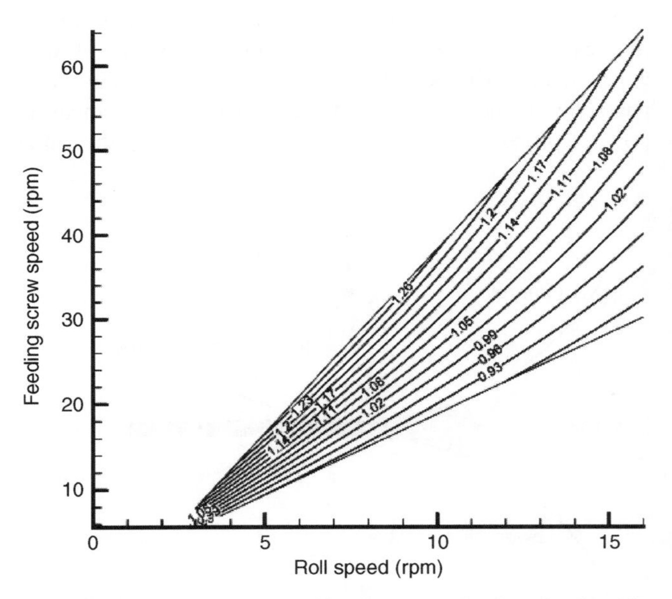

FIGURE 38.15 The effect of feed screw and roller speed on the gap opening at constant pressure. Reprinted from Ref. 44, Copyright (2003), with permission from Elsevier.

speed and lower roller speed resulting in a larger gap at constant feed speed.

Four parameters (pressure, gap, roller type, and sieve aperture) were studied for their effects on a buccal tablet formulation by Rambali et al. [11]. A partial factorial design showed that smooth rollers promoted larger granule size and higher tablet strength but resulted in slower dissolution compared to ribbed rollers. It was noted that smaller granule size typically gave higher tablet strength. The lower than expected tablet strength for small granules produced by ribbed rollers was attributed to the higher frictional force and powder rearrangement induced by the ribbed surface. Smaller gaps and higher pressures produced larger granules. However, the tablet strength was found to be inversely proportional to gap and pressure.

A factorial design was also used by Weyenberg et al. [36] to study the effects of roller speed, pressure, and horizontal feed speed on granule flow, granule size, granule friability, and the resulting tablet strength for a bioadhesive tablet formulation. The roller speed and the compaction force had the largest effects on the granule properties followed by the horizontal screw feed speed. Higher roller speeds combined with low compaction force had reduced granule flow, increased granule friability, and smaller granule size. The horizontal screw speed did not have a large influence on granule properties. The highest tablet strength was obtained from smaller granules prepared at high roller speed and low compaction force. Tablets from the smaller granules best matched the dissolution profile of tablets that were previously prepared by direct compression, tested *in vivo*, and used for comparison with tablet from a roller compaction granulation.

In a study by am Ende et al. [72], a two stage design was used to examine the effects of roller compaction on the content uniformity, granule properties, and tablet properties of a low active concentration granulation. The first stage of the experiment used a full factorial design to study the effects of compaction pressure, and gap width, where the feed screw was automatically adjusted to maintain gap. The study responses were the amount of uncompacted fines and the potency of the uncompacted fines. It was found that the amount of uncompacted fines was affected by the pressure and gap settings with a significant interaction between the pressure and gap variables. Higher levels of fines were present for low pressures and high gap widths. The potency of the fines, a measure of the uniformity of the granules, could not be predicted by these two variables alone.

The second stage of the experiment used a D-optimal design to examine the effects of roller speed, gap width, compaction pressure, and postcompaction mill sieve opening on granule and tablet properties. The granulation size increased with increasing pressure but had less of an effect as the mill sieve opening decreased showing the interaction between these two variables. Compaction pressure and sieve opening also had an influence on the uniformity of both the granulation and the tablets. As screen size increased the variability of the granulation potency increased. It was found that lower pressures and larger gaps during the roller compaction led to lower compaction forces to form a tablet of the desired strength. Tablet friability was unaffected by any of the roller compaction variables studied.

As discussed in this section a review of the mathematical descriptions and mechanistic understanding, combined with some knowledge of the materials being roller compacted and some studies of the system response to the parameter settings can be used to determine and even predict the settings of or system responses to the four main variables in the roller compactor: feed speed, roller speed, compaction pressure, and gap width.

38.4.3 Ribbon Breaking and Size Reduction

The roller compacted ribbon is typically broken and reduced in size to form the final granulation. The interparticle bonds in the ribbon need to be strong enough such that fractures form and lead to granules rather than a loosely compacted ribbon returning to powder. Ribbon of a consistent strength and density should give consistent granule particle size distributions. Bacher et al. [14] suggested that the ribbon breaks at the weakest interparticle bonds and that ribbons with uniform interparticle bonds give the better compactability.

The milling operation can be integral to the machine or a separate step and is usually accomplished with a screening mill. General considerations common to screening mills such as choosing the impeller type, screen type, speed (impeller or screen depending on type of mill), and spacing of the screen and impeller are steps to achieving the desired granulation particle size distribution. Proper selection, setup, and maintenance can eliminate the need for a metal detector at the mill exit. Effects of screen selection affecting the granule and tablet properties in a roller compaction operation have been reported by am Ende et al. [72].

Information on the operation of mills can be applied directly to the milling subunit operation of roller compaction. Rekhi and Sidwell [73] described how Kick's Law, Rittinger's Law, and Bond's Law have all been developed to relate the mill energy input to the size reduction process. It was reported that only a small part of the energy used by the mill is consumed in breaking the particles making the ability to predict performance limited in practice. Some experimentation is necessary to select the mill change parts (e.g., screen, impeller, spacers), and operating parameters due to the limitations of the predictive methods. These experiments can also identify common milling problems such as screen blinding, heat generation (with possible melting), and interactions with moisture either from the environment or liberated during the milling process.

38.4.4 Process Models for Roller Compaction

The area of the roller compaction operation that is unique to the unit operation and of most interest to model is the area where the powder is consolidated, begins to move with the rollers and is compacted into the ribbon. Several approaches to process predictions have been explored: mathematical models, multivariate regression of designed experiments, predictions from compaction properties, computational modeling of the material in the process, and artificial intelligence/neural network construction to predict outcomes from a set of training data.

The foundation of a mathematical description for the roller compaction process was described by Johanson [15] and has been further studied by many authors including Bindhumadhavan et al. [65], Yusof et al. [66], and Sommer and Hauser [67]. Alternate approaches were presented by Yehia [69] and reviewed by Dec et al. [70]. Several studies have used multivariate regressions to characterize the roller compaction operation such as Falzone et al. [19], Hervieu et al. [71], Rambali et al. [11], Weyenberg et al. [36], and am Ende et al. [72].

Although achieved by passing between rollers, the compaction of ribbon causes the same physical processes within the powders that are induced during compaction within a die. Borrowing or combining concepts from tablet compaction can be useful to understanding roller compaction. For example, Farber et al. [37] examined the loss in tensile strength when comparing tablets made via a roller compaction granulation to a direct compression prototype. The roller compaction process was considered as part of the overall compaction history of the materials. A "unified compaction curve" was constructed that described both the roller compaction and tablet forming processes. It was found that, for materials that bond primarily from plastic deformation, compaction information generated with a compaction simulator or single station press could be used with information about the pressure exerted by the rollers to estimate the tensile strength of the final tablets.

Hein et al. [74] used a three-dimensional model of compaction properties populated by data from a single station press to predict the change in final tablet properties as a result of roller compaction. A reduction in final tablet strength was shown for materials with plastic deformation. A minimal reduction in tablet strength was shown for the material with primarily brittle fracture. The model was considered effective at screening materials for use in roller compaction.

Computational methods can be used to understand and predict process behavior and can be applied to roller compaction. Dec et al. [70] reviewed several finite element method (FEM) applications as part of a review of modeling methods. However, most of the models reviewed were published from researchers in the metals industry. The simulations were dependent on estimates of the feed stress and friction to predict the process conditions and postcompacted material performance.

The use of neural networks and artificial intelligence approaches to modeling have been explored to correlate different types of inputs, process parameters, and granule or resulting tablet performance. Inghelbrecht et al. [75] studied a 60 experiment data set and then predicted the results for an additional 20 experiments. The speed of two feed screws (horizontal and vertical), roller pressure, and roller speed were used to predict the granule friability and particle size. It was found that the neural network was more effective than a quadratic mathematical model approach in predicting the granule performance results.

The binder type, binder concentration, number of compaction passes, and addition of microcrystalline cellulose extragranularly, were used as inputs to predict the performance of acetominophen tablets by Turkoglu et al. [76]. The results were poorly predicted using a typical neural network learning algorithm of adjusting the weighting of parameters relating model inputs and outputs. A second calculation method using a "genetic" algorithm that progressively selects best fit solutions in "generations" of calculations gave better predictions.

A variety of material inputs were used with projected process settings for roller speed, and roller gap to predict the roller pressure, nip angle, ribbon density, and ribbon porosity by Mansa et al. [77]. The commercial software package employed used neural network, genetic algorithms, and fuzzy logic in order to predict the outputs. Good agreement was found inside the training range. However, some rules generated by the system did not seem to correlate with the physical system (e.g., roller gap not having an effect on ribbon density) and predictions outside the training range did not compare well with experimental values.

38.4.5 Control Strategies for Roller Compaction

The overall roller compaction operation includes the powder preparation and handling, compaction by the rollers, and the subsequent breaking and sizing of the resulting ribbon. The variables affecting powder preparation and handling as well as sizing operations are not specific to the roller compaction and can be chosen for the roller compaction operation from a base of information devoted specifically to these fields.

For the compaction portion of the operation, several references such as Johanson [15], Yehia [69], and others have discussed how powder measurements with or without experiments can be used to design a compactor that will subject the material to a compression history resulting in the desired output ribbon and granule properties. The typical situation in many areas of manufacture, including pharmaceutical applications, is the reverse. An equipment vendor has already spent the effort to design a piece of equipment of general applicability for the typical range of powder properties encountered by their clients. The decisions regarding screw feeder arrangement, design, placement in proximity to the rollers, roller diameter, width, a mechanism to maintain position of the rollers, a system to apply pressure on the powder, and the milling arrangement have already been chosen.

The challenge in most practical applications of roller compaction technology is to find the appropriate settings for adjustable parameters given a set piece of equipment and control system previously engineered by the manufacturer. The four common variables discussed in most of the literature are the feeder speed, roller speed, gap, and pressure. These variables depend on one another and need to be set in combinations that are appropriately balanced. The experience of the vendor who configured the machinery and who has the benefit of the knowledge from working with many clients can help in designing the roller compaction process and selecting parameter values. Extreme settings of the controllers are usually not effective at making the best product unless there is something very unusual about the system under study. Several investigators, such as Rambali et al. [11], Weyenberg et al. [36], and am Ende et al. [72] used statistically designed studies to understand and even predict the granule and tablet properties as a function of the roller compactor variables and select the parameter values.

The simplest method is to choose a feed speed, roller speed, and pressure, which delivers the desired gap or ribbon thickness. The challenge for this method is to deliver the powder very consistently. A balance of the powder feed speed and the roller settings must be achieved to avoid over or under feeding the roller compactor. As discussed by Lecompte et al. [64] it is necessary to feed enough material to the nip region such that a compact is formed and the powder is being transported fast enough to encourage a uniform packing across the roller width with a uniform ribbon resulting. Guigon and Simon [44] discussed how the feed speed and compactor throughput should not be so fast that there is leakage of powder from the sealing mechanism or that air entrainment disrupts the powder flow or strength of the compacted ribbon.

In some studies, such as Guigon and Simon [44], a link between the gap and the resulting product quality is established. Given the difficulty of feeding powders without variation, control systems have been developed such that a roller speed, pressure, and gap distance can be set. The feeder speed is then adjusted in a feedback loop with the gap measurement in order to maintain a consistent ribbon thickness and strength.

More recent variations in control systems also recognize the importance of the feed and seek to monitor the feed screw output or the actual mass throughput of the roller compactor in order to make adjustments to the roller compactor settings.

Two different approaches to setting and scaling the operation of the roller compactor are discussed in the next section.

38.5 SCALE-UP OF ROLLER COMPACTION

38.5.1 Scale-Up Strategies

A common method for scaling up a roller compaction process from development equipment to commercial equipment is to use a parametric strategy. The parametric strategy focuses on determining the commercial equipment parameter values by using equivalency factors. Equivalency factors are based on aspects of the equipment. Some of the values used may be taken from the equipment manufacturer, who has assembled information from the design and testing of the unit as well as the collected experience of the client base in order to develop scale-up factors.

In selecting the roller pressure for the commercial equipment, the hydraulic pressure required to generate a needed force on the rolls can be estimated by considering the roller width and roller diameter. In selecting the roller speed for the commercial equipment, the rpm can be set to obtain a linear velocity of the rolls equivalent to that used in development. Alternatively, dwell time can be used as a metric for selecting the appropriate roller speed for the commercial equipment. Often, the selection of roller speed and roller gap is synchronized to obtain the desired throughput. This approach neglects the quality of the ribbons and granules produced, and the potential affects on downstream processes.

An alternative strategy focuses on the attributes of ribbons and calls for adjusting the ribbon production parameters to attain attribute values on the commercial equipment equivalent to that produced in development. The intent of controlling the quality of the ribbons produced is to control the downstream granule properties (size distribution, solid fraction, and compactability). The hypothesis underlying this approach is that under similar milling conditions (mill speed and screen opening) the output particle size distribution is determined by the ribbon input quality (Campbell et al. [78], and Morrison et al. [79]).

The roller compaction unit operation is composed of subprocesses focused on achieving two sequential, independent, but linked manipulations of the material: ribbon production and granule production by milling. The two quality attributes of interest for the ribbons are the thickness after recovery due to relaxation, and the solid fraction. As with the general consolidation theory, these two ribbon attributes control the breaking strength of the ribbon. Both of these attributes should, therefore, influence the behavior of the ribbons during milling. The indirect effects on ribbon recovery should be carefully studied, since extensive recovery could affect ribbon strength.

One crucial decision to be made when contemplating scaling up a roller compaction process is whether or not to use the automatic gap feedback control system or to proceed with a preset feed rate. This decision will determine what parameters need to be considered in scale-up. If the roller compactor is run under gap control, the effect on ribbon thickness is muted, unless purposely varied (by changing the gap). To determine what process parameter to alter to obtain the desired intermediate attribute requires a detailed knowledge of the interplay between process parameters and ribbon/granule attributes.

38.5.2 Case Study I: Parameteric-Based Scale-Up

Case Study I reports on the roller compaction of a model microcrystalline cellulose/lactose blend with 5% active using the manual operation. In manual operation, the three parameters for ribbon production (screw speed, roller speed, and roller pressure) were set at predetermined values that remained constant for the duration of the run. For reproducible ribbon production, this implied that the powder flow into the rollers had to be constant.

The behavior of the powder during roller compaction depends on the region it is in (Inghelbrecht and Remon [20, 22], Bindhumadhavan et al. [65], Zinchuk et al. [42], Yusof et al. [66]). The powder in the slip region is densified slightly by rearrangement as it travels toward the rollers. In this region there is slippage between the powder and the rollers. In the slip region the velocity of the powder is slower than the linear velocity of the rollers. In the nip region the powder undergoes densification by particle rearrangement and by deformation. In the compaction region further densification by deformation occurs proceeded by bonding to form the ribbon.

In the parameter optimization study three factors were evaluated: screw speed, ratio of roller speed to screw speed (powder delivery rate), and roller pressure. Based on the data set, various parameter combinations were identified that produced granules of the desired flow and particle size distribution. The process parameters (screw speed, roller pressure, and roller speed) for commercial-scale equipment were determined using equivalency factors.

38.5.2.1 ***Adjusting Pressure*** Several researchers have reported on the loss of powder compactability upon roller compaction (Kochhar et al. [80], Bultmann [31], Freitag et al. [24]) and ascribe it to a reduction in the binding potential due to the consolidation that occurs during roller compaction (Malkowska and Kahn [40], Falzone et al. [19], Kleinebudde [6]). To mitigate this reduction the authors state that only sufficient pressure should be applied during roller compaction to improve powder flow (the main benefit). The roller pressure for the commercial equipment was set considering the roller width and roller diameter. No modification to the calculated value was made during the run.

More recently, investigators have looked at instrumented rollers to gather data on the actual pressure curve on the rolls (Miguelez-Moran et al. [43]). Farber et al. [37] argued that the compression process (volume reduction) a powder undergoes during roller compaction is similar to that during tabletting. Therefore, instrumented rolls can be used in setting roller pressure on commercial equipment to achieve equivalent pressure on the commercial equipment (and hence similar powder behavior). The underlying principle being that similar ribbons would be produced on scale-up if the pressure is maintained.

38.5.2.2 ***Choosing Screw and Roller Speed*** The roller speed for the commercial equipment in Case Study I was set to obtain the desired process efficiency. In manual mode, the ratio of screw:roller speeds dictates, for a given material and roller pressure, the gap between the rollers. To attain a roller gap on the commercial equipment similar to that on the development equipment, the value of ratio of screw speed to roller speed obtained during development was used to set the screw speed on the commercial-scale roller compactor based on screw design and roller width and diameter. The force generated by the powder propelled by the feeder screw into the slip region is counter-balanced by the hydraulic pressure applied to the rollers. Based on the material properties, the balance between these two forces determines the gap between the rollers.

38.5.2.3 ***Mill Screens and Speeds*** The velocity at which the milling blade rotates affects the milling process in two ways: the force of the instantaneous impact on the ribbon and the residence time in the milling chamber. These two aspects have two different potential consequences. The higher impact could result in the ribbon shattering into smaller granules. Residence time in the chamber affects the amount of attrition the ribbon undergoes, which results in finer particles. Brittle materials are more susceptible to the impact, whereas pliable (viscoelastic) materials are more prone to attrition. The properties of the microcrystalline cellulose/lactose active blend in Case Study I suggested that the ribbons produced would posses both brittle as well as viscoelastic properties. Since ribbons are porous solid bodies, it is plausible that the ribbons fractured into smaller pieces that underwent size reduction by attrition. Hence, it was determined that residence time was the important factor to study and not the instantaneous impact.

A derived parameter was used for mill speed. The effect of milling speed is due to its effect on residence time of the ribbon in the milling chamber. To more accurately estimate the effect of residence time on granule properties, a ratio of mill speed to roller speed was studied as a derived parameter. Increasing the mill-to-roller speed ratio reduced residence time for the ribbons in the milling chamber by increasing the output rate for a given input rate. Increasing the mill-to-roller speed ratio (shorter residence time) increased the mass flow rate of the granules. This was due primarily to the increased mean particle diameter and reduced fines.

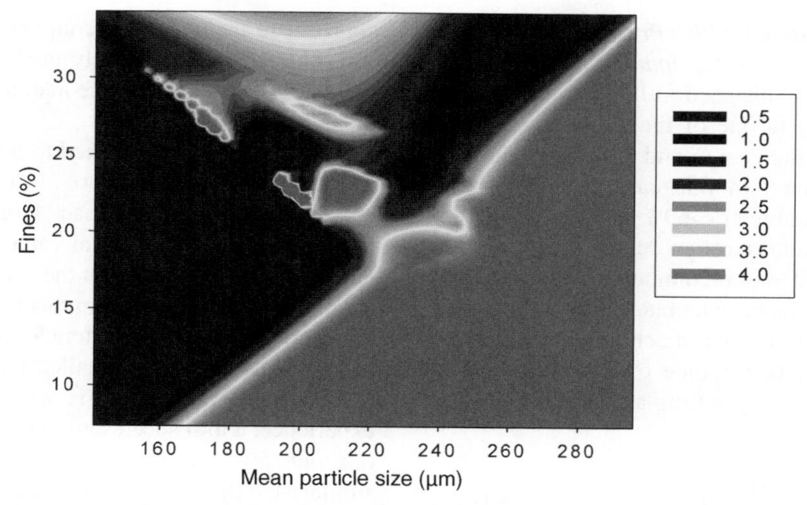

FIGURE 38.16 The interactive effect of mean particle size (µm) and amount of fines (%) on bulk mass flow (g/s).

The bulk powder property of interest for tablet compression is the mass flow. Powder mass flow should be governed by the properties of the particles making up the bulk. The two particle properties controlled by the roller compaction process are mean particle size and amount of fines. These two particle properties have an opposite effect on mass flow. Figure 38.16 is a contour plot showing the interactive effect of mean size and fines on flow. To maximize flow, the mean particle size would have to be increased to compensate for a higher amount of fines. The area of maximum flow is bounded approximately by the lower left to upper right diagonal in Figure 38.16.

Due to the statistical design used, the effect of screen size was nested within the effect of mill/roller speed ratio. The effect of the upper and lower screen could only be determined for a given mill to roller speed ratio. As expected, the effect of the lower screen was more pronounced than that for the upper screen. A maximum in mean particle size was observed with a 1 mm lower screen opening. This maximum mean particle size coincided with a similar maximum in mass flow with a 1 mm lower screen opening.

To understand the milling process in more depth, the difference between the upper screen opening and the lower screen opening was studied. It was theorized that this difference may be important since the upper screen controls the quality of the input for the lower screen, and because size reduction in this case is partially through attrition. Figure 38.17 depicts the relationship between mass flow and the difference in screen opening between the upper and lower screen. Equation 38.5 depicts the relationship, where M is the mass flow (g/s), M_{max} M is the maximum predicted flow (g/s), X is the difference in screen opening (mm) between the upper and lower screens, and X_0 is the difference at maximum flow.

$$M = M_{max}\, e^{-0.5\left(\frac{X-X_0}{b}\right)^2} \tag{38.5}$$

The fitted model strongly suggests an optimum for mass flow at 1.96 mm difference in screen opening. The maximum predicted flow is 3.99 g/s. The combination of the lower screen maximum at ~1 mm and the screen difference maximum at ~2 mm defines the upper screen optimum at ~3 mm. These same screen openings were used for scale-up in Case Study I.

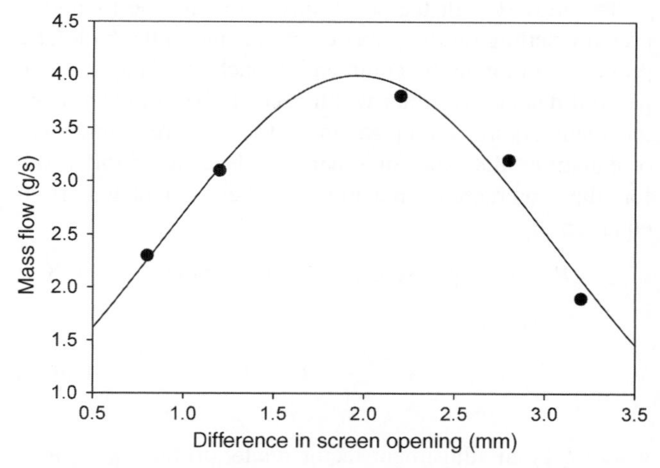

FIGURE 38.17 The effect of difference in opening between upper and lower screens on mass flow.

38.5.2.4 Parametric Scale-up With a Procedure and Parameters Recommended from the Equipment Manufacturer The example presented in this section illustrates scale-up recommendations similar to one of the major equipment manufacturers based on the design and experience developed with their equipment. In this case the pilot-scale equipment has a roller diameter ~60% of the size of the commercial equipment and a roller depth ~53% of the commercial equipment. It was recommended that the gap for the commercial machine be calculated from the acceptable values from the pilot-scale machine according to equation 38.6, which for this choice of equipment sizes leads to the larger machine gap setting about 1.7 times the pilot-scale machine.

$$G_2 = \frac{D_2}{D_1} \times G_1 \quad (38.6)$$

The screens on the postcompaction mill for the commercial machine are approximately five times the size of the pilot-scale machine so a throughput increase of approximately 4.5–5 times is expected in this example. The suggested scale-up procedure then uses machine parameters to calculate the roller speed based on equation 38.7. The throughput of the pilot-scale machine and the desired ribbon density are used as inputs with the remaining parameters coming from the machine geometry. The roller depth, B, may have an adjustment factor to be added for some machine designs. The roller speed calculation typically has a result with the commercial-scale machine about 1.7 times faster than the pilot-scale machine for the commonly available units this example was based on.

$$n_2 = \frac{T_1 \times 5}{D_2 \times \pi \times (B) \times G_2 \times \rho} \quad (38.7)$$

The next step in the scale-up is to relate the hydraulic pressure setting on the pilot-scale machine to the hydraulic pressure setting on the commercial machine. First the force per unit distance of roller width needs to be estimated from the manufacturer's conversion factor by a rearrangement of equation 38.8. The force per unit distance of roll width for the commercial machine is then calculated from equation 38.9.

$$P = F \times (\text{pressure conversion factor}) \quad (38.8)$$

$$F_2 = \frac{D_2}{D_1} \times F_1 \times t_c \quad (38.9)$$

where t_c is an adjustment factor related to the dwell time.

The hydraulic pressure setting for the commercial-scale machine is then calculated from equation 38.8 using the appropriate factor for the commercial-scale machine. Typical values are approximately an 8% increase in the hydraulic pressure as suggested by the manufacturer for the models in this example.

The screw feed is adjusted to achieve the calculated gap and the calculated pressure. It is recommended by this manufacturer to use a feedback loop to adjust the feed screw to maintain the gap. Typical values for screw speed on the larger unit are about 80% of the value for the pilot-scale unit for two typical models from this manufacturer but it should be noted that the larger system has twin screws compared to the single screw on the smaller machine.

Finally, the mill speed is set. From the manufacturer's experience, a mill speed setting on the commercial machine of about 75% of the pilot-scale machine will yield granules of similar size distribution when using screens with the same aperture.

38.5.3 Case Study II: Attribute-Based Scale-Up

Case Study II reports on the roller compaction of a model microcrystalline cellulose/lactose active blend with 2.5% active using the automatic operation. The roller speed for the commercial equipment was set at high speed to maximize operational efficiency. The roller gap and roller pressure were adjusted to produce ribbons with the attributes identified in development as optimal for downstream processing. The feed screw speed was allowed to float to maintain the powder feed rate sufficient to maintain an adequate flow into the nip area.

The optimum ribbon attributes were identified from development results by multiple constraint optimization of the granule properties such as powder flow and compactability. The variability in ribbon and granules properties depended on roller compaction process stability that is controlled by the powder flow to the feed screw during manufacture.

38.5.3.1 Ribbon Thickness and Density Roller speed had a marginal impact on the thickness, recovery, or density of ribbons produced. Increasing the roller pressure increased the ribbon density and reduced the ribbon recovery. The augmented load experienced by the powder during its transit through the rollers resulted in increased consolidation producing stronger ribbons. The increased interparticle bonding resisted the relaxation postcompaction, which explains the reduced recovery seen. Increasing the roller gap produced thicker ribbons, while reducing the recovery. The effect of gap on ribbon density was minimal.

There was little interactive effect between gap and pressure on ribbon thickness (Figure 38.18). The desired ribbon thickness could be obtained simply by setting the gap to the necessary setting (under gap control) allowing for the appropriate relaxation. Equation 38.10 depicts this relationship, where h is the ribbon thickness (mm), G is the gap setting

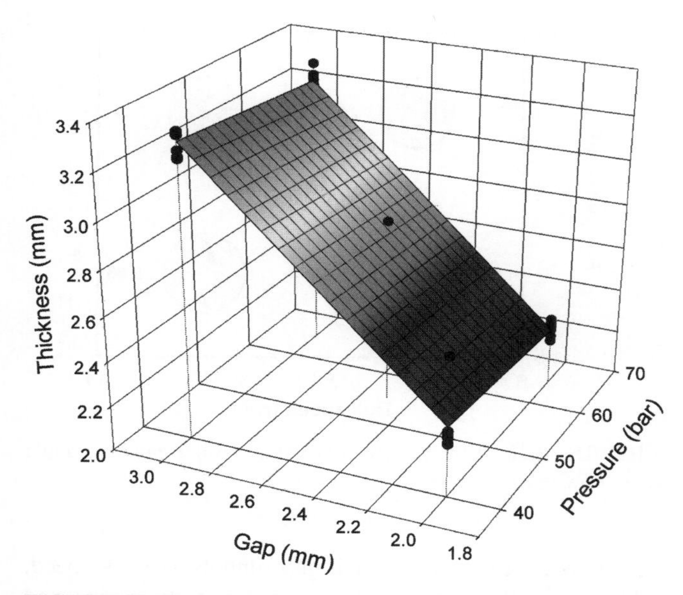

FIGURE 38.18 Effect of roller gap and roller pressure on ribbon thickness.

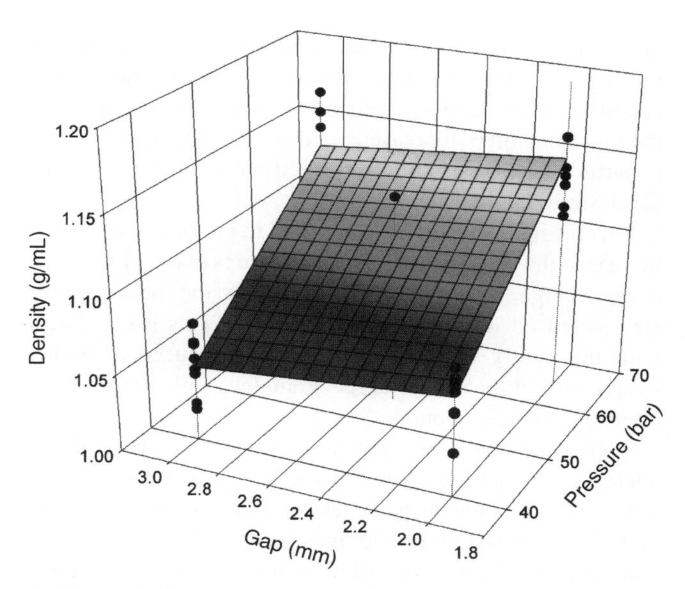

FIGURE 38.19 Effect of roller gap and roller pressure on ribbon density.

(mm), and P is the roll pressure (bar). The intercept accounts for the relaxation postcompaction.

$$h = 0.98 \times G - 0.002 \times P + 0.45 \qquad (38.10)$$

There was a marginal two-way interactive effect between gap and pressure on ribbon density (Figure 38.19). Equation 38.11 depicts this relationship, where ρ_a is the apparent ribbon density (g/mL), G is the gap setting (mm), and P is the roll pressure (bar). The intercept probably refers to the density in the powder bed just prior to significant consolidation.

$$\rho_a = 0.94 + 0.004 \times P - 0.008 \times G - 0.0001 \times P \times G \qquad (38.11)$$

Ribbon densification was more efficient at lower gap settings, which aligns with the common understanding of force transmission though a powder bed under load. The desired ribbon density could be obtained simply by setting the load to the necessary setting (under gap control).

Figure 38.20 shows that ribbon recovery is determined by roll pressure and roll gap. Equation 38.12 depicts this relationship, where R is the apparent ribbon relaxation (%), ε refers to the maximum elastic recovery (19.8%), G is the gap setting (mm), and P is the roll pressure (bar).

$$R = \varepsilon - 0.03 \times P - 3.298 \times G \qquad (38.12)$$

Both applied pressure and gap influenced the degree of recovery seen postcompaction, with gap being more dominant. The least recovery is seen with high pressure and wide gap, and most recovery is seen with low pressure and narrow gap. The effect of compression pressure on recovery is opposite that seen with tablets (Adolfsson and Nyström [81]). This is consistent with the view that weaker ribbons recover more.

38.5.3.2 Particle Size and Powder Flow Increasing roller speed reduced the powder flow. This reduction was mainly attributed to the change in particle size distribution; the mean particle size decreased while the fines increased. Potentially,

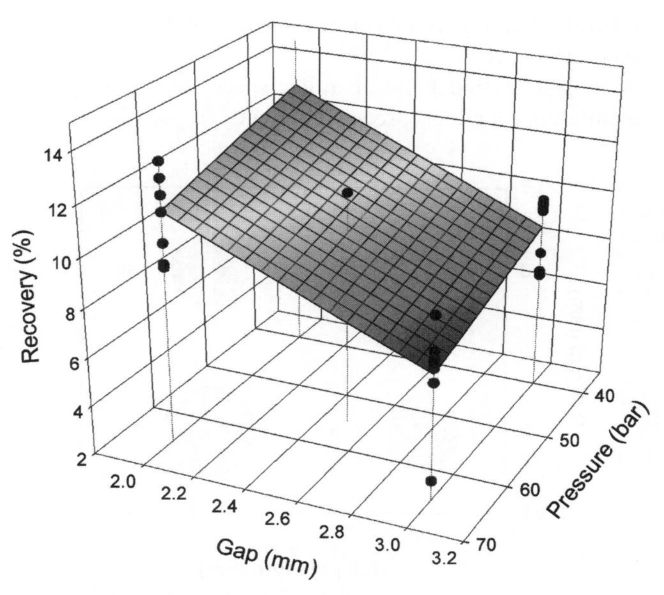

FIGURE 38.20 Effect of roller gap and roller pressure on ribbon recovery.

this can be the result of increased residence time in the milling chamber. Increasing the input rate into the milling chamber, while keeping the output constant, would increase the residence time. Increased residence time should produce a particle size distribution skewed to smaller sizes (see Case Study I).

Increasing the roller pressure used to produce the ribbons increased the powder flow. This increase is mainly attributed to the change in particle size distribution; the mean particle size increased while the fines decreased. This is consistent with the theory that stronger ribbon (produced at higher loads) would produce larger granules with fewer fines (Herting and Kleinebudde [26, 33]).

Increasing the roller gap used to produce the ribbons, decreased the powder flow. This reduction is mainly attributed to the change in particle size distribution; the mean particle decreased while the fines increased. This is consistent with the theory that ribbons have a density gradient decreasing from the surface to the center. The thicker the ribbon is the lower the density at the center. This low-density center will mimic ribbons made at lower loads and produce smaller particles during the milling process. Using principal component analysis, Soh et al. [16], showed that roller gap is an important predictor of granule properties. Figure 38.21 shows the interactive effect of roller pressure and roller gap on mass flow. The data shows that the ribbons made at opposite extremes of pressure and gap have minimum and maximum flow. Low pressure and large gap produced minimum flow; high pressure and narrow gap produced maximum flow. These powder properties are directly linked to the particle size distribution obtained from the ribbons. The data strongly suggests that pressure can be used with some degree of success to offset the effect of increased gap. This clearly illustrates the importance of control over ribbon quality as a scale-up metric.

The interaction between roller speed and roller gap is notable since it affects process efficiency (Figure 38.22). The

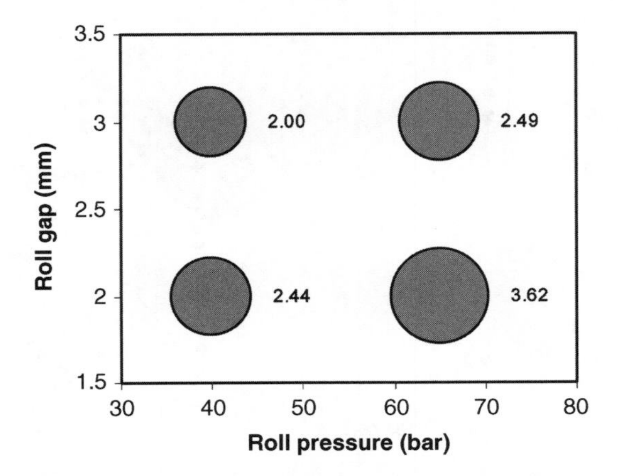

FIGURE 38.21 Effect of roller pressure and roller gap on granule mass (g/s).

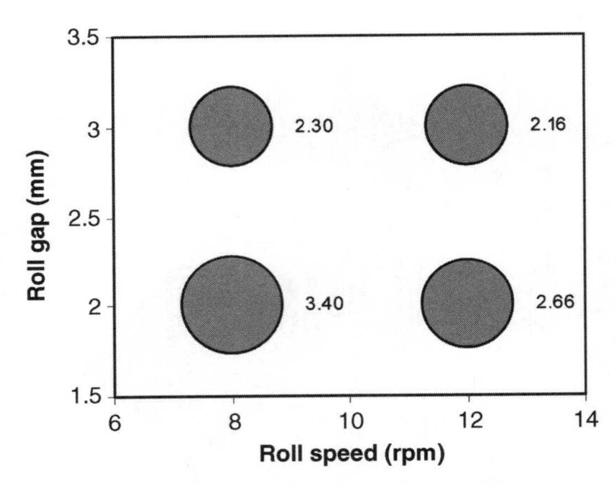

FIGURE 38.22 Effect of roller speed and roller gap on granule mass flow (g/s).

best flow is seen with a small gap running at slow speed, which coincides with the largest mean particle size and lowest amount of fines. These conditions would be the least efficient. Efficiency gains could be obtained with wider roller gaps at faster speeds. However, this combination yielded the slowest flow, which coincides with the smallest mean particle size and the highest amount of fines.

The effect of ribbon density and thickness on bulk mass flow is illustrated in Figure 38.23. Equation 38.13 depicts this relationship, where M is the mass flow (g/s), ρ_a is the apparent ribbon density (g/mL), h is the ribbon thickness (mm).

$$M = 9.49 \times \rho_a - 0.69 \times h - 0.582 \quad (38.13)$$

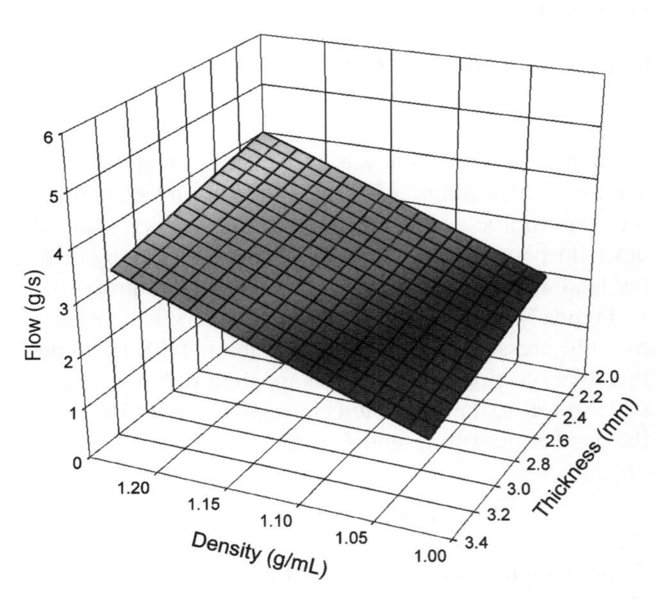

FIGURE 38.23 Effect of ribbon density and thickness on granule mass flow.

This coincided with their effect on granule mean size. Increased ribbon density produces larger granules with enhanced flow. Increased ribbon thickness leads to a smaller granules with reduced flow. The effect of ribbon thickness on the amount of fines produced after milling was marginal, but the effect of ribbon density was significant. Higher ribbon densities correlated with higher amounts of fines. This could be due to increased residence time in the mill leading to more attrition by shear. Herting and Kleinebudde [26, 33] reported a similar growth in particle size and improved flow with increased ribbon density.

Fragility of granules can be a main contributor to reduction on mass flow with processing postroller compaction. This was due to particle attrition during additional mixing of the granules with extragranular ingredients. To separate out the relative importance of changes in fines from changes in mean size on flow, the relative change in flow (final blend-granule) was evaluated against the relative change in mean size and fines (Figure 38.24). Equation 38.14 depicts this relationship, where ΔM is the change in mass flow (g/s), ms is the mean size and f is the amount of fines.

$$\Delta M = -0.24 - 0.04 \times \Delta ms + 0.004 \times \Delta t - 0.04 \times \Delta ms^2 + 0.0004 \times \Delta f^2 \quad (38.14)$$

Flow was relatively resistant to change mean size over the range observed. Flow was quite sensitive to changes in the amount of fines. This indicates that efforts to improve flow are better spent reducing the amount of fines rather than optimizing the mean size. The bulk powder property of interest for tablet compression is the mass flow. Powder mass flow is governed by the properties of the particles making up the bulk. The two particle properties controlled by the roller compaction process are mean particle size and amount of fines. These two particle properties have an opposite effect on mass flow. To maximize flow, the mean particle size would have to be increased to compensate for a higher amount of fines.

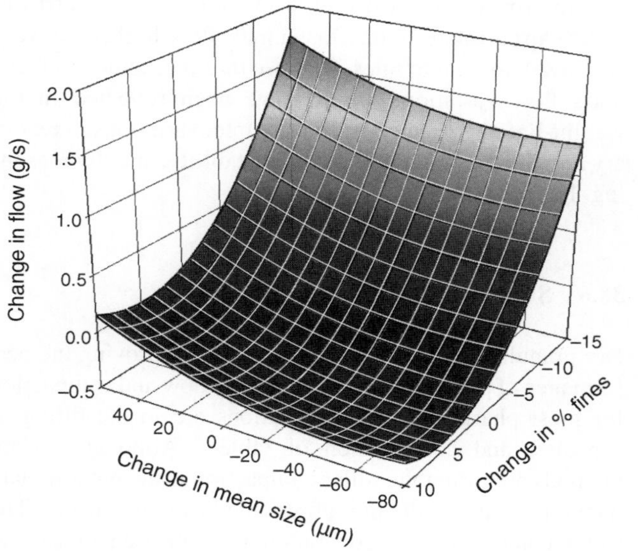

FIGURE 38.24 The interaction between change in mean diameter and fines on change in bulk mass flow.

*38.5.3.3 **Powder Compactability*** The performance of the compression operation depends on the characteristics of the input final blend. There are two main metrics that describe the performance of the process: compactability (how hard the press has to work to make tablets) and process stability (how variable the product properties are).

Ribbon production variables affected compactability, whereas ribbon milling conditions did not affect compactability appreciably. There were two populations of ribbons analyzed for compactability, one set corresponding to low roller pressure and one corresponding to high roller pressure. The ribbons compressed at low pressure had a lower apparent density and a higher compactability compared to ribbons compressed at high pressure that had a higher apparent density and a lower compactability. Malkowska and Khan [40] reported a similar observation of increased compactability with a reduction in ribbon density.

In addition to density, ribbons also had a characteristic thickness. The effect of density and thickness on compactability was evaluated. Figure 38.25 suggests that ribbon thickness had only a small impact, while ribbon density had a substantial impact on compactability. Equation 38.15 depicts this relationship, where K is the compactability

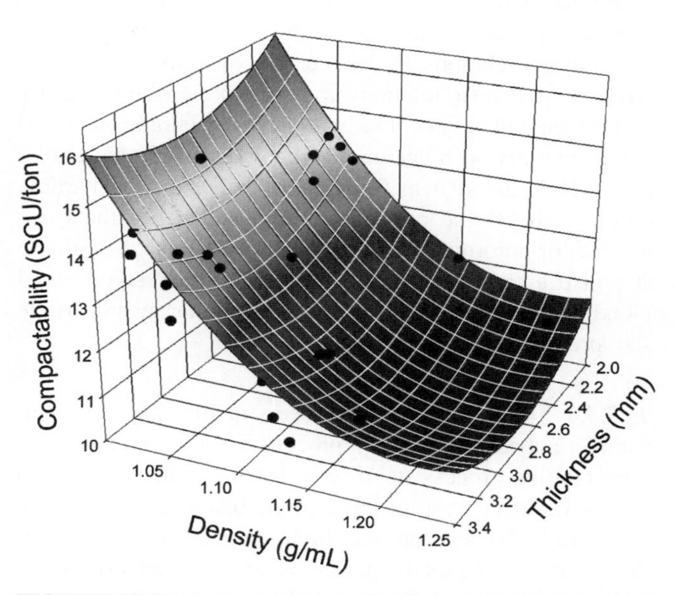

FIGURE 38.25 Effect of ribbon density and thickness on final blend compactability.

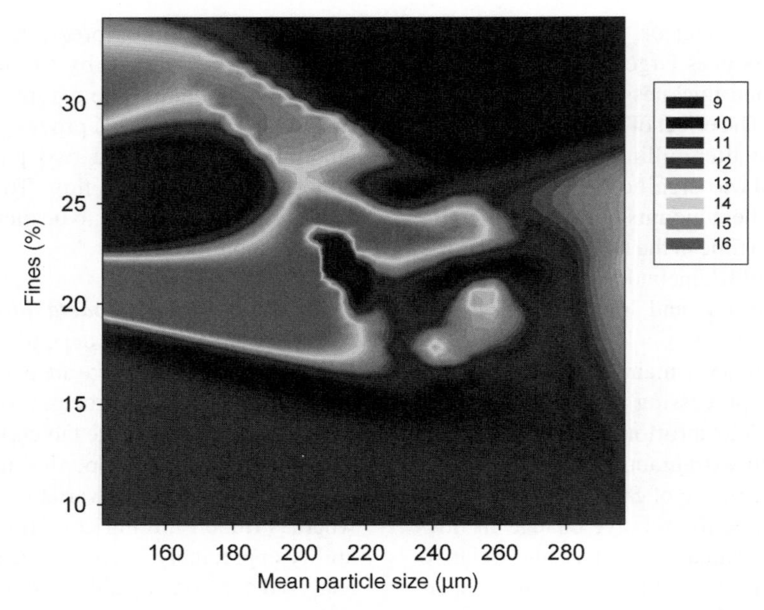

FIGURE 38.26 Effect of mean particle size and amount of fines on final blend compactability.

(SCU/ton), ρ_a is the apparent ribbon density (g/mL), and h is the ribbon thickness (mm).

$$K = 174 - 9.74 \times h - 245 \times \rho_a + 1.74 \times h^2 + 100 \times \rho_a \quad (38.15)$$

In addition to the direct affect of density on compression, the link between density and compactability may also be mediated through particle size (Figure 38.26).

Blends with lower amounts of fines or a small mean particle size had on average higher compactability. There was a direct relationship between ribbon density and mean size, but an inverse relationship between the ribbon density and the amount of fines. The hypothesis explaining increased compactability with decreased ribbon density includes two parts. The ribbon density (and by extension the particle density) affects the particle strength and hence the force needed for deformation. Ribbon density affects the particle size distribution (mean size and fines) of the resulting final blend. A higher amount of fines and a smaller mean particle size promotes compactability.

38.5.3.4 Process Quality Process quality has two measurable quantities, that is, process stability and process repeatability. Process stability is defined as the variance in the metric of interest during a single batch run (within run variability). Process repeatability or capability is defined as the variance in the metric of interest across a study (between run variability), also as the capability of a process to meet its purpose. This analysis can be done on unit operations for which the parameters are not varied, or for upstream operation variables under evaluation.

Process repeatability is a metric for the downstream effect of the roller compaction unit operation. Based on the proposed target core tablet attributes (tablet weight, hardness, and thickness), the process capability indices (*Cp*) [82] were calculated. A *Cp* value of 2 is generally expected for a process under control (6σ).

The *Cp* analysis based on tablet weight showed a compression operation unaffected by the roller compaction settings. Process capability (*Cp*) based on tablet hardness was improved with a decrease in roller gap. A narrower roller gap correlated with a faster mass flow, higher recovery and lower amount of fines. Of these three coincidental facts, mass flow variations would have been reflected in the weight-based *Cp* analysis. In principle, variations in recovery and fines amount affect compactability, and, by reasoning, hardness-based *Cp*.

38.6 SUMMARY

Dry granulation is an effective means of improving the performance of powders. Improvements in flow and compaction for most pharmaceutical applications aid in the filling of capsules, and compression of tablets. Although several methods are available, roller compaction is the predominant method for the production of granules without solvent. The roller compaction unit operation is comprised of several subprocesses. Powder feeding and milling, the first and last of these subprocesses, have large bodies of knowledge which can be adapted to roller compaction. The compaction step itself is of most interest when designing the roller compaction process. Material properties, in-process and final product measurements, and their interaction with the feed rate, roller speed, roller pressure, and gap have been reviewed and examples given for process development and scale-up. This information confirms the utility of roller compaction and provides tools for designing the roller compaction process.

SYMBOLS

B	roller width
Cp	process capability index
D	roller diameter
F	compaction force per unit distance of roller width
M	mass flow (g/s)
G	roller gap
G	gap setting (mm)
h	thickness of compacted ribbon (mm)
K	compactability (SCU/ton)
n	roller speed
P	hydraulic pressure (roller pressure) (bar)
R	apparent ribbon relaxation (%)
S	gap
T	throughput
α	nip angle
χ	position
δ	angle of internal friction
δ_E	effective angle of internal friction
ε	maximum elastic recovery (%)
ϕ, ϕ_w	angle of wall stress
μ	friction coefficient
ϑ	angular roll position
ϑ_h	angular position at which feed pressure is applied
ρ	density
ρ_a	apparent ribbon density (g/mL)
τ	shear stress
υ	acute angle roller tangent to normal stress
ms	mean size
f	amount of fines

REFERENCES

1. Nishii K, Horio M. Dry granulation. In: Salmon AD, Hounslow MJ, Seville JPK,editors. *Granulation*, Vol. 11, Elsevier, Amsterdam, 2007, pp. 289–322.
2. Horio M. Binderless granulation—its potential, achievements and future issues. *Powder Technol.* 2003; 130:1–7.
3. Pietsch W. *Agglomeration in Industry* Volume 1. Wiley-VCH, Weinheim, 2004.
4. Johnson CE. Production of granular products by roll compaction. In: Beestman GB, Vander Hooven DIB, editors. *Pesticide Formulations and Application Systems: Seventh Volume*, ASTM Special Technical Publication 968. American Society for Testing and Materials, Philadelphia, 1987, pp. 199–206.
5. Bakele W. New developments in the production of heavy soda-ash via compacting method. *Powder Technol.* 2003;130: 253–256.
6. Kleinebudde P. Roll compaction/dry granulation: pharmaceutical applications. *Eur. J. Pharmaceut. Biopharmaceut.* 2004;58 (2):317–326.
7. Mouro D, Noack R, Musico B, King H, Shah U. *Enhancement of Xcelodose Capsule-Filling Capabilities Using Roller Compaction*, Pharmaceutical Technology, February 2006. Available at http://license.icopyright.net/3.74587?icx_id=301460 Accessed 2009 April 18
8. Miller RW. Advances in pharmaceutical roller compactor feed system designs. *Pharmaceut. Technol.* 1994; March:154–162.
9. Miller RW. Roller compaction technology. In: Parikh D, editor. *Handbook of Pharmaceutical Granulation Technology*, 2nd edition, Taylor & Francis, New York, 2005, pp. 159–172.
10. Daugherity PD, Chu JH. Investigation of serrated roll surface differences on ribbon thickness during roll compaction. *Pharmaceut. Dev. Technol.* 2007;12:603–608.
11. Rambali B, Baert L, Jans E, Massart DL. Influence of the roll compactor parameter settings and the compression pressure on the buccal bio-adhesive tablet properties. *Int. J. Pharmaceut.* 2001;220:129–140.
12. Pietsch W. Granulate dry particulate solids by compaction and retain key powder particle properties. *Chem. Eng. Prog.* 1997; April:24–46.
13. Guangsheng H, Lingyun W, Zhongwei Z, Guangjie H, Fuhseng P. Manufacturing technique of magnesium alloy sheets by powder rolling. *Mater. Sci. Forum* 2005;488–489:445–448.
14. Bacher C, Olsen PM, Bertelsen P, Sonnergaard JM. Granule fraction inhomogeneity of calcium carbonate/sorbitol in roller compacted granules. *Int. J. Pharmaceut.* 2008;349:19–23.
15. Johanson JR. A rolling theory for granular solids. *Trans. Am. Soc. Mech. Eng.* 1965; December:842–848.
16. Soh JLP, Boersen N, Carvajal TM, Morris KR, Peck GE, Pinal R. Importance of raw material attributes for modeling ribbon and granule properties in roller compaction: multivariate analysis on roll gap and NIR spectral slope as process critical control parameters. *J. Pharmaceut. Innov.* 2007;2:106–124.
17. Soh JLP, Wang F, Boersen N, Pinal R, Peck GE, Carvajal TM, Cheney J, Valthorsson H, Pazdan J. Utility of multivariate analysis in modeling the effects of raw material properties and operating parameters on granule and ribbon properties prepared in roller compaction. *Drug Dev. Ind. Pharmacy* 2008;34: 1022–1035.
18. Sheskey PJ, Dasbach TP. Evaluation of various polymers as dry binders in the preparation of an immediate-release tablet formulation by roller compaction. *Pharmaceut. Technol.* 1995;19 (10):98–112.
19. Falzone AM, Peck GE, McCabe GP. Effects of changes in roller compactor parameters on granulation produced by compaction. *Drug Dev. Ind. Pharmacy* 1992;18:469–489.
20. Inghelbrecht S, Remon JP. Roller compaction and tableting of microcrystalline cellulose/drug mixtures. *Int. J. Pharmaceut.* 1998;161:215–224.

21. Riepma KA, Vromans H, Zuurman K, Lerk CF. The effect of dry granulation on the consolidation and compaction of crystalline lactose. *Int. J. Pharmaceut.* 1993;97:29–38.
22. Inghelbrecht S, Remon JP. The roller compaction of different types of lactose. *J. Int. Pharmaceut.* 1998;166:135–144.
23. Sheskey PJ, Cabelka TD, Robb RT, Boyce BM. Use of roller compaction in the preparation of controlled-release hydrophilic matrix tablets containing methylcellulose and hydroxypropyl methylcellulose polymers. *Pharmaceut. Technol.* 1994;18 (9):132–150.
24. Freitag F, Kleinebudde P. How do roll compaction/dry granulation affect the tableting behavior of inorganic materials? Comparison of four magnesium carbonates. *Eur. J. Pharmaceut. Sci.* 2003;19:281–189.
25. Freitag F, Runge J, Kleinebudde P. Coprocessing of powdered cellulose and magnesium carbonate: direct tableting versus tableting after roll compaction/dry granulation. *Pharmaceut. Dev. Technol.* 2005;10:353–362.
26. Herting MG, Kleinebudde P. Roll compaction/dry granulation: effect of raw material particle size on granule and tablet properties. *Int. J. Pharmacuet.* 2007;338:110–118.
27. Bacher C, Olsen PM, Bertelsen P, Kirstensen J. Sonnergaard JM. Improving the compaction of roller compacted calcium carbonate. *Int. J. Pharmaceut.* 2007;342:115–123.
28. He X, Secreast PJ, Amidon GE. Mechanistic Study of the effect of roller compaction and lubricant on tablet mechanical strength. *J. Pharmaceut. Sci.* 2007;96 (5):1342–1355.
29. Gupta A, Peck GE, Miller RW, Morris KR. Influence of ambient moisture on the compaction behavior of microcrystalline cellulose powder undergoing uni-axial compression and roller compaction: a comparative study using near-infrared spectroscopy. *J. Pharmaceut. Sci.* 2005;94 (10): 2301–2313.
30. Inghelbrecht S, Remon JP. Reducing dust and improving granule and tablet quality in the roller compaction process. *Int. J. Pharmaceut.* 1998;171:195–206.
31. Bultmann JM. Multiple compaction of microcrystalline cellulose in a roller compactor. *Eur. J. Pharmaceut. Biopharmaceut.* 2002;54:59–64.
32. Miguelez-Moran AM, Wu CY, Seville JPK. The effect of lubrication on density distributions on roller compacted ribbons. *Int. J. Pharmaceut.* 2008;362:52–59.
33. Herting MG, Kleinebudde P. Reduction of tensile strength of tablets after roll compaction/dry granulation. *Int. J. Pharmaceut.* 2008;70:372–378.
34. Herting MG, Klose K, Kleinebudde P. Comparison of different dry binders for roll compaction/dry granulation. *Pharmaceut. Dev. Technol.* 2007;12 (5):525–532.
35. Sheskey PJ, Hendren J. The effects of roll compaction equipment variables, granulation technique, and HPMC polymer level on a controlled-release matrix model drug formulation. *Pharmaceut. Technol.* 1999;223:90–106.
36. Weyenberg W, Vermeire A, Vandervoort J, Remon JP, Ludwig A. Effects of roller compaction settings on the preparation of bioadhesive granules and ocular minitablets. *Eur. J. Pharmaceut. Biopharmaceut.* 2005;59:527–536.
37. Farber L, Hapgood KP, Michaels JN, Fu X-Y, Meyer R, Johnson M-A, Li F. Unified compaction curve model for tensile strength of tablets made by roller compaction and direct compression. *Int. J. Pharmaceut.* 2008;346:17–24.
38. Patel C, Kaushal AM, Bansal AK. Compaction behavior of roller compacted ibuprofen. *Eur. J. Pharmaceut. Biopharmaceut.* 2008;69:743–749.
39. Sun CC, Himmilspach MW. Reduced tabletability of roller compacted granules as a result of granule size enlargement. *Int. J. Pharmaceut.* 2006;95:200–206.
40. Malkowska S, Khan KA. Effect of recompression on the properties of tablets prepared by dry granulation. *Drug Dev. Ind. Pharmacy* 1983;9:331–347.
41. Sheskey PJ, Cabelka TD. Re-workability of sustained release tablet formulation containing HPMC polymers. *Pharmaceut. Technol.* 1992;7:60–74.
42. Zinchuk AV, Mullarney MP, Hancock B. Simulation of roller compaction using a laboratory scale compaction simulator. *Int. J. Pharmaceut.* 2004;269:403–415.
43. Miguelez-Moran AM, Wu CY, Dong H, Seville JPK. *Eur. J. Pharmaceut. Biopharmaceut.* 2009;72:173–182.
44. Guigon P, Simon O. Roll press design—influence of force feed systems on compaction. *Powder Technol.* 2003;130:41–48.
45. Funakoshi Y, Asogawa T, Satake E. The use of a novel roller compactor with a concavo-convex roller pair to obtain uniform compacting pressure. *Drug Dev. Ind. Pharmacy* 1977;3 (6): 555–573.
46. Parrott EL. Densification of powders by concavo-convex roller compactor. *J. Pharmaceut. Sci.* 1981;70:288–291.
47. Guigon P, Simon O. Interaction between feeding and compaction during lactose compaction in a laboratory roll press. *KONA* 2000;18:131–138.
48. Zhang JQ, Yan Y. On-line continuous measurement of particle size using electrostatic sensors. *Powder Technol.* 2003;135: 164–168.
49. Frake P, Greenhalg D, Grierson SM, Hempenstall JM, RuddJr., Process control and endpoint determination of a fluid bed granulation by application of near infra-red spectroscopy. *Int. J. Pharmaceut.* 1997;151:75–80.
50. Rantanen J, Yliruusi J. Determination of particle size in a fluidized bed granulator with a near infrared (NIR) set-up. *Pharmaceut. Pharmacol. Commun.* 1998;4:73–75.
51. Gupta A, Peck GE, Miller RW, Morris KR. Nondestructive measurements of the compact strength and the particle-size distribution after milling of roller compacted powders by near-infrared spectroscopy. *J. Pharmaceut. Sci.* 2004;93: 1047–1053.
52. Rantanen J, Rasanen E, Tenhunen J, Kansakoski M, Mannermaa JP, Yliruusi J. In-line moisture measurement during granulation with a four-wavelength near infrared sensor: an evaluation of particle size and binder effects. *Eur. J. Pharmaceut. Biopharmaceut.* 2000;50:271–276.
53. Bordes C, Garcia F, Snabre P, Frances C. On-line characterization of particle size during an ultrafine wet grinding process. *Powder Technol.* 2002;128:218–228.

54. Liao CW, Tarng YS. On-line automatic optical inspection system for coarse particle size distribution. *Powder Technol.* 2009;189:508–513.

55. Sprockel OL, Yang E, Jayawickrama D, Li L. Ultrasound Propagation in Porous Materials, *AIChE Annual Meeting*, 2008, p. 294.

56. Feng T, Feng W, Pinal R, Wassgren C, Carvajal MT. Investigation of the variability of NIR in-line monitoring of roller compaction process by using fast fourier transform (FFT) analysis. *Pharmaceut. Sci. Technol.* 2008;9:419–424.

57. Ghorab MK, Chatlapalli R, Hasan S, Nagi A. Application of thermal effusivity as a process analytical technology tool for monitoring and control of the roller compaction process. *Pharmaceut. Sci. Technol.* 2007;8:1–7.

58. Alexander AW, Muzzio FJ. Batch size increase in dry blending and mixing. In: Levin M,editor. *Pharmaceutical Process Scale Up*, Informa Health Care, New York, 2002, pp. 115–132.

59. Prescott JK. Powder handling. In: Levin M,editor. *Pharmaceutical Process Scale Up*, Informa Health Care, New York, 2002, pp. 133–150.

60. Yang S, Evans JRG. Metering and dispensing of powder; the quest for new solid freeforming techniques. *Powder Technol.* 2007;178:56–72.

61. Sander U, Schonert K. Operation conditions of a screw-feeder-equipped high-pressure roller mill. *Powder Technol.* 1999;105: 282–287.

62. Simon O, Guigon P. Correlation between powder-packing properties and roll press compact heterogeneity. *Powder Technol.* 2003;130:257–264.

63. Lecompte T, Doremus P, Thomas G, Perrier-Camby L, LeThiesse L-C, Masteau J-C, Debove L. Dry granulation of organic powders—dependence of pressure 2D distribution on different process parameters. *Chem. Eng. Sci.* 2005;60: 3933–3940.

64. Podczeck F. *Particle-Particle Adhesion in Pharmaceutical Powder Handling*, Imperial College Press, London, 1998.

65. Bindhumadhavan G, Seville JPK, Adams MJ, Greenwood RW, Fitzpatrick S. Roll compaction of a pharmaceutical excipient: experimental validation of rolling theory for granular solids. *Chem. Eng. Sci.* 2005;60:3891–3897.

66. Yusof YA, Smith AC, Briscoe BJ. Roll compaction of maize powder. *Chem. Eng. Sci.* 2005;60:3919–3931.

67. Sommer K, Hauser G. Flow and compression properties of feed solids for roll-type presses and extrusion presses. *Powder Technol.* 2003; 272–276.

68. Schonert K, Sander U. Shear stresses and material slip in high pressure roller mills. *Powder Technol.* 2002;122:136–144.

69. Yehia KA. Estimation of roll press design parameters based on the assessment of a particular nip region. *Powder Technol.* 2007;177:148–153.

70. Dec RT, Zavaliangos A, Cunningham JC. Comparison of various modeling methods for analysis of powder compaction in roller press. *Powder Technol.* 2003;130:265–271.

71. Hervieu P, Dehont F, Jerome E, Delacourte A, Guyot JC. Granulation of pharmaceutical powders by compaction an experimental study. *Drug Dev. Ind. Pharmacy* 1994;20 (1): 65–74.

72. am Ende MT, Moses SK, Carella AJ, Gadkari RA, Graul TW, Otano AL, Timpano RJ. Improving the content uniformity of a low-dose tablet formulation through roller compaction optimization. *Pharmaceut. Dev. Technol.* 2007;12:391–404.

73. Rekhi GS, Sidwell R. Sizing of granulation. In: Parikh D,editor. *Handbook of Pharmaceutical Granulation Technology*, 2nd edition, Taylor & Francis, New York, 2005, pp. 159–172.

74. Hein S, Picker-Freyer KM, Langridge J. Simulation of roller compaction with subsequent tableting and characterization of lactose and microcrystalline cellulose. *Pharmaceut. Dev. Technol.* 2008;13:523–532.

75. Inghelbrecht S, Remon J-P, Fernandes de Aguiar P, Walczak B, Luc Massart D, Van De Velde F, De Baets P, Vermeersch H, DeBacker P. Instrumentation of a roll compactor and the evaluation of the parameter setting by neural networks. *Int. J. Pharmaceut.* 1997;148:103–115.

76. Turkoglu M, Aydin I, Murray M, Sakr A. Modeling of a roller-compaction process using neural networks and genetic algorithms. *Eur. J. Pharmaceut. Biopharmaceut.* 1999;48: 239–245.

77. Mansa RF, Bridson RH, Greenwood RW, Barker H, Seville JPK. Using intelligent software to predict the effects of formulation and processing parameters on roller compaction. *Powder Technol.* 2008;181:217–225.

78. Campbell GM, Bunn PJ, Webb C, Hook SCW. On predicting roller milling performance: Part II. The breakage function. *Powder Technol.* 2001;115:243–255.

79. Morrison RD, Shi F, Whyte R. Modeling of incremental rock breakage by impact—for use in DEM models. *Miner. Eng.* 2007;20:303–309.

80. Kochhar SK, Rubinstein MH, Barnes D. The effects of slugging and recompression on pharmaceutical excipients. *Int. J. Pharmaceut.* 1995;115:35–43.

81. Adolfsson A, Nyström C. Tablet strength, porosity, elasticity and solid state structure of tablets compressed at high loads, *Int. J. Pharmaceut.* 1996;132:95–106.

82. *NIST/Semateche-Handbook of Statistical Methods,* http://www.itl.nist.gov/div898/handbook, accessed November 1 2009.

39

WET GRANULATION PROCESSES

KAREN P. HAPGOOD
Department of Chemical Engineering, Monash University, Clayton, VIC, Australia
JAMES D. LITSTER
School of Chemical Engineering and Department of Industrial & Physical Pharmacy, Purdue University, West Layafette, IN, USA

39.1 INTRODUCTION

Granulation is a size enlargement process, where individual powder particles, usually of several different components, are aggregated together to form a larger structured particle where the original particles can still be distinguished. Granulation of one or more drugs and excipients is a common first step in the manufacture of tablets or capsules for pharmaceutical drug delivery. Although direct compression is possible, granulation usually ensures good flow properties and uniform bulk density essential for tabletting, and reduces the risk of segregation by creating multicomponent granules with more uniform composition than the dry blend. Dry granulation, also known as roller compaction, uses compressive forces to form the aggregates, and is covered in detail in Chapter 38.

In this chapter, we focus on wet granulation, where a liquid is used to form wet agglomerates, which are subsequently dried. We first provide an overview of the many advances in our understanding of the science underlying wet granulation. In the second half of the chapter, we discuss scale-up approaches and provide a case study of how the mechanistic knowledge can be applied to design the granulation process and scale-up with minimal trial-and-error. We conclude with some future directions for granulation process development and manufacturing in the pharmaceutical industry.

39.2 MECHANISMS IN WET GRANULATION

Wet granulation is the process of using liquid and a binder material (usually a polymer such as HPC, PVP, HMPC) to aggregate the individual particles in the dry mix into particle assemblies. The assemblies contain a mixture of drug and excipients, and have a porous structure, which provides improved compression properties due to the rearrangement of particles during the collapse of the granule structure. There are three main stages in granulation [1] (refer to Figure 39.1), which are as follows:

1. *Nucleation and Wetting*: In this stage, the spray drops form the initial granules or nuclei.
2. *Consolidation and Growth*: Agitation leads to granule–granule collisions and granule–particle collisions, resulting in larger granules—this is a size enlargement process. These collisions may also result in a reduction in internal pore space of the granule—this densification process is termed as consolidation.
3. *Attrition and Breakage*: Agitation forces exceeding granule strength will result in either fracture of the granule into several large pieces, or attrition of the outer layer of particles from the granule. These are both size reduction processes, and both may occur simultaneously.

Chemical Engineering in the Pharmaceutical Industry: R&D to Manufacturing Edited by David J. am Ende

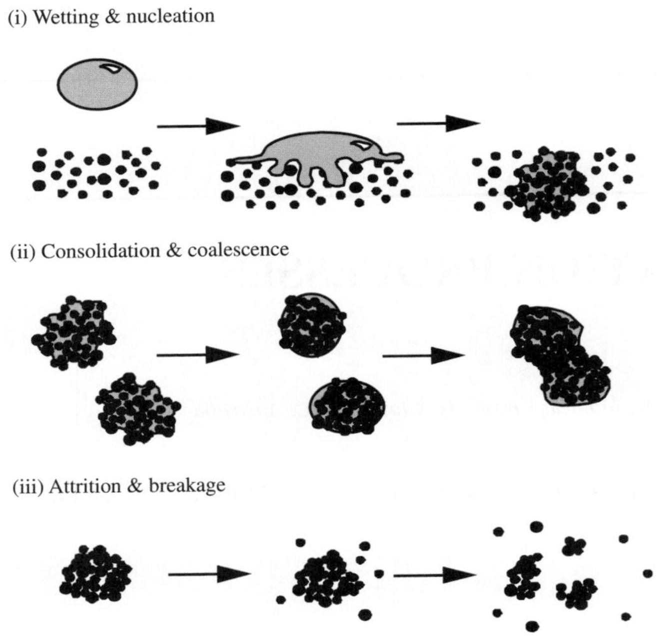

(i) Wetting & nucleation

(ii) Consolidation & coalescence

(iii) Attrition & breakage

FIGURE 39.1 Rate processes in granulation (i) wetting and nucleation; (ii) consolidation and (iii) growth; and attrition and breakage [1].

Each mechanism is discussed in more detail below.

39.2.1 Nucleation

Mixing and distribution of the liquid during the nucleation phase (spraying phase) is an important step, and poor initial liquid distribution leads to a heterogeneous nuclei size distribution and increased variability in the granulation process [2]. Nuclei formed from under conditions of poor liquid distribution will have broad distributions of size, porosity, and saturations [2]. This in turn will lead to different growth and breakage rates for each granule. Ensuring a controlled nucleation step is the first step toward a controlled granulation process.

Immersion nucleation is the step where fine powders are engulfed by larger drops to form nuclei [3–5]. There are five steps in immersion nucleation as shown in Figure 39.2. Initially, the drop must be formed at the nozzle. After landing on the powder surface, the drop may potentially shatter and break into fragments, as shown experimentally [6] or

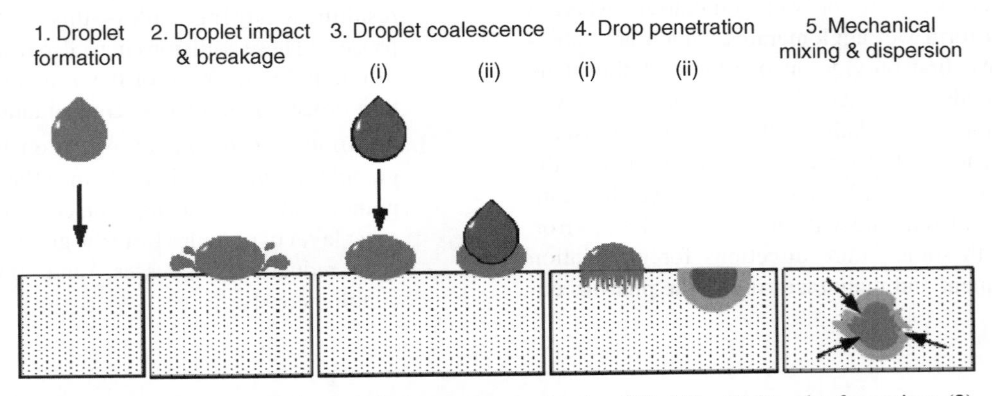

FIGURE 39.2 Five possible steps in immersion nucleation [10, 14]. (1) Droplet formation; (2) droplet impact and possible breakage on the powder bed; (3) droplet coalescence upon contact with other droplets at high spray flux; (4) drop penetration into the powder bed (i) and formation of a nucleus granule (ii); (5) mechanical mixing and dispersion of the liquid and powder.

coalesce with another drop already at the surface of the powder. Once it reaches the powder surface, the drop may penetrate into the powder bed via capillary action (step 4 in Figure 39.2), or it may require mechanical agitation and shear to disperse the fluid through the powder. A similar diagram for drops landing on a fluidized powder bed has also been published [7], which also depicts the nuclei passing through the spray zone multiple times. Distribution nucleation [3–5] can occur in fluid beds operated with very fine drops, which are of the same size or smaller than the particle size. The mechanisms of distribution nucleation are less well understood.

There are two parameters describing the immersion nucleation process in the spray zone of the granulator—the drop penetration time and the dimensionless spray flux [8, 9]. These parameters have been combined to form a nucleation regime map [10] and can be used for scale-up [11] and to quantitatively model the nuclei size distribution in the spray zone [12, 13].

39.2.1.1 Drop Penetration Time As the powder moves beneath the spray zone, the small droplets will land on the powder surface and begin to mix with the powder (see Figure 39.2). When the powder is easily wetted or hydrophilic, that is, the contact angle between the powder and fluid is less than 90°, penetration of the fluid into the powder pores will begin to occur. The rate at which a single drop of fluid, with volume V_d, viscosity μ, and surface tension γ_{lv} will penetrate into a static porous medium with cylindrical pores of radius R_{pore} and an overall porosity of ε is given by the drop penetration time t_p [15, 16]

$$t_p = 1.35 \frac{V_d^{2/3}}{\varepsilon^2 R_{pore}} \frac{\mu}{\gamma_{lv} \cos\theta} \tag{39.1}$$

In a loosely packed powder, similar to that found during agitation of the powder in the granulator, the voidage and pore size will be fairly heterogeneous. The powder will contain some combination of small pores and much larger pore spaces, or macrovoids. Liquid will flow through the small micropores, but there is no capillary driving force for the liquid to flow into the large macrovoids, as the rapid expansion of the pore radius dramatically reduces the Laplace pressure, which is the driving force for capillary flow. This means that the liquid does not flow through the macrovoids and must instead find a path around the macrovoid, slowing down the penetration [8] (Figure 39.3).

The effective voidage ε_{eff} is the first simple step toward including the effect of nonuniform pore structure on the penetration of a fluid into a real powder bed [8]. At the tap density ε_{tap}, the bed is assumed to contain no macrovoids. As the bed becomes less densely packed, the porosity of the bed increases, and the fraction of the voidage above the initial tapped bed voidage is assumed to form macrovoids.

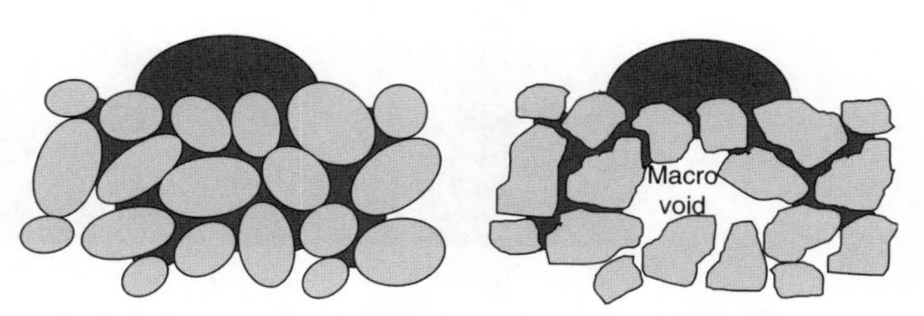

FIGURE 39.3 Liquid will penetrate the micropores which are driven by capillary action, but wherever a pore rapidly expands into a macrovoid, the fluid has to find a path around the flow obstruction [8].

The effective bed voidage thus estimates the amount of pore space that is actually available for capillary driven flow

$$\varepsilon_{eff} = \varepsilon_{tap}(1 - \varepsilon + \varepsilon_{tap}) \tag{39.2}$$

The effective porosity is used in the Kozeny equation in place of overall bed porosity ε to estimate the effective micropore size R_{eff}, which is pore size available for liquid flow

$$R_{eff} = \frac{\varphi d_{32}}{3} \frac{\varepsilon_{eff}}{1 - \varepsilon_{eff}} \tag{39.3}$$

Thus, the most appropriate equation for estimating the drop penetration time t_p into a loosely packed porous powder bed is [8]

$$t_p = 1.35 \frac{V_d^{2/3}}{\varepsilon_{eff}^2 R_{eff}} \frac{\mu}{\gamma_{lv} \cos\theta} \tag{39.4}$$

The drop penetration time is an indication of the kinetics of nucleus formation. Equation 39.4 shows that the penetration time depends on several factors, including the drop size V_d, the fluid properties (viscosity μ, surface tension γ_{lv}, and contact angle θ), and the powder packing structure (ε_{eff} and R_{eff}). However, the fluid viscosity has the largest effect, as the fluid viscosity can range over several orders of magnitude. This is also commonly found in pharmaceutical granulation, where low-viscosity fluids such as water or ethanol are used as frequently as high-viscosity fluids (e.g., a 7% HPC solution has a viscosity of 104 mPa s [9]). Figure 39.4 shows a water drop penetrating into a lactose powder bed over 2.3 s. In contrast, a similar drop of 7% HPC takes approximately 2 min to penetrate into the powder bed [8]. This timescale is clearly much longer than the timescale of agitation, and implies that nucleation via wetting and capillary action (step 4 in Figure 39.2) is not the dominant mechanism for high-viscosity systems. Instead, dispersion of the fluid through the powder will need to occur via mechanical mixing of the powder (see step 5 in Figure 39.2).

For a given powder, the drop penetration time is proportional to the liquid properties group $\mu/\gamma_{lv}\cos\theta$. For a given fluid, as the powder becomes finer (i.e., the d_{32} particle size decreases) the drop penetration time increases, primarily due

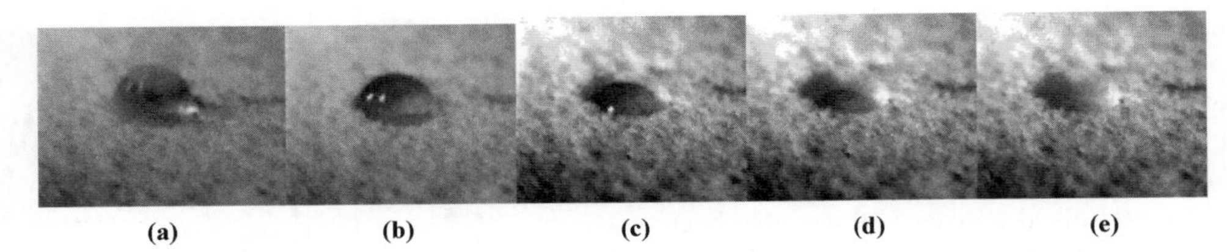

FIGURE 39.4 Water drop (~6 μL volume) penetrating into lactose powder. (a) Impact; (b) 0.23 s; (c) 0.9 s; (d) 1.4 s; and (e) 2.3 s (adapted from Ref. 8).

to a decrease in packing efficiency which creates a higher proportion of macrovoids [8]. Equation 39.4 can be used to estimate the drop penetration time from theory, but can have large errors for very fine, cohesive powders [8]. Experimental tests are recommended in this case, with equation 39.4 used to scale the experimental results to account for differences in drop size, etc. between the experiment and the actual manufacturing process (e.g., see Ref. 17).

Drop penetration time tends to decrease if the powder is already partially wet [14, 18], and the effect is more pronounced for viscous fluids with long penetration times [14]. For drops with long penetration times that land adjacent to a prewet patch of powder, the drop will tend to migrate over and penetrate into the preexisting wet patch, rather than penetrate into the dry powder directly underneath it (see Figure 39.5).

The above discussion is for drop penetration into a stationary powder bed. Drops impacting and penetrating into moving powders show more complex behavior. In fluid bed granulation, the fluid slowly flows outwards, and the powder layer is slowly built up by collisions between the wet outer surface of the drop or wet agglomerate and the agitated dry powder [19]. Fluid flow and nuclei formation kinetics are still controlled by the same fluid properties as shown in equation 39.4 but the rate of powder addition to the exterior surface is an additional factor in the kinetics [19]. Growth will continue until the saturation of the agglomerate decreases to the point that no further liquid can reach the powder surface. This saturation limit was defined as the "wetting saturation" S_w and determines whether additional growth will occur (see Figure 39.6).

39.2.1.2 Dimensionless Spray Flux Dimensionless spray flux [9, 13] considers the granulator spray zone and the flux of drops landing on the moving powder surface. The derivation of spray flux is straightforward [9] and is not intended to be equipment specific, although it has been most frequently applied in mixer granulation.

During the liquid addition stage of the granulation process, the powder surface is moving beneath the spray with a velocity v underneath a spray of width W at the powder surface. Therefore, the area of dry powder passing beneath the spray nozzle per second is vW. Each drop hitting the powder surface in the spray zone will leave a "footprint" as it wets into the powder. The number of drops hitting the powder surface per unit time can be calculated by dividing the total liquid flow rate by the volume of an individual drop. Assuming that each drop lands separately on the surface, without overlapping with any other drops, the total wetted area created per unit time can be estimated by multiplying the

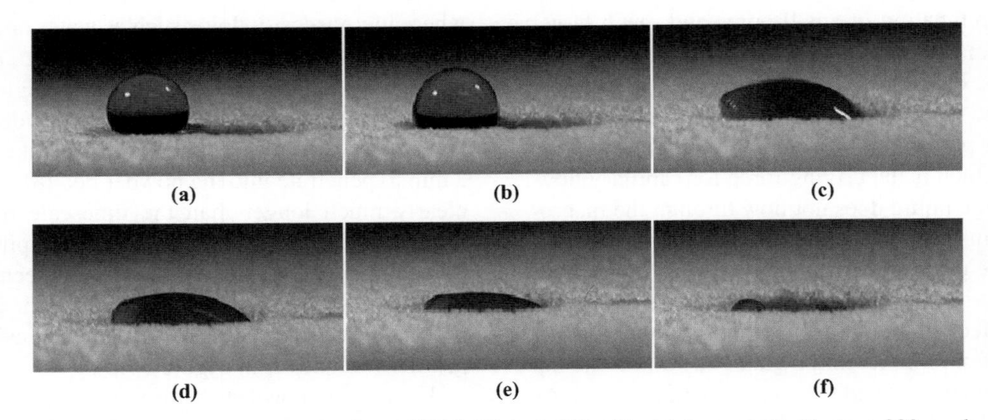

FIGURE 39.5 Penetration of a drop of PEG400 ($\mu = 120$ mPa s) into prewetted lactose 200 mesh powder. The footprint of the previous drop is visible to the right of the added drop (separation distance of 3 mm). (a)Impact; (b) 4.05 s after impact; (c) 5 s; (d) 5.02 s; (e) 12.06 s; and (f) 51.6 s [14].

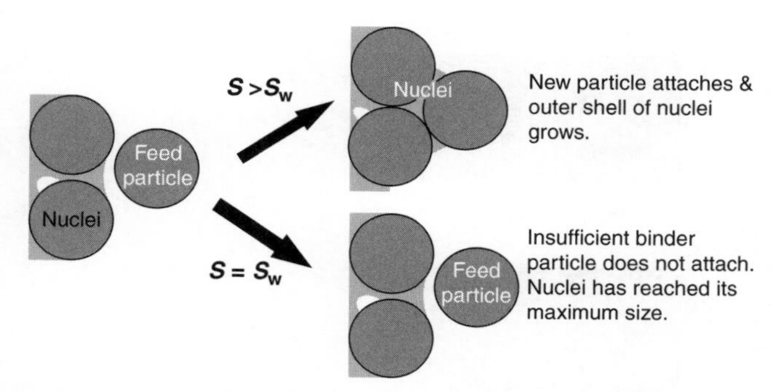

FIGURE 39.6 Nuclei growth in a fluid bed via growth of the outer nuclei shell continues, provided the agglomerate saturation is higher than the wetting saturation limit S_w (adapted from Ref. 19).

number of drops being added per second by the area each drop will occupy when it lands on the powder surface. Dimensionless spray flux is then defined as the ratio of the rate at which wetted area is created by the incoming droplets compared to the total area of dry powder passing through the spray zone [9]:

$$\Psi_a = \frac{\dot{a}}{\dot{A}} = \frac{3\dot{V}}{2\dot{A}d_d} \tag{39.5}$$

The dimensionless spray flux is a measure of the density of drops falling on the powder surface. Physically, $\Psi_a = 0.1$ means that ~10% of the powder surface will be wetted per pass through the spray zone. At low spray flux ($\Psi_a \ll 1$) drop footprints will not overlap and each drop will form a separate nucleus granule. The size of the nuclei will be directly proportional to the drop size [8, 20]. As Ψ_a increases, the fraction powder wetted also increases, although the relationship is linear only at low Ψ_a values. At high spray flux ($\Psi_a \sim 1$) the spray rate is high compared to the rate of dry powder entering the spray zone and the drops significantly overlap each other as they land on the powder bed. Spray flux $\Psi_a \geq 1$ means that the incoming droplets will theoretically cover 100% of the dry powder passing beneath the nozzle, assuming no drop overlap occurs [9]. At this value, the nuclei formed will be fragments of a much larger sheet of wet powder and will bear little relationship to original drop size. Good granulation can still be achieved provided that the shear forces during granulation are large enough and uniform enough to be effective. In pharmaceutical granulation, these conditions usually also consolidate the granules leading to lower porosity granules with slower dissolution. The process is illustrated schematically in Figure 39.1.

Note that $\dot{A}$ is a dynamic quantity, defined as the outer perimeter of powder sprayed per second (m²/s) and is not equivalent to static footprint spray area, A. The difference between the static footprint area of the spray and the dynamic area flux is illustrated schematically in Figure 39.7 for several cases of varying powder velocity.

The impact of Ψ_a on nuclei formation has been studied in both mixer granulators [11, 17, 21, 22] and externally on a simplified moving power bed [9, 23]. To eliminate the effect of granule growth, the nuclei size distribution was measured after a single pass of the powder through the spray zone [9]. The results clearly show that at low spray flux ($\Psi_a < 0.2$) the nuclei size distribution is quite narrow. As spray flux increases, the distribution broadens as agglomerates begin to form. At very high spray flux ($\Psi_a > 1$), the spray zone has become a continuous cake and the nuclei distribution bears no resemblance to the drop distribution. Further, when the spray flux is low, changes to the spray drop distribution are directly mapped onto the nuclei size distribution [9].

The dimensionless spray flux parameter is intended to capture the major effects of drop overlap in the spray zone on the nuclei distribution as simply as possible, to encourage its use as a scale-up parameter. The derivation contains several major simplifying assumptions. First, the spray is assumed to consist of mono-sized drops. Second, the spray flow rate is assumed to be uniformly distributed over the entire spray area. This is rarely true in industrial applications, as generally, the flow rate is higher in the center of the spray area than at the sides [7, 24]. The theory has been developed and validated for the case of a completely dry powder entering the spray zone. In practice, this is true only for the very initial stages of the granulation process. As spraying progresses, a mixture of dry powder and previously formed nuclei and granules will pass through the spray and will be rewetted, which will affect the final size distribution for some fluids [14].

Finally, the derivation ignores the fact that the nucleus size is always larger than the drop size. During spraying, two drops may land close to each other but without touching. As the liquid penetrates into the powder, the larger nuclei may grow and touch each other, causing coalesce of the nuclei even when the spray drops landed separately. Wildeboer et al. [13] extended the theory and modeling to account for the effects of drop size distribution, nonuniform spray density, and for probability of coalescence due to nucleus spreading. Nucleus spreading is described by the nucleation ratio K, which can be defined on either a volume basis—that is, the ratio of the nucleus volume to the drop volume, K_v [19]—or on an area basis where K_a is the ratio of the projected area of the drop (a_d) to the nucleus (a_n) [13].

Where a is the projected area of the nucleus (a_n) and drop (a_d). Typical values for K_v range between 3 and 30 or higher [18, 19, 23, 25, 26]. The probability of a single drop forming a single nucleus is, therefore, related to the dimensionless nucleation number, Ψ_n [13]:

$$\Psi_n = K_a \frac{3\dot{V}}{2wvd_d} \tag{39.6}$$

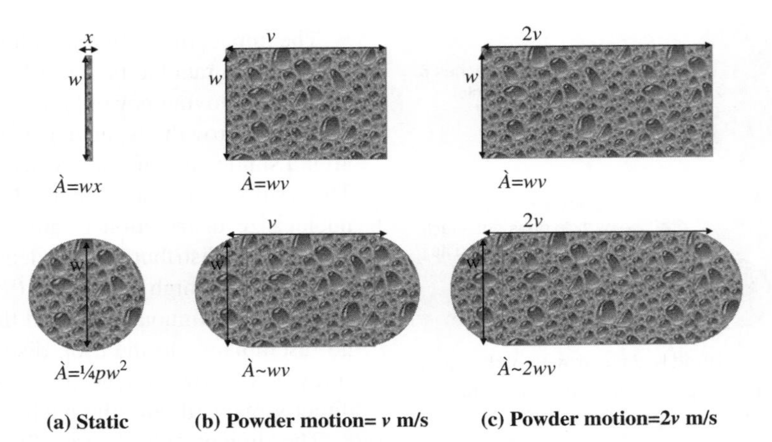

FIGURE 39.7 Relationship between the spray area of the nozzle, A, and the dynamic spray area $\dot{A}$ for a line spray. Upper case shows a line spray and lower case shows a circular spray. Case (a) represents the static spray case where there is no powder motion; case (b) shows the area flux of powder beneath the nozzle for powder moving at v m/s; and case (c) area wetted by the spray for powder moving at $2v$ m/s.

The dimensionless nucleation number Ψ_n differs from the original spray flux Ψ_a by the factor K_a, which accounts for the degree of nucleus spreading. Equation 39.6 has been used to model the nuclei distributions in the spray zone over a range of Ψ_n [13], accounting for nonuniform sprays and nucleus spreading and coalescence. At a given value of the dimensionless nuclei number Ψ_n, the density and size distribution of the nuclei formed on the surface is constant, that is, the final value of Ψ_n is the sole determinant of the final nuclei size distribution (Figure 39.8).

Assuming complete spatial randomness, spatial statistics can be used to derive an analytical solution for both the fraction surface coverage and fraction of agglomerates formed [12]. Under these conditions, the drops landing randomly on the target area are described by a Poisson distribution. The fraction of the surface covered by drops in a single pass through the spray zone is given by [12]

$$f_{\text{covered}} = 1 - \exp(-\Psi_a) \tag{39.7}$$

The fraction of nuclei formed from n drops is given by

$$f_n = \exp(-4\Psi_a)\left(\frac{(4\Psi_a)^{n-1}}{(n-1)!}\right) \tag{39.8}$$

Thus, we can calculate the number of single drops, not overlapping with any other drops, and by difference, the number of agglomerates [12]

$$f_{\text{single}} = \exp(-4\Psi_a) \tag{39.9}$$

Equation 39.8 can be used to estimate the initial nuclei size distribution as a function of spray flux for mono-sized drops. Figure 39.9 shows the nuclei distributions predicted by equation 39.8 at a range of spray flux values for 100 μm mono-size drops. As the spray flux increases, a higher percentage of nuclei are formed from multiple drops, creating a larger and broader nuclei size distribution. However, prediction of the bimodal nuclei distributions that form at

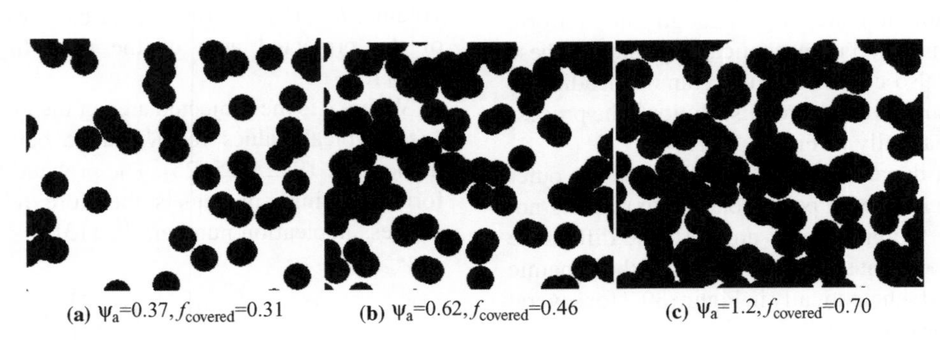

FIGURE 39.8 Monte-Carlo simulations of spray flux.

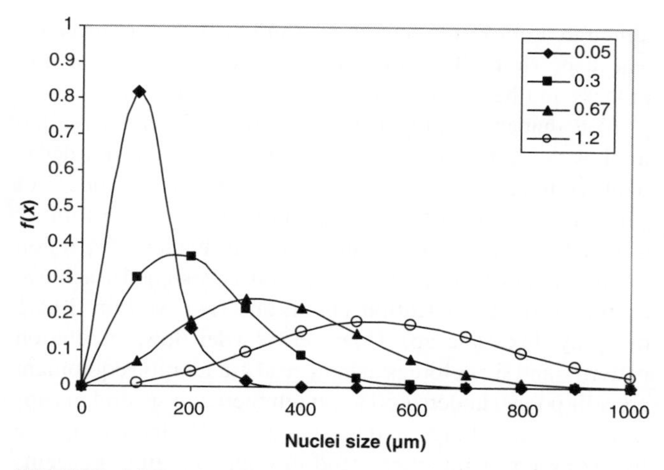

FIGURE 39.9 Nuclei size distributions predicted by equation 39.9 at a range of spray flux values assuming 100 μm mono-sized drops.

higher spray flux [27] appears to require a more sophisticated modeling approach incorporating overlapping drop size distributions.

Practical application of dimensionless spray flux requires measurements of the drop size distribution and powder velocity, and spray distribution for Ψ_a. Spray distribution is the easiest of the three parameters to measure (e.g., see Ref. 24), and laser diffraction and/or Doppler analysis of spray nozzles is widely available for drop size characterization. Currently, the most challenging parameter in Ψ_a to measure in real granulators is the powder velocity, and this is discussed separately in Section 39.3.2. Drop size in a spray is a strong function of the distance from the nozzle, so the measurements of drop size must be taken at the approximate distance where the spray intersects the bed. This distance usually increases as the granulator scale increases, and varies depending on spray lance design. Increase in the nozzle lance depth causes a reduction in the spray area and hence an increase in spray fluxes. Maintaining a constant nozzle height is important in reducing unwanted variation in granulator performance.

39.2.1.3 Nucleation Regime Map The drop penetration time describes the kinetics of nucleation from a single drop as a function of the material properties in the formulation. The spray flux describes the physical interactions of multiple drops in the spray zone. Together, these two parameters form the basis of the nucleation regime map [10], which describes the optimal conditions for uniform liquid distribution and suggests some ideal conditions for controlled nucleation.

When the drop penetration time is short, the fluid will sink quickly into the powder bed to form a nucleus granule. If no other drops land on top of the sinking drop as it passes through the spray zone, a single nucleus granule will be formed with a size equal to 2–3 times the drop volume (see equation 39.7). If this process occurs for all the drops, the nuclei size distribution will be directly proportional to the drop size distribution. This is known as "drop-controlled nucleation" [10] and occurs at low drop penetration time and low spray flux (i.e., low spray density). As a guide, the spray flux needs to be less than $\Psi_a < 01$ for approximately 2/3 of the nuclei to be formed from a single drop (see equation 39.11).

This "one drop produces one granule" mode of nucleation will not occur with a formulation with the same low drop penetration time, but granulated at a high spray flux. At higher spray flux, the spray density will be too high and the vast majority of the drops will coalesce with another drop on the powder surface. The surface of the powder will be wetted by an almost continuous sheet of liquid, rather than discrete drops. As the powder moves due to agitation, the "caked" powder will be broken and the fragments will be dispersed through the powder. This is the "mechanical dispersion" regime of nucleation [10], where the liquid is dispersed primarily due to powder agitation and shear, rather than by fluid flow and wetting.

If the drop penetration time is much longer, the liquid will remain on the surface of the powder for an extended period of time, in the order of minutes. The constant powder motion means that it is more likely to coalesce with other unpenetrated droplets (see step 3 in Figure 39.2), merge into a section of powder wetted by an earlier drop [14] or roll into depressions in the powder and form rivulets [23]. Even if the drop does not coalesce with other drops, powder agitation and shear will be required to disperse the fluid through the powder to form a nucleus (step 5 in Figure 39.2). Again, this is the "mechanical dispersion" nucleation regime.

These two regimes of nucleation can be summarized using a nucleation regime map (Figure 39.10). The axes of the map are the dimensionless spray flux Ψ_a and the dimensionless penetration time τ_p

$$\tau_p = \frac{t_p}{t_c} \tag{39.10}$$

where t_c is the circulation time for the droplet or nuclei to return to the spray zone. This is currently not quantified due to insufficient understanding of powder flow and circulation patterns in most industrial granulation equipment. In general, the drop penetration time needs to be much faster than the circulation time, and an arbitrary limit of 1/10th of the circulation time ($\tau_p < 0.1$) has been set as the upper limit for drop-controlled nucleation [10].

In the lower left-hand corner of the map is the drop-controlled regime, which occurs at low dimensionless drop penetration time ($\tau_p < 0.1$) and low spray flux ($\Psi_a < 0.1$). In this corner, each drop will generally land separately without touching any other droplets. As soon as either the spray flux or the penetration time increases slightly, the system enters an intermediate region where both wetting and agitation will be

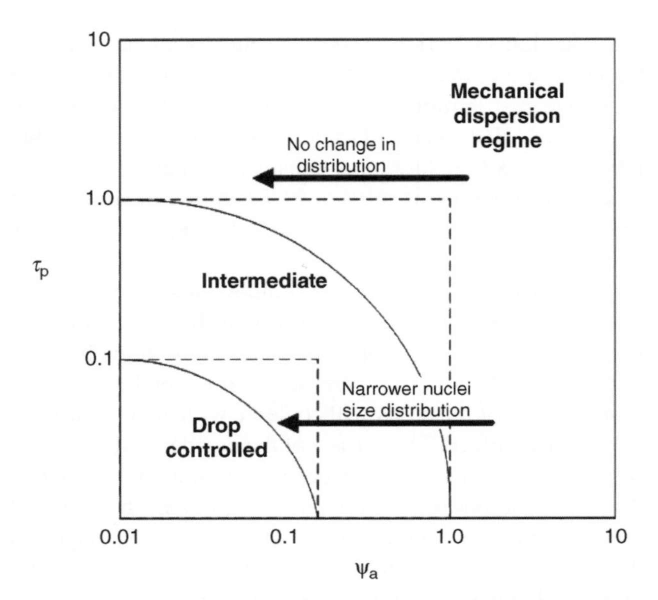

FIGURE 39.10 Nucleation regime map with adjusted boundaries incorporating the drop migration due to rewetting of powder during multiple passes through the spray zone [14]. Dotted lines represent original regime boundaries [10].

equally important in determining liquid distribution and nucleation. As spray flux or penetration time increase further, the mechanical mixing will become the dominant mechanism in nucleation, as the wetting kinetics occur on a much longer timescale.

Understanding which nucleation regime the process is operating in can be extremely useful for understanding how to optimize liquid dispersion and trouble shoot during manufacturing. For formulations with low penetration times, optimizing the spray flux by adjusting the fluid flow rate, the drop size, spray area, powder velocity, etc., will have a significant effect on the process, as it moves from the mechanical dispersion regime toward the drop-controlled regime, crossing the regime boundaries (see lower horizontal arrow in Figure 39.10). As the process nears the drop-controlled regime, the nuclei size distribution will approach the drop size distribution of the spray. Atomization of a fluid is fairly well understood (e.g., [28]) and is therefore much easier to optimize compared to a granulation process. In contrast, when working with a formulation with a long drop penetration time, optimizing the spray parameters to reduce spray flux will have little effect on the nucleation process. Once the drop penetration time is high, it is extremely difficult to achieve drop-controlled nucleation, as the liquid will form a "puddle" on the powder due to either slowly penetrating drops or high spray density. In addition, atomizing a highly viscous fluid can be difficult, and some industrial processes add the fluid as a steady unatomized stream or even scoop in extremely viscous pastes. There is little benefit in atomizing these extremely viscous fluids, as although atomization may assist fluid distribution, mechanical dispersion will still control the overall fluid dispersion process and hence control the nuclei distribution.

Note that granulating in the mechanical dispersion regime implies that efficient mixing and agitation of the powder is required to achieve effective liquid dispersion, and does not automatically mean "poor" liquid distribution. Most mixer granulation processes operate in the mechanical dispersion regime, unless they have been consciously designed to achieve drop-controlled nucleation conditions (see Section 39.3.1 on spray flux scale-up). However, powder flow, circulation patterns, and shear forces in industrial granulation equipment are still poorly understood although there are several groups actively researching in this area [29–33]. Increasing the fluidizing air or impeller speed to improve agitation seems a simple solution, but commonly cannot be done, either due to equipment or process limitations (e.g., fixed speed impellers in mixers, or maximum pressure drop limit in a fluid bed) or due to negative side effects in on the granulation process (e.g., change in granule density and/or growth regime—see Section 39.2.1.2). One way to improve mechanical dispersion is to position the spray in a highly agitated region of the granulator. In a mixer granulator, this can be achieved by placing the spray nozzle either directly over the chopper [34], or shortly before the chopper, so that the recently sprayed powder flows almost immediately into the turbulent chopper zone, rather than completing a 180–270° rotation before being agitated.

39.2.2 Consolidation and Growth

The second mechanism in wet granulation is granule growth (see Figure 39.1) and is inherently linked to both the liquid level and porosity of the granules. It is well known that different formulations show different growth behavior in the same equipment, and that the same formulation can demonstrate changed growth behavior when different granulation conditions or equipment are used. We can think of two extreme models for the process of granule growth, which are as follows:

1. *Growth of Deformable Porous Granules*: The starting force for granule growth and consolidation is a porous deformable nuclei formed by the processes described in Section 39.2.1. The nucleus contains substantial amounts of liquid in the pores but is not necessarily surface wet. This corresponds to a drop size in the spray zone larger than the primary particle size and this model is a reasonable picture of granule growth in high shear mixers.
2. *Near Elastic Granules*: Here we consider the granule to be a nearly elastic particle coated with a liquid binder after it leaves the spray zone. This model is well suited for processes where granules are dried before

they reenter the spray zone and where the liquid drop size is small compared to primary particle size. Thus, this model is often a reasonable description for fluidized bed granulation.

For both models, if we can answer the question: "Will two colliding granules coalesce or rebound?" we will have gone a long way to describing the granule growth behavior. Granule growth regime maps summarize the causes of much of this behavior. Before, describing these models, we first discuss granule strength and deformation, and granule consolidation.

39.2.2.1 Granule Strength and Deformation One way to experimentally study granule strength is through the use of wet granule pellets, which are small wet powder compacts with controlled size, porosity, and liquid content. Pellets are deformed under different conditions (in particular, strain rate $\dot{\varepsilon}$), and the effects of many variables, including binder fluid viscosity and surface tension, particle size, pellet porosity, etc. can be investigated [35–38]. The stress versus time is recorded and the peak stress σ_{pk} indicates the point of failure and converted to a dimensionless peak flow stress Str* [38]

$$\text{Str}^* = \frac{\sigma_{pk} d_p}{\gamma_{lv} \cos\theta} \tag{39.11}$$

The Str* data for all the experiments was plotted against the capillary number Ca, which is the ratio of viscous forces to surface tension forces, and proportional to the strain rate $\dot{\varepsilon}$

$$\text{Ca} = \frac{\mu d_p \dot{\varepsilon}}{\gamma_{lv} \cos\theta} \tag{39.12}$$

Figure 39.11 shows this data for spheres. A single relationship can be formed with two distinct regimes [37, 38]. In region 1, the strain rate is low (low Ca) and the peak flow stress is independent of strain rate. In region 2, the higher strain rates applied (high Ca) means that the viscous resistance forces begin to dominate, and the peak flow stress is proportional to the strain rate.

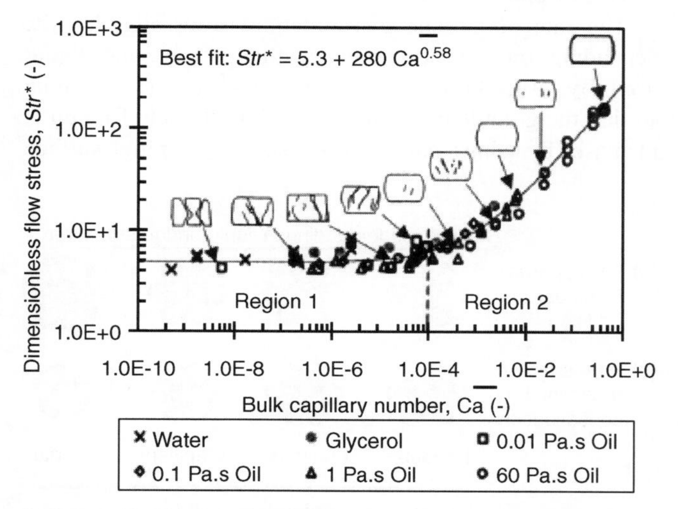

FIGURE 39.11 Dimensionless flow stress versus capillary number [37]. In region 1, the stress is independent of flow rate. At higher strain rates (region 2), viscous forces dominate and the stress is proportional to the strain rate.

Note that this approach confirms that at high enough values of Ca, the peak flow stress (granule dynamic strength) is a strong function of strain rate. As collisions in the granulator are dynamic and collision velocities may vary from approximately 0.1 ms^{-1} to approximately 1 ms^{-1}, this is important. The granule strength should not be measured under static conditions, but rather at strain rates similar to those experienced in the granulator. Granules made from nonspherical primary particles follow a similar behavior to that illustrated in Figure 39.11, but are generally significantly stronger than granules made from spherical model particles. In Section 39.2.3.4, an extended version of the granule strength model is presented which includes effects of primary particle shape and granule liquid saturation.

39.2.2.2 Granule Consolidation A granule is a three-dimensional composite of solids particles, liquid bridges (which covert to solid bridges after drying), and vacant pore space occupied by air. Consolidation is the increase in granule density that results when the primary particles are forced to move closer to each other as a result of collisions between particles. Consolidation can only occur while the binder is still liquid. Consolidation determines the porosity and density of the final granules. Factors influencing the rate and degree of consolidation include particle size, size distribution, and binder viscosity as well as the impeller speed or fluidizing velocity [39–41].

The structure of granules, particularly the proportion and arrangement of the pore space, plays an important role in downstream processing, particularly compaction of the granules into a tablet, and in product performance, especially dissolution of the final solid dosage form [42].

The structure of real granules is complex and has not been able to be studied until the relatively recent development of micro X-ray tomography (XRT) [43–47]. Some examples of the structure of some pharmaceutical granules made in the same equipment from the same formulation at two different mixing conditions are shown in Figure 39.12 and this is an area of continuing research.

Although detailed analysis of the pore size distribution via mercury porosimetry [41, 48, 49] or XRT is possible [42–45], the overall average porosity of the granule has been found to be a very useful parameter in granulation. The overall porosity of a granule ε is defined as the volume fraction of air *within* a granule. Care is needed to avoid confusion between the interparticle voidage between the granules and the intraporosity of the granules (i.e., the internal porosity), both of which affect the bulk and tap density of an assembly of granules. The porosity of the granules generally begins at a

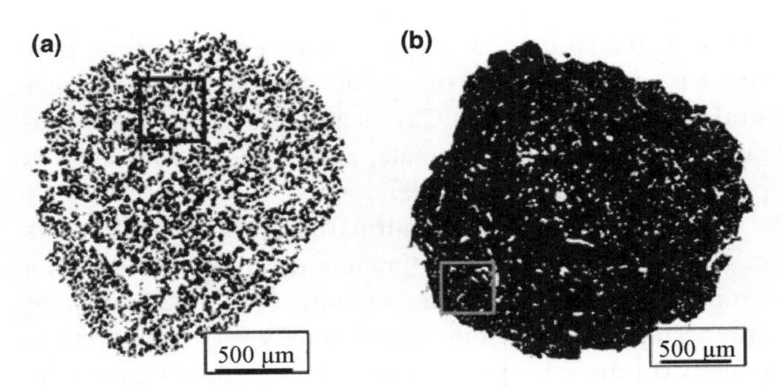

FIGURE 39.12 X-ray tomography images of internal structure of granules produced in a 2 L mixer at different shear conditions (a) 200 rpm impeller and 600 rpm chopper speed $\varepsilon \sim 58\%$ and (b) 600 rpm impeller and 1800 rpm chopper speed $\varepsilon \sim 15\%$ [43].

high value (approximately 50–60%) and decays exponentially as the granulation process proceeds [40] (see Figure 39.13), due to granule deformation and particle rearrangement as a result of collisions with other particles, granulator wall and the impeller or chopper in a mixer granulator. The rate of densification depends on several parameters, but large particle size, smooth round particles, and low-viscosity fluids allow rapid densification [40]. The final porosity reached by the granule, $\varepsilon_{\min}$, is often used to determine the yield strength of the granule. The rate at which granules densify can be related to the granule peak flow stress and the typical collision velocity in the granulator through the Stokes deformation number St_{def}

$$St_{\mathrm{def}} = \frac{\rho_g U_c^2}{2Y_g} \tag{39.13}$$

where ρ_g is the granule density, U_c is the collision velocity, and Y_g is the yield strength of the granule. The yield strength of the granule is a function of the formulation and the extent of consolidation, and is usually evaluated as the peak flow stress at the minimum porosity $\varepsilon_{\min}$ (see Figure 39.12). The granulation consolidation rate constant k_c is then given by

$$k_c = \beta_c \exp(a.St_{\mathrm{def}}) \tag{39.14}$$

where β_c and a are constant. k_c is the consolidation rate constant for a first-order consolidation equation of the form

$$\frac{\varepsilon - \varepsilon_{\min}}{\varepsilon_0 - \varepsilon_{\min}} = \exp(-k_c t) \tag{39.15}$$

Granule porosity is closely coupled with the granule saturation. The granule saturation s is defined as the proportion of pore space that is occupied by liquid

$$s = \frac{w\rho_s(1-\varepsilon)}{\rho_l \varepsilon} \tag{39.16}$$

where ε is the average granule porosity, w is the mass liquid/mass dry powder, ρ_l is the density of liquid, and ρ_s is the true density of the solid particles.

Four general saturation states have been defined—pendular, funicular, capillary, and droplet (Figure 39.14). There are two ways that the overall saturation of the granule can be increased—the amount of fluid added to the system can be increased, or the granule can be consolidated to reduce the pore space available [50, 51]. During the liquid addition phase of wet granulation, a combination of both processes is most likely occurring. For some formulations, consolidation of the granule will gradually decrease the porosity of the granule until the saturation reaches the droplet state, when the binder fluid will be squeezed to the exterior surface of the granule [24, 52–55]. The sudden presence of fluid at the outer granule surface often induces rapid coalescence and runaway growth of granules [40, 53, 56].

The rate of granule consolidation varies significantly, depending on the properties of the powder and liquid used [39, 40]. The interparticle friction must be overcome so that the granule can consolidate. Interparticle friction is increased by using smaller particle size, as their high surface

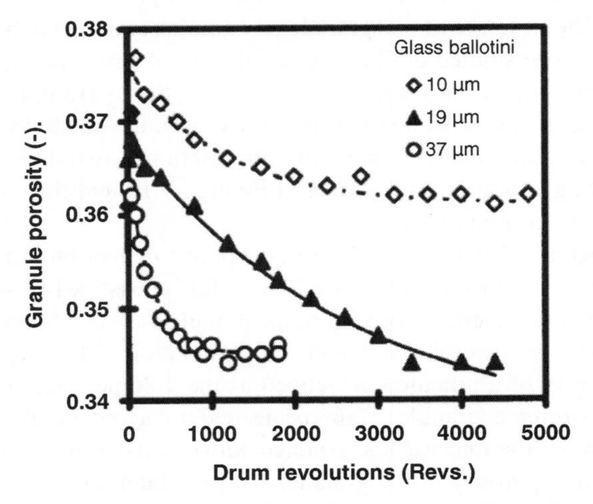

FIGURE 39.13 Exponential decay in granule porosity as granulation proceeds [40].

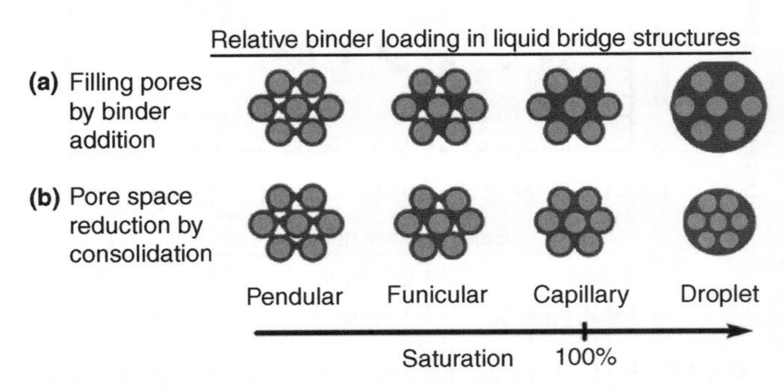

FIGURE 39.14 Granule saturation has four main states, and saturation increases as liquid content increases [50] and pore space decreases.

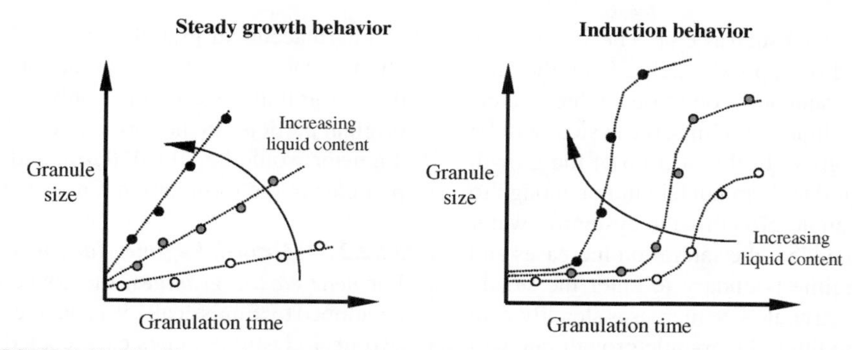

FIGURE 39.15 Two main types of granule growth behavior for deformable granules [52].

area and high number of interparticle contacts provide more resistance to consolidation [39, 40, 57]. Granules formed from coarser particles tend to consolidate more quickly (see Figure 39.12). Viscous binder fluids also reduce the rate of consolidation, as the liquid resists flowing during deformation. However, the presence of fluid also acts as a lubricant and higher liquid content (i.e., saturation) causes the rate of consolidation to increase.

39.2.2.3 Granule Growth Behavior for Deformable Porous Granules There are two main types of granule growth behavior for porous, deformable granules—steady growth and induction growth (Figure 39.15). "Steady growth" occurs when the rate of growth is constant (at a given liquid level). On a plot of granule size versus granulation time, steady growth behavior produces a linear trend. This type of growth occurs in formulations where the granules are easily deformed by the forces in the granulator, and tends to occur when the using coarse powders and low-viscosity fluids [52].

In contrast, "induction growth" occurs when the initial nuclei remain at a constant size for a long period, before very rapid granule growth occurs, resulting in a sudden increase in granule size (Figure 39.15). During the induction period, the granules consolidate and approach some minimum porosity (e.g., see Figure 39.13 for example data) but do not grow in size except through the layering of ungranulated fines. Eventually, the granule porosity can be reduced enough to squeeze liquid to the surface. If there are no ungranulated fines remaining, this excess free liquid on the granules causes sudden rapid coalescence of many granules, and results in the rapid increase in granule size characteristic of an induction formulation.

39.2.2.4 Granule Growth Regime Map for Deformable Granules It is well known that different formulations show different granulation behavior, such as induction versus steady growth. The links between formulation properties and granule growth behavior are summarized in the granule growth regime map [52, 54]. The horizontal axis indicates the granule saturation s, which is a function of the weight fraction of liquid w, and the granule porosity ε (see equation 39.13). On the vertical axis is the Stokes' deformation number, St_{def}, which is the ratio of the kinetic energy experienced by the granules during a collision compared to the yield stress or deformation of the granule (see equation 39.13).

To make effective use of the granulation regime map, we need reasonable estimates of the effective collision velocity U_c (controlled by process conditions) and dynamic yield stress Y (a function of formulation properties). Table 39.1 gives estimates of the average and maximum collision velocities for different process equipment. In high shear mixers, the difference between the average and maximum collision velocities can be very large and the estimates should be taken as indicative.

Let us consider the different growth regimes. First, we should consider a very weak formulation with a low granule yield strength Y_g. For example, coarse sand and water At the very beginning of the granulation process, the saturation s will be low and close to zero and the process in the "dry" free flowing powder section of the regime map, in the upper left-hand corner. As more fluid is added to the process, the saturation increases and we move from left to right across the map. However, in this case the impact forces from collisions are far exceeding the granule strength and the granules shatter as quickly as they are formed, forming a mixture reminiscent of damp sand. Increasing the amount of fluid does not help this, and eventually the system will end up as a slurry, in the upper right-hand corner of the map.

Considering now a formulation, which has a slightly higher granule yield stress Y_g, where granules deform rapidly, but are not shattered. For example, lactose granulated with

TABLE 39.1 Estimates of U_c for Different Granulation Processes

Type of Granulator	Average U_c	Maximum U_c
Fluidized beds	$6U_b d_p/d_b$	$6U_b d_p/d_b \delta^2$
Mixer granulators	$0.15\omega_i R$	$\omega_i R$

Modified from Ref. 58.

water or a low-viscosity binder such as PVP. The St_{def} for this type of formulation could be approximately 2/3 up the St_{def} axis (depending on the granulator conditions, which determine U_c). Initially, at low liquid amounts, the system will be in the nucleation only regime. In this section of the growth map, there is enough liquid to form nuclei but not enough to allow any significant granule growth—the system is water limited. As more fluid is added, the saturation increases and the process cross the regime boundary to enter the steady growth regime, where the granule size increases steadily with time. If we add still more fluid, the granule growth rate will increase further until we reach the rapid growth regime and eventually a slurry (Figure 39.16).

Finally, consider a formulation with a yield stress Y_g, which is able to resist the impact forces experienced in the granulator. A typical pharmaceutical example would be a formulation containing a very fine powder, granulated using a viscous binder fluid such as 5% HPC. After each collision, the granule consolidates slightly, and over time approaches a minimum porosity as shown in Figure 39.13. Initially, when only a small amount of fluid has been added, the saturation will be low and the process will be in the nucleation regime. As the amount of fluid increases, we cross into the induction regime. During the induction period, the granules densify and approach the minimum porosity. If the level of consolidation is enough for force the saturation to exceed 100% (see case (b) in Figure 39.14), the excess liquid that is squeezed to the surface of the granule will cause the granules to coalesce rapidly with the surrounding granules and extremely rapid granule growth occurs.

It is possible to switch the growth behavior from steady growth to induction growth or vice versa. For example, increasing the binder fluid viscosity or decreasing the particle size will slow the rate of consolidation, and move the system toward induction behavior.

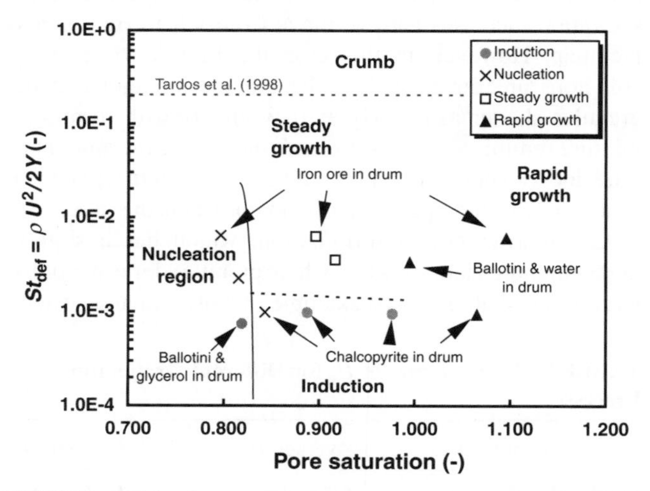

FIGURE 39.16 Granule growth regime map, summarizing the different types of granule growth and the conditions in which they will occur [52].

Pharmaceutical granulation mainly occurs in the nucleation, steady growth and induction regimes, as the final desired granule size is often only 2–4 times the size of the original particles in the formulation. A typical target granule diameter would be 200–400 μm, and the typical size of the particles in the formulation ranges between 50 and 200 μm.

39.2.2.5 Growth Regime Map for Nearly Elastic Granules

For *near elastic granules*, the conceptual model originally developed by Ennis et al. [59] considers the collision between two near elastic granules each coated with a layer of liquid (see Figure 39.17). This work has been summarized in several monographs [58, 60, 61]. In this case the key dimensionless group is the viscous Stokes number St_v

$$St_v = \frac{4\rho_g U_c d_p}{9\mu} \tag{39.17}$$

St_v is the ratio of the kinetic energy of the collision to the viscous dissipation in the liquid layer. Successful coalescence will occur, if St_v exceeds some critical value St^* and we can define three growth regimes, which are as follows:

1. *Noninertial Growth* $[St_{v,max} < St^*]$: The viscous Stokes number for all collisions in the granulator is less than the critical Stokes number. All collisions lead to sticking and growth by coalescence. In this regime, changes to process parameters will have little or no effect on the probability of coalescence.
2. *Inertial Growth* $[St_{v,av} \approx St^*]$: Some collisions cause coalescence while others lead to rebound. There will be steady granule growth by coalescence. The extent and rate of growth will be sensitive to process parameters that will determine the proportion of collisions that lead to coalescence. Varying process parameters and formulation properties can push the system into either the noninertial or coating regimes.
3. *Coating Regime* $[St_{v,min} > St^*]$: The kinetic energy in most or all collisions exceeds viscous dissipation in the

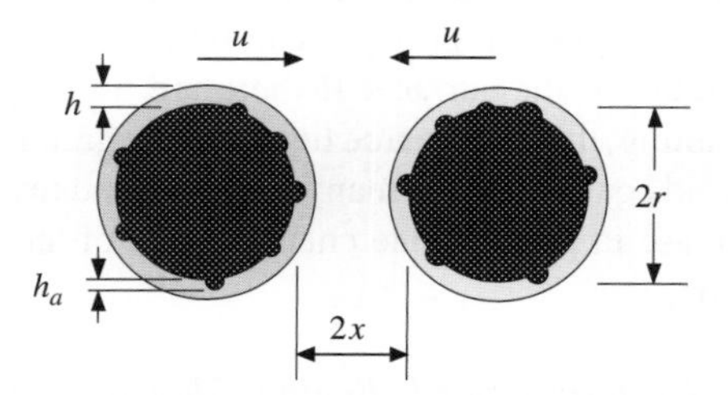

FIGURE 39.17 Two near elastic granules colliding—the basis for the coalescence/rebound criteria [9]. This simple model predicts that granules will grow to a maximum size by coalescence and such behavior is commonly seen in fluidized bed granulation. Note that the key formulation parameters are binder viscosity and particle size, as for deformable granule growth case.

liquid layer. There is no coalescence. Granule growth will only occur by the successive layering of new material in the liquid phase (melt, solution, or slurry) onto the granule.

This simple model predicts that granules will grow to a maximum size by coalescence and such behavior is commonly seen in fluidized bed granulation. Note that the key formulation parameters are binder viscosity and particle size, as for deformable granule growth case.

39.2.3 Breakage

Of the three granulation mechanisms, granule breakage is the least understood. There is no fully general regime map for breakage, although there is an active research effort to develop one. A thorough review of breakage research was published recently [62]. Breakage of wet granules is only important in mixer granulators. Attrition and breakage of dry granules can also occur during fluidized bed drying (and during fluid bed granulation) or during later handling.

The two main approaches to understanding wet granule breakage are to conduct breakage studies of single granules (both wet and dry), or to conduct studies of breakage during granulation within the granulator and granular motion in granulators (including the forces and velocities experienced by the granules). It is important to understand how an individual granule will deform and break under certain conditions. Therefore single granules studies, both experimental and theoretical, can be very useful. Breakage during granulation is usually studied by either analyzing the change in granule size distribution with time, or by using colored tracers.

Granule breakage is a function of the strength of the granule compared to the impact velocity and shear forces experienced by the granules within the granulator. Granules will respond differently under different conditions, and a given granule may break very differently, or not at all. Therefore, granular flow has a large impact on the breakage behavior in the granulating system, but is relatively poorly understood.

It is important to note that the breakage behavior of dry granules is completely different to the breakage behavior of wet granules. We discuss here the breakage of wet granules during the granulation process, which may also be applicable to a wet milling process, but cannot be extrapolated to a dry milling process or any other dry granule breakage process. Dry granule breakage is discussed in detail elsewhere [58].

39.2.3.1 Deformation and Breakage of Single Granules

Deformation and breakage of single wet granules has been studied with high velocity impacts [63], in controlled powder shear [64] and in unconfined compression at varying strain rates [65]. Tardos and coworkers were one of the first groups to study breakage of single wet granules [64]. They performed experiments with individual pellets in a fluidized coquette device, where the shear field applied to the granules was known and carefully controlled. Breakage occurs when St_{def} is greater than some critical value St_{def}^* of order 0.2. Here, the granule mechanics were modeled as a Herschel Bulkley fluid and the granule strength is taken as the plastic yield stress under shear. Smith and coworkers [65, 66] extended Iveson's work on granule strength under uniaxial and diametrical compression [35, 37, 38, 67] and showed that in fact the deformation and breakage behavior of single granules is complex. Some formulations show very plastic behavior, while other fail in semibrittle fashion, with propagation of single large cracks through the granule (see Figure 39.18). It is likely that plastic granules may fail

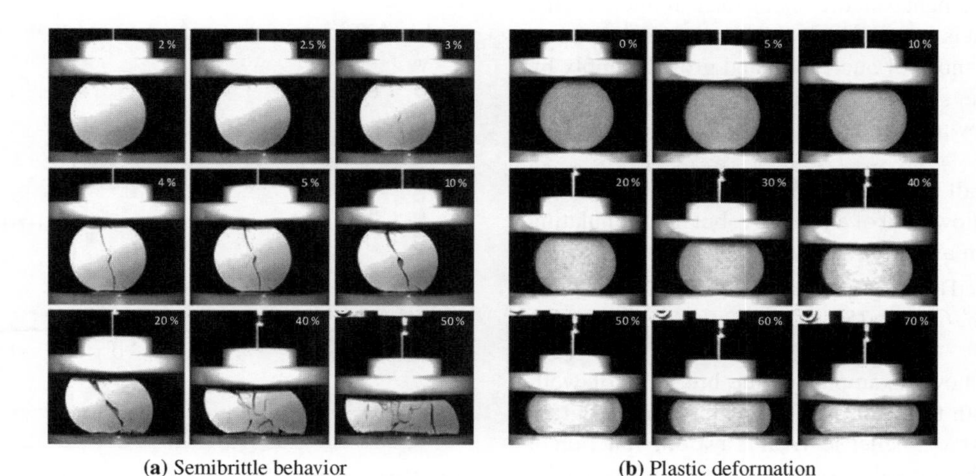

(a) Semibrittle behavior (b) Plastic deformation

FIGURE 39.18 Extremes of wet granule deformation and breakage behavior in diametrical compression [66].

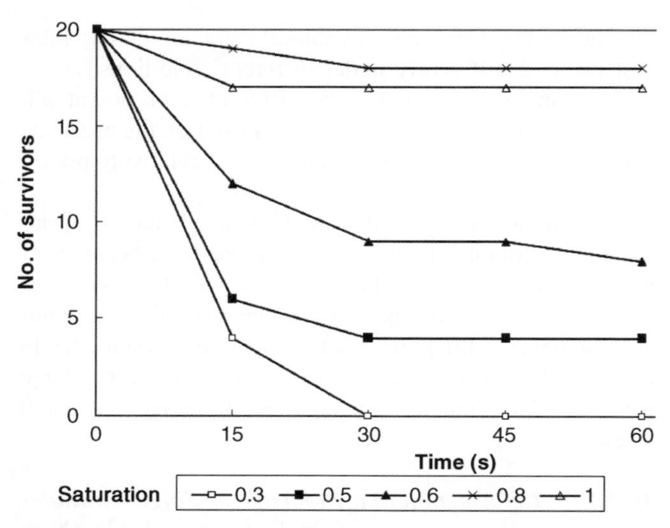

FIGURE 39.19 Breakage of wet granule pellets in a flow field of cohesive sand in a vertical axis mixer granulator (broad size distribution lactose with 1 Pa S silicone oil binder at different liquid saturation values) [70].

by shear and extensional flow in the granulator, while semi-brittle granules may break by impact in the impeller zone.

While plastic behavior is more likely to be seen at higher strain rate, there is no general relationship between capillary number Ca and breakage mode. However, granules made from nonspherical particles are both much stronger, and more likely to fail in semibrittle fashion by major crack propagation.

39.2.3.2 Breakage of Granules within the Granulator: Effect of Formulation Properties and Process Parameters

Wet granule breakage has only been identified as a significant mechanism in high shear granulators [26, 62, 65, 68–73]. The problem with fundamental studies using measurements in the granulator is that it is very difficult to deconvolute the effects of breakage from those of nucleation and growth simply by analyzing granule size distributions. To get around this problem, the following two approaches have been taken:

1. Using a small population of well-formed granules or pellets in a flow field of a cohesive, but nongranulating powder such as sticky sand [65, 69, 70] or
2. Using marked tracer granules in a flow field of the same material [26, 68, 72–75].

Figure 39.19 shows examples of the breakage of well-formed pellets with time within a vertical axis mixer granulator [70]. In this case, pellet survival is a strong function of liquid saturation. In general, strong granules (as measured by peak flow stress in uniaxial compression as described above) do not break as readily in the granulator. We can postulate that in the granulator, granules will break if they experience stresses, which exceed their peak flow stress. We can express this as a Stokes deformation number criterion. Breakage will occur if the Stokes deformation number on impact exceeds a critical value [70]

$$St_{\mathrm{def}} = \frac{\rho_{\mathrm{g}} v_{\mathrm{c}}^2}{2\sigma_{\mathrm{p}}} > St^* \tag{39.18}$$

This approach is similar to that used by Tardos and coworkers in the coquette flow rheometer [64]. Figure 39.20 shows that treating breakage data in this way leads to a surprisingly sharp transition from no breakage to breakage at $St^* = 0.2$ for a wide range of formulations.

In similar studies, van den Dries et al. [68] proposed a critical value of $St^* = 0.01$. Closer inspection of the analysis shows that differences in value for St^* are most likely due to differences in mixer geometry and in measuring or estimating the granule strength σ_{p} and the collision velocity v_{c}. In their work, Liu et al. [70] combined data from single granule strength measurements [65] with Rumpf's expression for granule strength where both viscous and capillary forces are important, to develop the following expression for granule strength

$$\sigma_{\mathrm{p}} = AR^{-4.3} S \left[6 \frac{1-\varepsilon}{\varepsilon} \frac{\gamma \cos\theta}{d_{3,2}} + \frac{9}{8} \frac{(1-\varepsilon)^2}{\varepsilon^2} \frac{9\pi\mu v_{\mathrm{p}}}{16 d_{3,2}} \right] \tag{39.19}$$

where AR is the aspect ratio of the primary particles, S is the granule pore saturation, ε is the porosity of the granule, $d_{3,2}$ is the specific surface area diameter of the particles, and v_{p} is the relative velocity of the moving particle inside a granule after impact. The expression explicitly accounts for the effects of primary particle size and shape, liquid binder properties, and liquid saturation. The collision velocity was assumed to be

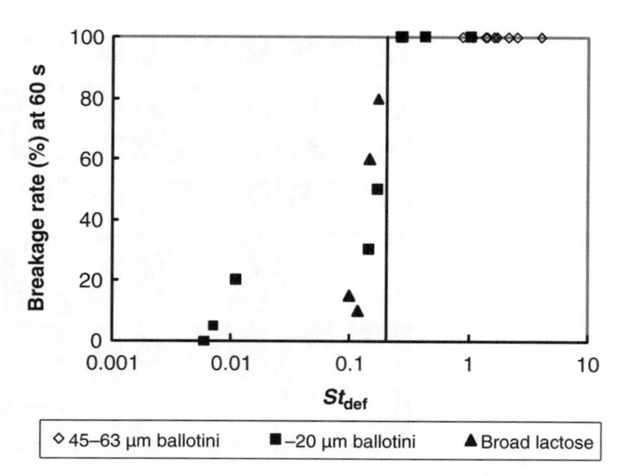

FIGURE 39.20 The extent of breakage at steady state versus Stokes deformation number St_{def} for all the formulations. The vertical line is the St_{def} value of 0.2 [70].

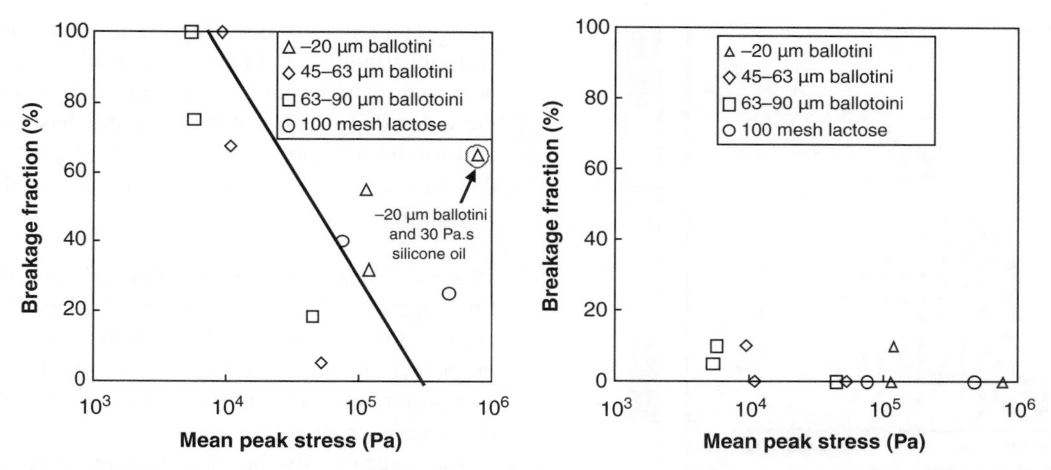

FIGURE 39.21 Breakage fraction of drop formed granules as a function of granule mean peak stress (a) 11° beveled blade at 500 rpm; (b) frictional flat plate impeller at 500 rpm. Modified from [69].

15% of the impeller tip speed. Note this correlation is similar to, although not identical to the Iveson correlation for granule strength (see Section 39.2.2.1). Both predict granule strength increases with increasing strain rate, increasing binder viscosity, and decreasing primary particle size.

Note that the mode of breakage as well as the granule strength may vary with formulation properties. Analysis of granule fragments from breakage shows evidence of significant elongation for some formulations (plastic deformation) and sharp angular fragments for others (semibrittle behavior) [69]. Breakage behavior is similar to that predicted from diametrical compression of single granules.

While this relatively simple approach is fairly general but able to predict the effect of formulation parameters, predicting the effect of changes in equipment parameters is more difficult. In general, increasing impeller speed increases the extent of breakage [26, 71, 73] and this is accounted for by the collision velocity term in the Stokes deformation number. However, breakage is very sensitive to changes in mixer and impeller geometry in ways we cannot yet predict *a priori*. Figure 39.21 shows an example of breakage of single drop formed granules in the same mixer with (a) a two-blade beveled edge impeller and (b) a frictional flat plate at the same impeller speed. For the beveled impeller, there is a reasonable correlation between breakage and granule strength consistent with equation 39.19. (The degree of scatter is due to the difficulty in keeping granule porosity and liquid saturation constant with this method of granule formation.) In contrast, there is *no* breakage using the frictional flat plate.

These results emphasize that breakage is not occurring uniformly in the bed, but rather in a narrow zone near the impeller. This is the reason that granules take time to break in the granulator. Figure 39.19 shows that the pellet granules take up to 1 min to break, even in a small granulator. Figure 39.22 shows an example of velocity fields in roping flow in a two-blade impeller vertical axis mixer. Powder velocity (and therefore applied stress) is very nonuniform with the highest velocities in a small zone near the impeller. As granules circulate in roping flow in the granulator, they will often bypass the impeller zone. Only those granules that enter this zone of high impact and shear stress are likely to break. The size of this breakage zone, and the maximum stresses and collision velocities seen by the granules in this zone will be a strong function of the impeller design.

In mixer granulators, a "chopper" mounted either in the side wall or in the granulator lid, rotates at high speed (e.g., 3000 rpm). This generates a small localized area with very intensive agitation, where breakage could be expected to dominate. The chopper is commonly thought to break up large lumps and granules, particularly at the powder surface (where they are generally located). Although this seems logical, there is no work to demonstrate this—the few pharmaceutical studies that have been performed [48, 77–79] find that the chopper had a very small effect, and that the overall granulation response (e.g., granule size and porosity) is dominated by the impeller speed, liquid level, etc.

39.2.3.3 *Breakage of Granules within the Granulator: Effect of Granule Size and Density* There are several studies, which show that the breakage probability is proportional to granule size [70, 72, 80]. These studies are consistent with the literature on particle size effects on crushing and grinding. Larger granules are more likely to have large pores or flaws which increases their probability of semibrittle fracture [80]. Consolidated, dense granules are well known to be stronger. An analogy between the granule growth regime map and breakage behavior has been postulated, where three "exchange mechanisms" have been proposed

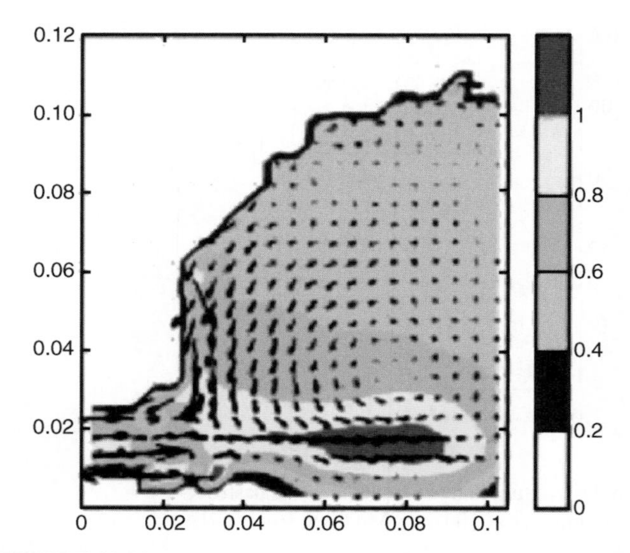

FIGURE 39.22 Powder velocity field during pseudoroping flow in as two-blade impeller vertical axis mixer measured using positron emission particle tracking [76].

to resemble three growth regimes, and linked to the type and rate of material transfer between granules [75].

Colored tracer granules have been used to follow breakage as a function of granule size. Three sizes of colored tracer granules (~200, 500, and 1000 μm) were added to a granulator while it was running. Samples were taken as a function of time, and the proportion of colored material in each size fraction was analyzed. The largest granules break at a higher rate than smaller granules, as shown in Figure 39.23 [72]. They also investigated the breakage rate of tracer granules that had been granulated for different times, before being removed and added to the running granulator. Younger, newly formed tracer granules broke at a faster rate than tracer granules that had been granulated for a longer period,

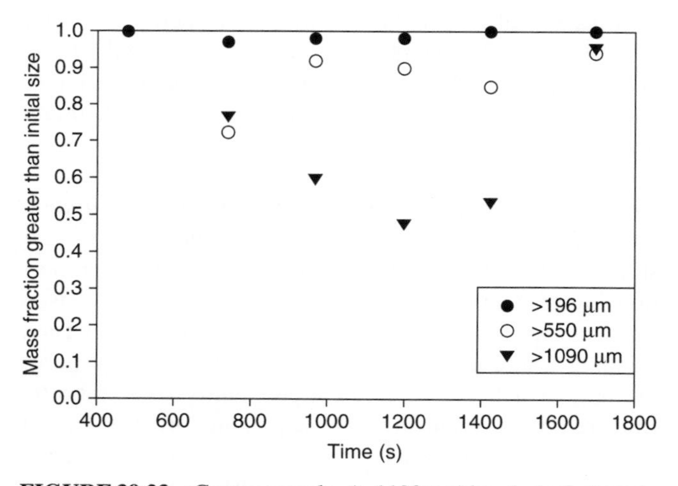

FIGURE 39.23 Coarse granules (>1190 μm) break at a faster rate than smaller granules [72].

allowing plenty of time for granule consolidation [72]. Granule strength, and hence granule breakage rates, have been shown to be quite heterogenous [74]. In some cases, the color distribution of the granules has become almost uniform within approximately 2 min [26, 68, 72], although this will not be true for all formulations or all granulation conditions.

39.2.3.4 Aiding Controlled Granulation via Breakage Breakage of wet granules during the granulation process is not necessarily detrimental. In pharmaceutical granulation, the formation of large granules is generally undesired, and size-preferential breakage of coarse granules [72] helps keep the proportion of coarse granules low.

Breakage can also occur early in the granulation process, in parallel with nucleation, and can assist in distributing the liquid evenly throughout the powder. The mechanical dispersion nucleation regime (see Section 39.2.1.3) requires breakage to disperse the wet clumps of binder fluid through the powder. Newly formed granules (i.e., nuclei) are easier to break than older granules, due to their relatively high porosity [72]. This mechanism of liquid dispersion via breakage of nuclei is called "destructive nucleation" [26] (Figure 39.24). The initial interaction of the drop either in a fluid bed (via layering of the powder on the exterior of the drop) or during the drop penetration process (more relevant to mixer granulation) forms a primary nuclei with a saturation gradient—the saturation decreases as you move from the inner core of the granule to the exterior surface. The large, low porosity, weak primary nuclei is broken into smaller and stronger secondary nuclei, which form starting materials for coalescence [26]. Tracer studies showed that the proportion of primary nuclei that survive decreases as the impeller speed increases [26, 68], and consequently the colored tracer fluid became more uniformly distributed both between granules and within granules. This effect may be smaller for other formulations and equipment (Figure 39.23).

The final stage in destructive nucleation shows a balance between coalescence and breakage, which implies a stable maximum granule size. This idea was applied, together with several other ideas, to produce a well controlled granulation which required only a single parameter—the impeller speed—to be adjusted on scale-up [81]. Note that because the kinetics of breakage in time scale to a typical pharmaceutical granulation (1–5 min), this "steady-state" approach will require longer granulation times for stable results. This work is described in more details in Section 39.3.4.

39.2.3.5 Summary Comments While much is still to be done in the area of wet granule breakage, we can draw some useful conclusions:

1. Approaches used to measure or estimate the effect of formulation properties on granule strength for the

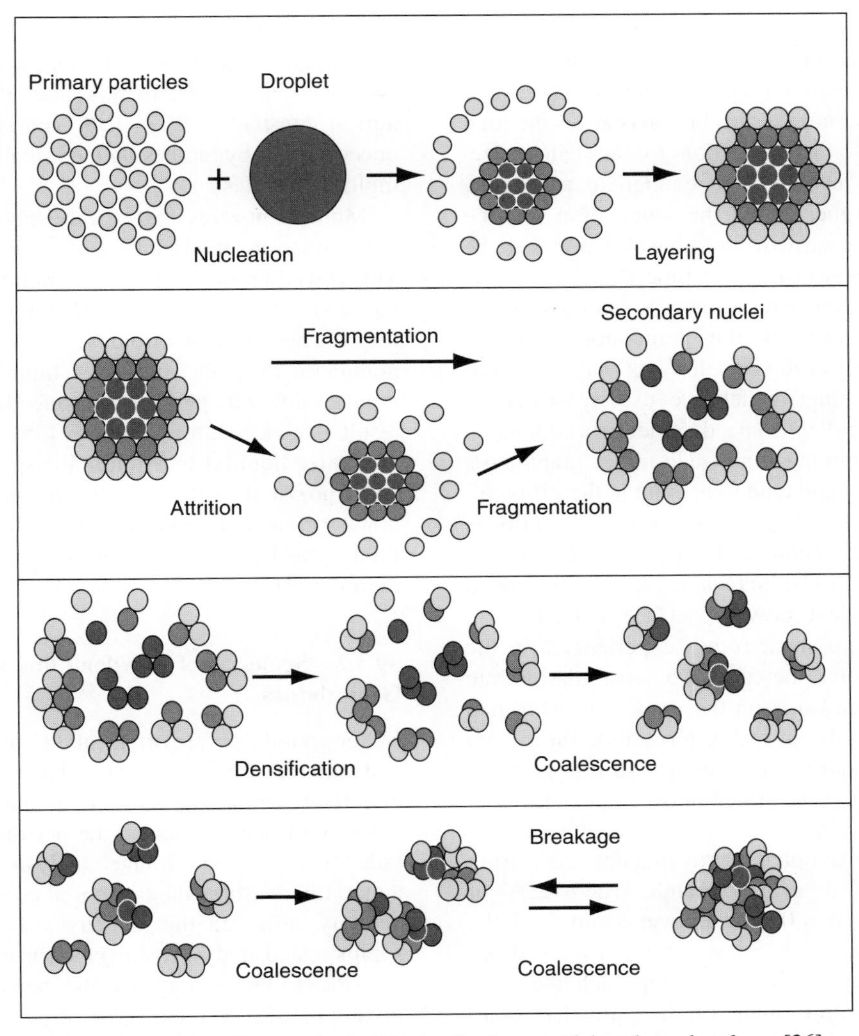

FIGURE 39.24 "Destructive nucleation" where nuclei undergo breakage [26].

purposes of understanding granule growth are also applicable to breakage studies. Remember, however, that changing the formulation can change the mode of breakage (plastic of semibrittle) as well as the yield stress.

2. In the granulator, granules will break due to some combination of shear and impact in a relatively small region near the impellers. Granulator and impeller geometry have a very strong, but difficult to predict effect on the rate and extent of breakage.
3. A simple Stokes deformation number criterion (equation 39.18), can be used to predict whether breakage will occur in a granulator of well defined geometry and to predict the effect of formulation effects and impeller speed on wet granule breakage in the granulator.
4. Granule breakage will occur more easily for large, low-density granules.
5. Breakage can sometimes have a positive effect in wet granulation to aid liquid distribution and limit maximum granule size.

39.3 SCALE-UP

39.3.1 Spray Flux Scale-Up

Dimensionless spray flux provides a good basis for scale up of the spray zone conditions to maintain good nucleation and equivalent liquid distribution conditions. However, maintaining constant spray flux on scale-up can be quite challenging (see also discussion in Ref. 50). The most common scale-up approach is to maintain the same solution addition time, which means that the flow rate through the nozzle increases and therefore spray flux also increases dramatically. The larger spray width and larger drop sizes at higher flow rates are not usually enough to compensate for this. An alternate approach of maintaining the same flow rate (and drop size, if an identical nozzle tip), but this causes the total batch time to increase in proportion to the batch size. Assuming that liquid level is high enough to produce a granule saturation higher than the nucleation regime limit [52, 54], then the growth and consolidation will be affected due to the longer processing time.

In addition, the powder surface velocity in a mixer granulator generally decreases as the granulator scale increases [21], particularly if the impeller has only two fixed "high" and "low" settings which are usually scaled to maintain equivalent impeller tip speed, rather than equivalent powder agitation and mixing. Lab scale mixer granulators display vigorous mixing and tend to operate in the "roping" regime, while at manufacturing scale the powder agitation is much less efficient, and spends at least part of the time in "bumping" flow [11, 21, 33]. Increasing the impeller speed to maintain an equivalent powder surface velocity also increases the impact and shear forces experienced by the granules. The higher force increases the consolidation rate and shifts the system upwards on the granule growth map (i.e., St_{def} increases). If this upward shift results in the system crossing a regime boundary, a fundamental shift in granulation behavior will occur as the system moves into a different granulation regime.

Table 39.2 gives an example of spray flux changes during scale-up from a 10 kg to a 50 kg batch, where 25% of granulating fluid is added at 0.5 L/min over 5 min. Initially, the calculated spray flux $\Psi_a = 0.36$, which is above the drop-controlled nucleation regime. In the first approach, the spray rate increases to 2.5 L/min to maintain the equivalent granulation time of 5 min and the impeller speed of 108 rpm is scaled using the common "tip speed" scaling rule [21]. The decrease in powder velocity and the increase in spray rate creates an increase in spray flux to $\Psi_a = 1.08$. In the second case, the spray rate is maintained at 0.5 L/min but the spray delivery time is increased from 5 min to 25 min. This is not commonly done, due to fears of significant changes in granule growth and consolidation, which may be unfounded (see Section 39.3.4). In this case, the spray flux remains almost constant at $\Psi_a = 0.38$. The spray flux could be reduced further by increasing the impeller speed, or by adding multiple nozzles.

TABLE 39.2 The Effect of Scale-Up on Spray Flux Ψ_a in Fielder Mixer Granulator

Scale-Up Approach	Base Case	Constant Spray Time	Constant Spray Rate
Batch size (kg)	10	50	50
Flow rate (L/min)	0.5	2.5	0.5
Spray time (min)	5	5	25
Drop size (μm)	200	350	200
Spray width (m)	0.25	0.3	0.3
Impeller speed (rpm)	216	108	108
Powder velocity (m/s)	0.7	0.55	0.55
Spray flux, Ψ_a	0.36	1.08	0.38

Multiple nozzles are the only way to maintain spray flux independently of granule growth and consolidation rates. This is well known in fluid bed granulation scale-up [82–84] but has not been applied to commercial mixer granulators. A four nozzle spray manifold was designed for a 300 L mixer granulator [85]. The increased liquid distribution area and reduced flow rate per nozzle reduced the spray flux and can result in reduced lump formation [85]. An alternate method to maximize liquid dispersion in the spray zone is to place the spray nozzle directly over the chopper, where the turbulent powder flow and strong localized shear forces disperse the fluid effectively, even in the mechanical dispersion regime [34].

39.3.2 Scale-Up of Powder Flow Patterns in Mixer Granulators

Mixer granulators are often called "high shear" granulators, and until recently it was assumed that the impeller was able to effectively agitate the powder bed during operation. Powder velocity measurements can be performed using high-speed video cameras and image analysis [11, 21, 22], which generally measures the tangential component of the powder velocity, although the velocity also varies radially. More sophisticated analysis using positron emission particle tracking (PEPT) technology has also been performed in several mixers [31, 86–89] and shows these trends in much greater detail although the data currently available is limited to lab-scale granulators. Experimental PEPT data for powder flow in larger scale equipment is currently being generated.

Two distinct types of powder flow have been observed in high shear mixer granulators [11]. At low impeller speeds, the powder surface remains horizontal and the bed "bumps" or "shunts" [90] up and down as the impeller passed underneath. The surface velocity was approximately an order of magnitude lower than the impeller tip velocity. As the impeller speed increases, the powder surface velocity increases linearly although there was still little vertical interchange of material. After increasing the impeller speeds above a critical point, a vortex appears and spiraling "roping" [11] or "torroidal" powder flow is observed [73]. In the roping regime, the surface velocity is independent of impeller speed [11, 73] and material from the bottom of the powder bed is forced up the vessel wall before tumbling down the vortex in the center of the powder flow. In all observed powder flows in mixer granulators, the powder velocity is at least one order of magnitude lower than the impeller tip speed [9, 21, 73, 88].

An industrial study of powder flow patterns and surface velocity was performed in a series of Fielder mixers (25, 65, and 300 L) as a function of impeller speed [21]. When running the 300 L granulator at the standard "low" speed setting of 180 rpm, the powder was stagnant approximately 1/3 of the time [21]. Changes in mixer geometry and fill level could significantly change the powder velocity at a given impeller speed. They also measured powder velocity during granulation, and found that the surface velocity gradually increases as granulation proceeded. The powder flow pattern also changed during granulation, shifting from bumping flow during dry mixing to roping flow during granulation [21]. Powder velocity during the dry mix stage has also been shown to vary between lots of API for a high drug load formulation [22]. The measured powder surface velocity for three batches containing different lots of the API varied between 0.64 and 0.95 m/s. This variation was presumed to be due to lot-to-lot differences in drug properties, although establishing the causal link between drug properties and powder flow is an area requiring further investigation.

In fluidized beds, the fluidizing airflow is always adjusted to maintain adequate fluidization of the powders. It would be unthinkable to attempt to scale-up a fluidized bed and select a set of operating conditions which did not fully fluidized the bed—yet we routinely scale-up mixer granulators in exactly this way. There are three main approaches to scaling powder flow in a mixer granulator which can be summarized by the following equation [33]:

$$ND^n = \text{constant} \qquad (39.20)$$

where N is the impeller speed (rpm), D is the impeller diameter (m), and n is a scaling index. The most common impeller scaling approach is to maintain tip speed, where $n = 0.5$ [47, 91–93]. An alternate approach is to use Froude number ($\mathrm{Fr} = N^2D/g$), which is commonly used to scale up fluid mixing by maintaining the ratio of centrifugal to gravitational forces [11, 91, 94, 95]. In this case, the scaling index in equation 39.13 is $n = 1$. More recent work used calibrated tracer pellets with a known yield stress to measure the average shear stress experienced by a granule during granulation. Scale-up studies showed that for a series of geometrically similar Fielder granulators, the scaling index n varied between 0.8 and 0.85 [33] depending on the height to diameter ratio, fill level and impeller style used in each case. The "equal shear" scale-up criterion is the subject of ongoing research [96, 97].

In addition to the impeller speed criteria outlined above, other scaling criteria include maintaining swept volume [79], constant energy per unit mass [98], and power number [99].

39.3.3 Granule Growth Scale-Up

It is important to stay within the same granule growth regime (see Figure 39.15) during scale-up to avoid dramatic changes in growth behavior. This is unfortunately easier said than done. The forces applied to the powder mass must remain similar; otherwise the system may shift vertically on the regime map, typically into or out of the induction growth regime. For fluid beds this implies maintaining similar fluidizing conditions and maintaining an equivalent excess gas velocity. For mixer granulators, the maximum impact (e.g., impeller and/or chopper tip speed) may need to be maintained if the granulation is controlled by direct impeller impacts, or perhaps the overall roping flow field to maintain equivalent shear. In some cases, these requirements may directly contradict the requirements needed to maintain equivalent nucleation conditions.

Shifting vertically on the regime map due to a change in the overall force applied to the granulation also implies that the granule porosity and/or structure will also shift, and this is usually undesirable as granule porosity is often shown to be directly linked to the dissolution rate of the granules, capsules, or tablets [42, 100]. Changing the porosity also changes the overall saturation of the granules, which also means that a change in the granule porosity moves the process both vertically and horizontally on the growth regime map. A change in porosity ε and the subsequent change in granule saturation s means that the granule size will change, even though the amount of liquid added to the batch (w in equation 39.11) remains constant. Typically, pharmaceutical process engineers concentrate on keeping the amount of liquid added to the batch constant. In actual fact, the granule saturation is the key factor in controlling granule growth but in the future we hope to see saturation being calculated at each stage of the scale-up process. Currently, the lack of knowledge of porosity changes meaning that the amount of liquid added to the batch is frequently adjusted from compensate for the changes in growth after scale-up.

39.3.4 Scale-Up Case Study: Steady-State Granulation

"Pseudosteady-state granulation" (also called one-dimensional granulation) is a recent approach to wet granulation which resulted in improved control of granule size, properties, and scale-up without any loss in product performance [81]. During a typical 5–25 min pharmaceutical granulation process, there are multiple dynamic subprocesses occurring including liquid distribution; dissolution and hydration of excipients such as lactose, MCC, and dry binders; granule growth; granule consolidation; granule breakage; and the overall granulator flow pattern and shear (particularly the transition between bumping and roping flow). Generally when the batch is stopped, each of these subprocesses is stopped abruptly, well before an equilibrium is reached. Each subprocess has its own characteristic timescale, and it is *impossible* to halt in *all* of these dynamic subprocesses at the *same* point at each scale, and this is why the granulation performance shifts as during scale-up.

An ideal granulation process would allow complete control of liquid sizes in the granulator, without the need for dry milling [81]. The ideal process would allow all the transient subprocesses to reach a repeatable, controllable equilibrium end point, and produce a narrow size distribution of granules between 200 and 500 μm with complete control of the granule size by adjusting only the liquid level. This would involve operating at a low spray flux by using a slow spray rate; a long granulation time to ensure that all complete dissolution/hydration of all the excipients; ensuring roping flow behavior during the entire process; and using some wet massing time to ensure that an equilibrium can be achieved. Vonk et al. [26] showed that an equilibrium granule size and saturation should exist where the rate of granule growth is exactly balanced by the rate of granule breakage (see lower diagram in Figure 39.18). As the entire granulation batch converges toward equilibrium point, the granule size distribution and saturation distribution will also converge. This "steady-state" granulation point should also be scale-independent [81].

To demonstrate steady-state granulation, Michaels et al. [81] granulated a standard lactose-MCC based formulation in a 2 L high shear mixer granulators using "conventional" conditions (40% fluid sprayed over 5 min) and "steady-state" conditions (28% fluid sprayed over 15 min plus up to 20 min wet massing time). The long granulation time caused the initial batches to heat up, creating new transients in evaporation and rheology, so the standard mixer cooling jackets were used to minimize the temperature rise in the batch. The granules produced by the steady-state process were typically 200–300 μm with a narrow size distribution with no granules larger than 1 mm. The final particle size distribution was a function only of the final liquid saturation and shear stress in the agitated wet mass. The granulating fluid level (ratio of liquid added to dry powder ingredients) became a material variable rather than a process variable, that is, scale-independent. In contrast, the conventional granules had a very broad distribution with more than a third of the granules above 1 mm, thus requiring the use of a dry mill.

Scale-up of the steady-state granulation process involves only one process variable: scale-up of the shear stress, controlled by the main impeller speed. Scale-up trials were conducted at 2, 25, and 300 L scale, using the "equal shear" scale-up correlation ($n = 0.8$ in equation 39.15) between shear stress and main impeller speed [33]. The granule size distributions were unimodal and centered at 200–300 μm depending on the liquid level, and the need for a milling step to control the granule size was eliminated [81]. The entire size distribution was matched "right first time" with only a single batch performed at 25 and 300 L (see Figure 39.25). This exact matching of the entire size distribution was repeated multiple times, over three different liquid levels (24%, 28%, 32%) and at two different impeller speed (shown as low, medium, and high shear on the x-axis) at each scale. Although the total granulation process time was far longer than normal (15 min solution delivery plus 20 min wet massing time), the lower liquid level (~28% compared to 45% for the same formulation granulated using a standard approach) and the small, uniformly size granules meant that the drying time was significantly shorter, resulting in the same

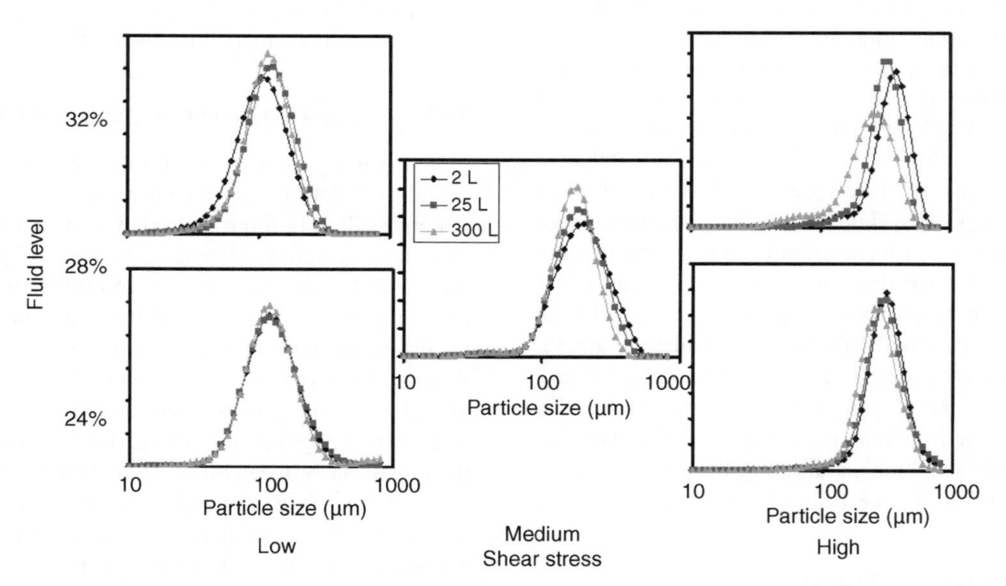

FIGURE 39.25 Particle size of dry, unmilled granules manufactured at 2, 25, and 300 L scale. Data shown for three fluid levels (24%, 28%, and 32%) and three shear stresses (low, medium, and high impeller speeds).

overall cycle time at production scale. In addition, the dissolution and compaction of these granules was unaffected, contrary to conventional opinion that such dense granules would fail to meet physical tablet specifications and drug release specifications. This highlights the surprisingly simple opportunity of using the steady-state granulation approach to scale up pharmaceutical formulations. The approach has also been successfully used for granulation of detergents in a fluid bed granulator [101].

39.4 FUTURE DIRECTIONS

There are several areas where the existing knowledge of granulation is currently insufficient—powder flow including the shear rates and collision rates experienced by the granules is one area, and a better understanding of the granule breakage mechanism is another. Both of these areas are the focus of several international groups. Areas particularly relevant to the pharmaceutical industry are the granulation behavior of multicomponent powder blends, and understanding how to obtain the most uniform drug distribution possible across all granule size fractions. The behavior of hydrophobic drugs in granulation can also be quite surprising [46, 102] and warrant further effort.

In future, the granulation process—that is, operating conditions for a given formulation to produce granules with a prespecified set of properties—will be designed entirely in advance using dimensionless groups and regime maps. Selection of robust process conditions based solely on theoretical considerations has already been demonstrated [81]. The existing knowledge provides valuable guidance for trouble-shooting process problem and estimating the process risk of atypical events.

Pharmaceutical granulation is also evolving—continuous granulation is now under serious development [103, 104] and is clearly a strong future direction of pharmaceutical granulation. Continuous granulation will also require a new effort to develop online process control and analytical technology—particularly for granule size and porosity. The technology for online and at-line granule size measurement already exists but is currently used only sparingly [81, 105], in part due to industry conservatism about applying a novel particle size measurement. The development of real process control will most likely also impact on more traditional batch granulation. Other improvements, including the use of foam to distribute the binder fluid instead of a spray [106–108] are also likely to expand the range of process options available during pharmaceutical wet granulation. The new engineering-focused approach to pharmaceutical granulation is leading to vast leaps in our understanding of the granulation mechanisms and the future direction is the gradual merging of granulation science and industrial know-how.

REFERENCES

1. Iveson SM, Litster JD, Hapgood KP, Ennis BJ. Nucleation, growth and breakage phenomena in agitated wet granulation processes: a review. *Powder Technol.* 2001;117(1–2): 3–39.
2. Knight PC, Instone T, Pearson JMK, Hounslow MJ. An investigation into the kinetics of liquid distribution and growth in high shear mixer agglomeration. *Powder Technol.* 1998;97:246–257.
3. Schæfer T, Mathiesen C. Melt pelletization in a high shear mixer IX. Effects of binder particle size. *Int. J. Pharmaceut.* 1996;139:139–148.
4. Schæfer T, Mathiesen C. Melt pelletization in a high shear mixer VIII. Effects of binder viscosity. *Int. J. Pharmaceut.* 1996;139:125–128.
5. Scott AC, Hounslow MJ, Instone T. Direct evidence of heterogeneity during high-shear granulation. *Powder Technol.* 1999;113:215–213.
6. Chouk V, Reynolds G, Hounslow M, Salman A. Single drop behaviour in a high shear granulator. *Powder Technol.* 2009;189(2):357–364.
7. Schaafsma SH, Vonk P, Kossen NWF, Hoffmann AC. A model for the spray zone in early-stage fluidized bed granulation. *AIChE J.* 2006;52(8):2736–2741.
8. Hapgood KP, Litster JD, Biggs SR, Howes T. Drop penetration into porous powder beds. *J. Colloid Interface Sci.* 2002; 253(2):353–366.
9. Litster JD, Hapgood KP, Michaels JN, Sims A, Roberts M, Kameneni SK, Hsu T. Liquid distribution in wet granulation: dimensionless spray flux. *Powder Technol.* 2001; 114(1–3):32–39.
10. Hapgood KP, Litster JD, Smith R. Nucleation regime map for liquid bound granules. *AIChE J.* 2003;49(2):350–361.
11. Litster JD, Hapgood KP, Michaels JN, Sims A, Roberts M, Kameneni SK. Scale-up of mixer granulators for effective liquid distribution. *Powder Technol.* 2002; 124(3):272–280.
12. Hapgood KP, Litster JD, White ET, Mort PR, Jones DG. Dimensionless spray flux in wet granulation: Monte-Carlo simulations and experimental validation. *Powder Technol.* 2004;141(1–2):20–30.
13. Wildeboer WJ, Litster JD, Cameron IT. Modelling nucleation in wet granulation. *Chem. Eng. Sci.* 2005;60:3751–3761.
14. Hapgood KP, Nguyen TH, Hauw S, Iveson SM, Shen W. Rewetting effects and droplet motion on partially wetted powder surfaces. *AIChE J.* 2009;55(6):1402–1415.
15. Denesuk M, Smith GL, Zelinski BJJ, Kreidl NJ, Uhlmann DR. Capillary penetration of liquid droplets into porous materials. *J. Colloid Interface Sci.* 1993;158:114–120.
16. Middleman S. *Modelling Axisymmetric Flows: Dynamics of Films, Jets, and Drops*, Chapter 8, Academic Press, San Diego, 1995, p. 299.
17. Ax K, Feise H, Sochon R, Hounslow M, Salman A. Influence of liquid binder dispersion on agglomeration in an intensive mixer. *Powder Technol.* 2008;179(3):190–194.

18. Hapgood KP. *Nucleation and Binder Dispersion in Wet Granulation*, PhD thesis, Department of Chemical Engineering, University of Queensland, Brisbane, Australia, 2000.
19. Schaafsma SH, Vonk P, Segers P, Kossen NWF. Description of agglomerate growth. *Powder Technol.* 1998;97:183–190.
20. Waldie B. Growth mechanism and the dependence of granule size on drop size in fluidised bed granulation. *Chem. Eng. Sci.* 1991;46(11):2781–2785.
21. Plank R, Diehl B, Grinstead H, Zega J. Quantifying liquid coverage and powder flux in high-shear granulators. *Powder Technol.* 2003;134(3):223–234.
22. Hapgood KP, Amelia R, Zaman MB, Merrett BK, Leslie P. Improving Liquid Distribution by Reducing Dimensionless Spray Flux in Wet Granulation—A Pharmaceutical Manufacturing Case Study. In: *9th International Symposium on Agglomeration*, Sheffield University, Sheffield, UK, 2009.
23. Wildeboer WJ, Koppendraaier E, Litster JD, Howes T, Meesters G. A novel nucleation apparatus for regime separated granulation. *Powder Technol.* 2007;171(2):96–105.
24. Wauters PAL, Jakobsen RB, Litster JD, Meesters GMH, Scarlett B. Liquid distribution as a means to describing the granule growth mechanism. *Powder Technol.* 2002;123:166–177.
25. Schaafsma SH, Kossen NWF, Mos MT, Blauw L, Hoffman AC. Effects and control of humidity and particle mixing in fluid-bed granulation. *AIChE J.* 1999;45(6):1202–1210.
26. Vonk P, Guillaume CPF, Ramaker JS, Vromans H, Kossen NWF. Growth mechanisms of high-shear pelletisation. *Int. J. Pharmaceut.* 1997;157:93–102.
27. Hapgood KP, Tan MXL, Chow DWY. Predicting nuclei size distributions in wet granulation using dimensionless spray flux. *Adv. Powder Technol.* 2009; 20(4): p. 293–297.
28. Marshall WRJ. Atomization and spray drying. *Chem. Eng. Progress,* Monograph Series 1954;50(2).
29. Stewart RL, Bridgwater J, Parker DJ. Granular flow over a flat-bladed stirrer. *Chem. Eng. Sci.* 2001;56:4257–4271.
30. Laurent BFC. Scaling factors in granular flow: analysis of experimental and simulations results. *Chem. Eng. Sci.* 2006;61(13):4138–4146.
31. Forrest S, Bridgwater J, Mort PR, Litster J, Parker DJ. Flow patterns in granulating systems. *Powder Technol.* 2003;130:91–96.
32. Laurent BFC, Bridgwater J, Parker DJ. Motion in a particle bed agitated by a single blade. *AIChE J.* 2000; 46(9):1723–1734.
33. Tardos GI, Hapgood KP, Ipadeola OO, Michaels JN. Stress measurements in high-shear granulators using calibrated test particles: application to scale-up. *Powder Technol.* 2004; 140(3):217–227.
34. Hapgood K, Jain S, Kline L, Moaddeb M, Zega J. Application of Spray Flux in Scale-Up of High Shear Wet Granulation Processes. In: *AIChE Annual Meeting*, AIChE, San Francisco, 2003.
35. Iveson SM, Beathe JA, Page NW. The dynamic strength of partially saturated powder compacts: the effect of liquid properties. *Powder Technol.* 2002;127:149–161.
36. Iveson SM, Page N. *Dynamic mechanical properties of liquid bound powder compacts, in 3rd Australasian Congress on Applied Mechanics (ACAM2002)*, Sydney, Australia: World Scientific, 2002.
37. Iveson SM, Page N. Brittle to plastic transition in the dynamic mechanical behavior of partially saturated granular materials. *Trans. ASME* 2004;71:470–475.
38. Iveson SM, Page N. Dynamic strength of liquid-bound granular materials: the effect of particle size and shape. *Powder Technol.* 2005;152:79–89.
39. Iveson SM, Litster JD, Ennis BJ. Fundamental studies of granule consolidation. Part 1: effects of binder content and binder viscosity. *Powder Technol.* 1996;88:15–20.
40. Iveson SM, Litster JD. Fundamental studies of granule consolidation. Part 2: quantifying the effects of particle and binder properties. *Powder Technol.* 1998;99:243–250.
41. Zoglio MA, Cartensen JT. Physical aspects of wet granulation III. Effect of wet granulation on granule porosity. *Drug Dev. Ind. Pharm.* 1983;9(8):1417–1434.
42. Ohno I, Hasegawa S, Yada S, Kusai A, Moribe K, Yamamoto K. Importance of evaluating the consolidation of granules manufactured by high shear mixer. *Int. J. Pharmaceut.* 2007;338(1–2):79–86.
43. Farber L, Tardos G, Michaels JN. Use of X-ray tomography to study the porosity and morphology of granules. *Powder Technol.* 2003;132(1):57–63.
44. Ansari MA, Stepanek F. Formation of hollow core granules by fluid bed *in situ* melt granulation: modelling and experiments. *Int. J. Pharmaceut.* 2006;321(1–2):108–116.
45. Ansari MA, Stepanek F. Design of granule structure: computational methods and experimental realization. *AIChE J.* 2006;52(11):3762–3774.
46. Hapgood KP, Farber L, Michaels JN. Agglomeration of hydrophobic powders via solid spreading nucleation. *Powder Technol.* 2009;188(3):248–254.
47. Rahmanian N, Ghadiri M, Jia X, Stepanek F. Characterisation of granule structure and strength made in a high shear granulator. *Powder Technol.* 2009;192(2):184–194.
48. Jægerskou A, Holm P, Schæfer T. Granulation in high speed mixers. Part 3: effects of process variables on the intragranular porosity. *Pharm. Ind.* 1984;46(3):310–314.
49. Berggren J, Alderborn G. Effect of drying rate on porosity and tabletting behaviour of cellulose pellets. *Int. J. Pharmaceut.* 2001;227:81–96.
50. Mort PR. Scale-up of binder agglomeration processes. *Powder Technol.*150(2): 2005; 86–103.
51. Iveson SM. *Fundamental Studies of Granulation: Granule Deformation and Consolidation*, University of Queensland, Brisbane, Australia, 1997.
52. Iveson SM, Litster JD. Growth regime map for liquid-bound granules. *AIChE J.* 1998;44(7):1510–1518.
53. Wauters PAL, van de Water R, Litster JD, Meesters GMH, Scarlett B. Growth and compaction behaviour of copper concentrate granules in a rotating drum. *Powder Technol.* 2002;124(3):230–237.

54. Iveson SM, Wauters PAL, Forrest S, Litster JD, Meesters GMH, Scarlett B. Growth regime map for liquid-bound granules: further development and experimental validation. *Powder Technol.* 2001;117:83–97.
55. Tu W-D, Hsiau S-S, Ingram A, Seville J. The effect of powder size on induction behaviour and binder distribution during high shear melt agglomeration of calcium carbonate. *Powder Technol.* 2008;184(3):298–312.
56. Hoornaert F, Wauters PAL, Meesters GMH, Pratsinis SE. Agglomeration behaviour of powders in a Lödige mixer granulator. *Powder Technol.* 1998;96:116–128.
57. Mackaplow MB, Rosen LA, Michaels JN. Effect of primary particle size on granule growth and endpoint determination in high-shear wet granulation. *Powder Technol.* 2000; 108(1):32–45.
58. Litster JD, Ennis BJ. *The Science and Engineering and Granulation Processes*, Kluwer Academic Publishers, Dordrecht, 2004.
59. Ennis BJ, Li J, Tardos GI, Pfeffer R. The influence of viscosity on the strength of an axially strained pendular liquid bridge. *Chem. Eng. Sci.* 1990;45(10):3071–3088.
60. He Y, Liu LX, Litster JD, Kayrak-Talay D. Scale up considerations in granulation. In: Parikh DM, editor, *Handbook of Pharmaceutical Granulation*, Taylor and Francis Group, Boca Raton, FL, 2009, pp. 459–490.
61. Ennis BJ, Litster JD. Size reduction and size enlargement., In: Green D, editor, *Perry's Chemical Engineers' Handbook*, McGraw-Hill, New York, 1997.
62. Reynolds GK, Fu JS, Cheong YS, Hounslow MJ, Salman AD. Breakage in granulation: a review. *Chem. Eng. Sci.* 2005; 60(14):3969–3992.
63. Fu JS, Reynolds GK, Adams MJ, Hounslow MJ, Salman AD. An experimental study of the impact breakage of wet granules. *Chem. Eng. Sci.* 2005;60(14):4005–4018.
64. Tardos GI, Khan MI, Mort PR. Critical parameters and limiting conditions in binder granulation of fine powders. *Powder Technol.* 1997;94:245–258.
65. Smith R. *Wet Granule Breakage in High Shear Mixer Granulators*, PhD Thesis, Department of Chemical Engineering, University of Queensland, Brisbane, Australia, 2008.
66. Smith RM, Litster JD, Page NW. Diametrical compression of wet granular materials. *Powder Technol.* 2010; Submitted.
67. Iveson SM, Page NW, Litster JD. The importance of wet-powder dynamic mechanical properties in understanding granulation. *Powder Technol.* 2003;130:97–101.
68. van den Dries K, Vegt OMd, Girard V, Vromans H. Granule breakage phenomena in a high shear mixer; influence of process and formulation variables and consequences on granule homogeneity. *Powder Technol.* 2003;113:228–236.
69. Smith RM, Liu LX, Litster JD. Breakage of drop nucleated granules in a breakage only high shear mixer. *Chem. Eng. Sci.* 2010; doi:10.1016/j.ces.2010.06.037.
70. Liu LX, Smith R, Litster JD. Wet granule breakage in a breakage only high-hear mixer: effect of formulation properties on breakage behaviour. *Powder Technol.* 2009; 189(2):158–164.
71. Knight PC, Johansen A, Kristensen HG, Schæfer T, Seville JPK. An investigation of the effects on agglomeration of changing the speed of a mechanical mixer. *Powder Technol.* 2000;110:204–209.
72. Pearson JKM, Hounslow MJ, Instone T. Tracer studies of high-shear granulation I: experimental results. *AIChE J.* 2001;47(9):1978–1983.
73. Ramaker JS, Jelgersma MA, Vonk P, Kossen NWF. Scale-down of a high shear pelletisation process: flow profile and growth kinetics. *Int. J. Pharmaceut.* 1998;166:89–97.
74. Hounslow MJ, Pearson JMK, Instone T. Tracer studies of high-shear granulation: II population balance modeling. *AIChE J.* 2001;47(9):1984–1999.
75. Bouwman AM, Visser MR, Meesters GMH, Frijlink HW. The use of Stokes deformation number as a predictive tool for material exchange behaviour of granules in the 'equilibrium phase' in high shear granulation. *Int. J. Pharmaceut.* 2006; 318(1–2):78–85.
76. Tran A. Powder flow in mixer granulators. In: *Chemical Engineering*, University of Queensland, Brisbane, Australia, 2010.
77. Holm P, Jungersen O, Schæfer T, Kristensen HG. Granulation in high speed mixers. Part 1 effects of process variables during kneading. *Pharm. Ind.* 1983;45(8):806–811.
78. Holm P, Jungersen O, Schæfer T, Kristensen HG. Granulation in high speed mixers. Part 2 effects of process variables during kneading. *Pharm. Ind.* 1984;46(1):97–101.
79. Holm P. Effect of impeller and chopper design on granulation in a high speed mixer. *Drug Dev. Ind. Pharm.* 1987; 13(9–11):1675–1701.
80. Johansen A, Schæfer T. Effects of interactions between powder particle size and binder viscosity on agglomerate growth mechanisms in a high shear mixer. *Eur. J. Pharmaceut. Sci.* 2001;12:297–309.
81. Michaels JN, Farber L, Wong GS, Hapgood K, Heidel SJ, Farabaugh J, Chou JH, Tardos GI. Steady states in granulation of pharmaceutical powders with application to scale-up. *Powder Technol.* 2009;189(2):295–303.
82. Aulton M, Banks M. The factors affecting fluidised bed granulation. *Manuf. Chemist Aer. N.* 1978;49:50–56.
83. Capes CE. *Particle Size Enlargement*, Elsevier Scientific Publishing Company, Amsterdam, 1980.
84. Rambali B, Baert L, Massart DL. Scaling up of the fluidized bed granulation process. *Int. J. Pharmaceut.* 2003;252:197–206.
85. Hapgood K, Plank R, Zega J. Use of Dimensionless Spray Flux to Scale Up a Wet Granulated Product. In: *World Congress on Particle Technology 4*, Sydney, Australia, 2002.
86. Stewart RL, Bridgwater J, Zhou YC, Yu AB. Simulated and measured flow of granules in a bladed mixer—a detailed comparison. *Chem. Eng. Sci.* 2001;56:5457–5471.
87. Laurent BFC, Bridgwater J. Performance of single and six-bladed powder mixers. *Chem. Eng. Sci.* 2002;57: 1695–1709.

88. Nilpawar AM, Reynolds GK, Salman AD, Hounslow MJ. Surface velocity measurement in a high shear mixer. *Chem. Eng. Sci.* 2006;61(13):4172–4178.
89. Reynolds GK, Nilpawar AM, Salman AD, Hounslow MJ. Direct measurement of surface granular temperature in a high shear granulator. *Powder Technol.* 2008;182(2):211–217.
90. Hiseman MJP, Bridgwater J, Wilson DI. Positron Emission Particle Tracking Studies of Powder Mixing in a Planetary Mixer. In: *Control of Particulate Processes IV*, Engineering Foundation, Delft, Netherlands, 1997.
91. Horsthuis GJB, Laarhoven JAHV, van Rooij RCMB, Vromans H. Studies on upscaling parameters of the Gral high shear granulation process. *Int. J. Pharmaceut.* 1993;92:143–150.
92. Litster JD, Hapgood KP, Michaels JN, Kamineni SK, Sims A, Roberts M, Hsu T. Scale-Up of Mixer Granulators for Effective Liquid Distribution. In: *Control of Particulate Processes 6*, Engineering Foundation, Fraser Island, Australia, 1999.
93. Rahmanian N, Ghadiri M, Ding Y. Effect of scale of operation on granule strength in high shear granulators. *Chem. Eng. Sci.* 2008;63(4):915–923.
94. Knight PC, Seville JPK, Wellm AB, Instone T. Prediction of impeller torque in high shear powder mixers. *Chem. Eng. Sci.* 2001;56:4457–4471.
95. Landin M, York P, Cliff MJ, Rowe RC, Wigmore AJ. Scale-up of pharmaceutical granulation in fixed bowl mixer-granulation. *Int. J. Pharmaceut.* 1996;133:127–131.
96. Hassanpour A, Antony SJ, Ghadiri M. Modeling of agglomerate behavior under shear deformation: effect of velocity field of a high shear mixer granulator on the structure of agglomerates. *Adv. Powder Technol.* 2007;18:803–811.
97. Fu J, Chan EL, Jones MR, Kemp IC, Gilmour CM, Hounslow MJ, Salman AD. Characterisation of impact stress from main impeller on granules during granulation processes in a high shear mixer. In: Salman AD, editor, *9th International Symposium on Agglomeration*, Sheffield University, Sheffield, UK, 2009.
98. Mort PR. Scale-up and control of binder agglomeration processes—flow and stress fields. *Powder Technol.* 2009;189(2):313–317.
99. Faure A, Grimsey IM, Rowe RC, York P, Cliff MJ. Applicability of a scale-up method for wet granulation processes in Collette Gral high shear mixer-granulators. *Eur. J. Pharmaceut. Sci.* 1999;8:85–93.
100. Stepanek F, Rajniak P, Mancinelli C, Chern R, Determination of the Coalescence Probability of Wet Granules by Mesoscale Modeling. In: *AIChE Annual Meeting*, AIChE, San Francisco, CA, 2006.
101. Boerefijn R, Juvin PY, Garzon P. A narrow size distribution on a high shear mixer by applying a flux number approach. *Powder Technol.* 2009;189(2):172–176.
102. Hapgood KP, Khanmohammadi B. Granulation of hydrophobic powders. *Powder Technol.* 2009;189(2):253–262.
103. Vervaet C, Remon JP. Continuous granulation in the pharmaceutical industry. *Chem. Eng. Sci.* 2005;60(14):3949–3957.
104. Djuric D, Kleinebudde P. Impact of screw elements on continuous granulation with a twin-screw extruder. *J. Pharmaceut. Sci.* 2009;97(11):4934–4942.
105. Hu X, Cunningham JC, Winstead D. Study growth kinetics in fluidized bed granulation with at-line FBRM. *Int. J. Pharmaceut.* 2008;347(1–2):54–61.
106. Cantor SL, Kothari S, Koo OMY. Evaluation of the physical and mechanical properties of high drug load formulations: wet granulation vs. novel foam granulation. *Powder Technol.* 2009;195(1):15–24.
107. Keary CM, Sheskey PJ. Preliminary report of the discovery of a new pharmaceutical granulation process using foamed aqueous binders. *Drug Dev. Ind. Pharm.* 2004;30(8):831–845.
108. Tan MXL, Wong LS, Lum KH, Hapgood KP. Foam and drop penetration kinetics into loosely packed powder beds. *Chem. Eng. Sci.* 2009;64(12):2826–2836.

40

SPRAY ATOMIZATION MODELING FOR TABLET FILM COATING PROCESSES

ALBERTO ALISEDA
Department of Mechanical Engineering, University of Washington, Seattle, WA, USA
ALFRED BERCHIELLI
Pharmaceutical Development, Pfizer Global Research & Development, Groton, CT, USA
PANKAJ DOSHI
Chemical Engineering and Process Division, National Chemical Laboratory, Pune, India
JUAN C. LASHERAS
Jacobs School of Engineering, University of California, San Diego, La Jolla, CA, USA

40.1 INTRODUCTION

Film coatings are often used to enhance pharmaceutical tablet products. Tablet film coatings can provide many benefits including improved appearance, added functionality (e.g., sustained release, delayed release, or coatings with active ingredients), brand identity, dose strength identification, ease of swallowing, mask bad taste, improved mechanical strength for improved handling (e.g., during production packaging), and reduce worker exposure (e.g., when dispensing in a pharmacy or hospital setting). There is also potential to improve chemical stability by separation of incompatible ingredients or through reduced oxygen or vapor transmission. Improvements in stability can be significant for maintaining potency, reducing or masking potential color changes in the tablet core, or reducing the development of odor due to chemical reactions. Coating ingredients may also cause instability, so selection of coating components and the coating process conditions are critical for successful product development.

The application of coating material to the tablets is carried out in a complex process that includes four key steps (spray atomization, droplet transport, droplet impact/spreading/drying, and tablet mixing). A schematic of the four elementary processes is shown in Figure 40.1. A photograph of the exterior and close-up view of the corresponding interior of an actual chamber where tablet coating is being performed can be seen in Figure 40.2. First, the coating formulation is atomized into small droplets. This is done by using two-fluid coaxial atomizers in which the liquid formulation is injected through the inner nozzle at low speed and a process gas, typically air, is injected through the outer nozzle at very high speed. The atomizing gas exerts shear and pressure forces on the liquid jet and breaks it into droplets. This type of atomizer allows independent control of the liquid mass flow rate and droplet size. This is because, for all practical liquid flow rates of interest, the liquid speed at the nozzle is low and the breakup process that determines the droplet size is a function of the velocity differential between the two fluids that can be modified by changing the gas speed (i.e., flow rate). The description and modeling of the physics behind this complex process is the focus of this chapter.

The second step in the process is the transport of the liquid droplets onto the tablets' surface. The droplets move under the velocity transferred to them by the atomizing gas and are dried by a secondary flow of hot dry gas that is introduced

Chemical Engineering in the Pharmaceutical Industry: R&D to Manufacturing Edited by David J. am Ende

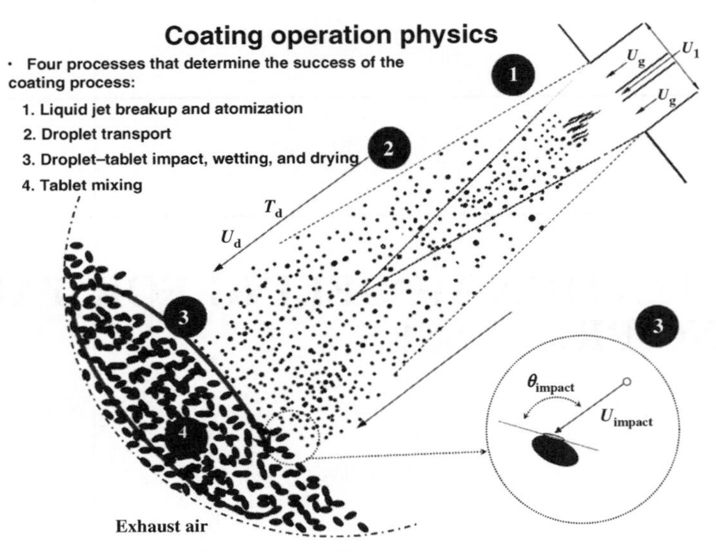

FIGURE 40.1 Schematic of the four elementary processes in the tablet coating operation.

with negligible momentum into the process chamber, so as not to disturb the droplet motion. This secondary gas flow plays the role of providing the energy source needed to balance the latent heat of evaporation of the solvent used in the coating formulation while keeping a sensitive balance between too little drying (droplets hitting the tablets with too much solvent) and too much drying (droplets drying out entirely during their transport). The third step is the impact and spreading of the droplets onto the tablets. The droplet spray hits the tablets at or near the surface of the bed. By rotating through the bed in a random manner, the individual tablets form a continuous film coating by integrating the effects of many tiny droplets that hit them at different times during the process. This last aspect forms the fourth elementary process that determines the success of the operation. For the droplet spreading on the individual tablets to lead to a uniform film coating, the tablets need to be adequately mixed in their recirculating porous bed.

While there are extensive studies in the pharmaceutical literature on the granular flow mechanics that dominate mixing in the tablet bed, evaluation of coating quality and properties, and the thermodynamics that set droplet and tablet drying [1–8], much less attention has been paid to the atomization process [9–12]. Mathematical modeling of all four aspects of tablet film coating is important and has far-reaching impact on the economics of pharmaceutical commercialization, potentially reducing both cost and time to market. Generally, this is the last unit process where a batch

FIGURE 40.2 *Left*: External view of Vector Corporation Laboratory Development Coating System (LDCS 20/30). *Right*: Internal close-up view of cascading tablets during a development coating run. *Note*: In the image above on right, a clear coating is added to white tablets. Numbers have been added to track individual tablets during the development coating run. The arrow indicates the direction of tablet movement.

may be rejected if process conditions are poorly selected or controlled resulting in an unacceptable defect rate. Tablets with surface defects may be removed from the batch after a complete visual quality inspection; however, this added step is time-consuming and should be avoided.

Tablet coating has evolved throughout history from sugar-based coatings, to solvent-based polymeric film coatings, to today's aqueous-based polymeric film coatings. Equipment preferences have also changed from nonperforated pans to partially perforated and fully perforated pans. Coating application has also drastically improved from ladling of sugar coatings to films applied with modern pneumatic or coaxial air blast spray guns. This chapter describes advances in mathematical modeling of the atomization process that enable droplet size predictions, based on models of the physical processes that control liquid jet breakup in two-fluid coaxial atomizers, for pharmaceutically relevant aqueous and organic polymeric film coatings.

In the process control and experimentation that is involved in the design and scale-up of coating operations, atomization is one of the most difficult elements to understand and optimize. Direct measurements of droplet size are typically not available during coating operations and scale-up studies. In addition, droplet size effects are often masked during development trials by the large variability in the results associated with the wide ranges of liquid and gas flow rates and velocities that need to be investigated. Predicting the size of the coating formulation droplets resulting from the atomization process is essential to the success of the coating process, as small droplets can dry and lead to reduced coating efficiency, or they may fail to coalesce on the tablet surface, not producing a smooth continuous film. Droplets that are too large, on the other hand, can see reduced solvent evaporation and lead to overwetting. This causes coating defects (e.g., sticking marks where tablets contact each other and stick together or rough coatings due to erosion of the tablet core components). Similarities between the effects of droplets below or above the valid size range make the diagnosis of these defects difficult.

As we will show in the following sections, droplet size is a function of fluid properties (e.g., surface tension, viscosity), atomizer geometry (e.g., liquid and gas nozzle diameters), and process conditions (e.g., liquid and gas flow rates). The selection of a coating formulation for a specific pharmaceutical tablet is influenced by many factors that may be unrelated to the physical processes involved. Once approved by regulatory agencies, however, changes to the formulation are not made unless necessary and often require justification and supporting regulatory data. The scale-up process, or changes in the industrial tablet manufacturing process that may occur during the lifetime of the product, may require modifications in atomizer geometry or process conditions. Under these circumstances, the atomization of the coating formulation will change. Unless a physics-based model is available to predict the direction and magnitude of those changes, a costly and time-consuming process is required to adapt the coating conditions to the new setup. These full-scale trials represent a major disruption to the manufacturing and commercialization, particularly if active ingredient costs are high, or the delays lead to reductions in product supply.

The atomization process is a key part of the tablet coating process used in the pharmaceutical industry. It is therefore of great importance to develop physics-based atomization models that can predict droplet size and volume density as a function of fluid properties, atomizer configuration, and process conditions. These models can play an important role in the design of the process, including selection of equipment and process parameters, and in the optimization at all different levels: lab, pilot, and full scale. In the following sections, we will discuss the rheology of coating formulations, characterization of coating suspensions and solutions, detailed physics of atomization and the associated equipment, droplet size and velocity measurement, a mathematical model for mean droplet size, insight into the scale-up of the coating operation, and some conclusions on the current state and future directions of this technology.

40.2 COATING FORMULATIONS, PHYSICAL PROPERTIES, AND RHEOLOGY CHARACTERIZATION

Coating formulations typically contain significant amounts of polymers, plasticizers, surfactants, pigments/colorants, antifoaming, antitack ingredients, film modifiers (i.e., sugars), and opacifiers. Because of this high content of large molecules in solution or of solids in colloidal dispersion (e.g., titanium dioxide, talc), coating fluids have complex rheology, exhibiting non-Newtonian behavior during the atomization process.

Tablet film coatings can be soluble (e.g., color coatings for immediate release formulations), insoluble (e.g., enteric coatings that are insoluble at gastric pH), or partially insoluble (e.g., porous or semipermeable membranes that may contain insoluble and soluble ingredients). Immediate release coating formulations are commonly based on polymers such as hypromellose (also known as hydroxypropyl methylcellulose or HPMC) and polyvinyl alcohol (PVA), but other polymers or natural products such as shellac have also been used. Functional coatings for controlled release are often based on polymers such as cellulose acetate, ethyl cellulose, or methacrylates. In this chapter, we will present experimental results obtained from two characteristic coating formulations for immediate release tablets: HPMC (Colorcon Codes: Y-30-18037, OY-LS-28914) and PVA (Colorcon Code: 85F18422), and a characteristic coating for controlled release that is composed of a mixture of cellulose acetate/polyethylene glycol and forms a semipermeable

TABLE 40.1 Example Coatings Discussed in This Chapter

Coating Number	Abbreviation	Vendor Code or Components	Composition
1	Opadry II White-HPMC	Colorcon Code: Y-30-18037	15% solids, 85% water
2	Opadry II White-PVA	Colorcon Code: 85F18422	20% solids, 80% water
3	Opadry II White-HPMC	Colorcon Code: OY-LS-28914	10%, 12%, 15% solids; 90%, 88%, 85% water
4	CA-PEG	Cellulose acetate 398-10	8%
		Polyethylene glycol 3350	2%
		Acetone	87%
		Water	3%

membrane allowing osmotic drug delivery [13, 14]. Table 40.1 lists these examples of coating formulations discussed in this chapter.

As we have pointed out in the previous sections and will show in the formulation of a mathematical model for the breakup of the liquid into individual droplets, the atomization process that constitutes the first stage of the pharmaceutical coating operation is strongly influenced by the physical properties of coating solutions. The two key nondimensional parameters that control this process are the Weber number $(We = \rho_g U_g^2 d/\sigma)$ and the Ohnesorge number $(Oh = \mu_l/\sqrt{\rho_l \sigma d})$, where ρ_g is the gas density, U_g is the gas velocity at the nozzle, d is the liquid jet diameter, μ_l is the liquid viscosity, ρ_l is the liquid density, and σ is the value of the surface tension between the liquid and the gas. The Weber number represents the relative importance of the inertia of the high-speed gas stream that disturbs the liquid jet and the surface tension stresses that minimize surface energy, bringing cohesion to the liquid. The Ohnesorge number characterizes the relative importance of viscous and surface tension stresses. Clearly, the characterization of the physical properties of the coating fluids is a key preliminary step in the understanding and modeling of the liquid atomization in the tablet coating processes.

40.2.1 Coating Solution Viscosity Measurement

A typical film coating formulation usually contains a mixture of solids comprising of polymer, plasticizer, pigment/opacifier, and other film modifiers dissolved in water or a nonaqueous organic solvent. The coating formulations could either be a solution if solids are soluble in the solvent or else they could form a colloidal suspension. Figure 40.3 shows the viscosity for four previously mentioned coating fluids at 25°C for different shear rates. These fluids show slight shear thinning behavior (i.e., viscosity decreases with increasing shear rate), up until shear rate reaches $200\,s^{-1}$. However, beyond this value of shear rate, viscosities are almost constant. In a typical atomization process, the shear rate is well above $200\,s^{-1}$ [9], so coating solution viscosity can be assumed to be constant for a given coating formulation.

40.2.2 Effect of Solids Content on Viscosity

Figure 40.4 shows the variation of viscosity for coating 3 with increasing solids content. For this formulation, the viscosity at low solids content, that is, up to 5%, is quite close to that of the solvent. However, with further increase in solids content, the viscosity increases in an exponential fashion, typical of this type of colloidal dispersion.

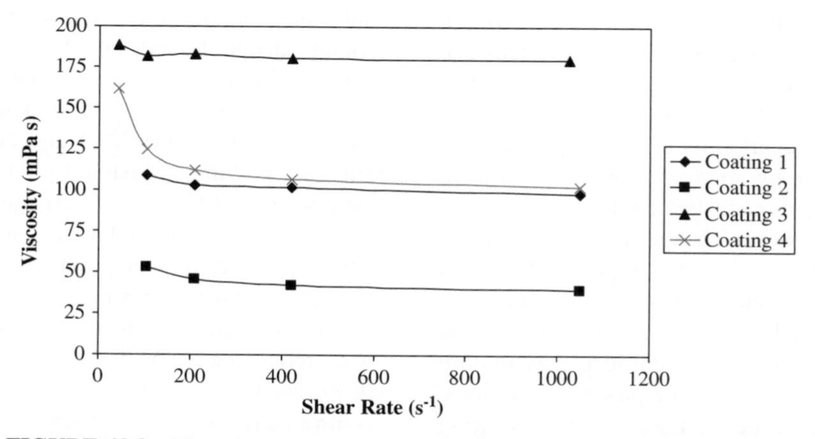

FIGURE 40.3 Viscosity versus shear rate for four coating formulations at 25°C.

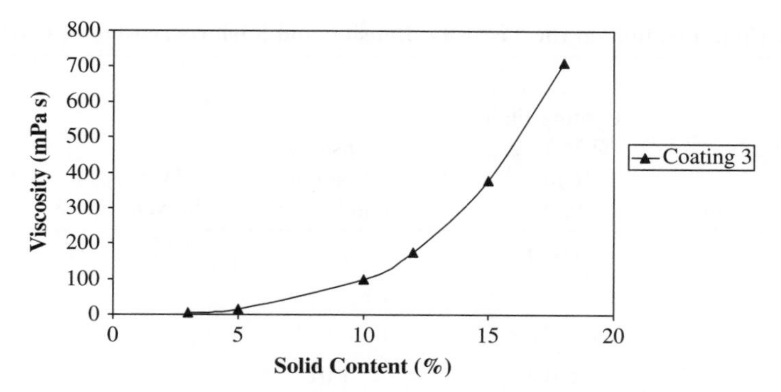

FIGURE 40.4 Solution viscosity versus solids content for coating 3 at 25°C and a shear rate of $1000\,s^{-1}$.

40.2.3 Effect of Temperature on Coating Solution Viscosity

Figure 40.5 shows the variation of viscosity of three different coating fluids with temperature. The viscosity of all three fluids decreases linearly with increasing temperature. The viscosity reduces to almost half their values when the fluids are heated from 20 to 35°C. This behavior can be taken advantage of, by heating very viscous coating formulation prior to injection in the atomizer, to facilitate tablet coatings with very high solids content, where the high viscosity makes atomization difficult. At the same time, this points out the need to monitor thermal gelation point of coating formulations, since operating conditions that include temperatures above that threshold will lead to semisolid and almost unsprayable coating fluids [15].

40.2.4 Coating Solution Surface Tension Measurement

Table 40.2 shows the surface tension data for different coating fluids and their respective solvents measured at approximately 23°C. The values of surface tension for all the aqueous coating fluids (1–3) are quite close to each other, irrespective of the solids concentration. However, they are quite low compared to the surface tension of the solvent (water). This observation is not true, however, for the coating formulation with organic solvent (acetone). These data are in accord with the finding in the literature that HPMC acts as a surfactant and reduces the surface tension of water for very low concentrations ($\sim 2 \times 10^{-5}$% w/w) [15]. After this point, further increase in HPMC concentration has little effect on reducing the surface tension. This behavior is similar to the critical micelle concentration shown by surface-active materials. The typical composition of coating formulations leads to surface concentrations that are always above this threshold, so the saturation effects in the values of surface tension found in our experiments are characteristic of these types of coatings.

40.3 SPRAYING PROCESS AND EQUIPMENT

As highlighted in previous sections, the atomization of coating liquid formulations into small droplets is a key step in the process of pharmaceutical tablet coating and the focus of this chapter. To contribute to success in the coating operation, atomization of the coating formulation must

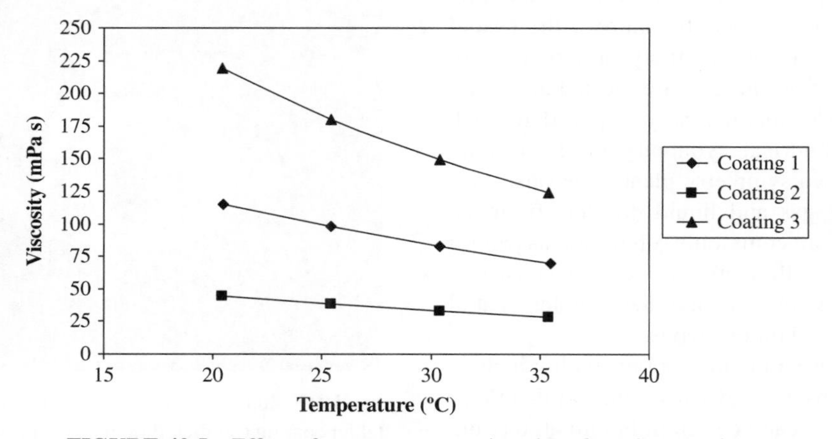

FIGURE 40.5 Effect of temperature on viscosity of coating solutions.

TABLE 40.2 Effect of Solids Concentration on the Viscosity, Density, and Surface Tension of Coating Fluids Measured

Coating No.	Coating Type	% Solids	Viscosity of Coating Fluid @25°C and 1048 s^{-1} (kg/m s)	Density of Coating Fluid (g/cc)	Density of Solvent (g/cc)	Surface Tension of Coating Fluid (dyn/cm)	Surface Tension of Solvent (dyn/cm)
1	Opadry II White (Y-30-18037), HPMC/lactose/ TiO_2/triacetin/water	15	0.098	1.05	1.00	46.98	72.8
2	Opadry II White (85F18422), PVA/PEG/water	20	0.039	1.07	1.00	43.93	72.8
3	Opadry II White (OY-LS-28914), HPMC/lactose/ TiO_2/triacetin	10	0.095	1.02	1.00	48.22	72.8
3	Opadry II White (OY-LS-28914), HPMC/lactose/ TiO_2/triacetin	12	0.181	1.03	1.00	47.66	72.8
3	Opadry II White (OY-LS-28914), HPMC/lactose/ TiO_2/triacetin	15	0.419	1.04	1.00	46.67	72.8
4	CA/PEG/acetone/ water	10	0.102	0.82	0.79/1.00	29.71	29.7

provide controllable droplet size distributions for a wide range of liquid flow rates. In addition, the concentration of droplets across the spray should be spatially uniform. These two requirements make two-fluid coaxial atomizers the most common in this setting. Unlike pressure atomizers, commonly used in diesel fuel injection, where the size of the liquid droplets depends on liquid flow rate, two-fluid atomizers allow control of the droplet size independent of liquid flow rates. In its simple configuration, they produce a very predictable spatial distribution of droplets within the spray, with a significant radial gradient of concentration, and high uniformity along concentric rings. To improve the spatial uniformity of the droplet flux, two auxiliary gas jets are used. In the atomizer design, these side jets are located at a small distance from the nozzle, diametrically opposed to each other, breaking the cylindrical symmetry of the coaxial atomizer and creating two symmetry planes. Because they help shape the flow of gas and liquid droplets from the atomizer, these side jets are commonly referred to as pattern air. The spray takes on an elliptical cross section, consistent with the two symmetry planes, and the droplet spatial distribution becomes very homogeneous.

Spray guns typically used in tablet spray coating include Schlick, Spraying Systems, Freund, Binks, and Walther Pilot. Typical spray gun designs can be seen in Figure 40.6 (front view of a Schlick fluid nozzle and air cap setup for creating an elliptical spray pattern), Figure 40.7 (side view of a Schlick fluid nozzle and air cap setup), Figure 40.8 (front view of a Spraying Systems fluid nozzle and air cap setup for creating an elliptical spray pattern), Figure 40.9 (side view of a Spraying Systems fluid nozzle and air cap setup), Figure 40.10 (front view of a Spraying Systems fluid nozzle

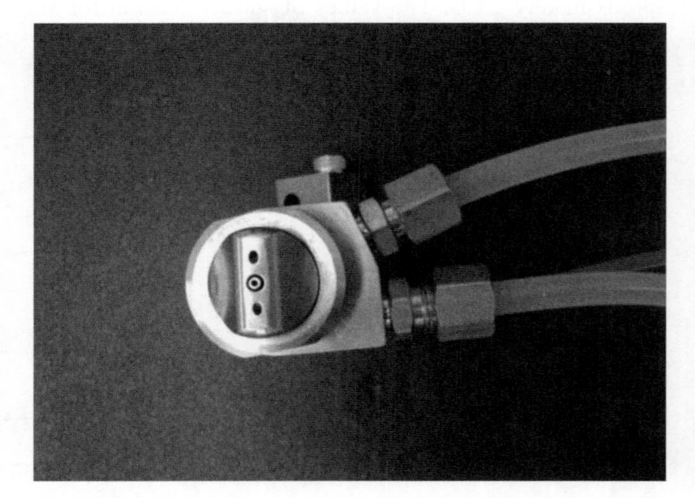

FIGURE 40.6 Front view of a laboratory-scale spray gun for tablet coating (Schlick fluid nozzle and air cap with spray gun body by Vector Inc.).

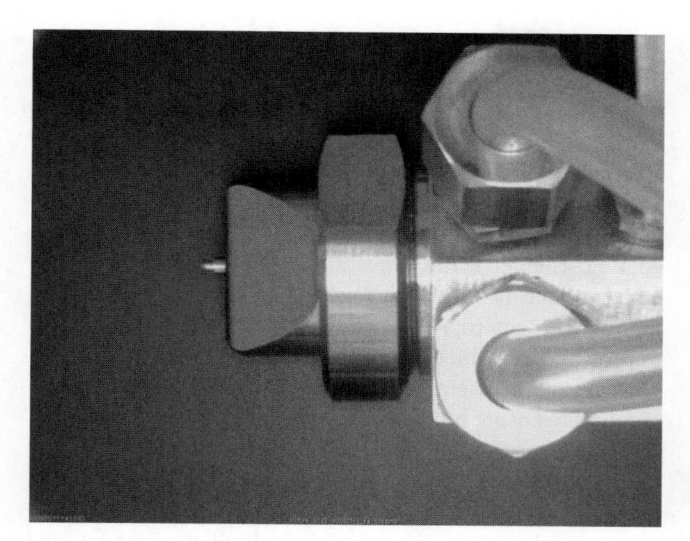

FIGURE 40.7 Side view of a laboratory-scale spray gun for tablet coating (Schlick fluid nozzle and air cap with spray gun body by Vector Inc.).

FIGURE 40.9 Side view of a laboratory-scale spray gun for tablet coating (Spraying Systems fluid nozzle and air cap with spray gun body by Vector Inc.).

and air cap setup for creating a round spray pattern), Figure 40.11 (side view of a Spraying Systems fluid nozzle and air cap setup). Figure 40.12 provides a sketch of the spray nozzle showing key geometric parameters that must be provided to successfully model the atomization process.

The use of side gas jets in pharmaceutical coating atomizers also improves spray coverage. This is defined as the cross-sectional area of the spray where it intersects with the tablet bed. This area, which is obviously determined by the distance from the spray gun to the tablet surface and the spray shape, is important because it determines the probability of a tablet moving along the surface of the tablet bed to collide with one or more coating formulation droplets. By stretching the spray in the direction perpendicular to the tablet motion (major axis of the spray elliptical cross section) and reducing the coverage in the perpendicular direction (minor axis of the ellipse), pattern air increases the number of tablets that will receive the impact of coating droplets as they are distributed on the tablet bed's surface, since it increases the width covered by the spray. At the same time, the shorter coverage along the trajectory of the tablets reduces the probability of a tablet being hit by too many droplets over a short time while it traverses the spray in its motion, thereby reducing the likelihood of overwetting or tablets sticking to each other.

Coating operations requires atomization to work efficiently over a wide range of operating conditions. Poor selection

FIGURE 40.8 Front view of a laboratory-scale spray gun for tablet coating (Spraying Systems fluid nozzle and air cap with spray gun body by Vector Inc.).

FIGURE 40.10 Front view of a laboratory-scale spray gun for tablet coating (Spraying Systems fluid nozzle and air cap with spray gun body by Vector Inc.).

FIGURE 40.11 Side view of a laboratory-scale spray gun for tablet coating (Spraying Systems fluid nozzle and air cap with spray gun body by Vector Inc.).

or poor control of process conditions can lead to nonoptimal results (i.e., poor coating appearance, tablet core erosion, tablet breakage, potency loss, or dosage form functionality changes) or potentially batch failure. Successful tablet coating requires selecting the appropriate tablet formulation components, tablet properties, coating formulation, coating properties, and coating process conditions. The liquid flow rate (i.e., spray rate) needs to be determined in combination with the appropriate thermodynamic [2] drying conditions (air flow, temperature, and humidity), so that evaporation of the solvent occurs at an adequate rate (too much evaporation leads to dry droplets not spreading over tablets, while too little leads to overwetting). Spray rate and coating formulation solid content determine the total flux of coating material onto the tablets, and, therefore, these two parameters need to be selected in conjunction to control the overall coating time. There is a minimum number of passes through the spray that a tablet needs to go through in order to develop a uniform coating. Longer spray times allow for increased passes of tablets under the spray zone and when combined with good mixing and appropriate tablet shape selection results in improved intratablet and intertablet coating uniformity; however, an efficient process should not be longer than is required to ensure product specifications are met. Changes in spray rate or coating fluid solids content will affect the droplet size of the atomized coating. Other atomization parameters need to be adjusted to maintain optimal droplet size under these changing conditions. Predictive models and accurate measurements of process conditions are essential to minimize the challenges presented by the wide range of needs within the coating operation. Manufacturers of atomization equipment typically provide values of the atomization and pattern air pressures as the key process settings because they are easy to measure in the industrial environment. These are, however, surrogates for the gas flow rates that play a key role in the atomization process and droplet redistribution within the spray. Different atomizer setups inside the coating equipment may translate to different pressure losses from the measurement to the atomizer head, introducing uncertainty about the actual flow rates in the process. It is, therefore, preferable to measure volumetric flow rates for both atomizing and pattern air. Air flow rates can economically be measured with rotameters (i.e., a graduated meter consisting

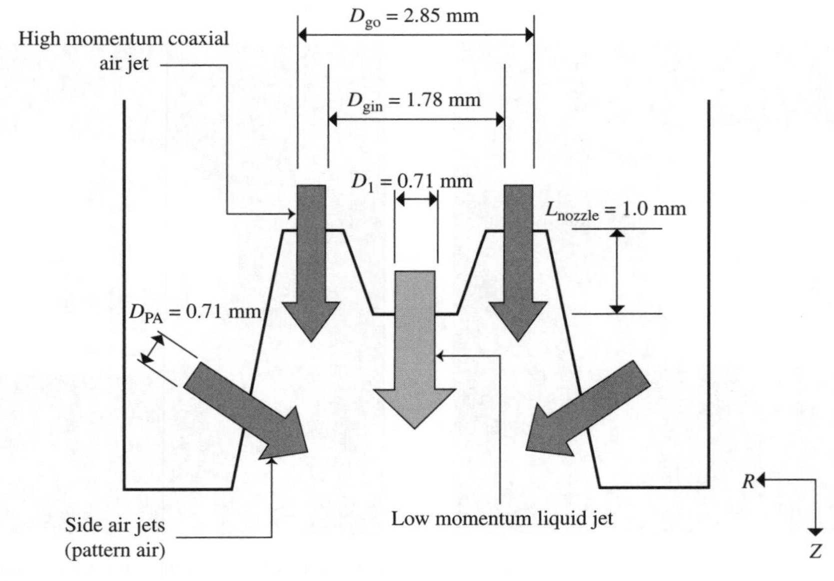

FIGURE 40.12 Side view sketch of a laboratory-scale spray gun for tablet coating (Spraying Systems fluid nozzle and air cap with corresponding measurements used in modeling).

of a tapered tube containing a free float for air flow measurement). The Key Instruments Flo-Rite Series is an example of a commercially available rotameter. Volumetric flow rates of air can also be measured with mass flow meters providing a digital output that is easy for the operator to read and allows coaters with data collection capabilities to store atomization and pattern air flow values electronically. The Sierra Top-Trak 824 series is an example of a mass flow meter. A combination of the appropriate monitoring equipment and accurate physics-based models greatly improves the effectiveness and efficiency of coating operations.

40.4 DROPLET SIZE AND VELOCITY MEASUREMENTS

The result of the atomization process described in previous sections, and modeled mathematically in the next section, is a distribution of droplets with a wide range of diameters and velocities. The combination of these two properties determines the droplet inertia and therefore how the droplets interact with the underlying high-speed turbulent air jet in which they are immersed. These interactions, together with the initial conditions for the motion of the droplets, will ultimately shape the spray coating process, setting the values of the droplet number density, droplet velocity at the impact with the tablet bed, and mass flux per unit area. These fluid variables determine the fate of individual droplets impacting on the tablets and, when ensemble over the whole droplet distribution, the success of the coating operation.

The droplet size distribution is given by the balance between cohesive forces, such as surface tension and viscosity, and disruptive forces, namely, pressure and shear forces exerted on the liquid surface by the atomizing gas stream. Because of the turbulent character of the high-speed gas stream and the nature of the instability with a wide range of unstable wavelengths with similar growth rates, the resulting droplets have a distribution of sizes, not a unique diameter as expected in other types of liquid breakup such as Rayleigh–Plateau instability [16]. Because of this, measurements and modeling of the droplet size are focused on characterizing the full distribution of diameters. One way to do this is to describe the different statistical moments of the distribution of diameters. Table 40.3 gives the mathematical definition and physical interpretation of the most used moments of the size distribution.

Droplet size distributions can be measured by a number of instruments that rely on light scattering to determine the droplet size unintrusively. The most widely used measurement technique is based on light interferometry to evaluate the phase shift introduced by a scattering spherical droplet on the fringe pattern created by the superposition of two laser beams with slightly different frequencies. This phase shift is proportional to the scatterer's diameter. A much more detailed description of the measurement principle was published by Bachalo [17]. This measurement technique has the added advantage that with some further processing of the information in the collected light scattered by the droplets, it also provides the droplet velocity distribution from the Doppler frequency shift. Although the physical principles

TABLE 40.3 Mathematical Definition and Physical Interpretation of the Most Used Moments of the Size Distribution

Statistical Moment	Mathematical Definition	Physical Meaning
D_{10}	$d_{10} = \frac{\sum_{i=1}^{N} n_i \cdot d_i}{\sum_{i=1}^{N} n_i}$	Diameter of the droplet that represents the arithmetic average of all droplet diameters. d_i is the diameter of droplets in size bin i and n_i is the number of droplets in that bin of the distribution
D_{20}	$d_{20} = \sqrt{\frac{\sum_{i=1}^{N} n_i \cdot d_i^2}{\sum_{i=1}^{N} n_i}}$	Characteristic diameter that if all droplets were this same size (monodisperse), they would contain the same surface area as the whole distribution under study
D_{30}	$d_{30} = \sqrt[3]{\frac{\sum_{i=1}^{N} n_i \cdot d_i^3}{\sum_{i=1}^{N} n_i}}$	Characteristic diameter that if all droplets were this same size (monodisperse), they would contain the same volume (or mass) as the whole distribution under study
D_{32}	$d_{32} = \frac{\sum_{i=1}^{N} n_i \cdot d_i^3}{\sum_{i=1}^{N} n_i \cdot d_i^2}$	Characteristic diameter that if all droplets were this same size (monodisperse), they would have the same volume to surface ratio as the whole distribution under study
D_{50}	$\frac{\sum_{i=1}^{M(D_{50})} n_i \cdot d_i^3}{\sum_{i=1}^{N} n_i \cdot d_i^3} = 0.5$	Characteristic diameter for which all droplets smaller than this contain 50% of the volume (or mass) of the whole distribution under study. M is the bin for which the cumulative volume summation reaches 50% of the total. The diameter corresponding to that bin is d_{50}
D_{90}	$\frac{\sum_{i=1}^{L(D_{90})} n_i \cdot d_i^3}{\sum_{i=1}^{N} n_i \cdot d_i^3} = 0.9$	Characteristic diameter for which all droplets smaller than this contain 90% of the volume (or mass) of the whole distribution under study. L is the bin for which the cumulative volume summation reaches 90% of the total. The diameter corresponding to that bin is d_{90}

behind these measurements are well established, the technological details involved in getting robust, accurate measurements in a wide variety of operating conditions are extremely complicated. There are currently only two instrument manufacturers that provide commercial solutions based on this technique: TSI Inc., Shoreline, MN (PDPA (phase Doppler particle analysis)) and Dantec Dynamics A/S, Skovlunde, Denmark (PDA (phase Doppler anemometry)).

The characteristics of the droplets in the spray vary spatially both radially and axially. The latter dependency is easy to describe and is universal for different spray types, with different phenomena having different relative weights in the resulting droplet distribution. An example of this axial evolution is shown in Figure 40.13. In the close proximity of the nozzle, droplets are very large or are growing to a large diameter. In this region, the liquid jet has been shattered by the atomizing air and large liquid masses, not necessarily spherical in shape, are being stripped from the liquid core. As these liquid masses get deformed by the high-speed gas stream, they take on shapes with low curvature that looks to the particle sizing measurements as increasingly large droplets. It is not until the deformation stage ends and the breakup into smaller droplets begins that the shape becomes truly spherical and accurate measurements are possible. This happens in the region between 5 and 10 diameters downstream. After this region, the breakup proceeds with the droplet diameter decreasing sharply before it plateaus in the region between 20 and 30 diameters. At this point, the breakup process has finished. The breakup of the liquid into smaller droplets makes surface tension a stronger cohesive force. Together with the dilution of the high momentum jet by expansion of its cross-sectional area due to entrainment, the balance between disruptive and cohesive forces tilts very soon toward the cohesive side, thereby stopping the breakup. From a practical point of view, this process usually takes place within the first 2 in. of the jet, giving rise to a minimum standoff distance between the coating gun and the tablet bed, since the coating is designed to impact spherical droplets that are small and do not extend beyond the boundaries of a single tablet. In the region beyond where breakup takes place, the droplet size is dependent on secondary processes that determine the dynamics of small spherical droplets immersed in a turbulent gas jet. At this stage, the evolution of the droplet size distribution can be evaluated with a population dynamics equation [18] where the sources on the right-hand side are turbulent transport, coalescence, and evaporation. Typically, these phenomena alter the droplet diameter very slowly in the range of interest for the coating operation, as seen by the constant mean diameter measured in the second half of the plot in Figure 40.13 (between 25 and 50 diameters downstream).

The evolution of the radial distribution of droplet sizes is determined by the different response of droplets to the high-speed gas flow due to their different inertia. The larger droplets are found along the centerline, where they maintain their strong axial momentum communicated by the high-speed jet near the nozzle. Smaller droplets have lower inertia

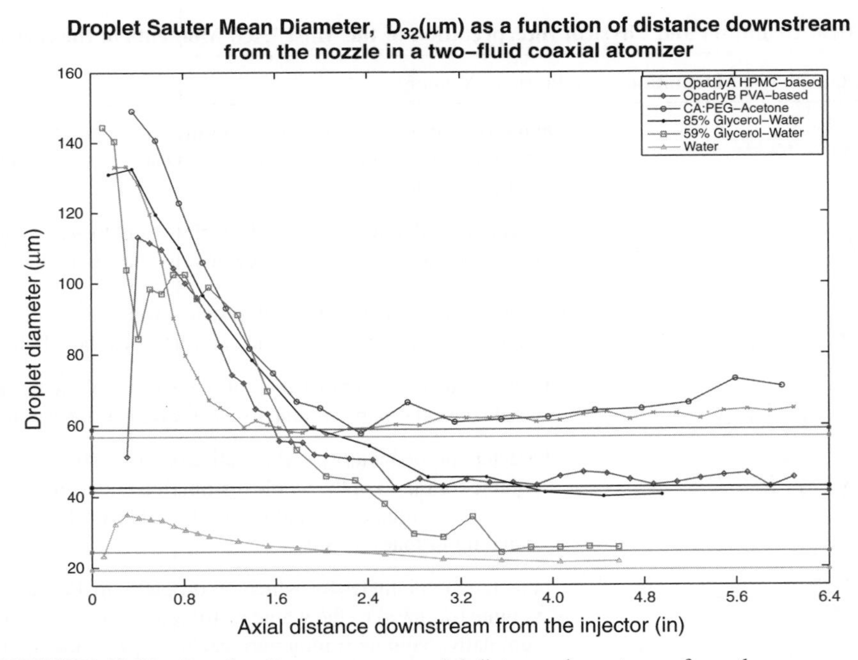

FIGURE 40.13 Droplet diameter versus axial distance downstream from the spray gun.

and are therefore more prone to radial excursions induced by small velocity fluctuations in that direction. Finally, there is typically an increase in the droplet size near the outer boundary of the jet. In swirled sprays, this is very acute and is caused by the centrifugal migration of large inertia droplets that sustain the initial azimuthal momentum longer and therefore travel farther distance radially. But even in non-swirled flows, the instability responsible for liquid breakup requires a radial velocity on the liquid that induces acceleration when it is exposed to the high-speed gas stream. Due to the natural stochastic distribution of this radial velocity, some liquid masses have high initial radial velocity and cross the gas jet quickly. Because of this, these droplets have reduced time of exposure to the high-speed gas stream, which decays exponentially in the radial direction, and have reduced breakup frequencies as a result. The tails of the radial velocity distribution are then responsible of those large droplets and their location in the outskirts of the spray.

The droplet sizes resulting from coaxial atomization of a low-speed liquid jet and a coaxial high-speed gas jet have been typically described in terms of lognormal or gamma distributions. If a specific distribution is used to characterize the spray, two or sometimes only one moment can be used to quantify the droplet size over its entire range.

40.4.1 Velocity Measurements

The velocity of the two phases present in the spray can be measured by the same phase Doppler technique as was used to determine the droplet diameter distribution. The basis for velocity measurements is the same as that for laser Doppler velocimetry. Taking advantage of the size measurements, the velocity data can be segregated for tracer droplets that represent the velocity of the carrier gas and the velocity of the larger droplets that, because of their inertia, do not track the gas velocity.

For a typical axisymmetric coaxial injector, the gas velocity behaves as a classical round jet. After a short development length, during which the flat top profile injected at the nozzle evolves to adjust to the boundary conditions, the flow dynamics become self-similar and the velocity profile Gaussian. In a coaxial atomizer, the gas jet flows through a ring-shaped nozzle that leaves inside space for the liquid nozzle. Given the small cross section necessary for the liquid injection, the gas coming out of the nozzle expands into the centerline so that at the end of the development length, the gas in the spray behaves exactly as a round jet. Figure 40.14 shows velocity measurements in an axisymmetric coaxial injector. The gas velocity can be fitted very accurately by a Gaussian profile. The velocity of the droplets tends to follow a very similar profile, but the magnitude of the velocity is slightly higher. Considering the inertia of the droplets and that the gas velocity is decreasing as it flows downstream from injection due to entrainment of ambient air, this is to be expected. Of particular interest is the strong deviation associated with the droplets near the edge of the jet. Those droplets, typically larger than the average across all radial positions, traverse the jet from the centerline to the edge with a significant radial velocity and therefore reach the region of very low gas velocity before they lose their forward

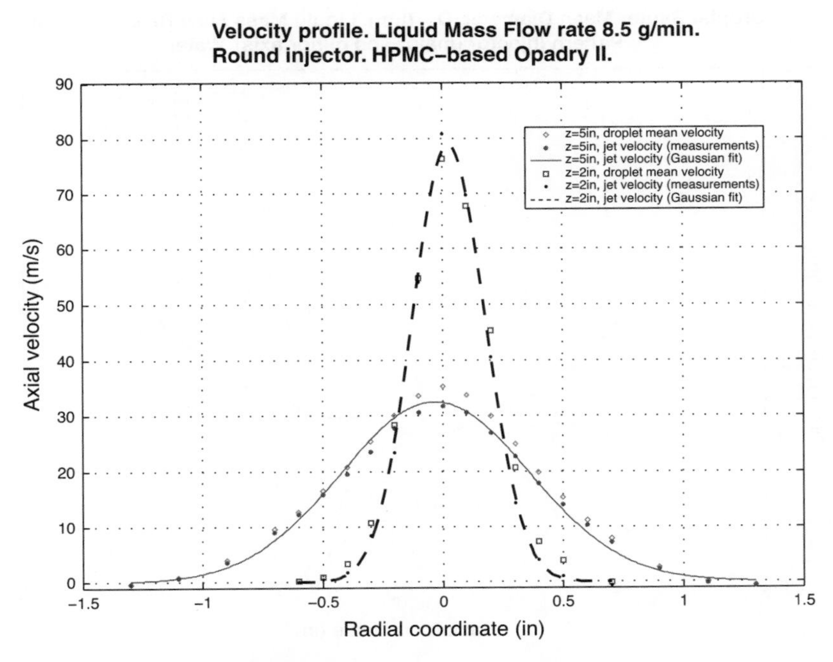

FIGURE 40.14 Droplet and gas jet velocity profile.

momentum. This produces the largest velocity difference between the two phases.

40.4.2 Effect of the Pattern Air on Turbulent Droplet Transport Within the Spray

The model for the breakup of the liquid jet into droplets does not take into account the existence of the two side jets that inject the pattern air into the spray.

The geometry of most pharmaceutical coating atomizers includes two auxiliary air jets that inject air from the edges of the liquid/air jet with a significant inward radial velocity component. Typical angles are between 30° and 45° with respect to the jet axial direction. Depending on the location and angle of the pattern air nozzles, these jets impinge on the droplet-laden jet after the liquid breakup process has finished or on the atomizing air as it is distorting the liquid jet. Pattern air plays an important role in the transport of the droplets of different sizes. Its effect on the liquid breakup is described in the next section. The goal of pattern air is to flatten the spray so that the spatial coverage of the atomizer in the coating pan is increased. To do this, the pattern air induces an asymmetry in the velocity field of the spray that makes its cross section become elliptical rather than circular. The spray is narrower along the axis of the side jets, where the pattern air momentum forces the spray inward (minor axis of the ellipse), and broader in the perpendicular axis (major axis of the ellipse). This is shown in Figure 40.15, where the size distribution is plotted in the radial direction along the minor and the major axis at two distances downstream of the injector. The effect of the pattern air on the droplet size distribution is rather complicated. First, the size distribution is concave along the major axis, and larger droplets are found as we move away from the axis due to the larger droplets conserving their radial momentum longer than the smaller ones. The contrary is true along the minor axis, where the effect of the side jets is more noticeable for the larger droplets that are subject to higher aerodynamic focusing toward the spray axis due to their larger inertia and lower diffusivity. The smallest droplets quickly adjust to the local gas velocity, minimizing any local perturbation inflicted by the pattern air. This explains the convex shape of the diameter distribution found along the cross-sectional minor axis. Furthermore, some very large droplets acquire a large radial velocity component as they interact with the side jets and, because of their large inertia, do not lose that momentum as they cross the spray axis. These droplets are then found in relatively high proportion in the outskirts of the jet, where the droplet number density is low and a small number of very large droplets have a significant impact on the average diameter of the distribution. This explains the change from convex to concave in the diameter radial distribution. Note that the inflexion point associated with this change in character moves outward from the spray axis with distance downstream, as one would expect those large droplets to keep moving in a radial direction, occupying the outer region of the spray. This inhomogeneity in the diameter distribution along the radial coordinate in the spray is a typical effect of pattern air that needs to be

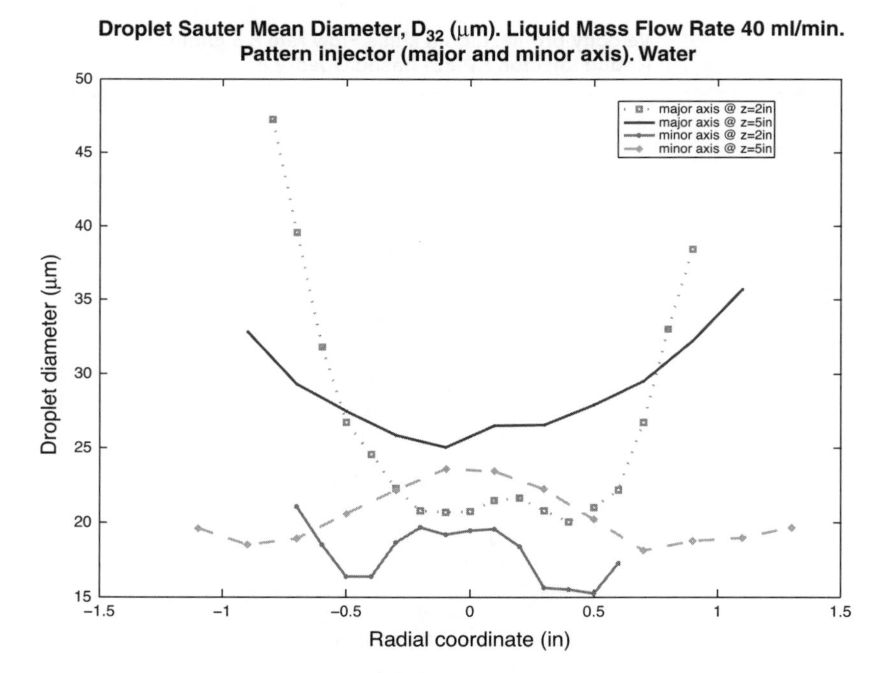

FIGURE 40.15 Droplet diameter versus radial distance from the center of the spray gun.

monitored to avoid large droplets accumulating at the edge of the spray.

40.5 MATHEMATICAL MODELING OF THE LIQUID BREAKUP: AVERAGE DROPLET SIZE

The atomization of a liquid jet by a coflowing, high-speed gas stream occurs via a series of fluid instabilities that lead to the disruption of the liquid jet and result in the liquid mass being broken into small spherical inclusions transported within a gas stream. Although the nature of the instability depends on the specific conditions present in the jet (Reynolds and Weber numbers, mass and dynamic pressure ratios between the air and liquid streams, etc.) and is the source of significant controversy, for the conditions typically present in pharmaceutical coating operations, it has been well documented ([9, 19]) that the breakup process follows a sequence of Kelvin–Helmholtz instability disrupting the liquid jet, followed by a Rayleigh–Taylor (RT) instability resulting in the breakup of the liquid cylinder into individual droplets. The Kelvin–Helmholtz instability develops in the annular shear layer that exists between the low-speed liquid injection and the high-speed coaxial gas jet. Once the liquid has been displaced from its axisymmetric position, it suffers a sudden acceleration as a result of the drag imposed by the gas flow and a Rayleigh–Taylor instability at the interface ensues. This process is shown in schematic form in Figure 40.16. The wavelength of the primary instability, λ_1, depends on the gas boundary layer thickness, δ_g, at the gas nozzle, as described by [20] and can be computed by the following equation:

$$\lambda_1 \approx 2\delta_g \sqrt{\frac{\rho_l}{\rho_g}} \tag{40.1}$$

where ρ_l and ρ_g are the liquid and gas densities, respectively. Spray atomization nozzles are typically designed with a strong convergence upstream of the nozzle so that the flow acceleration will keep the boundary layer laminar and, therefore, as thin as possible. The boundary layer thickness can then be calculated as

$$\delta_g = \frac{Cb_g}{\sqrt{Re_{bg}}} \tag{40.2}$$

where $Re_{bg} \equiv U_{Gas} b_g / \nu_{Gas}$ and the coefficient of proportionality C depends on nozzle design. Considering air as the working gas and typical flow rates used in coating operations (1–5 SCFM, 20–100 SLPM), the Reynolds number is of order 10^4. The disrupted liquid mass is accelerated by the surrounding gas and its resulting convective velocity is

$$U_c = \frac{\sqrt{\rho_l} U_{Liquid} + \sqrt{\rho_g} U_{Gas}}{\sqrt{\rho_l} + \sqrt{\rho_g}} \tag{40.3}$$

For the Kelvin–Helmholtz instability to develop quickly, before the gas jet velocity decays and becomes close to the liquid injection velocity, the Reynolds number of the liquid shear layer must be large:

$$Re_{\lambda 1} = \frac{(U_c - U_{Liquid})\lambda_1}{\nu_l} > 10 \tag{40.4}$$

This condition is necessary even though the instability is driven by the gas. In the case of coating solutions that may have non-Newtonian rheology, the applicable viscosity, ν_l, is the effective shear viscosity at the injection shear rate.

The liquid excursions resulting from the K-H instability, of thickness b_l, grow rapidly and are exposed to and accelerated by the high-speed gas stream. This liquid is subsequently subject to a Rayleigh–Taylor instability similar to

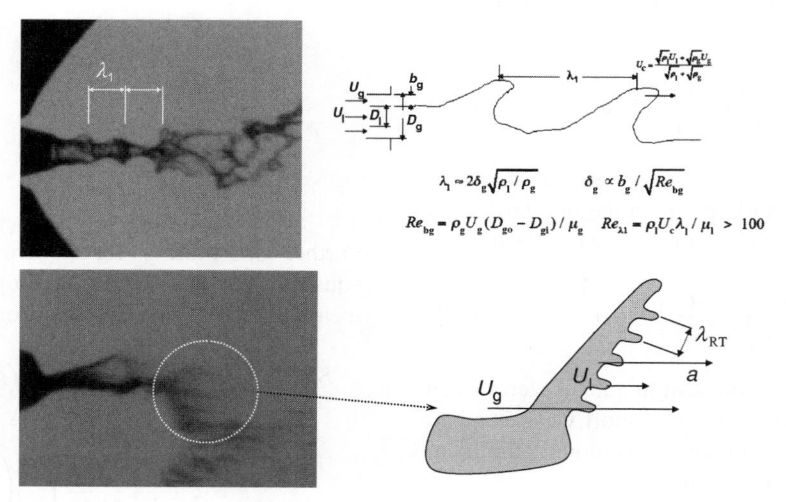

FIGURE 40.16 High-speed photograph (*top left*) and sketch (*top right*) of the primary Kelvin–-Helmholtz instability. High-speed photograph (*bottom left*) and sketch (*bottom right*) of the secondary Rayleigh–Taylor instability.

water placed over oil. Because oil is lighter than water, the interface between the two liquids is accelerated and creates fingers by which oil flows up and water goes down. The growth of a R-T instability in an accelerated drop suddenly injected into a high-speed gas stream is studied in [21]. The dispersion relation for the case $\rho_g << \rho_l$ is given by

$$-\left[1+\frac{1}{n^2}\left(-ak+\frac{\sigma k^3}{\rho_l}\right)\right]+4\frac{k^2}{n}\frac{\alpha_l}{\rho_l}+4\frac{k^3}{n^2}\left(\frac{\alpha_l}{\rho_l}\right)^2(q_l-k)=0 \quad (40.5)$$

where k is the magnitude of the wave vector, n the amplification rate, a the acceleration of the liquid tongue, σ the surface tension, α_l the effective shear viscosity of the liquid in $\tau_{ij}=2\alpha_l e_{ij}$, where τ_{ij} and e_{ij} are, respectively, the stress and the rate of strain tensors in the liquid, and q_l is given by

$$q_l=\sqrt{k^2+n\rho_l/\alpha_l} \quad (40.6)$$

For very low Ohnesorge numbers, when viscous effects are negligible, the wave number corresponding to maximum amplification is

$$k_\sigma=\sqrt{\frac{a\rho_l}{3\sigma}} \quad (40.7)$$

This is the dominant case in water atomization and in fuel injectors where linear chain hydrocarbons with very low viscosity values are used, but it is rarely the case in coating operations where the coating solutions are typically high solid content solutions or particle dispersions with complex rheology.

When viscous terms are important, as is the case for the tablet coating operations, α_l is large and it can be assumed that $(n\rho_l/k^2\alpha_l)<<1$ such that $(q_l-k)\approx(n\rho_l/2k\alpha_l)$ in equation 40.6. The simplified dispersion relation from equation 40.5 then reads

$$n=-\frac{k^2\alpha_l}{\rho_l}\pm\sqrt{\frac{k^4{\alpha_l}^2}{\rho_l^2}-\left(\frac{k^3\sigma}{\rho_l}-ka\right)} \quad (40.8)$$

Disturbances will grow when the second term in equation 40.8 is positive and larger than the first term. It is useful to rewrite equation 40.8 in the following form:

$$n=\frac{k^2\alpha_l}{\rho_l}\left[\left(1+\frac{a\rho_l^2}{k^3\alpha_l^2}-\frac{\sigma\rho_l}{k\alpha_l^2}\right)^{1/2}-1\right] \quad (40.9)$$

From equation 40.9, the amplification rate is zero when $k=\sqrt{(a\rho_l/\sigma)}$, which is the capillary cutoff wave number, and when $k=0$. The wave number of maximum amplification is given by the third-order equation

$$4\frac{\alpha_l^2}{\rho_l^2}k^3-\frac{3\sigma}{\rho_l}k^2+a=0 \quad (40.10)$$

The exact solution of this equation is too complex to be of practical interest. However, for the high viscosity fluids typical of tablet coating, the Ohnesorge number (which controls the relative importance of liquid viscosity and surface tension, $Oh=\mu_l/\sqrt{\rho_l\sigma D_l}$) based on the wavelength is large and the second term in equation 40.10 is small compared to the first one, so that the wave number of maximum amplification is

$$k_{\max}\approx\sqrt[3]{\frac{a\rho_l^2}{\alpha_l^2}} \quad (40.11)$$

The R-T wavelength is $\lambda_{RT}=2\pi/k_{\max}$ and ultimately the droplet diameter is a fraction of λ_{RT} [19]. Therefore, assuming viscous and surface tension effects are additive to the leading order according to the dispersion relation, we look for a correlation in the following form:

$$\lambda_{RT}=2\pi\left[\sqrt{\frac{3\sigma}{a\rho_l}}+C_2\sqrt[3]{\frac{\alpha_l^2}{a\rho_l^2}}\right] \quad (40.12)$$

The acceleration a in equation 40.12 is simply $a=\frac{F}{m}=\frac{F}{\rho_l V}$, where the force F is the drag force exerted by the gas stream on a liquid element, and here the liquid tongue of the primary instability

$$F=\frac{1}{2}C_D\rho_g(U_{Gas}-U_c)^2A_e \quad (40.13)$$

where $C_D\approx 2$ is the drag coefficient and A_e the projected area. The mass of the liquid to be accelerated is $m=\rho_l b_l A_e$ with $b_l\propto\lambda_1$. The expression for a is therefore given by

$$a\approx\frac{\rho_g(U_{Gas}-U_c)^2}{\rho_l b_l} \quad (40.14)$$

Substitution of equation 40.14 in equation 40.12 gives

$$\lambda_{RT}\propto\left(\frac{\sigma\lambda_1}{\rho_g(U_g-U_c)^2}\right)^{1/2}\times\left(1+C_2'\left\{\frac{\rho_g(U_g-U_c)^2}{\lambda_1\sigma}\right\}^{1/6}\left\{\frac{\alpha_l^2}{\rho_l\sigma}\right\}^{1/3}\right) \quad (40.15)$$

Further substituting for λ_1 from equation 40.1, using equation 40.2, and taking the drop diameter, say the Sauter mean diameter (SMD), proportional to λ_{RT} gives

$$\frac{\text{SMD}}{D_l}=C_1(1+m_r)\left(\frac{b_g}{D_l}\right)^{1/2}\left(\frac{\rho_l/\rho_g}{Re_{bg}}\right)^{1/4}\frac{1}{\sqrt{We_{D_l}}}\times\left\{1+C_2\left(\frac{D_l}{b_g}\right)^{1/6}\left(\frac{Re_{bg}}{\rho_l/\rho_g}\right)^{1/12}We_{D_l}^{1/6}Oh^{2/3}\right\} \quad (40.16)$$

In equation 40.16, the mass loading effect in the form $(1 + m_r)$ is obtained from energy arguments previously outlined by [22], where $m_r = m_l/m_g = (\rho_l U_{Liquid} A_l)/(\rho_g U_{Gas} A_g)$ and A_l and A_g are the areas of the liquid and gas nozzle exit sections, respectively. Furthermore, this equation indicates a dependency of the SMD on $U_{Gas}^{-5/4}$ and $\sigma^{-1/2}$. The drop diameter increases with $b_g^{1/4}$ if the coefficient of proportionality C in equation 40.3 remains constant when b_g is changed. As will be shown below, this would be the case only if the length of the gas jet potential cone is much larger than the liquid jet's intact length, which is not typical of atomizer designs.

The SMD in equation 40.16 has been made dimensionless by the liquid orifice diameter D_l and the Weber and Ohnesorge numbers are based on D_l following the usual convention. However, it should be emphasized that the drop diameter does not depend on the liquid orifice diameter but rather on the gas boundary layer thickness at the nozzle exit. This has been clearly demonstrated by [19], where the liquid orifice diameter was changed by a factor of 3 and the drop diameter remained practically identical for the same gas flow conditions.

For completeness, the various nondimensional parameters in equation 40.16 are defined as follows:

$$\begin{aligned} \text{Weber number}: We_{Dl} &= \frac{\rho_g (U_{Gas} - U_c)^2 D_l}{\sigma} \\ \text{Ohnesorge number}: Oh &= \frac{\alpha_l}{\sqrt{\rho_l \sigma D_l}} \\ \text{Reynolds number}: Re_{bg} &= \frac{U_{Gas} b_g}{\nu_g} \\ \text{Mass flux ratio}: m_r &= \frac{\rho_l U_{Liquid} A_l}{\rho_g U_{Gas} A_g} \end{aligned} \quad (40.17)$$

The coefficients C_1 and C_2 in equation 40.16 are of order 1 and values for both coefficients are determined from experiments. The value of C_1 depends on the gas nozzle geometry in general, and on the contraction ratio in particular, because for a given nozzle size the gas boundary layer thickness at the liquid nozzle discharge depends strongly on the contraction ratio. C_2 characterizes the viscosity dependence of the critical wave number in the R-T instability, compared to the surface tension dependence. This value is associated to the additivity and linearity of both cohesive effects, surface tension, and viscosity, which determine the growth rate of the instability. The validity of the linear theory for R-T instability has been confirmed for a wide parameter range via qualitative observation of the jet breakup process.

Another important parameter that does not appear explicitly in equation 40.16 is the dynamic pressure ratio M that determines the rate of atomization and hence the intact length of the liquid stream [23]. This ratio is defined as

$$M = \frac{\rho_g U_{Gas}^2}{\rho_l U_{Liquid}^2} \quad (40.18)$$

The dimensionless intact length of the liquid stream can be defined as $L/D_l \approx 6/\sqrt{M}$ and in coaxial injectors used under usual tablet coating parameters, M is typically large (of the order of 100). The gas potential cone length is approximately $6b_g$. For efficient atomization, it is desirable that the gas potential cone length be equal to or larger than the liquid intact length so that the primary atomization is completed before the gas velocity starts to decrease. This requirement is expressed by

$$\frac{b_g \sqrt{M}}{D_l} > 1 \quad (40.19)$$

It is worth noting that for the flow rates and atomizers utilized in pharmaceutical tablet coating, equation 40.19 is satisfied easily, with values typically exceeding 10, strongly suggesting that atomization is typically quite rapid and efficient. Finally, the fluid jets under the conditions of interest here are laminar at the injector nozzle but would potentially become turbulent if the flow rates are significantly increased or the gas contraction ratio decreased. Turbulent conditions of the liquid stream at the nozzle discharge plane would have little effect on the atomization process, while turbulent conditions in the high-speed gas stream would require altering the exponent of Re_{bg} in equation 40.16.

40.5.1 Pattern Air Effect on the Liquid Breakup

The previous model is derived for two coaxial streams of liquid and air, without taking into account the effect of pattern air on the liquid breakup. The effect of pattern air can be included in the model in two different ways.

In spray guns where the pattern air nozzles are oriented in a way that the jets impinge on the spray axis at a distance from the atomization nozzle at which liquid breakup is already underway, the effect of pattern air on droplet size is minor. Because the droplet size is set by the wavelength of the secondary R-T instability, pattern air does not modify it when the instability is already growing. Therefore, for this type of atomizers, only the gas flow through the atomization nozzle affects the breakup. In atomizers where the pattern air comes from the same supply line as the atomizer air, the effect of pattern air is simply to reroute some of the available gases from the atomizing nozzle to the side jets, and therefore reducing the atomizing air flow rate and the exit velocity. To account for that, in the case when pattern air does not have independent flow control and measurements, it is easiest to measure the cross-sectional area of both the atomizing nozzle and the pattern air and to prorate the total flow rate into the two air streams according to the pressure loss across the two nozzles (which is proportional to the diameter to the fourth power). The atomizing air is the value that is input into the breakup model, setting U_g and consequently the Weber number and the gas Reynolds number in the model.

In spray guns where the position and orientation of the pattern air nozzles result in the side jets impinging on the spray axis very close to the atomizing nozzle, in the region where the instability of the liquid jet is still developing, the impact of the pattern air jets on the droplet size resulting from the atomization is very strong. This effect can be modeled by considering that the atomizing air and the pattern air mix while the liquid jet is breaking up. The resulting air stream has axial momentum equal to the sum of all the jets but no radial momentum as the two pattern air side jets have equal but opposite values that cancel when the two streams mix. One can implement this into the previously derived model by computing the gas jet as the mean air velocity of the two streams, $U_g = (m_{atom}U_{atom} + m_{pattern}U_{pattern}\cos\theta)/(m_{atom} + m_{pattern})$, where θ is the angle between the pattern air jet axis and the atomizing air jet axis. Although there are a number of approximations underlying this model, most importantly that the air streams mix instantly and that the instability is not modified by the lack of axial symmetry induced by the pattern air, it has been proven to provide accurate predictions for the droplet size under typical pharmaceutical coating conditions.

40.6 SCALE-UP OF ATOMIZATION

The tablet coating process is often developed and optimized at a small scale and subsequently scaled up to larger equipment and batch size. The coating and atomization process are developed and optimized in a small lab coater, typically at 1–10 L capacity using a few kilograms of tablets, through detailed experimentation. The main goal of this experimentation is to explore the design space of the process parameters, such as the spray rate, atomization air flow rate, and pattern air flow rate, which will produce spray that will result in the desired film coat on the tablet surface. When this process is scaled up, process parameters should be selected such that coating spray will be similar to the one generated at the smaller scale.

An important spray characteristic that should be kept the same across different scale experiments is the SMD droplet size. In this example, equation 40.16 will be used to calculate the SMD for 1–2, 30, and 500 L coating pans based on the spray parameters given in Table 40.4. Furthermore, this

TABLE 40.4 Process Variables and Parameters for Spray Atomization for Three Different Scale Coaters Used in Scale-Up Example. Fluid Properties for Opadry II White (Colorcon Code: Y-30-18037)

Spray Parameters and Variables	1–2 L Scale	30 L Scale	500 L Scale
Input parameters			
Inner diameter of liquid nozzle (m)	1.00E − 03	1.50E − 03	1.50E − 03
Outer diameter of atomizing gas cap (m)	3.00E − 03	4.00E − 03	4.00E − 03
Inner diameter of atomizing gas cap (m)	2.00E − 03	3.00E − 03	3.00E − 03
Diameter of the pattern air side orifices (m)	1.0E − 03	2.0E − 03	2.0E − 03
Liquid surface tension (N/m)	0.047	0.047	0.047
Liquid density (kg/m^3)	1050.0	1050.0	1050.0
Infinite shear rate viscosity (kg/m s)	0.0981	0.0981	0.0981
Spray rate (g/min per spray gun)	10	60	120
Liquid mass flow rate (kg/s)	1.67E − 04	1.00E − 03	2.00E − 03
Air density (kg/m^3)	1.225	1.225	1.225
Air viscosity (kg/ms)	1.80E − 05	1.80E − 05	1.80E − 05
Gas volumetric flow rate (m^3/s)	1.18E − 03	2.92E − 03	3.41E − 03
Calculated values used in the model			
Liquid outlet velocity (m/s)	2.012E − 01	5.389E − 01	1.078E + 00
Gas outlet velocity (m/s)	214.61	247.97	289.63
Gas jet Reynolds number	7302.63	8437.946	9855.630
Convective velocity of the primary instability waves	7.284	8.711	10.608
Liquid Weber number	1120.311	2238.069	3043.828
Mass flux ratio	0.115	0.279	0.478
Dynamic pressure ratio	1315.525	246.988	84.239
Ohnesorge number	0.4416	0.3606	0.3606
Model output			
Sauter mean diameter (μm)	**48.4**	**48.4**	**48.4**

Note: Spray gun parameters are, for example purposes. They are not selected based on a specific spray gun model. Pattern air is not independent in this example.

TABLE 40.5 Impact of Coating Formulation Solids Content on the Mean Droplet Size (i.e., Sauter Mean Diameter) and Fluid Properties for Opadry II White (Colorcon Code: OY-LS-28914)

Spray Parameters and Variables	10% Solids	12% Solids	15% Solids
Input parameters			
Inner diameter of liquid nozzle (m)	1.20E − 03	1.20E − 03	1.20E − 03
Outer diameter of atomizing gas cap (m)	3.10E − 03	3.10E − 03	3.10E − 03
Inner diameter of atomizing gas cap (m)	2.60E − 03	2.60E − 03	2.60E − 03
Diameter of the pattern air side orifices (m)	1.50E − 03	1.50E − 03	1.50E − 03
Liquid surface tension (mN/m)	48.22	47.66	46.67
Liquid density (kg/m^3)	1020.0	1030.0	1040.0
Infinite shear rate viscosity (kg/m s)	0.095	0.181	0.419
Spray rate (g/min per spray gun)	60	60	60
Atomizing gas volumetric flow rate (m^3/s)	1.42E − 03	1.42E − 03	1.42E − 03
Calculated values used in the model			
Gas outlet velocity (m/s)	245.27	245.27	245.27
Model output			
Sauter mean diameter (μm)	**52.4**	**72.9**	**116.8**

Note: Spray gun parameters are, for example, purposes and are not selected based on a specific spray gun model. Pattern air is not independent in this example.

model will be used to set the atomization gas flow rate that will produce spray with same SMD for the given parameters.

In the example given in Table 40.5, the coating fluid solids content is increased from 10% to 12%, and then to 15%. This increase in % solids causes a dramatic change in fluid viscosity and also the droplet size as can be seen when equation 40.16 is used to predict droplet size. If atomization is left unchanged, the resulting droplet size of 116.8 μm from the 15% solids formulation would likely be too large. The model can then be used to determine the appropriate atomization air flow to achieve an acceptable droplet size.

TABLE 40.6 Impact of Coating Spray Gun Equipment Change on the Mean Droplet Size (i.e., Sauter Mean Diameter)

Spray Parameters And Variables	Spray Gun Setup 1 Process A	Spray Gun Setup 2 Process A	Spray Gun Setup 2 Process B
	Original Equipment and Process	Equipment Change/No Change in Process	Atomizing Gas Decreased to Match Droplet Size of Setup 1
Spray pattern type	Round	Ellipse	Ellipse
Fluid nozzle part #	PF2850	PF28100	PF28100
Air cap part #	120	PA110228-45	PA110228-45
Atomization air annulus (mm)	2.03	1.01	1.01
Input parameters			
Inner diameter of liquid nozzle (mm)	0.95	0.71	0.71
Outer diameter of the atomizing gas cap (mm)	3.33	2.79	2.79
Inner diameter of the atomizing gas cap (mm)	1.3	1.78	1.78
Diameter of the pattern air side orifices (m)	0	0.71	0.71
Liquid surface tension (mN/m)	43.9	43.9	43.9
Liquid density (kg/m^3)	1070.0	1070.0	1070.0
Infinite shear rate viscosity (kg/ms)	0.039	0.039	0.039
Spray rate (g/min per spray gun)	10	10	10
Atomizing gas volumetric flow rate (m^3/s)	1.18E − 03	1.18E − 03	6.65E − 04
Calculated values used in model			
Atomizing gas outlet velocity (m/s)	159.8	267.1	150.7
Model output			
Sauter mean diameter (μm)	**49.9**	**25.8**	**49.7**

Note: Fluid properties are for Opadry II White 20% solids (Colorcon Code: 85F18422). Example spray guns are available from Spraying Systems Co.

In the example given in Table 40.6, the coating process equipment is changed when product is processed with two coaters with different spray guns (i.e., setup 1 and setup 2). If the original process conditions (i.e., atomization gas volumetric flow rate) are kept and spray setup is changed from 1 to 2, the resulting droplet size is predicted to decrease from 49.9 to 25.8 μm due to differences in atomizer geometry. This reduction of droplet size may change coating appearance or influence efficiency of the coating process, and so the model can be used to predict the atomization air flow rate necessary to maintain the droplet size at about 50 μm (i.e., process B).

40.7 CONCLUSIONS

In this chapter, we have outlined the physical processes that are involved in the spray coating of pharmaceutical tablets. We have focused on the atomization of the complex rheology liquid formulations used in the coating because the size and spatial distribution of the droplets formed during the atomization process have a strong influence on the success or failure of the overall process. The essential physics of the problem have been described in detail and a mathematical model for the prediction of the average droplet size has been presented. The hydrodynamic stability analysis of the two parallel streams of gas and liquid characteristic of two-fluid atomizers, commonly used in the coating operations, provides quantification of the functional dependency of the resulting droplet size with the coating formulation properties (surface tension, viscosity) and the operational conditions (liquid and gas flow rates). We have also investigated the spatial distribution of droplets within the spray and the effect of the symmetry breaking side gas jets (commonly referred to in the industry as pattern air). We show the redistribution of droplets into an elliptical cross section spray due to these side jets and a more uniform spray density. The conditions under which these side jets need to be taken into account in the calculation of the droplet size have been identified and a simple way to include them in the breakup model is presented.

The availability of quantitative, physics-based models such as the one presented in this chapter can change the way pharmaceutical manufacturing operations, like tablet coating, is approached. The capability of predicting droplet size and distribution onto the coated material allows the optimization of the process with reduced input from experiments, as opposed to the traditional empirical approach that requires costly and time-consuming tests at each step of the process design (lab, pilot, and full scale). The improved understanding of the relationship between coating liquid rheology and coating outcomes can also provide more flexibility in the use of advanced coating formulations, reducing the barriers to apply novel coatings to improve a product or process.

ACKNOWLEDGMENTS

The authors would like to acknowledge Pfizer Inc. for supporting the research reported in this chapter. In addition, Douglas M. Kremer is thanked for his contributions to set up and develop this collaborative research effort. We also wish to thank Dr. Emil Hopfinger for his assistance in developing the atomization model described in Section 40.6. Katie Osterday is acknowledged for assisting in the viscosity measurements as well as in the measurements of the droplet size distributions using PDPA techniques. Daniel Bolleddula is acknowledged for helping set up the two-fluid coaxial atomizers for the independent pattern air experiments.

REFERENCES

1. Smith GW, Macleod GS, Fell JT. Mixing efficiency in side-vented coating equipment. *AAPS PharmSciTech*, 2003;4(3): E37.
2. am Ende MT, Berchielli A. A thermodynamic model for organic and aqueous tablet film coating. *Pharm. Dev. Technol.*, 2005;10(1):47–58.
3. Kalbag A, et al. Inter-tablet coating variability: residence times in a horizontal pan coater. *Chem. Eng. Sci.*, 2008;63(11): 2881–2894.
4. Ebey GC. A thermodynamic model for aqueous film-coating. *Pharm. Technol.*, 1987;11(4):40, 42–43, 46, 48, 50.
5. Porter SC, Felton LA. Techniques to assess film coatings and evaluate film-coated products. *Drug Dev. Ind. Pharm.* 36(2): 128–142.
6. McGinity JW, Felton LA, editors, *Aqueous Polymeric Coatings for Pharmaceutical Dosage Forms*, 3rd edition, Informa Healthcare, 2008, p. 488.
7. Turton R, Cheng XX. The scale-up of spray coating processes for granular solids and tablets. *Powder Technol.*, 2005;150(2): 78–85.
8. Gibson SHM, Rowe RC, White EFT. Mechanical properties of pigmented tablet coating formulations and their resistance to cracking. I. Static mechanical measurement. *Int. J. Pharm.*, 1988;48(1–3):63–77.
9. Aliseda A, et al. Atomization of viscous and non-Newtonian liquids by a coaxial, high-speed gas jet. Experiments and droplet size modeling. *Int. J. Multiphase Flow*, 2008;34 (2):161–175.
10. Mueller R, Kleinebudde P. Comparison of a laboratory and a production coating spray gun with respect to scale-up. *AAPS PharmSciTech*, 2007;8(1):3.
11. Mueller R, Kleinebudde P. Comparison study of laboratory and production spray guns in film coating: effect of pattern air and nozzle diameter. *Pharm. Dev. Technol.*, 2006. 11(4):425–433.
12. Tobiska S, Kleinebudde P. Coating uniformity: influence of atomizing air pressure. *Pharm. Dev. Technol.*, 2003;8(1): 39–46.

13. Thombre AG, et al. Osmotic drug delivery using swellable-core technology. *J. Control. Release*, 2004;94(1):75–89.
14. Thombre AG, et al. Asymmetric membrane capsules for osmotic drug delivery. II. In vitro and in vivo drug release performance. *J. Control. Release*, 1999;57(1):65–73.
15. Cole G, Hogan J, Aulton ME. *Pharmaceutical Coating Technology*, Informa Healthcare, 1995.
16. de Gennes PG, Brochard-Wyart F, Quéré D. *Capillary and Wetting Phenomena: Drops, Bubbles, Pearls, Waves*, Springer, 2002.
17. Bachalo WD. Experimental methods in multiphase flows. *Int. J. Multiphase Flow*, 1994;20 (Suppl.):233–259.
18. Williams FA. *Combustion Theory*, 2nd edition, Addison Wesley, 1985.
19. Varga CM, Lasheras JC, Hopfinger EJ. Initial breakup of a small-diameter liquid jet by a high speed gas stream. *J. Fluid Mech.*, 2003;497:405–434.
20. Marmottant P. *Atomisation d'un courant liquide dans un courant gazeux*, Institut National Polytechnique de Grenoble, Grenoble, 2001.
21. Joseph DD, Beaver GS, Funada T. Rayleigh–Taylor instability of viscoelastic drops at high Weber numbers. *J. Fluid Mech.*, 2002;453:109–132.
22. Mansour A, Chigier N. Air-blast atomization of non-Newtonian liquids. *J. Non-Newtonian Fluid Mech.*, 1995;58:161–194.
23. Lasheras JC, Hopfinger EJ. Liquid jet instability and atomization in a coaxial gas stream. *Ann. Rev. Fluid Mech.*, 2000;32: 275–308.

41

THE FREEZE-DRYING PROCESS: THE USE OF MATHEMATICAL MODELING IN PROCESS DESIGN, UNDERSTANDING, AND SCALE-UP

VENKAT KOGANTI AND SUMIT LUTHRA
Pharmaceutical Development, Pfizer Global Research and Development, Pfizer, Inc., Groton, CT, USA

MICHAEL J. PIKAL
University of Connecticut, Storrs, CT, USA

41.1 INTRODUCTION

Freeze-drying, also termed "lyophilization," is a drying process employed to convert solutions of labile materials into solids of sufficient stability for distribution and storage. A typical production scale freeze-dryer consists of a drying "chamber" containing temperature-controlled shelves that is connected to a "condenser" chamber via a large valve. The condenser chamber houses a series of plates or coils capable of being maintained at very low temperature (i.e., less than −50°C). One or more vacuum pumps in series are connected to the condenser chamber to achieve pressures in the range of 0.03–0.3 Torr in the entire system during operation. A commercial freeze-dryer may have 10–20 shelves with a total load on the order of 50,000 10 cc vials. The objective in a freeze-drying process is to convert most of the water into ice in the "freezing stage," remove the ice by direct sublimation in the "primary drying stage," and finally remove most of the unfrozen water in the "secondary drying" stage by desorption. The water removed from the product is reconverted into ice by the condenser.

In a typical freeze-drying process, an aqueous solution containing the drug and various formulation aids, or "excipients," is filled into glass vials, and the vials are loaded onto the temperature-controlled shelves. The shelf temperature is reduced, typically in several stages, to a temperature in the vicinity of −40°C, thereby converting nearly all of the water into ice. Some excipients, such as buffer salts and mannitol, may partially crystallize during freezing, but most "drugs," particularly proteins, remain amorphous. The drug and excipients are typically converted into an amorphous glass also containing large amounts of unfrozen water (15–30%) dissolved in the solid (i.e., glassy) amorphous phase. Thus, most of the desiccation actually occurs during the freezing stage of the freeze-drying process. After all water and solutes have been converted into solids, the entire system is evacuated by the vacuum pumps to the desired control pressure, the shelf temperature is increased to supply energy for sublimation, and primary drying begins. Due to the large heat flow required during primary drying, the product temperature runs much colder than the shelf temperature. The removal of ice crystals by sublimation creates an open network of "pores" that allows pathways for escape of water vapor out of the product. The ice–vapor boundary (i.e., the boundary between frozen and "dried" regions) generally moves from the top of the product toward the bottom of the vial in roughly planar fashion as primary drying proceeds. Primary drying is normally the longest part of the freeze-drying process. Primary drying times on the order of days are not uncommon, and in rare cases, weeks may be required for a combination of poor formulation and suboptimal process design. While some secondary drying does occur during primary drying (i.e., desorption of water from the amorphous phase occurs to a limited extent once the

Chemical Engineering in the Pharmaceutical Industry: R&D to Manufacturing Edited by David J. am Ende

ice is removed from that region), the start of secondary drying is normally defined, in an operational sense, as the end of primary drying (i.e., when ice is removed). Of course, since not all vials behave identically, some vials enter secondary drying while other vials are in the last stages of primary drying. When the judgment is made that all vials are devoid of ice, the shelf temperature is typically increased to provide the higher product temperature required for efficient removal of the unfrozen water. The final stages of secondary drying are normally carried out at shelf temperatures in the range of 25–50°C for several hours. Here, since the demand for heat is low, the shelf temperature and the product temperature are nearly identical.

Since freeze-drying plants are very expensive and process times are often long, a freeze-dried dosage form is relatively expensive to produce. Indeed, because of both cost and ease of use, a "ready-to-use" solution is the preferred option for a parenteral dosage form, particularly if the solution can withstand terminal heat sterilization. However, most parenteral drugs undergo excessive degradation during terminal sterilization. Even if sterility requirements may be satisfactorily met without terminal sterilization (i.e., sterile processing), many drugs do not have sufficient stability in the solution state to allow the long-term storage required for pharmaceutical products. Certainly, terminal sterilization is not an option for a protein product. Indeed, many proteins are insufficiently stable in aqueous solution, even when refrigerated, to allow storage for more than a few months without suffering significant degradation. Of course, some proteins are quite stable in aqueous solution, insulin being the classic example of a "solution stable" protein product [1]. When an aqueous solution does not have sufficient stability, the product must be produced in solid form. At least for small molecules, stability normally increases in the order: solution $\ll$ glassy solid $<$ crystalline solid [2–4], likely a result of restricted motion in solids with the high degree of order in the crystalline solid limiting reactivity even further. It should also be noted that the enhanced stability upon crystallization of a solid noted for small molecules may not extend to proteins. Although a direct experimental comparison is limited to one example, that of insulin, crystalline insulin is actually significantly less stable than amorphous freeze-dried insulin [5]. Since pharmaceutical proteins cannot generally be produced on a commercial scale by crystallization, a glassy solid is usually the only solid-state option.

Freeze-drying [6–9] and spray drying [10–12] are drying methodologies in common use in the pharmaceutical industry that are suitable for the production of glassy solids. Freeze-drying is basically a low temperature process. In general, a protein formulation can be dried to on the order of 1% water or less without any of the product exceeding 30°C. Thus, conventional wisdom states that freeze-drying is less likely to cause thermal degradation than a "high temperature" process, such as spray drying. However, it must be noted that due to self cooling as the water evaporates, the product temperature in a spray drying process is far less than the input air temperature, and residence times in the dryer are very short. Indeed, it has been shown that, suitably formulated, stable protein glasses may be produced by direct evaporative drying (i.e., drying without freezing) [13, 14]. Such direct evaporative drying process may involve spray drying or alternate new technologies [15]. While some of the factual material in reference 14 has been challenged [16], it must be admitted that freeze-drying is not the only process by which proteins solutions may be successfully converted into "stable" glasses.

Historically, freeze-drying is the method of choice for products intended for parenteral administration. Sterility and relative freedom from particulates are critical quality attributes for parenterals. Largely because the solution is sterile filtered immediately before filling into the final container, and further processing is relatively free of exposure to humans, a freeze-drying process maintains sterility and "particle free" characteristics of the product much easier than processes that must deal with dry powder handling issues, such as dry powder filling of a spray dried or bulk crystallized powder. Indeed, with modern robotics automatic loading systems [17], humans can be removed from the sterile processing area entirely, at least in principle. Furthermore, since the vials are sealed in the freeze-dryer, moisture and headspace gas can easily be controlled, an important advantage for products whose storage stability is adversely affected by residual moisture and/or oxygen. Since the critical heat and mass transfer characteristics for freeze-drying are nearly the same at the laboratory scale as in full production, resolution of scale-up problems tends to be easier for a freeze-drying process than for spray drying, at least in our experience. Also, development of a freeze-dried product requires less material for formulation and process development, a particularly important factor early in a project.

While freeze-drying has a long history in the pharmaceutical industry as a technique for stabilization of labile drugs, including proteins, many proteins suffer irreversible change, or degradation, during the freeze-drying process [18–22]. Even when the labile drug survives the freeze-drying process without degradation, the resulting product is rarely found perfectly stable during long-term storage, particularly when analytical techniques with a sensitivity to detect low levels of degradation (i.e., $\approx$0.1%) are employed. Both small molecules [2–4, 23] and proteins [24–27] show degradation during storage of the freeze-dried glass. In many cases, instability is serious enough to require refrigerated storage [24, 25, 28].

Stability problems are most often addressed by a combination of formulation optimization and attention to process control. Lyoprotectants are added to stabilize the protein during the freeze-drying process as well as to provide storage stability, and the level and type of buffer is optimized. Optimization of the freezing process may be critical, control of product temperature during drying is critical for products that tend to suffer cake "collapse" during primary drying, and control of residual moisture is nearly always critical for storage stability. Formulation and process are interrelated in that the process design depends on formulation, and process variations, particularly freezing variations, can change the physical state of the formulation. A bad formulation can be nearly impossible to freeze-dry, and even with a well designed formulation, a poorly designed process may require more than a week to produce material of suboptimal quality. While blind empiricism may, in time, yield an acceptable formulation and process, an appreciation for the materials science of amorphous systems and some understanding of heat and mass transfer relevant to freeze-drying are needed for efficient development of freeze-dried pharmaceuticals. Obviously, one also requires at least a phenomenological understanding of the major degradation pathways specific to the protein under consideration.

Once a suitable formulation and process are developed in the laboratory, hopefully at least close to optimized, the laboratory process needs to be transferred to manufacturing. While freeze-drying is, in many ways, a relatively easy process to scale-up since the volume and the nature of the primary container system is independent of scale (i.e., the same fill volume and vial in both laboratory and manufacturing), there do exist a number of differences between laboratory and manufacturing. First, the timescales are often far different for some stages of the process, the most notable being the time required for loading a manufacturing dryer being much longer than the corresponding time in the laboratory. Thus, particularly if the relative humidity in the vicinity of the freeze-dryer is high, one may experience condensation on the shelves during loading a production dryer but not during the corresponding operation in the laboratory. Such condensation may cause freezing variations that have consequences for the drying process as well. In addition, it is often possible to change shelf temperature very rapidly in the laboratory, but due to greater thermal mass in the manufacturing environment, the production dryer may be unable to match the laboratory dryer's performance. Secondly, there may exist heat and mass transfer differences between the laboratory dryer and the production dryer that require a slightly different shelf temperature profile with time for manufacturing than was used for the laboratory process [29]. Since the objective is to maintain the same product temperature history during the process in manufacturing that was validated in the laboratory, the shelf temperature versus time program in manufacturing may well be different than found optimal for the laboratory [29]. In addition, the maximum heat and mass transfer that the dryer can handle without loosing control varies with the dryer design, and one may find that a process that runs very well in the laboratory dryer may overload the manufacturing dryer and cause loss of chamber pressure control [29]. However, the most important and most troublesome scale-up issue is often the difference in freezing behavior that one experiences in the Class 100 "clean" environment of the production operation relative to the relatively high particulate content in the laboratory air. Thus, typically [30] supercooling is greater in manufacturing than in the laboratory, and since the size of the ice crystals (and resulting pores in the dry cake) decreases with increasing degree of supercooling, the laboratory produces larger ice crystals, larger pores, and less resistance to vapor transport. The net result is that the primary drying time is shorter in the laboratory than in manufacturing, and the product temperature runs colder in the laboratory than in manufacturing. The effect can be significant, with primary drying running from 10% to 30% longer in manufacturing [30–32]. Thus, the material being freeze-dried in the laboratory is typically not representative of the material being dried in production, simply because the freezing behavior is different. Frequently, the process is arbitrarily adjusted in an attempt to compensate for this anticipated difference. That is, the duration of primary drying is extended so that one may be reasonably confident that in the production batch all vials are devoid of ice when the shelf temperature is increased to facilitate secondary drying. However, one is never sure of how much extension is necessary, so there are better solutions to the problem, as discussed in more detail later in this chapter.

Freeze-drying is one of the few unit operations where the underlying physics is relatively well understood, and the theoretical models are not mostly empirical. Thus, modeling can be used very effectively for process design and scale-up using simple steady-state models [33] or more elaborate non-steady-state models that address both sublimation and desorption drying [34]. Use of the theoretical models, with appropriate input data to define heat and mass transfer parameters, allows greater insight into the impact of changes in operating conditions on product quality and brings much greater efficiency into robustness testing than possible with the empirical testing characteristic of pure statistical approaches.

This chapter begins with a detailed discussion of the freezing process, with suggestions for circumventing the problems arising from freezing differences between laboratory and manufacturing, and includes a brief discussion of the few attempts to model freezing behavior. We then continue with a discussion of drying behavior, which includes a detailed discussion of the utility of the various theoretical models that can be effectively utilized in process design and scale-up. We conclude with a brief discussion of modeling vapor flow within the freeze-dryer, a topic of importance but one that has received little attention in the literature.

41.2 FREEZING PROCESSES

Freezing is the first step in the lyophilization process where most of the water is separated from the solute as ice crystals.

It is also a very important step as the structure of the frozen matrix governs the rate of sublimation and desorption. Figure 41.1 shows the shelf and product temperature profiles of product frozen in vials on a laboratory-scale shelf freeze-dryer. The data shown are for 5% w/w sucrose solution. During the cooling step, the solution remains liquid well below the equilibrium freezing temperature (−13°C for the example shown in Figure 41.1). Ice nucleation occurs in the supercooled liquid in a stochastic manner and proceeds rapidly. Nucleation is quickly followed by crystal growth. Ice nucleation can occur by two mechanisms: homogeneous or heterogeneous. Homogeneous nucleation is defined as aggregation of pure material (in this case water, which undergoes homogenous nucleation at −40°C) [35]. Heterogeneous nucleation, on the other hand, occurs when aggregates form on foreign solids such as dirt and container wall. Homogeneous ice nucleation does not occur in pharmaceutical freeze-drying operations due to inevitable presence of foreign solids. The foreign surface acts as a catalyst thus reducing the surface free energy for formation of the nuclei. This lowering of free energy means that degree of supercooling is less as compared to homogeneous nucleation. The size of the ice crystals is dependent on the nucleation and growth rate, which are both governed by the degree of supercooling. A higher nucleation rate would result in smaller ice crystals and vice versa. The stochastic nature of nucleation would mean different degrees of supercooling and thus differences in ice crystal size and distribution across the batch and significant differences between batches, particularly when laboratory and production batches are compared [30]. In general, it is desirable to have larger ice crystals and narrow size distribution. Therefore, control over degree of supercooling is desirable. However, to date, the methods to control ice nucleation are only in the experimental stage and have not yet been applied in commercial setting. These issues are discussed in more detail later.

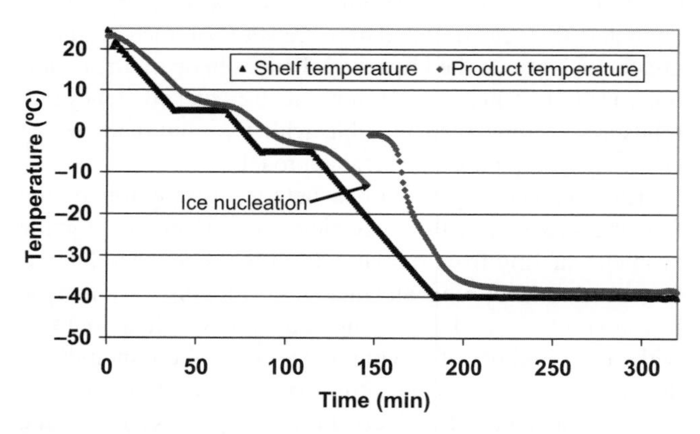

FIGURE 41.1 Typical shelf and product temperature profile during freezing for a formulation containing amorphous solute. The data are for 5% w/w sucrose solution lyophilized in a laboratory-scale freeze-dryer. The product temperature profile is recorded with a thermocouple located at the bottom center of the vial. The product temperature profile shows significant supercooling before ice nucleation. Ice nucleation causes an increase in product temperature up to approximately the equilibrium melting point. The product temperature stays high as ice crystallization proceeds and it eventually starts to drop and follow the shelf temperature.

41.2.1 Cooling Rate

In a typical freezing process the only variable that is under direct control is the cooling rate of the shelf fluid. Further, the range of cooling rates achievable is not particularly large. At most, the samples can be cooled at ≈2°C/min. It has been noted that a cooling rate of 1°C/min is generally optimal as it provides moderate supercooling and reasonably fast freezing rate [36].

41.2.2 Solute Concentration and Phase Changes

When liquid water is removed as ice crystals as freezing proceeds, the solutes (active pharmaceutical ingredient (API) and excipients) are concentrated in the unfrozen region between the ice crystals. The concentration continues until the solute crystallizes or converts into an amorphous glassy system. The physical nature of the solutes and their properties has a profound impact on the rest of the process and merits further discussion. Upon completion of freezing, the solute matrix may be completely amorphous, or crystalline, or a mixture of amorphous and crystalline phases. The importance of this matrix is in the mechanical structure it provides for efficient drying and formation of an elegant product. However, to maintain this solid matrix it is important that the product temperature during primary drying should not increase above a critical temperature known as the collapse temperature. In completely amorphous matrices, the collapse temperature (T_c) is related to the glass transition temperature of the frozen concentrate (T_g') and is generally about 2°C higher than T_g' [37]. In a completely crystalline matrix, T_c is equal to the temperature of the eutectic melt (T_{eu})[1]. The relationship in a mixed amorphous and crystalline system is governed by the relative ratio of amorphous and crystalline phase [38]. In most cases, mixed systems are dominated by the crystalline phase by design so the effective collapse temperature is close to the eutectic melt even though the collapse temperature (and T_g') of the amorphous phase may be much lower. These crystalline phases designed into the product are called bulking agents and are added to provide mechanical strength and/or raise the effective collapse temperature. Mannitol is the most commonly used bulking agent.

[1] A rigorous definition of "eutectic melting" in ternary systems requires all the components to exist in the crystalline state. This is not the case in the majority of pharmaceutical systems for freeze-drying. However, we have retained the term "eutectic" to describe solute + ice melting, as is the practice in the pharmaceutical community.

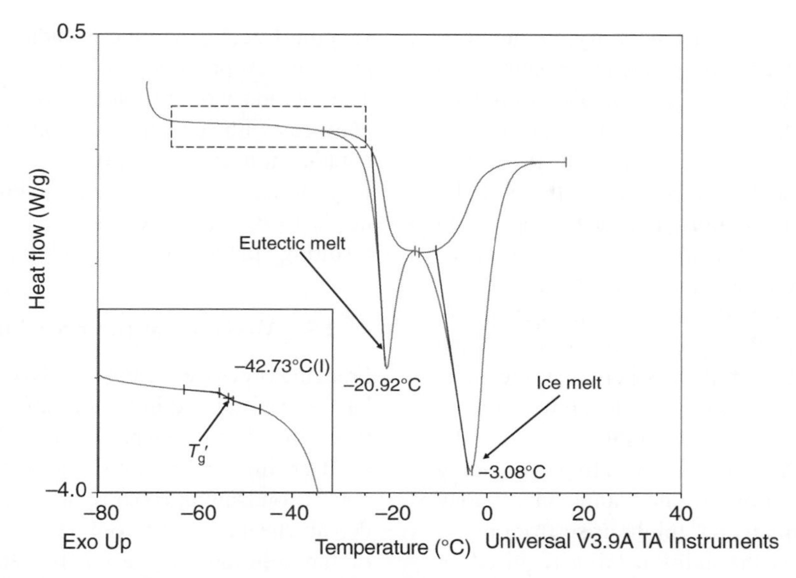

FIGURE 41.2 Thermogram of a frozen solution containing amorphous and crytalline solutes obtained by differential scanning calorimetry. The heating curve shown displays the glass transition temperature of the amorphous frozen concentrate (T_g'), Eutectic melt of crystalline component (T_{eu}) and the ice melt event. The data show that there is approximately 20°C window between T_g' and T_{eu} where annealing could be carried out.

Figure 41.2 shows the thermogram for a frozen solution containing amorphous and crystalline solutes. The heating curve shows the presence of T_g' and T_{eu} followed by the ice melting endotherm.

41.2.3 Annealing

Annealing is carried out by holding samples isothermal above the T_g' for few hours. It is relatively common to employ an annealing step to facilitate solute crystallization of API or bulking agents [39]. Figure 41.3 shows the shelf and product temperature profile for a solution containing solute that crystallized during freezing. The data shown are for 5% w/w mannitol solution freeze-dried on a laboratory scale freeze-dryer. The shelf temperature profile includes an annealing step at −20°C in this case. A distinct feature of the product temperature profile in this case is the appearance of bumps during the annealing step. These bumps are likely a result of heat released due to crystallization of solute and water during annealing. However, ice nucleation in adjacent vials could be another explanation for the observed bumps. Annealing is also currently the only commercially viable method to modify ice morphology. Annealing is believed to result in increased mean crystal size and narrower distribution due to Ostwald ripening [40]. When the frozen matrix is heated above T_g', the ice crystals below a critical size decrease in size and effectively "melt," and those larger than the critical size grow in a diffusion-dependent manner. The increase in mean size and narrower distribution are both advantageous as larger ice crystals means larger pores for water vapor to escape during primary drying and narrowing of size distribution results in batch homogeneity. The changes in primary drying time that have been reported are significant, such as a 3.5-fold increase in primary drying rate reported in one study [4]. This increase in primary drying rate is correlated with a decrease in mass flow resistance of the dry layer, as reported in another study [41]. However, there have been reported cases (and unpublished experience of the authors) where annealing has not led to a decrease in

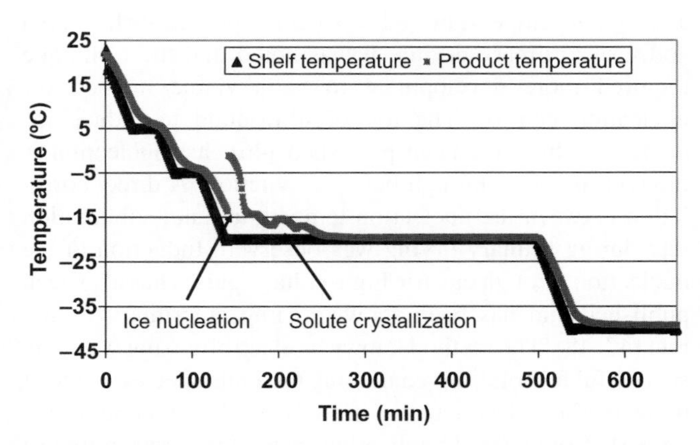

FIGURE 41.3 Typical shelf and product temperature profile for a formulation containing a crystallizable solute. The data are for 5% mannitol w/w solution freeze-dried in a laboratory-scale freeze-dryer. The shelf temperature profile incorporates an annealing step at -20°C. In addition to the ice nucleation event observed in Figure 41.1 thermal event related potentially to solute crystallization are apparent in the product temperature profile.

mass flow resistance or increase in primary drying rate [42]. There are other reasons why annealing may not be suitable in all cases. The increase in the size of ice crystals will reduce the surface area, which decreases desorption rate during secondary drying and/or prolong reconstitution time. The residual water content of the dried product may therefore be higher [43]. Annealing may also promote phase separation resulting in unintended crystallization of a solute such as a buffer, which would produce a large pH shift, or phase separation within the amorphous phase [44]. Such phase separations may adversely affect the in-process or storage stability of lyophilized protein products where lyoprotectants are used to improve physical and chemical stability and are required to be in the same phase as the protein to impart the stabilization effect. Thus, the benefit of annealing in reducing primary drying time and improving batch homogeneity may be offset by stability problems that might arise from phase separation and in some cases, the additional time required for annealing and secondary drying may negate the reduction in primary drying time. Certainly, optimization of the annealing step is required to achieve the greatest benefit.

41.2.4 Methods to Control Ice Nucleation

There are other methods that are currently under development to control the ice nucleation temperature. Ice fog technology is based on introducing cold nitrogen gas in the drying chamber to create an ice fog at the desired temperature of nucleation [45]. The concept was further [32] developed to evaluate the impact of nucleation temperature on mass flow resistance of the dry layer and primary drying time. An empirical direct correlation between specific surface area and mass flow resistance was also established. It was shown that a lower degree of supercooling, that is, nucleation occurring at a higher temperature led to smaller specific surface area and faster primary drying. It was noted that the technique required more development to be a viable method for nucleation control. The use of ultrasound to induce ice nucleation has also been published [46]. Ice nucleation is thought to occur through bubble cavitation. A direct correlation between ice nucleation temperature and sublimation rate during primary drying was observed. Induction of ice nucleation through electric high-voltage pulses has also been published and has been shown to impact primary drying rate [47, 48]. The method known as electrofreezing was only successful for solutions containing nonionic species. None of these methods has actually been applied to a commercial process. Commercial application may require retrofitting of existing units, which would be expensive in the least but may also be impractical in other cases. Nonetheless, these methods do demonstrate that that control of ice nucleation temperature has great potential benefit. An option that would not require any changes to the existing equipment would be preferable, such as induction of freezing by vacuum. A vacuum freezing method to induce top–down freezing arising from evaporative cooling has been published [49, 50]. The method would require further development to address concerns with secondary drying kinetics and residual water content similar to annealing. In addition, top–down freezing may increase the chances of vial breakage and the technique needs to be carefully controlled to avoid boiling and the resulting splattering of the solution.

41.2.5 Modeling of the Freezing Process

Freezing processes relevant to lyophilization, that is, in vials has been discussed in the literature [51]. For the purpose of modeling, the process has been divided into two parts: cooling and freezing. The cooling step has been modeled using Fourier's second law and is straightforward as the input parameters density, heat capacity, cooling rate, and thermal conductivity of the solution can be easily determined from published literature and experimentation. Modeling of the freezing step becomes more complicated as one would expect because the theories of nucleation and crystallization are less well defined than simple heat transfer. Also, the terms involving phase changes introduce additional parameters that at best are difficult to estimate or measure. Nonetheless, this approach was used to estimate mean ice crystal size and mass flow resistance of the dry layer. It is clear that modeling the freezing step is not as advanced as the drying steps and is not at a level where a scientist would determine the input parameters in the laboratory and plug the values into a simple model that produces useful quantitative results. However, the published data do confirm some general concepts that nucleation temperature and cooling rate both impact the mean ice crystal size and distribution and that increasing the nucleation temperature and lowering the cooling rate result in larger ice crystals, thus lower mass flow resistance of the dry layer. Additional progress in modeling the freezing process is needed so that the impact of process variations, for example, annealing, on the mean ice crystal size and distribution could be accessed and the mass flow resistance parameter for modeling the drying process could readily be obtained.

41.2.6 Scale-Up Issues

The manufacturing of clinical or commercial pharmaceutical products is conducted in sterile, particulate free, Class 100 areas whereas lyophilization cycle development is generally conducted in laboratories that are not at all "particle free." The particulate load, that is, heterogeneous nucleation sites, thus vary when scaling up from laboratory to clinical or commercial site. This means that the degree of supercooling in a production environment would be greater, leading to smaller ice crystals and higher resistance to mass flow through the dried layer. Therefore, primary drying parameters developed in the laboratory would generally lead to higher product temperature and longer drying time in the production environment [30].

41.2.7 Rational Freezing Process Development

A development scientist should take (essentially) the following approach to develop a rational freezing process.

1. *Selection of Cooling Rate*: The impact of cooling rate parameter has been discussed previously in this chapter. In summary, a cooling rate between 0.5 and 1°C/min would generally lead to moderate supercooling, which is uniform intravial and intervial, while providing a sufficiently fast freezing rate to avoid phase separation.
2. *Selection of Final Freezing Temperature and Time*: The lowest temperature during the freezing stage depends on the solidification temperature of the system. The temperature should be at least 2°C lower than the solidification temperature [36]. However, it is common to see from −40 to −50°C as empirical choice for lowest temperature, and while often not necessary also does not normally cause a significant problem. The hold time at the lowest temperature is a function of the fill volume. During development, the product temperature from thermocouple should be monitored as a function of time to determine the time required to reach the lowest desired temperature. In general, 1–2 h hold time is sufficient for fill volumes less than 1 cm and 2–4 h hold time is sufficient for fill volumes more than 1 cm. Fill volumes greater than 2 cm are generally not recommended.
3. *Selection of Annealing Conditions*: Choosing annealing conditions remains somewhat empirical and existing literature provides limited guidance on this aspect. As discussed above, annealing may be carried out with different objectives. If the formulation contains a component that must be crystallized, it has been suggested that annealing should be carried out at a temperature between T_g' and onset of the ice melt endotherm [36]. However, the temperature differential between the two may be quite large. The example shown in Figure 41.2 demonstrates this point where the difference between T_g' and T_{eu} is about 20°C. The exact choice of annealing temperature and time need to be optimized experimentally. Annealing must be carried out at a temperature that (a) is above the T_g' and (b) falls on the crystal growth curve. Therein lies the difficulty as the crystal growth curves are not readily available. For commonly used bulking agents such as mannitol and glycine annealing could be carried out at temperatures between −25 and −20°C for several hours to maximize crystallization, provided the fraction of crystallizing solute is high. If annealing is desired to bring about change in ice morphology and improving batch homogeneity for a totally amorphous system, one needs annealing temperatures relatively close to the onset of melting. It was shown for model amorphous systems such as sucrose and hydroxyethyl starch (HES) that annealing at temperatures between −10 and −2.4°C for 5–10 h was needed for maximum change in primary drying rate and ice morphology [40].
4. *Addressing Scale-Up Issues*: The difference in nucleation temperature between laboratory-scale and production-scale due to change in environmental particulate load could be eliminated if development work were carried out in the same environment. However, this is normally not practical. Therefore, currently, annealing is the only method that has been used to minimize or eliminate supercooling effects in a manufacturing environment. Further development of techniques to control ice nucleation may change this scenario in the future.

41.3 DRYING PROCESSES

After freezing, drying is the next step in freeze drying process. Mathematical representation of the drying problem can be described as follows: Once the solution is frozen, the vials are heated by raising the shelf temperature, resulting in sublimation of frozen ice initially and desorption of unfrozen water later. As drying proceeds from top to the bottom of the vial, the dried layers of the cake offer resistance to the water vapor flow due to sublimation of the ice from the layers underneath. The freeze-drying problem is hence a heat transfer (to the vial from shelf and surroundings) and mass transfer (transport of water vapor through porous dried layers and then from the main chamber to the condenser) problem that can be modeled utilizing the fundamentals of heat and mass transfer processes. Further details of the current state of knowledge in modeling this process are described below.

Mathematical modeling of the drying process provides methodology that streamlines experimental screening approaches for developing optimal freeze-drying cycles that produce a quality product in a robust process. A particularly important application of drying process modeling is in the area of freeze-drying process scale-up. A typical scale-up from laboratory-scale to commercial-scale freeze-drying will increase the shelf surface area available for freeze-dryer from 4.5 ft^2 to 220 ft^2 [52]. Figure 41.4 shows an image of a typical commercial-scale freeze-dryer. Heat transfer may differ, and differences in heterogeneous ice nucleation normally produce significant differences in ice structure and therefore in pore structure of the cake, which impacts mass transfer within the dry layer. Also, differences in dryer design may lead to differences in transport properties between laboratory and commercial dryers. A clear understanding of

FIGURE 41.4 Image of typical commercial-scale freeze-dryer. This freeze-dryer has a total shelf surface area of 39 m^2. It has 24 shelves and 6 trays per shelf. This image is obtained from Pfizer Kalamazoo manufacturing facility.

the dependence of drying kinetics on the heat and mass transfer characteristics of the vial and dried cake, respectively, and the impact of the differences between laboratory scale and commercial scale will facilitate rational scale-up of the freeze-drying process, avoiding expensive failures, and hence will result in efficient development of a robust process, thereby decreasing the cost and time of development [53].

41.3.1 Steady-State Heat and Mass Transfer Modeling

Although most reported modeling work uses nonsteady-state modeling techniques, there are some examples describing the use of simple steady-state theory to model the primary drying process. Using the pseudosteady-state approximation, solution of the heat and mass transfer equations has been obtained at several stages during primary drying phase, thus evaluating temperature and pressure profiles as a function of time [33]. Using this simple model, the authors studied the effect of the product temperature on drying time, effects of shelf temperature and chamber pressure were evaluated, and the optimum vial size to minimize primary drying time was identified. This simple model was also utilized to evaluate the effect of process nonuniformities (e.g., variability of vial heat transfer coefficient within the same freeze-drying run, nonuniform shelf temperature, product resistance variation) on the drying times and product temperature during primary drying.

Equations 41.1–41.4 describe a typical freeze-drying process where the solution to be freeze-dried is in a vial, which is placed on top of a temperature-controlled shelf. The steady-state approximation is used, meaning that all of the heat supplied from the shelf is utilized in subliming the ice from the interface [33]. See "Symbols" section for nomenclature details.

$$\frac{\mathrm{d}m}{\mathrm{d}t} = \frac{A_\mathrm{p} \times (P_0 - P_\mathrm{c})}{R_\mathrm{p}} \tag{41.1}$$

$$\Delta H_\mathrm{s} \times \frac{\mathrm{d}m}{\mathrm{d}t} = \frac{\mathrm{d}Q}{\mathrm{d}t} \tag{41.2}$$

$$\frac{\mathrm{d}Q}{\mathrm{d}t} = A_\mathrm{v} \times K_\mathrm{v} \times (T_\mathrm{s} - T_\mathrm{b}) \tag{41.3}$$

$$\Delta H_\mathrm{s} \times \frac{P_0 - P_\mathrm{c}}{R_\mathrm{p}} \times \frac{A_\mathrm{p}}{A_\mathrm{v} \times K_\mathrm{v}} = T_\mathrm{s} - T \tag{41.4}$$

EXAMPLE 41.1

Sucrose is common excipient used in parenteral formulations. Determine the product temperature during primary drying and length of primary drying time during lyophilizing a sucrose solution at the following given conditions. After building the model, utilize the model to perform an *in silico* robustness test at shelf temperatures that are ±3°C relative to shelf temperature set point and chamber pressure ±50 mTorr relative to the chamber pressure set point.

Conditions.

1. Shelf temperature set point (T_shelf): −25°C
2. Chamber pressure set point (P_chamber): 100 mTorr
3. Heat of sublimation of ice: 660 cal-g^{-1}
4. Average dry layer resistance (R_p): 3 cm^2-h-Torr-g^{-1}
5. Overall heat transfer coefficient between vial and surroundings (K_v): 0.00042 cal/(s-cm^2-K)
6. Vial dimensions: Inner cross-sectional area = 5.85 cm^2; Outer cross-sectional area = 7.08 cm^2
7. Formulation details: Solids concentration = 0.05 g solid/g liquid; fill volume = 5 mL

Solution

Part A. Equation 41.4 can be solved for the unknown interface temperature, *T*, using the given parameters. The solution can be obtained manually or by using the Solver feature in Excel. The interface temperature, *T*, and interface vapor pressure, P_0, are related as follows:

$$P_0 = 2.698 \times 10^{10} \times \exp(-6144.96/T)$$

where *T* is in Kelvin and P_0 is the vapor pressure of ice in Torr. Following the above procedure, one finds the average interface temperature is −34.5°C and the primary drying time is 31.5 h. Once P_0 is obtained, dm/dt can be calculated using equation 41.1. Total amount of ice to be sublimed (Δm)

can be calculated from the fill volume and solids concentration. Using these two quantities, time required to sublime all the ice (primary drying time) can be calculated as $\Delta m/(dm/dt)$. In a laboratory-scale freeze-drying run (same operating conditions described in this example) with 5 wt% sucrose as the solution (in same vial whose dimensions are described in this example), the product temperature was measured by placing a thermocouple at the bottom of the vial during drying. It was found that the product temperature was $-35°C$, and the primary drying time was 28 h. The vial was located in the center of the freeze-dryer and the end point of primary drying was considered to be the point when the product temperature reading starts to increase from the steady-state value to reach the shelf temperature set point. One simplification in the above equations is to neglect the difference between the sublimation interface temperature and bottom temperature difference. To obtain more accurate results, this difference can also be accounted for [54].

Part B. Now, we have a mathematical model that describes the primary drying process of sucrose. During a manufacturing process operation, set point deviations may occur due to a variety of reasons. The model can be used to predict the effect of set point deviations and thereby test process robustness. Table 41.1 lists the product temperatures and drying times for four extreme deviations from the set points. Also listed in the table are the changes in product temperature and drying time when compared to original set point ($T_{shelf} = -25°C$ and $P_{chamber} = 100$ mTorr). As noted above, the model predictions can differ from the experimental values. Therefore, understanding the relative changes in product temperature and drying time as a result of process deviations is one of the useful output from model predictions. Model predictions for the original set point are also listed in Table 41.1. This example demonstrates one of the several ways in which a successful model can be used in lyophilization cycle development to aid in choosing the operating conditions at the laboratory scale and also investigates the robustness of the process.

Recently an interactive modeling tool has been proposed, assuming quasisteady-state heat transfer in frozen layer and in dry product region as well as quasisteady-state mass transfer in the dried layer [55]. This software that is based on a one-dimensional heat and mass transfer model that describes both the primary and the secondary drying stages and also describes the transition region between primary drying and secondary drying. Using this interactive tool, the user can optimize the shelf temperature and/or chamber pressure profile to achieve desired product temperatures.

41.3.2 Nonsteady-State Heat and Mass Transfer Modeling

While the simple steady-state models quantitatively describe primary drying, desorption drying (i.e., secondary drying) cannot be accurately described by such models, and several researchers have developed nonsteady-state models of sublimation and desorption. Some advantages of nonsteady-state models include residual moisture prediction as a function of time, and describing the nonsteady-state parts of primary drying (immediately after a change in shelf temperature) [34]. Liapis et al. have presented a sorption–sublimation model to describe the effect of operating conditions on drying times [56, 57]. The initial model, which takes into account heat transfer only from the top surface of the frozen cake [56], has been extended to a more pharmaceutically representative case of heat transfer from both top and bottom surfaces of the frozen cake [57]. One-dimensional energy and material balance equations for frozen and dried layers were solved. Utilizing this mathematical model, the authors could calculate the sorbed water concentration profiles at the end of primary drying at different positions in the cake as a function of different operating conditions. A further comparison of the performance of different mathematical models for predicting sorbed water concentrations as a function of time was also provided [58]. The authors utilize the one-dimensional mathematical model to predict optimal operating conditions in a freeze-drying cycle. This mathematical model of sublimation and desorption drying was further extended to model the primary and secondary drying stages in vial lyophilization, which is a more realistic

TABLE 41.1 Evaluation of Effect of Change in Shelf Temperature and Chamber Pressure Changes on Product Temperature During Primary Drying and Primary Drying Time

T_{shelf} (°C)	$P_{chamber}$ (mTorr)	Product Temperature (°C)	ΔT (Product Temperature at New Condition - Product Temperature at Original Conditions) (°C)	Drying Time (h)	Δtime (Drying Time at New Condition - Drying Time at Original Conditions) (h)
−25	100	−34.5	0	31.5	0
−22	50	−35.5	−1	22	−9.5
−22	150	−32	2.5	30	−1.5
−28	50	−37.5	−3	31.5	0
−28	150	−33.5	1	53	21.5

representation of a typical pharmaceutical freeze-drying process [59, 60]. A further application of modeling was in understanding the mechanism of bound water removal utilizing one-dimensional nonsteady-state modeling of both primary and secondary drying stages [61]. Using the mathematical model coupled with experimental confirmation, they conclude that the removal of bound water during primary drying portion of freeze-drying is negligible.

A further improvement of the one-dimensional model was made by extending it to a two-dimensional system [62]. A finite element method was used to solve two-dimensional heat and mass transfer equations for the frozen and dried layers. Accounting for the removal of ice and sorbed water, this model predicts the position and geometric shape of the moving interface, thus modeling the entire primary and secondary drying stages. A physical rationale for the choice of boundary conditions used is provided elsewhere [34]. The authors further evaluate the model predictions using experimental results suggesting the usefulness of the utilizing mathematical modeling in freeze-drying process development [34]. An example for nonsteady-state model equations is shown below. Equations 41.5–41.11 are intended to give the reader a sense for the nonsteady-state formulation of the freeze-drying problem. Equation 41.5 describes the water vapor flow in the dry layer. This equation can be summarized as the sum of change in water concentration in the gas phase and the change in water concentration in the solid phase equals the flux of water out of the system. Equations 41.6 and 41.7 describe the change in water content of the solid phase and molar water flux in the dried region. Equation 41.6 assumes that the rate of change in water content of the solid phase is proportional to the difference between the water content and the equilibrium water content at the surrounding water activity, a_w, at temperature T, denoted $C^*(a_w,T)$ where k_g is a "rate constant" assumed to exhibit Arrhenius temperature dependence. The terms on the right-hand side of equation 41.7 represent the contributions of both diffusion and bulk fluid flow to the water flux. Bulk fluid flow is mostly Knudsen flow in usual pharmaceutical freeze-drying applications. Further simplification of equation 41.7 is possible by considering the fact that vapor composition in the cake and in the drying chamber during practical primary drying situation is nearly 100% water vapor, we can ignore diffusion in the dry layer during primary drying leaving only bulk flow or Knudsen flow as primary contributor to water vapor flux. These simplifications are described in Ref. 34.

Equations 41.9–41.11 describe the heat transfer in dried layer and frozen layer. Equation 41.9 describes the heat transfer in dry layer, which is conservation of energy in dry layer. Energy conservation for frozen layer leads to equation 41.10. Equation 41.11 is the conservation of energy at the sublimation interface, where rates of heat flow into the interface from dry and frozen layers is compensated by the rates of heat removed by gas flow and sublimation. As mentioned above, complete details and further simplifications of these equations can be found in Ref. 34.

41.3.2.1 Nonsteady-State Model Equations Representing Freeze-Drying (see "Symbols" Section)

$$\varepsilon\frac{\partial C_{w,g}}{\partial t}+\rho_I\frac{\partial C_{w,s}}{\partial t}=-\nabla\cdot N_w \tag{41.5}$$

$$\frac{\partial C_{w,s}}{\partial t}=-k_g\left(C_{w,s}-C^*(a_w,T)\right) \tag{41.6}$$

$$N_w=-k_1\nabla C_{w,g}-k_2C_{w,g}\nabla P \tag{41.7}$$

$$N_w=-\frac{M_w}{RT}\cdot K_w\left(1+\frac{C_{01}P_w}{K_w\mu_{mx}}\right)\nabla P_w\approx-\frac{M_w}{RT}\cdot K_w\nabla P_w \tag{41.8}$$

$$\rho_I C_{pI}\frac{\partial T}{\partial t}=k_I\nabla^2T+\rho_I\Delta H_v\frac{\partial C_{w,s}}{\partial T}-C_{p,g}\nabla(N_tT) \tag{41.9}$$

$$\rho_{II}C_{PII}\frac{\partial T}{\partial t}=k_{II}\nabla^2T \tag{41.10}$$

$$-k_I\left(\frac{\partial T}{\partial n}\right)_I+k_{II}\left(\frac{\partial T}{\partial n}\right)_{II}=-N_{t_n}C_{p,g}T-N_{w_n}\Delta H_s \tag{41.11}$$

Utilization of mathematical modeling techniques described above will result in greater understanding of the effect of process variables on the quality of the product. These modeling tools provide an excellent opportunity to apply the quality by design principles to ensure that quality of the freeze-dried product is built into the process. One such practical industrial application of mathematical modeling of freeze-drying is demonstrated in freeze-drying of Azithromycin solution [63]. The authors have confirmed the model predictions of the PASSAGE software [34, 62][2] and further utilized the mathematical model to predict the operating conditions at pilot scale to achieve a product temperature profile that is equivalent to the profile achieved for an optimized laboratory-scale lyophilization cycle. This work demonstrated the utility of the numerical models in the area of scale-up and optimization of lyophilization cycles at commercial scale [63].

Several other modeling approaches and applications have been reported in the literature that describe the freeze-drying process utilizing fundamental heat and mass transfer models [64–70]. These reports demonstrate that the fundamental understanding of the drying stage has evolved extensively

[2] PASSAGE is commercially available freeze-drying software capable to solving unsteady state mass and heat transfer equations. This is commercially available from Technalysis, Inc.

and scientists can utilize these techniques to rapidly develop optimized lyophilization cycles.

41.3.3 Determination of Modeling Parameters: Dry Layer Resistance and Heat Transfer Coefficient

The importance of heat transfer coefficient and mass transfer resistance during drying, especially for modeling purposes should now be evident to the reader. The manometric temperature measurement (MTM) procedure is one of the techniques that is useful in estimation of the mass transfer resistance offered by the dried layer and stopper.

In the MTM method, the flow of water vapor from the product chamber to condenser is momentarily interrupted during primary drying by quickly closing the valve separating the chamber and condenser, resulting in an increase in chamber pressure due to sublimation [71]. This transient increase in chamber pressure is modeled by considering the several factors contributing to the pressure rise. A curvilinear regression estimates the vapor pressure of ice and resistance offered by dried layer. The capabilities of this modeling technique have been further examined in estimation of product temperature [54], measurement of dry layer resistance [72], and heat and mass transfer measurements, including the vial heat transfer coefficient [73].

An "expert system" (SMART) that will allow development of an optimized freeze-drying process during laboratory-scale development in one experiment is another application of the MTM technique [74]. SMART is an excellent example of the utilization of heat and mass transfer theory and modeling to facilitate the development of freeze-drying cycle conditions. Measurement of product resistance and vapor pressure of ice by MTM allows calculation of the sublimation rate [71], which is then used to optimize the shelf temperature settings. Figure 41.5 summarizes the SMART concept. The expert systems algorithms control the different parts of the freeze-drying. Freezing conditions are chosen based on the input parameters regarding formulation details. Primary drying conditions are chosen based on the MTM feedback. Also, the algorithm chooses conditions for secondary drying based on input parameters. A detailed description can found in Ref. 74. A variation of the original MTM approach, denoted "pressure rise analysis (PRA)," was proposed as an improvement to the original MTM algorithm [75]. Using this PRA model, the authors estimate the same parameters as obtained by MTM, sublimation front temperature, resistance offered by dry layer, and the overall vial heat transfer coefficient. Utility of SMART and MTM during the lyophilization cycle development has been demonstrated in measuring product resistance, predicting product temperature and primary drying time [76].

Another method, based on an analysis of the normal product temperature history during primary drying, has been suggested as a method to obtain product resistance data without doing special experiments, such as the MTM experiment [77]. This technique once again utilizes the understanding of heat and mass transfer mechanisms to determine desired parameters (mass transfer resistance) of the formulation during freeze-drying. Other techniques utilize similar applications of heat and mass transfer fundamentals to estimate mass transfer resistances during freeze-drying [78–80]. Recently, another variation of the original MTM approach was suggested as a more rigorous model for describing the pressure rise during the valve closing procedure [81]. The stated final goal of using this advanced approach is to develop an online tool for controlling the heating strategy during freeze-drying, and the same group recently reported an online monitoring system (and hence control) for primary drying phase of lyophilization as a further application of the mathematical modeling [82]. This system provides in-line control signals (adjusting shelf heating fluid temperature throughout primary drying),

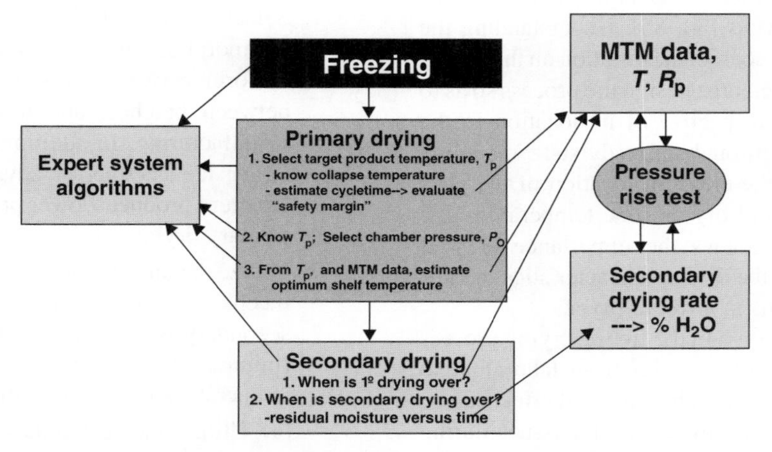

FIGURE 41.5 Summary of the "SMART" Freeze-Dryer concept. Reproduced from Ref. 74 with permission from Springer.

utilizing simple heat and mass transfer algorithm in conjunction with measurements made during the freeze-drying process to achieve desired product and process performance. While this general procedure is identical to that employed by the MTM procedure within the SMART Freeze-Dryer technology, the details of the heat and mass transfer analysis do differ. Further details can be found in Ref. 82.

Another important input parameter for models is the heat transfer coefficient of the vial. This parameter can be determined both by performing simple water sublimation experiments in the laboratory as a function of chamber pressure [83] and by MTM measurements (or variations on the MTM technique) as noted above.

Successful model development for the drying process depends on the accuracy of the modeling parameters. Dry layer resistance and heat transfer coefficient are two critical parameters. While one might estimate these quantities from an existing database, ideally these parameters should be measured for the application of interest, particularly with the dry layer resistance that varies considerably between products and is relatively difficult to quantitatively predict.

41.3.4 Issues in Scale-Up of Freeze-Drying Process

The heat transfer coefficient of a given vial depends not only on the bottom contour of the vial but also on the location of the vial within the vial array in a given freeze-dryer. Heat is transferred between shelf and vial due to three mechanisms: heat transfer from the contact area between the vial and the shelf, conduction of gas molecules between vial bottom and shelf surface, and due to radiation. While the first two modes of heat transfer can be considered independent of the location of the vial, the radiative heat transfer depends on the location of the vial. Edge vials have a greater area of exposure to radiation heat transfer (i.e., part of the side as well as top and bottom) and often the chamber surfaces that are responsible for side radiation are hotter than the shelf surface. These effects result in differences in heat transfer coefficients between center and edge vials [84, 85]. Understanding the impact of such heat transfer coefficient variation on the drying performance will help the scientist design the process so as to achieve optimal product irrespective of its location in the freeze-dryer. A multidimensional unsteady-state modeling was utilized to determine the effect of location of the vials in freeze-dryer on the overall drying time temperature distribution [86]. This study shows the importance of wall temperature in influencing the drying characteristics in vials located at different positions in a freeze-dryer.

Another important challenge in the freeze-drying process is the scale-up of lyophilization cycles from laboratory to pilot to commercial-scale dryers. The edge vial effect can be scale dependent. The heat and mass transfer issues during freeze-drying process development have been summarized in an excellent review, Ref. 87. In addition to providing a review of heat and mass transfer mechanisms to be considered during freeze-drying cycle development the authors also discuss the related scale-up issues. As mentioned above, radiative heat transfer varies dues to the location of the vial in the freeze-dryer. This radiation effect is also different from a laboratory dryer to pilot or a commercial-scale dryer. Differences in percentage of edge vials and differences in wall temperatures and differences in emissivities between laboratory scale and manufacturing scale introduce scale-up differences that can be significant. For example, a front vial in a laboratory freeze-dryer can receive $\approx$1.8 times greater heat transfer than a corresponding vial in a manufacturing freeze-dryer [29]. This additional heat input will directly affect the product temperature and drying time. Hence, understanding these differences between freeze-dryers is essential to proper scale-up. Another important scale based difference is the temperature distribution across the freeze-dryer shelf. Differences between the shelf temperature set point and measured shelf surface temperature as a function of the sublimation rate are reported [29]. These differences can be dryer specific in that dryer design and heat transfer characteristics of the fluid can cause shelf surface temperature differences between different freeze-dryers even at identical thermal loads. A properly designed shelf mapping study can determine the magnitude of the expected effects [87]. These differences between laboratory-scale and production-scale dryers have been highlighted and a step-by-step systematic approach to correlate dryers at two scales leading toward a successful scale-up are discussed in Ref. 52.

It should be obvious that the application of engineering principles of heat and mass transfer modeling and scale-up adjustments are essential to the successful design of a freeze-drying process for manufacturing. Utilizing these engineering principles will facilitate a rationale lyophilization cycle development and scale-up effort. The "general rules" for successful process design and scale-up can be summarized as follows:

1. Select the optimal freezing conditions that results in an ice morphology that is uniform within a given vial, uniform between vials in the same batch, uniform between batches, and uniform from laboratory to manufacturing. In addition, a structure composed of larger ice crystals is advantageous in that such a structure produces lower product resistance and faster primary drying.
2. Utilize the understanding of the heat and mass transfer mechanisms during drying to select the primary and secondary conditions to achieve maximal drying rate (minimal drying time) while maintaining the product temperature below the critical product temperature.
   (a) Utilize the techniques described in literature to measure heat transfer coefficient of the vial of interest.

(b) Determine the resistance offered by dry layer during the drying stage.
(c) Utilizing mathematical models obtain initial estimates for the first laboratory cycle and then optimize the cycle conditions using the SMART Freeze-Dryer methodology (or equivalent) and/or a few experiments to confirm results.

3. Understand the freeze-dryer differences with respect to the heat and mass transfer mechanisms during scale-up and utilize mathematical models to estimate the cycle conditions at larger scale.

Some examples of application of these concepts for rationale scale-up have been reported in literature [63, 88, 89]. Heat and mass transfer models in conjunction with limited lyophilization runs were utilized to successfully determine heat transfer coefficients and to evaluate the robustness of the lyophilization cycle at different operating conditions [88]. A systematic approach by utilizing laboratory-scale experiments to determine heat and mass transfer coefficients and mathematical modeling to predict operating conditions at pilot scale thus minimizing the number of pilot scale runs has also been reported [63]. Hopefully, increased use of theoretical modeling will be a norm in the future as increased emphasis is placed on "quality by design."

41.3.5 Mass Flow from Chamber to Condenser

Many advancements in modeling the sublimation and desorption of water vapor within the vial have been made as discussed in the Sections 41.3.1–41.3.4. The literature on modeling the flow of water vapor once it leaves the vial, that is, from chamber to condenser is, however, sparse. The lyophilization process can be limited by mass transfer within the freeze-dryer at high sublimation load [90]. In addition, freeze-dryer design differences at different scales may lead to different product temperature profiles that may not be captured in vial modeling. Therefore, it is important that these factors be captured in the models for freeze-drying. Some recent publications have described the use of various tools to model vapor flow in the freeze-dryer. Most of the work cited has used computational fluid dynamics (CFD) for this purpose. The effect of some geometrical parameters of the drying chamber such as clearances between the shelves and the position of the duct between the chamber and the condenser on the global fluid dynamics of the sublimated vapor in both small-scale and industrial-scale drying chambers were investigated as a function of the sublimation rate [91]. It was concluded that local pressure differentials existed in the freeze-dryer and contributed to heterogeneity in sublimation rate in addition to the commonly known effects such as radiation effects. These effects were more pronounced in larger scale freeze-dryers. However, the mag-

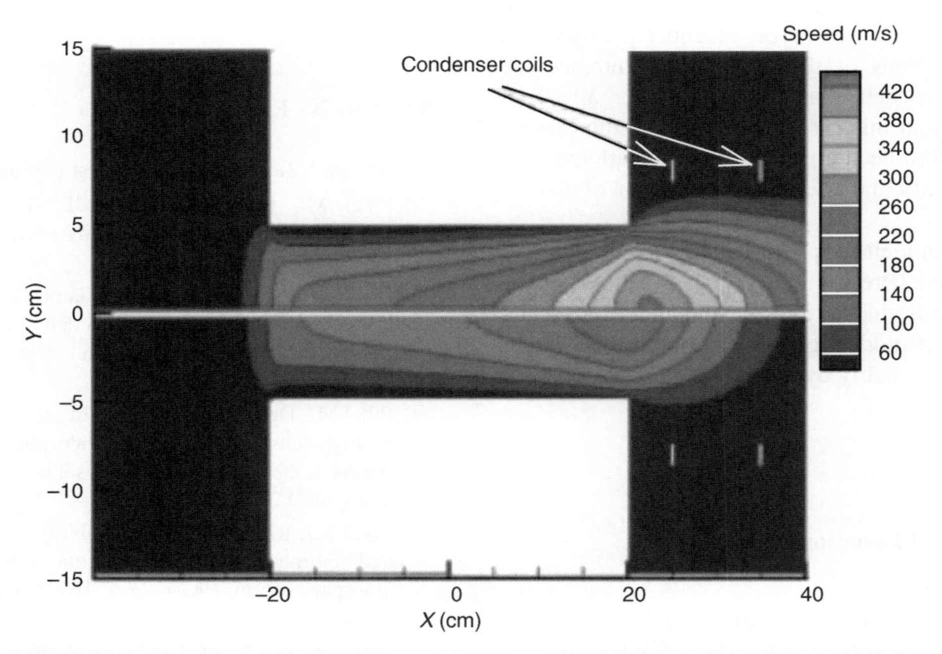

FIGURE 41.6 Velocity profile of water vapor in laboratory-scale freeze-dryer. The top half shows the results for 50 mTorr chamber pressure and choked flow conditions at the condenser entrance are apparent. The lower half shows simulation for 30 mTorr. The vapor velocity in this case remains below the Mach I limit. The data show potential choked flow conditions for some operating parameters. Reprinted from Ref. 91 with permission of Wiley-Liss, Inc., a subsidiary of John Wiley & Sons, Inc.

nitude of these effects was not explicitly discussed. It has been documented that the duct between the drying chamber and the condenser may be a bottleneck at high sublimation rates [90]. This effect has been called choked flow, and CFD has been employed to identify critical variables impacting choked flow [92]. It was determined that ratio of chamber to condenser pressure determined the onset of choked flow. Recently, a more comprehensive model that includes the drying chamber, duct, and condenser using CFD has been published [93]. The Knudsen number (*Kn*), a dimensionless parameter, has been used to define the flow regime and various equations that are valid in different flow regimes have been described. *Kn* is the ratio of the molecular mean free path (λ) and characteristic length scale of flow (L). For example, Navier–Stokes equation has been used where *Kn* is below 0.01 and continuum hypothesis is valid, that is, one is dealing with fluid flow or "viscous flow." The Boltzmann equation has been used in rarefied or "free molecular flow" regime ($Kn > 0.1$). A new approach to solving the Boltzmann equation using a statistical direct simulation Monte Carlo (DSMC) method has been described. Differences between a laboratory-scale and industrial-scale freeze- dryer, especially the impact of CIP/SIP (clean in place/steam in place) line in industrial freeze-dryer, on the fluid flow were presented. Simulation of water vapor flow in the freeze-dryer also showed choked flow conditions may exist at the point of maximum flow velocity, which is along the axis of the duct at the entrance to the condenser. Figure 41.6 shows the simulation for two different chamber pressures. The top half shows the results for 50 mTorr chamber pressure and choked flow conditions at the condenser entrance are apparent. The lower half shows simulation for 30 mTorr. The vapor velocity in this case remains below the Mach I limit. It has therefore been argued that computational fluid flow studies may also be useful in the design of freeze-dryers.

Following the guidelines presented in this chapter and the references cited herein, it should result in systematic approach to freeze-drying process development and scale-up. Increased use of modeling wherever applicable will be consistent with the "quality by design" philosophy.

SYMBOLS

Steady-State Model Equations

- dm/dt sublimation rate
- A_p inner cross-sectional area of the vial
- P_0 vapor pressure of ice at interface temperature
- P_c chamber pressure
- R_P dry layer resistance
- dQ/dt rate of heat transfer
- A_v outer cross-sectional area of the vial
- K_v heat transfer coefficient between the vial and the surroundings (includes heat transfer form shelf contact and also radiation)
- T_s shelf temperature
- T_b product temperature at the bottom of the vial
- T product temperature at the sublimation interface
- ΔH_s heat of sublimation of ice

Nonsteady-State Model Equations

- ε void fraction in the dried region
- ρ_I density of dry region
- $C_{w,s}$ concentration of sorbed water
- N_w molar flux of water
- k_g mass transfer coefficient for desorption
- C^* equilibrium concentration of sorbed water
- a_w water activity
- T temperature
- k_1 bulk diffusivity constant
- k_2 self diffusivity constant
- P total pressure
- C_{pI} heat capacity of dry layer
- k_I thermal conductivity of dry layer
- ΔH_v heat of vaporization of sorbed water
- ρ_{II} density of the frozen region
- C_{PII} heat capacity of frozen region
- k_{II} thermal conductivity of frozen layer
- N_{w_n} molar flux of water, normal component
- ΔH_s heat of sublimation of ice

REFERENCES

1. Brange J. *Galenics of Insulin*, 1st edition, Springer, 1987.
2. Pikal MJ, Lukes AL, Lang JE. Thermal decomposition of amorphous β-lactam antibacterials. *J. Pharm. Sci.* 1977;66(9):1312–1316.
3. Pikal MJ, Lukes AL, Lang JE, Gaines K. Quantitative crystallinity determinations for β-lactam antibiotics by solution calorimetry: correlations with stability. *J. Pharm. Sci.* 1978;67(6):767–773.
4. Pikal MJ, Dellerman KM. Stability testing of pharmaceuticals by high-sensitivity isothermal calorimetry at 25°C: cephalosporins in the solid and aqueous solution states. *Int. J. Pharm.* 1989;50(3):233–252.
5. Pikal MJ, Rigsbee DR. The stability of insulin in crystalline and amorphous solids: observation of greater stability for the amorphous form. *Pharm. Res.* 1997;14(10):1379–1387.
6. Pikal MJ. Freeze-drying of proteins. Part II. Formulation selection. *BioPharm* 1990;3(9):26–30.
7. Pikal MJ. Freeze-drying of proteins. Part I. Process design. *BioPharm* 1990;3(8):18–20, 22–24, 26–28.
8. Pikal MJ. Freeze-drying of proteins: process, formulation, and stability. *ACS Symp. Ser.* 1994;567:120–133.

9. MacKenzie AP. Factors affecting the transformation of ice into water vapor in the freeze-drying process. *Ann. N.Y. Acad. Sci.* 1965;125:522–547.
10. Broadhead J, Rouan SKE, Rhodes CT. The spray drying of pharmaceuticals. *Drug Dev. Ind. Pharm.* 1992;18(11–12): 1169–1206.
11. Mumenthaler M, Hsu CC, Pearlman R. Feasibility study on spray-drying protein pharmaceuticals: recombinant human growth hormone and tissue-type plasminogen activator. *Pharm. Res.* 1994;11(1):12–20.
12. Masters K. Applications of spray-drying in the food industry, in the pharmaceutical-biochemical industry. In: *Spray-Drying Handbook*, Longman Scientific and Technical, Essex, UK, 1991, pp. 491–676.
13. Franks F. Solid aqueous solutions. *Pure Appl. Chem.* 1993; 65(12):2527–2537.
14. Roser B. Trehalose drying: a novel replacement for freeze-drying. *BioPharm* 1991;4(8):47–53.
15. Abdul-Fattah AM, Kalonia DS, Pikal MJ. The challenge of drying method selection for protein pharmaceuticals: product quality implications. *J. Pharm. Sci.* 2007;96(8): 1886–1916.
16. Levine H, Slade L. Another view of trehalose for drying and stabilizing biological materials. *BioPharm* 1992;5(4):36–40.
17. Pregnolato FC. *Proceedings of the International Congress, Advanced Technologies for Manufacturing of Asceptic and Terminally Sterilized Pharmaceuticals and Biopharmaceuticals*, Basel, Switzerland, 1992, pp 4–30.
18. Carpenter JF, Crowe JH, Arakawa T. Comparison of solute-induced protein stabilization in aqueous solution and in the frozen and dried states. *J. Dairy Sci.* 1990; 73(12):3627–3636.
19. Tanaka K, Takeda T, Miyajima K. Cryoprotective effect of saccharides on denaturation of catalase by freeze-drying. *Chem. Pharm. Bull.* 1991;39(5):1091–1094.
20. Hellman K, Miller DS, Cammack KA. The effect of freeze-drying on the quaternary structure of L-asparaginase from *Erwinia carotovora. Biochim. Biophys. Acta, Protein Struct. Mol. Enzymol.* 1983;749(2):133–142.
21. Ressing ME, Jiskoot W, Talsma H, Van Ingen CW, Beuvery EC, Crommelin DJA. The influence of sucrose, dextran, and hydroxypropyl β-cyclodextrin as lyoprotectants for a freeze-dried mouse IgG2a monoclonal antibody (MN12). *Pharm. Res.* 1992;9(2):266–270.
22. Izutsu K-I, Yoshioka S. Stabilization of protein pharmaceuticals in freeze-dried formulations. *Drug Stab.* 1995;1 (1):11–21.
23. Bell LN, Hageman MJ. Differentiating between the effects of water activity and glass transition dependent mobility on a solid state chemical reaction: aspartame degradation. *J. Agric. Food Chem.* 1994;42(11):2398–2401.
24. Pikal MJ, Dellerman KM, Roy MI, Riggin RM. The effects of formulation variables on the stability of freeze-dried human growth hormone. *Pharm. Res.* 1991;8(4):427–436.
25. Pikal MJ, Dellerman K, Roy ML. Formulation and stability of freeze-dried proteins: effects of moisture and oxygen on the stability of freeze-dried formulations of human growth hormone. *Dev. Biol. Stand.* 1992;74:21–38.
26. Townsend MW, DeLuca P.P. Use of lyoprotectants in the freeze-drying of a model protein, ribonuclease A. *J. Parenter. Sci. Technol.* 1988;42(6):190–199.
27. Prestrelski SJ, Pikal KA, Arakawa T. Optimization of lyophilization conditions for recombinant human interleukin-2 by dried-state conformational analysis using Fourier-transform infrared spectroscopy. *Pharm. Res.* 1995;12(9): 1250–1259.
28. Roy ML, Pikal MJ, Rickard EC, Maloney AM. The effects of formulation and moisture on the stability of a freeze-dried monoclonal antibody-vinca conjugate: a test of the WLF glass transition theory. *Dev. Biol. Stand.* 1992;74:323–340.
29. Rambhatla S, Tchessalov S, Pikal Michael J. Heat and mass transfer scale-up issues during freeze-drying. III. Control and characterization of dryer differences via operational qualification tests. *AAPS PharmSciTech* 2006;7(2):E39.
30. Roy ML, Pikal MJ. Process control in freeze drying: determination of the end point of sublimation drying by an electronic moisture sensor. *J Parenter Sci Technol* 1989, 43, (2):60–6.
31. Searles, JA, Carpenter, JF, Randolph, TW. The ice nucleation temperature determines the primary drying rate of lyophilization for samples frozen on a temperature-controlled shelf, *J. Pharm. Sci.* 2001; 90(7):860–871.
32. Rambhatla S, Ramot R, Bhugra C, Pikal Michael J. Heat and mass transfer scale-up issues during freeze drying. II. Control and characterization of the degree of supercooling. *AAPS PharmSciTech* 2004;5(4):e58.
33. Pikal MJ. Use of laboratory data in freeze drying process design: heat and mass transfer coefficients and the computer simulation of freeze drying. *J. Parenter. Sci. Technol.* 1985; 39(3):115–139.
34. Pikal MJ, Cardon S, Bhugra C, Jameel F, Rambhatla S, Mascarenhas WJ, Akay HU. The nonsteady state modeling of freeze drying: in-process product temperature and moisture content mapping and pharmaceutical product quality applications. *Pharm. Dev. Technol.* 2005;10(1); 17–32.
35. Gilra NK. Homogeneous nucleation temperature of supercooled water. *Phys. Lett. A* 1968;28(1):51–52.
36. Tang X, Pikal MJ. Design of freeze-drying processes for pharmaceuticals: practical advice. *Pharm. Res.* 2004; 21(2):191–200.
37. Pikal MJ, Shah S. The collapse temperature in freeze drying: dependence on measurement methodology and rate of water removal from the glassy phase. *Int. J. Pharm.* 1990; 62(2–3):165–186.
38. Chatterjee K, Shalaev EY, Suryanarayanan R. Partially crystalline systems in lyophilization. II. Withstanding collapse at high primary drying temperatures and impact on protein activity recovery. *J. Pharm. Sci.* 2005;94(4); 809–820.
39. Milton N, Nail SL. The physical state of nafcillin sodium in frozen aqueous solutions and freeze-dried powders. *Pharm. Dev. Technol.* 1996;1(3):269–277.
40. Searles JA, Carpenter JF, Randolph TW. Annealing to optimize the primary drying rate, reduce freezing-induced drying rate

heterogeneity, and determine T_g' in pharmaceutical lyophilization. *J. Pharm. Sci.* 2001;90(7):872–887.

41. Chouvenc P, Vessot S, Andrieu J. Experimental study of the impact of annealing on ice structure and mass transfer parameters during freeze-drying of a pharmaceutical formulation. *PDA J. Pharm. Sci. Technol.* 2006;60(2):95–103.
42. Lu X, Pikal Michael J. Freeze-drying of mannitol-trehalose-sodium chloride-based formulations: the impact of annealing on dry layer resistance to mass transfer and cake structure. *Pharm. Dev. Technol.* 2004;9(1):85–95.
43. Pikal MJ, Shah S, Roy ML, Putman R. The secondary drying stage of freeze drying: drying kinetics as a function of temperature and chamber pressure. *Int. J. Pharm.* 1990; 60(3):203–217.
44. Heller MC, Carpenter JF, Randolph TW. Manipulation of lyophilization-induced phase separation: implications for pharmaceutical proteins. *Biotechnol. Prog.* 1997;13(5):590–596.
45. Rowe TD. *A technique for nucleation of ice.* International Symposium on Biological Product Freeze-Drying and Formulation, Geneva, Switzerland, 1990.
46. Nakagawa K, Hottot A, Vessot S, Andrieu J. Influence of controlled nucleation by ultrasounds on ice morphology of frozen formulations for pharmaceutical proteins freeze-drying. *Chem. Eng. Process* 2006;45(9):783–791.
47. Petersen A, Rau G, Glasmacher B. Reduction of primary freeze-drying time by electric field induced ice nucleus formation. *Heat Mass Transf.* 2006;42(10):929–938.
48. Petersen A, Schneider H, Rau G, Glasmacher B. A new approach for freezing of aqueous solutions under active control of the nucleation temperature. *Cryobiology* 2006;53 (2):248–257.
49. Liu J, Viverette T, Virgin M, Anderson M, Dalal P. A study of the impact of freezing on the lyophilization of a concentrated formulation with a high fill depth. *Pharm. Dev. Technol.* 2005;10(2):261–272.
50. Kramer M, Sennhenn B, Lee G. Freeze-drying using vacuum-induced surface freezing. *J. Pharm. Sci.* 2002;91(2):433–443.
51. Nakagawa K, Hottot A, Vessot S, Andrieu J. Modeling of freezing step during freeze-drying of drugs in vials. *AIChE J.* 2007;53(5):1362–1372.
52. Kuu WY, Hardwick LM, Akers MJ. Correlation of laboratory and production freeze drying cycles. *Int. J. Pharm.* 2005; 302(1–2):56–67.
53. Sadikoglu H, Ozdemir M, Seker M. Freeze-drying of pharmaceutical products: research and development needs. *Drying Technol.* 2006;24(7):849–861.
54. Tang X, Nail Steven L, Pikal Michael J. Evaluation of manometric temperature measurement, a process analytical technology tool for freeze-drying. Part I. Product temperature measurement. *AAPS PharmSciTech* 2006;7(1):E14.
55. Trelea IC, Passot S, Fonseca F, Marin M. An interactive tool for the optimization of freeze-drying cycles based on quality criteria. *Drying Technol.* 2007;25(5):741–751.
56. Litchfield RJ, Liapis AI. An adsorption–sublimation model for a freeze dryer. *Chem. Eng. Sci.* 1979;34(9):1085–1090.
57. Millman MJ, Liapis AI, Marchello JM. An analysis of the lyophilization process using a sorption–sublimation model and various operational policies. *AIChE J.* 1985; 31(10):1594–604.
58. Liapis AI, Marchello JM. Advances in the modeling and control of freeze-drying. *Adv. Drying* 1984;3:217–244.
59. Liapis AI, Bruttini R. Freeze-drying of pharmaceutical crystalline and amorphous solutes in vials: dynamic multi-dimensional models of the primary and secondary drying stages and qualitative features of the moving interface. *Drying Technol.* 1995;13(1):43–72.
60. Sheehan P, Liapis AI. Modeling of the primary and secondary drying stages of the freeze drying of pharmaceutical products in vials: numerical results obtained from the solution of a dynamic and spatially multi-dimensional lyophilization model for different operational policies. *Biotechnol. Bioeng.* 1998; 60(6):712–728.
61. Sadikoglu H, Liapis AI. Mathematical modeling of the primary and secondary drying stages of bulk solution freeze-drying in trays: parameter estimation and model discrimination by comparison of theoretical results with experimental data. *Drying Technol.* 1997;15(3–4):791–810.
62. Mascarenhas WJ, Akay HU, Pikal MJ. A computational model for finite element analysis of the freeze-drying process. *Comput. Mehtods Appl. Mecha. Eng.* 1997;148:105–124.
63. Kramer T, Kremer DM, Pikal MJ, Petre WJ, Shalaev EY, Gatlin LA. A procedure to optimize scale-up for the primary drying phase of lyophilization. *J. Pharm. Sci.* 2009;98(1):307–318.
64. Nastaj J, Witkiewicz K. Numerical model of freeze drying of random solids at two-region conductive–radiative heating. *Inz. Chem. Procesowa* 2004;25(1):109–121.
65. Boss EA, Maciel Filho R, II, Vasco de Toledo EC, III. Freeze drying process: real time model. *Tech. Pap. ISA* 20021;426:59–70.
66. Tu W, Chen M, Yang Z, Chen H. A mathematical model for freeze-drying. *Chin. J. Chem. Eng.* 2000;8(2):118–122.
67. Kisakurek B, Celiker H. A modified moving boundary model for freeze-dryers. *Proc. Int. Drying Symp.* 1984;2:420–424.
68. Hottot Al, Peczalski R, Vessot S, Andrieu J. Freeze-drying of pharmaceutical proteins in vials: modeling of freezing and sublimation steps. *Drying Technol.* 2006;24(5):561–570.
69. Boss EA, Filho RM, de Toledo ECV. Freeze drying process: real time model and optimization. *Chem. Eng. Process.* 2004; 43(12):1475–1485.
70. Velardi SA, Barresi AA. Development of simplified models for the freeze-drying process and investigation of the optimal operating conditions. *Chem. Eng. Res. Des.* 2008;86(1):9–22.
71. Milton N, Pikal MJ, Roy ML, Nail SL. Evaluation of manometric temperature measurement as a method of monitoring product temperature during lyophilization. *PDA J. Pharm. Sci. Technol.* 1997;51(1):7–16.
72. Tang Xiaolin C, Nail Steven L, Pikal Michael J. Evaluation of manometric temperature measurement, a process analytical technology tool for freeze-drying. Part II. Measurement of dry-layer resistance. *AAPS PharmSciTech* 2006;7(4):93.

73. Tang Xiaolin C, Nail Steven L, Pikal Michael J. Evaluation of manometric temperature measurement (MTM), a process analytical technology tool in freeze drying. Part III. Heat and mass transfer measurement. *AAPS PharmSciTech* 2006;7(4):97.
74. Tang X, Nail SL, Pikal MJ. Freeze-drying process design by manometric temperature measurement: design of a SMART Freeze-Dryer. *Pharm. Res.* 2005;22(4):685–700.
75. Chouvenc P, Vessot S, Andrieu J, Vacus P. Optimization of the freeze-drying cycle: a new model for pressure rise analysis. *Drying Technol.* 2004;22(7):1577–1601.
76. Gieseler H, Kramer T, Pikal MJ. Use of manometric temperature measurement (MTM) and SMART Freeze-Dryer technology for development of an optimized freeze-drying cycle. *J. Pharm. Sci.* 2007;96(12):3402–3418.
77. Kuu WY, Hardwick LM, Akers MJ. Rapid determination of dry layer mass transfer resistance for various pharmaceutical formulations during primary drying using product temperature profiles. *Int. J. Pharm.* 2006;313(1–2); 99–113.
78. Pikal MJ, Shah S, Senior D, Lang JE. Physical chemistry of freeze-drying: measurement of sublimation rates for frozen aqueous solutions by a microbalance technique. *J. Pharm. Sci.* 1983;72(6):635–650.
79. Kuu W-Y, McShane J, Wong J. Determination of mass transfer coefficients during freeze drying using modeling and parameter estimation techniques. *Int. J. Pharm.* 1995;124(2):241–252.
80. Zhai S, Su H, Taylor R, Slater NKH. Pure ice sublimation within vials in a laboratory lyophiliser; comparison of theory with experiment. *Chem. Eng. Sci.* 2005;60(4):1167–1176.
81. Velardi SA, Rasetto V, Barresi AA. Dynamic parameters estimation method: advanced manometric temperature measurement approach for freeze-drying monitoring of pharmaceutical solutions. *Ind. Eng. Chem. Res.* 2008;47(21):8445–8457.
82. Barresi AA, Pisano R, Fissore D, Rasetto V, Velardi SA, Vallan A, Parvis M, Galan M. Monitoring of the primary drying of a lyophilization process in vials. *Chem. Eng. Process.* 2009; 48(1):408–423.
83. Pikal MJ, Roy ML, Shah S. Mass and heat transfer in vial freeze-drying of pharmaceuticals: role of the vial. *J. Pharm. Sci.* 1984;73(9):1224–1237.
84. Rambhatla S, Pikal MJ. Heat and mass transfer scale-up issues during freeze-drying. I. Atypical radiation and the edge vial effect. *AAPS PharmSciTech* 2003;4(2):111–120.
85. Brülls M, Rasmuson A. Heat transfer in vial lyophilization. *Int. J. Pharm.* 2002;246(1–2):1–16.
86. Gan KH, Bruttini R, Crosser OK, Liapis AI. Freeze-drying of pharmaceuticals in vials on trays: effects of drying chamber wall temperature and tray side on lyophilization performance. *Int. J. Heat Mass Transf.* 2005;48(9):1675–1687.
87. Rambhatla S, Pikal MJ. Heat and mass transfer issues in freeze-drying process development. *Biotechnol. Pharm. Aspects* 2004;2:75–109.
88. Tsinontides SC, Rajniak P, Pham D, Hunke WA, Placek J, Reynolds SD. Freeze drying: principles and practice for successful scale-up to manufacturing. *Int. J. Pharm.* 2004;280 (1–2):1–16.
89. Tchessalov S, Dixon D, Nick W.Principles of lyophilization scale up. *Am. Pharm. Rev.* 2007; 10(3), 88–92.
90. Searles J. Observation and implications of sonic water vapor flow during freeze-drying. *Am. Pharm. Rev.* 2004;7(2):58, 60, 62, 64, 66–68, 75.
91. Rasetto VM, Daniele L, Baressi Antonello A.Computational fluid dynamics in freeze-drying technology. *Freeze Drying of pharmaceuticals and Biologicals*, Breckenridge, CO, 2008.
92. Patel SM, Chaudhuri S, Pikal MJ.Choked Flow and Importance of Mach I in Freeze-Drying Process Design. *Freeze Drying of Pharmaceutical and Biologicals,* Breckeridge, CO, 2008.
93. Alexeenko AA, Ganguly A, Nail SL. Computational analysis of fluid dynamics in pharmaceutical freeze-drying. *J. Pharm. Sci.* 2009;98(9):3483–3494.

42

ACHIEVING A HOT MELT EXTRUSION DESIGN SPACE FOR THE PRODUCTION OF SOLID SOLUTIONS

Luke Schenck
Merck & Co., Inc., Rahway, NJ, USA

Gregory M. Troup, Mike Lowinger, Li Li, and Craig McKelvey
Merck & Co., Inc., West Point, PA, USA

42.1 INTRODUCTION

Industrial applications of hot melt extrusion date back to the early 1900s. This process has since grown in use, and today is one of the most widely applied unit operations in the polymer industry. Recent increases in both patents and publications indicate it is rapidly becoming a key processing route for pharmaceutical dosage units as well [1].

Pharmaceutical extrusion has been used to process pastes (e.g., wet granulation) and polymer melts in which an active pharmaceutical ingredient (API) is present in an amorphous and/or crystalline state. The extrusion process has enabled a wide range of product applications including [1] solid solutions for the oral delivery of insoluble poorly soluble APIs, [2] implants, [3] intra oral delivery, [4] ophthalmic delivery, [5] controlled release (via matrix or multiparticulates), [6] conventional (i.e., API is in crystalline state) tablets by continuous wet granulation, and [7] nanocrystalline formulations.

Although many uses exist for hot melt extrusion, its application to produce solid solutions meeting pharmaceutical quality requirements remains limited. This chapter outlines design for six sigma (DFSS) led development activities used to define the design space for an extrusion process generating a solid solution. The use of DFSS methodology is a natural fit given its long standing application to new technologies, and added benefit of having significant overlap with the FDA's quality by design (QbD) initiative. The benefits of QbD development include added regulatory flexibility enabling continuous process improvement post filing, more prioritized development and an overall more effective management of risk. This text outlines how QbD methodology was reduced to a development roadmap using the DFSS tool set.

This chapter begins with an overview of solid solutions to provide a motivation for implementing hot melt extrusion in the pharmaceutical industry followed by a more detailed review of the extrusion process itself. The risk assessment activities that set the stage for more focused development efforts are then reviewed and the final section summarizes the use of process analytical technology (PAT) to facilitate definition of a multifactor design space. The chapter ends with a forward-looking vision of how this design space ultimately might be translated into a control strategy.

42.2 INTRODUCTION TO SOLID SOLUTIONS

Solid solutions have been developed largely to modulate undesirable drug properties—particularly, the poor aqueous solubility and/or wetting behavior of development candidates. Product pipelines in major pharmaceutical companies are increasingly composed of insoluble drug candidates driving the need for alternative oral delivery strategies including nanocrystalline, lipid, and solid solution based formulation technologies [2]. By some estimates the fraction

Chemical Engineering in the Pharmaceutical Industry: R&D to Manufacturing Edited by David J. am Ende

of development candidates considered very soluble has dropped below 10% [3]. A primary driver for deploying solid solution approaches in pharmaceutical development is to increase the exposure of orally administered poorly soluble active compounds and several reviews on this subject have been written [4–7].

Solid solutions are solutions of API in a glassy polymer often prepared by melt, solvent, and/or mechanical means by processes such as extrusion, spray drying, or mechanical activation, respectively. Increased oral absorption from solid solution formulations is achieved by supersaturation and/or *in situ* formation of nanoparticles [8, 9]. Increasing the apparent solubility of active compounds drives both potentially faster dissolution and permeability rates [10], making it possible to achieve dose proportional increases in exposure at higher doses and reduced potential for formulation related food effect. In addition to these benefits, solid solution formulations enable combination products in a solid format and have been used to bridge from liquid filled capsule type self-emulsifying and/or self-microemulsifying formulation approaches (Figure 42.1). Solid solutions have also found use in the preclinical setting, often in the form of suspensions [11]. The exposure increase compared to common alternative formulations can be dramatic (Figure 42.2). The benefit in exposure does come with a commensurate physical instability risk posed by these stabilized amorphous systems, although several compounds have been successfully launched.

Any process that reliably produces homogeneous glasses with consistent properties can be used to make solid solutions (Figure 42.3), however, there are relatively few papers that have discussed the relationship of process selection on product performance [12–16]. Those authors that have broached this subject have typically compared the performance of identical or unique formulations prepared at a single set of operating conditions using multiple preparation processes. It is not possible to make generalizations about superiority of a particular process for solid solution manufacture from the current body of published work. This is largely because an understanding of the operating space explored and associated characterization data used to determine if a homogeneous glass was produced in each case was not presented.

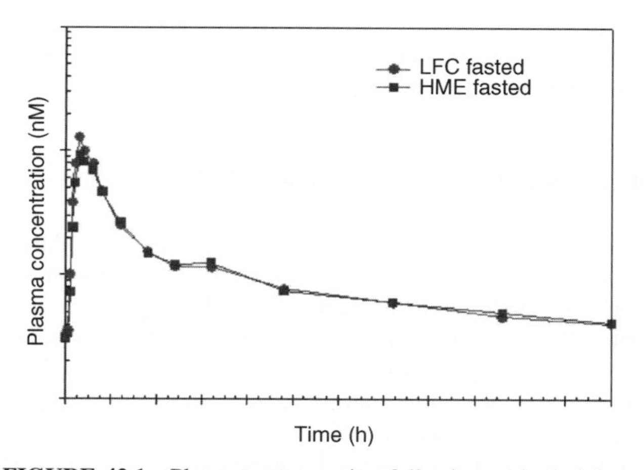

FIGURE 42.1 Plasma concentration following oral administration in fasted healthy adults of liquid filled capsule (circles) and a solid solution intermediate based tablet (squares) (Compound A; $n = 24$; 200 mg).

Each preparation process possesses advantages and disadvantages. Solvent-based processes are generally easier to scale down to mg-scale while extrusion approaches generally provide greater production rates per equipment volume. Access to multiple approaches is likely required to broadly enable a diverse portfolio of compounds with solid solution technology. The identification of drug candidate polymorphs may cause development challenges for spray drying process development in the identification of suitable solvent systems or extrusion processes in the case of thermal degradation or temperature-induced polymorphic transitions.

A particularly desirable aspect of extrusion in today's pharmaceutical development environment is the ability to continuously process in a direct-to-drug-product manner (Figure 42.4). Extrusion naturally lends itself to continuous processing from raw materials pneumatically conveyed into individual feed hoppers supplied by bulk containers to molded drug product fed into bulk product containers or even blister packages. This process has a small specific volume, making it amenable to real-time quality control and keeping the equipment footprint on a manufacturing floor exceptionally small. The product in an extrusion process naturally flows through a relatively narrow cross-section making continuous and direct process analytic interrogation straightforward (e.g., reducing issues of sampling).

42.2.1 Systematic Development Strategy

Prior to starting the design space definition process, efforts began by attempting to translate existing knowledge (i.e., experimental information, first principles understanding, models and best practices gleaned from peer reviewed literature) into a comprehensive view of the extrusion process. Process input parameters were summarized to ensure none of the process parameters were overlooked. This was facilitated with process mapping exercises following fishbone, or Ishikawa diagram methodology.

Subsequent risk assessment activities focused the parameters from the process map to a subset of potential critical process parameters (CPP's). These potential CPP's were identified as having a higher probability of impacting potential critical quality attributes (CQA's) as defined from an understanding of solid solution product requirements relevant to the patient. The risk-based evaluation of all parameters against the potential CQA's employed a quality function deployment (QFD) grid consistent with the house

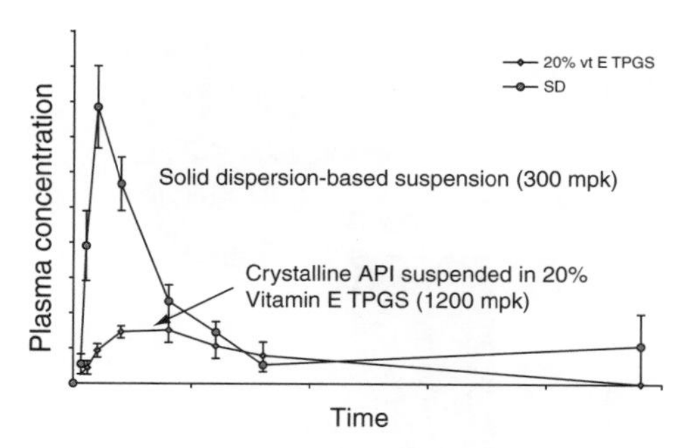

Cpd	SD (susp. dose, mpk)	Reference formulation	Exposure increase
A	300	1200 mpk 20% vitamin E TPGS	2–8X
B	100	100 mpk Imwitor 742:PS 80	67X
C	100	300 mpk methocel suspension	2–6X
D	200	750 mpk 10% PS80	4–16X

FIGURE 42.2 Preclinical impact of solid solution suspensions. *Left*: Plasma concentration profile following oral administration in male Sprague-Dawley rats (Compound B; $n = 4$). *Right*: Cross-project preclinical formulation comparison of solid solution based suspensions and a variety of reference formulations (mpk = milligrams per kilogram).

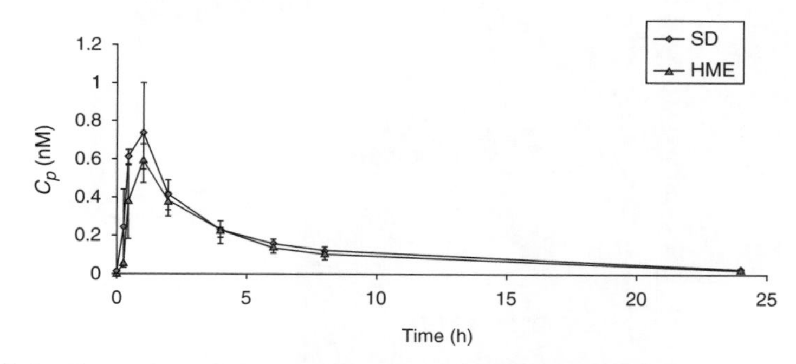

FIGURE 42.3 Comparison of identical solid solution formulations prepared by spray drying (SD) and extrusion (HME). Plasma concentration profile following oral administration of compound B in HPMC-AS-LF as tablets (male beagle dogs; $n = 6$; crossover; 50 mg dose).

of quality methodology [17]. This risk assessment exercise was not one of the tools explicitly outlined in the ICH Q9, however, it successfully managed the complexity of the hot melt extrusion process and delivered on the spirit of providing "transparent and reproducible methods to accomplish steps of the quality risk management process based on current knowledge about assessing the probability, severity, and, sometimes, detectability of the risk." [18] The outcome of this exercise was a summary of clearly prioritized development targets focused on a manageable set of potential CPP's.

Development efforts aimed to generate a fundamental understanding of the system recognizing that the degree of regulatory flexibility attainable was predicated by depth of knowledge. Enhancing this fundamental understanding was a focus on scale-independent parameters rather than scale-dependent parameters. An example of achieving this more fundamental understanding of the extrusion process by studying scale-independent parameters is demonstrated here for the case of shear stress in the extruder. Shear stress is a key parameter thought to influence quality attributes including degradation, and the capacity to achieve a molecularly dispersed product. Shear stress (the scale-independent parameter) is primarily manipulated via screw speed, or rpm (the scale-dependent parameter). While rpm is an easily accessible parameter to incorporate into design of experiments (DOE's), only studying the influence of rpm on product quality would have resulted in at best a correlative understanding of the system. This type of understanding does not capture the added impact of the degree of fill, scale, equipment manufacturer, or the result of wear in the extruder; all of which can change the shear stress at a given rpm. Focusing on scale-dependent parameters (i.e., rpm) rather than scale-independent parameters (i.e., shear stress) would have limited the potential to apply findings broadly to both anticipated and unanticipated process changes.

Initial small-scale experiments were conducted via DOE and successfully identified one scale-independent parameter that had a disproportionate impact on the key product quality attributes. This also yielded supportive data for more definitive CQA and CPP definitions. The next stage of development sought to build empirical models for the relationship between key scale-independent and scale-dependent parameters. This was achieved through response surface mapping, which was needed due to the multidimensional and quadratic means by which scale-dependent parameters (e.g., rpm, degree of fill, clearance) influenced scale-independent parameters (e.g., shear stress). This DOE was conducted on

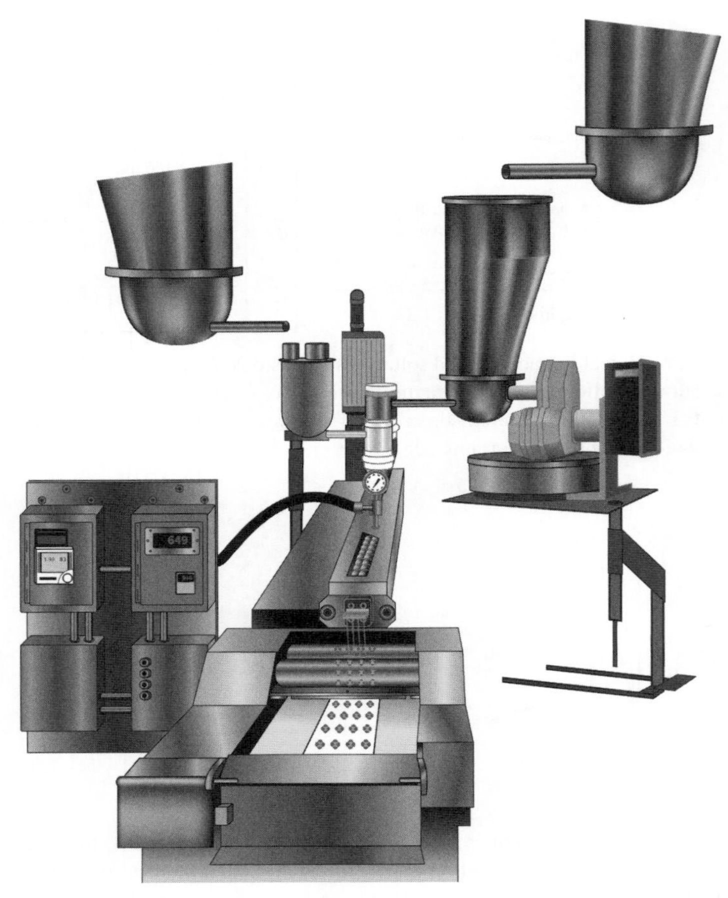

FIGURE 42.4 Direct-to-drug-product extrusion process train schematic, highlighting bulk feeders, precision loss in weight feeders, liquid injection, and calendaring of tablet between two chilled rolls.

commercial scale equipment incorporating in-line process analytical technology that facilitated evaluation of both product quality and process robustness. Robustness was assessed by introducing known perturbations at all processing conditions, and monitoring the system's capacity to dampen the upset. The combined understanding of quality attributes as a function of scale-independent parameters, the empirical models from response surface mapping and the PAT data on process robustness allowed definition of a design space that achieved balance between achieving target quality attributes and optimizing operating conditions including throughput and process stability.

42.2.2 HME Process Overview and Mapping

The intent of process mapping is to create an objective view of the extrusion process, and generate the set of potential process inputs to serve as the basis for risk assessment activities. Consider extrusion as a series of suboperations (1) material feeding, (2) powder conveying and degassing, (3) melting and mixing, (4) melt conveying and venting, and (5) pumping, shaping, and cooling. Each of these suboperations was mapped independently. The following sections include the map of process inputs and discuss some of the features of each suboperation.

42.2.2.1 Material Feeding Extruder feed systems in many ways ultimately control the content uniformity of product. While extruders offer some backmixing to dampen out high frequency feed rate perturbations, low frequency disturbances in material feed rates can result in compositional variations in the extrudate, and, hence, compromised product quality [19]. One route to decouple feeder performance from the extrudate compositional uniformity would be to preblend all of the feed streams. However, preblending requires an extra unit operation, presents a risk for segregation, and is generally complex when one or more components are liquids. Figure 42.5 is a process map outlining a subset of the potential process inputs for the material feeding process.

Solids feed to the extruder can be achieved via volumetric or gravimetric feeders. Volumetric feeders are best suited for applications where the materials flow well and control of the composition of the feed stream will not vary (i.e., for use with powder preblends). The applications presented here focus on

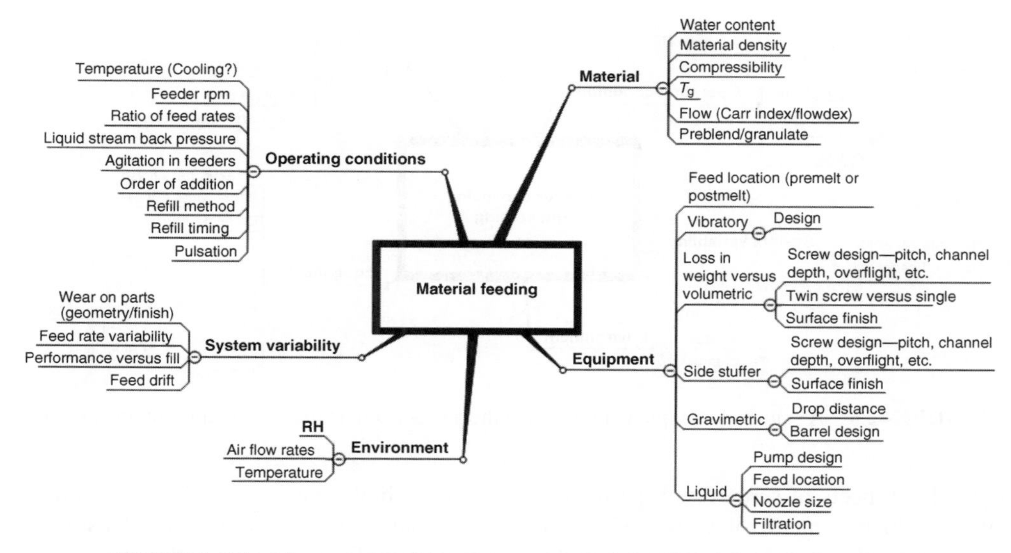

FIGURE 42.5 A summary of input parameters of the material feeding process.

gravimetric feeders. These feeders are often single or twin-screw feeders coupled to a load cell(s). Constant flow rates into the extruder are attained via loss in weight feedback control.

A key consideration when evaluating solids feeders is that flow rates from these units are never fully continuous. Intrinsic to powder conveying screws is some level of pulsing into the extruder, since the feed material is not a continuous media. For very poorly flowing material exhibiting strong propensity for avalanching, the frequency of the pulsing for a twin-screw feeder could be expressed by

$$\frac{\text{rpm}}{(2n-1)60} \tag{42.1}$$

where n is the number of flights in the feed screws. For example, for a twin-screw feeder with single flighted conveying screws feeding poorly flowing material, running at 60 rpm would result in a pulse every second. Similarly the concept of continual versus pulsed addition applies for liquids. At low liquid flow rates, any inconsistent flow, or dripping should be evaluated to ensure it is not occurring at a frequency undampened by the extruder's backmixing. This can be overcome by trying to achieve back pressure to deliver the liquid as a continuous stream.

Powder feeders can be coupled to the extruders such that material drops from the end of the feed screws into an open barrel section. A limitation of this approach is that incorporation of the powders into the extruder is limited to a fraction of the open barrel feed port where the extruder's down turning screw conveys material into the extruder [20]. The feeding of low bulk density powders can be particularly problematic and can significantly constrain the maximum attainable throughput. A common means to overcome flood feed limitations of low bulk density materials is via a side stuffer. Here, the loss in weight feeder delivers powder gravimetrically to volumetric (constant rpm) conveyors coupled to the side of the extruder. These conveying screws produce some predensification and force the powder into the extruder enabling higher throughputs for low bulk density powders. Successful densification in the side stuffer requires sufficient venting of entrapped air.

A consideration when designing feed locations are the disparities in shear stress, temperature, and mixing histories that materials experience. The highest viscosity that polymers transition through occurs at the melt onset, which is also the point of maximum shear stress. Active ingredients or liquids could be added upstream or downstream from this point. Adding the active downstream limits the total time at temperature and reduces the shear stress, however, it also potentially reduces the extent of mixing achieved. An additional consideration when determining the feed location are the difficulties in mixing materials having large viscosity differences. Lower viscosity materials can act as lubricants, reducing shear rates and mixing intensity [21]. In extreme cases, streaming of the low-viscosity material through the extruder may be observed. Special attention should be paid to the mixing sections of the screw profile when there is a need to incorporate materials having large viscosity differences.

42.2.2.2 Powder Conveying and Venting The next task that must be accomplished in the extruder is conveying the bulk powder to the melt zone. A list of process inputs to consider when feeding and conveying low bulk density powders is listed in Figure 42.6. In conveying sections, material passes through the extruder via drag flow with very little pressure generation. In an ideal conveying scenario, material would demonstrate perfect slip with the screw, and perfect friction with the barrel.

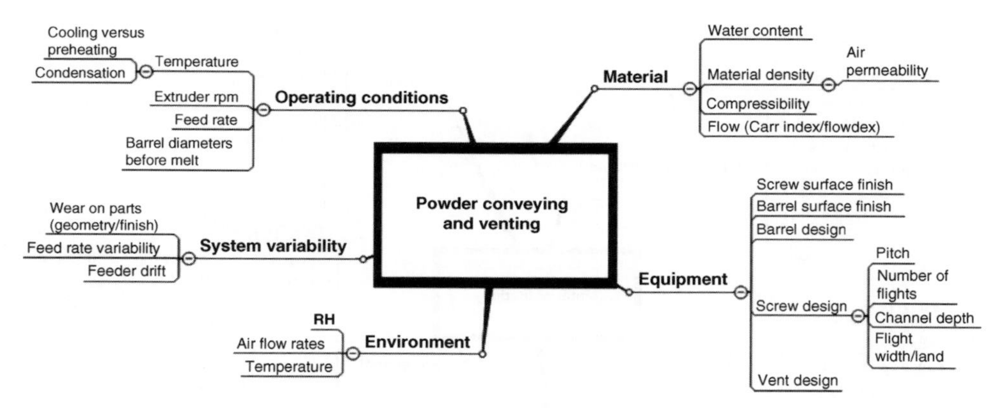

FIGURE 42.6 A summary of input parameters of the powder conveying and venting suboperations.

Figure 42.7 highlights aspects of screws with particular relevance to conveying. Increasing flight width results in greater pumping efficiency of the screws. Increasing flight pitch, or angle of the helix, generally results in faster material conveying per revolution, although low bulk density materials may not follow this trend. The internal (D_i) and external (D_o) diameters of the screw have numerous implications. The channel depth (D_o–D_i) sets the free volume of the extruder and largely determines the maximum feed rate (without the use of mechanical predensification via a side stuffer, etc.). The D_i generally limits available torque at small scales, although at larger scales the torque constraints of the drive motor may be limiting. Although a larger D_i enables greater torque, this generally reduces throughput by sacrificing free volume.

Conveying involves added complications for pharmaceutical applications because many of the feedstocks possess substantially lower bulk densities than commodity polymer extrusion operations based on 3 mm pellets. It is not uncommon for an API to have 0.2–0.4 g/cm^3 bulk density. Many of the pharmaceutical polymers and surfactants are also only available from manufacturers as powders (not pellets), having bulk densities of 0.4–0.6 g/cm^3. As noted in the previous section, this low bulk density largely impacts the maximum feed rate achieved in the extruder due to flooding of the feed port. Increasing screw speed (in some cases) [22], employing a side stuffer, or increasing the free volume of the screws can enable higher feed rates. One way to achieve a higher free volume in the feed section is to use undercut, square channel screws in the feed zone [23]. While these are not self wiping, this may not be a concern since there is no molten material in the feed section, and this would not extend the heat history of the material.

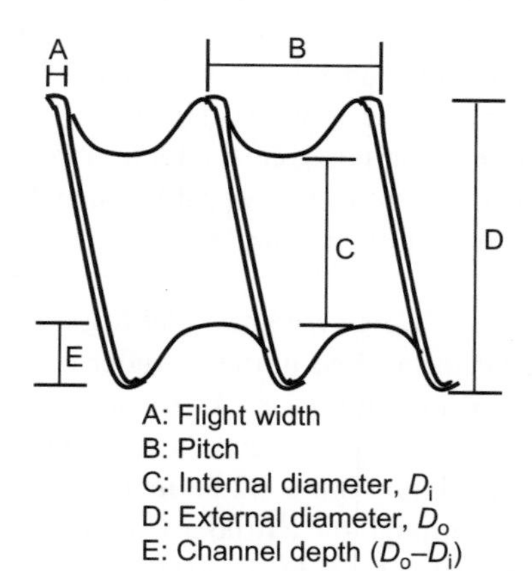

FIGURE 42.7 A subset of the critical design components of conveying screw elements.

A further challenge in feeding low bulk density powders involves the entrained air. Low bulk density feed materials undergo significant densification during extrusion because up to 80% of powder feed streams are entrained air. Effective venting mechanisms are essential to remove this air and maximize throughput [22]. In addition to venting air, moisture removed from the feedstock by heating in the extruder may condense on colder upstream barrel sections or feed powder in the feed zones [24]. Given these two considerations, proper venting in the feed section is critical.

42.2.2.3 Melting and Mixing

Achieving a melt is generally accomplished through the input of energy by the extruder into the formulation. A generic free energy diagram (Figure 42.8) illustrates this principle. The formulation components, whether crystalline API or amorphous polymer, undergo a transformation to a more mobile and deformable state at higher temperatures. The melt extrusion process enables energy input through both thermal and mechanical means. Thermal energy input is typically achieved through electric or oil heating of the barrel, which is transferred to the formulation via conduction. This method can be efficient in small-scale extruders, although conductive heating alone is generally not a sufficient source of energy to achieve a melt due to the poor thermal diffusivity of polymers [25]. Relying heavily on conductive heating also poses scale up challenges since heat transfer is a function of surface area, while scale up to preserve key material properties generally occurs on a

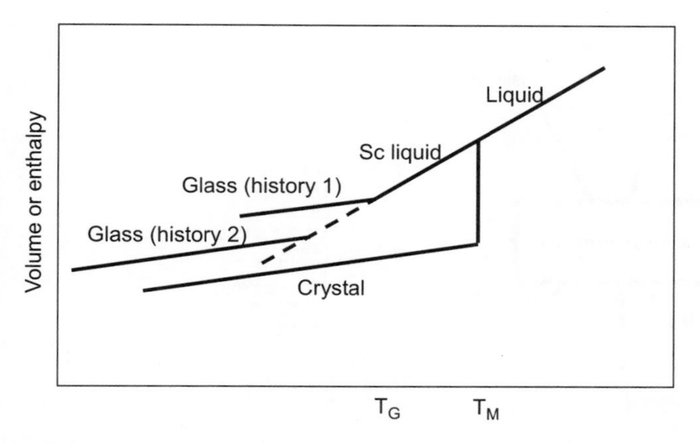

FIGURE 42.8 Behavior of glass forming single component material.

volumetric basis. Moreover, greater product temperature uniformity could be expected when minimal conductive heat is transferred from the barrel [26].

While conductive heat can contribute to the total energy input at smaller scales, melting is largely achieved by viscous dissipation via frictional forces (including interparticle, material/wall, and material/screw friction) [25, 27]. Some estimates suggest at least 80–90% of the energy to achieve a melt is supplied by the extruder's screws [28]. This energy is generally referred to as the specific energy, which is the ratio of mechanical energy (as measured by a wattmeter on the drive motor) to feed rate in units of kW h/kg. It is important to note that the wattmeter reading does not account for losses to the thrust bearing (which can be significant if high pressures are generated in the extrusion process), or drive motor efficiency at the given rpm, although these adjustments can be made. Mechanical energy input is achieved through deliberate design of the extruder screws to impart mechanical stress on the formulation, with specific energy often serving as a target to ensure process consistency upon scale up [29].

Extruder screws are generally modular and consequently allow for a number of different configurations, which have a direct impact on the specific energy, residence time distribution, and maximum shear stress among other process responses. The same formulation extruded under different conditions, therefore, may exhibit disparate levels of quality [30].

Pharmaceutical extrusion often features the use of double-flighted screws, creating three distinctly separate channels down the length of the extruder barrel. A configuration using only conveying elements would largely move material through the extruder in plug flow with minimal backmixing or material transfer between these three channels. The only mixing achieved in a pure conveying system would be laminar in nature, and would potentially be less than expected due to viscous polymers not following no-slip boundary layer conditions [31]. As such, under these conditions there is almost no high-frequency disturbance dampening, potentially leading to poor compositional uniformity. Compositional uniformity requires backmixing, and is achieved via screws designed to allow pressure flow to cause material movement between the channels. Here, the screw flights are opened, often taking the form of mixing blocks. Figure 42.9 attempts to show conceptually how conveying elements are discretized in a way that allows exchange of material via mixing blocks. While mixing blocks are the conventional means to achieve a melt and sufficient backmixing, other routes include blister rings [26], gear or turbine type elements [32], or simply staggering or offsetting conveying elements [20].

Mixing screw elements have been designed to ensure compositional uniformity. The axial width of a mixing paddle is an indicator of the magnitude of global pool capture (a region of the screw channel known for its high shear) and determines the extent to which mechanical energy will be imparted by a single mixing element. In plastics extrusion, wide mixing paddles reduce particle size through attrition (dispersive mixing), whereas narrow paddles and similar lower energy elements are utilized primarily to achieve compositional and thermal homogeneity (distributive mixing) [33]. Formulations consisting of miscible components benefit from both dispersive and distributive mixing to impart energy for melting, and create surface area for diffusion; thereby yielding uniformity at a molecular level [34, 35]. The number of mixing cycles (cycles of volume expansion and compression during screw rotation) may also serve as a measure of mixing intensity. This mechanism is analogous to kneading pizza dough as a method to incorporate and mix ingredients (Figure 42.10) [20, 33]. A full list of process parameters evaluated for the melting and mixing stage is highlighted in Figure 42.10.

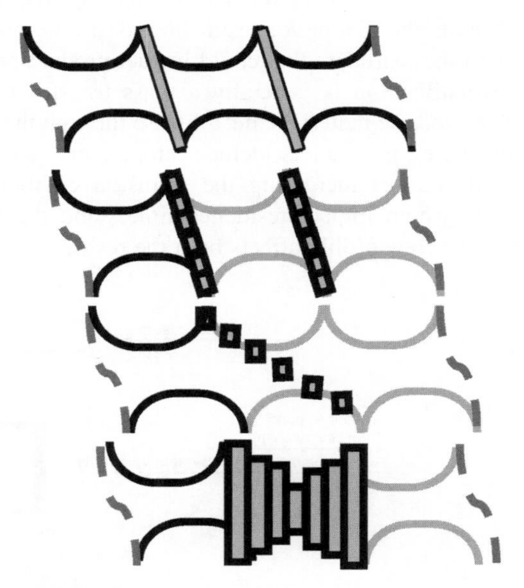

FIGURE 42.9 Illustration of the transition from conveying elements with a closed channel (*top*) to mixing elements with an open channel (*bottom*), to produce backmixing.

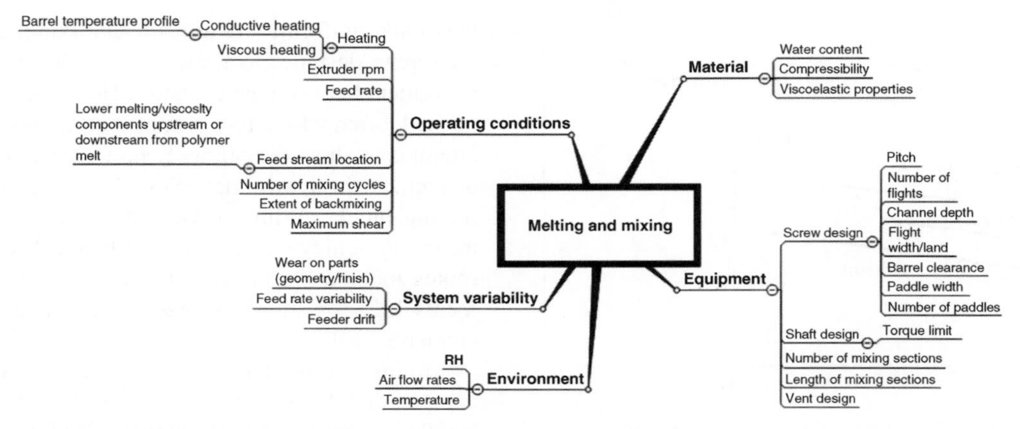

FIGURE 42.10 A summary of input parameters of the melting and mixing suboperations.

*42.2.2.4 **Melt Conveying and Venting*** Downstream of the mixing paddles, the aggregate material is in a molten state and any additional conveying before the die continues to constitute a formulation's time at temperature, which is the most relevant component of the residence time distribution as it relates to product quality (i.e., degradation versus homogeneity). The conveying screws are generally less than completely filled, and residence time is a function of screw speed. See Figure 42.11 for a list of process input parameters related to this stage. This is in contrast to residence time in full barrel sections (e.g., reverse conveying elements, neutral mixing sections) were residence time is a function of feedrate with minimal contributions from screw speed [23].

Venting is an important aspect of the melt conveying step. The melting of API and excipients liberates solvents (e.g., water) and without proper venting, even small quantities of water or other volatiles could result in bubbling/foaming as the melt exits the die and/or result in undesirable residual solvents in the extrudate that could impact product properties. Devolatilization is generally a mass transfer limited process, as volatile materials must diffuse through the melt. Key influencing parameters include temperature (with elevated temperatures increasing the diffusion coefficients), feed rates (which impact residence time), and the screw profile. The screw profile affects both the residence time in venting sections, and diffusion distances. Screws sections under the vent are typically designed to minimize local degree of fill using multiflight large pitch screws to maximize surface area and minimize diffusion distance [28]. Screw profiles can also be designed to enhance the surface renewal phenomena to further reduce the diffusion distances.

The extrusion of formulations containing API solvates is an intriguing application of venting to enable what has not been possible with conventional pharmaceutical processing. A given drug molecule can exist as any number of solvates, depending on its chemical synthesis route. Safety considerations prohibit the vast majority of solvates from moving into drug development despite their sometimes favorable physical properties (i.e., improved flow, enabling better control of feeding to the extruder and improved content uniformity of the extrudate). Using the melt extrusion process, an API solvate may be fed into the extruder, with the solvent removed upon melting. The final drug product should contain low levels of residual solvent, meeting ICH [36] specifications.

*42.2.2.5 **Pumping, Shaping, and Cooling*** The stage of extrusion most closely associated with its namesake is the pumping of molten extrudate through a die. Die geometry may play a role in the final product, such as in the production of transdermal films, which would require a slit die. The

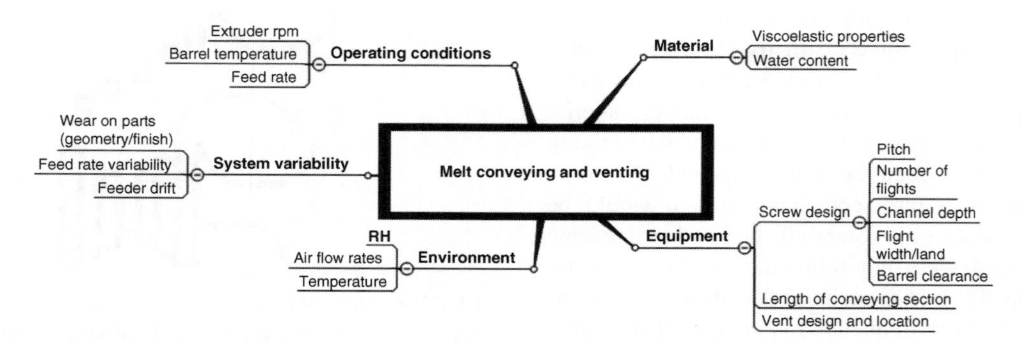

FIGURE 42.11 A summary of input parameters of the melt conveying and venting.

molten extrudate may also be processed downstream via conventional unit operations (i.e., milling and compression) and in this case precise die geometry is not critical. A common shape for pharmaceutical extrusion application is a circular die, with multiple strands facilitating more rapid quench cooling. In some cases, particularly for multiparticulate controlled release applications, the extrudate strand may be passed through a die face cutter resulting in small pellets [37]. Here, strand diameter plays a more central role in product quality, and is determined by the die geometry, viscoelastic properties of the formulation, process conditions (e.g., pressure) and material flow rates.

The extruder die can be roughly characterized as a resistance to material flow. Sufficient pressure is generated in the screw channels prior to the die to overcome this resistance. The pressure resistance can again be considered a function of the die geometry and the melt's viscoelastic properties. Significant pressure increases near the die may result in melt temperature increases due to viscous dissipation [38], resulting in the product's maximum melt temperature. For inviscid materials, additional heating during pumping out of the die may not be significant. In this case, the maximum product temperature may be achieved upstream of the die at some point after achieving a melt.

The extrudate can take many paths to a finished pharmaceutical dosage form following extrusion. The formulation is often quenched using such methods as passing along a conveyor belt with compressed air, or feeding through chilled stainless-steel rolls. Once cooled, the extrudate may be sized using conventional pharmaceutical mills, then compressed into tablets or filled into capsules. Alternative options for manufacturing finished pharmaceutical dosage forms include direct shaping methods of the extrudate. Directly formed tablets may be created by calendaring [39] or injection molding [40]. Figure 42.12 shows an example of a directly shaped dosage form. Molding enables production of complex shapes with features such as embossing to improve patient compliance, enable branding opportunities, and prevent counterfeiting. Molding also has the potential to reducing the need for fillers and compression aids. See Figure 42.13 for a list of input parameters related to this process stage.

42.3 RISK ASSESSMENT

42.3.1 Quality Attribute Definition

Multiple processes and process conditions can produce material with similar *in vivo* responses (as demonstrated in Figure 42.3) provided the process achieves a true solid solution. In the case of extrusion, heat generally drives miscibility unless the enthalpy of mixing is very unfavorable at high temperatures. Miscible formulations produced by extrusion will be homogeneous provided sufficient mixing, time, and heat.

FIGURE 42.12 Example of a directly shaped tablet following a melt extrusion process illustrating the ability to form complex shapes, such as the Merck corporate logo (image weight ca. 200 mg).

While the bulk of extrudate properties can be explained by the heat history and energy input, relaxation state and particle physical attributes (e.g., particle size) may also impact formulation performance. The relaxation state can be varied systematically by changing quench rate. Figure 42.8 is a common illustration of how different quench rates of a single component system can lead to materials of different relaxation states and properties.

HME processing conditions are generally constrained at one end by the maximum throughput. This operating limit is usually characterized by low specific energy, short residence time, low product temperature, and/or limited mixing (Figure 42.14). An extreme case for a low energy limit would be to employ conveying elements, no barrel heat, and no die restriction [41]. With more conventional compounding screw profiles, it is possible that the low energy limit resulting in unacceptable product quality will lie outside of the accessible operating space (e.g., a torque limit may be encountered before inhomogeneous product is produced).

The upper end of HME processing conditions as shown in Figure 42.15 may be constrained by thermal degradation. Efforts to circumvent thermal degradation in HME include the use of plasticizers with the goal to process at low barrel temperatures and/or with a less aggressive screw profile [42]. Incorporating antioxidants during extrusion [43] and nitrogen blanketing can effectively stabilize oxidatively sensitive drugs. Elucidating degradation mechanisms for the polymers and developing analytical characterization methodologies is more complex compared to thermal degradation of small drug molecules for which stability indication assay method and LC-MS are commonly used.

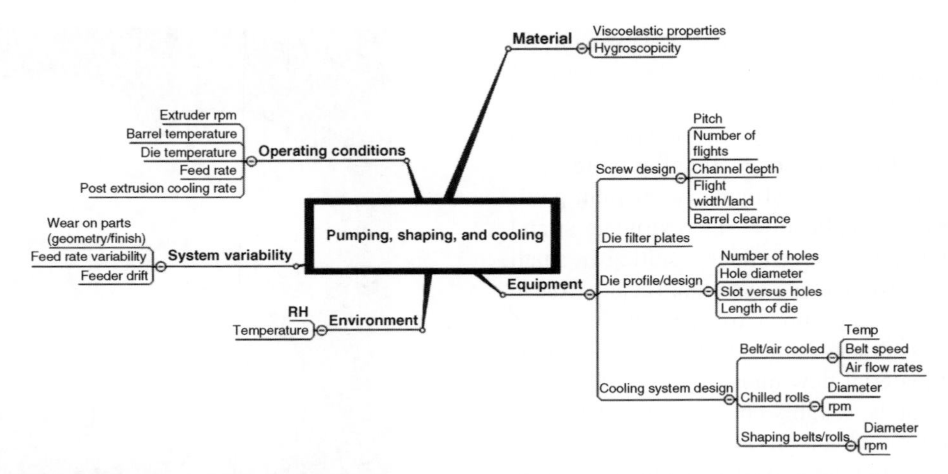

FIGURE 42.13 A summary of input parameters of melt pumping, shaping, and cooling.

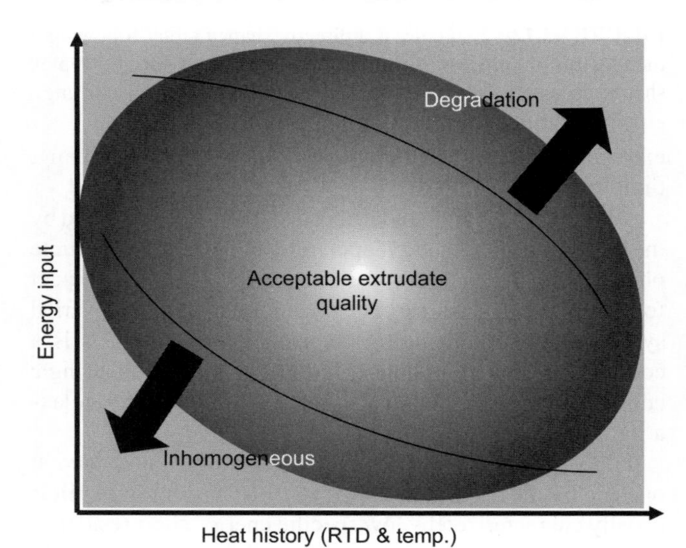

FIGURE 42.14 A schematic showing the balance of energy input and heat history, with the extreme combinations of these resulting in either degradation or inhomogeneous material.

The implication of polymer degradation is twofold (1) potential safety concerns of the degradation products at levels exceeding the ICH qualification threshold and (2) potential impact on polymer functionality (e.g., stabilizing high energy amorphous drug in solid state and/or enhancing solubilization of poorly soluble compounds during *in vitro* drug release).

Hydroxypropyl methylcellulose acetate succinate (HPMC-AS) is a widely used polymer for making solid solutions owing to its unique physical and chemical properties including high T_G, ample solubility in organic solvents, and potential for diversified molecular interactions with a broad range of poorly soluble drugs. HPMC-AS is a linear polymer consisting of $\beta(1 \rightarrow 4)$ linked substituted D-glucose units. Three possible reactions could occur after thermal stress: (1) the dissociation of substitution groups, (2) the breakage of glycosidic linkages, and (3) rearrangement of the polymer backbone via intramolecular reactions (Scheme 1). A detailed understanding in degradation chemistry involves polymer characterization using an array of analytical techniques such as TGA, GC-MS, size exclusion chromatography (SEC), NMR, IR or mass spectrometry.

TGA is an effective tool for monitoring polymer breakdown products as a function of temperature. It has been routinely used to derive activation energy associated with polymer decomposition [44]. TGA analysis of HPMC-AS showed a modest volatile formation (2%) at temperatures ranging from 175 to 250°C (linear ramp in 5 min). Headspace GC-MS analysis showed the volatiles mostly consist of CO_2, formic acid, acetic acid, succinic acid, and other unidentified small molecular species. The volatiles are likely the by-products of dissociation of side chains from the polymer backbone. The TGA/GC-MS results are consistent with those from variable temperature IR, where an increase in OH signal was observed, indicative of the loss of R groups. At representative HME processing temperature (e.g., <230°C), the loss of R groups is believed to dominate the reaction pathway leading to polymer degradation. TGA was also conducted for a series of other polymers including HPMC-phthalate, HPMC-trimellitate polyvinyl pyrrolidone, and polyvinyl pyrrolidone–polyvinyl acetate copolymer (PVP–PVAc). The polymer decomposition rate appears to increase in the order of PVP < PVP–PVAc < HPMC-AS < HPMC-T, and HPMC-P.

Thermally induced glycosidic bond breakage could lead to depolymerization or cross-linking of the polymer, which could potentially compromise polymer functionality. SEC results of HPMC-AS processed at several extrusion conditions suggest polymer breakdown or cross-linking is not likely to occur at typical operating temperatures (e.g., 170–230°C). This is consistent with biorelevant dissolution of HME extrudate samples in which typical process

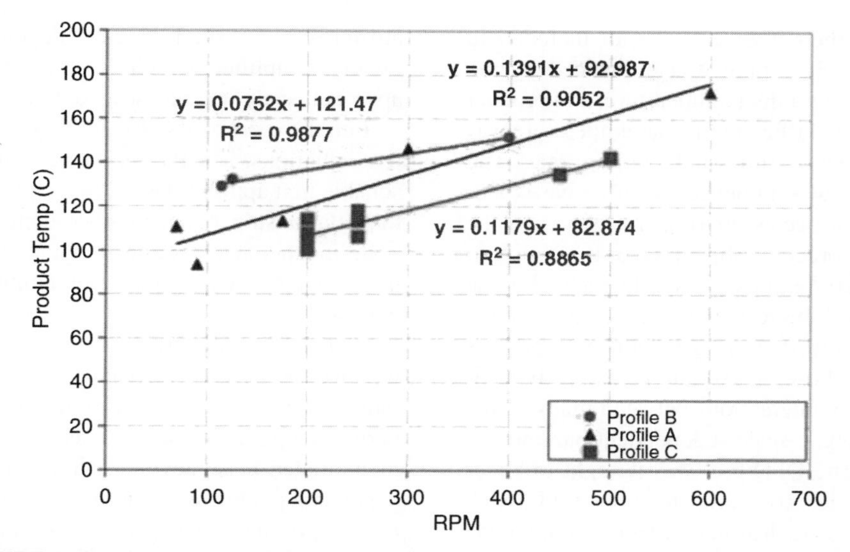

FIGURE 42.15 Impact of rpm and screw profile on the product temperature. Screw profiles varied from modest to aggressive: B → C → A.

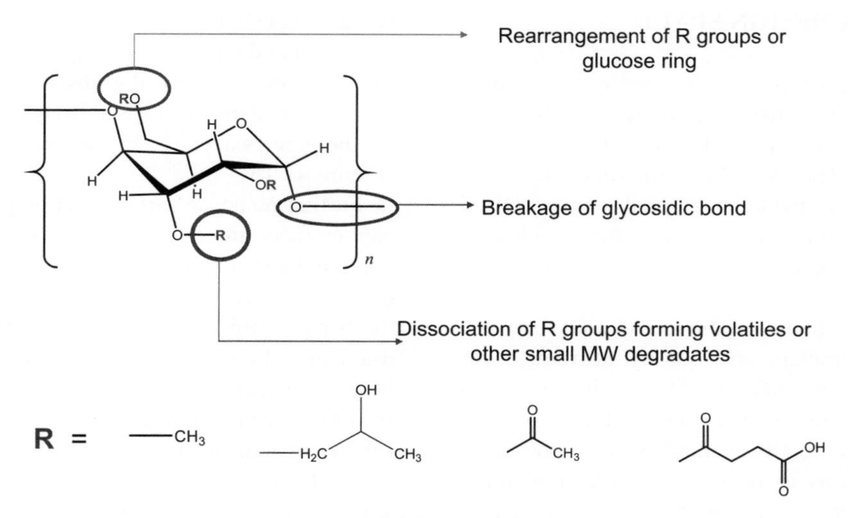

Scheme 42.1

conditions had minimal impact on dissolution rate and apparent drug concentration. Under extreme processing conditions (e.g., 220°C barrel temperature coupled with an aggressive screw profile), polymer cross-linking is evident, as shown by a significant drop in polymer aqueous solubility. The nature of the cross-linked polymer is not clear and remains the subject of further investigation. The polymer cross-linking could significantly compromise *in vitro* performance of solid dispersions due to the entrapment of drug molecules by cross-linked polymer networks.

42.3.2 Risk Assessment Tool

The foundation of risk assessment began with existing knowledge of how the extrusion process conditions would likely influence product quality. The quality attributes of the extrudate include achieving a molecularly dispersed glass having micro- and macroscale compositional uniformity and no degradate products. The critical process parameters to achieve this molecular dispersion include delivering appropriate thermal driving force without degrading any of the components, and coupling this thermal driving force with sufficient time and adequate surface area for molecular diffusion to occur. Ensuring macroscale compositional uniformity requires process conditions optimized to minimize sensitivity to feed rate disturbances.

A series of scale-independent parameters were understood to influence each critical quality attribute. Formulation dependent parameters such as viscosity and glass transition temperature are also important, however, due to the limited ability to adjust these; their discussion is not included in detail here. Thermal driving force was influenced by the specific energy, maximum product temperature, and product temperature distributions. Sufficient time at temperature was influenced by the residence time and cooling rate. The surface area for diffusion was influenced by effective expansion/compression mixing cycles, mixing intensity, and the shear stress profile. The process robustness, or the ability to dampen feed input perturbations, was influenced by the extent of backmixing and the residence time distribution.

The risk assessment efforts sought to prioritize process parameters and identify those that should be carefully evaluated. There are complex interactions between many of the scale-dependent parameters, scale-independent parameters, and quality attributes. The QFD tool was thought to better address these complexities compared to more traditional, linear risk assessment tools such as fault trees or failure mode effects and analysis.

42.4 ACHIEVING A DESIGN SPACE

Initial efforts focused on determining how scale-independent parameters impact quality attributes. Integral to this was identifying characterization methods indicative of patient relevant quality attributes. A battery of solid state and thermal characterization methods coupled with a DFSS driven measurement system analysis were used to identify a technique believed to be a good indicator of molecularly dispersed extrudate.

A quantitative assessment of the correlation between scale-independent parameters and quality attributes was evaluated within the framework of a DOE, with the scale-independent parameters set as the design factors. Constitutive equations and commercially available software based on one-dimensional solutions to heat, mass, and momentum balances were an integral part of this design. These models were used to determine how to adjust scale-independent parameters (screw profile, rpm, throughput, barrel temp, etc.) to achieve low and high factor level set points for scale-independent parameters (residence time distribution, specific energy, shear stress, mixing, etc.). Analysis of the result focused on quantitatively linking variations in the quality attributes to the scale-independent parameters.

While scale up of extrusion processes is well understood [23, 25, 28], and achieving comparable product quality attributes upon scale up is certainly feasible, it is unrealistic to expect all scale-independent parameters will be equivalent upon scale up. For this reason, it was particularly important to have a deep understanding of which scale-independent parameters most influence quality attributes, and where to focus scale up efforts.

Results from these experiments indeed suggested that variations in the key measurements of quality could be attributed to a single scale-independent parameter. This greatly simplified which scaling rules to apply, how to approach scale up, and how to define scale up success.

Initial experiments also highlighted that while constitutive equations, and the modeling software provided a reasonable first approximation, they did not sufficiently explain the relationship between scale-dependent parameters and scale-independent parameters. Statistically designed experiments were used to generate empirical models for this purpose.

Figure 42.15 illustrates one case of observed multifactor interactions between dependent variables used to describe independent variables. During these experiments, feed rate, barrel temperature, screw speed, and screw profile were manipulated to achieve targeted responses in the scale-independent parameters. Figure 42.15 shows how the response for the product temperature (scale-independent parameter) can be largely explained just by the screw profile and screw speed (scale-dependent parameters). Attempts were made to generate this correlation in one comprehensive empirical equation via DOE; however, the screw profile had too large of an impact and required a multiple linear regression approach to effectively describe the observed behavior. The product temperature dependence is a stronger function of screw speed as the aggressiveness of the screw profile is increased (Figure 42.15).

Further response surface mapping experiments sought to capture factor interactions and curvature. Figure 42.16 shows contour lines for the mean residence time as a function of barrel temperature and feed rate. This illustrates the nature of the barrel temperature feed rate interaction and that a quadratic model was necessary. Here, lower feed rates resulted in longer mean residence times. This is expected since lower feed rates would equate to more backmixing per unit mass. Longer residence times were observed at higher feed rates when the barrel temperature was decreased from 160 to 120°C. This could be due to the impact of lower barrel temperatures achieving lower product temperatures, and hence higher viscosity. This would increase backup lengths in filled sections, effectively extending residence time. Similar figures and empirical models were generated for the other key scale-independent parameters.

Understanding curvature and parameter interactions enabled a multifactored optimization of processing conditions. The resulting equations also made a detailed definition of the design space possible. The models provided insight into how quality attributes could be influenced in regions outside the space explored by the DOE. Several points were included outside the space of the DOE where a specific combination of operating conditions could achieve higher throughputs. This high throughput could not be universally achieved such that it could not have been a high factor setting in the DOE. Using the empirical model built via the DOE, the quality responses at these extreme points were predicted accurately.

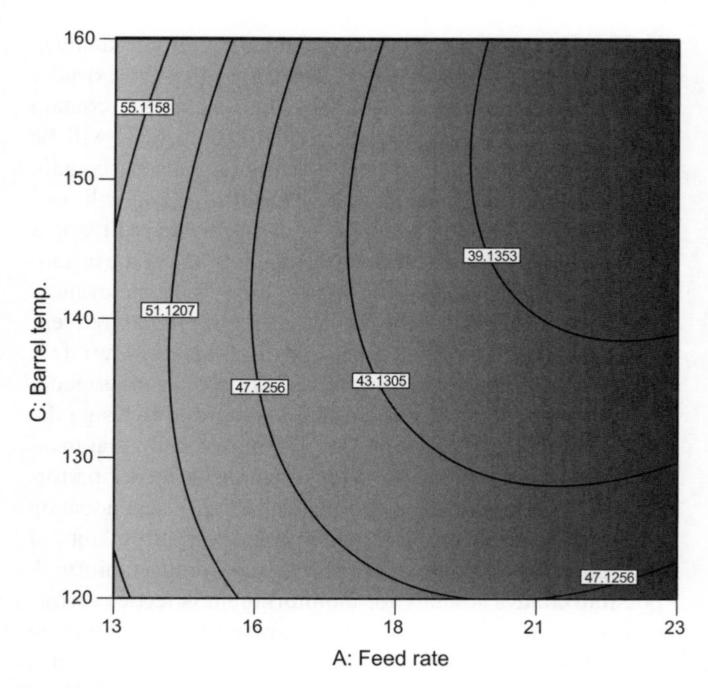

FIGURE 42.16 Impact of feed rate (kg/h) and barrel temperature (°C) on mean residence time.

42.4.1 Process Robustness

42.4.1.1 Introduction The extrusion process is the integration of the mass feeding systems with the extruder itself, and when operated with multiple feed streams, an understanding of the overall system dynamics is vital to ensuring product quality. The compositional homogeneity of the extrudate is dependent on performance and stability of each independent mass feeding system. Process upsets and perturbations as well as unstable mass feeder operation and degrading mass feeder performance threaten product quality.

In this work, PAT, was used as a key enabler for holistic process understanding and quality control. An in-line transmission near infrared measurement system was developed and implemented to monitor the composition of the multicomponent melt stream exiting the extruder in real time. Real time measurement of product composition during operation enables an operator to verify that the process is stable, identify process upsets, and ensure the targeted set point is realized. This PAT tool was also used to develop a dynamic model of the process to both pulsed and stepped inputs. The process model was then used to understand the disturbance dampening capabilities of the process and to inform performance requirements for the mass feeding systems.

42.4.1.2 System Identification A series of system identification tests were conducted, that consisted of both pulsed [45], and step change inputs. Pulsed inputs were achieved by rapidly introducing preweighed amounts of API into the feed throat of the extruder. Step change inputs were achieved by changing the mass flow set points on the feeder controllers. The dynamic response of the process to the input signal was measured in each test case. The process responses to the compositional step change identification tests were fit to a first order plus dead time (FOPDT) model equation (42.2), where $I(t)$ is the component concentration as a function of time, K is the process gain ($K = 1$, mass flow changes are fully realized), Δx is the magnitude and the direction of the step change, $u(t - t_0)$ is the unit step change with dead time, t is time, t_0 is dead time, τ is the process time constant, and $I(0)$ is the concentration of the component before the step change. More sophisticated models were considered [46, 47], but in this case the FOPDT model was found to adequately describe the data. The best-fit global FOPDT model parameters are shown in Table 42.1. The process time constants are similar for all three components in the formulation. This is consistent with expectations, because the extruder acts as a CSTR-PFR model in the sense that all mass elements will experience the same environment as they pass through the extruder. The processing environment is fixed by the extruder process parameters (screw profile, barrel temperature profile, mass flow rate, screw speed) and the formulation rheology. The process dead times are also similar, but their variation is consistent with the location in which they are introduced into the process. Example model fits are shown in Figure 42.17 (panels a, b, and c are the responses for each formulation component). The high speed of data acquisition of the spectrometer (one prediction every 1.31 s) was well suited for model parameter identification, in particular the process dead time.

$$I(t) = K \cdot \Delta x \cdot u(t - t_0) \cdot (1 - \exp(-(t - t_0)/\tau)) + I(0) \tag{42.2}$$

42.4.1.3 Process Disturbance Rejection Capability Disturbance rejection capability of the process can be assessed by calculating the periodogram of the time derivative of the identified process model. The periodogram [48] is defined as the absolute value of the square of the finite Fourier transform (FFT) versus a frequency vector as described by equation 42.3, where U is the FFT of the time series $u(t)$, N is the number of elements in the time series, ω is the frequency

TABLE 42.1 First Order Plus Dead Time Model Parameters

Model Parameter	API	Surfactant	Polymer
τ (time constant, s^{-1})	11.1	10.5	10.3
t_0 (dead time, s)	59.4	66.2	62.2

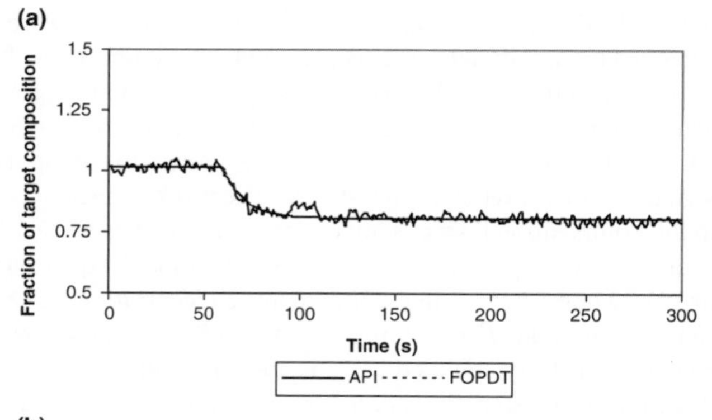

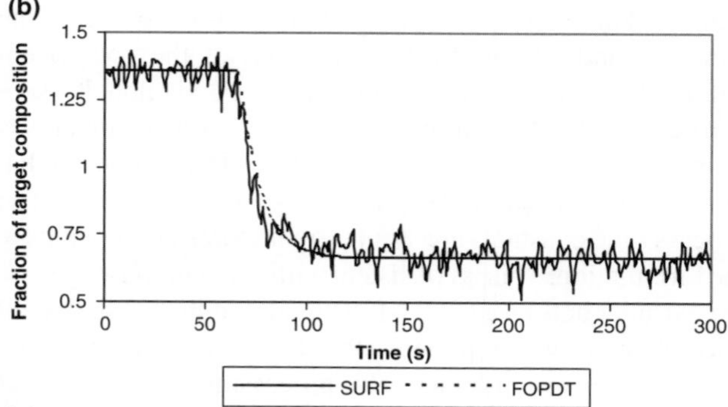

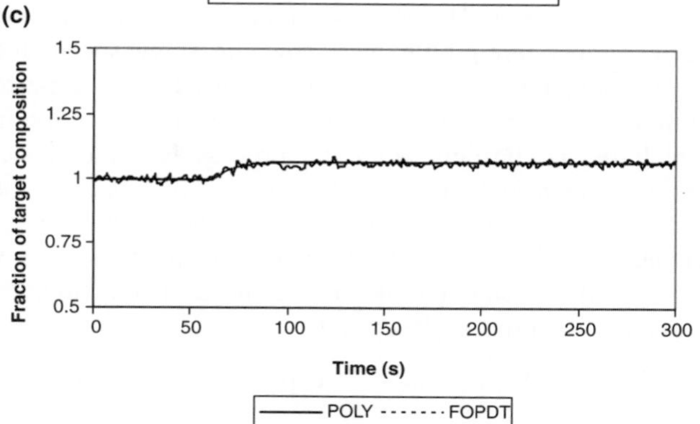

FIGURE 42.17 Example FOPDT model fits of the extruder process response to simultaneous mass feed rate step changes in API (upper), surfactant (middle), and polymer (lower) as measured by in-line transmission NIR.

in inverse time, i is the square root of negative one, and t is time.

$$|U_N(\omega)|^2 = \left|\frac{1}{\sqrt{N}}\sum_{t=1}^{N} u(t)e^{-i\omega t}\right|^2 \tag{42.3}$$

The time derivative of the identified process model is an estimate of the finite impulse response function for the process. The periodogram, which is a plot of signal power versus frequency, is the frequency response function of the process. Figure 42.18 shows the periodogram of the process model. This plot describes the extruder's ability to dampen mass feeder input disturbances and shows that the extruder acts as a low pass filter. Specifically, input signals that contain energy content at frequencies above 0.05 cycles/s will be nearly entirely damped. Conversely, input signals that contain energy content at frequencies below 0.05 cycles/s will pass through the extruder virtually undamped and will affect product uniformity. Additionally, the amplitude of the instantaneous variability in mass flow rate is essentially irrelevant as long as the variability frequency is above the critical frequency (ca. 0.05 cycles/s in this work) and that the mean flow rate is on target. This knowledge can be used to either set mass feeder performance specifications or could be used to redesign the extrusion process (screw profile, or screw speed, total mass flow rate) to be compatible with known mass feeder performance. The mass feeder performance is highly dependent on the feed material properties that could vary from lot–lot including particle size, Carr index, and compactability. A possible control strategy for monitoring mass feeder performance would be numerically calculating the periodogram from the real time mass flow rate data (from loss on weight) over a moving window (1 min, for example) that would trigger an alarm when the time series has frequency contributions below the critical frequency or a defined threshold value.

The process model can also be used for real time predictive process monitoring. This could be achieved by numerical convolution of the differential form of the process model with the time discretized mass flow rate data coming from the mass feeders calculated over a moving window. This enables predicting the extruder outlet composition one mean residence time in the future and triggering quality control actions in a feed forward manner.

42.4.1.4 The Future of Quality Control and Process Understanding for HME The future of quality control and process understanding is the integration of PAT tools and model based process knowledge. Large progress toward process understanding can be made during development, but additional information required to increase robustness and optimize the process can only be addressed in the supply phase. For example, during the product and process development cycle, it is not practical to study every factor with design of experiments. Additionally, the full range of raw material variability has not been experienced by the process, as development typically involves only a few lots of API. Lastly, unmeasured disturbances can pose a threat to product quality. PAT tools that include process sensors to measure physical and chemical information in real time together with the implementation of multivariate data analysis techniques can be used to enhance process understanding.

Multivariate statistical process control (MSPC) provides an efficient way of reducing all of the real time process data streams into one convenient control chart that also captures

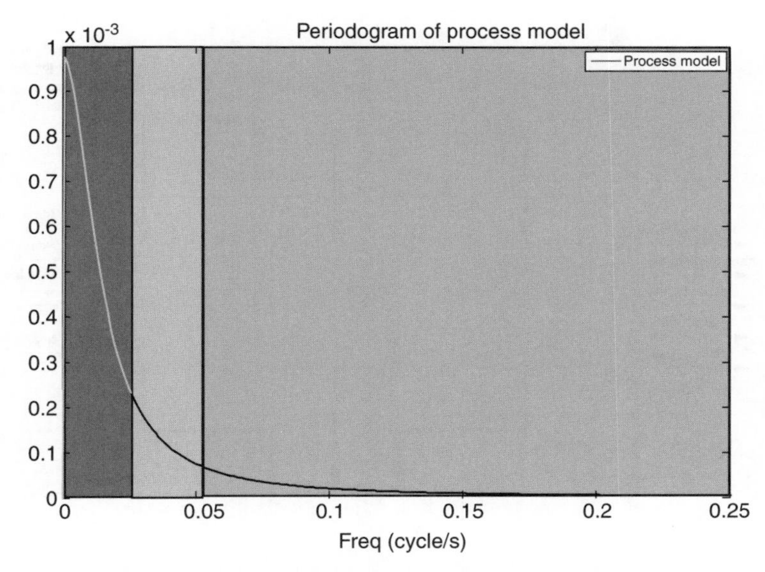

FIGURE 42.18 Periodogram of the process model showing that the process acts as a low pass filter capable of rejecting high frequency noise, but that input disturbances with a frequency of 0.05 cycles/s and lower will pass through the process potentially impacting product quality.

all of the variable interactions as well. MSPC models the covariance pattern of the real time process data and signals an alarm when the base covariance pattern is broken. When faults and deviations are detected, the variables causing the breakdown in the covariance pattern are identified. The implementation of MSPC to a manufacturing line can improve process robustness in several ways. MSPC can detect both sensor faults and deviations from normal process behavior. Early detection of sensor faults and process abnormalities improves process robustness by enabling

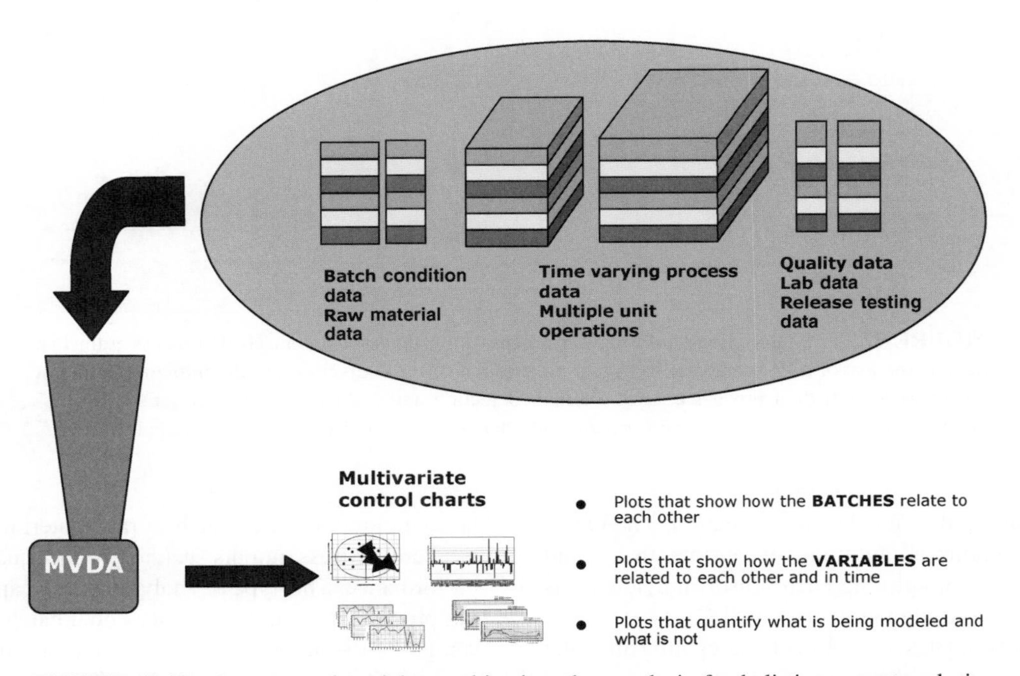

FIGURE 42.19 Overview of applying multivariate data analysis for holistic process analysis. MVDA techniques enable fast and efficient analysis of large and complex data sets.

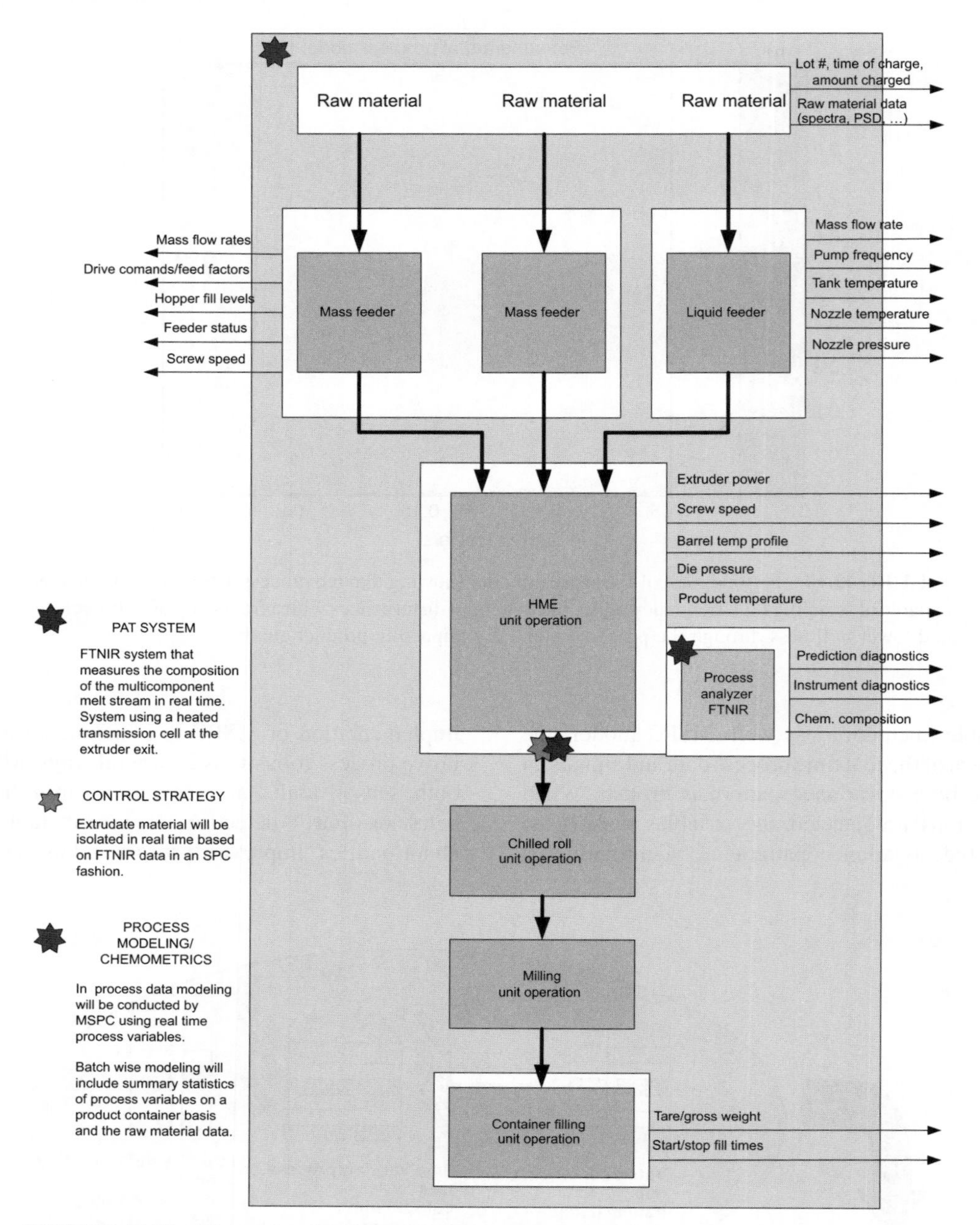

FIGURE 42.20 Material and information process flow diagram for an HME process train that depicts the use of a PAT analyzer to trigger an automated waste diverter and the implementation of multivariate statistical process control for fault detection and isolation as well as holistic process analysis that compares process performance on a batch-to-batch or campaign-to-campaign level.

intervention and rapid root cause identification. The extrusion process contains a multitude of sensors that could potentially fail, and monitoring each sensor individually is impractical, making it a good fit with MSPC.

Holistic process analysis consists of aggregating all of the data sources related to the process and the product and analyzing the data with multivariate tools. This type of data aggregation and analysis will detect any existing correlations between process inputs such as raw materials or equipment used and process outputs such as product quality or process performance. This type of analysis is also capable of detecting process drifts and differences on a batch-to-batch and a campaign-to-campaign basis. To facilitate holistic process analysis, it is necessary to have an IT infrastructure capable of aggregating raw material data, real time process data from multiple unit operations and process trains, in process testing, PAT data, equipment status and system suitability data, and quality testing and product release data. These IT systems need to be able to trace material and product genealogy as well. Figure 42.19 shows an overview of the holistic analysis process and Figure 42.20 shows a process flow diagram for an extrusion intermediate production process. The process depicted contains (1) a PAT system that measures the composition of the extrudate in real time and triggers a diverter system to isolate off-specification material, (2) a multivariate statistical process control system for sensor and process fault detection, and (3) a PAT-IT system that aggregates raw material data, lot/batch genealogy, process data, and release data for post batch holistic multivariate analysis. The implementation of these tools in manufacturing facilitates expanding the process knowledge space by detecting process upsets and deviations in real time and providing that data and information to rapidly identify correlations between process inputs and process performance. A logical progression of process understanding is identification of correlations, establishing causation, and model-based understanding. Ideally, model based process understanding will be based on first principles, but empirical, hybrid, and statistically based models can often be sufficient for improving product quality and rejecting process disturbances. Identified process models can be used to develop feed forward control systems that make processes robust to measured input variability.

This work describes the application of QbD principles and design space development to one unit operation in a pharmaceutical process train. A more complete process design space needs to be holistic, going from raw materials all the way to the final product image, with model based process knowledge describing the relationships between process inputs, process set points, and final product attributes.

42.5 CONCLUSION

Extrusion is a pharmaceutical process technology that meets a growing need to enable oral delivery of insoluble candidates and is particularly well suited to the quality by design approach. The application of DFSS methodology helped manage the technical complexity of developing the hot melt extrusion process and deliver on QbD. The knowledge gained from this development exercise facilitated process optimization and clear definition of a robust design space. The integration of PAT enabled added process robustness, process understanding, and derivation of a multifactor quadratic model of the process parameters most responsible for influencing the CQA's. The future of pharmaceutical extrusion will include fully integrated process analysis and control beyond single system PAT approaches and the continuous delivery of final drug product via a small footprint seamless process.

ACKNOWLEDGEMENTS

This work is the composite output of numerous functional areas and individuals within Merck. The authors would like to thank our colleagues for their inspiration, managerial support, analytical efforts, technical input, hard work, and discussions on quality by design and DFSS methodology.

REFERENCES

1. Repka M, Majumdar S, Battu SK, Srirangam R, Upadhye SB. Application of hot-melt extrusion for drug delivery. *Expert Opin. Drug Deliv.* 2008;5(12):1357–1376.
2. Lipinski C. Poor aqueous solubility—an industry wide problem in drug discovery. *Am. Pharm. Rev.* 2002;5:82–85.
3. Benet LZ, Wu CY. *Using a Biopharmaceutics Drug Disposition Classification System to Predict Bioavailability and Elimination Characteristics of New Molecular Entities.* NJDMDG, Somerset, NJ, 2006.
4. Leuner C, Dressman J. Improving drug solubility for oral delivery using solid dispersion. *Eur. J. Pharm. Biopharm. Sci.* 2000;50(1):47–60.
5. Serajuddin A.T.M. Solid dispersion of poorly water-soluble drugs: Early promises, subsequent problems, and recent breakthroughs. *J. Pharm. Sci.* 1999;88(10):1058–1066.
6. Chiou WL, Riegelman S. Pharmaceutical applications of solid dispersion systems. *J. Pharm. Sci.* 1971;60(9):1281–130.
7. Yu L. Amorphous pharmaceutical solids: Preparation, characterization and stabilization. *Adv. Drug Deliv. Rev.* 2001;48(1):27–42.
8. Friesen DT, Shanker R, Crew M, Smithey DT, Curatolo WJ, Nightingale JA. Hydroxypropyl methylcellulose acetate succinate-based spray-dried dispersions: An overview. *Mol. Pharmaceut.* 2008;5(6):1003–1019.
9. Curatolo W, Nightingale JA, Herbig SM. Utility of hydroxylpropylmethylcellulose acetate succinate (HPMCAS) for initiation and maintenance of drug supersaturation in the GI milieu. *Pharm. Res.* 2009;26(6):1419–1431.
10. Oh DM, Curl RL, Amidon GL. Estimating the fraction of dose absorbed from suspensions of poorly soluble compounds in humans: A mathematical model. *Pharm. Res.* 1993;10(2):264–270.
11. Moser JD, et al. Enhancing bioavailability of poorly soluble drugs using spray dried solid dispersions: Part I. *Am. Pharm. Rev.* (2008), 11(6):68, 70–71, 73.
12. Van den Mooter G, et al. Evaluation of Inutec SP1 as a new carrier in the formulation of solid dispersions for poorly soluble drugs. *Int. J. Pharm.* 2006;316(1-2):1–6.
13. Janssens S, et al. Characterization of ternary solid dispersions of itraconazole, PEG 6000, and HPMC 2910, *J. Pharm. Sci.* 2008;97(6):2110–2120.
14. Patterson JE, et al. Preparation of glass solutions of three poorly soluble drugs by spray drying, melt extrusion and ball milling. *Int. J. Pharm.* 2007;336(1):22–34.
15. Dong Z, et al. Evaluation of solid state properties of solid dispersions prepared by hot-melt extrusion and solvent coprecipitation. *Int. J. Pharm.* 2008;355(1-2):141–149.

16. Patterson JE, et al. Melt extrusion and spray drying of carbamazepine and dipyridamole with polyvinylpurrolidone/vinyl acetate copolymers. *Drug Dev. Ind. Pharm.* 2008;34(1):95–106.
17. Hauser JR. The house of quality. *Harvard Bus. Rev.* 1988; May–June:3–13.
18. ICH Q9. *Quality Risk Management*.
19. Kim EK, White JL. Transient compositional effects from feeders in a starved flow modular co-rotating twin-screw extruder. *Polym. Eng. Sci.* 2002;42 (Nov):2084–2093.
20. Todd D. Practical aspects of processing in intermeshing twin screw extruders. *J. Reinforc. Plast. Compos.* 1998; Vol. 17: 1607–1616.
21. Lee SH. Continuous mixing of low viscosity and high viscosity polymer melts in a modular co-rotating twin screw extruder. *Int. J. Polym. Proc.* 1997; Vol. 12:316–322.
22. Ishibashi J, Kikutani T. Experimental study of factors influencing throughput rate and process of polymer-mineral filler mixing in a twin screw extruder. *Int. Polym. Proc.* 2005; 20(4):388–397.
23. Todd D. *Plastics Compounding: Equipment & Processing*, Hanser, Munich, 1998.
24. Todd D. Practical aspects of processing in intermeshing twin screw extruders. *J. Reinforc. Plast. Compos.* 1998; Vol. 17: 1607–1616.
25. Todd D. Melting of plastics in kneading blocks. *Int. Polym. Proc.* 1993; 113–118.
26. Tadmor Z, Klein I. *Engineering Principles of Plasticating Extrusion*, Van Nostrand Reinhold, New York, 1970.
27. Jung J, White JL. Investigation of melting phenomena in modular co-rotating twin screw extrusion. *Int. Polym. Process.* 2003;XVIII:127–132.
28. Rauwendaal C. *Polymer Extrusion*, 4th edition, Hanser Gardner Publications, Inc., Cincinnati, OH, 2001, pp. 463–476.
29. Potluri R, Todd D, Gogos C. Mixing immiscible blends in an intermeshing counter-rotating twin screw extruder. *Adv. Polym. Technol.* 2006; Vol. 25 no. 2:81–89.
30. Lim S, White JL. Flow mechanisms, material distributions and phase morphology development in a modular intermeshing counter-rotating twin screw extruder of Leistritz design. *Int. Polym. Process.* 1994;IX:33–45.
31. Denn MM. Simulation of polymer melt processing. *AIChE J.* 2009; Vol. 55 1641–1647.
32. Brouwer T, Todd D, Janssen LPB. Flow characteristics of screws and special mixing enhancers in a co-rotating twin screw extruder. *Int. Polym. Process.* 2002;XVII:26–32.
33. Tadmor Z, Gogos C. *Principles of Polymer Processing*, 2nd edition, Wiley, 2006, pp. 322–354.
34. Sekiguchi K, Obi N. Studies on absorption of entectic mixtures. *Chem. Pharm. Bull.* 1961;9:866–872.
35. Breitenbach J. Melt extrusion: From process to drug delivery technology. *Eur. J. Pharm. Biopharm.* 2002;54(2): 107–117.
36. ICH Q3C (R3). *Impurities: Guidelines for Residual Solvents*.
37. Young CR, et al. Production of spherical pellets by a hot-melt extrusion and spheronization process. *Int. J. Pharm.* 2002;242:87–92.
38. Rauwendaal C, del Pilar Noriega M. *Troubleshooting the Extrusion Process*, Hanser Gardner Publications, Inc., Cincinnati, OH, 2001, pp. 67–70.
39. Chong J. Calendering thermoplastic materials. *J. Appl. Polym. Sci.* 1968;12(1):19.
40. Isayev A. *Injection and Compression Molding Fundamentals*, CRC Press, Boca Raton, FL, 1987.
41. Nakamichi K, Nakano T, Yasuura J, Izumi S, Kawashima Y. The role of the kneading paddle and the effects of screw revolution speed and water content on the preparation of solid dispersions using a twin-screw extruder. *Int. J. Pharm.* 2002;241:203–211.
42. Verreck G, et al. Hot stage extrusion of P-amino salicylic acid with EC using CO_2 as temporary plasticizer. *J. Supercrit. Fluid.* 2006;327:45–50.
43. Munjal M, Elsohly MA, Repka MA. Polymeric systems for amorphous Δ9-tetrahydrocannabinol produced by hot-melt method. Part II: Effect of oxidation mechanisms and chemical interactions on stability. *J. Pharm. Sci.* 2006; 95(11):2473–2485.
44. Villetti MA, et al. Thermal degradation of natural polymers. *J. Therm. Anal. Calorim.* 2002;67:295–303.
45. Chen H, Sundararaj U, Nandakumar K, Wetzel MD. Investigation of the melting mechanism in an twin-screw extruder using a pulse method and online measurement. *Ind. Eng. Chem. Res.* 2004;43:6822–6831.
46. Puaux JP, Bozga G, Ainser A. Residence time distribution in an corotating twin-screw extruder. *Chem. Eng. Sci.* 2000; 55:1641–1651.
47. Kim EK, White JL. Transient compositional effects from feeders in a starved flow modular co-rotating twin-screw extruder. *Polym. Eng. Sci.* Nov 2002; 2084–2093.
48. Zhu Y. *Multivariable System Identification for Process Control*, Pergamon, New York, 2001.

43

CONTINUOUS PROCESSING IN SECONDARY PRODUCTION

Martin Warman
Analytical Development, Vertex Pharmaceuticals Inc., Cambridge, MA, USA

43.1 INTRODUCTION

For the entire history of pharmaceutical manufacturing, secondary production processes have been carried out in batches. There are deep-rooted reasons for this [1], some historical (the ancestry of many pharmaceutical processes come from food or confectionary), some founded in the need to track the "history" of the dosage form (in terms of materials and processes, i.e., the "batch record"). However, even though pharmaceutical secondary production shares many unit operations with other industries, the reality is, many of those industries have already realized the commercial and operational benefits of continuous manufacturing and have started running those unit operations continuously, and have abandoned batch production.

It would be easy to simply justify the continued use of batch approaches for these reasons, however the pharmaceutical industry is coming under huge quality, efficiency, financial and business pressures (which has even reached the public's attention, e.g., in the article in Wall Street Journal [2]), but also the regulatory landscape has changed. The initial PAT Framework [3], the twenty-first century Initiative Final Report [4], the ICH Quality Trio (ICH Q8, Q9, and Q10) as well as the new guidance on Process Validation have all put a focus on the application of new technologies with a science-based approach. They also introduce a new term to the pharmaceutical industry; that of quality by design (QbD).

Running continuous processes during development and commercial manufacturing facilitates both the application of QbD approaches but also (as will be shown later in this text) the ability to implement these approaches in a highly efficient way.

43.2 DEFINITIONS

To be clear from the start, when describing Batch production, we are describing an overall production system whereby the entire mass/volume goes through the unit operation at same time, normal in one "container"; as an example, unit operations such as bin blending, where we start with individual components being added to a single processing unit and during the production process the entire batch changes to reach a single end point—the process is only complete when the entire mass of the blend is uniform. Whereas continuous production describes where the input materials continually enter into the unit operation and output materials continually exit, under a "first in/first out principle," taking the blending example, a continuous blend operation is where the input materials are continually being feed into a continuous mixing process. The outcome of the process is not only continuous but in a steady state, resulting in a uniform output where each unit dose mass is not only the same/having the correct concentration of each component (interdose uniformity) but also that those components are optimally dispersed within the unit dose, thus ensuring correct delivery performance. In the case of continuous blending (and many other continuous processes) the individual dosage is "generated" early on, and in some ways the process consists of a stream of individual unit doses, such that verification of performance of continuous systems has to be considered against this production

Chemical Engineering in the Pharmaceutical Industry: R&D to Manufacturing Edited by David J. am Ende

paradigm. It is also very important that continuous production processes are not be confused with flow production in which standard batch operations are simply linked together, for example it is very common for consecutive unit operations in secondary production to be "daisy chained" together, for example a bin blend discharging into high shear granulator, into a fluidized bed dryer, and so on—the bulk material flow maybe linked but the entire batch goes through a transformation at the same time within each unit operation.

43.3 REVIEW OF TYPICAL UNIT OPERATIONS

As previously indicated many pharmaceutical unit operations are shared with other industries, however, we also have to acknowledge that many of the individual unit operations used are themselves continuous operations—as an industry we simple chose to collect the output and form/maintain the batch. So let us first look at traditional solid dosage operations and consider if they are "batch," "continuous," or could be made continuous.

43.3.1 Typical Batch Process Operations

The vast majority of current pharmaceutical products are currently "solid oral doses," commonly known as tablets or capsules. Typically production of solid dosage forms is carried out in three types of process streams. The simplest is well described and commonly called direct compression (Figure 43.1). In a simple direct compression process, the Active Pharmaceutical Ingredient/excipients are dispensed (via a screen to de-lump) into a V-shell or bin blender. Post the blend operation the material is transferred to the feed hopper of the tablet press, postcompression the batch is coated "en-mass" in a pan coater, before packaging.

Small variations on this basic process workflow may occur, for example, replacing the compression/coating steps with an encapsulation step during capsule production, but the general workflow stays the same.

However, if the particle size and/or physical characteristics of the individual API/excipient powders are likely to cause segregation, between or during subsequent processing steps, it is common to introduction some form of granulation step postblending. This is often then followed by an additional (second) blend step when a lubricant excipient is needed to improve flowability within the process and to prevent sticking/chipping during compression.

FIGURE 43.1 Schematic of direct compression process.

If the granulation step is "dry" (better known as roller compaction or RC), the premixed materials are forced through two counter-rotating rollers that exert mechanical pressure on the powders during a high-pressure agglomeration or "compact." The compact can be of several forms however in each case the true granulate is formed by milling. Typically, the high-pressure compact formation, and subsequent milling are together described as a single unit operation (Figure 43.2).

In comparison, the alternate is wet granulation (WG), in which shear and compression forces are used along with the wet/massing forces during addition of a binder to firstly generate agglomerated particles using the three phases shown in Figure 43.3.

Leuenberger [5] identified the optimal granulation point as being the initiation of the capillary state. However, granulates in this phase then need to be dried. For this reason wet granulation is typically described as a two part process with the initial wet granulation being followed by drying process (even though the two steps can and often are carried out in the same vessel—commonly described as a "single pot" granulator (see Figure 43.4).

For more detail on the types of secondary production please refer to the specific sections of this publication, they are listed here purely for background before discussing the individual unit operations.

43.4 SOLID DOSAGE UNIT OPERATION

From the process workflows given above it is apparent that several unit operations are common, even repeated. But let us consider each of these "common steps" in turn and consider how they are run now, and what are the opportunities for continuous processing.

43.4.1 Dispensing

Under a traditional batch paradigm the production material is typically weighed in the pharmacy, verified and released, having already been individually bagged ready to be loaded into the production process. However, it is not uncommon for the toxicity (and so containment) of some materials to cause manual handling issues. These are often overcome by automated dispensing systems. In some cases these dispensing systems are connected directly to the production process. Although these systems are typically used to initiate a batch process, they are (in themselves) continuous systems—the first powder into the feeder is the first powder out. In reality pharmaceutical production may use the current range of volumetric and loss-in weight feeders to deliver in an automated way to ensure containment, they were actually designed for, and able to run, as part of continuous systems. Several of the feeder suppliers have even extended their

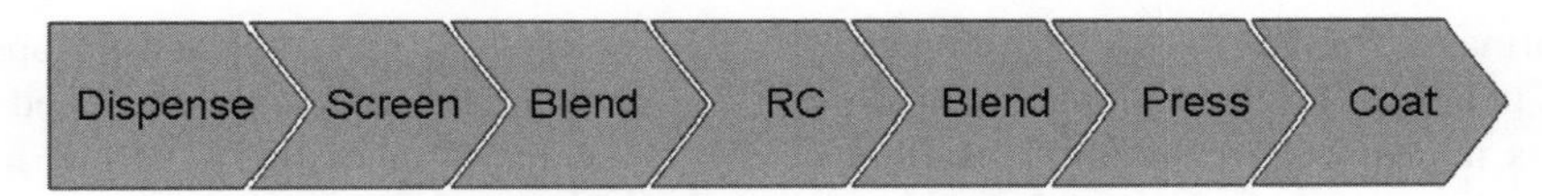

FIGURE 43.2 Schematic of dry granulation process.

FIGURE 43.3 Three phases of granulate wetting.

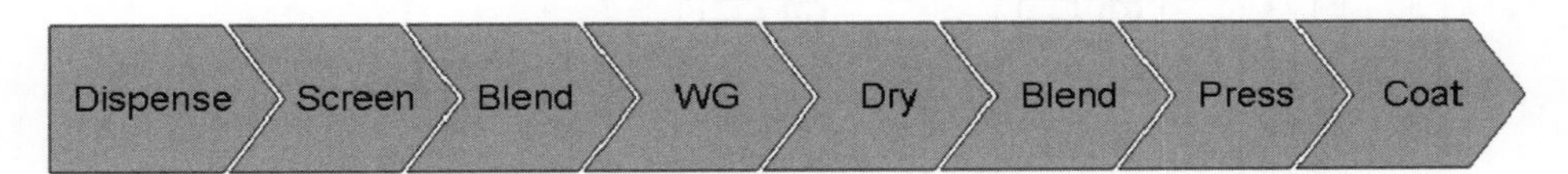

FIGURE 43.4 Schematic of wet granulation process.

product range forward, such that their control system becomes the basis for controlling material flow through the entire continuous system.

Although the basic principle of using existing automated feeders as part of a continuous processing system is sound, what we are trying to achieve is very different. A feeder on a batch dispensing line simply has to deliver, reproducibly, the correct mass of powder. The mass is simply what is needed for an entire batch. So, for example, a automatic feeder dispensing 20% API to a 1000 kg batch process simply has to, reproducibly deliver 200 kg of API. The automatic feeder is able to accelerate during bulk dispensing, slow as it approaches the end point and operate what can only be described as fine step control to end up at the predefined mass. Taking the same 20% API example, on a typical continuous process producing 40,000/500 mg image tablets per hour, the feeder is required to deliver not only exactly 4 kg/h of API but with adequate precision to ensure each of the approx 11 unit doses that will be generated every second, meet requirements around API uniformity. Using this simple example (and in the worst case) the API feeder needs to deliver approx 1.1 g of API every second, reproducibility, across the entire production run (see Figure 43.5). From a mechanical engineering perspective the two are very different challenges (and even worse if the API or excipient concentration is lower (e.g., a typical lubricant addition rate of 1% equates to 55 mg/s addition rate or rather 5 mg/90 ms).

In general the only additional consideration for continuous use is the maintenance of an acceptable level of materials in the feeder charge hopper.

43.4.2 Screening

Even when running under a continuous paradigm it is anticipated that many raw materials will be delivered in as drums/lots/batches. It is a straightforward logistic operation to track the use of the material to final dosage form and in effect allowing traceability of lots to whatever is defined as final batch integrity for compliance purposes. However, in

FIGURE 43.5 K-Tron MT12™ twin-screw microfeeder, capable of both batch and continuous operation.

many cases the performance of the feeder is impacted directly by fluctuations in hopper level, so simply, manually charging the hoppers is not an option. Material handling solutions have been developed for other industries such as food and food ingredients, which are even capable of receiving raw materials on rail or by road, transfer the local storage facilities before charging feed hoppers at local unit operations. These systems can also include flow aids, filtered venting/exhaust systems as well redundancy/parallel storage to ensure supply (see Figure 43.6).

The next step in all typical batch operations is to screen the input materials commonly described as delumping. This is often carried out using a screen mill but is in itself a continuous unit operation (it operates on a first in/first out principle) so although used with a batch paradigm could very easily be used as part of a continuous process. The main considerations when doing so are, does the performance of the screen mill change over time (e.g., does the screen become blocked, or does the mechanical action actually wear or cause the screen to break). These considerations are

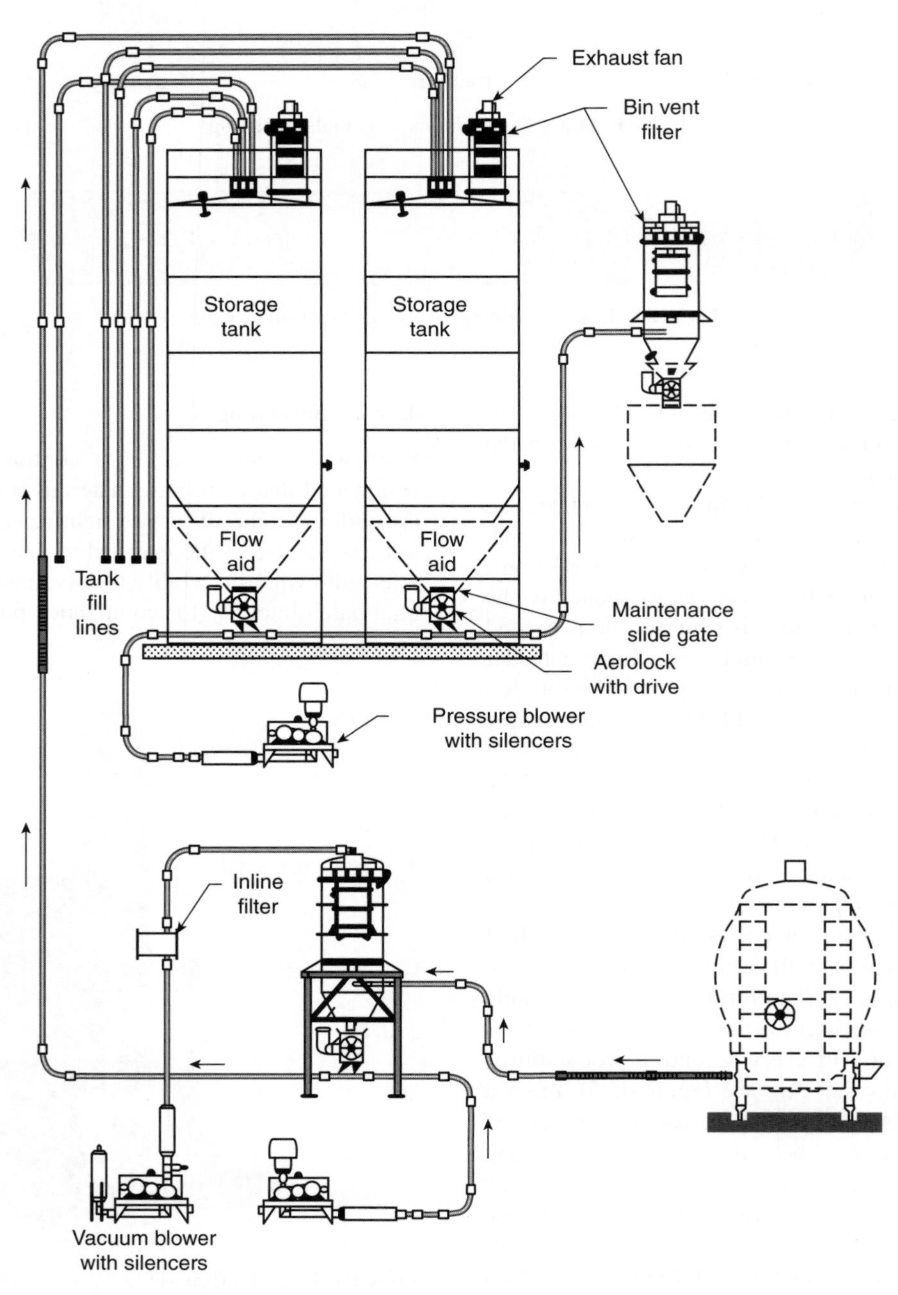

FIGURE 43.6 Example pneumatic conveying system courtesy of K-Tron Premier™.

often not important within a batch paradigm simply because the effects are not seen but critical to a continuous process.

However, a more elegant approach is to combine the screen into the dispensing operation. Many feeders can have mixers, microcentrifugal feeders, or even screens incorporated within their design. Combining the two unit operations in this way simplifies the production process but also has similarities with a dry granulation unit operation where milling is regarded as integral to the granulation process.

43.4.3 Blending

There can be little doubt the blending unit operation is the most common unit operation in pharmaceutical secondary production, it is also the operation with the fewest comparisons to continuous production (because of how the operation is carried out). However, continuous blending is common place in other industries and in the last few years continuous blenders that claim compliance to CFR and cGMP have also become available.

Because of the widespread use of this unit operation, but also because of the lack of experience in applying continuous blending to pharmaceutical processes, this processing step has been subject to the greatest intensity in academic and theoretical research in recent years. Laurent and Bridgwater [6–8] were one of the first to investigate the flow patterns within a continuous blender, using techniques such as tracking radioactive tracer; this allowed them to generate the axial and radial displacements as well as velocity fields with respect to time This was followed by Marikh et al. [9] where the focus is on the characterization and quantification of the stirring action, relating it empirically to the flow rate and the rotational speed of the continuous blender. In doing so it systematically investigates the effects of, operating conditions (such as rotational speed and processing angle) and design parameters (such as blade sign) on the mixing efficiency.

However, the key to the successful use of continuous blending is recognition that the blender actually has to fulfill more than one purpose. Its primary role is to take the variation in the disparate individual feeds (API and excipients) and generate a single uniform blend, such that each and every individual unit dose is of appropriate quality. However, in order to achieve the blender's primary role the continuous blender has to remove any variability remaining from the dosing operation. As such, a continuous blender has previously been described as "variability reduction ratio" (VRR) device. Williams and Rahman [10] proposed a mathematical approach to predict the VRR, utilizing data generated from a residence time distribution test for both and "ideal" and "nonideal" blender. The metric of "ideality" is defined by a mixing efficiency proposed by Beaudry [11]. In another publication, Williams and Rahman [12] investigated this mathematical methodology by using a salt/sand formulation of different compositional ratios. They verified the predicted VRR with experiments and suggested that the results where comparable. They also illustrated that (over at least typical conditions) the mixing speed and VRR were directly correlated. Harwood et al. [13] studied the performance of seven continuous mixers as well as the outflow sample size effect of sand and sugar mixtures. All of these activities was reviewed and then additionally verified by Portillo et al. [14], including experimental investigation of operation and design parameters such as processing angle, impeller rotation rate, and blade design are examined.

In summary all these investigations show that the powder's residence time and number of blade passes it experiences was affected not only by rotation rate but also by the processing angle, and that an upward processing angle and low impeller rotation rate are the optimal processing settings, when combined with optimal blade design. These generate a slight backflow between blade rotation and a turbulent flow within the linear flow of the process.

In Ref. 14 a new type of continuous/in-line blender (manufactured by GEA PharmaSystems) called the Continuous Dry Blender is used—this is the first dedicated, purpose designed for the pharmaceutical industry, continuous blender (see Figure 43.7).

This system is now commercialized and further details are included later in this chapter.

Fundamental research is still ongoing into continuous blending with the primary focus being the addition order of individual components as for the first time the dry blending process can be engineered to allow optimized mixing/interaction of components to effectively "build" the formulation in a structured way. One area where this is critical is around the addition of the lubricant component; or more accurately what type of effect is trying to be achieved by the addition of the lubricant, that is, do we want the lubricant to be in a

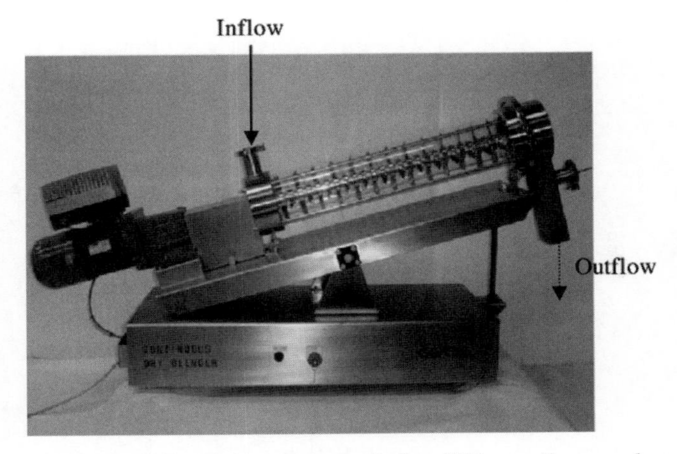

FIGURE 43.7 Early version of the GEA continuous dry blender [14].

distributed within the blend but remain as a discrete powder or do we want the lubricant to be smeared over particles of the other components? In the past, we could change blend time and rotational speed but little else.

Not only continuous dry blenders but also specialized PAT measurement systems, to monitor the blending process, are now commercially available. In the same way that on-line NIR systems are commonly used to monitor the batch blending process [15], ultra fast scanning (diode array) NIR linked to optimized sample presentation systems are available. Using these systems it is relatively easy the get comparative data on the trajectory and end point of both batch and continuous systems and therefore even compare the output from both. If we first look at a simple development scale batch process, an NIR prediction model can be generated trend how each components changes over the blend process over time (Figure 43.8).

In the example above (carried out at in a Paterson Kelly 4qt V-shell blender) we can see the batch process reaches uniformity after around 20 rotations, with one component taking longer to reach uniformity than the other two. If we look at exactly the same composition running from start-up, the plot is slightly different (Figure 43.9).

This data is from a "dry" start-up, that is, from when the feeders themselves are started and with the blender empty. In this case, the continuous blender has a volume of 500 ml and at the powder flow rate used (20 kg/h) so the blender has a residence time of around 90 s—meaning it therefore takes 90 s before powder starts to exit the blender. From this time point it then takes approx 3 min to reach a %RSD equivalent to the batch process.

The significant detail in this case is the start-up process used approx 2 kg of powder to reach this steady state (approx the same weight as used in the 4qt development batch) but there is no scale-up involved moving to commercial scale; the powder flow rate used (20 kg/h) is equivalent to 140,000 kg/year at expected equipment utilization rates (80–85%). We are able to go from development straight to commercial scale because unlike batch processes with scale in "space" continuous processes scale in "time."

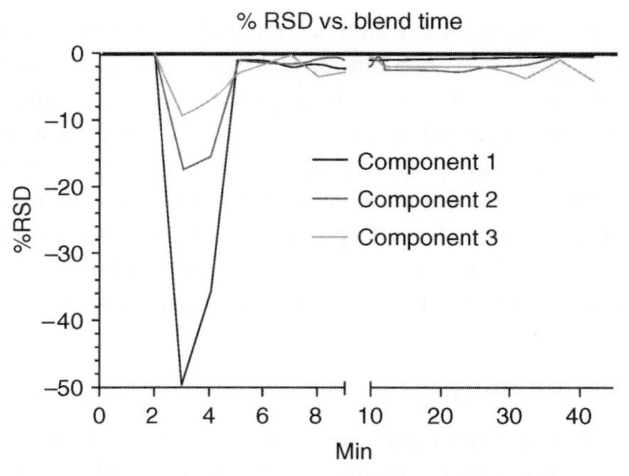

FIGURE 43.9 Typical continuous blend plot.

However, we also have to consider the scrutiny of scale when describing measurement systems. In a batch process we can statistically sample (spatially) the output of the blend process; in continuous blending we have a different scrutiny scale, effectively the uniformity of the individual/consecutive unit doses is generated here and simply doing a unit dose scale measurement at a fixed time interval across the batch can miss unit dose to unit dose variability (in much the same way that inadequate or poorly specified sampling will miss variability in a batch process). If we look at the unit dose (in this case the product is 500 mg image) to unit dose variance and calculate %RSD we get a very good demonstration of the high frequency variability in the system (Figure 43.10).

Although direct compression is the simplest form of solid dosage production it has significant restrictions in use.

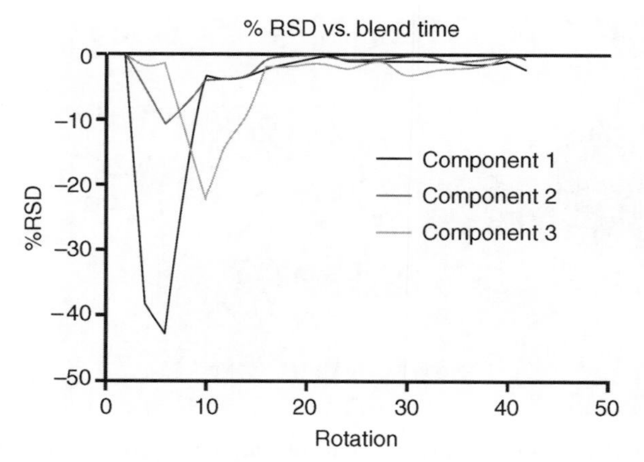

FIGURE 43.8 Typical batch blend plot.

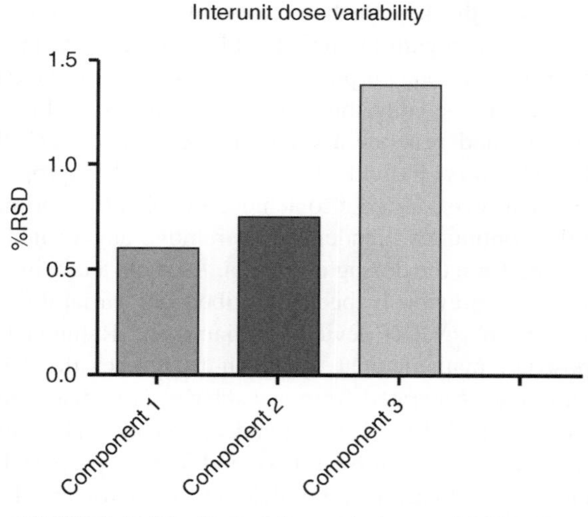

FIGURE 43.10 Typical interunit dose variability plot.

Typically solid dosage formulations do not use API and excipients with comparable physical properties such as particle size, which means there is a tendency to segregate postblending but before the final dosage form is made. Because if this it is very common to follow the first blend step with some form of granulation.

43.4.4 Dry Granulation

As previous described, in dry granulation the blending step is typically followed by another unit operation originally developed for another industry; roller compaction. In typical pharmaceutical processes the roller compaction step is carried out as part of a batch production line, however the activity follows a first in/first out principle and is inherently "continuous." Powder is fed from a feed or charge hopper into the RC unit and between two counter-rotating rolls. The compression force (and utilizing the elastic strength of the individual particles) causes the free flow blend to form solid compacts (sometimes ribbons, sometimes briquettes), see Figure 43.11.

Even though the actual RC activity is "continuous," variations in powder flow into the feed hopper will impact the uniformity of the compacts produced, both in terms of physical (i.e., tensile strength) and chemical (segregation). For this reason most commercial scale and many development scale RC units include a mechanical system (similar in design to an automated feeder) to deliver a constant feed rate at the rollers. Most RC units used in pharmaceutical were themselves developed/optimized with sophisticated feedback controls (for speed, press and even torque) to function with little variability.

Because the mechanism used when reducing variability in the output of the blending step may actually cause variability in the roller compaction step, when using RC as part of a continuous process specific consideration has to be given to changing this paradigm. It is possible that the RC process itself will need to be adjusted in order to cope with varying input and thus ensure a constant output. The fundamental

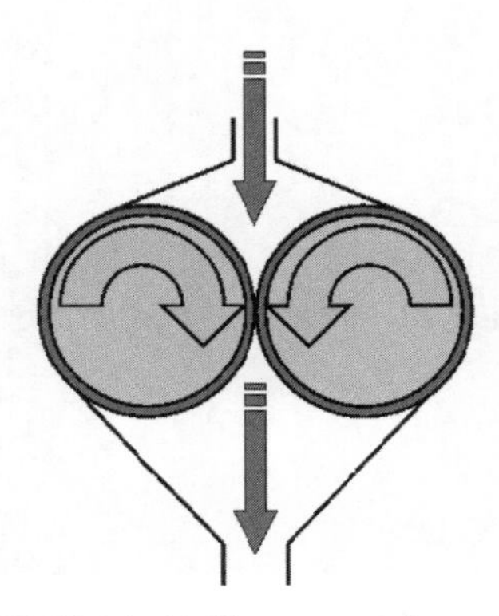

FIGURE 43.11 Roller compaction schematic.

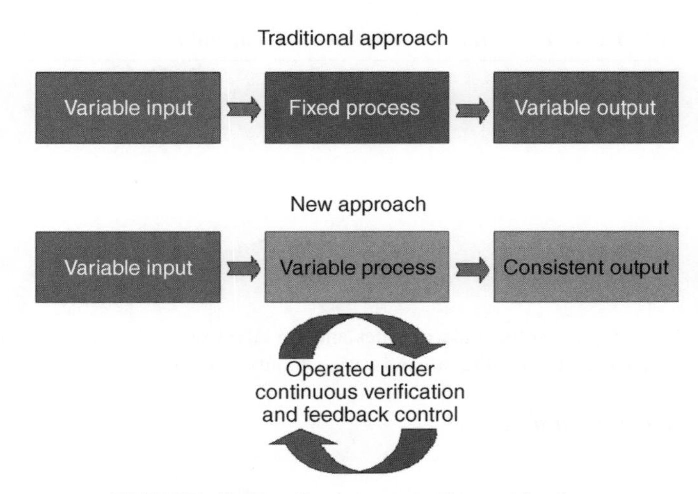

FIGURE 43.12 Continuous quality verification.

change to the way pharmaceutical processes are run is consistent with the current initiatives and is often represented graphically using a form of the diagram below, and is known as continuous quality verification (CQV) and is particularly important to highly constrained traditional processes such as roller compaction (Figure 43.12).

There are PAT measurement systems now available that have the capability to monitor both physical and chemical changes (e.g., density that in turn impacts tensile strength) of roller compaction ribbons. These allow continuous on-line measurement and real-time prediction (Table 43.1). Specific optical measuring heads have been developed that allows the use of NIR directly onto the compacts as the come-off the RC rollers (see Figure 43.13 and Table 43.1).

In this case study, intact ribbons are generated as a result of compaction. This is not always the case and it is essential that the capabilities of the measurement systems are match to the

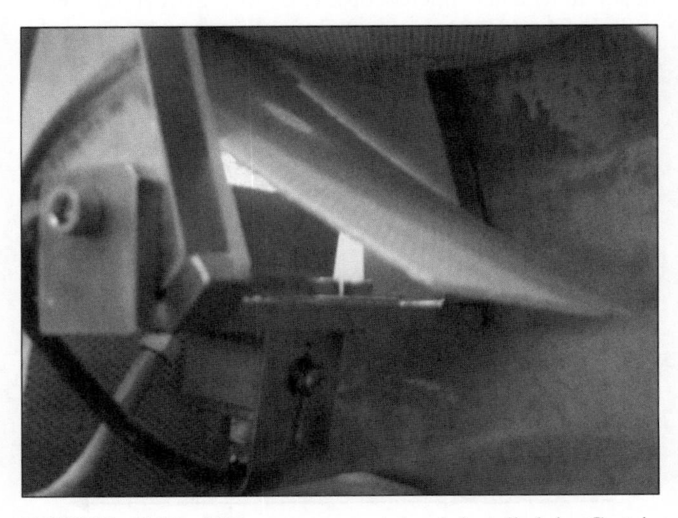

FIGURE 43.13 NIR measurement head installed in Gerteis Macropactor.

TABLE 43.1 Online Measurement of Variability

API trend	
Showing constituent alarm states and the effect seen when a process parameter changes to take the process variance outside expected norms	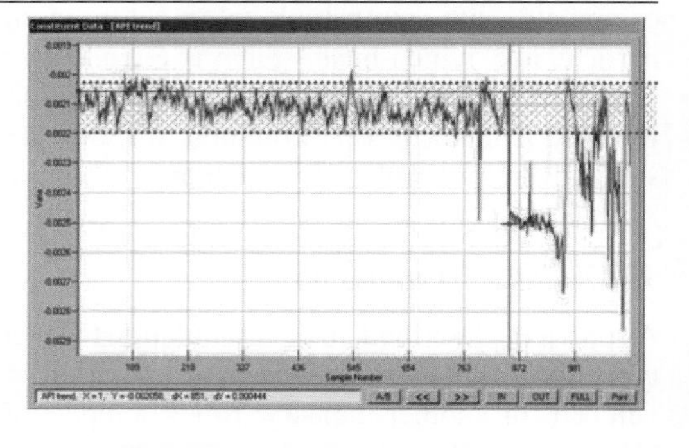
Physical variance	
Showing constituent alarm states and the effect seen when a process parameter changes to take the process variance outside expected norms	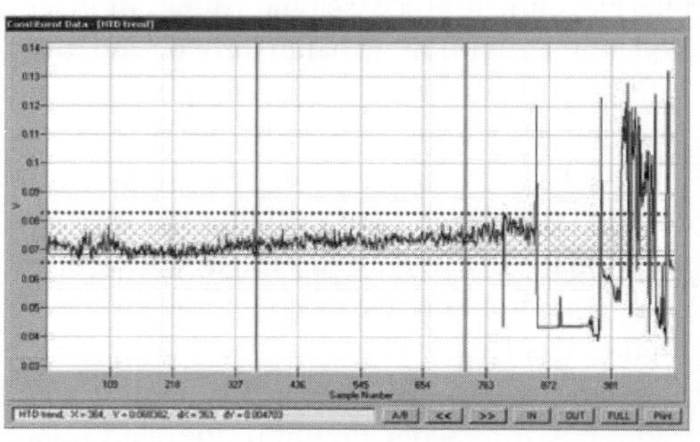

application in terms of speed of analysis, rate of analysis, reproducibility, and sample size.

The second component of dry granulation is the milling or granulation step. Here the compacts are typically put through a screen mill, again on a first in/first out basis so inherently continuous. When running this sort of mill under batch conditions (much like the example given for the screening process) the main considerations is around the performance of the screen mill and does it change over time. Particle characterization post the mill can be carried out using focal beam reflectance microscopy (FBRM). FBRM utilizes a spinning laser (of known rotational speed) to measure the chord length across any particle, by simply back calculating the duration of reflection of the laser off the particle (see Figure 43.14).

The laser light is delivered by fiber optic probe so is relatively easy to install in the output stream of the RC granulator, often fitted with optional gas purge to keep the tip clear (see Figures 43.15 and 43.16)

Laser beam
Scan direction
Typical particle
Chord length
I_R
Intensity profile
t_0 t_1 t

FIGURE 43.14 Theory of FBRM (courtesy of Mettler-Toldeo).

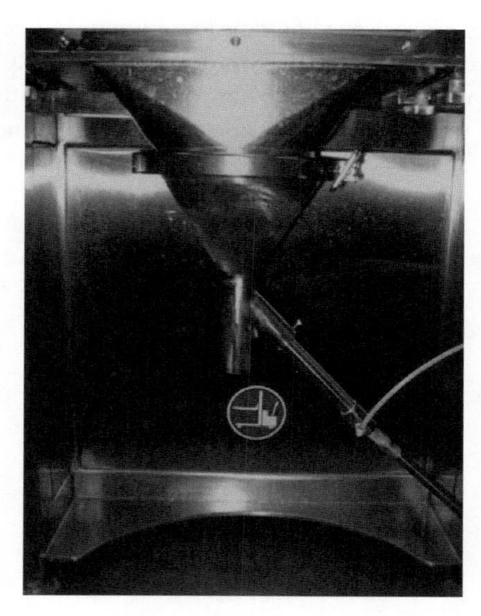

FIGURE 43.15 FBRM with purge tip installed on the Gerteis Macropactor.

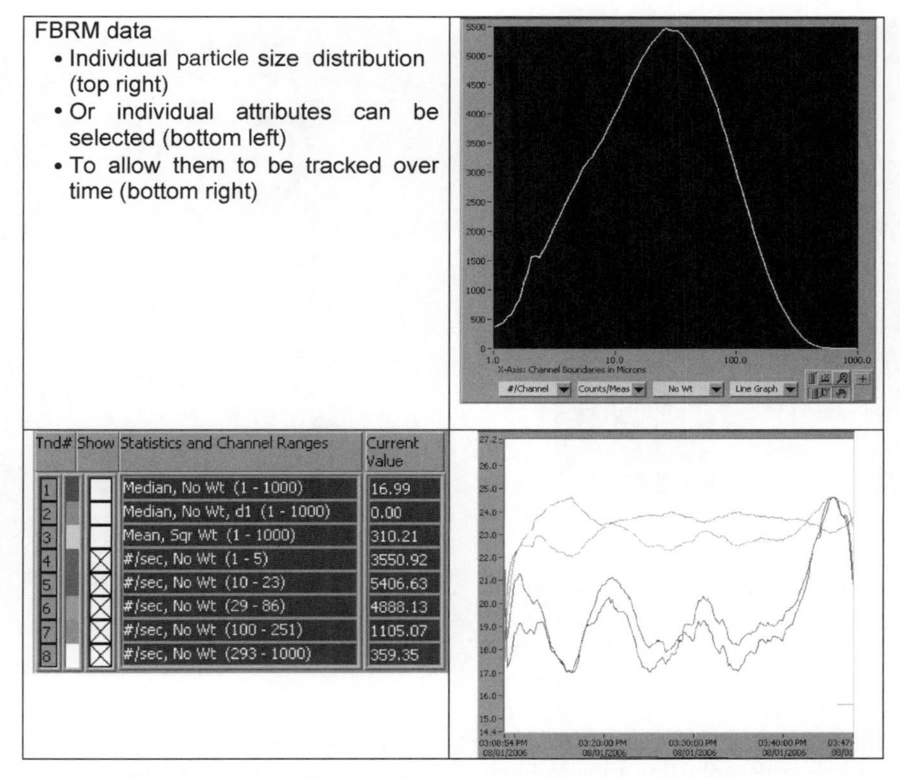

FIGURE 43.16 Monitoring the RC milling process using FBRM.

43.4.5 Wet Granulation

As previously described the need to granulate pharmaceutical powders is common however it is not always possible to dry granulate, possibly because the powders do not have sufficient elasticity (they are two brittle) or because the differences in particle size distribution or physical properties is too great. In these cases it is very common to wet granulate.

Unlike dry granulation the wet granulation is not inherently continuous however there are examples dating back to the mid 1980s, suggesting alternate approaches to traditional wet granulation that could be run continuously or semicontinuously, for example, Koblitz and Ehrhardt [16] published on wet granulation and using continuous variable frequency fluid bed drying.

A breakthrough approach came from Glatt with the launch of their Glatt Multicell (CMC) in the late 1990s. The technology has not seen widespread adoption but is well documented including several publications by Leuenberger [17].

The CMC 30 comprises of a 27 L High Speed Plough–Shear granulator, which equates to a 5–9 kg subbatch (Figure 43.17). The granulator "self-discharges" via a

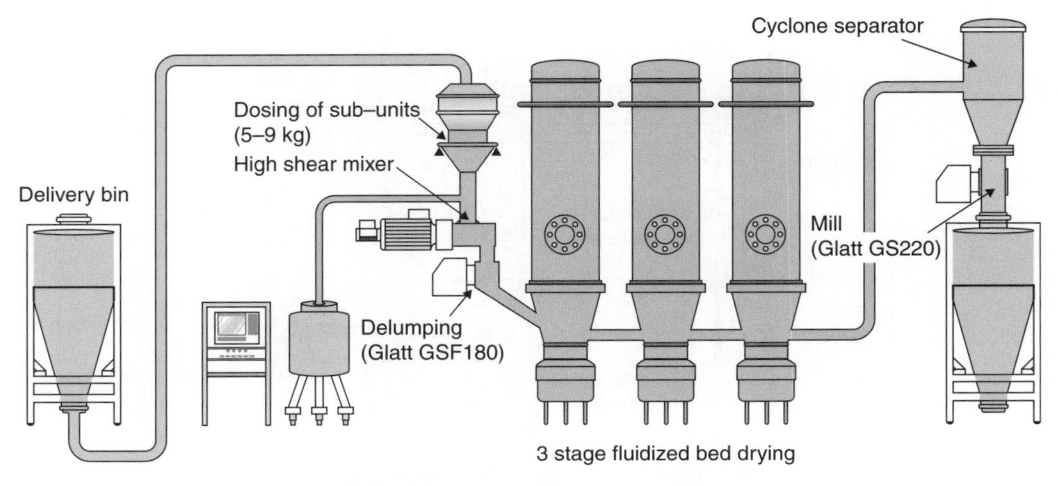

FIGURE 43.17 Glatt Multicell GMC 30.

FIGURE 43.18 HSP-S mixer.

delumping system into a multistage fluidized bed dryer. After granulation/wet massing, the material is conveyed sequentially through three stages of drying. In this way, four small batches (one in granulator and three in drying) are processed simultaneously and the cycle repeats for semi-continuous operation (Figures 43.18 and 43.19).

Although this system is best described as a micro-Flow system the technical significance of this system should be recognized, especially as it was one of the earliest examples of continuous verification and feedback control; each of the three fluidized bed towers can be fitted with a noninvasive NIR measurement system, which simply views the drying process through the preexisting inspection windows. A control strategy is then put in place balancing the subbatch throughput (Figure 43.20).

Once dried the subbatches could be discharged directly into a second continuous blender for lubricant addition although all known implementations currently collect the subbatches to form a single batch that moves forward.

FIGURE 43.19 GMC 30.

In their White Paper for PharmaManufacturing.com, Mollan and Lodaya [18] identified that a continuous fluidized bed granulation system would have five or more functional zones. These are product in-feed zone, product mixing and preheating zone, spraying zone, drying and cooling zone, and discharge zone, with a more detailed explanation being given by Paul et al. [19]. Continuous versions of some of these individual functions are available and have been published. Lindberg [20] used an Iversion mixer (where powders and liquid are metered into a narrow space at the periphery of the grooved disc, which rotates at high speed) to study wet granulation of placebo. Applegren et al. [21] used a similar system to study continuous melt granulation, and a system is commercially available that uses a Planetary Extruder to

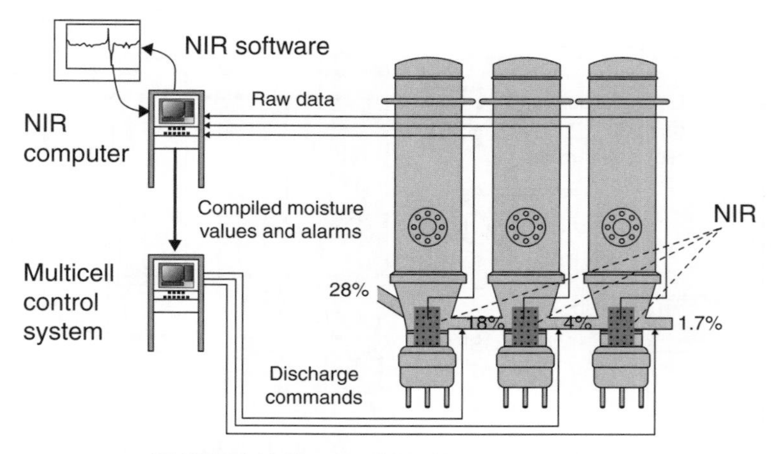

FIGURE 43.20 GMC 30 NIR control strategy.

granulate and Microwave tube to dry continuously. As an alternate to microwave energy there is also a commercial system based on using radiofrequency energy (using wavelengths specific for aqueous drying. Until recently, radiofrequency heating has been used mainly in other industries such as food, paper, and ceramic. Jones and Rowley [22] have reviewed several applications for drying where dielectric heating is used by itself or in combination with other methods.

However, the wet granulation that has the greatest potential for continuous operation is using a twin-screw granulator/mixer. This is a modified twin-screw extruder and relies on twin intermeshing screws that convey, mix, wet granulate, and wet mass the powder blend. These systems offer several advantages over traditional wet granulation processes, and the interchangeability of screw elements ensures flexibility. Twin-screw extruders themselves have been utilized for wet granulation since the 1980s [23, 24], and some aspects of the application are also covered by patent [25]. In addition, Ghebre-Sellassie et al. [26] have published on a continuous wet granulation and drying system that combines twin-screw mixer (for wet granulation and wet massing) with radiofrequency energy (for drying).

Very recently a commercial system has become available from GEA Pharma Systems called the CONSIGMA™. This system is at the center of GEA's philosophy of an integrated tablet line and starts with separate liquid (binder) and powder feed systems (the powder feed either being metered directly or coming from the outlet of one for their continuous dry blender systems). These feeds coming together in a twin-screw granulator, which continuously outputs into either a linear or segmented dryer (Figure 43.21).

FIGURE 43.21 GEA Pharma Systems CONSIGMA™.

Interestingly, currently the CONSIGMA™ does not utilize a second GEA continuous dry blender for addition of the lubricate postdrying. Instead, after drying the product is held in a dedicated discharge hopper/mill and small batch "lubricant" blender/feeder. All of this is under an integrated control system.

43.4.6 Compression/Encapsulation

No matter what the route taken to reach the compression or encapsulation unit operation, and no matter what vendor system is adopted in the step, it will be carried out under a "first in/first out" principle and as such is continuous. As an industry we choose to charge the feed hopper as a batch and collect the output as a single entity. If this step would run truly continuously some consideration needs to be given to dust buildup (even if preceded by a dedusting step), product build up on press tooling (which will cause issues with subsequent defects on tablet cores), and (in the case of encapsulation) the addition of additional raw materials (the empty capsule shells).

The addition of automated tablet testers and capsule checkers for feedback control also allows the compression/encapsulation process to be adjusted in real-time (for weight, thickness, hardness, and Cu)

43.4.7 Coating

When describing the coating process we have to remember there are two input streams. The first is the uncoated cores but we also have to consider the spray solution itself, so even if a continuous coated was feed directly from the press the spray solution preparation also needs to be addressed. That said, continuous coating is performed in food, flavor, and nutraceutical processing and the first commercially available continuous coater with claimed compliance to current CFR for use in pharmaceutical production has recently been launched by O'Hara Technologies. Their design resembles an extension (in depth) of a standard batch rotating drum coater and spray manifold but with the addition of inlet and outlet chutes to create a flow through process. There is still the same level of development needed and adjustments to tablet feed rate, pan RPM and residence amount and/or time all affect the coating uniformity. The unit is quite new to the market but details from the vendor indicate that tablet cooling, elevating and/or waxing can be added at the discharge end of the coater. The unit also includes some improvements to their spray manifold design, improving solution distribution and an update of their air caps, providing better antibearding properties (which are required for continuous operation). The coater is also designed to be run in batch mode during start-up and development. Fundamentally the continuous rotating drum O'Hara unit fills the same gap as the Glatt Multicell, providing continuous process by adaptation of the current batch approaches.

Another approach would be to investigate alternates to how the sample is presented to the coating spray. Fluidized bed coating is one possibility, and is the basic approach used in the SUPERCELL™ from GEA Pharma Systems. This is a small, modular, batch design where tablets are coated in batches ranging from 30 to 40 g, and even though the system was designed to have a linear scale-up to production it is unique as the tablets are coated, with the coating spray in the same direction as the drying gas (not orthogonal to the drying gas), resulting in a more efficient process. The process time is short, seconds or minutes as opposed to hours, and could be used on a semicontinuous mode. It would be relatively straightforward to make a semicontinuous (and possibly continuous) version.

There is also ongoing fundamental research and development ongoing in academia and industry to creating the radial movement of the tablet cores relative to the coating spray and also to provide an axial movement (to facilitate movement through the coater, so that the tablet cores follow a corkscrew, rather than circular motion, although it is expected that most of this work is/will be subject to IP (intellectual property).

43.4.8 Packaging (Including Printing of Final Dosage Form)

All current Pharmaceutical packaging processes are run as batch processes but are inherently continuous. However, much like encapsulation, consideration has to be given to the multiple input streams (product and packaging materials) and the biggest issue is maintaining a continuous, traceable, supply of package materials.

43.5 CREAMS, LIQUIDS, AND SUSPENSIONS

Continuous processing concepts have also been implemented in the area of sterilization, and solution manufacture. In addition, it is normal for the containers/bottles/ampoules/pouches used to be manufactured along side the actual product, even under the same sterile conditions. It is also common for the product strength or even alternate products to be run concurrently (with the appropriate changeover procedures being run automatically—including appropriate PAT measurement systems to provide verification of the change). This class of products comes as close as any to realizing true continuous processing in pharmaceutical production, as although they typically start and finish with a batch solution/suspension preparation, all over unit operations (including things like "blow fill" container formations, and dosing) run continuously until the product is collected into batches at the end.

43.6 LYOPHILIZATION

Currently lyophilization is carried out in very large batch sizes (based on number of individual samples and there are no indications/research ongoing to suggest this will change in the near future. However, Rey [27] proposed some very interesting concepts on continuous or semicontinuous lyophilization technology based on practices from the food industry where continuous freeze drying is deployed.

43.7 NOVEL UNIT OPERATIONS

43.7.1 Spray Dried Dispersion

These are common in food and other industries and becoming more so in Pharmaceutical, primarily as they provide a way to alter/control the bioavailability of certain API. They also provide a mechanism for holding the API in a something state/form. During the spray dried dispersion (SDD) process, the API and a waxy polymer are dissolved in solvent before the solution is sprayed under controlled conditions to generate a modified API, with defined particle characteristics (which actual make secondary formulation more straightforward—often direct compression). The actual spraying process is continuous (first in/first out) and even current manufacturing approaches could easily be adapted to flow production, but also modified to be truly continuous.

43.7.2 Melt Congeal Extrusion/Spinning Disk Extrusion

Much like SDD production, the melt congeal/spinning disk extrusion process is deployed to modify the availability of the API, however in this case normal to modify the rate of release; they often provide the basis of slow/sustained release formulations. The actual extrusion process is continuous and the batch nature of production comes not even from the initial feeder hopper but from the collection into batch postprocessing.

43.7.3 Webs/Oral Care Strips

Web-based products such as oral care strips bear more resemblance to screen printing than pharmaceutical manufacturing. Their production is continuous but the two input streams are both batch (the support/paper backing and gel like product suspension). Even though the suspension preparation could be made continuous this process is more easily adapted to flow production than continuous.

43.7.4 Transdermal Patches

Much like Web-based products; transdermal patches have more in common with printing than pharmaceuticals. Typical they are produced by deposition (sometimes spraying, more often roller deposition) into a permeable support medium, over which a protective coat is then applied—forming a sandwich. The support medium and protective coating comes on long rolls (much like the paper used in a cash register and used to provide a till receipt). The solution preparation (typically purely a dilution of the API in a carrier) is batch and normally highly toxic (e.g., nicotine solution used in nicotine patches is classified as an occupational exposure band (OEB) level 4/5 because in solution form it is not only toxic but readily absorbed. It is not likely that production of these types of product will become truly continuous in the near future.

43.8 WHY CONSIDER CONTINUOUS PROCESS FOR DRUG PRODUCT OPERATIONS?

43.8.1 Benefits of Continuous

The biggest advantage in developing continuous processes rather than batch is around scale-up, or rather, as has already been indicated in this chapter the lack of scale-up. Processes are developed at the same process flow rate as they will run in commercial manufacturing; it is purely that the process runs for a longer period of time in commercial production. This is key; the process performance changes with scale, and often development activities are not carried out on the same design of process equipment (e.g., a V-shell blender being used in development but a bin blender used in commercial manufacture). These types of dramatic changes equipment scale result in differences in physical characteristics just as surface area to volume, which lead to significant differences in the way the process to make the product performs.

Typically this goes hand in hand with a reduced equipment footprint, for example, a development scale blenders is around 3 ft tall, while a production size V-blender can be 1–2 stories high, and this is just the blend step—a complete direct compression equipment train with gravity flow between production steps, typically requires a building 3 stories high. The same annual output can be achieved from a self-contained, typically wall mounted, process suite occupying only one room.

Continuous processes also provide the ability to vary batch sizes based on product and demand—we simply run longer. Having a smaller footprint in a cGMP space is a huge cost saving, if the equipment could be "skid mounted" and pulled out of storage only when needed for use. This introduces the idea of the equipment being housed in a cGMP bubble that could (in theory) be dropped into any cGMP facility (e.g., a contract manufacturing organization (CMO)) and run under that facilities compliant processes.

Smaller equipment also typically means cheaper equipment; certainly comparing the cost of the large V-shell to a typical continuous blender has the V-shell costing around 10 times more.

Because these systems are designed to run continuously (with 80–85% availability) they have much higher equipment utilizations rates (a typical batch blender has 25–30% utilization). They are also (typically) highly automated, resulting in lower labor costs and higher operating efficiencies. Another advantage of continuous processing is a reduction in Work In Progress time and therefore inventory that needs to be held, leading to just-in-time manufacturing.

In commercial manufacturing there is also a significant advantage in running processes continually at steady state (rather than those that progress toward an end point); there is a reduction in variance but also it is simple to introduce PAT measurement systems to increase quality and reduce waste through continuous improvement. Especially when we also consider that these systems are typically contained, from start to finish and therefore more applicable to high potency products but also often include automated clean-in-place systems which allow automated changeover between products, which is particularly important when you consider the benefit of efficient start up and shut down.

This last statement is key, if we consider the benefit of continuous purely from a development viewpoint. Part of the twenty-first century quality initiatives is the principle of establishing Process understanding using tools such a design of experiments (DoE). To run a DoE even at development scale with take multiple small-scale batches. Whereas running the DoE (automated) on a continuous system simply means "driving" the continuous process around process space whilst tracking/isolating the product produced (so that the impact on the product performance can be determined). This could be carried out in two ways; the most basic is where the process simply drives to the next set of DoE conditions, waits for steady state, collects product, then moves again; the more complicated and more information rich is there the process trajectory is investigated between the points on the DoE, this allows for a more detailed surface response curve to be generated and the uncertainty within the process space to be lowered.

43.8.2 Cost Analysis

It is possible to quantify possible cost savings by comparing continuous to batch activities based on yield increases (a 2% yield improvement is common simply from start-up and shut savings). As an example a typical direct compression solid dosage formulation requiring 80,000 kg/year, could be achieved by running 100 × 800 kg batches (about the maximum number of batches possible through a single commercial blender). Start up and shut down of the 100 batches will account for approx 2% or 1600 kg of waste. Whereas the same volume could be delivered by running four separate 52-day production cycles (208 days in total) of a continuous system running 20 kg/h. The continuous process would waste only 64 kg. This could be further improved if production was carried out in a single production run, however, this would be product being held on inventory (impacting shelf life) for up to 5 months. In addition to the yield improvements, the improved equipment efficiencies would be the equipment that will be available for other use equivalent to an additional 50,000 kg of production.

43.9 IMPLEMENTATION OF CONTINUOUS PROCESSES

43.9.1 Regulatory Implications

One of the main reservations when considering developing and implementing continuous processes is regulatory burden. The first element often to be considered is the traceability provided by running a "batch." According to the CFR, the definition of a batch is

> A specific quantity of a drug or other material that is intended to have uniform character and quality, within specified limits, and is produced according to a single manufacturing order during the same cycle of manufacture [28].

It is clear that the regulatory definitions are already in place to support the concept of a batch being a period of time, whether that time period is very short (possibly even an individual dosage form), per day, or even a longer period if the process output can be adequately controlled using CQV.

"Continuous quality verification is described as an approach to Process Validation where manufacturing process (or supporting utility system) performance is continuously monitored, evaluated and adjusted as necessary."

More specifically "it is a science-based approach to verify that a process is capable and will consistently produce product meeting its predetermined critical quality attributes. With real time quality assurance (that CQV will provide), the desired quality attributes are ensured through continuous assessment during manufacture. Data from production batches can serve to validate the process and reflect the total system design concept, essentially supporting validation with each manufacturing batch" [29].

Under this paradigm the idea of a "batch" is becomes no more than form of tracking and quality assurance. Also of note; currently at the time of writting there is another ASTM activity (WK9192) as part of E55. This is a new Standard Guide for the Application of Continuous Processing Technology to the Manufacture of Pharmaceutical Products and is expected to clarify and give guidance around regulatory implications

43.9.2 Validation

One of the latest documents being drafted as part of the twenty-first century quality initiatives is a new Process Validation Guidance. At the time of writing this is only available in draft form but due to be issued in the very near future. This guidance divides Process Validation into three component parts. New Validation now includes establishing process understanding during development, followed by a performance qualification (PQ) of the process (this step replaces the old three batch validation activity) that is in turn followed by continued verification. Developing a continuous process under a QbD paradigm actually leads to a process with significantly higher level of process understanding (because we would have been able to investigate the impact on the product of many more process conditions (as we drive between points on the DoE). There is even a possibility that our confidence in how the process will run "in commercial manufacturing" (because there is no scale-up) will be so high that we could only carry out the PQ immediately before launch, reducing the financial burden of holding registration/validation material on inventor.

There is also an expectation that continuous processes will be adaptive and under continuous quality verification/feedback control that is aligned with the principle of Continued Verification. It also supports the principle of real-time release (RTR) where the process is under feedback control ensuring the output quality.

REFERENCES

1. Kossik J. *Think Small: Pharmaceutical Facility Could Boost Capacity and Slash Costs by Trading in Certain Batch Operations for Continuous Versions*. Pharmamag.com, article ID/ DDAS-SEX 52B.
2. Factory shift: new prescription for drug makers: update the plants. *Wall Street J.* September 3, 2003.
3. US Food and Drug Administration, Center for Drug Evaluation and Research, *Guidance for Industry PAT—A Framework for Innovative Pharmaceutical Manufacturing and Quality Assurance*, August 2003. http://www.fda.gov/cder/guidance/5815dft.htm
4. US, Food and Drug Administration, Center for Drug Evaluation and Research, *Pharmaceutical cGMPS for the 21st Century—A Risk-Based Approach: Second Progress Report and Implementation Plan*, September 2004. http://www.fda.gov/Drugs/DevelopmentApprovalProcess/Manufacturing/QuestionsandAnswersonCurrentGoodManufacturingPracticescGMPforDrugs/UCM071836
5. Leuenberger H. Moist agglomeration of pharmaceutical powders (size enlargement of particulate material)—the production of granules by moist agglomeration of powders in mixers/kneaders. In: Chulia D, Deleuil M, Pourcelot Y, editors. *Powder Technology and Pharmaceutical Processes, Handbook of*

Powder Technology, Vol. 9, Elsevier, Amsterdam, 1994, pp. 377–389.

6. Laurent BFC, Bridgwater J. Convection and segregation in a horizontal mixer. *Powder Technol.* 2002;123:9–18.
7. Laurent BFC, Bridgwater J. Performance of single and size-bladed powder mixers. *Chem. Eng. Sci.* 2002;57: 1695–1709.
8. Laurent BFC, Bridgwater J. Influence of agitator design on powder flow. *Chem. Eng. Sci.* 2002;57:3781–3793.
9. Marikh K, Berthiaux H, Mizonov V, Barantseva E. Experimental study of the stirring conditions taking place in a pilot plant continuous mixer of particulate solids. *Powder Technol.* 2005;157:138–143.
10. Williams J, Rahman M. Prediction of the performance of continuous mixers for particulate solids using residence time distributions, Part I: Theoretical. *Powder Technol.* 1971;5: 87–92.
11. Beaudry JP. Blender efficiency. *Chem. Eng.* 1948;55:112–113.
12. Williams J, Rahman M. Prediction of the performance of continuous mixers for particulate solids using residence time distributions, Part II: Experimental. *Powder Technol.* 1971;5: 307–316.
13. Harwood C, Walanski K, Luebcke E, Swanstrom C. The performance of continuous mixers for dry powders. *Powder Technol.* 1975;11:289–296.
14. Portillo PM, Ierapetritou MG, Muzzio FJ. Characterization of continuous convective powder mixing processes. *Powder Technol.* 2008;182:368–378.
15. Warman M. Using near infrared spectroscopy to unlock the pharmaceutical blending process. *Am. Pharm. Rev.* 2004; 7(2):54–57.
16. Koblitz T, Ehrhardt L. *Continuous Variable-Frequency Fluid Bed Drying of Pharmaceutical Granulations*, Pharmaceutical Technology, March 1985.
17. Leunberger H. New trends in the production of pharmaceutical granules: batch vs. continuous processing. *Eur. J. Pharm. Biopharm.* 2001;52:289–296.
18. Mollan MJ, Jr., Lodaya M. *Continuous Processing in Pharmaceutical Manufacturing (white paper),* PharmaManufacturing.com, http://www.pharmamanufacturing.com/whitepapers/2004/11.html
19. Paul S, Knoch A, Lee G. Continuous Granulation: Review. *PZ Prisma*, 1997;4:112–124.
20. Lindberg NO. Some experiences of continuous wet granulation. *Acta Pharm. Suec.* 1988;25:239–246.
21. Appelgren C, Eskilson C, Medical L. A novel method for the granulation and coating of pharmacologically active substances. *Drug Dev. Ind. Pharm.* 1990;16(15):2345–2351.
22. Jones PL, Rawley AT. Dielectric drying. *Drying Technol.* 1996;14(5):1063–1098.
23. Lindberg NO, Turfvesson C, Olbjer L. Extrusion of an effervescent granulation with a twin screw extruder. *Drug Dev Ind. Pharm.* 1987;13:1891–1913.
24. Gamlen M, Eardly C. Continuous Granulation using a Baker Perkins MP50 (multipurpose) extruder. *Drug Dev. Ind. Pharm.* 1986;12:1710–1713.
25. Ghebre-Sellassie I, et al. Continuous Production of Pharmaceutical Granulation. United States Patent US 6,499,984 B1 Dec. 31, 2002.
26. Lodaya M, Mollan M, Ghebre-Sellassie I. Twin screw wet granulation. In: Ghebre-Sellassie I, Martin C,editors. *Pharmaceutical Extrusion Technology, Drugs and the Pharmaceutical Sciences Series*, Vol. 133, Marcel Dekker, 2003.
27. Rey L. Some leading edge prospects in lyophilization. *Am. Pharm. Rev.* 2003;6(2):32–44.
28. US Food and Drug Administration, Center for Drug Evaluation and Research, *21 Code of Federal Regulations, Parts 210 and 211 Current Good Manufacturing Practice for Manufacturing, Processing, Packing, or Holding of Drugs*, http://www.fda.gov/cder/dmpq/cgmpregs.htm.
29. ASTM E2537: *Standard Guide for the Application of Continuous Quality Verification to Pharmaceutical and Biopharmaceutical Manufacturing.*

44

PHARMACEUTICAL MANUFACTURING: THE ROLE OF MULTIVARIATE ANALYSIS IN DESIGN SPACE, CONTROL STRATEGY, PROCESS UNDERSTANDING, TROUBLESHOOTING, AND OPTIMIZATION

Theodora Kourti

Pharma Launch and Global Supply, GlaxoSmithKline, Global Functions

44.1 INTRODUCTION

The goal of any industry (be it chemical, pharmaceutical, steel, pulp, and paper) is to produce a product satisfactory to the customer (i.e., within prescribed quality specifications) under safety and environmental regulations and at a minimum cost. Quality control and regulatory specifications (safety, environmental) will help the manufacturer achieve the first three objectives but sometimes at the expense of cost. Understanding the process and monitoring and controlling process performance will help meet all four targets (quality product, safety constraints, environment constraints, minimum cost) simultaneously.

Process analysis and understanding, monitoring, and control have been practiced by several industries (notably petrochemical) for several decades. These industries adopted the above practices gradually. First, they saw the need for real-time quality measurements and developed real-time analyzers; as an example, the first analytical and control instrument group of UOP (Universal Oil Products) was formed in 1959 with the mission to develop online analyzers for internal pilot plant applications. This first step made the industry capable of collecting real-time measurements of quality properties and other process variables. The second step was the development of automatic process control techniques. This required some form of modeling. Attempts were made to understand the fundamental mechanisms of processes and built sophisticated mechanistic (first-principles) or empirical (data-driven) models. Later, in the 1990s the industry made a third step, which was the use of multivariate statistical analysis methods. With these multivariate approaches, it became possible to analyze and understand the process by looking at historical data containing hundreds of variables, detect abnormal situations, diagnose the sources of the abnormalities, and make appropriate modifications. Furthermore, by utilizing multivariate statistical process control (MSPC), it became possible to monitor the wellness of the process and product in real time, by looking simultaneously at hundreds of variables as they are collected. As a result, several industries managed not only to assure acceptable end-product quality, but also to improve process performance and maintenance, and to significantly reduce cost. The quick adoption of the methodologies and the benefits becomes evident from a very impressive set of applications presented by industry in 2003 in the symposium of "abnormal situation detection and projection methods—industrial applications [1]."

However, the picture of the pharmaceutical industry was different. At 2003, an article in the Wall Street Journal [2] proclaimed that "The pharmaceutical industry has a little secret: Even as it invents futuristic new drugs, its manufacturing techniques lag far behind those of potato-chip and laundry-soap makers." The article went on to explain that "in other industries, manufacturers constantly fiddle with their production lines to find improvements" but "regulations leave drug-manufacturing processes virtually frozen in

Chemical Engineering in the Pharmaceutical Industry: R&D to Manufacturing Edited by David J. am Ende

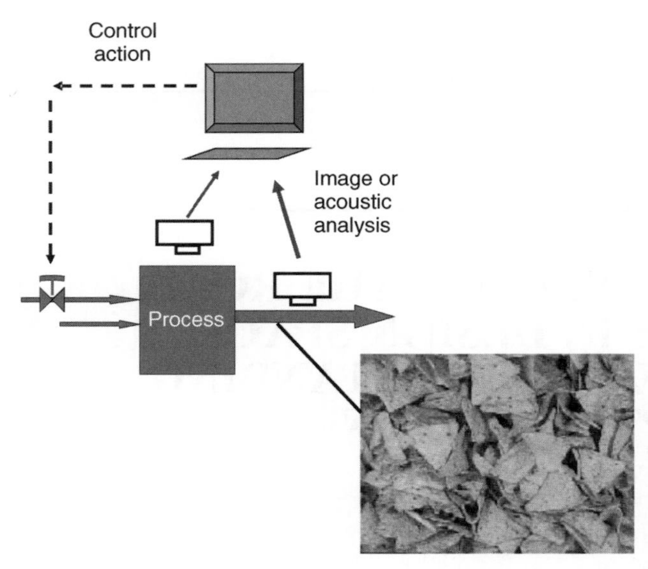

FIGURE 44.1 Monitoring and feedback control based on image and vibrational analysis. Example from snack food industry [3, 4].

time." Applications from the snack food industry (potato chip) that were published the same year [3, 4] were illustrating the use of inferential sensors based on a digital imaging system that had been developed for monitoring and control of the amount of coating applied to the base food product and the distribution of the coating among the individual product pieces, with real-time results from the implementation of such imaging system on snack food production lines. The imaging system was used to monitor product quality variables and to detect and diagnose operational problems in the plants. It was also used to implement closed loop feedback control over coating concentration. It was based on multivariate image analysis. Figure 44.1 shows the setup for feedback control based on Image Analysis. Images are collected by cameras from the process or from a stream exiting the process. This information then can be used in a feedback control loop.

The situation depicted in Figure 44.1 is the desired state in the pharmaceutical industry. That is, it is desired that we have the ability to measure or infer quality in real time, to assess the deviation from expected quality value, and to calculate a real-time control action for correction. Such ability stems from good models that provide process understanding, (different) models that convert spectral or other sensor data to quality, and (different) models that calculate the required control action. The word "different" was added on purpose in the previous phrase to indicate that in this endeavor different types and classes of models are employed and that there are several modeling activities required. The models required for all these classes of modeling may be first principles/mechanistic, data based/empirical, or hybrid. Multivariate projection methods or latent variable methods play an integral part in empirical and hybrid modeling. Such models can be developed to relate final quality properties to raw material attributes and process parameters and can be used for process understanding, process monitoring, and troubleshooting. Multivariate models can also be developed and utilized for process control, scale-up, and site transfer. There is a wealth of literature describing the theoretical foundation of the latent variable or projection methods [5–7], as well as the experiences from practitioners in industry [1, 8].

A lot of changes have happened since 2003 in the pharmaceutical industry that has entered a new era. The introduction of concepts such as quality by design, design space, and control strategy are also examples of such changes (Table 44.1). Multivariate methods are most suitable to address the requirements associated with these concepts. The pharmaceutical industry can learn from existing methodologies and from experiences from other industries and utilize multivariate technologies for fast process and product improvements.

In this chapter, the fundamentals behind latent variables modeling will be presented briefly together with references for in-depth presentations of methodologies and their use for process understanding, troubleshooting, monitoring, and control. Case studies are shown for such applications or are referenced. The chapter should be used by the reader as guidance for the types of problems that can be solved utilizing the methodology; the reader will use the detailed references to seek in-depth analysis and detailed solutions to specific problems.

TABLE 44.1 Terms Related to Quality by Design

Quality by design (*QbD*) is defined as a systematic approach to development that begins with predefined objectives and emphasizes product and process understanding and process control based on sound science and quality risk management

Design space is the multidimensional combination and interaction of input variables (e.g., material attributes) and process parameters that have been demonstrated to provide assurance of quality

Control strategy is a planned set of controls derived from current product and process understanding that ensures (good) process performance and product quality. The controls can include parameters and attributes related to drug substance and drug product materials and components, facility and equipment operating conditions, in-process controls, finished product specifications, and the associated methods and frequency of monitoring and control

Note: These terms are defined by the International Conference on Harmonisation of Technical Requirements for Registration of Pharmaceuticals for Human Use (ICH is a unique project that brings together the regulatory authorities of Europe, Japan, and the United States and experts from the pharmaceutical industry in the three regions to discuss scientific and technical aspects of product registration).

44.2 THE NATURE OF PROCESS AND QUALITY DATA

44.2.1 Multivariate Nature of Quality

Understanding the multivariate nature of quality is of great importance. Product quality is defined by the simultaneously correct values of all the measured properties; that is, product quality is a multivariate property. Most of the time, the property variables are not independent of one another, and none of them adequately defines product quality by itself; therefore, it is not a good practice to separately monitor key properties of the final product using univariate control charts.

Figure 44.2 is a classic illustration of the problem with using separate control charts for two quality variables (y_1, y_2). In this figure, the two variables are plotted against each other (upper left of the figure). The same observations are also plotted as individual (univariate) charts for y_1 (the horizontal plot) and y_2 (the vertical plot) with their corresponding upper and lower control limits. Suppose that when only common cause variation is present, y_1 and y_2 follow a multivariate normal distribution; the dots in the joint plot represent a set of observations from this distribution. Notice that y_1 and y_2 are correlated. The ellipse represents a $(1 - \alpha)\%$ joint confidence limit of the distribution (i.e., when the process is in control, $\alpha\%$ of the points will fall outside the ellipse).

The point indicated by the $\oplus$ symbol is clearly outside the joint confidence region, and it is different from the normal in-control population of the product. However, neither of the univariate charts gives any indication of a problem for point $\oplus$; it is within limits in both of the charts. The individual univariate charts effectively create a joint acceptance region shaped like a square (shown with the ellipse). This will lead to accepting wrong products as good (point $\oplus$), but also rejecting a good product as bad (point $\diamondsuit$). The problem worsens as the number of variables increases. It is clear that an efficient fault detection scheme should look at the variables together.

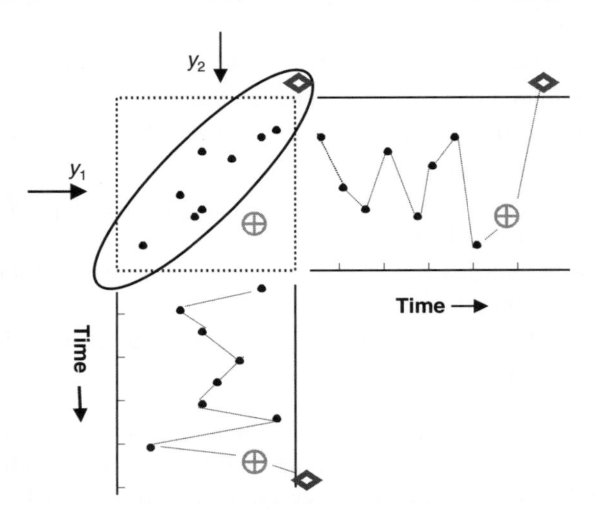

FIGURE 44.2 The multivariate nature of quality.

Multivariate charts are required to test quality [9] when it is described by many variables.

Recognizing the multivariate nature of quality should guide the procedures that will be used for the following cases:

- Raw material evaluation
- Intermediate quality evaluation
- Final quality evaluation
- Process control for quality
- Product transfer and scale-up (the multivariate nature of quality should be preserved for raw materials, intermediate qualities, and final qualities). This is a minimum requirement. Later, we will discuss the requirements on the multivariate space of the process variable trajectories as well.

44.2.2 Real-Time Monitoring and Process Signature: The Need to Utilize Information from Process Data

There is a widespread belief that the use of real-time quality measurements will help maintain the process "in control." Several practitioners are using real-time quality measurements to determine the "end point." The questions are the following:

- Does real-time "in-control" quality guarantee "in-control" process ?
- Is one or two final quality properties a sufficient "metric" of whether or not the process was in control?

Consider the example of Figure 44.3, where the trajectories of each one of the three process variables are plotted for three different batch runs, A, B, C. The final product at

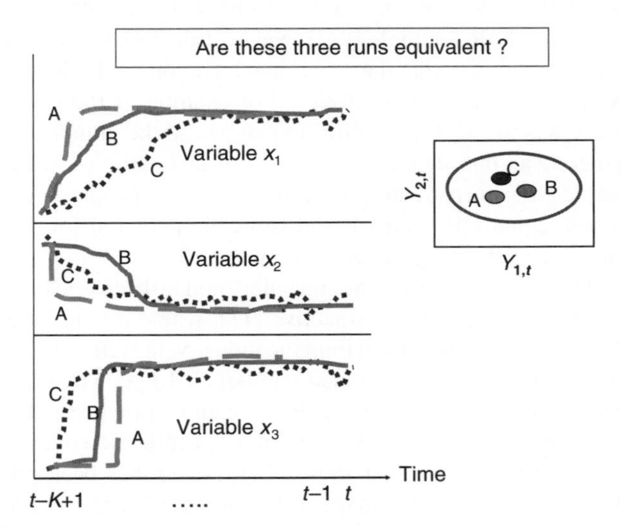

FIGURE 44.3 Different paths to the end point may result in similar values for some quality properties but may affect others in a different way. Consistent paths will assure that overall quality is consistent.

time t (end point quality) is determined from properties $y_{1,t}$ and $y_{2,t}$. These properties are plotted for each product produced by runs A, B, and C against the desired confidence limits of acceptable multivariate quality defined by the ellipse. Suppose that for all three runs the two "end properties" are on target in the multivariate control chart (points fall within the ellipse). However, the process trajectories follow different paths for each run. Are these runs equivalent?

It is well known in industry that the same measured quality properties can be achieved by taking different process paths. However, very frequently the few properties measured on a product are insufficient to define entirely the product quality. In polymer industry, for example, if only the viscosity of a polymer is measured and kept within specifications, any variation in end-use application (downstream processability) that arises due to variation in chemical structure (branching, composition, end-group concentration) will not be captured. To achieve consistency in all the product properties (measured quality and ability to process down the stream), the process conditions (path to end point) must also be kept in statistical control. When this is not the case, although the measured product properties may be on target, the properties that determined the processability of the product may not be within acceptable limits. Therefore, monitoring process data (temperatures, pressures, etc.,) together with real-time analyzers will give valuable information about events with special causes that may not only affect the final quality, but also give early warnings for potential equipment failure.

Another example that further corroborates this argument has been reported from the pharmaceutical industry [10]: "Conventional process control of drying of granulate in a fluidized bed drier would be to measure the loss on drying of a sample of powder, to determine water content. An advance on this may be to determine water content using an online NIR technology. However, true Process Understanding requires that the route by which you get to this end point be known and controlled. For example if the drying process is too vigorous, attrition may cause the granulate to generate an unacceptably high level of fine particles, which may cause downstream processing problems or dissolution issues; equally if drying is too slow, the potential for degradation of the drug molecule may exist."

This "process path to the end point" is also discussed in the European Regulatory Perspective [11], where it is reported that "during discussions within the industry, the term *process signature* has been mentioned regularly." To get a common understanding of this, the EU PAT Team had invited public comments on the following definition: "A collection of batch specific information that shows that a batch has been produced within a design space of the product." The EU PAT team mentions as examples of process signatures the amount of water added in relation to time (wet massing), air flow rate, and bed temperature during fall rate drying (fluidized bed drying). They concluded that their understanding is that there is no unique process signature, but instead a family of process signatures with common characteristics (salient features).

The above observations point to the importance of monitoring the process together with the product quality. By monitoring only the quality variables (in a univariate or multivariate chart), one performs statistical quality control (SQC). Real-time measurements on temperatures, pressure, pH, RPM, and so on combined with real-time measurements from analytical technology (spectroscopy, ultrasound, etc) will lead to online *process* monitoring and make MSPC, fault detection, and isolation possible. Combining information from the process measurements with the information from the analytical tools gives a very powerful tool to monitor the process. These two sets of measurements are not independent from one another but interrelated. As a matter of fact, these measurements "confirm" each other. This is the reason that process variables are sometimes used to assess the reliability of real-time analyzers. It will also be pointed out later that sometimes real-time process measurements may eliminate the need of some real-time analyzers. Information about the process may also include the vessels used for a specific run, the operators that were on shift, suppliers of raw material, and so on. Another advantage to using process measurements is that any abnormal events that occur will also have their fingerprints in the process data. Thus, once an abnormal situation is detected, it is easier to diagnose the source of the problem, as we are dealing directly with the process variables. For example, a pending equipment failure means that our production is not in control. However, there are situations that while there may be a pending equipment failure, real-time quality measurements may still be acceptable. By monitoring process variables, we have a very high probability to detect a pending problem.

Process data can be utilized together with appropriate models to

- Infer final product quality from process conditions during production
- Ease process understanding and troubleshooting
- Infer a quality in real time (soft sensors)
- Establish an overall "process signature" and monitor it
- Monitor analyzer reliability
- Check that the process is in a state of statistical process control (SPC)
- Decide on midcourse correction of variable trajectories to control final quality
- Establish operational knowledge that can be used for product transfer and scale-up

It should be emphasized here that although some relations between process operating conditions and final quality are known from the initial design of experiments (DOE), once we are in production, these relations may be influenced

TABLE 44.2 Some Multivariate Process Data Formats

Matrix Symbol	Dimensions	Explanation
X	$(n \times k)$; two-way matrix; n observations in time, or n batches; k process variable measurements	Data from a continuous process, at given instant in time, or summary data from a batch (max T, min T, length of batch run, etc.)
Y	$(n \times m)$; two-way matrix; n observations in time, or n batches; m product quality values	Quality data from a continuous process corresponding to the process measurements in **X**, properly lagged, or quality data at the end of a batch.
$\underline{\mathbf{X}}$	$(I \times J \times K)$; three-way matrix; I batches; J process variables measured at K time intervals for each batch	Data collected from batch process at several time intervals during production.
Z	$(n \times r)$; two-way matrix; n observations in time, or n batches; r other variable measurements	Raw material, total cycle times, length between processes, preprocessing information

by other factors and may change locally. By investigating production data, we can uncover the true relationships between process conditions and quality under the closed loop operations.

Finally, it is very frequently stated that critical process parameters should be identified and monitored together with critical quality attributes utilizing SPC. It should be emphasized here that the parameters that appear to be critical in the DOE are not necessarily the ones that will give information about the "wellness of the process" in SPC charts. The reason for this is that the parameters that are identified as important in DOE will be tightly controlled during production. SPC charts on routine data are noncausal; therefore, things that were important in DOE will not be important in SPC, unless something goes really wrong (i.e., the controller fails and cannot keep the desired target). As an example, suppose that temperature is important to the yield, as determined by DOE. This means that during production the desired temperature profile will be regulated by the controllers; in SPC monitoring, what is important (and in some types of processes will indicate the presence of excess impurities and other disturbances) is how much effort the controller is putting to maintain the temperature; that is, how much the valve to the cooling agent opened or closed during the reaction. Therefore, monitoring the controller action will provide much more information about abnormal situations than monitoring the temperature, although temperature was identified as critical process parameter by DOE.

44.2.3 Multivariate Nature, Structure, and Other Characteristics of Process Data

Databases containing measurements collected during production may become very large in size. The data are noncausal in nature (unless they come from designed experiments). They consist of highly correlated variables with many missing measurements and low content of information in any one variable (due to the low signal-to-noise ratios).

Multivariate Structure of Data: The convention that will be used throughout this chapter in expressing data is that of Table 44.2. Other formats that may appear in specific sections only will be defined in their corresponding sections.

44.2.4 Process Analysis and Process Understanding

Unfortunately, sometimes the term "process analysis" is wrongly being used as an equivalent term to process analytical chemistry. Collecting real-time measurements on a specific property may or may not reflect what is happening in the rest of process or the state of the process, unless the state of the process is completely observable in that quality, as discussed in Section 44.2.2; therefore, collecting a real-time quality measurement is not process analysis.

A process can be defined as a series of physical and/or chemical operations that converts input to output. Process analysis is a systematic examination of a process to understand it in order to develop ideas to improve it. Improvement could translate to better quality, lower cost, more efficient energy consumption, less pollutants to environment, and safer operation. One can perform process analysis utilizing both off-line and real-time measurements.

Process analysis leads to process understanding. Again, there may be several definitions of process understanding. There is a widespread belief that one gains process understanding only when they can describe the process by first principles, that is, by a theoretical or mechanistic model. However, one can gain tremendous insight into the process from empirical models derived from databases. These empirical models can lead to fast improvements that in several situations would have been impossible if people had been waiting for the development of theoretical first-principles models. Empirical models based on process data can be extremely valuable to diagnose abnormal operations such as pending equipment failure.

So while process understanding may by some definitions mean uncovering the mechanisms and path of a chemical reaction, or modeling a fermentation process, it may also mean uncovering production problems such as the examples reported by practitioners in several industries:

- Diagnosing that the production of abnormal batches followed a specific pattern and as a result uncovering an incorrect operator practice that led to an abnormal batch every time a routine maintenance task was taking place.
- Understanding why recent process data projected on a latent variable space seem to form two clusters indicating different operation practices and hence solving an important operation problem as a result. Examination revealed that the cooling agent valve was not capable of meeting the demands in hot days and the reactor temperature could not be controlled properly. The valve was resized.
- Assessing that an operational problem caused the readings of a specific thermocouple to be erroneous and appear as outliers; thermocouple was too close to entrance of cool reactant; erroneous readings fed to the controller inappropriately alter the reactor temperature.
- Understanding where is the maximum process variability; is this variability noise or is it assignable to a cause—can we reduce it?

Understanding the way the process behaves in real scale production is a tremendous asset to the effort of product quality improvement and sometimes it weighs equally importantly to understanding the detail mechanisms of the reaction that takes place during production.

44.3 LATENT VARIABLE METHODS FOR TWO-WAY MATRICES

Latent variables exploit the main characteristic of process databases, namely, that although they consist of measurements on a large number of variables (hundreds), these variables are highly correlated and the effective dimension of the space in which they move is very small (usually less than 10 and often as low as 2). Typically, only a few process disturbances or independent process changes routinely occur, and the hundreds of measurements on the process variables are only different reflections of these few underlying events. For a historical process data set consisting of a $(n \times k)$ matrix of process variable measurements $\mathbf{X}$ and a corresponding $(n \times m)$ matrix of product quality data $\mathbf{Y}$, for linear spaces, latent variable models have the following common framework [12]:

$$\mathbf{X} = \mathbf{TP}^{\mathrm{T}} + \mathbf{E} \tag{44.1}$$

$$\mathbf{Y} = \mathbf{TQ}^{\mathrm{T}} + \mathbf{F} \tag{44.2}$$

where $\mathbf{E}$ and $\mathbf{F}$ are error terms, $\mathbf{T}$ is an $(n \times A)$ matrix of latent variable scores, and $\mathbf{P}$ $(k \times A)$ and $\mathbf{Q}$ $(m \times A)$ are loading matrices that show how the latent variables are related to the original $\mathbf{X}$ and $\mathbf{Y}$ variables. The dimension A of the latent variable space is often quite small and determined by cross-validation or some other procedure.

Latent variable models assume that the data spaces $(\mathbf{X}, \mathbf{Y})$ are effectively of very low dimension (i.e., nonfull rank) and are observed with error. The dimension of the problem is reduced by these models through a projection of the high-dimensional $\mathbf{X}$ and $\mathbf{Y}$ spaces onto the low-dimensional latent variable space $\mathbf{T}$, which contains most of the important information. By working in this low-dimensional space of the latent variables $(t_1, t_2, \ldots, t_A)$, the problems of process analysis, monitoring, and optimization are greatly simplified. There are several latent variable methods. Principal component analysis (PCA) models only a single space ($\mathbf{X}$ or $\mathbf{Y}$) by finding the latent variables that explain the maximum variance. Principal components can then be used in regression (PCR). In PCR, there appears to be a misconception that the principal components (PC) with small eigenvalues will very rarely be of any use in regression. The author's personal experience is that these components can be as important as those with large variance. Projection to latent structures or partial least squares (PLS) maximizes the covariance of $\mathbf{X}$ and $\mathbf{Y}$ (i.e., the variance of $\mathbf{X}$ and $\mathbf{Y}$ explained, plus correlation between $\mathbf{X}$ and $\mathbf{Y}$). Reduced rank regression (RRR) maximizes the variance of $\mathbf{Y}$ and the correlation between $\mathbf{X}$ and $\mathbf{Y}$. Canonical variate analysis (CVA), or canonical correlation regression (CCR), maximizes only the correlation between $\mathbf{X}$ and $\mathbf{Y}$. A discussion of these latent variable models can be found elsewhere [12]. The choice of method depends on the objectives of the problem; however, all of them lead to a great reduction in the dimension of the problem. Some of them (PCR and PLS) model the variation both in the $\mathbf{X}$ space and in the $\mathbf{Y}$ space. This point is crucial in most of the applications related to PAT that are discussed in the following sections, as well as for the problem of treating missing data. The properties of PCA and PLS are discussed briefly below.

44.3.1 Principal Component Analysis

For a sample of mean centered and scaled measurements with n observations on k variables, $\mathbf{X}$, the principal components are derived as linear combinations $\mathbf{t}_i = \mathbf{X}\mathbf{p}_i$ in such a way that subject to $|\mathbf{p}_i| = 1$, the first PC has the maximum variance, the second PC has the next greatest variance and is subject to the condition that it is uncorrelated with (orthogonal to) the first PC, and so on. Up to k, PCs are similarly defined. The sample principal component loading vectors $\mathbf{p}_i$ are the eigenvectors of the covariance matrix of $\mathbf{X}$ (in practice, for mean centered data, the covariance matrix is estimated by $(n-1)^{-1}\mathbf{X}^{\mathrm{T}}\mathbf{X}$). The corresponding eigenvalues give the variance of the PCs (i.e., $\mathrm{var}(\mathbf{t}_i) = \lambda_i$). In practice, one rarely needs to compute all k eigenvectors, since most of the predictable variability in the data is captured in the first few PCs. By retaining only the first A PCs, the **X** matrix is approximated by equation 44.1.

44.3.2 Partial Least Squares

PLS can extract latent variables that explain the high variation in the process data, **X**, which is most predictive of the product quality data, **Y**. In the most common version of PLS, the first PLS latent variable $\mathbf{t}_1 = \mathbf{X}\mathbf{w}_1$ is the linear combination of the x variables that maximizes the covariance between $\mathbf{t}_1$ and the **Y** space. The first PLS weight vector $\mathbf{w}_1$ is the first eigenvector of the sample covariance matrix $\mathbf{X}^T\mathbf{Y}\mathbf{Y}^T\mathbf{X}$. Once the scores for the first component have been computed, the columns of **X** are regressed on $\mathbf{t}_1$ to give a regression vector, $\mathbf{p}_1 = \mathbf{X}\mathbf{t}_1/\mathbf{t}_1^T\mathbf{t}_1$; the **X** matrix is then deflated (the $\hat{\mathbf{X}}$ values predicted by the model formed by $\mathbf{p}_1$, $\mathbf{t}_1$, and $\mathbf{w}_1$ are subtracted from the original **X** values) to give residuals $\mathbf{X}_2 = \mathbf{X}-\mathbf{t}_1\mathbf{p}_1^T$. **Q** are the loadings in the **Y** space. In the so called NIPALS algorithm, $\mathbf{q}_1$ is obtained by regressing $\mathbf{t}_1$ on **Y**, and then **Y** is deflated $\mathbf{Y}_2 = \mathbf{Y}-\mathbf{t}_1\mathbf{q}_1^T$. The second latent variable is then computed from the residuals as $\mathbf{t}_2 = \mathbf{X}_2\mathbf{w}_2$, where $\mathbf{w}_2$ is the first eigenvector of $\mathbf{X}_2^T\mathbf{Y}_2\mathbf{Y}_2^T\mathbf{X}_2$, and so on. The new latent vectors or scores ($\mathbf{t}_1$, $\mathbf{t}_2$, ...) and the weight vectors ($\mathbf{w}_1$, $\mathbf{w}_2$, ...) are orthogonal. The final models for **X** and **Y** are given by equations 44.1 and 44.2 [13, 14].

44.3.3 Latent Variables for Process Understanding

Latent variable methods are excellent tools for data exploration to identify periods of unusual/abnormal process behavior and to diagnose possible causes for such behavior (troubleshooting). The scores and loadings calculated by PCA and PLS and the weights by PLS can be utilized for this purpose. By plotting the latent variables (t_1, t_2, ..., t_A) against each other, the behavior of the original data set (be it process **X**, or quality data **Y**) can be observed on the projection space. By examining the behavior in the projection spaces, regions of stable operation, sudden changes, or slow process drifts may be readily observed. Outlier and cluster detection also becomes easy, both for the process and for the quality space. An interpretation of the process movements in this reduced space can be found by examining the loading vectors ($\mathbf{p}_1$, $\mathbf{p}_2$, ..., $\mathbf{p}_A$) or ($\mathbf{w}_1$, $\mathbf{w}_2$, ..., $\mathbf{w}_A$) in the case of PLS and the contribution plots. For a PCA analysis on **X** or a PLS analysis on **X** and **Y**, each point on a t_1 versus t_2 plot is the summary of measurements on k variables.

Figure 44.4a gives a simplified schematic interpretation of the methods. Suppose that we have measurements from five variables in a process (here, we plot the variable deviations from their nominal trajectories) during a time period. Suppose that variables x_1, x_3, and x_4 are correlated with each other, while variable x_2 is correlated with x_5. With the multivariate projection methods, new variables (latent

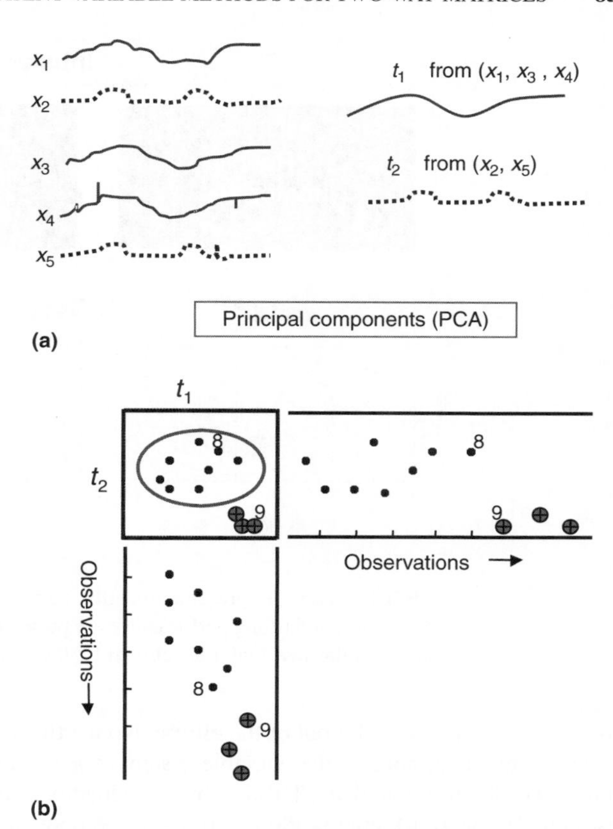

FIGURE 44.4 (a) Simple interpretation of PCA and dimensionality reduction. The principal components t_1 and t_2 use the correlation of five variables and break the process in two orthogonal events. The first principal component corresponds to the event that affects the largest number of variables, the second to the event that affects the next number of variables, and so on. (b) These components can be plotted against each other. A five-variable system is projected onto a two-dimensional plane.

variables) are calculated. In PCA, the first principal component t_1 is a weighted average of x_1, x_3, and x_4, while the second component, t_2, is a weighted average of x_2 and x_5. PCA can be seen as a classification of the main events that affect a process. The first principal component corresponds to the event that affects the largest number of variables, the second to the event that affects the next number of variables, and so on. The description here is a simplified explanation. It may be the case that two or more events affect the same variable, in which case this variable will contribute to the values of more than one component. We have reduced the number of the initial five raw variables to two principal components; we have *reduced the dimensionality* of the system. We can now plot these components against each other, as shown in Figure 44.4b. Each point on the plot summarizes the behavior of five raw variables. When the process is in statistical control, the points will be within the control limits, shown with an ellipse that is determined by statistical criteria (discussed later). If there is a problem in the

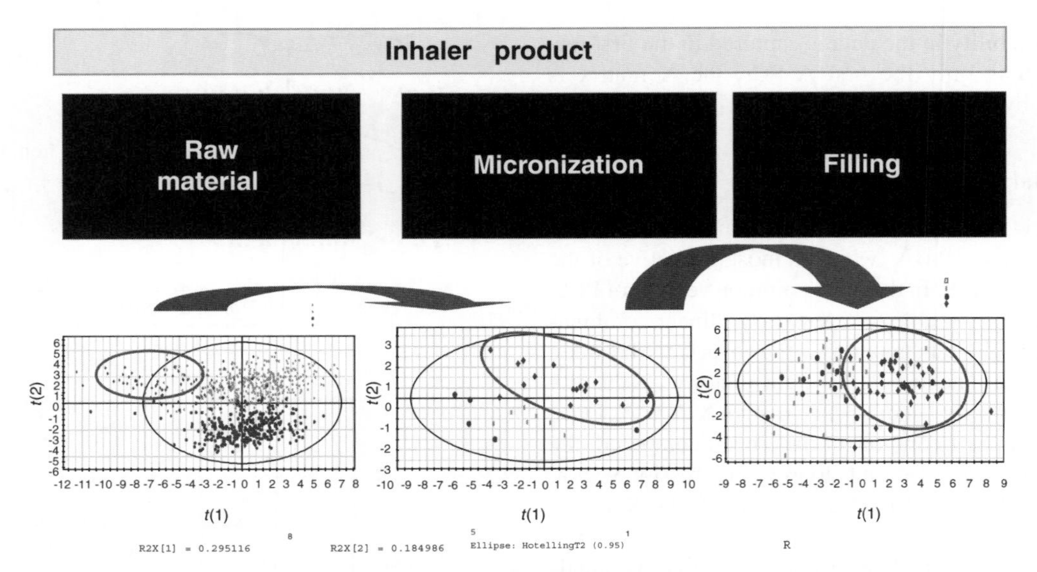

FIGURE 44.5 Representation utilizing projection methods. Raw material properties, micronization properties, and filling performance are projected on a latent variable space. Product batches produced from similar raw material (circled by the small ellipse) have similar filling performance.

process, the points will plot out of the ellipse. Notice that by using the latent variables, a five-variable system is *projected* onto a two-dimensional plane. This is why these methods are also called projection methods. Plotting principal components against each other is a good way to visualize process behavior and detect outliers and clusters. Typically, a small number of components are required to describe the main events in a unit (usually less than 10, and sometimes only 3 to 4).

44.3.3.1 Process Understanding Example 1: Relating Issues Across Unit Operations The power of projection methods in exploring large databases is demonstrated with the following example, shown in Figure 44.5, where we plot projections of the raw material quality, micronized material quality, and final quality. More than 25 measurements of physical and chemical properties are collected per lot of raw material. These variables are projected on a space defined by latent variables that allow us to visualize better the process behavior. In the particular example, raw material is produced at three supplier locations. The raw material properties are within univariate specifications at all locations. Projected on a multivariate space t_1 vs t_2, however, they form three clusters; one cluster projects at the low part of the t_1 vs t_2 plot (negative values of t_2) while the other two clusters at the upper part of the plot (positive values of t_2). One of these two clusters is marked by a small ellipse. This indicates that in a multivariate sense the material possesses slightly different characteristics depending on the location it was produced (covariance structure changes with location). A few of the batches of this raw material were subsequently used for a specific product, and we show its behaviour after micronization and filling. The material properties after micronization are also projected on principal components and it could be observed that the material corresponding to the batches of the small ellipse projects on a different location from the rest of the batches. The filling performance of the material originating from the batches of the small ellipse is different from the rest of the material. The conclusion for this example is that the raw material differences propagate in the final quality. Given the large number of variables involved, it is clear that the projection methods provided a very quick diagnostic of the problem. This could have not been possible by dealing with univariate charts. The reader may note here that although the control ellipses shown are set by default in the vendor software, they are not interpretable when there is clustering; the assumptions for the calculation of these ellipses are for process monitoring and not for process exploration where there is intentional variation such as that introduced by design of experiments.

44.3.3.2 Process Understanding Example 2: Quick Diagnosis of Effect of Raw Material Variability to Granule Characteristics The advantage of PLS is that many highly correlated quality responses can be analyzed simultaneously. The **Y** matrix can contain several parameters related to quality. In this example, we wanted to see the correlation between certain API physical properties and the granule properties. We chose to use as **Y** the entire tapped density profile and the entire PSD profile. As seven variables corresponded to each property profile, we did not have to use any special scaling discussed in the multiblock section. First we use as **X** matrix the API properties and used PLS, between **Y** and **X** to relate API variability to the variability of the granule properties. Figure 44.6 shows the variability explained for each variable in the tapped density profile, and granule size profile, in a cross-validated model. Light grey is the direct

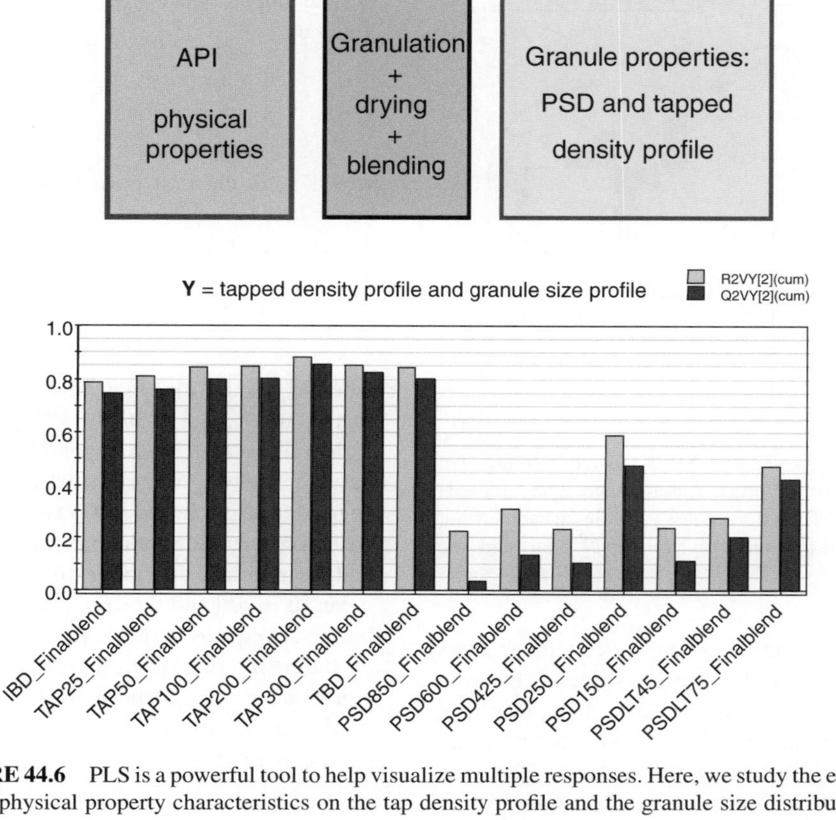

FIGURE 44.6 PLS is a powerful tool to help visualize multiple responses. Here, we study the effect of API physical property characteristics on the tap density profile and the granule size distribution.

fitting variability and dark the cross-validated variability. One can very clearly visualize that the tapped granule density is very highly correlated to the specific API physical property variability (as a high % of variability of the tapped density profile is explained by the variability of the API property values).

44.4 MULTIVARIATE STATISTICAL PROCESS CONTROL

From routine operation, we can establish acceptable limits of good process behavior. On a t_1 versus t_2 plane, such limits will take the form of an ellipse. When the process is in statistical control, the points will be within the ellipse. If there is a problem in the process, the points will plot out of the ellipse. In Figure 44.4b, the ellipse is calculated based on PCA on the data from good operation. Notice that while for raw correlated data the ellipse is tilted, indicating correlation (Figure 44.2), this is not the case when it is calculated for the principal components that are orthogonal.

To monitor the process in real time, however, it would have become cumbersome to have to plot all combinations of principal components (even if we had four components, we would need six charts). A statistic (Hotelling's T^2) can be calculated and the overall variability of the main events of the system can be monitored with a single chart, such as the one shown at the upper left corner of Figure 44.7. The line corresponds to acceptable performance. For the case of two components, this solid line corresponds to the perimeter of the ellipse of Figure 44.4b. For three components, it would correspond to the surface of an ellipsoid, and for four components the surface of a hyperellipsoid.

Hotelling's T^2 for scores is calculated as

$$T_A^2 = \sum_{i=1}^{A} \frac{t_i^2}{\lambda_i} = \sum_{i=1}^{A} \frac{t_i^2}{s_{t_i}^2} \tag{44.3}$$

where $s_{t_i}^2$ is the estimated variance of the corresponding latent variable t_i. This chart essentially checks if a new observation vector of measurements on k process variables projects on the hyperplane within the limits determined by the reference data.

As mentioned above, the A principal components explain the main variability of the system. The variability that cannot be explained forms the residuals (squared prediction error (SPE)). This residual variability is also monitored and a control limit for typical operation is being established. By monitoring the residuals (Figure 44.7, bottom left), we test that the unexplained disturbances of the system remain similar to the ones observed when we derived the model. For example, a model derived with data collected in the summer may not be valid in the winter when different

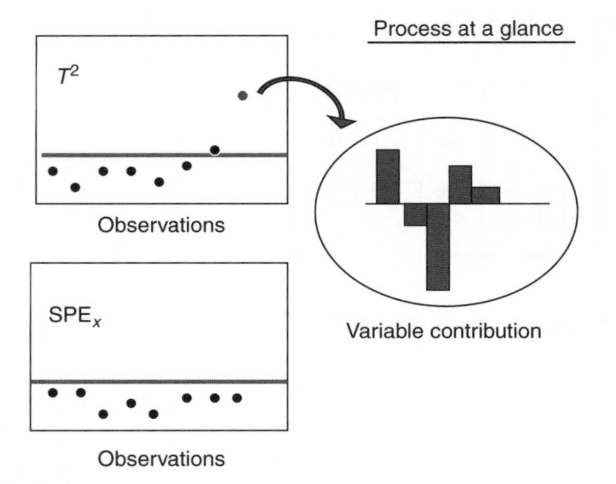

FIGURE 44.7 Two charts (T^2 and SPE) are required to give a picture of the wellness of the process at a glance. The hundreds of measurements collected from the process variables at each instant in real time are translated into one point for the T^2 chart and one point for the SPE chart.

disturbances affect the system (cooling water temperatures different, equipment walls colder, valves may reach limits in capacity of providing heating agent, etc). It is therefore important to check the validity of the model by checking the type of disturbances affecting the system. When the residual variability is out of limit, it is usually an indication that a new set of disturbances have entered the system; it is necessary to identify the reason for the deviation and it may become necessary to change the model.

SPE_X is calculated as

$$\text{SPE}_X = \sum_{i=1}^{k} (x_{\text{new},i} - \hat{x}_{\text{new},i})^2 \qquad (44.4)$$

where $\hat{x}_{\text{new}}$ is computed from the reference PLS or PCA model. Notice that SPE_X is the sum over the squared elements of a row in matrix **E** in equation 44.1. This latter plot will detect the occurrence of any new events that cause the process to move away from the hyperplane defined by the reference model.

44.4.1 Calculation of Chart Control Limits

For a Hotelling's T^2 chart (either for PCA or PLS), an upper control limit based on the A first PCs and derived from n observations is obtained using the F distribution and given by

$$T^2_{A,\text{UCL}} = \frac{(n^2-1)A}{n(n-A)} F_{\alpha(A,n-A)} \qquad (44.5)$$

where $F_{\alpha(A,n-A)}$ is the upper $100\alpha\%$ critical point of the F distribution with $(A, n-A)$ degrees of freedom.

For the SPE_X chart, limits can be computed using approximate results from the distribution of quadratic forms. The critical upper $100(1-\alpha)\%$ confidence interval on SPE is given as

$$\theta_1 \left[\frac{z_a \sqrt{2\theta_2 h_0^2}}{\theta_1} + \frac{\theta_2 h_0 (h_0 - 1)}{\theta_1^2} + 1 \right]^{1/h_0} \qquad (44.6)$$

where z_a is the unit normal deviate corresponding to the upper $100(1-\alpha)\%$, and α is the chance taken to incorrectly declare a fault because of the type I error,

$$\theta_i = \sum_{j=A+1}^{m} \lambda_j^i = Tr(\mathbf{E}^i) \quad \text{for } i = 1, 2, 3 \qquad (44.7)$$

where λ_i is the ith eigenvalue referring to the covariance matrix and $h_0 = 1-(2\theta_1\theta_3/3\theta_2^2)$.

Nomikos and MacGregor [15] used an approximation based on the weighted chi-square distribution $(g\chi^2(h))$. They suggested a simple and fast way to estimate the g and h that is based on matching moments between a $g\chi^2(h)$ distribution and the reference distribution of SPE at any time interval. The mean $[\mu = gh]$ and the variance $[\sigma^2 = g^2(2h)]$ of the distribution are equated to the sample mean (b) and variance (v) at each time interval. Therefore, g and h are estimated from the equations $\hat{g} = v/2b$ and $\hat{h} = 2b^2/v$.

Hence, the upper control limit on the SPE at significance level α is given by

$$\frac{v}{2b} \chi^2_\alpha \left(\frac{2b^2}{v} \right) \qquad (44.8)$$

It should be emphasized that the models built for process monitoring model only common cause variation and not causal variation. The main concepts behind the development and use of these multivariate SPC charts based on latent variables for monitoring continuous processes were laid out in early 1990s [9].

These two charts (T^2 and SPE) are two complementary indices; together they give a picture of the wellness of the system at a glance. With this methodology, the hundreds of measurements collected from the process variables at each instant in real time are translated into one point for the T^2 chart and one point for the SPE chart (these two points summarize the process at that instant). As long as the points are within their respective limits, everything is in order. Once a point is detected out of limit, then the so-called *contribution plots* can be utilized that give us a list of all the *process* variables that mainly contribute to the out of limit point and hence allow us to diagnose the process problem immediately. Contribution plots can be derived for out of limit points in both charts.

Contributions to SPE: When an out of control situation is detected on the *SPE* plot, the contribution of each variable of the original data set is simply given by $(x_{\text{new},j} - \hat{x}_{\text{new},j})^2$. Variables with high contributions are investigated.

Contributions to Hotelling's T2: Contributions to an out of limit value in Hotelling's T^2 chart are obtained as follows: a bar plot of the normalized scores $(t_i/s_{ti})^2$ is plotted and scores with high normalized values are further investigated by calculating variable contributions. A variable contribution plot indicates how each variable involved in the calculation of that score contributes to it. The contribution of each variable of the original data set to the score of component q is given by the following equation:

$$c_j = p_{q,j}(x_j - \bar{x}_j) \text{ for PCA}$$

$$c_j = w_{q,j}(x_j - \bar{x}_j) \text{ for PLS} \qquad (44.9)$$

where c_j is the contribution of the jth variable at the given observation, $p_{q,j}$ is the loading, and $w_{q,j}$ is the weight of this variable to the score of the principal component q and $\bar{x}_j$ is its mean value (which is zero from mean centered data). Variables on this plot that appear to have not only the largest contributions to it, but also the same sign as the score should be investigated (contributions of the opposite sign will make the score only smaller). When there are K scores with high values, an "overall average contribution" per variable is calculated over all the K scores [6].

As an example, consider Figure 44.4b that illustrates that two clusters of points were observed on a t_1 versus t_2 plot. The use of contribution plots may help to investigate which variables have contributed to the move from point 8 to 9. So equation 44.9 would give the contribution of variable j to the move of the score values between two observations (say, 8 and 9) for component q calculated as [14]

$$p_{jq} \times (x_{j,9} - x_{j,8}) \text{ for PCA}$$

Utilizing contribution plots, when an abnormal situation is detected, the source of the problem can be diagnosed such that corrective action is taken. Some actions can be taken immediately, in real time. Others may require interventions to the process. One such example of an abnormal situation appeared in a reactor, in which the reactor temperature should be controlled in an exothermic reaction to 50°C. On a very hot day, the charts indicated abnormalities. Contribution plots pointed to a break in the correlation of cooling water flow and reactor temperature. It turned out that although the cooling water valve was fully open, it could not cope with the demand, as the cooling water was warmer. The valve had to be resized. MSPC pointed to a problem that had to be corrected. Therefore, the contribution plots are very important tools in understanding factors influencing the process during production and help in an "ongoing process understanding" philosophy.

44.4.2 How to Utilize the Control Charts

Since latent variable-based control charts were introduced, their use in industry is increasing. The charts answer the need of process industries for a tool that allows them to utilize the massive amounts of data being collected on hundreds of process variables, as well as the spectral data collected from modern analyzers.

Latent variable control charts can be constructed to monitor either a group of response variables **Y** (e.g., product quality variables) or a group of predictor variables **X** (process variables). For example, multivariate charts can be constructed to assess the consistency of the multivariate quality of raw materials, **Z**, and to test the final product **Y** for consistent quality. If there is spectral analysis on some of the materials, then multiblock concepts, discussed later, can be used.

A very important advantage of latent variables is that they can be used to monitor predictor variables taking into account their effect on the response variables. A model is built to relate **X** and **Y** using available historical or specially collected data. Monitoring charts are then constructed for future values of **X**. This approach means that the process performance can be monitored even at times when the product quality measurements, **Y**, are not available.

The main approach of SQC methods developed throughout the statistical literature has been to monitor only product quality data (**Y**) and, in some cases, a few key process variables (**X**). However, often hundreds of process variables are measured much more frequently (and usually more accurately) than the product quality data. So monitoring the process data is expected to supply much more information on the state of the process and supply this information more frequently. Furthermore, any special events that occur will also have their fingerprints in the process data. So, once a special event is detected, it is easier to diagnose the source of the problem as we are dealing directly with the process variables. On the contrary, control charts on the product variables only indicate that the product properties are no longer consistent with specification and they do not point to the process variables responsible for this.

Control charts on process variables are useful in multistep operations when quality data are not available between successive steps. For example, if a catalyst is conditioned in a batch process before being used for polymer production, the quality of the catalyst (success of conditioning) is assessed by its performance in the subsequent polymer production. It would be useful to know if the catalyst will produce good product before using it; monitoring the batch process variables with a latent variable chart would give early detection of poor quality product. Similarly, the few properties measured on a product are sometimes not sufficient to define product performance for several different customers. For example, if only viscosity of a polymer is measured, end-use applications that depend on chemical structure (e.g. branching, composition, end-group concentration) are unlikely to receive good material. In these cases, the process data may contain much more information about events with special causes that affect the hidden product quality variables.

The philosophy applied in developing multivariate SPC procedures based on projection methods is the same as that used for the univariate or multivariate Shewhart charts. An appropriate reference set is chosen that defines the normal operating conditions for a particular process. Future values are compared against this set. A PCA or PLS model is built based on data collected from periods of plant operation when performance was good. Periods containing variations due to special events are omitted at this stage. The choice and quality of this reference set is critical to the successful application of the procedure.

44.5 BATCH PROCESS MONITORING

Figure 44.3 shows schematically the nature of batch process trajectories that are nonlinear and dynamic. Modeling batch operations requires taking into account their nonlinear dynamic nature. The methodology for developing multivariate control charts based on latent variables for batch process monitoring was initially presented by Nomikos and MacGregor [15–17] in a series of landmark papers. Any operation of finite duration, such as batch granulation, batch drying, blending/mixing additives for a finite time, fermentation, batch distillation, drying, and so on, can be modeled by the same methodology. This section will present the main issues that need be addressed in batch empirical modeling and will also give references to publications where these issues are discussed in detail.

44.5.1 Modeling of Batch Process Data

Most of the processes in the pharmaceutical industry are batch processes. Collecting real-time data during a batch process generates very large data sets. The top of Figure 44.8 gives the possible measurements that could be collected for a batch process. Information may be collected, at different time intervals for the duration of the batch, for several process variables such as agitation rate (RPM), pH, cooling agent flow (F), temperatures in different locations in the reactor (T_1, T_2, T_3). Data may also be collected in the form of spectra from real-time analyzers such as NIR. Finally, information may be available on raw material analysis, recipe, other preprocessing data, and even information on who was the operator on shift and which vessel was used.

Historical data collected from a batch process had traditionally been represented by a three-dimensional data array $\underline{\mathbf{X}}$ where a matrix $\underline{\mathbf{X}}$ $(I \times J \times K)$ indicates that J process variables are measured at K time intervals, or K aligned observation numbers (A.O.N.), for each one of I batches.

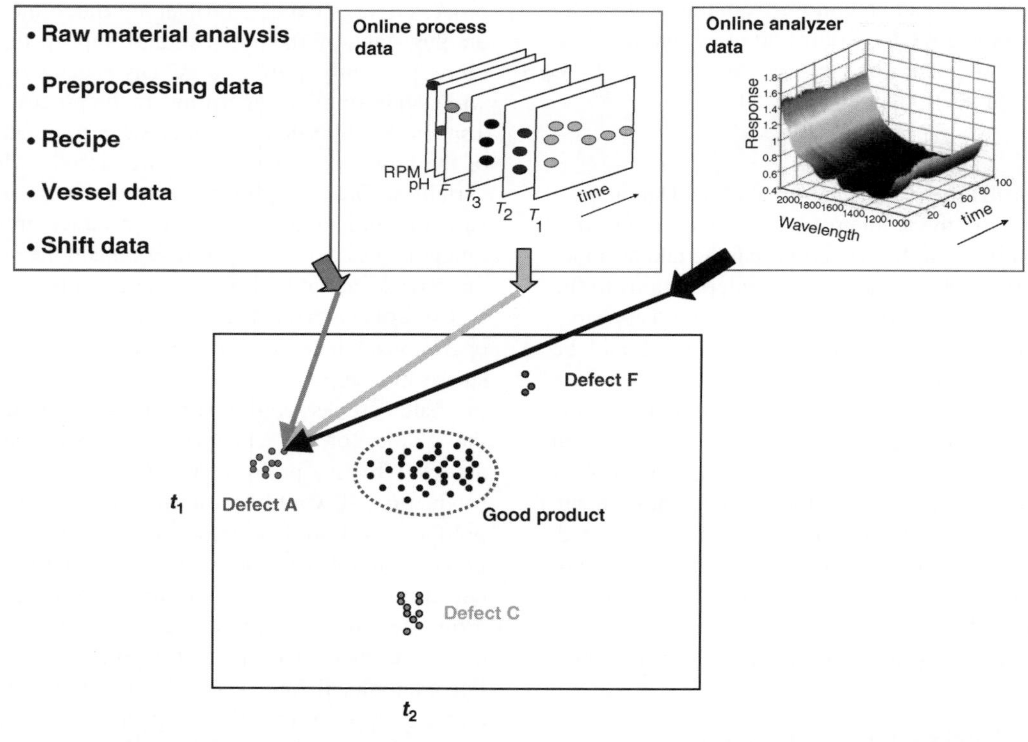

FIGURE 44.8 Data generated during batch runs are projected on a lower dimensional space, defined by two principal components t_1 and t_2. Each point on the plane corresponds to one batch run; that is, each point is the summary of the hundreds of measurements taken during the run.

Kourti [18,19] discussed that, in practice, it is not necessary that the same number of measurements are available for all the variables for the duration of the batch process. Some variables may not be present or measured for the full duration of the batch. Furthermore, the frequency of measurements may be different due to several reasons: (1) some variables may be measured more frequently than others (i.e., some every minute and others every 15 min); (2) certain phases in the process may be sampled more frequently to catch important phenomena (in emulsion polymerization, particle nucleation occurring at the very first few minutes in the reaction determines the number of particles and the particle size distribution; one may need to capture this with more frequent sampling at those stages). Therefore, Kourti argued the data set in such situations does not form a complete cube, but rather a cube where some columns are missing (Figure 44.9). Consequently, the methods used to model batch processes should be capable of modeling the structure of this incomplete cube. There are several methods for modeling three-way data. The choice of the method depends on the use of the model (i.e., prediction of final quality, monitoring, process control) and the types of the data sets available. Critical discussions on modeling procedures in batch processes for robust process monitoring, fault detection, and control can be found in selected publications [15–22].

The method presented by Nomikos and MacGregor [15] is termed in the literature as "batchwise unfolding" and is capable of modeling the incomplete cube structure. Furthermore, it is capable of modeling three-way structures generated when formulating the control problem of batch processes using latent variables that is discussed later. The method unfolds the three-dimensional structure into a two-dimensional array. In this new array, different time slices are arranged next to each other; variables observed at a given time interval are grouped in one time slice; the number of variables in each time slice may vary. Figure 44.9 shows an example of an unfolded matrix where variable x_4 is not measured at time $t = 1$, and variable x_2 is not measured at times $t = 3$–4. Once the three-way structure is unfolded to a two-way matrix **X**, equations 44.1 and 44.2 can be used to model **X** using PCA, or **X** and **Y** using PLS.

Multivariate control charts (Hotelling's T^2 and SPE) can be constructed for batch processes in a straightforward manner. Multivariate charts have superior detection capabilities to univariate charts for batch processes. In the words of a colleague from industry: "In most cases in practice, changes in the covariance structure precede detectable deviations from nominal trajectories. This was the problem that univariate monitoring approaches for batch processes could not address. In most process upsets it is the correlation among the monitored variables that changes first, and later, when the problem becomes more pronounced, the monitored variables deviate significantly from their nominal trajectories. There are cases where a process upset will change dramatically only the correlation among the variables without causing any of the variables involved to deviate significantly from its nominal trajectory. These particular cases, although rare, can result to significant cost to a company since they can go unnoticed for long periods of time (usually they are detected from a customer complaint)." (P. Nomikos, personal communication, 2002).

44.5.2 Alignment of Batch Processes of Different Time Duration

Sometimes batches have different time duration. In other words, using the same recipe, it may take different time to achieve the same conversion. This is due to the fact that time is not the deciding factor for the completion of the batch.

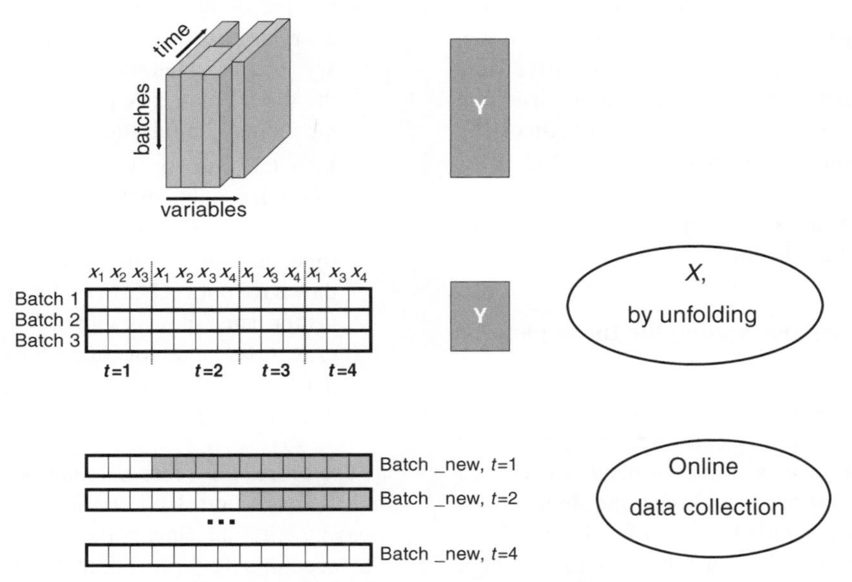

FIGURE 44.9 The structure of data in batch processes.

Sometimes the deciding factor may be the rate at which a certain reactant is added. One important stage before modeling batch process data is the alignment or synchronization of the data, such that they are expressed against the correct aligning factor (which may not be time). With alignment or synchronization, we must achieve the following: (1) Establish common start points at different phases of the run. For example, we could define that the first observation in all the runs for phase one of the reaction will correspond to the start of the monomer feed, for the second phase to the initiator injection, and so on; (2) match the shape of the trajectories of key variables. Once the shapes match, it is not necessary that the length of the batches match [19].

A critical discussion of the various synchronization approaches can be found in Kourti [19]. Attempts for batch data alignment involved the use of the cumulative amount of reactant added to the reactor as an indicator variable, where the variable trajectories were expressed as a function of the indicator variable, rather than time. The extent of the reaction was also used as an indicator variable later [23]. Dynamic time warping, based on speech recognition methods, was also suggested as well as the use of total time as a variable in the **Z** matrix, as extra information to describe the batch [24]. Taylor [25] suggested to include the cumulative warping, up to a given warped observation, as a new variable trajectory; his argument was that this would provide much richer information on the state of the batch by comparing it to the "typical batch" and would provide it in real time, rather than waiting for the batch to finish so that we can calculate the total time; the cumulative time spent could be used as extra trajectory in the case of alignment with an indicator variable. This suggestion was later used with excellent results and also provided the basis for designing batches with desired duration. The cumulative warp can be used as an extra variable to take into account time effects on batch quality when batches were synchronized by dynamic time warping, while García-Muñoz et al. [26] used the cumulative time when batches were synchronized by the indicator variable approach. Provided that an indicator variable exists (or can be constructed by nonlinear transformations from other variables and/or process knowledge), the indicator variable approach is usually chosen as the simplest and most convenient way for industrial applications.

44.5.3 Mean Centering and Scaling the Incomplete Cube

Mean centering the two-way matrix, formed by batchwise unfolding the three-way data, is equivalent to subtracting from each variable trajectory its average trajectory over the I batches, and thus converting a nonlinear problem to one that can be tackled with linear methods such as PCA and PLS.

When the three-way data form a full cube, that is an $\underline{\mathbf{X}}$ $(I \times J \times K)$ matrix, it is common practice to autoscale the two-way matrix formed by unfolding $\underline{\mathbf{X}}$ (i.e., divide each column by its standard deviation). This accomplishes two things: (1) gives an equal weight to all periods and consequently does not give high weights to noisy phases, or underweight a variable in tight control; after all, a variable is in tight control either because it is important to the product quality or because of safety and/or environmental concerns. (2) For the case of complete cube, it gives an equal weight to all the variables considered. However, in the case where variables are sampled less frequently, or are not present for the full run, the weights have to be adjusted accordingly, depending on the objective. To give equal weight to all variables, for example, after autoscaling, each column corresponding to variable j must be divided by $\sqrt{K_j}$, where K_j is the number of times that variable j was sampled in the run. In the example of Figure 44.9, after autoscaling the two-way matrix, all columns corresponding to x_1 and x_3 must be divided by $\sqrt{4}$, to x_2 by $\sqrt{2}$ and to x_4 by $\sqrt{3}$.

44.5.4 Online Monitoring of Batch Processes

Each batch run has a finite duration and the process variables exhibit a dynamic behavior during the run. This means that not only the autocorrelation structure of each variable changes during the run but also the cross-correlation of the variables changes. Models utilizing batchwise unfolding take into account this changing covariance structure of variables and time for the batch duration. For online monitoring, it is the structure of each evolving batch that it is compared against the typical behavior, as modeled by the training set of batches. The procedure for online monitoring of batch processes is slightly more complicated than that for continuous process because of the following reasons:

- For the online monitoring of continuous processes, at every time instant, we have a vector of new observations $\mathbf{x}_{\text{new}}^{\text{T}}$ that has a length equal to the number of columns in the model matrix **X** (and occasionally some measurements may be missing due to sensor failure, etc).
- In batch process monitoring, we have a vector with a length equal to the number of columns in the unfolded **X,** *only* when the batch run has finished. At any other time, data are missing from this vector, simply because they have not yet been collected. In Figure 44.9, the new vector batch_new is shown for different time intervals; gray areas have not been collected yet. Of course, the part with the collected data may also have missing data due to sensor failure, and so on.

Therefore, the score calculations and the limits for the multivariate control charts have to be developed in such a way that they take these "incomplete" measurement vectors into account. The procedure for the development of the multivariate control charts for the duration of the batch was outlined by Nomikos and MacGregor [15]. To deal with the incomplete measurement vector, several approaches have been suggested. García-Muñoz et al. [27] recently investigated these approaches and demonstrated that using the *missing data* option and solving the score estimation problem with an appropriate method is equivalent to the use of an accurate forecast for the future samples over the shrinking horizon of the remainder of the batch. As PCA can model the covariance structure of the process variables, it facilitates handling of missing data. In batch processes, a PCA model describes the variance–covariance structure between variables over the entire batch; in other words, there exists information over all combinations of x_{jk}, j being the variable number and k the time interval (e.g., one can find how variable 4 at time 6 is related to itself at time 15 and also to variable 8 at time 35). Because of the tremendous structural information built into these multivariate PCA models for batch processes, the missing data option for predicting the future trajectory is shown to yield the best performance by all measures, even from the beginning of the batch.

Provided that there are no faults for the prediction of the future process variable trajectories, the final scores, and the product quality, these missing data estimation methods are very powerful. They have also been proven critical to the success of the control methods using latent variables. However, for process monitoring and online detection of process faults, all the alternative "filling in" methods give similar results. When a fault occurs, the model structure is not valid anymore. In that case, the differences among the trajectory estimation methods appear to be much less critical since the control charts used in each case are tailored to the filling in mechanism employed. All the approaches appear to provide powerful charting methods for monitoring the progress of batch processes.

The calculation of monitoring charts and their limits for batch processes, discussed in Ref. 15, where Hotelling's T^2 statistic for the analysis of batch process data (called D statistic) is calculated as

$$D = \mathbf{t}_R^{\mathrm{T}} \mathbf{S}^{-1} \mathbf{t}_R \mathbf{I}/(\mathbf{I}-1)^2 \qquad (44.10)$$

where $\mathbf{t}_R$ is the vector containing the R retained components of the model, and $\mathbf{S}$ represents the covariance matrix of the R retained score vectors. It is mentioned that $\mathbf{S}$ is a diagonal matrix due to the orthogonality of the scores, which is true for the final score estimate (i.e., when the batch run is complete). When computing this statistic for the online monitoring of batches, one should consider that the covariance of the scores changes with time and the scores might become nonorthogonal; therefore, Hotelling's statistic should be computed using the correct and complete variance–covariance matrix that corresponds to each time sample. This time-varying variance–covariance is computed using the reference set of batches. Therefore, the estimate of the Hotelling statistic at time k for batch i (D_{ki}) is a function of the estimate of the score vector for the R retained components at time k ($\hat{\tau}_{Rki}$) for batch i and the covariance matrix of the scores at time k ($\mathbf{S}_k$). Notice that D_{ki} will change depending on the method used to solve the missing data problem (or the option selected to "fill in"), since it is a function of $\hat{\tau}_{Rki}$ that has been shown [27] to differ from method to method and option to option. Using this corrected version of Hotelling's statistic dramatically improves abnormality detection.

44.5.5 Industrial Practice

Industrial applications for batch analysis, monitoring, and fault diagnosis have been reported [26–31]. It should be noted here that several companies choose to use the methodology not for real-time monitoring but as a tool to *real-time release* of the batch product in the following way: the batch is not monitored as it evolves, but rather immediately after the batch finishes, the process data are passed through the model and the scores for the complete batch are investigated. If they are within control limits, the product is released. If there is a problem, the product is sent for analysis in the laboratory. This procedure saves the company time and money. The batch run may last 2–3 h but the product analysis may take many more hours. That means that they do not have to waste batches while they are waiting for the results from the laboratory. By checking the process data as soon as the batch is complete, they can detect problems before starting a new batch. Other applications of multiway methods to batch analysis, optimization, and control have been reported and will be discussed later in the corresponding sections. Multiway methods and design of experiments can be used [32] to determine optimal process variable trajectories in a batch process in order to obtain a desired quality property.

There is a great potential for applications of multivariate batch analysis in the pharmaceutical industry. It is a superb tool for achieving process understanding. Using it to analyze past historical data will provide the user with summaries of the process history such as the one shown in Figure 44.8. The figure shows the projection of several batch runs on a plane, defined by two principal components t_1 and t_2. Each point corresponds to one batch run; that is, each point is the summary of the hundreds of measurements taken during the run. For example, for a batch run that lasts 16 h and where 5 min averages are collected on six process variable trajectories, each point is the summary of 1152 process measurements plus all the spectra scans at all time intervals plus other information on recipe, and so on. In the figure, one can immediately detect not only the cluster of good operation, but also clusters corresponding to specific product problems. Notice that the problems in product quality are observable by the projection of the process data and the measurements taken by real-time analyzers; that is, the problem in the quality is observable without laboratory information. Utilizing contribution plots one can interrogate the multivariate model and determine combinations of variables and periods of operation that will drive a process away from producing a good product to producing defect A. Such an example is discussed in the literature [20] where information from the process and the raw materials was incorporated in the analysis. Based on this analysis, it was possible to determine raw material combinations and processing conditions that will result in a bad product.

An industrial example whereby monitoring the batch process variables a pending equipment failure was detected is reported in Ref. 31. A critical discussion on other batch processes modeling and monitoring procedures and other issues related to batch process analysis can be found in two recent studies [18, 19]. A discussion of the various approaches to synchronize runs of different durations can be found in Refs [13, 18, 19].

44.6 MULTISTAGE OPERATIONS: MULTIBLOCK ANALYSIS

There are multiple steps in pharmaceutical manufacturing and each step may involve multiple unit operations. Having a control chart for each unit rather than one for the whole process could be helpful to operators. However, building a model for each unit operation separately, does not consider interactions between unit operations. Such cases can also be addressed with latent variable models. Rather than building a model for each unit, one can build a model for the full process that will take into account the interactions between units and their relative importance to the final product quality by weighting them differently. Then, from this model, individual charts per unit operation can be derived. This way, interactions between unit operations are preserved. This is the approach of multiblock PLS (MB-PLS).

In the MB-PLS approach, large sets of process variables (**X**) are broken into meaningful blocks, with each block usually corresponding to a process unit or a section of a unit. MB-PLS is not simply a PLS between each block **X** and **Y**. The blocks are weighted in such a way that their combination is most predictive of **Y.** Several algorithms have been reported for multiblock modeling and for a good review, it is suggested that the reader consult Refs [33–35].

Multivariate monitoring charts for important subsections of the process, as well as for the entire process, can then be constructed, and contribution plots are used for fault diagnosis as before. In a multiblock analysis of a batch process, for example, one could have the combination of three blocks (**Z**, $\underline{\mathbf{X}}$, and **Y**); block **Z** could include information available on recipes, preprocessing times, hold times, as well as information of the shifts (which operator was in charge) or the vessels used (i.e. which reactor was utilized); $\underline{\mathbf{X}}$ would include process variable trajectories; and **Y** would be quality. Analysis of this type of data could even point to different ways the operators operate the units and relate product quality to operator, or different process behavior of vessels and identify faulty vessels, and so on. The reader is referred to the work of García-Muñoz et al. [20] for detailed examples where the multiblock analysis is utilized in batch processes for troubleshooting.

Several alternative ways to perform multiblock appear in commercial software. One approach that is being frequently used to deal with a data structure of several blocks involves two stages: PCA is performed for each one of the **Z** and **X** blocks and then the scores and/or residuals derived from these initial models are related to **Y** with a PLS. In an alternative version, PLS is performed between **Z** and **Y** and **X** and **Y,** and the resulting scores are related to **Y**. The users should exercise caution because these approaches may fail to take into account combinations of variables from different blocks that are most predictive of **Y**. For example, in situations where process parameters in **X** are modified to account for variability of raw material properties in **Z** (i.e., when **X** settings are calculated as a feedforward control to deviations of **Z**), a PLS between **Z** and **Y** will show that **Z** is not predictive of **Y** variability; similarly, a PLS between **X** and **Y** will show that **X** is not predictive of **Y**; a MB-PLS of **[Z, X]** and **Y** will identify the correct model. Finally, MB-PLS handles missing data in a very effective way.

As might be expected in multistage continuous processes, there can be significant time delays between the moment an event occurs in one unit (and therefore affects the variables of that unit) and the moment its effect will become obvious on a product variable at the end of the process. These delays significantly affect the interaction and correlation structures of the process variables and need to be handled by lagged variables created from the original process variables. Data can be time shifted to accommodate time delays between process units.

In some multistage operations, the path of the product through the various process units can be easily traced, and eventually one can relate a specific lot number to several process stages (via a multiblock PLS). In such cases, the process conditions of these units can be used to predict the quality of the product. There are situations, however, where a product (or the composition of the effluent stream of a process) is a result of a multistage operation but its path cannot be traced clearly due to mixing of streams from several parallel units in one vessel and then splitting to a number of other vessels. A discussion on monitoring difficult multistage operations can be found in Ref. 19. In those cases, the best alternative to achieve consistent operation is to monitor each unit, separately, by a PCA model. By assuring a consistent operation per unit, one hopes for a consistent product. Once an unusual event is detected in one unit, one may decide not to mix the product further, or investigate lab quality before proceeding to the next stage.

44.7 PROCESS CONTROL TO ACHIEVE DESIRED PRODUCT QUALITY

The term "control" currently appears in the pharmaceutical literature to describe a variety of concepts, such as end point determination, feedback control, statistical process control, or simply monitoring. Process control refers to a system of measurements and actions within a process intended to ensure that the output of the process conforms with pertinent specifications.

In this chapter, we use some terms related to process control with the following definition:

- *Feedback Control*: to indicate that we are reactive; that is, the corrective action is taken on the process based on information from the process output (e.g., measurements on product quality, at given time.)
- *Feedforward Control*: to indicate that we are proactive; that is, the process conditions are adjusted based on measured deviations of the input to the process (e.g., information on raw material)

44.7.1 Feedforward Estimation of Process Conditions

The concept of adjusting the process conditions of a unit based on measured disturbances (feedforward control) is a concept well known to the process systems engineering community for several decades. The methodology is also used in multistep (multiunit) processes where the process conditions of a unit are adjusted based on information of the intermediate quality achieved by the previous unit (or based on raw material information).

An example of a feedforward control scheme in the pharmaceutical industry, where multivariate analysis was involved, is described by Westerhuis et al. [36] The authors related crushing strength, disintegration time, and ejection force of the tablets with process variables from both the wet granulation and tableting steps and the composition variables of the powder mixture. They also included physical properties of the intermediate granules. The granule properties may differ from batch to batch due to uncontrolled sources such as humidity temperature, and so on. This model is then used for each new granulation batch. A feedforward control scheme was devised that can adjust the variables of the tableting step of the process based on the intermediate properties to achieve desirable final properties of the tablets.

To the author's knowledge, there are several unpublished examples in the chemical and other industries where information on the raw data $\mathbf{Z}$ is used to determine the process conditions $\mathbf{X}$ or $\underline{\mathbf{X}}$ in order to achieve the desired quality $\mathbf{Y}$, utilizing projection methods. Sometimes such information from $\mathbf{Z}$ may simply be used to determine the length of the run, while in other cases, it may be a multivariate sophisticated scheme that determines a multivariate combination of trajectories for the manipulated variables. To achieve this, historical databases can be used to develop multiblock models $\mathbf{Z}$, $\mathbf{X}$ (or $\underline{\mathbf{X}}$), and $\mathbf{Y}$.

44.7.2 End Point Determination

There have been reports in the literature where real-time analyzers are used for "end point detection" or "end point control." In most of these situations, a desired target concentration is sought, for example, % moisture in drying operations.

An example is described by Findlay et al. [37], where NIR spectroscopy is used to determine granulation end point. The moisture content and the particle size determined by the near-infrared monitor correlate well with off-line moisture content and particle size measurements. Given a known formulation, with predefined parameters for peak moisture content, final moisture content, and final granule size, the near-infrared monitoring system can be used to control a fluidized bed granulation by determining when binder addition should be stopped and when drying of the granules is complete.

44.7.3 Multivariate Manipulation of Process Variables

It was discussed in Section 44.2.2 that regulating only the final value of a property (or even several properties) is not sufficient. In other words, end point control may not be sufficient. The process signatures are equally important. These process signatures should be regulated in a correct, multivariate way, not simply on a univariate basis. It is possible that two batch runs produce products with different quality, even if the trajectory (path to end point) of one quality variable follows the same desired path in both the runs. This will happen if the covariance structure of the trajectory of this variable with the trajectories of the rest of the process variables (temperatures, agitation rate, reactant addition) is different for these two batches. This concept is very important both in control and in scale-up. Latent variable methodology allows taking into consideration the process variable trajectories in a multivariate way.

Control of batch product quality requires the online adjustment of several manipulated variable trajectories. Traditional approaches based on detailed theoretical models are based on either nonlinear differential geometric control or online optimization. Many of the schemes suggested in the 'literature require substantial model knowledge or are computationally intensive and therefore difficult to implement in practice. Empirical modeling offers the advantage of easy model building.

Lately, latent variable methods have found their way to control batch product quality and have been applied in industrial problems. Latent variable methodology allows taking into consideration the process signatures in a multivariate way for end point detection problems. Marjanovic et al. [38] describe a preliminary investigation into the development of a real-time monitoring system for a batch process. The process shares many similarities with other batch processes in that cycle times can vary considerably, instrumentation is limited, and inefficient laboratory assays are required to determine the end point of each batch. The aim of the work conducted in this study was to develop a data-based system able to accurately identify the end point of the batch. This information can then be used to reduce the overall cycle time of the process. Novel approaches based upon multivariate statistical techniques are shown to provide a soft sensor able to estimate the product quality throughout the batch and a prediction model able to provide a long-term estimate of the likely cycle time. This system has been implemented online and initial results indicate that it offers the potential to reduce operating costs.

In another application [39], latent variable methodology was used for soft sensor development that could be used to provide fault detection and isolation capabilities and can be integrated within a standard model predictive control framework to regulate the growth of biomass within a fermenter. This model predictive controller is shown to provide its own monitoring capabilities that can be used to identify faults within the process and also within the controller itself. Finally, it is demonstrated that the performance of the controller can be maintained in the presence of fault conditions within the process.

Work has also been reported for complicated control problems where adjustments are required for the full manipulated variable trajectories [40]. Control through complete trajectory manipulation using empirical models is possible by controlling the process in the reduce space (scores) of a latent variable model rather than in the real space of the manipulated variables. Model inversion and trajectory reconstruction are achieved by exploiting the correlation structure in the manipulated variable trajectories. Novel multivariate empirical model predictive control strategy (LV-MPC) for trajectory tracking and disturbance rejection for batch processes, based on dynamic PCA models of the batch processes, has been presented. The method presented by Nomikos and MacGregor [15] is capable of modeling three-way structures generated when formulating the control problem of batch processes using latent variables.

44.7.4 Setting Raw Material Multivariate Specifications as a Means to Control Quality

Dushesne and MacGregor [41] presented a methodology for establishing multivariate specification regions on raw/incoming materials or components. The thought process here is that if the process remains fixed, we should control the incoming material variability. PLS is used to extract information from databases and to relate the properties of the raw materials supplied to the plant and the process variables at the plant to the quality measures of the product exiting the plant. The specification regions are multivariate in nature and are defined in the latent variable space of the PLS model. The authors emphasize that although it is usually assumed that the raw material quality can be assessed univariately, that is, by setting specification limits on each variable separately, this is valid only when the raw material properties of interest are independent of one another. However, most of the times the properties of products are highly correlated. In other words, treating the raw material properties in a univariate way, for two properties, it would mean that (referring to Figure 44.1) while we can process only material that falls in the ellipse, we agree to buy material from the supplier with the specifications set in the square; that is, we agree to use material that we know in advance it will not perform well.

To develop models to address the problem, multiblock PLS is used for **Z**, **X,** and **Y**; **Z** contains measurements on N lots of raw material data from the past; **X** contains the steady-state processing conditions used to process each one of the N lots; **Y** contains final product quality for these N lots. The methodology could be easily extended to batch process $\underline{\mathbf{X}}$.

It should become one of the priorities in industries to express the raw material orders as a multivariate request to the supplier.

44.8 OTHER APPLICATIONS OF LATENT VARIABLE METHODS

44.8.1 Exploiting Databases for Causal Information

Recently, there has been a lot of interest in exploiting historical databases to derive empirical models (using tools such as Neural Networks regression or PLS) and use them for process optimization. The idea is to use already available data rather than collecting new through a design of experiments. The problem is that for process optimization causal information must be extracted from the data, so that a change in the operating variables can be made that will lead to a better quality product, or higher productivity and profit. However, databases obtained from routine operation contain mostly noncausal information. Inconsistent data, range of variables limited by control, noncausal relations, spurious relations due to feedback control, and dynamic relations are some of the problems the user will face using such happenstance data. These are discussed in detail in the section "Hazards of fitting regression equations to happenstance data" in Ref. 42 where the advantage of experimental designs as a means of obtaining causal information is emphasized. In fact, in a humorous way, the authors warn the young scientists that they need a strong character to resist the suggestion of their boss to use data from past plant operation every time they suggest performing designed experiments to collect data.

In spite of this, several authors have proposed approaches to optimization and control based on interpolating historical bases. However, in all these cases, their success was based on making strong assumptions that allowed the database to be reorganized and causal information to be extracted. One approach was referred to as "similarity optimization" that combined multivariate statistical methods for reconstructing unmeasured disturbances with nearest neighbor methods for finding similar conditions with better performance. However, it too was shown to fail for many of the same reasons. In general, it was concluded that one can optimize only the process if there exist manipulated variables that change independently of the disturbances and if disturbances are piecewise constant, a situation that would be rare in historical process operations.

The reader should therefore exercise caution about how historical databases are used when it comes to retrieving causal information. However, databases obtained from routine operation are great a source of data for building monitoring schemes.

44.8.2 Product Design

Given the reservations about the use of historical databases, one area where some success has been achieved is in identifying a range of process operating conditions for a new grade of product with a desired set of quality properties and in matching two different production plants to produce the same grade of product. If fundamental models of the process exist, then these problems are easily handled as constrained optimization problems. If not, optimization procedures based on response surface methodology can be used. However, even before one performs experiments, there exists information within the historical database on past operating conditions for a range of existing product grades.

In this case, the historical data used are selected from different grades and therefore contain information on variables for several levels of past operation (i.e., there is intentional variation in them, and they are not happenstance data). The key element in this empirical model approach is the use of latent variable models that both reduce the space of **X** and **Y** to a lower dimensional orthogonal set of latent variables and provide a model for both **X** and **Y**. This is essential in providing solutions that are consistent with past operating policies. In this sense, principal component regression and PLS are acceptable approaches, while MLR, neural networks, and reduced rank regression are not.

The major limitation of this approach is that one is restricted to finding solutions within the space and bounds of the process space **X** defined by previously produced grades. There may indeed be equivalent or better conditions in other regions where the process has never been operated before, and hence where no data exist. Fundamental models or more experimentation would be needed if one hopes to find such novel conditions.

A very good discussion on these issues can be found in García-Muñoz et al. [26]. The authors illustrate a methodology with an industrial batch emulsion polymerization process where the batch trajectories are designed to satisfy certain customer requirements in the final properties of the polymer while using the minimal amount of time for the batch run. The cumulative time, or used time, is added as an extra variable trajectory after the alignment of the batches.

44.8.3 Site Transfer and Scale-Up

Product transfer to different sites and scale-up falls into the same class of problems: one needs to estimate the process operating conditions of plant B to produce the same product that is currently produced in plant A.

Attempts have been made to solve such problems with latent variable methods, utilizing historical data from both locations for transferring other products.

The main points to keep in mind when addressing such a problem are as follows:

- The quality properties of the product should always be checked within a multivariate context because univariate charts may be deceiving. The multivariate quality space for both the sites should be the same. Correct product transfer cannot be achieved by comparing end point quality on univariate charts from the two sites (or from pilot scale and manufacturing). The product quality has to be mapped from site to site in a multivariate way (the products in both sites have to project on the same multivariate space).
- The end point quality may not be sufficient to characterize a product. The path to end product is important. Whenever full mechanistic models exist, these models describe the phenomena that are important for the process and therefore determine this path. When changing sites, the full mechanistic model will describe the desired path in the new site taking into account size, mass, and energy balances and/or other phenomena related to the process. When mechanistic models do not exist, this mapping of the "desired process paths" or "process signatures" has to happen with empirical data.

A methodology has been developed for product transfer and scale-up based on latent variables [43]. The methodology utilizes databases with information on previous products and their corresponding process conditions from both sites. The two sites may differ in equipment, number of process variables, locations of sensors, and history of products produced.

44.9 QUALITY BY DESIGN

44.9.1 Design Space: Expressing Quality as a Function of Input Material Attributes and Process Parameters

Several regulatory agencies participating in the International Conference of Harmonization have adopted the concept of design space. The definition of design space as "the multidimensional combination and interaction of input variables (e.g., material attributes) and process parameters that have been demonstrated to provide assurance of quality" [44] stems from a well known fact that if the variability in the raw material is not compensated by the process, it will be transferred to quality.

The effect of the raw material attributes on the process performance, if the process operating conditions remain fixed, is clear in the example depicted in Figure 44.5. Recall that in that example the raw material is characterized by more than 25 physical and chemical properties and that these variables are projected on a space defined by latent variables that allow us to visualize better the process behavior. Raw material is produced at three supplier locations. Although the raw material properties are within univariate specifications, at all locations, the projection in three clusters indicates that in a multivariate sense the material possesses slightly different characteristics depending on the location it was produced (covariance structure changes with location). Some batches of that raw material was used for a specific product. The material properties after micronization are projected on principal components. It can be observed that after micronization, the batches from a specific supplier location (circled by the small ellipse) project on a different location from the rest. The filling performance of the material originating from this location is also different from the rest of the material. *The raw material variability propagates to quality if the process remains fixed.* A note here that although the control ellipses (large ellipses) shown are set by default in the vendor software, they are not interpretable when there is clustering; the assumptions for the calculation of these ellipses are for process monitoring and not for process exploration where there is intentional variation such as that introduced by design of experiments.

The concept of the design space can be easily understood with the example below, depicted in Figures 44.10 and 44.11. In the figures, we have a process where the raw material is

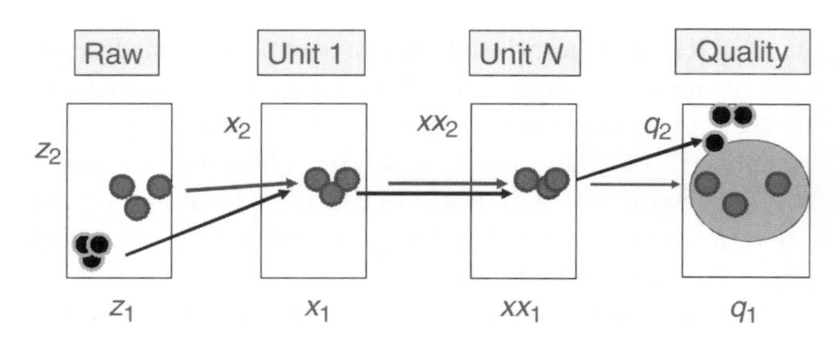

FIGURE 44.10 By maintaining fixed process conditions, we propagate raw material variability to quality.

described by two attributes z_1 and z_2, quality is described by q_1 and q_2, and unit operations described by process parameters x_1 and x_2 for unit 1 and xx_1 and xx_2 for unit N. Two attributes, two quality, and two process parameters per unit are used for illustration purposes, but this does not affect generalization of the following discussion. Each circle represents the values of these parameters for one batch. Figure 44.10 shows what happens when a fixed process is considered, depicted by the Grey circles. Suppose that we run the traditional three batches at a selected range of z_1–z_2 and selected range of process parameters, and we achieve the target quality (all grey circles fall on a multivariate target). The Dark circles represent raw material from, say, a different manufacturer, with attribute values different from the range initially examined. If we process the Dark material on the fixed process conditions (e.g., in the range of the grey circle values), chances are that the final quality will differ from that produced by the grey raw material. Figure 44.11 illustrates that if we carefully choose to operate at appropriate different process conditions for each different material then we can have quality on target. In other words, there is a multidimensional combination of raw material and process parameters that assures quality.

These appropriate process conditions (depicted by the paths that relate raw material and process parameters with quality) are the solutions to the equations of the model that relates raw material and process conditions to quality, and these solutions are obtained when we solve for the values of process conditions given the values of the raw material properties, such that quality falls in a desirable range.

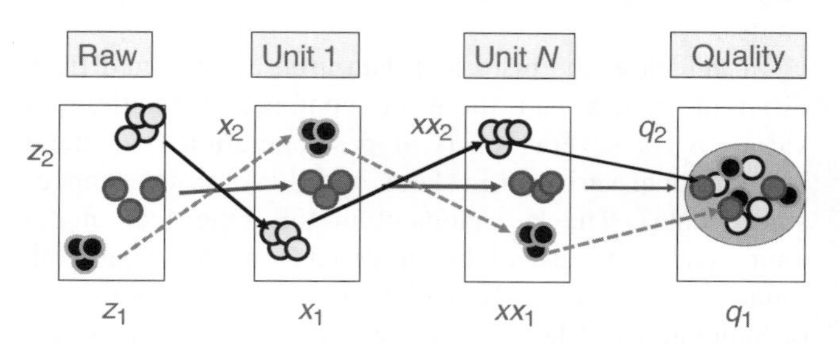

FIGURE 44.11 By taking a feedforward approach where the process conditions are flexible to account for raw material variability, we can maintain quality on target.

Sometimes optimization can be used to introduce constraints, such that the solution takes into account cost, duration of a process, and so on. The model may be theoretical, empirical, or hybrid.

44.9.2 Design Space Modeling

The design space can be established as a model that relates input material and process parameters to quality. The model may be theoretical (based on first principles), or empirical, derived from design of experiments, or a hybrid. Together with the model, one has to specify the range of parameters for which the model has been verified. The model may cover one unit operation or a series of unit operations.

The model will express quality as a function of the raw material attributes and process parameters

$$\text{Quality} = f(\text{raw material, process parameters})$$

or, more specifically, as

$$[q_1, q_2, \ldots, q_N] = f(z_1, z_2, \ldots, z_K, x_1, x_2, \ldots, x_M, \ldots, xx_1, xx_2, \ldots, xx_P)$$

and then solve for the combination of process parameters that will result in a desired $q1$, $q2$, q_N given the values of z_1, z_2, z_N. The function may be linear or nonlinear, and more than one models will in general be required to describe the behavior of a multiunit plant if we wish to be able to predict intermediate quality as well (i.e., granule properties). Multivariate projection methods can be used for empirical modeling. That is, the design space is a collection of models that relate (1) the final quality to all previous units, raw material, and intermediate quality; (2) intermediate quality to previous unit operations and raw material. The design space consists of models (relationships/paths) plus the range of parameters for which the models have been verified. Random combinations in the range of parameters will not work in general (i.e., in Figure 44.11, dark raw material operated in grey conditions may not result in acceptable quality). Therefore, the range without the paths or multidimensional combinations (the model) cannot describe the design space, unless it is the range selected with traditional approaches that describe a fixed process. This range (in this case, grey circle path) may however be only one of the acceptable solutions and therefore restricts our flexibility in dealing with a wide variety of raw materials and/or dealing with disturbances that affect the process.

The above function uses more than two attributes and process parameters in the multidimensional relationships to reflect a general case. Multivariate projection methods or latent variables are proven very useful to describe and solve for these relationships where many variables are present.

The effect of raw material on the quality as it propagates through different unit operations is shown for a tableting process in Figure 44.12. When the raw material properties have certain characteristics (marked with a small ellipse), the material projects on a different area. The properties of granules produced from raw material with such characteristics (black) are different from the rest, and the final quality also shows differences. The difference in the quality can be theoretically explained based on the physical phenomena that govern the whole process. The idea of the design space is to express these phenomena by a model.

Recognizing the continuum in drug production that spans from the drug substance to the drug product will help create a more versatile and robust design space. The final product that delivers the active pharmaceutical ingredient to a patient is indeed the result of a multidimensional combination of raw material attributes and process parameters that span several unit operations including those of the drug substance production (such as reaction and crystallization), the ones from drug product production (such as granulation and compression), and also packaging. Each one of them has an impact not

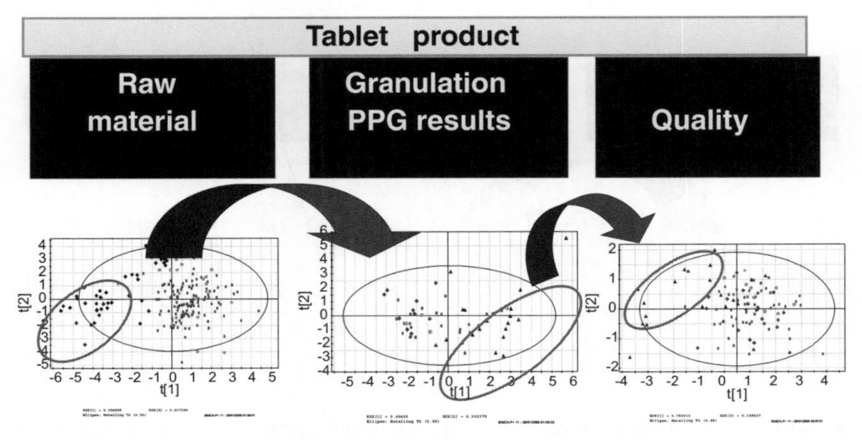

FIGURE 44.12 Projection space representation for a tablet product. Batches produced from raw material with similar characteristics have similar final quality.

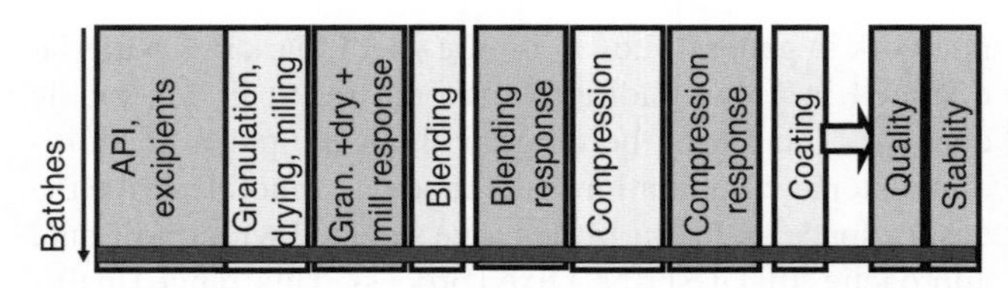

FIGURE 44.13 Both quality and stability profiles can be modeled as a function of input material, process variables, and intermediate attributes.

only on one or more final quality characteristics, but also on stability.

Incorporating stability into the quality by design framework has been discussed; it involves including stability time profiles in the model for the design space [45].

Therefore, if the design space is addressed in a holistic form, then the quality of the final product, including its stability, should be expressed as a function of raw material characteristics and process parameters. Also, by treating the design space in a holistic way, it would provide the manufacturer with the most cost-efficient operation and guarantee high yield and low operating costs because problems at later unit operations will be anticipated and corrected in earlier operations. In other words, the control strategy will be part of the design, such that it can be implemented in the most cost-effective way.

A model that describes the design space for the entire tableting process can be derived by relating quality to both the raw material properties and the process parameters of the unit operations (Figure 44.13). One row in the database depicted in Figure 44.13 would include the process conditions and quality experienced by the product as it is processed through the units. The empirical models derived are causal and based on carefully designed experiments (DOE). Some DOEs will also be necessary to estimate parameters even if mechanistic models are used.

The level of detail in the models varies depending on the objective of the model and the depth of process understanding one wishes to achieve. For example, the variable trajectories of a granulation may be described by carefully selected summary data or by the full variable trajectories aligned against time or another indicator variable.

44.9.3 Control Strategy

Based on the process understanding gained from the design space modeling, the control strategy can be derived to assure final quality. There are several ways of controlling a process, as discussed in Section 44.7. If we decide to keep the process fixed, we may apply a control strategy for the incoming material to reduce raw material variability (see Section 44.7.4).

If we wish to apply the principles depicted in Figure 44.11, then feedforward control should be applied (Section 44.7.1).

Figure 44.14 depicts action in feedforward control that would apply in the case of Figure 44.12. When a different raw material enters the process, we have the choice to adjust granulation conditions in a feedforward manner. However, we also have the choice to adjust compression, that is, to apply the feedforward action at a later stage. When a deviation in the granules is detected that may result in quality different from that typically observed if the compression operates at certain conditions, we may bring the quality on target by altering the compression settings. The choice of the process conditions at unit operation at which we will perform the action will be dictated by a model that takes into account the value of the properties of the input to that unit and calculates process conditions such that quality is on target. When the model is empirical, multivariate analysis can be used.

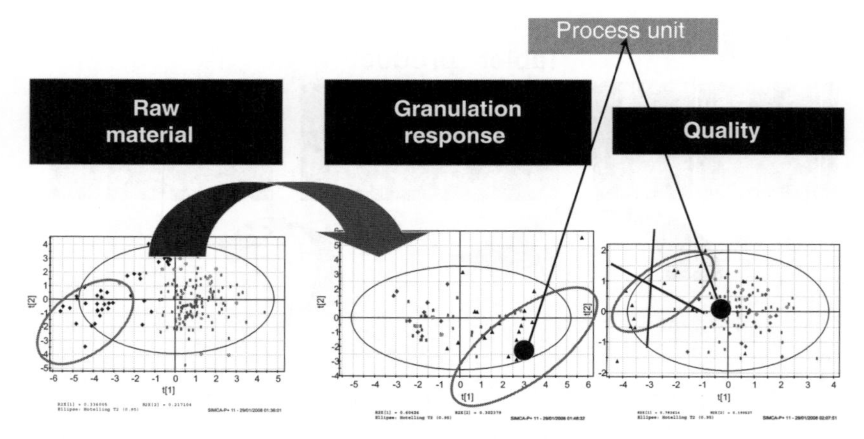

FIGURE 44.14 Feedforward control. When a deviation in the granules is detected that may result in quality different from that typically observed if the next process operates at given conditions, we may bring the quality on target by altering the process conditions.

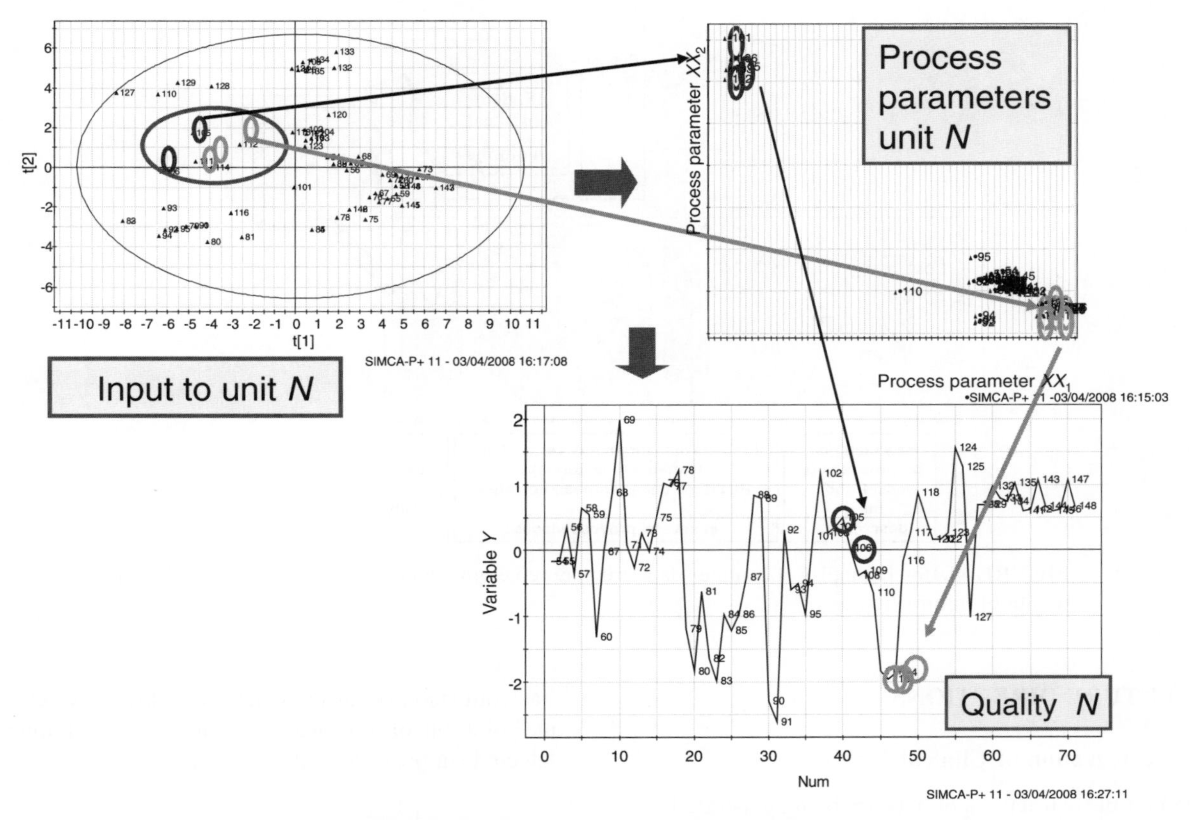

FIGURE 44.15 Control strategy using projection space.

Figure 44.15 illustrates such a case. The example here illustrates a feedforward control scheme for unit *N* based on input information on the "state of the intermediate product" from unit *N*−1. The settings are calculated and adjusted such that the target value for quality *Y* is met. A multivariate model was built (from batch data) to relate product quality to the process parameters of unit *N* and the state of the intermediate product from unit *N*−1, (i.e input to unit *N*). From this model, a quantitative understanding was developed showing how process parameters in *N* and the state of the intermediate product from *N*−1 interact to affect quality. Using multivariate analysis assures that the multivariate nature of quality is respected. In this case, the input to unit *N* is such that, the five batches that project in an area within the small circle (two dark batches and three grey) have the same state of intermediate product—meaning that up to that time the five batches experienced same raw material and processing conditions. The grey batches when processed with typical operating conditions in unit *N*, marked grey, resulted in quality below average. By taking a feedforward action and processing the dark batches with different operating conditions, in unit *N*, the quality improves with values above average.

For real-time monitoring and control of an individual unit operation, for example, batch granulation, the principles described in batch process monitoring (Section 44.5.4) and process control by manipulating multivariate trajectories (Section 44.7.3) apply.

44.9.4 Design Space Management

It is accepted that the design space will evolve after the initial submission, and therefore design space management is very important in the product life cycle. There are several issues to consider with design space management, beyond the obvious ones (i.e., beyond managing the design space at the current site to address issues not considered because of limited data available by the initial submission). These include situations where a larger scale is considered at the same site, as well as when there is site transfer. Other issues would be situations where there are different suppliers of API or excipients and when the raw material characteristics are altered slightly within the same supplier. Production changes, such as opportunities to use soft sensors instead of real-time analyzers or to expand the current process analytical technology capabilities, should also be considered. Solutions to such problems can be addressed under the framework of design space management.

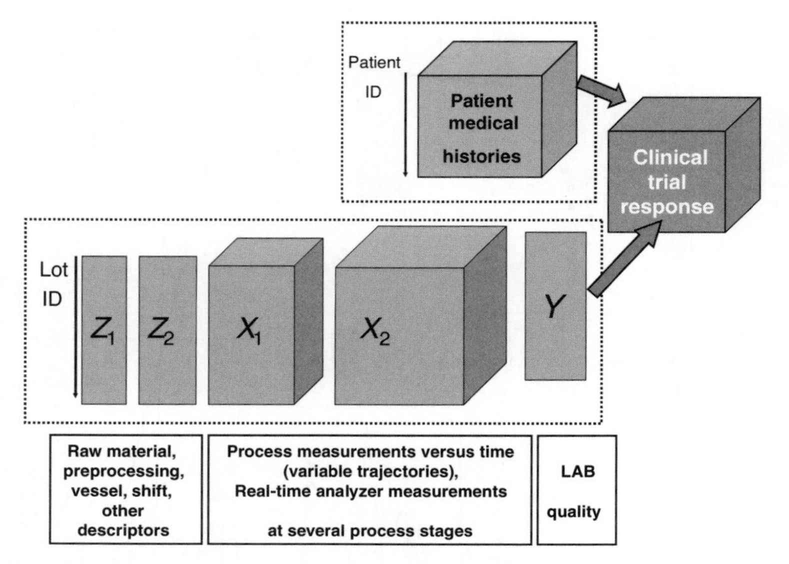

FIGURE 44.16 Examples of complex data structures emerging in industry that can be mined for a wealth of information.

44.10 FUTURE DIRECTIONS

44.10.1 Integration of Clinical Trials

As more complex structures of data are being generated, the multivariate analysis offers great opportunities for information integration and analysis. Both manufacturing data and patient histories can be integrated and then the clinical trial responses incorporated into design space.

Figure 44.16 shows an example of the possibilities that can be explored. Quality in product **Y** can be related to past information of raw materials, preprocessing and holding times, the type of the vessel used, the operator that run the process, and other recipe information, as well as process measurement trajectories and analyzer information. The quality **Y** (and details of manufacturing), as well as the patient medical histories and clinical responses, can be used to establish a better understanding of the design space.

44.10.2 Quality by Design in Analytical Methods

The methodology described for design space can be applied in analytical methods. Chromatography is not only a laboratory method but also a unit operation in biopharmaceuticals. Process transfer ideas can be also applied in method transfer ideas; in other words, method transfer and site transfer could be treated with similar principles.

MSPC for analytical methods has been reported. Multivariate monitoring of a chromatographic system has been carried out using a check sample containing five analytes to test column performance (Nijhuis et al. [46]). A T^2 chart and an SPE_X chart were used to monitor analyte peak area percent of the five analytes. The results indicated that false alarms that would have occurred with univariate charts were avoided and points out of control due to change in correlation could be detected (impossible with univariate charts).

ACKNOWLEDGMENT

The author would like to acknowledge Gordon Muirhead and Bernadette Doyle, GlaxoSmithKline, for their continuous mentoring and support.

REFERENCES

1. Kourti T. Symposium report on abnormal situation detection and projection methods. *Chemom. Intell. Lab. Syst.* 2005;76: 215–220.
2. Abboud L, Hensley S.Factory Shift: New Prescription For Drug Makers: Update the Plants; After Years of Neglect, Industry Focuses on Manufacturing; FDA Acts as a Catalyst; The Three-Story Blender, Wall Street J., Eastern edition, New York, NY, September 3, 2003, page A.1.
3. Yu H, MacGregor JF, Haarsma G, Bourg W. Digital imaging for on-line monitoring and control of industrial snack food processes. *Ind. Eng. Chem. Res.* 2003;42: 3036–3044.
4. Yu H, MacGregor JF. Multivariate image analysis and regression for prediction of coating content and distribution in the production of snack foods. *Chemom. Intell. Lab. Syst.* 2003;67:125–144.
5. Kourti T. Process analytical technology beyond real-time analyzers: the role of multivariate analysis critical reviews in analytical chemistry. *Crit. Rev. Anal. Chem.* 2006;36: 257–278.

6. Kourti T. Multivariate statistical process control and process control, using latent variables. In: Brown S, Tauler R, Walczak R, editors, *Comprehensive Chemometrics*, Vol. 4, Elsevier, Oxford, 2009, pp. 21–54.
7. Brown S, Tauler R, Walczak R, editors. *Comprehensive Chemometrics*, Vol. 1–4, Elsevier, Oxford, 2009.
8. Miletic I, Boudreau F, Dudzic M, Kotuza G, Ronholm L, Vaculik V, Zhang Y. Experiences in applying data-driven modelling technology to steelmaking processes. *Can. J. Chem. Eng.* 2008;86:937–946.
9. Kourti T, MacGregor JF. Process analysis, monitoring and diagnosis using multivariate projection methods—a tutorial. *Chemom. Intell. Lab. Syst.* 1995;28:3–21.
10. Muirhead GT. Process analytical technology at GlaxoSmithKline. Presented at the *19th IFPAC, International Forum Process Analytical Technology*, Arlington, VA, January 10–13, 2005.
11. Graffner C. PAT—European regulatory perspective. *J. Process Anal. Technol.* 2005;2(6):8–11.
12. Burnham AJ, Viveros R, MacGregor JF. Frameworks for latent variable multivariate regression. *J. Chemometrics* 1996;10: 31–45.
13. Kourti T. Application of latent variable methods to process control and multivariate statistical process control in industry. *Int. J. Adapt. Control Signal Process.* 2005;19: 213–246.
14. Kourti T. Process analysis and abnormal situation detection: from theory to practice. *IEEE Control Syst.* 2002;22(5):10–25.
15. Nomikos P, MacGregor JF. Multivariate SPC charts for monitoring batch processes. *Technometrics* 1995;37(1):41–59.
16. Nomikos P, MacGregor JF. Monitoring of batch processes using multi-way principal component analysis. *AIChE J.* 1994;40 (8):1361–1375.
17. Nomikos P, MacGregor JF. Multiway partial least squares in monitoring batch processes. *Chemom. Intell. Lab. Syst.* 1995;30:97–108.
18. Kourti T. Multivariate dynamic data modelling for analysis and statistical process control of batch processes, start-ups and grade transitions. *J. Chemom.* 2003;17:93–109.
19. Kourti T. Abnormal situation detection, three way data and projection methods—robust data archiving and modeling for industrial applications. *Annu. Rev. Control* 2003;27(2):131–138.
20. García-Muñoz S, Kourti T, MacGregor JF, Mateos AG, Murphy G. Troubleshooting of an Industrial Batch Process Using Multivariate Methods. *Ind. Eng. Chem. Res.* 2003;42:3592–3601.
21. Camacho J, Picó J, Ferrer A. Bilinear modelling of batch processes. Part I: theoretical discussion. *J. Chemom.* 2008;22(5):299–308.
22. Camacho J, Picó J, Ferrer A. Bilinear modelling of batch processes. Part II: a comparison of PLS soft-sensors. *J. Chemom.*, 2008;22(10):533–547.
23. Neogi D, Schlags CE. Multivariate statistical analysis of an emulsion batch process. *Ind. Eng. Chem. Res.* 1998;37:3971–3979.
24. Kassidas A, MacGregor JF, and Taylor PA, Synchronization of batch trajectories using dynamic time warping. *AIChE Journal*, 1998;44:864–875.
25. Taylor PA.Computing and Software Department, McMaster University, Hamilton, Ontario, Canada, personal communication, May 1998.
26. García-Muñoz S, MacGregor JF, Neogi D, Latshaw BE, Mehta S. Optimization of batch operating policies. Part II. Incorporating process constraints and industrial applications. *Ind. Eng. Chem. Res.* 2008;47(12):4202–4208.
27. García-Muñoz S, MacGregor JF, Kourti T. Model predictive monitoring for batch processes with multivariate methods. *Ind. Eng. Chem. Res.* 2004;43:5929–5941.
28. Kosanovich KA, Piovoso MJ, Dahl KS. Multi-way PCA applied to an industrial batch process. *Ind. Chem. Res.* 1995;35:138–146.
29. Lennox B, Montague GA, Hiden HG, Kornfeld G, Goulding PR. Process monitoring of an industrial fed-batch fermentation. *Biotechnol. Bioeng.* 2001;74(2):125–135.
30. Kourti T, Nomikos P, MacGregor JF. Analysis, monitoring and fault diagnosis of batch processes using multiblock and multiway PLS. *J. Process Control* 1995;5:277–284.
31. Nomikos P. Detection and diagnosis of abnormal batch operations based on multiway principal component analysis. *ISA Trans.* 1996;35:259–267.
32. Duchesne C, MacGregor JF. Multivariate analysis and optimization of process variable trajectories for batch processes. *Chemom. Intell. Lab. Syst.* 2000;51:125–137.
33. Westerhuis J, Kourti T, MacGregor JF. Analysis of multiblock and hierarchical PCA and PLS models. *J. Chemom.* 1998;12: 301–321.
34. Qin JS, Valle S, Piovoso MJ. On unifying multiblock analysis with application to decentralized process monitoring. *J. Chemom.* 2001;15:715–742.
35. Höskuldsson A. Multi-block and path modelling procedures. *J. Chemometrics* 2008;22(11–12): 571–579.
36. Westerhuis JA, Coenegracht PMJ, Coenraad FL. Multivariate modelling of the tablet manufacturing process with wet granulation for tablet optimization and in-process control. *Int. J. Pharm.* 1997;156:109–117.
37. Findlay P, Morris K, Kildsig D. PAT in fluid bed granulation. Presented at *AIChE*, San Francisco, CA, 2003.
38. Marjanovic O, Lennox B, Sandoz D, Smith K, Crofts M. Real-time monitoring of an industrial batch process. Presented at CPC7: *Chemical Process Control*, Lake Louise, Alberta, Canada, January 8–13, 2006.
39. Zhang H, Lennox B. Integrated condition monitoring and control of fed-batch fermentation processes. *J. Process Control* 2004;14:41–50.
40. Flores-Cerrillo J, MacGregor JF. Control of batch product quality by trajectory manipulation using latent variable models. *J. Process Control* 2004;14:539–553.
41. Dushesne C, MacGregor JF. Establishing Multivariate Specification Regions for incoming materials. *J. Qual. Technol.* 2004;36:78–94.
42. Box GEP, Hunter WG, Hunter JS, Statistics for Experimenters. An introduction to Design, Data Analysis and Model Building.

John Wiley & Sons, New York. Wiley Series in Probability and Mathematical Statistics, 1978.

43. García-Muñoz S, MacGregor JF, Kourti T. Product transfer between sites using Joint Y_PLS. *Chemom. Intell. Lab. Syst.* 2005;79:101–114.
44. International Conference on Harmonization, ICH Draft Step 4, Q8(R2) Pharmaceutical development August 2009.
45. Kourti T, Gonzalez S, Balaguer P, Frances C, Paris G. Stability in the QbD framework. Presented in IFPAC 2010, Baltimore, MD, February 2–4, 2010.
46. Nijhuis A, de Jong S, Vandeginste BGM. Multivariate statistical process control in chromatography. *Chemom. Intell. Lab. Syst.* 1997;38:51–62.

INDEX

Chemical Engineering in the Pharmaceutical Industry: R&D to Manufacturing Edited by David J. am Ende